McGraw-Hill LearnSmart™ Study Modules

Powered by Connect® Biology, McGraw-Hill LearnSmart™ provides students with a Guided Path to Success (GPS). It is an adaptive diagnostic tool that constantly assesses a student's knowledge of the course material. Sophisticated diagnostics adapt to each student's individual knowledge base, and vary the questions to determine what the student doesn't know, knows but has forgotten, and how best to match and improve their knowledge level. Students actively learn the required concepts, and instructors can get specific LearnSmart reports to monitor overall progress.

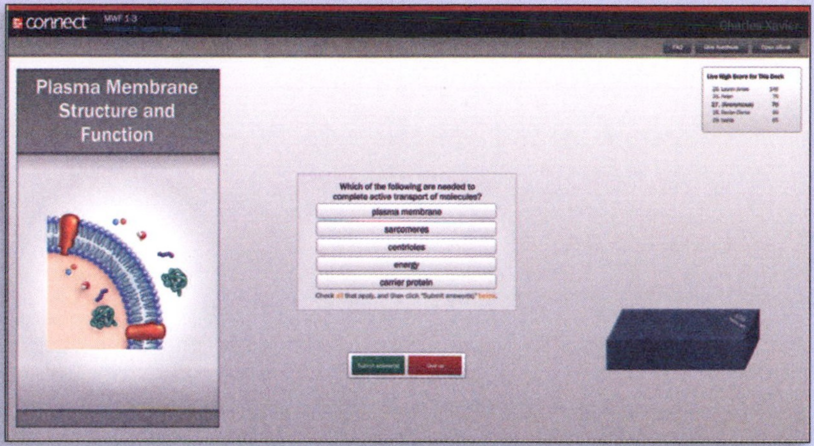
connect
Plasma Membrane Structure and Function
Which of the following are needed to complete active transport of molecules?
plasma membrane
sarcomeres
centrioles
energy
carrier protein

Learn Fast. Learn Easy. Learn Smart.

Learn more at **www.mhlearnsmart.com**

McGraw-Hill LabSmart™

Based on the same world-class adaptive technology as McGraw-Hill LearnSmart™, McGraw-Hill LabSmart is a must-see, outcomes-based lab simulation. It assesses a student's knowledge and adapts automatically to address deficiencies, allowing the student to learn faster and retain more knowledge with greater success. Whether your need is to overcome the logistical challenges of a traditional lab, provide better lap prep, improve student performance, or create an online experience that rivals the real world, LabSmart accomplishes it all.

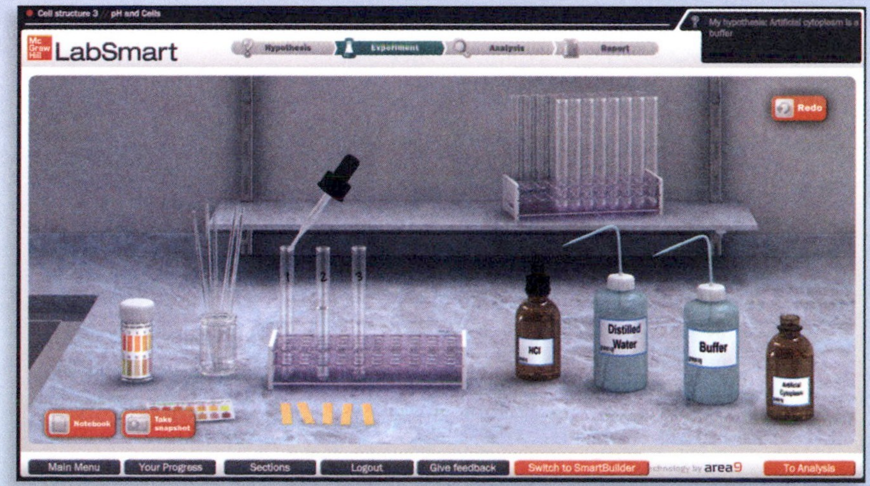
LabSmart
Hypothesis
Experiment
Analysis
Report
Redo
HCl
Distilled Water
Buffer
Notebook
Take snapshot
Main Menu
Your Progress
Sections
Logout
Give feedback
Switch to SmartBuilder
To Analysis

Learn more at **www.mhlabsmart.com**

Brief Contents

Fourteenth Edition

INQUIRY *into Life*

Sylvia S. Mader

Michael Windelspecht
Appalachian State University

With contributions by

David Cox
Lincoln Land Community College

Jeffrey Isaacson
Nebraska Wesleyan University

McGraw Hill
Connect Learn Succeed™

The McGraw·Hill Companies

INQUIRY INTO LIFE, FOURTEENTH EDITION

Published by McGraw-Hill, a business unit of The McGraw-Hill Companies, Inc., 1221 Avenue of the Americas, New York, NY 10020.

This book is printed on acid-free paper.

1 2 3 4 5 6 7 8 9 0 QVR/QVR 1 0 9 8 7 6 5 4 3

ISBN 978-0-07-352552-5
MHID 0-07-352552-9

Senior Vice President, Products & Markets: *Kurt L. Strand*
Vice President, General Manager, Products & Markets: *Marty Lange*
Vice President, Content Production & Technology Services: *Kimberly Meriwether David*
Director of Development: *Rose Koos*
Managing Director: *Michael S. Hackett*
Director: *Lynn Breithaupt*
Brand Manager: *Eric Weber*
Director of Digital Content: *Tod Duncan, Ph.D.*
Development Editor: *Liz Recker*
Marketing Manager: *Chris Loewenberg*
Senior Project Manager: *Jayne Klein*
Senior Buyer: *Sandy Ludovissy*
Senior Media Project Manager: *Tammy Juran*
Senior Designer: *Laurie B. Janssen*
Cover Designer: *Ron Bissell*
Cover Image: *Christian Beirle González, Getty Images*
Content Licensing Specialist: *Lori Hancock*
Photo Research: *Evelyn Jo Johnson*
Compositor and Art Studio: *Electronic Publishing Services Inc., NYC*
Typeface: *10/12 Utopia Std*
Printer: *Quad/Graphics*

Library of Congress Cataloging-in-Publication Data

Cataloging-in-Publication Data has been requested from the Library of Congress.

About the Authors

Sylvia S. Mader

Sylvia Mader has authored several nationally recognized biology texts published by McGraw-Hill. Educated at Bryn Mawr College, Harvard University, Tufts University, and Nova Southeastern University, she holds degrees in both Biology and Education. Over the years she has taught at University of Massachusetts, Lowell; Massachusetts Bay Community College; Suffolk University; and Nathan Mayhew Seminars. Her ability to reach out to science-shy students led to the writing of her first text, *Inquiry into Life*, that is now in its fourteenth edition. Highly acclaimed for her crisp and entertaining writing style, her books have become models for others who write in the field of biology.

Although her writing schedule is always quite demanding, Dr. Mader enjoys taking time to visit and explore the various ecosystems of the biosphere. Her several trips to the Florida Everglades and Caribbean coral reefs resulted in talks she has given to various groups around the country. She has visited the tundra in Alaska, the taiga in the Canadian Rockies, the Sonoran Desert in Arizona, and tropical rain forests in South America and Australia. A photo safari to the Serengeti in Kenya resulted in a number of photographs for her texts. She was thrilled to think of walking in Darwin's steps when she journeyed to the Galápagos Islands with a group of biology educators. Dr. Mader was also a member of a group of biology educators who traveled to China to meet with their Chinese counterparts and exchange ideas about the teaching of modern-day biology.

Michael Windelspecht

As an educator, Dr. Windelspecht has taught introductory biology, genetics, and human genetics in the online, traditional, and hybrid environments at community colleges, comprehensive universities, and military institutions. For over a decade he served as the Introductory Biology Coordinator at Appalachian State University, where he directed a program that enrolled over 4,500 students annually. He was educated at Michigan State University and the University of South Florida. Dr. Windelspecht is also active in promoting the scientific literacy of secondary school educators. He has led multiple workshops on integrating water quality research into the science curriculum, and has spent several summers teaching Pakistani middle school teachers.

As an author, Dr. Windelspecht has published five reference textbooks and multiple print and online lab manuals. He served as the series editor for a ten-volume work on the human body. For years Dr. Windelspecht has been active in the development of multimedia resources for online and hybrid science classrooms. Along with his wife, Sandra, he owns a multimedia production company, Ricochet Creative Productions, which actively develops and assesses new technologies for the science classroom.

Contributors

Dave Cox serves as Associate Professor of Biology at Lincoln Land Community College, in Springfield, Illinois. He was educated at Illinois College and Western Illinois University. As an educator, Professor Cox teaches introductory biology for nonmajors in the traditional classroom format as well as in a hybrid format. He also teaches biology for majors, and marine biology and biological field studies as study-abroad courses in Belize. He serves as the Educational Director for the Sibun Education and Adventure Lodge located in Belmopan, Belize. Dave served as a contributor to the eleventh edition of *Biology* and the upcoming thirteenth edition of *Human Biology*.

Jeffrey A. Isaacson is an Associate Professor of Biology at Nebraska Wesleyan University, a four-year, private, liberal arts institution. He currently serves as Chair of the Undergraduate Curriculum Committee and Co-Chair of the General Education Revision Team at NWU. His main teaching responsibilities include microbiology, immunology, and cell biology. He also supervises research projects for biology majors and conducts his own research on developing new antiviral drugs. Dr. Isaacson was educated at Nebraska Wesleyan University, Kansas State College of Veterinary Medicine, and Iowa State University. Prior to taking his current academic position, he worked as a small-animal veterinarian for several years, and he also completed a post-doctoral fellowship in the Department of Immunology at the Mayo Clinic.

As an author, Dr. Isaacson has written several scientific papers and served as a significant contributor and coauthor for the twelfth and thirteenth editions of Mader's *Inquiry into Life*.

Preface

Goals of the Fourteenth Edition

Dr. Sylvia Mader's text, *Inquiry into Life,* was originally developed to reach out to science-shy students. The text now represents one of the cornerstones of introductory biology education. *Inquiry into Life* was founded on the belief that teaching science from a human perspective, coupled with human applications, would make the material more relevant to the student. Interestingly, even though it has been thirty-six years since the first edition was published, this style of relevancy-based education remains the focus of the national efforts to increase scientific literacy in the general public.

Humans are a naturally inquisitive species. Like the child on the front cover, we express an interest in the living world from a very young age. We want to know not only why we are different, but how we are the same as the species we share the planet with. Students in the 21st century are being exposed, almost on a daily basis, to exciting new discoveries and insights that, in many cases, were beyond our predictions even a few short years ago. It is our task, as instructors, not only to make these findings available to our students, but to enlighten students as to why these discoveries are important to their lives and society. At the same time, we must provide students with a firm foundation in those core principles on which biology is founded, and in doing so, provide them with the background to keep up with the many discoveries still to come.

In this edition, we focused on the concept of inquiry and our inherent desire to learn. To do this, we integrated a tested, traditional learning system with modern digital and pedagogical approaches designed to stimulate and engage today's student.

The authors of the text identified several goals that guided them through the revision of *Inquiry into Life,* Fourteenth Edition:

1. build upon the strengths of the previous editions of the text,
2. enhance the learning process by integrating more human-related applications,
3. deploy new pedagogical elements, including multimedia assets, to increase student interaction with the text,
4. develop a new series of digital assets designed to engage the modern student and provide assessment of learning outcomes.

Integrating Featured Readings with Quantitative Analysis and Digital Content

To focus on the relevancy of biology to our everyday lives, the authors developed a series of new feature readings for this text. Each of these readings has a human focus, and in many cases represents topics that have recently been in the news. Each feature is supplemented by a series of questions that instructors may use to promote conversations in the classroom, or as the basis for one-minute opinion papers.

In addition, the majority of these features are now linked to a new style of digital questions. These multipart questions, all assignable within Connect, explore the quantitative aspects of the topic at a level which is comfortable for introductory biology students. In these questions, students analyze graphs and inquire into trends in the data.

SCIENCE IN YOUR LIFE ▶ HEALTH

How Safe Is Your Cell Phone?

The frequency of cell phone use in the United States has risen from 110 million users in 2000 to over 285 million users in 2009. The National Cancer Institute has raised concerns about the increased frequency of usage due to the fact that cell phones emit radiofrequency (RF) energy. Radiofrequency energy is a form of nonionizing electromagnetic radiation that is being studied to determine its effects on the human body.

At this point there is no conclusive evidence that the nonionizing radiation emitted by cell phones is associated with cancer risk. The radiation that is emitted by a cell phone comes from its antenna, which is housed within the handset. Due to the new design of today's cell phones, the handset is often held close to the side of the user's head. This leads to an increased absorption of nonionizing radiation to the side of the head, neck, and face. The level of exposure to RF depends on a variety of variables, such as the frequency and duration of calls as well as the quality of transmission and the size of the handset.

Numerous studies have been conducted to determine the relationship between cell phone use and the risk of developing malignant and benign brain tumors. Researchers reported that, overall, cell phone users did not show an increased risk for two of the most common brain tumors. However, the study group that had the greatest usage of cell phones showed an increased risk of glioma, a malignant type of brain tumor. The researchers considered this finding inconclusive.

Many of the other studies that have been conducted show a variety of discrepancies between their results. A Danish study showed no increased risk of acoustic neuroma in long-term cell phone users compared to short-term users. A Swedish study examined similar populations and discovered an elevated risk in the long-term users compared to the short-term users. So should the government ban the production and use of cell phones because of the potential for an increased risk of brain tumors? Or should people be allowed to freely use cell phones despite the potential risk?

Figure 1A Are cell phones safe? Several studies have not indicated any danger from radiation associated with cell phone use.

Questions to Consider

1. At this point, there is no conclusive evidence that the radiation emitted by cell phones is associated with cancer risk. How many studies do you think are necessary to prove cell phones are or are not safe to use?
2. What type of study would you design if you were trying to prove that cell phones are safe to use?
3. Do you think that cell phones can have a negative impact upon the various levels of biological organization? If so, which levels do you think are impacted the most and in what way?

connect Explore the concepts through a variety of multimedia assets, question types, and data interpretation.
www.mcgrawhillconnect.com

Media Integration

Students can improve the effectiveness of their learning by integrating the digital assets of today's courses into their study habits. LearnSmart™ and Connect®, McGraw-Hill's flagship digital tools, can increase student preparedness for class as well as increase retention of difficult topics. These assets may easily be uploaded into any course management system to provide your students with useful study tutorials.

As educators, the authors recognize that today's students are digital learners. Therefore, a significant new feature of this edition is the integration of media assets into the chapter content. Virtually every section of the textbook is now linked to MP3 files, animations of biological processes, and National Geographic and ScienCentral videos. In addition, McGraw-Hill's new 3D animations are integrated into the more difficult chapters of the text.

MP3 Files

These three- to five-minute audio files serve as a review of the material in the chapter, and they also assist the student in the pronunciation of scientific terms.

Animations

Drawing on McGraw-Hill's vast library of animations, the authors have selected animations that will enhance the student's understanding of complex biological processes.

3D Animations

For topics such as photosynthesis and cellular respiration, McGraw-Hill has produced a series of dynamic 3D animations that may be used both as presentation tools in the classroom, and as mini-tutorials that can be assigned within Connect or your course management system.

Videos

Two different types of movies are integrated into this edition of the text. The ScienCentral videos are short news clips on recent advances in the sciences. The National Geographic videos provide students with a glimpse of the complexity of life that normally would not be possible in the classroom.

Virtual Labs

These simulated experiments serve as excellent tutorials, allowing students to explore the topics covered in select chapters of the text.

Guided Tutorials

In addition to the assets listed above, the author of the textbook has prepared a series of short (2-3 minute) guided tutorials of some of the more difficult topics in the text. A complete list of these tutorials is provided on the inside back cover of the text.

A Student's Guide to Using This Textbook

Case Study The opening case study is designed to demonstrate how the chapter content is relevant to your life. The authors have not only chosen current-event topics, but have also included a series of two or three questions that you should be thinking of as you progress through the chapters.

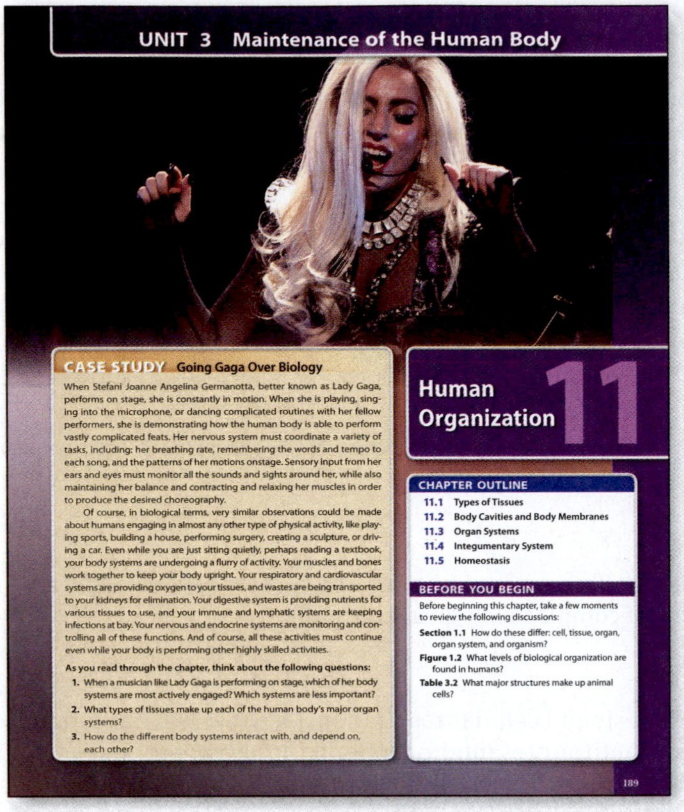

UNIT 3 Maintenance of the Human Body

CASE STUDY Going Gaga Over Biology

When Stefani Joanne Angelina Germanotta, better known as Lady Gaga, performs on stage, she is constantly in motion. When she is playing, singing into the microphone, or dancing complicated routines with her fellow performers, she is demonstrating how the human body is able to perform vastly complicated feats. Her nervous system must coordinate a variety of tasks, including: her breathing rate, remembering the words and tempo to each song, and the patterns of her motions onstage. Sensory input from her ears and eyes must monitor all the sounds and sights around her, while also maintaining her balance and contracting and relaxing her muscles in order to produce the desired choreography.

Of course, in biological terms, very similar observations could be made about humans engaging in almost any other type of physical activity, like playing sports, building a house, performing surgery, creating a sculpture, or driving a car. Even while you are just sitting quietly, perhaps reading a textbook, your body systems are undergoing a flurry of activity. Your muscles and bones work together to keep your body upright. Your respiratory and cardiovascular systems are providing oxygen to your tissues, and wastes are being transported to your kidneys for elimination. Your digestive system is providing nutrients for various tissues to use, and your immune and lymphatic systems are keeping infections at bay. Your nervous and endocrine systems are monitoring and controlling all of these functions. And of course, all these activities must continue even while your body is performing other highly skilled activities.

As you read through the chapter, think about the following questions:

1. When a musician like Lady Gaga is performing on stage, which of her body systems are most actively engaged? Which systems are less important?
2. What types of tissues make up each of the human body's major organ systems?
3. How do the different body systems interact with, and depend on, each other?

Human Organization 11

CHAPTER OUTLINE

BEFORE YOU BEGIN

Before beginning this chapter, take a few moments to review the following discussions:

Section 1.1 How do these differ: cell, tissue, organ, organ system, and organism?

Figure 1.2 What levels of biological organization are found in humans?

Table 3.2 What major structures make up animal cells?

189

Chapter Outline Lists the major sections that will be discussed in the chapter

Before You Begin Links the content of the chapter with material from earlier in the text. The questions designate important topics that you should understand before proceeding into the chapter.

Learning Outcomes Provide you with an overview of what you are to know. Your instructor can assign activities through Connect® to help you achieve these outcomes.

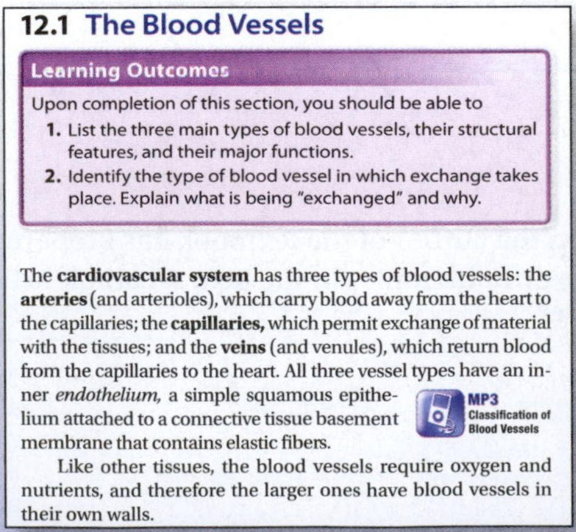

12.1 The Blood Vessels

Learning Outcomes

Upon completion of this section, you should be able to

1. List the three main types of blood vessels, their structural features, and their major functions.
2. Identify the type of blood vessel in which exchange takes place. Explain what is being "exchanged" and why.

The **cardiovascular system** has three types of blood vessels: the **arteries** (and arterioles), which carry blood away from the heart to the capillaries; the **capillaries,** which permit exchange of material with the tissues; and the **veins** (and venules), which return blood from the capillaries to the heart. All three vessel types have an inner *endothelium,* a simple squamous epithelium attached to a connective tissue basement membrane that contains elastic fibers.

MP3 Classification of Blood Vessels

Like other tissues, the blood vessels require oxygen and nutrients, and therefore the larger ones have blood vessels in their own walls.

Media Integration Enhances your study of biology with media. Go to www.mhhe.com/maderinquiry14 to access the animations, videos, and MP3 files referenced throughout this book. Ask your instructor about related quizzes that are available through Connect® Biology.

Check Your Progress
Questions at the end of each section help you assess and/or apply your understanding of the material in the section.

Media Study Tools
Provide a link to the *Inquiry into Life* website, which contains practice tests, animations, and videos organized and integrated by chapter to help you succeed in your study of biology. The ConnectPlus® platform provides a media-rich eBook, interactive learning tools, and access to the LearnSmart™ system for enhanced classroom performance.

Chapter 6 Metabolism: Energy and Enzymes 111

you would soon realize that certain substrates recur along the way. This allows the cell to switch metabolic gears and even switch to building up molecules rather than breaking them down. A good example is when you get fat from eating too much ice cream, cake, and pie. A particular breakdown product of glucose can be used to form fatty acids, which then join with glycerol and the product fat gets stored at various locations in your body.

Check Your Progress 6.4

1. Explain why overall equations for photosynthesis and cellular respiration represent redox reactions.
2. Summarize how the inputs and outputs of cellular respiration and photosynthesis are related.
3. Describe the three types of nutrients that can be used in mitochondria to build up ATP molecules.

Case Study Conclusion

The enzymes in the saliva of the vampire bat are interfering with a complicated biochemical pathway in the blood. The clotting cascade involves several enzymes that circulate in the blood. When a blood vessel is damaged, platelets release prothrombin activator, an enzyme that converts the plasma protein prothrombin into thrombin. Thrombin, in turn, is an enzyme that converts fibrinogen into fibrin, forming the basis of the clot. When the damage has been repaired, the plasma protein plasminogen is converted into plasmin, which destroys the fibrin. By releasing a potent plasminogen enzyme in their saliva, vampire bats are able to manipulate a biochemical pathway for their own needs.

MEDIA STUDY TOOLS

www.mhhe.com/maderinquiry14

Enhance your study of this chapter with study tools and practice tests. Also ask your instructor about the resources available through ConnectPlus, including LearnSmart, the media-rich eBook, interactive learning tools, and animations.

connect plus+ | BIOLOGY

SUMMARIZE

6.1 Life and the Flow of Energy

- **Energy** is the ability to do work. Energy may exist in several forms, including **potential energy** (**chemical energy**) and **kinetic energy** (**mechanical energy**). Two energy laws are basic to understanding energy-use patterns in cells and ecosystems. The first law states that energy cannot be created or destroyed but can only be changed from one form to another. The second law of thermodynamics states that one usable form of energy cannot be converted into another form without loss of usable energy. Therefore, every energy transformation makes the universe less organized. As a result of these laws, we know that the **entropy** of the universe is increasing and that only a constant input of energy maintains life's organization.

6.2 Energy Transformations and Metabolism

- The term **metabolism** encompasses all the chemical reactions occurring in a cell. **Catabolism** involves breaking down **reactants,** whereas **anabolism** involves building up **products. Free energy** is the amount of energy that is actually available to do work.
- Considering individual reactions, only those that result in products that have less usable energy than the reactants go forward. Such reactions, called **exergonic reactions,** release energy. **Endergonic reactions,** which require an input of energy, occur in cells as a **coupled reactions** with an exergonic process. For example, glucose breakdown is an exergonic metabolic pathway that drives the buildup of many **ATP** molecules. These ATP molecules then supply energy for cellular work. ATP goes through a cycle of constantly being built up from, and then broken down to, **ADP** + Ⓟ.

6.3 Enzymes and Metabolic Pathways

- A **metabolic pathway** is a series of reactions that proceed in an orderly, step-by-step manner. Each reaction requires a specific **enzyme.** Enzymes function by reducing the **energy of activation** (**E_a**) needed for the reaction to proceed.
- Reaction rates increase when enzymes form a complex with their **substrates,** called an **induced fit model.** Generally, enzyme activity increases as substrate concentration increases. Once all **active sites** are filled, the maximum rate has been achieved. Any environmental factor, such as temperature or pH, affects the shape of a protein. A **denatured** enzyme loses its ability to do its job.
- Cellular mechanisms regulate enzyme quantity and activity. The activity of most metabolic pathways is regulated by **enzyme inhibition,** usually associated with feedback. Many enzymes have **cofactors** or **coenzymes,** such as **vitamins,** that help them carry out a reaction.

6.4 Oxidation-Reduction Reactions and Metabolism

- The overall equation for photosynthesis is the opposite of that for cellular respiration. Both processes involve **oxidation-reduction reactions.** Redox reactions are a major way in which energy is transformed in cells.
- During photosynthesis, carbon dioxide is reduced to glucose, and water is oxidized. Glucose formation requires energy, and this energy comes from the sun. Chloroplasts capture solar energy and convert it to the chemical energy of ATP molecules, which are used along with hydrogen atoms to reduce carbon dioxide to glucose. During cellular respiration, glucose is oxidized to carbon dioxide, and oxygen is reduced to water. This reaction releases energy, which is used to synthesize ATP molecules in all types of cells.
- Energy flows through all living things. Photosynthesis is a metabolic pathway in chloroplasts that transforms solar energy to the chemical energy within carbohydrates, and cellular respiration is

Case Study Conclusion
The case study conclusion at the end of the chapter explains how the material you just learned relates to the scenario presented on the opening pages.

Summarize
Provides an excellent overview of the chapter concepts using concise, bulleted summaries, summary tables, and key illustrations. The boldfaced terms represent the key terms from each section of the chapter. These are placed within the summary to help you understand how the term relates to the content of the chapter. An expanded definition of each boldface term is provided in the Glossary.

ASSESS
Testing Yourself
Questions help you review the material and prepare for tests. (See Appendix A for answers.)

112 UNIT 1 Cell Biology

a metabolic pathway completed in mitochondria that transforms this energy into that of ATP molecules. Eventually, the energy within ATP molecules becomes heat.

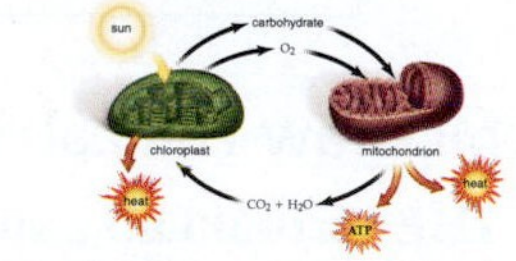

sun
carbohydrate
O_2
chloroplast
mitochondrion
heat
$CO_2 + H_2O$
ATP
heat

- Cellular respiration is an aerobic process that requires oxygen and gives off carbon dioxide. Cellular respiration involves oxidation. The air we inhale contains the oxygen, and the food we digest after eating contains the carbohydrate glucose needed for cellular respiration.
- Besides carbohydrates, fats and proteins can be used to provide ATP via cellular respiration.
- Anabolism can use molecules directly from our food or substrates from other metabolic pathways to synthesize new molecules.

ASSESS

Testing Yourself

Choose the best answer for each question.

1. The fact that energy transformation is never 100% efficient is stated in the ____ law of thermodynamics.
   a. first c. third
   b. second d. None of these are correct.

In questions 2–5, match the statements concerning enzyme action to the appropriate graph.

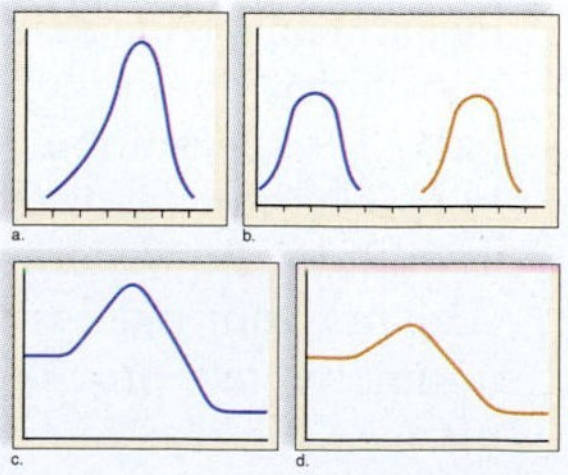

a.
b.
c.
d.

2. Progress of reaction in the absence of an enzyme
3. Enzymatic action with varying temperature
4. Progress of reaction in the presence of an enzyme
5. Enzymes with varying pH optima
6. Enzymatic action is sensitive to changes in
   a. temperature. c. substrate concentration.
   b. pH. d. All of these are correct.
7. Which has more potential energy: glucose or CO_2 and H_2O?
   a. glucose
   b. CO_2 and H_2O
   c. Both are the same because reactants and products are equal.
   d. Neither is correct.
8. ____ is the loss of electrons, and ____ is the gain of electrons.
   a. Oxidation, reduction
   b. Reduction, oxidation
   c. Neither a nor b is correct.
   d. Need more information to answer.
9. Entropy is a term used to indicate the level of
   a. usable energy. c. enzyme action.
   b. disorganization. d. None of these are correct.
10. Enzymes catalyze reactions by
    a. bringing the reactants together.
    b. lowering the activation energy.
    c. Both a and b are correct.
    d. Neither a nor b is correct.
11. The active site of an enzyme
    a. is identical to that of any other enzyme.
    b. is the part of the enzyme where its substrate can fit.
    c. can be used over and over again.
    d. is not affected by environmental factors such as pH and temperature.
    e. Both b and c are correct.

ENGAGE

Virtual Lab
Enzyme Controlled Reactions

The virtual lab "Enzyme Controlled Reactions" provides an interactive tutorial to explore many of the factors influencing enzyme activity.

Thinking Critically

1. Entropy is increased when nutrients break down, so why are enzymatic metabolic pathways required for cellular respiration? (*Hint:* See Fig. 6.5.)
2. Why would you expect glucose storage as glycogen to be an energy-requiring process?
3. If photosynthesis and cellular respiration are reverse equations, then why can't mitochondria carry on photosynthesis?

ENGAGE
Thinking Scientifically
Questions give you an opportunity to reason as a scientist.

Virtual Labs
For selected chapters, these online labs can serve as tutorials to help you better understand the content and provide you with the opportunity to investigate topics associated with the chapter from a scientific perspective.

Teaching and Learning Tools

McGraw-Hill Higher Education and Blackboard Have Teamed Up

Blackboard®, the Web-based course-management system, has partnered with McGraw-Hill to better allow students and faculty to use online materials and activities to complement face-to-face teaching. Blackboard features exciting social learning and teaching tools that foster active learning opportunities for students. You'll transform your closed-door classrooms into communities where students remain connected to their educational experience 24 hours a day.

This partnership allows you and your students access to McGraw-Hill Connect® and McGraw-Hill Create™ right from within your Blackboard course—all with one single sign-on.

Not only do you get single sign-on with Connect and Create, you also get deep integration of McGraw-Hill content and content engines right in Blackboard. Whether you're choosing a book for your course or building Connect assignments, all the tools you need are right where you want them—inside of Blackboard.

Gradebooks are now seamless. When a student completes an integrated Connect assignment, the grade for that assignment automatically (and instantly) feeds your Blackboard grade center.

McGraw-Hill and Blackboard can now offer you easy access to industry leading technology and content, whether your campus hosts it or we do. Be sure to ask your local McGraw-Hill representative for details.

McGraw-Hill LearnSmart™

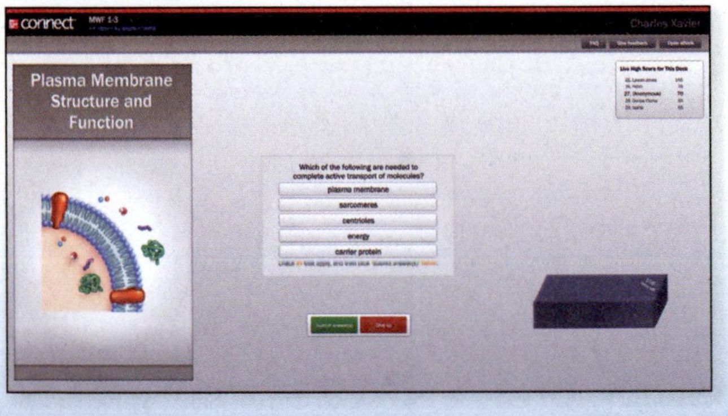
connect
Plasma Membrane Structure and Function

McGraw-Hill LearnSmart™ is available as an integrated feature of McGraw-Hill Connect® Biology and provides students with a GPS (Guided Path to Success) for your course. Using artificial intelligence, LearnSmart intelligently assesses a student's knowledge of course content through a series of adaptive questions. It pinpoints concepts the student does not understand and maps out a personalized study plan for success. This innovative study tool also has features that allow instructors to see exactly what students have accomplished and a built-in assessment tool for graded assignments. Visit the following site for a demonstration.

www.mhlearnsmart.com

McGraw-Hill Connect® Biology

McGraw-Hill Connect® Biology provides online presentation, assignment, and assessment solutions. It connects your students with the tools and resources they'll need to achieve success.

With Connect Biology, you can deliver assignments, quizzes, and tests online. A robust set of questions and activities are presented and aligned with the textbook's learning outcomes. As an instructor, you can edit existing questions and author entirely new problems. Track individual student performance—by question, assignment, or in relation to the class overall—with detailed grade reports. Integrate grade reports easily with Learning Management Systems (LMS), such as WebCT and Blackboard—and much more.

ConnectPlus Biology provides students with all the advantages of Connect Biology, plus 24/7 online access to an eBook. This media-rich version of the book is available through the McGraw-Hill Connect platform and allows seamless integration of text, media, and assessments. To learn more, visit

www.mcgrawhillconnect.com

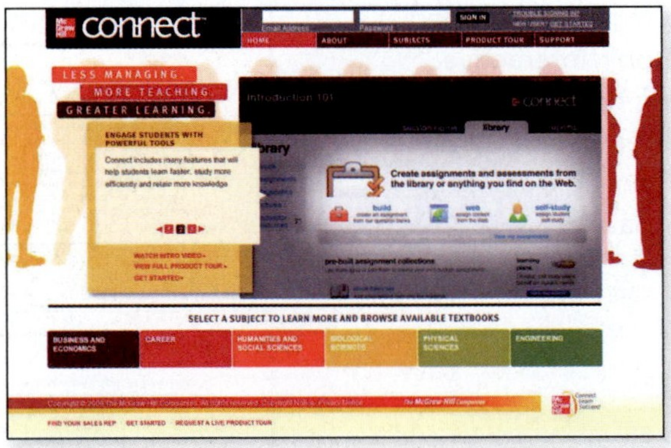
connect
LESS MANAGING.
MORE TEACHING.
GREATER LEARNING.

McGraw-Hill LabSmart™

THE Virtual Lab Experience

Based on the same world-class, super-adaptive technology as LearnSmart, McGraw-Hill LabSmart is a must-see, outcomes based lab simulation. It assesses a student's knowledge and adaptively corrects deficiencies, allowing the student to learn faster and retain more knowledge with greater success.

First, a student's knowledge is adaptively leveled on core learning outcomes: Questioning reveals knowledge deficiencies that are corrected by the delivery of content that is conditional on a student's response. Then, a simulated lab experience requires the student to think and act like a scientist: recording, interpreting, and analyzing data using simulated equipment found in labs and clinics. The student is allowed to make mistakes—a powerful part of the learning experience! A virtual coach provides subtle hints when needed, asks questions about the student's choices, and allows the student to reflect upon and correct those mistakes. Whether your need is to overcome the logistical challenges of a traditional lab, provide better lab prep, improve student performance, or make your online experience one that rivals the real world, LabSmart accomplishes it all. To learn more, visit

www.mhlabsmart.com

My Lectures—Tegrity®

McGraw-Hill Tegrity® records and distributes your class lecture with just a click of a button. Students can view anytime/anywhere via computer, iPod, or mobile device. It indexes as it records your PowerPoint® presentations and anything shown on your computer so students can use keywords to find exactly what they want to study. Tegrity is available as an integrated feature of McGraw-Hill Connect Biology and as a standalone.

Animations for a New Generation

Dynamic, 3D animations of key biological processes bring an unprecedented level of control to the classroom. Innovative features keep the emphasis on teaching rather than entertaining.

- An options menu lets you control the animation's level of detail, speed, length, and appearance, so you can create the experience you want.
- Draw on the animation using the whiteboard pen to highlight important areas.
- The scroll bar lets you fast forward and rewind while seeing what happens in the animation, so you can start at the exact moment you want.
- A scene menu lets you instantly jump to a specific point in the animation.
- Pop-ups add detail at important points and help students relate the animation back to concepts from the lecture and the textbook.
- A complete visual summary at the end of the animation reminds students of the big picture.
- Animation topics include: Cellular Respiration, Photosynthesis, Molecular Biology of the Gene, DNA Replication, Cell Cycle and Mitosis, Membrane Transport, and Plant Transport.

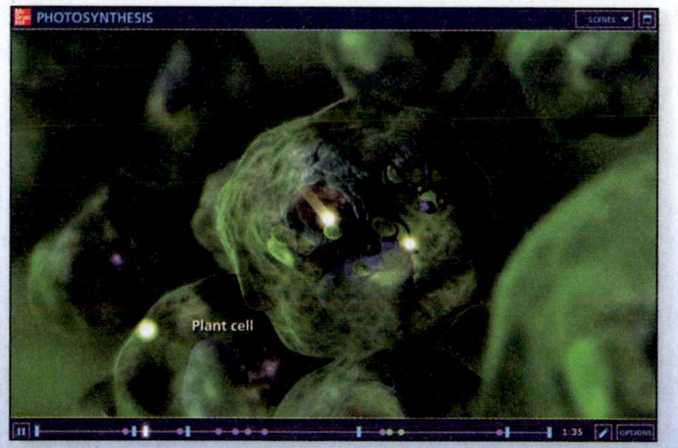
PHOTOSYNTHESIS
Plant cell
1:35

McGraw-Hill Create™

With **McGraw-Hill Create™**, you can easily rearrange chapters, combine material from other content sources, and quickly upload content you have written, like your course syllabus or teaching notes. Find the content you need in Create by searching through thousands of leading McGraw-Hill textbooks. Arrange your book to fit your teaching style. Create even allows you to personalize your book's appearance by selecting the cover and adding your name, school, and course information. Order a Create book and you'll receive a complimentary print review copy in 3–5 business days or a complimentary electronic review copy (eComp) via e-mail in minutes. Go to www.mcgrawhillcreate.com today and register to experience how McGraw-Hill Create empowers you to teach *your* students *your* way. To learn more visit

www.mcgrawhillcreate.com

Presentation Tools

Everything you need for outstanding presentations in one place.

www.mhhe.com/maderinquiry14

- *FlexArt Image Powerpoints*—including every piece of art that has been sized and cropped specifically for superior presentations, as well as labels that can be edited and flexible art that can be picked up and moved on key figures. Also included are tables, photographs, and unlabeled art pieces.
- *Lecture PowerPoints with Animations*—animations illustrating important processes are embedded in the lecture material.
- *Animation PowerPoints*—animations only are provided in PowerPoint.
- *Labeled JPEG Images*—full-color digital files of all illustrations that can be readily incorporated into presentations, exams, or custom-made classroom materials.
- *Base Art Image Files*—unlabeled digital files of all illustrations.

Presentation Center

In addition to the images from your book, this online digital library contains photos, artwork, animations, and other media from an array of McGraw-Hill textbooks.

Computerized Test Bank

A comprehensive bank of test questions is provided within a computerized test bank powered by McGraw-Hill's flexible electronic testing program, **EZ Test Online.** A new tagging scheme allows you to sort questions by Bloom's difficulty level, learning outcome, topic, and section. With EZ Test Online, instructors can select questions from multiple McGraw-Hill test banks or author their own, and then either print the test for paper distribution or give it online.

Instructor's Manual

The instructor's manual contains chapter outlines, lecture enrichment ideas, and discussion questions.

Laboratory Manual

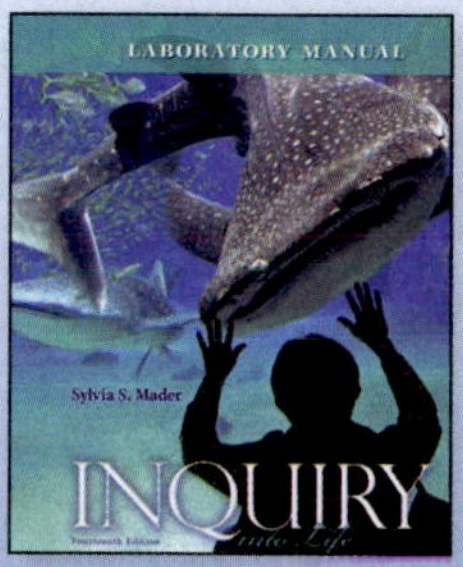
LABORATORY MANUAL
Sylvia S. Mader
INQUIRY
into Life

The *Inquiry into Life Laboratory Manual* is written by Dr. Sylvia Mader. Every laboratory has been written to help students learn the fundamental concepts of biology and the specific content of the chapter to which the lab relates, as well as gain a better understanding of the scientific method.

Companion Website

www.mhhe.com/maderinquiry14

The Mader *Inquiry into Life* companion website allows students to access a variety of free digital learning tools that include

- Chapter-level quizzing
- Vocabulary flashcards
- Animations and videos
- Virtual labs

Detailed List of Content Changes in *Inquiry into Life*, Fourteenth Edition

Every chapter now includes icons that direct the student to relevant movies, animations, and MP3 files. All of these are available from the companion site **www.mhhe.com/maderinquiry14**

Chapter 1: The Study of Life now opens with a feature on the recent nuclear crisis in Japan. A new health feature on the safety of cell phones enhances the discussion of the scientific method. A virtual lab on dependent and independent variables is now referenced at the end of the chapter.

Unit 1: Cell Biology

Chapter 2: The Molecules of Cells now begins with a feature on sports drinks and exercise. A new ecology feature on the fate of prescription medicines is included. An enhanced discussion of trans-fats is provided in the section on lipids. The material on a balanced diet has been updated to reflect the new MyPlate recommendations. **Chapter 3: Cell Structure and Function** includes a new opening feature on liposome disorders. The chapter features a new section (3.4) that focuses on the structure and function of the cytoskeleton. A new illustration is provided on endosymbiosis (Fig. 3.16). **Chapter 4: Membrane Structure and Function** contains a new section (4.3) on modifications to the cell surface and the junctions between cells. The 3D animation "Membrane Transport" supplements the chapter content. In **Chapter 5: Cell Division,** there is an enhanced discussion of control of the cell cycle. A new health feature examines the difference between benign and malignant tumors. The chapter is now supplemented by two 3D animations, "Cell Cycle and Mitosis" and "Meiosis" and the virtual lab "Cell Reproduction." **Chapter 6: Metabolism: Energy and Enzymes** includes a new bioethical feature on enzyme inhibitors such as Sarin. The virtual lab "Enzyme Controlled Reactions" is referenced at the end of the chapter. **Chapter 7: Cellular Respiration** now begins with a feature on the metabolic demands of athletes. The anaerobic processes of glycolysis and fermentation are now placed closer in the chapter. The chapter includes a new health feature on the metabolic fate of foods such as pizza. The 3D animation "Cellular Respiration" supplements the chapter content.

Unit 2: Plant Biology

Chapter 8: Photosynthesis is now supplemented by the 3D animation "Photosynthesis." **Chapter 9: Plant Organization and Function** includes a new ecology feature on the uses of bamboo, and a new bioethical feature on the use of plants for bioremediation. The 3D animation "Plant Transport" and the virtual lab "Plant Transpiration" supplement the chapter content. **Chapter 10: Plant Reproduction and Responses** contains a new figure (Fig.10.21) on the activation of phytochromes.

Unit 3: Maintenance of the Human Body

Chapter 11: Human Organization now begins with a focus on the physiological demands of Lady Gaga's performances. **Chapter 12: Cardiovascular System** was reorganized so that the discussion of blood follows the material on blood vessels. The virtual lab "Virtual Frog Dissection" supplements the chapter content. **Chapter 13: Lymphatic and Immune Systems** has been reorganized so that the innate and adaptive immune responses are covered in separate sections. The bioethical feature on the necessity of childhood vaccinations has been enhanced with recent research. A new figure on the production of monoclonal antibodies (Fig. 13.11) is included. **Chapter 14: Digestive System and Nutrition** references the new MyPlate dietary planning guidelines. Two virtual labs "Virtual Frog Dissection" and "Nutrition" supplement the chapter content. **Chapter 15: Respiratory System** contains a new scientific inquiry feature on artificial lung technology. **Chapter 16: Urinary System and Excretion** is supplemented by a virtual lab "Virtual Frog Dissection."

Unit 4: Integration and Control of the Human Body

Chapter 17: Nervous System contains a new section on the transmission of nerve impulses. The drug abuse section contains new material on "bath salts." **Chapter 19: Musculoskeletal System** includes a new scientific inquiry feature on forensic medicine. The 3D animation "Skeletal Muscle Contraction" and the virtual lab "Muscle Stimulation" supplement the chapter content. In **Chapter 20: Endocrine System,** the material on pheromones is now included with the general discussion of hormones. The health feature on melatonin now includes information on melatonin-containing soft drinks.

Unit 5: Continuance of the Species

Chapter 21: Reproductive System includes a new section dedicated to sexually-transmitted diseases. **Chapter 22: Development and Aging** begins with a new feature regarding pregnancy. The chapter has been reorganized to separate the material on development and pregnancy. A new bioethical feature on cloning has been added and the health feature on preventing birth defects has been revised to include examples of genetic testing of the fetus. The material on aging has been updated to reflect recent advances. **Chapter 23: Patterns of Gene Inheritance** includes a revised scientific inquiry feature on Mendel's

mathematical approach to his study of genetics. In addition, the material on polygenetic and multifactorial inheritance has been enhanced, including new figures on the genetics of skin color (Fig. 23.16). A virtual lab "Punnett Squares" supplements the chapter content. **Chapter 24: Chromosomal Basis of Inheritance** now begins with a discussion of gene linkage and the effects of crossing over on linkage (Figs. 24.1 and 24.2). The chapter content is supplemented by the virtual lab "Sex-Linked Traits." **Chapter 25: DNA Structure and Gene Expression** contains a new scientific inquiry feature that provides historical context on the search for the structure of DNA. The figures on DNA replication have been updated (Figs. 25.4 and 25.5) to provide more information on the molecular basis of replication. Two 3D animations, "DNA Replication" and "Molecular Biology of the Gene," and the virtual lab "DNA and Genes" supplement the chapter content. **Chapter 26: Biotechnology and Genomics** begins with a new article on biotechnology and agriculture. A new scientific inquiry feature explores the use of microchips in screening for genetic disorders. The virtual labs "Knocking Out Genes," "Gene Splicing," and "Classifying Using Biotechnology" supplement the chapter content.

Unit 6: Evolution and Diversity

The content on microbiology from the previous edition (Chapter 28) has been split into two chapters: Chapter 28: Microbiology and Chapter 29: Protists and Fungi. **Chapter 27: Evolution of Life** has been reorganized to focus on evolutionary theory, evidence of evolution, and the processes of microevolution and macroevolution (speciation). The illustrations and content regarding microevolution has been revised to provide a more visual representation of the process (Figs. 27.12 and 27.13). A new scientific inquiry feature demonstrates how microevolution occurs in human populations. The content on systematics has also been revised to reflect a better distinction between systematics and taxonomy. **Chapter 28: Microbiology** contains material on bacteria (eubacteria and archaebacteria) and viruses. The chapter now contains a more detailed examination of the origins of microbial life. **Chapter 29: Protists and Fungi** presents some of the new classification work on both protists and fungi, including an introduction to the supergroup level of classification for protists. A new health feature on African sleeping sickness is included. The content of **Chapter 30: Plants** has been supplemented by the virtual lab "Classifying Using Biotechnology." **Chapter 31: Animals: The Invertebrates** has been reorganized so that it examines the phyla of invertebrate animals using recent classification changes. The chapter opens with a new feature on Neglected Tropical Diseases. Two virtual labs, "Earthworm Dissection" and "Classifying Arthropods," supplement the chapter content. **Chapter 32: Animals: Chordates and Vertebrates** has been reorganized to provide more detail on the evolution of vertebrates and the process of human evolution. A new figure on the evolution of jaws (Fig. 32.6) and a scientific inquiry feature on biocultural evolution have been added.

Unit 7: Behavior and Ecology

Chapter 33: Behavioral Ecology now begins with a new opener on social living in African lions. The chapter content is supplemented by the virtual lab "Mealworm Behavior." **Chapter 34: Population and Community Ecology** contains a new opening feature on exotic species. A new ecology feature examines the impact of the loss of wolves to the Yellowstone ecosystem. The virtual lab "Model Ecosystems" supplements the chapter content. The content of **Chapter 35: Nature of Ecosystems** is supplemented by the virtual lab "Model Ecosystems." **Chapter 36: Major Ecosystems of the Biosphere** begins with a new article on DDT. **Chapter 37: Conservation Biology** starts with a new article on waste-to-energy projects. The chapter contains a new ecology feature on floodplain restoration in Illinois.

Acknowledgments

Dr. Sylvia Mader represents one of the icons of science education. Her dedication to her students, coupled to her clear, concise writing style, has benefited the education of thousands of students over the past three decades. It is an honor to continue her legacy, and to bring her message to the next generation of students. I have been privileged to work with a fantastic pair of coauthors, Dave Cox and Jeff Isaacson. I would also like to thank April Cognato for her assistance with the outline of the evolution chapters in this text, and her discussions on the digital assets for this edition. My coauthors are dedicated and talented individuals and have made a significant contribution to this text, and my abilities as a science educator have been enhanced by our teamwork. Together, we have striven to ensure that the material was written and illustrated in the familiar Mader style.

Many dedicated and talented individuals assisted in the development of *Inquiry into Life.* I am very grateful for the help of so many professionals at McGraw-Hill who were involved in bringing this book to fruition. In particular, let me thank my developmental editors Rose Koos and Liz Recker. The talents, project management skills, advice, and, most of all, patience of these individuals is greatly appreciated. The managing director, Michael Hackett, continues to support the development of innovative products. The desire of my editor, Eric Weber, to impact the lives of our students and assist those who instruct them is evident throughout this text. The project manager, Jayne Klein, faithfully and carefully steered the book through the publication process. Tamara Maury and Chris Loewenberg, the marketing managers, tirelessly promoted the text and educated the sales reps on its message.

The design of the book is the result of the creative talents of Laurie Janssen and many others who assisted in deciding the appearance of each element in the text. I was very lucky to have Bea Sussman as my copyeditor, and Dawnelle Krouse and Mary Schwartz as proofreaders. Electronic Publishing Services produced this textbook, in the process emphasizing pedagogy and beauty to arrive at the best presentation on the page. Lori Hancock and Evelyn Jo Johnson did a superb job of finding just the right photographs and micrographs.

Who I am, as an educator and an author, is a direct reflection of what I have learned from my students. Education is a two-way street, and it is my honest opinion that both my professional and personal life have been enriched by my interactions with my students. They have encouraged me to learn more, teach better, and never stop questioning the world around me. Last, but certainly not least, I would also like to acknowledge my wife, Sandra. She has never wavered in her patience and support of my endeavors.

Michael Windelspecht, Ph.D.
Blowing Rock, NC

The fourteenth edition of *Inquiry into Life* would not have the same excellent quality without the input of these contributors and those of the many contributors and reviewers listed below.

McGraw-Hill's 360° Development Process is an ongoing, never-ending, market-oriented approach to building accurate and innovative print and digital products. It is dedicated to continual large-scale and incremental improvement driven by multiple customer feedback loops and checkpoints. This is initiated during the early planning stages of our new products, and intensifies during the development and production stages, then begins again upon publication in anticipation of the next edition.

This process is designed to provide a broad, comprehensive spectrum of feedback for refinement and innovation of our learning tools, for both student and instructor. The 360° Development Process includes market research, content reviews, course- and product-specific symposia, accuracy checks, and art reviews. We appreciate the expertise of the many individuals involved in this process.

Ancillary Authors

Appendix Answer Bank: Betsy Harris, *Appalachian State University;* **Connect Question Bank:** Sandy Windelspecht, Krissy Johnson, and Alex James, *Ricochet Creative Productions;* **Test Bank:** Dave Cox, *Lincoln Land Community College;* **Lecture Outlines:** Lou Scala, *Montclair State University;* **Instructor's Manual:** Noelle Relles, *College of William & Mary;* **Website and eBook Quizzes:** Eric Rabitoy, *Citrus College*

LearnSmart Authors:

Daniel Matusiak, *St. Charles Community College*
Sylvester Allred, *Northern Arizona University*
Dena Berg, *Tarrant County College*
Joy Brookshire, *Kennesaw State University*
Patrick Galliart, *North Iowa Area Community College*
Jeffrey Isaacson, *Nebraska Wesleyan University*

Reviewers:

Jill Nugent, *University of North Texas*
MaryJo Witz, *Monroe Community College*
Stacy Zell, *Carroll Community College*

Fourteenth Edition Reviewers

Phyllis Ballard, *Rivercrest High School, Paris Junior College*
Michael Barnett, *Paris Junior College*
Wendy Billock, *Biola University*
Ruth E. Birch, *Fontboone University*
Robert Bogardus, *Mid-Plains Community College*
Carrie Brown, *Lincoln Trail College*
Xiaomei Cheng, *Mount St. Mary's College*
Jennifer Chotiner, *Mount St. Mary's College*
Carol Cleveland, *Northwest Mississippi Community College*
Yvonne Cole, *Fontbonne University*
Gary Corbin, *Columbia College–St. Louis College*
Matt Cruzen, *Biola University*
David Demezas, *University of Wisconsin at Fond du Lac*
Patrick Downie, *Pacific College of Oriental Medicine (Chicago)*
Frank Duroy, *Essex County College*
Michael Farabee, *Estrella Mountain Community College*
Dr. Angela M. Foster, *Wake Tech Community College*
Linda Grossman, *Wichita Area Technical College*
Irene Guttilla, *Saint Joseph College*
William Harness, *Three Rivers Community College*
Keith Hench, *Kirkwood Community College*
Brian Scott Hobson, *University of Arkansas Community College–Hope*
Vallie Holloway, *Columbia College*
Dan F. Ippolito, *Anderson University*
James R. Jacob, *Tompkins Cortland Community College*
Sandra Johnson, *Middle Tennessee State University*
Scott Johnson, *Waka Technical Community College*
Timothy Johnston, *Murray State University*
Ron Klepper, *Redlands Community College*
Travis Krehbiel, *Wichita Area Technical College*
Cynthia Littlejohn, *University of Southern Mississippi*
Ndiva Mongoh Mafany, *Sitting Bull College*
Michael Mendel, *Mount Vernon Nazarene University*
Mary Delores McCright, *Texarkana College*
Jeffery Miller, *Estrella Mountain Community College*
Victor A. Moss, *Three Rivers Community College*
Eric R. Myers, *South Suburban College*
Syed Naqui, *South Louisiana Community College*
Dr. Kathryn Stanley Podwell, *Nassau Community College*
David Riensche, *Las Positas College*
Todd Rimkus, *Marymount University*
Elizabeth Sharpe-Aparicio, *Blinn College*
Ronica Thomas, *Trendholm State Technical College*
Timothy Turner, *Allegany College of Maryland*
Beth Waters-Earhart, *Columbia College*
Jamie Welling, *South Suburban College*
James Wise, *Hampton University*
Kymbr Wright, *Columbia College*
Margaret Wright, *Baptist Bible College of Pennsylvania*
Michelle Young, *Three Rivers Community College*

Contents

Unit 2 Plant Biology

Unit 3 Maintenance of the Human Body

UNIT 4 Integration and Control of the Human Body

UNIT 5 Continuance of the Species

Unit 7 Behavior and Ecology

Science in Your Life

SCIENCE IN YOUR LIFE ▶ BIOETHICAL

SCIENCE IN YOUR LIFE ▶ HEALTH

SCIENCE IN YOUR LIFE ▶ ECOLOGY

SCIENCE IN YOUR LIFE ▶ SCIENTIFIC INQUIRY

1 The Study of Life

CHAPTER OUTLINE

CASE STUDY Japan Nuclear Power Plant Crisis

On March 11, 2011, a magnitude 8.9 earthquake struck the coast of Japan, causing the shutdown of 11 of Japan's nuclear power reactors, including reactors 1, 2, and 3 at the Fukushima Daiichi power plant. Shortly after the shutdown, the pressure within reactor 1 built up to twice that of normal levels. In an attempt to relieve pressure, steam-containing radiation was vented from the reactors. Over the next several days, the plant continued to have problems and the levels of radiation in the surrounding areas continued to rise. Shortly thereafter, Japanese officials formed a 10-kilometer (km) evacuation zone, which, over the next few days, was increased to 20 km, resulting in the relocation of close to 390,000 people.

The United States expressed concerns about the possibility of a complete meltdown of the reactor core and the potential release of radioactive cesium 137 (^{137}Cs) into the atmosphere. Since the prevailing jet streams blow from Japan across the Pacific Ocean to the West Coast of the United States, any release of ^{137}Cs had the potential of reaching North America. In fact, readings in San Francisco and Seattle detected small amounts of ^{137}Cs. Concerns were that individuals may be exposed to ^{137}Cs by inhalation or ingesting radioactive particles in food and water. Exposure to ^{137}Cs may lead to an increased risk of cancer, serious burns, and, if the level of exposure is significant enough, even death.

In this chapter, we will learn the basic characteristics of life and how all life is connected. We will explore the process scientists use to determine how the radiation from the Japanese power plant crisis will impact not only people in Japan, but the rest of the world as well.

As you read through the chapter, think about the following questions:

1. How would scientists design a test to determine the level of radiation exposure from the power plant fallout?
2. What levels of biological organization may this accident influence, both in Japan and the United States?

1.1 The Characteristics of Life

Learning Outcomes

Upon completion of this section, you should be able to
1. Identify the basic characteristics of life.
2. Distinguish between the levels of biological organization.

Life. Except for the most desolate and forbidding regions of the polar ice caps, planet Earth is teeming with life. Without life, our planet would be nothing but a barren rock hurtling through space. The variety of life on Earth is staggering, and humans are a part of it. So are butterflies, trees, birds, mushrooms, and giraffes (Fig. 1.1). The variety of living things ranges in size from bacteria, much too small to be seen by the naked eye, all the way up to giant sequoia trees that can reach heights of 100 meters (m) or more.

The diversity of life seems overwhelming, and yet all living things have certain characteristics in common. Taken together, these characteristics give us insight into the nature of life and help us distinguish living things from nonliving things. Living things generally (1) are organized, (2) acquire materials and energy, (3) reproduce, (4) respond to stimuli, (5) are homeostatic, (6) grow and develop, and (7) have the capacity to adapt to their environment. Let us consider each of these characteristics in order.

MP3 Life Characteristics

Organisms Are Organized

Organisms can be organized in a hierarchy of levels (Fig. 1.2). In trees, humans, and all other organisms, atoms join together to form molecules, such as DNA molecules that occur within cells. A **cell** is the smallest unit of life, and some organisms are single-celled. In multicellular organisms, a cell is the smallest

Masai giraffes
monarch butterfly feeding on nectar
giant sequoia
planet Earth
mushroom on northern forest floor
humans in a city
male peacock displaying feathers

Figure 1.1 Living things on planet Earth. If aliens ever visit our corner of the universe, they will be amazed at the diversity of life on our planet. They will also note that it is the only known place where oxygen is present in large quantities in the atmosphere. This is one of the criteria for the diversity of modern life as we know it.

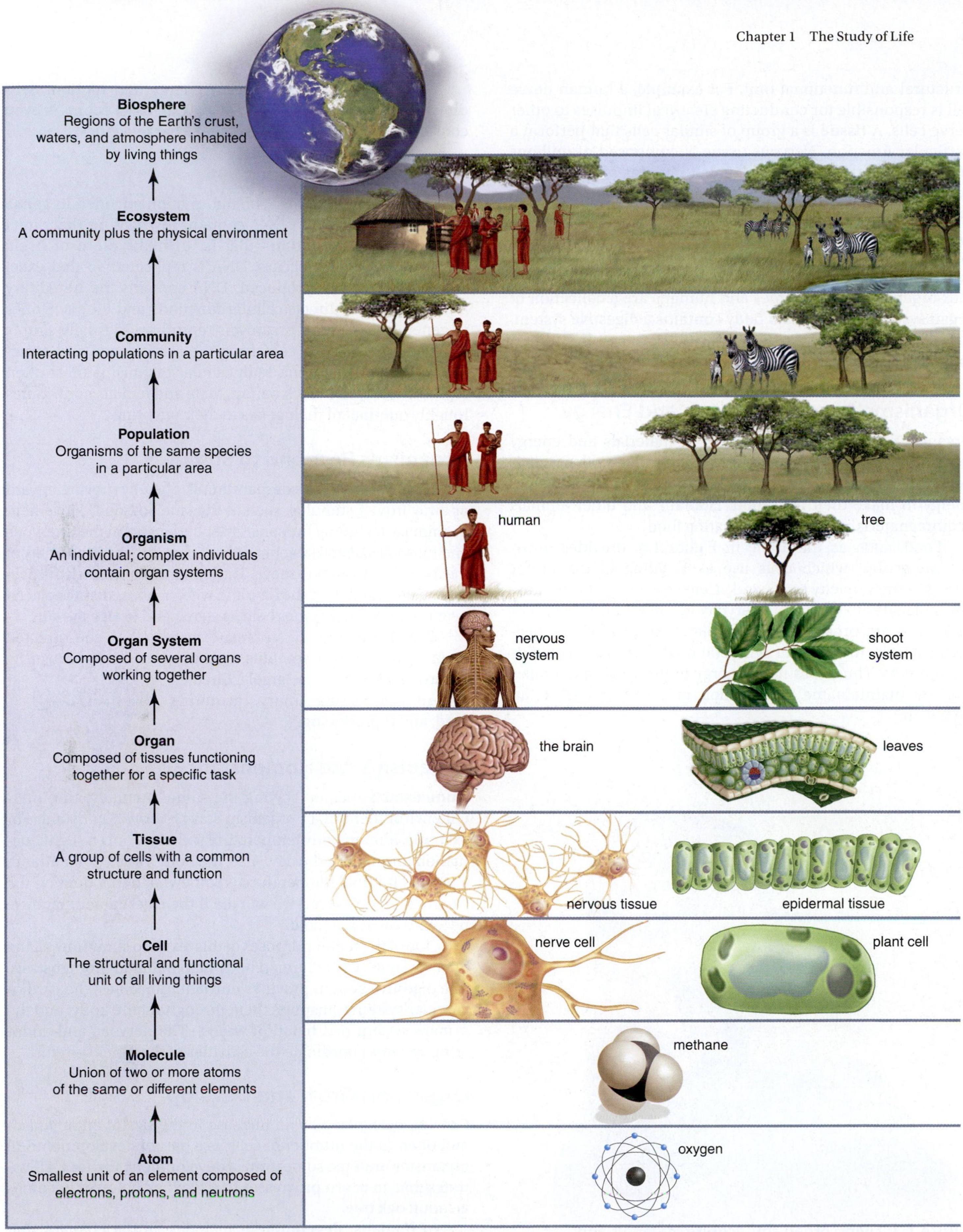
Biosphere
Regions of the Earth's crust, waters, and atmosphere inhabited by living things
Ecosystem
A community plus the physical environment
Community
Interacting populations in a particular area
Population
Organisms of the same species in a particular area
Organism
An individual; complex individuals contain organ systems
Organ System
Composed of several organs working together
Organ
Composed of tissues functioning together for a specific task
Tissue
A group of cells with a common structure and function
Cell
The structural and functional unit of all living things
Molecule
Union of two or more atoms of the same or different elements
Atom
Smallest unit of an element composed of electrons, protons, and neutrons
human
tree
nervous system
shoot system
the brain
leaves
nervous tissue
epidermal tissue
nerve cell
plant cell
methane
oxygen

Figure 1.2 Levels of biological organization. At the lowest level of organization, atoms join to form molecules (including DNA) that are found in cells. Cells, the smallest units of life, differ in size, shape, and function. Tissues, groups of cells that work together to serve the same function, may be arranged into organs. An organ system has a shared function in the body. Organs and organ systems perform functions that keep the organism alive. Populations consist of organisms of the same species. Communities represent interactions between populations. The community plus the environment forms the ecosystem.

structural and functional unit. For example, a human nerve cell is responsible for conducting electrical impulses to other nerve cells. A **tissue** is a group of similar cells that perform a particular function. Nervous tissue is composed of millions of nerve cells that transmit signals to all parts of the body. Several tissues then join together to form an **organ.** The main organ that receives signals from nerves is the brain. Organs then work together to form an **organ system.** In the nervous system, the brain sends messages to the spinal cord, which in turn sends them to body parts through spinal nerves. Complex organisms such as trees and humans are a collection of organ systems. The human body contains a digestive system, a cardiovascular system, and several other systems in addition to the nervous system.

Organisms Acquire Materials and Energy

Organisms need an outside source of materials and energy to maintain their organization and carry on life's other activities. Plants, such as trees, use carbon dioxide, water, and solar energy to make their own food. Humans and other animals acquire materials and energy by eating food.

Food, such as the apple in Figure 1.3, provides nutrient molecules, which cells use as building blocks or for **energy**—the capacity to do work. Cells use energy from nutrient molecules to carry out everyday activities. Some nutrient molecules are broken down completely to provide the necessary energy to carry out synthetic reactions, such as building proteins. Thus, cells need energy to do work, and it takes work to maintain the organization of a cell as well as an organism.

Figure 1.3 Living organisms acquire materials and energy. For example, people eat fruit as a source of energy.

Many organisms can convert energy into motion. Self-directed movement, as when we walk across a room, is even considered by some to be one of life's characteristics.

Organisms Reproduce

Life comes only from life. The information encoded in **genes** within each individual's DNA contains instructions that direct organismal **reproduction**—that is, to make more of itself. Before reproduction occurs, DNA is replicated so that exact copies of genes are produced. DNA contains the hereditary information that directs cellular functions and life's activities in general. Unicellular organisms reproduce asexually simply by dividing. The new cells have the same genes and structure as the single parent. Multicellular organisms most often reproduce sexually. Each parent, male and female, contributes roughly one-half of their genes to their offspring.

Organisms Respond to Stimuli

Organisms respond to external stimuli, often by moving toward or away from a stimulus, such as the smell of food. Movement in animals, including humans, is dependent upon their nervous and musculoskeletal systems. Other living things use a variety of mechanisms in order to move. The leaves of plants track the passage of the sun during the day, and when a houseplant is placed near a window, hormones help its stem bend to face the sun.

The movement of an organism, whether self-directed or in response to a stimulus, constitutes a large part of its **behavior.** Behavior is largely directed toward minimizing injury, acquiring food, and reproducing.

Video
Bat Echolocation

Organisms Are Homeostatic

Homeostasis means "staying the same." Actually, the internal environment of an organism stays *relatively* constant. For example, human body temperature will show only a slight fluctuation throughout the day. Also, the body's ability to maintain a normal internal temperature is somewhat dependent on the external temperature—we will die if the external temperature becomes too hot or cold.

One of the major purposes of this text is to show how all the organ systems of the human body help maintain homeostasis. The digestive system provides nutrient molecules; the cardiovascular system transports them throughout the body; and the urinary system rids blood of wastes. The nervous and endocrine systems coordinate the activities of the other systems.

Organisms Grow and Develop

Growth, recognized by an increase in the size of an organism and often in the number of cells, is a part of development. All organisms undergo some form of development. Figure 1.4 illustrates that an acorn progresses to a seedling before it becomes an adult oak tree.

In humans, **development** includes all the changes that take place between conception and death. First, the fertilized egg develops into a newborn, and then a human goes through childhood, adolescence, adulthood, and aging. Development also includes repair that takes place following an injury.

Organisms Have the Capacity to Adapt

Throughout the nearly 4 billion years that life has been on Earth, the environment has constantly been changing. For example, glaciers that once covered much of the world's surface 10,000–15,000 years ago have since receded, and many areas that were once covered by ice are now habitable. On a smaller scale, a hurricane or fire could drastically change the landscape in an area quite rapidly.

As the environment changes, some individuals of a species (a group of organisms that can successfully interbreed and produce fertile offspring) may possess certain features that make them better suited to the new environment. We call such features **adaptations.** For example, consider a hawk , which can catch and eat a rabbit. A hawk, like other birds, can fly because it has hollow bones, which is an adaptation. Similarly, its strong feet can take the shock of a landing after a hunting dive, and its sharp claws can grab and hold onto prey.

Video
Harris Hawk

Individuals of a species that are better adapted to their environment tend to live longer and produce more offspring than other individuals. This differential reproductive success, called **natural selection,** results in changes in the characteristics of a population (all the members of a species within a particular area) through time. That is, adaptations that result in higher reproductive success tend to increase in frequency in a population from one generation to the next. This change in the frequency of traits in populations and species is called **evolution.**

Evolution explains both the unity and diversity of life. As stated at the beginning of this chapter, all organisms share the same basic characteristics of life because we all share a common ancestor—the first cell or cells—that arose nearly 4 billion years ago. During the past 4 billion years, the Earth's environment has changed drastically, and the diversity of life has been shaped by the evolutionary responses of organisms to these changes.

Check Your Progress 1.1

1. List the common characteristics of all living organisms.
2. Explain how adaptations relate to evolutionary change.

Figure 1.4 Living things grow and develop. A small acorn gives rise to a very large oak tree through the process of growth and development.

1.2 The Classification of Organisms

Learning Outcomes

Upon completion of this section, you should be able to

1. Describe how living things are classified.
2. Distinguish between the three domains of life.

Because life is so diverse, it is helpful to have a classification system to group organisms based upon their similarities.

Systematics is the discipline of identifying and classifying organisms according to specific criteria. Each type of organism is placed in a **domain, kingdom, phylum, class, order, family, genus,** and finally **species.**

Domains

Domains are the largest classification category. Based upon biochemical and genetic evidence scientists have identified three domains: **domain Archaea, domain Bacteria,** and **domain Eukarya** (Fig. 1.5). Both domain Archaea and domain Bacteria contain unicellular **prokaryotes,** which lack the membrane-bounded nucleus found in the cells of **eukaryotes** in domain Eukarya. The genes of eukaryotes are found in the nucleus. Prokaryotes are structurally simple (Fig. 1.5*a, b*) but metabolically complex. The *Archaea* live in aquatic environments that lack oxygen or are too salty, too hot, or too acidic for most other organisms. Perhaps these environments are similar to those of the primitive Earth, and archaea represent the first cells that evolved. *Bacteria* are found almost anywhere—in the water, soil, and atmosphere, as well as on our skin and in our digestive tracts. Although some bacteria cause diseases, others are beneficial, both environmentally and commercially. For example, bacteria can be used to develop new medicines, to clean up oil spills, or to help purify water in sewage treatment plants.

Animation Three Domains

Kingdoms

Systematists are in the process of deciding how to categorize archaea and bacteria into kingdoms. The *eukaryotes* are currently classified into at least four kingdoms with which you may be familiar (Fig. 1.5*c*). **Protists** (kingdom Protista) range from unicellular to a few multicellular organisms. Some can make their own food (photosynthesizers), while others must ingest their food. Because of the great diversity among protists, however, there is debate among biologists on how to split the Protista into several kingdoms. Currently Protista is being split into various supergroups to more accurately represent their evolutionary relationships. Among the **fungi** (kingdom Fungi) are the familiar molds and mushrooms that, along with bacteria, help decompose dead organisms. **Plants** (kingdom Plantae) are well known as multicellular photosynthesizers, while **animals** (kingdom Animalia) are multicellular and ingest their food.

Other Categories

The other classification categories are phylum, class, order, family, genus, and species. Each classification category is more specific than the one preceding it. For example, the species within one genus share very similar characteristics, while those within the same kingdom share only general characteristics. Modern humans are the only living species in the genus *Homo,* but many different types of animals are in the animal kingdom (Table 1.1). To take another example, all species in the genus *Pisum* (pea plants) look quite similar, while the species in the plant kingdom can be quite different, as is evident when we compare grasses to trees.

TABLE 1.1 Classification of Humans

Classification Category	Characteristics
Domain Eukarya	Cells with nuclei
Kingdom Animalia	Multicellular, motile, ingestion of food
Phylum Chordata	Dorsal supporting rod and nerve cord
Class Mammalia	Hair, mammary glands
Order Primates	Adapted to climb trees
Family Hominidae	Adapted to walk erect
Genus *Homo*	Large brain, tool use
Species *Homo sapiens**	Body characteristics similar to modern humans

* To specify an organism, you must use the full binomial name, such as *Homo sapiens.*

Systematics helps biologists make sense out of the bewildering variety of life on Earth because organisms are classified according to their presumed evolutionary relationships. Organisms placed in the same genus are the most closely related, and those in separate domains are the most distantly related. Therefore, all eukaryotes are more closely related to one another than they are to bacteria or archaea. Similarly, all animals are more closely related to one another than they are to plants. As more is learned about evolutionary relationships among species, systematic relationships are changed. Systematists are even now making observations and performing experiments that will soon result in changes in the classification system adopted by this text.

Scientific Names

Within the broad field of systematics, **taxonomy** is the assignment of a binomial, or two-part name, to each species. For example, the scientific name for human beings is *Homo sapiens,* and for the garden pea, *Pisum sativum.* The first word is the genus to which the species belongs, and the second word is the specific epithet, or species name. (Note that both words are in italics, but only the genus name is capitalized.) The genus name can be used alone to refer to a group of related species. Also, a genus can be abbreviated to a single letter if used with the species name (e.g., *P. sativum*).

Scientific names are in a common language—Latin—and biologists use them universally to avoid confusion. Common names, by contrast, tend to overlap across multiple species.

Check Your Progress 1.2

1. Recognize the main criteria for classification of organisms into domains and kingdoms.
2. List the levels of taxonomic classification from most inclusive to least inclusive.
3. Explain why scientists assign species to a hierarchical classification system (e.g., kingdom, phylum, class).

DOMAIN ARCHAEA

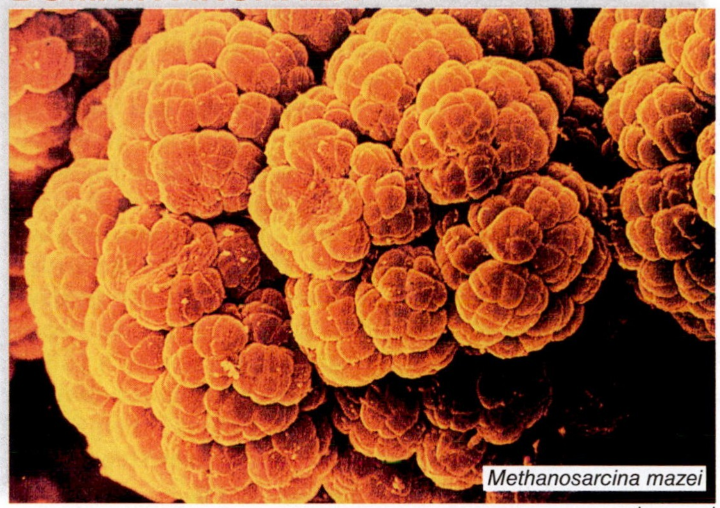
Methanosarcina mazei
1.6 μm

a. Archaea are capable of living in extreme environments.

DOMAIN BACTERIA

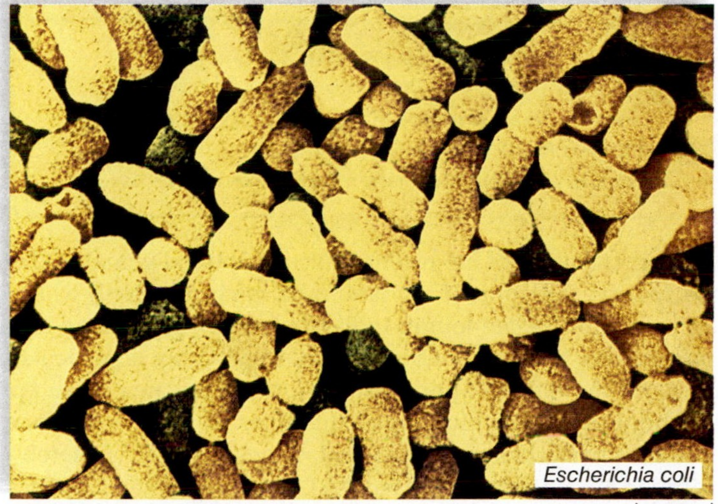
Escherichia coli
1.5 μm

b. Bacteria are found nearly everywhere.

DOMAIN EUKARYA

Kingdom	Organization	Type of Nutrition	Representative Organisms	
Protista	Complex single cell, some multicellular	Absorb, photosynthesize, or ingest food	 paramecium, euglenoid, slime mold, dinoflagellate	Protozoans, algae, water molds, and slime molds
Fungi	Some unicellular, most multicellular filamentous forms with specialized complex cells	Absorb food	 black bread mold, yeast, mushroom, bracket fungus	Molds, yeasts, and mushrooms
Plantae	Multicellular form with specialized complex cells	Photosynthesize food	 moss, fern, pine tree, nonwoody flowering plant	Mosses, ferns, nonwoody and woody flowering plants
Animalia	Multicellular form with specialized complex cells	Ingest food	 sea star, earthworm, finch, human	Invertebrates, fishes, reptiles, amphibians, birds, and mammals

c. Eukaryotes are divided into four kingdoms.

Figure 1.5 Pictorial representation of the three domains of life. Archaea **(a)** and bacteria **(b)** are both prokaryotes but are so biochemically different that they are not believed to be closely related. **(c)** Eukaryotes are biochemically similar but structurally dissimilar. Therefore, they have been categorized into four kingdoms. Many protists are unicellular, but the other three kingdoms are characterized by multicellular forms.

1.3 The Organization of the Biosphere

Learning Outcomes

Upon completion of this section, you should be able to

1. Distinguish among populations, communities, ecosystems, and the biosphere.
2. List ways in which humans influence ecosystems.

The organization of life extends beyond the individual to the **biosphere,** the zone of air, land, and water on Earth where living organisms are found. Individual organisms belong to a **population,** all the members of a species within a particular area. All the different populations in the same area make up a **community,** in which organisms interact with each other and the physical environment (soil, atmosphere, etc.), thereby forming an **ecosystem** (see Fig. 1.2).

Figure 1.6 depicts a grassland inhabited by populations of rabbits, mice, snakes, hawks, and various types of plants. These populations exchange gases with and give off heat to the atmosphere. They also take in and give off water to the physical environment. In addition, the populations interact with each other by forming food chains in which one population feeds on another. For example, mice feed on plants and seeds, snakes feed on mice, and hawks feed on rabbits and snakes. The interactions among the various food chains make up a food web.

Ecosystems are characterized by chemical cycling and energy flow, both of which begin when photosynthetic plants, algae, and some bacteria take in solar energy and inorganic nutrients to produce food in the form of organic nutrients. The gray arrows in Figure 1.6 represent chemical cycling—chemicals move from one species to another in a food chain, until with death and decomposition, inorganic nutrients are returned to living plants once again. The yellow to red arrows represent energy flow. Energy flows from the sun through plants and other members of the food chain as one species feeds on another. With each transfer, most energy is lost as heat. Eventually, all the energy taken in by photosynthesizers has dissipated into the atmosphere. Because energy flows instead of cycling, ecosystems could not stay in existence without a constant input of solar energy and the ability of photosynthesizers to absorb it.

Video Tallgrass Prairie Ecology

Climate largely determines where different ecosystems are found around the globe. For example, deserts are found in areas of almost no rain, grasslands require a minimum amount of rain, and forests generally require much rain. The two most biologically diverse ecosystems—tropical rain forests and coral reefs—occur where solar energy is most abundant. Coral reefs, which are found just offshore from the continents and islands of the Southern and Northern Hemispheres, are built up from the skeletons of sea animals called corals. Inside the tissues of

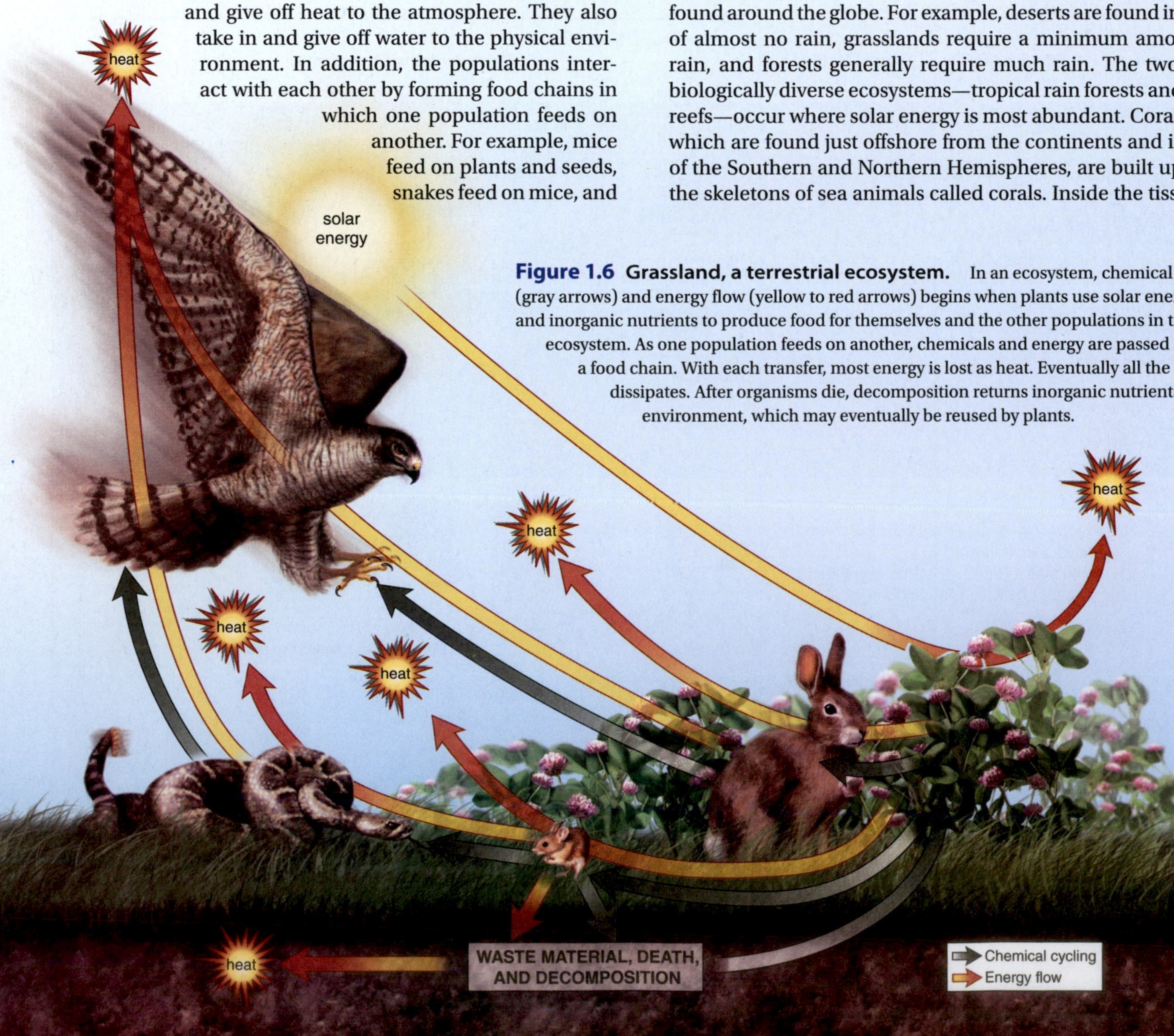
heat
solar energy
heat
heat
heat
heat
heat
WASTE MATERIAL, DEATH, AND DECOMPOSITION
Chemical cycling
Energy flow

Figure 1.6 Grassland, a terrestrial ecosystem. In an ecosystem, chemical cycling (gray arrows) and energy flow (yellow to red arrows) begins when plants use solar energy and inorganic nutrients to produce food for themselves and the other populations in the ecosystem. As one population feeds on another, chemicals and energy are passed along a food chain. With each transfer, most energy is lost as heat. Eventually all the energy dissipates. After organisms die, decomposition returns inorganic nutrients to the environment, which may eventually be reused by plants.

corals are tiny, one-celled protists that carry on photosynthesis and provide food to their hosts. Reefs provide a habitat for many other animals, including jellyfish, sponges, snails, crabs, lobsters, sea turtles, moray eels, and some of the world's most colorful fishes (Fig. 1.7).

The Human Species

The human species tends to modify existing ecosystems for its own purposes. Humans clear forests or grasslands in order to grow crops; later, they build houses on what was once farmland; and finally, they convert small towns into cities. As coasts are developed, humans send sediments, sewage, and other pollutants into the sea.

Like tropical rain forests, coral reefs are severely threatened as the global human population increases in size. Some reefs are 50 million years old, and yet in just a few decades, human activities have destroyed or severely degraded nearly 25% of all coral reefs and another two-thirds are threatened. At this rate, nearly three-quarters could be destroyed within 50 years. Similar levels of destruction have occurred in tropical rain forests.

Video Coral Reef Ecosystems

It has long been clear that humans depend on healthy ecosystems for food, medicines, and various raw materials. We are only now beginning to realize that we depend on them even more for the services they provide. Just as chemical cycling occurs within ecosystems, so do ecosystems keep chemicals cycling throughout the entire biosphere. The dynamics of ecosystems ensure that the environmental conditions of the biosphere are suitable for the continued existence of humans. In turn, many biologists believe that ecosystems cannot function properly unless they remain biologically diverse.

Biodiversity

Biodiversity encompasses the total number of species, the variability of their genes, and the ecosystems in which they live. The present biodiversity of our planet has been estimated at as high as 5–30 million species, and so far, about 1.8 million have been identified and named. Extinction is the permanent loss of a species or larger taxonomic group. Human activities are causing the extinction of about 400 species per day.

great barracuda
attached algae
reef shark
green moray
yellowtail snapper
yellow jack
Spanish hogfish
foureye butterfly fish
corals
bar jack
queen angelfish
sponges
parrotfish
sea urchin
Bermuda chub
spiny lobster
surgeonfish
phytoplankton
zooplankton
yellowtail damselfish
sea grass
sea star
detritus

Figure 1.7 Coral reef, a marine ecosystem. Coral reefs, a type of ecosystem found in tropical seas, contain many diverse forms of life, a few of which are shown here. Coral reefs are now threatened because of certain human activities. Saving such biodiversity is one of our greatest modern-day challenges.

For example, several species of fishes have all but disappeared from the coral reefs of Indonesia and along the African coast because of overfishing. Many biologists are alarmed about the present rate of extinction and believe it has surpassed the rates of the five mass extinctions that have occurred during our planet's history. The dinosaurs became extinct during the last mass extinction, 65 million years ago.

Most biologists agree that a primary bioethical issue of our time is conservation of biodiversity. Just as a native fisherman who assists in overfishing a reef is doing away with his own food source, so we as a society are contributing to the destruction of our home, the biosphere. If instead we adopt an ethic that conserves the biosphere, we are helping to ensure the continued existence of its species.

Check Your Progress 1.3

1. Explain the relationship between a population, a community, and an ecosystem.
2. Describe some unintentional ways in which human activities affect ecosystems.
3. Discuss why ecosystems with high biodiversity might be more vulnerable to destruction by human activities.

1.4 The Process of Science

Learning Outcomes

Upon completion of this section, you should be able to

1. Distinguish between a theory and a hypothesis.
2. Identify the components of the scientific method.
3. Analyze a scientific experiment and identify the hypothesis, experiment, control groups, and conclusions.

Biology is the scientific study of life. Biologists can be found almost anywhere studying life-forms. Often, they perform experiments, either in the field (outside the laboratory) or in the laboratory. Biologists—and all scientists—generally test hypotheses using the scientific method (Fig. 1.8).

Observation

Scientists tend to be curious about nature and how the world works. They believe that natural **phenomena** can be understood more fully by observing and studying them. Scientists use all their senses to make **observations.** We can observe with our noses that dinner is almost ready and our ears that a piano needs tuning. Scientists also extend the ability of their senses by using instruments. For example, a microscope enables us to see objects too small to view with the naked eye. Finally, scientists may expand their understanding even further by taking advantage of the knowledge and experiences of other scientists. For instance, they may look up previous studies in scientific journals, or they may attend scientific meetings and hear presentations from others researching similar topics.

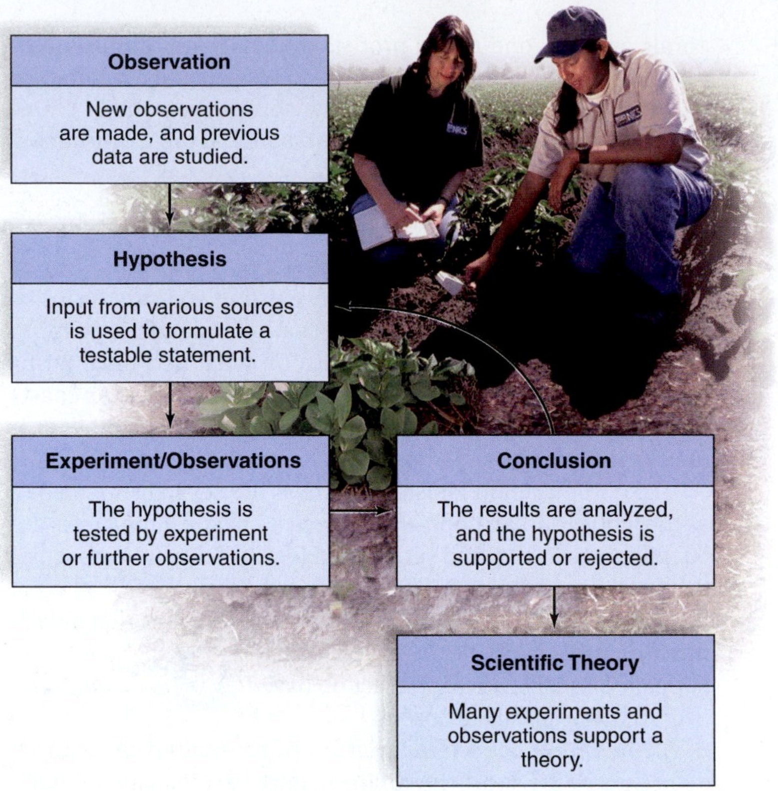
Observation
New observations are made, and previous data are studied.
Hypothesis
Input from various sources is used to formulate a testable statement.
Experiment/Observations
The hypothesis is tested by experiment or further observations.
Conclusion
The results are analyzed, and the hypothesis is supported or rejected.
Scientific Theory
Many experiments and observations support a theory.

Figure 1.8 Flow diagram for the scientific method. On the basis of new and/or previous observations, a scientist formulates a hypothesis. The scientist then tests the hypothesis by making further observations and/or performing experiments. The resulting new data either support or do not support the hypothesis. The return arrow indicates that a scientist often chooses to retest the same hypothesis or to test an alternative hypothesis. Conclusions from many different but related experiments may lead to the development of a scientific theory. For example, studies of development, anatomy, and fossil remains all support the theory of evolution.

Chance alone can help a scientist get an idea. A famous case pertains to penicillin. When examining a petri dish, Alexander Fleming observed an area around a mold that was free of bacteria. Upon investigating, Fleming found that the mold produced an antibacterial substance he called penicillin. This observation led Fleming to think that perhaps penicillin would be useful to humans.

The classic vision of the scientific method is that observations lead to hypotheses and that these, in turn, lead to experimentally testable predictions. In this way, we can evaluate new ideas in order to arrive at accurate conclusions about natural processes. However, it is important to realize many biologists study life from a descriptive point of view.

Hypothesis

After making observations and gathering knowledge about a phenomenon, a scientist uses inductive reasoning. **Inductive reasoning** occurs whenever a person uses creative thinking to combine isolated facts into a cohesive whole. In this case, the scientist comes up with a **hypothesis,** a tentative explanation for the natural event. The scientist presents this hypothesis as a falsifiable statement.

Past experiences and observations, no matter what they might be, most likely influence the formation of a hypothesis. But a scientist considers only hypotheses that can be tested. Moral and religious beliefs, while very important to our lives, differ between cultures and through time, and are not testable scientifically.

Experiment/Further Observations

Testing a hypothesis involves either conducting an experiment or making further observations. To determine how to test a hypothesis, a scientist uses deductive reasoning. **Deductive reasoning** involves "if, then" logic. For example, a scientist might reason, *if* organisms are composed of cells, *then* microscopic examination of any part of an organism should reveal cells. We can also say that the scientist has made a **prediction** that the hypothesis can be supported by doing microscopic studies. Making a prediction helps a scientist know what to do next.

Further observations may be used to test the hypothesis. Most often, however, scientists conduct experiments. The manner in which a scientist intends to conduct an experiment is called the **experimental design.** A good experimental design ensures testing is specific and that results will be meaningful. An experiment should always include a control group. A control group, or simply the **control,** goes through all the steps of an experiment but lacks the factor, or is not exposed to the factor, being tested. For example, in a drug trial the treatment group receives the actual drug being tested while the control group receives a placebo (sugar pill).

In their experiments, scientists often use a **model,** a representation of an actual subject. For example, scientists can use computer modeling to suggest future ways human activities may affect global climate change, or they may use mice as a model, instead of humans, when doing cancer research. When a medicine is effective in mice, researchers still need to test it in humans before making recommendations or using it in widespread treatment. Thus, results obtained using models are considered a hypothesis until the experiment can be performed using the actual subject.

Data

The results of an experiment are referred to as the **data.** Data should be observable and objective, rather than subjective or based on opinion. Mathematical data are often displayed in the form of a graph or table. Many studies rely on statistical data. Let's say an investigator wants to know if eating onions can prevent women from getting osteoporosis (weak, porous bones). The scientist conducts a survey asking women about their onion-eating habits and then correlates these data with the condition of their bones. Other scientists critiquing this study would want to know: How many women were surveyed? How old were the women? What were their exercise habits? What proportion of the diet consisted of onions? And what criteria were used to determine the condition of their bones? Should the investigators conclude that eating onions does protect a woman from osteoporosis, other scientists would want to know that the relationship did not occur simply by chance. The larger the sample size (in this case, the number of women studied) and the larger the effect (in this case, strength of bones in women eating more onions compared to those eating fewer onions), the less likely that the result is due to chance alone. Statistical tests are used to evaluate how likely it is that the results of an experiment are due to chance. And in the end, statistical data of this sort would only be suggestive until we know of some ingredient in onions that has a direct biochemical or physiological effect on bone strength. Statistical studies are correlation studies. The more data collected from a variety of sources allows scientists to assess the validity of their scientific method. Even so, a correlation does not necessarily translate to causation. Therefore, scientists are skeptics who always pressure one another to keep looking for additional information.

Conclusion

Scientists must analyze the data to reach a **conclusion** to determine if the hypothesis can be supported or not. Because science progresses, the conclusion of one experiment can lead to the hypothesis for another experiment, as represented by the return arrow in Figure 1.8. Sometimes, results that do not support one hypothesis can help a scientist formulate another hypothesis to be tested. Scientists report their findings in scientific journals, subject to critical review by their peers, so that their methodology and data are available to other scientists. Experiments and observations must be repeatable—that is, any scientist who repeats an experiment must get the same results, or else the data are suspect. This scientific process is continuous, and in biology scientists are discovering new things all the time.

Scientific Theory

The ultimate goal of science is to understand the natural world in terms of **scientific theories,** concepts that join together well-supported and related hypotheses. In ordinary speech, the word *theory* refers to a speculative idea. But in science, a theory is supported by a broad range of observations, experiments, and data and is thus similar to a law, such as the "theory" of gravity.

Some of the basic theories of biology are as follows:

Theory	Concept
Cell	All organisms are composed of cells, and new cells only come from preexisting cells.
Homeostasis	The internal environment of an organism stays relatively constant—within a range that is protective of life.
Evolution	A change in the frequency of traits that affect reproductive success in a population or species across generations.

When we use the word "theory" in everyday terms, we often mean a guess or a hunch. However, in scientific terms, a theory is supported by observation and repeated empirical investigation. So, a theory in science is akin to a "law," such as gravitational theory or atomic theory. The theory of evolution is the unifying concept of biology because it pertains to many different aspects of living things. The theory of evolution enables scientists to understand the history of life; the diversity of living things; and the anatomy, physiology, and development of organisms—even their behavior.

The theory of evolution has been a very fruitful theory scientifically, meaning that it has helped scientists generate new hypotheses. Because this theory has been supported by so many observations and experiments for over 100 years, some biologists refer to the **principle** of evolution, a term that refers to theories generally accepted by an overwhelming number of scientists. Other biologists even prefer to use the term **law** instead of principle.

nodules

Figure 1.9 Root nodules. Bacteria that live in nodules on the roots of legumes, such as pea plants, convert nitrogen in the soil to a form that plants can use to make proteins and other nitrogen-containing molecules.

A Controlled Study

Most investigators perform controlled studies in which some groups receive a treatment (experimental groups), while other groups receive no treatment (control). To ensure that the outcome is due to the **experimental variable** (independent variable) alone, all other conditions between the groups in the experiment are controlled to be identical. The result is called the **responding variable** (or dependent variable) because it is due to the experimental treatment.

Experimental Variable (Independent Variable)	**Response Variable** (Dependent Variable)
Factor of the experiment being tested	Result or change that occurs due to the experimental variable

In the following study, researchers performed an experiment in which nitrogen fertilizer is the experimental variable and crop yield is the response variable. Nitrogen fertilizer in the short run has been known to enhance yield and increase food supplies. However, excessive nitrogen application can cause pollution by adding toxic levels of nitrates to water supplies. Also, applying nitrogen fertilizer year after year may alter soil properties such that crop yields may decrease instead of increase. A solution is to let the land remain unplanted for several years until the soil recovers naturally.

An alternative to nitrogen fertilizers is the use of legumes, plants such as peas and beans, that provide a home for bacteria that convert atmospheric nitrogen to a form usable by plants. The bacteria live in nodules on the roots (Fig. 1.9). The products of photosynthesis move from the leaves to the root nodules. In turn, the nodules supply the plant with nitrogen compounds the plant can use to make proteins.

Numerous legume crops can be rotated (planted every other season) with any number of cereal crops. The nitrogen added to the soil by the legume crop increases cereal crop yield. The particular rotation used depends on the location, climate, and market demand.

The Experiment

The pigeon pea plant is a legume with a high rate of atmospheric nitrogen conversion. This plant is widely grown as a food crop in India, Kenya, Uganda, Pakistan, and other subtropical countries. Researchers formulated the hypothesis that rotating pigeon pea and winter wheat crops would be a reasonable alternative to using nitrogen fertilizer to increase the yield of winter wheat.

HYPOTHESIS: A pigeon pea/winter wheat rotation will cause winter wheat production to increase as well as or better than the use of nitrogen fertilizer.

PREDICTION: Wheat production (biomass) following the growth of pigeon peas will surpass wheat biomass following nitrogen fertilizer treatment.

In this study, the investigators decided on the following experimental design (Fig. 1.10*a*):

Control Pots

- Winter wheat was planted in pots of soil that received no fertilization treatment (i.e., no nitrogen fertilizer and no preplanting of pigeon peas).

Test Pots

- Winter wheat was grown in clay pots in soil treated with nitrogen fertilizer equivalent to 45 kilograms (kg)/hectare (ha).
- Winter wheat was grown in clay pots in soil treated with nitrogen fertilizer equivalent to 90 kg/ha.
- Pigeon pea plants were grown in clay pots in the summer. The pigeon pea plants were then tilled into the soil, and winter wheat was planted in the same pots.

Control pot
no fertilization treatment
Test pot
90 kg of nitrogen/ha
Test pot
45 kg of nitrogen/ha
a. Control pot and three types of test pots

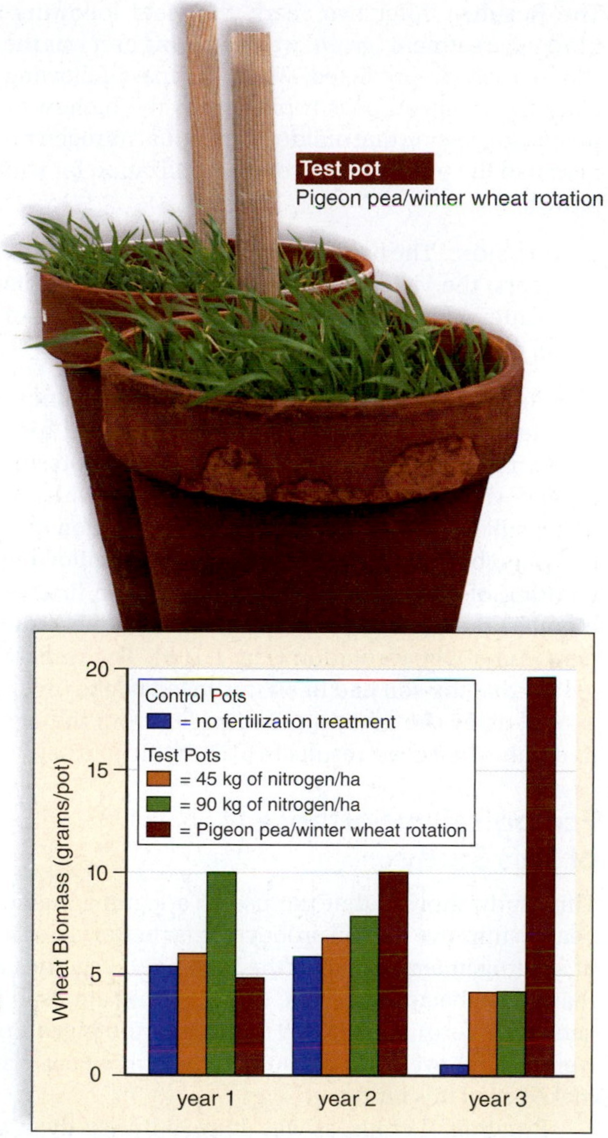
Test pot
Pigeon pea/winter wheat rotation
Control Pots
= no fertilization treatment
Test Pots
= 45 kg of nitrogen/ha
= 90 kg of nitrogen/ha
= Pigeon pea/winter wheat rotation
Wheat Biomass (grams/pot)
20
15
10
5
0
year 1
year 2
year 3
b. Results

Figure 1.10 **Pigeon pea/winter wheat rotation study.** **a.** The experiment involves control pots that receive no fertilization and three types of test pots: test pots that received 45 kg/ha of nitrogen; test pots that received 90 kg/ha of nitrogen; and test pots in which pigeon peas rotated with winter wheat. **b.** The graph compares wheat biomass for each of three years. Wheat biomass in test pots that received the most nitrogen fertilizer declined, while wheat biomass in test pots with pigeon pea/winter wheat rotation increased dramatically.

To ensure a controlled experiment, the conditions for the control pots and the test pots were identical with the exception of the treatments noted. The plants were exposed to the same environmental conditions and watered equally. During the following spring, the wheat plants were dried and weighed to determine total wheat production (biomass) in each of the pots.

The Results After the first year, wheat biomass was higher in certain test pots than in the control pots (Fig. 1.10*b*). Specifically, test pots with 45 kg/ha of nitrogen fertilizer (*orange*) had only slightly more wheat biomass production than the control pots, but test pots that received 90 kg/ha treatment (*green*) had nearly twice the biomass production of the control pots. Thus, application of nitrogen fertilizer had the potential to increase wheat biomass. To the surprise of investigators, wheat production following summer planting of pigeon peas did not result in as high a biomass production as the control pots.

CONCLUSION: The hypothesis is not supported. Wheat biomass following the growth of pigeon peas is not as high as that obtained with nitrogen fertilizer treatments.

Continuing the Experiment

The researchers decided to continue the experiment using the same design and the same pots as before, to see if the buildup of residual soil nitrogen from pigeon peas would eventually increase wheat biomass. They formed a new hypothesis.

HYPOTHESIS: A sustained pigeon pea/winter wheat rotation will eventually cause an increase in winter wheat production.

PREDICTION: Wheat biomass following two years of pigeon pea/winter wheat rotation will surpass wheat biomass following nitrogen fertilizer treatment.

The Results After two years, the yield following 90 kg/ha nitrogen treatment (*green*) was not as big as it was the first year (Fig. 1.10*b*). As predicted, wheat biomass following summer planting of pigeon peas (*brown*) was the highest of all treatments, suggesting that buildup of residual nitrogen from pigeon peas had the potential to provide fertilization for winter wheat growth.

CONCLUSION: The hypothesis was supported. At the end of two years, the yield of winter wheat following a pigeon pea/winter wheat rotation was better than for the other types of pots.

The researchers continued their experiment for still another year. After three years, winter wheat biomass production had decreased in both the control pots and the pots treated with nitrogen fertilizer. Pots treated with nitrogen fertilizer still had more wheat biomass production than the control pots, but not nearly as much as pots following summer planting of pigeon peas. Compared to the first year, wheat biomass increased almost fourfold in pots having a pigeon pea/winter wheat rotation (Fig. 1.10*b*). The researchers suggested that the soil had been improved by the organic matter, as well as by the addition of nitrogen from the pigeon peas, and published their results in a scientific journal.[1]

Ecological Importance of This Study

This study showed that the use of a legume, namely pigeon peas, to improve the soil produced a far better yield than the use of a nitrogen fertilizer over the long haul. Legumes are plants that house bacteria capable of converting atmospheric nitrogen into a form a plant can use. When the pigeon pea plants were turned over into the soil, the winter wheat plants could make use of this nitrogen.

Rotation of crops, as was done in this study, is an important part of organic farming. Sometimes there are adverse economic results when farmers switch from chemical-intensive to organic farming practices. But, because the benefit of making the transition may not be realized for several years, some researchers advocate that the process be gradual. This study suggests, however, that organic farming will eventually be beneficial both for farmers and the environment!

Check Your Progress 1.4

1. Distinguish between the roles of the test group and the control group in an experiment.
2. Identify the role of the experimental variable in an experiment.
3. Describe the process by which a scientist may test a hypothesis about an observation.

[1] Bidlack, J. E., Rao, S. C., and Demezas, D. H. 2001. Nodulation, nitrogenase activity, and dry weight of chickpea and pigeon pea cultivars using different *Bradyrhizobium* strains. *Journal of Plant Nutrition* 24:549–60.

1.5 Science and Social Responsibility

Learning Outcomes

Upon completion of this section, you should be able to

1. Identify the costs and benefits of technology.
2. Explain what could happen to the human population if we stopped using technology.

Many scientists are engaged in fields of study that sometimes seem remote from our everyday lives. Other scientists are interested in using the findings of past and present scientists to produce a product or develop a technique that affects our lives. The application of scientific knowledge for a practical purpose is called technology. For example, virology, the study of viruses and molecular chemistry, led to the discovery of new drugs that extend the life span of people who have HIV/AIDS. Cell biology research has allowed physicians to develop various cancer treatments.

Most technologies have benefits but also drawbacks. Research has led to modern agricultural practices that help to feed the burgeoning human population. However, the use of nitrogen fertilizers leads to water pollution, and the use of pesticides, as you may know, kills not only pests but also beneficial organisms.

Who should decide how, and even whether, a technology should be put to use? Making value judgments is not a part of science. Ethical and moral decisions must be made by all people. Therefore, the responsibility for how to use scientific technology must reside with people from all walks of life, not with scientists alone. Presently, we need to decide whether we want to stop producing bioengineered organisms that may be harmful to the environment. Also, through gene therapy, we are developing the ability to cure diseases and the possibility of altering the genes of our offspring. Perhaps one day we might even be able to clone ourselves. Should we do these things? So far, as a society, we continue to believe in the sacredness of human life, and therefore we have passed laws against human cloning. Even if the procedure is perfected, we may also continue to rule against human cloning.

The Health feature, "How Safe Is Your Cell Phone," discusses the safety concerns associated with cell phone usage. Each of us must wrestle with this and the other bioethical issues discussed in this text, in order to make better informed decisions that are beneficial to society overall.

Check Your Progress 1.5

1. Technological advances from scientific discoveries can help people but may also harm the environment. Identify an example of such a discovery and discuss the pros and cons.
2. Explain the potential consequences to the human population if we were to stop using technology.

SCIENCE IN YOUR LIFE ▶ HEALTH

How Safe Is Your Cell Phone?

The frequency of cell phone use in the United States has risen from 110 million users in 2000 to over 285 million users in 2009. The National Cancer Institute has raised concerns about the increased frequency of usage due to the fact that cell phones emit radiofrequency (RF) energy. Radiofrequency energy is a form of nonionizing electromagnetic radiation that is being studied to determine its effects on the human body.

At this point there is no conclusive evidence that the nonionizing radiation emitted by cell phones is associated with cancer risk. The radiation that is emitted by a cell phone comes from its antenna, which is housed within the handset. Due to the new design of today's cell phones, the handset is often held close to the side of the user's head. This leads to an increased absorption of nonionizing radiation to the side of the head, neck, and face. The level of exposure to RF depends on a variety of variables, such as the frequency and duration of calls as well as the quality of transmission and the size of the handset.

Figure 1A Are cell phones safe? Several studies have not indicated any danger from radiation associated with cell phone use.

Numerous studies have been conducted to determine the relationship between cell phone use and the risk of developing malignant and benign brain tumors. Researchers reported that, overall, cell phone users did not show an increased risk for two of the most common brain tumors. However, the study group that had the greatest usage of cell phones showed an increased risk of glioma, a malignant type of brain tumor. The researchers considered this finding inconclusive.

Many of the other studies that have been conducted show a variety of discrepancies between their results. A Danish study showed no increased risk of acoustic neuroma in long-term cell phone users compared to short-term users. A Swedish study examined similar populations and discovered an elevated risk in the long-term users compared to the short-term users. So should the government ban the production and use of cell phones because of the potential for an increased risk of brain tumors? Or should people be allowed to freely use cell phones despite the potential risk?

Questions to Consider

1. At this point, there is no conclusive evidence that the radiation emitted by cell phones is associated with cancer risk. How many studies do you think are necessary to prove cell phones are or are not safe to use?
2. What type of study would you design if you were trying to prove that cell phones are safe to use?
3. Do you think that cell phones can have a negative impact upon the various levels of biological organization? If so, which levels do you think are impacted the most and in what way?

connect BIOLOGY Explore the concepts through a variety of multimedia assets, question types, and data interpretation.
www.mcgrawhillconnect.com

Case Study Conclusion

The potential consequences of the Japanese nuclear plant crisis will continue to unfold for years to come. Scientists will continue to monitor the levels of radiation surrounding the power plants to determine if they have increased to dangerous levels. If they have, the inhabitants on the surrounding areas will be at an increased risk of radiation exposure. Exposure to radiation has been shown to increase the risk of mutations in cells that can lead to cancer. Even if an individual doesn't have direct exposure to the radiation, indirect exposure, through contaminated food and water, can lead to health problems.

If large enough concentrations of radioactive elements are released into the atmosphere, it is possible that the jet stream may blow it to the western shores of the United States. Even though we are thousands of miles away from the site of the problem, we will be impacted by what has happened in Japan for many years to come.

MEDIA STUDY TOOLS

www.mhhe.com/maderinquiry14

Enhance your study of this chapter with study tools and practice tests. Also ask your instructor about the resources available through ConnectPlus, including LearnSmart, the media-rich eBook, interactive learning tools, and animations.

SUMMARIZE

1.1 The Characteristics of Life

Evolution accounts for both the diversity and the unity of life we see about us. All organisms share the following characteristics of life:

- Organization: the levels of biological organization extend as follows: atoms and **molecules** ⟶ **cells** ⟶ **tissues** ⟶ **organs** ⟶ **organ systems** ⟶ organisms ⟶ populations ⟶ communities ⟶ ecosystems. In an ecosystem, populations interact with one another and with the physical environment.
- Acquisition of materials and **energy** from the environment: they need an outside source of nutrients.
- **Reproduction:** they produce offspring that resemble themselves.
- Respond to stimuli: they react to internal and external events.
- Internal **homeostasis:** means they stay just about the same despite changes in the external environment.
- Growth and **development:** during their lives, they change—most multicellular organisms undergo various stages from fertilization to death.
- **Adaptations** to a changing environment.

1.2 The Classification of Organisms

- Organisms are classified according to evolutionary relationships into ever-more-inclusive categories: **domain, kingdom, phylum, class, order, family, genus,** and **species.** Species in different domains are only distantly related; species in the same genus are very closely related.
- Each species has a Latin scientific name that consists of the genus and species name. Both names are italicized, but only the genus is capitalized, as in *Homo sapiens.*

1.3 The Organization of the Biosphere

- Organisms belong to a **population,** defined as all members of a single species in a particular area.
- Populations interact with each other within a **community** and with the physical environment, forming an **ecosystem.**
- Ecosystems are characterized by chemical cycling and energy flow, which begin when a photosynthesizer becomes food (organic nutrients) for an animal.
- Food chains tell who eats whom in an ecosystem. As one population feeds on another, the energy dissipates, but nutrients do not. Eventually, inorganic nutrient molecules are decomposed and return to photosynthesizers, which use them plus solar energy to produce more food.
- Human activities have totally altered many ecosystems and are putting stress on most of the others. Coral reefs and tropical rain forests, for example, are quickly disappearing.
- The health of the **biosphere** is essential to the future continuance of the human species.

1.4 The Process of Science

- When studying the natural world, scientists often use the scientific method. This process involves:
  1. **Observations,** along with previous **data,** are used to formulate a **hypothesis.**
  2. New observations and/or experiments are carried out to test the hypothesis. **Experimental designs** should include a control group.
  3. The experimental and observational results are analyzed, and the scientist decides whether the results support the hypothesis or prove it false.
  4. Hypotheses can be supported and modified based on new experiments because science is always open to change.
- Several related **conclusions** resulting from similar scientific experiments may allow scientists to arrive at a **scientific theory,** such as the cell theory, the gene theory, or the theory of evolution. The theory of evolution is a unifying theory of **biology.**

1.5 Science and Social Responsibility

- Science does not consider moral or ethical questions. It is up to all of us to decide how the various technologies that grow out of basic science should be used or regulated.

ASSESS

Testing Yourself

Choose the best answer for each question.

1. Which sequence represents the correct order of increasing complexity in living systems?
   a. cell, molecule, organ, tissue
   b. organ, tissue, cell, molecule
   c. molecule, cell, tissue, organ
   d. cell, organ, tissue, molecule
2. Classification of organisms reflects
   a. similarities.
   b. evolutionary history.
   c. Neither a nor b is correct.
   d. Both a and b are correct.
3. A population is defined as
   a. the number of species in a given geographic area.
   b. all of the individuals of a particular species in a given area.
   c. a group of communities.
   d. all of the organisms in an ecosystem.
4. The ultimate source of energy in any ecosystem is (are)
   a. the sun.
   b. green plants.
   c. fungi.
   d. animals.
5. Which are the most biologically diverse ecosystems?
   a. deserts and rain forests
   b. rain forests and coral reefs
   c. deserts and coral reefs
   d. None of these are correct.

6. What is the unifying theory in biology that explains the relationships of all living things?
   - **a.** ecology
   - **b.** evolution
   - **c.** biodiversity
   - **d.** taxonomy
7. Which sequence exhibits an increasingly more-inclusive scheme of classification?
   - **a.** kingdom, phylum, class, order
   - **b.** phylum, class, order, family
   - **c.** class, order, family, genus
   - **d.** genus, family, order, class
8. The best explanation as to why "big, fierce carnivores (animals that eat other animals) are rare" would be the fact that
   - **a.** energy flows and dissipates.
   - **b.** energy is stored in living systems.
   - **c.** nutrients are cycled.
   - **d.** nutrients flow.
9. Features that make an organism suited to its way of life are called
   - **a.** ecosystems.
   - **b.** populations.
   - **c.** adaptations.
   - **d.** None of these are correct.

For questions 10–13, match the statements with the correct terms.

- **a.** photosynthetic ability
- **b.** absorb food
- **c.** ingest food
- **d.** all populations within a given area
- **e.** all the interactions of all organisms with each other and with the environment in a given area

10. Animals
11. Plants
12. Ecosystem
13. Community
14. What's the difference between a test group and a control group?
    - **a.** There is no real difference because both are part of a study.
    - **b.** The test group is testing something, but the control group is not testing anything.
    - **c.** The test group is exposed to the factor being tested, but the control group is not exposed.
    - **d.** The control group proves the results true, but the test group does not.
    - **e.** All of these are correct.
15. In the controlled study of nitrogen fertilizers and legumes (see pages 12–14),
    - **a.** pigeon peas and winter wheat were treated similarly; for example, winter wheat was kept inside to stabilize the temperature.
    - **b.** no pigeon peas were grown in the control pots, and no fertilizer was added.
    - **c.** there were two test groups; therefore, there were two test pots.
    - **d.** the investigators were surprised that the control group didn't show better results.
    - **e.** All of these are correct.

ENGAGE

Virtual Lab
Dependent and Independent Variables

The virtual lab "Dependent and Independent Variables" provides an interactive examination of the scientific process.

Thinking Critically

1. How can evolution explain both the unity and diversity of life?
2. What is the definition of life? Do you think that computers could be considered living based upon the biological definition of life?
3. You are a scientist working at a pharmaceutical company and have developed a new cancer medication that has the potential for use in humans. Outline a series of experiments, including the use of a model, to test whether the cancer medication works.

2 The Molecules of Cells

CHAPTER OUTLINE

BEFORE YOU BEGIN

Before beginning this chapter, take a few moments to review the following discussions:

Section 1.1 Why must organisms acquire materials and energy?

Figure 1.2 Where are atoms and molecules located in the levels of biological organization?

Section 1.3 How does energy flow through an ecosystem?

CASE STUDY Sports Drinks and Exercise

Water is an essential molecule for life. In the human body (which is 60-70% water), water regulates temperature, as well as serving as a transport medium for nutrients and wastes. During exercise a person can lose up to 64 ounces (oz) of fluid per hour, even more during extremely hot and humid weather. Since a loss of even 10% can have profound influence on how our bodies function, it is essential that we maintain an adequate state of hydration.

To maintain adequate hydration, people need to be aware of their own body chemistry and how they respond to exercise. In general, you should drink 8–10 ounces of water for every 10–15 minutes of exercise. Only when exercise lasts longer than 90 minutes should you consider consuming 8–10 oz of a sports drink for every 15–30 minutes of exercise.

Exercise causes our bodies to not only lose water but also deplete our supply of electrolytes and carbohydrates. This results in a drop in blood volume, forcing the heart to work harder to circulate the blood. Muscle cramps, fatigue, dizziness, and an increase in core body temperature can also result. In an attempt to avoid these issues, many athletes will consume sports drinks during exercise. There are a variety of sports drinks, but most contain water, carbohydrates, and electrolytes (salts).

Issues arise when people consume sports drinks during workouts that are less than 90 minutes. For short duration exercise there is no need to replace electrolytes or carbohydrates. Each serving of a sports drink may contain up to 200 calories, thus reducing the effectiveness of exercise for weight loss. In this chapter, we will learn about the importance of water to our lives, as well as how elements combine to form the basic organic molecules that all life is built from.

As you read through the chapter, think about the following questions:

1. What properties of water make it so essential to our lives?
2. Which elements are most abundant in living organisms?

2.1 Basic Chemistry

Learning Outcomes

Upon completion of this section, you should be able to

1. Describe how protons, neutrons, and electrons relate to atomic structure.
2. Use the periodic table to determine relationships among atomic number and mass number.
3. Describe how variations in an atomic nucleus account for its physical properties.
4. Identify the beneficial and harmful uses of radiation.

Everything—including the book you're holding, the chair you're sitting on, the water you drink, and the air you breathe—is composed of matter. **Matter** refers to anything that takes up space and has mass. Matter has many diverse forms, but it can exist only in three distinct states: solid, liquid, and gas.

All matter, both nonliving and living, is composed of certain basic substances called **elements.** An element is a substance that cannot be broken down to simpler substances with different properties by ordinary chemical means. (A property is a physical or chemical characteristic, such as density, solubility, melting point, and reactivity.) Only 92 naturally occurring elements serve as the building blocks of all matter. Other elements have been "human-made" and are not biologically important.

The Earth's crust, as well as all organisms, are composed of elements, but they differ as to which elements are predominant (Fig. 2.1). Only six elements—carbon, hydrogen, nitrogen, oxygen, phosphorus, and sulfur—are basic to life and make up about 95% of the body weight of organisms. The acronym CHNOPS helps us remember these six elements. The properties of these elements are essential to the uniqueness of cells and organisms. Other elements, including sodium, potassium, calcium, iron, and magnesium are also important to all organisms.

Atomic Structure

In the early 1800s, the English scientist John Dalton championed the atomic theory, which says that elements consist of tiny particles called **atoms.** An atom is the smallest part of an element that displays the properties of the element. An element and its atom share the same name. The atomic symbol is composed of one or two letters, which stands for this name. For example, the symbol H means a hydrogen atom, the symbol Cl stands for chlorine, and the symbol Na (for *natrium* in Latin) is used for a sodium atom.

Physicists have identified a number of subatomic particles that make up atoms. The three best-known subatomic particles are positively charged **protons,** uncharged **neutrons,** and negatively charged **electrons.** Protons and neutrons are located within the nucleus of an atom, and electrons move around the nucleus. Figure 2.2 shows the arrangement of the subatomic particles in a helium atom, which has only two electrons. In Figure 2.2*a,* the stippling shows the probable location of electrons, and in Figure 2.2*b,* the circle represents an electron orbital, the location of electrons.

Animation Atomic Structure

Earth's crust
organisms
60
40
20
0
Percent by Weight
Fe
Ca
K
S
P
Si
Al
Mg
Na
O
N
C
H
Element

Figure 2.1 Elements that make up Earth's crust and its organisms. Humans are just one of the many organisms that exist on the Earth. The graph inset shows that Earth's crust primarily contains the elements silicon (Si), aluminum (Al), and oxygen (O). Organisms primarily contain the elements oxygen (O), nitrogen (N), carbon (C), and hydrogen (H). Along with sulfur (S) and phosphorus (P), these elements make up biological molecules.

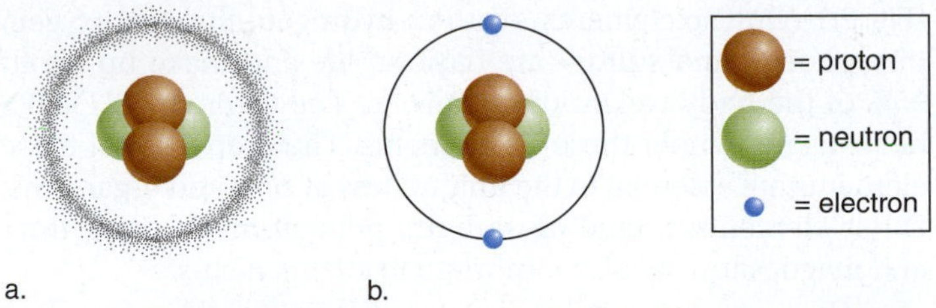
= proton
= neutron
= electron
a.
b.

Subatomic Particles			
Particle	**Electric Charge**	**Atomic Mass Unit (AMU)**	**Location**
Proton	+1	1	Nucleus
Neutron	0	1	Nucleus
Electron	–1	0	Electron shell

c.

Figure 2.2 Model of helium (He). Atoms contain subatomic particles called protons, neutrons, and electrons. Protons and neutrons are within the nucleus, and electrons are outside the nucleus. **a.** The stippling shows the probable location of the electrons in the helium atom. **b.** A circle termed an electron orbital represents the average location of an electron. **c.** The electric charge and the atomic mass units of the subatomic particles vary as shown.

Periods	I	II	III	IV	V	VI	VII	VIII
1	1 H 1.008							2 He 4.003
2	3 Li 6.941	4 Be 9.012	5 B 10.81	6 C 12.01	7 N 14.01	8 O 16.00	9 F 19.00	10 Ne 20.18
3	11 Na 22.99	12 Mg 24.31	13 Al 26.98	14 Si 28.09	15 P 30.97	16 S 32.07	17 Cl 35.45	18 Ar 39.95
4	19 K 39.10	20 Ca 40.08	31 Ga 69.72	32 Ge 72.59	33 As 74.92	34 Se 78.96	35 Br 79.90	36 Kr 83.60

1 — atomic number; H — atomic symbol; atomic mass — 4.003

Groups

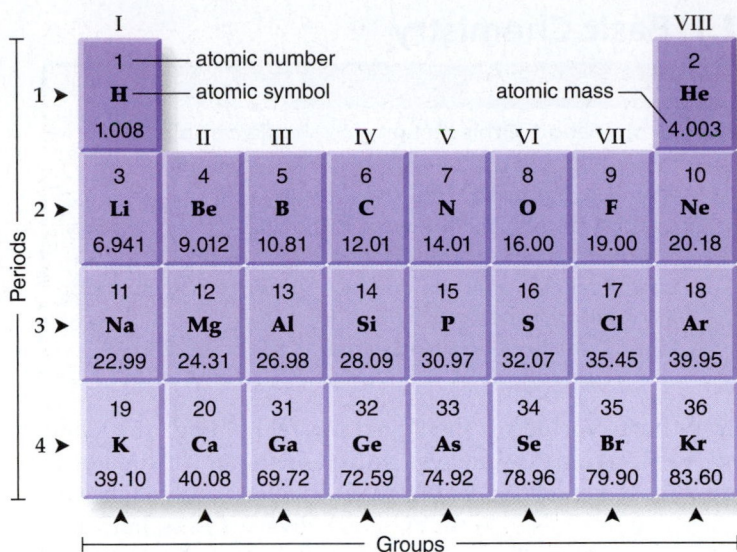

Figure 2.3 A portion of the periodic table. In the periodic table, the elements, and therefore atoms, are arranged in the order of their atomic numbers and placed in groups (vertical columns) and periods (horizontal rows). All the atoms in a particular group have certain chemical characteristics in common. These four periods contain the elements that are most important in biology. The complete periodic table is on the back endsheet.

The concept of an atom has changed greatly since Dalton's day. If an atom could be drawn the size of a football field, the nucleus would be like a gumball in the center of the field. The electrons would be tiny specks whirling about in the upper stands. Most of an atom is empty space. We can only indicate the orbital where the electrons are expected to be most of the time. In our analogy, the electrons might very well stray outside the stadium at times.

All atoms of an element have the same number of protons. This is called the **atomic number.** The number of protons in the nucleus makes each atom unique. The atomic number is often written as a subscript to the lower left of the atomic symbol.

Each atom also has its own **mass number,** which is dependent on the number of subatomic particles in that atom. Protons and neutrons are assigned one atomic mass unit (AMU) each. Electrons are so small that their AMU is considered zero in most calculations (Fig. 2.2*c*). Therefore, the mass number of an atom is the sum of protons and neutrons in the nucleus. The term *atomic mass* is used, rather than *atomic weight,* because mass is constant while weight changes according to the gravitational force of a body. The gravitational force of Earth is greater than that of the moon. Therefore, substances weigh less on the moon even though their mass has not changed. The atomic mass is often written as a superscript to the upper left of the atomic symbol. For example, the carbon atom can be noted in this way:

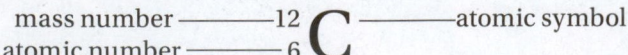

mass number — 12; atomic number — 6; C — atomic symbol

$$^{12}_{6}C$$

The Periodic Table

Once chemists discovered a number of the elements, they began to realize that even though each element consists of a different atom, certain chemical and physical characteristics recur, or showed periodicity. The periodic table was constructed as a way to group the elements, and therefore atoms, according to these characteristics. Notice in Figure 2.3 that the periodic table is arranged according to increasing atomic number. The vertical columns in the table are groups. The horizontal rows are periods, which cause each atom to be in a particular group. For example, all the atoms in group VII react with one atom at a time, for reasons we will soon explore. The atoms in group VIII are called the noble gases because they are inert and rarely react with another atom. Helium and krypton are examples of noble gases.

In Figure 2.3 the atomic number is above the atomic symbol and the atomic mass is below the atomic symbol. The atomic number tells you the number of positively charged protons. It therefore also tells you the number of negatively charged electrons if the atom is electrically neutral. The **atomic mass** is the average of the AMU for all the isotopes (discussed next) of that atom. To determine the number of neutrons, subtract the number of protons from the atomic mass, and take the closest whole number. The periodic table in Figure 2.3 shows only some of the elements. A complete table is available on the back endsheet of the book.

Isotopes

Isotopes are atoms of the same element that differ in the number of neutrons. Isotopes have the same number of protons, but they have different atomic masses. Because the number of protons gives an atom its identity, changing the number of neutrons affects the atomic mass but not the name of the atom. For example, the element carbon has three common isotopes:

$$^{12}_{6}\mathrm{C} \qquad ^{13}_{6}\mathrm{C} \qquad ^{14}_{6}\mathrm{C}$$

radioactive

Carbon 12 has six neutrons, carbon 13 has seven neutrons, and carbon 14 has eight neutrons. Unlike the other two isotopes of carbon, carbon 14 is unstable. It changes over time into nitrogen 14, which is a stable isotope of the element nitrogen. As carbon 14 decays, it releases various types of energy in the form of rays and subatomic particles, and therefore it is a **radioactive isotope.** Today, biologists use radiation to date objects, create images, and trace the movement of substances through the human body.

Animation Half-Life

Low Levels of Radiation

The chemical behavior of a radioactive isotope is essentially the same as that of the stable isotopes of an element. This means that you can put a small amount of radioactive isotope in a sample and it becomes a tracer or tag by which to detect molecular changes.

The importance of chemistry to medicine is nowhere more evident than in the many medical uses of radioactive isotopes. Specific tracers are used in imaging the body's organs and tissues. For example, after a patient drinks a solution containing a minute amount of radioactive ^{131}I, it becomes concentrated in the thyroid—the only organ to take up iodine. A subsequent image of the thyroid indicates whether it is healthy in structure and function (Fig. 2.4*a*). Positron emission tomography (PET) is a way to determine the comparative activity of tissues. Radioactively labeled glucose, which emits a subatomic particle known as a positron, is injected into the body. The radiation given off is detected by sensors and analyzed by a computer. The result is a color image that shows which tissues took up glucose and are metabolically active (Fig. 2.4*b*). A PET scan of the brain can help diagnose a brain tumor, Alzheimer disease, epilepsy, or whether a stroke has occurred.

Video Nuclear Medicine

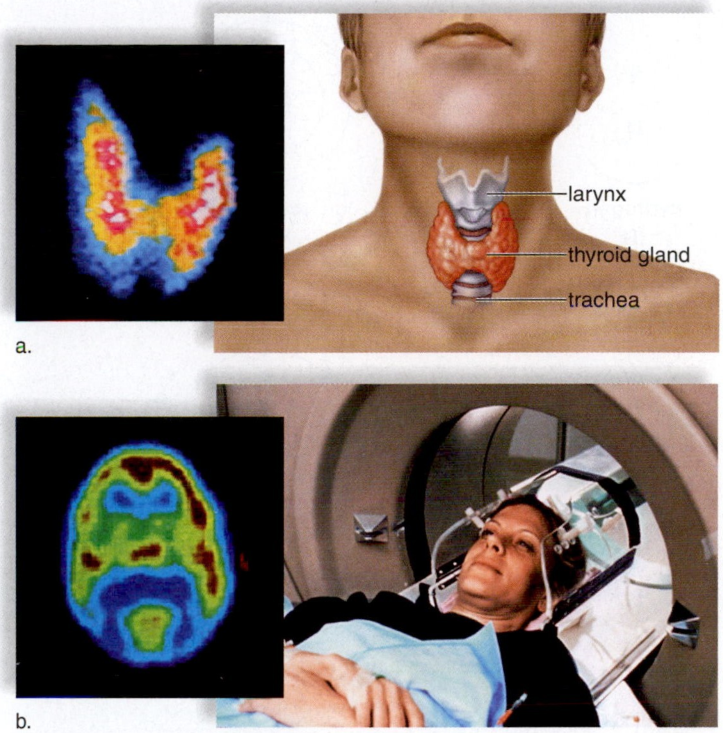
larynx
thyroid gland
trachea
a.
b.

Figure 2.4 Low levels of radiation. **a.** The missing area in this thyroid scan (*upper left*) indicates the presence of a tumor that does not take up the radioactive iodine. **b.** A PET (positron emission tomography) scan reveals which portions of the brain are most active (yellow and red colors).

High Levels of Radiation

Radioactive substances in the environment can harm cells, damage DNA, and cause cancer. When researchers, such as Marie Curie, began studying radiation in the nineteenth century, its harmful effects were not known, and many developed cancer. The release of radioactive particles following a nuclear power plant accident, such as occurred in Japan in 2011 following the tsunamis, can have far-reaching and long-lasting effects on human health. However, the effects of radiation can also be put to good use (Fig. 2.5). Radiation from radioactive isotopes has been used for many years to sterilize medical and dental products. Radiation is now used to sterilize the U.S. mail and other packages to free them of possible pathogens, such as anthrax spores. The ability of radiation to kill cells is often applied to cancer cells. Targeted radioisotopes can be introduced into the body so that the subatomic particles emitted destroy only cancer cells, with little risk to the rest of the body. X rays, another form of high-energy radiation, can be used for medical diagnosis.

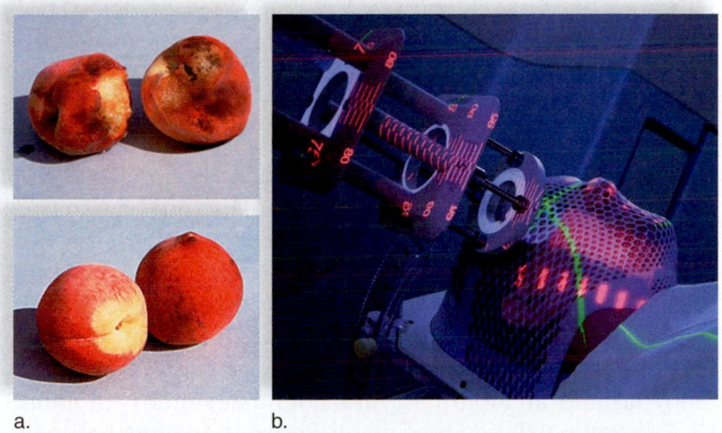
a.
b.

Figure 2.5 High levels of radiation. **a.** Radiation kills bacteria and fungi. Irradiated peaches (*bottom*) spoil less quickly and can be kept for a longer length of time. **b.** Physicians use radiation therapy to kill cancer cells.

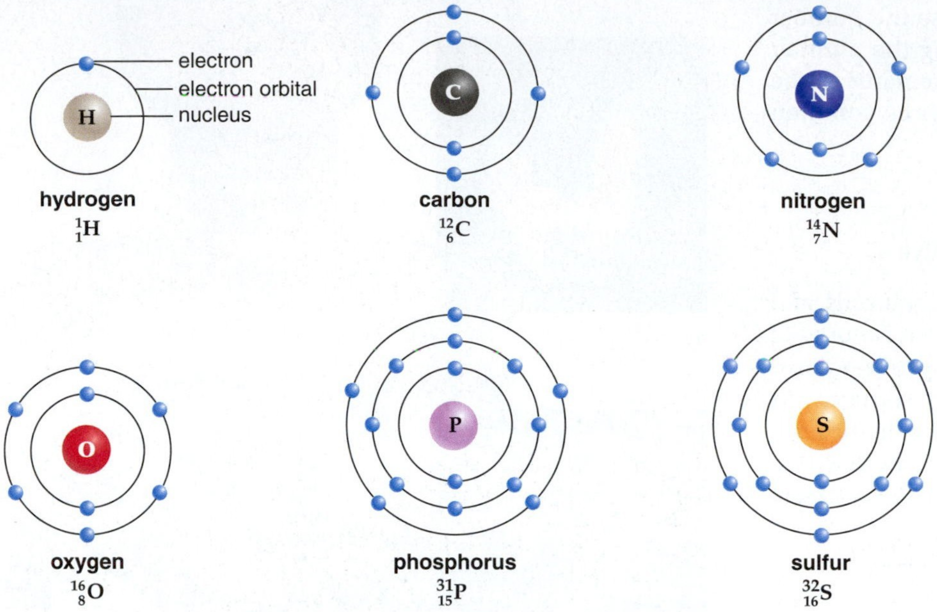
electron
electron orbital
nucleus
H
hydrogen
$^{1}_{1}H$
C
carbon
$^{12}_{6}C$
N
nitrogen
$^{14}_{7}N$
O
oxygen
$^{16}_{8}O$
P
phosphorus
$^{31}_{15}P$
S
sulfur
$^{32}_{16}S$

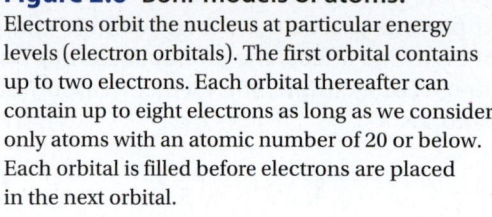

Figure 2.6 Bohr models of atoms. Electrons orbit the nucleus at particular energy levels (electron orbitals). The first orbital contains up to two electrons. Each orbital thereafter can contain up to eight electrons as long as we consider only atoms with an atomic number of 20 or below. Each orbital is filled before electrons are placed in the next orbital.

Electrons

In an electrically neutral atom, the positive charges of the protons in the nucleus are balanced by the negative charges of electrons moving about the nucleus. Various models in years past have attempted to illustrate the precise location of electrons. Figure 2.6 uses the Bohr model, which is named after the physicist Niels Bohr. The Bohr model is useful, but today's physicists tell us it is not possible to determine the exact location of individual electrons at any given moment.

In the diagrams in Figure 2.6, the energy levels of the electrons (also termed electron orbitals) are drawn as concentric rings about the nucleus. For atoms up through number 20 (i.e., calcium), the first orbital (closest to the nucleus) can contain two electrons. Thereafter, each additional orbital can contain eight electrons. For these atoms, each lower level is filled with electrons before the next higher level contains any electrons.

The sulfur atom, with an atomic number of 16, has two electrons in the first orbital, eight electrons in the second orbital, and six electrons in the third, or outer, orbital. Revisit the periodic table (see Fig. 2.3), and note that sulfur is in the third period. In other words, the horizontal row tells you how many orbitals an atom has. Also note that sulfur is in group VI. The group tells you how many electrons an atom has in its outer orbital.

If an atom has only one orbital, the outer orbital is complete when it has two electrons. Otherwise, atomic orbitals follow the **octet rule,** which states that the outer orbital is most stable when it has eight electrons. As mentioned previously, atoms in group VIII of the periodic table are called the noble gases because they do not ordinarily react. Atoms that have incomplete outer orbitals react with other atoms in such a way that after the reaction, each has a stable outer orbital. Atoms can give up, accept, or share electrons in order to have a complete outer orbital.

Check Your Progress 2.1

1. Identify why the key elements in the Earth's crust would differ from those present in living organisms.
2. Identify the protons, neutrons, and electrons within a calcium atom.
3. Explain how radiation can be both beneficial and harmful to humans.
4. Explain the differences between oxygen 16 and oxygen 18.

2.2 Molecules and Compounds

Learning Outcomes

Upon completion of this section, you should be able to

1. Describe how elements are combined into molecules and compounds.
2. List the different types of bonds that occur between elements.
3. Compare the relative strengths of ionic, covalent, and hydrogen bonds.

Atoms, except for noble gases, routinely bond with one another. A **molecule** is formed when two or more atoms bond together. For example, oxygen does not exist in nature as a single atom, O. Instead, two oxygen atoms are joined to form a molecule of oxygen, O_2. When atoms of two or more different elements bond together, the product is called a **compound.** Water (H_2O) is a compound that contains atoms of hydrogen and oxygen. We can also speak of molecules of water because a molecule is the smallest part of a compound that still has the properties of that compound.

Electrons possess energy, and the bonds that exist between atoms also contain energy. Organisms are directly dependent on chemical-bond energy to maintain their organization. When a chemical reaction occurs, electrons shift in their relationship to one another, and energy may be given off or absorbed. This same energy is used to carry on our daily lives.

MP3 Chemical Bonding

Ionic Bonding

Ions form when electrons are transferred from one atom to another. For example, sodium (Na), with only one electron in its third orbital, tends to be an electron donor (Fig. 2.7*a*). Once it gives up this electron, the second orbital, with eight electrons, becomes its outer orbital. Chlorine (Cl), on the other hand, tends to be an electron acceptor. Its outer orbital has seven electrons, so it requires one more electron to have a complete outer orbital. When a sodium atom and a chlorine atom come together, an electron is transferred from the sodium atom to the chlorine atom. Now both atoms have eight electrons in their outer orbitals.

This electron transfer, however, causes a charge imbalance in each atom. The sodium atom has one more proton than it has electrons. Therefore, it has a net charge of +1 (symbolized by Na^+). The chlorine atom has one more electron than it has protons. Therefore, it has a net charge of –1 (symbolized by Cl^-). Such charged particles are called **ions.** Sodium (Na^+) and chloride (Cl^-) are not the only biologically important ions. Some, such as potassium (K^+), are formed by the transfer of a single electron to another atom. Others, such as calcium (Ca^{2+}) and magnesium (Mg^{2+}), are formed by the transfer of two electrons.

Ionic compounds are held together by an attraction between negatively and positively charged ions called an **ionic bond.** When sodium reacts with chlorine, an ionic compound called sodium chloride (NaCl) results. Sodium chloride is a salt, commonly known as table salt because it is used to season our food (Fig. 2.7*b*). Salts can exist as a dry solid, but when salts are placed in water, they release ions as they dissolve. NaCl separates into Na^+ and Cl^-. In biological systems, because they are 70–90% water, ionic compounds tend to be dissociated (ionized) rather frequently.

Animation Ionic Bonds

Covalent Bonding

A **covalent bond** results when two atoms share electrons in such a way that each atom has a complete outer orbital. In a hydrogen atom, the outer orbital is complete when it contains two electrons. If hydrogen is in the presence of a strong electron acceptor, it gives up its electron to become a hydrogen ion (H^+). But if this is not possible, hydrogen can share with another atom and thereby have a completed outer orbital. For example, one hydrogen atom will share with another hydrogen atom. Their two orbitals overlap, and the electrons are shared between them (Fig. 2.8*a*). Each atom has a completed outer orbital due to the sharing of electrons.

Animation Ionic versus Covalent Bonding

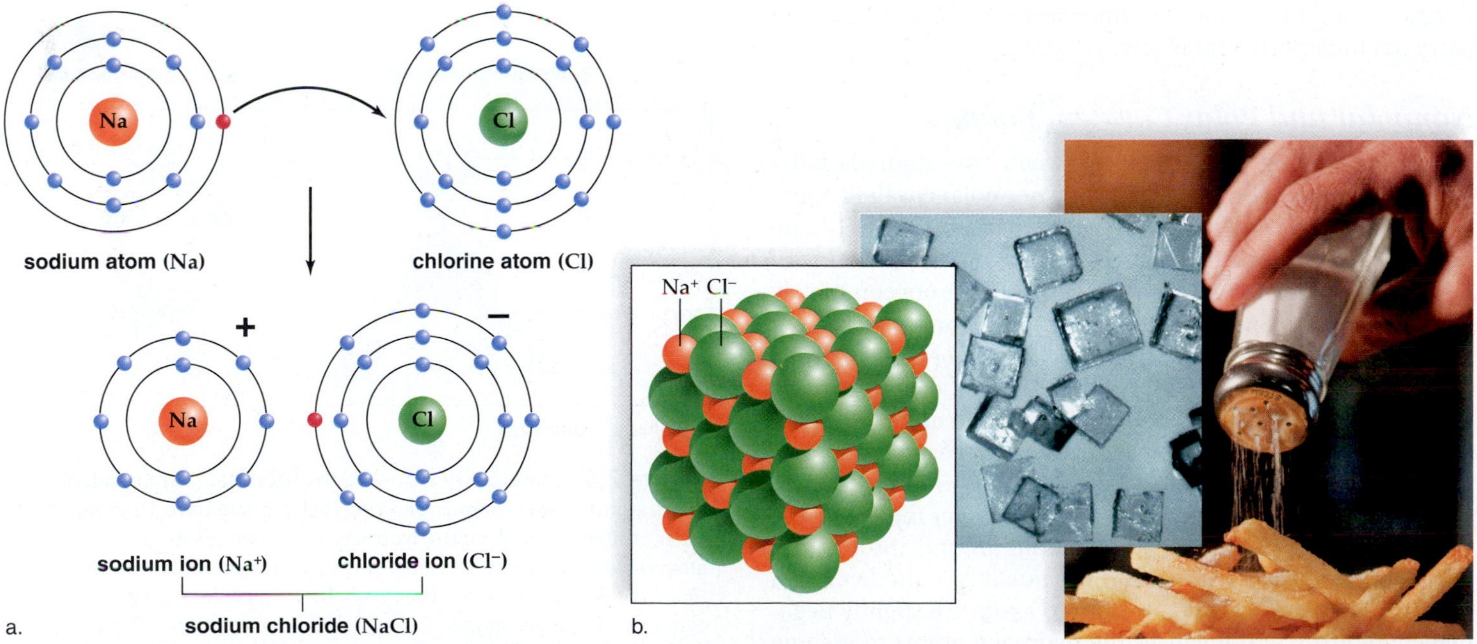
Na
Cl
sodium atom (Na)
chlorine atom (Cl)
+
−
Na
Cl
sodium ion (Na+)
chloride ion (Cl−)
sodium chloride (NaCl)
a.
Na+ Cl−
b.

Figure 2.7 Formation of sodium chloride (table salt). **a.** To form sodium chloride, an electron is transferred from the sodium atom to the chlorine atom. At the completion of the reaction, each atom is an ion containing eight electrons in its outer orbital. **b.** In a sodium chloride crystal, ionic bonding between Na^+ and Cl^- causes the atoms to assume a three-dimensional lattice in which each sodium ion is surrounded by six chloride ions, and each chloride ion is surrounded by six sodium ions. This forms crystals, such as in table salt.

A more common way to symbolize that atoms are sharing electrons is to draw a line between the two atoms, as in the structural formula H—H. In a molecular formula, the line is omitted and the molecule is simply written as H_2.

Sometimes, atoms share more than one pair of electrons to complete their octets. A double covalent bond occurs when two atoms share two pairs of electrons (Fig. 2.8*b*). To show that oxygen gas (O_2) contains a double bond, the molecule can be written as O═O.

It is also possible for atoms to form triple covalent bonds, as in nitrogen gas (N_2), which can be written as N≡N. Single covalent bonds between atoms are quite strong, but double and triple bonds are even stronger.

Shape of Molecules

Structural formulas make it seem as if molecules are one-dimensional, but actually molecules have a three-dimensional shape that often determines their biological function. Molecules consisting of only two atoms are always linear, but a molecule such as methane with five atoms (Fig. 2.8*c*) has a tetrahedral shape. Why? Because, as shown in the ball-and-stick model, each bond is pointing to the corners of a tetrahedron (Fig. 2.8*d, left*). The space-filling model comes closest to the actual shape of the molecule. In space-filling models, each type of atom is given a particular color—carbon is always black and hydrogen is always off-white (Fig. 2.8*d, right*).

The shapes of molecules are necessary to the structural and functional roles they play in organisms. For example, hormones have specific shapes that allow them to be recognized by the cells in the body. Antibodies combine with disease-causing agents, like a key fits a lock, to protect us. Similarly, homeostasis is maintained only when enzymes have the proper shape to carry out their particular reactions in cells.

Nonpolar and Polar Covalent Bonds

When the sharing of electrons between two atoms is fairly equal, the covalent bond is said to be a nonpolar covalent bond. All the molecules in Figure 2.8, including methane (CH_4), are nonpolar. In the case of water (H_2O), however, the sharing of electrons between oxygen and each hydrogen is not completely equal. The attraction of an atom for the electrons in a covalent bond is called its **electronegativity.** The larger oxygen atom, with the greater number of protons, is more electronegative than the hydrogen atom. The oxygen atom can attract the electron pair to a greater extent than each hydrogen atom can. It may help to think of electronegativity as where the electron pair chooses to "spend its time." In a water molecule, the shared electron pair spends more time around the nucleus of the oxygen atom than around the nucleus of the hydrogen atom. This causes the oxygen atom to assume a slightly negative charge (δ^-), and it causes the hydrogen atoms to assume slightly positive charges (δ^+). The unequal sharing of electrons in a covalent bond creates a polar covalent bond. In the case of water, the molecule itself is a polar molecule (Fig. 2.9*a*).

Animation
Electronegativity

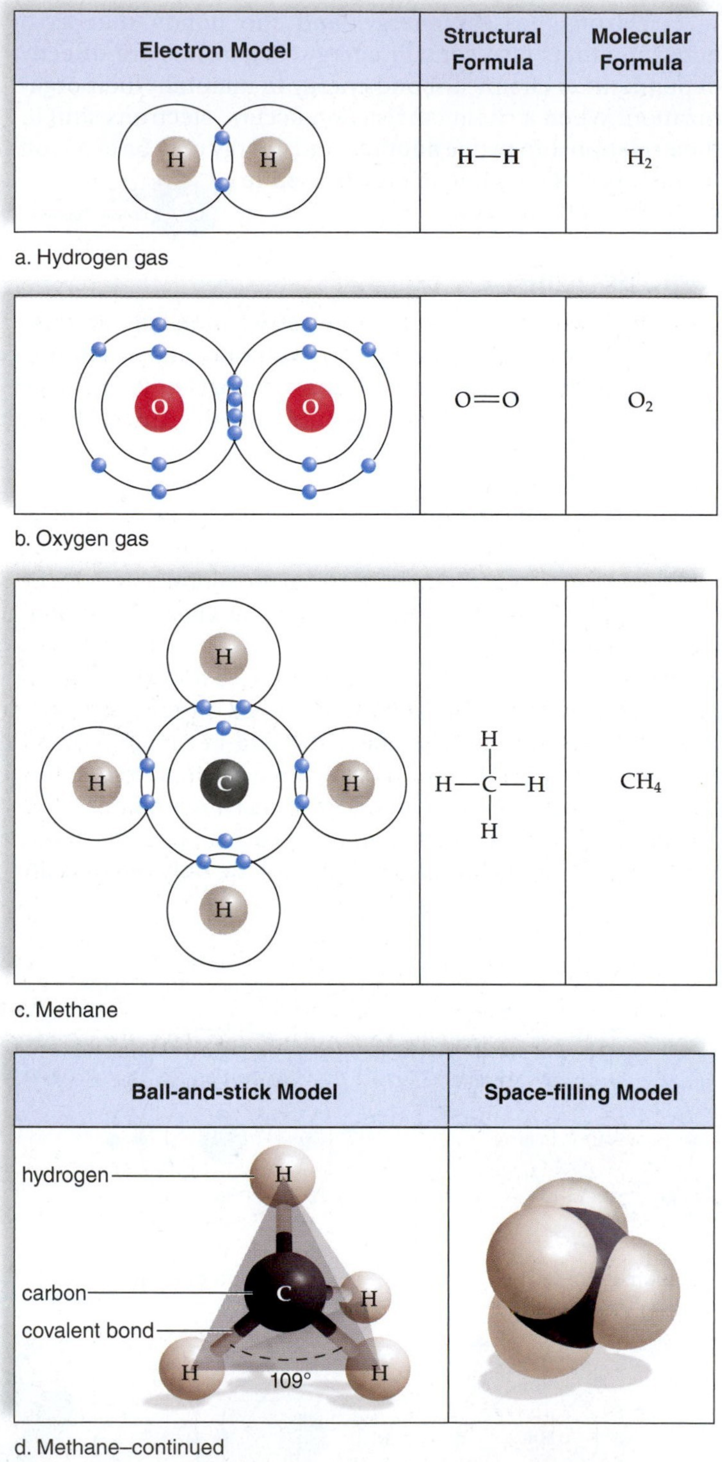
Electron Model
Structural Formula
Molecular Formula
H
H
H—H
H_2
a. Hydrogen gas
O
O
O═O
O_2
b. Oxygen gas
H
C
H
H
H
H—C—H
CH_4
c. Methane
Ball-and-stick Model
Space-filling Model
hydrogen
carbon
covalent bond
109°
d. Methane–continued

Figure 2.8 Covalently bonded molecules. In a covalent bond, atoms share electrons, allowing each atom to have a completed outer orbital. **a.** A molecule of hydrogen (H_2) contains two hydrogen atoms sharing a pair of electrons. This single covalent bond can be represented in any of these three ways. **b.** A molecule of oxygen (O_2) contains two oxygen atoms sharing two pairs of electrons. This results in a double covalent bond. **c.** A molecule of methane (CH_4) contains one carbon atom bonded to four hydrogen atoms. **d.** When carbon binds to four other atoms, as in methane, each bond actually points to one corner of a tetrahedron. Ball-and-stick models and space-filling models are three-dimensional representations of a molecule.

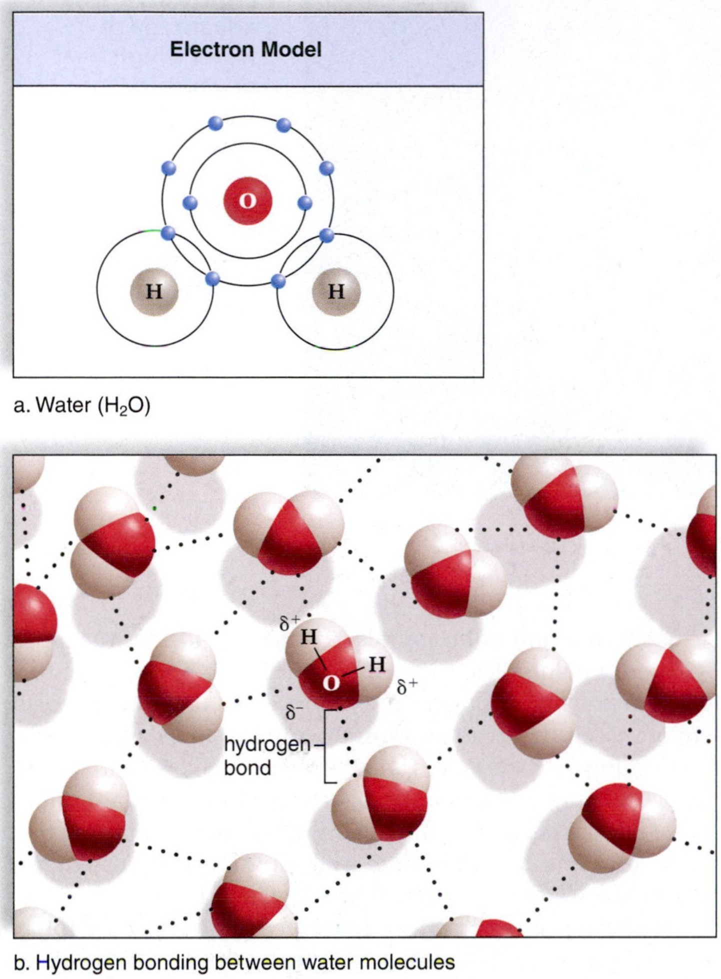
Electron Model
O
H
H
a. Water (H_2O)
δ+
H
O
H
δ+
δ−
hydrogen bond
b. Hydrogen bonding between water molecules

Figure 2.9 Water molecule. **a.** The electron model of water. **b.** Hydrogen bonding between water molecules. A hydrogen bond is the attraction of a slightly positive hydrogen to a slightly negative atom in the vicinity. Each water molecule can hydrogen-bond to four other molecules in this manner. When water is in its liquid state, some hydrogen bonds are forming and others are breaking at all times.

Hydrogen Bonding

Polarity within a water molecule causes the hydrogen atoms in one molecule to be attracted to the oxygen atoms in other water molecules (Fig. 2.9*b*). This attraction, although weaker than an ionic or covalent bond, is called a **hydrogen bond.** Because a hydrogen bond is easily broken, it is often represented by a dotted line. Hydrogen bonding is not unique to water. Many biological molecules have polar covalent bonds involving an electropositive hydrogen and usually an electronegative oxygen or nitrogen. In these instances, a hydrogen bond can occur within the same molecule or between different molecules.

Although a hydrogen bond is more easily broken than a covalent bond, many hydrogen bonds taken together are quite strong. Hydrogen bonds between cellular molecules help maintain their proper structure and function. For example, hydrogen bonds hold the two strands of DNA together. When DNA makes a copy of itself, each hydrogen bond easily breaks, allowing the DNA to unzip. On the other hand, the hydrogen bonds acting together add stability to the DNA molecule. As we shall see, many of the important properties of water are the result of hydrogen bonding.

Check Your Progress 2.2

1. Identify when carbon dioxide (CO_2) and nitrogen gas (N_2) are considered molecules, compounds, or both.
2. Describe why a carbon atom (C) is capable of forming four covalent bonds.
3. Explain why hydrogen ions form polar bonds that have a partially positive charge.

2.3 Chemistry of Water

Learning Outcomes

Upon completion of this section, you should be able to

1. Evaluate which properties of water are important for biological life.
2. Identify common acidic and basic substances.
3. Describe how buffers are important to living organisms.

The first cell(s) evolved in water, and organisms are composed of 70–90% water. Water is a polar molecule, and water molecules are hydrogen-bonded to one another (see Fig. 2.9*b*). Due to hydrogen bonding, water molecules cling together. Without hydrogen bonding between molecules, water would change from a solid to liquid state at −100°C and from a liquid to gaseous state at −91°C. This would make most of the water on Earth steam, and life unlikely. But because of hydrogen bonding, water is a liquid at temperatures typically found on Earth's surface. It melts at 0°C and boils at 100°C. These and other unique properties of water make it essential to the existence of life.

Properties of Water

Water has a high heat capacity. A **calorie** is the amount of heat energy needed to raise the temperature of 1 gram (g) of water by 1°C. In comparison, other covalently bonded liquids require input of only about half this amount of energy to rise 1°C in temperature. The many hydrogen bonds that link water molecules help water absorb heat without a great change in temperature.

Converting 1 g of the coldest liquid water to ice requires the loss of 80 calories of heat energy. Water holds onto its heat, and its temperature falls more slowly than that of other liquids. This property of water is important not only for aquatic organisms but also for all organisms. Because the temperature of water rises and falls slowly, organisms are better able to maintain their normal internal temperatures and are protected from rapid temperature changes.

Animation Properties of Water

Water has a high heat of vaporization. Converting 1 g of the hottest water to a gas requires an input of 540 calories of heat energy. Water has a high heat of vaporization because hydrogen bonds must be broken before water boils and changes to a vaporized state. Water's high heat of vaporization gives animals in a hot environment an efficient way to release excess body heat (Fig. 2.10). When an animal sweats, or gets splashed, body heat is used to vaporize the water, thus cooling the animal.

Temperatures along coasts are moderate due to water's high heat capacity and high heat of vaporization. During the summer, the ocean absorbs and stores solar heat, and during the winter, the ocean slowly releases it. In contrast, the interior regions of continents can experience severe changes in temperature.

Water is a solvent. Due to its polarity, water facilitates chemical reactions, inside and outside living organisms. It dissolves a great number of substances. A solution contains dissolved substances called **solutes.** When ionic salts—for example, sodium chloride (NaCl)—are put into water, the negative ends of the water molecules are attracted to the sodium ions, and the positive ends of the water molecules are attracted to the chloride ions. This causes the sodium ions and the chloride ions to separate, or dissociate, in water:

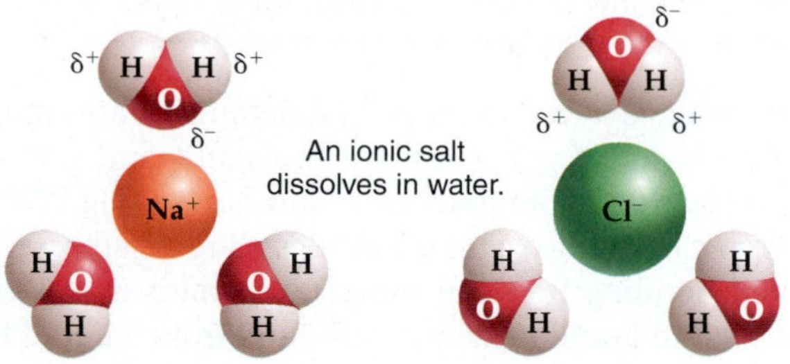
δ+
H
H
δ+
O
δ−
Na+
H
O
H
O
H
H
An ionic salt
dissolves in water.
δ−
O
H
H
δ+
δ+
Cl−
H
O
H
H
H
O

Water is also a solvent for larger molecules that contain ionized atoms or are polar molecules.

Molecules that can attract water are said to be **hydrophilic.** When ions and molecules disperse in water, they move about and collide, allowing reactions to occur. Nonionized and nonpolar molecules, such as oil, that cannot attract water are said to be **hydrophobic.**

Water molecules are cohesive and adhesive. Cohesion is apparent because water flows freely, and yet water molecules do not separate from each other. They cling together because of hydrogen bonding. Water exhibits adhesion because its positive and negative poles allow it to adhere to polar surfaces. Cohesion and adhesion allow water to fill a tubular vessel. Therefore, water is an excellent transport system, both inside and outside of living organisms. Unicellular organisms rely on external water to transport nutrient and waste molecules, but multicellular organisms often contain internal vessels through which water transports nutrients and wastes. For example, the liquid portion of our blood, which transports dissolved and suspended substances throughout the body, is 90% water.

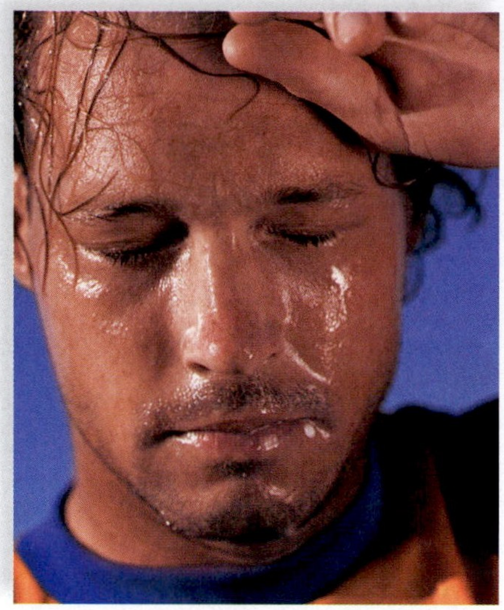

Figure 2.10 The advantage of water's high heat of vaporization. When body temperature increases, sweat is produced by glands in the dermal layer of the skin. The evaporation of the water off your skin aids in cooling your body.

Cohesion and adhesion also contribute to the transport of water in plants. The roots of plants absorb water while the leaves lose water through evaporation. A plant contains a system of vessels that reaches from the roots to the leaves. Water evaporating from the leaves is immediately replaced with water molecules from the vessels. Because water molecules are cohesive, a tension is created that pulls a water column up from the roots. Adhesion of water to the walls of the vessels also helps prevent the water column from breaking apart.

Water has a high surface tension. The stronger the force between molecules in a liquid, the greater the surface tension. As with cohesion, hydrogen bonding causes water to have a high surface tension. This property makes it possible for humans to skip rocks on water. The water strider, a common insect, can even walk on top of a pond without breaking the surface.

Video Basilisk Lizard

Frozen water (ice) is less dense than liquid water. As liquid water cools, the molecules come closer together. They are densest at 4°C, but they are still moving about, bumping into each other (Fig. 2.11). At temperatures below 4°C, including at 0°C when water is frozen, the water forms a regular crystal lattice that is rigid and has more open space between the water molecules. For this reason water expands as it freezes, which is why cans of soda burst when placed in a freezer or why frost heaves make northern roads bumpy in the winter. It also means that ice is less dense than liquid water, and floats on liquid water.

If ice did not float on water, it would sink, and ponds, lakes, and perhaps even regions of the ocean would freeze solid. This would make life impossible in the water and also on land. Instead, bodies of water always freeze from the top down. When a body of water freezes on the surface, the ice acts as an insulator to prevent the water below it from freezing. This protects aquatic organisms so that they can survive the winter. As ice melts in the spring, it draws heat from the environment, helping to prevent a sudden change in temperature that might be harmful to life.

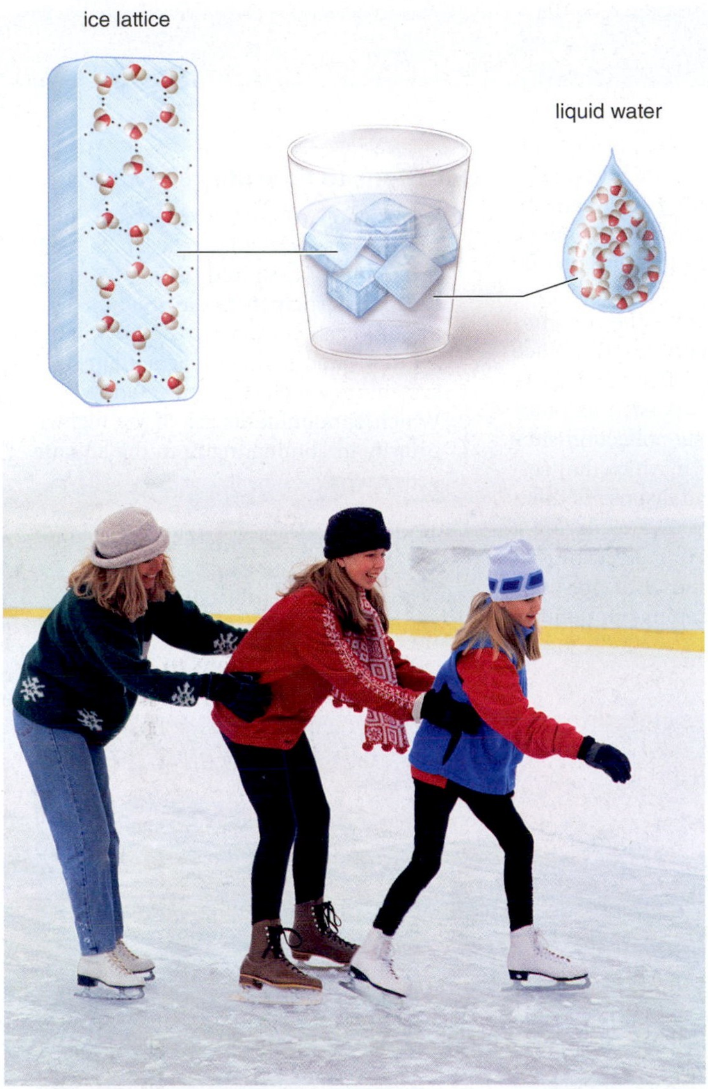
ice lattice
liquid water

Figure 2.11 Ice floats on water. Remarkably, water is more dense at 4°C than at 0°C. While most substances contract when they solidify, water expands when it freezes because the water molecules in ice form a lattice in which the hydrogen bonds are farther apart than in liquid water.

Acids and Bases

When water ionizes, it releases an equal number of hydrogen ions (H^+; also called protons[1]) and hydroxide ions (OH^-) into the solution:

$$\underset{\text{water}}{H{-}O{-}H} \rightleftharpoons \underset{\text{hydrogen ion}}{H^+} + \underset{\text{hydroxide ion}}{OH^-}$$

Only a few water molecules at a time dissociate, and the actual number of H^+ and OH^- is very small (1×10^{-7} moles/liter).[2]

[1] A hydrogen atom contains one electron and one proton. A hydrogen ion has only one proton, so it is often simply called a proton.

[2] In chemistry, a mole is defined as the amount of matter that contains as many objects (atoms, molecules, ions) as the number of atoms in exactly 12 g of ^{12}C.

Acidic Solutions (High H^+ Concentrations)

Lemon juice, vinegar, tomatoes, and coffee are all acidic solutions. **Acids** are substances that release hydrogen ions (H^+) when they dissociate in water. Therefore, they contain a higher concentration of H^+ than OH^-. For example, hydrochloric acid (HCl) is an important acid that dissociates in this manner:

$$HCl \longrightarrow H^+ + Cl^-$$

Because dissociation is almost complete, HCl is called a strong acid. If hydrochloric acid is added to a beaker of water, the number of hydrogen ions (H^+) increases greatly.

Basic Solutions (Low H^+ Concentrations)

Baking soda and antacids are common basic solutions familiar to most people. **Bases** are substances that either take up hydrogen ions (H^+) or release hydroxide ions (OH^-). They contain a higher concentration of OH^- than H^+. For example, sodium hydroxide (NaOH) is an important base that dissociates in this manner:

$$NaOH \longrightarrow Na^+ + OH^-$$

Because dissociation is almost complete, sodium hydroxide is called a strong base. If sodium hydroxide is added to a beaker of water, the number of hydroxide ions increases.

pH Scale

The **pH scale** is used to indicate the acidity or basicity (alkalinity) of solutions.[3] The pH scale (Fig. 2.12) ranges from 0 to 14. A pH of 7 represents a neutral state in which the hydrogen ion and hydroxide ion concentrations are equal. A pH below 7 is an acidic solution because the hydrogen ion concentration [H^+] is greater than

[3] pH is defined as the negative log of the hydrogen ion concentration [H^+]. A log is the power to which ten must be raised to produce a given number.

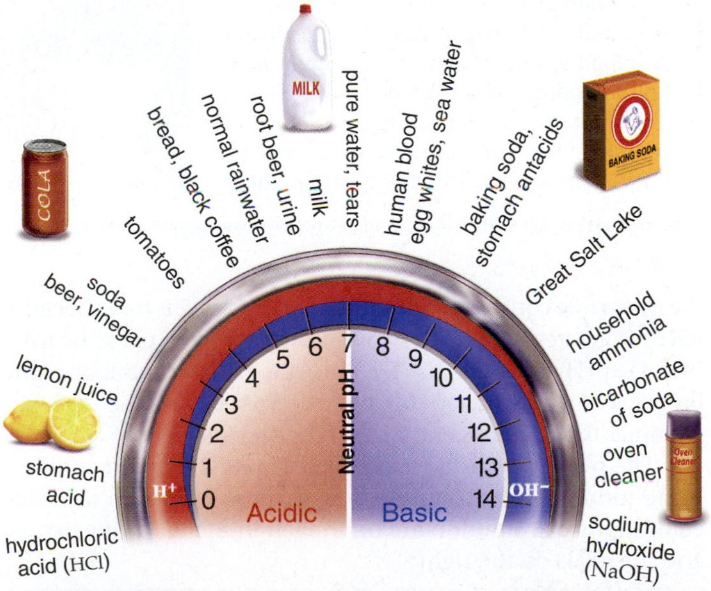
hydrochloric acid (HCl)
stomach acid
lemon juice
soda beer, vinegar
tomatoes
bread, black coffee
normal rainwater
root beer, urine
milk
pure water, tears
human blood egg whites, sea water
baking soda, stomach antacids
Great Salt Lake
household ammonia
bicarbonate of soda
oven cleaner
sodium hydroxide (NaOH)
COLA
MILK
BAKING SODA
0 1 2 3 4 5 6 7 8 9 10 11 12 13 14
H+
OH−
Neutral pH
Acidic
Basic

Figure 2.12 The pH scale. The dial of this pH meter indicates that pH ranges from 0 to 14, with 0 the most acidic and 14 the most basic. pH 7 (neutral pH) has equal amounts of hydrogen ions (H^+) and hydroxide ions (OH^-). An acidic pH has more H^+ than OH^-, and a basic pH has more OH^- than H^+.

SCIENCE IN YOUR LIFE ▶ ECOLOGY

The Fate of Prescription Medicine

What happens to prescriptions when they are flushed down the drain?

Recently, scientists have found detectable concentrations of pharmaceuticals and organic wastewater chemicals in rivers and streams across the United States (Fig. 2A). U.S. Geological Survey scientists found 12 of 22 pharmaceuticals and 32 of the 47 organic wastewater chemicals looked for in the streams of the Boulder Creek Watershed, Colorado. Some of these chemicals included antibiotics used to treat bacterial infections, antimicrobial agents used in soaps, endocrine-active chemicals (such as those found in birth control pills), as well as caffeine. While few of the detectable chemicals exceeded water-quality standards, the scientists did note that no standards exist for a number of these compounds, raising concerns about what represents acceptable levels in the water supply.

While difficult to fully assess the complete ecological impact of these chemicals, a wide range of problems have been attributed to them. Native fish populations were found to exhibit endocrine disruption due to exposure to these contaminants. Endocrine disruptions can produce adverse effects on the development of the nervous and reproductive systems as well as changes in the fish's response to stressors in the environment. A 2008 study showed fish populations with altered sex ratios, reduced gonad size, and disrupted ovarian and testicular development in streams containing a mixture of endocrine-active chemicals. Fish in contaminated streams were also found to contain both male and female reproductive organs (referred to as intersex). These results indicate that the reproductive potential of fish populations living in streams in which various pharmaceuticals and organic wastewater chemicals are present may be compromised.

In an attempt to decrease the contamination of the nation's waterways, the Office of National Drug Control Policy suggests working with local agencies such as pharmacies or hazardous waste collection sites that hold "take-back days" in which they collect unused medicines and dispose of them properly. Some states are also exploring a mail/ship back program to collect unused prescriptions. Proper disposal of prescriptions is the first step in keeping our nation's waterways safe.

Questions to Consider

1. Investigate whether or not your community has a pharmaceutical take-back program. If so, should people be mandated to participate in the program?
2. Should each community be held liable for the water quality of the aquatic ecosystems to which it is connected?
3. Which contaminants are of the highest priority in eliminating from the aquatic ecosystems?

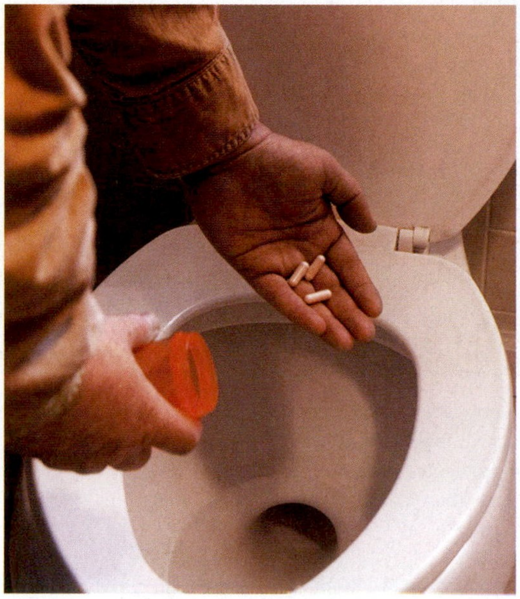

Figure 2A Effects of pharmaceuticals on fish. In some cases male fish have developed ovaries as a result of exposure to endocrine-active chemicals.

the hydroxide concentration [OH^-]. A pH above 7 is basic because [OH^-] is greater than [H^+]. As we move down the pH scale from pH 14 to pH 0, each unit has ten times the H^+ concentration of the previous unit. As we move up the scale from 0 to 14, each unit has ten times the OH^- concentration of the previous unit.

The pH scale was devised to eliminate the use of cumbersome numbers. For example, the possible hydrogen ion concentrations of a solution are on the left in the following listing, and the pH is on the right:

[H^+] (moles per liter)	pH
0.000001 = 1×10^{-6}	6
0.0000001 = 1×10^{-7}	7
0.00000001 = 1×10^{-8}	8

To further illustrate the relationship between hydrogen ion concentration and pH, consider the following question: Which of the pH values listed indicates a higher hydrogen ion concentration [H^+] than pH 7, and therefore would be an acidic solution? A number with a smaller negative exponent indicates a greater quantity of hydrogen ions than one with a larger negative exponent. Therefore, pH 6 is an acidic solution.

MP3 Water and pH

Buffers and pH

A **buffer** is a substance that keeps pH within normal limits. Many commercial products, such as Bufferin, shampoos, or deodorants, are buffered as an added incentive for us to buy them. Buffers resist pH changes because they can take up excess hydrogen ions (H^+) or hydroxide ions (OH^-).

In animals, the pH of body fluids is maintained within a narrow range, or else health suffers. The pH of our blood when we are healthy is always about 7.4—that is, just slightly basic (alkaline). If the blood pH drops to about 7, acidosis results. If the blood pH rises to about 7.8, alkalosis results. Both conditions can be life threatening. Normally, pH stability is possible because the body has built-in mechanisms to prevent pH changes. Buffers are the most important of these mechanisms. For example, carbonic acid (H_2CO_3) is a weak acid that minimally dissociates and then re-forms in the following manner:

$$\underset{\text{carbonic acid}}{H_2CO_3} \underset{\text{re-forms}}{\overset{\text{dissociates}}{\rightleftarrows}} H^+ + \underset{\text{bicarbonate ion}}{HCO_3^-}$$

Blood always contains a combination of carbonic acid and bicarbonate ions. When hydrogen ions (H^+) are added to blood, the following reaction occurs:

$$H^+ + HCO_3^- \longrightarrow H_2CO_3$$

When hydroxide ions (OH^-) are added to blood, this reaction occurs:

$$OH^- + H_2CO_3 \longrightarrow HCO_3^- + H_2O$$

These reactions prevent any significant change in blood pH.

Check Your Progress 2.3

1. Compare the difference between water's high heat capacity and high heat of vaporization.
2. Explain why solution with a pH of 6 contains more H^+ than a solution with a pH of 8.
3. Identify why a weakly dissociating acid/base is a better buffer than a strongly dissociating one.

2.4 Organic Molecules

Learning Outcomes

Upon completion of this section, you should be able to

1. Compare inorganic molecules to organic molecules.
2. Identify the role of a functional group.
3. Recognize how monomers are joined to form polymers.

Inorganic molecules constitute nonliving matter, but even so, inorganic molecules such as salts (e.g., NaCl) and water play important roles in living organisms. The molecules of life, however, are organic molecules. **Organic molecules** always contain carbon (C) and hydrogen (H). The chemistry of carbon accounts for the formation of the very large variety of organic molecules found in living organisms. Carbon atoms contain four electrons in their outer orbital. In order to achieve eight electrons in the outer orbital, a carbon atom shares electrons covalently with as many as four other atoms, as in methane (CH_4). Carbon atoms often share electrons with other carbon atoms, forming long hydrocarbon chains.

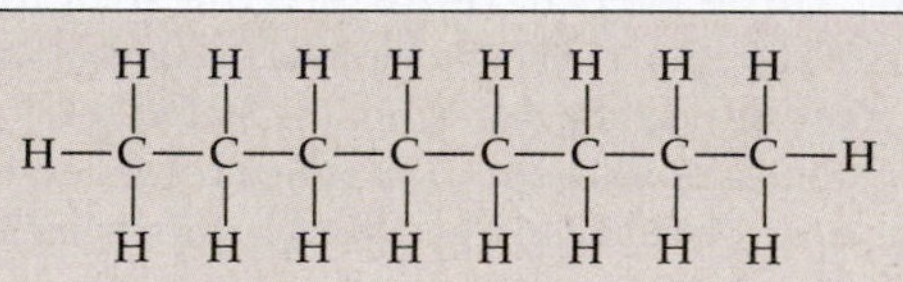

TABLE 2.1 Functional Groups

Functional Groups			
Group	***Structure***	***Compound***	***Significance***
Hydroxyl	*R*—OH	Alcohol as in ethanol	Polar, forms hydrogen bond Present in sugars, some amino acids
Carbonyl	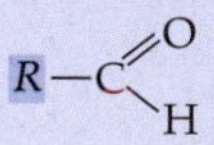	Aldehyde as in formaldehyde	Polar Present in sugars
	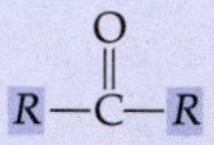	Ketone as in acetone	Polar Present in sugars
Carboxyl (acidic)	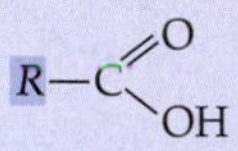	Carboxylic acid as in acetic acid	Polar, acidic Present in fatty acids, amino acids
Amino	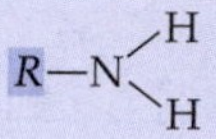	Amine as in tryptophan	Polar, basic, forms hydrogen bonds Present in amino acids
Sulfhydryl	*R*—SH	Thiol as in ethanethiol	Forms disulfide bonds Present in some amino acids
Phosphate	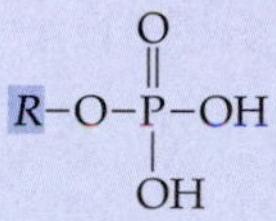	Organic phosphate as in phosphorylated molecules	Polar, acidic Present in nucleotides, phospholipids

R = remainder of molecule

Under certain conditions, a hydrocarbon chain can turn back on itself to form a ring compound. Attached to the carbon chains are **functional groups,** a specific combination of bonded atoms that always react in the same way. Table 2.1 lists some of the more common functional groups. Many molecules of life are macromolecules—that is, they contain many molecules joined together. A **monomer** (*mono,* one) is a simple organic molecule that exists individually or can link with other monomers to form a **polymer** (*poly,* many). The polymers in cells form from monomers as follows:

Polymer	Monomer
carbohydrate (e.g., starch)	monosaccharide
lipid	fatty acids
protein	amino acid
nucleic acid	nucleotide

Aside from carbohydrates, proteins, and nucleic acids, the other organic molecules in cells are lipids. You are very familiar with carbohydrates, lipids, and proteins, because certain foods are known to be rich in these molecules. The nucleic acid DNA makes up our genes, which are hereditary units that control our cells and the structure of our bodies.

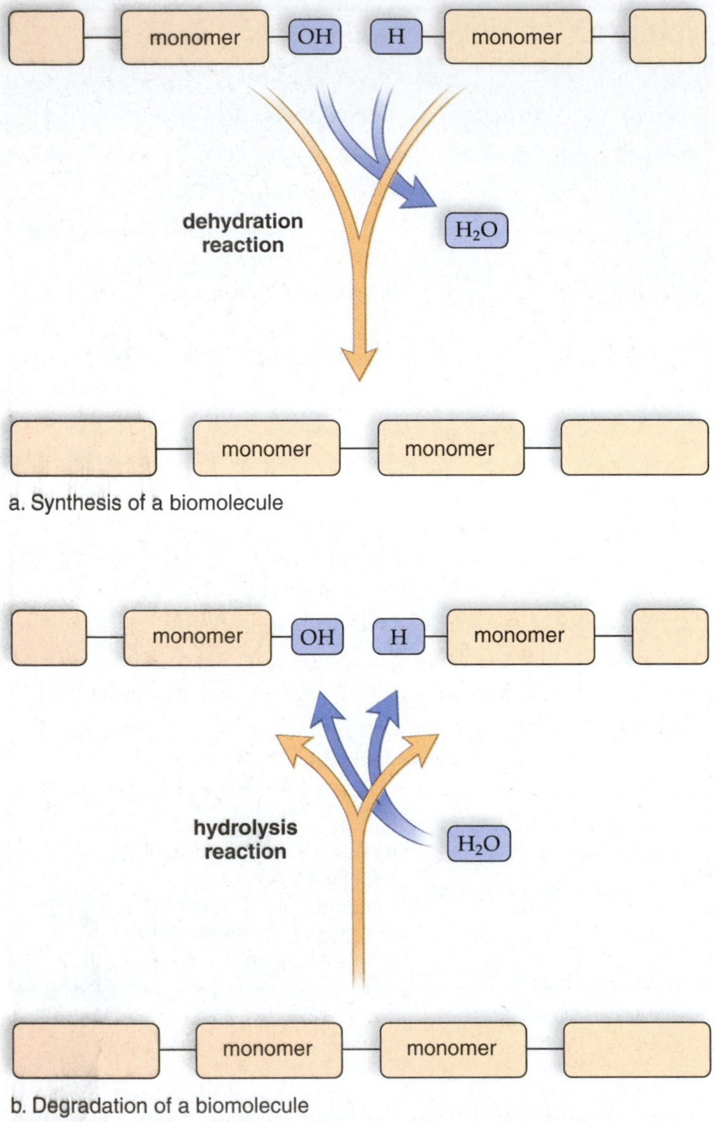

monomer
OH
H
monomer
dehydration reaction
H_2O
monomer
monomer
a. Synthesis of a biomolecule
monomer
OH
H
monomer
hydrolysis reaction
H_2O
monomer
monomer
b. Degradation of a biomolecule

Figure 2.13 Synthesis and degradation of polymers.
a. In cells, synthesis often occurs when monomers join (bond) during a dehydration reaction (removal of H_2O). **b.** Degradation occurs when the monomers in a polymer separate during a hydrolysis reaction (addition of H_2O).

Cells have a common way of joining monomers to build polymers. During a **dehydration reaction,** an —OH (hydroxyl group) and an —H (hydrogen atom), the equivalent of a water molecule, are removed as the reaction proceeds (Fig. 2.13*a*). To degrade polymers, the cell uses a **hydrolysis reaction,** in which the components of water are added (Fig. 2.13*b*).

MP3 Organic Molecules

Check Your Progress 2.4

1. Describe why organic molecules are considered the molecules of life.
2. Compare and contrast dehydration and hydrolysis reactions.

2.5 Carbohydrates

Learning Outcomes

Upon completion of this section, you should be able to

1. Identify the structural components of a carbohydrate.
2. List several examples of important monosaccharides and polysaccharides.

Carbohydrates first and foremost function for quick fuel and short-term energy storage in all organisms, including humans. Carbohydrates play a structural role in woody plants, bacteria, and animals such as insects. In addition, carbohydrates on cell surfaces are involved in cell-to-cell recognition, as we will learn in Chapter 4.

Carbohydrate molecules are characterized by the presence of the atomic grouping H—C—OH, in which the ratio of hydrogen atoms (H) to oxygen atoms (O) is approximately 2:1. Because this ratio is the same as the ratio in water, the term "hydrates of carbon" is often used.

Simple Carbohydrates

If the number of carbon atoms in a molecule is low (from three to seven), then the carbohydrate is a simple sugar, or **monosaccharide.** The designation **pentose** means a 5-carbon sugar, and the designation **hexose** means a 6-carbon sugar. **Glucose** is a hexose sugar found in our blood (Fig. 2.14). Our bodies use glucose as an immediate source of energy. Other common hexoses are fructose, found in fruits, and galactose, a constituent of milk. These three hexoses (glucose, fructose, and galactose) all occur as ring structures with the molecular formula $C_6H_{12}O_6$. The exact shape of the ring differs, as does the arrangement of the hydrogen (—H) and hydroxyl (—OH) groups attached to the ring.

A **disaccharide** (*di,* two; *saccharide,* sugar) contains two monosaccharides that have joined during a dehydration reaction. Figure 2.15 shows how the disaccharide maltose forms when two glucose molecules bond together. Note the position of this bond. Our hydrolytic digestive juices can break this bond, and the result is two glucose molecules. When glucose

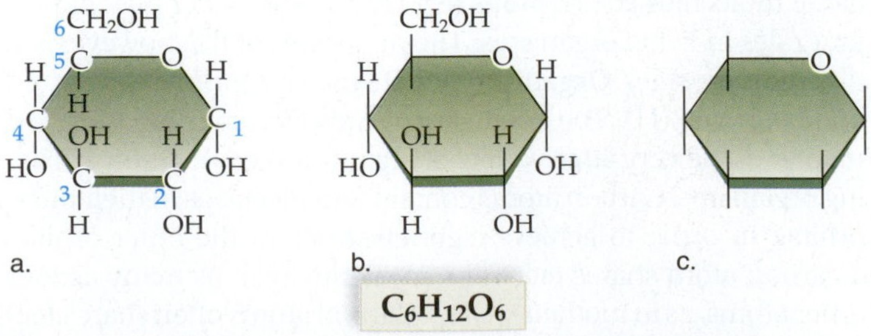

a.
b.
c.
$C_6H_{12}O_6$

Figure 2.14 Three ways to represent the structure of glucose. $C_6H_{12}O_6$ is the molecular formula for glucose. The *far left* structure **(a)** shows the carbon atoms, but the *middle* structure **(b)** does not show the carbon atoms. The *far right* structure **(c)** is the simplest way to represent glucose. Note that in **a** and **b,** each carbon has an attached H and OH group. Those groups are assumed in **c.**

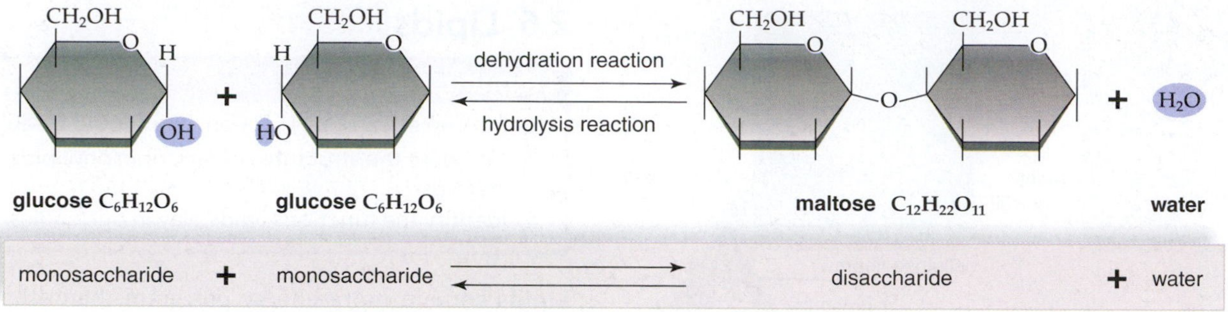

CH₂OH
H
OH
+
CH₂OH
H
HO
dehydration reaction
hydrolysis reaction
CH₂OH
CH₂OH
O
+
H₂O
glucose C₆H₁₂O₆
glucose C₆H₁₂O₆
maltose C₁₂H₂₂O₁₁
water
monosaccharide
+
monosaccharide
disaccharide
+
water

Figure 2.15 Synthesis and degradation of maltose, a disaccharide. Synthesis of maltose occurs following a dehydration reaction when a bond forms between two glucose molecules and water is removed. Degradation of maltose occurs following a hydrolysis reaction when this bond is broken by the addition of water.

and fructose join, the disaccharide sucrose forms. Sucrose is another disaccharide of special interest because we use it at the table to sweeten our food. We acquire sucrose from plants such as sugarcane and sugar beets. You may also have heard of lactose, a disaccharide found in milk. Lactose is glucose combined with galactose. Some people are lactose intolerant because they cannot break down lactose. This leads to unpleasant gastrointestinal symptoms when they consume dairy products.

Polysaccharides

Long polymers such as starch, glycogen, and cellulose are **polysaccharides** that contain many glucose subunits.

Starch and Glycogen

Starch and **glycogen** are large storage forms of glucose found in plants and animals. Some of the polymers in starch are long chains of up to 4,000 glucose units. Starch has fewer side branches, or chains of glucose that branch off from the main chain, than does glycogen, as shown in Figures 2.16 and 2.17. Flour, which we use for baking and usually acquire by grinding wheat, is high in starch, and so are potatoes.

After we eat starchy foods such as potatoes and bread, starch is hydrolyzed into glucose, which will then enter the bloodstream. The liver stores glucose as glycogen. In between meals, the liver releases glucose so that the blood glucose concentration is always about 0.1%.

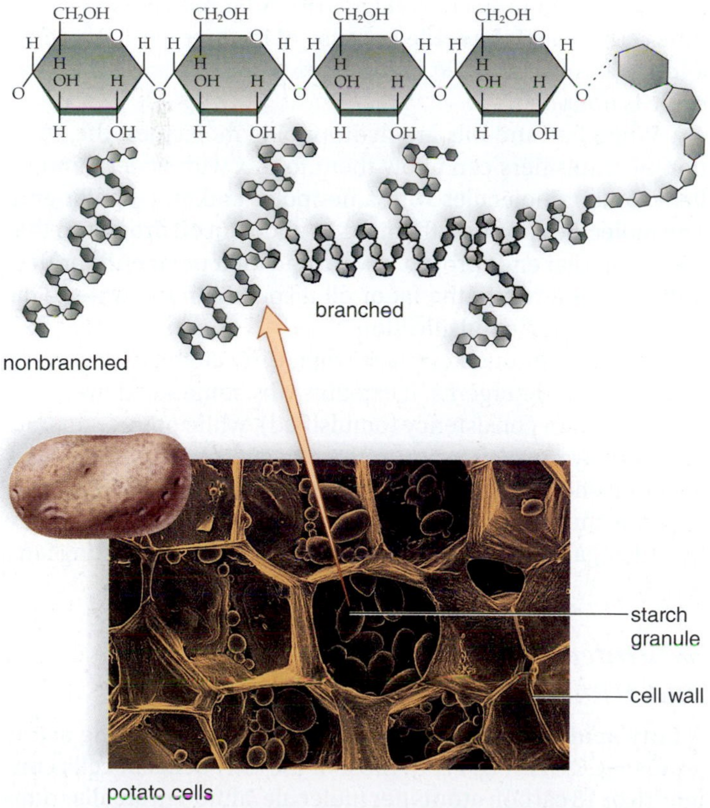

branched
nonbranched
starch granule
cell wall
potato cells

Figure 2.16 Starch structure and function. Starch is composed of chains of glucose molecules. Some chains are branched, as indicated. Starch is the storage form of glucose in plants. The electron micrograph shows starch granules in potato cells.

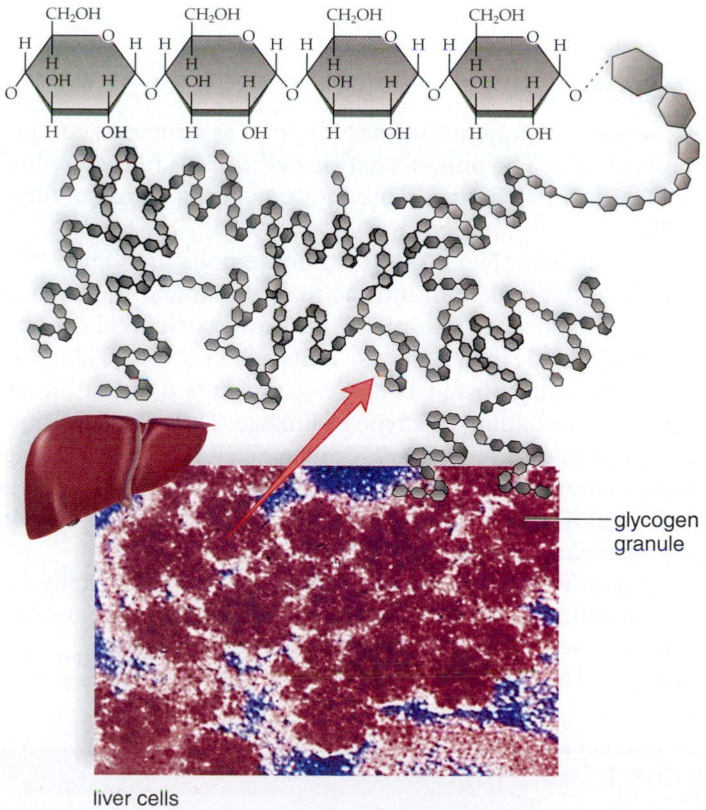

glycogen granule
liver cells

Figure 2.17 Glycogen structure and function. Glycogen is a highly branched polymer of glucose molecules that serves as the storage form of glucose in animals. The electron micrograph shows glycogen granules in liver cells.

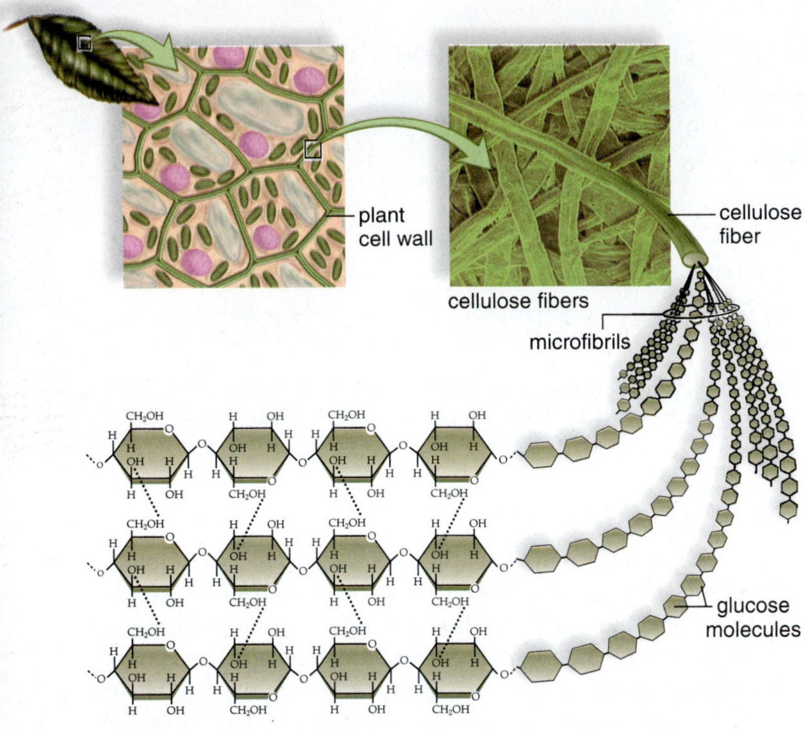
plant
cell wall
cellulose
fiber
cellulose fibers
microfibrils
glucose
molecules

Figure 2.18 Cellulose structure and function. In cellulose, the linkage between glucose molecules is slightly different from that in starch or glycogen. Plant cell walls contain cellulose, and the rigidity of these cell walls permits nonwoody plants to stand upright as long as they receive an adequate supply of water.

Cellulose

Some types of polysaccharides function as structural components of cells. The polysaccharide **cellulose** is found in plant cell walls, which helps account for the strong nature of these walls.

In cellulose (Fig. 2.18), the glucose units are joined by a slightly different type of linkage than that found in starch or glycogen. Notice the alternating up/down position of the oxygen atoms in the linked glucose units in Figure 2.18. This small difference is significant because it prevents us from digesting foods containing this type of linkage. Therefore, cellulose largely passes through our digestive tract as fiber, or roughage. Most doctors now recognize that fiber in the diet is necessary to good health, and some studies have suggested it may even help prevent colon cancer.

Chitin, which is found in the exoskeleton (shell) of crabs and related animals, is another structural polysaccharide. Scientists have discovered that chitin can be made into a thread and used as a suture material.

MP3 Carbohydrates

Check Your Progress 2.5

1. Identify the structural element that all carbohydrates have in common.
2. Explain why starch in plants is a source of glucose for our bodies but cellulose in plants is not.

2.6 Lipids

Learning Outcomes

Upon completion of this section, you should be able to

1. Compare the structures of fats, phospholipids, and steroids.
2. Identify the functions lipids play in our bodies.

Lipids contain more energy per gram than other biological molecules while fats and oils function as energy storage molecules in organisms. Phospholipids form a membrane that separates the cell from its environment and forms its inner compartments as well. The steroids are a large class of lipids that includes, among others, the sex hormones.

Lipids are diverse in structure and function, but they have a common characteristic: they do not dissolve in water. Lipids are hydrophobic.

Fats and Oils

The most familiar lipids are those found in fats and oils. **Fats** tend to be of animal origin (e.g., lard and butter), and are solid at room temperature. **Oils,** which are usually of plant origin (e.g., corn oil and soybean oil), are liquid at room temperature. Fat has several functions in the body: it is used for long-term energy storage, it insulates against heat loss, and it forms a protective cushion around major organs.

Fats and oils form when one glycerol molecule reacts with three fatty acid molecules (Fig. 2.19). A fat molecule is sometimes called a **triglyceride** because of its three-part structure, and the term *neutral fat* is sometimes used because the molecule is nonpolar.

While fats and oils are hydrophobic molecules, the addition of emulsifiers can allow them to mix with water. Emulsifiers contain molecules with a nonpolar end and a polar end. The molecules position themselves about an oil droplet so that their nonpolar ends project inward and their polar ends project outward. As a result, the fat or oil disperses in the water. This process is called emulsification.

Emulsification takes place when dirty clothes are washed with soaps or detergents. It explains why some salad dressings are uniform in consistency (emulsified), while others separate into two layers. Also, prior to the digestion of fatty foods, fats are emulsified by bile in the intestines. The liver produces bile, which is then stored in the gallbladder. Individuals who have had their gallbladder removed may have trouble digesting fatty foods.

Saturated, Unsaturated, and Trans-Fatty Acids

A **fatty acid** is a hydrocarbon chain that ends with the acidic group —COOH (Fig. 2.19). Most of the fatty acids in cells contain 16 or 18 carbon atoms per molecule, although smaller ones with fewer carbons are also known.

Fatty acids are either saturated or unsaturated. **Saturated fatty acids** have no double covalent bonds between carbon atoms. The carbon chain is saturated, so to speak, with all

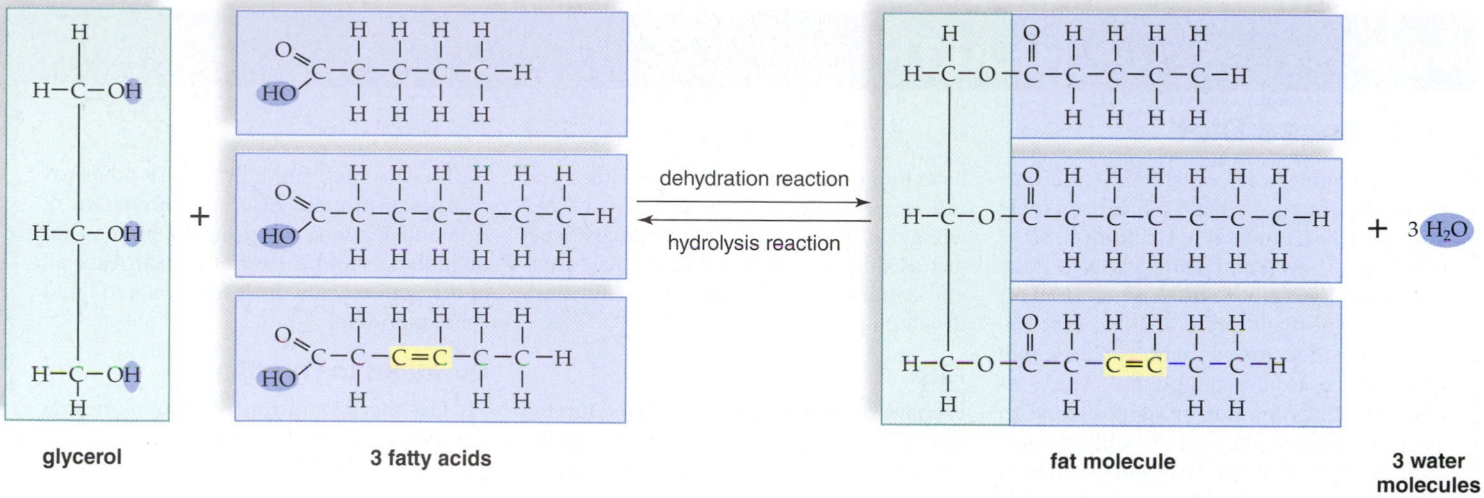
dehydration reaction
hydrolysis reaction
glycerol
3 fatty acids
fat molecule
3 water molecules

Figure 2.19 **Synthesis and degradation of a triglyceride.** Fatty acids can be saturated or unsaturated. Saturated fatty acids have no double bonds between carbon atoms, whereas unsaturated fatty acids have some double bonds (colored yellow) between carbon atoms. When a fat molecule (triglyceride) forms, three fatty acids combine with glycerol, and three water molecules are produced.

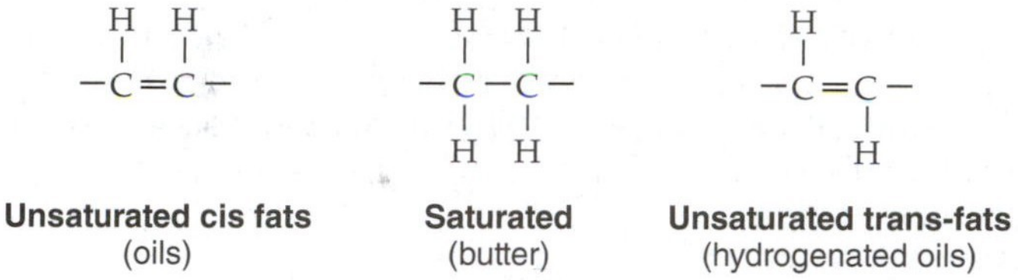
Unsaturated cis fats
(oils)
Saturated
(butter)
Unsaturated trans-fats
(hydrogenated oils)

Figure 2.20 **Comparison of saturated fats, unsaturated fats, and trans-fats.** Saturated fats have no double bonds between the carbon atoms, whereas unsaturated fats possess one or more double bonds. In a trans-fat, the hydrogen atoms are on opposite sides of the double bond.

the hydrogens it can hold. Saturated fatty acids account for the solid nature at room temperature of fats such as lard and butter. **Unsaturated fatty acids** have double bonds between carbon atoms wherever the number of hydrogens is less than two per carbon atom. Unsaturated fatty acids account for the liquid nature of vegetable oils at room temperature. Unsaturated fats are also often referred to as being "cis" or "trans." This terminology refers to the configuration of the hydrogen atoms in the double-bond of an unsaturated fat (Fig. 2.20). **Trans-fats** are often produced by hydrogenation, or the chemical addition of hydrogen to vegetable oils. This is done to convert the fat into a solid and is often found in processed foods.

Phospholipids

Phospholipids, as their name implies, contain a phosphate group. Essentially they are constructed like fats, except that in place of the third fatty acid, there is a polar phosphate group or a grouping that contains both phosphate and nitrogen. These molecules are not electrically neutral, as are fats, because the phosphate and nitrogen-containing groups are ionized. They form the polar (hydrophilic) head of the molecule, while the rest of the molecule becomes the nonpolar (hydrophobic) tails (Fig. 2.21).

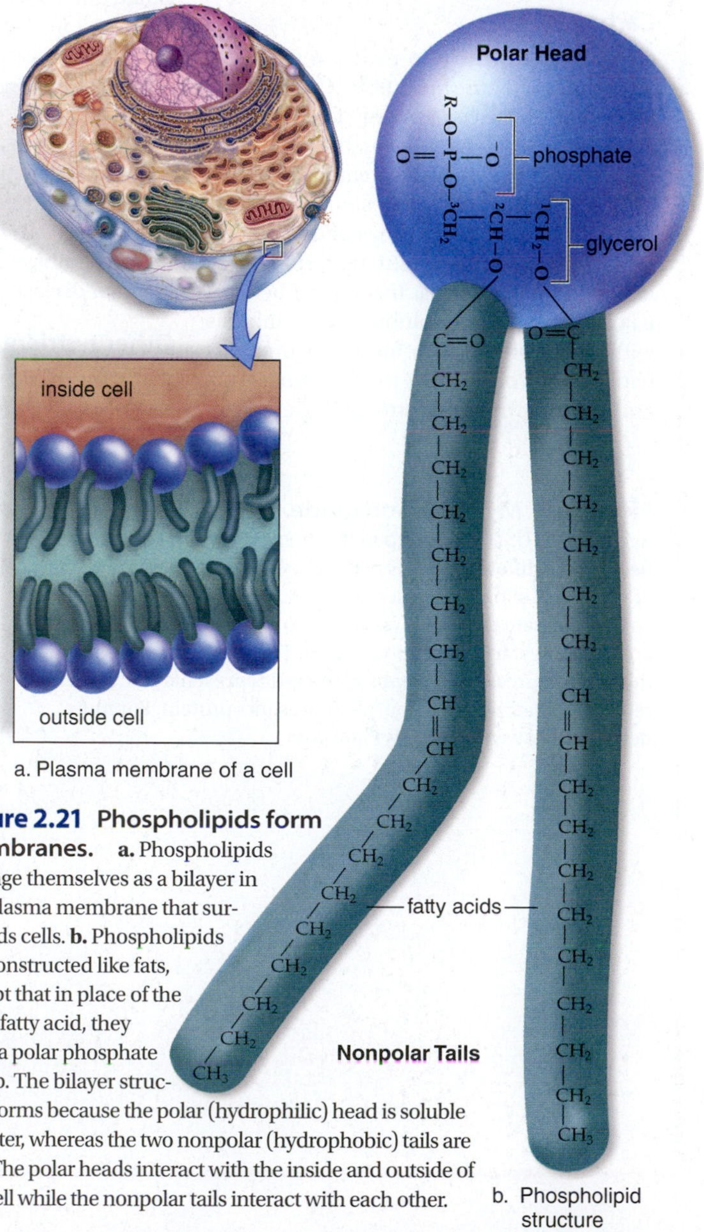
inside cell
outside cell
a. Plasma membrane of a cell
Polar Head
phosphate
glycerol
fatty acids
Nonpolar Tails
b. Phospholipid structure

Figure 2.21 **Phospholipids form membranes.** **a.** Phospholipids arrange themselves as a bilayer in the plasma membrane that surrounds cells. **b.** Phospholipids are constructed like fats, except that in place of the third fatty acid, they have a polar phosphate group. The bilayer structure forms because the polar (hydrophilic) head is soluble in water, whereas the two nonpolar (hydrophobic) tails are not. The polar heads interact with the inside and outside of the cell while the nonpolar tails interact with each other.

SCIENCE IN YOUR LIFE ▶ HEALTH

A Balanced Diet

Everyone agrees that we should eat a balanced diet, but just what is a balanced diet? The U.S. Department of Agriculture (USDA) released a new Food Plate in June 2011 (Fig. 2B). The new food plate contains a very colorful plate that is divided into the four basic food groups with a side circle representing a dairy component. The general guidelines are encouraging people to consume fewer calories by avoiding oversized portions. Half the plate should include fruits and vegetables, while making half of the grains consumed being whole grains. Reduction of foods that contain high levels of sodium is also recommended as is an increase in the amount of water consumed.

Carbohydrates

Carbohydrates include fruits, vegetables, and grains. They are the quickest, most readily available source of energy for the body. Complex carbohydrates, such as those in whole-grain breads and cereals, are preferable to simple carbohydrates, such as candy and ice cream, because they contain dietary fiber (nondigestible plant material), plus vitamins and minerals. Insoluble fiber has a laxative effect, and soluble fiber combines with the cholesterol in food and prevents cholesterol from exiting the intestinal tract and entering the bloodstream. Researchers have found that the starch in potatoes and processed foods, such as white bread and white rice, leads to a high blood glucose level just as simple carbohydrates do. Researchers wonder if this is why many adults are developing type 2 diabetes.

Fats

We have known for many years that saturated fats in animal products contribute to the formation of deposits called plaque, which clog arteries and lead to high blood pressure and heart attacks. Even more harmful than naturally occurring saturated fats are the so-called trans-fats, created artifically using vegetable oils. Trans-fats are partially hydrogenated to make them semisolid. Trans fats are found in shortenings, solid margarines, and some processed foods.

The new food plate advocates an intake of low-fat types of foods like 1% milk and low-fat yogurt instead of whole milk or regular yogurt. Polyunsaturated and monosaturated oils are still recommended in the Food Plate. These oils have been found to be protective against the development of cardiovascular disease.

Other nutrients

Red meat is rich in protein, but it is usually also high in saturated fat; therefore, lean meats, fish, and chicken are preferred sources of protein. Also, a combination of rice and *legumes* (a group of plants that includes peas and beans) can provide all of the amino acids the body needs to build cellular proteins.

Questions to Consider

1. List the proportion of your plate that should consist of fruits and vegetables. Are you getting the recommended amounts of these important foods? Do you have variety in the types of fruits and vegetables you consume?
2. Why are some fats "good" for you and some "bad" for you? What are ideal sources for these fats?
3. Chemically, what is the difference between a "whole-grain" carbohydrate and a simple carbohydrate?

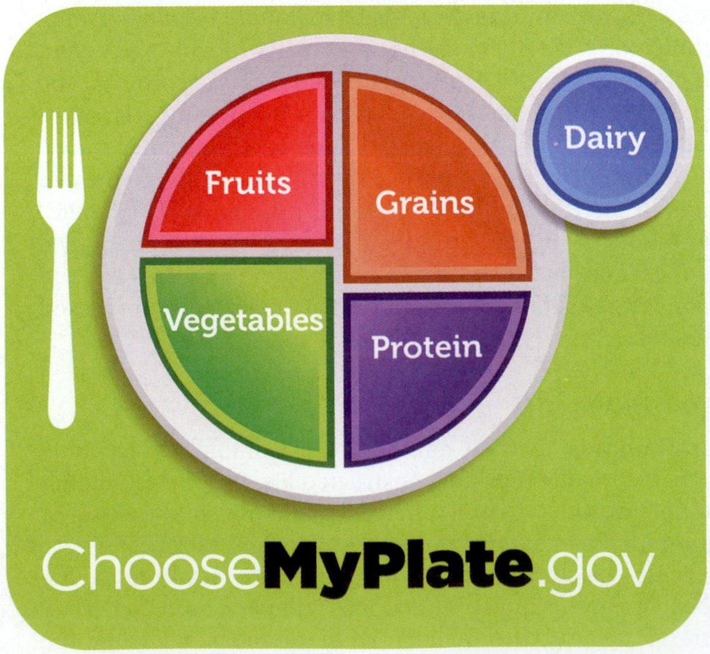
Fruits
Grains
Dairy
Vegetables
Protein
ChooseMyPlate.gov

Figure 2B **MyPlate food guide.** The U.S. Department of Agriculture (USDA) developed this Food Plate as a guide to better health. The different widths of the food groups suggest what portion of your meal should consist of each category. The five different colors illustrate that foods, in correct proportion from all groups, are needed each day for good health. The orange represents grains; the green represents vegetables; the red represents fruits; the blue represents dairy; and the purple represents protein. For more details visit www.ChooseMyPlate.gov.

OH
CH$_3$
CH$_3$
O
OH
CH$_3$
HO

a. Testosterone b. Estrogen

Figure 2.22 Steroids. All steroids have four adjacent rings, but their attached groups differ. The effects of **(a)** testosterone and **(b)** estrogen on the body largely depend on the difference in the attached groups (shown in blue).

Phospholipids illustrate that the chemistry of a molecule helps determine its function. Phospholipids are the primary components of cellular membranes. They spontaneously form a bilayer in which the hydrophilic heads face outward toward watery solutions and the tails form the hydrophobic interior. Plasma membranes separate extracellular from intracellular environments and are absolutely vital to the form and function of a cell.

Steroids

Steroids have a backbone of four fused carbon rings. Each one differs primarily by the arrangement of the atoms in the rings and the type of functional groups attached to them. Cholesterol is a steroid formed by the body that also enters the body as part of our diet. Cholesterol has several important functions. It is a component of an animal cell's plasma membrane and is the precursor of several other steroids, such as bile salts and the sex hormones testosterone and estrogen (Fig. 2.22).

We now know that a diet high in saturated fats, trans-fats, and cholesterol can cause fatty material to accumulate inside the lining of blood vessels thereby reducing blood flow. The Health feature, "A Balanced Diet," discusses which sources of carbohydrates, fats, and proteins are recommended for inclusion in the diet.

MP3 Lipids

Check Your Progress 2.6

1. List the two types of lipid molecules found in the plasma membranes of animal cells.
2. Explain how the presence of a double bond in an unsaturated fatty acid affects whether that substance is a solid or liquid.

2.7 Proteins

Learning Outcomes

Upon completion of this section, you should be able to

1. Describe the functions of proteins in cells.
2. Explain how a polypeptide is constructed from amino acids.
3. Compare the four levels of protein structure.

Proteins are polymers composed of amino acid monomers. An amino acid has a central carbon atom bonded to a hydrogen atom and three functional groups. The name of the molecule is appropriate because one of these groups is an amino group (—NH_2) and another is an acidic group (—COOH). The third group is called an *R* group which determines the uniqueness of each **amino acid.** The *R* group varies from having a single carbon to being a complicated ring structure (Fig. 2.23).

Proteins perform many functions. Proteins such as keratin, which makes up hair and nails, and collagen, which lends support to ligaments, tendons, and skin, are structural proteins. Some proteins are enzymes. Enzymes are necessary

Name	Structural Formula	*R* Group
alanine (Ala)	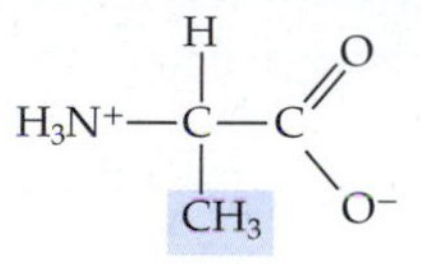	*R* group has a single carbon atom
valine (Val)	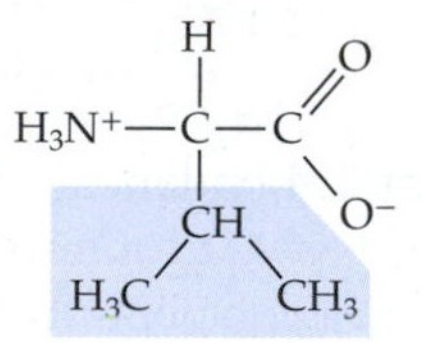	*R* group has a branched carbon chain
cysteine (Cys)	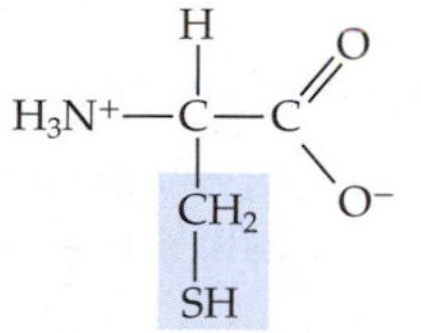	*R* group contains sulfur
phenylalanine (Phe)	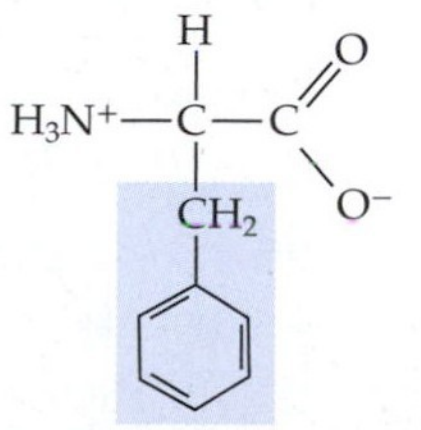	*R* group has a ring structure

Figure 2.23 Representative amino acids. Amino acids differ from one another by their *R* group. The simplest *R* group is a single hydrogen atom (H). *R* groups (blue) containing carbon vary as shown.

contributors to the chemical workings of the cell, and the body. **Enzymes** speed chemical reactions. They work so quickly that a reaction that normally takes several hours or days without an enzyme takes only a fraction of a second with an enzyme. Many hormones, messengers that influence cellular metabolism, are also proteins. The proteins actin and myosin account for the movement of cells and the ability of our muscles to contract. Some proteins transport molecules in the blood. Hemoglobin is a complex protein in our blood that transports oxygen. Antibodies in blood and other body fluids are proteins that combine with foreign substances, preventing them from destroying cells and upsetting homeostasis.

Proteins in the plasma membrane of cells have various functions: some form channels that allow substances to enter and exit cells; some are carriers that transport molecules into and out of the cell; and some are enzymes.

Peptides

Figure 2.24 shows that a synthesis reaction between two amino acids results in a dipeptide and a molecule of water. A **polypeptide** is a chain of amino acids that are joined to one another by a **peptide bond.** The atoms associated with a peptide bond unevenly because oxygen is more electronegative than nitrogen. Therefore, the hydrogen attached to the nitrogen has a slightly positive charge, while the oxygen has a slightly negative charge:

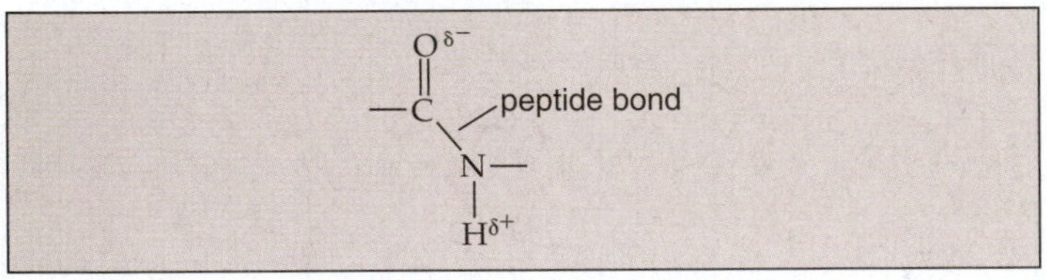
$O^{\delta-}$
peptide bond
$H^{\delta+}$

The polarity of the peptide bond means that hydrogen bonding is possible between the C=O of one amino acid and the N—H of another amino acid in a polypeptide. This hydrogen bonding influences the structure, or shape, of a protein.

Levels of Protein Organization

Proteins can have up to four levels of structural organization (Fig. 2.25). The first level, called the *primary structure*, is the linear sequence of the amino acids joined by peptide bonds. Polypeptides can be quite different from one another. A polypeptide chain is made from 20 different amino acids. Each particular polypeptide has its own sequence of amino acids and its own sequence of *R* groups.

The *secondary structure* of a protein comes about when the polypeptide takes on a certain orientation in space. A coiling of the chain results in an α (alpha) helix, or a right-handed spiral, similar to a spiral staircase. Or a folding of the chain results in a β (beta) pleated sheet similar to a hand-held fan. Hydrogen bonding between peptide bonds holds the shape in place.

The *tertiary structure* of a protein is its final three-dimensional shape. In muscles, myosin molecules have a rod shape ending in globular (globe-shaped) heads. In enzymes, the polypeptide bends and twists in different ways. Invariably, the hydrophobic portions are packed mostly on the inside, and the hydrophilic portions are on the outside where they can make contact with water. The tertiary shape of a polypeptide is maintained by various types of bonding between the *R* groups; covalent, ionic, and hydrogen bonding all occur. One common form of covalent bonding between *R* groups is a disulfide (S—S) linkage between two cysteine amino acids.

Some proteins have only one polypeptide, and others have more than one polypeptide, each with its own primary, secondary, and tertiary structures. In proteins with multiple polypeptide chains, these separate polypeptides are arranged to give such proteins a fourth level of structure, termed the *quaternary structure.* Hemoglobin is a complex protein having a quaternary structure. Most enzymes also have a quaternary structure. Thus, proteins can differ in many ways, such as in length, sequence, and structure. Each individual protein is chemically unique as well.

MP3 Protein

The final shape of a protein is very important to its function. As we will discuss in Chapter 6, enzymes cannot function unless they have their normal shape. When proteins are exposed to extremes in heat and pH, they undergo an irreversible change in shape. This is referred to as being **denatured.** For example, we are all aware that adding acid to milk causes curdling and that heating causes egg white, which contains a protein called albumin, to coagulate. Denaturation occurs because the normal bonding between the *R* groups has been

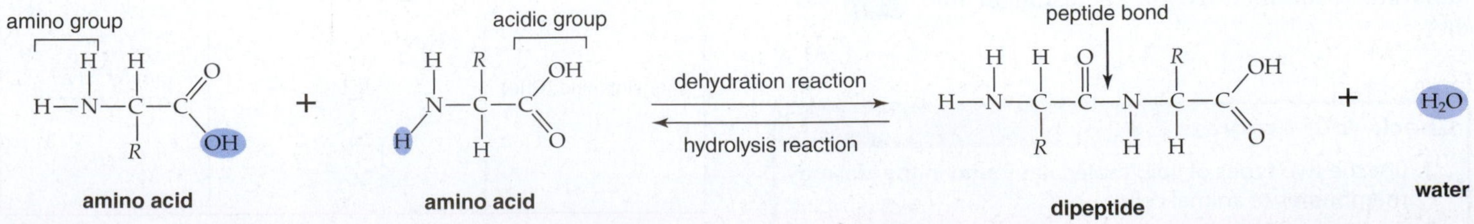
amino group
acidic group
peptide bond
dehydration reaction
hydrolysis reaction
H_2O
amino acid
amino acid
dipeptide
water

Figure 2.24 Synthesis and degradation of a dipeptide. Following a dehydration reaction, a peptide bond joins two amino acids, and a water molecule is released. Following a hydrolysis reaction, the bond is broken with the addition of water.

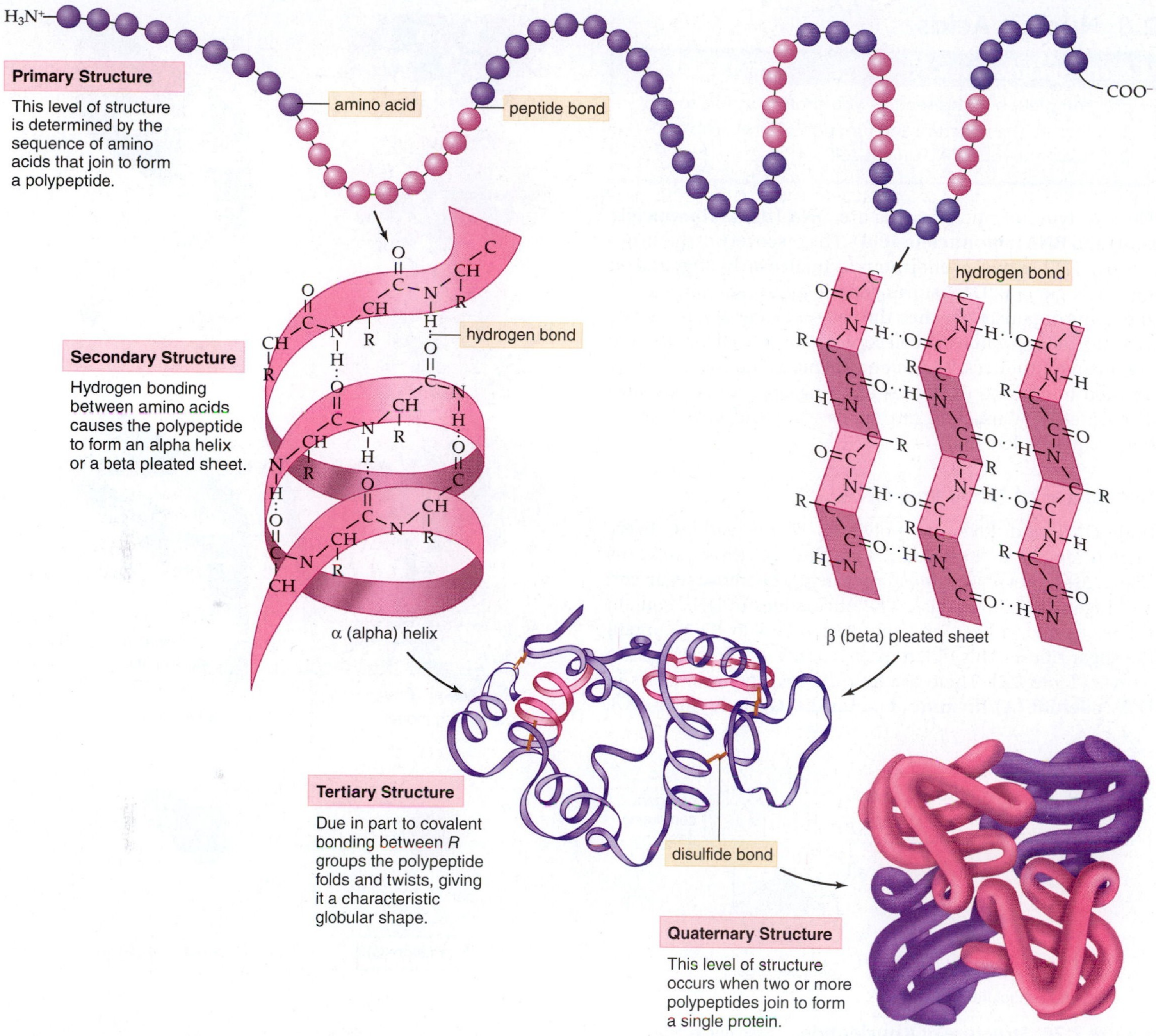
Primary Structure
This level of structure is determined by the sequence of amino acids that join to form a polypeptide.
amino acid
peptide bond
Secondary Structure
Hydrogen bonding between amino acids causes the polypeptide to form an alpha helix or a beta pleated sheet.
hydrogen bond
α (alpha) helix
β (beta) pleated sheet
Tertiary Structure
Due in part to covalent bonding between R groups the polypeptide folds and twists, giving it a characteristic globular shape.
disulfide bond
Quaternary Structure
This level of structure occurs when two or more polypeptides join to form a single protein.

Figure 2.25 Levels of protein organization. All proteins have a primary structure. Both fibrous and globular proteins have a secondary structure. They are either α (alpha) helices or β (beta) pleated sheets. Globular proteins always have a tertiary structure, and most have a quaternary structure.

disturbed. Once a protein loses its normal shape, it is no longer able to function normally. For example researchers hypothesize that an alteration in protein organization, forming structures called *prions*, is related to the development of Alzheimer disease and Creutzfeldt-Jakob disease (the human form of "mad cow" disease).

Animation Protein Denaturation

Animation How Prions Arise

Check Your Progress 2.7

1. List some of the functions of proteins.
2. Describe how amino acids are formed.
3. Compare and contrast the four levels of protein structure.

2.8 Nucleic Acids

Learning Outcomes

Upon completion of this section, you should be able to

1. Compare the structure and function of DNA and RNA.
2. Explain the role of ATP in the cell.

The two types of nucleic acids are **DNA** (**deoxyribonucleic acid**) and **RNA** (**ribonucleic acid**). The discovery of the structure of DNA has had an enormous influence on biology and on society in general. DNA stores genetic information in the cell and in the organism. Further, the cell replicates and transmits this information when the cell copies itself as well as when the organism reproduces. The science of biotechnology is largely devoted to altering the genes in living organisms. We have discovered how many of our genes work and have learned how to manipulate them.

Structure of DNA and RNA

Both DNA and RNA are polymers of nucleotides. Every **nucleotide** is a molecular complex of three subunits (Fig. 2.26)—phosphate (phosphoric acid), a pentose sugar, and a nitrogen-containing base. The nucleotides in DNA contain the sugar deoxyribose and the nucleotides in RNA contain the sugar ribose. This difference accounts for their respective names (Table 2.2). There are four different types of bases in DNA: **adenine** (**A**), **thymine** (**T**), **guanine** (**G**), and **cytosine** (**C**).

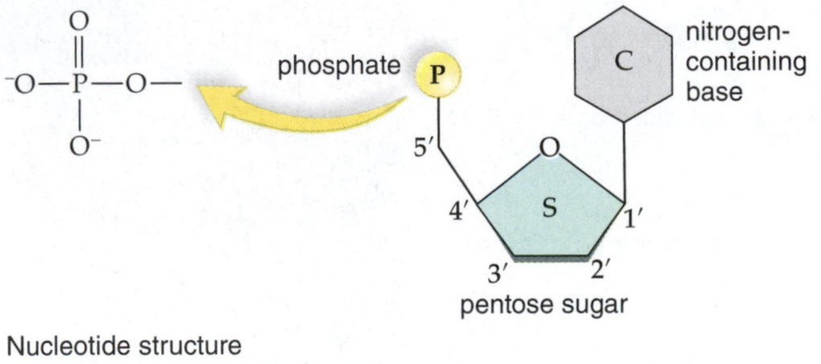
phosphate
P
nitrogen-containing base
C
S
5′
4′
3′
2′
1′
pentose sugar
Nucleotide structure

Figure 2.26 Structure of a nucleotide. Each nucleotide consists of a pentose sugar, a nitrogen-containing base, and a phosphate functional group.

TABLE 2.2 DNA Structure Compared With RNA Structure		
	DNA	**RNA**
Sugar	Deoxyribose	Ribose
Bases	Adenine, guanine, thymine, cytosine	Adenine, guanine, uracil, cytosine
Strands	Double stranded with base pairing	Single stranded
Helix	Yes	No

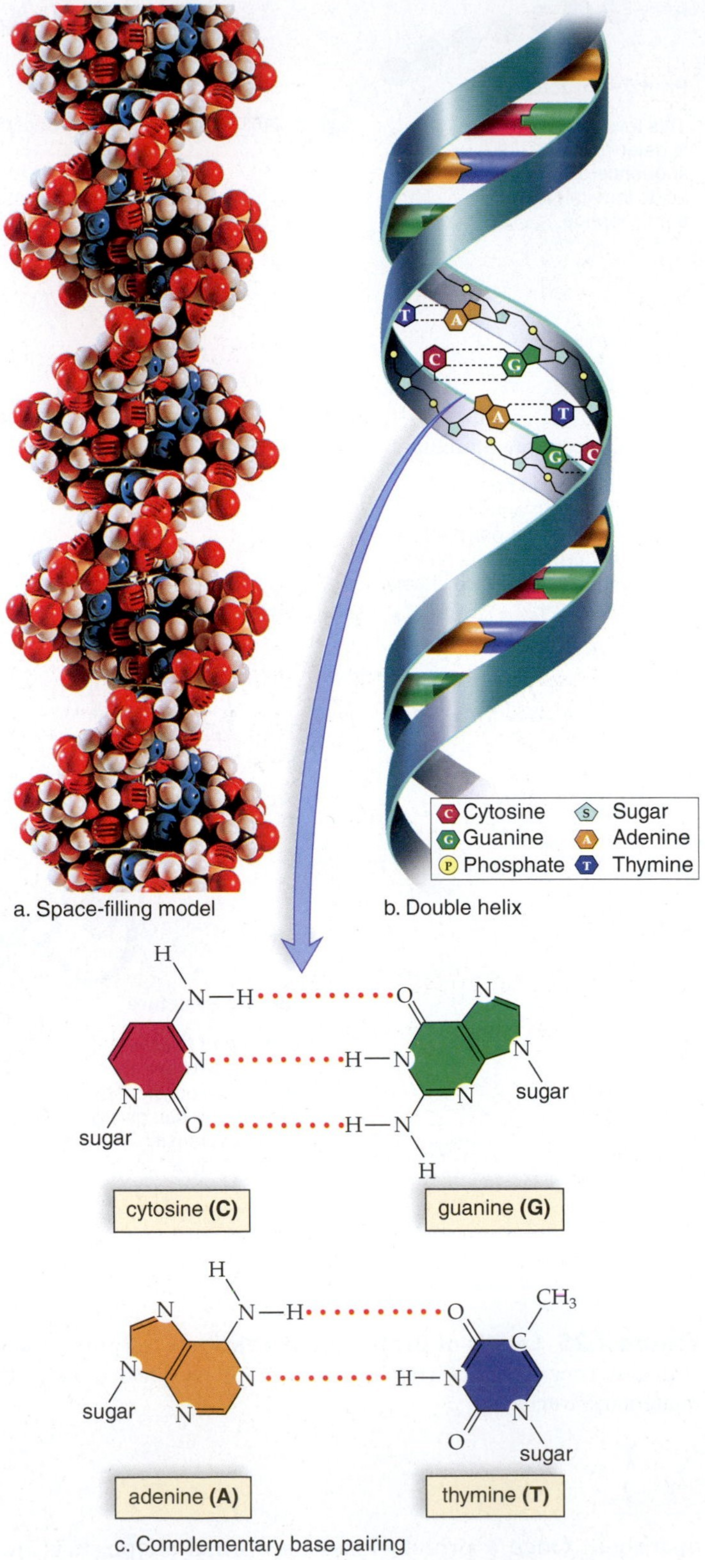
Cytosine
Guanine
Phosphate
Sugar
Adenine
Thymine
a. Space-filling model
b. Double helix
sugar
cytosine (C)
guanine (G)
sugar
adenine (A)
thymine (T)
c. Complementary base pairing

Figure 2.27 Overview of DNA structure. The structure of DNA is absolutely essential to its ability to replicate and to serve as the genetic material. **a.** Space-filling model of DNA's double helix. **b.** Complementary base pairing between strands. **c.** Ladder configuration. Notice that the uprights are composed of alternating phosphate and sugar molecules and that the rungs are complementary nitrogen-containing paired bases. The heredity information stored by DNA is the sequence of its bases, which determines the primary structure of the cell's proteins.

The base can have two rings (adenine or guanine) or one ring (thymine or cytosine). In RNA, the base **uracil (U)** replaces the base thymine. These structures are called bases because their presence raises the pH of a solution.

The nucleotides form a linear molecule called a strand, which has a backbone made up of alternating phosphates and sugars with the bases projecting to one side of the backbone. The nucleotides and their bases occur in a specific order. After many years of work, researchers now know the sequence of the bases in human DNA—the human genome. This breakthrough is expected to lead to improved genetic counseling, gene therapy, and medicines to treat the causes of many human illnesses.

DNA is double stranded, with the two strands twisted about each other in the form of a **double helix** (Fig. 2.27*a, b*). In DNA, the two strands are held together by hydrogen bonds between the bases. When unwound, DNA resembles a ladder. The uprights (sides) of the ladder are made entirely of the alternating phosphate and sugar molecules, and the rungs of the ladder are made only of complementary paired bases. Thymine (T) always pairs with adenine (A), and guanine (G) always pairs with cytosine (C). Complementary bases have shapes that fit together (Fig. 2.27*c*).

Complementary base pairing allows DNA to replicate in a way that ensures the sequence of bases will remain the same. The base sequence of specific sections of DNA contain a code that specifies the sequence of amino acids in the proteins of the cell.

Animation DNA Structure

RNA is single stranded and is formed by complementary base pairing with one DNA strand. There are several types of RNA. One type of RNA, mRNA or messenger RNA, carries the information from the DNA strand to the ribosome where it is translated into the sequence of amino acids specified by the DNA.

MP3 Nucleic Acids

ATP (Adenosine Triphosphate)

In addition to being the monomers of nucleic acids, nucleotides have other metabolic functions in cells. When adenosine (adenine plus ribose) is modified by the addition of three phosphate groups instead of one, it becomes **ATP (adenosine triphosphate)**, an energy carrier in cells. Glucose is broken down in a step-wise fashion so that the energy of glucose is converted to that of ATP molecules. ATP molecules serve as small "energy packets" suitable for supplying energy to a wide variety of a cell's chemical reactions. ATP can be said to be the energy "currency" of the cell. Reactions in the cell that need energy require ATP.

ATP is a high-energy molecule because the last two phosphate bonds are unstable and easily broken. In cells, the terminal phosphate bond usually is hydrolyzed, leaving the

SCIENCE IN YOUR LIFE ▶ **BIOETHICAL**

Blue Gold

Environmentalists believe that the world is running out of clean drinking water. Over 97% of the world's water is salt water found in the oceans. Salt water is unsuitable for drinking without expensive desalination. Of the fresh water in the world, most is locked in frozen form in the polar ice caps and glaciers and therefore unavailable. This leaves only a small percentage in groundwater, lakes, and rivers that could be available for drinking, industry, and irrigation. However, some of that water is polluted and unsuitable.

Water has always been the most valuable commodity in the Middle East, even more valuable than oil. But as fresh water becomes limited and the world's population grows, the lack of sufficient clean water is becoming a worldwide problem.

The combination of increasing demand and dwindling supply has attracted global corporations who want to sell water. Water is being called the "blue gold" of the twenty-first century, and an issue has arisen regarding whether the water industry should be privatized. That is, could water rights be turned over to private companies to deliver clean water and treat wastewater at a profit, similar to the way oil and electricity are handled? Private companies have the resources to upgrade and modernize water delivery and treatment systems, thereby conserving more water. However, opponents of this plan claim that water is a basic human right required for life, not a need to be supplied by the private sector. In addition, a corporation can certainly own the pipelines and treatment facilities, but who owns the rights to the water? For example, North America's largest underground aquifer, the Ogallala, covers 175,000 square miles under several states in the southern Great Plains. If water becomes a commodity, do we allow water to be taken away from people who cannot pay in order to give it to those who can?

Questions to Consider

1. Do you agree that the water industry should be privatized? Why or why not?
2. Is access to clean water a "need" or a "right"? If it is a right, who pays for that right?
3. Because water is a shared resource, everyone believes they can use water, but few people feel responsible for conserving it. What can you do to conserve water?

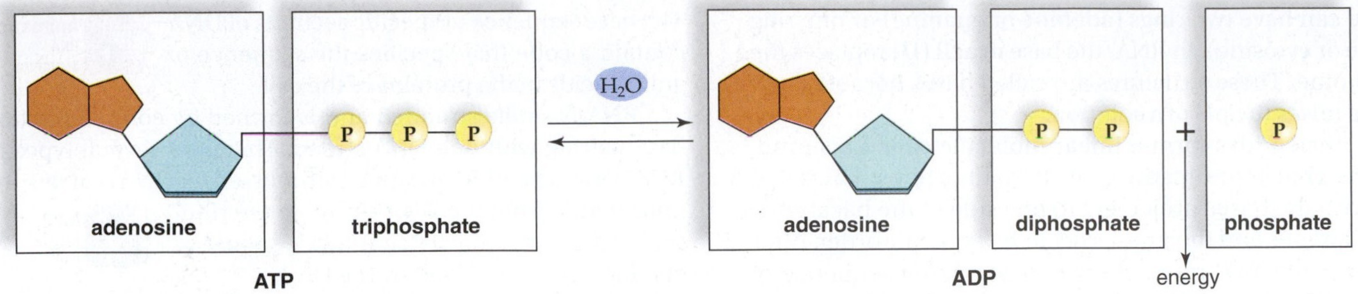

H_2O
P
P
P
adenosine
triphosphate
ATP
P
P
P
adenosine
diphosphate
phosphate
ADP
energy

Figure 2.28 ATP reaction. ATP, the universal energy "currency" of cells, is composed of adenosine and three phosphate groups (called a triphosphate). When cells require energy, ATP undergoes hydrolysis, producing ADP+Ⓟ, with the release of energy.

molecule **ADP** (**adenosine diphosphate**) and a molecule of inorganic phosphate Ⓟ (Fig. 2.28). The cell uses the energy released by ATP breakdown to synthesize macromolecules such as carbohydrates and proteins. Muscle cells use the energy for muscle contraction, while nerve cells use it for the conduction of nerve impulses. After ATP breaks down, it is rebuilt by the addition of Ⓟ to ADP. Notice in Figure 2.28 that an input of energy is required to re-form ATP.

MP3 ATP

Check Your Progress 2.8

1. Identify how information is stored in DNA.
2. Describe how energy is stored in ATP.

Case Study Conclusion

Most people realize that during and after exercising they need to replace their lost fluids and electrolytes. Each of us will be unique in how much fluid we lose based upon our individual body chemistry and the nature of our activity. What people are starting to realize is that it isn't necessary to replace lost fluids with sports drinks if the duration of our activity level is relatively short. While sports drinks are tasty and appear to be a growing fad, people need to pay more attention to the calories that they contain. The excessive consumption of sports drinks, when not necessary, can deter a person's ability to lose weight, which can lead to health problems. Our best bet to maintain hydration levels is to rely on water, the essential element of life.

MEDIA STUDY TOOLS

www.mhhe.com/maderinquiry14

Enhance your study of this chapter with study tools and practice tests. Also ask your instructor about the resources available through ConnectPlus, including LearnSmart, the media-rich eBook, interactive learning tools, and animations.

SUMMARIZE

2.1 Basic Chemistry

- Both living organisms and nonliving things are composed of **matter** consisting of **elements.** The acronym CHNOPS stands for the most significant elements found in all organisms: carbon, hydrogen, nitrogen, oxygen, phosphorus, and sulfur.
- Elements contain atoms, and **atoms** are composed of subatomic particles. **Protons** and **neutrons** in the nucleus determine the **atomic mass** of an atom. The **atomic number** indicates the number of protons and the number of **electrons** in electrically neutral atoms. Protons have positive charges, neutrons are uncharged, and electrons have negative charges. **Isotopes** are atoms of a single element that differ in their numbers of neutrons. **Radioactive isotopes** have many uses, including serving as tracers in biological experiments and medical procedures.
- The number of electrons in the outer orbital determines the reactivity of an atom. The first orbital is complete when it is occupied by two electrons. In atoms up through calcium, number 20, every orbital beyond the first one is complete with eight electrons. The **octet rule** states that atoms react with one another in order to have a completed outer orbital.

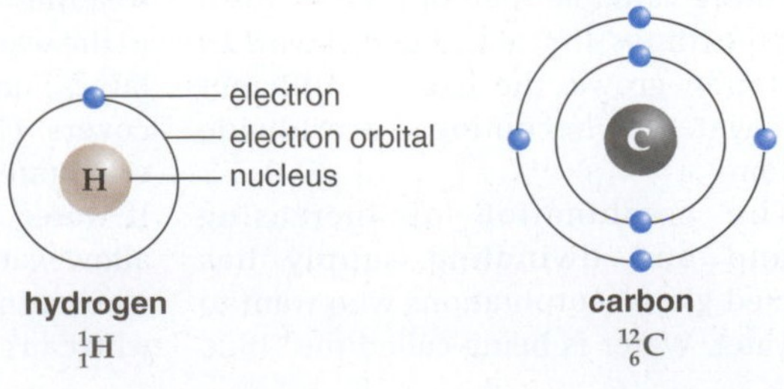

electron
electron orbital
nucleus
H
hydrogen
$^{1}_{1}H$
C
carbon
$^{12}_{6}C$

2.2 Molecules and Compounds

- **Ions** form when atoms lose or gain one or more electrons to achieve a completed outer orbital. An **ionic bond** is an attraction between oppositely charged ions. A **covalent bond** is one or more shared pairs of electrons.
- When carbon combines with four other atoms, the resulting **molecule** has a tetrahedral shape. The shape of a molecule is important to its biological role.
- In polar covalent bonds, the sharing of electrons is not equal. One of the atoms exerts greater attraction for the shared electrons than the other, resulting in a difference in **electronegativity.** This creates a slight charge on each atom. A **hydrogen bond** is a weak attraction between a slightly positive hydrogen atom and a slightly negative oxygen or nitrogen atom within the same or a different molecule. Hydrogen bonds help maintain the structure and function of cellular molecules.

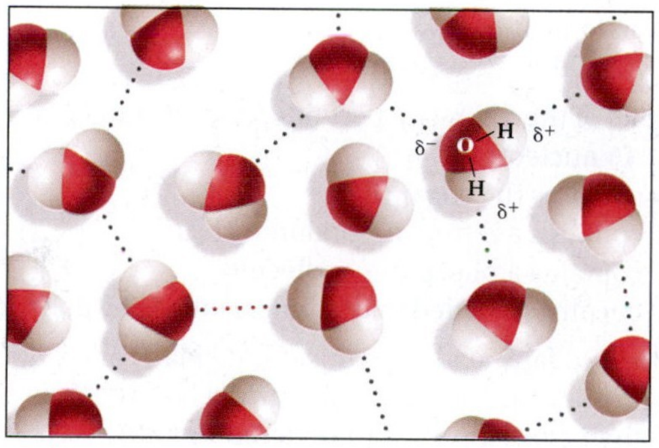
δ−
O
H
δ+
H
δ+

2.3 Chemistry of Water

- Water is a polar molecule. Its polarity allows hydrogen bonding to occur between water molecules. Water's polarity and hydrogen bonding account for its unique properties. These properties allow living things to exist and carry on cellular activities.
- A small fraction of water molecules dissociate to produce an equal number of hydrogen ions and hydroxide ions. Solutions with equal numbers of H^+ and OH^- are termed neutral. In **acidic** solutions, there are more hydrogen ions than hydroxide ions. These solutions have a pH less than 7. In **basic** solutions, there are more hydroxide ions than hydrogen ions. These solutions have a pH greater than 7. Cells are sensitive to pH changes. Biological systems often contain **buffers** that help keep the pH within a normal range.

$$\underset{\text{water}}{H{-}O{-}H} \rightleftharpoons \underset{\text{hydrogen ion}}{H^+} + \underset{\text{hydroxide ion}}{OH^-}$$

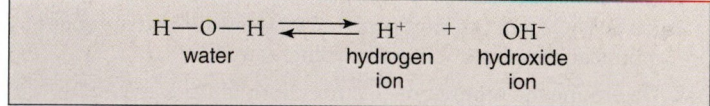

2.4 Organic Molecules

- The chemistry of carbon accounts for the chemistry of **organic molecules.** Carbohydrates, lipids, proteins, and nucleic acids are macromolecules with specific functions in cells.
- **Polymers** are formed by the joining together of **monomers.** Long-chain carbohydrates, proteins, and nucleic acids are all polymers. For each bond formed during a **dehydration reaction,** a molecule of water is removed. For each bond broken during a **hydrolysis** reaction, a molecule of water is added.

2.5 Carbohydrates

- Glucose is the 6-carbon sugar or **carbohydrate** most utilized by cells for "quick" energy. Sugar monomers join to form **polysaccharides.** Plants store glucose as **starch,** and animals store **glucose** as **glycogen.** Humans cannot digest **cellulose,** which is a structural component of plant cell walls.

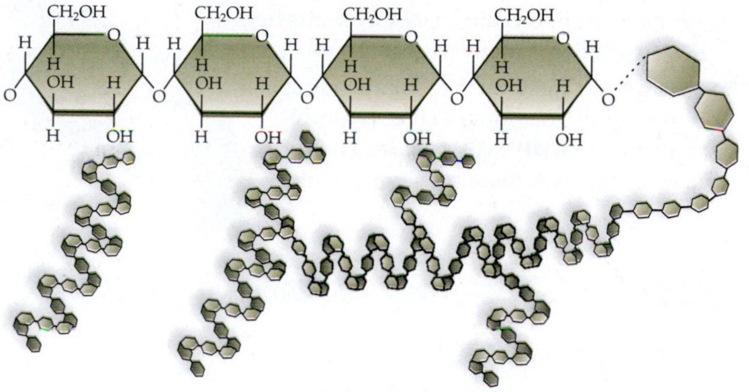
CH2OH
CH2OH
CH2OH
CH2OH
H
O
OH
H
OH

2.6 Lipids

- **Lipids** are varied in structure and function. **Fats** and **oils,** which function in long-term energy storage, contain glycerol and three **fatty acids.** Fatty acids can be **saturated** or **unsaturated.** Some unsaturated fats are called **trans-fats** due to the configuration of the atoms in the molecule. The **phospholipids** in the plasma membrane have polar heads and a nonpolar tail. Certain hormones are derived from cholesterol, a complex ring compound.

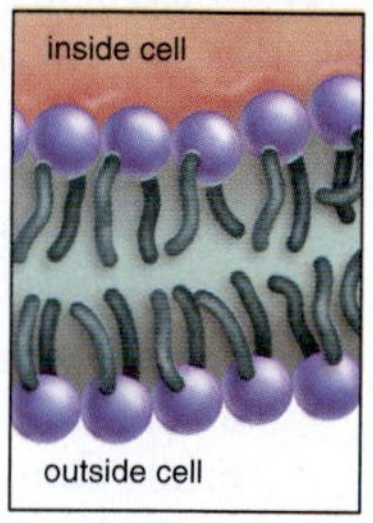
inside cell
outside cell

2.7 Proteins

- **Proteins** have numerous functions in cells. Some are **enzymes** that speed chemical reactions. The primary structure of a **polypeptide** is its own particular sequence of the possible 20 types of **amino acids.** The secondary structure is often an alpha (α) helix or a beta (β) pleated sheet. Tertiary structure occurs when a polypeptide bends and twists into a three-dimensional shape. A protein can contain several polypeptides, and this accounts for a possible quaternary structure.

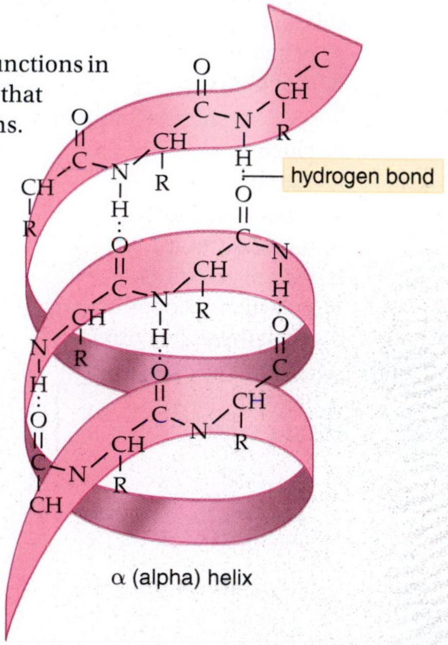
hydrogen bond
CH
R
C
N
H
O
α (alpha) helix

2.8 Nucleic Acids

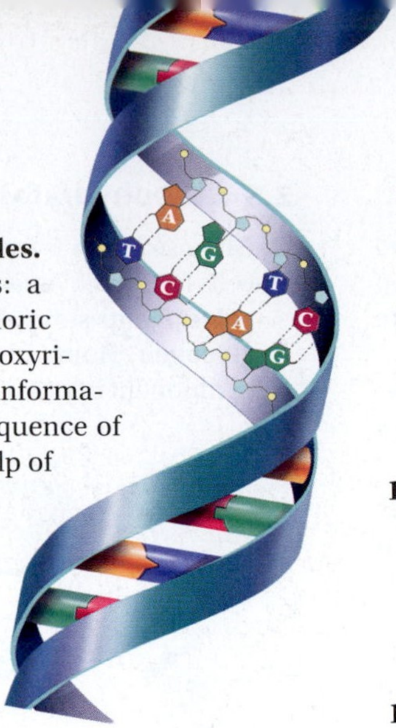
A
T
G
C
T
A
C
G

- Nucleic acids are polymers of **nucleotides.** Each nucleotide has three components: a sugar, a base, and phosphate (phosphoric acid). **DNA,** which contains the sugar deoxyribose, is the genetic material that stores information for its own replication and for the sequence of amino acids in proteins. DNA, with the help of **RNA,** specifies protein synthesis.
- **ATP,** with its unstable phosphate bonds, is the energy currency of cells. Hydrolysis of ATP to **ADP** + Ⓟ releases energy that the cell uses to do metabolic work.

ASSESS

Testing Yourself

Choose the best answer for each question.

1. Which of these ions is important for acid-base balance?
a. K^+ **b.** Na^+ **c.** Cl^- **d.** H^+

2. Which of these offers the most accurate information concerning the shape of a molecule?
a. electron-dot formula
b. structural formula
c. molecular formula
d. space-filling model

3. An example of a hydrogen bond would be
a. the bond between a carbon atom and a hydrogen atom.
b. the bond between two carbon atoms.
c. the bond between sodium and chlorine.
d. the bond between two water molecules.

4. Covalent bonding between *R* groups in proteins is associated with the ________________ structure.
a. primary **b.** secondary **c.** tertiary **d.** secondary and tertiary

5. The polysaccharide found in plant cell walls is
a. glucose. **b.** starch. **c.** cellulose. **d.** maltose.

6. Ionic bonding involves
a. loss of electrons.
b. gain of electrons.
c. attraction of opposite charges.
d. All of these are correct.

7. Phosphates can be found in
a. DNA. **b.** fatty acids. **c.** glucose. **d.** Both a and b are correct.

8. Which of these cannot be considered an organic molecule?
a. table salt **b.** glucose **c.** DNA **d.** enzyme

9. An example of a dehydration reaction would be
a. building amino acids from proteins.
b. building lipids from fatty acids.
c. breaking down glucose into starch.
d. breaking down starch into glucose.

10. Nucleotides
a. contain sugar, a nitrogen-containing base, and a phosphate molecule.
b. are the monomers for fats and polysaccharides.
c. join together by covalent bonding between the bases.
d. are found in DNA, RNA, and proteins.

11. The bond joining two adjacent amino acids is called
a. a peptide bond.
b. a hydrogen bond.
c. an ionic bond.
d. All of these are correct.

12. Which of the following pertains to an RNA nucleotide and not to a DNA nucleotide?
a. contains the sugar ribose
b. contains a nitrogen-containing base
c. contains a phosphate molecule
d. becomes bonded to other nucleotides by dehydration

13. Saturated fatty acids and unsaturated fatty acids differ in
a. the number of carbon-to-carbon bonds.
b. their consistency at room temperature.
c. the number of hydrogen atoms present.
d. All of these are correct.

14. Which of the following is not one of the elements commonly found in living organisms?
a. calcium
b. hydrogen
c. carbon
d. nitrogen

ENGAGE

Thinking Critically

1. Why can we use potatoes as a food source but not wood?

2. Because proteins are composed of the same limited number of amino acids, why are they all so different?

3. Determine the number of electrons normally present in the outer orbital of each atom.
a. K^+ **b.** Ca^{2+} **c.** F^- **d.** N^{3-}

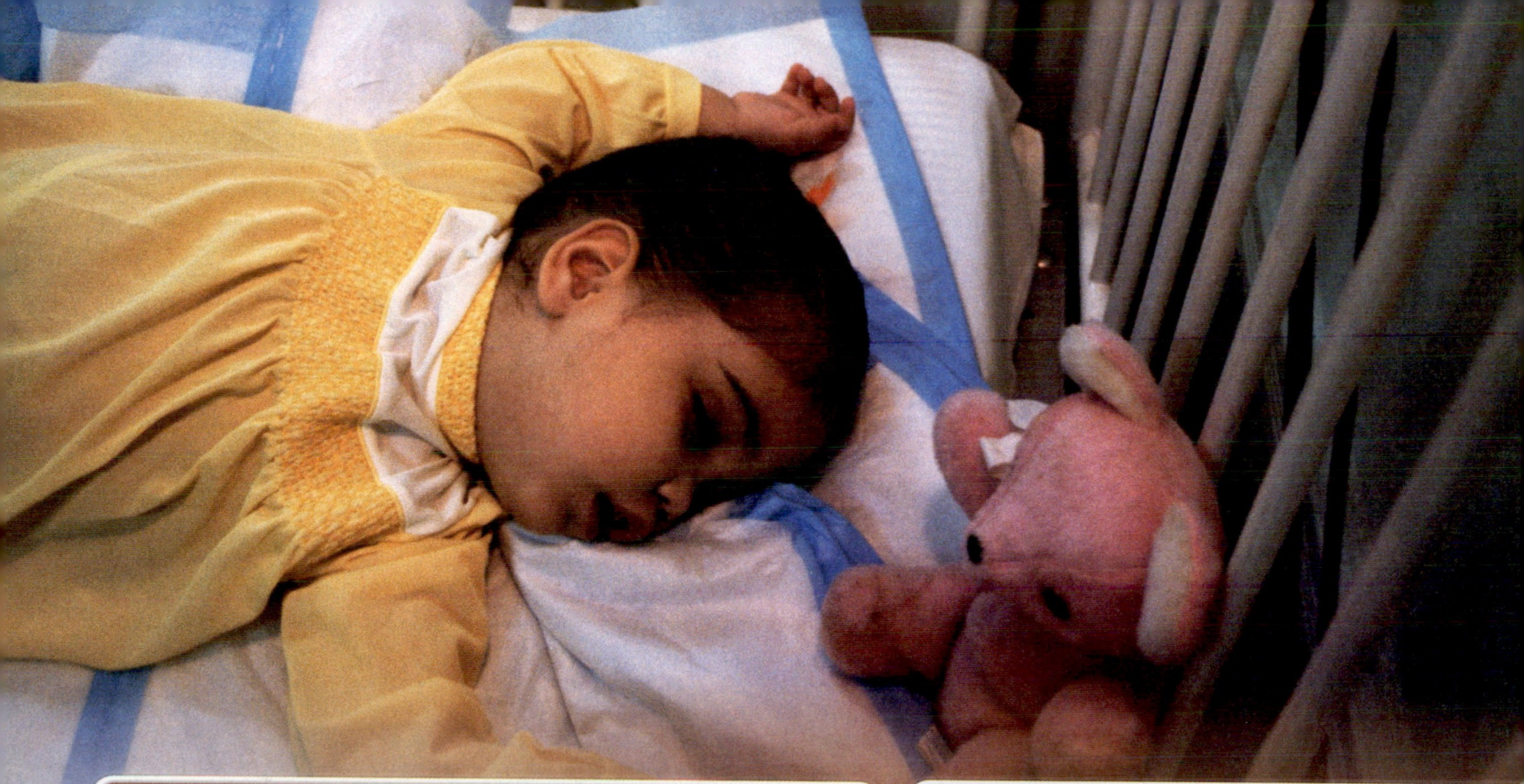

Cell Structure and Function

3

CASE STUDY Tay-Sachs: When Lysosomes Fail to Function

Tay-Sachs is a recessive neurological disease that is typically caused by the inheritance of a faulty gene from each of your parents. It is more common in individuals who are of eastern and central European Jewish heritage. Tay-Sachs disease is caused by the buildup of harmful quantities of a fatty substance called ganglioside GM2. This substance normally exists in the tissues and nerve cells in the brain. However, in Tay-Sachs patients, GM2 accumulates and the nerve cells begin to become malformed, resulting in a deterioration of mental and physical abilities. This deterioration leads to deafness, blindness, and atrophy of the muscles. Dementia and seizures often develop as well. A child who inherits Tay-Sachs will most likely die by the age of 4 due to severe neurological deterioration and recurring infections.

The root cause of Tay-Sachs disease is due to a mutation in the *HEXA* gene, which provides instructions for the production of an enzyme called beta-hexosaminidase A. This enzyme is produced in the lysosomes and is responsible for breaking down toxic substances and fatty acids that accumulate within the cell. The lysosome also acts as a recycling center within the cell. When the production of beta-hexosaminidase A is interrupted, the lysosome cannot perform its normal duties, leading to Tay-Sachs disease. Because Tay-Sachs disease prevents the normal functioning of the lysosomal enzymes, it is often referred to as a lysosomal storage disorder.

Currently, there is no treatment or cure for Tay-Sachs disease in humans. Clinical trials in gene therapy have been successful in curing Tay-Sachs in some lab animals. A variety of other clinical trials are under way with hopes of finding a cure in the near future.

As you read through the chapter, think about the following questions:

1. What other cellular organelles have a similar function to the lysosome?
2. Why doesn't the cell "clean up" the faulty lysosomes?

CHAPTER OUTLINE

BEFORE YOU BEGIN

Before beginning this chapter, take a few moments to review the following discussions:

Section 1.2 What are the main differences between prokaryotic and eukaryotic cells?

Section 2.3 Why is water so important to cells?

Section 2.6 What role do phospholipids play in forming the cell membrane?

3.1 The Cellular Level of Organization

Learning Outcomes

Upon completion of this section, you should be able to

1. Explain why cells are the basic unit of life and how the cell theory relates to this.
2. Explain the difference between the surface-area-to-volume ratios for large and small cells.

The cell marks the boundary between the nonliving and the living. The molecules that serve as food for a cell and the macromolecules that make up a cell are not alive, and yet the cell is alive. The cell is the structural and functional unit of an organism. It is the smallest structure capable of performing all the functions necessary for life. Thus, the answer to what life is must lie within the cell. The smallest living organisms are unicellular, while larger organisms are multicellular—that is, composed of many cells.

Fundamentally, three different types of cells exist in nature. Prokaryotic cells, commonly called bacteria, lack a membrane-enclosed nucleus. Another type of cell, called a eukaryotic cell, has a nucleus. A third group of cells, called archaeans, possess qualities of both prokaryotes and eukaryotes. Prokaryotic cells and archaeans were once thought to be closely related because of their similar size and shape. Comparisons of DNA and RNA sequences now show archaeans to be biochemically distinct from prokaryotes and eukaryotes. Although we have learned more about archaeans than was previously known, we have studied prokaryotes and eukaryotes in much greater detail.

The diversity of cells can be illustrated by the many types found in the human body. But despite a variety of forms and functions, human cells contain the same components. The basic components of eukaryotic cells are the subject of this chapter. Viewing these components requires a microscope. The Scientific Inquiry feature, "Microscopy Today," on page 51 introduces you to the microscopes most often used to study cells. Today, we are accustomed to thinking of living things as being constructed of cells. But the word "cell" didn't enter biology until the seventeenth century. The Dutch scientist Anton van Leeuwenhoek is recognized for making some of the earliest microscopes and observing things that no one had seen before. Robert Hooke, an English scientist, confirmed Leeuwenhoek's observations and was the first to use the term "cell." The tiny chambers he observed in the honeycomb structure of cork reminded him of the rooms, or cells, in a monastery.

Over 150 years later—in the 1830s—the German microscopist Matthias Schleiden stated that plants are composed of cells. His counterpart, Theodor Schwann, stated that animals are also made up of living units called cells. This was quite a feat, because aside from their own exhausting examination of tissues, both had to take into consideration the studies of many other microscopists. Rudolf Virchow, another German microscopist, later came to the conclusion that cells come from preexisting cells. Sperm fertilizing an egg and a human being developing from the zygote is just one example. These observations became the basis of the **cell theory,** which states that all organisms are made up of basic living units called cells, and that all cells come only from previously existing cells. Today, the cell theory is a basic theory of biology.

Cell Size

Cells are quite small. A frog's egg, at about 1 millimeter (mm) in diameter, is large enough to be seen by the human eye. But most cells are far smaller than 1 mm. Some are even as small

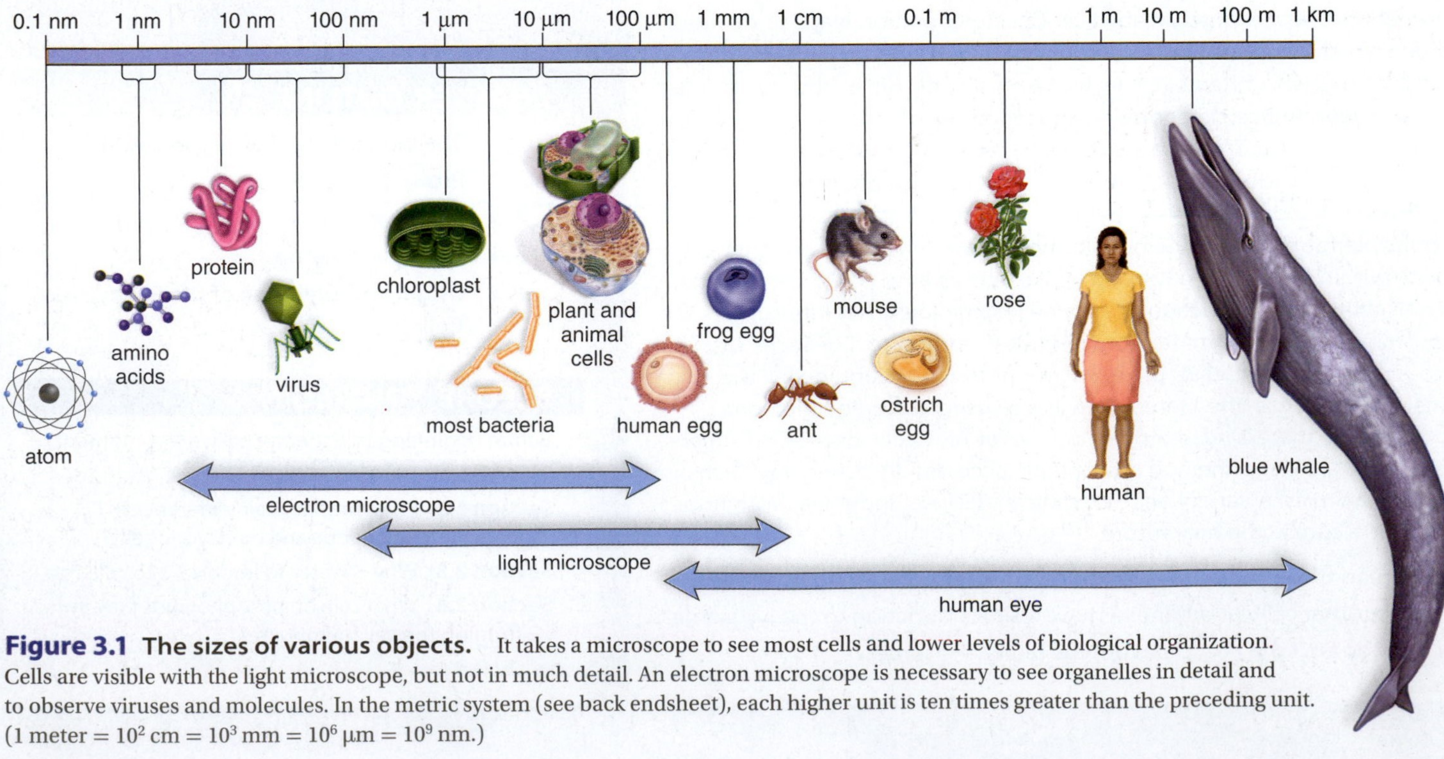
0.1 nm
1 nm
10 nm
100 nm
1 μm
10 μm
100 μm
1 mm
1 cm
0.1 m
1 m
10 m
100 m
1 km
protein
chloroplast
amino acids
virus
atom
plant and animal cells
most bacteria
frog egg
human egg
mouse
ant
ostrich egg
rose
human
blue whale
electron microscope
light microscope
human eye

Figure 3.1 The sizes of various objects. It takes a microscope to see most cells and lower levels of biological organization. Cells are visible with the light microscope, but not in much detail. An electron microscope is necessary to see organelles in detail and to observe viruses and molecules. In the metric system (see back endsheet), each higher unit is ten times greater than the preceding unit. (1 meter = 10^2 cm = 10^3 mm = 10^6 μm = 10^9 nm.)

as 1 micrometer (μm)—one-thousandth of a millimeter. Cell inclusions and macromolecules are smaller than a micrometer and are measured in terms of nanometers (nm). Figure 3.1 outlines the visual range of the eye, light microscope, and electron microscope. The discussion of microscopy in the Scientific Inquiry feature, "Microscopy Today," on page 51 explains why the electron microscope allows us to see so much more detail than the light microscope.

The fact that cells are so small is a great advantage for multicellular organisms. Nutrients, such as glucose and oxygen, enter a cell, and wastes, such as carbon dioxide, exit a cell at its surface. Therefore, the amount of surface area affects the ability to get material into and out of the cell. A large cell requires more nutrients and produces more wastes than a small cell. But, as cells get larger in volume, the proportionate amount of surface area actually decreases. For example, for a cube-shaped cell, the volume increases by the cube of the sides (height × width × depth), while the surface area increases by the square of the sides and number of sides (height × width × 6). If a cell doubles in size, its surface area increases only fourfold, while its volume increases eightfold. Therefore, small cells, not large cells, are likely to have a more adequate surface area for exchanging nutrients and wastes. As Figure 3.2 demonstrates, cutting a large cube into smaller cubes provides a lot more surface area per volume. Smaller cells tend to have higher metabolic rates

Most actively metabolizing cells are small. The frog's egg is not actively metabolizing. But once the egg is fertilized and metabolic activity begins, the egg divides repeatedly without growth. These cell divisions restore the amount of surface area needed for adequate exchange of materials. Further, cells that specialize in absorption have modifications that greatly increase their surface-area-to-volume ratio. For example, the columnar epithelial cells along the surface of the intestinal wall have surface foldings called microvilli (sing., microvillus) that increase their surface area.

Check Your Progress 3.1

1. Identify why humans are made up of trillions of cells instead of just one.
2. Explain how you would classify the cells in your body.

One 4-cm cube
Eight 2-cm cubes
Sixty-four 1-cm cubes

Figure 3.2 Surface-area-to-volume relationships. All three have the same volume, but the group on the right has four times the surface area.

3.2 Prokaryotic Cells

Learning Outcomes

Upon completion of this section, you should be able to

1. Describe the fundamental components of a bacterial cell.
2. Identify the key differences between the archaea and bacteria.

Cells can be categorized by the presence or absence of a nucleus. **Prokaryotic cells** lack a membrane-bound nucleus. The domains Archaea and Eubacteria consist of prokaryotic cells. Prokaryotes generally exist as unicellular organisms (single cells) or as simple strings and clusters. Many people think of germs when they hear the word bacteria, but not all bacteria cause disease. In fact, most bacteria are beneficial and are essential for other living organisms' survival.

Video *E.coli* Wars

Plasma Membrane and Cytoplasm

All cells are surrounded by a **plasma membrane** consisting of a phospholipid bilayer embedded with protein molecules.

The plasma membrane is a living boundary that separates the living contents of the cell from the surrounding environment. Inside the cell is a semifluid medium called the **cytoplasm.** The cytoplasm is composed of water, salts, and dissolved organic molecules. The plasma membrane regulates the entrance and exit of molecules into and out of the cytoplasm.

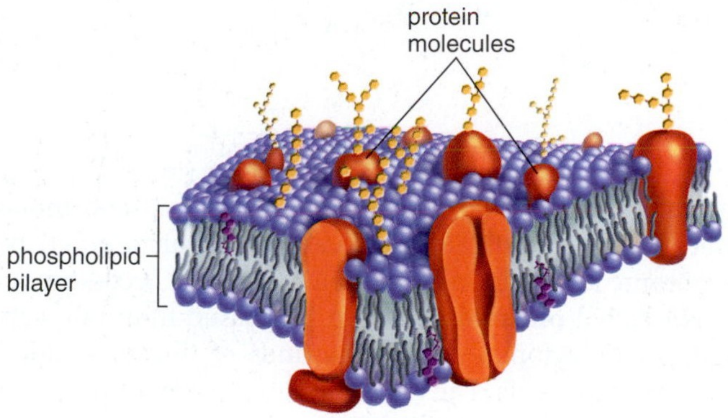
protein molecules
phospholipid bilayer

Bacterial Anatomy

Figure 3.3 illustrates the main features of prokaryotic cell anatomy. The **cell wall,** located outside of the plasma membrane, contains *peptidoglycan,* a complex molecule that is unique to bacteria and composed of chains of disaccharides joined together by peptide chains. The cell wall acts as a form of protection for bacteria. Some classes of antibiotics interfere with the synthesis of the peptidoglycans. In some bacteria, the cell wall is further surrounded by a gelatinous sheath called a **capsule**. Some bacteria have long, thin appendages called **flagella** (sing., **flagellum**), which are composed of subunits of the protein flagellin. The flagella rotate like propellers, allowing the bacterium to move rapidly in a fluid medium. Some bacteria also have **fimbriae,** which are short appendages that help them attach to an appropriate surface. The capsule and fimbriae often increase the ability of pathogenic bacteria to cause disease.

Animation Cell Wall Antibiotics

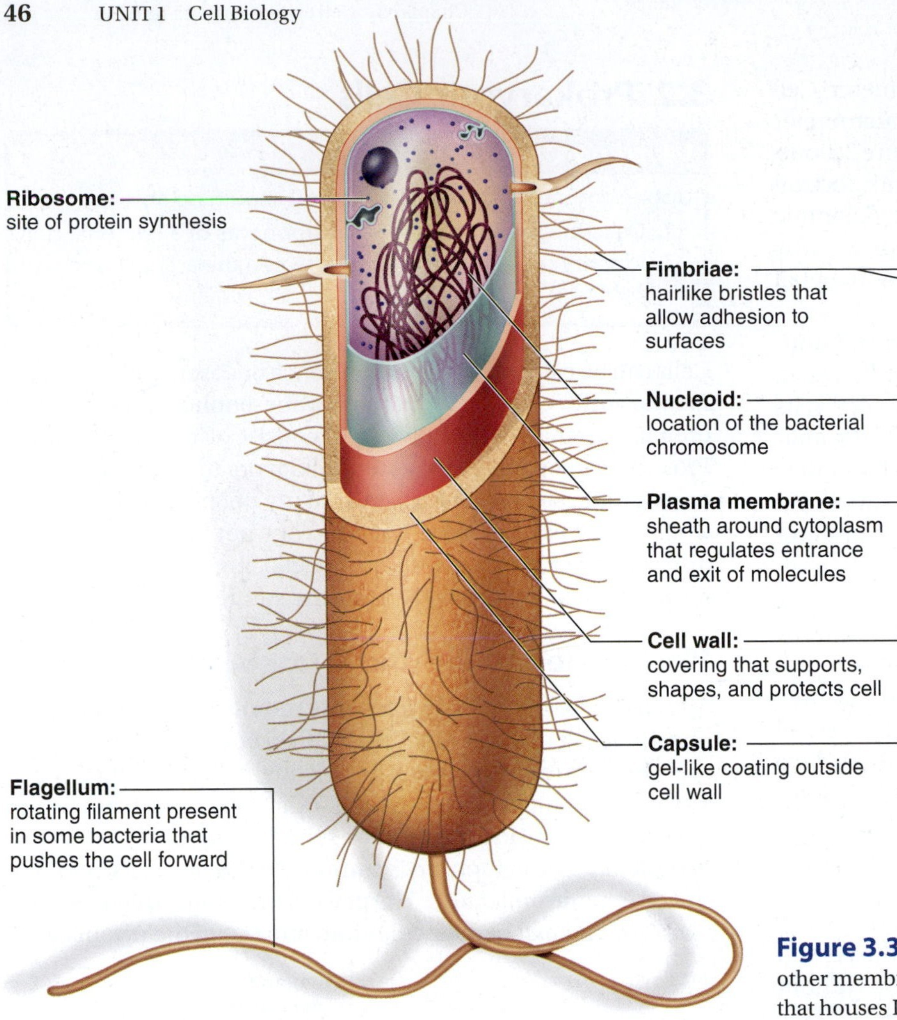
Ribosome:
site of protein synthesis
Fimbriae:
hairlike bristles that allow adhesion to surfaces
Nucleoid:
location of the bacterial chromosome
Plasma membrane:
sheath around cytoplasm that regulates entrance and exit of molecules
Cell wall:
covering that supports, shapes, and protects cell
Capsule:
gel-like coating outside cell wall
Flagellum:
rotating filament present in some bacteria that pushes the cell forward

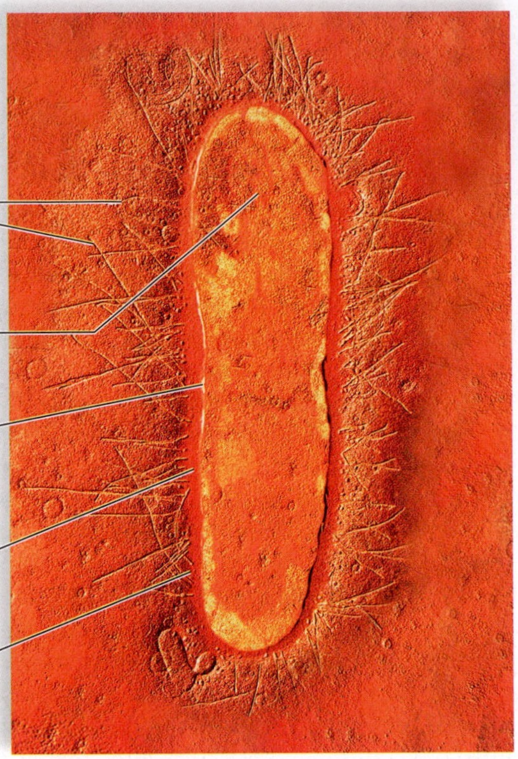
Escherichia coli

Figure 3.3 **Prokaryotic cell.** Prokaryotic cells lack a nucleus and other membrane-bound organelles, but they possess a nucleoid region that houses DNA.

Prokaryotes have a single chromosome (loop of DNA and associated proteins) located within a region of the cytoplasm called the **nucleoid.** The nucleoid is not surrounded by a membrane. Many prokaryotes also have small accessory rings of DNA called **plasmids** that can be passed from one cell to another. The cytoplasm has thousands of ribosomes for the synthesis of proteins. The ribosomes of prokaryotic organisms are smaller and structurally different from those of eukaryotic cells, which makes ribosomes a good target for antibacterial drugs. In addition, the photosynthetic cyanobacteria have light-sensitive pigments, usually within a series of internal membranes called **thylakoids.**

Although prokaryotes are structurally simple, they are much more metabolically diverse than eukaryotes. Many of them can synthesize all their structural components from very simple, even inorganic, molecules. Indeed, humans exploit the metabolic capability of bacteria by using them to produce a wide variety of chemicals and products. Prokaryotes also have adapted to living in almost every environment on Earth. In particular, archaeans have been found living under conditions that would not support any other form of life. Archaeal membranes have unique membrane-spanning lipids that help them survive in extremes of heat, pH, and salinity. Table 3.1 compares the major structures of prokaryotes (archaea and bacteria) with those of eukaryotes.

Check Your Progress 3.2

1. Recognize the function of the plasma membrane.
2. Identify the key bacterial structures and their function.
3. Identify which structures allow bacteria to survive as a unicellular organism.

TABLE 3.1 Comparison of Major Structural Features of Archaea, Bacteria, and Eukaryotes

	Archaea	Bacteria	Eukaryotes
Cell wall	Usually present, no peptidoglycan	Usually present, with peptidoglycan	Sometimes present, no peptidoglycan
Plasma membrane	Yes	Yes	Yes
Nucleus	No	No	Yes
Membrane-bound organelles	No	No*	Yes
Ribosomes	Yes	Yes	Yes, larger than prokaryotic

* Some possess internal membranes where chemical reactions may occur.

3.3 Eukaryotic Cells

Learning Outcomes

Upon completion of this section, you should be able to

1. Recognize the structure and function of the organelles within eukaryotic cells.
2. Identify the cellular structures unique to both plant and animal cells.

Eukaryotic cells are structurally very complex. The principal distinguishing feature of eukaryotic cells is the presence of a nucleus, which separates the genetic material from the cytoplasm of the cell. In addition, eukaryotic cells possess a variety of other organelles, many of which are surrounded by membranes. Animals, plants, fungi, and protists are all composed of eukaryotic cells.

MP3 Organelles

Cell Walls

Some eukaryotic cells have a permeable but protective cell wall in addition to a plasma membrane. Many plant cells have both a primary and a secondary cell wall. A main constituent of a primary cell wall is cellulose. Cellulose forms fibrils that lie at right angles to one another for added strength. The secondary cell wall, if present, forms inside the primary cell wall. Such secondary cell walls contain lignin, a substance that makes them even stronger than primary cell walls. The cell walls of some fungi are composed of cellulose and chitin, the same type of polysaccharide found in the exoskeleton of insects. Algae, members of the kingdom Protista, contain cell walls composed of cellulose.

Organelles of Eukaryotic Cells

Originally the term **organelle** referred to only membranous structures. However, it has come to mean any well-defined subcellular structure that performs a particular function (Table 3.2). By analogy, a eukaryotic cell can be thought of as a factory. Just as all the assembly lines of a factory operate at the same time, so do all the organelles of a cell. Raw materials enter a factory where different departments turn them into various products. In the same way, the cell takes in chemicals, and then the organelles process them. The factory must also get rid of waste, and the cell performs that function as well.

In this chapter, we will focus primarily on animal and plant cells (Figs. 3.4 and 3.5). While there are many similarities in these cell types, there are some important differences. For example, both contain mitochondria, but chloroplasts are primarily found in the cells of plants, while centrioles are found in the cells of animals. In the illustrations throughout this text, note that each of the organelles has an assigned color.

TABLE 3.2 Structures of Eukaryotic Cells

Structure	Composition	Function
Cell wall (absent in animal cells)	Contains polysacharrides	Support and protection
Plasma membrane	Phospholipid bilayer with embedded proteins	Defines cell boundary; regulates molecule passage into and out of cells
Nucleus	Nuclear envelope, nucleoplasm, chromatin, and nucleoli	Storage of genetic information; synthesis of DNA and RNA
Nucleoli	Concentrated area of chromatin, RNA, and proteins	Ribosomal subunit formation
Ribosomes	Protein and RNA in two subunits	Protein synthesis
Endoplasmic reticulum (ER)	Membranous flattened channels and tubular canals	Synthesis and/or modification of proteins and other substances, and distribution by vesicle formation
Rough ER	Network of folded membranes studded with ribosomes	Folding, modification, and transport of proteins
Smooth ER	Network of folded membranes having no ribosomes	Various; lipid synthesis in some cells
Golgi apparatus	Stack of small membranous sacs	Processing, packaging, and distribution of proteins and lipids
Lysosomes (animal cells only)	Membranous vesicle containing digestive enzymes	Intracellular digestion
Vacuoles and vesicles	Membranous sacs of various sizes	Storage of substances
Peroxisomes	Membranous vesicle containing specific enzymes	Various metabolic tasks
Mitochondria	Inner membrane (cristae) bounded by an outer membrane	Cellular respiration
Chloroplasts (primarily in plant cells)	Membranous grana bounded by two membranes	Photosynthesis
Cytoskeleton	Microtubules, intermediate filaments, actin filaments	Shape of cell and movement of its parts
Cilia and flagella (cilia are rare in plant cells)	9 + 2 pattern of microtubules	Movement of cell
Centriole (animal cells only)	9 + 0 pattern of microtubules	Formation of basal bodies

Figure 3.4 Animal cell anatomy. Micrograph of an insect cell and drawing of a generalized animal cell. See Table 3.2 for a description of these structures, along with a listing of their functions.

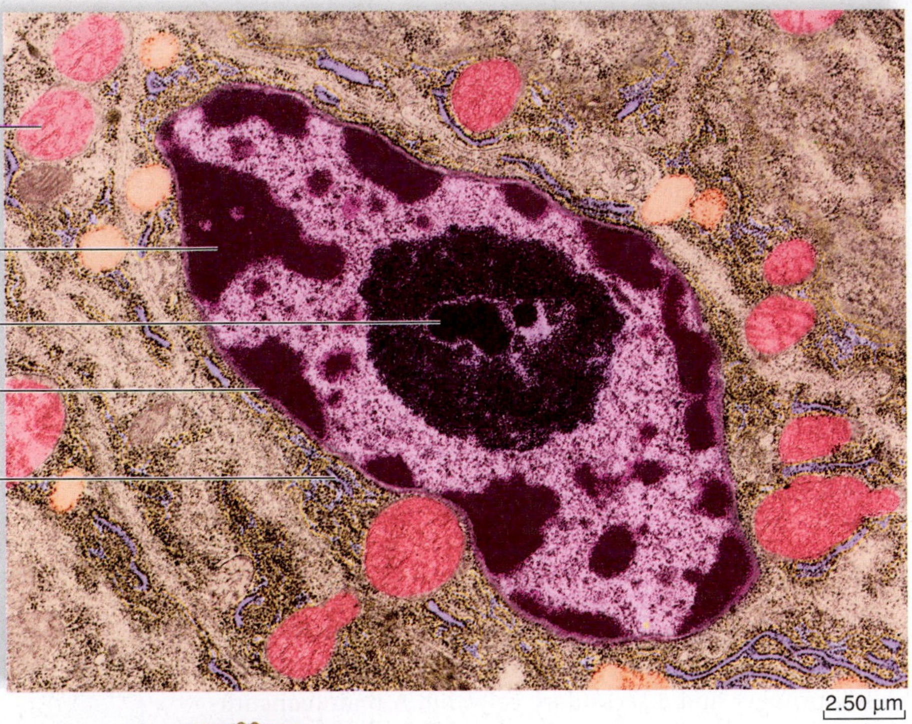
mitochondrion
chromatin
nucleolus
nuclear envelope
endoplasmic reticulum
2.50 μm

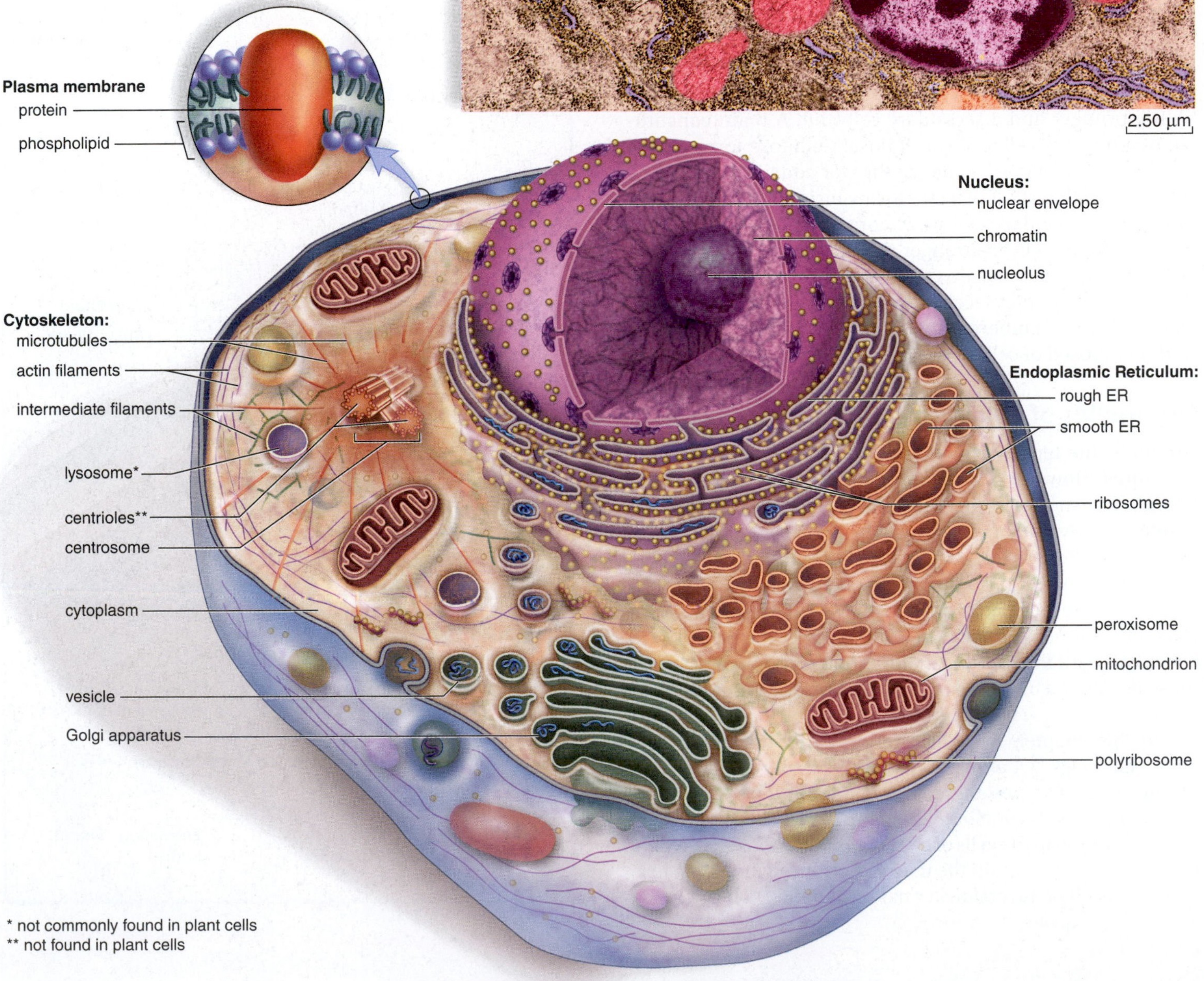
Plasma membrane
protein
phospholipid
Nucleus:
nuclear envelope
chromatin
nucleolus
Cytoskeleton:
microtubules
actin filaments
intermediate filaments
lysosome*
centrioles**
centrosome
cytoplasm
vesicle
Golgi apparatus
Endoplasmic Reticulum:
rough ER
smooth ER
ribosomes
peroxisome
mitochondrion
polyribosome
* not commonly found in plant cells
** not found in plant cells

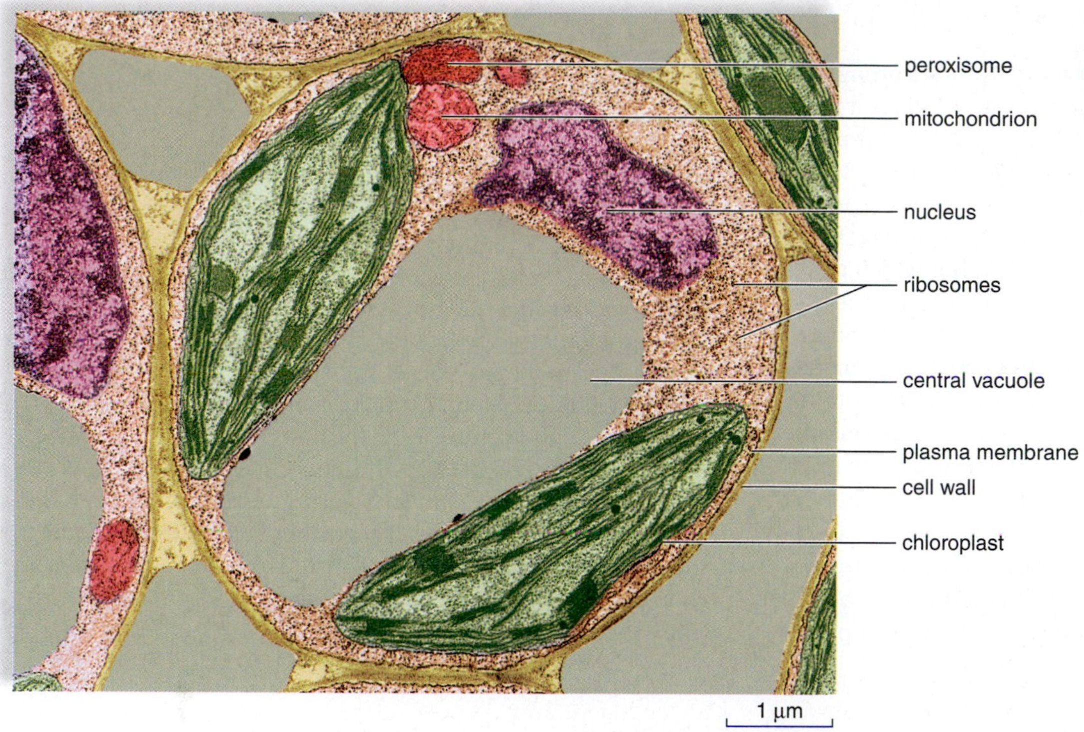
peroxisome
mitochondrion
nucleus
ribosomes
central vacuole
plasma membrane
cell wall
chloroplast
1 µm

Figure 3.5 Plant cell anatomy. Micrograph of a plant cell and drawing of a generalized plant cell. See Table 3.2 for a description of these structures, along with a listing of their functions.

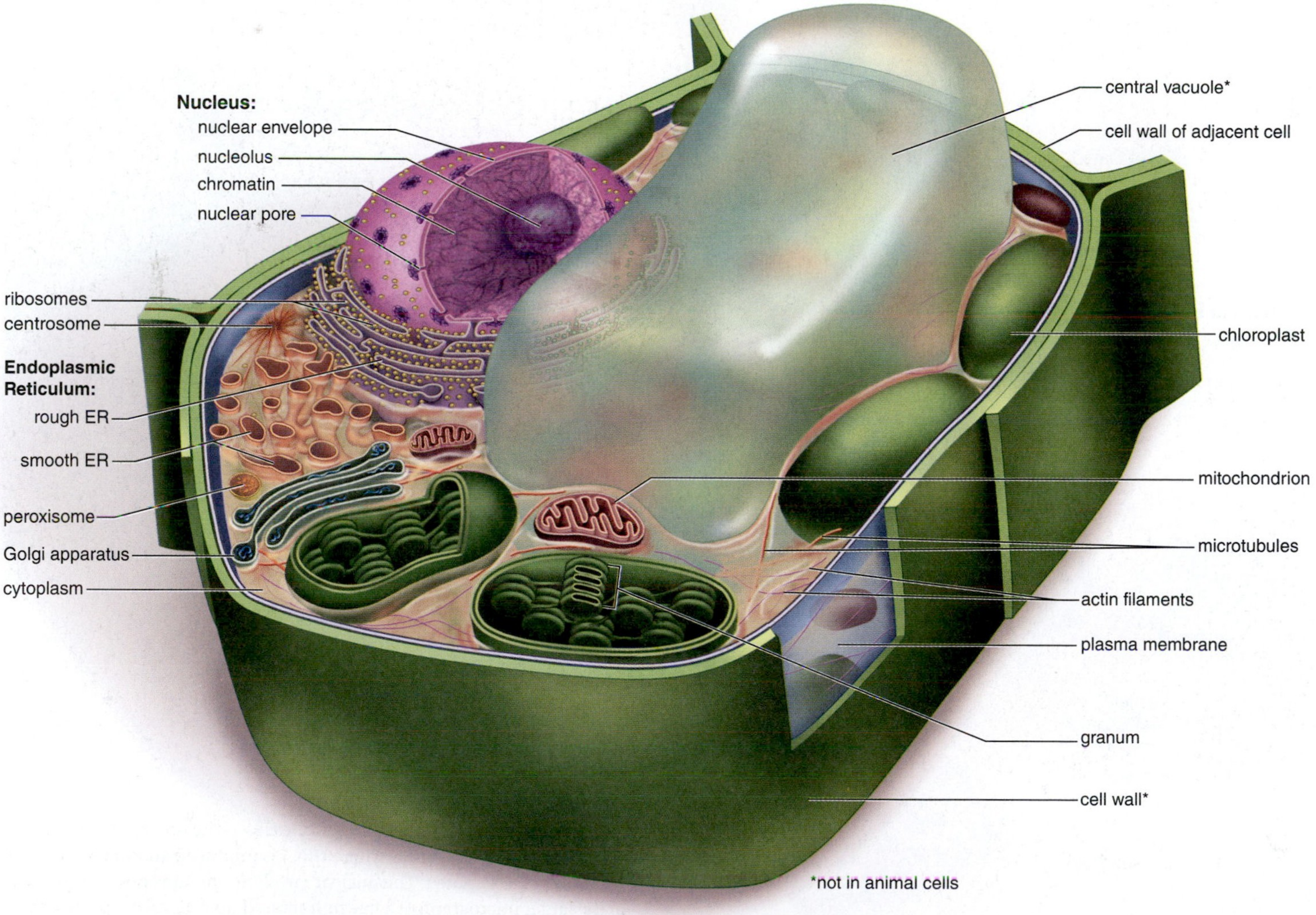
Nucleus:
nuclear envelope
nucleolus
chromatin
nuclear pore
ribosomes
centrosome
Endoplasmic Reticulum:
rough ER
smooth ER
peroxisome
Golgi apparatus
cytoplasm
central vacuole*
cell wall of adjacent cell
chloroplast
mitochondrion
microtubules
actin filaments
plasma membrane
granum
cell wall*
*not in animal cells

The Nucleus

The **nucleus,** which has a diameter of about 5 μm, is a prominent structure in the eukaryotic cell. In photographs or slides it looks like a dark golf ball within the cell. The nucleus is of primary importance because it stores the genetic material, DNA. In the factory analogy, the nucleus is the head office. DNA governs the characteristics of the cell and its metabolic functioning. Every cell in an individual contains the same DNA, but which genes are turned on and which are turned off will differ among cells.

When you look at the nucleus, even in an electron micrograph, you cannot see a DNA molecule (Fig. 3.6). You can see **chromatin,** which consists of DNA and associated proteins. Chromatin in most eukaryotic cells is not one continuous strand. During most of the cell's lifetime, chromatin is present, but when the cell is ready to undergo cell division, it will undergo coiling and become highly condensed structures called **chromosomes.** Human cells contain 46 chromosomes, which are immersed in a semifluid medium called the **nucleoplasm.** A difference in pH between the nucleoplasm and the cytoplasm suggests that the nucleoplasm has a different composition.

When you look at an electron micrograph of a nucleus, you may see one or more regions that look darker than the rest of the chromatin. These are nucleoli (sing., **nucleolus**), where another type of RNA, called ribosomal RNA (rRNA), is produced and where rRNA joins with proteins to form the subunits of ribosomes described in the next section.

The nucleus is separated from the cytoplasm by a double membrane known as the **nuclear envelope,** which is continuous with the endoplasmic reticulum (also discussed next). The nuclear envelope has **nuclear pores** of sufficient size (100 nm) to permit the bidirectional transport of proteins and ribosomal subunits.

Interestingly, during cell division, the nuclear envelope completely disappears and the contents of the nucleus are mixed with the cytoplasm. Following cell division, the nuclear envelope re-forms around the chromosomes, and the other contents of the nucleus are transported into the nucleus. Regulation of transport through the nuclear pores can control events within the nucleus or the entire cell.

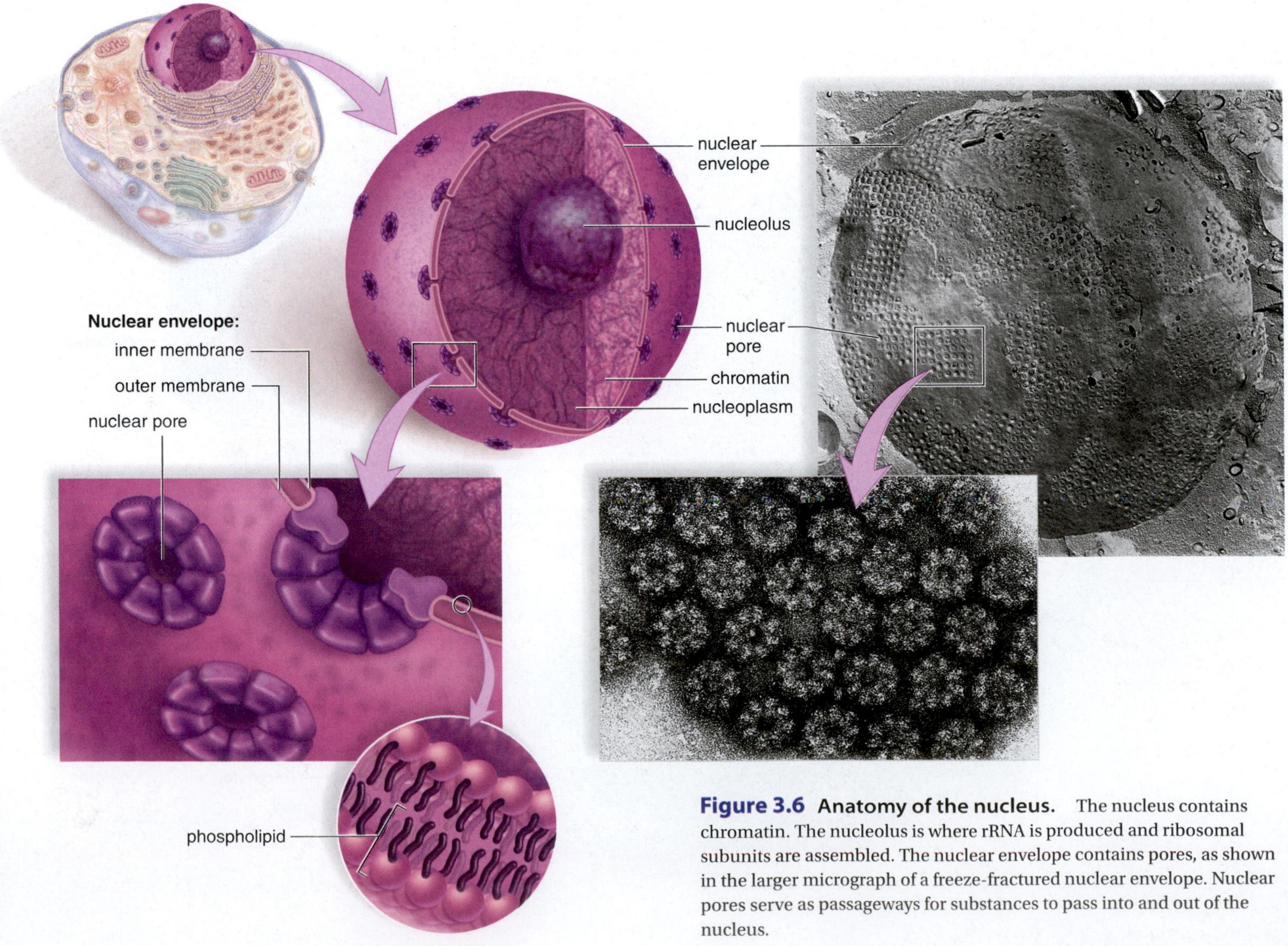
nuclear envelope
nucleolus
Nuclear envelope:
inner membrane
outer membrane
nuclear pore
nuclear pore
chromatin
nucleoplasm
phospholipid

Figure 3.6 Anatomy of the nucleus. The nucleus contains chromatin. The nucleolus is where rRNA is produced and ribosomal subunits are assembled. The nuclear envelope contains pores, as shown in the larger micrograph of a freeze-fractured nuclear envelope. Nuclear pores serve as passageways for substances to pass into and out of the nucleus.

SCIENCE IN YOUR LIFE ▶ **SCIENTIFIC INQUIRY**

Microscopy Today

Today, scientists use two main types of microscopes: light microscopes and electron microscopes. Figure 3A depicts one type of light microscope and two types of electron microscopes, along with micrographs obtained with each one.

The most commonly used light microscope is the compound light microscope. "Compound" refers to the presence of more than one lens. In a compound light microscope, light rays pass through a specimen such as a blood smear, a tissue section, or even living cells. The light rays are brought to a focus by a set of glass lenses and the resulting image is then viewed. Specimens are often stained to provide viewing contrast.

In an electron microscope, electrons rather than light rays are brought to focus by a set of electromagnetic lenses, and the resulting image is projected onto a viewing screen or photographic film. Electron microscopes produce images of much higher magnification than sometimes can be obtained with the best compound light microscope. The ability to distinguish two points as separate points (resolution) is much greater with electron microscopes. The greater resolving power is due to the fact that electrons travel at a much shorter wavelength than do light rays. Live specimens, however, cannot be viewed because the specimen must be dried out. Two commonly used types of electron microscopes are transmission and scanning.

A transmission electron microscope (TEM) may magnify an object's image over 200,000 times, compared to approximately 1,000 times with a compound light microscope. Also, a TEM has a much greater ability to make out detail. Much of our knowledge of the internal structures of cells comes from using a TEM.

The magnifying and resolving powers of a typical scanning electron microscope (SEM) are not as great as those of a TEM, although they are still greater than those of a compound light microscope. A SEM can provide a three-dimensional view of an object.

A picture obtained using a compound light microscope is sometimes called a photomicrograph, and a picture resulting from the use of an electron microscope is called a transmission or scanning electron micrograph. Color may be added to electron micrographs, such as those in Figure 3A, using a computer.

Questions to Consider

1. What distortions or artifacts might be seen in electron microscopy when viewing a dried, nonliving specimen?
2. What happens to an image if the magnification is increased without increasing the resolution?
3. Why are electron micrographs always in black and white (or artificially colored)?

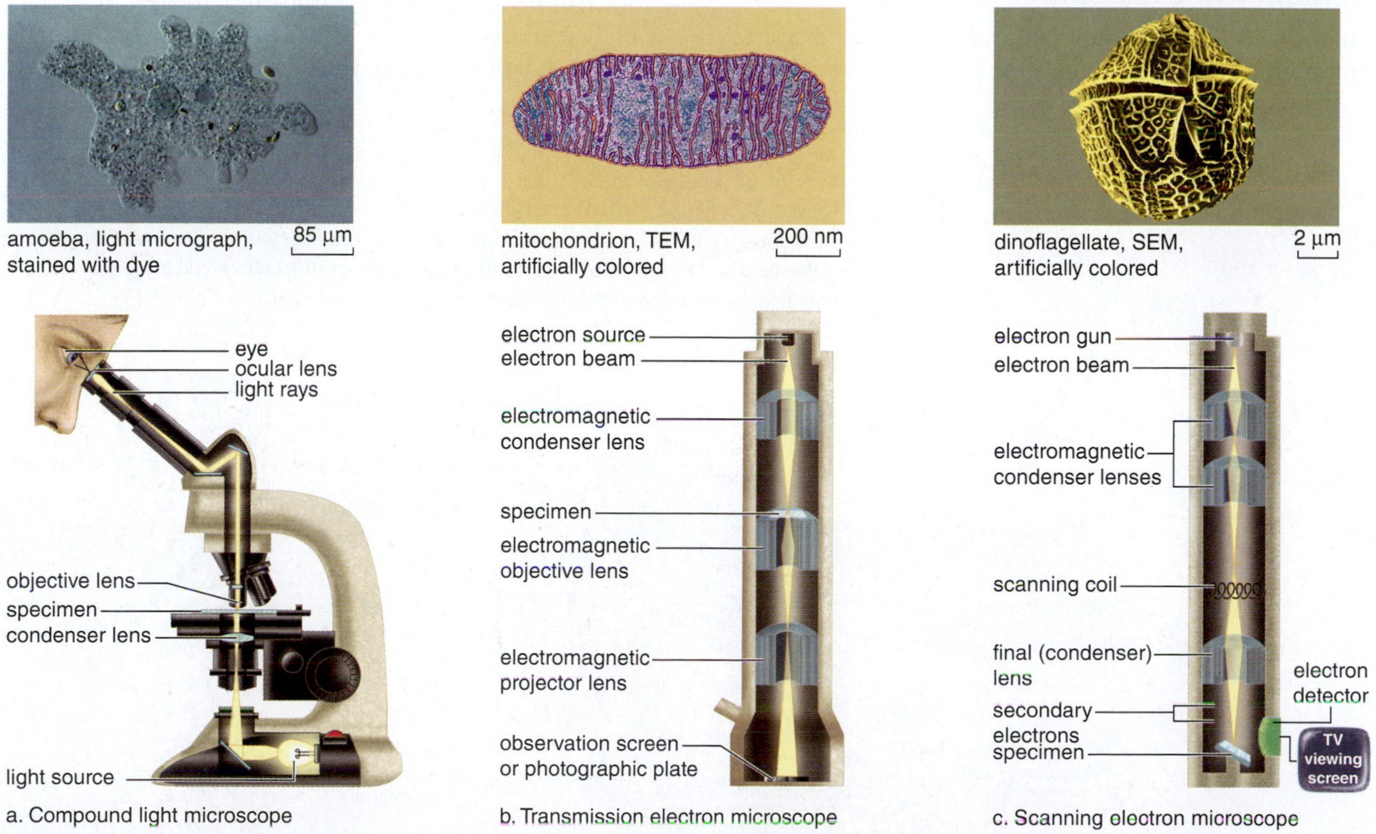
amoeba, light micrograph, stained with dye
85 µm
mitochondrion, TEM, artificially colored
200 nm
dinoflagellate, SEM, artificially colored
2 µm
eye
ocular lens
light rays
objective lens
specimen
condenser lens
light source
a. Compound light microscope
electron source
electron beam
electromagnetic condenser lens
specimen
electromagnetic objective lens
electromagnetic projector lens
observation screen or photographic plate
b. Transmission electron microscope
electron gun
electron beam
electromagnetic condenser lenses
scanning coil
final (condenser) lens
secondary electrons
specimen
electron detector
TV viewing screen
c. Scanning electron microscope

Figure 3A Diagram of microscopes with accompanying micrographs. The compound light microscope and the transmission electron microscope provide an internal view of an organism. The scanning electron microscope provides an external view of an organism.

Ribosomes

Ribosomes are responsible for the synthesis of proteins using messenger RNA as a template. Ribosomes are composed of two subunits, called "large" and "small" because of their relative sizes. Each subunit is a complex of unique ribosomal RNA (rRNA) and protein molecules. Ribosomes can be found individually in the cytoplasm, as well as in groups called **polyribosomes** (several ribosomes associated simultaneously with a single mRNA molecule). Ribosomes can also be found attached to the endoplasmic reticulum, a membranous system of sacs and channels discussed in the next section. Proteins synthesized at ribosomes attached to the endoplasmic reticulum have a different destination from that of proteins synthesized at ribosomes free in the cytoplasm.

The Endomembrane System

The endomembrane system consists of the nuclear envelope, the endoplasmic reticulum, the Golgi apparatus, and several **vesicles** (tiny membranous sacs). Continuing the factory analogy, the endomembrane system is essentially the transportation and product-processing section of the cell. This system compartmentalizes the cell so that particular enzymatic reactions are restricted to specific regions. Organelles that make up the endomembrane system are connected either directly or by transport vesicles.

The Endoplasmic Reticulum

The **endoplasmic reticulum (ER)**, a complicated system of membranous channels and sacs (flattened vesicles), is physically continuous with the outer membrane of the nuclear envelope. Rough ER is studded with ribosomes on the side of the membrane that faces the cytoplasm (Fig. 3.7). As proteins are synthesized on these ribosomes they pass into the interior of the ER, where processing and modification begin. Proteins synthesized here are destined for the membrane of the cell or to be secreted from the cell.

Proper folding, processing, and transport of proteins are critical to the functioning of the cell. For example, in cystic fibrosis a mutated plasma membrane channel protein is retained in the endoplasmic reticulum because it is folded incorrectly. Without this protein in its correct location, the cell is unable to regulate the transport of the chloride ion, resulting in the various symptoms of the disease.

Smooth ER, which is continuous with rough ER, does not have attached ribosomes. Smooth ER synthesizes the phospholipids found in cell membranes as well as those that perform various other functions. In the testes, it produces testosterone, and in the liver, it helps detoxify drugs. In muscle cells, the smooth ER stores calcium ions. Regardless of any specialized function, smooth ER also forms vesicles in which products are transported to the Golgi apparatus.

The Golgi Apparatus

The **Golgi apparatus** is named for Camillo Golgi, who discovered its presence in cells in 1898. The Golgi apparatus consists of a stack of three to twenty slightly curved sacs whose appearance can be compared to a stack of pancakes (Fig. 3.8). The Golgi apparatus is referred to as the shipping center of the cell because it collects, sorts, packages, and distributes materials

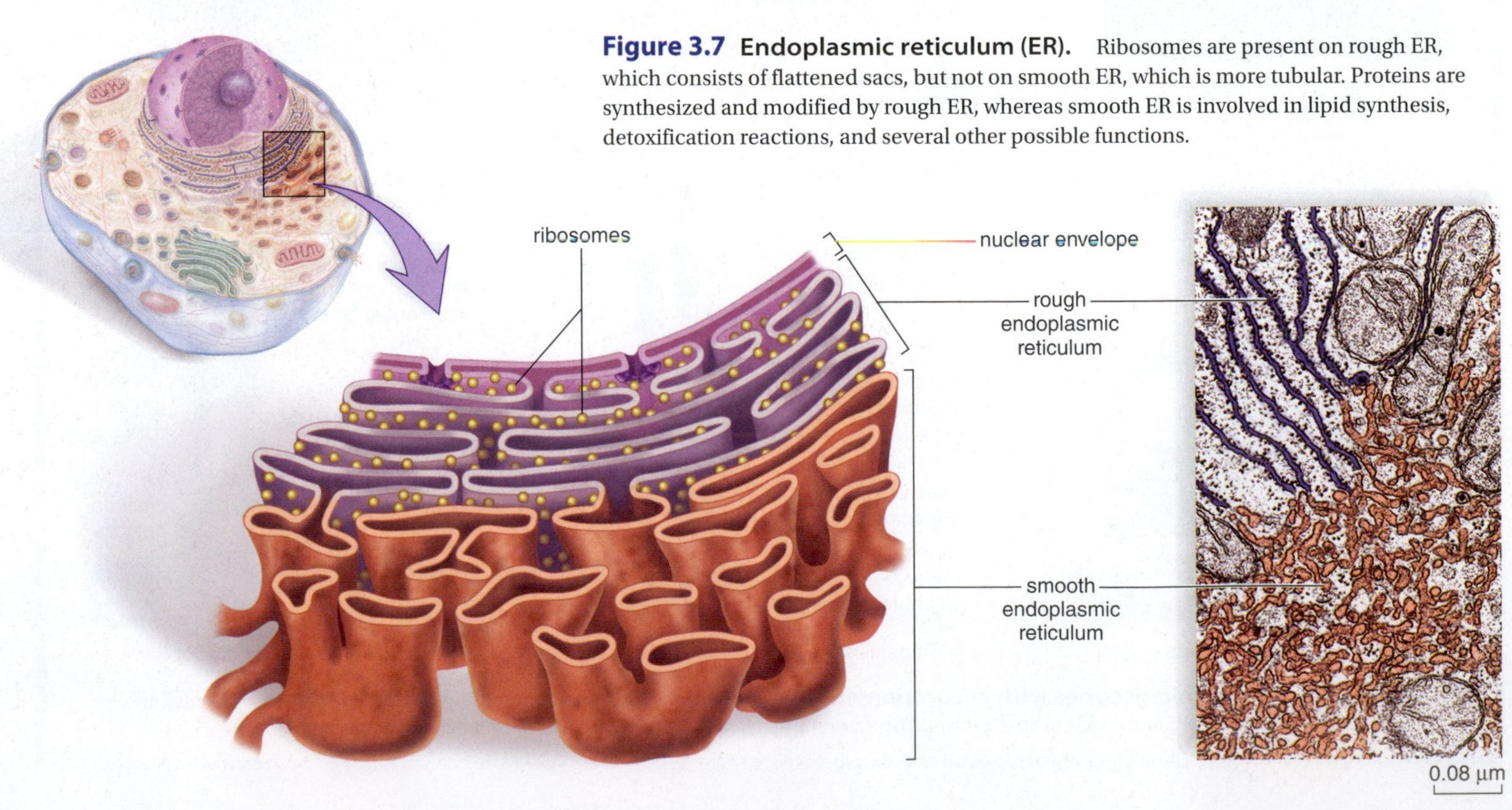
ribosomes
nuclear envelope
rough endoplasmic reticulum
smooth endoplasmic reticulum
0.08 µm

Figure 3.7 Endoplasmic reticulum (ER). Ribosomes are present on rough ER, which consists of flattened sacs, but not on smooth ER, which is more tubular. Proteins are synthesized and modified by rough ER, whereas smooth ER is involved in lipid synthesis, detoxification reactions, and several other possible functions.

such as proteins and lipids. In animal cells, one side of the stack (the inner face) is directed toward the ER, and the other side of the stack (the outer face) is directed toward the plasma membrane. Vesicles can frequently be seen at the edges of the sacs.

The Golgi apparatus receives proteins and also lipid-filled vesicles that bud from the ER. These molecules then move through the Golgi from the inner face to the outer face. During their passage through the Golgi apparatus, proteins and lipids can be modified before they are repackaged in secretory vesicles. Secretory vesicles proceed to the plasma membrane, where they discharge their contents. This action is termed **secretion.**

The Golgi apparatus is also involved in the formation of lysosomes, vesicles that contain enzymes that remain within the cell. It appears that proteins made at the rough ER have specific molecular tags that serve as "zip codes" to tell the Golgi apparatus whether they belong inside the cell in some membrane-bound organelle or in a secretory vesicle.

Lysosomes

Lysosomes are membrane-bound vesicles produced by the Golgi apparatus. Lysosomes contain hydrolytic digestive enzymes. Lysosomes are the garbage disposals of the factory.

Animation Lysosomes

Sometimes macromolecules are brought into a cell by vesicle formation at the plasma membrane (Fig. 3.8). When a lysosome fuses with such a vesicle, its contents are digested by lysosomal enzymes into simpler subunits that then enter the cytoplasm. Some white blood cells defend the body by engulfing pathogens via vesicle formation. When lysosomes fuse with these vesicles, the bacteria are digested. Even parts of a cell are digested by its own lysosomes (called autodigestion). For example, the finger webbing found in the human embryo is later dissolved by lysosomes so that the fingers are separated.

Lysosomes contain many enzymes for digesting all sorts of molecules. Occasionally, a child inherits the inability to

Figure 3.8 Endomembrane system. The organelles in the endomembrane system work together to carry out the functions noted.

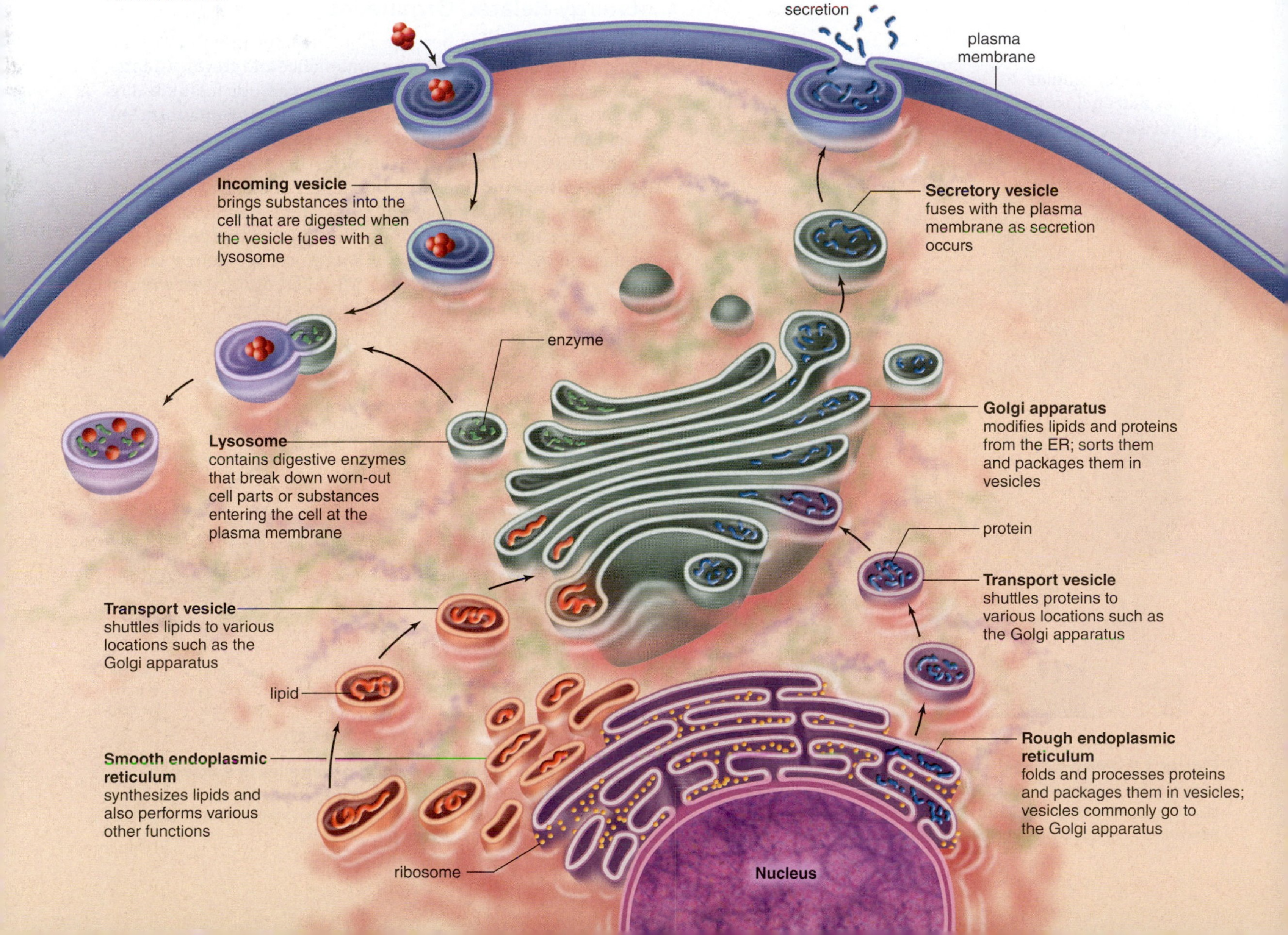
secretion
plasma membrane
Incoming vesicle brings substances into the cell that are digested when the vesicle fuses with a lysosome
Secretory vesicle fuses with the plasma membrane as secretion occurs
enzyme
Golgi apparatus modifies lipids and proteins from the ER; sorts them and packages them in vesicles
Lysosome contains digestive enzymes that break down worn-out cell parts or substances entering the cell at the plasma membrane
protein
Transport vesicle shuttles proteins to various locations such as the Golgi apparatus
Transport vesicle shuttles lipids to various locations such as the Golgi apparatus
lipid
Smooth endoplasmic reticulum synthesizes lipids and also performs various other functions
Rough endoplasmic reticulum folds and processes proteins and packages them in vesicles; vesicles commonly go to the Golgi apparatus
ribosome
Nucleus

make a lysosomal enzyme, and therefore has a lysosomal storage disease. For example, in Tay-Sachs disease, the cells that surround nerve cells cannot break down the lipid ganglioside GM2, which then accumulates inside lysosomes and affects the nervous system. At about six months, the infant can no longer see and, then, gradually loses hearing and even the ability to move. Death follows at about three years of age.

Vacuoles

A **vacuole** is a large membranous sac. A vacuole is larger than a vesicle. Although animal cells have vacuoles, they are much more prominent in plant cells. Typically, plant cells have a large central vacuole filled with a watery fluid that provides added support to the cell (see Fig. 3.5).

Vacuoles store substances. Plant vacuoles contain not only water, sugars, and salts, but also pigments and toxic molecules. The pigments are responsible for many of the red, blue, or purple colors of flowers and some leaves. The toxic substances help protect a plant from herbivorous animals. The vacuoles present in unicellular protozoans are quite specialized. They include contractile vacuoles for ridding the cell of excess water and digestive vacuoles for breaking down nutrients.

Peroxisomes

Peroxisomes, similar to lysosomes, are membrane-bound vesicles that enclose enzymes (Fig. 3.9). However, the enzymes in peroxisomes are synthesized by cytoplasmic ribosomes and transported into a peroxisome by carrier proteins. Typically, peroxisomes contain enzymes whose action results in hydrogen peroxide (H_2O_2), a toxic molecule:

$$RH_2 + O_2 \longrightarrow R + H_2O_2$$
$$(R = \text{remainder of molecule})$$

Hydrogen peroxide is immediately broken down to water and oxygen by another peroxisomal enzyme called catalase.

The enzymes present in a peroxisome depend on the function of the cell. Peroxisomes are especially prevalent in cells that are synthesizing and breaking down fats. In the liver, some peroxisomes break down fats and others produce bile salts from cholesterol. In a disease called adrenoleukodystrophy (ALD), cells lack a carrier protein to transport an enzyme into peroxisomes. As a result, long-chain fatty acids accumulate in the brain, resulting in neurological damage.

Plant cells also have peroxisomes. In germinating seeds, they oxidize fatty acids into molecules that can be converted to sugars needed by the growing plant. In leaves, peroxisomes can carry out a reaction that is opposite to photosynthesis—the reaction uses up oxygen and releases carbon dioxide.

Energy-Related Organelles

Life is possible only because of a constant input of energy. Organisms use this energy for maintenance and growth. Chloroplasts and mitochondria are the two eukaryotic membranous organelles that specialize in converting energy to a form the cell can use. **Chloroplasts** use solar energy to synthesize carbohydrates, which are broken down by the **mitochondria** (sing., **mitochondrion**) to produce ATP molecules, as shown in Figure 3.10. This diagram shows that chemicals recycle between chloroplasts and mitochondria in the presence of solar energy. When cells use ATP as an energy source, energy dissipates as heat. Most life cannot exist without a constant input of solar energy.

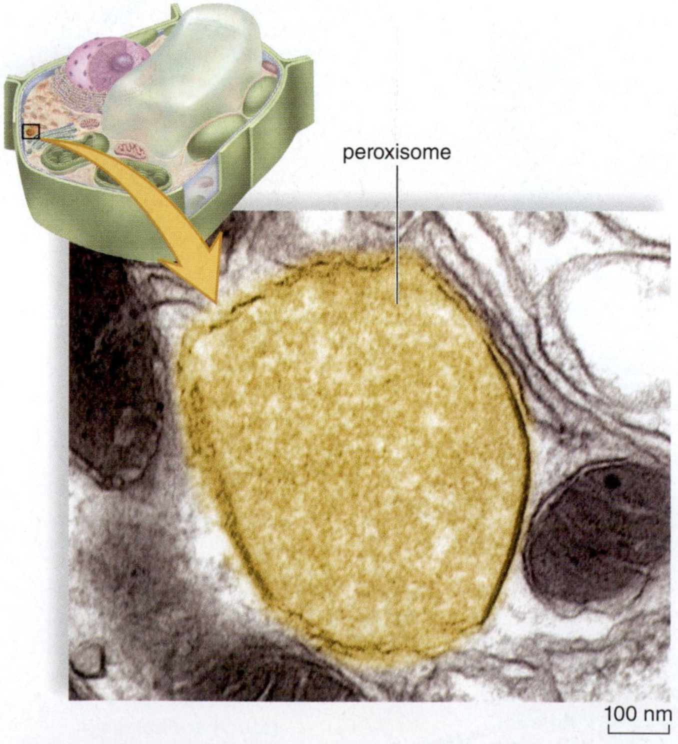
peroxisome
100 nm

Figure 3.9 Peroxisomes. Peroxisomes contain one or more enzymes that can oxidize various organic substances. Peroxisomes also contain the enzyme catalase, which breaks down the hydrogen peroxide (H_2O_2) that builds up after organic substances are oxidized.

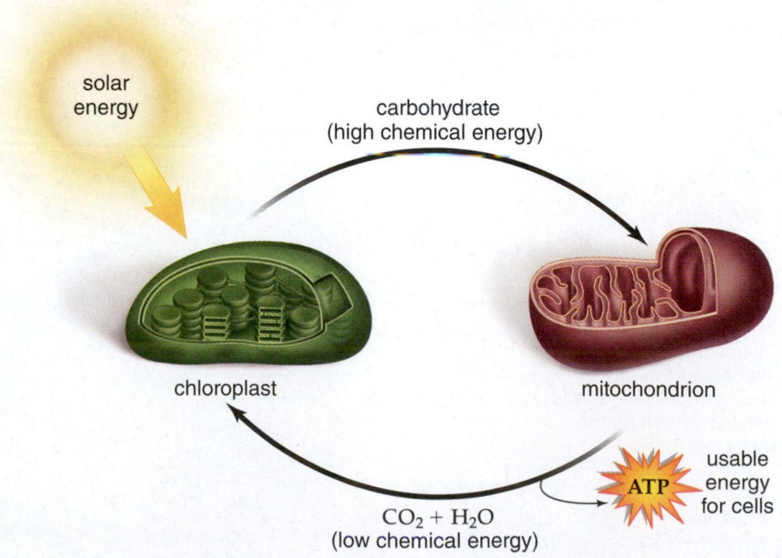
solar energy
carbohydrate (high chemical energy)
chloroplast
mitochondrion
ATP
usable energy for cells
CO2 + H2O (low chemical energy)

Figure 3.10 Energy-producing organelles in eukaryotic cells. Chloroplasts use sunlight to produce carbohydrates, which in turn are used by mitochondria. The mitochondria then produce carbon dioxide and water, which in turn is used by the chloroplasts.

Plants, algae, and cyanobacteria are all capable of carrying on **photosynthesis** in this manner:

solar energy + carbon dioxide + water ⟶ carbohydrate + oxygen

Plants and algae have chloroplasts (Fig. 3.11), while cyanobacteria carry on photosynthesis within independent thylakoids. Solar energy is the ultimate source of energy for most cells because nearly all organisms, either directly or indirectly, use the carbohydrates produced by photosynthesizers as an energy source.

Many organisms carry on **cellular respiration,** the process by which the chemical energy of carbohydrates is converted to that of ATP (adenosine triphosphate), the common energy carrier in cells. All organisms, except prokaryotes, complete the process of cellular respiration in mitochondria. Cellular respiration can be represented by this equation:

carbohydrate + oxygen ⟶ carbon dioxide + water + energy

Here, *energy* is in the form of ATP molecules. When a cell needs energy, ATP supplies it. The energy of ATP is used for all energy-requiring processes in cells.

Chloroplasts

Plant and algal cells contain chloroplasts, the organelles that allow them to use solar energy to produce organic molecules. Chloroplasts are about 4–6 μm in diameter and 1–5 μm in length. They belong to a group of organelles known as plastids. Among the plastids are also the *amyloplasts,* common in roots, which store starch, and the *chromoplasts,* common in leaves, which contain red and orange pigments. A chloroplast is green because it contains the green pigment chlorophyll. A typical plant cell in a leaf may contain approximately 50 chloroplasts. Chloroplasts divide by splitting in two in a manner similar to how bacteria divide.

A chloroplast is bounded by two membranes that enclose a fluid-filled space called the **stroma** (Fig. 3.11). The stroma contains a single circular DNA molecule as well as ribosomes. The chloroplast contains its own genetic material and makes most of its own proteins. The others are encoded by nuclear genes and imported from the cytoplasm. A membrane system within the stroma is organized into interconnected flattened sacs called thylakoids. In certain regions, the thylakoids are stacked up in structures called **grana** (sing., **granum**). There can be hundreds of grana within a single chloroplast (Fig. 3.11). Chlorophyll, which is located within the thylakoid membranes of grana, captures the solar energy needed to enable chloroplasts to produce carbohydrates. The solar energy excites an electron within the chlorophyll molecule. The energy from

Figure 3.11 Chloroplast structure. Chloroplasts carry out photosynthesis. **a.** Electron micrograph of a longitudinal section of a chloroplast. **b.** Generalized drawing of a chloroplast in which the outer and inner membranes have been cut away to reveal the grana, each of which is a stack of membranous sacs called thylakoids. In some grana, but not all, it is obvious that thylakoid spaces are interconnected.

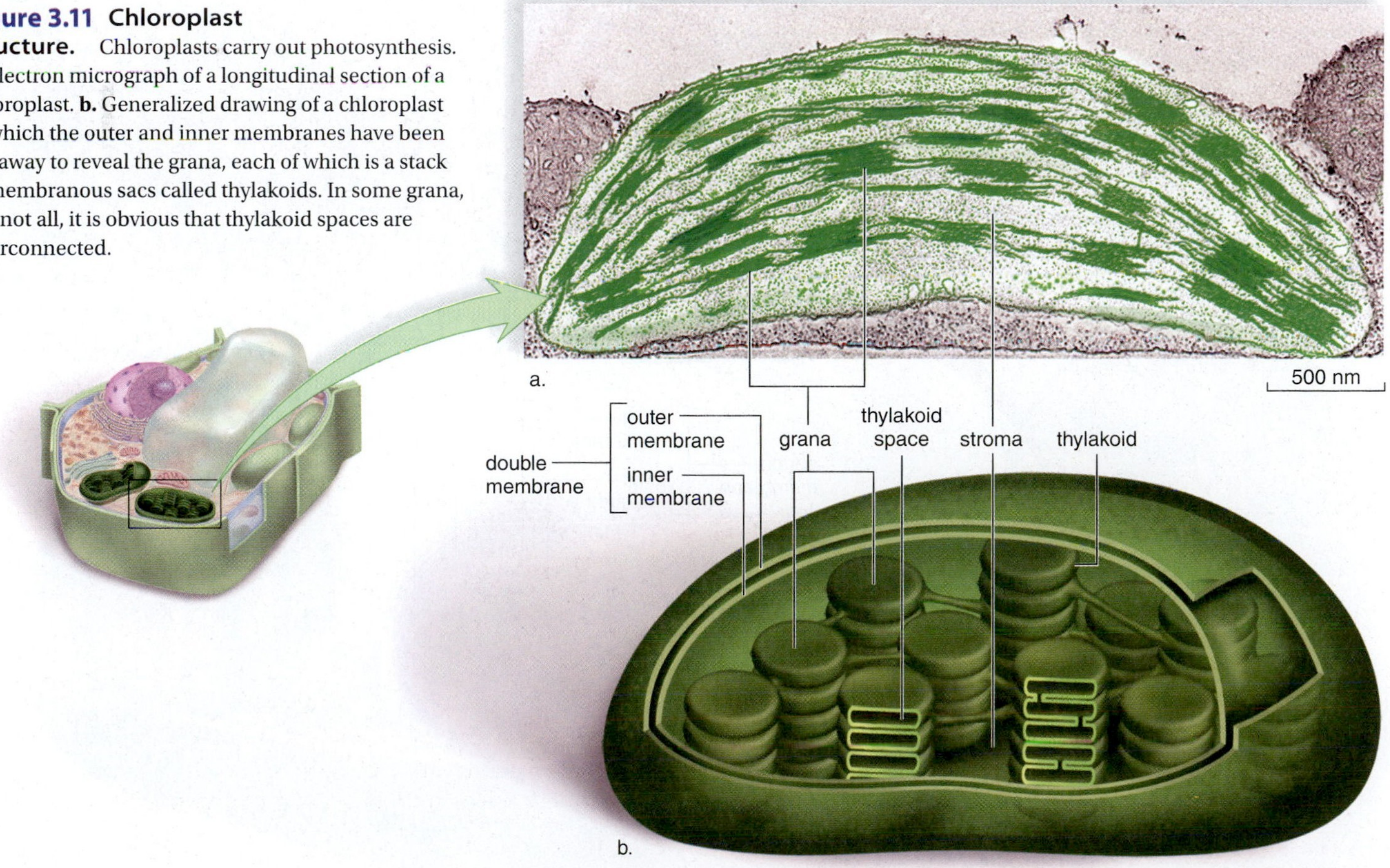
a.
500 nm
double membrane
outer membrane
inner membrane
grana
thylakoid space
stroma
thylakoid
b.

that excited electron is used to make high-energy compounds. The stroma contains the enzymes that synthesize carbohydrates from carbon dioxide and water using these high-energy compounds. The chloroplast is a tiny solar panel for collecting sunlight and a factory for making carbohydrates.

Mitochondria

All eukaryotic cells, including those of plants and algae, contain mitochondria. This means that plant cells contain *both* chloroplasts and mitochondria. Mitochondria are usually 0.5–1.0 μm in diameter and 2–5 μm in length. Mitochondria are the power plants of the cell. Substrates broken down in the cytoplasm are transported into the mitochondria and converted into ATP to be used by the cell for its various needs. Mitochondria are also involved in cellular differentiation and cell death. It is thought that mitochondria may play a role in aging.

Mitochondria, like chloroplasts, divide by splitting in two. Mitochondria are also bounded by a double membrane (Fig. 3.12). In mitochondria, the inner fluid-filled space is called the **matrix.** As with chloroplasts, mitochondria contain their own circular DNA chromosome and encode some, but not all, of their own proteins. The matrix contains ribosomes, and enzymes that break down carbohydrate products, releasing energy to be used for ATP production.

The inner membrane of a mitochondrion invaginates to form **cristae** (sing., **crista**). Cristae provide a much greater surface area to accommodate the protein complexes and other participants that produce ATP.

The number of mitochondria per cell can vary considerably. Some cells have only one mitochondrion, while other cells may have thousands. Tissues that need large amounts of energy have more mitochondria per cell than tissues with lower energy demand. Mutations in the mitochondrial DNA usually affect high-energy-demand tissues such as the eye, central nervous system, and muscles.

In humans, all mitochondria come from the mother's egg. The father's sperm does not contribute any mitochondria to the offspring. This has made mitochondrial DNA studies useful for population genetic studies. An interesting example of this is the search for "mitochondrial Eve," which examines different models on how early humans migrated from Africa (see section 32.5).

Check Your Progress 3.3

1. Explain the function of the cell wall in eukaryotes.
2. Define the role of select organelles within the cell.
3. Recognize the advantages that different compartments provide for the cell.

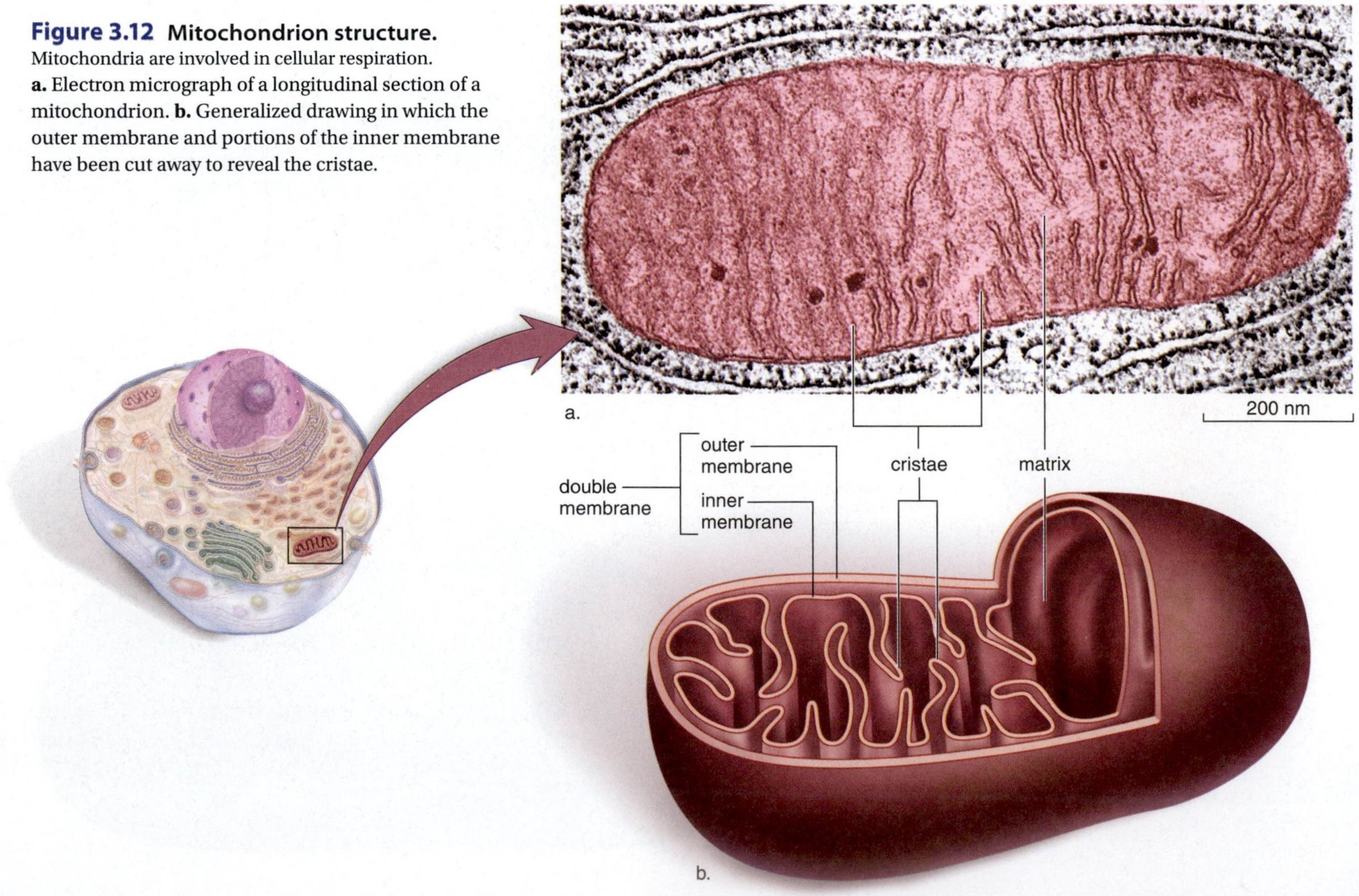
a.
200 nm
outer membrane
double membrane
inner membrane
cristae
matrix
b.

Figure 3.12 Mitochondrion structure. Mitochondria are involved in cellular respiration. **a.** Electron micrograph of a longitudinal section of a mitochondrion. **b.** Generalized drawing in which the outer membrane and portions of the inner membrane have been cut away to reveal the cristae.

3.4 The Cytoskeleton

Learning Outcomes

Upon completion of this section, you should be able to

1. Compare and contrast the structural differences between actin filaments, intermediate filaments, and microtubules.
2. Identify the cellular structures that are composed of the different cytoskeleton components.

The protein components of the **cytoskeleton** interconnect and extend from the nucleus to the plasma membrane in eukaryotic cells. Prior to the 1970s, scientists believed that the cytoplasm was an unorganized mixture of organic molecules. Then, high-voltage electron microscopes, which can penetrate thicker specimens, showed that the cytoplasm is highly organized. The technique of immunofluorescence microscopy identified the makeup of the protein components within the cytoskeletal network (Fig. 3.13).

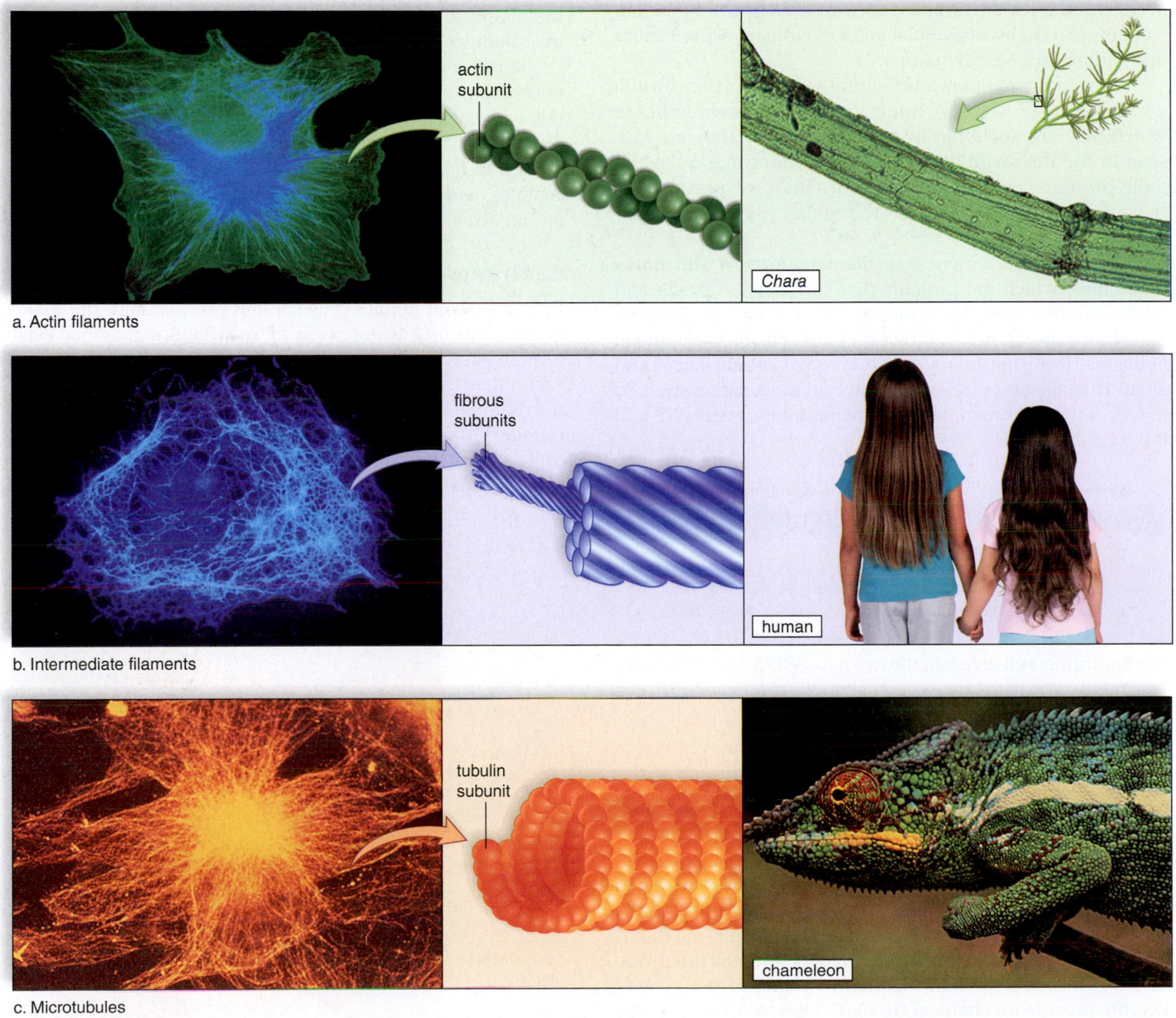
actin
subunit
Chara
a. Actin filaments
fibrous
subunits
human
b. Intermediate filaments
tubulin
subunit
chameleon
c. Microtubules

Figure 3.13 The cytoskeleton. The cytoskeleton maintains the shape of the cell and allows its parts to move. Three types of protein components make up the cytoskeleton. **a.** *Left to right:* Fibroblasts in animal tissue contain actin filaments. The drawing shows that actin filaments are composed of a twisted double chain of actin subunits. The giant cells of the green alga *Chara* rely on actin filaments to move organelles from one end of the cell to another. **b.** *Left to right:* Fibroblasts in an animal tissue contain intermediate filaments. The drawing shows that fibrous proteins account for the ropelike structure of intermediate filaments. Human hair is strengthened by the presence of intermediate filaments. **c.** *Left to right:* Fibroblasts in an animal tissue contain microtubules. The drawing shows that microtubules are hollow tubes composed of tubulin subunits. The skin cells of a chameleon rely on microtubules to move pigment granules around so that they can take on the color of their environment.

The cytoskeleton contains actin filaments, intermediate filaments, and microtubules, which maintain cell shape and allow the cell and its organelles to move. Therefore, the cytoskeleton is often compared with the bones and muscles of an animal. However, the cytoskeleton is dynamic, especially because its protein components can assemble and disassemble as needed.

Actin Filaments

Actin filaments (formerly called microfilaments) are long, extremely thin, flexible fibers (about 7 nm in diameter) that occur in bundles or meshlike networks. Each actin filament contains two chains of globular actin monomers twisted about one another in a helical manner.

Actin filaments play a structural role when they form a dense, complex web just under the plasma membrane, to which they are anchored by special proteins. They are also seen in the microvilli that project from intestinal cells, and their presence accounts for the formation of pseudopods (false feet), extensions that allow certain cells to move in an amoeboid fashion.

To produce movement, actin filaments interact with **motor molecules,** which are proteins that can attach, detach, and reattach farther along the actin filament. For example, in muscle cells the motor molecule myosin pulls actin filaments along in this way using the energy of ATP. Myosin has both a head and a tail. The tails of several muscle myosin molecules are joined to form a thick filament, while the heads interact with ATP and the actin filament.

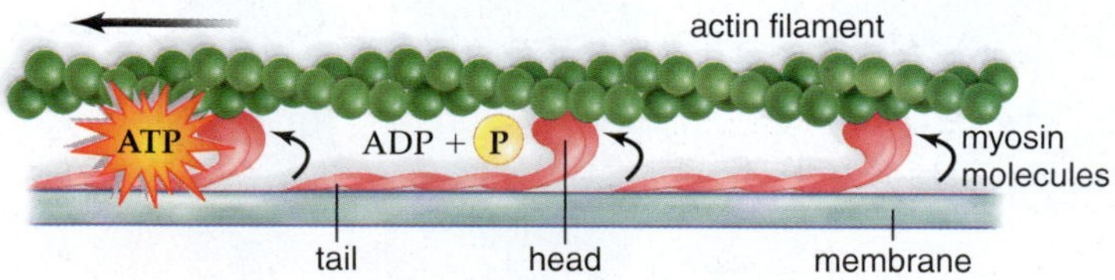
actin filament
ATP
ADP + P
myosin molecules
tail
head
membrane

During animal cell division, the two new cells form when actin, in conjunction with myosin, pinches off the cells from one another.

Intermediate Filaments

Intermediate filaments (8–11 nm in diameter) are intermediate in size between actin filaments and microtubules that will perform a structural role in the cell. They are a rope-like assembly of fibrous polypeptides that vary according to the type of tissue. Some intermediate filaments support the nuclear envelope, whereas others support the plasma membrane and take part in the formation of cell-to-cell junctions. In skin cells, intermediate filaments, made of the protein keratin, provide mechanical strength. They, too, are dynamic structures.

Microtubules

Microtubules are small, hollow cylinders about 25 nm in diameter and 0.2–25 μm in length.

Microtubules are made of the globular protein tubulin, which is of two types, called α and β. Microtubules have 13 rows of tubulin dimers, surrounding what appears to be an empty central core.

The regulation of microtubule assembly is controlled by a microtubule organizing center. In most eukaryotic cells, the main microtubule organizing center is in the **centrosome,** which lies near the nucleus. Microtubules radiate from the centrosome, helping to maintain the shape of the cell and acting as tracks along which organelles can move. Whereas the motor molecule myosin is associated with actin filaments, the motor molecules kinesin and dynein are associated with microtubules.

Before a cell divides, microtubules disassemble and then reassemble into a structure called a spindle. The spindle apparatus attaches to the chromosomes and ensures that they are distributed in an orderly manner. It also participates in dividing the cell in half. At the end of cell division, the spindle disassembles, and microtubules reassemble once again into their former array.

Centrioles

Both plant and animal cells contain centrosomes, the major microtubule organizing center for the cell. But in animal cells, a centrosome contains two centrioles lying at right angles to each other. The centrioles may be involved in the process of microtubule assembly and disassembly. **Centrioles** are short cylinders of microtubules with a 9 + 0 pattern of microtubule triplets—that is, a ring having nine sets of microtubule triplets, with none in the middle (Fig. 3.14).

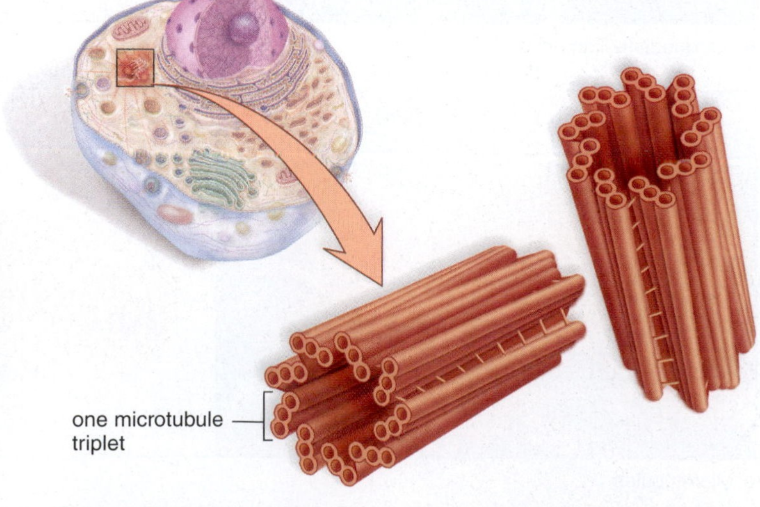
one microtubule triplet

Figure 3.14 Centrioles. In a nondividing animal cell, a single pair of centrioles lies in the centrosome located just outside the nucleus. Just before a cell divides, the centrioles replicate, producing two pairs of centrioles. During cell division, centrioles in their respective centrosomes separate so that each new cell has one centrosome containing one pair of centrioles.

Before an animal cell divides, the centrioles replicate such that the members of each pair are again at right angles to one another (Fig. 3.14). Then, each pair becomes part of a separate centrosome. During cell division, the centrosomes move apart and may function to organize the mitotic spindle.

Cilia and Flagella

Cilia (sing., **cilium**) and **flagella** (sing., **flagellum**) are hairlike projections that can move either in an undulating fashion, like a whip, or stiffly, like an oar. Some cells that have these organelles are capable of movement. For example, unicellular organisms called paramecia move by means of cilia, whereas sperm cells move by means of flagella. In the human body, the cells that line our upper respiratory tract have cilia that sweep debris trapped within mucus back up into the throat, where it can be swallowed or ejected. This action helps keep the lungs clean.

In eukaryotic cells, cilia are much shorter than flagella, but they have a similar construction. Both are membrane-bound cylinders. The cylinders are composed of nine microtubule doublets arranged in a circle around two central microtubules. Therefore, they have a 9 + 2 pattern of microtubules. Cilia and flagella move when the microtubule doublets slide past one another (Fig. 3.15).

Check Your Progress 3.4

1. Identify the structural makeup of actin filaments, intermediate filaments, and microtubules.
2. Contrast the structure of cilia and flagella to that of centrioles.
3. Explain how cilia and flagella move.

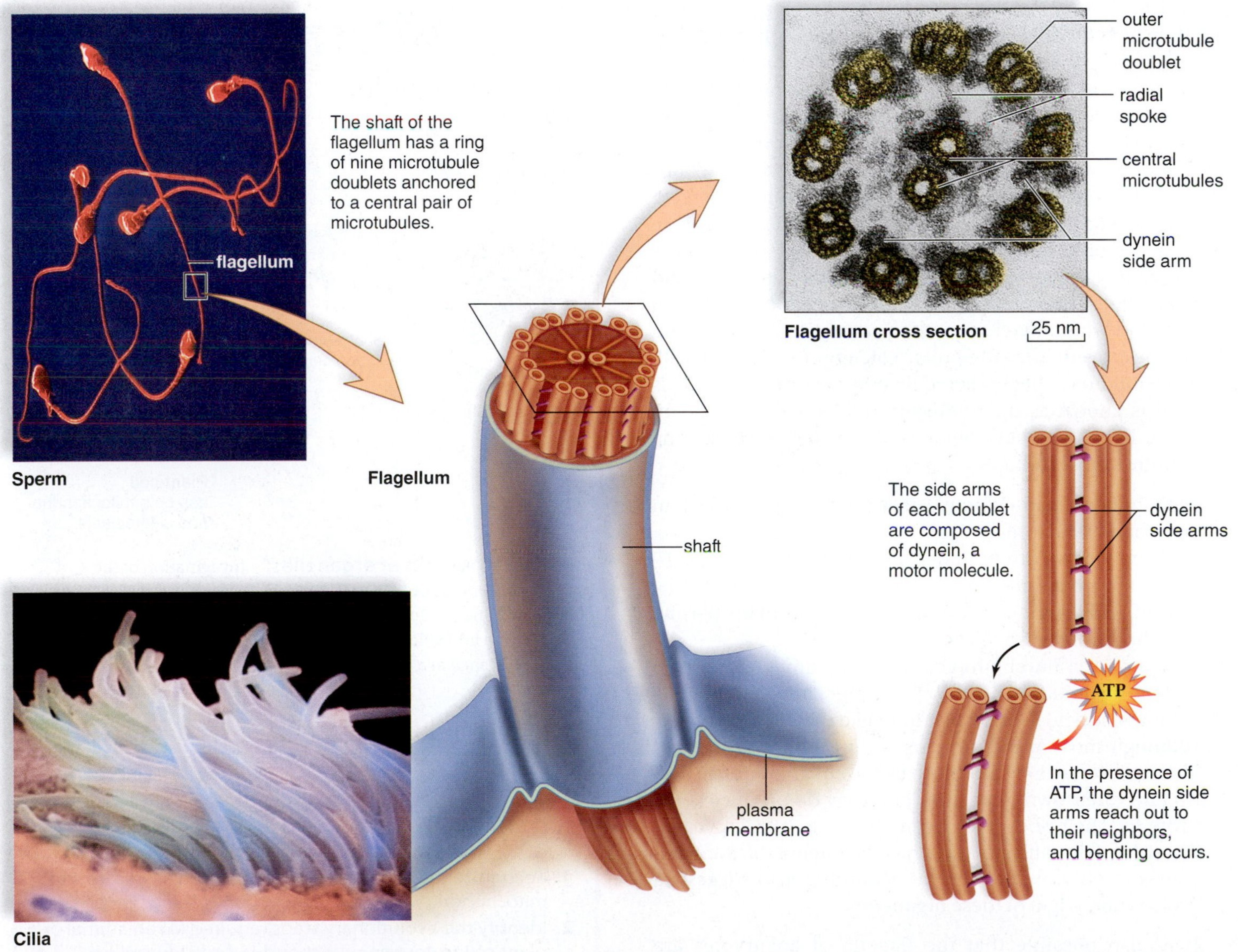

flagellum
Sperm
The shaft of the flagellum has a ring of nine microtubule doublets anchored to a central pair of microtubules.
Flagellum
shaft
plasma membrane
Cilia
outer microtubule doublet
radial spoke
central microtubules
dynein side arm
Flagellum cross section
25 nm
The side arms of each doublet are composed of dynein, a motor molecule.
dynein side arms
ATP
In the presence of ATP, the dynein side arms reach out to their neighbors, and bending occurs.

Figure 3.15 Structure of a flagellum or cilium. The shaft of a flagellum (or cilium) contains microtubule doublets whose side arms are motor molecules that cause the projection to move. Sperm have flagella. Without the ability of sperm to move to the egg, human reproduction would not be possible. Cilia cover the surface of the cells of the respiratory system where they beat upward to remove foreign matter.

3.5 Origin and Evolution of the Eukaryotic Cell

Learning Outcomes

Upon completion of this section, you should be able to

1. Identify the first cells and how eukaryotic cells came to be.
2. Describe the evidence that supports the endosymbiotic theory.

The fossil record, which is based on the remains of ancient life, suggests that the first cells were prokaryotes. Therefore, scientists believe that eukaryotic cells evolved from prokaryotic cells. Biochemical data suggest that eukaryotes are more closely related to the archaea than the bacteria. The eukaryotic cell probably evolved from a prokaryotic cell in stages. Invagination of the plasma membrane might explain the origin of the nuclear envelope and such organelles as the endoplasmic reticulum and the Golgi apparatus. Some believe that the other organelles could also have arisen in a similar fashion.

There is evidence that a similar process was involved in the origin of the energy organelles. Observations in the laboratory indicate that an amoeba infected with bacteria can become dependent upon them. Some investigators believe mitochondria and chloroplasts are derived from prokaryotes that were taken up by larger cells (Fig. 3.16). Perhaps mitochondria were originally aerobic heterotrophic bacteria, and chloroplasts were originally cyanobacteria. The eukaryotic host cell would have benefited from an ability to utilize oxygen or synthesize organic food when, by chance, the prokaryote was taken up and not destroyed. After the prokaryote entered the host cell, the two would have begun living together cooperatively. This proposal is known as the **endosymbiotic theory** (*endo-*, in; *symbiosis,* living together). Some of the evidence supporting this hypothesis is as follows:

1. Mitochondria and chloroplasts are similar to bacteria in size and in structure.
2. Both organelles are bounded by a double membrane—the outer membrane may be derived from the engulfing vesicle, and the inner one may be derived from the plasma membrane of the original prokaryote.
3. Mitochondria and chloroplasts contain a limited amount of genetic material and divide by splitting. Their DNA (deoxyribonucleic acid) is a circular loop like that of prokaryotes.
4. Although most of the proteins within mitochondria and chloroplasts are now produced by the eukaryotic host, they do have their own ribosomes and they do produce some proteins. Their ribosomes resemble those of prokaryotes.
5. The RNA (ribonucleic acid) base sequence of the ribosomes in chloroplasts and mitochondria also suggests a prokaryotic origin of these organelles.

It is also possible that the flagella of eukaryotes are derived from an elongated bacterium with a flagellum that became attached to a host cell. However, the flagella of eukaryotes are constructed differently from those of modern bacteria.

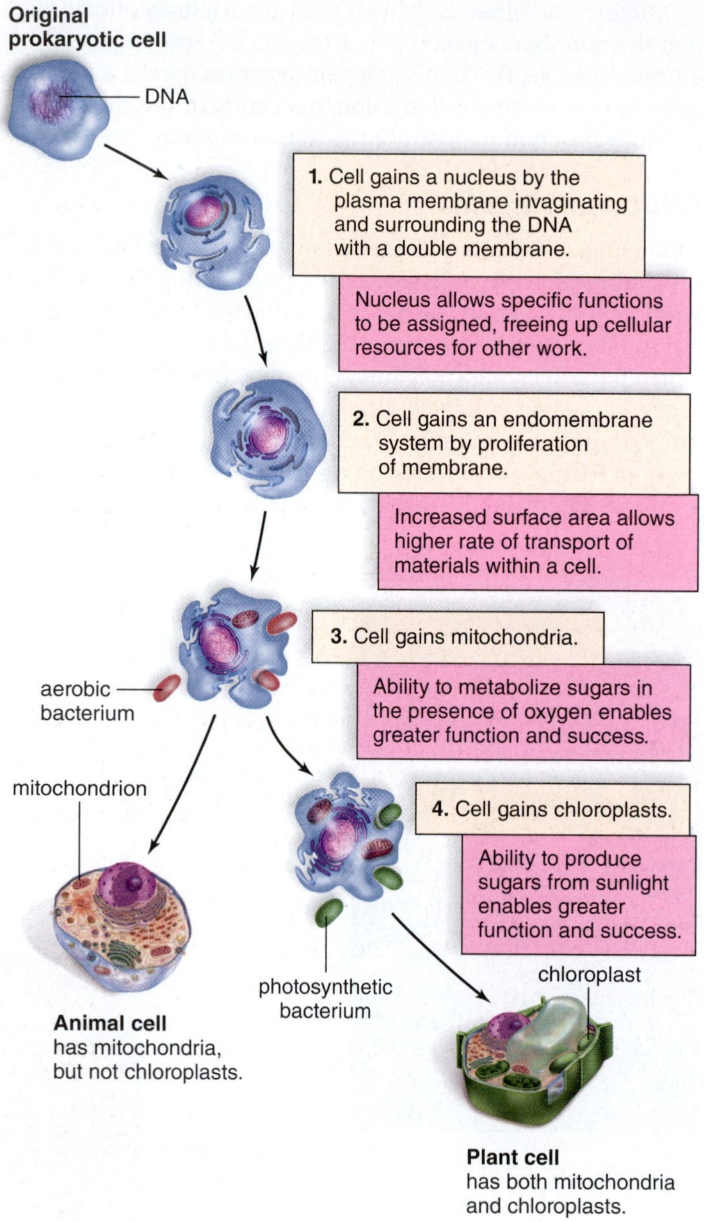
Original prokaryotic cell
DNA
1. Cell gains a nucleus by the plasma membrane invaginating and surrounding the DNA with a double membrane.
Nucleus allows specific functions to be assigned, freeing up cellular resources for other work.
2. Cell gains an endomembrane system by proliferation of membrane.
Increased surface area allows higher rate of transport of materials within a cell.
3. Cell gains mitochondria.
Ability to metabolize sugars in the presence of oxygen enables greater function and success.
aerobic bacterium
mitochondrion
4. Cell gains chloroplasts.
Ability to produce sugars from sunlight enables greater function and success.
chloroplast
photosynthetic bacterium
Animal cell
has mitochondria, but not chloroplasts.
Plant cell
has both mitochondria and chloroplasts.

Figure 3.16 Origin of organelles. Invagination of the plasma membrane could have created the nuclear envelope and an endomembrane system. The endosymbiotic theory suggests that mitochondria and chloroplasts were once independent prokaryotes that took up residence in a eukaryotic cell.

Animation Endosymbiosis

Check Your Progress 3.5

1. Recognize the evidence that suggests chloroplasts and mitochondria were once independently living prokaryotes.
2. Identify the evolutionary steps required for an animal or plant cell to contain mitochondria and chloroplasts.

Case Study Conclusion

Tay-Sachs disease is a recessive neurological disorder that is more common in individuals of eastern and central European Jewish heritage than the rest of the general population. A child who inherits Tay-Sachs will most likely die by the age of 4 due to neurological breakdown and recurring infections as the result of faulty lysosomes. The lysosomes do not produce the enzyme necessary to digest the fatty substance ganglioside GM2. Instead, GM2 accumulates within the lysosome, causing it to distend—which in turn impacts the function of neurological cells in the brain. Deafness, blindness, and seizures are some of the physical complications associated with the disease. Unfortunately, there is no cure or treatment available for individuals affected by Tay-Sachs disease. Clinical trials are under way with hopes of a cure in the near future.

MEDIA STUDY TOOLS

www.mhhe.com/maderinquiry14

Enhance your study of this chapter with study tools and practice tests. Also ask your instructor about the resources available through ConnectPlus, including LearnSmart, the media-rich eBook, interactive learning tools, and animations.

SUMMARIZE

3.1 The Cellular Level of Organization

- All organisms are composed of cells, the smallest units of living matter.
- Cells are capable of self-reproduction, and new cells come only from preexisting cells.
- Cells must remain small in order to have an adequate ratio of surface-area-to-volume for exchange of molecules with the environment.

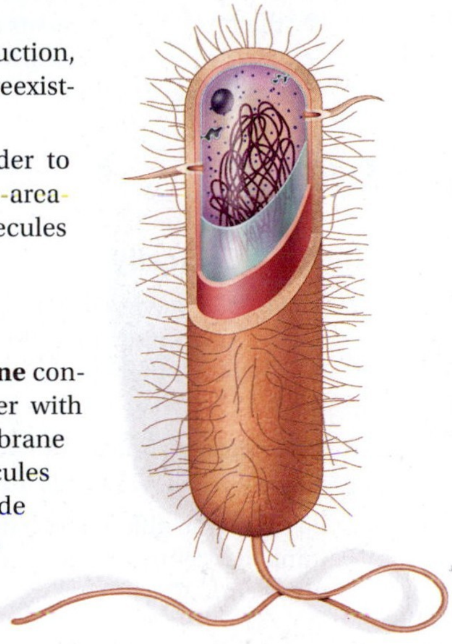

3.2 Prokaryotic Cells

- All cells have a **plasma membrane** consisting of a phospholipid bilayer with embedded proteins. The membrane regulates the movement of molecules into and out of the cell. The inside of the cell is filled with a fluid called **cytoplasm.** Some prokaryotic cells possess a gel-like sheath called a **capsule** and most possess a **cell wall** for protection.
- **Prokaryotic cells** do not have a nucleus, but they do have a **nucleoid** that is not bounded by a nuclear envelope. They also lack most of the other organelles that compartmentalize eukaryotic cells.
- Prokaryotic cells, as exemplified by the bacteria and archaea, are structurally less complex than eukaryotic cells but metabolically very diverse.

3.3 Eukaryotic Cells

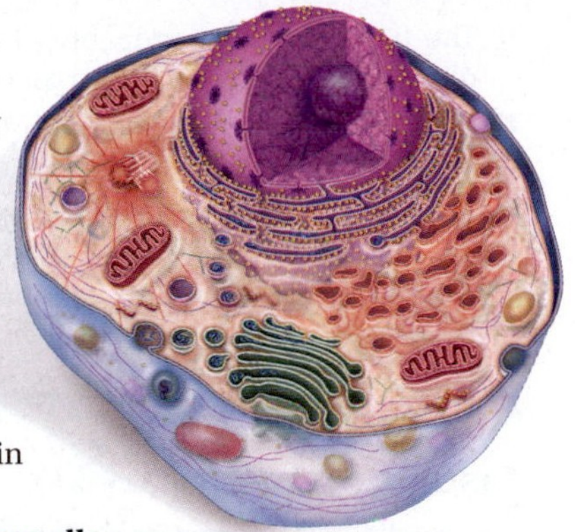

- **Eukaryotic cells** are very complex. Animals, plants, fungi, and protists are all examples of eukaryotes. The common feature is the presence of a **nuclear envelope** that contains the genetic material.
- Some eukaryotic cells contain cell walls for protection.
- Eukaryotic cells contain **organelles,** which are subcellular structures that perform a particular function.
- The **nucleus** of eukaryotic cells is bounded by a nuclear envelope containing pores. These pores serve as passageways between the cytoplasm and the **nucleoplasm.** Within the nucleus, the **chromatin** is a complex of DNA and protein. Chromatin is divided into separate structures called **chromosomes.**
- The **nucleolus** is a special region of the chromatin where rRNA is produced and where proteins from the cytoplasm gather to form ribosomal subunits.
- **Ribosomes** are organelles that function in protein synthesis. They can be bound to the **endoplasmic reticulum** (**ER**) or can exist within the cytoplasm singly or in groups called **polyribosomes.**
- The endomembrane system includes the ER, the **Golgi apparatus,** the **lysosomes,** and other types of **vesicles** and **vacuoles.** The endomembrane system compartmentalizes the cell. The rough endoplasmic reticulum (RER) is covered with ribosomes and is involved in the folding, modification, and transport of proteins. The smooth ER has various metabolic functions depending on the cell type, but it also forms vesicles that carry products to the Golgi apparatus.
- The Golgi apparatus processes proteins and repackages them into lysosomes, which carry out intracellular digestion, or into vesicles for transport to the plasma membrane or other organelles.

- Vacuoles are large storage sacs, and vesicles are smaller ones. The large single plant cell vacuole not only stores substances but also lends support to the plant cell.
- **Peroxisomes** contain enzymes that oxidize molecules by producing hydrogen peroxide, which is subsequently broken down.
- Cells require a constant input of energy to maintain their structure. **Chloroplasts** capture solar energy to carry on **photosynthesis,** which produces carbohydrates. Carbohydrate-derived products are broken down in **mitochondria** to produce ATP. The mitochondria are considered the energy powerhouse of the cell.

3.4 The Cytoskeleton

- The **cytoskeleton** contains **actin filaments, intermediate filaments,** and **microtubules.** These maintain cell shape and allow the cell and its organelles to move. Actin filaments interact with motor molecules such as myosin in muscle cells. Microtubules are present in **centrioles, cilia,** and **flagella.** In the cytoplasm they serve as tracks along which vesicles and other organelles move due to the action of specific motor molecules.

3.5 Origin and Evolution of the Eukaryotic Cell

- The first cells were probably prokaryotic cells. Eukaryotic cells most likely arose from prokaryotic cells in stages.
- Biochemical data suggest that eukaryotic cells are closer evolutionarily to the archaea than to the bacteria.
- The nuclear envelope most likely evolved through invagination of the plasma membrane, but mitochondria and chloroplasts may have arisen through endosymbiotic events.

ASSESS

Testing Yourself

Choose the best answer for each question.

1. Proteins are produced
   a. in the cytoplasm or the ER. **c.** in the Golgi apparatus.
   b. in the nucleus. **d.** None of these are correct.
2. Which structure is characteristic of prokaryotic cells?
   a. nucleus **c.** nucleoid
   b. mitochondria **d.** chloroplast
3. Ribosomes are found in what type of cells?
   a. animal **c.** bacterial
   b. plant **d.** All of these are correct.
4. Lysosomes function in
   a. protein synthesis. **c.** intracellular digestion.
   b. processing and packaging. **d.** All of these are correct.
5. Which of these could you see with a light microscope?
   a. atom **c.** amino acid
   b. proteins **d.** None of these are correct.

For questions 6–10, match the structure to the function in the key.

Key:
a. movement of cell
b. transport of proteins
c. photosynthesis
d. ribosome formation
e. folds and processes proteins

6. Chloroplasts
7. Flagella
8. Golgi apparatus
9. Rough ER
10. Nucleolus
11. Which of these sequences depicts a hypothesized evolutionary scenario?
    a. cyanobacteria ⟶ mitochondria
    b. Golgi ⟶ mitochondria
    c. mitochondria ⟶ cyanobacteria
    d. cyanobacteria ⟶ chloroplast
12. Label these parts of the cell that are involved in protein synthesis and modification. Give a function for each structure.

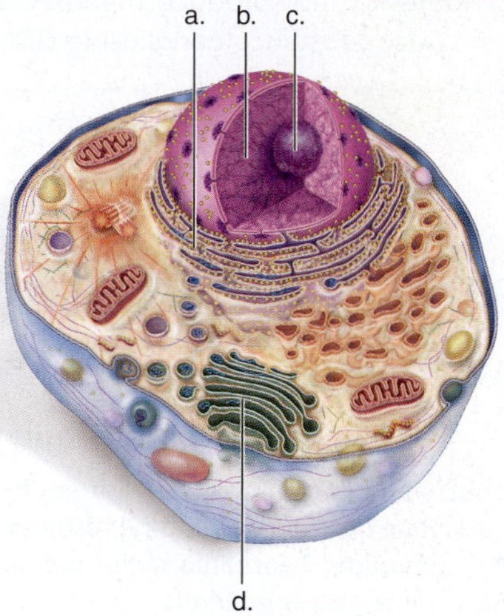
a.
b.
c.
d.

13. The cell theory states:
    a. Cells form as organelles and molecules become grouped together in an organized manner.
    b. The normal functioning of an organism does not depend on its individual cells.
    c. The cell is the basic unit of life.
    d. Only eukaryotic organisms are made of cells.
14. The small size of cells is best correlated with
    a. the fact that they are self-reproducing.
    b. their prokaryotic versus eukaryotic nature.
    c. an adequate surface area for exchange of materials.
    d. their vast versatility.

ENGAGE

Thinking Critically

1. In the 1958 movie *The Blob,* a giant, single-celled alien creeps and oozes around, attacking and devouring helpless humans. Why couldn't there be a real single-celled organism as large as the Blob?
2. Calculate the surface-area-to-volume ratio of a 1-mm cube and a 2-mm cube. Which has the smaller ratio?
3. Based on the differences between plant and animal cells, do you think that centrioles are required for microtubule assembly and disassembly?

Membrane Structure and Function 4

CASE STUDY Red Hot Chili Peppers

Have you ever bitten into a hot pepper and had the sensation that your mouth is on fire? Your eyes water and you are in real pain! The feelings of heat and pain are due to a membrane protein in your sensory nerves. Capsaicin is the chemical in chili peppers that binds to a channel protein in specialized sensory nerve cell endings called nociceptors (*noci-* means hurtful). One of the important functions of a membrane is to control what molecules move into and out of the cell and when they move. This particular channel protein, when activated, allows calcium ions to flow into the cell. In addition to capsaicin, other factors such as an acidic pH, heat, electrostatic charges, and a variety of chemical agents can activate this channel protein. Once activated by any of these signals, the response is the same. The channel opens, calcium ions flow into the cell, and the nociceptor sends a signal to the brain. The brain then interprets this signal as pain. As long as the capsaicin is present, this pathway will continue to send signals to the brain. So the quickest way to alleviate the pain is to remove the capsaicin and close the channel protein. While some people drink cold water, this does very little other than cool down their mouth because capsaicin is lipid-soluble and does not dissolve in water. However, drinking milk, or eating rice or bread, usually helps. If you are a true "chili head," you know that if you survive the first bite, the next bite is easier. That is because within minutes, the pathway becomes desensitized, or fails to respond, to the pain. However, other pathways, such as those in the eyes, may become activated if exposed to the capsaicin!

In this chapter, we will discuss the various functions of proteins embedded in the membranes of your cells and how the membrane controls what enters and leaves the cell. We will also describe how cells communicate with each other through signals sent to receptor proteins in the cell membrane.

As you read though this chapter, think about the following questions:

1. What are the roles of the proteins in the plasma membrane of cells?
2. What type of transport is the calcium channel in this story exhibiting?

CHAPTER OUTLINE

BEFORE YOU BEGIN

Before beginning this chapter, take a few moments to review the following discussions:

Section 2.6 How does the structure of a phospholipid make it an ideal molecule for the plasma membrane?

Section 2.7 How does a protein's shape relate to its function?

Figures 3.4 and 3.5 What are the key features of animal and plant cells?

4.1 Plasma Membrane Structure and Function

Learning Outcomes

Upon completion of this section, you should be able to

1. Describe the fluid-mosaic model of membrane structure.
2. Describe the diverse roles of proteins in membranes.

As we introduced in Chapter 3, the plasma membrane separates the internal environment of the cell from the external environment. In doing so, it regulates the entrance and exit of molecules from the cell. In this way, it helps the cell and the organism maintain a steady internal environment, a process called **homeostasis.** The plasma membrane is made primarily of phospholipids, a type of lipid with both hydrophobic and hydrophilic properties. The phospholipids of the membrane form a bilayer, with the hydrophilic (water-loving) polar heads of the phospholipid molecules facing the outside and inside of the cell (where water is found), and the hydrophobic (water-fearing) nonpolar tails facing each other (Fig. 4.1). The phospholipid bilayer has a fluid consistency, comparable to that of light oil. The fluidity of the membrane is regulated by steroids such as *cholesterol,* which serve to stiffen and strengthen the membrane.

Throughout the membrane are numerous proteins, in which protein molecules are either partially or wholly embedded. These proteins are scattered either just outside or within the membrane, and may be either partially or wholly embedded in the phospholipid bilayer. *Peripheral proteins* are associated with only one side of the plasma membrane. Peripheral proteins on the inside of the membrane are often held in place by cytoskeletal filaments. In contrast, *integral proteins* span the membrane, and can protrude from one or both sides. They are embedded in the membrane, but they can move laterally, changing their position in the membrane. The proteins in the membrane form a *mosaic* pattern, and this combination of proteins, steroids, and

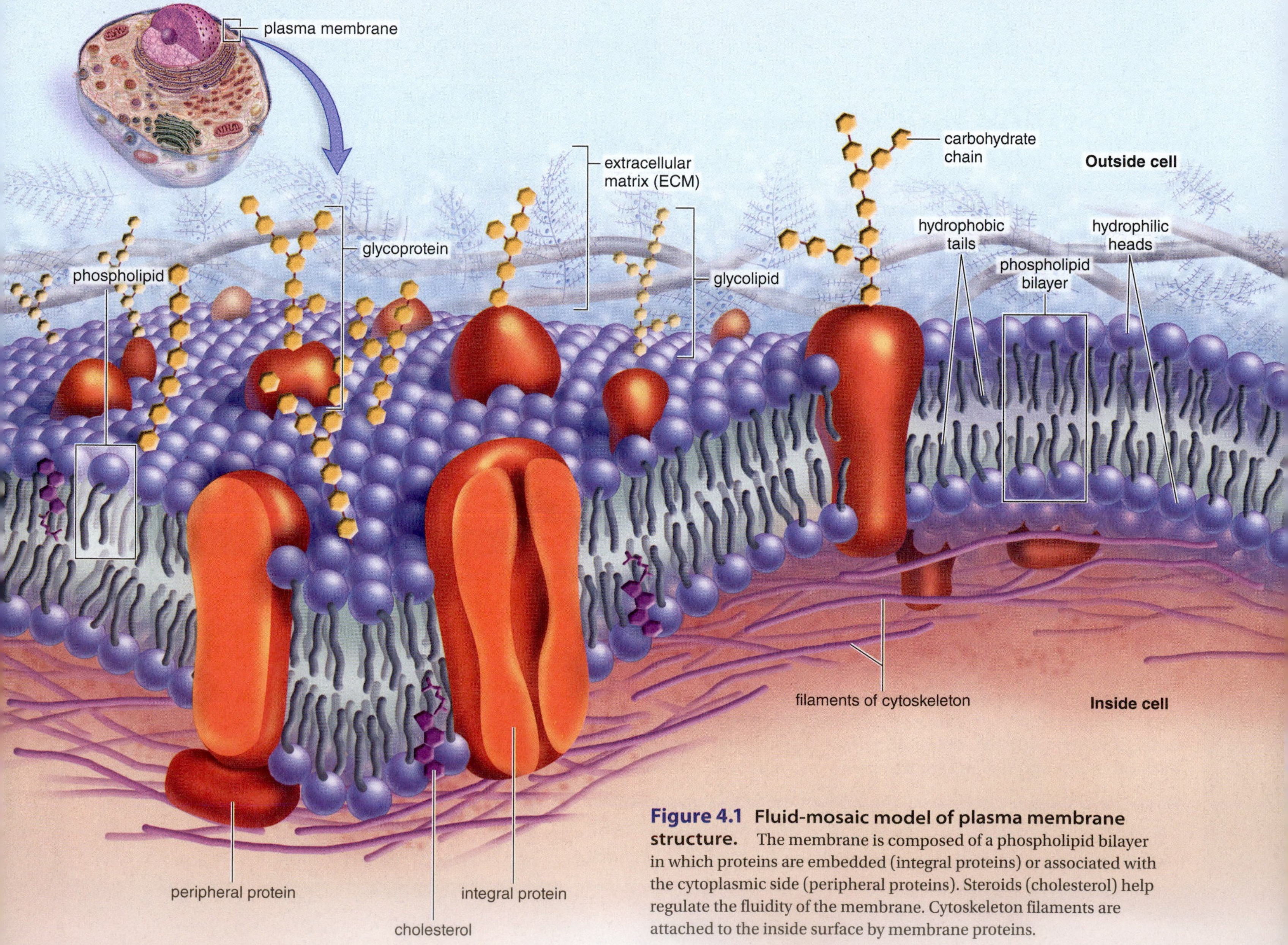
plasma membrane
extracellular matrix (ECM)
carbohydrate chain
Outside cell
glycoprotein
hydrophobic tails
hydrophilic heads
phospholipid
glycolipid
phospholipid bilayer
filaments of cytoskeleton
Inside cell
peripheral protein
integral protein
cholesterol

Figure 4.1 Fluid-mosaic model of plasma membrane structure. The membrane is composed of a phospholipid bilayer in which proteins are embedded (integral proteins) or associated with the cytoplasmic side (peripheral proteins). Steroids (cholesterol) help regulate the fluidity of the membrane. Cytoskeleton filaments are attached to the inside surface by membrane proteins.

phospholipids is called the **fluid-mosaic model** of membrane structure (Fig. 4.1).

3D Animation Lipid Bilayer

MP3 Membrane Structure

Both phospholipids and proteins can have attached carbohydrate (sugar) chains. If so, these molecules are called **glycolipids** and **glycoproteins,** respectively. Because the carbohydrate chains occur only on the outside surface and peripheral proteins occur asymmetrically on one surface or the other, the two halves of the membrane are not identical. These molecules play an important role in cellular identification.

Functions of the Membrane Proteins

The plasma membranes of various cells and the membranes of various organelles each have their own unique collections of proteins. The peripheral proteins often have a structural role in that they help stabilize and shape the plasma membrane (see Section 3.4). They may also function in signaling pathways. The integral proteins largely determine a membrane's specific functions. Integral proteins can be of the following types:

Channel proteins are involved in the passage of molecules through the membrane. They have a channel that allows a substance to simply move across the membrane (Fig. 4.2*a*). For example, a channel protein allows hydrogen ions to flow across the inner mitochondrial membrane. Without this movement of hydrogen ions, ATP would never be produced. Channel proteins may contain a gate that must be opened by the binding of a specific molecule to the channel.

Carrier proteins are also involved in the passage of molecules through the membrane. They combine with a substance and help it move across the membrane (Fig. 4.2*b*). For example, a carrier protein transports sodium and potassium ions across a nerve cell membrane. Without this carrier protein, nerve conduction would be impossible.

Cell recognition proteins are glycoproteins (Fig. 4.2*c*). Among other functions, these proteins help the body recognize when it is being invaded by pathogens so that an immune reaction can occur.

Receptor proteins have a shape that allows a specific molecule to bind to it (Fig. 4.2*d*). The binding of this molecule causes the protein to change its shape and thereby bring about a cellular response. The coordination of the body's organs is totally dependent on such signal molecules. For example, the liver stores glucose after it is signaled to do so by insulin.

Enzymatic proteins carry out metabolic reactions directly (Fig. 4.2*e*). The integral membrane proteins of the electron transport chain carry out the final steps of aerobic respiration. Without the presence of enzymes, some of which are attached to the various membranes of the cell, a cell would never be able to perform the metabolic reactions necessary to its proper function.

The peripheral proteins often have a structural role in that they help stabilize and shape the plasma membrane.

Check Your Progress 4.1

1. Describe the role of proteins, steroids, and phospholipids in the fluid-mosaic model.
2. Distinguish between the roles of the various integral proteins in the plasma membrane.

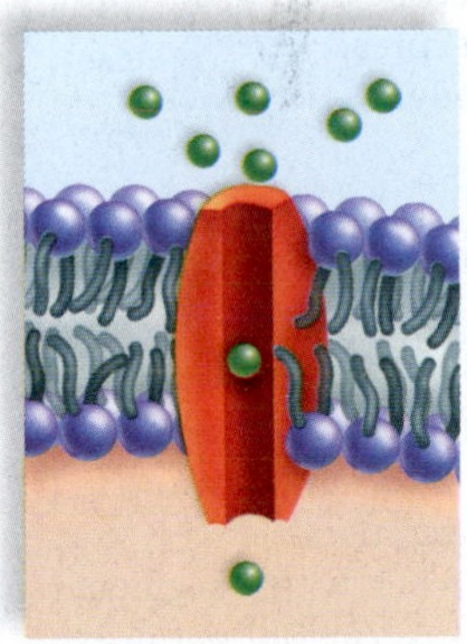

Allows a particular molecule or ion to cross the plasma membrane freely. Cystic fibrosis, an inherited disorder, is caused by a faulty chloride (Cl^-) channel; a thick mucus collects in airways and in pancreatic and liver ducts.

a. Channel protein

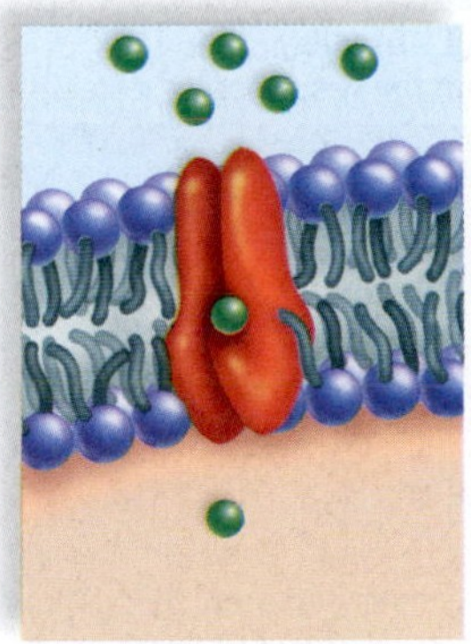

Selectively interacts with a specific molecule or ion so that it can cross the plasma membrane. The family of GLUT carriers transfers glucose in and out of the various cell types of the body. Different carriers respond differently to blood levels of glucose.

b. Carrier protein

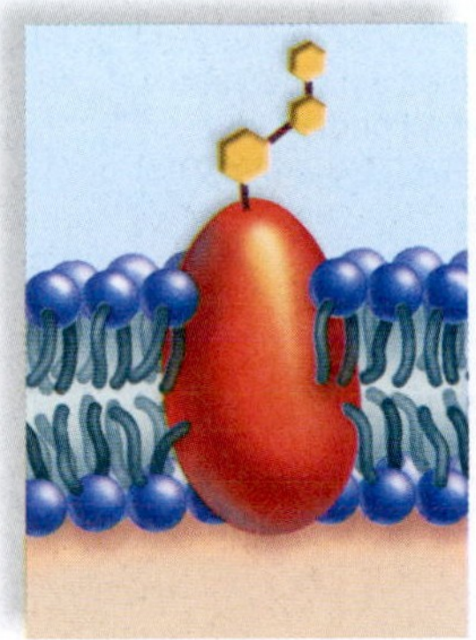

The MHC (major histocompatibility complex) glycoproteins are different for each person, so organ transplants are difficult to achieve. Cells with foreign MHC glycoproteins are attacked by white blood cells responsible for immunity.

c. Cell recognition protein

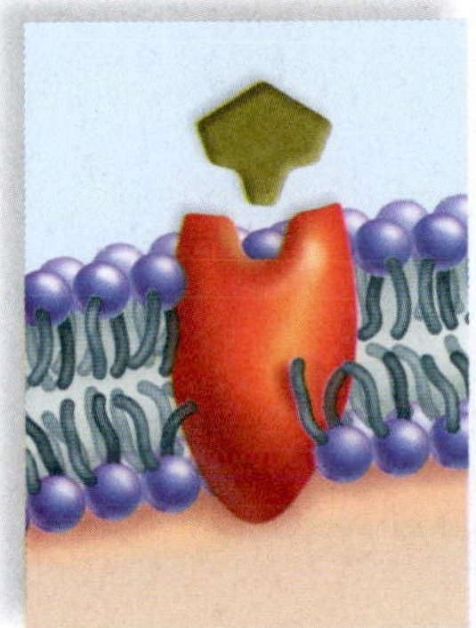

Shaped in such a way that a specific molecule can bind to it. Some types of dwarfism result not because the body does not produce enough growth hormone, but because the plasma membrane growth hormone receptors are faulty and cannot interact with growth hormone.

d. Receptor protein

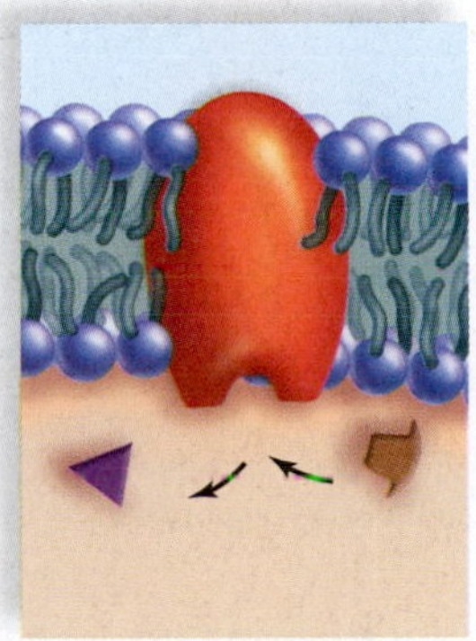

Catalyzes a specific reaction. The membrane protein, adenylate cyclase, is involved in ATP metabolism. Cholera bacteria release a toxin that interferes with the proper functioning of adenylate cyclase, which eventually leads to severe diarrhea.

e. Enzymatic protein

Figure 4.2 Examples of membrane protein diversity. These are some of the functions performed by integral proteins found in the plasma membrane.

SCIENCE IN YOUR LIFE ▶ HEALTH

How Cells Talk to One Another

All organisms are able to sense and respond to specific signals in their environment. A bacterium that has taken up residence in your body is responding to signaling molecules when it finds food and escapes immune cells in order to stay alive. Signaling helps bread mold on stale bread in your refrigerator detect the presence of an opposite mating strain and begin its sexual life cycle. Similarly, the cells of an embryo are responding to signaling molecules when they move to specific locations and assume the shape and perform the functions of specific tissues (Fig. 4A). In plants, external signals, such as a change in the amount of light, tells them when it is time to resume growth or flower. Internal signaling molecules enable plants to coordinate the activities of roots, stems, and leaves.

Cell Signaling

The cells of a multicellular organism "talk" to one another by using signaling molecules, called chemical messengers. Some messengers are produced at a distance from a target tissue and, in animals, are carried by the circulatory system to various sites around the body. For example, the pancreas releases a hormone called insulin, which is transported in blood vessels to the liver, and thereafter, the liver stores glucose as glycogen. Failure of the liver to respond appropriately results in a medical condition called diabetes. Growth factors act locally as signaling molecules and cause cells to divide. Overreacting to growth factors can result in a tumor characterized by unlimited cell division. We have learned that cells respond to only certain signaling molecules. Why? Because they must bind to a receptor protein, and cells have receptors for only certain signaling molecules. Each cell has receptors for numerous signaling molecules and often the final response is due to a summing up of all the various signals received. These molecules tell a cell what it should be doing at the moment, and without any signals, the cell dies.

Signaling not only involves a receptor protein, it also involves a pathway called a transduction pathway and a response. To understand the process, consider an analogy. When a TV camera (the receptor) is shooting a scene, the picture is converted to electrical signals (transduction pathway) that are understood by the TV in your house and are converted to a picture on your screen (the response). The process in cells is more complicated because each member of the pathway can turn on the activity of a number of other proteins. As shown in Figure 4A, the cell response to a transduction pathway can be a change in the shape or movement of a cell, the activation of a particular enzyme, or the activation of a specific gene.

Advances in Understanding Cell Communication

The importance of cell signaling causes much research to be directed toward understanding the intricacies of the process, and this research is starting to yield some significant results. Recently, researchers have discovered the basis of communication between melanoma cells and neighboring cells of the body. Melanoma is an aggressive form of cancer, in which the cells possess an enhanced ability to move (metastasis) by entering neighboring blood vessels. By identifying the signals generated by the melanoma cells, and their target receptors on the blood vessel cells, researchers have been able to develop chemicals that prevent the movement of the melanoma cells. These chemicals may someday be used to develop drugs that prevent the spread of melanoma in the body.

Video Bug Speak

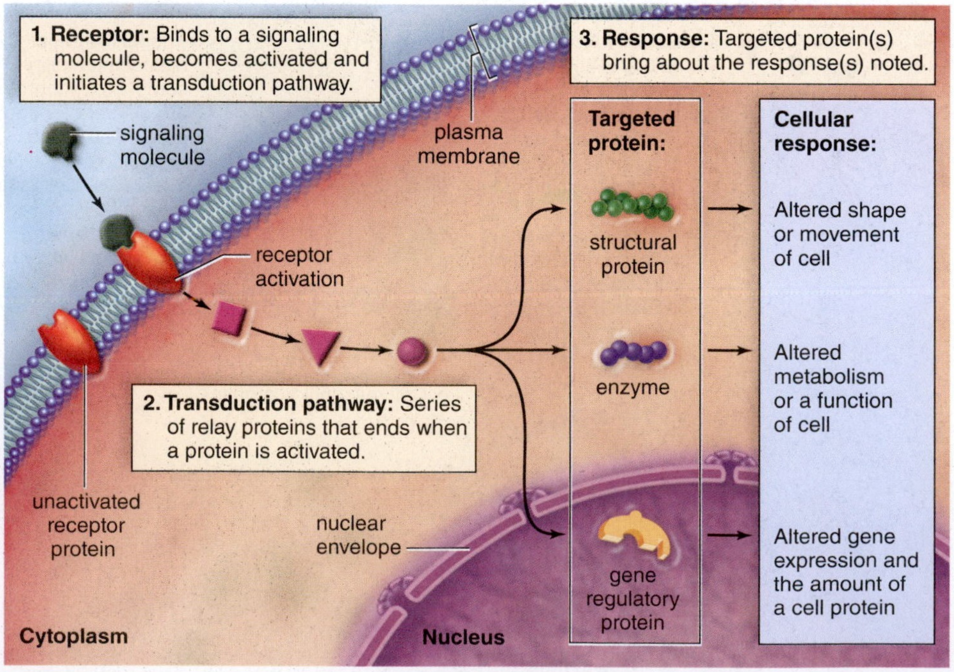

1. Receptor: Binds to a signaling molecule, becomes activated and initiates a transduction pathway.
signaling molecule
plasma membrane
receptor activation
2. Transduction pathway: Series of relay proteins that ends when a protein is activated.
unactivated receptor protein
nuclear envelope
Cytoplasm
Nucleus
3. Response: Targeted protein(s) bring about the response(s) noted.
Targeted protein:
structural protein
enzyme
gene regulatory protein
Cellular response:
Altered shape or movement of cell
Altered metabolism or a function of cell
Altered gene expression and the amount of a cell protein

Figure 4A Cell signaling and the transduction pathway. The process of signaling involves three steps: binding of the signaling molecule, transduction of the signal, and response of the cell depending on what type protein is targeted.

Questions to Consider

1. What happens if a cell is missing a receptor for a particular signaling molecule?
2. How does the binding of one type of signaling molecule sometimes result in multiple cellular responses?
3. As a cancer researcher, which segment of the signal transduction pathway would you target to prevent the spread of cancer and why?

4.2 The Permeability of the Plasma Membrane

Learning Outcomes

Upon completion of this section, you should be able to

1. Explain why a membrane is selectively permeable.
2. Predict the movement of molecules in diffusion and osmosis.
3. Describe the role of proteins in the movement of molecules across a membrane.

The plasma membrane regulates the passage of molecules into and out of the cell. This function is critical because the life of the cell depends on maintenance of its normal composition. The plasma membrane can carry out this function because it is **selectively permeable,** meaning that certain substances can move across the membrane while others cannot.

Table 4.1 lists, and Figure 4.3 illustrates, which types of molecules can freely (i.e., passively) cross a membrane and which may require transport by a carrier protein and/or an expenditure of energy. In general, small, noncharged molecules, such as carbon dioxide, oxygen, glycerol, and alcohol, can freely cross the membrane. They are able to slip between the hydrophilic heads of the phospholipids and pass through the hydrophobic tails of the membrane. These molecules are said to go "down" their **concentration gradient** as they move from an area where their concentration is high to an area where their concentration is low. Some molecules are able to go "up" their concentration gradient, or move from an area where their concentration is low to an area where their concentration is high, but this requires energy.

Water, a polar molecule, would not be expected to readily cross the primarily nonpolar membrane. While the small size of the water molecule may allow some water to diffuse across the plasma membrane, the majority of cells have special channel proteins called *aquaporins* that allow water to quickly cross the membrane.

Large molecules and some ions and charged molecules are unable to freely cross the membrane. They can cross the plasma membrane through channel proteins, with the assistance of carrier proteins, or in vesicles. A channel protein forms a pore through the membrane that allows molecules of a certain size and/or charge to pass. Carrier proteins are specific for the substances they transport across the plasma membrane—for example, sodium ions, amino acids, or glucose.

Vesicle formation is another way a molecule can exit a cell by exocytosis or enter a cell by endocytosis. This method of crossing a plasma membrane is reserved for macromolecules or even larger materials, such as a virus.

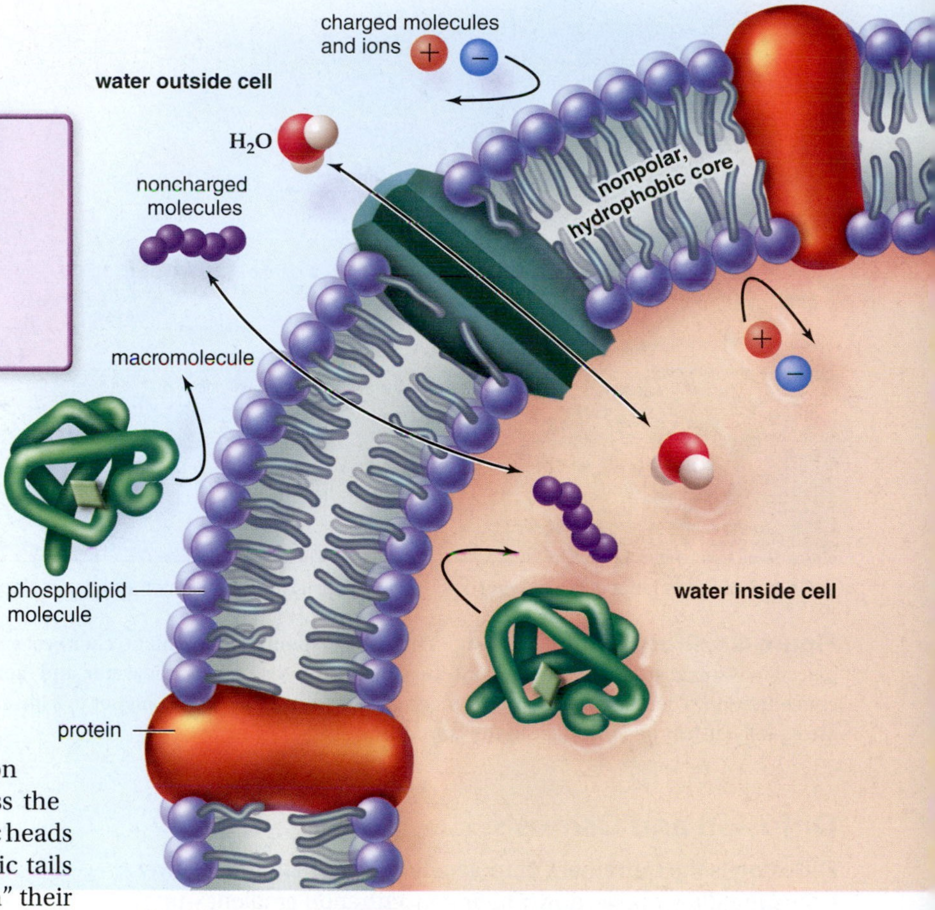
charged molecules and ions
water outside cell
H_2O
noncharged molecules
nonpolar, hydrophobic core
macromolecule
phospholipid molecule
water inside cell
protein

Figure 4.3 How molecules cross the plasma membrane. Molecules that can diffuse across the plasma membrane are shown with long back-and-forth arrows. Substances that cannot diffuse across the membrane are indicated by the curved arrows.

TABLE 4.1 Passage of Molecules Into and Out of the Cell

	Name	Direction	Requirement	Examples
Energy Not Required	Diffusion	Toward lower concentration	Concentration gradient	Lipid-soluble molecules and gases
	Facilitated transport	Toward lower concentration	Channels or carrier and concentration gradient	Some sugars and some amino acids
Energy Required	Active transport	Toward higher concentration	Carrier plus energy	Sugars, amino acids, and ions
	Exocytosis	Toward outside	Vesicle fuses with plasma membrane	Macromolecules
	Endocytosis	Toward inside	Vesicle formation	Macromolecules

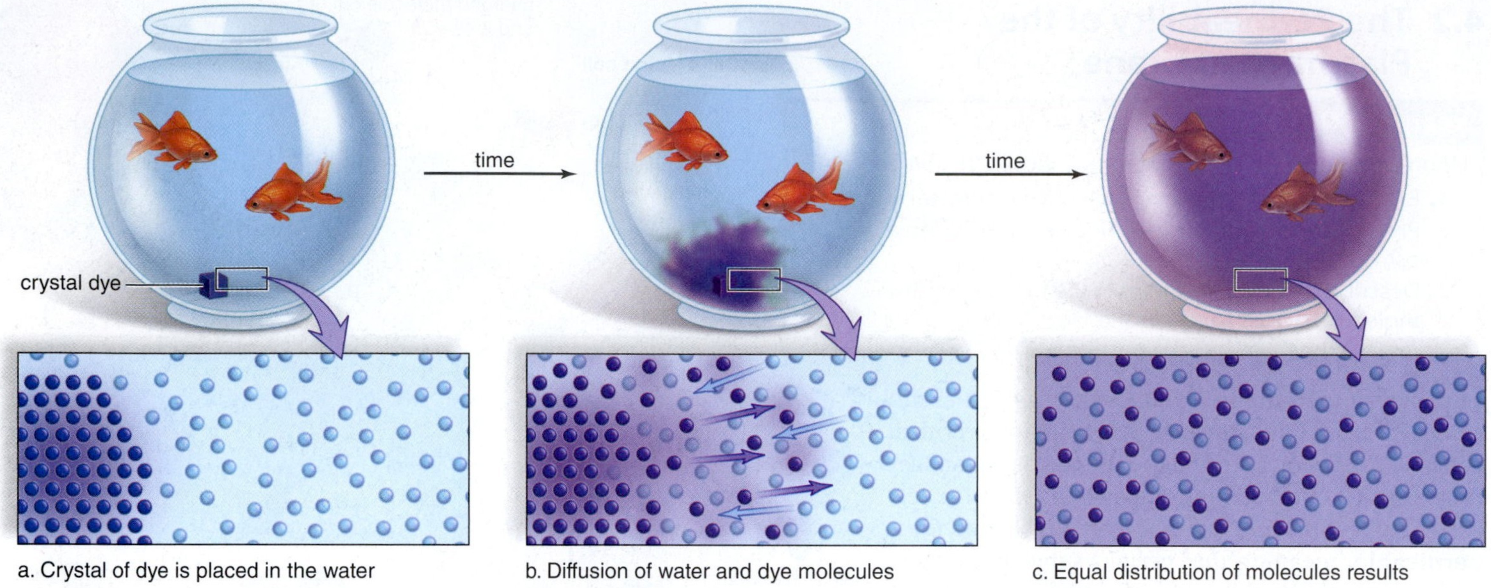
time
time
crystal dye
a. Crystal of dye is placed in the water
b. Diffusion of water and dye molecules
c. Equal distribution of molecules results

Figure 4.4 Process of diffusion. Diffusion is spontaneous, and no chemical energy is required to bring it about. **a.** When a dye crystal is placed in water, it is concentrated in one area. **b.** The dye dissolves in the water, and there is a net movement of dye molecules from a higher to a lower concentration. There is also a net movement of water molecules from a higher to a lower concentration. **c.** Eventually, the water and the dye molecules are equally distributed throughout the container.

Diffusion and Osmosis

Diffusion is the movement of molecules from a higher to a lower concentration—that is, down their concentration gradient—until equilibrium is achieved and they are distributed equally. Diffusion is a physical process that can be observed with any type of molecule. For example, when a crystal of dye is placed in water (Fig. 4.4), the dye and water molecules move in various directions, but their net movement, which is the sum of their motion, is toward the region of lower concentration. Eventually, the dye is dissolved in the water, resulting in equilibrium and a colored solution.

A solution contains both a **solute**, usually a solid, and a **solvent,** usually a liquid. In this case, the solute is the dye and the solvent is the water molecules. Once the solute and solvent are evenly distributed, their molecules continue to move about, but there is no net movement of either one in any direction.

The chemical and physical properties of the plasma membrane allow only a few types of molecules to enter and exit a cell simply by diffusion. Gases can diffuse through the lipid bilayer. This is the mechanism by which oxygen enters cells and carbon dioxide exits cells. Also, consider the movement of oxygen from the alveoli (air sacs) of the lungs to the blood in the lung capillaries (Fig. 4.5). After inhalation (breathing in), the concentration of oxygen in the alveoli is higher than that in the blood. Therefore, oxygen diffuses into the blood. Diffusion also plays an important role in maintaining the resting potential of neurons using gradients of potassium and sodium ions (see section 17.1)

Several factors influence the rate of diffusion. Among these factors are temperature, pressure, electrical currents, and molecular size. For example, as temperature increases, the rate of diffusion increases.

3D Animation Diffusion

Animation How Diffusion Works

MP3 Diffusion

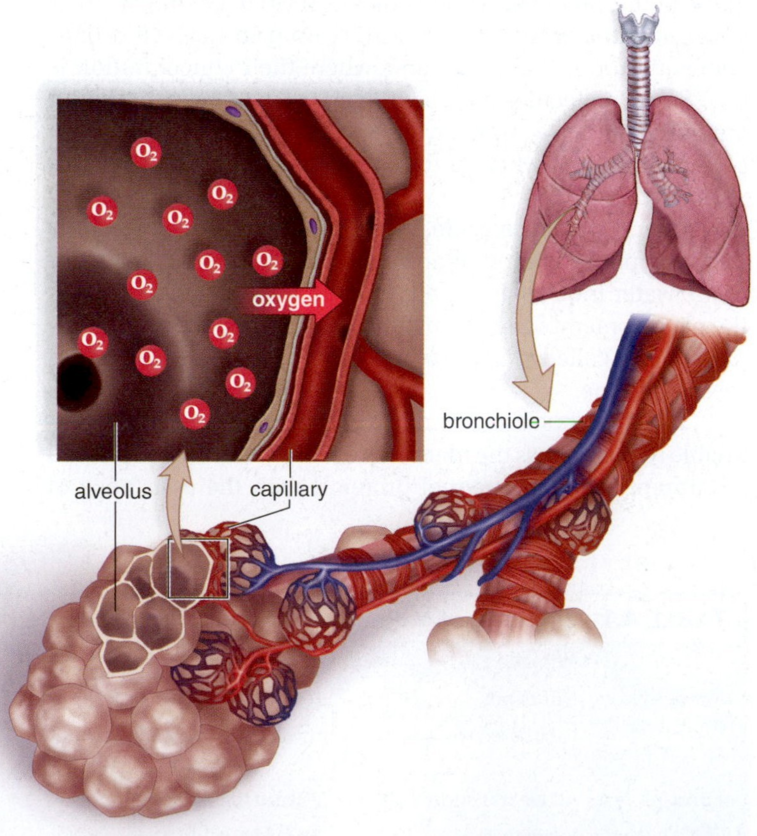
O_2
oxygen
alveolus
capillary
bronchiole

Figure 4.5 Gas exchange in lungs. Oxygen (O_2) diffuses into the capillaries of the lungs because there is a higher concentration of oxygen in the alveoli (air sacs) than in the capillaries.

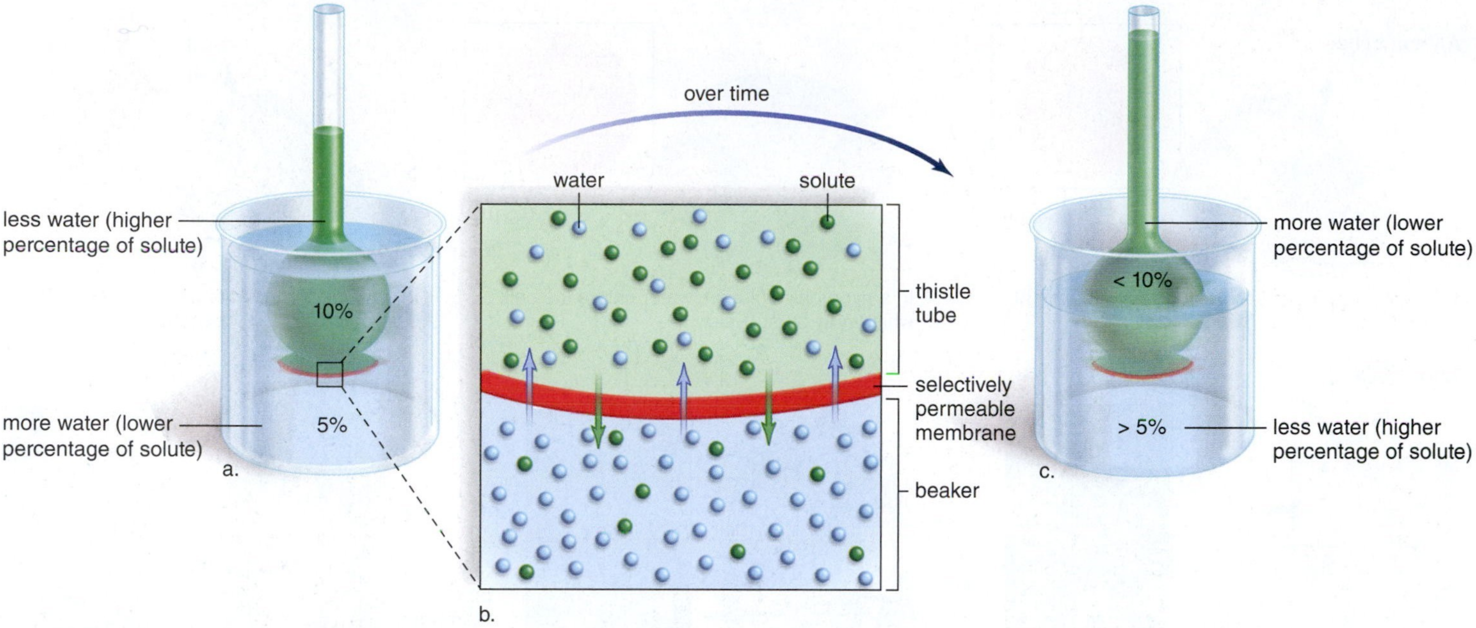
over time
water
solute
less water (higher percentage of solute)
10%
more water (lower percentage of solute)
5%
a.
thistle tube
selectively permeable membrane
beaker
b.
more water (lower percentage of solute)
< 10%
> 5%
less water (higher percentage of solute)
c.

Figure 4.6 **Osmosis demonstration.** **a.** A thistle tube, covered at the broad end by a differentially permeable membrane, contains a 10% solute solution. The beaker contains a 5% solute solution. **b.** The solute (green circles) is unable to pass through the membrane, but the water (blue circles) passes through in both directions. There is a net movement of water toward the inside of the thistle tube, where the percentage of water molecules is lower. **c.** Due to the incoming water molecules, the level of the solution rises in the thistle tube.

Osmosis

The diffusion of water across a selectively permeable membrane due to concentration differences is called **osmosis.** To illustrate osmosis, a thistle tube containing a 10% solute solution[1] is covered at one end by a selectively permeable membrane and then placed in a beaker containing a 5% solute solution (Fig. 4.6). The beaker has a higher concentration of water molecules (lower percentage of solute), and the thistle tube has a lower concentration of water molecules (higher percentage of solute). Diffusion always occurs from higher to lower concentration. Therefore, a net movement of water takes place across the membrane from the beaker to the inside of the thistle tube.

MP3 Osmosis

The solute does not diffuse out of the thistle tube. Why not? Because the membrane is not permeable to the solute. As water enters and the solute does not exit, the level of the solution within the thistle tube rises (Fig. 4.6*c*). In the end, the concentration of solute in the thistle tube is less than 10%. Why? Because there is now less solute per unit volume of solution. Furthermore, the concentration of solute in the beaker is greater than 5% because there is now more solute per unit volume.

Water enters the thistle tube due to the osmotic pressure of the solution within the thistle tube. **Osmotic pressure** is the pressure that develops in a system due to osmosis.[2] In other words, the greater the possible osmotic pressure, the more likely it is that water will diffuse in that direction. Due to osmotic pressure, water is absorbed by the kidneys and taken up by capillaries in the tissues. Osmosis also occurs across the plasma membrane.

Animation How Osmosis Works

Isotonic Solution In the laboratory, cells are normally placed in **isotonic solutions.** The prefix *iso-* means "the same as," and the term *tonicity* refers to the osmotic pressure or tension of the solution. In an isotonic solution, the solute concentration and the water concentration both inside and outside the cell are equal, and therefore there is no net gain or loss of water (Fig. 4.7). For example, a 0.9% solution of the salt sodium chloride (NaCl) is known to be isotonic to red blood cells. Therefore, intravenous solutions medically administered usually have this tonicity. Terrestrial animals can usually take in either water or salt as needed to maintain the tonicity of their internal environment. Many animals living in an estuary, such as oysters, blue crabs, and some fishes, are able to cope with changes in the salinity (salt concentrations) of their environment using specialized kidneys, gills, and other structures.

Hypotonic Solution Solutions that cause cells to swell, or even to burst, due to an intake of water are said to be **hypotonic solutions.** The prefix *hypo-* means "less than" and refers to a solution with a lower concentration of solute (higher concentration of water) than inside the cell. If a cell is placed in a hypotonic solution, water enters the cell. The net movement of water is from the outside to the inside of the cell.

---

[1] Percent solutions are grams of solute per 100 ml of solvent. Therefore, a 10% solution is 10g of sugar with water added to make 100 ml of solution.

[2] Osmotic pressure is measured by placing a solution in an osmometer and then immersing the osmometer in pure water. The pressure that develops is the osmotic pressure of a solution.

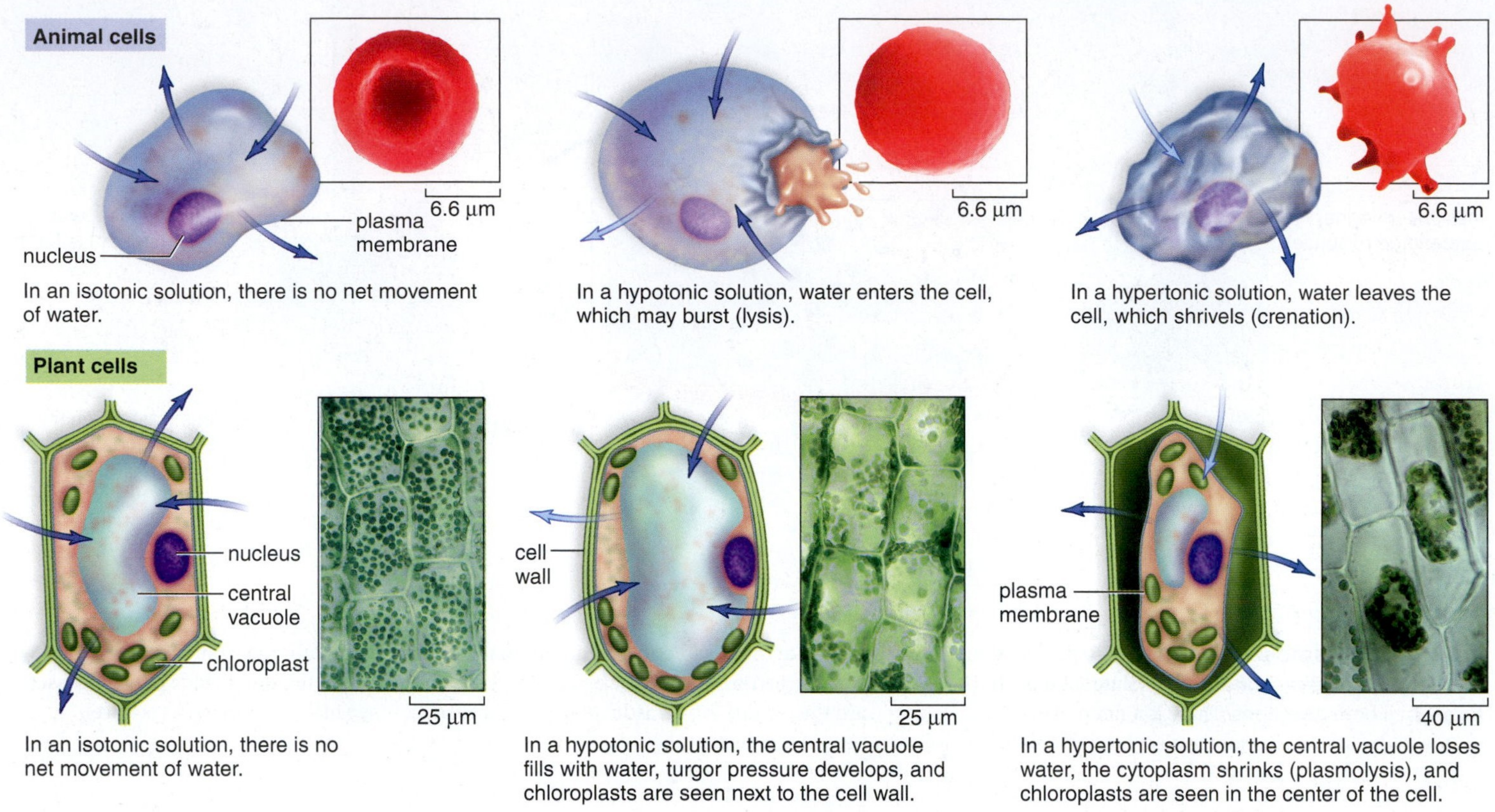
Animal cells
6.6 μm
plasma membrane
nucleus
In an isotonic solution, there is no net movement of water.
6.6 μm
In a hypotonic solution, water enters the cell, which may burst (lysis).
6.6 μm
In a hypertonic solution, water leaves the cell, which shrivels (crenation).
Plant cells
nucleus
central vacuole
chloroplast
25 μm
In an isotonic solution, there is no net movement of water.
cell wall
25 μm
In a hypotonic solution, the central vacuole fills with water, turgor pressure develops, and chloroplasts are seen next to the cell wall.
plasma membrane
40 μm
In a hypertonic solution, the central vacuole loses water, the cytoplasm shrinks (plasmolysis), and chloroplasts are seen in the center of the cell.

Figure 4.7 Osmosis in animal and plant cells. The arrows indicate the movement of water molecules. To determine the net movement of water, compare the number of arrows that are taking water molecules into the cell with the number that are taking water out of the cell. In an isotonic solution, a cell neither gains nor loses water; in a hypotonic solution, a cell gains water; and in a hypertonic solution, a cell loses water.

Any concentration of a salt solution lower than 0.9% is hypotonic to red blood cells. Animal cells placed in such a solution expand and sometimes burst or lyse due to the buildup of pressure. The term *cytolysis* is used to refer to disrupted cells. However, if the cell is a red blood cell, the term *hemolysis* is used.

The swelling of a plant cell in a hypotonic solution creates *turgor pressure*. When a plant cell is placed in a hypotonic solution, the cytoplasm expands because the large central vacuole gains water and the plasma membrane pushes against the rigid cell wall. The plant cell does not burst because the cell wall does not give way. Turgor pressure in plant cells is extremely important to the maintenance of the plant's erect position. If you forget to water your plants, they wilt due to decreased turgor pressure.

Organisms that live in fresh water have to prevent their internal environment from becoming hypotonic. Many protozoans, such as paramecia, have contractile vacuoles that rid the body of excess water. Freshwater fishes have well-developed kidneys that excrete a large volume of dilute urine. Even so, they have to take in salts at their gills. Even though freshwater fishes are good osmoregulators, they would not be able to survive in either distilled water or a marine environment.

Video Contractile Vacuoles

Hypertonic Solution Solutions that cause cells to shrink or shrivel due to loss of water are said to be **hypertonic solutions.** The prefix *hyper-* means "more than" and refers to a solution with a higher percentage of solute (lower concentration of water) than the cell. If a cell is placed in a hypertonic solution, water leaves the cell. The net movement of water is from the inside to the outside of the cell.

Any concentration of a salt solution higher than 0.9% is hypertonic to red blood cells. If animal cells are placed in this solution, they shrink. The term *crenation* refers to the shriveling of a cell in a hypertonic solution. Meats are sometimes preserved by salting them. The salt kills any bacteria present because it makes the meat a hypertonic environment.

Animation Hemolysis and Crenation

When a plant cell is placed in a hypertonic solution, the plasma membrane pulls away from the cell wall as the large central vacuole loses water. This is an example of *plasmolysis*, shrinking of the cytoplasm due to osmosis. The dead plants you may see along a salted roadside died because they were exposed to a hypertonic solution during the winter. Also, when salt water invades coastal marshes due to storms or human activities, coastal plants die. Without roots to hold the soil, it washes into the sea, thereby losing many acres of valuable wetlands.

3D Animation Osmosis

Animation Plasmolysis

Marine animals cope with their hypertonic environment in various ways that prevent them from losing water. Sharks increase or decrease the urea in their blood until their blood is isotonic with the environment. Marine fishes and other types of animals excrete salts across their gills. Have you ever seen a marine turtle cry? It is ridding its body of salt by means of glands near the eye.

Transport by Carrier Proteins

The plasma membrane impedes the passage of all but a few substances. Yet, biologically useful molecules are able to enter and exit the cell at a rapid rate because of carrier proteins in the membrane. Carrier proteins are specific. Each can combine with only a certain type of molecule or ion, which is then transported through the membrane. Scientists do not completely understand how carrier proteins function, but after a carrier combines with a molecule, the carrier is believed to undergo a change in shape that moves the molecule across the membrane. Carrier proteins are required for both facilitated transport and active transport (see Table 4.1).

Facilitated Transport

Facilitated transport explains the passage of such molecules as glucose and amino acids across the plasma membrane even though they are not lipid-soluble. The passage of glucose and amino acids is facilitated by their reversible combination with carrier proteins, which in some manner transport them through the plasma membrane. These carrier proteins are specific. For example, various sugar molecules of identical size might be present inside or outside the cell, but glucose can cross the membrane hundreds of times faster than the other sugars. Another example is provided by the aquaporins, which allow water to rapidly move across the plasma membrane of the cell.

A model for facilitated transport (Fig. 4.8) shows that after a carrier has assisted the movement of a molecule to the other side of the membrane, it is free to assist the passage of other similar molecules. Neither diffusion nor facilitated transport requires an expenditure of energy (use of ATP) because the molecules are moving down their concentration gradient in the same direction they tend to move anyway.

Animation
How Facilitated Diffusion Works

Active Transport

During **active transport,** molecules or ions move through the plasma membrane, accumulating either inside or outside the cell. For example, iodine collects in the cells of the thyroid gland; glucose is completely absorbed from the gut by the cells lining the digestive tract; and sodium can be almost completely withdrawn from urine by cells lining the kidney tubules. In these instances, molecules have moved to the region of higher concentration, exactly opposite to the process of diffusion.

3D Animation
Active Transport

Both carrier proteins and an expenditure of energy are needed to transport molecules against their concentration gradient. In this case, chemical energy, usually in the form of ATP, is required for the carrier to combine with the substance to be transported. Therefore, it is not surprising that cells involved primarily in active transport, such as kidney cells, have a large number of mitochondria near membranes where active transport is occurring.

Proteins involved in active transport are often called pumps because, just as a water pump uses energy to move water against the force of gravity, proteins use energy to move a substance against its concentration gradient. One type of pump that is active in all animal cells, but is especially associated with nerve and muscle cells, moves sodium ions (Na^+) to the outside of the cell and potassium ions (K^+) to the inside of the cell. These two events are linked, and the carrier protein is called a **sodium-potassium pump.** A change in carrier shape after the

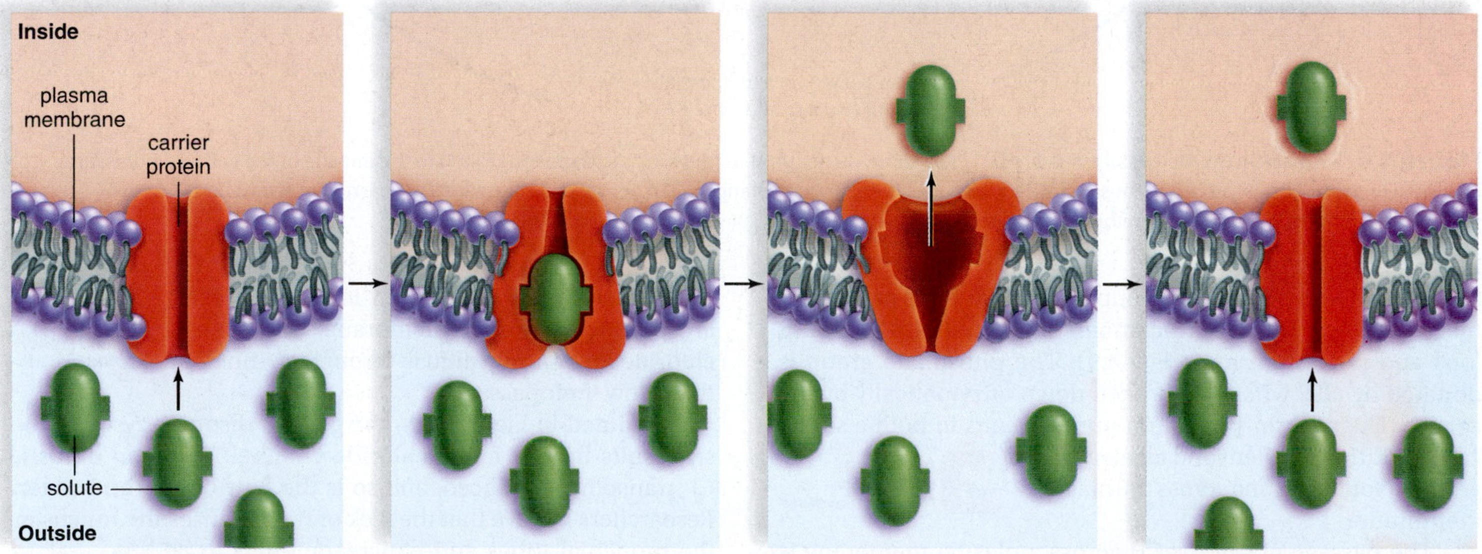
Inside
plasma membrane
carrier protein
solute
Outside

Figure 4.8 Facilitated transport. During facilitated transport, a carrier protein speeds the rate at which the solute crosses the plasma membrane toward a lower concentration. Note that the carrier protein undergoes a change in shape as it moves a solute across the membrane.

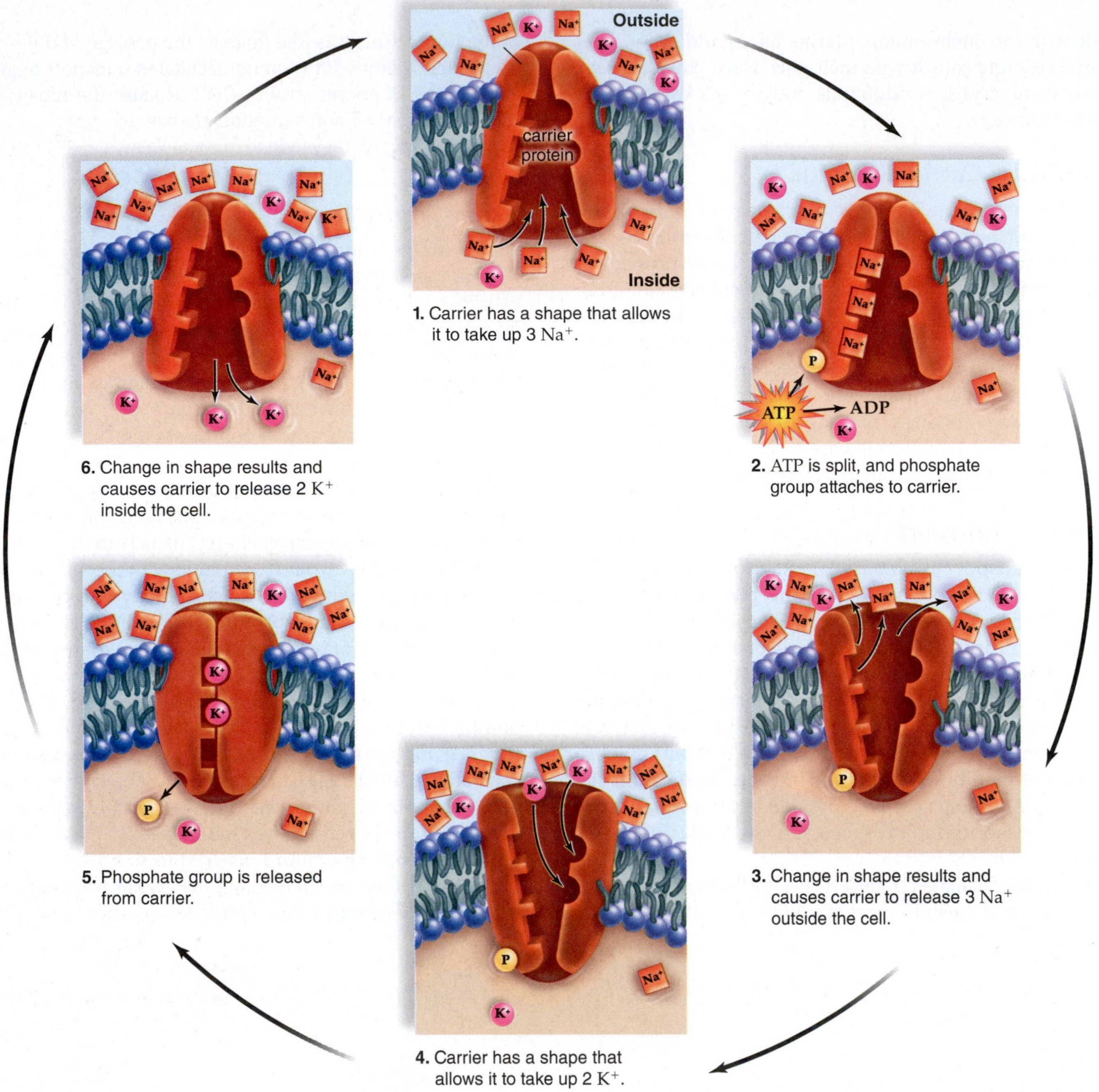
Outside
carrier protein
Inside
1. Carrier has a shape that allows it to take up 3 Na^+.
ATP
ADP
2. ATP is split, and phosphate group attaches to carrier.
3. Change in shape results and causes carrier to release 3 Na^+ outside the cell.
4. Carrier has a shape that allows it to take up 2 K^+.
5. Phosphate group is released from carrier.
6. Change in shape results and causes carrier to release 2 K^+ inside the cell.

Figure 4.9 The sodium-potassium pump. The same carrier protein transports sodium ions (Na^+) to the outside of the cell and potassium ions (K^+) to the inside of the cell because it undergoes an ATP-dependent change in shape. Three sodium ions are carried outward for every two potassium ions carried inward. Therefore, the inside of the cell is negatively charged compared to the outside.

attachment of a phosphate group, and again after its detachment, allows the carrier to combine alternately with sodium ions and potassium ions (Fig. 4.9). The phosphate group is donated by ATP when it is broken down enzymatically by the carrier. The sodium-potassium pump results in both a solute concentration gradient and an electrical gradient for these ions across the plasma membrane.

Animation How the Sodium-Potassium Pump Works

The passage of salt (NaCl) across a plasma membrane is of primary importance to most cells. The chloride ion (Cl^-) usually crosses the plasma membrane because it is attracted by positively charged sodium ions (Na^+). First sodium ions are pumped across a membrane, and then chloride ions simply diffuse through channels that allow their passage.

Animation Cotransport

As noted in Figure 4.2*a,* the genetic disorder cystic fibrosis results from a faulty chloride channel. In cystic fibrosis, Cl^- transport is reduced, and so is the flow of Na^+ and water. Researchers believe that the lack of water causes the mucus in the bronchial tubes and pancreatic ducts to be abnormally thick, thus interfering with the function of the lungs and pancreas.

Video Good Poison

Bulk Transport

How do macromolecules such as polypeptides, polysaccharides, or polynucleotides enter and exit a cell? Because they are too large to be transported by carrier proteins, macromolecules are transported into and out of the cell by vesicle formation. Vesicle formation is called membrane-assisted transport because membrane is needed to form the vesicle. Vesicle formation requires an expenditure of cellular energy, but an added benefit is that the vesicle membrane keeps the contained macromolecules from mixing with molecules within the cytoplasm. Exocytosis is a way substances can exit a cell, and endocytosis is a way substances can enter a cell.

Animation Endocytosis and Exocytosis

Exocytosis

During **exocytosis,** a vesicle fuses with the plasma membrane as secretion occurs (Fig. 4.10). Hormones, neurotransmitters, and digestive enzymes are secreted from cells in this manner. The Golgi apparatus often produces the vesicles that carry these cell products to the membrane. Notice that during exocytosis, the membrane of the vesicle becomes a part of the plasma membrane, which is thereby enlarged. For this reason, exocytosis can be a normal part of cell growth. The proteins released from the vesicle adhere to the cell surface or become incorporated in an extracellular matrix.

Cells of particular organs are specialized to produce and export molecules. For example, pancreatic cells produce digestive enzymes or insulin, and anterior pituitary cells produce growth hormone, among other hormones. In these cells, secretory vesicles accumulate near the plasma membrane, and the vesicles release their contents only when the cell is stimulated by a signal received at the plasma membrane.

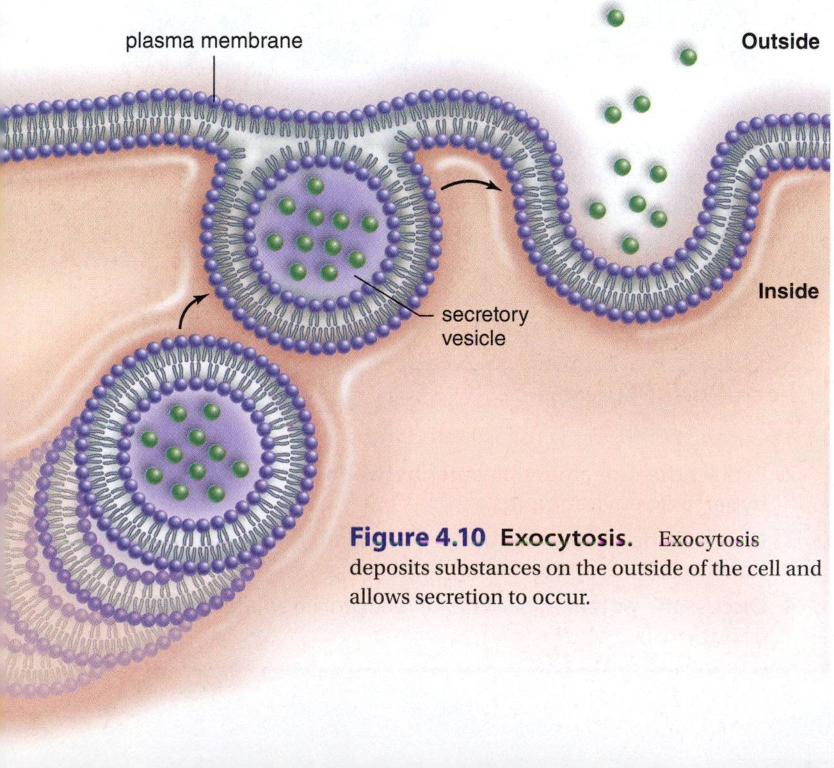
plasma membrane
Outside
Inside
secretory vesicle

Figure 4.10 Exocytosis. Exocytosis deposits substances on the outside of the cell and allows secretion to occur.

A rise in blood sugar, for example, signals pancreatic cells to release the hormone insulin. This is called regulated secretion, because vesicles fuse with the plasma membrane only when it is appropriate to the needs of the body.

Endocytosis

During **endocytosis,** cells take in substances by vesicle formation. A portion of the plasma membrane invaginates to envelop the substance, and then the membrane pinches off to form an intracellular vesicle. Endocytosis occurs in one of three ways, as illustrated in Figure 4.11. Phagocytosis transports large substances, such as viruses, and pinocytosis transports small substances, such as macromolecules, into cells. Receptor-mediated endocytosis is a special form of pinocytosis.

Phagocytosis When the material taken in by endocytosis is large, such as a food particle or another cell, the process is called **phagocytosis.** Phagocytosis is common in unicellular organisms such as amoebas (Fig. 4.11*a*). It also occurs in humans. Certain types of human white blood cells are amoeboid—that is, they are mobile like an amoeba, and they are able to engulf debris such as worn-out red blood cells or viruses. When an endocytic vesicle fuses with a lysosome, digestion occurs. We will see that this process is a necessary and preliminary step toward the development of immunity to bacterial diseases.

Pinocytosis **Pinocytosis** occurs when vesicles form around a liquid or around very small particles (Fig. 4.11*b*). Blood cells, cells that line the kidney tubules or the intestinal wall, and plant root cells all use pinocytosis to ingest substances.

Whereas phagocytosis can be seen with the light microscope, an electron microscope is required to observe pinocytic vesicles, which are no larger than 0.1–0.2 μm. Still, pinocytosis involves a significant amount of the plasma membrane because it occurs continuously. The loss of plasma membrane due to pinocytosis is balanced by the occurrence of exocytosis, however.

Receptor-Mediated Endocytosis **Receptor-mediated endocytosis** is a form of pinocytosis that is quite specific because it uses a receptor protein shaped so that a specific molecule, such as a vitamin, peptide hormone, or lipoprotein, can bind to it (Fig. 4.11*c*). The receptors for these substances are found at one location in the plasma membrane. This location is called a coated pit because there is a layer of protein on the cytoplasmic side of the pit. Once formed, the vesicle becomes uncoated and may fuse with a lysosome. When empty, used vesicles fuse with the plasma membrane, and the receptors return to their former location.

Receptor-mediated endocytosis is selective and much more efficient than ordinary pinocytosis. It is involved in uptake and also in the transfer and exchange of substances between cells. Such exchanges take place when substances move from maternal blood into fetal blood at the placenta, for example.

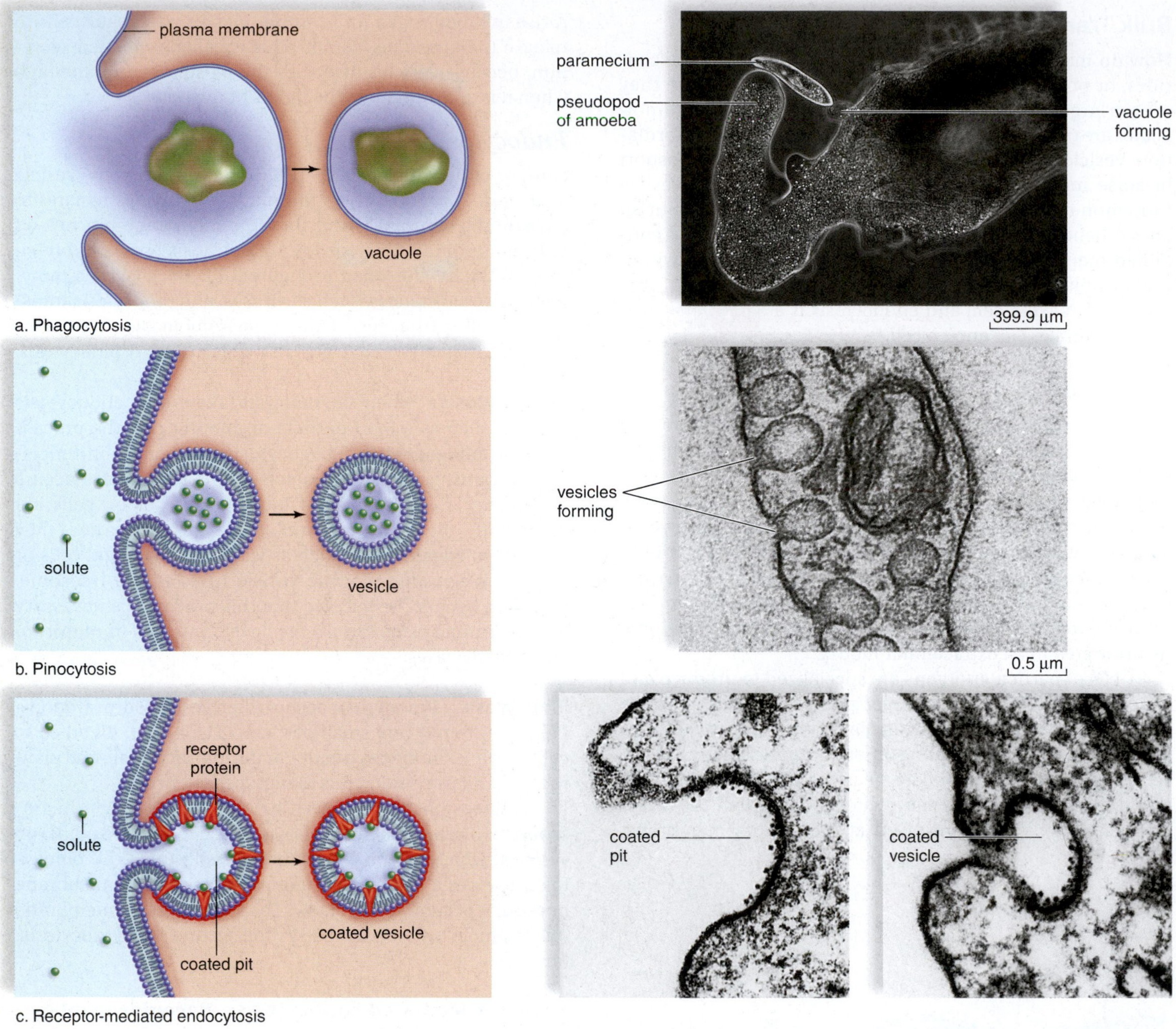
plasma membrane
vacuole
a. Phagocytosis
paramecium
pseudopod of amoeba
vacuole forming
399.9 µm
solute
vesicle
b. Pinocytosis
vesicles forming
0.5 µm
receptor protein
solute
coated vesicle
coated pit
c. Receptor-mediated endocytosis
coated pit
coated vesicle

Figure 4.11 Three methods of endocytosis. **a.** Phagocytosis occurs when the substance to be transported into the cell is large. Amoebas ingest by phagocytosis. Digestion occurs when the resulting vacuole fuses with a lysosome. **b.** Pinocytosis occurs when a macromolecule such as a polypeptide is transported into the cell. The result is a vesicle (small vacuole). **c.** Receptor-mediated endocytosis is a form of pinocytosis. Molecules first bind to specific receptor proteins, which migrate to or are already in a coated pit. The vesicle that forms contains the molecules and their receptors.

The importance of receptor-mediated endocytosis is demonstrated by a genetic disorder called familial hypercholesterolemia. Cholesterol is transported in the blood by a complex of lipids and proteins called low-density lipoprotein (LDL). Ordinarily, body cells take up LDL when LDL receptors gather in a coated pit. But in some individuals, the LDL receptor is unable to properly bind to the coated pit, and the cells are unable to take up cholesterol. Instead, cholesterol accumulates in the walls of arterial blood vessels, leading to high blood pressure, occluded (blocked) arteries, and heart attacks.

Check Your Progress 4.2

1. Contrast diffusion with facilitated transport.
2. Explain the movement of water between hypotonic and hypertonic environments.
3. Describe the differences between facilitated and active transport.
4. Discuss the potential benefits of receptor-mediated endocytosis.

4.3 Modifications of Cell Surfaces

Learning Outcomes

Upon completion of this section, you should be able to

1. Explain the role of the extracellular matrix in animal cells.
2. Compare the structure and function of adhesion, tight, and gap junctions.

Most cells do not live isolated from other cells. Rather, they live and interact within an external environment that can dramatically affect cell structure and function. This extracellular environment is made of large molecules produced by nearby cells and secreted from their membranes. In plants, prokaryotes, fungi, and most algae, the extracellular environment is a fairly rigid cell wall, which is consistent with a somewhat sedentary lifestyle. Animals, which tend to be more active, have a more varied extracellular environment that can change, depending on the tissue type.

Cell Surfaces in Animals

In this chapter, we will focus on two different types of animal cell surface features: (1) the extracellular matrix (ECM) that is observed outside cells, and (2) junctions that occur between some types of cells. Both of these can connect to the cytoskeleton and contribute to communication between cells, and therefore tissue formation.

Extracellular Matrix

A protective **extracellular matrix (ECM)** is a meshwork of proteins and polysaccharides in close association with the cell that produced them (Fig. 4.12). Collagen and elastin fibers are two well-known structural proteins in the ECM; collagen resists stretching and elastin gives the ECM resilience.

Fibronectin is an adhesive protein (colored green in Fig. 4.12) that binds to a protein in the plasma membrane called integrin. Integrins are integral membrane proteins that connect to fibronectin externally and to the actin cytoskeleton internally. Through its connections with both the ECM and the cytoskeleton, integrin plays a role in cell signaling, permitting the ECM to influence the activities of the cytoskeleton and, therefore, the shape and activities of the cell.

Amino sugars in the ECM form multiple polysaccharides that attach to a protein and are, therefore, called proteoglycans. Proteoglycans, in turn, attach to a very long, centrally placed polysaccharide. The entire structure, which looks like an enormous bottle brush, resists compression of the extracellular matrix. Proteoglycans assist cell signaling when they regulate the passage of molecules through the ECM to the plasma membrane, where receptors are located. Thus, the ECM has a dynamic role in all aspects of a cell's behavior.

In Chapter 11, during the discussion of tissues, you'll see that the extracellular matrix varies in quantity and in consistency from being quite flexible, as in loose connective tissue; semiflexible, as in cartilage; and rock solid, as in bone. The extracellular matrix of bone is hard because, in addition to the components mentioned, mineral salts, notably calcium salts, are deposited outside the cell.

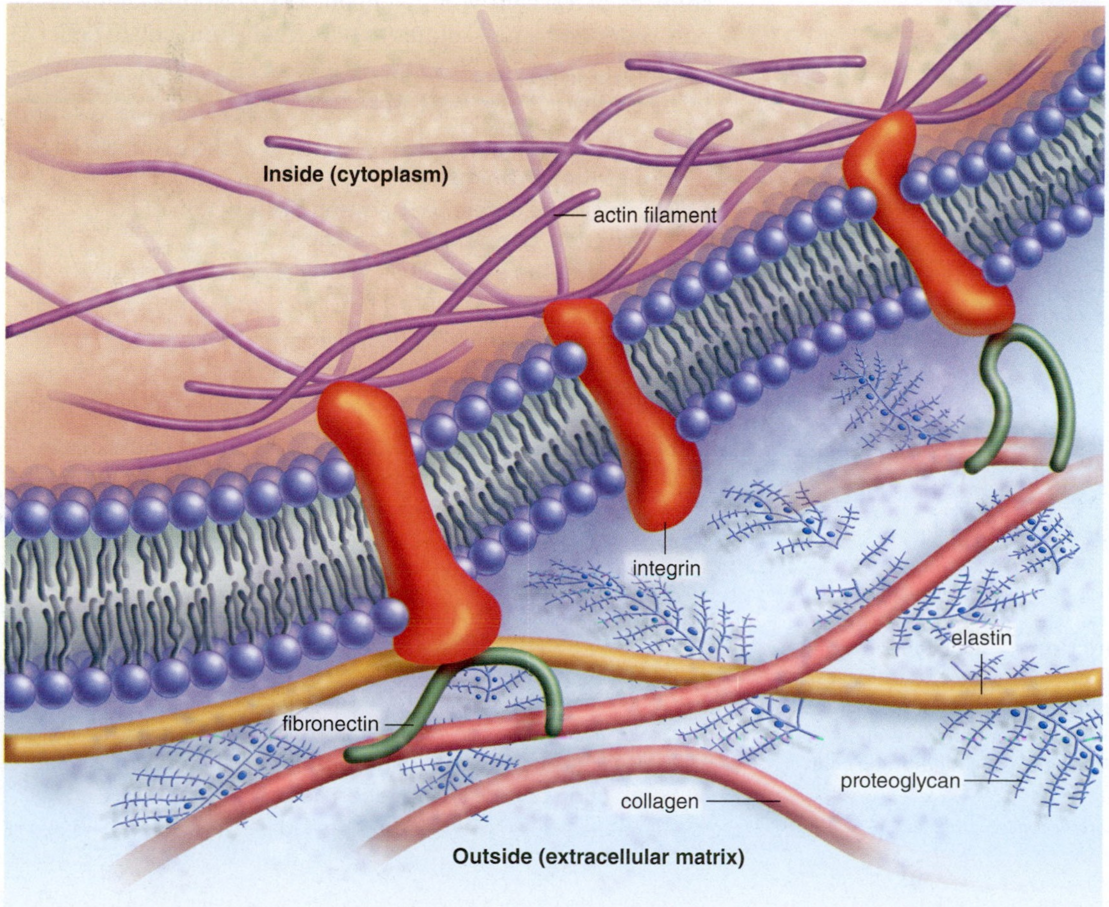
Inside (cytoplasm)
actin filament
integrin
elastin
fibronectin
proteoglycan
collagen
Outside (extracellular matrix)

Figure 4.12 Extracellular matrix of an animal cell. In the extracellular matrix (ECM), collagen and elastin have a support function, while fibronectins bind to integrin, thus assisting communication between the ECM and the cytoskeleton.

Junctions Between Cells

Certain tissues of vertebrate animals are known to have junctions between their cells that allow them to behave in a coordinated manner. Three types of junctions are shown in Figure 4.13. Adhesion junctions (Fig. 4.13*a*) serve to mechanically attach adjacent cells. **Desmosomes** are one form of adhesion junctions. In a desmosome, internal cytoplasmic plaques, firmly attached to the intermediate filament cytoskeleton within each cell, are joined by integral membrane proteins called cadherins between cells. The result is a sturdy but flexible sheet of cells. In some organs—such as the heart, stomach, and bladder, where tissues get stretched—desmosomes hold the cells together. Adhesion junctions are the most common type of intercellular junction between skin cells.

Another type of junction between adjacent cells are tight junctions (Fig. 4.13*b*), which bring cells even closer than desmosomes. Tight junction proteins actually connect plasma membranes between adjacent cells together, producing a zipperlike fastening. Tissues that serve as barriers are held together by tight junctions; in the intestine, the digestive juices stay out of the rest of the body, and in the kidneys, the urine stays within kidney tubules, because the cells are joined by tight junctions.

A gap junction (Fig. 4.13*c*) allows cells to communicate. A gap junction is formed when two identical plasma membrane channels join. The channel of each cell is lined by six plasma membrane proteins. A gap junction lends strength to the cells, but it also allows small molecules and ions to pass between them. Gap junctions are important in heart muscle and smooth muscle because they permit a flow of ions that is required for the cells to contract as a unit.

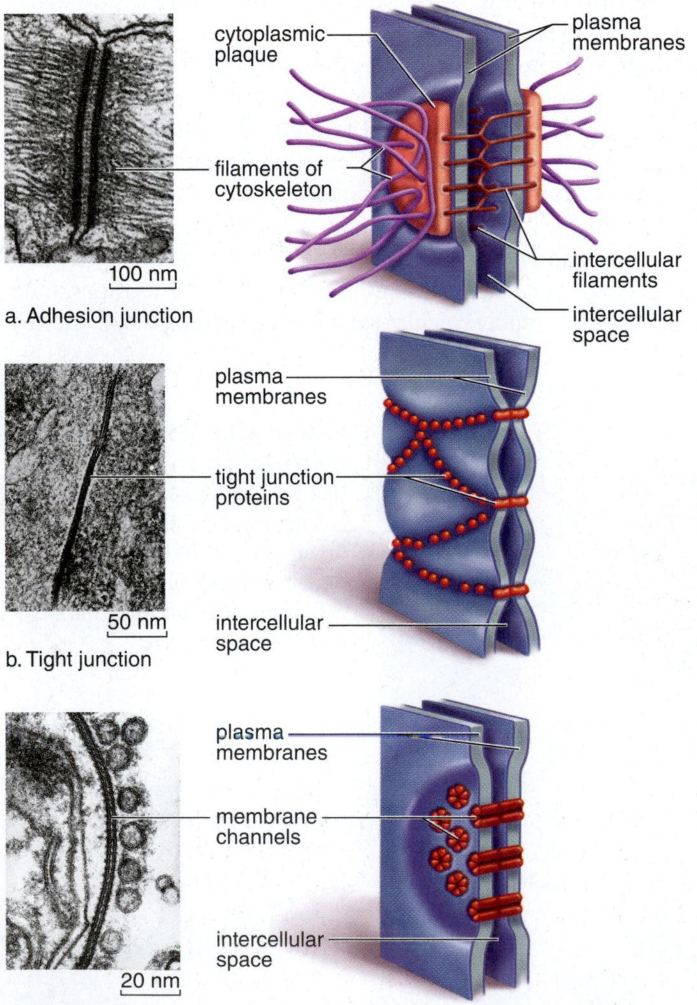
cytoplasmic plaque
plasma membranes
filaments of cytoskeleton
100 nm
intercellular filaments
intercellular space
a. Adhesion junction
plasma membranes
tight junction proteins
50 nm
intercellular space
b. Tight junction
plasma membranes
membrane channels
intercellular space
20 nm
c. Gap junction

Figure 4.13 Examples of cell junctions. **a.** In adhesion junctions, such as the desmosome, adhesive proteins connect two cells. **b.** Tight junctions between cells have joined their adjacent plasma membranes, forming an impermeable layer. **c.** Gap junctions allow communication between two cells by joining plasma membrane channels between the cells.

a. © Kelly, 1966. Originally published in *The Journal of Cell Biology*, 28:51–72.

Plant Cell Walls

In addition to a plasma membrane, plant cells are surrounded by a porous cell wall that varies in thickness, depending on the function of the cell.

All plant cells have a cell wall. The primary cell wall contains cellulose fibrils (very fine fibers) in which microfibrils are held together by noncellulose substances. Pectins allow the wall to stretch when the cell is growing, and noncellulose polysaccharides harden the wall when the cell is mature. Pectins are especially abundant in the **middle lamella,** which is a layer of adhesive substances that holds the cells together.

In a plant, the cytoplasm of living cells is connected by plasmodesmata (sing., plasmodesma), numerous narrow, membrane-lined channels that pass through the cell wall. Cytoplasmic strands within these channels allow direct exchange of some materials between adjacent plant cells and eventually connect all the cells within a plant. The plasmodesmata allow only water and small solutes to pass freely from cell to cell.

Check Your Progress 4.3

1. Describe the molecule composition of the extracellular matrix of an animal cell.
2. Explain the difference between the function of an adhesion, gap, and tight junction.
3. Contrast the extracellular matrix of an animal cell with the cell wall of a plant cell.

Case Study Conclusion

In the case of the particular channel described in the opening of this chapter, exposure to the capsaicin molecules in a chili pepper opens a channel protein in the membrane of the cells. This allows calcium ions to enter the cell and initiate a cascade of events that results in the sensation of pain. As we have seen, this is a type of facilitated diffusion, because the calcium ions are moving down their concentration gradient with the assistance of a channel protein. It is interesting to note that later in this text you will learn about a different type of voltage-gated ion channel that also allows calcium ions to enter the cell. In cells with this type of channel, muscular contraction or neuron excitation occurs.

MEDIA STUDY TOOLS

www.mhhe.com/maderinquiry14

Enhance your study of this chapter with study tools and practice tests. Also ask your instructor about the resources available through ConnectPlus, including LearnSmart, the media-rich eBook, interactive learning tools, and animations.

connect plus+ | BIOLOGY

3D Animation
Membrane Transport

For a detailed examination of the processes involved in the movement of molecules across the plasma membrane, watch McGraw-Hill's new 3D animation "Membrane Transport."

SUMMARIZE

4.1 Plasma Membrane Structure and Function

- The plasma membrane plays an important role in isolating the cell from the external environment and in maintaining **homeostasis** within the cell. According to the **fluid-mosaic model** of the plasma membrane, a lipid bilayer is fluid and has the consistency of light oil. The hydrophilic heads of phospholipids form the inner and outer surfaces, and the hydrophobic tails form the interior.
- Proteins within the membrane are the mosaic portion. The peripheral proteins often have a structural role in that they help stabilize and shape the plasma membrane. They may also function in signaling pathways. The integral proteins have a variety of functions, including acting as **channel proteins, carrier proteins, cell recognition proteins, receptor proteins,** and **enzymatic proteins.**
- Carbohydrate chains are attached to some of the lipids and proteins in the membrane. These are **glycolipids** and **glycoproteins.**

4.2 The Permeability of the Plasma Membrane

- Some substances, such as gases, freely cross a plasma membrane, while others—particularly ions, charged molecules, and macromolecules—have to be assisted across.
- Passive ways of crossing a plasma membrane (**diffusion** and **facilitated transport**) do not require an expenditure of chemical energy (ATP). Active ways of crossing a plasma membrane (**active transport** and vesicle formation) do require an expenditure of chemical energy.
- Lipid-soluble compounds, water, and gases simply diffuse across the plasma membrane by moving down their **concentration gradient** (high to low concentration)
- The diffusion of water across a **selectively permeable** membrane is called **osmosis.** Water (a **solvent**) moves across the membrane into the area of lower water (or higher **solute**) content. When cells are in an **isotonic solution,** they neither gain nor lose water; when they are in a **hypotonic solution,** they gain water; and when they are in a **hypertonic solution,** they lose water. **Osmotic pressure** occurs as a result of differences in tonicity.
- Some molecules are transported across the membrane by carrier proteins that span the membrane. During **facilitated transport,** a carrier protein assists the movement of a molecule down its concentration gradient. No energy is required.
- During **active transport,** a carrier protein acts as a pump that causes a substance to move against its concentration gradient. Energy in the form of ATP molecules is required for active transport to occur. The **sodium-potassium pump** is one example of active transport.
- Larger substances can exit and enter a membrane by **exocytosis** and **endocytosis.** Exocytosis involves secretion. Endocytosis includes **phagocytosis** and **pinocytosis. Receptor-mediated endocytosis,** a type of pinocytosis, makes use of receptor molecules in the plasma membrane and a coated pit, which pinches off to form a vesicle.

4.3 Modifications of Cell Surfaces

- Animal cells have an **extracellular matrix** (**ECM**) that influences their shape and behavior. The amount and character of the ECM varies by tissue type. Some animal cells have junction proteins that join them to other cells of the same tissue. **Adhesion junctions** and **tight junctions** help hold cells together; **gap junctions** allow passage of small molecules between cells.
- Plant cells have a freely permeable cell wall, with cellulose as its main component. Also, plant cells are joined by narrow, membrane-lined channels called plasmodesmata that span the cell wall and contain strands of cytoplasm that allow materials to pass from one cell to another.

ASSESS

Testing Yourself

Choose the best answer for each question.

1. Energy is used for
   a. diffusion.
   b. osmosis.
   c. active transport.
   d. None of these are correct.
2. Carrier proteins are _____ in their action.
   a. specific
   b. not specific
   c. involved in diffusion
   d. None of these are correct.
3. Which of these are methods of endocytosis?
   a. phagocytosis
   b. pinocytosis
   c. receptor-mediated
   d. All of these are correct.
4. An isotonic solution has the/a _____ concentration of water as the cell and the/a _____ concentration of solute as the cell.
   a. same, same
   b. same, different
   c. different, same
   d. different, different
5. The mosaic part of the fluid-mosaic model of membrane structure refers to
   a. the phospholipids.
   b. the proteins.
   c. the glycolipids.
   d. the cholesterol.
6. The carbohydrate chains on lipids and proteins are found
   a. on the inside of the membrane.
   b. on the outside of the membrane.
   c. on both sides of the membrane.

7. Integrin and elastin are ______ associated with the _______ of a animal cell.
   a. lipids, membrane
   b. proteins, ECM
   c. lipids, cell wall
   d. proteins, membrane
8. A coated pit is associated with
   a. diffusion.
   b. osmosis.
   c. receptor-mediated endocytosis.
   d. pinocytosis.
9. Exocytosis involves
   a. fusion with the plasma membrane.
   b. fusion with the nucleus.
   c. fusion with the mitochondria.
   d. fusion with a ribosome.
10. Which of these passes through a plasma membrane by way of diffusion?
    a. carbon dioxide
    b. oxygen
    c. lipid-soluble molecules
    d. All of these are correct.
11. This lipid helps regulate the fluidity of the plasma membrane.
    a. cholesterol
    b. glycogen
    c. triglycerides
    d. glycerol
12. When a cell is placed in a hypotonic solution,
    a. solute exits the cell to equalize the concentration on both sides of the plasma membrane.
    b. water exits the cell toward the area of lower solute concentration.
    c. water enters the cell toward the area of higher solute concentration.
    d. solute exits and water enters the cell.
    e. Both c and d are correct.
13. When a cell is placed in a hypertonic solution,
    a. solute exits the cell to equalize the concentration on both sides of the plasma membrane.
    b. water exits the cell toward the area of higher solute concentration.
    c. water enters the cell toward the area of higher solute concentration.
    d. solute exits and water enters the cell.
    e. Both a and c are correct.
14. ________ form communication channels between animal cells.
    a. Plasmodesmata
    b. Gap junctions
    c. Tight junctions
    d. Adhesion junctions
15. Which of the following is not found in the extracellular matrix of an animal cell?
    a. collagen
    b. proteoglycans
    c. cellulose
    d. All of these are found in animal cells.

ENGAGE

Thinking Critically

1. When a signal molecule such as a growth hormone binds to a receptor protein in the plasma membrane, it stays on the outside of the cell. How might the inside of the cell know that the signal has bound?
2. Some antibiotics interfere with the formation of the bacterial cell wall, thus weakening the cell wall. How might this cause a bacterium to be killed?
3. How could a channel protein regulate what enters the cell?

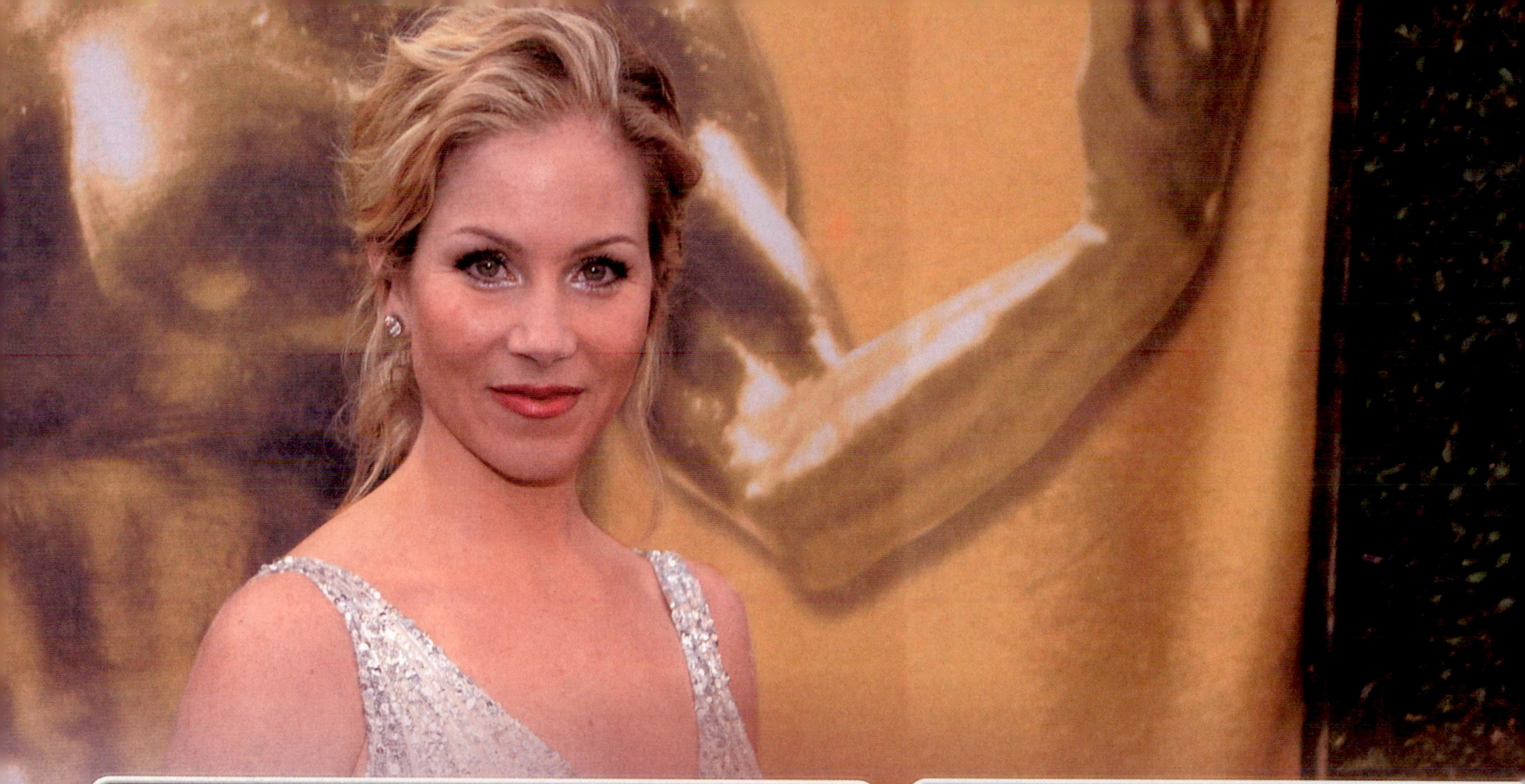

Cell Division 5

CHAPTER OUTLINE

BEFORE YOU BEGIN

Before beginning this chapter, take a few moments to review the following discussions:

Section 2.8 What is the role of DNA in a cell?

Section 3.1 What does the cell theory tell us about the need for cell division?

Section 3.3 What are microtubules and actin filaments?

CASE STUDY Genetics of Breast Cancer

In August 2008, actress Christina Applegate, then 36 years old, revealed to shocked fans that she had been diagnosed with breast cancer. Every year over 192,000 American women are diagnosed with breast cancer. Christina's mother is a breast and cervical cancer survivor, and Christina tested positive for a *BRCA1* (*br*east *ca*ncer susceptibility gene *1*) gene mutation, linked to both breast and ovarian cancer. Cancer results from a failure to control the cell cycle, a series of steps that all cells go through prior to initiating cell division. *BRCA1,* and a similar gene, *BRCA2,* are important components of that control mechanism. Both of these genes are tumor suppressor genes. At specific checkpoints in the cell cycle, the proteins encoded by these genes check the DNA for damage. In Christina's case, the version of *BRCA1* that she inherited from her parents was not acting as a gatekeeper, and instead was allowing cells to continuously divide, resulting in a tumor in one of her breasts.

It is estimated that 1 in 833 people possess the *BRCA1* mutation associated with breast cancer. Although it is not the only genetic contribution to breast and ovarian cancer, it does play a major role. Several medical approaches, including increased surveillance and surgery, are available to those that test positive for a mutated *BRCA1* gene.

As you read through this chapter, think about the following questions:

1. What are the roles of the checkpoints in the cell cycle?
2. At what checkpoint would you think that *BRCA1* would normally be active?
3. Why would a failure of the checkpoints in the cell cycle result in cancer?

5.1 The Cell Cycle

Learning Outcomes

Upon completion of this section, you should be able to

1. Distinguish between the two processes that change the number of cells in the body.
2. Describe the stages of the cell cycle and what occurs in each stage.

We start life as a single-celled fertilized egg. By the process of cell division, we grow into organisms that contain trillions of cells. Cell division serves to increase the number of **somatic cells,** or body cells. Cell division does not stop when you become an adult. Every day your body produces thousands of new red blood cells, skin cells, and cells that line your respiratory and digestive tracts. In addition to growth, cell division is involved in the repair of tissues after an injury.

Apoptosis, or programmed cell death, decreases the number of cells. Both cell division and apoptosis are normal parts of growth and development. Apoptosis occurs during development to remove unwanted tissue—for example, the tail of a tadpole disappears as it matures into a frog. In humans, the fingers and toes of the embryo are initially webbed, but are normally freed from one another later in development as a result of apoptosis. Apoptosis also plays an important role in preventing cancer. An abnormal cell that could become cancerous will often die via apoptosis, thus preventing a tumor from developing.

The Cell Cycle

Cell division is a part of the cell cycle. The **cell cycle** is an orderly set of stages that take place between the time a cell divides and the time the resulting cells also divide.

Animation Overview of Cell Division

The Stages of Interphase

As Figure 5.1 shows, most of the cell cycle is spent in **interphase.** This is the time when a cell carries on its usual functions, which are dependent on its location in the body. It also gets ready to divide: it grows larger, the number of organelles doubles, and the amount of DNA doubles. For mammalian cells, interphase lasts for about 20 hours, which is 90% of the cell cycle.

Interphase is divided into three stages: the G_1 stage occurs before DNA synthesis, the S stage includes DNA synthesis, and the G_2 stage occurs after DNA synthesis. Originally G stood for the "gaps" that occur before and after DNA synthesis during interphase. But now that we know growth occurs during these stages, the G can be thought of as standing for "growth."

Animation Cell Cycle and Mitosis: Interphase

During the G_1 stage, a cell doubles its organelles (such as mitochondria and ribosomes), and it accumulates the materials needed for DNA synthesis. Some cells, such as nerve and muscle cells, typically do not complete the cell cycle and are arrested. These cells are said to have entered a G_0 stage. Cells in G_0 phase continue to perform their normal functions, but no longer prepare for cell division.

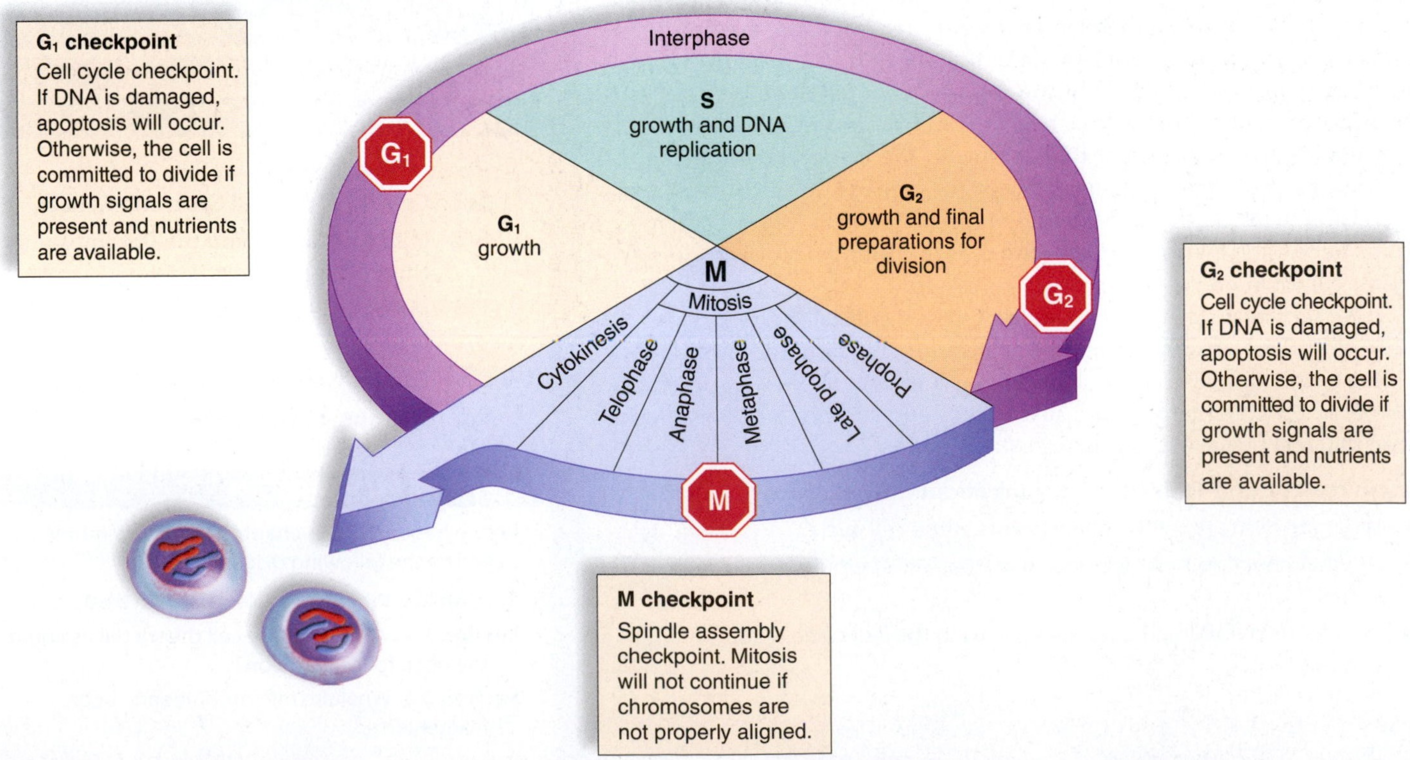
G1 checkpoint
Cell cycle checkpoint. If DNA is damaged, apoptosis will occur. Otherwise, the cell is committed to divide if growth signals are present and nutrients are available.
Interphase
S
growth and DNA replication
G1
G1
growth
G2
growth and final preparations for division
G2
M
Mitosis
Cytokinesis
Telophase
Anaphase
Metaphase
Late prophase
Prophase
M
G2 checkpoint
Cell cycle checkpoint. If DNA is damaged, apoptosis will occur. Otherwise, the cell is committed to divide if growth signals are present and nutrients are available.
M checkpoint
Spindle assembly checkpoint. Mitosis will not continue if chromosomes are not properly aligned.

Figure 5.1 The cell cycle. Cells go through a cycle that consists of four stages: G_1, S, G_2, and M. Checkpoints regulate the speed at which a cell moves through the cell cycle.

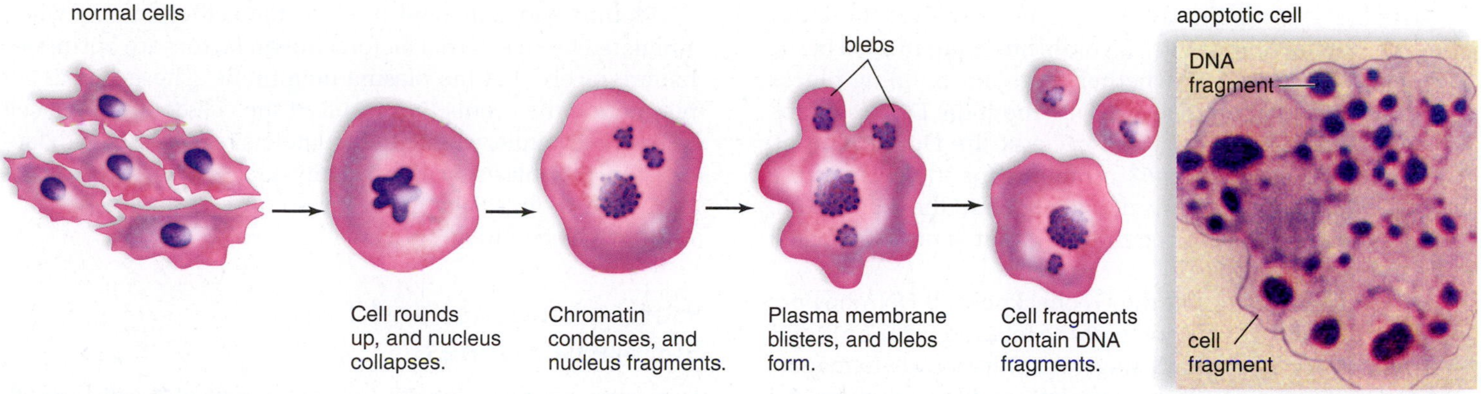
normal cells
blebs
apoptotic cell
DNA fragment
Cell rounds up, and nucleus collapses.
Chromatin condenses, and nucleus fragments.
Plasma membrane blisters, and blebs form.
Cell fragments contain DNA fragments.
cell fragment

Figure 5.2 Apoptosis. Apoptosis is a sequence of events that results in a fragmented cell. The fragments are engulfed by white blood cells and neighboring tissue cells.

During the S stage, DNA replication occurs. At the beginning of the S stage, each chromosome is composed of one DNA molecule, which is called a **chromatid.** At the end of this stage, each chromosome consists of two **sister chromatids** that have identical DNA sequences. Another way of expressing this is to say that DNA replication has resulted in duplicated chromosomes. The sister chromatids remain attached until they are separated during mitosis.

During the G_2 stage, the cell synthesizes the proteins needed for cell division, such as the proteins that make up the microtubules found in the spindle apparatus (see section 5.3).

The Mitotic Stage

Following interphase, the cell enters the M (for mitotic) stage. This stage not only includes **mitosis,** the division of the nucleus and genetic material, but also **cytokinesis,** the division of the cytoplasm, if it occurs. During mitosis, as we shall see, the sister chromatids of each chromosome separate, becoming daughter chromosomes that are distributed to two daughter nuclei. When cytokinesis is complete, two daughter cells that are identical to the mother cell are present. Mammalian cells usually require only about four hours to complete the mitotic stage.

Apoptosis

During apoptosis, the cell progresses through a typical series of events that bring about its destruction (Fig. 5.2). The cell rounds up and loses contact with its neighbors. The nucleus fragments, and the plasma membrane develops blisters. Finally, the cell fragments, and its bits and pieces are engulfed by white blood cells.

A remarkable finding of the past few years is that cells routinely harbor the enzymes, now called caspases, that bring about apoptosis. The enzymes are ordinarily held in check by inhibitors, but they can be unleashed either by internal or external signals. There are two sets of caspases. The first set, the "initiators," receive the signal to activate the second set, the "executioners," which then activate the enzymes that dismantle the cell. For example, executioners turn on enzymes that tear apart the cytoskeleton and enzymes that chop up DNA.

Check Your Progress 5.1

1. Illustrate the cell cycle, and explain the major events that occur during each stage.
2. Explain why apoptosis is a beneficial process.

5.2 Control of the Cell Cycle and Cancer

Learning Outcomes

Upon completion of this section, you should be able to

1. Distinguish between internal and external controls of the cell cycle.
2. Describe the checkpoints for the cell cycle.
3. Differentiate between the role of proto-oncogenes and tumor suppressor genes in regulating the cell cycle.

Eukaryotic cells have evolved a complex system for regulation of the cell cycle. The cell cycle is controlled by both internal and external signals. The internal signals ensure that the stages follow one another in the normal sequence and that each stage is properly completed before the next stage begins. The external signals tell the cell whether or not to divide.

The events of the cell cycle must occur in the correct order, even if the steps take longer than normal. The red stop signs in Figure 5.1 represent three checkpoints when the cell cycle possibly stops. Researchers have identified proteins called *cyclins* that increase and decrease as the cell cycle continues. The appropriate cyclin has to be present for the cell to proceed from the G_1 stage to the S stage and from the G_2 stage to the M stage.

The first checkpoint during G_1 allows the cell to determine whether conditions are favorable to begin the cell cycle. The cell needs to assess whether there are building blocks available for duplication of the DNA and if the DNA is intact. DNA damage can stop the cell cycle at the G_1 checkpoint. In mammalian cells, the p53 protein stops the cycle at the G_1 checkpoint when DNA is damaged. First, the p53 protein attempts to initiate DNA repair, but if that is not possible, it brings about apoptosis.

The cell cycle stops at the G_2 checkpoint if DNA has not finished replicating. This prevents the initiation of the M stage before completion of the S stage. Also, if DNA is damaged, stopping the cell cycle at this checkpoint allows time for the damage to be repaired. If repair is not possible, apoptosis occurs.

Another cell cycle checkpoint occurs during the mitotic (M) stage. The cycle stops if the chromosomes are not going to be distributed accurately to the daughter cells.

These checkpoints are critical for preventing cancer development. A damaged cell should not complete mitosis, but instead should undergo apoptosis.

3D Animation Cell Cycle and Mitosis: Checkpoints

Mammalian cells tend to enter the cell cycle only when stimulated by an external factor. Growth factors are hormones that are received at the plasma membrane. These signals set into motion the events that result in the cell entering the cell cycle. For example, when blood platelets release a growth factor, skin fibroblasts in the vicinity are stimulated to finish the cell cycle so an injury can be repaired.

Animation Cell Proliferation Signaling Pathway

Proto-oncogenes and Tumor Suppressor Genes

Two types of genes control the movement of a cell through the cell cycle: proto-oncogenes and tumor suppressor genes. **Proto-oncogenes** encode proteins that promote the cell cycle and prevent apoptosis. They are often likened to the gas pedal of a car because they cause cells to continue through the cell cycle. **Tumor suppressor genes** encode proteins that stop the cell cycle and promote apoptosis. They are often likened to the brakes of a car because they inhibit cells from progressing through the cell cycle (Fig. 5.3*a*).

Animation Control of the Cell Cycle

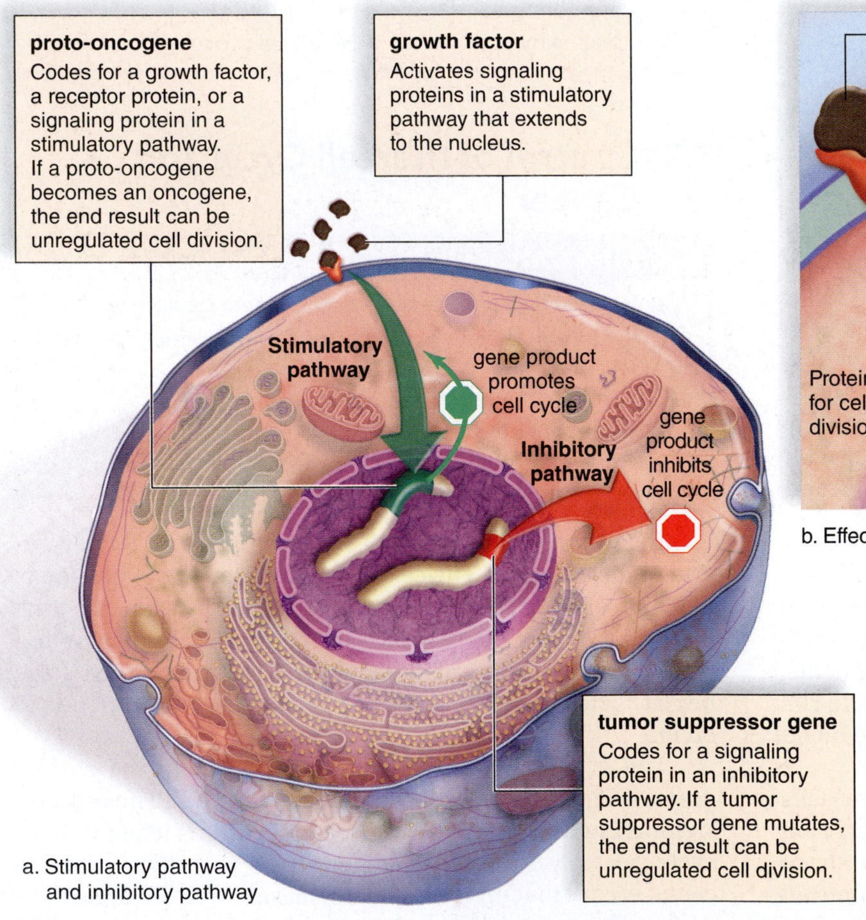
proto-oncogene
Codes for a growth factor, a receptor protein, or a signaling protein in a stimulatory pathway. If a proto-oncogene becomes an oncogene, the end result can be unregulated cell division.
growth factor
Activates signaling proteins in a stimulatory pathway that extends to the nucleus.
Stimulatory pathway
gene product promotes cell cycle
Inhibitory pathway
gene product inhibits cell cycle
tumor suppressor gene
Codes for a signaling protein in an inhibitory pathway. If a tumor suppressor gene mutates, the end result can be unregulated cell division.
a. Stimulatory pathway and inhibitory pathway

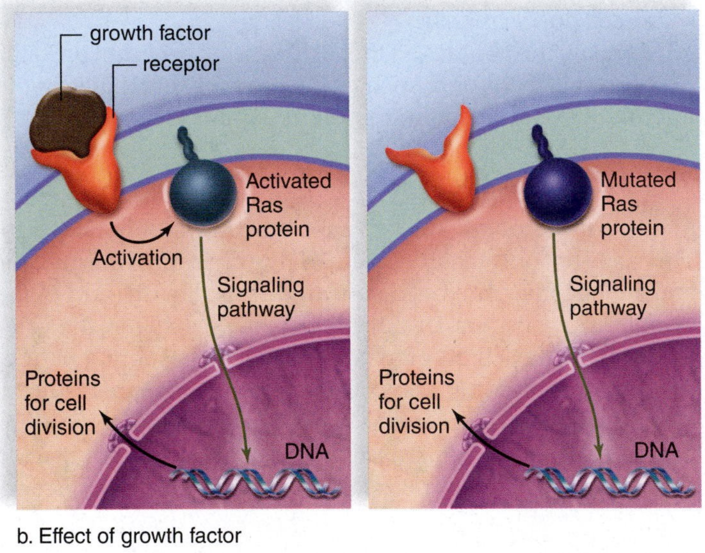
growth factor
receptor
Activated Ras protein
Activation
Signaling pathway
Proteins for cell division
DNA
Mutated Ras protein
Signaling pathway
Proteins for cell division
DNA
b. Effect of growth factor

Figure 5.3 Role of proto-oncogenes and tumor suppressor genes. **a.** Mutated proto-oncogenes and tumor suppressor genes may cause cancer by changing stimulatory and inhibitory pathways. **b.** *RAS* is a proto-oncogene that promotes cell division. When a growth factor binds to the Ras receptor on the cell surface, a signal transduction pathway is activated and relays a signal to the nucleus to express genes whose products promote cell division. Mutations in the *RAS* proto-oncogene cause the Ras protein to be always activated—thus sending a signal for cell division without the presence of a growth factor.

SCIENCE IN YOUR LIFE ▶ HEALTH

Benign Versus Malignant Tumors

Cancer is a disease that is characterized by unrestricted cell growth. As a cancer cell continues its unregulated division, it may form a population of cells called a tumor. There are two different types of tumors: benign tumors and malignant tumors.

A benign tumor is usually surrounded by a fibrous capsule (usually made of connective tissue). Because of this capsule, a benign tumor does not invade the adjacent tissue. The cells of a benign tumor resemble normal cells fairly closely. Many people have benign tumors. For example, a mole (or nevus) is a benign tumor of skin cells called melanocytes. However, not all benign tumors are harmless. Sometimes these tumors restrict a normal tissue's blood supply, or cause normal tissue to not function correctly. If this is in a critical area of the body, such as areas of the brain that control the heartbeat or breathing, the results may be fatal.

The cells of malignant tumors do not resemble normal cells, and are able to invade surrounding tissues. As the cells of the tumor invade nearby tissues, they may come in contact with either blood or lymph vessels, allowing the tumor to spread throughout the body. The tumors associated with lung cancer are often malignant, and may easily spread to other organs, such as the brain or liver. The more deformed a cancer cell, the greater the chance that it will be malignant, and thus dangerous to the individual.

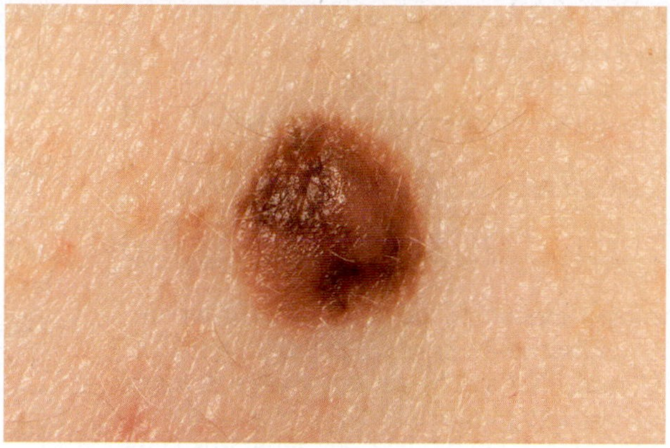

Figure 5A Benign tumor. A mole is a common example of a benign tumor.

Questions to Consider

1. Why might oncologists (cancer physicians) look for abnormal cells in the blood of a suspected cancer patient?
2. What differences in the cell cycle might occur between the cells of a benign and malignant tumor?

Cancer is unregulated cell growth. Carcinogenesis, or the development of cancer, is a multistage process involving disruption of normal cell division and behavior. This typically occurs due to mutations in either proto-oncogenes or tumor suppressor genes. These genetic changes may be inherited from a parent, the result of an error in replication, or induced by an environmental agent. Later, the altered cell grows and divides to become a population of cells, also called a tumor. The characteristics of cancer cells are discussed more fully in Chapter 25. When proto-oncogenes mutate, they become cancer-causing genes, called **oncogenes.** For example, the protein product of the *RAS* proto-oncogene is part of a signal transduction pathway, one of a series of relay proteins (Fig. 5.3*b*). When a signaling molecule, such as a growth factor, binds to a receptor on the cell surface, the Ras protein is activated. Activated Ras protein conveys a message to the nucleus, resulting in gene expression that brings about cell division. If the *RAS* proto-oncogene is mutated to become an oncogene in such a way that the Ras protein is always activated, no growth factor needs to bind to the cell for the altered Ras protein to send the signal to divide. An altered Ras protein is found in approximately 25% of all tumors.

When tumor suppressor genes mutate, their products no longer inhibit the cell cycle. *p53* and *BRCA1* are both examples of tumor suppressor genes. The p53 protein is such an important protein for regulating the cell cycle that almost half of all human cancers have a mutation in the *p53* gene. Mutations in *BRCA1* are often associated with breast and ovarian cancer.

Animation How Tumor Suppressor Genes Block Cell Division

Check Your Progress 5.2

1. Explain what factors would cause a cell to stop at each of the three cell cycle checkpoints.
2. Distinguish between the action of oncogenes and mutated tumor suppressor genes.

5.3 Mitosis: Maintaining the Chromosome Number

Learning Outcomes

Upon completion of this section, you should be able to

1. Explain the role of mitosis and how it maintains the chromosome number of a cell.
2. Summarize the major events that occur during mitosis and cytokinesis.
3. Compare and contrast mitosis and cytokinesis in plant and animal cells.

Eukaryotic chromosomes are composed of **chromatin,** a combination of both DNA and protein. Some of these proteins are concerned with DNA and RNA synthesis, but a large proportion, termed *histones,* seem to play primarily a structural role. A human cell contains at least 2 m of DNA. Yet all of this DNA is packed into a nucleus that is about 5 μm in diameter. The histones are responsible for packaging the DNA so that it can fit into such a small space. When a eukaryotic cell is not undergoing division, the chromatin is dispersed or extended. This makes the DNA available for RNA synthesis (see Chapter 25). At the time of cell division, chromatin coils, loops, and condenses into a highly compacted form.

Each species has a characteristic chromosome number. For instance, human cells contain 46 chromosomes, corn has 20 chromosomes, and the crayfish has 200! This number is called the **diploid** (**2n**) number because it contains two (a pair) of each type of chromosome. Humans have 23 pairs of chromosomes. One member of each pair originates from the mother, and the other member originates from the father. In Figure 5.4, the blue chromosomes are inherited from one parent, and the red are inherited from the other.

Half the diploid number, called the **haploid** (**n**) number of chromosomes, contains only one of each kind of chromosome. In the life cycle of humans, only sperm and eggs have the haploid number of chromosomes.

Overview of Mitosis

Mitosis is nuclear division in which the chromosome number stays constant. A 2n nucleus divides to produce daughter nuclei that are also 2n (Fig. 5.4). It would be possible to diagram this as 2n $\longrightarrow$ 2n. Before nuclear division takes place, DNA replication occurs, duplicating the chromosomes. A duplicated chromosome is composed of two sister chromatids held together in a region called the **centromere. Sister chromatids** are genetically identical—they contain the same DNA sequences. At the completion of mitosis, each chromosome consists of a single chromatid:

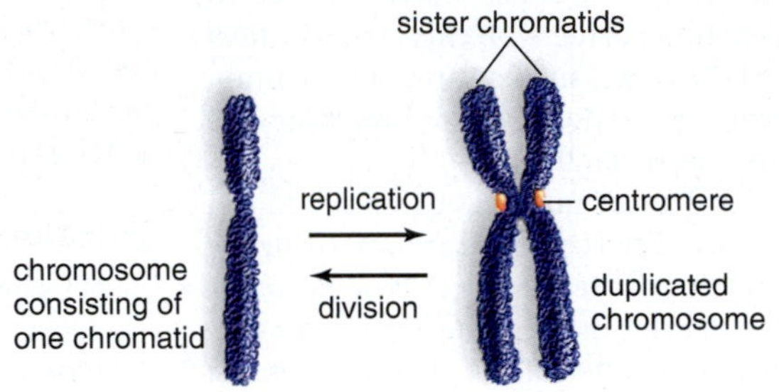
sister chromatids
replication
centromere
chromosome
consisting of
one chromatid
division
duplicated
chromosome

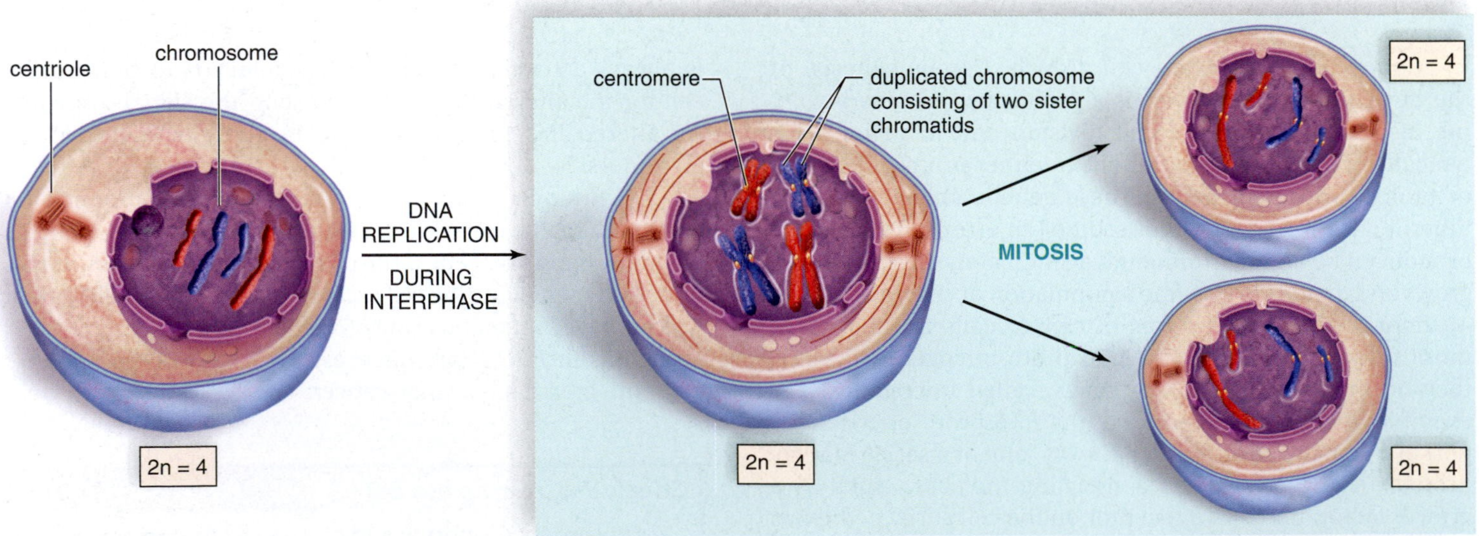
chromosome
centriole
DNA
REPLICATION
DURING
INTERPHASE
2n = 4
centromere
duplicated chromosome
consisting of two sister
chromatids
2n = 4
MITOSIS
2n = 4
2n = 4

Figure 5.4 Mitosis overview. Following DNA replication during interphase, each chromosome in the parental nucleus is duplicated and consists of two sister chromatids. During mitosis, the centromeres divide and the sister chromatids separate, becoming daughter chromosomes that move into the daughter nuclei. Therefore, daughter cells have the same number and kinds of chromosomes as the parental cell. (The blue chromosomes were inherited from one parent, and the red chromosomes were inherited from the other.)

Figure 5.4 gives an overview of mitosis; for simplicity, only four chromosomes are depicted. (In determining the number of chromosomes, it is only necessary to count the number of independent centromeres.) During mitosis, the centromeres divide and then the sister chromatids separate, becoming **daughter chromosomes.** Therefore, each daughter nucleus gets a complete set of chromosomes and has the same number of chromosomes as the parental cell. This makes the daughter cells genetically identical to each other and to the parental cell.

Animation Mitosis

MP3 Mitosis

Mitosis is the type of nuclear division that occurs when tissues grow or when repair occurs. Following fertilization, the zygote begins to divide mitotically, and mitosis continues during development and the life span of the individual.

Mitosis in Detail

Mitosis is nuclear division that produces *two daughter nuclei, each with the same number and kinds of chromosomes as the parental nucleus.*

During mitosis, a **spindle** brings about an orderly distribution of chromosomes to the daughter cell nuclei. The spindle contains many fibers, each composed of a bundle of microtubules. Microtubules are able to disassemble and assemble. The centrosome, which is the main microtubule organizing center of the cell, divides during late interphase. It is believed that centrosomes are responsible for organizing the spindle. In animal cells, each centrosome contains a pair of barrel-shaped organelles called *centrioles* and an *aster,* which is an array of short microtubules that radiate from the centrosome. The fact that plant cells lack centrioles suggests that centrioles are not required for spindle formation.

Mitosis in Animal Cells

Mitosis is a continuous process that is arbitrarily divided into four phases for convenience of description: prophase, metaphase, anaphase, and telophase (Fig. 5.5).

3D Animation Cell Cycle and Mitosis: Mitosis

Prophase It is apparent during early **prophase** that cell division is about to occur. The centrosomes begin moving away from each other toward opposite ends of the nucleus. Spindle fibers appear between the separating centrosomes as the nuclear envelope begins to fragment, and the nucleolus begins to disappear.

The chromatin condenses and the chromosomes are now visible. Each is duplicated and composed of sister chromatids held together at a centromere. The spindle begins forming during late prophase, and the chromosomes become attached to the spindle fibers. Their centromeres attach to fibers called kinetochore (or centromeric) fibers. As yet, the chromosomes have no particular orientation, moving first one way and then the other.

Metaphase By the time of **metaphase,** the fully formed spindle consists of poles, asters, and fibers. The *metaphase plate* is a plane perpendicular to the axis of the spindle and equidistant from the poles. The chromosomes attached to centromeric spindle fibers line up at the metaphase plate during metaphase. Polar spindle fibers reach beyond the metaphase plate and overlap.

Anaphase At the beginning of **anaphase,** the centromeres uniting the sister chromatids divide. Then the sister chromatids separate, becoming daughter chromosomes that move toward the opposite poles of the spindle. Daughter chromosomes have a centromere and a single chromatid.

What accounts for the movement of the daughter chromosomes? First, the kinetochore spindle fibers shorten, pulling the daughter chromosomes toward the poles. Second, the polar spindle fibers push the poles apart as they lengthen and slide past one another.

Telophase During **telophase,** the spindle disappears, and nuclear envelope components reassemble around the daughter chromosomes. Each daughter nucleus contains the same number and kinds of chromosomes as the original parental cell. Remnants of the polar spindle fibers are still visible between the two nuclei.

The chromosomes become more diffuse once again, and a nucleolus appears in each daughter nucleus. Cytokinesis is under way, and soon there will be two individual daughter cells, each with a nucleus that contains the diploid number of chromosomes.

Video Mitosis

Mitosis in Plant Cells

As with animal cells, mitosis in plant cells permits growth and repair. A particular plant tissue called meristematic tissue retains the ability to divide throughout the life of a plant. Meristematic tissue is found at the root tip and also at the shoot tip of stems. Lateral meristematic tissue accounts for the ability of trees to increase their girth each growing season.

Figure 5.5 illustrates mitosis in plant cells. Exactly the same phases occur in plant cells as in animal cells. Although plant cells have a centrosome and spindle, there are no centrioles or asters during cell division. The spindle still brings about the distribution of the chromosomes to each daughter cell.

Cytokinesis in Animal and Plant Cells

Cytokinesis, or cytoplasmic cleavage, usually accompanies mitosis, but they are separate processes. Division of the cytoplasm begins in anaphase and continues in telophase but does not reach completion until just before the next interphase. By that time, the newly forming cells have received a share of the cytoplasmic organelles that duplicated during the previous interphase.

3D Animation Cell Cycle and Mitosis: Cytokinesis

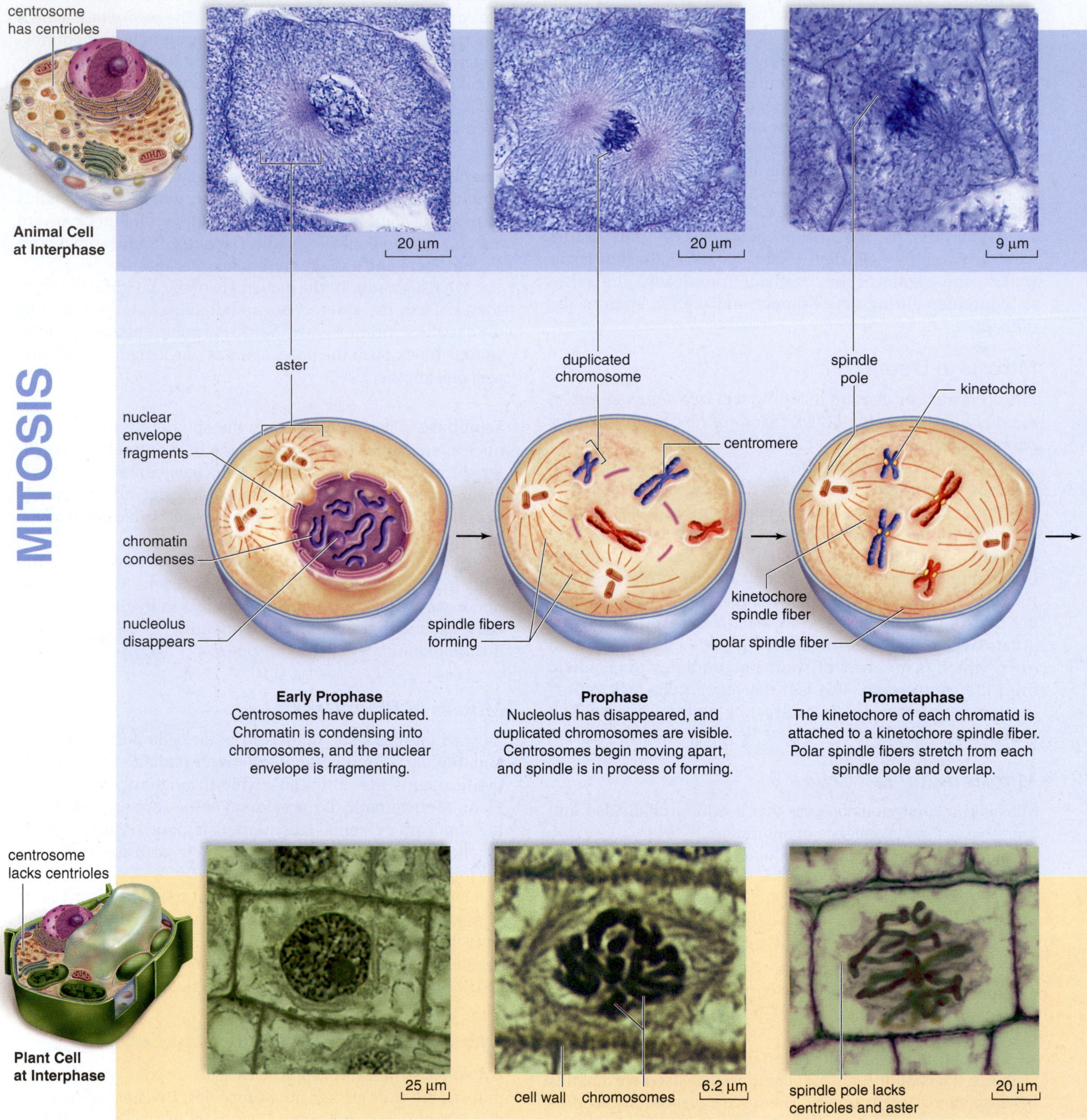
MITOSIS
centrosome
has centrioles
Animal Cell
at Interphase
20 μm
20 μm
9 μm
aster
duplicated
chromosome
spindle
pole
kinetochore
nuclear
envelope
fragments
centromere
chromatin
condenses
nucleolus
disappears
spindle fibers
forming
kinetochore
spindle fiber
polar spindle fiber
Early Prophase
Centrosomes have duplicated. Chromatin is condensing into chromosomes, and the nuclear envelope is fragmenting.
Prophase
Nucleolus has disappeared, and duplicated chromosomes are visible. Centrosomes begin moving apart, and spindle is in process of forming.
Prometaphase
The kinetochore of each chromatid is attached to a kinetochore spindle fiber. Polar spindle fibers stretch from each spindle pole and overlap.
centrosome
lacks centrioles
Plant Cell
at Interphase
25 μm
cell wall
chromosomes
6.2 μm
spindle pole lacks
centrioles and aster
20 μm

Figure 5.5 Phases of mitosis in animal and plant cells. The blue chromosomes were inherited from one parent, and the red from the other parent.

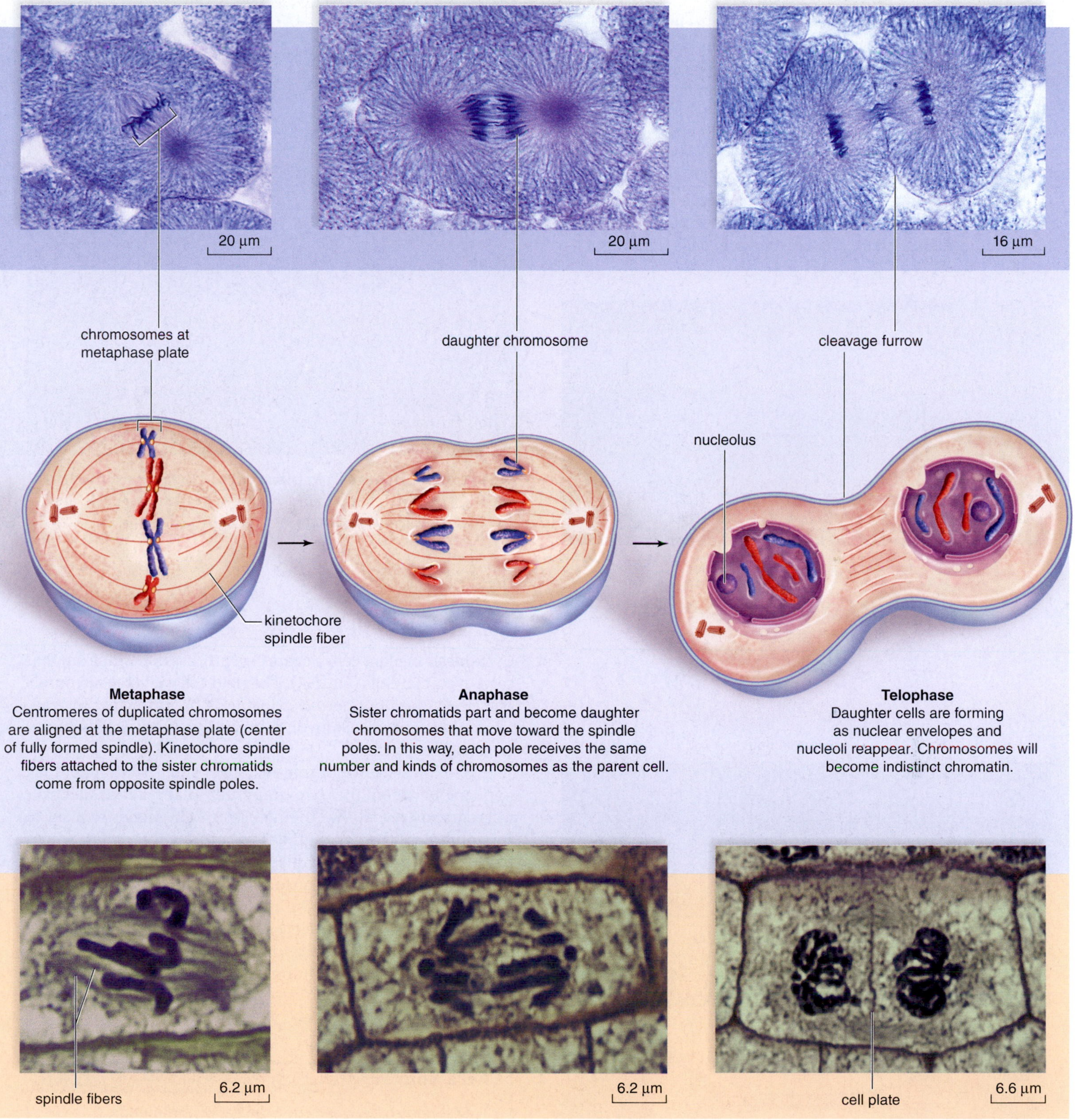
20 µm
20 µm
16 µm
chromosomes at metaphase plate
daughter chromosome
cleavage furrow
nucleolus
kinetochore spindle fiber
spindle fibers
6.2 µm
6.2 µm
cell plate
6.6 µm

Metaphase
Centromeres of duplicated chromosomes are aligned at the metaphase plate (center of fully formed spindle). Kinetochore spindle fibers attached to the sister chromatids come from opposite spindle poles.

Anaphase
Sister chromatids part and become daughter chromosomes that move toward the spindle poles. In this way, each pole receives the same number and kinds of chromosomes as the parent cell.

Telophase
Daughter cells are forming as nuclear envelopes and nucleoli reappear. Chromosomes will become indistinct chromatin.

Cytokinesis in Animal Cells

As anaphase draws to a close in animal cells, a **cleavage furrow,** which is an indentation of the membrane between the two daughter nuclei, begins to form. The cleavage furrow deepens when a band of actin filaments, called the contractile ring, slowly forms a constriction between the two daughter cells. The action of the contractile ring can be likened to pulling a drawstring ever tighter about the middle of a balloon, causing the balloon to constrict in the middle.

A narrow bridge between the two cells can be seen during telophase, and then the contractile ring continues to separate the cytoplasm until there are two daughter cells (Fig. 5.6).

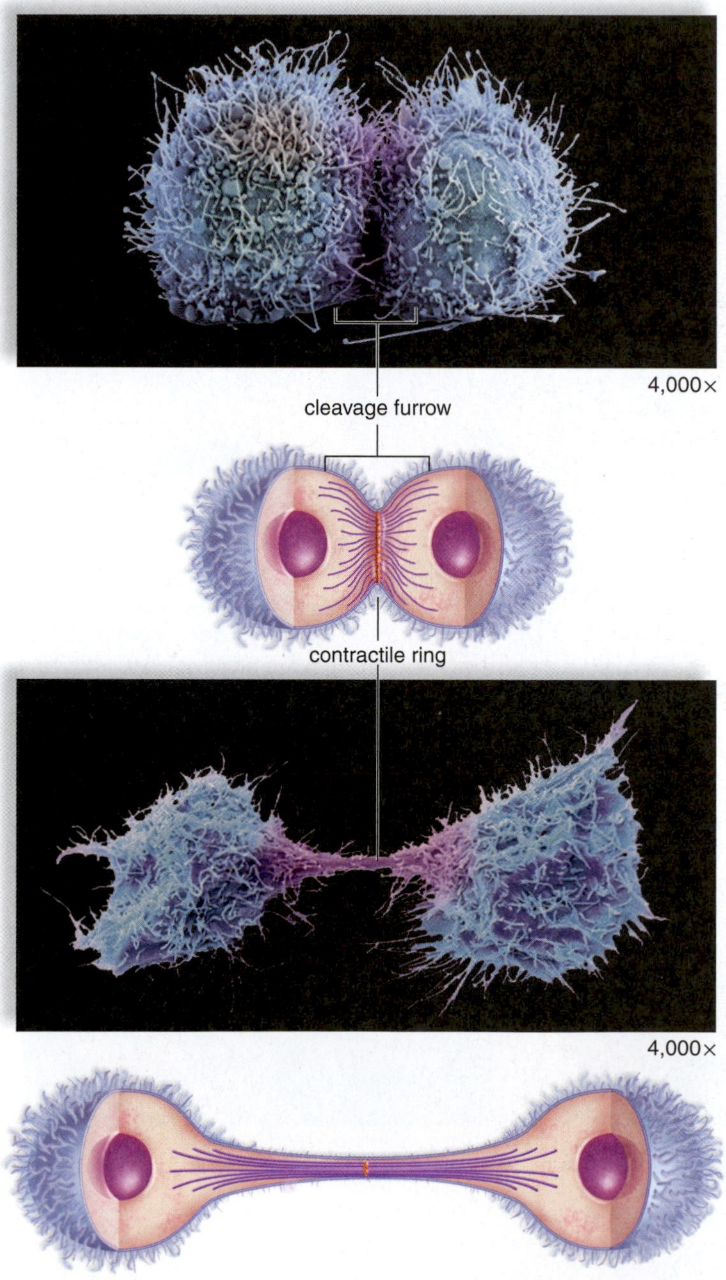
4,000×
cleavage furrow
contractile ring
4,000×

Figure 5.6 Cytokinesis in animal cells. A single cell becomes two cells by a furrowing process. A contractile ring composed of actin filaments gradually gets smaller, and the cleavage furrow pinches the cell into two cells.

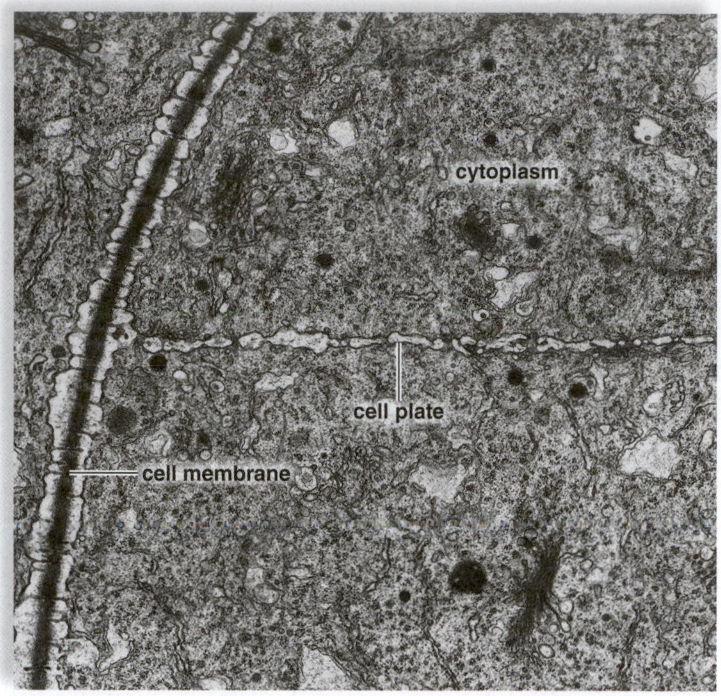
cytoplasm
cell plate
cell membrane

Figure 5.7 Cytokinesis in plant cells. During cytokinesis in a plant cell, a cell plate forms midway between two daughter nuclei and extends to the plasma membrane.

Cytokinesis in Plant Cells

Cytokinesis in plant cells occurs by a process different from that seen in animal cells (Fig. 5.7). The rigid cell wall that surrounds plant cells does not permit cytokinesis by furrowing. Instead, cytokinesis in plant cells involves building new cell walls between the daughter cells.

Cytokinesis is apparent when a small, flattened disk appears between the two daughter plant cells. Electron micrographs reveal that the disk is at right angles to a set of microtubules. The Golgi apparatus produces vesicles, which move along the microtubules to the region of the disk. As more vesicles arrive and fuse, a **cell plate** can be seen. The membrane of the vesicles completes the plasma membrane for both cells, and they release molecules that form the new plant cell walls. The cell walls are later strengthened by the addition of cellulose fibrils.

Animation
Cytokinesis

Check Your Progress 5.3

1. Distinguish between the number of chromatids each chromosome contains before and after replication.
2. Summarize the events that are occurring in each stage of mitosis.
3. Compare and contrast cytokinesis in plant and animal cells.

5.4 Meiosis: Reducing the Chromosome Number

Learning Outcomes

Upon completion of this section, you should be able to

1. Summarize the purpose of meiosis.
2. Explain what is meant by the term homologous chromosomes.
3. Describe the events of meiosis.

Meiosis occurs in any life cycle that involves sexual reproduction. **Meiosis** reduces the chromosome number in such a way that the daughter nuclei receive only one of each kind of chromosome. The process of meiosis ensures that the next generation of individuals will have only the diploid number of chromosomes and a combination of traits different from that of either parent.

Overview of Meiosis

At the start of meiosis, the parental cell has the diploid number of chromosomes. In Figure 5.8, the diploid number is 4, which you can verify by counting the number of centromeres. Notice that meiosis requires two cell divisions, meiosis I and meiosis II, and that the four daughter cells have the haploid number of chromosomes. DNA replication occurs prior to meiosis I.

Recall that when a cell is 2n, or diploid, the chromosomes occur in pairs. For example, the 46 chromosomes of humans occur in 23 pairs. In Figure 5.8, there are two pairs of chromosomes. The members of a pair are called **homologous chromosomes** (or **homologues**). In the diagrams in this text, homologues have the same size but are indicated by different colors—the blue chromosomes are inherited from one parent, and the red are inherited from the other.

Animation How Meiosis Works

MP3 Meiosis

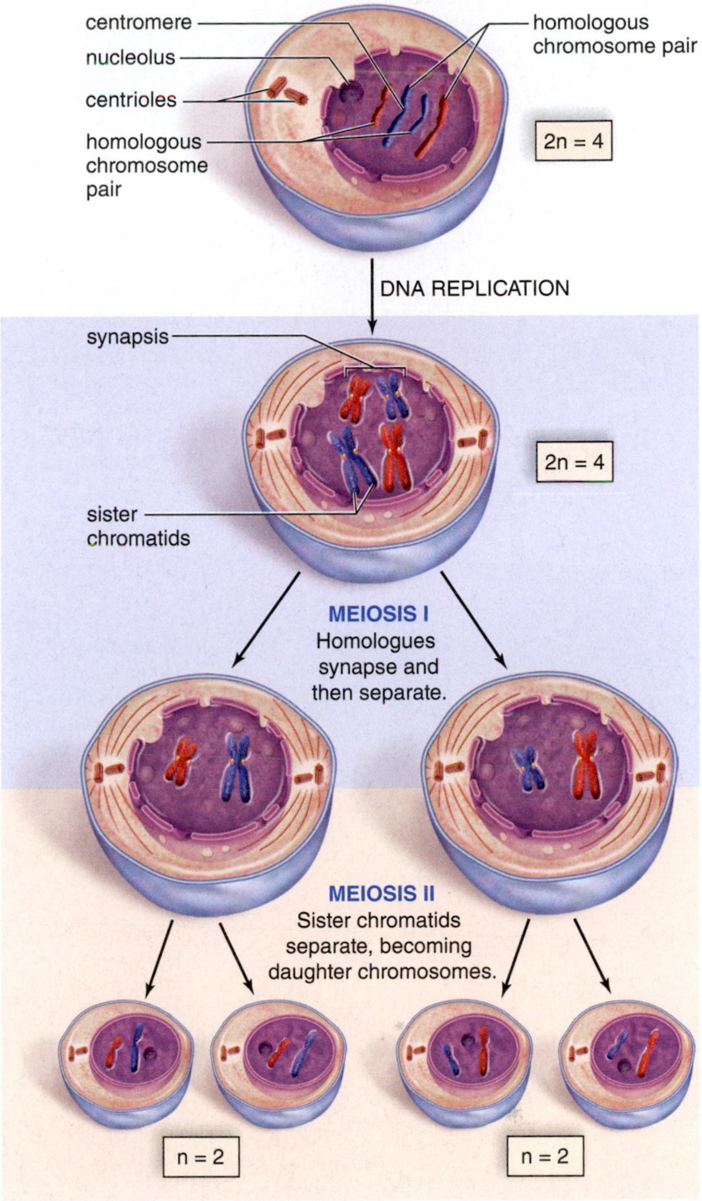
centromere
nucleolus
centrioles
homologous chromosome pair
homologous chromosome pair
2n = 4
DNA REPLICATION
synapsis
2n = 4
sister chromatids
MEIOSIS I
Homologues synapse and then separate.
MEIOSIS II
Sister chromatids separate, becoming daughter chromosomes.
n = 2
n = 2

Figure 5.8 Overview of meiosis. Following DNA replication, each chromosome is duplicated. During meiosis I, the homologous chromosomes pair during synapsis and then separate. During meiosis II, the centromeres divide and the sister chromatids separate, becoming daughter chromosomes that move into the daughter nuclei.

Meiosis I

During meiosis I, the homologous chromosomes come together and line up side by side. This process is called **synapsis** and it results in an association of four chromatids that stay in close proximity during the first two phases of meiosis I.

Because of synapsis, there are pairs of homologous chromosomes at the metaphase plate during meiosis I. Notice that only during meiosis I is it possible to observe paired chromosomes at the metaphase plate. When the members of these pairs separate, each daughter nucleus receives one member of each pair. Therefore, each daughter cell now has the haploid number of chromosomes, as you can verify by counting its centromeres.

Meiosis II and Fertilization

Replication of DNA does not occur between meiosis I and meiosis II because the chromosomes are still duplicated (composed of two sister chromatids). During meiosis II, the centromeres divide and the sister chromatids separate, becoming daughter chromosomes that are distributed to daughter nuclei. In the end, each of four daughter cells has the haploid number of chromosomes, and each chromosome consists of one chromatid.

In some life cycles, such as that of humans (see Fig. 5.14), the daughter cells mature into **gametes** (sex cells—sperm and egg) that fuse during fertilization. **Fertilization** restores the diploid number of chromosomes in a cell that will develop into a new individual. If the gametes carried the diploid instead of the haploid number of chromosomes, the chromosome number would double with each fertilization.

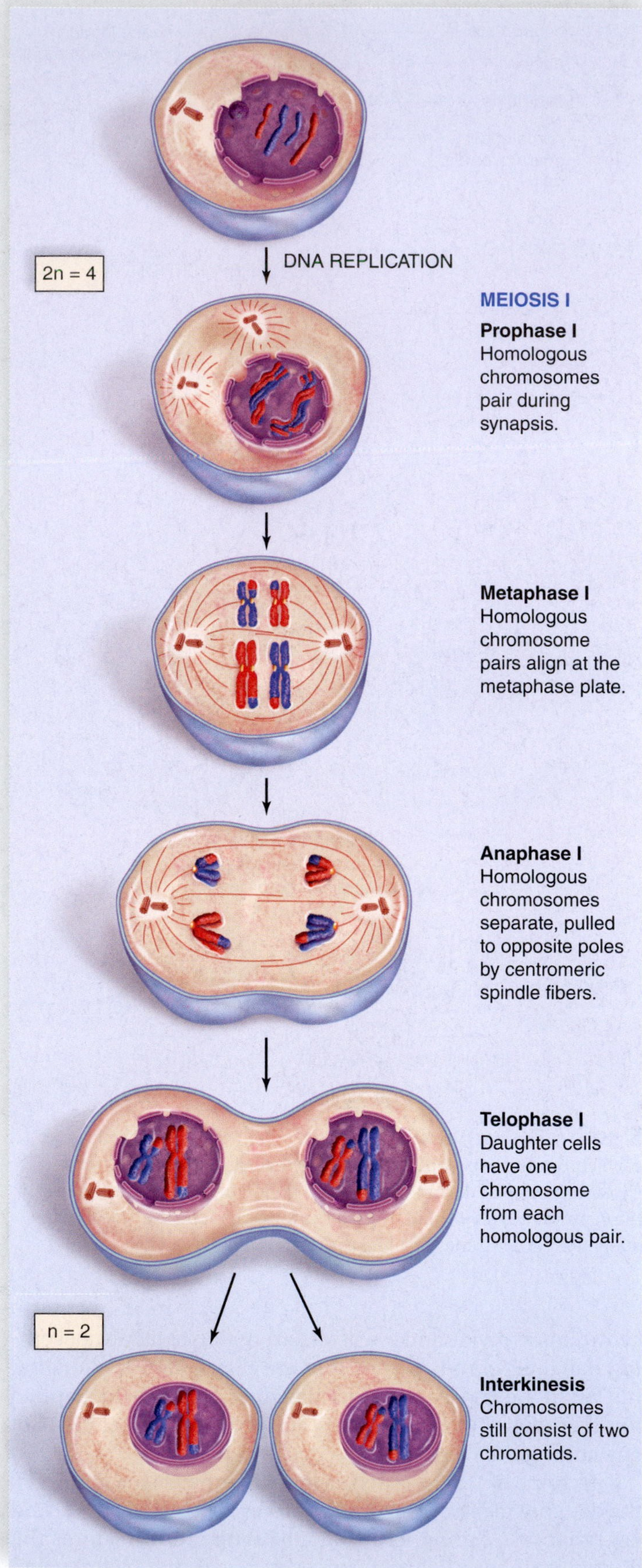
2n = 4
DNA REPLICATION
MEIOSIS I
Prophase I
Homologous chromosomes pair during synapsis.
Metaphase I
Homologous chromosome pairs align at the metaphase plate.
Anaphase I
Homologous chromosomes separate, pulled to opposite poles by centromeric spindle fibers.
Telophase I
Daughter cells have one chromosome from each homologous pair.
n = 2
Interkinesis
Chromosomes still consist of two chromatids.

Figure 5.9 Meiosis I in an animal cell. The exchange of color between nonsister chromatids represents crossing-over. Notice that the daughter cells have only 2 centromeres compared to the parental cell's 4.

Meiosis in Detail

Meiosis, which requires two nuclear divisions, results in four daughter nuclei, each having one of each kind of chromosome and therefore half the number of chromosomes as the parental cell.

First Division

Meiosis I has four phases, called prophase I, metaphase I, anaphase I, and telophase I. The phases of meiosis I for an animal cell are diagrammed in Figure 5.9. Meiosis helps ensure that **genetic variation** of the parental genes occurs through two key events: crossing-over and independent assortment of chromosome pairs. Because the members of a homologous pair can carry slightly different instructions for the same genetic trait, these events are significant. For example, one homologue may carry instructions for brown eyes, while the corresponding homologue may carry instructions for blue eyes. Fertilization ensures the offspring will have different combinations of genes from either parent.

Animation Meiosis I

Prophase I In prophase I, synapsis occurs, and then the spindle appears while the nuclear envelope fragments and the nucleolus disappear. Figure 5.10 shows that during synapsis, the homologous chromosomes come together and line up side by side. Now an exchange of genetic material may occur between the *nonsister* chromatids of the homologues. This exchange is called **crossing-over.** Crossing-over means that the chromatids held together by a centromere are no longer identical. As a result of crossing-over, the daughter cells receive chromosomes with recombined genetic material. In Figures 5.9 and 5.10, crossing-over is represented by an exchange of color. Because of crossing-over, first the nonsister chromatids, and then the chromosomes they give rise to, have a different combination of genes than the parental cell.

Animation Meiosis with Crossing-Over

Metaphase I and Anaphase I During metaphase I, the homologues align at the metaphase plate. Depending on how they align, the maternal or paternal member of each pair may be

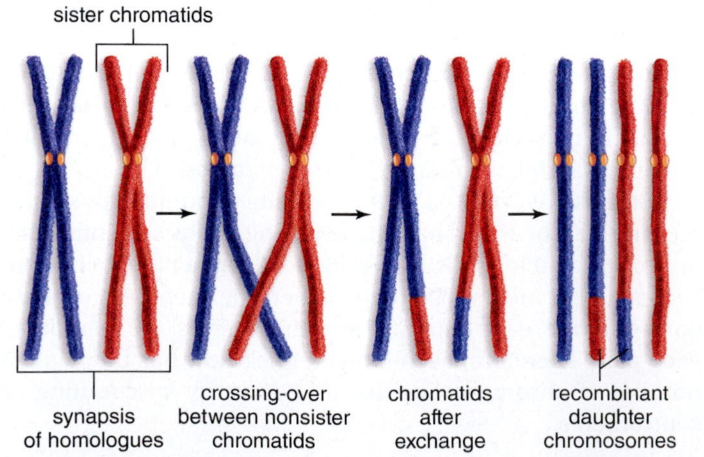
sister chromatids
synapsis of homologues
crossing-over between nonsister chromatids
chromatids after exchange
recombinant daughter chromosomes

Figure 5.10 Synapsis and crossing-over. During prophase I, *from left to right,* duplicated homologous chromosomes undergo synapsis when they line up with each other. During crossing-over, nonsister chromatids break and then rejoin. The two resulting daughter chromosomes will have a different combination of genes than they had before.

oriented toward either pole. **Independent assortment** occurs when these homologous pairs separate from each other during anaphase I, generating cells with different combinations of maternal and paternal chromosomes. Notice that each chromosome still consists of two sister chromatids. Figure 5.11 shows two possible orientations for a cell that contains only two pairs of chromosomes ($2n = 4$). Gametes from the first cell, but not the other, will have two paternal or two maternal chromosomes. The gametes from the other cell will have different combinations of maternal and paternal chromosomes. When all possible orientations are considered for a cell containing two pairs of chromosomes, the result will be 2^2, or four, possible combinations of maternal and paternal chromosomes in the resulting gametes from this cell. In humans, where there are 23 pairs of chromosomes, the number of possible chromosomal combinations in the gametes is a staggering 2^{23}, or 8,388,608. And this does not even consider the genetic recombinations that are introduced due to crossing-over. Together, prophase I, metaphase I, and anaphase I introduce the genetic variation that helps ensure that no two offspring have the same combination of genes as the parents.

Animation Random Orientation of Chromosomes During Meiosis

Telophase I In some species, including humans, telophase I occurs at the end of meiosis I. If so, the nuclear envelopes re-form, and nucleoli reappear. This phase may or may not be accompanied by cytokinesis, which is separation of the cytoplasm.

3D Animation Meiosis: Interphase and Meiosis I

Interkinesis The period of time between meiosis I and meiosis II is called **interkinesis.** No replication of DNA occurs during interkinesis. Why is this appropriate? Because the chromosomes are already duplicated.

Second Division

Phases of meiosis II for an animal cell are diagrammed in Figure 5.12. At the beginning of prophase II, a spindle appears while the nuclear envelope disassembles and the nucleolus disappears. Each duplicated chromosome attaches to the spindle, and then they align at the metaphase plate during metaphase II. During anaphase II, sister chromatids separate, becoming daughter chromosomes that move into the daughter nuclei. In telophase II, the spindle disappears as the nuclear envelope re-forms.

During the cytokinesis that can follow meiosis II, the plasma membrane furrows to produce two complete cells, each of which has the haploid number of chromosomes. Because each cell from meiosis I undergoes meiosis II, there are four daughter cells altogether.

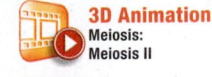

3D Animation Meiosis: Meiosis II

Animation Meiosis II

The Importance of Meiosis

Meiosis produces haploid cells. One of the hallmarks of meiosis is the possibility of genetic variation: the haploid cells produced are no longer identical to the diploid parent cell from which they came.

Variation occurs in two ways. First, during prophase I, crossing-over between nonsister chromatids of the paired homologous chromosomes rearranges genes, with the result that the sister chromatids of each homologue may no longer be identical. Second, the independent assortment of chromosomes during anaphase I means that the gametes produced by meiosis II may have different combinations of chromosomes than the parental cell.

Animation Genetic Diversity

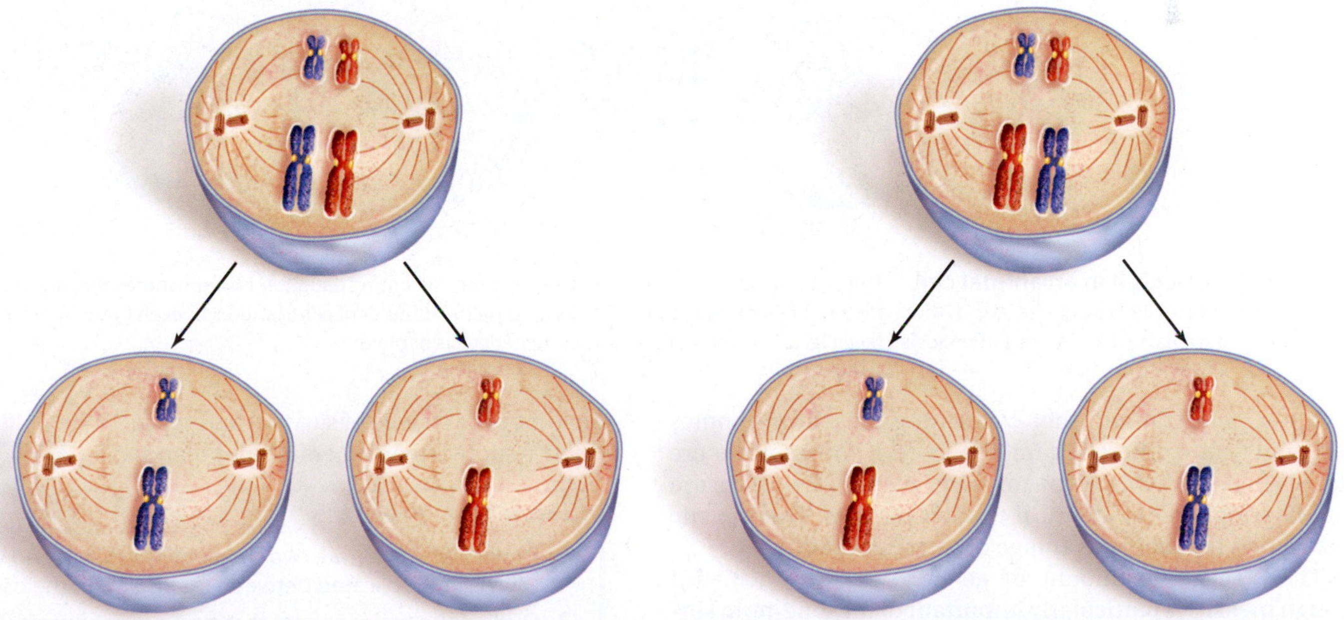

Figure 5.11 Independent assortment. Two possible orientations of homologous chromosome pairs at the metaphase plate are shown for metaphase I. Each of these will result in daughter nuclei with a different combination of parental chromosomes (independent assortment). In a cell with two pairs of homologous chromosomes, there are 2^2 possible combinations of parental chromosomes in the daughter nuclei.

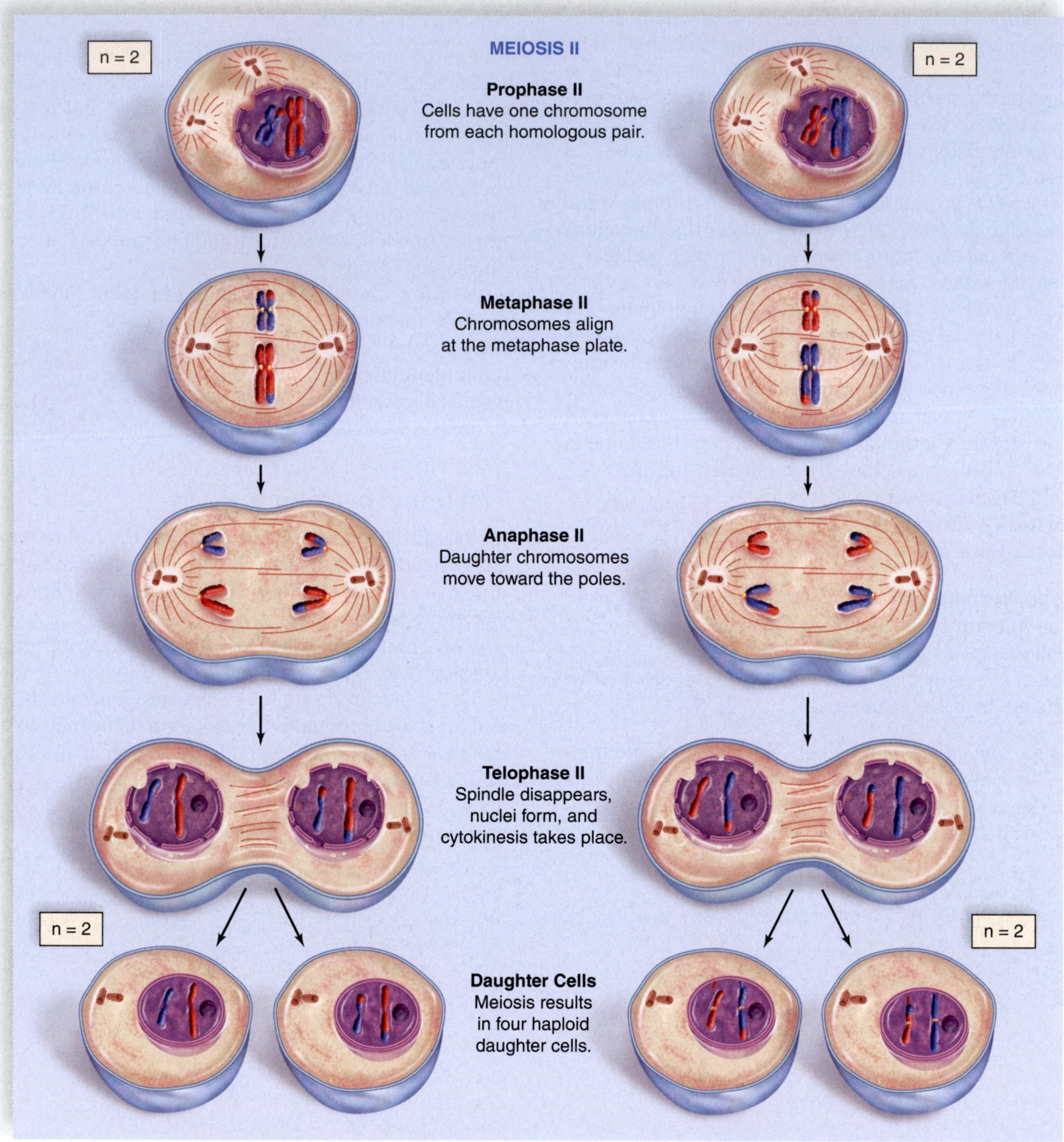
MEIOSIS II
n = 2
n = 2
Prophase II
Cells have one chromosome from each homologous pair.
Metaphase II
Chromosomes align at the metaphase plate.
Anaphase II
Daughter chromosomes move toward the poles.
Telophase II
Spindle disappears, nuclei form, and cytokinesis takes place.
n = 2
n = 2
Daughter Cells
Meiosis results in four haploid daughter cells.

Figure 5.12 Meiosis II in an animal cell. During meiosis II, sister chromatids separate, becoming daughter chromosomes that are distributed to the daughter nuclei. Following meiosis II, there are four haploid daughter cells. Comparing the number of centromeres in each daughter cell with the number in the parental cell at the start of meiosis I (see Fig. 5.10) verifies that each daughter cell is haploid.

Upon fertilization, the combining of chromosomes from genetically different gametes, even if the gametes are from a single individual (as in many plants), helps ensure that offspring are not identical to their parents. This genetic variability is the main advantage of sexual reproduction.

The staggering amount of genetic variation achieved through meiosis is particularly important to the long-term survival of a species because it increases genetic variation within a population. If the environment changes, genetic variability among offspring introduced by sexual reproduction may be advantageous. Under the new conditions, some offspring may have a better chance of survival and reproductive success than others in a population.

3D Animation Meiosis: Genetic Diversity

Check Your Progress 5.4

1. Explain why there are two divisions in meiosis and why the DNA is not replicated between meiosis I and meiosis II.
2. Outline the major events that occur during meiosis I and II, focusing on the activities of the chromosomes.
3. Explain how crossing-over and independent assortment introduce genetic variation.

5.5 Comparison of Meiosis with Mitosis

Learning Outcomes

Upon completion of this section, you should be able to

1. Compare and contrast the processes of meiosis and mitosis.
2. Identify the differences in the behavior of homologous chromosomes in meiosis and mitosis.

Figure 5.13 compares meiosis to mitosis. Notice the following differences:

- DNA replication takes place only once prior to either meiosis or mitosis. However, meiosis requires two nuclear divisions, whereas mitosis requires only one nuclear division.
- Meiosis produces four daughter nuclei, and following cytokinesis, there are four daughter cells. Mitosis followed by cytokinesis results in two daughter cells.
- The four daughter cells following meiosis are haploid and have half the chromosome number as the parental cell. The daughter cells following mitosis have the same chromosome number as the parental cell.

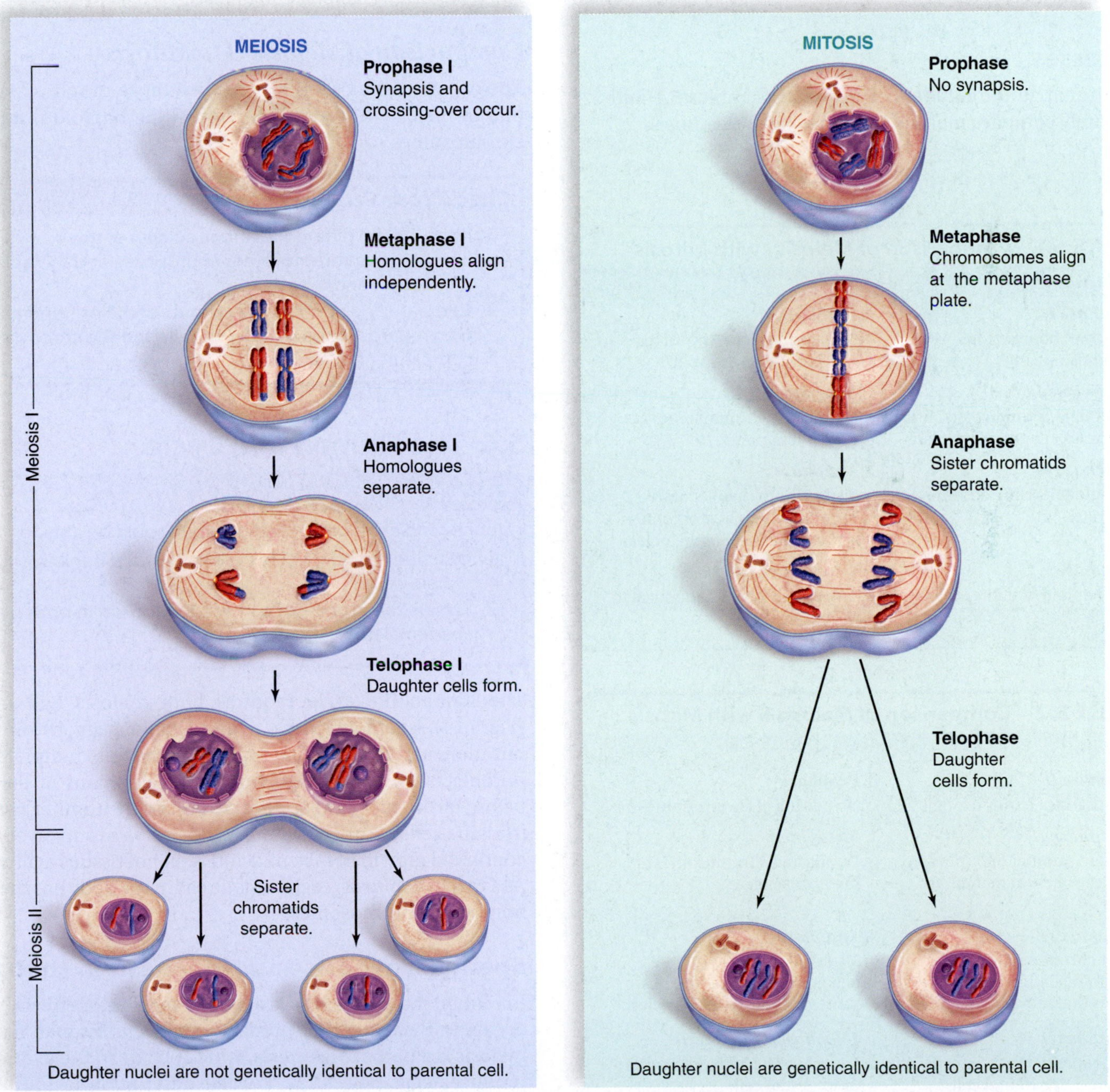
MEIOSIS
Prophase I
Synapsis and crossing-over occur.
Metaphase I
Homologues align independently.
Anaphase I
Homologues separate.
Telophase I
Daughter cells form.
Meiosis I
Meiosis II
Sister chromatids separate.
Daughter nuclei are not genetically identical to parental cell.
MITOSIS
Prophase
No synapsis.
Metaphase
Chromosomes align at the metaphase plate.
Anaphase
Sister chromatids separate.
Telophase
Daughter cells form.
Daughter nuclei are genetically identical to parental cell.

Figure 5.13 Meiosis I compared to mitosis. In meiosis metaphase I, homologues are paired at the metaphase plate, but in mitosis metaphase, all chromosomes align individually at the metaphase plate. Individual homologues separate during meiosis anaphase I, so that the daughter cells are haploid.

- The daughter cells resulting from meiosis are not genetically identical to each other or to the parental cell. The daughter cells resulting from mitosis are genetically identical to each other and to the parental cell.

The specific differences between these nuclear divisions can be categorized according to when they occur and the processes involved.

Occurrence

Meiosis occurs only at certain times in the life cycle of sexually reproducing organisms. In humans, meiosis occurs only in the reproductive organs and produces the gametes. Mitosis occurs almost continuously in all tissues during growth and repair.

Processes

To summarize the differences in process, Tables 5.1 and 5.2 separately compare meiosis I and meiosis II to mitosis.

TABLE 5.1 Comparison of Meiosis I with Mitosis

Meiosis I	Mitosis
Prophase I Pairing of homologous chromosomes	***Prophase*** No pairing of chromosomes
Metaphase I Homologous duplicated chromosomes at metaphase plate	***Metaphase*** Duplicated chromosomes at metaphase plate
Anaphase I Homologous chromosomes separate.	***Anaphase*** Sister chromatids separate, becoming daughter chromosomes that move to the poles.
Telophase I Two haploid daughter cells	***Telophase/Cytokinesis*** Two daughter cells, identical to the parental cell

TABLE 5.2 Comparison of Meiosis II with Mitosis

Meiosis II	Mitosis
Prophase II No pairing of chromosomes	***Prophase*** No pairing of chromosomes
Metaphase II Haploid number of duplicated chromosomes at metaphase plate	***Metaphase*** Duplicated chromosomes at metaphase plate
Anaphase II Sister chromatids separate, becoming daughter chromosomes that move to the poles.	***Anaphase*** Sister chromatids separate, becoming daughter chromosomes that move to the poles.
Telophase II Four haploid daughter cells, not identical to parental cell	***Telophase/Cytokinesis*** Two daughter cells, identical to the parental cell

Comparison of Meiosis I to Mitosis

Notice that these events distinguish meiosis I from mitosis:

- Homologous chromosomes pair and undergo crossing-over during prophase I of meiosis, but not during mitosis.
- Paired homologous chromosomes align at the metaphase plate during metaphase I in meiosis. These paired chromosomes have four chromatids altogether. Individual chromosomes align at the metaphase plate during metaphase in mitosis. They each have two chromatids.
- Homologous chromosomes (with centromeres intact) separate and move to opposite poles during anaphase I in meiosis. Centromeres split, and sister chromatids, now called daughter chromosomes, move to opposite poles during anaphase in mitosis.

Animation Comparison of Meiosis and Mitosis

Comparison of Meiosis II to Mitosis

The events of meiosis II are just like those of mitosis, except that in meiosis II, the nuclei contain the haploid number of chromosomes.

Check Your Progress 5.5

1. Describe the differences in the activity of the homologous chromosomes in prophase and metaphase of meiosis I and mitosis.
2. Compare and contrast the outputs of meiosis and mitosis with regards to the number of cells and the chromosome complement of each cell.

5.6 The Human Life Cycle

Learning Outcomes

Upon completion of this section, you should be able to

1. Describe the human life cycle in terms of haploid and diploid cells.
2. Explain the process of gamete production in both males and females.

The human life cycle requires both meiosis and mitosis (Fig. 5.14). A haploid sperm (n) and a haploid egg (n) join at fertilization, and the resulting **zygote** has the full, or diploid (2n), number of chromosomes. During development of the fetus before birth, mitosis keeps the chromosome number constant in all the cells of the body. After birth, mitosis is involved in the continued growth of the child and repair of tissues at any time. As a result of mitosis, each somatic cell in the body has the same number of chromosomes.

Spermatogenesis and Oogenesis in Humans

In human males, meiosis is a part of **spermatogenesis,** which occurs in the testes and produces sperm. In human females, meiosis is a part of **oogenesis,** which occurs in the ovaries and produces eggs (Fig. 5.15). Further description of the human reproductive system can be found in Chapter 21.

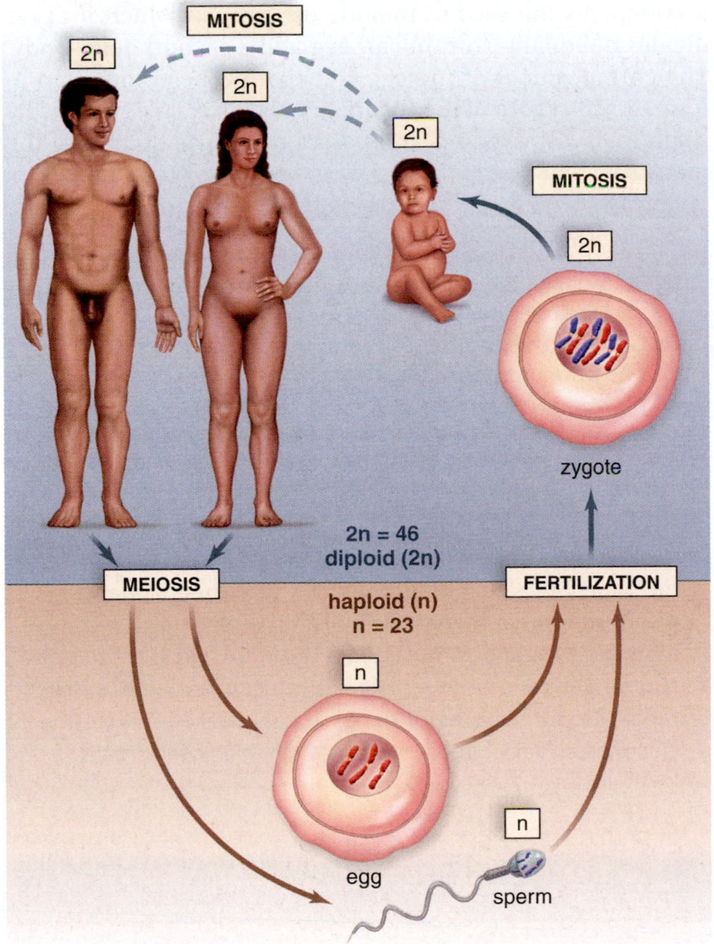
MITOSIS
2n
2n
2n
MITOSIS
2n
zygote
2n = 46
diploid (2n)
haploid (n)
n = 23
MEIOSIS
FERTILIZATION
n
egg
n
sperm

Figure 5.14 Life cycle of humans. Meiosis in males is a part of sperm production, and meiosis in females is a part of egg production. When a haploid sperm fertilizes a haploid egg, the zygote is diploid. The zygote undergoes mitosis as it develops into a newborn child. Mitosis continues throughout life during growth and repair.

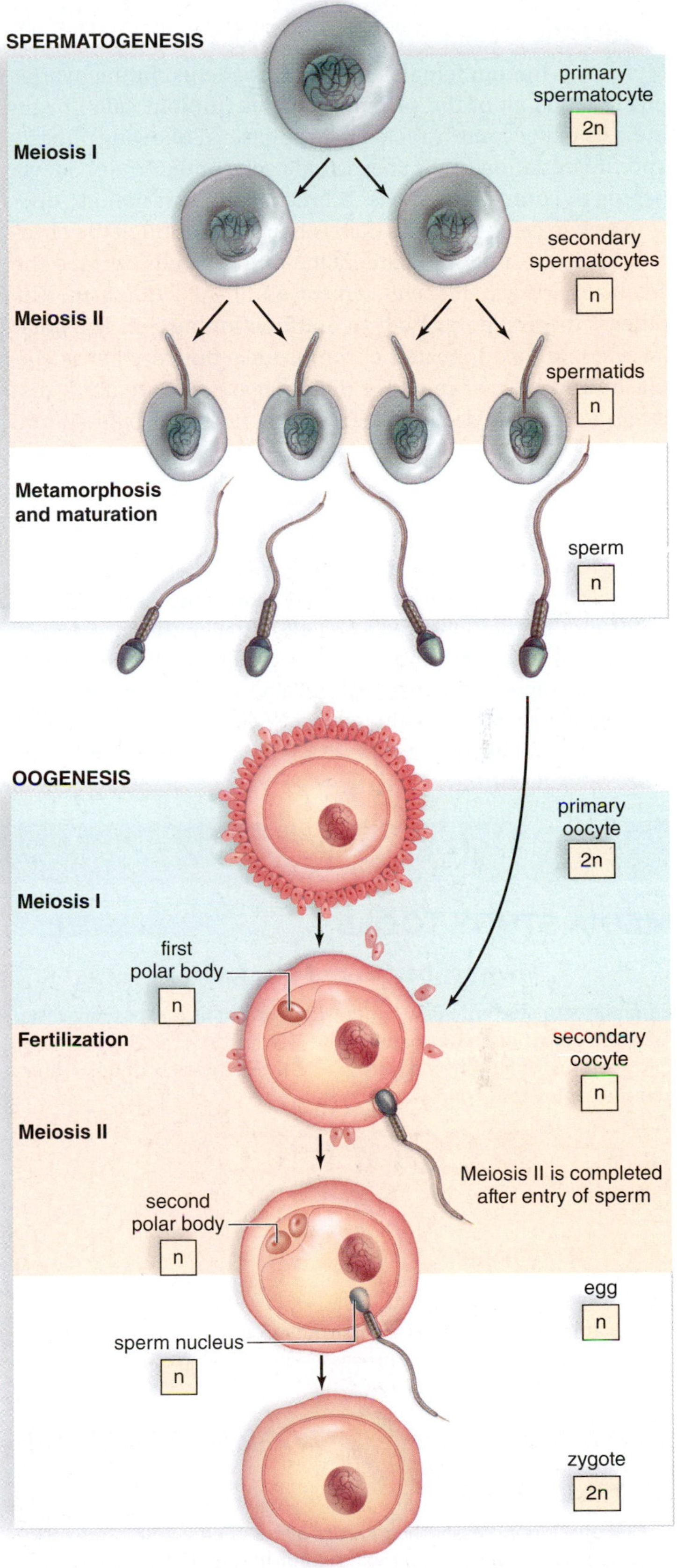
SPERMATOGENESIS
Meiosis I
primary spermatocyte
2n
secondary spermatocytes
n
Meiosis II
spermatids
n
Metamorphosis and maturation
sperm
n
OOGENESIS
primary oocyte
2n
Meiosis I
first polar body
n
Fertilization
secondary oocyte
n
Meiosis II
Meiosis II is completed after entry of sperm
second polar body
n
egg
n
sperm nucleus
n
zygote
2n

Figure 5.15 Spermatogenesis and oogenesis in mammals. Spermatogenesis produces four viable sperm, whereas oogenesis produces one egg and at least two polar bodies. In humans, both the sperm and the egg have 23 chromosomes each. Therefore, following fertilization, the zygote has 46 chromosomes.

Spermatogenesis

Following puberty, the time of life when the sex organs mature, spermatogenesis is continual in the testes of human males. As many as 300,000 sperm are produced per minute, or 400 million per day.

During spermatogenesis, primary spermatocytes, which are diploid, divide during the first meiotic division to form two *secondary spermatocytes,* which are haploid. Secondary spermatocytes divide during the second meiotic division to produce four *spermatids***,** which are also haploid. What is the difference between the chromosomes in haploid secondary spermatocytes and those in haploid spermatids? The chromosomes in secondary spermatocytes are duplicated and consist of two chromatids, whereas those in spermatids consist of only one chromatid. Spermatids then mature into **sperm** (spermatozoa).

Animation
Spermatogenesis

Oogenesis

Meiosis in human females begins in the fetus. In the ovaries of the fetus, all of the primary oocytes (diploid cells) begin meiosis but become arrested in prophase I. Following puberty and the initiation of the menstrual cycle, one primary oocyte begins to complete meiosis. It finishes the first meiotic division as two cells, each of which is haploid, although the chromosomes are still duplicated. One of these cells, termed the *secondary oocyte,* receives almost all of the cytoplasm. The other is the first **polar body,** a nonfunctioning cell. The polar body contains duplicated chromosomes but very little cytoplasm and may or may not divide again. Eventually it disintegrates. If the secondary oocyte is fertilized by a sperm, it completes the second meiotic division, in which it again divides unequally, forming an **egg** and a second polar body. The chromosomes of the egg and sperm nuclei then join to form the 2n **zygote.** If the secondary oocyte is not fertilized by a sperm, it disintegrates and passes out of the body with the menstrual flow.

Check Your Progress 5.6

1. Summarize the cells in the human body that are haploid and the role these cells play in the life cycle.
2. Contrast the number of sperm that are produced from one primary spermatocyte and the number of eggs produced from one oocyte.

Case Study Conclusion

Approximately 10–15% of the women who are diagnosed with breast cancer have a hereditary form of the disease. This means that they inherited a genetic mutation that increases their risk of developing cancer. A genetic mutation does not guarantee that they will develop cancer, nor does it determine when or where they may develop cancer, if they do. Many of these mutations occur in proto-oncogenes or tumor suppressor genes. In Christina Applegate's case, she inherited a mutated *BRCA1* gene. The *BRCA1* gene is a tumor suppressor gene whose protein product is involved in DNA repair. Tumor suppressor genes act as gatekeepers for the cell cycle and thus control the rate at which cells divide.

MEDIA STUDY TOOLS

www.mhhe.com/maderinquiry14

Enhance your study of this chapter with study tools and practice tests. Also ask your instructor about the resources available through ConnectPlus, including LearnSmart, the media-rich eBook, interactive learning tools, and animations.

Mc Graw Hill connect plus+ | BIOLOGY

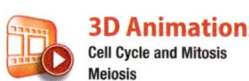

3D Animations
Cell Cycle and Mitosis
Meiosis

The 3D animations "Cell Cycle and Mitosis" and "Meiosis" provide an interactive exploration of the processes of cell division.

SUMMARIZE

5.1 The Cell Cycle

- Cell division increases the number of cells in the body, and **apoptosis** reduces this number when appropriate. Cells go through a **cell cycle** that includes (1) **interphase** and (2) cell division, consisting of **mitosis** and **cytokinesis.** Interphase, in turn, includes G_1 (growth as certain organelles double), S (DNA synthesis), and G_2 (growth as the cell prepares to divide). Cell division occurs during the mitotic stage (M) when daughter cells receive a full complement of chromosomes.

5.2 Control of the Cell Cycle and Cancer

- Internal and external signals control the cell cycle, which can stop at any of three checkpoints: in G_1 prior to the S stage, in G_2 prior to the M stage, and near the end of mitosis. DNA damage is one reason the cell cycle stops. The p53 protein is active at the G_1 checkpoint, and if DNA is damaged and can't be repaired, this protein initiates apoptosis.
- Both **proto-oncogenes** and **tumor suppressor genes** produce proteins that control the cell cycle. Proto-oncogenes encode proteins that promote the cell cycle and prevent apoptosis. *RAS* is an example of a proto-oncogene. Mutations in proto-oncogenes create **oncogenes,** which may cause cancer. When tumor suppressor genes mutate, their products no longer inhibit the cell cycle. *p53* and *BRCA1* are tumor suppressor genes.

5.3 Mitosis: Maintaining the Chromosome Number

- Each species has a characteristic number of chromosomes. The total number is the **diploid (2n)** number, and half this number is the **haploid (n)** number. Among eukaryotes, cell division involves nuclear division and division of the cytoplasm (cytokinesis).

- Replication of the **chromatids** precedes cell division. The duplicated chromosome is composed of two **sister chromatids** held together at a **centromere.** During mitosis, the centromeres divide, and **daughter chromosomes** go into each new nucleus.

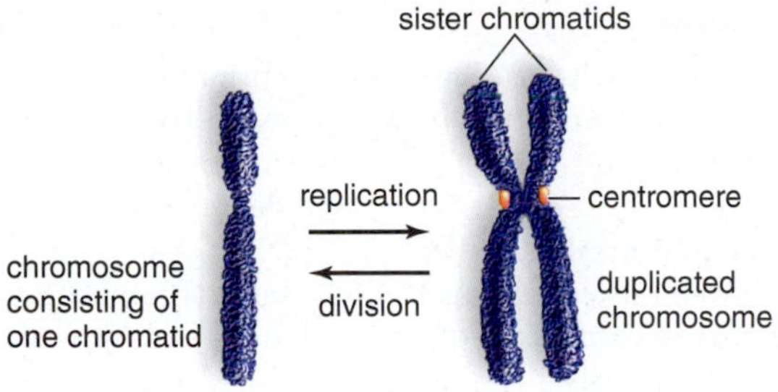
sister chromatids
replication
centromere
chromosome consisting of one chromatid
division
duplicated chromosome

- Mitosis has the following phases: **prophase,** which includes early prophase, when chromosomes have no particular arrangement, and late prophase, when the chromosomes are attached to **spindle** fibers; **metaphase,** when the chromosomes are aligned at the metaphase plate; **anaphase,** when the chromatids separate, becoming daughter chromosomes that move toward the poles; and **telophase,** when new nuclear envelopes form around the daughter chromosomes and cytokinesis is well under way.
- Cytokinesis differs between plant and animal cells. Plant cells form a **cell plate** between the daughter cells, while animal cells develop a **cleavage furrow** that separates the daughter cells.

5.4 Meiosis: Reducing the Chromosome Number

- **Meiosis** occurs in any life cycle that involves sexual reproduction. The end result of meiosis is daughter cells with the haploid number of **homologous chromosomes.** In some life cycles, the daughter cells become **gametes,** and upon **fertilization,** the offspring have the diploid number of chromosomes, the same as their parents. **Crossing-over** and **independent assortment** of chromosomes during meiosis I ensure **genetic variation** in daughter cells.
- Meiosis utilizes two nuclear divisions. During meiosis I, homologous chromosomes undergo **synapsis,** and crossing-over between nonsister chromatids occurs. When the homologous chromosomes separate during meiosis I, each daughter nucleus receives one member from each pair of chromosomes. Therefore, the daughter cells are haploid. Interkinesis during meiosis I and meiosis II does not include DNA replication. Distribution of daughter chromosomes derived from sister chromatids during meiosis II then leads to a total of four new cells, each with the haploid number of chromosomes.

5.5 Comparison of Meiosis with Mitosis

- While mitosis and meiosis differ in the number of cells produced, and the chromosome complement of each cell, there are many similarities in the processes. The primary difference is the behavior of the homologous chromosomes.

5.6 The Human Life Cycle

- The human life cycle involves both mitosis and meiosis. Mitosis ensures that each somatic cell has the diploid number of chromosomes. In humans, meiosis is a part of **spermatogenesis** and **oogenesis.** Spermatogenesis in males produces four viable **sperm,** whereas oogenesis in females produces one **egg** and **polar bodies.** Oogenesis does not go on to completion unless a sperm fertilizes the secondary oocyte. Fertilization produces a **zygote,** which grows by mitosis.

ASSESS

Testing Yourself

Choose the best answer for each question.

1. DNA synthesis occurs during
   a. interphase. **b.** prophase. **c.** metaphase. **d.** telophase.

In questions 2–4, match the part of the diagram of the cell cycle to the statements provided.

2. DNA replicates.
3. Organelles double.
4. Cytokinesis occurs.

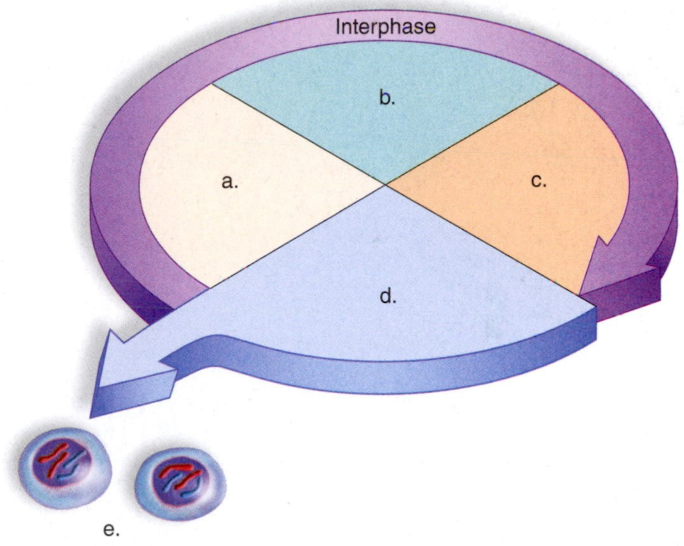
Interphase
b.
a.
c.
d.
e.

5. Which is not one of the stages of mitosis?
   a. anaphase **b.** telophase **c.** interphase **d.** metaphase
6. Where is the checkpoint for DNA damage?
   a. G_1 **b.** S **c.** G_2 **d.** M

In questions 7–11, match the part of the diagram that follows to the appropriate statement.

7. Meiosis II
8. Primary spermatocyte
9. Sperm
10. Secondary spermatocyte
11. Spermatids

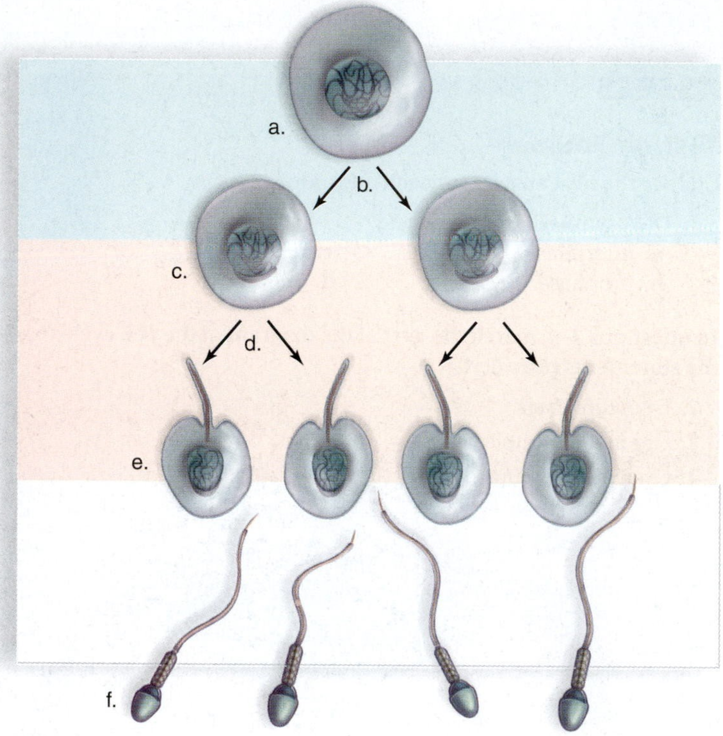
a.
b.
c.
d.
e.
f.

12. Programmed cell death is called
- **a.** mitosis.
- **b.** meiosis.
- **c.** cytokinesis.
- **d.** apoptosis.

13. Which of the following provide genetic diversity?
- **a.** independent assortment
- **b.** crossing-over
- **c.** genetic recombination
- **d.** All of these are correct.

14. If a parental cell has 14 chromosomes prior to mitosis, how many chromosomes will the daughter cells have?
- **a.** 28
- **b.** 14
- **c.** 7
- **d.** 196

15. Crossing-over occurs
- **a.** between sister chromatids of the same chromosome.
- **b.** between chromatids of homologous chromosomes.
- **c.** during mitosis.
- **d.** during meiosis II.

ENGAGE

Virtual Lab
Cell Reproduction

The virtual lab "Cell Reproduction" provides an interactive examination of the cell cycle and provides for a comparison of normal cells and cancer cells.

Thinking Critically

1. Why is crossing-over much more likely to occur during meiosis rather than mitosis?
2. How would a nonfunctional cell cycle checkpoint lead to carcinogenesis?
3. From an evolutionary standpoint, why is sexual reproduction advantageous for the continuation of a species?

Metabolism: Energy and Enzymes 6

CASE STUDY Vampire Bats

Vampire bats were inspiration for the many legends about vampires. Vampire bats do exist, but they are not as frightening as legends would have you believe. Vampire bats are not large animals and they do not normally attack humans. The body of the common vampire bat, *Desmodus rotundus*, is about the size of a human thumb with a wingspan of only eight inches. These bats normally feed on cattle, horses, and pigs. Also, rather than "suck" the blood of their animal host the bats make small cuts in the skin of the animal and lap up the blood that flows from the injury. Scientists have long known that vampire bat saliva has an amazing ability to dissolve blood clots, allowing the blood to continue to flow from the wound while the bat feeds. Normally, blood clots due to the action of fibrin, an insoluble protein in the blood plasma. Fibrin is dissolved by the enzyme plasmin, which circulates in the blood in an inactive form called plasminogen. Another enzyme activates plasminogen, which is converted to plasmin, and then dissolves the fibrin of the clot. Researchers have discovered an enzyme in the saliva of the vampire bat that is 150 times more potent at activating plasminogen—and thus dissolving clots—than any known drug. This enzyme may one day be used to treat victims of stroke, caused when a clot blocks blood supply to the brain.

This chapter describes the general characteristics and functions of enzymes, and how enzymes function in the flow of energy and metabolism. Each enzyme in our body is responsible for a unique metabolic reaction, as illustrated by the vampire bat enzyme, which is responsible for converting plasminogen into plasmin.

As you read through the chapter, think about the following questions:

1. What is the role of an enzyme in a metabolic pathway?
2. What environmental factors influence the rate of enzyme activity?

CHAPTER OUTLINE

BEFORE YOU BEGIN

Before beginning this chapter, take a few moments to review the following discussions:

Section 1.1 What is the relationship between energy and the basic characteristics of life?

Section 2.7 How does the shape of a protein relate to its function?

Section 4.2 In what ways does energy play a role in the movement of molecules across a plasma membrane?

6.1 Life and the Flow of Energy

Learning Outcomes

Upon completion of this section, you should be able to

1. Describe how and why energy flows through an ecosystem.
2. Summarize the two laws of thermodynamics and how these laws apply to cells.

Energy is the ability to do work or bring about a change. In order to maintain their organization and carry out metabolic activities, cells as well as organisms need a constant supply of energy. This energy allows living things to carry on the processes of life, including growth, development, locomotion, metabolism, and reproduction.

The majority of organisms get their energy from organic nutrients produced by photosynthesizers (algae, plants, and some bacteria). Therefore, life on Earth is ultimately dependent on solar energy.

Forms of Energy

Energy occurs in two forms: kinetic and potential. **Kinetic energy** is the energy of motion, as when a ball rolls down a hill or a moose walks through grass. **Potential energy** is stored energy—its capacity to accomplish work is not being used at the moment. The food we eat has potential energy because it can be converted into various types of kinetic energy. Food is specifically called **chemical energy** because it contains energy in the chemical bonds of organic molecules. When a moose walks, it has converted chemical energy into a type of kinetic energy called **mechanical energy** (Fig. 6.1).

Animation Energy Conversion

Two Laws of Thermodynamics

Figure 6.1 illustrates the flow of energy in a terrestrial ecosystem. Through photosynthesis, plants capture a small portion of the incoming solar energy. This is stored in the chemical bonds of organic molecules, such as carbohydrates. When plants use their carbohydrates (through cellular respiration), some of the energy dissipates as heat. In an ecosystem, plants serve as the food source for other organisms, such as moose. The cellular respiration pathways in the moose use these carbohydrates as energy, once again releasing energy as heat. All living things metabolize organic nutrient molecules, and eventually, all the captured solar energy dissipates as heat. Therefore, energy flows through an ecosystem. It does not recycle. Two laws of thermodynamics explain why energy flows in ecosystems and in cells.

> The first law of thermodynamics—the law of conservation of energy—states that energy cannot be created or destroyed, but it can be changed from one form to another.

During photosynthesis, plant cells use solar energy to covert energy-poor molecules such as carbon dioxide and water, to energy-rich molecules such as carbohydrates. As we have seen, not all of the captured solar energy is used to form carbohydrates; some becomes heat:

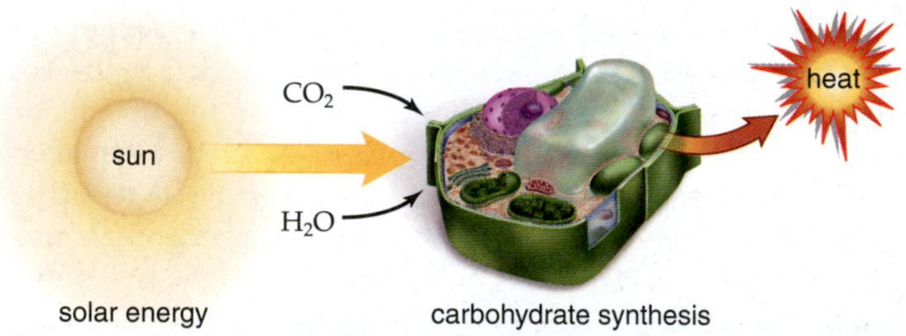

sun
CO_2
H_2O
heat
solar energy
carbohydrate synthesis

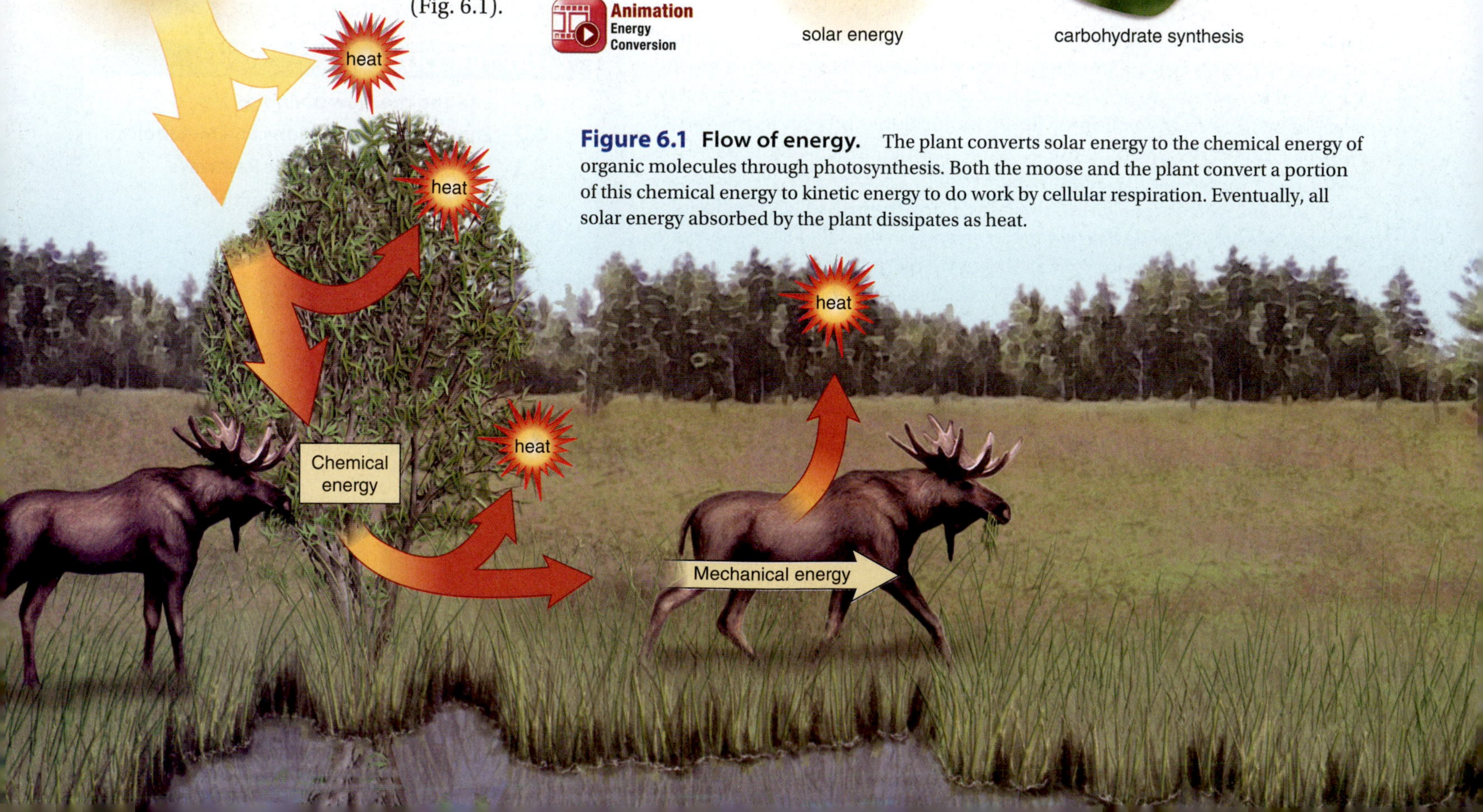

Solar energy
heat
heat
heat
heat
Chemical energy
Mechanical energy

Figure 6.1 Flow of energy. The plant converts solar energy to the chemical energy of organic molecules through photosynthesis. Both the moose and the plant convert a portion of this chemical energy to kinetic energy to do work by cellular respiration. Eventually, all solar energy absorbed by the plant dissipates as heat.

Obviously, plant cells do not create the energy they use to produce carbohydrate molecules. That energy comes from the sun. Is any energy destroyed? No, because heat is also a form of energy. Similarly, a moose uses the energy derived from carbohydrates to power its muscles. And as its cells use this energy, none is destroyed, but some becomes heat, which dissipates into the environment:

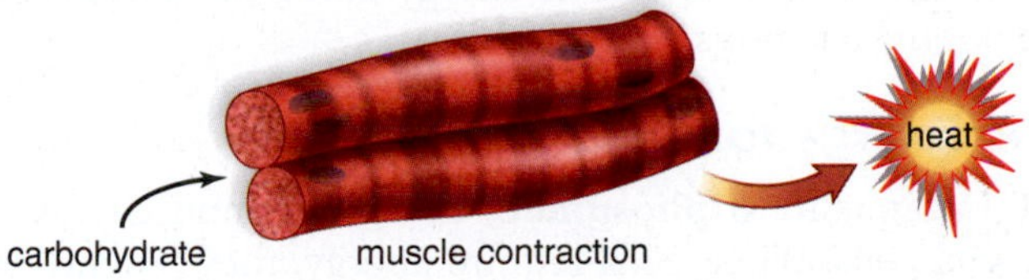
heat
carbohydrate
muscle contraction

The second law of thermodynamics, therefore, applies to living systems.

> The second law of thermodynamics states that energy cannot be changed from one form to another without a loss of usable energy.

The heat released during photosynthesis (from the plant) and cellular respiration (from both the plant and moose) dissipates into the environment. This energy is no longer usable—that is, it is not available to do work. With transformation upon transformation, eventually all usable forms of energy become heat that is lost to the environment. Heat that dissipates into the environment cannot be captured and converted to one of the other forms of energy.

As a result of the second law of thermodynamics, no process requiring a conversion of energy is ever 100% efficient. Much of the energy is lost in the form of heat. In automobiles, the gasoline engine is between 20% and 30% efficient in converting chemical energy into mechanical energy. The majority of energy is obviously lost as heat. Cells are capable of about 40% efficiency, with the remaining energy given off to their surroundings as heat.

MP3 Laws of Thermodynamics

Cells and Entropy

The second law of thermodynamics can be stated another way: Every energy transformation makes the universe less organized and more disordered. The term **entropy** is used to indicate the relative amount of disorganization. Because the processes that occur in cells are energy transformations, the second law means that every process that occurs in cells always does so in a way that increases the total entropy of the universe. Then, too, any one of these processes makes less energy available to do useful work in the future.

Figure 6.2 illustrates the second law of thermodynamics. The second law of thermodynamics tells us that glucose tends to break apart into carbon dioxide and water (Fig. 6.2*a*). Why? Because glucose is more organized, and therefore less stable, than its breakdown products. Also, hydrogen ions on one side of a membrane tend to move to the other side unless they are prevented from doing so (Fig. 6.2*b*). Why? Because when they are distributed randomly, entropy has increased. As an analogy, you know from experience that a neat room (Fig. 6.2*c*) is more

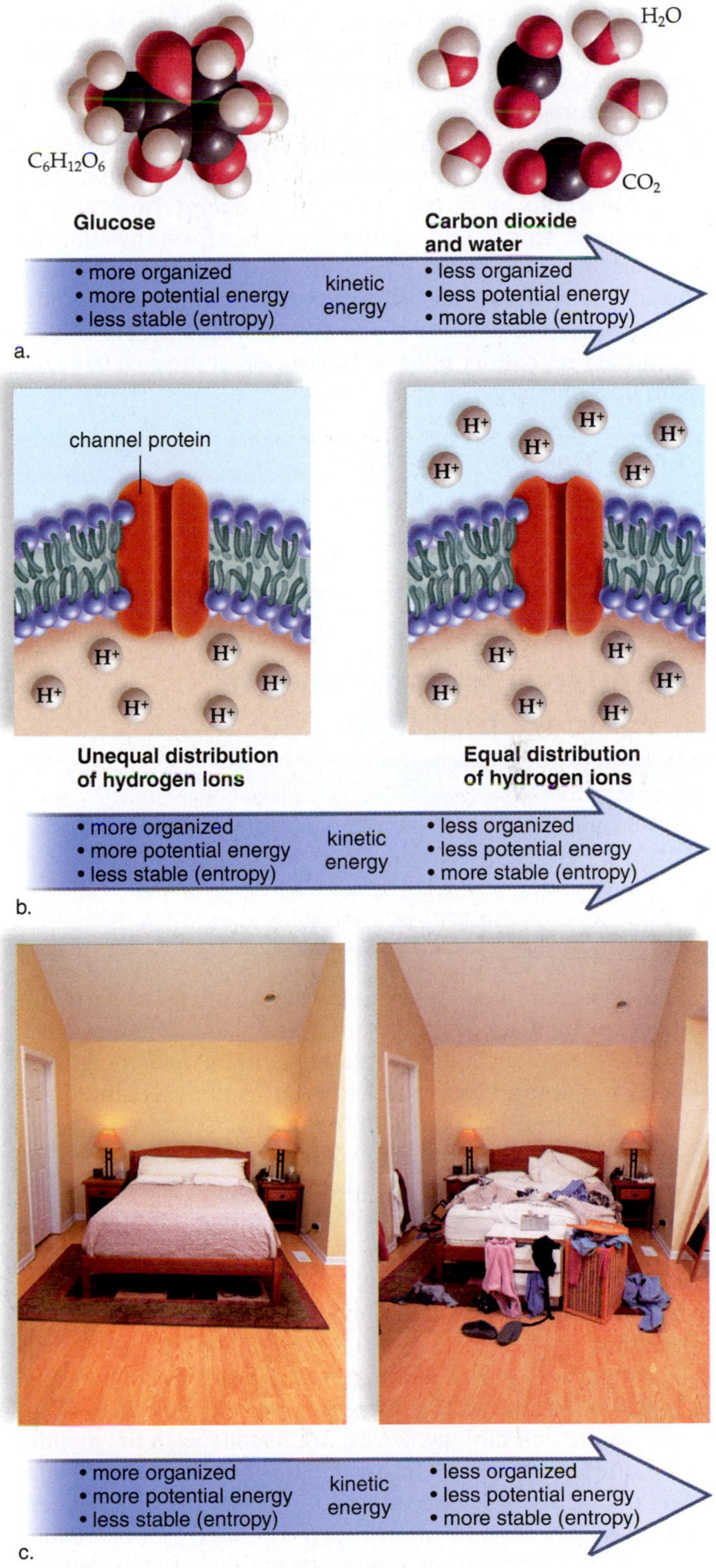
$C_6H_{12}O_6$
H_2O
CO_2
Glucose
Carbon dioxide and water
• more organized
• more potential energy
• less stable (entropy)
kinetic energy
• less organized
• less potential energy
• more stable (entropy)
a.
channel protein
H^+
Unequal distribution of hydrogen ions
Equal distribution of hydrogen ions
• more organized
• more potential energy
• less stable (entropy)
kinetic energy
• less organized
• less potential energy
• more stable (entropy)
b.
• more organized
• more potential energy
• less stable (entropy)
kinetic energy
• less organized
• less potential energy
• more stable (entropy)
c.

Figure 6.2 Cells and entropy. The second law of thermodynamics tells us that **(a)** glucose, which is more organized, tends to break down to carbon dioxide and water, which are less organized. **b.** Similarly, hydrogen ions (H^+) on one side of a membrane tend to move to the other side so that the ions are randomly distributed. **c.** Entropy also affects nonliving structures, such as this room. All of these are examples of a loss of potential energy and an increase in entropy.

organized but less stable than a messy room , which is disorganized but more stable. How do you know a neat room is less stable than a messy room? Consider that a neat room always tends to become more messy.

Cellular processes, such as the synthesis of glucose and ion transport across a membrane, are possible because cells have the ability to obtain an input of energy from an outside source. This energy ultimately comes from the sun. Life depends on a constant supply of energy from the sun because the ultimate fate of all solar energy in the biosphere is to become randomized in the universe as heat. A living cell is a temporary repository of order purchased at the cost of a constant flow of energy.

Check Your Progress 6.1

1. Explain why energy is not recycled in an ecosystem.
2. Summarize how the two laws of thermodynamics are involved as you consume your daily meals.

6.2 Energy Transformations and Metabolism

Learning Outcomes

Upon completion of this section, you should be able to

1. Distinguish between anabolic, catabolic, endergonic, and exergonic reactions.
2. Summarize the ATP cycle and the role of ATP in the cell.

Cellular **metabolism** is the sum of all the chemical reactions that occur in a cell. A significant part of cellular metabolism involves the breaking down and the building up of molecules. The term **catabolism** is used to refer to the breaking down of molecules, and the term **anabolism** is used to refer to the building up (synthesis) of molecules. In a chemical reaction, **reactants** are substances that participate in a reaction, while **products** are substances that form as a result of a reaction. For example, in the reaction $A + B \longrightarrow C + D$, A and B are the reactants while C and D are the products. How do you know whether this reaction will proceed in the indicated direction?

Using the concept of entropy, it is possible to state that a reaction will occur if it increases the entropy of the universe. But in cell biology, we do not usually wish to consider the entire universe; we simply want to consider a particular reaction. In such instances, cell biologists use the concept of free energy instead of entropy. **Free energy** (also called ΔG) is the amount of energy available—that is, energy that is still "free" to do work—after a chemical reaction has occurred. The change in free energy after a reaction occurs is calculated by subtracting the free energy content of the reactants from that of the products. A negative result (negative ΔG) means that the products have less free energy than the reactants, and the reaction will go forward. In our reaction, if C and D have less free energy than A and B, the reaction will occur.

Exergonic reactions are spontaneous and release energy, while **endergonic reactions** require an input of energy to occur. In the body, many reactions, such as protein synthesis, nerve impulse conduction, or muscle contraction, are endergonic, and they are driven by the energy released by exergonic reactions. ATP is a carrier of energy between exergonic and endergonic reactions.

ATP: Energy for Cells

ATP (**adenosine triphosphate**) is the common energy currency of cells. When cells require energy, they "spend" ATP. The more active the organism, the greater the demand for ATP. However, the amount on hand at any one moment is minimal because ATP is constantly being generated from **ADP** (**adenosine diphosphate**) and a molecule of inorganic phosphate Ⓟ (Fig. 6.3). A cell is assured of a supply of ATP,

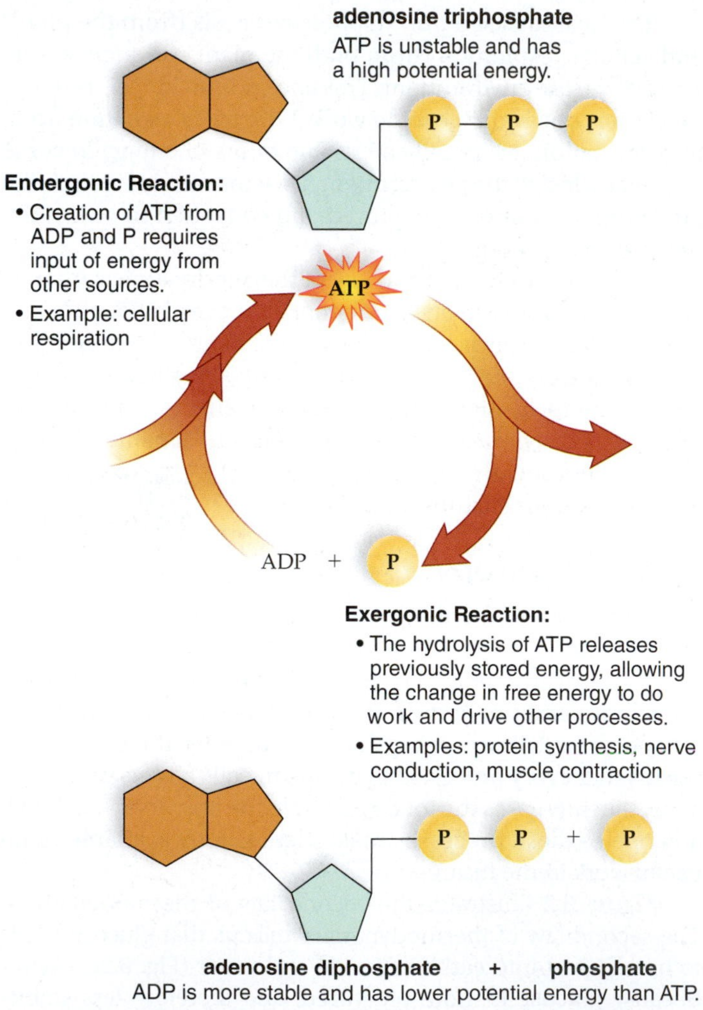
adenosine triphosphate
ATP is unstable and has a high potential energy.
P
P
P
Endergonic Reaction:
• Creation of ATP from ADP and P requires input of energy from other sources.
• Example: cellular respiration
ATP
ADP + P
Exergonic Reaction:
• The hydrolysis of ATP releases previously stored energy, allowing the change in free energy to do work and drive other processes.
• Examples: protein synthesis, nerve conduction, muscle contraction
P
P
+
P
adenosine diphosphate + phosphate
ADP is more stable and has lower potential energy than ATP.

Figure 6.3 The ATP cycle. In cells, ATP carries energy between exergonic reactions and endergonic reactions. When a phosphate group is removed by hydrolysis, ATP releases the appropriate amount of energy for most metabolic reactions.

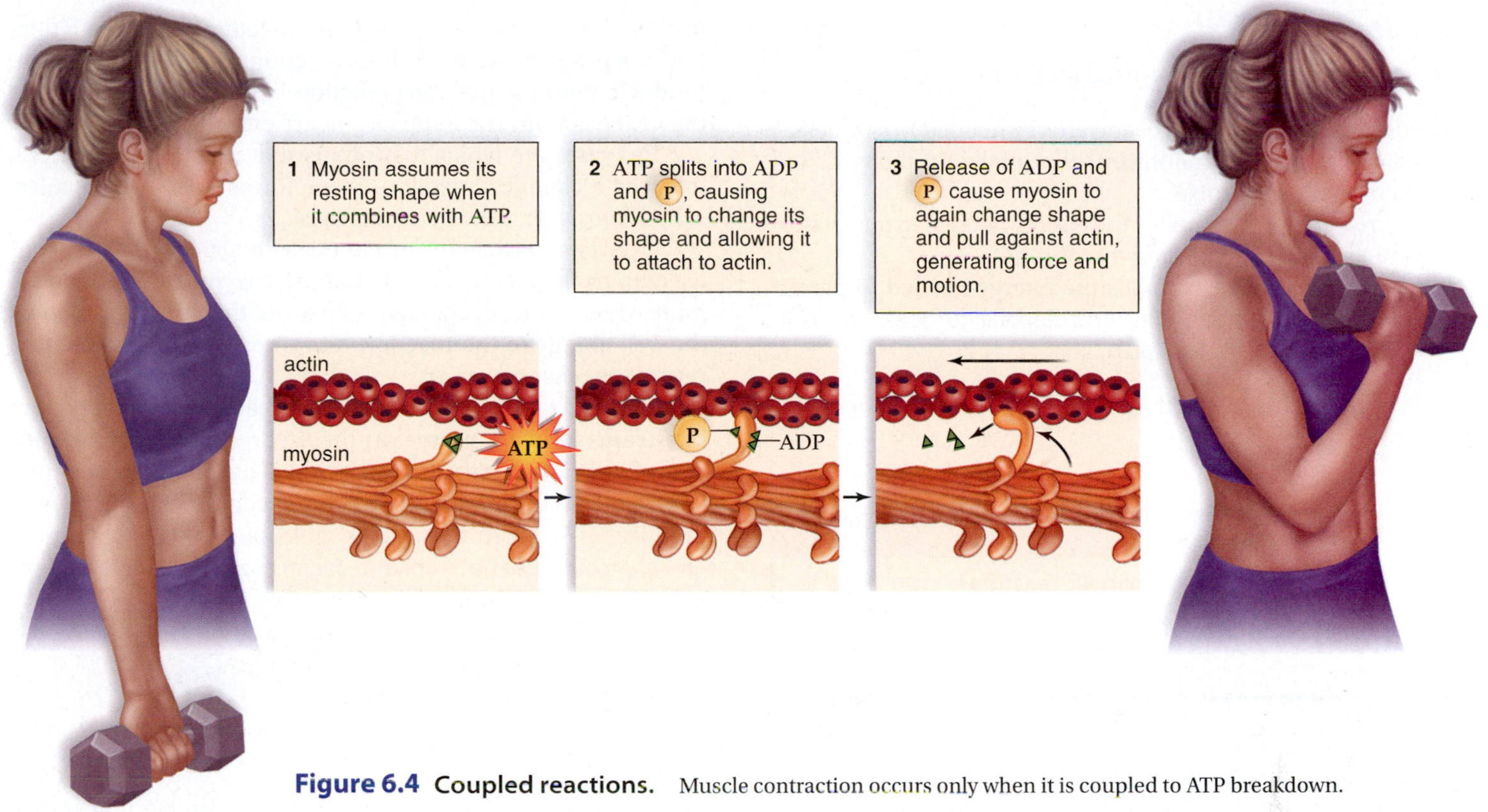
1 Myosin assumes its resting shape when it combines with ATP.
2 ATP splits into ADP and P, causing myosin to change its shape and allowing it to attach to actin.
3 Release of ADP and P cause myosin to again change shape and pull against actin, generating force and motion.
actin
myosin
ATP
P
ADP

Figure 6.4 Coupled reactions. Muscle contraction occurs only when it is coupled to ATP breakdown.

because glucose breakdown during cellular respiration provides the energy for the buildup of ATP in mitochondria. Only 39% of the free energy of glucose is transformed to ATP; the rest is lost as heat.

There are many biological advantages to the use of ATP as an energy carrier in living systems. ATP is a common and universal energy currency because it can be used in many different types of reactions. Also, when ATP is converted to energy, ADP, and Ⓟ, the amount of energy released is sufficient for a particular biological function, and little energy is wasted. In addition, ATP breakdown can be coupled to endergonic reactions in such a way that it minimizes energy loss.

Structure of ATP

ATP is a nucleotide composed of the nitrogen-containing base adenine and the 5-carbon sugar ribose (together called adenosine) and three phosphate groups (see Fig. 6.3). ATP is called a "high-energy" compound because of the energy stored in the chemical bonds of the phosphates. Under cellular conditions, the amount of energy released when ATP is hydrolyzed to ADP + Ⓟ is about 7.3 kcal per mole.[1]

MP3 ATP

Coupled Reactions

In **coupled reactions,** the energy released by an exergonic reaction is used to drive an endergonic reaction. ATP breakdown is often coupled to cellular reactions that require an input of energy. Coupling, which requires that the exergonic reaction and the endergonic reaction be closely tied, can be symbolized like this:

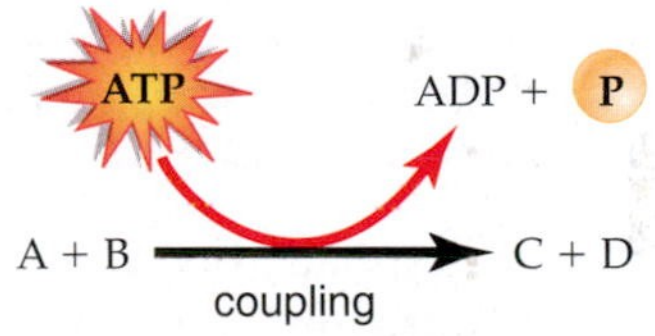
ATP
ADP + P
A + B
C + D
coupling

Notice that the word *energy* does not appear following ATP breakdown. Why not? Because this energy was used to drive forward the coupled reaction. Figure 6.4 shows that ATP breakdown provides the energy necessary for muscular contraction to occur. During muscle contraction, myosin filaments pull actin filaments to the center of the cell, and the muscle shortens. ① A myosin head combines with ATP (three connected green triangles) and takes on its resting shape. ② ATP breaks down to ADP (two connected green triangles) plus Ⓟ (one green triangle). A resulting change in shape allows myosin to attach to actin. ③ The release of ADP and from the myosin head causes it to change its shape again and pull on the actin filament. The cycle begins again ① when the myosin head combines with ATP and takes on its resting shape. During this cycle, chemical energy has been transformed to mechanical energy, heat has been released, and entropy has increased.

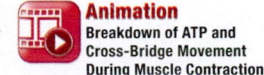

Animation Breakdown of ATP and Cross-Bridge Movement During Muscle Contraction

[1] A mole is the number of molecules present in the molecular weight of a substance (in grams).

Function of ATP

In living systems, ATP can be used for the following:

Chemical work. ATP supplies the energy needed to synthesize macromolecules (anabolism) that make up the cell, and therefore the organism.
Transport work. ATP supplies the energy needed to pump substances across the plasma membrane.
Mechanical work. ATP supplies the energy needed to permit muscles to contract, cilia and flagella to beat, chromosomes to move, and so forth.

In most cases, ATP is the immediate source of energy for these processes.

Check Your Progress 6.2

1. Determine whether an anabolic reaction is more likely to be exergonic or endergonic.
2. Explain how the ATP pathway is similar to a rechargeable battery.

6.3 Enzymes and Metabolic Pathways

Learning Outcomes

Upon completion of this section, you should be able to

1. Explain the purpose of a metabolic pathway and how enzymes help regulate it.
2. Describe how an enzyme binds its substrate.
3. Explain how an enzyme increases the rate of a reaction, and list factors that affect enzyme speed.

Reactions do not occur haphazardly in cells. They are usually part of a **metabolic pathway,** a series of linked reactions. Metabolic pathways begin with a particular reactant and terminate with an end product. Some metabolic pathways are cyclical, regenerating the starting material. While it is possible to write an overall equation for a pathway as if the beginning reactant went to the end product in one step, actually many specific steps occur in between. In the pathway, one reaction leads to the next reaction, which leads to the next reaction, and so forth in an organized, highly structured manner. This arrangement makes it possible for one pathway to lead to several others, because various pathways have several molecules in common. Also, metabolic energy is captured and utilized more easily if it is released in small increments rather than all at once.

A metabolic pathway can be represented by the following diagram:

$$A \xrightarrow{E_1} B \xrightarrow{E_2} C \xrightarrow{E_3} D \xrightarrow{E_4} E \xrightarrow{E_5} F \xrightarrow{E_6} G$$

In this diagram, the letters A–F are reactants, and the letters B–G are products in the various reactions. In other words, the products from the previous reaction become the reactants of the next reaction. The letters E_1–E_6 are enzymes.

Enzymes are typically proteins that function as catalysts to speed a chemical reaction. Some forms of RNA molecules, called *ribozymes,* can act as catalysts.

Catalysts participate in chemical reactions, but are not used up by the reaction. Note that the enzyme does not determine whether the reaction goes forward; that is determined by the free energy of the reaction. Enzymes simply increase the rate of the reaction.

The reactants in an enzymatic reaction are called the **substrates** for that enzyme. In the first reaction, A is the substrate for E_1, and B is the product. Now B becomes the substrate for E_2, and C is the product. This process continues until the final product (G) forms. Any one of the molecules (A–G) in this linear pathway could also be a substrate for an enzyme in another pathway. The presence or absence of an active enzyme determines which reaction takes place. A diagram showing all the possibilities would be highly branched.

Animation Biochemical Pathways

MP3 Enzymes

Energy of Activation

Molecules frequently do not react with one another unless they are activated in some way. In the lab, for example, in the absence of an enzyme, activation is very often achieved by heating a reaction flask to increase the number of effective collisions between molecules. The energy that must be added to cause molecules to react with one another is called the **energy of activation (E_a)** (Figure 6.5). Even though the reaction will proceed (ΔG is negative),

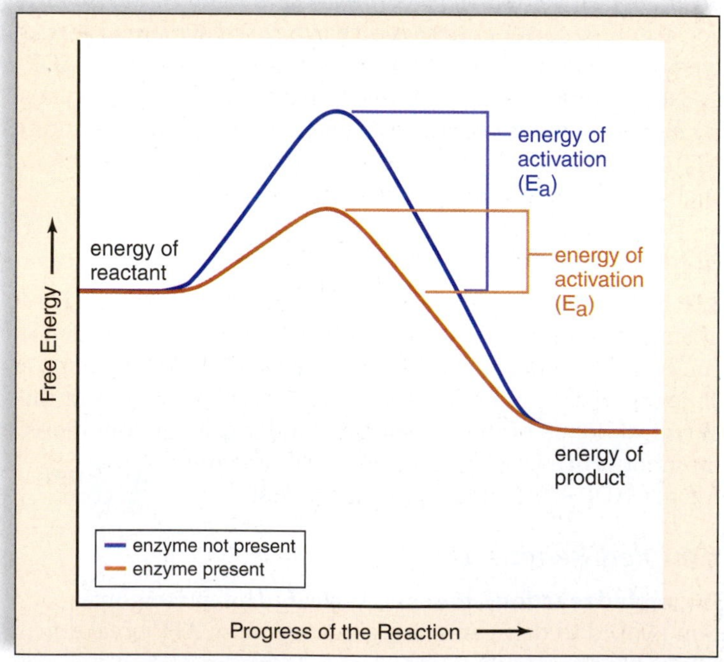
energy of activation (Ea)
energy of activation (Ea)
energy of reactant
energy of product
Free Energy
enzyme not present
enzyme present
Progress of the Reaction

Figure 6.5 Energy of activation (E_a). Enzymes speed the rate of reactions because they lower the amount of energy required for the reactants to react.

the energy of activation must be overcome. The burning of firewood is a very exergonic reaction, but firewood in a pile does not spontaneously combust. The input of some energy, perhaps a lit match, is required to overcome the energy of activation.

Animation Energy of Activation

Figure 6.5 shows E_a when an enzyme is not present compared to when an enzyme is present, illustrating that enzymes lower the amount of energy required for activation to occur. Nevertheless, the addition of the enzyme does not change the end result of the reaction. Notice that the energy of the products is less than the energy of the reactants. This results in a negative ΔG, so the reaction will proceed. But the reaction will not go at all unless the energy of activation is overcome. Without the enzyme, the reaction rate will be very slow. By lowering the energy of activation, the enzyme increases the rate of the reaction.

Sugar in your kitchen cupboard will break down to carbon dioxide and water because the energy of the products (carbon dioxide and water) is much less than the free energy of the reactant (sugar). However, the rate of this reaction is so slow that you never see it. If you eat the sugar, the enzymes in your digestive system greatly increase the speed at which the sugar is broken down. However, the end result (carbon dioxide and water) is still the same.

How Enzymes Function

The following equation, which is illustrated in Figure 6.6, is often used to indicate that an enzyme forms a complex with its substrate:

S	+	E	⟶	ES	⟶	E	+	P
substrate		enzyme		enzyme-substrate complex				product

In most instances, only one small part of the enzyme, called the **active site,** complexes with the substrate(s). It is here that the enzyme and substrate fit together, seemingly like a key fits a lock. However, cell biologists now know that the active site undergoes a slight change in shape in order to accommodate the substrate(s). This is called the **induced fit model** because the enzyme is induced to undergo a slight alteration to achieve optimum fit (Fig. 6.7).

The change in shape of the active site facilitates the reaction that now occurs. After the reaction has been completed, the product(s) is released, and the active site returns to its original state, ready to bind to another substrate molecule. Only a small amount of enzyme is actually needed in a cell because enzymes are not used up by the reaction.

Some enzymes do more than simply complex with their substrate(s); they participate in the reaction. For example, trypsin digests protein by breaking peptide bonds. The active site of trypsin contains three amino acids with *R* groups that actually interact with members of the peptide bond—first to break the bond and then to introduce the components of water. This illustrates that the formation of the enzyme-substrate complex is very important in speeding up the reaction.

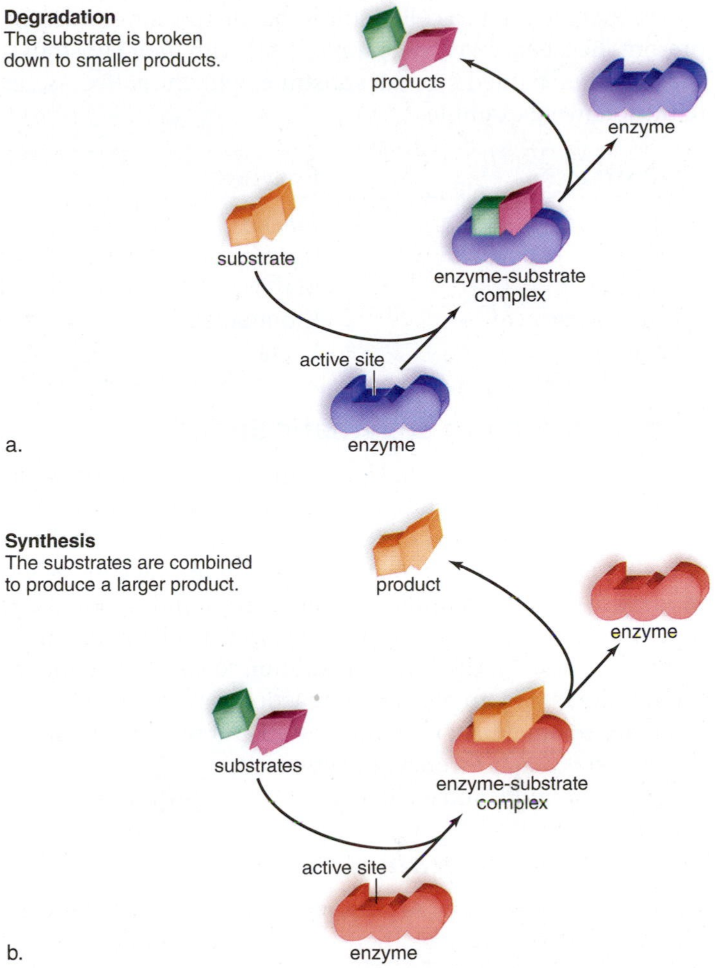
Degradation
The substrate is broken down to smaller products.
products
enzyme
substrate
enzyme-substrate complex
active site
enzyme
a.
Synthesis
The substrates are combined to produce a larger product.
product
enzyme
substrates
enzyme-substrate complex
active site
enzyme
b.

Figure 6.6 Enzymatic action. An enzyme has an active site where the substrates and enzyme fit together in such a way that the substrates react. Following the reaction, the products are released, and the enzyme is free to act again. **a.** The enzymatic reaction can result in the degradation of a substrate into multiple products (catabolism) or, **b.** the synthesis of a product from multiple substrates (anabolism).

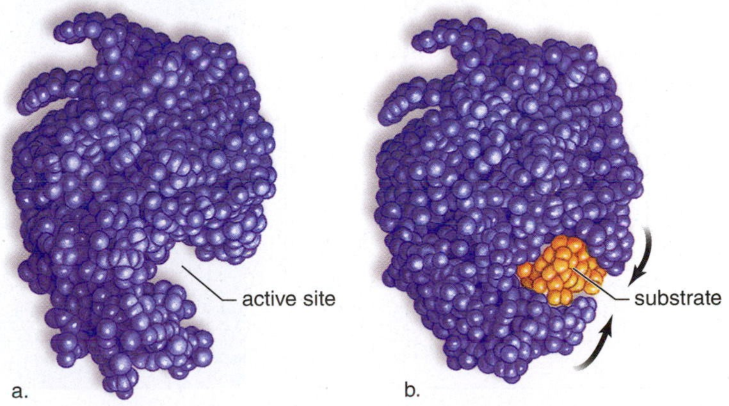
active site
substrate
a.
b.

Figure 6.7 Induced fit model. These computer-generated images show an enzyme called lysozyme that hydrolyzes its substrate, a polysaccharide that makes up bacterial cell walls. **a.** Shape of enzyme when no substrate is bound to it. **b.** After the substrate binds, the shape of the enzyme changes so that hydrolysis can better proceed.

Every reaction in a cell requires that its specific enzyme be present. Because enzymes complex only with their substrates, they are often named for their substrates with the suffix *-ase,* as in the following examples:

Substrate	Enzyme
Lipid	Lipase
Urea	Urease
Maltose	Maltase
Ribonucleic acid	Ribonuclease
Lactose	Lactase

Factors Affecting Enzymatic Speed

Enzymatic reactions proceed quite rapidly. Consider, for example, the breakdown of hydrogen peroxide (H_2O_2) as catalyzed by the enzyme catalase: $2\ H_2O_2 \longrightarrow 2\ H_2O + O_2$. The breakdown of hydrogen peroxide can occur 600,000 times a second when catalase is present! In order to achieve maximum product per unit time, there should be enough substrate to fill the enzyme's active sites most of the time. In addition to substrate concentration, the rate of an enzymatic reaction can also be affected by environmental factors (temperature and pH) or by cellular mechanisms, such as enzyme activation, enzyme inhibition, and cofactors.

Animation How Enzymes Work

Substrate Concentration

Generally, enzyme activity increases as substrate concentration increases because there are more collisions between substrate molecules and the enzyme. As more substrate molecules fill active sites, more product results per unit time. But when the enzyme's active sites are filled almost continuously with substrate, the enzyme's rate of activity cannot increase any more. The maximum rate has been reached.

Temperature and pH

As the temperature rises, enzyme activity increases (Fig. 6.8). This occurs because higher temperatures cause more effective collisions between enzyme and substrate. However, if the temperature rises beyond a certain point, enzyme activity eventually levels out and then declines rapidly because the enzyme is **denatured.** An enzyme's shape changes during denaturation, due to the loss of secondary and tertiary structure, and then it can no longer bind its substrate(s) efficiently. As the temperature decreases, enzyme activity decreases.

Each enzyme also has a preferred pH at which the rate of the reaction is highest. Figure 6.9 shows the preferred pH for the enzymes pepsin and trypsin. At this pH value, these enzymes have their normal configurations. The globular structure of an enzyme is dependent on interactions, such as hydrogen bonding, between *R* groups (see Fig. 2.23). A change in pH can alter the ionization of *R* groups and disrupt normal interactions, and under extreme pH conditions, denaturation eventually occurs. If the enzyme's shape is altered, it is then unable to combine efficiently with its substrate.

Enzyme Activation

Not all enzymes are needed by the cell all the time. Genes can be turned on to increase the concentration of an enzyme in a cell or turned off to decrease the concentration. But enzymes can also be present in the cell in an inactive form. Activation of enzymes occurs in many different ways. Some enzymes are covalently modified by the addition or removal of phosphate groups. An enzyme called a kinase adds phosphates to proteins,

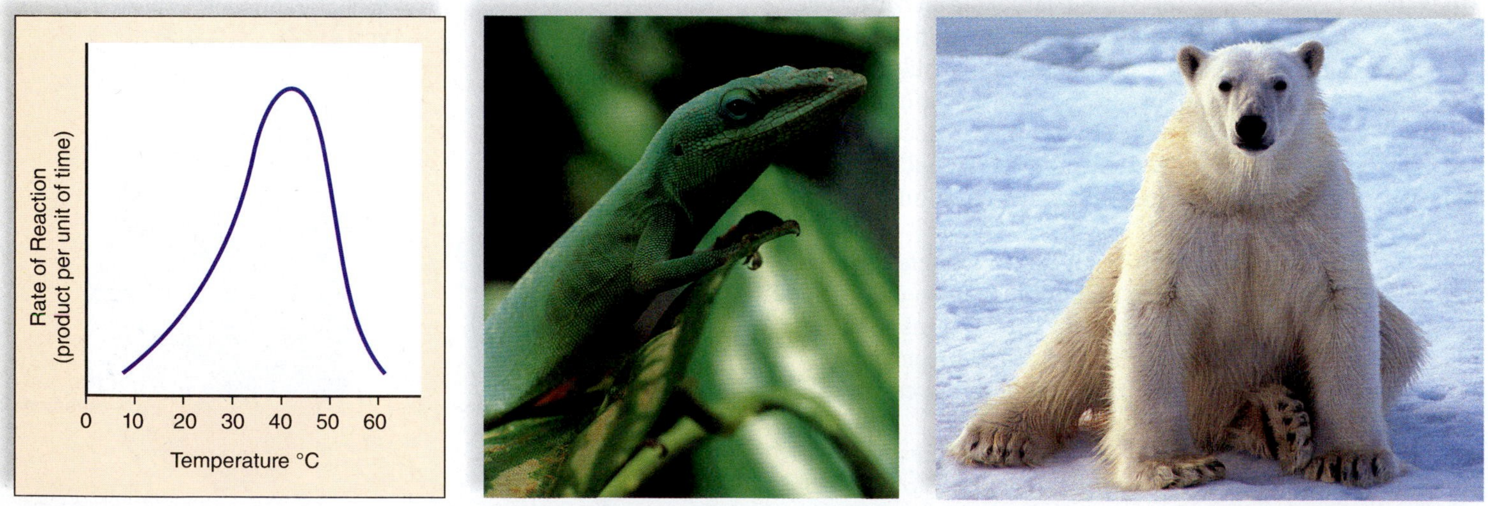
Rate of Reaction (product per unit of time)
0 10 20 30 40 50 60
Temperature °C

a. Rate of reaction as a function of temperature.

b. Body temperature of ectothermic animals often limits rates of reactions.

c. Body temperature of endothermic animals promotes rates of reactions.

Figure 6.8 The effect of temperature on rate of reaction. **a.** Usually, the rate of an enzymatic reaction doubles with every 10°C rise in temperature. This enzymatic reaction is maximum at about 40°C. Then it decreases until the reaction stops altogether, because the enzyme has become denatured. **b.** The body temperature of ectothermic animals, which require an environmental source of heat, often limits rates of reactions. **c.** The body temperature of endothermic animals, which generate heat through their own metabolism, promotes rates of reaction.

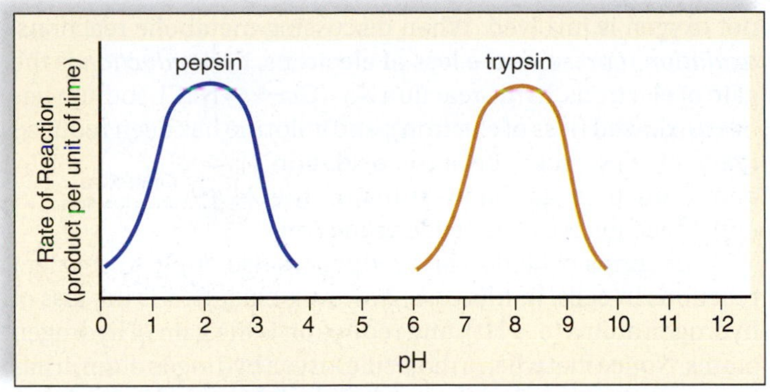

pepsin
trypsin
Rate of Reaction
(product per unit of time)
0 1 2 3 4 5 6 7 8 9 10 11 12
pH

Figure 6.9 The effect of pH on rate of reaction. The preferred pH for pepsin, an enzyme that acts in the stomach, is about 2, while the preferred pH for trypsin, an enzyme that acts in the small intestine, is about 8. At the preferred pH, an enzyme maintains its shape so that it can bind with its substrates.

as shown below, and an enzyme called a phosphatase removes them. In some proteins, adding phosphates activates them; in others, removing phosphates activates them. Enzymes can also be activated by cleaving or removing part of the protein, or by associating with another protein or cofactor.

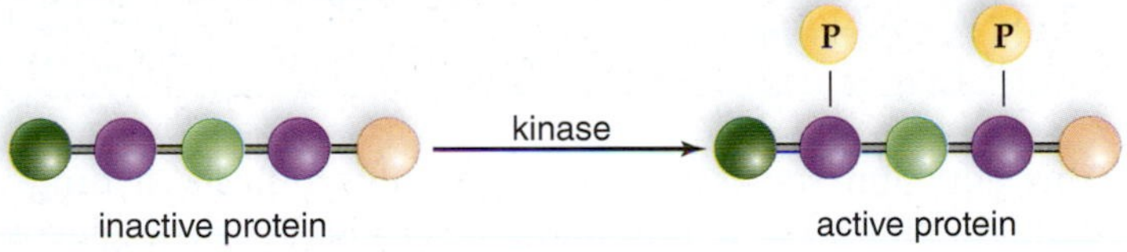

P
P
kinase
inactive protein
active protein

Enzyme Inhibition

Enzyme inhibition occurs when the substrate is unable to bind to the active site of an enzyme. The activity of almost every enzyme in a cell is regulated by feedback inhibition. In the simplest case, when there is plenty of product, it binds to the enzyme's active site, and then the substrate is unable to bind. As the product is used up, inhibition is reduced, and more product can be produced. In this way, the concentration of the product is always kept within a certain range.

Most metabolic pathways in cells are regulated by a more complicated type of feedback inhibition (Fig. 6.10). In these instances, the end product of an active pathway binds to a site other than the active site of the first enzyme. The binding changes the shape of the active site so that the substrate is unable to bind to the enzyme, and the pathway shuts down (inactive). Therefore, no more product is produced.

Animation Feedback Inhibition of Biochemical Pathways

Poisons are often enzyme inhibitors. Cyanide is an inhibitor for an essential enzyme (cytochrome *c* oxidase) in all cells, which accounts for its lethal effect on humans. Penicillin is an *antimicrobial agent* that blocks the active site of an enzyme unique to bacteria. Therefore, penicillin is a poison for bacteria.

Enzyme Cofactors

Many enzymes require an inorganic ion or an organic, but nonprotein, helper to function properly. The inorganic ions are metals such as copper, zinc, or iron. These helpers are called **cofactors.** The organic, nonprotein molecules are called **coenzymes.** These cofactors assist the enzyme and may even accept or contribute atoms to the reactions.

Vitamins are often components of coenzymes. **Vitamins** are relatively small organic molecules that are required in trace amounts in our diet and in the diets of other animals for synthesis of coenzymes that affect health and physical fitness. The vitamin becomes a part of the coenzyme's molecular structure. For example, the vitamin niacin is part of the coenzyme NAD, and riboflavin (B_2) is part of the coenzyme FAD.

A deficiency of any one of these vitamins results in a lack of the coenzyme and therefore a lack of certain enzymatic actions. In humans, this eventually results in vitamin-deficiency symptoms. For example, niacin deficiency results in a skin disease called pellagra, and riboflavin deficiency results in cracks at the corners of the mouth.

Animation B Vitamins

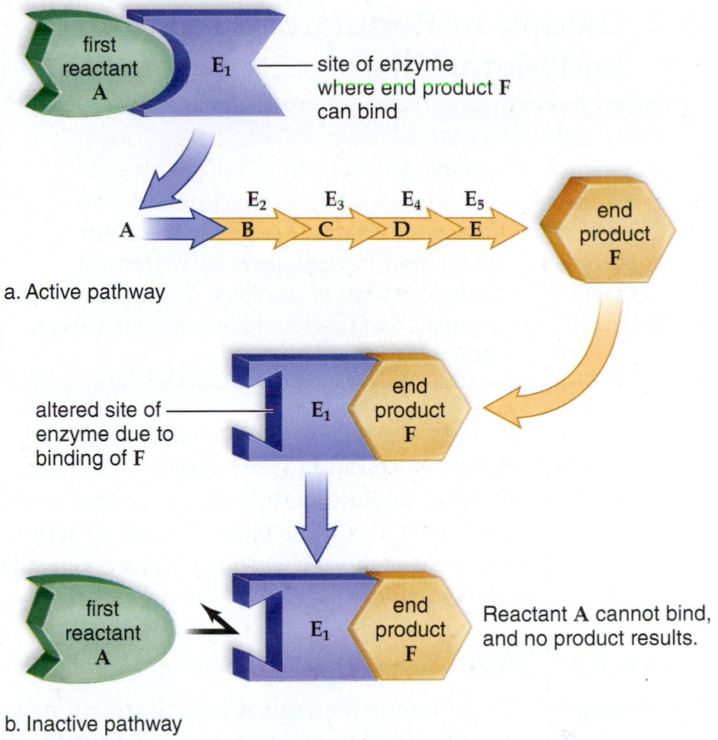

first reactant A
E1
site of enzyme where end product F can bind
A
E2 E3 E4 E5
B C D E
end product F
a. Active pathway
altered site of enzyme due to binding of F
E1
end product F
first reactant A
E1
end product F
Reactant A cannot bind, and no product results.
b. Inactive pathway

Figure 6.10 Feedback inhibition. **a.** In an active pathway, the first reactant (A) is able to bind to the active site of enzyme E_1. **b.** Feedback inhibition occurs when the end product (F) of the metabolic pathway binds to the first enzyme of the pathway—at a site other than the active site. This binding causes the active site to change its shape. Now reactant A is unable to bind to the enzyme's active site, and the whole pathway shuts down.

Check Your Progress 6.3

1. Summarize why enzymes are needed in biochemical pathways and how cells may regulate their activity.
2. Explain why denaturing an enzyme causes a change in its ability to act as a catalyst.
3. Discuss why the three-dimensional shape of an enzyme is important to its function.

6.4 Oxidation-Reduction Reactions and Metabolism

Learning Outcomes

Upon completion of this section, you should be able to

1. Explain how equations for photosynthesis and cellular respiration represent oxidation-reduction reactions.
2. Summarize the relationship between the metabolic reactions of photosynthesis and cellular respiration.
3. Explain how proteins and fats in addition to glucose can be used as mitochondrial fuel.

In the next two chapters, you will explore two important metabolic pathways: cellular respiration (see Chapter 7) and photosynthesis (see Chapter 8). Both of these pathways are based on the use of special enzymes to facilitate the movement of electrons. The movement of these electrons plays a major role in the energy-related reactions associated with these pathways.

Oxidation-Reduction Reactions

When oxygen (O) combines with a metal such as iron or magnesium (Mg), oxygen receives electrons and becomes an ion that is negatively charged. The metal loses electrons and becomes an ion that is positively charged. When magnesium oxide (MgO) forms, it is appropriate to say that magnesium has been oxidized. On the other hand, oxygen has been reduced because it has gained negative charges (i.e., electrons). Reactions that involve the gain and loss of electrons are called **oxidation-reduction reactions.** Sometimes, the terms oxidation and reduction are applied to other reactions, whether or not oxygen is involved. When discussing metabolic reactions, *oxidation* represents the loss of electrons, and *reduction* is the gain of electrons. In the reaction $Na + Cl \longrightarrow NaCl$, sodium has been oxidized (loss of electron), and chlorine has been reduced (gain of electrons). Because oxidation and reduction go hand-in-hand, the entire reaction is called a **redox reaction.**

Animation Redox Reactions

The terms oxidation and reduction also apply to covalent reactions in cells. In this case, however, oxidation is the loss of hydrogen atoms ($e^- + H^+$), and reduction is the gain of hydrogen atoms. Notice that when a molecule loses a hydrogen atom, it has lost an electron, and when a molecule gains a hydrogen atom, it has gained an electron. This form of oxidation-reduction is exemplified in the overall equations for photosynthesis and cellular respiration.

Chloroplasts and Photosynthesis

The chloroplasts in plants capture solar energy and use it to convert water and carbon dioxide into a carbohydrate. Oxygen is a by-product that is released (Fig. 6.11 *left*). The overall equation for photosynthesis can be written like this:

$$\underset{}{\text{energy}} + \underset{\text{carbon dioxide}}{6\,CO_2} + \underset{\text{water}}{6\,H_2O} \longrightarrow \underset{\text{glucose}}{C_6H_{12}O_6} + \underset{\text{oxygen}}{6\,O_2}$$

This equation shows that during photosynthesis, hydrogen atoms are transferred from water to carbon dioxide as glucose forms. In this reaction, therefore, carbon dioxide has been reduced and water has been oxidized. It takes energy to reduce

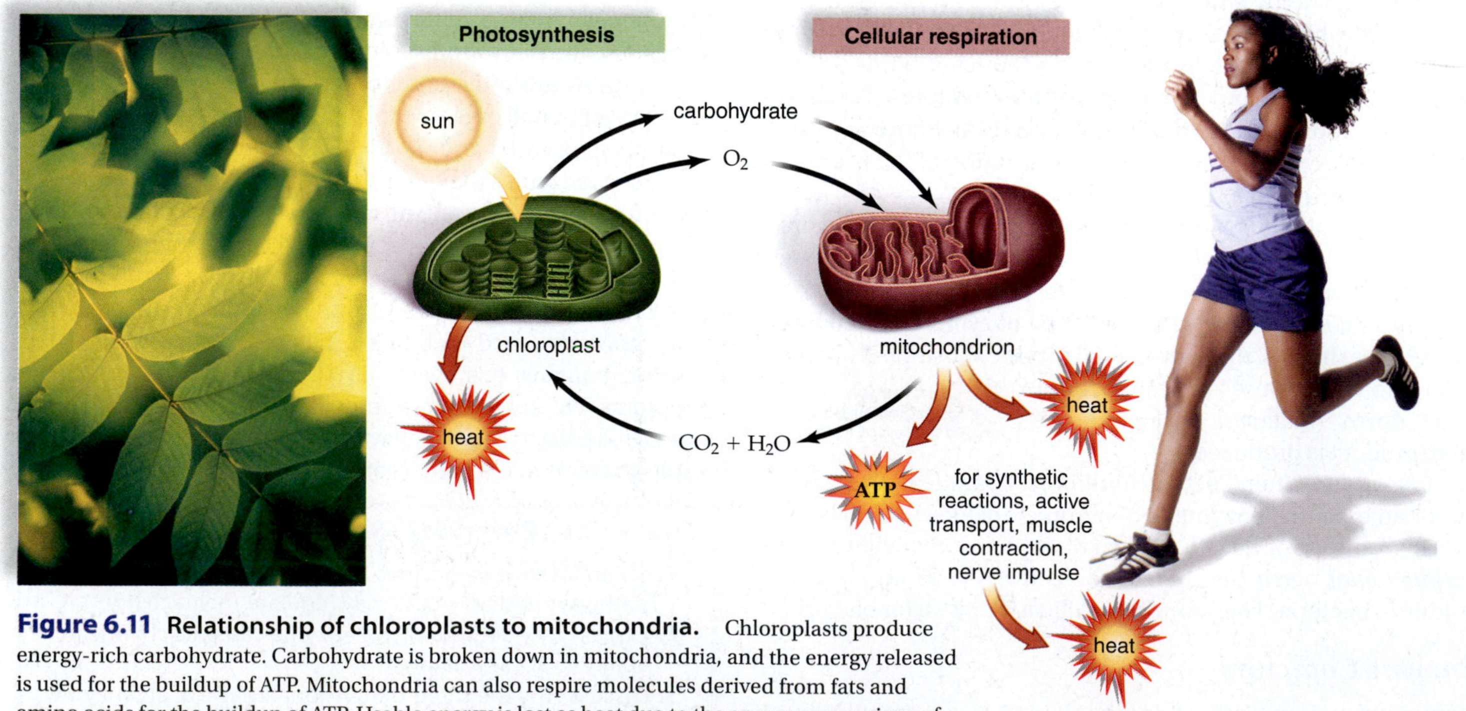

Photosynthesis
Cellular respiration
sun
carbohydrate
O2
chloroplast
mitochondrion
heat
CO2 + H2O
heat
ATP
for synthetic reactions, active transport, muscle contraction, nerve impulse
heat

Figure 6.11 Relationship of chloroplasts to mitochondria. Chloroplasts produce energy-rich carbohydrate. Carbohydrate is broken down in mitochondria, and the energy released is used for the buildup of ATP. Mitochondria can also respire molecules derived from fats and amino acids for the buildup of ATP. Usable energy is lost as heat due to the energy conversions of photosynthesis, cellular respiration, and the use of ATP in the body.

carbon dioxide to glucose, and this energy is supplied by solar energy. Chloroplasts are able to capture solar energy and convert it to the chemical energy of ATP, which is used along with hydrogen atoms to reduce carbon dioxide.

The reduction of carbon dioxide to form a mole of glucose stores 686 kcal in the chemical bonds of glucose. This is the energy that living organisms utilize to support themselves only because carbohydrates (and other nutrients) can be oxidized in mitochondria.

Mitochondria and Cellular Respiration

Mitochondria, present in both plants and animals, oxidize carbohydrates and use the released energy to build ATP molecules (Fig. 6.11 *right*). Cellular respiration therefore consumes oxygen and produces carbon dioxide and water, the very molecules taken up by chloroplasts. The overall equation for cellular respiration is the opposite of the one we used to represent photosynthesis:

$C_6H_{12}O_6$	+	$6\ O_2$	$\longrightarrow$	$6\ CO_2$	+	$6\ H_2O$	+	energy
glucose		oxygen		carbon dioxide		water		

In this reaction, glucose has lost hydrogen atoms (been oxidized), and oxygen has gained hydrogen atoms (been reduced). When oxygen gains electrons, it becomes water. The complete oxidation of a mole of glucose releases 686 kcal of energy, and some of this energy is used to synthesize ATP molecules. If the energy within glucose were released all at once, most of it would dissipate as heat instead of some of it being used to produce ATP. Instead, cells oxidize glucose step by step. The energy is gradually stored and then converted to that of ATP molecules, which is used in animals in the many ways listed in Figure 6.11.

Figure 6.11 shows us very well that chloroplasts and mitochondria are involved in a cycle. Carbohydrate produced within chloroplasts becomes a fuel for cellular respiration in mitochondria, while carbon dioxide released by mitochondria becomes a substrate during photosynthesis in chloroplasts. These organelles are involved in a redox cycle because carbon dioxide is reduced during photosynthesis and carbohydrate is oxidized during cellular respiration. Note that energy does not cycle between the two organelles; instead, it flows from the sun through each step of photosynthesis and cellular respiration until it eventually becomes unusable heat as ATP is used by the cell.

Cellular Respiration and Humans

Humans, like all eukaryotic organisms, are involved in the cycling of molecules between chloroplasts and mitochondria. Our food is derived from plants or we eat other animals that have eaten plants. Also, we take in oxygen released by plants. Nutrients from our food and oxygen enter our mitochondria, which produce ATP (Fig. 6.12). Without a supply of energy-rich molecules, ultimately derived from plants, we could not produce the ATP molecules needed to maintain our bodies.

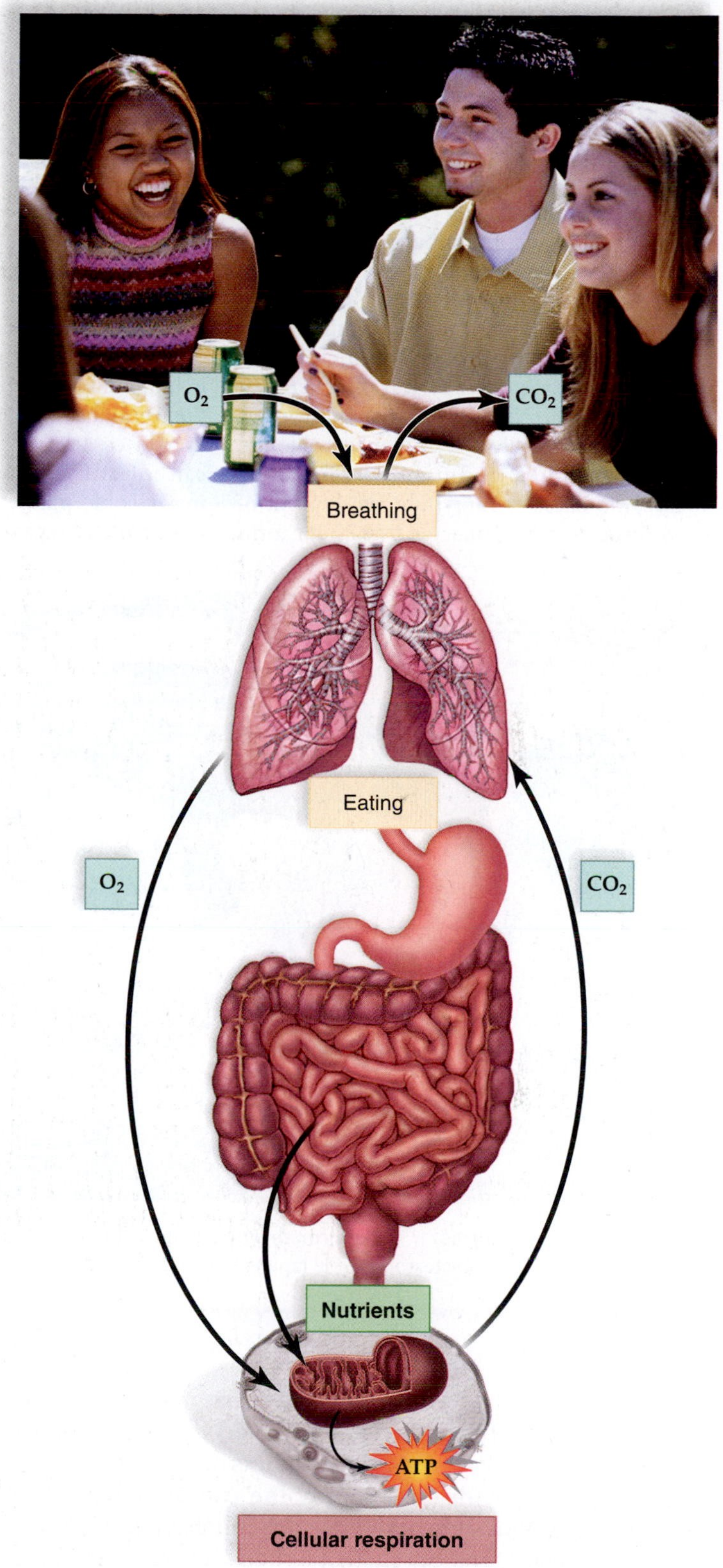
O2
CO2
Breathing
Eating
O2
CO2
Nutrients
ATP
Cellular respiration

Figure 6.12 Relationship between breathing, eating, and cellular respiration. The O_2 we inhale and the nutrients resulting from the digestion of our food are carried by the bloodstream to our cells, where they enter mitochondria and undergo cellular respiration. Following cellular respiration, the ATP stays in the cell, but the CO_2 is carried to the lungs for exhalation.

SCIENCE IN YOUR LIFE ▶ BIOETHICAL

Enzyme Inhibitors Can Spell Death

Cyanide gas was formerly used to execute people. How did it work? Cyanide can be fatal because it binds to a mitochondrial enzyme necessary for the production of ATP. MPTP (1-methyl-4-phenyl-1,2,3,6-tetrahydropyridine) is another enzyme inhibitor that stops mitochondria from producing ATP. The toxic nature of MPTP was discovered in the early 1980s, when a group of intravenous drug users in California suddenly developed symptoms of Parkinson disease, including uncontrollable tremors and rigidity. All of the drug users had injected a synthetic form of heroin that was contaminated with MPTP. Parkinson disease is characterized by the death of brain cells, the very ones that are also destroyed by MPTP.

Sarin is a chemical that inhibits an enzyme at neuromuscular junctions, where nerves stimulate muscles. When the enzyme is inhibited, the signal for muscle contraction cannot be turned off, so the muscles are unable to relax and become paralyzed. Sarin can be fatal if the muscles needed for breathing become paralyzed. In 1995, terrorists released sarin gas on a subway in Japan (Fig. 6A). Although many people developed symptoms, only 17 died.

A fungus that contaminates and causes spoilage of sweet clover produces a chemical called warfarin. Cattle that eat the spoiled feed die from internal bleeding because warfarin inhibits a crucial enzyme for blood clotting. Today, warfarin is widely used as a rat poison. Unfortunately, it is not uncommon for warfarin to be mistakenly eaten by pets and even very small children, with tragic results.

Many people are prescribed a medicine called warfarin (Coumadin), which acts as an anticoagulant to prevent inappropriate blood clotting. It is often prescribed for people who have received an artificial heart valve or have irregular heartbeats.

These examples all show how our understanding of science can have positive or negative consequences, and emphasize the role of ethics in scientific investigation.

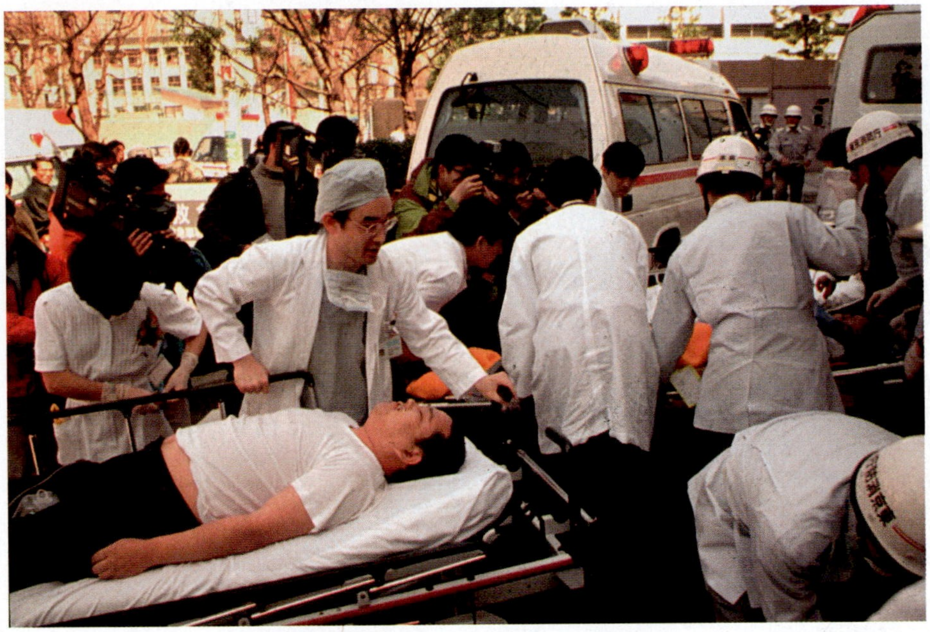

Figure 6A Sarin gas. The aftermath when sarin, a nerve gas that results in the inability to breathe, was released by terrorists in a Japanese subway in 1995.

Questions to Consider

1. Do you feel it is ethical to use dangerous chemicals to treat diseases?
2. Many drug companies are actively looking for species that might contain chemical compounds that are new to humans. How might this have both beneficial and harmful side-effects?

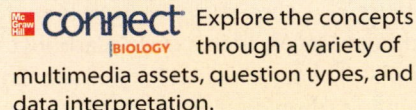

On the other hand, our mitochondria release carbon dioxide and water. The carbon dioxide is exhaled and it enters the atmosphere where it is accessible to plants once again.

So far we have discussed only the use of glucose as a mitochondrial fuel, but hopefully not much of your diet is actually glucose. Our food, such as a slice of pizza with cheese and pepperoni on top, consists of carbohydrates, fats, and proteins. These macromolecules are broken down to simpler molecules in our digestive tract. Starch in the crust of a pizza is digested to glucose, but the fat is broken down to glycerol and fatty acids, and proteins are hydrolyzed to amino acids. Glycerol, fatty acids, and amino acids enter cellular respiration at different places in the cellular respiration pathway. Fatty acids are a concentrated source of energy compared to a carbohydrate because each fatty acid has a long chain of reduced carbon atoms. Only when we are on a diet that restricts carbohydrates and fats do our bodies then use amino acids as an energy source. When amino acids are metabolized, the amino groups are removed in a process called *deamination*. The amino group becomes part of urea, the primary excretory product of humans. The remainder of the amino acid can enter cellular respiration at various places in the pathway, depending on the length of its carbon skeleton. When we are literally starving and have exhausted stored supplies of glycogen and fat, the body will withdraw protein from our muscles and use it as an energy source.

If we were to study all the various metabolic steps involved in the cellular respiration of glucose, fatty acids, and amino acids, you would soon realize that certain substrates recur along the way. This allows the cell to switch metabolic gears and even switch to building up molecules rather than breaking them down. A good example is when you get fat from eating too much ice cream, cake, and pie. A particular breakdown product of glucose can be used to form fatty acids, which then join with glycerol and the product fat gets stored at various locations in your body.

Check Your Progress 6.4

1. Explain why overall equations for photosynthesis and cellular respiration represent redox reactions.
2. Summarize how the inputs and outputs of cellular respiration and photosynthesis are related.
3. Describe the three types of nutrients that can be used in mitochondria to build up ATP molecules.

Case Study Conclusion

The enzymes in the saliva of the vampire bat are interfering with a complicated biochemical pathway in the blood. The clotting cascade involves several enzymes that circulate in the blood. When a blood vessel is damaged, platelets release prothrombin activator, an enzyme that converts the plasma protein prothrombin into thrombin. Thrombin, in turn, is an enzyme that converts fibrinogen into fibrin, forming the basis of the clot. When the damage has been repaired, the plasma protein plasminogen is converted into plasmin, which destroys the fibrin. By releasing a potent plasminogen enzyme in their saliva, vampire bats are able to manipulate a biochemical pathway for their own needs.

MEDIA STUDY TOOLS

www.mhhe.com/maderinquiry14

Enhance your study of this chapter with study tools and practice tests. Also ask your instructor about the resources available through ConnectPlus, including LearnSmart, the media-rich eBook, interactive learning tools, and animations.

Mc Graw Hill connect plus+ BIOLOGY

SUMMARIZE

6.1 Life and the Flow of Energy

- **Energy** is the ability to do work. Energy may exist in several forms, including **potential energy** (**chemical energy**) and **kinetic energy** (**mechanical energy**). Two energy laws are basic to understanding energy-use patterns in cells and ecosystems. The first law states that energy cannot be created or destroyed but can only be changed from one form to another. The second law of thermodynamics states that one usable form of energy cannot be converted into another form without loss of usable energy. Therefore, every energy transformation makes the universe less organized. As a result of these laws, we know that the **entropy** of the universe is increasing and that only a constant input of energy maintains life's organization.

6.2 Energy Transformations and Metabolism

- The term **metabolism** encompasses all the chemical reactions occurring in a cell. **Catabolism** involves breaking down **reactants,** whereas **anabolism** involves building up **products. Free energy** is the amount of energy that is actually available to do work.
- Considering individual reactions, only those that result in products that have less usable energy than the reactants go forward. Such reactions, called **exergonic reactions,** release energy. **Endergonic reactions,** which require an input of energy, occur in cells as a **coupled reactions** with an exergonic process. For example, glucose breakdown is an exergonic metabolic pathway that drives the buildup of many **ATP** molecules. These ATP molecules then supply energy for cellular work. ATP goes through a cycle of constantly being built up from, and then broken down to, **ADP** + Ⓟ.

6.3 Enzymes and Metabolic Pathways

- A **metabolic pathway** is a series of reactions that proceed in an orderly, step-by-step manner. Each reaction requires a specific **enzyme.** Enzymes function by reducing the **energy of activation** (**E_a**) needed for the reaction to proceed.
- Reaction rates increase when enzymes form a complex with their **substrates,** called an **induced fit model.** Generally, enzyme activity increases as substrate concentration increases. Once all **active sites** are filled, the maximum rate has been achieved. Any environmental factor, such as temperature or pH, affects the shape of a protein. A **denatured** enzyme loses its ability to do its job.
- Cellular mechanisms regulate enzyme quantity and activity. The activity of most metabolic pathways is regulated by **enzyme inhibition,** usually associated with feedback. Many enzymes have **cofactors** or **coenzymes,** such as **vitamins,** that help them carry out a reaction.

6.4 Oxidation-Reduction Reactions and Metabolism

- The overall equation for photosynthesis is the opposite of that for cellular respiration. Both processes involve **oxidation-reduction reactions.** Redox reactions are a major way in which energy is transformed in cells.
- During photosynthesis, carbon dioxide is reduced to glucose, and water is oxidized. Glucose formation requires energy, and this energy comes from the sun. Chloroplasts capture solar energy and convert it to the chemical energy of ATP molecules, which are used along with hydrogen atoms to reduce carbon dioxide to glucose. During cellular respiration, glucose is oxidized to carbon dioxide, and oxygen is reduced to water. This reaction releases energy, which is used to synthesize ATP molecules in all types of cells.
- Energy flows through all living things. Photosynthesis is a metabolic pathway in chloroplasts that transforms solar energy to the chemical energy within carbohydrates, and cellular respiration is

a metabolic pathway completed in mitochondria that transforms this energy into that of ATP molecules. Eventually, the energy within ATP molecules becomes heat.

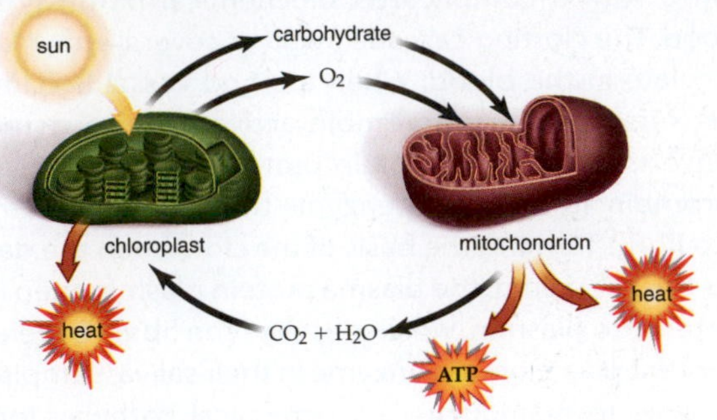
sun
carbohydrate
O_2
chloroplast
mitochondrion
heat
heat
heat
$CO_2 + H_2O$
ATP

- Cellular respiration is an aerobic process that requires oxygen and gives off carbon dioxide. Cellular respiration involves oxidation. The air we inhale contains the oxygen, and the food we digest after eating contains the carbohydrate glucose needed for cellular respiration.
- Besides carbohydrates, fats and proteins can be used to provide ATP via cellular respiration.
- Anabolism can use molecules directly from our food or substrates from other metabolic pathways to synthesize new molecules.

ASSESS

Testing Yourself

Choose the best answer for each question.

1. The fact that energy transformation is never 100% efficient is stated in the ____ law of thermodynamics.
   a. first **c.** third
   b. second **d.** None of these are correct.

In questions 2–5, match the statements concerning enzyme action to the appropriate graph.

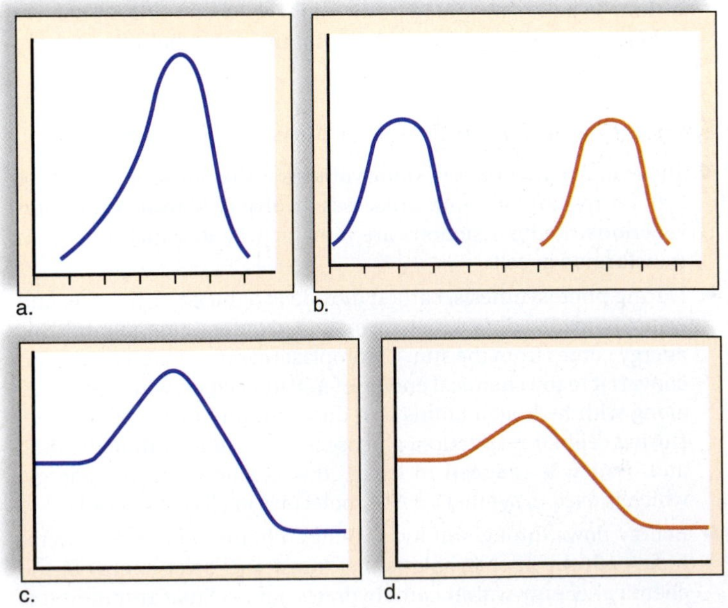
a.
b.
c.
d.

2. Progress of reaction in the absence of an enzyme
3. Enzymatic action with varying temperature
4. Progress of reaction in the presence of an enzyme
5. Enzymes with varying pH optima
6. Enzymatic action is sensitive to changes in
   a. temperature. **c.** substrate concentration.
   b. pH. **d.** All of these are correct.
7. Which has more potential energy: glucose or CO_2 and H_2O?
   a. glucose
   b. CO_2 and H_2O
   c. Both are the same because reactants and products are equal.
   d. Neither is correct.
8. ____ is the loss of electrons, and ____ is the gain of electrons.
   a. Oxidation, reduction
   b. Reduction, oxidation
   c. Neither a nor b is correct.
   d. Need more information to answer.
9. Entropy is a term used to indicate the level of
   a. usable energy. **c.** enzyme action.
   b. disorganization. **d.** None of these are correct.
10. Enzymes catalyze reactions by
    a. bringing the reactants together.
    b. lowering the activation energy.
    c. Both a and b are correct.
    d. Neither a nor b is correct.
11. The active site of an enzyme
    a. is identical to that of any other enzyme.
    b. is the part of the enzyme where its substrate can fit.
    c. can be used over and over again.
    d. is not affected by environmental factors such as pH and temperature.
    e. Both b and c are correct.

ENGAGE

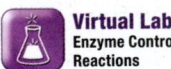

Virtual Lab
Enzyme Controlled Reactions

The virtual lab "Enzyme Controlled Reactions" provides an interactive tutorial to explore many of the factors influencing enzyme activity.

Thinking Critically

1. Entropy is increased when nutrients break down, so why are enzymatic metabolic pathways required for cellular respiration? (*Hint:* See Fig. 6.5.)
2. Why would you expect glucose storage as glycogen to be an energy-requiring process?
3. If photosynthesis and cellular respiration are reverse equations, then why can't mitochondria carry on photosynthesis?

CASE STUDY Metabolic Demands on Athletes

During a typical basketball game, such as the one shown above involving the 2011 NBA Champion Dallas Mavericks, the starting players run an average of 4–5 kilometers (around 3 miles) during a 40-minute game. However, unlike the endurance running experienced by marathoners, basketball players experience periods of intense activity (sprinting), followed by brief periods of rest. This start-and-stop nature of the game means that the muscles of the athlete are constantly switching between aerobic and anaerobic metabolism.

During aerobic metabolism, the muscle cells use oxygen in order to completely break down glucose, producing more ATP, a high-energy molecule used for muscle contraction, than otherwise. The breakdown of glucose with oxygen to produce carbon dioxide and water in the cytoplasm and mitochondria of the cell is called cellular respiration. However, running short, fast sprints quickly depletes oxygen levels and drives the muscles into anaerobic metabolism. Without oxygen, glucose cannot be broken down completely. It is changed into lactate, which is responsible for that muscle burn we sometimes feel after strenuous exercise. Once oxygen is restored to the muscles, the body is able to return to aerobic metabolism and dispose of the lactate.

In this chapter, we will discuss the metabolic pathways of cellular respiration that allow the energy within a glucose molecule, and other organic nutrients, to be converted into ATP.

As you read through the chapter, think about the following questions:

1. What are the differences between the aerobic and anaerobic pathways?
2. How is the energy of a glucose molecule harvested by a cell?
3. How are other organic nutrients, such as proteins and fats, used as energy?

Cellular Respiration 7

CHAPTER OUTLINE

BEFORE YOU BEGIN

Before beginning this chapter, take a few moments to review the following discussions:

Section 2.5 What is the role of carbohydrates in the body?

Section 3.3 What is the structure and function of the mitochondria in a cell?

Figure 6.3 How does the ATP cycle resemble a rechargeable battery?

7.1 Overview of Cellular Respiration

Learning Outcomes

Upon completion of this section, you should be able to

1. Describe the overall equation for cellular respiration.
2. Explain the role of electron carriers in respiration.
3. Summarize the phases of cellular respiration and indicate where they occur in a cell.

Cellular respiration is the release of energy from molecules such as glucose accompanied by the use of this energy to synthesize ATP molecules. Cellular respiration is an aerobic process that requires oxygen (O_2) and gives off carbon dioxide (CO_2). It usually involves the complete breakdown of glucose as shown here:

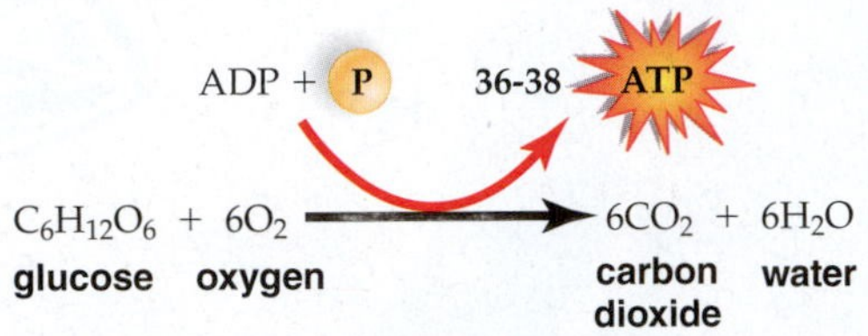
ADP + P
36-38
ATP
$C_6H_{12}O_6$ + $6O_2$
$6CO_2$ + $6H_2O$
glucose
oxygen
carbon dioxide
water

Glucose is a high-energy molecule, and its breakdown products, CO_2 and H_2O, are low-energy molecules. As glucose is broken down, energy is released. This energy is used to produce ATP molecules. The breakdown of one glucose molecule results in 36 to 38 ATP molecules. This represents about 39% of the potential energy within a glucose molecule. The rest of the energy dissipates. This conversion is more efficient than many others. For example, only about 25% of the energy within gasoline is converted to the motion of a car.

The pathways of cellular respiration allow the energy within a glucose molecule to be released slowly so that the ATP can be produced gradually. The energy in the ATP can be used for most cellular reactions, such as joining one amino acid to another during protein synthesis or joining actin to myosin during muscle contraction. This energy is released by a relatively simple procedure: the removal of a single phosphate group. The cell uses the reverse reaction for building up ATP again.

NAD$^+$ and FAD

Cellular respiration involves many individual reactions, each one catalyzed by its own enzyme. Some of these enzymes utilize the coenzyme **NAD$^+$ (nicotinamide adenine dinucleotide)** as an electron carrier. Coenzymes help an enzyme do its job and may even participate in the reaction. In this instance, NAD$^+$ receives two electrons (is reduced) as the substrate glucose is oxidized. Recall from Chapter 6 that each electron is received by NAD$^+$ as part of a hydrogen atom. A hydrogen atom consists of a hydrogen ion (H^+) and an electron (e^-). As shown in Figure 7.1, NAD$^+$ receives two e^- and two H^+ to give NADH + H^+.

Animation How the NAD$^+$ Works

FAD (flavin adenine dinucleotide) is another coenzyme frequently used as an electron carrier. When FAD accepts two e^- and two H^+, $FADH_2$ results. NAD$^+$ and FAD are analogous to electron shuttle buses. They pick up electrons at specific enzymatic reactions in either the cytoplasm or the matrix of the mitochondria and carry these high-energy electrons to an electron transport chain in the cristae of the mitochondria, where they drop them off. The empty NAD$^+$ or FAD is then free to go back and pick up more electrons.

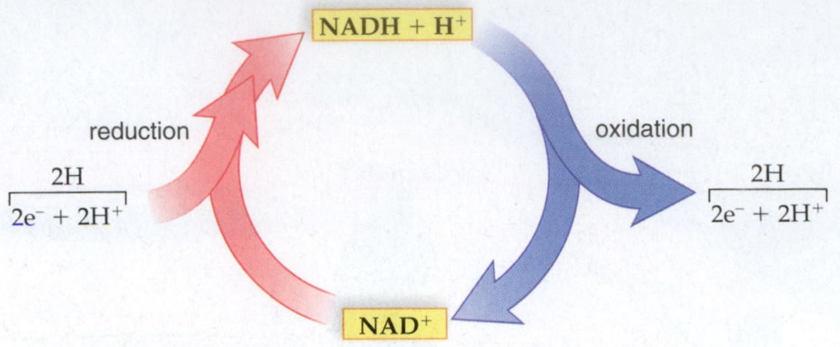
NADH + H+
reduction
oxidation
2H
2e− + 2H+
2H
2e− + 2H+
NAD+

Figure 7.1 The NAD$^+$ cycle. The coenzyme NAD$^+$ accepts two hydrogen atoms ($H^+ + e^-$), and NADH + H^+ results. When NADH passes on electrons, NAD$^+$ results. Only a small amount of NAD$^+$ need be present in a cell, because each NAD$^+$ molecule is used over and over again.

Phases of Cellular Respiration

The metabolic pathways of cellular respiration couple the release of energy within a glucose molecule to the production of ATP. The coupling of these reactions reduces the amount of energy lost as heat, which would be significant if glucose breakdown occurred all at once. Cellular respiration involves four phases (Fig. 7.2). The first phase, glycolysis, takes place outside the mitochondria and does not utilize oxygen. Therefore, glycolysis is **anaerobic.** The other phases take place inside the mitochondria, where oxygen is the final acceptor of electrons. These phases are **aerobic.**

- Glycolysis is the breakdown of glucose ($C_6H_{12}O_6$) to two molecules of pyruvate, a C_3 molecule. Oxidation by removal of electrons (e^-) and hydrogen ions (H^+) provides enough energy for the immediate buildup of two ATP.
- During the preparatory (prep) reaction, pyruvate is oxidized to a C_2 acetyl group carried by CoA (coenzyme A), and CO_2 is removed. Because glycolysis ends with two molecules of pyruvate, the prep reaction occurs twice per glucose molecule.
- The citric acid cycle is a cyclical series of oxidation reactions that give off CO_2 and produce one ATP. The citric acid cycle used to be called the Krebs cycle in honor of the man who worked out most of the steps. The cycle is now named for citric acid (or citrate), the first molecule in the cycle. The citric acid cycle turns twice because two acetyl CoA molecules enter the cycle per glucose molecule. Altogether, the citric acid cycle accounts for two immediate ATP molecules per glucose molecule.
- The electron transport chain is a series of membrane-bound carriers that pass electrons from one carrier to another. High-energy electrons are delivered to the chain, and low-energy electrons leave it (Fig. 7.3). The electron transport chain is like a flight of stairs. As something bounces down the stairs it loses potential energy. Similarly, as electrons pass through the carriers from a higher-energy to a lower-energy state, energy is released and used for ATP synthesis.

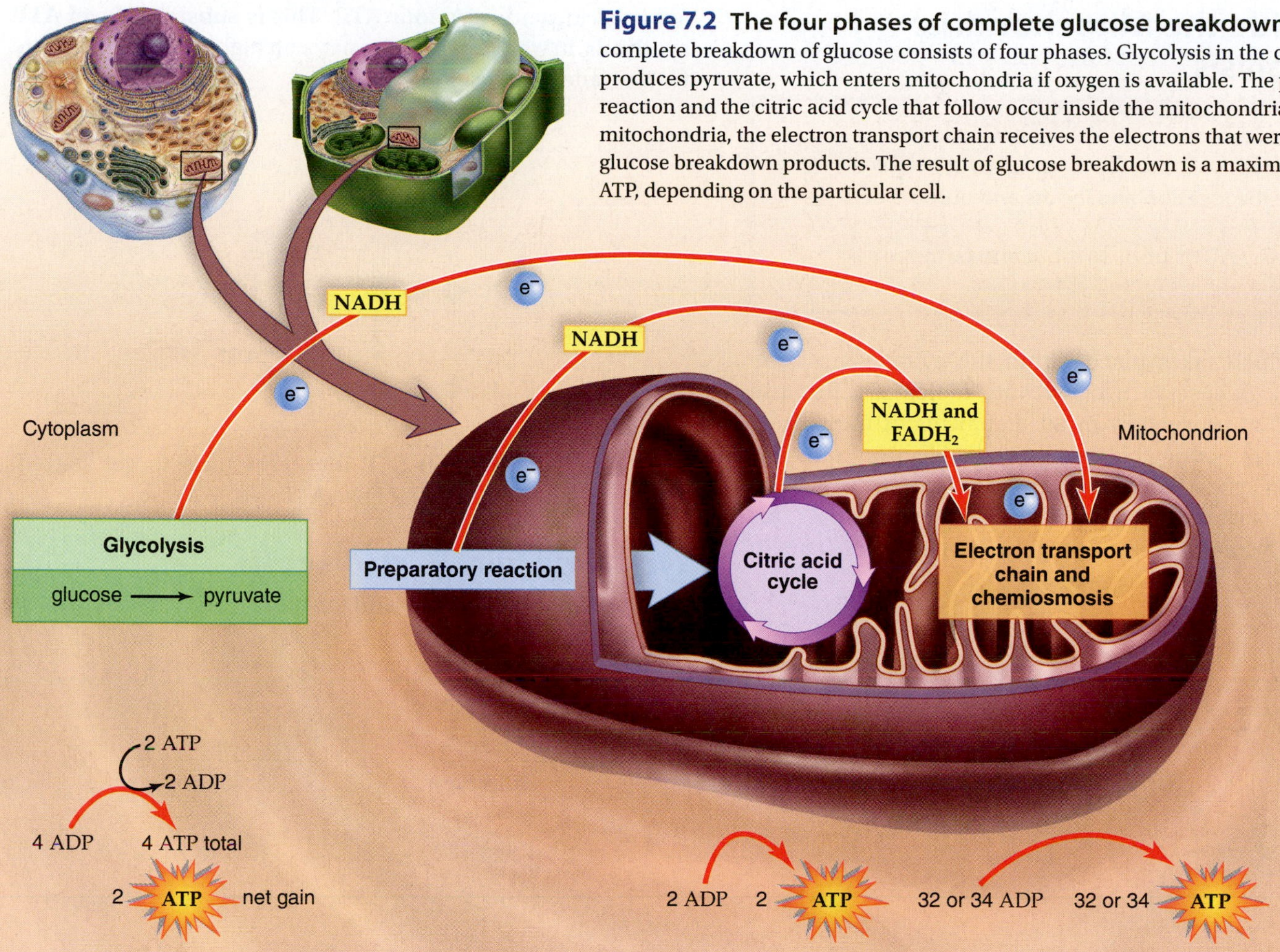
NADH
e⁻
NADH
e⁻
e⁻
e⁻
NADH and FADH$_2$
e⁻
Cytoplasm
Mitochondrion
e⁻
e⁻
Glycolysis
glucose ⟶ pyruvate
Preparatory reaction
Citric acid cycle
Electron transport chain and chemiosmosis
2 ATP
2 ADP
4 ADP
4 ATP total
2 ATP net gain
2 ADP
2 ATP
32 or 34 ADP
32 or 34 ATP

Figure 7.2 The four phases of complete glucose breakdown. The complete breakdown of glucose consists of four phases. Glycolysis in the cytoplasm produces pyruvate, which enters mitochondria if oxygen is available. The preparatory reaction and the citric acid cycle that follow occur inside the mitochondria. Also inside mitochondria, the electron transport chain receives the electrons that were removed from glucose breakdown products. The result of glucose breakdown is a maximum of 36 to 38 ATP, depending on the particular cell.

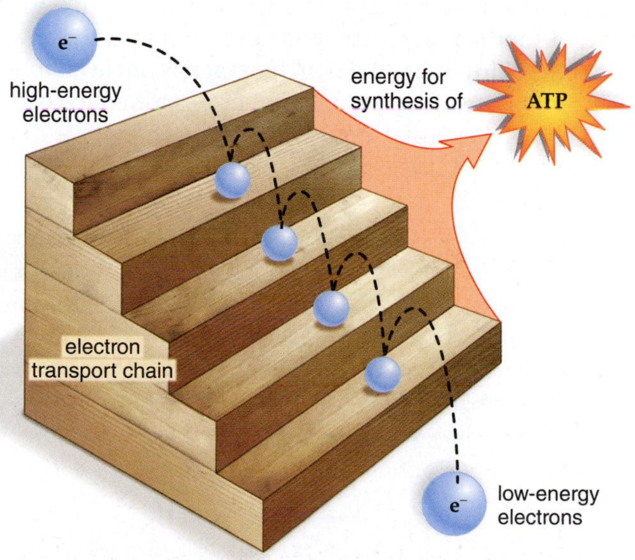
e⁻
high-energy electrons
energy for synthesis of
ATP
electron transport chain
e⁻
low-energy electrons

Figure 7.3 Electron transport chain. High-energy electrons are delivered to the chain. As they pass from carrier to carrier, energy is released and used for ATP production.

The electrons from one glucose molecule passing down the electron transport chain result in a maximum of 32 or 34 ATP, depending on certain conditions.

In cellular respiration, the low-energy electrons are finally received by O_2, which then combines with H^+ and becomes water.

Pyruvate is a pivotal metabolite in cellular respiration. If oxygen is not available to the cell, fermentation occurs in the cytoplasm instead of continued aerobic (with oxygen) cellular respiration. During fermentation, pyruvate is reduced to lactate or to carbon dioxide and alcohol, depending on the organism. As we shall see in section 7.3, fermentation results in a net gain of only two ATP per glucose molecule.

MP3 Cellular Respiration

Check Your Progress 7.1

1. Explain the role of NAD^+ and FAD in cellular respiration.
2. Distinguish between the aerobic and anaerobic phases of cellular respiration.
3. Summarize the location and function of each phase of cellular respiration.

7.2 Outside the Mitochondria: Glycolysis

Learning Outcomes

Upon completion of this section, you should be able to

1. Describe the location and inputs and outputs of glycolysis.
2. Explain why ATP is both an input and output of glycolysis.

Glycolysis, which takes place within the cytoplasm, is the breakdown of glucose to two pyruvate molecules. Because glycolysis occurs universally in all organisms, it most likely evolved before the citric acid cycle and the electron transport chain. Glycolysis most likely evolved when environmental conditions were anaerobic and before cells had mitochondria. This may be why glycolysis does not require oxygen and occurs in the cytoplasm.

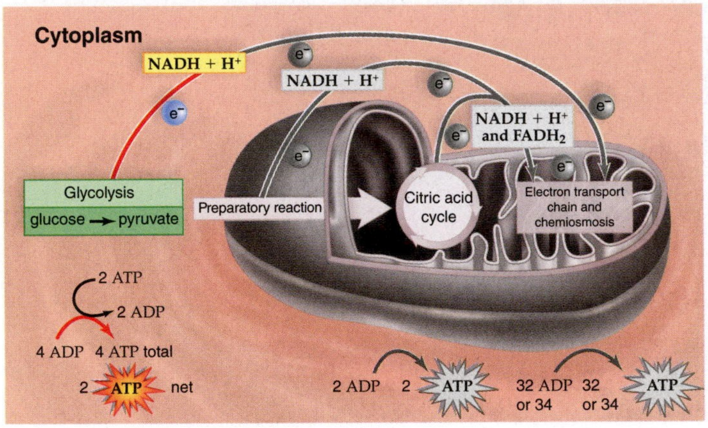

Cytoplasm
NADH + H+
NADH + H+
NADH + H+ and FADH2
e-
Glycolysis
glucose → pyruvate
Preparatory reaction
Citric acid cycle
Electron transport chain and chemiosmosis
2 ATP
2 ADP
4 ADP 4 ATP total
2 ATP net
2 ADP 2 ATP
32 ADP or 34
32 or 34 ATP

Energy-Investment Steps

As glycolysis begins, two ATP are used to activate glucose, and the molecule that results splits into two C_3 molecules (*G3P, glyceraldehyde 3-phosphate*), each of which has an attached phosphate group. From this point on, each C_3 molecule undergoes the same series of reactions (Fig. 7.4).

Energy-Harvesting Steps

Oxidation of G3P occurs by the removal of hydrogen atoms ($H^+ + e^-$). The hydrogen atoms are picked up by NAD^+, and $NADH + H^+$ results. Later, NADH will pass electrons on to the electron transport chain. Oxidation of G3P and subsequent substrates results in four high-energy phosphate groups, which are used to synthesize four ATP. This is **substrate-level ATP synthesis,** in which an enzyme passes a high-energy phosphate to ADP, and ATP results:

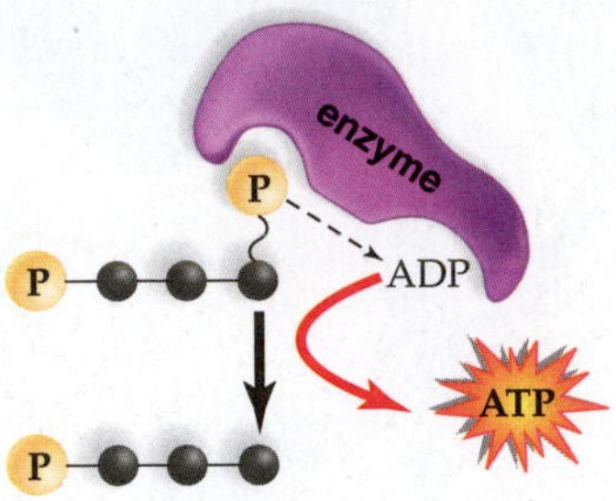

enzyme
P
ADP
ATP

Subtracting the two ATP that were used to get started, glycolysis yields a net gain of two ATP (Fig. 7.4).

Inputs and Outputs of Glycolysis

All together, the inputs and outputs of glycolysis are as follows:

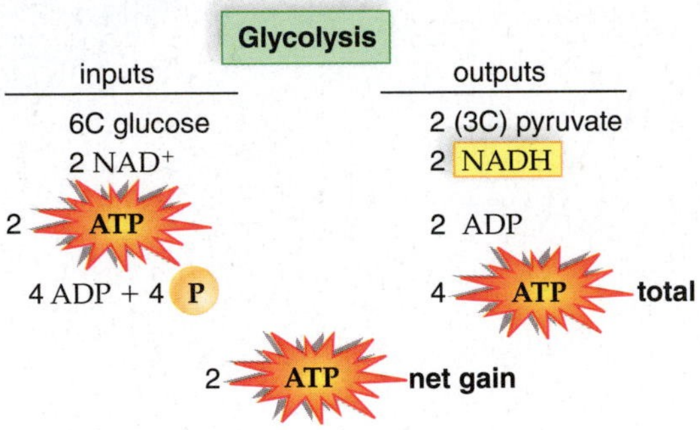

Glycolysis
inputs
outputs
6C glucose
2 NAD+
2 ATP
4 ADP + 4 P
2 (3C) pyruvate
2 NADH
2 ADP
4 ATP total
2 ATP net gain

Notice that, so far, we have accounted for only two of the 36 to 38 ATP molecules that are theoretically possible when glucose is completely broken down. When oxygen is available, the end product, pyruvate, enters the mitochondria, where it undergoes further breakdown. If oxygen is not available, pyruvate can be used for fermentation as described in section 7.3. For each glucose that enters glycolysis, two ATP, two $NADH + H^+$, and two pyruvate are formed.

3D Animation Cellular Respiration: Glycolysis

Check Your Progress 7.2

1. Explain why there is an energy-investment phase and energy-harvesting phase to glycolysis.
2. Summarize the inputs and outputs of glycolysis and state the net number of ATP that are produced.

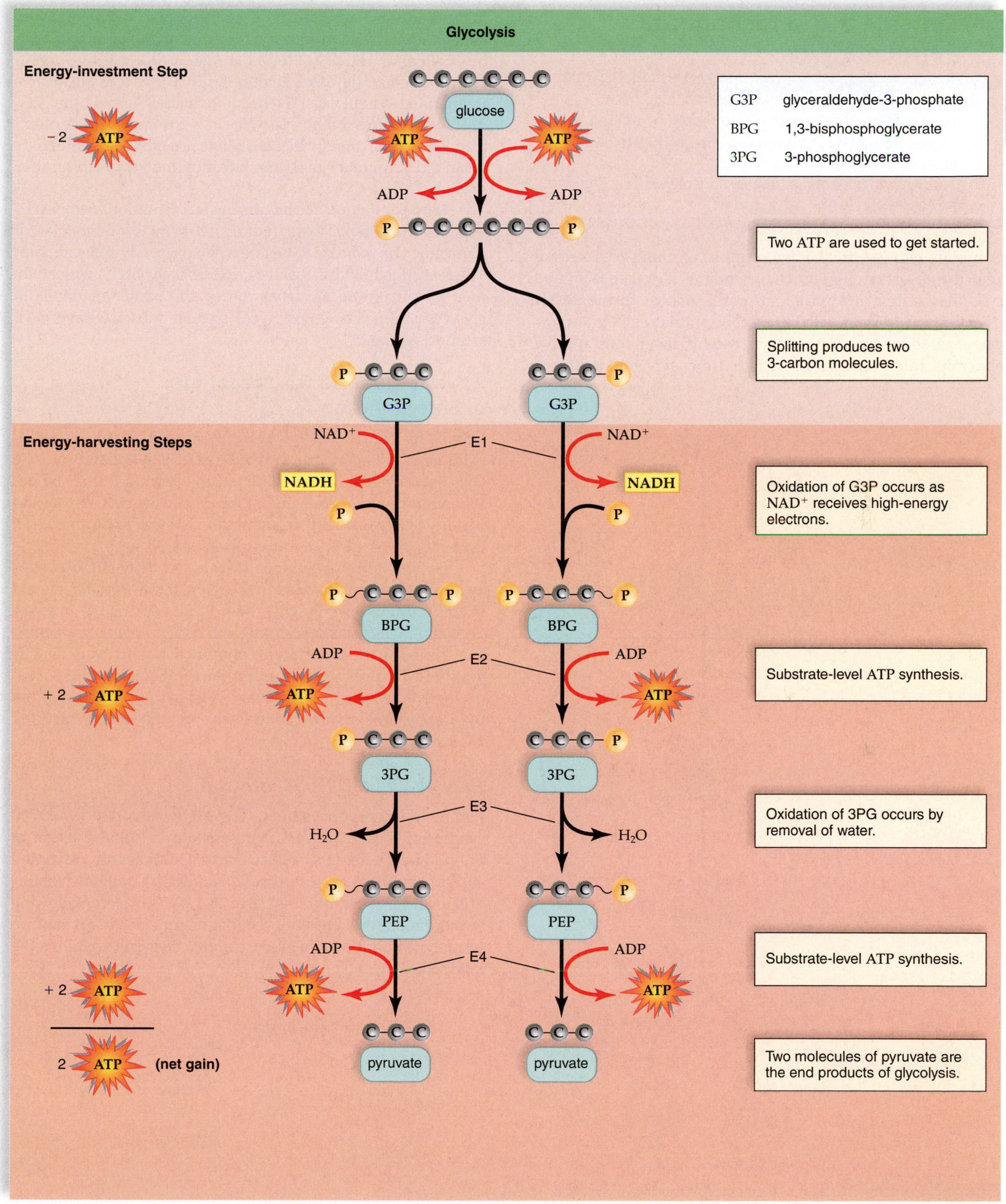

Glycolysis
Energy-investment Step
C-C-C-C-C-C
glucose
–2 ATP
ATP
ATP
ADP
ADP
P-C-C-C-C-C-C-P
G3P glyceraldehyde-3-phosphate
BPG 1,3-bisphosphoglycerate
3PG 3-phosphoglycerate
Two ATP are used to get started.
Splitting produces two 3-carbon molecules.
P-C-C-C
C-C-C-P
G3P
G3P
Energy-harvesting Steps
NAD+
NADH
P
E1
Oxidation of G3P occurs as NAD+ receives high-energy electrons.
BPG
ADP
ATP
E2
+2 ATP
Substrate-level ATP synthesis.
3PG
E3
H2O
Oxidation of 3PG occurs by removal of water.
PEP
E4
+2 ATP
Substrate-level ATP synthesis.
2 ATP (net gain)
pyruvate
Two molecules of pyruvate are the end products of glycolysis.

Figure 7.4 Glycolysis. Glycolysis begins with glucose and ends with two pyruvate molecules. There is a gain of two NADH + H$^+$ and a net gain of two ATP from glycolysis.

7.3 Outside the Mitochondria: Fermentation

Learning Outcomes

Upon completion of this section, you should be able to

1. Explain how ATP can continue to be produced in the absence of oxygen.
2. Describe the advantages and disadvantages of fermentation.

Complete glucose breakdown requires an input of oxygen to keep the electron transport chain working. If oxygen is limited, cells may utilize anaerobic pathways, such as **fermentation** (Fig. 7.5). In human cells, like other animal cells, the pyruvate formed by glycolysis accepts two hydrogen ions and two electrons and is reduced to lactate. Other types of organisms instead produce alcohol with the release of CO_2. Bacteria vary as to whether they produce an organic acid, such as lactate, or an alcohol and CO_2. Yeasts are good examples of organisms that generate ethyl alcohol and CO_2 as a result of fermentation. When yeast is used to leaven bread, the CO_2 produced makes bread rise. Yeast is also used to ferment fruit juice into wine. In that case, it is the ethyl alcohol that is desired. Eventually, yeasts are killed by the very alcohol they produce.

Notice in Figure 7.5 that during fermentation, two NADH pass electrons to pyruvate, reducing it. Why is it beneficial for pyruvate to be reduced to lactate when oxygen is not available? The reason is that this reaction regenerates NAD^+, which can then pick up more electrons during the earlier reactions of glycolysis, and this keeps glycolysis and substrate-level ATP synthesis going.

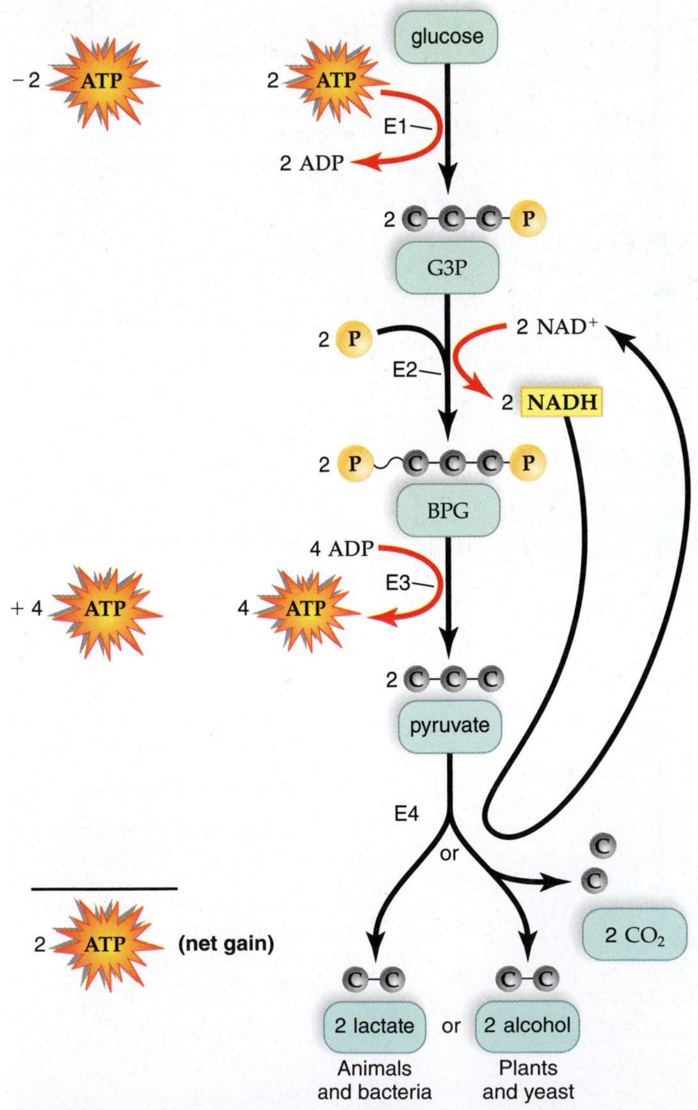

−2 ATP
2 ATP
glucose
E1
2 ADP
2 C-C-C-P
G3P
2 P
2 NAD+
E2
2 NADH
2 P~C-C-C-P
BPG
4 ADP
E3
+4 ATP
4 ATP
2 C-C-C
pyruvate
E4
or
C
C
2 CO2
2 ATP (net gain)
C-C
C-C
2 lactate
or
2 alcohol
Animals and bacteria
Plants and yeast

Figure 7.5 Fermentation. Fermentation consists of glycolysis followed by a reduction of pyruvate by NADH + H^+. The resulting NAD^+ returns to the glycolytic pathway to pick up more hydrogen atoms.

Advantages and Disadvantages of Fermentation

Despite its low yield of only two ATP, fermentation is essential to humans. It can provide a rapid burst of ATP, and thus muscle cells more than other cells are apt to carry on fermentation. When our muscles are working vigorously over a short period of time, as when we run, fermentation is a way to produce ATP even though oxygen is temporarily in limited supply.

Lactate, however, is toxic to cells. At first, blood carries away all the lactate formed in muscles. But eventually lactate begins to build up, changing the pH and causing the muscles to "burn." When we stop running, our bodies are in *oxygen debt,* as signified by the fact that we continue to breathe very heavily for a time. Recovery is complete when the lactate is transported to the liver, where it is reconverted to pyruvate. Some of the pyruvate is broken down completely, and the rest is converted back to glucose.

Energy Yield of Fermentation

Fermentation produces only two ATP by substrate-level ATP synthesis. These two ATP represent only a small fraction of the potential energy stored in a glucose molecule. As noted earlier, complete glucose breakdown during cellular respiration results in a maximum of 36 to 38 ATP. Therefore, following fermentation, most of the potential energy a cell can capture from the respiration of a glucose molecule is still waiting to be released.

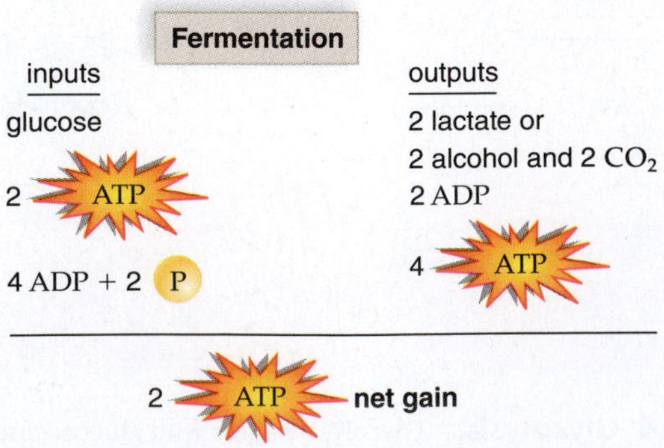

Fermentation
inputs
glucose
2 ATP
4 ADP + 2 P
outputs
2 lactate or
2 alcohol and 2 CO2
2 ADP
4 ATP
2 ATP net gain

Check Your Progress 7.3

1. Describe the environmental conditions that would cause a muscle cell to undergo fermentation.
2. Explain how fermentation acts as a NAD^+ recycling system.

7.4 Inside the Mitochondria

Learning Outcomes

Upon completion of this section, you should be able to

1. Recognize the importance of the mitochondria in cellular respiration.
2. Summarize the inputs and outputs of the preparatory reaction, the citric acid cycle, and the electron transport chain.
3. Identify how each stage of the aerobic pathway contributes to the generation of ATP in a cell.

The final reactions of cellular respiration, the preparatory reaction, the citric acid cycle, and the electron transport chain, all occur within the mitochondria.

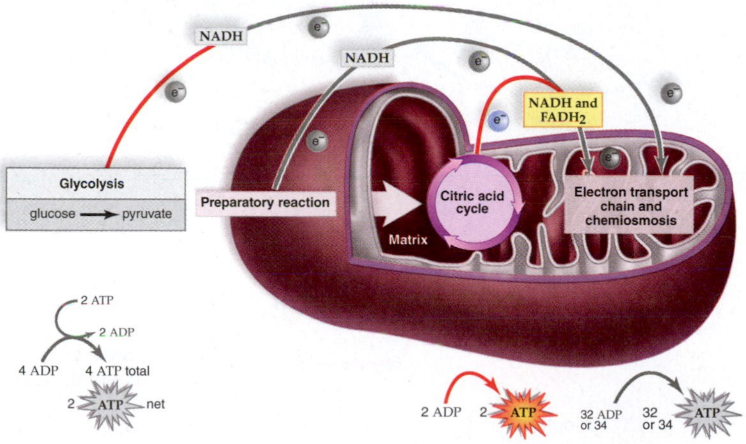
NADH
NADH
NADH and FADH2
Glycolysis
glucose → pyruvate
Preparatory reaction
Citric acid cycle
Electron transport chain and chemiosmosis
Matrix
2 ATP
2 ADP
4 ADP
4 ATP total
2 ATP net
2 ADP
2 ATP
32 ADP or 34
32 or 34 ATP

Preparatory Reaction

As stated, the preparatory reaction occurs inside the mitochondria. Where specifically does it occur? As you know, the *cristae* of a mitochondrion are folds of inner membrane that jut out into the *matrix,* an innermost compartment filled with a gel-like fluid. The preparatory reaction and the citric acid cycle are located in the matrix (see Fig. 7.2).

The **preparatory (prep) reaction** is so called because it produces the molecule that can enter the citric acid cycle. In this reaction, pyruvate is converted to a C_2 *acetyl group* attached to *coenzyme A* (*CoA*), and CO_2 is given off. This is an oxidation reaction in which hydrogen atoms ($H^+ + e^-$) are removed from pyruvate by NAD^+ and $NADH + H^+$ results. This reaction occurs twice per glucose molecule:

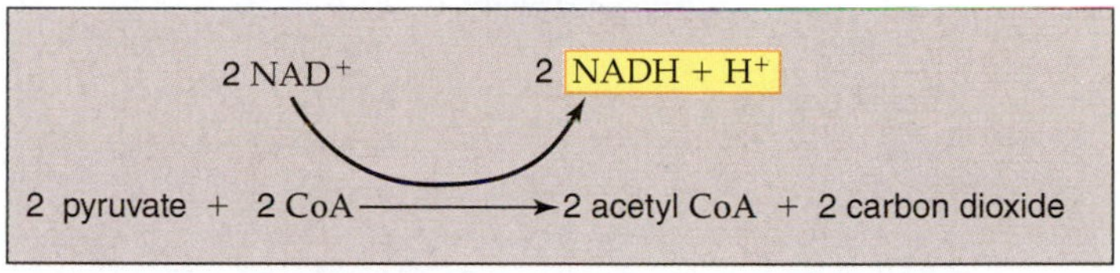
2 NAD+
2 NADH + H+
2 pyruvate + 2 CoA → 2 acetyl CoA + 2 carbon dioxide

Citric Acid Cycle

The **citric acid cycle** (also called the Krebs cycle) is a cyclical metabolic pathway located in the matrix of mitochondria (Fig. 7.6). At the start of the citric acid cycle, the C_2 acetyl group carried by CoA joins with a C_4 molecule, and a C_6 citrate molecule results. Note that because there are two acetyl CoA molecules entering the cycle for each glucose, the cycle will turn twice.

Animation How the Krebs Cycle Works

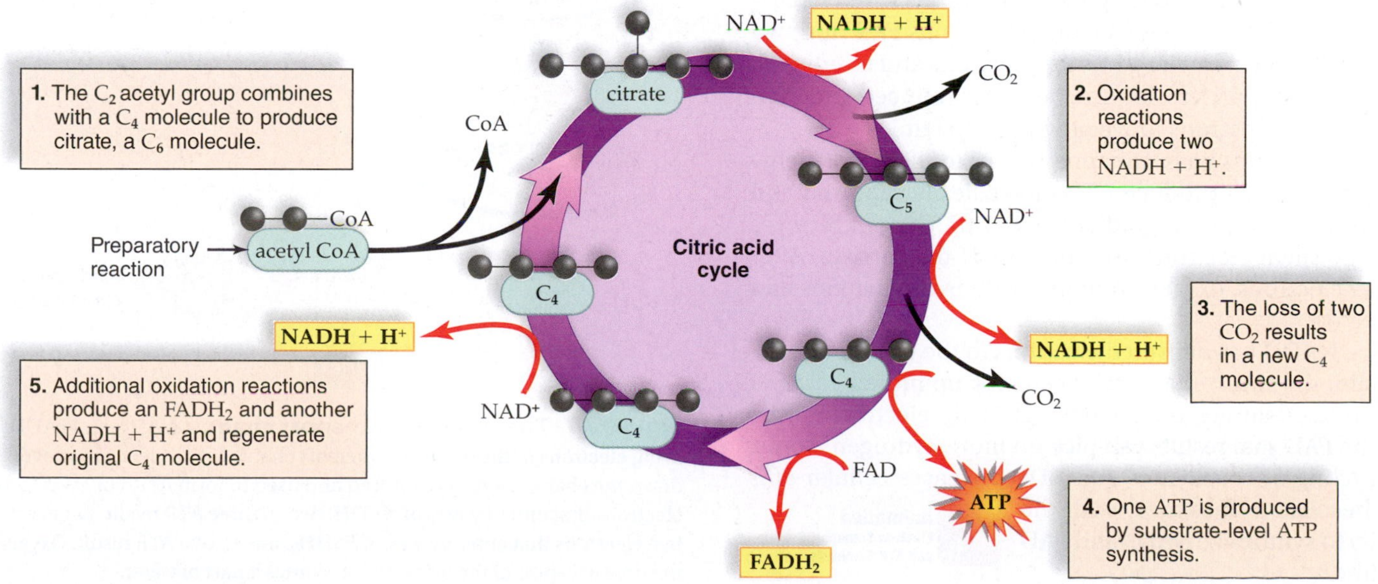
1. The C2 acetyl group combines with a C4 molecule to produce citrate, a C6 molecule.
CoA
Preparatory reaction
acetyl CoA
CoA
citrate
NAD+
NADH + H+
CO2
2. Oxidation reactions produce two NADH + H+.
C5
NAD+
Citric acid cycle
C4
NADH + H+
3. The loss of two CO2 results in a new C4 molecule.
C4
CO2
NADH + H+
NAD+
5. Additional oxidation reactions produce an FADH2 and another NADH + H+ and regenerate original C4 molecule.
C4
FAD
FADH2
ATP
4. One ATP is produced by substrate-level ATP synthesis.

Figure 7.6 Citric acid cycle. The net result of this cycle of reactions is the oxidation of an acetyl group to two molecules of CO_2, along with a transfer of electrons to NAD^+ and FAD and a gain of one ATP. The citric acid cycle turns twice per glucose molecule.

The CoA returns to the preparatory reaction to pick up another acetyl group. During the citric acid cycle, each acetyl group received from the preparatory reaction is oxidized to two CO_2 molecules. As the reactions of the cycle occur, oxidation is carried out by the removal of hydrogen atoms ($H^+ + e^-$). In three instances, NADH + H^+ results, and in one instance, $FADH_2$ is formed.

Substrate-level ATP synthesis is also an important event of the citric acid cycle. In substrate-level ATP synthesis, an enzyme passes a high-energy phosphate to ADP, and ATP results.

Because the citric acid cycle turns twice for each original glucose molecule, the inputs and outputs of the citric acid cycle per glucose molecule are as follows:

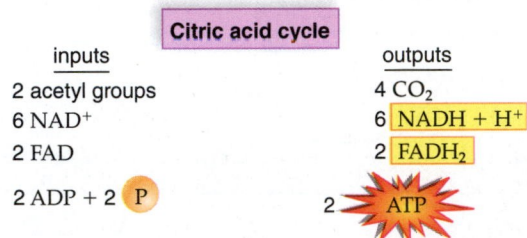

Citric acid cycle

inputs	outputs
2 acetyl groups	4 CO_2
6 NAD^+	6 NADH + H^+
2 FAD	2 $FADH_2$
2 ADP + 2 P	2 ATP

The six carbon atoms originally located in the glucose molecule have now become CO_2. The preparatory reaction produces two CO_2, and the citric acid cycle produces four CO_2 per glucose molecule.

3D Animation Cellular Respiration: Citric Acid Cycle

Electron Transport Chain

The **electron transport chain** located in the cristae of the mitochondria is a series of carriers that pass electrons from one to the other. Some of the electron carriers of the system are called *cytochrome* molecules. Cytochromes are a class of iron-containing proteins important in redox reactions.

Figure 7.7 is arranged to show that high-energy electrons enter the system and low-energy electrons leave the system. Notice that when NADH gives up its two electrons to the chain, it becomes NAD^+ and two H^+ remain. Similarly, when $FADH_2$ gives up two electrons to the chain, it becomes FAD and two H^+ remain. The next carrier gains the electrons and is reduced. Then each of the carriers, in turn, becomes reduced and then oxidized as the electrons move down the system.

As the electrons pass from one carrier to the next, energy that will be used to produce ATP molecules is captured and stored as a hydrogen ion gradient (Fig. 7.8). Oxygen receives the energy-spent electrons from the last of the carriers. After receiving electrons, oxygen combines with hydrogen ions and forms water.

Once NADH has delivered electrons to the electron transport chain, the NAD^+ that results can pick up more hydrogen atoms. In like manner, once $FADH_2$ gives up electrons to the chain, the FAD that results can pick up more hydrogen atoms. The recycling of coenzymes and ADP increases cellular efficiency because it does away with the necessity to synthesize NAD^+ and FAD every time.

Animation Electron Transport System and ATP Synthesis

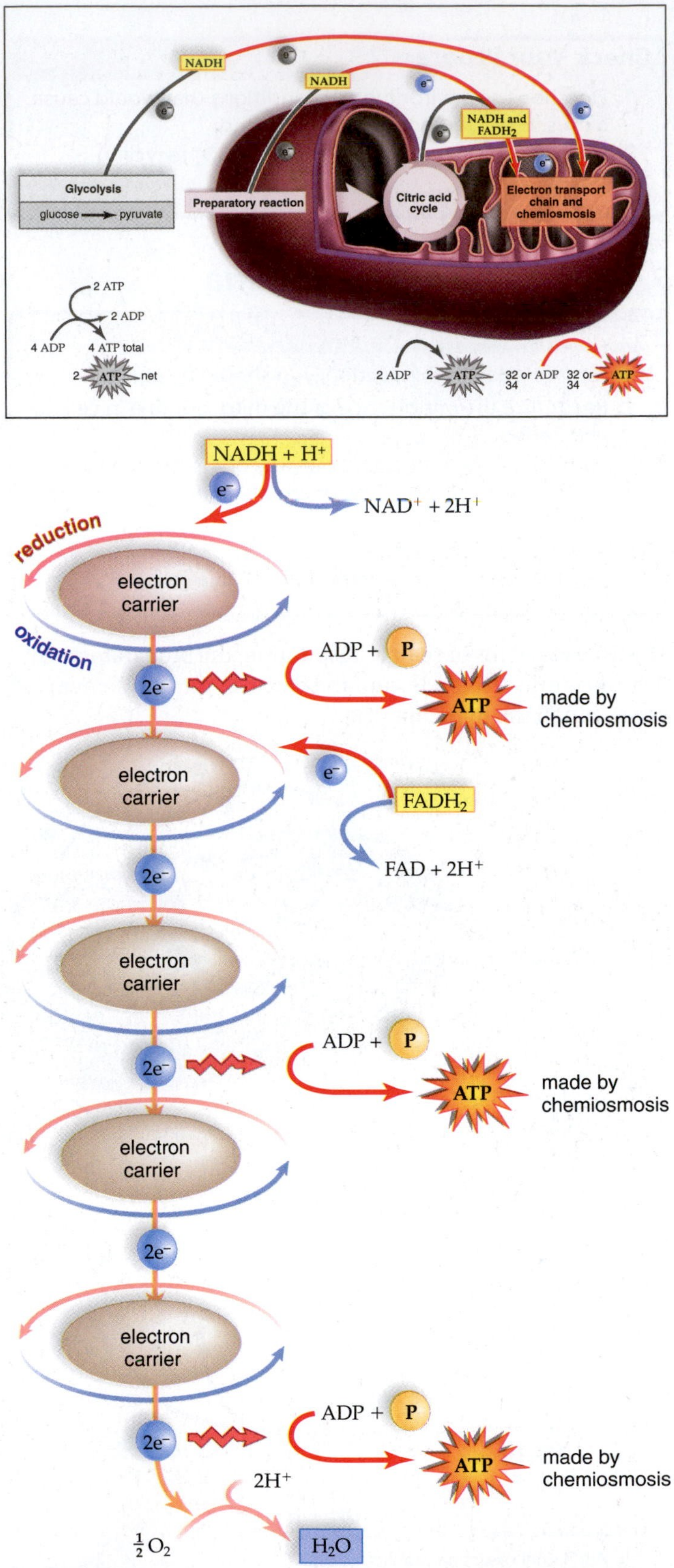
NADH
NADH
NADH and FADH2
e⁻
Glycolysis
glucose → pyruvate
Preparatory reaction
Citric acid cycle
Electron transport chain and chemiosmosis
2 ATP
2 ADP
4 ADP
4 ATP total
2 ATP net
2 ADP
2 ATP
32 or 34 ADP
32 or 34 ATP
NADH + H+
NAD+ + 2H+
reduction
oxidation
electron carrier
2e⁻
ADP + P
ATP
made by chemiosmosis
FADH2
FAD + 2H+
2H+
½ O2
H2O

Figure 7.7 The electron transport chain. NADH and $FADH_2$ bring electrons to the electron transport chain. As the electrons move down the chain, energy is captured and used to form ATP. For every two electrons that enter by way of NADH, two to three ATP result. For every two electrons that enter by way of $FADH_2$, one to two ATP result. Oxygen, the final acceptor of the electrons, becomes a part of water.

Figure 7.8 Organization and function of the electron transport chain. The electron transport chain is located in the cristae of the mitochondria. As electrons move from one protein complex to the other, hydrogen ions (H^+) are pumped from the mitochondrial matrix into the intermembrane space. As hydrogen ions flow down a concentration gradient from the intermembrane space into the matrix, ATP is synthesized by the enzyme ATP synthase. ATP leaves the matrix by way of a channel protein.

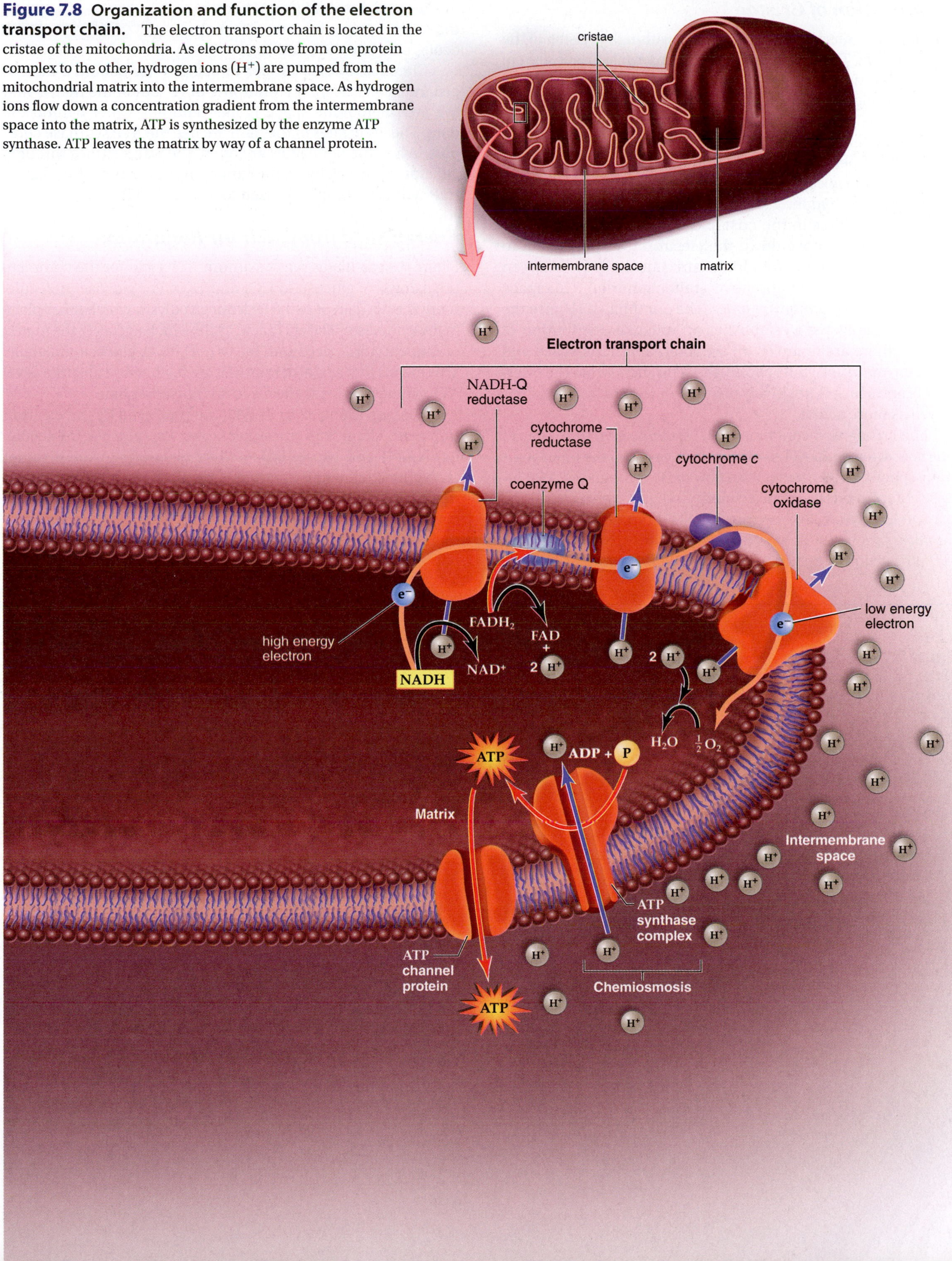

cristae
intermembrane space
matrix
Electron transport chain
NADH-Q reductase
cytochrome reductase
cytochrome c
coenzyme Q
cytochrome oxidase
high energy electron
low energy electron
NADH
NAD+
FADH2
FAD + 2 H+
2 H+
H2O
½ O2
ATP
ADP + P
Matrix
Intermembrane space
ATP synthase complex
ATP channel protein
Chemiosmosis

Organization of Cristae

The electron transport chain is located within the cristae of the mitochondria. The cristae increase the internal surface area of a mitochondrion, thereby increasing the area devoted to ATP formation. Figure 7.7 is a simplified overview of the electron transport chain. Figure 7.8 shows how the components of the electron transport chain are arranged in the cristae.

We have been stressing that the carriers of the electron transport chain accept electrons, which they pass from one to the other. What happens to the hydrogen ions from NADH + H^+? The complexes in the cristae use the energy released by electrons as they move down the electron transport chain to pump H^+ from the mitochondrial matrix into the space between the outer and inner membrane of a mitochondrion. This space is called the *intermembrane space.* The pumping of H^+ into the intermembrane space establishes an unequal distribution of H^+ ions; there are many H^+ in the intermembrane space and few in the matrix of a mitochondrion. This means that the intermembrane space is positively charged in relation to the matrix as well as more acidic.

The cristae also contain an *ATP synthase complex.* The H^+ ions flow through an ATP synthase complex from the intermembrane space into the matrix. The flow of H^+ through an ATP synthase complex brings about a change in shape, which causes the enzyme ATP synthase to synthesize ATP from ADP + Ⓟ. Mitochondria produce ATP by **chemiosmosis,** which indicates that ATP production is tied to an electrochemical gradient, namely the unequal distribution of H^+ across the cristae. Once formed, ATP molecules pass into the cytoplasm.

3D Animation
Cellular Respiration: Electron Transport Chain

Chemiosmosis is similar to using water behind a dam to generate electricity. The pumping of H^+ out of the matrix into the intermembrane space is like pumping water behind the dam. The floodgates of the dam are like the ATP synthase complex. When the floodgates are open, the water rushes through, generating electricity. In the same way, H^+ rushing through the ATP synthase complex is used to produce ATP.

Energy Yield from Cellular Respiration

Figure 7.9 calculates the maximum ATP yield for the complete breakdown of glucose to CO_2 and H_2O. Per glucose molecule, there is a net gain of two ATP from glycolysis, which takes place in the cytoplasm. The citric acid cycle, which occurs in the matrix of mitochondria, accounts for two ATP per glucose molecule. This means that a total of four ATP are formed by substrate-level ATP synthesis outside the electron transport chain.

The remaining 32 to 34 ATP are produced by the electron transport chain and chemiosmosis. Per glucose molecule, ten NADH and two $FADH_2$ deliver electrons to the electron transport chain. ATP synthesis can be measured by experiments where intact mitochondria are suspended in a solution containing oxygen. Such experiments yield between two and three ATP for the two electrons delivered to the electron transport chain by each NADH, and between one and two ATP for

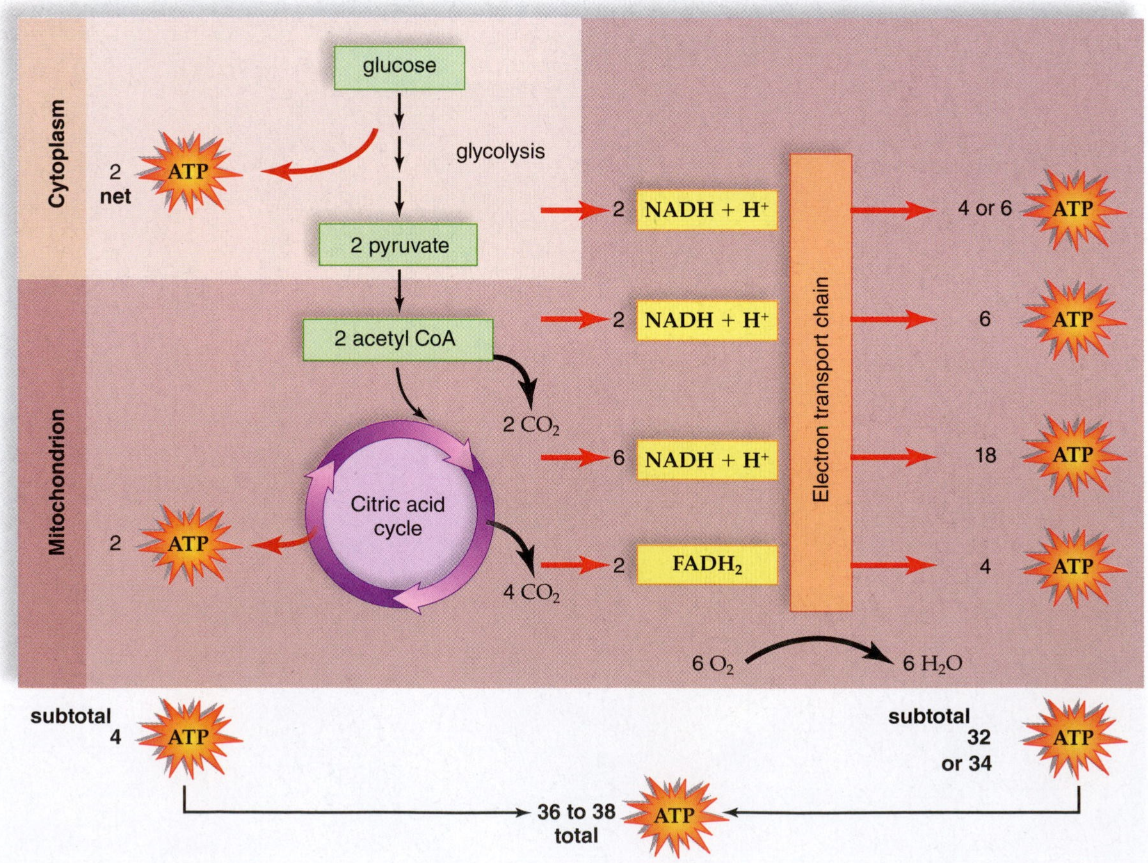
Cytoplasm
Mitochondrion
glucose
glycolysis
2 net ATP
2 pyruvate
2 acetyl CoA
2 CO₂
Citric acid cycle
2 ATP
4 CO₂
2 NADH + H⁺
2 NADH + H⁺
6 NADH + H⁺
2 FADH₂
Electron transport chain
4 or 6 ATP
6 ATP
18 ATP
4 ATP
6 O₂
6 H₂O
subtotal 4 ATP
subtotal 32 or 34 ATP
36 to 38 total ATP

Figure 7.9 Accounting of the maximum energy yield per glucose molecule breakdown. Substrate-level ATP synthesis during glycolysis and the citric acid cycle accounts for four ATP. Chemiosmosis accounts for a maximum of 32 or 34 ATP, depending on the shuttle mechanism involved for transporting cytoplasmic NADH + H^+ into the mitochondrion. The maximum total of ATP is therefore 36 to 38 ATP.

SCIENCE IN YOUR LIFE ▶ **HEALTH**

Metabolic Fate of Pizza

Obviously our diets do not solely consist of carbohydrates. Because fats and proteins are also organic nutrients, it makes sense that our bodies can utilize the energy found in the bonds of these molecules. In fact, the metabolic pathways we have discussed in this chapter are more than capable of accessing the energy of fats and proteins. For example, lets trace the fate of a pepperoni pizza, which contains carbohydrates (crust), fats (cheese), and protein (pepperoni).

We already know that the glucose in the carbohydrate crust is broken down during cellular respiration. When the cheese in the pizza (a fat) is used as an energy source, it breaks down to glycerol and three fatty acids. As Figure 7A indicates, glycerol can be converted to pyruvate and enter glycolysis. The fatty acids are converted to 2-carbon acetyl CoA that enters the citric acid cycle. An 18-carbon fatty acid results in nine acetyl CoA molecules. Calculation shows that respiration of these can produce a total of 108 ATP molecules. This is why fats are an efficient form of stored energy—the three long fatty acid chains per fat molecule can produce considerable ATP when needed.

Proteins are less frequently used as an energy source, but are available as necessary. The carbon skeleton of amino acids can enter glycolysis, be converted to acetyl groups, or enter the citric acid cycle at some other juncture. The carbon skeleton is produced in the liver when an amino acid undergoes deamination, or the removal of the amino group. The amino group becomes ammonia (NH_3), which enters the urea cycle and becomes part of urea, the primary excretory product of humans. Just where the carbon skeleton begins degradation depends on the length of the *R* group, because this determines the number of carbons left after deamination.

In Chapter 14, "Digestive System and Nutrition," we will take a more detailed look at the nutritional needs of humans, including discussions on how vitamins and minerals interact with metabolic pathways, and the dietary guidelines for proteins, fats, and carbohydrates.

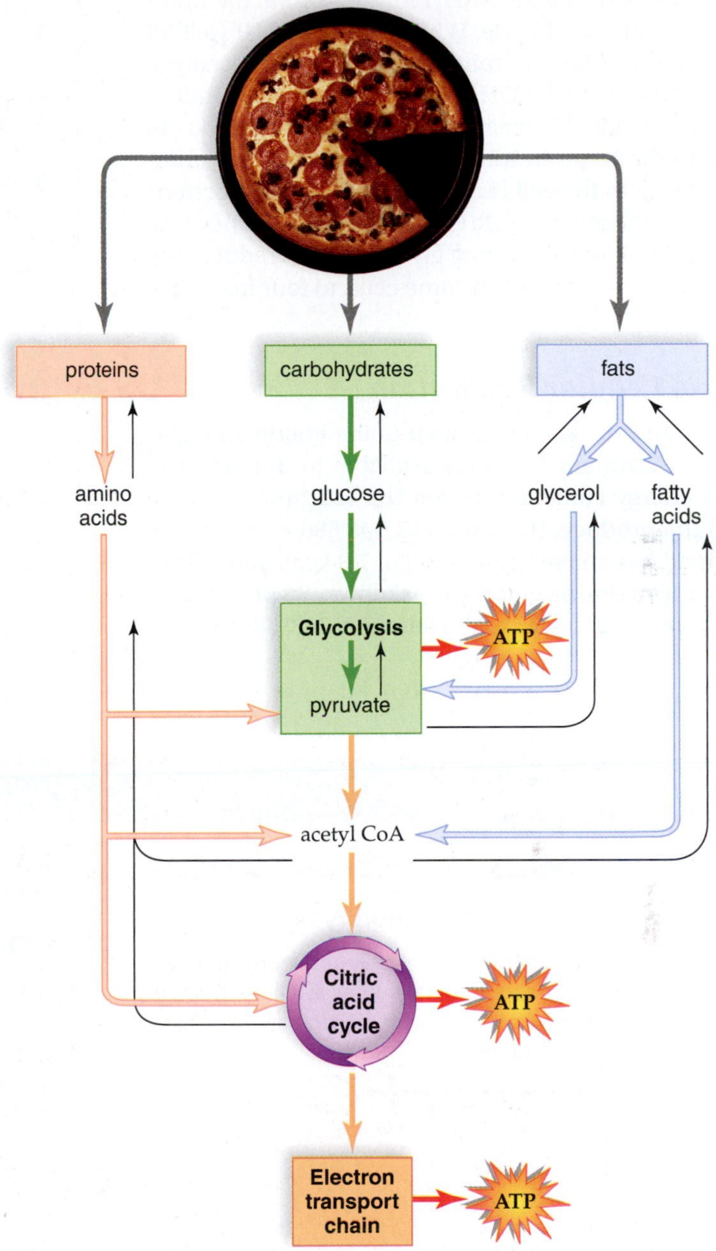
proteins
carbohydrates
fats
amino acids
glucose
glycerol
fatty acids
Glycolysis
pyruvate
ATP
acetyl CoA
Citric acid cycle
ATP
Electron transport chain
ATP

Figure 7A The metabolic pool concept. Carbohydrates, fats, and proteins can be used as energy sources, and their monomers (carbohydrates and proteins) or subunits (fats) enter degradative pathways at specific points.

Questions to Consider

1. How might a meal of a cheeseburger and fries be processed by the cellular respiration pathways?
2. While Figure 7A does not indicate the need for water, it is an important component of our diet. Where would water interact with these pathways?

the two electrons delivered by each $FADH_2$. Why the difference between NADH and $FADH_2$? Figure 7.7 shows the reason for this difference: $FADH_2$ delivers its electrons to the transport chain lower in the chain than does NADH, and therefore these electrons cannot account for as much ATP production.

These calculations are for NADH formed *inside* the mitochondria by the citric acid cycle. What about the ATP yield of NADH generated *outside* the mitochondria by the glycolytic pathway? In some cells, NADH cannot cross mitochondrial membranes, but a transfer mechanism allows its electrons to be delivered to the electron transport chain inside the mitochondria. The cost to the cell is one ATP for the two electrons transferred from outside to inside the mitochondrion. Because there are two NADH formed during glycolysis, this reduces the overall count of ATP produced, in some cells, to four instead of six ATP (Fig. 7.9).

Efficiency of Cellular Respiration

It is interesting to calculate how much of the energy in a glucose molecule eventually becomes available to the cell. The difference in energy content between the reactants (glucose and O_2) and the products (CO_2 and H_2O) is 686 kcal. An ATP phosphate bond has an energy content of 7.3 kcal, and if 36 of these are produced during glucose breakdown, 36 phosphates are equivalent to a total of 263 kcal. Therefore, 263/686, or 39%, of the available energy is usually transferred from glucose to ATP. The rest of the energy dissipates in the form of heat.

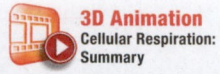

3D Animation
Cellular Respiration: Summary

Check Your Progress 7.4

1. Explain the relationship between the metabolic pathways within the mitochondria with glycolysis.
2. Calculate the number of NADH, $FADH_2$, and ATP molecules produced by each stage of cellular respiration per glucose molecule.
3. Discuss why there is variation in the number of ATP molecules produced per glucose.

Case Study Conclusion

As we have observed, the metabolic pathways have the ability to break down energy nutrients, primarily carbohydrates, and produce ATP. The specific pathway that is followed is dependent on the availability of oxygen. During aerobic exercise, oxygen is available for the muscles to completely break down glucose to carbon dioxide and water. During anaerobic exercise, oxygen is not available and the muscles produce lactic acid.

MEDIA STUDY TOOLS

www.mhhe.com/maderinquiry14

Enhance your study of this chapter with study tools and practice tests. Also ask your instructor about the resources available through ConnectPlus, including LearnSmart, the media-rich eBook, interactive learning tools, and animations.

Mc Graw Hill connect plus+ | BIOLOGY

3D Animation
Cellular Respiration

For an interactive exploration of the phases of cellular respiration, watch McGraw-Hill's 3D animation "Cellular Respiration."

SUMMARIZE

7.1 Overview of Cellular Respiration

- The metabolic pathways that produce ATP may be either **aerobic** or **anaerobic,** depending on the availability of oxygen. During aerobic **cellular respiration,** glucose is oxidized to CO_2 and H_2O. Four phases are required for glucose to be metabolized to carbon dioxide and water. These include glycolysis, the preparatory reactions, the citric acid cycle, and the electron transport chain. During oxidation of substrates, electrons are removed along with hydrogen ions (H^+). **NAD^+ (nicotinamide adenine dinucleotide)** and **FAD (flavin adenine dinucleotide)** act as electron carriers. In the absence of oxygen, fermentation reactions occur.

7.2 Outside the Mitochondria: Glycolysis

- **Glycolysis,** the breakdown of glucose to two pyruvates, is a series of anaerobic enzymatic reactions that occur in the cytoplasm.
- Oxidation by NAD^+ releases enough energy immediately to give a net gain of two ATP by **substrate-level ATP synthesis.** Two NADH + H^+ are formed.

7.3 Outside the Mitochondria: Fermentation

- **Fermentation** involves glycolysis, followed by the reduction of pyruvate by NADH + H^+ to either lactate or alcohol and CO_2. The reduction process regenerates NAD^+ so that it can accept more electrons during glycolysis.
- Although fermentation results in only two ATP, it still serves a purpose: in humans, it provides a quick burst of ATP energy for short-term, strenuous muscular activity. The accumulation of lactate puts the individual in oxygen debt because oxygen is needed to completely metabolize lactate to CO_2 and H_2O.

7.4 Inside the Mitochondria

- Pyruvate from glycolysis enters a mitochondrion, where the **preparatory (prep) reaction** takes place. During this reaction, oxidation occurs as CO_2 is removed. NAD^+ and FAD are reduced, and CoA receives the C_2 acetyl group that remains. Because the reaction must take place twice per glucose, two NADH + H^+ result.

- The acetyl group enters the **citric acid cycle,** a series of reactions located in the mitochondrial matrix. Complete oxidation follows, as two CO_2, three NADH + H^+, and one $FADH_2$ are formed. The cycle also produces one ATP. The entire cycle must turn twice per glucose molecule.
- The final stage of glucose breakdown involves the **electron transport chain** located in the cristae of the mitochondria. The electrons received from NADH and $FADH_2$ are passed down a chain of electron carriers until they are finally received by O_2, which combines with H^+ to produce H_2O. As the electrons pass down the chain, ATP is produced.
- The carriers of the electron transport chain are located in protein complexes on the cristae of the mitochondria. Each protein complex receives electrons and pumps H^+ into the intermembrane space, setting up an electrochemical gradient. When H^+ ions flow down this gradient through the ATP synthase complex, energy is released and used to form ATP molecules from ADP and Ⓟ. This is ATP synthesis by **chemiosmosis.**
- To calculate the total number of ATP produced per glucose molecule, consider that for each NADH + H^+ formed inside the mitochondrion, three ATP are produced. Each molecule of $FADH_2$ results in the formation of only two ATP because the electrons enter the electron transport chain at a lower energy level than the electrons delivered by NADH. In cells that cannot transfer NADH across the mitochondrial membrane, the ATP counts are reduced by two because the electrons from NADH generated in the cytoplasm during glycolysis must be transferred from outside to inside the mitochondria using two ATP.
- Of the maximum number of 36 or 38 ATP formed by cellular respiration, four are produced outside the electron transport chain: two by glycolysis and two by the citric acid cycle. The rest are produced by the electron transport chain.

ASSESS

Testing Yourself

Choose the best answer for each question.

For questions 1–5, match the letters in the diagram to the appropriate statement.

1. Preparatory reaction
2. Electron transport chain
3. Glycolysis
4. Citric acid cycle
5. NADH + H^+ and $FADH_2$

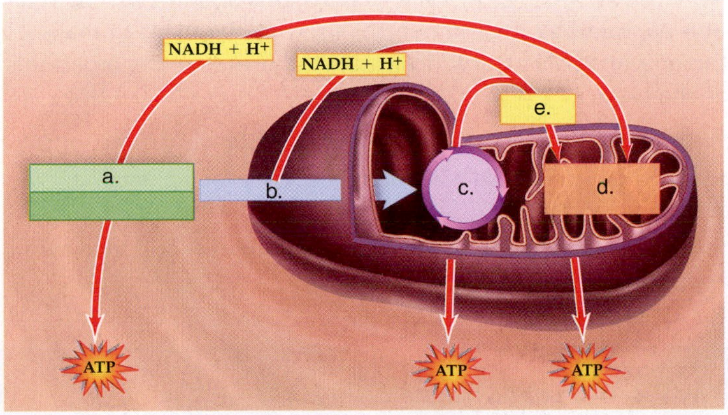
NADH + H+
NADH + H+
e.
a.
b.
c.
d.
ATP
ATP
ATP

6. Which phase of cellular respiration occurs in the cytoplasm?
   a. glycolysis
   b. citric acid cycle
   c. preparatory reaction
   d. electron transport chain
7. Cellular respiration cannot occur without
   a. sodium.
   b. oxygen.
   c. lactic acid.
   d. All of these are correct.
8. Mitochondria produce ATP by
   a. glycolysis.
   b. fermentation.
   c. chemiosmosis.
   d. All of these are correct.
9. An ATP synthase complex can be found in the
   a. cytoplasm.
   b. mitochondria.
   c. Both a and b are correct.
   d. Neither a nor b is correct.
10. Which of the following is needed for glycolysis to occur?
    a. pyruvate
    b. glucose
    c. NAD^+
    d. ATP
    e. All of these are needed except a.
11. Which of the following is *not* a product, or end result, of the citric acid cycle?
    a. carbon dioxide
    b. pyruvate
    c. NADH + H^+
    d. ATP
    e. $FADH_2$
12. How many ATP molecules are produced by the oxidation of one molecule of NADH via the electron transport chain?
    a. 1 to 2
    b. 2 to 3
    c. 36 to 38
    d. 8 to 10
13. Which are possible products of fermentation?
    a. lactic acid
    b. alcohol
    c. CO_2
    d. All of these are correct.
14. The metabolic process that produces the most ATP molecules is
    a. glycolysis.
    b. the citric acid cycle.
    c. the electron transport chain.
    d. fermentation.
15. Substrate-level ATP synthesis takes place during
    a. glycolysis and the citric acid cycle.
    b. the electron transport chain and the prep reaction.
    c. glycolysis and the electron transport chain.
    d. the citric acid cycle and the prep reaction.

ENGAGE

Thinking Critically

1. Bacteria do not have mitochondria, and yet they contain an electron transport chain. On what membrane could this be located?
2. Rotenone is a broad-spectrum insecticide that inhibits the electron transport chain. Why might it be toxic to humans?
3. Some fat-burning compounds accelerate the movement of fatty acids into the cellular respiration pathways. Explain how these compounds may work.

UNIT 2 Plant Biology

8 Photosynthesis

CHAPTER OUTLINE

BEFORE YOU BEGIN

Before beginning this chapter, take a few moments to review the following discussions:

Section 3.3 Which cellular structures are necessary for photosynthesis?

Figure 6.1 How does energy flow in biological systems?

Section 6.3 What role do enzymes play in regulating metabolic processes?

CASE STUDY Colors of Fall

Taking a walk in the woods in the fall when the leaves are turning colors can be an enjoyable and relaxing form of exercise. Interestingly, the same process that causes leaves to change colors in the fall is also involved in the ripening of fruits such as apples and pears—and the end product of the process may have health benefits as well! Leaves contain several types of pigments, including the green chlorophylls and the yellow to red carotenoids. These pigments absorb the solar energy that the plant utilizes to carry out photosynthesis. In the fall when lower temperatures signal a change in the seasons, the supply of water and nutrients to the leaves declines and chlorophyll begins to degrade. Now, the carotenoids, which were formerly masked by chlorophyll, become visible allowing us to enjoy the change of color. The breakdown of chlorophyll provides us with colorful woodlands and can also give us a healthy dose of antioxidants when we eat fruits. Researchers have discovered that when a fruit ripens and its skin changes color, chlorophyll is degraded in the same manner as in leaves. This produces antioxidants that become concentrated in the skin. Antioxidants stabilize free radicals, dangerous molecules that otherwise damage the DNA and proteins of a cell. Several health problems, including cancer and heart disease, are thought to be promoted by the accumulation of free radicals. In this chapter, we will see how pigments are involved in the process of photosynthesis. The solar energy that is captured by the pigments is used by the plant to make its food. It is this food that will be used to sustain the plant and the organisms that feed on the plant.

As you read through the chapter, think about the following questions:

1. Which pigments provide the maximum efficiency for a plant as it conducts photosynthesis?
2. Why do leaves appear green in the spring and summer and then turn to red or yellow in the fall?

8.1 Overview of Photosynthesis

Learning Outcomes

Upon completion of this section, you should be able to

1. Compare and contrast autotrophs and heterotrophs.
2. Explain the role of photosynthesis for all organisms on Earth.
3. Recognize the overall chemical equation for photosynthesis.
4. Describe the process of photosynthesis in terms of two sets of reactions that take place in a chloroplast.

Photosynthesis converts solar energy into the chemical energy of a carbohydrate. Photosynthetic organisms, including plants, algae, and cyanobacteria, are called **autotrophs** because they produce their own food (Fig. 8.1). Each year, photosynthesizing organisms produce approximately 170 billion metric tons of carbohydrates. No wonder photosynthetic organisms are able to sustain themselves and all other organisms on Earth.

With some exceptions, it is possible to trace the majority of food chains back to plants and algae. In other words, producers, which have the ability to synthesize carbohydrates, feed not only themselves but also consumers, which must take in preformed organic molecules. Collectively, consumers are called **heterotrophs.** Both autotrophs and heterotrophs use organic molecules produced by photosynthesis as a source of building blocks for growth and repair and as a source of chemical energy for cellular work.

Pigments allow photosynthetic organisms to capture solar energy, which acts as the "fuel" that makes photosynthesis possible. Most photosynthetic organisms contain **chlorophyll,** the pigment that gives them a green color. However, the green of chlorophyll can be masked by other pigments. The **carotenoids** give photosynthesizing cells a yellow to red color. In addition, the phycobilins give red algae their red color and cyanobacteria a bluish color.

Flowering Plants as Photosynthesizers

Although many different organisms can photosynthesize, we will focus our discussion on the flowering plants. Portions of the plant, particularly the leaves, contain chlorophyll and other pigments that enable the plant to carry on photosynthesis. The leaf of a flowering plant

Video Plants

a. *Oscillatoria* 100× b. Kelp c. Sequoias

Figure 8.1 Photosynthetic organisms. Photosynthetic organisms include **a.** cyanobacteria such as *Oscillatoria,* which are a type of bacterium; **b.** algae such as kelp, which typically live in water and can range in size from microscopic to macroscopic; and **c.** plants such as the sequoia, which typically live on land.

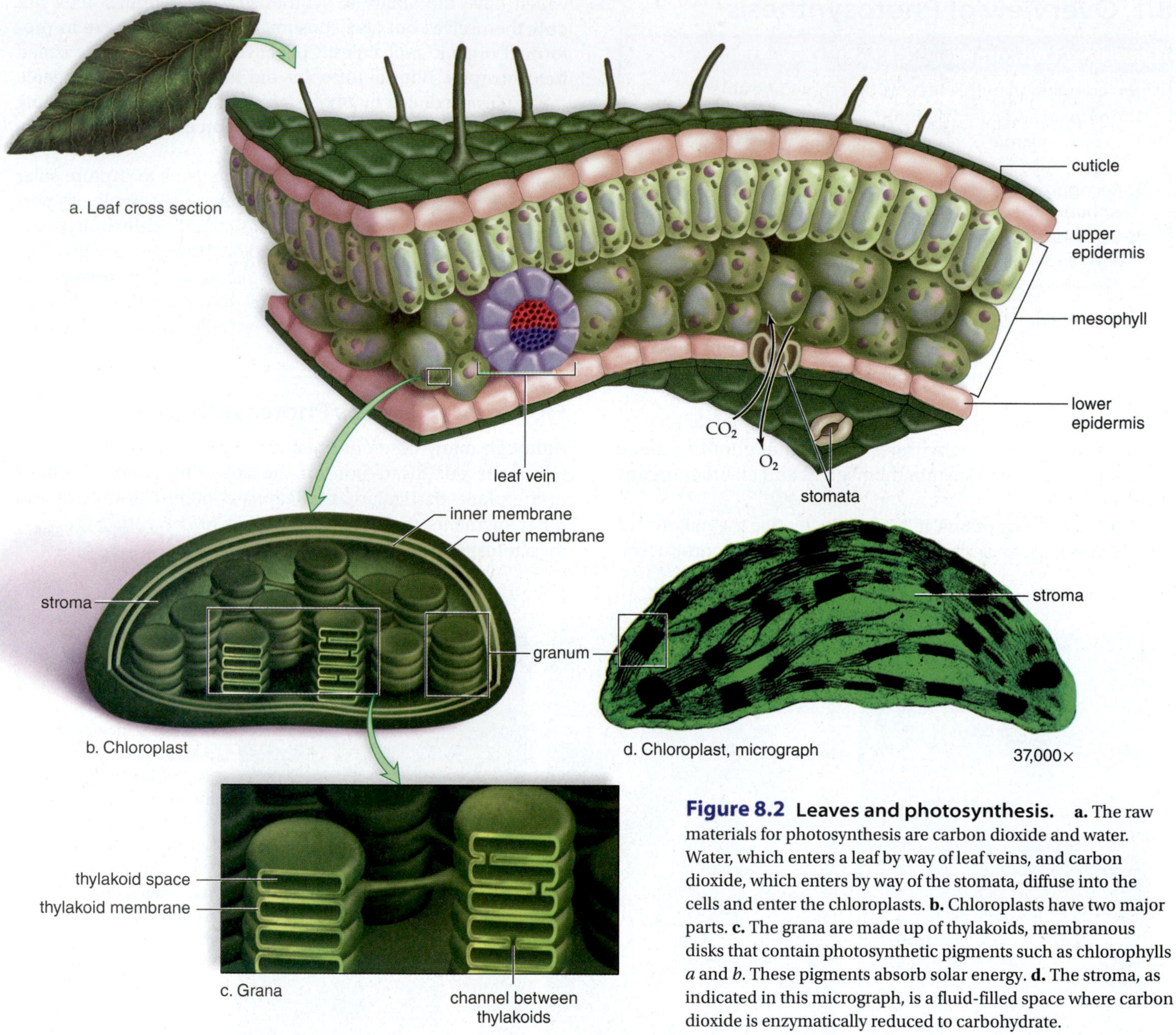
a. Leaf cross section
cuticle
upper epidermis
mesophyll
lower epidermis
leaf vein
CO_2
O_2
stomata
inner membrane
outer membrane
stroma
granum
b. Chloroplast
stroma
d. Chloroplast, micrograph
37,000×
thylakoid space
thylakoid membrane
c. Grana
channel between thylakoids

Figure 8.2 Leaves and photosynthesis. **a.** The raw materials for photosynthesis are carbon dioxide and water. Water, which enters a leaf by way of leaf veins, and carbon dioxide, which enters by way of the stomata, diffuse into the cells and enter the chloroplasts. **b.** Chloroplasts have two major parts. **c.** The grana are made up of thylakoids, membranous disks that contain photosynthetic pigments such as chlorophylls *a* and *b*. These pigments absorb solar energy. **d.** The stroma, as indicated in this micrograph, is a fluid-filled space where carbon dioxide is enzymatically reduced to carbohydrate.

contains mesophyll tissue, which contains cells that are specialized for photosynthesis (Fig. 8.2*a*). The raw materials for photosynthesis are water and carbon dioxide. The roots of a plant absorb water, which then moves through vascular tissue up the stem to a leaf by way of the leaf veins. Carbon dioxide in the air enters a leaf through small openings called **stomata** (sing., stoma). After entering a leaf, carbon dioxide and water diffuse into the cells and enter the chloroplasts (Fig. 8.2*b, d*), the organelles that carry on photosynthesis.

3D Animation Photosynthesis: Structure of a Chloroplast

A double membrane surrounds a chloroplast and its fluid-filled interior, which is called the stroma**.** A different membrane system within the stroma forms flattened sacs called thylakoids (Fig. 8.2*c*). In some places, thylakoids are stacked to form *grana* (sing., granum), so called because they looked like piles of seeds to early microscopists. The space of each thylakoid is connected to the space of every other thylakoid within a chloroplast, thereby forming a continuous inner compartment within the chloroplast called the thylakoid space.

Chlorophyll and other pigments that are part of a thylakoid membrane are capable of absorbing solar energy. This is the energy that drives photosynthesis. The stroma, filled with enzymes, is where carbon dioxide is first attached to an organic compound and is then reduced to a carbohydrate. Therefore, it is proper to associate the absorption of solar energy with the thylakoid membranes making up the grana, and to associate the reduction of carbon dioxide to a carbohydrate with the stroma of a chloroplast.

Humans, and indeed nearly all organisms, release carbon dioxide into the air as a waste product from metabolic processes such as cellular respiration. This is some of the same carbon dioxide that enters a leaf through the stomata and is converted to carbohydrate. Carbohydrate, in the form of glucose, is the chief energy source for most organisms.

Photosynthetic Reaction

For convenience, the overall equation for photosynthesis is sometimes simplified in this manner:

$$6\,CO_2 + 6\,H_2O \xrightarrow[\text{pigments}]{\text{solar energy}} C_6H_{12}O_6 + 6\,O_2$$

This equation shows glucose and oxygen as the products of photosynthesis.

The oxygen given off by photosynthesis comes from water. This was proven experimentally by exposing plants first to CO_2 and then to H_2O that contained an isotope of oxygen called heavy oxygen (^{18}O). Only when heavy oxygen was a part of water did this isotope appear in O_2 given off by the plant. Therefore, O_2 released by chloroplasts comes from H_2O, not from CO_2.

The equation also shows that during photosynthesis, CO_2 gains hydrogen atoms and becomes a carbohydrate. Reduction of any molecule requires energy and during photosynthesis this energy is ultimately provided by the sun.

The overall chemical equation for photosynthesis tells us why plants are so important to our lives. Plants provide us with food in the form of carbohydrates and much of the oxygen that we breathe and use for cellular respiration.

Two Sets of Reactions

The overall equation tells us H_2O is oxidized and CO_2 is reduced during photosynthesis. The word photosynthesis suggests that the process requires two sets of reactions: *photo,* which means light, refers to the reactions that capture solar energy, and *synthesis* refers to the reactions that produce carbohydrate.

The two sets of reactions are called the **light reactions** (light-dependent reactions) and the **Calvin cycle reactions** (light-independent reactions) (Fig. 8.3). We will see that the light reactions release oxygen and provide the molecules that allow the Calvin cycle reactions to reduce carbon dioxide to a carbohydrate. The coenzyme **NADP$^+$** (**nicotinamide adenine dinucleotide phosphate**) carries hydrogen atoms from the light reactions to the Calvin cycle reactions. When NADP$^+$ accepts hydrogen atoms, it becomes NADPH. Also, ATP carries energy from the light reactions to the Calvin cycle reactions.

The light reactions are sometimes called the light-dependent reactions because they cannot occur unless light is

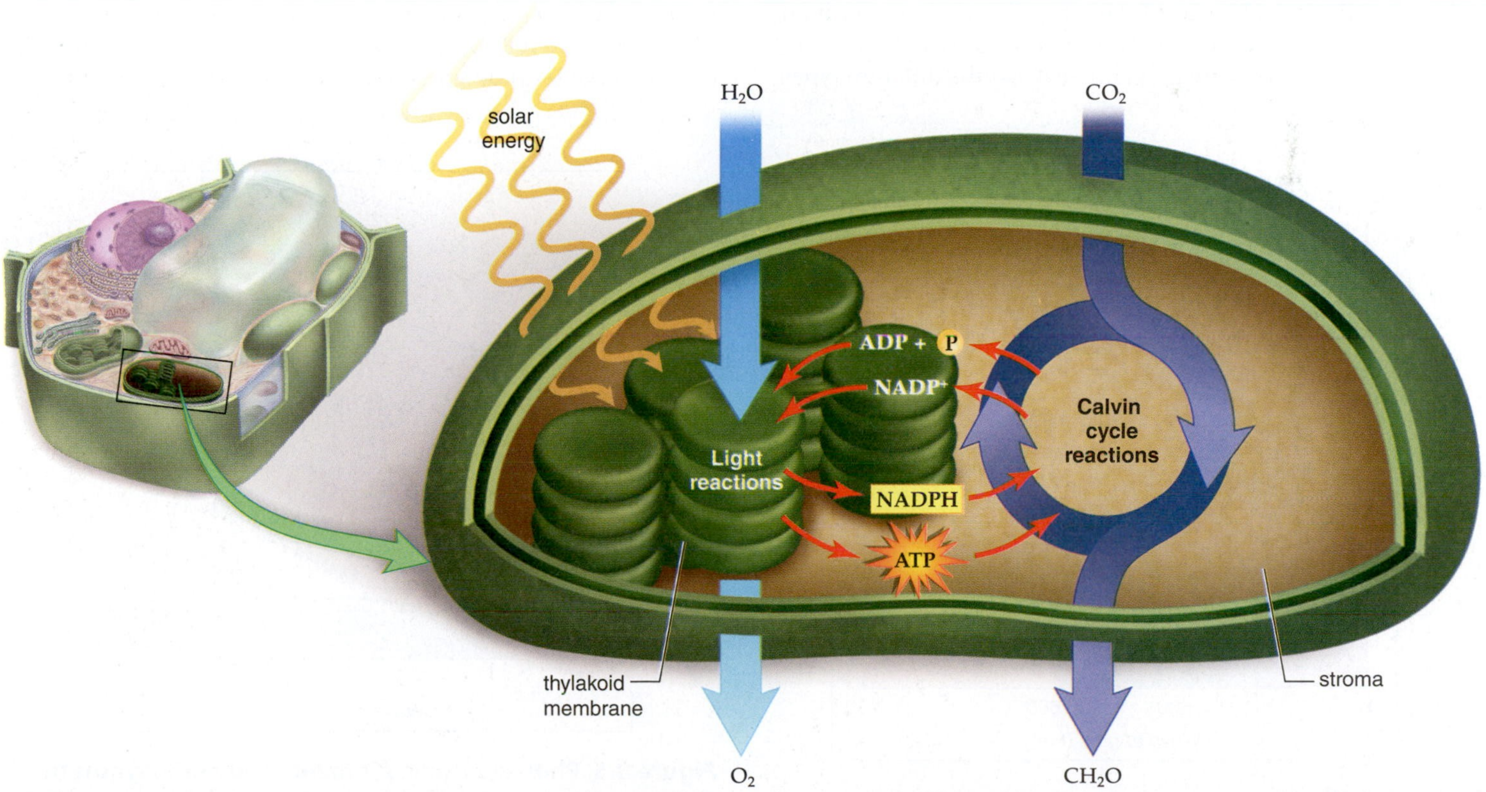
solar energy
H₂O
CO₂
ADP + P
NADP+
Light reactions
NADPH
ATP
Calvin cycle reactions
thylakoid membrane
stroma
O₂
CH₂O

Figure 8.3 Overview of photosynthesis. The process of photosynthesis consists of the light reactions and the Calvin cycle reactions. The light reactions, which produce ATP and NADPH, occur in the thylakoid membrane. These molecules are used in the Calvin cycle reactions in the stroma to reduce carbon dioxide to a carbohydrate.

present. The Calvin cycle reactions are sometimes called the light-independent reactions because they will occur whether light is present or not.

Check Your Progress 8.1

1. Recognize the primary source of energy for carbohydrate production in plants.
2. Identify the major components in the structure of a chloroplast.
3. Identify the overall equation for photosynthesis.
4. Describe the two sets of reactions involved in photosynthesis.

8.2 Plants as Solar Energy Converters

Learning Outcomes

Upon completion of this section, you should be able to

1. Identify the photosynthetic pigments required to absorb the various wavelengths of light necessary for photosynthesis.
2. Explain the role of the noncyclic electron pathway and the cyclic electron pathway.
3. Describe the organization of the thylakoid and how this organization is critical to the production of ATP during photosynthesis.

During the light reactions, the various pigments within the thylakoid membranes absorb solar energy. Solar energy (radiant energy from the sun) can be described in terms of its wavelength and its energy content. Figure 8.4 lists the different types of radiant energy, from the shortest wavelength, gamma rays, to the longest, radio waves. White, or *visible,* light is only a small portion of this spectrum.

3D Animation Photosynthesis: Properties of Light

Visible Light

Visible light itself contains various wavelengths of light. When it is passed through a prism, we see all the different colors that make up visible light. (Actually, it is our brains that interpret these wavelengths as colors.) The colors in visible light range from violet (the shortest wavelength) to indigo, blue, green, yellow, orange, and red (the longest wavelength). The energy content is highest for violet light and lowest for red light.

Only about 42% of the solar radiation that hits Earth's atmosphere ever reaches the surface of Earth, and most of this radiation is within the visible-light range. Higher-energy wavelengths are screened out by the ozone layer in the atmosphere, and lower-energy wavelengths are screened out by water vapor and carbon dioxide (CO_2) before they reach the Earth's surface. Both the organic molecules within organisms and certain life processes, such as vision and photosynthesis, are adapted to the solar radiation that is most prevalent in the environment.

The pigments found within most types of photosynthesizing cells, the chlorophylls *a* and *b* and the carotenoids, are capable of absorbing various portions of visible light. The absorption spectrum for these pigments is shown in Figure 8.5. Both chlorophyll *a* and chlorophyll *b* absorb violet, indigo, blue, and red light better than the light of other colors. Because green light is reflected and only minimally absorbed, leaves appear green to us. The yellow or orange carotenoids are able to absorb light in the violet-blue-green range. These pigments and others

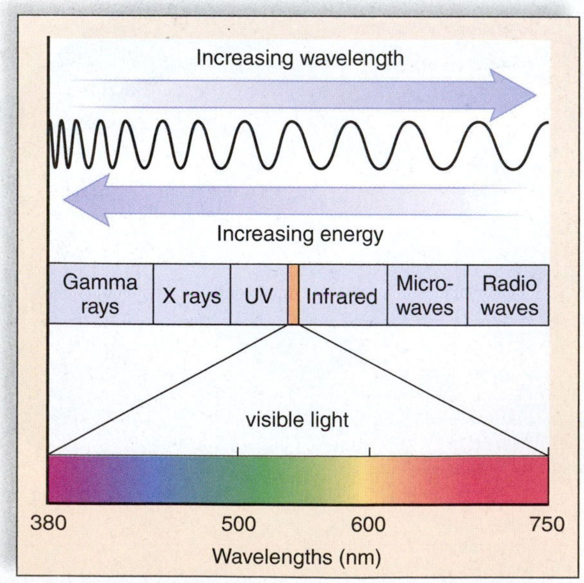
Increasing wavelength
Increasing energy
Gamma rays
X rays
UV
Infrared
Micro-waves
Radio waves
visible light
380
500
600
750
Wavelengths (nm)

Figure 8.4 The electromagnetic spectrum. The electromagnetic spectrum extends from the very short wavelengths of gamma rays through the very long wavelengths of radio waves. Visible light, which drives photosynthesis, is expanded to show its component colors. The components differ according to wavelength and energy content.

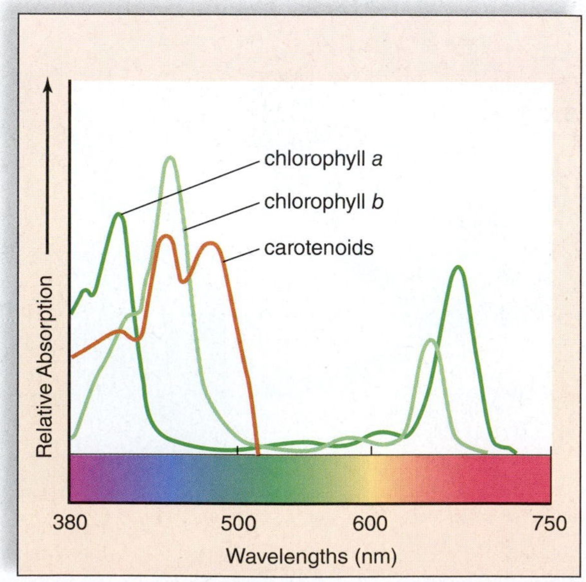
chlorophyll *a*
chlorophyll *b*
carotenoids
Relative Absorption
380
500
600
750
Wavelengths (nm)

Figure 8.5 Photosynthetic pigments and photosynthesis. The photosynthetic pigments in chlorophylls *a* and *b* and the carotenoids absorb certain wavelengths within visible light. This is their absorption spectrum. Notice that they do not absorb green light. That is why the leaves of plants appear green to us. You observe the color that is not absorbed by the leaf.

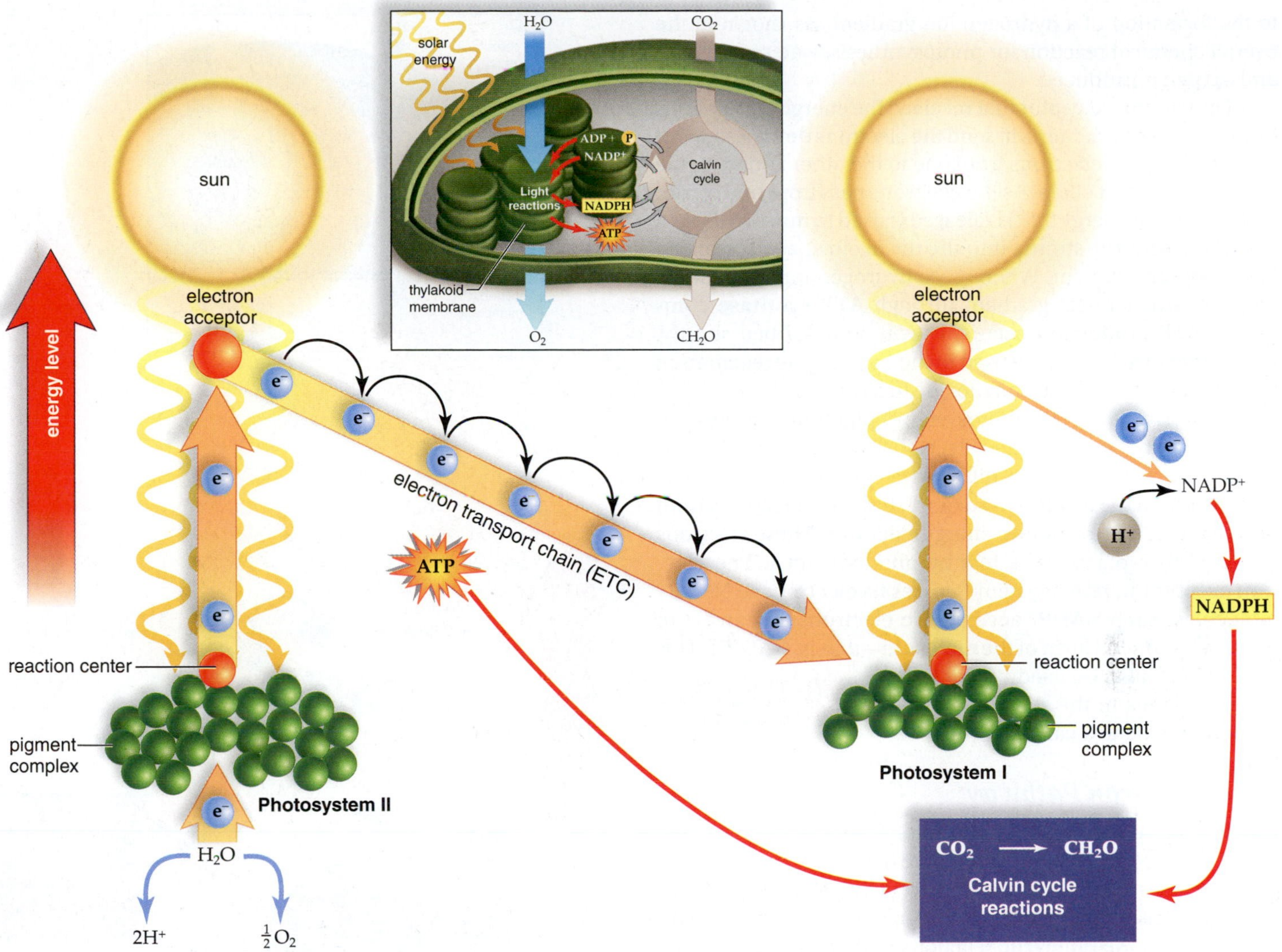
solar energy
H_2O
CO_2
ADP + P
NADP+
Calvin cycle
Light reactions
NADPH
ATP
thylakoid membrane
O_2
CH_2O
sun
energy level
electron acceptor
e−
electron transport chain (ETC)
ATP
reaction center
pigment complex
Photosystem II
H_2O
$2H^+$
$\frac{1}{2}O_2$
sun
electron acceptor
NADP+
H^+
NADPH
reaction center
pigment complex
Photosystem I
CO_2 → CH_2O
Calvin cycle reactions

Figure 8.6 Noncyclic electron pathway: Electrons move from water to NADP⁺. Energized electrons (taken from water, which splits, releasing oxygen) leave photosystem II and pass down an electron transport chain, leading to the formation of ATP. Energized electrons (replaced by photosystem II) leave photosystem I and pass to $NADP^+$, which then combines with H^+, becoming NADPH.

become noticeable in the fall when chlorophyll breaks down and the other pigments are uncovered. Photosynthesis begins when the pigments within thylakoid membranes absorb solar energy.

Animation Absorption of Light

Light Reactions

The light reactions that occur in the thylakoid membrane consist of two electron pathways called the noncyclic electron pathway and the cyclic electron pathway. Both electron pathways produce ATP, but only the noncyclic pathway also produces NADPH.

Noncyclic Electron Pathway

The **noncyclic electron pathway** is so named because the electron flow can be traced from water to a molecule of $NADP^+$ (Fig. 8.6). This pathway uses two photosystems, called photosystem I (PS I) and photosystem II (PS II). The photosystems are named for the order in which they were discovered, not for the order in which they participate in the photosynthetic process. A **photosystem** consists of a pigment complex (molecules of chlorophyll *a,* chlorophyll *b,* and the carotenoids) and an electron acceptor within the thylakoid membrane. The pigment complex serves as an "antenna" for gathering solar energy.

The noncyclic pathway begins with photosystem II. The pigment complex absorbs solar energy, which is then passed from one pigment molecule to another until it is concentrated in a particular pair of chlorophyll *a* molecules called the *reaction center.* Electrons (e^-) in the reaction center chlorophyll become so energized that they escape from the reaction center and move to a nearby electron acceptor.

Photosystem II would disintegrate without replacement electrons. These replacement electrons are provided by water, which splits, releasing oxygen to the atmosphere. Many organisms, including plants and even ourselves, use this oxygen. The hydrogen ions (H^+) stay in the thylakoid space and contribute to the formation of a hydrogen ion gradient. As shown in the overall chemical reaction for photosynthesis, water is used up and oxygen is produced.

The electron acceptor that received the energized electrons from the reaction center will send the electrons down a series of carriers that pass electrons from one to the other known as the electron transport chain. As the electrons pass from one carrier to the next, the energy that is released is used to move hydrogen ions (H^+) from the stroma into the thylakoid space, forming a hydrogen ion gradient. When these hydrogen ions flow down their electrochemical gradient through **ATP synthase** complexes, ATP production occurs, as will be described shortly. Notice that this ATP will be used in the Calvin cycle reactions in the stroma to reduce carbon dioxide to a carbohydrate.

Similarly, when the photosystem I pigment complex absorbs solar energy, energized electrons leave its reaction center and are captured by a different electron acceptor. After going through the electron transport chain, the electrons from photosystem II are now low-energy electrons. These electrons are used to replace those lost by photosystem I. The electron acceptor in photosystem I passes its electrons to $NADP^+$ molecules. Each $NADP^+$ accepts two electrons and an H^+ to become a reduced form of the molecule—that is, NADPH. This NADPH will also be used by the Calvin cycle reactions in the stroma to reduce carbon dioxide to a carbohydrate.

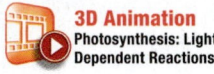

3D Animation Photosynthesis: Light-Dependent Reactions

Cyclic Electron Pathway

Under some environmental conditions, such as high oxygen levels, NADPH levels may accumulate in the cell, and therefore photosynthetic cells may enter the **cyclic electron pathway** (Fig. 8.7). This pathway is also found in many prokaryotic cells. It differs from the noncyclic reactions in that the electrons are recycled back to photosystem I. The pathway begins when the photosystem I pigment complex absorbs solar energy that is passed from one pigment to the next until it becomes concentrated in a reaction center. As with photosystem II, electrons (e^-) become so energized that they escape from the reaction center and move to nearby electron acceptor molecules.

This time, instead of the electrons moving on to $NADP^+$, energized electrons (e^-) taken up by an electron acceptor are sent down an electron transport chain. As the electrons pass from one carrier to the next, their energy becomes stored as a hydrogen (H^+) gradient. Again, the flow of hydrogen ions down their electrochemical gradient through ATP synthase complexes produces ATP.

The spent electrons return to photosystem I after the electron transport chain. This is how photosystem I receives replacement electrons and why this electron pathway is called cyclic. It is also why the cyclic pathway produces ATP but does not produce NADPH.

The Organization of the Thylakoid Membrane

As we have discussed, the following molecular complexes are present in the thylakoid membrane (Fig. 8.8):

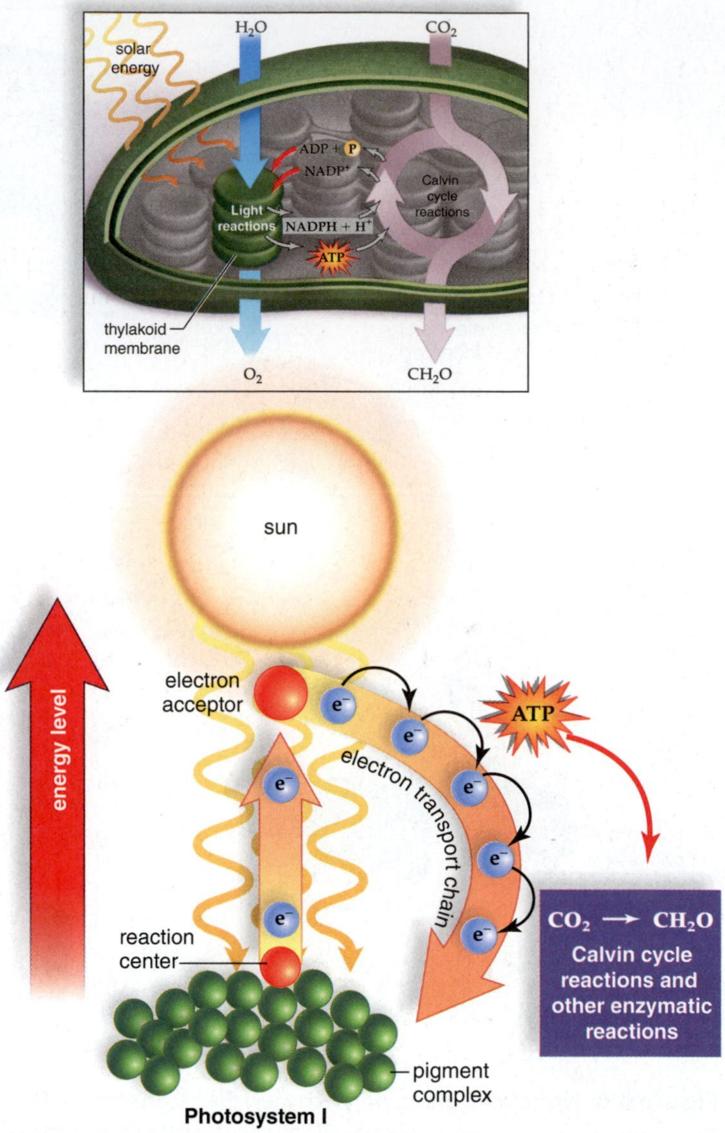

solar energy
H_2O
CO_2
ADP + P
$NADP^+$
Calvin cycle reactions
Light reactions
NADPH + H^+
ATP
thylakoid membrane
O_2
CH_2O
sun
energy level
electron acceptor
ATP
electron transport chain
$CO_2 \longrightarrow CH_2O$
Calvin cycle reactions and other enzymatic reactions
reaction center
pigment complex
Photosystem I

Figure 8.7 Cyclic electron pathway: Electrons leave and return to photosystem I. Energized electrons leave the photosystem I reaction center and are taken up by an electron acceptor, which passes them down an electron transport chain before they return to photosystem I. Only ATP production occurs as a result of this pathway.

- Photosystem II, which consists of a pigment complex and an electron acceptor molecule, receives electrons from water, which splits, releasing oxygen.
- The electron transport chain carries electrons from photosystem II to photosystem I, and pumps H^+ from the stroma into the thylakoid space.
- Photosystem I, which also consists of a pigment complex and an electron acceptor molecule, is adjacent to NADP reductase, which reduces $NADP^+$ to NADPH.
- The ATP synthase complex crosses the thylakoid membrane and contains an interior channel and a protruding ATP synthase, an enzyme that joins ADP+ Ⓟ.

Animation Photosynthetic Electron and ATP Synthesis

ATP Production

The thylakoid space acts as a reservoir for hydrogen ions (H^+). Each time oxygen is removed from water, two H^+ remain in the thylakoid space. As the electrons move from carrier to carrier along the electron transport chain, the electrons give up energy, which is used to pump H^+ from the stroma into the thylakoid space. This is like pumping water into a reservoir. The end result is that there are more H^+ in the thylakoid space than in the stroma. Because H^+ is charged, this is an electrochemical gradient. Not only are there more hydrogen ions, but there are more positive charges within the thylakoid space than the stroma. When a channel is opened in the ATP synthase complex, H^+ flows from the thylakoid space into the stroma, just like water flowing out of the reservoir. The flow of H^+ from high to low concentration across the thylakoid membrane provides the energy that allows the ATP synthase enzyme to enzymatically produce ATP from ADP + Ⓟ. This method of producing ATP is called chemiosmosis because ATP production is tied to the establishment of an H^+ gradient.

Animation Proton Pump

Video Spinach Battery

Check Your Progress 8.2

1. Explain why leaves appear green.
2. Compare the production of NADPH to ATP in noncyclic photosynthesis.
3. Identify which part of a thylakoid will contain the photosystems, electron transport chain, and the ATP synthase complex.
4. Describe why the H^+ gradient across a thylakoid membrane is referred to as a storage of energy.

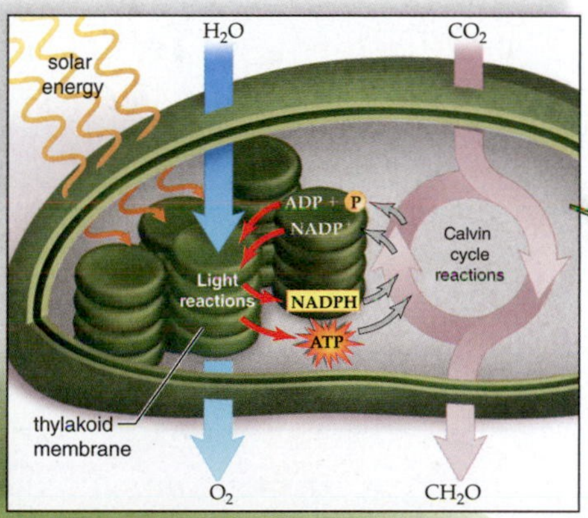

solar energy
H₂O
CO₂
ADP + P
NADP⁺
Calvin cycle reactions
Light reactions
NADPH
ATP
thylakoid membrane
O₂
CH₂O

Figure 8.8 Organization of a thylakoid. Each thylakoid membrane within a granum produces NADPH and ATP. Electrons move through photosystem II, photosystem I, and electron transport chains within the thylakoid membrane. Electrons pass to $NADP^+$, after which it becomes NADPH. A carrier at the start of the electron transport chain pumps hydrogen ions from the stroma into the thylakoid space. When hydrogen ions flow back out of the space into the stroma through the ATP synthase complex, ATP is produced from ADP + Ⓟ.

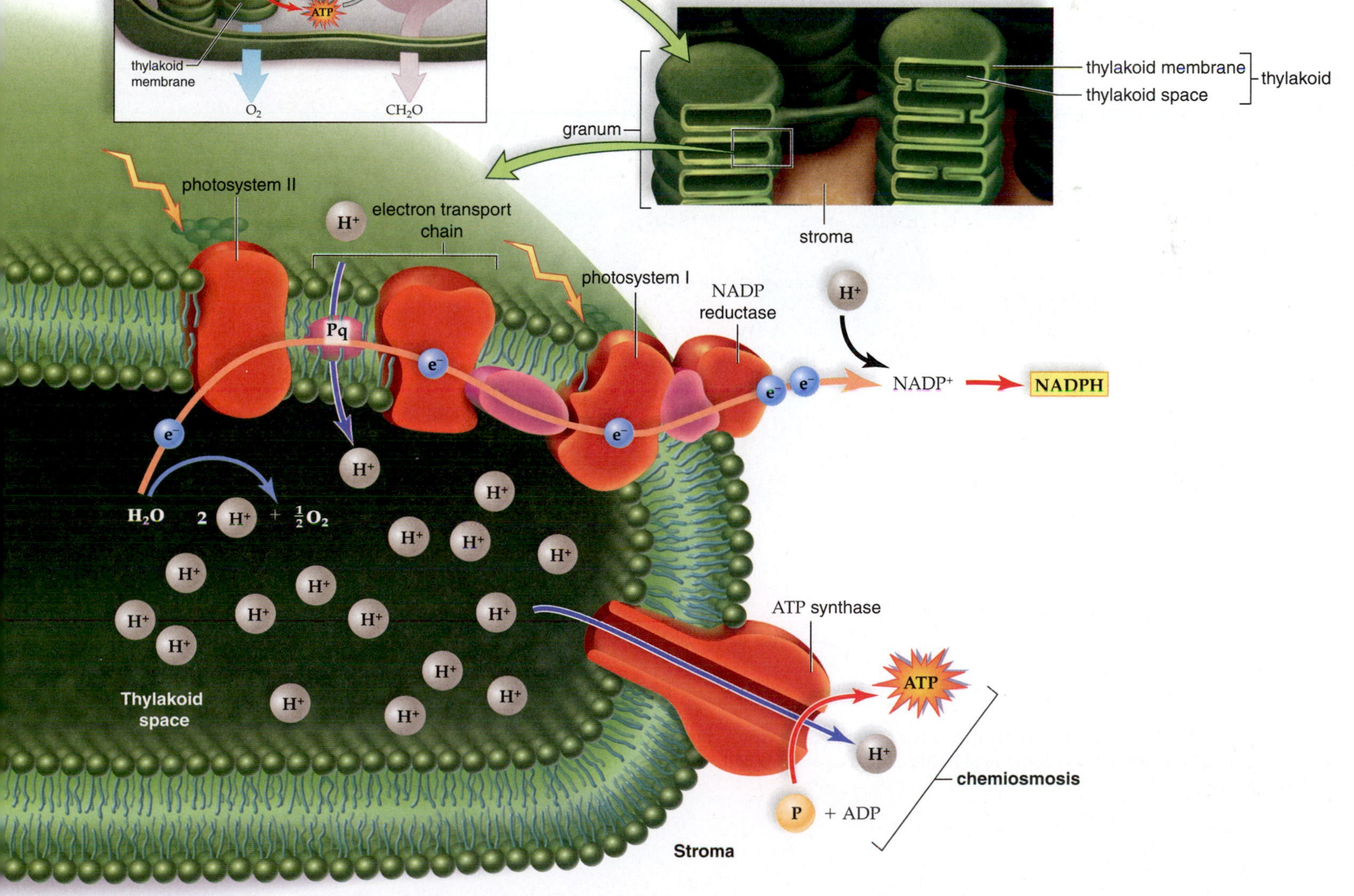

thylakoid membrane
thylakoid space
thylakoid
granum
stroma
photosystem II
electron transport chain
H⁺
Pq
e⁻
photosystem I
NADP reductase
NADP⁺
NADPH
H₂O
2 H⁺ + ½O₂
Thylakoid space
ATP synthase
ATP
chemiosmosis
P + ADP
Stroma

8.3 Plants as Carbon Dioxide Fixers

Learning Outcomes

Upon completion of this section, you should be able to

1. Describe the three phases of the Calvin cycle.
2. Explain how the products of the Calvin cycle are used to form the other molecules found in plants.

The Calvin cycle reactions take place after the light reactions. This is a series of reactions that produce a carbohydrate before returning to the starting point (Fig. 8.9). The cycle is named for Melvin Calvin, who, with colleagues, used the radioactive isotope ^{14}C as a tracer to discover the reactions making up the cycle.

This series of reactions uses atmospheric carbon dioxide to produce a carbohydrate. Humans and most other organisms

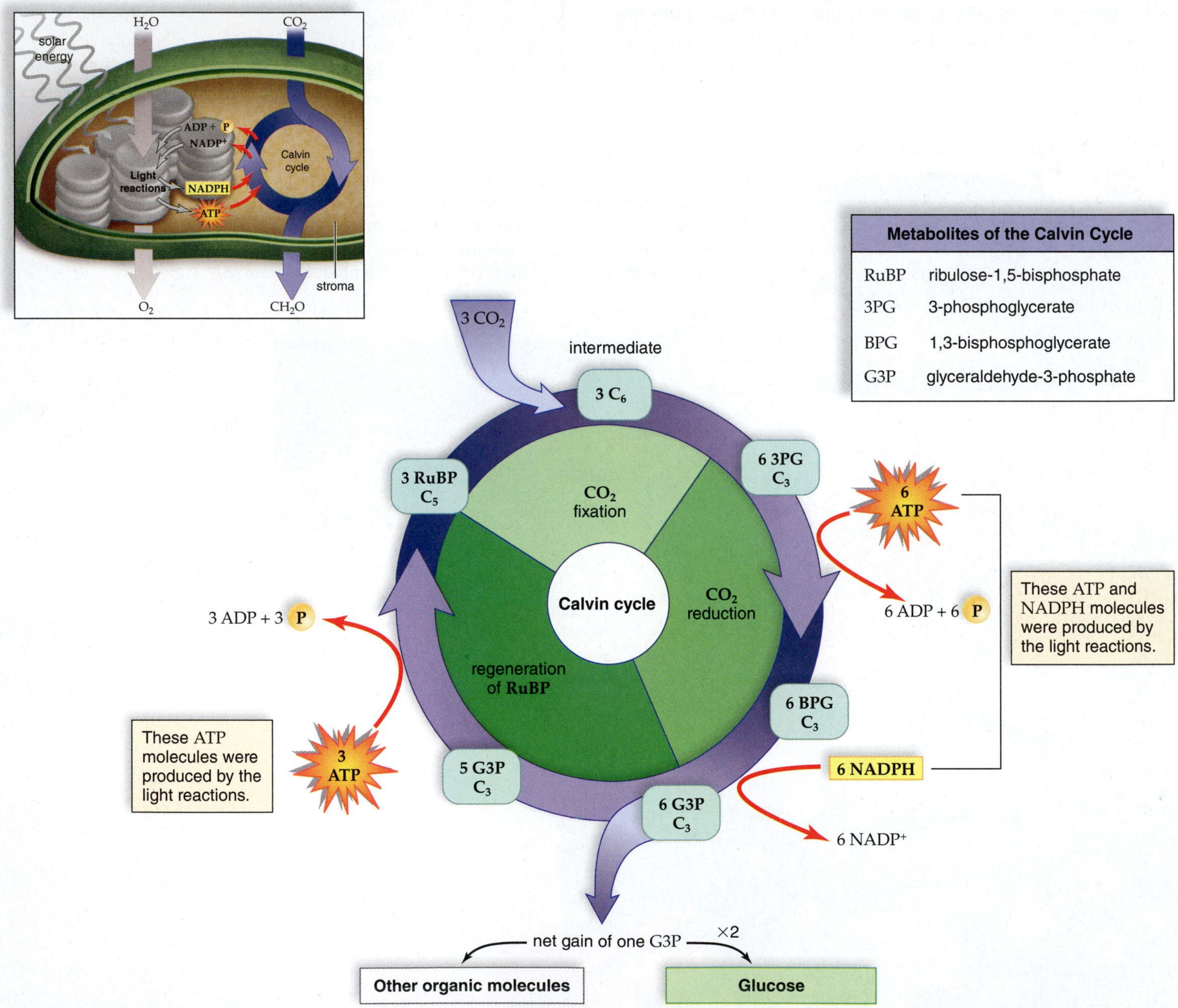

solar energy
H2O
CO2
ADP + P
NADP+
Light reactions
NADPH
ATP
Calvin cycle
stroma
O2
CH2O
Metabolites of the Calvin Cycle
RuBP ribulose-1,5-bisphosphate
3PG 3-phosphoglycerate
BPG 1,3-bisphosphoglycerate
G3P glyceraldehyde-3-phosphate
3 CO2
intermediate
3 C6
6 3PG C3
6 ATP
6 ADP + 6 P
These ATP and NADPH molecules were produced by the light reactions.
CO2 fixation
CO2 reduction
regeneration of RuBP
Calvin cycle
3 RuBP C5
3 ADP + 3 P
These ATP molecules were produced by the light reactions.
3 ATP
5 G3P C3
6 G3P C3
6 BPG C3
6 NADPH
6 NADP+
net gain of one G3P
×2
Other organic molecules
Glucose

Figure 8.9 The Calvin cycle reactions. The Calvin cycle is divided into three portions: CO_2 fixation, CO_2 reduction, and regeneration of RuBP. Because five G3P are needed to re-form three RuBP, it takes three turns of the cycle to have a net gain of one G3P. Two G3P molecules are needed to form glucose.

take in oxygen from the atmosphere and release carbon dioxide. The Calvin cycle includes (1) carbon dioxide fixation; (2) carbon dioxide reduction; and (3) regeneration of RuBP (ribulose-1, 5-bisphosphate), the starting material of the cycle.

3D Animation Photosynthesis: Calvin Cycle

Fixation of Carbon Dioxide

Carbon dioxide fixation is the first step of the Calvin cycle. During this reaction, three molecules of carbon dioxide from the atmosphere are attached to three molecules of RuBP, a 5-carbon molecule. The result is three 6-carbon molecules. Each 6-carbon molecule then splits in half, forming a total of six 3-carbon molecules. This first 3-carbon molecule is called 3-phosphoglycerate or 3PG.

The enzyme that speeds this reaction, called **RuBP carboxylase,** is a protein that makes up about 20% to 50% of the protein content in chloroplasts. The reason for its abundance may be that it is unusually slow (it processes only a few molecules of substrate per second compared to thousands per second for a typical enzyme), and so there has to be a lot of it to keep the Calvin cycle going.

Reduction of Carbon Dioxide

Each of two 3PG molecules undergoes reduction to G3P in two steps. The first step converts 3PG into 1,3-bisphosphoglycerate (BPG) using ATP, and the second reaction converts BPG into glyceraldehyde-3-phosphate (G3P) using NADPH.

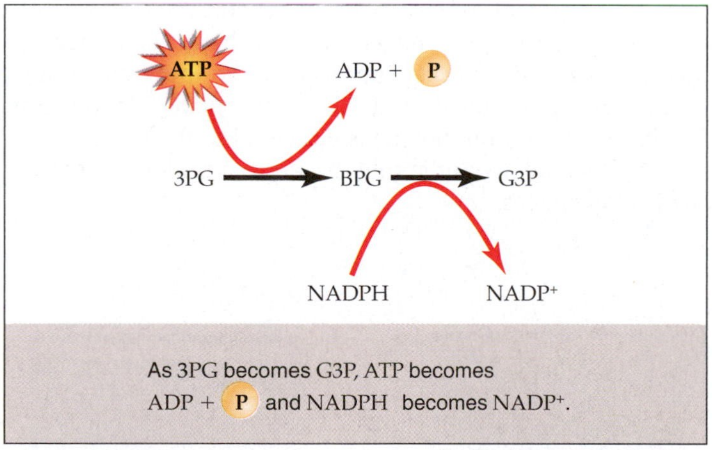
ATP
ADP + P
3PG
BPG
G3P
NADPH
NADP+

As 3PG becomes G3P, ATP becomes ADP + P and NADPH becomes $NADP^+$.

This sequence of reactions uses ATP and NADPH from the light reactions. These reactions also result in the reduction of carbon dioxide to a carbohydrate because R—CO_2 has become R—CH_2O. Energy and electrons are needed for this reduction reaction, and they are supplied by ATP and NADPH.

Regeneration of RuBP

Notice that the Calvin cycle reactions in Figure 8.9 are multiplied by three because it takes three turns of the Calvin cycle to allow one G3P to exit. For every three turns of the Calvin cycle, five molecules of G3P are used to re-form three molecules of RuBP and the cycle continues. Notice that 5 × 3 (carbons in G3P) = 3 × 5 (carbons in RuBP):

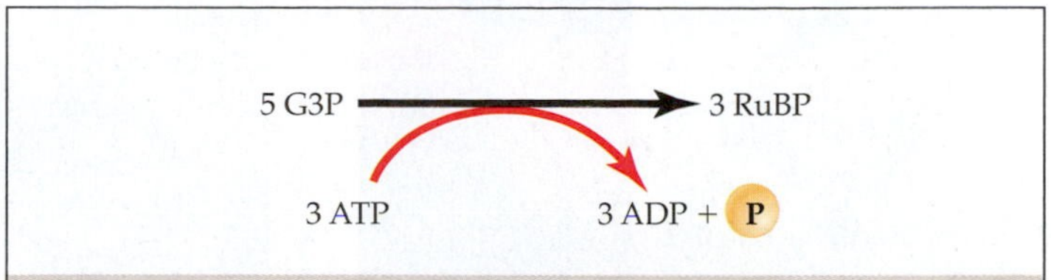
5 G3P
3 RuBP
3 ATP
3 ADP + P

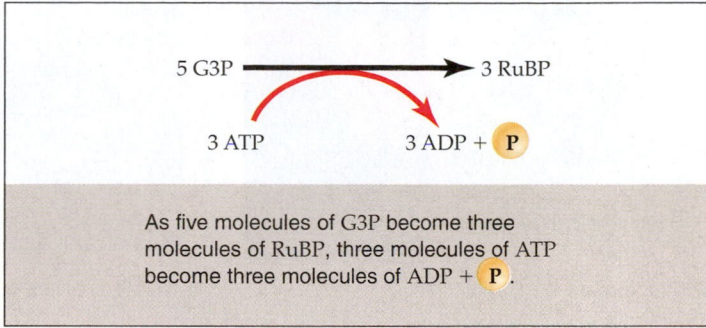

As five molecules of G3P become three molecules of RuBP, three molecules of ATP become three molecules of ADP + P.

This reaction also uses some of the ATP produced by the light reactions.

The Importance of the Calvin Cycle

G3P (glyceraldehyde-3-phosphate) is the product of the Calvin cycle that can be converted to other molecules a plant needs. Compared to animal cells, algae and plants have enormous biochemical capabilities. They use G3P for glucose, sucrose, starch, cellulose, fatty acid, and amino acid synthesis.

Glucose phosphate is one of the organic molecules that results from G3P metabolism. Glucose is the molecule that plants and animals most often metabolize to produce the ATP molecules they require for their energy needs.

Glucose phosphate can be combined with fructose (and the phosphate removed) to form sucrose, the molecule plants use to transport carbohydrates from one part of the plant to the other.

Glucose phosphate is also the starting point for the synthesis of starch and cellulose. Starch is the storage form of glucose. Some starch is stored in chloroplasts, but most starch is stored in roots. Cellulose is a structural component of plant cell walls and becomes fiber in our diet because we are unable to digest it.

A plant can use the hydrocarbon skeleton of G3P to form fatty acids and glycerol, which are combined in plant oils, such as corn oil, sunflower oil, or olive oil—commonly used in cooking. Also, when nitrogen is added to the hydrocarbon skeleton derived from G3P, amino acids are formed.

Check Your Progress 8.3

1. Describe how carbon dioxide is fixed and then reduced to a carbohydrate.
2. Illustrate why it takes three turns of the Calvin cycle to produce one glucose molecule.
3. Explain why G3P is an important molecule in plant metabolism.

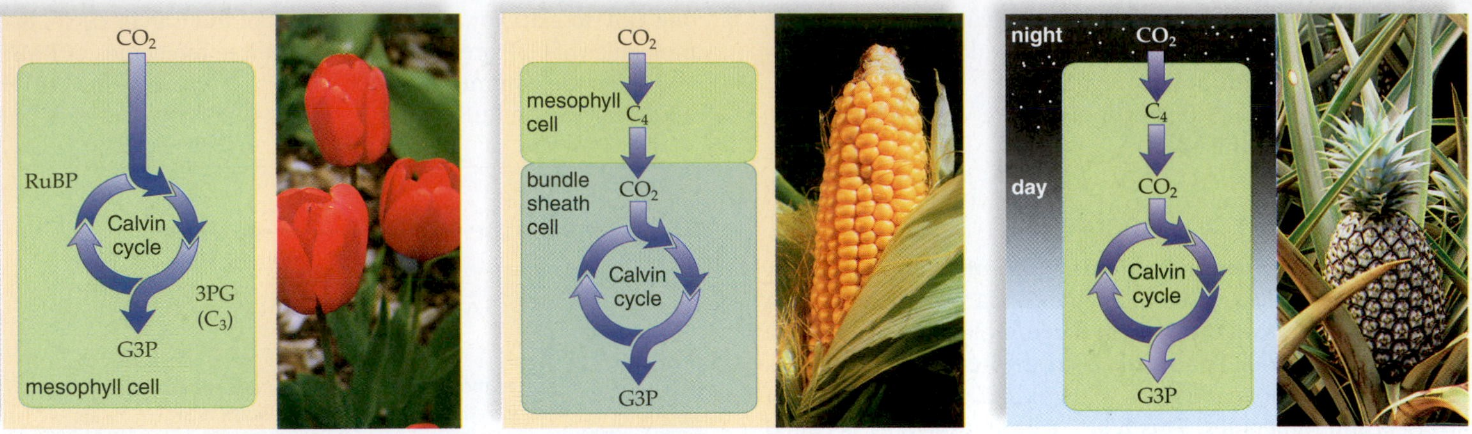
CO_2
RuBP
Calvin cycle
3PG (C_3)
G3P
mesophyll cell
CO_2
mesophyll cell
C_4
bundle sheath cell
CO_2
Calvin cycle
G3P
night
CO_2
C_4
day
CO_2
Calvin cycle
G3P

a. CO_2 fixation in a C_3 plant, tulip

b. CO_2 fixation in a C_4 plant, corn

c. CO_2 fixation in a CAM plant, pineapple

Figure 8.10 Alternative pathways for photosynthesis. Photosynthesis can be categorized according to how, where, and when CO_2 fixation occurs. **a.** In C_3 plants, CO_2 is taken up by the Calvin cycle directly in mesophyll cells. **b.** C_4 plants from a C_4 molecule in mesophyll cells prior to releasing CO_2 to the Calvin cycle in bundle sheath cells. **c.** CAM plants fix CO_2 at night, forming a C_4 molecule.

8.4 Alternative Pathways for Photosynthesis

Learning Outcomes

Upon completion of this section, you should be able to

1. Contrast C_3, C_4, and CAM photosynthesis.
2. Explain how different photosynthetic modes allow plants to adapt to a particular environment.

Plants are able to live under all sorts of environmental conditions, and one reason is that various modes of photosynthesis have evolved (Fig. 8.10).

C_3 Photosynthesis

The leaves of C_3 plants have a particular structure and a different means of fixing CO_2 compared with C_4 plants. In **C_3 plants,** such as wheat, rice, and oats, mesophyll cells are in parallel layers. The bundle sheath cells around the plant veins do not contain chloroplasts (Fig. 8.11*a*). This structure exposes the cells containing the Calvin cycle to the incoming CO_2. CO_2 is fixed by RuBP carboxylase of the Calvin cycle, and the first detectable molecule following fixation is a 3-carbon molecule (see Fig. 8.9). Unfortunately, RuBP carboxylase can not only bind CO_2, it can also bind with O_2. When the enzyme binds oxygen, it undergoes a nonproductive, wasteful reaction called photorespiration because it uses oxygen and releases carbon dioxide. Photorespiration makes C_3 photosynthesis an inefficient way to produce carbohydrate when oxygen concentrations rise in the leaf space. This can happen when drought conditions occur because this type of weather leads to the closing of stomata in order to conserve water.

C_4 Photosynthesis

In **C_4 plants,** such as sugarcane and corn, the mesophyll cells are arranged in concentric rings around the bundle sheath cells, which also contain chloroplasts (Fig. 8.11*b*). In the mesophyll cells, CO_2 is initially fixed by forming a 4-carbon molecule. (The formation of this molecule accounts for the terms C_4 plant and C_4 photosynthesis.) The 4-carbon molecule releases CO_2 to the Calvin cycle and carbohydrate production follows.

Because of the need to transport molecules from where CO_2 is fixed to where the Calvin cycle is located, it would appear that the C_4 pathway would be little utilized among plants. However, when the weather is hot and dry, C_4 plants have an advantage. This is because when the stomata close in order to conserve water and oxygen increases in leaf air space, RuBP carboxylase is not exposed to this oxygen in C_4 plants and photorespiration does not occur. Instead, in C_4 plants, the carbon dioxide is delivered to the Calvin cycle, which is located in bundle sheath cells that are sheltered from the leaf air spaces.

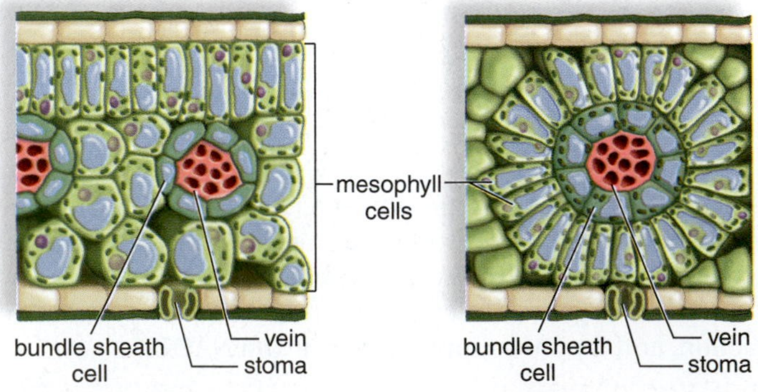
mesophyll cells
bundle sheath cell
vein
stoma
bundle sheath cell
vein
stoma

a. C_3 Plant

b. C_4 Plant

Figure 8.11 C_3 and C_4 plant leaf cell arrangement. **a.** C_3 plants contain mesophyll cells in parallel layers. The bundle sheath cells do not contain chloroplasts. **b.** C_4 plants contain mesophyll cells arranged in concentric rings around chlorophyll containing bundle sheath cells.

SCIENCE IN YOUR LIFE ▶ ECOLOGY

The New Rice

According to international government leaders who met in Japan in July of 2008, a global food crisis is developing. The world's population continues to grow, and grain is being diverted for biofuels and to feed livestock. The world's farmers are not keeping up with the demand for wheat, sorghum, maize, and rice. But according to the International Rice Research Institute (IRRI) in Los Baños, Philippines, two-thirds of the world's poorest people subsist primarily on rice (Fig. 8A). So a shortage of rice affects world hunger more than the other grains. More rice is being consumed than grown, and world rice stockpiles have been declining since 2001. In order to keep up with demand, the world must grow 50 million tons more rice per year than it did in 2005. This requires an increase globally of 1.2% per year.

Several environmental factors are causing these declines, including global climate change and regional flooding. For example, most varieties of rice will die if they are submerged for three or more days. However, one variety, called flood resistant (FR), could survive even if submerged for three weeks. A single gene, called the *Sub1A* gene, appeared to confer a significant degree of submergence tolerance to the FR variety. IRRI scientists were able to introduce the gene into commercial rice strains via hybridization, and four varieties are currently in field trials.

Figure 8A Rice to feed the world.
Rice provides food for two-thirds of the world's poorest people. New technologies may increase the production of rice.

China plants approximately 57% of its rice crop in hybrid varieties. Hybrid rice produces approximately 20% more per acre than the traditional inbred varieties. The hybrids developed by the Chinese are best suited for growth in the tropics and are not effective in temperate zones. In addition, the rice lacks in flavor compared to the inbred varieties.

One of the more long-term research programs with rice is trying to convert rice from a C_3 into a C_4 plant! C_4 plants are 50% more efficient at turning sunlight into food than C_3 plants. There are two aspects of a C_4 plant that would need to be introduced into rice. The first would be the actual enzymes involved in fixing carbon dioxide. The second would be the leaf anatomy of the mesophyll and bundle sheath cells. Researchers have succeeded in introducing the appropriate enzymes from maize, a C_4 plant, into rice plants. Without the proper leaf anatomy, however, there is no guarantee the cloned enzymes will increase the efficiency of photosynthesis.

Video
Global Warming Hurts Rice

Questions to Consider

1. If the poorest people in the world go hungry due to the lack of grain, how much difference would it make to solving world hunger if you reduced your meat and fuel consumption?
2. Are genetically modified organisms the only way to feed the world's growing population?

When the weather is moderate, C_3 plants ordinarily have the advantage, but when the weather becomes hot and dry, C_4 plants have the advantage, and we can expect them to predominate. In the early summer, C_3 plants such as Kentucky bluegrass and creeping bent grass predominate in lawns in the cooler parts of the United States, but by mid-summer, crabgrass, a C_4 plant, begins to take over.

CAM Photosynthesis

Another alternative pathway, called the **CAM** pathway, has also evolved because of environmental pressures. CAM stands for crassulacean-acid metabolism. Crassulaceae is a family of flowering succulent plants that live in warm, arid regions of the world. CAM photosynthesis is prevalent among most succulent plants that grow in desert environments, including the cacti, but it is now known to occur in a variety of other plants.

While the C_4 pathway separates components of photosynthesis by location, the CAM pathway separates them in time. CAM plants fix CO_2 into a 4-carbon molecule at night, when they can keep their stomata open without losing much water. The 4-carbon molecule is stored in large vacuoles in their mesophyll cells until the next day. During the day, the 4-carbon molecule releases the CO_2 to the Calvin cycle within the same cell.

Plants are capable of fixing carbon by more than one pathway. These pathways appear to be the result of adaptation to different climates.

Check Your Progress 8.4

1. Identify various plants that use a method of photosynthesis other than C_3 photosynthesis.
2. Explain why C_4 photosynthesis is advantageous in hot, dry conditions.

8.5 Photosynthesis Versus Cellular Respiration

Learning Outcomes

Upon completion of this section, you should be able to

1. Compare the overall chemical equation for photosynthesis and cellular respiration.
2. Describe the similarities and differences between cellular respiration and photosynthesis.

Both plant and animal cells carry on cellular respiration, but animal cells cannot photosynthesize. Plants, algae, and cyanobacteria are capable of photosynthesis. The organelle for cellular respiration is the mitochondrion, whereas the organelle for photosynthesis is the chloroplast. Note that in some bacteria, such as the cyanobacteria, photosynthesis occurs without chloroplasts—this is because these bacteria do have thylakoid-like structures in the cell. Photosynthesis is the formation of glucose, whereas cellular respiration is the breaking down of glucose. Figure 8.12 compares the two processes.

The following overall chemical equation for photosynthesis is the opposite of that for cellular respiration. The reaction in the forward direction represents photosynthesis, and the word *energy* stands for solar energy. The reaction in the opposite direction represents cellular respiration, and the word *energy* then stands for ATP:

$$\text{energy} + 6\,CO_2 + 6\,H_2O \underset{\text{cellular respiration}}{\overset{\text{photosynthesis}}{\rightleftharpoons}} C_6H_{12}O_6 + 6\,O_2$$

Both photosynthesis and cellular respiration are metabolic pathways within cells, and therefore consist of a series of reactions that the overall equation does not indicate. Both pathways, which use an electron transport chain located in membranes, produce ATP by chemiosmosis. Both also use an electron carrier; photosynthesis uses $NADP^+$, and cellular respiration uses NAD^+.

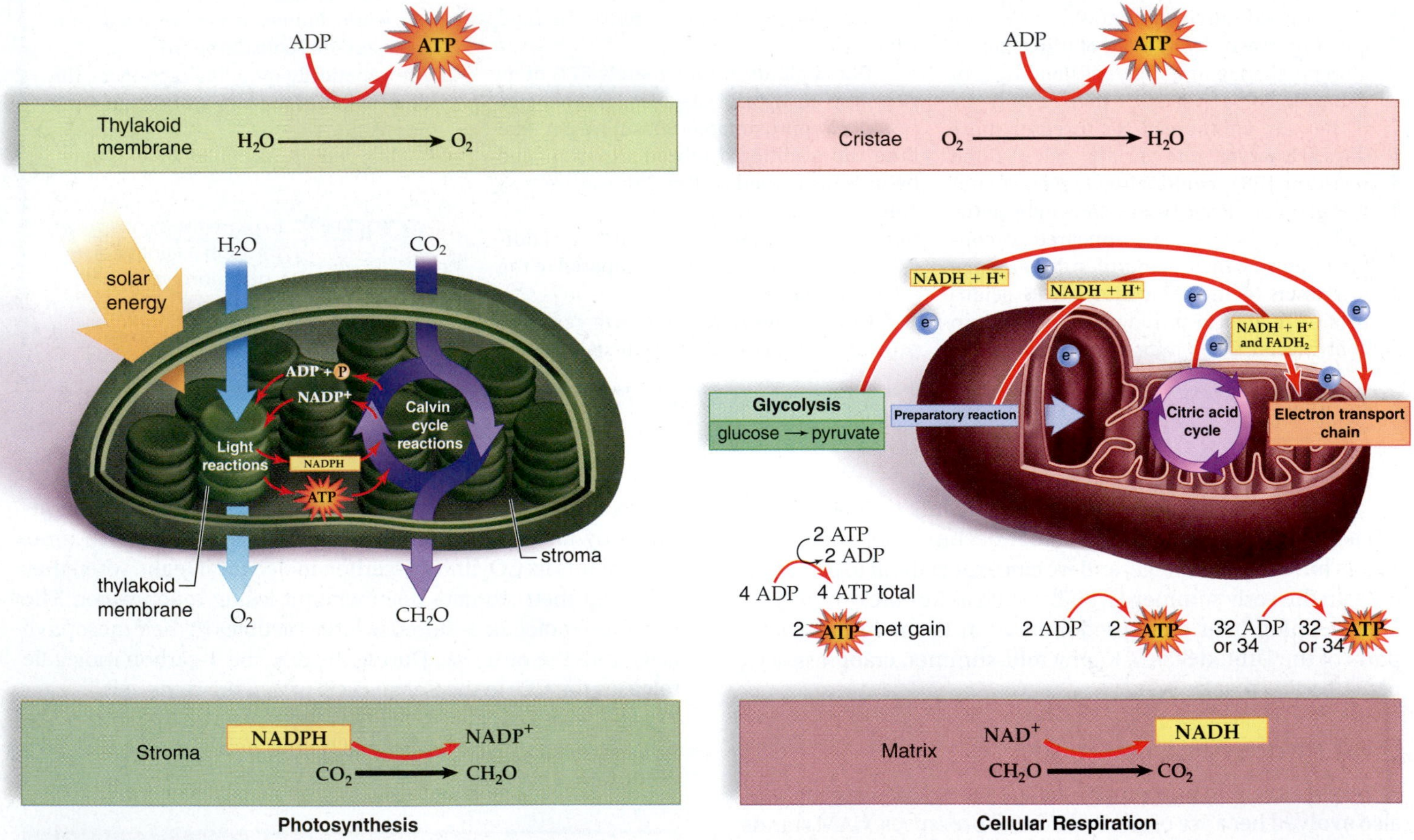

ADP
ATP
Thylakoid membrane
H2O → O2
Cristae
O2 → H2O
H2O
CO2
solar energy
ADP + P
NADP+
Calvin cycle reactions
Light reactions
NADPH
ATP
stroma
thylakoid membrane
O2
CH2O
NADH + H+
NADH + H+
NADH + H+ and FADH2
e−
Glycolysis
glucose → pyruvate
Preparatory reaction
Citric acid cycle
Electron transport chain
2 ATP
2 ADP
4 ADP
4 ATP total
2 ATP net gain
2 ADP
2 ATP
32 ADP or 34
32 or 34 ATP
Stroma
NADPH
NADP+
CO2 → CH2O
Matrix
NAD+
NADH
CH2O → CO2
Photosynthesis
Cellular Respiration

Figure 8.12 Photosynthesis versus cellular respiration. Both photosynthesis and cellular respiration have an electron transport chain located within membranes, where ATP is produced. Both processes have enzyme-catalyzed reactions located within the fluid interior of respective organelles. In photosynthesis, hydrogen atoms are donated by NADPH + H^+ when CO_2 is reduced in the stroma of a chloroplast. During cellular respiration, NADH forms when glucose is oxidized in the cytoplasm and glucose breakdown products are oxidized in the matrix of a mitochondrion.

SCIENCE IN YOUR LIFE ▶ **ECOLOGY**

Diesel Power from Algae

Algae are at the base of many food chains. But researchers are now using algae to produce a variety of different fuel types. Solazyme, a San Francisco–based company, has discovered the methods necessary to produce crude oil from various plant-based sugars with the help of microalgae. The process they have developed does not depend on the photosynthetic ability of the algae. It uses genetically modified algae grown in large vats. Instead of being exposed to sunlight, the algae are fed sugar that they convert into various types of oil. Different types of algae produce different types of oil. Because the algae are grown in the dark, the genes for photosynthesis have been switched off. With these genes turned off, the algae actually make more oil.

Algae-based biofuels are considered "green" fuels. After the oil is harvested, the remaining waste can be used as fertilizer. Algae can be grown on marginally productive, even desert, lands and do not require fresh water. And if the fuel is spilled—it is biodegradable!

In 2010, Solazyme delivered 80,000 liters of marine diesel and jet fuel to the U.S. Navy that was derived from algae-based production. They have also been awarded a contract to produce an additional 550,000 liters of marine diesel for the U.S. Department of Defense.

Some of the benefits of Solazyme-based fuels over traditional petroleum-based fuels are that they meet industry standards and can be used with factory-standard diesel engines without any type of modifications.

As diesel costs continue to rise, algae fuel may soon become economically feasible. Then the question will become whether companies produce enough fuel to meet the needs of a growing human population.

Questions to Consider

1. If algae-based biofuel reduces our dependence on foreign oil, how much cheaper does it need to be before we will switch?
2. What other costs might be involved in switching to biofuel besides just the production costs of the fuel?

connect BIOLOGY Explore the concepts through a variety of multimedia assets, question types, and data interpretation.
www.mcgrawhillconnect.com

Both pathways utilize the following reaction, but in opposite directions. For photosynthesis, read the reaction from left to right; for cellular respiration, read the reaction from right to left:

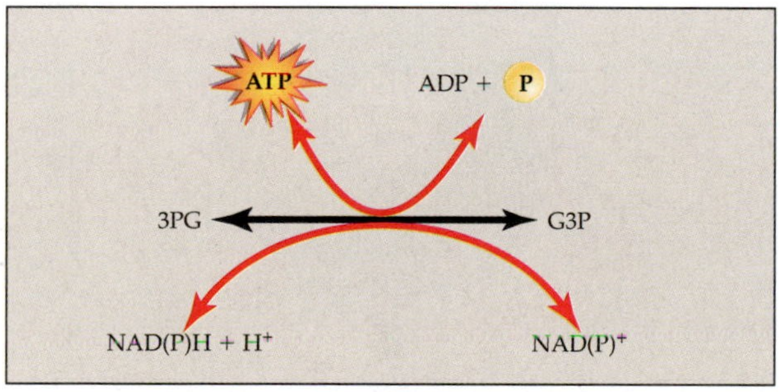
ATP
ADP + P
3PG
G3P
NAD(P)H + H+
NAD(P)+

Both photosynthesis and cellular respiration occur in plant cells. In plants, both processes occur during the daylight hours, whereas only cellular respiration occurs at night. During daylight hours, the rate of photosynthesis exceeds the rate of cellular respiration, resulting in a net increase and storage of glucose. The stored glucose is used for cellular metabolism, which continues during the night.

Check Your Progress 8.5

1. Explain the similarities, and differences, between cellular respiration and photosynthesis. Use the chemical equation for these reactions in your answer.
2. Explain why a plant cell must contain both chloroplasts and mitochondria.

Case Study Conclusion

Throughout the summer and fall every year, plants undergo a wide variety of color changes. During the summer, leaves and fruits are typically green due to the large amounts of chlorophyll present in them. As the season progresses, the chlorophyll degrades and the carotenoids, yellow to red pigments, become more visible. It is this wide range of pigments that enables the plants of the world to absorb the necessary solar energy to run photosynthesis. Once the solar energy is absorbed, a series of chemical reactions are set in place that will ultimately produce oxygen and glucose. The consumers of the world require this oxygen and glucose to run cellular respiration.

The additional benefit that scientists have discovered is that the degradation of chlorophyll in the skin of fruits results in the production of antioxidants. It is thought that antioxidants help stabilize free radicals, the molecules that can damage DNA and lead to cancer. Photosynthesis not only supplies the oxygen and glucose necessary for consumers, but it also has health benefits for us as well.

MEDIA STUDY TOOLS

www.mhhe.com/maderinquiry14

Enhance your study of this chapter with study tools and practice tests. Also ask your instructor about the resources available through ConnectPlus, including LearnSmart, the media-rich eBook, interactive learning tools, and animations.

connect plus+ | BIOLOGY

3D Animation
Photosynthesis

For an interactive exploration on the processes of photosynthesis, take a moment to view McGraw-Hill's 3D animation "Photosynthesis."

SUMMARIZE

8.1 Overview of Photosynthesis

- **Photosynthesis** is absolutely essential for the continuance of life because it supplies the biosphere with food, which is used as a source of building blocks and energy.
- Chloroplasts carry on photosynthesis. During photosynthesis, solar energy is converted to chemical energy within carbohydrates. A chloroplast contains two main portions: the stroma and membranous grana made up of thylakoid sacs. Chlorophyll and other pigments like the **carotenoids** that are located within the thylakoid membrane absorb solar energy, and enzymes in the fluid-filled stroma reduce CO_2. CO_2 enters the leaf through small openings called stomata.

- The overall equation for photosynthesis is $6\ CO_2 + 6\ H_2O \longrightarrow C_6H_{12}O_6 + 6\ O_2$. Photosynthesis consists of two reactions: the **light reactions** and the **Calvin cycle reactions.** The Calvin cycle reactions, located in the stroma, use NADPH and ATP to reduce carbon dioxide. These molecules are produced by the light reactions located in the thylakoid membranes of the grana, after chlorophyll captures the energy of sunlight.

8.2 Plants as Solar Energy Converters

- Photosynthesis uses solar energy in the visible-light range. Chlorophylls *a* and *b* and the carotenoids largely absorb violet, indigo, blue, and red wavelengths and reflect green wavelengths. This causes leaves to appear green to us.

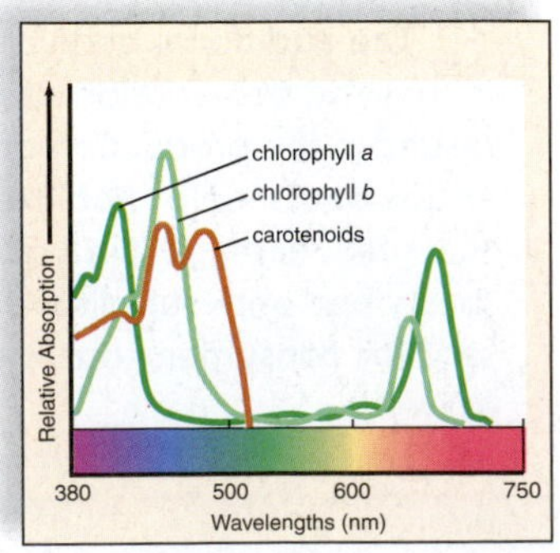

chlorophyll a
chlorophyll b
carotenoids
Relative Absorption
380
500
600
750
Wavelengths (nm)

- Photosynthesis begins when pigment complexes within **photosystem** I and photosystem II absorb radiant energy. In the **noncyclic electron pathway,** electrons are energized in photosystem II before they enter an electron transport chain. Electrons from H_2O replace those lost in photosystem II. As the electrons pass through the electron transport chain, they help establish a hydrogen ion gradient. The electrons energized in photosystem I pass to $NADP^+$, which becomes NADPH. Electrons from the electron transport chain replace those lost by photosystem I.
- In the **cyclic electron pathway** of the light reactions, electrons energized by the sun leave photosystem I and enter an electron transport chain that produces a hydrogen ion gradient. Then the energy-spent electrons return to photosystem I.
- The hydrogen ion gradient across the thylakoid membrane is used to synthesize ATP using an **ATP synthase** enzyme complex.

8.3 Plants as Carbon Dioxide Fixers

- The ATP and NADPH made in thylakoid membranes pass into the stroma, where carbon dioxide is reduced during the Calvin cycle reactions.
- **Carbon dioxide fixation** attaches CO_2 to a 5-carbon molecule named RuBP by the enzyme **RuBP carboxylase.** The resulting 6-carbon molecule splits into two molecules of 3-carbons, each called 3PG. ATP and NADPH from the light reactions are then used to reduce 3PG to G3P.
- G3P is used to synthesize various molecules, including carbohydrates such as glucose.

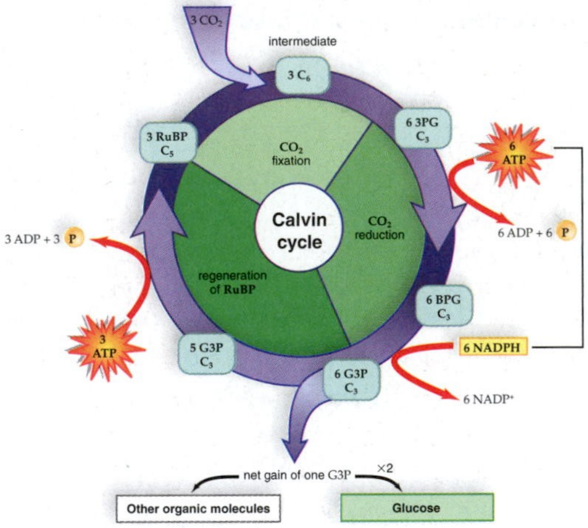

3 CO2
intermediate
3 C6
6 3PG C3
3 RuBP C5
CO2 fixation
Calvin cycle
CO2 reduction
regeneration of RuBP
6 ATP
6 ADP + 6 P
3 ADP + 3 P
3 ATP
6 BPG C3
6 NADPH
5 G3P C3
6 G3P C3
6 NADP+
net gain of one G3P
×2
Other organic molecules
Glucose

8.4 Alternative Pathways for Photosynthesis

- In C_3 **plants,** the first molecule observed following carbon dioxide fixation is a 3-carbon molecule. The structure of these plants allows RuBP carboxylase to bind with oxygen during photorespiration.
- Plants utilizing C_4 photosynthesis have a different construction than plants using C_3 photosynthesis. C_4 **plants** fix CO_2 in mesophyll cells and then deliver the CO_2 to the Calvin cycle in bundle sheath cells. Now, O_2 cannot compete for the active site of RuBP carboxylase when stomata are closed due to hot and dry weather. This represents a partitioning of pathways in space: carbon dioxide fixation occurs in mesophyll cells, and the Calvin cycle occurs in bundle sheath cells.

- **CAM** plants fix carbon dioxide at night when their stomata remain open. This represents a partitioning of pathways in time: carbon dioxide fixation occurs at night, and the Calvin cycle occurs during the day.

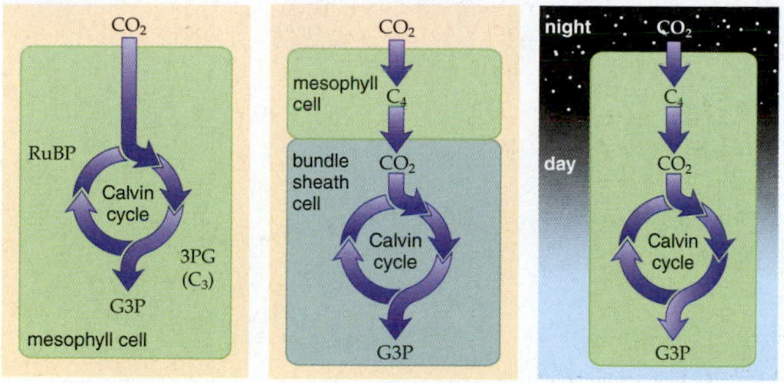
CO_2
RuBP
Calvin cycle
3PG (C_3)
G3P
mesophyll cell
CO_2
mesophyll cell
C_4
bundle sheath cell
CO_2
Calvin cycle
G3P
night
CO_2
C_4
day
CO_2
Calvin cycle
G3P

8.5 Photosynthesis Versus Cellular Respiration

- Both photosynthesis and cellular respiration utilize an electron transport chain and ATP synthesis. However, photosynthesis reduces CO_2 to a carbohydrate. The oxidation of H_2O releases O_2. Cellular respiration oxidizes carbohydrate, and CO_2 is given off. Oxygen is reduced to water.

ASSESS

Testing Yourself

Choose the best answer for each question.

1. Photosynthetic activity can be found in
   a. plants. **c.** algae.
   b. bacteria. **d.** All of these are correct.
2. Cellular respiration and photosynthesis both
   a. use oxygen.
   b. produce carbon dioxide.
   c. contain an electron transport chain.
   d. occur in the chloroplast.
3. The noncyclic electron pathway, but not the cyclic pathway, generates
   a. 3PG. **c.** ATP.
   b. chlorophyll. **d.** NADPH.

For questions 4–8, match each definition with a type of plant in the key.

Key:

a. C_3 plants **b.** C_4 plants

4. Contain chloroplasts in bundle sheath cells
5. Mesophyll cells arranged in parallel
6. Rice
7. Corn
8. Can be inefficient
9. In the absence of sunlight, plants are not able to engage in the Calvin cycle due to a lack of
   a. ATP. **c.** NADPH.
   b. oxygen. **d.** Both a and c are correct.
10. Which light rays are used for photosynthesis?
    a. light having long wavelengths
    b. light having short wavelengths
    c. violet, indigo, and blue light
    d. ultraviolet light
    e. Both b and c are correct.
11. Label a, b, c, d, and e in the following diagram of a chloroplast.

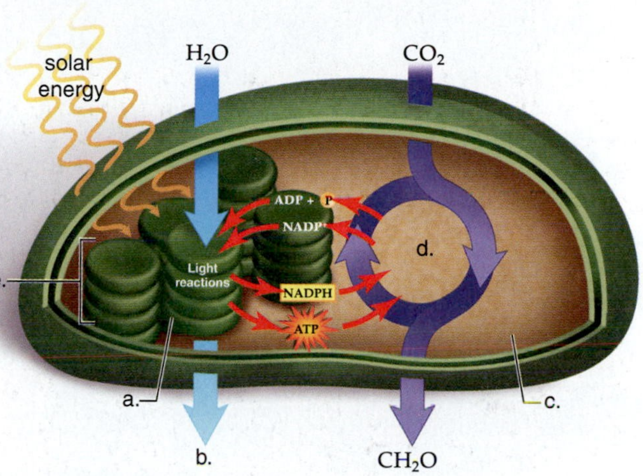
solar energy
H_2O
CO_2
ADP + P
NADP+
d.
e.
Light reactions
NADPH
ATP
a.
b.
c.
CH_2O

12. The oxygen given off by photosynthesis comes from
    a. H_2O. **c.** CO_2.
    b. glucose. **d.** RuBP.
13. Which of these should not be associated with an electron transport chain?
    a. chloroplasts
    b. protein complexes
    c. movement of H^+ into the thylakoid space
    d. formation of ATP
    e. the RuBP complex

ENGAGE

Thinking Critically

1. How do broad, thin leaves provide an advantage for photosynthesis?
2. Based on what you know about pigments, why do some people who live in very hot, sunny climates wear white?
3. How does the electron transport chain in a chloroplast pump H^+ into the thylakoid space?

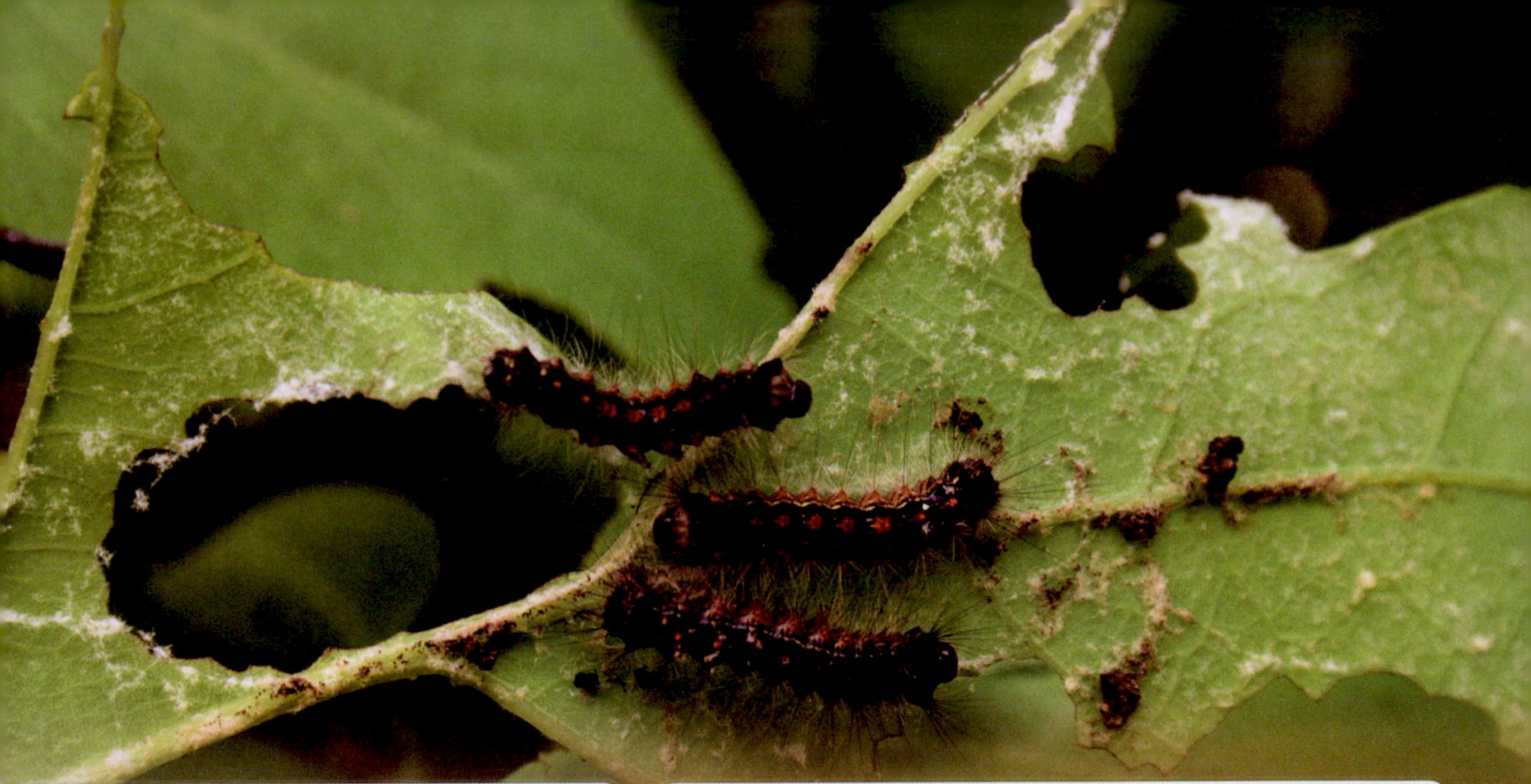

9 Plant Organization and Function

CHAPTER OUTLINE

BEFORE YOU BEGIN

Before beginning this chapter, take a few moments to review the following discussions:

Section 2.3 How do the properties of water make it possible for the movement of nutrients in plants?

Table 3.2 Which cell structures are unique to plants?

Section 8.1 Which parts of the plant are responsible for conducting photosynthesis?

CASE STUDY Plant Structures That Enable Survival

Plants have evolved a variety of mechanisms that help to protect themselves from the damaging effects of the environment and predators. In nonwoody plants, a thin, outer layer of epidermal cells secretes a waxy coating called a cuticle that is impermeable to water. The cuticle covers the surfaces of the leaves and aerial portions of the plant. Cacti have thick cuticles to protect against water loss in the dry desert. Seashore plants have thick cuticles to protect against salt in seawater spray. The epidermis of a plant can also produce hairs that help to reduce evaporation in windy habitats by breaking up the wind flow across the plant surface. In cold climates, plant hairs protect the living tissue from freezing. Plant hairs can also inhibit herbivores attempting to feed on the plant. This inhibition is a result of hair size, prickliness, and taste. Plants have also evolved antiherbivory chemical defenses against predators. For example, nicotine, made in the roots and stored in the leaves of tobacco plants, is toxic to insects. Caffeine from the leaves and fruits of some plants can kill certain insects feeding on the plant. In spite of these defenses, insect herbivory causes considerable damage to plants every year.

Overall, the roots, stems, and leaves of plants represent evolutionary adaptations that are specialized to carry out particular functions.

As you read through the chapter, think about the following questions:

1. What structural features are necessary for the function of flowering plants?
2. How do monocot and eudicot structural differences adapt them to their various roles in the environment?
3. Are adaptations against predators different from those that are used to help a plant survive physical trauma from environmental factors?

9.1 Plant Organs and Systems

Learning Outcomes

Upon completion of this section, you should be able to

1. Describe the vegetative and reproductive organs of plants and indicate their major functions.
2. Compare the structure and function of roots, stems, and leaves.

From cacti living in hot deserts to water lilies growing in a pond, the flowering plants, or **angiosperms,** are extremely diverse. But despite their great diversity in size and shape, flowering plants share many common structural features. Most flowering plants possess a root system and a shoot system (Fig. 9.1). The **root system** simply consists of the roots, whereas the **shoot system** consists of the stem and leaves. The structure of these systems is suitable to their functions. A typical plant features three vegetative organs (structures that contain different tissues and perform one or more specific functions). The roots, stems, and leaves are the *vegetative organs*. Vegetative organs are involved in growth and nutrition, but not reproduction. Flowers, seeds, and fruit are examples of reproductive structures.

Root System

Roots have various adaptations and associations to perform the functions of anchorage, absorption of water and minerals, and storage of carbohydrates. The root system in the majority of plants is located underground. The depth and distribution of plant **roots** depends on the type of plant, the timing and amount of rainfall, and the soil composition. For example, plants growing in deserts tend to have deeper roots than those growing in temperate grasslands. A common misconception is that the size and distribution of tree roots reflect the aboveground trunk and branches of the tree. In reality, 90% of a tree's roots are located within 1 meter (m) of the surface. However, a tree's roots extend out much farther than the crown of the tree. On average, a tree's roots will extend two to four times the diameter of the aboveground portion of the tree.

Extensive root systems of plants help anchor them in the soil and provide support (Fig. 9.2*a*). The root system absorbs water and minerals from the soil for the entire plant. The cylindrical shape of a root allows it to penetrate the soil as it grows and to absorb water from all sides. The absorptive capacity of a root is also increased by its many branch (lateral) roots and root hairs located in a special zone near a root tip. Root hairs, which are projections from epidermal root-hair cells, are so numerous that they increase the absorptive surface of a root tremendously. In a classic study completed in 1937, H. J. Dittmer counted 13,800,000 root hairs in a single, 20-inch-high rye plant. Since root-hair cells are constantly being replaced, a typical rye plant may form about 100 million new root-hair cells every day. Roughly pulling a plant out of the soil damages the small lateral roots and root hairs, and these plants do not fare well when transplanted. Transplantation is usually more successful if part of the surrounding soil is moved along with the plant, since this leaves many of the branch roots and the root hairs intact.

Roots also have other functions. In certain plants, the roots are modified for food storage. For example, branch roots form

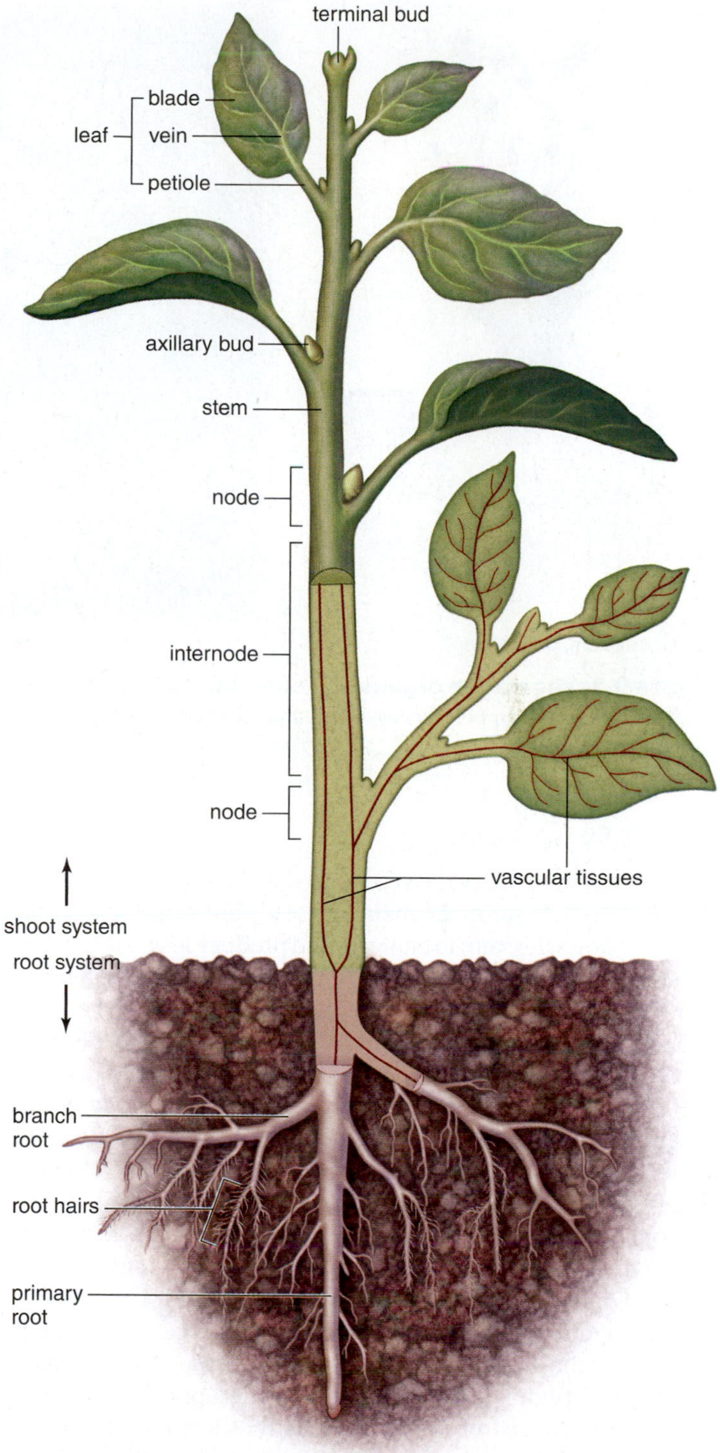
terminal bud
leaf
blade
vein
petiole
axillary bud
stem
node
internode
node
vascular tissues
shoot system
root system
branch root
root hairs
primary root

Figure 9.1 Organization of a plant body. The body of a plant consists of a root system and a shoot system. The shoot system contains the vegetative organs of the stem and leaves. Axillary buds can develop into branches or flowers, the reproductive structures of a plant. The root system and shoot system are connected by vascular tissue that extends from the roots to the leaves.

storage organs for carbohydrates in yams and sweet potatoes. Roots can also store water. Some plants in the pumpkin family store large amounts of water in their roots. Roots also produce hormones that stimulate the growth of stems and coordinate their size with the size of the root.

stem
roots
lateral root
stems
blade
petiole
a. Root system, dandelion
b. Shoot system, bean seedling
c. Leaves, pumpkin seedling

Figure 9.2 Vegetative organs of a eudicot. **a.** The root system anchors the plant and absorbs water and minerals. **b.** The shoot system, the stem and branches, support the leaves and transport water and organic nutrients. **c.** The leaves carry out photosynthesis.

Shoot System: Stems

The shoot system of a plant is composed of the stem, the branches, and the leaves. A **stem,** the main axis of a plant, terminates in tissue that allows the stem to elongate and produce leaves (Fig. 9.2*b*). A stem can sometimes expand in girth as well as length. As trees grow taller each year, they accumulate woody tissue that adds to their girth. The increase in girth strengthens the stem of the plant. Most stems are upright and support leaves in such a way that each leaf is exposed to as much sunlight as possible. A **node** occurs where leaves are attached to the stem, and an **internode** is the region between the nodes (see Fig. 9.1). The presence of nodes and internodes is the characteristic used to identify a stem, even if it happens to be an underground stem. Some stems are horizontal, and in these cases the nodes may give rise to new plants.

In addition to supporting the leaves, a stem has vascular tissue that transports water and minerals from the roots through the stem to the leaves. Vascular tissue also transports the organic products of photosynthesis, usually in the opposite direction. Xylem cells consist of nonliving cells that form a continuous pipeline for water and mineral transport, whereas phloem cells consist of living cells that join end to end for organic nutrient transport. Stems may have functions other than transport. In some plants (e.g., cactus), the stem is the primary photosynthetic organ. In succulent plants, the stem is a water reservoir. Tubers are underground horizontal stems that store nutrients.

Shoot System: Leaves

Leaves are the major component of the plant that require H_2O, CO_2, and sunlight to carry on photosynthesis. With few exceptions, their cells are living, and the bulk of a leaf contains tissue specialized to carry on photosynthesis. Leaves receive water from the root system by way of the stem. The position of the leaves on a plant maximizes the plant's photosynthetic efforts. The shape of the leaf also influences its function. For example, leaves that are broad and flat have the maximum surface area for the absorption of carbon dioxide and the collection of solar energy needed for photosynthesis. Also, unlike stems, leaves are almost never woody. When describing a leaf, the wide portion of a leaf is called the **blade,** and the stalk that attaches the blade to the stem is the **petiole** (Fig. 9.2*c*).

The size, shape, color, and texture of leaves can be highly variable. These characteristics are fundamental in plant identification. The leaves of some aquatic duckweeds may be less than 1 millimeter (mm) in diameter, while some palms may have leaves that exceed 6 m in length. The shapes of leaves can vary from cactus spines to deeply lobed oak leaves. Leaves can exhibit a variety of colors, from many shades of green to deep purple. The texture of leaves varies from smooth and waxy like a magnolia to coarse like a sycamore. Not all leaves are foliage leaves. Some are specialized to protect buds, attach to objects (tendrils), store food (bulbs), or even capture insects.

The upper acute angle between the petiole and stem is the leaf axil, where an **axillary bud** (or lateral bud) originates. This bud may become a branch or a flower. Plants that lose their leaves every year are called **deciduous.** This is in contrast to most of the gymnosperm plants (conifers), which usually retain their leaves for two to seven years. These plants are called **evergreens.**

Check Your Progress 9.1

1. Explain why it is more accurate to say that a stem has nodes and internodes than to define a stem as being aboveground.
2. Compare the structure and function of roots, stems, and leaves.

9.2 Cells and Tissues of Plants

Learning Outcomes

Upon completion of this section, you should be able to

1. Explain the function of meristematic tissue.
2. Describe the types of tissues found in plants.

Plants have levels of biological organization similar to animals (see Fig. 1.2). As in animals, a tissue is composed of specialized cells that perform a particular function, and an organ is a structure made up of multiple tissues. In plants, simple tissues are made up of a single cell type, whereas complex tissues contain a variety of cell types. All the tissue types in a plant arise from *meristematic tissue* (Fig. 9.3). Meristematic tissue allows a plant to grow its entire life because it retains cells that ever have the ability to divide and produce more tissues. Plants can grow their entire lives. Even a 5,000-year-old tree is still growing!

Video Seedling Growth

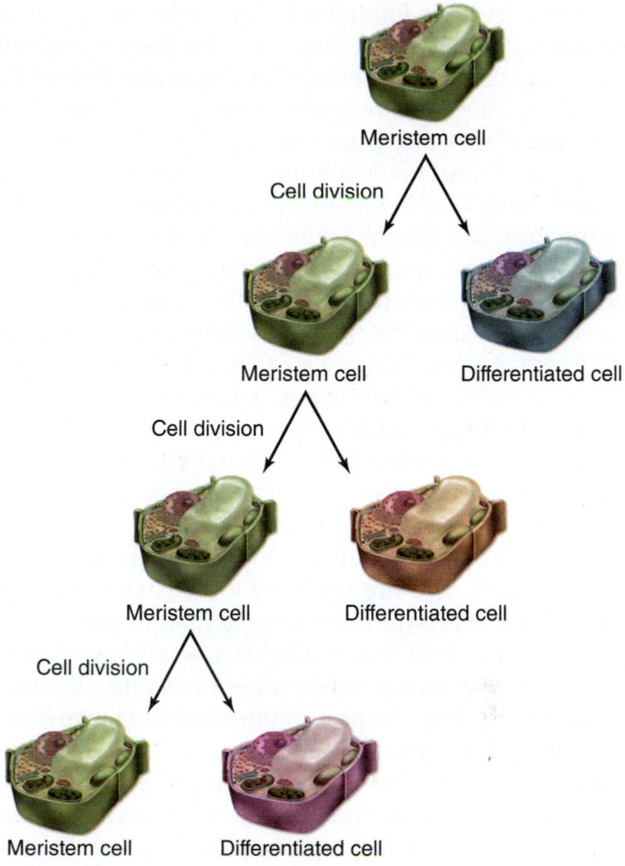
Meristem cell
Cell division
Meristem cell
Differentiated cell
Cell division
Meristem cell
Differentiated cell
Cell division
Meristem cell
Differentiated cell

Figure 9.3 Role of meristematic tissue. Meristematic tissue gives rise to a variety of different tissue types, including epidermal, ground, and vascular tissue.

Meristematic Tissue

Meristematic tissue is present in the shoot tip and root tip, where it is called the **apical meristem.** Apical meristem continually produces three types of primary meristem that develop into the three types of specialized primary tissues in the body of a plant: protoderm gives rise to epidermal tissue; ground meristem produces ground tissue; and procambium produces vascular tissue. The functions of these three specialized tissues include the following:

1. **Epidermal tissue** forms the outer protective covering of a plant.
2. **Ground tissue** fills the interior of a plant.
3. **Vascular tissue** transports water and nutrients in a plant and provides support.

A summary of each of these tissue types is presented in Table 9.1.

Epidermal Tissue

The entire body of both nonwoody (herbaceous) and young woody plants is covered by a layer of **epidermis.** The walls of epidermal cells that are exposed to air are covered with a waxy **cuticle** to minimize water loss. The cuticle also protects against bacteria and other organisms that might cause disease.

TABLE 9.1 Specialized Tissues in Plants

Tissue	Cell Types	Description	Function
Epidermal tissue	Epidermis	Tissue that covers roots, leaves, and stems	Protection, prevention of water loss
	Periderm (older woody plants)	Composed of waterproof cork cells	Protection against attacks by fungi, bacteria, and animals
Ground tissue	Parenchyma	Most abundant, corresponds to the typical plant cell	Gives rise to specialized, photosynthesis, and storage cells
	Colenchyma	Contains thick primary walls	Gives flexible support to immature regions of plant body
	Sclerenchyma	Contains thick secondary walls, contains lignin making the wall tough and hard	Supports the mature regions of the plant
Vascular: xylem	Tracheids	Elongated with tapered ends	Transport water and minerals from roots to leaves
	Vessel elements	Short and wide, contain perforated plates	Transport water and minerals from roots to leaves
Vascular: phloem	Sieve-tube members	Lack nuclei, associated with a companion cell, sieve plate	Transport sugar and organic compounds throughout the plant

In roots, certain epidermal cells have long, slender projections called **root hairs** (Fig. 9.4*a*). The hairs increase the surface area of the root for absorption of water and minerals.

On stems, leaves, and reproductive organs, epidermal cells produce small hairs that have two important functions: protecting the plant from too much sun and conserving moisture. Sometimes these hairs help protect a plant from herbivores by producing toxic substances. Under the slightest pressure, the stiff hairs of the stinging nettle lose their tips, forming "hypodermic needles" that inject an intruder with a stinging secretion.

A waxy cuticle reduces the gas exchange in leaves. Leaves will often contain specialized cells called *guard cells* and microscopic pores called *stomata* (sing., stoma) that assist in gas exchange. Guard cells, which are epidermal cells with chloroplasts, surround stomata (Fig. 9.4*b*). When the stomata are open, gas exchange is possible but water loss also occurs.

In older woody plants, the epidermis of the stem is replaced by boxlike **cork** cells. At maturity, cork cells can be sloughed off, and new cork cells are made by a meristem called cork cambium (Fig. 9.4*c*). As the new cork cells mature, they increase slightly in volume, and their walls become encrusted with a lipid material that makes them waterproof and chemically inert. These nonliving cells protect the plant and make it resistant to attack by fungi, bacteria, and animals. Some cork tissues are commercially used for bottle corks and other products.

The cork cambium overproduces cork in certain areas of the stem surface, causing ridges and cracks to appear. The areas where the cork layer is thin and the cells are loosely packed are called *lenticels*. Lenticels are important in gas exchange between the interior of a stem and the air.

Ground Tissue

Ground tissues form the bulk of a plant and fill the space between the epidermal and the vascular tissue. Most of the photosynthesis and carbohydrate storage takes place in ground tissue. Ground tissue is also responsible for producing hormones, toxins, pigments, and other specialized chemicals. The cell types present in ground tissue are parenchyma, collenchyma, and sclerenchyma cells.

Parenchyma cells are the most abundant and correspond best to the typical plant cell (Fig. 9.5*a*). These are the least specialized of the cell types and are found in all the organs of a plant. Parenchyma cells can divide and give rise to more specialized cells such as roots. They have relatively thin walls, are alive at maturity, and may contain chloroplasts and carry on photosynthesis, or they may contain colorless plastids that store the products of photosynthesis. The fleshy part of an apple is mostly storage parenchyma cells.

Collenchyma cells are like parenchyma cells except they have thicker primary walls (Fig. 9.5*b*). The thickness is irregular with the corners of the cell being thicker than other areas. Collenchyma cells often form bundles just beneath the epidermis and give flexible support to immature regions of a plant body. The familiar strands in celery stalks (leaf petioles) are composed mostly of collenchyma cells.

Sclerenchyma cells have a thick secondary cell wall that forms between the primary cell wall and the plasma membrane. The secondary cell wall is impregnated with **lignin,** a highly resistant organic substance that makes the walls tough and hard (Fig. 9.5*c*). If we compare a cell wall to reinforced concrete, cellulose fibrils would play the role of steel rods, and lignin would be analogous to the cement. Most sclerenchyma cells are dead at maturity. Their primary function is to support the mature regions of a plant. Tracheids and vessel elements are types of sclerenchyma cells that not only function in support but also transport water. Two other types of sclerenchyma cells are fibers and sclereids. Fibers are long and slender, and may be grouped in bundles that can be used commercially. Hemp fibers can be used to make rope, and flax fibers can be woven into linen. Flax fibers, however, are not lignified, which is why linen is soft. Sclereids, which are shorter than fibers and have variable shapes, are found in seed coats and nutshells. Sclereids, or "stone cells," are responsible for the gritty texture of pears as well as the hardness of nuts and peach pits.

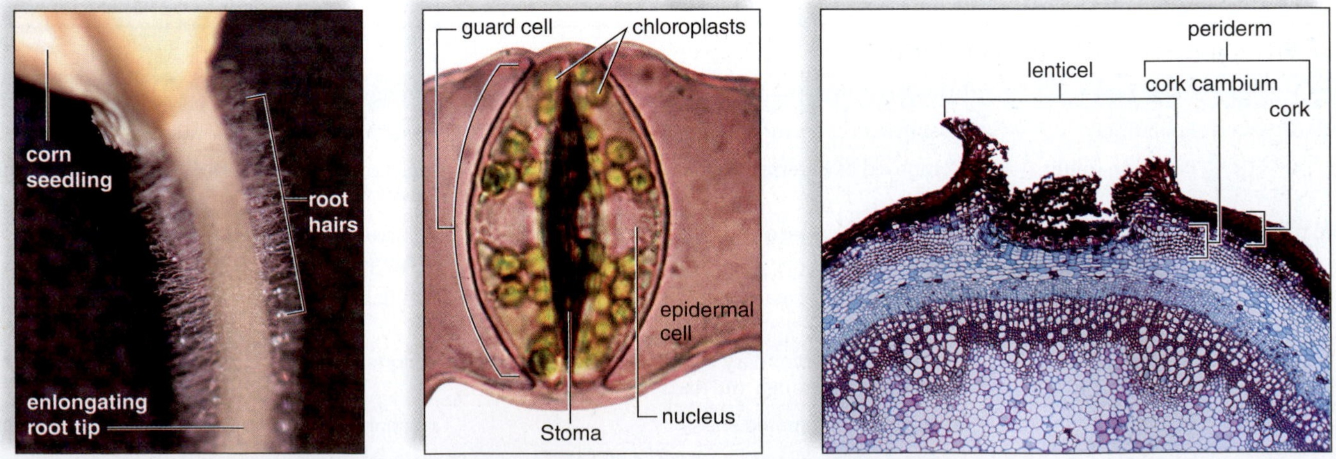
corn
seedling
root
hairs
enlongating
root tip
guard cell
chloroplasts
epidermal
cell
nucleus
Stoma
periderm
lenticel
cork cambium
cork

a. Root hairs   b. Stoma of leaf   c. Cork of older stem

Figure 9.4 Modifications of epidermal tissue. **a.** Root epidermis has root hairs to absorb water. **b.** Leaf epidermis contains stomata (sing., stoma) for gas exchange. **c.** Periderm includes cork and cork cambium. Lenticels in cork are important in gas exchange.

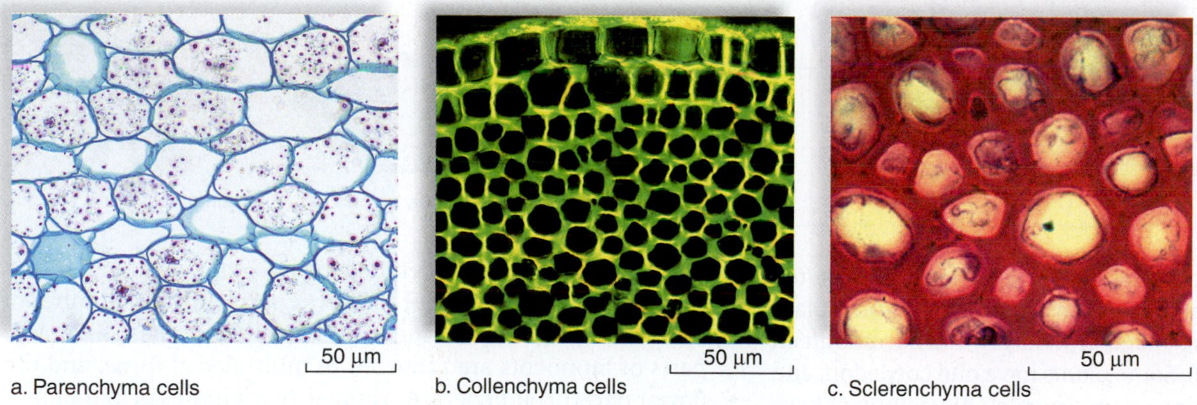
50 µm
a. Parenchyma cells
50 µm
b. Collenchyma cells
50 µm
c. Sclerenchyma cells

Figure 9.5 Ground tissue cells. **a.** Parenchyma cells are the least specialized of the plant cells. **b.** The walls of collenchyma cells are much thicker than those of parenchyma cells. **c.** Sclerenchyma cells have very thick walls and are nonliving at maturity—their primary function is to give strong support.

Vascular Tissue

There are two types of vascular (transport) tissue. **Xylem** transports water and minerals from the roots to the leaves, and **phloem** transports sugar and other organic compounds, including hormones, throughout the plant. Both xylem and phloem are considered complex tissues because they are composed of two or more kinds of cells. Xylem contains two types of conducting cells: tracheids and vessel elements (Fig. 9.6*a*). Both types of conducting cells are hollow and nonliving, but the **vessel elements** are shorter and wider. Vessel elements have plates with perforations in their end walls and are arranged to form a continuous vessel for water and mineral transport. The elongated **tracheids,** with tapered ends, form a less efficient means of transport, but water can move across the end walls and side walls because there are pits, or depressions, where the secondary wall does not form. In addition to vessel elements and tracheids, xylem contains parenchyma cells that store various substances and fibers that lend support.

The conducting cells of phloem are **sieve-tube members** arranged to form a continuous sieve tube (Fig. 9.6*b*). Sieve-tube members contain cytoplasm but no nuclei. The term *sieve* refers to a cluster of pores in the end walls, collectively called a sieve plate. Each sieve-tube member has a **companion cell,** which contains a nucleus. The two are connected by numerous **plasmodesmata,** strands of cytoplasm extending from one sieve-tube member to another, through the sieve plate. The nucleus of the companion cell controls and maintains the life of both cells. The companion cells are also believed to be involved in phloem's transport function.

It is important to realize that vascular tissue (xylem and phloem) extends from the root through stems to the leaves and vice versa (see Fig. 9.1). In the roots, the vascular tissue is located in the **vascular cylinder.** In the stem, it forms **vascular bundles,** and in the leaves, it is found in **leaf veins.**

Animation Vascular System of Plants

Check Your Progress 9.2

1. Describe the type of tissue in a plant that gives rise to all the other types and allows plants to grow their entire lives.
2. Identify the functions of epidermal tissue, ground tissue, and vascular tissue in plants.
3. Recognize the cell types found in ground tissue and vascular tissue and indicate their function.

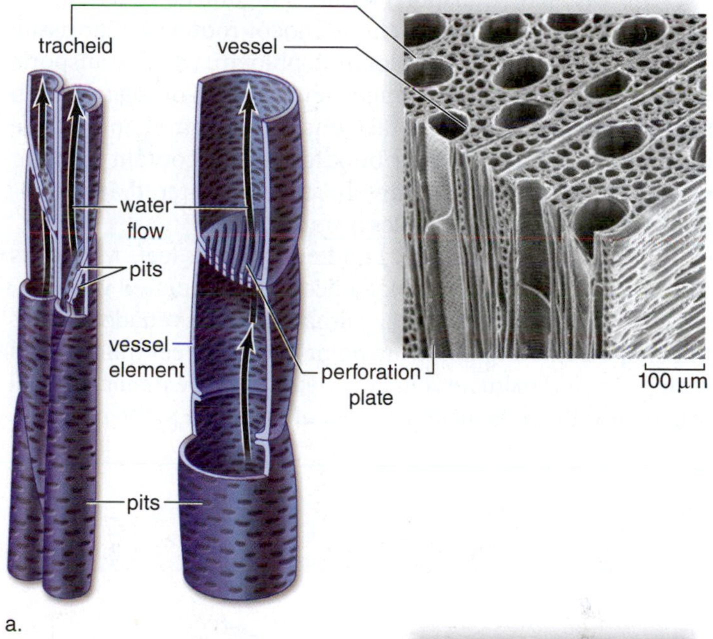
tracheid
vessel
water flow
pits
vessel element
perforation plate
pits
100 µm
a.

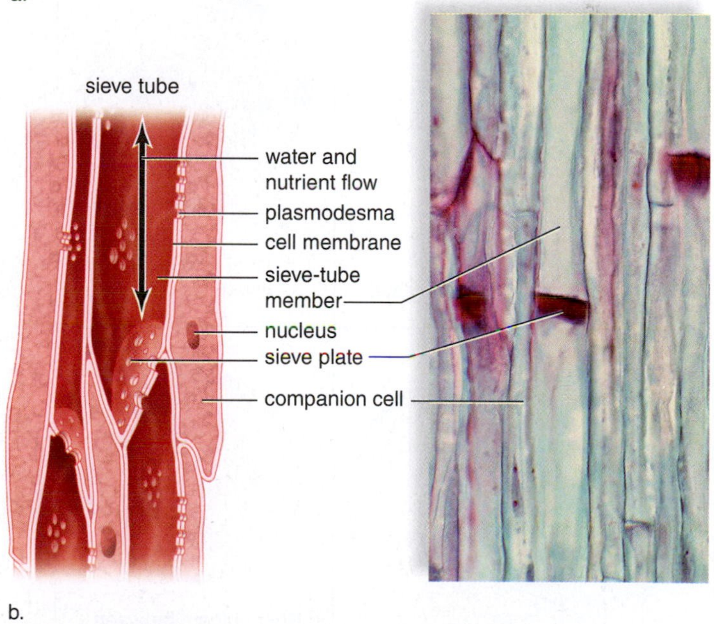
sieve tube
water and nutrient flow
plasmodesma
cell membrane
sieve-tube member
nucleus
sieve plate
companion cell
b.

Figure 9.6 Structure of xylem and phloem. **a.** Photomicrograph of xylem vascular tissue and drawing showing tracheids and vessel elements. **b.** Photomicrograph of phloem vascular tissue and drawing showing sieve tubes and companion cells.

9.3 Monocot Versus Eudicot Plants

Learning Outcome

Upon completion of this section, you should be able to

1. List and describe the key features of monocots and eudicots.

Flowering plants are divided into two groups, depending on the number of **cotyledons,** or seed leaves, in the embryonic plant (Fig. 9.7). Most cotyledons emerge, grow larger, and become green when the seed germinates. Some plants have one cotyledon, and are known as monocotyledons, or **monocots.** Other embryos have two cotyledons, and are known as eudicotyledons, or **eudicots.**

The vascular (transport) tissue is organized differently in monocots and eudicots. In the monocot root, vascular tissue occurs in a ring. In the eudicot root, phloem, which transports organic nutrients, is located between the arms of xylem, which transports water and minerals, and has a star shape. In the monocot stem, the vascular bundles, which contain vascular tissue surrounded by a bundle sheath, are scattered. In a eudicot stem, the vascular bundles occur in a ring.

Leaf veins are vascular bundles within a leaf. Monocots exhibit parallel venation, and eudicots exhibit netted venation, which may be either pinnate or palmate. Pinnate venation means that major veins originate from points along the centrally placed main vein, and palmate venation means that the major veins all originate at the point of attachment of the blade to the petiole:

Netted venation:
pinnately veined
palmately veined

Adult monocots and eudicots have other structural differences, such as the number of flower parts and the number of apertures (thin areas in the wall) of pollen grains. The flower parts of monocots are arranged in multiples of three, and the flower parts of eudicots are arranged in multiples of four or five. Eudicot pollen grains usually have three apertures, and monocot pollen grains usually have one aperture.

Although the division between monocots and eudicots may seem to be of limited importance, it does in fact affect many aspects of their structure. The eudicots are the larger group and include some of our most familiar flowering plants—from dandelions to oak trees. The monocots include grasses, lilies, orchids, and palm trees, among others. Some of our most significant food sources are monocots, such as rice, wheat, and corn.

Check Your Progress 9.3

1. Compare the following in monocots and eudicots: number of cotyledons, leaf venation, and flower parts.
2. List examples of plants that are monocots and eudicots.

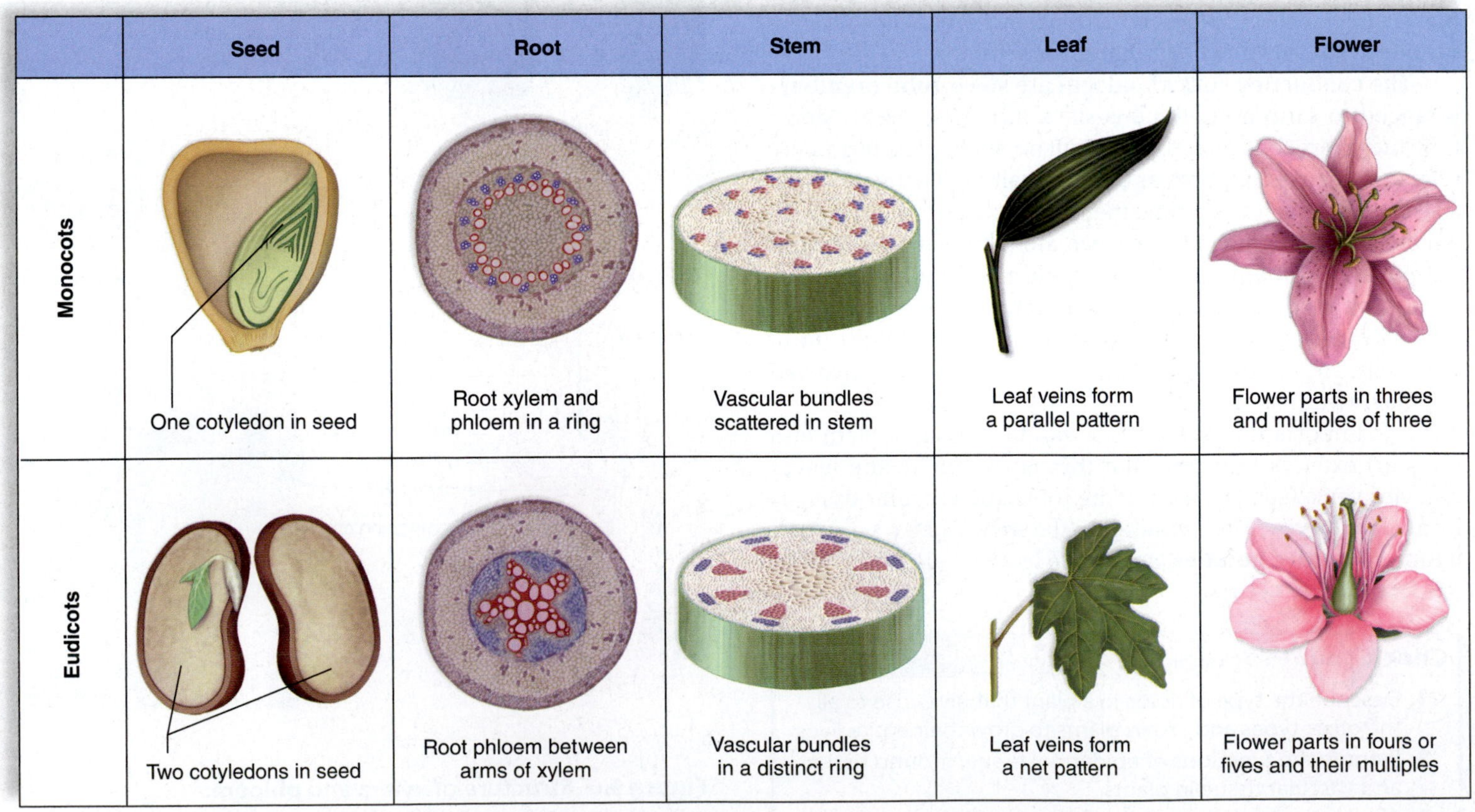
Seed
Root
Stem
Leaf
Flower
Monocots
One cotyledon in seed
Root xylem and phloem in a ring
Vascular bundles scattered in stem
Leaf veins form a parallel pattern
Flower parts in threes and multiples of three
Eudicots
Two cotyledons in seed
Root phloem between arms of xylem
Vascular bundles in a distinct ring
Leaf veins form a net pattern
Flower parts in fours or fives and their multiples

Figure 9.7 Flowering plants are either monocots or eudicots. Five features are used to distinguish monocots from eudicots: the number of cotyledons in the seed; the arrangement of vascular tissue in roots, stems, and leaves; and the number of flower parts.

9.4 Organization of Roots

Learning Outcomes

Upon completion of this section, you should be able to

1. List the zones in the root involved in primary growth.
2. Describe the anatomy of eudicot and monocot roots.
3. Identify different types of roots and root specializations.

Primary growth, which causes a plant to grow lengthwise, is centered in the apex (tip) of the shoot and the root systems. The growth of many roots is continuous, pausing only when temperatures become too cold or when water becomes too scarce. Figure 9.8*a*, a longitudinal section of a eudicot root, reveals zones where cells are in various stages of differentiation as primary growth occurs. These zones are called the *zone of cell division,* the *zone of elongation,* and the *zone of maturation.*

The zone of cell division is protected by the **root cap,** which is composed of parenchyma cells and protected by a slimy sheath. As the root grows, root cap cells are constantly removed by rough soil particles and replaced by new cells.

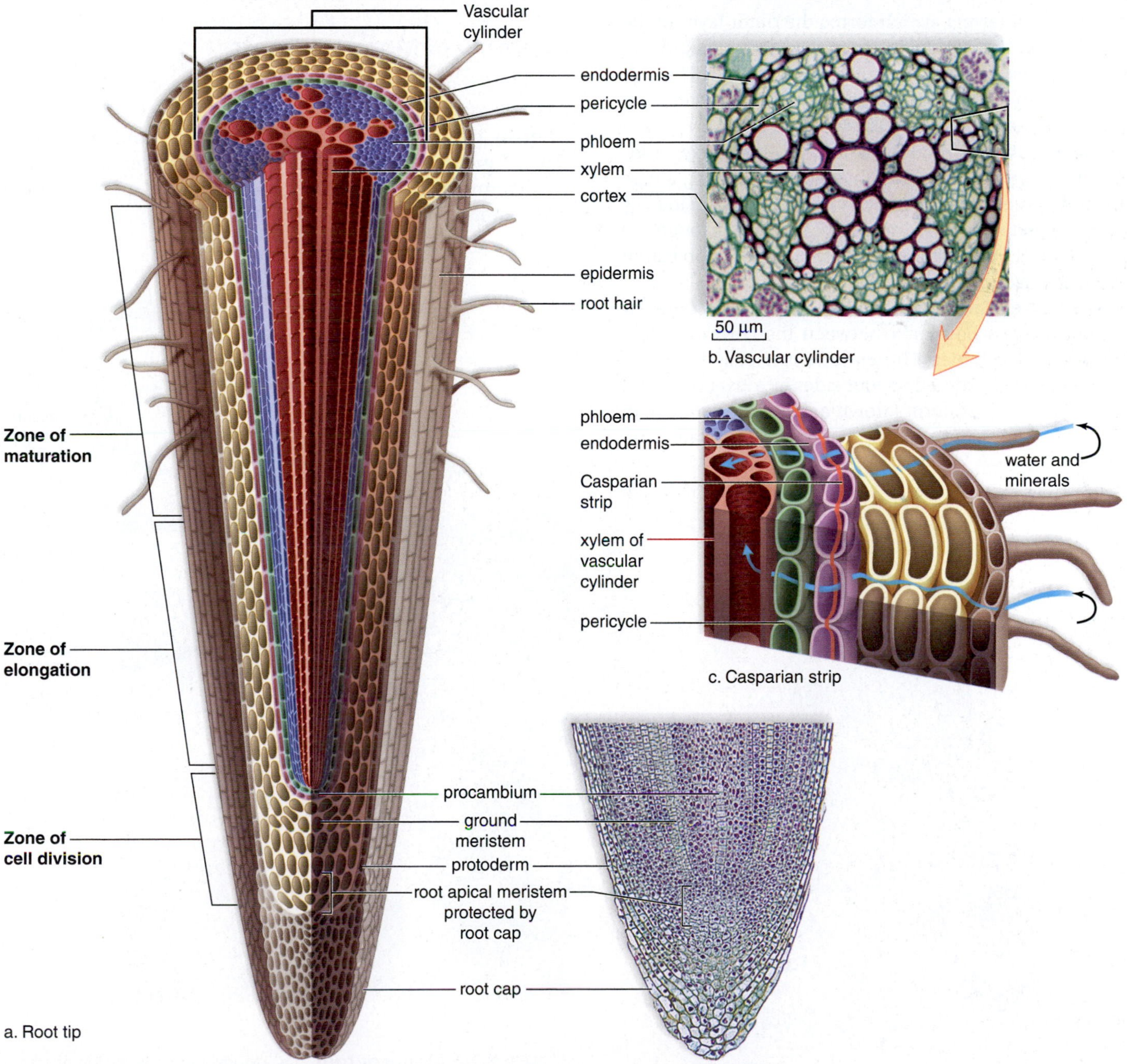

Vascular cylinder
endodermis
pericycle
phloem
xylem
cortex
epidermis
root hair
Zone of maturation
Zone of elongation
Zone of cell division
procambium
ground meristem
protoderm
root apical meristem protected by root cap
root cap
a. Root tip
50 µm
b. Vascular cylinder
phloem
endodermis
Casparian strip
xylem of vascular cylinder
pericycle
water and minerals
c. Casparian strip

Figure 9.8 Eudicot root tip. **a.** The root tip is divided into three zones, best seen in a longitudinal section such as this. **b.** The vascular cylinder of a eudicot root contains the vascular tissue. Xylem is typically star-shaped, and phloem lies between the points of the star. **c.** Water and minerals either pass between cells, or they enter root cells directly. In any case, because of the Casparian strip, water and minerals must pass through the cytoplasm of endodermal cells in order to enter the xylem. In this way, endodermal cells regulate the passage of minerals into the vascular cylinder.

The zone of cell division contains the root apical meristem, where mitosis produces relatively small, many-sided cells having dense cytoplasm and large nuclei. These meristematic cells give rise to the primary meristems, called protoderm, ground meristem, and procambium. Eventually, the primary meristems develop, respectively, into the three mature tissue types discussed previously: epidermis, ground tissue (i.e., cortex), and vascular tissue.

Anatomy of a Eudicot Root

Figure 9.8*a, b* also shows a cross section of a root at the zone of maturation. These specialized tissues are identifiable:

Epidermis. The epidermis, which forms the outer layer of the root, consists of only a single layer of cells. The majority of epidermal cells are thin-walled and rectangular. The epidermal cells in the zone of maturation have root hairs that project as far as 5–8 mm into the soil.

Cortex. Underneath the epidermis is a layer of large, thin-walled parenchyma cells that make up the **cortex,** a type of ground tissue. These irregularly shaped cells are loosely packed, so that water and minerals can move through the cortex without entering the cells. The cells contain starch granules, which function in food storage.

Endodermis. The **endodermis** is a single layer of rectangular cells that forms a boundary between the cortex and the inner vascular cylinder. The endodermal cells fit snugly together and are bordered on four sides by a layer of impermeable lignin and suberin known as the **Casparian strip** (Fig. 9.8*c*). This strip prevents the passage of water and mineral ions between adjacent cell walls. The two sides of each cell that contact the cortex and the vascular cylinder respectively remain permeable. Therefore, the only access to the vascular cylinder is through the endodermal cells, as shown by the arrows in Figure 9.8*c*.

Vascular tissue. The first layer of cells within the vascular cylinder is the **pericycle.** This tissue can become meristematic and start the development of branch roots (Fig. 9.9). The main portion of the vascular cylinder contains xylem and phloem. The xylem appears star-shaped in eudicots because several arms of tissue radiate from a common center (see Fig. 9.8). The phloem is found in separate regions between the arms of the xylem.

Anatomy of Monocot Roots

Monocot roots (Fig. 9.10) have the same growth zones as eudicot roots, but they do not undergo secondary growth as many eudicot roots do. A monocot root contains **pith,** a type of ground tissue, which is centrally located. The pith is surrounded by a vascular ring composed of alternating xylem and phloem bundles.

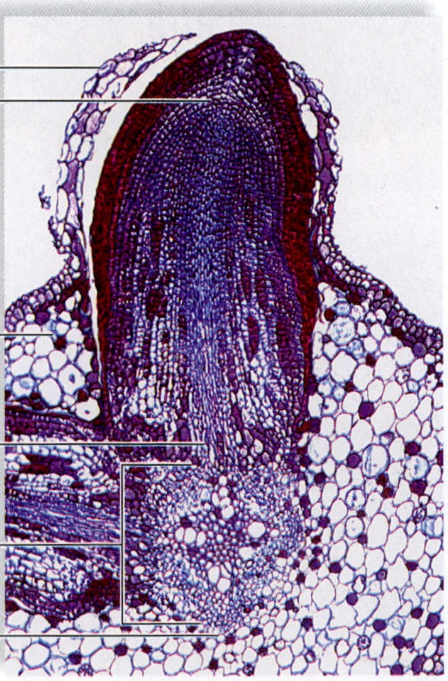
epidermis
emerging branch root
cortex
pericycle
vascular cylinder
endodermis

Figure 9.9 Branching of eudicot root. This cross section shows the origination and growth of a branch root from the pericycle.

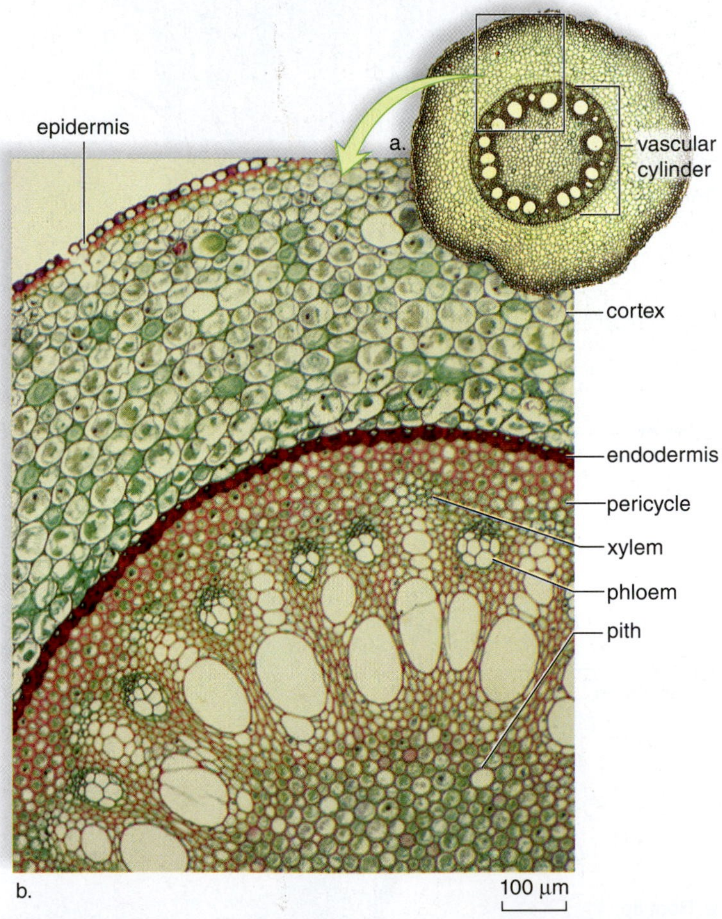
epidermis
a.
vascular cylinder
cortex
endodermis
pericycle
xylem
phloem
pith
b.
100 μm

Figure 9.10 Monocot root. **a.** This overall cross section of a monocot root shows that a vascular ring surrounds a central pith. **b.** The enlargement shows the exact placement of various tissues.

Root Diversity

In some plants, notably eudicots, the first or **primary root** grows straight down and remains the dominant root of the plant. This so-called **taproot** can penetrate the soil from a few centimeters (cm) to a reported 35 m for a mesquite tree. Lateral roots grow from the taproot to provide additional surface area for water absorption. The taproot is often fleshy and stores food (Fig. 9.11*a*). Carrots, beets, and turnips have taproots that we consume as vegetables. Sweet potato plants have roots that expand to store starch.

Other plants, notably monocots, lack a single, main root, but rather have a large number of slender roots. These grow from the lower nodes of the stem when the first (primary) root dies. These slender roots make up a **fibrous root system** (Fig. 9.11*b*). Fibrous root systems grow relatively close to the soil surface, forming dense mats. Many grasses have fibrous root systems that strongly anchor the plants to the soil.

Root Specializations

When roots develop from organs of the shoot system instead of the root system, they are known as *adventitious roots.* One style of adventitious root is typically found in corn plants (Fig. 9.11*c*). These are called *prop roots* because they emerge above the soil and act as anchors for the plant. Other examples of adventitious roots are those found on horizontal stems or the rootlets at the nodes of climbing English ivy. As the vines climb, the rootlets attach the plant to any available vertical structure (Fig 9.11*d, e*).

Black mangroves live in swampy water and have pneumatophores, root projections that rise above the water and acquire oxygen for cellular respiration (Fig. 9.11*f*). Some plants have poorly developed roots or no roots at all because minerals and water are supplied by other mechanisms. **Epiphytes** are "air plants." They do not grow in soil but on larger plants, which give them support, but no nutrients. Some epiphytes have roots that absorb moisture from the atmosphere, and many catch rain and minerals in special pockets at the base of their leaves.

Plants such as dodders and broomrapes are parasitic on other plants. Their stems have rootlike projections called haustoria (sing., haustorium) that grow into the host plant and make contact with vascular tissue from which they extract water and nutrients.

Two symbiotic relationships assist roots in taking up mineral nutrients. In the first type, legumes (soybeans and alfalfa) have roots infected by nitrogen-fixing *Rhizobium* bacteria. These bacteria can fix atmospheric nitrogen (N_2) by breaking the $N \equiv N$ bond and reducing nitrogen to NH_4^+ for incorporation into organic compounds. The bacteria live in **root nodules** and are supplied with carbohydrates by the host plant (Fig. 9.12*a*). The bacteria, in turn, furnish their host with nitrogen compounds.

Animation Root Nodule Formation

a. Taproot

b. Fibrous root system

c. Prop roots, a type of adventitious root

d. English ivy climbing up a tree trunk

e. Aerial roots

f. Pneumatophores of black mangrove trees

Figure 9.11 Root diversity and specialization. **a.** A taproot may have branch roots in addition to a main root. **b.** A fibrous root has many slender roots with no main root. **c.** Prop roots are specialized for support. **d.** English ivy climbs up the trunk because it has aerial roots. **e.** A closer look at aerial roots. **f.** The pneumatophores of a black mangrove tree allow it to acquire oxygen even though it lives in swampy water.

nodule
root
a. Root nodule
ectomycorrhizae
root
b. Mycorrhizae
c. Growth experiment

Figure 9.12 Root nodules and mycorrhizae. **a.** Nitrogen-fixing bacteria live in nodules on the roots of plants, particularly legumes. **b.** Ectomycorrhizae on red pine roots. **c.** Experimental results show that plants grown with mycorrhizae (two plants on right) grow much larger than a plant (left) grown without mycorrhizae.

The second type of symbiotic relationship, called a mycorrhizal association, involves fungi and almost all plant roots (Fig. 9.12*b*). Only a small minority of plants do not have mycorrhizae (sing., mycorrhiza). Ectomycorrhizae form a mantle that is exterior to the root that grows between the cell walls. Endomycorrhizae can penetrate cell walls. In any case, the fungus increases the surface area available for mineral and water uptake and breaks down organic matter, releasing nutrients for the plant to use. In return, the root furnishes the fungus with sugars and amino acids. Plants are extremely dependent on mycorrhizae for maximum growth (Fig. 9.12*c*). Orchid seeds, which are quite small and contain limited nutrients, do not germinate until a mycorrhizal fungus has invaded their cells. Nonphotosynthetic plants, such as Indian pipe, use their mycorrhizae to extract nutrients from nearby trees. Plants without mycorrhizae are usually limited as to the environment in which they can grow.

Check Your Progress 9.4

1. Explain how a plant controls what enters the vascular bundle.
2. Contrast eudicot and monocot roots.
3. List several specializations of roots and identify which of the three main functions of a root is involved in each specialization.

9.5 Organization of Stems

Learning Outcomes

Upon completion of this section, you should be able to

1. Identify the structures of a woody twig.
2. Describe the anatomy of eudicot and monocot stems.
3. Explain the secondary growth of stems.
4. Identify the adaptations that lead to stem diversity.

The anatomy of a woody twig helps us review the organization of a stem (Fig. 9.13). The **terminal bud** contains the shoot tip protected by bud scales, which are modified leaves. Leaf scars and bundle scars mark the location of leaves that have dropped. Dormant axillary buds that can give rise to branches or flowers are also found here. Each spring when growth resumes, bud scales fall off and leave a scar. You can tell the age of a stem by counting these groups of bud scale scars because there is one for each year's growth.

The terminal bud includes the shoot apical meristem and also leaf primordia (young leaves) that differentiate from cells produced by the shoot apical meristem (Fig. 9.14*a*). The activity of the terminal bud would be hindered by a protective covering comparable to the root cap. Instead, the leaf primordia fold over the apical meristem, providing protection. At the start of the season, the leaf primordia are, in turn, covered by terminal

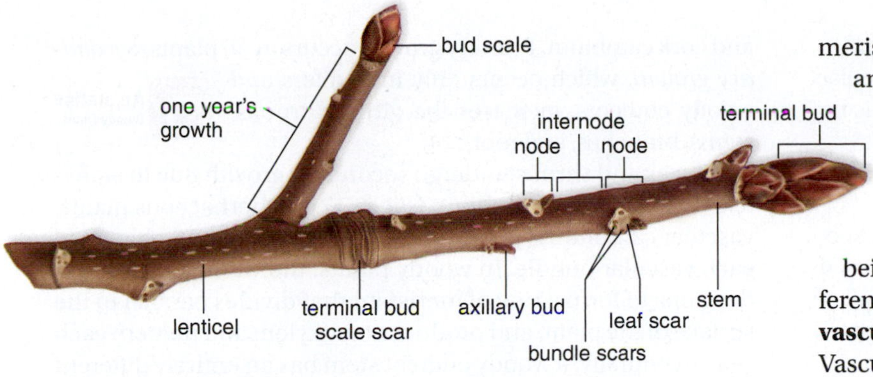
bud scale
one year's growth
internode
node
node
terminal bud
lenticel
terminal bud scale scar
axillary bud
bundle scars
leaf scar
stem

Figure 9.13 Woody twig. The major parts of a stem are illustrated by a woody twig collected in winter.

bud scales (scalelike leaves), but these drop off, or abscise, as growth continues.

The shoot apical meristem produces everything in a shoot: leaves, axillary buds, additional stem, and sometimes flowers. In the process, it gives rise to the same primary meristems as in the root. These primary meristems, in turn, develop into the differentiated tissues of a shoot system. The protoderm becomes the epidermis of the stem and leaves. Ground meristem produces parenchyma cells that become the cortex and pith in the stem and mesophyll in the leaves.

Procambium differentiates into the xylem and phloem of a vascular bundle. Certain cells become tracheids and others become vessel elements. The first sieve-tube members of a vascular bundle do not have companion cells and are short-lived (some live only a day before being replaced). Mature vascular bundles contain fully differentiated xylem and phloem, and a lateral meristem called **vascular cambium,** which is responsible for secondary growth. Vascular cambium is discussed more fully later in this section.

Herbaceous Stems

Mature nonwoody stems, called **herbaceous stems,** exhibit only primary growth. The outermost tissue of herbaceous stems is the epidermis, which is covered by a waxy cuticle to prevent water loss. In the vascular bundle, xylem is typically found toward the inside of the stem, and phloem is found toward the outside.

In a herbaceous eudicot stem, such as a sunflower, the vascular bundles are arranged in a distinct ring that separates the cortex from the central pith, where water and the products of photosynthesis

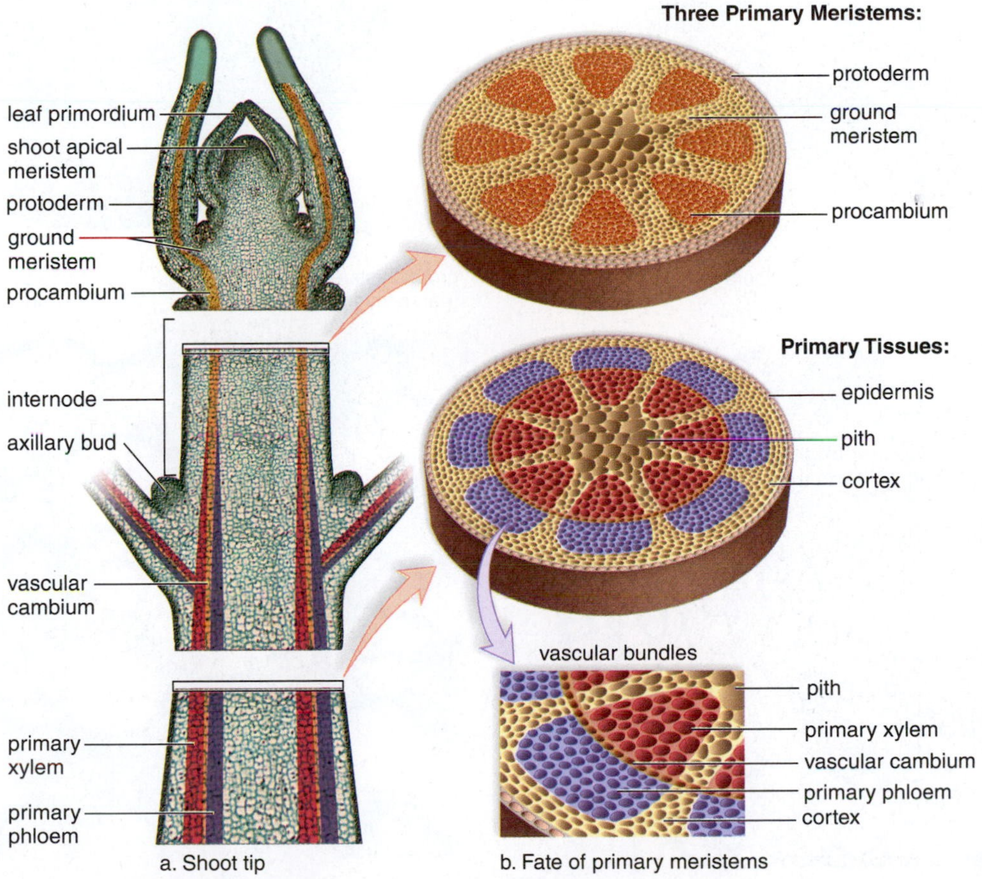
leaf primordium
shoot apical meristem
protoderm
ground meristem
procambium
internode
axillary bud
vascular cambium
primary xylem
primary phloem
a. Shoot tip
Three Primary Meristems:
protoderm
ground meristem
procambium
Primary Tissues:
epidermis
pith
cortex
vascular bundles
pith
primary xylem
vascular cambium
primary phloem
cortex
b. Fate of primary meristems

Figure 9.14 Shoot tip and primary meristems. **a.** The shoot apical meristem within a terminal bud is surrounded by leaf primordia. **b.** The shoot apical meristem produces the primary meristems. Protoderm gives rise to epidermis; ground meristem gives rise to pith and cortex; and procambium gives rise to vascular tissue, including primary xylem, primary phloem, and vascular cambium.

are stored (Fig. 9.15). The cortex is sometimes green and carries on photosynthesis. In a monocot stem such as corn, the vascular bundles are scattered throughout the stem, and often there is no well-defined cortex or well-defined pith (Fig. 9.16).

Woody Stems

A woody plant, such as an oak tree, has both primary and secondary tissues. Primary tissues are those new tissues formed each year from primary meristems right behind the shoot apical meristem. Secondary tissues develop during the first and subsequent years of growth from lateral meristems: vascular cambium and cork cambium. Primary growth occurs in all plants. *Secondary growth,* which occurs only in conifers and woody eudicots, increases the girth of trunks, stems, branches, and roots.

Animation Woody Dicot

Trees and shrubs undergo secondary growth due to activities of the vascular cambium (Fig. 9.17). In herbaceous plants, vascular cambium is present between the xylem and phloem of each vascular bundle. In woody plants, the vascular cambium develops to form a ring of meristem that divides parallel to the surface of the plant, and produces new xylem and phloem each year. Eventually, a woody eudicot stem has an entirely different

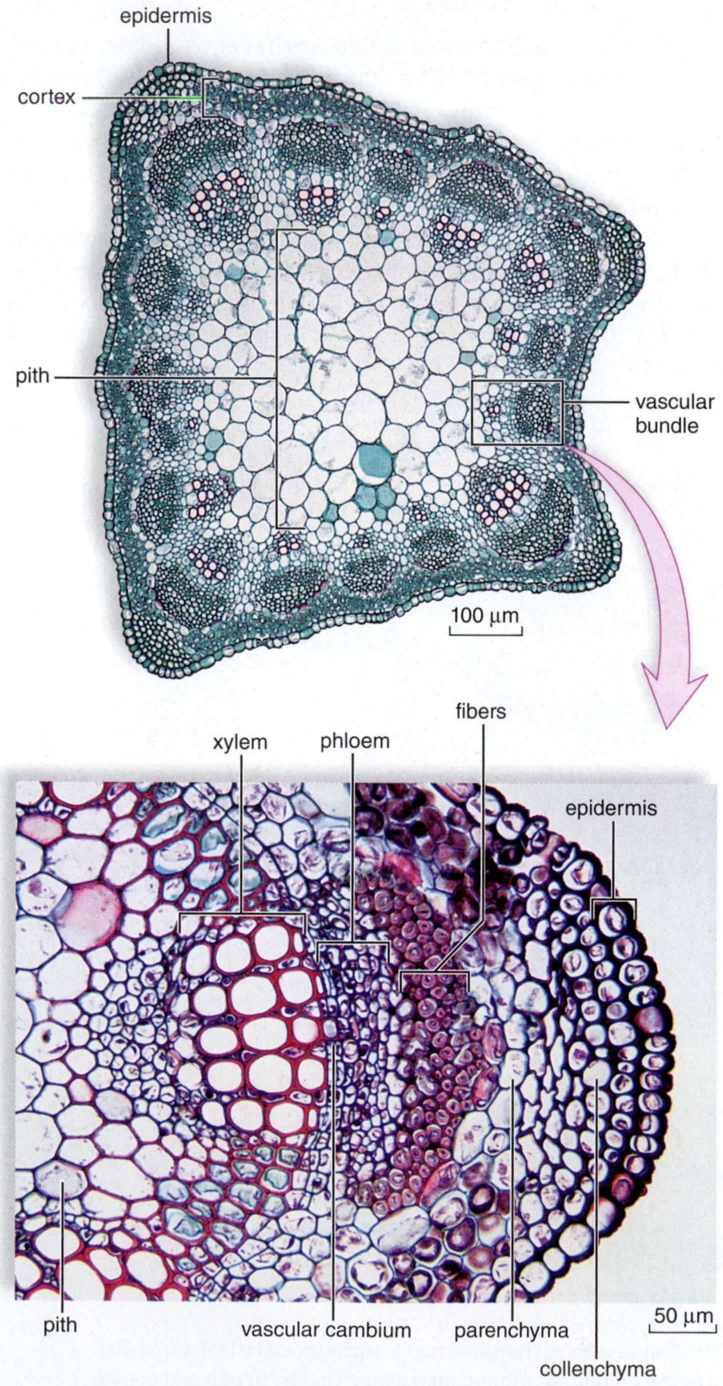
epidermis
cortex
pith
vascular bundle
100 μm
fibers
xylem
phloem
epidermis
pith
vascular cambium
parenchyma
50 μm
collenchyma

Figure 9.15 Herbaceous eudicot stem. In eudicot stems, the vascular bundles are arranged in a ring around the pith.

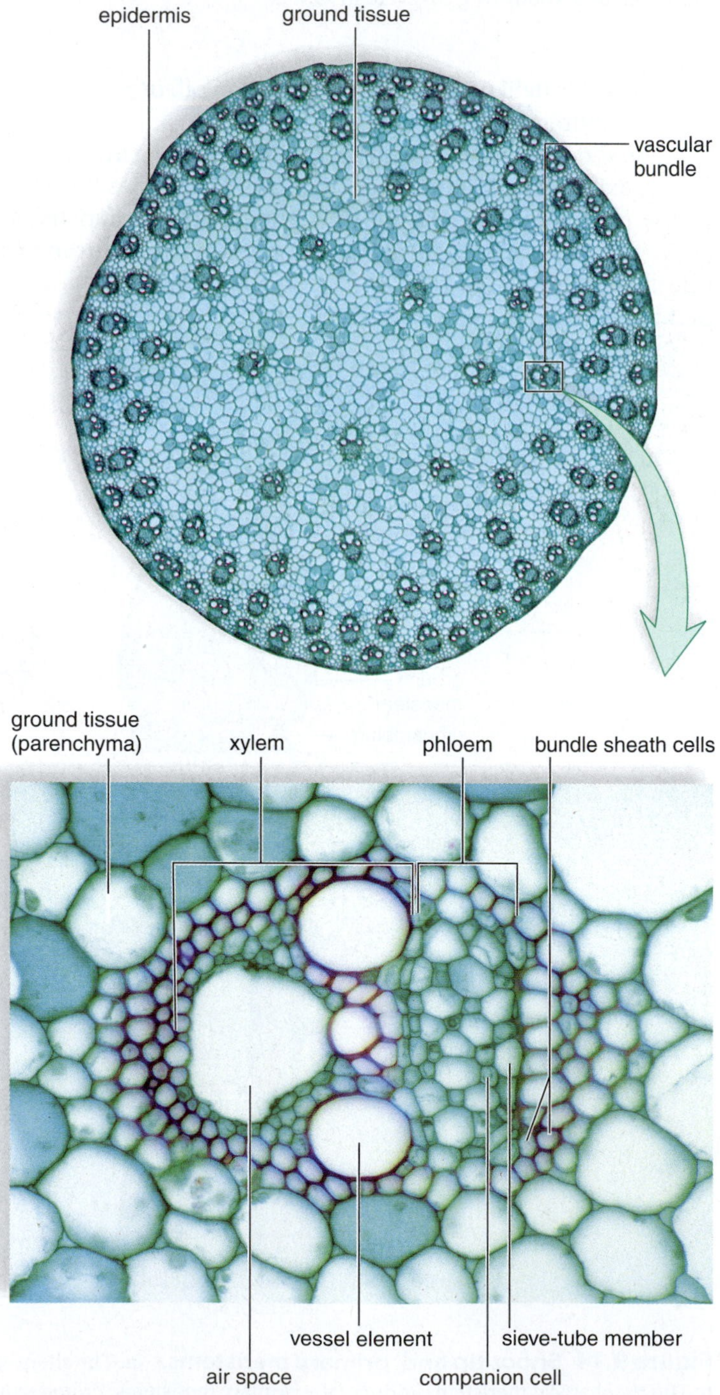
epidermis
ground tissue
vascular bundle
ground tissue (parenchyma)
xylem
phloem
bundle sheath cells
vessel element
sieve-tube member
air space
companion cell

Figure 9.16 Monocot stem. In the monocot stem, the vascular bundles are scattered throughout the stem.

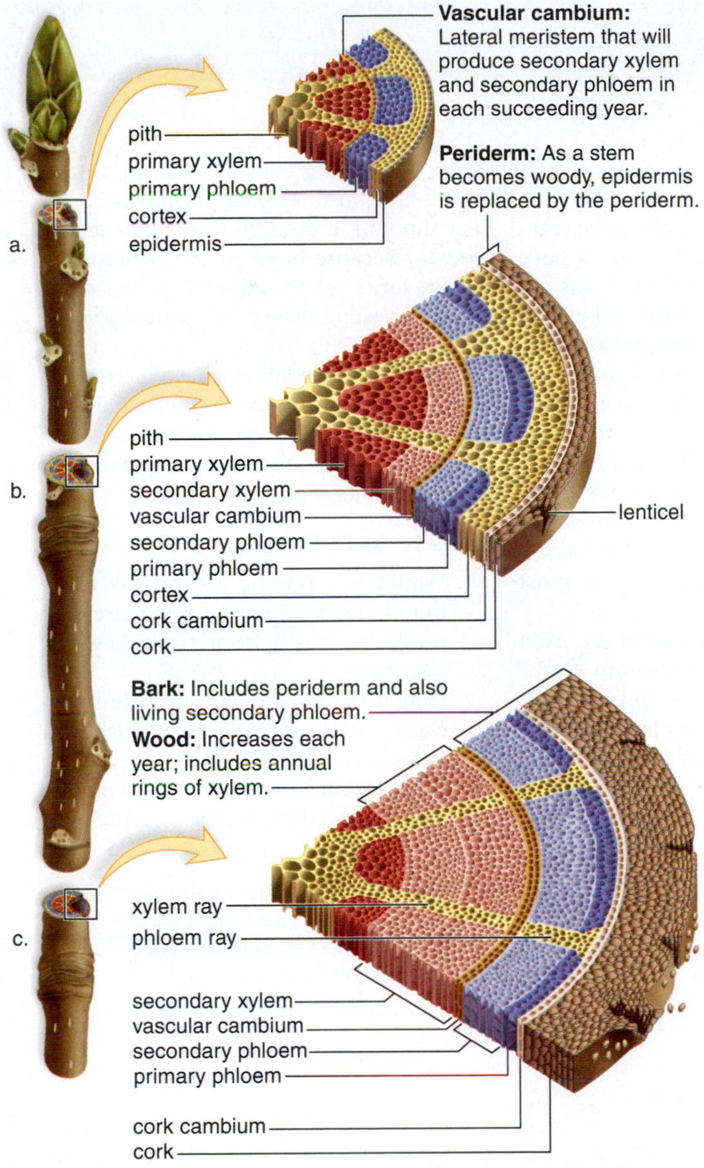
Vascular cambium: Lateral meristem that will produce secondary xylem and secondary phloem in each succeeding year.
pith
primary xylem
primary phloem
cortex
a.
epidermis
Periderm: As a stem becomes woody, epidermis is replaced by the periderm.
pith
primary xylem
b.
secondary xylem
vascular cambium
secondary phloem
primary phloem
cortex
cork cambium
cork
lenticel
Bark: Includes periderm and also living secondary phloem.
Wood: Increases each year; includes annual rings of xylem.
xylem ray
c.
phloem ray
secondary xylem
vascular cambium
secondary phloem
primary phloem
cork cambium
cork

Figure 9.17 Secondary growth of stems. **a.** Diagram showing a eudicot herbaceous stem just before secondary growth begins. **b.** Secondary growth has begun, and periderm has replaced the epidermis. Vascular cambium produces secondary xylem and secondary phloem each year. **c.** In a two-year-old stem, the primary phloem and cortex have disappeared, and only the secondary phloem (within the bark) produced by vascular cambium will be active that year. Secondary xylem builds up to become the annual rings of a woody stem.

organization from that of a herbaceous eudicot stem. A woody stem has no distinct vascular bundles and instead has three distinct areas: the bark, the wood, and the pith. The pith is a collection of parenchyma cells located at the center of the stem. Vascular cambium lies between the bark and the wood, which are discussed next.

You will also notice in Figure 9.17*c* the xylem rays and phloem rays that are visible in the cross section of a woody stem. Rays consist of parenchyma cells that permit lateral conduction of nutrients from the pith to the cortex as well as some storage of food. A phloem ray is actually a continuation of a xylem ray. Some phloem rays are much broader than other phloem rays.

Bark

The **bark** of a tree contains both periderm (cork, cork cambium, and a single layer of cork cells filled with suberin) and phloem. Although secondary phloem is produced each year by vascular cambium, phloem does not build up from season to season. The bark of a tree can be removed. However, doing so is very harmful because without phloem organic nutrients cannot be transported.

Cork cambium develops beneath the epidermis. When cork cambium first begins to divide, it produces tissue that disrupts the epidermis and replaces it with cork cells. Cork cells are impregnated with suberin, a waxy layer that makes them waterproof but also causes them to die. This is protective because now the stem is less edible. But an impervious barrier means that gas exchange is impeded except at lenticels, which are pockets of loosely arranged cork cells not impregnated with suberin.

Wood

Wood is secondary xylem that builds up year after year, thereby increasing the girth of trees. In trees that have a growing season, vascular cambium is dormant during the winter. In the spring, when moisture is plentiful and leaves require much water for growth, the secondary xylem contains wide vessel elements with thin walls. In this so-called *spring wood,* wide vessels transport sufficient water to the growing leaves. Later in the season, moisture is scarce, and the wood at this time, called *summer wood,* has a lower proportion of vessels (Fig. 9.18). Strength is required

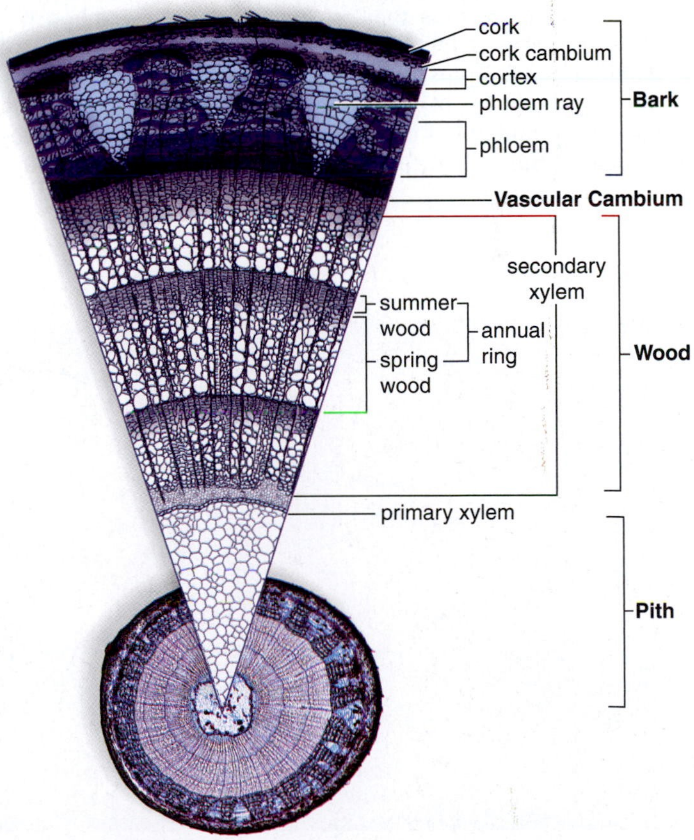
cork
cork cambium
cortex
phloem ray
phloem
Bark
Vascular Cambium
secondary xylem
summer wood
spring wood
annual ring
Wood
primary xylem
Pith

Figure 9.18 Three-year-old woody twig. The buildup of secondary xylem in a woody stem results in annual rings, which tell the age of the stem. The rings can be distinguished because each one begins with spring wood (large vessel elements) and ends with summer wood (smaller and fewer vessel elements).

SCIENCE IN YOUR LIFE ▶ ECOLOGY

The Many Uses of Bamboo

Due to its resilience, amazing rate of growth, and ability to grow in a variety of climates, bamboo is quickly becoming a valuable crop. Certain varieties of bamboo are capable of growing up to a foot per day and can reach their full height within one year in the right conditions.

Bamboo is classified as a grass that contains varieties ranging in height from one foot to over 100 feet. Globally there are over 1,400 species of bamboo. Approximately 900 can be found in tropical climates, whereas the remaining 500 are found in temperate environments. Several varieties of bamboo are even found in the eastern and southeastern United States.

Bamboo is recognized for its versatility as a building material, commercial food product, and clothing material. It can be processed into roofing material, flooring, and support beams, as well as a variety of construction materials. The mature stalks can be used as support columns in "Green" construction. It can also be used to replace steel reinforcing rods that are typically used in concrete-style construction. The bamboo goods industry started increasing in popularity in the United States during the mid-1990s and is expected to exceed $25 billion in the next several years. Bamboo products are three times harder than oak.

While not used extensively in the commercial food market, bamboo does have a variety of uses (Fig. 9A). The shoots are often used in a variety of Asian dishes as a vegetable. They can be boiled and added to a variety of dishes or eaten raw. In China, bamboo is made into a variety of alcoholic drinks; other Asian countries make soups, pancakes, and broths. The hollow bamboo stalk can be used to cook rice and soups within it. This will give the foods a subtle but distinct taste. Additional uses include modifying bamboo into cooking utensils, most notably chopsticks.

Clothing products are now being made out of bamboo. When made into clothing, bamboo is reported as being very light and extremely soft. It also has the ability to wick moisture away from the skin, making it ideal to wear during exercise. Several lines of baby clothing are being made from bamboo.

Bamboo is being recognized as one of the most eco-friendly crops. It requires lesser amounts of chemicals or pesticides to grow. It will remove nearly five times more greenhouse gases and produce nearly 35% more oxygen than an equivalent stand of trees. Harvesting can be done starting at the second to third year of growth all the way through to the fifth to seventh year of growth. Because bamboo is a perennial, it allows for a yearly regrowth of the stand after harvesting instead of requiring yearly replanting.

Versatility, hardiness, ease of growing, and greater awareness of environmentally friendly products are quickly making bamboo an ideal product that may someday replace metal, wood, and plastics.

Questions to Consider

1. Would you purchase products made from bamboo, even if it cost more than products made from nonrenewable materials?
2. Should the government provide financial incentives for farmers to switch their crops over to bamboo?
3. What are the negative consequences of growing and using bamboo products?

Figure 9A The many uses of bamboo. Bamboo is quickly becoming a multifunctional product in today's society. Uses range from building materials to food to clothing.

because the tree is growing larger and summer wood contains numerous thick-walled tracheids. At the end of the growing season, just before the cambium becomes dormant again, only heavy fibers with especially thick secondary walls may develop. When the trunk of a tree has spring wood followed by summer wood, the two together make up one year's growth, or an **annual ring.** You can tell the age of a tree by counting the annual rings (Fig. 9.19*a*). The outer annual rings, where transport occurs, are called sapwood.

In older trees, the inner annual rings, called the heartwood, no longer function in water transport. The cells become plugged with deposits such as resins, gums, and other substances that inhibit the growth of bacteria and fungi. Heartwood may help support a tree, although some trees stand erect and live for many years after the heartwood has rotted away. Figure 9.19*b* shows the layers of a woody stem in relation to one another.

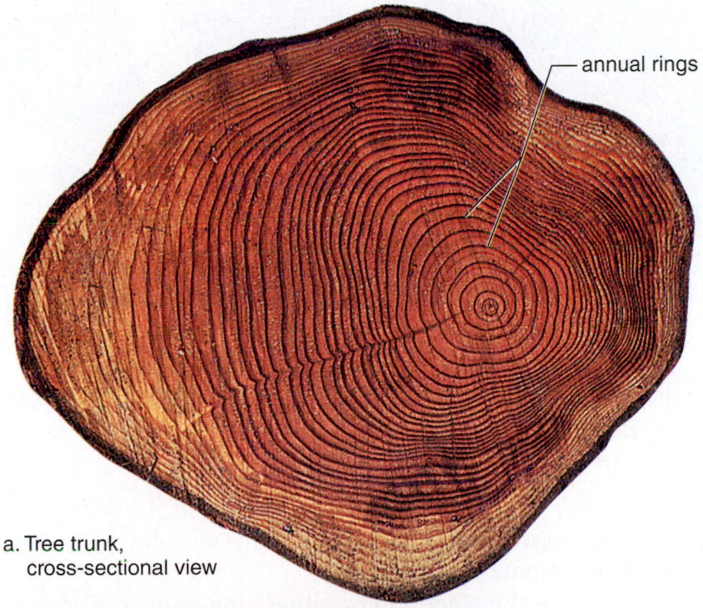
annual rings
a. Tree trunk,
cross-sectional view

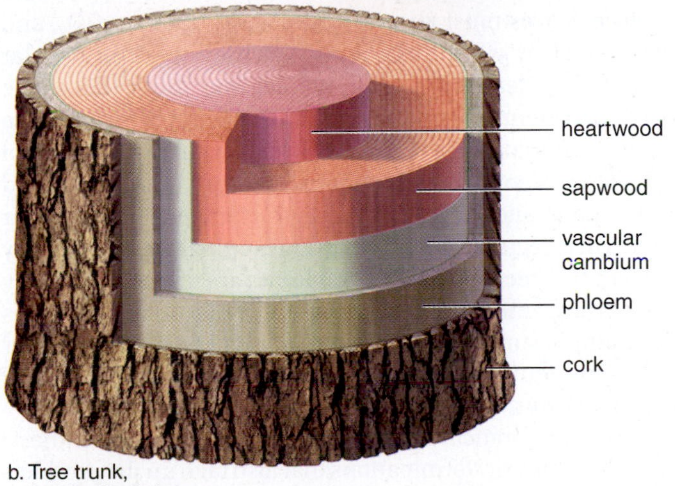
heartwood
sapwood
vascular cambium
phloem
cork
b. Tree trunk,
longitudinal view

Figure 9.19 Tree trunk. **a.** A cross section of a 39-year-old tree trunk. The xylem within the darker heartwood is inactive. The xylem within the lighter sapwood is active. **b.** The relationship of bark, vascular cambium, and wood is retained in a mature stem. The pith has been buried by the growth of layer after layer of new secondary xylem.

The annual rings are important not only in telling the age of a tree, but also in serving as a historical record of tree growth. For example, if rainfall and other conditions were extremely favorable during a season, the annual ring may be wider than usual.

Woody Plants

Is it advantageous to be woody? With adequate rainfall, woody plants can grow taller than herbaceous plants and increase in girth because they have adequate vascular tissue to support and service their leaves. However, it takes energy to produce secondary growth and prepare the body for winter if the plant lives in the temperate zone. Also, woody plants need more defense mechanisms because a long-lasting plant that stays in one spot is likely to be attacked by herbivores and parasites. Then, too, trees don't usually reproduce until they have grown for several seasons, by which time they may have succumbed to an accident or disease. In certain habitats, it is more advantageous for a plant to put most of its energy into simply reproducing rather than being woody.

Stem Diversity

Stems exist in diverse forms, or modifications (Fig. 9.20). Aboveground horizontal stems, called **stolons** or runners, produce new plants where the nodes touch the ground. The strawberry plant contains this type of stem, which functions in vegetative reproduction.

Aboveground vertical stems can also be modified. For example, cacti have succulent stems specialized for water storage, and the tendrils of grape plants (which are stem branches) allow them to climb. Morning glory and its relatives have stems that twine around support structures. Such tendrils and twining shoots help plants expose their leaves to the sun.

Underground horizontal stems, called **rhizomes,** may be long and thin, as in sod-forming grasses, or thick and fleshy, as in irises. Rhizomes survive the winter and contribute to asexual reproduction because each node bears a bud. Some rhizomes have enlarged portions called tubers, which function in food storage. Potatoes are an example of tubers.

Corms are bulbous underground stems that lie dormant during the winter, just as rhizomes do. They have thin, papery leaves and a thick stem. They also produce new plants the next growing season. Gladiolus corms are referred to as bulbs by laypersons, but the botanist reserves the term *bulb* for a structure composed of thick modified leaves attached to a short vertical stem. An onion is a bulb.

Humans make use of stems in many ways. The stem of the sugarcane plant is a primary source of table sugar. The spice cinnamon and the drug quinine are derived from the bark of *Cinnamomum verum* and various *Cinchona* species, respectively.

Check Your Progress 9.5

1. Compare and contrast eudicot and monocot stems.
2. Explain why there are two portions to every ring in a tree's annual rings.
3. Describe various stem modifications.

stolon
node
adventitious
roots
adventitious roots
rhizome
axillary
bud
rhizome
branch
tuber
axillary
bud
papery
leaves
corm
adventitious roots
a. Stolon
b. Rhizome
c. Tuber
d. Corm

Figure 9.20 Stem diversity. **a.** A strawberry plant has aboveground horizontal stems called stolons. Every other node produces a new shoot system. **b.** The underground horizontal stem of an iris is a fleshy rhizome. **c.** The underground stem of a potato plant has enlargements called tubers. We call the tubers potatoes. **d.** The corm of a gladiolus is a thick stem covered by papery leaves.

9.6 Organization of Leaves

Learning Outcomes

Upon completion of this section, you should be able to

1. Identify the structures of a leaf.
2. Describe some types of leaf modifications.

Figure 9.21 shows a cross section of a generalized foliage leaf of a temperate-zone eudicot plant. The functions of a foliage leaf are to carry on photosynthesis, regulate water loss, and be protective against parasites and predators.

Epidermal tissue is located on the leaf's upper and lower surfaces, where it is well situated to be protective. Epidermis contains trichomes, little hairs that are highly protective, especially when they secrete irritating substances against predators. Epidermal cells play a significant role in regulating water loss. Closely packed epidermal cells secrete an outer, waxy cuticle that helps prevent water loss but does not allow the uptake of CO_2 needed for photosynthesis. However, the stomata, located particularly in the lower epidermis, do allow the uptake of CO_2 and can also be closed by guard cells to prevent water loss. Trichomes can also help prevent parasites from entering stomata, and their presence cuts down on water loss when stomata are open.

Leaves are the chief photosynthesizing organs of a plant. Therefore, leaves must absorb solar energy, take up CO_2, and receive water by way of leaf veins. For the reception of solar energy, foliage leaves are generally flat and thin—this shape allows solar energy to penetrate the entire width of the leaf. The body of a leaf is composed of **mesophyll.** The elongated cells of the palisade mesophyll carry on most of the photosynthesis and are situated to allow its chloroplasts to efficiently absorb solar energy. The irregular cells of spongy mesophyll are bounded by air spaces that receive CO_2 when stomata are open. The loosely packed arrangement of the cells in spongy mesophyll increases the amount of surface area for absorption of CO_2 and for water loss. Evaporation of water is actually the means by which water rises in leaf veins, as discussed in section 9.7.

Leaf veins branch, and water needed for photosynthesis diffuses from narrow terminations that also take up the products of photosynthesis for distribution to other parts of the plant. **Bundle sheaths** are layers of cells surrounding vascular tissue. Most bundle sheaths are parenchyma cells, sclerenchyma cells, or a combination of the two. The parenchyma cells help regulate the entrance and exit of materials into and out of the leaf vein.

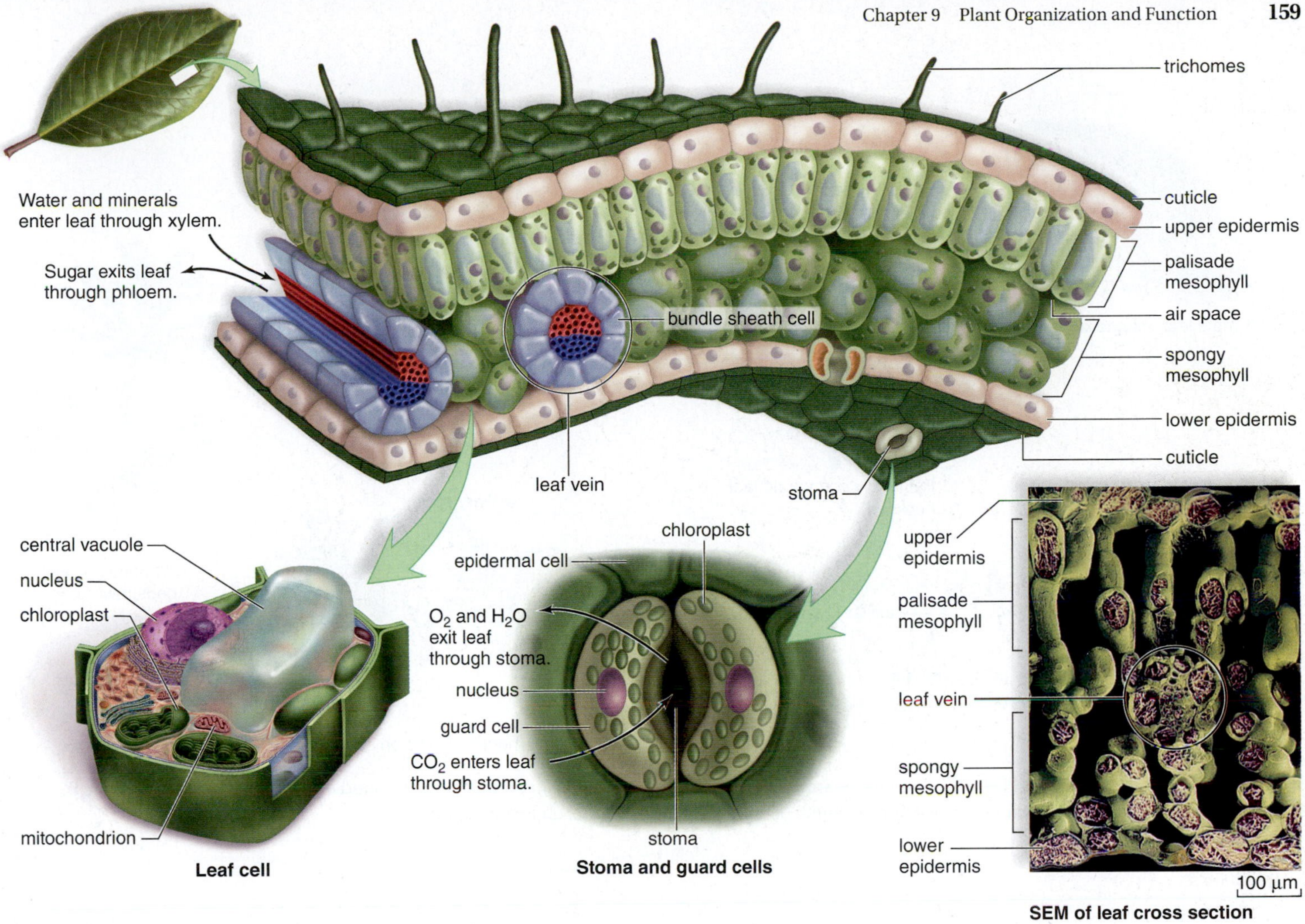

trichomes
cuticle
upper epidermis
palisade mesophyll
air space
spongy mesophyll
lower epidermis
cuticle
Water and minerals enter leaf through xylem.
Sugar exits leaf through phloem.
bundle sheath cell
leaf vein
stoma
central vacuole
nucleus
chloroplast
mitochondrion
Leaf cell
chloroplast
epidermal cell
O_2 and H_2O exit leaf through stoma.
nucleus
guard cell
CO_2 enters leaf through stoma.
stoma
Stoma and guard cells
upper epidermis
palisade mesophyll
leaf vein
spongy mesophyll
lower epidermis
100 μm
SEM of leaf cross section

Figure 9.21 Leaf structure. Photosynthesis takes place in the mesophyll tissue of leaves. The leaf is enclosed by epidermal cells covered with a waxy layer, the cuticle. The veins contain xylem and phloem for the transport of water and solutes. Stoma in the epidermis permit the exchange of gases.

Leaf Diversity

The blade of a leaf can be simple or compound (Fig. 9.22*a*). A simple leaf has a single blade in contrast to a compound leaf, which is divided into leaflets. For example, a magnolia tree has simple leaves, and a buckeye tree has compound leaves. In pinnately compound leaves, the leaflets occur in pairs, as in a black walnut tree, while in palmately compound leaves, all of the leaflets are attached to a single point, as in a buckeye tree. Plants such as the mimosa have bipinnately compound leaves, with leaflets subdivided into even smaller leaflets.

Leaves can be arranged on a stem in three ways: alternate, opposite, or whorled (Fig. 9.22*b*). The American beech has alternate leaves; the maple has opposite leaves, being attached to the same node; and bedstraw has a whorled leaf arrangement, with several leaves originating from the same node.

Leaves are adapted to environmental conditions. Shade plants tend to have broad, wide leaves, and desert plants tend to have reduced leaves with sunken stomata. The leaves of a cactus are the spines attached to the succulent (fleshy) stem (Fig. 9.23*a*). Other succulents have leaves adapted to hold moisture.

An onion bulb is composed of leaves surrounding a short stem. In a head of cabbage, large leaves overlap one another. The petiole of a leaf can be thick and fleshy, as in celery and rhubarb. Leaves of climbing plants, such as those of peas and cucumbers, are modified into tendrils that can attach to nearby objects (Fig. 9.23*b*).

Some plants have leaves that are specialized for catching insects. A sundew has sticky trichomes that trap insects and others that secrete digestive enzymes. The Venus flytrap has hinged leaves that snap shut and interlock when an insect triggers sensitive trichomes that project from inside the leaves (Fig. 9.23*c*). Insectivorous plants commonly grow in marshy regions, where the supply of soil nitrogen is severely limited. The digested insects provide the plants with a source of organic nitrogen.

Check Your Progress 9.6

1. Describe how the two regions found in the body of a leaf differ.
2. Identify how leaves are adapted to their environment.

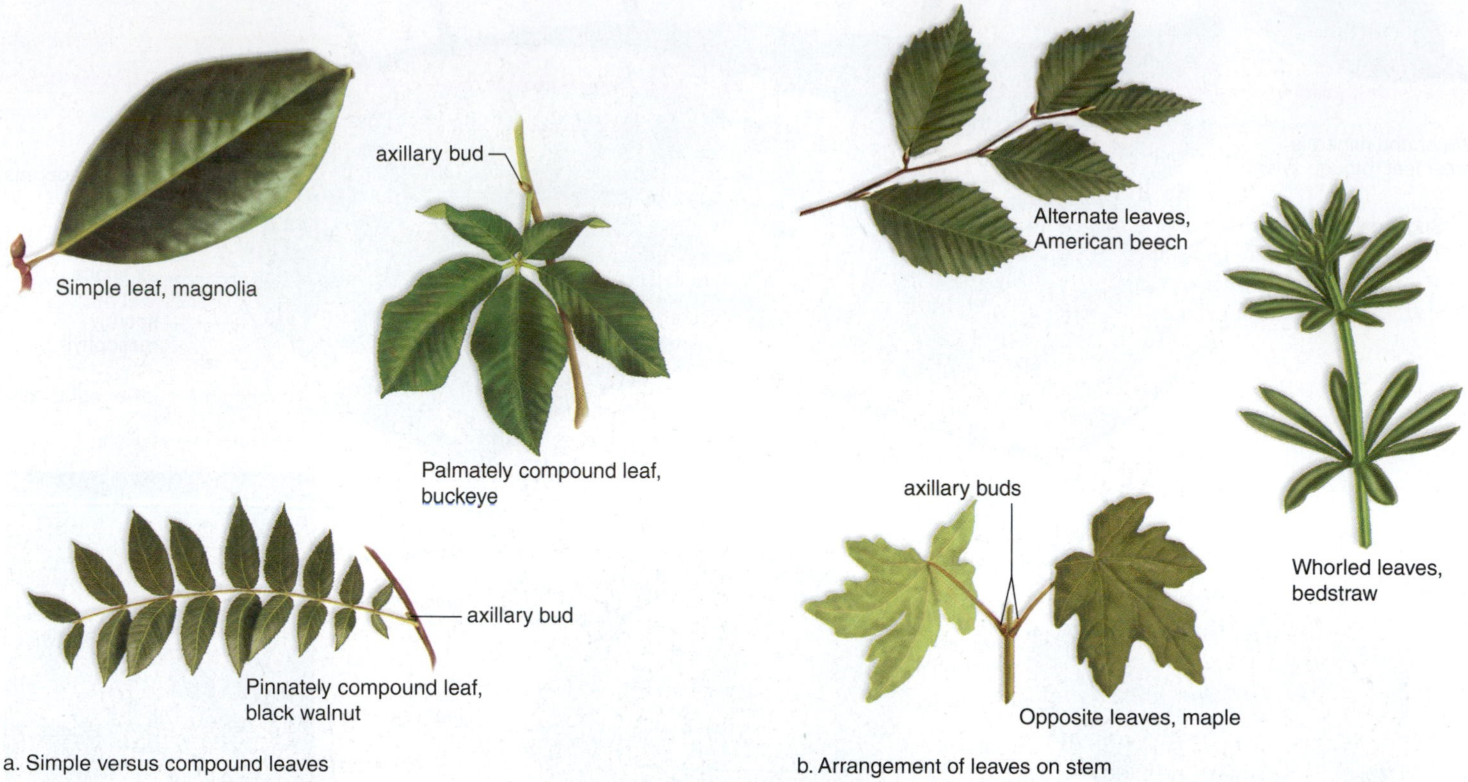

axillary bud
Simple leaf, magnolia
Palmately compound leaf, buckeye
axillary bud
Pinnately compound leaf, black walnut
a. Simple versus compound leaves
Alternate leaves, American beech
axillary buds
Whorled leaves, bedstraw
Opposite leaves, maple
b. Arrangement of leaves on stem

Figure 9.22 Classification of leaves. **a.** Leaves are simple or compound, being either pinnately compound or palmately compound. Note the one axillary bud per compound leaf. **b.** The leaf arrangement on a stem can be alternate, opposite, or whorled.

stem
spine
tendril
hinged leaves
a. Cactus
b. Cucumber
c. Venus flytrap

Figure 9.23 Leaf diversity. **a.** The spines of a cactus are modified leaves that protect the fleshy stem from animal predation. **b.** The tendrils of a cucumber are modified leaves that attach the plant to a physical support. **c.** The modified leaves of the Venus flytrap serve as a trap for insect prey. When triggered by an insect, the leaf snaps shut. Once shut, the leaf secretes digestive juices that break down the soft parts of the insect's body.

9.7 Uptake and Transport of Nutrients

Learning Outcomes

Upon completion of this section, you should be able to

1. Explain the movement of water in a plant according to the cohesion-tension model of xylem transport.
2. Explain the movement of organic nutrients in a plant according to the pressure-flow model of phloem transport.

In order to produce a carbohydrate, a plant requires carbon dioxide from the air and water from the soil. The transport system of the plant consists of the vascular tissue: the xylem and phloem. Water and minerals are transported through a plant in xylem; the products of photosynthesis are transported in phloem.

Water Uptake and Transport

Water and minerals enter the plant through the root, primarily through the root hairs. Osmosis of water and diffusion of minerals aid the entrance of water and minerals from the soil into the plant. Eventually, however, a plant uses active transport to bring minerals into root cells. From there, water, along with minerals, moves across the tissues of a root until it enters xylem.

Once water enters xylem, it must be transported upward to the leaves of the plant. This can be a daunting task. Water entering root cells creates a positive pressure called root pressure that tends to push xylem sap upward. Atmospheric pressure can support a column of water to a maximum height of approximately 10.3 m. However, some trees can exceed 90 m in height, so other factors must be involved in causing water to move from the roots to the leaves. Figure 9.24 illustrates the accepted model for the transport of water and, therefore, minerals in a plant. It is called the **cohesion-tension model.**

Cohesion-Tension Model of Xylem Transport

The vascular tissue of xylem contains the hollow conducting cells called tracheids and vessel elements (see Fig. 9.6). The vessel elements are larger than the tracheids, and they are stacked one on top of the other to form a pipeline that stretches from the roots to the leaves. It is an open pipeline because the vessel elements have no end walls, just perforation plates, separating one from the other. The tracheids, which are elongated with tapered ends, form a less obvious means of transport. Water can move across the end and side walls of tracheids because of pits, or depressions, where the secondary wall does not form.

Explanation of the Model Unlike animals, which rely on a pumping heart to move blood through their vessels, plants utilize a passive, not active, means of transport to move water in xylem. The cohesion-tension model of xylem transport relies on the properties of water (see section 2.3). The term *cohesion* refers to the tendency of water molecules to cling together. Because of hydrogen bonding, water molecules interact with one another, forming a continuous water column in xylem from the roots to the leaves that is not easily broken. *Adhesion* refers to the ability of water, a polar molecule, to interact with the molecules making up the walls of the vessels in xylem. Adhesion gives the water column extra strength and prevents it from slipping downward.

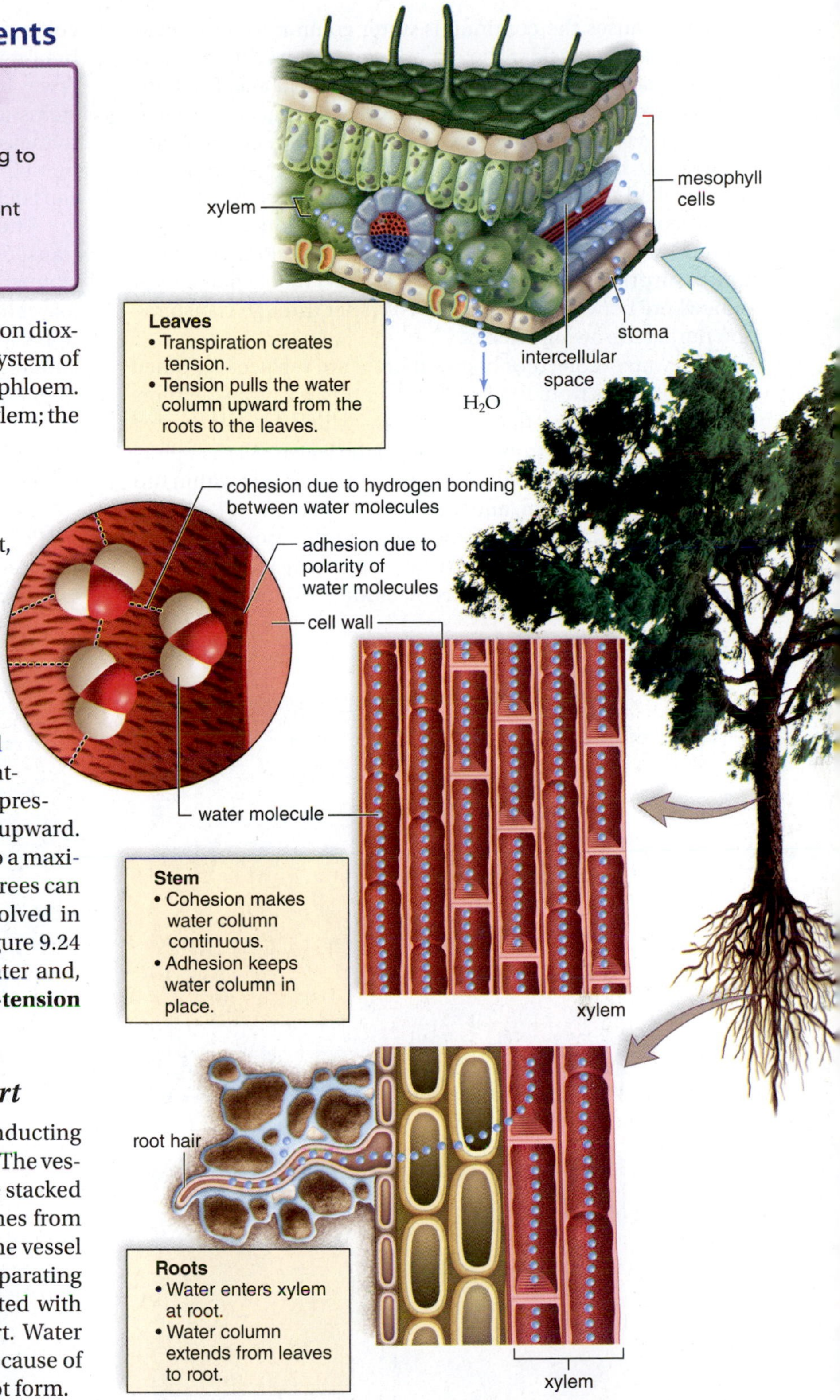

xylem
mesophyll cells
Leaves
• Transpiration creates tension.
• Tension pulls the water column upward from the roots to the leaves.
stoma
intercellular space
H_2O
cohesion due to hydrogen bonding between water molecules
adhesion due to polarity of water molecules
cell wall
water molecule
Stem
• Cohesion makes water column continuous.
• Adhesion keeps water column in place.
xylem
root hair
Roots
• Water enters xylem at root.
• Water column extends from leaves to root.
xylem

Figure 9.24 Cohesion-tension model of xylem transport. Tension created by evaporation (transpiration) at the leaves pulls water along the length of the xylem from the root hairs to the leaves.

3D Animation Plant Transport: Water Transport in Xylem

What causes the continuous water column to move passively upward? Consider the structure of a leaf. When the sun rises, stomata open, and carbon dioxide enters a leaf. Within the leaf, the mesophyll cells—particularly the spongy layer—are exposed to the air, which can be quite dry. Water now evaporates from mesophyll cells. Evaporation of water from leaf cells is called **transpiration.** At least 90% of the water taken up by the roots is eventually lost by transpiration. This means that the total amount of water lost by a plant over a long period of time is surprisingly large. A single *Zea mays* (corn) plant loses somewhere between 135 and 200 l of water through transpiration during a growing season.

The water molecules that evaporate are replaced by other water molecules from the leaf veins. In this way, transpiration exerts a driving force—that is, a *tension*—which draws the water column up in vessels from the roots to the leaves. As transpiration occurs, the water column is pulled upward, first within the leaf, then from the stem, and finally from the roots.

Tension can reach from the leaves to the root only if the water column is continuous. What happens if the water column within xylem is broken, as by cutting a stem? The water column "snaps back" in the xylem vessel, away from the site of breakage, making it more difficult for conduction to occur. This is why it is best to maximize water conduction by cutting flower stems under water. This effect has also allowed investigators to measure the tension in stems. A device called the pressure bomb measures how much pressure it takes to push the xylem sap back to the cut surface of the stem.

There is an important consequence to the way water is transported in plants. When a plant is under water stress, the stomata close. Now the plant loses little water because the leaves are protected against water loss by the waxy cuticle of the upper and lower epidermis. When stomata are closed, however, carbon dioxide cannot enter the leaves, and plants are unable to photosynthesize. (CAM plants are a notable exception. See page 137.) Photosynthesis, therefore, requires an abundant supply of water so that the stomata remain open and allow carbon dioxide to enter.

Opening and Closing of Stomata

Each stoma, a small pore in the leaf epidermis, is bordered by two **guard cells** (Fig. 9.25). When water enters the guard cells

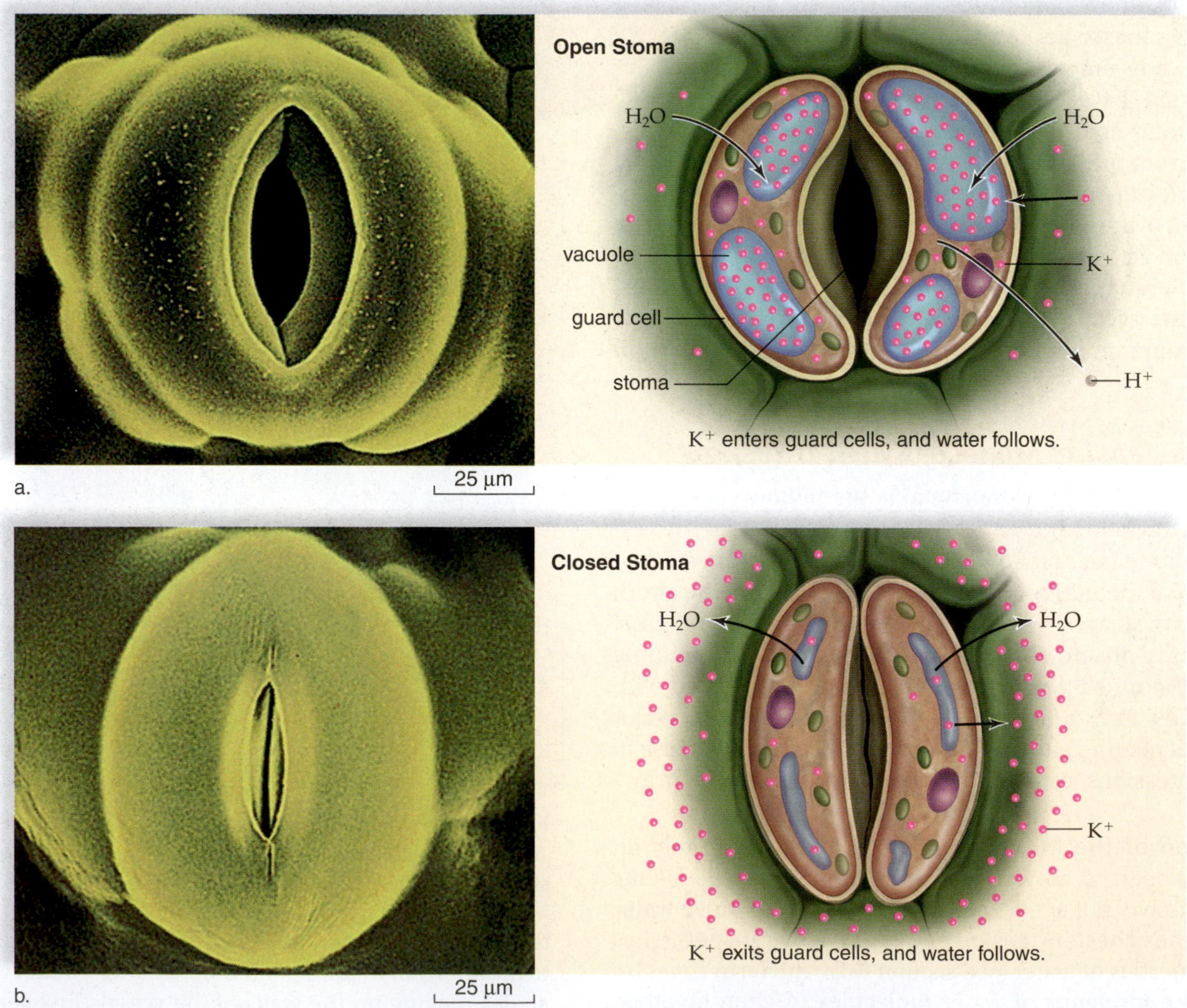
Open Stoma
H_2O
H_2O
vacuole
guard cell
stoma
K^+
H^+
K^+ enters guard cells, and water follows.
a.
25 µm
Closed Stoma
H_2O
H_2O
K^+
K^+ exits guard cells, and water follows.
b.
25 µm

Figure 9.25 Opening and closing of stomata. **a.** A stoma opens when turgor pressure increases in guard cells due to the entrance of K^+ followed by the entrance of water. **b.** A stoma closes when turgor pressure decreases due to the exit of K^+ followed by the exit of water.

and turgor pressure increases, the stoma opens. When water exits the guard cells and turgor pressure decreases, the stoma closes. Notice in Figure 9.25 that the guard cells are attached to each other at their ends and that the inner walls are thicker than the outer walls. When water enters, a guard cell's radial expansion is restricted because of cellulose microfibrils in the walls, but lengthwise expansion of the outer walls is possible. When the outer walls expand lengthwise, they buckle out from the region of their attachment, and the stoma opens.

Since about 1968, it has been clear that potassium ions (K^+) accumulate within guard cells when stomata open. In other words, active transport of K^+ into guard cells causes water to follow by osmosis and stomata to open.

If plants are kept in the dark, stomata open and close about every 24 hours, just as if they were responding to the presence of sunlight in the daytime and the absence of sunlight at night. This means that some sort of internal biological clock must be keeping time. Circadian rhythms (behaviors that occur nearly every 24 hours) and biological clocks are areas of intense investigation at this time. Other factors that influence the opening and closing of stomata include temperature, humidity, and stress.

Organic Nutrient Transport

Plants transport water and minerals from the roots to the leaves. They also transport organic nutrients from the leaves to the parts of plants that need them. This includes young leaves that have not yet reached their full photosynthetic potential, flowers that are in the process of making seeds and fruits, and the roots, whose location in the soil prohibits them from carrying on photosynthesis.

Role of Phloem

As long ago as 1679, Marcello Malpighi suggested that bark is involved in moving sugars from leaves to roots. He observed that if a strip of bark is removed below the level of the majority of the tree's leaves, the bark swells just above the cut, and sugar accumulates in the swollen tissue. We know today that this is because the phloem is being removed, but the xylem is left intact. Therefore, the results suggest that phloem is the tissue that transports sugars.

Radioactive tracer studies with carbon 14 (^{14}C) have confirmed that phloem transports organic nutrients. When ^{14}C-labeled carbon dioxide (CO_2) is supplied to mature leaves, radioactively labeled sugar is soon found moving down the stem into the roots. It's difficult to get samples of sap from phloem without injuring the phloem, but this problem is solved by using aphids, small insects that are phloem feeders. The aphid drives its stylet, a sharp mouthpart that functions like a hypodermic needle, between the epidermal cells, and sap enters its body from a sieve-tube member (Fig. 9.26). If the aphid is anesthetized using ether, its body can be carefully cut away, leaving the stylet. Phloem can then be collected and analyzed. The use of radioactive tracers and aphids has revealed that the movement of nutrients through phloem can be as fast as 60–100 cm per hour and possibly up to 300 cm per hour.

Pressure-Flow Model of Phloem Transport

The **pressure-flow model** is the current explanation for the movement of organic materials in phloem. Consider the

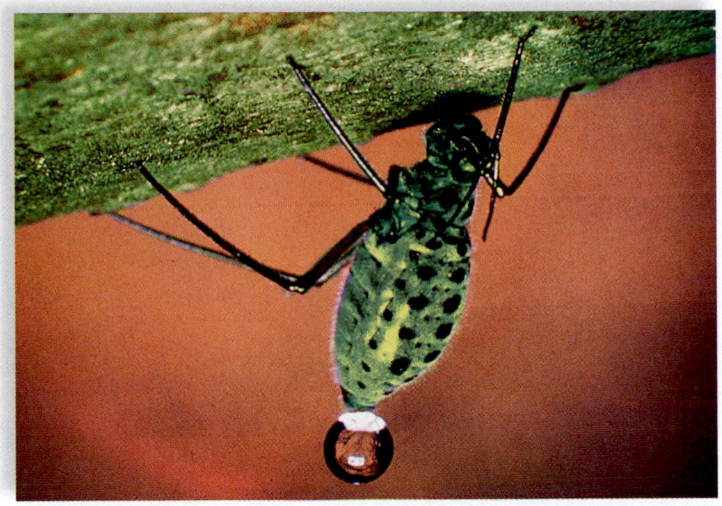

a. An aphid feeding on a plant stem

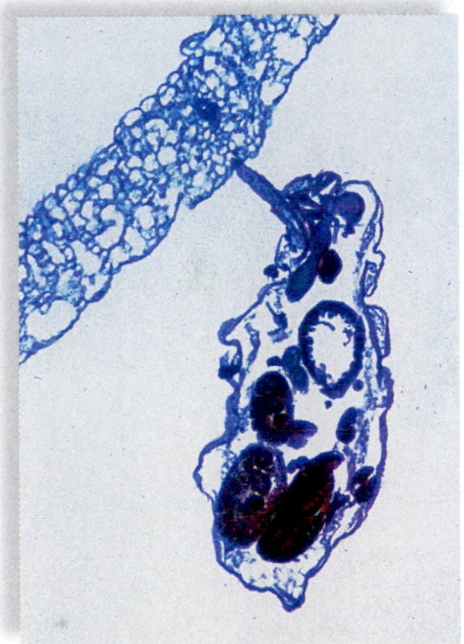

b. Aphid stylet in place

Figure 9.26 Acquiring phloem sap. Aphids are small insects that remove nutrients from phloem by means of a needlelike mouthpart called a stylet. **a.** Excess phloem sap appears as a droplet after passing through the aphid's body. **b.** Micrograph of a stylet in plant tissue. When an aphid is cut away from its stylet, phloem sap becomes available for collection and analysis.

following experiment in which two bulbs are connected by a glass tube. The left-hand bulb contains solute at a higher concentration than the right-hand bulb. Each bulb is bounded by a differentially permeable membrane, and the entire apparatus is submerged in distilled water:

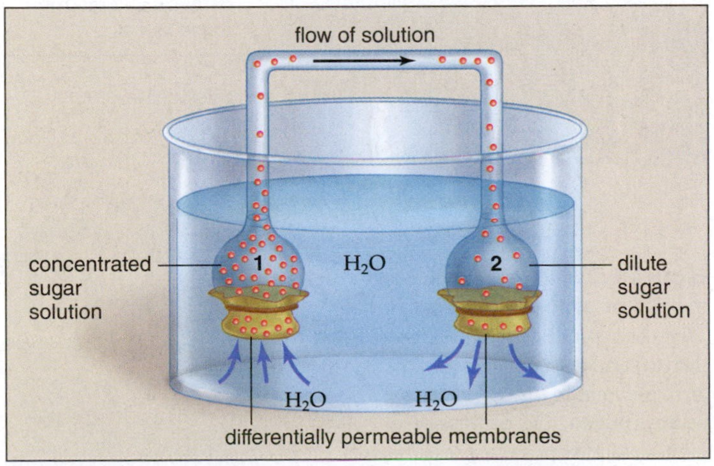
flow of solution
concentrated sugar solution
1
H_2O
2
dilute sugar solution
H_2O
H_2O
differentially permeable membranes

Distilled water flows into the left-hand bulb to a greater extent because it has the higher solute concentration. The

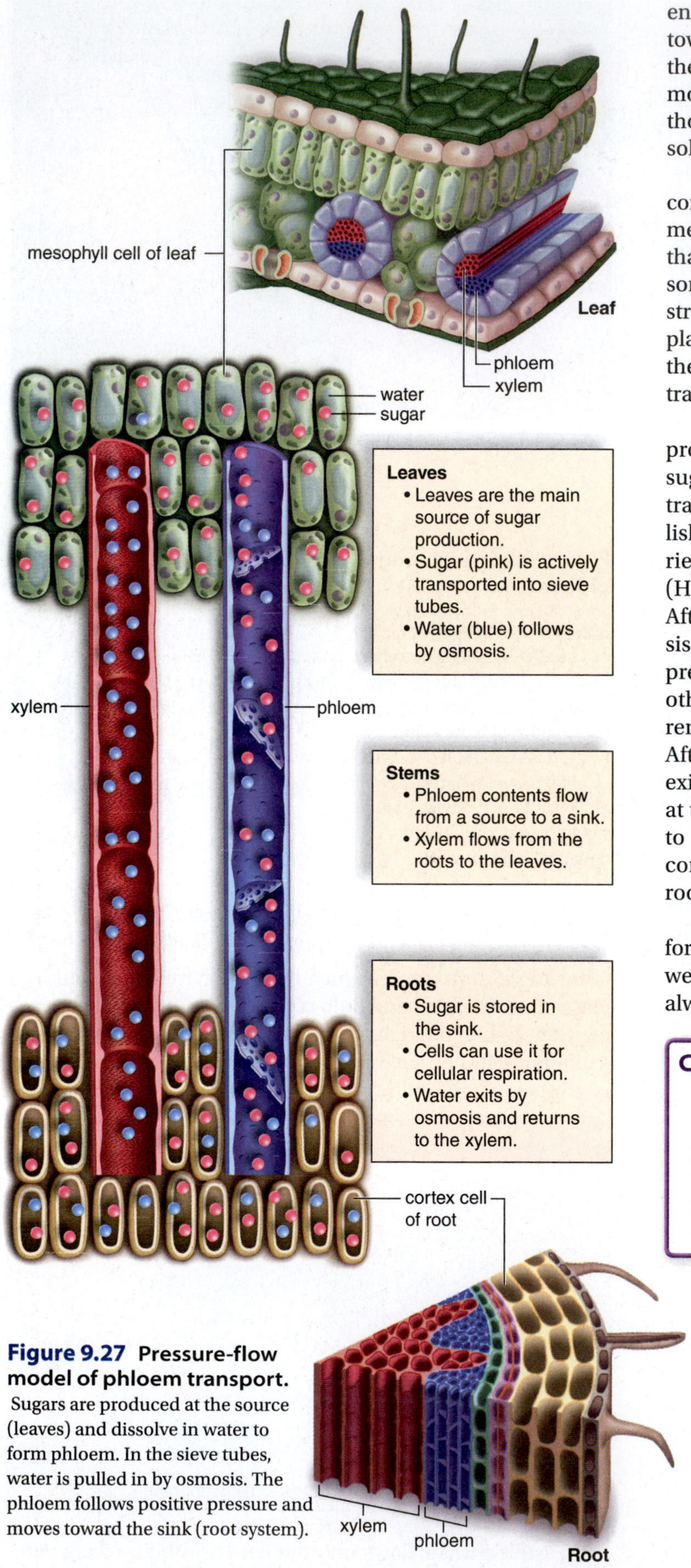
mesophyll cell of leaf
Leaf
phloem
xylem
water
sugar
Leaves
• Leaves are the main source of sugar production.
• Sugar (pink) is actively transported into sieve tubes.
• Water (blue) follows by osmosis.
xylem
phloem
Stems
• Phloem contents flow from a source to a sink.
• Xylem flows from the roots to the leaves.
Roots
• Sugar is stored in the sink.
• Cells can use it for cellular respiration.
• Water exits by osmosis and returns to the xylem.
cortex cell of root
xylem
phloem
Root

Figure 9.27 Pressure-flow model of phloem transport. Sugars are produced at the source (leaves) and dissolve in water to form phloem. In the sieve tubes, water is pulled in by osmosis. The phloem follows positive pressure and moves toward the sink (root system).

entrance of water creates a positive *pressure,* and water *flows* toward the second bulb. This flow not only drives water toward the second bulb, but it also provides enough force for water to move out through the membrane of the second bulb—even though the second bulb contains a higher concentration of solute than the distilled water.

In plants, sieve tubes are analogous to the glass tube that connects the two bulbs. Sieve tubes are composed of sieve-tube members, each of which has a companion cell. It is believed that the companion cells assist the sieve-tube members in some way. The sieve-tube members align end to end, and strands of plasmodesmata (cytoplasm) extend through sieve plates from one sieve-tube member to the other. Sieve tubes, therefore, form a continuous pathway for organic nutrient transport throughout a plant.

During the growing season, photosynthesizing leaves are producing sugar (Fig. 9.27). Therefore, they are a **source** of sugar. This sugar is actively transported into phloem. Again, transport is dependent on an electrochemical gradient established by a proton pump, a form of active transport. Sugar is carried across the membrane in conjunction with hydrogen ions (H^+), which are moving down their concentration gradient. After sugar enters sieve tubes, water follows passively by osmosis. The buildup of water within sieve tubes creates the positive pressure that starts a flow of phloem contents. The roots (and other growth areas) are a **sink** for sugar, meaning that they are removing sugar, storing it, or using it for cellular respiration. After sugar is actively transported out of sieve tubes, water exits phloem passively by osmosis (which reduces pressure at the sink) and is taken up by xylem, which transports water to leaves, where it is used for photosynthesis. Now, phloem contents continue to flow from the leaves (source) to the roots (sink).

The pressure-flow model of phloem transport can account for any direction of flow in sieve tubes if we consider that the direction of flow is always from source to sink.

3D Animation
Plant Transport: Translocation in Phloem

Check Your Progress 9.7

1. Describe cohesion and adhesion and how they are involved in moving water up a plant.
2. Explain how transpiration is involved in moving water throughout a plant.
3. Explain how osmosis is involved in moving organic nutrients in the plant.

SCIENCE IN YOUR LIFE ▶ BIOETHICAL

Using Plants to Clean Up Toxic Messes

Phytoremediation uses plants like poplar, mustard, and mulberry to clean up lead, uranium, and other environmental pollutants. The genetic makeup of these plants allows them to absorb, store, degrade, or transform substances that normally kill or harm other plants and animals. "It's an elegantly simple solution to pollution problems," says Louis Licht, who runs Ecolotree, an Iowa City phytoremediation company. The idea behind phytoremediation is not new, but the idea of using these plants on contaminated sites has just gained support in the last decade. Different plants work on different contaminants. The mulberry bush, for instance, is effective on industrial sludge; some grasses attack petroleum wastes; and sunflowers (together with soil additives) remove lead. The plants clean up sites in different ways, depending on the substance involved. The plants or microbes around their roots break down organic substances, allowing the remains to be absorbed by the plant or left in the soil or water. Inorganic contaminants are trapped within the body of the plant, which must then be harvested and disposed of, or processed to reclaim the trapped contaminant.

Poplars Take Up Excess Nitrates

The poplars act like vacuum cleaners, sucking up nitrate-laden runoff from a fertilized cornfield before this runoff reaches nearby waterways. Nitrate runoff into the Mississippi River from Midwest farms, after all, is a major cause of the large "dead zone" of oxygen-depleted water that develops each summer in the Gulf of Mexico. Poplars planted along the edge of farm fields have been shown to reduce the nitrate levels of streams by more than 90% (Fig. 9B).

Figure 9B **Phytoremediation of nitrates.** Poplars and various other plants can be used to absorb excess nitrates that run off from farm fields.

Mustard Plants Take Up Uranium

Phytoremediation has also helped clean up badly polluted sites, in some cases at a fraction of the usual cost. Uranium was removed from a Superfund site on an Army firing range in Aberdeen, Maryland. Mustard plants removed the uranium at as little as 10% of the cost of traditional cleanup methods.

Limitations of Phytoremediation

One of the main limitations to phytoremediation is the pace. Depending on the contaminant, it can take several growing seasons to clean a site—much longer than conventional methods. Phytoremediation is also only effective at depths that plant roots can reach, making it useless against deep-lying contamination unless the soil is excavated. Phytoremediation will not work on lead and other metals unless chemicals are added to the soil. Plants will also vary in the amount of pollutant that they can absorb.

Despite its shortcomings, experts see a bright future for this technology because, for one reason, the costs are relatively small compared to those of traditional remediation technologies. Traditional methods of cleanup require much energy input and therefore have higher cost. In general, phytoremediation is a low-cost alternative to traditional methods because less energy is required for operation and maintenance.

Questions to Consider

1. What happens to the pollutants when the plant dies?
2. Why would one plant be more adapted to absorbing a particular nutrient than another?
3. Should farmers be required by law to border their fields with plants that can phytoremediate pollutants?

connect BIOLOGY Explore the concepts through a variety of multimedia assets, question types, and data interpretation.
www.mcgrawhillconnect.com

Case Study Conclusion

The roots, stems, and leaves have structural and chemical adaptations to protect the plant from the physical elements and attack by other organisms such as insects and bacteria. The cuticle acts as a barrier against water loss; the epidermis serves as another line of defense by trying to deter predators with prickliness and an unpleasant taste. Toxins are present in a variety of plants that will also help reduce the chance of predation. The wide diversity of adaptations employed by plants have enabled them to become very successful.

MEDIA STUDY TOOLS

www.mhhe.com/maderinquiry14

Enhance your study of this chapter with study tools and practice tests. Also ask your instructor about the resources available through ConnectPlus, including LearnSmart, the media-rich eBook, interactive learning tools, and animations.

3D Animation
Plant Transport

For an interactive exploration of how nutrients are moved within a plant, watch the McGraw-Hill 3D animation "Plant Transport."

SUMMARIZE

9.1 Plant Organs and Systems

- A flowering plant has three vegetative organs. **Roots** anchor a plant, absorb water and minerals, and store the products of photosynthesis. **Stems** support leaves, conduct materials to and from roots and leaves, and help store plant products. **Leaves** are specialized for gas exchange, and they carry on most of the photosynthesis in the plant.

9.2 Cells and Tissues of Plants

- A plant has the ability to grow its entire life because it possesses meristematic (embryonic) tissue.
- **Apical meristems** are located at or near the tips of stems and roots, where they increase the length of these structures. This increase in length is called *primary growth.* The apical meristems continually produce three types of meristem: protoderm, ground meristem, and procambium.
- The entire body of both nonwoody (herbaceous) and young woody plants is covered by a layer of **epidermis,** which in most plants contains a single layer of closely packed epidermal cells. The walls of epidermal cells that are exposed to air are covered with a waxy **cuticle** to minimize water loss.
- **Ground tissue** forms the bulk of a plant and contains **parenchyma**, **collenchyma**, and **sclerenchyma** cells. Parenchyma cells are thin-walled and capable of photosynthesis when they contain chloroplasts. Collenchyma cells have thicker walls for flexible support. Sclerenchyma cells are hollow, nonliving support cells with secondary walls fortified by **lignin.**
- **Vascular tissue** consists of **xylem** and **phloem.** Xylem contains two types of conducting cells: **vessel elements** and **tracheids.** Xylem transports water and minerals. In phloem, sieve tubes are composed of **sieve-tube members,** each of which has a **companion cell.** Phloem transports sugar and other organic compounds, including hormones.

9.3 Monocot Versus Eudicot Plants

- Flowering plants are divided into **monocots** and **eudicots** according to the number of **cotyledons** in the seed; the arrangement of vascular tissue in roots, stems, and leaves; and the number of flower parts.

9.4 Organization of Roots

- A root tip has a zone of cell division (containing the primary meristems), a zone of elongation, and a zone of maturation.
- A cross section of a herbaceous eudicot root reveals the epidermis, which protects; the **cortex,** which stores food; the **endodermis,** which regulates the movement of minerals; and the vascular cylinder, which is composed of vascular tissue.
- In the vascular cylinder of a eudicot, the xylem appears star-shaped, and the phloem is found in separate regions, between the arms of the xylem. In contrast, a monocot root has a ring of vascular tissue with alternating bundles of xylem and phloem surrounding the **pith.**
- Roots are diversified. **Taproots** are specialized to store the products of photosynthesis. A **fibrous root system** covers a wider area. Prop roots are adventitious roots specialized to provide increased anchorage.
- Roots may enter into symbiotic relationships with bacteria and fungi.

9.5 Organization of Stems

- The activity of the shoot apical meristem within a **terminal bud** accounts for the primary growth of a stem. A terminal bud contains internodes and leaf primordia at the nodes. When stems grow, the internodes lengthen.
- In a cross section of a nonwoody eudicot stem, epidermis is the outermost layer of cells, followed by cortex tissue, vascular bundles in a ring, and an inner pith. Monocot stems have scattered vascular bundles, and the cortex and pith are not well defined.
- Secondary growth of a woody stem is due to activities of the **vascular cambium,** which produces new xylem and phloem every year, and **cork cambium,** which produces new cork cells when needed. Cork, a part of the **bark,** replaces epidermis in woody plants.
- In a cross section of a woody stem, the bark is composed of all the tissues outside the vascular cambium: secondary phloem, cork cambium, and cork. **Wood** consists of secondary xylem, which builds up year after year and forms **annual rings.**
- Stems are diverse. Modified stems include horizontal aboveground and underground stems, corms, and some tendrils.

9.6 Organization of Leaves

- The bulk of a leaf is **mesophyll** tissue bordered by an upper and lower layer of epidermis. The epidermis is covered by a cuticle and may bear trichomes. Stomata tend to be in the lower layer. Vascular tissue is present within **leaf veins.** The leaf veins conduct water and the products of photosynthesis to various parts of the plant body.
- Leaves are diverse. The spines of a cactus are leaves. Other succulents have fleshy leaves. An onion is a bulb with fleshy leaves, and the tendrils of peas are leaves. The Venus flytrap has leaves that trap and digest insects.

9.7 Uptake and Transport of Nutrients

- Water transport in plants occurs within xylem. The **cohesion-tension model** of xylem transport states that transpiration (evaporation of water at stomata) creates tension, which pulls water upward in xylem. This method works only because water molecules are cohesive. **Transpiration** and carbon dioxide uptake occur when stomata are open.
- Stomata open when **guard cells** take up potassium (K^+) ions and water follows by osmosis. Stomata open because the entrance of water causes the guard cells to buckle out.
- Transport of organic nutrients in plants occurs within phloem. The **pressure-flow model** of phloem transport states that sugar is actively transported into phloem at a **source,** and water follows by osmosis. The resulting increase in pressure creates a flow, which moves water and sugar to a **sink.**

ASSESS

Testing Yourself

Choose the best answer for each question.

1. _____ forms the outer protective covering of a plant.
- **a.** Ground tissue
- **b.** Root tissue
- **c.** Epidermal tissue
- **d.** Vascular tissue

2. Which of these is an incorrect contrast between monocots (stated first) and eudicots (stated second)?
- **a.** one cotyledon—two cotyledons
- **b.** leaf veins parallel—net veined
- **c.** vascular bundles in a ring—vascular bundles scattered
- **d.** flower parts in threes—flower parts in fours or fives
- **e.** All of these are correct contrasts.

3. Which of these cells in a plant is apt to be nonliving?
- **a.** parenchyma
- **b.** collenchyma
- **c.** sclerenchyma
- **d.** epidermal
- **e.** guard cells

4. Root hairs are found in the zone of
- **a.** cell division.
- **b.** elongation.
- **c.** maturation.
- **d.** apical meristem.
- **e.** All of these are correct.

5. Insect-eating plants are common in regions where _____ is in limited supply.
- **a.** water
- **b.** CO_2
- **c.** nitrogen
- **d.** oxygen

6. Stomata are open when
- **a.** guard cells are filled with water.
- **b.** potassium ions move into the guard cells.
- **c.** turgor pressure increases.
- **d.** All of these are correct.

7. Annual rings are the number of
- **a.** internodes in a stem.
- **b.** rings of vascular bundles in a monocot stem.
- **c.** layers of xylem in a stem.
- **d.** bark layers in a woody stem.
- **e.** Both b and c are correct.

8. What role do cohesion and adhesion play in xylem transport?
- **a.** Like transpiration, they create a tension.
- **b.** Like root pressure, they create a positive pressure.
- **c.** Like sugars, they cause water to enter xylem.
- **d.** They create a continuous water column in xylem.

9. The sugar produced by mature leaves moves into sieve tubes by way of _____, while water follows by _____.
- **a.** osmosis, osmosis
- **b.** active transport, active transport
- **c.** osmosis, active transport
- **d.** active transport, osmosis

10. In a cross section of which of the following structures would you find scattered vascular bundles?
- **a.** monocot root
- **b.** monocot stem
- **c.** eudicot root
- **d.** eudicot stem

11. Which of the following is considered a stem?
- **a.** potato
- **b.** sweet potato
- **c.** cabbage
- **d.** carrot

For questions 12–15, match the definitions with the structure in the key.

Key:
- **a.** stoma
- **b.** node
- **c.** xylem
- **d.** parenchyma cell

12. Major constituent of ground tissue

13. Region where leaves attach to stem

14. Opens when transpiration occurs

15. Contains tracheids

ENGAGE

Virtual Lab
Plant Transpiration

The virtual lab "Plant Transpiration" provides a more detailed look at the environmental conditions associated with the transpiration of water from the leaves of a plant.

Thinking Critically

1. Contrast epidermal, ground, and vascular tissue on the basis of function.
2. What are some of the potential evolutionary advantages of being woody?
3. Compare and contrast the transport of water and organic nutrients in plants with the transport of blood in humans.

10 Plant Reproduction and Responses

CHAPTER OUTLINE

BEFORE YOU BEGIN

Before beginning this chapter, take a few moments to review the following discussions:

Section 5.4 What role does meiosis play in sexual reproduction?

Section 6.3 What role do membrane proteins play in the control of plant growth?

Figure 6.11 What are the basic requirements for photosynthesis and cellular respiration?

CASE STUDY Seedless Plants

It is nice to be able to eat a cold watermelon on a hot summer day without being bothered by the seeds. Seedless watermelons, once a novelty, are now a common convenience. But how are seedless watermelons produced?

Seeded watermelons have a diploid number (2n) of 22 chromosomes. Genetic manipulation of a 2n plant can produce a 4n plant with 44 chromosomes. When a 2n plant is crossed with a 4n plant , the resulting fruit has 3n seeds. The plant that germinates from this 3n seed is sterile (unable to produce seeds) because meiosis cannot proceed as usual. Hence the seedless watermelon!

Although seedless watermelons do not produce viable seeds, they develop small white "seeds." These are actually soft, tasteless seed coats that can be eaten with the watermelon. Since these seed coats can't be used to start next year's crop, the 3n seeds needed for planting must be produced each year by crossing 2n and 4n plants. Expense, time, and energy are required in order to save us the inconvenience of spitting out the seeds!

Seedless watermelons are just one example of the capability we have to genetically modify plants and other organisms, often simply called GMOs. Unlike seedless watermelons, there is controversy surrounding the potential health effects of eating GM (genetically modified) foods. Nonetheless, humans modify plants all the time.

The fundamentals of agriculture and supplying plant-based products to the human population include a better understanding of how flowering plant reproduction occurs, seed production and development, and how plants operate under natural conditions.

As you read through the chapter, think about the following questions:

1. What hormones are associated with the normal growth and development of flowering plants?
2. What are some examples of how humans have manipulated plants to better serve our needs?

10.1 Sexual Reproduction in Flowering Plants

Learning Outcomes

Upon completion of this section, you should be able to

1. Recognize that flowering plants exhibit an alternation of generations even though they produce two types of spores and two types of gametophytes.
2. Identify the reproductive parts of a flower and describe the function of each part.
3. Diagram and describe the development of male and female gametophytes and the development of the sporophyte of flowering plants.

Like the animals, sexual reproduction in plants is advantageous because it generates variation among the offspring through the process of meiosis and random fertilization. In contrast to animals, plants have evolved a two-stage life cycle, called an **alternation of generations,** as an adaptation to life on the land environment. These evolutionary adaptations are covered in more detail in section 30.1. For now, it is important to recognize that this life cycle is composed of a diploid (2n) sporophyte stage and a haploid (n) gametophyte stage.

As an overview (Fig. 10.1), notice how the **sporophyte** stage produces haploid **spores** by meiosis. These spores then divide by mitosis to become gametophytes. The **gametophyte** stage produces gametes by mitosis. Upon fertilization, the cycle returns to the sporophyte stage. The next section will examine how this life cycle pertains to a flowering plant.

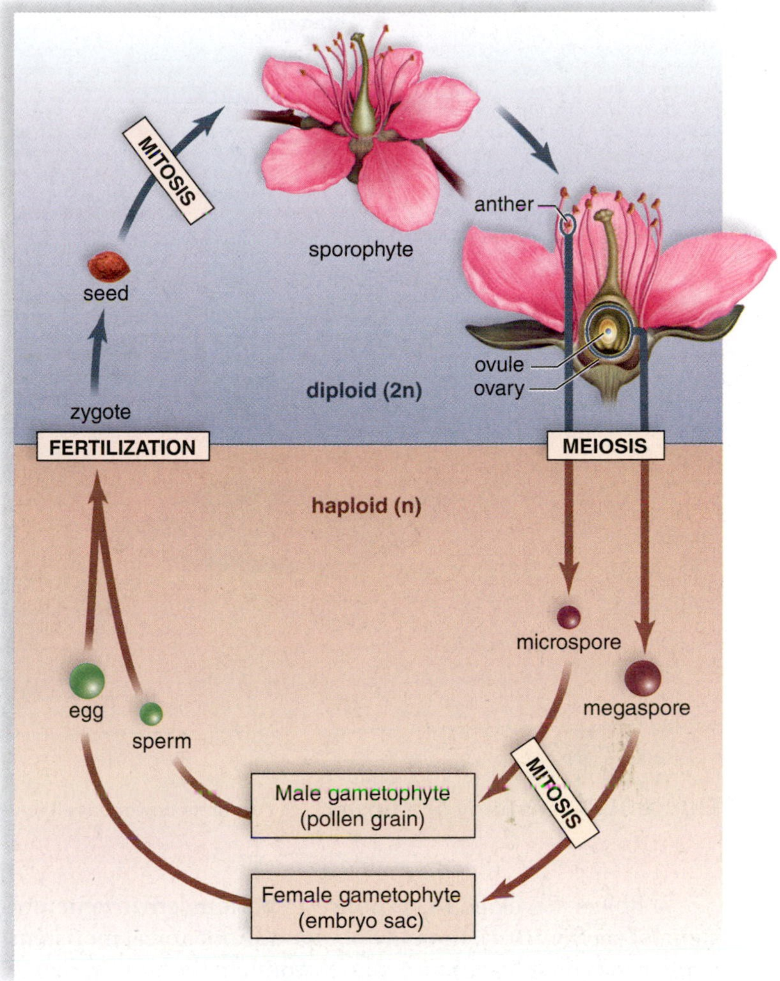
MITOSIS
sporophyte
anther
seed
ovule
ovary
diploid (2n)
zygote
FERTILIZATION
MEIOSIS
haploid (n)
microspore
megaspore
egg
sperm
Male gametophyte
(pollen grain)
MITOSIS
Female gametophyte
(embryo sac)

Figure 10.1 Alternation of generations in flowering plants. The sporophyte bears flowers. The flower produces microspores within anthers and megaspores within ovules by meiosis. A megaspore becomes a female gametophyte, which produces an egg within an embryo sac, and a microspore becomes a male gametophyte (pollen grain), which produces sperm. Fertilization results in a seed-enclosed zygote and stored food.

Flowering Plant Life Cycle

In flowering plants, the sporophyte is dominant and it is the generation that bears flowers (Fig. 10.1). A **flower** is a reproductive structure and it produces two types of spores: microspores and megaspores. A **microspore** undergoes mitosis and becomes a pollen grain, which is the male gametophyte. Meanwhile, the **megaspore** has undergone mitosis to become a microscopic embryo sac, which is the female gametophyte. The female gametophyte is retained within the flower.

A pollen grain is either windblown or carried by an animal to the vicinity of the embryo sac. At maturity, a pollen grain contains two nonflagellated sperm that travel down a pollen tube to the embryo sac. Upon fertilization, the structure surrounding the embryo sac develops into a **seed.** The seeds are enclosed by a **fruit,** which aids in dispersing the seeds. When a seed germinates, a new sporophyte emerges and develops into a mature organism.

Adaptation to a Land Environment

The life cycle of flowering plants is adapted to a land existence because all stages of the life cycle are protected from drying out (see section 30.1). For example, the microscopic gametophytes develop within the sporophyte. Pollen grains are not released until they have a thick wall, and they are carried by wind or an animal (usually an insect) to another flower, where they develop a pollen tube that carries the sperm to the egg. Following fertilization, the seed coat protects the embryo until conditions favor regrowth.

Flowers

The flower is unique to the group of plants known as the angiosperms. The evolution of the flower was a major factor leading to the success of angiosperms, which now number over 240,000 described species. Aside from producing the spores and protecting the gametophytes, flowers often attract pollinators. Flowers also produce the fruits that enclose the seeds.

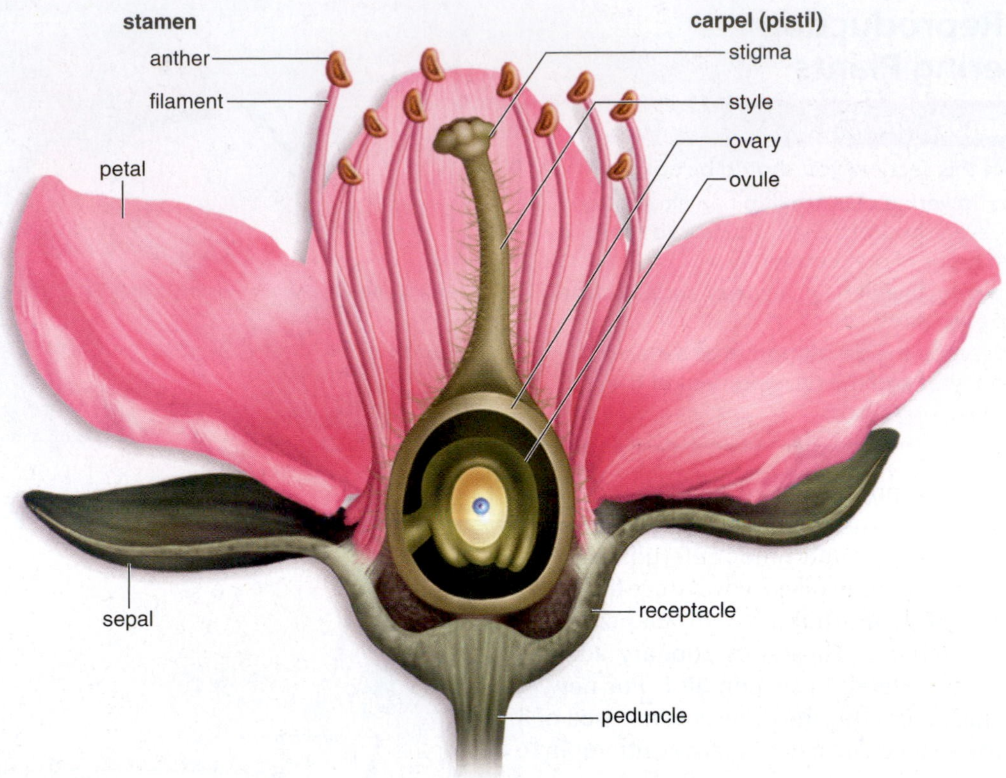
stamen
anther
filament
petal
sepal
carpel (pistil)
stigma
style
ovary
ovule
receptacle
peduncle

Figure 10.2 Anatomy of a flower. A complete flower has all flower parts: sepals, petals, stamens, and at least one carpel.

A flower develops in response to various environmental signals (section 10.4). In many plants, a shoot apical meristem stops producing leaves and starts producing a bud that contains a flower. In other plants, axillary buds develop directly into flowers.

A typical flower has four whorls of modified leaves attached to a receptacle at the end of a flower stalk. The receptacle bearing a single flower is attached to a structure called a peduncle, while the structure that bears one of several flowers is called a pedicel. Figure 10.2 shows the following basic floral structures:

1. **Sepals** are the most leaflike of all the flower parts, are usually green, and protect the bud as the flower develops within. Sepals can also be the same color as the flower petals.
2. **Petals** are the structures used to attract a variety of pollinators. Wind-pollinated flowers may have no petals at all.
3. **Stamens** are the "male" portion of the flower. Each stamen has two parts: the **anther,** a saclike container, and the **filament,** a slender stalk. Pollen grains develop from the microspores produced in the anther.
4. At the very center of a flower is the **carpel** (also called the pistil), a vaselike structure that represents the "female" portion of the flower. A carpel usually has three parts: the **stigma,** an enlarged sticky knob; the **style,** a slender stalk; and the **ovary,** an enlarged base that encloses one or more ovules.

In monocots, flower parts occur in threes and multiples of three. In eudicots, flower parts are in fours or fives and multiples of four or five (Fig. 10.3).

A flower can have a single carpel or multiple carpels. Sometimes several carpels are fused into a single structure, in which case this compound ovary has several chambers containing **ovules,** each of which will house a megaspore. For example, an orange develops from a compound ovary, and every section of the orange is a chamber.

Not all flowers have sepals, petals, stamens, and carpels. Those that do, such as the flower in Figure 10.2, are said to be *complete,* and those that do not are said to be *incomplete.* Flowers that have both stamens and carpels are called perfect (bisexual) flowers. Those with only stamens or only carpels are imperfect (unisexual) flowers. If staminate and carpellate (pistillate) flowers are on one plant, the plant is monoecious (one house) (Fig. 10.4). If staminate and carpellate flowers are on separate plants, the plant is dioecious (two houses).

Life Cycle of Flowering Plants

In plants, the sporophyte produces haploid spores by meiosis. The haploid spores grow and develop into haploid gametophytes, which produce gametes by mitotic division.

Flowering plants are heterosporous, meaning that they produce microspores and megaspores. Microspores become mature male gametophytes (sperm-bearing pollen grains), and megaspores become mature female gametophytes (egg-bearing embryo sacs).

stamen
carpel
petal
sepal
s1
p2
s2
p1
p3
s3

a. Daylily

carpel
stamen
petal
p2
p3
p4
p1
p5

b. Festive azalea

Figure 10.3 Monocot versus eudicot flowers. **a.** Monocots, such as daylilies, have flower parts usually in threes. In particular, note the three petals and three sepals. **b.** Azaleas are eudicots. They have flower parts in fours or fives; note the five petals of this flower. (p = petal; s = sepal)

a. Staminate flowers b. Carpellate flowers

Figure 10.4 Corn plants are monoecious. A corn plant has (**a**) clusters of staminate flowers and (**b**) clusters of carpellate flowers. Staminate flowers produce the pollen that is carried by wind to the carpellate flowers, where an ear of corn develops.

Development of Male Gametophyte

Microspores are produced in the anthers of flowers (Fig. 10.5). An anther has four pollen sacs, each containing many **microsporocytes.** A microsporocyte undergoes meiosis to produce four haploid microspores. In each, the haploid nucleus divides mitotically, followed by unequal cytokinesis, and the result is two cells enclosed by a finely sculptured wall. This structure, called the **pollen grain,** is at first an immature male gametophyte that consists of a tube cell and a generative cell. The larger tube cell will eventually produce a *pollen tube.* The smaller generative cell divides mitotically either now or later to produce two sperm. Once these events take place, the pollen grain has become the mature male gametophyte.

Development of Female Gametophyte

The ovary contains one or more ovules. An ovule has a central mass of parenchyma cells almost completely covered by layers of tissue called integuments, except for the opening called the micropyle. One parenchyma cell enlarges to become a **megasporocyte,** which undergoes meiosis, producing four haploid megaspores (Fig. 10.5). Three of these megaspores are nonfunctional. In a typical pattern, the nucleus of the functional megaspore divides mitotically to form eight nuclei in the female gametophyte**.** When cell walls form later, there are seven cells, one of which is binucleate. The female gametophyte, also called the **embryo sac,** consists of these seven cells, as follows:

1 egg cell;
2 synergid cells;
1 central cell containing two polar nuclei; and
3 antipodal cells.

Pollination and Fertilization

The walls separating the pollen sacs in the anther break down when the pollen grains are ready to be released (Fig. 10.6). **Pollination** is simply the transfer of pollen from an anther to the stigma of a carpel. Self-pollination occurs if the pollen is from the same plant, and cross-pollination occurs if the pollen is from a different plant of the same species.

Angiosperms often have adaptations to foster cross-pollination. For example, the carpels may mature only after the anthers have released their pollen. Cross-pollination may also be brought about with the assistance of a particular pollinator. The Ecology feature, "Pollinators and You," on page 174 explains the importance of insects and other pollinators to the life cycle of a flowering plant. Plants have evolved mechanisms, such as color, odors, or production of nectar, to attract specific pollinators to their flowers. If a pollinator goes from flower to flower of only one type of plant, cross-pollination is more likely to occur in an efficient manner. Over time, certain pollinators have become adapted to reach the nectar of only one type of flower. The process by which the evolution of two species is connected is called **coevolution**, and it is considered to be a major factor in the success of the angiosperms. When a pollen grain lands on the stigma of the same species, it germinates, forming a pollen tube (see Fig. 10.5). The germinated pollen grain, containing a tube cell and two sperm, is the mature male gametophyte. As it grows, the pollen tube passes between the cells of the stigma and the style to reach the micropyle, a pore of the ovule. Now **double fertilization** occurs. One sperm nucleus unites with the egg nucleus, forming a 2n zygote, and the other sperm nucleus migrates and unites with the polar nuclei of the central cell, forming a 3n endosperm

Figure 10.5 Life cycle of flowering plants. *Development of gametophytes (far page):* A pollen sac in the anther contains a microsporocyte, which produces microspores by meiosis. A microspore develops into a pollen grain, which germinates and has two sperm. An ovule in an ovary contains a megasporocyte, which produces a megaspore by meiosis. A megaspore develops into an embryo sac containing seven cells, one of which is an egg. *Development of sporophyte (this page):* A pollen grain contains two sperm by the time it germinates and forms a pollen tube. During double fertilization, one sperm fertilizes the egg to form a diploid zygote, and the other fuses with the polar nuclei to form a triploid (3n) endosperm cell. A seed contains the developing sporophyte embryo plus stored food.

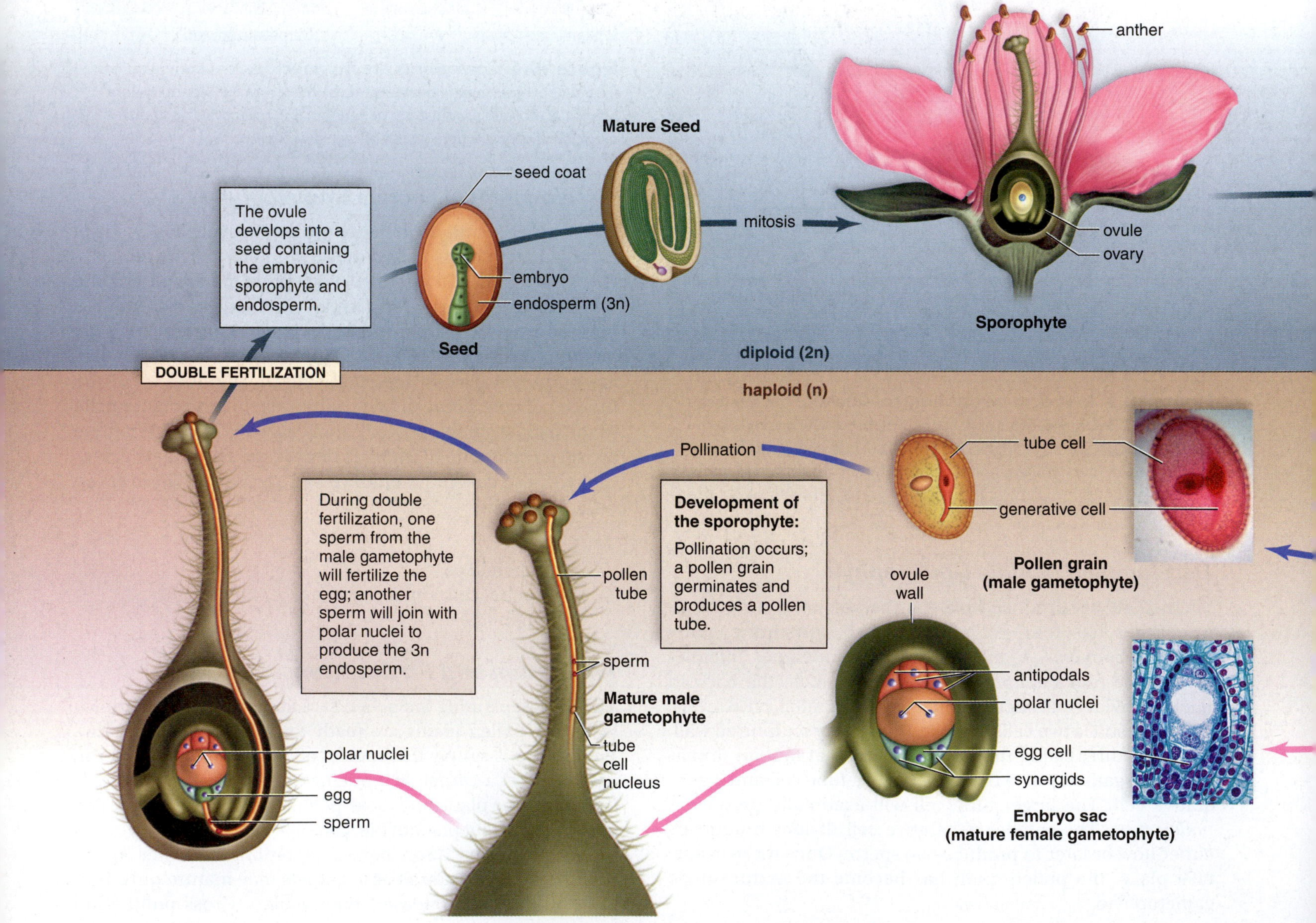
anther
Mature Seed
seed coat
mitosis
ovule
ovary
The ovule develops into a seed containing the embryonic sporophyte and endosperm.
embryo
endosperm (3n)
Seed
Sporophyte
diploid (2n)
DOUBLE FERTILIZATION
haploid (n)
Pollination
tube cell
generative cell
During double fertilization, one sperm from the male gametophyte will fertilize the egg; another sperm will join with polar nuclei to produce the 3n endosperm.
Development of the sporophyte:
Pollination occurs; a pollen grain germinates and produces a pollen tube.
pollen tube
ovule wall
Pollen grain (male gametophyte)
sperm
Mature male gametophyte
antipodals
polar nuclei
tube cell nucleus
egg cell
synergids
polar nuclei
egg
sperm
Embryo sac (mature female gametophyte)

cell. The zygote divides mitotically to become the **embryo,** a young sporophyte, and the endosperm cell divides mitotically to become the endosperm. **Endosperm** is the tissue that will nourish the embryo and seedling as they develop.

Video
Pollinators

Check Your Progress 10.1

1. Identify which generation produces two types of spores and what will happen to the spores.
2. List the reproductive parts of a flower and give a function for each part.
3. Describe what is meant by the term heterosporous.
4. Describe the process of double fertilization.

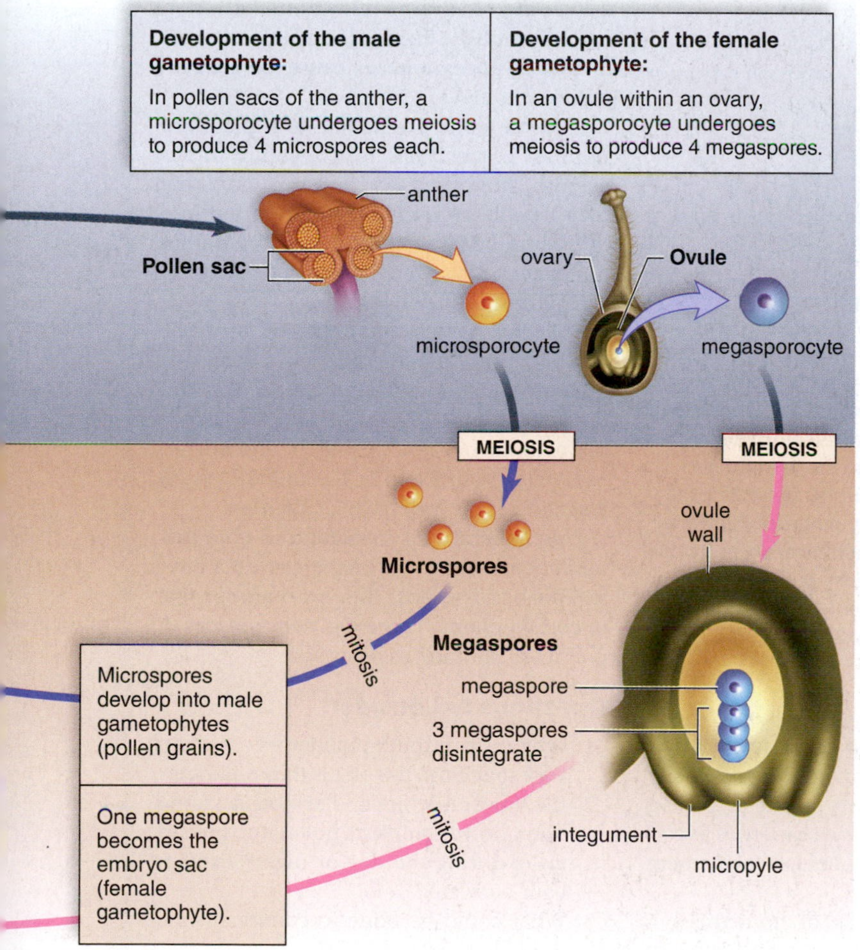
Development of the male gametophyte:
In pollen sacs of the anther, a microsporocyte undergoes meiosis to produce 4 microspores each.
Development of the female gametophyte:
In an ovule within an ovary, a megasporocyte undergoes meiosis to produce 4 megaspores.
anther
Pollen sac
ovary
Ovule
microsporocyte
megasporocyte
MEIOSIS
MEIOSIS
Microspores
ovule wall
mitosis
Megaspores
megaspore
3 megaspores disintegrate
Microspores develop into male gametophytes (pollen grains).
One megaspore becomes the embryo sac (female gametophyte).
mitosis
integument
micropyle

a.
b.
118 μm
c.
8 μm

Figure 10.6 Pollen grains. **a.** Cocksfoot grass releasing pollen. **b.** Pollen grains of Canadian goldenrod. **c.** Pollen grains of pussy willow. The shape and pattern of pollen grain walls are quite distinctive, and experts can use them to identify the genus, and sometimes even the species, that produced a particular pollen grain. Pollen grains have strong walls resistant to chemical and mechanical damage. Therefore, they frequently become fossils.

Pollinators and You

Bees, butterflies, and other pollinators play a critical role in our world. Many agricultural crops are dependent upon insects to carry their pollen from one flower to another. Pollination is required for these plants to produce a fruit. Pollinators are often necessary for the production of fruits, vegetables, and even the coffee we enjoy every day.

Plants and their pollinators have a symbiotic relationship. In return for pollination and fertilization of flowers, the flower provides the pollinator with food, called nectar. Flowers are often colors that attract a specific pollinator species and open at times the pollinators are active. In some cases, the structure of the flower is an exact match to the mouthparts of a particular species of pollinator!

Important pollinators include a variety of insects as well as birds and mammals, particularly bats. The best-known pollinator is the honeybee (Fig. 10A*a*). Beekeepers maintain around 2.4 million colonies of the European honeybee in the United States. Wild honeybee colonies also occur. Current estimates value honeybee pollination at approximately $5 billion in Canada and $15 billion in the United States per year. Unlike humans, bees can see ultraviolet light. Bee-pollinated flowers often have ultraviolet shadings, called nectar guides, which highlight the reproductive portion of the flower. The nectar guides point to a narrow floral tube that fits the bee's feeding apparatus. Bee-pollinated flowers are sturdy and irregular in shape because of the landing platform, where the bee can alight. Here, the bee brushes up against the anther and stigma as it moves toward the floral tube. Pollen clings to its hairy body, and the bee then stores the pollen in pollen baskets on the third pair of its legs. Bees feed pollen to their larvae.

Butterflies have a weak sense of smell, so their flowers tend to be odorless. The flowers are often brightly colored. Many butterfly-pollinated flowers are red because red can be seen by butterflies. Composite flowers (composed of a compact head of numerous individual flowers) are especially favored by butterflies because these flowers provide a flat platform for the butterflies to land on. Each flower has a long, slender floral tube, accessible to the long, thin butterfly proboscis (Fig. 10A*b*).

Today, there are more than 240,000 described species of flowering plants and over 900,000 described species of insects. This diversity suggests that the success of angiosperms and insects is due to their mutual relationship.

In recent years, however, populations of bees and other pollinators have been declining worldwide. Consequently, some plants are endangered because they have lost their normal pollinator. Research suggests that the decline in pollinator populations has been caused by a variety of factors, including pollution, habitat loss, and emerging diseases.

Recently, declines of honeybees in particular have become a global concern. In 2006, commercial beekeepers in the United States began reporting rapid declines in their honeybee colonies. Scientists later named this phenomenon "colony collapse disorder" (CCD) due to its severity. Although honeybee losses have been common in the past, CCD has alarmed beekeepers and farmers alike due to the combination of rapid and widespread losses, along with failure of bees to return to the hive. Current reports suggest CCD affects bee colonies in 35 of the 50 U.S. states. A survey of U.S. honeybee colonies suggests losses increased by 14% between 2007 and 2008.

a.

b.

Figure 10A Two types of pollinators. **a.** A bee-pollinated flower is a color other than red (bees can't see this color) and has a landing platform where the reproductive structures of the flower brush up against the bee's body. **b.** A butterfly-pollinated flower is often a composite, containing many individual flowers. The broad expanse provides room for the butterfly to land, after which it lowers its proboscis into each flower, in turn.

Possible causes of CCD include pesticides, parasites and diseases, stress, poor nutrition, and a lack of genetic diversity among bees. These factors may act together, with impacts magnified relative to single causes.

Although the actual cause of CCD remains a mystery, two serious emerging diseases of honeybees are caused by mites. The tracheal mite, a stowaway from Europe, lives in the breathing tubes of adult honeybees and sucks their blood, causing adult bees to become disoriented and weak. The other mite, called the varoa mite, from Asia, is an external parasite that feeds on pupae, larvae, and adult stages, and can cause the death of entire colonies.

Mites alone are unable to explain the number of declines. Also recently implicated in honeybee declines is a virus that may have been accidentally imported into the United States via infected Australian honeybees. However, the role of this virus in CCD is currently debated.

Today, professional farmers rent bee colonies and have them trucked long distances to ensure crop pollination. You can do your part to help sustain honeybee populations by planting pollen- and nectar-producing plants and by limiting the use of insecticides on your lawn and garden.

In general, it is important to note that pollinators cannot be exchanged on a one-for-one basis because they are often specific to certain plants. Therefore, each species of pollinator should be conserved.

Questions to Consider

1. Why do you think pollinators are often very specific to the plants they pollinate?
2. If a flower is white and exhibits a strong odor, do you think its pollinator species visits during the day or night? Explain your answer.
3. What could be done to counteract the possible causes of CCD?

10.2 Growth and Development

Learning Outcomes

Upon completion of this section, you should be able to

1. Recognize the developmental steps of a eudicot embryo and compare the function of its cotyledons to that of a cotyledon in monocots.
2. Identify different types of fruits.
3. Label seed structure and describe germination and dispersal.

Development of the Eudicot Embryo

The endosperm cell, shown in Figure 10.7*a,* divides to produce the endosperm tissue, shown in Figure 10.7*b*. The zygote also divides, but asymmetrically. One of the resulting cells is small, with dense cytoplasm. This cell is destined to become the embryo, and it divides repeatedly in different planes, forming a ball of cells. The other, larger cell also divides repeatedly, but it forms an elongated structure called a suspensor, which has a basal cell. The suspensor pushes the embryo deep into the endosperm tissue. The suspensor then disintegrates as the seed matures.

During the globular stage, the embryo is a ball of cells (Fig. 10.7*c*). The root-shoot axis of the embryo is already established at this stage because the embryonic cells near the suspensor will become a root, whereas those at the other end will ultimately become a shoot.

The embryo has a heart shape when the **cotyledons,** or seed leaves, appear (Fig. 10.7*d*). As the embryo continues to enlarge and elongate, it takes on a torpedo shape (Fig. 10.7*e*). Now the root tip and shoot tip are distinguishable. The shoot apical meristem in the shoot tip is responsible for aboveground growth, and the root apical meristem in the root tip is responsible for underground growth. In a mature embryo, the epicotyl is the portion between the cotyledon(s) that contributes to shoot development (Fig. 10.7*f*). The hypocotyl is that portion below the cotyledon(s) that contributes to stem development. The radicle contributes to root development. Now, the embryo is ready to develop the two main parts of a plant: the shoot system and the root system.

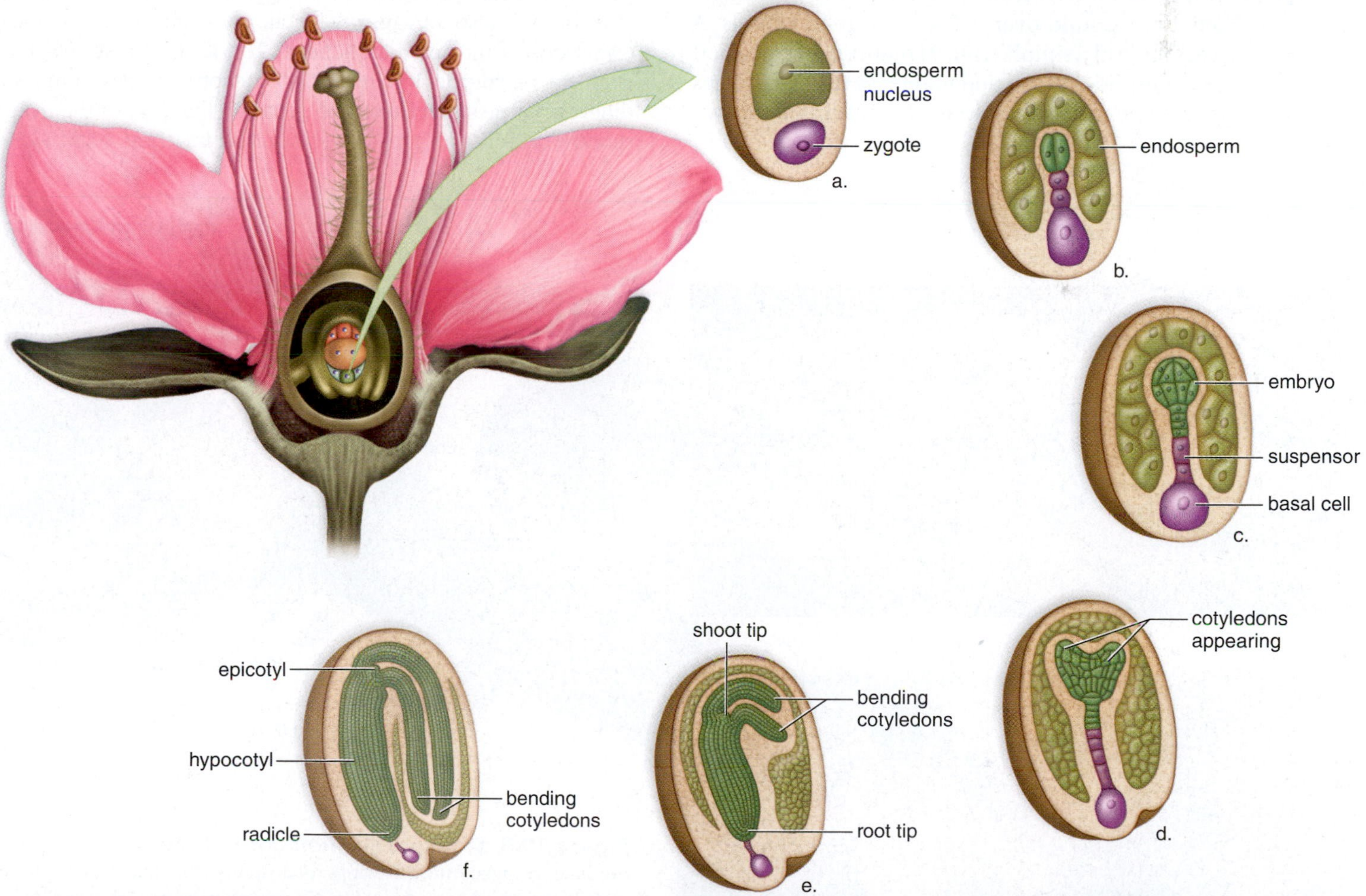
endosperm nucleus
zygote
a.
endosperm
b.
embryo
suspensor
basal cell
c.
cotyledons appearing
d.
shoot tip
bending cotyledons
root tip
e.
epicotyl
hypocotyl
radicle
bending cotyledons
f.

Figure 10.7 Development of a eudicot embryo. **a.** The unicellular zygote lies beneath the endosperm nucleus. **b., c.** The endosperm is a mass of tissue surrounding the embryo. The embryo is located above the suspensor. **d.** The embryo becomes heart-shaped as the cotyledons begin to appear. **e.** There is progressively less endosperm as the embryo differentiates and enlarges. As the cotyledons bend, the embryo takes on a torpedo shape. **f.** The embryo consists of the epicotyl, the hypocotyl, and the radicle.

The cotyledons are quite noticeable in a eudicot embryo and may fold over. As the embryo develops, the wall of the ovule becomes the seed coat.

Monocot Versus Eudicot Embryos

Eudicots have two cotyledons; monocots have only one. In monocots, the cotyledon rarely stores food. Instead, it absorbs food molecules from the endosperm and passes them to the embryo. In eudicots, the cotyledons usually store the nutrient molecules that the embryo uses. Therefore, the endosperm disappears because it has been taken up by the two cotyledons.

Figures 10.9 and 10.10 contrast the structure of a bean seed (eudicot) and a corn kernel (monocot).

Fruit Types and Seed Dispersal

A fruit is derived from an ovary and sometimes other flower parts. It serves to protect and help disperse the offspring. As a fruit develops, the ovary wall thickens to become the pericarp (Fig. 10.8*a*). The pericarp can contain up to three layers: exocarp, mesocarp, and endocarp.

Simple fruits are derived from a simple ovary of a single carpel or from a compound ovary of several fused carpels. A pea pod is an example of a simple fruit. At maturity, the pea pod breaks open on both sides of the pod to release the seeds. Peas and beans are legumes. A *legume* is a fruit that splits along two sides when mature.

Like legumes, cereal grains of wheat, rice, and corn are *dry fruits*. Sometimes, these grainlike fruits are mistaken for seeds because a dry pericarp adheres to the seed within. These dry fruits are indehiscent, meaning that they don't split open. Humans gather grains before they are released from the plant and then process them to acquire their nutrients.

In some simple fruits, the mesocarp becomes fleshy. When ripe, *fleshy fruits* often attract animals and provide them with food (Fig. 10.8*b*). Peaches and cherries are examples of fleshy fruits that have a hard endocarp. This type of endocarp protects the seed so it can pass through the digestive system of an animal and remain unharmed. In a tomato, the entire pericarp is fleshy. If you cut open a tomato, you see several chambers because the flower's carpel is composed of several fused carpels.

Among fruits, apples are an example of an *accessory fruit* because the bulk of the fruit is not from the ovary, but from the receptacle. Only the core of an apple is derived from the ovary. If you cut an apple crosswise, it is obvious that an apple, like a tomato, came from a compound ovary with several chambers.

Compound fruits develop from several individual ovaries. For example, each little part of a raspberry or blackberry is derived from a separate ovary. Because the flower had many separate carpels, the resulting fruit is called an *aggregate fruit* (Fig. 10.8*c*). The strawberry is also an aggregate fruit, but each ovary becomes a one-seeded fruit called an achene. The flesh of a strawberry is from the receptacle. In contrast, a pineapple comes from many different carpels that belong to separate flowers. As the ovaries mature, they fuse to form a large, *multiple fruit* (Fig. 10.8*d*).

seed covered by pericarp
wing
a.
one fruit
b.
one fruit
c.
fruits from ovaries of one flower
d. fruits from ovaries of many flowers

Figure 10.8 Structure and function of fruits. **a.** Maple trees produce a winged fruit. The wings rotate in the wind and keep the fruit aloft. **b.** Strawberry plants produce an accessory fruit. Each "seed" is actually a fruit on a fleshy, expanded receptacle. **c.** Like strawberries, blackberries are aggregate fruits. Each "berry" is derived from an ovary within the same flower. **d.** A pineapple is a multiple fruit derived from the ovaries of many flowers.

Dispersal of Seeds

To increase a plant's chance of success, it needs to be widely distributed. In order to achieve success, plant seeds need to be dispersed away from the parent plant.

Plants have various means of ensuring that dispersal takes place. The hooks and spines of clover, bur, and cocklebur attach to the fur of animals and the clothing of humans. Birds and mammals sometimes eat fruits, including the seeds, which are then defecated (passed through the digestive tract with the feces) some distance from the parent plant. Squirrels and other animals gather seeds and fruits, which they bury some distance away.

Some plants have fruits with trapped air or seeds with inflated sacs that help them float in water. The fruit of the coconut palm, for example, which can be dispersed by ocean currents, may land hundreds of kilometers away from the parent plant. Many plants have seeds with structures such as woolly hairs, plumes, and wings that adapt them for dispersal by wind. The seeds of an orchid, however, are so small and light that they need no special adaptation to carry them far away. The somewhat heavier dandelion fruit uses a tiny "parachute" for dispersal. The winged fruit of a maple tree, which contains two seeds, has been known to travel up to 10 kilometers (km) from its parent (see Fig. 10.8*a*). A touch-me-not plant has seed pods that swell as they mature. When the pods finally burst, the ripe seeds are hurled out.

Video
Fruit Bat Seed Dispersal

Germination of Seeds

Provided the environmental conditions are right seeds **germinate**—that is, they begin to grow so that a seedling appears (Fig. 10.9). Some seeds do not germinate until they have been dormant for a specific period of time. For seeds, **dormancy** is the time during which no growth occurs, even though conditions may be favorable for growth. In the temperate zone, seeds often have to be exposed to a period of cold weather before dormancy

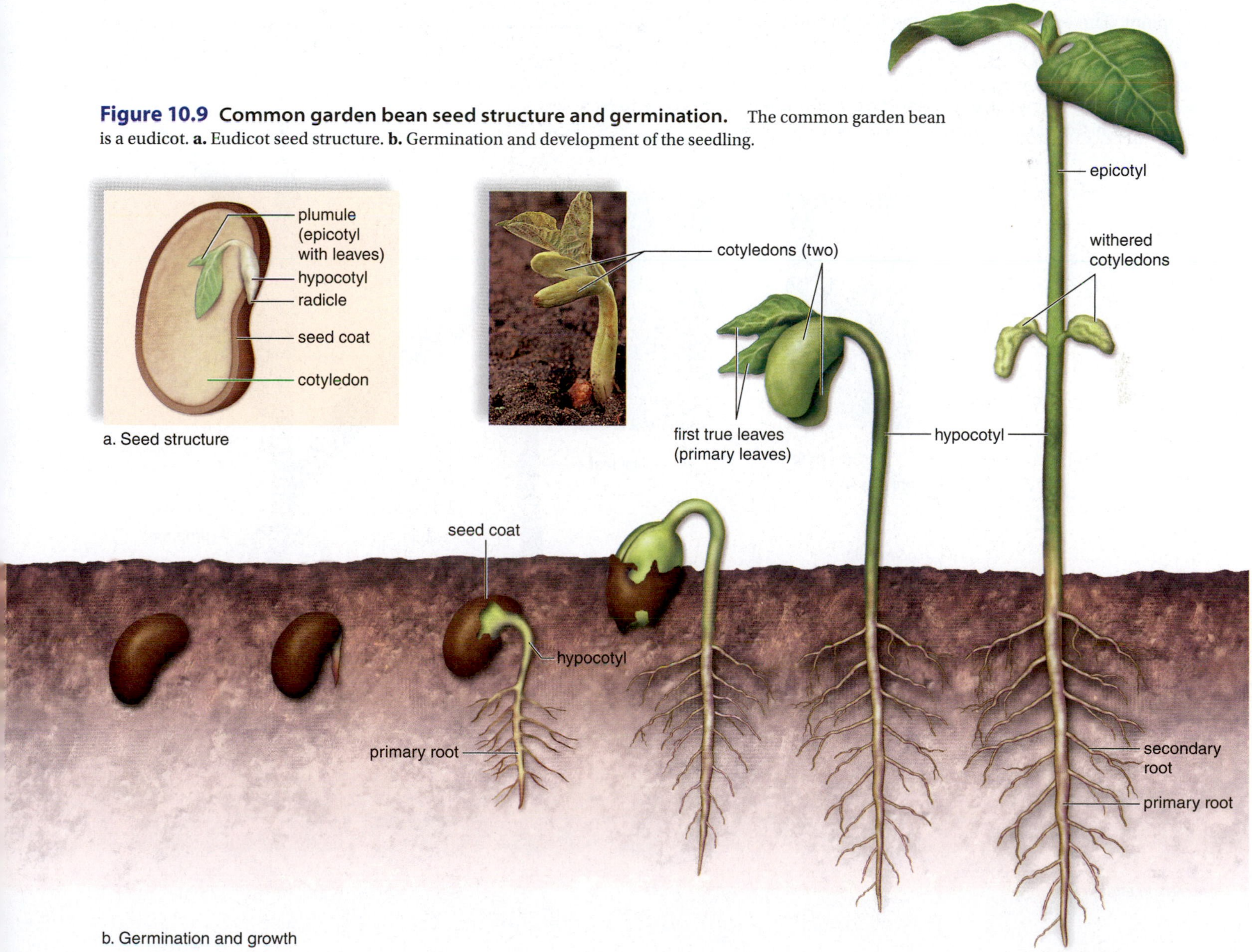
plumule
(epicotyl
with leaves)
hypocotyl
radicle
seed coat
cotyledon
a. Seed structure
cotyledons (two)
first true leaves
(primary leaves)
hypocotyl
epicotyl
withered
cotyledons
seed coat
hypocotyl
primary root
secondary
root
primary root
b. Germination and growth

Figure 10.9 Common garden bean seed structure and germination. The common garden bean is a eudicot. **a.** Eudicot seed structure. **b.** Germination and development of the seedling.

is broken. In deserts, germination does not occur until there is adequate moisture. This requirement helps ensure that seeds do not germinate until the most favorable growing conditions are present. Germination takes place if there is sufficient water, warmth, and oxygen to sustain growth.

Germination requires regulation, and both inhibitors and stimulators are known to exist. Fleshy fruits (e.g., apples and tomatoes) contain inhibitors so that germination does not occur until the seeds are removed and washed. In contrast, stimulators are present in the seeds of some temperate-zone woody plants. Mechanical action may also be required. Water, bacterial action, and even fire can act on the seed coat, allowing it to become permeable to water. The uptake of water causes the seed coat to burst.

Eudicot Versus Monocot Seed Germination

Eudicot embryos have two cotyledons that have absorbed the endosperm (Fig. 10.9). The cotyledons supply nutrients to the embryo and seedling and eventually shrivel and disappear. If the two cotyledons of a bean seed are parted, you can see a rudimentary plant. The epicotyl bears young leaves and is called a *plumule*. As the eudicot seedling emerges from the soil, the shoot is hook-shaped to protect the delicate plumule. The hypocotyl becomes part of the stem, and the radicle develops into the roots.

In monocots, the endosperm is the food-storage tissue, and the single cotyledon does not have a storage role (Fig. 10.10). Corn is a monocot, and its kernels are actually fruits, therefore, the outer covering is the pericarp. The plumule and radicle are enclosed in protective sheaths called the coleoptile and the coleorhiza, respectively. The plumule and the radicle burst through these coverings during germination.

Check Your Progress 10.2

1. Compare the function of endosperm in monocots versus eudicots.
2. Identify examples of plant adaptations for seed dispersal by wind.
3. Explain why seeds don't germinate immediately after dispersal.
4. Describe the differences between eudicot and monocot seed germination.

Figure 10.10 Corn kernel structure and germination. A corn plant is a monocot. **a.** Grain structure. **b.** Germination and development of the seedling.

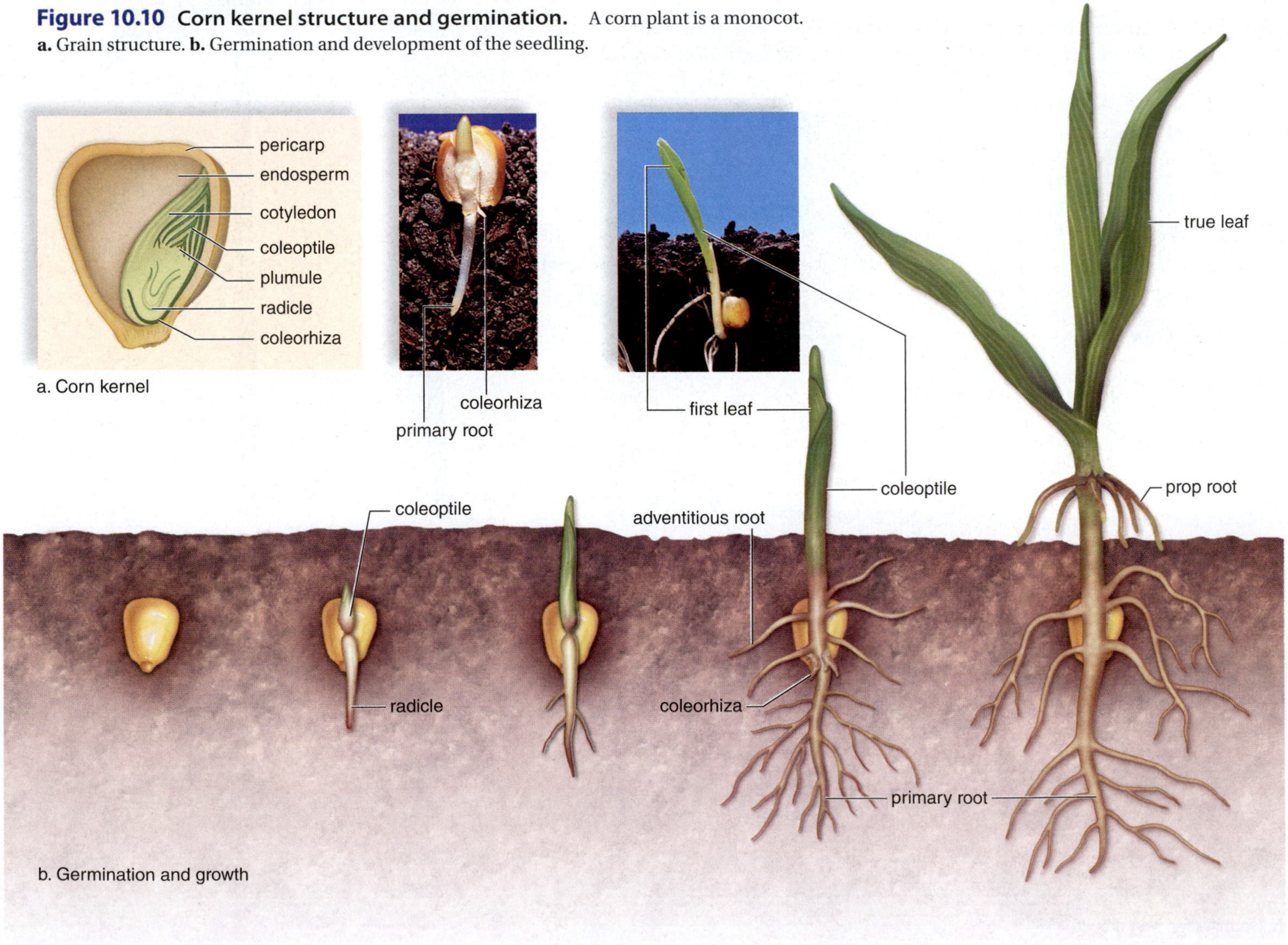
pericarp
endosperm
cotyledon
coleoptile
plumule
radicle
coleorhiza
a. Corn kernel
coleorhiza
primary root
first leaf
coleoptile
true leaf
prop root
coleoptile
adventitious root
radicle
coleorhiza
primary root
b. Germination and growth

10.3 Asexual Reproduction and Genetic Engineering in Plants

Learning Outcomes

Upon completion of this section, you should be able to

1. Recognize how asexual reproduction in plants differs from sexual reproduction.
2. Describe how plants are propagated in tissue culture.
3. Explain how genetic engineering can be used to alter plant traits.

In asexual reproduction, there is only one parent involved. Because plants contain nondifferentiated meristem tissue, they routinely reproduce asexually by vegetative propagation. For example, complete strawberry plants will grow from the nodes of stolons (see Fig. 9.20*a*) and irises will grow from the nodes of rhizomes (see Fig. 9.20*b*). White potatoes are actually portions of underground stems, and each eye is a bud that will produce a new potato plant if it is planted with a portion of the swollen tuber. Sweet potatoes are modified roots. They can be propagated by planting sections of the root. You may have noticed that the roots of some fruit trees, such as cherry and apple trees, produce "suckers," small plants that can grow into new trees.

Many types of plants are now propagated from stem cuttings. The discovery that the plant hormone auxin can cause roots to develop from a stem, as discussed in section 10.4, has expanded the list of plants that can be propagated from stem cuttings.

Propagation of Plants in Tissue Culture

Tissue culture is the growth of a *tissue* in an artificial liquid or solid *culture* medium. Tissue culture now allows botanists to breed a large number of plants from somatic tissue and to select any that have superior hereditary characteristics. These and any plants with desired genotypes can be propagated rapidly in tissue culture.

Plant cells are **totipotent,** which means that an entire plant can be produced from most plant cells. The only exceptions are the plant cells that lose their nuclei (sieve-tube members) or that are dead at maturity (xylem, sclerenchyma, and cork cells). Commercial micropropagation methods now produce thousands, even millions, of identical seedlings in a small vessel. One favorite such method is meristem culture. If the correct proportions of hormones are added to a liquid medium, many new shoots will develop from a single shoot tip. When these are removed, more shoots form. Because the shoots are genetically identical, the adult plants that develop from them, called clonal plants, all have the same traits. Another advantage to meristem culture is that meristem, unlike other portions of a plant, is virus-free. Therefore, the plants produced are also virus-free. When mature plant cells are used to grow entire plants, enzymes are often used to digest the cell walls of mesophyll tissue from a leaf. The result is cells without walls, called *protoplasts* (Fig. 10.11*a*). The protoplasts regenerate a new cell wall (Fig. 10.11*b*) and begin to divide, forming aggregates of cells and then a callus (Fig. 10.11*c, d*). These clumps of cells can be manipulated to produce *somatic embryos* (asexually produced embryos) (Fig. 10.11*e*). It's possible to produce millions of somatic embryos at once in large tanks called bioreactors. This is done for certain vegetables, such as tomatoes, celery, and asparagus, and for ornamental plants, such as lilies, begonias, and African violets. Somatic embryos that are encapsulated in a protective hydrated gel (and sometimes called artificial seeds) can be shipped anywhere. A mature plant develops from each somatic embryo (Fig. 10.11*f*). Plants generated from the somatic embryos vary somewhat because of mutations that

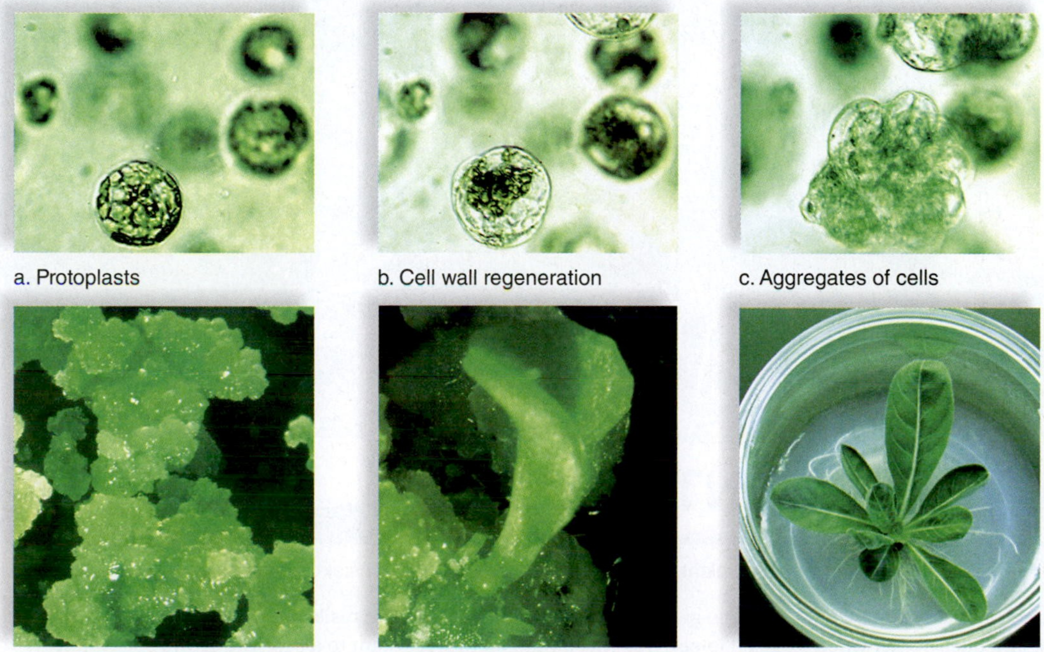
a. Protoplasts
b. Cell wall regeneration
c. Aggregates of cells
d. Callus (undifferentiated mass)
e. Somatic embryo
f. Plantlet

Figure 10.11 Tissue culture of plants. **a.** When plant cell walls are removed by digestive enzyme action, the resulting cells are called protoplasts. **b.** Cell walls regenerate, and cell division begins. **c.** Cell division produces an aggregate of cells. **d.** An undifferentiated mass, called a callus, forms. **e.** Somatic cell embryos appear. **f.** The embryos develop into plantlets that can be transferred to soil for growth into adult plants.

arise during the production process. These are called somaclonal variations and represent another way to produce new plants with desirable traits.

Recall that pollen grains are produced in the anthers of a flower. Anther culture is a direct way to produce a line of plants whose homologous chromosomes have the same genes. Anther culture involves mature anthers that are cultured in a medium containing vitamins and growth regulators. The haploid tube cells within the pollen grains divide, producing proembryos consisting of as many as 20 to 40 cells. Finally, the pollen grains rupture, releasing haploid embryos. The experimenter can now generate a haploid plant, or chemical agents can be added that encourage chromosomal doubling. The resulting plants are diploid, and the homologous chromosomes carry the same genes.

The culturing of plant tissues has led to a technique called cell suspension culture. Rapidly growing calluses are cut into small pieces and shaken in a liquid nutrient medium so that single cells or small clumps of cells break off and form a suspension. These cells will produce the same chemicals as the entire plant. For example, cell suspension cultures of *Cinchona ledgeriana* produce quinine, and those of *Digitalis lanata* produce digitoxin.

Genetic Engineering of Plants

Traditionally, **hybridization,** the crossing of different varieties of plants or even species, was used to produce plants with desirable traits. Hybridization, followed by vegetative propagation of the mature plants, generated a large number of identical plants with these traits. Today, it is possible to directly alter the genes of organisms and, in that way, produce new varieties with desirable traits.

Tissue Culture and Genetic Engineering

Protoplasts growing in a tissue culture medium can be genetically altered. A foreign gene isolated from any type of organism—plant, animal, or bacteria—is placed in the tissue culture medium. High-voltage electric pulses are then used to create pores in the plasma membrane so that the foreign gene enters the cells. In one such procedure, a gene for the production of the firefly enzyme luciferase was inserted into tobacco protoplasts, and the adult plants glowed when sprayed with the substrate luciferin.

Unfortunately, the regeneration of cereal grains from protoplasts has been difficult. As a result, other methods are used to introduce DNA into plant cells with intact cell walls. Today, it is possible to use a gene gun to bombard a callus (undifferentiated mass of cells) with DNA-coated microscopic metal particles. Then genetically altered somatic embryos develop into genetically altered adult plants.

Many plants, including corn and wheat varieties, have been genetically engineered by these methods. Such plants are called **transgenic organisms** (or **genetically modified organisms**) because they carry a foreign gene and have new and different traits. Figure 10.12 shows two types of transgenic plants. In another technique, foreign DNA is inserted into the plasmid of the bacterium *Agrobacterium,* which normally infects plant cells. The plasmid then contains *recombinant DNA* because it has genes from different sources—namely, those of the plasmid and also the foreign genes of interest. When a bacterium with a recombinant plasmid infects a plant, the recombinant plasmid enters the cells of the plant and can incorporate into the plant's DNA.

Agricultural Plants with Improved Traits

Corn and cotton plants, in addition to soybean and potato plants, have been engineered to be resistant to either herbicides

CONTROL
B.t.t. +

a. Herbicide-resistant soybean plants b. Nonresistant potato plant c. Pest-resistant potato plant

Figure 10.12 Transgenic plants. **a.** These soybean plants have been given a gene that causes them to be resistant to a particular herbicide. **b.** This potato plant is not resistant to a pest and does poorly when exposed to pests. **c.** This potato plant is resistant to a pest and grows beautifully.

Transgenic Crops of the Future	
Improved Agricultural Traits	
Herbicide resistant	Wheat, rice, sugar beets, canola
Salt tolerant	Cereals, rice, sugarcane, canola
Drought tolerant	Cereals, rice, sugarcane
Cold tolerant	Cereals, rice, sugarcane
Improved yield	Cereals, rice, corn, cotton
Modified wood pulp	Trees
Disease protected	Wheat, corn, potatoes
Improved Food-Quality Traits	
Fatty acid/oil content	Corn, soybeans
Protein/starch content	Cereals, potatoes, soybeans, rice, corn
Amino acid content	Corn, soybeans

a. Desirable traits

b. WT X10E

Wild-type (WT) and transgenic tomato (X1OE) grown in the presence of 200 mM NaCl

Figure 10.13 Transgenic crops of the future. **a.** Transgenic crops of the future include those with improved agricultural or food quality traits. **b.** Salt-tolerant tomato plants have been engineered. The plant to the left does poorly when watered with a salty solution, but the engineered plant to the right is tolerant of the solution. The development of additional salt-tolerant crops would help increase food production in the future.

or insect pests. Some corn and cotton plants have been developed that are both insect- and herbicide-resistant. According to the not-for-profit group International Service for the Acquisition of Agri-biotech Applications (ISAAA), over 16.7 million farmers in 29 countries grew biotech crops in 2011. Biotech crops represent one of the most significant crop technologies in the history of modern agriculture. If crops are resistant to a broad-spectrum herbicide and weeds are not, then the herbicide can be used to kill the weeds. When herbicide-resistant plants were planted, weeds were easily controlled, less tillage was needed, and soil erosion was minimized. However, wildlife and native plants declined because increased amounts of herbicides were used. Another concern is the potential transfer of herbicide-resistant genes from the modified crop plant to a related wild, "weedy" species through natural hybridization.

Crops with other improved agricultural and food-quality traits are desirable (Fig. 10.13). Irrigation, even with fresh water, inevitably leads to salinization of the soil, which reduces crop yields. Thus, the development of salt-tolerant crops would increase yields on such land. Salt-tolerant tomatoes have been developed to address this situation. First, scientists identified a gene coding for a channel protein that transports Na^+ into a vacuole, preventing it from interfering with plant metabolism. Then the scientists used the gene to engineer tomato plants that overproduce the channel protein. The modified plants thrived despite being watered with a salty solution. Salt- and also drought- and cold-tolerant cereals, rice, and sugarcane might help provide enough food for a growing world population.

Potato blight is the most serious potato disease in the world. In the mid-1840s, it was responsible for the Irish potato famine that caused the death of millions of people. By placing a gene from a naturally blight-resistant wild potato into a farmed variety, researchers have now made potato plants that are no longer vulnerable to a range of blight strains.

Some progress has also been made in increasing the food quality of crops. Soybeans have been developed that mainly produce monounsaturated fatty acid, a change that may improve human health. These altered plants also produce acids that can be used as hardeners in paints and plastics. The necessary genes were derived from *Vernonia* and castor bean seeds and were transferred into the soybean DNA.

Other types of genetically engineered plants are also expected to increase productivity. Stomata might be altered to take in more carbon dioxide or lose less water. A team of Japanese scientists is working on introducing the C_4 photosynthetic cycle into rice (see the Ecology feature, "The New Rice," in Chapter 8). Unlike C_3 plants, C_4 plants do well in hot, dry weather. These modifications would require a more complete reengineering of plant cells than the single-gene transfers that have been done so far.

Some people have expressed health and environmental concerns regarding the growing of transgenic crops, as discussed in the Bioethical feature, "Transgenic Crops," on page 187.

Commercial Products

Single-gene transfers have allowed plants to produce various products, including human hormones, clotting factors, and antibodies. One type of antibody made by corn can deliver radioisotopes to tumor cells, and another made by soybeans may be developed to treat genital herpes.

The tobacco mosaic virus has been used as a vector to introduce a human gene into adult tobacco plants in the field.

(Note that this technology bypasses the need for tissue culture completely.) Tens of grams of α-galactosidase, an enzyme needed for the treatment of a human lysosomal storage disease, were harvested per acre of tobacco plants. And it took only 30 days to get tobacco plants to produce antibodies to treat non-Hodgkin's lymphoma after being sprayed with a genetically engineered virus.

Check Your Progress 10.3

1. List two ways in which plants can reproduce asexually.
2. Identify how plants are grown in tissue culture.
3. Discuss two examples of how and why plants are genetically engineered.

10.4 Control of Growth and Responses

Learning Outcomes

Upon completion of this section, you should be able to

1. Explain the importance of plant hormones.
2. Identify the various types of plant hormones and their function.
3. Recognize how plants respond to stimuli.

Plants respond to environmental stimuli such as light, gravity, and seasonal changes, usually by a change in their pattern of growth. Hormones are involved in these responses.

Plant Hormones

Plant **hormones** are small organic molecules produced by the plant that serve as chemical signals between cells and tissues. Currently, the five commonly recognized groups of plant hormones are auxins, gibberellins, cytokinins, abscisic acid, and ethylene. Other chemicals, some of which differ only slightly from the natural hormones, also affect the growth of plants. These and the naturally occurring hormones are sometimes grouped together and called plant growth regulators.

Plant hormones bring about a physiological response in target cells after binding to a specific receptor protein in the plasma membrane. Figure 10.14 shows how the hormone auxin brings about elongation of a plant cell, a necessary step toward differentiation and maturation.

Auxins

The most common naturally occurring **auxin** is indoleacetic acid (IAA). It is produced in shoot apical meristem and is found in young leaves, flowers, and fruits. Therefore, you would expect auxin to affect many aspects of plant growth and development.

Effects of Auxin Auxin is present in the apical meristem of a plant shoot, where it causes the shoot to grow from the top, a phenomenon called **apical dominance.** Only when the terminal bud is removed, deliberately or accidentally, are the axillary buds able to grow, allowing the plant to branch (Fig. 10.15). Pruning the top (apical meristem) of a plant generally achieves a fuller look because of increased branching of the main body of the plant.

The application of a weak solution of auxin to a woody cutting causes adventitious roots to develop more quickly than they would otherwise. Auxin production by seeds also promotes the growth of fruit. As long as auxin is concentrated in

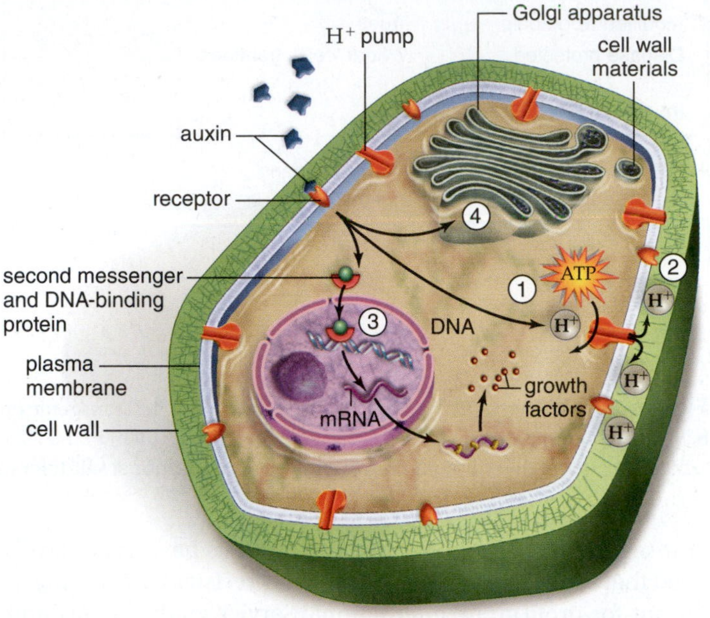
Golgi apparatus
H^+ pump
cell wall materials
auxin
receptor
4
second messenger and DNA-binding protein
1
ATP
2
H^+
3
DNA
plasma membrane
growth factors
cell wall
mRNA

Figure 10.14 Auxin mode of action. (**1**) The binding of auxin to a receptor stimulates an H^+ pump, moving hydrogen ions into the cell wall. (**2**) The resulting acidity causes the cell wall to weaken, and water enters the cell. (**3**) At the same time, a second messenger stimulates production of growth factors, and (**4**) new cell wall materials are synthesized. The result of all these actions is that the cell elongates, causing the stem to elongate.

axillary buds
apical bud removed
lateral branches
a. Plant with apical bud intact
b. Plant with apical bud removed

Figure 10.15 Apical dominance. **a.** Auxin in the apical bud inhibits axillary bud development, and the plant exhibits apical dominance. **b.** When the apical bud is removed, lateral branches develop.

leaves or fruits rather than in the stem, leaves and fruits do not fall off. Therefore, trees can be sprayed with auxin to keep mature fruit from falling to the ground.

How Auxins Work When a plant is exposed to unidirectional light, auxin moves to the shady side, where it binds to receptors and activates an ATP-driven pump that transports hydrogen ions (H^+) out of the cell (see Fig. 10.14). The acid environment weakens cellulose fibrils, and activated enzymes further degrade the cell wall. Water now enters the cell, and the resulting increase in turgor pressure causes the cell to elongate and the stem to bend toward the light. This growth response to unidirectional light is called **phototropism.** Besides phototropism, auxins are also involved in **gravitropism,** in which roots curve downward and stems curve upward in response to gravity, as discussed later in this section.

Gibberellins

Gibberellins were discovered in 1926 while a Japanese scientist was investigating a fungal disease of rice plants called "foolish seedling disease." The plants elongated too quickly, causing the stem to weaken and the plant to collapse. The fungus infecting the plants produced an excess of a chemical called gibberellin, named after the fungus *Gibberella fujikuroi.* It was not until 1956 that gibberellic acid was able to be isolated from a flowering plant. Sources of gibberellin in flowering plant parts are young leaves, roots, embryos, seeds, and fruits.

Gibberellins are growth-promoting hormones that bring about elongation of the resulting cells. We know of about 70 gibberellins that differ slightly in their chemical makeup. The most common of these is gibberellic acid, GA_3 (the subscript designation distinguishes it from other gibberellins).

Effects of Gibberellins When gibberellins are applied externally to plants, the most obvious effect is stem elongation (Fig. 10.16). Gibberellins can cause dwarf plants to grow, cabbage plants to become 2 m tall, and bush beans to become pole beans.

Figure 10.16 Effect of gibberellins. Stems elongate and grape size increases.

During dormancy a plant does not grow, even though conditions may be favorable for growth. The dormancy of seeds and buds can be broken by applying gibberellins, and research with barley seeds has shown how GA_3 influences the germination of seeds. Barley seeds have a large, starchy endosperm that must be broken down into sugars to provide energy for the embryo to grow. It is hypothesized that after GA_3 attaches to a receptor in the plasma membrane, calcium ions (Ca^{2+}) combine with a protein. This complex is believed to activate the gene that codes for amylase. Amylase then acts on starch to release the sugars needed for the seed to germinate. Gibberellins have been used to promote barley seed germination for the beer brewing industry.

Cytokinins

Cytokinins were discovered as a result of attempts to grow plant tissue and organs in culture vessels in the 1940s. It was found that cell division occurred when coconut milk (a liquid endosperm) and yeast extract were added to the culture medium. Although the effective agent or agents could not be isolated, they were collectively called cytokinins because cytokinesis refers to division of the cytoplasm. A naturally occurring cytokinin was not isolated until 1967. Because it came from the kernels of maize (*Zea*), it was called zeatin.

Effects of Cytokinins The cytokinins promote cell division. These substances, which are derivatives of adenine, one of the purine bases in DNA and RNA, have been isolated from actively dividing tissues of roots and also from seeds and fruits. A synthetic cytokinin, called kinetin, also promotes cell division.

Researchers have found that **senescence** (aging) of leaves can be prevented by applying cytokinins. When a plant organ, such as a leaf, loses its natural color, it is most likely undergoing senescence. During senescence, large molecules within the leaf are broken down and transported to other parts of the plant. Senescence does not always affect the entire plant at once. For example, as some plants grow taller, they naturally lose their lower leaves. Not only can cytokinins prevent the death of leaves, but they can also initiate leaf growth. Axillary buds begin to grow despite apical dominance when cytokinin is applied to them.

Researchers are well aware that the ratio of auxin to cytokinin and the acidity of the culture medium determine whether a plant tissue forms an undifferentiated mass, called a callus, or differentiates to form roots, vegetative shoots, leaves, or floral shoots. Researchers have reported that chemicals called oligosaccharins (chemical fragments released from the cell wall) are effective in directing differentiation. They hypothesize that reception of auxin and cytokinins, which leads to the activation of enzymes, releases these fragments from the cell wall.

Abscisic Acid

Abscisic acid (**ABA**) is produced by any "green tissue" with chloroplasts, monocot endosperm, and roots. Abscisic acid was once thought to function in **abscission,** the dropping of leaves, fruits, and flowers from a plant. But even though the external

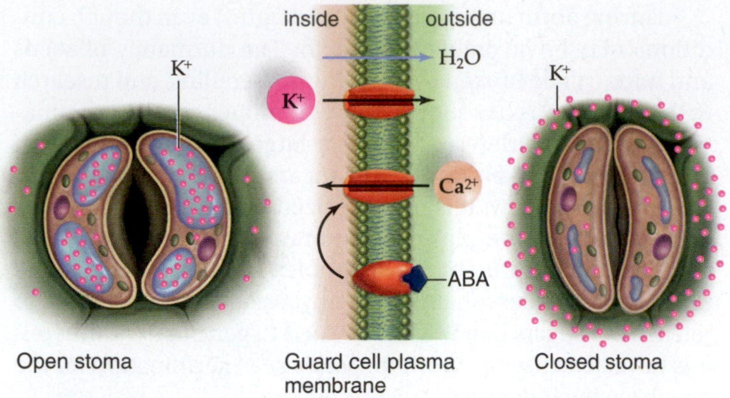
inside
outside
H_2O
K^+
K^+
K^+
Ca^{2+}
ABA
Open stoma
Guard cell plasma membrane
Closed stoma

Figure 10.17 Abscisic acid control of stoma closing. K^+ is concentrated inside the guard cells, and the stoma is open (*left*). When ABA binds to its receptor in the guard cell plasma membrane, calcium ions ($Ca2^+$) enter (*middle*). Then, potassium (K^+) channels open and K^+ exits the cell. Water follows and the stoma closes (*right*).

application of abscisic acid promotes abscission, researchers no longer believe this hormone functions naturally in this process. Instead, they think the hormone ethylene, discussed next, brings about abscission.

Effects of Abscisic Acid Abscisic acid is sometimes called the stress hormone because it initiates and maintains seed and bud dormancy and brings about the closure of stomata when a plant is under water stress (Fig. 10.17). Dormancy has begun when a plant stops growing and prepares for adverse conditions (even though conditions at the time are favorable for growth). For example, researchers believe that abscisic acid moves from leaves to vegetative buds in the fall, and thereafter, these buds are converted to winter buds. A winter bud is covered by thick, hardened scales. A reduction in the level of abscisic acid and an increase in the level of gibberellins are believed to break seed and bud dormancy. Then seeds germinate, and buds send forth leaves.

Ultimately abscisic acid causes potassium ions (K^+) to leave guard cells, causing the guard cells to lose water, and the stomata to close.

Ethylene

Ethylene is a gas that works with other hormones to bring about certain effects.

Effects of Ethylene Ethylene is involved in abscission. Low levels of auxin and perhaps gibberellin in the leaf, compared to the stem, probably initiate abscission. But once the process of abscission has begun, ethylene stimulates certain enzymes, such as cellulase, which cause leaf, fruit, or flower drop (Fig. 10.18*a, b*). Cellulase hydrolyzes cellulose in plant cell walls.

In the early 1900s, it was common practice to prepare citrus fruits for market by placing them in a room with a kerosene stove. Only later did researchers realize that ethylene, an incomplete combustion product of kerosene, was ripening the fruit (Fig. 10.18*c*). Because it is a gas, ethylene can act from a distance. A barrel of ripening apples can induce the ripening

a. No abscission
b. Abscission
c. Ripening

Figure 10.18 Functions of ethylene. **a.** Normally, there is no abscission when a holly twig is placed under a glass jar for a week. **b.** When an ethylene-producing ripe apple is also under the jar, abscission of the holly leaves occurs. **c.** Similarly, ethylene given off by this one ripe tomato will cause the others to ripen.

of a bunch of bananas, even if they are in different containers. Ethylene is released at the site of a plant wound due to physical damage or infection (which is why one rotten apple spoils the whole bushel).

Plant Responses to Environmental Stimuli

Environmental signals determine the seasonality of growth, reproduction, and dormancy in plants. Plant responses are strongly influenced by such environmental stimuli as light, day length, gravity, and touch. The ability of a plant to respond to environmental signals fosters the survival of the plant and the species in a particular environment.

Plant responses to environmental signals can be rapid, as when stomata open in the presence of light, or they can take some time, as when a plant flowers in season. Despite their variety, most plant responses to environmental signals are due to growth and sometimes differentiation, brought about at least in part by particular hormones.

Plant Tropisms

Plant growth toward or away from a directional stimulus is called a **tropism.** Tropisms are due to differential growth—one side of an organ elongates faster than the other, and the result is a curving toward or away from the stimulus. The following three well-known tropisms were each named for the stimulus that causes the response:

phototropism	growth in response to a light stimulus
gravitropism	growth in response to gravity
thigmotropism	growth in response to touch

Growth toward a stimulus is called a positive tropism, and growth away from a stimulus is called a negative tropism. For example, in positive phototropism, stems curve toward the light, and in negative gravitropism, stems curve away from the direction of gravity (Fig. 10.19). Roots, of course, exhibit positive gravitropism.

The role of auxin in the positive phototropism of stems has been studied for quite some time. Because blue light, in particular, causes phototropism to occur, researchers believe that a yellow pigment related to the vitamin riboflavin acts as a photoreceptor for light. Following reception, auxin migrates from the bright side to the shady side of a stem. The cells on that side elongate faster than those on the bright side, causing the stem to curve toward the light. Negative gravitropism of stems occurs when auxin moves to the lower part of a stem when a plant is placed on its side. Figure 10.14 explains how auxin brings about elongation.

Video
Plant Tactile Response

Figure 10.19 Tropisms. **a.** In positive phototropism, the stem of a plant curves toward the light. This response is due to the accumulation of auxin on the shady side of the stem. **b.** In negative gravitropism, the stem of a plant curves away from the direction of gravity 24 hours after the plant was placed on its side. This response is due to the accumulation of auxin on the lower side of the stem. **c.** In thigmotropism, contact directs the growth of the plant, as is shown in this morning glory plant.

Flowering

The flowering of angiosperms is a striking response to environmental seasonal changes. In some plants, flowering occurs according to the **photoperiod,** which is the ratio of the length of day to the length of night over a 24-hour period. Plants can be divided into the following three groups:

1. **Short-day plants** flower when the day length is shorter than a critical length. (Examples are cocklebur, poinsettia, and chrysanthemum.)
2. **Long-day plants** flower when the day length is longer than a critical length. (Examples are wheat, barley, clover, and spinach.)
3. **Day-neutral plants** do not depend on day length for flowering. (Examples are tomato and cucumber.)

Experiments have shown that the length of continuous darkness, not light, controls flowering. For example, the cocklebur does not flower if a suitable length of darkness is interrupted by a flash of light. In contrast, clover does flower when an unsuitable length of darkness is interrupted by a flash of light (Fig. 10.20). (Interrupting the light period with darkness has no effect on flowering.)

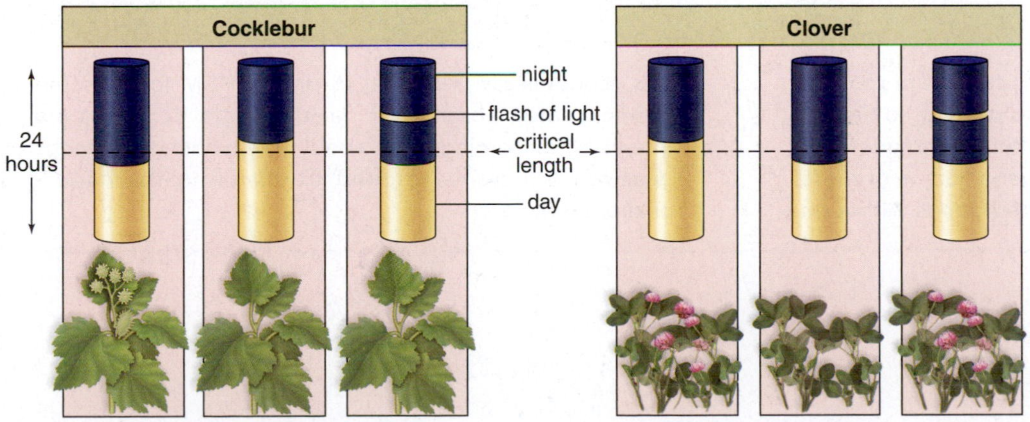
Cocklebur
Clover
24 hours
night
flash of light
critical length
day

a. Short-day (long-night) plant
b. Long-day (short-night) plant

Figure 10.20 Photoperiodism and flowering. **a.** Short-day plant. When the day is shorter than a critical length, this type of plant flowers. The plant does not flower when the day is longer than the critical length. It also does not flower if the longer-than-critical-length night is interrupted by a flash of light. **b.** Long-day plant. The plant flowers when the day is longer than a critical length. When the day is shorter than a critical length, this type of plant does not flower. However, it does flower if the slightly longer-than-critical-length night is interrupted by a flash of light.

Figure 10.21 Phytochrome control of growth pattern. **a.** The inactive form of phytochrome (P_r) is converted to the active form (P_{fr}) in the presence of red light. **b.** In sunlight, phytochrome is active and normal growth occurs. In the shade, etiolation occurs.

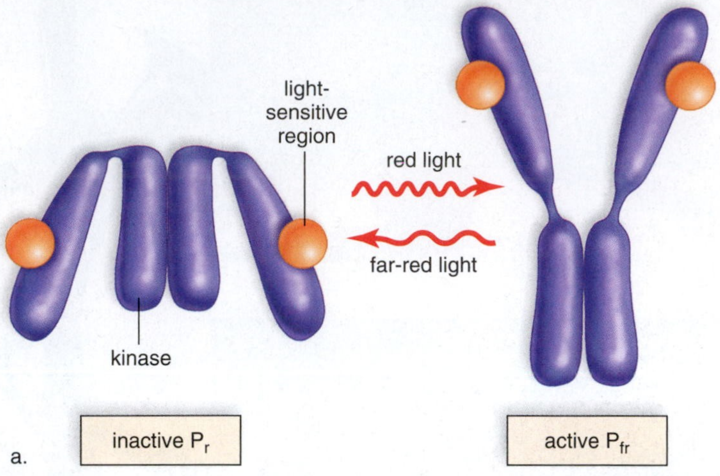
light-sensitive region
red light
far-red light
kinase
inactive P_r
active P_{fr}
a.

b.
Normal growth
Etiolation

Phytochrome and Plant Flowering

If flowering is dependent on day and night length, plants must have some way to detect these periods. This appears to be the role of **phytochrome,** a blue-green leaf pigment that alternately exists in two forms—P_r and P_{fr} (Fig. 10.21*a*).

Direct sunlight contains more red light than far-red light; therefore, P_{fr} (far red) is apt to be present in plant leaves during the day. In the shade and at sunset, there is more far-red light than red light. Therefore, P_{fr} is converted to P_r (red) as night approaches. There is also a slow metabolic replacement of P_{fr} by P_r during the night.

Animation Phytochrome Signaling

It is possible that phytochrome conversion is the first step in a signaling pathway that results in flowering. A flowering hormone has never been discovered.

Other Functions of Phytochrome

The $P_r \longrightarrow P_{fr}$ conversion cycle has other functions. The presence of P_{fr} indicates to seeds of some plants that sunlight is present and conditions are favorable for germination. Such seeds must be only partly covered with soil when planted. Following germination, the presence of P_r indicates that stem elongation may be needed to reach sunlight. Seedlings that are grown in the dark etiolate—that is, the stem increases in length, and the leaves remain small (Fig. 10.21*b*). Once the seedling is exposed to sunlight and P_r is converted to P_{fr}, the seedling begins to grow normally—the leaves expand, and the stem branches.

Check Your Progress 10.4

1. Identify the functions of auxins, gibberellins, cytokinins, abscisic acid, and ethylene.
2. Recognize the different types of plant growth tropism.
3. Compare the differences between short-day and long-day plants.
4. Identify the potential role of phytochrome in plant flowering.

Case Study Conclusion

Genetic manipulation of watermelon is capable of producing a 3n seedless watermelon. It is also possible to genetically modify plants by using tissue culture techniques and various ways of giving them foreign genes. Some of these plants become GM (genetically modified) foods. Although it is economically beneficial to produce GM foods, some are concerned about their health effects. Scientists are interested in learning more about how plants grow, seeds develop, and plants respond to stimuli in the natural environment.

SCIENCE IN YOUR LIFE ▶ **BIOETHICAL**

Transgenic Crops

Transgenic plants (genetically modified plants, or genetically modified organisms [GMOs]) are helping to improve crop yields in order to keep up with the ever-increasing human population. Some of these plants have the added benefit of requiring less fertilizer and/or pesticides, which has benefits to both people and the environment.

Some scientists are concerned that transgenic crops do pose a threat to the environment, and many activists believe that GMOs are dangerous to our health. Studies have shown that wind-carried pollen can cause transgenic crops to hybridize with nearby weedy relatives. This is leading to the production of herbicide-resistant weeds. There is also concern that the toxin produced by transgenic crops could hurt other species. Many researchers are conducting tests to see if this might occur. Also, although transgenic crops have not caused any illnesses in humans, some scientists concede the possibility that people could be allergic to a transgene's protein product. Such concerns led to a massive recall in September 2000 that pulled about 2.8 million boxes of Taco Bell taco shells from grocery stores after unapproved genetically modified corn was found in the product.

Already, transgenic plants must be approved by the Food and Drug Administration before they are considered safe for human consumption, and they must meet certain Environmental Protection Agency standards. Some people believe safety standards for transgenic crops should be further strengthened, whereas others fear stricter standards will compromise food production. Another possibility is to retain the current standards but require clear labeling so buyers can choose whether or not to eat genetically modified foods.

Questions to Consider

1. If plant pollen is transferred by an insect rather than by the wind, do you think there would still be a risk of GMOs hybridizing with weedy relatives?
2. What are the advantages and disadvantages of genetically engineering a plant versus using artificial breeding methods to develop new traits?
3. If a food label states that the food has been genetically modified, do you think most people would still buy the product? Why or why not?

MEDIA STUDY TOOLS

www.mhhe.com/maderinquiry14

Enhance your study of this chapter with study tools and practice tests. Also ask your instructor about the resources available through ConnectPlus, including LearnSmart, the media-rich eBook, interactive learning tools, and animations.

SUMMARIZE

10.1 Sexual Reproduction in Flowering Plants

- Flowering plants have an **alternation-of-generations** life cycle.
- **Flowers** borne by the **sporophyte** produce **microspores** and **megaspores** by meiosis. Microspores develop into a male gametophyte, and megaspores develop into the female gametophyte. **Gametophytes** produce gametes by mitosis.
- Flowers contain **sepals** that form an outer whorl; **petals,** which are the next whorl; and **stamens** that form a whorl around the base of at least one **carpel.** The carpel, in the flower's center, consists of a **stigma, style,** and **ovary,** which contains **ovules.**
- **Anthers** contain **microsporocytes,** each of which divides meiotically to produce four haploid microspores. Each microspore divides mitotically to produce a two-celled **pollen grain**—a tube cell and a generative cell. The generative cell later divides mitotically to produce two sperm cells.
- The pollen grain is the male gametophyte. After **pollination,** the pollen grain germinates, and as the pollen tube grows, sperm cells travel to the **embryo sac.** Plants have undergone **coevolution** with many of their pollinators to ensure fertilization occurs.
- Each ovule contains a **megasporocyte,** which divides meiotically to produce four haploid **megaspores,** only one of which survives. This megaspore divides mitotically to produce the female gametophyte (embryo sac), which usually has seven cells.
- Flowering plants undergo **double fertilization.** One sperm nucleus unites with the egg nucleus, forming a 2n zygote, and the other unites with the polar nuclei of the central cell, forming a 3n **endosperm** cell. The zygote becomes the sporophyte **embryo,** and the endosperm cell divides to become endosperm tissue. Collectively these structures are known as the **seed,** which is often found within a **fruit.**

10.2 Growth and Development

- Prior to seed formation, the zygote undergoes growth and development to become an embryo.
- The seeds enclosed by a fruit contain the embryo (hypocotyl, epicotyl, plumule, radicle) and stored food (endosperm and/or **cotyledons**).
- Following dispersal, a seed **germinates.**

10.3 Asexual Reproduction and Genetic Engineering in Plants

- Many flowering plants reproduce asexually (e.g., buds give rise to entire plants, or roots produce new shoots).
- Asexual reproduction allows for clonal propagation in tissue culture. This micropropagation is now commercialized—cell suspension

cultures can produce chemicals of medical importance, and plant **tissue culture** allows genetic engineering. **Transgenic organisms** are produced by inserting genes from other species into the host organism.

10.4 Control of Growth and Responses

- Plant **hormones** are chemical signals involved in plant growth and responses to environmental stimuli.
- There are five commonly recognized groups of plant hormones:
  - **Auxins** cause **apical dominance,** the growth of adventitious roots, and two types of **tropisms—phototropism** and **gravitropism.**
  - **Gibberellins** promote stem elongation and break seed dormancy so that germination occurs.
  - **Cytokinins** promote cell division, prevent **senescence** of leaves, and along with auxin, influence differentiation of plant tissues.
  - **Abscisic acid** (ABA) initiates and maintains seed and bud **dormancy** and closes stomata under water stress.
  - **Ethylene** causes **abscission** of leaves, fruits, and flowers. It also causes some fruits to ripen.
- Environmental signals play a significant role in plant growth and development.
- Tropisms are growth responses toward or away from unidirectional stimuli. When a plant is exposed to light, auxin moves laterally from the bright to the shady side of a stem. Then, cells on the shady side elongate, and the stem bends toward the light. Similarly, auxin is responsible for negative gravitropism and growth upward, opposite gravity.
- Flowering is a striking response to seasonal changes. **Phytochrome,** a plant pigment that responds to daylight, is believed to be part of a biological clock system that in some unknown way brings about flowering. Phytochrome has various other functions, such as seed germination, leaf expansion, and stem branching.
- **Short-day plants** flower when days are shorter (nights are longer) than a critical length, and **long-day plants** flower when days are longer than a critical length. Some plants are **day-neutral**.

ASSESS

Testing Yourself

Choose the best answer for each question.

1. Stigma is to carpel as anther is to
   a. sepal. **b.** stamen. **c.** ovary. **d.** style.
2. _____ always promotes cell division.
   a. Auxin
   b. Phytochrome
   c. Cytokinin
   d. None of these are correct.

For questions 3–6, match the definitions with the structure in the key.

Key:

a. flower **b.** megaspore **c.** gibberellin **d.** style **e.** senescence

3. Develops into the female gametophyte
4. Growth hormone that promotes elongation
5. Aging in plants
6. Produces seeds enclosed by fruits
7. Double fertilization refers to the formation of a _____ and a _____.
   a. zygote, zygote
   b. zygote, pollen grain
   c. zygote, megaspore
   d. zygote, endosperm
8. Which is the correct order of the following events: (1) megaspore becomes embryo sac, (2) embryo formed, (3) double fertilization, (4) meiosis?
   a. 1, 2, 3, 4 **b.** 4, 1, 3, 2 **c.** 4, 3, 2, 1 **d.** 2, 3, 4, 1
9. In an accessory fruit, such as an apple, the bulk of the fruit is from the
   a. ovary.
   b. style.
   c. pollen.
   d. None of these are correct.
10. Ethylene stimulates the action of _____ to produce _____.
    a. auxins, root formation
    b. gibberellins, flower formation
    c. cellulase, flower formation
    d. cellulase, flower dropping
11. Label the following diagram of a flower.

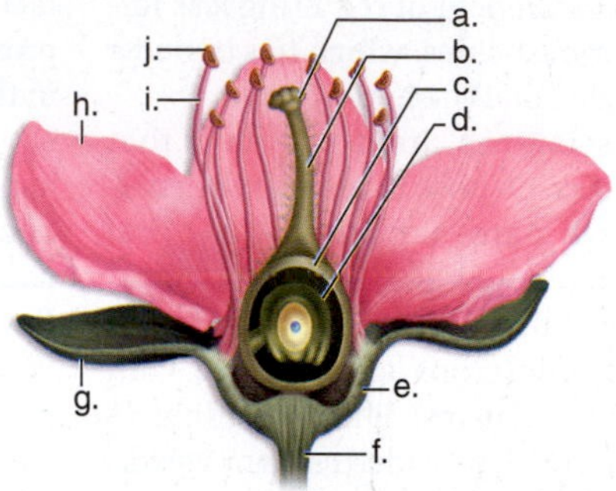
a.
b.
c.
d.
j.
i.
h.
g.
e.
f.

12. Label the diagram to the right of alternation of generations in flowering plants.

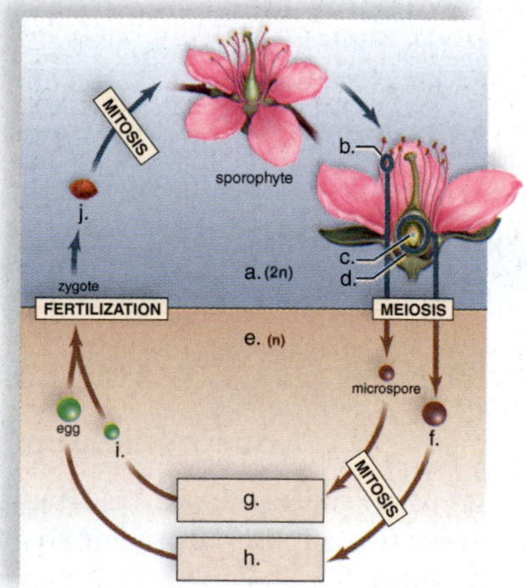
MITOSIS
sporophyte
b.
j.
a. (2n)
c.
d.
zygote
FERTILIZATION
MEIOSIS
e. (n)
microspore
egg
i.
f.
MITOSIS
g.
h.

13. Short-day plants
    a. are the same as long-day plants.
    b. are apt to flower in the fall.
    c. do not have a critical photoperiod.
    d. will not flower if a short day is interrupted by bright light.
    e. All of these are correct.
14. Which of the following plant hormones causes plants to grow in an upright position?
    a. auxin
    b. gibberellins
    c. cytokinins
    d. abscisic acid
    e. ethylene
15. How is the megaspore in the plant life cycle similar to the microspore? Both
    a. have the diploid number of chromosomes.
    b. become an embryo sac.
    c. become a gametophyte that produces a gamete.
    d. are necessary for seed production.
    e. Both c and d are correct.

ENGAGE

Thinking Critically

1. Compare sexual reproduction in humans and plants to show how they are both adapted to reproduction on land.
2. How might a plant benefit from having just one type of pollinator?
3. What feature of plant reproduction allows plants to be genetically modified utilizing asexual reproduction?

UNIT 3 Maintenance of the Human Body

CASE STUDY Going Gaga Over Biology

When Stefani Joanne Angelina Germanotta, better known as Lady Gaga, performs on stage, she is constantly in motion. When she is playing, singing into the microphone, or dancing complicated routines with her fellow performers, she is demonstrating how the human body is able to perform vastly complicated feats. Her nervous system must coordinate a variety of tasks, including: her breathing rate, remembering the words and tempo to each song, and the patterns of her motions onstage. Sensory input from her ears and eyes must monitor all the sounds and sights around her, while also maintaining her balance and contracting and relaxing her muscles in order to produce the desired choreography.

Of course, in biological terms, very similar observations could be made about humans engaging in almost any other type of physical activity, like playing sports, building a house, performing surgery, creating a sculpture, or driving a car. Even while you are just sitting quietly, perhaps reading a textbook, your body systems are undergoing a flurry of activity. Your muscles and bones work together to keep your body upright. Your respiratory and cardiovascular systems are providing oxygen to your tissues, and wastes are being transported to your kidneys for elimination. Your digestive system is providing nutrients for various tissues to use, and your immune and lymphatic systems are keeping infections at bay. Your nervous and endocrine systems are monitoring and controlling all of these functions. And of course, all these activities must continue even while your body is performing other highly skilled activities.

As you read through the chapter, think about the following questions:

1. When a musician like Lady Gaga is performing on stage, which of her body systems are most actively engaged? Which systems are less important?
2. What types of tissues make up each of the human body's major organ systems?
3. How do the different body systems interact with, and depend on, each other?

Human Organization 11

CHAPTER OUTLINE

BEFORE YOU BEGIN

Before beginning this chapter, take a few moments to review the following discussions:

Section 1.1 How do these differ: cell, tissue, organ, organ system, and organism?

Figure 1.2 What levels of biological organization are found in humans?

Table 3.2 What major structures make up animal cells?

11.1 Types of Tissues

Learning Outcomes

Upon completion of this section, you should be able to

1. List the four major types of tissues found in the human body, and describe the structural features of each.
2. Identify the common locations of each tissue type in the human body.
3. Explain how the cells that make up various types of tissues are specialized to perform their functions.

Recall the biological levels of organization described in Chapter 1. Cells are composed of molecules; a tissue has similar types of cells; an organ contains several types of tissues; and several organs make up an organ system. In this chapter, we consider the tissue, organ, and organ system levels of organization.

Tissues are composed of similarly specialized cells that perform a common function in the body. The tissues of the human body can be categorized into the following four major types:

Epithelial tissue covers body surfaces and lines body cavities.
Connective tissue binds and supports body parts.
Muscular tissue moves the body and its parts.
Nervous tissue receives stimuli, processes that information, and conducts nerve impulses.

We will now examine the structure and function of each of these four types of tissues.

MP3 Overview of Tissues

Epithelial Tissue

Epithelial tissue, also called *epithelium,* consists of tightly packed cells that form a continuous layer. Epithelial tissue covers surfaces and lines body cavities. Epithelial tissue has numerous functions in the body. Usually, it has a protective function, but it can also be modified to carry out secretion, absorption, excretion, and filtration.

On the external surface, epithelial tissue protects the body from injury, drying out, and possible invasion by microbes such as bacteria and viruses. On internal surfaces, modifications help epithelial tissue carry out both its protective and specific functions. Epithelial tissue secretes mucus along the digestive tract and sweeps up impurities from the lungs by means of cilia (sing., cilium). It efficiently absorbs molecules from kidney tubules and from the intestine because of minute cellular extensions called *microvilli* (sing., microvillus).

A *basement membrane* usually joins an epithelium to underlying connective tissue. We now know that the basement membrane consists of glycoprotein secreted by epithelial cells and collagen fibers that belong to the connective tissue.

Epithelial tissue is classified according to the shape of cell it is composed of (squamous, cuboidal, or columnar) and on the number of layers in the tissue (Fig. 11.1). **Squamous epithelium** is characterized by flattened cells and is found lining the lungs and blood vessels. **Cuboidal epithelium** contains cube-shaped cells and lines the kidney tubules. **Columnar epithelium** has cells resembling rectangular pillars or columns, with nuclei usually located near the bottom of each cell. This epithelium lines the digestive tract. Ciliated columnar epithelium lines the oviducts, where it propels the egg toward the uterus, or womb.

Classification is also based on the number of layers in the tissue. *Simple epithelium* has a single layer of cells, whereas *stratified epithelium* has layers of cells piled one on top of another. The walls of the smallest blood vessels, called *capillaries,* are composed of simple squamous epithelium. The permeability of this single layer of cells allows exchange of substances between the blood and tissue cells. Simple cuboidal epithelium lines kidney tubules and the cavities of many internal organs. Stratified squamous epithelium lines the nose, mouth, esophagus, anal canal, and vagina. The outer layer of skin is also stratified squamous epithelium, but the cells have been reinforced by keratin, a protein that provides strength.

Pseudostratified epithelium appears to be layered, but true layers do not exist because each cell touches the basement membrane. The lining of the windpipe, or trachea, is pseudostratified ciliated columnar epithelium. A secreted covering of mucus traps foreign particles, and the upward motion of the cilia carries the mucus to the back of the throat, where it either may be swallowed or expelled. Smoking can cause a change in mucous secretion and inhibit ciliary action, and the result is a chronic inflammatory condition called bronchitis.

When an epithelium secretes a product, it is said to be glandular. A **gland** can be a single epithelial cell, as are the mucus-secreting goblet cells within the columnar epithelium lining the digestive tract, or a gland can contain many cells. Glands that secrete their product into ducts are called exocrine glands, and those that secrete their product into the bloodstream are called endocrine glands. The pancreas is both an exocrine gland, because it secretes digestive juices into the small intestine via ducts, and an endocrine gland, because it secretes insulin into the bloodstream.

MP3 Epithelial Tissue

Junctions Between Epithelial Cells

The cells of a tissue can function in a coordinated manner when the plasma membranes of adjoining cells interact. Three common types of junctions link epithelial cells (see section 4.3). These are the *tight junctions,* which form an impermeable barrier between the cells; *gap junctions,* which serve to strengthen connections while allowing small molecules to pass; and *adhesion junctions,* which act like rivets or "spot welds" to anchor tissues in place, increasing their overall strength.

Connective Tissue

Connective tissue binds organs together, provides support and protection, fills spaces, produces blood cells, and stores fat. As a rule, connective tissue cells are widely separated by a *matrix,* consisting of a noncellular material that varies in consistency from solid to jellylike to fluid. A nonfluid matrix may have fibers of three possible types. White **collagen fibers** contain collagen, a protein that gives them flexibility and strength. **Reticular fibers** are very thin collagen fibers that are highly branched and form delicate supporting networks. Yellow **elastic fibers** contain elastin, a protein that is not as strong as collagen but more elastic.

MP3 Connective Tissue

Figure 11.1 Epithelial tissue.
Certain types of epithelial tissue—squamous, cuboidal, and columnar—are named for the shapes of their cells. They all have a protective function in addition to other specific functions.

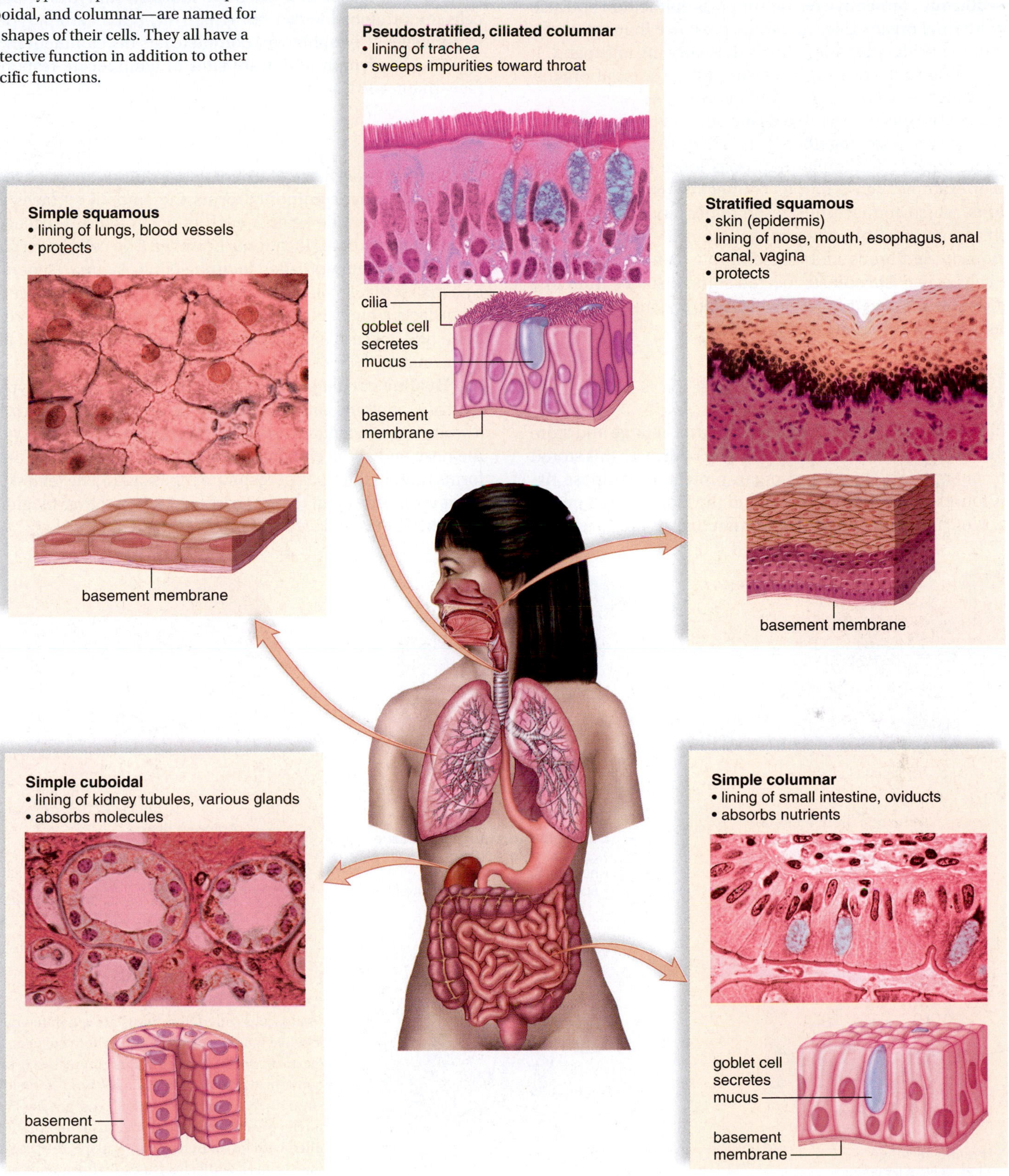
Pseudostratified, ciliated columnar
• lining of trachea
• sweeps impurities toward throat
cilia
goblet cell secretes mucus
basement membrane
Simple squamous
• lining of lungs, blood vessels
• protects
basement membrane
Stratified squamous
• skin (epidermis)
• lining of nose, mouth, esophagus, anal canal, vagina
• protects
basement membrane
Simple cuboidal
• lining of kidney tubules, various glands
• absorbs molecules
basement membrane
Simple columnar
• lining of small intestine, oviducts
• absorbs nutrients
goblet cell secretes mucus
basement membrane

Loose Fibrous and Dense Fibrous Tissues

Loose fibrous connective tissue supports epithelium and also many internal organs (Fig. 11.2*a*). Its presence in lungs, arteries, and the urinary bladder allows these organs to expand. It forms a protective covering enclosing many internal organs, such as muscles, blood vessels, and nerves.

Dense fibrous connective tissue contains many collagen fibers that are packed together (Fig. 11.2*b*). This type of tissue has more specific functions than does loose connective tissue. For example, dense fibrous connective tissue is found in **tendons,** which connect muscles to bones, and in **ligaments,** which connect bones to other bones at joints.

Both loose fibrous and dense fibrous connective tissues have cells called **fibroblasts** located some distance from one another and separated by a jellylike matrix containing white collagen fibers and yellow elastic fibers.

Adipose Tissue and Reticular Connective Tissue

In **adipose tissue** (Fig. 11.2*c*), the fibroblasts enlarge and store fat (thus, they are called adipocytes). The body uses this stored fat for energy, insulation, and organ protection. Adipose tissue is found beneath the skin, around the kidneys, and on the surface of the heart. **Reticular connective tissue** forms the supporting meshwork of lymphoid tissue in lymph nodes, the spleen, the thymus, and the bone marrow. All types of blood cells are produced in red bone marrow, but a certain type of lymphocyte (T lymphocyte) completes its development in the thymus. The lymph nodes are sites of lymphocyte responses (see Chapter 13).

Cartilage

Cartilage is a specialized form of dense fibrous connective tissue, which most commonly forms the smooth surfaces that allow bones to slide against each other in joints. Cartilage is found in many other locations, however. The cells in cartilage lie in small chambers called **lacunae** (sing., lacuna), separated by a matrix that is solid yet flexible. Unfortunately, because this tissue lacks a direct blood supply, it heals very slowly. There are three types of cartilage, distinguished by the main type of fiber in the matrix.

Hyaline cartilage (Fig. 11.2*d*), the most common type of cartilage, contains only very fine collagen fibers. The matrix has a white, translucent appearance. Hyaline cartilage is found in the nose and at the ends of the long bones and the ribs, and it forms rings in the walls of respiratory passages. The fetal skeleton is also made of this type of cartilage. Later, the cartilaginous fetal skeleton is replaced by bone.

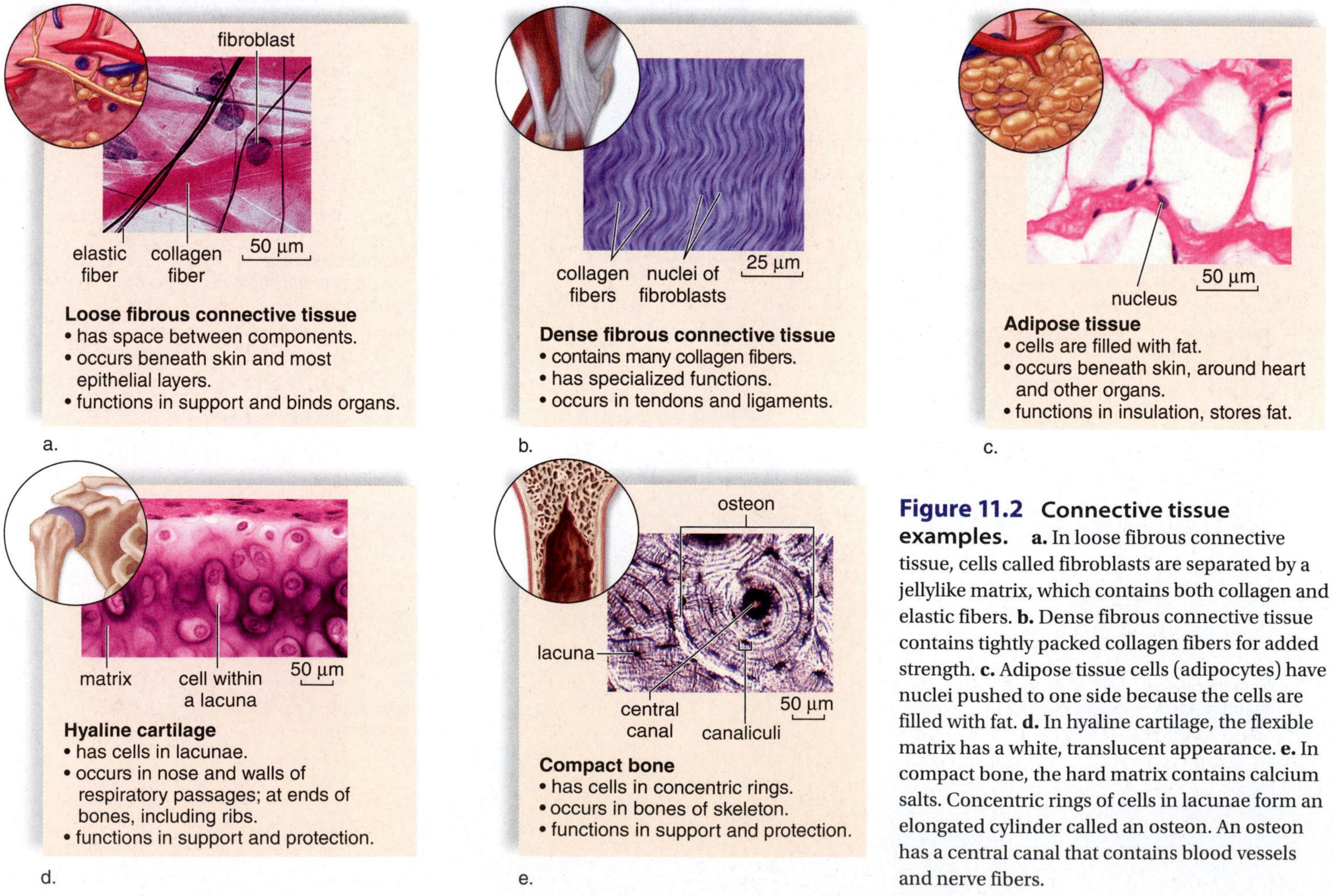
fibroblast
elastic fiber
collagen fiber
50 µm
Loose fibrous connective tissue
• has space between components.
• occurs beneath skin and most epithelial layers.
• functions in support and binds organs.
a.
collagen fibers
nuclei of fibroblasts
25 µm
Dense fibrous connective tissue
• contains many collagen fibers.
• has specialized functions.
• occurs in tendons and ligaments.
b.
nucleus
50 µm
Adipose tissue
• cells are filled with fat.
• occurs beneath skin, around heart and other organs.
• functions in insulation, stores fat.
c.
matrix
cell within a lacuna
50 µm
Hyaline cartilage
• has cells in lacunae.
• occurs in nose and walls of respiratory passages; at ends of bones, including ribs.
• functions in support and protection.
d.
osteon
lacuna
central canal
canaliculi
50 µm
Compact bone
• has cells in concentric rings.
• occurs in bones of skeleton.
• functions in support and protection.
e.

Figure 11.2 Connective tissue examples. **a.** In loose fibrous connective tissue, cells called fibroblasts are separated by a jellylike matrix, which contains both collagen and elastic fibers. **b.** Dense fibrous connective tissue contains tightly packed collagen fibers for added strength. **c.** Adipose tissue cells (adipocytes) have nuclei pushed to one side because the cells are filled with fat. **d.** In hyaline cartilage, the flexible matrix has a white, translucent appearance. **e.** In compact bone, the hard matrix contains calcium salts. Concentric rings of cells in lacunae form an elongated cylinder called an osteon. An osteon has a central canal that contains blood vessels and nerve fibers.

Elastic cartilage has more elastic fibers than hyaline cartilage. For this reason, it is more flexible and is found, for example, in the framework of the outer ear.

Fibrocartilage has a matrix containing strong collagen fibers. Fibrocartilage is found in structures that withstand tension and pressure, such as the pads between the vertebrae in the backbone and the wedges in the knee joint.

Bone

Bone is the most rigid connective tissue. It consists of an extremely hard matrix of inorganic salts, notably calcium salts, deposited around protein fibers, especially collagen fibers. The inorganic salts give bone rigidity, and the protein fibers provide elasticity and strength, much as steel rods do in reinforced concrete.

Compact bone (Fig. 11.2*e*) makes up the shaft of a long bone. It consists of cylindrical structural units called osteons. The central canal of each osteon is surrounded by rings of hard matrix. Bone cells are located in spaces called lacunae between the rings of matrix. In the central canal, nerve fibers carry nerve impulses and blood vessels carry nutrients that allow bone to renew itself. Nutrients can reach all of the bone cells because they are connected by thin processes within canaliculi (minute canals) that also reach to the central canal. The shaft of long bones such as the femur (thigh bone) has a hollow chamber filled with bone marrow, a site where blood cells develop.

The ends of a long bone contain **spongy bone,** which contains numerous bony bars and plates, separated by irregular spaces. Although lighter than compact bone, spongy bone is still designed for strength. Just as braces are used for support in buildings, the solid portions of spongy bone follow lines of stress. Section 19.1 provides more information about bone and cartilage.

Blood

Blood is unlike other types of connective tissue in that the matrix (i.e., plasma) is not made by the cells. In fact, some scientists do not classify blood as connective tissue, suggesting instead a separate category called vascular tissue.

The internal environment of the body consists of blood and the fluid between the body's cells. The systems of the body help keep the composition and chemistry of blood within normal limits, and blood, in turn, creates tissue fluid. Blood transports nutrients and oxygen to tissue fluid and removes carbon dioxide and other wastes. It helps distribute heat and also plays a role in fluid, ion, and pH balance. Various components of blood help protect us from disease, and blood's ability to clot prevents fluid loss.

If blood is transferred from a person's vein to a test tube and prevented from clotting, it separates into two layers (Fig. 11.3*a*). The upper, liquid layer, called **plasma,** represents about 55% of the volume of whole blood and contains a variety of inorganic and organic substances dissolved or suspended in water (Table 11.1). The lower layer consists of red blood cells (erythrocytes), white blood cells (leukocytes), and blood platelets (thrombocytes) (Fig. 11.3*b*). Collectively, these are called the formed elements, and they represent about 45% of the volume of whole blood. Formed elements are manufactured in the red bone marrow of the skull, ribs, vertebrae, and ends of the long bones.

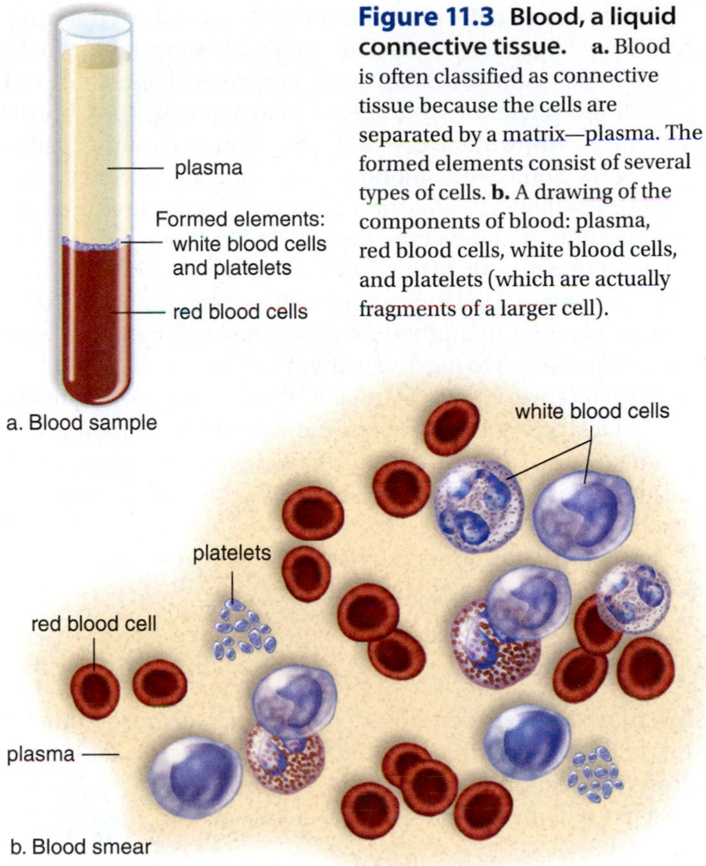
plasma
Formed elements:
white blood cells
and platelets
red blood cells
a. Blood sample
white blood cells
platelets
red blood cell
plasma
b. Blood smear

Figure 11.3 Blood, a liquid connective tissue. **a.** Blood is often classified as connective tissue because the cells are separated by a matrix—plasma. The formed elements consist of several types of cells. **b.** A drawing of the components of blood: plasma, red blood cells, white blood cells, and platelets (which are actually fragments of a larger cell).

TABLE 11.1 Components of Blood Plasma

Water (90–92% of total)	
Solutes (8–10% of total)	
Inorganic ions (electrolytes)	Na^+, Ca^{2+}, K^+, Mg^{2+}, Cl^-, HCO_3^-, HPO_4^{2-}, SO_4^{2-}
Gases	O_2, CO_2
Plasma proteins	Albumin, globulins, fibrinogen, transport proteins
Organic nutrients	Glucose, lipids, phospholipids, amino acids, etc.
Nitrogen-containing waste products	Urea, ammonia, uric acid
Regulatory substances	Hormones, enzymes

The **red blood cells** are small, biconcave, disk-shaped cells without nuclei. The presence of the red pigment hemoglobin makes the cells red and, in turn, makes the blood red. Hemoglobin is composed of four units. Each contains the protein globin and a complex, iron-containing structure called heme. The iron forms a loose association with oxygen, and in this way, red blood cells transport oxygen.

White blood cells differ from red blood cells in that they are usually larger, have a nucleus, and without staining would appear translucent. When smeared onto microscope slides and stained, the nuclei of white blood cells typically look bluish (Fig. 11.3*b*). White blood cells fight infection in a number of different ways. Some white blood cells are phagocytic and engulf infectious pathogens, while others are responsible for the adaptive immunity that develops after an individual is exposed to various pathogens or toxins, either through natural infection or by vaccination. One important feature of acquired immunity is the production of antibodies, molecules that combine with foreign substances to inactivate them.

Platelets are not complete cells. Rather, they are fragments of large cells present only in bone marrow. When a blood vessel is damaged, platelets help to form a plug that seals the vessel, and injured tissues release molecules that stimulate the clotting process.

For more details on red blood cells, white blood cells, and platelets, see section 12.2 and Chapter 13.

Muscular Tissue

Muscular tissue is composed of cells called muscle fibers. Muscle fibers contain actin filaments and myosin filaments, whose interaction accounts for movement. The three types of muscle tissue are skeletal, smooth, and cardiac.

MP3 Muscle Tissue

Skeletal muscle (Fig. 11.4*a*) is usually attached by tendons to the bones of the skeleton, and when it contracts, body parts move. Contraction of skeletal muscle is under voluntary control. Skeletal muscle fibers are cylindrical and quite long—sometimes they run the length of the muscle. They arise from fusion of several cells, resulting in one fiber with multiple nuclei. The fibers have alternating light and dark bands that give them a *striated* appearance.

Smooth muscle is so named because the cells lack striations. The spindle-shaped fibers, each with a single nucleus, form layers in which the thick middle portion of one cell is opposite the thin ends of adjacent cells. Consequently, the nuclei form an irregular pattern in the tissue (Fig. 11.4*b*). Smooth muscle is not under voluntary control, and therefore is said to be involuntary. Smooth muscle is found in the walls of viscera (intestine, stomach, and other internal organs) and blood vessels.

Cardiac muscle (Fig. 11.4*c*) is found only in the walls of the heart. Cardiac muscle combines features of both smooth muscle and skeletal muscle. Like skeletal muscle, it has striations, but the contraction of the heart is involuntary. Cardiac muscle cells also differ from skeletal muscle cells in that they usually have a single, centrally placed nucleus. The cells are branched and seemingly fused one with the other, and the heart appears to be composed of one large interconnecting mass of muscle cells. Actually, cardiac muscle cells are separate and individual, but they are bound end to end at *intercalated disks,* areas where folded plasma membranes between two cells contain adhesion junctions and gap junctions. These areas promote the flow of electrical current when the heart muscle contracts by allowing ions to flow freely between cells. See section 19.4 to learn more about the mechanism of muscle contraction.

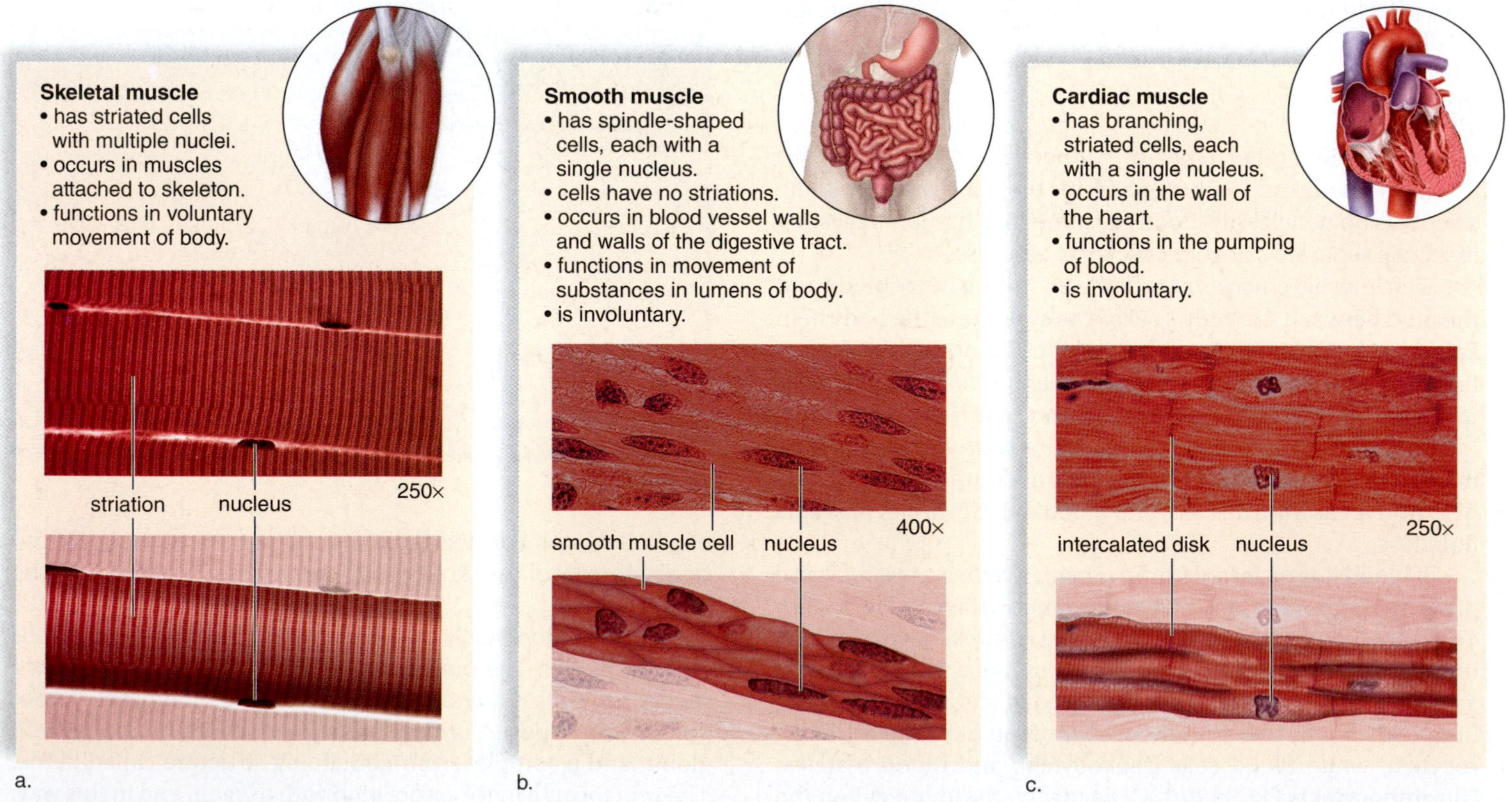
Skeletal muscle
• has striated cells with multiple nuclei.
• occurs in muscles attached to skeleton.
• functions in voluntary movement of body.
striation
nucleus
250×
a.
Smooth muscle
• has spindle-shaped cells, each with a single nucleus.
• cells have no striations.
• occurs in blood vessel walls and walls of the digestive tract.
• functions in movement of substances in lumens of body.
• is involuntary.
smooth muscle cell
nucleus
400×
b.
Cardiac muscle
• has branching, striated cells, each with a single nucleus.
• occurs in the wall of the heart.
• functions in the pumping of blood.
• is involuntary.
intercalated disk
nucleus
250×
c.

Figure 11.4 Muscular tissue. **a.** Skeletal muscle is voluntary and striated. **b.** Smooth muscle is involuntary and nonstriated. **c.** Cardiac muscle is involuntary and striated. Cardiac muscle cells branch and fit together at intercalated disks.

Nervous Tissue

Nervous tissue, which contains nerve cells called neurons (Fig. 11.5*a*), is present in the brain and spinal cord. A **neuron** is a specialized cell that has three parts: a cell body, dendrites, and an axon (Fig. 11.5*a, b*). The cell body contains the major concentration of the cytoplasm and the nucleus of the neuron. A dendrite is a process that conducts signals toward the cell body. An axon is a process that typically conducts nerve impulses away from the cell body. Long axons are covered by myelin, a white fatty substance, which increases the speed of nerve impulses. The term *fiber*[1] is used here to refer to an axon along with its myelin sheath if it has one. Outside the brain and spinal cord, fibers bound by connective tissue form **nerves.**

The nervous system has just three functions: sensory input, integration of data, and motor output. Nerves conduct impulses from sensory receptors to the spinal cord and the brain where integration occurs. The phenomenon called sensation occurs only in the brain, however. Nerves also conduct nerve impulses away from the spinal cord and brain to the muscles and glands, causing them to contract and secrete, respectively. In this way, a coordinated response to the stimulus is achieved.

MP3 Nervous Tissue

Neuroglia

In addition to neurons, nervous tissue contains neuroglia. **Neuroglia** are cells that outnumber neurons nine to one and take up more than half the volume of the brain (Fig. 11.5*a*). Although the primary function of neuroglia is to support and nourish neurons, research is currently being conducted to determine how much they directly contribute to brain function. The four types of neuroglia in the brain are microglia, astrocytes, oligodendrocytes, and ependymal cells. Microglia, in addition to supporting neurons, engulf bacterial and cellular debris. Astrocytes provide nutrients to neurons and produce a hormone known as glia-derived growth factor, which has potential as a treatment for Parkinson disease and other diseases caused by neuron degeneration. Oligodendrocytes form myelin sheaths. Ependymal cells line the fluid-filled spaces of the brain and spinal cord. Neuroglia don't have long processes, but even so, researchers are now beginning to gather evidence that they do communicate among themselves and with neurons. To learn more about the nervous system, see Chapter 17.

Check Your Progress 11.1

1. List five types of epithelium, and identify a location where each could be found in the human body.
2. Compare and contrast the three types of connective tissue.
3. Describe the structure and function of skeletal, smooth, and cardiac muscle.
4. Recall the three parts of a neuron, and define nerve fiber.

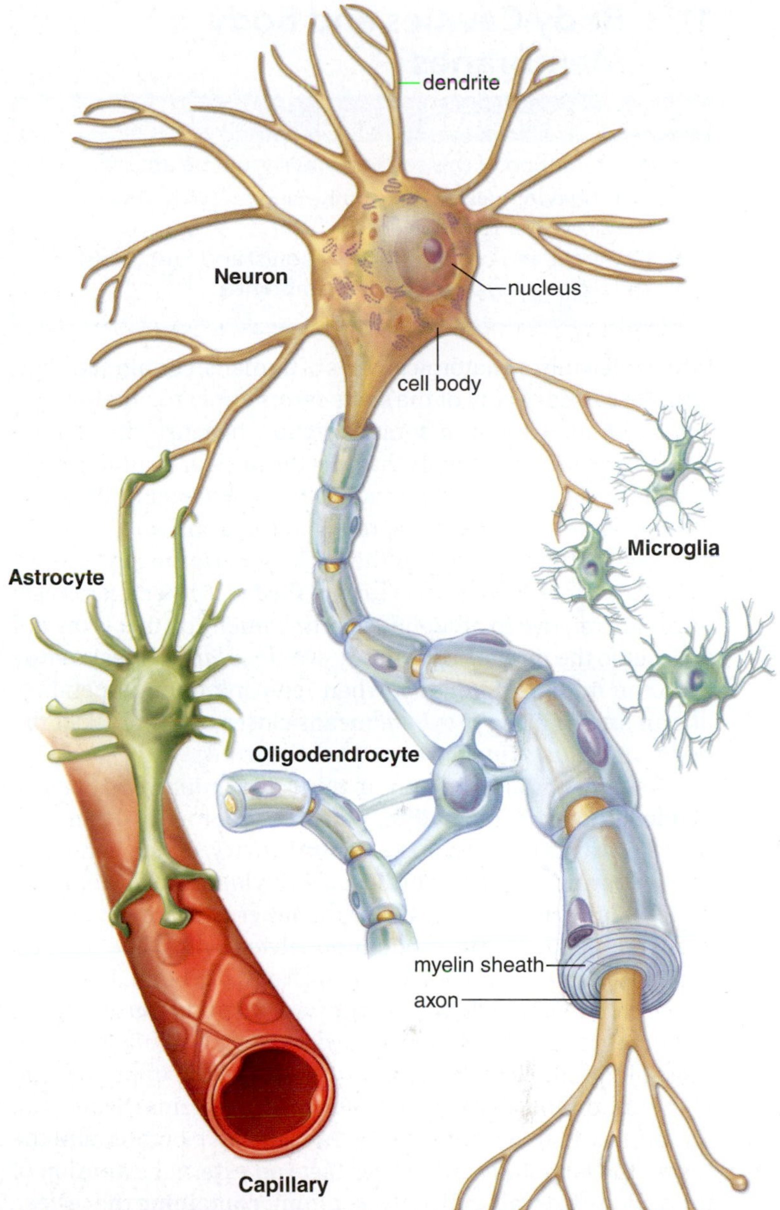
dendrite
Neuron
nucleus
cell body
Microglia
Astrocyte
Oligodendrocyte
myelin sheath
axon
Capillary
a. Neuron and neuroglia

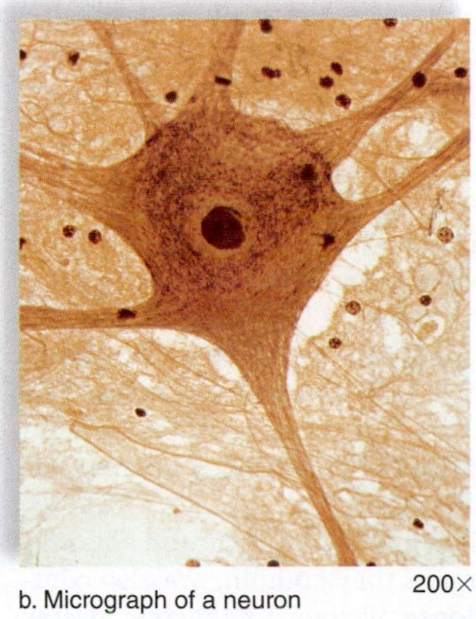
200×
b. Micrograph of a neuron

Figure 11.5 Neurons and neuroglia. **a.** Neurons conduct nerve impulses. Neuroglia consist of cells that support and nourish neurons and have various functions. Microglia become mobile in response to inflammation and phagocytize debris. Astrocytes lie between neurons and a capillary. Therefore, nutrients entering neurons from the blood must first pass through astrocytes. Oligodendrocytes form the myelin sheaths around fibers in the brain and spinal cord. **b.** A neuron cell body as seen with a light microscope.

[1]In connective tissue, a fiber is a component of the matrix; in muscle tissue, a fiber is a muscle cell; in nervous tissue, a fiber is an axon and its myelin sheath.

11.2 Body Cavities and Body Membranes

Learning Outcomes

Upon completion of this section, you should be able to

1. List the two main cavities of the human body, and the major organs found in each.
2. Compare and contrast the location(s) and function(s) of the different types of body membranes.

When referring to anatomical parts of humans, certain standard terms are used. Many of the same terms apply to other organisms, although there is some variation because the human terms always refer to a body that is in the upright, standing position, as in Figure 11.6. The term *ventral,* which means the same thing as *anterior* in humans, refers to the front, and *dorsal* or *posterior* both mean toward the back. *Superior* means toward the head, and *inferior* means toward the feet. Other anatomical terms are relative to other body parts; something that is *medial* is closer to the midline of the body, whereas *lateral* means away from the midline. Similarly, when referring to an appendage like an arm or a leg, *proximal* means closer to the trunk of the body, while *distal* means away from the trunk.

Keeping these definitions in mind, the human body can be divided into two main **body cavities:** the ventral cavity and the dorsal cavity (Fig. 11.6*a*). The **ventral cavity,** which is called a *coelom* during development, becomes divided into the thoracic, abdominal, and pelvic cavities (the latter two are sometimes grouped together as the abdominopelvic cavity). The *thoracic cavity* contains the right and left lungs and the heart. It is separated from the *abdominal cavity* by a horizontal muscle called the diaphragm. The stomach, liver, spleen, gallbladder, and most of the small and large intestines are in the upper portion of the abdominal cavity. The *pelvic cavity* contains the rectum, the urinary bladder, the internal reproductive organs, and the rest of the large intestine. Males have an external extension of the abdominal wall, called the scrotum, containing the testes.

The **dorsal cavity** also has two parts: the *cranial cavity* within the skull contains the brain; the *vertebral canal,* formed by the vertebrae, contains the spinal cord.

MP3 Body Cavities

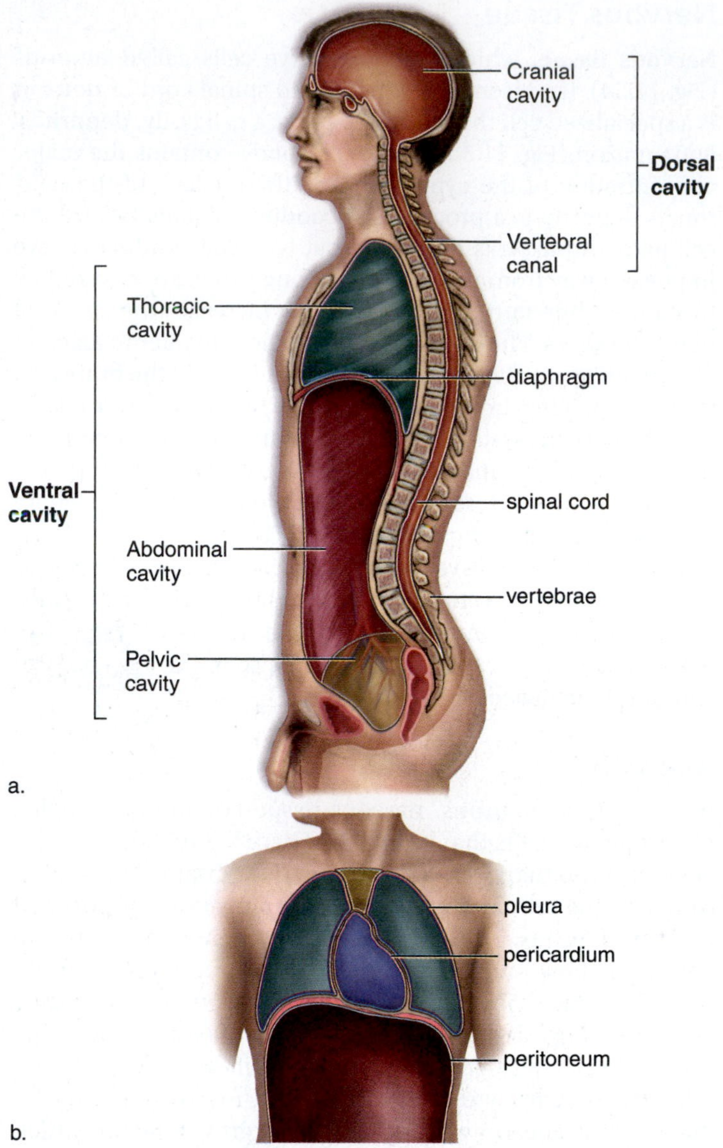

Cranial cavity
Dorsal cavity
Vertebral canal
Thoracic cavity
diaphragm
Ventral cavity
spinal cord
Abdominal cavity
vertebrae
Pelvic cavity
a.
pleura
pericardium
peritoneum
b.

Figure 11.6 Mammalian body cavities. **a.** Side view. The dorsal (toward the back) cavity contains the cranial cavity and the vertebral canal. The brain is in the cranial cavity, and the spinal cord is in the vertebral canal. In the ventral (toward the front) cavity, the diaphragm separates the thoracic cavity and the abdominal cavity. The heart and lungs are in the thoracic cavity, and most other internal organs are in the abdominal cavity. **b.** Types of serous membranes.

Body Membranes

Body membranes line cavities and the internal spaces of organs and tubes that open to the outside.

Mucous membranes line the tubes of the digestive, respiratory, urinary, and reproductive systems. They are composed of an epithelium overlying a loose fibrous connective tissue layer. The epithelium contains goblet cells that secrete mucus. This mucus ordinarily protects the body from invasion by bacteria and viruses. More mucus is secreted and expelled when a person has a cold and has to blow her or his nose.

Serous membranes, which line the thoracic and abdominal cavities and cover the organs they contain, are also composed of epithelium and loose fibrous connective tissue. They secrete a watery fluid that keeps the membranes lubricated. Serous membranes support the internal organs and compartmentalize the large thoracic and abdominal cavities. This helps hinder the spread of any infection.

Serous membranes have specific names according to their location (Fig. 11.6*b*). The *pleurae* (sing., pleura) line the thoracic cavity and cover the lungs; the *pericardium* covers the heart; the *peritoneum* lines the abdominal cavity and covers its organs. A double layer of peritoneum, called mesentery, supports the abdominal organs and attaches them to the abdominal wall. Peritonitis is a potentially life-threatening infection of the peritoneum that may occur if an inflamed appendix bursts before it is removed, or if the digestive tract is perforated for any other reason.

Synovial membranes, composed only of loose connective tissue, line freely movable joint cavities. They secrete synovial fluid that lubricates the cartilage at the ends of the bones so that they can move smoothly in the joint cavity.

The **meninges** are membranes within the dorsal cavity. They are composed only of connective tissue and serve as a protective covering for the brain and spinal cord.

Check Your Progress 11.2

1. Identify the two cavities that are separated by the diaphragm.
2. Describe the function of the fluids produced by various body membranes.

11.3 Organ Systems

Learning Outcomes

Upon completion of this section, you should be able to

1. List the major organs that make up each organ system.
2. Describe the general function(s) of each organ system.

Figure 11.7 illustrates the organ systems of the human body. Just as organs work together in an organ system, organ systems work together in the body. Therefore, while a particular organ may be assigned to one system, it may assist in the function of other organ systems.

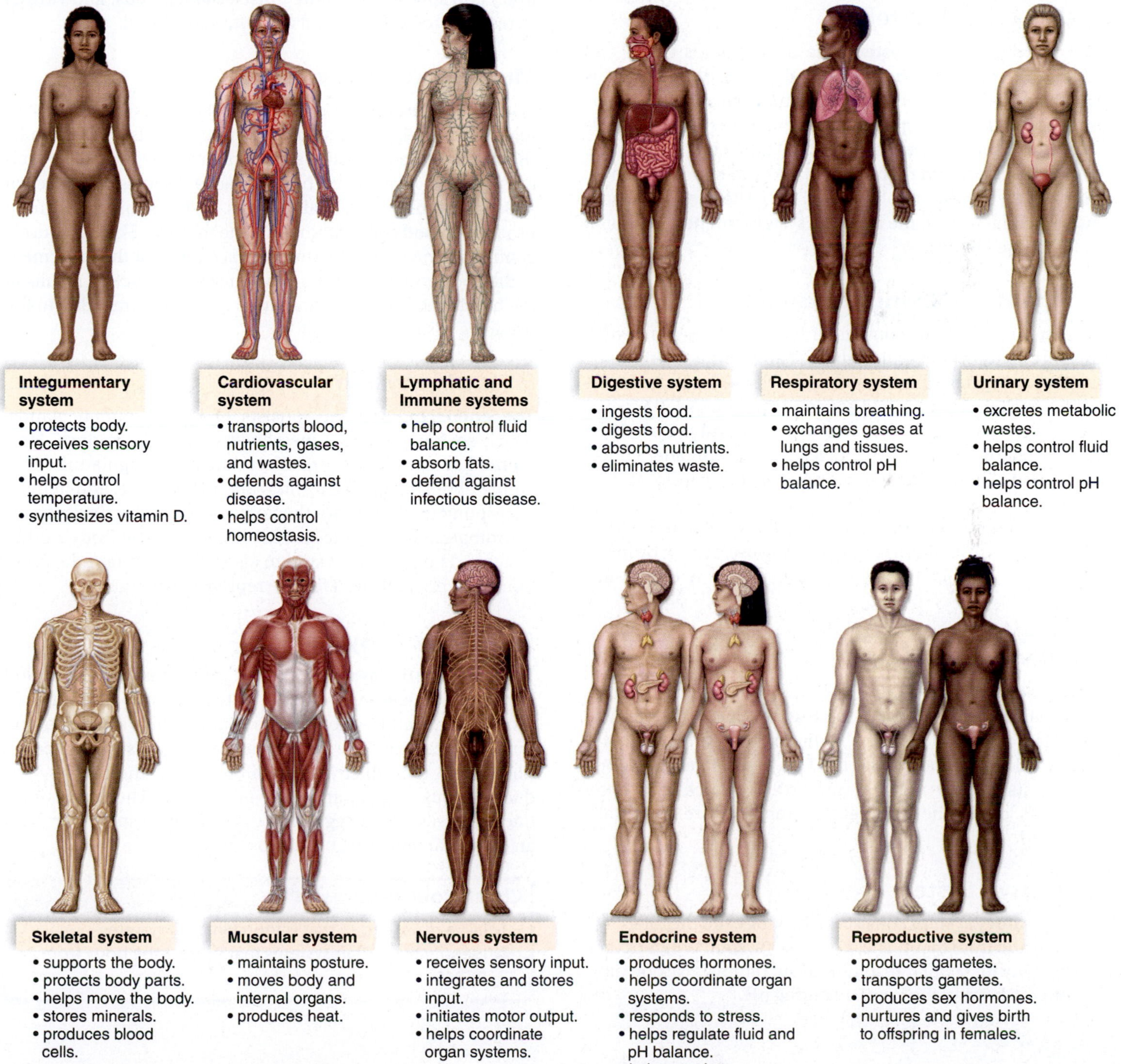
Integumentary system
• protects body.
• receives sensory input.
• helps control temperature.
• synthesizes vitamin D.
Cardiovascular system
• transports blood, nutrients, gases, and wastes.
• defends against disease.
• helps control homeostasis.
Lymphatic and Immune systems
• help control fluid balance.
• absorb fats.
• defend against infectious disease.
Digestive system
• ingests food.
• digests food.
• absorbs nutrients.
• eliminates waste.
Respiratory system
• maintains breathing.
• exchanges gases at lungs and tissues.
• helps control pH balance.
Urinary system
• excretes metabolic wastes.
• helps control fluid balance.
• helps control pH balance.
Skeletal system
• supports the body.
• protects body parts.
• helps move the body.
• stores minerals.
• produces blood cells.
Muscular system
• maintains posture.
• moves body and internal organs.
• produces heat.
Nervous system
• receives sensory input.
• integrates and stores input.
• initiates motor output.
• helps coordinate organ systems.
Endocrine system
• produces hormones.
• helps coordinate organ systems.
• responds to stress.
• helps regulate fluid and pH balance.
• helps regulate metabolism.
Reproductive system
• produces gametes.
• transports gametes.
• produces sex hormones.
• nurtures and gives birth to offspring in females.

Figure 11.7 Organ systems of the body.

Integumentary System

The **integumentary system** contains skin, which is made up of two main types of tissue: the epidermis is composed of stratified squamous epithelium, and the dermis is composed of fibrous connective tissue. The integumentary system also includes nails, located at the ends of the fingers and toes (collectively called the digits), hairs, muscles that move hairs, the oil and sweat glands, blood vessels, and nerves leading to sensory receptors. Besides having a protective function, skin also synthesizes vitamin D, collects sensory data, and helps regulate body temperature.

Cardiovascular System

In the **cardiovascular system,** the heart pumps blood and sends it under pressure into the blood vessels. While blood is moving throughout the body, it distributes heat produced by the muscles. Blood transports nutrients and oxygen to the cells and removes their waste molecules, including carbon dioxide. Despite the movement of molecules into and out of the blood, it has a fairly constant volume and pH. The blood is also a route by which cells of the immune system can be distributed throughout the body.

Lymphatic and Immune Systems

The **lymphatic system** consists of lymphatic vessels (which transport lymph), lymph nodes, and other lymphatic (lymphoid) organs. This system protects the body from disease by purifying lymph and storing lymphocytes, the white blood cells responsible for adaptive immunity. Lymphatic vessels absorb fat from the digestive system and collect excess tissue fluid, which is returned to the cardiovascular system.

The **immune system** consists of all the cells in the body that protect us from disease, especially those caused by infectious agents. The lymphocytes, in particular, belong to this system.

Digestive System

The **digestive system** includes the mouth, esophagus, stomach, small intestine, and large intestine (colon), along with associated organs such as the teeth, tongue, salivary glands, liver, gallbladder, and pancreas. This system receives food and digests it into nutrient molecules, which can enter the body's cells. The nondigested remains are eventually eliminated.

Respiratory System

The **respiratory system** consists of the lungs and the tubes that take air to and from them. The respiratory system brings oxygen into the body for cellular respiration and removes carbon dioxide from the body at the lungs, restoring pH.

Urinary System

The **urinary system** contains the kidneys, the urinary bladder, and the tubes that carry urine. This system rids the body of metabolic wastes, and helps regulate the fluid balance and pH of the blood.

Nervous System

The **nervous system** consists of the brain, spinal cord, and associated nerves. The nerves conduct nerve impulses from sensory receptors to the brain and spinal cord where integration occurs. Nerves also conduct nerve impulses from the brain and spinal cord to the muscles and glands, allowing us to respond to both external and internal stimuli.

Musculoskeletal System

Within the **musculoskeletal system,** the bones provide a scaffolding that helps hold and protect body parts. For example, the skull forms a protective encasement for the brain. The skeleton also helps move the body because it serves as a place of attachment for the skeletal muscles. It stores minerals, and it produces blood cells within the bone marrow. Skeletal muscle contraction maintains posture and accounts for the movement of the body and its parts. Cardiac muscle contraction results in the heartbeat. The walls of internal organs contract due to the presence of smooth muscle.

Endocrine System

The **endocrine system** consists of the hormonal glands, which secrete chemical messengers called hormones. Hormones have a wide range of effects, including regulating cellular metabolism, regulating fluid and pH balance, and helping us respond to stress. Both the nervous and endocrine systems coordinate and regulate the functioning of the body's other systems. The endocrine system also helps maintain the functioning of the male and female reproductive organs.

Reproductive System

The **reproductive system** has different organs in the male and female. The male reproductive system consists of the testes, other glands, and various ducts that conduct semen to and through the penis. The testes produce sex cells called sperm. The female reproductive system consists of the ovaries, oviducts, uterus, vagina, and external genitals. The ovaries produce sex cells called eggs, or oocytes. When a sperm fertilizes an oocyte, an offspring begins to develop.

Check Your Progress 11.3

1. List a major organ that is found in each system.
2. Identify two organ systems that protect the body from disease.

wrinkling.) The dermis also contains blood vessels that nourish the skin. When blood rushes into these vessels, a person blushes, and when there is minimal blood in them, the skin turns "blue."

Sensory receptors are specialized free nerve endings in the dermis that respond to external stimuli. There are sensory receptors for touch, pressure, pain, and temperature. The fingertips contain the most touch receptors, and these add to our ability to use our fingers for delicate tasks.

The *subcutaneous layer* is composed of loose connective tissue and adipose tissue, which stores fat. Fat is a stored source of energy in the body. Adipose tissue also helps thermally insulate the body from either gaining heat from the outside or losing heat from the inside.

MP3 Human Skin

Accessory Organs of the Skin

Nails, hair, and glands are structures of epidermal origin, even though some parts of hair and glands are largely found in the dermis.

Nails are a protective covering of the distal part of the digits. Nails can help pry things open or pick up small objects. Nails grow from special epithelial cells at the base of the nail in the portion called the nail root. These cells become keratinized as they grow out over the nail bed. The visible portion of the nail is called the nail body. The cuticle is a fold of skin that hides the nail root. The whitish color of the half-moon-shaped base, or lunula, results from the thick layer of cells in this area (Fig. 11.9).

Hair follicles are in the dermis and continue through the epidermis where the hair shaft extends beyond the skin. Epidermal cells form the root of hair, and their division causes a hair to grow. As the cells become keratinized and die, they are pushed farther from the root. Interestingly, chemical substances in the body such as illicit drugs and by-products of alcohol metabolism are incorporated into growing hair shafts, where they can be detected by laboratory tests (see the Bioethical feature, "Just a Snip, Please: Testing Hair for Drugs," on page 202).

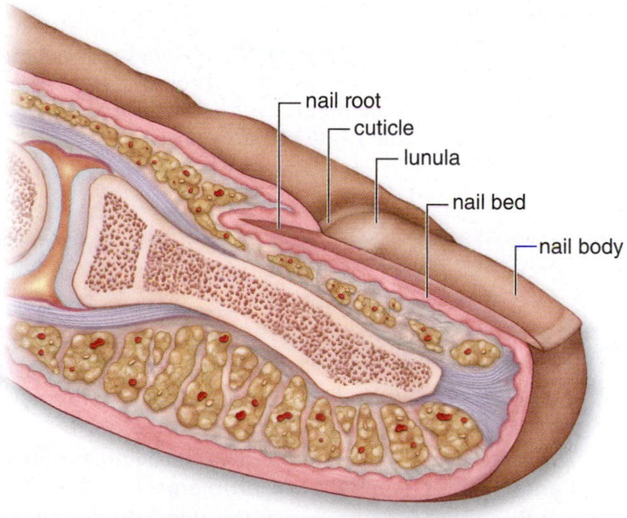
nail root
cuticle
lunula
nail bed
nail body

Figure 11.9 Nail anatomy. Cells produced by the nail root become keratinized, forming the nail body.

When we are scared or cold, contraction of the arrector pili muscles attached to hair follicles may cause the hairs to "stand on end" and goose bumps to develop. The purpose of this phenomenon in humans is unknown, although it may help fur-covered mammals to look larger, as well as to keep warm by trapping air within the fur. The color of hair is due to the variable presence of different forms of melanin. In general, the more melanin, the darker the hair. A decreasing melanin content with age results in the various shades of gray hair. White hair contains little or no melanin. Hair loss is most often age-related, but also may occur following many types of illnesses.

Each hair follicle has one or more **oil glands** (sebaceous glands), which secrete *sebum,* an oily substance that lubricates the hair within the follicle and the skin itself. With the exception of the palms of the hands and soles of the feet, all areas of human skin have oil glands.

The skin of an average adult also has about three thousand **sweat glands** per square inch. A sweat gland is a coiled tubule within the dermis that straightens out near its opening, or pore. Some sweat glands open into hair follicles, but most open onto the surface of the skin. Sweat glands play a role in modifying body temperature. When body temperature starts to rise, sweat glands become active. Sweat absorbs body heat as it evaporates. Once body temperature lowers, sweat glands are no longer active.

Disorders of the Skin

The integumentary system is susceptible to a number of diseases. Perhaps more than animals with skin covered by tough scales or a thick fur, human skin can be easily traumatized. It is also prone to certain infections, although its dryness, mildly acidic pH, and the presence of dead cells on its outermost layers renders skin resistant to many pathogens. Several types of cancer can arise from the skin, often secondary to damage from UV rays (see the Health feature, "UV Rays: Too Much Exposure or Too Little?"). Allergic reactions and various irritating chemicals can cause *dermatitis,* or inflammation of the skin.

Acne is a common skin disorder that usually first appears during adolescence. It occurs mostly on the face, shoulders, chest, and back—areas where the skin has the highest density of sebaceous glands. Around the time of puberty, increasing levels of certain hormones cause an increased production of sebum. This, in turn, can lead to a blockage of the exit of the sebum from the gland, and an increase in the growth of *Propionibacterium acnes,* a bacterium that is almost universally found on human skin. This causes an inflammatory response, commonly known as a pimple. In many cases white blood cells called neutrophils will migrate into the area, forming pus (see Chapter 13), and sometimes deeper tissues become involved, which increases the possibility of permanent scarring.

Most medications for acne work by drying out and unclogging blocked pores, killing the bacteria, and/or reducing inflammation. Over-the-counter medications can be useful for mild acne, but stronger, prescription medications may be needed for more severe cases. In 2005, the FDA approved Zeno, a hand-held medical device that kills *P. acnes* by heating the skin to 120 degrees for 2.5 minutes. Other available treatments rely on lasers or light to kill the bacteria and/or reduce the production of sebum.

SCIENCE IN YOUR LIFE ▶ **BIOETHICAL**

Just a Snip, Please: Testing Hair for Drugs

Imagine you have just started your first real full-time job, with health insurance, a pension plan, and other benefits. The first day, after you settle into your cubicle, a manager comes by and asks for a sample of your hair. "We normally just snip a bit off in the back, where it will hardly show," he says, smiling. When you ask why, he replies that it is company policy to screen employees for drug use every six months.

Tests that detect illicit drugs or their metabolic products in urine or saliva are of limited value, because most of these chemicals are secreted by the body for only a few days. In contrast, many illicit drugs and/or their metabolites are incorporated into growing hair shafts throughout the body, where they remain indefinitely. Especially in the first 1.5 inches of hair growth from the scalp, each 0.5 inch is considered to represent 30 days' worth of growth (and thus, potential drug use). Hair from anywhere on the body can be used, although the growth of body hair is usually slower, so the time of any drug use cannot be determined. It generally takes four to five days from the time a drug is taken into the body until it begins to appear in hair.

Many commercial laboratories now offer hair testing (Fig. 11B). In general, these labs contend that they can distinguish between environmental exposure to a particular drug—from being in the vicinity of someone smoking marijuana, for example—and actual drug use by an individual. In recent years, several court decisions have supported the idea that hair testing can accurately distinguish actual drug use from such "passive" exposure. This has led to the marketing of several shampoos for cleaning or "detoxifying" the hair shafts, but these may not be effective. And even if they were, a lab could conceivably test hair for common contaminants expected to be found in everyone's hair. If these contaminants were not found, this could be used as evidence that a person has attempted to hide prior drug usage.

Even if one accepts that hair testing is an accurate way to prove that a person has used drugs, is it ethical for businesses to require their employees to undergo such tests? Does it matter what type of job a person has? For example, would it be more difficult to argue against mandatory drug testing for school bus drivers than for stockbrokers? And for those of you still living with your parents, how upset would you be to find out that your mom or dad had slipped into your bedroom at night, snipped off a small bit of hair from the back of your head, and mailed it to one of several companies that now offer hair testing to the general public?

Questions to Consider

1. The Fourth Amendment to the U.S. Constitution guards against "unreasonable searches and seizures" by the government. Do you believe that the types of drug testing described here are unconstitutional?
2. In which of the following additional situations would you support mandatory drug testing: To test airline pilots for hallucinogens? To test high school athletes for steroids? To test NBA players for marijuana?
3. Could you imagine any circumstance where you might want to test your own children for drug use?

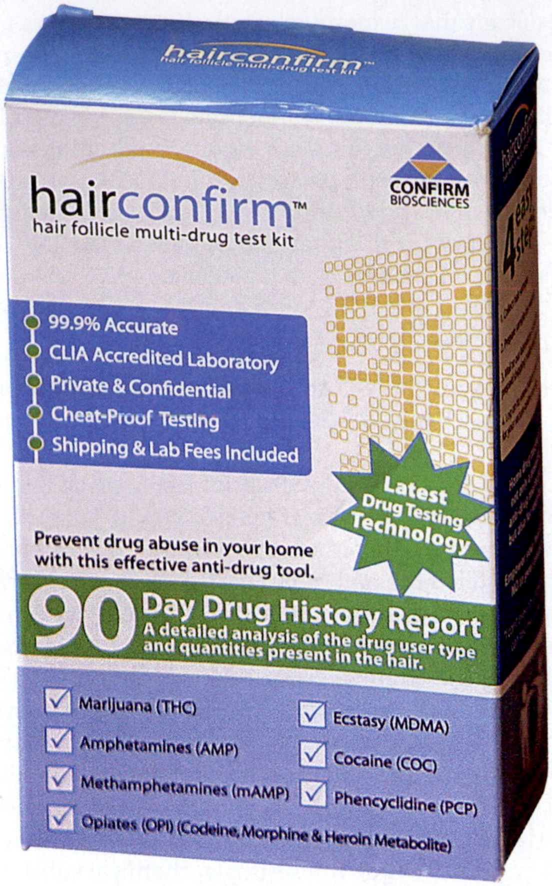
hairconfirm™
hair follicle multi-drug test kit
CONFIRM BIOSCIENCES
99.9% Accurate
CLIA Accredited Laboratory
Private & Confidential
Cheat-Proof Testing
Shipping & Lab Fees Included
Latest Drug Testing Technology
Prevent drug abuse in your home with this effective anti-drug tool.
90 Day Drug History Report
A detailed analysis of the drug user type and quantities present in the hair.
Marijuana (THC)
Ecstasy (MDMA)
Amphetamines (AMP)
Cocaine (COC)
Methamphetamines (mAMP)
Phencyclidine (PCP)
Opiates (OPI) (Codeine, Morphine & Heroin Metabolite)

Figure 11B An invasion of privacy? Kits for testing hair for illicit or prescription drugs can be purchased at retail stores or online.

connect BIOLOGY Explore the concepts through a variety of multimedia assets, question types, and data interpretation.
www.mcgrawhillconnect.com

Warts are small areas of skin proliferation caused by the human papillomavirus. Though they can occur at any age, nongenital warts most commonly occur between the ages of 12 and 16. Though sometimes embarrassing, these are generally harmless and disappear without treatment—more than half do so within two years. Warts that cause cosmetic disfigurement or are painful (such as plantar warts on the foot) can be surgically removed, frozen with liquid nitrogen, or treated with various pharmaceutical compounds. One example is cantharidin, an extract from the blister beetle. Genital warts are discussed in section 21.5.

Check Your Progress 11.4

1. Compare the structure of the epidermal and dermal layers of the skin.
2. Discuss why a dark-skinned individual living in northern Canada might develop bone problems.
3. List three accessory organs of the skin, and describe the major function(s) of each.

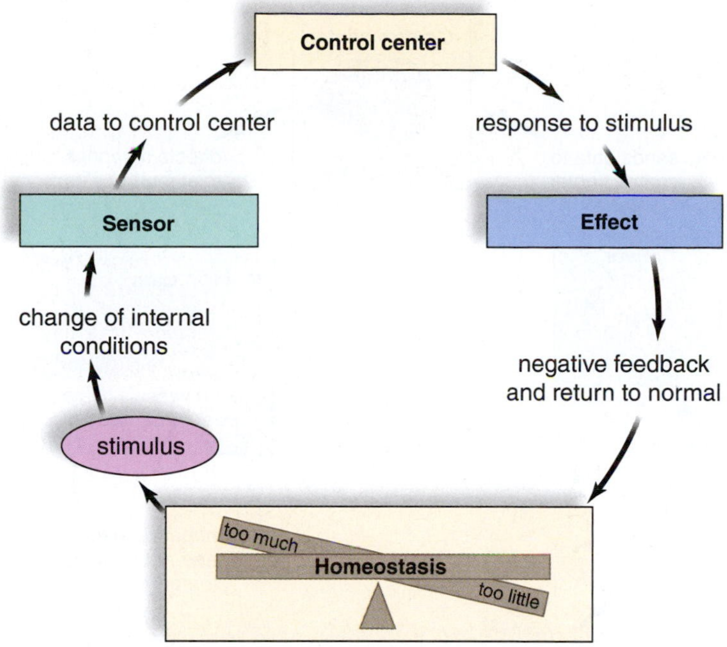
Control center
data to control center
response to stimulus
Sensor
Effect
change of internal conditions
negative feedback and return to normal
stimulus
too much
Homeostasis
too little

Figure 11.10 Negative feedback mechanism. The sensor and control center of a feedback mechanism work together to keep a variable close to a particular value.

11.5 Homeostasis

Learning Outcomes

Upon completion of this section, you should be able to

1. Define homeostasis, and describe why it is essential to living organisms.
2. Differentiate between positive and negative feedback mechanisms, and list specific examples of each.
3. Discuss the roles of specific organ systems in homeostasis, and the health consequences if homeostasis is disrupted.

Homeostasis is the maintenance of a relatively constant internal environment by an organism, or even by a single cell. Even though external conditions may change dramatically, internal conditions stay within a narrow range. For example, regardless of how cold or hot it gets, the temperature of the body stays around 98.6°F (37°C). Even if you consume an acidic food such as yogurt (with a pH around 4.5), the pH of your blood is usually about 7.4, and even if you eat a candy bar, the amount of sugar in your blood stays between 0.05% and 0.08%.

It is important to realize that internal conditions are not absolutely constant. They tend to fluctuate above and below a particular value. Therefore, the internal state of the body is often described as one of *dynamic* equilibrium. If internal conditions change to any great degree, illness results. This makes the study of homeostatic mechanisms medically important.

MP3 Homeostasis

Negative Feedback

Negative feedback is the primary homeostatic mechanism that keeps a variable close to a particular value, or set point. A homeostatic mechanism has at least two components: a sensor and a control center (Fig. 11.10). The sensor detects a change in internal conditions. The control center then directs a response that brings conditions back to normal again. Now, the sensor is no longer activated. In other words, a negative feedback mechanism is present when the output of the system dampens the original stimulus.

Let's take a simple example. When the pancreas detects that the blood glucose level is too high, it secretes insulin, a hormone that causes cells to take up glucose. Now the blood sugar level returns to normal, and the pancreas is no longer stimulated to secrete insulin.

Animation Feedback Mechanisms

Mechanical Example

A home heating system is often used to illustrate how a more complicated negative feedback mechanism works (Fig. 11.11).

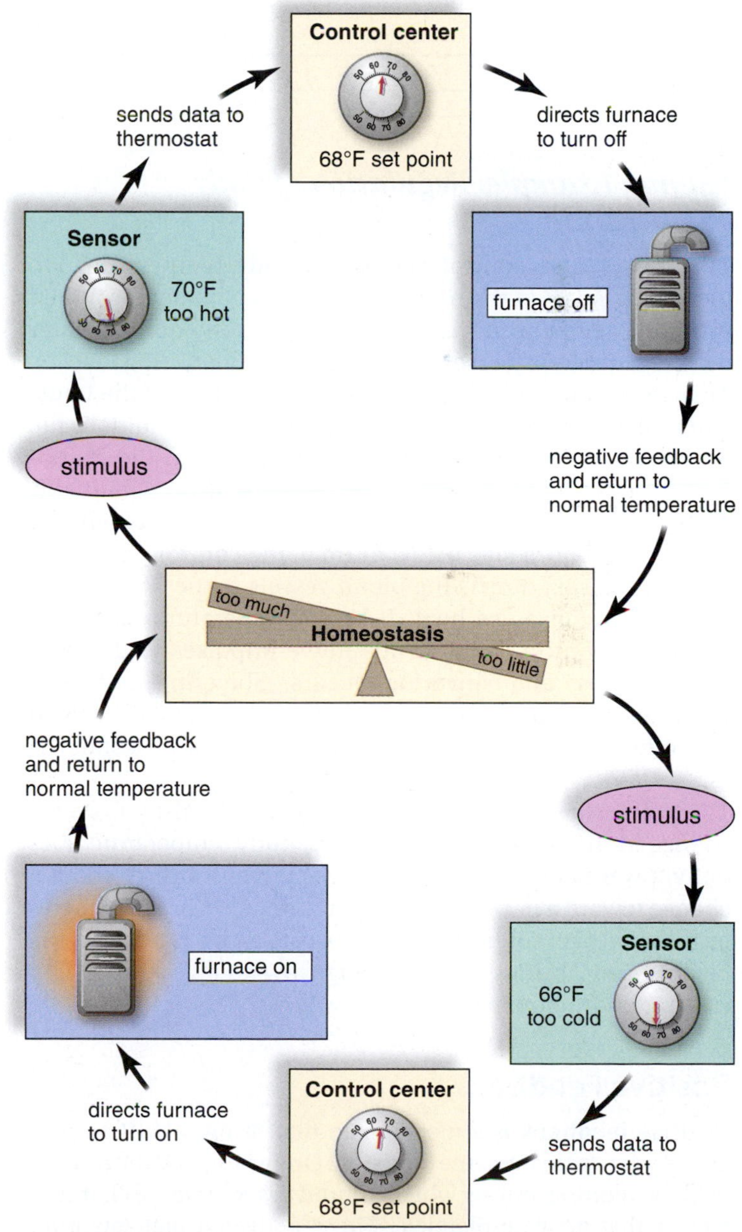
Control center
68°F set point
sends data to thermostat
directs furnace to turn off
Sensor
70°F too hot
furnace off
stimulus
negative feedback and return to normal temperature
too much
Homeostasis
too little
negative feedback and return to normal temperature
stimulus
furnace on
Sensor
66°F too cold
directs furnace to turn on
Control center
68°F set point
sends data to thermostat

Figure 11.11 Complex negative feedback mechanism. When a room becomes too warm, negative feedback allows the temperature to return to normal. A contrary cycle, in which the furnace turns on and gives off heat, returns the room temperature to normal when the room becomes too cool.

You set the thermostat at, say, 68°F. This is the *set point.* The thermostat contains a thermometer, a sensor that detects when the room temperature is above or below the set point. The thermostat also contains a control center. It turns the furnace off when the room is too hot and turns it on when the room is too cold. When the furnace is off, the room cools, and when the furnace is on, the room warms. In other words, typical of negative feedback mechanisms, there is a fluctuation above and below normal.

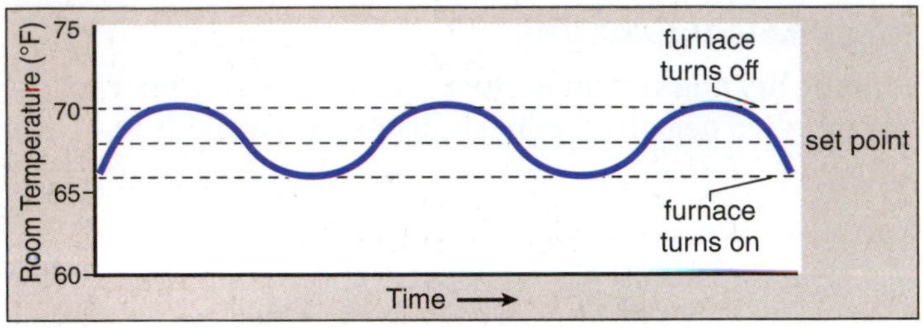
Room Temperature (°F)
75
70
65
60
furnace turns off
set point
furnace turns on
Time

Human Example: Regulation of Body Temperature

The sensor and control center for body temperature are located in a part of the brain called the hypothalamus. When the body temperature is above normal, the control center directs the blood vessels of the skin to dilate (Fig. 11.12). This allows more blood to flow near the surface of the body, where heat can be lost to the environment. In addition, the nervous system activates the sweat glands, and the evaporation of sweat helps lower body temperature. Gradually, body temperature decreases to 98.6°F (37°C). When the body temperature falls below normal, the control center (via nerve impulses) directs the blood vessels of the skin to constrict. This conserves heat. If body temperature falls even lower, the control center sends nerve impulses to the skeletal muscles, and shivering occurs. Shivering generates heat, and gradually body temperature rises to 98.6°F. When the temperature rises to normal, the control center is inactivated.

Notice that a negative feedback mechanism prevents change in the same direction; that is, body temperature does not get warmer and warmer because warmth brings about a change toward a lower body temperature. Also, body temperature does not get colder and colder because a body temperature below normal brings about a change toward a warmer body temperature.

MP3 Temperature Regulation

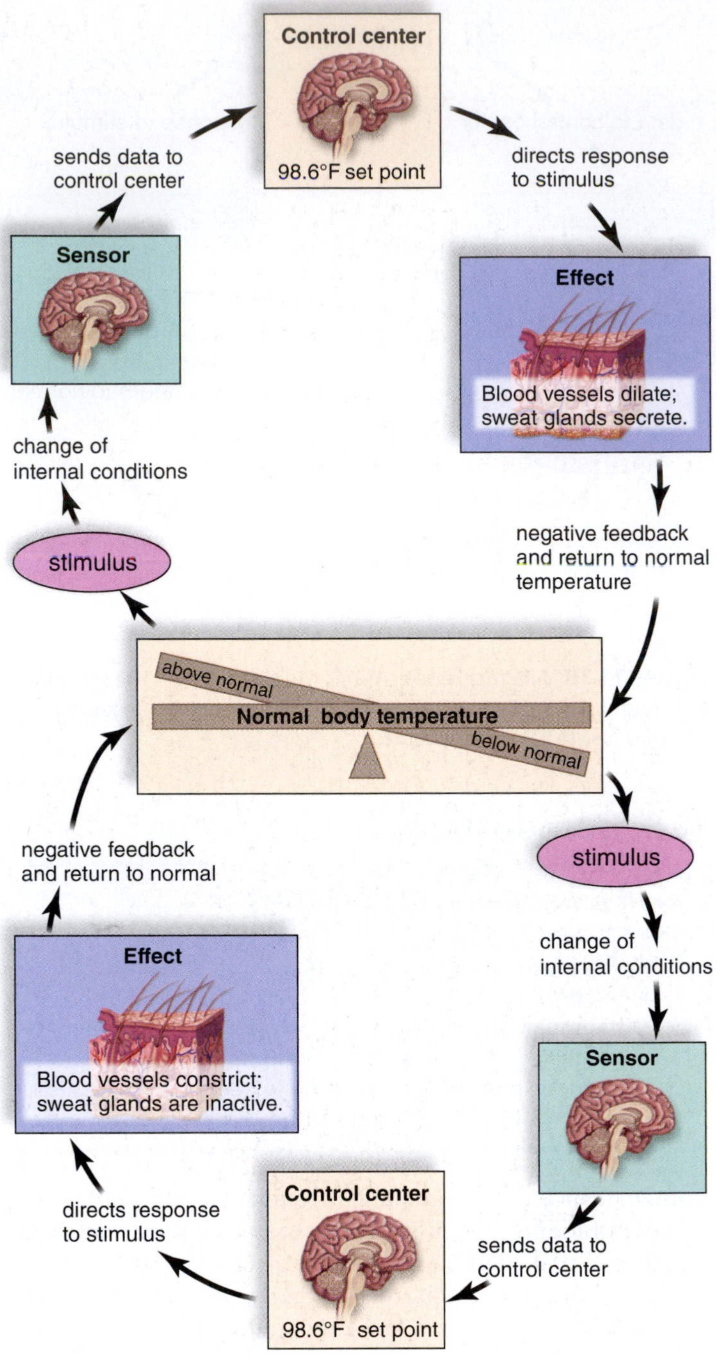
Control center
98.6°F set point
sends data to control center
directs response to stimulus
Sensor
Effect
Blood vessels dilate; sweat glands secrete.
change of internal conditions
stimulus
negative feedback and return to normal temperature
above normal
Normal body temperature
below normal
negative feedback and return to normal
stimulus
change of internal conditions
Effect
Blood vessels constrict; sweat glands are inactive.
Sensor
Control center
directs response to stimulus
sends data to control center
98.6°F set point

Figure 11.12 Regulation of body temperature. Normal body temperature is maintained by a negative feedback system.

Positive Feedback

Positive feedback is a mechanism that brings about an ever greater change in the same direction. One example is the process of blood clotting, during which injured tissues release chemical factors that activate platelets. These activated platelets initiate the clotting process, and also release factors that stimulate further clotting (see section 12.2).

Positive feedback loops tend to be involved in processes that have a definite cutoff point. Consider that when a woman is giving birth, the head of the baby begins to press against the cervix, stimulating sensory receptors there. When nerve impulses reach the brain, the brain causes the pituitary gland to secrete the hormone oxytocin. Oxytocin travels in the blood and causes the uterus to contract. As labor continues, the cervix is ever more stimulated, and uterine contractions become ever stronger until birth occurs.

Homeostasis and Body Systems

All systems of the body contribute toward maintaining homeostasis.

The Transport Systems

The cardiovascular system conducts blood to and away from capillaries, where exchange of gases, nutrients, and wastes occurs. Tissue fluid, which bathes all the cells of the body, is refreshed when molecules such as oxygen and nutrients move into tissue fluid from the blood and when carbon dioxide and wastes move from tissue fluid into the blood (Fig. 11.13).

The lymphatic system is an accessory to the cardiovascular system. Lymphatic capillaries collect excess tissue fluid, which is returned via lymphatic vessels to the cardiovascular system. Lymph nodes are sites where the immune system responds to invading microorganisms.

The Maintenance Systems

The respiratory system adds oxygen to and removes carbon dioxide from the blood. It also plays a role in regulating blood pH because removal of CO_2 causes the pH to rise, just as CO_2 retention helps to lower the pH. The digestive system takes in and digests food, providing nutrient molecules that enter the blood to replace the nutrients that are constantly being used by the body cells. The liver, an organ that assists the digestive process by producing bile, also plays a significant role in regulating blood composition. Immediately after glucose enters the blood, any excess is removed by the liver and stored as glycogen. Later, the glycogen can be broken down to replace the glucose used by the body cells. In this way, the glucose composition of the blood remains relatively constant. The liver also removes toxic chemicals, such as ingested alcohol and other drugs, from the blood. The liver makes urea, a nitrogenous end product of protein metabolism. Urea and other metabolic waste molecules are excreted by the kidneys, which are a part of the urinary system. Urine formation by the kidneys is extremely critical to the body, not only because it rids the body of unwanted substances, but also because urine formation offers an opportunity to carefully regulate blood volume, salt balance, and pH.

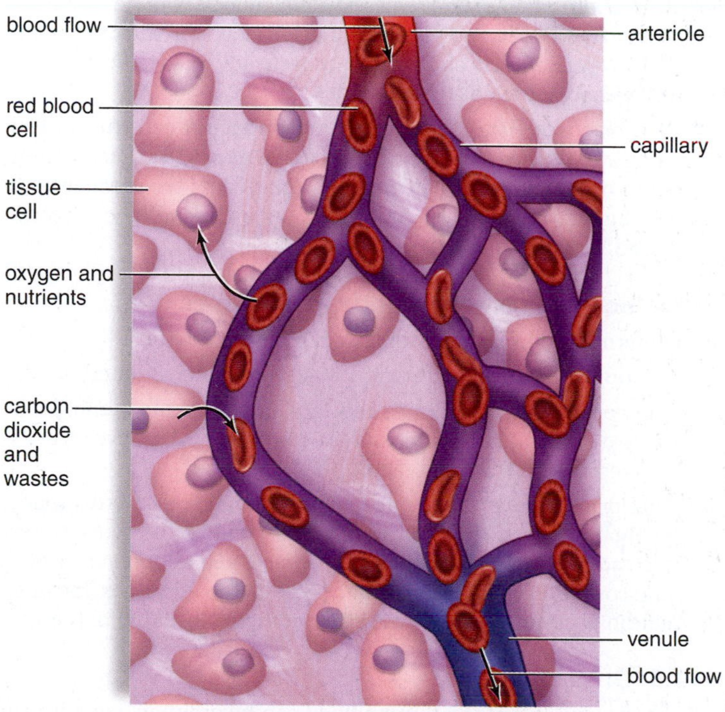
blood flow
arteriole
red blood cell
capillary
tissue cell
oxygen and nutrients
carbon dioxide and wastes
venule
blood flow

Figure 11.13 Regulation of tissue fluid composition. Cells are surrounded by tissue fluid, which is continually refreshed because oxygen and nutrient molecules constantly exit, and carbon dioxide and waste molecules continually enter the bloodstream.

The Support Systems

The integumentary and musculoskeletal systems protect the internal organs. In addition, the integumentary system produces vitamin D, while the skeleton stores minerals and produces the blood cells.

The Control Systems

The nervous system and the endocrine system work together to control other body systems so that homeostasis is maintained. We have already seen that in negative feedback mechanisms, sensory receptors send nerve impulses to control centers in the brain, which then rapidly direct effectors to become active. Effectors can be muscles or glands. Muscles bring about an immediate change. Endocrine glands secrete hormones that bring about a slower, more lasting change that keeps the internal environment relatively stable.

Disease

A disease is an abnormality in the body's normal processes that significantly impairs homeostasis. As will be seen in the remaining chapters in this part of the book, diseases (disorders) can affect virtually every part of the human body. The major causes of human diseases include blood vessel problems, cancers, infections, and inflammatory conditions.

A particular disease may be described as *systemic,* meaning that it affects the entire body or at least several organ systems. Other diseases, including many infections or inflammatory diseases such as dermatitis or arthritis, are more *localized* to a specific part of the body. Diseases may also be classified on the basis of their severity and duration. *Acute* diseases, such as poison ivy dermatitis or influenza, tend to occur suddenly and generally last a short time, although some can be life threatening. *Chronic* diseases, such as multiple sclerosis, AIDS, and most cancers, tend to develop slowly and last a long time, even for the rest of a person's life, unless an effective cure is available.

The term *cancer* refers to a group of disorders in which the usual controls on cell division fail, resulting in the production of abnormal cells that invade and destroy healthy body tissue. Cancers are classified according to the type of tissue from which they arise. *Carcinomas,* the most common type, are cancers of epithelial tissue. The Health feature,

"UV Rays: Too Much Exposure or Too Little," on page 200 describes one type of carcinoma, the melanoma. *Sarcomas* are cancers arising in muscle or connective tissue, especially bone or cartilage. *Leukemias* are cancers of the blood cells, and *lymphomas* are cancers that originate in lymph nodes. The chance of developing cancer in a particular tissue shows a positive correlation to the rate of cell division. Epithelial cells reproduce at a high rate, and 2,500,000 new blood cells appear each second. Therefore, carcinomas and leukemias are common types of cancers.

Check Your Progress 11.5

1. Explain how positive feedback differs from negative feedback.
2. Describe how several body systems can interact to maintain homeostasis.
3. List several specific diseases that result when a particular body system fails to perform its function.

Case Study Conclusion

The human body consists of an incredibly complex system of molecules, cells, tissues, organs, and organ systems functioning together. Ultimately, the interaction of these components provides humans with the ability to achieve the highest levels of physical and mental functions, as well as maintaining general homeostasis. We are aware of only a few of these processes, such as the activity of our muscles when we are trying to learn a new skill or perform a challenging physical task. When these homeostatic processes fail, various types of diseases can result.

MEDIA STUDY TOOLS

www.mhhe.com/maderinquiry14

Enhance your study of this chapter with study tools and practice tests. Also ask your instructor about the resources available through ConnectPlus, including LearnSmart, the media-rich eBook, interactive learning tools, and animations.

SUMMARIZE

11.1 Types of Tissues

In the human body, most cells are organized into **tissues.** Human tissues are categorized into four groups:

- **Epithelial tissue** covers the body and lines its cavities. **Squamous epithelium, cuboidal epithelium,** and **columnar epithelium** can be simple, stratified, or pseudostratified and may also have cilia or microvilli. Epithelial cells sometimes form **glands** that secrete either into ducts or into the blood.
- **Connective tissue,** in which cells are separated by a matrix, often binds body parts together. In most cases, the matrix contains **collagen fibers, reticular fibers,** and **elastic fibers.**
    - **Loose fibrous connective tissue** supports epithelium and encloses organs. **Dense fibrous connective tissue,** such as that of tendons and ligaments, contains a closely packed matrix. **Cartilage,** a type of dense fibrous connective tissue in which cells are found in **lacunae,** can be of three types: **hyaline cartilage, elastic cartilage,** or **fibrocartilage.**
    - **Bone** has a hard matrix due to the presence of calcium salts. **Compact bone** is found in the shafts of long bones; **spongy bone** is found on the ends.
    - **Adipose tissue** stores fat, while **reticular connective tissue** forms a supportive mesh in some organs.
    - **Blood** is a connective tissue in which the matrix is a liquid. Blood consists of the liquid part, called **plasma,** and the formed elements, which include the **red blood cells, white blood cells,** and **platelets.**
- **Muscular tissue** is of three types. **Skeletal muscle** is found in muscles attached to bones, and **smooth muscle** is found in internal organs. Both skeletal muscle and **cardiac muscle** are striated. Both cardiac and smooth muscle are involuntary.
- **Nervous tissue** has one main type of conducting cell, the **neuron,** and several types of supporting cells, the **neuroglia.** Each neuron has a cell body, dendrites, and an axon. Axons are specialized to conduct nerve impulses. Bundles of axons bound by connective tissue are called **nerves.**

11.2 Body Cavities and Body Membranes

The internal organs occur within two main **body cavities.**

- The **dorsal cavity** includes the cranial cavity and vertebral canal.
- The **ventral cavity** contains the thoracic, abdominal and pelvic cavities, which house organs of the respiratory, digestive, urinary, and reproductive systems, among others.
- Four types of membranes line body cavities and the internal spaces of organs. **Mucous membranes** line the tubes of the digestive system; **serous membranes** line the thoracic and abdominal cavities and cover the organs they contain; **synovial membranes** line movable joint cavities; and **meninges** cover the brain and spinal cord.

11.3 Organ Systems

The organ systems of the human body can be grouped according to their major function(s):

- The **integumentary system** protects deeper tissues and also makes vitamin D, collects sensory data, and helps regulate body temperature.

- The **cardiovascular, lymphatic, digestive, respiratory,** and **urinary systems** perform processing and transport functions that maintain the normal conditions of the body.
- The **immune system** defends against infectious disease.
- The **musculoskeletal system** supports and protects the body and permits movement.
- The **nervous system** receives sensory input and directs muscles and glands to respond.
- The **endocrine system** produces hormones, which influence the functions of the other systems.
- The **reproductive system** produces sex cells (gametes), and ultimately, offspring.

11.4 Integumentary System

- The integumentary system is composed of **skin** and the accessory organs.
- The main functions of the integumentary system are to protect underlying tissues while providing a barrier to pathogen invasion and water loss. Skin also functions in the sense of touch.
- Human skin has two regions: the **epidermis** contains basal cells that produce new epithelial cells, which become keratinized as they move toward the surface; and (2) the **dermis,** contains glands, hair follicles, blood vessels, and sensory receptors.
- The accessory organs of skin include the **nails, hair follicles, oil glands** (sebaceous glands), and **sweat glands.**
- The skin is susceptible to several types of disorders, including infections, cancers, and various inflammatory conditions.

11.5 Homeostasis

- The body's internal environment consists of blood and tissue fluid.
- **Homeostasis** is the maintenance of a relatively constant internal environment, mainly by two mechanisms: (1) **negative feedback** mechanisms keep the environment relatively stable (when a sensor detects a change above or below a set point, a control center brings about an effect that reverses the change and brings conditions back to normal again); and (2) **positive feedback** mechanisms bring about rapid change in the same direction as the stimulus. **Disease** results when one or more body systems fail.

ASSESS

Testing Yourself

Choose the best answer for each question.

1. A grouping of similar cells that perform a specific function is called a
   a. sarcoma.
   b. membrane.
   c. tissue.
   d. None of these are correct.
2. Tight junctions are associated with
   a. connective tissue.
   b. adipose tissue.
   c. cartilage.
   d. epithelium.
3. A reduction in red blood cells would cause problems with
   a. fighting infection.
   b. carrying oxygen.
   c. blood clotting.
   d. None of these are correct.
4. Which choice is true of both cardiac and skeletal muscle?
   a. striated
   b. single nucleus per cell
   c. multinucleated cells
   d. involuntary control
5. The skeletal system functions in
   a. blood cell production.
   b. mineral storage.
   c. movement.
   d. All of these are correct.
6. Which system plays the biggest role in fluid balance?
   a. cardiovascular
   b. urinary
   c. digestive
   d. integumentary
7. Label the diagram of body cavities below.

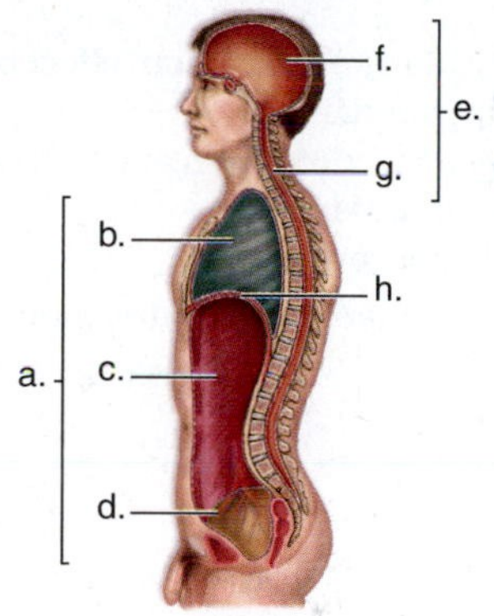

f.
e.
g.
b.
h.
a.
c.
d.

8. Which of the following is a function of skin?
   a. temperature regulation
   b. manufacture of vitamin D
   c. collection of sensory input
   d. protection from invading pathogens
   e. All of these are correct.
9. Which of these is involved in storing energy?
   a. epidermis
   b. dermis
   c. subcutaneous layer
   d. None of these are correct.
10. Which of these correctly describes a layer of the skin?
   a. The epidermis is simple squamous epithelium in which hair follicles develop and blood vessels expand when we are hot.
   b. The subcutaneous layer lies between the epidermis and the dermis. It contains adipose tissue, which keeps us warm.
   c. The dermis is a region of connective tissue that contains sensory receptors, nerve endings, and blood vessels.
   d. The skin has a special layer, still unnamed, in which there are all the accessory structures such as nails, hair, and various glands.

11. Without melanocytes, skin would
- **a.** be too thin.
- **b.** lack nerves.
- **c.** lack color.
- **d.** None of these are correct.

12. Label the diagram of human skin below.

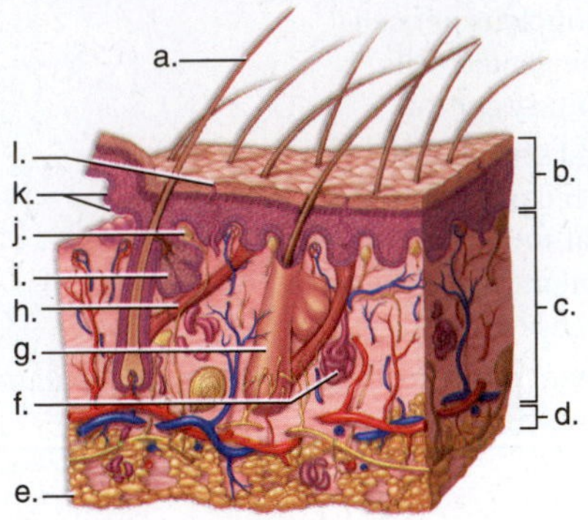
a.
b.
c.
d.
e.
f.
g.
h.
i.
j.
k.
l.

13. The correct order for homeostatic processing is
- **a.** sensory detection, control center, effect brings about change.
- **b.** control center, sensory detection, effect brings about change in environment.
- **c.** sensory detection, control center, effect causes no change in environment.
- **d.** None of these are correct.

14. Which allows rapid change in one direction and does not achieve stability?
- **a.** homeostasis
- **b.** positive feedback
- **c.** negative feedback
- **d.** All of these are correct.

15. Which of the following is an example of negative feedback?
- **a.** Air conditioning goes off when room temperature lowers.
- **b.** Insulin decreases blood sugar levels after eating a meal.
- **c.** Heart rate increases when blood pressure drops.
- **d.** All of these are examples of negative feedback.

ENGAGE

Thinking Critically

1. In what way(s) is blood like a tissue?

2. Which of these homeostatic mechanisms in the body are examples of positive feedback, and which are examples of negative feedback? Why?
- **a.** The adrenal glands produce epinephrine in response to a hormone produced by the pituitary gland in times of stress; the pituitary gland senses the epinephrine in the blood and stops producing the hormone.
- **b.** As the bladder fills with urine, pressure sensors send messages to the brain with increasing frequency signaling that the bladder must be emptied. The more the bladder fills, the more messages are sent.
- **c.** When you drink an excess of water, specialized cells in your brain as well as stretch receptors in your heart detect the increase in blood volume. Both signals are transmitted to the kidneys, which increase the production of urine.

3. Homeostatic systems in the body can be categorized as transport systems, maintenance systems, support systems, and control systems. What organ systems (from section 11.3) could be placed into more than one of these categories?

Cardiovascular System 12

CHAPTER OUTLINE

BEFORE YOU BEGIN

Before beginning this chapter, take a few moments to review the following discussions:

Section 4.2 What molecules must diffuse across the plasma membrane of cells lining capillaries?

Figure 11.4 How does the structure of cardiac muscle differ from skeletal and smooth muscle?

Section 11.5 The cardiovascular system interacts with which other body systems?

CASE STUDY The Deadliest Disease

On June 13, 2008, Tim Russert, the longtime host of NBC's *Meet The Press* TV program, collapsed in his office. Co-workers called 911, and paramedics arrived four minutes later. After determining that 58-year-old Russert's heart wasn't beating, they began CPR and attempted to defibrillate his heart three times, with no success. Less than an hour later, the much-respected television personality and journalist was pronounced dead at a local hospital.

The cardiovascular system, which includes the heart and blood vessels, transports oxygen, nutrients, and wastes to and from the tissues. Diseases affecting this system, such as heart attack and stroke, are a major cause of death in the more developed countries of the world. A person currently living in the United States has about a one-in-three chance of dying of heart disease, and if you add in all the other conditions that can affect the blood vessels, your odds are greater than 50:50 of eventually developing some type of cardiovascular disease (CVD).

Fortunately, there are steps that we can take to reduce the risk of CVD. Eating a heathy diet, exercising regularly, and avoiding smoking are all effective approaches to improving heart health. New technologies are also being developed to treat existing CVD, as discussed in this chapter. The ideal approach, however, is to do all you can to prevent CVD from affecting yourself or your loved ones. Perhaps the first step is to understand why the cardiovascular system is so essential to life.

As you read through the chapter, think about the following questions:

1. When the heart stops functioning, what is the actual cause of death?
2. Of the many functions of blood, which are the most essential to life?
3. What are the major factors that lead to cardiovascular disease, and which can be prevented?

12.1 The Blood Vessels

Learning Outcomes

Upon completion of this section, you should be able to

1. List the three main types of blood vessels, their structural features, and their major functions.
2. Identify the type of blood vessel in which exchange takes place. Explain what is being "exchanged" and why.

The **cardiovascular system** has three types of blood vessels: the **arteries** (and arterioles), which carry blood away from the heart to the capillaries; the **capillaries,** which permit exchange of material with the tissues; and the **veins** (and venules), which return blood from the capillaries to the heart. All three vessel types have an inner *endothelium,* a simple squamous epithelium attached to a connective tissue basement membrane that contains elastic fibers.

MP3 Classification of Blood Vessels

Like other tissues, the blood vessels require oxygen and nutrients, and therefore the larger ones have blood vessels in their own walls.

The Arteries

An arterial wall has three layers (Fig. 12.1*a*). As mentioned, the innermost layer is endothelium. The middle layer is the thickest layer and consists of smooth muscle that can contract to regulate blood flow and blood pressure. The outer layer is fibrous connective tissue near the middle layer, but it becomes loose connective tissue at its periphery. The largest artery in the human body is the **aorta.** It is approximately 25 mm wide and carries O_2-rich blood from the heart to other parts of the body.

Smaller arteries branch off from the aorta, eventually forming a large number of arterioles. **Arterioles** are small arteries just visible to the naked eye, averaging about 0.5 mm in diameter. The inner layer of arterioles is endothelium. The middle layer is composed of some elastic tissue but mostly of smooth muscle with fibers that encircle the arteriole. When these muscle fibers are contracted, the vessel has a smaller diameter (is constricted). When these muscle fibers are relaxed, the vessel has a larger diameter (is dilated). Whether arterioles are constricted or dilated affects blood pressure. The greater the number of vessels dilated, the lower the blood pressure.

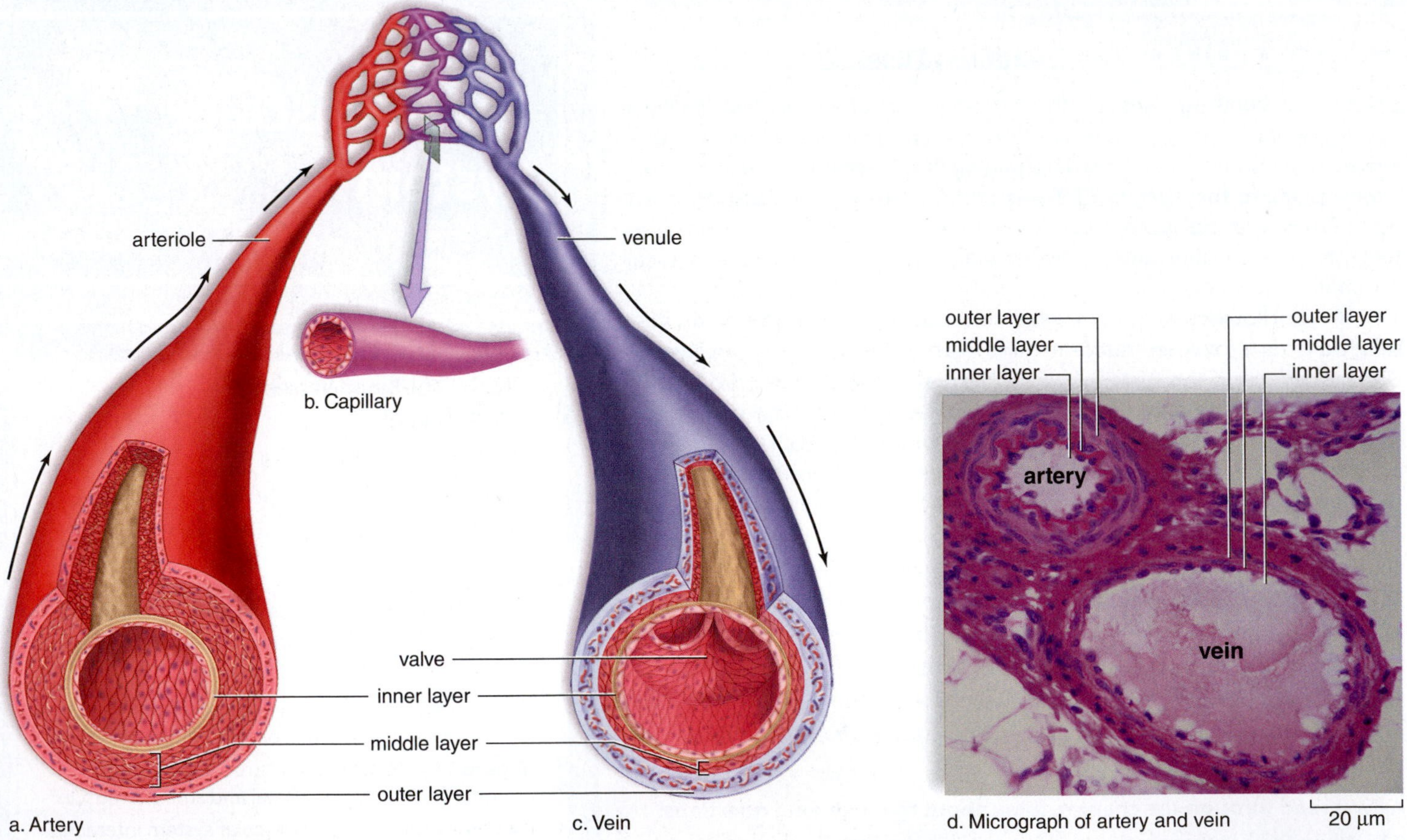
arteriole
venule
b. Capillary
valve
inner layer
middle layer
outer layer
a. Artery
c. Vein
outer layer
middle layer
inner layer
artery
vein
outer layer
middle layer
inner layer
d. Micrograph of artery and vein
20 µm

Figure 12.1 Blood vessels. The walls of arteries and veins have three layers. The inner layer is composed largely of endothelium, with a basement membrane that has elastic fibers; the middle layer is smooth muscle tissue; the outer layer is connective tissue (largely collagen fibers). **a.** Arteries have a thicker wall than veins because they have a larger middle layer than veins. **b.** Capillary walls are one-cell-thick endothelium. **c.** Veins are generally larger in diameter than arteries, so collectively, veins have a larger holding capacity than arteries. **d.** Light micrograph of an artery and a vein.

The Capillaries

Capillaries join arterioles to venules (Fig. 12.1*b*). Capillaries are extremely narrow—only 8–10 μm wide—and have thin walls composed only of a single layer of endothelium with a basement membrane. Although each capillary is small, they form vast networks. Their total surface area in a human body is about 6,000 square meters (m^2). Capillary beds (networks of many capillaries) are present in nearly all regions of the body. Consequently, a cut to almost any body tissue draws blood. One region of the body that is nearly capillary-free is the cornea of the eye, so that light can pass through. Therefore, the cells of the cornea must obtain nutrients by diffusion from the tears on the outside surface, and from the aqueous humor on the inside surface (see Chapter 18).

Capillaries play a very important role in homeostasis because an exchange of substances takes place across their thin walls. Oxygen and nutrients, such as glucose, diffuse out of a capillary into the tissue fluid that surrounds cells. Wastes, such as carbon dioxide, diffuse into the capillary. Some water also leaves a capillary. Any excess is picked up by lymphatic vessels, as shown in Figure 12.9 and discussed further in Chapter 13. The relative constancy of tissue fluid is absolutely dependent upon capillary exchange.

Because capillaries serve the cells, the heart and the other vessels of the cardiovascular system can be thought of as the means by which blood is conducted to and from the capillaries. Only certain capillary beds are completely open at any given time. For example, after eating, the capillary beds that serve the digestive system are mostly open, and those that serve the muscles are mostly closed. Each capillary bed has arteriovenous shunts, also called *anastomoses,* that allow blood to go directly from arterioles to venules, bypassing the bed (Fig. 12.2). Contracted precapillary sphincter muscles prevent the blood from entering the capillary vessels.

A decreased blood flow to the brain after meals has also been offered as an explanation for "postprandial somnolence," or the sleepiness that many people feel after eating. However, recent evidence suggests that the blood supply to the brain is maintained under most physiological conditions, including ingestion of a heavy meal, and that hormones released by the digestive tract may instead be the culprit.

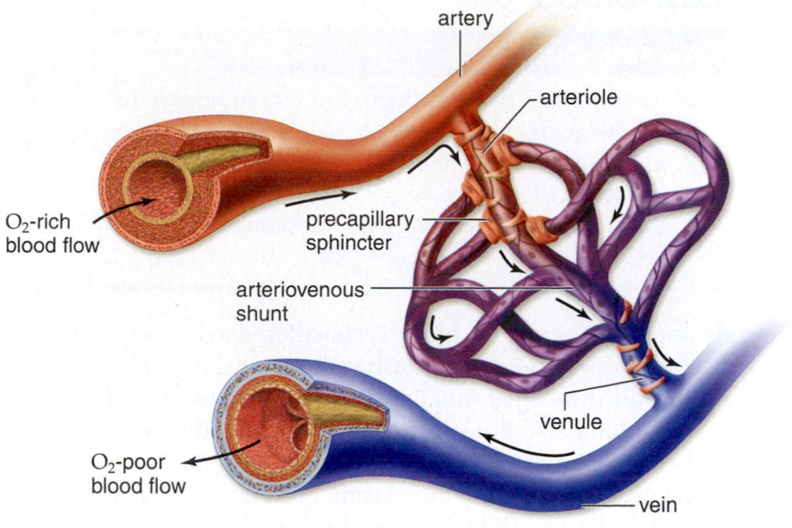
artery
arteriole
O_2-rich blood flow
precapillary sphincter
arteriovenous shunt
venule
O_2-poor blood flow
vein

Figure 12.2 Anatomy of a capillary bed. A capillary bed forms a maze of capillary vessels that lies between an arteriole and a venule. When precapillary sphincter muscles are relaxed, the capillary bed is open, and blood flows through the capillaries. When sphincter muscles are contracted, blood flows through a shunt (anastomosis) that carries blood directly from an arteriole to a venule. As blood passes through a capillary in the tissues, it gives up its oxygen (O_2). Therefore, blood goes from being O_2-rich in the arteriole (red color) to being O_2-poor in the vein (blue color).

The Veins

Veins and venules take blood from the capillary beds to the heart. First, the **venules** (small veins) drain blood from the capillaries and then join to form a vein. The walls of veins (and venules) have the same three layers as arteries, but there is less smooth muscle and connective tissue (see Fig. 12.1*c*). Therefore, the wall of a vein is thinner than that of an artery.

Veins often have *valves,* which allow blood to flow only toward the heart when open and prevent blood from flowing backward when closed. Valves are found in the veins that carry blood against the force of gravity, especially the veins of the lower limbs. Unlike blood flow in the arteries and arterioles, which is kept moving by the pumping of the heart, blood flow in veins is primarily due to skeletal muscle contraction. If these valves become damaged by disease or through the normal wear-and-tear of aging, blood may begin pooling in the veins, causing them to enlarge and be visible as *varicose veins.* These most commonly occur in the lower legs of older individuals. Varicose veins of the anal canal are known as hemorrhoids.

Because the walls of veins are thinner (see Fig. 12.1*d*), they can expand to a greater extent. At any one time, about 70% of the blood is in the veins. In this way, the veins act as a blood reservoir. If blood is lost due to hemorrhaging, nervous stimulation causes the veins to constrict, providing more blood to the rest of the body. The largest veins are the **venae cavae,** which include the *superior vena cava* (20 mm wide) and the *inferior vena cava* (35 mm wide)**.** Both veins deliver O_2-poor blood into the heart.

Check Your Progress 12.1

1. Describe how blood flow is controlled in each of the three major types of blood vessels.
2. List several specific substances that diffuse across capillary walls.

12.2 Blood

Learning Outcomes

Upon completion of this section, you should be able to

1. List the major types of blood cells, and their functions.
2. Identify the major molecular and cellular events that result in a blood clot.
3. Define capillary exchange, and describe the two major forces involved.

As noted in Chapter 11, blood is considered to be a liquid connective tissue. **Blood** has transport functions, regulatory functions, and protective functions. In addition to nutrients and wastes, blood also transports hormones. Blood helps regulate body temperature by dispersing body heat and helps regulate blood pressure because the plasma proteins contribute to the osmotic pressure of blood. Buffers in blood help maintain blood pH at about 7.4. Blood helps protect the body against invasion by disease-causing pathogens. Clotting mechanisms protect the body against potentially life-threatening loss of blood.

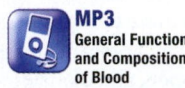

MP3 General Functions and Composition of Blood

If blood is collected from a person's vein into a test tube and prevented from clotting, it separates into three layers (Fig. 12.3*a*). The upper layer is **plasma,** the liquid portion of blood. The two lower layers together are the **formed elements,** which include the white blood cells and platelets (middle layer), and the red blood cells (bottom layer).

Plasma

Plasma contains a variety of inorganic and organic substances dissolved or suspended in water (Fig. 12.3*b*). Plasma proteins, which make up 7–8% of plasma, assist in transporting large organic molecules in blood. For example, *albumin* transports bilirubin, a breakdown product of hemoglobin. Lipoproteins transport cholesterol. Certain plasma proteins have specific functions. As discussed later in this section, fibrinogen is necessary to blood clotting, and immunoglobulins are antibodies that help fight infection. Plasma proteins also maintain blood volume because they are too large to leave capillaries. Therefore, blood in capillaries normally has a higher solute concentration than does tissue fluid, and water automatically diffuses into them.

The Red Blood Cells

Red blood cells (erythrocytes) are continuously manufactured in the red bone marrow of the skull, the ribs, the vertebrae, and the ends of the long bones. Normally, there are 4 to 6 million red blood cells per cubic millimeter (mm^3) of whole blood (Fig. 12.3*d*).

Mature red blood cells don't have a nucleus and are biconcave disks (Fig. 12.4*a, b*). Their shape increases their flexibility for moving through capillary beds and their surface area for diffusion of gases. Red blood cells carry oxygen because they contain **hemoglobin,** the respiratory pigment. Because hemoglobin is a red pigment, the cells are red. A hemoglobin molecule contains four polypeptide chains (Fig. 12.4*c*). Each chain is associated with heme, a complex iron-containing group. The iron portion of hemoglobin acquires oxygen in the lungs and gives it up in the tissues.

Possibly because they lack nuclei, red blood cells live only about 120 days. They are destroyed chiefly in the liver and the spleen, where they are engulfed by large phagocytic cells. The iron is mostly salvaged and reused, while the heme portion undergoes chemical degradation, and the liver excretes it into the bile as bile pigments.

When the body has an insufficient number of red blood cells or the red blood cells do not contain enough hemoglobin, an individual suffers from **anemia** and has a tired, rundown feeling. There are three basic causes of anemia: (1) decreased production of red blood cells, (2) loss of red blood cells from the body, and (3) destruction of red blood cells within the body. In the most common type of anemia, iron-deficiency anemia, red blood cell production is decreased, most often due to a diet that does not contain enough iron.

Whenever arterial blood carries a reduced amount of oxygen, as in chronic anemia or when an individual first takes up residence at a high altitude, the kidneys increase their production of a hormone called *erythropoietin,* which speeds the maturation of red blood cells in the bone marrow.

The White Blood Cells

White blood cells (leukocytes) differ from red blood cells in that they are usually larger, have a nucleus, lack hemoglobin, and without staining appear translucent (see Fig. 12.3*c, d*). White blood cells are not as numerous as red blood cells, with only 5,000–11,000 cells per mm^3. White blood cells fight infection and play a role in the development of immunity, the ability to resist disease.

On the basis of structure, it is possible to divide white blood cells into *granular leukocytes* and *agranular leukocytes,* the latter also being known as *mononuclear cells.* Granular leukocytes (neutrophils, basophils, and eosinophils) are filled with spheres that contain enzymes and proteins, which help white blood cells defend the body against microbes. **Neutrophils** are granular leukocytes with a multilobed nucleus joined by nuclear threads. They are the most abundant of the white blood cells and are able to phagocytize and digest bacteria. **Basophils** stain a deep blue and release histamine, which can cause inflammation. **Eosinophils** stain a deep red and are thought to fight parasitic worms, although they are also involved in some allergies.

Agranular leukocytes (monocytes and lymphocytes) typically have a kidney-shaped or spherical nucleus. **Monocytes** are the largest of the white blood cells, and they differentiate into phagocytic dendritic cells and macrophages. **Dendritic cells** are present in tissues that are in contact with the environment: skin, nose, lungs, and intestines. Once they have captured a microbe with their long, spiky arms, called dendrites, they stimulate other white blood cells to defend the body. **Macrophages,** well known

Figure 12.3 Composition of blood. **a.** When blood is collected into a test tube containing an anticoagulant to prevent clotting and then centrifuged, it consists of three layers. The transparent straw-colored or yellow top layer is the plasma, the liquid portion of the blood. The thin, middle buffy coat layer consists of leukocytes and platelets. The bottom layer contains the erythrocytes. **b.** Breakdown of the components of plasma. **c.** Micrograph of the formed elements, which are also listed in **(d).**

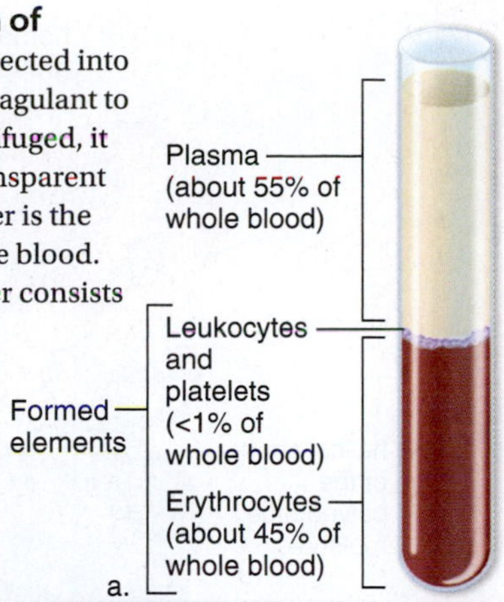
Plasma (about 55% of whole blood)
Formed elements
Leukocytes and platelets (<1% of whole blood)
Erythrocytes (about 45% of whole blood)

a.

Plasma		
Type	**Function**	**Source**
Water (90–92% of plasma)	Maintains blood volume; transports molecules	Absorbed from intestine
Plasma proteins (7–8% of plasma)	Maintain blood osmotic pressure and pH	
Albumin	Maintains blood volume and pressure, transport	Liver
Antibodies	Fight infection	B lymphocytes
Fibrinogen	Clotting	Liver
Salts (less than 1% of plasma)	Maintain blood osmotic pressure and pH; aid metabolism	Absorbed from intestine
Gases		
Oxygen	Cellular respiration	Lungs
Carbon dioxide	End product of metabolism	Tissues
Nutrients Lipids, Glucose, Amino acids	Food for cells	Absorbed from intestine
Nitrogenous wastes Urea, Uric acid	Excretion by kidneys	Liver
Other Hormones, vitamins, etc.	Aid metabolism	Varied

b.

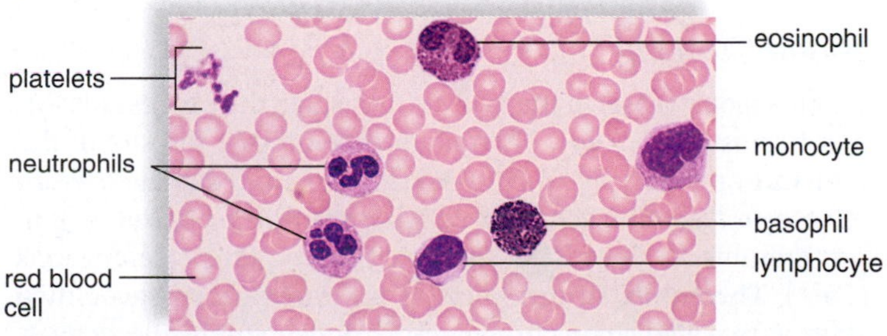
platelets
neutrophils
red blood cell
eosinophil
monocyte
basophil
lymphocyte

c. 250×

Formed Elements		
Type	**Function and Description**	**Source**
Red blood cells (erythrocytes)	Transport O_2 and help transport CO_2	Red bone marrow
 4 million–6 million per mm^3 blood	7–8 μm in diameter; Bright-red to dark-purple biconcave disks without nuclei	
White blood cells (leukocytes) 5,000–11,000 per mm^3 blood	Fight infection	Red bone marrow
Granular leukocytes		
Neutrophils * 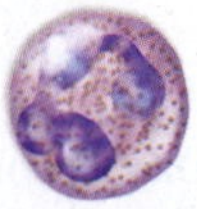 40–70%	10–14 μm in diameter; Spherical cells with multilobed nuclei; fine, pink granules in cytoplasm; phagocytize pathogens	
Eosinophils * 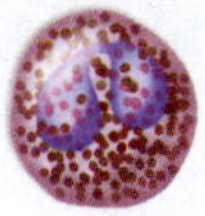 1–4%	10–14 μm in diameter; Spherical cells with bilobed nuclei; coarse, deep-red, uniformly sized granules in cytoplasm; phagocytize antigen-antibody complexes and allergens	
Basophils *  0–1%	10–12 μm in diameter; Spherical cells with lobed nuclei; large, irregularly shaped, deep-blue granules in cytoplasm; release histamine, which promotes blood flow to injured tissues	
Agranular leukocytes		
Lymphocytes * 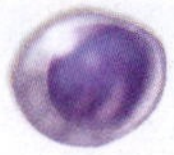 20–45%	5–17 μm in diameter (average 9–10 μm); Spherical cells with large, round nuclei; responsible for specific immunity	
Monocytes * 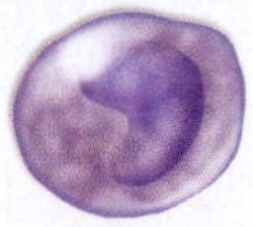 4–8%	10–24 μm in diameter; Large spherical cells with kidney-shaped, round, or lobed nuclei; become macrophages that phagocytize pathogens and cellular debris	
Platelets (thrombocytes)	Aid clotting	Red bone marrow
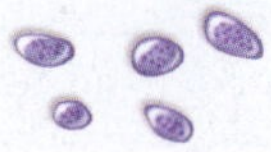 150,000–300,000 per mm^3 blood	2–4 μm in diameter; Disk-shaped cell fragments with no nuclei; purple granules in cytoplasm	

d. * Appearance with Wright's stain.

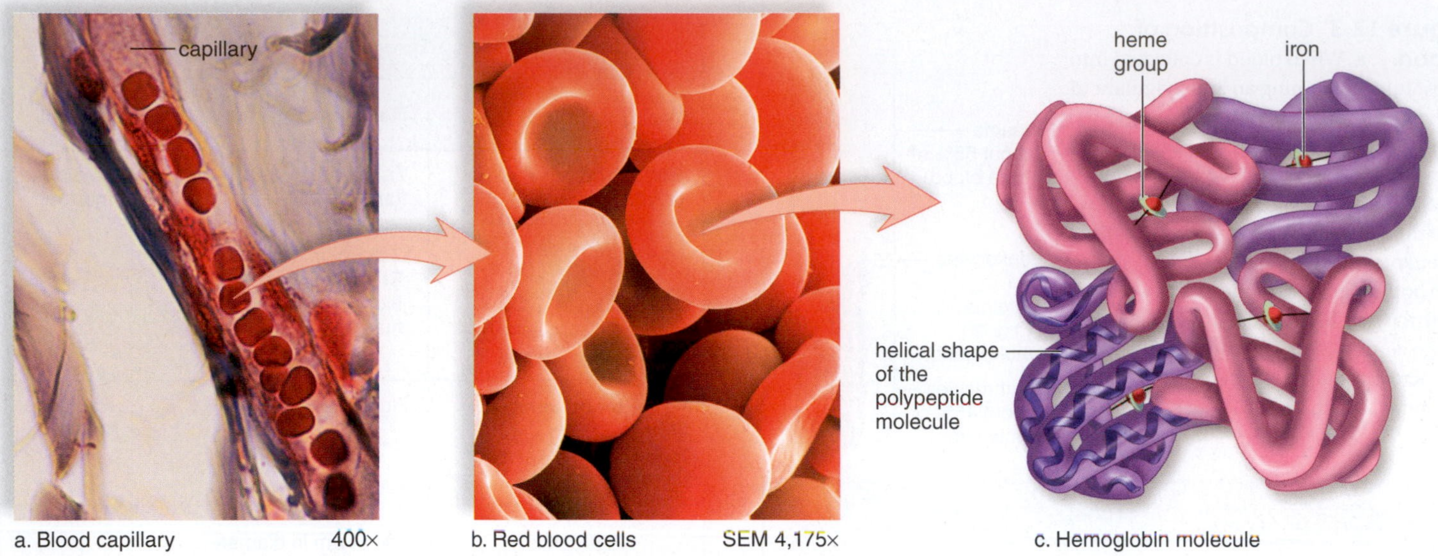

capillary
heme group
iron
helical shape of the polypeptide molecule
a. Blood capillary
400×
b. Red blood cells
SEM 4,175×
c. Hemoglobin molecule

Figure 12.4 Physiology of red blood cells. **a.** Red blood cells move in single file through the capillaries. **b.** Each red blood cell is a biconcave disk containing many molecules of hemoglobin, the respiratory pigment. **c.** Hemoglobin contains four polypeptide chains (purple). There is an iron-containing heme group in the center of each chain. Oxygen combines loosely with iron when hemoglobin is oxygenated. Oxyhemoglobin is bright red, and deoxyhemoglobin is a dark maroon color.

as ferocious phagocytes (Fig. 12.5), play a similar role in other organs, such as the liver, kidney, and spleen. The **lymphocytes** are of two major types: B lymphocytes (B cells) and T lymphocytes (T cells). B cells produce antibodies, whereas T cells occur as two main types: helper T cells that regulate the responses of other cells, and cytotoxic T cells that are able to kill other cells. To learn more about the immune system, see Chapter 13.

If the number of any type of white blood cell increases or decreases beyond normal, disease may be present. For example, neutrophil numbers tend to increase in response to many different bacterial infections. A person with *infectious mononucleosis*, caused by the Epstein-Barr virus, often has an increased number of B cells. In an HIV-infected person, AIDS is defined, in part, by having an abnormally low number of T cells. *Leukemia* is a form of cancer characterized by uncontrolled production of abnormal white blood cells.

White blood cells live different lengths of time. Many live only a few days or may die combating invading pathogens. Others live for months or even years. White blood cells are discussed more extensively in Chapter 13.

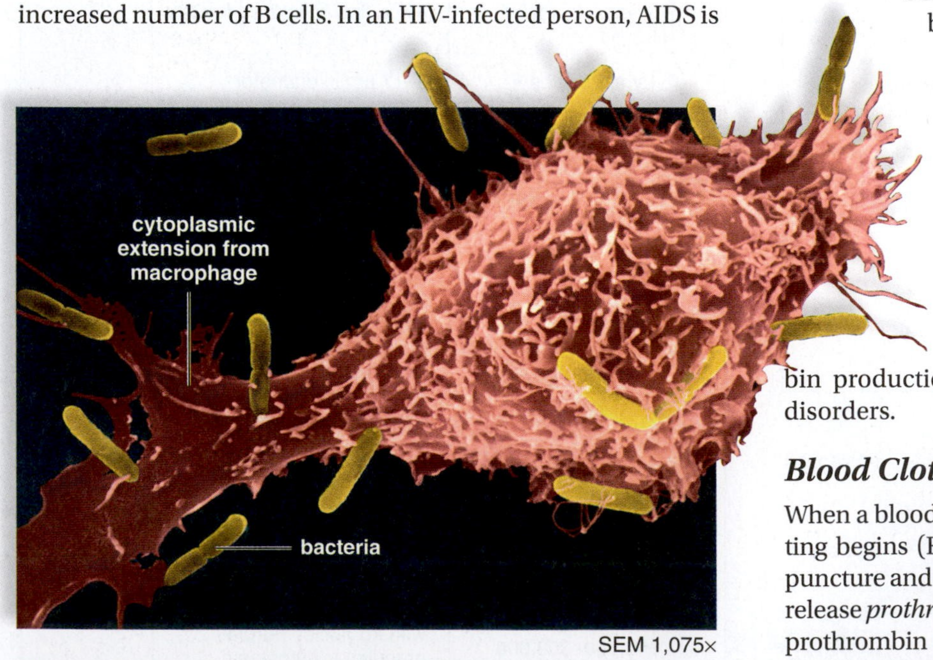

cytoplasmic extension from macrophage
bacteria
SEM 1,075×

Figure 12.5 Macrophage (red) engulfing bacteria. Monocyte-derived macrophages are the body's scavengers. They engulf microbes and debris in the body's fluids and tissues, as illustrated in this colorized scanning electron micrograph.

The Platelets and Blood Clotting

Platelets (also called *thrombocytes*) result from fragmentation of certain large cells, called *megakaryocytes,* in the red bone marrow. Platelets are produced at a rate of 200 billion a day, and the blood contains 150,000–300,000 per mm^3. These formed elements are involved in the process of blood **clotting,** or coagulation.

There are at least 12 clotting factors in the blood that participate with platelets in the formation of a blood clot. We will discuss the roles played by *fibrinogen* and *prothrombin,* which are proteins manufactured by the liver. Vitamin K, found in green vegetables and also formed by intestinal bacteria, is needed for prothrombin production. Vitamin K deficiency can result in clotting disorders.

Blood Clotting

When a blood vessel in the body is damaged, the process of clotting begins (Fig. 12.6*a*). First, platelets clump at the site of the puncture and partially seal the leak. Platelets and damaged tissue release *prothrombin activator,* which converts the plasma protein prothrombin to thrombin. This reaction requires calcium ions (Ca^{2+}). **Thrombin,** in turn, acts as an enzyme that severs two short amino acid chains from each fibrinogen molecule, activating it. The activated fragments then join end to end, forming long threads of **fibrin.** Fibrin threads wind around the platelet plug in

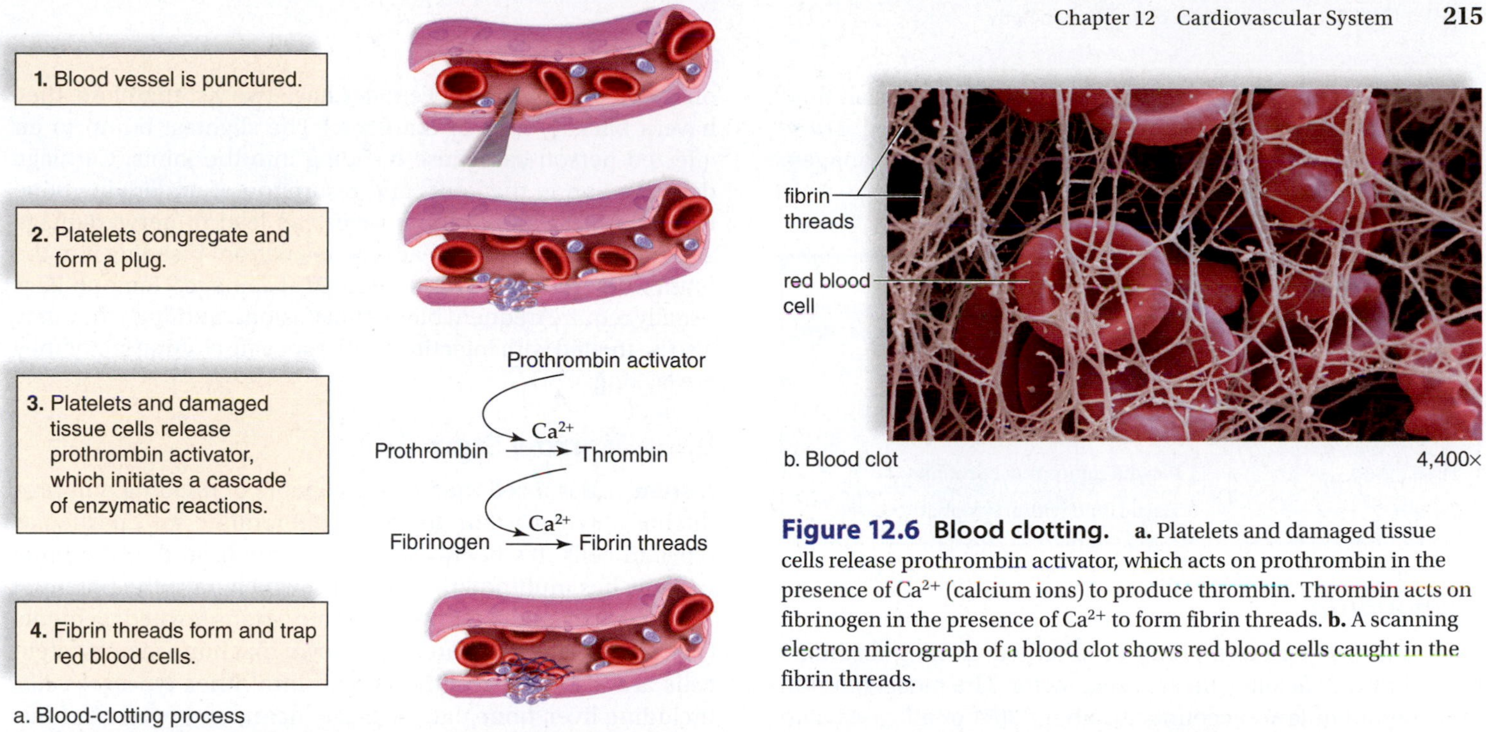
1. Blood vessel is punctured.
2. Platelets congregate and form a plug.
3. Platelets and damaged tissue cells release prothrombin activator, which initiates a cascade of enzymatic reactions.
Prothrombin activator
Prothrombin
Ca2+
Thrombin
Fibrinogen
Ca2+
Fibrin threads
4. Fibrin threads form and trap red blood cells.
a. Blood-clotting process
fibrin threads
red blood cell
b. Blood clot
4,400×

Figure 12.6 Blood clotting. **a.** Platelets and damaged tissue cells release prothrombin activator, which acts on prothrombin in the presence of Ca^{2+} (calcium ions) to produce thrombin. Thrombin acts on fibrinogen in the presence of Ca^{2+} to form fibrin threads. **b.** A scanning electron micrograph of a blood clot shows red blood cells caught in the fibrin threads.

SCIENCE IN YOUR LIFE ▶ HEALTH

Medicinal Leeches: Medicine Meets *Fear Factor*

Although it may seem more like an episode of a popular TV show than a real-life medical treatment, the U.S. Food and Drug Administration has approved the use of leeches as medical "devices" for treating conditions involving poor blood supply to various tissues.

Leeches are bloodsucking, aquatic creatures, whose closest living relatives are earthworms (Fig. 12A). Prior to modern times, medical practitioners frequently applied leeches to patients, mainly in an attempt to remove the bad "humors" that they thought were responsible for many diseases. This practice was abandoned, thankfully, in the nineteenth century when it was realized that the "treatment" often harmed the patient.

True to their tenacious nature, however, leeches are making a comeback in twenty-first-century medicine. By applying leeches to tissues that have been injured by trauma or disease, blood supply can be improved. When reattaching a finger, for example, it is easier to suture together the thicker-walled arteries than the veins. Poorly draining blood from the veins can pool in the appendage and threaten its survival. It turns out that leech saliva contains chemicals that dilate blood vessels and prevent blood from clotting by blocking the activity of thrombin. These effects can improve the circulation to the body part. Another substance in leech saliva actually anesthetizes the bite wound. In a natural setting, this allows the leech to feast on the blood supply of its victim undetected, but in a medical setting, it makes the whole experience more tolerable, at least physically. Mentally, however, the application of leeches can still be a rather unsettling experience, and patient acceptance is a major factor limiting their more widespread use.

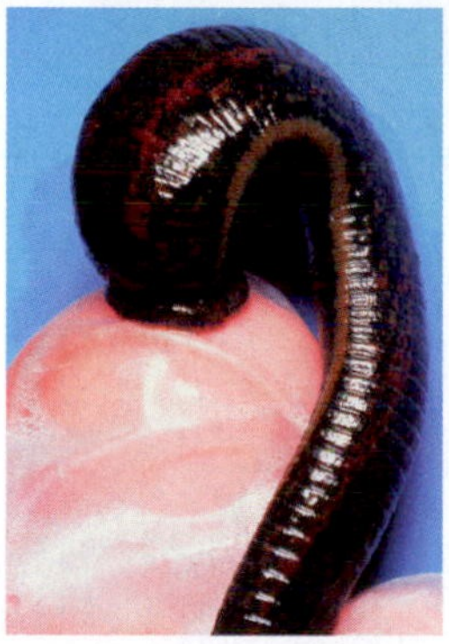

Figure 12A Leeches. Leeches can attach to human skin and suck blood out. Their use in medicine seems to be making a comeback.

Questions to Consider

1. If you had an injury and your doctor said it might help, would you be willing to let leeches feast for a few minutes, say, on your hand? On your face?
2. Leeches were historically used for bloodletting, one of the oldest medical practices. Can you think of several reasons why, prior to the mid-nineteenth century, early physicians might have gotten the idea that removing blood could help cure various diseases?
3. The use of actual living organisms in medical treatments is sometimes called "biotherapy." Can you think of other situations in which living creatures might be used for human health benefits, either directly or indirectly?

the damaged area of the blood vessel and provide the framework for the clot. Red blood cells are also trapped within the fibrin threads. These cells make a clot appear red (Fig. 12.6*b*).

A fibrin clot is only temporary. As soon as blood vessel repair is initiated, an enzyme called plasmin destroys the fibrin network and restores the fluidity of the plasma. Sometimes it is desirable to inhibit the clotting process. This can be done with medications such as heparin, or even with medicinal leeches (see the Health feature, "Medicinal Leeches: Medicine Meets *Fear Factor*.").

If blood is allowed to clot in a test tube, a yellowish fluid develops above the clotted material. This fluid is called *serum,* and it contains all the components of plasma except fibrinogen. Table 12.1 reviews the body fluids related to blood.

TABLE 12.1 Body Fluids Related to Blood

Name	Composition
Blood	Formed elements and plasma
Plasma	Liquid portion of blood
Serum	Plasma minus fibrinogen
Tissue fluid	Plasma minus most proteins
Lymph	Tissue fluid within lymphatic vessels

Hemophilia

Hemophilia refers to a group of inherited clotting disorders caused by a deficiency in a clotting factor. The most common type, hemophilia A, accounts for about 90% of all cases, and almost always occurs in males because the faulty gene is found on the X chromosome. (Females have two Xs; therefore, they have a backup copy of the gene.) The slightest bump to an affected person can cause bleeding into the joints. Cartilage degeneration in the joints and resorption of underlying bone can follow. Bleeding into muscles can lead to nerve damage and muscular atrophy. Death can result from bleeding into the brain with accompanying neurological damage. Hemophiliacs usually require frequent blood transfusions, although they may also be treated with injections of the specific clotting factor they are lacking.

Bone Marrow Stem Cells

A **stem cell** is a cell that is ever capable of dividing and producing new cells that go on to differentiate into particular types of cells. It's been known for some time that the bone marrow has multipotent stem cells, which have the potential to give rise to other stem cells for the various formed elements (Fig. 12.7). Recent research has shown that bone marrow stem cells are also able to differentiate into other types of cells, including liver, bone, fat, cartilage, heart, and even neurons. The possibility exists that a patient's own bone marrow stem

Figure 12.7 Blood cell formation in red bone marrow. Multipotent stem cells give rise to two specialized types of stem cells. The myeloid stem cells give rise to still other cells, which become red blood cells, platelets, and all the white blood cells except lymphocytes. The lymphoid stem cells give rise to lymphoblasts, which become lymphocytes.

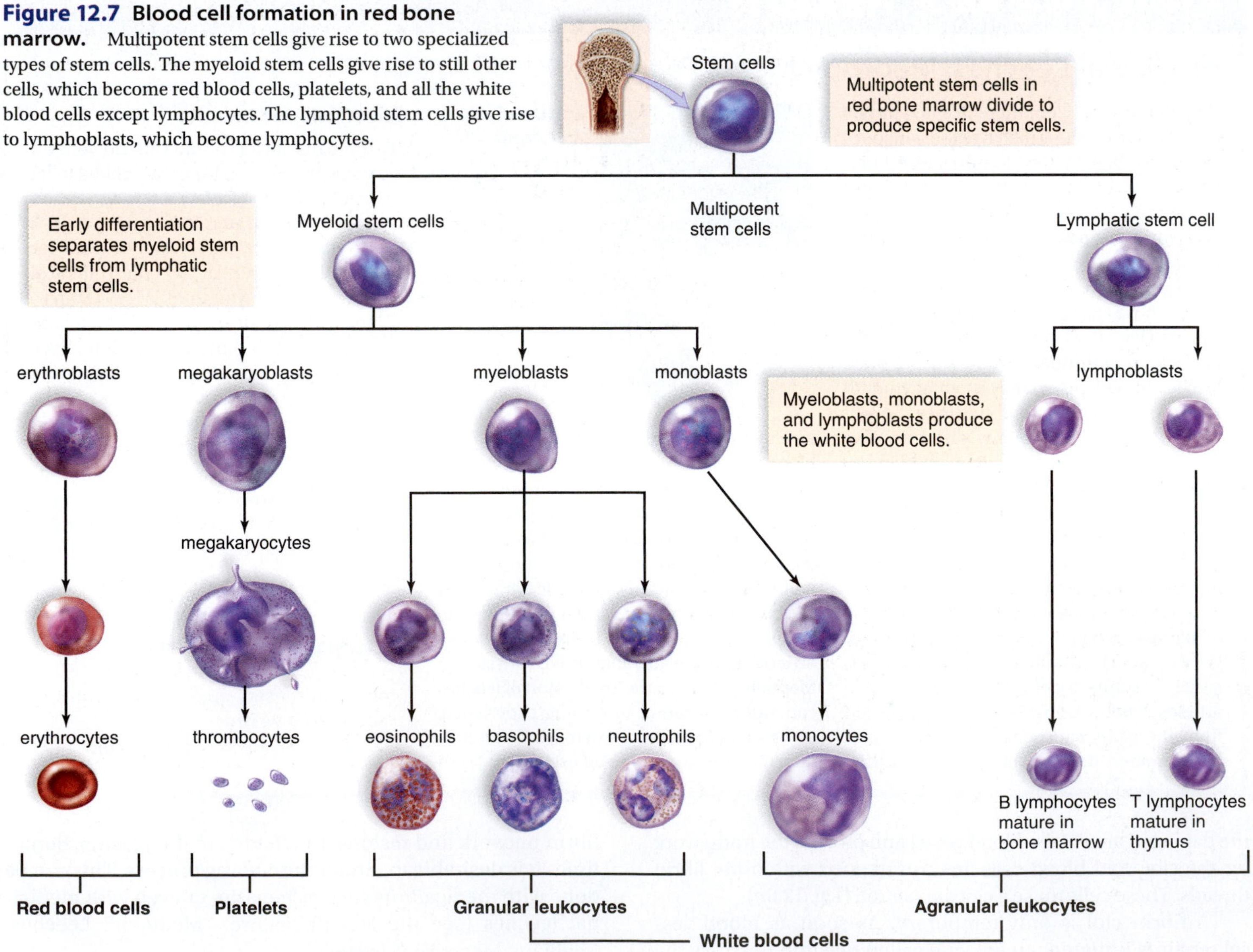
Stem cells
Multipotent stem cells in red bone marrow divide to produce specific stem cells.
Early differentiation separates myeloid stem cells from lymphatic stem cells.
Myeloid stem cells
Multipotent stem cells
Lymphatic stem cell
erythroblasts
megakaryoblasts
myeloblasts
monoblasts
Myeloblasts, monoblasts, and lymphoblasts produce the white blood cells.
lymphoblasts
megakaryocytes
erythrocytes
thrombocytes
eosinophils
basophils
neutrophils
monocytes
B lymphocytes mature in bone marrow
T lymphocytes mature in thymus
Red blood cells
Platelets
Granular leukocytes
Agranular leukocytes
White blood cells

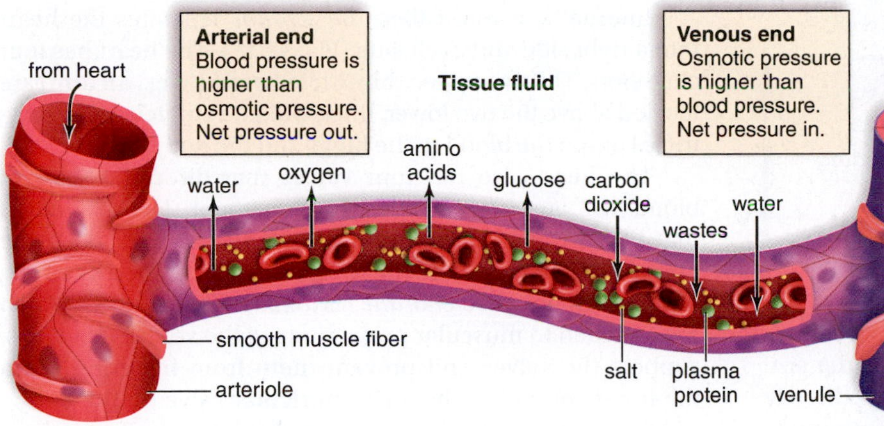
Arterial end
Blood pressure is higher than osmotic pressure. Net pressure out.
Venous end
Osmotic pressure is higher than blood pressure. Net pressure in.
from heart
Tissue fluid
to heart
water
oxygen
amino acids
glucose
carbon dioxide
wastes
water
smooth muscle fiber
arteriole
salt
plasma protein
venule

Figure 12.8 Capillary exchange in the systemic circuit. At the arterial end of a capillary (*left*) the blood pressure is higher than the osmotic pressure; therefore, water tends to leave the bloodstream. In the midsection, molecules, including oxygen and carbon dioxide, follow their concentration gradients. At the venous end of a capillary (*right*), the osmotic pressure is higher than the blood pressure; therefore, water tends to enter the bloodstream. Notice that the red blood cells and the plasma proteins are too large to exit a capillary.

cells could be used for curing certain conditions that might develop, such as diabetes, heart disease, liver disease, or even brain disorders (e.g., Alzheimer disease or Parkinson disease). In one study, researchers examined the brains of women who had received bone marrow stem cells from male donors as a part of their treatment for leukemia. All recipients had neurons in their brains that contained a Y chromosome! It seems that some of the bone marrow stem cells derived from male donors had traveled to the brain, where they differentiated into neurons.

The use of a person's own stem cells does away with the problem of possible rejection. Among all the various types of *adult stem cells* in the body, bone marrow stem cells are particularly attractive because they are the most accessible. Some researchers prefer to work with *embryonic stem cells,* thinking that they are more likely to become any type of cell. Embryonic stem cells are available because many early-stage embryos remain unused in fertility clinics, although this issue has become controversial.

Capillary Exchange

Two forces primarily control movement of fluid through the capillary wall: (1) osmotic pressure, created by salts and plasma proteins, which tends to cause water to move from the tissue fluid to the blood; and (2) blood pressure, which tends to cause water to move in the opposite direction. At the arterial end of a capillary, blood pressure is higher than the osmotic pressure of blood (Fig. 12.8), so water exits a capillary at this end.

Midway along the capillary, where blood pressure is lower, the two forces essentially cancel each other, and there is no net movement of water. Solutes now diffuse according to their concentration gradient—nutrients (glucose and amino acids) and oxygen diffuse out of the capillary, and wastes (carbon dioxide) diffuse into the capillary. In the pulmonary circuit, where the O_2 concentration is higher in the lung tissues, and the CO_2 concentration is lower, the movement of these gases is reversed.

Red blood cells and almost all plasma proteins remain in the capillaries, but small substances leave, contributing to **tissue fluid** (also called interstitial fluid)**,** the fluid between the body's cells. Tissue fluid tends to contain all the components of plasma but lesser amounts of protein, which generally stays in the capillaries.

At the venous end of a capillary, where blood pressure has fallen even more, osmotic pressure is greater than blood pressure, and water tends to move into the capillary. Almost the same amount of fluid that left the capillary returns to it, although some excess tissue fluid is always collected by the **lymphatic capillaries** (Fig. 12.9). Tissue fluid contained within lymphatic vessels is called **lymph.** Lymph is returned to the systemic venous blood when the major lymphatic vessels enter the subclavian veins in the shoulder region.

MP3 Capillary Exchange and Bulk Flow

Animation Fluid Exchange Across the Walls of Capillaries

Check Your Progress 12.2

1. List the major components of blood, along with their functions.
2. Describe the cellular and molecular events that lead to blood clotting.
3. Identify some types of diseases that would be expected to be treatable with stem cell therapy.
4. Predict the effect on tissue fluid if the protein content of the blood was greatly reduced, i.e., by malnutrition.

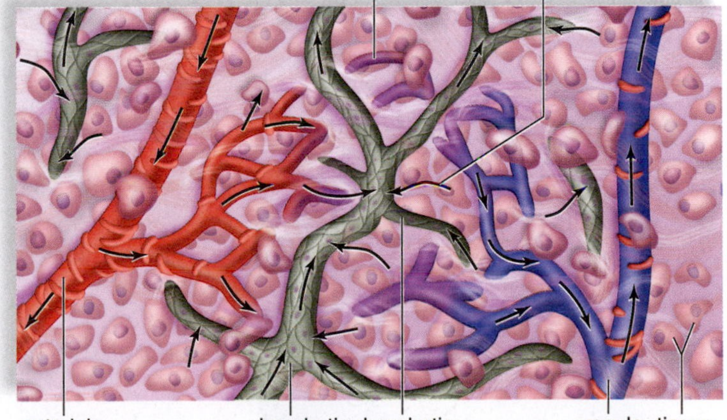
blood capillary
excess tissue fluid
arteriole
lymphatic duct
lymphatic capillaries
venule
tissue cells

Figure 12.9 Lymphatic capillaries. A lymphatic capillary bed (shown here in green) lies near a blood capillary bed. When lymphatic capillaries take up excess tissue fluid, it becomes lymph.

12.3 The Human Heart

Learning Outcomes

Upon completion of this section, you should be able to

1. Identify the major components of the heart, including the four chambers and four valves.
2. Trace the path of blood through the heart and lungs.
3. Describe the intrinsic and extrinsic control of the heartbeat.

The **heart** is a cone-shaped, muscular organ about the size of a fist (Fig. 12.10). It is located between the lungs directly behind the sternum (breastbone) and is tilted so that the apex (the pointed end) is oriented to the body's left. The major portion of the heart, called the *myocardium,* consists largely of cardiac muscle tissue. The muscle fibers of the myocardium are branched and tightly joined to one another. The heart lies within the *pericardium,* a thick, serous membrane that secretes a small quantity of lubricating liquid. The inner surface of the heart is lined with endocardium, a membrane composed of connective tissue and endothelial tissue.

Internally, a wall called the *septum* separates the heart into a right side and a left side (Fig. 12.11). The heart has four chambers. The two upper, thin-walled **atria** (sing., **atrium**) are located above the two lower, thick-walled **ventricles.** The ventricles pump the blood to the lungs and the body.

The heart also has four valves that direct the flow of blood and prevent its backward movement. The two valves that lie between the atria and the ventricles are called the **atrioventricular valves.** These valves are supported by strong fibrous strings called *chordae tendineae.* The chordae, which are attached to muscular projections of the ventricular walls, support the valves and prevent them from inverting when the heart contracts. The atrioventricular valve on the body's right side is called the *tricuspid valve* because it has three flaps, or cusps. The valve on the left side is called the *bicuspid (mitral) valve* because it has two flaps. The remaining two valves, between the ventricles and their attached vessels, are the **semilunar valves,** whose flaps resemble half-moons. The *pulmonary* semilunar valve lies between the right ventricle and the pulmonary trunk. The *aortic* semilunar valve lies between the left ventricle and the aorta.

MP3 Heart Structure and Function

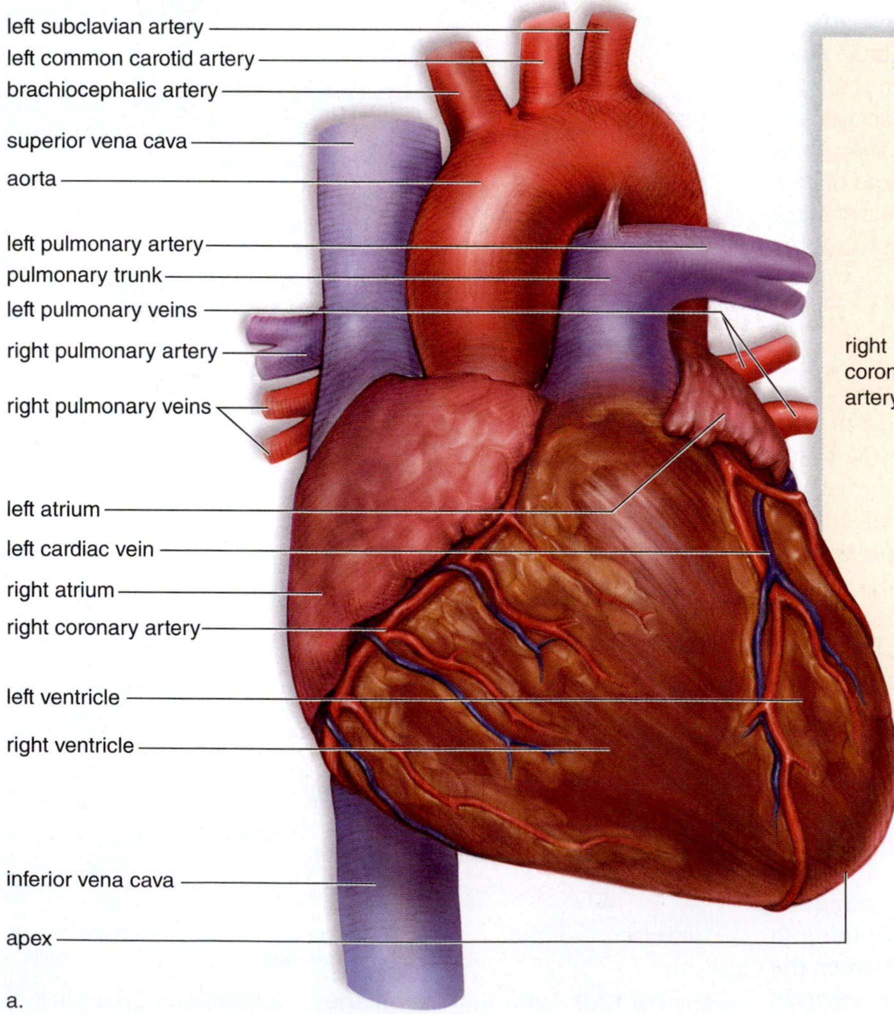
left subclavian artery
left common carotid artery
brachiocephalic artery
superior vena cava
aorta
left pulmonary artery
pulmonary trunk
left pulmonary veins
right pulmonary artery
right pulmonary veins
left atrium
left cardiac vein
right atrium
right coronary artery
left ventricle
right ventricle
inferior vena cava
apex

a.

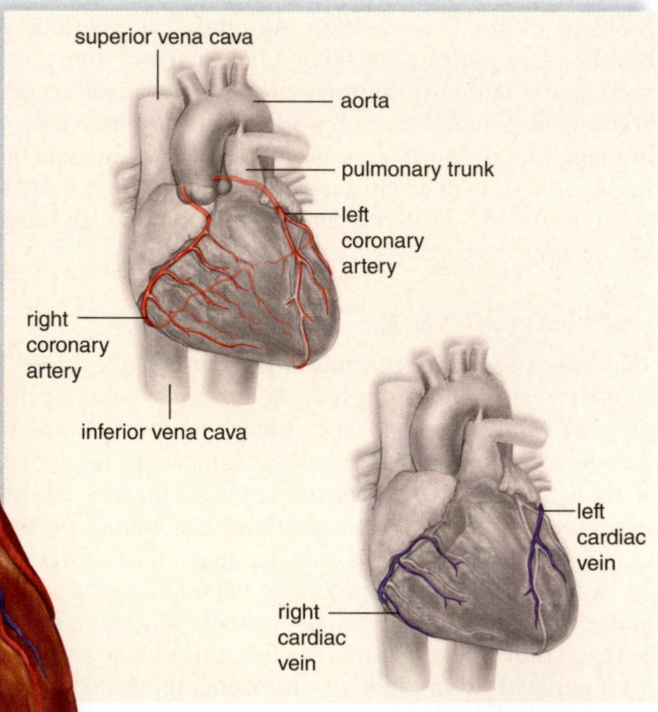
superior vena cava
aorta
pulmonary trunk
left coronary artery
right coronary artery
inferior vena cava
left cardiac vein
right cardiac vein

b.

Figure 12.10 External heart anatomy. **a.** The venae cavae and the pulmonary trunk are attached to the right side of the heart. The aorta and the pulmonary veins are attached to the left side of the heart. **b.** The coronary arteries and cardiac veins pervade cardiac muscle. The coronary arteries bring oxygen and nutrients to cardiac cells, which derive no benefit from blood coursing through the heart. The cardiac veins drain blood into the right atrium.

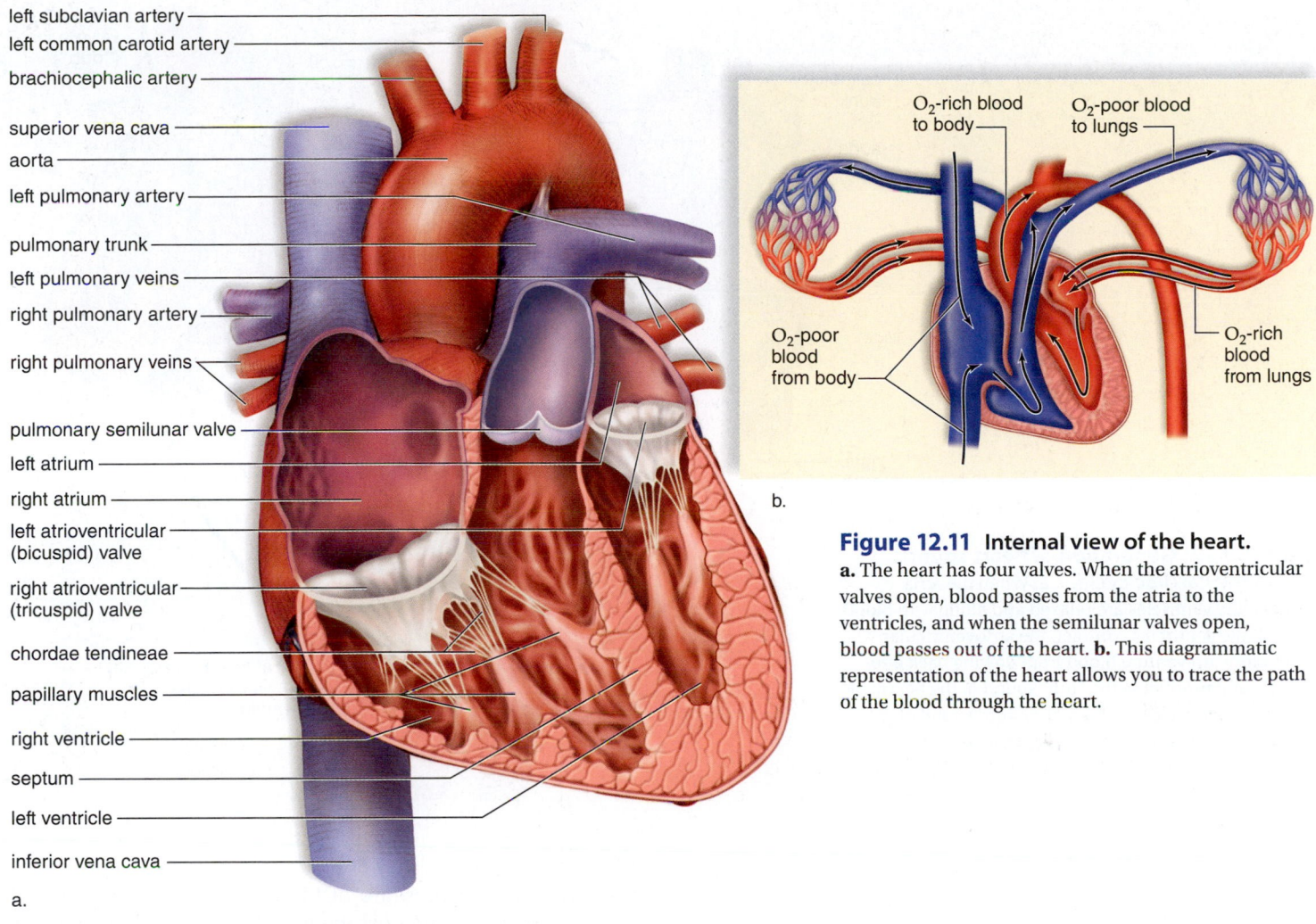

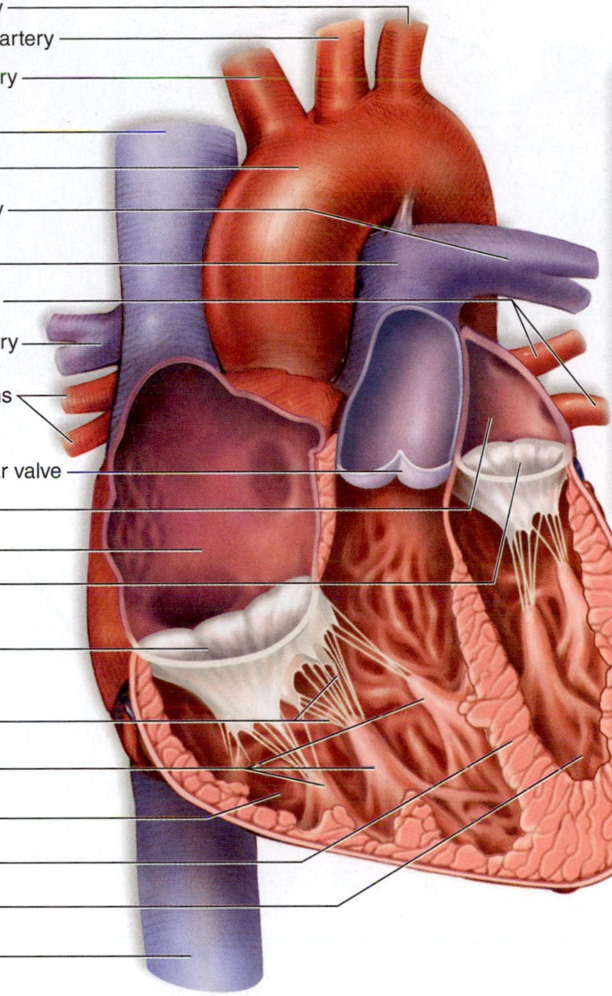
left subclavian artery
left common carotid artery
brachiocephalic artery
superior vena cava
aorta
left pulmonary artery
pulmonary trunk
left pulmonary veins
right pulmonary artery
right pulmonary veins
pulmonary semilunar valve
left atrium
right atrium
left atrioventricular (bicuspid) valve
right atrioventricular (tricuspid) valve
chordae tendineae
papillary muscles
right ventricle
septum
left ventricle
inferior vena cava

a.

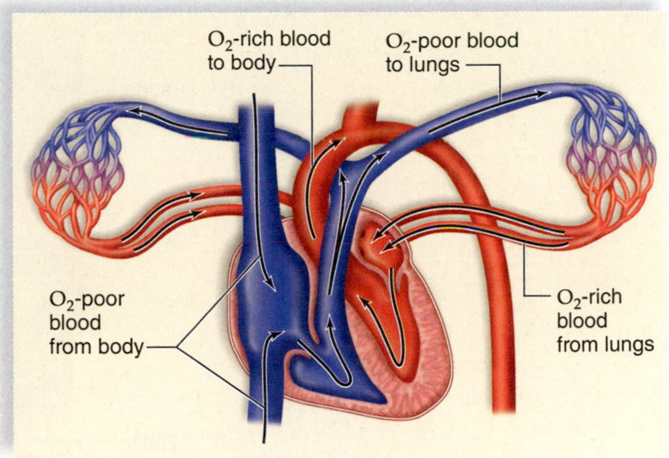
O_2-rich blood to body
O_2-poor blood to lungs
O_2-poor blood from body
O_2-rich blood from lungs

b.

Figure 12.11 Internal view of the heart.
a. The heart has four valves. When the atrioventricular valves open, blood passes from the atria to the ventricles, and when the semilunar valves open, blood passes out of the heart. **b.** This diagrammatic representation of the heart allows you to trace the path of the blood through the heart.

Path of Blood Through the Heart

By referring to Figure 12.11, we can trace the path of blood through the heart in the following manner:

- The superior vena cava and the inferior vena cava, which carry O_2-poor blood that is relatively high in carbon dioxide, enter the right atrium.
- The right atrium sends blood through the tricuspid valve to the right ventricle.
- The right ventricle sends blood through the pulmonary semilunar valve into the pulmonary trunk and through the two **pulmonary arteries** to the lungs.
- Four **pulmonary veins,** which carry O_2-rich blood, enter the left atrium.
- The left atrium sends blood through the bicuspid (mitral) valve to the left ventricle.
- The left ventricle sends blood through the aortic semilunar valve into the aorta to the rest of the body.

From this description, it is obvious that O_2-poor blood never mixes with O_2-rich blood, and that blood must go through the lungs in order to pass from the right side to the left side of the heart. In fact, the heart is a double pump because the right ventricle sends blood into the lungs and the left ventricle sends blood into the rest of the body. Because the left ventricle has the harder job of pumping blood to the entire body, its walls are thicker than those of the right ventricle, which pumps blood a relatively short distance to the lungs. The volume of blood that the left ventricle pumps per minute is called the **cardiac output.** In a person with an average heart rate of 70 beats per minute, the cardiac output is about 5.25 l of blood per minute. This is about equal to the total amount of blood in the body, and adds up to about 750 l (2,000 gallons) of blood per day. During heavy exercise, the cardiac output can increase as much as fivefold, or perhaps more in trained athletes.

The pumping of the heart sends blood out under pressure into the arteries. Because the left side of the heart is the stronger pump, blood pressure is greatest in the aorta. Blood pressure then decreases as the cross-sectional area of arteries and then arterioles increases. The **pulse** is a wave effect that passes down the walls of arteries when the aorta expands and then recoils with each ventricular contraction. The arterial pulse can be used to determine the heart rate, and a weak or "thready" pulse may indicate a weak heart or low blood pressure.

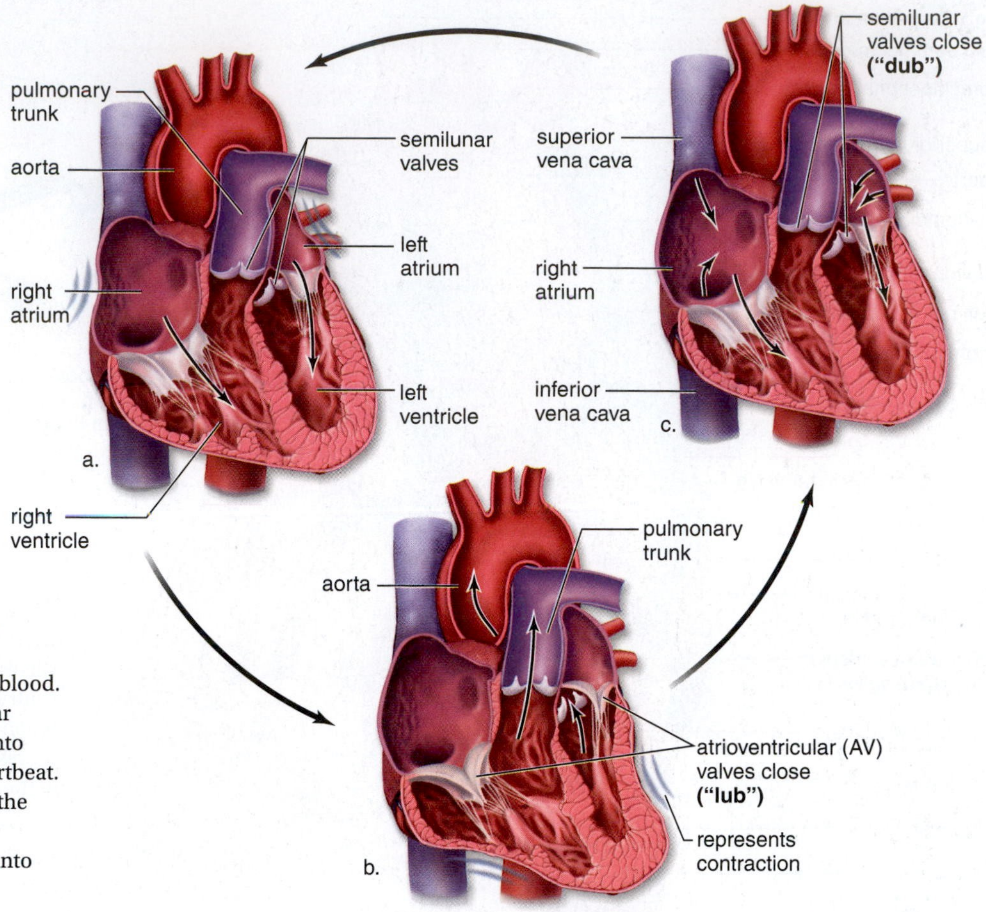
pulmonary trunk
aorta
right atrium
right ventricle
a.
semilunar valves
left atrium
left ventricle
superior vena cava
right atrium
inferior vena cava
c.
semilunar valves close ("dub")
aorta
pulmonary trunk
atrioventricular (AV) valves close ("lub")
represents contraction
b.

Figure 12.12 **Generation of heart sounds during the cardiac cycle.** **a.** When the atria contract, the ventricles are relaxed and filling with blood. **b.** When the ventricles contract, the atrioventricular valves close, preventing blood from flowing back into the atria and producing the "lub" sound of the heartbeat. **c.** After the ventricles contract, the "dub" sound of the heartbeat results from the closing of the semilunar valves to prevent arterial blood from flowing back into the ventricles.

The Heartbeat

Each heartbeat is called a **cardiac cycle** (Fig. 12.12). When the heart beats, first the two atria contract at the same time; then the two ventricles contract at the same time. Then all of the chambers relax. The word **systole** refers to contraction of heart muscle, and the word **diastole** refers to relaxation of heart muscle. The heart contracts, or beats, about 70 times a minute, and each heartbeat lasts about 0.85 seconds, apportioned as follows:

Cardiac Cycle		
Time	**Atria**	**Ventricles**
0.15 sec	Systole	Diastole
0.30 sec	Diastole	Systole
0.40 sec	Diastole	Diastole

A normal adult rate at rest can vary from 60 to 80 beats per minute, which adds up to about 2.5 million beats in a lifetime!

When the heart beats, the familiar "lub-dub" sound occurs. This is best heard using a *stethoscope,* an instrument used to isolate and amplify the body's internal sounds. The longer and lower-pitched "lub" is caused by vibrations occurring when the atrioventricular valves close due to ventricular contraction. The shorter and sharper "dub" results when the semilunar valves close due to back pressure of blood in the arteries. A heart murmur, or a slight whooshing sound after the "lub," is most commonly due to blood flowing back through an ineffective mitral valve (see section 12.5).

MP3
Cardiac
Cycle

Intrinsic Control of Heartbeat

The rhythmic contraction of the atria and ventricles is due to the intrinsic (or internal) conduction system of the heart that is made possible by the presence of nodal tissue, a unique type of cardiac muscle. Nodal tissue, which has both muscular and nervous characteristics, is located in two regions of the heart. The *SA (sinoatrial) node* is located in the upper dorsal wall of the right atrium. The *AV (atrioventricular) node* is located in the base of the right atrium very near the septum (Fig. 12.13*a*). The SA node initiates the heartbeat and sends out an excitation impulse every 0.85 seconds. This causes the atria to contract. When impulses reach the AV node, a slight delay allows the atria to finish contraction before the ventricles begin to contract. The signal for the ventricles to contract travels from the AV node through specialized cardiac muscle fibers called the *atrioventricular bundle* (AV bundle) before reaching the numerous and smaller *Purkinje fibers.*

The SA node is also called the **pacemaker** because it usually keeps the heartbeat regular. If the SA node fails to work properly, the heart still beats due to impulses generated by the AV node. But the beat is slower (40 to 60 beats per minute).

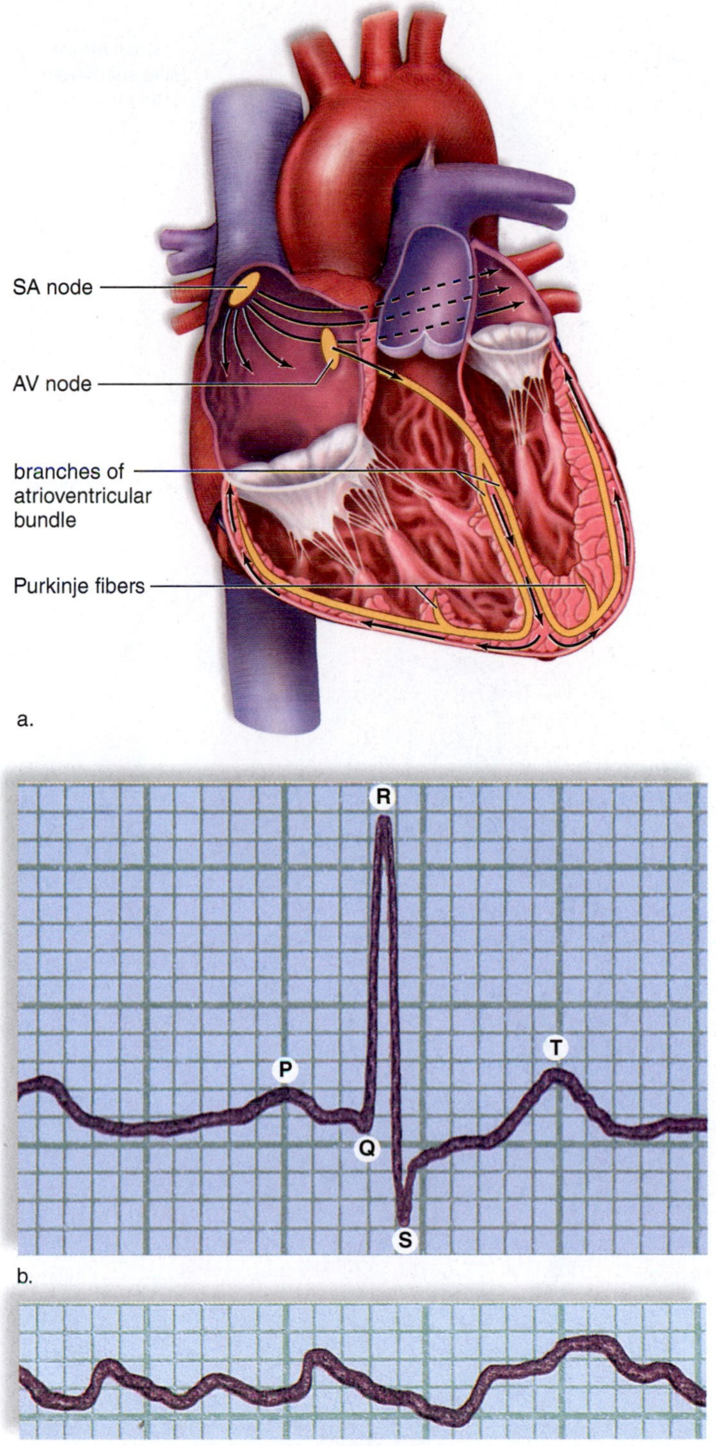
SA node
AV node
branches of atrioventricular bundle
Purkinje fibers
a.
R
P
T
Q
S
b.
c.

Figure 12.13 Conduction system of the heart. **a.** The SA node sends out a stimulus (black arrows), which causes the atria to contract. When this stimulus reaches the AV node, it signals the ventricles to contract. Impulses pass down the two branches of the atrioventricular bundle to the Purkinje fibers, and thereafter the ventricles contract. **b.** A normal ECG usually indicates that the heart is functioning properly. The P wave occurs just prior to atrial contraction; the QRS complex occurs just prior to ventricular contraction; and the T wave occurs when the ventricles are recovering from contraction. **c.** Ventricular fibrillation produces an irregular electrocardiogram due to irregular stimulation of the ventricles.

To correct this condition, an artificial pacemaker may be implanted that gives an electrical stimulus to the heart every 0.85 seconds.

Extrinsic Control of Heartbeat

The body also has extrinsic (or external) ways to regulate the heartbeat. A cardiac control center in the medulla oblongata, a portion of the brain that controls internal organs, can alter the beat of the heart by way of the autonomic system, a portion of the nervous system. The autonomic system has two subdivisions: the parasympathetic division, which promotes functions of a resting state, and the sympathetic division, which brings about responses to increased activity or stress. The parasympathetic division decreases SA and AV nodal activity when we are inactive, and the sympathetic system increases these nodes' activity when we are active or excited.

The hormones epinephrine and norepinephrine, which are released by the adrenal medulla (inner portion of the adrenal gland), also stimulate the heart. When we are frightened, for example, the heart pumps faster and stronger due to sympathetic stimulation and because of the release of epinephrine and norepinephrine.

The Electrocardiogram

An **electrocardiogram** (**ECG**) is a recording of the electrical changes that occur in the myocardium during a cardiac cycle. Body fluids contain ions that conduct electrical currents, and therefore the electrical changes in the myocardium can be detected on the skin's surface. Electrodes placed on or near the chest are connected by wires to an instrument that detects and records the myocardium's electrical changes. Figure 12.13*b* depicts the electrical changes during a normal cardiac cycle.

Animation Cardiac Cycle

When the SA node triggers an impulse, the atrial fibers produce an electrical change called the P wave. The P wave indicates that the atria are about to contract. After that, the QRS complex signals that the ventricles are about to contract. The electrical changes that occur as the ventricular muscle fibers recover produce the T wave.

Various types of abnormalities, known as *arrhythmias,* can be detected by an electrocardiogram. The most common type is called *atrial fibrillation* (*AF*). Instead of the heart generating single, regular impulses from the AV node, multiple, chaotic impulses are generated, resulting in an irregular, fast heartbeat. Symptoms of AF may include *palpitations,* meaning a fluttering sensation in the heart, along with dizziness, weakness, or chest pain. Sometime AF occurs infrequently and resolves on its own; other cases may require medication.

A more serious medical emergency is *ventricular fibrillation* (*VF*), an uncoordinated contraction of the ventricles (Fig. 12.13*c*). VF most commonly follows a heart attack, but it also can be caused by an injury or drug overdose. Because a heart in VF is not pumping blood, it must be defibrillated by applying a strong electrical current for a short period of time. Then the SA node may be able to reestablish a coordinated beat. Many public places now have *automatic external* *defibrillators* (*AEDs*), which are small devices that can be used to determine whether a person is suffering from ventricular fibrillation—and if so, administer an appropriate electrical shock to the chest.

Check Your Progress 12.3

1. Name each blood vessel and heart chamber that blood passes through on its journey through the heart and lungs.
2. Predict what might happen to the lungs if the left ventricle was not able to pump blood properly.
3. Describe the specific cause of each sound of the heartbeat.
4. Explain why it is important for the speed and strength of heart contractions to be regulated both intrinsically and extrinsically.

12.4 The Vascular Pathways

Learning Outcomes

Upon completion of this section, you should be able to

1. Describe the flow of blood from the heart through all major parts of the body.
2. Explain the factors that affect blood pressure in arteries, capillaries, and veins.

The cardiovascular system includes two circuits (Fig. 12.14). The **pulmonary circuit** circulates blood through the lungs, and the **systemic circuit** serves the needs of body tissues.

The Pulmonary Circuit

Blood from all regions of the body first collects in the right atrium and then passes into the right ventricle, which pumps it into the pulmonary trunk. The pulmonary trunk divides into the right and left pulmonary arteries, which branch as they approach the lungs. The arterioles take blood to the pulmonary capillaries, where gas exchange occurs. Blood then passes through the pulmonary venules, which lead to the four pulmonary veins (two from each lung) that enter the left atrium. Because blood in the pulmonary arteries is O_2-poor but blood in the pulmonary veins is O_2-rich, it is not correct to say that all arteries carry blood high in oxygen and all veins carry blood low in oxygen. It is just the reverse in the pulmonary circuit, as well as in the umbilical arteries and vein of the developing fetus (see Chapter 22).

The Systemic Circuit

The systemic circuit includes the major arteries and veins shown in Figure 12.15. As mentioned, the largest artery in the systemic circuit is the aorta; the largest veins are the venae cavae. The superior vena cava collects blood from the head, the chest, and the arms, and the inferior vena cava collects blood from the lower body regions. Both enter the right atrium. The aorta and the venae cavae serve as the major pathways for blood in the systemic circuit.

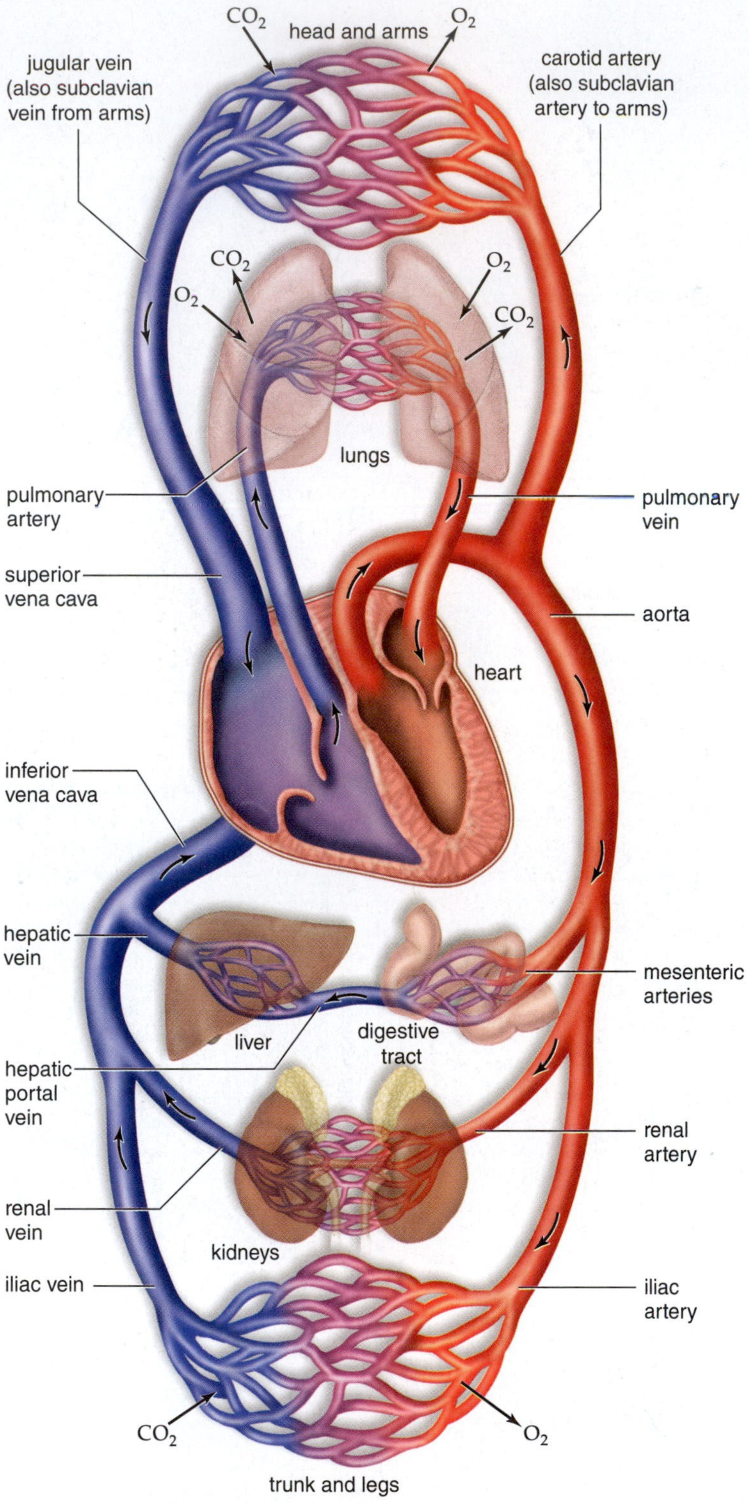

CO2
head and arms
O2
jugular vein (also subclavian vein from arms)
carotid artery (also subclavian artery to arms)
CO2
O2
O2
CO2
lungs
pulmonary artery
pulmonary vein
superior vena cava
aorta
heart
inferior vena cava
hepatic vein
mesenteric arteries
liver
digestive tract
hepatic portal vein
renal artery
renal vein
kidneys
iliac vein
iliac artery
CO2
O2
trunk and legs

Figure 12.14 Path of blood. This symbolic and not-to-scale drawing shows the path of blood in the pulmonary and systemic circuits. The pulmonary arteries and veins take blood from the right (blue) to the left (red) side of the heart. Tracing blood from the digestive tract to the right atrium in the systemic circuit involves the hepatic portal vein, the hepatic vein, and the inferior vena cava. The blue-colored vessels carry O_2-poor blood, and the red-colored vessels carry O_2-rich blood; the arrows indicate the direction of blood flow.

The path of systemic blood to any organ in the body begins in the left ventricle. For example, trace the path of blood to and from the legs in Figure 12.15:

left ventricle→aorta→common iliac artery→femoral artery→leg capillaries→femoral vein→common iliac vein→inferior vena cava→right atrium

Notice that, when tracing blood, you need only mention the aorta, the proper branch of the aorta, the region, and the vein returning blood to the vena cava. In most instances, the artery and the vein that serve the same region are given the same name (Fig. 12.15).

The *coronary arteries* (see Fig. 12.10) serve the heart muscle itself. (The heart is not nourished by the blood in its own chambers.) The coronary arteries are the first branches off the aorta. They originate just above the aortic semilunar valve, and they lie on the exterior surface of the heart, where they divide into diverse arterioles. Because they have a very small diameter, the coronary arteries may become clogged, as discussed in section 12.5. The coronary capillary beds join to form venules. The venules converge to form the cardiac veins, which empty into the right atrium.

A **portal system** in blood circulation begins and ends in capillaries. One such system, the *hepatic portal system,* is associated with the liver. Capillaries that occur in the villi of the small intestine pass into venules that join to form the *hepatic portal vein.* This vein carries the blood to a set of capillaries in the liver, an organ that monitors the makeup of the blood (see Fig. 12.14). The *hepatic vein* leaves the liver and enters the inferior vena cava.

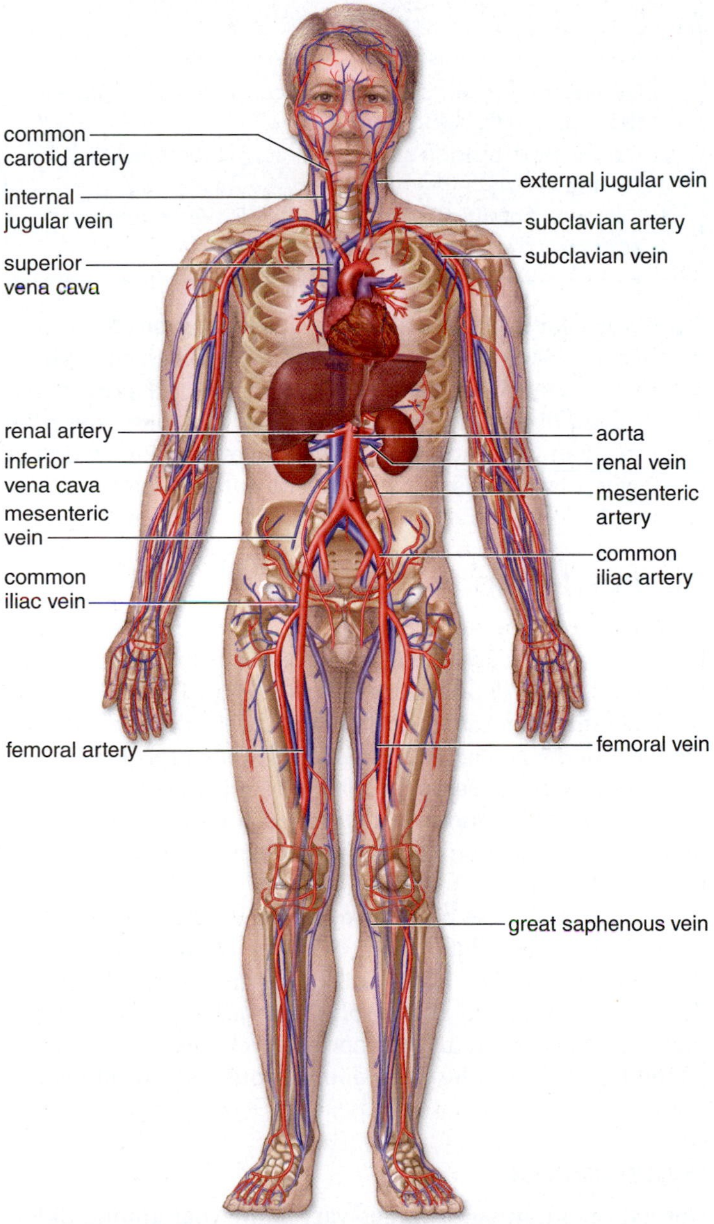
common carotid artery
internal jugular vein
superior vena cava
renal artery
inferior vena cava
mesenteric vein
common iliac vein
femoral artery
external jugular vein
subclavian artery
subclavian vein
aorta
renal vein
mesenteric artery
common iliac artery
femoral vein
great saphenous vein

Figure 12.15 Major arteries (red) and veins (blue) of the systemic circuit. This representation of the major blood vessels of the systemic circuit shows how the systemic arteries and veins are arranged in the body. The superior and inferior venae cavae take their names from their relationship to which organ?

Blood Pressure

When the left ventricle contracts, blood is forced into the aorta and then into other systemic arteries under pressure. *Systolic pressure* results from blood being forced into the arteries during ventricular systole, and *diastolic pressure* is the pressure in the arteries during ventricular diastole.

As blood flows from the aorta into the arteries and arterioles, blood pressure falls. Also, the difference between systolic and diastolic pressure gradually diminishes. In the capillaries, blood flow is slow and fairly even. This may be related to the very high total cross-sectional area of the capillaries (Fig. 12.16). It has

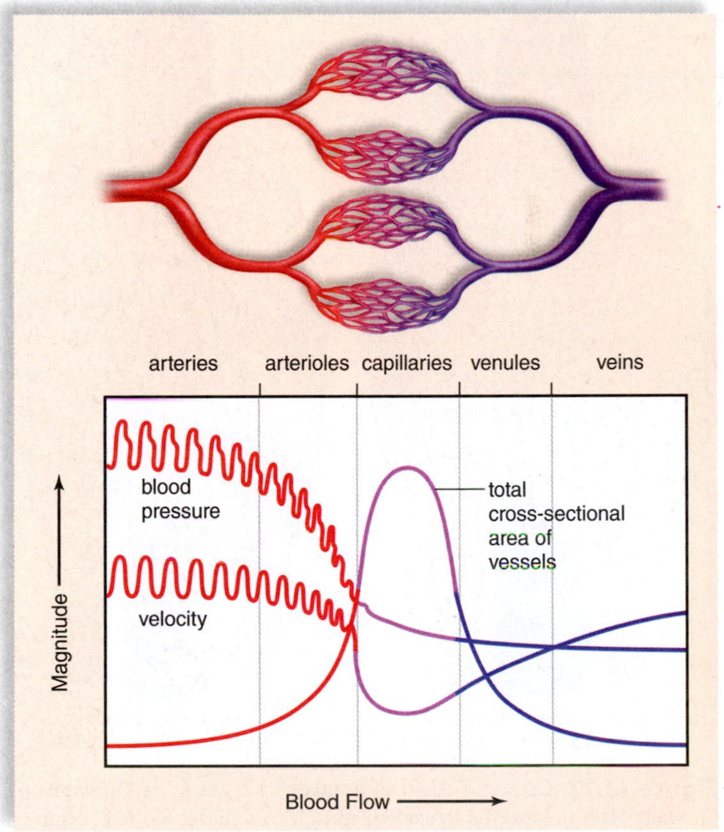
arteries
arterioles
capillaries
venules
veins
blood pressure
total cross-sectional area of vessels
velocity
Magnitude
Blood Flow

Figure 12.16 Blood velocity and blood pressure. In capillaries, blood is under minimal pressure and has the least velocity. Blood pressure and velocity drop off because capillaries have a greater total cross-sectional area than arterioles.

been calculated that if all the blood vessels in a human were connected end to end, the total distance would reach around Earth at the equator two times! Most of this distance would be due to the large number of capillaries.

Blood pressure can be measured with a *sphygmomanometer,* which has a pressure cuff that determines the amount of pressure required to stop the flow of blood through an artery. Blood pressure is normally measured on the *brachial artery,* an artery in the upper arm. Today, automated manometers are often used to take one's blood pressure instead. Blood pressure is expressed in millimeters of mercury (mm Hg). A blood pressure reading consists of two numbers that represent systolic and diastolic pressures, respectively—for example, 120/80 mm Hg. Blood pressure that is too low or—more commonly—too high can be a significant health problem (see section 12.5).

MP3 Blood Flow and Blood Pressure

Blood pressure in the veins is low and by itself is an inefficient means of moving blood back to the heart, especially from the limbs. When skeletal muscles near veins contract, they put pressure on the veins and the blood they contain. Valves prevent the backward flow of blood in veins, and therefore muscle contraction is sufficient to move blood toward the heart (Fig. 12.17). During long periods of sitting or other inactivity, clots may form in the deep veins of the legs. Because these can cause serious problems, especially if they break free and become lodged in the lungs, take frequent stretch breaks on long car or plane trips.

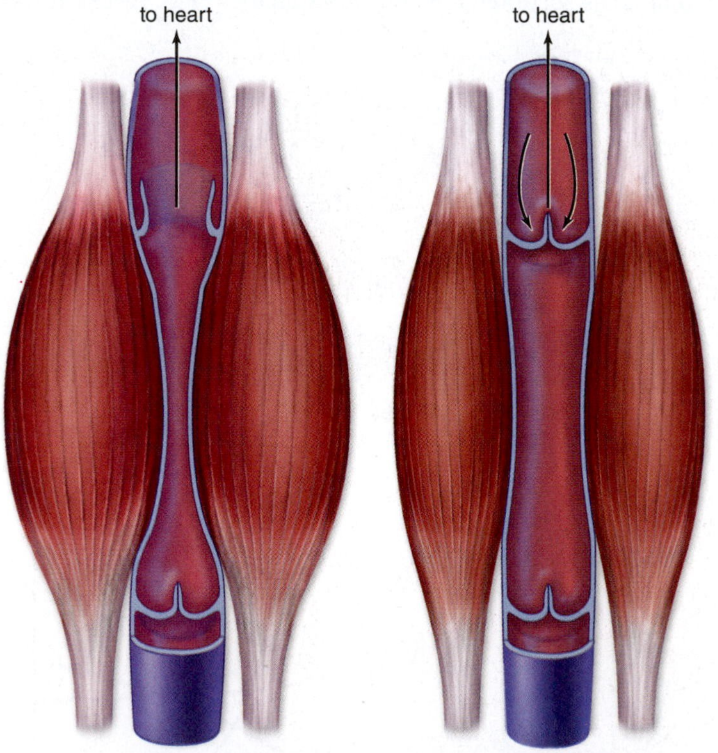
to heart
to heart

a. Contracted skeletal muscle pushes blood past open valve.

b. Closed valve prevents backward flow of blood.

Figure 12.17 Cross section of a valve in a vein. **a.** Pressure on the walls of a vein, exerted by skeletal muscles, increases blood pressure within the vein and forces the valve open. **b.** When external pressure is no longer applied to the vein, blood pressure decreases, and back pressure forces the valve closed. Closure of the valve prevents the blood from flowing in the opposite direction.

Check Your Progress 12.4

1. Identify which arteries carry O_2-poor blood.
2. Name the major blood vessels through which blood flows to the lungs and other parts of the body.
3. Describe why blood flows in one direction in an artery versus a vein.

12.5 Cardiovascular Disorders

Learning Outcomes

Upon completion of this section, you should be able to

1. Describe the major categories of cardiovascular disease that occur in the United States.
2. Define hypertension and explain its most common causes.
3. List three possible treatments for a blocked coronary artery.

Cardiovascular disease (CVD) is the leading cause of untimely death in most Western countries. Modern research efforts have resulted in improved diagnosis, treatment, and prevention. This section discusses some of the major advances that have been made in these areas. The Health feature, "Prevention of Cardiovascular Disease, describes ways to maintain cardiovascular health.

Atherosclerosis

Atherosclerosis is an accumulation of soft masses of fatty materials, particularly cholesterol, beneath the inner linings of arteries. Such deposits are called *plaque.* Plaque tends to protrude into the lumen of the vessel and interfere with the flow of blood (see Fig. 12B in the Health feature, "Prevention of Cardiovascular Disease"). In most instances, atherosclerosis begins in early adulthood and develops progressively through middle age, but symptoms may not appear until an individual is 50 or older.

Plaque can cause platelets to adhere to the irregular arterial wall, forming a clot. As long as the clot remains stationary, it is called a *thrombus,* but if it dislodges and moves along with the blood, it is called an *embolus. Thromboembolism,* a clot that has been carried in the bloodstream but is now lodged in a blood vessel, must be treated, or serious complications can arise.

Hypertension

Normal blood pressure values vary somewhat among different age groups, body sizes, and levels of athletic conditioning. However, it is estimated that about 33% of American adults have **hypertension,** which is high blood pressure. Another 30% are thought to have a condition called *prehypertension,* often

SCIENCE IN YOUR LIFE ▶ HEALTH

Prevention of Cardiovascular Disease

Many of us are predisposed to cardiovascular disease (CVD) due to factors beyond our control. Having a family history of heart attacks under age 55, being male or African American increases the risk. Other risk factors for CVD are related to our behavior.

The Don'ts

Smoking

When a person smokes, the drug nicotine, present in cigarette smoke, enters the bloodstream. Nicotine causes arterioles to constrict, including those that supply the heart itself. It also increases blood pressure, and the tendency of blood to clot. These factors may explain why about 20% of deaths from CVD are directly related to cigarette smoking.

Drug Abuse

Stimulants, such as cocaine and amphetamines, can cause an irregular heartbeat and lead to heart attacks even when using the drugs for the first time.

Obesity

People who are obese have more tissues that need to be supplied with blood. To meet this demand, the heart pumps blood out under greater pressure. Being overweight also increases the risk of type 2 diabetes, in which glucose damages blood vessels and makes them more prone to the development of plaque.

The Do's

Healthy Diet

Diet influences the amount of cholesterol in the blood. Cholesterol is ferried by two types of plasma proteins, called LDL (low-density lipoprotein) and HDL (high-density lipoprotein). LDL (the "bad" lipoprotein) takes cholesterol from the liver to the tissues, and HDL (the "good" lipoprotein) transports cholesterol out of the tissues to the liver. When the LDL level in blood is high or the HDL level is too low, plaque accumulates on arterial walls (Fig. 12B).

Eating foods high in saturated fat (red meat, cream, and butter) or trans-fats (most margarines, commercially baked goods, and deep-fried foods) increases LDL cholesterol. Replacement of these harmful fats with monounsaturated fats (olive and canola oil) and polyunsaturated fats (corn, safflower, and soybean oil) is recommended. Most nutritionists also suggest eating at least five servings of antioxidant-rich fruits and vegetables a day to protect against cardiovascular disease. Antioxidants protect the body from free radicals that can damage blood vessels.

The American Heart Association (AHA) recommends eating at least two servings of cold-water fish (e.g., halibut, sardines, tuna, and salmon) each week. These fish contain omega-3 polyunsaturated fatty acids that can reduce plaque. However, a 2008 study suggested that farm-raised tilapia contain low levels of these beneficial fatty acids and may, in fact, be detrimental to heart health.

Resveratrol

The "French paradox" refers to the observation that levels of CVD are relatively low in France, despite the common consumption of a high-fat diet. One possible explanation is that wine is frequently consumed with meals. In addition to its alcohol content, red wine contains an antioxidant called resveratrol. This chemical is mainly produced in the skin of grapes, so it is also found in grape juice, and supplements are available at health food stores.

Cholesterol Profile

Starting at age 20, all adults are advised to have their cholesterol levels tested at least every five years. Even in healthy individuals, an LDL level above 160 mg/100 ml and an HDL level below 40 mg/100 ml are matters of concern. Cholesterol-lowering medications are available for those who do not meet these minimum guidelines.

Exercise

People who exercise are less apt to have cardiovascular disease. Exercise not only helps keep weight under control, but may also help minimize stress and reduce hypertension. And short bursts of exercise may be superior to longer sessions. In one study, as few as three 10-minute workout sessions a day reduced triglyceride levels in blood better than one 30-minute session.

Anxiety and Stress

Mental stress can increase the odds of a heart attack. Within an hour of a strong earthquake that struck near Los Angeles in 1994, 16 people died of sudden heart failure (compared to the average of about 4 per day). Over the next several days, the number of heart-related deaths declined, suggesting that emotional stress had triggered fatal complications in those who were already predisposed to them. Obviously, it is difficult to avoid earthquakes, but we can learn healthy ways to avoid and manage stress.

Can Alcohol Benefit the Heart?

Alcohol abuse can destroy just about every organ in the body, the heart included. Based on several large research studies, however, the AHA notes that people who consume one or two drinks per day have a 30% to 50% reduced risk of CVD compared to nondrinkers. Importantly, because of the potential downsides of alcohol consumption, the AHA does not recommend that nondrinkers start using alcohol.

Questions to Consider

1. Review the risk factors discussed in this reading. Which, if any, could be affecting you right now?
2. Some scientists suspect that infections with certain types of bacteria and viruses may contribute to cardiovascular disease. What are some specific ways that infections might do this?
3. What are other possible explanations for the "French paradox"?

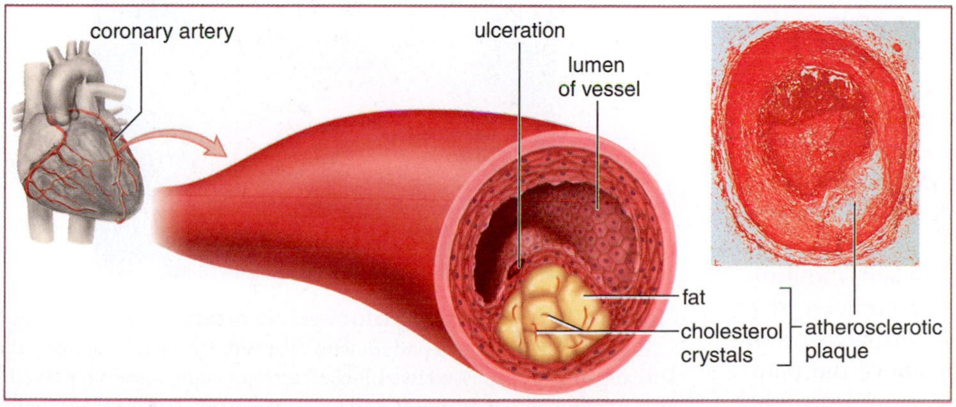

coronary artery
ulceration
lumen
of vessel
fat
cholesterol
crystals
atherosclerotic
plaque

Figure 12B Coronary arteries and plaque. Atherosclerotic plaque is an irregular accumulation of cholesterol and fat. When plaque is present in a coronary artery, a heart attack is more likely to occur because of restricted blood flow.

a precursor to hypertension. Under age 45, a blood pressure reading above 130/90 mm Hg is considered to be abnormally high, as is a reading above 140/95 mm Hg in older people.

Hypertension is sometimes called "the silent killer" because it may not be detected until a stroke or heart attack occurs. It most often occurs secondary to a narrowing of a person's arteries from atherosclerosis. Consider trying to force a gallon of water through a straw, versus a garden hose, in the same amount of time. You would need to apply much more force to get the water through the straw. In the same way, forcing blood through narrowed arteries over time creates additional pressure on the cardiovascular system that can damage the blood vessels, heart, and other organs.

Many medications are used to treat hypertension, including diuretics (which reduce the blood volume by increasing urine output), vasodilators (which dilate the blood vessels), and various drugs that improve heart function. Only about two-thirds of people with hypertension seek medical help for their condition, and it is likely that many people with high blood pressure are unaware of it.

Heart Valve Disease

Each year in the United States, around 90,000 people undergo surgery to have faulty heart valves repaired or replaced. In some patients, the heart valves are malformed at birth, but more commonly they degenerate due to age or infections, to the point where they no longer prevent the backflow of blood. A narrowing (stenosis) of the aortic valve opening is the most common, followed by mitral valve prolapse, in which abnormally thickened "leaflets" of the mitral valve protrude back into the left ventricle. In some cases, the faulty valves can be repaired during open-heart surgery, but more commonly they are replaced, using either artificial valves or valves removed from an animal (usually a pig) or a deceased human. Surgeons can also replace a defective heart valve by threading a compact artificial valve through an artery in the leg, thus avoiding open-heart surgery.

Stroke, Heart Attack, and Aneurysm

A cerebrovascular accident, also called a **stroke,** often results when an arteriole in the brain bursts or is blocked by an embolus. Lack of oxygen causes a portion of the brain to die, and paralysis or death can result. A person is sometimes forewarned of a stroke by a feeling of numbness in the hands or the face, difficulty in speaking, or temporary blindness in one eye. Cerebrovascular accidents become more common with age, but are certainly not limited to the elderly: in 2006, former Minnesota Twins baseball star Kirby Puckett died of a stroke at age 45.

If a coronary artery becomes partially blocked, the individual may suffer from **angina pectoris,** characterized by a squeezing or burning sensation in the chest. Nitroglycerin or related drugs dilate blood vessels and help relieve the pain. When a coronary artery is completely blocked, a portion of the heart muscle dies due to a lack of oxygen, and a myocardial infarction, or **heart attack,** occurs.

An **aneurysm** is the ballooning of a blood vessel, most often the abdominal aorta or the arteries leading to the brain. Atherosclerosis and high blood pressure can weaken the wall of an artery to the point that an aneurysm develops. If a major vessel such as the aorta should burst, about half of victims die before reaching a hospital. It is often possible to replace a damaged or diseased portion of a vessel, such as an artery, with a synthetic graft. At least two companies are developing wireless sensors that can be implanted along with these grafts, to help doctors monitor changes in pressure that may signal an aneurysm that is about to burst.

Coronary Bypass Operations

Since the late 1960s, coronary bypass surgery has been a common way to treat an obstructed coronary artery. During this operation, a surgeon usually takes a segment from another blood vessel and stitches one end to the aorta and the other end to a coronary artery past the point of obstruction. Figure 12.18 shows a heart in which three blocked coronary arteries have been bypassed using grafted arteries. Note that if veins are used as replacements, they must be placed in reverse orientation, due to the presence of valves. Even after successful bypass surgery, it is common for blockages to recur in the transplanted veins within five to seven years.

Some recent research suggests that heart muscle that has been damaged by a heart attack can be regenerated using stem cells. Studies using models of heart attacks in mice and rats have shown that stem cells injected directly into the damaged heart muscle will differentiate into new heart muscle cells, as well as new blood vessels. A few clinical trials are in progress,

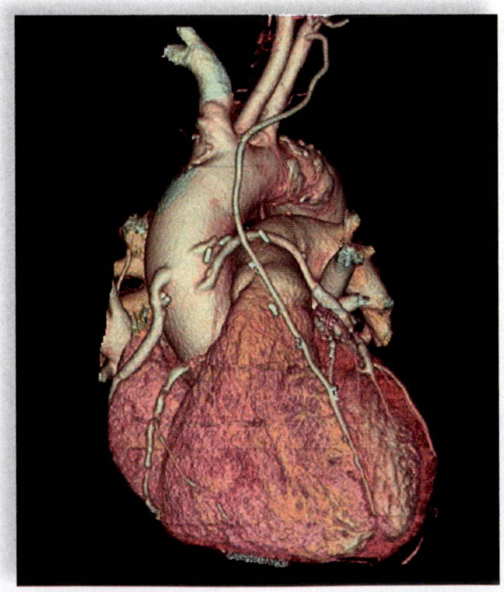

Figure 12.18 Bypassing blocked coronary arteries. This is a 3D scan of the heart of a patient who received a triple bypass operation. The surgeon has bypassed two blocked arteries using vessels removed from another part of the body, and used an existing artery that branches off of the left subclavian artery (see Figs. 12.10 and 12.11) to bypass a third blocked artery.

in which human patients have been injected with stem cells isolated from their own body, but the long-term results are still unknown. Scientists at the University of Minnesota went a step further in April 2011, when they announced that they had grown a human heart by starting with a cadaver heart, removing the cardiac muscle cells, then exposing this "scaffold" of blood vessels and connective tissues to human stem cells. Although these researchers are working to show that the resulting "bioartificial" heart can function normally, the ability to grow such organs would be a breakthrough for transplant medicine.

Video Heart Stem Cells

Clearing Clogged Arteries

Over the past decade, the proportion of heart attack patients receiving bypass surgery has decreased, mainly due to the increased usage of angioplasty and stent placement. In *angioplasty,* a cardiologist threads a catheter into an artery in the groin or upper part of the arm and guides it through a major blood vessel toward the heart. When the tube arrives at the region of plaque in an artery, a balloon attached to the end of the tube is inflated, forcing the vessel open, followed by removal of the tube. However, the artery may not always remain open because the trauma can cause smooth muscle cells in the wall of the artery to proliferate and close it.

More often, the same operation utilizes a cylinder of expandable metal mesh called a *stent.* Once the stent is in place, a balloon inside the stent is inflated, expanding it, and locking it in place (Fig. 12.19). In recent years, the most common types of stents implanted in patient's arteries are drug-eluting stents, because they have been coated with medications designed to inhibit inflammation and scar formation. Since 2003, these drug-coated stents have been implanted in over 6 million people worldwide. Compared to coronary bypass surgery, patients receiving stents generally have an easier recovery and shorter hospitalization, although there is some controversy about whether stents are now being overused.

Video Cardiac Repair

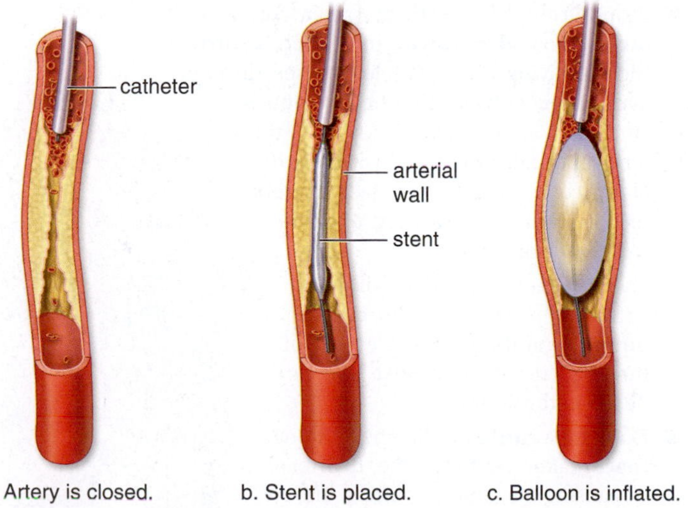
catheter
arterial wall
stent
a. Artery is closed.
b. Stent is placed.
c. Balloon is inflated.

Figure 12.19 Angioplasty with stent placement. **a.** A plastic tube (catheter) is inserted into the coronary artery until it reaches the clogged area. **b.** A metal stent with a balloon inside it is pushed out the end of the plastic tube into the clogged area. **c.** When the balloon is inflated, the vessel opens, and the stent is left in place to keep the vessel open.

Dissolving Blood Clots

Medical treatment for thromboembolism often includes the use of *tissue plasminogen activator* (*tPA*). This drug converts plasminogen into plasmin, an enzyme that dissolves blood clots, as mentioned in section 12.2. In fact, tPA is the body's own way of converting plasminogen to plasmin. tPA is also being used for thrombolytic stroke patients but with limited success because some patients experience life-threatening bleeding in the brain.

If a person has symptoms of angina or a stroke, aspirin may be prescribed. Aspirin reduces the stickiness of platelets, and thereby lowers the probability that a clot will form. Evidence indicates that aspirin protects against first heart attacks, but there is no clear support for taking aspirin every day to prevent strokes in symptom-free people. Even so, some physicians recommend taking a low dosage of aspirin every day.

Heart Transplants and Artificial Hearts

Heart transplants are usually successful today, but unfortunately, the need for hearts to transplant is greater than the supply because of a shortage of human donors. Therefore, a left ventricular assist device (LVAD), implanted in the abdomen, may help patients waiting for a transplant. A tube passes blood from the left ventricle to the device, which pumps it to the aorta. A cable passes from the device through the skin to an external battery the patient totes around.

Only a few patients have ever received a *total artificial heart.* One available model is called an AbioCor (Fig. 12.20). An internal battery and controller regulate the pumping speed, and an external battery powers the device by passing electricity through the skin. A rotating centrifugal pump moves silicon hydraulic fluid between left and right sacs to force blood out of the heart into the pulmonary trunk and the aorta. The exterior is made mainly of titanium, but the valves and membranes inside the ventricles are made of plastic, which has held up in testing to beating 100,000 times a day for years. So far, the longest a patient has lived after receiving an AbioCor heart is 512 days.

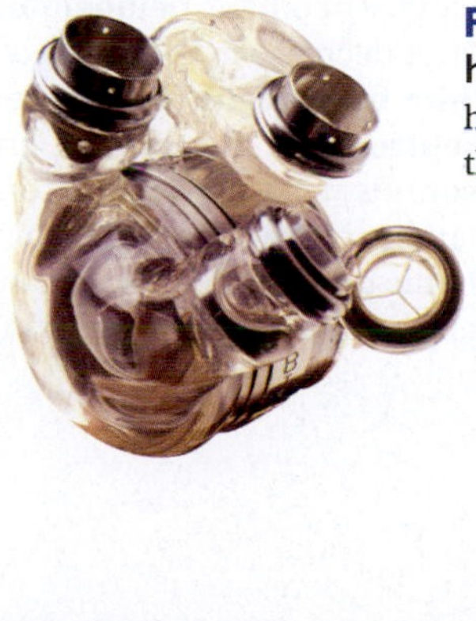

Figure 12.20 A total artificial heart. The AbioCor replacement heart is designed to be implanted within the chest cavity.

Check Your Progress 12.5

1. Define the following: thrombus, embolus, and aneurysm.
2. Describe some possible complications with stents placed in blocked arteries (i.e., what could go wrong).
3. Identify some treatment options, as well as potential limitations, for patients who need their heart replaced.

Case Study Conclusion

TV personality Tim Russert was a relatively famous recent victim of cardiovascular disease, but it is very likely that everyone reading this knows someone who has suffered a similar fate. Thus, in terms of life and death applications, the cardiovascular system is in some ways the most important system of all. Fortunately, medical science has identified many steps that can be taken to preserve the essential functions of our cardiovascular system.

MEDIA STUDY TOOLS

www.mhhe.com/maderinquiry14

Enhance your study of this chapter with study tools and practice tests. Also ask your instructor about the resources available through ConnectPlus, including LearnSmart, the media-rich eBook, interactive learning tools, and animations.

SUMMARIZE

12.1 The Blood Vessels

There are three types of blood vessels in the **circulatory system:**

- **Arteries** (and **arterioles**) take blood away from the heart. The largest artery is the **aorta.**
- **Capillaries** are the smallest vessels, where exchange of substances with the tissues occurs.
- **Veins** (and **venules**) take blood to the heart. Of these three types of vessels, only veins have valves. The largest veins are the **venae cavae.**

12.2 Blood

Blood is a connective tissue that functions in transport, homeostasis, protection, and temperature regulation. It is composed of the plasma and the formed elements:

- **Plasma** contains mostly water and proteins, as well as nutrients and wastes. The plasma proteins prothrombin and fibrinogen are well known for their contribution to the **clotting** process.
- The **formed elements** are red blood cells, white blood cells, and platelets. **Red blood cells (erythrocytes)** contain **hemoglobin,** which functions in oxygen transport. A decrease in red blood cell numbers is known as **anemia. White blood cells** (**leukocytes**) help defend against infections. **Neutrophils, monocytes,** and **dendritic cells** are phagocytic. **Basophils** release histamine, and **eosinophils** fight parasitic infestations. **Lymphocytes,** which include B and T cells, are involved in the development of adaptive immunity to infections. **Platelets** are involved in blood clotting, along with proteins like **thrombin** and **fibrin. Hemophilia** is a clotting deficiency caused by a genetic defect.
- **Stem cells** can give rise to many types of cells. The bone marrow contains multipotent cells that not only produce all types of blood cells, but may also be useful in treating many types of diseases.
- When blood reaches a capillary, water moves out at the arterial end due to blood pressure, forming **tissue fluid.** Some of this fluid moves back into capillaries due to osmotic pressure. Excess fluid is collected by **lymphatic capillaries,** forming **lymph.**

12.3 The Human Heart

The **heart** is made up largely of myocardium, and it is surrounded by a membrane called the pericardium. It has a right and left side and four chambers:

- On the right side, the venae cavae deliver O_2-poor blood from the body into the right atrium, and the right ventricle pumps it via the **pulmonary arteries** into the pulmonary circuit.
- On the left side, the **pulmonary veins** bring O_2-rich blood from the lungs into the left atrium, and the left ventricle pumps it through the aorta into the systemic circuit.
- Blood from each **atrium** (pl. **atria**) passes through **atrioventricular valves** into each **ventricle.** Blood leaving the ventricles passes through **semilunar valves.** The heart sounds, "lub-dub," are due to the closing of the atrioventricular valves, followed by the closing of the semilunar valves. **Systole** refers to heart contraction; **diastole** refers to relaxation.
- The volume of blood leaving the heart each minute is the **cardiac output,** and the pumping action of the heart results in a **pulse** that can be felt in the arteries.
- During the **cardiac cycle,** the SA node (**pacemaker**) initiates the heartbeat by causing the atria to contract. The AV node conveys the stimulus to the ventricles, causing them to contract. These electrical impulses can be monitored using an **electrocardiogram** (**ECG**).

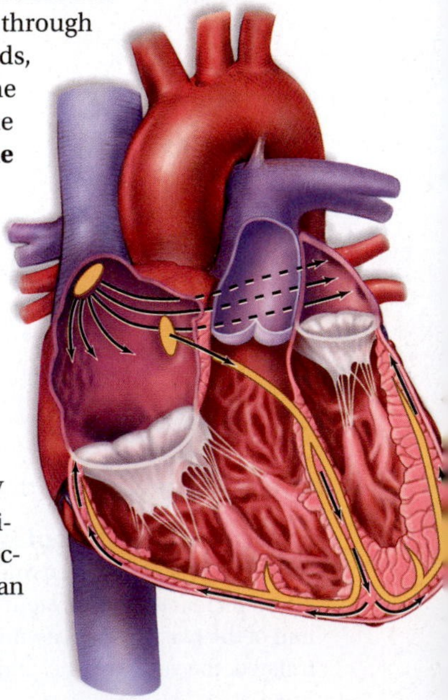

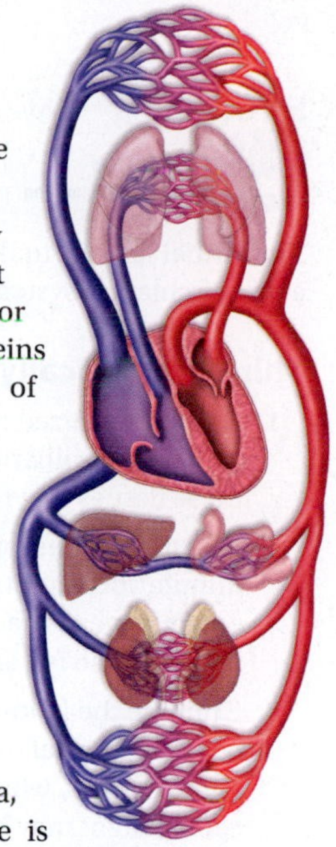

12.4 The Vascular Pathways

The cardiovascular system is divided into the pulmonary circuit and the systemic circuit:

- In the **pulmonary circuit,** the pulmonary trunk from the right ventricle of the heart and the two pulmonary arteries take O_2-poor blood to the lungs, and four pulmonary veins return O_2-rich blood to the left atrium of the heart.
- In the **systemic circuit,** O_2-rich blood is pumped from the left ventricle into the aorta, which branches off to form arteries going to specific organs.
- **Portal systems** begin and end in capillaries. For example, the hepatic portal vein originates from intestinal capillaries and finishes in hepatic capillaries.
- The left ventricle forces blood into the aorta, creating **blood pressure.** Blood pressure is higher in arteries than in veins, where muscular movement and the presence of one-way valves help return venous blood to the heart.

12.5 Cardiovascular Disorders

Although disorders of the cardiovascular system are the leading cause of death in many countries, medical science has developed many effective ways to treat, and preferably to prevent them:

- **Atherosclerosis** is an acculmulation of cholesterol and other substances in arteries. This condition can lead to **hypertension,** or high blood pressure.
- Diseases of the heart valves are also common, and usually treatable by surgery.
- Strokes, heart attacks, and aneurysms are very common disorders of the blood vessels. A **stroke** results when the blood supply to the brain is disrupted. An **aneurysm** is a ballooning of any blood/vessel, but aneurysms of the aorta or blood vessels supplying the brain are especially significant. A **heart attack** usually occurs when one or more of the coronary arteries supplying the heart becomes obstructed.

ASSESS

Testing Yourself

Choose the best answer for each question.

1. ____ lie between ____ and ____.
   a. Arteries, veins, capillaries
   b. Capillaries, arteries, veins
   c. Veins, arteries, capillaries
   d. None of these are correct.
2. The myocardium is made of
   a. muscle.
   b. epithelium.
   c. connective tissue.
   d. None of these are correct.
3. The average adult heart rate is about ____ beats per minute.
   a. 120
   b. 100
   c. 45
   d. 70
4. The "lub," or first heart sound, is produced by closing of the
   a. aortic semilunar valve.
   b. pulmonary semilunar valve.
   c. tricuspid valve.
   d. bicuspid valve.
   e. both AV valves.
5. Label the following ECG wave chart.

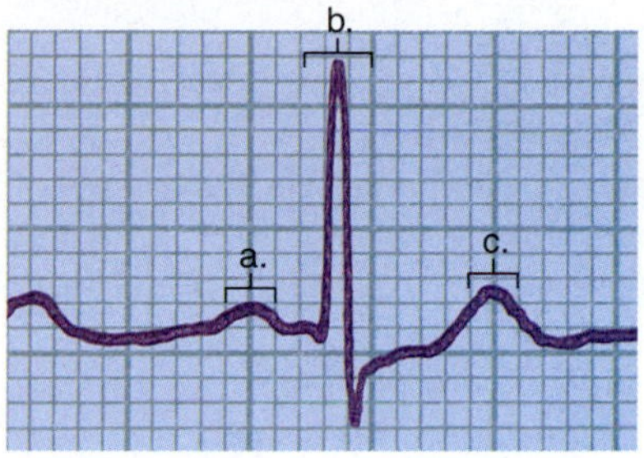

a.
b.
c.

6. All arteries in the body contain O_2-rich blood with the exception of the
   a. aorta.
   b. pulmonary artery.
   c. renal artery.
   d. coronary arteries.
7. The best explanation for the slow movement of blood in capillaries is
   a. skeletal muscles press on veins, not capillaries.
   b. capillaries have much thinner walls than arteries.
   c. there are many more capillaries than arterioles.
   d. venules are not prepared to receive so much blood from the capillaries.
   e. All of these are correct.
8. Label the following diagram of the systemic circuit.

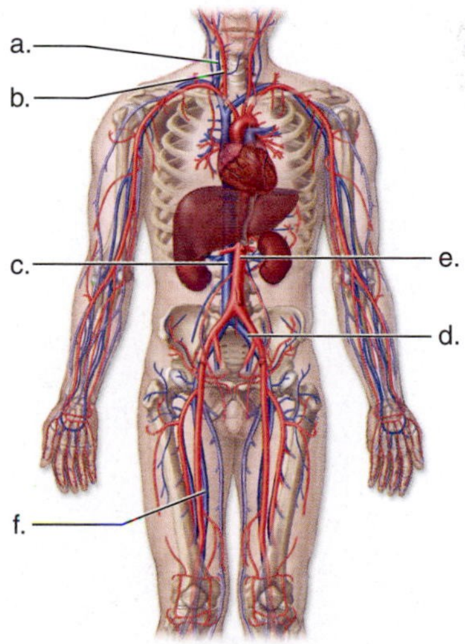

a.
b.
c.
e.
d.
f.

9. Which association is incorrect?
   a. white blood cells—fight infection
   b. red blood cells—inflammation
   c. red blood cells—contain hemoglobin
   d. platelets—blood clotting

10. A decrease in lymphocytes would result in problems associated with
    a. clotting.
    b. immunity.
    c. oxygen transport.
    d. All of these are correct.
11. Anemia can result from
    a. an iron-deficient diet.
    b. a loss of blood.
    c. destruction of red blood cells by a parasite.
    d. All of these are correct.
12. In the tissues, the amount of carbon dioxide is _____ than in the capillaries.
    a. lower
    b. higher
    c. neither higher nor lower
    d. None of these are correct.
13. Lymph is formed from
    a. urine.
    b. whole plasma.
    c. excess extracellular tissue fluid.
    d. All of these are correct.
14. Water enters the venous side of capillaries because of
    a. active transport from tissue fluid.
    b. higher osmotic pressure than blood pressure.
    c. higher blood pressure on the venous side.
    d. higher blood pressure than osmotic pressure.
    e. higher red blood cell concentration on the venous side.
15. A stent would most likely be used to treat
    a. hemophilia.
    b. atherosclerosis.
    c. mitral valve prolapse.
    d. hypertension.

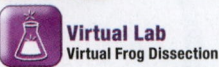

ENGAGE

Virtual Lab
Virtual Frog Dissection

The virtual lab "Virtual Frog Dissection" provides an interactive look at the circulatory system of a vertebrate animal.

Thinking Critically

1. A few specialized tissues do not contain any blood vessels, including capillaries. Can you think of two or three? How might these tissues survive without a direct blood supply?
2. Assume your heart rate is 70 beats per minute (bpm), and each minute your heart pumps 5.25 l of blood to your body. Based on your age to the nearest day, about how many times has your heart beat so far, and what volume of blood has it pumped?
3. Examine the following abnormal ECG tracings. Compared with the normal tracing in Testing Yourself Question 5 or Figure 12.13*b*, what is your best guess as to what type of problem each patient may have? What type of device can be implanted in a patient's heart to correct some abnormalities of the heart, and how does it generally work?

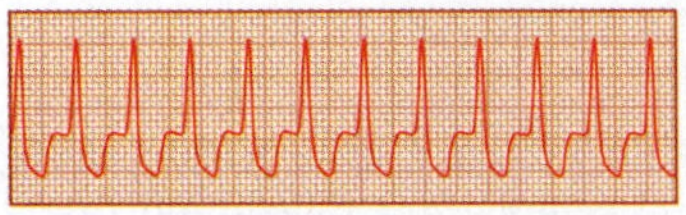
a.

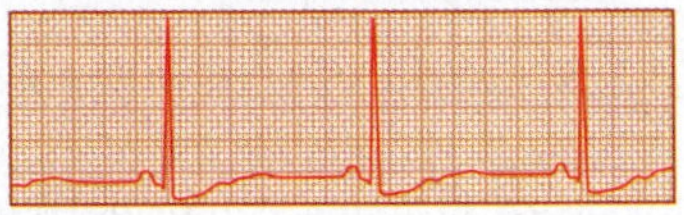
b.

4. There are not enough living hearts available to meet the need for heart transplants. What are some of the major problems a manufacturer would have to overcome when attempting to design an artificial heart that will last for years inside a patient's body?

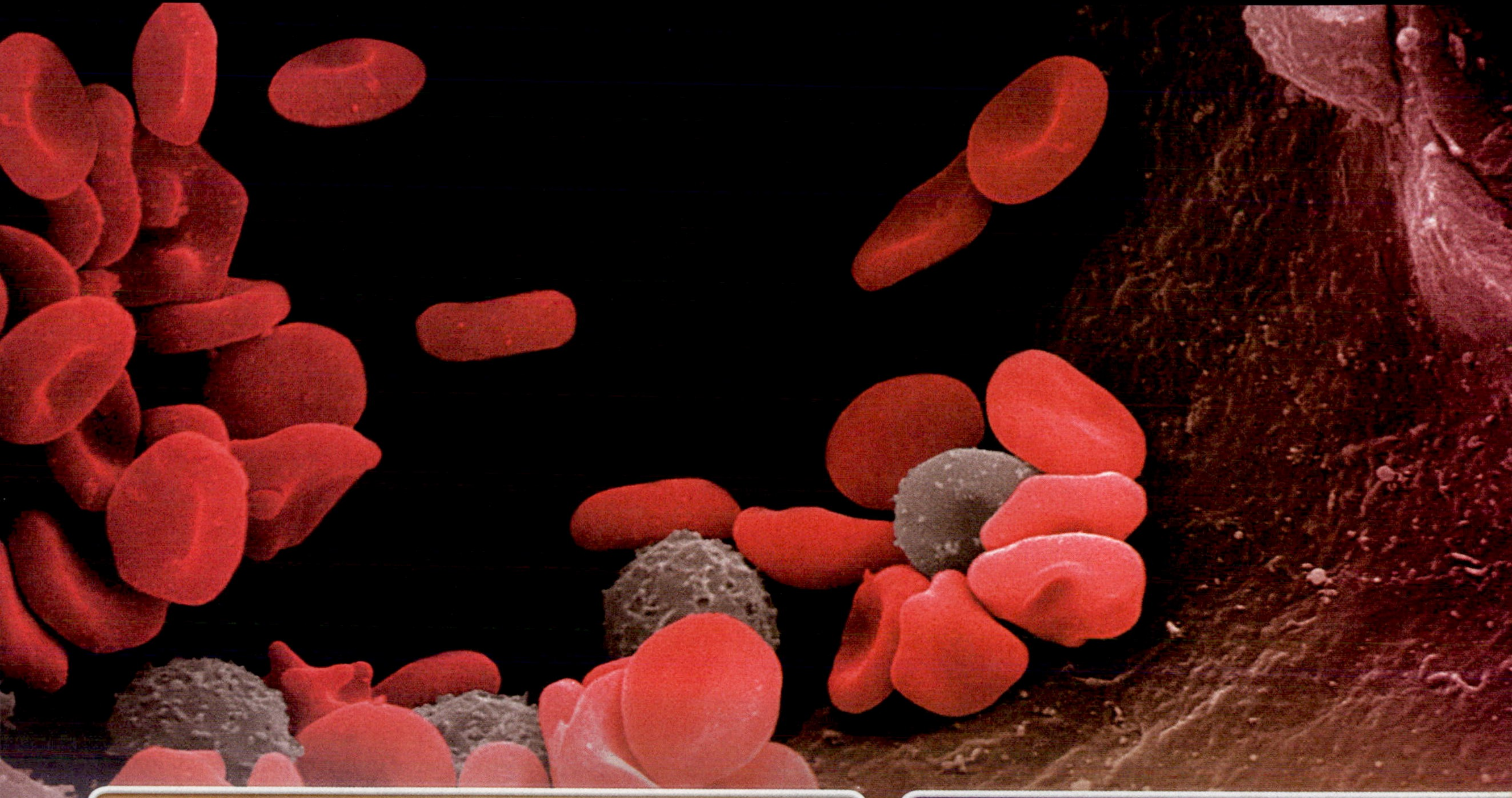

Lymphatic and Immune Systems 13

CASE STUDY When Antibodies Are Lacking

Jason is an active three-year-old boy, but his parents are worried. They have been making frequent visits to the pediatrician's office since their son was about six months old. He seems to be having one ear infection after another, and they can't remember him going a month without a cough or runny nose. These chronic infections led the doctor to suspect that Jason might be suffering from some kind of inherited defect in his immune system. After initial blood tests revealed that the boy had very low levels of antibodies in his blood, more specialized testing resulted in a diagnosis of X-linked agammaglobulinemia (XLA).

XLA is one of the most common inherited immunodeficiencies, affecting about 1 in 100,000 males (but extremely rare in females). Infant boys with XLA usually appear normal until about six months of age, when they start developing many infections. XLA is due to a mutated gene on the X chromosome that is needed for proper development of B lymphocytes, the type of white blood cell that produces antibodies (the ruffled-looking cells in the photo are white blood cells adhering to a blood vessel wall). As we will see in this chapter, antibodies are an important component of the human immune system, and without them, serious or even fatal infections often occur. In this case, Jason will need to be treated with intravenous injections of human antibodies every three to four weeks, for the rest of his life. His parents' hope, however, is that within their son's lifetime, doctors will be able to cure the genetic deficiency that caused his problem.

As you read through the chapter, think about the following questions:

1. Why do you think that individuals born with XLA usually appear to be normal until they reach about six months of age?
2. Why do patients with XLA tend to contract extracellular infections caused by bacteria, more than viral infections?

CHAPTER OUTLINE

BEFORE YOU BEGIN

Before beginning this chapter, take a few moments to review the following discussions:

Figure 2.25 What are the four levels of protein organization?

Section 4.1 What are the roles of proteins in the plasma membrane?

Section 11.5 What is the role of the immune system in maintaining homeostasis?

Photo: © Dr. Richard Kessel and Dr. Randy Kardon/Tissues and Organs/Visuals Unlimited

13.1 The Lymphatic System

Learning Outcomes

Upon completion of this section, you should be able to

1. Describe three major functions of the lymphatic system.
2. Distinguish between the roles of primary and secondary lymphoid organs, and list examples of each.

The **lymphatic system** consists of lymphatic vessels and the lymphoid organs (Fig. 13.1). This system, which is closely associated with the cardiovascular system, has three main functions that contribute to homeostasis: (1) lymphatic capillaries absorb excess tissue fluid and return it to the bloodstream; (2) lymphatic capillaries absorb fats from the digestive tract and transport them to the bloodstream; and (3) lymphoid organs help to defend the body against disease. This last function is mainly carried out by the white blood cells present in lymphatic vessels and lymphoid organs, as well as in the blood.

MP3 Lymphatic System

Lymphatic Vessels

Lymphatic vessels form a one-way system that begins with lymphatic capillaries. Most regions of the body are richly supplied with **lymphatic capillaries,** tiny, closed-ended vessels whose walls consist of simple squamous epithelium (Fig. 13.1). Lymphatic capillaries absorb excess tissue fluid. Tissue fluid is mostly water, but it also contains solutes (i.e., nutrients, electrolytes, and oxygen) derived from plasma and cellular products (i.e., hormones, enzymes, and wastes) secreted by cells. The fluid inside lymphatic vessels is called **lymph.**

The lymphatic capillaries join to form lymphatic vessels that merge before entering one of two ducts: the thoracic duct or the right lymphatic duct. The larger, thoracic duct

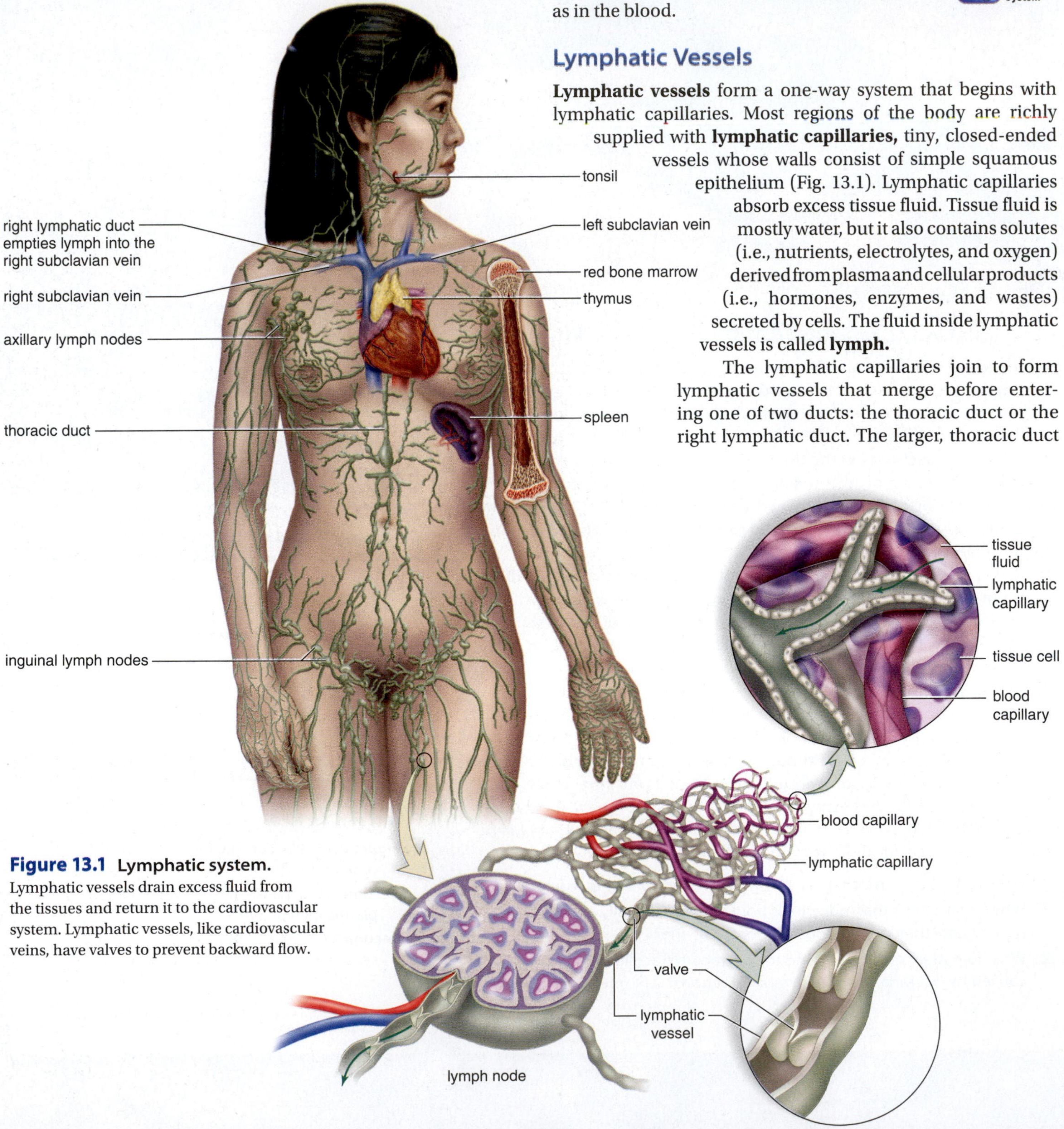
tonsil
right lymphatic duct empties lymph into the right subclavian vein
left subclavian vein
red bone marrow
right subclavian vein
thymus
axillary lymph nodes
spleen
thoracic duct
inguinal lymph nodes
tissue fluid
lymphatic capillary
tissue cell
blood capillary
blood capillary
lymphatic capillary
valve
lymphatic vessel
lymph node

Figure 13.1 Lymphatic system. Lymphatic vessels drain excess fluid from the tissues and return it to the cardiovascular system. Lymphatic vessels, like cardiovascular veins, have valves to prevent backward flow.

returns lymph collected from the body below the thorax and the left arm and the left side of the head and neck, into the left subclavian vein. The right lymphatic duct returns lymph from the right arm and right side of the head and neck into the right subclavian vein. On its way back to the cardiovascular system, lymph percolates through various lymph nodes, where any foreign material present can be recognized by the immune system (see "Secondary Lymphoid Organs" on page 234).

The structure of the larger lymphatic vessels is similar to that of cardiovascular veins. The movement of lymph within lymphatic capillaries is largely dependent upon skeletal muscle contraction. Lymph forced through lymphatic vessels as a result of muscular compression is prevented from flowing backward by one-way valves.

Edema is a localized swelling caused by the accumulation of tissue fluid that has not been collected by the lymphatic system. It occurs if too much tissue fluid is made and/or if not enough of it is drained away. Edema can lead to tissue damage and even death, illustrating the importance of the fluid-absorbing function of the lymphatic system.

Lymphoid Organs

Lymphoid organs contain large numbers of **lymphocytes,** the type of white blood cell that is mainly responsible for adaptive immunity. There are two types of lymphoid organs (Fig. 13.2). The red bone marrow and thymus are **primary lymphoid organs,** where lymphocytes develop and mature. The lymph nodes and spleen are examples of **secondary lymphoid organs**—sites where some lymphocytes become activated.

Primary Lymphoid Organs

Red bone marrow (Fig. 13.2*b*) contains a network of connective tissue fibers, along with stem cells that are ever capable of dividing and producing blood cells. Some of these cells become the various types of white blood cells: neutrophils, eosinophils, basophils, lymphocytes, and monocytes (see Fig. 12.3). The cells are packed around thin-walled sinuses filled with venous blood. Differentiated blood cells enter the bloodstream at these sinuses.

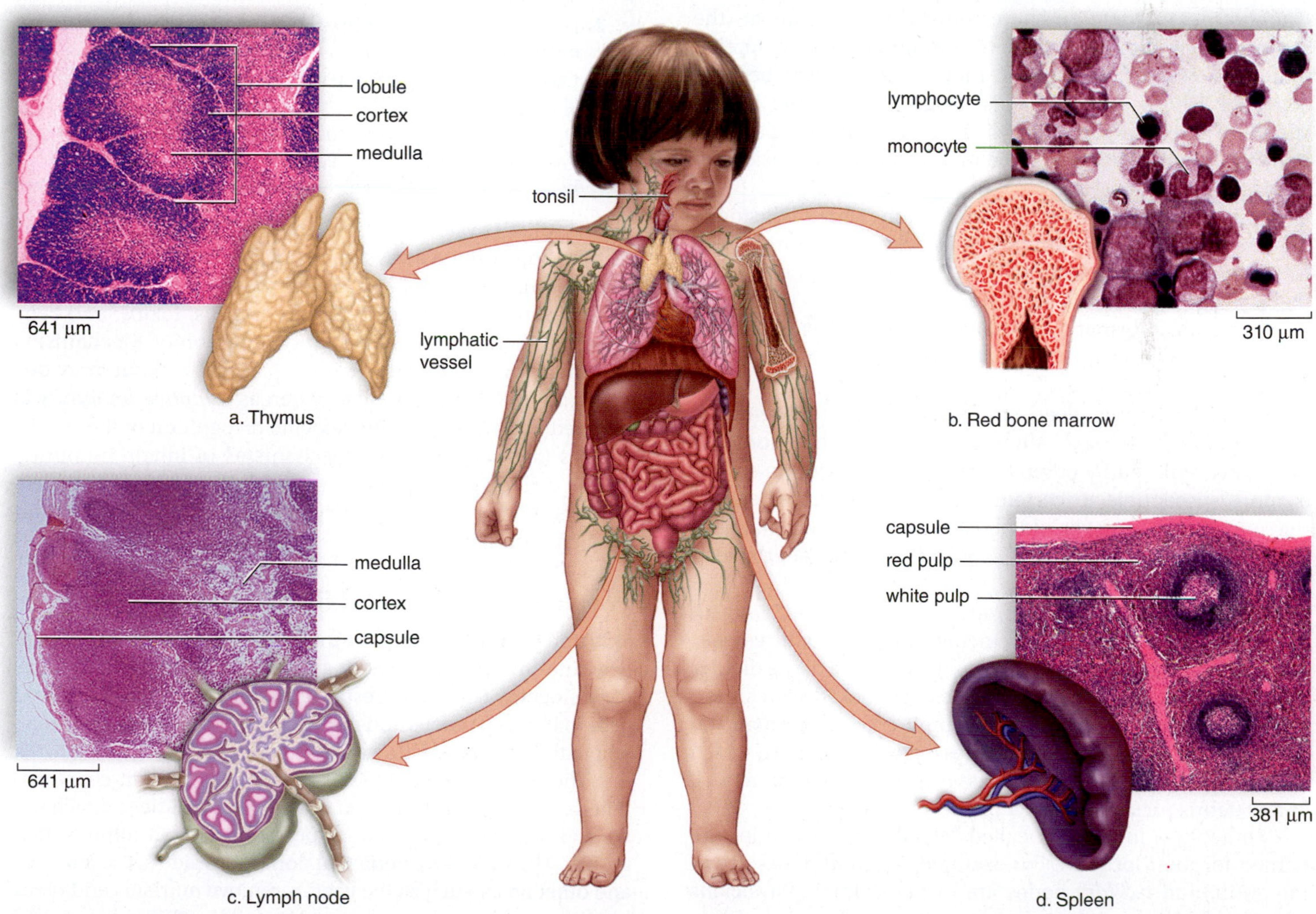
lobule
cortex
medulla
641 µm
a. Thymus
tonsil
lymphatic vessel
lymphocyte
monocyte
310 µm
b. Red bone marrow
medulla
cortex
capsule
641 µm
c. Lymph node
capsule
red pulp
white pulp
381 µm
d. Spleen

Figure 13.2 The lymphoid organs. The thymus (**a**) and red bone marrow (**b**) are the primary lymphoid organs. Blood cells, including lymphocytes, are produced in red bone marrow. B cells mature in the bone marrow. T cells mature in the thymus. The lymph nodes (**c**) and the spleen (**d**) are secondary lymphoid organs. Lymph is cleansed in the nodes, and blood is cleansed in the spleen.

In children, most bones contain red bone marrow, but in adults it is present mainly in the bones of the skull, sternum, ribs, clavicle, pelvis, and vertebral column. Although all lymphocytes begin their development in the bone marrow, the two major types of lymphocytes finish maturing in different locations. B lymphocytes, also called **B cells,** remain in the bone marrow until mature. Immature T lymphocytes (**T cells**) migrate from the bone marrow through the bloodstream to the thymus, where they mature.

The soft, bilobed **thymus** (Fig. 13.2*a*) is located in the thoracic cavity between the trachea and the sternum above the heart. The thymus grows largest around the time of puberty, and shrinks as we get older. Connective tissue divides the thymus into lobules, which are filled with T cells and supporting cells. Only about 5% of T cells ever exit the thymus. T cells that are capable of reacting to the body's own cells undergo *apoptosis,* which involves a cascade of specific cellular events leading to the death and destruction of the cell (see section 5.1). Those that leave the thymus are capable of reacting to foreign molecules (see section 13.4).

Secondary Lymphoid Organs

Lymphocytes frequently migrate from the bloodstream into the secondary lymphoid organs. Here, they may encounter foreign molecules or cells, after which they proliferate and become activated. These activated cells usually reenter the bloodstream, where they search for sites of infection or inflammation, like a team of highly trained military personnel seeking to destroy a specific enemy.

The **spleen** (Fig. 13.2*d*) is located in the upper left side of the abdominal cavity behind the stomach. Most of the spleen is *red pulp,* which consists of blood vessels and sinuses where macrophages remove old and defective blood cells. The spleen also has *white pulp* that consists of small areas of lymphoid tissue, where lymphocytes can react to foreign invaders that have gained access to the blood.

Occasionally, the spleen may need to be surgically removed due to trauma or disease. Although the spleen's functions can be mostly replaced by other organs, a person with no spleen is more susceptible to certain types of infections. Because the spleen also removes older red blood cells and platelets, the numbers of both types of cells may increase after spleen removal.

Lymph nodes (Fig. 13.2*c*) are small (from 1–25 mm in diameter), ovoid structures occurring along lymphatic vessels, through which the lymph must pass. Connective tissue divides the organ into nodules. Each of these is packed with B and T cells and contains a sinus. As lymph courses through the many sinuses, resident macrophages engulf **pathogens,** which are disease-causing agents such as viruses and bacteria, as well as any debris present in the lymph.

Sometimes incorrectly called "glands," lymph nodes are named for their location. For example, *inguinal* nodes are in the groin, and *axillary* nodes are in the armpits. Physicians often feel for the presence of swollen or tender lymph nodes as evidence that the body is fighting an infection.

Unfortunately, cancer cells sometimes enter lymphatic vessels and move undetected to other regions of the body, where they may produce secondary tumors. In this way, the lymphatic system sometimes inadvertently assists *metastasis,* the spreading of cancer far from its place of origin.

Animation Lymphatic System

Check Your Progress 13.1

1. Summarize the three functions of the lymphatic system.
2. Define edema. What can cause edema?
3. Identify two primary and two secondary lymphoid organs, and compare their specific functions.

13.2 Innate Immunity

Learning Outcomes

Upon completion of this section, you should be able to

1. List examples of physical and chemical barriers to infection.
2. Describe how an inflammatory response can be initiated.
3. Explain the major activities of phagocytes and natural killer cells.
4. Discuss the three main functions of the complement system.

The lymphatic system works with the **immune system** to protect the body from pathogens, toxins, and other invaders. The term **immunity** refers to a condition where the body is protected from various threats, like pathogens, toxins, and cancer cells. There are two main types of immunity. Mechanisms of **innate immunity** are fully functional without previous exposure to these substances, whereas *adaptive immunity* is initiated and amplified after specific recognition of these substances (see section 13.3). Mechanisms of innate immunity can be divided into at least four types: physical and chemical barriers, inflammation, phagocytes and natural killer cells, and protective proteins.

Physical and Chemical Barriers

Skin and the mucous membranes lining the respiratory, digestive, and urinary tracts serve as mechanical barriers to entry of pathogens. The upper respiratory tract is lined with ciliated cells that sweep mucus and trapped particles up into the throat, where they can be swallowed or expelled.

The secretions of oil glands in the skin contain chemicals that weaken or kill certain bacteria on the skin. The stomach has an acid pH, which kills many types of bacteria or inhibits their growth. The various bacteria that normally reside in the intestine and other areas, such as the vagina, remove nutrients and block binding sites that potentially could be used by pathogens.

MP3 Barriers and Nonspecific Defenses

Inflammatory Response

Damage to tissues—whether by physical trauma, chemical agents, or pathogens—initiates a series of events known as the **inflammatory response.** Inflammation tends to wall off infections and increase the exposure and access of the immune system to the inciting agent. Figure 13.3 illustrates the main participants in an inflammatory response.

An inflamed area has four "cardinal" signs: redness, heat, swelling, and pain. Most of these signs are due to capillary changes in the damaged area. At least three types of cells that reside in the skin and connective tissues promote inflammatory responses to various types of damage. **Mast cells,** a type of immune cell found especially in the skin, lungs, and intestinal tract, respond to damage by releasing **histamine,** a chemical that binds to receptors present on endothelial cells lining the blood vessels. This causes capillaries in the area to dilate and become more permeable, allowing fluids to escape into the tissues, resulting in swelling. The swollen area may stimulate free nerve endings, causing the sensation of pain. The increased blood flow also causes the skin to redden and feel warm.

Two other cell types that play an important role in inflammation are **macrophages** and **dendritic cells.** Both are phagocytic cells that reside in almost all types of tissues. Especially when they are exposed to invading microbes, both cell types can release various proinflammatory **cytokines,** chemical messengers that influence the activities of other immune cells. For example, after ingesting a bacterial pathogen, macrophages release a cytokine called *tumor necrosis factor alpha,* which acts on endothelial cells, and another known as *interleukin-8,* which attracts other types of immune cells to the scene. Some cytokines secreted by macrophages and dendritic cells act on the brain to induce a fever response; others, called *colony-stimulating factors,* cause the bone marrow to produce more white cells.

Animation Inflammatory Response

If the cause of inflammation cannot be eliminated, as occurs in tuberculosis and some types of arthritis, the inflammatory reaction may persist and become harmful rather than helpful. In these cases, anti-inflammatory medications such as aspirin, ibuprofen, or cortisone may be used to minimize the detrimental effects of chronic inflammation.

Phagocytes and Natural Killer Cells

At sites of inflammation, various types of white blood cells migrate through the walls of dilated capillaries, into the inflamed area. Among these are two types of **phagocytes,** which literally means "eating cell." **Neutrophils** are usually the first white blood cells to enter an inflamed area from the blood, and they may accumulate to form *pus,* which has a whitish appearance mainly due to their presence. If the inflammatory reaction continues, **monocytes** will migrate from the blood to the tissues, where they are then called macrophages (meaning, literally, "large eaters"). When a neutrophil or a macrophage encounters a pathogen, especially a bacterial cell, it will engulf (phagocytose) the pathogen into an endocytic vesicle (see Fig. 4.11), which fuses with a lysosome inside the cell. The lysosome has an acid pH that activates hydrolytic enzymes and various reactive oxygen compounds, which can usually destroy the pathogen.

Video Neutrophils

Natural killer (**NK**) **cells** are large, granular, lymphocyte-like cells that kill some virus-infected and cancer cells by

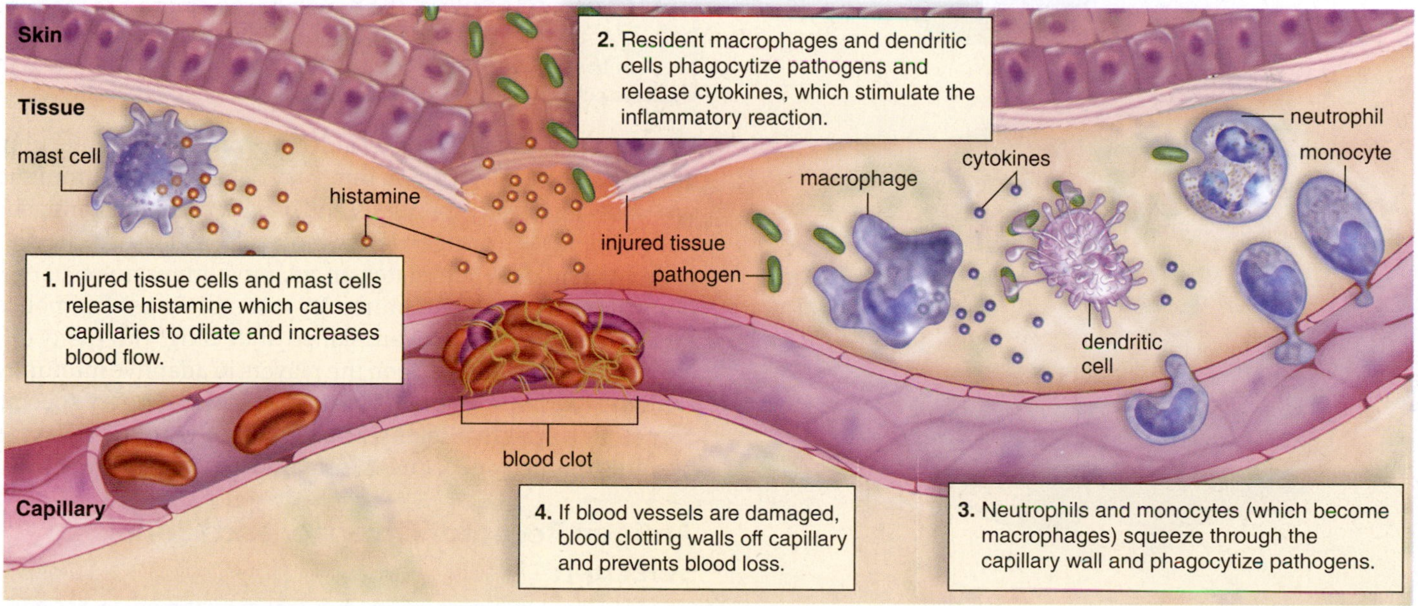
Skin
Tissue
mast cell
histamine
2. Resident macrophages and dendritic cells phagocytize pathogens and release cytokines, which stimulate the inflammatory reaction.
injured tissue
pathogen
macrophage
cytokines
neutrophil
monocyte
dendritic cell
1. Injured tissue cells and mast cells release histamine which causes capillaries to dilate and increases blood flow.
blood clot
Capillary
4. If blood vessels are damaged, blood clotting walls off capillary and prevents blood loss.
3. Neutrophils and monocytes (which become macrophages) squeeze through the capillary wall and phagocytize pathogens.

Figure 13.3 Inflammatory response. Due to capillary changes in a damaged area and the release of chemical mediators, such as histamine by mast cells, an inflamed area exhibits redness, heat, swelling, and pain. The inflammatory reaction can be accompanied by other reactions to the injury. Macrophages and dendritic cells, present in the tissues, phagocytize pathogens, as do neutrophils, which squeeze through capillary walls from the blood. Macrophages and dendritic cells release cytokines, which stimulate the inflammatory and other immune reactions. A blood clot can form to seal a break in a blood vessel.

cell-to-cell contact. Interestingly, NK cells and cytotoxic T cells (see next section) use very similar mechanisms to kill their target cells, by inducing them to undergo apoptosis. NK cells, however, seek out and kill cells that lack a particular type of "self" molecule, called MHC-I (major histocompatibility class I), on their surface. Because some virus-infected and cancer cells may lack these MHC-I molecules, they often become susceptible to being killed by NK cells. However, because NK cells do not recognize specific viral or tumor antigens, and do not proliferate when exposed to a particular antigen, they are considered a part of the innate immune system.

Protective Proteins

The complement system, often simply called **complement,** is composed of a number of blood plasma proteins designated by the letter C and a number (i.e., C3). The complement proteins "complement" certain immune responses, which accounts for their name. For example, they are involved in and amplify the inflammatory reaction because certain complement proteins can bind to mast cells and trigger histamine release, and others can attract phagocytes to the scene. Some complement proteins bind to the surface of pathogens, as described later in this section, which ensures that the pathogens will be phagocytized by a neutrophil, dendritic cell, or macrophage.

Animation Activation of Complement

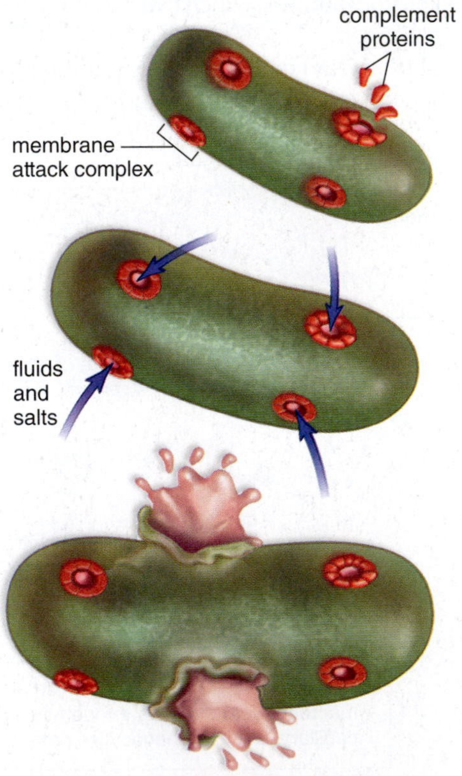
complement proteins
membrane attack complex
fluids and salts

Figure 13.4 Action of the complement system against a bacterium. When complement proteins in the blood plasma are activated in an immune response, they form a membrane attack complex. This complex makes holes in bacterial cell walls and plasma membranes, allowing fluids and salts to enter until the cell eventually bursts.

Certain other complement proteins join to form a *membrane attack complex* that produces holes in the surface of bacteria and some viruses. Fluids and salts then enter the bacterial cell or virus to the point that they burst (Fig. 13.4).

Interferons are proteins produced by virus-infected cells as a warning to noninfected cells in the area. Interferons can bind to receptors of noninfected cells, causing them to prepare for possible attack by producing substances that interfere with viral replication. Interferons may be administered to treat certain viral infections, such as hepatitis C.

Animation Antiviral Activity of Interferons

Check Your Progress 13.2

1. Name one physical and one chemical barrier to infection.
2. List the four cardinal signs of inflammation, and explain why each one occurs.
3. Contrast the way that macrophages typically kill pathogens with the method used by natural killer cells.
4. Summarize three major functions of the complement system.

13.3 Adaptive Immunity

Learning Outcomes

Upon completion of this section, you should be able to

1. Describe the major steps in the generation of antigen receptor diversity, and clonal selection.
2. Compare and contrast the activities of B cells versus T cells.
3. Describe the basic structure of an antibody molecule, and explain the different functions of IgG, IgA, IgM, and IgE.
4. Distinguish between the functions of helper T cells and cytotoxic T cells, and explain the role of MHC molecules in these responses.

When innate defenses have failed to prevent an infection, **adaptive immunity** comes into play. The adaptive immune system recognizes, responds to, and usually eliminates antigens from the body. An **antigen** is any molecule, usually a protein or carbohydrate, that stimulates an adaptive immune response. If mechanisms of innate immunity resemble a general police force, with members who are skilled at dealing with many different types of criminals, then the players in adaptive immunity are more like bounty hunters trained to seek out a very specific perpetrator. Adaptive defenses usually take five to seven days to become fully activated, and they may last for many years.

Antigen Receptor Diversity and Clonal Selection

The adaptive immune system depends primarily on the activity of B cells and T cells. Both cell types are capable of recognizing antigens because they have specific *antigen receptors*—plasma membrane proteins whose shape allows them to bind to particular antigens. Each lymphocyte has only one type of receptor. Because we could potentially encounter millions of different antigens during our lifetime, our B and T cells need a diversity of antigen receptors to protect us against them. Remarkably, diversification occurs to such an extent during the maturation process that there are specific B cells and/or T cells for almost any possible antigen.

Although the mechanisms accounting for the generation of antigen receptor diversity are complex, four major mechanisms account for this phenomenon as follows:

1. The genes that code for antigen receptors of B and T cells contain a variety of *gene segments*; small DNA sequences that each code for the portion of the antigen receptor that will eventually bind to the antigen.
2. In developing B and T cells, an enzyme cuts out some of these gene segments and links them back together in different combinations.
3. As the gene segments are being linked, another enzyme introduces mutations, thereby greatly increasing the variety in these genes.
4. Each antigen receptor is made up of at least two protein chains, each of which undergoes steps 1–3 independently, resulting in various combinations of the chains.

The **clonal selection theory** states that in a large population of B and T cells, only a few cells will have antigen receptors that can react with any particular antigen in the body (be it an infection, toxin, cancer cell, etc.). However, when an antigen binds to receptors on a particular lymphocyte, that cell is "selected," (i.e., it divides, forming "clones" of itself). This can be seen in Figure 13.5, where each B cell has a specific antigen receptor, or **B-cell receptor** (**BCR**), represented by shape. Only the B cell with a BCR that fits the antigen (green circles) undergoes clonal expansion.

Because our various immune defenses do not ordinarily react to our own normal cells, it is said that the immune system is able to distinguish "self" from "nonself." However, because the generation of antigen receptors is a random genetic process, some of these receptors will be capable of reacting with self-antigens (i.e., molecules normally found in the body). Such cells could attack our own tissues, a situation that would be incompatible with life. Fortunately this disaster is avoided, because any lymphocytes that react with self-antigens early in their development (in the primary lymphoid tissues) undergo apoptosis.

B Cells and Antibody-Mediated Immunity

B cells are usually activated in a lymph node or the spleen, when their BCRs bind to specific antigens. During clonal expansion, cytokines secreted by helper T cells stimulate B cells to divide (Fig. 13.5). Most of the resulting clones become **plasma cells,** which are specialized for the secretion of antibodies. **Antibodies** are the secreted form of the BCR of the B cell that was activated. Some cloned B cells become **memory B cells,** which are the means by which long-term immunity is possible. If the same antigen enters the system again, memory B cells quickly divide and give rise to more plasma cells capable of rapidly producing the correct type of antibody. Once an infection has been

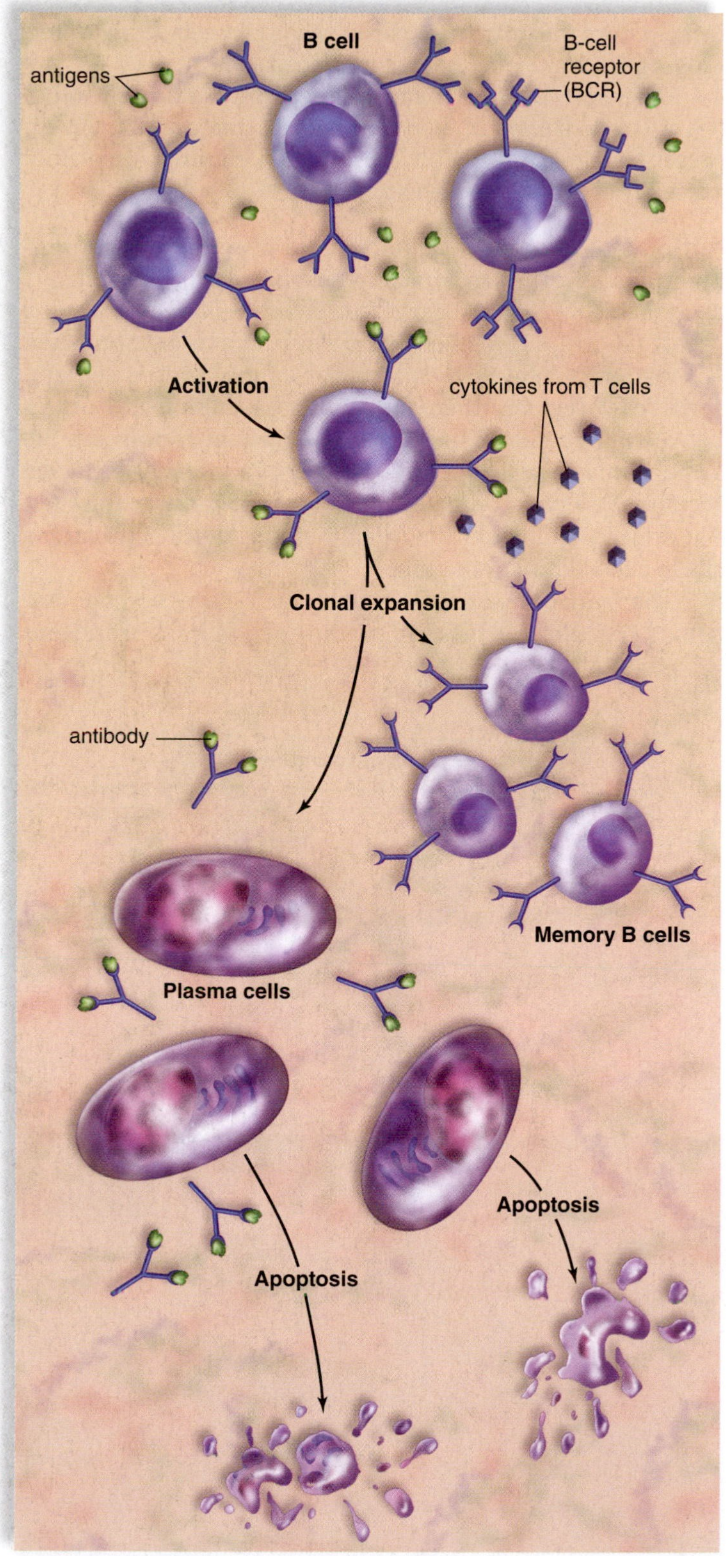
B cell
B-cell receptor (BCR)
antigens
Activation
cytokines from T cells
Clonal expansion
antibody
Memory B cells
Plasma cells
Apoptosis
Apoptosis

Figure 13.5 Clonal selection theory as it applies to B cells. Each B cell has a B-cell receptor (BCR) designated by shape that will combine with a specific antigen. Activation of a B cell occurs when its BCR can combine with an antigen (colored green). In the presence of cytokines, the B cell undergoes clonal expansion, producing many plasma cells and memory B cells. These plasma cells secrete antibodies specific to the antigen, and memory B cells immediately recognize the antigen in the future. After the infection passes, plasma cells undergo apoptosis, also called programmed cell death.

cleared from the body, new plasma cells cease to develop, and most of those present undergo apoptosis.

Defense by B cells is called **antibody-mediated immunity.** It is also called *humoral immunity* because these antibodies are present in blood and lymph. (Historically, a humor is any fluid normally occurring in the body.)

Antibody Structure Antibodies are also called **immunoglobulins (Ig).** They are typically Y-shaped molecules with two arms. Each arm has a "heavy" (long) polypeptide chain and a "light" (short) polypeptide chain. These chains have constant (C) regions, where the sequence of amino acids is set, and variable (V) regions, where the sequence of amino acids varies between antibodies (Fig. 13.6). The constant regions are the same within a particular type or class of antibodies (see next section). The variable regions form the antigen-binding sites. The antigen combines with the antibody at the antigen-binding site in a lock-and-key manner.

The antigen-antibody reaction can have several outcomes, but often the reaction produces complexes of antigens combined with antibodies. These complexes, also called *immune complexes,* may mark the antigens for destruction. For example, an immune complex may be engulfed by a neutrophil or macrophage, or it may activate complement. Antibodies may also "neutralize" viruses or toxins by preventing their binding to cells.

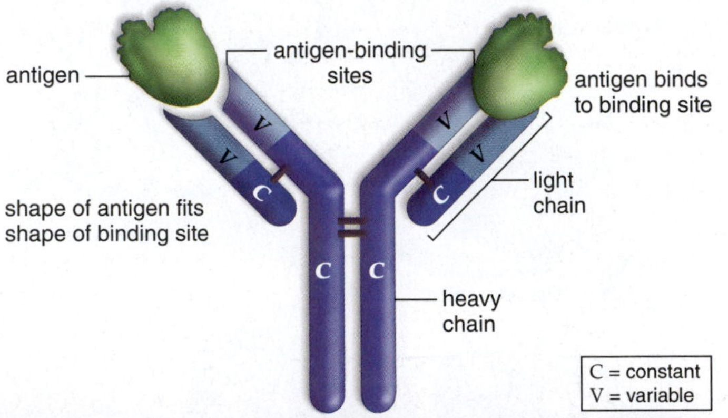
antigen
antigen-binding sites
antigen binds to binding site
V
V
V
V
C
C
C
C
light chain
shape of antigen fits shape of binding site
heavy chain
C = constant
V = variable

a.

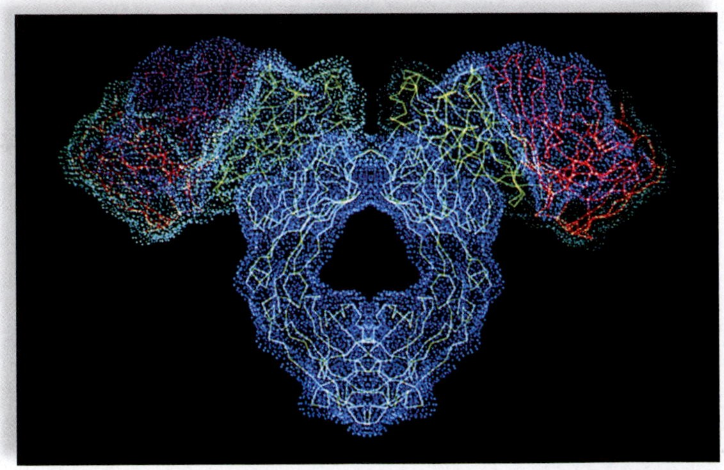

b.

Figure 13.6 Structure of an antibody. **a.** An antibody contains two heavy (long) polypeptide chains and two light (short) chains arranged so that there are two variable regions, where a particular antigen is capable of binding with an antibody (V = variable region, C = constant region). The shape of the antigen fits the shape of the binding site. **b.** Computer model of an antibody molecule. The antigen combines with the two side branches.

TABLE 13.1 Classes of Human Antibodies

Class	Distribution	Function
IgG	Main antibody type in circulation	Binds to pathogens, activates complement, and enhances phagocytosis
IgM	Antibody type found in circulation; largest antibody	Activates complement; clumps cells
IgA	Main antibody type in secretions such as saliva and milk	Prevents pathogens from attaching to epithelial cells in digestive and respiratory tract
IgD	Antibody type found on surface of immature B cells	Presence signifies readiness of B cell to respond to antigens
IgE	Antibody type found as antigen receptors on eosinophils in blood and on mast cells in tissues	Responsible for immediate allergic response and protection against certain parasitic worms

Types of Antibodies There are five major classes of antibodies in humans (Table 13.1). Immunoglobulin G (IgG) antibodies are the major type in blood, lymph, and tissue fluid. IgG antibodies bind to pathogens and their toxins and can activate complement. IgG is the only antibody class that can cross the placenta. Along with IgA, IgG is found in breast milk. IgM antibodies are pentamers, meaning that they contain five of the Y-shaped structures shown in Figure 13.6*a*. These antibodies appear in blood soon after a vaccination or infection and disappear relatively quickly. They are good complement activators. IgA antibodies are monomers in blood and lymph or dimers in tears, saliva, gastric juice, and mucous secretions. They are the main type of antibody found in body secretions. IgD molecules appear on the surface of B cells when they are ready to be activated. IgE antibodies are important in the immune system's response to parasites (e.g., parasitic worms). They are best known, however, for immediate allergic responses, including anaphylactic shock (discussed in section 13.5). IgD antibodies are found mainly on the surface of immature B cells.

T Cells and Cell-Mediated Immunity

As a T cell leaves the thymus, it has a unique **T-cell receptor (TCR)** similar to the BCR on B cells. Unlike B cells, however, there is no secreted form of the TCR, and T cells are unable to recognize an antigen without help. The antigen must be displayed, or "presented," to the TCR by an **MHC (major histocompatibility complex)** protein on the surface of another cell.

There are two major types of T cells: **helper T cells** (T_H cells) and **cytotoxic T cells** (T_C cells, or CTLs). Each type has a TCR

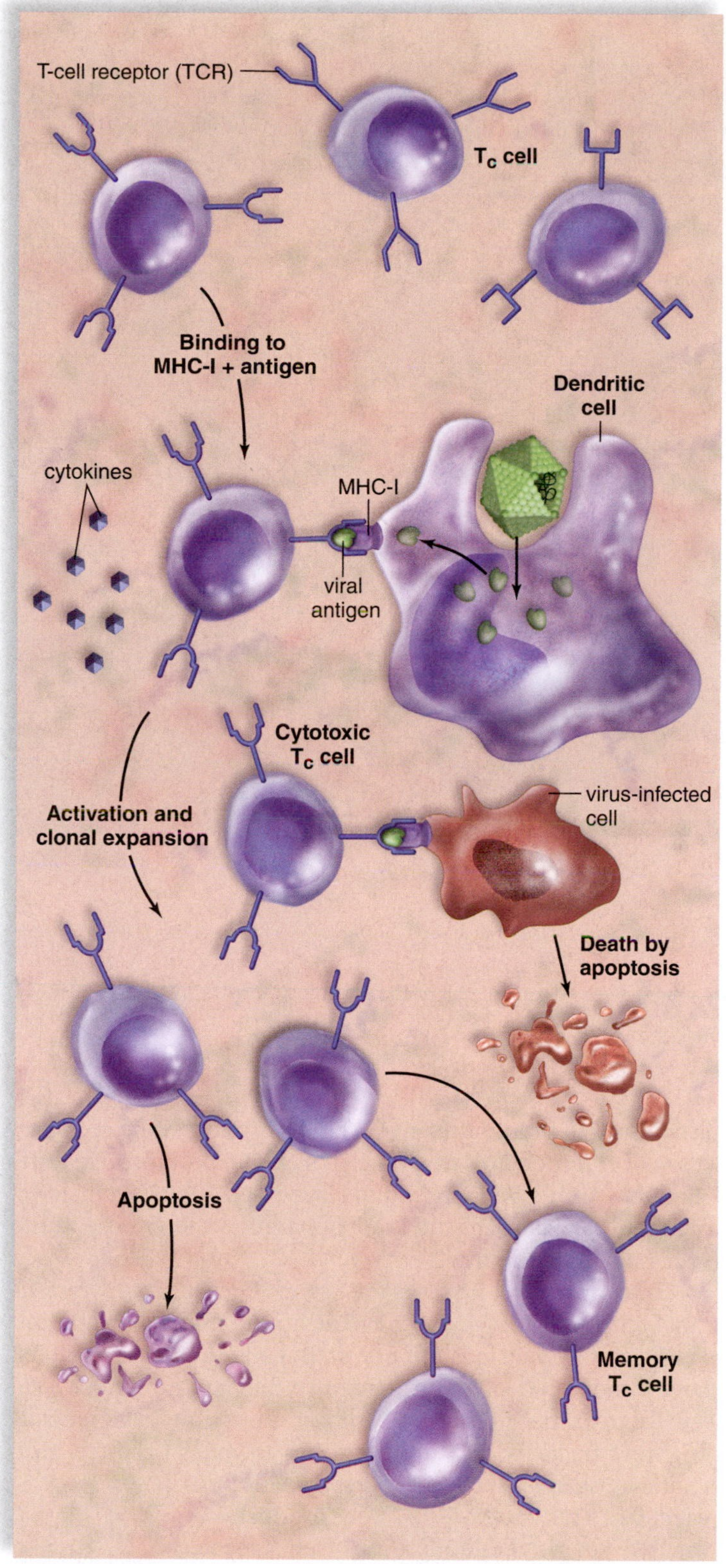
T-cell receptor (TCR)
T_C cell
Binding to
MHC-I + antigen
Dendritic
cell
cytokines
MHC-I
viral
antigen
Cytotoxic
T_C cell
virus-infected
cell
Activation and
clonal expansion
Death by
apoptosis
Apoptosis
Memory
T_C cell

Figure 13.7 Clonal selection model as it applies to T_C cells. Each T cell has a T-cell receptor (TCR) designated by a shape that will combine only with a specific antigen. Activation of a T_C cell occurs when its TCR can combine with an antigen such as a virus (colored green). A dendritic cell presents fragments of the antigen in the groove of its MHC class I (MHC-I) molecules. Thereafter, the T_C cell undergoes clonal expansion, producing many T_C cells with the same antigen specificity. After the antigen has been removed, the majority of T_C cells undergo apoptosis, but a small number of memory T_C cells remain. Memory T_C cells provide protection should the same antigen enter the body again at a future time.

that can recognize an antigen fragment in combination with an MHC molecule. However, T_H cells only recognize antigen presented by specialized **antigen-presenting cells (APCs)** with MHC class II proteins on their surface, whereas the T_C cells only recognize antigen presented by various cell types with MHC class I proteins on their surface (Fig. 13.7).

Video T Lymphocyte

APCs such as dendritic cells or macrophages play an important role in initiating most adaptive immune responses. After ingesting and destroying a pathogen, APCs travel to a lymph node or the spleen, where T cells also congregate. After breaking the pathogen apart with lysosomal enzymes, the APC displays a piece of the pathogen on its surface, in the binding groove of an MHC class II protein. After they are stimulated by APCs, T_H cells proliferate and secrete various cytokines. Cytokines produced by T_H cells help activate T_C cells, as well as B cells. Other cytokines, produced by a type of T_H cell called suppressor T cells, inhibit certain responses, thereby helping to control or regulate adaptive immunity.

In contrast to T_H cells, T_C cells often kill the cells they recognize. One method of killing used by T_C cells depends on storage vacuoles that contain many molecules of *perforin* and enzymes called *granzymes.* After an activated T_C cell binds to a virus-infected or cancer cell that is presenting foreign antigen on its MHC class I proteins, the T_C cell releases perforin, which forms pores in the plasma membrane of the abnormal cell. This allows the granzymes to enter the target cell, inducing it to undergo apoptosis (Fig. 13.8*a*). The T_C cell is then capable of moving on to kill another target cell (Fig. 13.8*b*).

Notice also in Figure 13.7 that, similar to B cells, some of the clonally expanded T cells become **memory T cells.** These may live for many years and can quickly jump-start an immune response to an antigen previously present in the body. Immunity mediated by T_H cells and T_C cells is sometimes referred to as **cell-mediated immunity.**

Animation Cytotoxic T-cell Activity Against Target Cells

Because HIV, the virus that causes AIDS, infects helper T cells and other cells of the immune system, it suppresses many components of acquired immune responses and makes HIV-infected individuals susceptible to opportunistic infections (see the Health feature, "Opportunistic Infections and HIV," on page 241). Infected macrophages and dendritic cells also serve as reservoirs for HIV.

Check Your Progress 13.3

1. Describe four mechanisms that result in the generation of antigen receptor diversity.
2. Discuss the major tenets (principles) of the clonal selection theory.
3. List the five classes of human antibodies and their main functions.
4. Contrast the specific functions of helper T cells, suppressor T cells, and cytotoxic T cells.

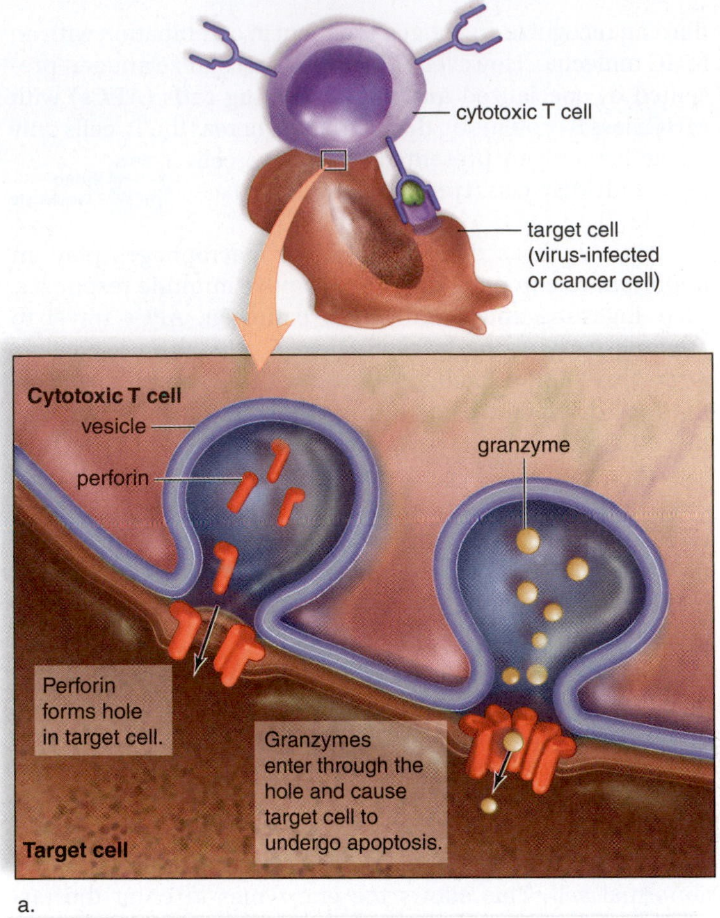

cytotoxic T cell
target cell
(virus-infected
or cancer cell)
Cytotoxic T cell
vesicle
perforin
granzyme
Perforin
forms hole
in target cell.
Granzymes
enter through the
hole and cause
target cell to
undergo apoptosis.
Target cell
a.

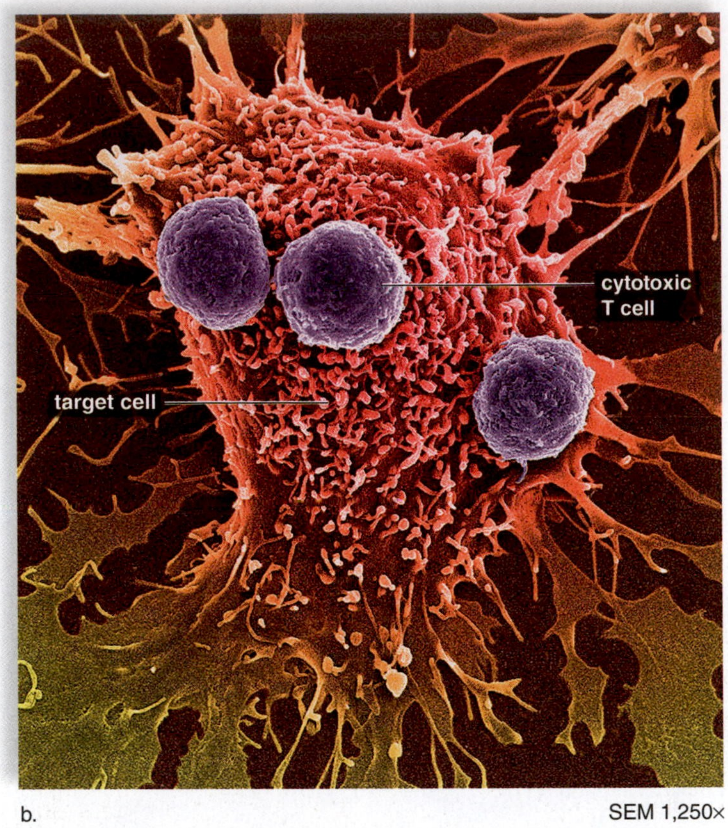

cytotoxic
T cell
target cell
b.
SEM 1,250×

Figure 13.8 Cell-mediated immunity. **a.** How a T_C cell destroys a virus-infected cell or cancer cell. **b.** The scanning electron micrograph shows cytotoxic T cells attacking and destroying a cancer cell (target cell).

13.4 Active Versus Passive Immunity

Learning Outcomes

Upon completion of this section, you should be able to

1. Distinguish between active and passive immunity.
2. Describe some specific clinical applications of cytokine therapies.
3. Explain the major steps in the production of monoclonal antibodies, and some of their practical uses.

Immunity is the condition where the body is protected from various threats—usually by the adaptive immune system. Adaptive immunity can be induced actively or passively. In **active immunity,** the individual alone produces an immune response against an antigen. In **passive immunity,** the individual receives antibodies or cells from another individual.

Active Immunity

Active immunity usually develops naturally after a person is infected with a pathogen. However, active immunity can also be induced artificially when a person is well, so that future infection will not take place. To prevent infections, people can be immunized (vaccinated) against them. The United States is committed to immunizing all children against the common types of childhood disease (Fig. 13.9*a*). The Bioethical feature, "Should Parents Be Required to Have Their Kids Vaccinated?" on page 243 examines the issue of parents who choose not to vaccinate their children.

Immunization involves the use of **vaccines,** preparations that contain an antigen to which the immune system responds. Traditionally, vaccines are the pathogens themselves, which have either been killed, or treated so they are no longer *virulent* (able to cause disease). Today, it is possible to genetically engineer bacteria or other microbes to mass-produce individual antigens, which can be used as *subunit* vaccines. This method has now produced a vaccine against hepatitis B, a viral disease, and is being used to develop a vaccine against malaria, a protozoal disease.

Video Potato Vaccine

Animation Constructing Vaccines

Most recently, animals have been successfully immunized by injecting them with DNA coding for a foreign protein. The DNA is taken up by the animal's cells, which produce the protein, to which the immune system responds. Several of these *DNA vaccines* are already approved for use in veterinary medicine, and DNA vaccines against human diseases are likely to follow.

After a vaccine is given, it is possible to monitor the amount of antibody (also called the *antibody titer*) present in samples of

SCIENCE IN YOUR LIFE ▶ **HEALTH**

Opportunistic Infections and HIV

AIDS (acquired immunodeficiency syndrome) is caused by the destruction of the immune system, especially the helper T cells, following an HIV (human immunodeficiency virus) infection. Then, the individual succumbs to many unusual types of infections that would not cause disease in a person with a healthy immune system. Such infections are known as opportunistic infections (OIs).

HIV not only kills helper T cells by directly infecting them, it also causes many uninfected T cells to die by a variety of mechanisms, including apoptosis. Many helper T cells are also killed by the person's own immune system as it tries to overcome the HIV infection. After initial infection with HIV, it may take up to ten years for an individual's helper T cells to become so depleted that the immune system can no longer organize a specific response to OIs (Fig. 13A). In a healthy individual, the number of helper T cells typically ranges from 800 to 1,000 cells per cubic millimeter (mm^3) of blood. The appearance of the following specific OIs can be associated with the helper T-cell count:

Animation How the HIV Infection Cycle Works

- Shingles. Painful infection with varicella-zoster (chicken pox) virus. Helper T-cell count of less than $500/mm^3$.
- Candidiasis. Yeast infection of the mouth, throat, or vagina. Helper T-cell count of about $350/mm^3$.
- *Pneumocystis* pneumonia. Fungal infection causing the lungs to fill with fluid and debris. Helper T-cell count of less than $200/mm^3$.
- Kaposi sarcoma. Cancer of blood vessels due to human herpesvirus 8; gives rise to reddish-purple, coin-sized spots and lesions on the skin. Helper T-cell count of less than $200/mm^3$.
- Toxoplasmic encephalitis. Protozoan infection characterized by severe headaches, fever, seizures, and coma. Helper T-cell count of less than $100/mm^3$.
- *Mycobacterium avium* complex (MAC). Bacterial infection resulting in persistent fever, night sweats, fatigue, weight loss, and anemia. Helper T-cell count of less than $75/mm^3$.
- Cytomegalovirus. Viral infection that leads to blindness, inflammation of the brain, throat ulcerations. Helper T-cell count of less than $50/mm^3$.

Due to development of powerful drug therapies that slow the progression of AIDS, people infected with HIV in the United States are suffering a lower incidence of OIs than was common in the 1980s and 1990s.

Questions to Consider

1. People who don't have HIV also sometimes get OIs, such as shingles or candidiasis. How would you expect the outcome of, for example, a yeast infection to be different in a healthy person compared to an HIV-infected person?
2. Kaposi sarcoma occurs relatively rarely in people who are infected with human herpesvirus 8, but are HIV-negative. How might an HIV infection increase the odds of getting Kaposi sarcoma?
3. Why do you suppose that the amount of HIV in the blood increases rapidly during the later stages of AIDS, as shown in Figure 13A?

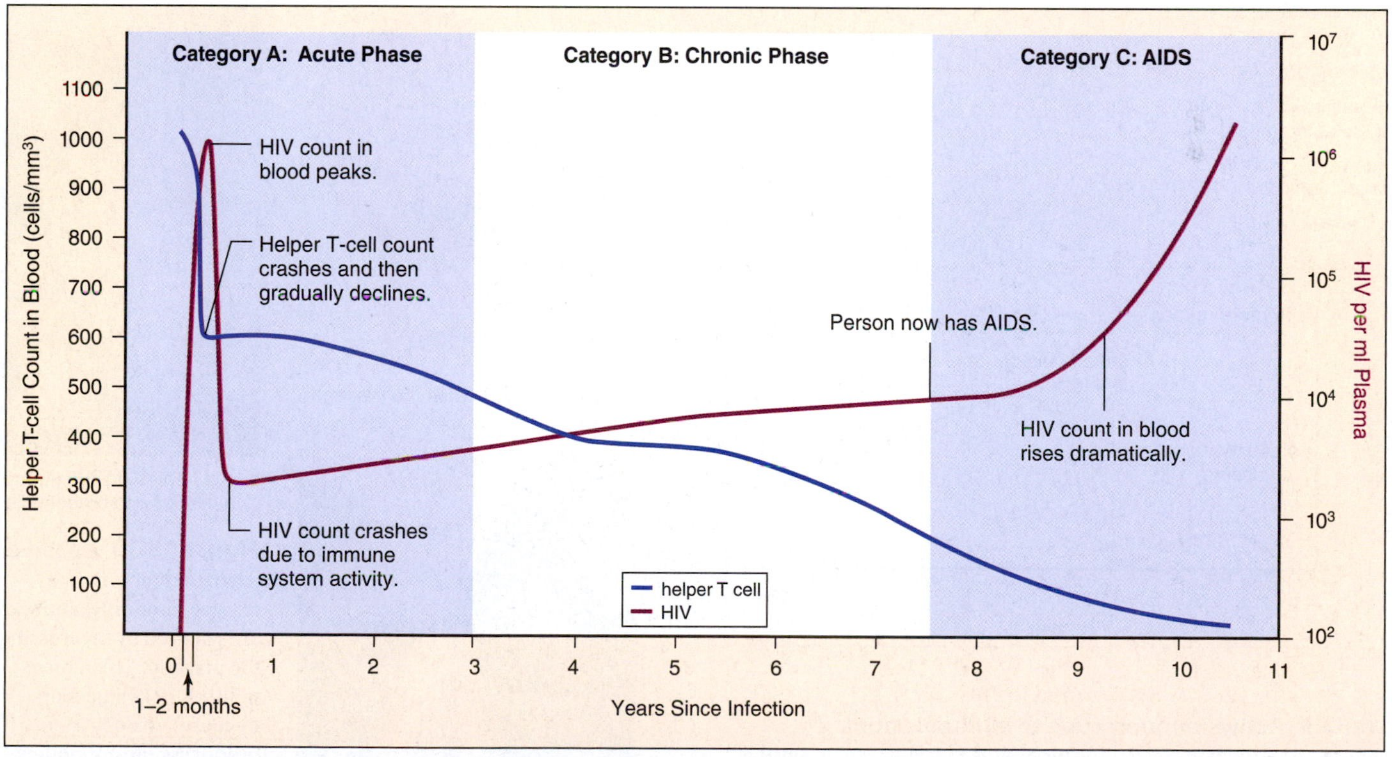
Category A: Acute Phase
Category B: Chronic Phase
Category C: AIDS
Helper T-cell Count in Blood (cells/mm³)
1100
1000
900
800
700
600
500
400
300
200
100
HIV count in blood peaks.
Helper T-cell count crashes and then gradually declines.
HIV count crashes due to immune system activity.
Person now has AIDS.
HIV count in blood rises dramatically.
helper T cell
HIV
10⁷
10⁶
10⁵
10⁴
10³
10²
HIV per ml Plasma
0
1
2
3
4
5
6
7
8
9
10
11
1–2 months
Years Since Infection

Figure 13A Progression of HIV infection during its three stages. In category A, the individual may have no symptoms or very mild symptoms associated with the infection. By category B, opportunistic infections have begun to occur, such as candidiasis, shingles, and diarrhea. Category C is characterized by more severe opportunistic infections and is clinically described as AIDS.

plasma over time. After the first vaccination, a *primary response* occurs. Typically, the antibody titer rises slowly, levels off, and gradually declines as the antibodies bind to the antigen or simply break down (Fig. 13.9*b*). Following a second vaccination, a *secondary response* occurs. The titer rises more rapidly, to a level much greater than before, then slowly declines. The second exposure is called a "booster" vaccination because it boosts the immune response to a high level. The high levels of antigen-specific antibodies and T cells usually prevent disease symptoms even if the individual is exposed to the disease-causing agent.

After exposure to a particular antigen, whether naturally or by immunization, active immunity is dependent upon the generation of a population of memory B cells and memory T cells that are capable of responding quickly to the same antigen, if encountered later. As to the question of how long this immunological memory can last, a 2008 study revealed that most survivors of the 1918 influenza epidemic had memory B cells that could produce antibodies that reacted with the 90-year-old virus strain, but not against more recent strains. The researchers also showed that these antibodies could protect mice from a lethal dose of that same, nearly century-old virus. While the mechanisms of immunological memory are not completely understood, it is possible that certain memory cells can last a lifetime.

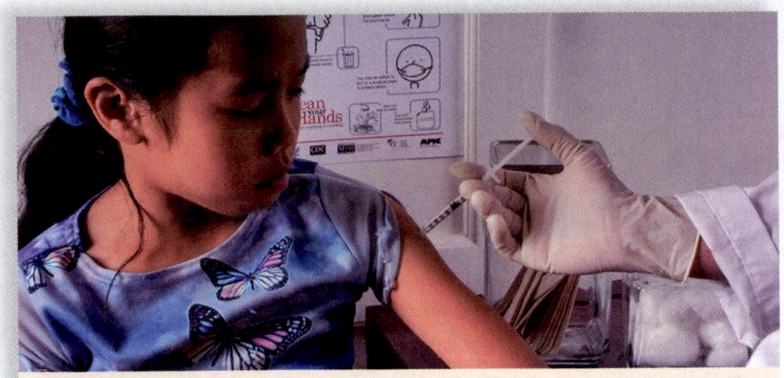

Suggested Immunization Schedule

Vaccine	Age (months)	Age (years)
Hepatitis B	Birth, 1–2, 6–18	
Diphtheria, tetanus, pertussis (DTP)	2, 4, 6, 15–18	4–6
Tetanus only		11–12, 13–18
Haemophilus influenzae, type b	2, 4, 6, 12–15	
Polio (IPV)	2, 4, 6–18	4–6
Pneumococcal	2, 4, 6, 12–15	
Measles, mumps, rubella (MMR)	12–15	4–6, 11–12
Varicella (chicken pox)	12–18	2–18
Hepatitis A (in selected areas)	12–18	2–18
Human papillomavirus, types 6, 11, 16, 18	–	11–12

a.

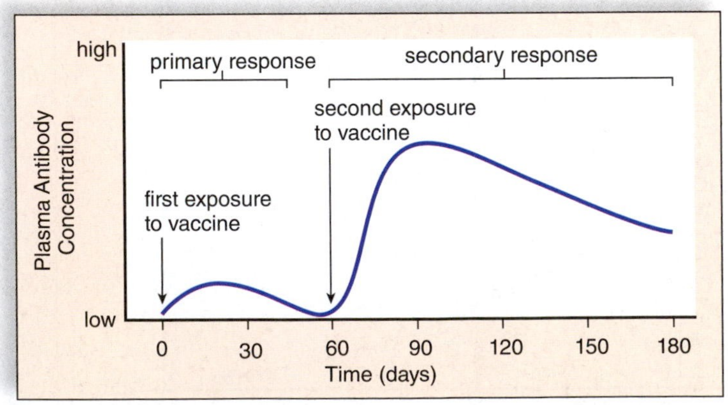

high
low
Plasma Antibody Concentration
primary response
secondary response
second exposure to vaccine
first exposure to vaccine
0
30
60
90
120
150
180
Time (days)

b.

Figure 13.9 Active immunity due to immunizations.
a. Suggested immunization schedule for infants and children. **b.** During immunization, the primary response, after the first exposure to a vaccine, is minimal. The secondary response, which occurs after the second exposure, shows a dramatic rise in the amount of antibody present in plasma.

Passive Immunity

Passive immunity occurs when an individual receives another individual's antibodies or immune cells. The passive transfer of antibodies is a common natural process. For example, newborn infants are passively immune to some diseases because IgG antibodies have crossed the placenta from the mother's blood (Fig. 13.10*a*). These antibodies soon disappear, however, so

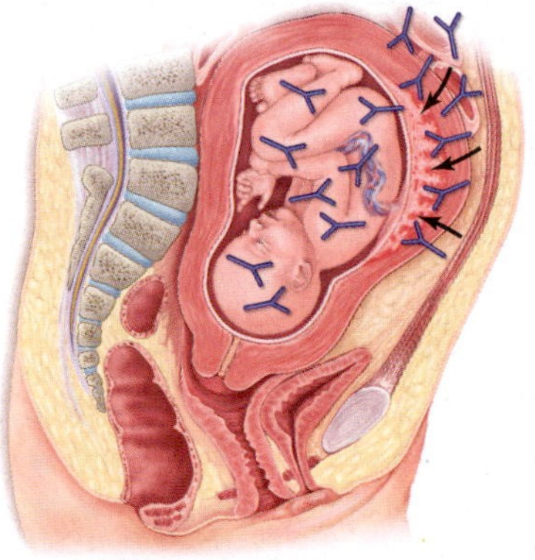

a. Antibodies (IgG) cross the placenta.

b. Antibodies (IgG, IgA) are secreted into breast milk.

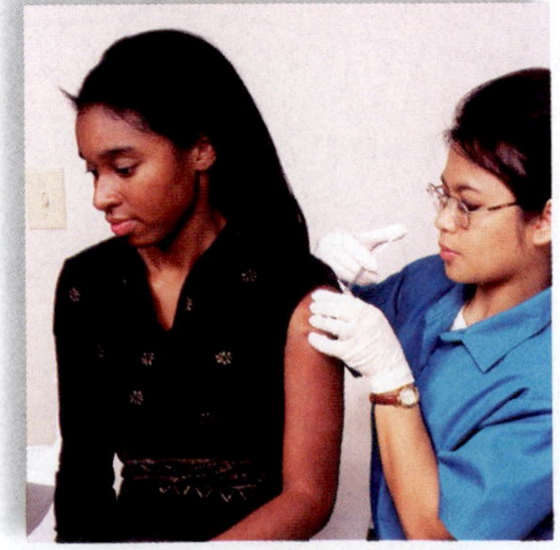

c. Antibodies can be injected by a physician.

Figure 13.10 Passive immunity. During passive immunity, antibodies are received by (**a**) crossing the placenta, (**b**) in breast milk, or (**c**) by injection. Because the body is not producing the antibodies, passive immunity is short-lived.

SCIENCE IN YOUR LIFE ► BIOETHICAL

Should Parents Be Required to Have Their Kids Vaccinated?

The development of vaccines has been one of the greatest medical achievements in the last 200 years. The World Health Organization estimates that vaccination saves about 3 million lives each year. Vaccination against smallpox eradicated that dreaded disease from the planet, and other feared diseases like measles, polio, diphtheria, whooping cough, and tetanus are now rare in countries where vaccination is common. Paradoxically, because vaccination is so effective, most parents of young children today have never seen a child suffering from these diseases, leading some of them to question whether their children need to be vaccinated at all.

Most often, these parents are worried about the possible side effects of vaccination. Indeed, a Google search on "vaccine risks" will turn up some reasonable medical advice, mixed in with a good dose of overhyped hysteria. As with any other medical treatment, vaccines can produce unintended side effects. In one example, a rotavirus vaccine called RotaShield was withdrawn from the U.S. market in 1999, less than a year after it had been approved by the FDA for use in protecting children from a serious intestinal infection. Out of about 1.5 million doses administered, 15 cases of life-threatening bowel obstructions were reported within 1–2 weeks of vaccination. This resulted in the manufacturer removing the product from the market.

A famous incidence of alleged vaccine side effects occurred in 1976, after an outbreak of "swine flu" at Fort Dix military base. Concerned about a nationwide influenza outbreak, President Gerald Ford recommended that every U.S. citizen should be vaccinated for this new, and possibly fatal, strain of virus. However, about two months after the mass vaccination was initiated, the government halted the program, in response to hundreds of lawsuits alleging the vaccine caused side effects ranging from a rare neurological syndrome, to sudden death. In all, after about 40 million people received the swine flu vaccine, 25 deaths, and hundreds of illnesses, were blamed on the vaccine. The feared influenza outbreak never occurred, and only one person died from the virus (a soldier at Fort Dix). Experts still debate whether the vaccine was to blame for all the medical issues. Regardless, some antivaccine activists point to the "swine flu fiasco" as a general argument against vaccination.

In February 1998, a paper was published in the British medical journal *The Lancet,* claiming that an autism-type condition developed in 8 or 12 children who received the measles-mumps-rubella (MMR) vaccine. The article was immediately controversial, and dozens of subsequent studies failed to turn up any vaccine-autism link. In February 2010, *The Lancet* officially retracted the article, accusing the paper's lead author, Andrew Wakefield, of several types of scientific fraud. However, by then considerable damage had been done to the reputation of vaccines in general, and the MMR in particular. In the years following the Wakefield paper, vaccination rates in Britain dropped from 92% to 73%. In the United States, more cases of measles were reported in 2008 than any year since 1997, and most of these cases were in unvaccinated children.

Ironically, even parents who don't vaccinate their children benefit from the fact that a large percentage of children are vaccinated, and thus are less likely to transmit infectious diseases to their unvaccinated child. This concept of "herd immunity" has been shown to be relevant in recent outbreaks of measles and whooping cough in communities where increasing numbers of parents are not vaccinating their children. On the front lines of the vaccine issue are pediatricians, many of whom feel so strongly about the need for vaccination that in a 2011 survey, 25% of them had turned away kids whose parents refused vaccination (compared to 18% in 2005).

The concept that vaccinating a high percentage of children can protect entire communities has led to immunization requirements. Most states require that a parent provide written proof of a child's immunizations prior to registering for school. If the child is not immunized, he or she will not be allowed to attend public school, unless the parent can demonstrate a medical reason, or a religious objection, that explains why the child is not vaccinated. As mentioned, there is no shortage of Internet sites voicing objections to any vaccine requirements, with some going so far as to question the benefit of all vaccinations. While vaccines are not risk-free, these risks seem to be greatly outweighed by the benefits, especially if one considers how infectious diseases were a major cause of death and suffering in our very recent past.

Questions to Consider

1. Do you agree that the benefits of vaccination outweigh the risks? What level of risk would seem acceptable if you were having your own child vaccinated against polio, for example?
2. Do you think it is appropriate to allow parents to forgo the required vaccinations for their children because of religious objections?
3. What is your opinion of pediatricians who refuse to see children if they are not vaccinated? Can you clearly articulate both sides of this issue?

connect BIOLOGY Explore the concepts through a variety of multimedia assets, question types, and data interpretation.
www.mcgrawhillconnect.com

within a few months, infants become more susceptible to infections. Breast-feeding may prolong the natural passive immunity an infant receives from the mother because IgG and IgA antibodies are present in the mother's milk (Fig. 13.10*b*). The "first milk," or *colostrum,* produced within the first few days after giving birth, is particularly rich in antibodies.

Even though passive immunity does not last, it is sometimes used to prevent illness in a patient who has been exposed to an infectious disease. Usually, the patient receives a gamma globulin injection (serum that contains antibodies), perhaps taken from individuals who have recovered from the illness (Fig. 13.10*c*). In the past, horses were immunized, and serum was taken from them to provide the needed antibodies against such diseases as diphtheria, botulism, and tetanus. Unfortunately, about 50% of patients who received these antibodies became ill because their immune system recognized horse protein as foreign. This is called *serum sickness.* Passive immunity using horse antibodies against poisonous snake and spider venom is still used today, but human gamma globulin is used to treat certain infections and immunodeficiencies.

Instead of antibodies, cells of the immune system may be passively transferred into a patient, although this practice is less common. The best example is a bone marrow transplant, in which stem cells that produce blood cells are replenished after a cancer patient's own bone marrow has been intentionally destroyed by radiation or chemotherapy.

MP3 Specific Immunity

Immune Therapies

Cytokines and Immunity

As mentioned, *cytokines* are chemical messengers produced by T cells, macrophages, and other cells. Because cytokines regulate white blood cell formation and/or function, they have multiple uses in medicine. Two of the most common applications of cytokine therapy are in the stimulation of blood cell production, and the treatment of cancer.

Several cytokines stimulate the bone marrow to produce more blood cells. One example is granulocyte-macrophage colony-stimulating factor, a cytokine produced by lymphocytes, macrophages, and other cells. Under the generic name sargramostim, it is mainly used to stimulate the production of white blood cells in patients receiving chemotherapy for cancer, as well as following bone marrow transplantation. It can also be useful in treating certain genetic deficiencies in white blood cell production.

Certain cytokines are used in treating cancer. Although surgery is the main treatment for melanoma, interferon is used to decrease the risk of the cancer recurring after surgery. Interferon may work by directly inhibiting the growth of cancer cells, and/or by stimulating the immune response to the tumor. Due to its antiviral activity, interferon may inhibit viruses that are involved in causing some cancers. The FDA has also approved another cytokine, called interleukin-2 (IL-2) or aldesleukin, for the treatment of kidney cancer, melanoma, and certain immunodeficiencies. IL-2 works mainly by stimulating activities of T cells and NK cells.

One serious drawback with cytokine therapy is the risk of side effects, which can be life-threatening. One approach to this problem is to block the activity of the cytokine until it is needed. In a study published in March 2011, researchers at the University of Rochester (New York) described how they attached another protein to IL-2, effectively blocking it from attaching to its receptor. They designed the protein, however, so that it would be cleaved by enzymes that are commonly present in a particular type of cancer. That way, the cytokine would be active only in the immediate vicinity of the cancer cells, not throughout the body, resulting in fewer side effects. Preliminary studies indicated that the approach works well in mice, but more work must be done before human trials can begin.

Monoclonal Antibodies

Every plasma cell derived from the same B cell secretes antibodies against a specific antigen. These are **monoclonal antibodies** because all of them are the same type and because they are produced by plasma cells derived from the same B cell. One method of producing monoclonal antibodies in vitro (outside the body in glassware) is depicted in Figure 13.11. Following immunization of an animal (usually lab mice are used) with an antigen of interest, spleen cells are removed and fused with myeloma cells (malignant plasma cells that live and divide indefinitely). The fused cells are called hybridomas—*hybrid-* because they result from the fusion of two

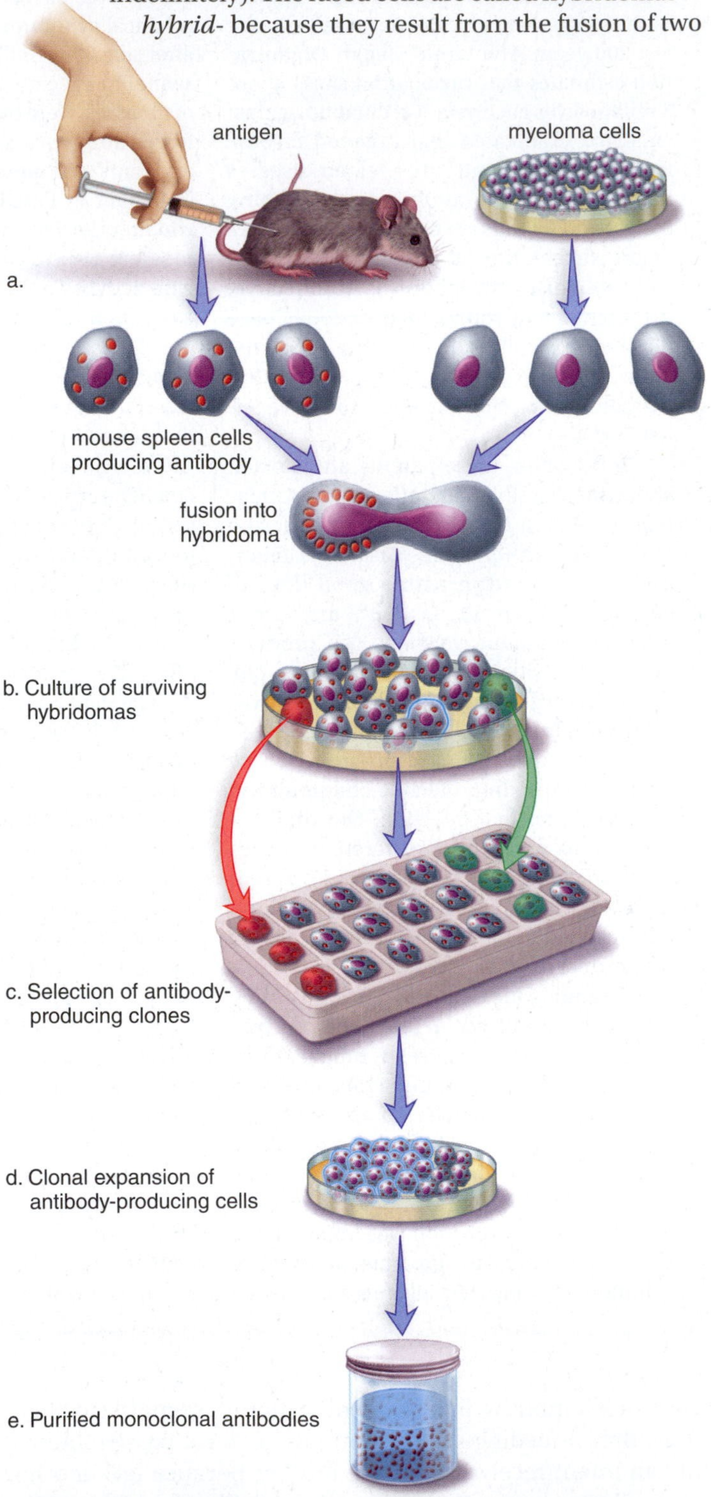
antigen
myeloma cells
a.
mouse spleen cells producing antibody
fusion into hybridoma
b. Culture of surviving hybridomas
c. Selection of antibody-producing clones
d. Clonal expansion of antibody-producing cells
e. Purified monoclonal antibodies

Figure 13.11 Production of monoclonal antibodies. **a.** Spleen cells derived from mice immunized with a specific antigen are fused with myeloma (cancerous) cells. **b.** This produces hybridoma cells that are "immortal." **c.** Hybridoma cells are cultured and clones producing the desired antibody are selected and expanded (**d**), producing monoclonal antibodies (**e**).

different cells, and *-oma* because one of the cells is a cancer cell. The cells are then cultured in a medium that kills any unfused myeloma cells, while unfused spleen cells die naturally after a few days. The cell cultures are then screened to see if any contain hybridomas that are producing the desired antibody. If so, these specific hybridomas are clonally expanded to produce a large number of cells that secrete the desired antibody.

Because they react very specifically with only one particular antigen, monoclonal antibodies can be used for quick and accurate diagnosis of various conditions. For example, a particular hormone called human chorionic gonadotropin (HCG) is present in the urine of a pregnant woman. A monoclonal antibody can be used to detect this hormone. If it is present, the woman is pregnant. Monoclonal antibodies are also used to identify infections. And because they can distinguish between cancerous and normal tissue cells, they are used to carry radioactive isotopes or toxic drugs to tumors, which can then be selectively destroyed. Trastuzumab (Herceptin) is a monoclonal antibody used to treat breast cancer. Given intravenously, it binds to a protein receptor found on some breast cancer cells and prevents the cancer cells from dividing so fast. Antibodies that bind to cancer cells also can activate complement, increase phagocytosis by macrophages and neutrophils, and attract NK cells.

Animation Monoclonal Antibody Production

Check Your Progress 13.4

1. Describe three different types of vaccines.
2. Explain why the passive transfer of antibodies is of great importance for the newborn.
3. Explain how the interaction of cytokines with receptors on target cells may be blocked, or encouraged.
4. Summarize the advantage of using a monoclonal antibody for diagnosis or treatment, versus using antibodies produced by injecting an animal with an antigen.

13.5 Adverse Effects of Immune Responses

Learning Outcomes

Upon completion of this section, you should be able to

1. Discuss the most common immunological mechanisms responsible for allergies.
2. Compare the adverse reactions involving the ABO blood system with those involving the Rh system.
3. Explain the types of precautions that must be taken when transplanting organs.

Sometimes the immune system responds to harmless antigens in a manner that damages the body, as when individuals develop allergies, receive an incompatible blood type, or suffer tissue rejection.

Allergies

Allergies are hypersensitivities to substances, such as pollen, food, or animal hair, that ordinarily would do no harm to the body. The response to these antigens, called *allergens,* usually includes some degree of tissue damage.

An **immediate allergic response** can occur within seconds of contact with the antigen. The response is caused by antibodies known as IgE (see Table 13.1). IgE antibodies are attached to receptors on the plasma membrane of mast cells in the tissues and also to eosinophils and basophils in the blood. When an allergen attaches to the IgE antibodies, these cells release histamine and other substances that bring about the allergic symptoms. When an allergen such as pollen is inhaled, histamine stimulates the mucous membranes of the nose and eyes, causing the runny nose and watery eyes typical of *hay fever.* In **asthma,** the airways leading to the lungs constrict, resulting in difficult breathing accompanied by wheezing. When food contains an allergen, nausea, vomiting, and diarrhea often result.

Anaphylactic shock is an immediate allergic response that occurs because the allergen has entered the bloodstream. Bee stings and penicillin shots are known to cause this reaction. Anaphylactic shock is characterized by a sudden and life-threatening drop in blood pressure due to increased permeability of the capillaries caused by histamine. Injecting epinephrine can counteract this reaction until medical help is available.

Treating mild to moderate allergies usually involves antihistamines to relieve the symptoms. In more serious cases, injections of the allergen can be given so that the body will build up high quantities of IgG antibodies (see the Health feature, "Immediate Allergic Responses," on page 246). The hope is that IgG antibodies will then combine with allergens before they have a chance to reach the IgE antibodies. A monoclonal antibody called Xolair is also available, which binds to and blocks the binding of IgE to the receptor found on inflammatory cells.

A **delayed allergic response** is initiated by memory T cells at the site of allergen contact in the body. The allergic response is regulated by the cytokines secreted by "sensitized" T cells at the site. A classic example of a delayed allergic response is the skin test for tuberculosis (TB). When the test result is positive, the tissue where the antigen was injected becomes red and hardened (Fig. 13.12), indicating prior exposure to tubercle bacilli, the cause of TB.

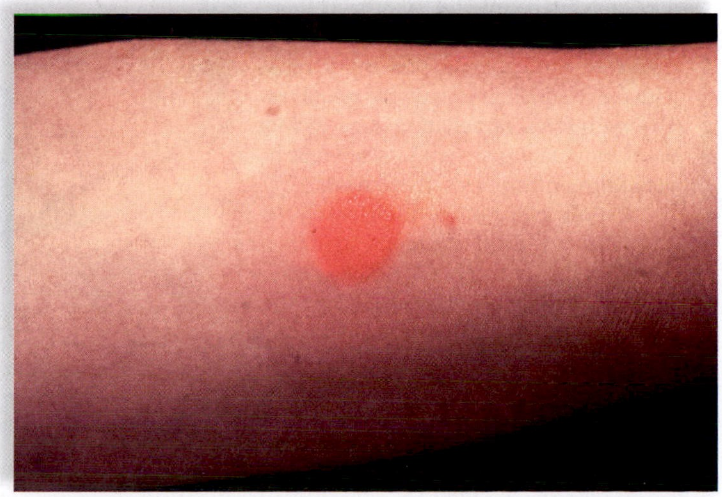

Figure 13.12 Skin testing for tuberculosis. The formation of a swollen, red area two to three days after injection of *Mycobacterium tuberculosis* antigens into the skin indicates previous exposure.

SCIENCE IN YOUR LIFE ▶ HEALTH

Immediate Allergic Responses

The runny nose and watery eyes of hay fever are often caused by an allergic reaction to the windblown pollen of trees, grasses, and ragweed, particularly in the spring and fall (Fig. 13B). Worse, if a person has asthma, the airways leading to the lungs constrict, resulting in difficult breathing characterized by wheezing. Most people can inhale pollen with no ill effects. But others have developed a hypersensitivity, meaning that their immune system responds in a deleterious manner. The problem stems from a type of antibody called immunoglobulin E (IgE), which causes the release of histamine from mast cells and also basophils whenever they are exposed to an allergen. Histamine causes the mucous membranes of the nose and eyes to release fluid as a defense against pathogen invasion. But in the case of allergies, copious fluid is released even though no real danger is present.

Many food allergies are also due to the presence of IgE antibodies, which usually bind to a protein in the food. The symptoms, such as nausea, vomiting, and diarrhea, are due to the mode of entry of the allergen. Skin symptoms may also occur, however. Adults are often allergic to shellfish, nuts, eggs, cows' milk, fish, and soybeans. Peanut allergy is a common food allergy in the United States, possibly because peanut butter is a staple in the diet. People seem to outgrow allergies to cows' milk and eggs more often than allergies to peanuts and soybeans.

Celiac disease occurs in people who are allergic to wheat, rye, barley, and sometimes oats—in short, any grain that contains gluten proteins. It is thought that the gluten proteins elicit a delayed cell-mediated immune response by T cells with the resultant production of cytokines. The symptoms of celiac disease include diarrhea, bloating, weight loss, anemia, bone pain, chronic fatigue, and weakness.

People can reduce their reactions to airborne and food allergens by avoiding the offending substances. Airlines are now required to provide a peanut-free zone in their planes for those who are allergic. Taking antihistamines can also be helpful.

If these precautions are inadequate, patients can be tested to measure their susceptibility to any number of possible allergens. A small quantity of a suspected allergen is injected just beneath the skin, and the strength of the subsequent reaction is noted. The appearance of a raised, red, so-called wheal-and-flare response at the skin prick site within a few minutes demonstrates that IgE antibodies attached to mast cells have reacted to an allergen. In an immunotherapy called hyposensitization, ever-increasing doses of the allergen are periodically injected subcutaneously with the hope that the body will produce an IgG response against the allergen. IgG, in contrast to IgE, does not cause the release of histamine after it combines with the allergen. If IgG combines first, the allergic response does not occur. Patients know they are cured when the allergic symptoms go away. Therapy may have to continue for as long as two to three years.

Allergic-type reactions can occur without involving the immune system. Wasp and bee stings contain substances that cause swelling, even in those whose immune system is not sensitized to substances in the sting. Also, jellyfish tentacles and certain foods (e.g., fish that is not fresh and strawberries) contain histamine or closely related substances that can cause a reaction. Furthermore, immunotherapy is not possible in people who are allergic to penicillin and bee stings. Upon the first exposure, high sensitivity has built up so that when reexposed, anaphylactic shock can occur. Among its many effects, histamine causes increased permeability of the capillaries, the smallest blood vessels. When this reaction occurs throughout the body, these individuals experience a drastic decrease in blood pressure that can be fatal within a few minutes. People who know they are allergic to bee stings can obtain a syringe of epinephrine to carry with them. This medication can counteract the symptoms of anaphylactic shock until medical help is reached.

Questions to Consider

1. Based on what you have read in this essay and the chapter, what are some factors that might determine a person's tendency to develop an immediate allergic response to an allergen?
2. An increase in allergic diseases has been reported in developed countries, compared to lower rates in less-developed countries. What are some possible reasons for this difference?
3. What are the specific effects of epinephrine that can save the life of a person who is in anaphylactic shock?

connect BIOLOGY Explore the concepts through a variety of multimedia assets, question types, and data interpretation.
www.mcgrawhillconnect.com

Figure 13B Pollen and allergies. When people are allergic to pollen, they develop symptoms that include watery eyes, sinus headaches, increased production of mucus, labored breathing, and sneezing.

Blood-Type Reactions

Several blood typing systems are currently in use. The most important is the ABO system. People often carry information about their ABO type in their wallets in case an accident requires them to need blood.

ABO System

In the ABO system, the presence or absence of type A and type B antigens on red blood cells determines a person's blood type. For example, a person with type A blood has the A antigen on his or her red blood cells. Because it is considered "self," this molecule is not recognized as an antigen by this individual's body, although it can be an antigen to a recipient who does not have type A blood.

The ABO system identifies four types of blood: A, B, AB, and O. Because the A and B antigens are commonly found on microbes present in and on our bodies, a person's plasma contains antibodies to the antigens that are *not* present on the red blood cells. These antibodies are called anti-A and anti-B. The antibodies present in the plasma of each blood type are shown here:

Blood Type	Antigen on Red Blood Cells	Antibody in Plasma
A	A	Anti-B
B	B	Anti-A
AB	A, B	None
O	None	Anti-A and anti-B

Because type A blood has anti-B and not anti-A antibodies in the plasma, a donor with type A blood can give blood to a recipient with type A blood (Fig. 13.13*a*). But giving type A blood to a type B recipient causes agglutination (Fig. 13.13*b*). *Agglutination,* or clumping of red blood cells, can stop blood from circulating in small blood vessels, leading to organ damage. It may be followed by *hemolysis,* or bursting of red blood cells, which if extensive can be fatal.

Theoretically, which blood type would be accepted by all other blood types? Type O blood has no A or B antigens on the red blood cells and is sometimes called the universal donor. Which blood type could receive blood from any other blood type? Type AB blood has no anti-A or anti-B antibodies in the plasma and is sometimes called the universal recipient. In practice, however, it is not safe to rely solely on the ABO system when matching blood, and instead, samples of the two types of blood are physically mixed, and the result is microscopically examined before blood transfusions are done.

Today, blood transfusions are a matter of concern not only because blood types should match, but also because each person wants to receive blood that is free of infectious agents. Blood is tested for the more serious agents, such as those that cause AIDS, hepatitis, and syphilis.

Rh System

Another important antigen in matching blood types is the Rh factor. Eighty-five percent of the U.S. population have this particular antigen on their red blood cells and are Rh-positive. Fifteen percent do not have this antigen and are Rh-negative. Rh-negative individuals normally do not have antibodies to the Rh factor, but they may make them when exposed to the Rh factor. The designation of blood type usually also includes whether the person has or does not have the Rh factor on the red blood cells (i.e., type A–positive).

During pregnancy, if the mother is Rh-negative and the father is Rh-positive, the fetus may be Rh-positive. The fetal Rh-positive red blood cells may leak across the placenta into the mother's cardiovascular system, especially as placental

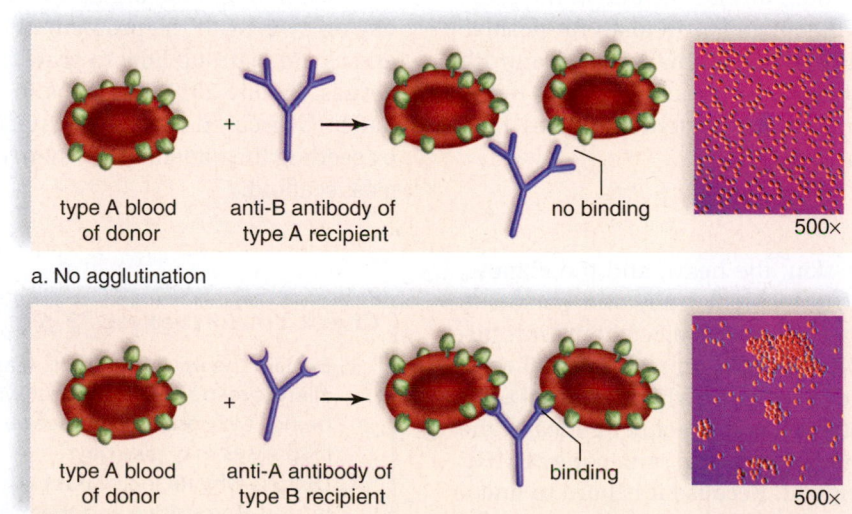

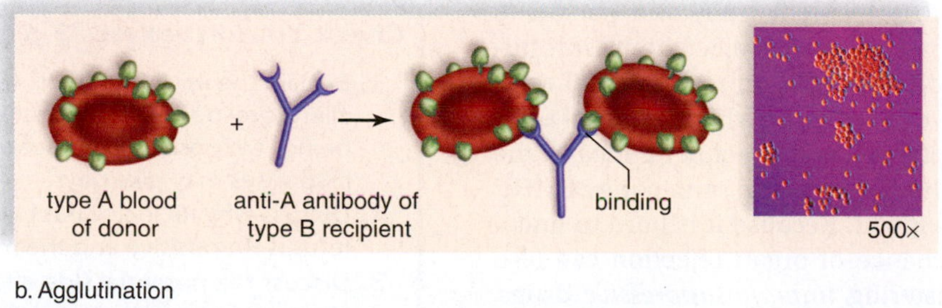
type A blood of donor
anti-B antibody of type A recipient
no binding
500×
a. No agglutination
type A blood of donor
anti-A antibody of type B recipient
binding
500×
b. Agglutination

Figure 13.13 Blood transfusions. No agglutination (**a**) versus agglutination (**b**) is determined by whether the recipient has antibodies in the plasma that can combine with antigens on the donor's red blood cells. The photos on the right show how the red blood cells appear under the microscope.

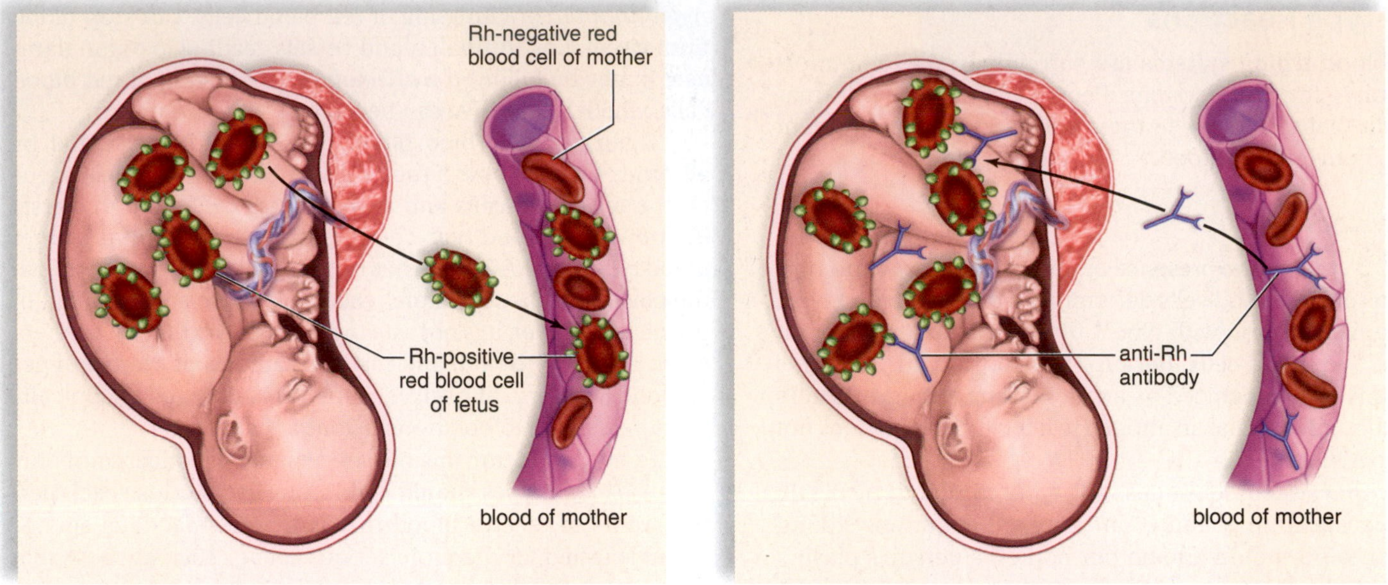
Rh-negative red blood cell of mother
Rh-positive red blood cell of fetus
blood of mother
a. Fetal Rh-positive red blood cells leak across placenta into mother's bloodstream.
anti-Rh antibody
blood of mother
b. Mother forms anti-Rh antibodies that cross the placenta and attack fetal Rh-positive red blood cells.

Figure 13.14 Hemolytic disease of the newborn. **a.** Due to a pregnancy in which the child is Rh-positive, an Rh-negative mother can begin to produce antibodies against Rh-positive red blood cells. **b.** In the same, but more likely a subsequent, pregnancy these antibodies can cross the placenta and cause hemolysis of an Rh-positive child's red blood cells.

tissues normally break down during birth (Fig. 13.14*a*). Now, the mother produces anti-Rh antibodies. In a subsequent pregnancy with another Rh-positive child, these antibodies may cross the placenta and destroy the fetal red blood cells (Fig. 13.14*b*). The resulting condition, called hemolytic disease of the newborn, can lead to brain damage or even death due to anemia and excess bilirubin in the blood.

The Rh problem is prevented by giving Rh-negative women an Rh immunoglobulin injection, toward the end of the pregnancy and within 72 hours after giving birth to an Rh-positive child. This injection contains relatively low levels of anti-Rh antibodies that help destroy any baby's red blood cells in the mother's blood before her immune system produces high levels of anti-Rh antibodies.

Tissue Rejection

Certain organs, such as the skin, the heart, and the kidneys, would be relatively easy to transplant from one person to another if the recipient did not attempt to reject them. Unfortunately, in addition to their role in antigen presentation to T cells, MHC proteins can also serve as antigens, when organs of a different MHC type are transplanted into a recipient. Ideally, the transplanted organ would have exactly the same type of MHC proteins as those of the recipient. Because it is hard to find a perfect MHC match, the chance of organ rejection can also be diminished by administering *immunosuppressive* drugs. Two well-known immunosuppressive drugs, cyclosporine and tacrolimus, both act by inhibiting the production of certain cytokines by T cells.

Xenotransplantation is the use of animal organs instead of human organs in transplant patients. Scientists have chosen to use the pig because pigs are raised as a meat source and are prolific. Genetic engineering can make pig organs less antigenic by removing the MHC antigens. The ultimate goal is to make pig organs as widely accepted as type O blood. Although advances in xenotransplantation are encouraging, many researchers hope that tissue engineering, including the production of human organs that lack MHC antigens, will one day do away with the problem of rejection. Alternatively, scientists are learning how to grow organs from a patient's own tissues. In July 2011, surgeons in Sweden replaced a cancer patient's diseased trachea with a lab-grown version, produced by seeding the patient's own stem cells onto a trachea-shaped glass scaffold.

Check Your Progress 13.5

1. Explain the immunological reason why immediate allergic responses can occur within seconds after being reexposed to an allergen, whereas delayed-type responses may take days.
2. Discuss why Rh incompatibility is only a problem when a fetus is Rh-positive and the mother is Rh-negative.
3. Discuss the potential side effects of the drugs cyclosporine and tacrolimus, which function by inhibiting cytokine production (rather than inhibiting specific responses to transplanted organs).

13.6 Disorders of the Immune System

Learning Outcomes

Upon completion of this section, you should be able to
1. Define autoimmune disease, and list several specific examples of these diseases.
2. Differentiate between acquired and congenital immunodeficiencies.

When a person has an **autoimmune disease,** cytotoxic T cells or antibodies mistakenly recognize the body's own cells or molecules as if they are foreign antigens. Exactly what causes autoimmune diseases is not known. However, sometimes they occur after an individual has recovered from an infection.

In the autoimmune disease **myasthenia gravis** (**MG**), antibodies attach to and interfere with neuromuscular junctions, and muscular weakness results. In **multiple sclerosis** (**MS**), T cells attack the myelin sheath of nerve fibers, and this causes various neuromuscular disorders. A person with **systemic lupus erythematosus** (**SLE**) has various symptoms prior to death due to kidney damage caused by the deposition of excessive antigen/antibody complexes. In **rheumatoid arthritis,** the joints are affected. As yet, there are no cures for autoimmune diseases, but they can sometimes be controlled with immunosuppressive drugs.

When a person has an **immunodeficiency disease,** the immune system is unable to protect the body against disease. AIDS is an example of an *acquired* immunodeficiency (see the Health feature, "Opportunistic Infections and HIV," on page 241). Immunodeficiencies may also be *primary* or congenital (i.e., inherited). As mentioned in the chapter-opening story, children may be born lacking the ability to produce antibodies. Less commonly, a child may be born with an impaired immune system caused by a defect in T cells. In **severe combined immunodeficiency** (**SCID**), both antibody- and cell-mediated immunity are lacking or inadequate. Without treatment, even common infections can be fatal. Gene therapy has been successful in some SCID patients.

Check Your Progress 13.6

1. Describe four examples of autoimmune diseases.
2. List two immunodeficiency diseases, and identify them as either congenital or acquired.

Case Study Conclusion

Because the immune system is of critical importance in protecting an individual from infectious disease, disorders affecting the immune system can have serious or even fatal consequences. Inherited immune deficiencies can affect a very limited part of the immune system, or target some of the more general forms of immune responses. In the story that opened this chapter, we learned about Jason, a three-year-old boy with an immunodeficiency called X-linked agammaglobulinemia (XLA), which mainly affects B-cell responses. Without antibodies, most people with XLA would die in childhood, typically from bacterial infections. With regular intravenous infusions of antibodies from a pool of human donors, these individuals can live relatively normal lives.

MEDIA STUDY TOOLS

www.mhhe.com/maderinquiry14

Enhance your study of this chapter with study tools and practice tests. Also ask your instructor about the resources available through ConnectPlus, including LearnSmart, the media-rich eBook, interactive learning tools, and animations.

McGraw Hill connect plus+ BIOLOGY

SUMMARIZE

13.1 The Lymphatic System

The **lymphatic system** consists of lymphatic vessels and lymphoid organs:

- **Lymphatic vessels** begin as **lymphatic capillaries** that absorb fats from the digestive tract and excess tissue fluid at blood capillaries, forming a fluid called **lymph.** The lymphatic vessels eventually return lymph to the bloodstream.
- **Lymphoid organs** include **primary lymphoid organs,** where **lymphocytes** are produced, and **secondary lymphoid organs,** where adaptive immune responses can be initiated by the **B cells** and **T cells.** The primary lymphoid organs are the **red bone marrow** and **thymus.** Secondary lymphoid organs include the **lymph nodes** and **spleen.** Disease-causing microbes known as **pathogens** may be present in these lymphoid tissues.

13.2 Innate Immunity

The lymphatic system works with the **immune system** to protect the body with **immunity** against pathogens, toxins, and other dangers. **Innate immunity** is fully functional without previous exposure to the threatening substance. Mechanisms of innate immunity include:

- Physical and chemical barriers, such as the ciliated cells and mucus of the respiratory tract, and stomach acid.
- The **inflammatory response,** which is the redness, heat, swelling, and pain that occur due to capillary changes in response to chemicals released by tissue-dwelling cells such as **mast cells, dendritic cells,** and **macrophages.** Mast cells release **histamine,** while the latter two cell types release **cytokines** that promote inflammation.
- **Phagocytes** and **natural killer** (**NK**) **cells,** which either ingest and kill microorganisms (phagocytes), or kill virus-infected cells (NK cells). The

first phagocytes to enter inflamed tissues from the bloodstream are usually the **neutrophils,** followed by **monocytes,** which become macrophages in the tissues.

- Protective proteins, such as **complement** and **interferons,** which enhance the effectiveness of the immune system.

13.3 Adaptive Immunity

Adaptive immunity, which is dependent on B cells and T cells, is enhanced after exposure to **antigens;** specific molecular configurations that can bind to antigen receptors:

- During their development, B and T cells undergo a rearrangement of gene segments that will code for the antigen-binding regions of their plasma membrane antigen receptors.
- B cells are responsible for **antibody-mediated immunity.** As predicted by the **clonal selection theory,** upon exposure to an antigen that binds to its **B-cell receptors,** a B cell replicates and differentiates, producing plasma cells and memory B cells. **Plasma cells** secrete antibodies and eventually undergo apoptosis. **Memory B cells** are long-lived and produce antibodies if the same antigen enters the body at a later date. **Antibodies,** also called **immunoglobulins (Ig),** are typically Y-shaped proteins with at least two binding sites for a specific antigen. The main classes of human antibodies are IgG, IgM, IgA, IgE, and IgD.
- T cells are responsible for **cell-mediated immunity.** Each T cell bears **T-cell receptors.** However, for a T cell to recognize an antigen, the antigen must be processed and presented on a cell surface by an **MHC (major histocompatibility complex) protein.** The two main types of T cells are helper T (T_H) cells and cytotoxic T (T_C) cells. **Helper T cells** respond to antigens presented by MHC class II proteins found on **antigen-presenting cells (APCs)** such as dendritic cells or macrophages. The activated T_H cells produce cytokines that stimulate other immune cells. **Cytotoxic T cells** kill virus-infected or cancer cells that bear a "nonself" protein on their MHC class I proteins. Thereafter, the activated T cell undergoes clonal expansion until the illness has been stemmed. Then, most of the activated T cells undergo apoptosis. A few cells remain, however, as **memory T cells.**

13.4 Active Versus Passive Immunity

Immunity is the condition where the body is protected from various threats—usually by the adaptive immune system. Active and passive immunity are induced in different ways:

- **Active immunity** can be induced through infection or by **immunization.** A person can be immunized with a **vaccine,** which can be a killed or modified form of a pathogen, a subunit, or even pure DNA.
- **Passive immunity** is acquired from another individual, whether naturally as antibodies passed from a mother to child, or artificially via an injection.

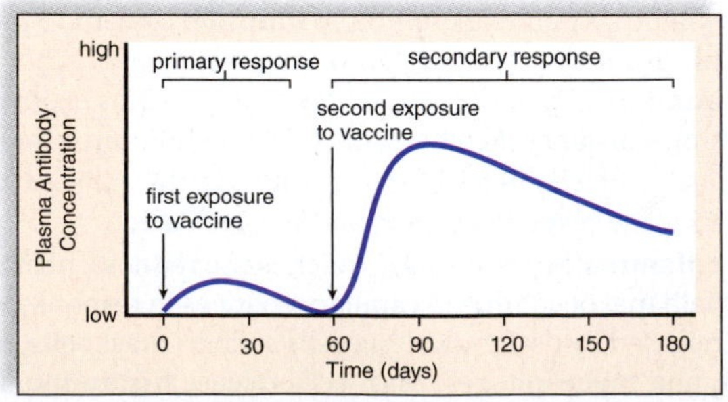
high
low
Plasma Antibody Concentration
primary response
secondary response
second exposure to vaccine
first exposure to vaccine
0
30
60
90
120
150
180
Time (days)

Immune therapies may include cytokines and monoclonal antibodies:

- Cytokines are chemical messengers that are used to stimulate blood cell production, and to treat cancer and other diseases.
- **Monoclonal antibodies,** which are produced by cells derived from the same plasma cell, are used for various purposes, from detecting infections to treating cancer.

13.5 Adverse Effects of Immune Responses

Immune responses to harmless compounds (called allergens) are known as **allergies** or hypersensitivities:

- **Immediate allergic responses** occur due to an inappropriate production of IgE antibody, which results in an inflammatory response to subsequent exposures to the allergen. The most extreme form of this is **anaphylactic shock,** which can be fatal. **Asthma** also has an immediate allergic component.
- **Delayed allergic responses,** such as a positive TB skin test, are due to the activity of T cells.
- The reactions that occur following improperly matched blood types may involve the ABO system, the Rh system, or other blood group antigens. Also, if an Rh-negative woman becomes pregnant with an Rh-positive fetus, hemolytic disease of the newborn may result.
- The rejection of transplanted tissues is usually due to an imperfect match of MHC proteins between the donor and recipient.

13.6 Disorders of the Immune System

In **autoimmune diseases,** a person's own T cells or antibodies target the body's own cells or molecules. Many types of immunodeficiencies also occur:

- Examples of **autoimmune diseases** include **myasthenia gravis (MG), multiple sclerosis (MS), systemic lupus erythematosus (SLE),** and **rheumatoid arthritis.**
- **Immunodeficiency diseases** can be acquired, as occurs with AIDS, or inherited, as is seen in children with **severe combined immunodeficiency (SCID),** who are born unable to produce their own antibody- and cell-mediated immunity. Depending on the severity, immunodeficiencies can be fatal, usually as a result of other, common infections.

ASSESS

Testing Yourself

Choose the best answer for each question.

1. What is the term for localized swelling caused by fluid accumulation?
   a. node
   b. edema
   c. valve
   d. None of these are correct.

2. Label a–d as either primary or secondary lymphoid organs.

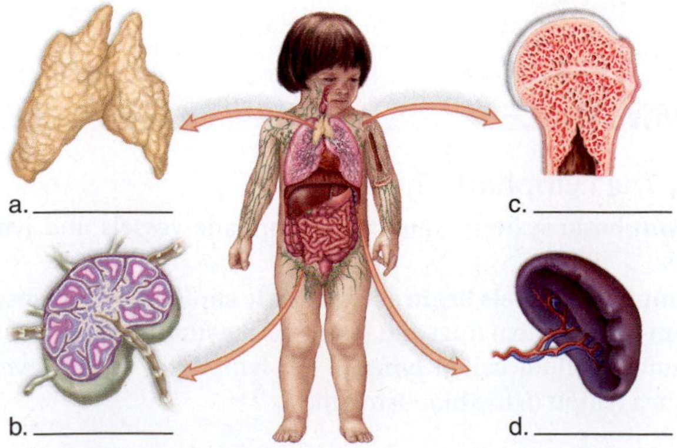
a. ________
b. ________
c. ________
d. ________

3. Which of the following is not a function of the lymphatic system?
   a. produces red blood cells
   b. returns excess fluid to the blood
   c. transports lipids absorbed from the digestive system
   d. defends the body against pathogens
4. The _____ filters _____, and in its absence, problems associated with _____ may occur.
   a. spleen, lymph, oxygen transport
   b. liver, lymph, infection
   c. spleen, blood, infection
   d. spleen, blood, oxygen transport
5. Which cell(s) phagocytize pathogens?
   a. neutrophils
   b. macrophages
   c. lymphocytes
   d. Both a and b are correct.
6. _____ can initiate an inflammatory response.
   a. Mast cells
   b. Dendritic cells
   c. Macrophages
   d. All of these are correct.
7. Complement
   a. is an innate defense mechanism.
   b. is involved in the inflammatory reaction.
   c. is a series of proteins present in the plasma.
   d. can enhance phagocytosis.
   e. All of these are correct.
8. Antibody-mediated immunity is most directly associated with
   a. T cells.
   b. monocytes.
   c. basophils.
   d. B cells.
9. Plasma cells are
   a. the same as memory cells.
   b. formed from blood plasma.
   c. B cells that are actively secreting antibody.
   d. inactive T cells carried in the plasma.
10. Which applies to T cells?
   a. begin development in the bone marrow
   b. mature in thymus
   c. Both a and b are correct.
   d. None of these are correct.
11. MHC proteins play a role in
   a. active immunity.
   b. presentation of antigens to T cells.
   c. tissue transplant rejection.
   d. All of these are correct.
12. Type O blood has _____ antibodies in the plasma.
   a. anti-A and anti-B
   b. anti-O
   c. no
   d. All of these are correct.
13. Label a–c on this IgG molecule using these terms: antigen-binding sites, light chain, heavy chain. What do V and C stand for in the diagram?

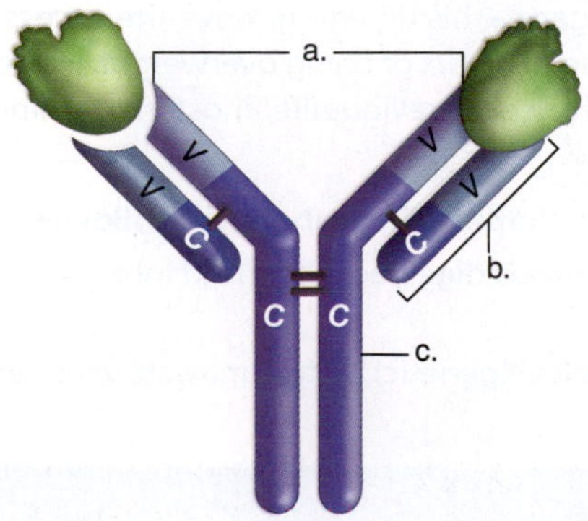

a.
b.
c.
V
V
V
V
C
C
C
C

14. Monoclonal antibodies are
   a. the same thing as vaccines.
   b. very specific in binding to a single antigen.
   c. usually isolated from the serum of immunized animals.
   d. All of these are correct.
15. Which of the following is not an example of an autoimmune disease?
   a. multiple sclerosis
   b. rheumatic fever
   c. hemolytic disease of the newborn
   d. systemic lupus erythematosus

ENGAGE

Thinking Critically

1. Some primitive organisms, such as invertebrates, have no lymphocytes and thus lack an adaptive immune system, but they have some components of an innate immune system, including phagocytes and certain protective proteins. What are some general features of innate immunity that make it very valuable to organisms lacking more specific antibody- and cell-mediated responses? What are some disadvantages to having only an innate immune system?
2. Compare the ways in which natural killer cells and cytotoxic T cells recognize the cells they are going to kill. Why does it make good biological sense for NK cells and cytotoxic T cells to recognize, for example, virus-infected cells in these different ways?
3. Why is it that Rh incompatibility can be a serious problem when an Rh-negative mother is carrying an Rh-positive fetus, but ABO incompatibility between mother and fetus is usually no problem? That is, a type A mother can usually safely carry a type B fetus. (*Hint:* The antibodies produced by an Rh-negative mother against the Rh antigen are usually IgG, whereas the antibodies produced against the A or B antigens are IgM.) Because the Rh antigen obviously serves no vital function (most humans lack it), why do you think it hasn't been completely eliminated during human evolution?

the brain (see Chapter 18). The *tongue* is composed of skeletal muscle that contracts to change the shape of the tongue. A fold of mucous membrane on the underside of the tongue attaches it to the floor of the mouth.

The roof of the mouth separates the *nasal cavity* from the mouth, and prevents ingested food from entering the nasal cavity. The roof has two parts: an anterior (toward the front) *hard palate* and a posterior (toward the back) *soft palate* (Fig. 14.2*a*). The hard palate contains several bones, but the soft palate is composed of muscle and connective tissue. The soft palate ends in a finger-shaped projection called the *uvula*.

In about one out of 700 newborns, the bones of the hard palate have not fused during fetal development, leaving a gap, or *cleft palate.* If not surgically repaired, such a defect can result in problems with feeding, speech, and hearing (due to ear infections).

The *tonsils* are in the back of the mouth, on either side of the tongue, and in the nasopharynx, where they are called *adenoids.* The tonsils contain lymphoid tissue that helps protect the body against infections (see Chapter 13). If the tonsils become inflamed, the person has *tonsillitis.* If tonsillitis recurs repeatedly, the tonsils may be surgically removed in a procedure called a *tonsillectomy.*

Three pairs of **salivary glands** (see Fig. 14.1) produce saliva. Saliva keeps the mouth moist and contains an enzyme that begins the process of digesting starch. One pair of salivary glands lies on either side of the face immediately below and in front of the ears. These glands swell when a person has the *mumps,* a disease caused by a viral infection. Another pair of salivary glands lies beneath the tongue, and still another pair lies beneath the floor of the mouth. The ducts from these salivary glands open under the tongue. You may be able to locate the openings if you use your tongue to feel for small flaps on the inside of your cheek and under your tongue.

The Teeth

With our *teeth,* we chew food into pieces convenient for swallowing. During the first two years of life, the smaller 20 *deciduous,* or baby, teeth appear. These are eventually replaced by 32 adult teeth (Fig. 14.2*a*). The third pair of molars, called the *wisdom teeth,* sometimes fail to erupt. If they push on the other teeth and/or cause pain, they can be removed by a dentist or oral surgeon.

Each tooth has two main divisions, a crown and a root (Fig. 14.2*b*). The *crown* has an outer layer of *enamel,* an extremely hard covering made of calcium compounds; *dentin,* a thick layer of bonelike material; and an inner *pulp,* which contains the nerves and the blood vessels. Dentin and pulp are also found in the root. The gum tissue, or *gingiva,* surrounds the teeth and normally forms a tight seal around them.

Bacteria that adhere to the teeth metabolize sugar and give off acids, which may erode the enamel. This can result in *dental caries,* or cavities, in the teeth. The enamel decays very slowly, but once the dentin layer is reached, the damage spreads more rapidly. Three measures can help to prevent tooth decay: limiting sugar intake, daily brushing and flossing, and regular visits to the dentist. Fluoride treatments, particularly in children, can make the enamel stronger and decay-resistant. Gum disease is more apt to occur with aging. Inflammation of the gums (*gingivitis*) can spread to the periodontal membrane, which lines the tooth socket. A person then has *periodontitis,* characterized by loss of bone and loosening of the teeth so that extensive dental work may be required. Stimulation of the gums in a manner advised by your dentist is helpful in controlling this condition.

Over time, and especially with exposure to cigarette smoke, soft drinks, or coffee, the enamel layer can become stained. In recent years, various tooth-whitening treatments, most of which contain a bleaching agent such as hydrogen peroxide, have become a popular way to remove these stains.

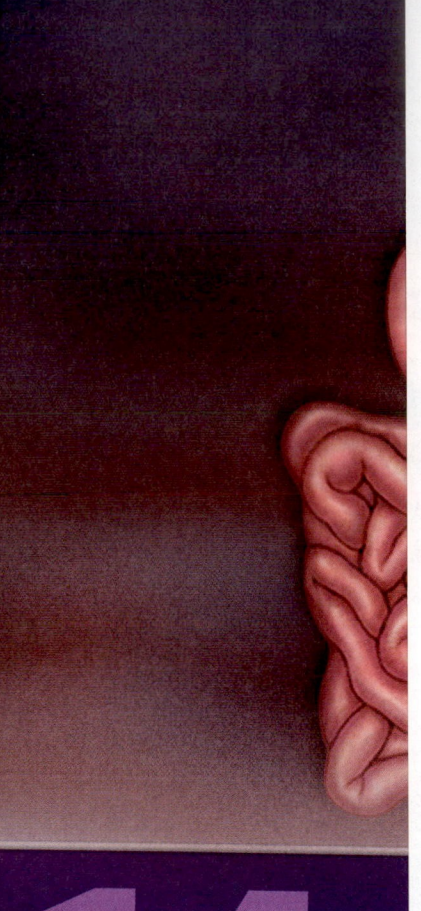

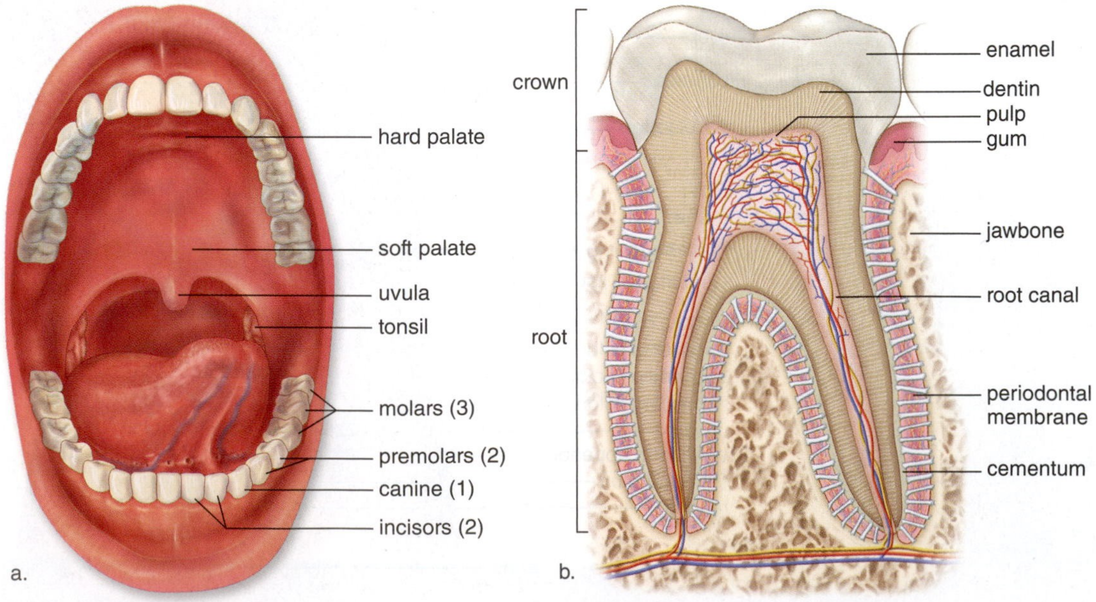

hard palate
soft palate
uvula
tonsil
molars (3)
premolars (2)
canine (1)
incisors (2)
a.
crown
root
enamel
dentin
pulp
gum
jawbone
root canal
periodontal membrane
cementum
b.

Figure 14.2 Adult mouth and teeth. **a.** The chisel-shaped incisors bite; the pointed canines tear; the fairly flat premolars grind; and the flattened molars crush food. The last molars, called wisdom teeth, may fail to erupt, or if they do, they are sometimes crooked and useless. **b.** Longitudinal section of a tooth. The crown is the portion that projects above the gum line and can be replaced by a dentist if damaged. When a "root canal" is done, the nerves are removed. When the periodontal membrane is inflamed, the teeth can loosen.

The Pharynx

The **pharynx** is a region that receives air from the nasal cavities and food from the mouth (see Fig. 14.1).

Table 14.1 lists some specific functions of the digestive organs. From the mouth, food (the bolus) passes through the pharynx and esophagus to the stomach, small intestine, and large intestine. The food passage and air passage cross in the pharynx because the trachea (windpipe) is ventral to (in front of) the esophagus, which takes food to the stomach. Swallowing, a process that occurs in the pharynx (Fig. 14.3), is a *reflex action* performed automatically, without conscious thought. During swallowing, the soft palate moves back to close off the *nasopharynx,* and the trachea moves up under the *epiglottis* to cover the glottis. The *glottis* is the opening to the larynx (voice box), and therefore the air passage. The up-and-down movement of the Adam's apple, the front part of the larynx, is easy to observe when a person swallows. During swallowing, food normally enters the esophagus because the air passages are blocked. We do not breathe when we swallow.

Unfortunately, we have all had the unpleasant experience of having food "go the wrong way." The wrong way may be either into the nasal cavities or into the trachea. If it is the latter, coughing will most likely force the food up out of the trachea and into the pharynx again.

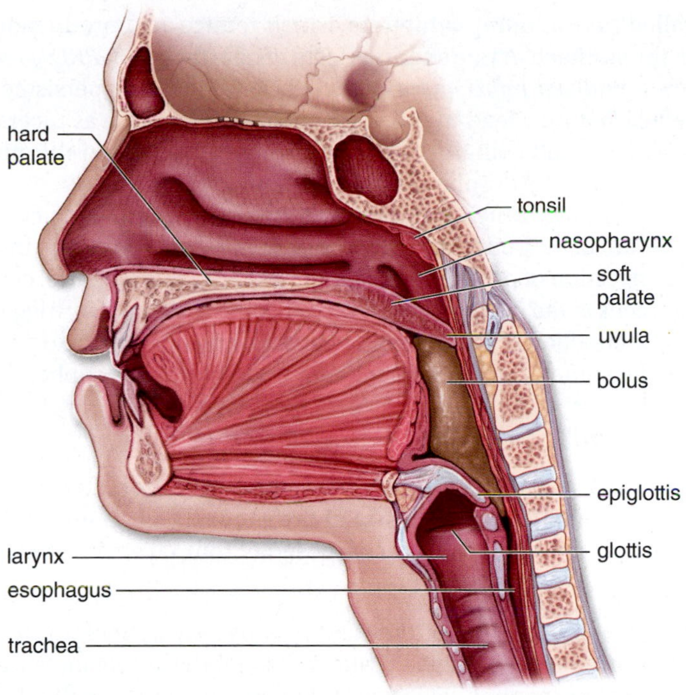
hard palate
tonsil
nasopharynx
soft palate
uvula
bolus
epiglottis
glottis
larynx
esophagus
trachea

Figure 14.3 Swallowing. When food is swallowed, the soft palate closes off the nasopharynx, and the epiglottis covers the glottis, forcing the bolus to pass down the esophagus. Therefore, a person does not breathe while swallowing.

The Esophagus

The **esophagus** is a long muscular tube that passes from the pharynx, through the thoracic cavity and diaphragm, and into the abdominal cavity, where it joins the stomach. The esophagus is ordinarily collapsed, but it opens and receives the bolus when swallowing occurs.

Rhythmic muscular contractions, collectively called **peristalsis,** push the food along the digestive tract. Peristalsis begins in the esophagus and continues in all the organs of the digestive tract. Occasionally, peristalsis begins even though there is no food in the esophagus. This produces the sensation of a lump in the throat.

The esophagus plays no role in the chemical digestion of food. Its sole purpose is to conduct the food bolus from the mouth to the stomach. *Sphincters* are muscles that encircle tubes in the body, acting as valves. The tubes close when sphincters contract, and they open when sphincters relax. The entrance of the esophagus to the stomach is marked by a constriction that is often called a sphincter, although the muscle is not as developed as it would be in a true sphincter. Relaxation of the sphincter allows the bolus to pass into the stomach, while contraction prevents the acidic contents of the stomach from backing up into the esophagus.

MP3 Oral Cavity, Esophagus, and the Swallowing Reflex

Heartburn, which feels like a burning pain rising into the throat, occurs when some of the stomach contents escape into the esophagus. Heartburn is a very common condition: according to Drugs.com, the top-selling drug in the United States in 2010 was Nexium, with total U.S. sales over $5 billion. Nexium (generic name, esomeprazole) belongs to a group of drugs

TABLE 14.1 Path of Food

Organ	General Function(s)	Special Feature(s)	Function of Special Feature(s)
Mouth	Receives food; starts digestion of starch	Teeth Tongue	Chew food Forms bolus
Pharynx	Passageway	—	—
Esophagus	Passageway	—	—
Stomach	Storage of food; acidity kills bacteria; starts digestion of protein	Gastric glands	Release gastric juices
Small intestine	Digestion of all foods; absorption of nutrients	Intestinal glands Villi	Release intestinal juices Absorb nutrients
Large intestine	Absorption of water; storage of indigestible remains	—	—

called proton pump inhibitors, which reduce acid production in the stomach. A more serious form of heartburn is *GERD* (*gastroesophageal reflux disease*), which is frequent or persistent reflux that may lead to more serious problems such as ulcers, difficulty swallowing, or even esophageal cancer due to chronic irritation of the esophageal wall by stomach acid.

Vomiting, which is the forceful expulsion of stomach contents out of the body through the mouth, can help protect against the ingestion of potentially harmful agents. In this scenario, certain cells in the intestinal tract, as well as in the brainstem, trigger the contraction of the abdominal muscles and diaphragm to propel the contents of the stomach upward through the esophagus.

The Wall of the Digestive Tract

The structure of the esophageal wall in the abdominal cavity is representative of that found in the stomach, small intestine, and large intestine. All are composed of four layers, as shown in Figure 14.4 and listed here:

Mucosa (mucous membrane layer) A layer of epithelium supported by connective tissue and smooth muscle lines the **lumen** (central cavity) and contains glandular epithelial cells that secrete digestive enzymes and goblet cells that secrete mucus.

Submucosa (submucosal layer) Beneath the mucosa lies the submucosa, a broad band of loose connective tissue that contains blood vessels. Lymph nodules, including some called Peyer's patches, are in the submucosa. Like the tonsils, they help protect us from disease.

Muscularis (smooth muscle layer) Two layers of smooth muscle make up this section. The inner, circular layer encircles the gut; the outer, longitudinal layer lies perpendicular to the circular layer. (The stomach also has oblique muscles.)

Serosa (serous membrane layer) Most of the digestive tract has a serosa, a very thin, outermost layer of squamous epithelium that secretes a serous fluid that keeps the outer surface of the intestines moist so that the organs of the abdominal cavity slide against one another.

The Stomach

The **stomach** (Fig. 14.5*a*) is a thick-walled, J-shaped organ that lies on the left side of the abdominal cavity below the liver and diaphragm. It is continuous with the esophagus above and the duodenum of the small intestine below. The stomach receives food from the esophagus, starts the digestion of proteins, and moves food into the small intestine. The human stomach is about 25 cm (10 in.) long, regardless of the amount of food it holds, but the diameter varies, depending on how full it is. As the stomach expands, deep folds in its wall, called *rugae,* gradually disappear. When full, it can hold about 4 l (1 gal).

The columnar epithelium lining the stomach has millions of gastric pits, which lead into *gastric glands* (Fig. 14.5*b*). The term "gastric" always refers to the stomach. The gastric glands produce gastric juice, which contains pepsinogen, hydrochloric acid (HCl), and mucus. As discussed in section 14.3, pepsinogen becomes the enzyme pepsin when exposed to HCl. The high acidity of the stomach (pH of about 2) is also beneficial because it kills most of the bacteria and other microbes present in food.

The stomach acts both physically and chemically on food. Its wall contains three muscle layers: one layer is longitudinal and another is circular, as shown in Figure 14.4*a*; the third is

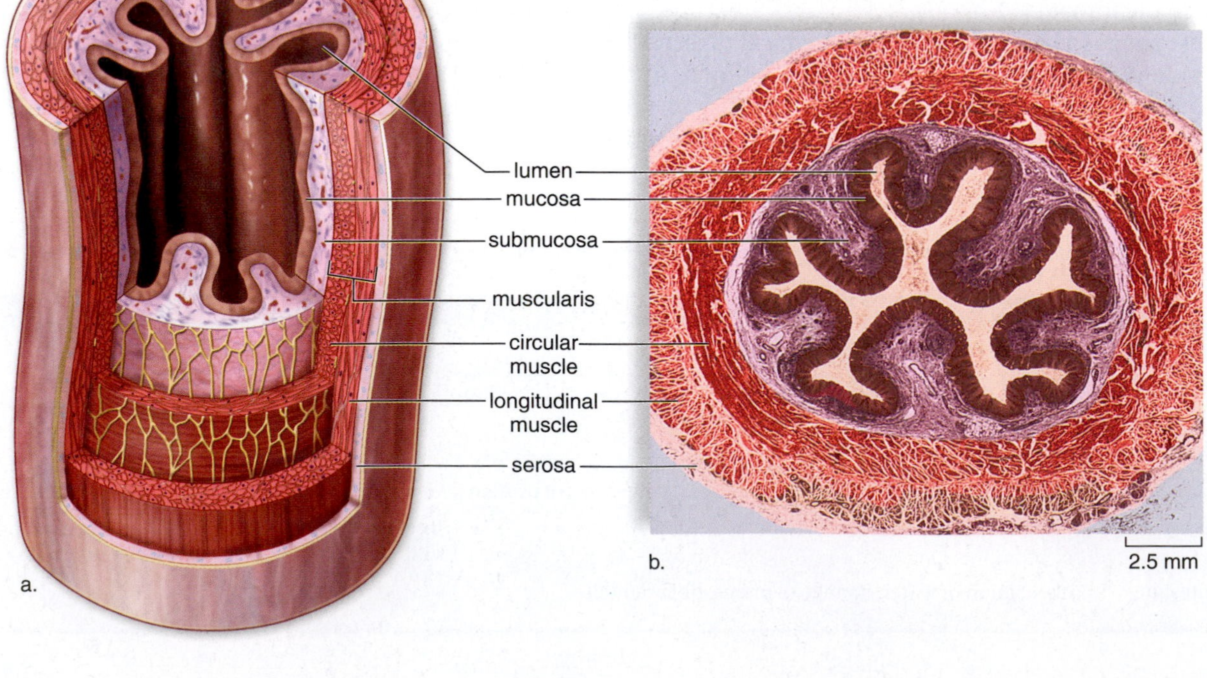
lumen
mucosa
submucosa
muscularis
circular muscle
longitudinal muscle
serosa
a.
b.
2.5 mm

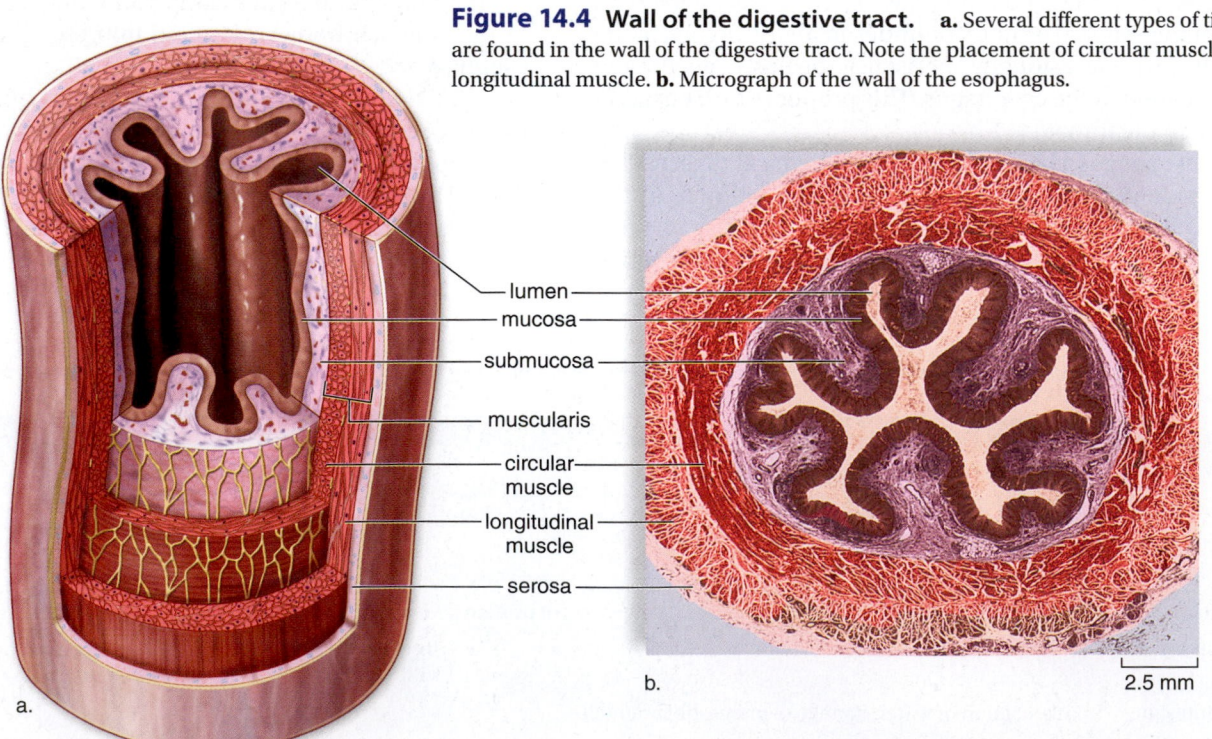

Figure 14.4 Wall of the digestive tract. **a.** Several different types of tissues are found in the wall of the digestive tract. Note the placement of circular muscle inside longitudinal muscle. **b.** Micrograph of the wall of the esophagus.

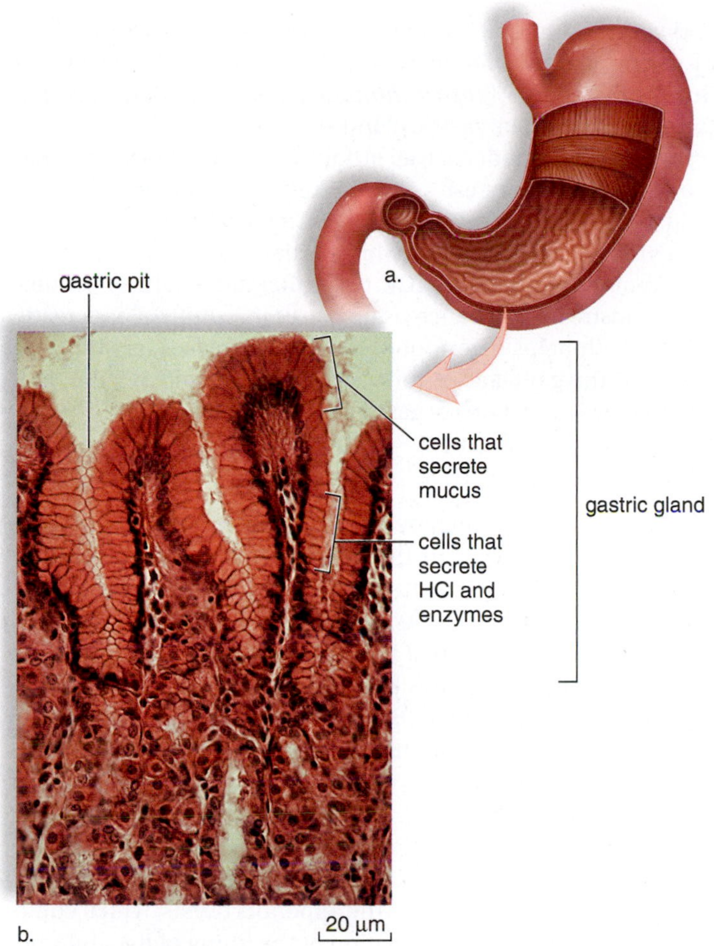

a.
gastric pit
cells that secrete mucus
gastric gland
cells that secrete HCl and enzymes
b.
20 µm

Figure 14.5 Anatomy of the stomach. **a.** The stomach has a thick wall with folds that allow it to expand and fill with food. **b.** The mucosa contains gastric glands, which secrete gastric juice containing mucus and digestive enzymes.

obliquely arranged. This muscular wall not only moves the food along, but it also churns, mixing the food with gastric juice and breaking it down into small pieces.

Alcohol and other liquids are absorbed in the stomach, but most solid food substances are not. Normally, the stomach empties in two to six hours. When food leaves the stomach, it is a thick, soupy liquid called **chyme.** Chyme enters the small intestine in squirts by way of the pyloric sphincter, which acts like a valve, repeatedly opening and closing.

Animation Three Phases of Gastric Secretion

MP3 Stomach

The Small Intestine

The **small intestine** is named for its small diameter (compared to that of the large intestine), but perhaps it should be called the long intestine. It averages about 6 meters (almost 20 ft) in length, compared to the large intestine, which is about 1.5 meters (5 ft) in length.

The first 25 cm (1 ft) of the small intestine is called the **duodenum.** Ducts from the liver and pancreas join to form one common bile duct that enters the duodenum (see Fig. 14.10). The small intestine receives bile from the liver and pancreatic juice from the pancreas via this duct. The intestine has a slightly basic pH because pancreatic juice contains sodium bicarbonate ($NaHCO_3$), which neutralizes the acid in chyme. The enzymes in pancreatic juice and enzymes produced by the intestinal wall complete the process of food digestion.

The middle part of the small intestine is called the **jejunum,** and the remainder is the **ileum.** The submucosal layer of the ileum contains aggregations of lymphoid tissues called *Peyer's patches,* which are involved in generating immune responses to intestinal pathogens.

The surface area of the small intestine has been compared to that of a tennis court. What factors contribute to such a large surface area? The wall of the small intestine contains fingerlike projections called **villi** (sing. villus), which give the intestinal wall a soft, velvety appearance (Fig. 14.6). A villus has an outer

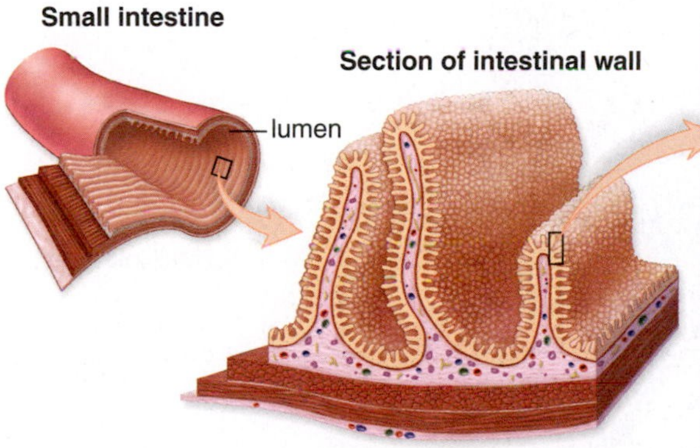

Small intestine
lumen
Section of intestinal wall

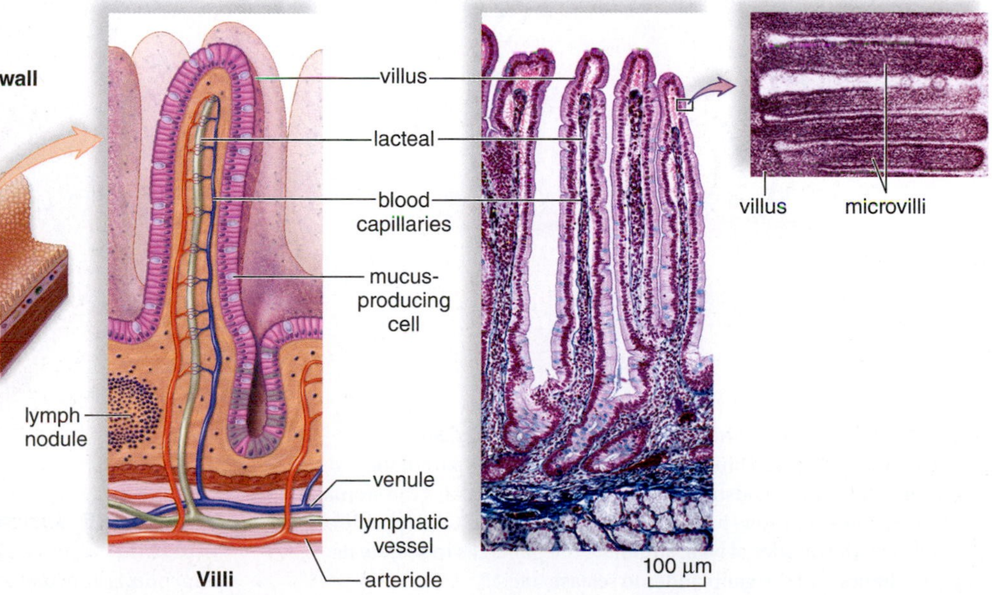

villus
lacteal
blood capillaries
mucus-producing cell
lymph nodule
venule
lymphatic vessel
arteriole
Villi
100 µm
villus
microvilli

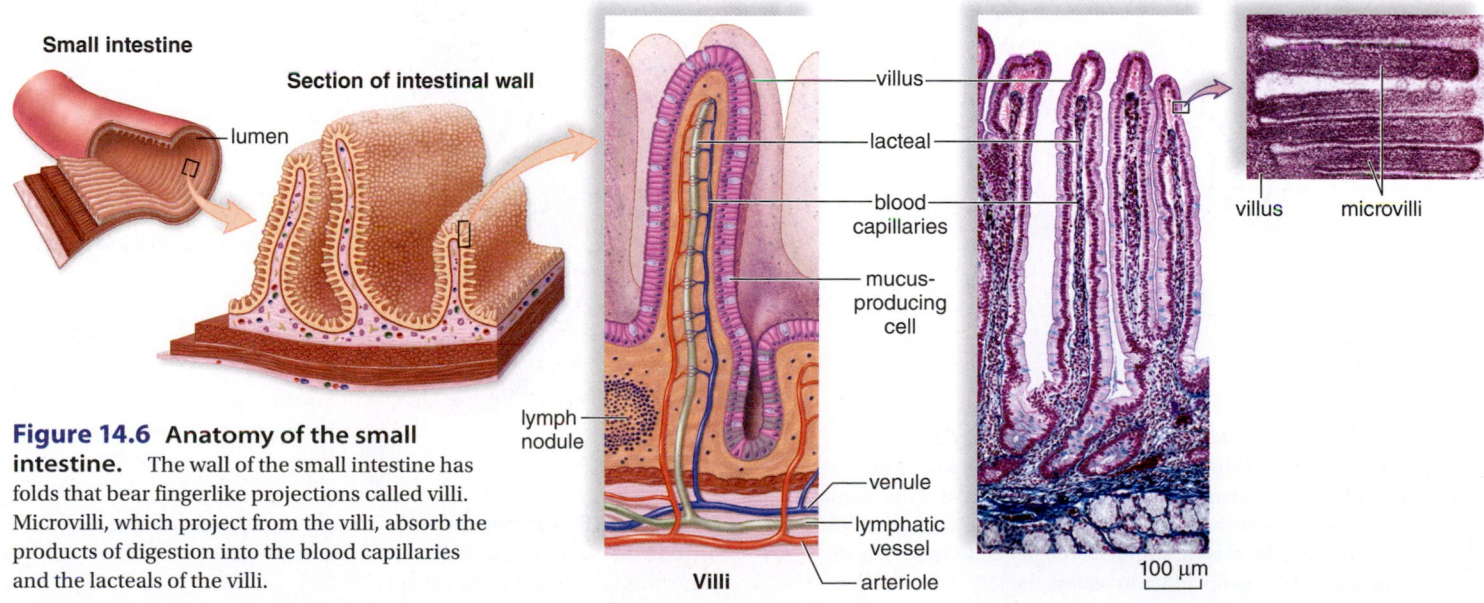

Figure 14.6 Anatomy of the small intestine. The wall of the small intestine has folds that bear fingerlike projections called villi. Microvilli, which project from the villi, absorb the products of digestion into the blood capillaries and the lacteals of the villi.

layer of columnar epithelial cells, and each of these cells has thousands of microscopic extensions called microvilli. Microvilli greatly increase the surface area of the villus for the absorption of nutrients.

Nutrients are absorbed into the vessels of a villus. Each villus contains blood capillaries and a small lymphatic capillary, called a **lacteal.** Glycerol and fatty acids (digested from fats) enter the epithelial cells of the villi, where they are joined and packaged as lipoprotein droplets that enter a lacteal. Lacteals are a part of the lymphatic system (see Chapter 13). The lymphatic system is an adjunct to the cardiovascular system; that is, its vessels carry a fluid called lymph to the cardiovascular veins. Sugars and amino acids directly enter the blood capillaries of a villus. After nutrients are absorbed, the bloodstream eventually carries them to all the cells of the body.

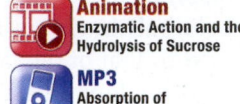

Animation Enzymatic Action and the Hydrolysis of Sucrose

MP3 Absorption of Nutrients and Water

Regulation of Digestive Secretions

The secretion of digestive juices is promoted by the nervous system and by hormones. A *hormone* is a substance produced by one set of cells that affects a different set of cells (for more on the endocrine system, see Chapter 20). Hormones are usually transported by the bloodstream. For example, when a person has eaten a meal particularly rich in protein, the stomach produces the hormone *gastrin.* Gastrin enters the bloodstream, and soon the stomach is churning, and the secretory activity of gastric glands is increasing. A hormone produced by the duodenal wall, *GIP* (*gastric inhibitory peptide*), works opposite to gastrin: it inhibits gastric gland secretion.

Cells of the duodenal wall produce two other hormones that are of particular interest—secretin and CCK (cholecystokinin). Acid, especially hydrochloric acid (HCl) present in chyme, stimulates the release of secretin, while partially digested protein and fat stimulate the release of CCK. Soon after these hormones enter the bloodstream, the pancreas increases its output of pancreatic juice, which helps digest food; the liver increases its output of bile; and the gallbladder contracts to release bile. Figure 14.7 summarizes the actions of gastrin, secretin, and CCK.

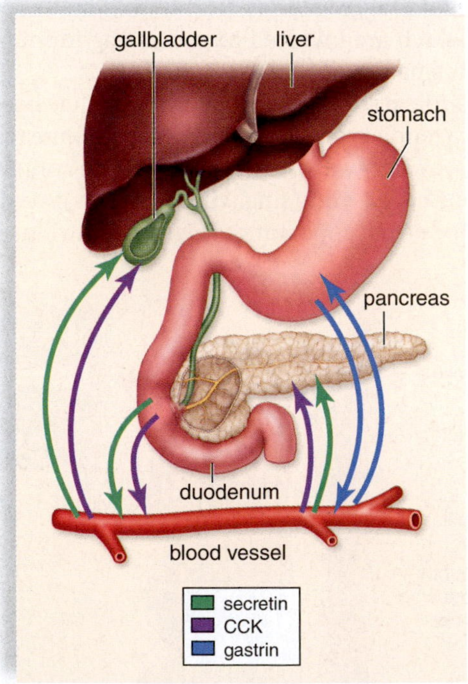
gallbladder
liver
stomach
pancreas
duodenum
blood vessel
secretin
CCK
gastrin

Figure 14.7 Hormonal control of digestive gland secretions. Gastrin (blue), produced by the lower part of the stomach, enters the bloodstream and thereafter stimulates the stomach to produce more digestive juices. Secretin (green) and CCK (purple), produced by the duodenal wall, stimulate the pancreas to secrete its digestive juices and the gallbladder to release bile.

The Large Intestine

The **large intestine** includes the cecum, colon, rectum, and anal canal. It is larger in diameter than the small intestine (6.5 cm compared to 2.5 cm), but shorter in length. The large intestine absorbs water, salts, and some vitamins. It also stores indigestible material until it is eliminated as feces.

The **cecum,** which lies below the junction with the small intestine, is a small pouch (6 cm long) that forms the first part of the large intestine. The human cecum has a small projection called the **appendix** or **vermiform appendix** (*vermiform* means wormlike) (Fig. 14.8). Like the tonsils, the appendix may play a role in fighting infection. This organ is subject to inflammation, a condition called appendicitis. If inflamed, the appendix should be removed before the appendix bursts, which could cause *peritonitis,* an inflammation of the lining of the abdominal cavity. Peritonitis can lead to death.

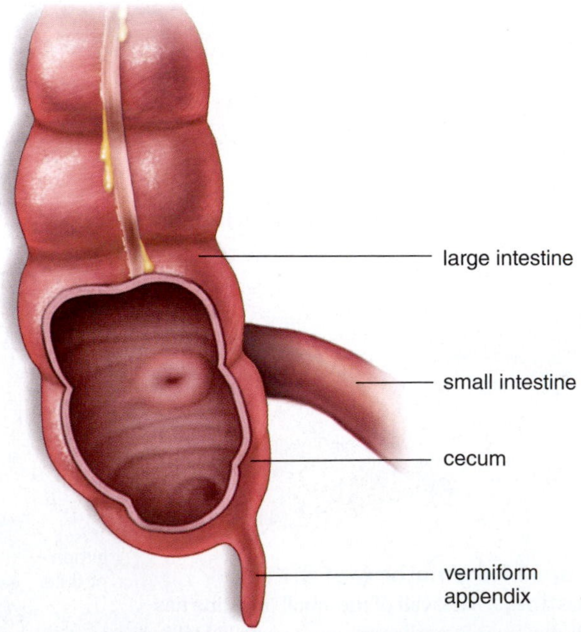
large intestine
small intestine
cecum
vermiform appendix

Figure 14.8 Junction of the small intestine and the large intestine. The cecum is the blind end of the large intestine. The appendix is attached to the cecum.

The *colon* includes the ascending colon, which goes up the right side of the body to the level of the liver; the transverse colon, which crosses the abdominal cavity just below the liver and the stomach; the descending colon, which passes down the left side of the body; and the sigmoid colon, which enters the *rectum,* the last 20 cm of the large intestine. The rectum opens at the **anus,** where *defecation,* the expulsion of feces, occurs. When feces are forced into the rectum by peristalsis, a defecation reflex occurs. The stretching of the rectal wall initiates nerve impulses to the spinal cord, and shortly thereafter, the rectal muscles contract, and the anal sphincters relax (Fig. 14.9). Ridding the body of indigestible remains is another way the digestive system helps maintain homeostasis. Feces are normally about three-quarters water and one-quarter solids. Bacteria, **fiber** (indigestible plant material), and other indigestible materials are in the solid portion. Bacterial action on undigested materials causes the odor of feces and also accounts for the presence of gas. A breakdown product of bilirubin (described in section 14.2) and the presence of oxidized iron cause the brown color of feces.

About 40–50% of the fecal mass consists of bacteria and other microbes; in fact, there are about 100 billion bacteria per gram of feces. These include facultative bacteria (which can live with or without oxygen) like *Escherichia coli,* as well as many obligate anaerobes (which die in the presence of oxygen). These bacteria break down some indigestible material, and produce some vitamins that our bodies can absorb and use. In this way, they perform a service for us, and we provide a good environment for them to live in.

Video Fat Microbes

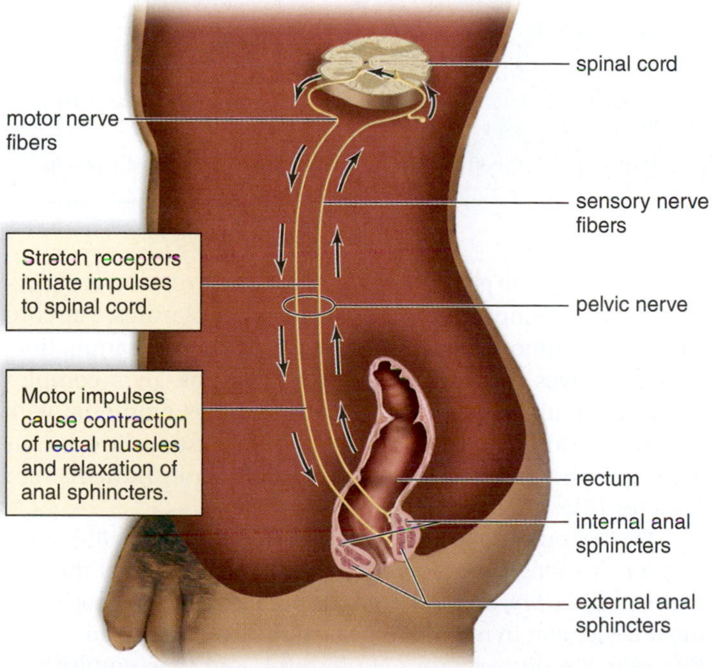
spinal cord
motor nerve fibers
sensory nerve fibers
Stretch receptors initiate impulses to spinal cord.
pelvic nerve
Motor impulses cause contraction of rectal muscles and relaxation of anal sphincters.
rectum
internal anal sphincters
external anal sphincters

Figure 14.9 Defecation reflex. The accumulation of feces in the rectum causes it to stretch, which initiates a reflex action resulting in rectal contraction and expulsion of the fecal material.

Water is considered unsafe for drinking when the coliform (intestinal) bacterial count reaches a certain number. A high count indicates that a significant amount of feces has entered the water. The more feces present, the greater the possibility that disease-causing bacteria are also present.

Check Your Progress 14.1

1. List the parts of the human digestive tract that food passes through, and note whether chemical and/or mechanical digestion occurs in each.
2. Identify the four layers that make up the wall of the esophagus, stomach, and intestines.
3. Compare the major functions of the small intestine and the large intestine.
4. Discuss strategies that might help obese people lose weight by inhibiting the secretion and/or activity of certain hormones.

14.2 Accessory Organs of Digestion

Learning Outcomes

Upon completion of this section, you should be able to

1. Summarize the major functions of the pancreas, the liver, and the gallbladder.
2. Describe the structure and function of the hepatic portal system.

The pancreas, liver, and gallbladder are accessory digestive organs. The salivary glands, discussed earlier, are also accessory organs. Figure 14.10*a* shows how the pancreatic duct from the pancreas and the common bile duct from the liver and gallbladder enter the duodenum.

The Pancreas

The **pancreas** lies deep in the abdominal cavity, resting on the posterior abdominal wall. It is an elongated and somewhat flattened organ that has both endocrine and exocrine functions. As an endocrine gland, it secretes insulin and glucagon, hormones that help keep the blood glucose level within normal limits. In this chapter, however, we are interested in its exocrine function. Most pancreatic cells produce pancreatic juice, which contains sodium bicarbonate ($NaHCO_3$) to help neutralize the stomach acid, and digestive enzymes for all types of food (see next section).

The Liver

The **liver,** which is the largest gland in the body, lies mainly in the upper right section of the abdominal cavity, under the diaphragm (see Fig. 14.1). The liver contains approximately 100,000 lobules that serve as its structural and functional units (Fig. 14.10*b*). Three structures are located between the lobules: a bile duct that takes bile away from the liver; a branch of the hepatic artery that brings O_2-rich blood to the liver; and a branch of the hepatic portal vein that transports nutrients

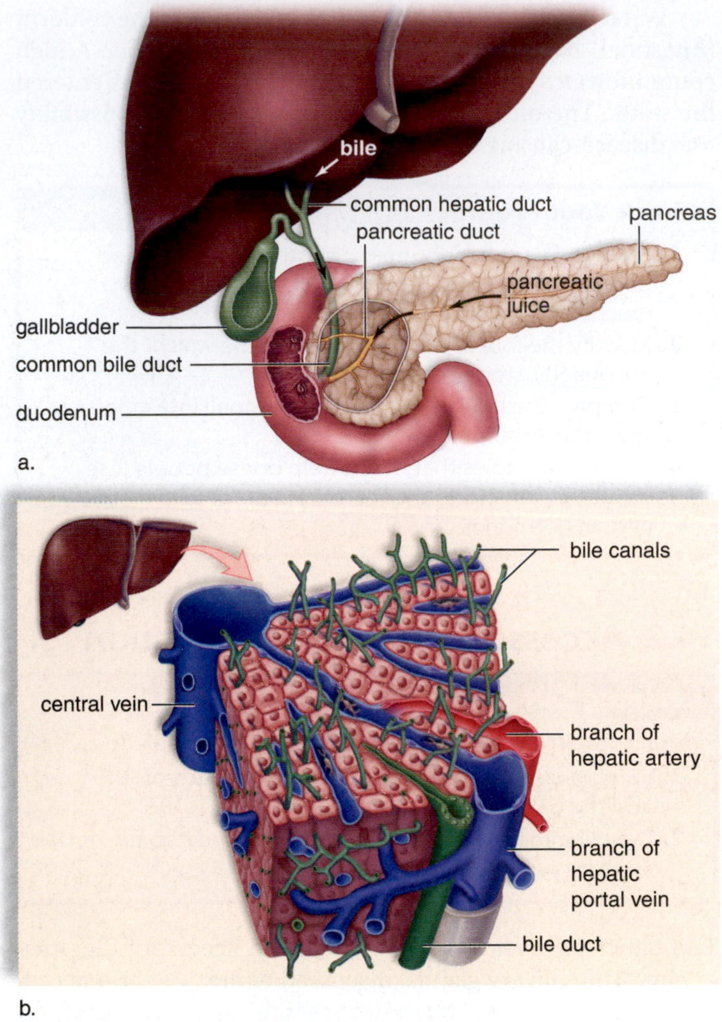
bile
common hepatic duct
pancreatic duct
pancreas
pancreatic juice
gallbladder
common bile duct
duodenum
a.
bile canals
central vein
branch of hepatic artery
branch of hepatic portal vein
bile duct
b.

Figure 14.10 Liver, gallbladder, and pancreas. **a.** The liver makes bile, which is stored in the gallbladder and sent (black arrow) to the small intestine by way of the common bile duct. The pancreas produces digestive enzymes that are sent (black arrows) to the small intestine by way of the pancreatic duct. **b.** A hepatic lobule. The liver contains over 100,000 lobules, each lobule composed of many cells that perform the various functions of the liver. They remove and add materials to the blood and deposit bile in a duct.

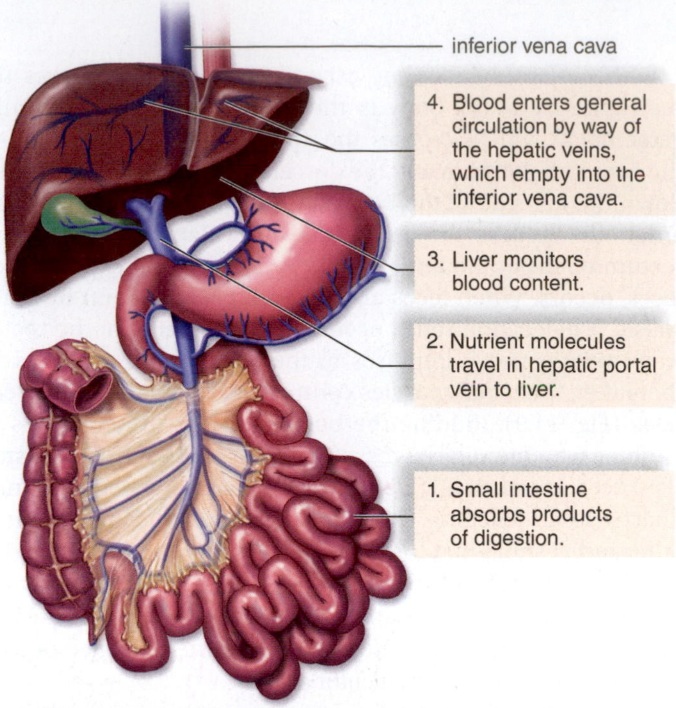
inferior vena cava
4. Blood enters general circulation by way of the hepatic veins, which empty into the inferior vena cava.
3. Liver monitors blood content.
2. Nutrient molecules travel in hepatic portal vein to liver.
1. Small intestine absorbs products of digestion.

Figure 14.11 Hepatic portal system. The hepatic portal vein takes the products of digestion from the digestive system to the liver, where they are processed before entering a hepatic vein.

TABLE 14.2 Functions of the Liver

1.	Detoxifies blood by removing and metabolizing poisonous substances
2.	Stores iron (Fe^{2+}) and vitamins A, D, E, K, and B_{12}
3.	Makes many plasma proteins, such as albumins and fibrinogen, from amino acids
4.	Stores glucose as glycogen after a meal, and breaks down glycogen to glucose to maintain the glucose concentration of blood between eating periods
5.	Produces urea after breaking down amino acids
6.	Removes bilirubin, a breakdown product of hemoglobin from the blood, and excretes it in bile, a liver product
7.	Helps regulate blood cholesterol level, converting some to bile salts

from the intestines. Each lobule has a central vein that enters a hepatic vein. In Figure 14.11, trace the path of blood from the intestines to the liver via the hepatic portal vein and from the liver to the inferior vena cava via the hepatic veins.

In some ways, the liver acts as the gatekeeper to the blood (Table 14.2). As blood from the hepatic portal vein passes through the liver, it removes poisonous substances and detoxifies them. The liver also removes and stores iron and vitamins A, D, E, K, and B_{12}. The liver makes many of the plasma proteins and helps regulate the quantity of cholesterol in the blood.

The liver maintains the blood glucose level at 50–80 mg/100 ml, even though a person eats intermittently. When insulin is present, any excess glucose in the blood is removed and stored by the liver (and muscles) as glycogen. Between meals, glycogen is broken down to glucose, which enters the hepatic veins, and in this way, the blood glucose level remains relatively constant.

If the supply of glycogen is depleted, the liver converts glycerol (from fats) and amino acids to glucose molecules. The conversion of amino acids to glucose necessitates deamination, which involves the removal of amino groups. By a complex metabolic pathway, the liver then combines ammonia with carbon dioxide to form *urea.* Urea is the usual nitrogenous waste product from amino acid breakdown in humans (see Chapter 16).

The liver produces **bile,** which is stored in the gallbladder. Bile has a yellowish-green color because it contains the bile pigment *bilirubin,* derived from the breakdown of hemoglobin, the red pigment in red blood cells. Bile also contains bile salts. Bile salts are derived from cholesterol, and they emulsify fat in the small intestine. When fat is emulsified, it breaks up into droplets, providing a much larger surface area that can be acted upon by a pancreatic lipase (see section 14.3).

The Gallbladder

The **gallbladder** is a pear-shaped, muscular sac attached to the surface of the liver (see Fig. 14.1). The liver produces about 400–800 ml of bile each day, and any excess is stored in the gallbladder. Water is reabsorbed by the gallbladder so that bile becomes a thick, mucuslike material. When needed, bile leaves the gallbladder and proceeds to the duodenum via the common bile duct.

Check Your Progress 14.2

1. Review what it means to say that the pancreas has both endocrine and exocrine functions.
2. Explain why the liver is necessary for life, whereas the gallbladder can be removed with few consequences.

14.3 Digestive Enzymes

Learning Outcomes

Upon completion of this section, you should be able to

1. Describe the overall function of digestive enzymes.
2. Compare the specific types of chemical digestion that take place in the mouth, stomach, and small intestine.

Digestive enzymes, like other enzymes, are proteins that speed up specific chemical reactions. Table 14.3 lists some of the major enzymes produced by the digestive tract. These enzymes help break down carbohydrates, proteins, nucleic acids, and fats, the major components of food.

Starch is a carbohydrate, and its digestion begins in the mouth. Saliva from the salivary glands has a neutral pH and contains **salivary amylase,** the first enzyme to act on starch:

$$\text{starch} + H_2O \xrightarrow{\text{salivary amylase}} \text{maltose}$$

In this equation, salivary amylase is written above the arrow to indicate that it is neither a reactant nor a product in the reaction. It merely speeds the reaction in which its substrate, starch, is digested to many molecules of maltose, a disaccharide. Maltose molecules cannot be absorbed by the intestine. Additional digestive action in the small intestine converts maltose to glucose, a monosaccharide. Glucose can be absorbed.

Protein digestion begins in the stomach. Gastric juice secreted by gastric glands is very acidic (pH of about 2) because it contains hydrochloric acid (HCl). Pepsinogen, a precursor that is converted to the enzyme **pepsin** when exposed to HCl, is also present in gastric juice. Pepsin acts on proteins, which are polymers of amino acids, to produce peptides:

$$\text{protein} + H_2O \xrightarrow{\text{pepsin}} \text{peptides}$$

Peptides vary in length, but they always consist of a number of linked amino acids. Peptides from the stomach are usually too large to be absorbed by the intestinal lining, but later in the small intestine, they are broken down to amino acids, the monomer of a protein molecule.

Pancreatic juice, which enters the duodenum, has a basic pH because it contains sodium bicarbonate ($NaHCO_3$). Sodium bicarbonate neutralizes the acid in chyme, producing the slightly basic pH that is optimum for pancreatic enzymes, including **pancreatic amylase,** which digests starch, and **trypsin,** which digests protein. Trypsin is secreted as trypsinogen, which is converted to trypsin in the duodenum.

Lipase, a third pancreatic enzyme, digests fat molecules in the fat droplets after they have been emulsified by bile salts:

$$\text{fat} \xrightarrow{\text{bile salts}} \text{fat droplets}$$

$$\text{fat droplets} + H_2O \xrightarrow{\text{lipase}} \text{glycerol} + 3 \text{ fatty acids}$$

Fats are triglycerides, which means that each one is composed of glycerol and three fatty acids. The end products of lipase digestion, glycerol and fatty acid molecules, are small enough to cross the cells of the intestinal villi, where absorption takes place. As mentioned in section 14.1, glycerol and fatty acids enter the cells

TABLE 14.3 Major Digestive Enzymes

Enzyme	Produced By	Site of Action	Optimum pH	Digestion
Salivary amylase	Salivary glands	Mouth	Neutral	Starch + H_2O ⟶ maltose
Pancreatic amylase	Pancreas	Small intestine	Basic	Starch + H_2O ⟶ maltose
Maltase	Small intestine	Small intestine	Basic	Maltose + H_2O ⟶ glucose + glucose
Pepsin	Gastric glands	Stomach	Acidic	Protein + H_2O ⟶ peptides
Trypsin	Pancreas	Small intestine	Basic	Protein + H_2O ⟶ peptides
Peptidases	Small intestine	Small intestine	Basic	Peptide + H_2O ⟶ amino acids
Nuclease	Pancreas	Small intestine	Basic	RNA and DNA + H_2O ⟶ nucleotides
Nucleosidases	Small intestine	Small intestine	Basic	Nucleotide + H_2O ⟶ base + sugar + phosphate
Lipase	Pancreas	Small intestine	Basic	Fat droplet + H_2O ⟶ glycerol + fatty acids

of the villi, and within these cells, they are rejoined and packaged as lipoprotein droplets before entering the lacteals (see Fig. 14.6).

Peptidases and **maltase,** enzymes secreted by the surface cells of the small intestinal villi, complete the digestion of protein to amino acids and starch to glucose, respectively. These small molecules can then be absorbed by the cells of the villi. Peptides, which result from the first step in protein digestion, are digested to amino acids by peptidases:

$$\text{peptides} + H_2O \xrightarrow{\text{peptidases}} \text{amino acids}$$

Maltose, a disaccharide that results from the first step in starch digestion, is digested to glucose by maltase:

$$\text{maltose} + H_2O \xrightarrow{\text{maltase}} \text{glucose} + \text{glucose}$$

Other disaccharides, each of which has its own enzyme, are digested in the small intestine. The absence of any one of these enzymes can cause illness. The most common situation involves a deficiency of lactase, the enzyme that digests lactose, the sugar found in milk. Interestingly, the rate of this enzyme deficiency varies between human races. For example, only 25% of adult Caucasians, but 75–90% of adult African Americans, are lactase-deficient. Consuming dairy products often gives these individuals the symptoms of *lactose intolerance* (diarrhea, gas, cramps), caused by the fermentation of nondigested lactose by intestinal bacteria. However, lactose-reduced products are available for most kinds of dairy products, including milk, cheese, and ice cream. It is important to include foods that are high in calcium in the diet. Otherwise, osteoporosis, a condition characterized by weak and fragile bones, may develop (see section 19.6).

As mentioned in Chapter 6, enzymes function best at an optimum temperature and pH, which helps to maintain the proper shape to fit their substrate. The compartmentalization of the digestive system assists in the regulation of enzymatic activity. Because the entire digestive system of humans is maintained at a constant temperature of 37°C (98.6°F), enzymatic activity is largely controlled by pH. As discussed in section 14.1, the pH of the stomach is between 1 and 2, allowing the activation of pepsinogen enzyme. However, shortly after chyme proceeds into the duodenum (small intestine), the sodium bicarbonate released from the pancreas returns the pH to around 7.4 to 7.8. This "switch" serves to activate digestive enzymes released by the pancreas, thus allowing the digestion of carbohydrates, fats, and proteins (Fig. 14.12).

Check Your Progress 14.3

1. Describe where in the digestive tract the chemical digestion of each of the following types of nutrients occurs: carbohydrates, proteins, and fats.
2. Identify the final molecule (monomer) resulting from the digestion of carbohydrates, proteins, and fats.
3. Explain how the structure of the digestive system assists in regulating the digestive enzymes.

14.4 Human Nutrition

Learning Outcomes

Upon completion of this section, you should be able to

1. List the six major classes of nutrients and their major functions in the body.
2. Describe the relationship between diet and obesity, type 2 diabetes, and cardiovascular disease.
3. Distinguish between vitamins, coenzymes, and minerals.

Every aspect of your body's functioning depends on proper *nutrition,* the intake of nutrients. A *nutrient* is a component of food that is utilized by the body. The six major classes of nutrients are carbohydrates, proteins, lipids, vitamins, minerals, and water. All of these serve essential roles in the body, but overconsumption of almost any nutrient can also lead to health problems.

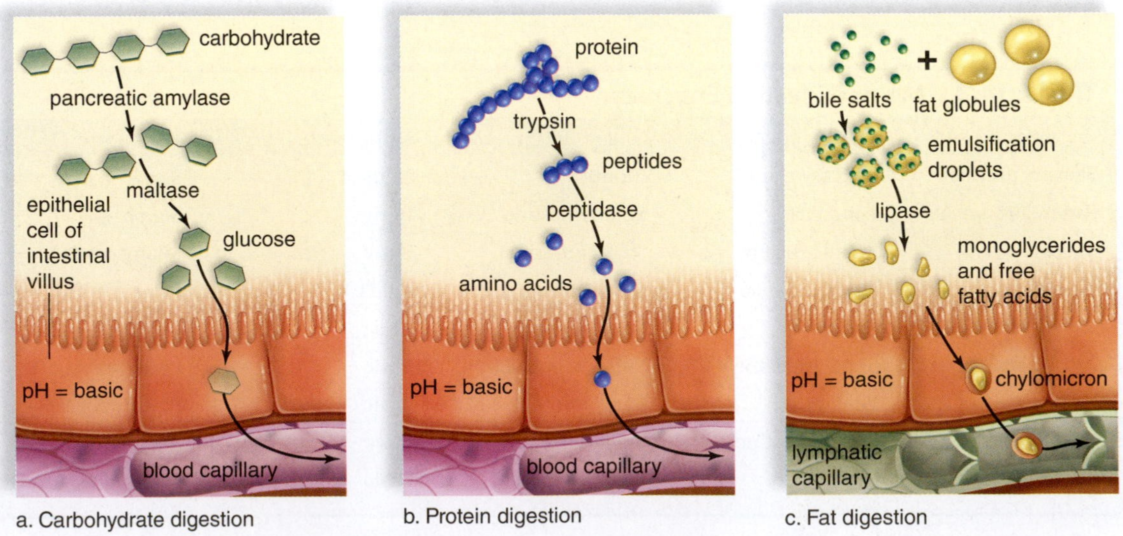
carbohydrate
pancreatic amylase
maltase
epithelial cell of intestinal villus
glucose
pH = basic
blood capillary
a. Carbohydrate digestion
protein
trypsin
peptides
peptidase
amino acids
pH = basic
blood capillary
b. Protein digestion
bile salts
fat globules
emulsification droplets
lipase
monoglycerides and free fatty acids
pH = basic
chylomicron
lymphatic capillary
c. Fat digestion

Figure 14.12 Digestion and absorption of nutrients. **a.** The breakdown of carbohydrates, such as starch, involves amylase enzymes. **b.** Protein digestion involves the action of protease enzymes. **c.** For fat digestion, bile salts emulsify the fats so that lipase enzymes can digest the particles.

Food choices to fulfill nutrient needs constitute the *diet.* In June 2011, the U.S. Department of Agriculture (USDA) replaced its 20-year-old symbol of a healthy diet, the food pyramid, with the MyPlate symbol (Fig. 14.13). As a part of First Lady Michelle Obama's campaign against obesity, the MyPlate replaced numerical recommendations of how much of different types of foods (grains, vegetables, fruits, milk, meat, and beans) with a simpler image of a plate symbolizing the need to eat a well-balanced diet.

Obesity

The most common definition of **obesity** in use today is having a body mass index (BMI) of 30 or greater. You can determine your own BMI using Figure 14.14. In general, the taller you are, the more you can weigh without being overweight or obese. Keep in mind, however, the BMI is only a general guide. It doesn't take into account factors like fitness, bone structure, or gender. For example, a weight lifter might have a high BMI, not because of excess body fat, but because of increased bone and muscle weight.

Animation Body Mass Index

According to the Centers for Disease Control and Prevention, obesity rates in the United States have doubled in the last 20 years. As of 2010, about one-third of U.S. adults, and 17% of children and adolescents aged 2 to 19 years, were obese.

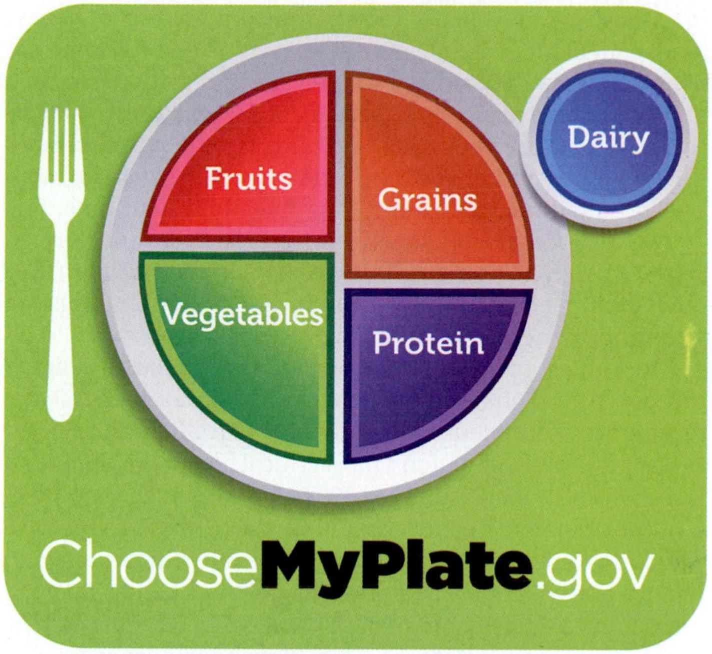
Fruits
Grains
Dairy
Vegetables
Protein
ChooseMyPlate.gov

Figure 14.13 The MyPlate icon. U.S. Department of Agriculture (USDA) released this symbol in 2011 as a guide to better health. It replaces the more complicated food pyramid models, and simply suggests eating more balanced portions. Additional information and recommendations are available at ChooseMyPlate.gov.

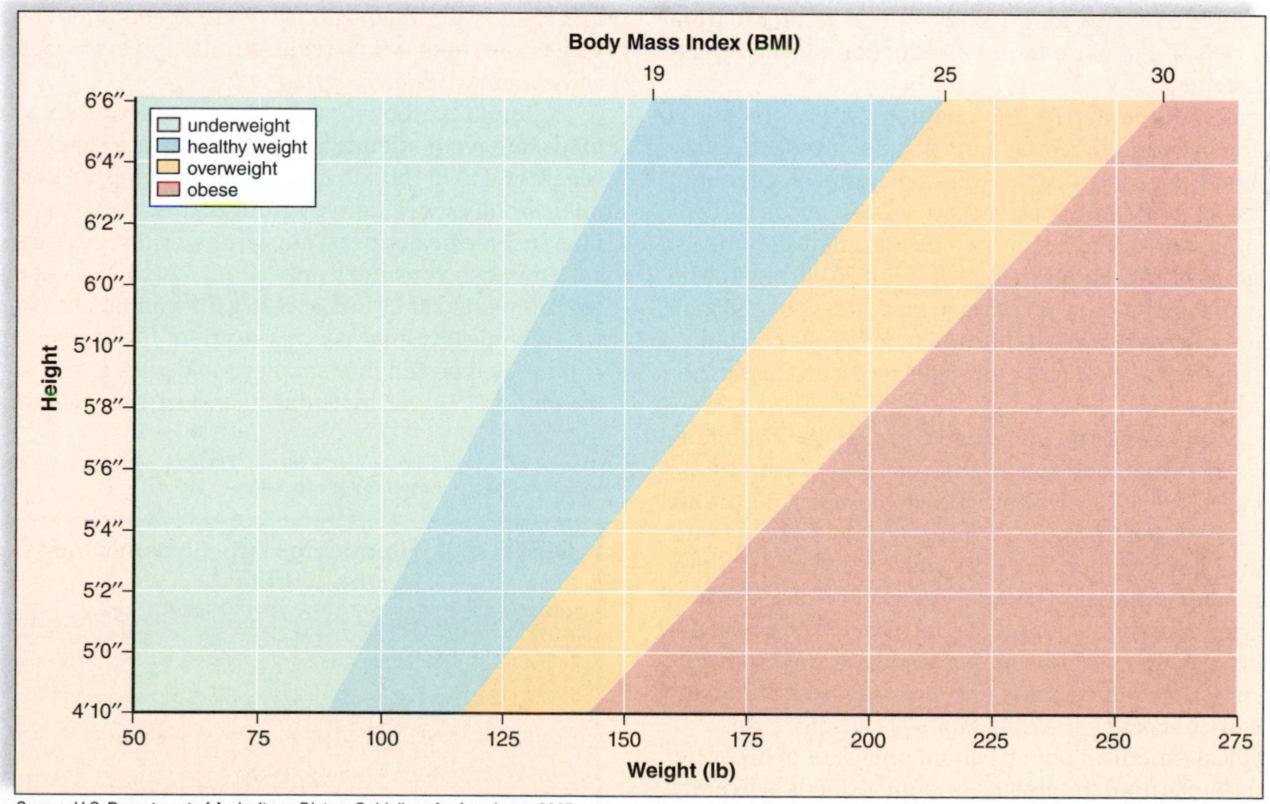
Body Mass Index (BMI)
19
25
30
underweight
healthy weight
overweight
obese
Height
6′6″
6′4″
6′2″
6′0″
5′10″
5′8″
5′6″
5′4″
5′2″
5′0″
4′10″
50
75
100
125
150
175
200
225
250
275
Weight (lb)

Source: U.S. Department of Agriculture: *Dietary Guidelines for Americans, 2005*

Figure 14.14 Body mass index (BMI). To calculate your BMI, match your weight with your height, then determine your body mass index (BMI). Healthy BMI = 18.5 to 24.9; overweight BMI = 25 to 29.9; obese BMI = 30 to 39.9; morbidly obese BMI = 40 or more.

Obesity is most likely caused by a combination of behavioral, genetic, metabolic, and social factors. Some people are obese because they suffer from an eating disorder (see next section); others may simply have a long-term pattern of overeating. People are certainly more sedentary than they used to be and they eat more processed foods, which can contain more calories than meals prepared from fresh foods at home. It is known that obese individuals have more fat cells than normal, and when they lose weight, the fat cells simply get smaller; they don't disappear. An inability to produce sufficient amounts of a satiety hormone, called leptin, might also be involved.

Regardless of the cause, obesity is associated with a higher risk of type 2 diabetes, hypertension, cardiovascular disease, respiratory dysfunction, osteoarthritis, and other conditions. Obesity is one of our nation's most critical health problems and is responsible for an estimated 300,000 deaths each year.

The treatment depends on the degree of obesity. For those who are severely obese, possible treatments include surgery to remove body fat, or even (as described in the chapter-opening case study) to bypass part of the stomach. But for most people, a lifelong commitment to a healthy diet and regular exercise is the best way to prevent obesity.

Major Classes of Nutrients

Carbohydrates

Carbohydrates are present in food in the form of complex polysaccharides like starch and fiber, and simple sugars like glucose and sucrose (table sugar). Glucose is the quickest, most readily available source of energy for the body. As you will recall from Chapter 7, cells use glucose for the production of ATP by cellular respiration.

As mentioned earlier in this chapter, excess glucose is stored by the liver and muscles in the form of glycogen. However, the human liver and muscles can store a total of only about 600 g of glucose in the form of glycogen; excess glucose is converted into fat and stored in adipose tissues. Between meals, the blood glucose level is maintained at about 50–80 mg/100 ml of blood by the breakdown of glycogen or by the conversion of fat or amino acids to glucose. While most body cells can utilize fatty acids as an energy source, brain cells require a steady supply of glucose.

If you, or someone you know, have lost weight by following a "low-carb" diet, you may think that carbohydrates are unhealthy and should always be avoided. However, according to the American Dietetic Association, some low-carbohydrate, high-fat diets have no benefits over well-balanced diets that include the same number of calories. In fact, a recent study of over 440 Canadian adults found the lowest risk of obesity in people who consumed about half of their calories from carbohydrates.[1] On the other hand, evidence suggests that many Americans are not eating the right kinds of carbohydrates.

The typical American obtains about one-sixth of his or her daily caloric intake from simple sugars found in foods like candy, cookies, and ice cream. High fructose corn syrup, the most commonly used sweetening agent, is found in soft drinks and a huge variety of other foods. These sugars immediately enter the bloodstream, as do those derived from the digestion of the starch in white bread and potatoes. These foods are said to have a high **glycemic index,** because the blood glucose response to these foods is high. When the blood glucose level rises rapidly, the pancreas produces an overload of insulin to bring the level under control. (Cells take up glucose in response to insulin.) Hunger quickly returns after eating high-glycemic index foods. A chronically high insulin level may have many harmful effects, such as increased fat deposition and a high blood fatty acid level. Over the years, the body's cells can become insulin-resistant, resulting in type 2 diabetes and other problems.

Table 14.4 gives suggestions on how to reduce your intake of high-glycemic-index carbohydrates. These recommendations are reinforced by a research study published in June 2011[2], which combined data from three separate studies of 120,877 U.S. women and men over a 20-year period. The researchers assessed the affects of various dietary and lifestyle choices on body weight. Instead of following people who were already obese and trying to lose weight, this study focused on the weight gain that often occurs gradually over decades. After adjusting for lifestyle factors like physical activity, alcohol use, and smoking, those dietary factors that had the greatest effect on body weight over time are shown in Figure 14.15. The biggest cause of weight gain was potatoes—especially potato chips and french fries. Specifically, adding just a single serving of potato chips a day was associated with a weight gain of 1.69 pounds over four years. Meats, whether processed or not, were also associated with weight gain, as were sweets and refined grains. Vegetables, nuts, whole grains, fruits, and even yogurt were all associated with weight loss.

Complex carbohydrates, like those found in whole-grain foods, are recommended because they are slowly digested to sugars and they contain fiber. **Fiber** includes various undigestible carbohydrates derived from plants. Food sources rich in fiber include beans, peas, nuts, fruits, and vegetables. The typical American consumes only about 15 g of fiber each day; the recommended daily intake is 25 g for women and 38 g for men.

Technically, fiber is not a nutrient for humans because it cannot be digested into smaller molecules and absorbed into the blood. However, insoluble fiber, such as that found in wheat

[1]Merchant, A. T., et al. 2009. Carbohydrate intake and overweight-obesity among healthy adults. *Journal of the American Dietetic Association* 109: 1165–72.

[2]Mozaffaria, D., et al. 2011. Changes in Diet and Lifestyle and Long-Term Weight Gain in Women and Men. *N Engl J Med.* 364:2392-2404.

TABLE 14.4 Reducing High-Glycemic Index Carbohydrates

To reduce dietary sugar:
1. Eat fewer sweets, such as candy, soft drinks, ice cream, and pastry.
2. Eat fresh fruits or fruits canned without heavy syrup.
3. Use less sugar—white, brown, or raw—and less honey and syrups.
4. Avoid sweetened breakfast cereals.
5. Eat less jelly, jam, and preserves.
6. Avoid potatoes and processed foods made from refined carbohydrates, such as white bread, rice, and pasta.

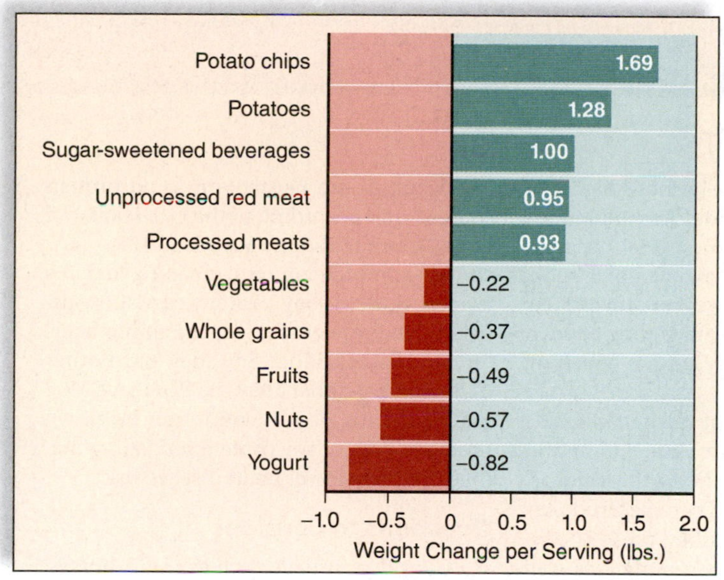

Potato chips
Potatoes
Sugar-sweetened beverages
Unprocessed red meat
Processed meats
Vegetables
Whole grains
Fruits
Nuts
Yogurt
1.69
1.28
1.00
0.95
0.93
−0.22
−0.37
−0.49
−0.57
−0.82
−1.0 −0.5 0 0.5 1.0 1.5 2.0
Weight Change per Serving (lbs.)

Figure 14.15 Dietary factors influencing body weight. Values indicate the average weight gain, or loss, over time by adding a single daily serving of each item.

bran, stimulates the movement of food through the large intestine, and may possibly guard against colon cancer by limiting the amount of time cancer-causing substances are in contact with the intestinal wall. Soluble fiber, such as that found in oat bran, combines with cholesterol in the small intestine and prevents it from being absorbed. While the benefits of fiber are clear, some evidence suggests that a diet too high in fiber can be detrimental, possibly impairing the body's ability to absorb iron, zinc, and calcium.

Proteins

Dietary proteins are digested to amino acids, which cells use to synthesize many different types of proteins. Foods rich in protein include red meat, fish, poultry, dairy products, legumes (i.e., peas and beans), nuts, and cereals. Following digestion of protein, amino acids enter the bloodstream and are transported to the tissues. Ordinarily, amino acids are not used as an energy source. Most are incorporated into structural proteins found in muscles, skin, hair, and nails. Others are used to synthesize such proteins as hemoglobin, plasma proteins, enzymes, and hormones.

Adequate protein formation requires 20 different types of amino acids. Of these, eight are required from the diet in adults (nine in children) because the body is unable to produce them. These are termed the **essential amino acids.** The body is capable of producing the other amino acids by simply transforming one type into another type.

Some protein sources, such as meat, milk, and eggs, are complete; they provide all 20 types of amino acids in the approximate amounts needed by the human body. The amino acid composition of other dietary protein sources varies considerably. For example, corn is a poor source of the amino acids lysine and tryptophan. (However, high-lysine corn can now be produced through genetic modification of the plant.) Legumes (beans and peas), seeds, nuts, and grains can be good protein sources but some contain insufficient amounts of at least one essential amino acid. Therefore, vegetarians may be advised to combine two or more incomplete types of plant products to acquire all the essential amino acids (see the Health feature, "Vegetarians: Where Do You Get Your Protein?" on page 266).

Amino acids are not stored in the body; thus a regular supply is needed. However, it does not take very much protein to meet the daily requirement. Two servings of meat a day (one serving is equal in size to a deck of cards) are usually plenty. While the body is harmed if the amount of protein in the diet is severely limited, it is also likely to be harmed by an overabundance of protein. Deamination of excess amino acids in the liver results in urea, our main nitrogenous excretion product. The water needed for excretion of urea can cause dehydration when a person is exercising and losing water by sweating. High-protein diets can also increase calcium loss in urine, which may increase the risk of kidney stones and osteoporosis.

Certain types of meat, especially red meat, are known to be high in saturated fats, while other sources, such as chicken, fish, nuts, and plant proteins, are more likely to be low in saturated fats. As discussed next, a high intake of saturated fats is associated with a greater risk of cardiovascular disease.

Lipids

Lipids include fats, oils, and cholesterol. Fats are a source of energy for the body; 1 gram (g) of fat contains nine calories, which is more than twice the amount found in carbohydrates or protein. *Saturated fats,* which are solid at room temperature, usually have an animal origin. Two well-known exceptions are palm oil and coconut oil, vegetable oils that contain mostly saturated fats. Butter and fats associated with meats (like the fat on steak and bacon) are also high in saturated fats. Saturated fats are particularly associated with cardiovascular disease.

Oils contain *unsaturated fatty acids,* which can be further characterized as monounsaturated and polyunsaturated fatty acids. Corn oil and safflower oil are high in polyunsaturated fatty acids. Polyunsaturated oils are nutritionally essential because they are the only type of fat that contains *linoleic acid* and *linolenic acid,* two fatty acids the body cannot make. Because these fatty acids must be supplied by diet, they are called **essential fatty acids.**

Olive or canola oils contain a larger percentage of monounsaturated fatty acids than other types of cooking oils. In particular, the omega-3 fatty acids are believed to be especially protective against heart disease. Some cold-water fishes, such as salmon, sardines, and trout, are rich sources of omega-3, but they contain only about half as much as flaxseed oil, the best source from plants.

Perhaps the unhealthiest type of saturated fats are the *trans-fatty acids* (often called *trans-fats*), which arise when unsaturated fatty acids are hydrogenated to produce a solid fat. Found largely in commercially produced products, trans-fats may reduce the function of the plasma membrane receptors that clear cholesterol from the bloodstream. Trans-fatty acids are commonly found in cookies and crackers, commercially

SCIENCE IN YOUR LIFE ▶ HEALTH

Vegetarians: Where Do You Get Your Protein?

There are approximately 5 million adult vegetarians in the United States. Although definitions vary, most abstain from eating red meat, poultry, and fish. About a third of vegetarians are vegans, who avoid all animal products, while less stringent vegetarians may include eggs and/or dairy products as a part of their regular diet. Although vegetarianism is an important part of certain Eastern religions, most American vegetarians cite ethical or health reasons. Ethical concerns may include the treatment of animals raised as food, and/or the effects of animal agriculture on the environment. On the health side, most nutritional research seems to support the benefits of a low-fat, well-balanced vegetarian diet in reducing the incidence of cardiovascular disease, obesity, diabetes, and certain cancers, especially colon and prostate cancer.

There are certain health risks associated with vegetarian diets. Highly restricted vegetarian diets are probably not appropriate for pregnant or lactating women, or for very young children. Vegans, especially, have to be careful to obtain enough iron, vitamin D, vitamin B_{12}, and certain fatty acids. However, most American vegetarians would likely agree that the most common question they are asked about their dietary habits is, "Where do you get your protein?" Many people believe that humans require meat in their diet in order to obtain sufficient protein. Adding to the confusion, in 1971, Frances Moore Lappé published *Diet for a Small Planet,* which advocated vegetarianism as a more ecologically sustainable diet. That book also introduced the concept of protein complementation, the idea that foods with insufficient levels of one or more essential amino acids needed to be ingested at the same time as foods higher in those amino acids. This idea has influenced American nutritionists for several decades, but in a later edition of her famous book, Lappé noted that with the exception of a few extreme diets, "if people are getting enough calories, they are virtually certain of getting enough protein."

A position paper published in 2003 by the American Dietetic Association and Dietitians of Canada echoed these ideas, stating that "plant protein can meet requirements when a variety of plant foods is consumed and energy needs are met," as well as "complementary proteins do not need to be consumed at the same meal." Because proteins in plant food such as cereals may be harder to digest than animal proteins, vegetarians may need to include a higher percentage of protein in their diet. Beans, nuts, and legumes are particularly good protein sources, and according to the U.S. Food and Drug Administration, soybeans (Fig. 14A) contain complete protein, meaning that like animal protein, soy protein contains sufficient amounts of all essential amino acids for human nutrition. Soy may have other benefits. In October 1999 the FDA gave food manufacturers permission to put labels on products high in soy protein indicating that they may help lower heart disease risk.

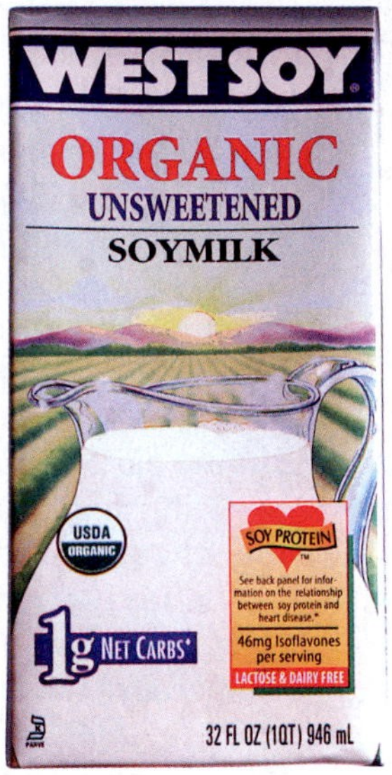
WESTSOY
ORGANIC
UNSWEETENED
SOYMILK
USDA ORGANIC
SOY PROTEIN
1g NET CARBS
46mg Isoflavones per serving
LACTOSE & DAIRY FREE
32 FL OZ (1QT) 946 mL

Figure 14A Vegetable proteins. For most adults, protein needs can be met by eating vegetables, such as soybeans, that contain high-quality proteins.

Questions to Consider

1. Consider animals such as cattle, horses, or even elephants. All are very large and powerful, while consuming a predominantly (or exclusively) vegetarian diet. In what way(s) is this a convincing argument that human vegetarians can obtain sufficient protein? Can you think of any potential flaws in this reasoning?
2. In general, proteins are not stored by the body in the way that carbohydrates or fats can be stored. Considering this idea, why do you think the idea of complementary proteins became so well established?
3. Most proteins contain varying amounts of all 20 amino acids. During the process of protein synthesis, what would specifically happen if one required amino acid was missing? What specific step in translation would not occur?

fried foods, such as french fries from some fast-food chains, and packaged snacks, such as microwave popcorn, as well as in vegetable shortening and some margarines. Any packaged goods that list "partially hydrogenated vegetable oils" or "shortening" on their label likely contain trans-fats.

As noted in Chapter 12, cardiovascular disease is often due to blockage of arteries by *plaque,* which contains saturated fats and cholesterol. Cholesterol is carried in the blood by two types of lipoproteins: low-density lipoprotein (LDL) and high-density lipoprotein (HDL). LDL is thought of as "bad" because it carries cholesterol from the liver to the cells, while HDL is thought of as "good" because it carries cholesterol to the liver, which takes it up and converts it to bile salts. Saturated fats tend to raise LDL cholesterol levels, whereas unsaturated fats lower LDL cholesterol levels. Figure 14.16 illustrates the amount of saturated fat, monounsaturated fat, polyunsaturated fat, and cholesterol found in some common cooking items.

Table 14.5 suggests ways to reduce dietary saturated fat and cholesterol. It is not a good idea to rely on commercially produced low-fat foods to reduce total fat intake. In some low-fat products, sugars have replaced the fat, meaning the calorie content can still be high. And even if the caloric content is low, consumers may believe they can eat more of a product that is low fat. Manufacturers have also developed fake fats (e.g.,

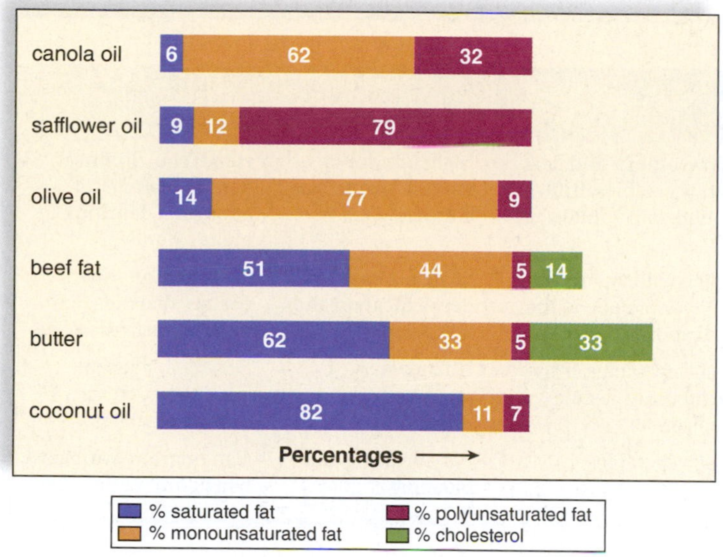
canola oil 6 62 32
safflower oil 9 12 79
olive oil 14 77 9
beef fat 51 44 5 14
butter 62 33 5 33
coconut oil 82 11 7
Percentages
% saturated fat
% monounsaturated fat
% polyunsaturated fat
% cholesterol

Figure 14.16 Lipid content of selected oils and fats. The consumption of a diet high in saturated fats and cholesterol has been associated with disorders of the cardiovascular and other systems.

TABLE 14.5 Reducing Certain Lipids

To reduce saturated fats and trans fats in the diet:
1. Choose poultry, fish, or dry beans and peas as a protein source.
2. Remove skin from poultry and trim fat from red meats before cooking.
3. Broil, boil, or bake rather than fry.
4. Limit your intake of butter, cream, trans-fats, shortenings, and tropical oils (coconut and palm oils).*
5. Use herbs and spices to season foods instead of butter or margarine.
6. Drink skim milk instead of whole milk, and use skim milk in cooking and baking.
To reduce dietary cholesterol:
1. Limit cheese, liver, and certain shellfish (shrimp and lobster). Preferably, eat white fish and poultry.
2. Substitute egg whites for egg yolks in both cooking and eating.
3. Include soluble fiber in the diet. Good sources are oat bran, oatmeal, beans, corn, and fruits, such as apples, citrus fruits, and cranberries.

*Although coconut and palm oils are from plant sources, they are mostly saturated fats.

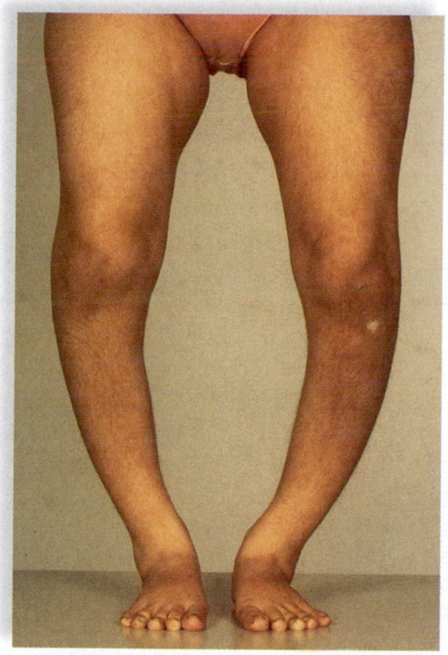

a. Rickets

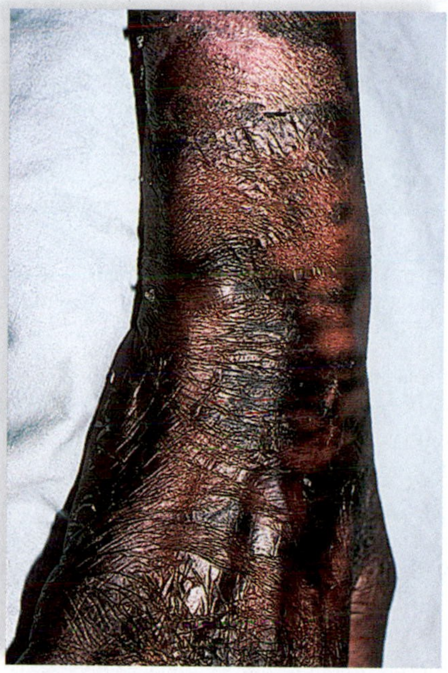

b. Pellagra

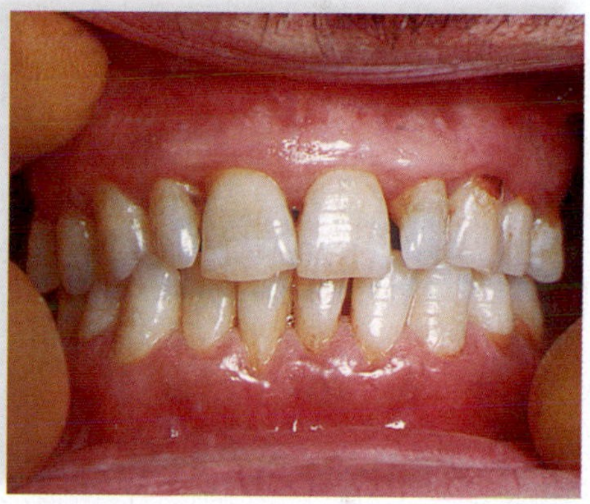

c. Scurvy

Figure 14.17 Illnesses due to vitamin deficiency. **a.** Bowing of bones (rickets) due to vitamin D deficiency. **b.** Dermatitis (pellagra) of areas exposed to light due to niacin deficiency. **c.** Bleeding of gums (scurvy) due to vitamin C deficiency.

olestra), which are not absorbed by the intestine. Nutritionists also express concern about these fake fats, however, because high levels of intake can lead to abdominal cramping, diarrhea, and reduced absorption of fat-soluble vitamins. In addition, recent research suggests that fat substitutes can interfere with the body's ability to regulate the intake of fat, which can result in overeating.

Vitamins

Vitamins are organic compounds (other than carbohydrates, fats, and proteins) that the body needs for metabolic purposes but is unable to produce. Many vitamins serve as coenzymes, which are enzyme helpers. For example, niacin is part of the coenzyme NAD, and riboflavin is part of the coenzyme FAD. Our bodies need coenzymes in only small amounts because each can be used over and over again. Not all vitamins are coenzymes; vitamin A, for example, is a precursor for the visual pigment that prevents night blindness. A variety of diseases may occur if vitamins are lacking in the diet (Fig. 14.17). Altogether, there are 13 vitamins, which are divided into those that are fat-soluble (Table 14.6) and those that are water-soluble (Table 14.7).

Animation B Vitamins

Cellular metabolism generates free radicals—unstable molecules that carry an extra electron. The most common free

TABLE 14.6 Fat-Soluble Vitamins

Vitamin	Functions	Food Sources	Conditions With: *Too Little*	Conditions With: *Too Much*
Vitamin A	Antioxidant synthesized from beta-carotene; needed for healthy eyes, skin, hair, and mucous membranes, and for proper bone growth	Deep yellow/orange and dark, leafy green vegetables, fruits, cheese, whole milk, butter, eggs	Night blindness, impaired growth of bones and teeth	Headache, dizziness, nausea, hair loss, abnormal development of fetus
Vitamin D	A group of steroids needed for development and maintenance of bones and teeth	Milk fortified with vitamin D, fish liver oil; also made in the skin when exposed to sunlight	Rickets, bone decalcification and weakening	Calcification of soft tissues, diarrhea, possible renal damage
Vitamin E	Antioxidant that prevents oxidation of vitamin A and polyunsaturated fatty acids	Leafy green vegetables, fruits, vegetable oils, nuts, whole-grain breads and cereals	Unknown	Diarrhea, nausea, headaches, fatigue, muscle weakness
Vitamin K	Needed for synthesis of substances active in clotting of blood	Leafy green vegetables, cabbage, cauliflower	Easy bruising and bleeding	Can interfere with anticoagulant medication

TABLE 14.7 Water-Soluble Vitamins

Vitamin	Functions	Food Sources	Conditions With: *Too Little*	Conditions With: *Too Much*
Vitamin C	Antioxidant; needed for forming collagen; helps maintain capillaries, bones, and teeth	Citrus fruits, leafy green vegetables, tomatoes, potatoes, cabbage	Scurvy, delayed wound healing, infections	Gout, kidney stones, diarrhea, decreased copper
Thiamine (vitamin B_1)	Part of coenzyme needed for cellular respiration; also promotes activity of the nervous system	Whole-grain cereals, dried beans and peas, sunflower seeds, nuts	Beriberi, muscular weakness, enlarged heart	Can interfere with absorption of other vitamins
Riboflavin (vitamin B_2)	Part of coenzymes, such as FAD; aids cellular respiration, including oxidation of protein and fat	Nuts, dairy products, whole-grain cereals, poultry, leafy green vegetables	Dermatitis, blurred vision, growth failure	Unknown
Niacin (nicotinic acid)	Part of coenzymes NAD and NADP; needed for cellular respiration, including oxidation of protein and fat	Peanuts, poultry, whole-grain cereals, leafy green vegetables, beans	Pellagra, diarrhea, mental disorders	High blood sugar and uric acid, vasodilation, etc.
Folacin (folic acid)	Coenzyme needed for production of hemoglobin and formation of DNA	Dark, leafy green vegetables, nuts, beans, whole-grain cereals	Megaloblastic anemia, spina bifida	May mask B_{12} deficiency
Vitamin B_6	Coenzyme needed for synthesis of hormones and hemoglobin; CNS control	Whole-grain cereals, bananas, beans, poultry, nuts, leafy green vegetables	Rarely, convulsions, vomiting, seborrhea, muscular weakness	Insomnia, neuropathy
Pantothenic acid	Part of coenzyme A needed for oxidation of carbohydrates and fats; aids in the formation of hormones and certain neurotransmitters	Nuts, beans, dark, leafy green vegetables, poultry, fruits, milk	Rarely, loss of appetite, mental depression, numbness	Unknown
Vitamin B_{12}	Complex, cobalt-containing compound; part of the coenzyme needed for synthesis of nucleic acids and myelin	Dairy products, fish, poultry, eggs, fortified cereals	Pernicious anemia	Unknown
Biotin	Coenzyme needed for metabolism of amino acids and fatty acids	Generally in foods, especially eggs	Skin rash, nausea, fatigue	Unknown

radicals in cells are superoxide (O_2^-) and hydroxide (OH^-). Because they are very reactive, free radicals combine with and damage DNA, proteins (including enzymes), or lipids, which are found in plasma membranes. Vitamins C, E, and A are believed to defend the body against free radicals, and therefore, they are termed *antioxidants.* These vitamins are especially abundant in fruits and vegetables.

The vitamin that has received the most attention recently in the popular media is vitamin D (cholecalciferol). Skin cells contain a precursor cholesterol molecule that is converted to vitamin D after UV exposure. Vitamin D leaves the skin and is modified first in the kidneys and then in the liver until finally it becomes calcitriol. Calcitriol promotes the absorption of calcium by the intestines. The lack of vitamin D leads to rickets in children (Fig. 14.17*a*). Rickets, characterized by bowing of the legs, is caused by defective mineralization of the skeleton. Since the 1930s, most milk has been fortified with vitamin D, which helps prevent the occurrence of rickets.

Minerals

In addition to vitamins, the body requires various inorganic chemical elements known as **minerals.** Minerals required as nutrients are divided into major minerals and trace minerals. The body contains more than 5 g of each major mineral and less than 5 g of each trace mineral (Fig. 14.18). The major minerals are constituents of cells and body fluids and are structural components of tissues. For example, calcium (present as Ca^{2+}) is needed for the construction of bones and teeth and for nerve conduction and muscle contraction. Phosphorus (present as PO_4^{3-}) is stored in the bones and teeth and is a part of phospholipids, ATP, and the nucleic acids. Potassium (K^+) is the major positive ion inside cells and is important in nerve conduction and muscle contraction, as is sodium (Na^+). Sodium also plays a major role in regulating the body's water balance, as does chloride (Cl^-). Magnesium (Mg^{2+}) is critical to the functioning of hundreds of enzymes.

The trace minerals are parts of larger molecules. For example, iron (Fe^{2+}) is present in hemoglobin, and iodine (I^-) is a part of thyroxine and triiodothyronine, hormones produced by the thyroid gland. Zinc (Zn^{2+}), copper (Cu^{2+}), and manganese (Mn^{2+}) are present in enzymes that catalyze a variety of reactions. Proteins called zinc-finger proteins, some of which contain zinc, bind to DNA when a particular gene is to be activated. As research continues, more and more elements are added to the list of trace minerals considered essential. During the past three decades, for example, very small amounts of selenium, molybdenum, chromium, nickel, vanadium, silicon, and even arsenic have been found to be essential to good health. Table 14.8 lists the functions of various minerals and gives their food sources and signs of deficiency and toxicity.

Occasionally, individuals do not receive enough iron, calcium, magnesium, or zinc in their diets. Adult females need more iron in the diet than males (18 mg compared to 10 mg) because they lose hemoglobin each month during menstruation. Due to its high-fiber content, a vegetarian diet may make zinc less available to the body. However, a varied and complete diet usually supplies enough of each type of mineral.

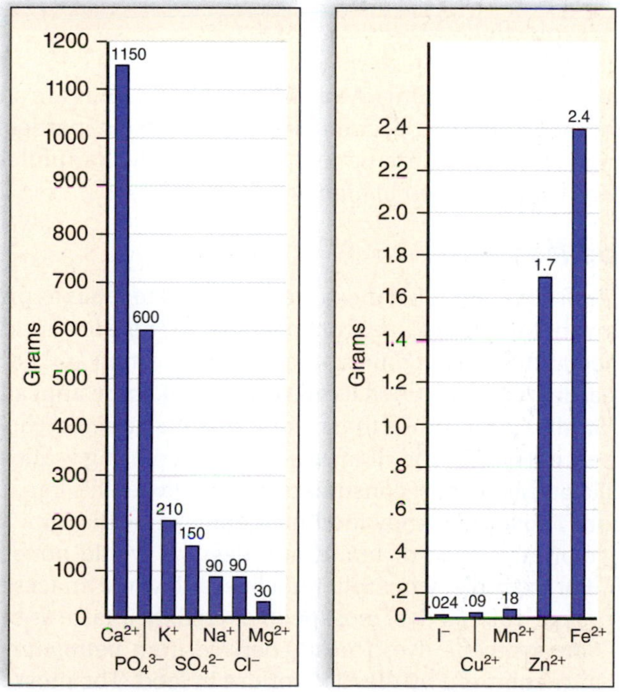

Grams
1200
1100
1000
900
800
700
600
500
400
300
200
100
0
1150
600
210
150
90 90
30
Ca2+ K+ Na+ Mg2+
PO43– SO42– Cl–
2.4
2.2
2.0
1.8
1.6
1.4
1.2
1.0
.8
.6
.4
.2
0
.024 .09 .18
1.7
2.4
I– Mn2+ Fe2+
Cu2+ Zn2+

a. Major minerals b. Trace minerals

Figure 14.18 Minerals in the body. These charts show the usual amount of certain minerals in a 60-kg (135-lb) person. The functions of minerals are given in Table 14.8. **a.** The major minerals are present in amounts larger than 5 g (about a teaspoon). **b.** Trace minerals are present in lesser amounts (notice the smaller scale).

Diet and Osteoporosis

Many people take calcium supplements to counteract **osteoporosis,** a degenerative bone disease that afflicts an estimated one-fourth of older men and one-half of older women in the United States. Osteoporosis develops because bone-eating cells called osteoclasts become more active than bone-forming cells called osteoblasts. Therefore, the bones become porous, and they break easily because they lack sufficient calcium. Due to recent studies that show consuming more calcium does slow bone loss in elderly people, the guidelines have been revised. A calcium intake of 1,000 mg a day is recommended for men and for women who are premenopausal or who use estrogen replacement therapy; 1,300–1,500 mg a day is recommended for postmenopausal women who do not use estrogen replacement therapy. To achieve this amount, supplemental calcium is most likely necessary.

Because it promotes calcium absorption, vitamin D is an essential companion to calcium in preventing osteoporosis. Other nutrients may also be helpful. For example, magnesium has been found to suppress the cycle that leads to bone loss. Along with adequate calcium and vitamin intake, regular exercise also helps to prevent osteoporosis. Factors that increase the risk of osteoporosis include smoking and a high daily caffeine intake. Medications are also available that slow bone loss while increasing skeletal mass. These are still being studied for their effectiveness and possible side effects.

Animation Osteoporosis

Sodium and Hypertension

The recommended amount of sodium intake per day is less than 2,400 mg, although the average American takes in 4,000–4,700 mg every day. High sodium intake has been linked to hypertension (high blood pressure) in some people. About one-third of the sodium we consume occurs naturally in foods; another one-third is added during commercial processing; and we add the last one-third either during home cooking or at the table in the form of table salt. Table 14.9 gives recommendations for reducing sodium intake.

Dietary Supplements

Dietary supplements are nutrients and plant products (such as herbal teas) that are used to enhance health. The U.S. government does not require dietary supplements to undergo the same safety and effectiveness testing that new prescription drugs must complete before they are approved. Therefore,

TABLE 14.8 Minerals

Mineral	Functions	Food Sources	Conditions With: *Too Little*	Conditions With: *Too Much*
Major minerals (more than 100 mg/day needed)				
Calcium (Ca^{2+})	Strong bones and teeth, nerve conduction, muscle contraction	Dairy products, leafy green vegetables	Stunted growth in children, low bone density in adults	Kidney stones; interferes with iron and zinc absorption
Phosphorus (PO_4^{3-})	Bone and soft tissue growth; part of phospholipids, ATP, and nucleic acids	Meat, dairy products, sunflower seeds, food additives	Weakness, confusion, pain in bones and joints	Low blood and bone calcium levels
Potassium (K^+)	Nerve conduction, muscle contraction	Many fruits and vegetables, bran	Paralysis, irregular heartbeat, eventual death	Vomiting, heart attack, death
Sodium (Na^+)	Nerve conduction, pH and water balance	Table salt	Lethargy, muscle cramps, loss of appetite	Edema, high blood pressure
Chloride (Cl^-)	Water balance	Table salt	Not likely	Vomiting, dehydration
Magnesium (Mg^{2+})	Part of various enzymes for nerve and muscle contraction, protein synthesis	Whole grains, leafy green vegetables	Muscle spasm, irregular heartbeat, convulsions, confusion, personality changes	Diarrhea
Trace minerals (less than 20 mg/day needed)				
Zinc (Zn^{2+})	Protein synthesis, wound healing, fetal development and growth, immune function	Meats, legumes, whole grains	Delayed wound healing, night blindness, diarrhea, mental lethargy	Anemia, diarrhea, vomiting, renal failure, abnormal cholesterol levels
Iron (Fe^{2+})	Hemoglobin synthesis	Whole grains, meats, prune juice	Anemia, physical and mental sluggishness	Iron toxicity disease, organ failure, eventual death
Copper (Cu^{2+})	Hemoglobin synthesis	Meat, nuts, legumes	Anemia, stunted growth in children	Damage to internal organs if not excreted
Iodine (I^-)	Thyroid hormone synthesis	Iodized table salt, seafood	Thyroid deficiency	Depressed thyroid function, anxiety
Selenium (SeO_4^{2-})	Part of antioxidant enzyme	Seafood, meats, eggs	Vascular collapse, possible cancer development	Hair and fingernail loss, discolored skin

TABLE 14.9 Reducing Dietary Sodium

To reduce dietary sodium:
1. Use spices instead of salt to flavor foods.
2. Add little or no salt to foods at the table, and add only small amounts of salt when you cook.
3. Eat unsalted crackers, pretzels, potato chips, nuts, and popcorn.
4. Avoid hot dogs, ham, bacon, luncheon meats, smoked salmon, sardines, and anchovies.
5. Avoid processed cheese and canned or dehydrated soups.
6. Avoid brine-soaked foods, such as pickles or olives.
7. Read labels to avoid high-salt products.

many of these products have not been tested scientifically to determine their benefits. People often think herbal products are safe because they are "natural," but many plants, such as lobelia, comfrey, and kava kava, can be poisonous.

Supplements can be useful to correct a deficiency, but most people in the United States take them to greatly increase the dietary intake of a particular nutrient. However, ingesting high levels of certain nutrients can cause harm. As mentioned earlier in this section, ingesting too much protein can be harmful, so most nutritionists do not recommend taking protein or amino acid supplements. Most fat-soluble vitamins are stored in the body and can accumulate to toxic levels, particularly vitamins A and D. Certain minerals can also be harmful, even deadly, when ingested in high doses.

Food Additives

Food additives are substances that are added to foods as preservatives, or to enhance flavor or appearance. There are hundreds of food additives, but among the most interesting and controversial are the dyes added to make food look more appealing.

Studies have shown that color is an important psychological part of tasting food. If tasteless yellow coloring is added to vanilla pudding, most consumers report it tastes like banana or lemon. Those who study food marketing know that gray pickles, colorless candy, or tan gelatin desserts would not sell as well. For example, when offered cheese-flavored snacks lacking orange-colored dye, most people rated their taste as bland.

Nine synthetic dyes (mostly derived from petroleum) are currently approved by the FDA for use in food. The most common of these is "Allura Red," also called Red No. 40. It is added to all sorts of candies, flavored drinks, frostings, and jellies. "Brilliant Blue" (Blue No. 1), is found in frozen treats, baked goods, sweets, drinks, and even soaps, mouthwash, and cosmetics.

While there is no doubt about the psychological effect of food coloring, the influence on health is less certain. In March

2011, and FDA advisory panel met to determine whether food dyes may contribute to hyperactivity in children. Many European countries have banned certain dyes, including Blue No. 1, because of this possibility. However, after reviewing the available studies, the panel concluded that the link between food color additives and hyperactivity has not been established. Other experts, however, have reached a different conclusion. For example, in 2008 the American Academy of Pediatrics suggested that a low-additive diet is a valid intervention for children with ADHD.

Check Your Progress 14.4

1. Discuss what it means to say that a particular food has a high glycemic index.
2. Predict some specific consequences of consuming a diet that is either deficient in protein, or too high in protein.
3. Describe the relationship between saturated fat intake, LDL, HDL, and heart disease. What other factors might influence your risk of having your arteries blocked by plaque?
4. Define vitamin. How do the functions of vitamins A and D differ from those of most other vitamins?

14.5 Eating Disorders

Learning Outcomes

Upon completion of this section, you should be able to

1. Describe the three main categories of eating disorders.
2. Discuss some of the psychological and social factors that can lead to eating disorders.
3. Distinguish between compulsive eating, binge eating, and pica.

People with *eating disorders* have attitudes and behaviors toward food that are outside the norm, which causes a serious disturbance to their diet. Anyone can develop an eating disorder, regardless of ethnicity, socioeconomic status, or intelligence level. Eating disorders are usually treated with a combination of medication and psychological counseling. If untreated, the most severe cases can be fatal.

According to the National Institute of Mental Health (NIMH), types of eating disorders include anorexia nervosa, bulimia nervosa, and eating disorders not otherwise specified (EDNOS). The latter category, which includes all eating disorders not meeting the criteria for anorexia nervosa or bulimia nervosa, is the most common diagnosis among people who seek treatment.

Anorexia Nervosa

Anorexia nervosa is a severe psychological disorder characterized by an irrational fear of getting fat. No matter how thin they become, people with this condition think they are overweight. This may be a form of body dysmorphic disorder, in which a person becomes preoccupied with a perceived body defect (Fig. 14.19*a*). A self-imposed starvation diet is often accompanied by purging episodes involving self-induced vomiting and laxative abuse. Anorexics may also engage in extreme physical activity to avoid weight gain. About 90% of people suffering from anorexia nervosa are young women; an estimated 1 in 200 teenage girls is affected. Athletes such as distance runners, wrestlers, and dancers are at an increased risk because they believe that being thin gives them a competitive edge.

A person with anorexia nervosa has symptoms of starvation, including low blood pressure, irregular heartbeat, constipation,

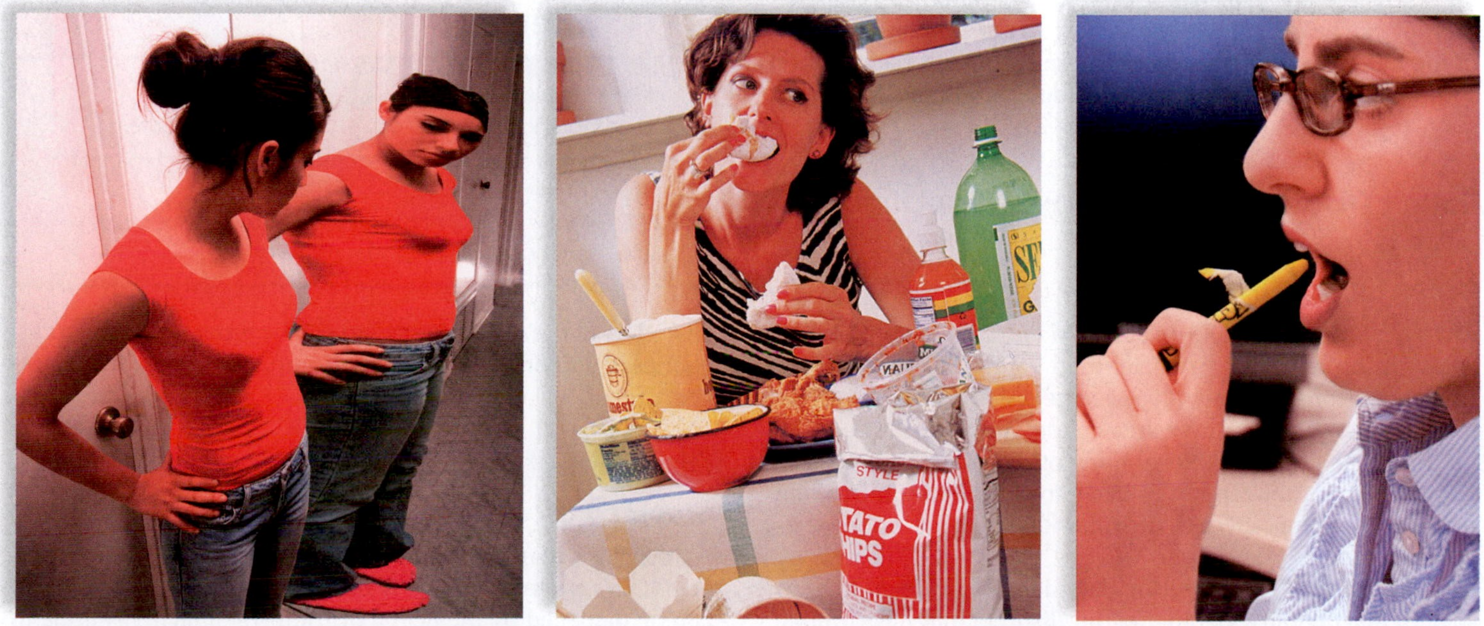

a. Anorexia nervosa b. Bulimia nervosa c. Pica

Figure 14.19 Characteristics of some eating disorders. Some common types of eating disorders. **a.** People with anorexia nervosa often think they are overweight, even if they are quite thin. **b.** Those with bulimia nervosa overeat and then purge their bodies of the food they have eaten. **c.** People suffering from pica eat unusual, sometimes bizarre nonfood items.

and constantly feeling cold. Decreased bone density, with associated stress fractures, may occur. The body begins to shut down; menstruation ceases in females; the internal organs, including the brain, don't function well; and the skin becomes dry. Impairment of the pancreas and digestive tract may interfere with proper utilization of any food that is consumed. Death may be imminent. If so, the only recourse may be hospitalization and force-feeding. Psychological counseling may be necessary to convince the person to eat properly.

Bulimia Nervosa

A person with **bulimia nervosa** binge-eats and then purges (vomits) to avoid gaining weight (Fig. 14.19*b*). Bulimia can coexist with either obesity or anorexia nervosa. According to the NIMH, about 4% of young women, and 1% of males, suffer from bulimia. The desire for a slender body in females or a manly physique in males may lead to excessive dieting, which can be a precursor to bulimia nervosa and anorexia nervosa.

As with anorexia, bulimic individuals are overly concerned about their body shape and weight, and therefore, they may be on a very restrictive diet. In bulimia, the restrictive diet brings on the desire to binge, typically on high-calorie foods like cakes, cookies, and ice cream. Then, a feeling of guilt most likely brings on the next phase, a purging of all the calories that have been taken in.

Bulimia is dangerous to a person's health. The chronic vomiting can alter blood, leading to an abnormal heart rhythm, and damage to the kidneys can even result in death. At the very least, repeated vomiting leads to inflammation of the pharynx and esophagus, and stomach acids can cause the teeth to erode. The esophagus and stomach may even rupture and tear due to strong contractions during repeated vomiting.

The most important aspect of treatment is to get the patient on a sensible and consistent diet. As with anorexia, psychotherapy can help the patient understand the emotional causes of the behavior. Medications, including antidepressants, have sometimes been effective in reducing the bulimic cycle and restoring normal appetite.

Eating Disorders Not Otherwise Specified (EDNOS)

Binge Eating Disorder

Affecting an estimated 3% of adults in the United States, binge eating disorder is the most common eating disorder. People with this disorder frequently eat large amounts of food while feeling a loss of control over their eating. However, they usually do not purge afterward by vomiting or using laxatives. Therefore, although they are not as likely to suffer from immediate medical problems, people with this disorder are often overweight or obese.

The cause of binge eating disorder is unknown. Studies suggest that people with binge eating disorder may have trouble handling some of their emotions. Many people who are binge eaters say that being angry, sad, bored, worried, or stressed can cause them to binge eat. As many as half of all people with binge eating disorder are depressed or have been depressed in the past. Whether depression causes binge eating disorder, or whether binge eating disorder causes depression, is not known. Other research suggests that genes may be involved in binge eating, since the disorder often occurs in several members of the same family. This research is still in the early stages.

Pica

Pica is defined as repeated ingestion of nonfood items, such as ash, charcoal, cotton, soap, soil, wool, or many other items (Fig. 14.19*c*). In most cases, pica affects young children, and stops before adolescence. Pica is often associated with intellectual disability, lack of proper parental care, and neglect. Children with pica are at greater risk of various poisoning (especially lead), as well as parasitic infections.

Adults can also suffer from pica, as documented on the TV show *My Strange Addiction.* On one episode called "Plastic Snacks," an 18-year-old female describes eating plastic forks, cocktail swords, water bottles, CD cases, and other items. She has consumed over 150 pounds of plastic, and prefers it over food. Other episodes have described people who eat chalk, rocks, dirt, their own hair, or even glass. In adults, pica may be a result of nutritional deficiencies, or a behavioral response to stress.

Check Your Progress 14.5

1. Compare and contrast anorexia nervosa, bulimia nervosa, binge eating disorder, and pica.
2. Predict some physiological factors that could contribute to an eating disorder.

14.6 Disorders of the Digestive System

Learning Outcomes

Upon completion of this section, you should be able to

1. Describe a common disorder that affects each part of the digestive system.
2. Classify the digestive disorders by the type of cause (i.e., infectious, cancer, inflammatory, etc.).

Disorders of the digestive system can be grouped into two categories: disorders of the tract itself, and disorders of the accessory organs.

Disorders of the Digestive Tract

Stomach Ulcers

A thick layer of mucus normally protects the wall of the stomach. If this protective layer is broken down, the stomach wall can be damaged by the acidic pH of the stomach, resulting in a **stomach ulcer,** or open sore in the wall caused by a gradual disintegration of the tissue. It now appears that most stomach ulcers are initiated by infection of the stomach by a bacterium called *Helicobacter pylori* (see the Scientific Inquiry feature, "A Bacterial Culprit for Stomach Ulcers"), which can impair the ability of the mucous cells to produce protective mucus. Thus, treatment of stomach ulcers now typically includes antibiotics to kill these bacteria, along with medication to reduce the amount of acid in the stomach. Other potential causes of stomach ulcers include certain viral infections, as well as the overuse of anti-inflammatory medications, which can have the side effect of damaging the stomach lining. Duodenal ulcers are also common because the duodenum receives acidic chyme from the stomach.

Intestinal Disorders

The most common problems associated with the small and large intestines are diarrhea and constipation. **Diarrhea,** or loose, watery feces, can be acute or chronic. Most cases of acute (sudden-onset) diarrhea are due to infections of the small or large intestines with any of several bacterial, viral, or protozoal agents. Collectively known as food poisoning, symptoms of these infections can range from a few hours of stomach and bowel discomfort, to life-threatening infections such as cholera, salmonellosis, and the more toxic forms of *E. coli.* In all cases, the intestinal wall becomes irritated by the infection, or by toxins produced by the organism, and peristalsis increases. Less water is absorbed, and the resulting diarrhea helps rid the body of the infectious organisms. Severe diarrhea can lead to dehydration because of water loss and to disturbances in the heart's contraction due to an imbalance of salts in the blood.

As opposed to intestinal infections that come and go, some people are afflicted with chronic diarrhea. About 7 in 100,000 people in the United States suffer from *Crohn's disease,* which is a persistent inflammation of the intestine that results in recurrent bouts of abdominal cramping (which may be severe) and bloody diarrhea. Usually diagnosed between ages 15 and 30, it seems to be caused by a misdirected immune response against one's own intestinal tissues, or perhaps against some of the normal bacteria that live in the gut. The incidence of Crohn's disease is higher among Caucasians compared to other races, and among wealthier, urban populations compared to poorer,

SCIENCE IN YOUR LIFE ▶ **SCIENTIFIC INQUIRY**

A Bacterial Culprit for Stomach Ulcers

A stomach (gastric) ulcer is a gaping, sometimes bloody, indentation in the stomach lining (Fig. 14B). Australian physician Barry Marshall didn't believe the accepted explanation that the major causes of stomach ulcers were stress or spicy foods. Instead, he and his colleague, Dr. Robin Warren, thought most stomach ulcers were due to infection with a bacterial species called *Helicobacter pylori,* which they had discovered in 1982. At that time, most physicians disagreed with their hypothesis, because they didn't believe any bacteria could survive in the very acidic conditions of the stomach. However, Marshall and Warren, a pathologist, had observed that nearly every biopsy of stomach ulcer tissue contained the corkscrew-shaped *H. pylori* bacteria.

To get the data they needed, Dr. Marshall decided to use himself as a guinea pig. One day in July 1984, he drank a culture containing *H. pylori.* About a week later, he started vomiting and suffering the painful symptoms of an ulcer. Medical tests confirmed that his stomach lining was teeming with bacteria. In a few weeks, his ulcer had disappeared, but the experiment had served its purpose. In recognition of his efforts, in 2005, Marshall and Warren were awarded the Nobel Prize in Physiology or Medicine for their discovery of *H. pylori* and its role in ulcers and other stomach ailments.

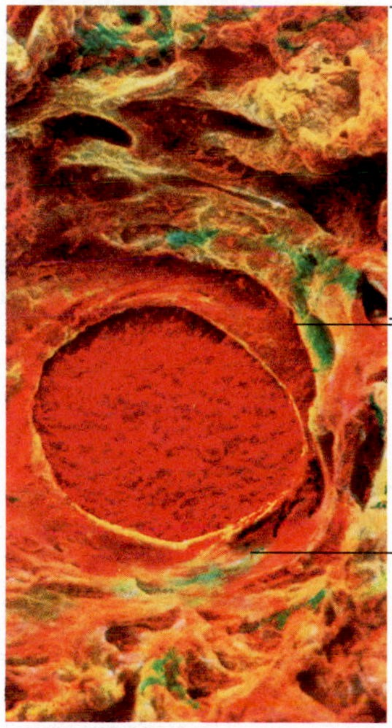
healthy gastric mucosa
gastric ulcer

Figure 14B Stomach ulcer. Drs. Barry Marshall and Robin Warren won a Nobel Prize in 2005 for discovering that most stomach ulcers are caused by a bacterial infection.

Prior to Dr. Marshall's bold experiment, ulcers were usually treated mainly with drugs to reduce the acidity of the stomach. Patients got temporary relief, but the ulcers tended to return, and serious complications could arise. Some patients had bleeding ulcers and underwent an operation to remove a portion of the stomach. Thanks to the work of Marshall and Warren, in 1994 the National Institutes of Health endorsed antibiotics as a standard treatment for stomach ulcers. An international effort is under way among scientists to develop a vaccine.

Questions to Consider

1. What kinds of adaptations would *H. pylori* need in order to survive in the acid pH of the stomach?
2. Why do you suppose that even after Dr. Marshall's experiment in 1984, it took ten years for his findings to be widely accepted by the medical community?
3. Do you think Dr. Marshall's experiment was ethical? If so, would it always be okay for scientists to experiment on themselves? If not, what would be a more ethical way to prove that *H. pylori* could cause ulcers?

rural ones. Genetic predisposition is certainly a factor, as are several environmental triggers, such as smoking and diet.

The opposite of diarrhea is **constipation.** When a person is constipated, the feces are dry and hard. One reason for this condition is that some people have learned to inhibit the defecation reflex to the point that the urge to defecate is ignored. Chronic constipation can lead to the development of *hemorrhoids,* which are enlarged and inflamed blood vessels of the anus. Increasing the amount of water and fiber in the diet can help prevent constipation. Laxatives and enemas can also be used, but frequent use is discouraged because they can be irritating to the colon.

Polyps and Colon Cancer

The colon is subject to the development of **polyps,** which are small growths arising from the epithelial lining. They may be benign or cancerous. Polyps are usually detected using a procedure known as colonoscopy, during which a long, flexible endoscope is inserted into the colon to visualize the wall of the large intestine. If colon cancer is detected and surgically removed while it is still confined to a polyp, a complete cure is expected.

Some investigators believe that dietary fat increases the likelihood of colon cancer because dietary fat causes an increase in bile secretion. Certain intestinal bacteria may convert bile salts to substances that promote the development of cancer. On the other hand, fiber in the diet seems to inhibit the development of colon cancer, possibly by diluting the concentration of bile salts and facilitating the movement of cancer-inducing substances through the intestine.

Disorders of the Accessory Organs

Disorders of the Pancreas

Pancreatitis is an inflammation of the pancreas. It can be caused by drinking too much alcohol, by gallstones that block the pancreatic duct, or by other unknown factors. In chronic pancreatitis, the digestive enzymes normally secreted by the pancreas can damage the pancreas and surrounding tissues. Eventually, chronic pancreatitis can decrease the ability of the pancreas to secrete insulin, potentially leading to diabetes.

Unfortunately, **pancreatic cancer** is almost always fatal. Only about 20% of patients are alive one year after this diagnosis. Reasons for this include the essential functions of the pancreas, the resistance of many forms of the disease to treatment, and the tendency of the cancer cells to spread to other organs before any symptoms appear. Steve Jobs, Apple CEO, died in 2011 at age 56 after a seven-year battle with pancreatic cancer.

Disorders of the Liver and Gallbladder

The liver performs many functions that are essential to life; therefore, diseases that affect the liver can be life threatening. A person who has a liver ailment may develop *jaundice,* which is a yellowish coloring in the whites of the eyes, as well as in the skin of light-pigmented persons. Jaundice can be caused by an abnormally large amount of bilirubin in the blood. **Hepatitis,** or inflammation of the liver, is most commonly caused by one of several viruses. Hepatitis A virus is usually acquired by consuming water or food that has been contaminated by sewage. Hepatitis B virus is usually spread by sexual contact, but can also be acquired through blood transfusions or contaminated needles. The hepatitis B virus is more contagious than the AIDS virus, which is spread in similar ways. Thankfully, however, a vaccine is available to prevent hepatitis B. Hepatitis C virus is spread by the same routes as hepatitis B, and it can lead to chronic hepatitis, liver cancer, and death. Unfortunately, no vaccine is available for hepatitis C, although antiviral drugs can be effective.

Video Halting Hepatitis

Cirrhosis is another chronic disease of the liver. It is often seen in alcoholics due to malnutrition and to the toxic effects of consuming excessive amounts of alcohol, which acts as a toxin on the liver. In cirrhosis, the liver first becomes infiltrated with fat, and then the fatty liver tissue is replaced by fibrous scar tissue. Although the liver has amazing regenerative powers, if the rate of damage exceeds the rate of regeneration, there may not be enough time for the liver to repair itself. The preferred treatment in most cases of liver failure is liver transplantation, but the supply of livers currently cannot meet the demand. Artificial livers have been developed, but with very limited success so far.

In some individuals, the cholesterol present in bile can come out of solution and form crystals. If the crystals grow in size, they form **gallstones.** The passage of the stones from the gallbladder may block the common bile duct and cause obstructive jaundice. If this happens, the obstruction can sometimes be relieved; otherwise, the gallbladder must be removed. Interestingly, gallstone formation is particularly common in individuals who have lost a significant amount of weight in a short time, such as those who have undergone gastric bypass procedures.

Check Your Progress 14.6

1. Identify the specific cause of most stomach ulcers.
2. Explain why the pancreas and liver are such important organs to the health of an individual.

Case Study Conclusion

The story that opened this chapter introduced Dennis, who struggled with obesity for most of his life until he had gastric bypass surgery. Considering the history of obesity in his family, it is likely that Dennis had a genetic predisposition to become overweight, but it is likely that this tendency was exacerbated by unhealthy lifestyle and dietary choices. Insufficient exercise, along with consistently consuming more calories than are needed to support one's daily activities, adds up to obesity for a significant percentage of people. Obesity, in turn, leads to a higher risk of serious diseases, including type 2 diabetes and cardiovascular disorders. Dennis's pattern of overeating large amounts of food at night indicated that he might have been suffering from an eating disorder, as well. The gastric bypass surgery, along with counseling to resolve underlying psychological issues, helped Dennis to maintain a healthy weight.

MEDIA STUDY TOOLS

www.mhhe.com/maderinquiry14

Enhance your study of this chapter with study tools and practice tests. Also ask your instructor about the resources available through ConnectPlus, including LearnSmart, the media-rich eBook, interactive learning tools, and animations.

SUMMARIZE

14.1 The Digestive Tract

The **digestive system** is involved in the ingestion and digestion of food and elimination of indigestible material. It includes the following parts:

- The **mouth** takes food into the body. **Salivary glands** send saliva into the mouth, where the teeth chew the food and the tongue forms a bolus for swallowing.
- The **pharynx** is where the air passage and food passage cross. When a person swallows, the air passage is usually blocked off by the epiglottis.
- The **esophagus** is a muscular tube that conducts food from the pharynx to the stomach.
- Beginning with the esophagus, each component of the digestive tract surrounding the central cavity, or **lumen,** has four layers: an inner **mucosa,** a **submucosa,** a **muscularis,** and a **serosa.** Muscular contractions called **peristalsis** move food through the tract.
- The **stomach** expands and stores food. When food enters the stomach, the stomach churns, mixing food with the acidic gastric juice. At this point, the food is known as **chyme.**
- The **small intestine** is made up of the **duodenum,** the **jejunum,** and the **ileum.** The duodenum receives bile from the liver and pancreatic juice from the pancreas. The walls of the small intestine have fingerlike projections called **villi,** where small nutrient molecules are absorbed. Amino acids and glucose enter the blood vessels of a villus. Glycerol and fatty acids are joined and packaged as lipoproteins before entering lymphatic vessels called **lacteals** in the villi.
- The **large intestine** begins at the **cecum,** which has a pouch called the **appendix** (**vermiform appendix**) attached. The large intestine continues as the colon and rectum, and ends at the **anus.** The large intestine absorbs water, salts, and some vitamins.

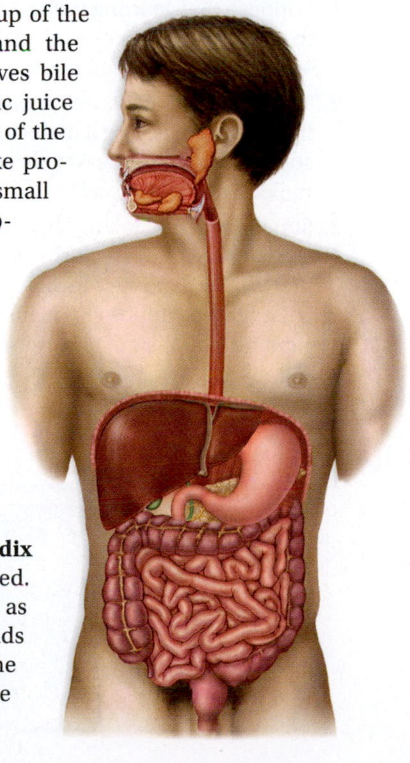

14.2 Accessory Organs of Digestion

Three accessory organs of digestion send secretions to the duodenum via ducts:

- The **pancreas** produces pancreatic juice, which contains sodium bicarbonate and enzymes that chemically digest starch, protein, and fat.
- The **liver** has many important functions. It receives blood from the small intestine by way of the hepatic portal vein and helps to detoxify any harmful substances present. It monitors blood glucose concentrations and helps eliminate excess nitrogen from the body. It also produces many plasma proteins, as well as **bile,** which helps to emulsify (break apart) fats in the intestine.
- The **gallbladder** stores bile produced by the liver, until it is secreted into the duodenum through the common bile duct. Bile emulsifies fat and readies it for digestion by lipase.

14.3 Digestive Enzymes

Like all enzymes, digestive enzymes are specific to their substrate and speed up specific reactions at optimum body temperature and pH. They break down food into smaller molecules like glucose, amino acids, fatty acids, and glycerol, which can be absorbed. As food passes through the digestive tract:

- In the mouth, **salivary amylase** begins digestion of starch.
- In the stomach, **pepsin** begins digesting protein to peptides.
- In the small intestine, **pancreatic amylase, trypsin,** and **lipase** digest starch, protein, and fat. **Peptidases** and **maltase** finish the digestion of protein and starch, respectively.

14.4 Human Nutrition

The nutrients released by the digestive process should provide us with an adequate amount of energy, essential amino acids and fatty acids, and all necessary vitamins and minerals. However, some recommendations can be given, based on scientific studies:

- Minimize your intake of carbohydrates with a high **glycemic index** (simple sugars and refined starches) because they cause a rapid release of insulin that can lead to health problems. An adequate intake of undigestible carbohydrates in the form of **fiber** also promotes health.
- Proteins supply us with **essential amino acids,** but it is wise to avoid meats that are fatty because fats from animal sources are saturated. While unsaturated fatty acids, particularly the omega-3 fatty acids, are protective against cardiovascular disease, the saturated fatty acids lead to plaque, which occludes blood vessels. The body cannot synthesize **essential fatty acids,** which must be supplied by the diet.
- Aside from carbohydrates, proteins, and fats, the body requires vitamins and minerals. **Vitamins** A, E, and C are antioxidants that protect cell contents from damage due to free radicals. The **mineral** calcium is needed to prevent the bone disease **osteoporosis.**

14.5 Eating Disorders

- An eating disorder is an abnormal behavior that causes significant changes in the diet. These changes often have severe health consequences for the individual. **Anorexia nervosa** and **bulimia nervosa** are examples of common eating disorders.

14.6 Disorders of the Digestive System

Various conditions can affect the function of the digestive tract, as well as the accessory organs.

- Most **stomach ulcers** are now thought to be caused by a bacterial infection and thus can be cured with antibiotics.

- **Diarrhea** is an increased frequency and looseness of the feces. Acute diarrhea is most frequently caused by intestinal infections. Chronic (longer-lasting) diarrhea may be due to an immune-mediated condition such as Crohn's disease. Conversely, with **constipation,** the feces become dry and hard.
- Disorders that affect the accessory digestive organs include **pancreatitis, pancreatic cancer, hepatitis,** and **cirrhosis.**

ASSESS

Testing Yourself

Choose the best answer for each question.

1. The digestive system
   a. breaks down food into usable nutrients.
   b. absorbs nutrients.
   c. eliminates waste.
   d. All of these are correct.
2. Tooth decay is caused by bacteria metabolizing _____ and giving off _____.
   a. sugar, protein
   b. protein, sugar
   c. acids, sugar
   d. sugar, acids
3. Chemical digestion begins in the
   a. esophagus.
   b. stomach.
   c. pharynx.
   d. None of these are correct.
4. Which organ has both an exocrine and endocrine function?
   a. liver
   b. esophagus
   c. pancreas
   d. cecum
5. Which of these is not a function of the liver in adults?
   a. produces bile
   b. detoxifies alcohol
   c. stores glucose
   d. produces urea
   e. makes red blood cells
6. Bile
   a. is an important enzyme for the digestion of fats.
   b. cannot be stored.
   c. is made by the gallbladder.
   d. emulsifies fat.
   e. All of these are correct.
7. Label the following diagram of the large and small intestine junction.

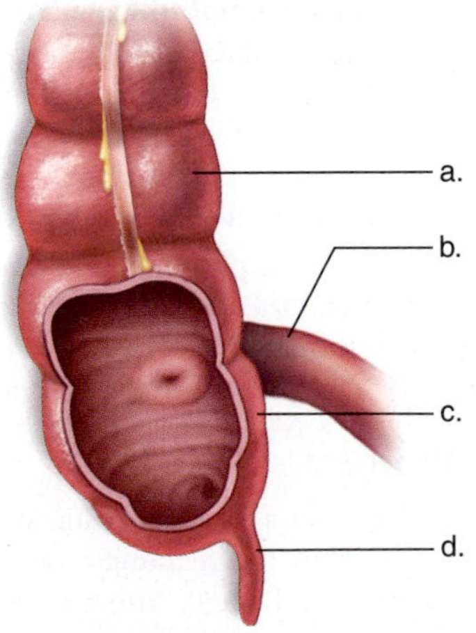
a.
b.
c.
d.

8. Most of the products of digestion are absorbed across the
   a. squamous epithelium of the esophagus.
   b. striated walls of the trachea.
   c. convoluted walls of the stomach.
   d. fingerlike villi of the small intestine.
   e. smooth wall of the large intestine.
9. The amino acids that must be consumed in the diet are called essential. Nonessential amino acids
   a. can be produced by the body.
   b. are needed only occasionally.
   c. are not needed for protein synthesis.
   d. are required in smaller amounts.
10. Which of the following are often organic portions of important coenzymes?
    a. minerals
    b. vitamins
    c. proteins
    d. carbohydrates
11. A vitamin A deficiency would most likely result in
    a. anemia.
    b. kidney stones.
    c. night blindness.
    d. osteoporosis.
12. Compared to anorexia nervosa, bulimia nervosa is *less* likely to be characterized by
    a. a restrictive diet often with excessive exercise.
    b. binge eating followed by purging.
    c. an obsession with body shape and weight.
    d. a distorted body image, so that the person feels fat even when emaciated.
    e. a health risk due to this complex.
13. Pancreatic cancer is one of the most deadly forms of cancer because
    a. the pancreas is essential for life.
    b. the cancer usually spreads before it is detected.
    c. it is often resistant to treatment.
    d. All of these are correct.

ENGAGE

Virtual Labs
Virtual Frog Dissection
Nutrition

The virtual lab "Virtual Frog Dissection" provides you with a more detailed look at a vertebrate digestive system. The virtual lab "Nutrition" allows you to input your personal information to assess the health of your diet.

Thinking Critically

1. Suppose that, starting today, your body would lose the ability to perform either mechanical digestion or chemical digestion. Which type would you choose to keep and why?
2. Trace *all* of the types and locations of enzymes that might be involved in digesting a molecule of a complex carbohydrate such as starch, all the way through to the utilization of a glucose molecule by a body cell.
3. Suppose you are taking large doses of creatine, an amino acid supplement that is advertised for its ability to enhance muscle growth. Because your muscles can grow only at a limited rate, what do you suppose happens to the excess creatine that is not used for synthesis of new muscle?
4. Predict and explain the expected digestive results per test tube for this experiment.

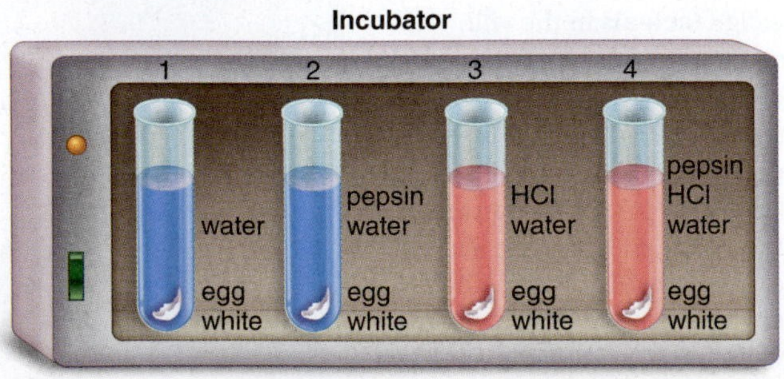
Incubator
1
2
3
4
water
egg white
pepsin
water
egg white
HCl
water
egg white
pepsin
HCl
water
egg white

Respiratory System 15

CASE STUDY Every Breath You Take

Growing up living in a neighborhood next to a freeway in downtown Los Angeles, 13-year-old Juan had plenty of challenges to deal with. His parents wanted him to be the first in his family to attend college. That meant avoiding local gangs that he often heard about. Fortunately, Juan loved playing sports, and he knew if he stayed focused on playing for school teams, it was easier to stay out of trouble. Basketball was his favorite—but unfortunately, over the past few weeks, whenever he ran up and down the court, he would begin coughing and wheezing. Sometimes he couldn't catch his breath for several minutes. The school nurse diagnosed Juan with asthma.

Asthma is a disease in which the airways become constricted (narrowed) and inflamed (swollen), both of which can result in difficulty breathing. Over 20 million children and adults in the United States have asthma, and the incidence seems to be increasing. Experts offer various explanations for this. One hypothesis is that we may be "too clean," in the sense that we are not exposed to enough common bacteria, viruses, and even worm parasites as children. As a result, our immune system may react to harmless material we inhale. In Juan's case, living next to a freeway may be another significant factor. There are many harmful forms of air pollution, but scientists are becoming increasingly concerned about tiny "ultrafine" particles, which are produced at high levels by diesel engines. At no more than 0.1 micrometers (μm) across, these particles can bypass the normal defenses of the upper respiratory tract and end up lodging deep in the lungs, with damaging effects. Though incurable, asthma usually can be treated. By using an asthma inhaler and other medications, Juan was able to play the sports he enjoyed.

As you read through the chapter, think about the following questions:

1. How would a narrowing and swelling of the airways affect the respiratory volumes?
2. How do the typical treatments for asthma work to reduce the symptoms?
3. Why is it so difficult to develop a cure for asthma?

CHAPTER OUTLINE

BEFORE YOU BEGIN

Before beginning this chapter, take a few moments to review the following discussions:

Section 7.1 During which stage(s) of cellular respiration are oxygen used and carbon dioxide produced?

Section 11.5 How does the respiratory system contribute to homeostasis in the body?

Section 12.4 What is the path of blood through the pulmonary and systemic circuits?

15.1 The Respiratory System

Learning Outcomes

Upon completion of this section, you should be able to

1. Summarize the major functions of the respiratory system.
2. Trace the path of air from the human nose to the lungs.
3. Describe the layers of cells a molecule of oxygen must pass through when diffusing from a lung alveolus into a red blood cell, and from the red blood cell into the tissues.

The primary function of the **respiratory system** is to allow oxygen from the air to enter the blood and carbon dioxide from the blood to exit into the air. During inspiration, or inhalation (breathing in), and expiration, or exhalation (breathing out), air is conducted toward or away from the lungs by a series of cavities, tubes, and openings (Fig. 15.1). **Ventilation,** another term for breathing, encompasses both inspiration and expiration.

The respiratory system works with the cardiovascular system to accomplish these homeostatic functions:

1. External respiration, the exchange of gases (oxygen and carbon dioxide) between air and the blood
2. Transport of gases to and from the lungs and the tissues
3. Internal respiration, the exchange of gases between the blood and tissue fluid

As was described in Chapter 7, cellular respiration, which produces ATP, uses the oxygen and gives off the carbon dioxide that makes gas exchange with the environment necessary.

The Respiratory Tract

Table 15.1 traces the path of air from the nose to the lungs. As air moves in along the airways, it is cleansed, warmed, and moistened. Cleansing is accomplished by coarse hairs just inside the nostrils and by cilia and mucus in the nasal cavities and the other airways of the respiratory tract. In the nose, the hairs and the cilia act as screening devices. In the trachea

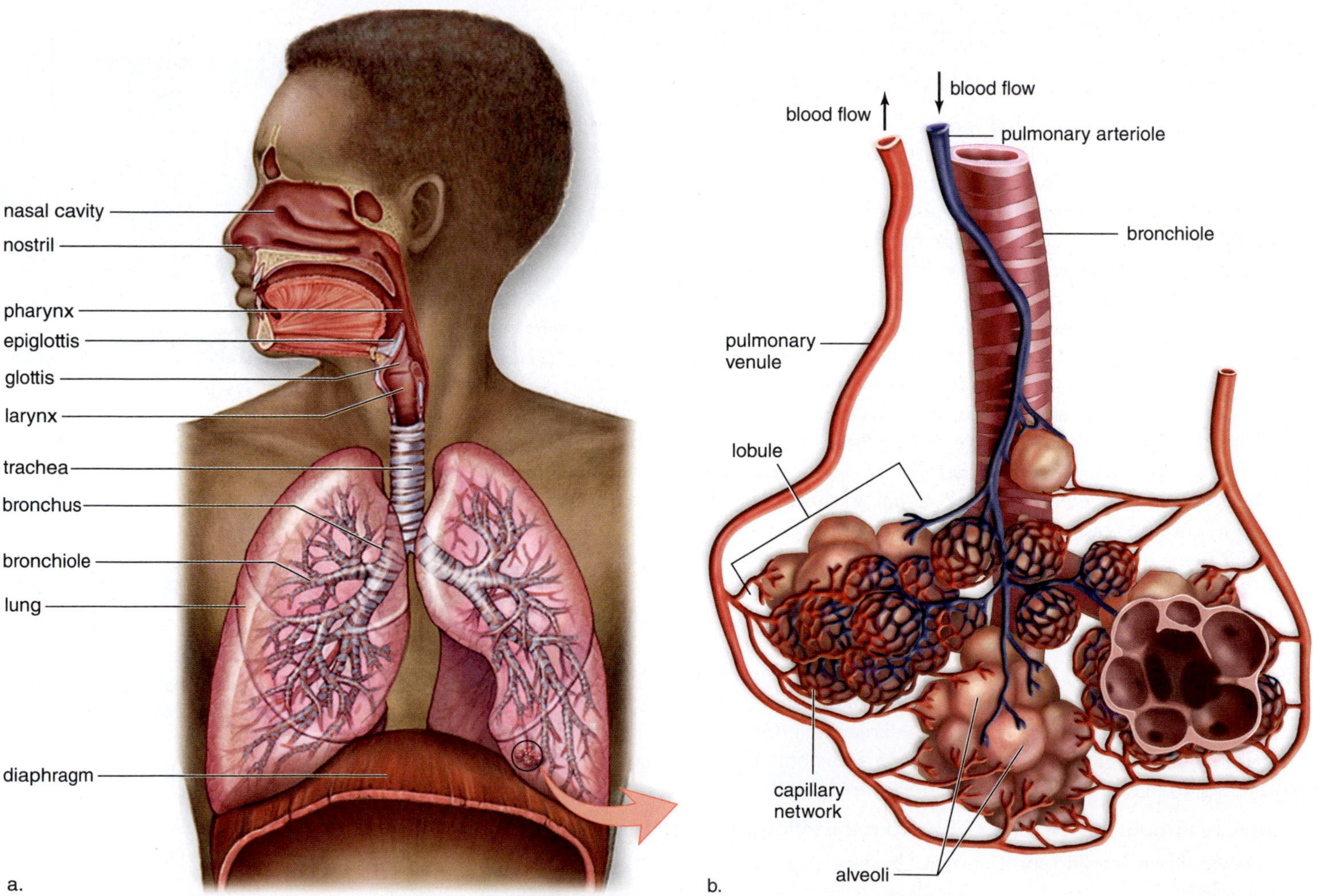
nasal cavity
nostril
pharynx
epiglottis
glottis
larynx
trachea
bronchus
bronchiole
lung
diaphragm
a.
blood flow
blood flow
pulmonary arteriole
bronchiole
pulmonary venule
lobule
capillary network
alveoli
b.

Figure 15.1 The respiratory tract. **a.** The respiratory tract extends from the nose to the lungs, which are composed of air sacs called alveoli (arrow). **b.** Gas exchange occurs between air in the alveoli and blood within a capillary network that surrounds the alveoli. Notice that the pulmonary arteriole is colored blue—it carries O_2-poor blood away from the heart to the alveoli. Then carbon dioxide leaves the blood, and oxygen enters the blood. The pulmonary venule is colored red—it carries O_2-rich blood from the alveoli toward the heart.

TABLE 15.1 Path of Air

Structure	Description	Function
Upper Respiratory Tract		
Nasal cavities	Hollow spaces in nose	Filter, warm, and moisten air
Pharynx	Chamber posterior to oral cavity; lies between nasal cavity and larynx	Connection to surrounding regions
Glottis	Opening into larynx	Passage of air into larynx
Larynx	Cartilaginous organ that houses the vocal cords; voice box	Sound production
Lower Respiratory Tract		
Trachea	Flexible tube that connects larynx with bronchi	Passage of air to bronchi
Bronchi	Paired tubes inferior to the trachea that enter the lungs	Passage of air to lungs
Bronchioles	Branched tubes that lead from bronchi to alveoli	Passage of air to each alveolus
Lungs	Soft, cone-shaped organs that occupy lateral portions of thoracic cavity	Contain alveoli and blood vessels
Alveoli	Thin-walled microscopic air sacs in lungs	Gas exchange between air and blood

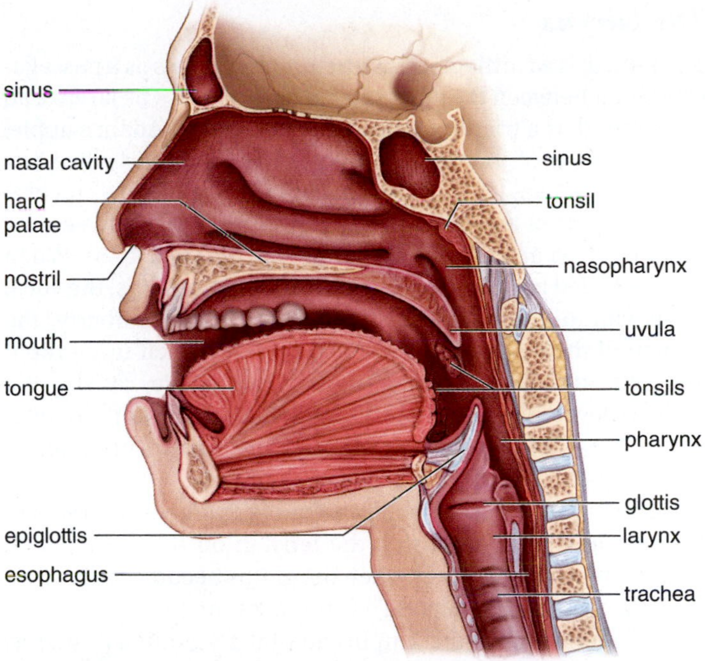
sinus
nasal cavity
hard palate
nostril
mouth
tongue
epiglottis
esophagus
sinus
tonsil
nasopharynx
uvula
tonsils
pharynx
glottis
larynx
trachea

Figure 15.2 The path of air. Air passes through the nasal cavities and mouth to and from the upper and lower respiratory tracts. The trachea is part of the lower respiratory tract; the other organs are in the upper respiratory tract.

and other airways, the cilia beat upward, carrying mucus, dust, and occasional bits of food that "went down the wrong way" into the pharynx, where the accumulation can be swallowed or expelled. The air is warmed by heat given off by the blood vessels lying close to the surface of the lining of the airways, and it is moistened by the wet surface of these passages.

Conversely, as air moves out during expiration, it cools and loses its moisture. As the air cools, it deposits its moisture on the lining of the trachea and the nose, and the nose may even drip as a result of this condensation. The air still retains so much moisture, however, that upon expiration on a cold day, it condenses and can be seen as a small cloud.

The Nose

The nose, a prominent feature of the face, is the only external portion of the respiratory system. It is a part of the upper respiratory tract, which contains the nasal cavities, the pharynx, and the larynx (Fig. 15.2). Air enters the nose through external openings called nostrils. The nose contains two *nasal cavities,* which are narrow canals separated from one another by a septum composed of bone and cartilage. Mucous membranes line the nasal cavities. Bony ridges that project laterally into the nasal cavity increase the surface area for moistening and warming air during inhalation and for trapping water droplets during exhalation. Odor receptors are on the cilia of cells located high in the recesses of the nasal cavities (see Fig. 18.5).

The tear (lacrimal) glands drain into the nasal cavities by way of tear ducts. For this reason, crying often produces a runny nose. The nasal cavities also communicate with *sinuses,* air-filled spaces that reduce the weight of the skull and act as resonating chambers for the voice. The nasal cavities are separated from the mouth by a partition called the *palate,* which has two portions. Anteriorly, the *hard palate* is supported by bone, while posteriorly the *soft palate* is made of muscle tissue surrounded by a mucous membrane.

The Pharynx

The **pharynx** is a funnel-shaped passageway that connects the nasal and oral cavities to the larynx. Consequently, the pharynx, commonly referred to as the "throat," has three parts: the *nasopharynx,* where the nasal cavities open posterior to the soft palate; the *oropharynx,* where the mouth opens; and the *laryngopharynx,* which opens into the larynx. The soft palate has a soft extension called the *uvula* projecting into the oropharynx, which you can see by looking into your throat using a mirror.

The *tonsils* form a protective ring at the junction of the mouth and the pharynx. The tonsils are lymphatic tissue containing lymphocytes that protect against invasion by inhaled bacteria and viruses.

In the pharynx, the air passage and the food passage cross because the larynx, which receives air, lies above and in front of the esophagus, which receives food. The larynx leads to the trachea. Both the larynx and the trachea are normally open, allowing air to pass, but the esophagus is normally closed and opens only when a person swallows.

The Larynx

The **larynx** is a cartilaginous structure that serves as a passageway for air between the pharynx and the trachea. The larynx can be pictured as a triangular box whose apex, the Adam's apple, is at the front of the neck. The larynx is called the voice box because it houses the vocal cords. The **vocal cords** are mucosal folds supported by elastic ligaments, and the slit between the vocal cords is an opening called the **glottis** (Fig. 15.3). When air is expelled past the vocal cords through the glottis, the vocal cords vibrate, producing sound. At the time of puberty, the growth of the larynx and the vocal cords is much more rapid and accentuated in the male than in the female, causing the male to develop a more prominent Adam's apple and a deeper voice. The voice "breaks" in the young male due to his inability to control the longer vocal cords.

The high or low pitch of the voice is regulated when speaking and singing by changing the tension on the vocal cords. The greater the tension, as when the glottis becomes more narrow, the higher the pitch. The loudness, or intensity, of the voice depends upon the amplitude of the vibrations—that is, the degree to which the vocal cords vibrate.

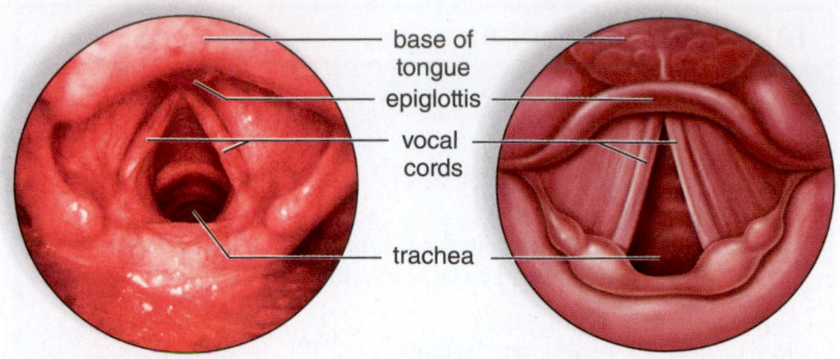
base of
tongue
epiglottis
vocal
cords
trachea

Figure 15.3 Placement of the vocal cords. Viewed from above, the vocal cords stretch across the glottis, the opening to the trachea. When air is expelled through the glottis, the vocal cords vibrate, producing sound. The glottis is narrow when we make a high-pitched sound, and it widens as the pitch deepens.

When food is swallowed, the larynx moves upward against the **epiglottis,** a flap of tissue that prevents food from passing into the larynx (Fig. 15.3). You can detect this movement by placing your hand gently on your larynx and swallowing.

SCIENCE IN YOUR LIFE ▶ **HEALTH**

Frequently Asked Questions About Tobacco and Health

Is there a safe way to smoke?

No. All forms of tobacco can cause damage, and smoking even a small amount is dangerous. Tobacco is perhaps the only legal product whose advertised and intended use—that is, smoking it—will hurt the body.

Does smoking cause cancer?

Yes, and not only lung cancer. Besides lung cancer, smoking a pipe, cigarettes, or cigars is also a major cause of cancers of the mouth, larynx (voice box), and esophagus. In addition, smoking increases the risk of cancer of the bladder, kidney, pancreas, stomach, and uterine cervix.

What are the chances of being cured of lung cancer?

Very low; the combined five-year survival rate is only 15%. Fortunately, lung cancer is a largely preventable disease. In other words, by not smoking, it can probably be prevented.

Does smoking cause other lung diseases?

Yes. Smoking leads to chronic bronchitis, a disease in which the airways produce excess mucus, forcing the smoker to cough frequently. Smoking is also the major cause of emphysema, a disease that slowly destroys a person's ability to breathe.

Why do smokers have "smoker's cough "?

Normally, cilia (tiny hairlike formations that line the airways) beat outward and "sweep" harmful material out of the lungs. Smoke, however, decreases this sweeping action, so some of the poisons in the smoke remain in the lungs.

If you smoke but don't inhale, is there any danger?

Yes. Wherever smoke touches living cells, it does harm. So, even if smokers of pipes, cigarettes, and cigars don't inhale, they are at an increased risk for lip, mouth, and tongue cancer.

Does smoking affect the heart?

Yes. Smoking doubles the risk of heart disease, which is the United States' number-one killer. Smoking, high blood pressure, high cholesterol, and lack of exercise are all risk factors for heart disease.

Is there any risk for pregnant women and their babies?

Pregnant women who smoke endanger the health and lives of their unborn babies. When a pregnant woman smokes, she really is smoking for two because the nicotine, carbon monoxide, and other dangerous chemicals in smoke enter her bloodstream and then pass into the baby's body. Smoking mothers have more stillbirths and babies of low birth weight than nonsmoking mothers.

Does smoking cause any special health problems for women?

Yes. Women who smoke and use the birth control pill have an increased risk of stroke and blood clots in the legs. In addition, women who smoke increase their chances of getting cancer of the uterine cervix.

What are some of the short-term effects of smoking cigarettes?

Almost immediately, smoking can make it hard to breathe. Within a short time, it can also worsen asthma and allergies. Only seven seconds after a smoker takes a puff, nicotine reaches the brain, where it produces a morphinelike effect.

Are there any other risks to the smoker?

Yes, there are many more risks. Smoking is a cause of stroke, which is the third leading cause of death in the United States.

The Trachea

The **trachea,** commonly called the *windpipe,* is a tube connecting the larynx to the primary bronchi. The trachea lies in front of the esophagus and is held open by C-shaped cartilaginous rings. The open part of the C-shaped rings faces the esophagus, and this allows the esophagus to expand when swallowing. The mucosa that lines the trachea has a layer of pseudostratified ciliated columnar epithelium (see Fig. 11.1). The cilia that project from the epithelium keep the lungs clean by sweeping mucus (produced by goblet cells) and debris toward the pharynx (Fig. 15.4). Smoking is known to destroy these cilia, and consequently, the toxins in cigarette smoke collect in the lungs. Smoking is discussed more fully in the Health feature, "Frequently Asked Questions About Tobacco and Health."

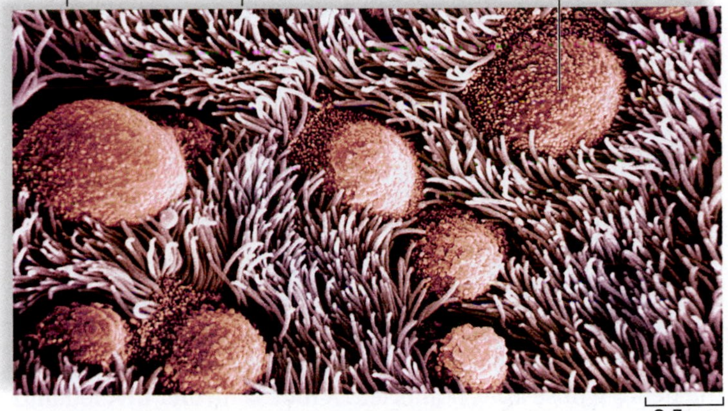
cilia
goblet cell
3.5 µm

Figure 15.4 The surface of the trachea. A scanning electron micrograph shows that the surface of the mucous membrane lining the trachea consists of goblet cells and ciliated cells. The cilia sweep mucus and the debris embedded in it toward the pharynx, where they are swallowed or expectorated. Smoking causes the cilia to disappear, allowing debris to enter the bronchi and lungs.

The Bronchial Tree

The trachea divides into right and left primary **bronchi** (sing., bronchus), which lead into the right and left lungs (see Fig. 15.1). The bronchi branch into a great number of secondary bronchi that eventually lead to **bronchioles.** The bronchi

Smokers are more likely to have and die from stomach ulcers than nonsmokers. Smokers have a higher incidence of cancer in general. If a person smokes and is exposed to radon or asbestos, the risk for lung cancer increases dramatically.

What are the dangers of passive smoking?

Passive smoking causes lung cancer in healthy nonsmokers. Children whose parents smoke are more likely to suffer from pneumonia or bronchitis in the first two years of life than children who come from smoke-free households. Passive smokers have a 30% greater risk of developing lung cancer than nonsmokers who live in a smoke-free house.

Are chewing tobacco and snuff safe alternatives to cigarette smoking?

No, they are not. Smokeless tobacco contains nicotine, the same addicting drug found in cigarettes and cigars. The juice from smokeless tobacco is absorbed through the lining of the mouth. There it can cause sores and white patches, which often lead to cancer of the mouth. Snuff dippers actually take in an average of over ten times more cancer-causing substances than cigarette smokers.

Lung Cancer Statistics

Lung cancer is the leading cause of death due to cancer for both men and women.

About 87% of lung cancer cases are linked to smoking. In the United States, an estimated 221,130 new cases and 156,940 deaths from lung cancer were expected to occur in 2011.

This means that lung cancer accounts for 28% of all cancer deaths. More people die of lung cancer than of colon, breast, and prostate cancers combined.

- Women who smoke are 1.5 times more likely to get lung cancer than are men who smoke.
- About 3,400 lung cancer deaths per year are attributed to secondhand smoke.
- Usually, 42% of people diagnosed with lung cancer die within the year.
- The five-year combined survival rate (the number of people alive after five years) after being diagnosed is only 15%. Even for those who are diagnosed early, the survival rate after five years is only about 49%.

Questions to Consider

1. Before reading this Health feature, were you aware of all the risks associated with smoking? If you are a smoker, how does knowing these risks affect the chances that you will quit? If you are a nonsmoker, do you think you might ever start?
2. Smoking is clearly addictive, because many people who want to quit are unable to do so. Whether you are a smoker or not (and it might be interesting to compare the responses of these two groups), what do you think is the percentage of smoking addiction that is physical versus psychological?
3. Suppose you became addicted to smoking two packs a day. If you decided to quit, how would you go about it? All at once, or a little at a time? Would you use any aids, such as nicotine gum or patches, or drugs? Perhaps a device that administers a mild electrical shock every time you smoke a cigarette? (Such a device exists!)

resemble the trachea in structure, but as the bronchial tubes divide and subdivide, their walls become thinner, and the small rings of cartilage are no longer present. Each bronchiole leads to an elongated space enclosed by a multitude of air pockets, or sacs, called **alveoli** (sing., alveolus). The components of the bronchial tree beyond the primary bronchi compose the lungs.

The Lungs

The **lungs** are paired, cone-shaped organs that occupy the thoracic cavity except for a central area that contains the trachea, the thymus, the heart, and the esophagus. The right lung has three lobes, and the left lung has two lobes, allowing room for the heart, whose apex points left. A lobe is further divided into lobules, and each lobule has a bronchiole serving many alveoli. The apex of a lung is narrow, while the base is broad and curves to fit the dome-shaped **diaphragm,** the muscle that separates the thoracic cavity from the abdominal cavity.

Each lung is covered by a very thin serous membrane called a *pleura* (see Fig. 15.7*a*). Another pleura covers the internal chest wall and diaphragm. Both membranes produce a lubricating serous fluid that helps the pleurae slide freely against each other during inspiration and expiration. Surface tension is the tendency for water molecules to cling to each other due to hydrogen bonding between the molecules. Surface tension holds the two pleural layers together when the lungs recoil during expiration.

With each inhalation, air passes by way of the respiratory passageways to the alveoli. Each alveolus is made up of simple squamous epithelium surrounded by blood capillaries. Gas exchange occurs between the air in an alveolus and the blood in the capillaries (Fig. 15.5). Oxygen diffuses across the alveolar and capillary walls to enter the bloodstream, while carbon dioxide diffuses from the blood across these walls to enter the alveoli.

If gas exchange is to occur, the alveoli must stay open to receive the inhaled air. The surface tension of fluid coating the alveoli is capable of causing them to close up. To prevent this, the alveoli are coated with *pulmonary surfactant,* a film of lipoprotein that lowers the surface tension and prevents them from closing. The lungs collapse in some newborn babies, especially premature infants, who lack this film. The condition, called *infant respiratory distress syndrome,* is the leading cause of death in babies born prematurely. It is usually treated by inserting a

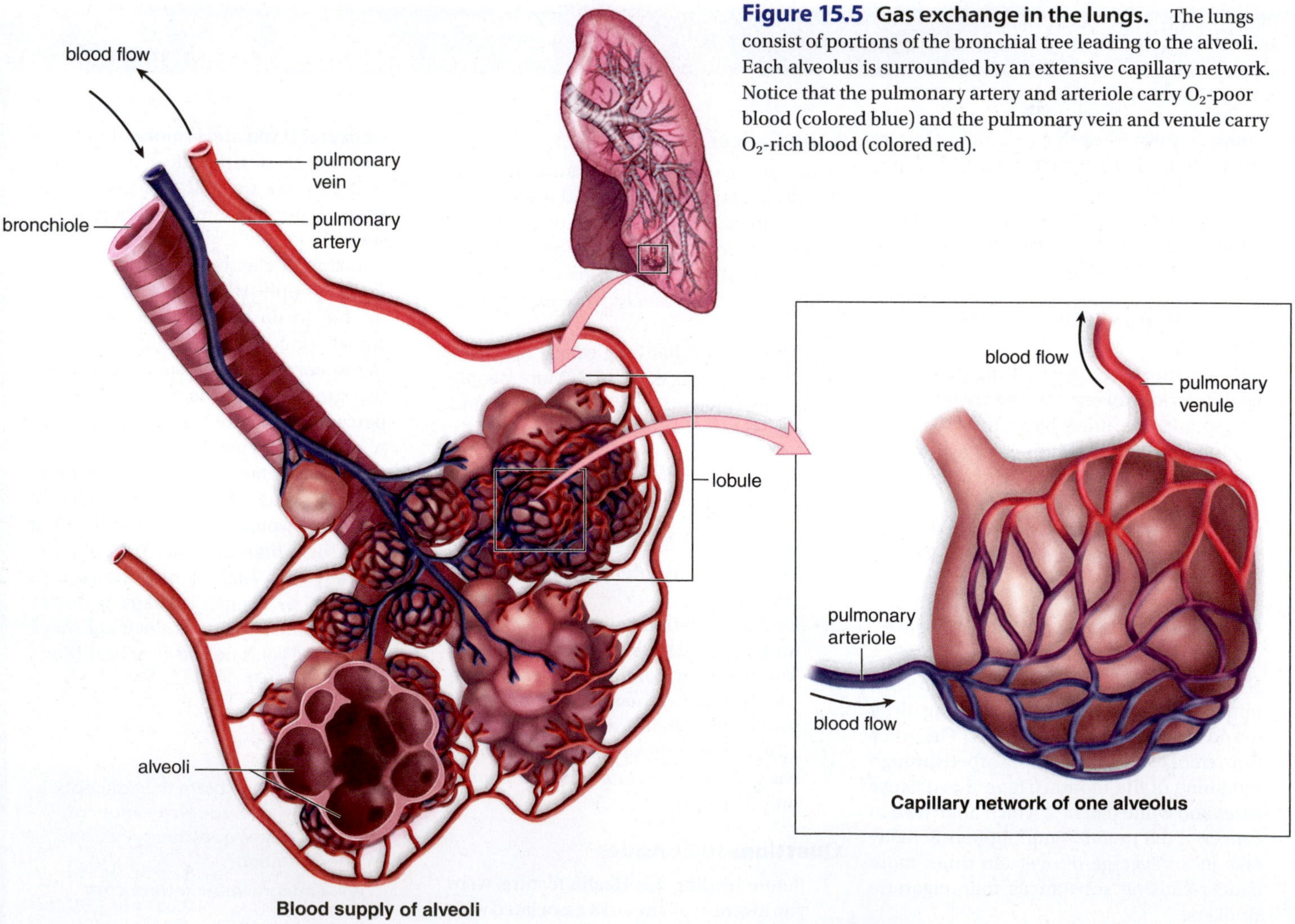
blood flow
pulmonary vein
bronchiole
pulmonary artery
lobule
alveoli
Blood supply of alveoli
blood flow
pulmonary venule
pulmonary arteriole
blood flow
Capillary network of one alveolus

Figure 15.5 Gas exchange in the lungs. The lungs consist of portions of the bronchial tree leading to the alveoli. Each alveolus is surrounded by an extensive capillary network. Notice that the pulmonary artery and arteriole carry O_2-poor blood (colored blue) and the pulmonary vein and venule carry O_2-rich blood (colored red).

plastic tube into the trachea and administering oxygen as well as surfactant that is either synthetic or extracted from cows' lungs. Severely ill patients may also receive *extracorporeal membrane oxygenation,* during which a cannula is installed in a large blood vessel, through which blood flows into a machine that adds oxygen and removes carbon dioxide prior to being returned to the patient.

MP3 Respiratory Structure and Function

Check Your Progress 15.1

1. Identify the structures that comprise the upper and lower respiratory tracts.
2. Discuss the arrangement of the trachea, esophagus, larynx, and epiglottis, which normally prevents food from entering the trachea.
3. Summarize the path of an oxygen molecule from the nose to the bloodstream.
4. Describe the function of pulmonary surfactant.

15.2 Mechanism of Breathing

Learning Outcomes

Upon completion of this section, you should be able to

1. Identify the different respiratory volumes that can be measured with a spirometer.
2. Contrast the mechanisms of inspiration and expiration.
3. Indicate how ventilation is controlled by nervous and chemical mechanisms.

During ventilation (breathing), a free flow of air is vitally important. Medical professionals use a device called a *spirometer* to record the volume of air exchanged during both normal and deep breathing. Examining these breathing patterns is also a useful way to increase our understanding of normal inspiration and expiration.

Respiratory Volumes

As shown in Figure 15.6, the amount of air inhaled and exhaled at rest (normally about 0.5 liters) is called the *tidal volume.* It is possible to increase the amount of air inhaled, and therefore the amount exhaled, by deep breathing. The maximum volume of air that can be moved in and out during a single breath is called the *vital capacity,* because breathing is essential to life. A number of the respiratory disorders discussed in section 15.4 can decrease vital capacity.

When taking a very deep breath, a healthy person can increase the volume of inhaled air beyond the tidal volume by about 3.0 liters, an amount called the *inspiratory reserve volume.* Similarly, one can forcefully exhale well beyond the normal tidal volume. This *expiratory reserve volume* is usually about 1.5 liters of air. Note from Figure 15.6 that vital capacity is the sum of the tidal, inspiratory reserve, and expiratory reserve volumes.

These respiratory volumes depend on various factors, such as age (decreasing after age 30), gender (10–20% lower in women), and physical activity (20–30% higher in conditioned athletes). During his bicycling career, Lance Armstrong's vital capacity was over 7 liters, which is 2–3 liters higher than the average adult male's.

During normal breathing, only about 70% of the tidal volume actually reaches the alveoli; 30% remains in the airways. Also, note from Figure 15.6 that even after a very deep exhalation, some air (about 1,000 ml) remains in the lungs; this is called the *residual volume.* This air is not as useful for gas exchange because it has been depleted of oxygen. In some lung diseases, the residual volume increases because the individual has difficulty emptying the lungs.

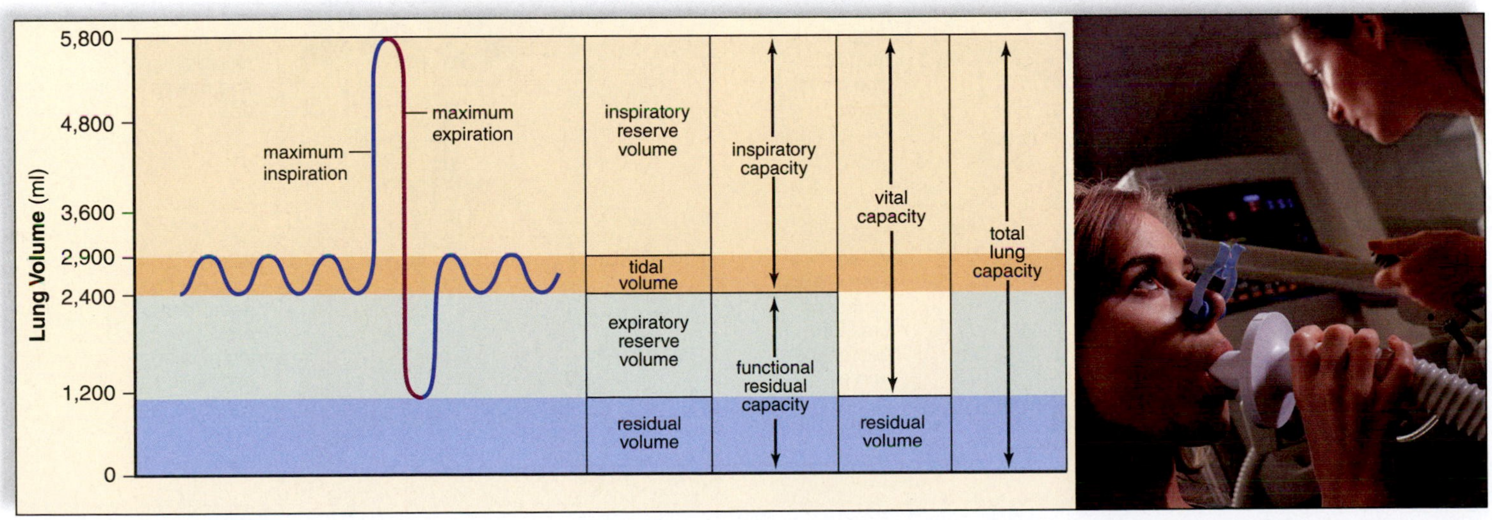
Lung Volume (ml)
5,800
4,800
3,600
2,900
2,400
1,200
0
maximum inspiration
maximum expiration
inspiratory reserve volume
tidal volume
expiratory reserve volume
residual volume
inspiratory capacity
functional residual capacity
vital capacity
residual volume
total lung capacity

Figure 15.6 Measuring ventilation. A spirometer measures the amount of air inhaled and exhaled with each breath. During inspiration, the pen moves up, and during expiration, the pen moves down. Vital capacity (red) is the maximum amount of air a person can exhale after taking the deepest inhalation possible.

Inspiration and Expiration

To understand ventilation, the manner in which air enters and exits the lungs, it is necessary to remember the following facts:

1. Normally, there is a continuous column of air from the pharynx to the alveoli of the lungs.
2. The lungs lie within the sealed-off thoracic cavity. The rib cage, consisting of the ribs joined to the vertebral column posteriorly and to the sternum anteriorly, forms the top and sides of the thoracic cavity. The intercostal muscles lie between the ribs. The diaphragm and connective tissue form the floor of the thoracic cavity.
3. The lungs adhere to the thoracic wall by way of the pleura. Normally, any space between the two pleurae is minimal due to the surface tension of the fluid between them.

Inspiration

Inspiration is the active phase of ventilation because this is the phase in which the diaphragm and the external intercostal muscles contract (Fig. 15.7*a*). In its relaxed state, the diaphragm is dome-shaped. During deep inspiration, it contracts and lowers. Also, the external intercostal muscles contract, and the rib cage moves upward and outward.

Following this contraction, the volume of the thoracic cavity is larger than it was before. As the thoracic volume increases, the lungs expand. Now the air pressure within the alveoli decreases, creating a partial vacuum. Because alveolar pressure is now less than atmospheric pressure outside the lungs, air naturally flows from outside the body into the respiratory passages and into the alveoli.

Notice that air comes into the lungs because they have already opened up. Air does not force the lungs open. This is why it is sometimes said that *humans inhale by negative pressure.* While inspiration is the active phase of breathing, the actual flow of air into the alveoli is passive.

Expiration

Usually, **expiration** is the passive phase of breathing, and no effort is required to bring it about. During expiration, the elastic properties of the thoracic wall and lungs cause them to recoil. In addition, the surface tension of the fluid lining the alveoli tends to draw them closed. During expiration, the abdominal organs press up against the diaphragm, and the rib cage moves down and inward (Fig. 15.7*b*).

The diaphragm and external intercostal muscles are usually relaxed as expiration occurs. However, when breathing is

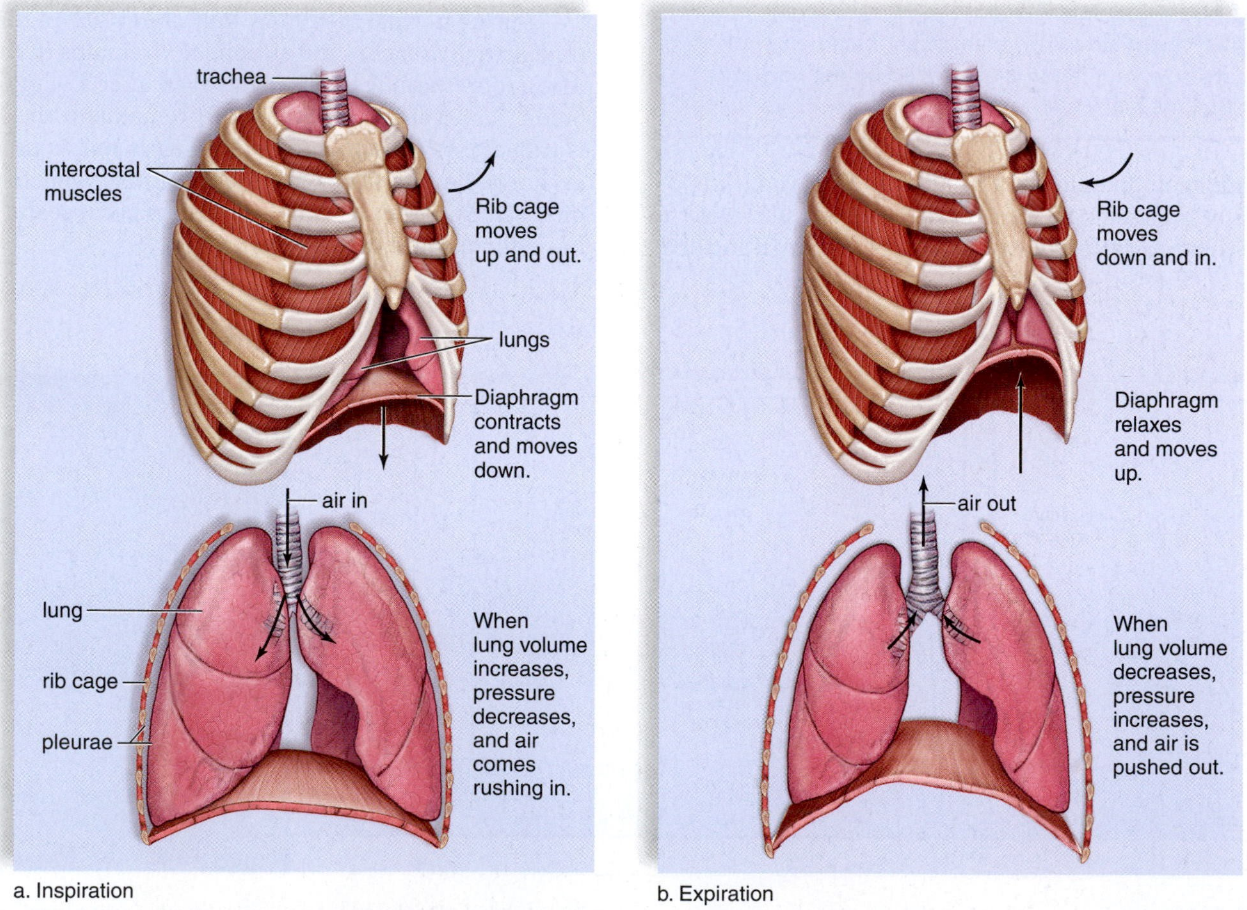
trachea
intercostal muscles
Rib cage moves up and out.
lungs
Diaphragm contracts and moves down.
air in
lung
rib cage
pleurae
When lung volume increases, pressure decreases, and air comes rushing in.
a. Inspiration
Rib cage moves down and in.
Diaphragm relaxes and moves up.
air out
When lung volume decreases, pressure increases, and air is pushed out.
b. Expiration

Figure 15.7 Inspiration and expiration compared. **a.** During inspiration, the thoracic cavity and lungs expand so that air is drawn in. **b.** During expiration, the thoracic cavity and lungs resume their original positions and pressures. Now, air is forced out.

deeper and/or more rapid, expiration can be active. Contraction of the internal intercostal muscles can force the rib cage to move downward and inward, and the increased pressure in the thoracic cavity helps to expel air.

Control of Breathing

Normally, adults have a resting breathing rate of 12 to 20 ventilations per minute. The rhythm of ventilation is controlled by a **respiratory center** located in the medulla oblongata of the brain.

The respiratory center stimulates inspiration by automatically sending impulses to the diaphragm by way of the phrenic nerve, and to the intercostal muscles by way of the intercostal nerves (Fig. 15.8). When the respiratory center temporarily stops sending neuronal signals to the diaphragm and the rib cage, the diaphragm relaxes and resumes its dome shape. Now expiration occurs.

Although the respiratory center automatically controls the rate and depth of breathing, its activity can also be influenced by nervous input and chemical input. Following forced inhalation, *stretch receptors* in the alveolar walls send inhibitory nerve impulses via the vagus nerve to the respiratory center, which temporarily inhibits the respiratory center from sending out nerve impulses.

MP3 Control of Respiration

Chemical Input As explained in more detail in the next section, most of the carbon dioxide (CO_2) that enters the blood combines with water, forming an acid, which breaks down and gives off hydrogen ions (H^+). The respiratory center contains cells that are sensitive to the levels of both CO_2 and H^+ in the blood. When either rises, the respiratory center increases the rate and depth of breathing. Respiration rate is also influenced by a group of cells called the **carotid bodies,** located in the carotid arteries, and the **aortic bodies,** located in the aorta, which are sensitive mainly to blood oxygen levels. When the concentration of oxygen decreases, these bodies communicate with the respiratory center, and the rate and depth of breathing increase.

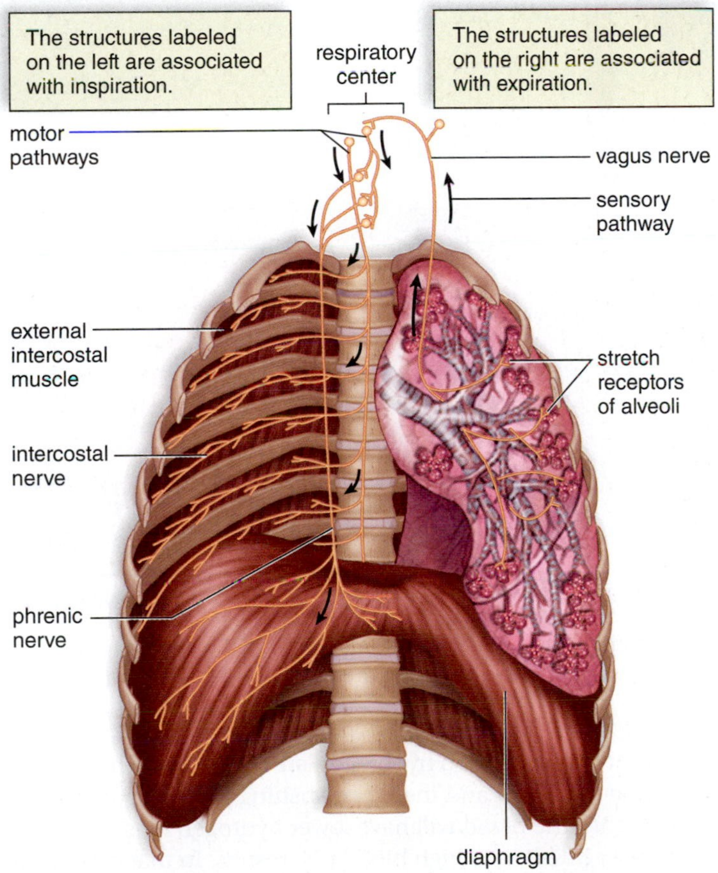
The structures labeled on the left are associated with inspiration.
respiratory center
The structures labeled on the right are associated with expiration.
motor pathways
vagus nerve
sensory pathway
external intercostal muscle
stretch receptors of alveoli
intercostal nerve
phrenic nerve
diaphragm

Figure 15.8 Nervous control of breathing. The respiratory center automatically stimulates the external intercostal (rib) muscles and diaphragm to contract via the phrenic nerve. After forced inhalation, stretch receptors send inhibitory nerve impulses to the respiratory center via the vagus nerve. Usually, expiration automatically occurs due to lack of stimulation from the respiratory center to the diaphragm and intercostal muscles.

Check Your Progress 15.2

1. Compare and contrast tidal volume, vital capacity, expiratory reserve volume, and residual volume.
2. Explain why inspiration is considered the active phase of ventilation, and expiration the passive phase.
3. Discuss the roles of the following in controlling respiration: respiratory center, vagus nerve, and carotid bodies.

15.3 Gas Exchanges in the Body

Learning Outcomes

Upon completion of this section, you should be able to

1. Distinguish between external and internal respiration.
2. Interpret the chemical reaction discussed in this section to explain the effect of hyperventilation or hypoventilation on blood pH.
3. Review how the differences in P_{O_2} and P_{CO_2} in arterial and venous blood determine how gases are exchanged in the lungs versus the tissues.

As mentioned in section 15.1, respiration includes the exchange of gases in the lungs (external respiration), and the exchange of gases in the tissues (internal respiration) (Fig. 15.9). Most of the O_2 carried in the blood is attached to the iron-containing *heme* portion of the protein **hemoglobin (Hb),** found in red blood cells.

MP3 Gas Exchange

External Respiration

External respiration refers to the exchange of gases between air in the alveoli and blood in the pulmonary capillaries (see Fig. 15.5). Gases exert pressure, and the amount of pressure each gas exerts is called its partial pressure, symbolized as P_{O_2} and P_{CO_2}. Blood in the pulmonary capillaries has a higher P_{CO_2} than atmospheric air does. Therefore, *CO_2 diffuses out of the*

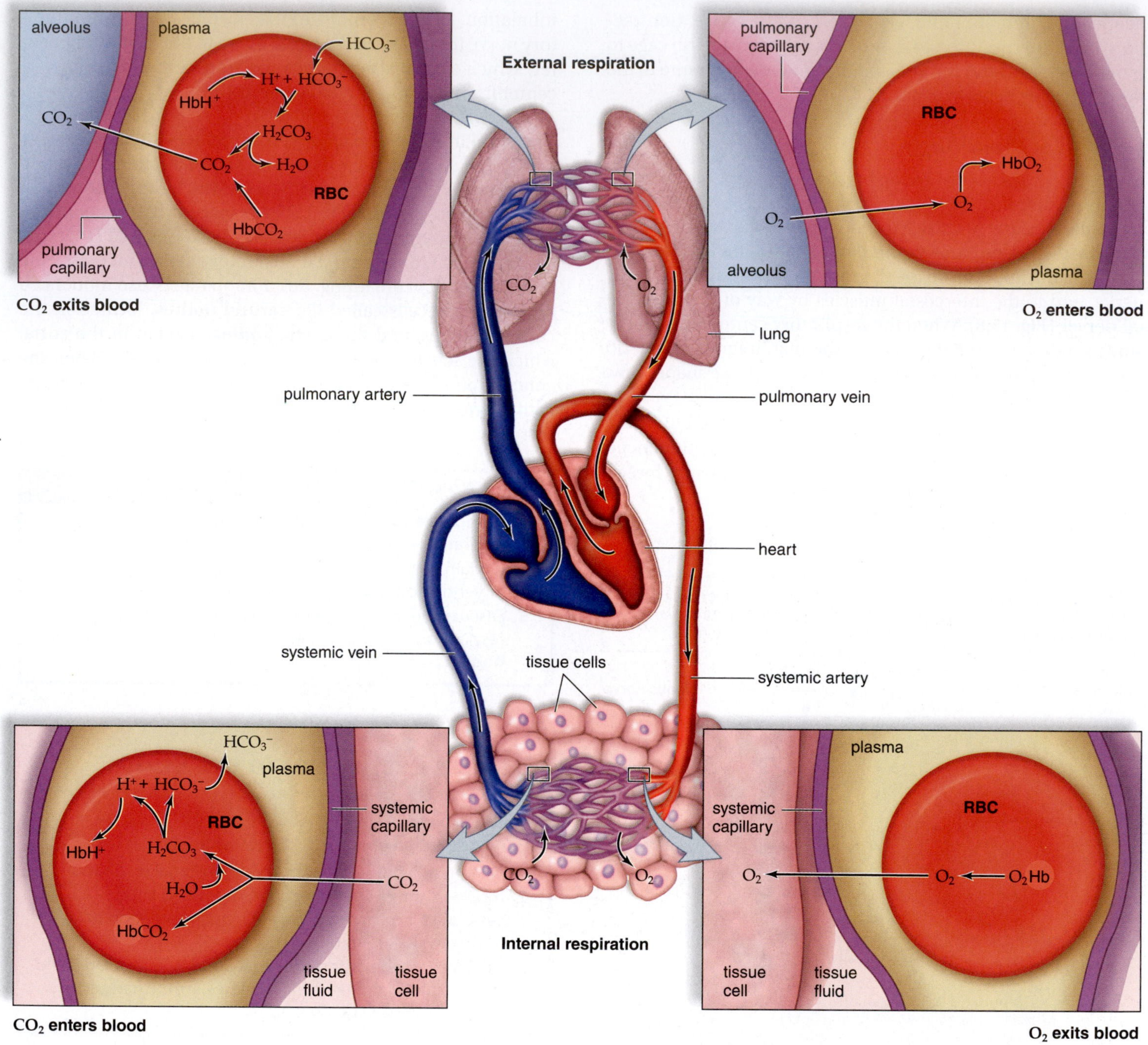
alveolus
plasma
HCO_3^-
$H^+ + HCO_3^-$
HbH^+
CO_2
H_2CO_3
CO_2
H_2O
RBC
$HbCO_2$
pulmonary capillary
CO_2 exits blood
External respiration
pulmonary capillary
RBC
HbO_2
O_2
O_2
alveolus
plasma
O_2 enters blood
CO_2
O_2
lung
pulmonary artery
pulmonary vein
heart
systemic vein
tissue cells
systemic artery
HCO_3^-
plasma
$H^+ + HCO_3^-$
RBC
systemic capillary
HbH^+
H_2CO_3
H_2O
CO_2
$HbCO_2$
tissue fluid
tissue cell
CO_2
O_2
Internal respiration
plasma
systemic capillary
RBC
O_2
O_2
O_2Hb
tissue cell
tissue fluid
CO_2 enters blood
O_2 exits blood

Figure 15.9 External and internal respiration. During external respiration in the lungs, carbon dioxide (CO_2) leaves the blood, and oxygen (O_2) enters the blood. During internal respiration in the tissues, oxygen leaves the blood, and carbon dioxide enters the blood.

plasma into the lungs. Most of the CO_2 is carried in the plasma as **bicarbonate ions** (HCO_3^-). As free CO_2 begins to diffuse out of the plasma, the enzyme **carbonic anhydrase,** present in red blood cells, speeds the breakdown of carbonic acid (H_2CO_3), driving this reaction to the right:

$$H^+ + HCO_3^- \longrightarrow H_2CO_3 \xrightarrow{\text{carbonic anhydrase}} H_2O + CO_2$$

H^+	+	HCO_3^-	→	H_2CO_3	carbonic anhydrase →	H_2O	+	CO_2
hydrogen ion		bicarbonate ion		carbonic acid		water		carbon dioxide

What happens if you hyperventilate (breathe at a high rate), removing more CO_2 and therefore pushing this reaction further to the right? The blood will have fewer hydrogen ions (H^+), and *respiratory alkalosis,* a high blood pH, results. In that case, your breathing will be inhibited, but in the meantime, you might feel dizzy and disoriented. Conversely, what happens if you hypoventilate (hold your breath), driving this reaction to the left? H^+ builds up in the blood, and *respiratory acidosis* occurs. As previously mentioned, increased H^+ and CO_2 levels will stimulate your respiratory center, and your breathing rate will most likely increase. If not, you might become confused, sleepy, and even comatose.

The pressure pattern for O_2 during external respiration is the reverse of that for CO_2. Blood returning from the systemic capillaries has a lower P_{O_2} than alveolar air does. Therefore, *in the lungs O_2 diffuses into plasma and then into red blood cells.* Hemoglobin takes up this oxygen and becomes **oxyhemoglobin (HbO_2)**:

$$\underset{\text{hemoglobin}}{Hb} + \underset{\text{oxygen}}{O_2} \longrightarrow \underset{\text{oxyhemoglobin}}{HbO_2}$$

Interestingly, only a relatively small percentage of the oxygen present in atmospheric air is utilized during normal external respiration. At sea level, air contains about 21% O_2, and exhaled air still contains 16–17% O_2. While this system may not seem very efficient, it does explain how a person whose heart and lungs have failed can sometimes be revived by mouth-to-mouth resuscitation, which involves filling the person's lungs with exhaled air.

Animation Changes in the Partial Pressure of Oxygen and Carbon Dioxide

Internal Respiration

Internal respiration refers to the exchange of gases between the blood in systemic capillaries and the tissue fluid. Therefore, internal respiration services tissue cells and without internal respiration, cells could not continue to produce the ATP that allows them to exist. Blood in the systemic capillaries is a bright red color because red blood cells contain oxyhemoglobin. Oxyhemoglobin gives up O_2, which diffuses out of the blood into the tissues:

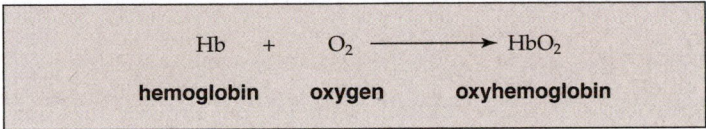

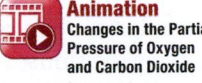

$$\underset{\text{oxyhemoglobin}}{HbO_2} \longrightarrow \underset{\text{hemoglobin}}{Hb} + \underset{\text{oxygen}}{O_2}$$

Oxygen diffuses out of the blood into the tissues because the P_{O_2} of tissue fluid is lower than that of blood. The lower P_{O_2} is due to cells continuously using up oxygen in cellular respiration. *Carbon dioxide diffuses into the blood from the tissues* because the P_{CO_2} of tissue fluid is higher than that of blood. Carbon dioxide, produced continuously by cells, collects in tissue fluid.

After CO_2 diffuses into the blood, it enters the red blood cells, where about 10% is taken up by hemoglobin, forming **carbaminohemoglobin ($HbCO_2$)**. Most of the remaining CO_2 combines with water in the blood plasma, forming carbonic acid (H_2CO_3), which dissociates to hydrogen ions (H^+) and bicarbonate ions (HCO_3^-). The increased concentration of CO_2 in the blood drives the reaction to the right:

$$\underset{\text{carbon dioxide}}{CO_2} + \underset{\text{water}}{H_2O} \xrightarrow{\text{carbonic anhydrase}} \underset{\text{carbonic acid}}{H_2CO_3} \longrightarrow \underset{\text{hydrogen ion}}{H^+} + \underset{\text{bicarbonate ion}}{HCO_3^-}$$

The enzyme carbonic anhydrase, mentioned previously, speeds up the reaction. Bicarbonate ions diffuse out of red blood cells and are carried in the plasma. The globin portion of hemoglobin combines with excess hydrogen ions produced by the overall reaction, and becomes **reduced hemoglobin (HHb)**. In this way, the pH of blood remains fairly constant. Blood that leaves the systemic capillaries is a dark maroon color because red blood cells contain reduced hemoglobin (though veins may actually look blue when seen under the skin, due to the bluish color of carbaminohemoglobin).

In either case, when blood reaches the lungs, the carbon dioxide readily diffuses out of the blood and is exhaled. In contrast, carbon monoxide, a gas that comes primarily from the incomplete combustion of natural gas and gasoline, has a much greater affinity for hemoglobin than oxygen does. Therefore, it stays combined for several hours, making hemoglobin unavailable for oxygen transport. Carbon monoxide detectors in homes can help prevent accidental death due to a malfunctioning furnace or inadequate exhaust system.

MP3 Gas Transport

Animation Gas Exchange During Respiration

Check Your Progress 15.3

1. Explain the role of hemoglobin in carrying O_2, CO_2, and hydrogen ions. Which is hemoglobin's most essential function?
2. Discuss why arterial blood is bright red in color, but venous blood is darker. This being the case, why does blood oozing from a cut always appear to be bright red?
3. Describe, as specifically as possible, why carbon monoxide poisoning can be rapidly fatal.

15.4 Disorders of the Respiratory System

Learning Outcomes

Upon completion of this section, you should be able to

1. Summarize several disorders that affect the upper respiratory tract and several that occur in the lower respiratory tract.
2. Classify disorders of the respiratory system according to whether they are caused by allergies, infections, a genetic defect, or exposure to toxins.

The respiratory tract is constantly exposed to the air in our environment, and thus it is susceptible to various infectious agents, as well as to pollution and, in some individuals, tobacco smoke. Disorders of the upper respiratory tract will be discussed before disorders of the lower respiratory tract.

Disorders of the Upper Respiratory Tract

The upper respiratory tract consists of the nasal cavities, the pharynx, and the larynx. Because it is responsible for filtering out many of the pathogens and other materials that may be present in the air, the upper respiratory tract is susceptible to a variety of viral and bacterial infections. Upper respiratory infections can also spread from these areas to the middle ear or the sinuses.

The Common Cold

Most "colds" are relatively mild viral infections of the upper respiratory tract characterized by sneezing, a runny nose, and perhaps a mild fever. Many different viruses can cause colds, the most common being a group called the rhinoviruses (*rhin* is Greek for nose). Most colds last from a few days to a week, when the immune response is able to eliminate the virus. Because colds are caused by viruses, antibiotics do not help, although decongestants and anti-inflammatory medications can ease the symptoms.

Pharyngitis, Tonsillitis, and Laryngitis

Pharyngitis is an inflammation of the throat, usually because of an infection. If the tonsils have not been removed previously, they often become involved as well. What we call "strep throat" is a pharyngitis caused by the bacterium *Streptococcus pyogenes* that can lead to a generalized upper respiratory infection and even a systemic (affecting the body as a whole) infection. The symptoms of strep throat are severe sore throat, high fever, and white patches on a dark-red pharyngeal or tonsillar area (Fig. 15.10). Most cases of strep throat can be successfully treated with antibiotics.

Tonsillitis occurs when the *tonsils*, aggregates of lymphoid tissue in the pharynx, become inflamed and enlarged. The tonsils in the posterior wall of the nasopharynx are often called adenoids. If tonsillitis occurs frequently and the enlarged tonsils make breathing difficult, the tonsils can be removed surgically in a *tonsillectomy*. Fewer tonsillectomies are performed today than in the past because we now know that the tonsils help initiate immune responses to many of the pathogens that enter the pharynx. Therefore, they are an important component of the body's immune system.

Laryngitis is an inflammation of the larynx with accompanying hoarseness, often leading to the inability to talk in an audible voice. Usually, laryngitis disappears after resting the vocal cords and treating any infection present. Sometimes benign growths, or polyps, can develop on the vocal cords. Laryngeal polyps are more likely to occur in people who put their vocal cords through excessive wear and tear, such as professional singers. Steven Tyler, lead singer of the rock band Aerosmith, as well as pop music superstar Adele, are two recent examples of singers who had to have surgery to remove polyps that were interfering with their ability to control their vocal cords.

Sinusitis

Sinusitis is an inflammation of the cranial sinuses, the cavities within the facial skeleton that drain into the nasal cavities. Sinusitis develops when nasal congestion blocks the tiny openings leading to the sinuses. Up to 10% of upper respiratory infections are accompanied by sinusitis, and allergies may also play a role. Symptoms may include postnasal discharge, headache, and facial pain that worsens when the patient bends forward. Successful treatment depends on addressing the cause of the inflammation, and restoring proper drainage of the sinuses. Rinsing the sinuses by instilling a warm saline solution into one nostril, and out the other, though somewhat uncomfortable, may help remove irritants and rinse out mucus.

Otitis Media

Otitis media is an inflammation of the middle ear. The middle ear is not a part of the respiratory tract, but this condition is considered here because, especially in children, nasal infections can spread to the ear by way of the *auditory (eustachian) tubes* that lead from the nasopharynx to the middle ear. Pain is the primary symptom of otitis media. A sense of fullness, hearing loss, vertigo (dizziness), and fever may also be present. If the cause is bacterial, antibiotic therapy is usually very effective. However, if viruses or allergies are the culprit, antibiotics are not effective. Tubes (called tympanostomy tubes) are sometimes surgically placed in the eardrums of children with multiple recurrences to help prevent the buildup of pressure in the middle ear and the possibility of hearing loss (Fig. 15.11). Normally, these tubes fall out with time.

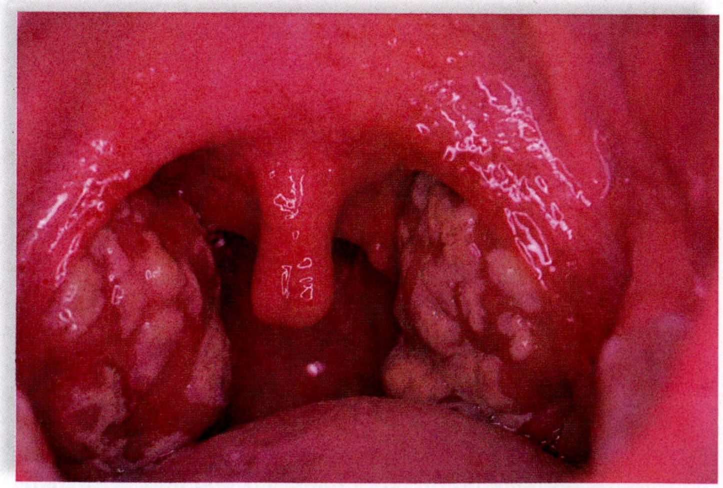

Figure 15.10 Strep throat. Pharyngitis, often caused by the bacterium *S. pyogenes,* can lead to swollen, inflamed tonsils, sometimes with whitish patches.

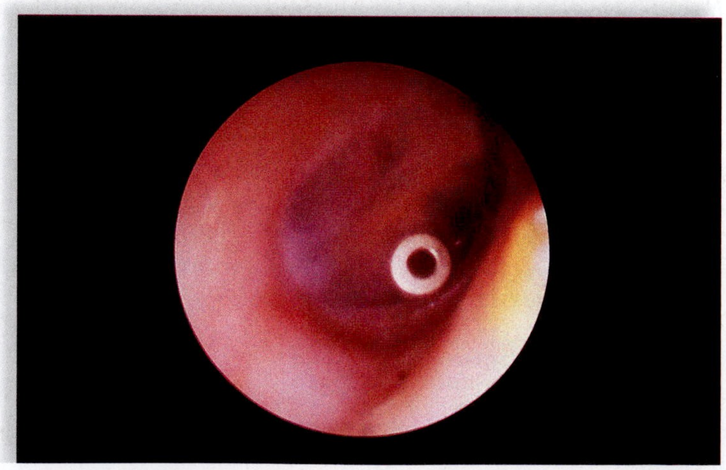

Figure 15.11 A common treatment for middle ear infections. Tympanostomy tubes may be surgically implanted in the eardrums to drain fluid from the middle ears of children with chronic otitis media.

Disorders of the Lower Respiratory Tract

Several disorders of the lower respiratory tract cause problems by obstructing normal airflow. Their causes range from a foreign object lodged in the trachea to excessive mucus in the bronchi and bronchioles. Other conditions tend to restrict the normal elasticity of the lung tissue itself (Fig. 15.12).

Disorders of the Trachea and Bronchi

One of the simplest—but most life-threatening—disorders that affects the trachea is choking. The best way for a person without extensive medical training to help someone who is choking is to perform the Heimlich maneuver, which involves grabbing the choking person around the waist from behind, and forcefully pulling both hands into their upper abdomen to expel whatever is lodged. Trained medical personnel may also insert a breathing tube by way of an incision made in the trachea. The operation is called a tracheotomy, and the opening is a *tracheostomy***.** Sometimes people whose larynx or trachea has been damaged or destroyed, usually as a result of smoking, must have a permanent tracheostomy tube installed.

Acute bronchitis is an inflammation of the primary and secondary bronchi. Usually it is preceded by a viral infection that has led to a secondary bacterial infection. Most likely, a nonproductive cough has become a deep cough that produces more mucus and perhaps pus (Fig. 15.12*a*). Typically, the bacterial infection can be successfully treated with antibiotics.

In **chronic bronchitis,** the airways are inflamed and filled with mucus. A cough that brings up mucus is common. The bronchi have undergone degenerative changes, including the loss of cilia and their normal cleansing action. Under these conditions, an infection is more likely to occur. The most frequent cause of chronic bronchitis is smoking, although exposure to environmental pollutants can also be a contributing factor.

Asthma is a disease of the bronchi and bronchioles that is marked by wheezing, breathlessness, and sometimes a cough and expectoration of mucus. The airways are unusually sensitive to specific irritants, which can include a wide range of allergens such as pollen, animal dander, dust, cigarette smoke, and industrial fumes. Even cold air can be an irritant. When exposed to an irritant, the smooth muscle in the bronchioles undergoes spasms (Fig. 15.12*b*). Most asthma patients have some degree of bronchial inflammation that reduces the diameter of the

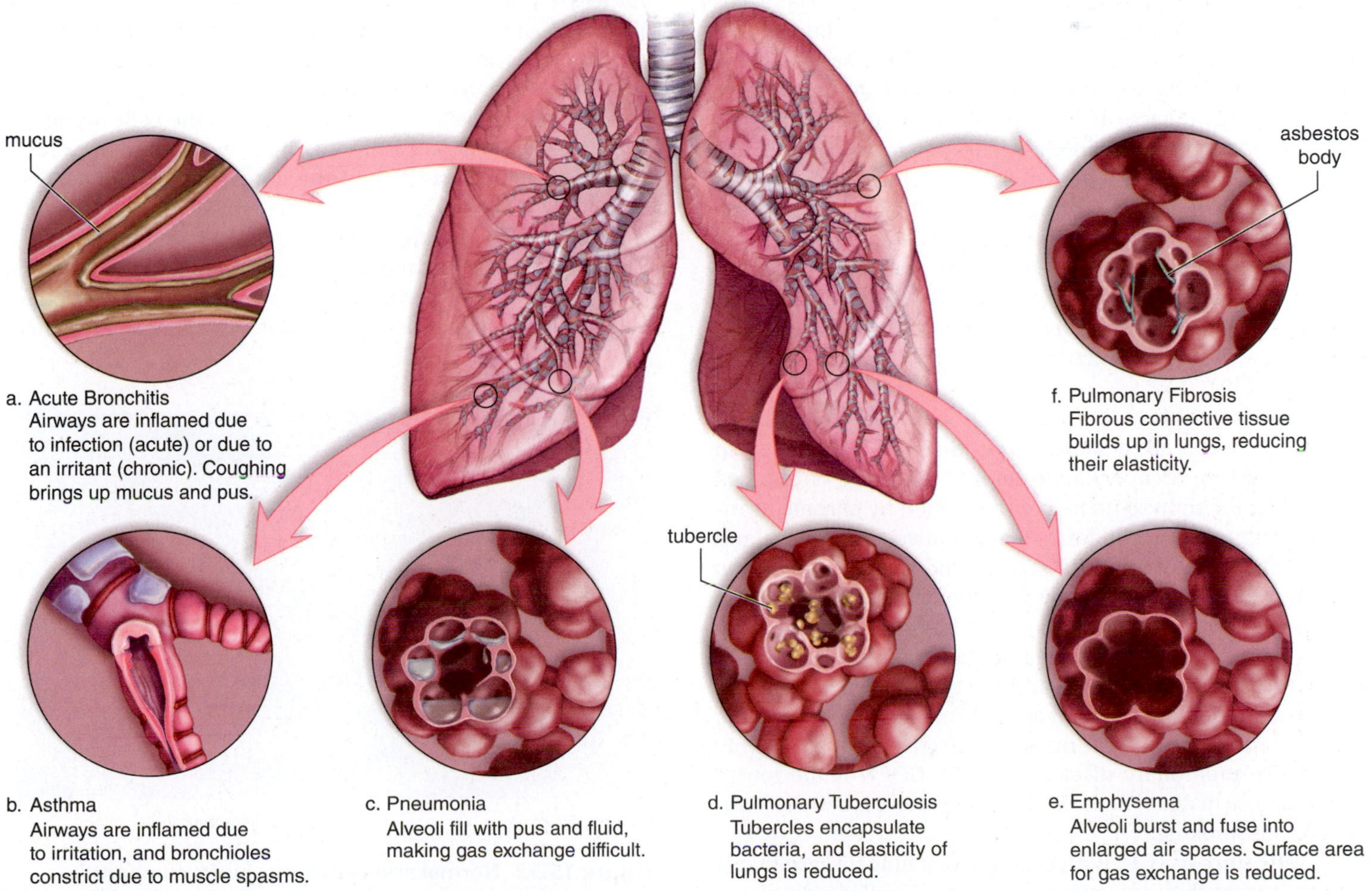
mucus
asbestos body
tuberculosis
a. Acute Bronchitis
Airways are inflamed due to infection (acute) or due to an irritant (chronic). Coughing brings up mucus and pus.
f. Pulmonary Fibrosis
Fibrous connective tissue builds up in lungs, reducing their elasticity.
tubercle
b. Asthma
Airways are inflamed due to irritation, and bronchioles constrict due to muscle spasms.
c. Pneumonia
Alveoli fill with pus and fluid, making gas exchange difficult.
d. Pulmonary Tuberculosis
Tubercles encapsulate bacteria, and elasticity of lungs is reduced.
e. Emphysema
Alveoli burst and fuse into enlarged air spaces. Surface area for gas exchange is reduced.

Figure 15.12 Common bronchial and pulmonary diseases. Exposure to infectious pathogens and/or polluted air, including tobacco smoke, causes various diseases and disorders.

airways and contributes to the seriousness of an attack. Asthma is not curable, but several types of drugs can prevent or treat asthma attacks. One group of drugs, called the beta-agonists, works by dilating the bronchioles. These drugs are usually administered using an asthma inhaler, as mentioned in the story that opened this chapter. Corticosteroids can help control the inflammation and hopefully prevent an attack.

Diseases of the Lungs

Pneumonia is an infection of the lungs in which the bronchi or alveoli fill with thick fluid (Fig. 15.12*c*). High fever and chills, with headache and chest pain, are symptoms of pneumonia. Rather than being a generalized lung infection, pneumonia may be localized in specific lobules of the lungs. Obviously, the more lobules involved, the more serious is the infection. Pneumonia can be caused by several types of bacteria, viruses, and other infectious agents. Certain types of pneumonia mainly strike individuals with reduced immunity. For example, AIDS patients are subject to a particularly rare form of pneumonia caused by the fungus *Pneumocystis carinii.*

Pulmonary tuberculosis is caused by the bacterium *Mycobacterium tuberculosis.* When *M. tuberculosis* bacteria invade the lung tissue, the cells accumulate around the invading bacteria, isolating them from the rest of the body. This accumulation of cells is called a tubercle (Fig. 15.12*d*). If the body's resistance is high, the imprisoned organisms die, but if the resistance is low, the organisms eventually escape and spread. If a chest X ray detects active tubercles, the individual is put on appropriate drug therapy to ensure the localization of the disease and the eventual destruction of any live bacteria. It is possible to tell if a person has ever been infected with tuberculosis bacteria with a TB skin test, in which a highly diluted extract of the bacteria is injected into the skin of the patient. A person who has never been exposed to *M. tuberculosis* shows no reaction, but one who has had or is fighting an infection shows an area of inflammation that peaks in about 48 hours (see Fig. 13.12).

Emphysema is a chronic and incurable disorder in which the alveoli are distended and their walls damaged so that the surface area available for gas exchange is reduced (Fig. 15.12*e*). Emphysema is often preceded by chronic bronchitis. Air trapped in the lungs leads to alveolar damage and a noticeable ballooning of the chest. The elastic recoil of the lungs is reduced, so the driving force behind expiration is also reduced. The victim often feels out of breath and may have a cough. Because the surface area for gas exchange is reduced, less oxygen reaches the heart and the brain. Because they are often seen in the same patient and tend to recur, emphysema, chronic bronchitis, and asthma are collectively called chronic obstructive pulmonary disease (COPD). COPD is the fourth leading cause of death in the United States, and is usually associated with smoking.

Cystic fibrosis (CF) is an example of a lung disease that is genetic, rather than infectious, although infections do play a role in the disease. One in 31 Americans carries the defective gene, but a child must inherit two copies of the faulty gene to have the disease. Still, CF is the most common inherited disease in the U.S. white population. The gene that is defective in CF codes for cystic fibrosis conductance transmembrane regulator (CFTR), a protein needed for proper transport of Cl^- ions out of the epithelial cells of the lung. Because this also reduces the amount of water transported out of the lung cells, the mucus secretions become very sticky and can form plugs that interfere with breathing. Symptoms of CF include coughing and shortness of breath, and part of the treatment involves clearing mucus from the airways by vigorously slapping the patient on the back as well as by administering mucus-thinning drugs. None of these treatments is curative, however, and because the lungs can be severely affected, the median survival age for people with CF is only 30 years. Researchers are attempting to develop gene therapy strategies to replace the faulty *CFTR* gene.

Another common lung disease is **pulmonary fibrosis,** in which fibrous connective tissue builds up in the lungs, causing a loss of elasticity (Fig. 15.12*f*). This restricts the ability of the lungs to expand during inhalation, and so reduces the vital capacity and other lung volumes. Pulmonary fibrosis most commonly occurs in elderly persons, and the risk is increased by environmental exposure to silica (sand), various types of dust, and asbestos.

Video
Good Poison

Lung cancer is more prevalent in men than in women, but it is the leading cause of cancer death in both genders. About 87% of lung cancers are associated with cigarette smoking. Autopsies on smokers have revealed the progressive steps by which the most common form of lung cancer develops. The first step appears to be thickening of the cells lining the bronchi. Then cilia are lost, making it impossible to prevent dust and dirt from settling in the lungs. Following this, cells with atypical nuclei appear, followed by a tumor consisting of disordered cells with atypical nuclei. A normal lung and a lung with cancerous tumors are shown in Figure 15.13. A final step occurs when some of these cells break loose and penetrate other tissues, a process called metastasis. Now the cancer

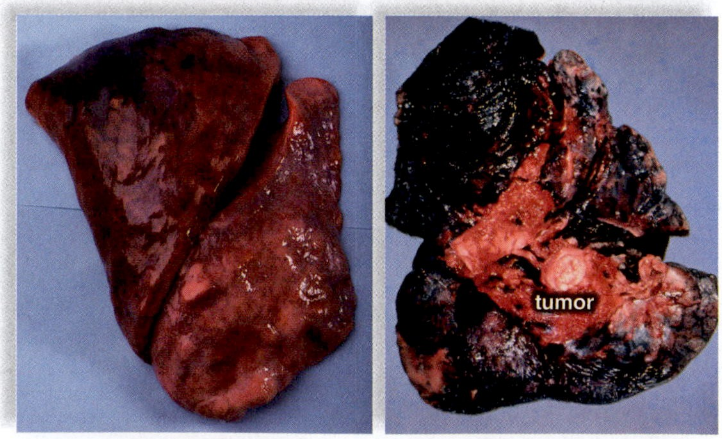
tumor

a. Normal lung   b. Lung cancer

Figure 15.13 Normal and cancerous lungs compared.
a. Normal lung. Note the healthy red color in both lobes. **b.** Lung of a heavy smoker. Notice that the lung is black except where cancerous tumors have formed.

SCIENCE IN YOUR LIFE ▶ SCIENTIFIC INQUIRY

Artificial Lung Technology

Some organs, such as the kidneys, are relatively easy to transplant from one well-matched individual to another, with a high rate of success. Lungs are more difficult to transplant, however, with a ten-year survival rate of only 10–20%. Some very recent research is showing how one day it may be possible to replace diseased lungs with laboratory-grown versions.

In early 2010, a group of scientists at Yale University anesthetized a group of adult rats, surgically removed their left lungs, and replaced them with tissue-engineered lungs that had been produced in the laboratory. For periods of up to two hours, the implanted lung tissue exchanged oxygen and carbon dioxide at similar rates as the natural lungs.[1]

In order to build the artificial rat lungs, the Yale researchers began by removing lungs from adult rats, and treating the organs to remove most of the cellular components, while preserving the airways and supporting connective tissue matrix. Next, they placed these decellularized structures, along with various types of cultured cells, into a sterile container designed to mimic some aspects of the fetal environment in which the lungs normally first develop. The scientists found that the cells were able to form much of the lung tissues as well as the blood vessels needed to supply the tissue with blood and transport gases.

Although these results are an important early step toward growing replacement lungs in the lab, there is a long way to go before anyone will contemplate implanting engineered lungs into humans.

In a separate study published on the same day as the Yale study, a Harvard group announced that they had developed a pea-sized device that mimics human lung tissues. Made of human lung cells, a permeable membrane, plus blood capillary cells, all mounted on a microchip (Fig. 15A), the device is able to mimic the function of alveoli. When the researchers placed bacteria on the alveolar side of the device, and white blood cells on the capillary side, the blood cells crossed the membrane, mimicking an immune response.[2]

At a minimum, the researchers hope that their "lung-on-a-chip" device can be used for testing certain drugs or the effects of various toxins on the lungs, which might replace much of the animal testing that is currently performed.

Questions to Consider

1. With regard to producing tissue-engineered human lungs, what is the major drawback in the procedure developed by the Yale group?
2. What are some of the aspects of lung structure that make it a more difficult organ to grow in the lab than, for example, a urinary bladder?
3. Besides increased availability, what are two other potential advantages of laboratory-grown lungs (or other tissues) compared to regular donor tissues?

[1]Peterson, T. H. et al. "Tissue-Engineered Lungs for in Vivo Implantation," *Science* 329:538–541 (2010). DOI: 10.1126/science.1189345.

[2]Huh, D. et al. "Reconstituting Organ-Level Lung Functions on a Chip," *Science* 328:1662–1668 (2010). DOI: 10.1126/science.1188302.

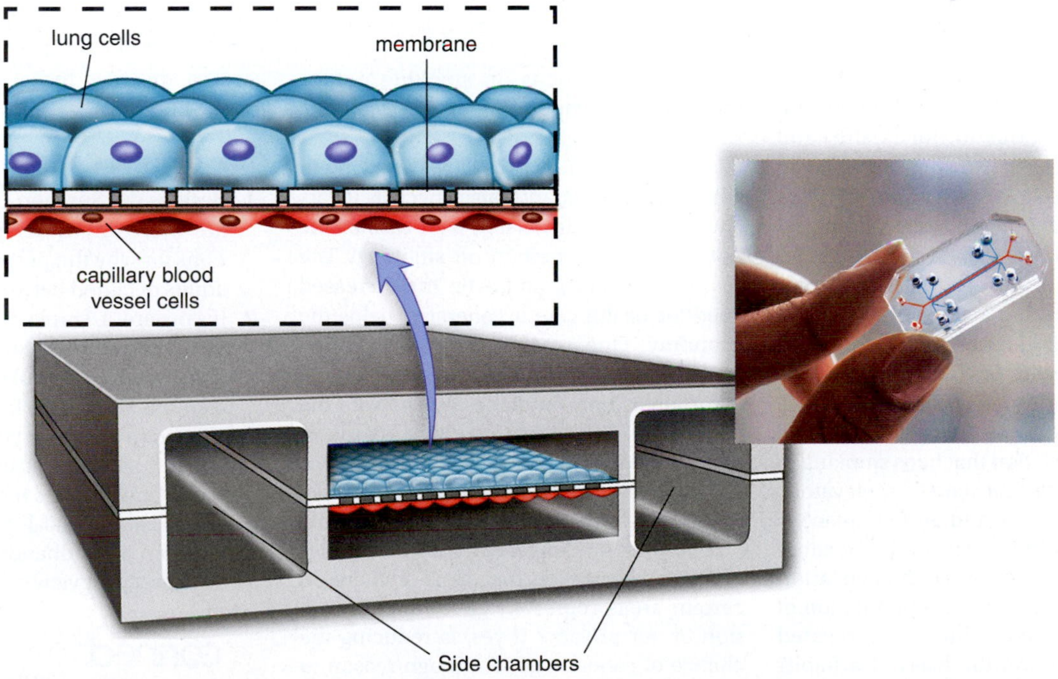
lung cells
membrane
capillary blood vessel cells
Side chambers

Figure 15A Lung on a chip. About the size of a credit card, the device consists of a semipermeable membrane with lung cells on one side and blood vessel cells on the other. Liquid medium flows in a channel on the blood cell side, while air flows in and out on the lung cell side.

has spread. The original tumor may grow until a bronchus is blocked, cutting off the supply of air to that lung. The entire lung then collapses, the secretions trapped in the lung spaces become infected, and pneumonia or a lung abscess (localized area of pus) results. The only treatment that offers a possibility of cure is *pneumonectomy,* in which a lobe or the whole lung is removed before metastasis has had time to occur. If the cancer has spread, chemotherapy and radiation are also required.

The Health feature, "Frequently Asked Questions About Tobacco and Health," on page 280 discusses the various illnesses, including cancer, that are apt to occur when a person smokes. The Bioethical feature, "Bans on Smoking," below raises questions of fairness regarding attempts to cut back on passive smoking by issuing smoking bans. If a person stops voluntary smoking and avoids passive smoking, and if the body tissues are not already cancerous, they may return to normal over time.

Check Your Progress 15.4

1. Describe a disorder of the lower respiratory tract that tends to cause a narrowing of the airways and one that restricts the lungs' ability to expand normally.
2. Explain how, in addition to the directly harmful effects of cigarette smoke, smoking might predispose the lung to damage by other harmful substances.
3. Describe three common disorders of the upper respiratory tract, along with the most common cause of each.
4. Explain how infections can spread from the respiratory system to the middle ear.

Case Study Conclusion

Thirteen-year-old Juan was having difficulty playing basketball due to the airway constriction and inflammation characteristic of asthma. Common symptoms of asthma include wheezing and feeling short of breath. Asthma can be brought on by a variety of different irritants in the environment, including pollens, dust, and smoke. In this case, heavy levels of air pollutants present near a busy freeway may have been a factor.

While the symptoms of asthma are treatable, there is no cure, nor is it clear why some people are more prone to develop asthma. Researchers are continuing to develop and test many new treatments for asthma, however, so until the cause of, and potentially the cure for, asthma can be found, people who suffer from this disease should be able to find relief.

SCIENCE IN YOUR LIFE ▶ BIOETHICAL

Bans on Smoking

In 1964, the Surgeon General of the United States announced to the general public that smoking is hazardous to our health, and thereafter, a health warning was placed on packs of cigarettes. At that time, 40.4% of adults smoked, but by 2009, only about 24% of adult males, and 18% of adult females were smokers. In the meantime, however, the public became aware that passive smoking—that is, just being in the vicinity of someone who is smoking—can also lead to cancer and other health problems. By now, many state and local governments have passed legislation that bans smoking in public places such as restaurants, elevators, public meeting rooms, and the workplace.

Is legislation that restricts the freedom to smoke ethical? Or is such legislation akin to racism and creating a population of second-class citizens who are segregated from the majority on the basis of a habit? Are the desires of nonsmokers being allowed to infringe on the rights of smokers? Or is this legislation one way to help smokers become nonsmokers? One study showed that workplace bans on smoking reduce the daily consumption of cigarettes among smokers by 10%.

Is legislation that disallows smoking in family-style restaurants fair, especially if bars and restaurants associated with casinos are not included in the ban on smoking? The selling of tobacco, and even the increased need for health care it generates, helps the economy. One smoker writes, "Smoking causes people to drink more, eat more, and leave larger tips. Smoking also powers the economy of Wall Street." Is this a reason to allow smoking to continue? Or should we simply require all places of business to put in improved air filtration systems? Would that do away with the dangers of passive smoking?

Does legislation that bans smoking in certain areas represent government invasion of our privacy? If yes, is reducing the chance of cancer a good enough reason to allow the government to invade our privacy? Some people are more prone to cancer than others. Should we all be regulated by the same legislation? Are we our brothers' keepers, meaning that we have to look out for one another?

Questions to Consider

1. Besides the Surgeon General's warning, are there any other possible explanations for why the percentage of smoking adults declined between 1964 and 2009?
2. If you are a smoker, do you smoke in your house? Your car? If not, why not? If you are a nonsmoker, how would you respond if a good friend or a relative wanted to smoke in your house?
3. It is estimated that 90 million nonsmokers in the United States have measurable levels of toxic chemicals in their body resulting from secondhand smoke. How does this affect your view of smoking bans?

MEDIA STUDY TOOLS

www.mhhe.com/maderinquiry14

Enhance your study of this chapter with study tools and practice tests. Also ask your instructor about the resources available through ConnectPlus, including LearnSmart, the media-rich eBook, interactive learning tools, and animations.

SUMMARIZE

15.1 The Respiratory System

The major functions of the **respiratory system** are to allow O_2 from the air to enter the blood, and CO_2 from the blood to exit the body. This occurs during **ventilation,** which includes inspiration and expiration. The respiratory tract can be divided into two major components:

- The upper respiratory tract, consisting of the nose (with nasal cavities), **pharynx, glottis, epiglottis,** and **larynx** (which contains the **vocal cords**).
- The lower respiratory tract, made up of the **trachea, bronchi, bronchioles,** and **lungs.** The lungs contain many **alveoli,** which are air sacs surrounded by a capillary network. The base of each lung contacts the **diaphragm,** a muscle that separates the thoracic and abdominal cavities.

15.2 Mechanism of Breathing

Ventilation includes both inspiration and expiration:

- The volumes of air that move in and out of the respiratory system during ventilation can be studied using a spirometer. The maximum volume of air that can be moved in and out during ventilation is called the vital capacity, which can be affected by many factors, including disease.
- **Inspiration** begins when the **respiratory center** in the brain sends excitatory nerve impulses to the diaphragm and the muscles of the rib cage. As they contract, the diaphragm lowers, and the rib cage moves upward and outward; the lungs expand, creating a partial vacuum, which causes air to rush in. The activity of the respiratory center is influenced by stretch receptors in the alveoli, as well as by the **aortic bodies** and **carotid bodies,** which are mainly sensitive to blood oxygen levels.

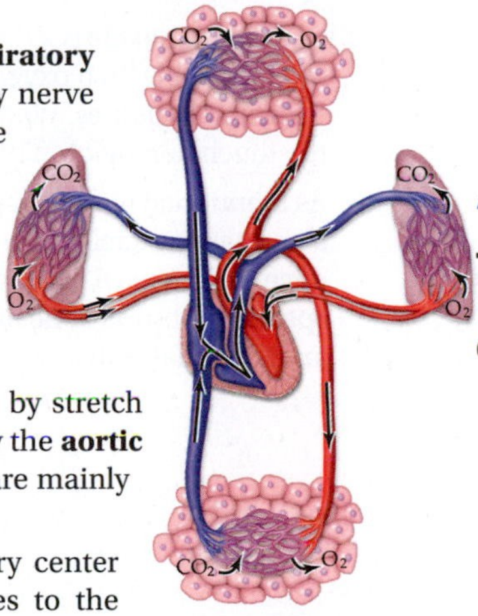

CO2
O2
CO2
O2
CO2
O2
CO2
O2

- **Expiration** occurs as the respiratory center temporarily stops sending impulses to the diaphragm and the muscles of the rib cage. As the diaphragm relaxes, it resumes its dome shape, and as the rib cage retracts, air is pushed out of the lungs.

15.3 Gas Exchanges in the Body

Oxygen and carbon dioxide are exchanged between the blood and tissues during external respiration and internal respiration:

- **External respiration** occurs in the lungs when CO_2 leaves the blood and O_2 enters the blood. In the lungs, the P_{CO_2} is higher in the blood than in alveoli, so CO_2 diffuses from the blood into the alveoli. Because carbon dioxide is present in the blood mainly as **bicarbonate ion** (HCO_3^-), carbonic acid forms first, which is then broken down by the enzyme **carbonic anhydrase** to carbon dioxide and water. Because the P_{O_2} is higher in the alveoli than in blood, O_2 diffuses from the alveoli into the blood where it combines with **hemoglobin (Hb)**, forming **oxyhemoglobin (HbO_2)**.
- **Internal respiration** occurs in the tissues, when O_2 leaves and CO_2 enters the blood. This occurs because the P_{O_2} is lower in the tissues compared to arterial blood, and the P_{CO_2} is higher. About 10% of this CO_2 enters red blood cells and binds to hemoglobin, forming **carbaminohemoglobin ($HbCO_2$)**. Most of the CO_2 combines with water in the plasma, forming carbonic acid that is broken down to bicarbonate and hydrogen ions. Hemoglobin can also combine with hydrogen ions to form **reduced hemoglobin (HHb)**.

15.4 Disorders of the Respiratory System

Disorders of the respiratory system can be divided into those that affect the upper respiratory tract and those that affect the lower respiratory tract:

- In the upper respiratory tract, the common cold, as well as **pharyngitis, tonsillitis,** and **laryngitis** are all conditions that most people experience at some time. In addition, these infections can spread to the sinuses (causing **sinusitis**) and middle ear (causing **otitis media**).
- The lower respiratory tract is subject to foreign bodies that can block the trachea and bronchi. In addition, infections can cause **acute bronchitis, chronic bronchitis, pneumonia,** and **pulmonary tuberculosis.** Other common disorders include **asthma, emphysema, cystic fibrosis** (CF), **pulmonary fibrosis,** and **lung cancer.** Smoking is a major risk factor associated with many of these.

ASSESS

Testing Yourself

Choose the best answer for each question.

1. Label this diagram of the human respiratory system.

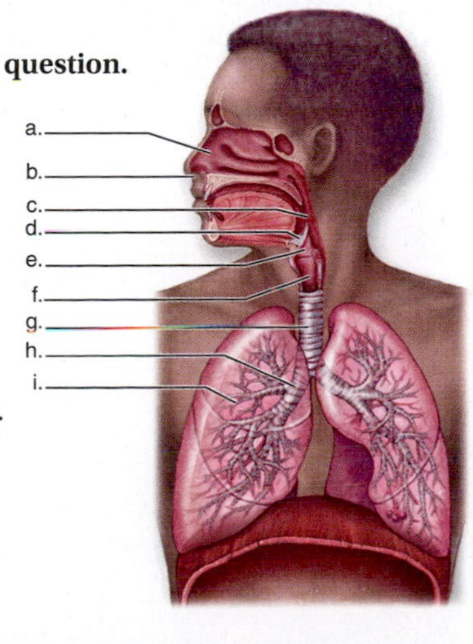

a.
b.
c.
d.
e.
f.
g.
h.
i.

2. The pulmonary _____ are rich in oxygen.
   - **a.** veins
   - **b.** arteries
   - **c.** Both a and b are correct.
   - **d.** Neither a nor b is correct.

3. Food and air both travel through the
  a. lungs.
  b. pharynx.
  c. larynx.
  d. trachea.

4. Which of these statements is anatomically incorrect?
  a. The nose has two nasal cavities.
  b. The pharynx connects the nasal cavity and mouth to the larynx.
  c. The larynx contains the vocal cords.
  d. The trachea enters the lungs.
  e. The lungs contain many alveoli.

5. If the digestive and respiratory tracts were completely separate in humans, there would be no need for
  a. swallowing.
  b. a nose.
  c. an epiglottis.
  d. a diaphragm.
  e. All of these are correct.

6. Label the following diagram of the path of air.

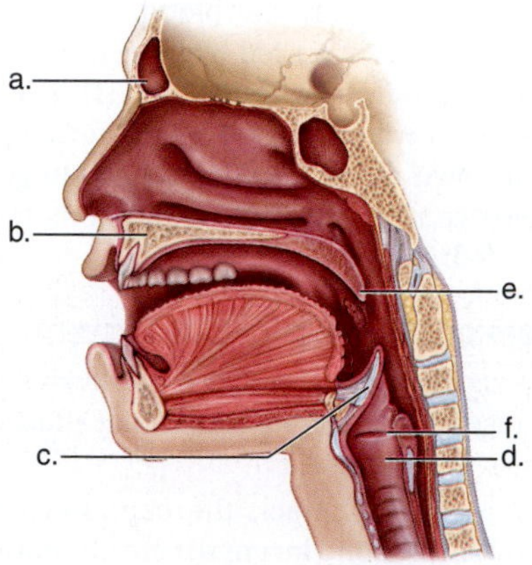
a.
b.
c.
d.
e.
f.

7. The amount of air moved into and out of the lungs during regular breathing is known as the
  a. vital capacity.
  b. tidal volume.
  c. residual volume.
  d. None of these are correct.

8. Which of these statements correctly goes with expiration rather than inspiration?
  a. Rib cage moves up and out.
  b. Diaphragm relaxes and moves up.
  c. Pressure in lungs decreases, and air comes rushing out.
  d. Diaphragm contracts and lowers.

9. The respiratory center is directly sensitive to
  a. carbon dioxide and hydrogen ions.
  b. oxygen and carbon dioxide.
  c. the amount of air in the alveoli.
  d. the breathing rate.

10. Carbon dioxide is carried in the plasma
  a. in combination with hemoglobin.
  b. as the bicarbonate ion.
  c. combined with carbonic anhydrase.
  d. only as a part of tissue fluid.
  e. All of these are correct.

11. The presence of carbon dioxide in the blood
  a. affects the respiratory volumes.
  b. makes blood more acidic.
  c. makes respiratory exchanges more difficult.
  d. can increase the breathing rate.
  e. Both b and d are correct.

12. Internal respiration refers to
  a. the exchange of gases between the air and the blood in the lungs.
  b. the movement of air into the lungs.
  c. the exchange of gases between the blood and tissue fluid.
  d. cellular respiration, resulting in the production of ATP.

13. Reduced hemoglobin is carrying
  a. oxygen.
  b. CO_2.
  c. hydrogen ions.
  d. sodium.

14. Smoking is associated with an increased risk of
  a. emphysema.
  b. several types of cancer.
  c. heart disease.
  d. All of these are correct.

ENGAGE

Thinking Critically

1. The respiratory system of birds is quite different from that of mammals. Unlike mammalian lungs, the lungs of birds are relatively rigid and do not expand much during ventilation. Most birds also have thin-walled air sacs that fill most of the body cavity not occupied by other organs. Inspired air passes through the bird's lungs, into these air sacs, and back through the lungs again on expiration. In what ways might this system be more efficient than the mammalian lung?
2. Carbon monoxide (CO) binds to hemoglobin 230 to 270 times more strongly than oxygen does, which means less O_2 is being delivered to tissues. What would be the specific cause of death if too much hemoglobin became bound to CO instead of to O_2?
3. As mentioned in the chapter, sometimes it is necessary to install a permanent tracheostomy—for example, in smokers who develop laryngeal cancer. Besides interfering with the ability to speak, what sorts of health problems would you expect to see in an individual with a permanent tracheostomy, and why?

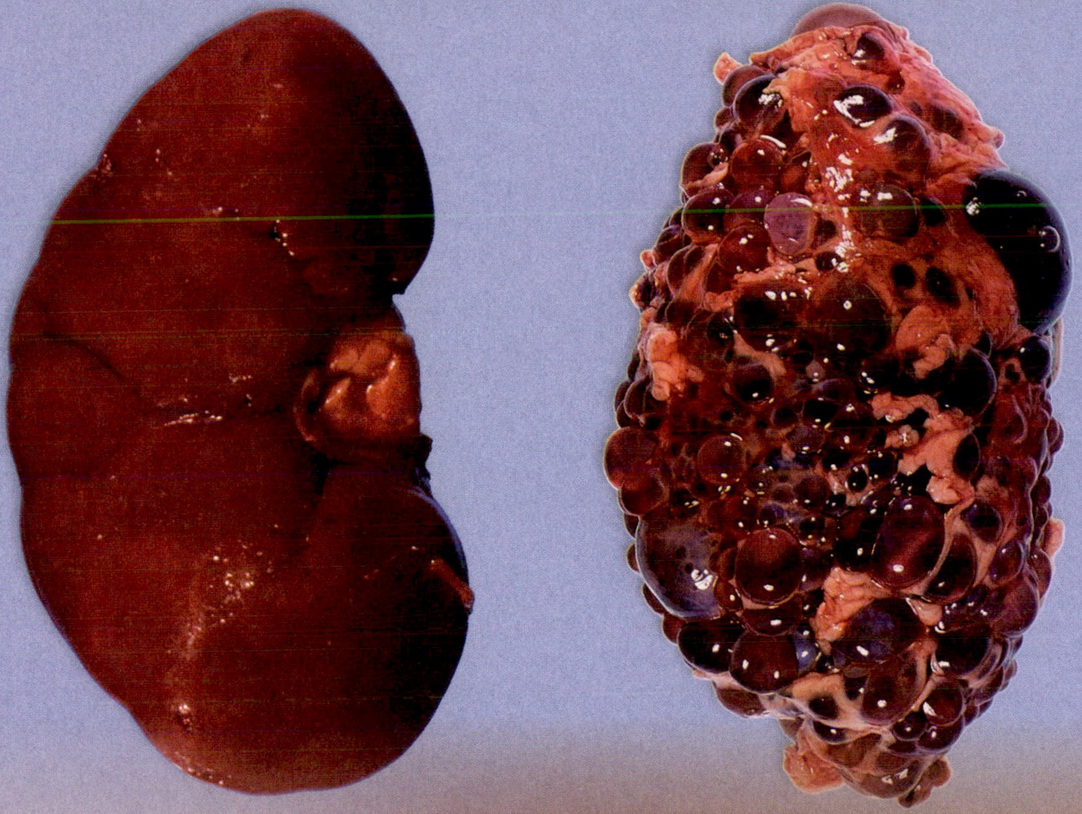

Urinary System and Excretion 16

CHAPTER OUTLINE

BEFORE YOU BEGIN

Before beginning this chapter, take a few moments to review the following discussions:

Section 2.3 What determines whether a solution is acidic or basic?

Figure 4.3 What types of molecules can diffuse across the plasma membrane?

Section 11.5 What are the major homeostatic functions of the urinary system?

CASE STUDY Born with Bad Kidneys

Michael and Jada were excited about the birth of their second baby. Married for three years, they already had a healthy daughter, and they were happy that Amber would have a younger sister to play with. After Aiesha was born, however, it soon became clear that something was wrong. She weighed only 5 lb, 4 oz at birth. The first time she urinated, there was an obvious tinge of blood in her urine. She also seemed to urinate much more frequently than normal. When her doctors performed ultrasound and magnetic resonance imaging (MRI) scans of her abdominal organs, they found that Aiesha had signs of polycystic kidney disease (PKD). Aiesha's doctors explained that in PKD, cysts (small, fluid-filled sacs) form within the collecting ducts of the nephrons in the interior of the kidneys. The ultrasound results indicated that both of Aiesha's kidneys were covered in cysts (as shown by the kidney on the right, above), which usually meant that the cysts were present inside the kidneys as well. Michael and Jada were informed that the presence of these cysts explained Aiesha's symptoms. The doctors also told them that PKD would most likely cause Aiesha's kidneys to fail and that she should immediately be placed on a wait list for a kidney transplant. Because PKD is a genetic disorder, the doctors suggested that both parents undergo genetic testing to see if they were carriers for PKD.

As you read through the chapter, think about the following questions:

1. What is the role of the kidneys in the body?
2. How would problems in the collecting ducts of the nephrons cause kidney failure?
3. Polycystic kidney disease seems to cause more serious problems in African Americans, especially those who have sickle-cell disease. Because sickle-cell is mainly a disease of the red blood cells, what does this have to do with the kidneys?

16.1 The Urinary System

Learning Outcomes

Upon completion of this section, you should be able to

1. Describe four major functions of the urinary system.
2. Trace the path of urine from its formation to its exit from the body.
3. Compare and contrast the structure and functions of the male versus female urethra.

The **urinary system** is involved in **excretion,** the removal of metabolic wastes from the body. People sometimes confuse the terms excretion and defecation, but they do not refer to the same process. Defecation, the elimination of feces from the body, is a function of the digestive system. The undigested food and bacteria that make up feces have never been a part of the functioning of the body, while most substances excreted in urine were once metabolites in the body.

Functions of the Urinary System

The urinary system produces **urine** and conducts it to outside the body. As the kidneys produce urine, they carry out the following four functions that contribute to homeostasis:

1. *Excretion of Metabolic Wastes* The kidneys excrete metabolic wastes, notably nitrogenous wastes. Urea is the primary nitrogenous end product of metabolism in human beings, but we also excrete some ammonium, creatinine, and uric acid.

   Urea is a by-product of amino acid metabolism. The breakdown of amino acids in the liver releases **ammonia,** which the liver rapidly combines with carbon dioxide to produce urea. Ammonia is very toxic to cells, but urea is much less toxic. Some ammonia (NH_3) is also excreted as ammonium ion (NH_4^+).

   Creatine phosphate is a high-energy phosphate reserve molecule in muscles. The metabolic breakdown of creatine phosphate results in *creatinine,* which is excreted.

   The breakdown of nucleotides, such as those containing adenine and thymine, produces **uric acid.** Uric acid is rather insoluble. If too much uric acid is present in blood, crystals may form and precipitate out in the joints, producing a painful ailment called **gout.**
2. *Osmoregulation* A principal function of the kidneys is **osmoregulation,** the maintenance of the appropriate balance of water and salt in the blood. Blood volume is intimately associated with the salt balance of the body. Salts, such as NaCl, have the ability to cause osmosis, the diffusion of water—in this case, into the blood. The more salts there are in the blood, the greater the blood volume and the greater the blood pressure. In this way, the kidneys are also involved in regulating blood pressure.

   The kidneys also help maintain the appropriate level of other ions, such as potassium (K^+), bicarbonate (HCO_3^-), and calcium (Ca^{2+}), in the blood.
3. *Regulation of Acid-Base Balance* Along with the respiratory system, the kidneys regulate the acid-base balance of the blood. The kidneys monitor and help keep the blood pH at about 7.4, mainly by excreting hydrogen ions (H^+) and reabsorbing the bicarbonate ions (HCO_3^-) as needed. Human urine usually has a pH of 6 or lower because our diet often contains acidic foods.
4. *Secretion of Hormones* The kidneys assist the endocrine system in hormone secretion. The kidneys release renin, an enzyme that leads to the secretion of the hormone aldosterone from the adrenal cortex, the outer portion of the adrenal glands, which lie atop the kidneys. As described in section 16.3, aldosterone promotes the reabsorption of sodium ions (Na^+) by the kidneys.

   Whenever the oxygen-carrying capacity of the blood is reduced, or oxygen demand increases, the kidneys secrete the hormone **erythropoietin (EPO)**, which stimulates red blood cell production.

   The kidneys also help activate vitamin D from the skin. Vitamin D is a hormonelike molecule that promotes calcium (Ca^{2+}) absorption from the digestive tract.

Organs of the Urinary System

The urinary system consists of the kidneys, ureters, urinary bladder, and urethra. Figure 16.1 shows these organs and also traces the path of urine.

Kidneys

The **kidneys** are paired, bean-shaped, reddish-brown organs located near the small of the back in the lumbar region on either side of the vertebral column. About the size of a fist, they lie in depressions dorsal to (behind) the peritoneum, where they receive some protection from the lower rib cage.

Each kidney is covered by a tough fibrous connective tissue layer called a *renal capsule.* The concave side of each kidney has a depression called the hilum where a *renal artery* enters and a *renal vein* and a ureter exit the kidney.

Ureters

The **ureters** conduct urine from the kidneys to the bladder. They are small, muscular tubes about 25 cm long and 5 mm in diameter. Each descends behind the peritoneum, from the hilum of a kidney, to enter the bladder at its dorsal surface.

The wall of each ureter has three layers: an inner mucosa, a smooth muscle layer, and an outer fibrous coat of connective tissue. Peristaltic contractions cause urine to enter the bladder in spurts, about one to five times per minute. Because a healthy adult human produces 1–2 l of urine a day, an average of 40–80 milliliters (ml) per hour enters the bladder.

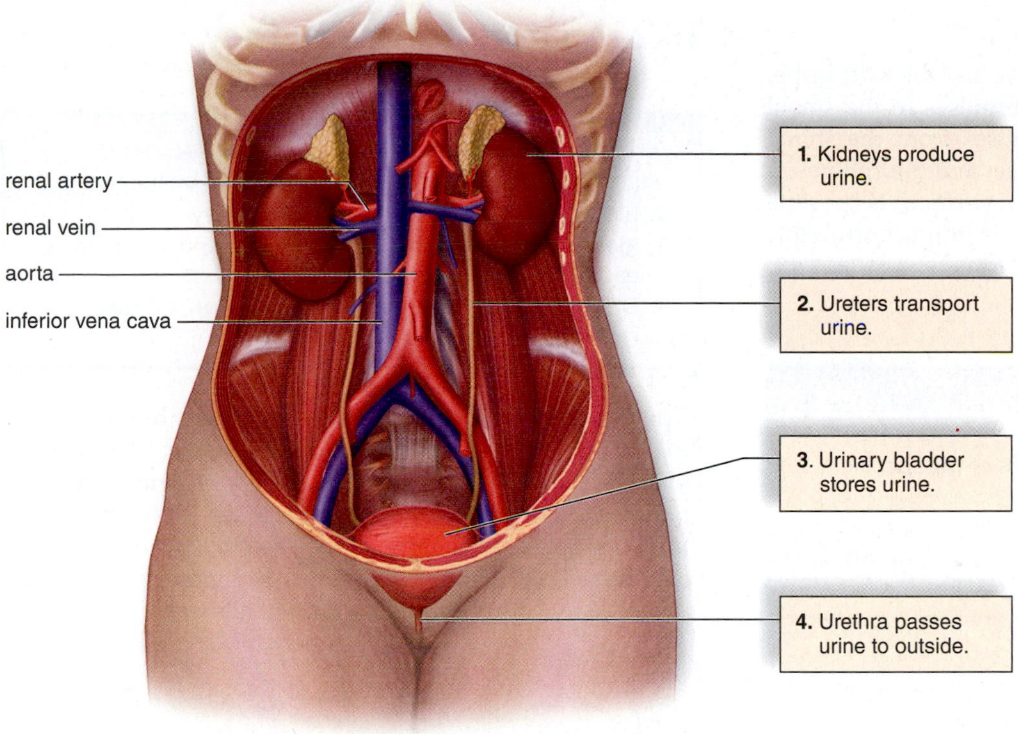
renal artery
renal vein
aorta
inferior vena cava
1. Kidneys produce urine.
2. Ureters transport urine.
3. Urinary bladder stores urine.
4. Urethra passes urine to outside.

Figure 16.1 The urinary system. Urine is found only within the kidneys, the ureters, the urinary bladder, and the urethra. The kidneys are important organs of homeostasis because they excrete metabolic wastes and adjust both the water-salt and acid-base balance of the blood.

Urinary Bladder

The **urinary bladder** stores urine until it is expelled from the body. The bladder is located in the pelvic cavity, behind the pubic symphysis and behind the peritoneum. The urinary bladder has three openings (orifices)—two for the ureters and one for the urethra, which drains the bladder (Fig. 16.2).

The bladder wall is expandable because it contains a middle layer of circular fiber and two layers of longitudinal muscle. The transitional epithelium of the mucosa becomes thinner, and folds in the mucosa called *rugae* disappear as the bladder enlarges.

After urine enters the bladder from a ureter, small folds of bladder mucosa act like a valve to prevent backward flow. Two sphincters lie in close proximity where the urethra exits the bladder. The internal sphincter occurs around the opening to the urethra. An external sphincter is composed of skeletal muscle that can be voluntarily controlled. Many people have some degree of *incontinence,* the involuntary loss of urine. At least one in ten people over age 65 have bladder control problems due to an age-related loss of muscle tone in the urinary bladder. Other causes of incontinence include enlarged prostate in men, pregnancy, and nervous system diseases that affect the control of urination.

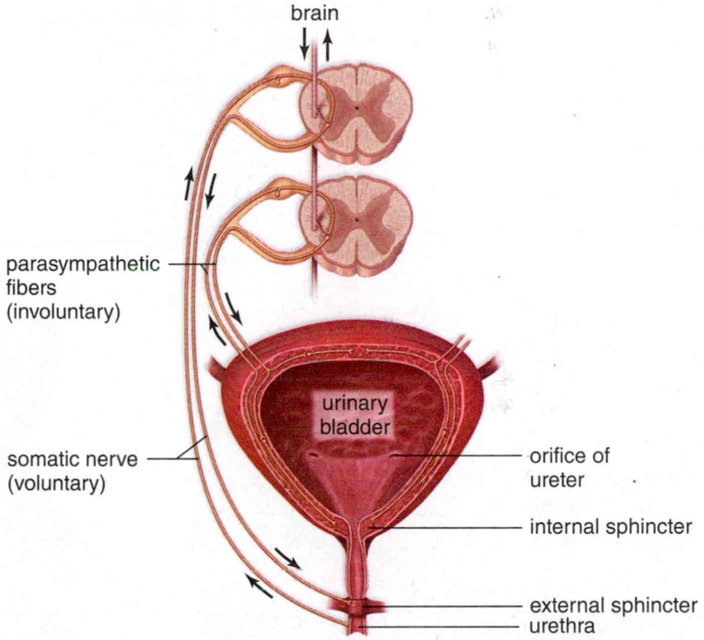
brain
parasympathetic fibers (involuntary)
somatic nerve (voluntary)
urinary bladder
orifice of ureter
internal sphincter
external sphincter
urethra

Figure 16.2 Urination. As the bladder fills with urine, sensory impulses go to the spinal cord and then to the brain. When urination occurs, motor nerve impulses cause the bladder to contract and an internal sphincter to relax. Motor nerve impulses also cause an external sphincter to relax.

Urethra

The **urethra** is a small tube that extends from the urinary bladder to an external opening. Therefore, its function is to remove urine from the body. In females, the urethra is only about 4 cm long, whereas in males, the urethra averages 20 cm when the penis is flaccid (nonerect). In females, the reproductive and urinary systems are not connected. In males, the urethra carries urine during urination and semen during ejaculation. As the urethra leaves the male urinary bladder, it is encircled by the prostate gland. In men over age 40, the prostate gland is subject to enlargement, as well as to prostate cancer. Either of these conditions can restrict urination, and may require treatment (see Fig. 21.18).

Urination

When the urinary bladder fills to about 250 ml with urine, stretch receptors send sensory nerve impulses to the spinal cord. Subsequently, motor nerve impulses from the spinal cord cause the urinary bladder to contract and the sphincters to relax so that urination, also called *micturition,* is possible (Fig. 16.2). In older children and adults, the brain controls this reflex, delaying urination until a suitable time.

Check Your Progress 16.1

1. Define excretion.
2. List and briefly describe the functions of the specific components of the urinary system.
3. Of the four major functions of the urinary system, which is/are exclusively performed by the kidneys, and which is/are shared with other body systems?

16.2 Anatomy of the Kidney and Excretion

Learning Outcomes

Upon completion of this section, you should be able to

1. Identify the structures of a human kidney.
2. Identify the parts of a nephron and state the function of each.
3. Distinguish between glomerular filtration, tubular reabsorption, and tubular secretion.

A lengthwise section of a kidney shows that many branches of the renal artery and renal vein reach inside a kidney (Fig. 16.3*a*). A kidney has three regions (Fig. 16.3*c*). The *renal cortex* is an outer, granulated layer that dips down in between a radially striated inner layer called the renal medulla. The *renal medulla* contains cone-shaped tissue masses called renal pyramids. The *renal pelvis* is a central space, or cavity, that is continuous with the ureter.

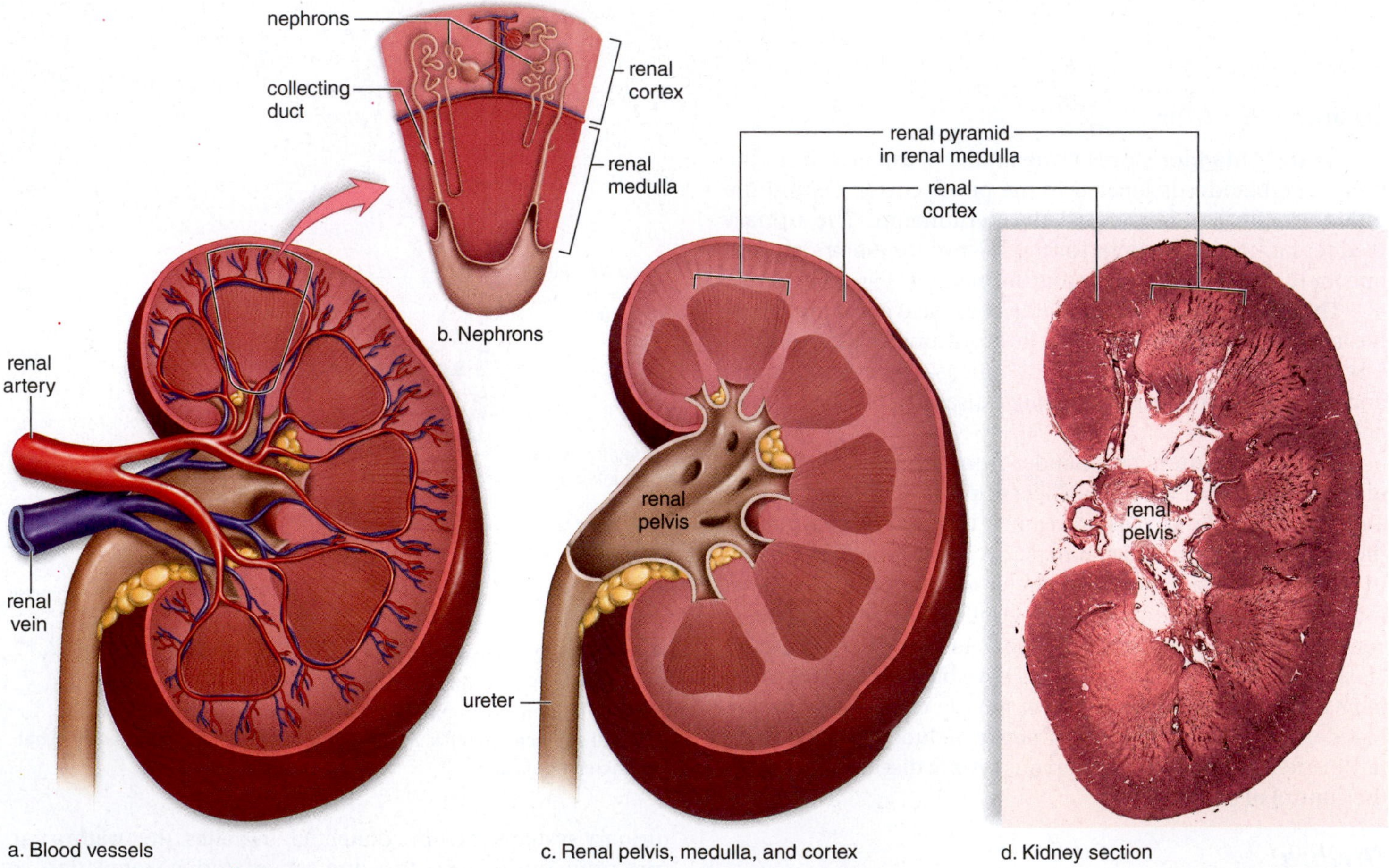
nephrons
collecting duct
renal cortex
renal medulla
b. Nephrons
renal artery
renal vein
a. Blood vessels
renal pyramid in renal medulla
renal cortex
renal pelvis
ureter
c. Renal pelvis, medulla, and cortex
renal pelvis
d. Kidney section

Figure 16.3 Anatomy of the kidney. **a.** A lengthwise section of the kidney showing the blood supply. Note that the renal artery divides into smaller arteries, and these divide into arterioles. Venules join to form small veins, which form the renal vein. **b.** An enlargement showing the placement of nephrons. **c.** The same section as **(a)** without the blood supply. Now it is easier to distinguish the renal pelvis , medulla, and cortex, which connects with the ureter. **d.** A lengthwise section of an actual kidney.

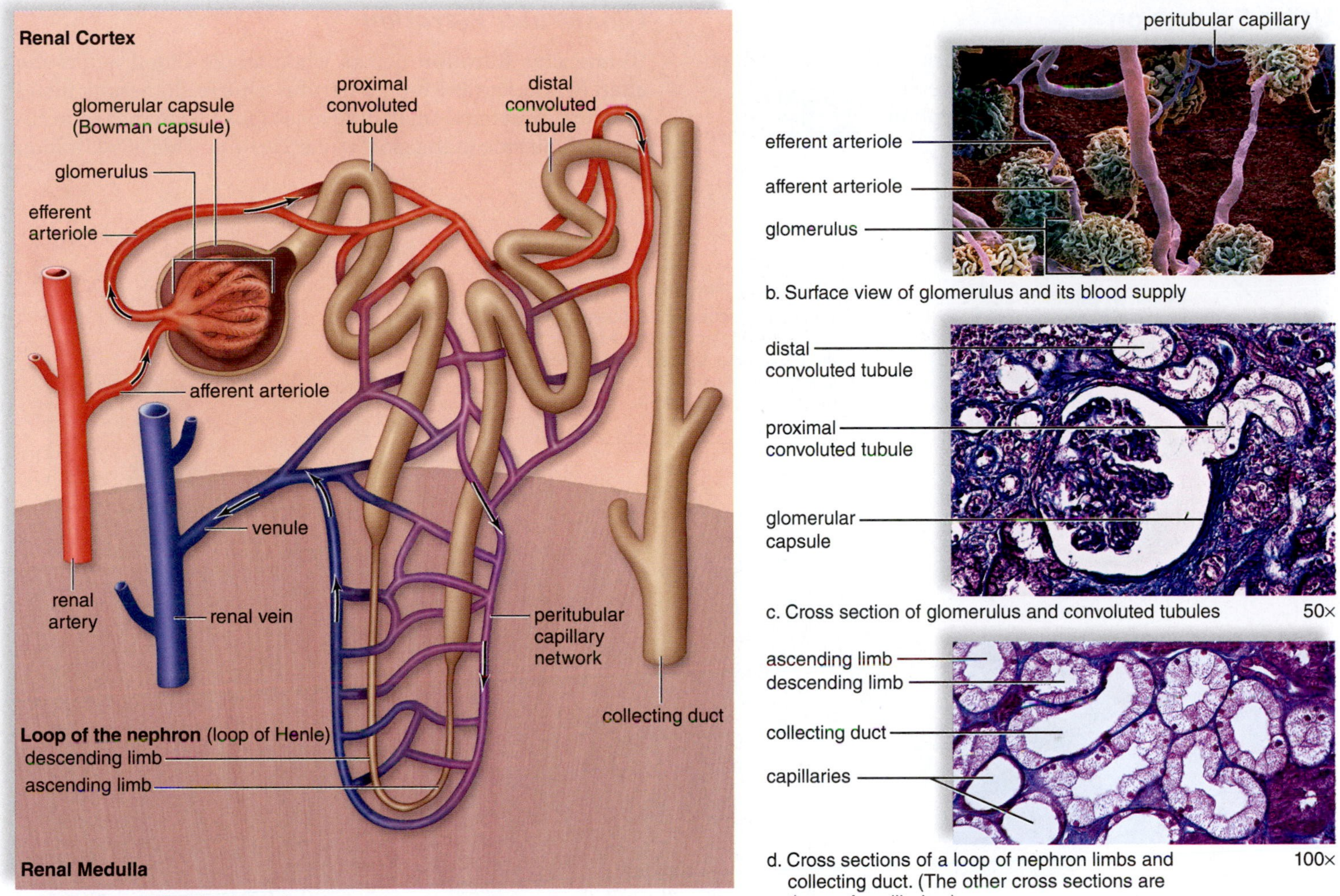
Renal Cortex
glomerular capsule (Bowman capsule)
proximal convoluted tubule
distal convoluted tubule
glomerulus
efferent arteriole
afferent arteriole
venule
renal artery
renal vein
peritubular capillary network
collecting duct
Loop of the nephron (loop of Henle)
descending limb
ascending limb
Renal Medulla
peritubular capillary
efferent arteriole
afferent arteriole
glomerulus
distal convoluted tubule
proximal convoluted tubule
glomerular capsule
50×
ascending limb
descending limb
collecting duct
capillaries
100×

a. A nephron and its blood supply

b. Surface view of glomerulus and its blood supply

c. Cross section of glomerulus and convoluted tubules

d. Cross sections of a loop of nephron limbs and collecting duct. (The other cross sections are those of capillaries.)

Figure 16.4 Nephron anatomy. **a.** You can trace the path of blood through a nephron by following the black arrows. **b, c, d.** A nephron is made up of a glomerular capsule, the proximal convoluted tubule, the loop of the nephron, the distal convoluted tubule, and the collecting duct.

Anatomy of a Nephron

Under higher magnification, the kidney is composed of over 1 million **nephrons,** sometimes called renal or kidney tubules (Fig. 16.3*b*). Each nephron has its own blood supply, including two capillary regions (Fig. 16.4). From the renal artery, an afferent arteriole leads to the **glomerulus,** a knot of capillaries inside the glomerular capsule. Blood leaving the glomerulus enters the efferent arteriole (the term afferent means going toward, while efferent means going away). The efferent arteriole takes blood to the *peritubular capillary network,* which surrounds the rest of the nephron. From there, the blood goes into a venule that joins the renal vein.

Parts of a Nephron

Each nephron is made up of several parts. First, the closed end of the nephron is pushed in on itself to form a cuplike structure called the **glomerular capsule** (Bowman capsule). The inner layer of the glomerular capsule is composed of *podocytes* (see Fig. 16.6) that have long cytoplasmic extensions. The podocytes cling to the capillary walls of the glomerulus and leave pores that allow easy passage of small molecules from the glomerulus to the inside of the glomerular capsule.

The glomerular capsule connects to a **proximal convoluted tubule** (PCT). The cuboidal epithelial cells lining this part of the nephron have numerous tightly packed microvilli that form a brush border, increasing the surface area for reabsorption. The tube then narrows and makes a U-turn called the **loop of the nephron** (loop of Henle), which is lined with simple squamous epithelium.

Following the loop, the tube becomes the **distal convoluted tubule** (DCT), which is composed of cuboidal epithelial cells that have numerous mitochondria but lack microvilli. This means the DCT is not specialized for reabsorption and is more likely to help move molecules from the blood into the tubule, a process called tubular secretion. The distal convoluted tubules of several nephrons enter one collecting duct. Many **collecting ducts** carry urine to the renal pelvis.

As shown in Figures 16.3*b* and 16.4*a,* the glomerular capsule and the convoluted tubules always lie within the renal cortex. The loop of the nephron dips down into the renal medulla; a few nephrons have a very long loop of the nephron, which penetrates deep into the renal medulla. Collecting ducts are also located in the renal medulla, and they help give the renal pyramids their lined appearance (see Fig. 16.3*d*).

Urine Formation

Figure 16.5 gives an overview of urine formation, which is divided into the following processes: glomerular filtration, tubular reabsorption, and tubular secretion.

MP3 Overview of Urine Formation

Glomerular Filtration

Glomerular filtration occurs when whole blood enters the afferent arteriole and the glomerulus. Due to glomerular blood pressure, water and small molecules move from the glomerulus to the inside of the glomerular capsule. This is a filtration

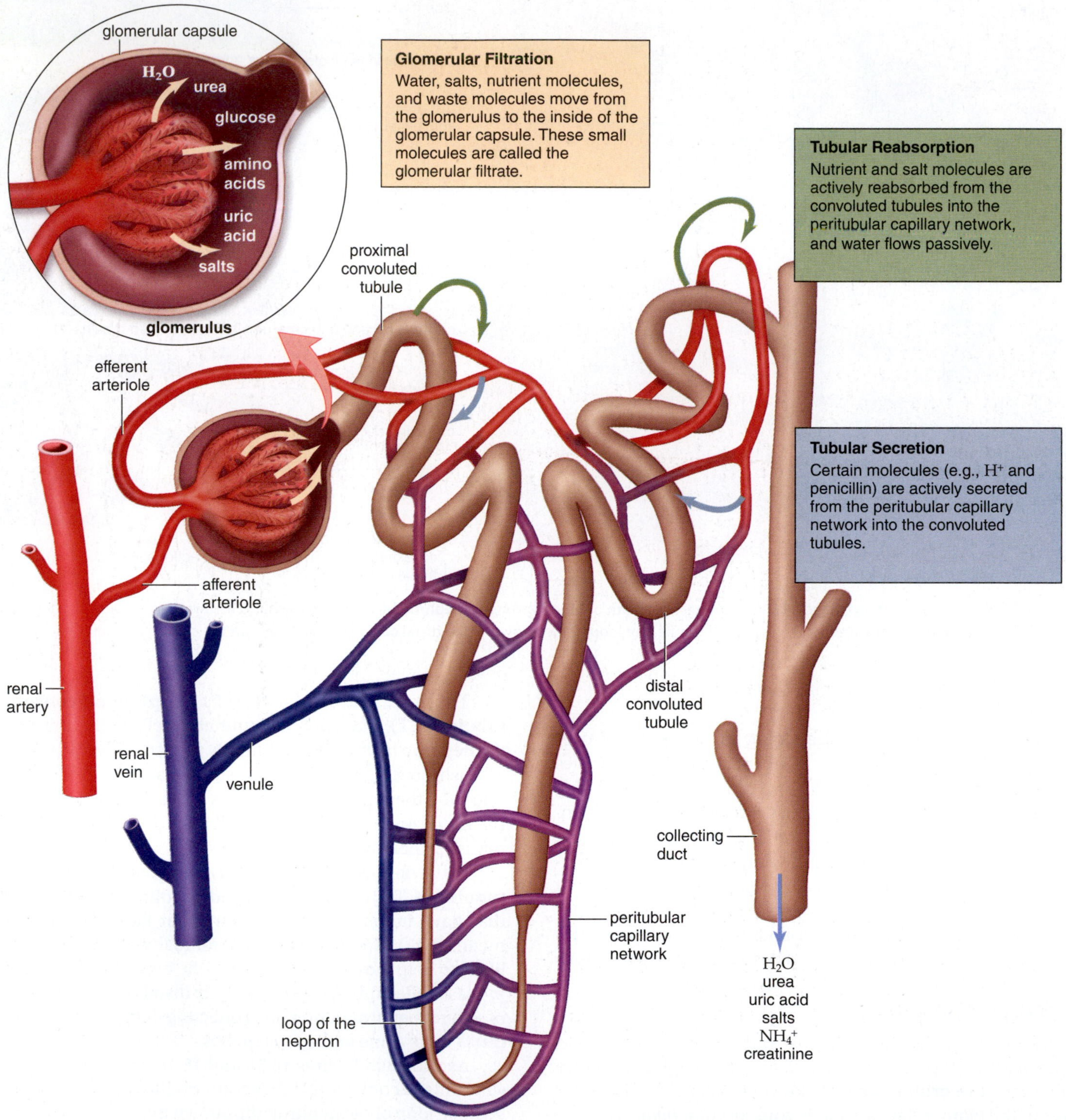

glomerular capsule
H_2O
urea
glucose
amino acids
uric acid
salts
glomerulus
Glomerular Filtration
Water, salts, nutrient molecules, and waste molecules move from the glomerulus to the inside of the glomerular capsule. These small molecules are called the glomerular filtrate.
Tubular Reabsorption
Nutrient and salt molecules are actively reabsorbed from the convoluted tubules into the peritubular capillary network, and water flows passively.
Tubular Secretion
Certain molecules (e.g., H^+ and penicillin) are actively secreted from the peritubular capillary network into the convoluted tubules.
proximal convoluted tubule
efferent arteriole
afferent arteriole
renal artery
renal vein
venule
distal convoluted tubule
collecting duct
peritubular capillary network
loop of the nephron
H_2O
urea
uric acid
salts
NH_4^+
creatinine

Figure 16.5 Processes in urine formation. The three main processes in urine formation are described in boxes and color-coded to arrows that show the movement of molecules into or out of the nephron at specific locations. In the end, urine is composed of the substances within the collecting duct (see blue arrow).

process because large molecules and formed elements are unable to pass through the capillary wall. In effect, then, blood in the glomerulus has two portions—the filterable components and the nonfilterable components:

Filterable Blood Components	Nonfilterable Blood Components
Water	Formed elements (blood cells and platelets)
Nitrogenous wastes	Plasma proteins
Nutrients	
Salts (ions)	

The *glomerular filtrate* contains small dissolved molecules in approximately the same concentration as plasma. Small molecules that escape being filtered and the nonfilterable components leave the glomerulus by way of the efferent arteriole.

As indicated in Table 16.1, nephrons in the kidneys filter 180 liters of water per day, along with a considerable amount of small molecules (such as glucose) and ions (such as sodium). If the composition of urine were the same as that of the glomerular filtrate, the body would continually lose water, salts, and nutrients. Therefore, we can conclude that the composition of the filtrate must be altered as this fluid passes through the remainder of the tubule.

Tubular Reabsorption

Tubular reabsorption occurs as molecules and ions are both passively and actively reabsorbed from the nephron into the blood of the peritubular capillary network. The osmolarity of the blood is maintained by the presence of both plasma proteins and salt. When sodium ions (Na^+) are actively reabsorbed, chloride ions (Cl^-) follow passively. The reabsorption of salt (NaCl) increases the osmolarity of the blood compared to the filtrate, and therefore water moves passively from the tubule into the blood. About 65% of Na^+ is reabsorbed at the proximal convoluted tubule.

Nutrients, such as glucose and amino acids, also return to the blood almost exclusively at the proximal convoluted tubule. This is a selective process because only molecules recognized by carrier proteins are actively reabsorbed. Glucose is an example of a molecule that ordinarily is completely reabsorbed because there is a plentiful supply of carrier proteins for it. However, every substance has a maximum rate of transport, and after all its carriers are in use, any excess in the filtrate will appear in the urine. For example, as reabsorbed levels of glucose approach 1.8–2 mg/ml of plasma, the rest appears in the urine.

In diabetes mellitus, excess glucose appears in the blood, because the liver and muscles have failed to store glucose as glycogen. The kidneys cannot reabsorb all the glucose in the filtrate, and it appears in the urine. The presence of glucose in the filtrate increases its osmolarity, and therefore, less water is reabsorbed into the peritubular capillary network. The frequent urination and increased thirst experienced by untreated diabetics are due to the fact that less water is being reabsorbed.

We have seen that the filtrate that enters the proximal convoluted tubule is divided into two portions—components that are reabsorbed from the tubule into blood and components that are not reabsorbed and continue to pass through the nephron to be further processed into urine:

Reabsorbed Filtrate Components	Nonreabsorbed Filtrate Components
Most water	Some water
Nutrients	Most nitrogenous waste
Required salts (ions)	Excess salts (ions)

The substances that are not reabsorbed become the tubular fluid, which enters the loop of the nephron.

Tubular Secretion

Tubular secretion is a second way by which substances are removed from blood in the peritubular network and added to the tubular fluid. Hydrogen ions, potassium ions, creatinine, and drugs, such as penicillin, are some of the substances that are removed by active transport from the blood. In the end, urine contains (1) substances that have undergone glomerular filtration but have not been reabsorbed and (2) substances that have undergone tubular secretion.

Animation
Kidney Function

TABLE 16.1 Reabsorption from Nephrons

Substance	Amount Filtered (per day)	Amount Excreted (per day)	Reabsorption (%)
Water, l	180	1.8	99.0
Sodium, g	630	3.2	99.5
Glucose, g	180	0.0	100.0
Urea, g	54	30.0	44.0

l = liters, g = grams

Check Your Progress 16.2

1. Identify the parts of a nephron that are found in the renal cortex and the renal medulla.
2. Explain the difference in microscopic structure which suggests that the PCT, but not the DCT, is specialized for reabsorption.
3. Summarize how the kidney uses the processes of diffusion, passive transport, and active transport.
4. Explain where in the nephron, and by what process, glucose is normally returned from the glomerular filtrate to the blood.

16.3 Regulatory Functions of the Kidneys

Learning Outcomes

Upon completion of this section, you should be able to

1. Summarize how the kidneys maintain the water-salt balance in the body.
2. Explain the functions of ADH, ANH, and aldosterone in homeostasis.
3. Describe two ways that the kidneys regulate blood pH.

The kidneys maintain the water-salt balance of the blood within normal limits. In this way, they also maintain the blood volume and blood pressure. Most of the water and salt (NaCl) present in the filtrate is reabsorbed across the wall of the proximal convoluted tubule.

Osmoregulation

The excretion of a hypertonic urine (one that is more concentrated than blood) is dependent upon the reabsorption of water from the loop of the nephron and the collecting duct. We can think of reabsorption of water as requiring (1) reabsorption of salt and (2) establishment of a solute gradient dependent on salt and urea before (3) water is reabsorbed. During the process of reabsorption, water passes through water channels, called **aquaporins,** which are proteins embedded in the plasma membrane.

Reabsorption of Salt

The kidneys regulate the salt balance in blood by controlling the excretion and the reabsorption of various ions. Sodium (Na^+) is an important ion in plasma that must be regulated, but the kidneys also excrete or reabsorb other ions, such as potassium ions (K^+), bicarbonate ions (HCO_3^-), and magnesium ions (Mg^{2+}), as needed.

Usually, more than 99% of sodium (Na^+) filtered at the glomerulus is returned to the blood. Most of this sodium (67%) is reabsorbed at the proximal tubule, and a sizable amount (25%) is reabsorbed by the ascending limb of the loop of the nephron. The rest is reabsorbed from the distal convoluted tubule and collecting duct.

Hormones regulate the reabsorption of sodium at the distal convoluted tubule. **Aldosterone,** a hormone secreted by the adrenal cortex, promotes the excretion of potassium ions (K^+) and the reabsorption of sodium ions (Na^+). The release of aldosterone is set in motion by the kidneys themselves. The *juxtaglomerular apparatus* is a region of contact between the afferent arteriole and the distal convoluted tubule (Fig. 16.6). When blood volume, and therefore blood pressure, is not sufficient to promote glomerular filtration, the juxtaglomerular apparatus secretes renin. **Renin** is an enzyme that changes angiotensinogen (a large plasma protein produced by the liver) into angiotensin I. Later, angiotensin I is converted to angiotensin II, a powerful vasoconstrictor that also stimulates the adrenal cortex to release aldosterone. The reabsorption of sodium

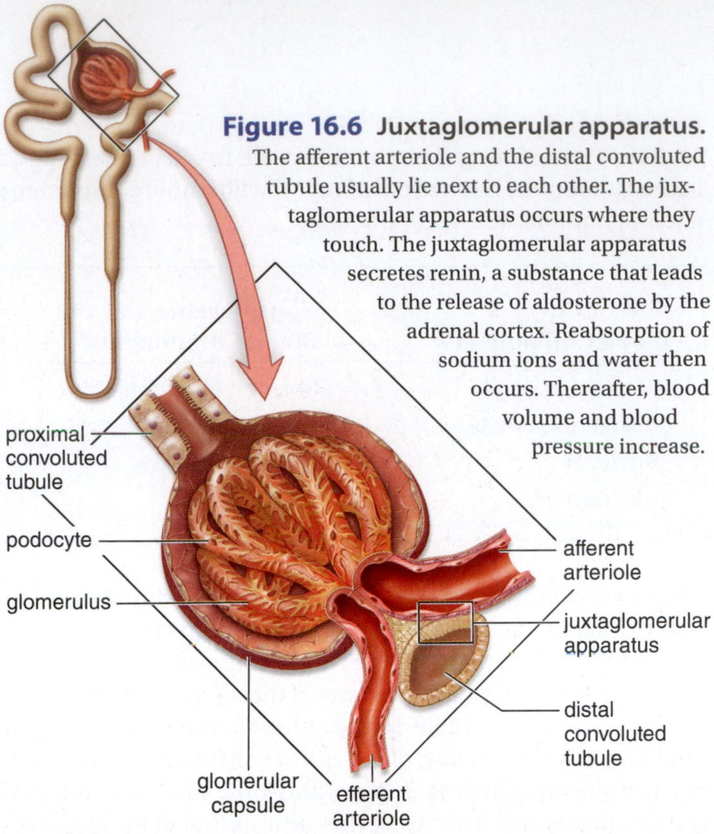
proximal convoluted tubule
podocyte
glomerulus
glomerular capsule
efferent arteriole
afferent arteriole
juxtaglomerular apparatus
distal convoluted tubule

Figure 16.6 Juxtaglomerular apparatus. The afferent arteriole and the distal convoluted tubule usually lie next to each other. The juxtaglomerular apparatus occurs where they touch. The juxtaglomerular apparatus secretes renin, a substance that leads to the release of aldosterone by the adrenal cortex. Reabsorption of sodium ions and water then occurs. Thereafter, blood volume and blood pressure increase.

ions is followed by the reabsorption of water. Therefore, blood volume and blood pressure increase.

Atrial natriuretic hormone (**ANH**) is a hormone secreted by the atria of the heart when cardiac cells are stretched due to increased blood volume. ANH inhibits the secretion of renin by the juxtaglomerular apparatus and the secretion of aldosterone by the adrenal cortex. Its effect, therefore, is to promote the excretion of Na^+, called natriuresis.

Establishment of a Solute Gradient

The long loop of a nephron, which typically penetrates deep into the renal medulla, is made up of a descending limb and an ascending limb. Salt (NaCl) passively diffuses out of the lower portion of the ascending limb, but the upper, thick portion of the limb actively extrudes salt out into the tissue of the outer renal medulla (Fig. 16.7). Less and less salt is available for transport as fluid moves up the thick portion of the ascending limb. Because of these circumstances, there is an osmotic gradient within the tissues of the renal medulla: the concentration of salt is greater in the direction of the inner medulla. (Note that water cannot leave the ascending limb because the limb is impermeable to water.)

The large arrow at the far left in Figure 16.7 indicates that the innermost portion of the inner medulla has the highest concentration of solutes. This cannot be due to salt because active transport of salt does not start until fluid reaches the thick portion of the ascending limb. Urea is believed to leak from the lower portion of the collecting duct, and it is this molecule that contributes to the high solute concentration of the inner medulla.

Reabsorption of Water

Because of the osmotic gradient within the renal medulla, water leaves the descending limb along its entire length. This is a countercurrent mechanism: as water diffuses out of the descending

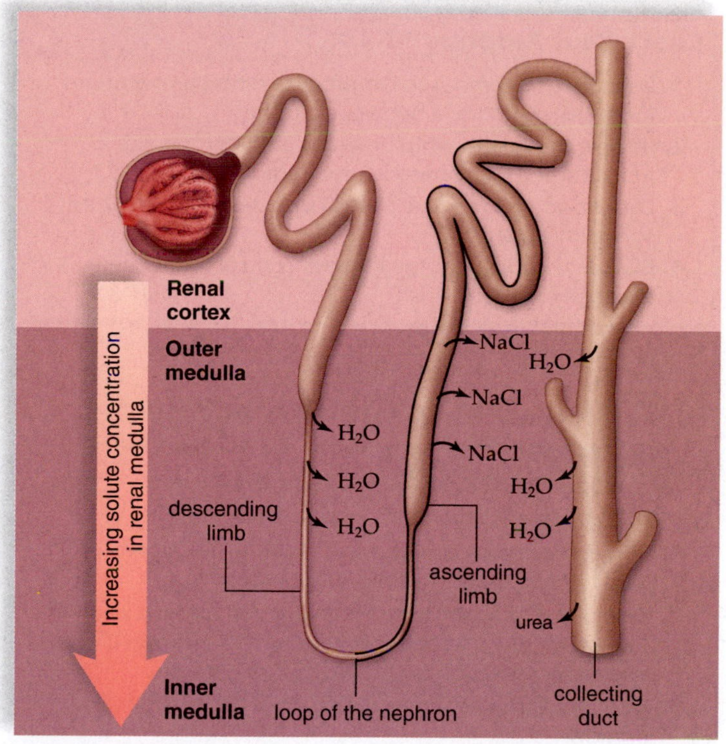
Renal cortex
Outer medulla
Inner medulla
Increasing solute concentration in renal medulla
NaCl
NaCl
NaCl
H2O
H2O
H2O
H2O
H2O
H2O
urea
descending limb
ascending limb
loop of the nephron
collecting duct

Figure 16.7 Reabsorption of water at the loop of the nephron and the collecting duct. Salt (NaCl) diffuses and is actively transported out of the ascending limb of the loop of the nephron into the renal medulla. Also, urea is believed to leak from the collecting duct and to enter the tissues of the renal medulla. This creates a hypertonic environment, which draws water out of the descending limb and the collecting duct. This water is returned to the cardiovascular system. (The thick black outline of the ascending limb means that it is impermeable to water.)

limb, the remaining fluid within the limb encounters an even greater osmotic concentration of solute. Therefore, water continues to leave the descending limb from the top to the bottom.

Fluid enters the collecting duct from the distal convoluted tubule. This fluid is hypotonic to the renal cortex, so to this point, the active transport of salt out of the ascending limb has decreased the osmolarity of the filtrate (see Fig. 16.7). This allows the production of urine that is hypotonic to the general body fluids—for example, when excess water needs to be excreted. When the urine needs to be hypertonic to body fluids (when the body is dehydrated, for example), the posterior pituitary gland releases **antidiuretic hormone** (**ADH**). In order to understand the action of this hormone, consider its name. *Diuresis* means an increased amount of urine, and *antidiuresis* means a decreased amount of urine. In the absence of ADH, the collecting duct is impermeable to water, and a dilute urine is produced. ADH increases the number of aquaporins (water channels) inserted in the cells of the collecting duct, so it becomes permeable to water. Because the filtrate within the collecting duct encounters the same osmotic gradient mentioned earlier, water can diffuse out of the collecting duct into the renal medulla. Usually, more ADH is produced at night, presumably so that we don't have to interrupt our sleep as often to urinate. This may also help to explain why the first urine of the day is more concentrated.

MP3 Water Conservation

Diuretics

Diuretics are chemicals that increase the flow of urine. Drinking alcohol causes diuresis because it inhibits the secretion of ADH. The dehydration that follows is believed to contribute to the symptoms of a hangover. Caffeine is a diuretic because it increases the glomerular filtration rate and decreases the tubular reabsorption of Na^+. Diuretic drugs developed to counteract high blood pressure inhibit active transport of Na^+ at the loop of the nephron or at the distal convoluted tubule. A decrease in water reabsorption and a decrease in blood volume follow.

Besides their medical uses, diuretics can also be abused. Bodybuilders and other athletes may take them for a more "cut" look, or to lose weight quickly, in the form of water. Also, because it may be more difficult to detect drugs in dilute urine, drug users who are attempting to pass a urine drug test may use furosemide (Lasix) or other diuretics. While this may be effective, it is also risky. Electrolyte imbalances resulting from inappropriate use of furosemide or other diuretic drugs can result in dehydration, irregular heartbeat, and even death.

Acid-Base Balance

The normal pH for body fluids is about 7.4. This is the pH at which our proteins, such as cellular enzymes, function best. If the blood pH rises above 7.4, a person is said to have *alkalosis,* and if the blood pH falls below 7.4, a person is said to have *acidosis.* Alkalosis and acidosis are abnormal conditions that may need medical attention.

The foods we eat add basic or acidic substances to the blood, and so does metabolism. For example, cellular respiration adds carbon dioxide that combines with water to form carbonic acid, and fermentation adds lactic acid. The pH of body fluids stays at just about 7.4 via several mechanisms, primarily acid-base buffer systems, the respiratory center, and the kidneys.

Acid-Base Buffer Systems

The pH of the blood stays near 7.4 because the blood is buffered. A *buffer* is a chemical or a combination of chemicals that can take up excess hydrogen ions (H^+) or excess hydroxide ions (OH^-). One of the most important buffers in the blood is a combination of carbonic acid (H_2CO_3) and bicarbonate ions (HCO_3^-). When hydrogen ions (H^+) are added to blood, the following reaction occurs:

$$H^+ + HCO_3^- \longrightarrow H_2CO_3$$

When hydroxide ions (OH^-) are added to blood, this reaction occurs:

$$OH^- + H_2CO_3 \longrightarrow HCO_3^- + H_2O$$

These reactions temporarily prevent any significant change in blood pH. A blood buffer, however, can be overwhelmed unless some more permanent adjustment is made. The next adjustment to keep the pH of the blood constant occurs at pulmonary capillaries.

Respiratory Center

As discussed in section 15.2, the respiratory center in the medulla oblongata increases the breathing rate if the hydrogen ion concentration of the blood rises. Increasing the breathing rate rids the body of hydrogen ions because the following reaction takes place in pulmonary capillaries:

$$H^+ + HCO_3^- \rightleftharpoons H_2CO_3 \rightleftharpoons H_2O + CO_2$$

In other words, when carbon dioxide is exhaled, this reaction shifts to the right, and the amount of hydrogen ions is reduced.

It is important to have the correct proportion of carbonic acid and bicarbonate ions in the blood. Breathing readjusts this proportion so that this particular acid-base buffer system can continue to absorb both H^+ and OH^- as needed.

The Kidneys

As powerful as the acid-base buffer and the respiratory center mechanisms are, only the kidneys can rid the body of a wide range of acidic and basic substances and otherwise adjust the pH. The kidneys are slower acting than the other two mechanisms, but they have a more powerful effect on pH. For the sake of simplicity, we can think of the kidneys as reabsorbing bicarbonate ions and excreting hydrogen ions as needed to maintain the normal pH of the blood (Fig. 16.8). If the blood is acidic, hydrogen ions are excreted, and bicarbonate ions are reabsorbed. If the blood is basic, hydrogen ions are not excreted, and bicarbonate ions are not reabsorbed. Because the urine is usually acidic, it follows that an excess of hydrogen ions is usually excreted. Ammonia (NH_3) provides another means of buffering and removing the hydrogen ions in urine: ($NH_3 + H^+ \longrightarrow NH_4^+$). Ammonia (whose presence is quite obvious in the diaper pail or kitty litter box) is produced in tubule cells by the deamination of amino acids. Phosphate provides another means of buffering hydrogen ions in urine.

The importance of the kidneys' ultimate control over the pH of the blood cannot be overemphasized. As mentioned, the enzymes of cells cannot continue to function if the internal environment does not have near-normal pH.

MP3 Acid-Base Balance

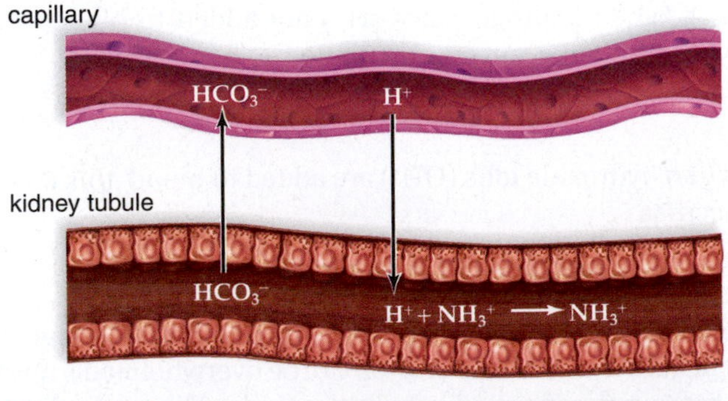
capillary
HCO3−
H+
kidney tubule
HCO3−
H+ + NH3+ ⟶ NH3+

Figure 16.8 Acid-base balance. In the kidneys, bicarbonate ions (HCO_3^-) are reabsorbed and hydrogen ions (H^+) are excreted as needed to maintain the pH of the blood. Excess hydrogen ions are buffered, for example, by ammonia (NH_3).

Check Your Progress 16.3

1. Describe the three steps or processes that are required for the excretion of a hypertonic urine.
2. Review the relationship between aldosterone, the juxtaglomerular apparatus, and renin.
3. Describe some medical uses of diuretics. Why are they sometimes abused?
4. Explain the ions that the kidneys control to maintain blood pH homeostasis.

16.4 Disorders of the Urinary System

Learning Outcomes

Upon completion of this section, you should be able to

1. Summarize how damage to the glomeruli/nephrons can lead to uremia and edema.
2. Describe one advantage and one disadvantage each of hemodialysis, continuous ambulatory peritoneal dialysis, and transplantation for kidney failure patients.
3. Describe three common disorders of the bladder and urethra.

Various disease conditions can affect the kidneys, ureters, urinary bladder, and urethra. The kidneys are essential for life, so if they are damaged sufficiently by disease, their function must somehow be restored if the patient is to survive.

Disorders of the Kidneys

Many major illnesses that affect other parts of the body, especially diabetes, hypertension, and certain autoimmune diseases, can also cause serious kidney disease. Most of these conditions tend to damage the glomeruli, resulting in a decreased glomerular filtration rate, and eventually kidney failure.

Infection of the kidneys is called **pyelonephritis.** Kidney infections usually result from infections of the urinary bladder that spread via the ureter(s) to the kidney(s). Kidney infections are usually curable with antibiotics if diagnosed in time. However, an infection called hemolytic uremic syndrome, caused by various bacteria but most famously by a strain of *Escherichia coli* known as O157:H7, can cause serious kidney damage in children even when antibiotics are administered.

Video *E. coli* Wars

Kidney stones are hard granules that can form in the renal pelvis. They may be composed of various materials, such as calcium, phosphate, uric acid, and protein. An estimated 5–15% of people will develop kidney stones at some time in their lives. Factors that contribute to the formation of these stones include too much animal protein in the diet, an imbalance in urinary pH, and/or urinary tract infection. Kidney stones may pass unnoticed in the urine flow. However, when a large kidney stone passes, strong contractions within a ureter can be excruciatingly painful. If a kidney stone grows large enough to block the renal pelvis or ureter, a reverse pressure builds up that may destroy the nephrons. To prevent this, it is sometimes necessary to break up a stone using a technique called lithotripsy, in which powerful ultrasound waves shatter the stone into smaller pieces that can be passed.

One of the first signs of kidney damage is the presence of albumin, white blood cells, and/or red blood cells in the urine. Once more than two-thirds of the nephrons have been destroyed by a disease process, urea and other waste products accumulate in the blood, a condition known as *uremia.* Although nitrogenous wastes can cause serious damage, the retention of water and salts by diseased kidneys is of even greater concern. The latter causes *edema,* the accumulation of fluid in body tissues. Imbalances in the ionic composition of body fluids can lead to loss of consciousness and heart failure.

Treatment Options for Kidney Failure

Patients whose kidneys are failing can undergo **hemodialysis,** continuous ambulatory peritoneal dialysis, or kidney transplantation. *Dialysis* is the diffusion of dissolved molecules through a semipermeable natural or synthetic membrane having pore sizes that allow only small molecules to pass through. In hemodialysis, the patient's blood is passed through an artificial kidney machine (Fig. 16.9). This machine contains a membranous tube, which is in contact with a dialysis solution, or *dialysate.* Substances more concentrated in the blood diffuse into the dialysate, and substances more common in the dialysate diffuse into the blood. The dialysate is continuously replaced to maintain favorable concentration gradients. In this way, the artificial kidney can be utilized either to extract substances from the blood, including waste products or toxic chemicals and drugs, or to add substances to the blood—for example, bicarbonate ions (HCO_3^-) if the blood is acidic. In the course of a three- to six-hour hemodialysis, 50–250 g of urea can be removed from a patient, which greatly exceeds the amount excreted by normal kidneys. Therefore, a patient needs to undergo treatment only about two or three times a week. Although about 98% of patients opt to be treated at dialysis centers, a growing number are purchasing portable units that can be used in their own homes.

Hemodialysis centers may not be available in rural areas, where continuous ambulatory peritoneal dialysis, or CAPD, is more commonly used. In CAPD, the peritoneum serves as the dialysis membrane. A fresh amount of dialysate is introduced directly into the abdominal cavity from a bag that is temporarily attached to a permanently implanted plastic tube. The dialysate flows into the peritoneal cavity by gravity. Waste and salt molecules pass from blood vessels in the abdominal wall into the dialysate before the fluid is collected several hours later. The solution is drained by gravity from the abdominal cavity into a bag, which is then discarded. One advantage of CAPD over an artificial kidney machine is that the individual can go about many of his or her normal activities during CAPD. A disadvantage, however, is an increased likelihood of infections.

Patients with renal failure may undergo a kidney transplant, which is the surgical replacement of a defective kidney with a functioning kidney from a donor. As with all organ transplants, organ rejection is a possibility (see the Scientific Inquiry feature, "Matching Organs for Transplantation," on page 306). Receiving a kidney from a close relative decreases the chances of rejection. The current one-year survival rate following a kidney transplant is 95–98%, but donor organs are in short supply. In the future, it may be possible to use kidneys from pigs or kidneys created in the laboratory.

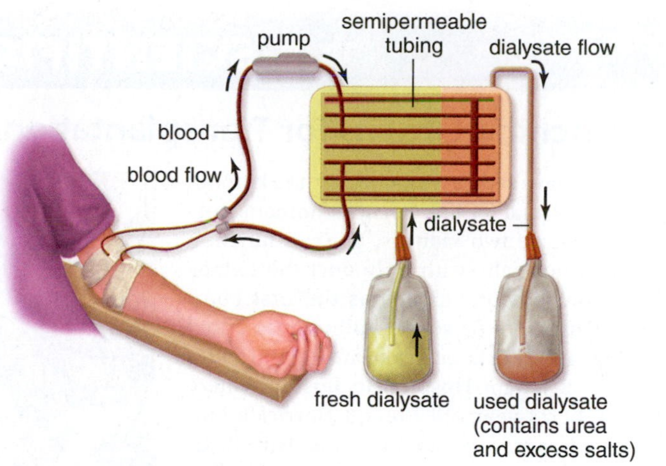
semipermeable tubing
pump
dialysate flow
blood
blood flow
dialysate
fresh dialysate
used dialysate (contains urea and excess salts)

Figure 16.9 An artificial kidney machine. As the patient's blood is pumped through dialysis tubing, the tubing is exposed to a dialysate (dialysis solution). Wastes exit the blood into the solution because of a preestablished concentration gradient. In this way, blood is not only cleansed, but its water-salt and acid-base balances are also adjusted.

Disorders of the Urinary Bladder and Urethra

Infections are probably the most common cause of problems in the urinary bladder and urethra. Urine leaving the kidneys is normally free of bacteria. The distal parts of the urethra, however, are normally colonized with bacteria. Most of the time, the flow of urine should be one-way from the bladder to the urethra. However, sometimes harmful bacteria from the urethra are able to gain access to the bladder, especially in females, whose urethra is shorter and broader than that of males. This helps explain why females are relatively more prone to bladder infections (Fig. 16.10).

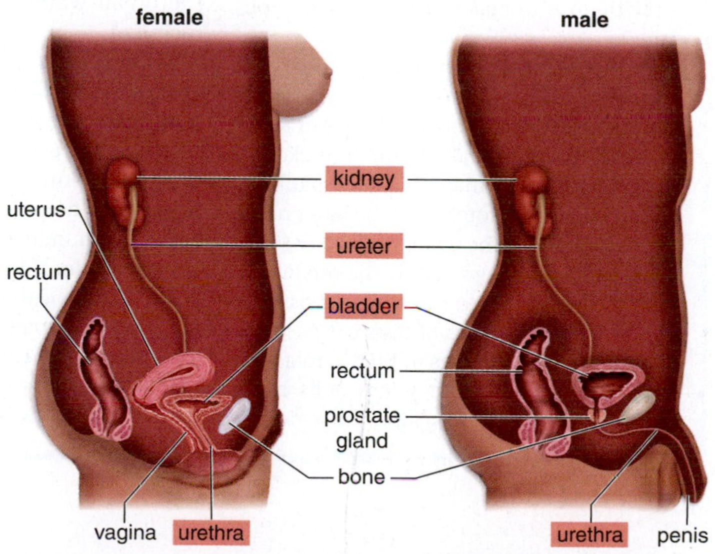
female
male
kidney
uterus
ureter
rectum
bladder
rectum
prostate gland
bone
vagina
urethra
urethra
penis

Figure 16.10 Female and male urinary tracts compared. Females have a short urinary tract compared to that of males. This makes it easier for bacteria to invade the urethra and helps explain why females are more likely than males to get a urinary tract infection.

SCIENCE IN YOUR LIFE ▶ SCIENTIFIC INQUIRY

Matching Organs for Transplantation

As we have seen in this chapter, the kidneys are essential for life. Because humans normally have two kidneys, but can function quite normally with only one, the kidney was an obvious choice as the first complex organ to be successfully transplanted (Fig. 16A). This occurred in 1954 at Peter Bent Brigham Hospital in Boston, where one of 23-year-old Ronald Herrick's kidneys was transplanted into his twin, Richard, who was dying of kidney disease. The operation was successful enough to allow Richard to live another eight years, while his brother lived with one kidney for over 50 years.

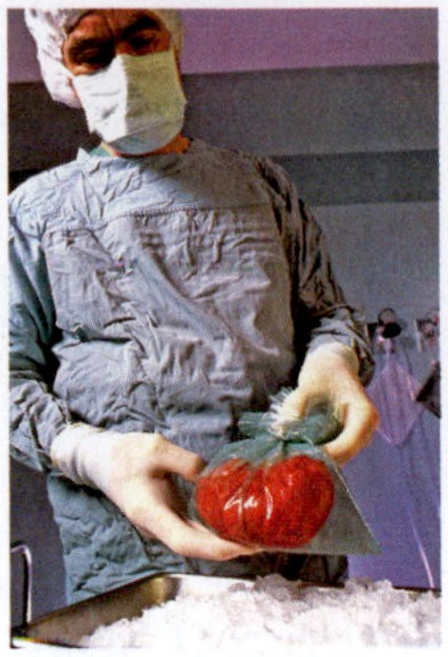

Figure 16A Preparing for transplantation. A donor kidney, kept moist and sterile inside a plastic bag, arrives in an operating room.

It was no accident, however, that the first successful transplant was performed between twins. Most cells and tissues of the body have molecules on their surface that can vary between individuals; that is, they are polymorphic. The major blood groups are a familiar example. Some individuals have only type A antigen on their red blood cells, some have only type B, some have both (blood type AB), and some have neither (blood type O). Because an individual naturally has antibodies against the A or B antigens their own cells lack, it is essential to make sure that donor blood is compatible before giving someone a blood transfusion. Moreover, because other cells besides red blood cells also have these antigens, the first requirement when attempting to match a potential organ donor with a recipient is to make sure the blood types match.

Another step in matching organs for transplantation involves determining whether the donor and the recipient share the same proteins, called human leukocyte antigens (HLAs) or major histocompatibility complex (MHC) proteins. As the latter name implies, these are the major molecules (antigens) that are recognized by the recipient's immune system when transplanted tissue is rejected. Various types of tests can be done to determine what types of MHC molecules the donor cells and recipient cells have, but the ideal situation would be a perfect MHC match between the donor organ and the recipient. This occurs 25% of the time between siblings who have the same mother and father, and also may occur by chance between unrelated individuals.

A final test that is usually performed prior to kidney transplantation is a cross-matching procedure in which the potential donor's cells are mixed with the blood (serum) of the recipient. This is to detect whether the recipient has any antibodies that could damage the cells of the donor kidney. Such antibodies could be present in patients who have been exposed to mismatched MHC molecules through a previous transplantation, blood transfusion, or even pregnancy. Even if the donor and recipient seem closely matched, if the recipient has preformed antibodies that could react with the donor organ, the organ would probably go to another recipient.

Because a perfect match (such as between two identical twins) is rare, transplantations would not have become as commonplace as they are today without the development of immunosuppressive drugs that help prevent transplant rejection. Unfortunately, at the present time all drugs used for this purpose cause a general suppression of the immune system, leaving transplant patients more susceptible to infections. Because transplant patients usually need to take these drugs for the rest of their lives, there is a need for new drugs that specifically target the cells responsible for the transplant rejection reaction, but leave the rest of the immune system intact.

In order to increase the odds of finding matching kidney donors, the National Kidney Registry has begun using a computerized matching system to find matches. Often, a patient who needs a transplant has one or more potential donors who are willing, but incompatible. The "donor chain" concept allows one of that patient's donors to provide a kidney to a compatible stranger, who in turn has potential donors who could donate to other patients. In an example from 2010, a father in New Jersey donated a kidney to a man in Los Angeles who was a part of a chain of five transplants that resulted in a successful kidney transplant for the New Jersey man's 22-month-old child.

Questions to Consider

1. As explained in section 13.5, why do you suppose all humans with blood type A, for example, have antibodies against the type B antigen? Where have we all been exposed?
2. Even if two individuals are tested and found to have a perfect match of all of their MHC-I and MHC-II molecules, a kidney transplanted between them still has a higher chance of failing compared to a transplant between identical twins. Can you explain why?
3. Besides the matching criteria described here, what other medical issues should be taken into consideration before a decision is made to transplant an organ into a recipient?

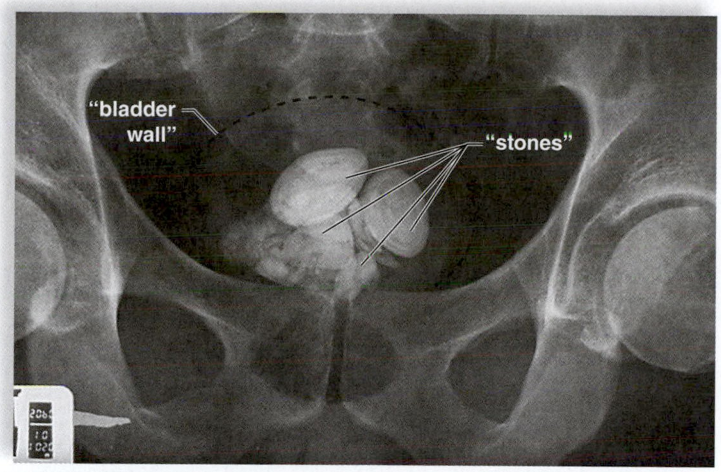
"bladder wall"
"stones"

Figure 16.11 Bladder stones. Hard granules such as those seen nearly filling the bladder in this X ray can form and lodge virtually anywhere in the urinary tract, including the kidneys, ureters, and urethra.

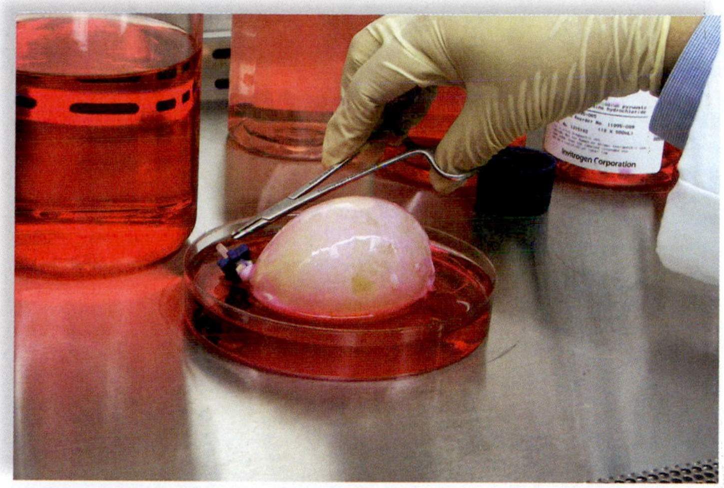

Figure 16.12 Organs grown in the lab. Researchers at Children's Hospital, Boston, have cultured bladder cells on biodegradable bladder-shaped molds to produce an organ that can be implanted in individuals with diseased or defective bladders.

In the United States, approximately 25–40% of women aged 20–40 years have had a urinary tract infection. Infections of the bladder usually result in inflammation of the bladder, or **cystitis,** and infection of the urethra may result in *urethritis.* Almost all urinary tract infections can be cured with appropriate antibiotics.

Bladder stones (Fig. 16.11) can form in people of any age. They most commonly occur as a result of another condition that interferes with normal urine flow, such as a bladder infection with associated inflammation or prostate enlargement in men. They can also form from kidney stones that pass into the urinary bladder and become larger. The most common symptoms of bladder stones are pain, difficulty urinating or increased frequency of urination, and blood in the urine. Similar to kidney stones, bladder stones may be removed surgically or, preferably, by lithotripsy.

The most common type of cancer affecting the urinary system is cancer of the bladder. In the United States, bladder cancer is the fourth most common type of cancer in men and the tenth most common in women. Because several toxic byproducts of tobacco are secreted in the urine, it is estimated that smoking quintuples the risk of bladder cancer. Certain types of bladder cancer are very malignant. In these cases, the bladder must be removed. With no bladder, the ureters must either be diverted into the large intestine or attached to an opening (or stoma) that is created in the skin near the navel. In the latter case, the patient typically must wear a removable plastic bag to collect the urine. Recently, however, researchers have had success in growing entire human bladders in the lab, and implanting them into a limited number of patients. Clinical trials of such replacement bladders are ongoing (Fig. 16.12).

Check Your Progress 16.4

1. Explain the specific mechanisms by which kidney failure may cause death.
2. Discuss how smoking may affect the urinary bladder.
3. Summarize the anatomical differences that make females more prone to cystitis. What unique structure may cause difficult urination in males?

Case Study Conclusion

Michael and Jada learned that there are two forms of polycystic kidney disease (PKD) and that both are genetic disorders. Over 90% of cases belong to a class called autosomal dominant PKD, named after the type of mutation that causes the disease. Individuals with this form of PKD typically develop symptoms in their 30s and 40s. The other, rarer form, is called autosomal recessive PKD, and it is caused by a defect in a gene called *PKHD1*. This gene contains the instructions for the manufacture of a protein called fibrocystin, an important molecule in ensuring that the kidneys function correctly. Unfortunately, genetic tests revealed that Michael and Jada were both carriers of a defective copy of *PKHD1*, and that Aiesha had inherited a defective copy from both parents. Without functioning fibrocystin protein, Aiesha's kidneys—and also possibly her liver—would eventually fail as the nephrons inside her kidneys struggled to perform their functions, and fluid accumulated inside the cysts. Besides dialysis, the only effective long-term treatment for Aiesha would be a kidney transplant. Fortunately, kidney transplants have a 90% success rate, so if a compatible donor could be found, Aiesha would have a good chance to have a normal, healthy childhood.

MEDIA STUDY TOOLS

www.mhhe.com/maderinquiry14

Enhance your study of this chapter with study tools and practice tests. Also ask your instructor about the resources available through ConnectPlus, including LearnSmart, the media-rich eBook, interactive learning tools, and animations.

Mc Graw Hill connect plus+ | BIOLOGY

SUMMARIZE

16.1 The Urinary System

The **urinary system** has several components, including the kidneys, ureters, urinary bladder, and urethra:

- The **kidneys** produce **urine.**
- The **ureters** take urine to the bladder.
- The **urinary bladder** stores urine.
- The **urethra** releases urine to the outside.

During urine formation, the kidney serves three homeostatic functions:

- **Excretion** of metabolic waste, especially nitrogenous waste from the breakdown of amino acids. This process produces **ammonia,** which is excreted mainly as **urea.** Nucleic acids are metabolized to **uric acid,** which may accumulate, causing the disease **gout.**
- **Osmoregulation,** the maintenance of normal water-salt balance.
- Helping to regulate the acid-base balance of the blood.

A fourth major function of the kidneys, which is not directly related to urine formation, is the production of the following hormones:

- **Erythropoietin (EPO),** which stimulates red blood cell production.
- Renin, an enzyme that leads to the secretion of aldosterone by the adrenal cortex.
- The kidneys also convert vitamin D to its biologically active form.

16.2 Anatomy of the Kidney and Excretion

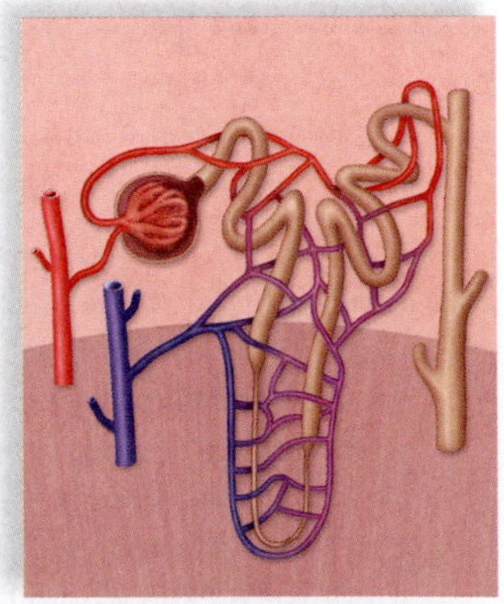

Macroscopically, the kidneys are divided into the renal cortex, renal medulla, and renal pelvis. Microscopically, they contain **nephrons.**

- Each nephron has its own blood supply. The afferent arteriole approaches the **glomerular capsule** and divides to become the **glomerulus,** a capillary tuft. The efferent arteriole leaves the capsule and immediately branches into the peritubular capillary network.
- In addition to the glomerular capsule, a nephron consists of the **proximal convoluted tubule,** the **loop of the nephron,** the **distal convoluted tubule,** and the **collecting duct.** The spaces between the podocytes of the glomerular capsule allow small molecules to enter the capsule from the glomerulus. The cuboidal epithelial cells of the proximal convoluted tubule have many mitochondria and microvilli to carry out active transport (and allow passive transport) from the tubule to the blood. The cuboidal epithelial cells of the distal convoluted tubule have numerous mitochondria but lack microvilli. They carry out active transport from the blood to the tubule.

Urine is composed primarily of nitrogenous waste products and salts in water. The following steps occur during urine formation:

- **Glomerular filtration,** during which water and other small molecules consisting of both nutrients and wastes move from the blood to the inside of the glomerular capsule.
- **Tubular reabsorption,** during which water and nutrients move back into the blood, mostly at the proximal convoluted tubule.
- **Tubular secretion,** where certain substances (e.g., hydrogen ions) are transported from the blood into the distal convoluted tubule.

16.3 Regulatory Functions of the Kidneys

Reabsorption of water and the production of a hypertonic urine involves three steps:

- The reabsorption of salt increases blood volume and pressure because more water is also reabsorbed. Two hormones control the kidneys' reabsorption of sodium (Na^+). Following the secretion of **renin** by the kidneys, **aldosterone,** a hormone secreted by the adrenal cortex, causes resorption of sodium ions, and excretion of potassium ions, by the distal convoluted tubules. **Atrial natriuretic hormone (ANH),** secreted by the heart, inhibits the secretion of renin by the kidney, and this inhibits aldosterone. Na^+ is actively transported out of the ascending limb of the loop of the nephron, establishing a solute gradient that increases toward the inner medulla.
- This gradient draws water from the descending limb of the loop of the nephron and also from the collecting duct. The permeability of the collecting duct is controlled by **antidiuretic hormone (ADH)** from the posterior pituitary gland, which increases the number of plasma membrane channels called **aquaporins** in the collecting ducts.

The kidneys also keep blood pH within normal limits:

- They reabsorb HCO_3^- and excrete H^+ as needed to maintain the pH at about 7.4. Ammonia also buffers H^+ in the urine.

16.4 Disorders of the Urinary System

These disorders can be divided into those that affect the kidney itself and those that affect the rest of the urinary tract:

- Disorders specifically affecting the kidneys include **pyelonephritis** (kidney infection) and **kidney stones.**
- Other major medical conditions, such as diabetes, hypertension, and autoimmune diseases, can lead to kidney failure, which necessitates undergoing **hemodialysis,** continuous ambulatory peritoneal dialysis, or kidney transplantation.
- The urinary bladder is susceptible to infections (called **cystitis**), **bladder stones,** and cancer.

ASSESS

Testing Yourself

Choose the best answer for each question.

1. _____ is the elimination of metabolic waste, and _____ is the elimination of undigested food.
   - **a.** Excretion, absorption
   - **b.** Defecation, excretion
   - **c.** Excretion, defecation
   - **d.** None of these are correct.
2. Urea is a by-product of _____ metabolism.
   a. glucose **b.** salt **c.** fat **d.** amino acid

3. Breakdown of _____ produces _____, which can cause _____.
 a. proteins, fats, gout
 b. nucleotides, ammonia, high blood pressure
 c. nucleotides, uric acid, gout
 d. fats, acids, diabetes

4. Kidneys control blood pH by excreting
 a. HCO_3^-.
 b. H^+.
 c. Both a and b are correct.
 d. Neither a nor b is correct.

5. The function of erythropoietin is
 a. reabsorption of sodium ions.
 b. excretion of potassium ions.
 c. reabsorption of water.
 d. stimulation of red blood cell production.
 e. to increase blood pressure.

6. Label the following diagram of the urinary system and nearby blood vessels.

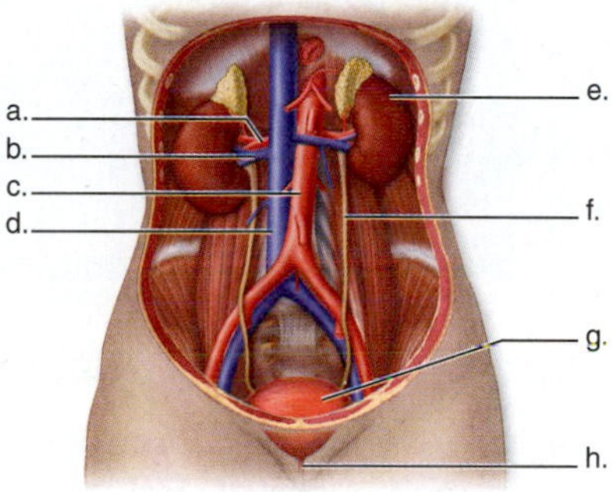
a.
b.
c.
d.
e.
f.
g.
h.

7. Label the following diagram of the kidneys.

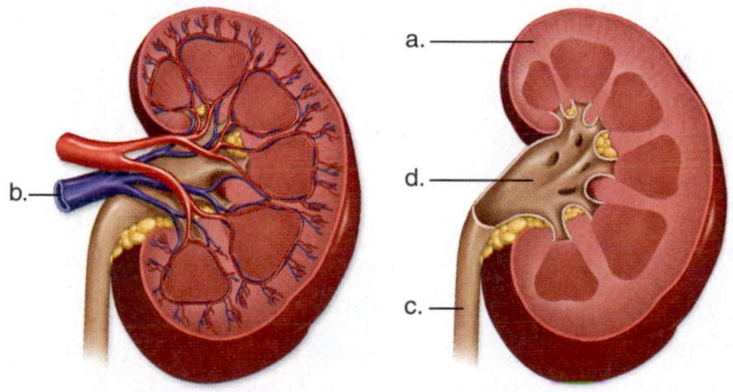
a.
b.
c.
d.

8. Which of the following is a difference between the urinary systems of males and females?
 a. Males have a longer urethra than females.
 b. In males, the urethra passes through the prostate.
 c. In males, the urethra serves both the urinary and reproductive systems.
 d. All of these are correct.

9. Put these parts of the nephron in correct order from start (nearest to glomerulus) to finish (nearest to the collecting duct).
 a. distal convoluted tubule
 b. proximal convoluted tubule
 c. glomerular capsule
 d. loop of the nephron tubule

10. Filtration is associated with the
 a. glomerular capsule.
 b. distal convoluted tubule.
 c. collecting duct.
 d. All of these are correct.

11. An _____ artery exits the glomerulus.
 a. afferent
 b. efferent
 c. Both a and b are correct.
 d. Neither a nor b is correct.

12. Which of the following materials would not be filtered from the blood at the glomerulus?
 a. water
 b. urea
 c. protein
 d. glucose
 e. sodium ions

13. _____ reabsorption from the nephrons is, ordinarily, 100%.
 a. Sodium
 b. Glucose
 c. Water
 d. Urea

14. The purpose of the loop of the nephron in the process of urine formation is
 a. reabsorption of water.
 b. production of filtrate.
 c. reabsorption of solutes.
 d. secretion of solutes.

15. By what transport process are most molecules secreted from the blood into the tubule?
 a. osmosis
 b. diffusion
 c. active transport
 d. facilitated diffusion

16. The cells of the distal convoluted tubule
 a. have many mitochondria.
 b. lack microvilli.
 c. can move molecules from the blood to the tubules.
 d. All of these are correct.

17. Excretion of a hypertonic urine in humans is most associated with the
 a. glomerular capsule and the tubules.
 b. proximal convoluted tubule only.
 c. loop of the nephron and collecting duct.
 d. distal convoluted tubule and peritubular capillary.

18. Which of these associations is mismatched?
 a. atrial natriuretic hormone (ANH)—secreted by the heart
 b. renin—secreted by the anterior pituitary
 c. aldosterone—secreted by the adrenal cortex
 d. antidiuretic hormone (ADH)—secreted by the posterior pituitary

19. Which hormone causes a rise in blood pressure?
 a. aldosterone
 b. oxytocin
 c. erythropoietin
 d. atrial natriuretic hormone

20. Urinary tract infections are usually caused by
 a. viruses.
 b. bacteria.
 c. fungi.
 d. None of these are correct.

ENGAGE

Virtual Lab
Virtual Frog Dissection

The virtual lab "Virtual Frog Dissection" provides an interactive opportunity to explore the urinary system of a vertebrate.

Thinking Critically

1. High blood pressure often is accompanied by kidney damage. In some people, the kidney damage is subsequent to the high blood pressure, but in others the kidney damage is what caused the high blood pressure. Explain how a low-salt diet would enable you to determine whether the high blood pressure or the kidney damage came first?
2. The renin-angiotensin-aldosterone system can be inhibited in order to reduce high blood pressure. Usually, the angiotensin-converting enzyme, which converts angiotensin I to angiotensin II, is inhibited by drug therapy. Why would this enzyme be an effective point to disrupt the system?

17 Nervous System

CHAPTER OUTLINE

BEFORE YOU BEGIN

Before beginning this chapter, take a few moments to review the following discussions:

Figure 4.9 How does the sodium-potassium pump move ions across the cell membrane?

Section 11.1 What are the major components of nervous tissue?

Section 11.5 The nervous system interacts with which other body systems?

CASE STUDY An Autoimmune Attack

As she sat in her car at the red light, Sarah noticed that the colors didn't seem quite right. "I must be stressed out," she thought to herself, but as she squinted at the traffic signal, the top light definitely seemed more orange than red. "Maybe there's something wrong with this light," she thought, but the next one was the same. At work later that day, she noticed that she was having trouble reading her e-mail, and by the end of the day she had a splitting headache. She tried telling herself that she had just been working too hard, but she had a feeling it might be more serious. Within a few weeks, she was almost completely blind in one eye, and the sensations in her feet felt muffled, like they were wrapped in gauze. Her doctors diagnosed multiple sclerosis (MS), and they were able to treat the inflammation in her optic nerves and spinal cord with high doses of immunosuppressive medications. Over the next few years, Sarah was able to keep her symptoms at bay by injecting herself daily with a drug called beta interferon, but she knew that someday she might need a wheelchair to get around.

MS is an inflammatory disease that affects the myelin sheaths, which wrap parts of some nerve cells like insulation around an electrical cord. As these sheaths deteriorate, the nerves no longer conduct impulses normally. For unknown reasons, MS often attacks the optic nerves first, before spreading to other areas of the brain. Most researchers think MS results from a misdirected attack on myelin by the body's immune system, although other factors may be involved. Like many diseases that affect the nervous system, there is no cure, so affected individuals must deal with their condition for the rest of their lives.

As you read through the chapter, think about the following questions:

1. What is the function of myelin? What specific type of neurological process will be affected if myelin is damaged?
2. MS affects the CNS only, not the PNS. What cells synthesize myelin in these two systems, and how might this help to explain why MS affects myelin in the CNS only?
3. Would MS cause more of a problem with the white matter or gray matter of the brain?

17.1 Nervous Tissue

Learning Outcomes

Upon completion of this section, you should be able to

1. Compare the location and functions of the central nervous system versus the peripheral nervous system with regard to location and function.
2. Describe the basic structure of a neuron, and compare the functions of the three classes of neurons.
3. Explain the structure, function, and locations of myelin.

Along with the endocrine system, the **nervous system** coordinates and regulates the functioning of the body's other systems. In most animals, including humans, the nervous system has two major anatomical divisions. The central nervous system (CNS) consists of the brain and spinal cord, which are located in the midline of the body. The peripheral nervous system (PNS) consists of nerves that carry sensory messages to the CNS and motor commands from the CNS to the muscles and glands (Fig. 17.1). Although the CNS and PNS are anatomically and functionally distinct, the two systems work together and are connected to one another.

The nervous system contains two types of cells: neurons and neuroglia (neuroglial cells). **Neurons** are the cells that transmit nerve impulses between parts of the nervous system. **Neuroglia,** which were described in section 11.1, support and nourish neurons, maintain homeostasis, form myelin, and may aid in signal transmission. This section discusses the structure and function of neurons and the myelin sheath that surrounds cell extensions called axons.

MP3 Cells of the Nervous System

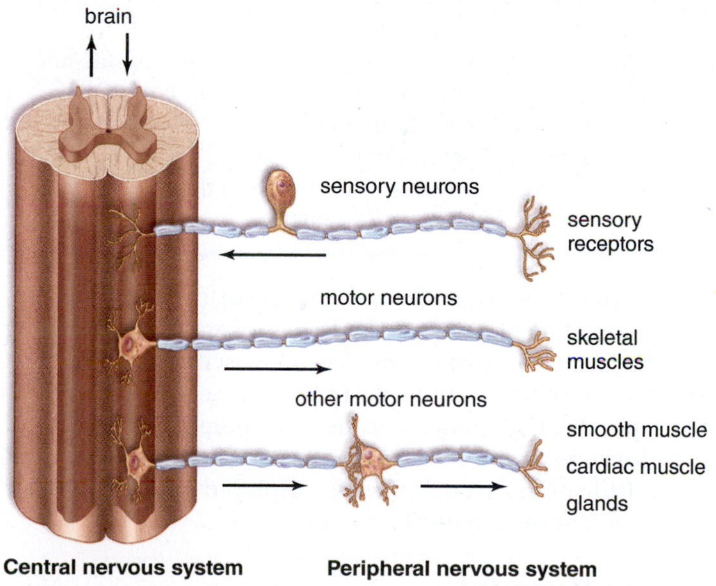

brain
sensory neurons
sensory receptors
motor neurons
skeletal muscles
other motor neurons
smooth muscle
cardiac muscle
glands
Central nervous system
Peripheral nervous system

Figure 17.1 Organization of the nervous system. The sensory neurons of the peripheral nervous system take nerve impulses from sensory receptors to the central nervous system (CNS), and motor neurons take nerve impulses from the CNS to muscles and glands.

Types of Neurons and Neuron Structure

Although the structure of the nervous system is extremely complex, the principles of its operation are simple. There are three classes of neurons: sensory neurons, interneurons, and motor neurons (Fig. 17.2). Their functions are best described in relation to the CNS. A **sensory neuron** takes messages to the CNS. Sensory neurons may be equipped with specialized endings called sensory receptors that detect changes in the environment. An **interneuron** lies entirely within the CNS. Interneurons can

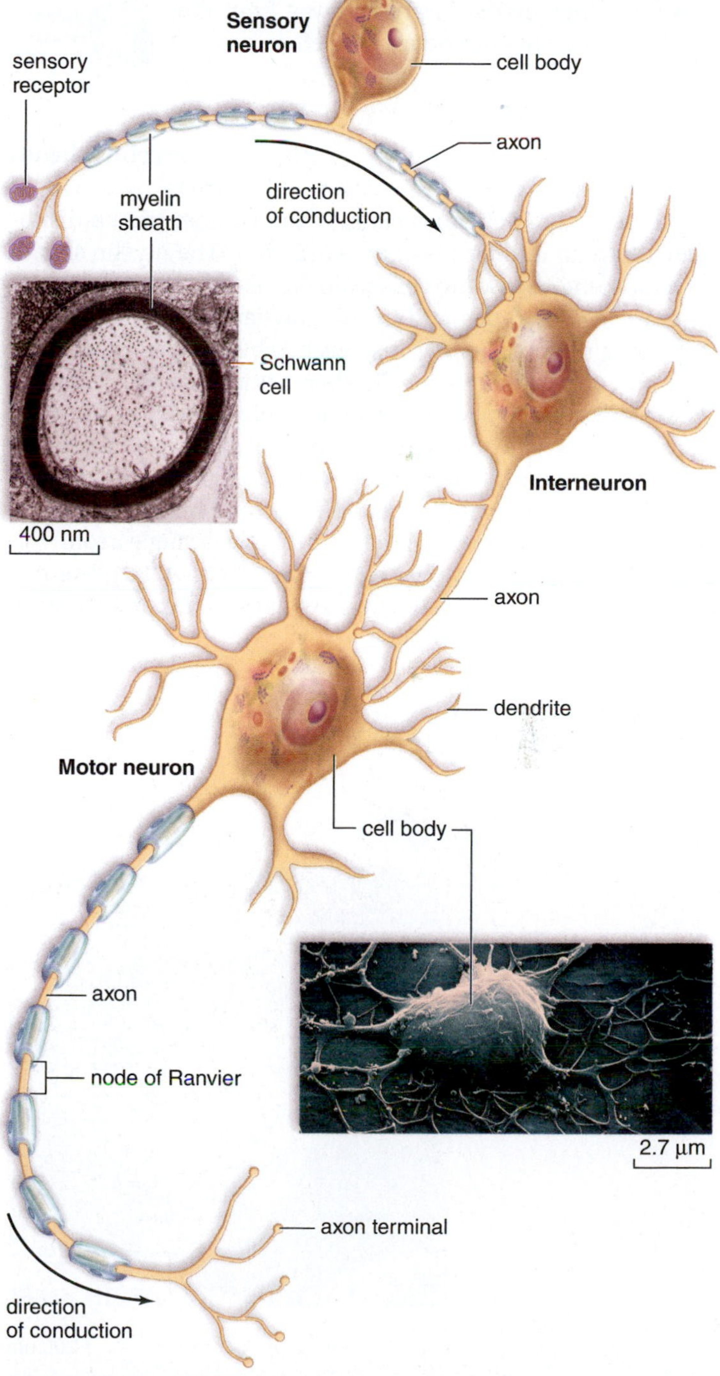

Sensory neuron
sensory receptor
cell body
axon
myelin sheath
direction of conduction
Schwann cell
400 nm
Interneuron
axon
dendrite
Motor neuron
cell body
axon
node of Ranvier
2.7 µm
axon terminal
direction of conduction

Figure 17.2 Types of neurons. Sensory neurons, interneurons, and motor neurons have different arrangements in the body. How does each neuron's arrangement correlate with its function?

receive input from sensory neurons and also from other interneurons in the CNS. Thereafter, they sum up all the messages received from these neurons before they communicate with motor neurons. A **motor neuron** takes messages away from the CNS to an effector (an organ, muscle fiber, or gland). Effectors carry out our responses to environmental changes.

Neurons vary in appearance, but most of them have just three parts: a cell body, dendrites, and an axon. The **cell body** contains the nucleus, as well as other organelles. **Dendrites** are extensions leading toward the cell body that receive signals from other neurons and send them on to the cell body. An **axon** conducts nerve impulses away from the cell body toward other neurons or effectors.

Video Making Brain Cells

Myelin Sheath

Some axons are covered by a protective **myelin sheath** (Fig. 17.3). In the PNS, this covering is formed by a type of neuroglia called **Schwann cells,** which contain the lipid substance myelin in their plasma membranes. The myelin sheath develops when Schwann cells wrap themselves around an axon as many as 100 times and in this way lay down many layers of plasma membrane. Because each Schwann cell myelinates only part of an axon, the myelin sheath is interrupted. The gaps where there is no myelin sheath are called **nodes of Ranvier.** Each Schwann cell only covers about one millimeter of an axon. Because a single axon may be several feet long (if it runs from the spinal cord to the foot, for example), several hundred or more Schwann cells may be required to myelinate a single axon.

In the PNS, myelin gives nerve fibers their white, glistening appearance. The myelin sheath also plays an important role in nerve regeneration within the PNS. If an axon is accidentally severed, the myelin sheath may remain and serve as a passageway for new fiber growth. In the CNS, myelin is produced by a type of neuroglia called **oligodendrocytes.** Unlike in the PNS, nerve regeneration does not seem to occur to any significant degree in the CNS.

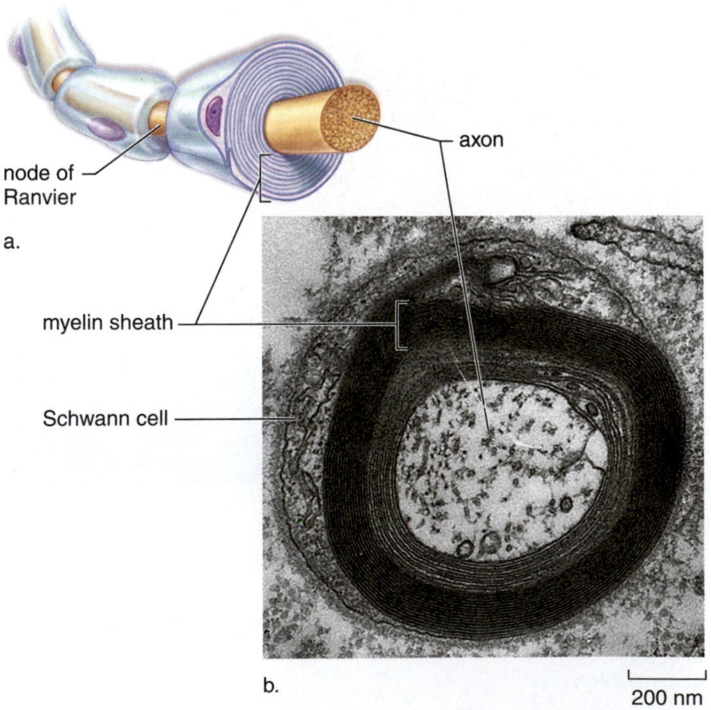
axon
node of Ranvier
a.
myelin sheath
Schwann cell
b.
200 nm

Figure 17.3 Myelin sheath. **a.** In the PNS, a myelin sheath forms when Schwann cells wrap themselves around an axon. **b.** Electron micrograph of a cross section of an axon surrounded by a myelin sheath.

The CNS is composed of two types of nervous tissue—gray matter and white matter. **Gray matter** is gray because it contains neurons with short, nonmyelinated axons. **White matter** is white because it contains myelinated axons that run together in bundles called **tracts.** In the CNS, neurons with short axons make up the gray matter because no part of these neurons has a myelin sheath. However, some neurons in the CNS have long axons with a myelin sheath. These long axons, which make up the white matter of the brain, are carrying messages from one part of the CNS to another. The surface layer of the brain is gray matter, and the white matter lies deep within the gray matter. The central part of the spinal cord consists of gray matter, and the white matter surrounds the gray matter.

Check Your Progress 17.1

1. Identify the three classes of neurons, and describe their relationship to each other.
2. Describe the three parts of a neuron.
3. Distinguish the cell types that form the myelin in the PNS versus the CNS.
4. Review the structure of gray matter and white matter, and describe where each is found in the CNS and the PNS.

17.2 Transmission of Nerve Impulses

Learning Outcomes

Upon completion of this section, you should be able to

1. Summarize the changes in ion concentrations inside and outside of a neuron that result in an action potential.
2. Describe saltatory conduction.
3. Define synaptic integration.
4. Explain the role of neurotransmitters in the transmission of a nerve impulse across a synapse.

The nervous system uses the **nerve impulse** to convey information. The nature of a nerve impulse has been studied by using excised axons and a voltmeter called an oscilloscope. Voltage, which is expressed in millivolts (mV), is a measure of the electrical potential difference between two points. In the case of a neuron, the two points are the inside (the *axoplasm*) and the outside of the axon. Voltage is displayed on the voltmeter screen as a trace, or pattern, over time.

Resting Potential

In the experimental setup shown in Figure 17.4, a voltmeter is wired to two electrodes: one electrode is placed inside an axon, and the other electrode is placed outside. When the axon is not conducting an impulse, the voltmeter records a potential difference across an axonal membrane (membrane of axon) equal to about –70 mV. This reading indicates that the inside of the axon is negative compared to the outside. This is called the **resting potential** because the axon is not conducting an impulse.

The existence of the polarity (charge difference) correlates with a difference in ion distribution on either side of the axonal membrane. As Figure 17.4*a* shows, the concentration of sodium ions (Na^+) is greater outside the axon than inside, and the concentration of potassium ions (K^+) is greater inside the axon than outside. The unequal distribution of these ions is maintained by the action of carrier proteins in the membrane called **sodium-potassium pumps** (see Fig. 4.9), which actively transport Na^+ out of and K^+ into the axon.

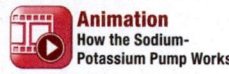

Animation How the Sodium-Potassium Pump Works

These pumps are always working because the membrane is somewhat permeable to these ions, and they tend to diffuse toward their lesser concentration. Because the membrane is more permeable to K^+ than to Na^+, there are always more positive ions outside the membrane than inside. This accounts for the polarity recorded by the voltmeter. Large, negatively charged organic ions in the axoplasm also contribute to the polarity across a resting axonal membrane.

Action Potential

An **action potential** is a rapid change in polarity across an axonal membrane as the nerve impulse occurs. An action potential is an all-or-none phenomenon. If a stimulus causes the axonal

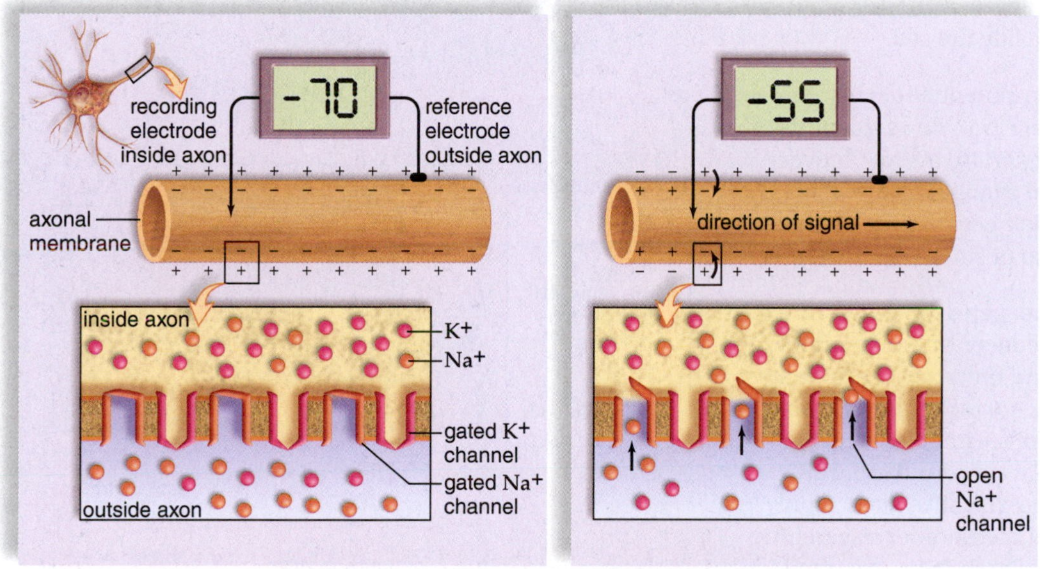
-70
recording electrode inside axon
reference electrode outside axon
axonal membrane
inside axon
K+
Na+
gated K+ channel
gated Na+ channel
outside axon
-55
direction of signal
open Na+ channel

a. Resting potential: Na^+ outside the axon, K^+ and large anions inside the axon. Separation of charges polarizes the cell and causes the resting potential.

b. Stimulus causes the axon to reach its threshold; the axon potential increases from –70 to –55. The action potential has begun.

Figure 17.4 Resting and action potentials of the axonal membrane. **a.** Resting potential. A voltmeter indicates the axonal membrane has a resting potential of –70 mV. There is a preponderance of Na^+ outside the axon and a preponderance of K^+ inside the axon. The permeability of the membrane to K^+ compared to Na^+ causes the inside to be negative compared to the outside. **b.** During an action potential, depolarization occurs when Na^+ gates open and Na^+ begins to move inside the axon. **c.** Depolarization continues until a potential of +35 mV is reached. **d.** Repolarization occurs when K^+ gates open and K^+ moves outside the axon. **e.** Graph of an action potential.

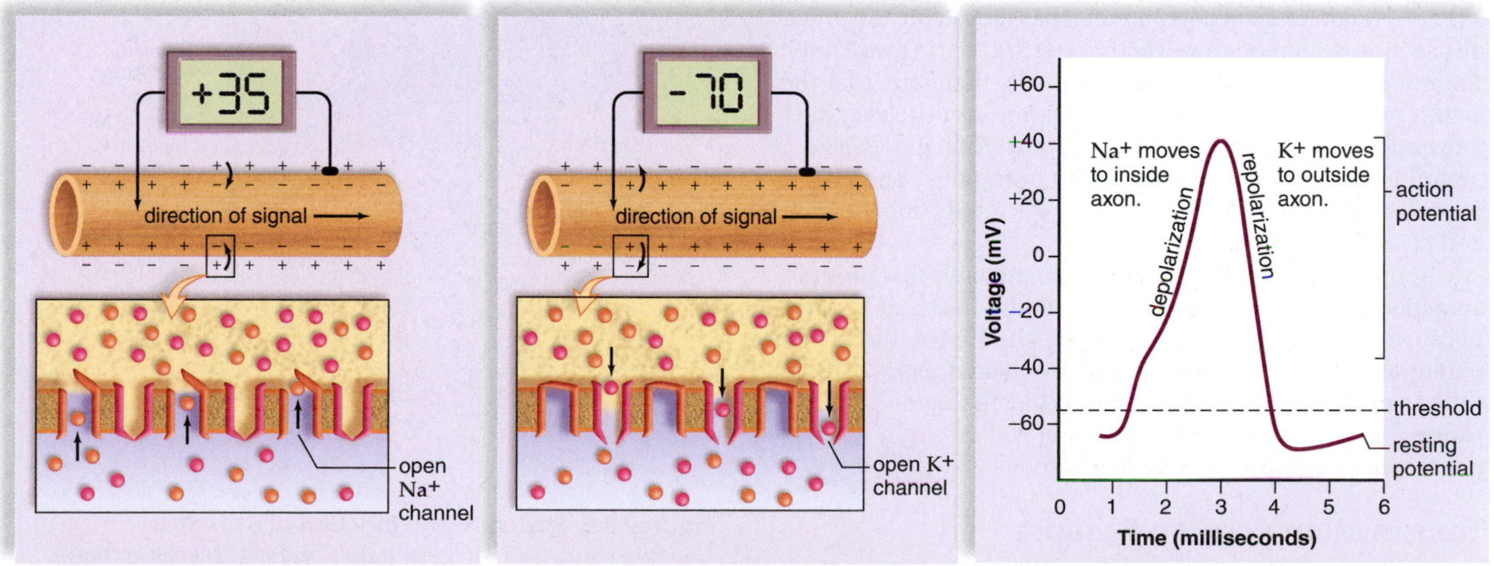
+35
direction of signal
open Na+ channel
-70
direction of signal
open K+ channel
Voltage (mV)
+60
+40
+20
0
-20
-40
-60
Na+ moves to inside axon.
K+ moves to outside axon.
depolarization
repolarization
action potential
threshold
resting potential
0 1 2 3 4 5 6
Time (milliseconds)

c. Depolarization continues as Na^+ gates open and Na^+ moves inside the axon.

d. Action potential ends: repolarization occurs when K^+ gates open and K^+ moves to outside the axon. The sodium-potassium pump returns the ions to their resting positions.

e. An action potential can be visualized if voltage changes are graphed over time.

membrane to depolarize to a certain level, called *threshold,* an action potential occurs. The strength of an action potential does not change, but an intense stimulus can cause an axon to fire (start an axon potential) more often in a given time interval than a weak stimulus.

The action potential requires two types of gated channel proteins in the membrane (Fig. 17.4*a*). One gated channel protein opens to allow Na^+ to pass through the membrane to inside the cell, and the second channel opens to allow K^+ to pass through the membrane to outside the cell.

Sodium Gates Open When an action potential begins, the gates of sodium channels open first, and Na^+ flows down its concentration gradient into the axon (Fig 17.4*b*). As Na^+ moves inside the axon, the membrane potential changes from –70 mV to +35 mV. This is called a *depolarization* because the charge inside the axon changes from negative to positive (Fig. 17.4*c*).

Potassium Gates Open Second, the gates of potassium channels open, and K^+ flows down its concentration gradient to outside the axon. As K^+ moves outside the axon, the action potential changes from +35 mV back to –70 mV. This is a *repolarization* because the inside of the axon resumes a negative charge as K^+ exits the axon (Fig. 17.4*d*). An action potential only takes 2 milliseconds (msec). In order to visualize such rapid fluctuations in voltage across the axonal membrane, researchers generally find it useful to plot the voltage changes over time.

Conduction of an Action Potential In nonmyelinated axons, the action potential travels down an axon one small section at a time. As soon as an action potential has moved on, the previous section undergoes a **refractory period,** during which the sodium gates are unable to open. Notice, therefore, that the action potential cannot move backward and instead always moves down an axon toward its terminals. When the refractory period is over, the sodium-potassium pump has restored the previous ion distribution by pumping Na^+ to outside the axon and K^+ to inside the axon.

In myelinated axons, the gated ion channels that produce an action potential are concentrated at the nodes of Ranvier. Because ion exchange occurs only at the nodes, the action potential travels faster than in nonmyelinated axons. This is called **saltatory conduction,** meaning that the action potential "jumps" from node to node. Speeds of 200 m per second (450 mph) have been recorded.

Animation Nerve Impulse

Transmission Across a Synapse

Every axon branches into many fine endings, each tipped by a small swelling called an *axon terminal.* Each terminal lies very close to either the dendrite or the cell body of another neuron (or a muscle cell). This region of close proximity is called a **synapse** or chemical synapse (Fig. 17.5). It is important to note

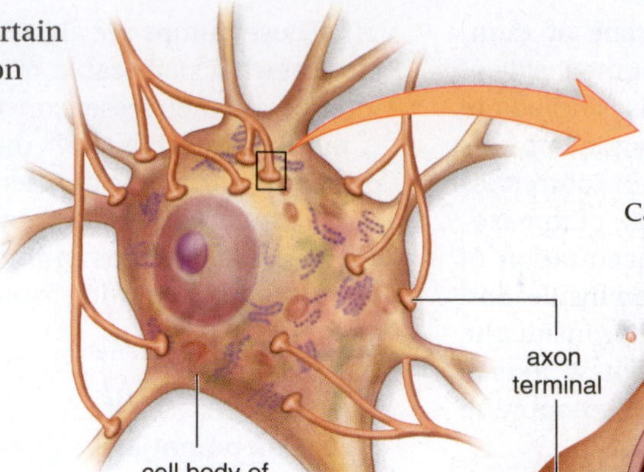
cell body of
postsynaptic
neuron

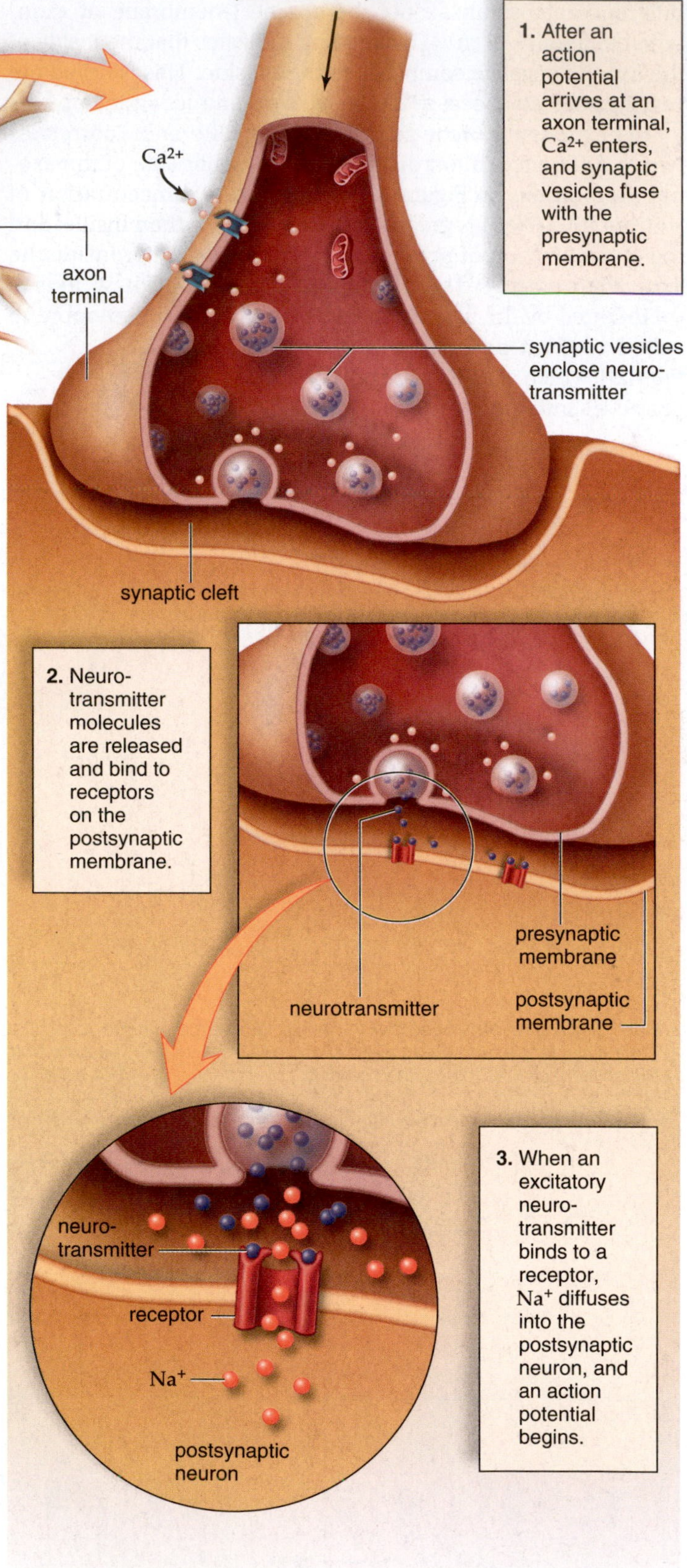
path of action potential
Ca^{2+}
axon
terminal
1. After an action potential arrives at an axon terminal, Ca^{2+} enters, and synaptic vesicles fuse with the presynaptic membrane.
synaptic vesicles
enclose neuro-
transmitter
synaptic cleft
2. Neurotransmitter molecules are released and bind to receptors on the postsynaptic membrane.
presynaptic
membrane
neurotransmitter
postsynaptic
membrane
neuro-
transmitter
receptor
Na^+
postsynaptic
neuron
3. When an excitatory neurotransmitter binds to a receptor, Na^+ diffuses into the postsynaptic neuron, and an action potential begins.

Figure 17.5 Structure and function of a synapse. Transmission across a synapse from one neuron to another occurs when an action potential causes a neurotransmitter to be released at the presynaptic membrane. The neurotransmitter diffuses across a synaptic cleft and binds to a receptor in the postsynaptic membrane. An action potential may begin in the postsynaptic membrane if the concentration of incoming Na^+ reaches threshold.

that the two neurons at a synapse don't ever physically touch each other. Instead, they are separated by a tiny gap called the **synaptic cleft.** At a synapse, the membrane of the first neuron is called the *pre*synaptic membrane, and the membrane of the next neuron is called the *post*synaptic membrane.

Animation Action Potential Propagation

An action potential cannot cross a synapse. Communication between the two neurons at a chemical synapse is carried out by molecules called **neurotransmitters,** which are stored in synaptic vesicles in the axon terminals. When nerve impulses traveling along an axon reach an axon terminal, gated channels for calcium ions (Ca^{2+}) open, and calcium enters the terminal. This sudden rise in Ca^{2+} stimulates synaptic vesicles to merge with the presynaptic membrane, and neurotransmitter molecules are released into the synaptic cleft. They diffuse across the cleft to the postsynaptic membrane, where they bind with specific receptor proteins.

MP3 Synapses

Depending on the type of neurotransmitter and receptor, the response of the postsynaptic neuron can be toward excitation (causing an action potential to happen) or toward inhibition (stopping an action potential from happening).

Synaptic Integration

The dendrites and cell body of a neuron can have synapses with many other neurons, thus a single neuron may receive many excitatory and inhibitory signals. Excitatory signals have a depolarizing effect, causing the charge across the neuron membrane to move closer to the threshold needed to trigger an action potential. Inhibitory signals usually have a hyperpolarizing effect; that is, they increase the potential difference across the axonal membrane. **Integration** is the summing up of excitatory and inhibitory signals by a neuron (Fig. 17.6). If a neuron receives many excitatory signals (either from different synapses or at a rapid rate from one synapse), the chances are that the neuron's axon will transmit a nerve impulse. On the other hand, if a neuron receives both inhibitory and excitatory signals, the summing up of these signals may prohibit the axon from firing.

Neurotransmitters

At least 25 different neurotransmitters have been identified, but two very well-known ones are *acetylcholine* (*ACh*) and *norepinephrine* (*NE*).

Once a neurotransmitter has been released into a synaptic cleft and has initiated a response, it is removed from the cleft. In some synapses, the postsynaptic membrane contains enzymes that rapidly inactivate the neurotransmitter. For example, the enzyme *acetylcholinesterase* (*AChE*) breaks down acetylcholine. In other synapses, the presynaptic membrane rapidly reabsorbs the neurotransmitter, possibly for repackaging in synaptic vesicles. The short existence of neurotransmitters at a synapse prevents continuous stimulation (or inhibition) of postsynaptic membranes.

Animation Chemical Synapse

It is of interest to note here that many drugs that affect the nervous system act by either interfering with or potentiating (enhancing) the action of neurotransmitters. As shown

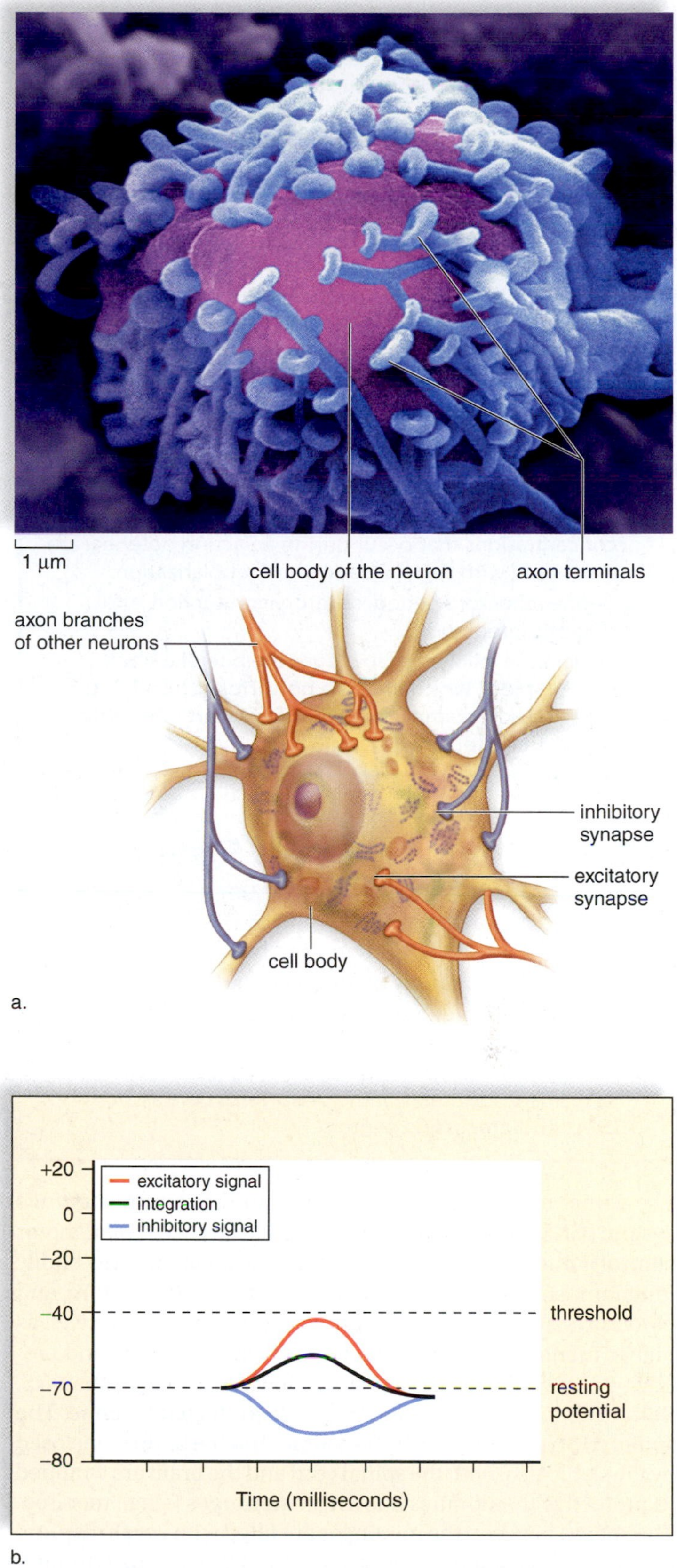
1 µm
cell body of the neuron
axon terminals
axon branches of other neurons
inhibitory synapse
excitatory synapse
cell body
a.
+20
0
–20
–40
–70
–80
excitatory signal
integration
inhibitory signal
threshold
resting potential
Time (milliseconds)
b.

Figure 17.6 Synaptic integration. **a.** Inhibitory signals and excitatory signals are summed up in the dendrites and cell body of the postsynaptic neuron. Only if the combined signals cause the membrane potential to rise above threshold does an action potential occur. **b.** In this example, threshold was not reached.

in Figure 17.18 on page 329, drugs can enhance or block the release of a neurotransmitter, mimic the action of a neurotransmitter or block the receptor, or interfere with the removal of a neurotransmitter from a synaptic cleft. In addition, several naturally occurring venoms and poisons, organophosphate insecticides, and nerve agents such as sarin gas, all inhibit the AChE enzyme, thus prolonging the activity of ACh. In contrast, the bacterium responsible for botulism produces a toxin that inhibits ACh release, which can paralyze muscles. Botox, a highly diluted preparation of this toxin, is now being used to treat a variety of conditions, from eyelid spasms and back pain to the wrinkles that appear on our faces with age.

Check Your Progress 17.2

1. Describe the activity of the sodium-potassium pump present in neurons.
2. Explain how the changes in Na^+ and K^+ ion concentrations that occur during an action potential are associated with depolarization and repolarization.
3. Define refractory period, saltatory conduction, and synaptic integration.
4. Explain the following: How does the bite of a black widow spider, which injects a powerful AChE inhibitor, cause muscle cramps, salivation, fast heart rate, and high blood pressure?

17.3 The Central Nervous System

Learning Outcomes

Upon completion of this section, you should be able to

1. Describe the anatomy of the spinal cord.
2. Identify the major regions of the brain and the main functions of each region.
3. Compare the functions of the primary motor and somatosensory areas of the brain versus the association areas and processing centers.

The spinal cord and the brain make up the **central nervous system** (**CNS**), where sensory information is received and motor control is initiated. The brain controls or influences many bodily functions, such as breathing, heart rate, body temperature, and blood pressure. It is also the source of our emotions, as well as higher mental functions such as reasoning, memory, and creativity. Figure 17.7 illustrates how the CNS relates to the PNS. Both the spinal cord and the brain are protected by bone. The spinal cord is surrounded by vertebrae, and the brain is enclosed by the skull. Also, both the spinal cord and the brain are wrapped in protective membranes known as **meninges** (sing., meninx). The spaces between the meninges are filled with **cerebrospinal fluid,** which cushions and protects the CNS. A small amount of this fluid is sometimes withdrawn from around the cord for laboratory testing when a *spinal tap* is performed.

MP3 Organization of the Nervous System

The brain has hollow interconnecting cavities termed **ventricles,** which also connect with the hollow *central canal* of

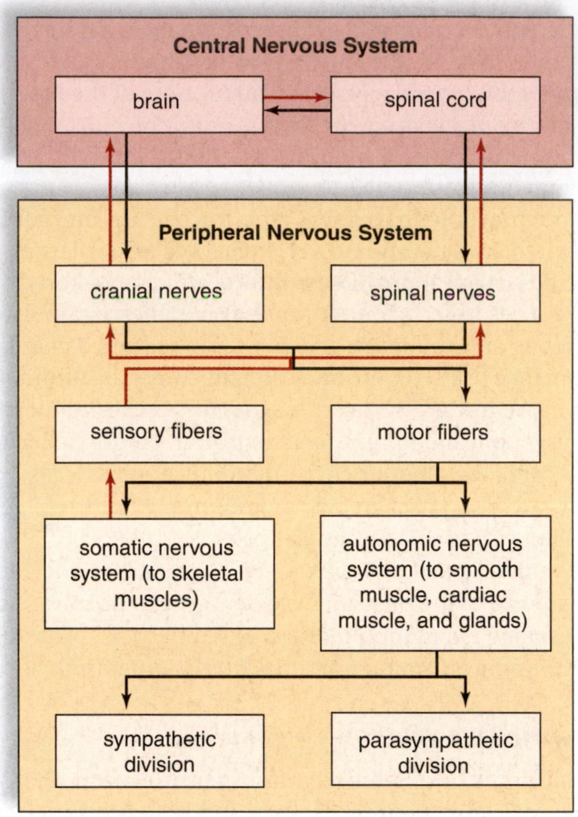
Central Nervous System
brain
spinal cord
Peripheral Nervous System
cranial nerves
spinal nerves
sensory fibers
motor fibers
somatic nervous system (to skeletal muscles)
autonomic nervous system (to smooth muscle, cardiac muscle, and glands)
sympathetic division
parasympathetic division

Figure 17.7 Organization of the nervous system. The CNS is composed of the spinal cord and brain. The PNS is composed of the cranial and spinal nerves. Nerves contain both sensory and motor fibers. In the somatic system, nerves conduct impulses from sensory receptors to the CNS and motor impulses from the CNS to the skeletal muscles. In the autonomic system, consisting of the sympathetic and parasympathetic divisions, motor impulses travel to smooth muscle, cardiac muscle, and glands.

the spinal cord. The ventricles produce and serve as a reservoir for cerebrospinal fluid, as does the central canal. Normally, any excess cerebrospinal fluid drains away into the cardiovascular system. However, blockages can occur. In an infant, the brain and skull can enlarge due to cerebrospinal fluid accumulation, resulting in a condition called hydrocephalus ("water on the brain"). If cerebrospinal fluid collects in an adult, the brain cannot enlarge and instead is pushed against the skull. Such situations can cause brain damage if not quickly corrected.

The Spinal Cord

The **spinal cord** extends from the base of the brain through a large opening in the skull called the foramen magnum and into the vertebral canal formed by openings in the vertebrae.

Structure of the Spinal Cord

Figure 17.8*a* shows how an individual vertebra protects the spinal cord. The spinal nerves project from the cord between the vertebrae that make up the vertebral column. Fluid-filled *intervertebral disks* cushion and separate the vertebrae. If a disk ruptures, the vertebrae press on the spinal cord and spinal nerves, causing pain and loss of motor function.

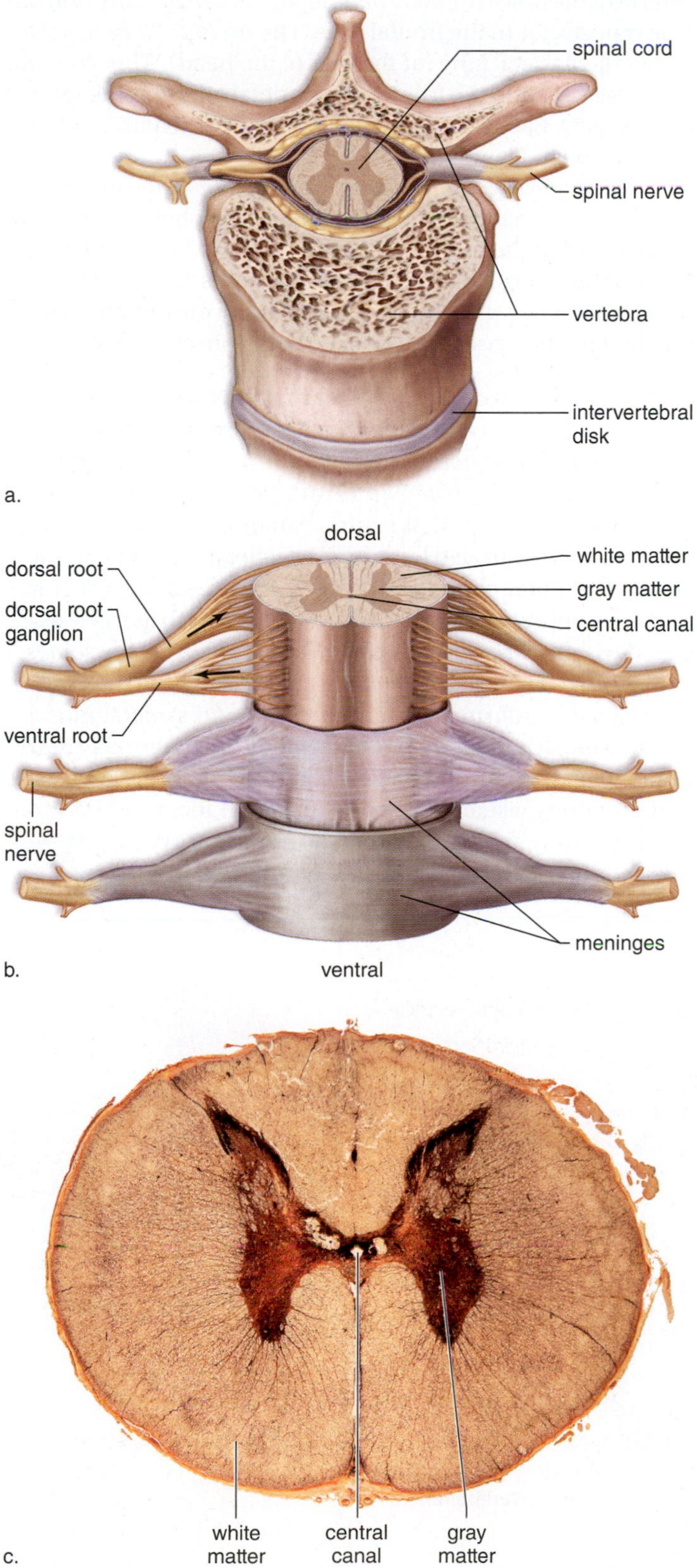
spinal cord
spinal nerve
vertebra
intervertebral disk
a.
dorsal
dorsal root
dorsal root ganglion
ventral root
spinal nerve
white matter
gray matter
central canal
meninges
b.
ventral
c.
white matter
central canal
gray matter

Figure 17.8 Spinal cord. **a.** The spinal cord passes through the vertebral canal formed by the vertebrae. **b.** The spinal cord has a central canal filled with cerebrospinal fluid, gray matter in an H-shaped configuration, and white matter. The white matter contains tracts that take nerve impulses to and from the brain. **c.** Photomicrograph of a cross section of the spinal cord.

A cross section of the spinal cord shows a central canal, gray matter, and white matter (Fig. 17.8*b, c*). The central canal contains cerebrospinal fluid, as do the meninges that protect the spinal cord. The gray matter is centrally located and shaped like the letter H. Portions of sensory neurons and motor neurons are found there, as are interneurons that communicate with these two types of neurons. The dorsal root of a spinal nerve contains sensory fibers entering the gray matter, and the ventral root of a spinal nerve contains motor fibers exiting the gray matter. The dorsal and ventral roots join before the spinal nerve leaves the vertebral canal. Spinal nerves are a part of the PNS.

The white matter of the spinal cord surrounds the gray matter. The white matter contains ascending tracts taking information to the brain (primarily located dorsally) and descending tracts taking information from the brain (primarily located ventrally). Because the tracts cross after they enter and exit the CNS, the left side of the brain controls the right side of the body, and the right side of the brain controls the left side of the body.

Functions of the Spinal Cord

The spinal cord provides a means of communication between the brain and the peripheral nerves that leave the cord. When someone touches your hand, sensory receptors generate nerve impulses that pass through sensory fibers to the spinal cord and up ascending tracts to the brain. When we voluntarily move our limbs, motor impulses originating in the brain pass down descending tracts to the spinal cord and out to our muscles by way of motor fibers. Therefore, if the spinal cord is severed, we suffer a loss of sensation and a loss of voluntary control—that is, paralysis (see section 17.7).

In section 17.5 we will see that the spinal cord is also the center for thousands of reflex arcs, which allow the nerves and muscles to respond very quickly to potentially dangerous stimuli. A stimulus causes sensory receptors to generate nerve impulses that travel in sensory axons to the spinal cord. Interneurons integrate the incoming data and relay signals to motor neurons. The reflex is complete when motor axons cause skeletal muscles to contract or a gland or an organ to respond. Each interneuron in the spinal cord has synapses with many other neurons, and therefore, they send signals to several other interneurons and motor neurons.

The spinal cord plays a similar role for the internal organs. For example, when blood pressure falls, sensory receptors in the carotid arteries and aorta generate nerve impulses that pass through sensory fibers to the cord and then up an ascending tract to a cardiovascular center in the brain. The brain then sends nerve impulses down a descending tract to the spinal cord. Motor impulses then travel via spinal nerves to smooth muscle in the walls of arteries and arterioles, causing their constriction so that the blood pressure rises.

The Brain

The human **brain** has been called the last great frontier of biology. The goal of modern neuroscience is to understand the structure and function of the brain's various parts well enough that it will be possible to prevent or correct the many disorders that can afflict it, robbing many people of a normal life. This section provides only a glimpse of what is known about the brain and the modern avenues of research.

Video Brain Bank

We will discuss the brain with reference to its four major parts: the cerebrum, the diencephalon, the cerebellum, and the brain stem. It may be helpful for you to associate these four regions with the brain's ventricles (fluid-filled spaces): the cerebrum with the two lateral ventricles, the diencephalon with the third ventricle, and the brain stem and the cerebellum with the fourth ventricle (Fig. 17.9*a*).

MP3 The Brain

The Cerebrum

The **cerebrum** is the largest portion of the brain in humans. The cerebrum is the last center to receive sensory input and carry out integration before commanding voluntary motor responses. It communicates with and coordinates the activities of the other parts of the brain.

MP3 The Cerebrum

The Cerebral Hemispheres Just as the human body has two halves, so does the cerebrum. These halves are called the left and right **cerebral hemispheres** (Fig. 17.9*b*). A deep groove, the *longitudinal fissure,* divides the left and right cerebral hemispheres. The two cerebral hemispheres communicate via the *corpus callosum,* an extensive bridge of nerve tracts.

Shallow grooves called *sulci* (sing., sulcus) divide each hemisphere into lobes (Fig. 17.10). The *frontal lobe* is the most ventral of the lobes (directly behind the forehead). The *parietal lobe* is posterior to the frontal lobe. The *occipital lobe* is posterior to the parietal lobe (at the rear of the head). The *temporal lobe* lies inferior to the frontal and parietal lobes (at the temple and the ear). Each lobe is associated with particular functions, as indicated in Figure 17.10.

The **cerebral cortex** is a thin, highly convoluted outer layer of gray matter that covers the cerebral hemispheres. Each fold, or convolution, in the cortex is called a *gyrus.* The cerebral cortex contains over 1 billion cell bodies and is the region of the brain that accounts for sensation, voluntary movement, and all the thought processes we associate with consciousness.

Primary Motor and Sensory Areas of the Cortex The cerebral cortex contains motor areas, sensory areas, and association areas. The *primary motor area* is in the frontal lobe just anterior to the central sulcus (see Fig. 17.10). Voluntary commands to skeletal muscles begin in the primary motor area, and each part of the body is controlled by a certain section (Fig. 17.11*a*). Large areas of cerebral cortex are devoted to controlling structures that carry out very fine, precise movements, such as those associated with the movements of the hand. The *primary somatosensory area* is just posterior to the central sulcus in the parietal lobe. Sensory information from the skin and skeletal muscles arrives here, where each part of the body is sequentially represented (Fig. 17.11*b*). As with the primary motor area, large parts of the primary visual area are dedicated to receiving information from body areas generating the most sensory information, such as the face and hands.

Video Brain Surgery

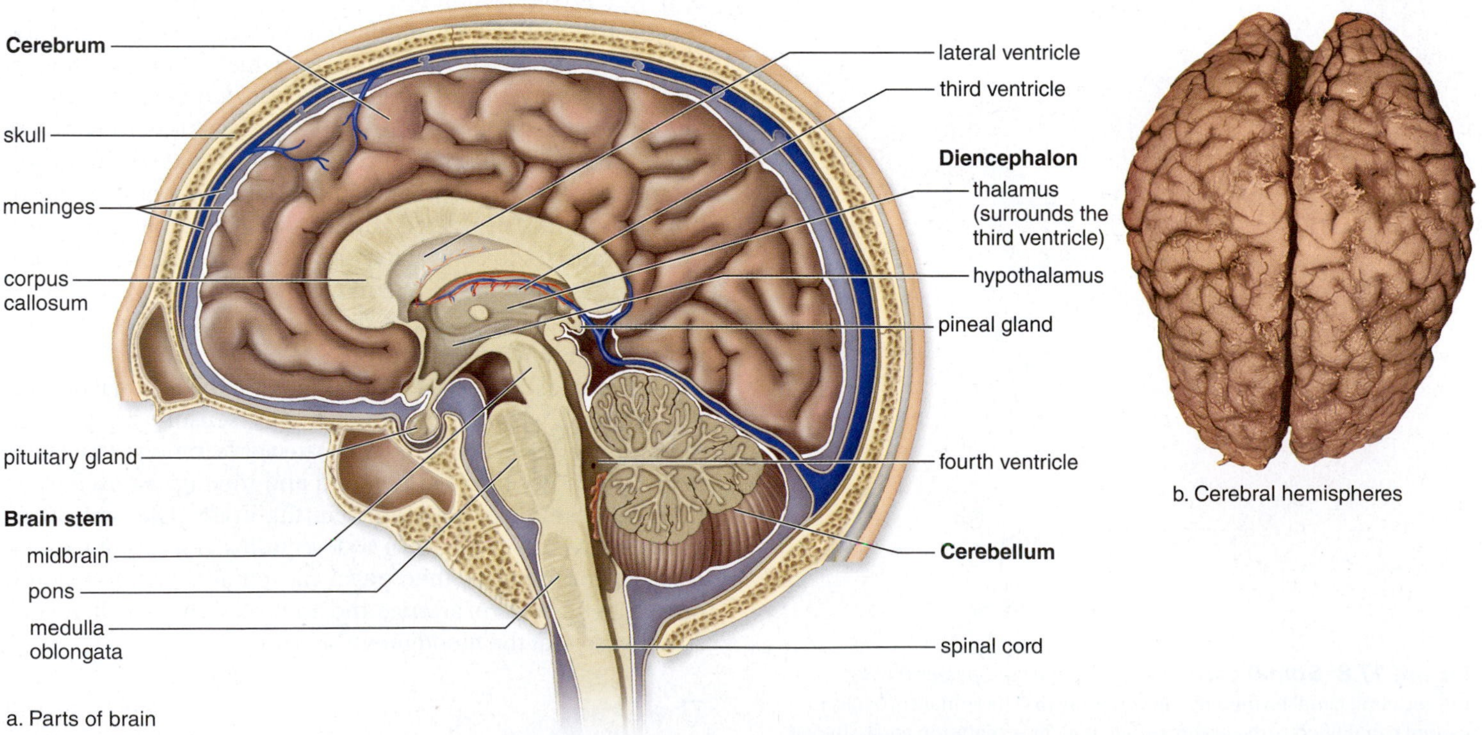
Cerebrum
skull
meninges
corpus callosum
pituitary gland
Brain stem
midbrain
pons
medulla oblongata
lateral ventricle
third ventricle
Diencephalon
thalamus (surrounds the third ventricle)
hypothalamus
pineal gland
fourth ventricle
Cerebellum
spinal cord
a. Parts of brain
b. Cerebral hemispheres

Figure 17.9 The human brain. **a.** The cerebrum, seen here in longitudinal section, is the largest part of the brain in humans. The right cerebral hemisphere is shown here. **b.** The cerebrum has left and right cerebral hemispheres, which are connected by the corpus callosum.

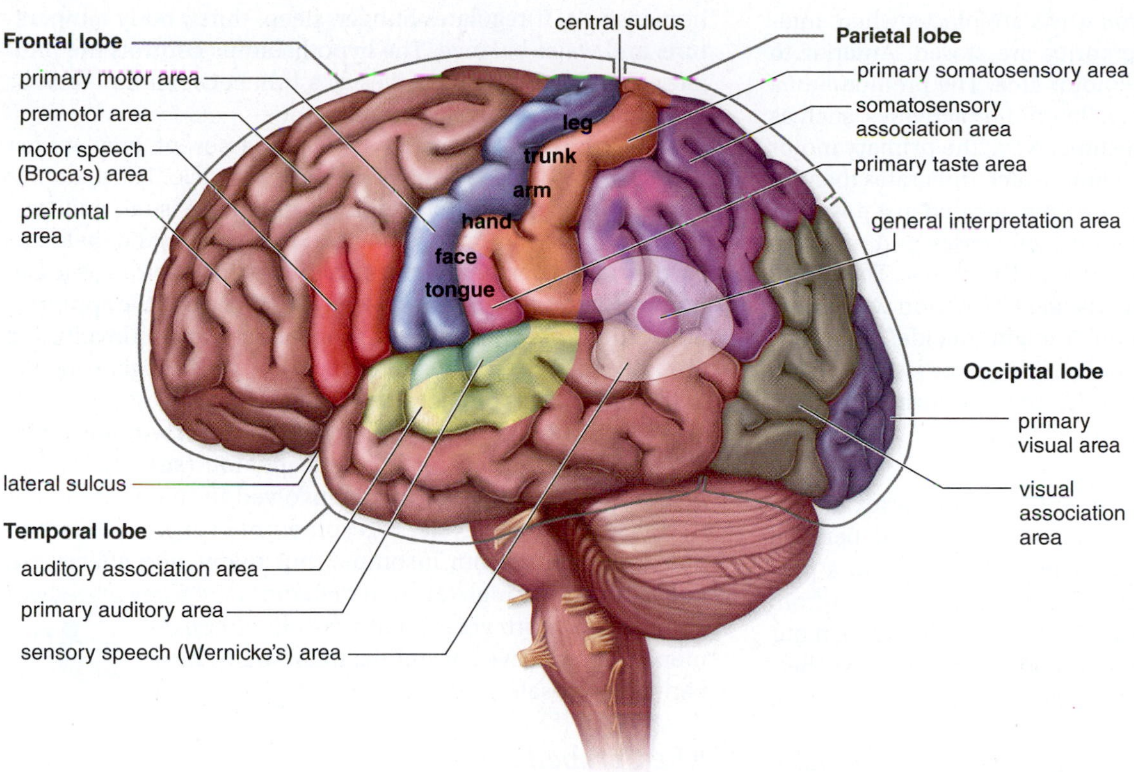
Frontal lobe
primary motor area
premotor area
motor speech
(Broca's) area
prefrontal
area
central sulcus
Parietal lobe
primary somatosensory area
somatosensory
association area
primary taste area
general interpretation area
leg
trunk
arm
hand
face
tongue
Occipital lobe
primary
visual area
visual
association
area
lateral sulcus
Temporal lobe
auditory association area
primary auditory area
sensory speech (Wernicke's) area

Figure 17.10 The lobes of a cerebral hemisphere. Each cerebral hemisphere is divided into four lobes: frontal, parietal, temporal, and occipital. The frontal lobe contains centers for reasoning and movement, the parietal lobe for somatic sensing and taste, the temporal lobe for hearing, and the occipital lobe for vision.

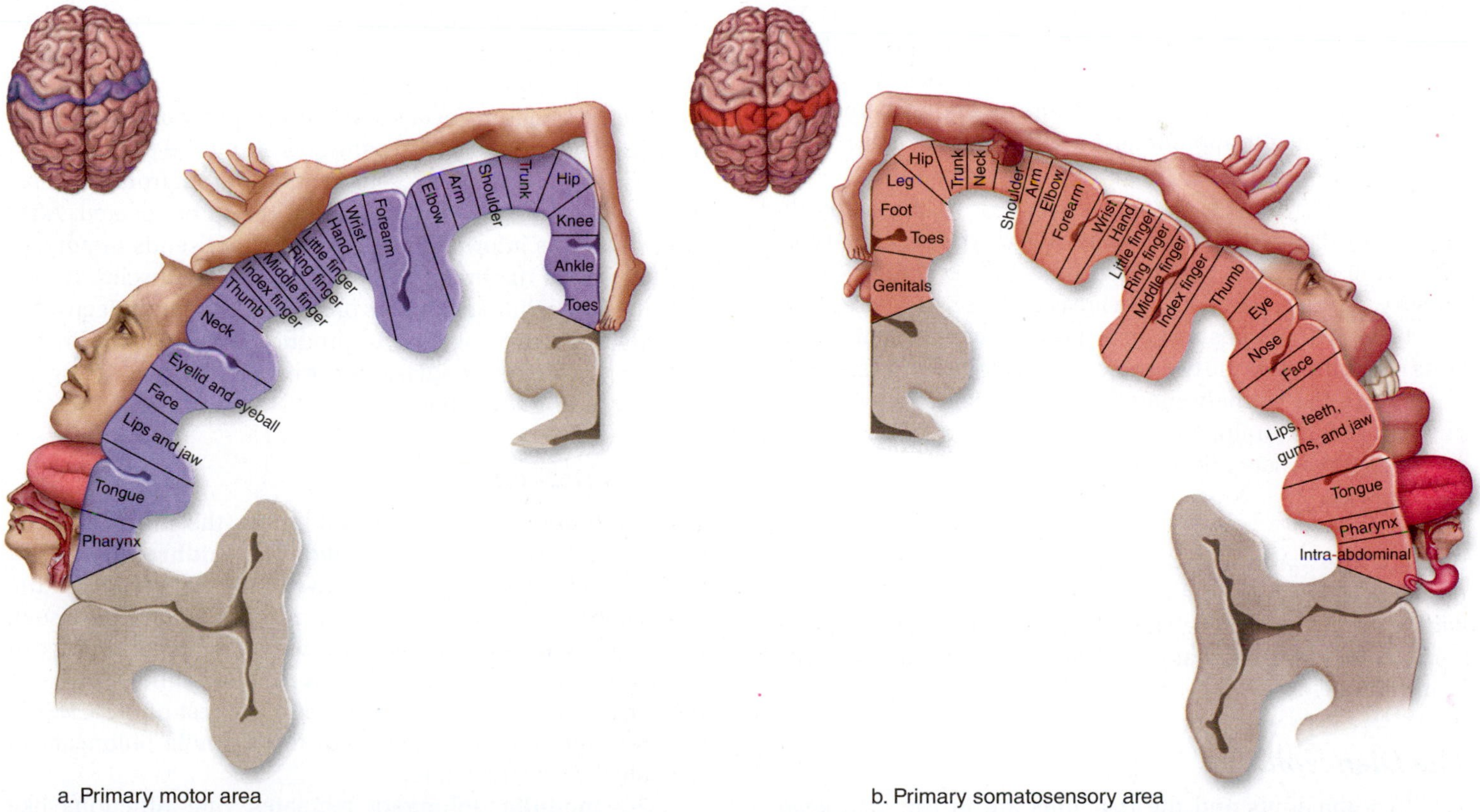
Hip
Trunk
Shoulder
Arm
Elbow
Forearm
Wrist
Hand
Little finger
Ring finger
Middle finger
Index finger
Thumb
Neck
Eyelid and eyeball
Face
Lips and jaw
Tongue
Pharynx
Knee
Ankle
Toes
a. Primary motor area
Hip
Leg
Foot
Toes
Genitals
Trunk
Neck
Shoulder
Arm
Elbow
Forearm
Wrist
Hand
Little finger
Ring finger
Middle finger
Index finger
Thumb
Eye
Nose
Face
Lips, teeth, gums, and jaw
Tongue
Pharynx
Intra-abdominal
b. Primary somatosensory area

Figure 17.11 The primary motor and somatosensory areas. In these drawings, the size of the body part reflects the amount of cerebral cortex devoted to that body part. For example, the amount of primary motor cortex (**a**) and somatosensory cortex (**b**) devoted to the thumb, fingers, and hand is greater than that for the foot and toes.

Association Areas *Association areas* are places where integration occurs and where memories are stored. Anterior to the primary motor area is a premotor area. The premotor area organizes motor functions for skilled motor activities, such as walking and talking at the same time. Next, the primary motor area sends signals to the cerebellum, which integrates them.

The *somatosensory association area,* located just posterior to the primary somatosensory area, processes and analyzes sensory information from the skin and muscles. The *visual association area* associates new visual information with previously received visual information. It might "decide," for example, whether we have seen this face or scene or symbol in the past before. The *auditory association area* performs the same functions with regard to sounds.

Processing Centers Processing centers of the cortex receive information from the other association areas and perform higher-level analytical functions. The *prefrontal area,* a processing center in the frontal lobe, receives information from the other association areas and uses it to reason and plan our actions. Integration in this center accounts for our most cherished human abilities (e.g., to think critically and to formulate appropriate behaviors).

The unique ability of humans to speak is partially dependent upon two processing centers found only in the left cerebral cortex: *Wernicke's area* in the posterior part of the left temporal lobe and *Broca's area* in the left frontal lobe. Broca's area is located just anterior to the portion of the primary motor area for speech musculature (lips, tongue, larynx, and so forth) (see Fig. 17.14). Wernicke's area helps us understand both the written and spoken word and sends the information to Broca's area. Broca's area adds grammatical refinements and directs the primary motor area to stimulate the appropriate muscles for speaking.

Central White Matter Much of the rest of the cerebrum beneath the cerebral cortex is composed of white matter. Tracts within the cerebrum take information between the different sensory, motor, and association areas pictured in Figures 17.10 and 17.11. The corpus callosum, previously mentioned, contains tracts that join the two cerebral hemispheres. Descending tracts from the primary motor area communicate with various parts of the brain, and ascending tracts from lower brain centers send sensory information up to the primary somatosensory area.

Basal Nuclei Although the majority of the cerebrum is composed of tracts, there are masses of gray matter located deep within the white matter. These **basal nuclei** integrate motor commands, ensuring that proper muscle groups are activated or inhibited.

The Diencephalon

The hypothalamus and the thalamus are in the *diencephalon,* a region that encircles the third ventricle (see Fig. 17.9*a*). The **hypothalamus** forms the floor of the third ventricle. The hypothalamus is an integrating center that helps maintain homeostasis. It regulates hunger, sleep, thirst, body temperature, and water balance. The hypothalamus controls the pituitary gland and thereby serves as a link between the nervous and endocrine systems.

The **thalamus** consists of two masses of gray matter that form the sides and roof of the third ventricle. The thalamus receives all sensory input except for smell. Visual, auditory, taste, and somatosensory information arrives at the thalamus via the cranial nerves and tracts from the spinal cord. The thalamus integrates this information and sends it on to the appropriate portions of the cerebrum. The thalamus is also involved in arousal of the cerebrum, and it participates in higher mental functions, such as memory and emotions.

The **pineal gland,** which secretes the hormone melatonin, is also located in the diencephalon (see Fig. 17.9*a*). *Melatonin* is a hormone that is involved in maintaining our normal sleep-wake cycle. It is sometimes recommended for people suffering from insomnia, but many side effects can occur. Relatively high levels of melatonin are a key ingredient in "relaxation brownies" that are sold at convenience stores as well as online; many states have banned their sale, however.

Video Winter Mood

The Cerebellum

The **cerebellum** lies under the occipital lobe of the cerebrum and is separated from the brain stem by the fourth ventricle (see Fig. 17.9*a*). The cerebellum has two portions that are joined by a narrow median portion. Each portion is primarily composed of white matter, which, in longitudinal section, has a treelike pattern. Overlying the white matter is a thin layer of gray matter that forms a series of complex folds.

The cerebellum receives sensory input from the joints, muscles, and other sensory pathways about the present position of body parts. It also receives motor output from the cerebral cortex about where these parts should be located. After integrating this information, the cerebellum sends motor signals by way of the brain stem to the skeletal muscles. In this way, the cerebellum maintains posture and balance. It ensures that muscles work together to produce smooth, coordinated voluntary movements, such as those required when playing the piano or hitting a baseball.

The Brain Stem

The **brain stem** contains the midbrain, the pons, and the medulla oblongata (see Fig. 17.9*a*). The **midbrain** acts as a relay station for tracts passing between the cerebrum and the spinal cord or cerebellum. It also has reflex centers for visual, auditory, and tactile responses. The word *pons* means bridge in Latin, and true to its name, the **pons** contains bundles of axons traveling between the cerebellum and the rest of the CNS. In addition, the pons functions with the medulla oblongata to regulate the breathing rate.

The **medulla oblongata** regulates vital functions like heartbeat, breathing, and blood pressure. It also contains reflex centers for vomiting, coughing, sneezing, hiccuping, and swallowing. The medulla oblongata lies just superior to the spinal cord. It contains tracts that ascend or descend between the spinal cord and higher brain centers.

The **reticular activating system (RAS)** contains the *reticular formation,* a complex network of nuclei and nerve fibers that extend the length of the brain stem. The reticular formation receives sensory signals that it sends up to higher centers, and motor signals that it sends to the spinal cord.

The RAS arouses the cerebrum via the thalamus and causes a person to be alert. Apparently, the RAS can filter out unnecessary sensory stimuli, explaining why you can study with the TV on. If you want to awaken the RAS, surprise it with a sudden stimulus, like an alarm clock, smelling salts, or splashing your face with cold water; if you want to deactivate it, remove visual and auditory stimuli. A severe injury to the RAS can cause a person to be comatose, from which recovery may be impossible.

Electroencephalograms

The electrical activity of the brain can be recorded in the form of an *electroencephalogram* (*EEG*). Electrodes are taped to different parts of the scalp, and an instrument records the so-called brain waves. The EEG is a diagnostic tool. For example, an irregular pattern can signify epilepsy or a brain tumor. A flat EEG is often used as a clinical and legal criterion for brain death.

Check Your Progress 17.3

1. Summarize the functions of the spinal cord.
2. Identify the four major parts of the brain and describe the general functions of each.
3. Describe the types of symptoms you would expect to see in a person who has sustained damage to their cerebellum, medulla oblongata, or RAS.

17.4 The Limbic System and Higher Mental Functions

Learning Outcomes

Upon completion of this section, you should be able to

1. Identify the structures of the limbic system.
2. Summarize how the limbic system is involved in memory, language, and speech.
3. Distinguish between short-term, long-term, semantic, episodic, and skill memory.
4. Explain how certain diseases, accidents, and experiments have helped scientists understand some basic components of how memories are made.

Emotions and higher mental functions are associated with the limbic system in the brain. The limbic system blends primitive emotions (rage, fear, joy, sadness) and higher mental functions (reason, memory) into a united whole. It accounts for why activities such as sexual behavior and eating seem pleasurable and also for why, say, mental stress can cause high blood pressure.

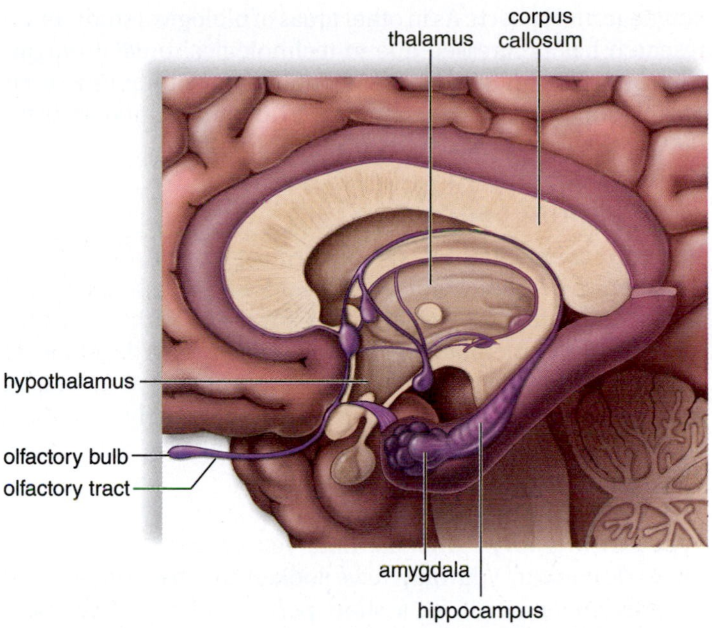
thalamus
corpus callosum
hypothalamus
olfactory bulb
olfactory tract
amygdala
hippocampus

Figure 17.12 The limbic system. Structures deep within the cerebral hemispheres and surrounding the diencephalon join higher mental functions such as reasoning with more primitive feelings such as fear and pleasure.

Anatomy of the Limbic System

The **limbic system** is a complex network of tracts and nuclei that includes portions of the cerebral lobes, the basal nuclei, and the diencephalon. Two significant structures within the limbic system are the hippocampus and the amygdala, which are essential for learning and memory (Fig. 17.12). The **hippocampus** is a seahorse-shaped structure deep in the temporal lobe that is well situated to communicate with the prefrontal area of the brain, which is also involved in learning and memory. The **amygdala** is an almond-shaped structure in the limbic system that allows us to respond to and display anger, avoidance, defensiveness, and fear. The amygdala prompts release of adrenaline and other hormones into the bloodstream.

A case report published in 2010 described a woman identified as S.M., who lacked an amygdala due to a rare genetic disease. Not only did S.M. seem to exhibit no response to typically fearful stimuli like snakes, tarantulas, and haunted houses, but she also could not recognize fearful facial expressions of others. She retained her ablity to feel other emotions like surprise, happiness, and disgust. While acknowledging they need to study more subjects, the researchers suggest this case shows that the amygdala is a critical brain region for triggering a state of fear.[1]

Higher Mental Functions

The well-developed human cerebrum is responsible for higher mental functions such as memory and learning, as well as

[1]Feinstein, J. S. et al. "The Human Amygdala and the Induction and Experience of Fear: *Current Biology* 21(1): 34–38 (2011). Epub 2010 December 16.

language and speech. As in other areas of biological study, brain research has progressed due to technological breakthroughs. Neuroscientists now have a wide range of techniques at their disposal for studying the human brain, including modern technologies that allow us to record its functioning.

Memory and Learning

Just as the connecting tracts of the corpus callosum are evidence that the two cerebral hemispheres work together, so the limbic system indicates that cortical areas may work with lower centers to produce memory and learning. **Memory** is the ability to hold a thought in mind or to recall events from the past, ranging from a word we learned only yesterday to an early emotional experience that has shaped our lives. **Learning** takes place when we retain and utilize past memories.

Types of Memory The last time you decided to try a new pizza restaurant, you may have looked up the number and tried to remember it for a short period of time. If you said you were trying to keep it in the forefront of your brain, you were exactly correct. The prefrontal area, which is active during *short-term memory,* lies just behind our forehead! On the other hand, there are probably some telephone numbers that you have memorized. These have gone into *long-term memory.* Think of the phone number of someone close to you whom you call frequently. Can you bring the number to mind without also thinking about the place or person associated with that number? Most likely you cannot, because long-term memory is typically a mixture of *semantic memory* (of ideas, concepts, and meanings) and *episodic memory* (of specific facts, persons, events, etc.). Due to brain damage, some people lose one type of memory but not the other. For example, without a working episodic memory, they can carry on a conversation but have no recollection of recent events. If you are talking to them and then leave the room, they don't remember you when you come back!

Skill memory is another type of memory that can exist independent of episodic memory. Skill memory is involved in performing motor activities, such as riding a bike, playing ice hockey, or typing a letter using a computer keyboard. When a person first learns a skill, more areas of the cerebral cortex are involved than after the skill is perfected. In other words, you have to think about what you are doing when you learn a skill, but later the actions become automatic. Skill memory involves all the motor areas of the cerebrum below the level of consciousness.

Long-Term Memory Storage and Retrieval The first step toward curing memory disorders is to know what parts of the brain are functioning when we remember something. Our long-term memories are stored in bits and pieces throughout the sensory association areas of the cerebral cortex. Visions are stored in the visual association area, sounds are stored in the auditory association area, and so forth. The hippocampus serves as a bridge between the sensory association areas and the prefrontal area. A classic case that demonstrated this involved a physicist identified as S.S., whose hippocampus was selectively destroyed by a herpesvirus infection. Even though he still had a high I.Q. and could remember childhood events and physics equations, S.S. forgot recent experiences within a few minutes. By observing S.S. and patients like him, we can surmise that the hippocampus must be involved in the conversion of short-term memories to long-term memories. As discussed in the Health feature, "Why Do We Sleep?," different stages of sleep may be involved in the formation of memory connections in the brain.

So why are some memories so emotionally charged? The amygdala is responsible for fear conditioning and associating danger with sensory stimuli received from both the diencephalon and the cortical sensory areas. Figure 17.13 diagrammatically illustrates the long-term memory circuits of the brain.

In addition to studying what regions of the brain are involved in memory formation, neurobiologists want to know what exactly neurons are doing when we store memories and bring them back. They have discovered a chemical process

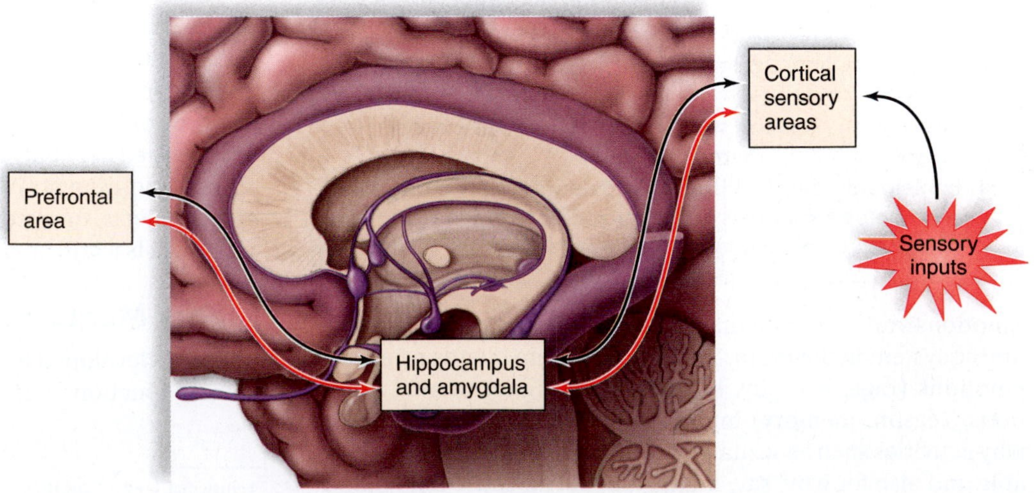
Cortical sensory areas
Prefrontal area
Sensory inputs
Hippocampus and amygdala

Figure 17.13 Long-term memory circuits. The hippocampus and the amygdala are believed to be involved in the storage and retrieval of memories. Semantic memory (red arrows) and episodic memory (black arrows) are stored separately, and therefore, you can lose one without losing the other.

SCIENCE IN YOUR LIFE ▶ **HEALTH**

Why Do We Sleep?

When was the last time you went without enough sleep? Studying for an exam, writing a paper, socializing with friends? How did you feel the next day? What do you think would happen if you only slept three or four hours a night for a month? But if you sleep eight hours a night, that's about 122 days a year. Couldn't you do something productive or fun with all that extra time?

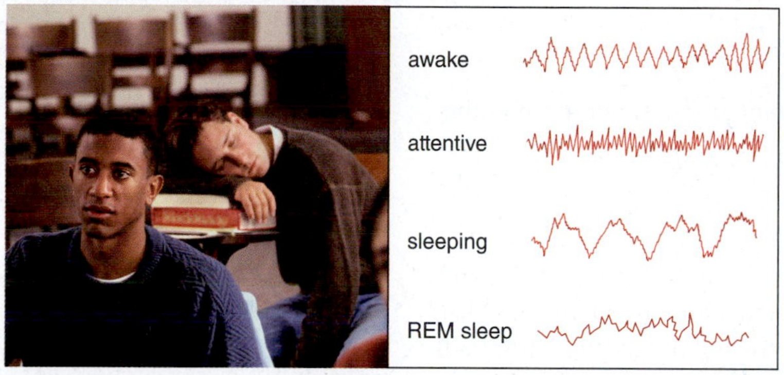
awake
attentive
sleeping
REM sleep

Figure 17A The sleeping brain. In addition to an increased need for daytime naps, chronic sleep deprivation may lead to serious health problems. When we are awake, attentive, or in different stages of sleep, the brain produces characteristic types of electrical waves that can be measured with an EEG.

Our bodies need sleep as much as they need food. Sleep deprivation affects us both mentally and physically. Without enough sleep we become irritable and have trouble concentrating. A 2008 study showed that adults deprived of sleep for 35 hours had increased activity in the amygdala, the brain's emotional center. Sleep-deprived bodies lose coordination and agility. Our reaction times are slower. Studies are also showing links between too little sleep and obesity, heart disease, and diabetes.

Virtually all mammals and birds sleep. Most fish and insects do, too, or at least they enter a sleeplike state. Even dolphins, which need to keep surfacing to breathe, seem to sleep in one-half of their brain at a time! And yet, scientists are still debating exactly why we need to sleep at all. Considering that we spend around a third of our lives in this inactive, seemingly unproductive state, the essential function of sleep, as one sleep researcher noted, "may be the biggest open question in biology."

Most people probably believe that sleep is simply a time for the body to rest. However, no matter how quietly you may sit or lie down, watching TV or just meditating all day, you will still need to sleep. Another idea is that sleeping serves to protect various creatures by keeping them still and quiet at times when predators are more common. However, this does not explain the severe effects of sleep deprivation—and besides, a sleeping animal cannot run away or defend itself as well as an awake one.

Prior to 1953, most scientists believed that during sleep the brain was completely inactive. That was the year that a graduate student at the University of Chicago studied his sleeping son and noticed there were periods when his eyes darted rapidly back and forth. Subsequent research has shown that during this rapid eye movement (REM) sleep the brain is very active. REM sleep may be important for brain development; infants spend about 50% of their sleeping time in this state compared to 15–20% for adults. REM sleep may also be a time when the brain is building neurological connections to consolidate memories learned from experiences during that day. A 2004 study showed that subjects who were asked to work on a math problem that contained a "hidden solution" were about twice as likely to realize that solution when they returned after a night's sleep compared to subjects who spent an equal amount of time awake.

If a person is hooked up to an EEG during REM sleep (Fig. 17A), the brain appears to be awake, but the muscles are very still. If people are awakened during REM sleep, they are very likely to report that they have been dreaming, and dreams may also play a role in sorting through our experiences, saving some as memories and deleting others.

Another, often overlooked, task the brain must perform is forgetting. During the activities of an average day, the brain is taking in mountains of information by building new connections between neurons. Because maintaining these new connections requires energy, if some were not broken down, the brain would soon require more energy than the body could supply. Studies show that this reduction of connections in the brain occurs during non-REM sleep. Perhaps sleep allows the brain time to "clean out" a large percentage of these memories, while solidifying others.

So if you have trouble sleeping, also known as insomnia, how can you get a better night's rest? Experts note that caffeine, nicotine, and even alcohol can disrupt sleep patterns. Exercising at night can raise metabolism when it needs to be slowing down for sleep. Taking naps longer than 15 minutes can make it harder to fall asleep later, as can exercising within three hours of bedtime. Regardless of the true purpose of sleep, all sleep researchers would agree that we can't rob ourselves of sleep for long, without serious consequences.

Questions to Consider

1. Briefly, how would you define what sleep is?
2. Do you agree that the necessity of sleep for mammals and birds is one of the biggest mysteries in biology? Why do you think that the answer has eluded scientists for so long?
3. Of the several ideas presented here regarding the purpose of sleep, which one makes the most sense to you?

called *long-term potentiation* (*LTP*), which is an enhanced response at synapses within the hippocampus. LTP seems to mainly involve glutamate, an amino acid that can function as a neurotransmitter. To prove this, investigators at the Massachusetts Institute of Technology produced mice that lacked the receptor for glutamate only in the hippocampus. Unlike control mice, the defective mice could not learn to run the maze, demonstrating an important role for this chemical in memory.

Language and Speech

Language is obviously dependent upon semantic memory. Therefore, we would expect some of the same areas in the brain to be involved in both memory and language. Seeing words and hearing words depend on the primary visual cortex in the occipital lobe and the primary auditory cortex in the temporal lobe, respectively (Fig. 17.14*a*). And speaking words depends on motor centers in the frontal lobe. Functional imaging by a technique known as PET (positron emission tomography) supports these suppositions (Fig. 17.14*b*).

From studies of patients with speech disorders, it has been known for some time that damage to the motor speech (Broca's) area results in an inability to speak, while damage to the sensory speech (Wernicke's) area results in the inability to comprehend speech. Actually, any disruption of pathways between various centers of the brain can contribute to an inability to comprehend our environment and use speech correctly.

An additional observation pertaining to language and speech is the recognition that the left brain and right brain have different functions. Only the left hemisphere, not the right, contains a Broca's area and Wernicke's area. In an attempt to cure epilepsy in the early 1940s, the corpus callosum was surgically severed in some patients. Later studies showed that these split-brain patients could only name objects seen by the left hemisphere. If objects were viewed only by the right hemisphere, a split-brain patient could choose the proper object for a particular use but was unable to name it. Based on these and various other studies, the popular idea developed that the left brain can be contrasted with the right brain along these lines:

Left Hemisphere	Right Hemisphere
Verbal	Nonverbal, visual, spatial
Logical, analytical	Intuitive
Rational	Creative

Further, researchers generally came to believe that one hemisphere was dominant in each person, accounting in part for personality traits. However, recent studies suggest that the hemispheres simply process the same information differently. The right hemisphere is more global in its approach, whereas the left hemisphere is more specific. For example, if only the left cerebral cortex is functional, a person has difficulty communicating the emotions involved with language. Therefore, both sides of the brain play a role in language use.

Figure 17.14 Brain regions involved in language and speech. **a.** The labeled areas are thought to be involved in speech comprehension and use. **b.** These PET images show the cortical pathway for reading words and then speaking them. Red indicates the most active areas of the brain, and blue indicates the least active areas.

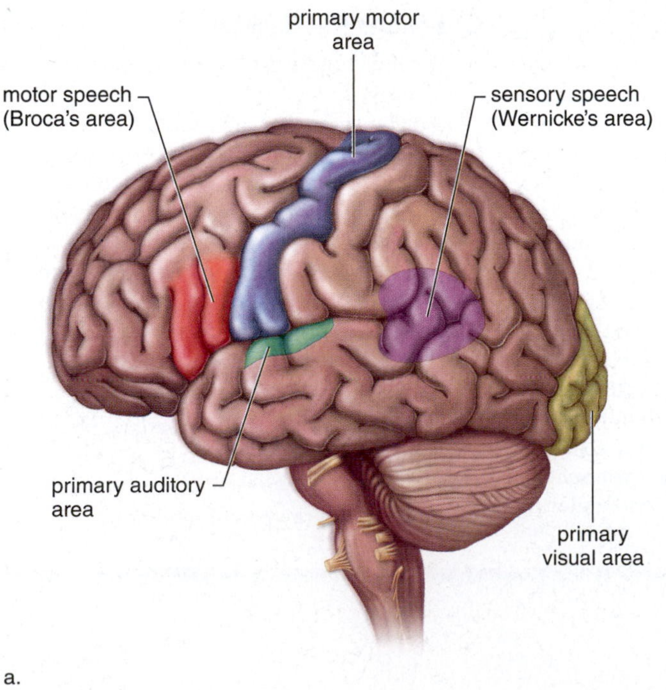
primary motor area
motor speech (Broca's area)
sensory speech (Wernicke's area)
primary auditory area
primary visual area
a.

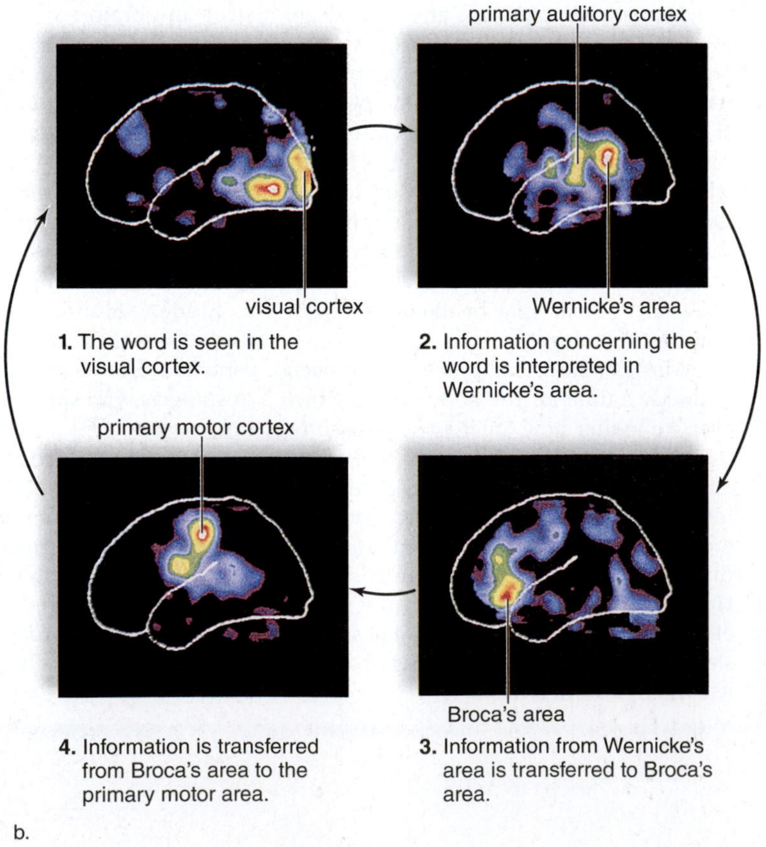
primary auditory cortex
visual cortex
Wernicke's area
1. The word is seen in the visual cortex.
2. Information concerning the word is interpreted in Wernicke's area.
primary motor cortex
Broca's area
4. Information is transferred from Broca's area to the primary motor area.
3. Information from Wernicke's area is transferred to Broca's area.
b.

Check Your Progress 17.4

1. Describe some possible behavioral changes that may be exhibited by a person who developed a tumor in his or her amygdala.
2. Discuss how the unfortunate case of S.M. helped prove the role of the hippocampus in long-term memory.
3. Summarize the evidence from the mouse study that showed the role of glutamate in memory formation.
4. Describe the specific functions of Broca's and Wernicke's areas.

17.5 The Peripheral Nervous System

Learning Outcomes

Upon completion of this section, you should be able to

1. Describe the overall anatomy of the peripheral nervous system, including the cranial nerves and spinal nerves.
2. Distinguish between the somatic and autonomic divisions of the peripheral nervous system.
3. Contrast the overall functions of the sympathetic and parasympathetic divisions of the autonomic nervous system.

The **peripheral nervous system (PNS)** lies outside the central nervous system and is composed of nerves and ganglia. In the PNS, **nerves** are bundles of axons. The axons that occur in nerves are called **nerve fibers.**

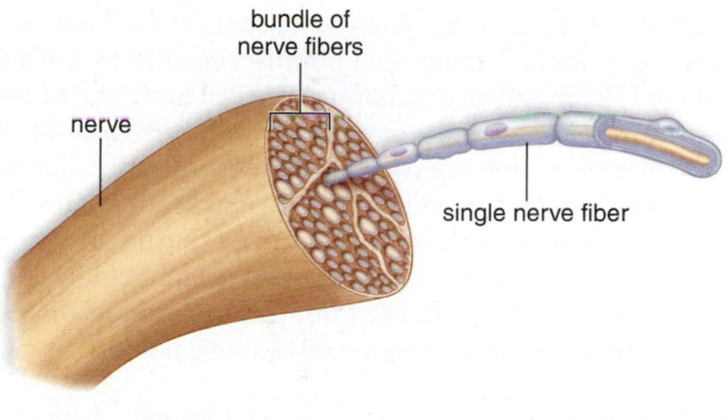
bundle of
nerve fibers
nerve
single nerve fiber

Sensory fibers carry information to the CNS, and motor fibers carry information away from the CNS. **Ganglia** (sing., ganglion) are swellings associated with nerves that contain collections of cell bodies.

MP3 Organization of the Nervous System

Humans have 12 pairs of **cranial nerves** attached to the brain. By convention, the pairs of cranial nerves are referred to by Roman numerals (Fig. 17.15*a*). Some of the cranial nerves contain only sensory input fibers, while others contain only motor output fibers. The remaining are mixed nerves that contain both sensory and motor fibers. Cranial nerves are largely concerned with the head, neck, and facial regions of the body. However, the vagus nerve (X) has branches not only to the pharynx and larynx, but also to most of the internal organs.

The **spinal nerves** of humans emerge in 31 pairs from between openings in the vertebral column of the spinal cord. Each spinal nerve originates when two short branches, or roots,

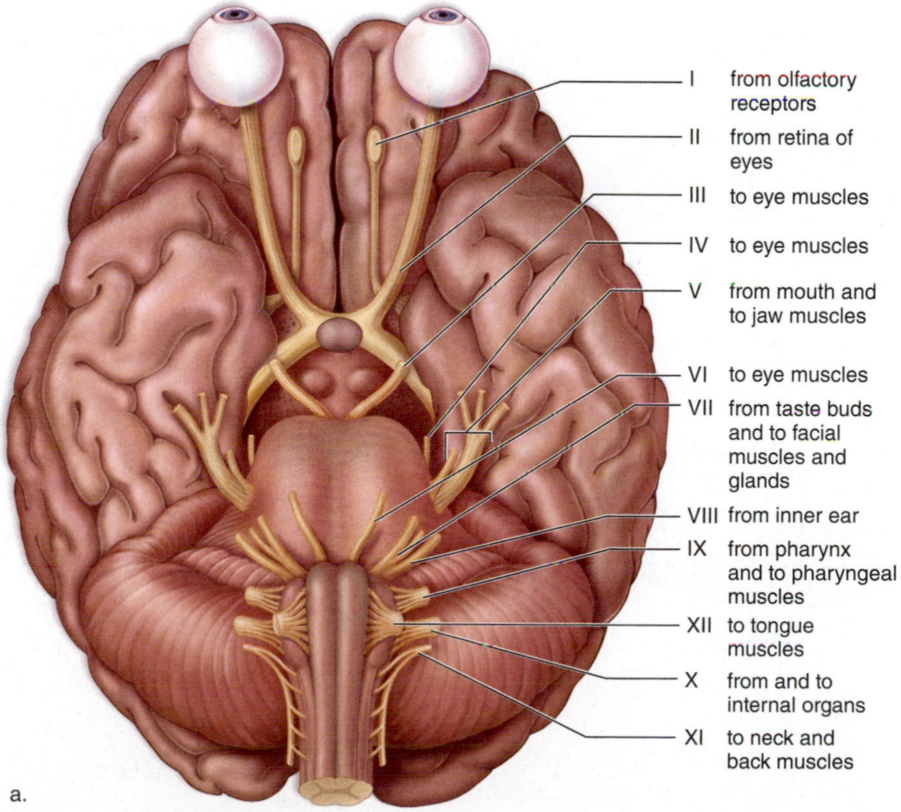
I from olfactory receptors
II from retina of eyes
III to eye muscles
IV to eye muscles
V from mouth and to jaw muscles
VI to eye muscles
VII from taste buds and to facial muscles and glands
VIII from inner ear
IX from pharynx and to pharyngeal muscles
XII to tongue muscles
X from and to internal organs
XI to neck and back muscles

a.

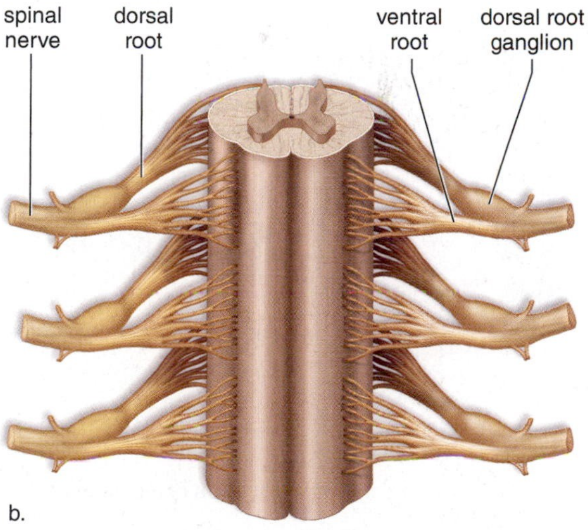
spinal
nerve
dorsal
root
ventral
root
dorsal root
ganglion

b.

Figure 17.15 Cranial and spinal nerves. **a.** Ventral surface of the brain showing the attachments of the 12 pairs of cranial nerves. **b.** Cross section of the spinal cord showing three pairs of spinal nerves. The human body has 31 pairs of spinal nerves altogether, and each spinal nerve has a dorsal root and a ventral root attached to the spinal cord.

join together (Fig. 17.15*b*). The dorsal root (at the back) contains sensory fibers that conduct impulses inward (toward the spinal cord) from sensory receptors. The cell body of a sensory neuron is in a **dorsal root ganglion.** The ventral root (at the front) contains motor fibers that conduct impulses outward (away from the cord) to effectors. Notice, then, that all spinal nerves are mixed nerves that contain many sensory and motor fibers. Each spinal nerve serves the particular region of the body in which it is located. For example, the intercostal muscles of the rib cage are innervated by thoracic nerves.

Somatic System

The PNS is subdivided into the somatic system and the autonomic system. The **somatic system** serves the skin, skeletal muscles, and tendons. It includes nerves that take sensory information from external sensory receptors to the CNS and motor commands away from the CNS to the skeletal muscles. Some actions in the somatic system are due to **reflex actions,** which are automatic responses to a stimulus. A reflex occurs quickly, without our even having to think about it. Other actions are voluntary, and these always originate in the cerebral cortex, as when we decide to move a limb.

The Reflex Arc

A reflex arc is a nerve pathway that carries out a reflex. Reflexes are programmed, built-in circuits that allow for protection and survival. They enable the body to react swiftly to stimuli that could disrupt homeostasis. They are present at birth and require no conscious thought to take place. Figure 17.16 illustrates the path of a withdrawal reflex that involves only the spinal cord. If your hand touches a sharp pin, sensory receptors in the skin generate nerve impulses that move along sensory fibers through the dorsal root ganglia toward the spinal cord. Sensory neurons that enter the cord dorsally pass signals on to many interneurons. Some of these interneurons synapse with motor neurons whose short dendrites and cell bodies are in the spinal cord. Nerve impulses travel along these motor fibers to an effector, which brings about a response to the stimulus. In this case, the effector is a muscle, which contracts so that you withdraw your hand from the pin. Various other reactions are also possible—you will most likely look at the pin, wince, and cry out in pain. This whole series of responses occurs because some of the interneurons involved carry nerve impulses to the brain. The brain makes you aware of the stimulus and directs these other reactions to it. You do not feel the pain until your brain receives the information and interprets it.

Figure 17.16 A somatic reflex arc showing the path of a spinal reflex. A stimulus (e.g., sharp pin) causes sensory receptors in the skin to generate nerve impulses that travel in sensory axons to the spinal cord. Interneurons integrate data from sensory neurons and then relay signals to motor axons. Motor axons convey nerve impulses from the spinal cord to a sketetal muscle, which contracts. Movement of the hand away from the pin is the response to the stimulus.

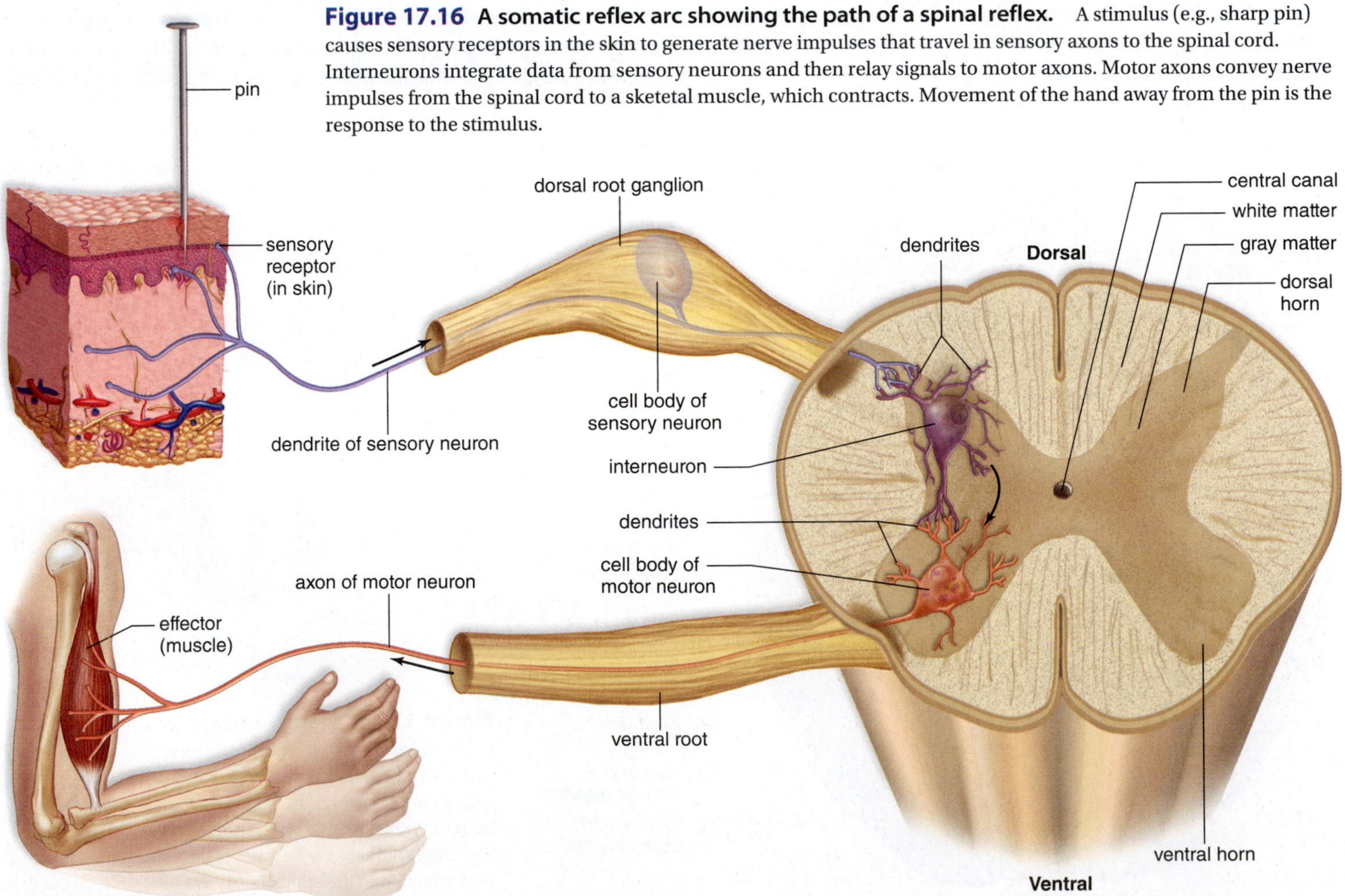

pin
sensory receptor (in skin)
dendrite of sensory neuron
dorsal root ganglion
cell body of sensory neuron
interneuron
dendrites
cell body of motor neuron
axon of motor neuron
effector (muscle)
ventral root
dendrites
Dorsal
central canal
white matter
gray matter
dorsal horn
ventral horn
Ventral

Autonomic System

The **autonomic system** of the PNS regulates the activity of cardiac and smooth muscle and glands. The system is composed of the sympathetic and parasympathetic divisions (Fig. 17.17). These two divisions have several features in common: (1) they function automatically and usually in an involuntary manner; (2) they innervate all internal organs; and (3) for each signal, they utilize two motor neurons that synapse at a ganglion. The first neuron has a cell body within the CNS, and its axon is called the preganglionic fiber. The second neuron has a cell body within the ganglion, and its axon is termed the postganglionic fiber.

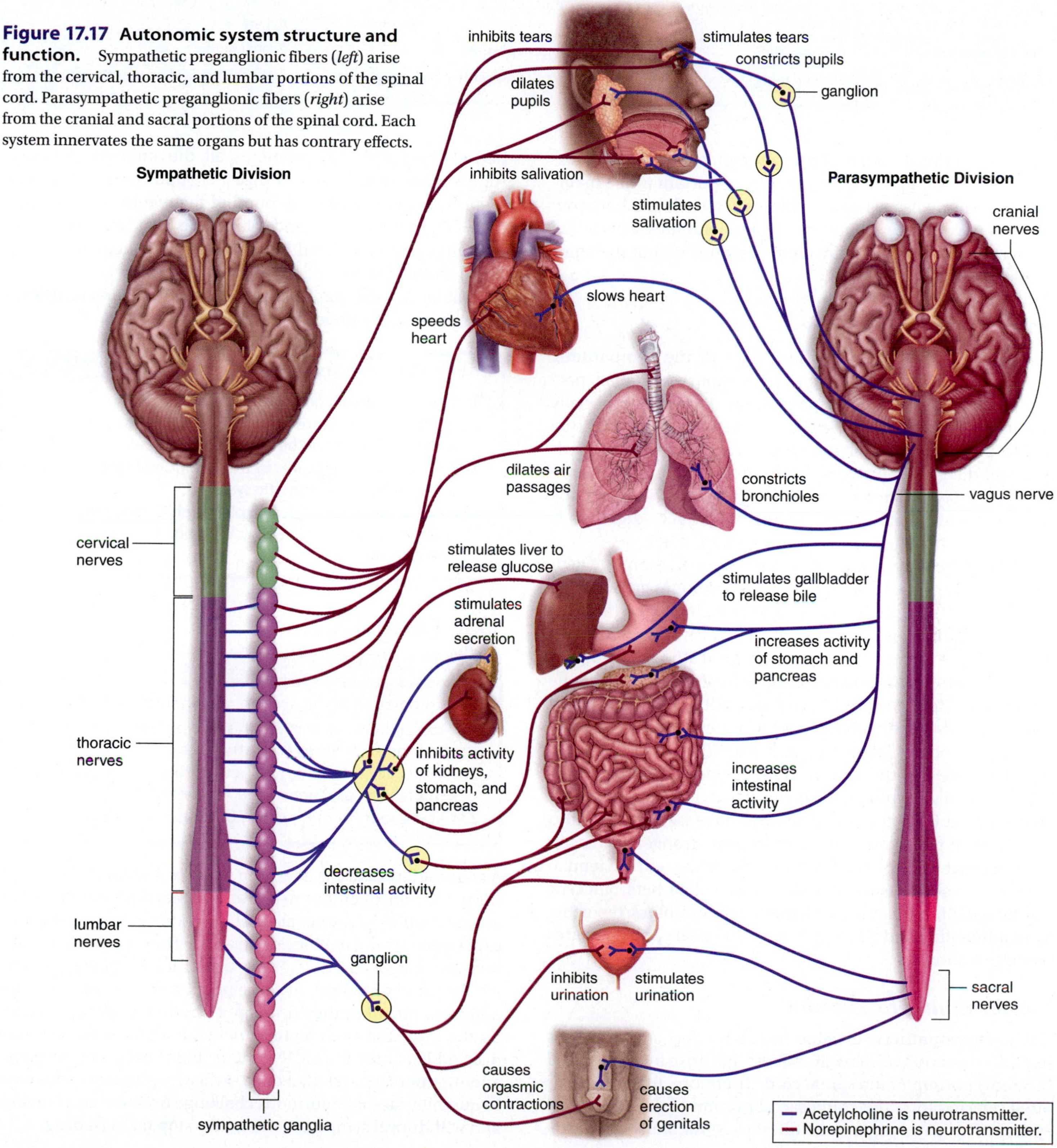
Sympathetic Division
Parasympathetic Division
inhibits tears
stimulates tears
constricts pupils
ganglion
dilates pupils
inhibits salivation
stimulates salivation
cranial nerves
slows heart
speeds heart
dilates air passages
constricts bronchioles
vagus nerve
cervical nerves
stimulates liver to release glucose
stimulates gallbladder to release bile
stimulates adrenal secretion
increases activity of stomach and pancreas
thoracic nerves
inhibits activity of kidneys, stomach, and pancreas
increases intestinal activity
decreases intestinal activity
lumbar nerves
ganglion
inhibits urination
stimulates urination
sacral nerves
causes orgasmic contractions
causes erection of genitals
sympathetic ganglia
Acetylcholine is neurotransmitter.
Norepinephrine is neurotransmitter.

Figure 17.17 Autonomic system structure and function. Sympathetic preganglionic fibers (*left*) arise from the cervical, thoracic, and lumbar portions of the spinal cord. Parasympathetic preganglionic fibers (*right*) arise from the cranial and sacral portions of the spinal cord. Each system innervates the same organs but has contrary effects.

TABLE 17.1 Comparison of Somatic Motor and Autonomic Motor Pathways

		Autonomic Motor Pathways	
	Somatic Motor Pathway	***Sympathetic***	***Parasympathetic***
Type of control	Voluntary/involuntary	Involuntary	Involuntary
Number of neurons per message	One	Two (preganglionic shorter than postganglionic)	Two (preganglionic longer than postganglionic)
Location of motor fiber	Most cranial nerves and all spinal nerves	Thoracolumbar spinal nerves	Cranial (e.g., vagus) and sacral spinal nerves
Neurotransmitter	Acetylcholine	Norepinephrine	Acetylcholine
Effectors	Skeletal muscles	Smooth and cardiac muscle, glands	Smooth and cardiac muscle, glands

Reflex actions, such as those that regulate the blood pressure and breathing rate, are especially important to the maintenance of homeostasis. These reflexes begin when the sensory neurons in contact with internal organs send messages to the CNS. They are completed by motor neurons within the autonomic system.

Sympathetic Division

The cell bodies of preganglionic fibers in the **sympathetic division** are located in the middle, or thoracolumbar, portion of the spinal cord. The sympathetic division is especially important during emergency situations when you might be required to fight back or run away. It accelerates the heartbeat and dilates the bronchi. Active muscles, after all, require a ready supply of glucose and oxygen. On the other hand, the sympathetic division inhibits the digestive tract—digestion is not an immediate necessity if you are under attack.

One of the functions of the sympathetic division is to activate the adrenal medulla to secrete the hormones epinephrine (adrenaline) and norepinephrine (NE) into the blood (also see Fig. 20.7). In fact, one special preganglionic sympathetic neuron goes right to the adrenal gland without stopping at a ganglion. The adrenaline and NE released by the adrenal medulla bind to receptors on various cell types, adding to the "fight or flight" response. Interestingly, the neurotransmitter released by sympathetic postganglionic axons is also NE. A small amount of the NE released by these axons spills over into the blood. During times of high sympathetic nerve activity, the amount of NE entering the blood from these neurons increases significantly.

Because they cause the heart to beat stronger and faster and constrict certain blood vessels, adrenaline and NE tend to increase blood pressure. Certain drugs, called beta blockers, can be used in patients with hypertension to block the activity of adrenaline and NE, which slows the heart and decreases blood pressure.

Parasympathetic Division

The **parasympathetic division** includes a few cranial nerves (e.g., the vagus nerve), as well as fibers that arise from the sacral (bottom) portion of the spinal cord. Therefore, this division is often referred to as the craniosacral portion of the autonomic system. The parasympathetic division, sometimes called the housekeeper division, promotes all the internal responses we associate with "rest and digest." Opposite to the sympathetic system, it causes the pupil of the eye to contract, promotes digestion of food, and retards the heartbeat. The major neurotransmitter utilized by the parasympathetic division is acetylcholine (ACh).

Table 17.1 compares and contrasts the characteristics of the somatic and autonomic systems.

Check Your Progress 17.5

1. Contrast cranial and spinal nerves.
2. Explain why, if you touch a hot stove, you usually withdraw your hand before feeling the pain.
3. Summarize the features of the autonomic system that are different from the somatic system.
4. Provide the neurological explanation for the following: You eat a big lunch, then go for a jog, during which your stomach starts to ache.

17.6 Drug Abuse

Learning Outcomes

Upon completion of this section, you should be able to

1. Explain the specific mechanisms whereby the most common illicit drugs affect the brain.
2. Classify illicit drugs as to whether they have a depressant, stimulant, or psychoactive effect on the body.
3. List the long-term effects of drug use on the body.

A wide variety of drugs affect the nervous system and can alter the mood and/or emotional state. The first time someone tries a drug such as nicotine, alcohol, or cocaine, the person may experience strong feelings of pleasure. However, with continued use, larger amounts of the drug may be needed to create the same sensation. At some point, the users may become addicted, meaning they develop a physiological dependence on the drug and can't feel normal without it. At the extreme, drug addicts may prioritize getting "high" over food, hygiene, and interpersonal relationships. Even drug abusers who want to quit may face an enormous challenge from intense cravings and withdrawal symptoms when they stop taking a drug.

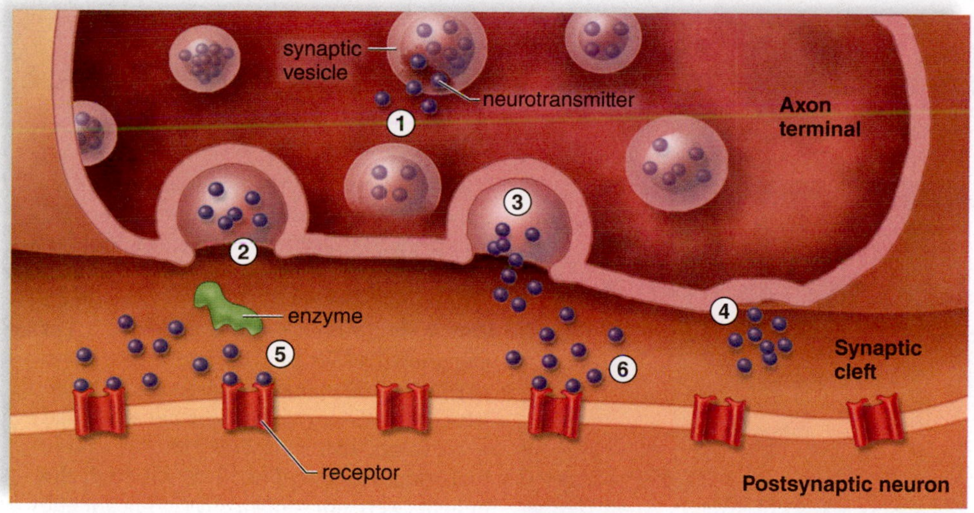
synaptic vesicle
neurotransmitter
Axon terminal
1
2
3
4
5
6
enzyme
Synaptic cleft
receptor
Postsynaptic neuron

Figure 17.18 Drug actions at a synapse. A drug can ① cause the neurotransmitter (NT) to leak out of a synaptic vesicle into the axon terminal; ② prevent release of NT into the synaptic cleft; ③ promote release of NT into the synaptic cleft; ④ prevent reuptake of NT by the presynaptic membrane; ⑤ block the enzyme that causes breakdown of the NT; or ⑥ bind to a receptor, mimicking the action of an NT.

Most illicit drugs affect the action of a particular neurotransmitter at synapses in the brain (Fig. 17.18). Stimulants are drugs that increase the likelihood of neuron excitation, and depressants decrease the likelihood of excitation. Increasingly, researchers believe that dopamine is one of the main neurotransmitters in the brain that is responsible for mood. Cocaine is known to potentiate the effects of dopamine by interfering with its uptake from synaptic clefts (Fig. 17.19). Many of the new medications developed to counter drug dependence and mental illness affect the release, reception, or breakdown of dopamine.

Some Specific Drugs of Abuse

Nicotine

In 2009, about 19% of U.S. high school students, and 5% of middle school students, were cigarette smokers. A 2010 survey of more than 14,000 students at 119 colleges found that about one-third of college students smoke or use some other type of tobacco product, up from 22% in the 1990s. When a person smokes a cigarette or cigar, or chews tobacco, nicotine is quickly distributed throughout the body. This causes a release of epinephrine from the adrenal cortex, which increases blood sugar levels and causes an initial feeling of stimulation. Then, as blood sugar falls, depression and fatigue set in, leading the smoker to seek more nicotine. In the CNS, nicotine causes neurons to release excess dopamine, which has a reinforcing effect that leads to dependence on the drug. Nicotine induces both physiological and psychological dependence, and once they start, many tobacco users find it extremely difficult to give up the habit. Withdrawal symptoms include headache, stomach pain, irritability, and insomnia. Pregnant women who smoke are more likely to have stillborn or premature infants, or babies born with other health problems.

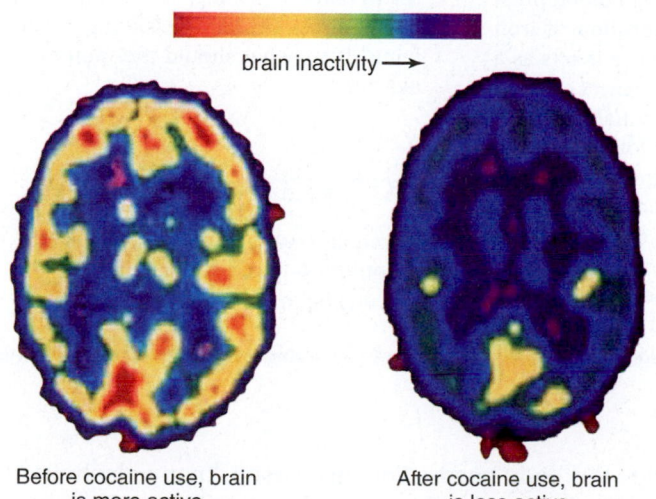

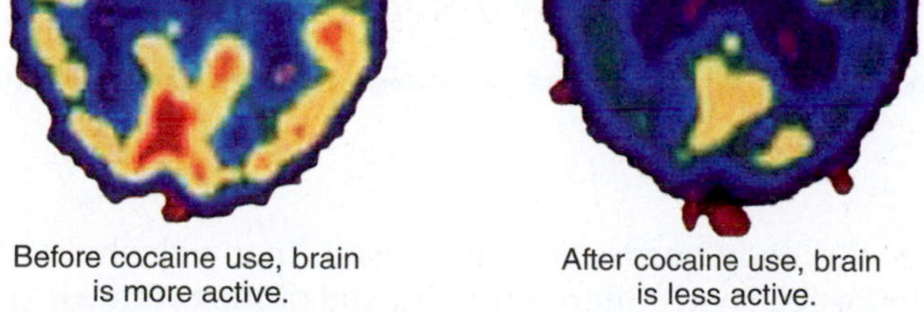
brain inactivity →
Before cocaine use, brain is more active.
After cocaine use, brain is less active.

Figure 17.19 Drug abuse. PET scans show that the usual activity of the brain is reduced and many areas become inactive after a person injects cocaine.

Alcohol (Ethanol)

With the possible exception of caffeine (see the Health feature, "Caffeine: Good or Bad for You?" on page 330), alcohol is the most used and abused drug among America's teenagers. According to a 2009 national survey, 42% of high school students reported drinking some alcohol, and 24% binge drank (five-plus drinks in one setting) during the 30 days preceding the survey. Among adults, 51% were regular drinkers, and 15% were binge drinkers (more than five drinks for men, or four drinks for women, within about a two-hour period).

When ingested, ethanol acts as a drug in the brain by influencing the action of gamma aminobutyrate (GABA), an inhibitory neurotransmitter, and other neurotransmitters like serotonin and dopamine. Alcohol is primarily metabolized in the liver, where it disrupts the normal workings of this organ so that fats cannot be broken down. Fat accumulation, the first stage of liver deterioration, begins after only a single night of heavy drinking. If heavy drinking continues, fibrous scar tissue is deposited during the next stage of liver damage. If heavy drinking stops at this point, the liver may still recover and become normal again. If not, the final and irrevocable stage,

SCIENCE IN YOUR LIFE ▶ **HEALTH**

Caffeine: Good or Bad for You?

Red Bull. Rockstar. Monster. Full Throttle. Walk down the beverage aisle of most U.S. grocery stores, and the colorful labels of these and other "energy drinks" compete for your attention (Fig. 17B). That's no surprise—the United States is the world's largest consumer of energy drinks, totaling roughly 290 million gallons in 2007.

Around 90% of American adults report using caffeine every day. Caffeine, or 1,3,7-trimethylxanthine, occurs naturally in over 60 plant species worldwide. In plants it is thought to function as a natural pesticide, killing certain insects that attempt to feed on the plant. It is naturally present in coffee, tea, and chocolate. It is also added to soft drinks and energy drinks, possibly to add flavor, though studies seem to prove that most people cannot taste the difference when caffeine is added to a caffeine-free beverage.

NOS
amp

Figure 17B A sampling of energy drinks found in most U.S. grocery stores. The caffeine content of these beverages varies widely.

For humans, caffeine is a central nervous system stimulant. It readily crosses the blood-brain barrier, where it seems to block the effects of adenosine, which it resembles chemically. In the brain, adenosine normally seems to suppress brain activity, and by blocking this effect, caffeine can increase alertness. However, in response to continuing doses of caffeine, brain cells adjust by producing more adenosine receptors, so that increased doses of caffeine are needed for the same effect. If a regular user suddenly stops ingesting caffeine, withdrawal symptoms—such as headache, inability to concentrate, and insomnia—may ensue within 12 to 24 hours and last from one to five days.

Americans still get most of their caffeine from that tried-and-true source, the cup of coffee, which averages about 100 mg of caffeine. By comparison, a 12-oz can of Diet Coke contains about 38 mg (about 10 mg more than regular Coca-Cola), Mountain Dew has 55 mg, and Diet Pepsi Max has 69 mg. Energy drinks vary even more, with Red Bull at 80 mg, Full Throttle and Monster doubling that to around 160. At the extreme end of the spectrum, a 20-oz bottle of Fixx Energy Drink contains 500 mg of caffeine! In 2008, 100 scientists and physicians wrote a letter to the U.S. Food and Drug Administration asking for more regulation and better labeling of increasingly popular energy drinks because their high caffeine content puts young drinkers at possible risk for caffeine intoxication.

As a stimulant, high doses of caffeine can have serious consequences. Adverse effects seen with high doses of caffeine include nervousness, irritability, tremors, insomnia, headaches, and high blood pressure. Caffeine reduces absorption of iron from the intestine, and because it acts as a diuretic (see section 16.3), increases calcium loss in the urine. Some individuals, perhaps 20%, are more sensitive to lower levels of caffeine, so even a few milligrams may cause problems for them. Pregnant and nursing women should also be careful to limit their caffeine intake. Fortunately, caffeine-induced deaths are rare. The lethal dose for adults has been estimated at 5,000 mg.

So there is definitely a downside to caffeine use (or abuse). But might there be an upside? In 2008, researchers at Harvard published results of an 18-year study of thousands of men and women. They found that women who drank two to three cups of caffeinated coffee per day had a 25% lower risk of death from heart disease, and women who drank five to seven cups per week had a 7% lower risk of death from all causes than non-drinkers. For men, although the benefits were not statistically significant, the scientists found no health risk to drinking up to seven cups of coffee per day. Other studies have suggested a link between caffeine and increased athletic and academic performance, as well as a reduced risk of Parkinson disease, migraine headaches, and even diabetes. So although the debate over caffeine's risks versus benefits may never be completely resolved, for many people, it is possible that consumption of a moderate amount of caffeine may end up doing more good than harm.

Video Caffeine Withdrawal

Questions to Consider

1. Estimate how much caffeine you consume per day. What effects can you notice after consuming caffeine?
2. Would you agree that manufacturers should be required to display the caffeine content of soft drinks and energy drinks on the label? Why or why not?
3. Do you believe that there should be any restrictions on selling, for example, an energy drink containing 500 mg of caffeine? If so, what should the minimum age be?

cirrhosis of the liver, occurs. Liver cells die, harden, and turn orange (cirrhosis means orange).

Alcohol is a carbohydrate and as such, can be converted into energy for the body to use. However, alcoholic beverages lack the vitamins, minerals, essential amino acids, and fatty acids the body needs. For this reason, many alcoholics are vitamin-deficient, undernourished, and prone to illness.

Based upon extensive research, the Surgeon General of the United States recommends that pregnant women drink no alcohol at all. Alcohol crosses the placenta freely and causes fetal alcohol syndrome in newborns, which is characterized by intellectual disability and various physical defects.

Marijuana

Marijuana is the most commonly used illegal drug in the United States. Surveys vary, but in 2006 about 16% of young adults aged 18 to 25 reported using marijuana in the last month, and 40% of the U.S. population has tried it at least once. Marijuana, or "pot," refers to the dried flowering tops, leaves, and stems of the Indian hemp plant *Cannabis sativa,* which contain and are covered by a resin that is rich in THC (tetrahydrocannabinol). Researchers have found that THC binds to a receptor in the brain for anandamide, a naturally occurring neurotransmitter that is important for short-term memory processing, and perhaps for feelings of contentment. Usually, marijuana is smoked in a cigarette form called a "joint," or in pipes or other devices, but it can also be consumed. The occasional user experiences a mild euphoria along with alterations in vision and judgment, which result in distortions of space and time. Motor incoordination, including the inability to speak coherently, takes place. Heavy use can result in hallucinations, anxiety, depression, body image distortions, paranoid reactions, and similar psychotic symptoms. Some researchers believe that long-term marijuana use leads to brain impairment. Fetal cannabis syndrome, which resembles fetal alcohol syndrome, has also been reported.

Cocaine and Crack

Cocaine is an alkaloid derived from the shrub *Erythroxylon coca.* It is sold in powder form and as crack, a more potent extract. Crack is cocaine that comes in a rocklike crystal that can be heated and its vapors inhaled. The term "crack" refers to the crackling sound heard when it is heated. Cocaine prevents the synaptic uptake of dopamine, and this causes the user to experience a rush sensation. The epinephrine-like effects of dopamine account for the state of arousal that lasts for several minutes after the rush experience.

A cocaine binge can go on for days, after which the user suffers a crash. During the binge period, the user is hyperactive and has little desire for food or sleep but an increased sex drive. During the crash period, the user is fatigued, depressed, and irritable, has memory and concentration problems, and displays no interest in sex. Indeed, men who use "coke" often become impotent.

Cocaine causes extreme physical dependence. With continued use, the body begins to make less dopamine to compensate for a seeming excess supply. The user subsequently experiences tolerance, withdrawal symptoms, and an intense craving for cocaine. Overdosing on cocaine can cause seizures and cardiac and respiratory arrest. It is possible that long-term cocaine abuse causes brain damage, and babies born to addicts suffer withdrawal symptoms and may have neurological and behavioral problems.

Heroin

Heroin is derived from morphine, an alkaloid of opium. Once heroin is injected into a vein, a feeling of euphoria, along with relief of any pain, occurs within three to six minutes. Heroin binds to receptors meant for the endorphins, naturally occurring neurotransmitters that kill pain and produce a feeling of tranquility. With repeated heroin use, the body's production of endorphins decreases. Tolerance develops so that the user needs to take more of the drug just to prevent withdrawal symptoms, and the original euphoria is no longer felt. Heroin withdrawal symptoms include perspiration, dilation of pupils, tremors, restlessness, abdominal cramps, vomiting, and increased blood pressure and respiratory rate. People who are excessively dependent may experience convulsions, respiratory failure, and death. Infants born to women who are physically dependent also experience those withdrawal symptoms.

"Club" Drugs

Ecstasy (MDMA [3,4-methylenedioxymethamphetamine]), Rohypnol (flunitrazepam), and ketamine are examples of drugs that are abused by many teens and young adults who attend night-long dances called raves or trances. Of course, these drugs may also be abused by others as well. MDMA is chemically similar to methamphetamine. Many users say that "XTC" increases their feelings of well-being and friendliness, even love, for other people. However, it can cause many of the same side effects as other stimulants, such as increase in heart rate and blood pressure, muscle tension, and blurred vision. In high doses, MDMA can interfere with temperature regulation, leading to hyperthermia followed by damage to the liver, kidneys, and heart. Chronic MDMA use can also damage memory and lead to depression. Perhaps because of these side effects, several recent surveys have reported that MDMA use has been decreasing among youths aged 12 to 17.

Rohypnol is in the same class of drugs as diazepam (Valium), a popular sedative. When mixed with alcohol, Rohypnol, or "roofies," can render victims incapable of resisting, for example, sexual assault. It can also produce anterograde amnesia, which means victims may not remember events they experienced while under the influence of the drug. Ketamine, or "Special K," is an anesthetic that veterinarians use to perform surgery on animals. When administered to humans, it typically induces a dreamlike state, sometimes with hallucinations. Because it can render the user unable to move, ketamine has also been used as a date rape drug. At higher doses, ketamine can cause dangerous reductions in heart and respiratory functions.

Methamphetamine

Methamphetamine, also called "meth" or "crank," is a powerful CNS stimulant. Meth is often synthesized or "cooked" in makeshift home laboratories, usually starting with either ephedrine or pseudoephedrine, common ingredients in many cold and asthma medicines. Many states have passed laws making these medications more difficult to purchase. Meth is usually produced as a powder (speed) that can be snorted, or as crystals (crystal meth) that can be smoked. The most immediate effect is an initial "rush" of euphoria, due to high amounts of dopamine released in the brain. The user may also experience an increased energy level, alertness, and elevated mood. After the initial rush, a state of high agitation typically occurs that, in some individuals, leads to violent behavior. Chronic use can lead to paranoia, irritability, hallucinations, insomnia, tremors, hyperthermia, cardiovascular collapse, and death. Similar to cocaine, long-term meth abuse renders the brain less able to produce normal levels of dopamine. The user may have strong cravings for more meth, without being able to reach a satisfactory high.

Video Meth and the Brain

"Bath Salts"

One of the newer illicit drugs to become available are "bath salts," synthetic powders sold in colorful containers under innocent-sounding names like "Blue Wave" and "Vanilla Sky." However, unlike the innocuous sodium salts and fragrances that have been added to bathwater for centuries, these new products often contain synthetic amphetamine-like chemicals such as methylenedioxypyrovalerone and mephedrone. Although they are often labeled "Not For Human Consumption," when ingested, injected, or snorted, these chemicals inhibit the reuptake of several neurotransmitters, producing a euphoric sensation. Unfortunately, a long list of side effects includes severe chest pain, hallucinations, extreme paranoia, and violent episodes often leading to suicide. In response to dramatic increases in bath salts–related emergency-room visits and deaths, in October 2011 the U.S. Drug Enforcement Administration banned these chemicals for at least one year. Unfortunately, illicit drugmakers are presumably working to develop new "designer drugs" to circumvent these bans.

Check Your Progress 17.6

1. Examine Figure 17.18 and hypothesize why, since illicit drugs affect activity of neurotransmitters, it might take increasing amounts of a drug to obtain the same effect over time.
2. Detail the specific modes of action of several illicit drugs.

17.7 Disorders of the Nervous System

Learning Outcomes

Upon completion of this section, you should be able to

1. Describe two abnormalities seen in the brain of Alzheimer disease patients.
2. Compare the mechanism causing the following diseases: Parkinson disease, MS, stroke, meningitis, and prion disease.
3. Describe the symptoms seen in several disorders of the brain and spinal cord and discuss why it is difficult to find cures for each.

A myriad of abnormal conditions can affect the nervous system. We will consider only a few of the most serious, common, and/or interesting ones here.

Disorders of the Brain

Alzheimer disease (AD) is the most common cause of dementia, an impairment of brain function significant enough to interfere with the patient's ability to carry on daily activities. Signs of AD may appear before the age of 50, but the condition is usually seen in individuals past the age of 65. It is estimated that at least 40% of individuals older than 80 develop AD. One of the early symptoms is loss of memory, particularly for recent events. Characteristically, the person asks the same question repeatedly and becomes disoriented even in familiar places. Gradually, the person loses the ability to perform any type of daily activity and becomes bedridden. Death is often due to pneumonia resulting from the patient's debilitated state.

In AD patients, abnormal neurons are present throughout the brain but especially in the hippocampus and amygdala. These neurons have two abnormalities: (1) plaques, containing a protein called beta amyloid, envelop the axons, and (2) neurofibrillary tangles are in the axons (Fig. 17.20) and may extend upward to surround the nucleus. The neurofibrillary tangles arise because a protein called tau no longer has the correct shape. Tau holds microtubules in place so that they support the structure of neurons. When tau changes shape, it grabs onto other tau molecules, and the tangles result.

Video Apples for Alzheimer

Researchers are working to discover the cause of beta amyloid plaques and neurofibrillary tangles. Several genes

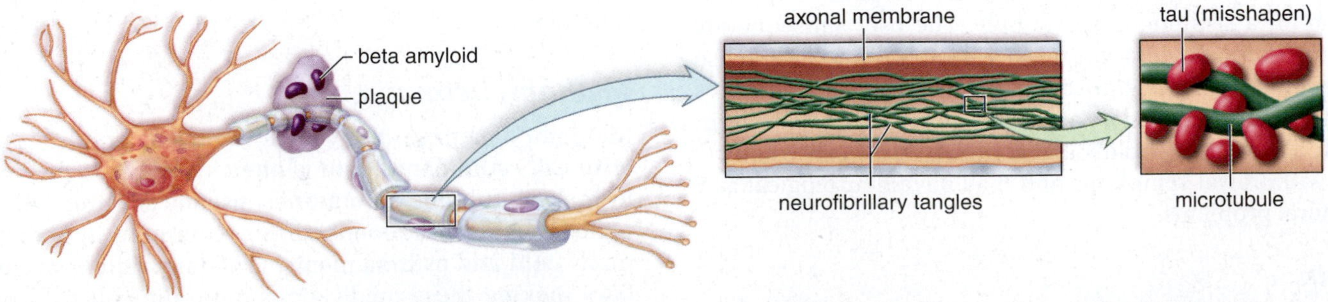
beta amyloid
plaque
axonal membrane
neurofibrillary tangles
tau (misshapen)
microtubule

Figure 17.20 Alzheimer disease. Some of the neurons of patients with AD have beta amyloid plaques and neurofibrillary tangles. AD neurons are present throughout the brain but concentrated in the hippocampus and amygdala.

that predispose a person to AD have been identified, but it is unknown why inheritance of these genes leads to AD. Moreover, although several treatments have been tried to date, none has been shown to prevent or cure AD. A variety of medications can help ease symptoms such as memory loss, agitation, and depression, but brain cells still die. Most drugs currently available for treating symptoms of AD are acetylcholinesterase inhibitors, which increase the amount of time that acetylcholine can reside in synapses to transmit nerve impulses.

Parkinson disease (PD) is characterized by a gradual loss of motor control, typically beginning between the ages of 50 and 60. Eventually the person develops a wide-eyed, unblinking expression, involuntary tremors of the fingers and thumbs, muscular rigidity, and a shuffling gait. Speaking and performing ordinary daily tasks becomes laborious. In Parkinson patients, the basal nuclei (see section 17.3) function improperly because of a degeneration of the dopamine-releasing neurons in the brain. Without dopamine, the excessive excitatory signals from the motor cortex result in the symptoms of PD.

Unfortunately, it is not possible to give PD patients dopamine directly because of the impermeability of the capillaries serving the brain. However, symptoms can be alleviated by giving patients L-dopa, a chemical that can be changed to dopamine in the body, until too few cells are left to do the job. Then patients must turn to a number of controversial surgical procedures. In 1998 the actor Michael J. Fox, who had been diagnosed with PD seven years earlier at the age of 30, had a surgical procedure called a thalamotomy, which has been successful in reducing the symptoms of some PD patients. However, in his biography *Lucky Man: A Memoir,* Fox admits that after a temporary improvement, his PD symptoms returned in full force. In 2007, the first report was published on treating PD with gene therapy, in which a virus engineered to produce the required neurotransmitter was injected into the brains of 12 PD patients. In one year, motor function had improved in these patients from 25–65% of normal.

Multiple sclerosis (**MS**) is the most common neurological disease that afflicts young adults. Nearly 350,000 people in the United States have MS, with approximately 10,000 new cases diagnosed each year. MS affects the myelin sheath of neurons in the white matter of the brain. Although the initiating cause remains unknown, MS is considered an autoimmune disease in which the patient's own white blood cells attack the myelin, oligodendrocytes, and eventually, neurons in the CNS. Specifically, MS is characterized by an infiltration of lymphocytes and macrophages into the brain, brain stem, optic nerves, and spinal cord. The most common symptoms of MS include fatigue, vision problems, limb weakness and/or pain, and abnormal sensations such as numbness or tingling. These symptoms, however, can be similar to those of several other brain disorders, so a technique called magnetic resonance imaging (MRI) can be very helpful in both diagnosing and monitoring the disease (Fig. 17.21). MS can also take several clinical forms, from the milder relapsing-remitting form to more relentlessly progressive forms. Although MS is not generally considered a fatal disease, people with MS die an average of seven years earlier than the general population. No treatment has been shown to reverse the damage to the myelin of MS patients, but several medications are effective in slowing the progression of the disease, especially in the early stages.

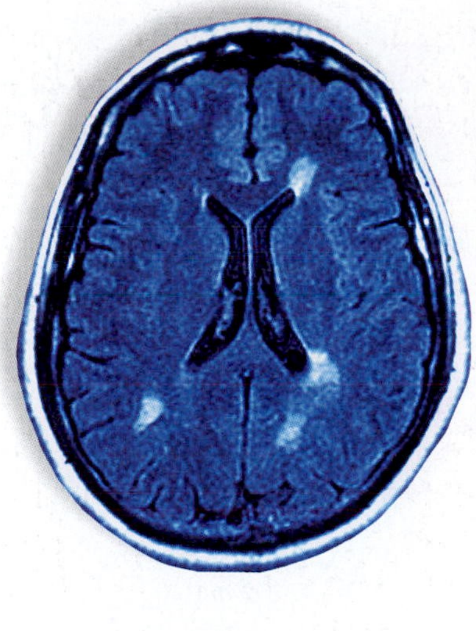

Figure 17.21 Multiple sclerosis. This MRI scan shows the brain of an MS patient. The white areas, also known as MS plaques, are sites of active inflammation and myelin destruction.

A **stroke** results in disruption of the blood supply to the brain. There are two major forms of stroke, hemorrhagic and ischemic. In hemorrhagic stroke, bleeding occurs into the brain due to leakage from small arteries, often after years of damage by high blood pressure. Ischemic stroke occurs when there is a sudden loss of blood supply to an area of the brain, usually due to a thrombus or thromboembolism (see section 12.5). In both types of stroke, the symptoms depend on the amount and specific area of brain tissue affected. However, some degree of paralysis and aphasia (loss of speech) are common signs. The risk of stroke increases with age, and African Americans have a higher incidence than other races in the United States. Physicians treating stroke patients must be careful to determine which type they are dealing with. For example, tissue plasminogen activator (tPA) can be used to help dissolve clots when treating ischemic stroke, but could have disastrous consequences if administered to a hemorrhagic stroke patient.

Video Brain Healing

Meningitis is an infection of the meninges that surround the brain and spinal cord. Most cases of meningitis are caused by bacteria or viruses that have usually gained access to the meninges after first infecting the blood, middle ear, or sinuses. Certain viruses are also capable of infecting the brain and meninges by infecting peripheral nerve cells and being transported back to the brain. Because cells of the immune system have somewhat limited access to the brain, the infection may spread into the brain tissue. Bacterial meningitis is especially serious, even life-threatening if untreated. Physicians may diagnose meningitis based on clinical signs such as a stiff neck, fever, and headache, and may confirm the diagnosis by examining a sample of cerebrospinal fluid, usually obtained by a lumbar puncture. Most cases of bacterial meningitis are caused by organisms called *Haemophilus influenzae* or *Neisseria meningitidis.* Outbreaks of *N. meningitidis,* also called meningococcus, have occurred recently among college students. Fortunately, vaccines are available to prevent infection with both organisms.

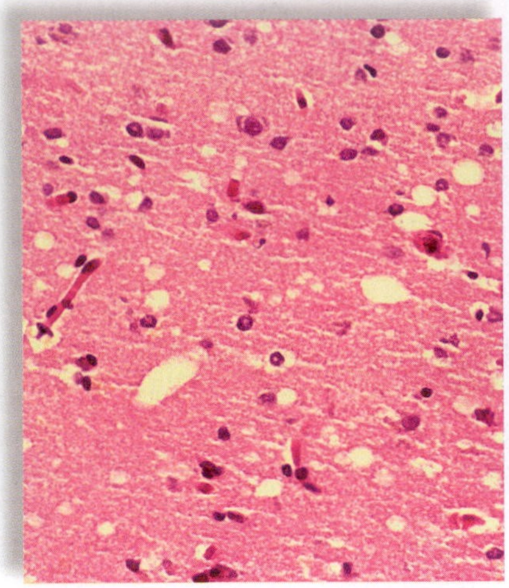

Figure 17.22 Effect of prions on the brain. In this stained section of brain tissue from a patient with a prion disease, the spongiform changes (holes) in the tissue are due to accumulation of the abnormal prion protein.

Several brain diseases of humans and animals are caused by **prions,** which are infectious agents that most researchers believe are composed of protein only (see Chapter 28). Examples of human diseases include kuru, Creutzfeldt-Jakob disease, and fatal familial insomnia. Prion diseases of animals include scrapie in sheep, mad cow disease in cattle, and chronic wasting disease in deer. Mad cow disease seems able to cross the species barrier and infect humans who consume infected tissues; about 200 human cases have been confirmed, mostly in the United Kingdom. All prion diseases are characterized by long incubation periods between initial exposure to the agent and onset of clinical signs, followed by rapid progress of the disease and, uniformly, death. Prion diseases are thought to result from the accumulation in the tissues of an abnormal form of a normal host (prion) protein. When this occurs in the brain, nerve cells die, forming areas that look like holes, which pathologists refer to as spongiform changes (Fig. 17.22). These diseases are also unique in that they can be inherited or can be spread by either eating infected tissues (as in mad cow disease) or through transplantation of certain organs or tissues (such as corneas) from an infected individual. Therefore, prion diseases are also called transmissible spongiform encephalopathies (TSEs). No treatment has yet been shown to delay the inevitably fatal progression of these diseases.

Disorders of the Spinal Cord

Spinal cord injuries may result from car accidents or other trauma. Because little or no nerve regeneration is possible in the CNS, any resulting disability is usually permanent. The cord may be completely cut across (called a transection) or only partially severed (a partial section). The location and extent of the damage produce a variety of effects, depending on the partial or complete stoppage of impulses passing up and down the spinal cord. If the spinal cord is completely transected, no sensations or somatic motor impulses are able to pass the point of injury. If the injury is between the first thoracic vertebra (T1) and the second lumbar vertebra (L2), paralysis of the lower body and legs occurs, a condition known as *paraplegia* (see Fig. 19.6 for the specific locations of these vertebrae). If the injury is in the neck region, the entire body below the neck is usually affected, a condition called *quadriplegia.* If the injury is above the second or third cervical vertebrae (C2 or C3), nervous control of respiration is also affected. In 1995, actor Christopher Reeve, best known for his movie role as "Superman," was thrown from his horse, crushing the top two cervical vertebrae. He immediately lost control of his arms and legs, and had to be placed on a respirator. Through an intense exercise and physical therapy program, Reeve regained the ability to move slightly (e.g., taking tiny steps while being held upright in a pool). Unfortunately, he died of cardiac arrest in 2004.

Amyotrophic lateral sclerosis (ALS), also known as Lou Gehrig's disease, is a rare but devastating condition that affects the motor nerve cells of the spinal cord. ALS is incurable, and most people die within five years of being diagnosed, mainly due to failure of the respiratory muscles. However, about 10% of ALS patients survive for ten or more years. Typically the first signs of ALS occur between the ages of 40 and 60 in the form of mild weakness or tremors in one or more limbs, starting in the fingers or toes and gradually involving the larger muscles nearer to the body. The cause(s) of nerve cell death in ALS are unclear. About 5–10% of all ALS cases are genetic in nature, but the majority of victims have no known risk factors. The cerebrospinal fluid of people with ALS often contains abnormally high levels of the neurotransmitter *glutamate,* which can be toxic to neurons. Autoimmune reactions may also play a role, as may environmental toxins.

Other than Lou Gehrig, the most famous person with ALS is probably physicist Steven Hawking, who has lived for over 30 years with the disease (Fig. 17.23). The fact that Hawking has remained an active scientist and author shows that ALS has very little effect on brain function.

Figure 17.23 Physicist Steven Hawking. Dr. Hawking has been a productive scientist and author for over 30 years, even though he has amyotrophic lateral sclerosis (ALS).

Disorders of the Peripheral Nerves

Guillain-Barré syndrome (GBS) is an inflammatory disease that causes demyelination of peripheral nerve axons. It is thought to result from an abnormal immune reaction to one of several types of infectious agents. In GBS patients, antibodies formed against these bacteria or viruses seem to cross-react with the myelin. Symptoms usually begin with weakness or unsteadiness two to four weeks after an infection or immunization. The muscle weakness typically starts in the lower limbs and ascends to the upper limbs over a period of days to weeks. The disease process may eventually affect the respiratory muscles, so that mechanical ventilation is required. In most cases, however, the paralysis begins to improve in a few weeks, and most patients make a full recovery within six to twelve months.

Myasthenia gravis (MG) is an autoimmune disorder in which antibodies are formed that react against the acetylcholine receptor (AChR) at the neuromuscular junction of the skeletal muscles. Normally, when the action potential arrives at the presynaptic terminal of peripheral motor nerves, acetylcholine (ACh) is released into the synaptic cleft and binds to the AChR on the muscle cell, stimulating contraction. In MG, antibodies bind to the AChR and block binding of ACh, preventing muscle stimulation.

Interestingly, the most common symptom of MG is weakness of the muscles of the eye or eyelid, although over time the weakness usually spreads to the face, trunk, and then the limbs. In severe cases, the respiratory muscles can be affected, leading to death if untreated. Even though there is no cure, MG patients often respond very well to treatment with immunosuppressive drugs and inhibitors of acetylcholinesterase (AChE), an enzyme that normally destroys the ACh after it is released. A technique called *plasmapheresis* can also be used in both GBS and MG to remove the harmful antibodies from the patient's blood.

Check Your Progress 17.7

1. Summarize the major symptoms of Alzheimer disease and explain the age at which these symptoms may appear.
2. Describe how L-dopa treats Parkinson disease.
3. Explain why MS is an autoimmune disease.
4. Discuss the following: One cause of bacterial meningitis, *Haemophilus influenzae*, can also be found in the throat (pharynx) of healthy people. What factors might determine when *H. influenzae* causes disease?

Case Study Conclusion

Years after her initial diagnosis, Sarah's condition continues to slowly deteriorate. Although she is still able to drive, she must walk with a cane and has occasional issues with her vision, plus fatigue, weakness, and other problems.

Multiple sclerosis (MS) is an autoimmune disease affecting the myelin. The initiating cause of MS is unknown, but most researchers agree that there are most likely many contributing factors, including infections and other environmental influences. Studies have shown that the risk of contracting MS depends in part on where in the world a person lives, although the specific factor related to geographical location has not been identified.

Although there is no cure, many individuals with MS are able to control their symptoms with the help of immunosuppressive medications such as corticosteroids and beta interferon. However, in most cases the disease does progress slowly, often over a period of decades.

MEDIA STUDY TOOLS

www.mhhe.com/maderinquiry14

Enhance your study of this chapter with study tools and practice tests. Also ask your instructor about the resources available through ConnectPlus, including LearnSmart, the media-rich eBook, interactive learning tools, and animations.

Mc Graw Hill connect plus+ BIOLOGY

SUMMARIZE

17.1 Nervous Tissue

The **nervous system** coordinates and regulates the body's other systems:

- In most animals, the nervous system is divided into a central nervous system (CNS) and peripheral nervous system (PNS).
- Nervous tissue contains **neurons** and **neuroglia.** There are three types of neurons. **Sensory neurons** take information from sensory receptors to the CNS; **interneurons** occur within the CNS; and **motor neurons** take information from the CNS to muscles or glands.

- A neuron is composed of **dendrites,** a **cell body,** and an **axon.** Some axons are covered by a **myelin sheath.** In the CNS, myelin is synthesized by neuroglia called **oligodendrocytes;** in the PNS, myelin is produced by **Schwann cells.** Gaps in the myelin covering of axons are called **nodes of Ranvier.**
- In the CNS, areas called **gray matter** contain short, nonmyelinated nerves, while **white matter** has neurons with longer, myelinated axons that run in **tracts.**

7.2 Transmission of Nerve Impulses

The nervous system transmits information using **nerve impulses:**

- In order to transmit a nerve impulse, an **action potential** must be generated. When an axon is not conducting a nerve impulse, the inside of the axon is negative (–70 mV) compared to the outside. **Sodium-potassium pumps** actively transport Na^+ out of an axon and K^+ to the inside of an axon. The **resting potential** is due to the leakage of K^+ to the outside of the neuron. When an axon is conducting a nerve impulse, Na^+ first moves into the axoplasm, causing membrane depolarization, until the inside of the axon is positive (+35 mV) compared to the outside. K^+ then moves out of the axoplasm, causing repolarization. Immediately following an action potential, there is a **refractory period,** during which the sodium gates are unable to open.
- In myelinated axons, action potentials are generated only at the unmyelinated nodes of Ranvier. Therefore, the action potentials travel much faster as they "jump" from node to node, a phenomenon known as **saltatory conduction.**
- At the junction of an axon with another nerve cell, or a muscle cell, a **synapse** forms, where the cell membranes are very close but don't touch. In order to send messages, chemicals called **neurotransmitters** are released into this gap, or **synaptic cleft.**
- After being released by the presynaptic neuron, binding of the neurotransmitter to receptors in the postsynaptic membrane can cause excitation or inhibition.
- **Integration** is the summing of excitatory and inhibitory signals.

17.3 The Central Nervous System

The **central nervous system (CNS)**, consisting of the spinal cord and brain, receives and integrates sensory input and formulates motor output:

- The spinal cord and brain are protected by membranes called **meninges.** Spaces between the meninges are filled with **cerebrospinal fluid,** which also fills hollow cavities in the brain called **ventricles.**
- The **spinal cord** sends sensory information to the brain, receives motor output from the brain, and carries out reflex actions.
- The **brain** can be divided into several functional and anatomical regions. Sensation, reasoning, learning and memory, and language and speech take place in the **cerebrum.** The **cerebral cortex** is a thin layer of gray matter covering the cerebrum. The cerebral cortex of each **cerebral hemisphere** has four lobes: a frontal, parietal, occipital, and temporal lobe. The primary motor area, in the frontal lobe, sends out motor commands to lower brain centers, which pass them on to motor neurons. The primary somatosensory area, in the parietal lobe, receives sensory information from lower brain centers in communication with sensory neurons. The cortex also contains processing centers for reasoning and speech. Deep within the cerebrum, the **basal nuclei** integrate motor activity.
- Other regions of the brain include the **hypothalamus,** which controls homeostasis, and the **thalamus,** which specializes in sending sensory input on to the cerebrum. The **cerebellum** primarily coordinates skeletal muscle contractions. The **brain stem** includes the **midbrain, medulla oblongata,** and **pons,** which relay messages to and from the cerebrum, and have centers for vital functions, such as breathing and the heartbeat. Also found in the brain stem is the **reticular activating system (RAS)**, a relay center that is also involved in alertness.

17.4 The Limbic System and Higher Mental Functions

The **limbic system** is associated with emotions as well as higher brain functions like **memory** and **learning:**

- The limbic system contains the **hippocampus,** which acts as a conduit for sending messages to long-term memory and retrieving them once again, and the **amygdala,** which adds emotional overtones to memories. In addition, it has various connections between the cerebral cortex and the hypothalamus, the thalamus, and the basal nuclei.
- Two basic types of memory are short-term and long-term memory. The hippocampus is essential for formation of long-term memories. At the chemical level, this process involves glutamate and is known as long-term potentiation.
- Language and speech involve many areas of the brain, but Broca's area and Wernicke's area are indispensable. The right and left hemispheres play different, though overlapping, roles.

17.5 The Peripheral Nervous System

The **peripheral nervous system (PNS)** lies outside of the CNS, where it conducts nerve impulses to and from the CNS.

- The PNS contains only nerves and ganglia. **Nerves** are bundles of axons; **ganglia** are collections of cell bodies. Each axon in a nerve is called a **nerve fiber.**
- Humans have 12 paired **cranial nerves,** and 31 paired **spinal nerves,** that emerge from the CNS. These may carry sensory or motor impulses, or both. The cell bodies of spinal nerves carrying sensory information lie in the **dorsal root ganglia.**
- The PNS can be subdivided into the somatic system and the autonomic system. The **somatic system** takes information from external sensory receptors to the CNS and voluntary motor commands from the cerebrum to skeletal muscles. It also carries out **reflex actions** in which interneurons in the spine quickly send messages to motor nerves without CNS input.
- The **autonomic system** involuntarily controls the smooth muscle of the internal organs and glands. The **sympathetic division** is associated with responses that occur during times of stress, and the **parasympathetic system** is associated with responses that occur during times of relaxation.

17.6 Drug Abuse

Most illicit drugs affect neurotransmitters in the brain, either promoting or inhibiting the action of a particular neurotransmitter. Examples include the following:

- Nicotine, which increases dopamine release in the brain.

- Cocaine, which inhibits uptake of dopamine.
- Heroin, which binds to receptors for natural painkillers called endorphins.

17.7 Disorders of the Nervous System

The precise cause of many of the disorders affecting the CNS remains unclear, and many are untreatable. Examples of brain disorders include the following:

- **Alzheimer disease** seems to be due to an accumulation of abnormal proteins in the brain.
- **Parkinson disease** results from a degeneration of dopamine-secreting neurons.
- **Multiple sclerosis** (**MS**) is the result of autoimmune destruction of myelin and myelin-producing cells.
- **Stroke,** whether hemorrhagic or ischemic, occurs when the blood supply to the brain is disrupted.
- Infections of the brain include **meningitis** and **prion** diseases.
- Similar to brain disorders, disorders of the spine and peripheral nerves can be serious and difficult to treat. Injuries to the spinal cord can cause a spectrum of problems, from weakness in one limb to complete paralysis.
- **Amyotrophic lateral sclerosis** (**ALS**) is a fatal disease resulting from a loss of motor neurons in the spinal cord.
- **Guillain-Barré syndrome** and **myasthenia gravis** are both autoimmune disorders that attack the peripheral nerves.

ASSESS

Testing Yourself

Choose the best answer for each question.

1. An interneuron can relay information
   a. from a sensory neuron to a motor neuron.
   b. only from a motor neuron to another motor neuron.
   c. only from a sensory neuron to another sensory neuron.
   d. None of these are correct.
2. In the PNS, myelin is formed by
   a. oligodendroglial cells.
   b. Schwann cells.
   c. axons.
   d. motor neurons.
3. Gray matter is found on the _____ of the brain and the _____ of the spinal cord.
   a. inside, inside
   b. outside, outside
   c. outside, inside
   d. inside, outside
4. Which of these correctly describes the distribution of ions on either side of an axon when it is not conducting a nerve impulse?
   a. more sodium ions (Na^+) outside and more potassium ions (K^+) inside
   b. more K^+ outside and less Na^+ inside
   c. charged protein outside and Na^+ and K^+ inside
   d. Na^+ and K^+ outside and water only inside
   e. chloride ions (Cl^-) outside and K^+ and Na^+ inside
5. Label this diagram of the action potential.

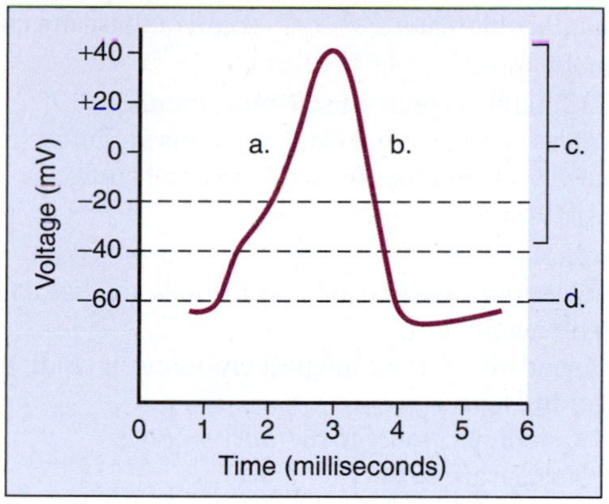

+40
+20
0
−20
−40
−60
Voltage (mV)
a.
b.
c.
d.
0 1 2 3 4 5 6
Time (milliseconds)

6. During the refractory period,
   a. sodium gates are open.
   b. sodium gates are closed.
   c. sodium and potassium gates are open.
   d. None of these are correct.
7. Transmission of an action potential across a synapse is accomplished by the
   a. movement of Na^+ and K^+.
   b. release of a neurotransmitter by a dendrite.
   c. release of a neurotransmitter by an axon.
   d. release of a neurotransmitter by a cell body.
   e. All of these are correct.
8. The summing up of inhibitory and excitatory signals is
   a. excitation.
   b. integration.
   c. depolarization.
   d. All of these are correct.
9. A spinal nerve takes nerve impulses
   a. to the CNS.
   b. away from the CNS.
   c. both to and away from the CNS.
   d. from the CNS to the spinal cord.
10. Label this diagram of the brain.

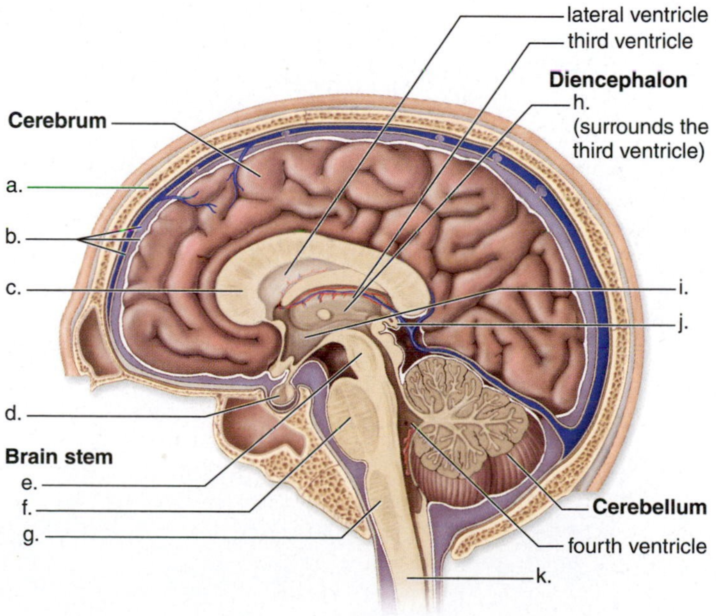

lateral ventricle
third ventricle
Diencephalon
h.
(surrounds the third ventricle)
Cerebrum
a.
b.
c.
i.
j.
d.
Brain stem
e.
f.
g.
Cerebellum
fourth ventricle
k.

11. Homeostasis is dependent on which of these?
   a. hypothalamus
   b. medulla oblongata
   c. parasympathetic division
   d. All of these are correct.
12. The cerebellum
   a. coordinates skeletal muscle movements.
   b. receives sensory input from the joints and muscles.
   c. receives motor input from the cerebral cortex.
   d. All of these are correct.
13. The limbic system
   a. involves portions of the cerebral lobes, the basal nuclei, and the diencephalon.
   b. is responsible for our deepest emotions, including pleasure, rage, and fear.
   c. is a system necessary to memory storage.
   d. All of these are correct.
14. The sympathetic division of the autonomic system will
   a. increase heart rate and digestive activity.
   b. decrease heart rate and digestive activity.
   c. cause pupils to constrict.
   d. None of these are correct.
15. Somatic is to skeletal muscle as autonomic is to
   a. cardiac muscle.
   b. smooth muscle.
   c. gland.
   d. All of these are correct.
16. A damaged hippocampus can cause problems associated with
   a. long-term memory.
   b. episodic memory.
   c. consolidating short-term into long-term memory.
   d. All of these are correct.
17. Damage to the Wernicke's area of the brain could cause problems in
   a. vision.
   b. taste.
   c. smell.
   d. speech.
18. Effects of drugs can include
   a. prevention of neurotransmitter release.
   b. prevention of reuptake by the presynaptic membrane.
   c. blockages to a receptor.
   d. All of these are correct.
   e. None of these are correct.
19. A stroke results from
   a. hemorrhage in the brain.
   b. a blockage of a blood vessel in the brain.
   c. a deficiency of tissue plasminogen activator.
   d. Either a or b is correct.
20. What do Creutzfeldt-Jakob and mad cow disease have in common?
   a. Both are fatal diseases.
   b. Both are caused by prions.
   c. Both cause spongiform changes in the brain.
   d. All of these are correct.

ENGAGE

Thinking Critically

1. How do you suppose scientists and physicians have determined which areas of the brain are responsible for different activities, as shown in Figures 17.10, 17.11, and 17.14?
2. Dopamine is a key neurotransmitter in the brain. Patients with Parkinson disease suffer because of a lack of this chemical, while abusers of drugs such as nicotine, cocaine, and methamphetamine enjoy an enhancement of dopamine activity. Would these drugs be a possible treatment, or even a cure, for Parkinson disease? Why or why not?
3. What are some of the factors that make brain diseases such as Alzheimer disease, Parkinson disease, and MS so difficult to treat?

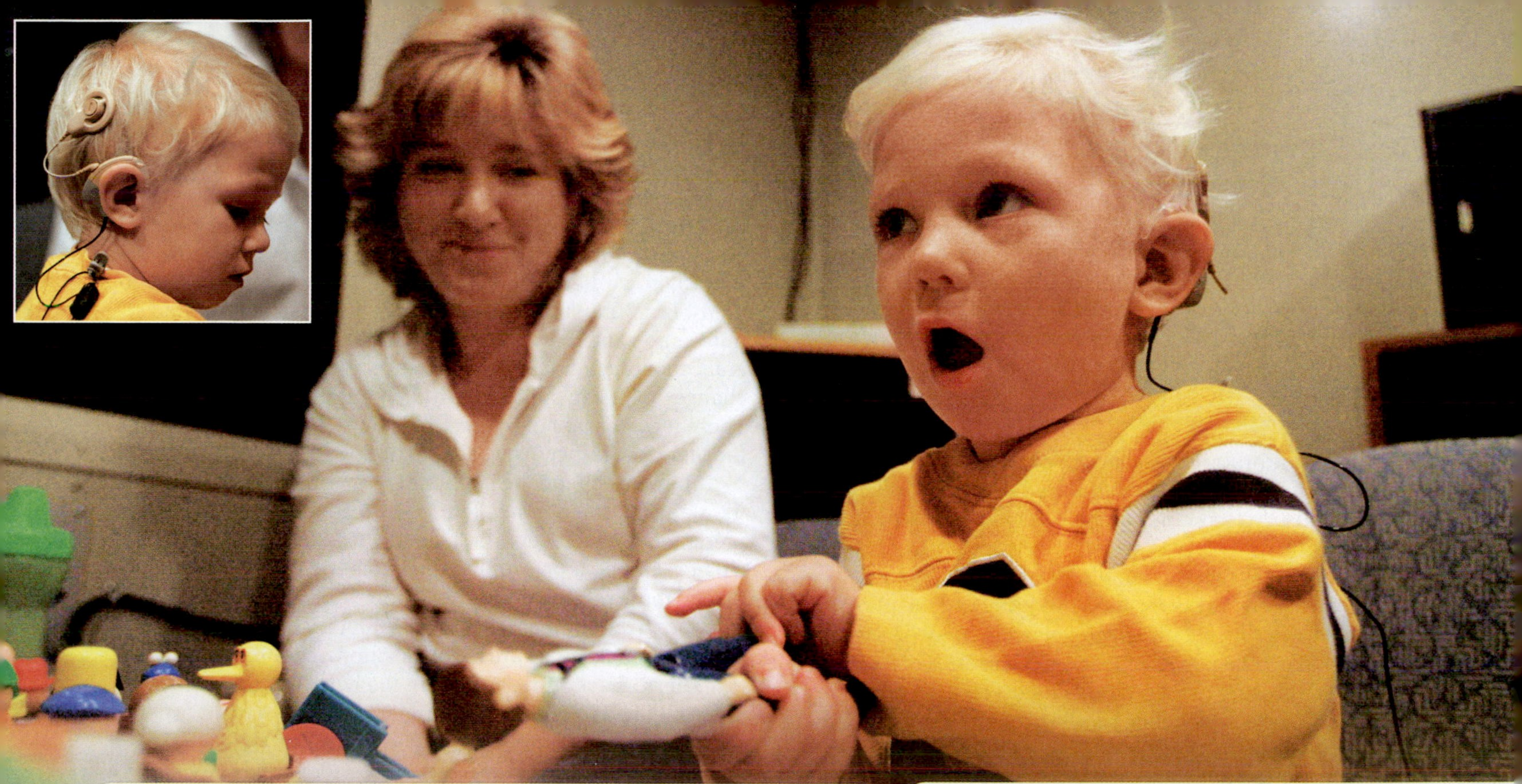

Senses 18

CASE STUDY Hearing for the Deaf

Jacob and Marlene were married for almost a year when they learned Marlene was pregnant with their first child. Almost immediately, they began discussing what they would do if their baby was born with hearing problems. Marlene had been born deaf, due to a genetic condition in which her inner ear did not develop properly. She had learned sign language and even became a teacher at a school for the deaf. However, because of technical advances since Marlene was a child, it was now possible to provide hearing for children with many forms of deafness.

One treatment option is cochlear implants. Cochlear implants are different from hearing aids, which simply amplify sounds. A cochlear implant consists of an external device that sits behind the ear, and an internal device that is surgically implanted under the skin (see inset). The external part picks up sounds from the environment and converts them into electrical impulses, which the internal implant sends directly to different regions of the cochlear nerve and then to the brain. As of 2010, about 219,000 people worldwide had received cochlear implants, including 28,400 children and 42,600 adults in the United States. Marlene and Jacob decided to have implants placed in their son, Tony, shortly after his first birthday, the earliest age approved by the FDA.

As you read through the chapter, think about the following questions:

1. Where is the cochlea located, and what is its specific function?
2. Some types of deafness are caused by abnormalities affecting the cochlea, whereas others affect only the auditory nerves. Which type(s) would be most amenable to treatment with cochlear implants, and why?
3. Compare how sound and light are detected. How could the technology associated with cochlear implants be used to help people with vision problems?

CHAPTER OUTLINE

BEFORE YOU BEGIN

Before beginning this chapter, take a few moments to review the following discussions:

Section 11.4 What specific types of receptors in the skin provide information about the external environment?

Figure 17.1 What are the roles of the central and peripheral nervous systems in the body?

Section 17.3 What are the functions of the primary somatosensory area and somatosensory association area of the cerebral cortex?

18.1 Sensory Receptors and Sensations

Learning Outcomes

Upon completion of this section, you should be able to

1. Distinguish between interoceptors and exteroceptors.
2. Describe the four categories of sensory receptors.
3. Explain the significance of sensory transduction and sensory adaptation.

Sensory receptors are specialized cells that detect certain types of *stimuli* (sing., stimulus). Based on the source of the stimulus, sensory receptors can be classified as interoceptors or exteroceptors.

Interoceptors receive stimuli from inside the body, such as changes in blood pressure, blood volume, and the pH of the blood. Interoceptors are directly involved in homeostasis and are regulated by a negative feedback mechanism. For example, when blood pressure rises, pressoreceptors signal a regulatory center in the brain, which then sends out nerve impulses to the arterial walls, causing them to relax so that the blood pressure falls. Once the pressoreceptors are no longer stimulated, the system shuts down.

Exteroceptors are sensory receptors that detect stimuli from outside the body, such as those related to touch, taste, smell, vision, hearing, and equilibrium (Table 18.1). Their function is to send messages to the central nervous system to report changes in environmental conditions.

Types of Sensory Receptors

Based on the nature of the stimulus, interoceptors and exteroceptors can be further classified into four categories: chemoreceptors, photoreceptors, mechanoreceptors, and thermoreceptors.

Chemoreceptors respond to chemical substances in the immediate vicinity. As Table 18.1 indicates, taste and smell are dependent on this type of sensory receptor, but certain chemoreceptors in various other organs are sensitive to internal conditions. For example, chemoreceptors that monitor blood oxygen (O_2) levels are located in the carotid arteries and aorta. If O_2 levels fall, the breathing rate increases.

Photoreceptors respond to light energy. Our eyes contain photoreceptors that are sensitive to light rays, and thereby provide us with the sense of vision. Stimulation of the photoreceptors known as rod cells results in black-and-white vision, whereas stimulation of the photoreceptors known as cone cells results in color vision (see section 18.4).

Mechanoreceptors are stimulated by mechanical forces, which most often result in pressure of some sort. For example, when we hear, airborne sound waves are converted to fluid-borne pressure waves that can be detected by mechanoreceptors in the inner ear. Similarly, mechanoreceptors are responding to fluid-borne pressure waves when we detect changes in gravity and motion, helping us keep our balance. These receptors are in the vestibule and semicircular canals of the inner ear, respectively (see section 18.6).

The sense of touch is dependent on pressure receptors in the skin and other organs that are sensitive to strong or slight pressures. Pressoreceptors located in certain arteries detect changes in blood pressure, and stretch receptors in the lungs detect the degree of lung inflation. Proprioceptors, which respond to the stretching of muscle fibers, tendons, joints, and ligaments, make us aware of the position of our limbs.

Thermoreceptors, located in the hypothalamus and skin, are stimulated by changes in temperature. Those that respond when temperatures rise are called warmth receptors, and those that respond when temperatures fall are called cold receptors.

MP3 Sensations and Receptors

How Sensation Occurs

Sensory receptors respond to environmental stimuli by generating nerve impulses. **Detection** occurs when environmental changes, such as pressure to the fingertips or light to the eye, stimulate sensory receptors. **Sensation** occurs when nerve impulses arrive at the cerebral cortex of the brain. **Perception** occurs when the brain interprets the meaning of stimuli.

As we discussed in section 17.5, sensory receptors are the first element in a reflex arc. We are only aware of a reflex action when sensory information reaches the brain. At that time, the brain integrates this information with other information received from other sensory receptors. After all, if you burn yourself and quickly remove your hand from a hot stove, the brain receives information not only from your skin, but also from your eyes, nose, and all sorts of sensory receptors.

Some sensory receptors are free nerve endings or encapsulated nerve endings, while others are specialized cells closely associated with neurons. The plasma membrane of a sensory receptor contains proteins that react to the stimulus. For example,

TABLE 18.1 Examples of Exteroceptors

Sensory Receptor	Stimulus	Category	Sense	Sensory Organ
Taste cells	Chemicals	Chemoreceptor	Taste	Taste buds
Olfactory cells	Chemicals	Chemoreceptor	Smell	Olfactory epithelium
Rod cells and cone cells in retina	Light rays	Photoreceptor	Vision	Eye
Hair cells in spiral organ	Sound waves	Mechanoreceptor	Hearing	Ear
Hair cells in semicircular canals	Motion	Mechanoreceptor	Rotational equilibrium	Ear
Hair cells in vestibule	Gravity	Mechanoreceptor	Gravitational equilibrium	Ear

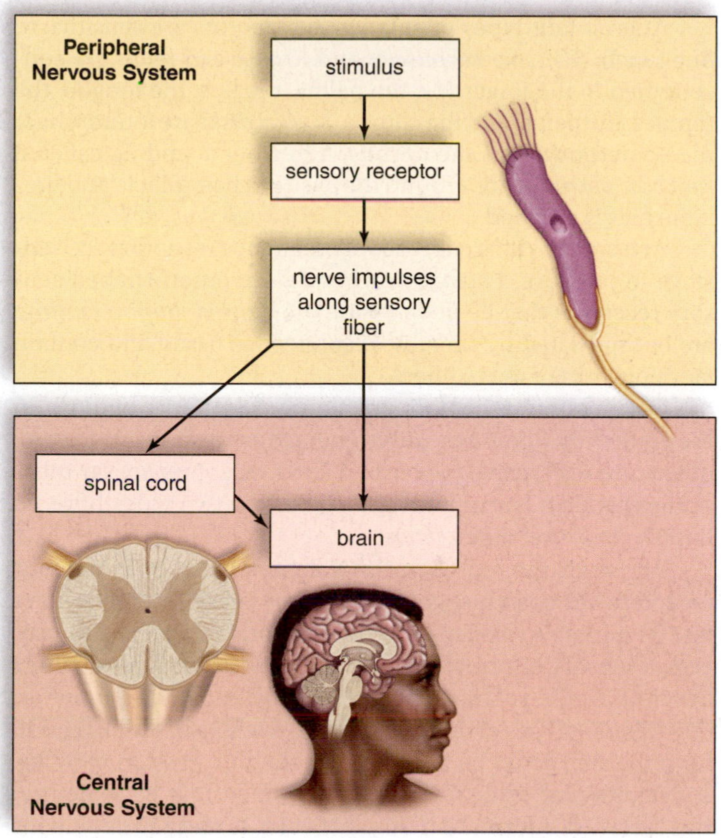
Peripheral
Nervous System
stimulus
sensory receptor
nerve impulses
along sensory
fiber
spinal cord
brain
Central
Nervous System

Figure 18.1 Sensation. The stimulus is received by a sensory receptor, which generates nerve impulses (action potentials). Nerve impulses are conducted to the CNS by sensory nerve fibers within the PNS, and only those impulses that reach the cerebral cortex result in sensation and perception.

the receptor proteins in the plasma membrane of chemoreceptors bind to certain molecules. When this happens, ion channels open, and ions flow across the plasma membrane. If the stimulus is sufficient, nerve impulses begin and are carried by a sensory nerve fiber within the PNS to the CNS (Fig. 18.1). In this process, known as **sensory transduction,** energy from the chemical or physical stimulus is transduced by the sensory receptors into an electrical signal. This electrical signal is in the form of nerve impulses (action potentials) that travel along the sensory nerve fiber. The stronger the stimulus, the greater the frequency of nerve impulses. Nerve impulses that reach the spinal cord first are conveyed to the brain by ascending tracts. If nerve impulses finally reach the cerebral cortex, sensation and perception occur.

All sensory receptors initiate nerve impulses; the sensation that results depends on the part of the brain receiving the nerve impulses. Nerve impulses that begin in the optic nerve reach the visual cortex, and thereafter, we see objects. Nerve impulses that begin in the auditory nerve reach the auditory cortex, and thereafter, we hear sounds.

Before sensory receptors initiate nerve impulses, they carry out **integration,** the summing up of signals. One type of integration is called *sensory adaptation,* a decrease in the response to a stimulus. For example, as you sit reading this, you are probably not aware of the feeling of the chair beneath you, until now as you are currently reminded. Some authorities believe that when sensory adaptation occurs, sensory receptors have stopped sending impulses to the brain. Others believe that the thalamus has filtered out the ongoing stimuli. As mentioned in section 17.3, most sensory information is conveyed from the brain stem through the thalamus to the cerebral cortex. The thalamus acts as a gatekeeper and only passes on information of immediate importance. Just as we gradually become unaware of particular environmental stimuli, we can suddenly become aware of stimuli that may have been present for some time. This can be attributed to the workings of the thalamus, which has synapses with all the major ascending sensory tracts.

The functioning of our sensory receptors makes a significant contribution to homeostasis. Without sensory input, we would not receive information about our internal and external environment. This information leads to appropriate reflex and voluntary actions to keep the internal environment constant.

Check Your Progress 18.1

1. Describe how interoceptors are often involved in homeostasis.
2. List a specific example of each of the four types of sensory receptors.
3. Explain why sensory adaptation is important to an organism.

18.2 Somatic Senses

Learning Outcomes

Upon completion of this section, you should be able to

1. Compare and contrast the functions of proprioceptors, cutaneous receptors, and pain receptors.
2. List specific types of cutaneous receptors that are sensitive to fine touch, pressure, and temperature.
3. Explain the importance of pain receptors to the survival of an organism.

It is logical to consider those senses whose receptors are associated with the skin, muscles, joints, and viscera together as the *somatic senses.* Somatic sensory receptors can be categorized into three types: proprioceptors, cutaneous receptors, and pain receptors. Some of these are interoceptors, while others are exteroceptors. Regardless, these receptors send nerve impulses via the spinal cord to the somatosensory areas of the cerebral cortex (see Fig. 17.11).

Proprioceptors

Proprioceptors are mechanoreceptors involved in reflex actions that maintain muscle tone, and thereby the body's equilibrium and posture. They help us know the position of our limbs in space by detecting the degree of muscle relaxation, the stretch of tendons, and the movement of ligaments. Two main types of proprioceptors help to maintain the amount of tension, or tone, in skeletal muscles. Muscle spindles act to increase the degree of muscle contraction, and Golgi tendon organs act to decrease it.

Figure 18.2 illustrates the activity of muscle spindles and Golgi tendon organs. In a muscle spindle, sensory nerve endings are wrapped around thin muscle cells within a connective tissue sheath. When the muscle relaxes and stretching of the muscle spindle occurs, nerve impulses are generated and travel to the spinal cord. The rapidity of these nerve impulses is proportional to the stretching of a muscle. A reflex action then occurs, which results in contraction of muscle fibers adjoining the muscle spindle. The knee-jerk reflex, which involves muscle spindles, offers an opportunity for physicians to test a reflex action.

Golgi tendon organs are sensory receptors buried in the collagen fibers of tendons. When skeletal muscles, and therefore tendons, are stretched excessively, these proprioceptors generate nerve impulses that travel via sensory neurons to the spinal cord, inhibiting further muscle contraction. In addition to maintaining proper muscle tone, Golgi tendon organs also protect muscles and tendons from injury due to overstretching.

Cutaneous Receptors

As was described in section 11.4, the skin is composed of two layers: the epidermis and the dermis (Fig. 18.3). The epidermis is stratified squamous epithelium in which cells become keratinized as they rise to the surface, where they are sloughed off. The dermis is a thick connective tissue layer. Both layers of the skin contain **cutaneous receptors,** which make the skin sensitive to touch, pressure, pain, and temperature (warmth and cold).

At least four types of cutaneous receptors are sensitive to fine touch. *Meissner corpuscles* and *Krause end bulbs* are concentrated in the fingertips, the palms, the lips, the tongue, the nipples, the penis, and the clitoris. *Merkel disks* are found where the epidermis meets the dermis. A free nerve ending called a *root hair plexus* winds around the base of a hair follicle and fires if the hair is touched.

At least two different types of cutaneous receptors are sensitive to pressure. *Pacinian corpuscles* are onion-shaped sensory receptors that lie deep inside the dermis. *Ruffini endings* are encapsulated by sheaths of connective tissue and contain lacy networks of nerve fibers.

Temperature receptors are simply free nerve endings in the epidermis. Some free nerve endings are responsive to cold; others are responsive to warmth. Cold receptors are far more numerous than warmth receptors, but the two types have no known structural differences.

Although the existence of different types of cutaneous receptors has been recognized for many years, researchers are just beginning to understand how the brain collects and processes sensory information from the skin. In 2011, researchers at Johns Hopkins University genetically engineered mice so that different types of neurons that receive sensory information from the skin could be visualized by their expression of different fluorescent proteins. They determined that each neuron was receiving information from between 30 and 150 cutaneous receptors, and that different types of neurons receive information from different types of receptors. Moreover, they found that nerves bringing sensory information from small areas of skin line up in small columns in the spinal cord (about 3,000 to 5,000 columns per mouse). The researchers admit they don't fully

Figure 18.2 Muscle spindles and Golgi tendon organs. ① When a muscle is stretched, muscle spindles send sensory nerve impulses to the spinal cord. ② Motor nerve impulses from the spinal cord result in muscle fiber contraction so that muscle tone is maintained. ③ When tendons are stretched excessively, Golgi tendon organs cause muscle relaxation.

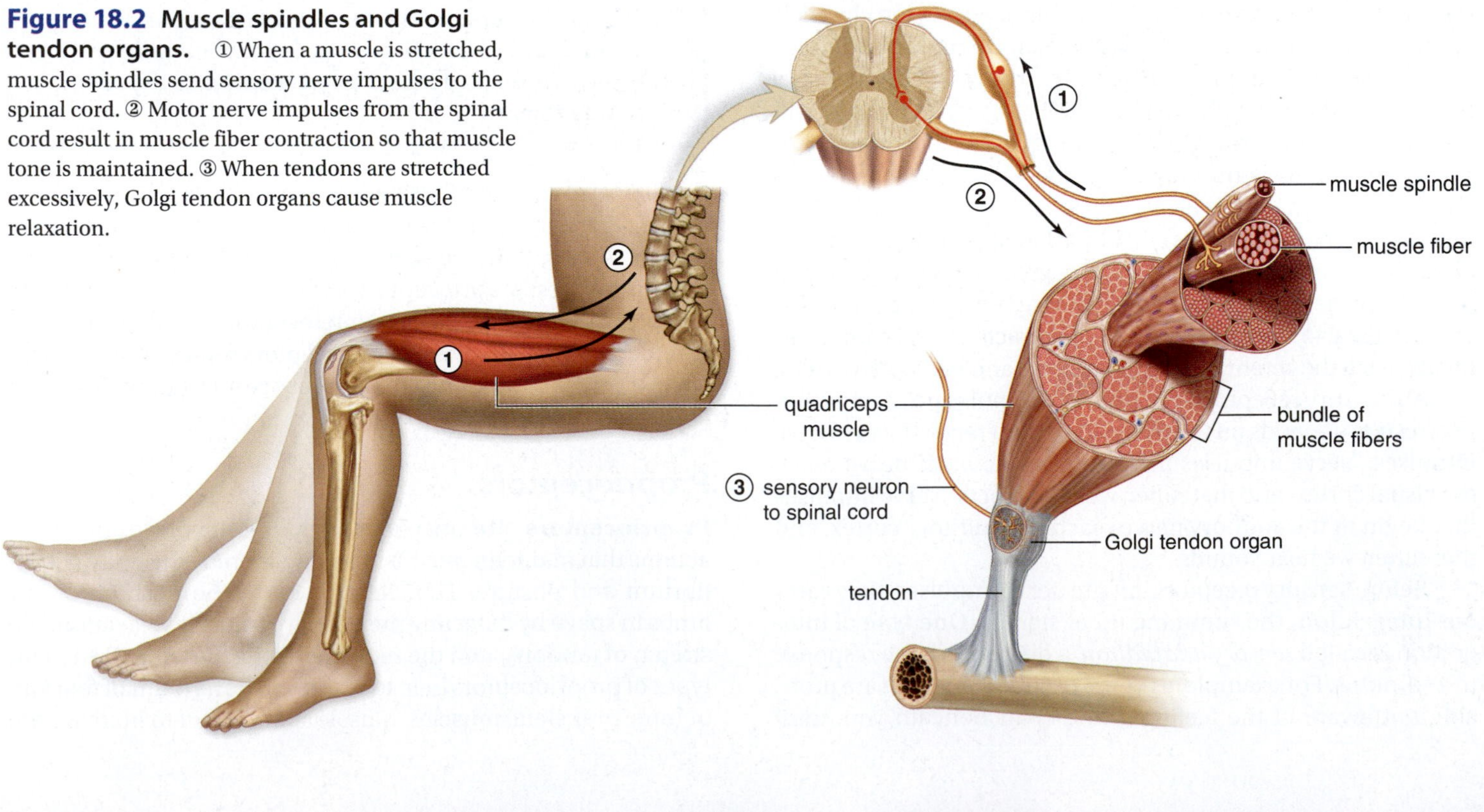

1
2
muscle spindle
muscle fiber
2
1
quadriceps muscle
bundle of muscle fibers
3 sensory neuron to spinal cord
Golgi tendon organ
tendon

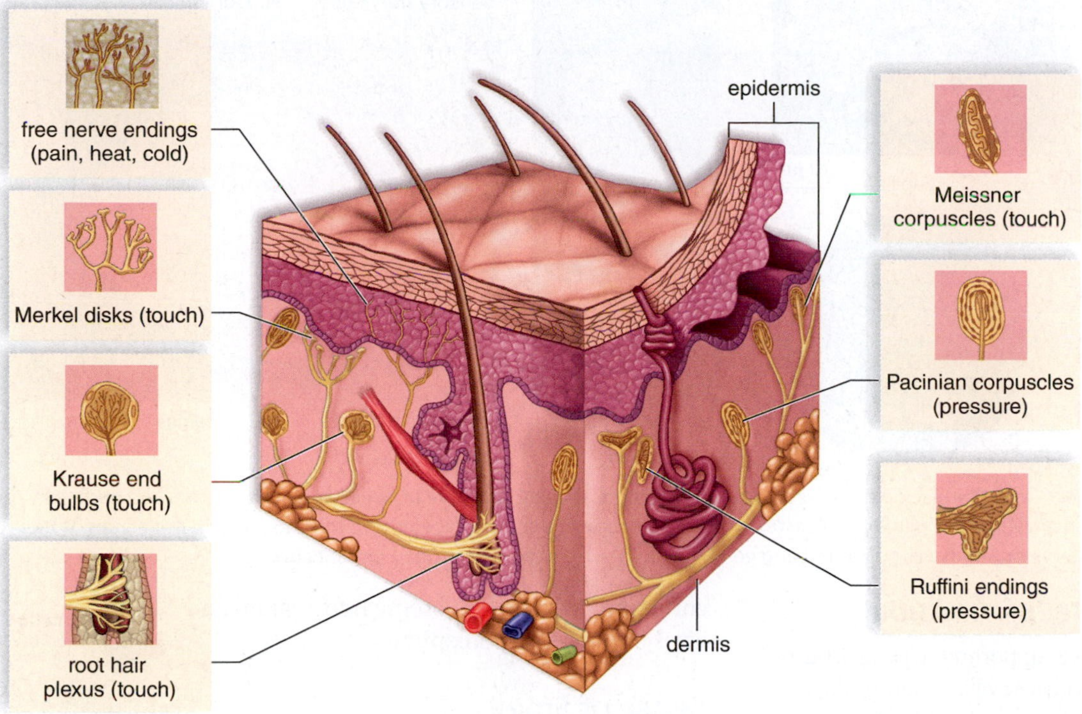
free nerve endings
(pain, heat, cold)
Merkel disks (touch)
Krause end
bulbs (touch)
root hair
plexus (touch)
epidermis
dermis
Meissner
corpuscles (touch)
Pacinian corpuscles
(pressure)
Ruffini endings
(pressure)

Figure 18.3 Cutaneous receptors in human skin. The classical view is that each cutaneous receptor has the main function shown here. However, investigators report that matters are not so clear-cut. For example, microscopic examination of the skin of the ear shows only free nerve endings (pain receptors), and yet the skin of the ear is sensitive to all sensations. Therefore, it appears that the receptors of the skin are somewhat, but not completely, specialized.

understand how this large amount of information is processed by the brain, but they suspect that the way these columns are arranged in the spinal cord is a key factor.

Pain Receptors

The skin and many internal organs and tissues have **pain receptors** (free nerve endings, also called *nociceptors*), which are sensitive to chemicals released by damaged cells. When damage occurs due to mechanical, thermal, or electrical stimulation, or a toxic substance, cells release chemicals that stimulate pain receptors. Aspirin and ibuprofen reduce pain by inhibiting the synthesis of one class of these chemicals.

Sometimes, stimulation of internal pain receptors is felt as pain from the skin, as well as the internal organs. This is called *referred pain.* Some internal organs have a referred pain relationship with areas located in the skin of the back, groin, and abdomen. For example, pain from the heart is typically felt in the left shoulder and arm. This most likely happens when nerve impulses from the pain receptors of internal organs travel to the spinal cord and synapse with neurons also receiving impulses from the skin. Pain receptors are protective because they alert us to possible danger. For example, without the pain of appendicitis, we might never seek the medical help needed to avoid a ruptured appendix.

Check Your Progress 18.2

1. Classify which specific types of somatic sensory receptors are interoceptors and which are exteroceptors.
2. Classify the cutaneous receptors by their location (i.e., in the dermis or epidermis).
3. Predict the likely outcome if an individual lacked muscle spindles, or pain receptors.

18.3 Senses of Taste and Smell

Learning Outcomes

Upon completion of this section, you should be able to

1. Identify the structures that are involved in taste, including the five types of taste receptors.
2. Describe the sensory receptors involved in smell.
3. Compare and contrast how the brain receives information about taste versus smell.

Taste and smell are called *chemical senses* because their receptors are sensitive to molecules in the food we eat and the air we breathe. Taste cells and olfactory cells are classified as chemoreceptors.

Sense of Taste

The sensory receptors for the sense of taste, the taste cells, are located in about 10,000 **taste buds** located primarily on the tongue (Fig. 18.4). Many lie along the walls of the *papillae,* the small elevations on the tongue that are visible to the naked eye. Isolated taste buds are also present on the hard palate, the pharynx, and the epiglottis. Different taste cells can detect at least five primary types of taste: salty, sour, bitter, sweet, and umami (Japanese; savory, delicious). Umami receptors are capable of detecting certain amino acids. In particular, this receptor detects the amino acid glutamate, which is part of the flavor enhancer monosodium glutamate (MSG). Taste buds containing sensory receptors for each type of taste are located throughout the tongue.

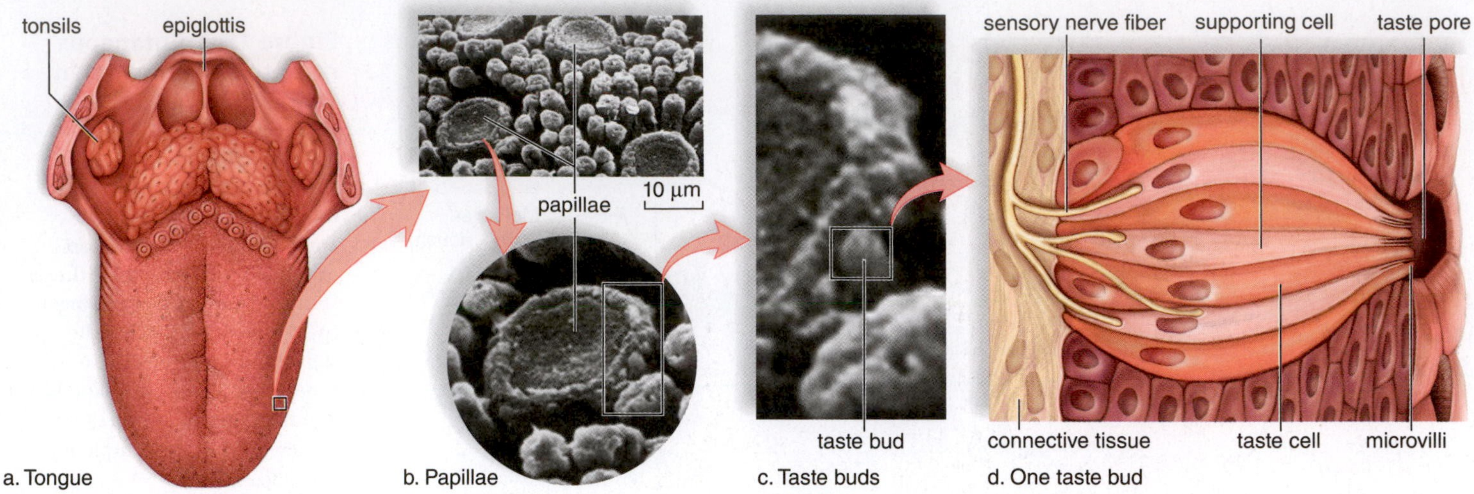
tonsils
epiglottis
10 µm
papillae
sensory nerve fiber
supporting cell
taste pore
taste bud
connective tissue
taste cell
microvilli
a. Tongue
b. Papillae
c. Taste buds
d. One taste bud

Figure 18.4 Taste buds in humans. **a, b.** Papillae on the tongue contain taste buds that are sensitive to sweet, sour, salty, bitter, and perhaps umami. **c.** Taste buds occur mainly along the walls of the papillae. **d.** Taste cells end in microvilli that bear receptor proteins for certain molecules. When molecules bind to the receptor proteins, nerve impulses are generated and go to the brain, where the sensation of taste occurs.

How the Brain Receives Taste Information

Taste buds open at a taste pore. They have supporting cells and a number of elongated taste cells that end in microvilli. When molecules bind to receptor proteins of the microvilli, nerve impulses are generated in sensory nerve fibers that go to the brain. When they reach the gustatory (taste) cortex, they are interpreted as particular tastes.

Because we can respond to a range of salty, sour, bitter, sweet, and umami tastes, the brain appears to survey the overall pattern of incoming sensory impulses and to take a "weighted average" of their taste messages as the perceived taste. Note that even though our senses are dependent on sensory receptors, the cortex integrates the incoming information and gives us our sense perceptions.

Animation Taste

Sense of Smell

Approximately 80–90% of what we perceive as "taste" is actually due to the sense of smell, which explains why food tastes dull when we have a head cold or a stuffed-up nose. Our sense of smell is dependent on **olfactory cells** located within olfactory epithelium high in the roof of the nasal cavity (Fig. 18.5). Olfactory cells are modified neurons. Each cell ends in a tuft of about five olfactory cilia, which bear receptor proteins for odor molecules.

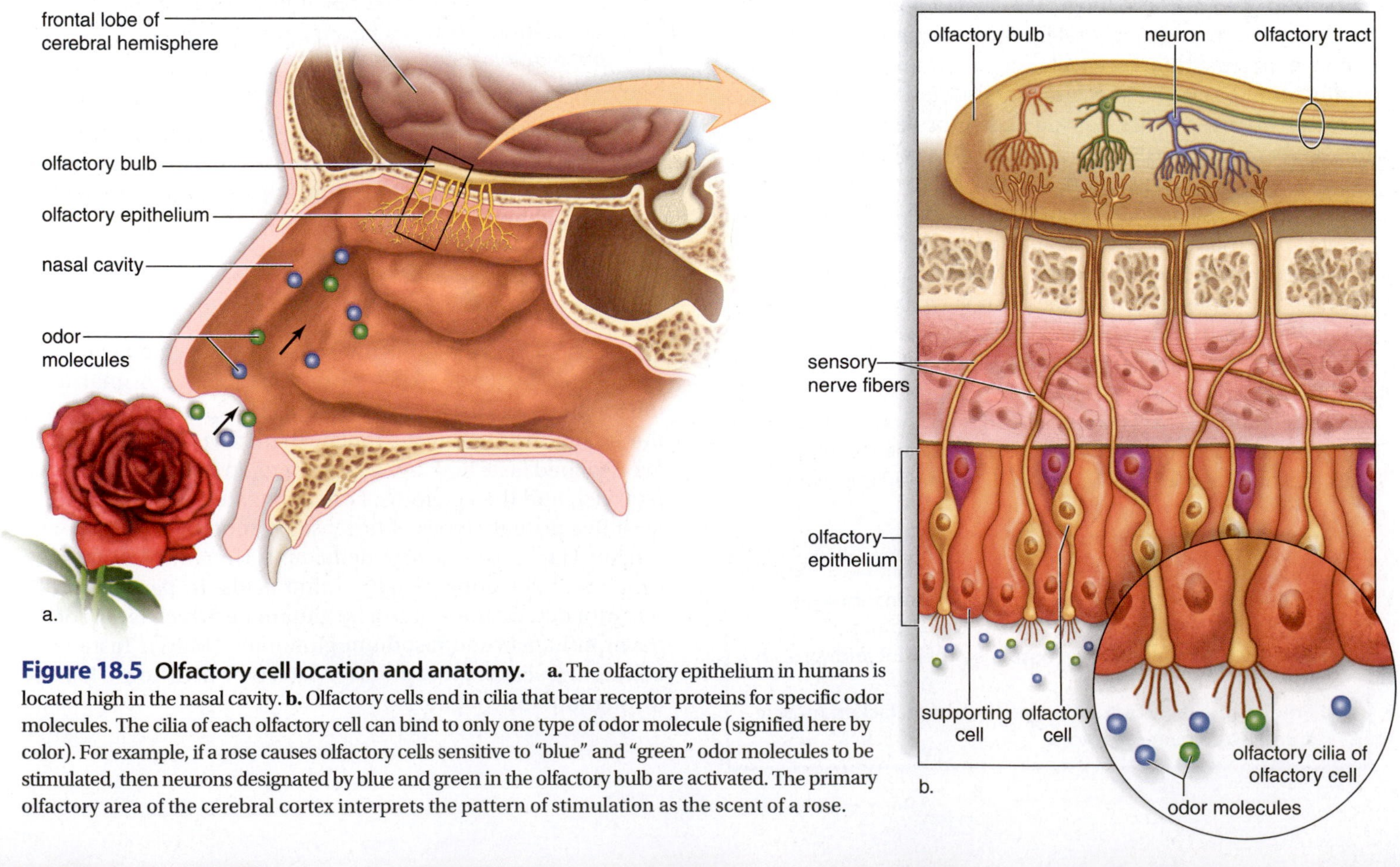
frontal lobe of cerebral hemisphere
olfactory bulb
olfactory epithelium
nasal cavity
odor molecules
a.
olfactory bulb
neuron
olfactory tract
sensory nerve fibers
olfactory epithelium
supporting cell
olfactory cell
olfactory cilia of olfactory cell
odor molecules
b.

Figure 18.5 Olfactory cell location and anatomy. **a.** The olfactory epithelium in humans is located high in the nasal cavity. **b.** Olfactory cells end in cilia that bear receptor proteins for specific odor molecules. The cilia of each olfactory cell can bind to only one type of odor molecule (signified here by color). For example, if a rose causes olfactory cells sensitive to "blue" and "green" odor molecules to be stimulated, then neurons designated by blue and green in the olfactory bulb are activated. The primary olfactory area of the cerebral cortex interprets the pattern of stimulation as the scent of a rose.

How the Brain Receives Odor Information

Each olfactory cell has only one out of about 1,000 different types of receptor proteins. Nerve fibers from like olfactory cells lead to the same neuron in the olfactory bulb, an extension of the brain. An odor contains many odor molecules, which activate a characteristic combination of receptor proteins. For example, a rose might stimulate olfactory cells, designated by blue and green in Figure 18.5*b*, while a hyacinth might stimulate a different combination. An odor's signature in the olfactory bulb is determined by which neurons are stimulated. The neurons communicate this information via the olfactory tract to the brain. When the information reaches the olfactory cortex, we know we have smelled a rose or a hyacinth.

Have you ever noticed that a certain aroma vividly brings to mind a certain person or place? A person's perfume may remind you of someone else, or the smell of boxwood may remind you of your grandfather's farm. The olfactory bulbs (see Fig. 17.12) have direct connections to the limbic system and its centers for emotions and memory. One investigator showed that when subjects smelled an orange while viewing a painting, they not only remembered the painting when asked about it later, but they also had many deep feelings about it.

MP3 Taste and Smell

Video Mosquitoes and Sweat

Check Your Progress 18.3

1. Compare and contrast the functions of the chemoreceptors on the tongue and in the nasal cavity.
2. Predict some structural variations that could provide certain animals, such as dogs, with a sense of smell thousands of times more sensitive than that of humans.
3. Describe the anatomical connections that exist in the brain between smells and memories.

18.4 Sense of Vision

Learning Outcomes

Upon completion of this section, you should be able to

1. Identify the structures of the human eye.
2. Explain how the eye focuses on near and far objects.
3. Discuss the function and distribution of rods and cones.
4. Trace the path of a nerve impulse from the retina to the visual cortex.

Vision requires the work of the eyes and the brain. Researchers estimate that at least a third of the cerebral cortex takes part in processing visual information.

Anatomy and Physiology of the Eye

The eye, an elongated sphere about 2.5 cm in diameter, has three layers, or coats: the sclera, the choroid, and the retina (Fig. 18.6). Only the retina contains photoreceptors for light energy. Table 18.2 lists the functions of the parts of the eye.

The outer layer, called the **sclera,** is white and fibrous except for the **cornea,** which is made of transparent collagen fibers. The cornea is the window of the eye. Beginning at the edge of the cornea, a thin layer of epithelial cells forms a mucous membrane called the **conjunctiva** that covers the sclera and inner eyelids, keeping them moist. An inflammation of this membrane, called *pink eye* or *conjunctivitis,* can be caused by viruses, bacteria, irritants, or allergic reactions. The middle, thin, darkly pigmented layer, the **choroid,** is vascular and absorbs stray light rays that photoreceptors have not absorbed. Toward the front, the choroid becomes the donut-shaped **iris.**

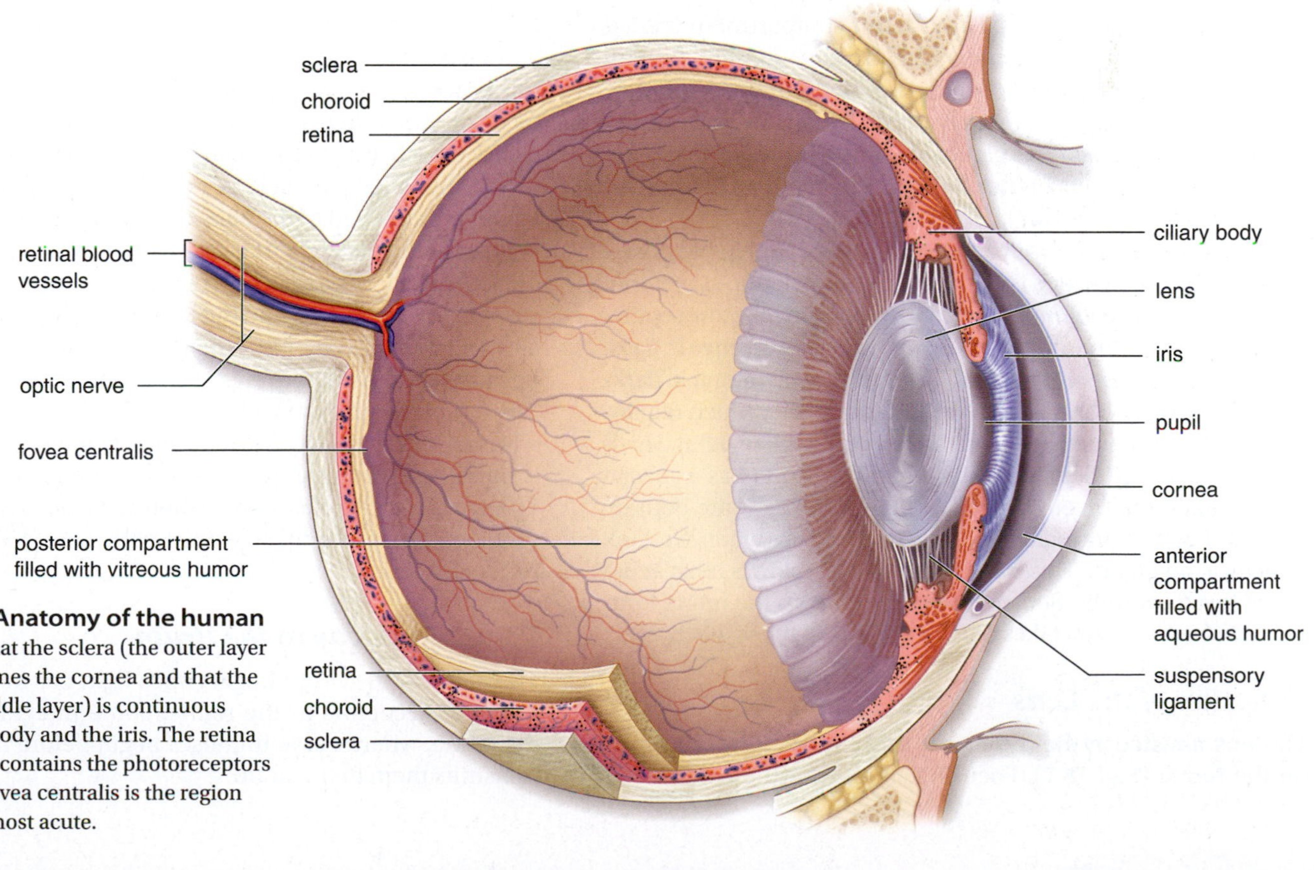
sclera
choroid
retina
retinal blood vessels
optic nerve
fovea centralis
posterior compartment filled with vitreous humor
retina
choroid
sclera
ciliary body
lens
iris
pupil
cornea
anterior compartment filled with aqueous humor
suspensory ligament

Figure 18.6 Anatomy of the human eye. Notice that the sclera (the outer layer of the eye) becomes the cornea and that the choroid (the middle layer) is continuous with the ciliary body and the iris. The retina (the inner layer) contains the photoreceptors for vision. The fovea centralis is the region where vision is most acute.

TABLE 18.2 Functions of the Parts of the Eye

Part	Function
Sclera	Protects and supports the eye
Cornea	Refracts light rays
Choroid	Absorbs stray light
Ciliary body	Holds lens in place; aids accommodation
Iris	Regulates light entrance
Pupil	Admits light
Retina	Contains sensory receptors for sight
Rods	Make black-and-white vision possible
Cones	Make color vision possible
Fovea centralis	Makes acute vision possible
Other	
Lens	Refracts and focuses light rays
Humors	Transmit light rays and support the eye
Optic nerve	Transmits impulse to brain

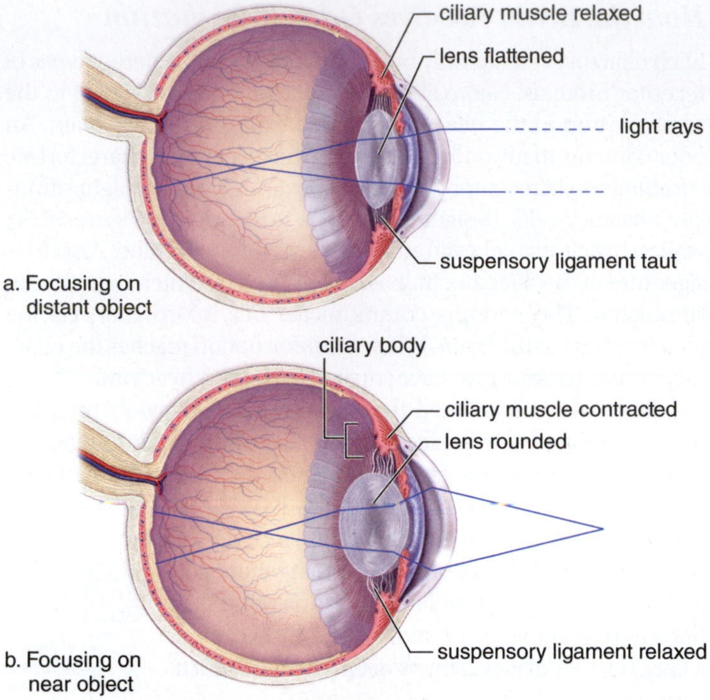
ciliary muscle relaxed
lens flattened
light rays
suspensory ligament taut
a. Focusing on distant object
ciliary body
ciliary muscle contracted
lens rounded
suspensory ligament relaxed
b. Focusing on near object

Figure 18.7 Focusing the human eye. Light rays from each point on an object are bent by the cornea and the lens in such a way that an inverted and reversed image forms on the retina. **a.** When focusing on a distant object, the lens is flat because the ciliary muscle is relaxed and the suspensory ligament is taut. **b.** When focusing on a near object, the lens accommodates; that is, it becomes rounded because the ciliary muscle contracts, causing the suspensory ligament to relax.

The iris contains smooth muscle that regulates the size of the **pupil,** a hole in the center of the iris through which light enters the eye. The color of the iris (and therefore, the color of your eyes) correlates with its pigmentation. Heavily pigmented eyes are brown, while lightly pigmented eyes are green or blue. Behind the iris, the choroid thickens and forms the circular ciliary body. The *ciliary body* contains the **ciliary muscle,** which controls the shape of the lens for near and far vision.

The **lens,** attached to the ciliary body by ligaments, divides the eye into two compartments. The one in front of the lens is the anterior compartment, and the one behind the lens is the posterior compartment. The anterior compartment is filled with a clear, watery fluid called the *aqueous humor.* A small amount of aqueous humor is continually produced each day. Normally, it leaves the anterior compartment by way of tiny ducts. When a person has *glaucoma,* these drainage ducts are blocked, and aqueous humor builds up, causing an increase in pressure (see section 18.7).

The third layer of the eye, the **retina,** is located in the posterior compartment, which is filled with a clear, gelatinous material called the *vitreous humor.* The retina contains photoreceptors called rod cells and cone cells. The rods are very sensitive to light, but they do not respond to color. Therefore, at night or in a darkened room, we see only shades of gray. The cones, which require bright light, are sensitive to different wavelengths of light, which give us the ability to distinguish colors. The retina has a special region called the **fovea centralis,** where cone cells are densely packed. Light is normally focused on the fovea when we look directly at an object. This is helpful because vision is most acute in the fovea centralis. Sensory nerve fibers from the retina form the *optic nerve,* which takes nerve impulses to the visual cortex.

Function of the Lens

The lens, assisted by the cornea and the humors, focuses images on the retina (Fig. 18.7). Focusing starts with the cornea and continues as the rays pass through the lens and the humors. The image produced is much smaller than the object because light rays are bent (refracted) when they are brought into focus.

Visual accommodation occurs for close vision. During this process, the lens rounds up in order to bring the image to focus on the retina. The shape of the lens is controlled by the ciliary muscle within the ciliary body. When you view a distant object, the ciliary muscle is relaxed, causing the suspensory ligaments attached to the ciliary body to be taut. Therefore, the lens remains relatively flat (Fig. 18.7*a*). When you view a near object, the ciliary muscle contracts, releasing the tension on the suspensory ligaments, and the lens rounds up due to its natural elasticity (Fig. 18.7*b*). Now the image is focused on the retina. Because close work requires contraction of the ciliary muscle, it very often causes muscle fatigue, known as *eyestrain.* Usually after the age of 40, the lens loses some of its elasticity and is unable to accommodate, a condition known as *presbyopia.* The use of reading glasses may then be necessary, or bifocal lenses for those who already have corrective lenses.

Video Artificial Eye

Visual Pathway to the Brain

The pathway for vision begins once light has been focused on the photoreceptors in the retina. Some integration occurs in the retina, where nerve impulses begin before the optic nerve transmits them to the brain.

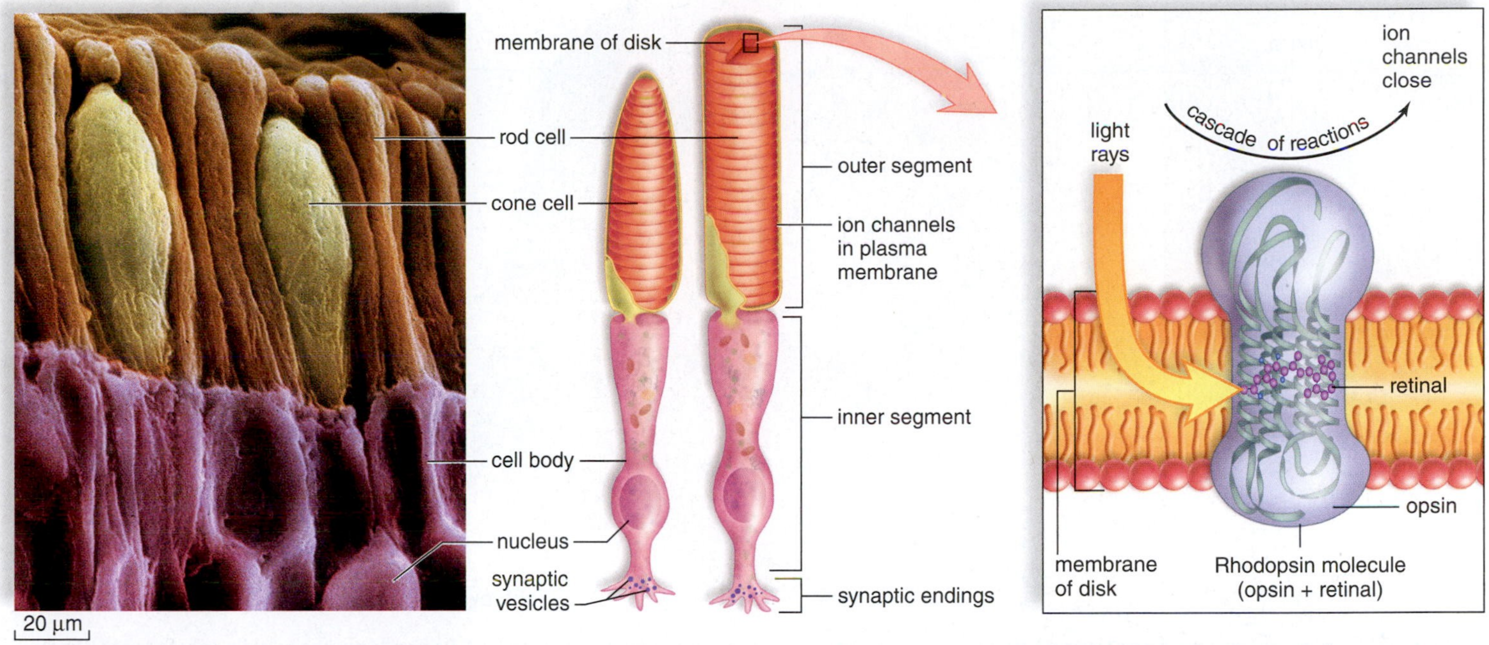
membrane of disk
rod cell
cone cell
cell body
nucleus
synaptic vesicles
outer segment
ion channels in plasma membrane
inner segment
synaptic endings
20 µm
ion channels close
cascade of reactions
light rays
retinal
opsin
membrane of disk
Rhodopsin molecule (opsin + retinal)

Figure 18.8 Photoreceptors in the eye. The outer segment of rods and cones contains stacks of membranous disks that contain visual pigments. In rods, the membrane of each disk contains rhodopsin, a complex molecule composed of the protein opsin and the pigment retinal. When rhodopsin absorbs light energy, it splits, releasing opsin, which sets in motion a cascade of reactions that cause ion channels in the plasma membrane to close. Thereafter, nerve impulses go to the brain.

Function of Photoreceptors The photoreceptors in the eye are of two types—**rod cells** and **cone cells.** Figure 18.8 illustrates their structure. Both rods and cones have an outer segment joined to an inner segment by a stalk. Pigment molecules are embedded in the membrane of the many disks present in the outer segment. Synaptic vesicles are located at the synaptic endings of the inner segment.

The visual pigment in rods is a deep purple pigment called rhodopsin. **Rhodopsin** is a complex molecule made up of the protein opsin and a light-absorbing molecule called **retinal,** which is a derivative of vitamin A. When a rod absorbs light, rhodopsin splits into opsin and retinal, leading to a cascade of reactions and the closure of ion channels in the rod cell's plasma membrane. The release of inhibitory transmitter molecules from the rod's synaptic vesicles ceases. Thereafter, signals go to other neurons in the retina. Rods are very sensitive to light, and are therefore suited to night vision. (Because carrots are rich in vitamin A, it is true that eating carrots can improve your night vision.) With the exception of the fovea centralis, rod cells are plentiful throughout the entire retina. Therefore, they also provide us with peripheral vision and perception of motion.

The cones, on the other hand, are located primarily in the fovea centralis and are activated by bright light. They allow us to detect the fine detail and the color of an object. **Color vision** depends on three different kinds of cones, which contain pigments called the B (blue), G (green), and R (red) pigments. Each pigment is an iodopsin made up of retinal and an opsin. There is a slight difference in the opsin structure of each, which accounts for the ability of the pigments to absorb different wavelengths (colors) of visible light. Various combinations of cones are believed to be stimulated by in-between shades of color.

Function of the Retina The retina has three layers of neurons (Fig. 18.9). The layer closest to the choroid (at the back of the eye) contains the rod cells and cone cells; the middle layer contains bipolar cells; and the innermost layer contains ganglion cells, whose sensory fibers become the optic nerve. Only the rod cells and the cone cells are sensitive to light, and therefore, light must penetrate to the back of the retina before they are stimulated.

The rod cells and the cone cells synapse with the bipolar cells, which in turn synapse with ganglion cells whose axons become the optic nerve. Notice in Figure 18.9 that there are many more rod cells and cone cells than ganglion cells. In fact, the retina has as many as 150 million rod cells and 6 million cone cells but only one million ganglion cells. The sensitivity of cones versus rods is due in part to how directly they connect to ganglion cells. As many as 150 rods, covering about one square millimeter of retina (about the size of a thumbtack hole) may synapse on the same ganglion cell. No wonder stimulation of rods results in vision that is blurred and indistinct. In contrast, some cone cells in the fovea centralis activate only one ganglion cell. This explains why cones, especially in the fovea, provide us with a sharper, more detailed image of an object.

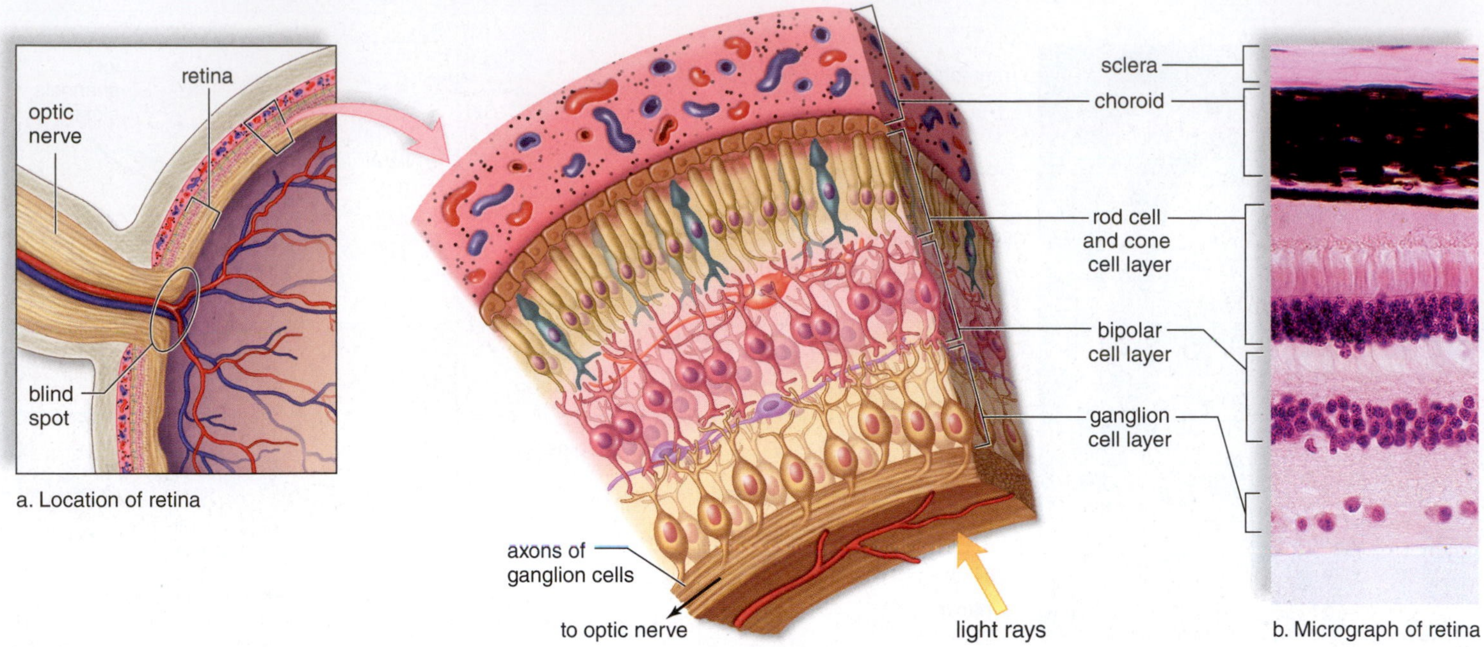
optic
nerve
retina
blind
spot
a. Location of retina
sclera
choroid
rod cell
and cone
cell layer
bipolar
cell layer
ganglion
cell layer
axons of
ganglion cells
to optic nerve
light rays
b. Micrograph of retina

Figure 18.9 Structure and function of the retina. **a.** The retina is the inner layer of the eyeball. Rod cells and cone cells, located at the back of the retina nearest the choroid, synapse with bipolar cells, which synapse with ganglion cells. Integration of signals occurs at these synapses. Therefore, much processing occurs in bipolar and ganglion cells. Notice also that many rod cells share one bipolar cell, but cone cells do not. Certain cone cells synapse with only one ganglion cell. Cone cells, in general, distinguish more detail than do rod cells. **b.** Micrograph shows that the sclera and choroid are relatively thin compared to the retina, which has several layers of cells.

Blind Spot Figure 18.9*a* provides an opportunity to point out that there are no rods and cones where the optic nerve exits the retina. Therefore, no vision is possible in this area. You can prove this to yourself by putting a dot to the right of center on a piece of paper. Use your right hand to move the paper slowly toward your right eye, while you look straight ahead. The dot will disappear at one point—this is your right eye's **blind spot.**

From the Retina to the Visual Cortex As stated previously, the axons of ganglion cells in the retina assemble to form the optic nerves. The optic nerves carry nerve impulses from the eyes to the optic chiasma. The *optic chiasma* has an X shape, formed by a crossing-over of optic nerve fibers. Fibers from the right half of each retina converge and continue on together in the *right optic tract,* and fibers from the left half of each retina converge and continue on together in the *left optic tract* (Fig. 18.10).

The optic tracts sweep around the hypothalamus, and most fibers synapse with neurons in nuclei (masses of neuron cell bodies) within the thalamus. Axons from the thalamic nuclei form optic radiations that take nerve impulses to the *visual area* within the occipital lobe. Notice in Figure 18.10 that the image arriving at the thalamus, and therefore the visual area, has been split because the left optic tract carries information about the right portion of the visual field (shown in green) and the right optic tract carries information about the left portion of the visual field (shown in red). Therefore, the right and left

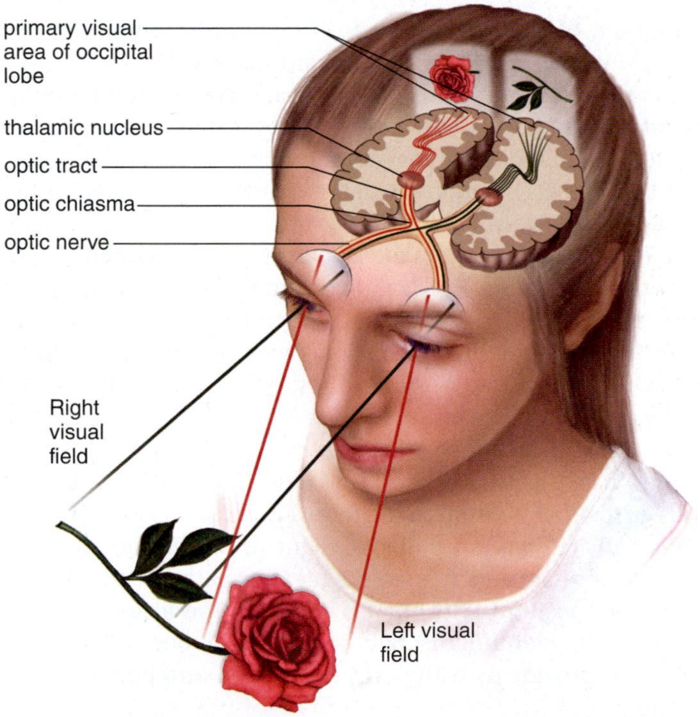
primary visual
area of occipital
lobe
thalamic nucleus
optic tract
optic chiasma
optic nerve
Right
visual
field
Left visual
field

Figure 18.10 Optic chiasma. Both eyes "see" the entire visual field. Because of the optic chiasma, data from the right half of each retina (red lines) go to the right visual cortex, and data from the left half of each retina (green lines) go to the left visual cortex. These data are then combined to allow us to see the entire visual field. Note that the visual pathway to the brain includes the thalamus, which has the ability to filter sensory stimuli.

visual areas must communicate with each other for us to see the entire visual field. Also, because the image is inverted and reversed, it must be righted in the brain for us to correctly perceive the visual field.

The most surprising finding has been that the primary visual area acts like a post office, parceling out information regarding color, form, motion, and possibly other attributes to different portions of the adjoining visual association area. Therefore, the brain has taken the visual field apart, even though we see a unified visual field. The visual association areas are believed to rebuild the field and give us an understanding of it at the same time.

MP3 Sense of Vision

Animation Vision

Check Your Progress 18.4

1. List all tissues that light passes through from when it enters the eye until it is converted into a nerve impulse.
2. Describe how the ciliary muscle controls the shape of the lens.
3. Explain how the number of rod or cone cells synapsing on a single ganglion cell allows cones to produce a clearer image.
4. Describe how the anatomy of the optic chiasma implies that the brain interprets visual information that has been "split" in half.

18.5 Sense of Hearing

Learning Outcomes

Upon completion of this section you should be able to

1. List the parts of the ear that make up the outer ear, middle ear, and inner ear.
2. Describe the mechanism by which sound waves in the outer ear are converted into nerve impulses in the inner ear.
3. Explain how the inner ear is able to distinguish sounds of different pitches and volumes.

The ear functions in hearing as well as balance (equilibrium). The sensory receptors for both senses are mechanoreceptors located in the inner ear, each consisting of **hair cells** with stereocilia (long microvilli) that are sensitive to mechanical stimulation.

Anatomy of the Ear

Figure 18.11 shows that the ear has three divisions: outer, middle, and inner. The **outer ear** consists of the **pinna** (external flap), which collects and funnels sound into the **auditory canal.** The opening of the auditory canal is lined with fine hairs and

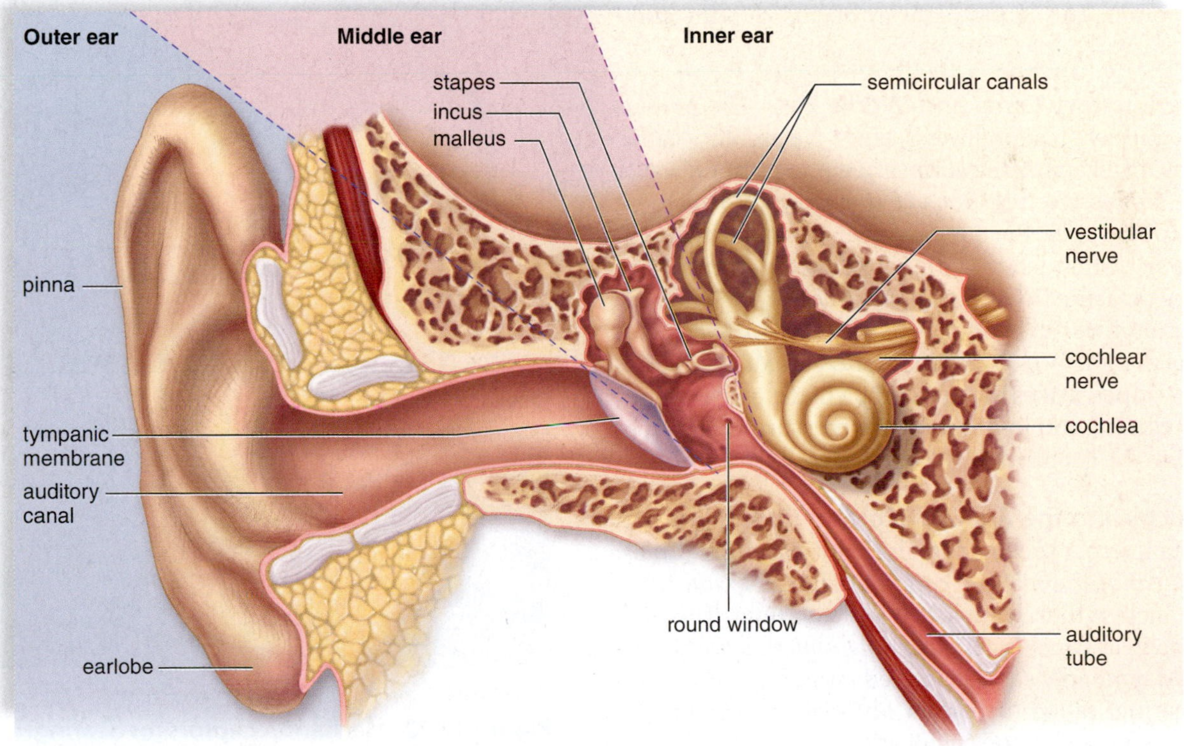

Outer ear
Middle ear
Inner ear
stapes
incus
malleus
semicircular canals
vestibular nerve
pinna
cochlear nerve
tympanic membrane
cochlea
auditory canal
round window
auditory tube
earlobe

Figure 18.11 Anatomy of the human ear. In the middle ear, the malleus (hammer), the incus (anvil), and the stapes (stirrup) amplify sound waves. In the inner ear, the mechanoreceptors for equilibrium are in the semicircular canals and the vestibule, and the mechanoreceptors for hearing are in the cochlea.

sweat glands. Modified sweat glands are located in the upper wall of the canal. They secrete *earwax,* a substance that helps guard the ear against the entrance of foreign materials, such as air pollutants and microorganisms.

The **middle ear** begins at the **tympanic membrane** (eardrum) and ends at a bony wall containing two small openings covered by membranes. These openings are called the **oval window** and the **round window.** Between the tympanic membrane and the oval window are three small bones, collectively called the **ossicles.** Individually, they are the **malleus** (hammer), the **incus** (anvil), and the **stapes** (stirrup) because their shapes resemble these objects. The malleus adheres to the tympanic membrane, and the stapes touches the oval window. An **auditory tube** (eustachian tube), which extends from the middle ear to the nasopharynx, permits equalization of air pressure. Chewing gum, yawning, and swallowing in elevators and airplanes help move air through the auditory tubes upon ascent and descent. As this occurs, we often hear the ears "pop."

Whereas the outer ear and the middle ear contain air, the inner ear is filled with fluid. Anatomically speaking, the **inner ear** has three areas: the **semicircular canals** and the **vestibule** function in equilibrium; the **cochlea** functions in hearing. The cochlea resembles the shell of a snail because it spirals.

Auditory Pathway to the Brain

Hearing requires the ear, the cochlear nerve, and the auditory cortex of the brain. The sound pathway begins with the auditory canal.

Through the Auditory Canal and Middle Ear The process of hearing begins when sound waves enter the auditory canal. Just as ripples travel across the surface of a pond, sound waves travel by the successive vibrations of molecules. When these vibrations impinge upon the tympanic membrane, it begins to vibrate at the same frequency as the sound waves. The malleus then takes the pressure from the inner surface of the tympanic membrane and passes it, by means of the incus, to the stapes in such a way that the pressure is multiplied about 20 times as it moves. The stapes strikes the membrane of the oval window, causing it to vibrate, and in this way, the pressure is passed to the fluid within the cochlea.

From the Cochlea to the Auditory Cortex Examining part of the cochlea in cross section reveals that it has three canals: the *vestibular canal,* the *cochlear canal,* and the *tympanic canal* (Fig. 18.12). Each of these canals is filled with fluid. The fluid within the cochlear canal is called *endolymph;* that within the vestibular and tympanic canals is called *perilymph.* The sense organ for hearing, called the **organ of Corti** (also called the *spiral organ*), is located in the cochlear canal. The spiral organ consists of little hair cells and a gelatinous material called the *tectorial membrane.* The hair cells sit on the basilar membrane, and their stereocilia are embedded in the tectorial membrane.

When the stapes strikes the membrane of the oval window, pressure waves move from the vestibular canal to the tympanic canal across the basilar membrane. The basilar

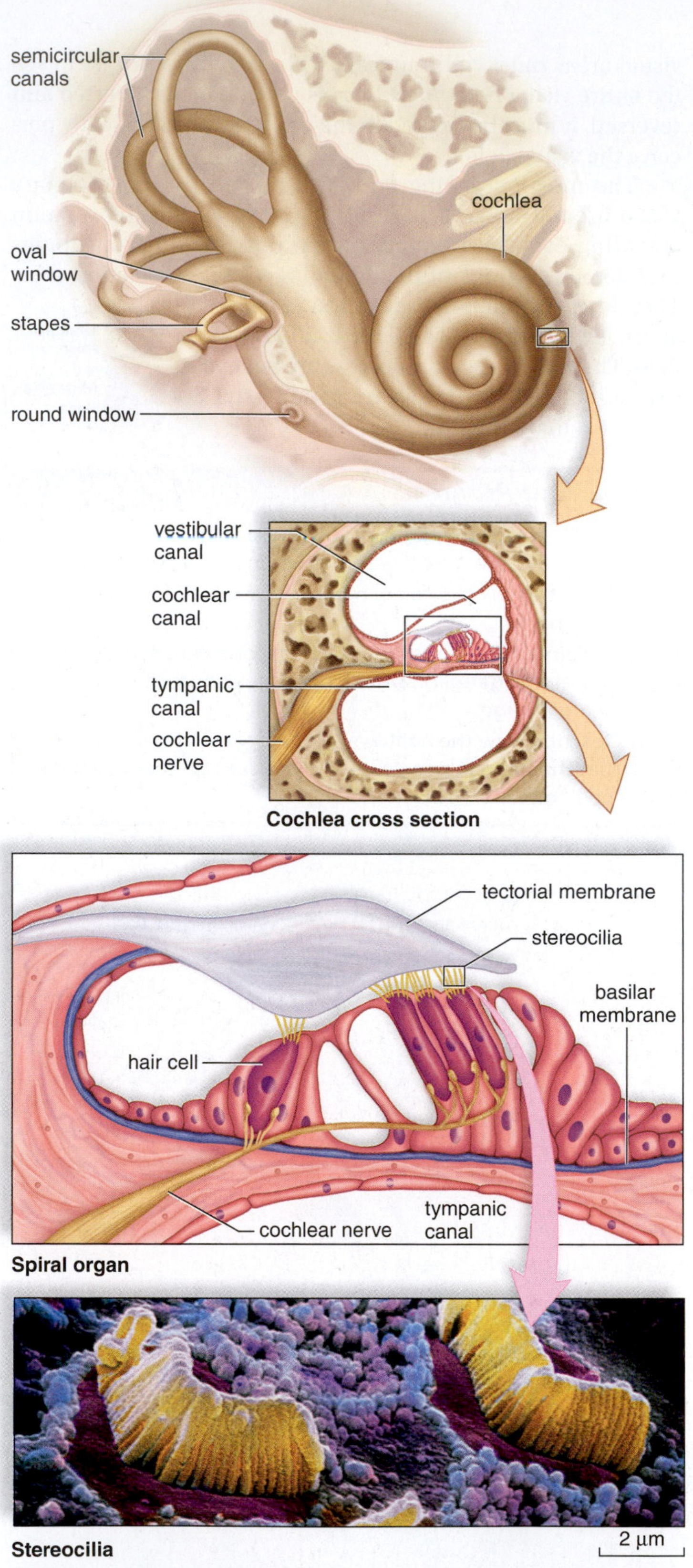

semicircular canals
cochlea
oval window
stapes
round window
vestibular canal
cochlear canal
tympanic canal
cochlear nerve
Cochlea cross section
tectorial membrane
stereocilia
basilar membrane
hair cell
cochlear nerve
tympanic canal
Spiral organ
Stereocilia
2 µm

Figure 18.12 Mechanoreceptors for hearing. The spiral organ is located within the cochlea. In the uncoiled cochlea, note that the spiral organ consists of hair cells resting on the basilar membrane, with the tectorial membrane above. Pressure waves move from the vestibular canal to the tympanic canal, causing the basilar membrane to vibrate. This causes the stereocilia (or at least a portion of the more than 20,000 hair cells) embedded in the tectorial membrane to bend. Nerve impulses traveling in the cochlear nerve to the auditory cortex result in hearing.

membrane moves up and down, and the stereocilia of the hair cells embedded in the tectorial membrane bend. Then nerve impulses begin in the *cochlear nerve* (also called the *auditory nerve*) and travel to the brain. When they reach the auditory cortex in the temporal lobe, they are interpreted as a sound.

Animation Effects of Sound Waves on Cochlear Structure

Each part of the organ of Corti is sensitive to different wave frequencies, or pitch. Near the tip, it responds to low pitches, such as a tuba, and near the base, it responds to higher pitches, such as a bell or a whistle. The nerve fibers from each region along the length of the organ of Corti lead to slightly different areas in the auditory cortex. The pitch sensation we experience depends upon which region of the basilar membrane vibrates and which area of the auditory cortex is stimulated.

Volume is a function of the amplitude of sound waves. Loud noises cause the fluid within the vestibular canal to exert more pressure and the basilar membrane to vibrate to a greater extent. The brain interprets the resulting increased stimulation as volume. Researchers believe that the brain interprets the tone of a sound based on the distribution of the hair cells stimulated. Exposure to loud noises over a long period of time is one of the causes of hearing loss. For more on hearing loss, see the Health feature, "Preventing a Loss of Hearing," on page 353.

Animation Hearing

MP3 Sense of Hearing and Equilibrium

Check Your Progress 18.5

1. Identify all structures of the ear that are involved with hearing, and describe their functions.
2. Explain how mechanoreceptors in the inner ear transduce sound waves into nerve impulses.
3. Discuss how the distribution of stereocilia along the spiral organ allows the brain to perceive different pitches.
4. Explain how the brain distinguishes the volume of a sound.

18.6 Sense of Equilibrium

Learning Outcomes

Upon completion of this section, you should be able to

1. Explain how mechanoreceptors are involved in the sense of equilibrium.
2. Review the structures involved in rotational and gravitational equilibrium.
3. Compare the role of the fluid found in the semicircular canals with the otoliths present in the utricle and saccule.

In addition to detecting sound waves, mechanoreceptors in the inner ear detect rotational and/or angular movement of the head, or **rotational equilibrium,** as well as movement of the head in the vertical or horizontal planes, or **gravitational equilibrium** (Fig. 18.13).

Through their communication with the brain, these mechanoreceptors help us maintain balance and equilibrium, but other structures are also involved. For example, we already mentioned that proprioceptors are necessary for maintaining our equilibrium. Vision, if available, also provides extremely helpful input the brain can act upon.

Rotational Equilibrium Pathway

Rotational equilibrium involves the three semicircular canals, which are arranged so that there is one in each dimension of space. The base of each of the three canals, called an ampulla, is slightly enlarged. Little hair cells, whose stereocilia are embedded within a gelatinous material called a cupula, are found within the ampullae. Because of the way the semicircular canals are arranged, each ampulla responds to head rotation in a different plane of space. As fluid within a semicircular canal flows over and displaces a cupula, the stereocilia of the hair cells bend, and the pattern of impulses carried by the vestibular nerve to the brain stem and cerebellum changes. The brain uses information from the hair cells within the ampullae of the semicircular canals to maintain rotational equilibrium through appropriate motor output to various skeletal muscles that can right our present position in space as need be.

Gravitational Equilibrium Pathway

Gravitational equilibrium depends on the **utricle** and the **saccule,** two membranous sacs located in the vestibule. Both of these sacs contain little hair cells, whose stereocilia are embedded within a gelatinous material called an otolithic membrane. Calcium carbonate ($CaCO_3$) granules, or **otoliths,** rest on this membrane. The utricle is especially sensitive to horizontal (back-and-forth) movements and the bending of the head, while the saccule responds best to vertical (up-and-down) movements.

When the body is still, the otoliths in the utricle and the saccule rest on the otolithic membrane above the hair cells. When the head bends or the body moves in the horizontal and vertical planes, the otoliths are displaced and the otolithic membrane sags, bending the stereocilia of the hair cells beneath. If the stereocilia move toward the largest stereocilium, called the kinocilium, nerve impulses increase in the vestibular nerve. If the stereocilia move away from the kinocilium, nerve impulses decrease in the vestibular nerve. If you are upside down, nerve impulses in the vestibular nerve cease. These data reach the vestibular cortex, which uses them to determine the direction of the movement of the head at the moment. The brain uses this information to maintain gravitational equilibrium through appropriate motor output to various skeletal muscles that can right our present position in space as need be.

MP3 The Sense of Hearing and Equilibrium

Table 18.3 summarizes how the various parts of the ear function in both hearing and equilibrium.

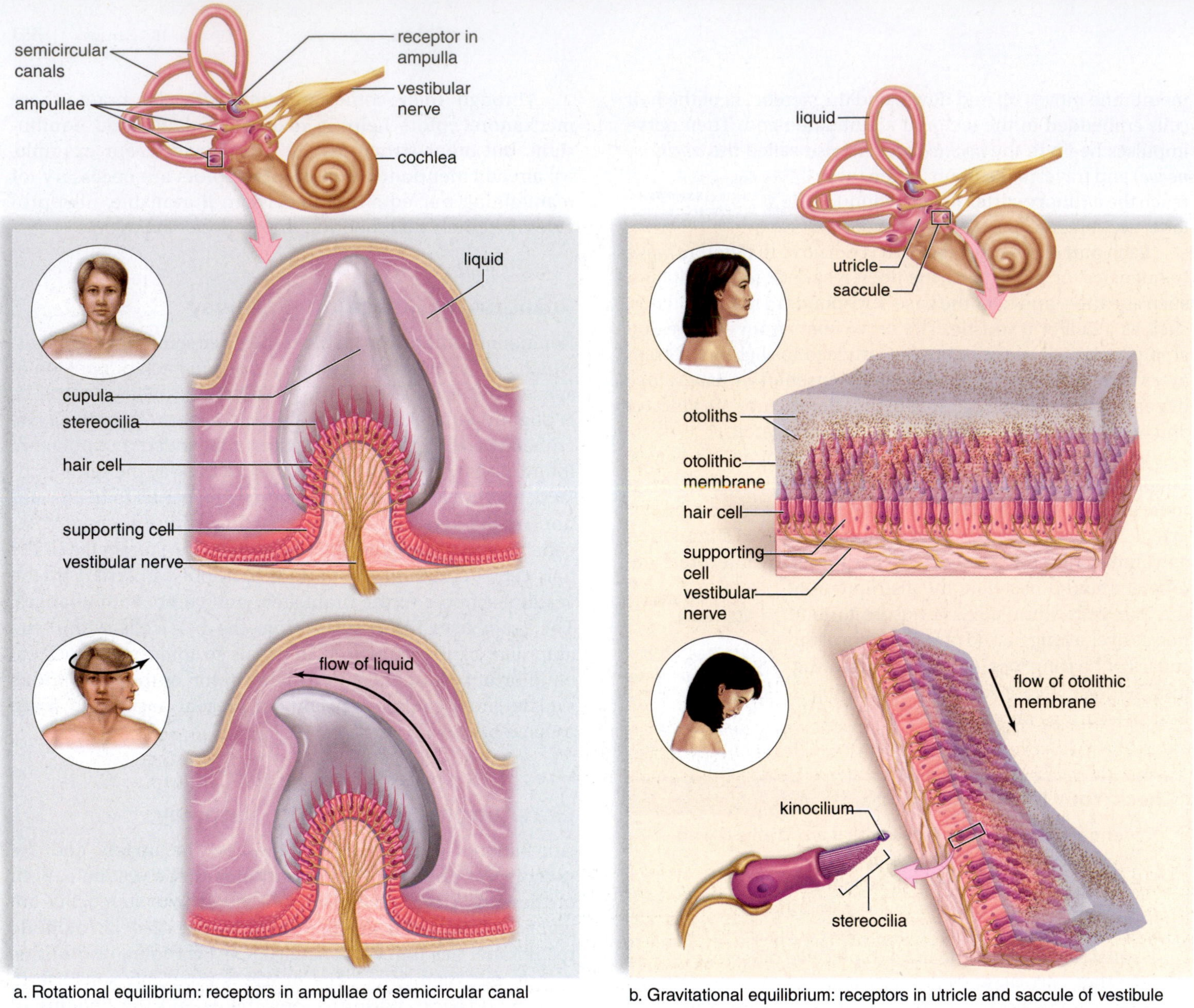

semicircular canals
ampullae
receptor in ampulla
vestibular nerve
cochlea
liquid
cupula
stereocilia
hair cell
supporting cell
vestibular nerve
flow of liquid
liquid
utricle
saccule
otoliths
otolithic membrane
hair cell
supporting cell
vestibular nerve
flow of otolithic membrane
kinocilium
stereocilia
a. Rotational equilibrium: receptors in ampullae of semicircular canal
b. Gravitational equilibrium: receptors in utricle and saccule of vestibule

Figure 18.13 Mechanoreceptors for equilibrium. **a.** Rotational equilibrium. The ampullae of the semicircular canals contain hair cells with stereocilia embedded in a cupula. When the head rotates, the cupula is displaced, bending the stereocilia. Thereafter, nerve impulses travel in the vestibular nerve to the brain. **b.** Gravitational equilibrium. The utricle and the saccule contain hair cells with stereocilia embedded in an otolithic membrane. When the head bends, otoliths are displaced, causing the membrane to sag and the stereocilia to bend. If the stereocilia bend toward the kinocilium, the longest of the stereocilia, nerve impulses increase in the vestibular nerve. If the stereocilia bend away from the kinocilium, nerve impulses decrease in the vestibular nerve. This difference tells the brain in which direction the head moved.

TABLE 18.3 Functions of the Parts of the Ear

Part	Medium	Function	Mechanoreceptor
Outer Ear	*Air*		
Pinna		Collects sound waves	—
Auditory canal		Filters air	—
Middle Ear	*Air*		
Tympanic membrane and ossicles		Amplify sound waves	—
Auditory tube		Equalizes air pressure	—
Inner Ear	*Fluid*		
Semicircular canals		Rotational equilibrium	Stereocilia embedded in cupula
Vestibule (contains utricle and saccule)		Gravitational equilibrium	Stereocilia embedded in otolithic membrane
Cochlea (contains spiral organ)		Hearing	Stereocilia embedded in tectorial membrane

SCIENCE IN YOUR LIFE ▶ **HEALTH**

Preventing a Loss of Hearing

Some degree of age-associated hearing loss is very common (see section 18.7), but deafness due to stereocilia damage from exposure to loud noises is preventable. Hospitals are now aware that even the ears of newborns need to be protected from noise and are taking steps to make sure neonatal intensive care units and nurseries are as quiet as possible.

In today's society, exposure to the types of noises listed in Table 18A is common. Noise is measured in decibels (db), and any noise above a level of 80 db could result in damage to the hair cells of the spiral organ. Hearing loss occurs once 25–30% of these cells are damaged. Eventually, the stereocilia and then the hair cells disappear completely (Fig. 18A).

Legendary rock musician Pete Townsend of The Who recently admitted that long-term exposure to loud music has caused him to experience hearing loss. Digital music players may pose an additional risk because they can produce music as loud as 120 db, plus hold thousands of songs and play for hours without recharging. Like all the sensory systems, the ears can become desensitized to loud music, so that it seems less loud. The response of many music enthusiasts, of course, is to turn up the volume. In 2006, an iPod user filed a lawsuit against Apple, claiming the company did not warn users sufficiently about the dangers of loud music, and asking the company to offer a software update to limit the device's output to 100 db, as it has already done in France.

The first hint of danger could be temporary hearing loss, a "full" feeling in the ears, muffled hearing, or tinnitus (e.g., ringing in the ears). If you have any of these symptoms, modify your listening habits immediately to prevent further damage. If exposure to noise is unavoidable, specially designed noise-reduction earmuffs are available, and it is also possible to purchase earplugs made from a compressible, sponge-like material at the drugstore or sporting-goods store. These earplugs are not the same as those worn for swimming, and they should not be used interchangeably.

In addition to loud music, noisy indoor or outdoor equipment can also be harmful to hearing if hearing protection is not used. Even motorcycles and recreational vehicles such as snowmobiles can contribute to a gradual loss of hearing. Exposure to intense sounds of short duration, such as a burst of gunfire, can result in an immediate hearing loss. Hunters may have a significant hearing reduction in the ear opposite the shoulder where the gun is held. (The butt of the rifle offers some protection to the ear nearest the gun when it is shot.)

Finally, people need to be aware that some medicines are ototoxic. Anticancer drugs, most notably cisplatin, and certain antibiotics (e.g., streptomycin, kanamycin, and gentamicin) make ears especially susceptible to a hearing loss. Anyone taking such medications needs to be especially careful to protect the ears from loud noises.

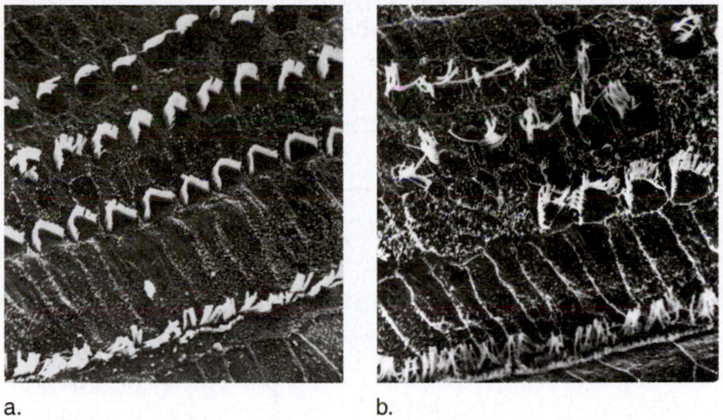

a.
b.

Figure 18A Hearing loss. **a.** Normal hair cells in the spiral organ of a guinea pig. **b.** Damaged hair cells in the spiral organ of a guinea pig. This damage occurred after 24-hour exposure to a noise-level typical of a rock concert.

Questions to Consider

1. Do you think that manufacturers of digital music players should be required to restrict the output of these devices to 100 db in the United States?
2. If you listen to loud music or are exposed to other noise, are you concerned that the hair cells in the spiral organs of your inner ears may become damaged? Why or why not?
3. Assume that you suddenly became deaf today, and list three to five of your everyday activities that would be most affected. If your deafness could not be corrected medically, what specific steps could you take to help you communicate with hearing people? With other deaf people?

TABLE 18A Noises That Affect Hearing

Type of Noise	Sound Level (decibels)	Effect
Jet engine, shotgun, rock concert	Over 125	Beyond threshold of pain; potential for hearing loss high
Music on ear buds (maximum), thunderclap	Over 120	Hearing loss likely
Chain saw, pneumatic drill, jackhammer, snowmobile, cement mixer	100–200	Regular exposure of more than 1 minute risks permanent hearing loss
Farm tractor, newspaper press, subway, motorcycle	90–100	Fifteen minutes of unprotected exposure potentially harmful
Lawn mower, food blender	85–90	Continuous daily exposure for more than eight hours can cause hearing damage
Diesel truck, average city traffic noise	80–85	Annoying; constant exposure may cause hearing damage

Source: National Institute on Deafness and Other Communication Disorders, National Institutes of Health.

Check Your Progress 18.6

1. Explain the role, and specific location, of mechanoreceptors involved in maintaining balance and equilibrium.
2. Classify each of the following as being associated with either the semicircular canals or the vestibule: ampulla, cupula, otoliths, saccule, and utricle.
3. Explain how having two separate systems for equilibrium provides more complete information than one system could.

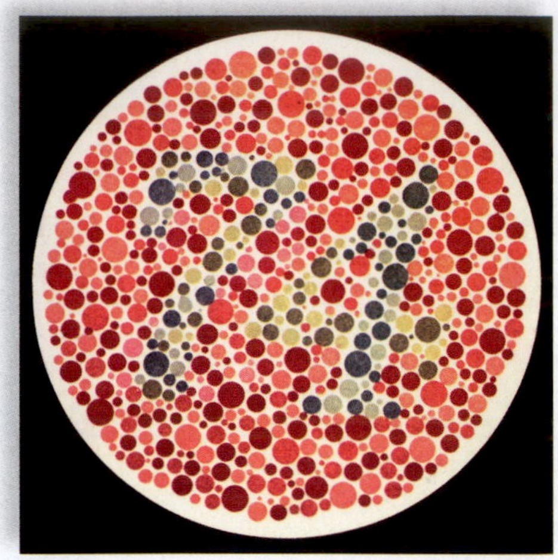

Figure 18.14 Testing for color blindness. This is one type of test plate used to diagnose color blindness. There are many different types of color blindness, and thus many different tests are used. In this case, you should be able to see the number "74" in green among the dots.

18.7 Disorders That Affect the Senses

Learning Outcomes

Upon completion of this section, you should be able to

1. List three common disorders that can affect the senses of taste and smell.
2. Explain the anatomical abnormalities that cause color blindness, nearsightedness, farsightedness, and astigmatism.
3. Characterize the three most common causes of blindness in the United States.
4. Describe some common disorders of hearing and equilibrium.

Disorders of Taste and Smell

It is hard for most of us to imagine our world without the ability to taste and smell. Disorders that affect these senses may not sound very serious, but these senses contribute substantially to our enjoyment of life. In addition, unpleasant smells and tastes can warn us about dangers such as fire, poisonous fumes, and spoiled food.

In most people, the sense of smell begins to decline after age 60, and a large proportion of elderly people lose some of their olfactory sense. Some people are born with a poor sense of smell or taste, and others completely lack a sense of smell, a condition known as *anosmia*. However, the most common conditions that affect our sense of taste and smell are upper respiratory infections, allergies, exposure to certain drugs and chemicals, and certain types of brain tumors. Brain injury due to trauma can also cause smell or taste problems, and smoking impairs the ability to identify odors and diminishes the sense of taste.

The extent of a loss of smell or taste can be tested using the lowest concentration of a chemical that a person can detect and recognize. Scientists have developed an easily administered "scratch-and-sniff" test to evaluate the sense of smell. For taste, patients are asked to evaluate different chemical concentrations, sometimes applied directly to specific areas of the tongue.

Disorders of the Eye

Color blindness and problems with visual focus are two common abnormalities of the eye. More serious disorders can result in blindness.

Color Blindness

Complete **color blindness** is extremely rare. Instead, in most instances a particular type of cone is lacking or deficient in number. The most common mutation is a lack of either red or green cones, resulting in various types of color blindness (Fig. 18.14). If the eye lacks red cones, the green colors are accentuated, and vice versa. Most genetic mutations affecting color vision are X-linked, recessive traits. Because females have two copies of the X chromosome, they are more likely to have a normal copy of the gene. Therefore, color blindness is much more common in males. Approximately 5–8% of males and 0.5% of females have some type of deficiency in color vision. See section 24.1 for additional discussion of the genetics of color blindness.

Visual Focus

The majority of people can see what is designated as a size 20 letter 20 feet away, and so are said to have 20/20 vision. Persons who can see close objects but cannot see the letter from this distance are said to be **nearsighted.** These individuals usually have an elongated shape to the eye, so that when they attempt to look at a distant object, the lens of the eye brings the image into focus in front of the retina (Fig. 18.15*a*). They can see close objects because the lens can compensate for the long eyeball. To see distant objects, these people can wear concave lenses, which diverge the light rays so that the image focuses on the retina.

Persons who can easily see the optometrist's chart but cannot see close objects well are **farsighted.** These individuals can see distant objects better than they can see close objects. Their eyes have a shortened shape, and when they try to see close objects, the image is focused behind the retina. When the object is distant, the lens can compensate for the shorter

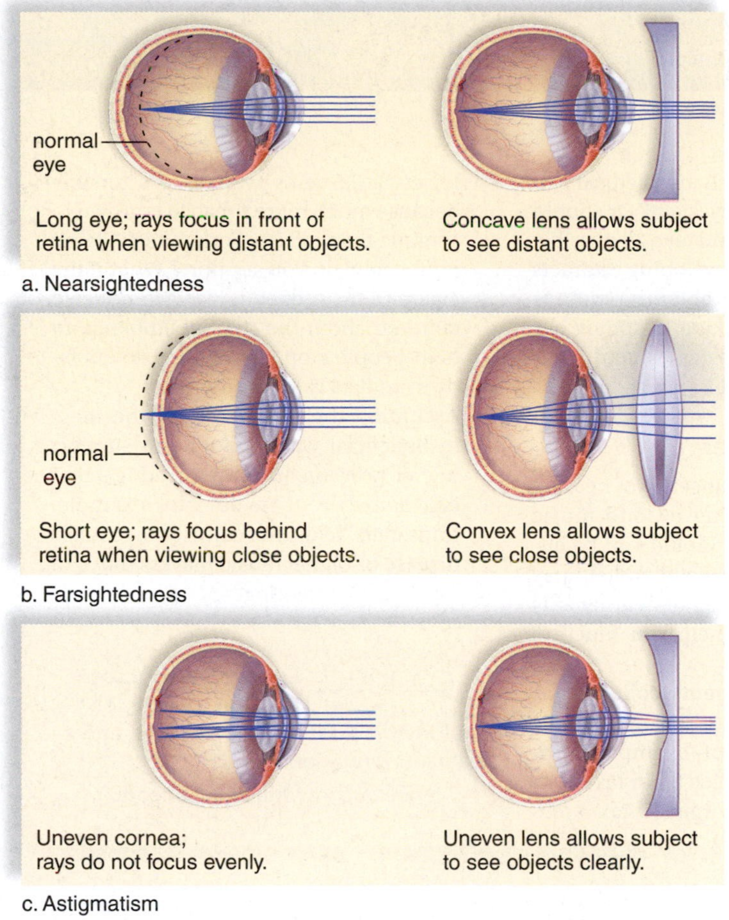
normal eye
Long eye; rays focus in front of retina when viewing distant objects.
Concave lens allows subject to see distant objects.
a. Nearsightedness
normal eye
Short eye; rays focus behind retina when viewing close objects.
Convex lens allows subject to see close objects.
b. Farsightedness
Uneven cornea; rays do not focus evenly.
Uneven lens allows subject to see objects clearly.
c. Astigmatism

Figure 18.15 Common abnormalities of the eye and possible corrective lenses. **a.** A concave lens in nearsighted persons focuses light rays on the retina. **b.** A convex lens in farsighted persons focuses light rays on the retina. **c.** An uneven lens in persons with astigmatism focuses light rays on the retina.

eye. When the object is close, these persons must wear convex lenses to increase the bending of light rays so that the image can be focused on the retina (Fig. 18.15*b*). People over the age of 40 usually begin to have trouble seeing objects close up due to a condition called presbyopia, in which the lens of the eye begins to lose its elasticity and thus cannot focus on close objects.

When the cornea or lens is uneven, the image is fuzzy. The light rays cannot be evenly focused on the retina. This condition, known as **astigmatism,** can be corrected by wearing an unevenly ground lens to compensate for the uneven cornea (Fig. 18.15*c*). Rather than wearing glasses or contact lenses, many people with problems of visual focus are now choosing to undergo a laser surgery procedure called LASIK (laser-assisted in situ keratomileusis).

Common Causes of Blindness

The most frequent causes of blindness in adults are retinal disorders, glaucoma, and cataracts, in that order. Retinal disorders are varied. In **diabetic retinopathy,** the leading cause of blindness in the United States for people under the age of 65, capillaries to the retina become damaged, and hemorrhages and blocked vessels can occur. Surgical lasers can be used to seal the leaking blood vessels, but eventually scarring can lead to blindness. A common condition in people over 50 years of age is **macular degeneration,** in which the cones are destroyed because thickened choroid vessels no longer function as they should. With **retinal detachment,** the retina becomes torn or separated from the choroid, often following trauma to the eye or head. If diagnosed quickly enough, the retina can usually be reattached with a laser.

Glaucoma occurs when the drainage system of the eyes fails, so fluid builds up and the resulting increase in pressure destroys the nerve fibers responsible for peripheral vision. If glaucoma is untreated, total blindness can result. Eye doctors usually measure the pressure inside the eye during routine exams, but it is advisable for everyone to be aware of the disorder because it can come on quickly, and damage to the nerves can be permanent. Those who have experienced acute glaucoma report that the eye feels as heavy as a stone. Rarely, the abnormalities that lead to glaucoma can be present at birth, even though the disease symptoms may not appear for a few months or years. Regular visits to an eye-care specialist, especially by the elderly, are a necessity in order to catch conditions such as glaucoma early enough to allow effective treatment.

With aging, the eye is increasingly subject to **cataracts,** cloudy spots on the lens of the eye. Eventually these pervade the whole lens, which takes on a milky-white to yellowish color and becomes incapable of transmitting light (Fig. 18.16). Factors that increase the risk of cataracts include exposure to ultraviolet or other types of radiation, diabetes, and heavy alcohol consumption. In addition, smoking can double the risk of cataracts, possibly by reducing the delivery of blood, and therefore nutrients, to the lens. In most cases vision can be restored by surgical removal of the unhealthy lens and replacement with a clear plastic artificial lens. Cataract surgery is the

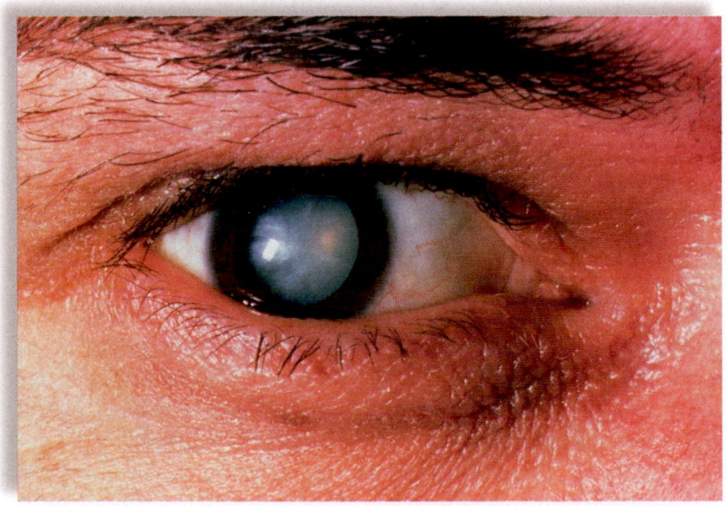

Figure 18.16 Cataract in a human eye. A cataract is a lens that has become opaque or cloudy. Risk factors that increase the chances of cataract formation include UV exposure, diabetes, and smoking. Cataracts can usually be surgically removed and replaced with an artificial lens.

SCIENCE IN YOUR LIFE ▶ HEALTH

Taking Personal Responsibility for Health

A cataract is a cloudiness of the lens that usually occurs in the elderly. A dense, centrally placed cataract causes severe blurring of vision. Certain factors have been identified as contributing to the chances of having a cataract:

- Smoking 20 or more cigarettes a day doubles the risk of cataracts in men, and smoking more than 30 cigarettes a day increases the chances of cataracts in women.
- Exposure to the ultraviolet radiation in sunlight can more than double the risk of cataracts.
- As many as one-third of cataracts may be caused by being overweight. Diet, rather than exercise, seems to reduce cataract formation.
- Most cataract operations today are performed on an outpatient basis, with minimal postoperative discomfort and a high expectation of sight restoration.

Should we simply rely on medical science to restore our eyesight? Or should we act responsibly and take all possible steps to keep from developing cataracts or other medical problems—such as refraining from smoking, watching our weight, wearing sunglasses, and turning down the volume?

Questions to Consider

1. Suppose you lose your sense of hearing after years of listening to loud music on your MP3 player. Should you be eligible for insurance benefits to help pay for thousands of dollars of hearing aids or other devices to help you function normally?
2. Some state governments collect relatively large amounts of revenue by taxing cigarettes. For example, New York City currently adds $3 in taxes per pack. Interestingly, Virginia's tax is the lowest, at 2.5 cents per pack, presumably because much more tobacco is grown in Virginia than in New York. Which state's philosophy do you support? What if the cigarette tax money is used to educate children about the risks of smoking, to help people stop smoking, or to subsidize medical care for smokers?
3. Assuming the money would be used in beneficial ways, would you support higher taxes on any other products that can damage our health—for example, trans-fatty acids that contribute to heart disease or digital music players that can damage your hearing?

a clear plastic artificial lens. Cataract surgery is the most frequently performed surgery in the United States. As of 2009, an estimated 6.1 million Americans over age 40 have had cataract surgery. Surgeons can now replace cataracts with multifocal lenses that allow patients to focus at different distances, even eliminating the need for glasses in some patients. The Health feature, "Taking Personal Responsibility for Health," explores the question of whether we should keep relying on surgery to solve this problem or take more preventive steps to keep cataracts from developing in the first place.

In 2008, two groups of researchers published studies showing that a certain form of hereditary blindness could be successfully treated with gene therapy. Both groups injected a genetically modified virus into one eye of patients with Leber's congenital amaurosis (LCA), a rare condition in which retinal degeneration begins in childhood and eventually causes total blindness. Starting two weeks after the injections, vision in the injected eyes improved from detecting only hand movements to reading lines on an eye chart. Equally significant, the injected eyes showed no signs of inflammation in the retina or other toxic side effects. If these positive results are confirmed, gene therapy might be useful for other forms of blindness.

Although it may sound like science fiction, researchers are currently testing an artificial eye of sorts, which may allow blind people to see. For blind patients who still have intact connections between their retinas and brain, an array of electrodes can be surgically implanted in the retina. These electrodes receive input from a tiny digital camera mounted on eyeglass frames worn by the patient. This input is converted into a nerve impulse, which the individual's brain interprets as sight. Clinical trials have shown that by using such a device, some formerly blind patients can detect light or even distinguish between objects such as a cup or plate, and more advanced devices may become available for testing soon.

Since several disorders of the eye have been associated with exposure to ultraviolet irradiation, it is recommended that everyone, especially those who live in sunny climates or work outdoors, wear sunglasses that absorb ultraviolet light. The Sunglass Association of America has adopted the following system for categorizing sunglasses:

- Cosmetic lenses, which absorb at least 70% of UVB (the more harmful type of ultraviolet radiation), 20% of UVA, and 60% of visible light. Such lenses are worn for comfort or appearance rather than protection.
- General-purpose lenses that absorb at least 95% of UVB, 60% of UVA, and 60–92% of visible light. These lenses are good for outdoor activities in temperate regions.
- Special-purpose lenses, which block at least 99% of UVB, 60% of UVA, and 60–92% of visible light. They are good for bright sun combined with sand, snow, or water.

Disorders of Hearing and Equilibrium

Hearing loss can develop gradually or suddenly and has many potential causes. Especially when we are young, the middle ear is subject to infections that can lead to hearing impairment if not treated promptly by a physician (see section 15.4). Hearing loss is also one of the most common conditions of older adults, affecting one in three people over age 60. Age-associated hearing loss usually develops gradually. The first sign may be problems understanding conversations when there is noise in the background. Actually, in most individuals age-related hearing loss probably begins at around age 20, with the ability to hear high-pitched sounds affected first. With age, the mobility of ossicles decreases, and in otosclerosis, new filamentous bone grows over the stirrup, impeding its movement. Surgical treatment is the only remedy for this type of conduction deafness. Some types of hearing loss seem to run in families, but most can be attributed to the effect of years of frequent exposure to loud noise (see the Health feature, "Preventing a Loss of Hearing," on page 353). As noted in the story that opened this chapter, some types of deafness can be treated with cochlear implants.

Sudden deafness, which may occur all at once or over a period of up to three days, most commonly affects people between the ages of 30 and 50, usually in only one ear. Potential causes include infections, head trauma, and the side effects of certain drugs. Depending on the cause, sudden deafness often resolves itself within a few days.

Deafness can also be present at birth. Several genetic disorders are associated with deafness at birth, as are infections with German measles (rubella) and mumps virus when contracted by a woman during her pregnancy. For this reason, every female should be immunized against these viruses before she reaches childbearing age.

Disorders of equilibrium often manifest as **vertigo,** the feeling that a person or the environment is moving when no motion is occurring. It is possible to simulate a feeling of vertigo by spinning your body rapidly and then stopping suddenly. Vertigo can be caused by problems in the brain as well as the inner ear. An estimated 20% of those who experience symptoms of vertigo have benign positional vertigo (BPV), which may result from the formation of particles in the semicircular canals. When individuals with BPV move their heads suddenly, especially when lying down, these particles shift like pebbles inside a tire. After the head movement, there is an initial delay, but when the movement stops, the particles tumble down with gravity, stimulating the stereocilia and resulting in the sensation of movement. Interestingly, most cases of BPV either resolve on their own or can be cured using a series of precise head movements designed to cause the particles to settle into a position where they can no longer stimulate the stereocilia.

Because the senses of hearing and equilibrium are anatomically linked, certain disorders can affect both. An example of this phenomenon is *Meniere's disease,* which is usually characterized by vertigo, a feeling of fullness in the affected ear(s), tinnitus (ringing in the ears), and hearing loss. In about 80% of patients, only one ear is affected, and the disease usually strikes between the ages of 20 and 50. The exact cause of Meniere's disease is unknown, but it seems related to an increased volume of fluid in the semicircular canals, vestibule, and/or cochlea. For this reason, it has been called "glaucoma of the ear." There is no cure for Meniere's disease, but the vertigo and feeling of pressure can often be managed by adhering to a low-salt diet, which is thought to decrease the volume of fluid produced. Any hearing loss that occurs, however, is usually permanent.

Check Your Progress 18.7

1. Describe how a lack of smell can be life threatening.
2. Explain how the shape of the eyeball can affect vision.
3. Distinguish how blindness can occur in macular degeneration, glaucoma, and cataracts.
4. List some common causes of hearing loss.

Case Study Conclusion

The sounds people hear with cochlear implants are different from what hearing people are familiar with. With postimplantation therapy and experience, however, implant recipients can learn to understand speech, and function very much like a person with normal hearing. Now three years old, Tony is learning to say his ABCs and sing nursery rhymes, but Marlene is making sure he learns sign language as well.

Cochlear implants were developed as a result of decades of research into understanding how the ear functions. An essential aspect was the collaboration of scientists from many different disciplines, including biomedical researchers, engineers, and speech-language pathologists. Scientists continue to work on improving the existing technology, including designing implants that can send auditory information directly to the brain of people who lack a functioning cochlear nerve.

MEDIA STUDY TOOLS

www.mhhe.com/maderinquiry14

Enhance your study of this chapter with study tools and practice tests. Also ask your instructor about the resources available through ConnectPlus, including LearnSmart, the media-rich eBook, interactive learning tools, and animations.

McGraw Hill connect plus+ BIOLOGY

SUMMARIZE

18.1 Sensory Receptors and Sensations

Sensory receptors are cells capable of converting many types of stimuli into nerve impulses, a process called **sensory transduction.** They can be classified in several ways:

- According to whether they receive stimuli from inside or outside the body—that is, as **interoceptors** or **exteroceptors.**
- According to the nature of the stimulus they receive—that is, as **chemoreceptors, photoreceptors, mechanoreceptors,** or **thermoreceptors.**

The conscious perception of stimuli is the result of several processes:

- **Detection** occurs when environmental changes stimulate sensory receptors, generating nerve impulses;
- **Sensation** occurs when these nerve impulses reach the cerebral cortex;
- **Perception** occurs when the brain interprets the meaning of sensations.
- **Integration** occurs when the incoming signals are summed together.

18.2 Somatic Senses

The somatic sensory receptors include proprioceptors, sensory receptors, and pain receptors:

- **Proprioceptors** help maintain equilibrium and posture. Proprioception is illustrated by the action of muscle spindles that are stimulated when muscle fibers stretch and Golgi tendon organs that are stimulated when muscles contract.
- **Cutaneous receptors** in the skin detect touch, pressure, pain, and temperature (warmth and cold).
- **Pain receptors** are free nerve endings that respond to chemicals released by damaged cells. The pain of internal organs is sometimes felt in the skin and is called referred pain.

18.3 Senses of Taste and Smell

Taste and smell are due to the detection of molecules in food and in the air by chemoreceptors:

- Located mainly on the tongue, **taste buds** contain taste cells. After food molecules bind to plasma membrane receptor proteins on the microvilli of taste cells, as well as on the cilia of **olfactory cells** in the nasal cavity, nerve impulses eventually reach the cerebral cortex, which determines the taste and odor according to the pattern of stimulation. The gustatory (taste) cortex responds to a range of salty, sour, bitter, sweet, and perhaps "umami" tastes.

18.4 Sense of Vision

Vision requires the eye, the optic nerves, and the visual cortex in the occipital lobe:

- The eye has three layers. The outer layer, called the **sclera,** can be seen as the white of the eye. It also forms the transparent portion of the front of the eye called the **cornea.** The exposed surface of the sclera is covered by the **conjunctiva.** The middle pigmented layer, called the **choroid,** absorbs stray light rays. At the front of the eye the choroid becomes the **iris,** which forms the **pupil,** or opening through which light passes. As light passes through the eye it is focused by the **lens** on its way to the **retina,** the inner layer of the eyeball. The choroid also forms the **ciliary muscle,** which controls the shape of the lens. To see a close object, **visual accommodation** occurs as the lens rounds up.
- Light is normally focused on the **fovea centralis,** an area of densely packed cone cells. Photoreceptors called **rod cells** (sensory receptors for dim light) and the **cone cells** (sensory receptors for bright light and color) are located in the retina. The visual pathway begins when light strikes **rhodopsin** within the membranous disks of rod cells. A cascade of reactions leads to the closing of ion channels in a rod cell's plasma membrane, and signals are passed to other neurons in the retina. From the rod and cone cell layer, impulses pass to the bipolar cell layer and the ganglion cell layer. Many rods may synapse on the same ganglion cell, but some cone cells may activate only one ganglion cell. There are no photoreceptors in the area of the retina known as the **blind spot,** where the optic nerve exits.
- The axons of ganglion cells become the optic nerve, which takes nerve impulses to the visual cortex in the occipital lobe. The primary visual area parcels out signals for color, form, and motion to the visual association area. Then the cortex rebuilds the field.

18.5 Sense of Hearing

Hearing is dependent on the ear, the cochlear nerve, and the auditory cortex of the brain:

- The ear is divided into three parts. The **outer ear** consists of the **pinna** and the **auditory canal,** which directs waves to the middle ear. The **middle ear** begins with the **tympanic membrane** and contains the **ossicles** (**malleus, incus,** and **stapes**). The malleus is attached to the **tympanic membrane,** and the stapes is attached to the **oval window,** which is covered by a membrane. Within the **inner ear,** the **cochlea** functions in hearing, while the **vestibule** and **semicircular canals** function in equilibrium.
- The auditory pathway begins when the outer ear receives and the middle ear amplifies the sound waves, which then strike the oval window membrane. Its vibrations set up pressure waves across the cochlear canal, which contains the **organ of Corti** (also called the spiral organ), consisting of hair cells whose stereocilia are embedded within the tectorial membrane. When the basilar membrane vibrates, the stereocilia of the hair cells bend. This movement is transduced into nerve impulses that are carried by the cochlear nerve to the primary auditory area in the temporal lobe of the cerebral cortex.

18.6 Sense of Equilibrium

The ear also contains mechanoreceptors for our sense of equilibrium, or balance:

- **Rotational equilibrium,** which is the sensing of head movements, is dependent on the stimulation of hair cells within the ampullae of the semicircular canals.
- **Gravitational equilibrium,** which is the sensing of the position of the head in space, relies on the stimulation of hair cells within the **utricle** and the **saccule** by small granules called **otoliths.**

18.7 Disorders That Affect the Senses

Most conditions that affect the sensory organs are not life threatening, but can affect the quality of life:

- Disorders of taste and smell may result from infections, allergies, certain drugs, or tumors.
- Disorders of the eye include **color blindness** and problems with visual focus such as **astigmatism** and being **nearsighted** or **farsighted.** The most common causes of blindness are retinal disorders like **diabetic retinopathy, macular degeneration,** and **retinal detachment,** as well as **glaucoma** (increased intraocular pressure) and **cataracts** (cloudy lens).
- Disorders of hearing are very common, especially in the aging population. Possible causes include excessive noise exposure, infections, trauma, and certain drugs. **Vertigo** is an equilibrium disorder characterized by a feeling of motion or dizziness.

ASSESS

Testing Yourself

Choose the best answer for each question.

1. Chemoreceptors are involved in
   - **a.** hearing.
   - **b.** taste.
   - **c.** smell.
   - **d.** vision.
   - **e.** Both b and c are correct.
2. Free nerve endings sense temperature and
   - **a.** pressure.
   - **b.** pain.
   - **c.** vision.
   - **d.** hearing.

3. Tasting "sweet" versus "salty" is a result of
   a. activating different sensory receptors.
   b. activating many versus few sensory receptors.
   c. activating no sensory receptors.
   d. None of these are correct.
4. Our sense of smell
   a. is completely separate from our sense of taste.
   b. is dependent on olfactory cells, which are modified neurons.
   c. requires millions of different types of odor receptors.
   d. None of these are correct.
5. Label the following diagram of the human eye.

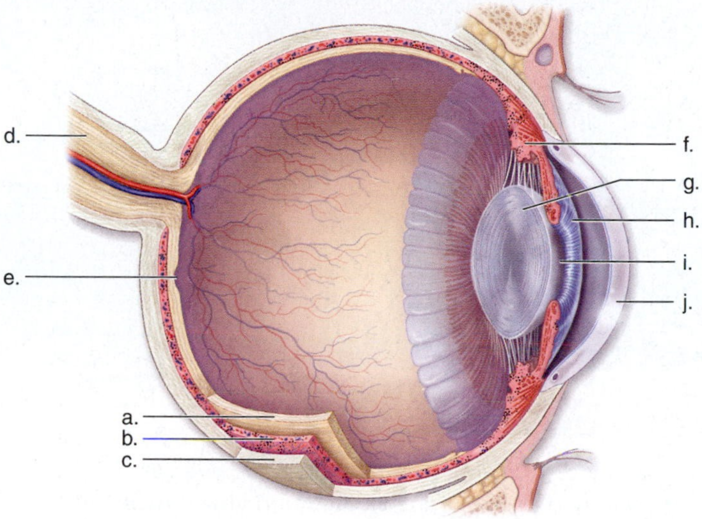
d.
e.
a.
b.
c.
f.
g.
h.
i.
j.

6. Which of the following structure-function associations is incorrect?
   a. lens—focusing
   b. cones—color vision
   c. iris—regulation of amount of light
   d. choroid—location of cones
   e. sclera—protection
7. Retinal is
   a. a derivative of vitamin A.
   b. sensitive to light energy.
   c. found in both rods and cones.
   d. a part of rhodopsin.
   e. All of these are correct.
8. Label the following diagram of the human ear.

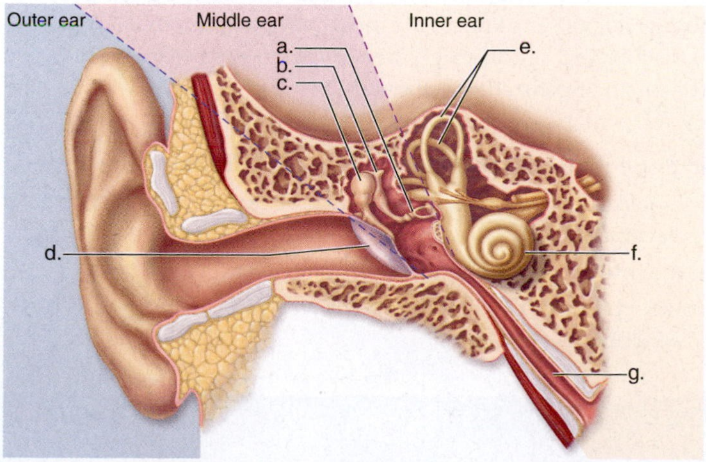
Outer ear
Middle ear
Inner ear
a.
b.
c.
e.
d.
f.
g.

9. Which is the correct location of the spiral organ?
   a. between the tympanic membrane and the oval window in the inner ear
   b. in the utricle and saccule within the vestibule
   c. between the tectorial membrane and the basilar membrane in the cochlear canal
   d. between the nasal cavities and the throat
   e. between the outer and inner ear within the semicircular canals
10. Our perception of pitch is dependent upon the region of the _____ vibrated and the regions of the _____ stimulated.
    a. cochlea, spinal cord
    b. basilar membrane, spinal cord
    c. basilar membrane, auditory cortex
    d. cochlea, cerebellum
11. The ossicle that articulates with the tympanic membrane is the
    a. malleus.
    b. stapes.
    c. incus.
    d. All of these are correct.
12. Which of these associations is mismatched?
    a. semicircular canals—inner ear
    b. utricle and saccule—outer ear
    c. auditory canal—outer ear
    d. cochlea—inner ear
    e. ossicles—middle ear
13. Both olfactory receptors and sound receptors have cilia, and both
    a. are chemoreceptors.
    b. are a part of the brain.
    c. are mechanoreceptors.
    d. initiate nerve impulses.
    e. All of these are correct.
14. A color-blind person has (an) abnormal
    a. rods.
    b. cochlea.
    c. cornea.
    d. cones.
    e. None of these are correct.
15. Which pair of terms is properly matched?
    a. nearsightedness—light rays focus in front of the retina
    b. nearsightedness—light rays focus behind the retina
    c. farsightedness—light rays focus in front of the retina
    d. None of these are correct.

ENGAGE

Thinking Critically

1. Some sensory receptors, such as those for taste, smell, and pressure, readily undergo the process of sensory adaptation, or decreased response to a stimulus. In contrast, receptors for pain are less prone to adaptation. Why does this make good biological sense? What do you think happens to children who are born without the ability to feel pain normally?
2. Trace the path of a photon of light through all the layers of the retina discussed in this chapter. Then trace the nerve impulse from a photoreceptor located on the right side of the left eyeball through the visual cortex.
3. The density of taste buds on the tongue can vary. Some obese individuals have a lower density of taste buds than usual. Assume that taste perception is related to taste bud density. If so, what hypothesis would you test to see if there is a relationship between taste bud density and obesity?

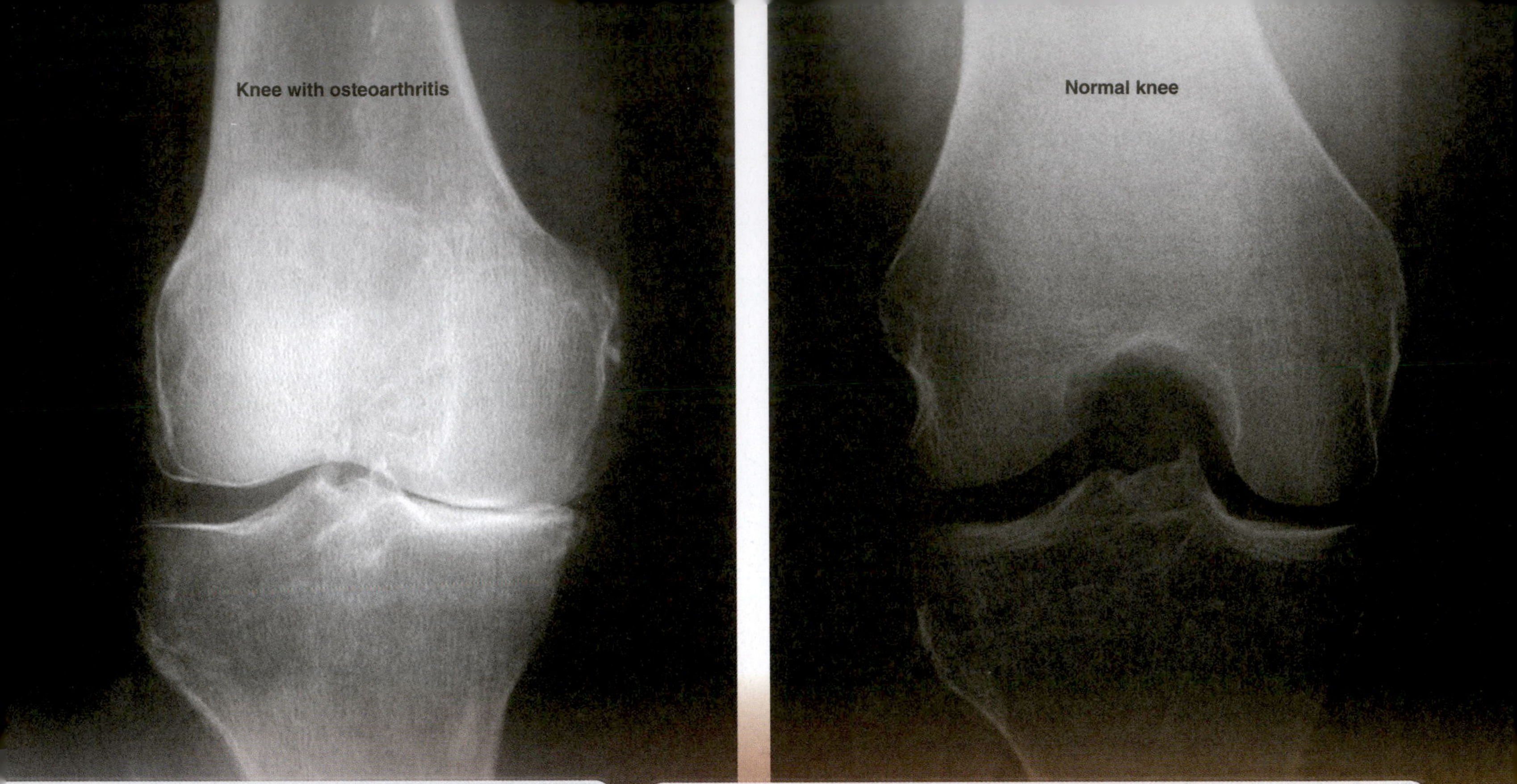
Knee with osteoarthritis
Normal knee

19 Musculoskeletal System

CHAPTER OUTLINE

BEFORE YOU BEGIN

Before beginning this chapter, take a few moments to review the following discussions:

Chapter 7 What biochemical pathways do cells use to produce ATP?

Section 11.1 What is the role of connective tissue in the body?

Section 11.5 How does the musculoskeletal system contribute to homeostasis?

CASE STUDY Replacing Joints

Jackie was an outstanding athlete in high school, and even as she reached her early fifties, she tried to stay in shape. But during her customary 3-mile jogs, she was having an increasingly hard time ignoring the pain in her left knee. She had torn some ligaments there playing intramural softball in college, and her knee had never quite felt the same. In her forties, she was able to control the pain by taking over-the-counter medications, but two years ago she had arthroscopic surgery to remove some torn cartilage and calcium deposits from the knee joint. Now that the pain was getting worse, she knew her best option might be a total knee replacement.

Although it sounds drastic, replacing old, arthritic joints with shiny new artificial ones is becoming increasingly routine. According to a 2012 estimate, almost 1 in 20 Americans over the age of 50 have an artificial knee, and the numbers are increasing as the population ages. During the procedure, a surgeon removes bone from the bottom of the femur and the top of the tibia, and replaces each with caps made of metal or ceramic, held in place with bone cement. A plastic plate is installed to allow the femur and tibia to move smoothly against each other, and a smaller plate is attached to the kneecap (patella) so it can function properly. The average cost for knee replacement surgery is about $40,000. The knee is the most common joint that is replaced, followed by the hip. And although it is more difficult, each year about 10,000 artificial ankles are being installed worldwide. In this chapter, we will examine the anatomy and physiology of bones, joints, and muscles.

As you read through the chapter, think about the following questions:

1. Why does Jackie's physical condition make her an ideal candidate for a knee replacement?
2. What is the role of cartilage in the knee joint?
3. Why would it be more difficult to replace an ankle than a knee or hip, or even an elbow or shoulder?

19.1 Overview of Bone and Cartilage

Learning Outcomes

Upon completion of this section, you should be able to

1. Explain the macroscopic and microscopic structure of bone.
2. Distinguish the structure and function of the three types of cartilage.
3. Describe how bones change during a person's lifetime.

The *skeleton* provides attachment sites for the muscles, whose contraction makes the bones move so that we can walk, play sports, hold this book (or eBook), and perform all sorts of other activities. Together, the bones and muscles make up the **musculoskeletal system.** In this chapter, we discuss the structures and functions of first the skeleton and then the muscles. Bones and the tissues associated with them are connective tissues, whereas muscles are, of course, composed of muscular tissue.

Organization of Tissues in the Skeleton

A **bone** is classified by its shape: long, short, flat, or irregular. Long bones do not have to be terribly long—the bones in your fingers are classified as long. But a long bone must be longer than it is wide. The bones of the arms and legs are long bones except for the kneecap and the wrist and ankle bones.

A long bone of the arm or leg can be used to illustrate the organization of tissues of the skeleton (Fig. 19.1). A bone is enclosed by a tough, fibrous, connective tissue covering called the **periosteum,** which is continuous with the ligaments that go across a joint. The periosteum contains blood vessels that enter the bone and give off branches that service various types of cells found in bone. Each expanded end of a long bone is termed an epiphysis; the portion between the epiphyses is called the diaphysis. Where the diaphysis of a long bone contacts (articulates with) another bone at a **joint,** it will be covered by a layer of hyaline cartilage, also called **articular cartilage.**

Structure of Bone and Associated Tissues

The primary connective tissues of the skeleton are bone, cartilage, and dense fibrous connective tissue. All connective tissues contain cells separated by a matrix that contains fibers. Bone tissue is particularly strong because the matrix contains collagen fibers plus mineral salts, notably calcium phosphate.

Bone

There are two types of bone tissue, compact bone and spongy bone. **Compact bone** is highly organized and composed of tubular units called *osteons.* Each bone cell, or **osteocyte,** lies in a *lacuna,* or tiny chamber. Lacunae are arranged in concentric circles around a *central canal* (Fig. 19.1). Central canals contain blood vessels, lymphatic vessels, and nerves. Tiny canals called *canaliculi* run through the matrix, connecting the lacunae with each other and with the central canal. In this way, the canaliculi bring nutrients from the blood vessel in the central canal to the cells in the lacunae.

The *epiphyses* (ends) of a long bone are composed largely of spongy bone. Compared to compact bone, **spongy bone** has an unorganized appearance (Fig. 19.1). Cells called osteocytes are found in *trabeculae,* numerous thin plates separated by unequal spaces. Although the spaces make spongy bone lighter than compact bone, spongy bone is still very strong. Just as braces are used for support in buildings, the plates of spongy bone follow lines of stress.

Splitting a bone open, as in Figure 19.1, shows that the *diaphysis* (shaft) is not solid but has a medullary cavity containing *yellow bone marrow.* Yellow bone marrow contains a large amount of fat.

The spaces of spongy bone are often filled with **red bone marrow,** a specialized tissue that produces all types of blood cells. In infants, red bone marrow is found in the cavities of most bones. In adults, red bone marrow occurs in a more limited number of bones (see section 19.2).

MP3 Bone Structure

Cartilage

Cartilage is not as strong as bone, but it is more flexible because the matrix is gel-like and contains many collagenous and elastic fibers. The cells lie within lacunae, which are irregularly grouped. Cartilage has no blood vessels, and therefore, injured cartilage is slow to heal.

All three types of cartilage—hyaline cartilage, fibrocartilage, and elastic cartilage—are associated with bones. **Hyaline cartilage** is firm and somewhat flexible. The matrix appears uniform and glassy (Fig. 19.1), but actually it contains a generous supply of collagen fibers. Hyaline cartilage is found in the joints at the ends of long bones and in the nose, at the ends of the ribs, and in the larynx and trachea.

Fibrocartilage is stronger than hyaline cartilage because the matrix contains wide rows of thick collagen fibers. Fibrocartilage is able to withstand both tension and pressure. This type of cartilage is found where support is of prime importance—in the disks between the vertebrae and in the cartilaginous disks between the bones of the knee.

Elastic cartilage is more flexible than hyaline cartilage because the matrix contains mostly elastin fibers. This type of cartilage is found in the ear flaps and epiglottis.

Dense Fibrous Connective Tissue

Dense fibrous connective tissue contains rows of cells called fibroblasts separated by bundles of collagen fibers. Dense fibrous connective tissue forms the flared sides of the nose. **Ligaments,** which bind bone to bone, are dense fibrous connective tissue, and so are **tendons,** which connect muscle to bone at joints.

Bone Growth and Remodeling

Bones are composed of living tissues, as exemplified by their ability to undergo growth and remodeling.

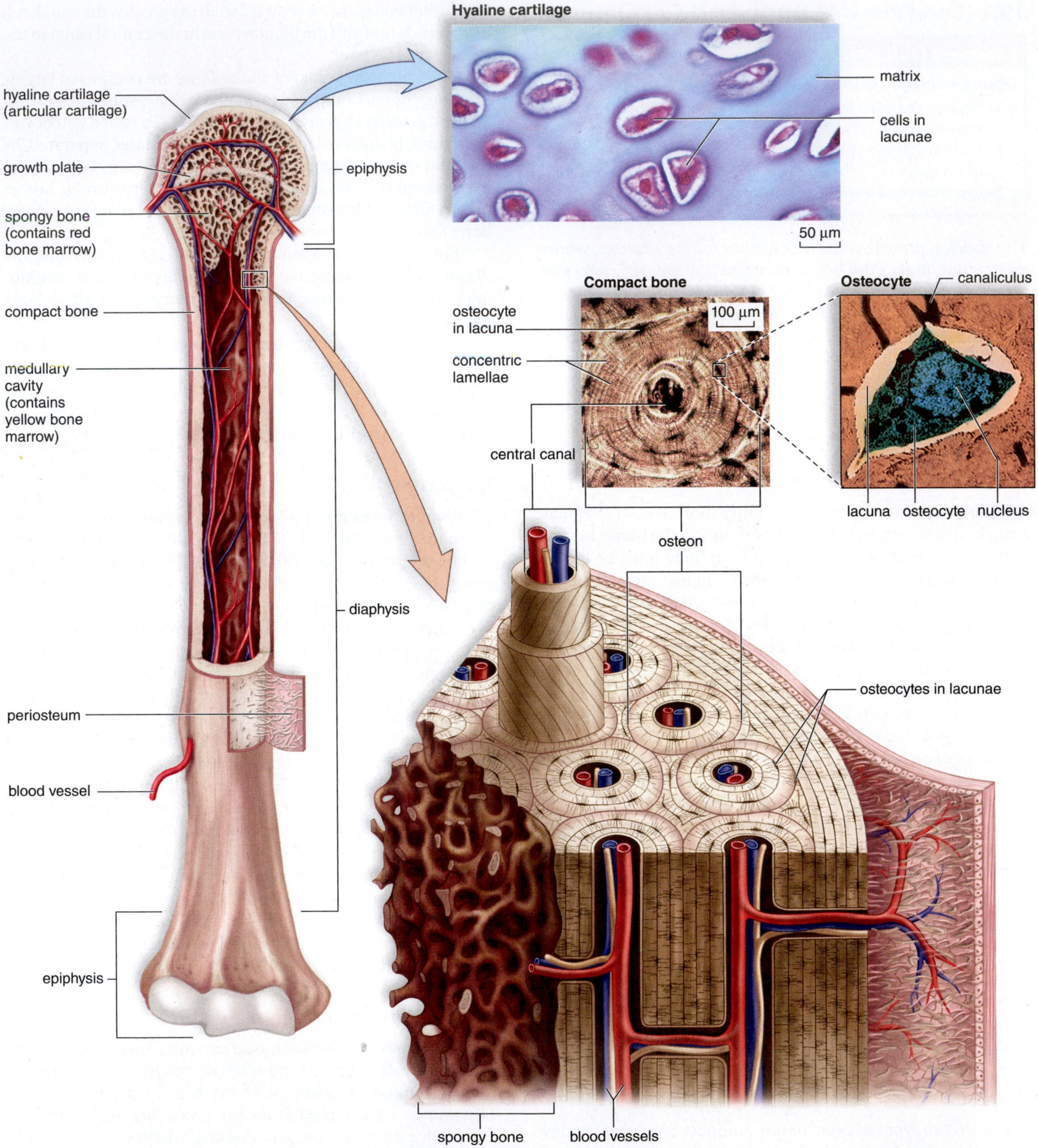

Hyaline cartilage
matrix
cells in lacunae
50 µm
hyaline cartilage (articular cartilage)
growth plate
spongy bone (contains red bone marrow)
epiphysis
compact bone
medullary cavity (contains yellow bone marrow)
Compact bone
100 µm
osteocyte in lacuna
concentric lamellae
central canal
Osteocyte
canaliculus
lacuna
osteocyte
nucleus
osteon
diaphysis
osteocytes in lacunae
periosteum
blood vessel
epiphysis
spongy bone
blood vessels

Figure 19.1 Anatomy of a bone from the macroscopic to the microscopic level. A long bone is encased by fibrous membrane (periosteum) except where it is covered at the ends by hyaline cartilage, shown on the micrograph (*right, top*). Spongy bone, located beneath the cartilage, may contain red bone marrow. The central shaft contains yellow bone marrow and is bordered by compact bone, which is shown in the enlargement and micrograph (*right, center*).

Bone Development and Growth

The bones of the human skeleton, except those of the skull, first appear during embryonic development as hyaline cartilage. These cartilaginous structures are then gradually replaced by bone, a process called *endochondral ossification* (Fig. 19.2).

During endochondral ossification, the cartilage begins to break down in the center of a long bone, which is now covered by a periosteum. **Osteoblasts** invade the region and begin to lay down spongy bone in what is called a primary ossification center. Other osteoblasts lay down compact bone beneath the periosteum. As the compact bone thickens, **osteoclasts,** which are large, multinucleated, macrophage-like cells, break down the spongy bone, and the cavity created becomes the medullary cavity.

The ends of developing bone continue to grow, but soon, secondary ossification centers appear in these regions. Here, spongy bone forms and does not break down. Also, a band of cartilage called a **growth plate** remains between the primary ossification center and each secondary center. The limbs keep increasing in length as long as the growth plates are present. The rate of growth is mainly controlled by growth hormone and sex hormones. Eventually, the growth plates become ossified, and the bone stops growing.

Animation Bone Growth

Video Nano Bones

Remodeling of Bones

In the adult, bone is continually being broken down and built up again, a process called remodeling. Osteoclasts break down bone, resulting in increased calcium levels in the blood. Simultaneously, the bone is repaired by the work of osteoblasts. As they form new bone, osteoblasts take calcium from the blood. Eventually, some of these cells get caught in the matrix they secrete and are converted to osteocytes, the cells found within the lacunae of osteons. Because of continual remodeling, the thickness of bones can change due to excessive use or disuse. Hormones such as calcitonin and parathyroid hormone (see section 20.3) can also affect the thickness of bones. When women reach menopause, the likelihood of a bone disease called osteoporosis increases, in part due to reduced estrogen levels (see section 19.6).

Bone remodeling also occurs when a fractured bone is healing. Bone healing can be divided into three major phases: (1) the *reactive phase,* which includes the body's inflammatory response to the injury; (2) the *reparative phase,* when the body generates a bone callus, which is a temporary connective tissue bridge that is gradually replaced in the last phase; and (3) the *remodeling phase,* during which osteoclasts and osteoblasts gradually replace the temporary bone with compact bone. Depending on factors such as the type of fracture, the age of the patient, and medical intervention, some fractures are healed in as little as three or four weeks, while others may not be properly healed after a year or more.

Check Your Progress 19.1

1. Describe the typical anatomy of compact and spongy bone.
2. List the three types of cartilage and where they are found in the body.
3. Compare and contrast ligaments and tendons.
4. Describe the role of osteoclasts and osteoblasts in remodeling bone.

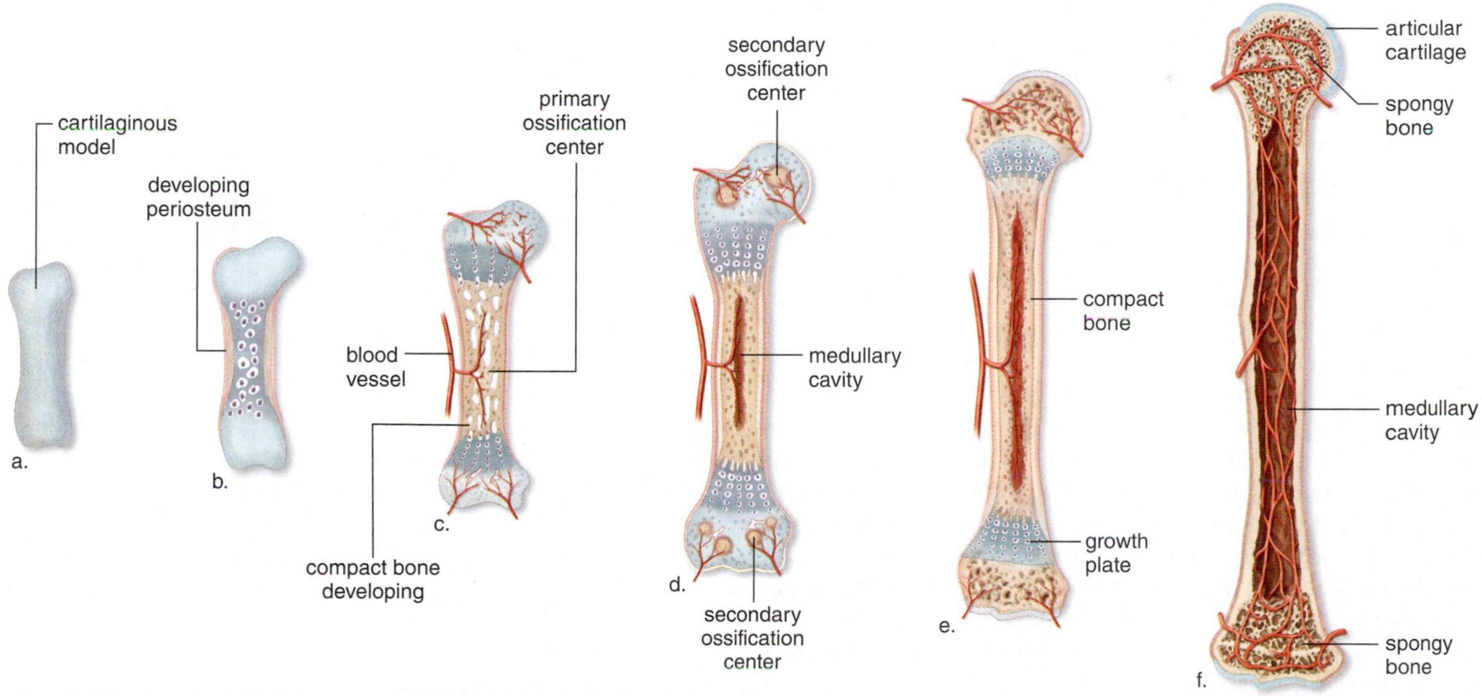
cartilaginous model
a.
developing periosteum
b.
blood vessel
primary ossification center
compact bone developing
c.
secondary ossification center
medullary cavity
secondary ossification center
d.
compact bone
growth plate
e.
articular cartilage
spongy bone
medullary cavity
spongy bone
f.

Figure 19.2 Endochondral ossification of a long bone. **a.** A cartilaginous model develops during embryonic development. **b.** A periosteum develops. **c.** A primary ossification center contains spongy bone surrounded by compact bone. **d.** The medullary cavity forms in the shaft, and secondary ossification centers develop at the ends of the long bone. **e.** Growth is still possible as long as cartilage remains at the growth plate. **f.** When the bone is fully formed, the growth plate disappears.

19.2 Bones of the Skeleton

Learning Outcomes

Upon completion of this section, you should be able to
1. Summarize the five major functions of the skeleton.
2. List the major bones that comprise the axial and appendicular skeletons.
3. Compare the structure and function of fibrous joints, cartilaginous joints, and synovial joints.

The functions of the skeleton pertain to particular bones:

The skeleton supports the body. The bones of the lower limbs (the femur in particular and also the tibia) support the entire body when we are standing, and the coxal bones of the pelvic girdle support the abdominal cavity.

The skeleton protects soft body parts. The bones of the skull protect the brain; the rib cage, composed of the ribs, thoracic vertebrae, and sternum, protects the heart and lungs.

The skeleton produces blood cells. All bones in the fetus have spongy bone containing red bone marrow that produces blood cells. In the adult, the flat bones of the skull, ribs, sternum, clavicles, and also the vertebrae and pelvis produce blood cells.

The skeleton stores minerals and fat. All bones have an extracellular substance that contains calcium phosphate. When bones are remodeled, osteoclasts break down bone and return calcium ions and phosphate ions to the bloodstream. Fat is stored in the yellow bone marrow (see Fig. 19.1).

The skeleton, along with the muscles, permits flexible body movement. While articulations (joints) occur between all the bones, we associate body movement, in particular, with the bones of the lower limbs (especially the femur and tibia) and the feet (tarsals, metatarsals, and phalanges) because we use them when walking.

Classification of the Bones

The approximately 206 bones of the skeleton are classified according to whether they occur in the axial skeleton or the appendicular skeleton. The **axial skeleton** is in the midline of the body, and the **appendicular skeleton** consists of the limbs along with their girdles (Fig. 19.3).

As mentioned, long bones, exemplified by the humerus and femur, are longer than they are wide. In contrast, short bones, such as the carpals and tarsals, are cube shaped—that is, their lengths and widths are about equal. Flat bones, such as those of the skull, are platelike, with broad surfaces. Irregular bones, such as the vertebrae and facial bones, have varied shapes that permit connections with other bones. Some authorities recognize another category, the round bones, which are circular in shape and usually embedded in a tendon. A good example of a round bone is the patella, or kneecap.

None of the bones of the skeleton are smooth. They have articulating depressions and protuberances at various joints. They also have projections, often called *processes*, where the muscles attach, as well as openings for nerves and/or blood vessels to pass through.

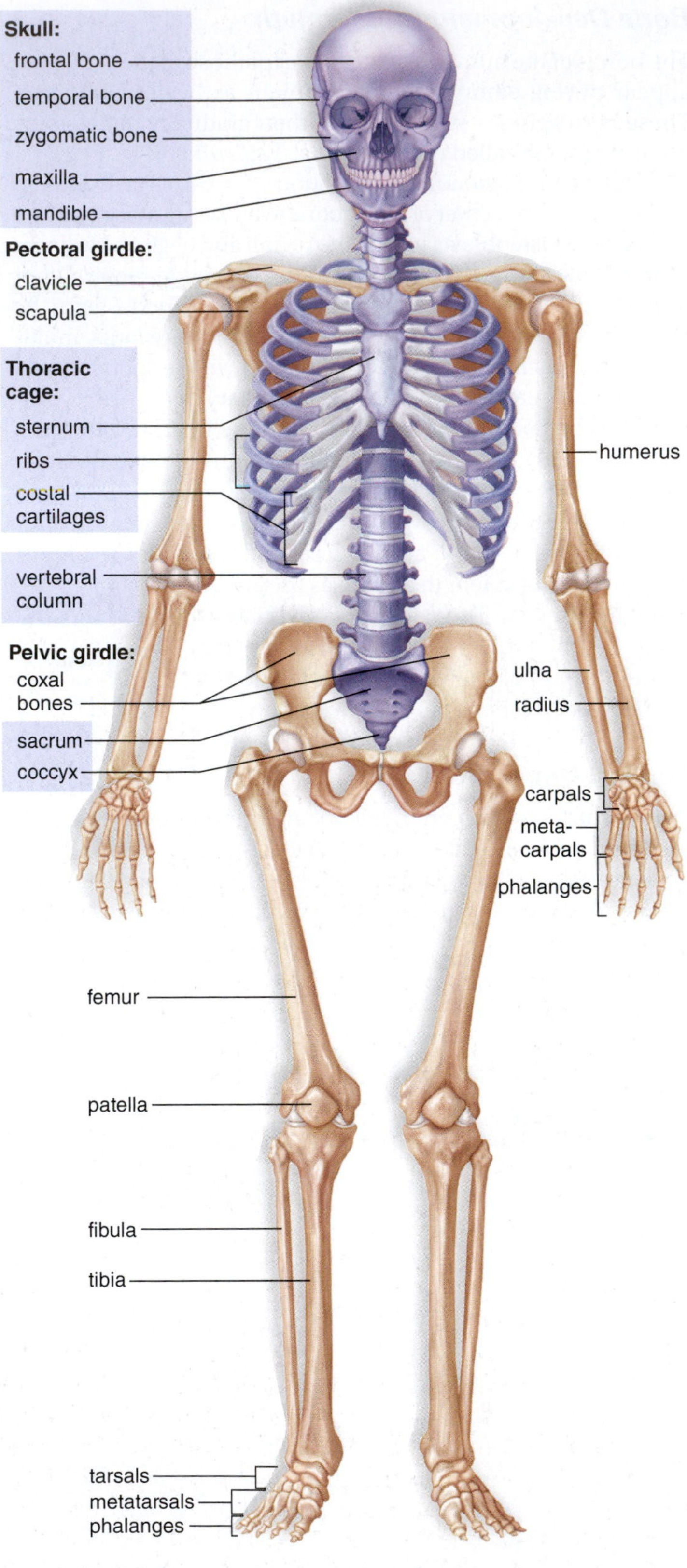
Skull:
frontal bone
temporal bone
zygomatic bone
maxilla
mandible
Pectoral girdle:
clavicle
scapula
Thoracic cage:
sternum
ribs
costal cartilages
vertebral column
Pelvic girdle:
coxal bones
sacrum
coccyx
humerus
ulna
radius
carpals
meta-carpals
phalanges
femur
patella
fibula
tibia
tarsals
metatarsals
phalanges

Figure 19.3 The skeleton. The skeleton of a human adult contains bones that belong to the axial skeleton (shaded in blue) and those that belong to the appendicular skeleton (unshaded). The bones of the axial skeleton are located along the body's axis, and the bones of the appendicular skeleton are located in the girdles and appendages.

The Axial Skeleton

The axial skeleton lies in the midline of the body and consists of the skull, hyoid bone, vertebral column, rib cage, and ossicles.

The Skull

The **skull** is formed by the cranium (braincase) and the facial bones. It should be noted, however, that some cranial bones also help to form the face.

The Cranium The *cranium* protects the brain. In adults, it is composed of eight bones fitted tightly together. In newborns, certain cranial bones are joined by membranous regions called **fontanels,** which allow the cranium to be compressed somewhat during birth. The fontanels usually close by the age of 24 months by the process of intramembranous ossification.

Some of the bones of the cranium contain **sinuses,** air spaces lined by mucous membrane. The sinuses reduce the weight of the skull and give a resonant sound to the voice. Two sinuses, called the *mastoid sinuses,* drain into the middle ear. Infections of the upper respiratory tract can spread into various sinuses, sometimes resulting in chronic sinusitis that is difficult to treat (see section 15.4).

The major bones of the cranium have the same names as the lobes of the brain: frontal, parietal, occipital, and temporal. On the top of the cranium (Fig. 19.4*a*), the *frontal bone* forms the forehead, the *parietal bones* extend to the sides, and the *occipital bone* curves to form the base of the skull. At the base, there is a large opening, the **foramen magnum** (Fig. 19.4*b*), through which the spinal cord passes and connects with the brain stem. Below the much larger parietal bones, each *temporal bone* has an opening (external auditory canal) that leads to the middle ear.

The *sphenoid bone,* which is shaped like a bat with wings outstretched, extends across the floor of the cranium from one side to the other. The sphenoid is the keystone of the cranial bones because all the other bones articulate with it. The sphenoid completes the sides of the skull and also helps form the eye sockets. The eye sockets are called *orbits* because of our ability to rotate our eyes. The *ethmoid bone,* which lies in front of the sphenoid, also helps form the orbits and the nasal septum. The orbits are completed by various facial bones.

The Facial Bones The most prominent facial bones are the mandible, the maxillae (maxillary bones), the zygomatic bones, and the nasal bones.

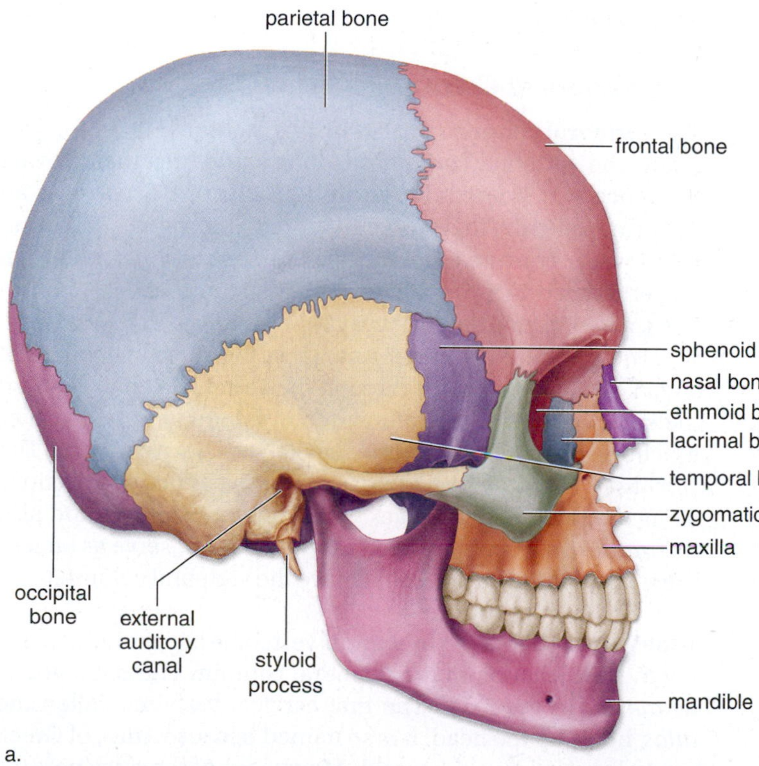
parietal bone
frontal bone
sphenoid bone
nasal bone
ethmoid bone
lacrimal bone
temporal bone
zygomatic bone
maxilla
occipital bone
external auditory canal
styloid process
mandible
a.

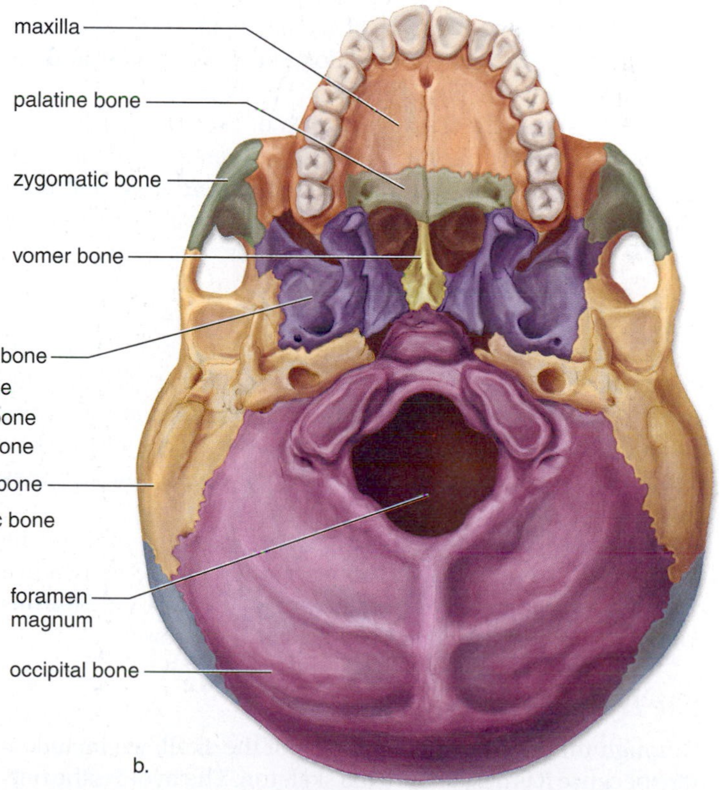
maxilla
palatine bone
zygomatic bone
vomer bone
foramen magnum
occipital bone
b.

Figure 19.4 Bones of the skull. **a.** Lateral view. **b.** Inferior view.

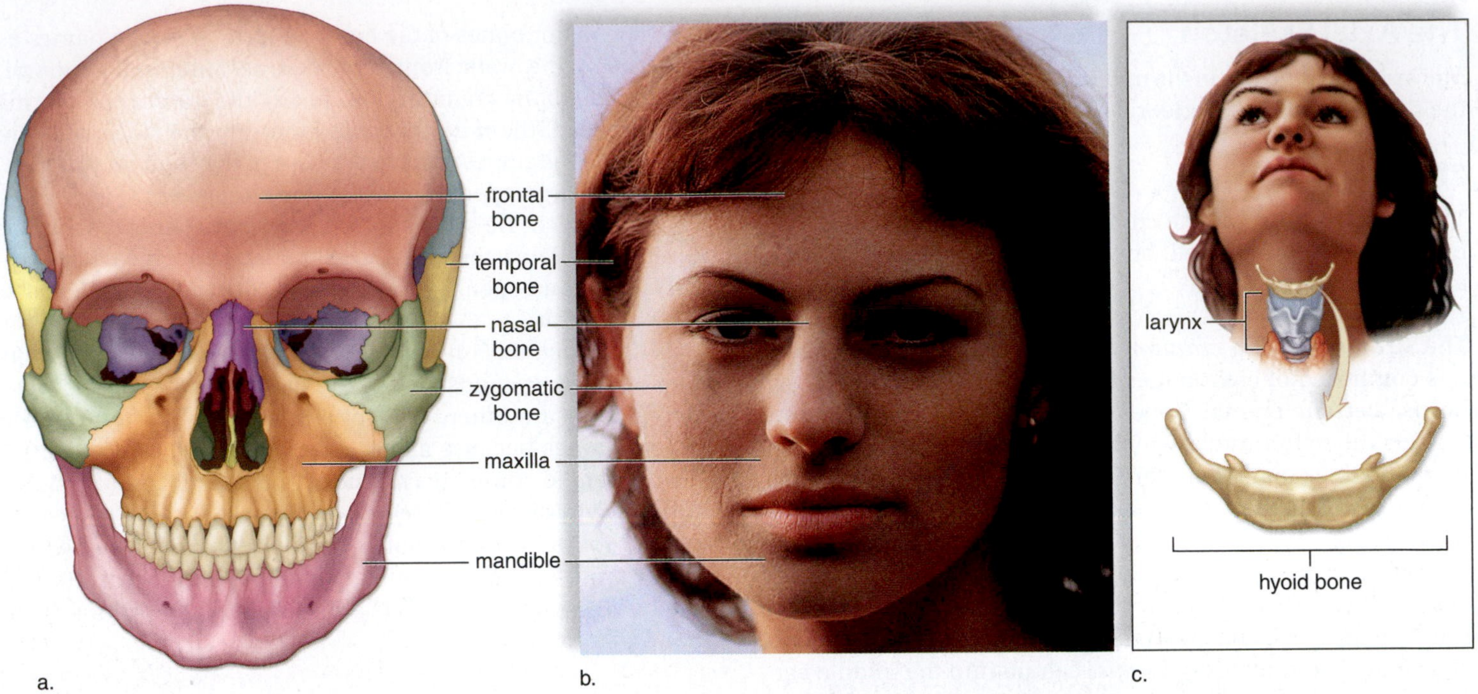
frontal bone
temporal bone
nasal bone
zygomatic bone
maxilla
mandible
larynx
hyoid bone
a.
b.
c.

Figure 19.5 Bones of the face and the hyoid bone. **a.** The frontal bone forms the forehead and eyebrow ridges, and the zygomatic bones form the cheekbones. The maxillae have numerous functions: assisting in the formation of the eye sockets and the nasal cavity; forming the upper jaw; and containing sockets for the upper teeth. The mandible is the lower jaw with sockets for the lower teeth. The mandible has a projection we call the chin. **b.** The maxillae, frontal bones, and nasal bones help form the external nose. **c.** The hyoid bone attaches to the larynx as shown.

The *mandible,* or lower jaw, is the only movable portion of the skull (Fig. 19.5*a*), and its action permits us to chew our food. It also forms the chin. Tooth sockets are located on the mandible and on the two *maxillae,* which form the upper jaw and also the anterior portion of the hard palate. The *palatine bones* make up the posterior portion of the hard palate and the floor of the nose (see Fig. 19.4*b*).

The lips and cheeks have a core of skeletal muscle. The *zygomatic bones* are the cheekbone prominences, and the nasal bones form the bridge of the nose. Other bones (e.g., ethmoid and vomer) are a part of the nasal septum, which divides the interior of the nose into two nasal cavities. The *lacrimal bone* (see Fig. 19.4*a*) contains the opening for the *nasolacrimal canal,* which brings tears from the eyes to the nose.

The temporal and frontal bones are cranial bones that contribute to the face. The temporal bones account for the flattened areas we call the temples. The frontal bone forms the forehead and has supraorbital ridges where the eyebrows are located. Glasses sit where the frontal bone joins the nasal bones (Fig. 19.5*b*).

While the ears are formed only by cartilage and not by bone, the nose is a mixture of bones, cartilages, and connective tissues. The cartilages complete the tip of the nose, and fibrous connective tissue forms the flared sides of the nose.

The Hyoid Bone

Although the *hyoid bone* is not part of the skull, we include it here because it is part of the axial skeleton. The hyoid is the only bone in the body that does not articulate with another bone. It is attached to processes of the temporal bones by muscles and ligaments and to the larynx by a membrane (Fig. 19.5*c*). The larynx is the voice box at the top of the trachea in the neck region. The hyoid bone anchors the tongue and serves as the site of attachment for the muscles associated with swallowing.

MP3 The Skull

The Vertebral Column

The **vertebral column** consists of 33 vertebrae (Fig. 19.6). Normally, the vertebral column has four curvatures that provide resilience and strength for an upright posture. *Scoliosis* is an abnormal lateral (sideways) curvature of the spine. Two other well-known abnormal curvatures are *kyphosis,* an abnormal posterior curvature that often results in a hunchback, and *lordosis,* an abnormal anterior curvature resulting in a swayback.

The vertebral column forms when the vertebrae join. The spinal cord, which passes through the vertebral canal, gives off the spinal nerves at the intervertebral foramina. Among other functions, spinal nerves control skeletal muscle contraction. The *spinous processes* of the vertebrae can be felt as bony projections along the midline of the back. The spinous processes and also the *transverse processes,* which extend laterally, serve as attachment sites for the muscles that move the vertebral column.

Types of Vertebrae The various vertebrae are named according to their location in the vertebral column. The *cervical vertebrae* are in the neck. The first cervical vertebra, called the *atlas,* holds up the head. It is so named because Atlas, of Greek mythology, held up the world. Movement of the atlas permits the "yes" motion of the head. It also allows the head to tilt from side to side. The second cervical vertebra is called the *axis*

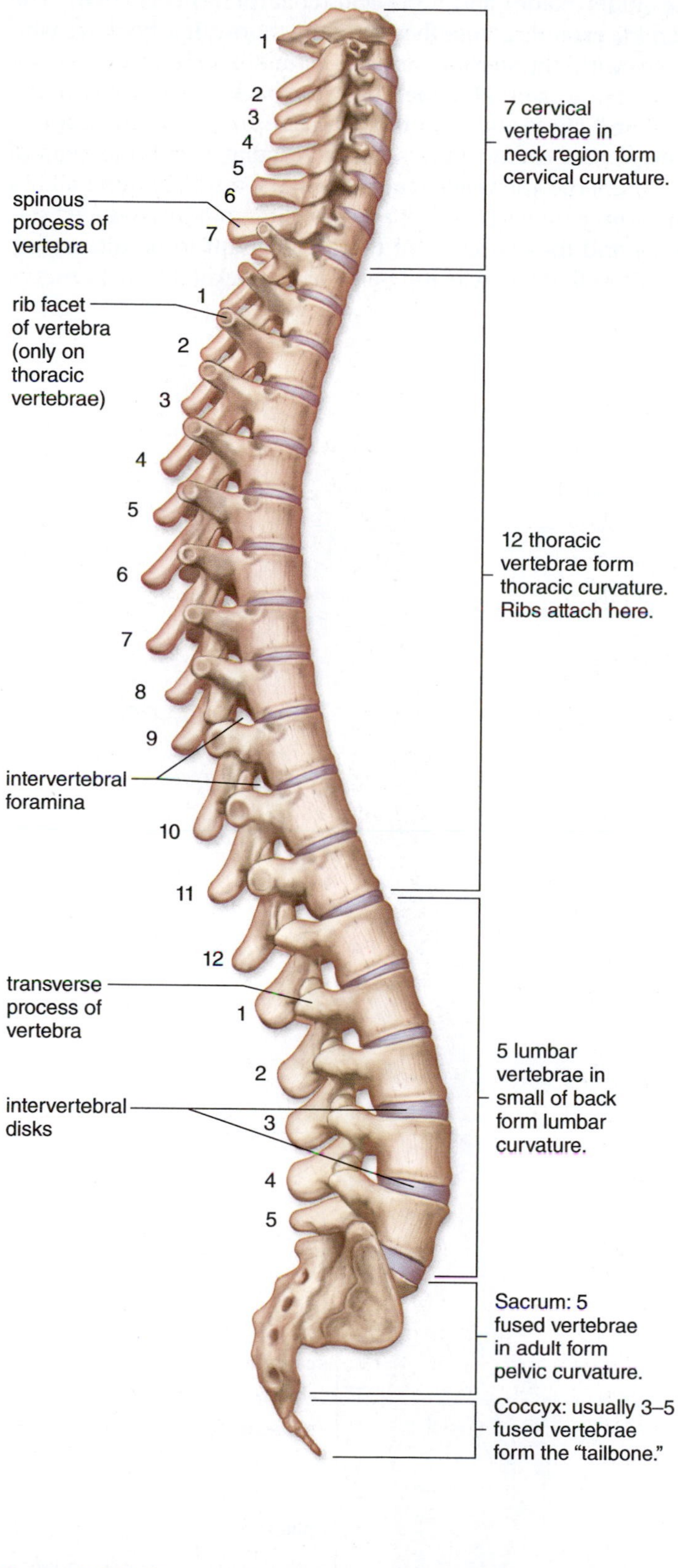
1
2
3
4
5
6
7
spinous process of vertebra
7 cervical vertebrae in neck region form cervical curvature.
rib facet of vertebra (only on thoracic vertebrae)
1
2
3
4
5
6
7
8
9
10
11
12
intervertebral foramina
12 thoracic vertebrae form thoracic curvature. Ribs attach here.
transverse process of vertebra
intervertebral disks
1
2
3
4
5
5 lumbar vertebrae in small of back form lumbar curvature.
Sacrum: 5 fused vertebrae in adult form pelvic curvature.
Coccyx: usually 3–5 fused vertebrae form the "tailbone."

Figure 19.6 The vertebral column. The vertebral column is flexible because the vertebrae are separated by intervertebral disks. The vertebrae are named for their location in the vertebral column. For example, the thoracic vertebrae are in the thorax. Note that humans have a coccyx, which is also called a tailbone.

because it allows a degree of rotation, as when someone shakes his or her head "no." The *thoracic vertebrae* have long, thin, spinous processes and articular facets for the attachment of the ribs (Fig. 19.7*a*). *Lumbar vertebrae* have a large body and thick processes. The five *sacral vertebrae* are fused together in the *sacrum*. The *coccyx*, or tailbone, is usually composed of three to five fused vertebrae.

Intervertebral Disks Between the vertebrae are **intervertebral disks** composed of fibrocartilage that act as a kind of padding. They prevent the vertebrae from grinding against one another and absorb shock caused by movements such as running, jumping, and even walking. Unfortunately, these disks become weakened with age and can even herniate and rupture. A herniated or "slipped" disk may press against the spinal cord and/or spinal nerves, resulting in pain, numbness, or even loss of function in the areas innervated by the spinal nerves at or below that region. Herniated disks are most common in the lumbar region; about 25% of people with lower back pain actually have a herniated disk. Although rest and treatment of symptoms usually results in much improvement, surgical treatment may be necessary, especially if there is chronic pain or loss of nerve function.

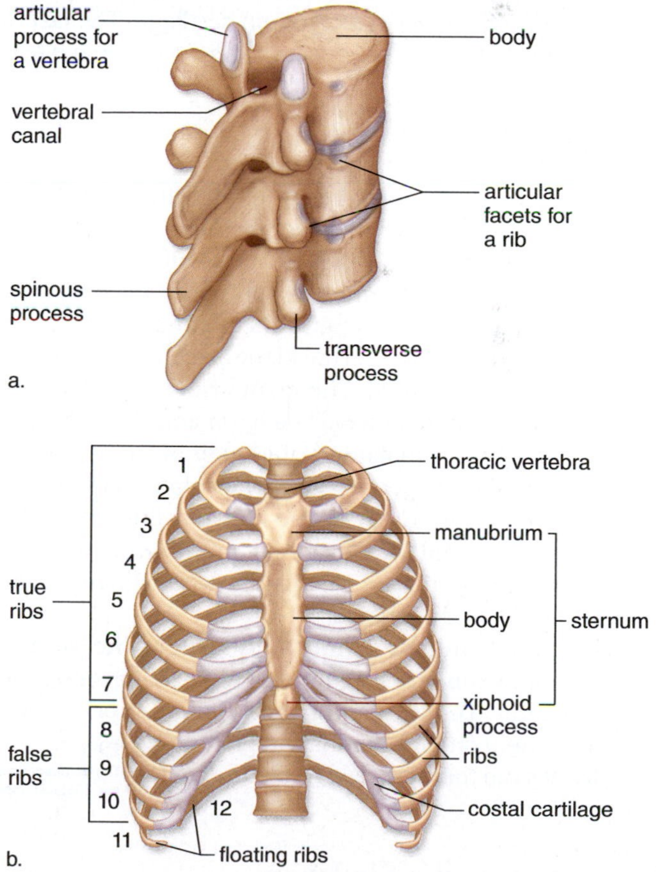
articular process for a vertebra
vertebral canal
body
articular facets for a rib
spinous process
transverse process
a.
thoracic vertebra
manubrium
body
sternum
xiphoid process
ribs
costal cartilage
floating ribs
true ribs
false ribs
1
2
3
4
5
6
7
8
9
10
11
12
b.

Figure 19.7 Thoracic vertebrae and the rib cage. **a.** The thoracic vertebrae articulate with each other at the articular processes and with the ribs at the articular facets. A thoracic vertebra has two facets for articulation with a rib; one is on the body, and the other is on the transverse process. **b.** The rib cage consists of the 12 thoracic vertebrae, the 12 pairs of ribs, the costal cartilages, and the sternum. The rib cage protects the lungs and the heart.

The Rib Cage

The *rib cage,* also called the thoracic cage, is composed of the thoracic vertebrae, the ribs and their associated cartilages, and the sternum (Fig. 19.7*b*). The rib cage is part of the axial skeleton.

The rib cage demonstrates how the skeleton is protective but also flexible. The rib cage protects the heart and lungs. Yet it swings outward and upward upon inspiration, and then downward and inward upon expiration. Because of their proximity to vital organs, however, the ribs also have the potential to cause harm. A fractured rib can puncture a lung.

The Ribs

There are twelve pairs of ribs, and all twelve connect directly to the thoracic vertebrae in the back. Each rib is a flattened bone that originates at a particular thoracic vertebra and proceeds toward the ventral thoracic wall. A rib articulates with the body and transverse process of its corresponding thoracic vertebra. It curves outward and then forward and downward.

The upper seven pairs of ribs connect directly to the sternum by means of costal cartilages. These are called the "true ribs." The next three pairs of ribs do not connect directly to the sternum, and they are called the "false ribs." They attach to the sternum by means of a common cartilage. The last two pairs are called "floating ribs" because they do not attach to the sternum at all.

The Sternum

The *sternum,* or breastbone, lies in the midline of the body. Along with the ribs, it helps protect the heart and lungs. The sternum is a flat bone that is shaped like a knife.

The sword-shaped sternum is composed of three bones that fuse during fetal development. These bones are the *manubrium* (the handle), the *body* (the blade), and the *xiphoid process* (the point of the blade). The manubrium articulates with the clavicles of the appendicular skeleton and the first pair of ribs. The manubrium joins with the body of the sternum at an angle. This is an important anatomical landmark because it occurs at the level of the second rib, and therefore allows the ribs to be counted. Medical personnel sometimes count the ribs to locate the apex of a patient's heart, which is usually between the fifth and sixth ribs.

The xiphoid process is the third part of the sternum. Composed of hyaline cartilage in the child, it becomes ossified in the adult. The variably shaped xiphoid process serves as an attachment site for the diaphragm, which divides the thoracic cavity from the abdominal cavity.

MP3 **The Vertebral Column and Thoracic Cage**

The Appendicular Skeleton

The appendicular skeleton consists of the bones within the pectoral and pelvic girdles and their attached limbs. The two (left and right) pectoral girdles and upper limbs are specialized for flexibility. The pelvic girdle and lower limbs are specialized for strength.

The Pectoral Girdles and Upper Limbs

Each **pectoral girdle** (shoulder girdle) consists of a scapula (shoulder blade) and a clavicle (collarbone) (Fig. 19.8). The *clavicle* extends across the top of the thorax. It articulates with (joins with) the sternum and the *acromion process* of the *scapula,* a bone external to the ribs in the back. The muscles of the arm and chest attach to the *coracoid process* of the scapula. The *glenoid cavity* of the scapula articulates with the head of the humerus, the single long bone in the arm. This joint allows the arm to move in almost any direction, but has reduced stability and thus is the joint that is most apt to be dislocated. Tendons that encircle and help form a socket for the humerus

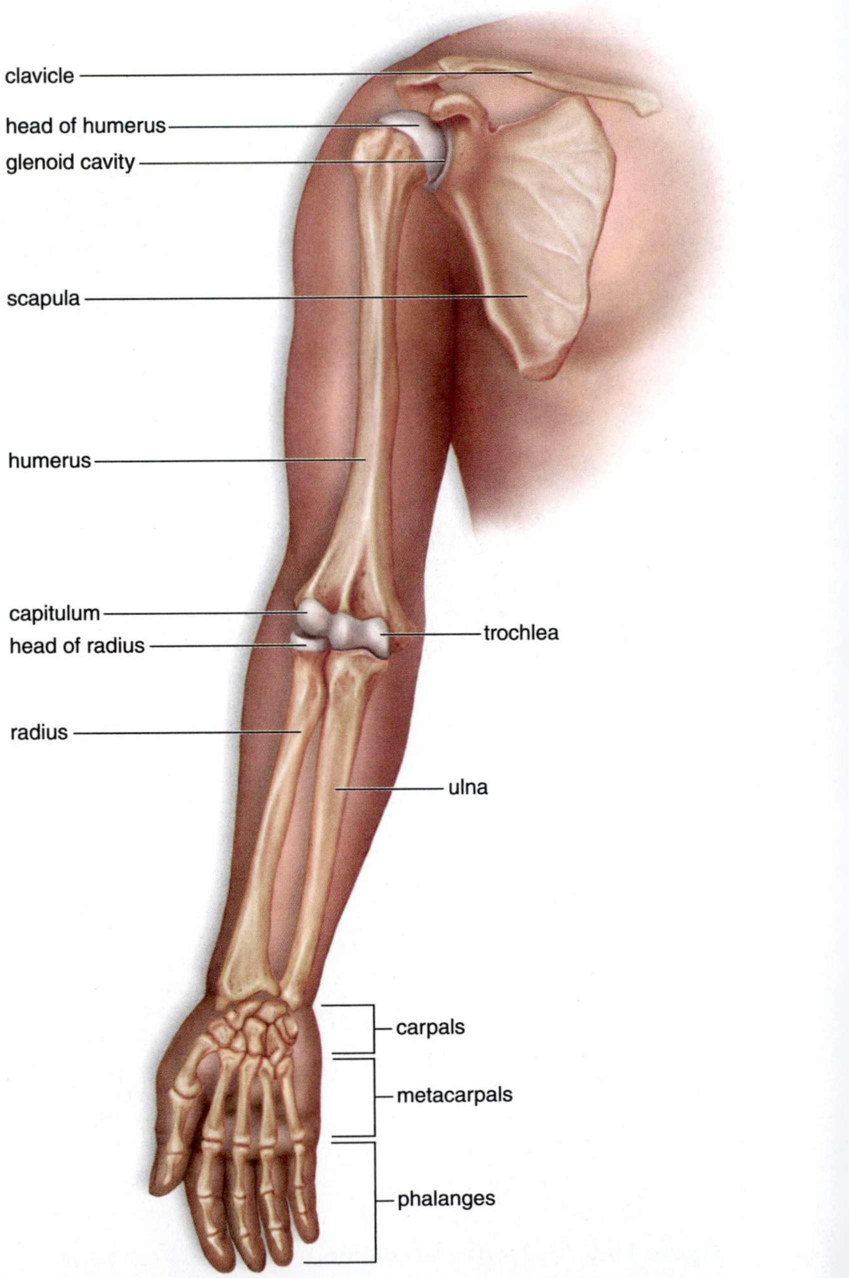
clavicle
head of humerus
glenoid cavity
scapula
humerus
capitulum
head of radius
trochlea
radius
ulna
carpals
metacarpals
phalanges

Figure 19.8 **Bones of the pectoral girdle and upper limb (arm, forearm, and hand).**

are collectively called the *rotator cuff.* Vigorous circular movements of the arm (such as pitching a baseball) can lead to rotator cuff injuries.

The upper limb consists of the *humerus* in the arm and the *radius* and *ulna* in the forearm. The shaft of the humerus has a tuberosity (protuberance) where the deltoid, a prominent muscle of the shoulder, attaches. Enlargement of this tuberosity occurs in people who do a lot of heavy lifting.

The far end of the humerus has two protuberances, called the *capitulum* and the *trochlea,* which articulate respectively with the radius and the ulna at the elbow. The bump at the back of the elbow is the *olecranon process* of the ulna. This area is sometimes called the "funny bone" because the ulnar nerve passes behind it with only skin for protection—thus it is susceptible to being bumped, resulting in pain or numbness.

When the arm is held with the palm turned forward, the radius and ulna are about parallel to one another. When the arm is turned so that the palm faces backward, the radius crosses in front of the ulna, a feature that contributes to the easy twisting motion of the forearm.

The hand has many bones, and this increases its flexibility. The wrist has eight *carpal bones,* which look like small pebbles. From these, five *metacarpal bones* fan out to form a framework for the palm. The metacarpal bone that leads to the thumb is opposable to the other digits. (The term *digits* refers to either fingers or toes.) The knuckles are the enlarged distal ends of the metacarpals. Beyond the metacarpals are the *phalanges,* the bones of the fingers and the thumb. The phalanges of the hand are long, slender, and lightweight.

The Pelvic Girdle and Lower Limb

Figure 19.9 shows how the lower limb is attached to the pelvic girdle. The **pelvic girdle** consists of two heavy, large coxal bones (hipbones). The pelvis is a basin composed of the pelvic girdle, sacrum, and coccyx. The pelvis bears the weight of the body, protects the organs within the pelvic cavity, and serves as the site of attachment for the lower limbs.

Each *coxal bone* has three parts: the ilium, the ischium, and the pubis, which are fused in the adult (Fig. 19.9). The hip socket, called the acetabulum, occurs where these three bones meet. We sit on the *ischium,* which has a posterior spine called the ischial spine. The *pubis,* from which the term pubic hair is derived, is the anterior part of a coxal bone. The two pubic bones are joined together by a fibrocartilage disk at the pubic symphysis.

The male and female pelves differ from one another. In the female, the ilia are more flared, the pelvic cavity is broader, and the outlet is wider. These adaptations facilitate giving birth.

The *femur* (thighbone) is the longest and strongest bone in the body. The head of the femur articulates with the coxal bones at the *acetabulum,* and the short neck better positions the legs for walking. The femur has two large processes, the greater and lesser trochanters, which are places of attachment for the thigh muscles and the muscles of the buttocks. At its distal end, the femur has medial and lateral condyles that articulate with the *tibia.* This is the region of the knee and the patella, or kneecap.

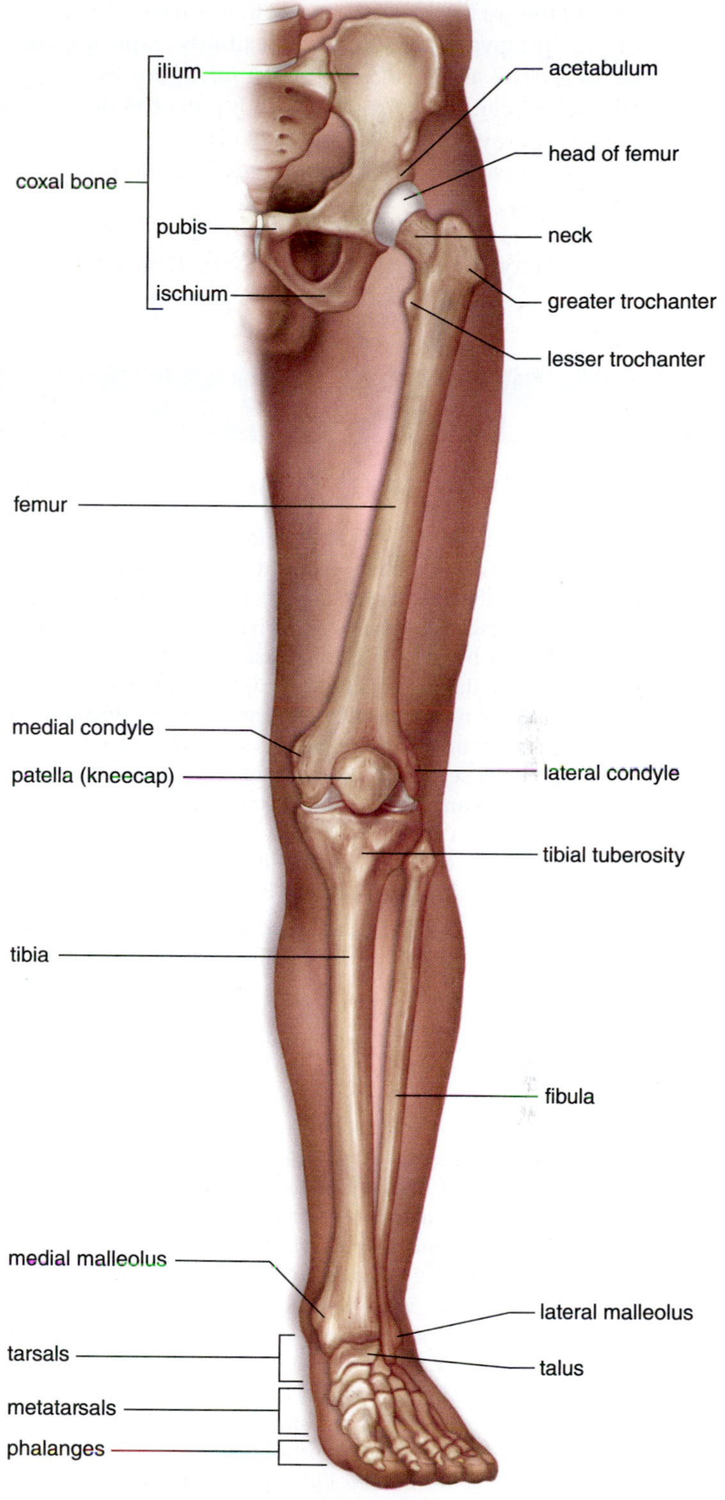

ilium
coxal bone
pubis
ischium
acetabulum
head of femur
neck
greater trochanter
lesser trochanter
femur
medial condyle
patella (kneecap)
lateral condyle
tibial tuberosity
tibia
fibula
medial malleolus
lateral malleolus
tarsals
talus
metatarsals
phalanges

Figure 19.9 **Bones of the pelvic girdle and lower limb.**

The *patella* is held in place by the quadriceps tendon, which continues as a ligament that attaches to the tibial tuberosity. At the distal end, the medial malleolus of the tibia causes the inner bulge of the ankle. The *fibula* is the more slender bone in the leg. The fibula has a head that articulates with the tibia and a prominence (lateral malleolus) that forms the outer bulge of the ankle.

Each foot has an ankle, an instep, and five toes. The many bones of the foot give it considerable flexibility, especially on rough surfaces. The ankle contains seven *tarsal* bones, one of which (the talus) can move freely where it joins the tibia and fibula. Strange to say, the calcaneus, or heel bone, is also considered part of the ankle. The *talus* and *calcaneus* support the weight of the body.

The instep has five elongated *metatarsal bones.* The distal end of the metatarsals forms the ball of the foot. If the ligaments that bind the metatarsals together become weakened, flat feet are apt to result. The bones of the toes are called phalanges, just like those of the fingers, but in the foot, the phalanges are stout and extremely sturdy.

Joints

Bones are joined at the joints, which are classified as fibrous, cartilaginous, or synovial based on their structure and their ability

SCIENCE IN YOUR LIFE ▶ **SCIENTIFIC INQUIRY**

Dead on the "Farm"—The Science of *Bones*

A 2009 episode of Fox TV's *Bones* begins with a scene of a wine-tasting event at a fancy winery. As the well-dressed participants begin tasting wine from an aged barrel, they immediately become visibly nauseated. Some spit out their wine; others nearly vomit. Someone then pulls a badly decomposed human finger out of a wineglass.

Soon after that scene, forensic anthropologist Dr. Temperance "Bones" Brennan and FBI Special Agent Seeley Booth receive a call from the vineyard. They arrive to discover the remains of a human body in a barrel of wine. The winery owner is, of course, horrified. To these experienced investigators, however, it's just another day in the office.

Back at the lab, Brennan and Booth get busy removing wine stains from the victim's bones using more than 700 tablets of denture cleanser. Once the bones are clean, Brennan notices a circular pattern of cuts on the victim's rib cage measuring about 9 inches in circumference. This turns out to match the circumference of a wine bottle, indicating the victim may have been stabbed with a broken wine bottle. This information, of course, is an essential clue to solving the case.

Bones is inspired by the work of a real-life forensic anthropologist and author, Kathy Reichs, who also works as a producer on the show. Reichs has trained FBI agents on how to properly recover and identify human remains. She has helped to exhume and identify war dead from World War II, the Korean War, and the Vietnam War, and she assisted with identification of remains at Ground Zero after the World Trade Center attacks on 9/11. Reichs has also written many best-selling books based on her experiences.

Clearly, the thought of studying dead bodies for a living does not appeal to everyone. However, it is a critical process in the investigation of many crimes and missing person cases. Forensic Science is a broad field that includes the application of a variety of scientific methods to answer legal questions. This may include an analysis of a diverse array of evidence such as DNA, bloodstain patterns, and even the types of insects that occupy a decomposing body.

In cases when a body is badly decomposed, the hard, mineralized tissues of the bones may provide the most useful clues. Indeed, bones may not entirely decompose for hundreds of years. This is the realm of forensic anthropologists, who tend to be called in to help identify the remains of bodies that are so decomposed or otherwise damaged as to be unrecognizable. Fortunately these situations are relatively uncommon, and in fact there are fewer than 100 board-certified forensic anthropologists in the United States and Canada.

But how do these scientists learn what they need to know about decomposing bodies? And how can students be trained in this unusual discipline? That's where decomposition research facilities (popularly called "body farms") come in. At a few locations around the country, students pursuing degrees in various aspects of forensic science can examine the decomposition process up close.

In 1971, local police asked Dr. William Bass of the University of Tennessee for help in analyzing bodies in criminal cases. Recognizing the need for research into human decomposition, Dr. Bass eventually established a 3-acre complex that now contains around 40 decomposing bodies at any one time. The obvious need for dead human bodies was initially met by using unclaimed corpses from medical examiners' offices. Later, people began donating their bodies to help with forensic studies. If requested, remains can be returned to families for burial after studies are completed, or the bones may be kept in collections at the facilities.

As of 2012, there were four decomposition facilities using human remains operating in the United States. Each facility has a different focus. The facility in Tennessee studies decomposition under a broad range of conditions, including buried, exposed, and underwater. Those left exposed may be covered with wire cages, to keep coyotes or other wildlife from removing parts of the body. A 7-acre facility operated by Texas State University studies decomposition processes that occur in a very dry climate. In addition to studying biological processes, students must learn to keep careful records about any manipulations of the bodies, similar to what would be required in a criminal case.

In addition to looking for clues as to how and when a person died, a forensic anthropologist can estimate a person's age, sex, ethnicity, and body type from skeletal remains. Age may be estimated by dentition, or the structure of the teeth. For example, infants up to age four months will have no teeth present; children aged approximately six to ten years will have missing deciduous teeth (or "baby teeth"); young adults acquire their last molars (or "wisdom teeth") around age 20. In older adults, a general wear and tear of the skeleton provides additional information about age. With age, the areas where the ribs meet the sternum become more ragged and less flat.

to move. Most **fibrous joints,** such as the **sutures** between the cranial bones, are immovable. **Cartilaginous joints** are connected by hyaline cartilage, as in the costal cartilages that join the ribs to the sternum, or by fibrocartilage, as in the intervertebral disks. Cartilaginous joints tend to be slightly movable.

In **synovial joints,** which are freely movable, the two bones are separated by a cavity (Fig. 19.10). Ligaments hold the two bones in place and help form a capsule around the joint. Tendons also help stabilize the joint. The joint capsule is lined by a **synovial membrane,** which produces a lubricating *synovial fluid.* Major synovial joints include the shoulder, elbow, hip, and knee. Aside from articular cartilage, the knee contains *menisci* (sing., meniscus), crescent-shaped pieces of hyaline cartilage between the bones (Fig. 19.10*b*). These give added stability and act as shock absorbers. Unfortunately, athletes often suffer injury to the menisci, known as torn cartilage. The anterior and posterior cruciate ligaments, which help stabilize the knee by attaching the femur to the tibia,

In addition, cartilage generally becomes worn, yellowed, and brittle with age, and the hyaline cartilages covering bone ends wear down over time.

It can be difficult to determine the gender of a child from the bones alone. In adults, however, the long bones will be thicker and denser in males, and points of muscle attachment will be bigger and more prominent. The skull of a male will have a square chin and more prominent ridges above the eye sockets or orbits. In adult females, the pelvis is shallower and wider than in males, and the space inside the pelvic bone is larger to accommodate the birthing process.

Because of an increasing demand for forensic scientists, training programs are growing across the country. With such a limited number of human decomposition facilities available, however, some forensic science programs have come up with innovative solutions to allow their students to study the decomposition process. Such is the case at Nebraska Wesleyan University (NWU), where Dr. Melissa Connor maintains a 1-acre facility with one major difference—the "bodies" in this case are of the porcine variety. In many midwestern states, where pigs outnumber people, this seems like a good solution, especially considering the many similarities in body structure between pigs and people (Fig. 19A). Students in NWU's Forensic Science program learn about the decomposition process and methods of recovering remains, both buried and on the surface. NWU's Master of Science in Forensic Science Program recently received accreditation from the Forensic Education Program Accreditation Commission (FEPAC).

Questions to Consider

1. If you discovered that a university was installing a human "body farm" within a mile of your house, what concerns would you have, if any?
2. Suppose investigators are examining the body of someone who died of natural causes. What types of diseases might leave evidence in the bones?
3. Besides studying decomposition processes, how might the investigation of criminal homicides involve disciplines such as botany, chemistry, geology, and psychology?

a.

b.

Figure 19A Studying decomposition using pigs. **a.** NWU forensic science students excavating a buried pig uncover a void in the soil that represents the pig's body cavity. **b.** Pig remains recovered in the excavation.

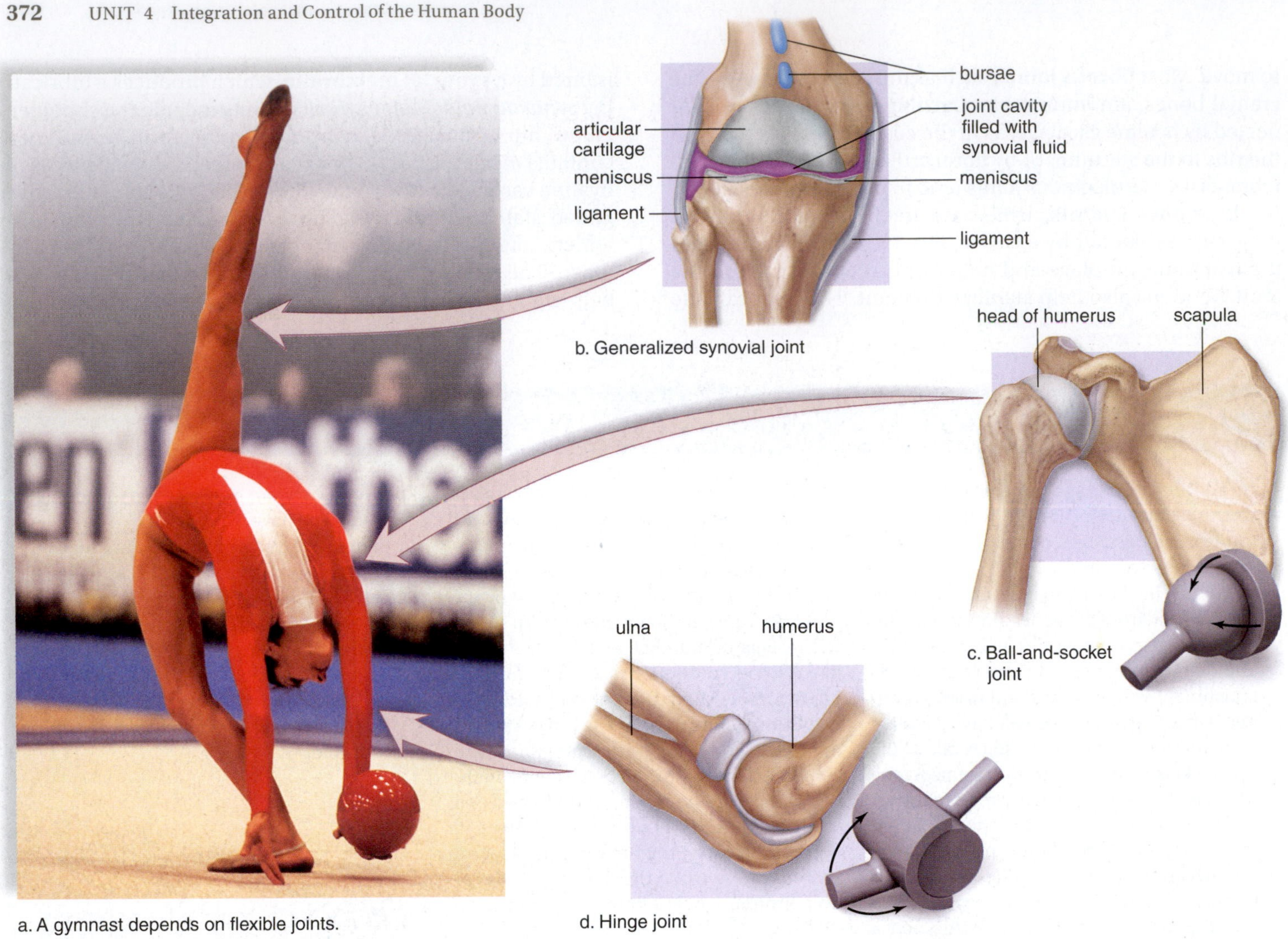
bursae
joint cavity filled with synovial fluid
articular cartilage
meniscus
meniscus
ligament
ligament
head of humerus
scapula
ulna
humerus

a. A gymnast depends on flexible joints.

b. Generalized synovial joint

c. Ball-and-socket joint

d. Hinge joint

Figure 19.10 The structure of a synovial joint. **a.** Synovial joints like the hips and shoulders are the most flexible type of joint. **b.** The ends of bones in synovial joints are capped by articular cartilage. The joint capsule is lined with synovial membrane, which gives off synovial fluid as a lubricant. Menisci (sing., meniscus) cushion and stabilize the joint, and bursae are fluid-filled sacs that reduce friction. **c.** Ball-and-socket joints form the shoulder and hip. **d.** Hinge joints construct the knee and elbow.

are also prone to being torn (sometimes called "ruptured"). The knee joint also contains 13 fluid-filled sacs called **bursae** (sing., bursa), which ease friction between the tendons and ligaments. Inflammation of the bursae is called **bursitis.** *Tennis elbow* is a form of bursitis.

There are different types of synovial joints. The shoulder is a *ball-and-socket joint,* which allows movement in all planes, even rotational movement (Fig. 19.10*c*). The knee and elbow joints are *hinge joints* because, like a hinged door, they largely permit movement in one direction only (Fig. 19.10*d*). The joint between the radius and ulna is a *pivot joint* in which only rotation is possible. Some of the movements at synovial joints are also listed in Table 19.1.

MP3 The Appendicular Skeleton

TABLE 19.1 Examples of Movements at Synovial Joints

Type	Example
Flexion	Forearm toward the arm
Extension	Forearm away from the arm
Abduction	Arms sideways, away from body
Adduction	Arms back to the body
Rotation	Head to answer "no"

Check Your Progress 19.2

1. List the major bones of the axial skeleton, including the normal numbers of cervical, thoracic, and lumbar vertebrae.
2. Identify the bones of the cranium and face.
3. Define the difference between true ribs, false ribs, and floating ribs.
4. List the three major types of joints and give a specific example of each.

19.3 Skeletal Muscles

Learning Outcomes

Upon completion of this section, you should be able to

1. Explain how antagonistic muscle groups can move bones in different directions.
2. Identify the major muscles in the human body.
3. Describe six characteristics that are used to name skeletal muscles.

Humans have three types of muscle tissue: smooth, cardiac, and skeletal (see Fig. 11.4). **Skeletal muscle** makes up the greatest percentage of muscle tissue in the body. Skeletal muscle is voluntary because its contraction can be consciously stimulated and controlled by the nervous system.

MP3 Muscle Tissue

Skeletal Muscles Work in Pairs

Skeletal muscles are covered by several layers of fibrous connective tissue called fascia, which extends beyond the muscle to become its tendon. Tendons are attached to the skeleton, and the contraction of skeletal muscle causes the bones at a joint to move. When skeletal muscles contract, one bone remains fairly stationary, and the other one moves. The **origin** of a muscle is on the stationary bone, and the **insertion** of a muscle is on the bone that moves. Table 19.1 lists some terms that are used to describe these movements, and gives examples. *Flexion* means bending bones at a joint so that the angle between the bones decreases, and the involved bones move closer together, while *extension* increases that angle and moves the parts farther apart. *Abduction* at a joint moves the involved part (usually a limb) away from the middle of the body, while *adduction* brings it back toward the midline. *Rotation* moves the part around an axis, due to a twisting motion at the joint. Note that not all of these movements can occur at all joints.

Most muscles have antagonists, and antagonistic pairs bring about movement in opposite directions. For example, the biceps brachii and the triceps brachii are antagonists; one flexes the forearm, and the other extends the forearm (Fig. 19.11).

Major Skeletal Muscles

There are approximately 650 skeletal muscles in the human body. The major ones are shown in Figure 19.12. As listed in Tables 19.2 and 19.3, the muscles of the head are involved in facial expression and chewing; those of the neck participate in head movements; the muscles of the trunk and upper limb move the arm, the forearm, and the fingers; and the muscles of the buttocks and lower limb move the thigh, the leg, and the toes.

Nomenclature

Skeletal muscles are named based on the following characteristics:

1. *Size.* The gluteus maximus that makes up the buttocks is the largest muscle (*maximus* means greatest). Other terms used to indicate size are vastus (huge) and longus (long).
2. *Shape.* The deltoid is shaped like a triangle. (The Greek letter delta has this appearance: Δ.) The trapezius is shaped like a trapezoid. Another term used to indicate shape is latissimus (wide).
3. *Location.* The frontalis overlies the frontal bone. The external obliques are located outside the internal obliques. Another term used to indicate location is pectoralis (chest).
4. *Direction of muscle fibers.* The rectus abdominis is a longitudinal muscle of the abdomen (*rectus* means straight). The orbicularis oculi is a circular muscle around the eye.
5. *Number of attachments.* The biceps brachii has two attachments, or origins (*bi-* means two).
6. *Action.* The extensor digitorum extends the fingers (digits). The adductor longus is a long muscle that adducts the thigh. Another term used to indicate action is masseter (to chew).

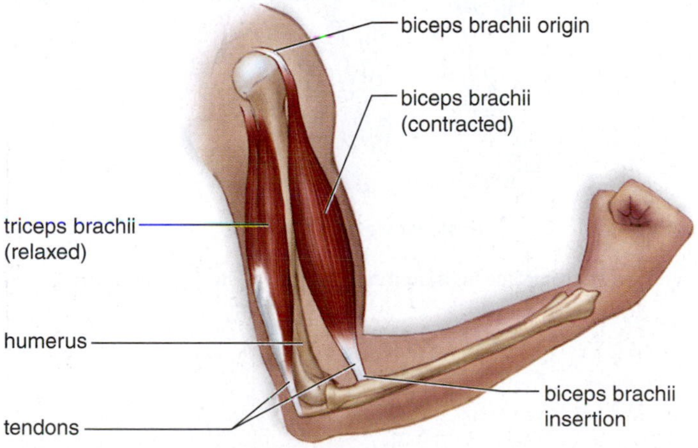
biceps brachii origin
biceps brachii (contracted)
triceps brachii (relaxed)
humerus
tendons
biceps brachii insertion

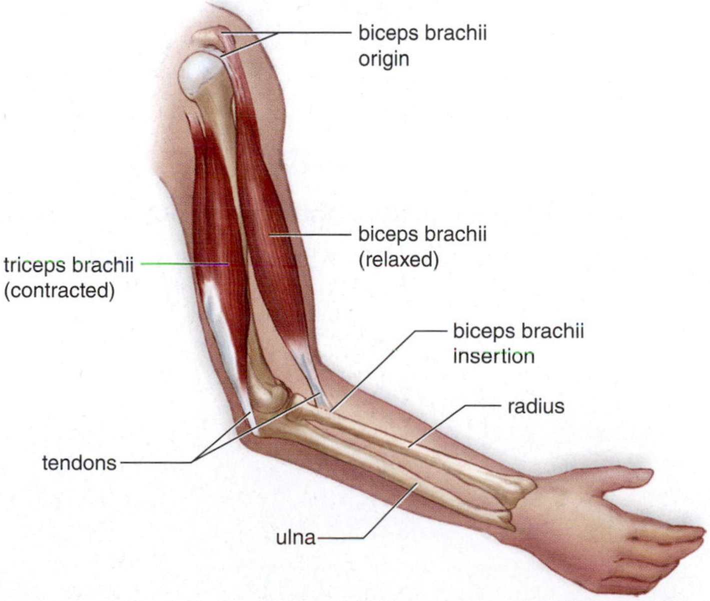
biceps brachii origin
biceps brachii (relaxed)
triceps brachii (contracted)
biceps brachii insertion
radius
tendons
ulna

Figure 19.11 Antagonistic muscles. Muscles can exert force only by shortening; therefore, they often work as antagonistic pairs. The biceps and triceps brachii exemplify an antagonistic pair of muscles that act opposite to one another. The biceps brachii raises the forearm, and the triceps brachii lowers the forearm.

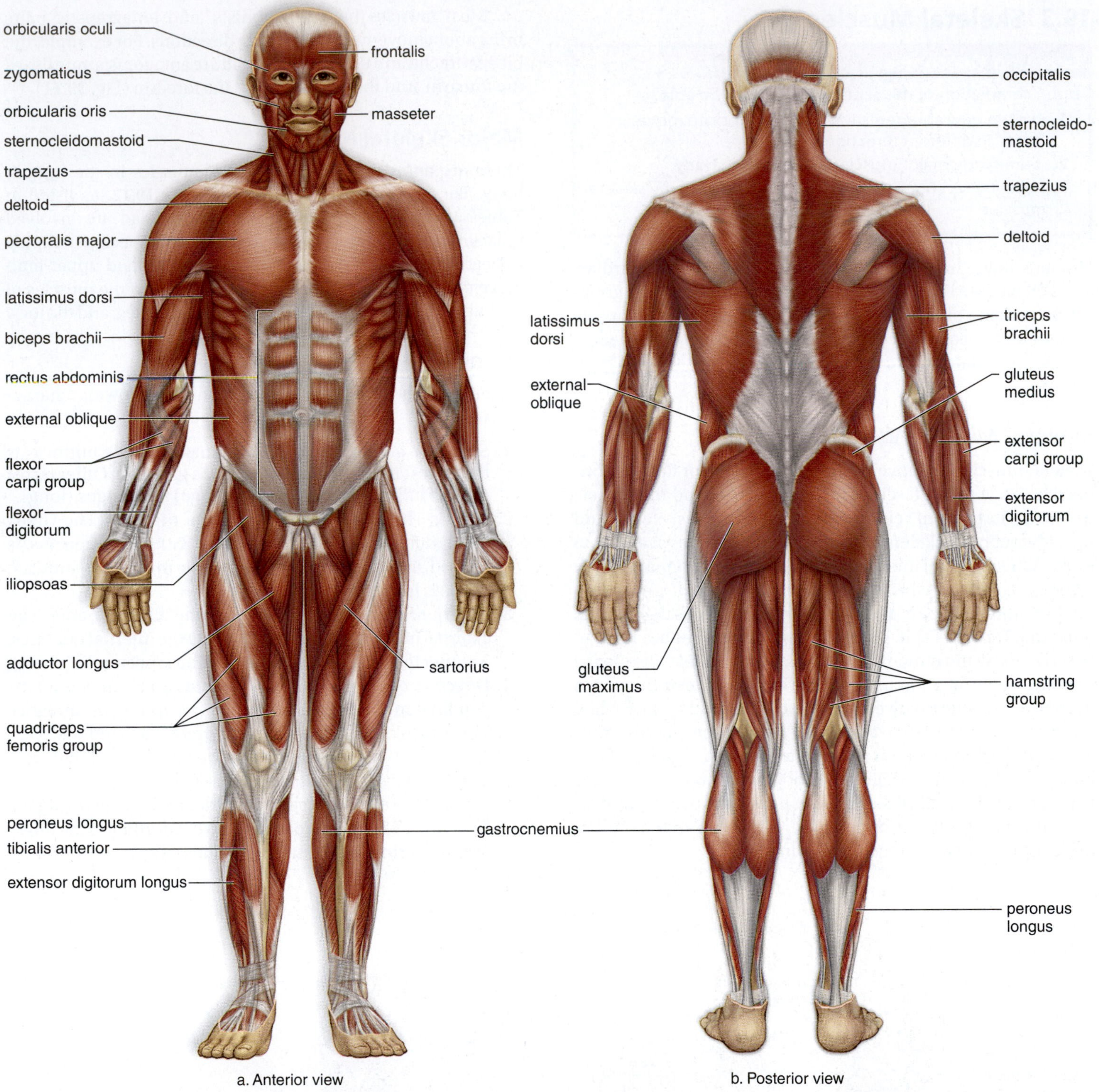
orbicularis oculi
frontalis
zygomaticus
orbicularis oris
masseter
sternocleidomastoid
trapezius
deltoid
pectoralis major
latissimus dorsi
biceps brachii
rectus abdominis
external oblique
flexor carpi group
flexor digitorum
iliopsoas
adductor longus
sartorius
quadriceps femoris group
peroneus longus
tibialis anterior
extensor digitorum longus
gastrocnemius
occipitalis
sternocleido-mastoid
trapezius
deltoid
latissimus dorsi
triceps brachii
external oblique
gluteus medius
extensor carpi group
extensor digitorum
gluteus maximus
hamstring group
peroneus longus
a. Anterior view
b. Posterior view

Figure 19.12 Major muscles of the human body. Skeletal muscles located near the surface in (**a**) anterior and (**b**) posterior views.

TABLE 19.2 Major Human Muscles (Fig. 19.12*a*: Anterior View)

Name	Action
Head and neck	
Frontalis	Wrinkles forehead and lifts eyebrows
Orbicularis oculi	Closes eye (winking)
Zygomaticus	Raises corner of mouth (smiling)
Masseter	Closes jaw, chewing
Orbicularis oris	Closes and protrudes lips (kissing)
Upper limb and trunk	
External oblique	Compresses abdomen; rotates trunk
Rectus abdominis	Flexes spine; compresses abdomen
Pectoralis major	Flexes and adducts arm ventrally (pulls arm across chest)
Deltoid	Abducts and moves arm up and down in front
Biceps brachii	Flexes forearm and rotates hand outward
Lower limb	
Adductor longus	Adducts and flexes thigh
Iliopsoas	Flexes thigh at hip joint
Sartorius	Raises and rotates thigh and leg
Quadriceps femoris group	Extends leg at knee; flexes thigh
Peroneus longus	Everts foot
Tibialis anterior	Dorsiflexes and inverts foot
Flexor digitorum longus	Flexes toes
Extensor digitorum longus	Extends toes

TABLE 19.3 Major Human Muscles (Fig, 19.12*b*: Posterior View)

Name	Action
Head and neck	
Occipitalis	Moves scalp backward
Sternocleidomastoid	Turns head to side; flexes neck and head
Trapezius	Extends head; raises scapula as when shrugging shoulders
Upper limb and trunk	
Latissimus dorsi	Extends and adducts arm dorsally (pulls arm across back)
Deltoid	Abducts and moves arm up and down in front
External oblique	Rotates trunk
Triceps brachii	Extends forearm
Flexor carpi group	Flexes hand
Extensor carpi group	Extends hand
Flexor digitorum	Flexes fingers
Extensor digitorum	Extends fingers
Buttocks and lower limb	
Gluteus medius	Abducts thigh
Gluteus maximus	Extends thigh back (forms buttocks)
Hamstring group	Flexes leg and extends thigh
Gastrocnemius	Plantar flexes foot (tiptoeing)

Check Your Progress 19.3

1. State the three types of muscle tissue in the human body.
2. Describe the difference between a muscle's origin and its insertion.
3. Identify the major skeletal muscles of the body.
4. Define flexion, extension, abduction, adduction, and rotation.

19.4 Mechanism of Muscle Fiber Contraction

Learning Outcomes

Upon completion of this section, you should be able to

1. Describe the microscopic structure of a muscle fiber.
2. Review the sliding filament model of muscle contraction.
3. Summarize how activities within the neuromuscular junction control muscle fiber contraction.
4. Indicate three ways that muscle cells can generate ATP.

We have already examined the structure of skeletal muscle as seen with the light microscope. Skeletal muscle tissue has alternating light and dark bands, giving it a striated appearance. These bands are due to the arrangement of myofilaments in a muscle fiber.

Muscle Fiber

A muscle fiber is a cell containing the usual cellular components, but special names have been assigned to some of these components (Table 19.4). Figure 19.13 shows the microscopic structure of a muscle fiber. The plasma membrane is called the **sarcolemma;** the cytoplasm is the sarcoplasm; and the endoplasmic reticulum is the **sarcoplasmic reticulum.** A muscle fiber also has some unique anatomical characteristics. One feature is its T (for transverse) system; the sarcolemma forms **T (transverse) tubules** that penetrate, or dip down, into the cell so that they come in contact—but do not fuse—with expanded portions of the sarcoplasmic reticulum. The expanded portions of the sarcoplasmic reticulum are calcium storage sites. Calcium ions (Ca^{2+}), as we shall see, are essential for muscle contraction.

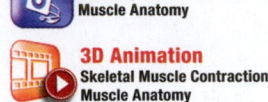

MP3 Muscle Anatomy

3D Animation Skeletal Muscle Contraction: Muscle Anatomy

The sarcoplasmic reticulum encases hundreds and sometimes even thousands of **myofibrils,** each about 1 μm in diameter, which are the contractile portions of the muscle fibers. Any other organelles, such as mitochondria, are located in the sarcoplasm between the myofibrils. The sarcoplasm also contains glycogen, which provides stored energy for muscle contraction, and the red pigment myoglobin, which binds oxygen until it is needed for muscle contraction.

TABLE 19.4 Microscopic Anatomy of a Muscle

Name	Function
Sarcolemma	Plasma membrane of a muscle fiber that forms T tubules
Sarcoplasm	Cytoplasm of a muscle fiber that contains the organelles, including myofibrils
Glycogen	A polysaccharide that stores energy for muscle contraction
Myoglobin	A red pigment that stores oxygen for muscle contraction
T tubule	Extension of the sarcolemma that extends into the muscle fiber and conveys impulses that cause Ca^{2+} to be released from the sarcoplasmic reticulum
Sarcoplasmic reticulum	The smooth ER of a muscle fiber that stores Ca^{2+}
Myofibril	A bundle of myofilaments that contracts
Myofilament	Actin filaments and myosin filaments whose structure and functions account for muscle striations and contractions

Myofibrils and Sarcomeres

Myofibrils are cylindrical in shape and run the length of the muscle fiber. The light microscope shows that skeletal muscle fibers have light and dark bands called *striations.* The electron microscope shows that the striations of skeletal muscle fibers are formed by the placement of myofilaments within units of myofibrils called **sarcomeres.** A sarcomere extends between two dark lines, called the *Z lines.* A sarcomere contains two types of protein myofilaments. The thick filaments are made up of a protein

Figure 19.13 Skeletal muscle fiber structure and function. A muscle fiber contains many myofibrils, divided into sarcomeres, which are contractile. When the myofibrils of a muscle fiber contract, the sarcomeres shorten. The actin (thin) filaments slide past the myosin (thick) filaments toward the center so that the H zone gets smaller, to the point of disappearing.

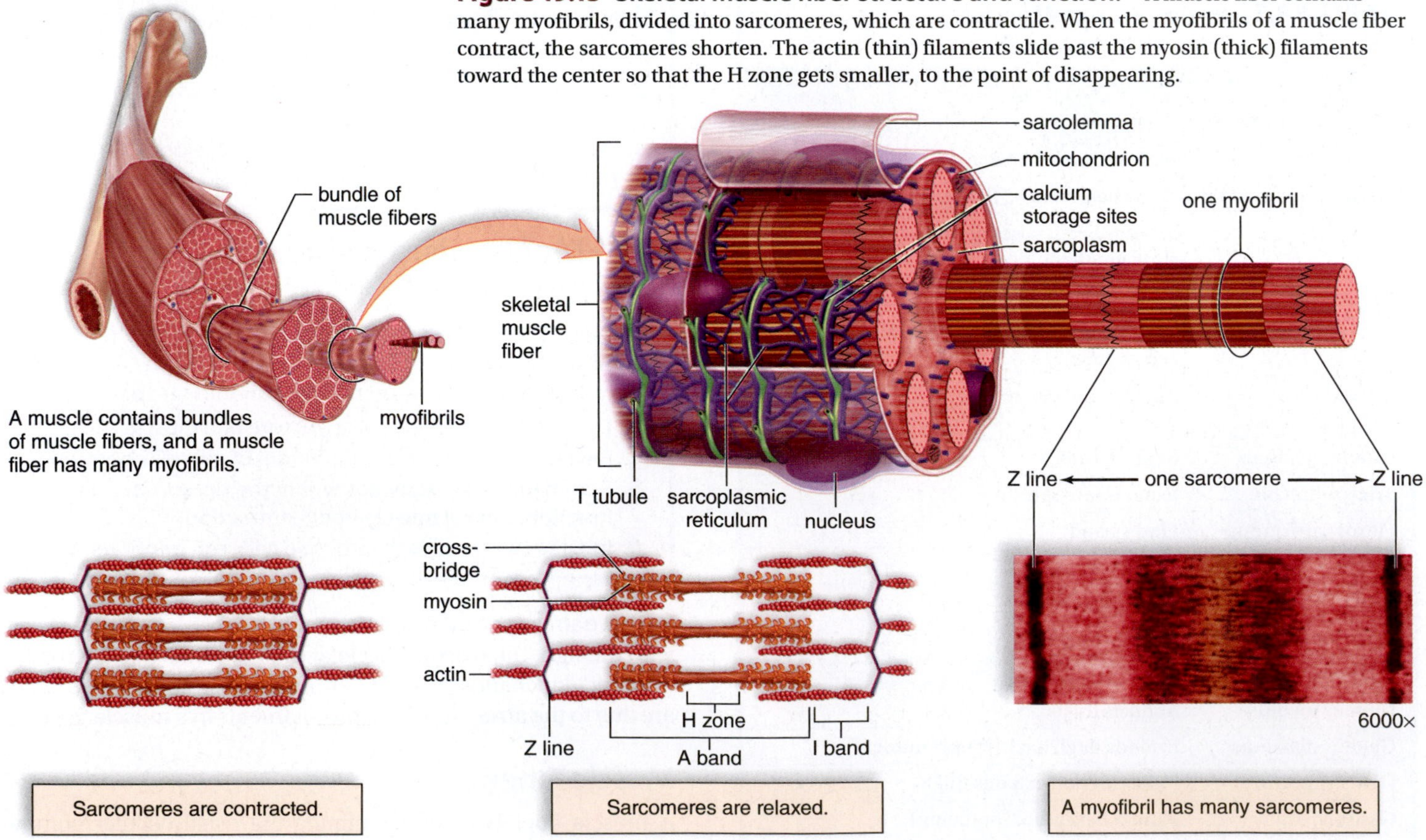

bundle of muscle fibers
A muscle contains bundles of muscle fibers, and a muscle fiber has many myofibrils.
myofibrils
skeletal muscle fiber
sarcolemma
mitochondrion
calcium storage sites
sarcoplasm
one myofibril
T tubule
sarcoplasmic reticulum
nucleus
Z line
one sarcomere
Z line
cross-bridge
myosin
actin
H zone
Z line
A band
I band
6000×
Sarcomeres are contracted.
Sarcomeres are relaxed.
A myofibril has many sarcomeres.

called **myosin,** and the thin filaments are made up of a protein called **actin.** Other proteins are also present. The I band is light colored because it contains only actin filaments attached to a Z line. The dark regions of the A band contain overlapping actin and myosin filaments, and its H zone has only myosin filaments.

Myofilaments

The thick and thin filaments differ in the following ways:

Thick Filaments A thick filament is composed of several hundred molecules of myosin. Each myosin molecule is shaped like a golf club, with the straight portion of the molecule ending in a double globular head. The myosin heads occur on each side of a sarcomere but not in the middle.

Thin Filaments A thin filament primarily consists of two intertwining strands of actin. Two other proteins, called tropomyosin and troponin, also play a role, as we will discuss later in this section.

Sliding Filaments We will also see that when muscles are stimulated, impulses travel down a T tubule, and calcium is released from the sarcoplasmic reticulum. Calcium triggers muscle fiber contraction as the sarcomeres within the myofibrils shorten. When a sarcomere shortens, the actin (thin) filaments slide past the myosin (thick) filaments and approach one another. This causes the I band to shorten and the H zone to almost or completely disappear. The movement of actin filaments in relation to myosin filaments is called the **sliding filament model** of muscle contraction. During the sliding process, the sarcomere shortens, even though the filaments themselves remain the same length. ATP supplies the energy for muscle contraction. Although the actin filaments slide past the myosin filaments, it is the myosin filaments that do the work. Myosin filaments break down ATP and form cross-bridges that pull the actin filaments toward the center of the sarcomere.

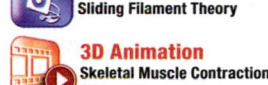

MP3 Sliding Filament Theory

3D Animation Skeletal Muscle Contraction: Sliding Filament Model

SCIENCE IN YOUR LIFE ▶ BIOETHICAL

What Constitutes an Unfair Advantage in Sports?

Sports TV news shows are full of reports of the latest scandals involving professional athletes using performance-enhancing drugs. Baseball players either accused or proven to have used anabolic steroids or human growth hormone (HGH) may be banned from the game, or at least from the Hall of Fame. Floyd Landis, the American cyclist who won the Tour de France in 2006, had to give up his title when he tested positive for synthetic testosterone. This list seems to go on and on.

But even if there is general agreement that performance-enhancing drugs can give athletes an "unfair advantage," what about athletes with other types of advantages? We wouldn't tend to think that being born with malformed bones below the knee as an advantage, but for Oscar Pistorius, that may be exactly the case. Pistorius, who won a gold medal at the Athens Paralympics in 2004, was banned from the Beijing Olympics in 2008, because examinations showed that his blade-shaped prosthetic legs (Fig. 19B) are more efficient than normal human legs. Or consider the case of Casey Martin, who suffers from a painful leg defect, and who in 2001 sued the Professional Golf Association for the right to use a golf cart, which is normally banned in tournaments. Martin was granted the right to use a cart under the Americans with Disabilities Act, but did this give him an unfair advantage over able-bodied golfers?

One of the major reasons performance-enhancing drugs are banned is that they pose clear health risks to users. Even so, shouldn't individuals be allowed to take these risks if they want to? And on what basis can we outlaw adaptations that are actually designed to increase a person's health? In other words, how can you justify allowing some strategies that enhance performance and not others?

Figure 19B Oscar Pistorius was banned from competing in the 2008 Olympics.

Questions to Consider

1. Are there any similarities between taking anabolic steroids and taking legal drugs, such as aspirin, to reduce inflammation in the body?
2. Do you see a significant difference in the advantage gained by taking drugs, using a prosthesis, or otherwise compensating for a disability?
3. Suppose a new pay-per-view sports network allowed athletes to take any drug, or undergo any surgical modification they wanted, in order to see what the human body could achieve. Would you watch such a show? What factors would then limit the athlete's performance?

Skeletal Muscle Contraction

Muscle fibers are stimulated to contract by motor neurons whose axons are in nerves (Fig. 19.14). The axon of one motor neuron can stimulate from a few to several muscle fibers of a muscle because each axon has several branches. A branch of an axon ends in an axon terminal that lies in close proximity to the sarcolemma of a muscle fiber. A small gap, called a synaptic cleft, separates the axon terminal from the sarcolemma. This entire region is called a **neuromuscular junction** (Fig. 19.14).

Animation Function of the Neuromuscular Junction

Axon terminals contain synaptic vesicles that are filled with the neurotransmitter acetylcholine (ACh). When nerve impulses traveling down a motor neuron arrive at an axon terminal, the synaptic vesicles release ACh into the synaptic cleft. ACh quickly diffuses across the cleft and binds to receptors in the sarcolemma. Now the sarcolemma generates impulses that spread over itself and down T tubules to the sarcoplasmic reticulum. The release of Ca^{2+} ions from the sarcoplasmic reticulum causes the filaments within the sarcomeres to slide past one another. Sarcomere contraction causes myofibril contraction, which in turn results in the contraction of a muscle fiber and, eventually, a whole muscle.

Animation Sarcomere Contraction

Botox is a trade name for botulinum toxin A, a neurotoxin produced by a bacterium. Botox, which can be injected by a physician to treat several medical conditions as well as to prevent wrinkling of the brow and skin around the eyes, prevents the release of ACh, and therefore, the contraction of muscles in these areas. Tetanus toxin, produced by a different bacterial species, interferes with muscle relaxation. Thus, it causes excessive contraction, which can be fatal if it affects the respiratory muscles. Fortunately, the tetanus vaccine is very effective at preventing this.

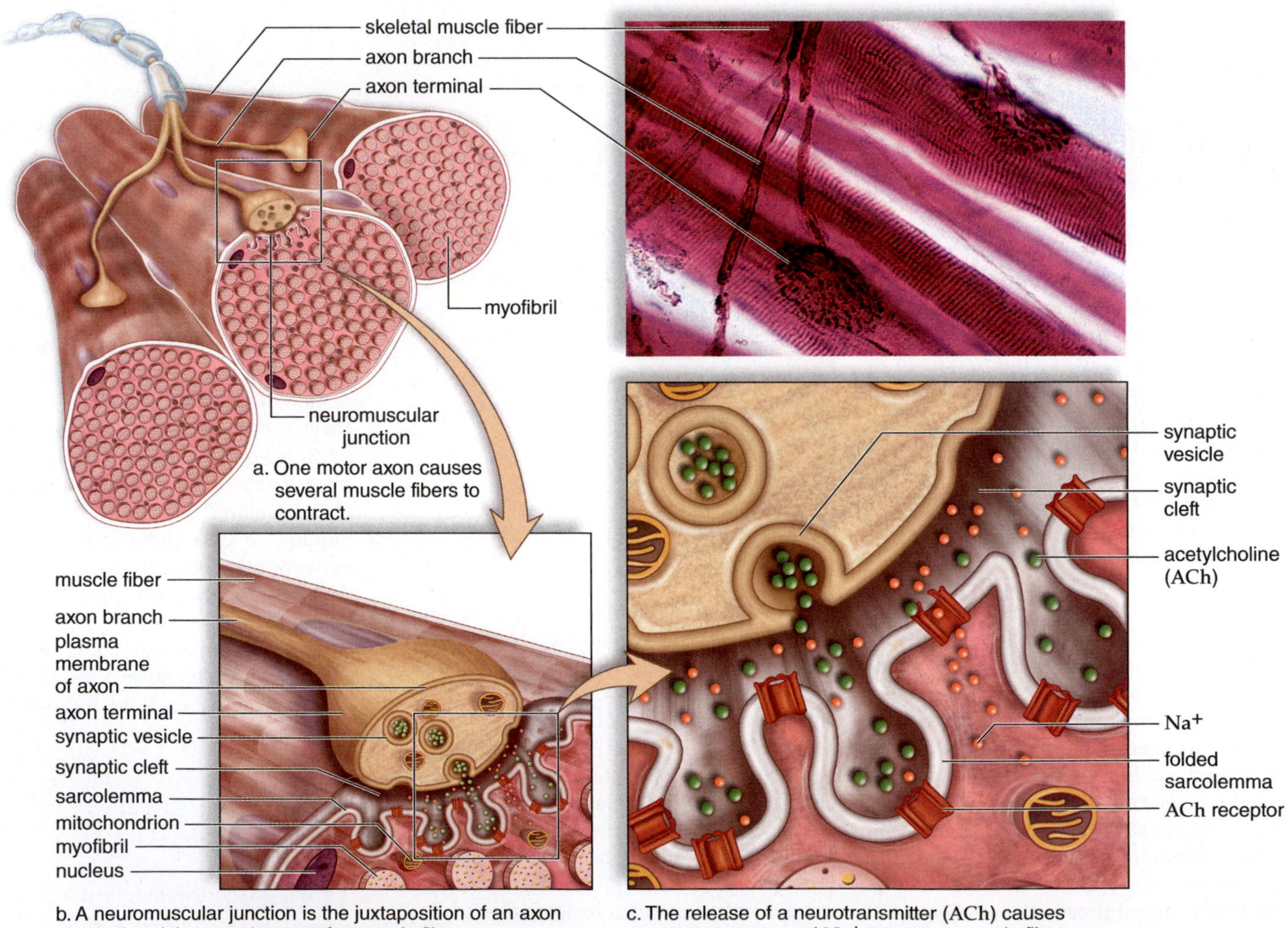
skeletal muscle fiber
axon branch
axon terminal
myofibril
neuromuscular junction
muscle fiber
axon branch
plasma membrane of axon
axon terminal
synaptic vesicle
synaptic cleft
sarcolemma
mitochondrion
myofibril
nucleus
synaptic vesicle
synaptic cleft
acetylcholine (ACh)
Na+
folded sarcolemma
ACh receptor

a. One motor axon causes several muscle fibers to contract.

b. A neuromuscular junction is the juxtaposition of an axon terminal and the sarcolemma of a muscle fiber.

c. The release of a neurotransmitter (ACh) causes receptors to open and Na^+ to enter a muscle fiber.

Figure 19.14 Neuromuscular junction. **a.** The branch of a motor nerve fiber ends in an axon terminal that meets but does not touch a muscle fiber. **b.** A synaptic cleft separates the axon terminal from the sarcolemma of the muscle fiber. **c.** Nerve impulses traveling down a motor fiber cause synaptic vesicles to discharge a neurotransmitter (ACh) that diffuses across the synaptic cleft. When the neurotransmitter is received by the sarcolemma of a muscle fiber, impulses begin that lead to muscle fiber contraction.

The Molecular Mechanism of Contraction

As shown in Figure 19.15, two other proteins are associated with actin filaments. Threads of *tropomyosin* wind about an actin filament, and *troponin* occurs at intervals along the threads. When Ca^{2+} ions are released from the sarcoplasmic reticulum, they combine with troponin, and this causes the tropomyosin threads to shift their position.

The double globular heads of a myosin filament have ATP binding sites, where an ATPase splits ATP into ADP and Ⓟ. The ADP and Ⓟ remain on the myosin heads until the heads attach to an actin filament, forming cross-bridges. Now, ADP and Ⓟ are released, and the cross-bridges change their positions. This is the power stroke that pulls the actin filament toward the center of the sarcomere. When ATP molecules again bind to the myosin heads, the cross-bridges are broken, and heads detach from the actin filament. Actin filaments move nearer the center of the sarcomere each time the cycle is repeated. When nerve impulses cease, the sarcoplasmic reticulum actively transports Ca^{2+} ions back into the sarcoplasmic reticulum, and the muscle relaxes.

3D Animation
Skeletal Muscle Contraction: Regulation by Calcium Ions

Energy for Muscle Contraction

ATP produced previous to strenuous exercise lasts a few seconds, and then muscles acquire new ATP in three different ways: creatine phosphate breakdown, cellular respiration, and fermentation. Creatine phosphate breakdown

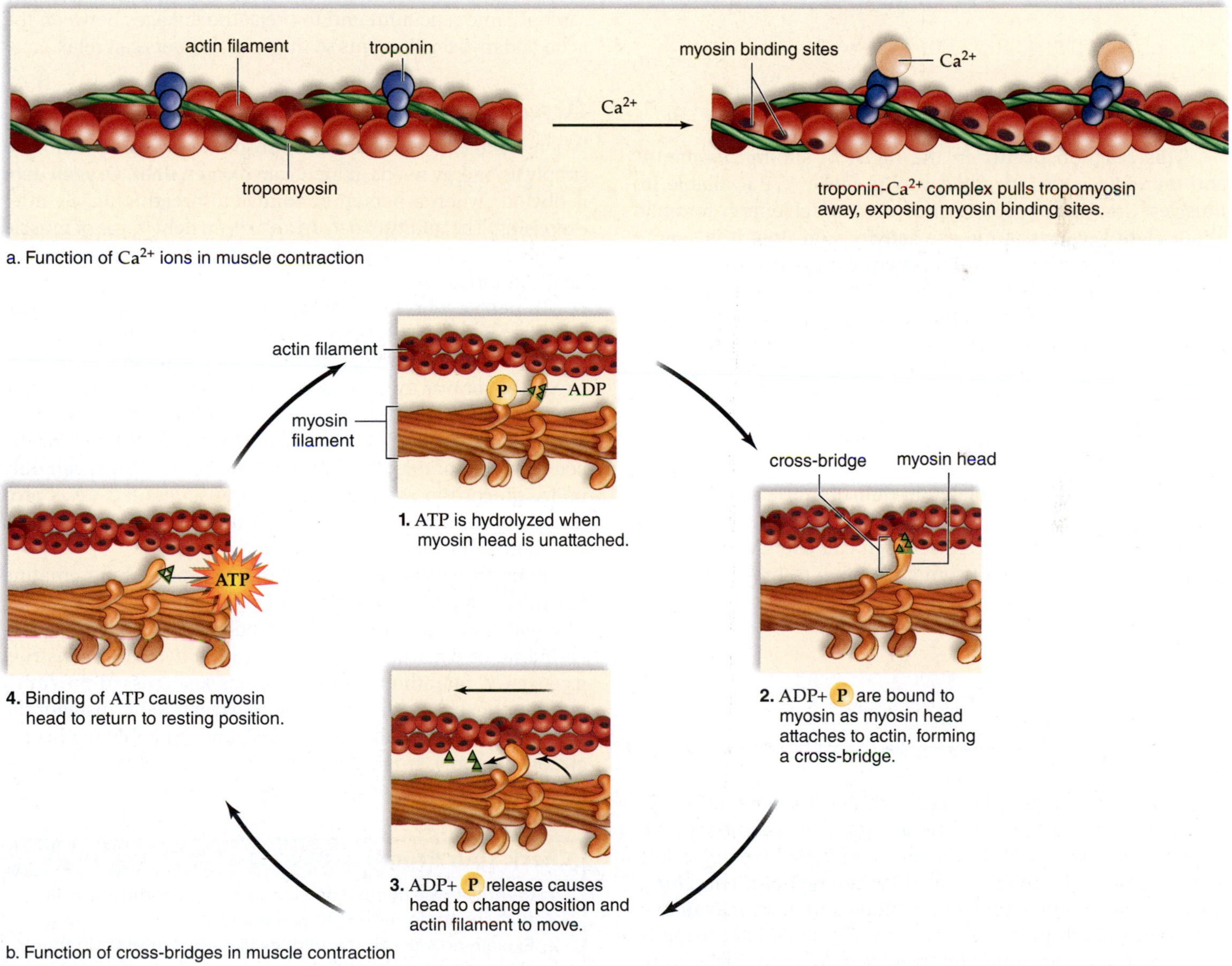
actin filament
troponin
tropomyosin
Ca2+
myosin binding sites
Ca2+
troponin-Ca2+ complex pulls tropomyosin away, exposing myosin binding sites.
a. Function of Ca2+ ions in muscle contraction
actin filament
myosin filament
P
ADP
1. ATP is hydrolyzed when myosin head is unattached.
cross-bridge
myosin head
2. ADP+ P are bound to myosin as myosin head attaches to actin, forming a cross-bridge.
3. ADP+ P release causes head to change position and actin filament to move.
ATP
4. Binding of ATP causes myosin head to return to resting position.
b. Function of cross-bridges in muscle contraction

Figure 19.15 The role of specific proteins and calcium in muscle contraction. **a.** Upon release, calcium binds to troponin, exposing myosin binding sites. **b.** After breaking down ATP (1), myosin heads bind to an actin filament (2), and later, a power stroke causes the actin filament to move (3). When another ATP binds to myosin, the head detaches from actin (4), and the cycle begins again. Although only one myosin head is featured, many heads are active at the same time.

and fermentation are anaerobic, meaning that they do not require oxygen. Creatine phosphate breakdown, used first, is a way to acquire ATP before oxygen starts entering mitochondria. Cellular respiration is aerobic and takes place only when oxygen is available. If exercise is vigorous to the point that oxygen cannot be delivered fast enough to working muscles, fermentation occurs. Fermentation brings on oxygen debt.

Creatine Phosphate Breakdown

Creatine phosphate is a high-energy compound built up when a muscle is resting. Creatine phosphate cannot participate directly in muscle contraction. Instead, it can regenerate ATP by the following reaction:

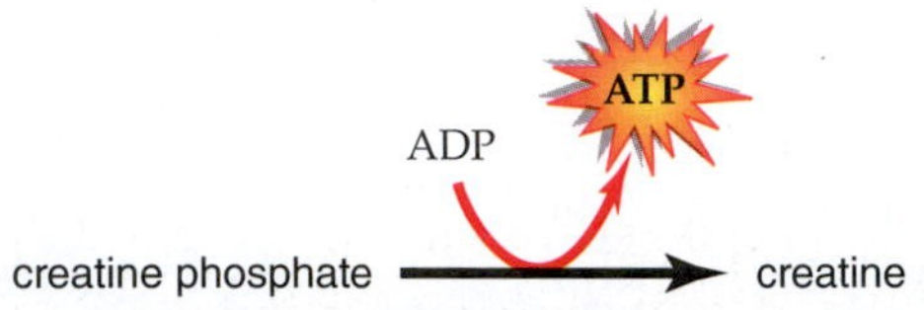
ATP
ADP
creatine phosphate
creatine

This reaction occurs in the midst of sliding filaments, and therefore is the speediest way to make ATP available to muscles. Creatine phosphate provides enough energy for only about eight seconds of intense activity, and then it is spent. Creatine phosphate is rebuilt when a muscle is resting by transferring a phosphate group from ATP to creatine. Creatine phosphate is also a very popular dietary supplement for training athletes, and there is evidence that consuming creatine in moderate doses is safe and may increase some types of athletic performance.

Cellular Respiration

Cellular respiration, completed in mitochondria, usually provides most of a muscle's ATP. Glycogen and fat are stored in muscle cells. Therefore, a muscle cell can use glucose (from glycogen) and fatty acids (from fat) as fuel to produce ATP if oxygen is available:

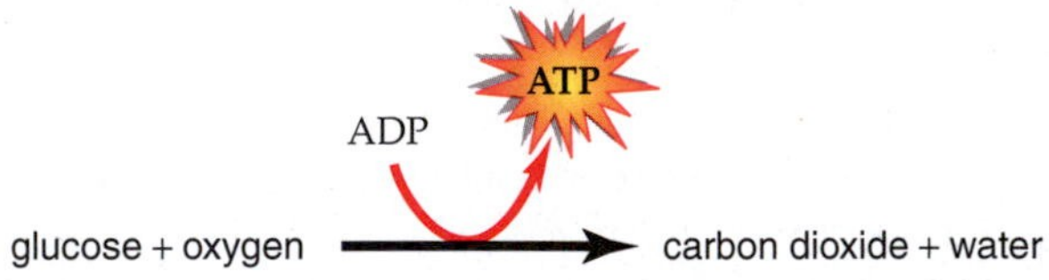
ATP
ADP
glucose + oxygen
carbon dioxide + water

Myoglobin, an oxygen carrier similar to hemoglobin, is synthesized in muscle cells, and its presence accounts for the reddish-brown color of skeletal muscle fibers. Myoglobin has a higher affinity for oxygen than does hemoglobin. Therefore, myoglobin can pull oxygen out of blood and make it available to muscle mitochondria that are carrying on cellular respiration. Then, too, the ability of myoglobin to temporarily store oxygen reduces a muscle's immediate need for oxygen when cellular respiration begins. The end products (carbon dioxide and water) are usually no problem. Carbon dioxide leaves the body at the lungs, and water simply enters the extracellular space. The by-product, heat, keeps the entire body warm.

Fermentation

Like creatine phosphate breakdown, fermentation supplies ATP without consuming oxygen. During fermentation, glucose is broken down to lactate (lactic acid):

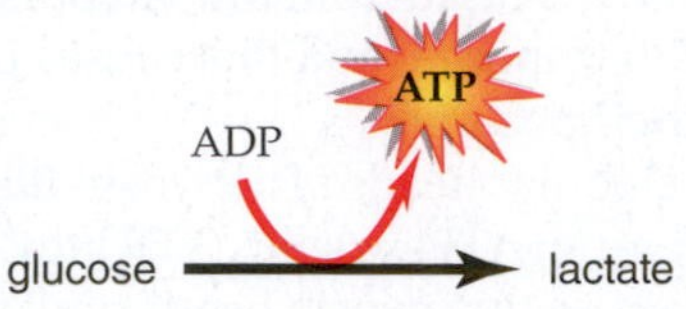
ATP
ADP
glucose
lactate

The accumulation of lactate in a muscle fiber makes the cytoplasm more acidic, and eventually enzymes cease to function well. If fermentation continues longer than two or three minutes, cramping and fatigue set in. Cramping seems to be caused by lack of the ATP needed to pump calcium ions back into the sarcoplasmic reticulum and to break the linkages between the actin and myosin filaments so that muscle fibers can relax.

Oxygen Debt

When a muscle uses creatine phosphate or fermentation to supply its energy needs, it incurs an **oxygen debt.** Oxygen debt is obvious when a person continues to breathe heavily after exercising. The ability to run up an oxygen debt is one of muscle tissue's greatest assets. Without oxygen, brain tissue cannot last nearly as long as muscles can.

In people who exercise regularly, the number of muscle mitochondria increases, and so fermentation is not needed to produce ATP. Their bodies' mitochondria can start consuming oxygen as soon as the ADP concentration begins to rise during muscle contraction. Because mitochondria can break down fatty acids instead of glucose, blood glucose is spared for the activity of the brain. (The brain, unlike other organs, can only utilize glucose to produce ATP.) Because less lactate is produced in people who train, the pH of the blood remains steady, and there is less of an oxygen debt.

Repaying an oxygen debt requires replenishing creatine phosphate supplies and disposing of lactate. Lactate can be changed back to pyruvate and metabolized completely in mitochondria, or pyruvate can be sent to the liver to reconstruct glycogen. A marathon runner who has just crossed the finish line is not usually exhausted due to oxygen debt. Instead, the runner has used up all the muscles', and probably the liver's, glycogen supply. It takes about two days to replace glycogen stores on a high-carbohydrate diet.

Check Your Progress 19.4

1. Identify three microscopic components of muscle cells that are not found in other types of cells.
2. Explain how the thin and thick filaments interact in the sliding filament model.
3. Describe how neurons specifically control muscle contraction.
4. List one specific advantage of each mechanism of generating ATP for muscle contraction.

19.5 Whole Muscle Contraction

Learning Outcomes

Upon completion of this section, you should be able to

1. Explain what is happening to an isolated muscle fiber during summation, tetanus, and fatigue.
2. Describe how muscles generate different levels of contractive force.
3. Explain the types of exercise that require more fast-twitch versus slow-twitch muscle fibers.

Researchers sometimes study muscles in the laboratory in an effort to understand whole muscle contraction in the body.

In the Laboratory

When a muscle fiber is isolated, placed on a microscope slide, and provided with ATP plus various salts, it contracts completely along its entire length. This observation has resulted in the *all-or-none law,* which states that a muscle fiber contracts completely or not at all. In contrast, a whole muscle shows degrees of contraction. To study whole muscle contraction in the laboratory, an isolated muscle is stimulated electrically, and the mechanical force of contraction is recorded as a visual pattern called a *myogram.* When the strength of the stimulus is above a threshold level, the muscle contracts and then relaxes. This action—a single contraction that lasts only a fraction of a second—is called a **muscle twitch.**

Figure 19.16*a* is a myogram of a muscle twitch, which is customarily divided into three stages: the *latent period,* or the period of time between stimulation and initiation of contraction; the *contraction period,* when the muscle shortens; and the *relaxation period,* when the muscle returns to its former length. It's interesting to use our knowledge of muscle fiber contraction to understand these events. From our study thus far, we know that a muscle fiber in an intact muscle contracts when calcium leaves storage sacs and relaxes when calcium returns to storage sacs.

But unlike the contraction of a muscle fiber, a whole muscle has degrees of contraction, and a twitch can vary in height (strength) depending on the degree of stimulation. Why? Obviously, a stronger stimulation causes more individual fibers to contract than before.

If a whole muscle is given a rapid series of stimuli, it can respond to the next stimulus without relaxing completely. Summation is increased muscle contraction until maximal sustained contraction, called **tetanus,** is achieved (Fig. 19.16*b*). The myogram no longer shows individual twitches; rather, the twitches are fused and blended completely into a straight line. Tetanus continues until the muscle becomes fatigued due to depletion of energy reserves. Fatigue is apparent when a muscle relaxes even though stimulation continues.

In the Body

In the body, nerves cause muscles to contract. As mentioned, each axon within a nerve stimulates a number of muscle fibers. A nerve fiber, together with all of the muscle fibers it innervates,

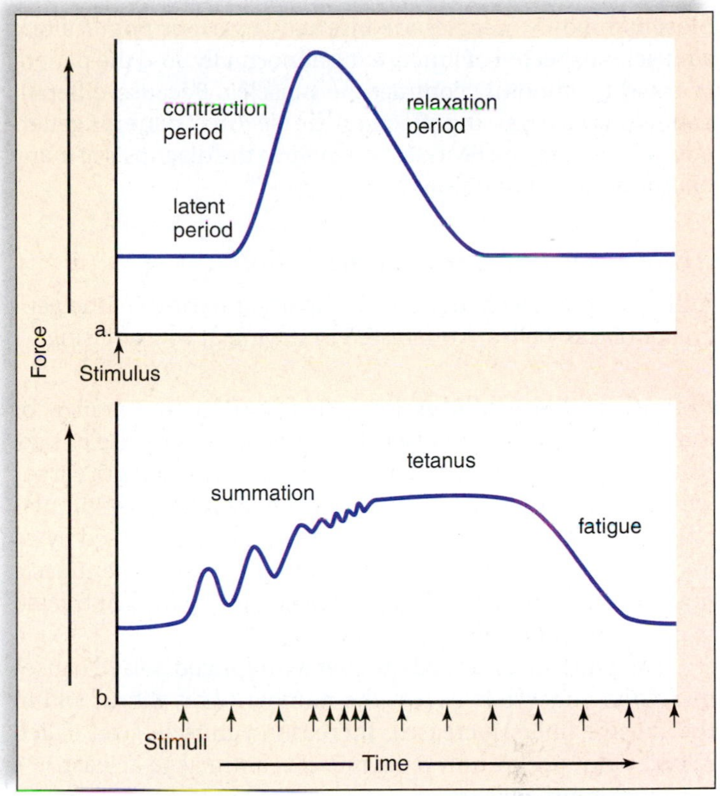
contraction period
relaxation period
latent period
a.
Stimulus
Force
tetanus
summation
fatigue
b.
Stimuli
Time

Figure 19.16 Physiology of skeletal muscle contraction. Stimulation of a muscle dissected from a frog resulted in these myograms. **a.** A simple muscle twitch has three periods: latent, contraction, and relaxation. **b.** Summation and tetanus. When the muscle is not permitted to relax completely between stimuli, the contraction gradually increases in intensity. The muscle becomes maximally contracted until it fatigues.

is called a **motor unit.** A motor unit obeys the all-or-none law. Why? Because all the muscle fibers in a motor unit are stimulated at once, and they all either contract or do not contract. A variable of interest is the number of muscle fibers within a motor unit. For example, in the ocular muscles that move the eyes, the innervation ratio is one motor axon per 23 muscle fibers, while in the gastrocnemius muscle of the leg, the ratio is about one motor axon per 1,000 muscle fibers. Thus, moving the eyes allows finer control than moving the legs.

Tetanic contractions ordinarily occur in the body because, as the intensity of nervous stimulation increases, more and more motor units are activated. This phenomenon, known as recruitment, results in stronger and stronger muscle contractions. But while some muscle fibers are contracting, others are relaxing. Because of this, intact muscles rarely fatigue completely. Even when muscles appear to be at rest, they exhibit **muscle tone,** in which some of their fibers are always contracting. Muscle tone is particularly important in maintaining posture. If all the fibers within the muscles of the neck, trunk, and legs were to suddenly relax, the body would collapse.

Medical professionals can measure the electrical activity of muscles using a test called an *electromyogram (EMG).*

Most commonly, needles are inserted at various points along a muscle suspected of functioning abnormally, and the patient is asked to smoothly contract the muscles. Because different conditions have varying effects on the electrical patterns generated by muscles, the EMG test can help in the diagnosis of many neuromuscular conditions.

Athletics and Muscle Contraction

Athletes who excel in a particular sport, and much of the general public as well, are interested in staying fit by exercising.

Exercise and Size of Muscles Muscles that are not used or that are used for only very weak contractions decrease in size, or *atrophy*. Atrophy can occur when a limb is placed in a cast or when the nerve serving a muscle is damaged. If nerve stimulation is not restored, muscle fibers are gradually replaced by fat and fibrous tissue. Unfortunately, atrophy can cause muscle fibers to shorten progressively, leaving body parts contracted in contorted positions.

Forceful muscular activity over a prolonged period causes muscle to increase in size as the number of myofibrils within the muscle fibers increases. Increase in muscle size, called *hypertrophy*, occurs only if the muscle contracts to at least 75% of its maximum tension.

Slow-Twitch and Fast-Twitch Muscle Fibers We have seen that all muscle fibers metabolize both aerobically and anaerobically. Some muscle fibers, however, utilize one method more than the other to provide myofibrils with ATP. Slow-twitch fibers tend to be aerobic, and fast-twitch fibers tend to be anaerobic (Fig. 19.17).

Slow-twitch fibers have a steadier tug and more endurance, despite having motor units with a smaller number of fibers. These muscle fibers are most helpful in sports such as long-distance running, biking, jogging, and swimming. Because they produce most of their energy aerobically, they tire only when their fuel supply is gone. Slow-twitch fibers have many mitochondria and are dark in color because they contain myoglobin, the respiratory pigment found in muscles. They are also surrounded by dense capillary beds and draw more blood and oxygen than fast-twitch fibers. Slow-twitch fibers have a low maximum tension, which develops slowly, but these muscle fibers are highly resistant to fatigue. Because slow-twitch fibers have a substantial reserve of glycogen and fat, their abundant mitochondria can maintain a steady, prolonged production of ATP when oxygen is available.

Fast-twitch fibers tend to be anaerobic and seem designed for strength because their motor units contain many fibers. They provide explosions of energy and are most helpful in sports activities such as sprinting, weight lifting, swinging a golf club, or throwing a shot. Fast-twitch fibers are light in color because they have fewer mitochondria, little or no myoglobin, and fewer blood vessels than slow-twitch fibers do. Fast-twitch fibers can develop maximum tension more rapidly than slow-twitch fibers can, and their maximum tension is greater. However, their dependence on anaerobic energy leaves them vulnerable to an accumulation of lactate that causes them to fatigue quickly.

Check Your Progress 19.5

1. List the three typical stages of a muscle twitch, as seen in a myogram.
2. Estimate which would have a greater ratio of motor nerve axons per muscle fibers: a muscle in your index finger or one in your back.
3. Explain why some domesticated birds, like turkeys and chickens, have mostly fast-twitch fibers in their breast (wing) muscles, whereas wild birds like geese and ducks have mostly slow-twitch fibers in those muscles.

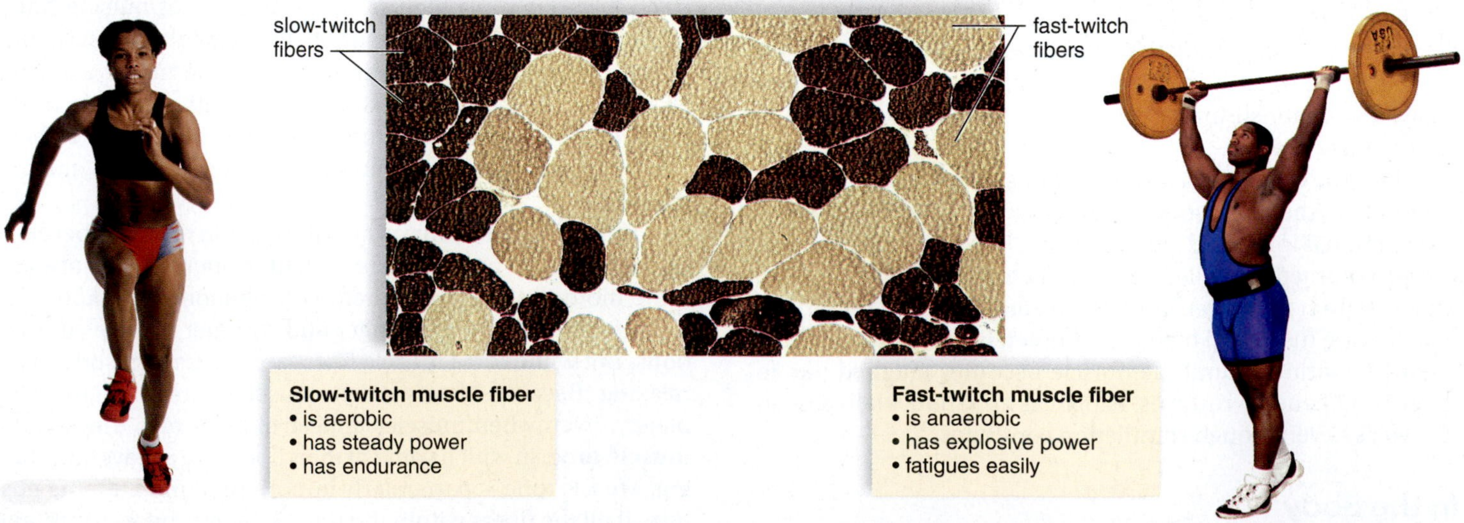
slow-twitch fibers
fast-twitch fibers
Slow-twitch muscle fiber
• is aerobic
• has steady power
• has endurance
Fast-twitch muscle fiber
• is anaerobic
• has explosive power
• fatigues easily

Figure 19.17 Slow- and fast-twitch muscle fibers. If your muscles contain many slow-twitch fibers (dark color), you probably do better at a sport such as cross-country running. But if your muscles contain many fast-twitch fibers (light color), you probably do better at a sport such as weight lifting.

19.6 Disorders of the Musculoskeletal System

Learning Outcomes

Upon completion of this section, you should be able to

1. List three common disorders that affect the bones and joints and three that affect the muscles.
2. Identify some of the most common steps that can be taken to help prevent osteoporosis.

Disorders of the Skeleton and Joints

Because bones are rigid, they are susceptible to being broken, or fractured. The most common cause of bone fractures is trauma, although pathological fractures can also occur during normal everyday activities if the bones are diseased. If the skin is punctured by the injury, the fracture is generally termed an "open" fracture; otherwise, it is "closed." Some common types of fractures include greenstick fractures, which do not pass all the way through the bone; comminuted fractures, which result in three or more bone fragments (Fig. 19.18); and stress fractures, one of the most common injuries in athletes. Stress fractures are small cracks in the bone that occur most commonly on the lower leg when muscles become fatigued and unable to absorb the stress of athletic activities. These small fractures usually heal with rest—that is, by avoiding the activity that produced them. Other types of uncomplicated fractures may be treated with casts or splints. Compound and comminuted fractures, however, often require surgery to realign bones or replace bone fragments. In children, the growth plate is the last part of the bone to ossify, leaving it more susceptible to fracture. Fracture of the growth plate may cause the affected bone to stop growing, resulting in limbs that are crooked or uneven in length.

Video Bear Bones

Osteoporosis is a condition in which bone loses mass and mineral content. This leads to an increased risk of fractures, especially of the pelvis, vertebrae, and wrist. As the bone softens affected individuals often develop a characteristic curvature of the spine (Fig. 19.19). Eighty percent of those affected by osteoporosis are women. About one in four women will have an osteoporosis-related fracture in their lifetime. Women generally have about 30% less bone mass than men to begin with, and after menopause women typically lose about 2% of their bone mass per year. Estrogen replacement therapy (ERT) has been shown to increase bone mass and reduce fractures, but because it also can increase the risk of cardiovascular disease and certain types of cancer, patients must be well informed before beginning ERT. Other factors that decrease the risk of osteoporosis for everyone include consuming 1,000–1,500 mg of calcium per day and engaging in regular physical exercise.

Animation Osteoporosis

There are many different types of **arthritis,** or inflammation of the joints. The Arthritis Foundation estimates that nearly one in three adult Americans has some degree of chronic joint pain. Arthritis is second only to heart disease as a cause of work disability in the United States. The most common form of arthritis is **osteoarthritis,** or degenerative joint disease, which results from the deterioration of the cartilage in one or more synovial joints. As the cartilage continues to soften, crack, and wear away, the resulting bone-on-bone contact causes pain and a decreased range of motion. Symptoms of OA, such as pain or stiffness in the joints after periods of inactivity, typically begin after age 40 and progress slowly. The hands, hips, knees, lower back, and neck are most often affected (Fig. 19.20). **Rheumatoid arthritis** is considered an autoimmune disease, in which the body's immune system attacks the joints as well as other tissues. RA affects more women than men, and often

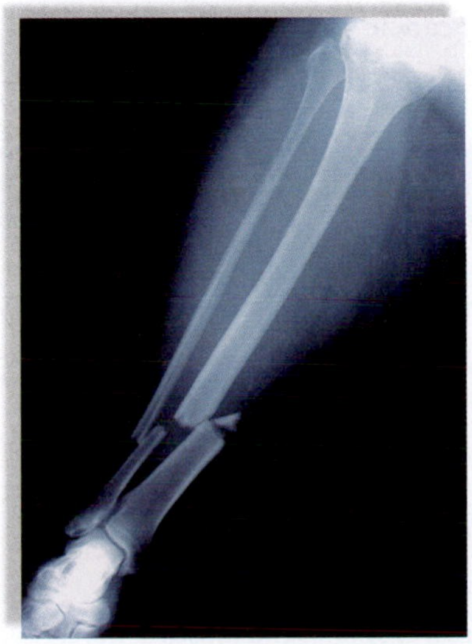

Figure 19.18 A comminuted fracture. This X ray shows a tibia that has fractured into at least three pieces. The fibula (left) is also fractured.

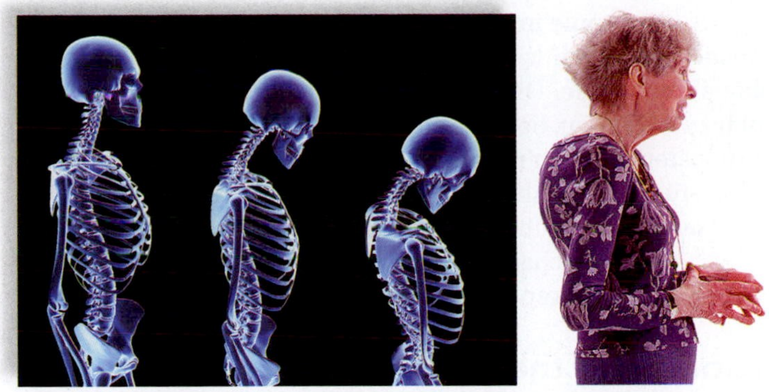

Figure 19.19 Osteoporosis. Osteoporosis can eventually result in the slumped posture and loss of height characteristically seen in some elderly people.

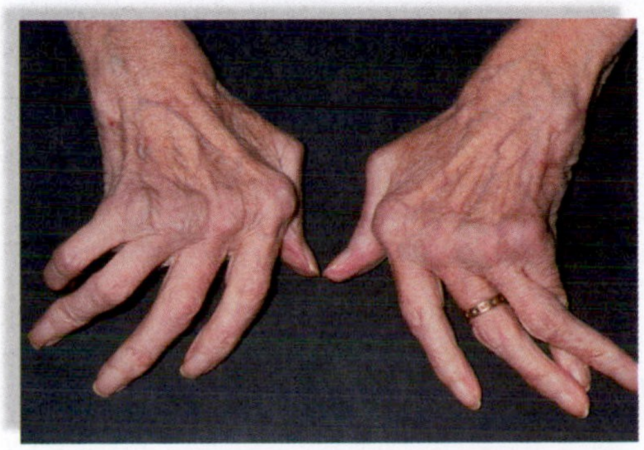

Figure 19.20 The effects of arthritis. Arthritis often affects the hands, resulting in swelling, pain, and loss of the ability to perform normal functions.

first strikes people in their twenties and thirties. The initiating cause of RA remains to be discovered, but genetic factors may play a major role. Over-the-counter anti-inflammatory drugs such as ibuprofen, or more powerful prescription drugs such as corticosteroids, can reduce the pain of most types of arthritis. Some studies have shown that dietary supplements containing glucosamine and chondroitin sulfate are also beneficial. As mentioned in the chapter-opening story, the damaged joint can also be surgically replaced with a prosthesis (artificial joint).

Disorders of the Muscles

Muscle cramps and twitches don't usually rise to the level of a "disorder," but almost everyone has experienced them. A true muscle cramp doesn't originate in the muscle itself, but rather is caused by abnormal nerve activity. This type of cramp is more common at night or after exercise, and can usually be ended by stretching the muscle. Most of us have also experienced occasional muscle twitches, or *fasciculations.* These also result from abnormal nerve activity, but only involve a small part of a muscle. Twitches most commonly occur in the eyelids, between the thumb and forefinger, and in the feet. They may occur more frequently due to stress, caffeine, or lack of sleep.

Fibromyalgia is a disorder that causes chronic pain in the muscles and ligaments. It affects about 4 million Americans, mostly women in their mid-30s to late-50s. The pain is usually most severe in the neck, shoulders, back, and hips. Although the cause of fibromyalgia is unknown, it may be due to low levels of serotonin or other neurotransmitters involved in pain perception and sleep regulation. As a result, patients with fibromyalgia may be more sensitive to pain, and they usually are unable to get enough deep sleep. As a result, chronic fatigue is also a major symptom. There are no specific treatments, but most people can reduce their symptoms by combining medications to reduce pain and improve sleep with regular exercise and a healthy diet.

Muscular dystrophy (**MD**) refers to a group of genetic diseases that affect the muscles. Most people are probably most familiar with these diseases from the annual Labor Day Telethon sponsored by the Muscular Dystrophy Association. These disorders vary considerably in their degree of severity. The more severe forms may cause problems beginning at a very young age, while other forms may not cause problems until middle age or later. The most common type of MD, called Duchenne MD, is due to an abnormality in the gene coding for a muscle-associated protein called dystrophin. Because it is an X-linked recessive disorder, Duchenne MD mainly affects boys. Muscle weakness is usually noticeable by age 3 to 5. By age 12, most affected children are unable to walk, and will eventually need a respirator to breathe. Corticosteroids have been shown to slow the degenerative process of Duchenne MD, but these drugs often have side effects. Gene therapy, in which the correct form of the faulty gene is inserted into the patient's cells, is being investigated as a potential cure for MD.

Check Your Progress 19.6

1. Describe the likely outcome if a bone fracture was not treated.
2. Predict why astronauts working in space would be likely to lose bone density.
3. Differentiate between the cause of rheumatoid arthritis and the cause of osteoarthritis.

Case Study Conclusion

For the first painful month or so after having her knee replaced, Jackie wondered if she had made the right decision. Just walking down the hall or up stairs was excruciating at first. Within two months, however, she was walking and swimming. Her physical therapist attributed her rapid recovery to her previous habit of staying in shape.

Jackie knows that without twenty-first-century medicine, she might have had a difficult time walking by the time she was 60. And she is reminded of the new metallic part of her every time she sets off an airport metal detector. Still, her doctor has reminded her that even in adults, bones are dynamic structures. Whereas her natural bone could be replaced by bone remodeling, the plastic and ceramic parts of her artificial knee would eventually wear out. So there was a very good chance that she would have to undergo a repeat replacement of her knee within 10 to 20 years. For Jackie, the ability to lead an active lifestyle would be worth going through the whole procedure again.

MEDIA STUDY TOOLS

www.mhhe.com/maderinquiry14

Enhance your study of this chapter with study tools and practice tests. Also ask your instructor about the resources available through ConnectPlus, including LearnSmart, the media-rich eBook, interactive learning tools, and animations.

Mc Graw Hill connect plus+ BIOLOGY

3D Animation
Skeletal Muscle Contraction

For an interactive exploration of the phases of muscle anatomy, the sliding filament model, and the moelcular mechanisms of muscle contraction, watch McGraw-Hill's 3D animation "Skeletal Muscle Contraction."

SUMMARIZE

19.1 Overview of Bone and Cartilage

The **musculoskeletal system** includes the bones and muscles. The skeleton is made up of bone, cartilage, and dense fibrous connective tissue:

- **Bone** is a connective tissue that contains mineral salts, which increase its strength. In a long bone, the wall of the shaft is **compact bone,** which is made up of units called osteons, consisting of a central canal surrounded by tiny canals called canaliculi and chambers called lacunae. Within the lacunae are **osteocytes.** The outermost layer of a bone is a tough connective tissue layer called **periosteum.** The ends of a long bone contain **spongy bone,** which is less organized, although still very strong. The spaces in spongy bones often contain **red bone marrow,** which produces all types of blood cells.
- Bone is a living tissue that is continuously being remodeled by **osteoblasts,** which build bone, and **osteoclasts,** which break it down. In children, the growth of long bones occurs at a band of cartilage called the **growth plate.**
- **Cartilage** is not as strong as bone, but is more flexible because it contains many collagenous and elastic fibers. **Hyaline cartilage** covers the ends of long bones, whereas **fibrocartilage** is mainly used for support. **Elastic cartilage** is the most flexible type.
- **Dense fibrous connective tissue,** which contains fibroblasts and bundles of collagen fibers, forms the ligaments and tendons. **Ligaments** connect bone to bone, and **tendons** connect muscle to bone at **joints.** Joints are covered by **articular cartilage.**

19.2 Bones of the Skeleton

The skeleton supports and protects the body, produces blood cells, serves as a storehouse for fats and mineral salts, and permits flexible movement. It can be divided into two major parts, plus the associated joints:

- The **axial skeleton** lies in the midline of the body and consists of the skull, hyoid bone, vertebral column, and rib cage. The **skull** contains the cranium, which protects the brain, and the facial bones. In newborns, bones of the cranium are joined by membranous **fontanels.** Certain cranial bones contain air spaces called **sinuses.** The spinal cord exits the base of the skull through a large opening called the **foramen magnum.** The **vertebral column** protects the spinal cord. Individual vertebrae are separated by **intervertebral disks** made of fibrocartilage.
- The **appendicular skeleton** consists of the bones of the **pectoral girdles,** upper limbs, **pelvic girdle,** and lower limbs.
- There are three types of joints. **Fibrous joints** tend to be immobile, like the **sutures** between the cranial bones. **Cartilaginous joints,** which are slightly movable, include the joints between the ribs and sternum. Flexible joints like the shoulder and knees are the **synovial joints,** in which two bones are separated by a cavity. This cavity, or joint capsule, is lined with a **synovial membrane** that is lubricated by synovial fluid. Many synovial joints contain fluid-filled sacs, or **bursae;** inflammation of these is called **bursitis.**

19.3 Skeletal Muscles

Skeletal muscles are striated and under voluntary control. Most work in antagonistic pairs, causing bones to move at a joint:

- The **origin** of a muscle attaches to a stationary bone; the **insertion** is on the bone that moves.
- Terms used to describe the types of movements due to muscle contraction include flexion, extension, abduction, and adduction.
- Muscles are grouped by location and named for characteristics such as size, shape, location, and action.

19.4 Mechanism of Muscle Fiber Contraction

Examining the microscopic anatomy of a muscle fiber helps explain the molecular mechanism of muscle contraction:

- The **sarcolemma** of a muscle fiber forms **T (transverse) tubules** that extend into the fiber and almost touch the **sarcoplasmic reticulum,** which stores calcium ions. Each muscle fiber contains many **myofibrils,** which contain myofilaments made of **actin** and **myosin.** These are arranged in subunits called **sarcomeres,** which are the contractile units of a muscle fiber.
- At a **neuromuscular junction,** synaptic vesicles release acetylcholine (ACh), which binds to protein receptors on the sarcolemma, causing impulses to travel down T tubules and calcium to leave the sarcoplasmic reticulum. Calcium ions bind to troponin and cause tropomyosin threads to shift their position, revealing myosin binding sites on actin. Myosin is an ATPase, and once it breaks down ATP, the myosin head is ready to attach to actin. The release of ADP + Ⓟ causes the head to change its position. This is the power stroke that causes the actin filament to slide toward the center of a sarcomere. This is the **sliding filament model** of muscle contraction.

A muscle fiber has three ways to acquire ATP after muscle contraction begins:

- **Creatine phosphate,** built up when a muscle is resting, can donate phosphates to ADP, forming ATP.
- Oxygen-dependent cellular respiration can occur within the mitochondria of the muscle cell. Myoglobin, an oxygen carrier that has a higher affinity for oxygen than does hemoglobin, provides oxygen for this process.
- Fermentation of glucose, with the concomitant production of lactate, quickly produces ATP. Fermentation can result in **oxygen debt,** and lactate accumulation may cause fatigue and cramping.

19.5 Whole Muscle Contraction

The contraction of whole muscles can be studied in the laboratory as well as in the body:

- In the laboratory, a single all-or-none contraction of a muscle fiber is called a **muscle twitch.** In whole muscle contraction, summation is followed by **tetanus,** or maximum sustained contractions, until the muscle must relax from fatigue.
- A **motor unit** consists of a single nerve fiber and all the muscle fibers it innervates. Muscles exhibit **muscle tone** because some muscle fibers are contracting at any time. Two major types of muscle fibers can be identified, with respect to their predominant method of ATP production. Slow-twitch fibers, specialized for endurance, are darker in color and have more mitochondria and myoglobin for aerobic cellular respiration. Fast-twitch fibers have more muscle fibers in their motor units and rely more on an anaerobic means of acquiring ATP. These fibers provide quick bursts of power, but they fatigue quickly.

19.6 Disorders of the Musculoskeletal System

Disorders affecting the skeletal system can be grouped into those affecting the bones and joints, and those affecting the muscles:

- Disorders of the skeleton and joints include fractures, osteoporosis, and arthritis. **Osteoporosis** is a softening of bone due to loss of mass and mineral content, which affects women more commonly than men. Among the many types of **arthritis,** or joint inflammation, are **osteoarthritis,** and **rheumatoid arthritis,** which is an autoimmune disease.
- Muscles can be affected by cramps and twitches, which are very common but usually just an annoyance. More serious muscle disorders include **fibromyalgia,** which causes intense pain and may be due to low levels of serotonin, and **muscular dystrophy,** a group of genetic disorders that affect muscle development.

ASSESS

Testing Yourself

Choose the best answer for each question.

1. _____ are involved in the breakdown of bone.
   a. Monocytes
   b. Osteoblasts
   c. Lymphocytes
   d. Osteoclasts
2. _____ connect bone to bone, and _____ connect muscle to bone.
   a. Ligaments, ligaments
   b. Tendons, ligaments
   c. Ligaments, tendons
   d. None of these are correct.
3. Label the following diagram of the skull.

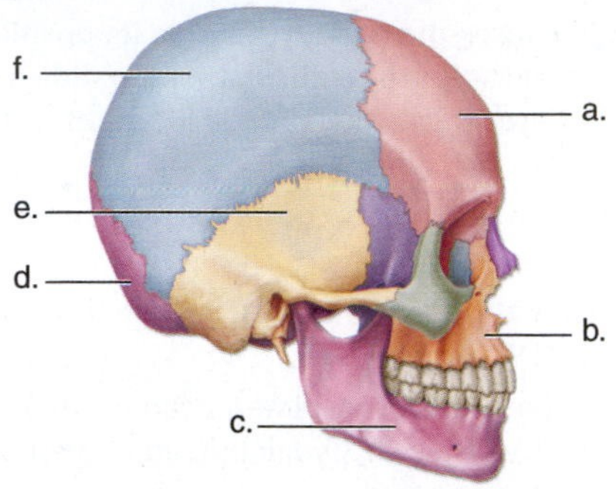
f.
a.
e.
d.
b.
c.

4. The knee joint contains
   a. cartilage.
   b. cruciate ligaments.
   c. synovial fluid.
   d. bursae.
   e. All of these are correct.
5. The _____ of the muscle attaches to the bone that is stationary during movement.
   a. origin
   b. insertion
   c. belly
   d. None of these are correct.
6. The biceps and triceps are considered
   a. synergists.
   b. antagonists.
   c. protagonists.
   d. None of these are correct.
7. Muscles can be named based on
   a. size.
   b. shape.
   c. location.
   d. action.
   e. All of these are correct.
8. Nervous stimulation of muscles
   a. occurs at a neuromuscular junction.
   b. results in an action potential that travels down the T system.
   c. causes calcium to be released from expanded regions of the sarcoplasmic reticulum.
   d. All of these are correct.
9. When muscles contract,
   a. sarcomeres increase in length.
   b. actin breaks down ATP.
   c. myosin slides past actin.
   d. the H zone disappears.
   e. calcium is taken up by the sarcoplasmic reticulum.

For questions 10–15, label the diagram of a muscle fiber, using these terms: myofibril, Z line, T tubule, sarcomere, sarcolemma, and sarcoplasmic reticulum.

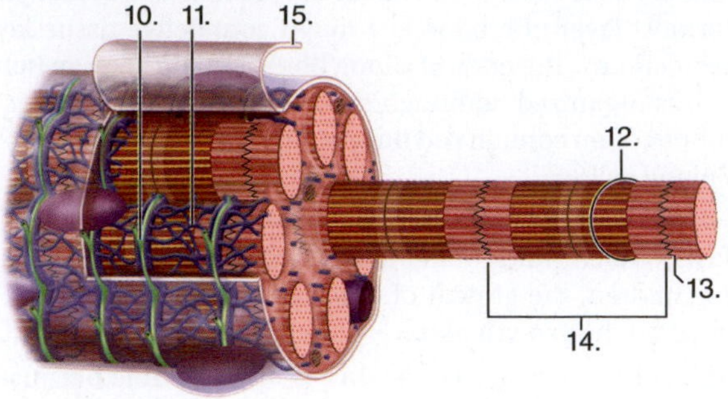
10.
11.
15.
12.
13.
14.

ENGAGE

Virtual Lab
Muscle Stimulation

The virtual lab "Muscle Stimulation" allows you to visualize how the load on a muscle influences the structures of the muscle at the microscopic level.

Thinking Critically

1. Determine whether each of the following joints is capable of these movements: flexion, extension, abduction, adduction, and rotation. You may "cheat" by using your own body.
   a. elbow
   b. neck
   c. knee
   d. phalanges
   e. hip
2. Similar to the activity in our muscle cells, certain bacteria can utilize sugars by aerobic cellular respiration (which requires oxygen) or by fermentation (which does not). However, like muscle cells, these bacteria only use fermentation when oxygen is not available. Why is aerobic respiration preferable? What is the specific role of oxygen in this process? (You may want to refer back to Chapter 7.)
3. Most authorities emphasize the importance of calcium consumption for the prevention of osteoporosis. However, surveys have shown that osteoporosis is rare in some countries, such as China, where calcium intake is lower than in the United States. These studies often point to the higher level of animal protein consumed in the United States as having a detrimental effect on bone mass. If this were true, what physiological mechanism might explain it? Can you think of any other differences between the lifestyles in China and the United States that also might explain the difference?

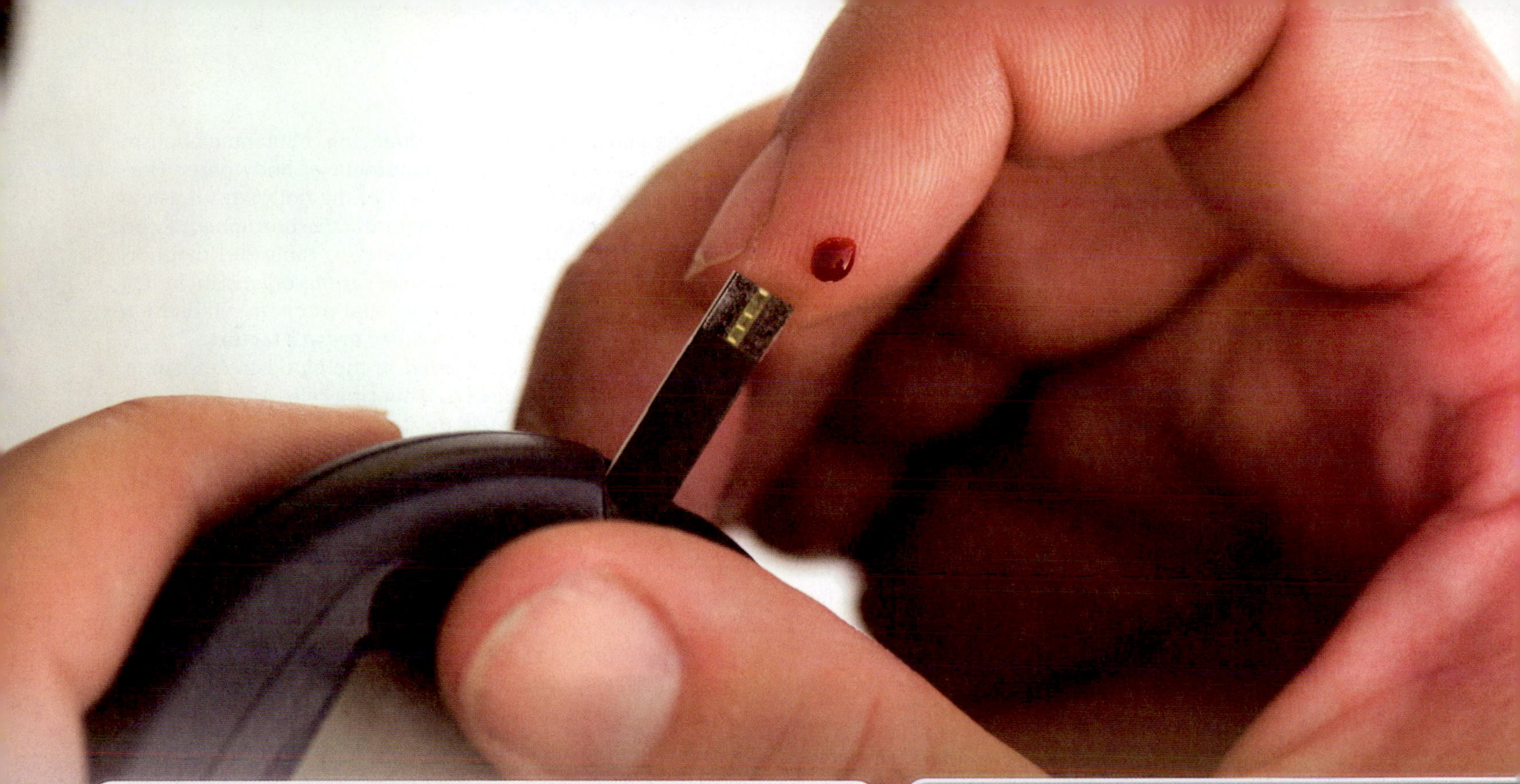

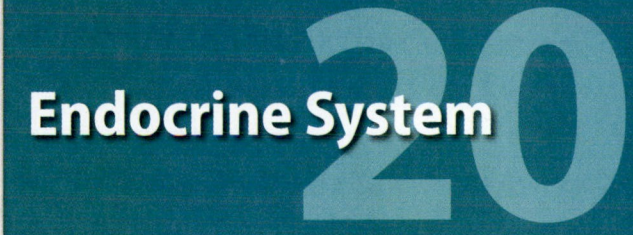

20 Endocrine System

CASE STUDY Grade School Diabetic

Hank is only in the second grade, but if you were around him for very long you would probably notice that he has some extra responsibilities compared to most other kids his age. Several times a day, Hank gets a small black kit out of his backpack, loads a plastic-covered needle onto an instrument, places it against his finger, and pushes a button. Once a drop of blood appears, he neatly and efficiently adds it to a strip, which he then inserts into a reader. If the number is too high, Hank knows it is time to go to the school nurse for an injection of insulin. If it is too low, he needs to eat a snack to restore his blood glucose level. Because Hank has type 1 (formerly called juvenile-onset) diabetes, he doesn't remember a time when he didn't have to perform these extra steps throughout his day.

In addition to the development of technology that allows a second-grader to monitor his blood sugar level, scientific advances are increasing the options available for treating diabetes. As we will see in this chapter, insulin is needed for the body to properly utilize glucose. Prior to the development of recombinant DNA technology, which now allows human insulin to be produced in large quantities, insulin was derived from the pancreases of pigs or cows. This required laborious purification, and because the animal insulins were not identical to the human form, sometimes immunological reactions occurred. Another recent advance is the insulin pump, a device a little bigger than a cell phone, which can deliver precise amounts of insulin via a small plastic catheter, more accurately mimicking the pancreas's natural release of the correct amount of insulin needed. According to recent surveys, about 375,000 diabetics in the United States are using an insulin pump.

As you read through the chapter, think about the following questions:

1. What hormones control the level of glucose in the blood?
2. How do feedback mechanisms help control blood glucose levels?
3. The symptoms of low versus high blood sugar (hypoglycemia versus hyperglycemia) can be similar, but severe hypoglycemia is considered to be a more serious emergency. Why would this be the case?

CHAPTER OUTLINE

BEFORE YOU BEGIN

Before beginning this chapter, take a few moments to review the following discussions:

Figure 2.22 What is the chemical structure of a steroid?

Section 11.5 How do negative feedback mechanisms function in homeostasis?

Section 17.2 Where is the hypothalamus located, and what are its functions?

20.1 Overview of the Endocrine System

Learning Outcomes

Upon completion of this section, you should be able to

1. Identify the endocrine glands of the human body.
2. Compare the mechanism of action of peptide hormones and steroid hormones.
3. Differentiate between first messengers and second messengers.
4. Define pheromones, and provide an example of pheromone activity in humans.

The **endocrine system** consists of glands and tissues that secrete hormones. **Hormones** are chemicals that affect the behavior of other glands or tissues, which can be located far away from the sites of hormone production. The endocrine system is rivaled only by the nervous system with respect to the degree that both influence the body's other systems. Hormones help maintain homeostasis by influencing cellular metabolism as well as the growth and development of body parts. Hormones are an essential component of the body's response to stress, and some (especially the reproductive hormones) act on the brain to influence behavior. Notice in Table 20.1 that hormones are involved in the function of various organs, including other endocrine organs. Hormones that promote cell division and mitosis are also sometimes called **growth factors.**

Not all hormones are carried in the bloodstream. Some, called *local hormones,* affect neighboring cells. As we shall see, prostaglandins, which mediate pain and inflammation, are good examples of local hormones.

Endocrine glands have no ducts. Instead, they usually secrete their hormones into tissue fluid. From there, hormones diffuse into the bloodstream for distribution throughout the body. Figure 20.1 depicts the locations of the major endocrine glands in the body, and Table 20.1 lists the hormones they release. In contrast, *exocrine glands* secrete their products through ducts. For example, the salivary glands send saliva into the mouth by way of the salivary ducts.

MP3 Endocrine System

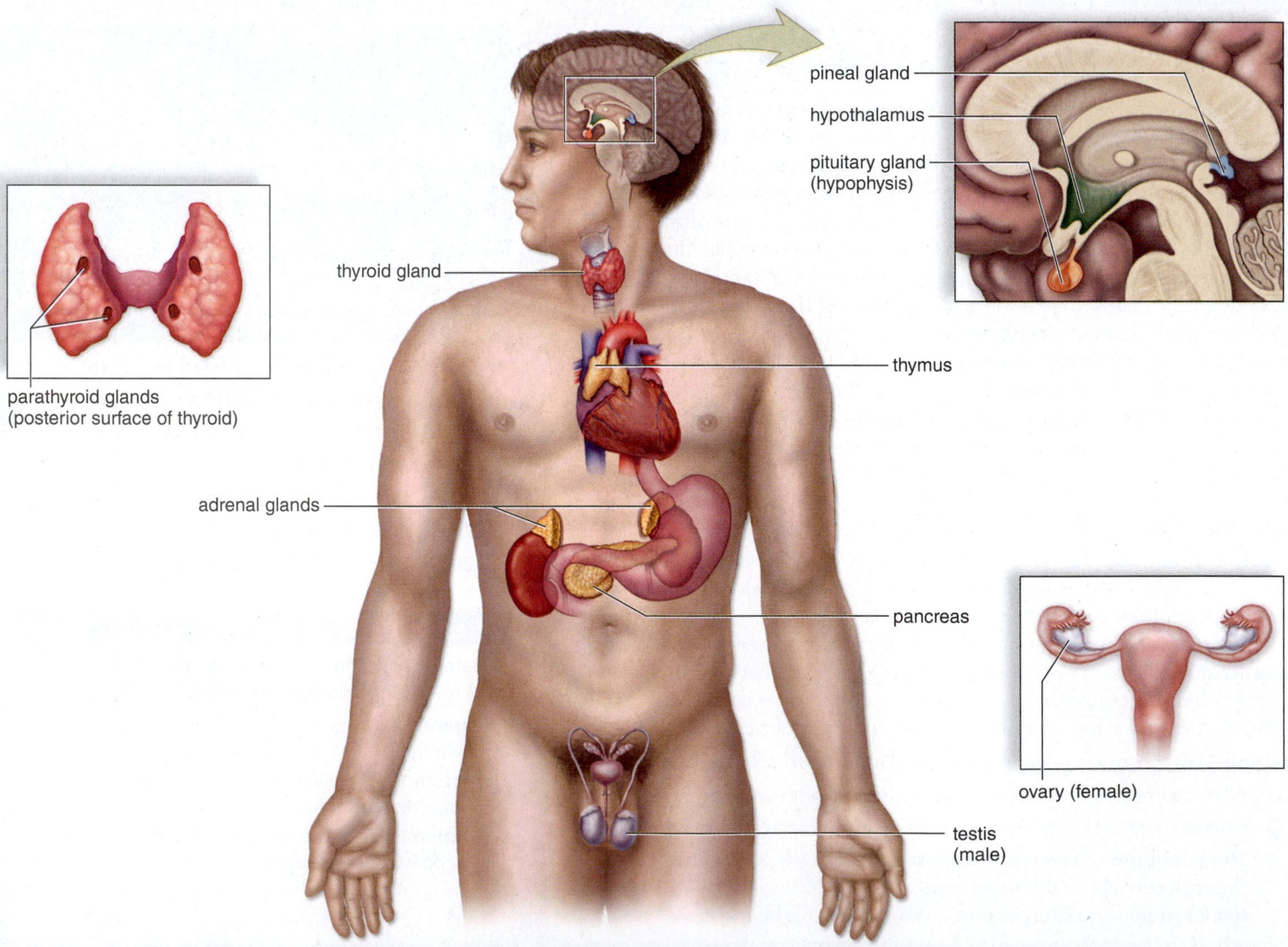
pineal gland
hypothalamus
pituitary gland
(hypophysis)
thyroid gland
parathyroid glands
(posterior surface of thyroid)
thymus
adrenal glands
pancreas
ovary (female)
testis
(male)

Figure 20.1 The endocrine system. Anatomical location of major endocrine glands in the body. Other organs, such as the kidneys and heart, produce hormones. However, hormone production is not a primary function of these organs.

TABLE 20.1 Principal Endocrine Glands and Hormones

Endocrine Gland	Hormone Released	Chemical Class	Target Tissues/Organs	Chief Function(s) of Hormone
Hypothalamus	Hypothalamic-releasing and hypothalamic-inhibiting	Peptide	Anterior pituitary	Regulate anterior pituitary hormones
Pituitary gland				
Posterior pituitary	Antidiuretic (ADH)	Peptide	Kidneys	Stimulates water reabsorption by kidneys
	Oxytocin	Peptide	Uterus, mammary glands	Stimulates uterine muscle contraction, release of milk by mammary glands
Anterior pituitary	Thyroid-stimulating (TSH)	Glycoprotein	Thyroid	Stimulates thyroid
	Adrenocorticotropic (ACTH)	Peptide	Adrenal cortex	Stimulates adrenal cortex
	Gonadotropic (FSH, LH)	Glycoprotein	Gonads	Egg and sperm production; sex hormone production
	Prolactin (PRL)	Protein	Mammary glands	Milk production
	Growth (GH)	Protein	Soft tissues, bones	Cell division, protein synthesis, and bone growth
	Melanocyte-stimulating (MSH)	Peptide	Melanocytes in skin	Unknown function in humans; regulates skin color in lower vertebrates
Thyroid	Thyroxine (T_4) and triiodothyronine (T_3)	Iodinated amino acid	All tissues	Increases metabolic rate; regulates growth and development
	Calcitonin	Peptide	Bones, kidneys, intestine	Lowers blood calcium level
Parathyroids	Parathyroid (PTH)	Peptide	Bones, kidneys, intestine	Raises blood calcium level
Adrenal gland				
Adrenal cortex	Glucocorticoids (cortisol)	Steroid	All tissues	Raise blood glucose level; stimulate breakdown of protein
	Mineralocorticoids (aldosterone)	Steroid	Kidneys	Reabsorb sodium and excrete potassium
	Sex hormones	Steroid	Gonads, skin, muscles, bones	Stimulate reproductive organs and bring about sex characteristics
Adrenal medulla	Epinephrine and norepinephrine	Modified amino acid	Cardiac and other muscles	Released in emergency situations; raise blood glucose level
Pancreas	Insulin	Protein	Liver, muscles, adipose tissue	Lowers blood glucose level; promotes formation of glycogen
	Glucagon	Protein	Liver, muscles, adipose tissue	Raises blood glucose level
Gonads				
Testes	Androgens (testosterone)	Steroid	Gonads, skin, muscles, bones	Stimulate male sex characteristics
Ovaries	Estrogens and progesterone	Steroid	Gonads, skin, muscles, bones	Stimulate female sex characteristics
Thymus	Thymosins	Peptide	T lymphocytes	Stimulate production and maturation of T lymphocytes
Pineal gland	Melatonin	Modified amino acid	Brain	Controls circadian and circannual rhythms; possibly involved in maturation of sexual organs

Hormones and Homeostasis

Like the nervous system, the endocrine system is intimately involved in homeostasis. However, while the nervous system is organized to respond rapidly to stimuli by transmitting nerve impulses, hormones secreted by the endocrine system must reach their target organs via the blood, resulting in a slower but often a more prolonged response. The blood concentration of a substance often prompts an endocrine gland to secrete its hormone. For example, the parathyroid glands secrete a hormone when the blood calcium level falls below normal. This hormone causes osteoclasts to slowly release calcium from bone. It takes time for the cells to respond, but the effect is longer lasting.

The production of most hormones is controlled in two ways: by negative feedback, and by the action of other hormones. When controlled by **negative feedback,** an endocrine gland can be sensitive to either the condition it is regulating or the blood level of the hormone it is producing. For example, when the blood glucose level rises, the pancreas produces insulin. Insulin causes the liver to store glucose, and glucose is removed from the blood for liver storage. The stimulus for the production of insulin is, thereby, inhibited, and the pancreas stops producing insulin. Or, the anterior pituitary produces a thyroid-stimulating hormone that causes the thyroid to produce two hormones. When these hormones rise, the anterior pituitary stops producing thyroid-stimulating hormone.

The effect of a hormone also can be controlled by the release of an *antagonistic* hormone. The effect of insulin, for example, is offset by the production of glucagon by the pancreas. Insulin lowers the blood sugar level, while glucagon raises it. Similarly, calcitonin produced by the thyroid gland lowers the blood calcium level, but parathyroid hormone increases blood calcium.

Animation Hormonal Communication

The Action of Hormones

Hormones have a wide range of effects on cells. Some of these effects induce a target cell to increase its uptake of particular substances (e.g., glucose) or ions (e.g., calcium). Other hormones bring about an alteration of the target cell's structure in some way.

Most hormones fall into two basic chemical classes: (1) **peptide hormones** are either peptides, proteins, glycoproteins, or modified amino acids; (2) **steroid hormones** always have the same complex of four carbon rings, but each specific hormone has different side chains. The chemical composition of a hormone also determines how it must be administered as a medication. Peptide hormones, such as insulin, must be administered by injection. If these hormones were taken orally, they would be digested in the stomach and small intestine. Steroid hormones, such as those in birth control pills, can often be taken orally.

Hormones are a type of chemical signal. **Chemical signals** are a means of communication between cells, between body parts, or even between individuals. They typically affect the metabolism of target cells that have receptors to receive them (Fig. 20.2). In the case of peptide hormones, these receptors are usually found on the cell surface (Fig. 20.3), whereas receptors for steroid hormones are usually in the nucleus but sometimes in the cytoplasm (Fig. 20.4). Epinephrine, a typical peptide hormone, binds to a receptor on muscle cells. This leads to a conversion of ATP to cyclic AMP (adenosine monophosphate), or cAMP, inside the cell. In this chemical signaling process, the peptide hormone is called the **first messenger,** and cAMP is called the **second messenger.**

Animation Peptide Hormone Action

The second messenger typically sets in motion an enzymatic pathway sometimes termed an enzyme cascade, because each enzyme, in turn, activates another. Because enzymes work over and over, every step in an enzyme cascade leads to more reactions—the binding of even a single peptide hormone molecule can result in a thousandfold response. In muscle cells, the activation of these cellular signaling pathways by epinephrine leads to the breakdown of glycogen to glucose (see Fig. 20.3).

Animation Second Messengers

To better understand the terms "first messenger" and "second messenger," imagine that the adrenal medulla, which

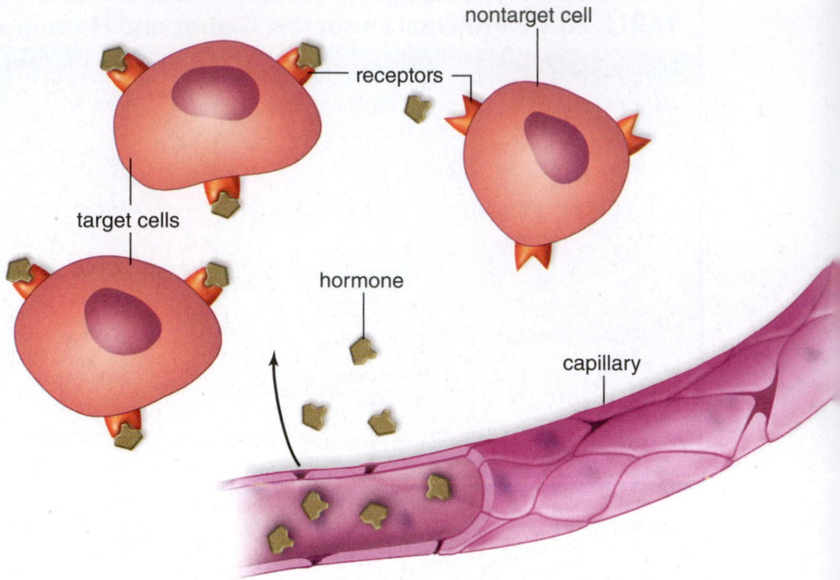
nontarget cell
receptors
target cells
hormone
capillary

Figure 20.2 Target cell concept. Most hormones are distributed by the bloodstream to target cells. Target cells have receptors for the hormone, and the hormone combines with the receptor as a key fits a lock.

produces epinephrine, is the home office that sends out a courier (i.e., the first messenger, epinephrine) to a factory (the cell). The courier doesn't have clearance to enter the factory, so when he arrives there, he tells a supervisor through a screen door that the home office wants the factory to produce a particular product. The supervisor (i.e., cAMP, the second messenger) walks over and flips a switch (the enzymatic pathway), and a product is made.

Steroid hormones act in a somewhat different way. After diffusing through the plasma membrane, these hormones typically bind to receptors in the cytoplasm or nucleus (see Fig. 20.4). Either way, the hormone-receptor complex usually binds to DNA and activates transcription of certain genes. Translation of messenger RNA (mRNA) transcripts results in enzymes and other proteins that can carry out a response to the hormonal signal. Because it takes longer to synthesize new proteins than to activate enzymes already present in cells, steroid hormones generally act more slowly than peptide hormones. Their actions usually last longer, however.

Animation Mechanism of Steroid Hormone Action

Pheromones

Chemical signals that act between individuals of the same species are called **pheromones.** The effects of pheromones are better documented in other animals than in humans. For example, female moths release a sex attractant that is received by male moth antennae even several miles away. Humans do produce a rich supply of airborne chemicals from a variety of areas, including the scalp, oral cavity, armpit (axilla), genital areas, and feet. Studies suggest that women prefer the axillary odors of men who are a different major histocompatibility complex (MHC) type than themselves. Choosing a mate of a different

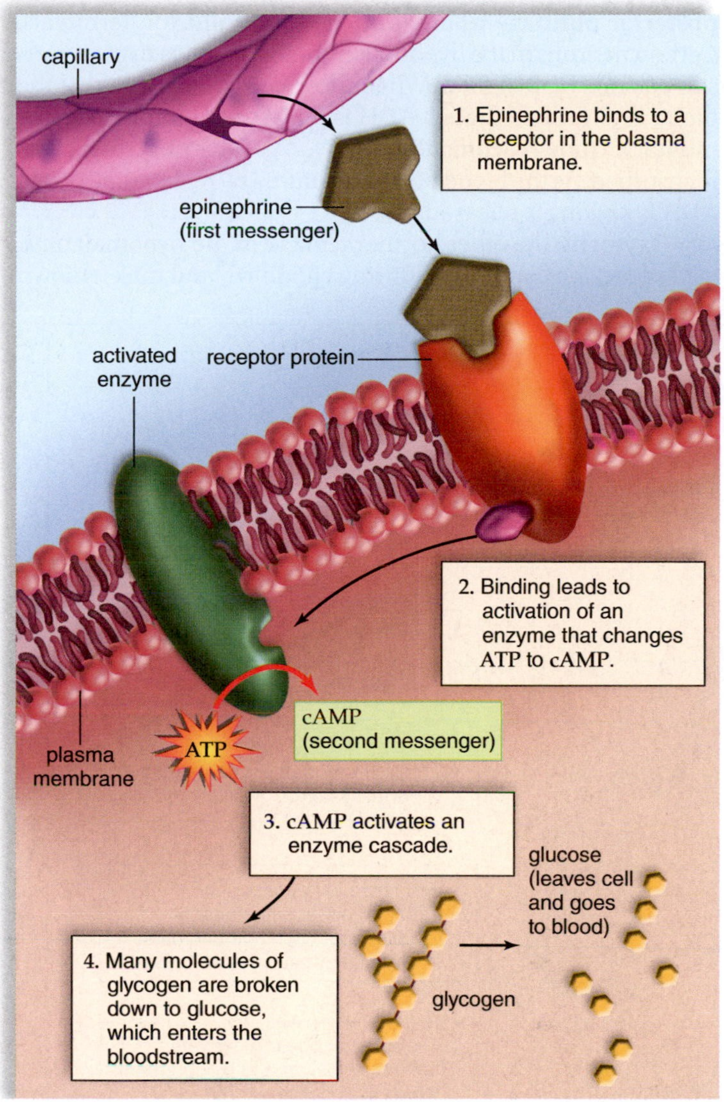
capillary
1. Epinephrine binds to a receptor in the plasma membrane.
epinephrine (first messenger)
activated enzyme
receptor protein
2. Binding leads to activation of an enzyme that changes ATP to cAMP.
plasma membrane
ATP
cAMP (second messenger)
3. cAMP activates an enzyme cascade.
glucose (leaves cell and goes to blood)
4. Many molecules of glycogen are broken down to glucose, which enters the bloodstream.
glycogen

Figure 20.3 Action of epinephrine, a peptide hormone. Epinephrine (first messenger) binds to a receptor in the plasma membrane of a muscle cell. Thereafter, a second messenger (cyclic AMP in this case) forms and activates an enzyme cascade.

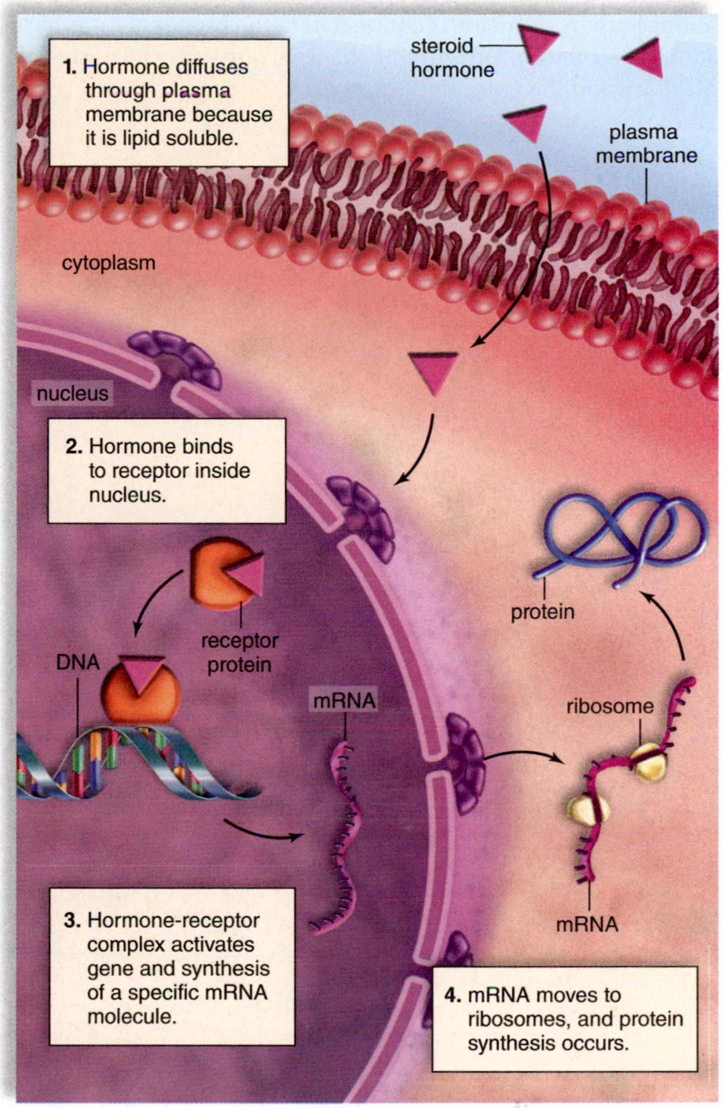
1. Hormone diffuses through plasma membrane because it is lipid soluble.
steroid hormone
plasma membrane
cytoplasm
nucleus
2. Hormone binds to receptor inside nucleus.
receptor protein
DNA
mRNA
protein
ribosome
mRNA
3. Hormone-receptor complex activates gene and synthesis of a specific mRNA molecule.
4. mRNA moves to ribosomes, and protein synthesis occurs.

Figure 20.4 Steroid hormone action. A steroid hormone passes directly through the target cell's plasma membrane before binding to a receptor in the nucleus or cytoplasm. The hormone-receptor complex binds to DNA, and gene expression follows.

MHC type could conceivably improve the immune response of offspring (see section 13.3 for an explanation of MHC functions). Several studies indicate that human pheromones can affect the menstrual cycle. Women who live in the same household often have menstrual cycles in synchrony. Researchers have found that a woman's axillary extract can alter another woman's cycle by a few days.

Video Sex and the Senses

Check Your Progress 20.1

1. List six major human endocrine organs.
2. Explain what it means to say that secretion of a hormone is regulated by negative feedback.
3. Explain why second messengers are needed for peptide hormones.
4. Provide a specific example that suggests human behavior may be influenced by pheromones.

20.2 Hypothalamus and Pituitary Gland

Learning Outcomes

Upon completion of this section, you should be able to

1. Describe the relationship between the hypothalamus and the pituitary gland.
2. List two hormones released by the posterior pituitary and six produced by the anterior pituitary.
3. Recognize which three pituitary hormones act on other endocrine glands.

The **hypothalamus** helps regulate the internal environment in two ways. Through the autonomic system, it modulates heartbeat, blood pressure, hunger and appetite, body temperature, and water balance. It also controls the glandular secretions of the **pituitary gland** (*hypophysis*). The pituitary, a small gland about 1 cm in diameter, is connected to the hypothalamus by a stalklike structure. The pituitary has two portions: the posterior pituitary and the anterior pituitary.

Posterior Pituitary

Neurons in the hypothalamus called *neurosecretory cells* produce the hormones **antidiuretic hormone** (**ADH**) and oxytocin (Fig. 20.5, *left*). These hormones pass through axons into the **posterior pituitary** where they are stored in axon terminals. Certain neurons in the hypothalamus are sensitive to the water-salt balance of the blood. When these cells determine that the blood is too concentrated, ADH is released from the posterior pituitary. Upon reaching the kidneys, ADH causes water to be reabsorbed. As the blood becomes diluted by reabsorbed water, ADH is no longer released.

Oxytocin, the other hormone made in the hypothalamus, causes uterine contraction during childbirth and milk letdown

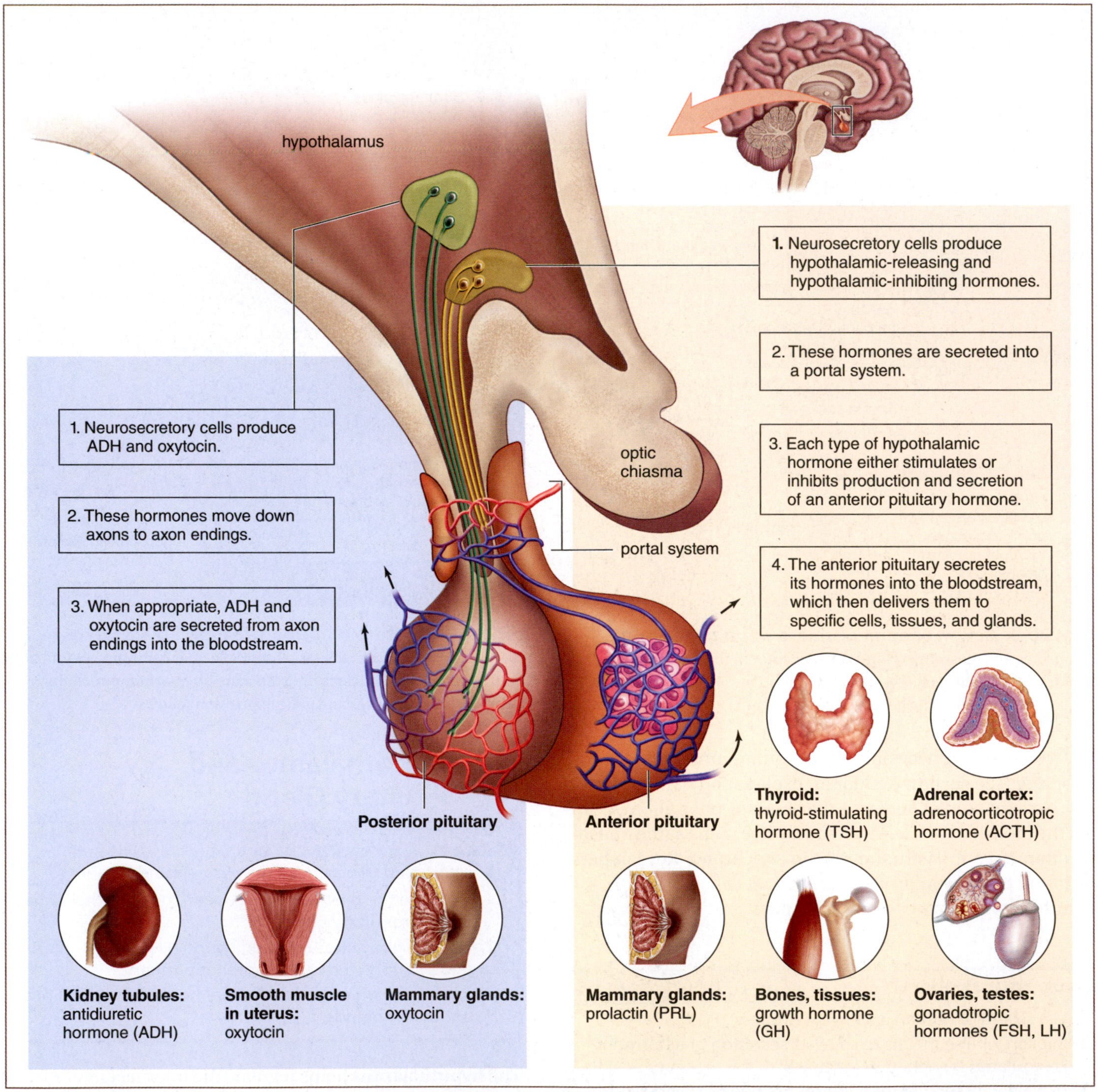
hypothalamus
1. Neurosecretory cells produce ADH and oxytocin.
2. These hormones move down axons to axon endings.
3. When appropriate, ADH and oxytocin are secreted from axon endings into the bloodstream.
1. Neurosecretory cells produce hypothalamic-releasing and hypothalamic-inhibiting hormones.
2. These hormones are secreted into a portal system.
3. Each type of hypothalamic hormone either stimulates or inhibits production and secretion of an anterior pituitary hormone.
4. The anterior pituitary secretes its hormones into the bloodstream, which then delivers them to specific cells, tissues, and glands.
optic chiasma
portal system
Posterior pituitary
Anterior pituitary
Thyroid: thyroid-stimulating hormone (TSH)
Adrenal cortex: adrenocorticotropic hormone (ACTH)
Kidney tubules: antidiuretic hormone (ADH)
Smooth muscle in uterus: oxytocin
Mammary glands: oxytocin
Mammary glands: prolactin (PRL)
Bones, tissues: growth hormone (GH)
Ovaries, testes: gonadotropic hormones (FSH, LH)

Figure 20.5 Hypothalamus and the pituitary. *Left:* The hypothalamus produces two hormones, ADH and oxytocin, which are stored and secreted by the posterior pituitary. *Right:* The hypothalamus controls the secretions of the anterior pituitary, and the anterior pituitary controls the secretions of the thyroid gland, adrenal cortex, and gonads, which are also endocrine glands. These activities of the hypothalamus illustrate that the nervous system exerts control over the endocrine system and is intimately involved in maintaining the constancy of the internal environment.

when a baby is nursing. The more the uterus contracts during labor, the more nerve impulses reach the hypothalamus, causing oxytocin to be released. Similarly, the more a baby suckles, the more oxytocin is released. In both instances, the release of oxytocin from the posterior pituitary is controlled by **positive feedback**—that is, the stimulus continues to bring about an effect that ever increases in intensity.

Anterior Pituitary

A *portal system,* consisting of two capillary networks connected by a vein, lies between the hypothalamus and the anterior pituitary (Fig. 20.5, *right*). The hypothalamus controls the anterior pituitary by producing **hypothalamic-releasing hormones** and, in some instances, **hypothalamic-inhibiting hormones.** For example, one hypothalamic-releasing hormone stimulates the anterior pituitary to secrete a thyroid-stimulating hormone, and a particular hypothalamic-inhibitory hormone prevents the anterior pituitary from secreting prolactin.

Three of the six hormones produced by the **anterior pituitary** have an effect on other glands: **thyroid-stimulating hormone (TSH)** stimulates the thyroid to produce the thyroid hormones; **adrenocorticotropic hormone (ACTH)** stimulates the adrenal cortex to produce cortisol; and **gonadotropic hormones** (FSH and LH) stimulate the gonads—the testes in males and the ovaries in females—to produce gametes and sex hormones. In each instance, the blood level of the last hormone in the hypothalamus-anterior pituitary-target gland control system exerts negative feedback over the secretions of the first two structures:

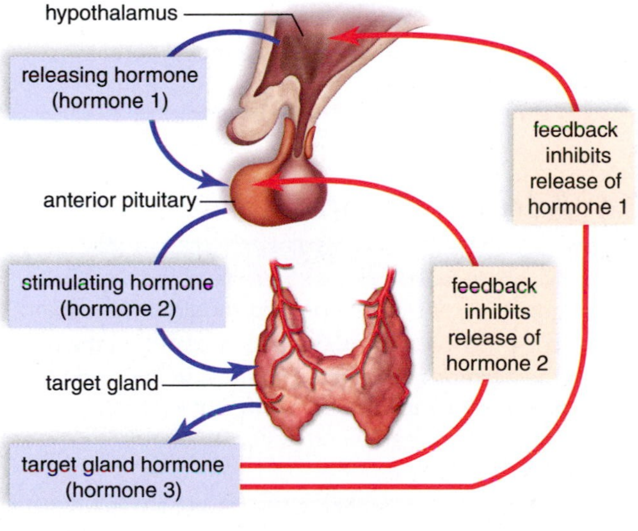
hypothalamus
releasing hormone (hormone 1)
anterior pituitary
stimulating hormone (hormone 2)
target gland
target gland hormone (hormone 3)
feedback inhibits release of hormone 1
feedback inhibits release of hormone 2

The other three hormones produced by the anterior pituitary do not affect other endocrine glands. **Prolactin (PRL)** is produced in quantity only during pregnancy and after childbirth. It causes the mammary glands in the breasts to develop and produce milk and suppress ovulation for a time in the nursing mother. It also plays a role in carbohydrate and fat metabolism.

Melanocyte-stimulating hormone (MSH) causes skin-color changes in many fishes, amphibians, and reptiles that have melanophores, special skin cells that produce color variations. The concentration of this hormone in humans is very low.

Growth hormone (GH) or somatotropic hormone, promotes skeletal and muscular growth. It increases the rate at which amino acids enter cells and causes increased protein synthesis. It also promotes fat metabolism as opposed to glucose metabolism.

The anterior pituitary normally produces higher levels of GH during childhood and adolescence, when most body growth is occurring. The amount of growth hormone produced during childhood is clearly one factor that influences height (see section 20.7), although other factors, such as the genetic predisposition to attain a certain height, are also important.

Check Your Progress 20.2

1. Explain how the posterior and anterior pituitary glands differ in the way that they are controlled by the hypothalamus.
2. List the hormones produced or released by the pituitary gland, and provide a function for each.
3. Describe how hormone secretion can be regulated by positive feedback, and provide an example.
4. Explain why cancer of the pituitary gland, which may affect any of its normal functions, might be difficult to diagnose.

20.3 Thyroid and Parathyroid Glands

Learning Outcomes

Upon completion of this section, you should be able to

1. Identify the anatomical location of the thyroid and parathyroid glands.
2. Distinguish between the functions of T_3, T_4, calcitonin, and parathyroid hormone.

The **thyroid gland** is a large gland located in the neck, where it is attached to the trachea just below the larynx (see Fig. 20.1). The **parathyroid glands** are embedded in the posterior surface of the thyroid gland.

Thyroid Gland

The thyroid gland contains a large number of *follicles,* each a small spherical structure made of thyroid cells that produce **triiodothyronine (T_3)**, which contains three iodine atoms, and **thyroxine (T_4)**, which contains four iodine atoms. To produce T_3 and T_4, the thyroid gland actively acquires iodine from the bloodstream. The concentration of iodine in the thyroid gland is approximately 25 times that found in the blood. The primary source for iodine in our diets is iodized salt.

Animation Mechanism of Thyroxine Action

The thyroid hormones T_3 and T_4 increase the metabolic rate. They do not have a single target organ; instead, they stimulate most cells of the body to metabolize more glucose and utilize more energy. Although the thyroid releases about ten times more T_4 than T_3, T_3 has a much more potent activity. Fortunately, the liver is able to convert most of the T_4 produced by the thyroid gland to T_3. Interestingly, even though T_3 and T_4 are considered peptide hormones because they are synthesized from the amino acid tyrosine, their receptor is actually located inside cells, more like a steroid hormone receptor.

The thyroid gland also produces a hormone called **calcitonin,** which helps control blood calcium levels (Fig. 20.6).

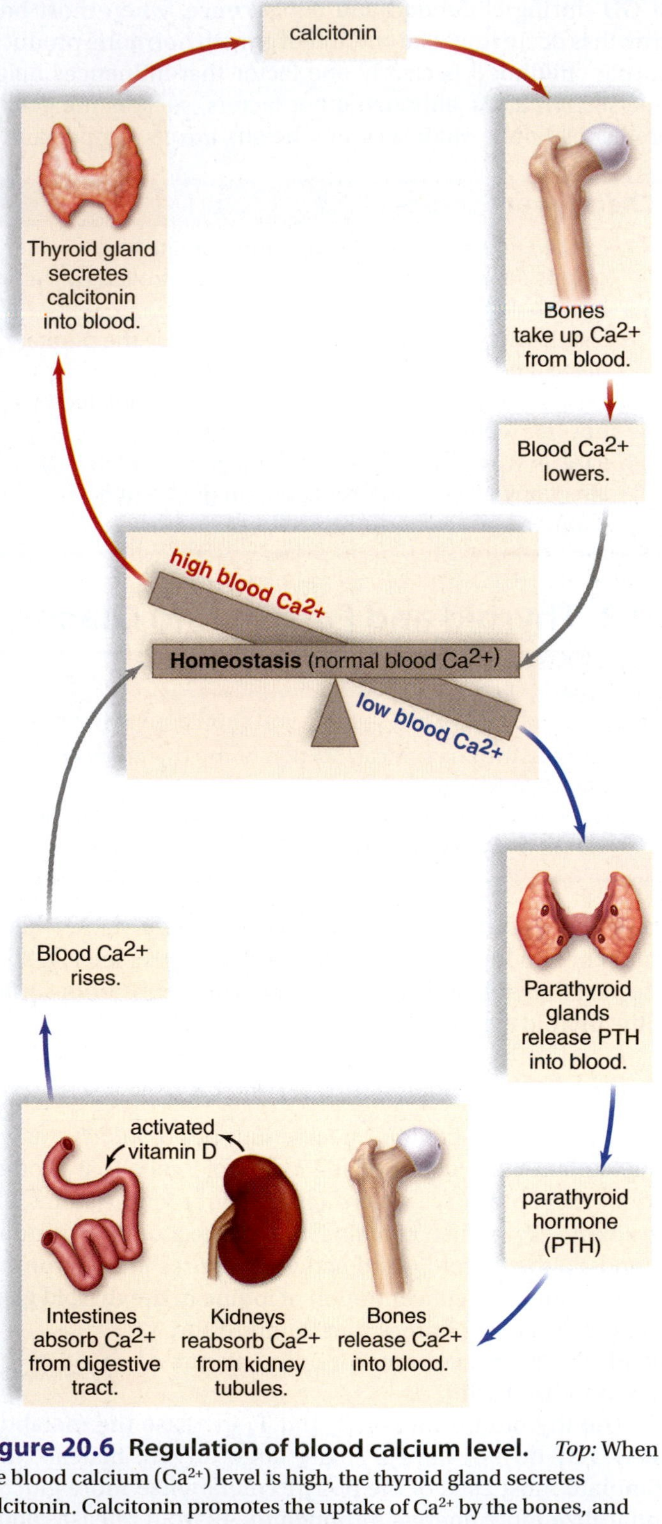
calcitonin
Thyroid gland secretes calcitonin into blood.
Bones take up Ca2+ from blood.
Blood Ca2+ lowers.
high blood Ca2+
Homeostasis (normal blood Ca2+)
low blood Ca2+
Parathyroid glands release PTH into blood.
parathyroid hormone (PTH)
activated vitamin D
Intestines absorb Ca2+ from digestive tract.
Kidneys reabsorb Ca2+ from kidney tubules.
Bones release Ca2+ into blood.
Blood Ca2+ rises.

Figure 20.6 Regulation of blood calcium level. *Top:* When the blood calcium (Ca^{2+}) level is high, the thyroid gland secretes calcitonin. Calcitonin promotes the uptake of Ca^{2+} by the bones, and therefore the blood Ca^{2+} level returns to normal. *Bottom:* When the blood Ca^{2+} level is low, the parathyroid glands release parathyroid hormone (PTH). PTH causes the bones to release Ca^{2+} and the kidneys to reabsorb Ca^{2+} and activate vitamin D. Thereafter, the intestines absorb Ca^{2+}. Therefore, the blood Ca^{2+} level returns to normal.

Calcium (Ca^{2+}) plays a significant role in many processes, including nerve conduction, muscle contraction, and blood clotting. The thyroid gland secretes calcitonin when the blood calcium level rises. The primary effect of calcitonin is to bring about the deposition of calcium in the bones. It does this by temporarily reducing the activity and number of osteoclasts, the cells that break down bone (see section 19.1). When the blood calcium level decreases to normal, the thyroid stops releasing calcitonin.

Parathyroid Glands

Many years ago, the four parathyroid glands were sometimes mistakenly removed during thyroid surgery because they are so small. **Parathyroid hormone** (**PTH**), the hormone produced by the parathyroid glands, causes the blood phosphate (HPO_4^{2-}) level to decrease and the blood calcium level to increase.

PTH promotes the activity of osteoclasts and the release of calcium from the bones. PTH also promotes the reabsorption of calcium by the kidneys, where it helps activate vitamin D. Vitamin D, in turn, stimulates the absorption of calcium from the intestine. These effects bring the blood calcium level back to the normal range. As soon as blood calcium is back to normal, the parathyroid glands no longer secrete PTH.

MP3 Calcium Homeostasis

Check Your Progress 20.3

1. Explain how the amount of T_3 and T_4 secreted by the thyroid gland is controlled.
2. Describe the symptoms that might be expected in a person with a thyroid tumor that was producing excess calcitonin.
3. Predict the expected effect(s) of removal of the parathyroid glands.

20.4 Adrenal Glands

Learning Outcomes

Upon completion of this section, you should be able to

1. Describe the location of the adrenal glands in the body.
2. List the hormones produced by the adrenal medulla and adrenal cortex and provide a function for each.
3. Distinguish between mineralocorticoid and glucocorticoid hormones.

The **adrenal glands** sit atop the kidneys (see Fig. 20.1). Each adrenal gland consists of an inner portion called the **adrenal medulla** and an outer portion called the **adrenal cortex.** These portions, like the anterior pituitary and the posterior pituitary, are two functionally distinct endocrine glands. The adrenal medulla is under nervous control, and a portion of the adrenal cortex is under the control of adrenocorticotropic hormone (ACTH), an anterior pituitary hormone. Stress of all types, including emotional and physical trauma, prompts the hypothalamus to stimulate a portion of the adrenal glands (Fig. 20.7).

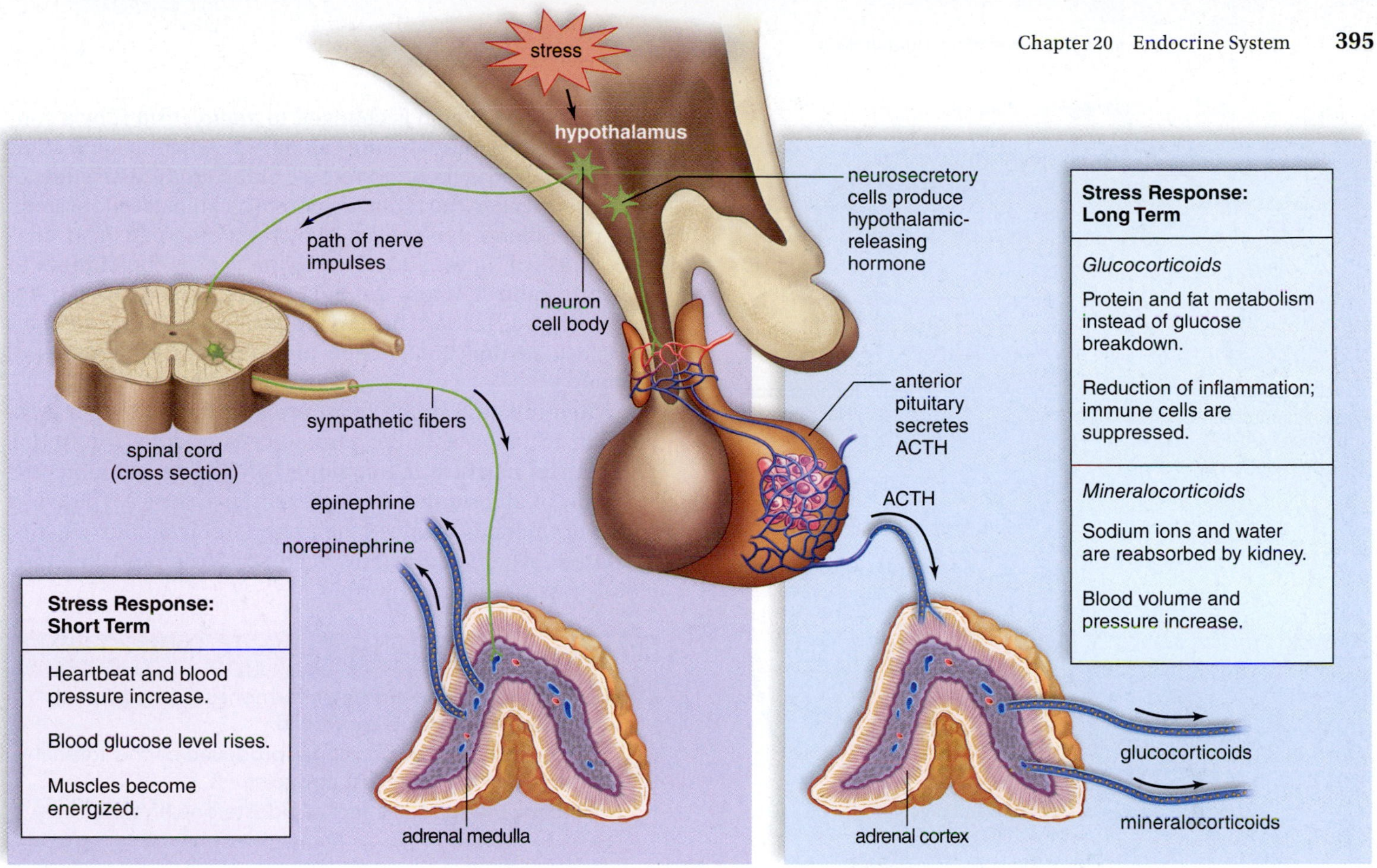
stress
hypothalamus
path of nerve impulses
neuron cell body
neurosecretory cells produce hypothalamic-releasing hormone
spinal cord (cross section)
sympathetic fibers
anterior pituitary secretes ACTH
ACTH
epinephrine
norepinephrine
Stress Response: Short Term
Heartbeat and blood pressure increase.
Blood glucose level rises.
Muscles become energized.
adrenal medulla
adrenal cortex
Stress Response: Long Term
Glucocorticoids
Protein and fat metabolism instead of glucose breakdown.
Reduction of inflammation; immune cells are suppressed.
Mineralocorticoids
Sodium ions and water are reabsorbed by kidney.
Blood volume and pressure increase.
glucocorticoids
mineralocorticoids

Figure 20.7 The stress response. Both the adrenal cortex and the adrenal medulla are under the control of the hypothalamus when they help us respond to stress. *Left:* Nervous stimulation causes the adrenal medulla to provide a rapid, but short-term, stress response. *Right:* The adrenal cortex provides a slower, but long-term, stress response. ACTH causes the adrenal cortex to release glucocorticoids. Independently, the adrenal cortex releases mineralocorticoids.

Adrenal Medulla

The hypothalamus initiates nerve impulses that travel by way of the brain stem, spinal cord, and sympathetic nerve fibers to the adrenal medulla, which then secretes its hormones.

Epinephrine (adrenaline) and **norepinephrine** (noradrenaline) produced by the adrenal medulla rapidly bring about all the body changes that occur when an individual reacts to an emergency situation in a fight-or-flight manner (Fig. 20.7, *left*). The effects of these hormones provide a short-term response to stress.

Adrenal Cortex

The hormones produced by the adrenal cortex provide a longer-term response to stress (Fig. 20.7, *right*). The two major types of hormones produced by the adrenal cortex are the mineralocorticoids and the glucocorticoids. **Mineralocorticoids** regulate salt and water balance, leading to increases in blood volume and blood pressure. **Glucocorticoids** regulate carbohydrate, protein, and fat metabolism, leading to an increase in the blood glucose level. The adrenal cortex also secretes small amounts of male and female sex hormones in both sexes.

Glucocorticoids

ACTH stimulates the portion of the adrenal cortex that secretes the glucocorticoids, of which **cortisol** is the most significant. Cortisol raises the blood glucose level in at least two ways: (1) it promotes the breakdown of muscle proteins to amino acids, which are taken up from the bloodstream and converted into glucose by the liver; (2) Cortisol promotes the metabolism of fatty acids, rather than carbohydrates, and this spares glucose. The rise in blood glucose is beneficial to a person under stress because glucose is the preferred energy source for neurons.

Cortisol also counteracts the inflammatory response that leads to the pain and swelling of joints in arthritis and bursitis. *Cortisone,* the medication often administered for inflammation of joints, is also a glucocorticoid. Very high levels of glucocorticoids in the blood can suppress the body's defense system, including the inflammatory response that occurs at infection sites. This can make a person more susceptible to injury and infection.

Animation Action of Glucocorticoid

Mineralocorticoids

Aldosterone is the most important of the mineralocorticoids. Aldosterone primarily targets the kidney, where it promotes renal absorption of sodium (Na^+) and renal excretion of potassium (K^+), and thereby helps regulate blood volume and blood pressure (Fig. 20.8).

The secretion of mineralocorticoids is not controlled by the anterior pituitary. When the blood sodium level and, therefore, blood pressure are low, the kidneys secrete **renin.** Renin is an enzyme that converts the plasma protein *angiotensinogen*

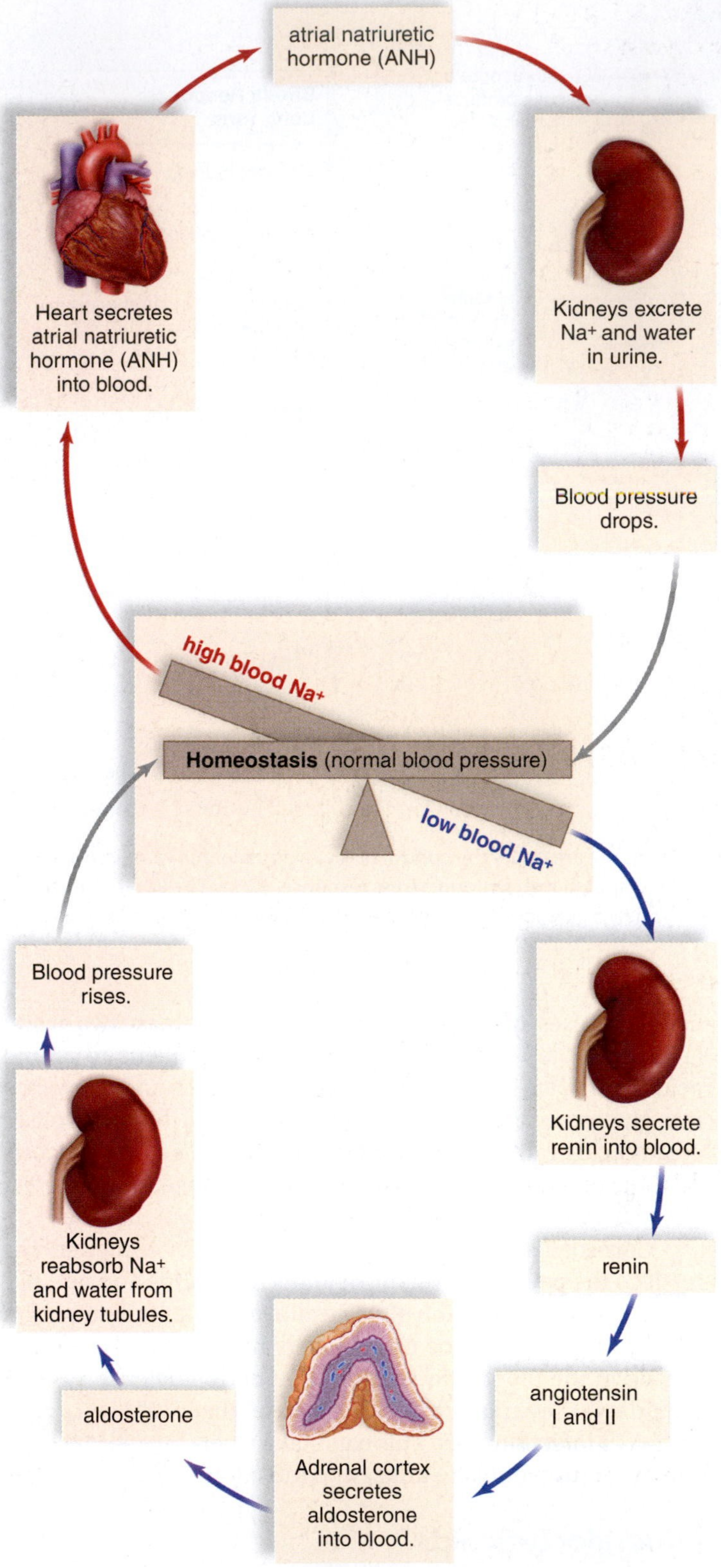

atrial natriuretic hormone (ANH)
Heart secretes atrial natriuretic hormone (ANH) into blood.
Kidneys excrete Na+ and water in urine.
Blood pressure drops.
high blood Na+
Homeostasis (normal blood pressure)
low blood Na+
Blood pressure rises.
Kidneys secrete renin into blood.
Kidneys reabsorb Na+ and water from kidney tubules.
renin
angiotensin I and II
aldosterone
Adrenal cortex secretes aldosterone into blood.

Figure 20.8 Regulation of blood pressure and volume. *Top:* When the blood Na^+ is high, high blood volume causes the heart to secrete atrial natriuretic hormone (ANH). ANH causes the kidneys to excrete Na^+, and water follows. The blood volume and pressure then return to normal. *Bottom:* When the blood sodium (Na^+) level is low, low blood pressure causes the kidneys to secrete renin. Renin leads to the secretion of aldosterone from the adrenal cortex. Aldosterone causes the kidneys to reabsorb Na^+, and water follows, so that blood volume and pressure return to normal.

to *angiotensin I,* which is changed to *angiotensin II* by a converting enzyme found in lung capillaries. Angiotensin II stimulates the adrenal cortex to release aldosterone. The effect of this system, called the renin-angiotensin-aldosterone system, is to raise blood pressure in two ways: angiotensin II constricts the arterioles, and aldosterone causes the kidneys to reabsorb sodium. When the blood sodium level rises, water is reabsorbed, in part because the hypothalamus secretes ADH (see section 20.2). Then blood pressure increases to normal.

When the atria of the heart are stretched due to a great increase in blood volume, cardiac cells release a hormone called **atrial natriuretic hormone** (**ANH**) which inhibits the secretion of aldosterone from the adrenal cortex. The effect of this hormone is to cause the excretion of sodium—that is, *natriuresis.* When sodium is excreted, so is water, and therefore, blood pressure lowers to normal.

Check Your Progress 20.4

1. Explain how the secretion of hormones by the adrenal medulla and cortex is controlled.
2. Summarize how the hormones produced by the adrenal glands help the body respond to stress.
3. Describe the specific effect of aldosterone on the kidney.

20.5 Pancreas

Learning Outcomes

Upon completion of this section, you should be able to

1. Explain how the pancreas has both endocrine and exocrine functions.
2. Describe the functions of insulin, glucagon, and somatostatin.

The **pancreas** is a long organ that lies in the abdomen between the kidneys and near the duodenum of the small intestine (see Fig. 20.1). It is composed of two types of tissue. Exocrine tissue produces and secretes digestive juices by way of ducts to the small intestine. The **pancreatic islets** (islets of Langerhans) are clusters of at least three types of endocrine cells: (1) *alpha cells* that produce glucagon, (2) *beta cells* that produce insulin, and (3) *delta cells* that produce somatostatin. Insulin and glucagon are important in regulating the blood glucose level (Fig. 20.9).

Insulin is secreted when the blood glucose level is high, which usually occurs just after eating. Insulin stimulates the uptake of glucose by cells, especially liver cells, muscle cells, and adipose tissue cells. In liver and muscle cells, glucose is then stored as glycogen. In muscle cells, the breakdown of glucose supplies energy for protein metabolism, and in fat cells, the breakdown of glucose supplies glycerol for the formation of fat. In these various ways, insulin lowers the blood glucose level.

Glucagon is secreted from the pancreas, usually before eating, when the blood glucose level is low. The major target tissues of glucagon are the liver and adipose tissue. Glucagon stimulates the liver to break down glycogen to glucose. Fat and

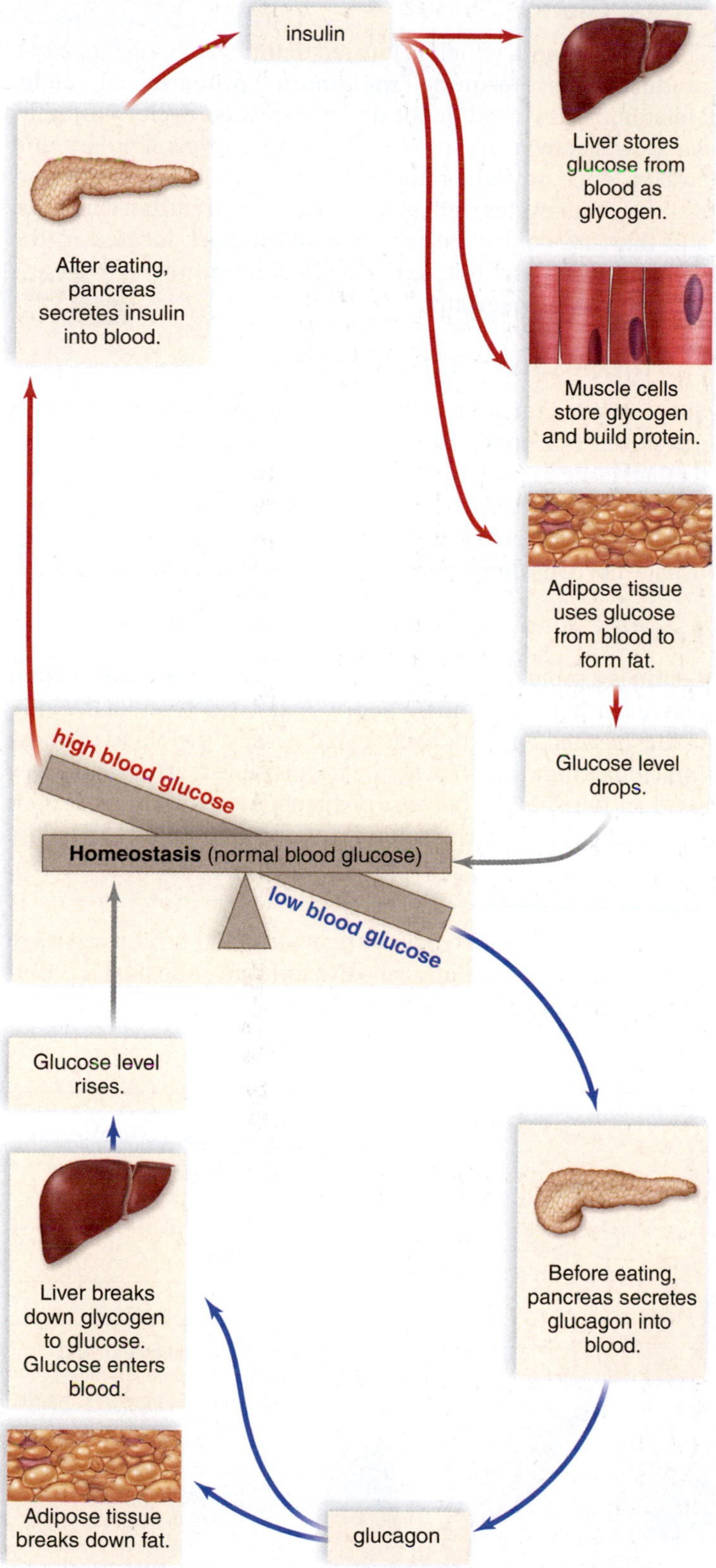
insulin
After eating, pancreas secretes insulin into blood.
Liver stores glucose from blood as glycogen.
Muscle cells store glycogen and build protein.
Adipose tissue uses glucose from blood to form fat.
Glucose level drops.
high blood glucose
Homeostasis (normal blood glucose)
low blood glucose
Glucose level rises.
Liver breaks down glycogen to glucose. Glucose enters blood.
Before eating, pancreas secretes glucagon into blood.
Adipose tissue breaks down fat.
glucagon

Figure 20.9 Regulation of blood glucose level. *Top:* When the blood glucose level is high, the pancreas secretes insulin. Insulin promotes the storage of glucose as glycogen and the synthesis of proteins and fats (as opposed to their use as energy sources). Therefore, insulin lowers the blood glucose level to normal. *Bottom:* When the blood glucose level is low, the pancreas secretes glucagon. Glucagon acts opposite to insulin. Therefore, glucagon raises the blood glucose level to normal.

protein are also used as energy sources by the liver, thus sparing glucose. Adipose tissue cells break down fat to glycerol and fatty acids. The liver takes these up and uses them as substrates for glucose formation. In these various ways, glucagon raises the blood glucose level.

Somatostatin is also known as a growth hormone inhibiting hormone. Besides the pancreas, somatostatin is also produced by cells in the stomach and small intestine. Its main effects are to inhibit the release of growth hormone by the anterior pituitary, as well as to suppress the release of various hormones produced by the digestive system, including insulin and glucagon. The overall effect of somatostatin on the digestive tract is to generally decrease the absorption of nutrients.

Check Your Progress 20.5

1. Distinguish between the endocrine and exocrine functions of the pancreas.
2. Describe how pancreatic hormones interact to regulate blood glucose levels.
3. Review the effect of somatostatin on other hormones, such as insulin and glucagon.

20.6 Other Endocrine Glands

Learning Outcomes

Upon completion of this section, you should be able to

1. Discuss the role of the testes and ovaries in the development of male and female secondary sex characteristics.
2. Describe the functions of thymosin, melatonin, leptin, and prostaglandins.
3. Define and list two examples of growth factors.

The **gonads** are the testes in males and the ovaries in females. The gonads are endocrine glands. Other lesser known glands and certain tissues also produce hormones.

Testes and Ovaries

The testes are located in the scrotum, and the ovaries are located in the pelvic cavity. Under the influence of the gonadotropic hormones secreted by the anterior pituitary, the **testes** produce sperm and **androgens** (e.g., **testosterone**), which are the male sex hormones. Under the influence of the gonadotropic hormones, the **ovaries** produce eggs as well as estrogen and progesterone, the female sex hormones. The hypothalamus and the pituitary gland control the hormonal secretions of these organs in a similar manner as previously described for the thyroid gland.

Greatly increased testosterone secretion at the time of puberty stimulates the growth of the penis and the testes. Testosterone also brings about and maintains the male secondary sex characteristics that develop during puberty, such as the growth of a beard, axillary (underarm) hair, and pubic hair. Testosterone also prompts the larynx and the vocal cords to enlarge, causing the voice to change. It is partially responsible for the muscular strength of males. This is the reason some athletes take supplemental amounts of illegal **anabolic steroids,** which are either testosterone or related chemicals. The side effects of taking anabolic steroids are listed in Figure 20.10. Testosterone also stimulates oil and sweat glands in the skin. Therefore, it is largely responsible for acne and body odor. Another side effect of testosterone is baldness. Genes for baldness are probably inherited by both sexes, but baldness occurs more often in males because of the presence of testosterone.

Estrogen and **progesterone** have many effects on the body. In particular, estrogen secreted at the time of puberty stimulates the growth of the uterus and the vagina. Estrogen is necessary for egg maturation and is largely responsible for the secondary sex characteristics in females, including female body hair and fat distribution. In general, females have a more rounded appearance than males because of a greater accumulation of fat beneath the skin. Also, the pelvic girdle is wider in females than in males, resulting in a larger pelvic cavity. Both estrogen and progesterone are required for breast development and regulation of the uterine cycle, which includes monthly menstruation (see section 21.3).

Thymus

The **thymus** is a lobular organ that lies just beneath the sternum (see Fig. 20.1). This organ reaches its largest size and is most active during childhood. With aging, the organ gets smaller and becomes fatty. Lymphocytes that originate in the bone marrow and then pass through the thymus differentiate into T lymphocytes (T cells). The lobules of the thymus are lined by epithelial cells that secrete hormones called *thymosins.* These hormones aid in the differentiation of T cells packed inside the lobules.

Pineal Gland

The **pineal gland,** which is located in the brain (see Fig. 20.1), produces the hormone **melatonin,** primarily at night. Melatonin is involved in our daily sleep-wake cycle. Normally, we grow sleepy at night when melatonin levels increase and awaken once daylight returns and melatonin levels are low. Daily 24-hour cycles such as this are called **circadian rhythms,** and they are controlled by a biological clock located in the hypothalamus, as discussed in the Health feature, "Melatonin," on page 400.

Video Winter Moods

Hormones from Other Tissues

Some organs that are usually not considered endocrine glands can secrete hormones. We have already mentioned in this chapter that the heart produces atrial natriuretic hormone. Figure 14.7 showed that the stomach and small intestine produce peptide hormones that regulate digestive secretions. Other tissues that are usually not considered endocrine glands also secrete hormones.

Leptin

Leptin is a protein hormone produced by adipose tissue. Leptin acts on the hypothalamus, where it signals satiety—that the individual has had enough to eat. Paradoxically, the blood of obese individuals may be rich in leptin. It is possible that the leptin they produce is ineffective because of a genetic mutation, or else their hypothalamic cells lack a suitable number of receptors for leptin.

Growth Factors

A number of different types of organs and cells produce **growth factors,** which stimulate cell division and mitosis. Like hormones, they act on cell types with specific receptors to receive

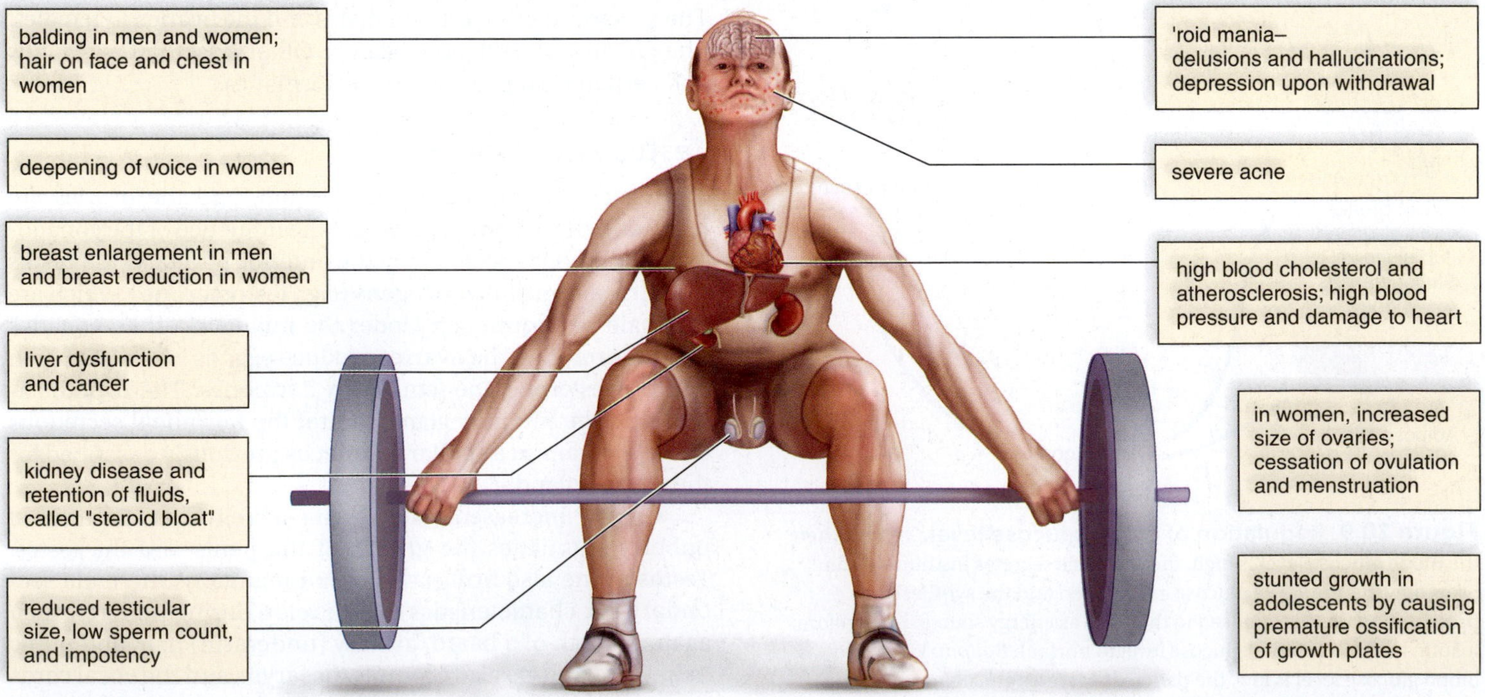
balding in men and women; hair on face and chest in women
deepening of voice in women
breast enlargement in men and breast reduction in women
liver dysfunction and cancer
kidney disease and retention of fluids, called "steroid bloat"
reduced testicular size, low sperm count, and impotency
'roid mania–delusions and hallucinations; depression upon withdrawal
severe acne
high blood cholesterol and atherosclerosis; high blood pressure and damage to heart
in women, increased size of ovaries; cessation of ovulation and menstruation
stunted growth in adolescents by causing premature ossification of growth plates

Figure 20.10 The side effects of anabolic steroid use.

them. Some, such as lymphokines, are released into the blood. Others diffuse to nearby cells. Growth factors of particular interest include:

Granulocyte-macrophage colony-stimulating factor is secreted by many different tissues. It causes bone marrow stem cells to form various types of white blood cells.

Platelet-derived growth factor is released from platelets and from many other cell types. It helps in wound healing and causes an increase in the number of fibroblasts, smooth muscle cells, and certain cells of the nervous system.

Epidermal growth factor and nerve growth factor stimulate the cells indicated by their names, as well as many others. These growth factors are also important in wound healing.

Prostaglandins

Prostaglandins are potent chemical signals produced within cells from *arachidonate,* a fatty acid. Prostaglandins are generally not distributed in the blood. Instead, they act locally, quite close to where they were produced. In the uterus, prostaglandins cause muscles to contract. Therefore, they are implicated in the pain and discomfort of menstruation in some women. Also, prostaglandins mediate the effects of *pyrogens,* chemicals that are believed to reset the temperature regulatory center in the brain. Aspirin reduces body temperature and controls pain because it inhibits the synthesis of prostaglandins.

Check Your Progress 20.6

1. Summarize the role of testosterone and estrogen in the body.
2. List some of the undesirable side effects of anabolic steroid use.
3. Describe the normal function of leptin.
4. Explain why prostaglandins are considered to be local hormones.

20.7 Disorders of the Endocrine System

Learning Outcomes

Upon completion of this section, you should be able to

1. Identify the cause of each of the following conditions: diabetes insipidus, pituitary dwarfism, gigantism, acromegaly, Cushing syndrome, and Addison disease.
2. Discuss the specific causes and likely outcomes of hypothyroidism and hyperthyroidism.
3. Compare and contrast type 1 versus type 2 diabetes mellitus.

The endocrine glands play a major role in regulating the development and function of many body systems. Therefore, an increase or decrease in the production of most hormones can cause significant disease. Increased production of a hormone is often due to cancer affecting the endocrine gland, while decreased production can be due to various conditions that result in destruction of the gland.

Disorders of the Pituitary Gland

Because it controls the secretions of several other endocrine glands, the pituitary has been called the "master gland" of the endocrine system. As such, disorders of the pituitary gland can have dramatic effects on the body. As discussed in section 20.2, the posterior pituitary produces antidiuretic hormone (ADH), which reduces the amount of urine formed by increasing water reabsorption by the collecting ducts of the kidney (see section 16.3). If too little ADH is secreted, a condition known as **diabetes insipidus** (DI) results. Patients with DI are usually very thirsty, produce large volumes of urine, and can become severely dehydrated if the condition is untreated. Another type of DI occurs when the kidneys are unable to respond to ADH produced by the pituitary. Consumption of alcohol also inhibits ADH production, such that the symptoms of a hangover are, in part, caused by dehydration.

The anterior pituitary produces several important hormones, including growth hormone (GH) and adrenocorticotropic hormone (ACTH). If too little GH is produced during childhood, the individual will have **pituitary dwarfism,** characterized by normal proportions but small stature. Conversely, if too much GH is secreted during childhood, the result may be **gigantism** (Fig. 20.11). People with gigantism usually have poor health, primarily because GH has a secondary effect on the blood glucose level, promoting diabetes mellitus (discussed later in this section).

Figure 20.11 Effect of growth hormone on height. Too much growth hormone can lead to gigantism, whereas an insufficient amount results in limited stature and even pituitary dwarfism.

SCIENCE IN YOUR LIFE ▶ HEALTH

Melatonin

The hormone melatonin is now sold as a nutritional supplement. The popular press promotes its use in pill form for sleep, aging, cancer treatment, sexuality, and more. At best, melatonin may have some benefits in certain sleep disorders. But most physicians do not yet recommend it for that use because so little is known about its dosage requirements and possible side effects.

Melatonin is produced by the pineal gland in greatest quantity at night and smallest quantity during the day. Notice in Figure 20A that melatonin's production cycle accompanies our natural sleep-wake cycle. Rhythms with a period of about 24 hours are called circadian ("around the day") rhythms. All circadian rhythms seem to be controlled by an internal biological clock because they are free-running—that is, they have a regular cycle even in the absence of environmental cues. In scientific experiments, humans have lived in underground bunkers where they never see the light of day. In a few people, the sleep-wake cycle drifts badly, but in most, the daily activity schedule is just about 25 hours.

An individual's internal biological clock is reset each day by the environmental day-night cycle. Characteristically, biological clocks that control circadian rhythms are reset by environmental cues, or else they drift out of phase with the environmental day-night cycle.

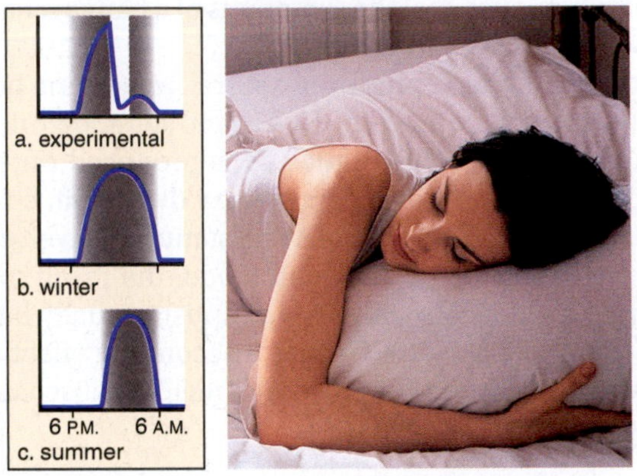
a. experimental
b. winter
6 P.M.
6 A.M.
c. summer

Figure 20A Melatonin production.
Melatonin production is greatest at night when we are sleeping. Light suppresses melatonin production (**a**) so its duration is longer in the winter (**b**) than in the summer (**c**).

Recent research suggests that our biological clock lies in a cluster of neurons within the hypothalamus called the suprachiasmatic nucleus (SCN). The SCN undergoes spontaneous cyclical changes in activity, and therefore, it can act as a pacemaker for circadian rhythms. Neural connections between the retina and the SCN indicate that reception of light by the eyes most likely resets the SCN and keeps our biological rhythms on a 24-hour cycle. Some people suffer from seasonal affective disorder, or SAD. As the days get darker and darker during the fall and winter, they become depressed, sometimes severely. They find it difficult to keep up because their biological clock has fallen behind without early morning light to reset it. If so, a half-hour dose of simulated daylight from a portable light box first thing in the morning makes them feel operational again. The SCN also controls the secretion of melatonin by the pineal gland, and, in turn, melatonin may quiet the operation of the neurons in the SCN.

Research is still going forward to see if melatonin will be effective for circadian rhythm disorders such as SAD, jet lag, sleep phase problems, recurrent insomnia in the totally blind, and some other less

Age 9
Age 16
Age 33
Age 52

Figure 20.12 Acromegaly. Acromegaly is caused by overproduction of GH in the adult. It is characterized by enlargement of the bones in the face, the hands, and the feet as a person ages.

common disorders. So-called jet lag occurs when you travel across several time zones and your biological clock is out of phase with local time. Jet-lag symptoms gradually disappear as your biological clock adjusts to the environmental signals of the local time zone.

Many young people have a sleep phase problem because their circadian cycle lasts 25 to 26 hours. As they lengthen their day, they get out of sync with normal times for sleep and activity. Medical researchers have speculated that melatonin could be used to adjust biological clocks; however, there is still much concern about possible side effects of melatonin supplementation.

An issue that highlights this concern is the widespread availability of melatonin-laced products like "relaxation brownies," as well as beverages with names like "Slowtivate" and "Marley's Mellow Mood." Although many such products bear label warnings such as "For adults only; not suitable for children," there is very little regulation of who can purchase these products, or even the levels of melatonin and other chemicals that are added.

An increasing number of states are moving to ban these products. In Arizona, after a two-year-old was rushed to a hospital in 2011 after eating part of a brownie and falling into a deep sleep, the state department of health issued a mass recall of the product. Parents in other states have also complained about the type of packaging used for these products, which may use cartoon characters or brightly colored designs that some say are trying to target children in a not-so-subtle way (Fig. 20B).

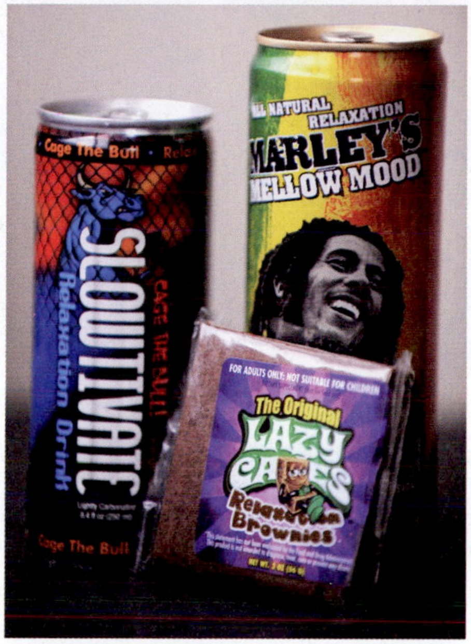
SLOWTIVATE
Relaxation Drink
Cage The Bull
ALL NATURAL RELAXATION
MARLEY'S MELLOW MOOD
FOR ADULTS ONLY; NOT SUITABLE FOR CHILDREN
The Original
LAZY CAKES
Relaxation Brownies

Figure 20B Packaging of products containing melatonin.

Questions to Consider

1. Have you or someone you know ever experienced seasonal affective disorder? If so, do you think it was at least partly psychological, or did you feel it was a "purely" chemical disorder over which you had no control?
2. Are you more of a "morning person" or a "night person"? Can you think of any way that variations in melatonin secretion could account for how different people feel about, for example, getting up early in the morning?
3. Currently melatonin is sold without a prescription in health food stores and drugstores in the United States, but a typical recommended dose is two to three times lower than that found in most "relaxation brownies." Do you see a difference between selling melatonin over the counter in health food stores versus selling melatonin-laced products in convenience stores? Should both be banned? Why or why not?

On occasion, GH is overproduced in adulthood, resulting in a condition known as **acromegaly.** Because long bone growth is no longer possible in adults, only the feet, hands, and face (particularly the chin, nose, and eyebrow ridges) can grow, and these bones become overly large (Fig. 20.12)

Some pituitary tumors produce large amounts of ACTH, which is a common cause of **Cushing syndrome.** The excess ACTH stimulates the adrenal cortex to overproduce cortisol. The increased amount of cortisol causes muscle protein to be metabolized and subcutaneous fat to be deposited in the face and midsection (Fig. 20.13). Depending on the cause and duration of Cushing syndrome, some people may experience more dramatic changes, including masculinization, high blood pressure, and weight gain. Cushing syndrome can also be caused by tumors of the adrenal gland or by taking high or prolonged doses of cortisone or similar drugs.

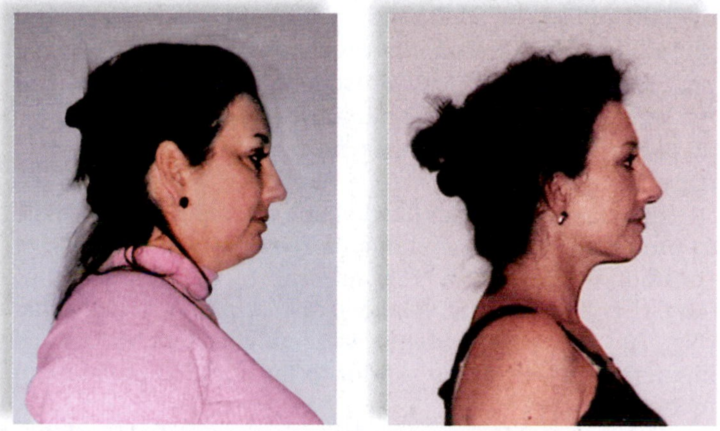

Figure 20.13 Cushing syndrome. This 40-year-old woman was diagnosed with a small tumor in her pituitary gland, which secreted large amounts of ACTH. The high ACTH levels stimulated the adrenal glands to produce excessive amounts of cortisol. *Left:* Patient at the time of surgery to remove her pituitary tumor. *Right:* Patient's appearance one year later.

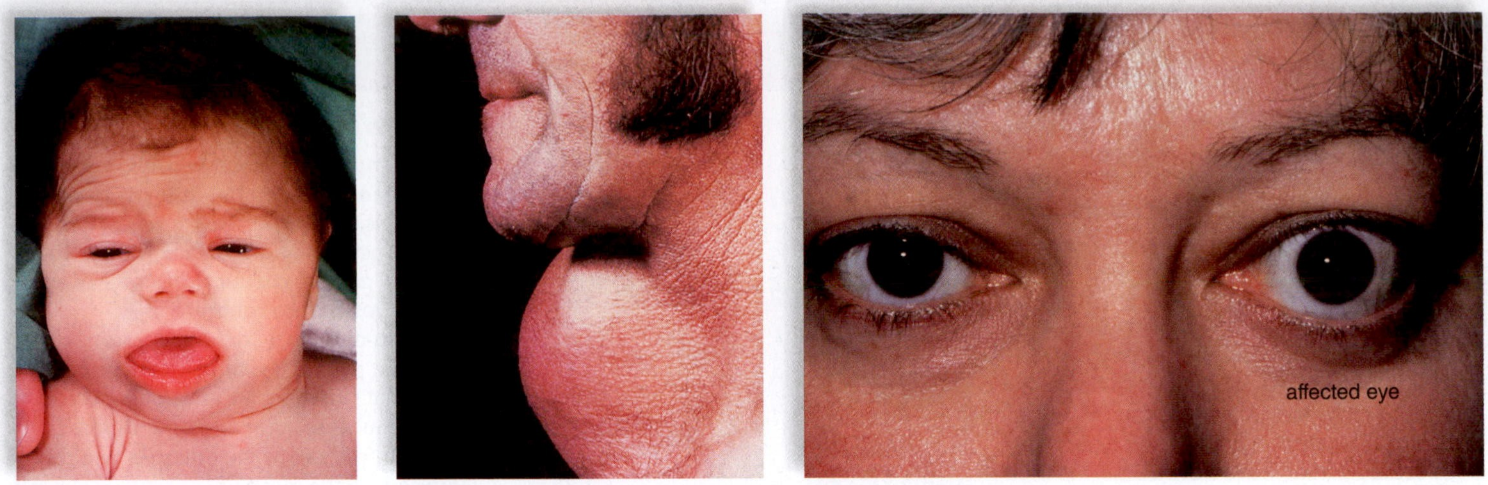
affected eye

a. Congenital hypothyroidism b. Simple goiter c. Exophthalmic goiter

Figure 20.14 Abnormalities of the thyroid gland. **a.** Individuals who develop hypothyroidism during infancy or childhood do not grow and develop as others do. **b.** An enlarged thyroid gland is often caused by a lack of iodine in the diet. Without iodine, the thyroid is unable to produce its hormones, and increased release of TSH from the anterior pituitary stimulation causes the gland to enlarge and form a simple goiter. **c.** In hyperthyroidism, accumulation of fluid behind the eyes can cause the eyes to protrude (exophthalmia). This individual's left eye was affected.

Disorders of the Thyroid, Parathyroid, and Adrenal Glands

If the thyroid gland does not produce enough thyroid hormones for any reason, **hypothyroidism** occurs. Failure of the thyroid to function properly in infancy or childhood results in *congenital hypothyroidism* (Fig. 20.14*a*). Individuals with this condition are short and stocky. Thyroid hormone therapy can initiate growth, but unless treatment is begun relatively quickly, stunted growth and intellectual disability may result. Hypothyroidism can also occur in adults, most often when the immune system produces antibodies that destroy the gland, a condition known as *Hashimoto thyroiditis.* Untreated hypothyroidism in adults may produce a group of clinical symptoms called *myxedema,* which is characterized by lethargy, weight gain, hair loss, constipation, slow heart rate, and thickened or puffy skin. The administration of adequate doses of thyroid hormones restores normal body functions and appearance.

If iodine is lacking in the diet, the thyroid gland is unable to produce sufficient amounts of T_3 and T_4. In response to constant stimulation by TSH released by the anterior pituitary, the thyroid gland enlarges, resulting in a *simple goiter* (Fig. 20.14*b*). The incidence of simple goiters was much higher in the United States prior to the addition of iodine to most salt.

Hyperthyroidism results from the oversecretion of thyroid hormones. In **Graves disease,** antibodies are produced that react with the TSH receptor on thyroid follicular cells, mimicking the effect of TSH and causing the production of too much T_3 and T_4. One typical sign of Graves disease is *exophthalmia,* or excessive protrusion of the eyes (Fig. 20.14*c*). One or both eyes protrude because of the edema in eye socket tissues and swelling of the muscles that move the eyes. The patient with Graves disease usually becomes hyperactive, nervous, and irritable, and may suffer from insomnia, diarrhea, and abnormal heart rhythms. Drugs that inhibit the synthesis or release of thyroid hormones, or their effects on body tissues, are used to manage Graves disease. Hyperthyroidism can also be caused by a thyroid tumor, which is usually detected as a lump during physical examination. Removal or destruction of a portion of the thyroid gland, either by surgery or by administration of radioactive iodine, is often curative.

When insufficient parathyroid hormone production leads to a dramatic drop in the blood calcium level, tetany results. In *tetany,* the body shakes from continuous muscle contraction, brought about by increased excitability of the nerves.

As mentioned previously, an increased production of cortisol by the adrenal glands is known as Cushing syndrome. In contrast, adrenal gland insufficiency is called **Addison disease.** Perhaps the most famous person to suffer from Addison disease was President John F. Kennedy. The most common cause of Addison disease in the United States is destruction of the adrenal cortex by the immune system. Clinical symptoms do not appear until about 90% of both adrenal cortexes have been destroyed. These symptoms are fairly nonspecific and may include weakness, weight loss, abdominal pain, mood disturbances, and hyperpigmentation of the skin (Fig. 20.15). Because the decreased production of mineralocorticoids can affect the balance of sodium and potassium, which in turn can have dramatic effects on the heart, Addison disease can be fatal unless properly treated by replacing the missing hormones.

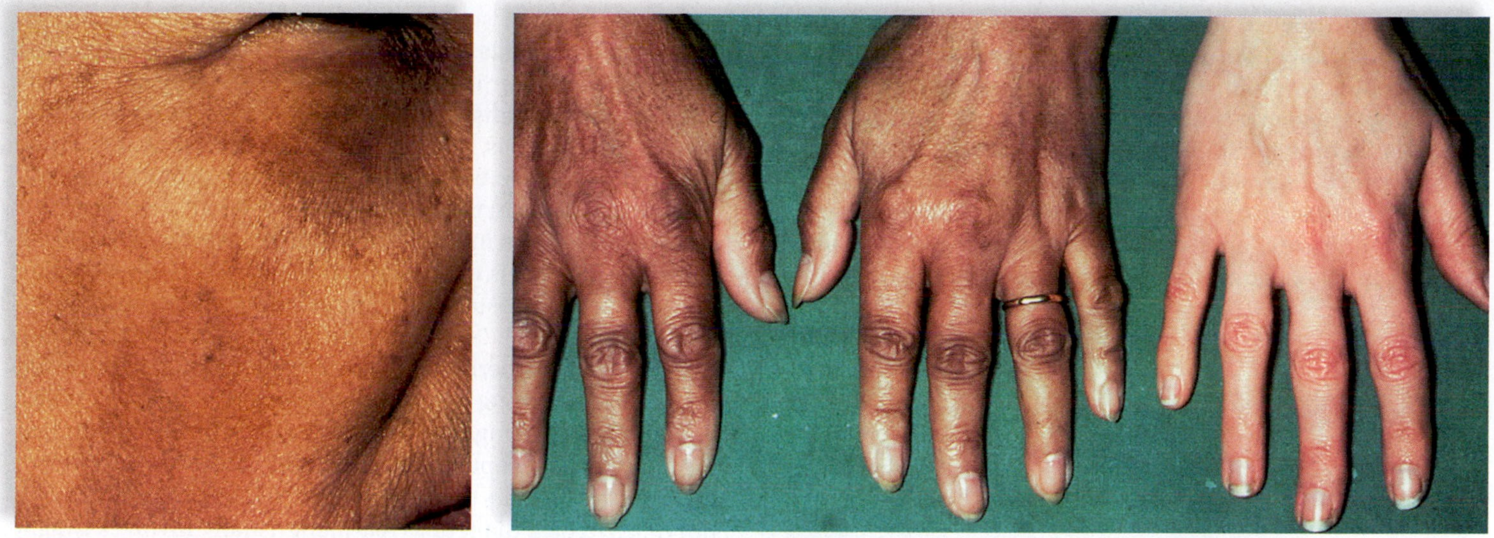

a. b.

Figure 20.15 Addison disease. Addison disease is characterized by a peculiar bronzing of the skin, particularly noticeable in light-skinned individuals. Note the color of (**a**) the face and (**b**) the hands compared to the hand of an individual without the disease.

Diabetes Mellitus

As of 2010, it is estimated that about 25.8 million Americans, or 8.3% of the population, have **diabetes mellitus,** often referred to simply as diabetes, a condition that affects their ability to regulate their glucose metabolism. People with diabetes either do not produce enough insulin (type 1) or cannot properly use the insulin they produce (type 2). In either case, although blood glucose levels rise, cellular famine exists in the midst of plenty. Some of the excess glucose in the blood is excreted into the urine, and water follows, causing the volume of urine to increase. Because the glucose in the blood cannot be used, the body turns to the metabolism of fat, which leads to the buildup of ketones in the blood. In turn, these ketones are metabolized to form various acids, which can build up in the blood (acidosis) and lead to coma and death.

The glucose tolerance test is often used to diagnose diabetes mellitus. After a person ingests a known amount of glucose, the blood glucose concentration is measured at intervals. In a diabetic, the blood glucose level usually rises greatly and remains elevated for hours (Fig. 20.16). In the meantime, glucose appears in the urine. In a nondiabetic, the blood glucose level rises somewhat, and then returns to normal after about two hours.

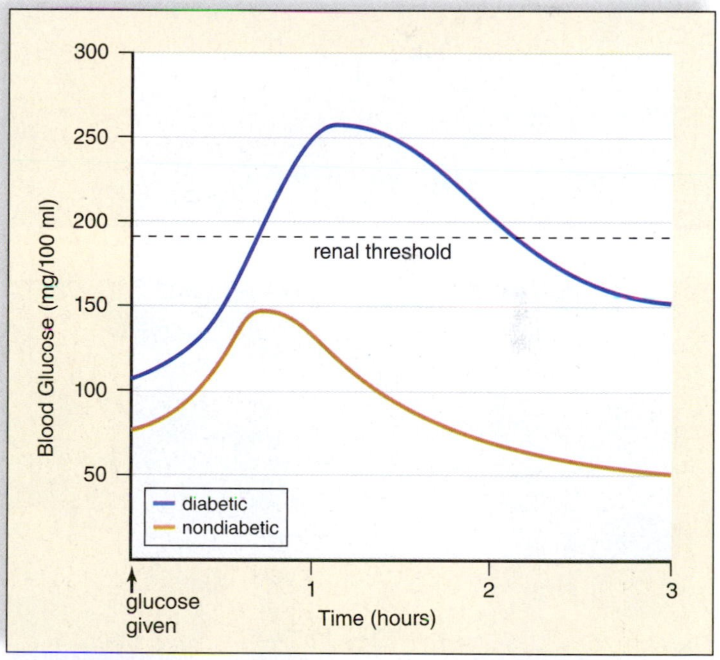

Blood Glucose (mg/100 ml)
300
250
200
150
100
50
renal threshold
diabetic
nondiabetic
glucose given
1
2
3
Time (hours)

Figure 20.16 Glucose tolerance test. Following the administration of 100 g of glucose, the blood glucose level rises dramatically in the diabetic and glucose appears in the urine. Also, the blood glucose level at 2 hours is equal to or more than 200 mg/100 dl.

About 10% of diabetics in the United States have *type 1 diabetes.* This condition usually begins in childhood or adolescence, and thus it is sometimes called juvenile-onset diabetes. However, type 1 diabetes can also occur later in life, following a viral infection, an autoimmune reaction, or an environmental agent that leads to destruction of the pancreatic islets. Because individuals with type 1 diabetes are suffering from an insulin shortage, treatment simply consists of providing the needed insulin through daily injections. These injections usually control the diabetic symptoms but if too much insulin is administered, or a meal is missed, the result can be hypoglycemia (low blood sugar). Symptoms of hypoglycemia include perspiration, pale skin, shallow breathing, and anxiety. Because the brain requires a constant supply of glucose, unconsciousness can result. The treatment is quite simple: immediate ingestion of a sugary snack or fruit juice can very quickly counteract hypoglycemia. Better control of glucose levels can sometimes be achieved with an insulin pump, a small device worn outside the body that is connected to a plastic catheter inserted under the skin (Fig. 20.17). It is also possible to transplant a working pancreas, or even fetal pancreatic islet cells, into patients with type 1 diabetes. Some researchers believe that some form of artificial pancreas will become available in the early 2010s.

Most diabetics in the United States have *type 2 diabetes.* Often, the patient is overweight or obese, and adipose tissue may produce a substance that impairs insulin receptor function. Because type 2 diabetes usually doesn't occur before age 40, it was formerly called adult-onset diabetes. However, due to the increasing prevalence of obesity in children, type 2 diabetes is occurring at younger ages. Normally, the binding of insulin to its receptor on cell surfaces causes the number of glucose transporters to increase in the plasma membrane, but not in the type 2 diabetic. Treatment usually involves weight loss, which can sometimes control symptoms in the type 2 diabetic. However, many type 2 diabetics also have low insulin levels, so they may require injections of insulin, along with medications to increase the effectiveness of the insulin they produce.

Regardless of the type of diabetes they have, diabetics usually need to monitor their blood glucose levels several times a day. As described in the story that opened this chapter, this is usually done by poking a finger to obtain a drop of blood, which is tested with an external device. A U.S. company is currently testing a special tattoo ink that changes colors in response to glucose levels in the skin. The FDA has already approved a watchlike device that measures glucose levels in tiny amounts of fluid it painlessly extracts from the skin every 20 minutes. For insulin administration, insulin pumps are now replacing the need for injecting insulin with a needle and syringe (Fig. 20.17).

Animation Blood Sugar Regulation in Diabetics

Long-term complications of both types of diabetes are blindness, kidney disease, and cardiovascular disorders, including reduced circulation. The latter can lead to gangrene in the arms and legs. Diabetic women who become pregnant have an increased risk of diabetic coma, and the child of a diabetic is more likely to be stillborn or to die shortly after birth. These complications are uncommon, however, if the mother's blood glucose level is carefully regulated during pregnancy.

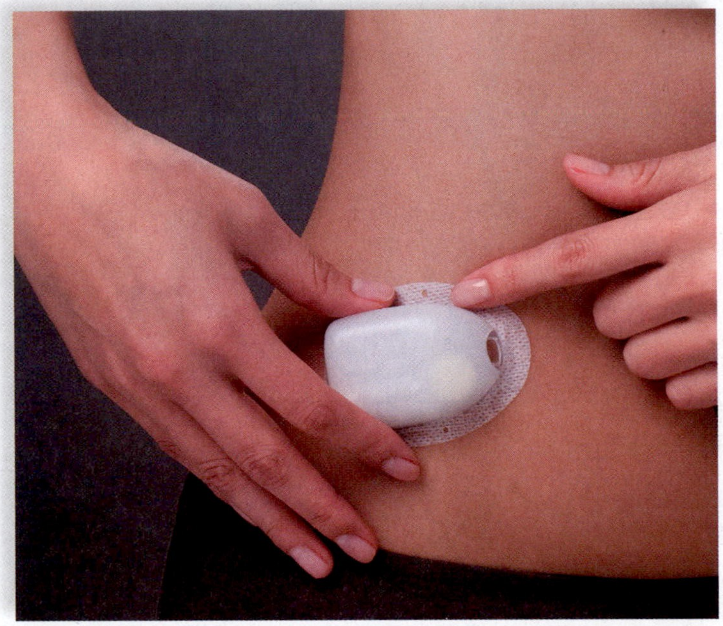

a.

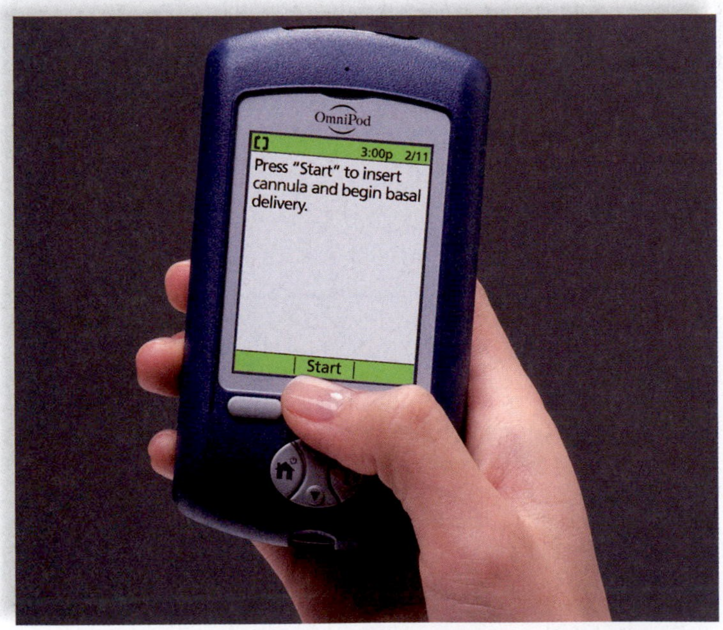
OmniPod
3:00p 2/11
Press "Start" to insert cannula and begin basal delivery.
Start

b.

Figure 20.17 An insulin pump. Insulin pumps administer preprogrammed small doses of insulin throughout the day via an implanted catheter. Most insulin pumps can be worn under clothing.

Check Your Progress 20.7

1. Explain how ADH is related to an increased urge to urinate when one drinks alcohol.
2. Describe why a goiter forms in iodine deficiency.
3. Differentiate the role of antibodies in causing Graves disease versus Addison disease.
4. Describe the causes and treatment of type 1 and type 2 diabetes.

SCIENCE IN YOUR LIFE ▶ BIOETHICAL

Human Growth Hormone: Does Rambo Know Something We Don't?

As discussed in the chapter, abnormally low levels of growth hormone production by the anterior pituitary can result in short stature. Suppose you have a son or daughter who is considerably shorter than most kids his or her age. The average adult American male is 5 ft 9 in. tall. If you knew your son would only reach 5 ft, do you think it would be ethical to treat him with human growth hormone (HGH), which could easily add 3 or 4 in. to his height? And because HGH injections can cost between $10,000 and $30,000 per year, should that be covered by health insurance, even if it raises insurance rates for other people?

Now suppose you are 55 years old. You are finding that even though you run and lift weights three days a week, you are slowly losing muscle mass, and you don't feel nearly as energetic as you once felt, even in your forties. Then you hear about how some professional athletes and movie stars are swearing by HGH as a way to recapture their youth. You find a website touting "legal" HGH at affordable prices, and claiming HGH will help you lose weight, tone your muscles, and give you younger-looking skin in a matter of weeks. Assuming it was legal and you could afford it, would you give it a try?

Prior to the development of recombinant DNA techniques, treatment of children who failed to grow due to insufficient HGH production required isolation of HGH from the pituitary glands of human cadavers. In 1985, the FDA approved the first recombinant HGH, produced by inserting the HGH gene into *E. coli* bacteria. In 2003, the FDA approved a new use for HGH, in the treatment of children of short stature due to unknown causes. However, HGH must still be administered by a physician, and it is illegal to use it for fitness purposes.

The production of recombinant HGH has been very beneficial in specific medical cases, but its use for other purposes remains controversial. Prior to filming his fourth Rambo movie, actor Sylvester Stallone admitted to using HGH as well as anabolic steroids to help tone his body (Fig. 20C). In 2007, Stallone was fined over $10,000 for attempting to smuggle 48 vials of HGH and 4 vials of testosterone into Australia. At age 61, he seems to be in incredible condition, but it is hard to say how much of that is due to hard workouts, genetics, or pharmaceuticals. Studies show that HGH can cut down on body fat and increase muscle mass, but no antiaging effects have been proven, and the list of scary side effects is long. Still, it's worth asking: Would you be willing to take the risk?

Figure 20C
At age 61, actor Sylvester Stallone admitted to using human growth hormone and anabolic steroids to help him get in shape for filming his latest *Rambo* movie.

Questions to Consider

1. Would you ever consider using HGH under any circumstances? Why or why not?
2. Do you believe that HGH should be illegal for personal use? If so, do you believe that taking HGH is worse than taking currently legal athletic supplements, such as creatine? Would it make a difference if someone was taking it to grow taller instead of for other perceived benefits?
3. Do you think that celebrities have an increased responsibility to avoid promoting behaviors that could be harmful to others?

Case Study Conclusion

As Hank matures, he is learning to manage his condition better, even though he feels a sense of resigned frustration that he may always have to deal with type 1 diabetes. When he was 10 years old, he began using an insulin pump, which eliminated the need for daily insulin injections, although he has to insert his infusion catheter (through which the pump dispenses insulin as needed) at a different site every couple of days. He also must remember to input his daily intake of carbohydrates into the machine, and he still must check his glucose levels in a drop of blood extracted from a finger-stick before eating, exercising, or going to bed at night.

Beginning in 2006, continuous glucose monitors have been available. Most of these devices include a probe that must be inserted under the skin, and a receiver that can display blood glucose readings taken hundreds of times a day.

However, most of these machines still require checking blood from a finger-stick several times a day. In January 2012, Hank's parents were excited to learn that the FDA had approved a remote glucose monitor, which would allow parents or other caregivers to monitor the blood glucose readings of a diabetic from another room. This would be especially reassuring at night, when many parents still wake up to make sure that a diabetic child's blood glucose isn't too low or high. Although current technology still would not allow them to monitor their son's glucose while he was away from home, they are hoping that it won't be long before automatic glucose monitors can communicate directly with an insulin pump, thereby serving as a sort of artificial pancreas that would allow their son to have more freedom from worrying about his blood sugar levels than he has ever had.

MEDIA STUDY TOOLS

www.mhhe.com/maderinquiry14

Enhance your study of this chapter with study tools and practice tests. Also ask your instructor about the resources available through ConnectPlus, including LearnSmart, the media-rich eBook, interactive learning tools, and animations.

SUMMARIZE

20.1 Overview of the Endocrine System

The **endocrine system** consists of **endocrine glands** that produce hormones. **Hormones** are **chemical signals** that affect other tissues, often at a distance. **Negative feedback** and antagonistic hormonal actions usually control the secretion of hormones. **Growth factors** are chemical signals that affect cell growth.

- Hormones are either peptides or steroids. Binding of a **peptide hormone** to a receptor at the plasma membrane typically activates an enzyme cascade inside the cell. A **steroid hormone** combines with a receptor in the cell, and that complex activates transcription of certain genes.
- A peptide hormone typically acts as a **first messenger.** This means the hormone binds to a cellular receptor, triggering activation of a **second messenger** inside the cell.
- A **pheromone** is a chemical signal that affects other individuals. There is some evidence for pheromone activity in humans.

20.2 Hypothalamus and Pituitary Gland

The **hypothalamus** controls hormone secretion by the **pituitary gland** through two distinct mechanisms:

- The hypothalamus itself synthesizes **antidiuretic hormone** (**ADH**) and **oxytocin,** which are stored in the **posterior pituitary** until they are released. In contrast to most hormones, the release of oxytocin is regulated by **positive feedback.**
- The hypothalamus communicates with the **anterior pituitary** via a portal system, through which **hypothalamic-releasing hormones** and **hypothalamic-inhibiting hormones** pass.
- Several hormones produced by the anterior pituitary stimulate other hormonal glands: **thyroid-stimulating hormone** (**TSH**) acts on the thyroid gland; **adrenocorticotropic hormone** (**ACTH**) stimulates the adrenal cortex; and the **gonadotropic hormones** FSH and LH stimulate the gonads.
- Three other hormones are produced by the anterior pituitary. **Prolactin** (**PRL**) causes mammary gland development after childbirth; **melanocyte-stimulating hormone** (**MSH**) affects skin color in many animals; **growth hormone** (**GH**) promotes skeletal and muscular growth.

20.3 Thyroid and Parathyroid Glands

The **thyroid gland** produces three hormones:

- **Thyroxine** (**T_4**) and **triiodothyronine** (**T_3**) increase the metabolic rate of most cells in the body.
- **Calcitonin** helps lower the blood calcium level by reducing the activity of osteoclasts in the bone.

The **parathyroid glands** secrete **parathyroid hormone** (**PTH**), which raises the blood calcium and decreases the blood phosphate levels.

20.4 Adrenal Glands

The paired **adrenal glands** respond to stress. The adrenal medulla is controlled directly by the nervous system, allowing an almost immediate response. The adrenal cortex responds on a longer-term basis to ACTH released by the anterior pituitary:

- The **adrenal medulla** secretes **epinephrine** and **norepinephrine,** which bring about fight-or-flight responses we associate with emergency situations.
- The **adrenal cortex** produces the **glucocorticoids** (e.g., cortisol) and the **mineralocorticoids** (e.g., aldosterone). **Cortisol** stimulates the conversion of amino acids to glucose, raising the blood glucose level. It can also suppress inflammatory responses. **Aldosterone** causes the kidneys to reabsorb sodium ions (Na^+) and to excrete potassium ions (K^+), which raises the blood pressure. The release of aldosterone is inhibited by **atrial natriuretic hormone** (**ANH**), released by cells in the atrium of the heart.

20.5 Pancreas

The **pancreas** has both exocrine and endocrine functions. The **pancreatic islets** secrete three hormones:

- **Insulin** lowers the blood glucose level by stimulating the uptake of glucose by various types of cells.
- **Glucagon** raises the blood glucose level.
- **Somatostatin** inhibits growth hormone release and decreases absorption of nutrients from the digestive tract.

20.6 Other Endocrine Glands

Other organs and tissues produce hormones:

- The male **gonads** (**testes**) produce **androgens** (male sex hormones) like **testosterone,** while in the female the **ovaries** produce **estrogen** and **progesterone.** Athletes who abuse **anabolic steroids** due to their muscle-building effects risk serious side effects.
- The **thymus** secretes thymosins, which stimulate T-lymphocyte production and maturation.
- The **pineal gland** produces **melatonin,** which may be involved in **circadian rhythms** and reproductive organ development.
- Tissues also produce hormones. Adipose tissue produces **leptin,** which acts on the hypothalamus, and various tissues produce **growth factors** that stimulate cell division. **Prostaglandins** act locally, near where they are produced.

20.7 Disorders of the Endocrine System

Increases or decreases in the amounts of hormones synthesized by endocrine glands cause several common disorders:

- Disorders caused by pituitary gland dysfunction include **diabetes insipidus** (insufficient ADH), **pituitary dwarfism** (GH deficiency), **gigantism** (too much GH in child), **acromegaly** (too much GH in adult), and **Cushing syndrome** (too much ACTH).
- **Hypothyroidism** (insufficient thyroid hormones) results in cretinism in childhood and myxedema in adults. **Graves disease** is a type of **hyperthyroidism.**
- **Addison disease** is a serious condition in which the adrenal cortexes have been damaged, often by an autoimmune process.

- **Diabetes mellitus** is a common condition affecting the body's ability to metabolize glucose. In type 1 diabetes, the pancreas doesn't produce enough insulin. In type 2 diabetes, insulin receptors do not function properly. Insulin injections may be required for both types of diabetes, but type 2 diabetes can be more difficult to treat, resulting in long-term complications.

ASSESS

Testing Yourself

Choose the best answer for each question.

1. Which hormones typically cross the plasma membrane?
   a. peptide hormones
   b. steroid hormones
   c. Both a and b are correct.
   d. Neither a nor b is correct.
2. Name the endocrine glands in the following diagram.

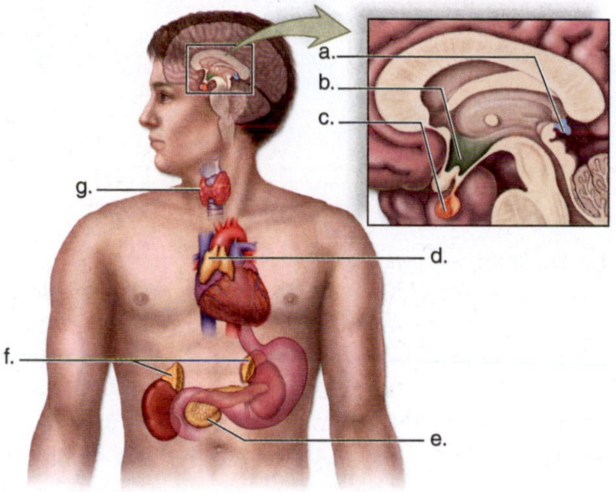
a.
b.
c.
d.
e.
f.
g.

3. The anterior pituitary controls the secretion(s) of
   a. both the adrenal medulla and the adrenal cortex.
   b. both the pancreas and the adrenal cortex.
   c. both the ovaries and the testes.
   d. Both b and c are correct.
4. Growth hormone is produced by the
   a. posterior adrenal gland.
   b. posterior pituitary gland.
   c. anterior pituitary gland.
   d. kidneys.
   e. None of these are correct.
5. PTH causes the blood levels of _____ to increase and _____ to decrease.
   a. calcium, sodium
   b. calcium, phosphate
   c. phosphate, sodium
   d. phosphate, calcium
   e. None of these are correct.
6. The body's response to stress includes
   a. water reabsorption by the kidneys.
   b. blood pressure increase.
   c. increase in the blood glucose level.
   d. heart rate increase.
   e. All of these are correct.
7. Lack of aldosterone will cause a blood imbalance of
   a. sodium.
   b. potassium.
   c. water.
   d. All of these are correct.
   e. None of these are correct.

Match the hormones in questions 8–12 to the correct gland in the key.

Key:

a. pancreas
b. anterior pituitary
c. posterior pituitary
d. thyroid gland
e. adrenal medulla
f. adrenal cortex

8. Cortisol
9. Growth hormone (GH)
10. Oxytocin storage
11. Insulin
12. Epinephrine
13. Name the hormones in the following diagram.

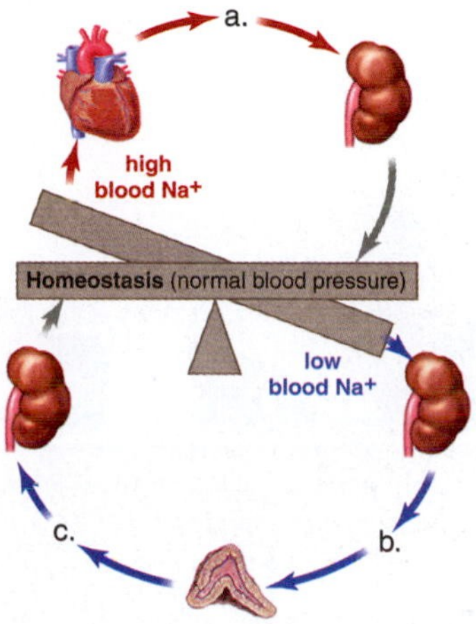
a.
high
blood Na+
Homeostasis (normal blood pressure)
low
blood Na+
b.
c.

14. Which hormone and condition are mismatched?
    a. growth hormone—acromegaly
    b. thyroxine—goiter
    c. parathyroid hormone—tetany
    d. cortisol—myxedema
    e. insulin—diabetes
15. Long-term complications of diabetes include
    a. blindness.
    b. kidney disease.
    c. circulatory disorders.
    d. All of these are correct.

ENGAGE

Thinking Critically

1. Because some of their functions overlap, why is it necessary to have both a nervous system and an endocrine system?
2. Certain endocrine disorders, such as Cushing syndrome, can be caused by excessive secretion of a hormone (in this case, ACTH) by the pituitary gland, or by a problem with the adrenal gland itself. If you were able to measure the ACTH levels of a Cushing patient, how could you tell the difference between a pituitary problem and a primary adrenal gland problem?
3. In animals, pheromones can affect many different behaviors. Because humans do produce a number of airborne chemicals, what potential human behaviors could be affected by these chemicals? How would these pheromones be received, and would we necessarily be conscious of their effects on us?

UNIT 5 Continuance of the Species

21 Reproductive System

CHAPTER OUTLINE

BEFORE YOU BEGIN

Before beginning this chapter, take a few moments to review the following discussions:

Section 5.4 How are human gametes produced?

Figure 20.5 How is the secretion of sex hormones controlled?

Section 20.6 Where are the male and female sex hormones produced, and what are their functions?

CASE STUDY In Vitro Fertilization

The case of Nadya Suleman, who gave birth to eight babies in January 2009, made headlines around the world. Cable TV news channels spent countless hours speculating about the motivation of this divorced single mother, also dubbed the "Octomom," who had six embryos (two of which became twins) implanted into her uterus at a fertility clinic in Beverly Hills, California. The embryos were left over from a previous in vitro fertilization (IVF) procedure using sperm from a donor. Complicating the situation was the fact that Ms. Suleman already had six children, she was unemployed and on government assistance, and she may have been interested in benefiting financially from the attention her story received.

Reproduction is a basic necessity for the survival of a species. However, the ethics of human reproduction can become quite complicated. In the great majority of cases, IVF and the other assisted reproductive technologies described in this chapter are used to help couples who are unable to conceive on their own. Guidelines from the American Society for Reproductive Medicine state that no more than three embryos should be implanted at a time. In June 2011, the Medical Board of California revoked the medical license of the doctor who implanted the embryos into Ms. Suleman. Other infertility doctors, however, argue that decisions on how many embryos to transfer should be left up to medical experts familiar with a patient's individual circumstances.

As you read through the chapter, think about the following questions:

1. What specific physical factors limit the number of live children a mother can gestate and deliver?
2. Suppose these limitations could be overcome—for example, by incubating embryos in an artificial womb. Assuming a couple was able to afford it, should they be allowed to have, say, 100 children? Why or why not?
3. Besides IVF, what other forms of assisted reproductive technology are available for infertile couples?

21.1 Male Reproductive System

Learning Outcomes

Upon completion of this section, you should be able to

1. Describe the path of sperm from the testes to the urethra.
2. List the organs that produce components of seminal fluid.
3. Analyze the role of gonadotropin-releasing hormone, luteinizing hormone, follicle-stimulating hormone, and testosterone in male sexual reproduction.

As described in Chapter 5, organisms that carry out sexual reproduction must produce **gametes**, haploid sex cells that become united during fertilization. Unlike other human organ systems, the **reproductive system** is quite different in males and females. The male reproductive system includes the organs depicted in Figure 21.1 and listed in Table 21.1. The male gonads are paired **testes** (sing., testis).

MP3 Male Reproductive Anatomy and Physiology

Genital Tract

Sperm produced by the testes mature within the epididymides (sing., **epididymis**), which are tightly coiled ducts lying outside the testes. Maturation seems to be required in order for sperm to swim to the egg. When sperm leave an epididymis, they enter a **vas deferens** (pl., vasa deferentia), where they may also be stored for a time. Each vas deferens passes into the abdominal cavity, where it curves around the urinary bladder and empties into an ejaculatory duct. The ejaculatory ducts connect to the **urethra.**

At the time of ejaculation, sperm leave the penis in a fluid called **semen** (**seminal fluid**). Three glands add secretions to semen. The paired **seminal vesicles** lie at the base of the bladder, and each has a duct that joins with a vas deferens. The **prostate gland** is a single, donut-shaped gland that surrounds the upper portion of the urethra just below the urinary bladder. **Bulbourethral glands** (Cowper glands) are pea-sized organs that lie underneath the prostate on either side of the urethra.

TABLE 21.1 Male Reproductive Structures

Organ	Function
Testes	Produce sperm and sex hormones
Epididymides	Ducts where sperm mature and are stored
Vasa deferentia	Conduct and also store sperm
Seminal vesicles	Contribute nutrients and fluid to semen
Prostate gland	Contributes basic fluid to semen
Urethra	Conducts sperm
Bulbourethral glands	Contribute viscous fluid to semen
Penis	Organ of sexual intercourse

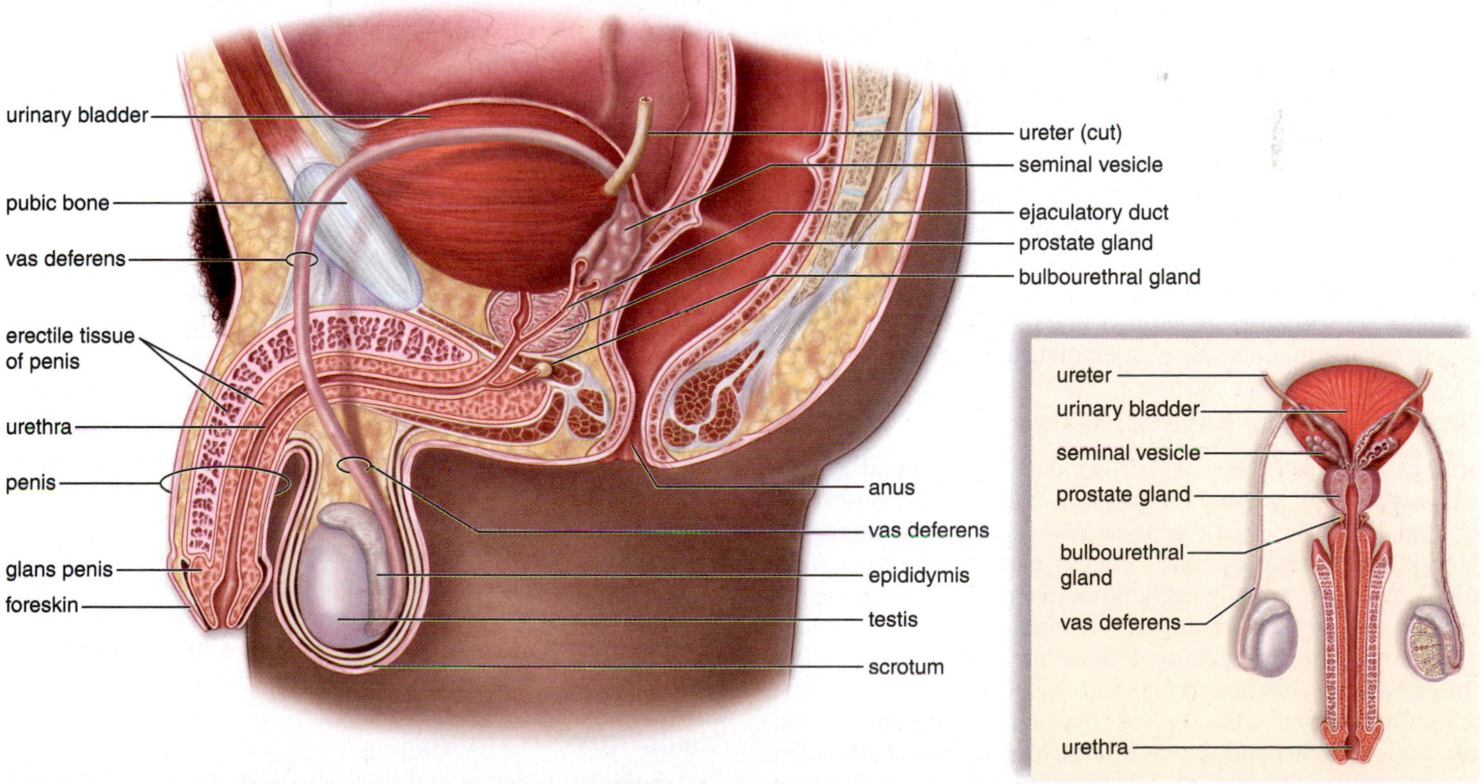

urinary bladder
pubic bone
vas deferens
erectile tissue of penis
urethra
penis
glans penis
foreskin
ureter (cut)
seminal vesicle
ejaculatory duct
prostate gland
bulbourethral gland
anus
vas deferens
epididymis
testis
scrotum
ureter
urinary bladder
seminal vesicle
prostate gland
bulbourethral gland
vas deferens
urethra

Figure 21.1 The male reproductive system. The testes produce sperm. The seminal vesicles, the prostate gland, and the bulbourethral glands provide a fluid medium for the sperm, which move from the vasa deferentia through the ejaculatory ducts to the urethra in the penis. The foreskin (prepuce) is removed when a penis is circumcised.

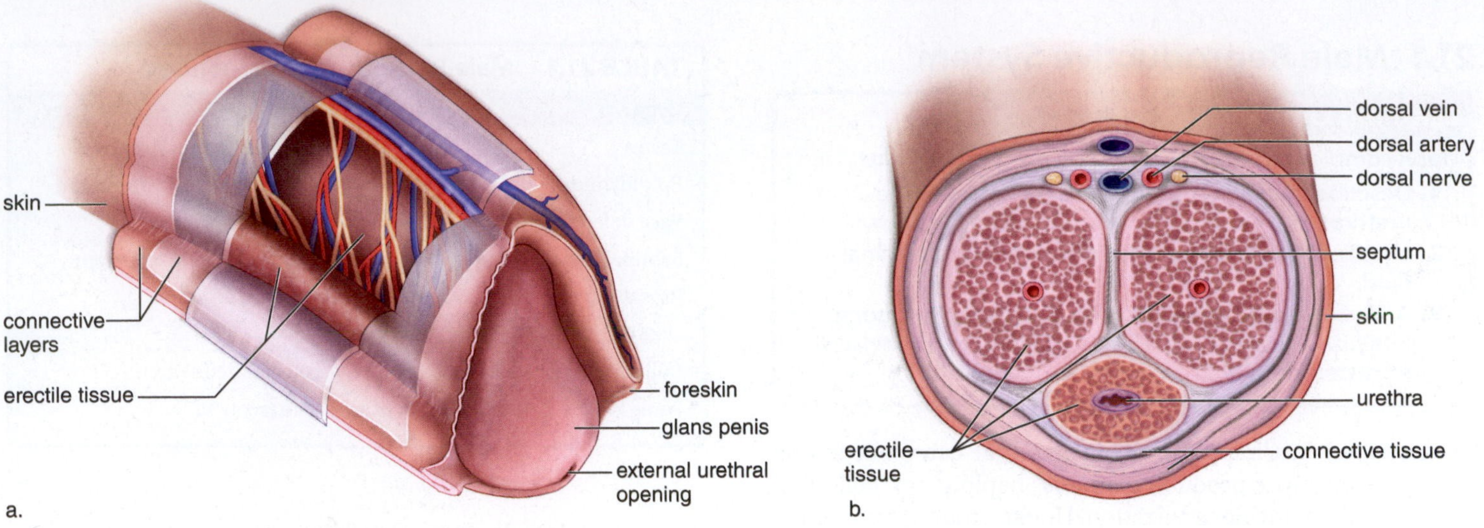
skin
connective layers
erectile tissue
foreskin
glans penis
external urethral opening
a.
dorsal vein
dorsal artery
dorsal nerve
septum
skin
urethra
connective tissue
erectile tissue
b.

Figure 21.2 Penis anatomy. **a.** Beneath the skin and the connective tissue lies the urethra, surrounded by erectile tissue. This tissue expands to form the glans penis, which in uncircumcised males is partially covered by the foreskin (prepuce). **b.** Two other columns of erectile tissue in the penis are located dorsally.

Each component of semen seems to have a particular function. Sperm are more viable in a basic solution, and semen, which is milky in appearance, has a slightly basic pH (about 7.5). Swimming sperm require energy, and semen contains the sugar fructose, which serves as an energy source. Semen also contains *prostaglandins,* local hormones that cause the uterus to contract. Some investigators believe that uterine contractions help propel the sperm toward the egg.

The **penis** (Fig. 21.2) is the male organ of sexual intercourse. The penis has a long shaft and an enlarged tip called the *glans penis.* At birth, the glans penis is covered by a layer of skin called the *foreskin.* **Circumcision,** the surgical removal of the foreskin, is usually done for religious reasons or perceived health benefits, including easier hygiene, decreased urinary tract infections, and reduced risk of penile cancer. A 2009 study of 3,000 Ugandan men showed that compared to uncircumcised men, circumcised men were 25–35% less likely to contract HIV, human papillomavirus, or genital herpes. However, opponents of the procedure focus on the risks, including pain and potential loss of sensitivity in the penis, and the American Academy of Pediatrics' position is that the benefits are not sufficient to recommend routine neonatal circumcision.

Erection and Orgasm in Males

Spongy, erectile tissue containing distensible blood spaces extends through the shaft of the penis. When a man is sexually excited, the arteries in the penis relax and widen. Increased blood flow causes the penis to enlarge and become erect. Also, the veins that normally carry blood away from the penis get compressed, and this maintains an erection.

When sexual stimulation intensifies, sperm enter the urethra from the vasa deferentia, and the accessory glands contribute secretions to the semen. Once semen is in the urethra, rhythmic muscle contractions cause it to be ejaculated from the penis in spurts. During *ejaculation,* a sphincter normally closes off the urinary bladder so that no urine enters the urethra, and no semen enters the bladder. (Notice that the urethra carries urine or semen at different times.)

The contractions that expel semen from the penis are a part of male *orgasm,* the physiological and psychological sensations that occur at the climax of sexual stimulation. The psychological sensation of pleasure is centered in the brain, but the physiological reactions involve the genital (reproductive) organs and associated muscles, as well as the entire body.

Following ejaculation and/or loss of sexual arousal, the penis returns to its normal flaccid state. After ejaculation, a male typically experiences a period of time, called the *refractory period,* during which stimulation does not bring about an erection. The length of the refractory period may be from minutes to hours, although it tends to increase with age.

There may be in excess of 400 million sperm in the approximately 2–6 ml of semen expelled during ejaculation. However, the sperm count can be much lower than this, and fertilization of the egg by a sperm still can take place.

Male Gonads, the Testes

The testes, which produce sperm as well as the male sex hormones, lie outside the abdominal cavity of the human male, within the saclike **scrotum.** The testes begin their development inside the abdominal cavity but descend into the scrotal sacs during the last two months of fetal development. If the testes fail to descend properly, and thus remain in the abdomen, male infertility may result. This is because the internal temperature of the body is too high to produce viable sperm. This condition can usually be surgically corrected, however. The scrotum helps regulate the temperature of the testes by holding them closer or farther away from the body. In fact, any activity that increases testicular temperature, such as taking hot baths, using a laptop computer, or even sitting for extended periods of time, may decrease sperm production.

Seminiferous Tubules

A longitudinal section of a testis shows that it is composed of compartments called *lobules,* each of which contains one to three tightly coiled **seminiferous tubules** (Fig. 21.3*a*). Altogether, these tubules have a combined length of approximately 250 meters. A

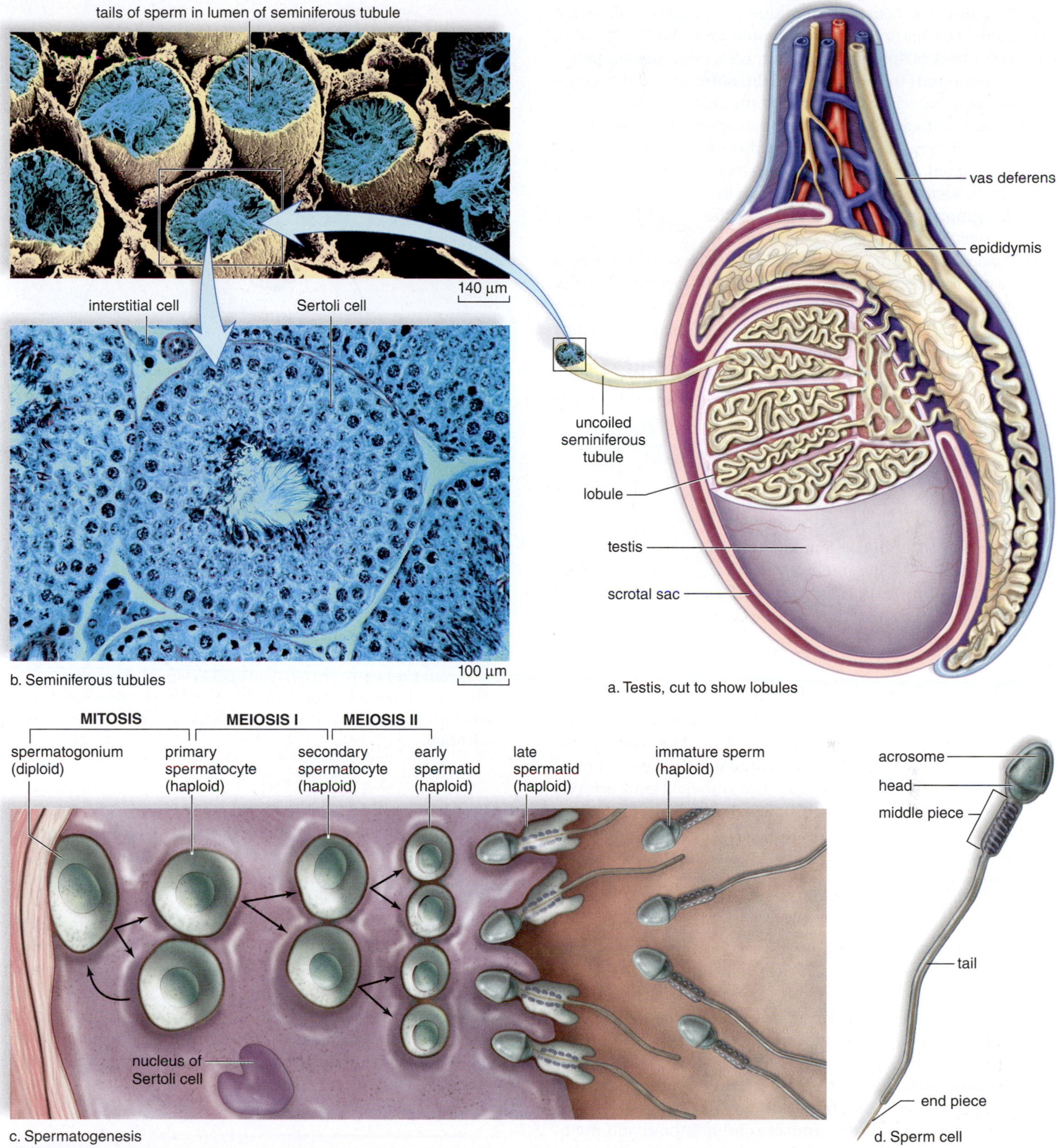
tails of sperm in lumen of seminiferous tubule
140 µm
interstitial cell
Sertoli cell
100 µm
b. Seminiferous tubules
vas deferens
epididymis
uncoiled seminiferous tubule
lobule
testis
scrotal sac
a. Testis, cut to show lobules
MITOSIS
MEIOSIS I
MEIOSIS II
spermatogonium (diploid)
primary spermatocyte (haploid)
secondary spermatocyte (haploid)
early spermatid (haploid)
late spermatid (haploid)
immature sperm (haploid)
nucleus of Sertoli cell
c. Spermatogenesis
acrosome
head
middle piece
tail
end piece
d. Sperm cell

Figure 21.3 Testis and sperm. **a.** The lobules of a testis contain seminiferous tubules. **b.** Light micrographs of a cross section of the seminiferous tubules, where spermatogenesis occurs. Note the location of interstitial cells in clumps among the seminiferous tubules. **c.** Diagrammatic representation of spermatogenesis, which occurs in the walls of the tubules. **d.** A sperm has a head, a middle piece, tail, and an end piece. The nucleus is in the head, which is capped by the enzyme-containing acrosome.

microscopic cross section of a seminiferous tubule reveals that it is packed with cells undergoing **spermatogenesis** (Fig. 21.3*b, c*), the production of sperm. Newly formed *spermatogonia* (sing., spermatogonium) move away from the outer wall and become *primary spermatocytes* that undergo meiosis I to produce *secondary spermatocytes* with 23 chromosomes. Secondary spermatocytes undergo meiosis II to produce four *spermatids* that are also haploid. Spermatids then differentiate into sperm. Also present are **Sertoli cells** (sustentacular cells), which support, nourish, and regulate the spermatogenic cells.

Animation Spermatogenesis

Mature **sperm,** or spermatozoa, have three main parts: a head, a middle piece, and a tail (Fig. 21.3*d*). Mitochondria in the middle piece provide energy for the movement of the tail, which has the structure of a flagellum (see Fig. 3.15). The head contains a nucleus covered by a cap called the *acrosome,* which stores enzymes needed to penetrate the egg. The ejaculated semen of a normal human male contains several hundred million sperm, but only one sperm normally enters an egg. Sperm usually do not live more than 48 hours in the female genital tract.

Video Human Sperm

Interstitial Cells

The male sex hormones, the androgens, are secreted by cells that lie between the seminiferous tubules. Therefore, they are called **interstitial cells** (Fig. 21.3*b*). The most important of the androgens is testosterone, whose functions are discussed next.

Hormonal Regulation in Males

The hypothalamus has ultimate control of the testes' sexual function because it secretes a hormone called **gonadotropin-releasing hormone** (**GnRH**) that stimulates the anterior pituitary to secrete the gonadotropic hormones. There are two gonadotropic hormones—**follicle-stimulating hormone** (**FSH**) and **luteinizing hormone** (**LH**)—in both males and females. In males, FSH promotes the production of sperm in the seminiferous tubules, which also release the hormone inhibin. *Inhibin,* as its name suggests, inhibits further FSH synthesis.

LH in males is sometimes given the name **interstitial cell-stimulating hormone** (**ICSH**) because it controls the production of testosterone by the interstitial cells. All these hormones are involved in a negative feedback relationship that maintains the fairly constant production of sperm and testosterone (Fig. 21.4).

Testosterone, the main sex hormone in males, is essential for the normal development and functioning of the organs listed in Table 21.1. Testosterone also brings about and maintains the male secondary sex characteristics that develop at the time of puberty. Males are generally taller than females and have broader shoulders and longer legs relative to trunk length. The deeper voices of males compared to those of females are due to a larger larynx with longer vocal cords. Because the so-called Adam's apple is a part of the larynx, it is usually more prominent in males than in females. Testosterone causes males to develop noticeable hair on the face, chest, and occasionally other regions of the body, such as the back. Testosterone also leads to the receding hairline and pattern baldness that occur in aging males.

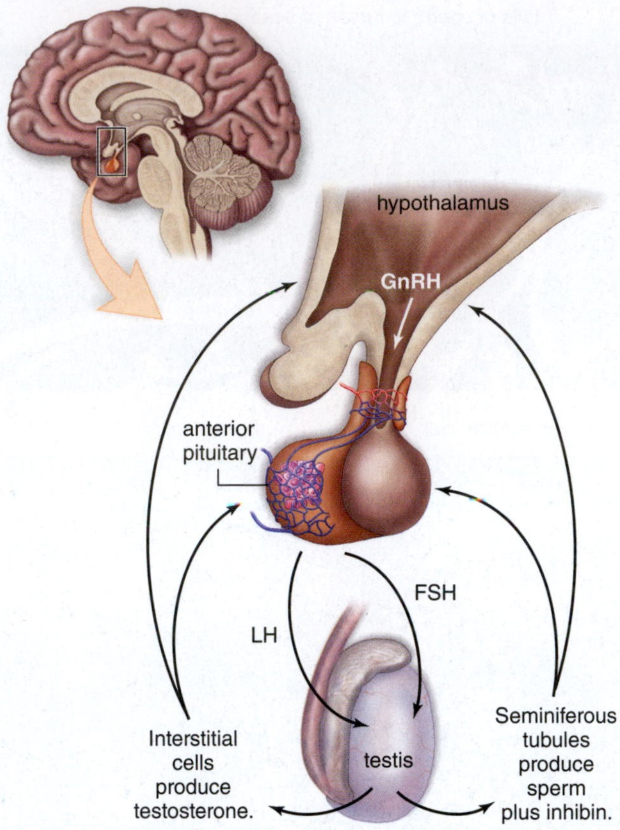
hypothalamus
GnRH
anterior pituitary
FSH
LH
Interstitial cells produce testosterone.
testis
Seminiferous tubules produce sperm plus inhibin.

Figure 21.4 Hormonal control of testes. GnRH stimulates the anterior pituitary to produce FSH and LH. FSH stimulates the testes to produce sperm, and LH stimulates the testes to produce testosterone. Testosterone from interstitial cells and inhibin from the seminiferous tubules exert negative feedback control over the hypothalamus and the anterior pituitary, and this ultimately regulates the level of testosterone in the blood.

Testosterone is responsible for the greater muscular development of males. Knowing this, both males and females sometimes take *anabolic steroids,* which are either testosterone or related steroid hormones resembling testosterone. Health problems involving the kidneys, the cardiovascular system, and hormonal imbalances can arise from such use. In males, the testes shrink in size, and feminization of other male traits occurs (see Fig. 20.10).

Check Your Progress 21.1

1. Compare the functions of the seminiferous tubules, interstitial cells, epididymis, vasa deferentia, and urethra.
2. List the three glands that add secretions to semen.
3. Explain the role of the following hormones in male reproduction: GnRH, FSH, ICSH, and testosterone.

21.2 Female Reproductive System

Learning Outcomes

Upon completion of this section, you should be able to
1. Identify where an oocyte is produced and how it is transported to the uterus.
2. Describe the major components of the female external genitalia.
3. Describe the events that occur during a female orgasm.

The female reproductive system includes the organs depicted in Figure 21.5 and listed in Table 21.2. The female gonads are paired **ovaries** that lie in shallow depressions, one on each side of the upper pelvic cavity. **Oogenesis** is the production of an egg, or **oocyte,** the female gamete. The ovaries usually alternate in producing one oocyte per month. **Ovulation** is the process by which an oocyte bursts from an ovary and usually enters an oviduct.

Animation Female Reproductive System

MP3 Female Reproductive Anatomy and Physiology

The Genital Tract

The **uterine tubes,** also called oviducts or fallopian tubes, extend from the uterus to the ovaries. However, the uterine tubes are not attached to the ovaries. Instead, they have fingerlike projections called **fimbriae** (sing., fimbria) that sweep over the ovaries. When an oocyte bursts from an ovary during ovulation, it usually is swept into a uterine tube by the combined action of the fimbriae and the beating of cilia that line the uterine tubes.

TABLE 21.2 Female Reproductive Structures

Organ	Function
Ovaries	Produce oocyte and sex hormones
Uterine tubes (oviducts or fallopian tubes)	Conduct oocyte; location of fertilization; transport early zygote
Uterus (womb)	Houses developing fetus
Cervix	Contains opening to uterus
Vagina	Receives penis during sexual intercourse; serves as birth canal and as the exit for menstrual flow

Once in the uterine tube, the oocyte is propelled slowly by ciliary movement and tubular muscle contraction toward the uterus. An oocyte lives only approximately 6 to 24 hours unless fertilization occurs. Fertilization, and therefore formation of a **zygote,** usually takes place in the uterine tube. The developing embryo normally arrives at the uterus after several days and then embeds, or *implants,* itself in the uterine lining, which has been prepared to receive it. Occasionally however, the embryo may implant in the uterine tube or elsewhere, resulting in an *ectopic pregnancy.*

The **uterus** is a thick-walled, muscular organ about the size and shape of an inverted pear. The oviducts join the uterus at its upper end, while at its lower end, the **cervix** connects with the vagina nearly at a right angle (Fig. 21.5).

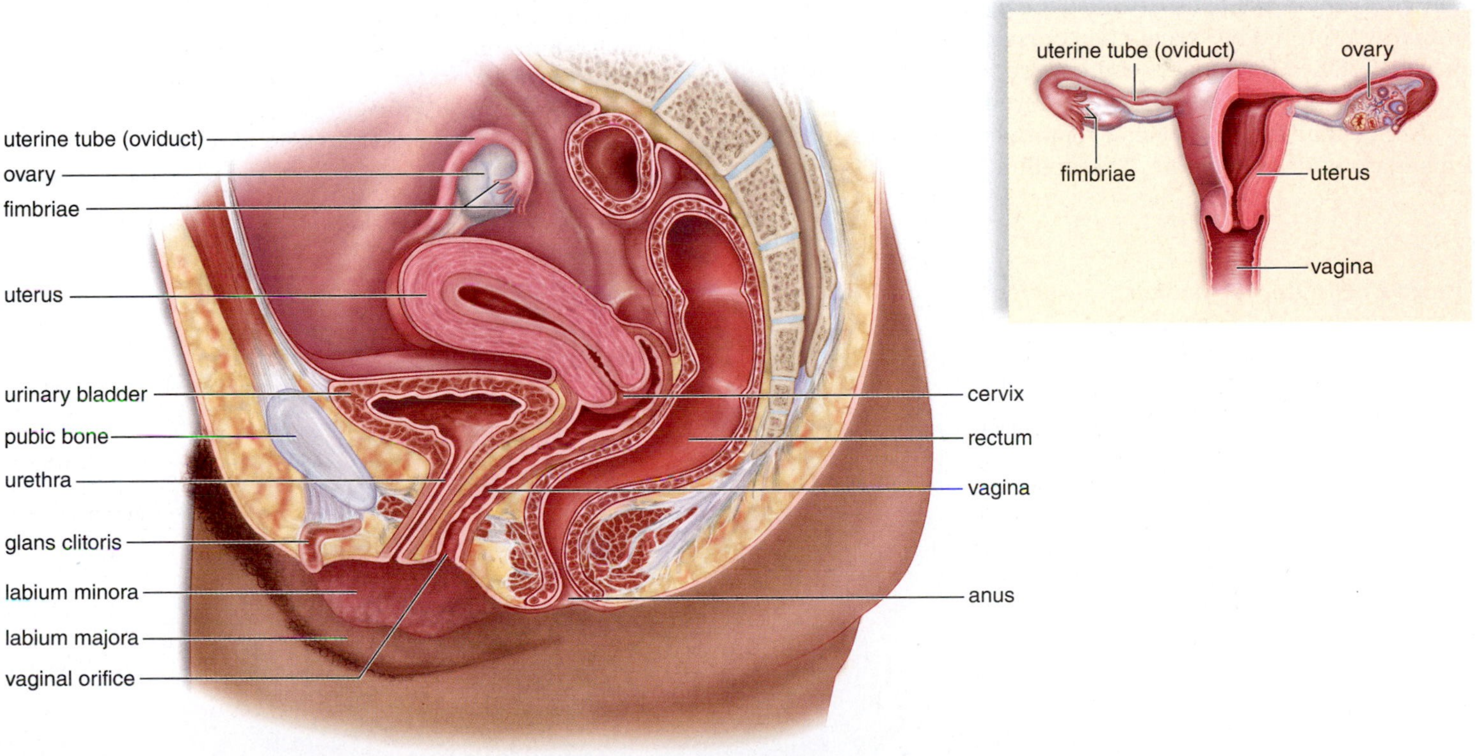
uterine tube (oviduct)
ovary
fimbriae
uterus
urinary bladder
pubic bone
urethra
glans clitoris
labium minora
labium majora
vaginal orifice
cervix
rectum
vagina
anus
uterine tube (oviduct)
ovary
fimbriae
uterus
vagina

Figure 21.5 The female reproductive system. The ovaries release one egg a month; fertilization occurs in the uterine tube, and development occurs in the uterus. The vagina is the birth canal, as well as the organ of sexual intercourse and the outlet for menstrual flow.

Development of the embryo normally takes place in the uterus. This organ, sometimes called the womb, is approximately 5 cm wide in its usual state but is capable of stretching to over 30 cm wide to accommodate a growing baby. The lining of the uterus, called the **endometrium,** participates in the formation of the placenta (see section 21.3), which supplies nutrients needed for embryonic and fetal development. In the nonpregnant female, the functional layer of the endometrium varies in thickness according to a monthly cycle of events called the uterine cycle.

A small opening in the cervix leads to the vaginal canal. The **vagina** is a tube that lies at a 45-degree angle to the small of the back. The mucosal lining of the vagina lies in folds and can extend. This is especially important when the vagina serves as the birth canal, and it facilitates sexual intercourse, when the vagina receives the penis. The vagina also acts as the exit for menstrual flow.

External Genitals

The external genital organs of the female are known collectively as the **vulva** (Fig. 21.6). The vulva includes two large, hair-covered folds of skin called the *labia majora* (sing., labium major). The labia majora extend backward from the *mons pubis,* a fatty prominence underlying the pubic hair. The *labia minora* are two small folds lying just inside the labia majora. The *glans clitoris* is the external portion of the **clitoris,** the organ of sexual arousal in females. Like the penis, the clitoris contains a shaft of erectile tissue that becomes engorged with blood during sexual stimulation.

The cleft between the labia minora contains the openings of the urethra and the vagina. The vagina may be partially closed by a ring of tissue called the *hymen.* The hymen is ordinarily ruptured by sexual intercourse or by other types of physical activity. If remnants of the hymen persist after sexual intercourse, they can be surgically removed.

Notice that the urinary and reproductive systems in the female are entirely separate. The urethra carries only urine, and the vagina serves only as the birth canal and the organ for sexual intercourse.

Orgasm in Females

Upon sexual stimulation, the labia minora, the vaginal wall, and the clitoris become engorged with blood. The breasts also swell, and the nipples become erect. The labia majora enlarge, redden, and spread away from the vaginal opening.

The vagina expands and elongates. Blood vessels in the vaginal wall release small droplets of fluid that seep into the vagina and lubricate it. Mucus-secreting glands beneath the labia minora on either side of the vagina also provide lubrication for entry of the penis into the vagina. Although the vagina is the organ of sexual intercourse in females, the extremely sensitive clitoris plays a significant role in the female sexual response. The thrusting of the penis and the pressure of the pubic symphyses of the partners acts to stimulate the clitoris, which may swell to two or three times its usual size.

Orgasm occurs at the height of the sexual response. Blood pressure and pulse rate rise, breathing quickens, and the walls of the vagina, uterus, and uterine tubes contract rhythmically. A sensation of intense pleasure is followed by relaxation when organs return to their normal size. Females have little or no refractory period between orgasms, and multiple orgasms can occur during a single sexual experience.

Check Your Progress 21.2

1. Explain how a tubal ligation, or cutting and sealing off the uterine tubes, prevents pregnancy.
2. Explain how males and females differ in the specialization of their genital tract for reproduction versus urination.
3. Describe the event(s) that occur during a female orgasm that may increase the chances of fertilization.

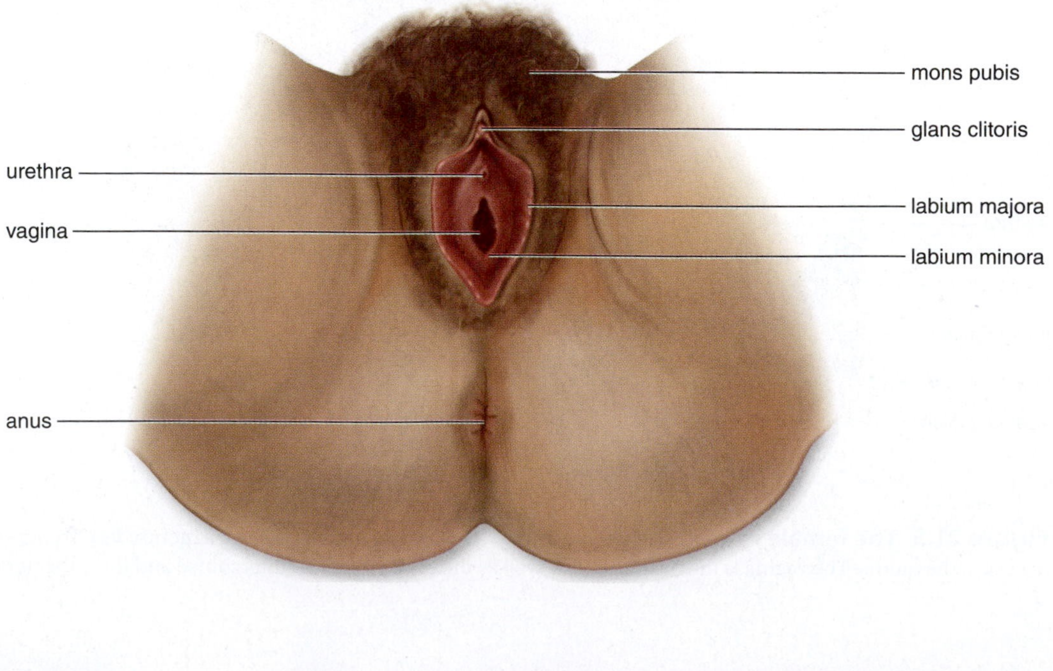
mons pubis
glans clitoris
urethra
labium majora
vagina
labium minora
anus

Figure 21.6 External genitals of the female. The opening of the vagina is partially blocked by a membrane called the hymen. Physical activities and sexual intercourse disrupt the hymen.

21.3 Ovarian and Uterine Cycles

Learning Outcomes

Upon completion of this section, you should be able to

1. List the stages of the ovarian cycle and explain what is occurring in each stage.
2. Describe the process of oogenesis.
3. Summarize how estrogen and progesterone affect the uterus.
4. Describe changes that occur in the uterus during menstruation, pregnancy, and menopause.

Hormone levels cycle in the female on a monthly basis, and the ovarian cycle drives the uterine cycle.

The Ovarian Cycle

A longitudinal section through an ovary shows that it is made up of an outer *cortex* and an inner *medulla* (Fig. 21.7*a*). In the cortex are many **follicles,** each one containing an immature oocyte. A female is born with all the ovarian follicles she will ever have, an estimated 700,000. However, only about 400 of these follicles will ever mature because a female usually produces only one oocyte per month during her reproductive years. Because oocytes are present at birth, they age as the woman ages. This may be one reason older women are more likely to produce children with genetic defects.

The **ovarian cycle** occurs as a follicle changes from a primary to a secondary to a vesicular (Graafian) follicle (Fig. 21.7*a*). Epithelial cells of a *primary follicle* surround a primary oocyte.

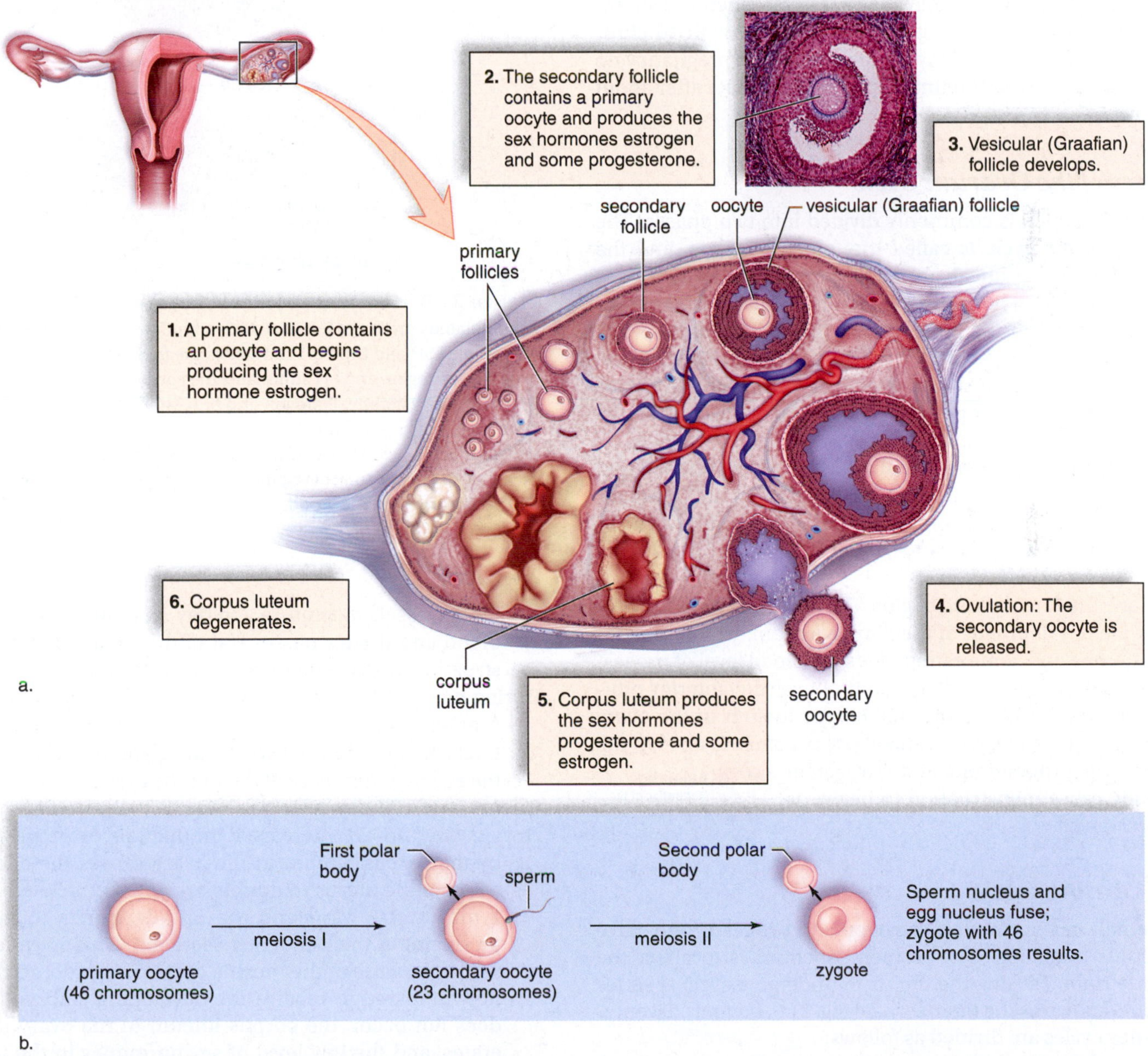

2. The secondary follicle contains a primary oocyte and produces the sex hormones estrogen and some progesterone.
3. Vesicular (Graafian) follicle develops.
secondary follicle
oocyte
vesicular (Graafian) follicle
primary follicles
1. A primary follicle contains an oocyte and begins producing the sex hormone estrogen.
6. Corpus luteum degenerates.
4. Ovulation: The secondary oocyte is released.
a.
corpus luteum
5. Corpus luteum produces the sex hormones progesterone and some estrogen.
secondary oocyte
First polar body
sperm
Second polar body
Sperm nucleus and egg nucleus fuse; zygote with 46 chromosomes results.
meiosis I
meiosis II
primary oocyte (46 chromosomes)
secondary oocyte (23 chromosomes)
zygote
b.

Figure 21.7 Ovarian cycle. **a.** A single follicle goes through six stages in one place within the ovary. As a follicle matures, layers of follicle cells surround the developing oocyte. Eventually, the mature follicle ruptures, and the secondary oocyte is released. The follicle then becomes the corpus luteum, which eventually disintegrates. **b.** During oogenesis, the chromosome number is reduced from 46 to 23. Fertilization restores the full number of chromosomes.

Pools of follicular fluid surround the oocyte in a *secondary follicle.* In a *vesicular follicle,* a fluid-filled cavity increases to the point that the follicle wall balloons out on the surface of the ovary.

As a follicle matures, oogenesis, depicted in Figure 21.7*b,* is initiated and continues. The *primary oocyte* divides, producing two haploid cells. One cell is a *secondary oocyte,* and the other is a *polar body.* The vesicular follicle bursts, releasing the secondary oocyte. This process is referred to as ovulation. Once a vesicular follicle has lost the secondary oocyte, it develops into a **corpus luteum,** a glandlike structure that produces progesterone.

The secondary oocyte enters a uterine tube. If a sperm enters the secondary oocyte, fertilization occurs, and the secondary oocyte completes meiosis. An egg with 23 chromosomes and a second polar body results. When the sperm nucleus unites with the egg nucleus, a zygote with 46 chromosomes is produced. If zygote formation and pregnancy do not occur, the corpus luteum begins to degenerate after about ten days.

Phases of the Ovarian Cycle

The ovarian cycle is commonly divided into two phases. The first half of the cycle is called the follicular phase, and the second half is the luteal phase. During the *follicular phase,* follicle-stimulating hormone (FSH), produced by the anterior pituitary, promotes the development of a follicle in the ovary, which secretes estrogen and some progesterone (Fig. 21.8). As the estrogen level in the blood rises, it exerts negative feedback control over the anterior pituitary secretion of FSH so that the follicular phase comes to an end.

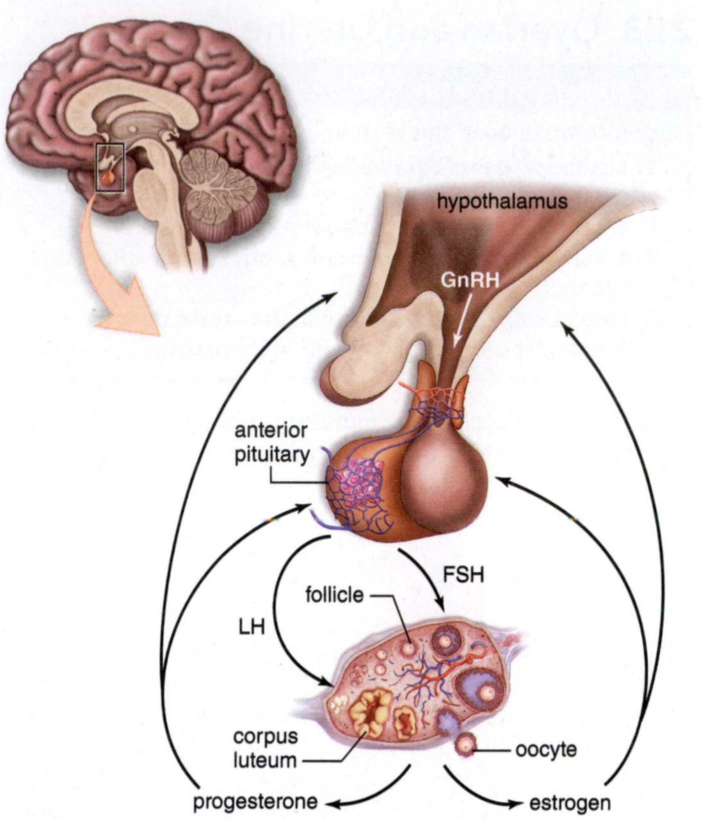

hypothalamus
GnRH
anterior pituitary
FSH
follicle
LH
corpus luteum
oocyte
progesterone
estrogen

Figure 21.8 Hormonal control of ovaries. The hypothalamus produces GnRH, which stimulates the anterior pituitary to produce FSH and LH. FSH stimulates the follicle to produce primarily estrogen, and LH stimulates the corpus luteum to produce primarily progesterone. Estrogen and progesterone maintain the sex organs (e.g., uterus) and the secondary sex characteristics, and they exert feedback control over the hypothalamus and the anterior pituitary. Feedback control regulates the relative amounts of estrogen and progesterone in the blood.

Presumably, an estrogen spike causes a sudden secretion of a large amount of GnRH from the hypothalamus. This leads to a surge of luteinizing hormone (LH) from the anterior pituitary, which causes ovulation at about the 14th day of a 28-day cycle.

Now, the *luteal phase* begins. During this phase, LH promotes the development of the corpus luteum, which secretes progesterone and some estrogen. As the blood level of progesterone rises, it exerts feedback control over the anterior pituitary secretion of LH so that the corpus luteum in the ovary begins to degenerate. As the luteal phase comes to an end, the low levels of progesterone and estrogen in the body cause menstruation to begin, as discussed next.

Animation Maturation of the Follicle and Oocyte

The Uterine Cycle

The female sex hormones, **estrogen** and **progesterone,** have numerous functions. One of their functions is to affect the endometrium, causing the uterus to undergo a cyclical series of events known as the *uterine cycle* (Fig. 21.9, *bottom*). Twenty-eight-day cycles are divided as follows:

During *days 1–5,* a low level of female sex hormones in the body causes the endometrium to disintegrate and its blood vessels to rupture. On day 1 of the cycle, a flow of blood and tissues passes out of the vagina during menstruation, also called the menstrual period.

During *days 6–13,* increased production of estrogen by a new ovarian follicle in the ovary causes the endometrium to thicken and become vascular and glandular. This is called the *proliferative phase* of the uterine cycle.

On *day 14* of a 28-day cycle, ovulation usually occurs.

During *days 15–28,* increased production of progesterone by the corpus luteum in the ovary causes the endometrium of the uterus to double or triple in thickness (from 1 mm to 2–3 mm) and the uterine glands to mature, producing a thick mucous secretion. This is called the *secretory phase* of the uterine cycle. The endometrium is now prepared to receive the developing embryo. If this does not occur, the corpus luteum in the ovary degenerates, and the low level of sex hormones in the female body results in the endometrium breaking down during menstruation.

TABLE 21.3 Ovarian and Uterine Cycles

Ovarian Cycle	Events	Uterine Cycle	Events
Follicular phase—Days 1-13*	FSH secretion begins.	Menstruation—Days 1-5	Endometrium breaks down.
	Follicle maturation occurs. Estrogen secretion is prominent.	Proliferative phase—Days 6-13	Endometrium rebuilds.
Ovulation—Day 14*	LH spike occurs.		
Luteal phase—Days 15-28*	LH secretion continues. Corpus luteum forms.	Secretory phase—Days 15-28	Endometrium thickens, and glands are secretory.

*Assuming a 28-day cycle.

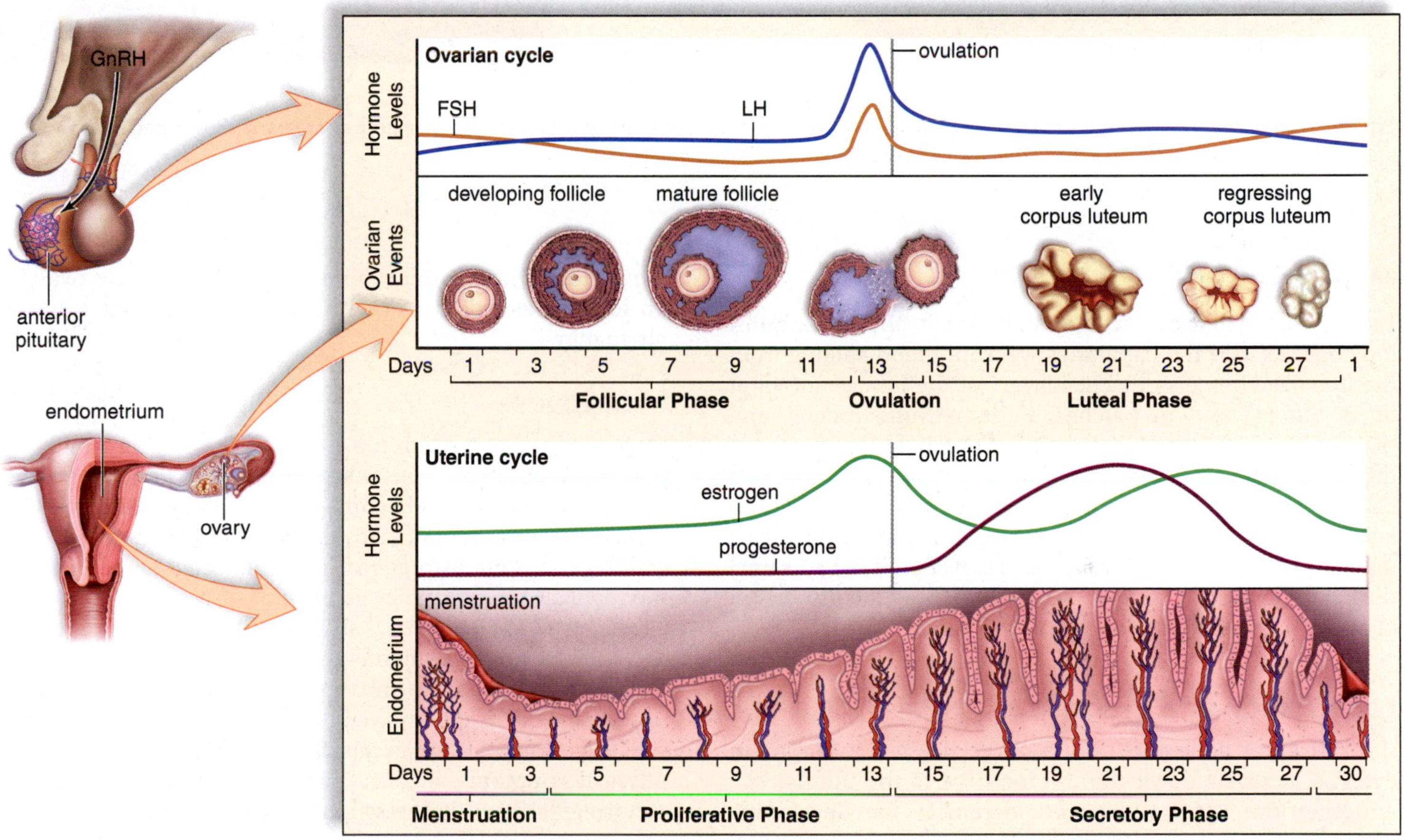
GnRH
anterior pituitary
endometrium
ovary
Ovarian cycle
Hormone Levels
FSH
LH
ovulation
Ovarian Events
developing follicle
mature follicle
early corpus luteum
regressing corpus luteum
Days 1 3 5 7 9 11 13 15 17 19 21 23 25 27 1
Follicular Phase
Ovulation
Luteal Phase
Uterine cycle
estrogen
progesterone
ovulation
menstruation
Endometrium
Days 1 3 5 7 9 11 13 15 17 19 21 23 25 27 30
Menstruation
Proliferative Phase
Secretory Phase

Figure 21.9 Female hormone levels during the ovarian and uterine cycles. During the follicular phase of the ovarian cycle (*top*), FSH released by the anterior pituitary promotes the maturation of a follicle in the ovary. The ovarian follicle produces increasing levels of estrogen, which causes the endometrium to thicken during the proliferative phase of the uterine cycle (*bottom*). After ovulation and during the luteal phase of the ovarian cycle, LH promotes the development of the corpus luteum. This structure produces increasing levels of progesterone, which causes the endometrium to become secretory. Menstruation and the proliferative phase begin when progesterone production declines to a low level.

Table 21.3 and Figure 21.9 compare the stages of the uterine cycle with those of the ovarian cycle.

Estrogen and progesterone affect not only the uterus but other parts of the body as well. Estrogen is largely responsible for the secondary sex characteristics in females, including body hair and fat distribution. In general, females have a more rounded appearance than males because of a greater accumulation of fat beneath the skin. The pelvic girdle becomes wider and deeper in females, so the pelvic cavity usually has a larger relative size in women compared to men. Like males, females develop axillary and pubic hair during puberty. Both estrogen and progesterone are also required for breast development.

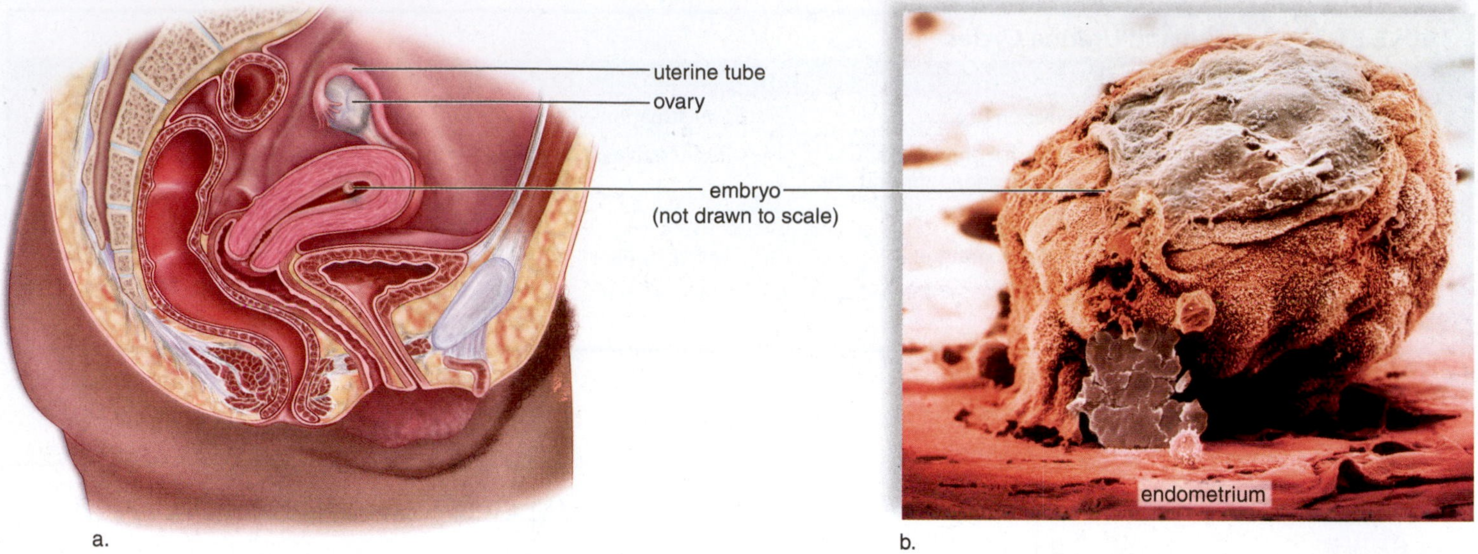
uterine tube
ovary
embryo
(not drawn to scale)
endometrium
a.
b.

Figure 21.10 Implantation. **a.** Site of implantation of an embryo in the uterine wall. **b.** A scanning electron micrograph showing an embryo implanted in the endometrium on day 12 following fertilization.

Menstruation

During **menstruation,** arteries that supply the uterine lining constrict and the capillaries weaken. Blood spilling from the damaged vessels detaches layers of the lining, not all at once, but in random patches. Endometrium, mucus, and blood descend from the uterus, and through the vagina, creating menstrual flow. *Fibrinolysin,* an enzyme released by dying cells, prevents the blood from clotting. Normal menstruation lasts from three to ten days. Some abdominal cramping, breast tenderness, and even moodiness are considered to be normal during the menstrual period. See section 21.6 for a discussion of abnormal menstruation.

Fertilization and Pregnancy

If fertilization does occur, an embryo begins development even as it travels down the uterine tube to the uterus. The endometrium is now prepared to receive the developing embryo, which becomes implanted in the lining several days following fertilization (Fig. 21.10). Pregnancy has now begun.

The **placenta,** which sustains the developing embryo and later fetus, originates from both maternal and fetal tissues. It is the region of exchange of molecules between fetal and maternal blood, although the two rarely mix. At first, the placenta produces **human chorionic gonadotropin (HCG)**, which maintains the corpus luteum in the ovary until the placenta begins its own production of progesterone and estrogen. A pregnancy test can usually detect the presence of HCG in the blood or urine by 10 days after fertilization.

Progesterone and estrogen produced by the placenta have two effects: they shut down the anterior pituitary so that no new follicle in the ovaries matures, and they maintain the endometrium so that the corpus luteum in the ovary is no longer needed. Usually, no menstruation occurs during pregnancy.

Menopause

Menopause, the period in a woman's life during which the ovarian and uterine cycles cease, usually occurs between ages 45 and 55. The ovaries become unresponsive to the gonadotropic hormones produced by the anterior pituitary, and they no longer secrete estrogen or progesterone. At the onset of menopause, the uterine cycle becomes irregular, but as long as menstruation occurs, it is still possible for a woman to conceive. Therefore, a woman is usually not considered to have completed menopause (and thus be infertile) until menstruation has been absent for a year.

The hormonal changes during menopause often produce physical symptoms, such as "hot flashes" (caused by circulatory irregularities), dizziness, headaches, insomnia, sleepiness, and depression. Until recently, many women took combined estrogen-progestin drugs to ease menopausal symptoms. However, accumulating data show long-term use of the combined drugs by postmenopausal women may cause increases in breast cancer, heart attacks, strokes, and blood clots.

Check Your Progress 21.3

1. Summarize the events that occur during the two phases of the ovarian cycle.
2. Distinguish between the proliferative phase and the secretory phase of the uterine cycle and the hormones that promote each.
3. Describe the changes that occur in the ovarian and uterine cycles during menstruation, pregnancy, and menopause.

21.4 Control of Reproduction

Learning Outcomes

Upon completion of this section, you should be able to

1. Distinguish several birth control methods that are designed to prevent ovulation from those that prevent fertilization.
2. Summarize the birth control methods that are effective at preventing the spread of sexually transmitted diseases.
3. Describe two medications used as emergency contraceptives.

Birth control methods are used to regulate the number of children an individual or couple will have.

Birth Control Methods

The most reliable method of birth control is *abstinence*—that is, not engaging in sexual contact. This form of birth control has the added advantage of preventing the spread of sexually transmitted diseases (STDs). Nearly as effective in preventing pregnancy are the surgical methods of sterilization: vasectomy in males and tubal ligation in females. A **vasectomy** involves cutting and sealing off each vas deferens, so that sperm are unable to reach the seminal fluid. **Tubal ligation** consists of cutting and sealing the uterine tubes.

Table 21.4 lists other means of birth control, and rates their effectiveness when used correctly. For example, the birth control pill is nearly 100% effective, meaning that very few sexually active women will get pregnant while taking the pill. On the other hand, with natural family planning, we expect that 70% will not get pregnant and 30% will get pregnant within the year.

TABLE 21.4 Common Birth Control Methods*

Name	Procedure	Methodology	Effectiveness	Risk
Abstinence	Refrain from sexual intercourse	No sperm in vagina	100%	None
Vasectomy	Vasa deferentia cut and tied	No sperm in semen	Almost 100%	Irreversible sterility
Tubal ligation	Uterine tubes cut and tied	No oocytes in uterine tube	Almost 100%	Irreversible sterility
Oral contraception	Hormone medication taken daily	Anterior pituitary does not release FSH and LH	Almost 100%	Increased risk of heart attack, stroke, and thromboembolism
Contraceptive implants	Tubes of progestin (form of progesterone) implanted under skin	Anterior pituitary does not release FSH and LH	More than 90%	Similar to other hormone treatments; rare problems with insertion/removal
Contraceptive injections/patch	Injections of hormones	Anterior pituitary does not release FSH and LH	About 99%	Similar to other hormone treatments
Vaginal contraceptive ring	Ring inserted into vagina each month	Anterior pituitary does not release FSH and LH	92–99%	May be somewhat lower than other hormone treatments
Intrauterine device (IUD)	Plastic coil inserted into uterus by physician	Prevents implantation	More than 90%	PID†, rare problems with insertion/removal
Diaphragm	Latex cup inserted into vagina to cover cervix before intercourse	Blocks entrance of sperm to uterus	With jelly, about 90%	Latex, spermicide allergy
Cervical cap	Latex cap held by suction over cervix	Delivers spermicide near cervix	Almost 85%	UTI‡, latex or spermicide allergy
Male condom	Latex sheath fitted over erect penis	Traps sperm and prevents STDs	About 85%	Latex allergy
Female condom	Polyurethane liner fitted inside vagina	Blocks entrance of sperm to uterus and prevents STDs	About 85%	Latex allergy
Coitus interruptus	Penis withdrawn before ejaculation	Prevents sperm from entering	About 75%	None
Spermicidal jellies, creams, or foams	Spermicidal products inserted before intercourse	Kill a large number of sperm	About 75%	UTI‡, vaginitis, allergy
Natural family planning	Day of ovulation determined by record keeping, various methods of testing	Intercourse avoided on certain days of the month	About 70%	None
Vaginal film	Thin waferlike substance inserted into vagina, near cervix	Kills a large number of sperm	74–94%	UTI‡, vaginitis, allergy

*Assuming a 28-day cycle; † = pelvic inflammatory disease; ‡ = urinary tract infection.

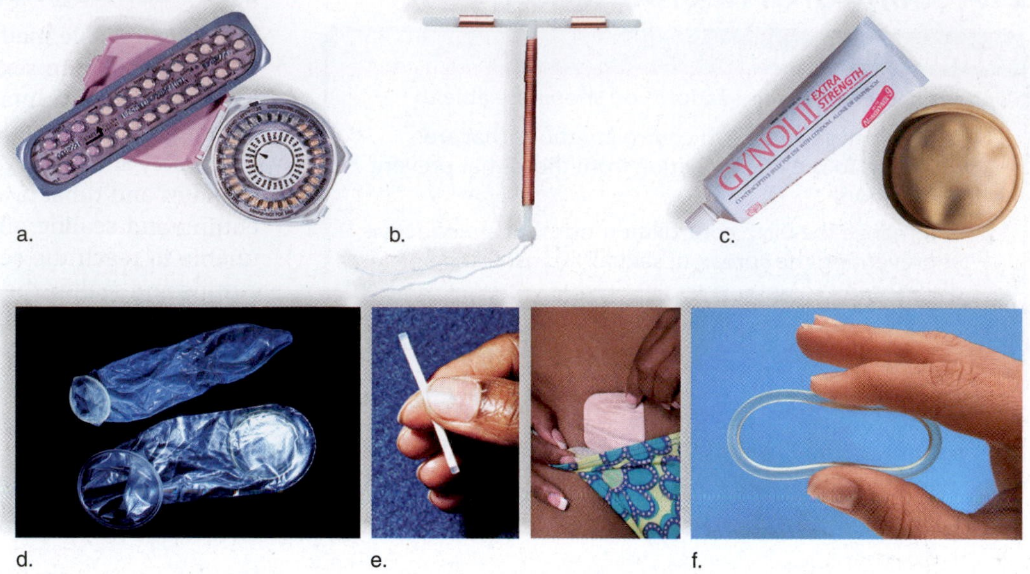
a.
b.
GYNOL II
EXTRA STRENGTH
c.
d.
e.
f.

Figure 21.11 Various methods of birth control. **a.** Oral contraception (birth control pills). **b.** Intrauterine device. **c.** Spermicidal jelly and diaphragm. **d.** Male and female condoms. **e.** Contraceptive implant and patch. **f.** A vaginal contraceptive ring.

Figure 21.11 shows some of the most commonly used means of birth control. **Contraceptives** are medications and devices that reduce the chance of pregnancy. Oral contraception involves taking **birth control pills** on a daily basis (Fig. 21.11*a*). Most of these contain a combination of estrogen and progesterone that mimics the hormonal condition of pregnancy, thus shutting down the pituitary production of both FSH and LH so that no follicle begins to develop in the ovary. Because ovulation does not occur, pregnancy cannot take place. A male birth control pill, or *male hormonal contraception,* will likely be available within a few years. Most approaches use a combination of a progesterone-like hormone to inhibit sperm production, along with testosterone to combat side effects. Concerns persist, however, about whether most men would be willing to consistently take any contraceptive, whether in the form of a pill, an injection, or an implant.

An **intrauterine device** (**IUD**) (Fig. 21.11*b*) is a small device made of molded plastic or copper that is inserted into the uterus by a physician. IUDs are believed to alter the environment of the uterus and oviducts so that fertilization probably will not occur—but if it does, implantation cannot take place. Another type of IUD has a hormone cylinder that releases a synthetic progesterone.

The **diaphragm** is a soft latex cup with a flexible rim that lodges behind the pubic bone and fits over the cervix (Fig. 21.11*c*). Each woman must be properly fitted by a physician, and the diaphragm can be inserted into the vagina no more than two hours before sexual relations. Also, it must be used with spermicidal jelly or cream and should be left in place for at least six hours after sexual relations. The *cervical cap* is a mini-diaphragm.

In addition to helping prevent pregnancy, both male and female condoms provide some degree of protection against sexually transmitted diseases (Fig. 21.11*d*). However, condoms are not 100% effective in preventing pregnancy or STDs. A **male condom** is most often a latex sheath that fits over the erect penis. The ejaculate is trapped inside the sheath, and thus does not enter the vagina. A **female condom** consists of a large polyurethane tube with a flexible ring that fits onto the cervix. The open end of the tube has a ring that covers the external genitals. Better protection is achieved when a condom is used in conjunction with a spermicide.

Contraceptive implants (Fig. 21.11*e*) use synthetic progesterone to prevent ovulation by disrupting the ovarian cycle. The newest version consists of a single capsule that remains effective for about three years. A **contraceptive patch** is also available. Known by the brand name Ortho Evra, the patch is applied to the skin once a week for three weeks, then no patch is worn during the menstrual period. **Contraceptive injections** are available as progesterone only or as a combination of estrogen and progesterone. The time between injections can vary from one to several months.

One of the newer developments in hormonal contraceptives is the **vaginal contraceptive ring** (Fig. 21.11*f*). Known by the brand name NuvaRing, this device is a flexible vinyl ring about two inches in diameter, that contains estrogen and progesterone. Currently available by prescription only, it is inserted into the vagina on about the fifth day after menstruation ceases, and left in place for 21 days. Used properly, the NuvaRing prevents ovulation for that month's cycle. It also thickens the cervical mucus, preventing sperm from entering the uterus.

Contraceptive vaccines are now being developed. For example, a vaccine intended to immunize women against HCG, the hormone so necessary to maintaining the implantation of the embryo, was successful in a limited clinical trial. Because HCG is not normally present in the body, no autoimmune reaction is expected, but the immunization does wear off with time. Similar approaches are being investigated to develop a safe antisperm vaccine for use in men or women.

Emergency contraception includes approaches to birth control that can prevent pregnancy after unprotected intercourse. The expression "morning-after pill" is a misnomer in that a woman can use these medications up to several days after unprotected intercourse.

One type, called the Preven Emergency Contraceptive Kit, contains four synthetic progesterone pills; two are taken up to 72 hours after unprotected intercourse, and two more are taken 12 hours later. The medication upsets the normal uterine cycle, preventing an embryo from implanting in the endometrium. Used correctly, Preven is about 85% effective in preventing unintended pregnancies. Currently Preven is available only when prescribed by a doctor.

Another type of emergency contraception is Plan B One-Step, which is up to 89% effective in preventing pregnancy if taken within 72 hours after unprotected sex. It is available to women age 17 and older without a prescription. A similar medication called ella is more effective up to day four or five after unprotected sex, but a prescription is required. In some cases, these medications can be combined with the insertion of an IUD.

Check Your Progress 21.4

1. Classify which birth control methods in Table 21.4 prevent ovulation, and which prevent fertilization.
2. List the birth control methods discussed in this section that physically block sperm from entering the uterus.
3. Summarize which birth control methods reduce the risk of STDs.
4. Define "emergency contraception."

21.5 Sexually Transmitted Diseases

Learning Outcomes

Upon completion of this section, you should be able to

1. List the specific cause and common symptoms of three STDs that are caused by viruses and three caused by bacteria.
2. Describe how HIV specifically affects the immune system and how this explains the three categories of HIV infection.
3. Discuss four types of antiretroviral drugs.
4. Explain why antibiotics should not be used to treat viral infections.

Many infectious diseases are transmitted by sexual contact and are therefore called sexually transmitted diseases (STDs). Note that these diseases may or may not affect the reproductive organs.

STDs Caused by Viruses

AIDS, genital herpes, genital warts, and hepatitis B are common STDs caused by viruses. Therefore, they do not respond to traditional antibiotics. Antiviral drugs have been developed to treat some of these, but in many cases infection is lifelong.

AIDS

Acquired immunodeficiency syndrome (AIDS) is caused by the *human immunodeficiency virus (HIV)*. AIDS is considered a pandemic because the disease is prevalent in the entire human population around the globe. As of the end of 2010, the World Health Organization reports that AIDS has killed nearly 30 million people worldwide, and about 34 million more are currently estimated to be infected with HIV. The epidemic has been particularly devastating in sub-Saharan Africa, where 30–40% of the population is infected in some countries. It is also estimated that AIDS could kill 31 million people in India and 18 million in China by the year 2025. In the United States, an estimated 1 million people are infected, and over 500,000 have died. Homosexual men make up the largest proportion of people with AIDS in the United States, but the fastest rate of increase is now seen among heterosexuals, and over half of all new HIV infections are occurring in people under the age of 25.

HIV is transmitted by sexual contact with an infected person, including vaginal or anal intercourse and oral/genital contact. Also, needle-sharing among intravenous drug users is a very efficient way to transmit HIV. The Health feature, "Preventing Transmission of STDs," on page 422 suggests ways to protect yourself against HIV and other STDs. Babies born to HIV-infected mothers may become infected before or during birth or through breast-feeding after birth. Cultural factors may also play a very important role in the transmission of HIV. For example, polygamy is common in many African countries, as are misconceptions about the effectiveness of birth control, and, indeed, about the fact that HIV is the cause of AIDS. Such practices and beliefs may significantly increase the risk of HIV infection.

Stages of an HIV Infection When a person initially becomes infected with HIV, the virus replicates at a high level in the macrophages and $CD4^+$ helper T cells, and spreads throughout the lymphoid tissues of the body (Fig. 21.12). Even though the person's blood contains a high "viral load" at this point, most experience only mild symptoms such as fever, chills, aches, and

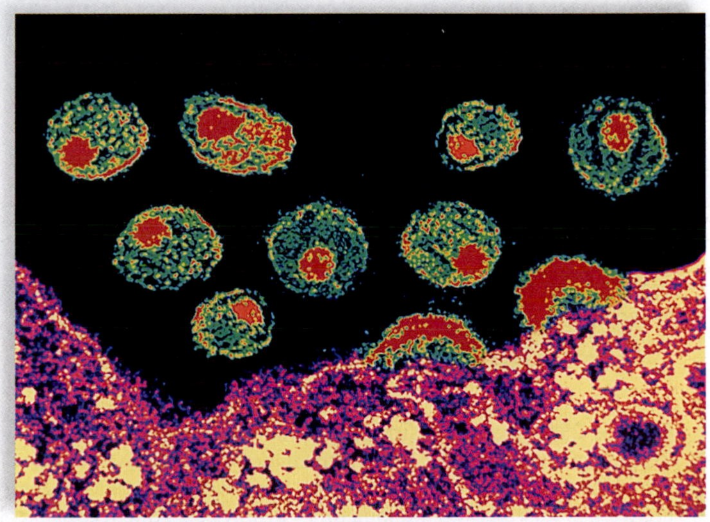

Figure 21.12 HIV, the AIDS virus. False-colored micrographs show HIV particles budding from an infected helper T cell. These viruses can infect helper T cells and also macrophages, which work with helper T cells to stem the infection.

SCIENCE IN YOUR LIFE ▶ HEALTH

Preventing Transmission of STDs

Being aware of how STDs are spread (Fig. 21A) and then observing the following guidelines will greatly help prevent the transmission of STDs.

Sexual Activities Transmit STDs

Abstain from sexual intercourse or develop a long-term monogamous (always the same partner) *sexual relationship* with a partner who is free of STDs.

Refrain from multiple sex partners or having relations with someone who has multiple sex partners. If you have sex with two other people and each of these has sex with two people and so forth, the number of people having sexual contact is quite large.

Remember that, although the prevalence of AIDS is presently higher among homosexuals and bisexuals, the highest rate of increase is now occurring among heterosexuals. The lining of the uterus is only one cell thick, and thus it can allow infected cells from a sexual partner to enter the blood.

Be aware that having relations with an intravenous drug user is risky because the behavior of this group risks hepatitis and an HIV infection. Be aware that anyone who already has another sexually transmitted disease is more susceptible to an HIV infection.

Avoid anal-rectal intercourse (in which the penis is inserted into the rectum) because the lining of the rectum is thin and cells infected with HIV can easily enter the body there.

Unsafe Sexual Practices Transmit STDs

Always use a latex condom during sexual intercourse if you do not know for certain that your partner has been free of STDs for some time. Be sure to follow the directions supplied by the manufacturer.

Avoid fellatio (kissing and insertion of the penis into a partner's mouth) *and cunnilingus* (kissing and insertion of the tongue into the vagina) because they may be a means of transmission. The mouth and gums often have cuts and sores that facilitate the entrance of infected cells.

Be cautious about using alcohol or any drug that may prevent you from being able to control your behavior.

Drug Use Transmits Hepatitis and HIV

Stop, if necessary, or do not start the habit of injecting drugs into your veins. Be aware that hepatitis and HIV can be spread by blood-to-blood contact.

Always use a new, sterile needle for injection or one that has been cleaned in bleach if you are a drug user and cannot stop your behavior.

Questions to Consider

1. How can you be certain that your partner doesn't have any STDs? Do people necessarily know they have an STD?
2. What does it mean if a person tells you he or she just had an AIDS test, and it was negative? What does this test detect in the blood?
3. Which methods of birth control are most effective in preventing transmission of HIV? Which are least effective?

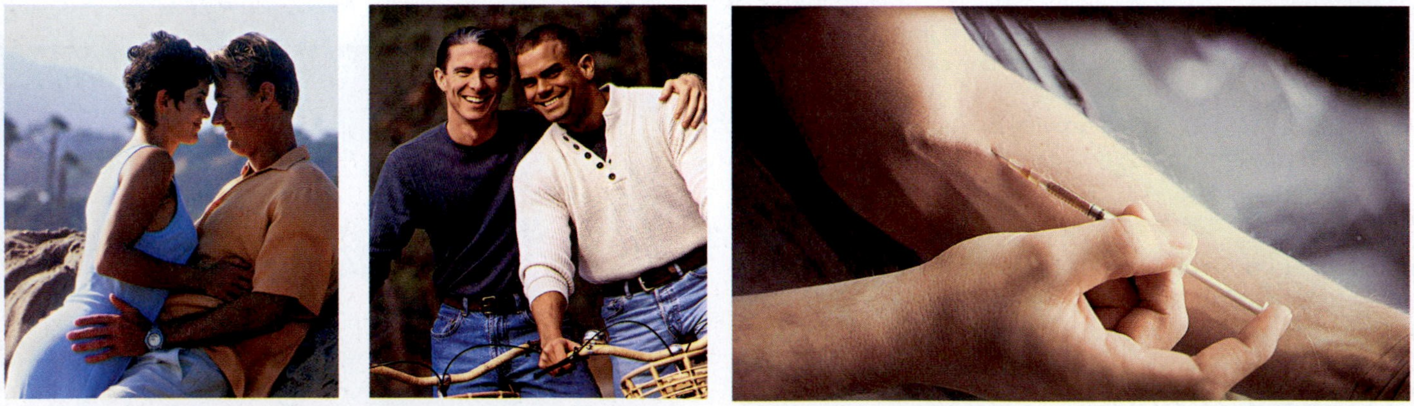

a. b.

Figure 21A Transmission of STDs. **a.** Sexual activities transmit STDs. **b.** Sharing needles transmits STDs. In the United States, more than 65 million people are living with an incurable STD. An additional 15 million people become infected with one or more STDs each year. Approximately one-fourth of these new infections occur in teenagers. Despite the fact that STDs are quite widespread, many people remain unaware of the risks and the sometimes deadly consequences of becoming infected.

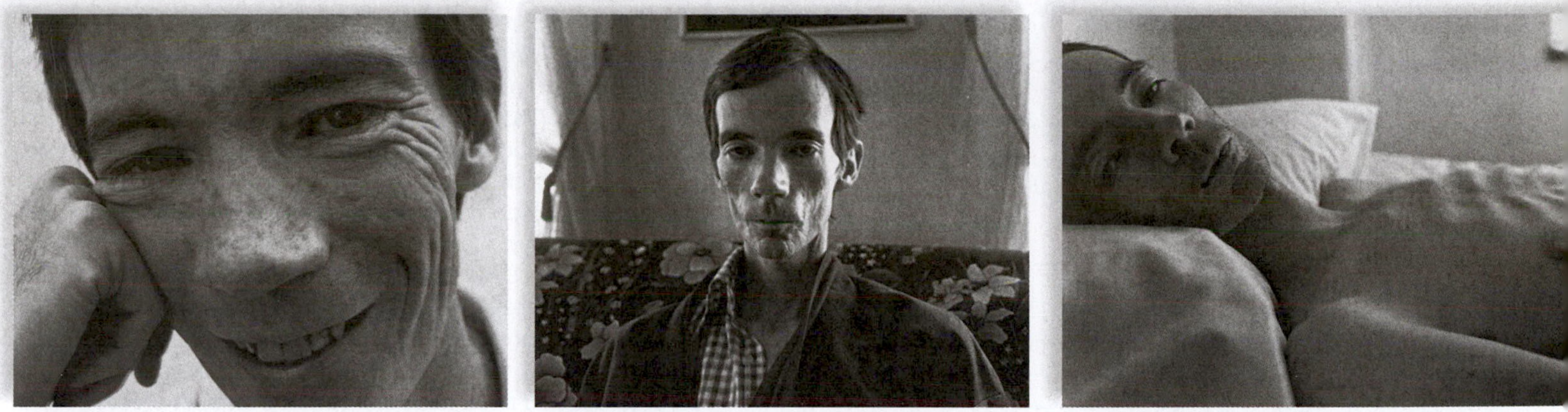

a. AIDS patient Tom Moran, at diagnosis b. AIDS patient Tom Moran, 6 months later c. AIDS patient Tom Moran, in category C of AIDS

Figure 21.13 The course of an AIDS infection. These photos show the effect of an HIV infection in one individual who progressed through all the phases of HIV infection. At first, the body can fight off the effects of the virus, but as the infection takes over the number of $CD4^+$ T cells drops from thousands to hundreds. Eventually, an AIDS patient will die from the opportunistic diseases that flourish because the immune system has been destroyed.

swollen lymph nodes, and soon these symptoms disappear. Even if a person had an HIV test at this time, it would probably be negative because it tests for the presence of anti-HIV antibodies, which are not detectable for an average of 25 days. This means that early in the infection, a person can be highly infectious, even though the HIV antibody test is negative.

Usually no other HIV-related symptoms occur for several years. Even so, the virus is beginning to cause a loss of $CD4^+$ T cells, which is one of the hallmarks of an HIV infection that is progressing to AIDS. HIV is thought to kill these cells directly by infecting them and also indirectly by inducing apoptosis, even of uninfected cells. Cytotoxic T cells are also thought to kill large numbers of HIV-infected cells. During this stage of infection, the body is able to maintain a reasonably healthy immune system by producing as many as one to two billion new $CD4^+$ helper T cells per day. As long as the body's production of new $CD4^+$ T cells is able to keep pace with the loss of $CD4^+$ T cells (much like leaving the water running in a sink with the drain open), the person stays relatively healthy. However, even during this asymptomatic stage, evidence indicates that the virus continues to replicate at a high rate. Eventually, the number of $CD4^+$ T cells falls so low that the patient begins to suffer from **opportunistic infections,** which have the *opportunity* to occur because the immune system is weakened.

Animation How the HIV Infection Works

In order to help physicians assess the prognosis of individuals at different stages of HIV infection, the U.S. Centers for Disease Control and Prevention have defined three categories of HIV infection, based mainly on the type of clinical disease seen in the patient (Fig. 21.13). Within each category, the number of $CD4^+$ T cells is also monitored. Patients in *category A* usually have $CD4^+$ T-cell counts of 500 or greater cells per cubic millimeter (cells/mm^3) of blood, and either lack symptoms completely or have persistently enlarged lymph nodes. Patients in *category B* have symptoms that are indicative of an unhealthy immune system. Their $CD4^+$ T-cell count is usually between 200 and 499 cells/mm^3 of blood. Examples of category B conditions include thrush, a yeast infection of the oral cavity, and repeated episodes of shingles, a reactivation of a childhood chickenpox virus infection. Patients in *category C* usually have $CD4^+$ T-cell counts less than 200 cells/mm^3 of blood, and have one or more AIDS-indicator conditions, such as recurrent bacterial pneumonia, various fungal infections, Kaposi sarcoma (caused by a human herpesvirus), mycobacterial infections, and toxoplasmosis of the brain (see the Health feature, "Opportunistic Infections and HIV," in Chapter 13).

Although not composing an official category of HIV infection, another small, yet very interesting group of HIV-infected people remain healthy for at least ten years, even without any antiviral treatments. Often called *long-term nonprogressors*, these people have been very interesting for HIV researchers to study because their viral loads remain low and their $CD4^+$ T-cell counts usually remain normal. As of this writing, no single factor has been identified that explains how some people are able to remain relatively healthy in the face of HIV infection, but their existence continues to provide hope that effective host resistance to HIV is possible.

Treatments for HIV Infection Prior to the development of antiviral drugs, most HIV-infected individuals eventually progressed to category C or full-blown AIDS, which was almost inevitably fatal. There is still no cure for HIV infection, but since 1995 an increasing number of drugs have become available to help control the replication and spread of HIV, and the life span of many HIV-infected patients has increased to the point where HIV is no longer considered an automatic death sentence. Unfortunately, this is not the case in developing nations where most people cannot afford these drugs.

In order to understand how antiretroviral drugs work, a brief review of the life cycle of HIV is necessary (Fig. 21.14). HIV is a type of RNA virus called a retrovirus. HIV has a protein on its outer surface called gp120, which attaches to a CD4 molecule,

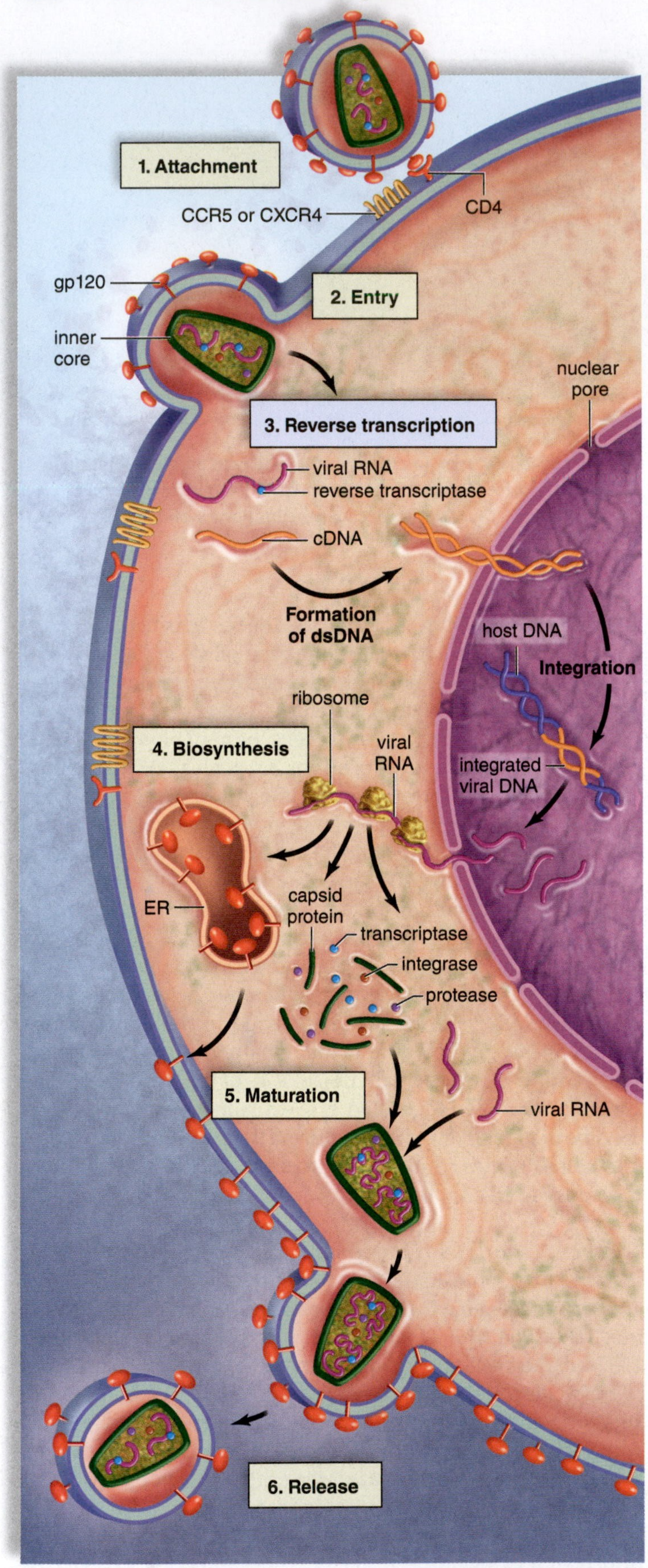

1. Attachment
CCR5 or CXCR4
CD4
gp120
2. Entry
inner core
nuclear pore
3. Reverse transcription
viral RNA
reverse transcriptase
cDNA
Formation of dsDNA
host DNA
Integration
ribosome
4. Biosynthesis
viral RNA
integrated viral DNA
ER
capsid protein
transcriptase
integrase
protease
5. Maturation
viral RNA
6. Release

Figure 21.14 Reproduction of the retrovirus HIV. HIV attaches to CD4 plus CCR5 or CXCR4, then uses reverse transcription to produce a double-stranded DNA copy of viral RNA. The viral dsDNA integrates into the cell's chromosomes before the virus reproduces and buds from the cell.

plus another molecule called CCR5 or CXCR4 on a host T cell, macrophage, or dendritic cell. This attachment causes *fusion* of HIV with the host plasma membrane, releasing HIV's inner core inside the cell. This core contains the viral RNA, plus several viral enzymes. One such enzyme, called *reverse transcriptase,* makes a DNA copy of the viral RNA, called cDNA. Duplication occurs and the result is double-stranded DNA (dsDNA). During *integration,* the viral enzyme integrase inserts the viral dsDNA into host DNA, where it will remain for the life of the cell. Following integration, viral DNA serves as a template for production of more viral RNA. Some of the viral RNA proceeds to the ribosomes, where it brings about the synthesis of long polypeptides that need to be cleaved into smaller pieces before they can be assembled into new viruses. This cleavage process depends on a third viral enzyme, called *protease.* Finally, the new virus particles are assembled and then released from the host plasma membrane by a process called budding.

The National Institute of Allergy and Infectious Disease lists five classes of currently available antiretroviral drugs: (1) *reverse transcriptase inhibitors,* which block the production of viral DNA from RNA; (2) *protease inhibitors,* which block the viral enzyme that cleaves long viral polypeptides prior to assembly; (3) *fusion/entry inhibitors,* which interfere with HIV's ability to fuse with the host plasma membrane during initial infection; (4) *integrase inhibitors,* which inhibit the insertion of viral DNA into host DNA; and (5) *multidrug combination products* that incorporate drugs from more than one class into a single product. When the first antiretroviral drug, known as AZT, became available in 1995, it was initially very effective at reducing the viral load in many patients, but the virus tended to become resistant to the drug very quickly. This is because the HIV reverse transcriptase enzyme is very error-prone, which means the virus mutates at a very high rate, resulting in drug-resistant mutants. However, today HIV-infected patients are typically treated with three or four antiretroviral drugs at the same time, a combination called *highly active antiretroviral therapy* (*HAART*). Because it is unlikely that a particular mutant virus will be resistant to several drugs all at once, HAART is usually able to inhibit HIV replication to the point that no virus is detectable in the blood. It is important to note, however, that these patients presumably still have large numbers of HIV-infected cells, and thus they can still transmit the virus. Other potential problems with HAART include various side effects, which can be severe, the high cost, and the complexity of dosing schedules for the different medications.

Investigators have found that when HAART is discontinued, the virus rebounds. Therefore, therapy is usually continued indefinitely. An HIV-positive pregnant woman who takes reverse transcriptase inhibitors during her pregnancy reduces the chances of HIV transmission to her newborn. If possible, drug therapy should be delayed until the 10th to 12th weeks of her pregnancy to minimize any adverse effects of antiretroviral drugs on fetal development.

Animation Treatment of HIV Infection

HIV Vaccines It is unlikely that the global HIV pandemic will ever be controlled unless an effective vaccine against HIV is developed. Some of the obstacles to developing an effective vaccine include the existence of many strains of HIV, the tendency of HIV to mutate its proteins, thus making it a "moving target" for the immune system, and the ability of HIV to "hide" inside cells. One type of vaccine that has been studied is a recombinant vaccine, in which a piece of DNA encoding an HIV protein is inserted into another organism (usually another type of virus), which can then infect cells and elicit a cytotoxic T-cell response, without any danger of causing an infection with HIV. Other types of vaccines focus more on inducing an antibody response, especially targeted against the gp120 protein (Fig. 21.14). In a 2009 study involving 16,000 participants in Thailand, people who received an HIV vaccine were 31% less likely to become HIV-positive than those who received a placebo. Other vaccine trials are currently being conducted in several countries.

Genital Herpes

Genital herpes is caused by the *herpes simplex virus,* of which there are two types: type 1, which usually causes cold sores and fever blisters, and type 2, which more often causes genital herpes. In the United States, it is estimated that approximately one in four women, and one in five men, have had a genital herpes infection, but 20% may never have symptoms. Even without symptoms, however, the virus can be transmitted. After becoming infected, most individuals experience a tingling or itching sensation before blisters appear at the infected site within 2 to 20 days (Fig. 21.15*a*). Once the blisters rupture, they leave painful ulcers, which take from five days to three weeks to heal. These symptoms can be accompanied by fever, pain upon urination, and swollen lymph nodes. Several antiviral medications are available to treat herpesvirus infections. These can be used topically in ointment form, taken orally, or in severe cases, injected. All of these medications inhibit the replication of viral DNA.

After the blisters heal, the virus becomes latent (dormant), and blisters can recur repeatedly at variable intervals. Sunlight, sexual intercourse, menstruation, and stress seem to cause the symptoms of genital herpes to recur. While the virus is latent, it primarily resides in the ganglia of the sensory nerves associated with the affected skin. The ability of the virus to "hide" in this manner has made it more difficult to develop an effective vaccine.

Infection of the newborn can occur if the child comes in contact with a lesion in the birth canal. Within one to three weeks, the infant may be gravely ill and can become blind, have neurological disorders, including brain damage, or die. Birth by cesarean section prevents these adverse developments.

Human Papillomavirus

An estimated 20 million Americans are currently infected with the *human papillomavirus* (*HPV*). There are over 100 types of HPV. Most cause warts, and about 30 types cause **genital warts,** which are sexually transmitted. Genital warts may appear as flat or raised warts on the penis and/or foreskin of males (Fig. 21.15*b*), and on the vulva, vagina, and/or cervix of females. Note that if the wart(s) are only on the cervix, there may be no outward signs or symptoms of the disease. Newborns can also be infected with HPV during passage through the birth canal.

About ten types of HPV can cause **cervical cancer,** the second leading cause of cancer death in women in the United States (approximately 500,000 deaths per year). These HPV types produce a viral protein that inactivates a host protein called p53, which normally acts as a "brake" on cell division. Once p53 has been inactivated in a particular cell, that cell is more prone to the uncontrolled cell division characteristic of cancer. Early detection of cervical cancer is possible by means of a **Pap test,** in which a few cells are removed from the region

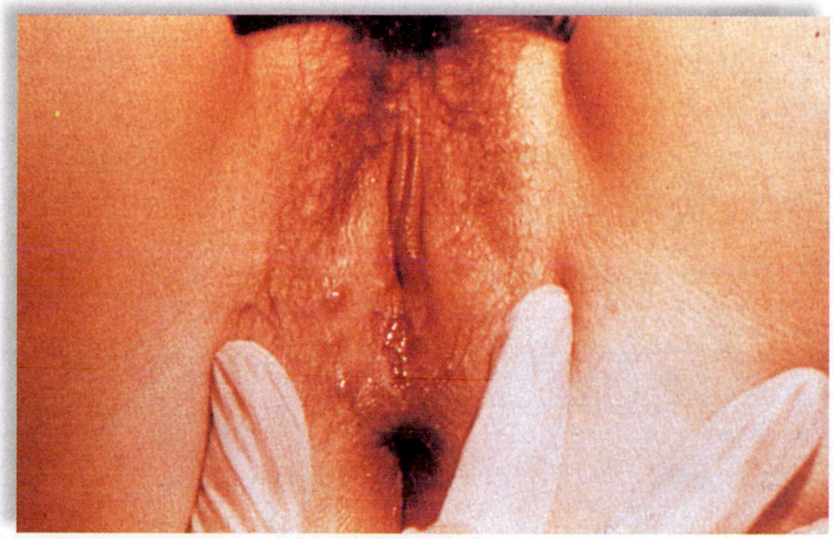

a.

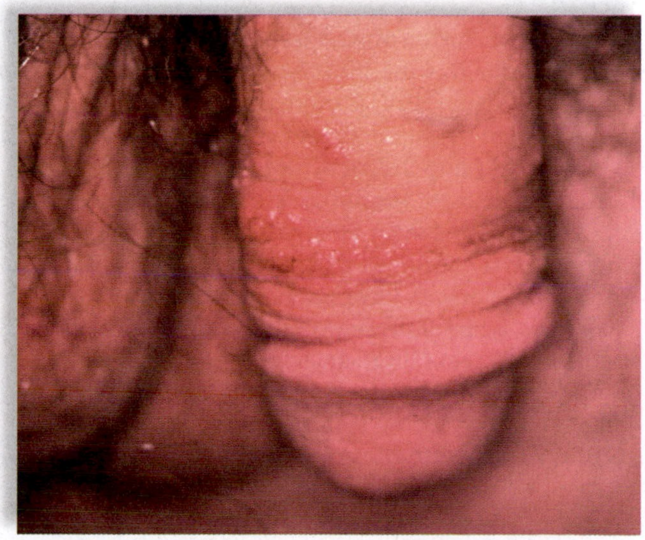

b.

Figure 21.15 Two STDs causes by viruses. **a.** There are several types of herpesviruses, and usually the type called herpes simplex 2 causes genital herpes. Symptoms of genital herpes are due to an outbreak of blisters, which can be present on the labia of females (seen here) or on the penis of males. **b.** Several types of human papillomaviruses (HPV) can cause genital warts, seen here on the penis. In the female, genital warts may occur externally on the vulva, or internally in the vagina or cervix. Certain HPV types can also cause cancer.

of the cervix for microscopic examination. If the cells are cancerous, a hysterectomy (removal of the uterus) may be recommended. In males, HPV can cause cancers of the penis, anus, and other areas. According to a 2008 study, HPV now causes as many cancers of the mouth and throat in U.S. males as does tobacco, perhaps due to an increase in oral sex as well as a decline in tobacco use.

Currently, there is no cure for an HPV infection, but the warts can be treated effectively by surgery, freezing, application of an acid, or laser burning. However, even after treatment, the virus can sometimes be transmitted. Therefore, once someone has been diagnosed with genital warts, abstinence or the use of a condom is recommended to help prevent transmission of the virus.

In June 2006, the U.S. Food and Drug Administration licensed Gardasil, an HPV vaccine that is effective against the four most common types of HPV found in the United States, including the two types that cause about 70% of cervical cancers. Because the vaccine doesn't protect those who are already infected, ideally children should be vaccinated before they become sexually active. This vaccine has been controversial, however. For example, in February 2007, Texas governor Rick Perry issued an executive order requiring that all girls entering the sixth grade be vaccinated for HPV. Within two months, however, the Texas legislature passed a bill blocking public officials from carrying out the mandatory vaccination.

In 2009, the U.S. Food and Drug Administration approved Gardasil for use in males. The CDC recommends that 11- to 12-year-old girls and boys receive three doses of the vaccine. Nonpregnant females between ages 13 and 26, and males from age 13 to 21, can also be vaccinated if they did not receive any or all of the three recommended doses when they were younger. Older individuals should speak with their doctor to find out if getting vaccinated is right for them.

Hepatitis B

Of the several forms of viral **hepatitis,** or inflammation of the liver, *hepatitis B* is the most likely to be transmitted sexually. Hepatitis B virus (HBV) is more contagious than HIV, and HBV is transmitted by similar routes. About 50% of persons infected with HBV develop flulike symptoms, including fatigue, fever, headache, nausea, vomiting, and muscle aches. As the virus replicates in the liver, a dull pain in the upper right half of the abdomen may occur, as may jaundice, a yellowish cast to the skin. Some persons have an acute infection that only lasts a few weeks, while others develop a chronic form of the disease that leads to liver failure and the need for a liver transplant. A safe and effective vaccine is available for the prevention of HBV infection. This vaccine is now on the list of recommended immunizations for children.

Video Halting Hepatitis

STDs Caused by Bacteria

Chlamydia, gonorrhea, and syphilis are bacterial diseases that are usually curable with appropriate antibiotic therapy if diagnosed early enough. However, an increasing number of these bacteria are becoming resistant to antibiotics.

Chlamydia

Chlamydia is named for the tiny bacterium that causes it, *Chlamydia trachomatis.* New chlamydial infections are more numerous than any other sexually transmitted disease. Some estimate that the actual incidence could be as high as 6 million new cases per year. For every reported case in men, more than five cases are detected in women. The low rates in men suggest that many of the sex partners of women with chlamydia are not diagnosed or reported.

Chlamydial infections of the lower reproductive tract are usually mild or asymptomatic. About 8 to 21 days after infection, men experience *urethritis,* an inflammation of the urethra, resulting in a mild burning sensation on urination and a mucous discharge. The infection can spread to the prostate gland and epididymides in males.

Women may have a vaginal discharge, along with the symptoms of a urinary tract infection. If not properly treated, the infection can eventually cause **pelvic inflammatory disease (PID)**. This condition is characterized by inflamed oviducts that become partially or completely blocked by scar tissue. As a result, the woman may become sterile or subject to ectopic pregnancy, which can be a medical emergency. Chlamydial infections may also increase the possibility of premature and stillborn births. If a newborn is exposed to chlamydia during delivery, pneumonia or infection of the eyes can result (Fig. 21.16).

Detection and Treatment of Chlamydia New and faster laboratory tests are now available for detecting a chlamydial infection. Their expense sometimes prevents public clinics from testing for chlamydia. Thus, the following criteria have been suggested to help physicians decide which women should be tested: no more than 24 years old; a new sex partner within the preceding two months; cervical discharge; bleeding during parts of the vaginal exam; and use of a nonbarrier method of contraception. Some doctors, however, are routinely prescribing additional antibiotics appropriate to treating chlamydia for anyone who has gonorrhea, because 40% of females and 20% of males with gonorrhea also have chlamydia.

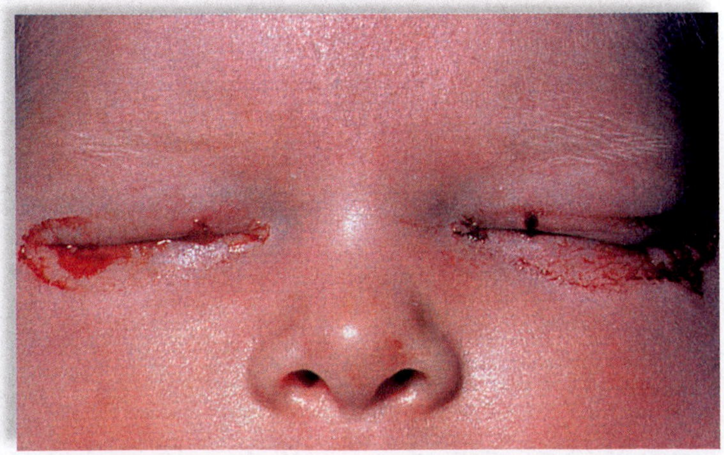

Figure 21.16 Chlamydial eye infection. This newborn's eyes were infected after passing through the birth canal of an infected mother.

Gonorrhea

Gonorrhea is caused by the bacterium *Neisseria gonorrhoeae.* Although as many as 20% of infected males are asymptomatic, most have urethritis, and experience pain at urination with a thick, greenish-yellow urethral discharge three to five days after contact with an infected partner. In contrast, 60–80% of infected females are asymptomatic until they develop severe PID. Similarly, scarring of each vas deferens may occur in untreated males. There is also strong evidence that infection with *N. gonorrhoeae* increases the risk of transmitting HIV.

Gonorrhea proctitis, or infection of the anus, with symptoms that may include pain in the anus and blood or pus in the feces, is spread by anal intercourse. Oral sex can cause infection of the throat and the tonsils. Gonorrhea can also spread to other parts of the body, causing heart damage or arthritis. If, by chance, the person touches infected genitals and then his or her eyes, a severe eye infection can result.

Eye infection leading to blindness can occur as a baby passes through the birth canal. Because of this, all vaginally delivered infants receive eyedrops containing antibacterial agents, such as silver nitrate, tetracycline, or penicillin, as a protective measure.

Reported gonorrhea rates declined steadily until the late 1990s, and since then, they have stayed fairly constant. In 2010, approximately 309,000 cases of gonorrhea were reported in the United States, or about 1 case per 1,000 people. About 75% of all reported cases are found in people between 15 and 29 years of age. Gonorrhea rates among African Americans are about 20 times greater than among Caucasians. Antibiotic-resistant strains of *N. gonorrhoeae* are also becoming more common. As with AIDS, condoms can help protect against gonorrheal and chlamydial infections.

Syphilis

Syphilis is caused by the bacterium *Treponema pallidum,* an actively motile corkscrewlike organism (Fig. 21.17*a*). Rates of infection have mostly been decreasing since reporting began in the 1940s, largely because of the development of effective diagnostic and treatment methods. In 2010, only about 14,000 cases were reported in the United States. However, syphilis remains a serious health problem in certain areas, mainly in the South and in large metropolitan areas. About 67% of new cases occur in men who have sex with men, with rates rising fastest among males aged 20 to 29 years. Worldwide, an estimated 10 million new infections occur each year.

Syphilis has three stages, which can be separated by intervening latent periods during which the bacteria are not multiplying. During the *primary stage,* a hard chancre (ulcerated sore with hard edges) indicates the site of infection (Fig. 21.17*b*). The chancre can go unnoticed, especially because it usually heals spontaneously, leaving little scarring. During the *secondary stage,* the individual breaks out in a rash, coinciding with the replication and spread of bacteria throughout the body. Curiously, the rash does not itch and is even seen on the palms of the hands and the soles of the feet (Fig. 21.17*c*). Hair loss can occur, and infectious gray patches may develop on the mucous membranes, including the mouth. These symptoms disappear of their own accord.

Not all cases of secondary syphilis go on to the *tertiary stage.* During this stage, which lasts until the patient dies, syphilis may affect the cardiovascular system by weakening arterial walls (particularly in the aorta), allowing aneurysms to develop. The disease may also affect the nervous system, causing psychological problems. Sometimes during the tertiary stage, gummas, large destructive ulcers, may develop on the skin or within the internal organs (Fig. 21.17*d*).

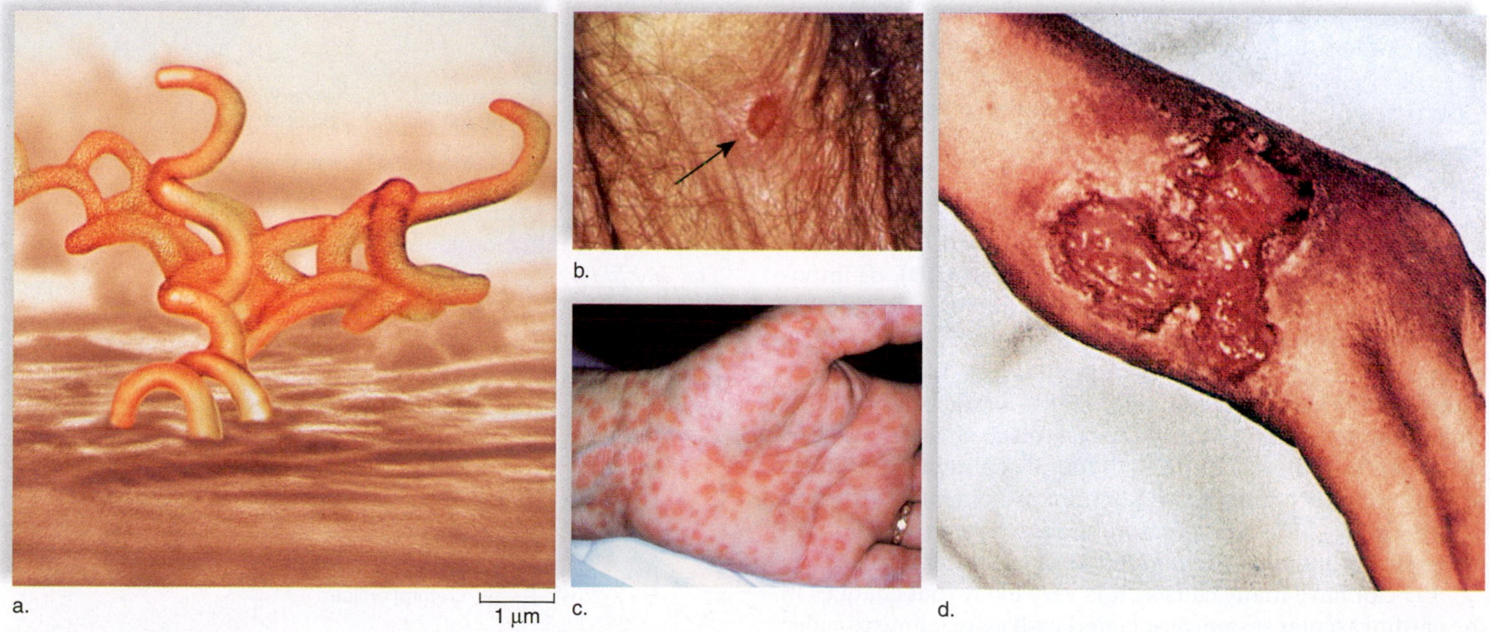
1 µm

a. b. c. d.

Figure 21.17 Syphilis. **a.** Scanning electron micrograph of *Treponema pallidum,* the cause of syphilis. **b.** The primary stage of syphilis is a chancre at the site where the bacterium enters the body. **c.** The secondary stage is a body rash that occurs even on the palms of the hands and soles of the feet. **d.** In the tertiary stage, gummas may appear on the skin or internal organs.

Congenital syphilis, which is present at birth, is caused by *T. pallidum* bacteria crossing the placenta. The child is born blind and/or with numerous anatomical malformations.

As with most bacterial diseases, antibiotics are effective in treating syphilis, especially in the early stages. Therefore, it is crucial for all sexual contacts to be identified and treated.

Check Your Progress 21.5

1. List three STDs that are caused by bacteria and three that are caused by viruses.
2. Describe three ways that HIV may destroy $CD4^+$ T cells.
3. List four classes of antiretroviral drugs and the specific stage of the HIV life cycle that each class targets.
4. Explain why viral diseases, in general, are more difficult to treat than bacterial diseases.

21.6 Other Disorders of the Reproductive System

Learning Outcomes

Upon completion of this section, you should be able to

1. Describe some common disorders of the male reproductive system.
2. Describe some common disorders of the female reproductive system.
3. Discuss four assisted reproductive technologies that can help infertile couples conceive a child.

Several common disorders affect the reproductive system. While some of these affect the ability of young adult males or females to reproduce, others tend to occur in older individuals, and may affect their overall health.

Common Conditions Affecting the Male Reproductive System

Erectile Dysfunction

It is estimated that about 50% of men aged 40 to 75 have experienced some degree of **erectile dysfunction** (**ED**), or impotence, which is the inability to produce or maintain an erection sufficient to perform sexual intercourse. Beginning in 1998, several new drugs have been developed to treat ED. All of these drugs—like Viagra, Levitra, and Cialis—work by increasing blood flow to the penis during sexual arousal. Specifically, they inhibit an enzyme called *PDE-5,* found mainly in the penis, which normally breaks down some of the chemicals responsible for an erection. Surveys have shown these medications to be helpful in about 65% of men who are experiencing ED.

ED can have many causes. It is very likely that changes in the cardiovascular system associated with aging alone can also produce some degree of ED. A normal erection is dependent on blood flow, so almost any condition that affects the cardiovascular system, such as atherosclerosis, high blood pressure, or diabetes mellitus, can lead to ED. Certain medications can cause ED, as can smoking, alcohol, and some illicit drugs. Psychological factors, especially depression, can also be very important contributors to ED. Depending on the cause, other therapies besides the Viagra-type drugs may be indicated. For example, chemicals that dilate blood vessels can be injected directly into the penis, applied as a gel or a patch, or inserted into the urethra as a suppository. If chemical treatments do not work, it is possible to surgically implant a device into the penis that either causes it to remain semirigid all the time or that must be inflated with a small pump, which is usually implanted in the scrotum. Obviously, such devices are usually a last resort in men with ED for whom other approaches have been unsuccessful.

Disorders of the Prostate

Beginning around age 40, most men have some enlargement of the prostate gland, which may eventually grow from its normal size of a walnut to that of a lime or even a lemon. This condition is called **benign prostatic hyperplasia** (**BPH**). Because the prostate gland wraps around the urethra, this enlargement initially may force the bladder muscle to work harder to empty the urethra, leading to irritation of the bladder and an urge to urinate more frequently (Fig. 21.18). As the prostate grows larger, this increased pressure may result in bladder or kidney damage, and may even completely block urination, a condition that is fatal if untreated.

Early signs of an enlarged prostate include a more frequent urge to urinate, a weak stream of urine, and/or a sense of not fully emptying the bladder after urination. The presence of an enlarged prostate can usually be confirmed by a simple digital

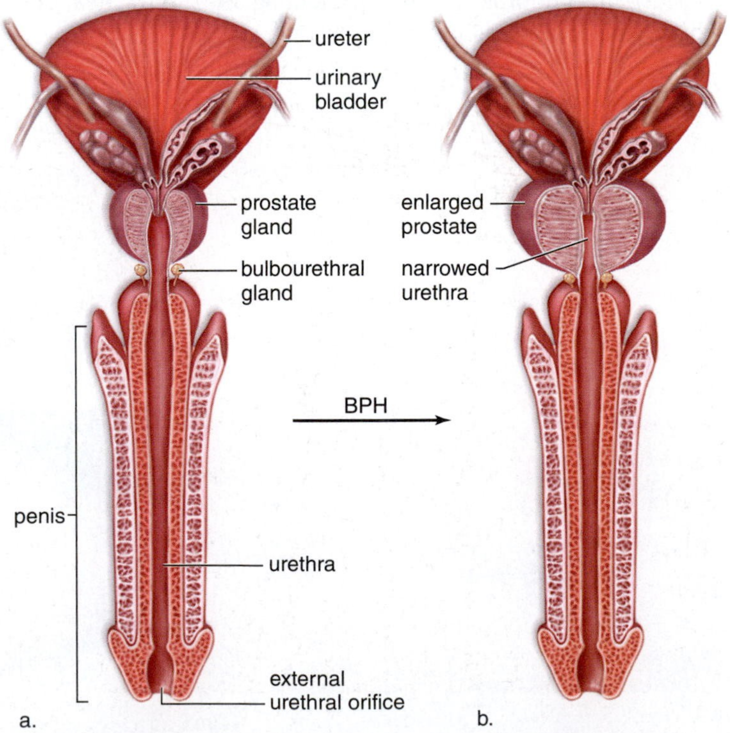
ureter
urinary bladder
prostate gland
bulbourethral gland
enlarged prostate
narrowed urethra
BPH
penis
urethra
external urethral orifice
a.
b.

Figure 21.18 Benign prostatic hyperplasia (BPH). **a.** This longitudinal section shows how the prostate gland normally surrounds the urethra. **b.** When the prostate gland enlarges, it can constrict the urethra as it exits the bladder, making it difficult to fully empty the bladder.

exam of the prostate, which a physician performs by inserting the finger of a gloved hand into the rectum. Depending on the severity of the symptoms, treatments for BPH range from avoiding liquids in the evening (to reduce the need to urinate during the night), to taking medications that shrink the prostate and/or improve urine flow, to having surgery.

Prostate enlargement is often due to an enzyme (5-alpha reductase) that acts on the male sex hormone testosterone, converting it to a substance that promotes prostate growth. That growth is needed during puberty, but continued growth in the adult is undesirable. Several chemical substances have been shown to interfere with the action of this enzyme. Saw palmetto, which is sold in tablet form as an over-the-counter nutritional supplement, has been shown to be effective. Prescription drugs such as finasteride (Proscar) are more powerful inhibitors of the enzyme, but men taking these drugs may experience erectile dysfunction and loss of libido.

If drugs don't control symptoms or the symptoms are severe, surgery may be necessary to reduce the pressure on the urethra and/or reduce the size of the prostate. In the most common operation, called *transurethral resection* of the prostate, the surgeon scrapes away the innermost core of the gland using a small instrument inserted into the urethra. In a more limited operation, called transurethral incision of the prostate, the doctor simply makes one or two small cuts in the prostate to relieve pressure. Alternatively, a newer procedure called *transurethral microwave thermotherapy* uses microwaves to produce heat that destroys, and hopefully shrinks, the enlarged gland.

Prostate cancer is the most commonly diagnosed cancer in American males, with one in ten men expected to develop the disease in their lifetime. After lung cancer, prostate cancer is the second leading cancer-related cause of death in men in the United States, because it has a tendency to metastasize and spread throughout the body. Many men are concerned that BPH may be associated with prostate cancer, but the two conditions are not necessarily related. If prostate cancer is detected early, usually by a blood test for prostate-specific antigen (PSA), it can usually be successfully treated. However, some experts now recommend against the routine screening of PSA levels, mainly because this test often detects prostate cancers that are small or slow growing and thus may not require treatment.

Testicular Cancer

Testicular cancer is the most common type of cancer in males aged 15 to 35. There are several types of testicular cancer, and some of them have a greater tendency to metastasize to other tissues. If discovered and treated early, testicular cancer is one of the most curable of all cancers. Therefore, experts recommend that men perform a monthly testicular self-examination after a hot shower or bath, when the scrotal skin is looser. Men should examine each testicle for any lumps or abnormal tenderness and for any change in the size of one testicle compared to the other. Any such findings should be reported to a doctor as soon as possible. If testicular cancer is confirmed, treatment usually includes surgery to remove the affected testicle, plus radiation or chemotherapy for types of cancer that have a high risk of spreading. However, if diagnosed and treated before it has spread, the cure rate for testicular cancer approaches 100%.

Common Conditions Affecting the Female Reproductive System

Endometriosis

Endometriosis, the presence of endometrial-like tissue at locations outside the uterine cavity, is the most common cause of chronic pelvic pain in women. Estimates indicate that between 10% and 33% of women aged 25 to 35 years will have some degree of this condition. The abnormal tissue often grows on the ovaries, the uterine tubes, the outside of the uterus, or on virtually any other abdominal organ. Endometriosis can induce a chronic inflammatory reaction, which often results in pain, abnormal menstrual cycles, or infertility. The exact cause of endometriosis is unclear, but various theories have been suggested. During menstruation, it is possible that live cells shed from the endometrium actually flow backward, through the oviducts and into the abdomen, where they are able to form endometrial tissue. Another hypothesis suggests that under certain hormonal conditions, the cells that normally line the abdominal organs can become endometrial tissue.

Endometriosis is often diagnosed when a surgeon sees the abnormal tissue during a surgical procedure to investigate the cause of abdominal pain (Fig. 21.19). The initial treatment usually focuses on controlling the pain, but may also involve drugs, such as oral contraceptives, which inhibit ovulation, and thus reduce the normal hormonally induced changes in the endometrial tissue. If these approaches do not work, surgery may be performed to either remove the abnormal endometrial tissue or to remove the uterus and ovaries in women who do not wish to have children.

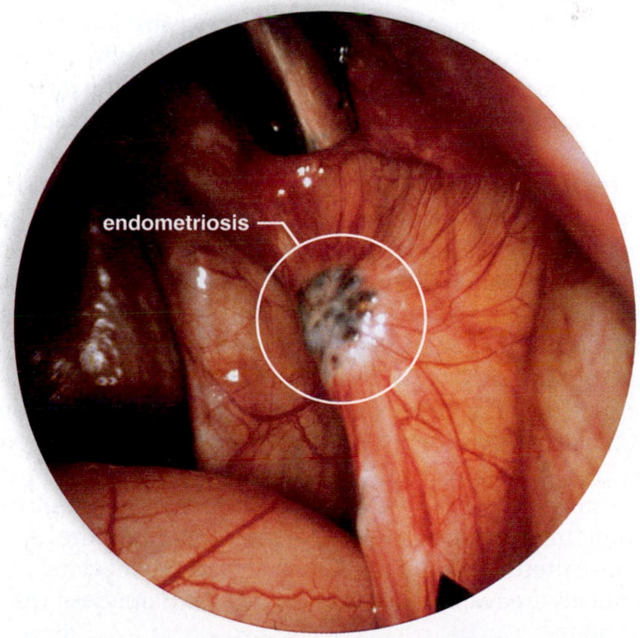
endometriosis

Figure 21.19 Endometriosis. Endometriosis is a common cause of pelvic pain, abnormal menstruation, and infertility in women. The condition occurs when endometrial tissue, which is often darkly pigmented, grows outside the uterus (usually in the abdominal cavity).

Ovarian Cancer

Ovarian cancer accounts for about 4% of all women's cancers in the United States, causing about 15,000 deaths a year. Unlike testicular cancer, ovarian cancer typically occurs in individuals over the age of 40. However, it also has a worse prognosis than testicular cancer, with only 50% of women surviving five years after diagnosis. The main reason for this may be that ovarian cancer is often "silent," showing no obvious signs or symptoms until late in its development. The most common symptoms include abdominal pain, feeling bloated or full, and frequent urge to urinate.

Women over age 40 should have a cancer-related checkup every year. Early detection requires periodic, thorough pelvic examinations. The Pap test, useful in detecting cervical cancer, does not reveal ovarian cancer. Testing for the level of tumor marker CA-125, a protein antigen, is helpful. Surgery, radiation therapy, and drug therapy are treatment options. Surgery usually involves removing one or both ovaries, the uterus, and the oviducts.

Women who have never had children are twice as likely to develop ovarian cancer as those who have. Early age at first pregnancy, early menopause, and the use of oral contraceptives, which reduces ovulation frequency, can help protect against ovarian cancer. If a woman has had breast cancer, her chances of developing ovarian cancer double.

Disorders of Menstruation

Most menstrual cycles occur every 21 to 35 days, with 3 to 10 days of bleeding during which 30–40 ml of blood are lost. Abnormalities include too little bleeding (or a complete lack of menstruation) or too much bleeding. The former usually indicates a problem with the control of menstruation by the hypothalamus or pituitary gland. This may be seen, for example, in young women who are undergoing strenuous athletic training, and/or who have very low levels of body fat. Excessive bleeding may be due to a large number of other disorders, including various STDs, certain medications, or problems with blood clotting.

The most common disorder of menstruation is *dysmenorrhea,* or painful menstruation. About half of all women experience this condition at some point, which includes such symptoms as agonizing cramps, headaches, backaches, and nausea. Painful menstruation is the leading cause of lost time from school and work among young women. There are many potential causes of dysmenorrhea. One common factor is a high level of prostaglandins released by the endometrium during the luteal and menstrual phases of the cycle. Research has shown that prostaglandin levels are four or five times higher than normal in women who experience painful menstruation. Because prostaglandins are known to cause uterine contractions, investigators suspect that these chemicals are responsible for much of the pain these women experience.

Other potential causes of painful menstruation include endometriosis (discussed previously) and various types of *ovarian cysts.* These fluid-filled sacs can develop on the ovaries of females of all ages, but are more common in the childbearing years. Two major types of ovarian cysts are follicular cysts, usually caused by the failure of a developing follicle to burst, and corpus luteum cysts, which occur when the corpus luteum doesn't break down normally. On rare occasions, surgery may be needed to remove ovarian cysts.

Premenstrual syndrome (*PMS*) is a group of symptoms related to the menstrual cycle. Anywhere from two weeks to a few days before menstruation, a significant number of women experience such symptoms as irrational mood swings, headaches, joint pain, digestive upsets, and sore breasts. The mood swings can be particularly troublesome if they lead to behaviors uncharacteristic of the individual.

Self-help measures for painful menstruation include applying a heating pad to the lower abdomen, drinking warm beverages, and taking warm showers or baths. Over-the-counter anti-inflammatory medications may be helpful; if not, a physician can prescribe other medications. For PMS, avoiding salt, sugar, caffeine, and alcohol may also help, especially when symptoms are present.

Infertility

Infertility is the failure of a couple to achieve pregnancy after one year of regular, unprotected intercourse. The American Medical Association estimates that 15% of all couples are infertile. The cause of infertility can be attributed to the male (40%), the female (40%), or both (20%).

Causes of Infertility

The most frequent cause of infertility in males is low sperm count and/or a large proportion of abnormal sperm, which can have many causes. The public is particularly concerned about endocrine-disrupting contaminants, which are discussed in the Ecology feature, "Endocrine-Disrupting Contaminants." Other possible causes of male infertility include various STDs, abnormal hormone levels, smoking, and alcohol. As noted in section 21.1, activities that increase testicular temperature can lower sperm count.

Common causes of infertility in females include blocked oviducts due to pelvic inflammatory disease and endometriosis, as discussed earlier in this section. Infertility problems also begin to increase sharply after age 35 in women, more so than in men. Several studies have also linked obesity and infertility in both genders, but the exact mechanism(s) involved remain unclear.

Sometimes the cause of infertility can be corrected by medical intervention so that couples can have children. It is possible to give females fertility drugs, which are gonadotropic hormones that stimulate the ovaries and bring about ovulation, although this may cause multiple births.

When reproduction does not occur in the usual manner, many couples adopt a child. Others sometimes try one of the assisted reproductive technologies discussed on page 432.

SCIENCE IN YOUR LIFE ▶ ECOLOGY

Endocrine-Disrupting Contaminants

Rachel Carson's book *Silent Spring*, published in 1962, predicted that pesticides would have a deleterious effect on animal life. Soon thereafter, it was found that pesticides caused the eggshells of bald eagles to become so thin that their eggs broke and the chicks died. Additionally, populations of terns, gulls, cormorants, and lake trout declined after they ate fish contaminated by high levels of environmental toxins. The concern was so great that the U.S. Environmental Protection Agency (EPA) came into existence. This agency, together with civilian environmental groups, has brought about a reduction in pollution release and a cleaning up of emissions. Even so, we are now aware of more subtle effects of pollutants.

Hormones influence nearly all aspects of physiology and behavior in animals. Therefore, when wildlife in contaminated areas began to exhibit certain abnormalities, researchers began to think that certain pollutants could affect the endocrine system. In England, male fish exposed to sewage developed ovarian tissue and produced a metabolite normally found only in females during egg formation. In California, western gulls displayed abnormalities in gonad structure and nesting behaviors. Hatchling alligators in Florida possessed abnormal gonads and hormone concentrations linked to nesting.

Figure 21B Exposure to endocrine-disrupting contaminants. Various types of wildlife, as well as humans, are exposed to endocrine-disrupting contaminants that can seriously affect their health and reproductive abilities.

At first, such effects seemed to involve only the female hormone estrogen, and researchers therefore called the contaminants ecoestrogens. Many of the contaminants interact with hormone receptors, and in that way cause developmental effects. Others bind directly with sex hormones, such as testosterone and estradiol (a potent estrogen). Still others alter the physiology of growth hormones and neurotransmitters responsible for brain development and behavior. Therefore, the preferred term today for these pollutants is endocrine-disrupting contaminants (EDCs).

Many EDCs are chemicals used as pesticides and herbicides in agriculture, and some are associated with the manufacture of various other organic molecules, such as PCBs (polychlorinated biphenyls). PCBs have been banned in the United States since 1979, but because they are chemically stable, they are still present in the environment. Other chemicals shown to influence hormones are found in plastics, food additives, and personal hygiene products. In mice, phthalate esters, which are plastic components, affect neonatal development when present in the part-per-trillion range. It is of great concern that EDCs have been found at levels comparable to functional hormone levels in the human body. Therefore, it is not surprising that EDCs are affecting the endocrine systems of a wide range of organisms (Fig. 21B).

Scientists and representatives of industrial manufacturers continue to debate whether EDCs pose a health risk to humans. Some suspect that EDCs lower sperm counts, reduce male and female fertility, and increase rates of certain cancers (breast, ovarian, testicular, and prostate). Additionally, some studies suggest that EDCs contribute to learning deficits and behavioral problems in children.

Laboratory and field research continues to identify chemicals that have the ability to influence the endocrine system. Millions of tons of potential EDCs are produced annually in the United States, and the EPA is under pressure to certify these compounds as safe. The European Economic Community has already restricted the use of certain EDCs and has banned the production of specific plastic components intended for use by children. Only through continued scientific research and the cooperation of industry can we identify the risks the EDCs pose to the environment, wildlife, and humans.

Questions to Consider

1. There are undeniable benefits to the use of pesticides, one of the major ones being more attractive, insect-free food. How would you personally weigh these benefits compared to the risks of ingesting endocrine-disrupting contaminants?
2. How might endocrine-disrupting chemicals specifically affect cells? (Review how peptide and steroid hormones work in section 20.1.)
3. What would be some implications for agricultural and manufacturing industries if the use of all potential endocrine-disrupting chemicals were banned? How much extra would you be willing to pay for your monthly food (and other products) if their costs increased?

Assisted Reproductive Technologies

Assisted reproductive technologies (*ART*) are techniques that increase the chances of pregnancy.

Artificial Insemination by Donor (AID) During artificial insemination, sperm are placed in the vagina by a physician. Sometimes a woman is artificially inseminated by her partner's sperm. This is especially helpful if the partner has a low sperm count, because the sperm can be collected over a period of time and concentrated so that the sperm count is sufficient to result in fertilization. Often, however, a woman is inseminated by sperm acquired from a donor who is a complete stranger to her. At times, a combination of partner and donor sperm is used.

A variation of AID is *intrauterine insemination* (*IUI*). In IUI, fertility drugs are given to stimulate the ovaries, and then the donor's sperm is placed in the uterus, rather than in the vagina.

If the prospective parents wish, sperm can be sorted into those that are believed to be X chromosome–bearing or Y chromosome–bearing to increase the chances of having a child of the desired sex.

In Vitro Fertilization (IVF) During IVF, conception occurs in laboratory glassware. Ultrasound machines can now spot follicles in the ovaries that hold immature oocytes. Therefore, the latest method is to forgo the administration of fertility drugs and retrieve immature oocytes by using a needle. The immature oocytes are then brought to maturity in glassware before concentrated sperm are added. After about two to four days, the embryos are ready to be transferred to the uterus of the woman, who is now in the secretory phase of her uterine cycle. If desired, the embryos can be tested for a genetic disease, so that only those free of disease are used. If implantation is successful, development is normal and continues to term.

Video
One Healthy Baby

Video
In vitro Fertilization

Gamete Intrafallopian Transfer (GIFT) The term *gamete* refers to a sex cell, either a sperm or an oocyte. GIFT was devised to overcome the low success rate (15–20%) of in vitro fertilization. The method is exactly the same as in vitro fertilization, except the oocytes and the sperm are placed in the uterine tubes immediately after they have been brought together. GIFT has the advantage of being a one-step procedure for the woman—the oocytes are removed and reintroduced all in the same time period. A variation on this procedure is to fertilize the oocytes in the laboratory and then place the zygotes in the uterine tubes.

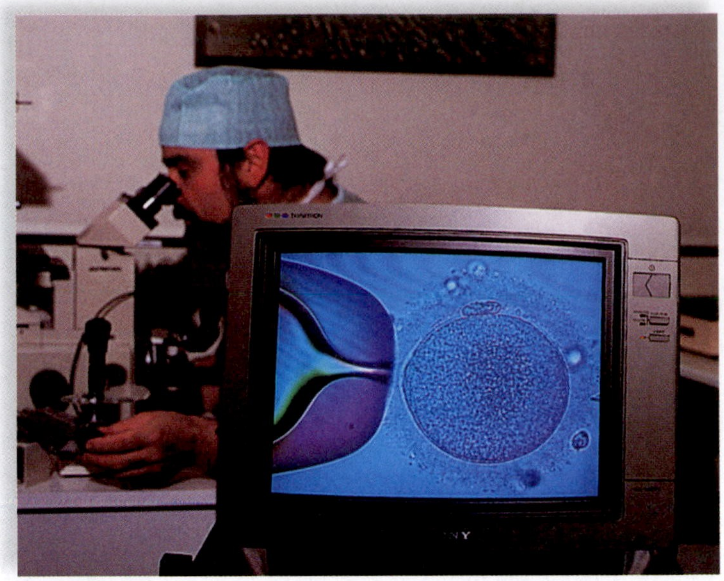

Figure 21.20 Intracytoplasmic sperm injection. This procedure uses a microscope connected to a monitor to indicate the injection of sperm into the egg using a microscopic needle.

Surrogate Mothers In some instances, women are contracted and paid to have babies. These women are called surrogate mothers. The sperm and even the oocyte can be contributed by the contracting parents.

Intracytoplasmic Sperm Injection (ICSI) In this highly sophisticated procedure, a single sperm is injected into an oocyte (Fig. 21.20). It is used effectively when a man has severe infertility problems.

Check Your Progress 21.6

1. Explain how sildenafil (Viagra) and similar drugs work to enhance erections.
2. Describe how endometriosis causes infertility.
3. List some factors that explain why ovarian cancer is more frequently fatal compared to testicular cancer.
4. Compare the degree of medical intervention required for artificial insemination by donor (AID), in vitro fertilization (IVF), gamete intrafallopian transfer (GIFT), and intracytoplasmic sperm injection (ICSI).

Case Study Conclusion

The Suleman octuplets—Noah, Maliyah, Isaiah, Nariyah, Jonah, Josiah, Jeremiah, and Makai—celebrated their third birthday on January 26, 2012. Their initial weeks in the hospital cost taxpayers hundreds of thousands of dollars, and their mother has estimated the cost of taking care of her 14 children to be $15,000 a month. That may be an underestimate, because children of multiple births have a higher rate of medical problems and learning disabilities, compared to other children. Ms. Suleman has attempted to benefit from her celebrity in several ways, from starring in a documentary to participating in a celebrity boxing match (she won) to filming a weekly Internet program called *Octo-TV,* during which she provides parenting advice.

Appearing in 2011 on *The Today Show* with nine of her kids, Suleman noted that she had fallen behind on mortgage payments and was planning on selling her house. She also noted that she feels guilty for putting her children under a media microscope, and that she suffers from panic attacks due to frequent death threats that she receives. She also said, "I personally cannot waste my energy fixating on the past and my past choices, regardless if they were good or bad choices. . . ." Unfortunately, it seems likely that the choices she and her fertility doctors made will present a series of difficult consequences for her 14 children in the future.

MEDIA STUDY TOOLS

www.mhhe.com/maderinquiry14

Enhance your study of this chapter with study tools and practice tests. Also ask your instructor about the resources available through ConnectPlus, including LearnSmart, the media-rich eBook, interactive learning tools, and animations.

SUMMARIZE

21.1 Male Reproductive System

The **reproductive system** produces **gametes,** haploid sex cells that join during fertilization. The male genital tract consists of **testes** that produce **sperm,** accessory organs that add secretions to the **semen** (**seminal fluid**), and the external genitalia:

- **Spermatogenesis** occurs in the **seminiferous tubules** of the testes, producing sperm that mature and are stored in each **epididymis.** As they develop, the sperm are supported by **Sertoli cells.** Sperm enter each **vas deferens** before entering the **urethra** along with seminal fluid produced by the **seminal vesicles, prostate gland,** and **bulbourethral glands.**
- The external genitals of males are the **penis,** the organ of sexual intercourse, and the testes, which are within the **scrotum.** Erection of the penis occurs when expandable tissue in the penis fills with blood. Orgasm in males is a physical and emotional climax that results in ejaculation of semen from the penis. **Circumcision** is the surgical removal of the foreskin of the penis.

The hypothalamus controls the male reproductive system through the secretion of hormones:

- **Gonadotropin-releasing hormone** (**GnRH**) from the hypothalamus causes the anterior pituitary to secrete **follicle-stimulating hormone** (**FSH**) and **luteinizing hormone** (**LH**). FSH promotes spermatogenesis in the seminiferous tubules of the testes. LH, also called **interstitial cell-stimulating hormone** (**ICSH**) in males, promotes testosterone production by **interstitial cells. Testosterone** is required for proper functioning of the male sex organs, as well as for the development of male secondary sex characteristics.

21.2 Female Reproductive System

The female genital tract includes the internal organs that produce the female sex cells and allow fertilization to occur, plus the external genitalia:

- **Oogenesis** is the production of an egg, or **oocyte,** by an **ovary.** After **ovulation,** the release of an oocyte, it is transported by **fimbriae** through a **uterine tube** (oviduct) to the **uterus,** where a fertilized oocyte, or **zygote,** can implant in the **endometrium** and develop. The lower part of the uterus is called the **cervix.**
- The external genitals of the female include the **vagina,** the **clitoris,** and the **vulva** (labia majora and labia minora). The vagina is the organ of sexual intercourse, the birth canal, and the exit for menstrual fluids. The external genitals, especially the clitoris, play an active role in orgasm, which culminates in vaginal, uterine, and oviduct contractions.

21.3 Ovarian and Uterine Cycles

The **ovarian cycle** is under the control of the hypothalamus and anterior pituitary:

- During the follicular phase, FSH causes maturation of a **follicle** that secretes estrogen and some progesterone. This phase ends with ovulation.
- During the luteal phase, which occurs during the second half of the cycle, LH converts the follicle into the **corpus luteum,** which secretes **progesterone** and **estrogen.**

The **uterine cycle** varies depending on whether fertilization occurs:

- After ovulation, estrogen produced by a new ovarian follicle causes the endometrium to thicken. Progesterone from the corpus luteum causes the endometrium to become secretory, in preparation to receive a fertilized egg.

- If no pregnancy occurs, the low level of hormones causes the endometrium to break down, resulting in **menstruation.** If a fertilized embryo implants itself in the thickened endometrium, pregnancy occurs. Initially, production of **human chorionic gonadotropin** (HCG) by the **placenta** maintains the corpus luteum. As the placenta develops, it takes over production of estrogen and progesterone.
- During **menopause,** the ovaries become unresponsive to gonadotropic hormones, and the ovarian and uterine cycles cease.

21.4 Control of Reproduction

Only abstinence provides 100% protection against STDs, although condoms are somewhat effective. Common methods of preventing pregnancy are summarized here:

- Abstaining from sex is the most reliable birth control method.
- Surgical approaches to birth control—that is, **vasectomy** and **tubal ligation**—are virtually 100% effective.
- Oral **contraceptives,** or **birth control pills,** are nearly 100% effective in preventing ovulation.
- An **intrauterine device** (**IUD**) is inserted into the uterus to prevent either fertilization or implantation of a fertilized embryo.
- A **diaphragm** is inserted to cover the cervix, to prevent sperm from entering the uterus.
- **Male condoms** and **female condoms** both form a barrier that prevents sperm from entering the uterus, while providing some protection against STDs.
- **Contraceptive implants, contraceptive patches,** and **contraceptive injections** all utilize estrogen and/or progesterone to prevent ovulation.
- A **vaginal contraceptive ring** can be inserted into the vagina every month to prevent ovulation.
- **Emergency contraception,** such as Preven and Plan B One-Step, can be taken after sexual activity (and fertilization) has occurred. They work by preventing implantation of the embryo.

21.5 Sexually Transmitted Diseases

Many infectious diseases can be transmitted sexually. Several common STDs are caused by viruses:

- **Acquired immunodeficiency syndrome** (**AIDS**), caused by the human immunodeficiency virus (HIV), devastates the immune system, resulting in **opportunistic infections.** Inflammation of the liver, or **hepatitis,** can be caused by several viruses, including hepatitis B virus, which can be transmitted sexually.
- **Genital herpes** may cause blisters that flare up repeatedly.
- **Cervical cancer, genital warts,** and cancers of the anus and other tissues are caused by human papillomaviruses. Cervical cancer is often detected using the **Pap test.**
- Hepatitis B can lead to liver failure.

Several common STDs are caused by bacteria:

- **Chlamydia** and **gonorrhea** can cause urethritis, and possibly **pelvic inflammatory disease** (**PID**) in females.
- **Syphilis** may cause cardiovascular and neurological complications if untreated.

21.6 Other Disorders of the Reproductive System

The reproductive system is affected by several other common conditions:

- Common disorders of the male reproductive tract include **erectile dysfunction** (**ED**), **benign prostatic hyperplasia** (**BPH**), **prostate cancer,** and **testicular cancer.**
- Common disorders of the female reproductive tract include **endometriosis, ovarian cancer,** and abnormal menstruation.
- Couples experiencing **infertility** may use assisted reproductive technologies in order to have a child. Such technologies include artificial insemination, in vitro fertilization, gamete intrafallopian transfer, and intracytoplasmic sperm injection.

ASSESS

Testing Yourself

Choose the best answer for each question.

1. Which of these structures contributes fluid to semen?
   - **a.** prostate
   - **b.** seminal vesicles
   - **c.** bulbourethral gland
   - **d.** All of these are correct.
   - **e.** None of these are correct.
2. Production of testosterone in the interstitial cells is controlled by
   - **a.** LH.
   - **b.** ICSH.
   - **c.** GnRH.
   - **d.** All of these are correct.
   - **e.** Both a and b are correct.
3. Label the anatomical structures in this diagram of the male reproductive system.

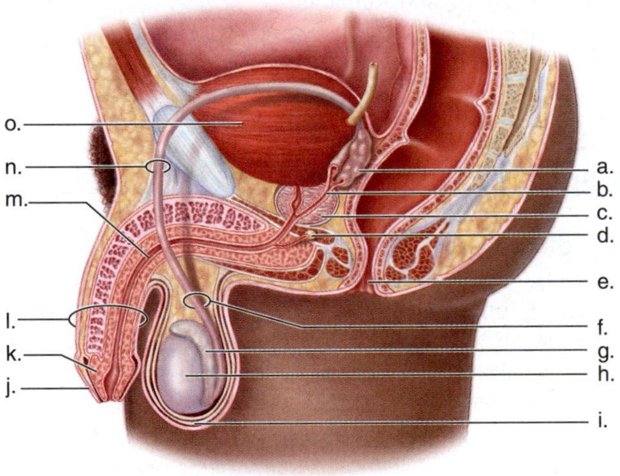

o.
n.
m.
l.
k.
j.
a.
b.
c.
d.
e.
f.
g.
h.
i.

4. An oocyte is fertilized in the
   - **a.** vagina.
   - **b.** uterus.
   - **c.** uterine tube.
   - **d.** ovary.
5. During pregnancy,
   - **a.** the ovarian and uterine cycles occur more quickly than before.
   - **b.** GnRH is produced at a higher level than before.
   - **c.** the ovarian and uterine cycles do not occur.
   - **d.** the female secondary sex characteristics are not maintained.
6. Home pregnancy tests are based on the presence of
   - **a.** estrogen.
   - **b.** progesterone.
   - **c.** follicle-stimulating hormone.
   - **d.** human chorionic gonadotropin.

7. Label the anatomical structures in this diagram of the female reproductive system.

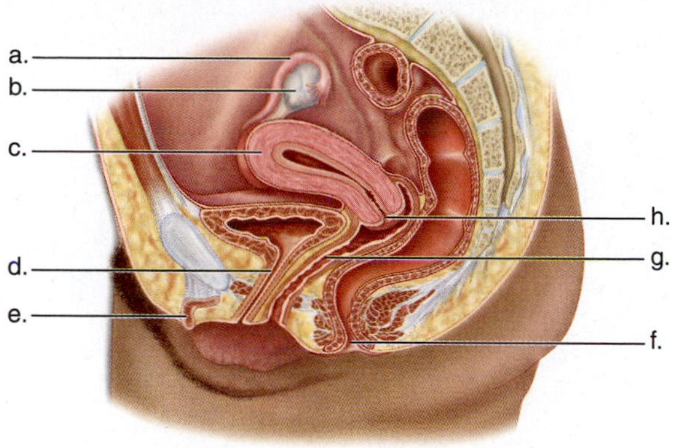
a.
b.
c.
d.
e.
h.
g.
f.

8. Female oral contraceptives prevent pregnancy because
   a. the pill inhibits the release of luteinizing hormone.
   b. oral contraceptives prevent the release of an egg.
   c. follicle-stimulating hormone is not released.
   d. All of these are correct.

For questions 9–15, match the terms in the key with the following graph of the female reproductive cycles.

Key:

a. estrogen
b. ovulation
c. menstruation
d. LH
e. follicular phase
f. luteal phase
g. progesterone

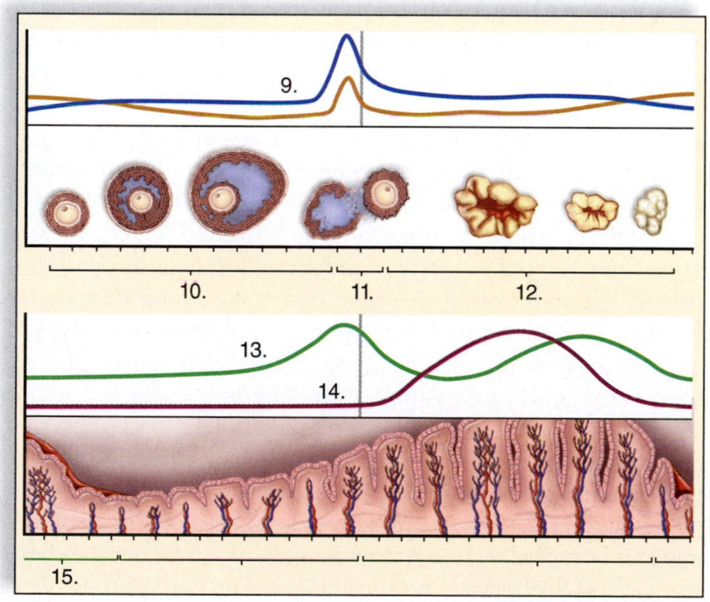
9.
10.
11.
12.
13.
14.
15.

16. Which of these diseases is caused by a virus?
   a. gonorrhea
   b. genital warts
   c. chlamydia
   d. syphilis
   e. None of these are correct.

ENGAGE

Thinking Critically

1. The average sperm count in males is now lower than it was several decades ago. The reasons for the lower sperm count usually seen today are not known. What data might be helpful in order to formulate a testable hypothesis?
2. In the animal kingdom, only primates menstruate. Other mammals come into season, or "heat," during certain times of the year, while still others only ovulate after having sex. What would be some possible advantages of a monthly menstrual cycle?
3. The condition called benign prostatic hyperplasia is usually not life threatening, but prostate cancer can be. Because the prostate gland is typically enlarged in both conditions, why is one condition benign and the other potentially life threatening?

22 Development and Aging

CHAPTER OUTLINE

BEFORE YOU BEGIN

Before beginning this chapter, take a few moments to review the following discussions:

Section 5.4 How does meiosis in animals result in the production of haploid gametes?

Figures 21.3 and 21.7 What are the differences between spermatogenesis and oogenesis?

Sections 21.2 and 21.3 Where does fertilization usually occur, and what hormonal changes result?

CASE STUDY A Healthy Pregnancy

For several months, Amber and Kent had been trying to conceive a child. The couple had put off having children for several years while they pursued their careers. Now, as they both approached the age of 40, they were feeling the pressure of time. After she stopped taking birth control pills, Amber had begun taking prenatal vitamins; additionally, she became more conscientious about eating a healthy diet. Although neither of them was ever really into physical exercise, both began walking several times a week. After four months of trying to conceive, Amber announced that the home pregnancy test was positive! They immediately scheduled an appointment with their regular physician to begin preparing for the next stage of their lives.

At the first visit following the positive results of the home pregnancy test, their physician performed a complete physical of Amber, as well as a blood test to confirm pregnancy. The physician informed the couple that a blood test was much more accurate in detecting levels of the hormone human chorionic gonadotropin (HCG) than over-the-counter (OTC) urine tests. The results of the blood test confirmed what Amber and Kent suspected—that in a period of just 40 weeks, they would be parents.

Their physician promptly gave Amber a list of items to avoid. In addition to increasing her level of exercise, she needed to drink eight to ten glasses of water per day and eat plenty of fruits and vegetables. The doctor informed Amber that it was crucial for her to inform her physician of any OTC drugs she may want to take, especially in the first trimester. He noted that the first trimester was a period when critical organ systems developed in their baby and that most medications and alcohol were now forbidden. Both parents-to-be were up to the challenge and excited about the prospects of adding to their family.

As you read through the chapter, think about the following questions:

1. What is the source of the HCG hormone in pregnancy?
2. What physiological changes should Amber expect over the course of her pregnancy?

22.1 Fertilization and Early Stages of Development

Learning Outcomes

Upon completion of this section, you should be able to

1. Identify the major structures of a sperm cell and an oocyte, and provide a function for each.
2. Describe the steps involved in fertilization.
3. Describe the cellular, tissue, and organ stages of development.

Animal development begins with a single cell that multiplies and changes to form a complete organism.

Fertilization

Fertilization is the union of a sperm and an oocyte, resulting in a **zygote.** A sperm has three distinct parts: a *head,* a *middle piece,* and a *tail.* The head contains a nucleus and is capped by a membrane-bound *acrosome.*

Video Human Sperm

In humans, the plasma membrane of the oocyte is surrounded by an extracellular matrix termed the *zona pellucida.* In turn, the zona pellucida is surrounded by a few layers of adhering follicle cells called the *corona radiata.* These cells nourished the oocyte when it was in a follicle of the ovary. During fertilization, which usually occurs in a uterine tube, a sperm cell must make its way to the inside of the oocyte. As illustrated in Figure 22.1, the stages of fertilization are: (1) the sperm squeezes through the corona radiata; (2) the sperm releases acrosomal enzymes that allow it to penetrate the zona pellucida; (3) the plasma membrane of the sperm head fuses with the plasma membrane of the oocyte; (4) the sperm nucleus enters the oocyte and releases its chromatin; (5) oocyte *cortical granules* secrete enzymes that turn the zona pellucida into a *fertilization membrane;* and (6) the sperm chromatin reorganizes into a sperm pronucleus that can fuse with the egg pronucleus, forming a single nucleus.

Normally, out of perhaps hundreds of sperm that reach an oocyte, only one is able to penetrate the egg. If more than one sperm enters an oocyte, an event called *polyspermy,* the zygote would have too many chromosomes, and development would be abnormal. Prevention of polyspermy depends on changes in the oocyte's plasma membrane and in the zona pellucida. As soon as a sperm touches an oocyte, the oocyte's plasma membrane depolarizes (from –65 mV to 10 mV), and

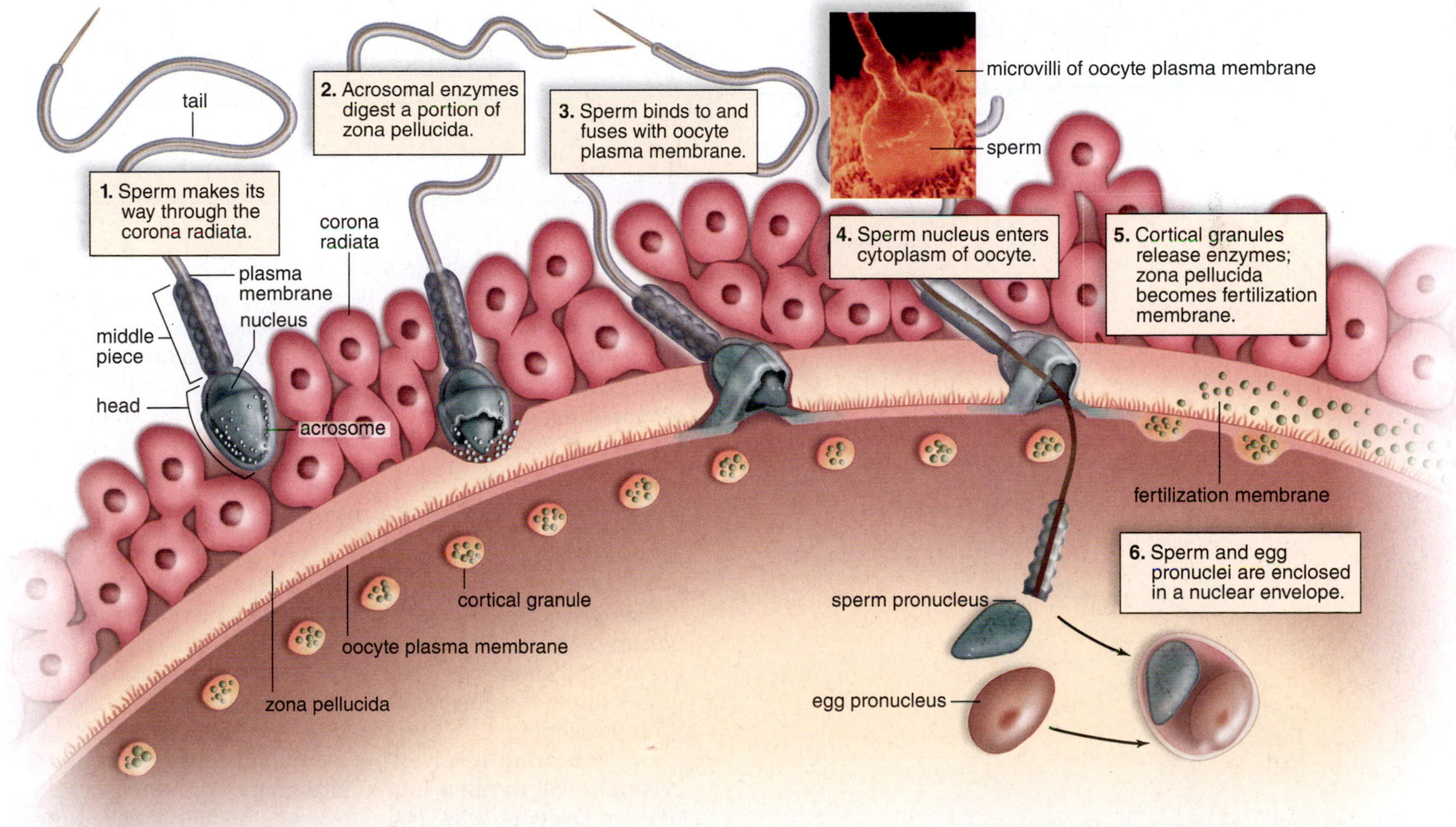
tail
1. Sperm makes its way through the corona radiata.
2. Acrosomal enzymes digest a portion of zona pellucida.
3. Sperm binds to and fuses with oocyte plasma membrane.
microvilli of oocyte plasma membrane
sperm
4. Sperm nucleus enters cytoplasm of oocyte.
5. Cortical granules release enzymes; zona pellucida becomes fertilization membrane.
corona radiata
plasma membrane
nucleus
middle piece
head
acrosome
fertilization membrane
6. Sperm and egg pronuclei are enclosed in a nuclear envelope.
cortical granule
sperm pronucleus
oocyte plasma membrane
egg pronucleus
zona pellucida

Figure 22.1 Fertilization. During fertilization, a single sperm enters the oocyte. A sperm makes its way through the corona radiata surrounding the oocyte. The head of a sperm has a membrane-bound acrosome filled with enzymes. When released, these enzymes digest a pathway for the sperm through the zona pellucida. After it binds to the plasma membrane of the oocyte, a sperm enters the oocyte. When the sperm pronucleus joins with the oocyte pronucleus, fertilization is complete.

this prevents the binding of any other sperm. In addition, the enzymes secreted by cortical granules during formation of the fertilization membrane cause the zona pellucida to lift away from the surface of the oocyte. Now, other sperm cannot bind to the zona pellucida either.

As described in the Bioethical feature, "Cloning," techniques have been developed that allow an individual animal to be "cloned" by implanting the nucleus from a cell of one animal into the egg of another. However, many ethical issues related to this procedure have yet to be resolved, especially as they apply to human cloning.

Early Stages of Animal Development

The early stages of animal development occur at the cellular, tissue, and organ levels of organization.

Cellular Stages of Development

The cellular stages of development are (1) cleavage resulting in a multicellular embryo, and (2) formation of the blastula. **Cleavage** is cell division without growth. DNA replication and mitotic cell division occur repeatedly, and the cells get smaller with each division. In other words, cleavage increases only the number of cells. It does not change the original volume of the egg cytoplasm.

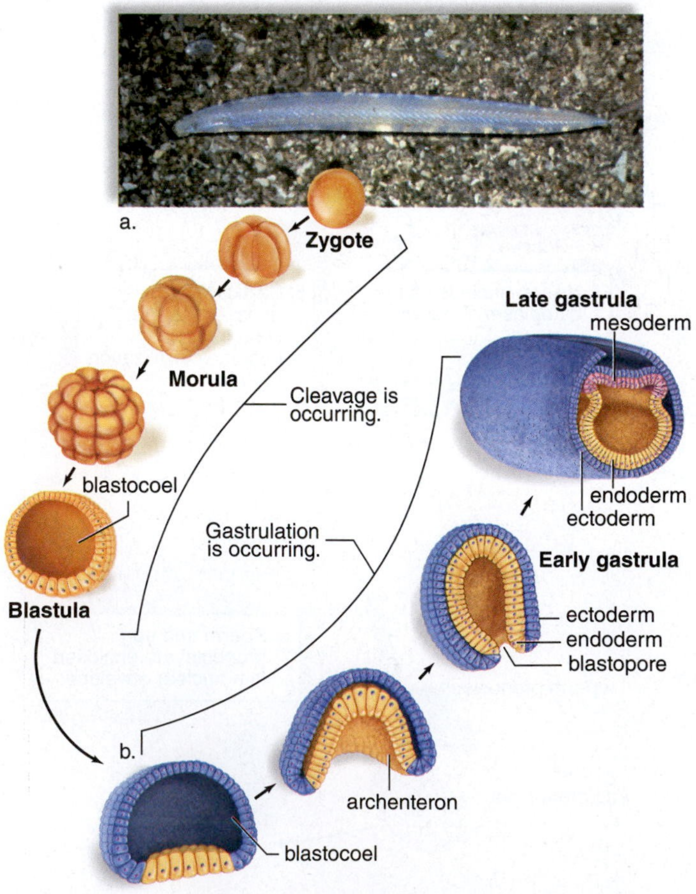

a.
Zygote
Morula
Cleavage is occurring.
blastocoel
Blastula
Gastrulation is occurring.
Late gastrula
mesoderm
endoderm
ectoderm
Early gastrula
ectoderm
endoderm
blastopore
b.
archenteron
blastocoel

Figure 22.2 Lancelet early development. **a.** A lancelet. **b.** The early stages of development are exemplified in the lancelet. Cleavage produces a number of cells that form a ball called a morula. The blastula then develops a cavity, the blastocoel. Invagination during gastrulation produces the germ layers ectoderm and endoderm. Then the mesoderm arises.

As shown in Figure 22.2, cleavage in a lancelet (Fig. 22.2*a*), a primitive chordate, is *equal* (cells of uniform size) and results in a **morula,** which is a ball of cells. The next cellular stage in lancelet development is formation of a **blastula,** which is a hollow ball of cells having a fluid-filled cavity called a **blastocoel.** The blastocoel forms when the cells of the morula pump Na^+ into extracellular spaces and water follows by osmosis. The water collects in the center, and the result is a hollow ball of cells.

The zygotes of other animals also undergo cleavage and form a morula. In frogs, cleavage is *not equal* because of the presence of yolk, a dense nutrient material. When yolk is present, the zygote and embryo are said to exhibit polarity because the embryo has an *animal pole* (smaller cells) and a *vegetal pole* (larger cells containing yolk). A chicken develops on land and lays a hard-shelled egg containing a large amount of yolk. In the developing chick, the yolk does not participate in cleavage. All vertebrates have a blastula stage, but the appearance of the blastula can be different from that of a lancelet. In the chick, the blastula forms as a layer of cells that spreads out over the yolk. The blastocoel is a space that separates these cells from the yolk:

Video Blastocyst Formation

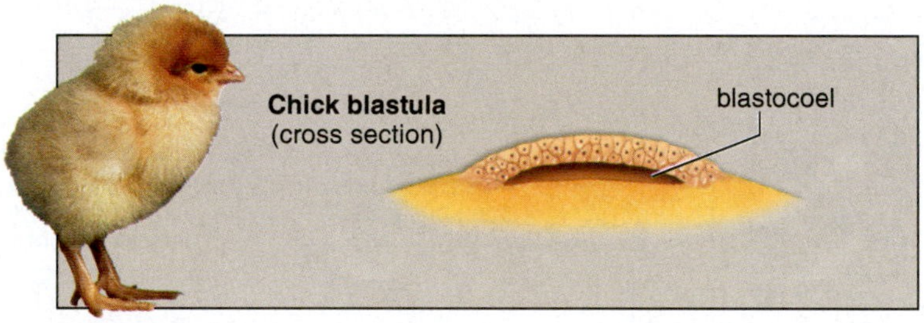

Chick blastula
(cross section)
blastocoel

The blastula of humans resembles that of the chick embryo, but this resemblance cannot be related to the amount of yolk because the human egg contains little yolk. Rather, the evolutionary history of these two animals can explain this similarity. Both birds and mammals are related to reptiles, and all three groups develop similarly, despite a difference in the amount of yolk in their eggs.

Tissue Stages of Development

The tissue stages of development are: (1) the early gastrula and (2) the late gastrula. The early gastrula stage begins when certain cells begin to push, or invaginate, into the blastocoel, creating a double layer of cells (Fig. 22.2*b*). Cells migrate during this and other stages of development, sometimes traveling quite a distance before reaching a destination, where they continue developing.

Gastrulation involves the formation of three layers of cells that will develop into adult organs. An early gastrula has two layers of cells. The outer layer of cells is called the **ectoderm,** and the inner layer is called the **endoderm.** The endoderm borders the gut, which at this point is termed either the *archenteron* or the primitive gut. The pore, or hole, created by invagination is the **blastopore,** and in a lancelet, the

SCIENCE IN YOUR LIFE ▶ BIOETHICAL

Cloning

In March 1997, Scottish investigators announced that they had cloned a sheep called Dolly, and their procedure is now routinely used (Fig. 22A). A donor 2n nucleus is substituted for the n nucleus of an egg. A stimulus is applied that triggers cell division, and the resulting embryo is implanted into a surrogate mother where it develops to term. Using the procedure developed by the Scottish researchers, it is now common practice to clone all sorts of farm animals (horses, cows, sheep, goats, pigs) and also cats and monkeys.

Success of Cloning

Even so, cloning of animals is still relatively new, and although advances in the technology are rapidly developing, many problems still exist. (1) The vast majority of pregnancies involving clones are not successful. To clone Dolly the sheep, it took 247 tries before one was successful. In many cases, clone pregnancies spontaneously abort. (2) Of the small number of animal clones born, many have severe abnormalities, including: malfunctioning livers, abnormal blood vessels and heart problems, underdeveloped lungs, diabetes, and immune system deficiencies. (3) Even if the newborn clone appears healthy, it may develop diseases seen in older animals. Dolly was euthanized in 2003 because she was suffering from lung cancer and crippling arthritis. She had lived only half the normal life span for a Dorset sheep.

Reproductive Cloning Versus Therapeutic Cloning

Animal cloning is a form of reproductive cloning. In reproductive cloning, the desired end is to create an individual (Fig. 22A). Even if such a feat were accomplished, a clone would not be identical to the person being cloned. Recall that mitochondria have genes, and these genes are contributed by the egg even if the egg nucleus is removed. The clone could be identical to its genetic parent only if the donor nucleus and the donor egg came from the same female. Further, the clone would be subject to different environmental factors and a different upbringing from his or her genetic parent. In therapeutic cloning, the desired end is not an individual. Rather, it is the embryonic stem cells that could possibly be "coaxed" into becoming other types of cells. The goal is twofold. Stem cell research will undoubtedly yield useful information about how cells develop and become specialized. However, the ultimate goal of therapeutic cloning is to provide cells and tissues that could treat human illnesses: insulin-secreting cells for diabetics, nerve cells for stroke patients or those with Parkinson disease, cardiac cells for those with heart disease, and so forth. Yet, ethical concerns about therapeutic cloning remain—after all, any pre-embryo is potentially a living, breathing human.

Anticipating intense interest in therapeutic cloning, the U.S. National Academy of Sciences proposed strict new guidelines for federally funded research in 2005. As a result of the Academy's recommendations, Embryonic Stem Cell Research Oversight (ESCRO) committees must approve embryonic stem cell research before it is begun. Thus, this type of research is subject to two reviews: (1) an ESCRO committee analysis and (2) reviews already required by an Institutional Review Board.

ESCRO committees consist of bioethicists and legal experts, as well as members of the general public. All research requires informed consent from the donors of ova or sperm prior to beginning the research. Stem cells created for therapeutic cloning can never be used for reproductive purposes under the proposed guidelines. The guidelines also require that the embryos created cannot be grown in culture for longer than 14 days. At that point, the primitive streak of the developing nervous system begins to form.

Questions to Consider

1. Do you approve of the restrictions currently in place for therapeutic cloning? If not, how would you see these restrictions changed?
2. Scientific researchers have made great strides in stem cell research, successfully converting both adult cells and umbilical cord cells back to stem cell forms. In light of these successes, should therapeutic cloning be funded at all?
3. Commercial companies in South Korea currently offer the opportunity to clone people's dogs—for around $50,000. Should the cloning of pets be allowed when there are so many unwanted pets?
4. Should scientists be allowed to bring extinct species back to life (as in the movie *Jurassic Park*) using frozen tissue samples and reproductive cloning?

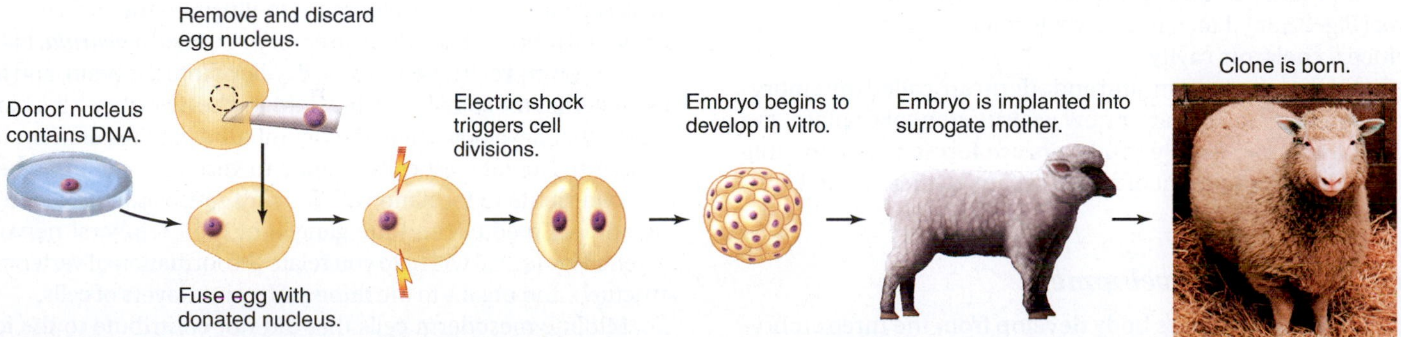
Remove and discard egg nucleus.
Donor nucleus contains DNA.
Electric shock triggers cell divisions.
Embryo begins to develop in vitro.
Embryo is implanted into surrogate mother.
Clone is born.
Fuse egg with donated nucleus.

Figure 22A The process of reproductive cloning.

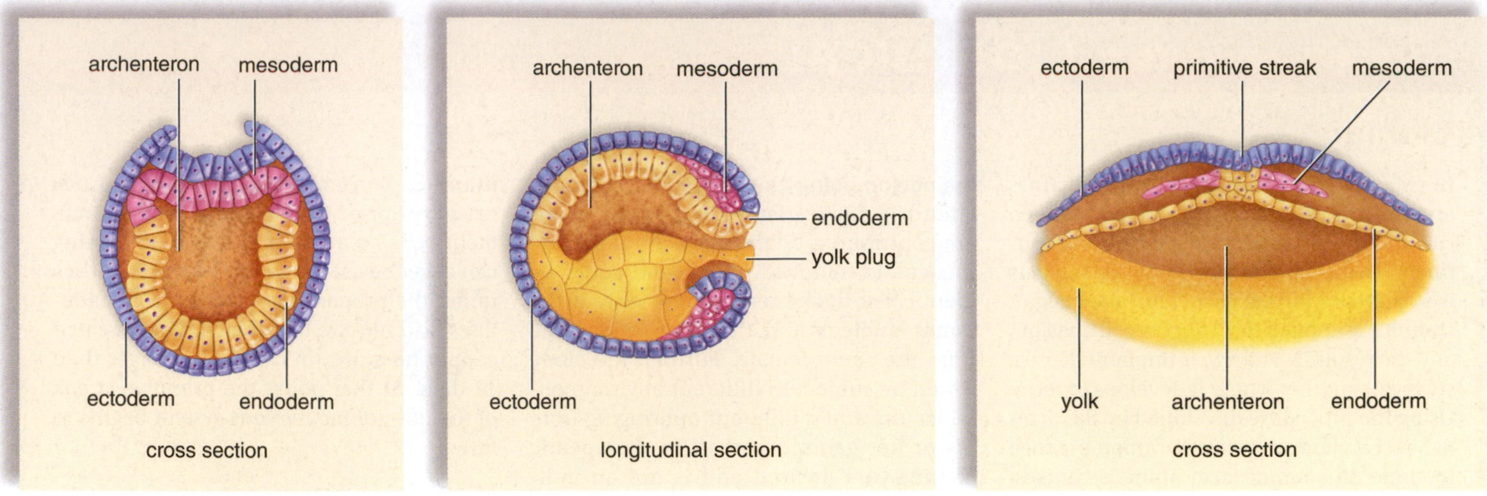
archenteron
mesoderm
ectoderm
endoderm
cross section
a. Lancelet late gastrula
archenteron
mesoderm
endoderm
yolk plug
ectoderm
longitudinal section
b. Frog late gastrula
ectoderm
primitive streak
mesoderm
yolk
archenteron
endoderm
cross section
c. Chick late gastrula

Figure 22.3 Comparative development of mesoderm. **a.** In the lancelet, mesoderm forms by an outpocketing of the archenteron. **b.** In the frog, mesoderm forms by migration of cells between the ectoderm and endoderm. **c.** In the chick, mesoderm also forms by invagination of cells.

blastopore eventually becomes the anus. In addition to ectoderm and endoderm, the late gastrula has a middle layer of cells called the **mesoderm.**

Figure 22.2 illustrates gastrulation in a lancelet, and Figure 22.3 compares the lancelet, frog, and chick late gastrula stages. In the lancelet, mesoderm formation begins as outpocketings from the archenteron (Fig. 22.3*a*). These outpocketings will grow in size until they meet and fuse, forming two layers of mesoderm. The space between them is the coelom. The *coelom* is a body cavity lined by mesoderm that contains internal organs. In humans, the coelom becomes the thoracic and abdominal cavities.

In the frog, the cells containing yolk do not participate in gastrulation, and therefore, they do not invaginate. Instead, a slitlike blastopore is formed when the animal pole cells begin to invaginate from above, forming endoderm. Animal pole cells also move down over the yolk, to invaginate from below. Some yolk cells, which remain temporarily in the region of the blastopore, are called the *yolk plug.* Mesoderm forms when cells migrate between the ectoderm and endoderm (Fig. 22.3*b*). Later, a splitting of the mesoderm creates the coelom.

The chicken egg contains so much yolk that endoderm formation does not occur by invagination. Instead, an upper layer of cells becomes ectoderm, and a lower layer becomes endoderm. Mesoderm arises by an invagination of cells along the edges of a longitudinal furrow in the midline of the embryo. Because of its appearance, this furrow is called the *primitive streak* (Fig. 22.3*c*). Later, the newly formed mesoderm splits to produce a coelomic cavity.

Ectoderm, mesoderm, and endoderm are called the embryonic **germ layers.** No matter how gastrulation takes place, the result is the same: three germ layers are formed. It is possible to relate the development of future organs to these germ layers (Table 22.1).

Organ Stages of Development

The organs of an animal's body develop from the three embryonic germ layers. We will concentrate on how one organ system, the nervous system, develops.

TABLE 22.1 Embryonic Germ Layers

Embryonic Germ Layer	Vertebrate Adult Structures
Ectoderm (outer layer)	Nervous system; epidermis of skin; epithelial lining of oral cavity and rectum
Mesoderm (middle layer)	Musculoskeletal system; dermis of skin; cardiovascular system; urinary system; reproductive system—including most epithelial linings; outer layers of respiratory and digestive systems
Endoderm (inner layer)	Epithelial lining of digestive tract and respiratory tract, associated glands of these systems, epithelial lining of urinary bladder

The newly formed mesoderm cells lying along the main longitudinal axis of the animal coalesce to form a dorsal supporting rod called the **notochord.** The notochord persists in lancelets, but in frogs, chicks, and humans, it is later replaced by the vertebral column. Therefore, these animals are called vertebrates.

The nervous system develops from midline ectoderm located just above the notochord. At first, a thickening of cells, called the *neural plate,* is seen along the dorsal surface of the embryo. Then, neural folds develop on either side of a neural groove, which becomes the *neural tube* when these folds fuse. Figure 22.4 shows cross sections of frog development to illustrate the formation of the neural tube. At this point, the embryo is called a *neurula.* Later, the anterior end of the neural tube develops into the brain, and the rest becomes the spinal cord. In addition, the *neural crest* is a band of cells that develops where the neural tube pinches off from the ectoderm. Neural crest cells migrate to various locations, where they contribute to formation of skin and muscles, in addition to the adrenal medulla and the ganglia of the peripheral nervous system. Figure 22.5 will help you relate the formation of vertebrate structures and organs to the three embryonic layers of cells.

Midline mesoderm cells that did not contribute to the formation of the notochord now become two longitudinal masses of tissue. These two masses become blocked off into *somites,* which

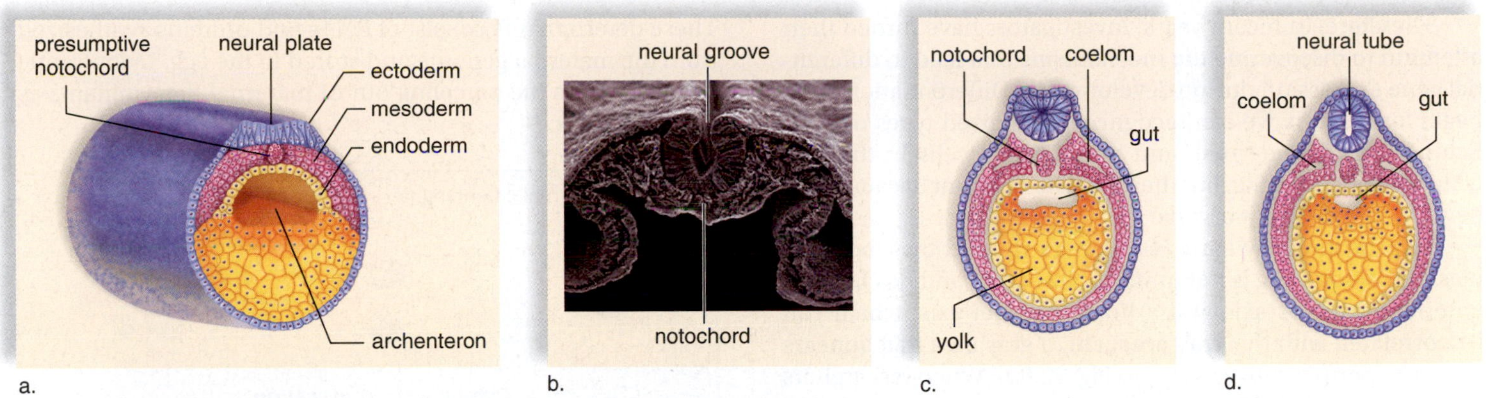
presumptive notochord
neural plate
ectoderm
mesoderm
endoderm
archenteron
a.
neural groove
notochord
b.
notochord
coelom
gut
yolk
c.
neural tube
coelom
gut
d.

Figure 22.4 Development of neural tube and coelom in a frog embryo. **a.** Ectodermal cells that lie above the future notochord (called presumptive notochord) thicken to form a neural plate. **b.** The neural groove and folds are noticeable as the neural tube begins to form. **c.** Splitting of the mesoderm produces a coelom, which is completely lined by mesoderm. **d.** A neural tube and a coelom have now developed.

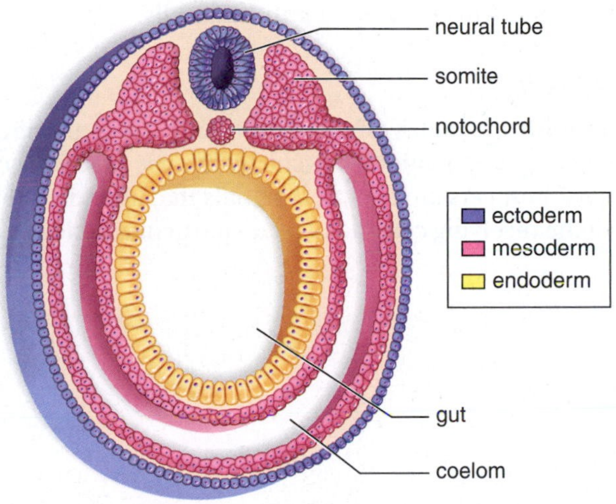
neural tube
somite
notochord
ectoderm
mesoderm
endoderm
gut
coelom

Figure 22.5 Vertebrate embryo, cross section. At the neurula stage, each of the germ layers, indicated by color (see key), can be associated with the later development of particular parts. The somites give rise to the muscles of each segment and to the vertebrae, which replace the notochord in vertebrates.

are serially arranged along both sides for the length of the notochord. Somites give rise to muscles associated with the axial skeleton and to the vertebrae. The serial origin of axial muscles and the vertebrae testifies that vertebrates are segmented animals.

A primitive gut tube is formed by endoderm as the body itself folds into a tube. The heart, too, begins as a simple tubular pump. Organ formation continues until the germ layers have given rise to the specific organs listed in Table 22.1.

MP3 Fertilization and Pre-embryonic Development

Check Your Progress 22.1

1. Outline the steps of fertilization, beginning with sperm in the uterine tube.
2. Describe two ways that an oocyte avoids polyspermy.
3. Compare and contrast the cellular and tissue stages of development in the lancelet, frog, and chick.
4. Identify the organ systems that are formed from each of the primary germ layers.

22.2 Processes of Development

Learning Outcomes

Upon completion of this section, you should be able to

1. Explain how cellular differentiation and morphogenesis play a role in development.
2. Explain how cytoplasmic segregation and induction are able to influence cellular differentiation.
3. Describe how morphogen genes, homeotic genes, and apoptosis help determine pattern formation during development.

Aside from growth, the process of development requires (1) cellular differentiation and (2) morphogenesis. **Cellular differentiation** occurs when cells become specialized in structure and function. For example, a muscle cell looks different and acts differently than a nerve cell. **Morphogenesis** produces the shape and form of the body. One of the earliest indications of morphogenesis is cell migration to form the germ layers. Later, morphogenesis includes the process of **pattern formation,** which directs how tissues and organs are arranged in the body. Apoptosis, or programmed cell death, which was first discussed in section 5.1, is an important part of pattern formation.

Cellular Differentiation

At one time, investigators mistakenly believed that irreversible genetic changes must account for cellular differentiation. Perhaps the genes are parceled out as development occurs, they thought, and that's why cells of the body have different structures and functions. However, the process of somatic cell cloning shows that every cell in the body contains all the genes necessary to develop into an organism. In other words, all an organism's genes are in each cell of the body.

The process by which different tissues come about becomes clearer when we consider that specialized cells produce only certain proteins. In other words, we now know that specialization is not due to the parceling out of genes; rather, it is due to differential gene expression. Certain genes and not others are turned on in differentiated cells.

Therefore, in recent years, investigators have turned their attention to discovering the mechanisms that lead to differential gene expression during development. Differentiation must begin long before we can recognize specialized types of cells. Ectodermal, endodermal, and mesodermal cells in the gastrula look quite similar, but they must be different because they develop into different organs.

Investigations by early researchers showed that the cytoplasm of a frog's egg is not uniform. It is polar and has both an anterior/posterior axis and a dorsal/ventral axis, which can be correlated with the **gray crescent,** a gray area that appears after the sperm fertilizes the egg (Fig. 22.6*a*). When researchers in the lab divided the fertilized egg such that each half contained gray crescent, each experimentally separated daughter cell developed into a complete embryo (Fig. 22.6*b*). However, if the egg was divided so that only one daughter cell received the gray crescent, only that cell became a complete embryo (Fig. 22.6*c*). This experiment allowed investigators to hypothesize that the gray crescent contains particular chemical signals that are needed for development to proceed normally.

Cytoplasmic Segregation

The oocyte is now known to contain substances called *maternal determinants* that influence the course of development. These determinants consist of RNAs and proteins synthesized from the maternal genome and stored in the egg. Cytoplasmic **segregation** is the parceling out of maternal determinants as mitosis occurs:

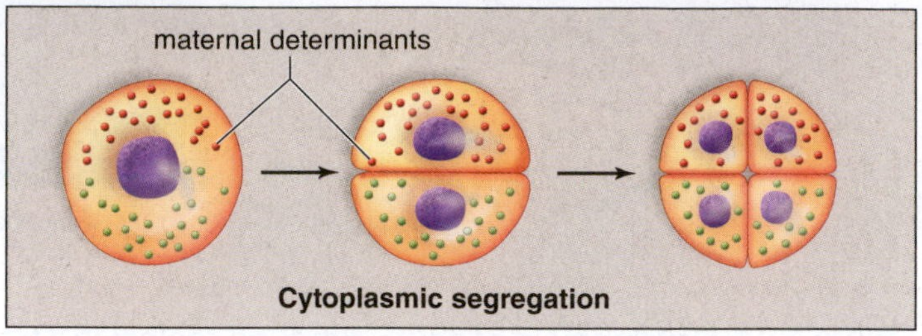
maternal determinants
Cytoplasmic segregation

Cytoplasmic segregation helps determine how the various cells of the morula will later develop. As development proceeds, specialization of cells is influenced by maternal determinants and the signals given off by neighboring cells.

Induction

Induction is the ability of one embryonic tissue to influence the development of another tissue by the use of signals called inducers. Inducers are chemical signals that alter the metabolism of the receiving cell and activate particular genes.

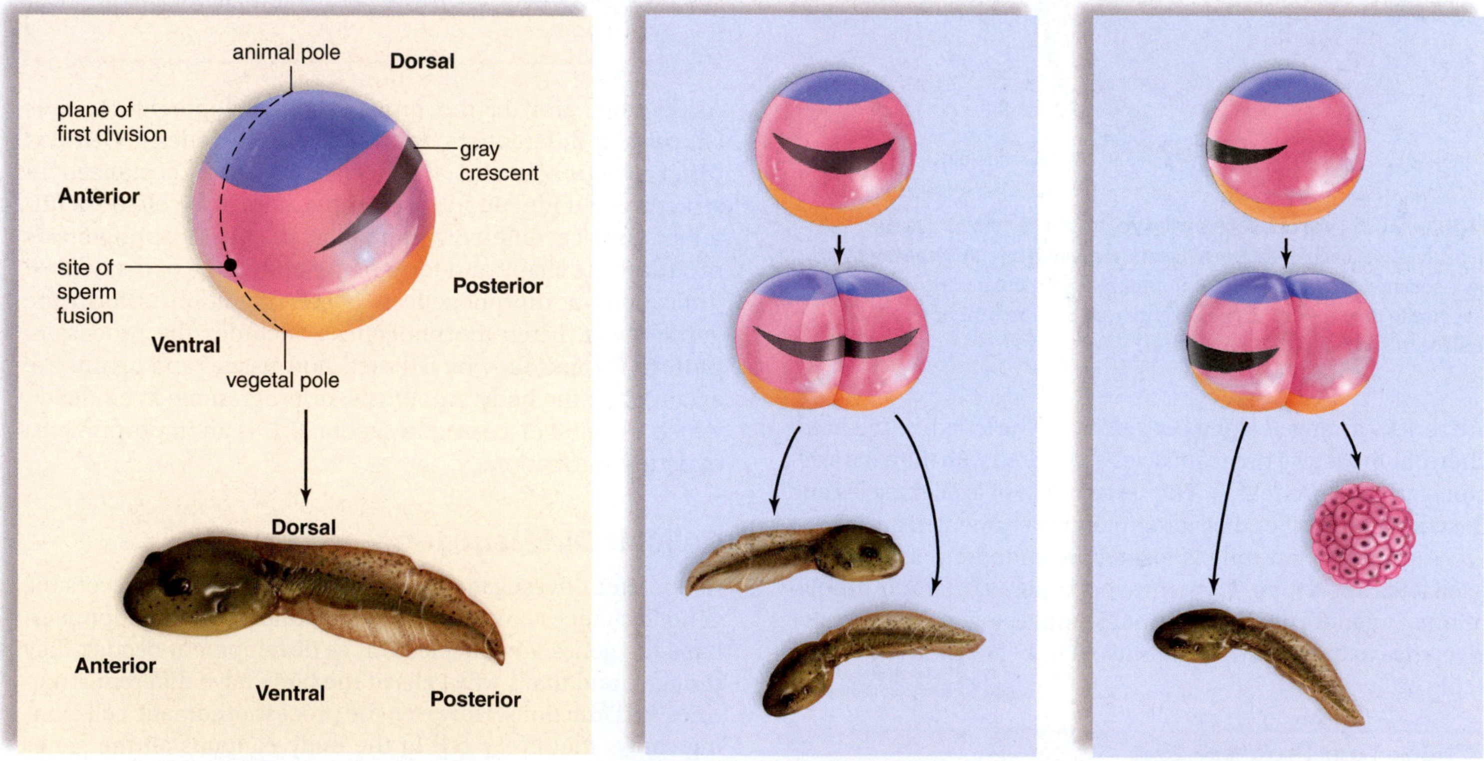
animal pole
Dorsal
plane of first division
gray crescent
Anterior
site of sperm fusion
Posterior
Ventral
vegetal pole
Dorsal
Anterior
Ventral
Posterior

a. Frog's egg is polar and has axes.
b. Each cell receives a part of the gray crescent.
c. Only the cell on the left receives the gray crescent.

Figure 22.6 Experimental determination of cytoplasmic influence on development. **a.** A frog's egg has anterior/posterior and dorsal/ventral axes that correlate with the position of the gray crescent. **b.** When researchers cut the fertilized egg such that the gray crescent is divided in half, each daughter cell is capable of developing into a complete tadpole. **c.** If the fertilized egg is cut such that only one daughter cell receives the gray crescent, then only that cell can become a complete embryo.

Induction and Frog Experiments A frog embryo's gray crescent becomes the dorsal lip of the blastopore, where gastrulation begins. Because this region is necessary for complete development, early investigators called the dorsal lip of the blastopore the *primary organizer.* The cells closest to the primary organizer become endoderm, those farther away become mesoderm, and those farthest away become ectoderm. This suggests that a molecular concentration gradient may act as a chemical signal (inducer) to cause germ layer differentiation.

The gray crescent of a frog's egg marks the dorsal side of the embryo where the mesoderm becomes notochord and ectoderm becomes nervous system. Experiments have shown that presumptive (potential) notochord tissue induces the formation of the nervous system. If presumptive nervous system tissue, located just above the presumptive notochord, is cut out and transplanted to another region of an embryo, nervous tissue does not develop in that region. The presumptive nervous system is missing some sort of signal to differentiate. On the other hand, if presumptive notochord tissue is cut out and transplanted beneath what would be ectoderm in another region, this other ectoderm does differentiate into nervous tissue. This experiment shows that signals from the presumptive notochord system cause ectoderm to differentiate into nervous tissue.

A well-known series of inductions accounts for the development of the vertebrate eye. An optic vesicle, which is a lateral outgrowth of a developing brain, induces the overlying ectoderm to thicken and become the lens of the eye. The developing lens, in turn, induces an optic vesicle to form an optic cup, where the retina forms.

Induction and Roundworm Experiments More recently, work with the roundworm *Caenorhabditis elegans* has also shown that induction is necessary to the process of differentiation. *C. elegans* is only 1 mm long, and vast numbers can be raised in the laboratory in either petri dishes or a liquid medium. The worm is hermaphroditic, and self-fertilization is the rule. Development of *C. elegans* takes three days, and the adult worm contains only 959 cells. Investigators have been able to watch the developmental process from beginning to end, especially because the worm is transparent. Its entire genome has been sequenced, and many modern genetic studies have been done utilizing this worm.

Fate maps, diagrams that trace the differentiation of developing cells, have been developed that show the destiny of each cell of *C. elegans* as it arises following successive cell divisions (Fig. 22.7). Some investigators have studied in detail the development of the vulva, a pore through which eggs are laid. A cell called the *anchor cell* induces the vulva to form. The cell closest to the anchor cell receives the most inducer and becomes the inner vulva. This cell, in turn, produces another inducer, which acts on its two neighboring cells, and they become the outer vulva. Additional work with *C. elegans* has shown that a series of inductions occurs as development proceeds.

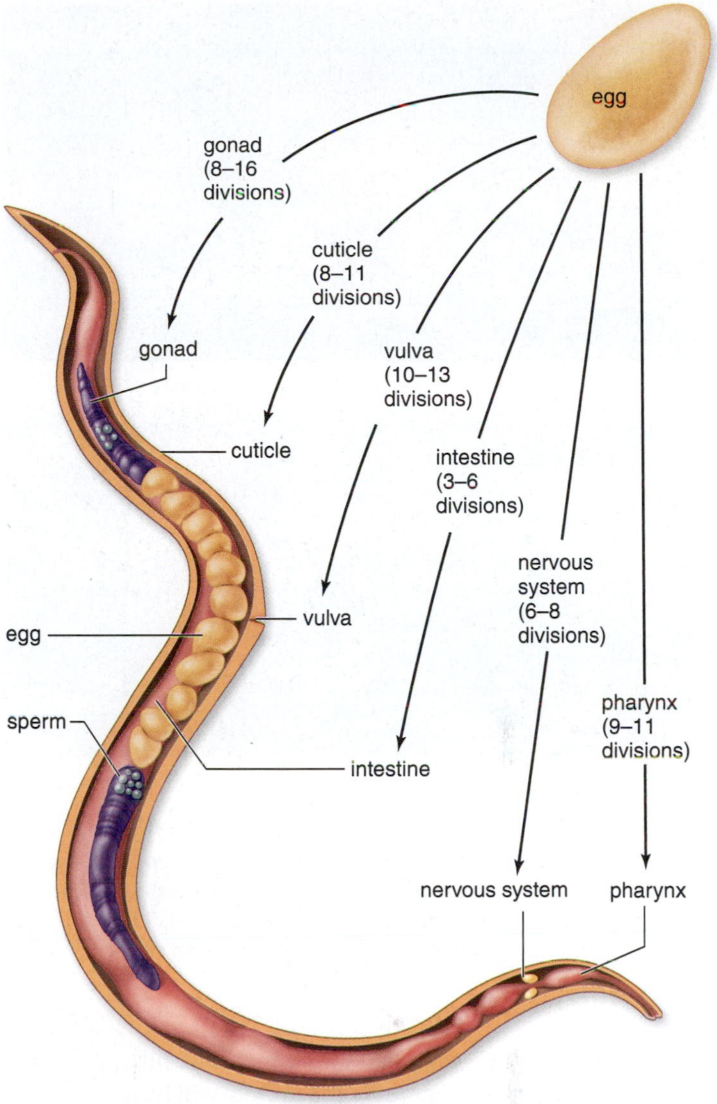
egg
gonad (8–16 divisions)
cuticle (8–11 divisions)
vulva (10–13 divisions)
intestine (3–6 divisions)
nervous system (6–8 divisions)
pharynx (9–11 divisions)
gonad
cuticle
vulva
egg
sperm
intestine
nervous system
pharynx

Figure 22.7 Development of *C. elegans,* a nematode.
A fate map of the worm showing that as cells arise by cell division, they are destined to become particular structures.

Morphogenesis

Pattern formation is the ultimate in morphogenesis. To understand the concept of pattern formation, think about the pattern of your own body. Your vertebral column runs along the main axis, and your arms and legs occur in certain locations. Pattern formation refers to how this arrangement of body parts comes about during development. Fruit fly experiments, in particular, have contributed to our knowledge of pattern formation. A few pairs of fruit flies can produce hundreds of offspring in a couple of weeks, all within a small bottle kept on a laboratory bench.

Morphogen Genes

In fruit flies, investigators have discovered certain genes, now called **morphogen genes,** that determine the relationship of individual parts. For example, some genes control which end

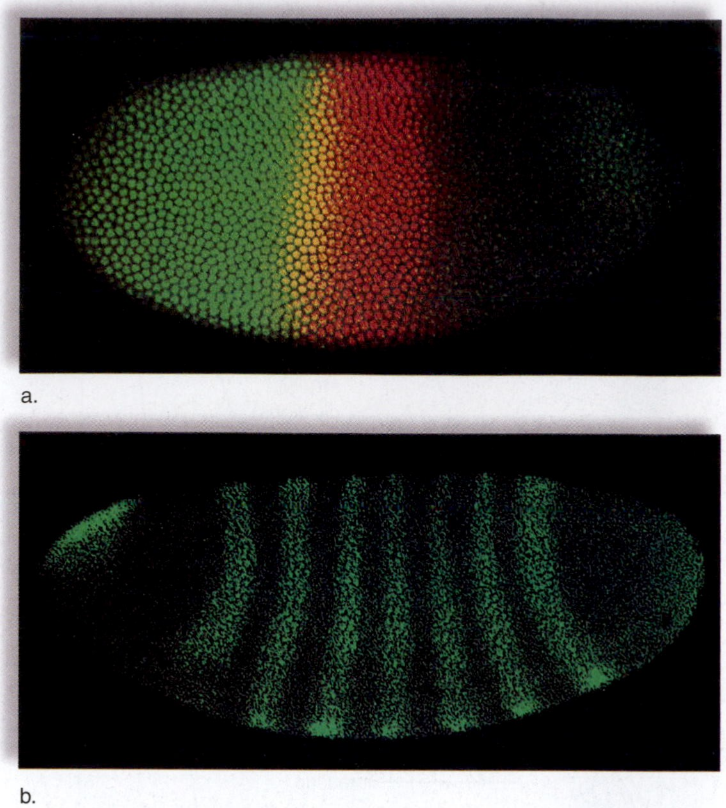
a.
b.

Figure 22.8 Morphogen gradients in the fruit fly. Different morphogen gradients appear as development proceeds. **a.** This early gradient determines which end is the head and which is the tail. The different colors show the concentration gradients of two different proteins. **b.** Another gradient determines the number of segments.

of the animal will be the head and which the tail, and others determine how many segments the animal will have.

Each of these genes codes for a protein that is present in a gradient—cells at the start of the gradient contain high levels of the protein, and cells at the end of the gradient contain low levels of the protein. These gradients are called morphogen gradients because they determine the shape of the organism (Fig. 22.8). A morphogen gradient is efficient because it has a range of effects, depending on the particular concentration in a portion of the animal. Sequential sets of master genes code for morphogen gradients that activate the next set of master genes in turn. This is one reason development turns out to be so orderly.

Homeotic Genes

Homeotic genes act by controlling the identity of each segment. They encode master regulatory proteins that control the expression of other genes, which in turn are responsible for the development of segment-specific structures. Mutations in homeotic genes cause particular segments to develop as if they were located elsewhere in the body. For example, the second thoracic segment of the fruit fly should develop a pair of wings.

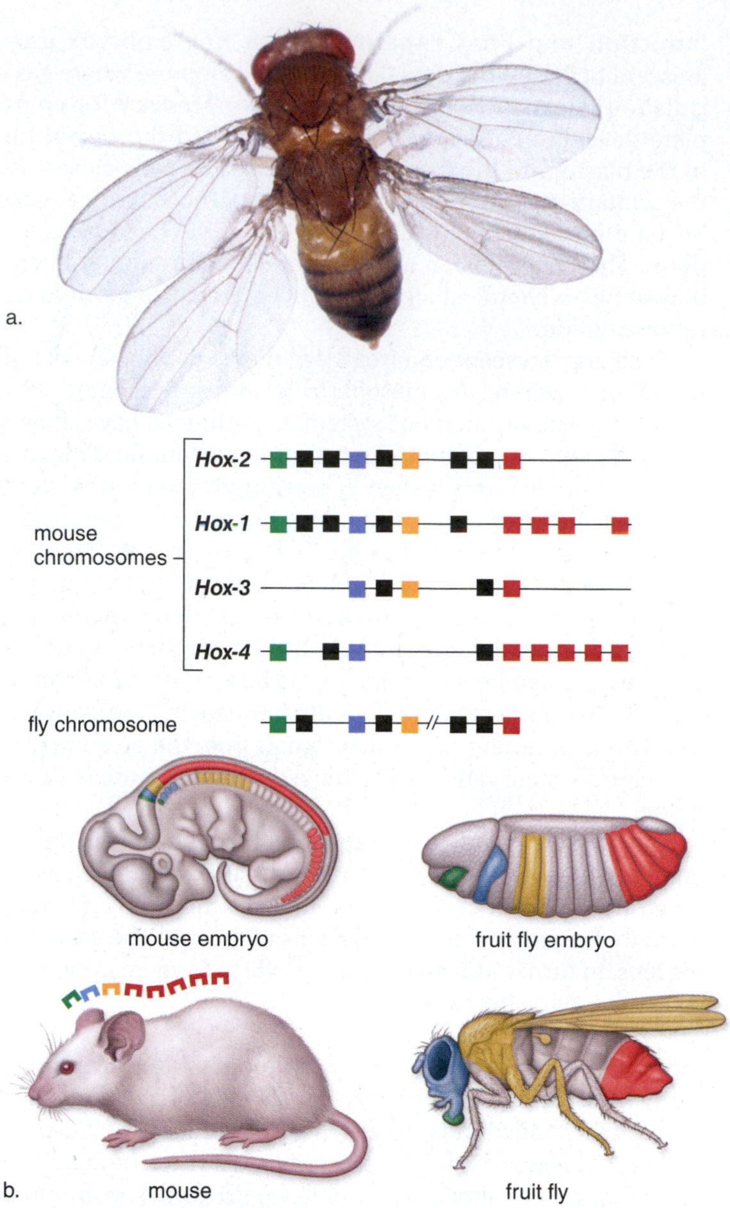
a.
Hox-2
Hox-1
mouse chromosomes
Hox-3
Hox-4
fly chromosome
mouse embryo
fruit fly embryo
b.
mouse
fruit fly

Figure 22.9 Pattern formation in *Drosophila*. Homeotic genes control pattern formation, an aspect of morphogenesis. **a.** If homeotic genes are activated at inappropriate times, abnormalities such as a fly with four wings occur. **b.** The green, blue, yellow, and red colors show that homologous homeotic genes occur on four mouse chromosomes and on a fly chromosome in the same order. These genes are color-coded to the region of the embryo, and therefore the adult, where they regulate pattern formation. The black boxes are homeotic genes that are not identical between the two animals. In mammals, homeotic genes are called *Hox* genes.

However, a mutation in the homeotic gene that controls the development of the third thoracic segment allows that segment to develop wings when it should not, resulting in a four-winged fly (Fig. 22.9*a*).

Homeotic genes have now been found in many other organisms, and surprisingly, they all contain the same particular sequence of nucleotides, called a *homeobox*. The homeobox codes for a particular sequence of 60 amino acids called a **homeodomain.** A homeodomain protein binds to DNA and helps determine which genes are turned on:

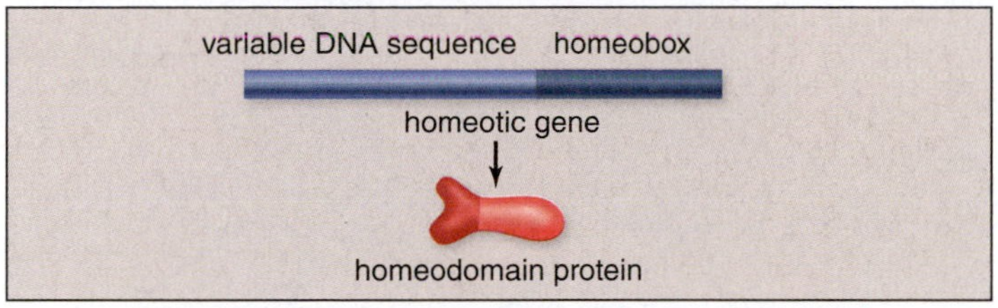
variable DNA sequence
homeobox
homeotic gene
homeodomain protein

In *Drosophila,* homeotic genes are located on a single chromosome. In mice and also humans, the same four clusters of homeotic genes are located on four different chromosomes (Fig. 22.9*b*). In all three types of animals, homeotic genes are expressed from anterior to posterior in the same order. The first clusters determine the final development of anterior segments of the embryo, while those later in the sequence determine the final development of posterior segments of the embryo.

Because the homeotic genes of so many different organisms contain the same homeodomain, we know that this nucleotide sequence arose early in the history of life and that it has been largely conserved as evolution occurred. In general, it has been intriguing to learn that developmental processes are similar in organisms ranging from yeasts to plants to a wide variety of animals, including humans.

Apoptosis

We have already discussed the importance of **apoptosis** (programmed cell death) in the normal day-to-day operation of the body (see Fig. 5.2). Apoptosis is also an important part of pattern formation in all organisms. In humans, for example, we know that apoptosis is necessary for the shaping of the hands and feet. If it does not occur, the child is born with webbing between the fingers and toes.

The fate maps of *C. elegans* indicate that apoptosis occurs in 131 cells as development takes place. When a cell-death signal is received, an inhibiting protein becomes inactive, allowing a cell-death cascade to proceed, which ends in enzymes destroying the cell.

Check Your Progress 22.2

1. Describe the two processes that are associated with cellular differentiation.
2. Define the term "morphogen."
3. Explain the function of the homeobox sequence in a homeotic gene.

22.3 Human Embryonic and Fetal Development

Learning Outcomes

Upon completion of this section, you should be able to

1. Identify the extraembryonic membranes and provide a function for each.
2. Summarize, in chronological order, the major events that occur during fetal development from three to nine months.
3. Describe the flow of blood in a fetus and explain the role of the placenta.

In humans, the length of the time from *conception* (fertilization followed by implantation) to birth is approximately nine months. It is customary to calculate a baby's "due date" by adding 280 days to the start of the last menstrual period because this date is usually known, whereas the day of fertilization is usually unknown. Because the time of birth is influenced by so many variables, only about 5% of babies actually arrive on the forecasted date.

Human development before birth is often divided into embryonic development (months one and two) and fetal development (months three to nine). **Embryonic development** consists of early formation of the major organs, and fetal development is the refinement of these structures.

Extraembryonic Membranes

The **extraembryonic membranes** lie outside of the embryo. The names of these membranes are derived from their function in reptiles and birds. In reptiles, these membranes made development on land possible. If an embryo develops in the water, the water supplies oxygen for the embryo and takes away waste products. The surrounding water prevents desiccation, or drying out, and provides a protective cushion. For animals that live on land, all these functions are performed by the extraembryonic membranes.

In the chick, the extraembryonic membranes develop from extensions of the germ layers, which spread out over the yolk. Figure 22.10*a* shows the chick surrounded by the membranes. The **chorion** lies next to the shell and carries on gas exchange. The **amnion** contains the protective amniotic fluid, which bathes the developing embryo. The **allantois** collects nitrogenous wastes, and the **yolk sac** surrounds the remaining yolk, which provides nourishment.

Humans (and other mammals) also have these extraembryonic membranes (Fig. 22.10*b*). As will be discussed further in this section, the functions of the extraembryonic membranes in humans have been modified to suit internal development. However, their very presence indicates our relationship to birds and to reptiles. It is interesting to note that all chordate animals develop in water, either in bodies of water or within amniotic fluid.

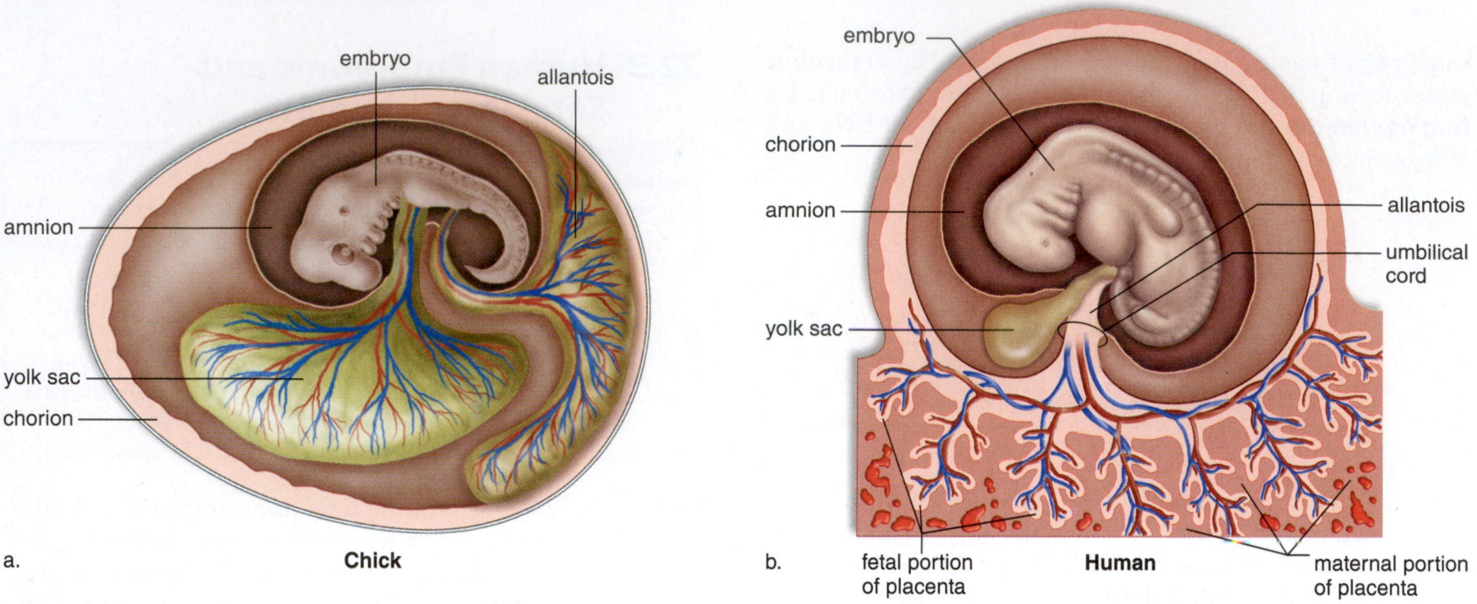
embryo
allantois
amnion
yolk sac
chorion
a.
Chick
embryo
chorion
amnion
allantois
umbilical cord
yolk sac
b.
fetal portion of placenta
Human
maternal portion of placenta

Figure 22.10 Extraembryonic membranes. Extraembryonic membranes, which are not part of the embryo, are found during the development of (**a**) chicks and (**b**) humans. Each has a specific function.

Embryonic Development

Embryonic development encompasses the first two months of development following fertilization.

The First Week

Fertilization usually occurs in the distal third (closest to the ovary) of a uterine tube (Fig. 22.11), and cleavage begins even as the embryo passes down this duct to the uterus. By the time the embryo reaches the uterus on the third day, it is a morula. By about the fifth day, the morula is transformed into the blastocyst. The **blastocyst** has a fluid-filled cavity, a single layer of outer cells called the **trophoblast,** and an inner cell mass. Later, the trophoblast, reinforced by a layer of mesoderm, gives rise to the chorion, one of the extraembryonic membranes (see Fig. 22.10). The chorion develops into the fetal half of the placenta. The inner cell mass eventually becomes the embryo, which develops into a fetus.

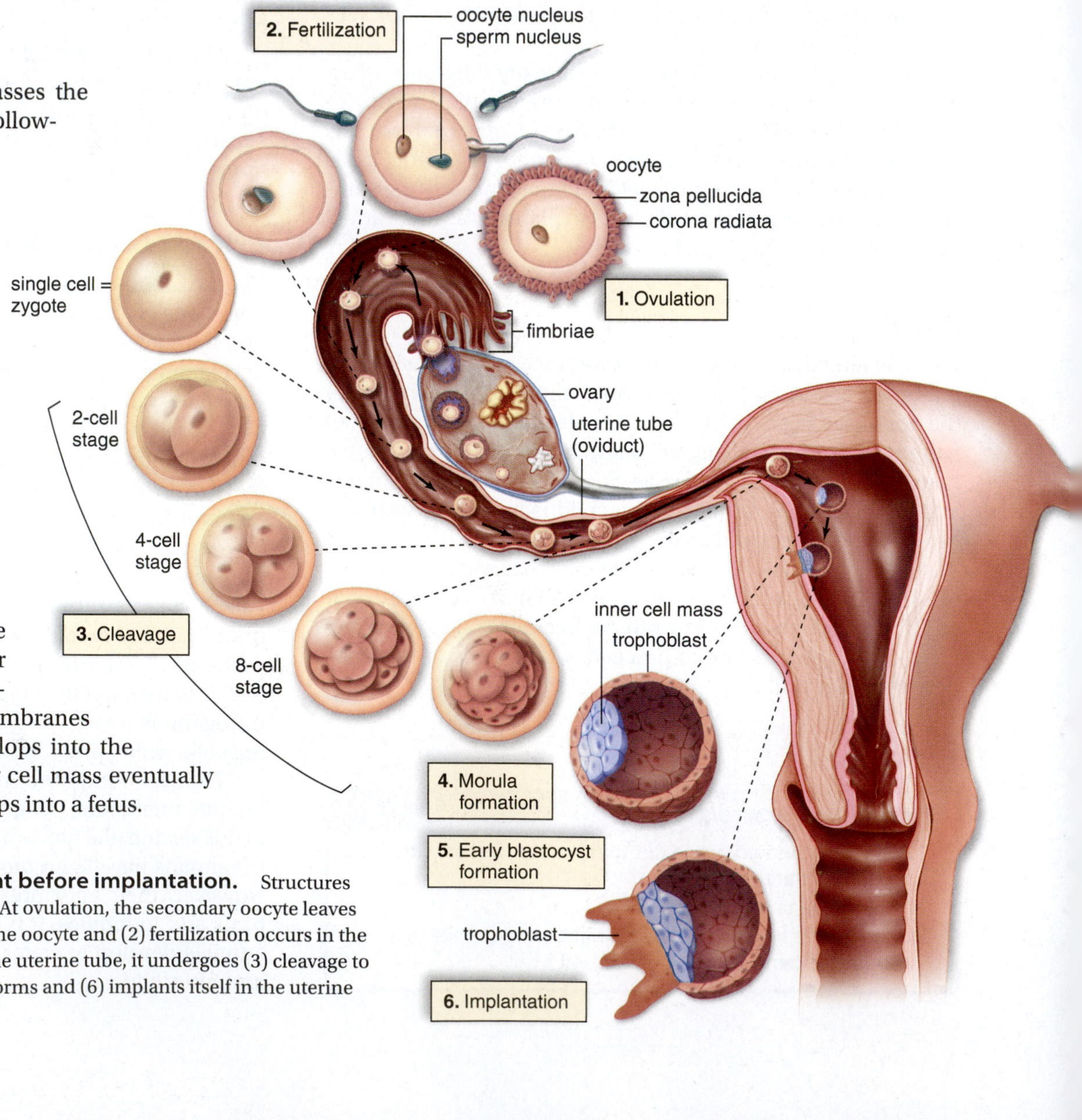
2. Fertilization
oocyte nucleus
sperm nucleus
oocyte
zona pellucida
corona radiata
single cell = zygote
1. Ovulation
fimbriae
ovary
uterine tube (oviduct)
2-cell stage
4-cell stage
3. Cleavage
8-cell stage
inner cell mass
trophoblast
4. Morula formation
5. Early blastocyst formation
trophoblast
6. Implantation

Figure 22.11 Human development before implantation. Structures and events proceed counterclockwise. (1) At ovulation, the secondary oocyte leaves the ovary. A single sperm nucleus enters the oocyte and (2) fertilization occurs in the uterine tube. As the zygote moves along the uterine tube, it undergoes (3) cleavage to produce (4) a morula. (5) The blastocyst forms and (6) implants itself in the uterine lining.

The Second Week

At the end of the first week, the embryo begins the process of implanting in the wall of the uterus. The trophoblast secretes enzymes to digest away some of the tissue and blood vessels of the endometrium of the uterus (Fig. 22.11). The embryo is now about the size of the period at the end of this sentence. The trophoblast begins to secrete the hormone **human chorionic gonadotropin** (**HCG**), which is the basis for the pregnancy test and serves to maintain the corpus luteum past the time it normally disintegrates (see section 21.3). Because of this, the endometrium is maintained and menstruation does not occur.

As the week progresses, the inner cell mass detaches itself from the trophoblast, and two more extraembryonic membranes form (Fig. 22.12*a*). The yolk sac, which forms below the embryonic disk, has no nutritive function as it does in chicks, but it is the first site of blood cell formation. The amnion has a cavity where the embryo (and then the fetus) develops. In humans, amniotic fluid acts as an insulator against cold and heat and also absorbs shock, such as that caused by the mother exercising.

Gastrulation occurs during the second week. The inner cell mass now has flattened into the **embryonic disk,** composed of two layers of cells: ectoderm above and endoderm below. Once the embryonic disk elongates to form the primitive streak, the third germ layer, mesoderm, forms by invagination of cells along the streak. The trophoblast is reinforced by mesoderm and becomes the chorion (Fig. 22.12*b*). It is possible to relate the development of future organs to these germ layers (see Table 22.1).

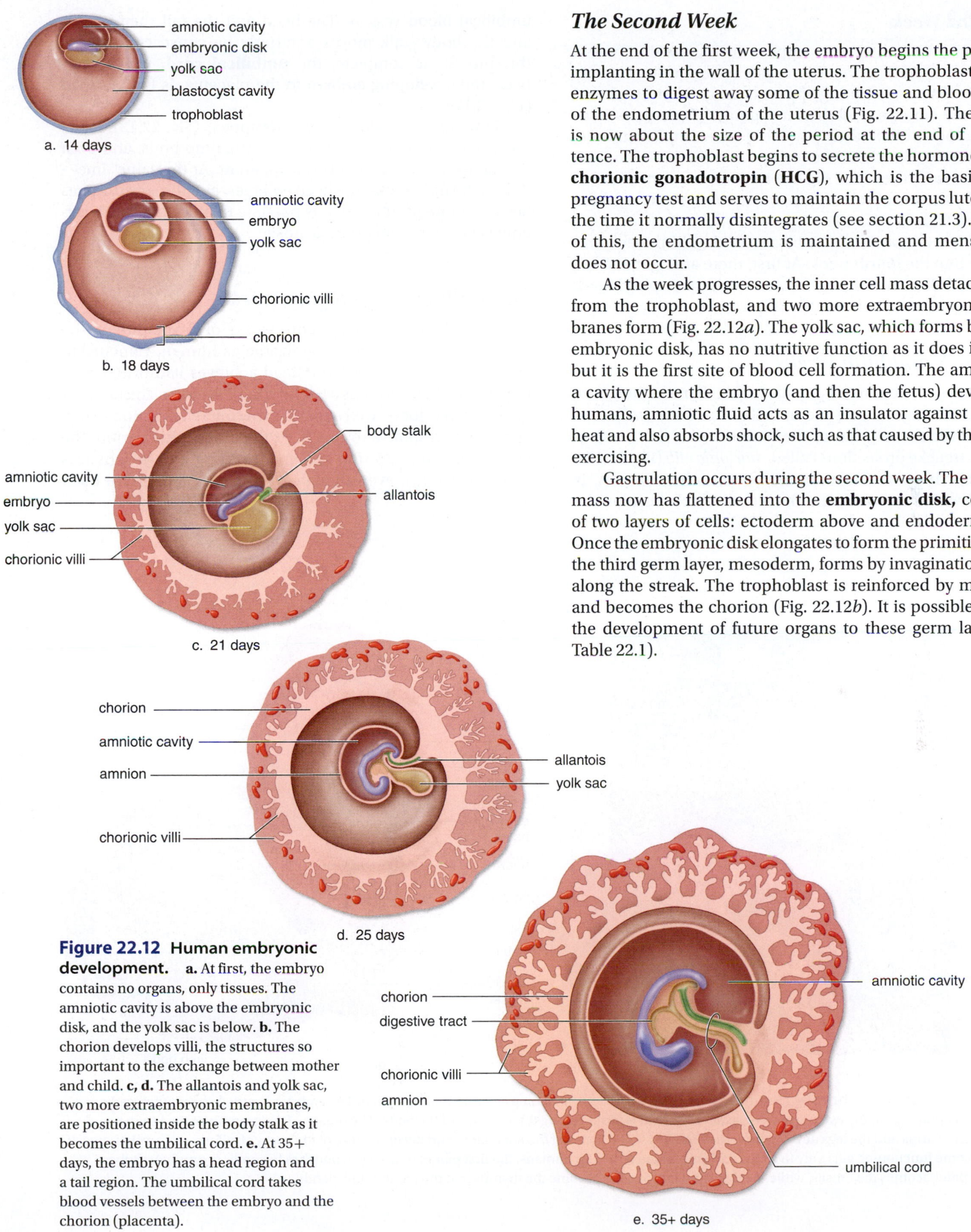
amniotic cavity
embryonic disk
yolk sac
blastocyst cavity
trophoblast
a. 14 days
amniotic cavity
embryo
yolk sac
chorionic villi
chorion
b. 18 days
body stalk
amniotic cavity
embryo
yolk sac
chorionic villi
allantois
c. 21 days
chorion
amniotic cavity
amnion
chorionic villi
allantois
yolk sac
d. 25 days
amniotic cavity
chorion
digestive tract
chorionic villi
amnion
umbilical cord
e. 35+ days

Figure 22.12 Human embryonic development. **a.** At first, the embryo contains no organs, only tissues. The amniotic cavity is above the embryonic disk, and the yolk sac is below. **b.** The chorion develops villi, the structures so important to the exchange between mother and child. **c, d.** The allantois and yolk sac, two more extraembryonic membranes, are positioned inside the body stalk as it becomes the umbilical cord. **e.** At 35+ days, the embryo has a head region and a tail region. The umbilical cord takes blood vessels between the embryo and the chorion (placenta).

The Third Week

Two important organ systems appear during the third week. The nervous system is the first organ system to be visually evident. At first, a thickening appears along the entire dorsal length of the embryo, and then invagination occurs as neural folds appear. When the neural folds meet at the midline, the neural tube, which later develops into the brain and the nerve cord, is formed (see Fig. 22.4). After the notochord is replaced by the vertebral column, the nerve cord is called the spinal cord.

Development of the heart begins in the third week and continues into the fourth week. At first, there are right and left heart tubes. When these fuse, the heart begins pumping blood, even though its chambers are not fully formed.

The Fourth and Fifth Weeks

At four weeks, the embryo is barely larger than the height of this print. A bridge of mesoderm called the *body stalk* connects the caudal (tail) end of the embryo with the chorion, which has treelike projections called *chorionic villi* (Fig. 22.12*c*, *d*). The fourth extraembryonic membrane, the allantois, is contained within this stalk, and its blood vessels become the umbilical blood vessels. The head and the tail then lift up, and the body stalk moves anteriorly by constriction. Once this process is complete, the **umbilical cord,** which connects the developing embryo to the placenta, is fully formed (Fig. 22.12*e*).

Little flippers called *limb buds* appear (Fig. 22.13). Later, the arms and the legs develop from the limb buds, and even the hands and the feet become apparent. At the same time—during the fifth week—the head enlarges, and the sense organs become more prominent. It is possible to make out the developing eyes, ears, and even the nose.

The Sixth Through Eighth Weeks

During the sixth through eighth weeks of development, the embryo becomes more recognizable as human. Concurrent with brain development, the head achieves its normal relationship with the body as a neck region develops. The nervous system is developed well enough to permit reflex actions, such as a startle response to touch. At the end of this period, the embryo is about 38 mm (1.5 inches) long and weighs less than 1 gram (g), even though all organ systems have been established.

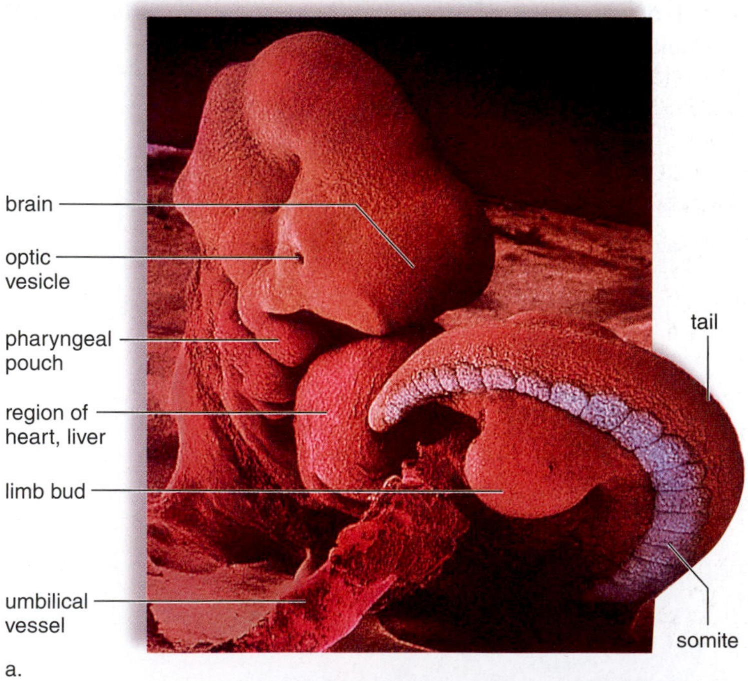
brain
optic vesicle
pharyngeal pouch
region of heart, liver
limb bud
umbilical vessel
tail
somite
a.

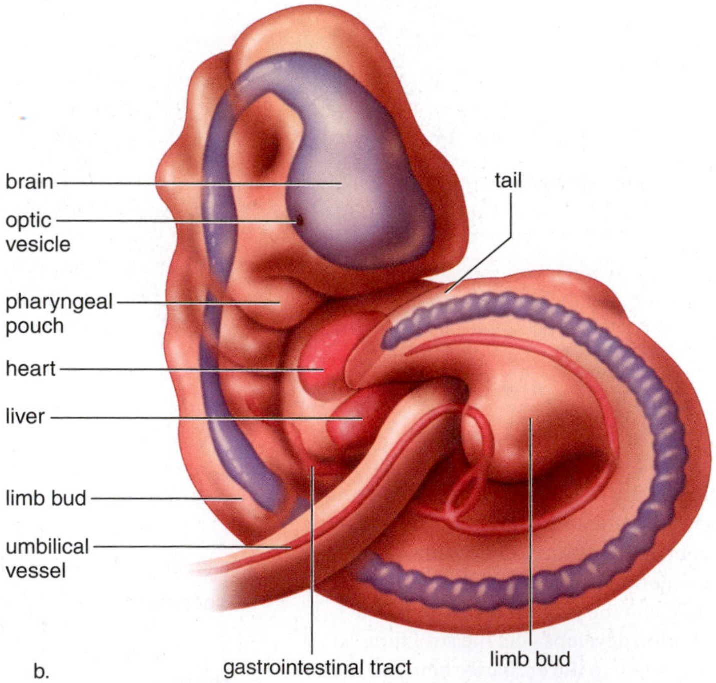
brain
optic vesicle
pharyngeal pouch
heart
liver
limb bud
umbilical vessel
tail
gastrointestinal tract
limb bud
b.

Figure 22.13 Human embryo at beginning of fifth week. **a.** Scanning electron micrograph. **b.** The embryo is curled so that the head touches the heart and liver, the two organs whose development is farther along than the rest of the body. The organs of the gastrointestinal tract are forming, and the arms and the legs develop from the limb buds. The bones of the tail regress and become those of the coccyx (tailbone). The pharyngeal pouches become functioning gills only in fishes and amphibian larvae. In humans, the first pair of pharyngeal pouches becomes the auditory tubes. The second pair becomes the tonsils, while the third and fourth pairs become the thymus and the parathyroid glands.

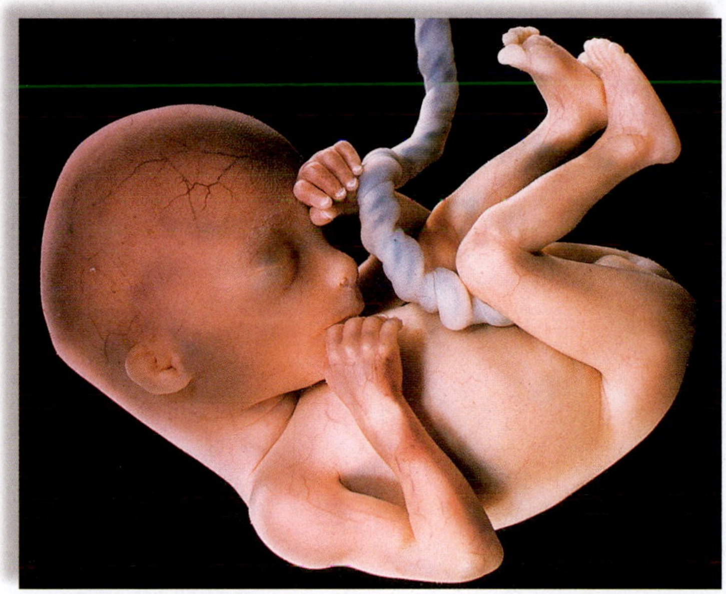

Figure 22.14 Three- to four-month-old fetus. At this stage, the fetus looks human, with its face, hands, and fingers well defined.

Fetal Development

Fetal development includes the third through ninth months of development. At this time, the fetus looks human (Fig. 22.14).

The Third and Fourth Months

At the beginning of the third month, the fetal head is still relatively large, the nose is flat, the eyes are far apart, and the ears are well formed. Head growth now begins to slow down as the rest of the body increases in length. Epidermal refinements, such as eyelashes, eyebrows, hair on head, fingernails, and nipples, appear.

Cartilage begins to be replaced by bone as ossification centers appear in most of the bones. Cartilage remains at the ends of the long bones, and ossification is not complete until age 18 or 20 years. The skull has six large membranous areas called **fontanels,** which permit a certain amount of flexibility as the head passes through the birth canal and allow rapid growth of the brain during infancy.

Sometime during the third month, it is possible to distinguish males from females. The presence or absence of the *SRY* gene on the X chromosome determines whether the gonads will differentiate into testes or ovaries. Once these have differentiated, they produce the sex hormones (see section 20.6) that influence the differentiation of the genital tract. At this time, either testes or ovaries are located within the abdominal cavity, but later, in the last trimester of fetal development, the testes descend into the scrotal sacs (see section 21.1).

During the fourth month, the fetal heartbeat is loud enough to be heard when a physician applies a stethoscope to the mother's abdomen. By the end of this month, the fetus is about 152 mm (6 inches) in length and weighs about 170 g (6 oz).

The Fifth Through Seventh Months

During the fifth through seventh months, the mother begins to feel movement. At first, there is only a fluttering sensation, but as the fetal legs grow and develop, kicks and jabs are felt. The fetus, though, is in the fetal position, with the head bent down and in contact with the flexed knees.

The wrinkled, translucent skin is covered by a fine down called **lanugo.** This, in turn, is coated with a white, greasy, cheeselike substance called **vernix caseosa,** which probably protects the delicate skin from the amniotic fluid. The eyelids are now fully open.

At the end of this period, the length has increased to about 300 mm (12 inches), and the weight is about 1,380 g (3 lb). If born now, it is likely that the baby will survive.

Fetal Circulation

The fetus has circulatory features that are not present in the adult circulation (Fig. 22.15). All of these features are necessary because the fetus does not use its lungs for gas exchange. For example, much of the blood entering the right atrium is shunted into the left atrium through the **foramen ovale** between the two atria. Also, any blood that does enter the right ventricle and is pumped into the pulmonary trunk is shunted into the aorta by way of the **ductus arteriosus.**

Blood within the aorta travels to the various branches, including the iliac arteries, which connect to the *umbilical arteries* leading to the placenta. Exchange of gases and nutrients between maternal blood and fetal blood takes place at the placenta. The *umbilical vein* carries blood rich in nutrients and oxygen to the fetus. The umbilical vein enters the liver and then joins the *ductus venosus,* which merges with the inferior vena cava, a vessel that returns blood to the heart. The umbilical arteries and vein run alongside one another in the umbilical cord, which is tied off and cut at birth, leaving only the umbilicus (navel).

The most common of all cardiac defects in the newborn is the persistence of the foramen ovale. Once the umbilical cord has been severed and the lungs have expanded, blood enters the lungs in quantity. The return of this blood to the left side of the heart usually causes a flap to cover the opening. Incomplete closure occurs in nearly one out of four individuals, but even so, passage of the blood from the right atrium to the left atrium rarely occurs because either the opening is small or it closes when the atria contract. In a small number of cases, the passage of impure blood from the right side to the left side of the heart is sufficient to cause a "blue baby." Such a condition can usually be corrected by threading a catheter into the heart, and sealing the defect.

The ductus arteriosus normally closes because endothelial cells divide and block off this duct. Remains of the ductus arteriosus and parts of the umbilical arteries and vein are later transformed into connective tissue.

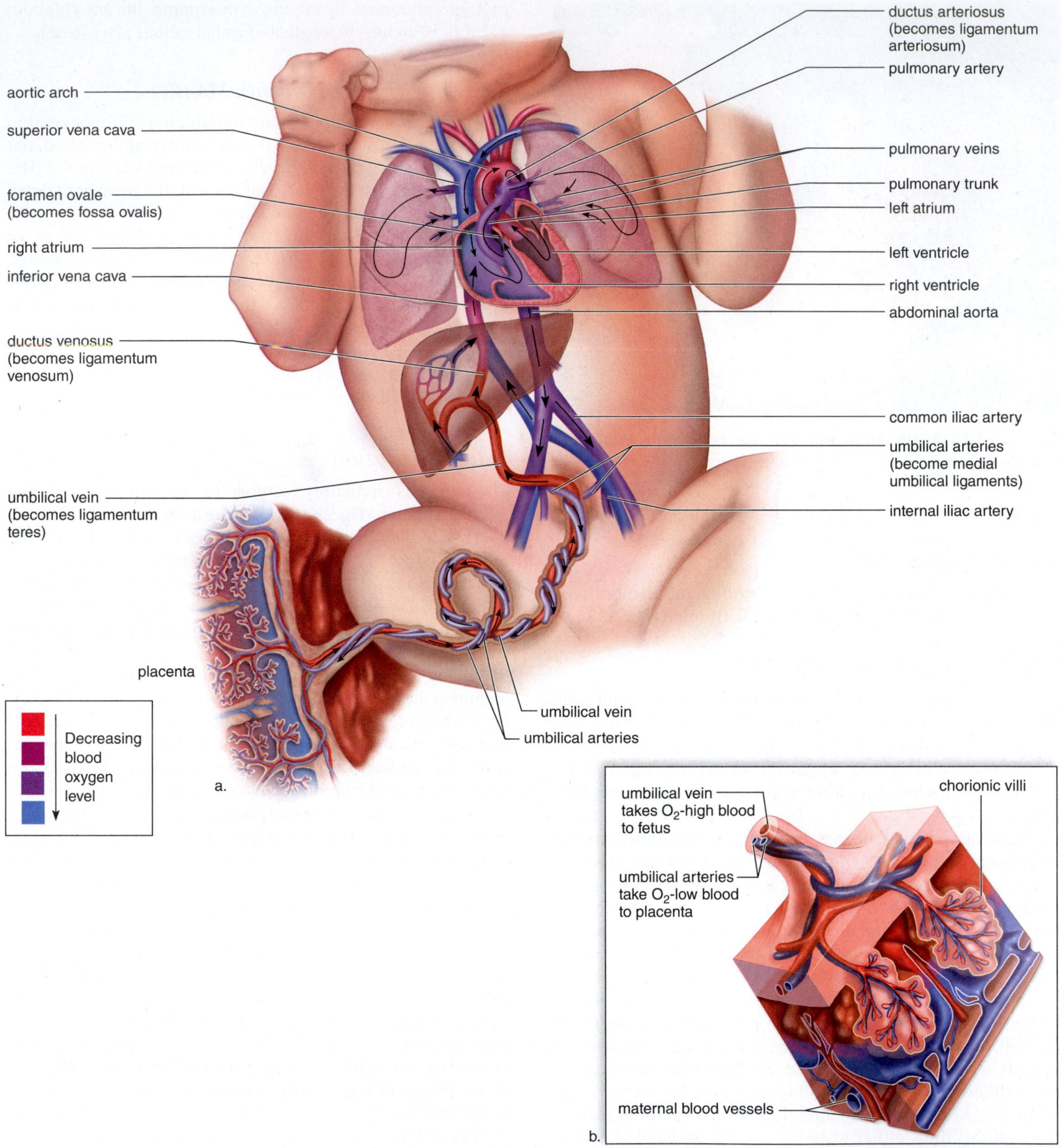
aortic arch
superior vena cava
foramen ovale
(becomes fossa ovalis)
right atrium
inferior vena cava
ductus venosus
(becomes ligamentum
venosum)
umbilical vein
(becomes ligamentum
teres)
placenta
Decreasing
blood
oxygen
level
a.
ductus arteriosus
(becomes ligamentum
arteriosum)
pulmonary artery
pulmonary veins
pulmonary trunk
left atrium
left ventricle
right ventricle
abdominal aorta
common iliac artery
umbilical arteries
(become medial
umbilical ligaments)
internal iliac artery
umbilical vein
umbilical arteries
umbilical vein
takes O_2-high blood
to fetus
chorionic villi
umbilical arteries
take O_2-low blood
to placenta
maternal blood vessels
b.

Figure 22.15 Fetal circulation and the placenta. **a.** The lungs are not functional in the fetus, and the blood passes directly from the right atrium to the left atrium or from the right ventricle to the aorta. The umbilical arteries take fetal blood to the placenta, and the umbilical vein returns fetal blood from the placenta. **b.** At the placenta, an exchange of molecules between fetal and maternal blood takes place across the walls of the chorionic villi. Oxygen and nutrient molecules diffuse into the fetal blood, and carbon dioxide and urea diffuse out of the fetal blood.

Structure and Function of the Placenta Humans belong to the group of mammals called placental mammals. The **placenta** is firmly attached to the uterine wall by the allantois and by fingerlike projections called the chorionic villi. The placenta is a structure that functions only before birth. When the child is born, the placenta becomes part of the *afterbirth.*

The placenta functions in gas, nutrient, and waste exchange between the embryonic (later fetal) and maternal circulatory systems. The placenta begins to form once the embryo is fully implanted. At first, the entire chorion has chorionic villi that project into the endometrium. Later, these disappear in all areas except where the placenta develops. By the tenth week, the placenta is fully formed and is producing progesterone and estrogen. These hormones have two effects: (1) due to their negative feedback control of the hypothalamus and the anterior pituitary, they prevent any new follicles from maturing, and (2) they maintain the lining of the uterus so that the corpus luteum is no longer needed. Usually no menstruation occurs during pregnancy.

The placenta has a fetal side contributed by the chorion and a maternal side consisting of uterine tissues. Notice in Figure 22.15 how the chorionic villi are surrounded by maternal blood. Normally, maternal and fetal blood never mix, because exchange always takes place across the walls of the chorionic villi.

The umbilical cord stretches between the placenta and the fetus. The umbilical cord is the lifeline of the fetus because it contains the umbilical arteries and vein, which transport waste molecules (carbon dioxide and urea) to the placenta for disposal and take oxygen and nutrient molecules from the placenta to the rest of the fetal circulatory system.

As discussed in the Health feature, "Preventing and Testing for Birth Defects" on page 454, harmful chemicals, such as alcohol, can also cross the placenta. This is of particular concern during the embryonic period, when various structures are first forming. Each organ or part seems to have a sensitive period during which a substance can alter its normal development. Even over-the-counter drugs should not be taken during pregnancy without permission from your physician.

Check Your Progress 22.3

1. Name the extraembryonic membrane that gives rise to each of the following: umbilical blood vessels, the first blood cells, and the fetal half of the placenta.
2. Summarize the major events by month during fetal development.
3. Describe the path of blood flow in the fetus starting with the placenta, and name the structures unique to fetal circulation.
4. Explain the function of the placenta.

22.4 Human Pregnancy, Birth, and Lactation

Learning Outcomes

Upon completion of this section, you should be able to

1. Describe changes that occur in female physiology during pregnancy.
2. Outline the stages of birth.
3. Summarize the advantages of breast-feeding.

Pregnancy

Many changes that take place in the mother's body during pregnancy are due to the hormones progesterone and estrogen. Others are due to the increasing size of the uterus, from a nonpregnant weight of 60–80 g to 900–1,200 g at term.

Morning Sickness and Energy Level

About six weeks into her pregnancy, the mother may experience nausea and vomiting, loss of appetite, and fatigue. These symptoms usually subside by around the 12th week. Subsequently, some mothers then report increased energy levels and a general sense of well-being despite gaining weight. The weight gain of pregnancy is due to breast and uterine enlargement; weight of the fetus; amount of amniotic fluid; size of the placenta; the mother's own increase in total body fluid; and an increase in storage of proteins, fats, and minerals.

Effects on Smooth Muscle

Progesterone decreases uterine motility by relaxing smooth muscle, including the smooth muscle in the walls of arteries. The arteries expand, and this leads to a low blood pressure that sets in motion the renin-angiotensin-aldosterone mechanism (see section 20.4). Aldosterone promotes sodium and water retention, and blood volume increases until it reaches its peak sometime during weeks 28 to 32 of pregnancy. Altogether, blood volume increases from the normal 5 liters up to 7 liters—a 40% rise. An increase in the number of red blood cells follows. With the rise in blood volume, cardiac output increases by 20–30%. Blood flow to the kidneys, placenta, skin, and breasts rises significantly. Smooth muscle relaxation also explains the common gastrointestinal effects of pregnancy. The heartburn experienced by many is due to relaxation of the esophageal sphincter and reflux of stomach contents into the esophagus. Constipation can be a result of decreased intestinal tract motility.

Other Effects

Compression of the ureters and urinary bladder by an enlarged uterus can result in incontinence, or the involuntary leakage of urine from the bladder. Compression of the inferior vena cava, especially when lying down, decreases venous return, and the result is edema (fluid accumulation) and varicose veins.

In addition to the steroid hormones progesterone and estrogen, the placenta also produces some peptide hormones. Several of these may make cells resistant (unresponsive) to insulin, and the result can be gestational diabetes. Hormones released by the placenta, rather than stretching of the skin, may also be the cause of *striae gravidarum,* commonly called "stretch marks," which form over the abdomen and lower breasts during pregnancy. Melanocyte activity also increases during pregnancy. Darkening of certain areas of the skin, including the face, neck, and breast areolae, is common.

Birth

The uterus has contractions during the last trimester of the pregnancy. At first, these are light, lasting about 20 to 30 seconds and occurring every 15 to 20 minutes. Near the end of pregnancy, the contractions may become stronger and more frequent. However, the onset of true labor is marked by uterine contractions that occur regularly every 10–15 minutes and last for 40 seconds or longer. These contractions are generally called "labor pains," although, when they first appear, they are not usually painful and only become painful at a later stage.

A positive feedback mechanism regulates the onset and continuation of labor. Uterine contractions are induced by stretching of the cervix, which also brings about the release of oxytocin from the posterior pituitary. Oxytocin stimulates uterine contractions, which push the fetus downward, and the cervix stretches even more. This cycle keeps repeating itself until the baby is born. Three events, in any order, indicate that delivery will soon occur: (1) The uterine contractions are occurring about every five minutes and becoming stronger. (2) The amnion, which contains amniotic fluid, ruptures, causing water to flow out of the vagina. This event is sometimes referred to as "breaking water." (3) A plug of mucus from the cervical canal leaves the vagina. This plug prevents bacteria and sperm from entering the uterus during pregnancy. Its expulsion may be the least noticeable of the events, unless it is mixed with blood.

Stage 1

Parturition, the process of giving birth to an offspring, can be divided into three stages (Fig. 22.16). Prior to the first stage, there can be a "bloody show" caused by expulsion of the mucous plug. At first, the uterine contractions of labor occur in such a way

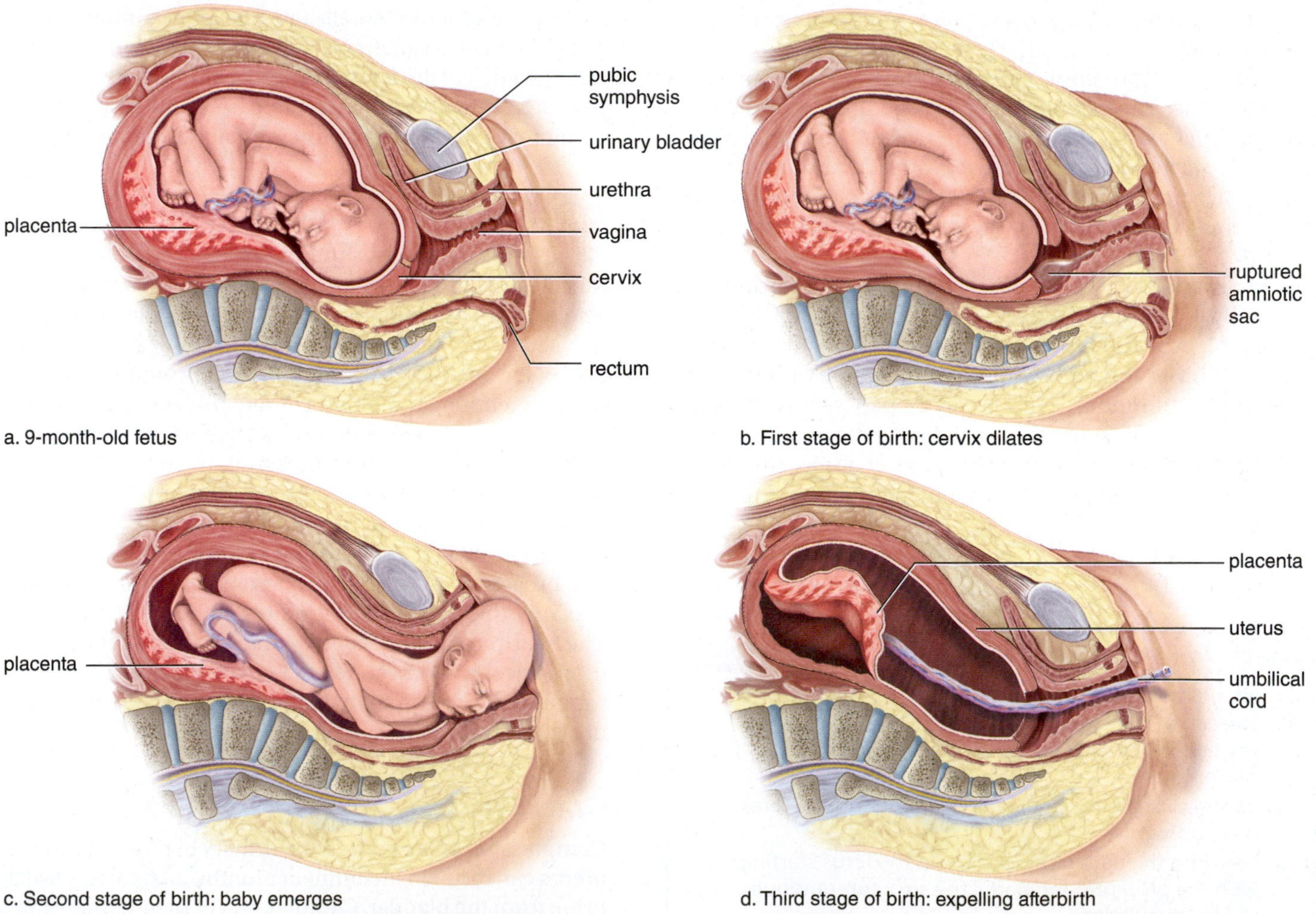
pubic symphysis
urinary bladder
urethra
placenta
vagina
cervix
rectum
a. 9-month-old fetus
ruptured amniotic sac
b. First stage of birth: cervix dilates
placenta
c. Second stage of birth: baby emerges
placenta
uterus
umbilical cord
d. Third stage of birth: expelling afterbirth

Figure 22.16 **Three stages of parturition (birth).** **a.** Position of fetus just before birth begins. **b.** Dilation of cervix. **c.** Birth of baby. **d.** Expulsion of afterbirth.

that the cervical canal slowly disappears as the lower part of the uterus is pulled upward toward the baby's head (Fig. 22.16*b*). This process is called effacement, or "taking up the cervix." With further contractions, the baby's head acts as a wedge to assist cervical dilation. If the amniotic membrane has not already ruptured, it is apt to do so during this stage, releasing the amniotic fluid, which leaks out the vagina. The first stage of parturition ends once the cervix is dilated completely.

Stage 2

During the second stage of parturition, the uterine contractions occur every 1 to 2 minutes and last about one minute each. They are accompanied by a desire to push, or bear down. As the baby's head gradually descends into the vagina, the desire to push becomes greater. When the baby's head reaches the exterior, it turns so that the back of the head is uppermost (Fig. 22.16*c*). If the vaginal orifice does not expand enough to allow passage of the head, an **episiotomy** may be performed. This incision, which enlarges the opening, is sewn together later. As soon as the head is delivered, the baby's shoulders rotate so that the baby faces either to the right or the left. At this time, the physician may hold the head and guide it downward, while one shoulder and then the other emerges. The rest of the baby follows easily.

Once the baby is breathing normally, the umbilical cord is cut and tied, severing the child from the placenta. The stump of the cord shrivels and leaves a scar, which is the umbilicus.

Stage 3

The placenta, or **afterbirth,** is delivered during the third stage of parturition (Fig. 22.16*d*). About 15 minutes after delivery of the baby, uterine muscular contractions shrink the uterus and dislodge the placenta. The placenta is then expelled into the vagina. As soon as the placenta and its membranes are delivered, the third stage of parturition is complete.

Female Breast and Lactation

A female breast contains 15 to 25 lobules, each with a milk duct. Each milk duct begins at the nipple and divides into numerous other ducts that end in blind sacs called alveoli (Fig. 22.17).

During pregnancy, the breasts enlarge as the ducts and alveoli increase in number and size. The same hormones that affect the mother's breasts can also affect the child's. Some newborns, including males, even secrete a small amount of milk for a few days.

Usually, no milk is produced during pregnancy. The hormone prolactin is needed for lactation to begin, and the production of this hormone is suppressed because of the feedback control that the increased amount of estrogen and progesterone during pregnancy has on the pituitary. Once the baby is delivered, however, the pituitary begins secreting prolactin. It takes a couple of days for milk production to begin, and in the meantime, the breasts produce **colostrum,** a thin, yellow, milky fluid rich in protein, including antibodies.

The continued production of milk requires a suckling child. When a breast is suckled, the nerve endings in the areola are stimulated, and a nerve impulse travels along neural pathways from the nipples to the hypothalamus, which directs the pituitary gland to release the hormone oxytocin. When this hormone arrives at the breast, it causes contraction of the lobules so that milk flows into the ducts (called milk letdown), where it may be drawn out of the nipple by the suckling child.

Whether to breast-feed or not is a private decision based in part on a woman's particular circumstances. However, it is well known that breast milk contains antibodies produced by the mother that can help a baby survive. Babies have immature immune systems, less stomach acid to destroy foreign antigens, and also unsanitary habits. Breast-fed babies are less likely to develop stomach and intestinal illnesses, including diarrhea, during the first 13 weeks of life. Breast-feeding also has physiological benefits for the mother. Suckling causes uterine contractions that can help the uterus return to its normal size, and breast-feeding uses up calories and thus can help a woman return to her normal weight.

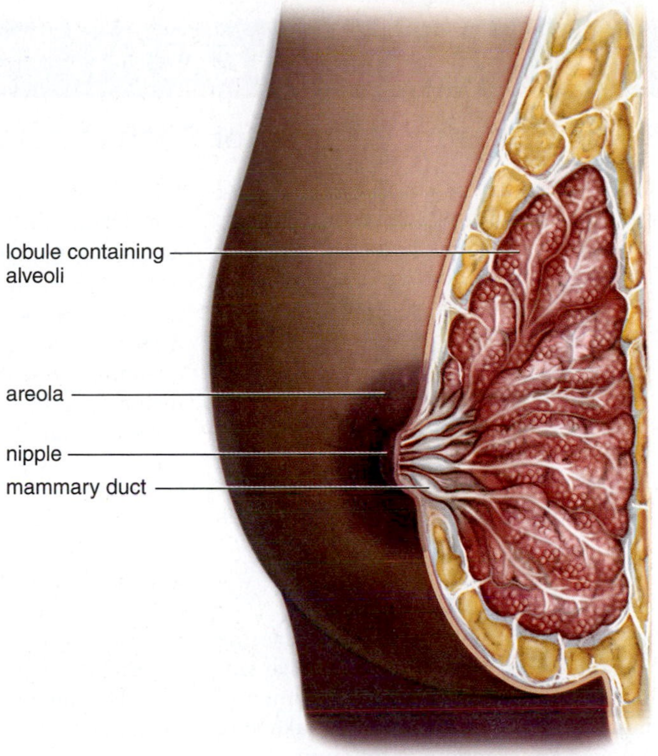
lobule containing alveoli
areola
nipple
mammary duct

Figure 22.17 Female breast anatomy. The female breast contains lobules consisting of ducts and alveoli. The alveoli are lined by milk-producing cells in the lactating (milk-producing) breast. Milk droplets congregate in the ducts that enter sinuses just behind the nipple. When a baby suckles, the sinuses are squeezed.

Check Your Progress 22.4

1. Describe the physiological changes that occur in a pregnant woman's circulatory system.
2. Identify the stage of parturition in which the baby is delivered.
3. Define colostrum, and list three advantages of breast-feeding.

SCIENCE IN YOUR LIFE ▶ HEALTH

Preventing and Testing for Birth Defects

Birth defects, or congenital disorders, are abnormal conditions that are present at birth. According to the Centers for Disease Control and Prevention (CDC), about 1 in 33 babies born in the United States has a birth defect. Those due to genetics can sometimes be detected before birth by one of the methods shown in Figure 22B; however, these methods are not without risk. Some birth defects are not serious, and not all can be prevented. But women can take steps to increase their chances of delivering a healthy baby.

Eat a Healthy Diet

Certain birth defects occur because the developing embryo does not receive sufficient nutrition. For example, women of childbearing age are urged to make sure they consume adequate amounts of the vitamin folic acid in order to prevent neural tube defects such as spina bifida and anencephaly. In spina bifida, part of the vertebral column fails to develop properly and cannot adequately protect the delicate spinal cord. With anencephaly, most of the fetal brain fails to develop. Anencephalic infants are stillborn or survive for only a few days after birth. Fortunately, folic acid is plentiful in leafy green vegetables, nuts, and citrus fruits. It is also present in multivitamins, and many breads and cereals are fortified with it. The CDC recommends that all women of childbearing age get at least 400 micrograms (μg) of folic acid every day through supplements, in addition to eating a healthy, folic acid–rich diet. Unfortunately, neural tube birth defects can occur just a few weeks after conception, when many women are still unaware that they are pregnant—especially if the pregnancy is unplanned.

Avoid Alcohol, Smoking, and Drug Abuse

As shown in Figure 22.15, the placenta is the region for exchange between the fetal and maternal circulatory systems. Alcohol consumption during pregnancy is a leading cause of birth defects. In severe instances, the baby is born with fetal alcohol syndrome (FAS), estimated to occur in 0.2 to 1.5 of every 1,000 live births in the United States. Children with FAS are frequently underweight, have an abnormally small head, abnormal facial development, and intellectual disability. As they mature, children with FAS often exhibit short attention span, impulsiveness, and poor judgment, as well as serious difficulties with learning and memory. Heavy alcohol use also reduces a woman's folic acid level, increasing the risk of the neural tube defects just described. Cigarette smoking causes many birth defects. Babies born to smoking mothers typically have low birth weight, and are more likely to have defects of the face, heart, and brain than those born to nonsmokers. Illegal drugs should also be avoided. For example, cocaine causes blood pressure fluctuations that deprive the fetus of oxygen. Cocaine-exposed babies may have problems with vision and coordination, and may be intellectually disabled.

Alert Medical Personnel If You Are or May Be Pregnant

Several medications that are safe for healthy adults may pose a risk to a developing fetus. If pregnant women need to be immunized, they are usually given killed or inactivated forms of the vaccine, because live forms often present a danger to the fetus. Because the rapidly dividing cells of a developing embryo or fetus are very susceptible to damage from radiation, pregnant women should avoid unnecessary X rays. If a medical X ray is unavoidable, the woman should notify the X-ray technician so that her fetus can be protected as much as possible, for example, by draping a lead apron over her abdomen if this area is not targeted for X-ray examination.

Avoid Infections That Cause Birth Defects

Certain pathogens such as the virus that causes rubella (German measles) can result in birth defects. In the past, this virus caused many birth defects, specifically mental disability, deafness, blindness, and heart defects. Rubella is much less of a danger in developed countries today because most women have been vaccinated against rubella as children. Other infections that can cause birth defects include toxoplasmosis, herpes simplex, and cytomegalovirus.

Questions to Consider

1. Why is it likely that the actual rate of birth defects is higher than what is reported?
2. Besides tobacco, alcohol, and illegal drugs, what are some other potential risks that a pregnant woman should avoid? Why?
3. Should a woman be prosecuted for causing damage to her unborn child by drinking or illegal drug use?

22.5 Human Development After Birth

Learning Outcomes

Upon completion of this section, you should be able to

1. Discuss the effects of aging on various body systems.
2. Compare and contrast the preprogrammed theories to the damage accumulation theories of aging.

Development does not cease once birth has occurred but continues throughout the stages of life: infancy, childhood, adolescence, and adulthood. Infancy, the toddler years, and preschool years are times of remarkable growth. During the birth to five-year-old stage, humans acquire gross motor and fine motor skills. These include the ability to sit up and then to walk, as well as being able to hold a spoon and manipulate small objects. Language usage begins during this time and will become increasingly sophisticated

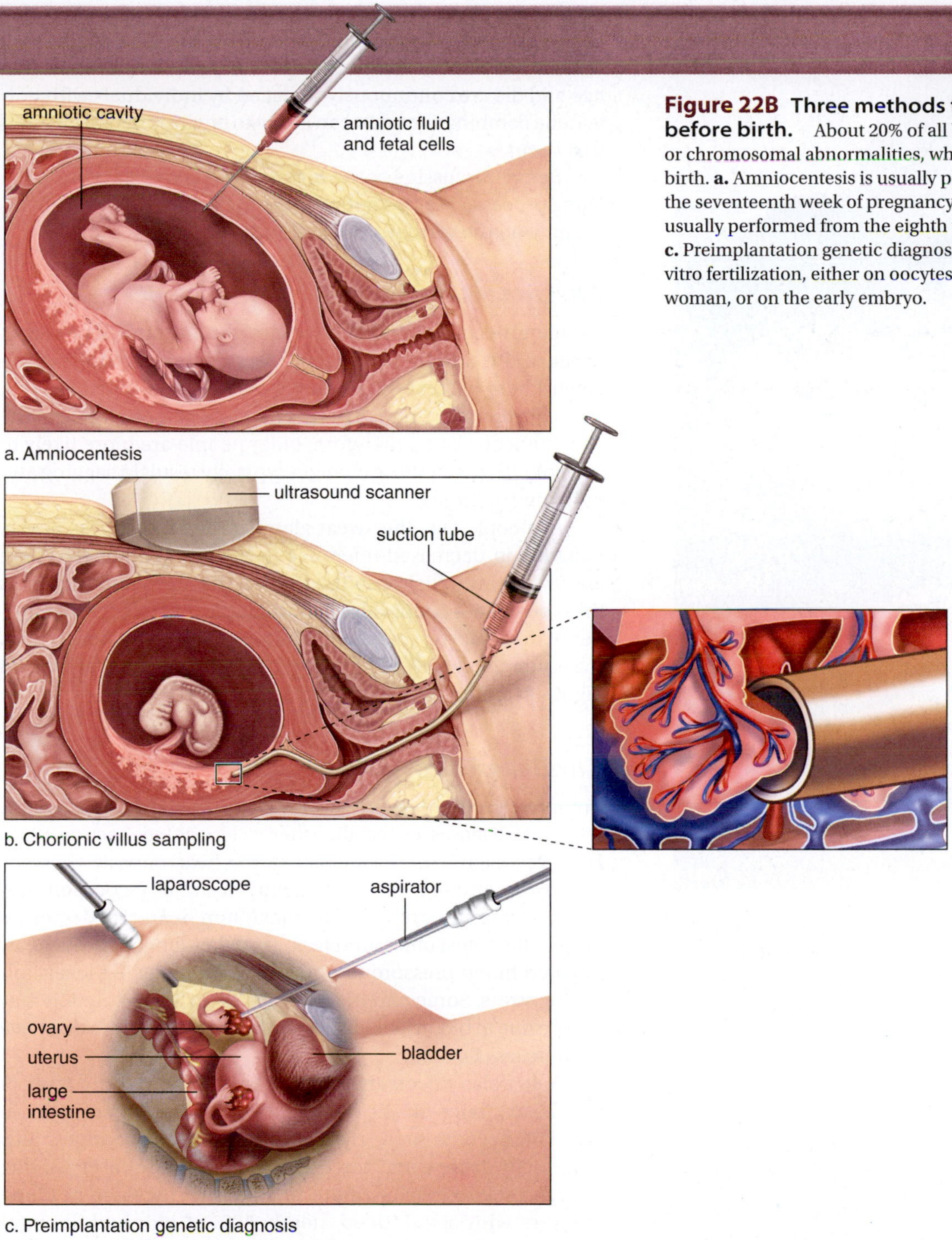
amniotic cavity
amniotic fluid and fetal cells
a. Amniocentesis
ultrasound scanner
suction tube
b. Chorionic villus sampling
laparoscope
aspirator
ovary
uterus
large intestine
bladder
c. Preimplantation genetic diagnosis

Figure 22B Three methods for genetic defect testing before birth. About 20% of all birth defects are due to genetic or chromosomal abnormalities, which may be detected before birth. **a.** Amniocentesis is usually performed from the fifteenth to the seventeenth week of pregnancy. **b.** Chorionic villus sampling is usually performed from the eighth to the twelfth week of pregnancy. **c.** Preimplantation genetic diagnosis can be performed prior to in vitro fertilization, either on oocytes that have been collected from the woman, or on the early embryo.

throughout childhood. As infants and toddlers explore their environment, their senses—vision, taste, hearing, smell, and touch—mature dramatically. Socialization is very important, as a child forms emotional ties with its caregivers and learns to separate self from others. Babies do not all develop at the same rate, and there is a large variation in what is considered normal.

The preadolescent years, from 6 to 12 years of age, are a time of continued rapid growth and learning. Preadolescents form identities apart from parents, and peer approval becomes very important. Adolescence begins with the onset of *puberty,* as the young person achieves sexual maturity. For girls, puberty begins between 10 and 14 years of age, whereas for boys it generally occurs between ages 12 and 16. During this time, the sex-specific hormones cause the secondary sexual characteristics to appear. Profound social and psychological changes are also associated with the transition from childhood to adulthood.

Figure 22.18 Aging. Aging is a slow process during which the body undergoes changes that eventually bring about death, even if no marked disease or disorder is present. Medical science is trying to extend the human life span as well as the health span, the length of time the body functions normally.

Aging encompasses these progressive changes, which contribute to an increased risk of infirmity, disease, and death. Today, there is great interest in **gerontology,** the study of aging, because our society includes more older individuals than ever before. In the next half-century, the number of people over age 65 will increase by 147%. The human *life span* is judged to be a maximum of 120 to 125 years. The goal of gerontology is not necessarily to increase the life span, but to increase the *health span,* the number of years an individual enjoys the full functions of all body parts and processes (Fig. 22.18).

MP3 Aging

The Effects of Aging on Organ Systems

Aging and death are as much a part of biology as are embryonic development and birth. If animals—and indeed, all forms of life—did not age and eventually die, their exponentially increasing numbers would presumably overwhelm the ability of the Earth to support life. Moreover, aging and death are a critical part of the evolutionary process, because animals that age and die are continuously replaced by individuals with new genetic combinations that can adapt to environmental changes that occur.

Before discussing some possible mechanisms behind the aging process, it seems appropriate to first consider the effects of aging on the various body systems.

Integumentary System

As aging occurs, the skin becomes thinner and less elastic because the number of elastic fibers decreases and the collagen fibers become increasingly cross-linked to each other, reducing their flexibility. There is also less adipose tissue in the subcutaneous layer; therefore, older people are more likely to feel cold. Together these changes typically result in sagging and wrinkling of the skin.

As people age, the sweat glands also become less active, resulting in decreased tolerance to high temperatures. There are fewer hair follicles, so the hair on the scalp and the limbs thins out. Older people also experience a decrease in the number of melanocytes, making their hair gray and their skin pale. In contrast, some of the remaining pigment cells are larger, and pigmented blotches ("age spots") appear on the skin.

Cardiovascular System

Common problems with cardiovascular function are usually related to diseases, especially atherosclerosis (see section 12.5). However, even with normal aging, the heart muscle weakens somewhat, and may increase slightly in size as it compensates for its decreasing strength. The maximum heart rate decreases even in the fittest older athlete, and it takes longer for the heart rate and blood pressure to return to normal resting levels following stress. Some part of this decrease in heart function may also be due to blood leaking back through heart valves that have become less flexible.

Aging also affects the blood vessels. The middle layer of arteries contains elastic fibers, which, like collagen fibers in the skin, become more rigid with time. These changes, plus a frequent decrease in the internal diameter of arteries due to atherosclerosis, contribute to a gradual increase in blood pressure with age. Indeed, nearly 50% of older adults have chronic hypertension. Such changes are common in individuals living in Western industrialized countries but not in agricultural societies. This indicates that a diet low in cholesterol and saturated fatty acids, along with a sensible exercise program, may help to prevent age-related cardiovascular disease.

Immune System

As people age, many of their immune system functions become compromised. Because a healthy immune system normally protects the entire body from infections, toxins, and at least some types of cancer, some investigators believe that decreasing immune function can play a major role in the aging process.

As noted in Chapter 13, the thymus is an important site for T-cell maturation. Beginning in adolescence, the thymus begins to involute, gradually decreasing in size, and eventually becoming replaced by fat and connective tissue. The thymus of a 60-year-old adult is about 5% of the size of the thymus of a newborn, resulting in a decrease in the ability of older people to generate T-cell responses to new antigens. The evolutionary rationale for this may be that the thymus is energetically expensive for an organism to maintain, and that compared to younger animals that must respond to a high number of new infections and other antigens, older animals have already responded to most of the antigens to which they will be exposed in their life.

Aging also affects other immune functions. Because most B-cell responses are dependent on T cells, antibody responses also begin to decline. This, in turn, may explain why the elderly do not respond as well to vaccinations as young people do. This presents challenges in protecting older people against diseases like influenza and pneumonia, which can otherwise be prevented by annual vaccination.

Not all immune functions decrease in older animals. The activity of *natural killer cells,* which are a part of the innate immune system, seems to change very little with age. Perhaps by investigating how these cells remain active throughout a normal human life span, researchers can learn to preserve other aspects of immunity in the elderly.

Digestive System

The digestive system is perhaps less affected by the aging process than other systems. Because secretion of saliva decreases, more bacteria tend to adhere to the teeth, causing more decay and periodontal disease. Blood flow to the liver is reduced, resulting in less efficient metabolism of drugs or toxins. This means that, as a person gets older, less medication is needed to maintain the same level in the bloodstream.

Respiratory System

Cardiovascular problems are often accompanied by respiratory disorders, and vice versa. Decreasing elasticity of lung tissues means that ventilation is reduced. Because we rarely use the entire vital capacity, these effects may not be noticed unless the demand for oxygen increases, such as during exercise.

Excretory System

Blood supply to the kidneys is also reduced. The kidneys become smaller and less efficient at filtering wastes. Salt and water balance are difficult to maintain, and the elderly dehydrate faster than young people. Urinary incontinence (lack of bladder control) increases with age, especially in women. In men, an enlarged prostate gland may reduce the diameter of the urethra, causing frequent or difficult urination.

Nervous System

Between the ages of 20 and 90, the brain loses about 20% of its weight and volume. Neurons are extremely sensitive to oxygen deficiency, and neuron death may be due not to aging itself but to reduced blood flow in narrowed blood vessels. However, contrary to previous scientific opinion, recent studies using advanced imaging techniques show that most age-related loss in brain function is not due to whatever loss of neurons is occurring. Instead, decreased function may occur due to alterations in complex chemical reactions, or increased inflammation in the brain. For example, an age-associated decline in levels of dopamine can affect brain regions involved in complex thinking.

Perhaps more important than the molecular details, recent studies have confirmed that lifestyle factors can affect the aging brain. For example, animals on restricted calorie diets developed fewer Alzheimer-like changes in their brains. Other positive factors that may help maintain a healthy brain include attending college (the "use it or lose it" idea), regular exercise, and sufficient sleep.

Sensory Systems

In general, with aging more stimulation is needed for taste, smell, and hearing receptors to function as before. A majority of people over age 80 have a significant decline in their sense of smell, and about 15% suffer from anosmia, or a total inability to smell. The latter condition can be a serious health hazard, due to the inability to detect smoke, gas leaks, or spoiled food. After age 50, most people gradually begin to lose the ability to hear tones at higher frequencies, and this can make it difficult to identify individual voices and to understand conversation in a group.

Starting at about age 40, the lens of the eye does not accommodate as well, resulting in *presbyopia,* or difficulty focusing on near objects, which causes many people to require reading glasses as they reach middle age. Finally, as noted in section 18.7, cataracts and other eye disorders become much more common in the aged.

Musculoskeletal System

For the average person, muscle mass peaks between age 16 to 19 for females and 18 to 24 for males. Beginning in the twenties or thirties, but accelerating with increasing age, muscle mass generally decreases, due to decreases in both the size and number of muscle fibers. Most people who reach age 90 have 50% less muscle mass compared to when they were 20. Although some of this loss may be inevitable, regular exercise can slow this decline (Fig. 22.19).

Like muscles, bones tend to shrink in size and density with age. Due to compression of the vertebrae, along with changes

Figure 22.19 Remaining active. The aim of gerontology is to allow seniors to enjoy living.

in posture, most of us lose height as we age. Those who reach age 80 will be about 2 inches shorter than they were in their twenties. Women lose bone mass more rapidly than men do, especially after menopause. As discussed in section 19.6, osteoporosis is a common disease in the elderly. Although some decline in bone mass is a normal result of aging, certain extrinsic factors are also important. A proper diet and moderate exercise program has been found to slow the progressive loss of bone mass.

Endocrine System

As with the immune system, aging of the hormonal system can affect many different organs of the body. These changes are complex, however, with some hormone levels tending to decrease with age, while others increase. The activity of the thyroid gland generally declines, resulting in a lower basal metabolic rate. The production of insulin by the pancreas may remain stable, but cells become less sensitive to its effects, resulting in a rise in fasting glucose levels of about 10 mg/dL each decade after age 50.

Human growth hormone (HGH) levels also decline with age, but it is very unlikely that taking HGH injections will "cure" aging, despite Internet claims. In fact, one study found that people with lower levels of HGH actually lived longer than those with higher levels.

Reproductive System

Testosterone levels are highest in men in their twenties. After age 30, testosterone levels decrease by about 1% per year. Extremely low testosterone levels have been linked to a decreased sex drive, excessive weight gain, loss of muscle mass, osteoporosis, general fatigue, and depression. However, the levels below which testosterone treatment should be initiated remain controversial. Testosterone replacement therapy, whether through injection, patches, or gels, is associated with side effects like enlargement of the prostate, acne or other skin reactions, and the production of too many red blood cells.

Menopause, the period in a woman's life during which the ovarian and uterine cycles cease, usually occurs between ages 45 and 55. The ovaries become unresponsive to the gonadotropic hormones produced by the anterior pituitary, and they no longer secrete estrogen or progesterone. At the onset of menopause, the uterine cycle becomes irregular, but as long as menstruation occurs, it is still possible for a woman to conceive. Therefore, a woman is usually not considered to have completed menopause (and thus be infertile) until menstruation has been absent for a year.

The hormonal changes during menopause often produce physical symptoms such as "hot flashes" (caused by circulatory irregularities), dizziness, headaches, insomnia, sleepiness, and depression. To ease these symptoms, female hormone replacement therapy (HRT) was routinely prescribed until 2002, when a large clinical study showed that in the long term, HRT caused more health problems than it prevented. For this reason, most doctors no longer routinely recommend long-term HRT for the prevention of postmenopausal conditions.

It is also of interest that, as a group, females live longer than males. It is likely that estrogen offers women some protection against cardiovascular disorders when they are younger. Males suffer a marked increase in heart disease in their forties, but an increase is not noted in females until after menopause, when women lead men in the incidence of stroke. Men remain more likely than women to have a heart attack at any age, however.

Hypotheses About Why We Age

Aging is a complex process, and multiple factors can affect the aging process. The many hypotheses about why aging occurs can be grouped into two major categories: (1) theories suggesting that aging is mainly due to preprogrammed genetic events and (2) theories that aging is mainly due to the accumulation of cellular damage.

Preprogrammed Theories

Most scientists who study gerontology believe that aging is partly genetically preprogrammed. This idea is supported by the observation that longevity runs in families—that is, the children of long-lived parents tend to live longer than those of short-lived parents. As would also be expected, studies show that identical twins have a more similar life span than nonidentical twins.

Many laboratory studies of aging have been performed in the nematode *C. elegans,* in which single-gene mutations have been shown to influence the life span. For example, mutations that decrease the activity of a hormone receptor similar to the insulin receptor more than double the life span of the worms, which also behave and look like much younger worms throughout their prolonged lives. Interestingly, small-breed dogs like poodles or terriers, which may live 15 to 20 years, have lower levels of an analogous receptor, compared to large dog breeds that live 6 to 8 years.

Studies of the behavior of cells grown in the lab also suggest a genetic influence on aging. Most types of differentiated cells can divide only a limited number of times. One factor that may control the number of cell divisions is the length of the telomeres, sequences of DNA at the ends of chromosomes. As discussed further in section 25.5, telomeres protect the ends of chromosomes from deteriorating or fusing with other chromosomes. Each time a cell divides, the telomeres normally shorten, and cells with shorter telomeres tend to undergo fewer divisions. Perhaps as we grow older, more and more cells are unable to divide, and instead, they undergo degenerative changes and die.

If aging were mainly a function of genes, however, we would expect much less variation in life span to be observed among individuals of a given species than is actually seen. For this reason, experts have estimated that in most cases, genes account only for about 25% of what determines the length of life.

Damage Accumulation Theories

A second group of hypotheses postulate that aging involves the accumulation of damage over time. In 1900, the average human life expectancy was around 45 years. A baby born in the United States in 2009 is expected to live an average of about 78 years. Because human genes have presumably not changed much in such a short time, most of this gain in life span is due to better medical care, along with the application of scientific knowledge about how to prolong our lives.

There are two basic types of cellular damage that can accumulate over time. The first type can be thought of as agents that are unavoidable; for example, the accumulation of harmful DNA mutations, or the buildup of harmful metabolites. For some time, medical researchers have known that proteins—such as the collagen fibers present in many support tissues—become increasingly cross-linked as people age. This cross-linking may account for the inability of such organs as the blood vessels, the heart, and the lungs to function as they once did. Some researchers have now found that glucose has the tendency to attach to any type of protein, which is the first step in a cross-linking process. They are currently experimenting with drugs that can prevent cross-linking.

Other sources of cellular damage, however, may be avoidable, such as a poor diet or exposure to the sun. Recent work suggests that when an animal produces fewer free radicals, it lives longer. Free radicals are unstable molecules that have an unpaired electron. In order to stabilize themselves, free radicals react with another molecule, such as DNA or proteins (e.g., enzymes) or lipids found in plasma membranes. Eventually, these molecules are unable to function, and the cell is destroyed. By increasing one's consumption of natural antioxidants, such as those present in brightly colored and dark green leafy vegetables, citrus fruits, nuts, fish, shellfish, and red wine, we can reduce our exposure to free radicals, and slow the aging process.

The longest-lived human whose age has been verified by modern methods was Jeanne Calment (Fig. 22.19), who lived in France her entire life, and died at the age of 122 years, 164 days (she had lived on her own until age 110, when she entered a nursing home). This is currently considered to be the longest potential human life span. However, our understanding of the aging process could be considered, ironically, to be in its infancy, and animal models and other studies suggest that it may be possible to extend human life far beyond that.

Check Your Progress 22.5

1. List the changes that occur from infancy to adulthood.
2. Describe how involution of the thymus can lead to decreased responses to vaccines in the elderly.
3. Define menopause, and explain why it occurs.
4. Distinguish between the two hypotheses regarding aging, and give examples of each.

Case Study Conclusion

Over the course of the next few office visits, Amber's doctor suggested a variety of tests. Because Amber was over 35 years old when she conceived her first child, the doctor recommended that a chorionic villus sampling test be performed to check for birth defects, such as Down syndrome. The doctor also performed a glucose tolerance test to determine if Amber was experiencing gestational diabetes. An additional blood test, called an alpha-fetoprotein (AFP) test, screened for neural tube defects and additional evidence of Down syndrome. Much to the relief of Amber and Kent, all of these tests came back normal. Then, around week 20 of Amber's pregnancy, the doctor made an appointment for Amber to get an ultrasound exam, which is frequently used to confirm the due date of the baby but can also be used to look for birth defects such as cleft palate. Everything once again appeared normal. Finally, the technician asked the parents the question that they had been waiting for: were they interested in knowing the sex of their baby? Within a few minutes, Amber and Kent found out that, in just a few short months, they would be parents of a baby girl.

MEDIA STUDY TOOLS

www.mhhe.com/maderinquiry14

Enhance your study of this chapter with study tools and practice tests. Also ask your instructor about the resources available through ConnectPlus, including LearnSmart, the media-rich eBook, interactive learning tools, and animations.

McGraw Hill connect plus+ BIOLOGY

SUMMARIZE

22.1 Fertilization and Early Stages of Development

Animal development begins with a single cell that develops into an embryo. Development proceeds though cellular, organ, and tissue stages:

- Development begins at **fertilization.** Only one sperm enters the oocyte, and its nucleus fuses with the oocyte nucleus, forming a **zygote.**
- During the cellular stages, **cleavage** (cell division) occurs, without overall growth. The result is a **morula,** which becomes the **blastula** when an internal cavity (the **blastocoel**) appears.
- During the tissue stages, **gastrulation** results in formation of the **germ layers: ectoderm, mesoderm,** and **endoderm.** Invagination of the gastrula forms a **blastopore.** Both the cellular and tissue stages can be affected by the amount of yolk.
- Organ formation can be related to germ layers. For example, the nervous system develops from midline ectoderm, just above the **notochord.**

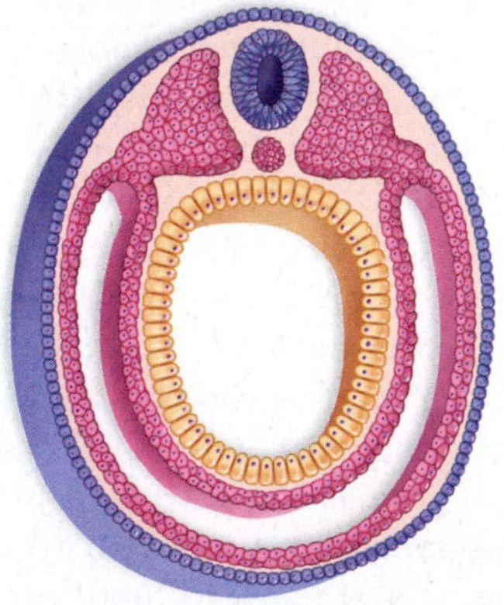
ectoderm
mesoderm
endoderm

22.2 Processes of Development

Development involves an increasing specialization of cells, as well as **morphogenesis,** the formation of the shape of the body. Morphogenesis progresses to **pattern formation,** the positioning of tissues and organs.

- **Cellular differentiation** begins with **cytoplasmic segregation** in the egg. An example of this is the **gray crescent** of a frog's egg.
- Some cellular differentiation is due to **induction,** an ongoing process in which one tissue regulates the development of another tissue through chemical signals. In simple animals, **fate maps** have been constructed to map the destiny of individual cells.
- During morphogenesis, sequential sets of **morphogen genes** code for morphogen gradients that, in turn, activate the next set of morphogen genes. **Homeotic genes** code for proteins that contain a **homeodomain,** a sequence of 60 amino acids that is highly conserved among a wide variety of organisms. **Apoptosis,** or programmed cell death, also plays an important role in development.

22.3 Human Embryonic and Fetal Development

Human development before birth can be divided into **embryonic development** (months one to two) and **fetal development** (months three to nine).

- The **extraembryonic membranes** appear early in human development. The **trophoblast** of the **blastocyst** is the first sign of the **chorion,** which becomes the fetal part of the **placenta.** This structure also produces **human chorionic gonadotropin** (**HCG**), the basis of the pregnancy test. The **amnion** contains amniotic fluid, which cushions and protects the embryo. The **yolk sac** serves to produce blood cells. The **allantois** gives rise to the **umbilical cord.**
- Gastrulation is marked by the appearance of the **embryonic disk,** followed by a steady progression of organ formation during embryonic development. During fetal development, features are refined, and the fetus adds weight. The skin is covered by fine hairs called **lanugo,** as well as by **vernix caseosa.**
- Because the fetus receives oxygen and nutrients from the **placenta,** the fetal heart contains a **foramen ovale** and a **ductus arteriosus.**

22.4 Human Pregnancy, Birth, and Lactation

Many physiological changes occur in a woman's body during pregnancy. Birth, or **parturition,** requires three stages, and occurs about 280 days after the start of the mother's last menstruation.

- Many pregnant women experience morning sickness early in their pregnancy. Due to a general relaxation of smooth muscles, a pregnant woman's body generally takes on fluid, and blood flow to many organs increases. Stretch marks, or striae gravidarum, may appear on the skin.
- During stage one of parturition, uterine contractions begin to push the baby's head against the cervix. Uterine contractions increase during stage two, as the baby's head enters the birth canal and the rest of the body usually follows. An **episiotomy** may be required to enlarge the vaginal opening. The placenta, or **afterbirth,** is delivered during stage 3.
- After a baby is born, prolactin from the pituitary gland stimulates milk production. For the first day or two after birth, the breasts produce **colostrum,** an antibody- and nutrient-rich fluid. Breastfeeding is usually beneficial to the newborn and the mother.

22.5 Human Development After Birth

Development after birth consists of infancy, childhood, adolescence, and adulthood.

- **Gerontology** is the study of **aging.** The aging process affects virtually all the body systems.
- The biological processes behind aging include changes that are genetically preprogrammed, as well as those that follow from an accumulation of cellular and tissue damage over time.

ASSESS

Testing Yourself

Choose the best answer for each question.

1. Which occurs in the egg to prevent fertilization by multiple sperm cells?
   a. Plasma membrane depolarizes.
   b. Zona pellucida moves away from the egg.
   c. Killer enzymes are secreted.
   d. Both a and b are correct.
2. The inner lining of the digestive tract forms from which germ layer?
   a. endoderm
   b. ectoderm
   c. mesoderm
   d. blastula
3. In some animals, the notochord is replaced by the
   a. neural plate.
   b. vertebral column.
   c. neurula.
   d. None of these are correct.
4. The umbilical vein carries blood rich in
   a. nutrients.
   b. oxygen.
   c. Both a and b are correct.
   d. Neither a nor b is correct.
5. The archenteron will develop into the
   a. heart.
   b. vertebral column.
   c. gut.
   d. None of these are correct.
6. The outer cell layer in the blastocyst is called the
   a. ectoderm.
   b. endoderm.
   c. trophoblast.
   d. embryo.
   e. None of these are correct.
7. Implantation occurs at the _____ stage.
   a. gastrula
   b. blastula
   c. morula
   d. None of these are correct.
8. The cavity in the gastrula is the
   a. blastocoel.
   b. gastrocoel.
   c. archenteron.
   d. embryo cavity.
   e. None of these are correct.
9. The ability to experimentally cause a second neural plate to develop above notochord donor tissue is due to
   a. segregation.
   b. induction.
   c. reduction.
   d. gastrulation.
   e. None of these are correct.
10. Identify the stages of animal development in the following diagram.

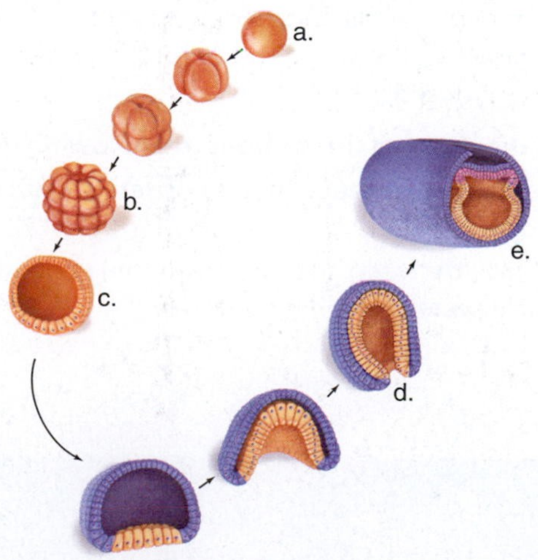
a.
b.
c.
d.
e.

11. Identify the regions of a cross section of an animal embryo in the following diagram.

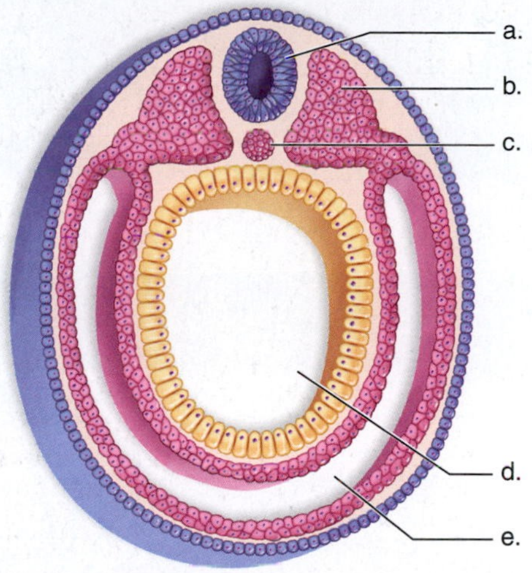
a.
b.
c.
d.
e.

12. In human development, which part of the blastocyst develops into a fetus?
   a. morula
   b. trophoblast
   c. inner cell mass
   d. chorion
   e. yolk sac
13. Genes that control the structural pattern of an animal are
   a. homeotic genes.
   b. morphogen genes.
   c. induction genes.
   d. All of these are correct.
   e. Both a and b are correct.
14. Morphogenesis is associated with
   a. protein gradients.
   b. pattern formation.
   c. homeotic genes.
   d. All of these are correct.
15. Which of the germ layers is best associated with development of the heart?
   a. ectoderm
   b. mesoderm
   c. endoderm
   d. neurula
   e. All of these are correct.

ENGAGE

Thinking Critically

1. Mitochondria contain their own genetic material. Mitochondrial mutations are inherited from mother to child, without any contribution from the father. Why?
2. In one form of prenatal testing for fetal genetic abnormalities, chorionic villus samples are taken and analyzed. Why is this possible?
3. The Health feature, "Preventing and Testing for Birth Defects," on page 454 suggests that women of childbearing age begin taking precautions to prevent birth defects (good health habits, rubella vaccination, etc.). Why should these precautions begin before a woman is pregnant?

USA
7
VOLLEYBALL

23 Patterns of Gene Inheritance

CHAPTER OUTLINE

BEFORE YOU BEGIN

Before beginning this chapter, take a few moments to review the following discussions:

Chapter 2 What are the roles of DNA and proteins in a cell?

Section 5.4 How are chromosomes segregated during the process of meiosis?

Section 5.6 How does the process meiosis relate to the human life cycle?

CASE STUDY Genetics of Marfan Syndrome

Flo Hyman was considered the best volleyball player of her time. Born in 1954, Hyman was athletically talented and very tall—over 6 ft tall in junior high school. Her final height was 6 ft 5 in. At the 1981 World Cup in Tokyo, she was named to the six-member All-World Cup Team. Hyman went on to lead the U.S. Olympic team to a silver medal in the 1984 Summer Olympic Games in Los Angeles, California. She earned the nickname "Flying Clutchman" and could serve a volleyball at speeds of up to 100 miles per hour. After the Olympics, Hyman lobbied for the Civil Rights Restoration Act and testified before Congress in 1985 on behalf of Title IX legislation, which prohibits sex discrimination in sports by schools receiving federal funds. In 1982, Hyman returned to Japan to play professional women's volleyball. Her height, however, was a symptom of the condition that would kill her. Hyman collapsed during a match in January 1986 and died later that evening of a ruptured aorta, a complication of Marfan syndrome. As described in this chapter, Marfan syndrome is an autosomal dominant genetic disorder that results in a defect in the gene (*FBN1*) for an elastic connective tissue protein called fibrillin. Connective tissue is abundant throughout the body and fibrillin plays a special role in strengthening the wall of the aorta. Hyman did not suffer from other abnormalities common to those with Marfan syndrome, including curvature of the spine and breastbone, and this is probably why she went undiagnosed.

As you read through the chapter, think about the following questions:

1. How do the patterns of inheritance that we observe relate to the principles of meiosis?
2. What is the difference between an autosomal dominant and autosomal recessive pattern of inheritance?

23.1 Mendel's Laws

Learning Outcomes

Upon completion of this section, you should be able to

1. Describe how Mendel's law of segregation and law of independent assortment are related to the movements of chromosomes during gamete formation and fertilization.
2. Define the term allele, and explain what it means for an allele to be dominant or recessive.
3. Contrast genotype and phenotype, and use appropriate terms for describing genotypes.
4. Predict outcome ratios and probabilities for problems based on Mendelian patterns of inheritance.

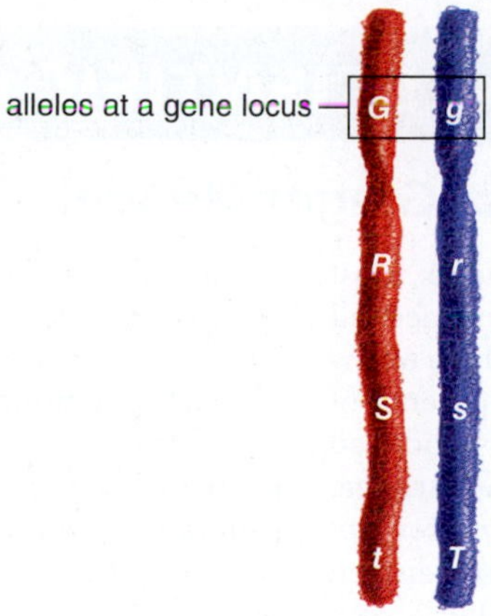
alleles at a gene locus
G
g
R
r
S
s
t
T

Figure 23.2 Homologous chromosomes. The letters represent alleles—that is, alternate forms of a gene. Each allelic pair, such as *Gg* or *Tt*, is located on homologous pairs of chromosomes at a particular gene locus. The different colors represent the maternal or paternal origin of the chromosome.

The science of genetics explains the process of inheritance (why you are human, as are your parents) and also why there are variations between offspring from one generation to the next (why you have a different combination of traits than your parents). Virtually every culture in history has attempted to explain observed inheritance patterns. An understanding of these patterns has always been important to agriculture, animal husbandry (the science of breeding animals), and medicine.

Gregor Mendel

Our modern understanding of genetics is based largely on the work of Gregor Mendel (Fig. 23.1), an Austrian monk who developed the basic principles of inheritance after performing a series of ingenious experiments in the 1860s. Mendel's parents were farmers, which provided him with the practical experience he needed to grow his experimental organism, the pea plant. Mendel was also a mathematician. He kept careful and complete records, even though he crossed and catalogued some 24,034 plants through several generations. He concluded that the plants transmitted distinct factors to offspring. We know now that the factors that control traits are called genes, and that genes are found on chromosomes.

Figure 23.1 Gregor Mendel, 1822–84. Mendel was a skilled mathematician with excellent skills of observation. His work established the early principles of inheritance that serve as the basis for modern genetic studies.

In pea plants and in humans, the chromosomes are found in pairs, called *homologous chromosomes.* One member of the pair is inherited from the mother, while the other member is inherited from the father. Homologous chromosomes have the same length and centromere location and are similar in the types of genes that they contain. For example, one trait in humans is finger length. One member of the homologous pair of chromosomes could contain the gene for long fingers, whereas the other member could contain the gene for short fingers. These alternate forms of a gene for the same trait are called **alleles.** Alleles are always at the same spot, or **locus,** on each member of the homologous pair. In Figure 23.2, the letters on the homologous chromosomes stand for alleles. *G* is an allele of *g*, and vice versa; *R* is an allele of *r*, and vice versa. *G* could never be an allele of *R* because *G* and *R* are at different loci and govern different traits.

Mendel knew nothing about homologous pairs of chromosomes, but his work with pea plants—described in the Scientific Inquiry feature, "The Investigations of Gregor Mendel," on page 464—caused him to decide that pea plants have two factors for every trait, such as stem length. He observed that one of the factors controlling the same trait can be dominant over the other, which is recessive. Therefore, a tall pea plant could have one factor for long stem and another for short stem. While most of the offspring from two such tall pea plants were tall, some were short. Mendel reasoned that this would be possible if the gametes (i.e., sperm and oocyte) contained only one factor for each trait. Only in that way could an offspring receive two factors for short stem. On the basis of such studies, Mendel formulated his first law, the law of segregation. The law of segregation states the following:

Each individual has two factors for each trait.
The factors segregate (separate) during the formation of the gametes.
Each gamete contains only one factor from each pair of factors.
Fertilization gives each new individual two factors for each trait.

SCIENCE IN YOUR LIFE ▶ **SCIENTIFIC INQUIRY**

The Investigations of Gregor Mendel

When Gregor Mendel began his work, most plant and animal breeders acknowledged that both sexes contribute equally to a new individual. They thought that parents of contrasting appearance always produced offspring of intermediate appearance. This concept, called the *blending concept of inheritance,* meant that a cross between plants with red flowers and plants with white flowers would yield only plants with pink flowers. When red and white flowers reappeared in future generations, the breeders mistakenly attributed this to instability in the genetic material.

Mendel chose to work with the garden pea, *Pisum sativum* (Fig. 23A). The garden pea was a good choice for several reasons, including the fact that they were easy to cultivate and had a short generation time. Although peas normally self-pollinate (pollen only goes to the same flower), they could be cross-pollinated by hand by transferring pollen from an anther to a stigma. Many varieties of peas were available, and Mendel chose 22 for his experiments. When these varieties self-pollinated, they were *true-breeding.* A true-breeding line always produced individuals of the same phenotype. In contrast to his predecessors, Mendel studied the inheritance of relatively simple and discrete traits, such as seed shape, seed color, and flower color.

Mendel utilized both one-trait and two-trait crosses in his research. He cross-pollinated the plants for the F_1 (first filial) generation. None of the offspring of this generation were intermediate between the parents. All of the offspring exhibited the trait of one or the other parent, which Mendel termed the dominant trait. The trait that disappeared in the offspring he called the recessive trait. He then allowed the F_1 plants to self-pollinate to obtain an F_2 generation. He counted many plants and always found approximately the same ratios of plants exhibiting dominant and recessive traits in this generation (Fig. 23B).

As Mendel followed the inheritance of individual traits, he kept careful records, and he used his understanding of the mathematical laws of probability and statistics to interpret his results. Mendel recognized that discrete units of heredity, which he called "elementens," were being passed from generation to generation. We know that elementens are actually genes. His results became the basis of the *particulate theory of inheritance.* Inheritance involves reshuffling the same genes from generation to generation.

Mendel achieved his success in genetics by studying large numbers of offspring, keeping careful records, and treating his data quantitatively. However, a certain amount of luck was involved in the traits Mendel chose. None of the traits he analyzed exhibited the more complicated patterns of inheritance.

Questions to Consider

1. From a statistical perspective, why are large numbers of offspring needed?
2. Besides the reasons listed in this feature, what other reasons would make the pea plant a better object for genetic study than humans?
3. What "more complicated patterns of inheritance" could have confused Mendel's results?

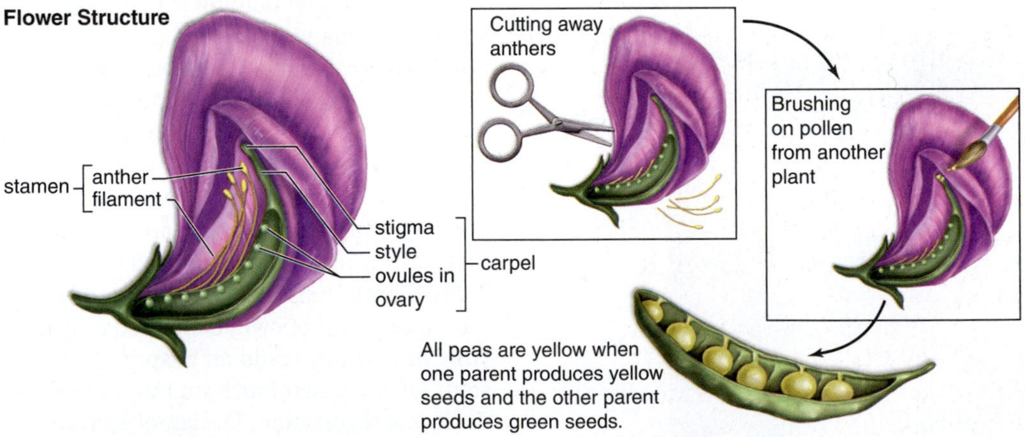
Flower Structure
stamen
anther
filament
stigma
style
ovules in ovary
carpel
Cutting away anthers
Brushing on pollen from another plant
All peas are yellow when one parent produces yellow seeds and the other parent produces green seeds.

Figure 23A Garden pea anatomy and a few traits. In the garden pea, *Pisum sativum,* pollen grains produced in the anther contain sperm, and ovules in the ovary contain oocytes. Mendel brushed pollen from one plant onto the stigma of another plant. After fertilization, the ovules developed into seeds (peas). The open pod shows the results of a cross between plants with yellow seeds and plants with green seeds.

Trait	Characteristics		F$_2$ Results		
	Dominant	Recessive	Dominant	Recessive	Ratio
Stem length	Tall	Short	787	277	2.84:1
Pod shape	Inflated	Constricted	882	299	2.95:1
Seed shape	Round	Wrinkled	5,474	1,850	2.96:1
Seed color	Yellow	Green	6,022	2,001	3.01:1
Flower position	Axial	Terminal	651	207	3.14:1
Flower color	Purple	White	705	224	3.15:1
Pod color	Green	Yellow	428	152	2.82:1
		Totals:	14,949	5,010	2.98:1

Figure 23B The traits Mendel selected for study. The parent (P generation) plants were true-breeding and were cross-pollinated to produce an F$_1$ (first filial) generation. These plants always resembled the parent with the dominant characteristic. The F$_1$ plants were allowed to self-pollinate. In the F$_2$ (second filial) generation, he always achieved a 3:1 (dominant to recessive) phenotypic ratio. The results of all of Mendel's crosses are shown here.

The Inheritance of a Single Trait

The **phenotype** of an organism refers to the individual's actual appearance. Phenotype may describe either an individual's physical characteristics, or microscopic and metabolic characteristics. In this chapter, we will use the traits of hairline and finger length in our examples (Fig. 23.3). Both of these traits have variations in the phenotype, which are determined by the combination of alleles that the individual inherited. The **genotype** refers to the alleles the chromosomes carry that are responsible for that trait. In diploid organisms, such as humans, a pair of homologous chromosomes contains two alleles for each trait, one on each member of the homologous pair. Mendel suggested using letters to indicate these factors, which we now call alleles. A capital letter indicates a **dominant allele** and a lowercase letter indicates a **recessive allele.** The word *dominant* is not meant to imply that the dominant allele is better or stronger than the recessive allele. Dominant means that this allele will mask the expression of the recessive allele when they are together in the same organism.

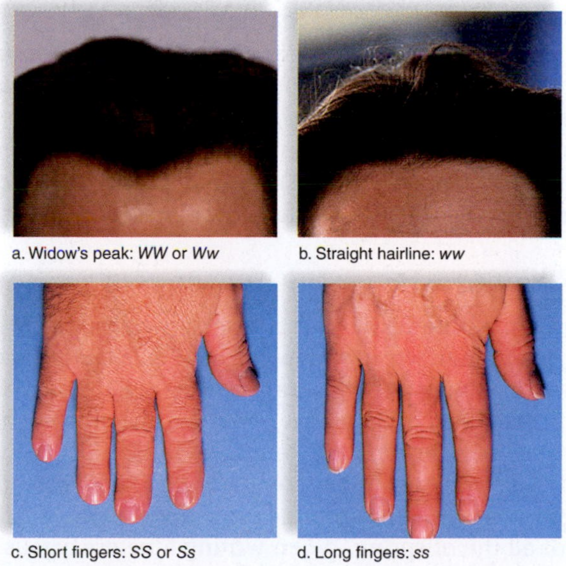
a. Widow's peak: *WW* or *Ww*
b. Straight hairline: *ww*
c. Short fingers: *SS* or *Ss*
d. Long fingers: *ss*

Figure 23.3 Dominance of human traits. In the hairline trait of humans, (**a**) widow's peak is dominant over (**b**) straight hairline. For finger length, short fingers (**c**) is dominant over long fingers (**d**).

When working with the alleles for hairline, we will represent the dominant allele of widow's peak with a *W*, and the recessive allele of a straight hairline with a *w*. Likewise, for finger length, short fingers is the dominant allele (represented by *S*), whereas long fingers is the recessive allele (represented by *s*). In the case of a single trait, such as hairline, there are three possible combinations of the two alleles: *WW, Ww,* and *ww*. These are considered the possible genotypes for this trait. If the two members of the allelic pair are the same (*homo*), the organism is said to be **homozygous.** Both *WW* and *ww* are homozygous, with *WW* homozygous dominant, and *ww* homozygous recessive. If the two members of the allelic pair are different (*hetero*), the organism is said to be **heterozygous.** Only *Ww* is a heterozygous genotype. But what about the phenotype? When an organism has a *W*, it exhibits the dominant phenotype regardless of whether the other allele is *W* or *w*. Both *WW* and *Ww* exhibit the dominant phenotype, a widow's peak. The recessive allele *w* is masked by the expression of the dominant allele *W* in the heterozygous genotype. The only time the recessive phenotype, straight hairline, is expressed, is when the genotype is homozygous recessive, or *ww*. The phenotypes and genotypes for this trait are summarized in Table 23.1.

TABLE 23.1 Genotype Related to Phenotype

Genotype	Genotype	Phenotype
WW	Homozygous dominant	Widow's peak
Ww	Heterozygous	Widow's peak
ww	Homozygous recessive	Straight hairline

Gamete Formation

The genotype has two alleles for each trait, whereas the gametes (i.e., sperm and oocyte) have only one allele for each trait in accordance with Mendel's law of segregation. During meiosis, the homologous chromosomes separate (see Chapter 5), resulting in only one member of each pair of chromosomes in each haploid cell. In humans, these haploid cells go on to form gametes. Therefore, there is only one allele for each trait, such as type of hairline, in each gamete. When working problems in genetics *no two letters in a gamete can be the same letter of the alphabet.* If the genotype of an individual is *Ww*, the gametes for this individual will contain either a *W* or a *w* but not both. If the genotype is *WwSs*, all combinations of any two different letters can be present.

One-Trait Crosses

It is now possible for us to consider a particular cross. If a homozygous man with a widow's peak (Fig. 23.3*a*) reproduces with a woman with a straight hairline (Fig. 23.3*b*), what kind of hairline will their children have? To solve this problem, (1) use Table 23.1 to determine the genotype of each parent; (2) list the possible gametes from each parent; (3) combine all possible gametes; and (4) determine the genotypes and the phenotypes of all the offspring. When writing a heterozygous genotype (*Ww*), always put the capital letter first to avoid confusion.

In the following diagram, the letters in the first row are the genotypes of the parents. Each parent has only one type of gamete in regard to hairline, and therefore all the offspring have similar genotypes and phenotypes. The children are heterozygous (*Ww*) and have a widow's peak:

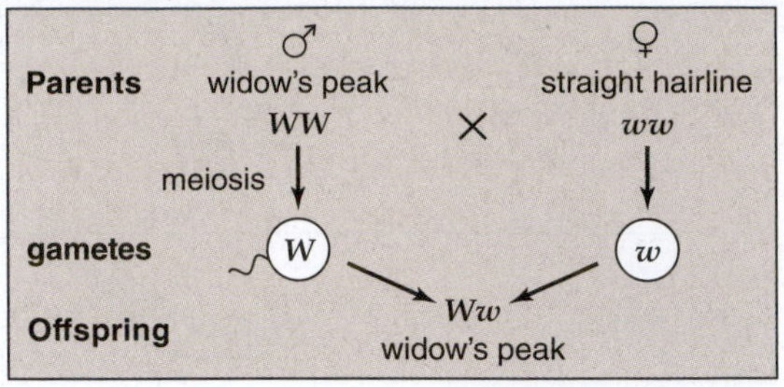

♂
♀
Parents
widow's peak
straight hairline
WW
×
ww
meiosis
gametes
W
w
Offspring
Ww
widow's peak

These children are monohybrids; that is, they are heterozygous for only one pair of alleles. If they reproduce with someone else of the same genotype, what type of hairline will their children have? In this problem (*Ww* × *Ww*), also called a **monohybrid cross,** each parent has two possible types of gametes (*W* or *w*), and we must ensure that all types of sperm have an equal chance to fertilize all possible types of oocytes. One way to do this is to use a **Punnett square** (Fig. 23.4), in which all possible types of sperm are lined up vertically and all possible types of oocytes are lined up horizontally (or vice versa), and every possible combination of gametes occurs within the squares.

After we determine the genotypes and the phenotypes of the offspring, we can determine the genotypic and the phenotypic ratios. The genotypic ratio is 1 *WW*: 2 *Ww*: 1 *ww*, or simply 1:2:1, but the phenotypic ratio is 3:1. Why? Because three individuals have a widow's peak and one has a straight hairline. This 3:1 phenotypic ratio is always expected for a monohybrid cross when one allele is completely dominant over the other. The exact ratio is more likely to be observed if a large number of matings take place and if a large number of offspring result. Only then do all possible kinds of sperm have an equal chance of fertilizing all possible kinds of oocytes. Naturally, in humans, we do not routinely observe hundreds of offspring from a single type of cross. The best interpretation of Figure 23.4 in humans is to say

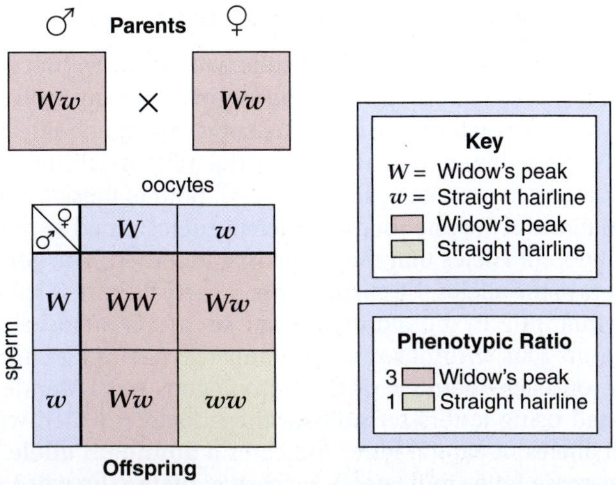

♂
Parents
♀
Ww
×
Ww
oocytes
♂
♀
W
w
sperm
W
WW
Ww
w
Ww
ww
Offspring
Key
W = Widow's peak
w = Straight hairline
Widow's peak
Straight hairline
Phenotypic Ratio
3 Widow's peak
1 Straight hairline

Figure 23.4 Monohybrid cross. A Punnett square diagrams the results of a cross. When the parents are heterozygous, each child has a 75% chance of having the dominant phenotype and a 25% chance of having the recessive phenotype.

that each child has three chances out of four to have a widow's peak, or one chance out of four to have a straight hairline. It is important to realize that *chance has no memory;* for example, if two heterozygous parents already have three children with a widow's peak and are expecting a fourth child, this child still has a 75% chance of having a widow's peak and a 25% chance of having a straight hairline.

One-Trait Crosses and Probability

Another method of calculating the expected ratios uses the rules of probability or chance. First, we must know the *product rule of probability,* which states that the chance of two or more independent events occurring together is the product (multiplication) of their chance of occurring separately. For example, if you flip a coin, the chance of getting "heads" is $1/2$ (one head out of two possibilities, or $1/2$). If you flip two coins, the chance of getting two heads is $1/2 \times 1/2$, or $1/4$. You must multiply together the probability of each head occurring separately ($1/2$).

In the cross just considered ($Ww \times Ww$), what is the chance of obtaining either a W or a w from a parent?

The chance of $W = 1/2$, *and the chance of* $w = 1/2$

Therefore, the probability of having these genotypes is as follows:

The chance of $WW = 1/2 \times 1/2 = 1/4$

The chance of $Ww = 1/2 \times 1/2 = 1/4$

The chance of $wW = 1/2 \times 1/2 = 1/4$

The chance of $ww = 1/2 \times 1/2 = 1/4$

Now we must consider the *sum rule of probability,* which states that the chance of an event that can occur in more than one way is the sum (addition) of the individual chances. Therefore, to calculate the chance of an offspring having a widow's peak, add the chances of WW, Ww, or wW from the preceding list to arrive at $3/4$, or 75%. The chance of offspring with a straight hairline (only ww from the preceding) is $1/4$, or 25%.

The One-Trait Testcross

It is not possible to tell by observation if an individual expressing a dominant allele is homozygous dominant or heterozygous. Therefore, breeders sometimes do a testcross. In a **testcross,** an organism showing the dominant phenotype, but unknown genotype, is crossed with a homozygous recessive individual. Why use a homozygous recessive individual? Because this is the one case where the phenotype indicates a known genotype. The results of a testcross should indicate the genotype of an organism with the dominant phenotype (Fig. 23.5).

The results of crosses between people also tell us the genotype of the parents. For example, Figure 23.5 shows two possible results when a man with a widow's peak reproduces with a woman who has a straight hairline. If the man is homozygous dominant, all his children will have a widow's peak. If the man is heterozygous, each child has a 50% chance of having a straight hairline. The birth of just one child with a straight hairline indicates that the man is heterozygous.

The Inheritance of Two Traits

Figure 23.6 represents two pairs of homologues undergoing meiosis. The homologues are distinguished by length—that is, one pair of homologues is short and the other is long. In Figure 23.6, the color signifies which parent the chromosome is inherited from: one homologue of each pair is the "paternal" chromosome, and the other is the "maternal" chromosome. When the homologues separate (segregate) during meiosis, each gamete receives one member from each pair of homologues. The homologues separate independently. It does not matter which member of a pair goes into which gamete. In the

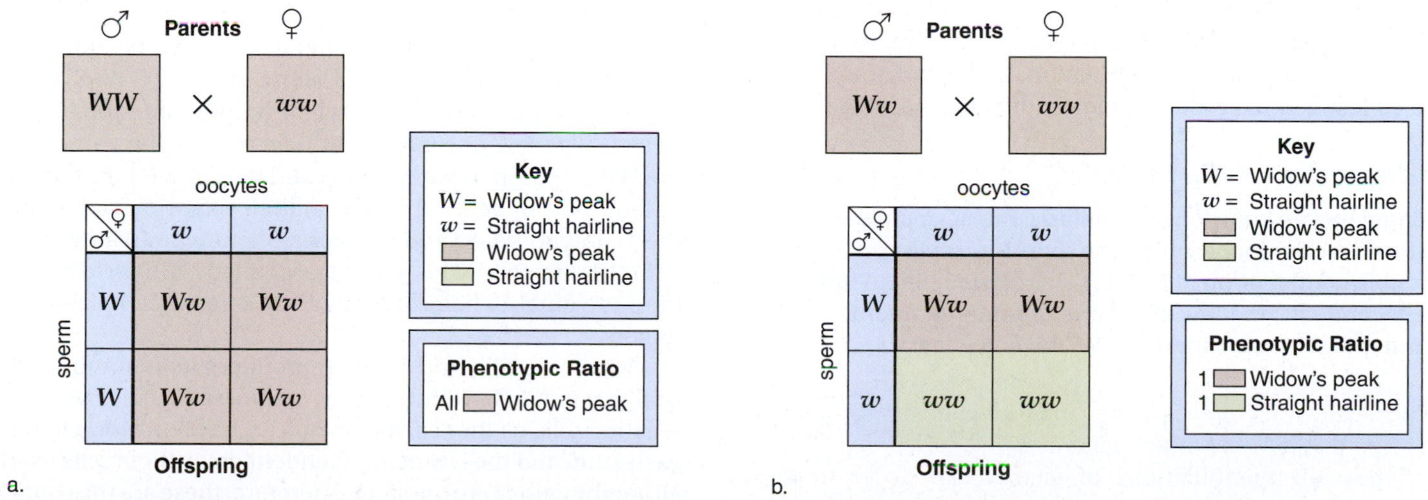

Parents
WW
ww
oocytes
w
w
sperm
W
W
Ww
Ww
Ww
Ww
Offspring
a.
Key
W = Widow's peak
w = Straight hairline
Widow's peak
Straight hairline
Phenotypic Ratio
All Widow's peak
Parents
Ww
ww
oocytes
w
w
sperm
W
w
Ww
Ww
ww
ww
Offspring
b.
Key
W = Widow's peak
w = Straight hairline
Widow's peak
Straight hairline
Phenotypic Ratio
1 Widow's peak
1 Straight hairline

Figure 23.5 One-trait testcross. A testcross determines if an individual with the dominant phenotype is homozygous or heterozygous. **a.** Because all offspring show the dominant characteristic, the individual is most likely homozygous, as shown. **b.** Because the offspring show a 1:1 phenotypic ratio, the individual is heterozygous, as shown.

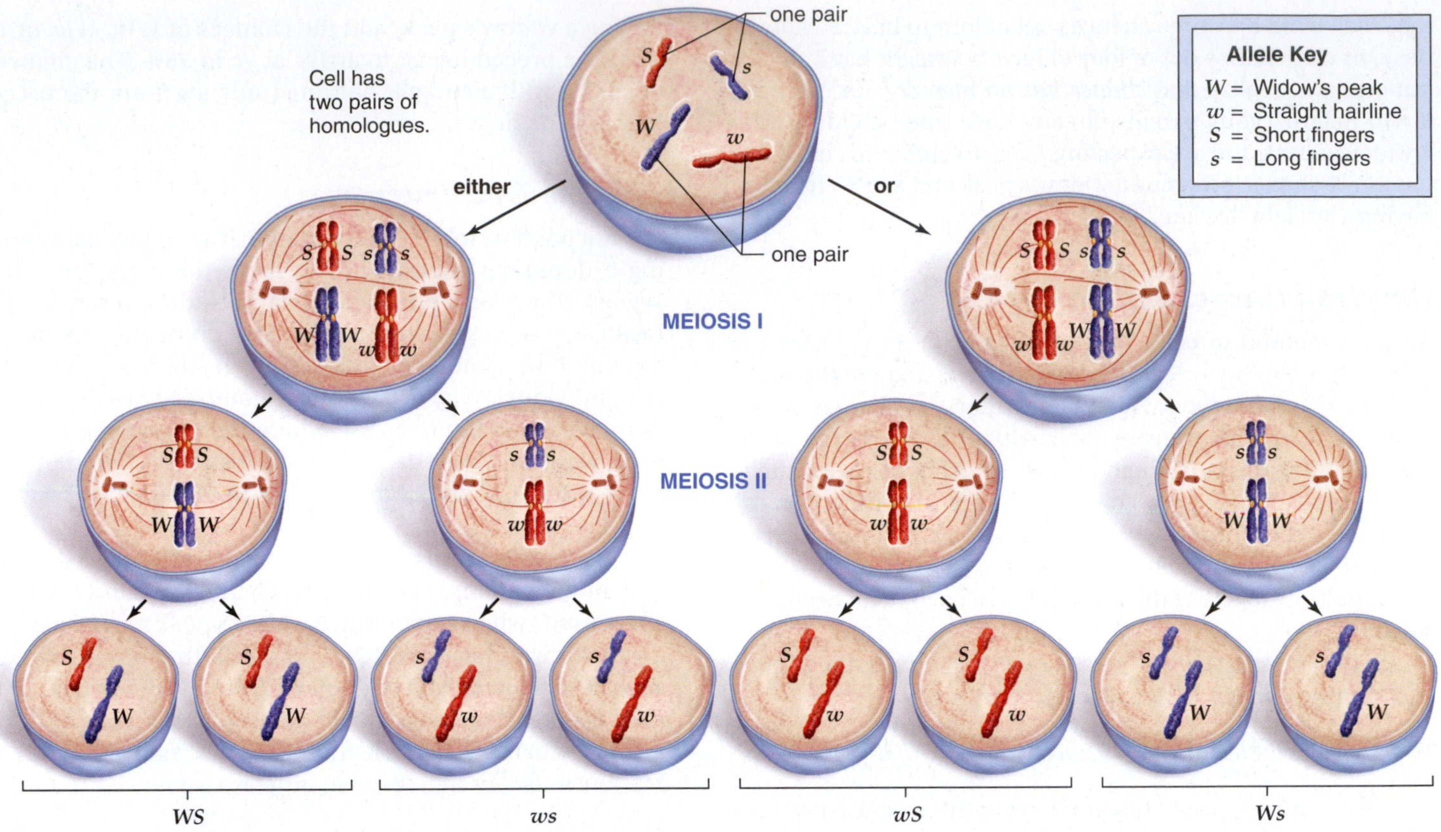
one pair
Cell has two pairs of homologues.
Allele Key
W = Widow's peak
w = Straight hairline
S = Short fingers
s = Long fingers
either
or
one pair
MEIOSIS I
MEIOSIS II
WS
ws
wS
Ws

Figure 23.6 Homologous pairs of chromosomes during meiosis. A cell has two pairs of homologous chromosomes (homologues), a short pair and a long pair. The short pair of homologues carries alleles for finger length, and the long pair of homologues carries alleles for type of hairline. The homologues, and the alleles they carry, segregate independently during meiosis I. This means that either a red or blue chromosome from each pair can go into either of the daughter cells. Therefore, all possible combinations of chromosomes and alleles occur in the gametes. The last line of the illustration lists the possible allele combinations for this particular cross.

simplest terms, a gamete in Figure 23.6 will *receive one short and one long chromosome of either color.* Therefore, all possible combinations of chromosomes are in the gametes.

We will assume that the alleles for two genes are on these homologues. The alleles *S* and *s* are on one pair of homologues, and the alleles *W* and *w* are on the other pair of homologues. Because there are no restrictions as to which homologue goes into which gamete, a gamete can receive either an *S* or an *s* and either a *W* or a *w* in any combination. In the end, the gametes will collectively have all possible combinations of alleles.

Independent Assortment

Mendel had no knowledge of meiosis, but he could see that his results were attainable only if the sperm and eggs contain every possible combination of factors. This caused him to formulate his second law, the **law of independent assortment.** The law of independent assortment states the following:

> Each pair of factors separates independently (without regard to how the others separate).
> All possible combinations of factors can occur in the gametes.

Figure 23.6 illustrates the law of segregation and the law of independent assortment. Because it does not matter which homologue of a pair faces which spindle pole, a daughter cell can receive either homologue and either allele.

Animation Random Orientation of Chromosomes During Meiosis

Two-Trait Crosses

In the two-trait cross depicted in Figure 23.7, a person homozygous for widow's peak and short fingers (*WWSS*) reproduces with one who has a straight hairline and long fingers (*wwss*). The law of segregation tells us that the gametes for the *WWSS* parent have to be *WS* and the gametes for the *wwss* parent have to be *ws.* Therefore, their offspring will all have the genotype *WwSs* and the same phenotype (widow's peak with short fingers). This genotype is called a **dihybrid** because the individual is heterozygous in two regards: hairline and fingers.

When the dihybrid *WwSs* reproduces with another dihybrid that is *WwSs,* what gametes are possible? The law of segregation tells us that each gamete can have only one letter of each kind, and the law of independent assortment tells us that all combinations are possible. Therefore, these are the gametes for both dihybrids: *WS, Ws, wS,* and *ws.*

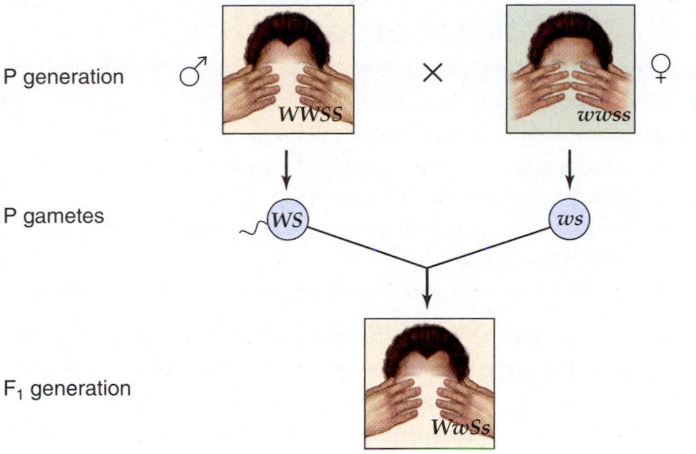
P generation
WWSS
wwss
P gametes
WS
ws
F1 generation
WwSs

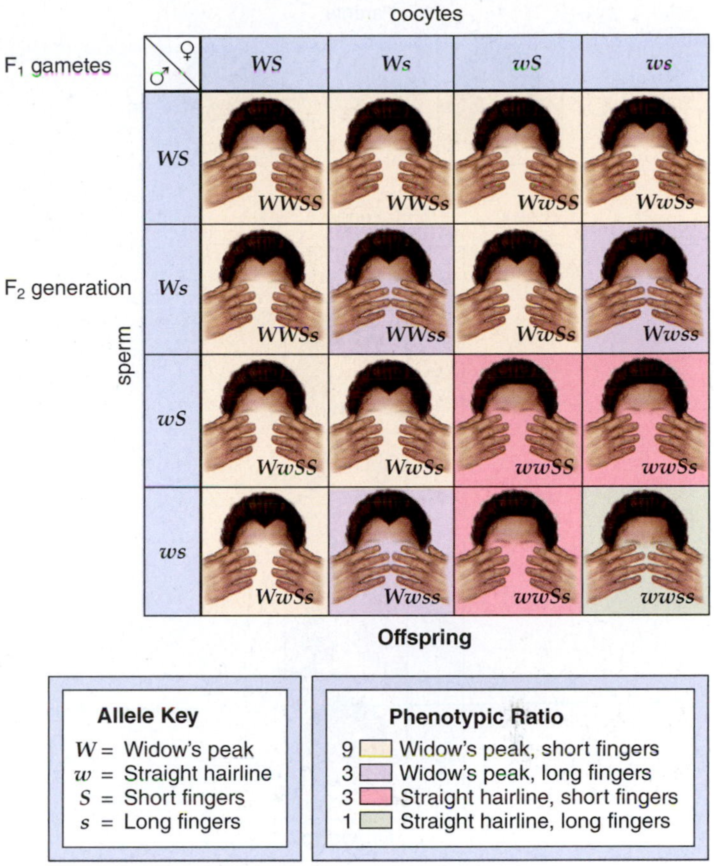
oocytes
F1 gametes
WS
Ws
wS
ws
F2 generation
sperm
WWSS
WWSs
WwSS
WwSs
WWss
Wwss
wwSS
wwSs
wwss
Offspring
Allele Key
W = Widow's peak
w = Straight hairline
S = Short fingers
s = Long fingers
Phenotypic Ratio
9 Widow's peak, short fingers
3 Widow's peak, long fingers
3 Straight hairline, short fingers
1 Straight hairline, long fingers

Figure 23.7 Dihybrid cross. Each dihybrid can form four possible types of gametes, so four different phenotypes occur among the offspring in the proportions shown.

A Punnett square represents all possible sperm fertilizing all possible oocytes, so following are the expected phenotypic results for a dihybrid cross:

9 widow's peak and short fingers
3 widow's peak and long fingers
3 straight hairline and short fingers
1 straight hairline and long fingers

This 9:3:3:1 phenotypic ratio is always expected for a dihybrid cross when simple dominance is present. We can use this expected ratio to predict the chances of each child receiving a certain phenotype. For example, the chance of getting the two dominant phenotypes together is 9 out of 16, and the chance of getting the two recessive phenotypes together is 1 out of 16.

Two-Trait Crosses and Probability

Instead of using a Punnett square, it is also possible to use the product rule and the sum rule of probability to predict the results of a dihybrid cross. For example, we know that the probable results for two separate monohybrid crosses are as follows:

Probability of widow's peak = $3/4$
Probability of short fingers = $3/4$
Probability of straight hairline = $1/4$
Probability of long fingers = $1/4$

Using the product rule, we can calculate the probable outcome of a dihybrid cross as follows:

Probability of widow's peak and short fingers =
$3/4 \times 3/4 = 9/16$

Probability of widow's peak and long fingers =
$3/4 \times 1/4 = 3/16$

Probability of straight hairline and short fingers =
$1/4 \times 3/4 = 3/16$

Probability of straight hairline and long fingers =
$1/4 \times 1/4 = 1/16$

In this way, the rules of probability tell us that the expected phenotypic ratio when all possible sperm fertilize all possible oocytes is 9:3:3:1.

The Two-Trait Testcross

It is impossible to tell by inspection whether an individual expressing the dominant allele for two traits is homozygous dominant or heterozygous in regard to these traits. A two-trait testcross occurs when this individual is crossed with a homozygous recessive for both traits. The recessive phenotype is used because it has a *known* phenotype.

For example, if a man who is homozygous dominant for widow's peak and short fingers reproduces with a woman who is homozygous recessive for both traits, then all his children will

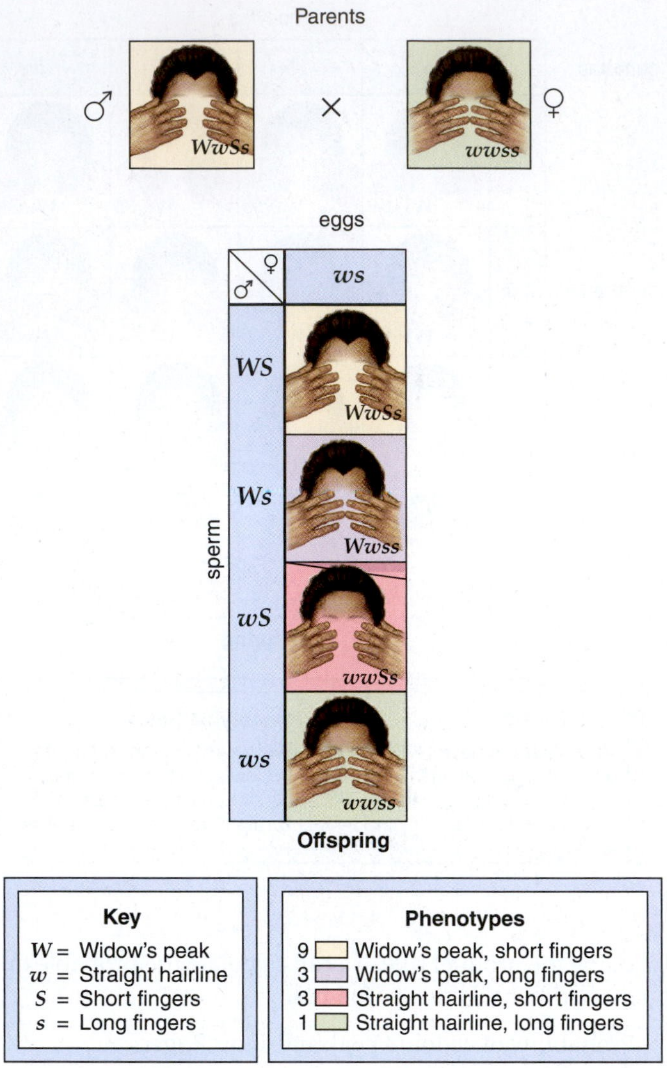
Parents
♂
WwSs
×
wwss
♀
eggs
♀
♂
ws
sperm
WS
WwSs
Ws
Wwss
wS
wwSs
ws
wwss
Offspring
Key
W = Widow's peak
w = Straight hairline
S = Short fingers
s = Long fingers
Phenotypes
9 Widow's peak, short fingers
3 Widow's peak, long fingers
3 Straight hairline, short fingers
1 Straight hairline, long fingers

Figure 23.8 Two-trait testcross. A testcross determines if the individual with a dominant phenotype is homozygous or heterozygous. If the individual is heterozygous as shown, there is a 25% chance for each possible phenotype. If the individual in question was *WWSs,* could any of the children have a straight hairline? If the individual in question was *WwSS,* could any of the children have long fingers?

have the dominant phenotypes. However, if a man is heterozygous for both traits, then each child has a 25% chance of showing either one or both recessive traits. A Punnett square (Fig. 23.8) shows that the expected ratio is 1 widow's peak with short fingers: 1 widow's peak with long fingers: 1 straight hairline with short fingers: 1 straight hairline with long fingers, or 1:1:1:1.

Check Your Progress 23.1

1. Distinguish between the terms genotype and phenotype.
2. Explain the purpose of a testcross.
3. Summarize Mendel's law of segregation and law of independent assortment.
4. Describe the types of crosses that result in a 3:1 and 9:3:3:1 ratio.

23.2 Pedigree Analysis and Genetic Disorders

Learning Outcomes

Upon completion of this section, you should be able to

1. Recognize autosomal dominant and autosomal recessive patterns of inheritance when examining a pedigree.
2. Discuss certain genetic disorders and do problems that involve these disorders.

Many traits and disorders in humans, and other organisms also, are genetic in origin and follow Mendel's laws. These traits are often controlled by a single pair of alleles on the autosomal chromosomes. An autosome is any chromosome other than a sex (X or Y) chromosome. The patterns of inheritance associated with the sex chromosomes are discussed in Chapter 24.

Patterns of Inheritance

In humans, it is unethical or impossible to carry out some of the crosses that can be done in plants, such as a testcross or self-pollination. Instead, human inheritance is diagrammed in a pedigree. A **pedigree** is a chart of a family's history with regard to a particular genetic trait. Pedigrees can be done for humans as well as other animals, such as dogs and livestock. In the chart, males are designated by squares and females by circles. A line between a square and a circle represents a mating. A vertical line going downward leads directly to a single child. If there are more children, they are placed off a horizontal line. If the purpose of the pedigree is to follow the inheritance of a particular trait, those individuals with the trait, called the affected individuals, are usually shaded. A pedigree is often used to determine the inheritance of a genetic disease.

Pedigrees for Autosomal Disorders

It is possible to decide if an inherited condition is due to an autosomal dominant or an autosomal recessive allele by studying a pedigree. Does the following pattern of inheritance represent an autosomal dominant or an autosomal recessive characteristic?

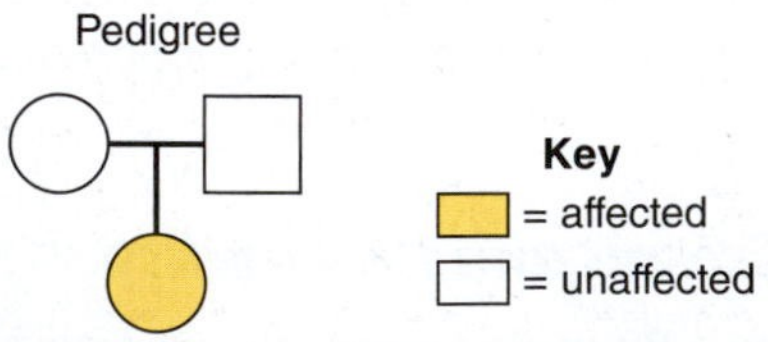
Pedigree
Key
= affected
= unaffected

In this pattern, the child is affected, but neither parent is. This can happen only if the disorder is recessive and the parents are heterozygotes. Notice that the parents are carriers because they are unaffected but are capable of having a child with the genetic disorder. If the family pedigree suggests that the parents are carriers for an autosomal recessive disorder, it would be possible to do prenatal testing for the genetic disorder (see Chapter 26).

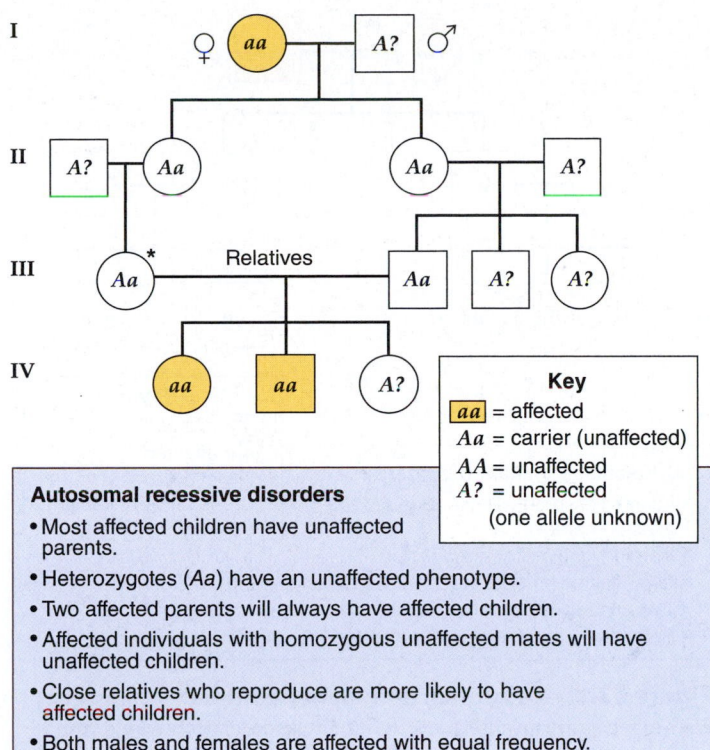

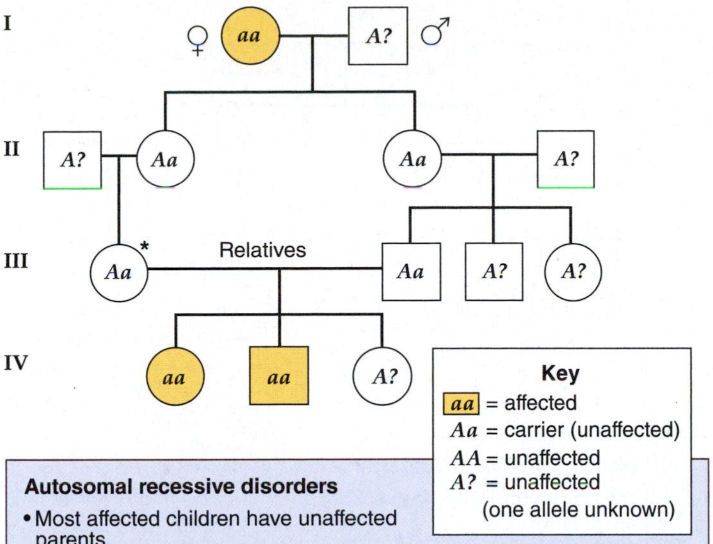
I
II
III
IV
aa
A?
A?
Aa
Aa
A?
Aa
Relatives
Aa
A?
A?
*
aa
aa
A?
Key
aa = affected
Aa = carrier (unaffected)
AA = unaffected
A? = unaffected
(one allele unknown)

Autosomal recessive disorders

- Most affected children have unaffected parents.
- Heterozygotes (*Aa*) have an unaffected phenotype.
- Two affected parents will always have affected children.
- Affected individuals with homozygous unaffected mates will have unaffected children.
- Close relatives who reproduce are more likely to have affected children.
- Both males and females are affected with equal frequency.

Figure 23.9 Autosomal recessive pedigree. The list gives ways to recognize an autosomal recessive disorder. How would you know the individual at the asterisk is heterozygous?

*See Appendix A for answers.

Notice that in the pedigree in Figure 23.9, cousins are the parents of three children, two of whom have the disorder. Aside from illustrating that reproduction between cousins is more likely to bring out recessive traits, this pedigree also shows that "chance has no memory." Each child born to heterozygous parents has a 25% chance of having the disorder. In other words, it is possible that if a heterozygous couple has four children, each child might have the condition.

Autosomal Recessive Disorders

Inheritance of two recessive alleles is required before an autosomal recessive disorder will appear.

Tay-Sachs disease is a well-known autosomal recessive disorder that usually occurs among Jewish people in the United States, most of whom are of central and eastern European descent. Tay-Sachs disease results from a lack of the enzyme hexosaminidase A (Hex A) and the subsequent storage of its substrate, a glycosphingolipid, in lysosomes. Lysosomes build up in many body cells, but the primary sites of storage are the cells of the brain, which accounts for the onset of symptoms and the progressive deterioration of psychomotor functions (Fig. 23.10*a*).

At first, it is not apparent that a baby has Tay-Sachs disease. However, development begins to slow down between four and eight months of age, and neurological impairment and psychomotor difficulties then become apparent. The child gradually becomes blind and helpless, develops uncontrollable seizures, and eventually becomes paralyzed.

Cystic fibrosis (CF) is an autosomal recessive disorder that occurs among all ethnic groups, but it is the most common lethal genetic disorder among Caucasians in the United States. Research has demonstrated that chloride ions (Cl^-) fail to pass

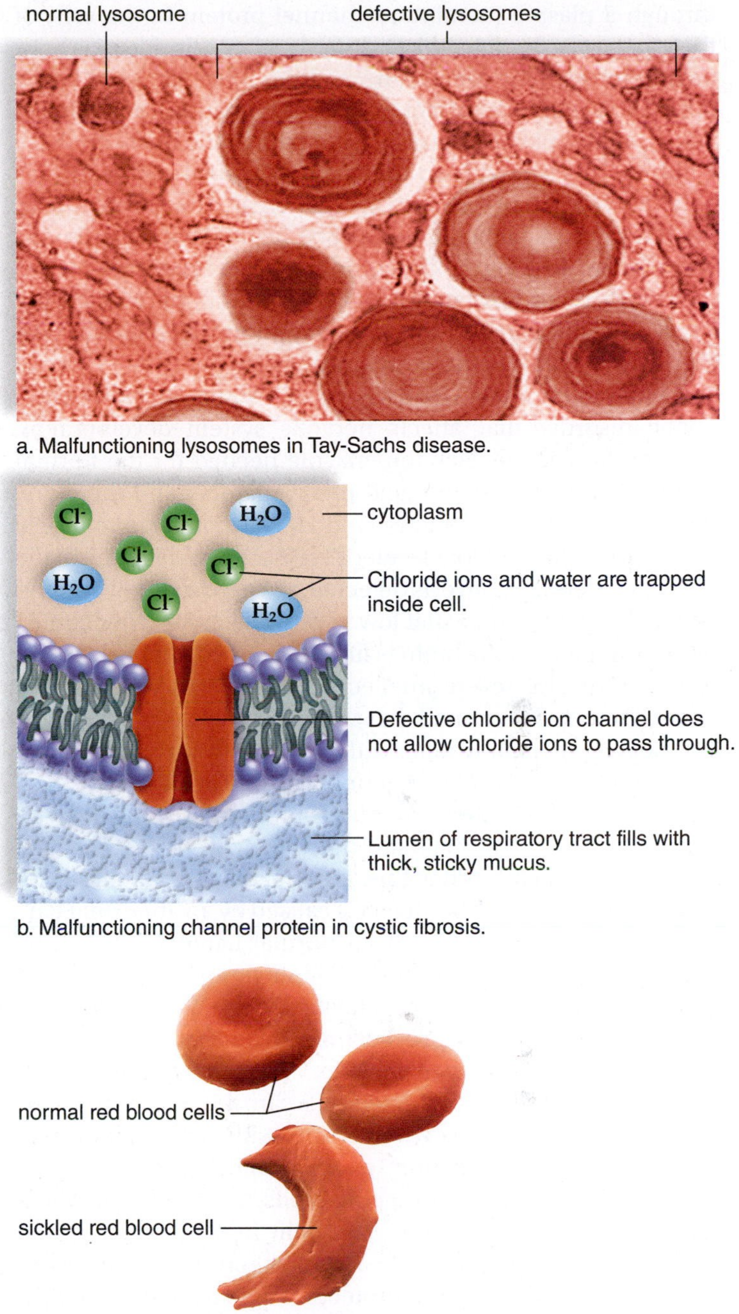
normal lysosome
defective lysosomes
Cl^-
H_2O
cytoplasm
Chloride ions and water are trapped inside cell.
Defective chloride ion channel does not allow chloride ions to pass through.
Lumen of respiratory tract fills with thick, sticky mucus.
normal red blood cells
sickled red blood cell

a. Malfunctioning lysosomes in Tay-Sachs disease.

b. Malfunctioning channel protein in cystic fibrosis.

c. Abnormally shaped red blood cells in sickle cell disease.

Figure 23.10 Autosomal recessive genetic disorders. **a.** A neuron in Tay-Sachs disease contains many defective lysosomes because of a deficient lysosomal enzyme. **b.** Cystic fibrosis is due to a faulty protein that is supposed to regulate the flow of chloride ions into and out of cells through a channel protein. **c.** Individuals with sickle cell disease have sickle-shaped red blood cells because of an abnormal hemoglobin A molecule.

through a plasma membrane channel protein in the cells of these patients (Fig. 23.10*b*). Ordinarily, after chloride ions have passed through the membrane, sodium ions (Na^+) and water follow. It is believed that lack of water is the cause of abnormally thick mucus in bronchial tubes and pancreatic ducts. In these children, this mucus is so thick and viscous that it interferes with the function of the lungs and pancreas. To ease breathing, the thick mucus in the lungs must be manually loosened periodically, but the lungs still become infected frequently. Clogged pancreatic ducts prevent digestive enzymes from reaching the small intestine, and to improve digestion, CF patients take digestive enzymes before every meal.

Phenylketonuria (PKU) is an autosomal recessive metabolic disorder that affects nervous system development. Affected individuals lack the enzyme needed for the normal metabolism of the amino acid phenylalanine, and therefore, it appears in the urine and the blood. Newborns are routinely tested in the hospital for elevated levels of phenylalanine in the blood. If an elevated level is detected, the newborn will develop normally if placed on a diet low in phenylalanine, which must be continued until the brain is fully developed, around the age of seven, or else severe intellectual disabilities will develop. Some doctors recommend that the diet continue for life, but in any case, a pregnant woman with phenylketonuria must be on the diet in order to protect her unborn child from harm.

Sickle cell disease is an autosomal recessive disorder in which the red blood cells are not biconcave disks like normal red blood cells; rather, they are irregular, and often sickle-shaped (Fig. 23.10*c*). This defect is caused by an abnormal type of hemoglobin that differs from normal hemoglobin by one amino acid in the protein globin. The single amino acid change causes hemoglobin molecules to stack up and form insoluble rods, and the red blood cells become sickle-shaped.

Because sickle-shaped cells can't pass along narrow capillary passageways as disk-shaped cells can, they clog the vessels and break down. This is why persons with sickle cell disease experience poor circulation, anemia, and low resistance to infection. Internal hemorrhaging leads to further complications, such as jaundice, episodic pain in the abdomen and joints, and damage to internal organs. Interestingly, sickle cell disease does provide some protection against malaria, which is why the allele still persists in the human population. This is covered in greater detail in section 27.3.

Sickle cell heterozygotes have sickle cell traits in which the blood cells are normal unless they experience dehydration or mild oxygen deprivation. Still, at present, most experts believe that persons with the sickle cell trait do not need to restrict their physical activity.

Autosomal Dominant Disorders

Now consider this pattern of inheritance:

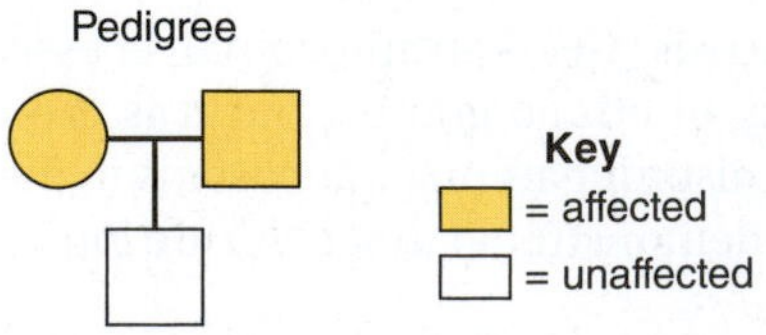
Pedigree
Key
= affected
= unaffected

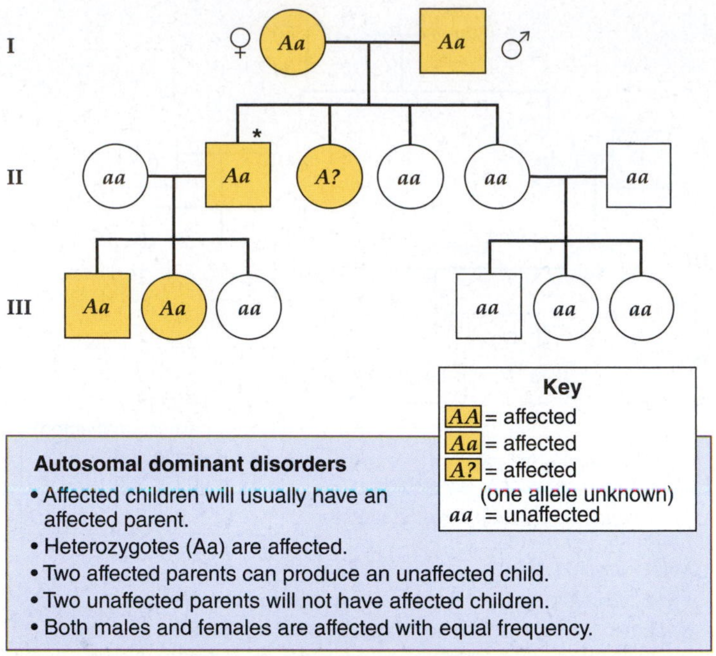
I
II
III
Aa
Aa
aa
Aa
A?
aa
aa
aa
Aa
Aa
aa
aa
aa
aa
*
Key
AA = affected
Aa = affected
A? = affected
(one allele unknown)
aa = unaffected
Autosomal dominant disorders
• Affected children will usually have an affected parent.
• Heterozygotes (Aa) are affected.
• Two affected parents can produce an unaffected child.
• Two unaffected parents will not have affected children.
• Both males and females are affected with equal frequency.

Figure 23.11 Autosomal dominant pedigree. The list gives ways to recognize an autosomal dominant disorder. How would you know the individual at the asterisk is heterozygous?

* See Appendix A for answers.

In this pattern, the child is unaffected, but the parents are both affected. This can happen if the condition is autosomal dominant and the parents are heterozygotes. Figure 23.11 lists other ways to recognize an autosomal dominant pattern of inheritance. This pedigree also illustrates that when both parents are unaffected, all their children are unaffected. Why? Because neither parent has a dominant gene that passes the condition on. Inheritance of only one dominant allele is necessary for an autosomal dominant genetic disorder to appear. The following describe some examples of autosomal dominant disorders in humans.

Marfan syndrome, an autosomal dominant disorder, is caused by a defect in an elastic connective tissue protein called fibrillin. This protein is normally abundant in the lens of the eye, the bones of the limbs, fingers, and ribs, and also in the wall of the aorta. The affected person often has a dislocated lens, long limbs and fingers, and a caved-in chest. The aorta wall is weak and may burst without warning. A tissue graft can strengthen the aorta, but people with Marfan syndrome still should not overexert themselves.

Huntington disease is a neurological disorder that leads to progressive degeneration of brain cells (Fig. 23.12). The disease is caused by a mutated copy of the gene for a protein called huntingtin. Most patients appear normal until they are of middle age and have already had children, who may later also be stricken. Occasionally, the first sign of the disease appears during the teen years or even earlier. There is no effective treatment, and death comes 10 to 15 years after the onset of symptoms.

Several years ago, researchers found that the gene for Huntington disease was located on chromosome 4. They developed a test to detect the presence of the gene. However,

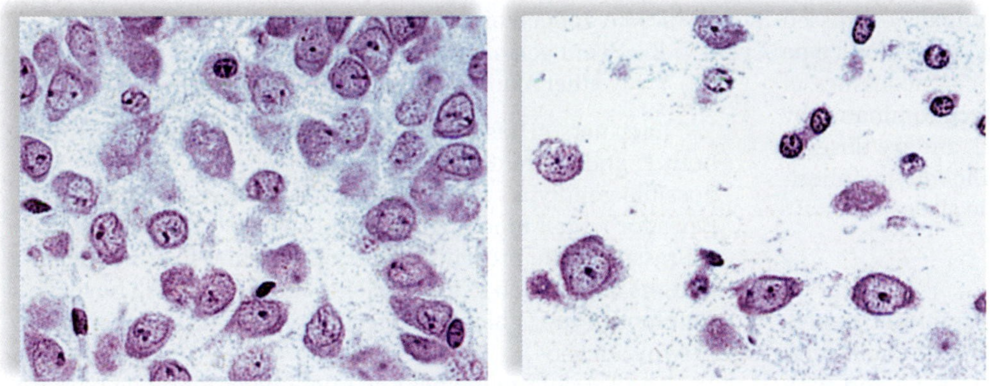
Many neurons in normal brain.
Loss of neurons in Huntington brain.

Figure 23.12 Autosomal dominant genetic disorder. Huntington disease is characterized by increasingly serious psychomotor and mental disturbances because of the loss of nerve cells.

few people want to know they have inherited the gene because there is no cure. At least now we know that the disease stems from a mutation that causes the huntingtin protein to have too many copies of the amino acid glutamine. The normal version of huntingtin has stretches of between 10 and 25 glutamines. If huntingtin has more than 36 glutamines, it changes shape and forms large clumps inside neurons. Even worse, it attracts and causes other proteins to clump with it. One of these proteins, called CBP, which helps nerve cells survive, is inactivated when it clumps with huntingtin. Researchers hope to combat the disease by boosting CBP levels.

Check Your Progress 23.2

1. Describe the characteristics of a pedigree for an autosomal dominant and an autosomal recessive trait.
2. Distinguish between autosomal recessive and autosomal dominant genetic disorders.

23.3 Beyond Simple Inheritance Patterns

Learning Outcomes

Upon completion of this section, you should be able to

1. Explain how the phenotypes for incompletely dominant and codominant traits differ from dominant/recessive trait phenotypes.
2. Explain how ABO blood types are an example of multiple allele inheritance.
3. Describe polygenic inheritance.

Certain traits, such as those studied in sections 23.1 and 23.2, are controlled by one set of alleles that follows a simple dominant or recessive inheritance. But we now know of many other types of inheritance patterns.

Incomplete Dominance and Codominance

Incomplete dominance occurs when the heterozygote is intermediate between the two homozygotes. For example, when a curly-haired Caucasian reproduces with a straight-haired Caucasian, their children have wavy hair. When two wavy-haired persons reproduce, the expected phenotypic ratio among the offspring is 1:2:1—that is, one curly-haired child to two with wavy hair to one with straight hair (Fig. 23.13). We can explain incomplete dominance by assuming that only one allele codes for a product and the single dose of the product gives the intermediate result. Notice in Figure 23.13 that we do not use the same capital and lowercase letters as we used with dominant/recessive traits.

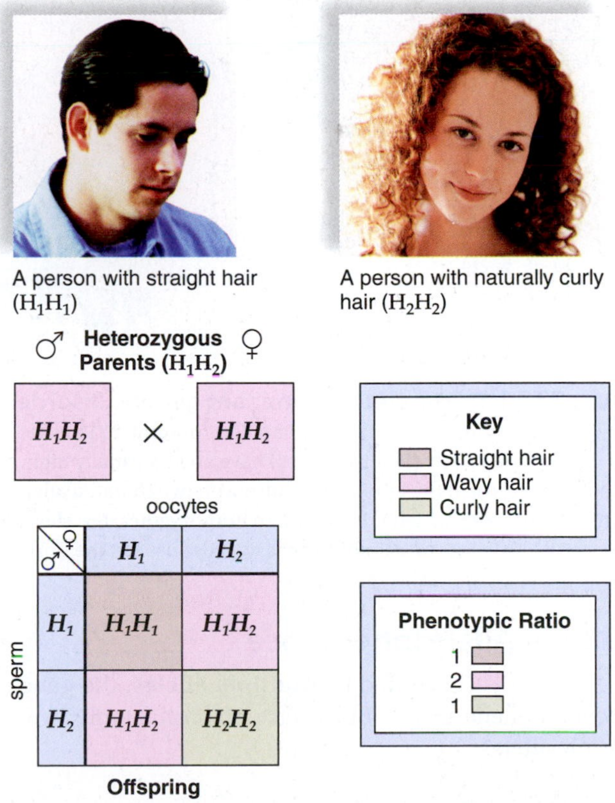
A person with straight hair (H_1H_1)
A person with naturally curly hair (H_2H_2)
♂ Heterozygous Parents (H_1H_2) ♀
H_1H_2 × H_1H_2
oocytes
sperm
H_1
H_2
H_1H_1
H_1H_2
H_1H_2
H_2H_2
Offspring
Key
Straight hair
Wavy hair
Curly hair
Phenotypic Ratio
1
2
1

Figure 23.13 Incomplete dominance. Among Caucasians, neither straight nor curly hair is dominant. When two wavy-haired individuals reproduce, each offspring has a 25% chance of having either straight or curly hair and a 50% chance of having wavy hair, the intermediate phenotype.

Codominance occurs when alleles are equally expressed in a heterozygote. A familiar example is the human blood type AB, in which the red blood cells have the characteristics of both type A and type B blood. We can explain codominance by assuming that both genes code for a product, and we observe the results of both products being present. Blood type inheritance is also said to be an example of multiple alleles.

Incompletely Dominant Disorders

The severity of symptoms exhibited in **familial hypercholesterolemia** parallels the number of LDL-cholesterol receptor proteins in the plasma membrane. A person with two mutated alleles lacks LDL-cholesterol receptors; a person with only one mutated allele has half the normal number of receptors; and a person with two normal alleles has the usual number of receptors. The presence of excessive cholesterol in the blood causes cardiovascular disease. Therefore, people with no receptors die of cardiovascular disease before 30 years of age. These individuals develop cholesterol deposits in the skin, tendons, and cornea. Individuals with one-half the number of receptors may die while still young or after they have reached middle age. People with the full number of cholesterol receptors do not have familial hypercholesterolemia (Fig. 23.14).

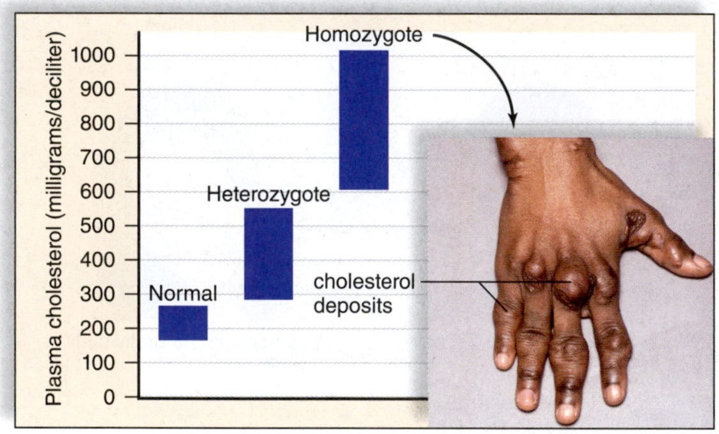
Plasma cholesterol (milligrams/deciliter)
1000
900
800
700
600
500
400
300
200
100
0
Homozygote
Heterozygote
Normal
cholesterol deposits

Figure 23.14 Incompletely dominant genetic disorder. Familial hypercholesterolemia is incompletely dominant. Persons with one mutated allele (heterozygotes) have an abnormally high level of cholesterol in the blood, and those with two mutated alleles (homozygotes) have a higher level still. In homozygotes, the elevated cholesterol level may sometimes form visible deposits in the skin.

Multiple Allele Inheritance

When a trait is controlled by **multiple alleles,** the gene exists in several allelic forms. But each person has only two of the possible alleles.

ABO Blood Types

Three alleles for the same gene control the inheritance of ABO blood types. These alleles determine the presence or absence of antigens on red blood cells:

I^A = A antigen on red blood cells
I^B = B antigen on red blood cells
i = Neither A nor B antigen on red blood cells

Each person has only two of the three possible alleles, and both I^A and I^B are dominant over i. Therefore, there are two possible genotypes for type A blood and two possible genotypes for type B blood. On the other hand, I^A and I^B are fully expressed in the presence of the other. Therefore, if a person inherits one of each of these alleles, that person will have type AB blood. Type O blood can result only from the inheritance of two i alleles.

The possible genotypes and phenotypes for blood type are as follows:

Phenotype	Genotype
A	I^AI^A, I^Ai
B	I^BI^B, I^Bi
AB	I^AI^B
O	ii

In the past, blood typing was used in paternity suits. However, the results of a blood test can only suggest that the tested individual *might* be the father, not that he definitely *is* the father. For example, it is possible, but not definite, that a man with type A blood (genotype I^Ai) is the father of a child with type O blood. On the other hand, a blood test can sometimes definitely prove that a man is *not* the father. For example, a man with type AB blood cannot possibly be the father of a child with type O blood. Therefore, blood tests can be used in legal cases only to exclude a man from possible paternity.

As a point of interest, the Rh factor is inherited separately from A, B, AB, or O blood types. When you are Rh-positive, your red blood cells have a particular antigen, and when you are Rh-negative, that antigen is absent. There are multiple recessive alleles for Rh-negative, but they are all recessive to Rh-positive. More information on the interactions of the ABO and Rh blood types with the immune system is provided in section 13.4.

Figure 23.15 shows that matings between certain genotypes can have surprising results in terms of blood type.

Polygenic Inheritance

Many traits, such as size or height, shape, weight, skin color, metabolic rate, and behavior, are governed by several genes (each with sets of alleles). **Polygenic inheritance** occurs when a trait is governed by two or more genes (sets of alleles). The individual has a copy of all allelic pairs, possibly located on many different pairs of chromosomes. Each dominant allele codes for a product, and therefore the dominant alleles have a quantitative effect on the phenotype, and these effects are additive. The result is a *continuous variation* of phenotypes, resulting in a distribution of these phenotypes that resembles a bell-shaped curve. The more genes involved, the more continuous are the variations and distribution of the phenotypes.

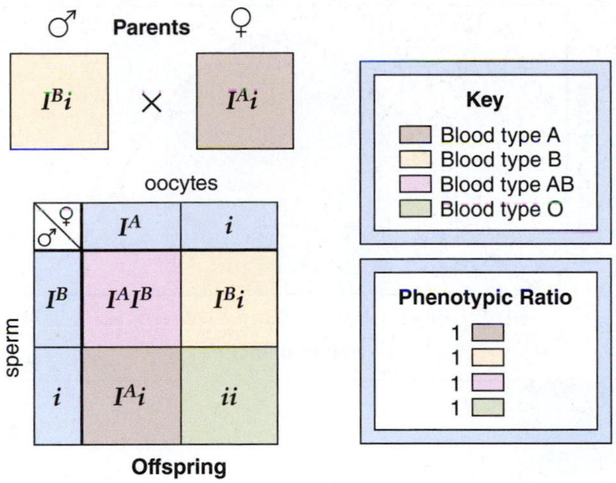

♂ Parents ♀
I^Bi
×
I^Ai
oocytes
♀
♂
I^A
i
I^B
I^AI^B
I^Bi
i
I^Ai
ii
sperm
Offspring
Key
Blood type A
Blood type B
Blood type AB
Blood type O
Phenotypic Ratio
1
1
1
1

Figure 23.15 Inheritance of blood types. Blood type exemplifies multiple allele inheritance. A mating between individuals with type A blood and type B blood can result in any one of the four blood types. Why? Because the parents are I^Ai and I^Bi. If both parents had type AB blood, predict the blood types of their potential offspring.

Skin Color

Skin color is the result of pigmentation produced by cells called melanocytes in the skin (see section 11.4). We now know that there are over 100 different genes that influence skin color. For simplicity, we will assume that skin has only three pairs of alleles (*Aa, Bb,* and *Cc*) and that each capital letter contributes pigment to the skin. When a very dark person reproduces with a very light person, the children have medium-brown skin. When two people with the genotype *AaBbCc* reproduce with one another, the individuals may range in skin color from very dark to very light. The distribution of these phenotypes typically follows a bell-shaped curve, meaning that few people have the extreme phenotypes and most people have the phenotype that lies in the middle. A bell-shaped curve is a common identifying characteristic of a polygenic trait (Fig. 23.16).

However, skin color is also influenced by the sunlight in the environment. Notice how a range of phenotypes exists for each genotype. For example, individuals who are *AaBbCc* may vary in their skin color, even though they possess the same genotype, and several possible phenotypes fall between the two extremes. The interaction of the environment with polygenic traits is discussed in the next section.

Check Your Progress 23.3

1. Interpret the genotype of the heterozygote if the inheritance pattern for a genetic disorder is shown to be incompletely dominant.
2. Determine the genotype of the child, the mother, and the possible genotypes of the father for a child with type O blood who is born to a mother with type A blood.

23.4 Environmental Influences

Learning Outcomes

Upon completion of this section, you should be able to

1. Describe ways in which the environment can influence genetic traits.
2. Understand how scientists determine the effect of nature versus nurture.

Environmental factors, such as nutrition or temperature, can also influence the expression of genetic traits. Polygenic traits seem to be particularly influenced by the environment. For example, in the case of height, differences in nutrition are one of the factors that bring about a bell-shaped curve (Fig. 23.17).

Temperature can also affect the phenotypes of plants and animals. For example, primroses have white flowers when grown above 32°C and red flowers when grown at 24°C. The coats of Siamese cats and Himalayan rabbits are darker in color at the ears, nose, paws, and tail. Himalayan rabbits are known to be homozygous for the allele *ch,* which is involved in the production of melanin. Experimental evidence suggests that the enzyme coded for by this gene is active only at low temperatures, and therefore black fur occurs only at the extremities, where body heat is lost to the environment (Fig. 23.18).

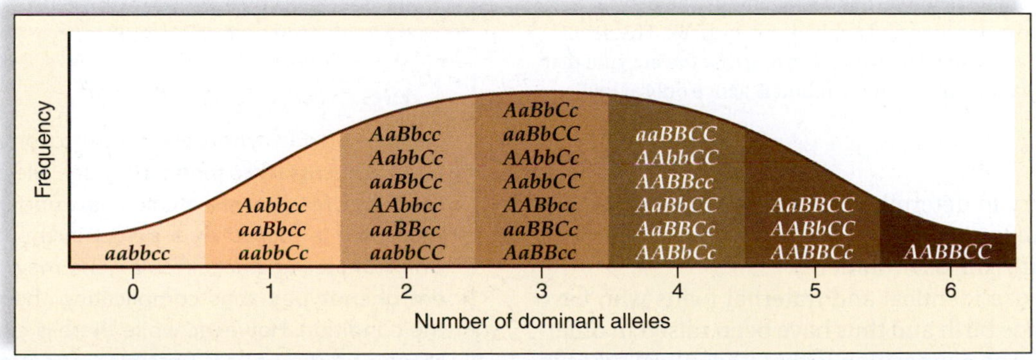

Frequency
aabbcc
Aabbcc
aaBbcc
aabbCc
AaBbcc
AabbCc
aaBbCc
AAbbcc
aaBBcc
aabbCC
AaBbCc
aaBbCC
AAbbCc
AabbCC
AABbcc
aaBBCc
AaBBcc
aaBBCC
AAbbCC
AABBcc
AaBbCC
AaBBCc
AABbCc
AaBBCC
AABbCC
AABBCc
AABBCC
0
1
2
3
4
5
6
Number of dominant alleles

Figure 23.16 Polygenic inheritance. Skin color is controlled by many pairs of alleles, which results in a range of phenotypes. The vast majority of people have skin colors in the middle range, while fewer people have skin colors in the extreme range.

Figure 23.17 Polygenic inheritance and environmental effects. A person's height is determined by both environmental and genetic factors. This results in a characteristic bell-shaped distribution curve. Notice how there are two distinct curves in this photo, one for women (white) and one for men (blue).

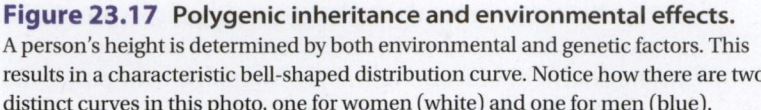

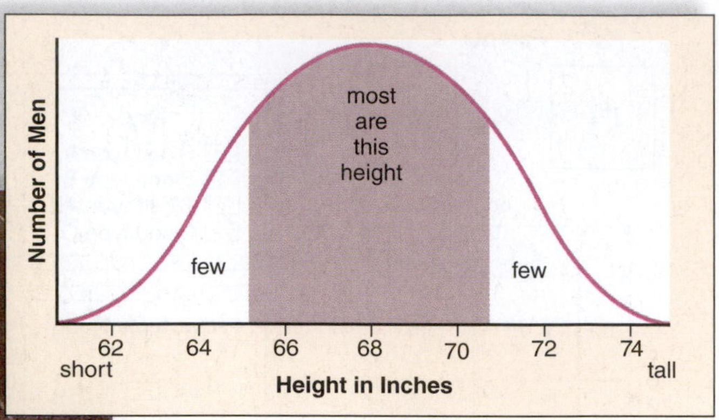
Number of Men
most
are
this
height
few
few
62
short
64
66
68
70
72
74
tall
Height in Inches

5:0
5:1
5:2
5:3
5:4
5:5
5:6
5:7
5:8
5:9
5:10
5:11
6:0
6:1
6:2
6:3
6:4
6:5

Figure 23.18 Coat color in Himalayan rabbits. The influence of the environment on the phenotype has been demonstrated by plucking out the fur from one area and applying an ice pack. The new fur that grew in was black instead of white, showing that the enzyme that produces melanin (a dark pigment) in the rabbit is active only at low temperatures.

Investigators try to determine what percentage of various human traits is due to nature (inheritance) and what percentage is due to nurture (the environment).

Some studies use identical and fraternal twins who have been separated since birth and thus have been raised in different environments. The supposition is that if identical twins in different environments share the same trait, that trait is most likely inherited. These studies have found that identical twins are more similar in their intellectual talents, personality traits, and levels of lifelong happiness than are fraternal twins when both groups have been separated since birth. Biologists conclude that all behavioral traits are partly inheritable and that genes exert their effects by acting together in complex combinations susceptible to environmental influences.

Scientists have also learned how to manipulate the environment in order to prevent the damaging effects of some genetic disorders. For example, the bodies of people who have the genetic disorder phenylketonuria (PKU) lack an enzyme that breaks down the amino acid phenylalanine. The accumulation of phenylalanine leads to neurological damage in these patients. But scientists have discovered that if these patients modify their diet so that they consume very little protein containing phenylalanine, they can avoid many of the damaging effects of this disease.

Check Your Progress 23.4

1. Summarize the environmental factors that may influence polygenic traits such as height and skin color.
2. Discuss how studies using identical twins may help determine how much of a phenotype is based on genetics and how much is environmental.

Case Study Conclusion

Because Marfan syndrome is an autosomal dominant disorder, it is necessary to inherit only one allele to suffer from the condition. Medical researchers now understand that Marfan syndrome is caused by a defect in the fibrin gene (*FBN1*). Unfortunately, defects in this gene may cause many different phenotypes, thus complicating the early detection of the condition. However, while there is currently no cure for Marfan syndrome, it is possible to treat many of the symptoms so that the individuals may lead longer, healthier lives.

SCIENCE IN YOUR LIFE ▶ **SCIENTIFIC INQUIRY**

Genetics of Taste

People differ in how they experience the taste of food. As children, we all refused to eat certain foods. Sometimes we were just being stubborn, but other times it was because that food tasted horrible! Researchers are looking at the genetics of taste. On the tongue are approximately 10,000 taste buds and within each taste bud, there are 50 to 100 taste receptor cells. There are five different types of taste receptors: sweet, sour, salty, bitter, and umami (savory). When a food molecule binds to a taste receptor on the outside of the cell, the cell triggers a nerve impulse that travels to the brain where it is interpreted as a particular taste. There is no taste receptor for spicy food. Spicy food causes pain and is registered by the brain as that!

Humans contain about 30 different genes for bitter taste receptors. Natural plant toxins are generally bitter so it makes sense that humans would evolve an aversion to bitter-tasting compounds. In one study, children who possess two positive alleles (+ +) for a particular bitter taste receptor were very sensitive to any bitterness in their food. Interestingly, the mothers of those children who were also + + did not share that same degree of sensitivity. Researchers think that as we age our sense of taste may lessen and so the mothers' tastes were not quite as sensitive. So a child who refuses to eat a particularly bitter-tasting vegetable may find it much more offensive than mom does.

A common genetics laboratory experiment is to taste PTC (phenylthiocarbamide), a compound not found in nature. A single gene with at least three different alleles is responsible for the ability to taste PTC. A PTC taster finds low concentrations extremely bitter, while a nontaster cannot taste anything even at higher concentrations. For those who can taste PTC, the bitter taste is unforgettable! Researchers have found that there is a correlation between the ability to taste PTC and food preferences. Those who taste PTC tend to dislike bitter foods such as dark chocolate and black coffee. And broccoli. Or that could just be stubbornness.

Questions to Consider

1. How might the environment influence the polygenic trait of taste?
2. If something tastes very bitter, it is more likely that a human or animal will not try to eat it (at least after the first try). Explain why it is advantageous for plants to produce bitter chemicals.

MEDIA STUDY TOOLS

www.mhhe.com/maderinquiry14

Enhance your study of this chapter with study tools and practice tests. Also ask your instructor about the resources available through ConnectPlus, including LearnSmart, the media-rich eBook, interactive learning tools, and animations.

SUMMARIZE

23.1 Mendel's Laws

- Genes carry instructions for a trait. Alternative forms of a gene (producing differences in the trait) are called **alleles.** The location of a gene on a chromosome is called its **locus.** In genetics, **phenotype** refers to the physical appearance of the individual, whereas the **genotype** lists the individual's combination of alleles. For a trait, **dominant alleles** mask the expression of **recessive alleles.** It is customary to use letters to represent the genotypes of individuals. **Homozygous** dominant is indicated by two capital letters, and homozygous recessive is indicated by two lowercase letters. **Heterozygous** is indicated by a capital letter and a lowercase letter.
- An individual has two alleles for each trait, whereas the gametes have one allele for each trait. Therefore, in keeping with Mendel's **law of segregation,** each heterozygous individual can form two types of gametes.
- When performing an actual cross, it is assumed that all possible types of sperm fertilize all possible types of oocytes. The Punnett square is based on this assumption. The results may be expressed as a probable phenotypic ratio. It is also possible to state the chance of an offspring having a particular phenotype. When a heterozygote reproduces with a heterozygote (called a **monohybrid cross**), the phenotypic ratio is 3:1; that is, each child has a 75% chance of having the dominant phenotype and a 25% chance of having the recessive phenotype. **Punnett squares** are often used to indicate the potential offspring of a genetic cross.

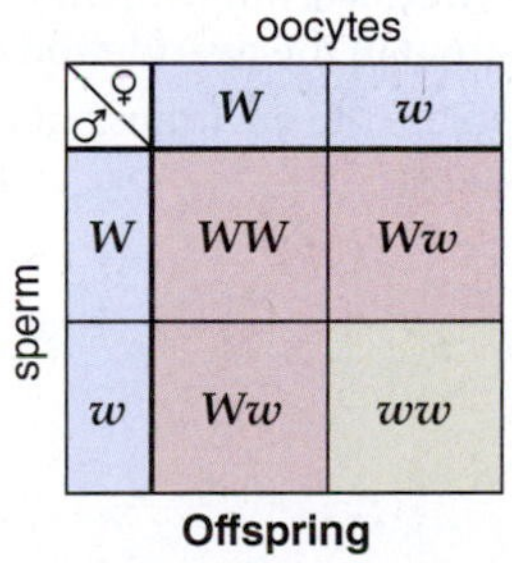
oocytes
♀
♂
W
w
W
WW
Ww
sperm
w
Ww
ww
Offspring

- A **testcross** is sometimes used to determine the genotype of an individual with a dominant phenotype.

- In keeping with Mendel's **law of independent assortment,** an individual heterozygous for two traits (**dihybrid**) can form four types of gametes. Therefore, the phenotypic ratio for a dihybrid cross is 9:3:3:1; in other words, 9/16 of the offspring have the two dominant traits, 3/16 have one of the dominant traits with one of the recessive traits, 3/16 have the other dominant trait with the other recessive trait, and 1/16 have both recessive traits.

23.2 Pedigree Analysis and Genetic Disorders

- A **pedigree** is a chart of a family's history with regard to a particular genetic trait. It may be possible to decide if an inherited condition is due to an autosomal dominant or an autosomal recessive allele by studying a pedigree.
- **Tay-Sachs disease** (a lysosomal storage disease), **cystic fibrosis (CF)** (faulty regulator of chloride channel), **phenylketonuria (PKU)** (inability to metabolize phenylalanine), and **sickle cell disease** (sickle-shaped red blood cells) are all autosomal recessive disorders.
- **Marfan syndrome** (defective elastic connective tissue) and **Huntington disease** (abnormal huntingtin protein) are both autosomal dominant disorders.

23.3 Beyond Simple Inheritance Patterns

- There are many exceptions to Mendel's laws, including **incomplete dominance** (curly hair), **codominance,** and **multiple alleles** (ABO blood types).
- **Familial hypercholesterolemia** (liver cells lack cholesterol receptors) is an incompletely dominant autosomal disorder.
- In **polygenic inheritance,** such as height and skin color in humans, several genes each contribute to the overall phenotype in equal, small degrees. If the environment is involved, the result is a bell-shaped curve of phenotypes.

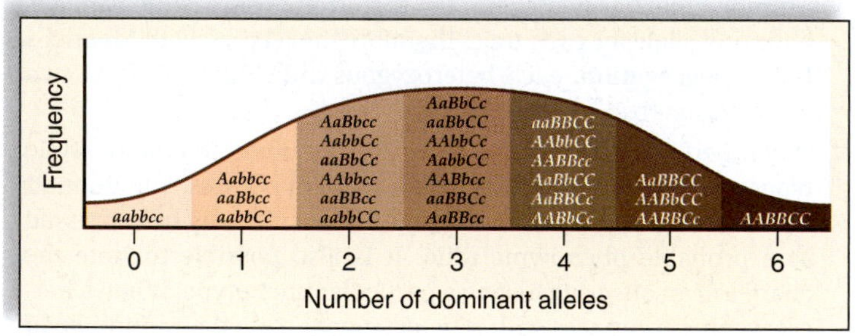
Frequency
aabbcc
Aabbcc
aaBbcc
aabbCc
AaBbcc
AabbCc
aaBbCc
AAbbcc
aaBBcc
aabbCC
AaBbCc
aaBbCC
AAbbCc
AabbCC
AABbcc
aaBBCc
AaBBcc
aaBBCC
AAbbCC
AABBcc
AaBbCC
AaBBCc
AABbCc
AaBBCC
AABbCC
AABBCc
AABBCC
0
1
2
3
4
5
6
Number of dominant alleles

23.4 Environmental Influences

- Environmental factors, such as nutrition and temperature, affect the expression of certain traits. In humans, nutrition helps determine height. In Siamese cats and Himalayan rabbits, the effect of temperature is apparent in the distribution of fur color.
- Humans can manipulate the environment to lessen the damaging effects of genetic defects.

ASSESS

Testing Yourself

Choose the best answer for each question.

1. According to the law of segregation,
   a. each gamete contains two copies of each factor for each trait.
   b. each individual has one copy of each factor for each trait.
   c. fertilization gives each new individual one factor for each trait.
   d. All of these are correct.
   e. None of these are correct.
2. Peas are good for genetic studies because they
   a. cannot self-pollinate.
   b. have a long generation time.
   c. are easy to grow.
   d. are more genetically variable than most plants.
3. Which of the following genotypes indicates a heterozygous individual?
   a. *AB*
   b. *AA*
   c. *Aa*
   d. *aa*
4. List the gametes produced by *AaBb*.
   a. *Aa, Bb*
   b. *A, a, B, b*
   c. *AB, ab*
   d. *AB, Ab, aB, ab*
5. Homologous chromosomes contain identical
   a. genes.
   b. alleles.
   c. DNA sequences.
   d. All of these are correct.
6. The genotypic ratio from the F_2 of a monohybrid cross is
   a. 1:1.
   b. 3:1.
   c. 1:2:1.
   d. 9:3:3:1.
   e. 1:1:1:1.
7. If two parents with short fingers (dominant) have a child with long fingers, what is the chance their next child will have long fingers?
   a. There is no chance.
   b. 1/2
   c. 1/4
   d. 1/16
   e. 3/16
8. Using the product rule, determine the probability that an *Aa* individual will be produced from an *Aa* × *Aa* cross.
   a. 50%
   b. 25%
   c. 75%
   d. 100%
   e. 0%
9. According to the law of independent assortment,
   a. all possible combinations of factors can occur in the gametes.
   b. only the parental combinations of gametes can occur in the gametes.
   c. only the nonparental combinations of gametes can occur in the gametes.

10. Which of the following is *not* a feature of polygenic inheritance?
    a. Effects of dominant alleles are additive.
    b. Genes affecting the trait may be on multiple chromosomes.
    c. Environment influences phenotype.
    d. Recessive alleles are harmful.
11. A distribution graph of a trait is bell-shaped. What pattern of inheritance does this suggest?
    a. incomplete dominance
    b. single-gene, autosomal recessive
    c. polygenic with environmental influences
    d. a testcross
12. Which is a late-onset neuromuscular genetic disorder?
    a. cystic fibrosis
    b. Huntington disease
    c. phenylketonuria
    d. Tay-Sachs disease

Additional Genetics Problems

1. For each of the following genotypes, give all possible gametes.
    a. *WW*
    b. *WWSs*
    c. *Tt*
    d. *Ttgg*
    e. *AaBb*
2. For each of the following, state whether a genotype or a gamete is represented.
    a. *D*
    b. *Ll*
    c. *Pw*
    d. *LlGg*
3. Both a man and a woman are heterozygous for freckles. Freckles (*F*) are dominant over no freckles (*f*). What is the chance that their child will have freckles?
4. Both you and your sister or brother have attached earlobes, but your parents have unattached earlobes. Unattached earlobes (*E*) are dominant over attached earlobes (*e*). What are the genotypes of your parents?
5. A father has dimples, the mother does not have dimples, and all five of their children have dimples. Dimples (*D*) are dominant over no dimples (*d*). Give the probable genotypes of all persons concerned.
6. Attached earlobes are recessive. What genotype do children have if one parent is homozygous recessive for attached earlobes and homozygous dominant for widow's peak, and the other is homozygous dominant for unattached earlobes and homozygous recessive for straight hairline?
7. If an individual from this cross reproduces with another of the same genotype, what are the chances that they will have a child with a straight hairline and attached earlobes?
8. A child who does not have dimples or freckles is born to a man who has dimples and freckles (both dominant) and a woman who does not. What are the genotypes of all persons concerned?

ENGAGE

Virtual Lab
Punnett Squares

The virtual lab "Punnett Squares" allows you the ability to test your knowledge of Mendelian patterns of inheritance.

Thinking Critically

1. You are working with the fruit fly *Drosophila* and want to determine whether a newly found characteristic is dominant or recessive. Design an experiment that will help you determine the dominance of the trait.
2. How would you determine whether a disease in humans is simply polygenic or has an environmental influence?
3. If identical twins have exactly the same genetic makeup, why are there any differences at all between them?

24 Chromosomal Basis of Inheritance

CHAPTER OUTLINE

BEFORE YOU BEGIN

Before beginning this chapter, take a few moments to review the following discussions:

Section 5.4 What is meant by the term homologous chromosomes?

Section 23.1 What does Mendel's law of segregation tell us about chromosomal inheritance?

Figure 23.4 How does a Punnett square help predict the genotypes and phenotypes of a genetic cross?

CASE STUDY Down Syndrome

Sarah's youngest son, Brandon, acted like any other six-year-old boy. He loved to eat, and play, and enjoyed spending time with his family. However, Brandon was born with Down syndrome, a genetic disorder that affects over 400,000 Americans. Down syndrome occurs when an individual inherits an extra copy of chromosome 21. It is the most common chromosomal anomaly in humans and the incidence of the condition is slowly rising in Western cultures. From 1983 to 2003, the total number of cases of Down syndrome increased by 24.2%. People with Down syndrome have similar characteristics, including an upward slant to the eyes, low muscle tone, a deep crease across the center of the palm, and an increased risk of cardiac defects. There is some degree of intellectual disability, but most individuals with Down syndrome have IQs in the mild to moderate disability range. Despite these health concerns, the life expectancy of Down syndrome individuals has increased tremendously in recent years, from 25 years in 1983 to over 60 years today.

The National Down Syndrome Society tells us that individuals with Down syndrome can be integrated into all aspects of society. They attend school, work, and participate in community activities. Some individuals graduate from high school and attend college. Some socialize, date, and even marry.

In this chapter, we will examine genetic conditions that are caused by changes in chromosomal number and structure.

As you read through the chapter, think about the following questions:

1. What causes a change in the number of chromosomes?
2. What is the genetic basis that causes the symptoms of Down syndrome?
3. How can the chromosome number of an individual be tested before they are born?

24.1 Gene Linkage

Learning Outcomes

Upon completion of this section, you should be able to

1. Describe what is meant by the term linkage group.
2. Predict which alleles in a linkage group will cross over more frequently.

Not long after Mendel's study of inheritance (see section 23.1) investigators began to realize that many different types of alleles are on a single chromosome. A chromosome doesn't contain just one or two alleles; it contains a long series of alleles in a definite sequence. The sequence is fixed because each allele has its own particular locus on a chromosome. All the alleles on one chromosome form a **linkage group** because they tend to be inherited together. Figure 24.1 shows some of the traits in the chromosome 19 linkage group of humans. When we do two-trait crosses (see Fig. 23.7), we are assuming that the alleles are on nonhomologous chromosomes and therefore are not linked. Alleles that are linked do not show independent assortment. Figure 24.2*a* shows the results of a cross when linkage is complete: a dihybrid produces only two types of gametes in equal proportion.

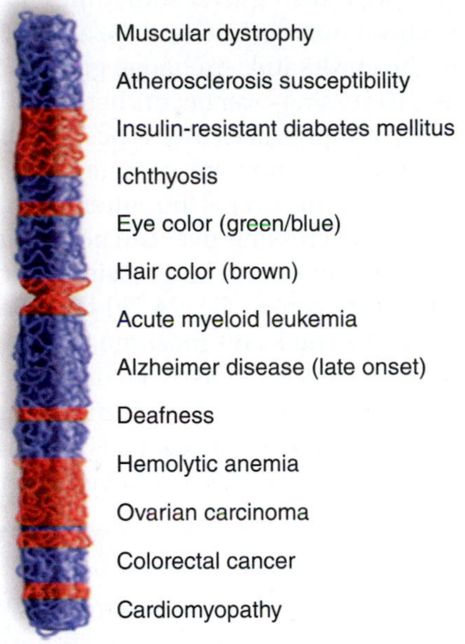

Muscular dystrophy
Atherosclerosis susceptibility
Insulin-resistant diabetes mellitus
Ichthyosis
Eye color (green/blue)
Hair color (brown)
Acute myeloid leukemia
Alzheimer disease (late onset)
Deafness
Hemolytic anemia
Ovarian carcinoma
Colorectal cancer
Cardiomyopathy

Figure 24.1 Human chromosome 19. A selection of traits located on human chromosome 19 (not to scale). These traits form a single linkage group. In human genetics, genes are sometimes named after the disease that they are associated with, not their function in the body.

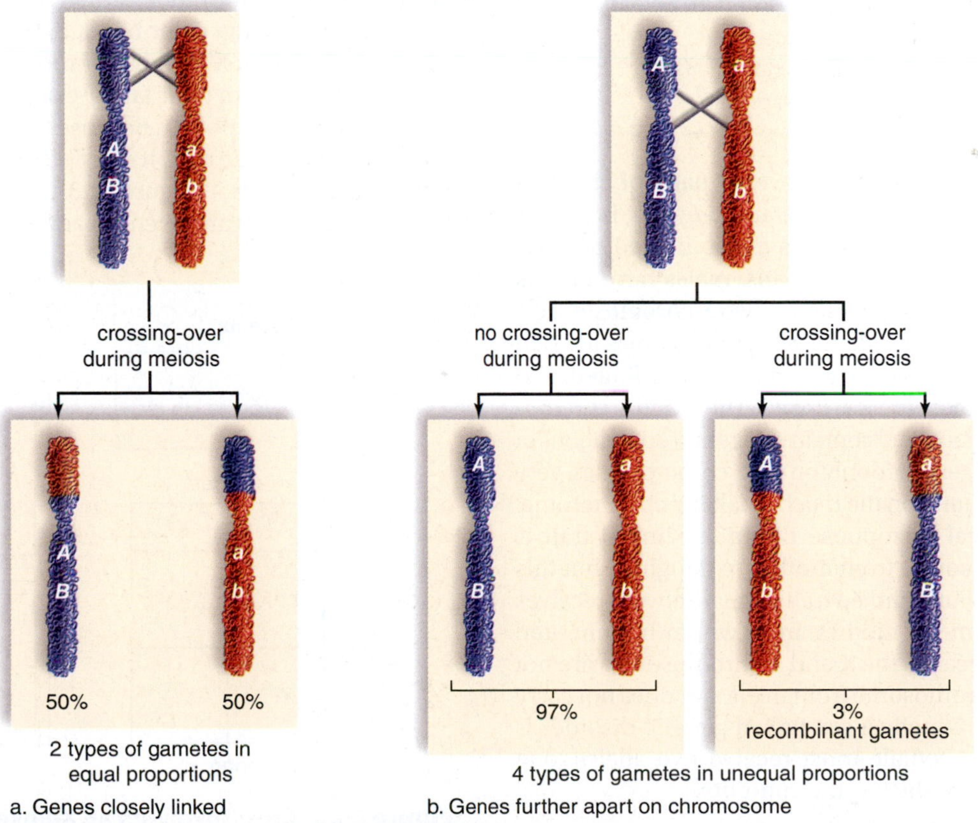

A
B
a
b
crossing-over during meiosis
50%
50%
2 types of gametes in equal proportions
a. Genes closely linked
no crossing-over during meiosis
crossing-over during meiosis
97%
3% recombinant gametes
4 types of gametes in unequal proportions
b. Genes further apart on chromosome

Figure 24.2 Recombination and linkage groups. In this example, alleles *A* and *B* are on one member of a homologous pair, and alleles *a* and *b* are on the other member. **a.** When the genes for *A* and *B* are located closely together, recombination between the alleles is rare and only two types of gametes are produced in equal proportion. **b.** When the genes are farther apart, crossing-over occurs more frequently between the alleles, producing four different types of gametes. The recombinant gametes occur in reduced proportion because crossing-over occurs infrequently.

During meiosis, crossing-over sometimes occurs between the nonsister chromatids of a tetrad (see Fig. 5.11). During crossing-over, the chromatids exchange genetic material, and therefore genes. If crossing-over occurs between the two alleles of interest, a dihybrid produces four types of gametes instead of two (Fig. 24.2*b*). Recombinant gametes occur in reduced number because crossing-over is infrequent.

The occurrence of crossing-over can help tell the sequence of genes on a chromosome because crossing-over occurs more often between distant genes (Fig. 24.2*b*) than between genes that are close together on a chromosome (Fig. 24.2*a*). Historically, maps of chromosomes were based on the recombination frequencies between alleles, with each map unit equal to an observed 1% of crossing-over. Thus, in Figure 24.2*b*, the distance between the *A* and *B* genes would be 3 map units.

Check Your Progress 24.1

1. Explain why linked alleles do not show independent assortment.
2. Explain what a distance of 10 map units tells you about the distance between two genes.

24.2 Sex-Linked Inheritance

Learning Outcomes

Upon completion of this section, you should be able to

1. Explain X-linked genetic inheritance.
2. Solve crosses involving sex-linked traits.
3. Describe four X-linked recessive disorders.

Normally, both males and females have 23 pairs of chromosomes; 22 pairs are called **autosomes,** and one pair is the sex chromosomes. The **sex chromosomes** are so named because they differ between the sexes. In humans, males have the sex chromosomes X and Y, and females have two X chromosomes.

Traits controlled by genes on the sex chromosomes are said to be **sex-linked;** an allele on an X chromosome is **X-linked,** and an allele on the Y chromosome is Y-linked. The X and Y chromosomes are not homologous, and contain different combinations of genes. Most sex-linked genes are only on the X chromosomes. Very few alleles have been found on the much smaller Y chromosome.

It would be logical to suppose that a sex-linked trait is passed from father to son or from mother to daughter, but this is not always the case. For X-linked traits, a male always receives an X-linked allele from his mother, from whom he inherited an X chromosome. Because the X and Y chromosomes are not homologous, the Y chromosome from the father does not carry an allele for the trait. Usually a sex-linked genetic disorder is recessive. Therefore, a female must receive two alleles, one from each parent, before she has the condition.

Sex-Linked Alleles

When considering X-linked traits, the allele on the X chromosome is shown as a letter attached to the X chromosome. For example, following is the key for red-green color blindness, a well-known X-linked recessive disorder:

X^B = normal vision X^b = color blindness

The possible genotypes and phenotypes in both males and females are:

Genotypes	Phenotypes
X^BX^B	Female who has normal color vision
X^BX^b	Carrier female with normal color vision
X^bX^b	Female who is color blind
X^BY	Male who has normal vision
X^bY	Male who is color blind

The second genotype is a *carrier* female because, although a female with this genotype has normal color vision, she is capable of passing on an allele for color blindness. Color-blind females are rare because they must receive the allele from both parents. Color-blind males are more common because they need only one recessive allele to be color blind. The allele for color blindness has to be inherited from their mother because it is on the X chromosome. Males only inherit the Y chromosome from their father.

Now let us consider a mating between a man with normal vision and a heterozygous woman (Fig. 24.3). What is the chance that this couple will have a color-blind daughter? A color-blind son? All daughters will have normal color vision because they all receive an X^B from their father. The sons, however, have a 50% chance of being color blind, depending on whether they receive an X^B or an X^b from their mother. The inheritance of a Y chromosome from their father cannot offset the inheritance of an X^b from their mother. Because the Y chromosome doesn't have an allele for the trait, it can't possibly prevent color blindness in a son. Note in Figure 24.3 that the phenotypic results for sex-linked traits are given separately for males and females.

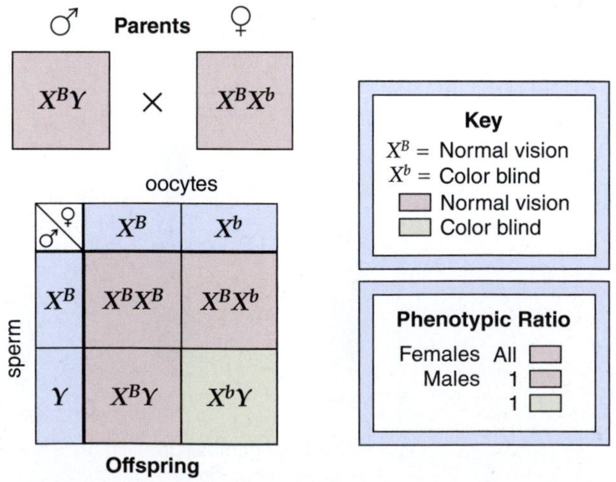

♂ Parents ♀
X^B Y
×
X^B X^b
oocytes
♀ ♂
X^B
X^b
sperm
X^B
Y
X^B X^B
X^B X^b
X^B Y
X^b Y
Offspring
Key
X^B = Normal vision
X^b = Color blind
Normal vision
Color blind
Phenotypic Ratio
Females All
Males 1
1

Figure 24.3 Cross involving an X-linked allele. The male parent is normal, but the female parent is a carrier—that is, an allele for color blindness is located on one of her X chromosomes. Therefore, each son has a 50% chance of being color blind. The daughters will have normal color vision, but each one has a 50% chance of being a carrier.

Pedigree for X-Linked Disorders

Like color blindness, most sex-linked disorders are carried on the X chromosome. Figure 24.4 gives a pedigree for an *X-linked recessive disorder.* More males than females have the disorder because recessive alleles on the X chromosome are always expressed in males. The Y chromosome lacks an allele for the disorder. X-linked recessive conditions often pass from grandfather to grandson because the daughters of a male with the disorder are carriers. Figure 24.4 lists various ways to recognize a recessive X-linked disorder.

Only a few known traits are *X-linked dominant*. One example is incontinentia pigmenti, which causes discoloration of the skin. If a disorder is X-linked dominant, affected males pass the trait *only* to daughters, who have a 100% chance of having the condition. Females can pass an X-linked dominant allele to both sons and daughters. If a female is heterozygous and her partner is normal, each child has a 50% chance of escaping an X-linked dominant disorder. This depends on which maternal X chromosome is inherited.

X-Linked Recessive Disorders of Interest

Color blindness, an X-linked recessive disorder, does not prevent males from leading a normal life. About 8% of Caucasian men have red-green color blindness. Most of them see brighter greens as tans, olive greens as browns, and reds as reddish browns. A few cannot tell reds from greens at all. They see only yellows, blues, blacks, whites, and grays.

Duchenne muscular dystrophy is an X-linked recessive disorder characterized by wasting away of the muscles. Symptoms, which include a waddling gait, toe walking, frequent falls, and difficulty in rising, may appear as soon as the child starts to walk. Muscle weakness intensifies until the individual is confined to a wheelchair. Death usually occurs by age 20; therefore, affected males are rarely fathers. The recessive allele remains in the population by passing from carrier mother to carrier daughter.

The absence of a protein, now called dystrophin, is the cause of Duchenne muscular dystrophy. Much investigative work determined that the lack of dystrophin causes calcium to leak into the cell, which promotes the action of an enzyme that dissolves muscle fibers. When the body attempts to repair the tissue, fibrous tissue forms (Fig. 24.5), and this cuts off the blood supply so that more and more cells die. Immature muscle cells can be injected into muscles, but it takes 100,000 cells for dystrophin production to increase by 30–40%.

Fragile X syndrome is caused by an abnormal number of repeat sequences in the genome and is the most common cause of inherited mental impairment. These impairments can range from mild learning disabilities to more severe intellectual disabilities. It is also associated with some forms of *autism*, a class of social, behavioral, and communication disorders. Males with full symptoms of the condition have characteristic physical abnormalities. A long face, prominent jaw, and large ears are facial features often seen in fragile X males. Excessively flexible joints and genital abnormalities are common as well. Females with the condition have variable symptoms. Most of the same traits seen in males with fragile X have been reported in females as well, but often the symptoms are milder in females and present with lower frequency. The Health feature, "Trinucleotide Repeat Expansion Disorders (TREDs)," on page 484 provides additional information on fragile X syndrome.

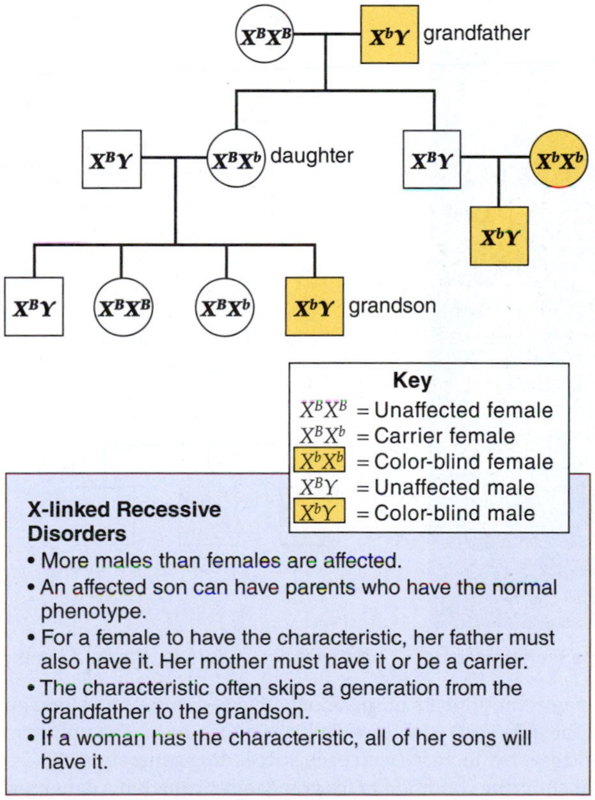

X^BX^B
X^bY grandfather
X^BY
X^BX^b daughter
X^BY
X^bX^b
X^bY
X^BY
X^BX^B
X^BX^b
X^bY grandson
Key
X^BX^B = Unaffected female
X^BX^b = Carrier female
X^bX^b = Color-blind female
X^BY = Unaffected male
X^bY = Color-blind male
X-linked Recessive Disorders
• More males than females are affected.
• An affected son can have parents who have the normal phenotype.
• For a female to have the characteristic, her father must also have it. Her mother must have it or be a carrier.
• The characteristic often skips a generation from the grandfather to the grandson.
• If a woman has the characteristic, all of her sons will have it.

Figure 24.4 X-linked recessive pedigree. This pedigree for color blindness exemplifies the inheritance pattern of an X-linked recessive disorder. The list gives various ways of recognizing the X-linked recessive pattern of inheritance.

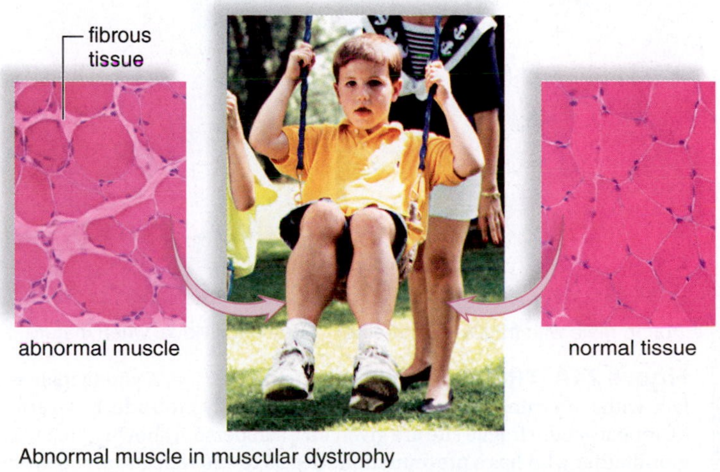

fibrous tissue
abnormal muscle
normal tissue
Abnormal muscle in muscular dystrophy

Figure 24.5 Duchenne muscular dystrophy. In Duchenne muscular dystrophy, an X-linked recessive disorder, the calves of the legs enlarge because fibrous tissue develops as muscles waste away due to lack of the protein dystrophin.

SCIENCE IN YOUR LIFE ▶ **HEALTH**

Trinucleotide Repeat Expansion Disorders (TREDs)

Several of the genetic diseases we have studied result from a single change in the DNA that results in a single change in an amino acid in a particular protein. For example, the most common form of sickle cell disease results from a one-nucleotide change in the DNA that causes a one-amino-acid difference in the protein hemoglobin. However, in at least 15 human diseases, including Huntington disease and fragile X syndrome, the problem lies in a DNA sequence that is repeated too many times, called a *trinucleotide repeat.*

As described in Chapter 23, Huntington disease is a neurological disorder in which the huntingtin protein is altered. The trinucleotide sequence CAG normally appears several times at the beginning of the gene that encodes the huntingtin protein. The presence of multiple CAGs results in multiple glutamines in the protein. People with 10 to 35 copies of the CAG display no signs of Huntington disease. However, people with more than 40 copies have the disease.

Fragile X syndrome is one of the most common genetic causes of mental impairment. As children, fragile X individuals may be hyperactive or autistic. Their speech is delayed in development and often repetitive in nature. As adults, males have large testes and big, usually protruding ears. They are short in stature, but the jaw is prominent, and the face is long and narrow (Fig. 24A*a*). The term *fragile X syndrome* originated because diagnosis used to be dependent upon observation of an X chromosome whose tip is attached to the rest of the chromosome by only a thin thread (Fig. 24A*b*). This disease results from repeats of the trinucleotide CGG. People with 6 to 50 copies appear normal, whereas those with over 230 copies have the disease. Those with between 50 and 230 copies are carriers of fragile X, with either no symptoms or very mild symptoms.

People who have the intermediate number of repeats are said to have a pre-mutation. Pre-mutations can lead to full-blown mutations in future generations by expansion of the number of repeats. This is sometimes called genetic anticipation. The number of repeats increases with each generation, as does the severity of the disease (Fig. 24A*c*). It is not known why the repeats expand in number. Using the current molecular technology, it is possible to accurately determine the number of trinucleotide repeats present in an individual.

Questions to Consider

1. The repeat expansion appears to occur during meiosis but not mitosis. Why might this be true?
2. If a male has fragile X syndrome, is his mother or father more likely to be a carrier for the disease? Why?
3. Why is fragile X, but not Huntington disease, considered a syndrome?

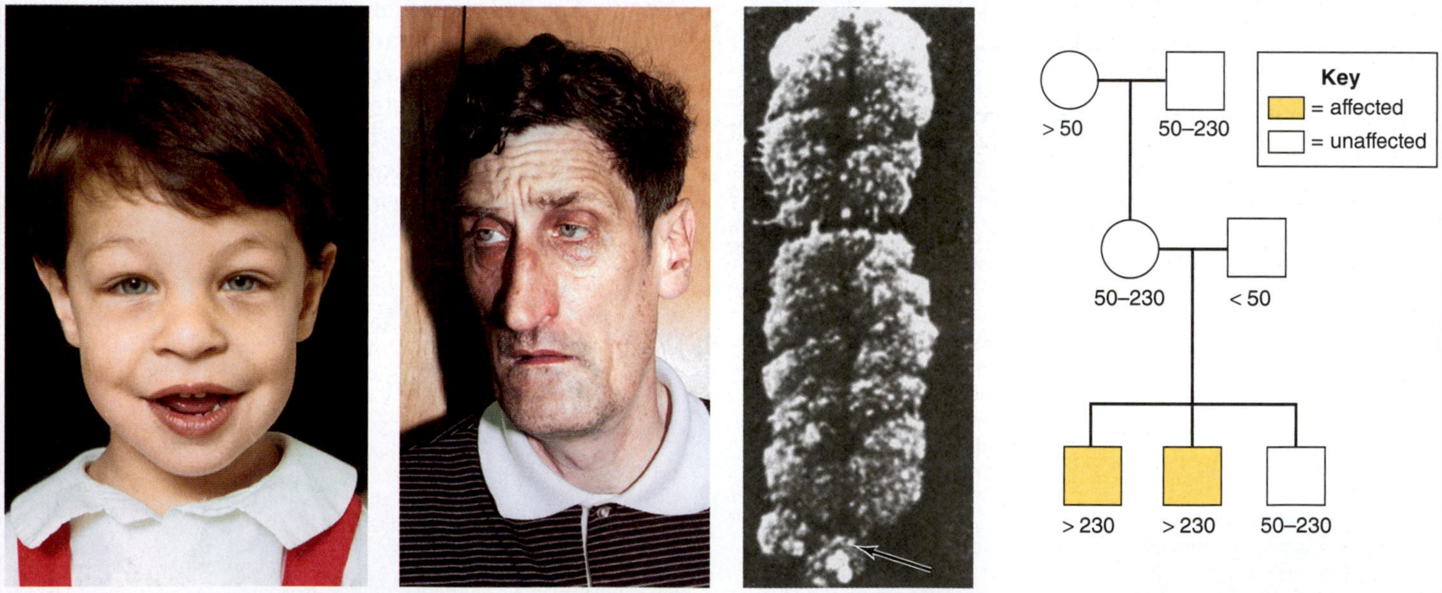
Key
= affected
= unaffected
> 50
50–230
50–230
< 50
> 230
> 230
50–230

a. Young male with fragile X syndrome   Same individual when mature   b. Fragile X chromosome   c. Inheritance pattern for fragile X syndrome

Figure 24A TRED in fragile X syndrome. **a.** A young male with fragile X syndrome appears normal but with age develops an elongated face with a prominent jaw and ears that noticeably protrude. **b.** An arrow points out the fragile site of this fragile X chromosome. **c.** The number of repeats at the fragile site are given for each person, showing that the incidence and severity increase with each succeeding generation. A grandfather who has a pre-mutation with 50 to 230 repeats has no symptoms but transmits the condition to his grandsons through his daughter. Two grandsons have full-blown mutations with more than 230 repeats.

Hemophilia, of which there are two common types, is another X-linked recessive disorder. Hemophilia A is due to the absence or minimal presence of a clotting factor known as factor VIII, and hemophilia B is due to the absence of clotting factor IX. Hemophilia is called the bleeder's disease because the affected person's blood either does not clot or clots very slowly. Although hemophiliacs bleed externally after an injury, they also bleed internally, particularly around joints. Hemorrhages can be stopped with transfusions of fresh blood (or plasma) or concentrates of the missing clotting protein. Factors VIII and IX are also now available as biotechnology products.

At the turn of the century, hemophilia was prevalent among the royal families of Europe, and all of the affected males could trace their ancestry to Queen Victoria of England (Fig. 24.6). Of Queen Victoria's 26 grandchildren, four grandsons had hemophilia, and four granddaughters were carriers. Because none of Queen Victoria's ancestors were affected, it seems that the faulty allele she carried arose by mutation, either in Victoria or in one of her parents. Her carrier daughters, Alice and Beatrice, introduced the allele into the ruling houses of Russia and Spain, respectively. Alexei, the last heir to the Russian throne before the Russian Revolution, was a hemophiliac. There are no hemophiliacs in the present British royal family because Victoria's eldest son, King Edward VII, did not receive the allele.

Check Your Progress 24.2

1. Explain how sex-linked patterns of inheritance differ from autosomal inheritance.
2. Describe the phenotypic ratio that is expected for a cross in which both parents have one X-linked recessive allele.
3. Summarize the causes of the X-linked recessive disorders discussed in this chapter.

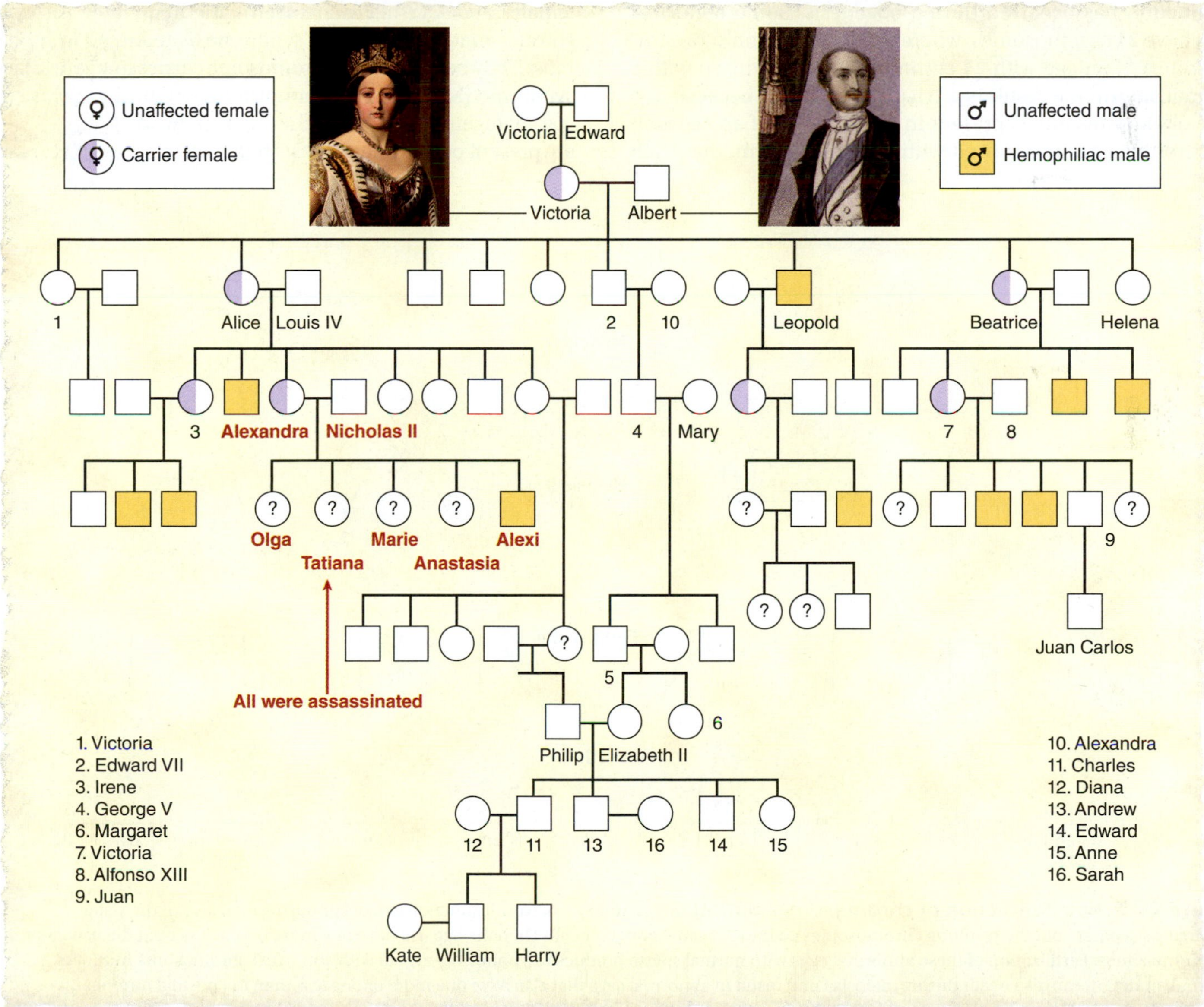
Unaffected female
Carrier female
Unaffected male
Hemophiliac male
Victoria
Edward
Victoria
Albert
1
Alice
Louis IV
2
10
Leopold
Beatrice
Helena
3
Alexandra
Nicholas II
4
Mary
7
8
Olga
Tatiana
Marie
Anastasia
Alexi
9
All were assassinated
Juan Carlos
5
6
Philip
Elizabeth II
12
11
13
16
14
15
Kate
William
Harry
1. Victoria
2. Edward VII
3. Irene
4. George V
6. Margaret
7. Victoria
8. Alfonso XIII
9. Juan
10. Alexandra
11. Charles
12. Diana
13. Andrew
14. Edward
15. Anne
16. Sarah

Figure 24.6 A pedigree showing X-linked inheritance of hemophilia in European royal families. Because Queen Victoria was a carrier, each of her sons had a 50% chance of having the disorder, and each of her daughters had a 50% chance of being a carrier. This pedigree shows only a portion of the European royal families.

24.3 Changes in Chromosome Number

Learning Outcomes

Upon completion of this section, you should be able to

1. Describe how chromosome number disorders arise.
2. Give several examples of chromosome number disorders.

Normally, a human receives 22 pairs of autosomes and two sex chromosomes. Sometimes individuals are born with either too many or too few autosomes or sex chromosomes, most likely due to nondisjunction during meiosis. **Nondisjunction** occurs during meiosis I, when both members of a homologous pair go into the same daughter cell, or during meiosis II, when the sister chromatids fail to separate and both daughter chromosomes go into the same gamete. Figure 24.7 assumes that nondisjunction has occurred during oogenesis. Some abnormal eggs have 24 chromosomes, whereas others have only 22 chromosomes. If an egg with 24 chromosomes is fertilized with a normal sperm, the result is a **trisomy,** so called because one type of chromosome is present in three copies. If an egg with 22 chromosomes is fertilized with a normal sperm, the result is a **monosomy,** so called because one type of chromosome is present in a single copy.

Normal development depends on the presence of exactly two of each kind of chromosome. Too many chromosomes are tolerated better than a deficiency of chromosomes, and several trisomies are known to occur in humans. Among autosomal trisomies, only trisomy 21 (Down syndrome) has a reasonable chance of survival after birth. This is most likely due to the fact that chromosome 21 is one of the smallest chromosomes.

The chances of survival are greater when trisomy or monosomy involves the sex chromosomes. In normal XX females, one of the X chromosomes becomes a darkly stained mass of chromatin called a **Barr body.** A Barr body is an inactive X chromosome. Therefore, we now know that the cells of females function with a single X chromosome just as those of males do. This is most likely the reason that a zygote with one X chromosome (Turner syndrome) can survive. Then, too, all extra X chromosomes beyond a single one become Barr bodies, and this explains why poly-X females and XXY males are seen fairly frequently. An extra Y chromosome, called Jacobs syndrome, is tolerated in humans, most likely because the Y chromosome carries few genes. Jacobs syndrome (XYY) is due to nondisjunction during meiosis II of spermatogenesis. We know this because two Ys are present only during meiosis II in males.

Animation X Inactivation

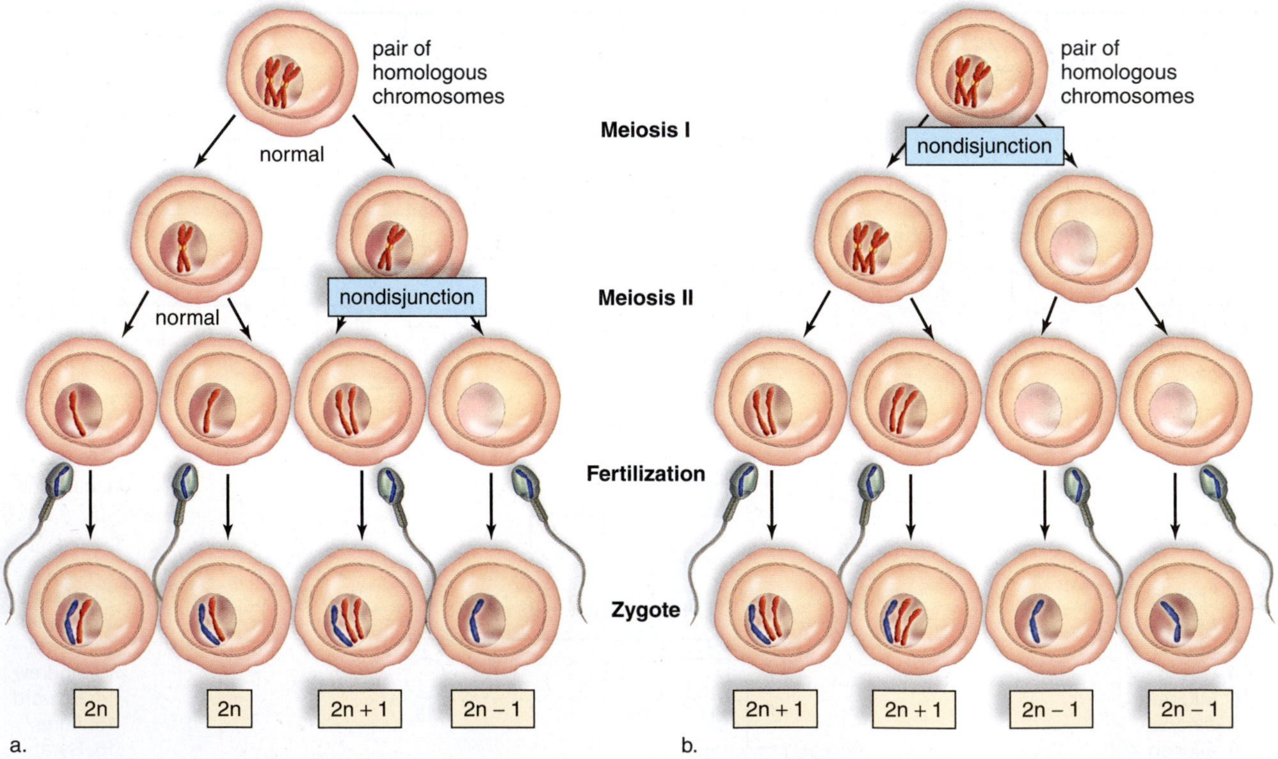
pair of homologous chromosomes
normal
nondisjunction
normal
Meiosis I
Meiosis II
Fertilization
Zygote
pair of homologous chromosomes
nondisjunction
2n
2n
2n + 1
2n – 1
2n + 1
2n + 1
2n – 1
2n – 1
a.
b.

Figure 24.7 Nondisjunction of chromosomes during oogenesis. **a.** Nondisjunction can occur during meiosis II if the sister chromatids separate but the resulting chromosomes go into the same daughter cell. Then the egg will have one more or one less than the usual number of chromosomes. Fertilization of these abnormal eggs with normal sperm produces an abnormal zygote with abnormal chromosome numbers. **b.** Nondisjunction can also occur during meiosis I and result in abnormal eggs that also have one more or one less than the normal number of chromosomes. Fertilization of these abnormal eggs with normal sperm results in a zygote with an abnormal chromosome number. In humans (2n), the normal diploid number of chromosomes is 46. An individual who is 2n + 1 would have 47 chromosomes, whereas an individual who is 2n – 1 would have 45 chromosomes.

Down Syndrome

The most common autosomal trisomy seen among humans is trisomy 21, also called Down syndrome (Fig. 24.8). This syndrome is easily recognized by the following characteristics: short stature; an eyelid fold; a flat face; stubby fingers; a wide gap between the first and second toes; a large, fissured tongue; a round head; a palm crease (called the simian line); and mental impairment, which may range from mild to severe.

Persons with Down syndrome usually have three copies of chromosome 21 because the egg had two copies instead of one. However in around 23% of the cases the extra chromosome 21 came from the sperm. The chances of a woman having a Down syndrome child increase rapidly with age, starting at about age 40, and the reasons for this are still being determined.

Although an older woman is more likely to have a Down syndrome child, most babies with Down syndrome are born to women younger than age 40 because this is the age group having the most babies. A **karyotype,** or a visual inspection of the chromosomes, can detect changes in chromosome number such as occurs with Down syndrome. However, young women are not routinely encouraged to undergo the procedures necessary to get a sample of fetal cells (i.e., amniocentesis or chorionic villus sampling) because the risk of complications is greater than the risk of having a Down syndrome child. Fortunately, a test based on substances in maternal blood can help identify fetuses that may need to be karyotyped. The Scientific Inquiry feature, "Viewing the Chromosomes," on page 488 describes karyotyping, amniocentesis, and chorionic villus sampling.

The genes that cause Down syndrome are located on the long arm of chromosome 21 (Fig. 24.8*b*), and extensive investigative work has been directed toward discovering the specific genes responsible for the characteristics of the syndrome. Thus far, investigators have discovered several genes that may account for various conditions seen in persons with Down syndrome. For example, they have located genes most likely responsible for the increased tendency toward leukemia, cataracts, and accelerated rate of aging. Researchers have also discovered that an extra copy of the *Gart* gene causes an increased level of purines in the blood, a finding associated with mental impairment. One day, it may be possible to control the expression of the *Gart* gene even before birth so that at least this symptom of Down syndrome does not appear.

Changes in Sex Chromosome Number

An abnormal sex chromosome number is the result of inheriting too many or too few X or Y chromosomes. Figure 24.7 can be used to illustrate nondisjunction of the sex chromosomes during oogenesis, if you assume that the chromosomes shown represent X chromosomes. Nondisjunction during oogenesis or spermatogenesis can result in gametes that have too few or too many X or Y chromosomes. After fertilization, the syndromes listed in Table 24A (other than Down syndrome, which is autosomal) are possibilities.

A person with only one sex chromosome, an X (Turner syndrome), is a female, and a person with more than one X chromosome plus a Y (Klinefelter syndrome) is a male. This shows that in humans, the presence of a Y chromosome, not the number of X chromosomes, determines maleness. The *SRY* (*s*ex-determining *r*egion of *Y*) gene, on the short arm of the Y chromosome, produces a hormone called testis-determining factor, which plays a critical role in the development of male genitals.

Video
Why a Guy Is a Guy

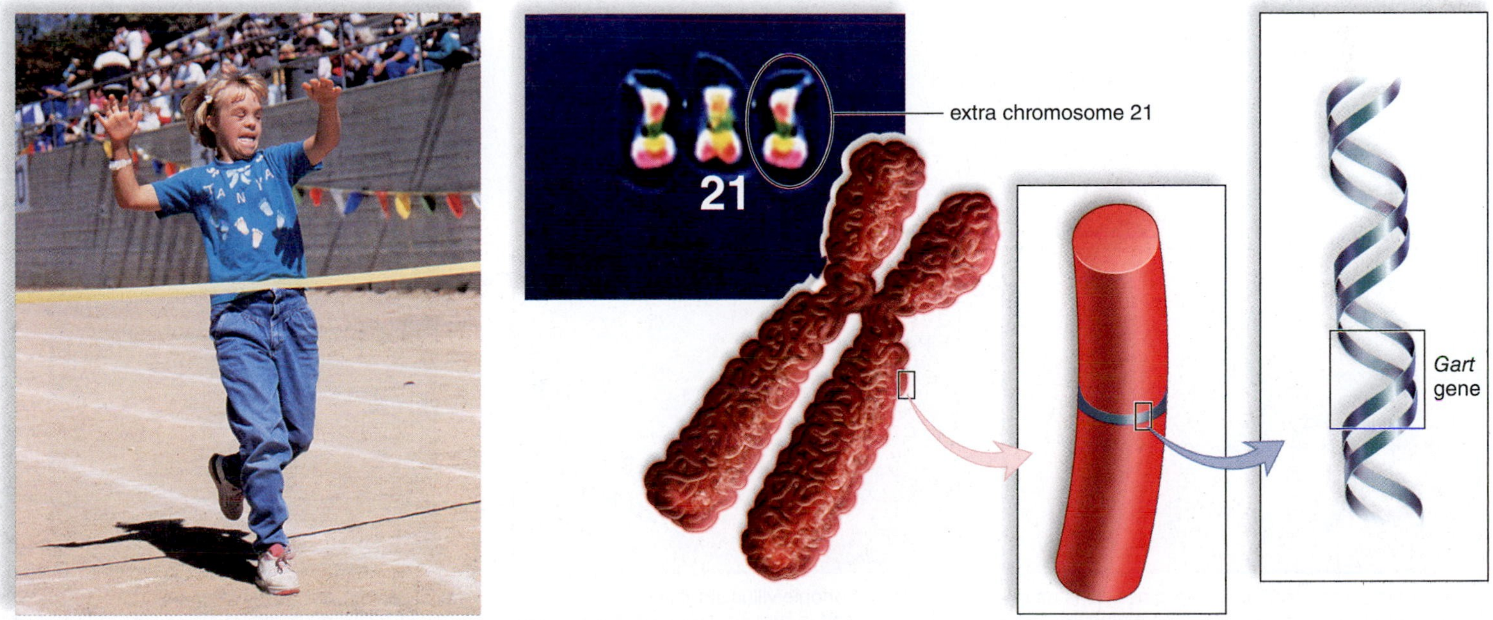
extra chromosome 21
21
Gart gene

a. b.

Figure 24.8 Abnormal autosomal chromosome number. **a.** Persons with Down syndrome have an extra chromosome 21. **b.** Karyotype of an individual with Down syndrome shows an extra chromosome 21. More sophisticated technologies allow investigators to pinpoint the location of specific genes associated with the syndrome. An extra copy of the *Gart* gene, which leads to a high level of purines in the blood, may account for the mental impairment seen in persons with Down syndrome.

SCIENCE IN YOUR LIFE ▶ **SCIENTIFIC INQUIRY**

Viewing the Chromosomes

A karyotype displays the chromosomes by size, shape, and banding pattern. Usually, both human males and females have 23 pairs of chromosomes: 22 pairs are autosomes, and 1 pair is the sex chromosomes. Males have the sex chromosomes X and Y, and females have two X chromosomes.

Various human disorders, or syndromes, result from abnormal chromosome number or structure. Table 24A lists several syndromes that are due to an abnormal chromosome number. A karyotype analysis will reveal such abnormalities.

Any cell in the body, except red blood cells, which lack a nucleus, can be a source of chromosomes for karyotyping. In adults, it is easiest to use white blood cells separated from a blood sample for this purpose. For fetuses, cells can be obtained by either amniocentesis or chorionic villus sampling. These cells can be used for karyotype analysis or for DNA analysis, as described in Chapter 26.

Amniocentesis

Amniocentesis is a procedure for obtaining a sample of amniotic fluid from the uterus of a pregnant woman (Fig. 24B*a*). Amniocentesis is not usually performed until after the fifteenth week of pregnancy. Blood tests and the mother's age are used to determine whether the procedure should be done. The risk of spontaneous abortion increases by about 0.25–0.5% due to amniocentesis, and doctors use the procedure only if it is medically warranted.

Chorionic Villus Sampling

Chorionic villus sampling (*CVS*) is a procedure for obtaining chorionic cells in the region where the placenta will develop (Fig. 24B*b*). This procedure is usually done between the tenth and fourteenth weeks of pregnancy. An ultrasound assists in the sampling of chorionic cells by suction. CVS carries a greater risk of spontaneous abortion (0.5–1%), but it does provide karyotype results at an earlier date.

Video Safer Fetal Test

Karyotyping

For a karyotype analysis, the cell sample is stimulated to divide in a culture medium. Mitosis is stopped during metaphase when chromosomes are the most highly compacted and condensed. The cells are then spread on a microscope slide and dried. Stains, or fluorescent markers, are applied to the slides, and the cells are photographed. The stains and markers produce patterns that can be used, in addition to size and shape, to pair up the chromosomes. Today, a computer is used to arrange the chromosomes in pairs. The karyotype of a person who has Down syndrome usually has three chromosomes 21, instead of the usual two (Fig. 24A*c*).

Questions to Consider

1. Why are cells at metaphase of mitosis used for a karyotype analysis?
2. What other information may be obtained from a karyotype analysis?

TABLE 24A Syndromes from Abnormal Chromosome Numbers

Syndrome	Sex	Chromosomes	Chromosome Number	Frequency	
				Spontaneous Abortions	***Live Births***
Down	M or F	Trisomy 21	47	1/40	1/733
Poly-X	F	XXX (or XXXX)	47 or 48	0	1/1,500
Klinefelter	M	XXY (or XXXY)	47 or 48	1/300	1/800
Jacobs	M	XYY	47	?	1/1,000
Turner	F	X	45	1/18	1/2,500

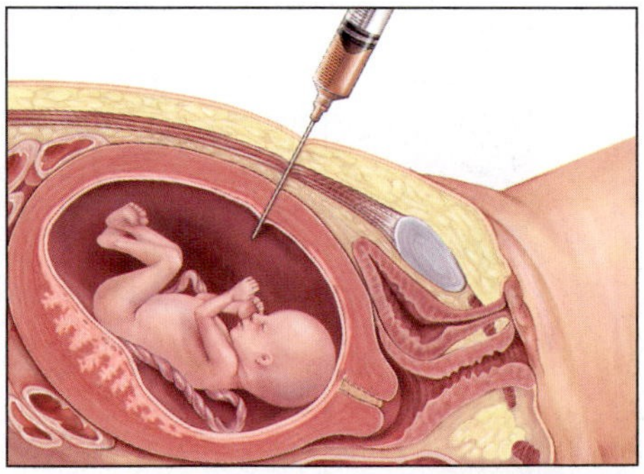

a. During amniocentesis, a long needle is used to withdraw amniotic fluid containing fetal cells.

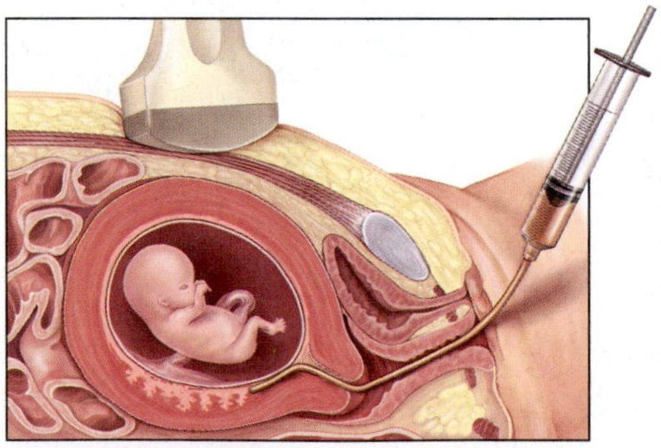

b. During chorionic villus sampling, a suction tube is used to remove cells from the chorion, where the placenta will develop.

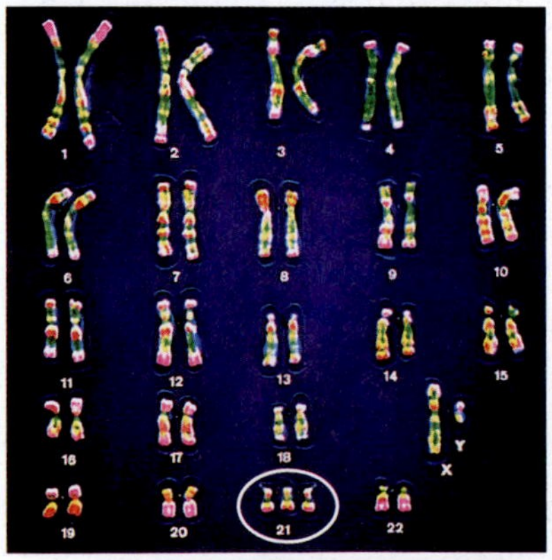
1
2
3
4
5
6
7
8
9
10
11
12
13
14
15
16
17
18
X
Y
19
20
21
22

c. Down syndrome karyotype with an extra chromosome 21

Figure 24B Human karyotype preparation. A karyotype presents chromosomes according to their size, shape, and banding pattern. **a.** Amniocentesis and (**b**) chorionic villus sampling provide cells for karyotyping to determine if the unborn child has a chromosomal abnormality. **c.** Karyotype of a male with Down syndrome, showing three chromosomes 21.

Turner Syndrome Females with Turner syndrome have only one sex chromosome, an X. They are usually short and may have malformed features, such as a webbed neck, high palate, and small jaw. Many have congenital heart and kidney defects. Most have ovarian failure and do not undergo puberty or menstruate without sex hormone replacement therapy. However, pregnancy has been achieved through in vitro fertilization using donor eggs. Women with Turner syndrome have a normal range of intelligence but often have a nonverbal learning disability. They can lead very successful, fulfilling lives if they receive appropriate care.

Klinefelter Syndrome About 1 in 650 males is born with two X chromosomes and one Y chromosome. The symptoms of this condition (referred to as "47, XXY") are often so subtle that only 25% are ever diagnosed, and those are usually not diagnosed until after age 15. Earlier diagnosis opens the possibility for educational accommodations and other interventions that can help mitigate common symptoms, which include speech and language delays. Those 47, XXY males who develop more severe symptoms as adults are referred to as having "Klinefelter syndrome." All 47, XXY adults will require assisted reproduction in order to father children. Affected individuals commonly receive testosterone supplementation beginning at puberty.

Poly-X Females A poly-X female has more than two X chromosomes and thus extra Barr bodies in the nucleus. Females with three X chromosomes (sometimes called triplo-X) have no distinctive phenotype, aside from a tendency to be tall and thin. Although some have delayed motor and language development, most poly-X females are not mentally impaired. Some may have menstrual difficulties, but many menstruate regularly and are fertile. Their children usually have a normal karyotype.

Females with more than three X chromosomes occur rarely. Unlike XXX females, XXXX females are tall and more likely to be mentally impaired. They exhibit various physical abnormalities, but may menstruate normally.

Jacobs Syndrome XYY males, who have Jacobs syndrome, can only result from nondisjunction during spermatogenesis. Affected males are usually taller than average, suffer from persistent acne, and tend to have speech and reading problems. Despite the extra Y chromosome, there is no difference in behavior between XYY males and XY males.

Check Your Progress 24.3

1. Explain why the chances of survival are greater for a trisomy or monosomy of the sex chromosomes than for autosomes.
2. Explain the nondisjunction event that would cause a Turner or Klinefelter syndrome individual.

24.4 Changes in Chromosome Structure

Learning Outcomes

Upon completion of this section, you should be able to

1. Describe a chromosomal deletion, duplication, translocation, and inversion.
2. Summarize the disorders caused by changes in chromosome structure.

Chromosomal mutations occur when chromosomes break. Various environmental agents—radiation, certain organic chemicals, or even viruses—can cause chromosomes to break apart. Ordinarily, when breaks occur, the segments reunite to give the same sequence of genes. But their failure to reunite correctly can result in one of several types of mutations: deletion, duplication, translocation, or inversion. Chromosomal mutations can occur during meiosis, and if the offspring inherits the abnormal chromosome, a syndrome may develop.

Deletions and Duplications

A **deletion** occurs when a single break causes a chromosome to lose an end piece or when two simultaneous breaks lead to the loss of an internal chromosomal segment. An individual who inherits a normal chromosome from one parent and a chromosome with a deletion from the other parent no longer has a pair of alleles for each trait, and a syndrome can result.

Williams syndrome occurs when chromosome 7 loses a tiny end piece (Fig. 24.9). Children who have this syndrome

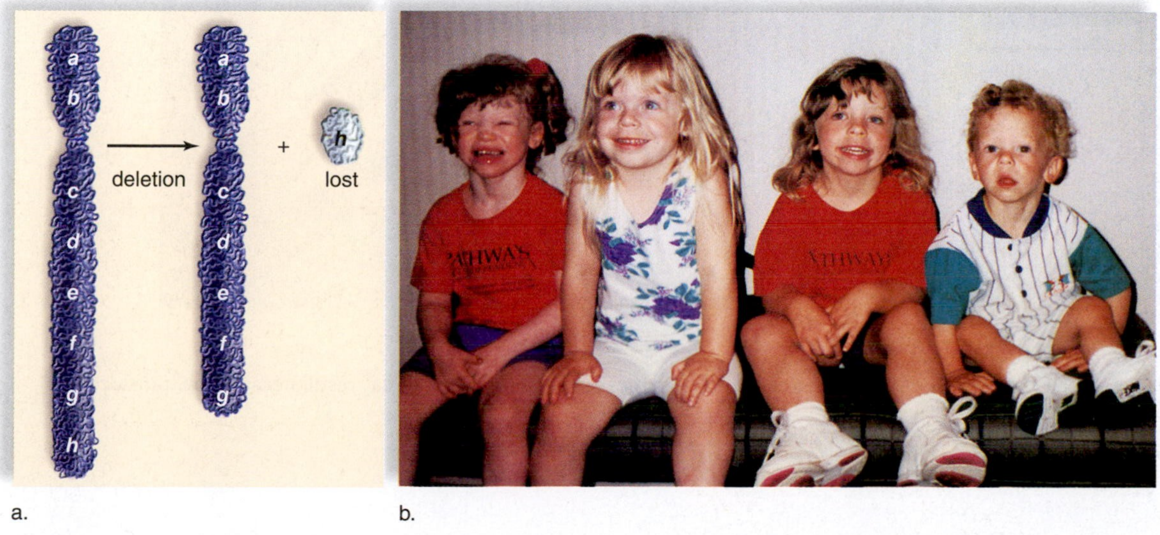
a
b
c
d
e
f
g
h
deletion
+
lost
a.
b.

Figure 24.9 Deletion. **a.** When chromosome 7 loses an end piece, the result is Williams syndrome. **b.** These children, although unrelated, have similar appearance, health, and behavioral problems.

look like pixies because they have turned-up noses, wide mouths, small chins and large ears. Although their academic skills are poor, they exhibit excellent verbal and musical abilities. The gene that governs the production of the protein elastin is missing, which affects the health of the cardiovascular system and causes their skin to age prematurely. Such individuals are very friendly but need an ordered life, perhaps because of the loss of a gene for a protein that is normally active in the brain.

Cri du chat (cat's cry) syndrome is seen when chromosome 5 is missing an end piece. The affected individual has a small head, is mentally impaired, and has facial abnormalities. Abnormal development of the glottis and larynx results in the most characteristic symptom—the infant's cry resembles that of a cat.

In a **duplication,** a chromosomal segment is repeated in the same chromosome or in a nonhomologous chromosome. In any case, the individual has more than two alleles for certain traits. An inverted duplication is known to occur in chromosome 15. **Inversion** means that a segment joins in the direction opposite from normal. Children with this syndrome, called inv dup 15 syndrome, have poor muscle tone, mental impairment, seizures, a curved spine, and autistic characteristics, including poor speech, hand flapping, and lack of eye contact (Fig. 24.10).

The Health feature, "Trinucleotide Repeat Expansion Disorders (TREDs)," on page 484 describes a type of duplication called a trinucleotide repeat expansion. In the case of Huntington disease, the trinucleotide repeat consists of the sequence CAG repeated over and over again. In the case of fragile X syndrome, the trinucleotide repeat is CGG. Persons with duplications over a particular number, which is specific for each disease, exhibit symptoms of the disease. In the case of Huntington disease, sufferers have over 40 copies of the CAG sequence within the huntingtin protein gene.

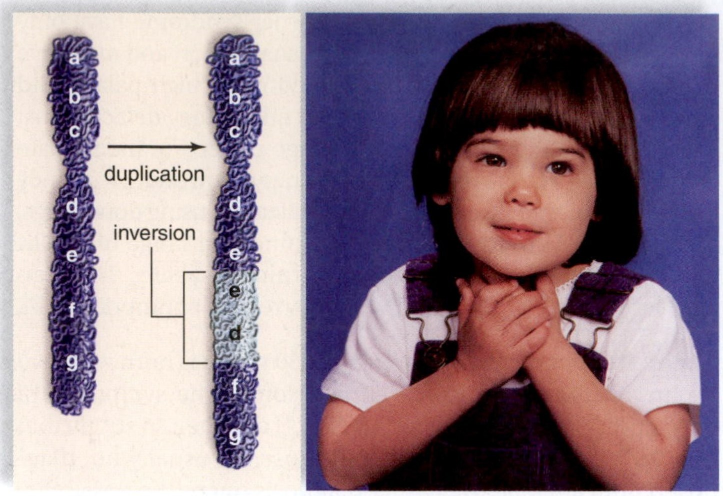
a
b
c
d
e
f
g
duplication
inversion
a
b
c
d
e
e
d
f
g

a. b.

Figure 24.10 Duplication. **a.** When a piece of chromosome 15 is duplicated and inverted, (**b**) a syndrome results that is characterized by poor muscle tone and autistic characteristics.

Translocation

A **translocation** is the exchange of chromosomal segments between two nonhomologous chromosomes. A person who has both of the involved chromosomes has the normal amount of genetic material and is healthy, unless the chromosome exchange breaks an allele into two pieces. The person who inherits only one of the translocated chromosomes will have only one copy of certain alleles and three copies of certain other alleles.

In 5% of cases, a translocation that occurred in a previous generation between chromosomes 21 and 14 is the cause of one type of Down syndrome. The affected person inherits two normal chromosomes 21 and an abnormal chromosome

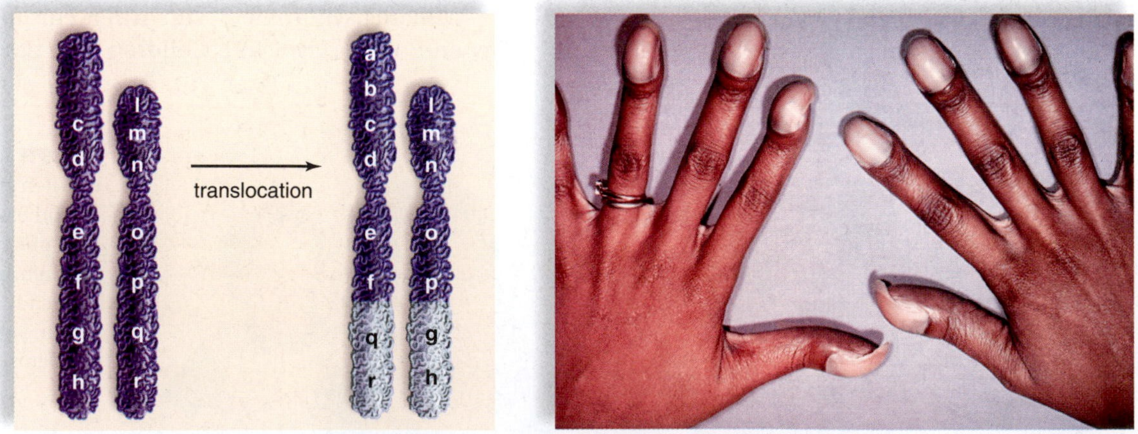
translocation

a. b.

Figure 24.11 Translocation. **a.** When chromosomes 2 and 20 exchange segments, (**b**) Alagille syndrome may result because the translocation disrupts an allele on chromosome 20.

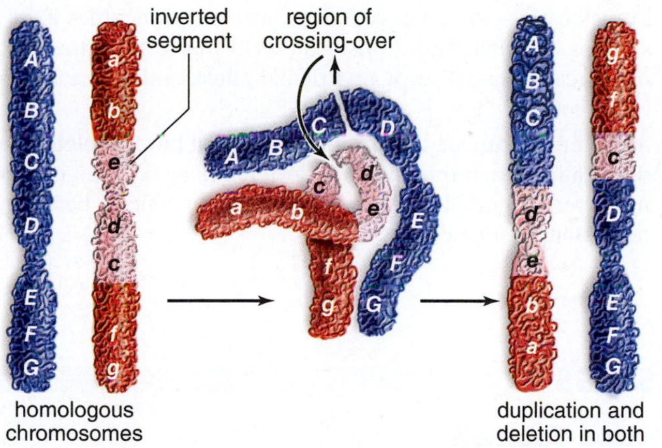
inverted segment
region of crossing-over
homologous chromosomes
duplication and deletion in both

Figure 24.12 Inversion. *Left:* A segment of one homologue is inverted. Notice that in the inverted segment *edc* occurs instead of *cde*. *Middle:* The two homologues can pair only when the inverted sequence forms an internal loop. After crossing-over, a duplication and a deletion can occur. *Right:* The homologue on the left has *AB* and *ba* sequences and neither *gf* nor *FG* genes. The homologue on the right has *gf* and *FG* sequences and neither *AB* nor *ba* genes.

14 that contains a segment of chromosome 21. In these cases, Down syndrome is not related to the age of the mother, but instead tends to run in the family of either the father or the mother.

Figure 24.11 shows the phenotype of an individual with a translocation between chromosomes 2 and 20. Although they have the normal amount of genetic material, the rearrangement of the material produces the phenotypic changes associated with Alagille syndrome. People with this syndrome ordinarily have a deletion on chromosome 20. Therefore, it can be deduced that the translocation disrupted an allele on chromosome 20. The symptoms of Alagille syndrome range from mild to severe, so some people may not be aware they have the syndrome.

Inversion

An inversion occurs when a segment of a chromosome is turned 180 degrees. You might think this is not a problem because the same genes are present, but the reverse sequence of alleles can lead to altered gene activity.

Crossing-over between an inverted chromosome and the noninverted homologue can lead to recombinant chromosomes that have both duplicated and deleted segments. This happens because alignment between the two homologues is only possible when the inverted chromosome forms a loop (Fig. 24.12).

Check Your Progress 24.4

1. Explain why a duplication on one chromosome is usually associated with a deletion on the corresponding homologous chromosome.
2. Explain how it is possible for a person with a translocation or an inversion to be phenotypically normal.

Case Study Conclusion

Research on the causes of Down syndrome has indicated that 95% of Down syndrome cases are a result of nondisjunction of chromosome 21, resulting in three copies of this chromosome. However, we now also know that some cases are caused by translocations. Because of this research, scientists have been able to identify specific genes that are associated with the symptoms of the syndrome. For Down syndrome individuals, such as Brandon, these medical advances have ensured that he can lead a long, productive life.

MEDIA STUDY TOOLS

www.mhhe.com/maderinquiry14

Enhance your study of this chapter with study tools and practice tests. Also ask your instructor about the resources available through ConnectPlus, including LearnSmart, the media-rich eBook, interactive learning tools, and animations.

SUMMARIZE

24.1 Gene Linkage

- All the genes on one chromosome form a **linkage group,** which is broken only when crossing-over occurs. Genes that are linked tend to go together into the same gamete.
- If crossing-over occurs, a dihybrid cross yields all possible phenotypes among the offspring, but the expected ratio is greatly changed because recombinant phenotypes are reduced in number.

24.2 Sex-Linked Inheritance

In humans, there are 22 pairs of **autosomes** (1 to 22), and one pair of **sex chromosomes** (X and Y). Some traits are **sex-linked,** meaning that although they do not determine gender, they are carried on the sex chromosomes. Most of the alleles for these traits are carried on the X chromosome (**X-linked**), whereas the Y does not bear alleles for those same traits.

- The phenotypic results of sex-linked crosses are given for females and males separately.
- An X-linked recessive pedigree shows why more males than females have these disorders. Females are carriers of the disease.

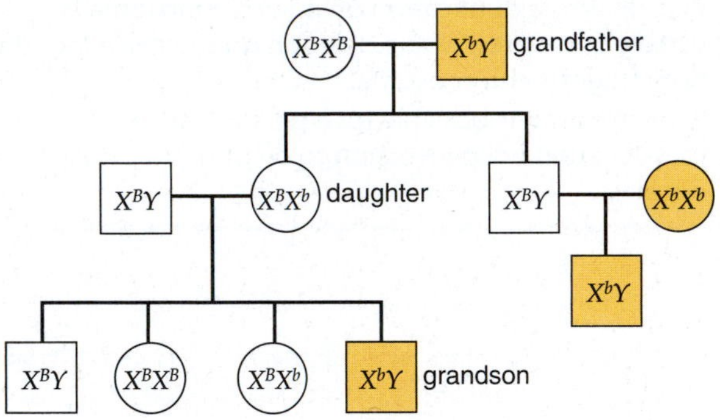
X^BX^B
X^bY grandfather
X^BY
X^BX^b daughter
X^BY
X^bX^b
X^bY
X^BY
X^BX^B
X^BX^b
X^bY grandson

- **Color blindness, Duchenne muscular dystrophy, fragile X syndrome,** and **hemophilia** are all X-linked recessive disorders.

24.3 Changes in Chromosome Number

- **Nondisjunction** during meiosis can result in an abnormal number of autosomes or sex chromosomes in the gametes. In a **trisomy,** individuals have an extra chromosome, whereas in a **monosomy** they are missing a chromosome. If nondisjunction occurs with the X chromosome, extra X chromosomes may be inactivated, forming **Barr bodies.** Monosomies and trisomies are often detected by **karyotype** analysis.
- Down syndrome results when an individual inherits three copies of chromosome 21.
- Females who possess a single X have Turner syndrome, and those who are XXX are poly-X females. Males with Klinefelter syndrome are XXY. Males who are XYY have Jacobs syndrome.

24.4 Changes in Chromosome Structure

- Changes in chromosomal structure also affect the phenotype. **Chromosomal mutations** include **deletions, duplications, translocations,** and **inversions.**
- Translocations do not necessarily cause any difficulties if the person has inherited both translocated chromosomes. However, the translocation can disrupt a particular allele, and then a syndrome will follow.
- An inversion can lead to chromosomes that have a deletion and a duplication when the inverted piece loops back to align with the noninverted homologue, and crossing-over follows between the nonsister chromatids.

ASSESS

Testing Yourself

Choose the best answer for each question.

1. The following pedigree pertains to color blindness. Using the letter *B* for the normal allele, what is the genotype of the individual with the asterisk?

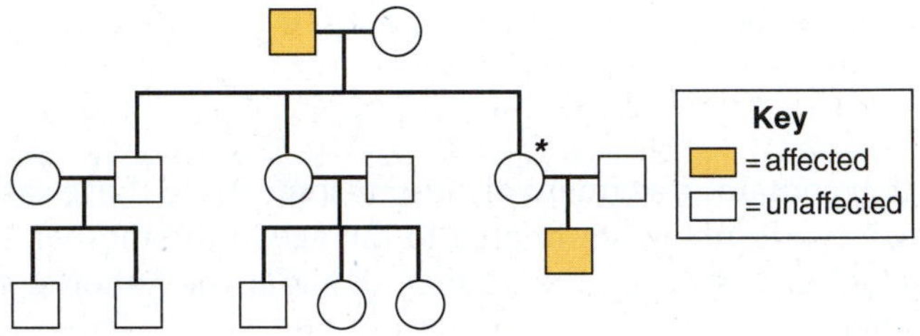
*
Key
= affected
= unaffected

   a. $X^B X^B$
   b. $X^B X^b$
   c. $X^b X^b$

2. All the genes on one chromosome are said to form a
   a. chromosomal group.
   b. recombination group.
   c. linkage group.
   d. crossing-over group.
3. If alleles *R* and *s* are linked on one chromosome and *r* and *S* are linked on the homologous chromosome, what gametes will be produced? (Assume that no crossing-over occurs.)
   a. *RS, Rs, rS, rs*
   b. *RS, rs*
   c. *Rs, rS*
   d. *R, S, r, s*
4. An individual can have too many or too few chromosomes as a result of
   a. nondisjunction.
   b. Barr bodies.
   c. trisomy.
   d. amniocentesis.
   e. monosomy.
5. Assume that two parents with normal vision have a color-blind son. Which parent is responsible for the son's color blindness?
   a. the mother
   b. the father
   c. either parent
   d. None of these are correct because two normal parents cannot have a color-blind son.

For questions 6–10, match the chromosome disorder to its description in the key.

Key:

- **a.** female with undeveloped ovaries and uterus, unable to undergo puberty, normal intelligence, can live normally with hormone replacement
- **b.** XXY male, can inherit more than two X chromosomes
- **c.** male or female, mentally impaired, short stature, flat face, stubby fingers, large tongue, simian palm crease
- **d.** XXX or XXXX female
- **e.** caused by nondisjunction during spermatogenesis

6. Klinefelter syndrome
7. Poly-X female
8. Down syndrome
9. Turner syndrome
10. Jacobs syndrome

For questions 11–14, match the chromosomal mutation to its description in the key.

Key:

- **a.** turned-up nose, wide mouth, small chin, large ears, poor academic skills, excellent verbal and musical abilities, prematurely aging cardiovascular system
- **b.** deletion in chromosome 5
- **c.** poor muscle tone, mental impairment, seizures, curved spine, autistic characteristics, poor speech, hand flapping, lack of eye contact
- **d.** translocation between chromosomes 2 and 20

11. Alagille syndrome
12. Inv dup 15 syndrome
13. Williams syndrome
14. Cri du chat syndrome
15. A karyotype is prepared
    - **a.** for an unborn child through amniocentesis or chorionic villus sampling.
    - **b.** by arranging chromosomes into pairs by size, shape, and banding pattern.
    - **c.** as a means of diagnosing an abnormal chromosomal disorder.
    - **d.** using a photograph of a cell sample arrested during metaphase.
    - **e.** All of these are correct.

Additional Genetics Problems

1. What phenotypic ratio is expected for a cross in which both parents have one X-linked dominant allele?
2. Both the mother and the father of a son with hemophilia appear to be normal. From whom did the son inherit the allele for hemophilia? What are the genotypes of the mother, the father, and the son?
3. A woman is color blind. What are the chances that her sons will be color blind? If she is married to a man with normal vision, what are the chances that her daughters will be color blind? Will be carriers?
4. Both the husband and wife have normal vision. The wife gives birth to a color-blind daughter. Is it more likely the father had normal vision or is color blind? What does this lead you to deduce about the girl's parentage?

ENGAGE

Virtual Lab
Sex-Linked Traits

The virtual lab "Sex-Linked Traits" provides you with the ability to test your knowledge of sex-linked patterns of inheritance.

Thinking Critically

1. Why might expectant parents want to undergo fetal genetic testing if they know that, regardless of the findings, they plan to have the baby?
2. Why do you think there are no viable trisomies of chromosome 1?
3. Why can a person carrying a translocation be normal except for the inability to have children?

25 DNA Structure and Gene Expression

CHAPTER OUTLINE

BEFORE YOU BEGIN

Before beginning this chapter, take a few moments to review the following discussions:

Figure 2.26 What are the three components of a nucleotide?

Section 2.8 How does the structure of DNA differ from that of RNA?

Section 23.1 What is the relationship between the genotype and phenotype?

CASE STUDY Xeroderma Pigmentosum

Sometimes they are called "children of the night" because they cannot play outside in the sunlight. They even have a summer camp named Camp Sundown in which they do all the things that other children do at summer camp such as hiking, horseback riding, and swimming, except they do it after sundown. Whenever these children are outside during daylight, they wear protective clothing and eyewear, use sunblock, and travel in cars with tinted windows. These are children that suffer from a genetic disease called xeroderma pigmentosum (XP). This is a genetic disease in which the enzymes that are needed to repair DNA damage due to ultraviolet (UV) light are defective. Therefore, those who suffer with XP cannot be exposed to UV light. XP is very rare, only about 1 in a million individuals in the United States have XP, although those with Japanese descent have a higher incidence. There is no cure. DNA damage cannot be repaired and mutations accumulate throughout the lifetime of the patient. These individuals have a 1,000-fold higher risk of skin cancer than those with normal DNA repair enzymes. Exposure to UV light results in blistering and freckling, and individuals suffer from premature aging of the skin along with eye tumors. More severe cases of XP may result in progressive neurological complications such as intellectual disability and hearing loss. Fewer than 40% of individuals with XP survive beyond the age of 20, although those with milder cases may survive into middle age. The most common cause of death is skin cancer, either squamous cell carcinoma or metastatic melanoma. This chapter describes the structure of DNA and how mutations could affect the function of RNA and proteins. It also describes the progression of cancer from a single mutation to a metastatic tumor.

As you read through the chapter, think about the following questions:

1. How is the information in the DNA interpreted into a functional protein, such as an enzyme?
2. How might a mutation in the DNA result in the formation of cancer?

25.1 DNA Structure

Learning Outcomes

Upon completion of this section, you should be able to
1. Summarize the experiments that enabled scientists to determine that DNA made up the genetic material.
2. Describe the structure of a DNA molecule.

The middle of the twentieth century was an exciting period of scientific discovery. On one hand, geneticists were busy determining that **DNA (deoxyribonucleic acid)** is the genetic material of life. On the other hand, biochemists were in a frantic race to describe the structure of DNA. The classic experiments performed during this era set the stage for an explosion in our knowledge of modern molecular biology.

When researchers began their work, they knew that the genetic material must be (1) able to store information that pertains to the development, structure, and metabolic activities of the cell or organism; and (2) stable so that it can be replicated with high accuracy during cell division and be transmitted from generation to generation.

In this section, we will explore how the structure of a DNA molecule satisfies both of these requirements.

The Nature of the Genetic Material

During the late 1920s, the bacteriologist Frederick Griffith was attempting to develop a vaccine against a form of bacteria (*Streptococcus pneumoniae*) that causes pneumonia in mammals. In 1931, he performed a classic experiment with the bacterium. He noticed that when these bacteria are grown on culture plates, some, called S strain bacteria, produce shiny, smooth colonies, and others, called R strain bacteria, produce colonies that have a rough appearance. Under the microscope, S strain bacteria have a capsule (mucous coat) that makes them smooth, but R strain bacteria do not.

When Griffith injected mice with the S strain of bacteria, the mice died (Fig. 25.1*a*), and when he injected mice with the R strain, the mice did not die (Fig. 25.1*b*). In an effort to determine whether the capsule alone was responsible for the virulence (ability to kill) of the S strain bacteria, he injected mice with heat-killed S strain bacteria. The mice did not die (Fig. 25.1*c*).

Finally, Griffith injected the mice with a mixture of heat-killed S strain and live R strain bacteria. Most unexpectedly, the mice died—and living S strain bacteria were recovered from the bodies! Griffith concluded that some substance necessary for the bacteria to produce a capsule and be virulent must have passed from the dead S strain bacteria to the living R strain bacteria, thus transforming the R strain bacteria (Fig. 25.1*d*). This change in the phenotype of the R strain bacteria must be due to a corresponding change in their genotype. Thus, the transforming substance was potentially the genetic material. Reasoning such as this prompted investigators at the time to begin looking for the transforming substance to determine the chemical nature of the genetic material.

By the 1940s, scientists recognized that genes are on chromosomes and that chromosomes contain both proteins and nucleic acids. However, there was some debate about whether protein or DNA was the genetic material. Many thought that the protein component of chromosomes must be the genetic material because proteins contain up to 20 different amino acids that can be sequenced in any particular way. On the other hand, nucleic acids—DNA and RNA—contain only four types of nucleotides as basic building blocks. Some argued that DNA did not have enough variability to be able to store information and be the genetic material.

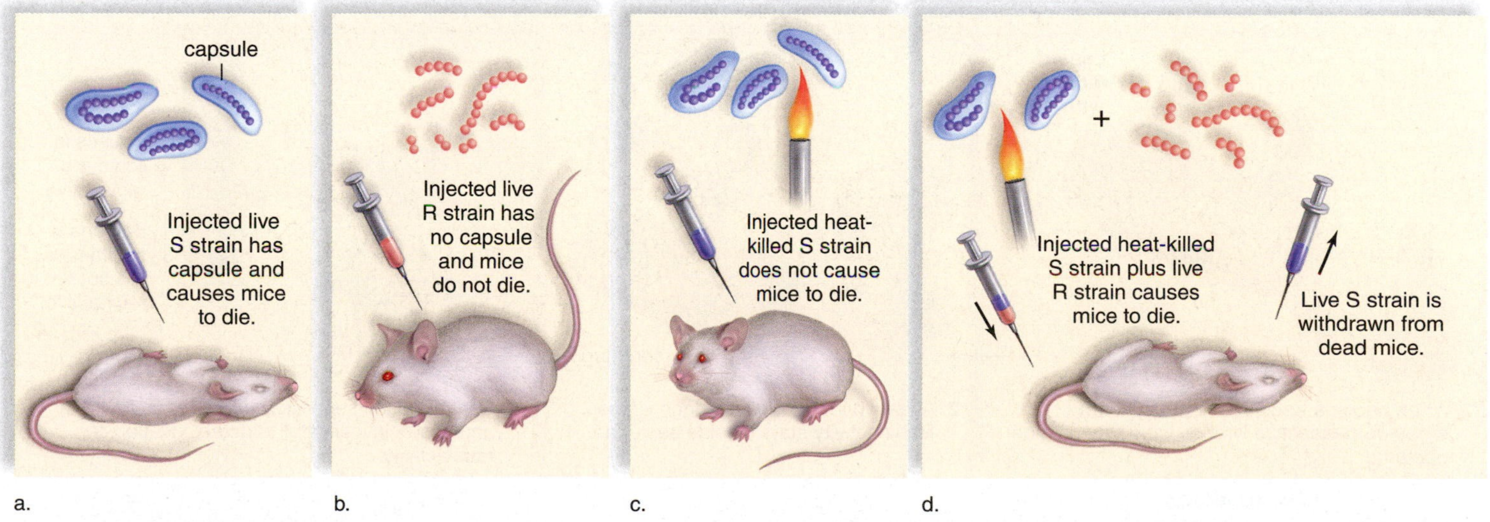
capsule
Injected live S strain has capsule and causes mice to die.
Injected live R strain has no capsule and mice do not die.
Injected heat-killed S strain does not cause mice to die.
+
Injected heat-killed S strain plus live R strain causes mice to die.
Live S strain is withdrawn from dead mice.
a.
b.
c.
d.

Figure 25.1 Griffith's experiment. **a.** Encapsulated S strain is virulent and kills mice. **b.** Nonencapsulated R strain is not virulent and does not kill mice. **c.** Heat-killed S strain bacteria do not kill mice. **d.** If heat-killed S strain and R strain are both injected into mice, they die because the R strain bacteria have been transformed into the virulent S strain.

In 1944, after 16 years of research, Oswald Avery and his coinvestigators, Colin MacLeod and Maclyn McCarty, published a paper demonstrating that the transforming substance that allows *S. pneumoniae* to produce a capsule and be virulent is DNA. This meant that DNA is the genetic material. Here is what they found out:

1. DNA from S strain bacteria causes R strain bacteria to be transformed so that they can produce a capsule and be virulent.
2. The addition of DNase, an enzyme that digests DNA, prevents transformation from occurring. This supports the hypothesis that DNA is the genetic material.
3. The molecular weight of the transforming substance is large. This suggests the possibility of genetic variability.
4. The addition of enzymes that degrade proteins has no effect on the transforming substance nor does RNase, an enzyme that digests RNA. This shows that neither protein nor RNA is the genetic material.

These experiments showed that DNA is the transforming substance and, therefore, the genetic material. Although some scientists remained skeptical, many felt that the evidence for DNA being the genetic material was overwhelming.

An experiment by Alfred Hershey and Martha Chase in the early 1950s helped to firmly establish DNA as the genetic material. Hershey and Chase used a virus called a T phage, composed of radioactively labeled DNA and capsid coat proteins, to infect *Escherichia coli* bacteria (Fig 25.2). They discovered that the radioactive tracers for DNA, but not protein, ended up inside the bacterial cells, causing them to become transformed. As only the genetic material could have caused this transformation, Hershey and Chase determined that DNA must be the genetic material.

Animation Hershey-Chase Experiment

Structure of DNA

The structure of DNA was determined by James Watson and Francis Crick in the early 1950s. The data they used and

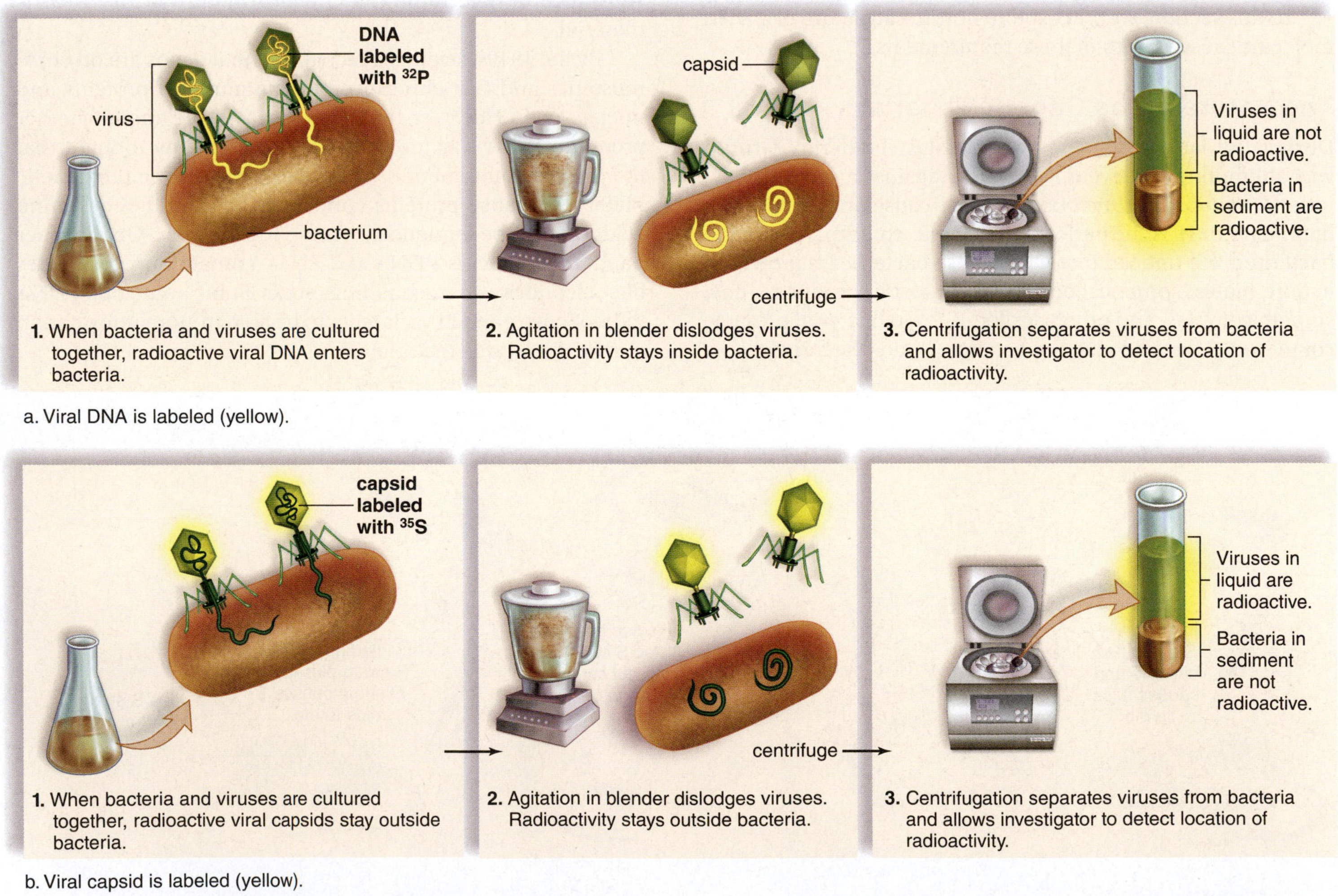
virus
DNA labeled with ^{32}P
bacterium
capsid
centrifuge
Viruses in liquid are not radioactive.
Bacteria in sediment are radioactive.
1. When bacteria and viruses are cultured together, radioactive viral DNA enters bacteria.
2. Agitation in blender dislodges viruses. Radioactivity stays inside bacteria.
3. Centrifugation separates viruses from bacteria and allows investigator to detect location of radioactivity.
a. Viral DNA is labeled (yellow).
capsid labeled with ^{35}S
centrifuge
Viruses in liquid are radioactive.
Bacteria in sediment are not radioactive.
1. When bacteria and viruses are cultured together, radioactive viral capsids stay outside bacteria.
2. Agitation in blender dislodges viruses. Radioactivity stays outside bacteria.
3. Centrifugation separates viruses from bacteria and allows investigator to detect location of radioactivity.
b. Viral capsid is labeled (yellow).

Figure 25.2 Hershey-Chase experiments. These experiments concluded that viral DNA, not protein, was responsible for directing the production of new viruses.

how they interpreted the data to deduce DNA's structure are reviewed in the Scientific Inquiry feature, "Finding the Structure of DNA," on page 498.

DNA is a chain of nucleotides. Each nucleotide is a complex of three subunits—phosphoric acid (phosphate), a pentose sugar (deoxyribose), and a nitrogen-containing base. There are four possible bases: two are **purines** with a double ring, and two are **pyrimidines** with a single ring. Adenine (A) and guanine (G) are purines; thymine (T) and cytosine (C) are pyrimidines.

A DNA polynucleotide *strand* has a backbone made up of alternating phosphate and sugar molecules. The bases are attached to the sugar but project to one side. DNA has two such strands, and the two strands twist about one another in the form of a **double helix** (Fig. 25.3*a*). The strands are held together by hydrogen bonding between the bases: A always pairs with T by forming two hydrogen bonds, and G always pairs with C by forming three hydrogen bonds. Notice that a purine is always bonded to a pyrimidine. This is called **complementary base pairing.** When the DNA helix unwinds, it resembles a ladder (Fig. 25.3*b*). The sides of the ladder are the sugar-phosphate backbones, and the rungs of the ladder are the complementary paired bases.

The two DNA strands are *antiparallel,* meaning that they are oriented in opposite directions, which you can verify by noticing that the sugar molecules are oriented differently. The carbon atoms in a sugar molecule are numbered, and the fifth carbon atom (5′) is uppermost in the strand on the left, while the third carbon atom (3′) is uppermost in the strand on the right (Fig. 25.3*c*).

Animation DNA Structure

Check Your Progress 25.1

1. Summarize the significance of the Griffith and Avery experiments.
2. Explain how, at the completion of the Hershey-Chase experiment, the results suggested that DNA was the genetic material.
3. Describe the structure of the DNA molecule.

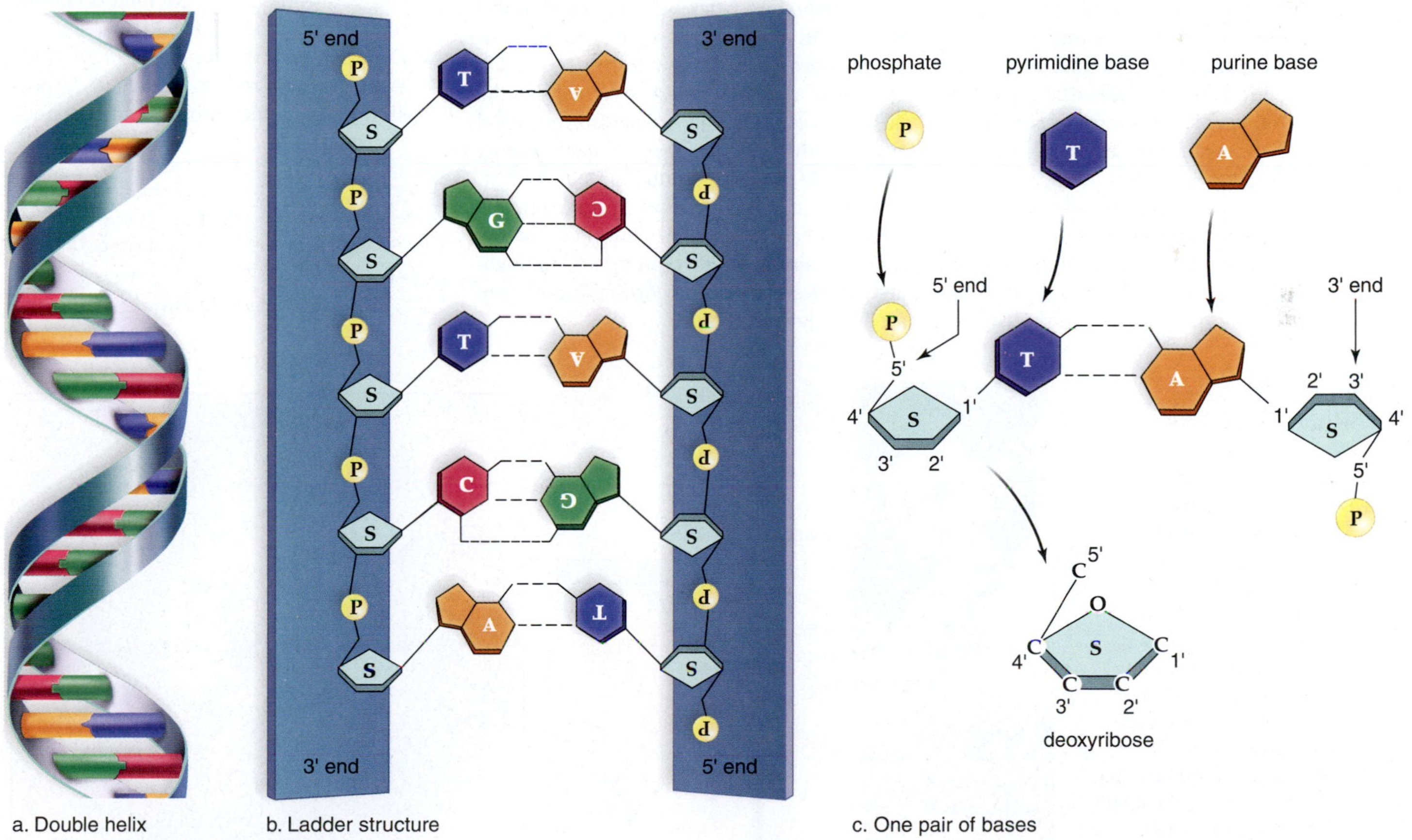
5' end
3' end
3' end
5' end
phosphate
pyrimidine base
purine base
P
S
T
A
G
C
5'
4'
3'
2'
1'
O
deoxyribose
a. Double helix
b. Ladder structure
c. One pair of bases

Figure 25.3 Overview of DNA structure. **a.** DNA double helix. **b.** Unwinding the helix reveals a ladder configuration in which the uprights are composed of sugar and phosphate molecules and the rungs are complementary bases. The bases in DNA pair in such a way that the sugar-phosphate backbones are oriented in different directions. **c.** Notice that 3′ and 5′ are part of the system for numbering the carbon atoms that make up the sugar.

SCIENCE IN YOUR LIFE ▶ SCIENTIFIC INQUIRY

Finding the Structure of DNA

In 1953, James Watson, an American biologist, and Francis Crick, a British physicist, set out not only to determine the structure of DNA, but also to build a model that would explain how DNA, the genetic material, can vary from species to species and even from individual to individual. In the process, they also proposed how DNA replicates (makes a copy of itself) so that daughter cells receive an identical copy.

The bits and pieces of data available to Watson and Crick were like puzzle pieces they had to fit together. *This is what they knew from the research of others:*

1. DNA is a polymer of nucleotides, each one having a phosphate group, the sugar deoxyribose, and a nitrogen-containing base. There are four types of nucleotides because there are four different bases: adenine (A) and guanine (G) are purines, while cytosine (C) and thymine (T) are pyrimidines.
2. A chemist, Erwin Chargaff, had determined in the late 1940s that regardless of the species under consideration, the number of purines in DNA always equals the number of pyrimidines. Further, the amount of adenine equals the amount of thymine (A = T), and the amount of guanine equals the amount of cytosine (G = C). These findings came to be known as Chargaff's rules.
3. Rosalind Franklin (Fig. 25A*a*), working with Maurice Wilkins at King's College, London, had just prepared an X-ray diffraction photograph of DNA (Fig. 25A*b*, *c*). It showed that DNA is a double helix of constant diameter and that the bases are regularly stacked on top of one another.

Video DNA Dark Lady

Using these data, Watson and Crick deduced that DNA has a twisted, ladderlike structure. The sugar-phosphate molecules make up the sides of the ladder, and the bases make up the rungs. The double helices of the DNA molecule are antiparallel, meaning that they are oriented in opposite directions. Further, they determined that if A is normally hydrogen-bonded with T, and G is normally hydrogen-bonded with C (Chargaff's rules), then the rungs always have a constant width, consistent with the X-ray photograph.

Watson and Crick built an actual model of DNA out of wire and tin. This double-helix model does indeed allow for differences in DNA structure between species because the base pairs can be in any order. Also, the model suggests that complementary base pairing plays a role in the replication of DNA. As Watson and Crick pointed out in their original paper, "It has not escaped our notice that the specific pairing we have postulated immediately suggests a possible copying mechanism for the genetic material."

3D Animation DNA Replication: DNA Structure

When the Nobel Prize for the discovery of the double helix was awarded in 1962, the honor went to James Watson, Francis Crick, and Maurice Wilkins. Rosalind Franklin was not listed as one of the recipients. Tragically, Franklin developed ovarian cancer in 1956, and she died in 1958 at the age of 37. Franklin had not been nominated for the award, and according to the rules at that time, was ineligible for the Nobel Prize.

Questions to Consider

1. Based on Chargaff's rules, if a segment of DNA is composed of 20% adenine (A) bases, what is the percentage of guanine (G)?
2. Watson and Crick's discovery of DNA is clearly one of the most important biological discoveries in the last century. What advances in medicine and science can you think of that are built on knowing the structure of DNA?
3. Describe why the structure of DNA led Watson and Crick to point out "a possible copying mechanism for the genetic material."

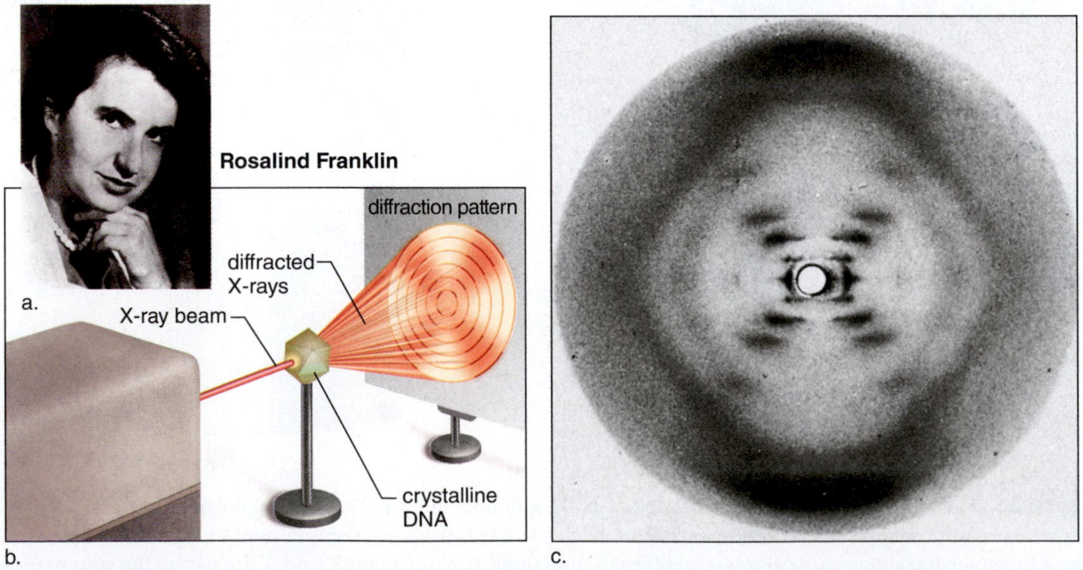
Rosalind Franklin
a.
diffraction pattern
diffracted X-rays
X-ray beam
crystalline DNA
b.
c.

Figure 25A X-ray diffraction of DNA. **a.** Rosalind Franklin, 1920–1958. **b.** The X-ray diffraction system used by Franklin. **c.** The diffraction pattern of DNA produced by Rosalind Franklin. The crossed (X) pattern in the center told investigators that DNA is a helix, and the dark portions at the top and the bottom told them that some feature is repeated over and over. Watson and Crick determined that this feature was the result of the hydrogen-bonded bases.

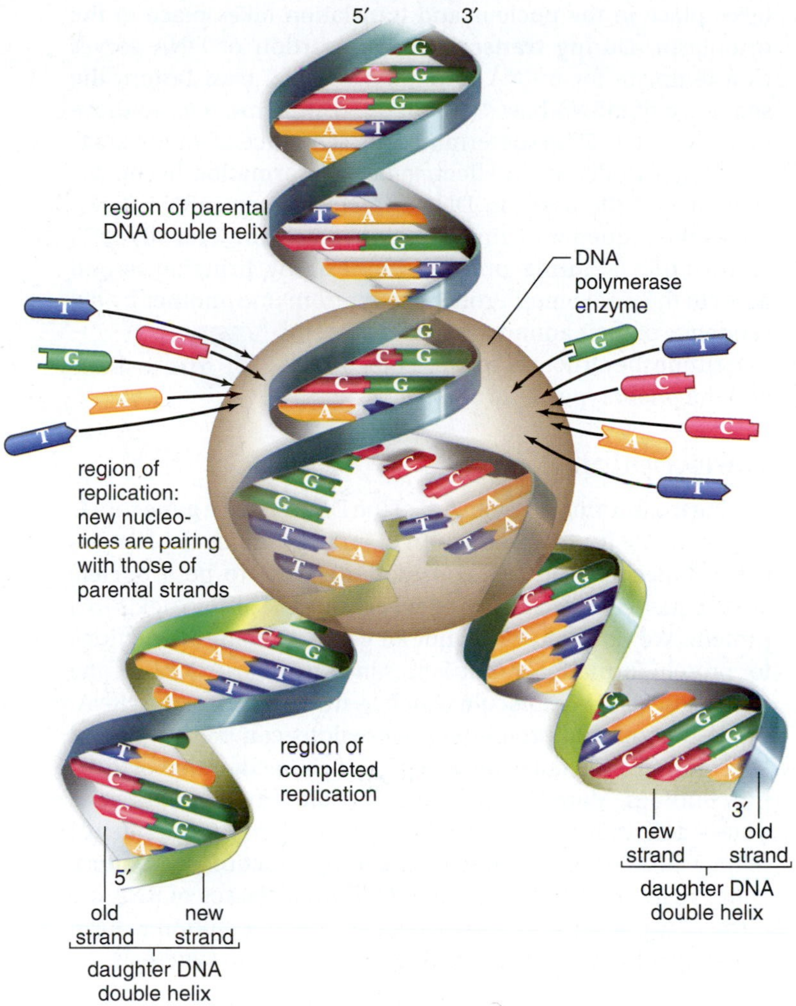
5′
3′
region of parental DNA double helix
DNA polymerase enzyme
region of replication: new nucleotides are pairing with those of parental strands
region of completed replication
3′
new strand
old strand
daughter DNA double helix
5′
old strand
new strand
daughter DNA double helix

Figure 25.4 Overview of DNA replication. Replication is called semiconservative because each new double helix is composed of an old (parental) strand and a new (daughter) strand.

25.2 DNA Replication

Learning Outcomes

Upon completion of this section, you should be able to

1. Explain why DNA replication is semiconservative.
2. Summarize the events that occur during the process of DNA replication.

When the body grows or heals itself, cells divide. Each new cell requires an exact copy of the DNA contained in the chromosomes. The process of copying one DNA double helix into two identical double helices is called **DNA replication.** The process of DNA replication is carried out by an enzyme called DNA polymerase (Fig. 25.4). The DNA polymerase uses each original strand as a template for the formation of a complementary new strand. As a result, DNA replication is termed *semiconservative* because a new double helix has one conserved old strand and one new strand. Replication results in two DNA helices that are identical to each other and to the original molecule.

Animation Meselson and Stahl Experiment

Animation DNA Replication

At the molecular level, several enzymes and proteins participate in the synthesis of the new DNA strands. This process is summarized in Figure 25.5:

1. The enzyme *DNA helicase* unwinds and "unzips" the double-stranded DNA by breaking the weak hydrogen bonds between the paired bases.
2. New complementary DNA nucleotides, always present in the nucleus, fit into place by the process of complementary base pairing. These are positioned and joined by the enzyme *DNA polymerase.*
3. Because the strands of DNA are oriented in an antiparallel configuration, and the DNA polymerase may add new nucleotides only to one end of the chain, DNA synthesis

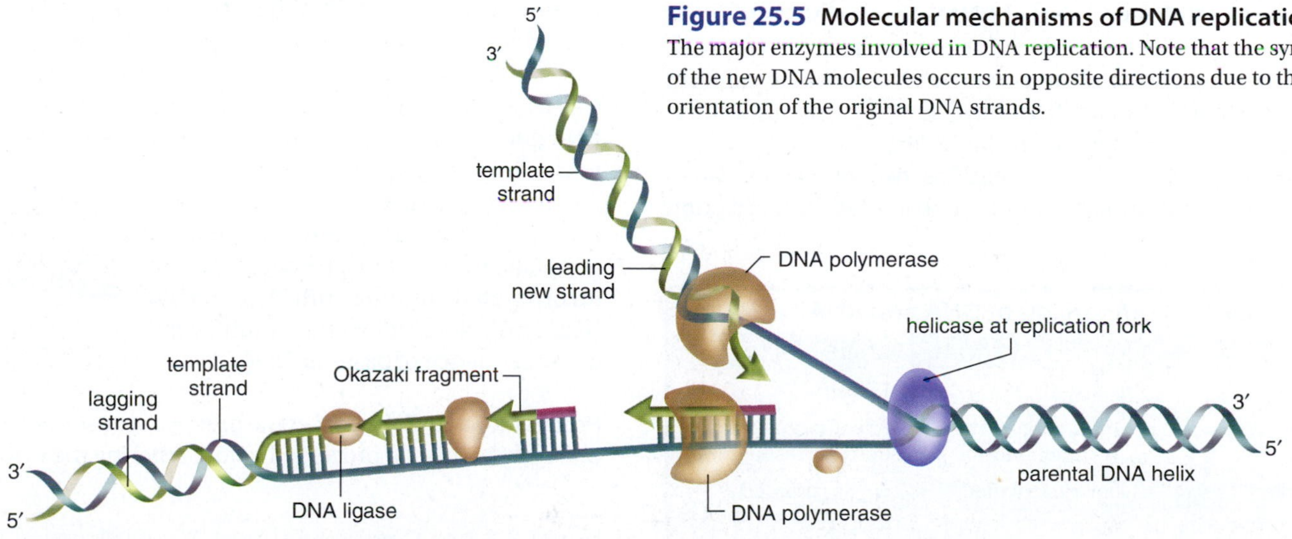
5′
3′
template strand
leading new strand
DNA polymerase
helicase at replication fork
3′
5′
parental DNA helix
template strand
lagging strand
Okazaki fragment
3′
5′
DNA ligase
DNA polymerase

Figure 25.5 Molecular mechanisms of DNA replication. The major enzymes involved in DNA replication. Note that the synthesis of the new DNA molecules occurs in opposite directions due to the orientation of the original DNA strands.

occurs in opposite directions. The *leading strand* follows the helicase enzyme, while synthesis on the *lagging strand* results in the formation of short segments of DNA called *Okazaki fragments.*

4. To complete replication, the enzyme *DNA ligase* connects the Okazaki fragments and seals any breaks in the sugar-phosphate backbone.
5. The two double helix molecules are identical to each other and to the original DNA molecule.

Chemotherapeutic drugs for cancer treatment stop replication and, therefore, cell division. Some chemotherapeutic drugs are analogs that have a similar, but not identical, structure to one of the four nucleotides in DNA. When these are mistakenly used by the cancer cells to synthesize DNA, replication stops, and the cancer cells die off.

3D Animation DNA Replication

> **Check Your Progress 25.2**
>
> 1. Explain why DNA replication is said to be semiconservative.
> 2. Summarize the sequence of events that occur during DNA replication.

25.3 Gene Expression

> **Learning Outcomes**
>
> Upon completion of this section, you should be able to
>
> 1. Describe the roles of RNA molecules in gene expression.
> 2. Summarize the sequence of events that occurs during gene expression.
> 3. Determine the sequence of amino acids in a peptide, given the messenger RNA sequence.
> 4. Explain the purpose of mRNA processing.

The process of using a gene sequence to synthesize a protein is called gene expression. Gene expression relies on the participation of several different forms of **RNA (ribonucleic acid)** molecules, most important of which are messenger RNA (mRNA), transfer RNA (tRNA), and ribosomal RNA (rRNA). Recall from section 2.8 that DNA and RNA have several structural differences. These are summarized in Table 25.1.

Overall, gene expression requires two processes called transcription and translation. In eukaryotes, transcription takes place in the nucleus and translation takes place in the cytoplasm. During **transcription,** a portion of DNA serves as a template for mRNA formation. During **translation,** the sequence of mRNA bases (which are complementary to those in the template DNA) determines the sequence of amino acids in a polypeptide. So, in effect, genetic information lies in the sequence of the bases in DNA, which through mRNA determines the sequence of amino acids in a protein. Transfer RNA assists mRNA during protein synthesis by bringing amino acids to the ribosomes. Proteins differ from one another by the sequence of their amino acids, and proteins determine the structure and function of cells and the phenotype of the organism.

MP3 Protein Synthesis

TABLE 25.1 Comparison of DNA and RNA

	DNA	RNA
Sugar	Deoxyribose	Ribose
Bases	Adenine, guanine, thymine, cytosine	Adenine, guanine, uracil, cytosine
Strands	Double stranded	Single stranded
Helix	Yes	No

Transcription

During transcription, a segment of the DNA called a **gene** serves as a template for the production of an RNA molecule. Historically, molecular genetics considered a gene to be a nucleic acid sequence that codes for the sequence of amino acids in a protein. We now know that not all genes contain instructions for protein formation. Some genes include instructions for the formation of DNA molecules, such as mRNA, tRNA, and rRNA. We also know that protein-coding regions can be interrupted by regions that do not code for a protein. In recognition of these new findings, Mark Gerstein and associates in 2007 suggested a new definition for a gene: "A gene is a genomic sequence (either DNA or RNA) directly encoding functional products, either RNA or protein."[1] Although all three classes of RNA are formed by transcription, we will focus on transcription to form messenger RNA (mRNA), the first step in protein synthesis.

Messenger RNA

The purpose of **messenger RNA (mRNA)** is to carry genetic information from the DNA to the ribosomes for protein synthesis. Messenger RNA is formed by the process of transcription which, in eukaryotes, occurs in the nucleus. Transcription begins when the enzyme **RNA polymerase** binds tightly to a **promoter,** a region of DNA that contains a special sequence of nucleotides. This enzyme opens up the DNA helix just in front of it so that complementary base pairing can occur in the same way as in DNA replication. Then, RNA polymerase inserts the RNA nucleotides, and an mRNA molecule results. When mRNA forms, it has a sequence of bases complementary to that of the DNA; wherever A, T, G, or C are present in the DNA template, U, A, C, or G, respectively, are incorporated into the mRNA molecule (Fig. 25.6). Now, mRNA is a faithful copy of the sequence of bases in DNA.

Animation Stages of Transcription

MP3 Transcription

3D Animation Molecular Biology of the Gene: Transcription

Processing of mRNA After the mRNA is transcribed in eukaryotic cells, it must be processed before entering the cytoplasm.

[1]Gerstein, M. B., Bruce, C., and Rozowsky, J. S. et al. "What is a gene. post-ENCODE? History and updated definition," *Genomic Research* 17:669–681 (2007).

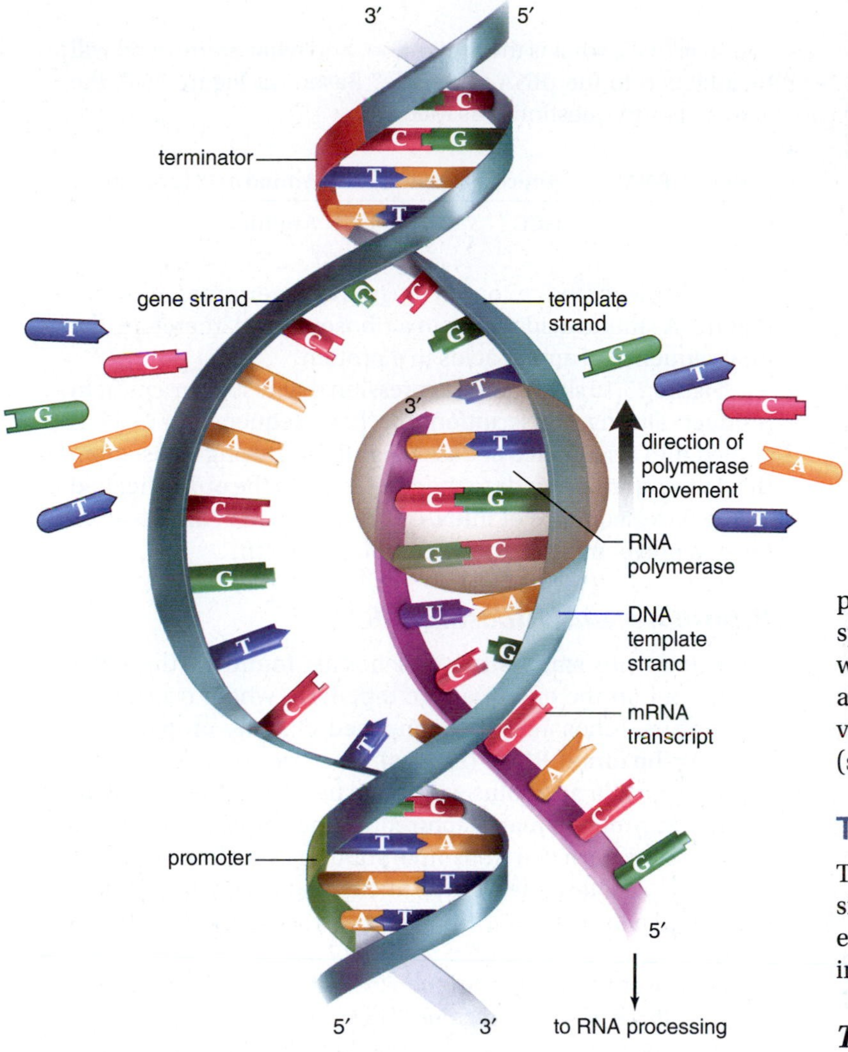
3′
5′
terminator
gene strand
template strand
direction of polymerase movement
RNA polymerase
DNA template strand
mRNA transcript
promoter
5′
5′
3′
to RNA processing

Figure 25.6 Transcription of DNA to form mRNA. During transcription, complementary RNA is made from a DNA template. At the point of attachment of RNA polymerase, the DNA helix unwinds and unzips, and complementary RNA nucleotides are joined together. After RNA polymerase has passed by, the DNA strands rejoin and the mRNA transcript is released.

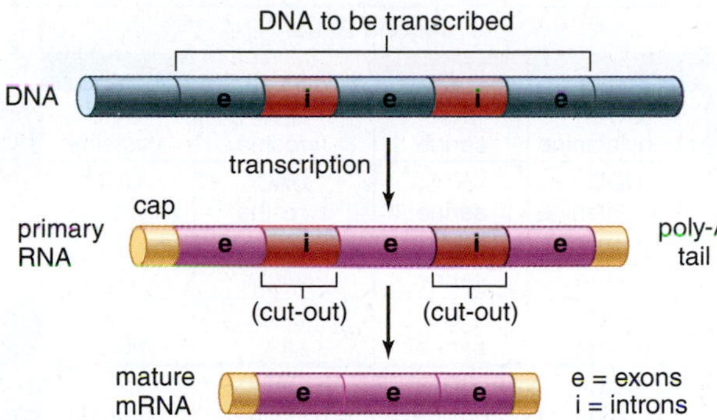
DNA to be transcribed
DNA
transcription
cap
primary RNA
poly-A tail
(cut-out)
(cut-out)
mature mRNA
e = exons
i = introns

Figure 25.7 mRNA processing. During processing, a cap and tail are added to mRNA, and the introns (i) are removed so that only exons (e) remain.

The newly synthesized *primary mRNA* molecule becomes a *mature mRNA* molecule after processing. Most genes in humans are interrupted by segments of DNA that are not part of the gene. These portions are called **introns** because they are intragene segments. The other portions of the gene are called **exons** because they are ultimately expressed (Fig. 25.7). Only exons result in a protein product.

Primary mRNA contains bases that are complementary to both exons and introns, but during processing, (1) one end of the mRNA is modified by the addition of a cap, composed of an altered guanine nucleotide, and the other end is modified by the addition of a poly-A tail, a series of adenosine nucleotides. (2) The introns are removed, and the exons are joined to form a mature mRNA molecule consisting of continuous exons. Ordinarily, processing brings together all the exons of a gene. In some instances, cells use only certain exons to form a mature RNA transcript. The result can be a different protein product in each cell. Alternate mRNA splicing accounts for the ability of white blood cells to produce a specific antibody for each type of bacteria and virus we encounter on a daily basis (see section 13.3).

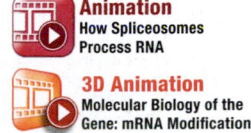

Animation How Spliceosomes Process RNA

3D Animation Molecular Biology of the Gene: mRNA Modifications

Translation

Translation is the second process by which gene expression leads to protein synthesis. Translation requires several enzymes, and several different types of RNA molecules, including mRNA, tRNA, and rRNA.

The Genetic Code

The sequence of bases in DNA is transcribed into mRNA, which ultimately codes for a particular sequence of amino acids to form a polypeptide. Can four mRNA bases (A, C, G, U) provide enough combinations to code for 20 amino acids? If only one base stood for an amino acid (i.e., a "singlet code"), then only four amino acids would be possible. If two bases stood for one amino acid, there would only be 16 possible combinations (4×4). If the code is a triplet, then there are 64 possible triplets of the four bases ($4 \times 4 \times 4$). Each triplet of nucleotides is called a **codon.** The code is also degenerate, meaning that most amino acids are coded for by more than one codon (Fig. 25.8). For example, leucine has six codons and serine has four codons. This degeneracy offers some protection against possibly harmful mutations that change the sequence of bases. Of the 64 codons, 61 code for amino acids, whereas the remaining three are *stop codons* (UAA, UGA, UAG), codons that do not code for amino acids but instead signal polypeptide termination.

The genetic code is just about universal in all living organisms. This means that a codon in a fruit fly codes for the same amino acid as in a bird, a fern, or a human. The universal nature of the genetic code suggests that it dates back to the very first organisms on Earth and that all living organisms have a common evolutionary history.

First Base	Second Base: U	Second Base: C	Second Base: A	Second Base: G	Third Base
U	UUU phenylalanine	UCU serine	UAU tyrosine	UGU cysteine	U
U	UUC phenylalanine	UCC serine	UAC tyrosine	UGC cysteine	C
U	UUA leucine	UCA serine	UAA *stop*	UGA *stop*	A
U	UUG leucine	UCG serine	UAG *stop*	UGG tryptophan	G
C	CUU leucine	CCU proline	CAU histidine	CGU arginine	U
C	CUC leucine	CCC proline	CAC histidine	CGC arginine	C
C	CUA leucine	CCA proline	CAA glutamine	CGA arginine	A
C	CUG leucine	CCG proline	CAG glutamine	CGG arginine	G
A	AUU isoleucine	ACU threonine	AAU asparagine	AGU serine	U
A	AUC isoleucine	ACC threonine	AAC asparagine	AGC serine	C
A	AUA isoleucine	ACA threonine	AAA lysine	AGA arginine	A
A	**AUG *(start)* methionine**	ACG threonine	AAG lysine	AGG arginine	G
G	GUU valine	GCU alanine	GAU aspartic acid	GGU glycine	U
G	GUC valine	GCC alanine	GAC aspartic acid	GGC glycine	C
G	GUA valine	GCA alanine	GAA glutamic acid	GGA glycine	A
G	GUG valine	GCG alanine	GAG glutamic acid	GGG glycine	G

Figure 25.8 Messenger RNA codons. Notice that in this chart, each of the codons (in boxes) is composed of three letters representing the first base, second base, and third base. For example, find the box where C for the first base and A for the second base intersect. You will see that U, C, A, or G can be the third base. The bases CAU and CAC are codons for histidine.

Transfer RNA

Transfer RNA (tRNA) molecules bring amino acids to the ribosomes, the site of protein synthesis. Each tRNA molecule is a single-stranded polynucleotide that doubles back on itself such that complementary base pairing creates a bootlike shape. On one end is an amino acid, and on the other end is an **anticodon,** a triplet of three bases complementary to a codon of mRNA (Fig. 25.9). Although there are 64 possible codons, there are only 40 different tRNA molecules. This is because of the *wobble effect,* which states that for some tRNAs, the third nucleotide in the mRNA codon may vary. This is believed to provide additional degeneracy to the genetic code, and help protect against mutations that may alter the amino acid sequence of a protein.

When a tRNA–amino acid complex comes to the ribosome, its anticodon pairs with an mRNA codon. For example, if the codon is CGG, what is the anticodon, and what amino acid will be attached to the tRNA molecule? Based on Figure 25.8, the answer to this question is as follows:

Codon (mRNA)	Anticodon (tRNA)	Amino Acid (protein)
CGG	GCC	Arginine

The order of the codons of the mRNA determines the order that tRNA–amino acids come to a ribosome and, therefore, the final sequence of amino acids in a protein.

Figure 25.10 shows gene expression that results in a protein product. During transcription, the base sequence in DNA is copied into a sequence of bases in mRNA. During translation, tRNAs bring amino acids to the ribosomes in the order dictated by the base sequence of mRNA. The sequence of amino acids forms a polypeptide chain, or complete protein.

Ribosomes and Ribosomal RNA

Ribosomes are small structural bodies found in the cytoplasm and on the endoplasmic reticulum where translation also occurs. Ribosomes are composed of many proteins and several **ribosomal RNAs (rRNAs)**. In eukaryotic cells, rRNA is produced in a nucleolus within the nucleus. Then the rRNA joins with proteins manufactured in and imported from the cytoplasm to form two ribosomal subunits, one large and one small. The subunits leave the nucleus and join together in the cytoplasm to form a ribosome just as protein synthesis begins.

A ribosome has a binding site for mRNA as well as binding sites for three tRNA molecules. These binding sites facilitate

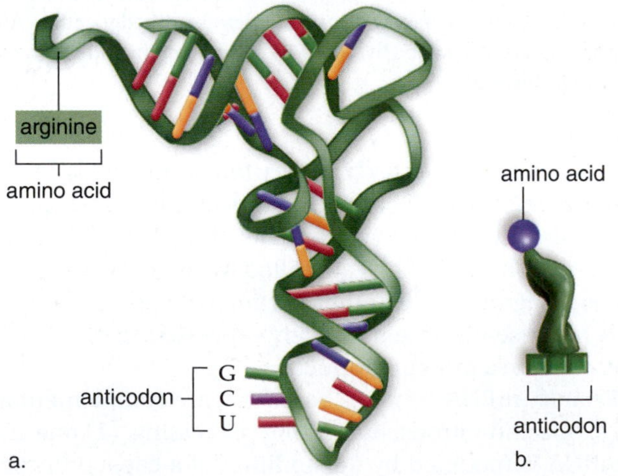
arginine
amino acid
anticodon
G
C
U
a.
amino acid
anticodon
b.

Figure 25.9 Transfer RNA: amino acid carrier. **a.** A tRNA is a polynucleotide that folds into a bootlike shape because of complementary base pairing. At one end of the molecule is its specific anticodon—in this case, GCU (which hybridizes to the codon CGA). At the other end, an amino acid attaches that corresponds to this anticodon—in this case, arginine. **b.** tRNA is represented like this in the illustrations that follow.

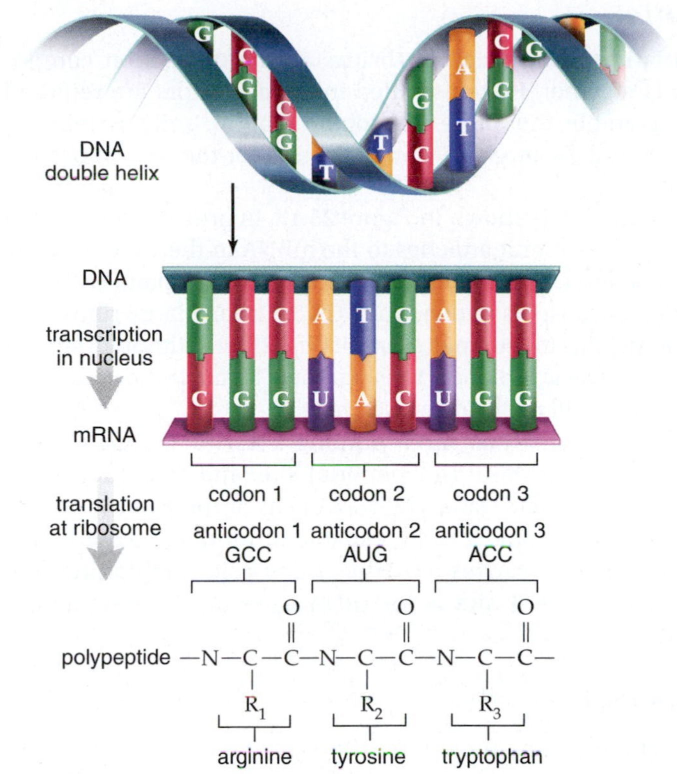
DNA
double helix
DNA
transcription
in nucleus
mRNA
translation
at ribosome
codon 1
codon 2
codon 3
anticodon 1
GCC
anticodon 2
AUG
anticodon 3
ACC
polypeptide
arginine
tyrosine
tryptophan

Figure 25.10 Overview of gene expression. One strand of DNA acts as a template for mRNA synthesis, and the sequence of bases in mRNA determines the sequence of amino acids in a polypeptide.

complementary base pairing between tRNA anticodons and mRNA codons. As the ribosome moves down the mRNA molecule, new tRNAs arrive, and a polypeptide forms and grows longer. Translation terminates once the polypeptide is fully formed and an mRNA stop codon is reached. The ribosome then dissociates into its two subunits and falls off the mRNA molecule.

As soon as the initial portion of mRNA has been translated by one ribosome and the ribosome has begun to move down the mRNA, another ribosome attaches to the same mRNA. Therefore, several ribosomes are often attached to and translating a single mRNA, thus forming several copies of a polypeptide simultaneously. The entire complex is called a *polyribosome* (Fig. 25.11).

Translation Requires Three Steps

During translation, the codons of an mRNA base pair with the anticodons of tRNA molecules carrying specific amino acids. The order of the codons determines the order of the tRNA molecules at a ribosome and the sequence of amino acids in a polypeptide. The process of translation must be extremely orderly so that the amino acids of a polypeptide are sequenced correctly.

Protein synthesis involves three steps: initiation, elongation, and termination. Enzymes are required for each of the three steps to function properly. The first two steps, initiation and elongation, require energy.

Animation How Translation Works

Figure 25.11 Polyribosome structure and function. **a.** Structure of a ribosome. **b.** Internal view of a ribosome showing the tRNA binding sites. **c.** Several ribosomes, collectively called a polyribosome, move along an mRNA at one time. Therefore, several polypeptides can be made at the same time. **d.** Electron micrograph of a polyribosome.

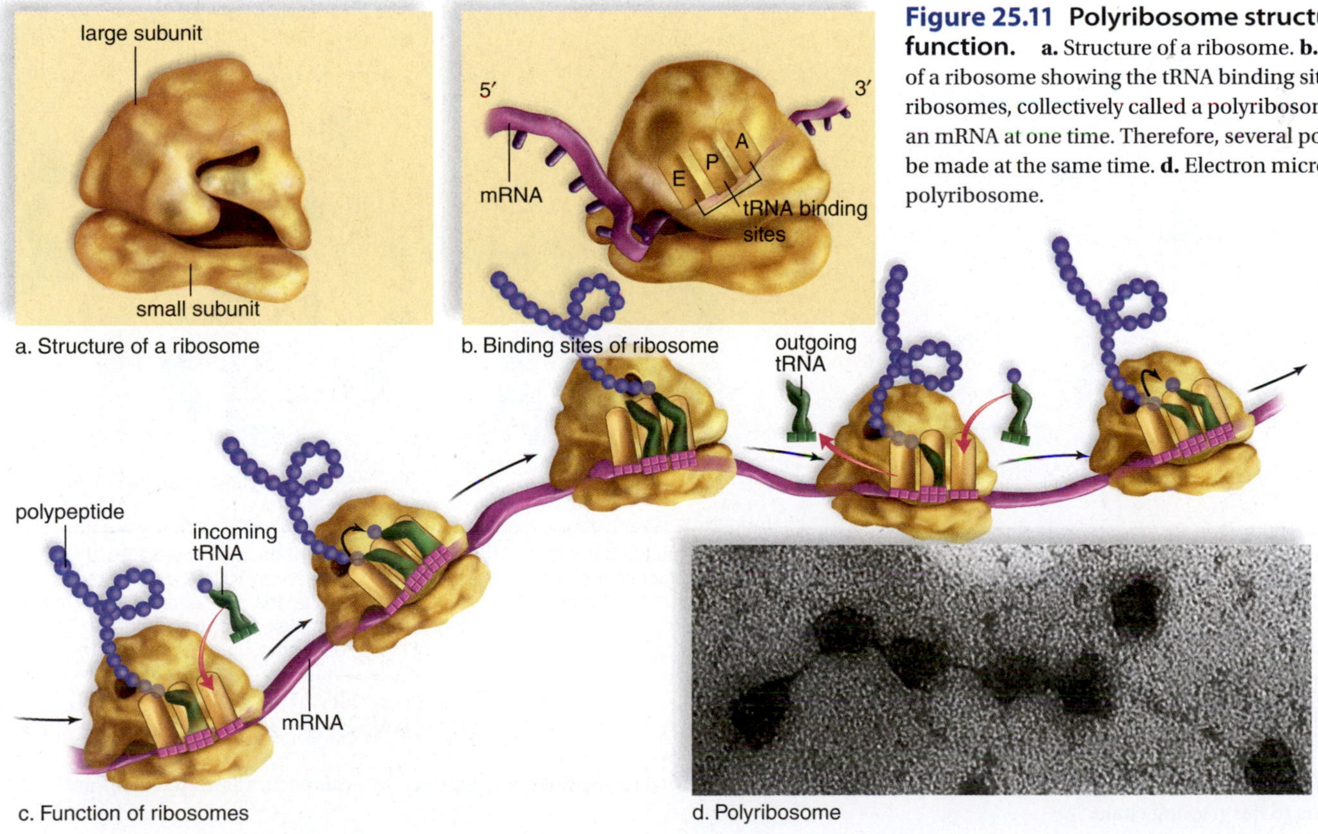
large subunit
small subunit
a. Structure of a ribosome
5′
3′
mRNA
E
P
A
tRNA binding
sites
b. Binding sites of ribosome
outgoing
tRNA
polypeptide
incoming
tRNA
mRNA
c. Function of ribosomes
d. Polyribosome

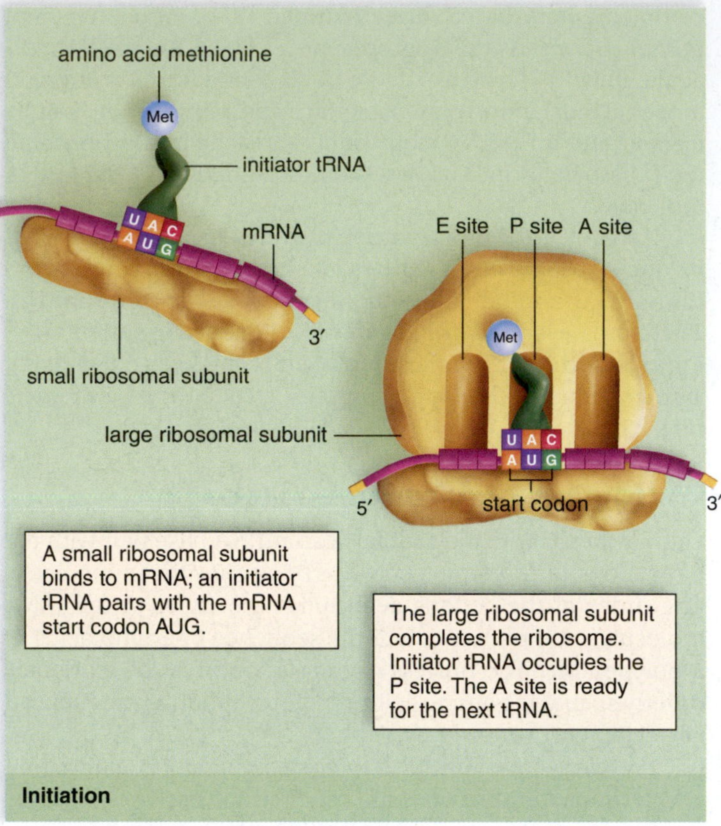
amino acid methionine
Met
initiator tRNA
mRNA
5′
3′
small ribosomal subunit
large ribosomal subunit
E site
P site
A site
Met
start codon
5′
3′
A small ribosomal subunit binds to mRNA; an initiator tRNA pairs with the mRNA start codon AUG.
The large ribosomal subunit completes the ribosome. Initiator tRNA occupies the P site. The A site is ready for the next tRNA.
Initiation

Figure 25.12 Initiation. During initiation, participants in the translation process assemble as shown. The start codon, AUG, also codes for the first amino acid, methionine.

Initiation

Initiation is the step that brings all the translation components together. Proteins called initiation factors are required to assemble the small ribosomal subunit, mRNA, initiator tRNA, and the large ribosomal subunit for the start of protein synthesis.

Initiation is shown in Figure 25.12. In prokaryotes, a small ribosomal subunit attaches to the mRNA in the vicinity of the *start codon* (AUG). The first or initiator tRNA pairs with this codon because its anticodon is UAC. Then, a large ribosomal subunit joins to the small subunit (Fig. 25.12). Although similar in many ways, initiation in eukaryotes is much more complicated and will not be discussed here.

A ribosome has three binding sites for tRNAs. One of these is called the P (for peptide) site, and the other is the A (for amino acid) site. The tRNA exits at the E site. The initiator tRNA happens to be capable of binding to the P site, even though it carries only the amino acid methionine (see Fig. 25.8). The A site is for tRNA carrying the next amino acid.

Elongation

Elongation is the protein synthesis step in which a polypeptide increases in length one amino acid at a time (Fig. 25.13). In addition to the participation of tRNAs, elongation requires elongation factors, which facilitate the binding of tRNA anticodons to mRNA codons at a ribosome.

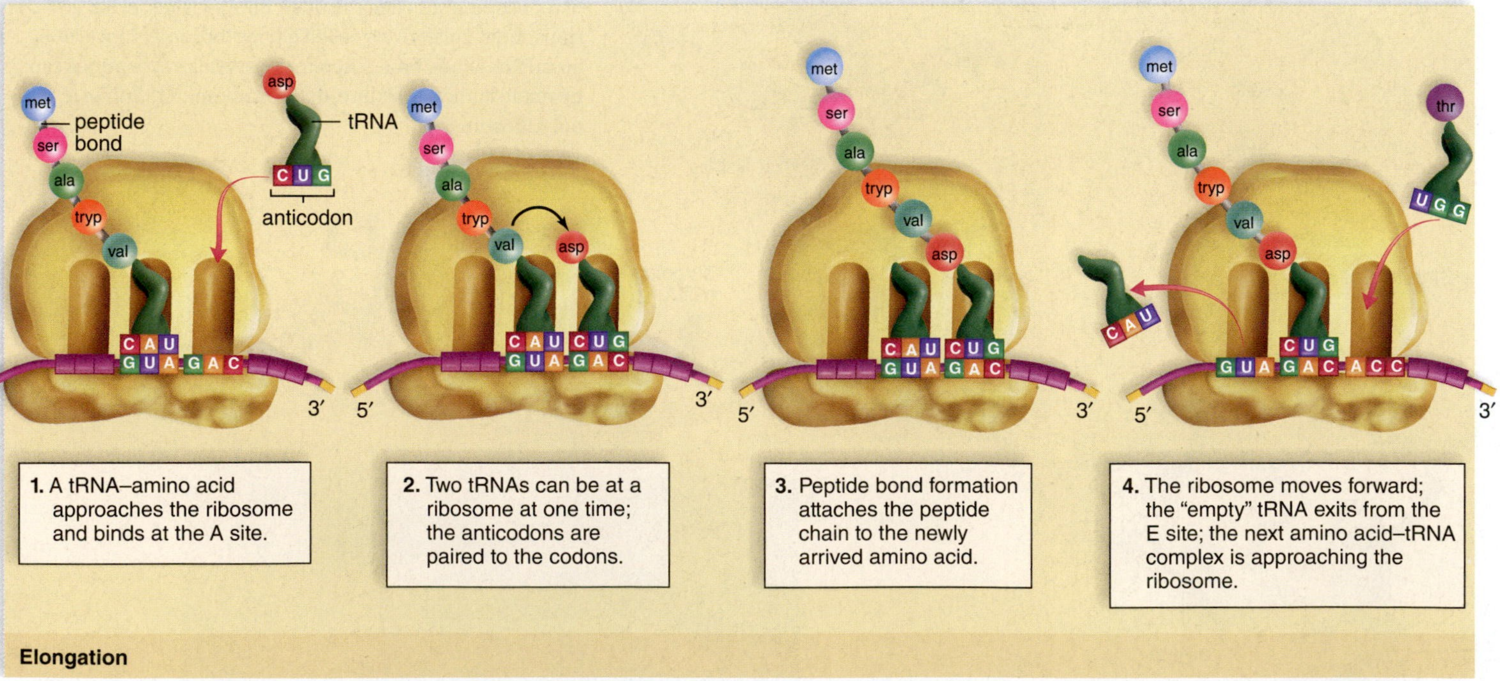
met
ser
ala
tryp
val
asp
thr
peptide bond
tRNA
anticodon
5′
3′
1. A tRNA–amino acid approaches the ribosome and binds at the A site.
2. Two tRNAs can be at a ribosome at one time; the anticodons are paired to the codons.
3. Peptide bond formation attaches the peptide chain to the newly arrived amino acid.
4. The ribosome moves forward; the "empty" tRNA exits from the E site; the next amino acid–tRNA complex is approaching the ribosome.
Elongation

Figure 25.13 Elongation. Note that a polypeptide is already at the P site. During elongation, polypeptide synthesis occurs as amino acids are added one at a time to the growing chain.

Elongation consists of a series of four steps (Fig. 25.13):

1. A tRNA with an attached peptide is already at the P site, and a tRNA carrying the next amino acid in the chain is just arriving at the A site.
2. Once the next tRNA is in place at the A site, the peptide chain will be transferred to this tRNA.
3. Energy and part of the ribosomal subunit are needed to bring about this transfer. The energy contributes to peptide bond formation, which makes the peptide one amino acid longer by adding the peptide from the A site.
4. Next, translocation occurs—the mRNA moves forward one codon length, and the peptide-bearing tRNA is now at the ribosome P site. The "spent" tRNA now exits. The new codon is at the A site and is ready to receive the next complementary tRNA.

The complete cycle in steps 1–4 is repeated at a rapid rate (e.g., about 15 times each second in *E. coli*).

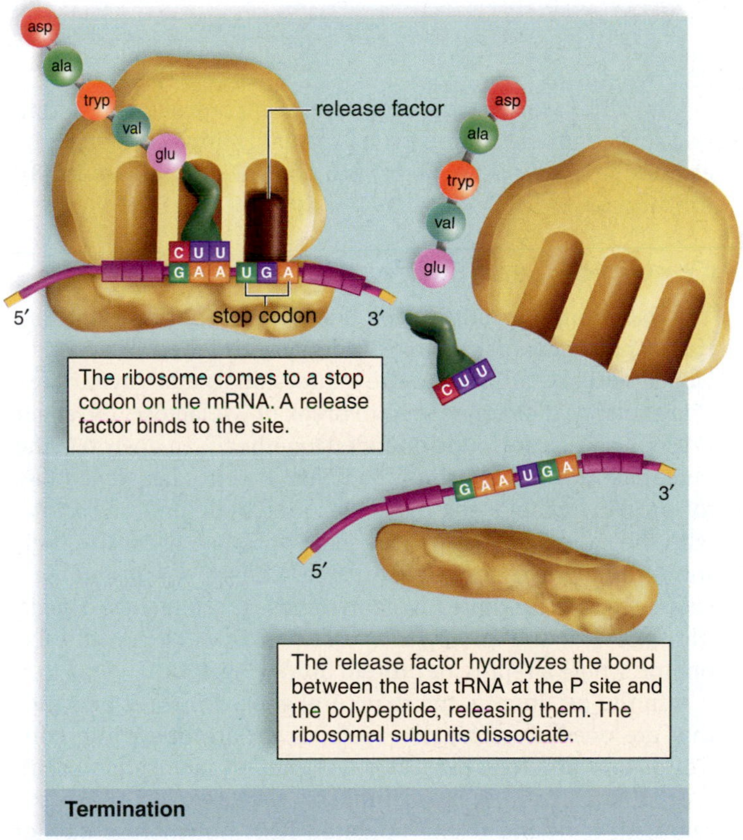
asp
ala
tryp
val
glu
release factor
C U U
G A A U G A
stop codon
5′
3′
The ribosome comes to a stop codon on the mRNA. A release factor binds to the site.
C U U
G A A U G A
3′
5′
The release factor hydrolyzes the bond between the last tRNA at the P site and the polypeptide, releasing them. The ribosomal subunits dissociate.
Termination

Figure 25.14 Termination. During termination, the finished polypeptide is released, as are the mRNA and the last tRNA.

Termination

Termination is the final step in protein synthesis. During termination, as shown in Figure 25.14, the polypeptide and the assembled components that carried out protein synthesis are separated from one another.

Termination of polypeptide synthesis occurs at a stop codon. Termination requires a protein called a release factor, which cleaves the polypeptide from the last tRNA. After this occurs, the polypeptide is set free and begins to take on its three-dimensional shape. The ribosome dissociates into its two subunits.

Properly functioning proteins are of paramount importance to the cell and to the organism. For example, if an organism inherits a faulty gene, the result can be a genetic disorder (such as Huntington disease) caused by a malfunctioning protein or a propensity toward cancer. Proteins are the link between genotype and phenotype. The DNA sequence underlying these proteins distinguishes different types of organisms. In addition to accounting for the difference between cell types, proteins account for the differences between organisms.

MP3 Translation

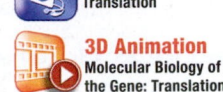

3D Animation Molecular Biology of the Gene: Translation

Review of Gene Expression

A gene is expressed when its protein product has been synthesized. Protein synthesis requires the process of transcription and translation (Fig. 25.15). During transcription, a segment of a DNA strand serves as a template for the formation of messenger RNA (mRNA). The bases in mRNA are complementary to those in DNA. Every three mRNA bases is a *codon* (a triplet code) for a certain amino acid. Messenger RNA is processed before it leaves the nucleus, during which time the introns are removed and the ends are modified. Messenger RNA carries a sequence of codons to the *ribosomes.* During translation, tRNAs bring attached amino acids to the ribosomes. Because tRNA anticodons pair with codons, the amino acids become sequenced in the order originally specified by DNA. The genes we receive from our parents determine the proteins in our cells and these proteins are responsible for our inherited traits!

Check Your Progress 25.3

1. Explain the role of mRNA, tRNA, and rRNA in gene expression.
2. Describe the movement of information from the nucleus to the formation of a functional protein.
3. Discuss why the genetic code is said to be degenerate.

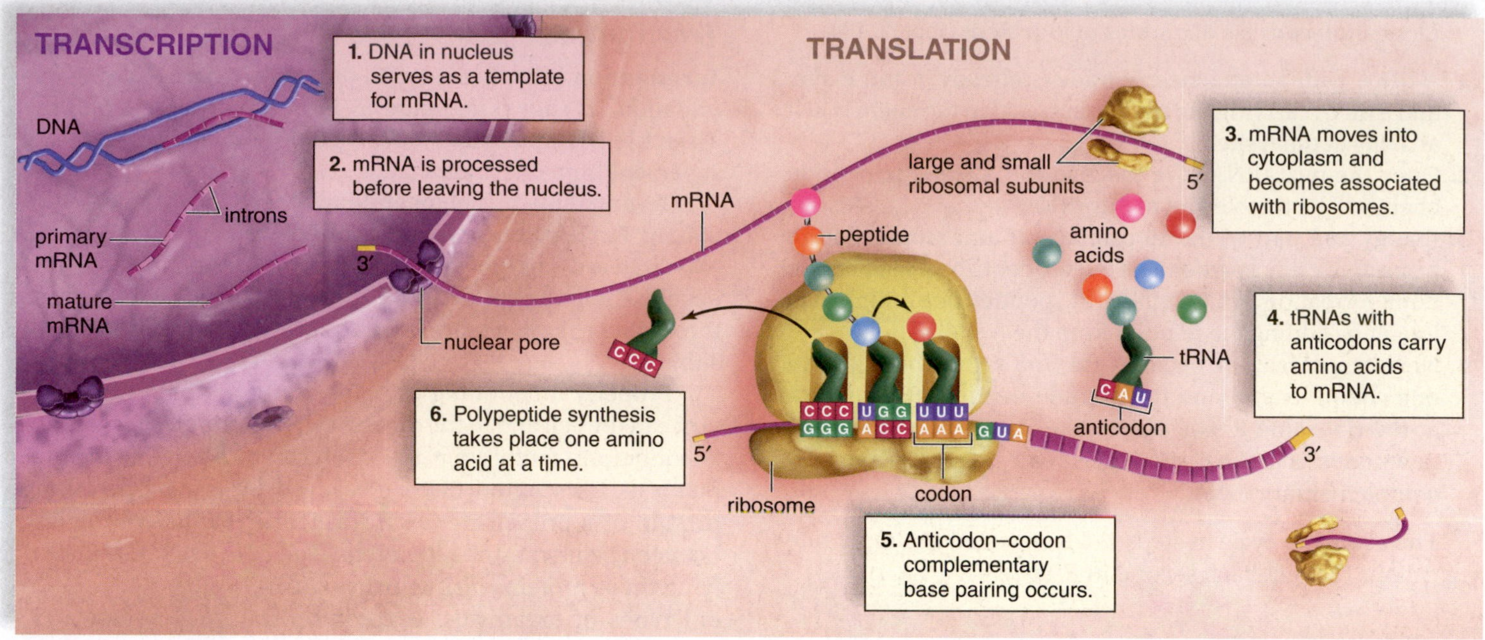
TRANSCRIPTION
1. DNA in nucleus serves as a template for mRNA.
TRANSLATION
DNA
2. mRNA is processed before leaving the nucleus.
introns
primary mRNA
3′
mature mRNA
nuclear pore
mRNA
peptide
large and small ribosomal subunits
5′
3. mRNA moves into cytoplasm and becomes associated with ribosomes.
amino acids
4. tRNAs with anticodons carry amino acids to mRNA.
tRNA
C C C
C A U
anticodon
C C C U G G U U U
G G G A C C A A A G U A
6. Polypeptide synthesis takes place one amino acid at a time.
5′
3′
ribosome
codon
5. Anticodon–codon complementary base pairing occurs.

Figure 25.15 Review of gene expression. Messenger RNA is produced and processed in the nucleus during transcription, and protein synthesis occurs at the ribosomes (in cytoplasm and rough ER) during translation.

25.4 Control of Gene Expression

Learning Outcomes

Upon completion of this section, you should be able to

1. Describe the relationship between gene regulation and gene activity in a cell.
2. Compare regulation in prokaryotes to regulation in eukaryotes.
3. Discuss the many ways genes are regulated in eukaryotes.

The human body contains many types of cells that differ in structure and function. Each cell type must contain its own mix of proteins that make it different from all other cell types. Therefore, only certain genes are active in cells that perform specialized functions, such as nerve, muscle, gland, and blood cells.

Some of these active genes are called housekeeping genes because they govern functions that are common to many types of cells, such as glucose metabolism. But otherwise, the activity of selected genes accounts for the specialization of cells:

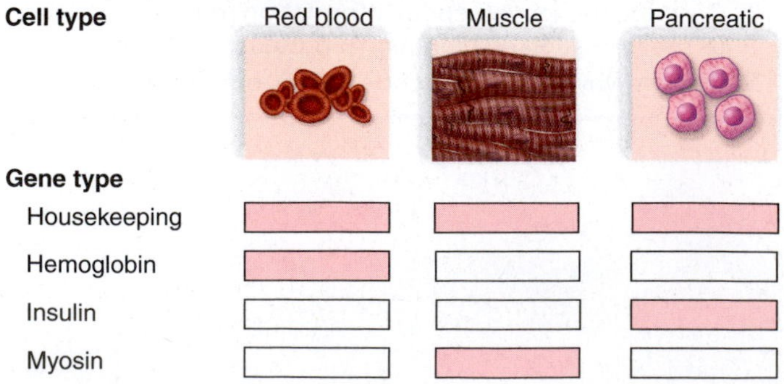
Cell type
Red blood
Muscle
Pancreatic
Gene type
Housekeeping
Hemoglobin
Insulin
Myosin

In other words, gene expression is controlled in a cell, and this control accounts for its specialization. Let's begin by examining a simpler system—the control of transcription in prokaryotes.

Control of Gene Expression in Prokaryotes

The bacterium *Escherichia coli* that lives in your intestine can use various sugars as a source of energy and carbon. This organism can quickly adjust its gene expression to match your diet. The enzymes that are needed to break down lactose, the sugar present in milk, are found encoded together in an operon in the bacterial DNA. An **operon** is a cluster of genes usually coding for proteins related to a particular metabolic pathway, along with the short DNA sequences that coordinately control their transcription. The control sequences consist of a *promoter,* a sequence of DNA where RNA polymerase first attaches to begin transcription, and an *operator,* a sequence of DNA in the *lac* operon where a repressor protein binds (Fig. 25.16).

In the *lac* operon, the structural genes for three enzymes that are needed for lactose metabolism are under the control of one promoter/operator complex. If lactose is absent, a protein called a *repressor* binds to the operator. When the repressor is bound to the operator, RNA polymerase cannot transcribe the three structural genes of the operon. The *lac* repressor is encoded by a regulatory gene located outside of the operon.

When lactose is present, the lactose binds with the *lac* repressor so that the repressor is unable to bind to the operator. Then RNA polymerase is able to transcribe the structural genes into a single mRNA, which is then translated into the three different enzymes. In this way, the enzymes needed to break down lactose are only synthesized when lactose is present.

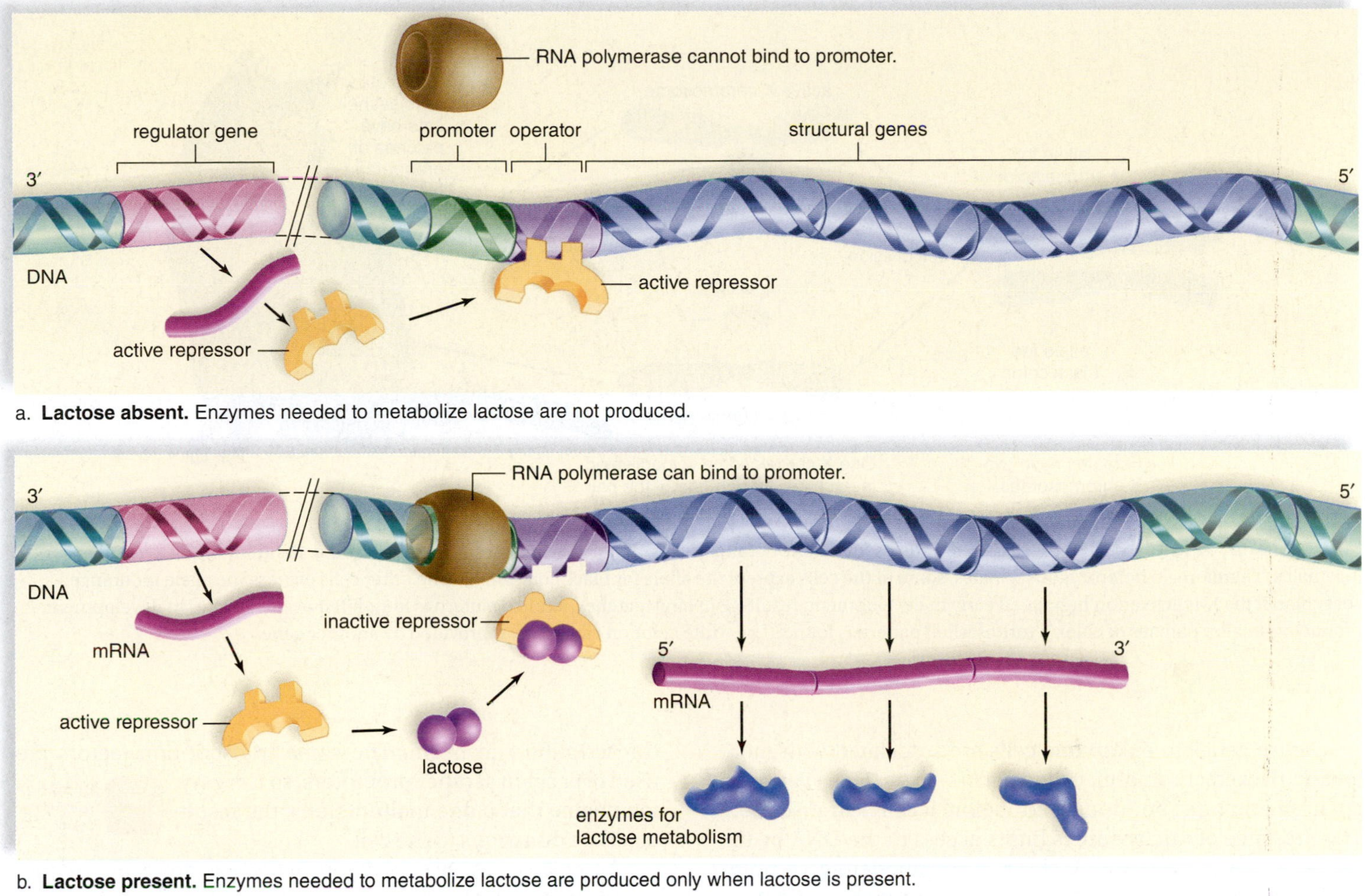
RNA polymerase cannot bind to promoter.
regulator gene
promoter
operator
structural genes
3′
5′
DNA
active repressor
active repressor
a. Lactose absent. Enzymes needed to metabolize lactose are not produced.
RNA polymerase can bind to promoter.
3′
5′
DNA
mRNA
inactive repressor
active repressor
lactose
5′
3′
mRNA
enzymes for lactose metabolism
b. Lactose present. Enzymes needed to metabolize lactose are produced only when lactose is present.

Figure 25.16 The *lac* operon. **a.** When lactose is absent, the regulator gene codes for a repressor that is normally active. When it binds to the operator, RNA polymerase cannot attach to the promoter, and structural genes are not expressed. **b.** When lactose is present, it binds to the repressor, changing its shape so that it is inactive and cannot bind to the operator. Now, RNA polymerase binds to the promoter, and the structural genes are expressed.

The *lac* operon is considered an inducible operon. It is only activated when lactose induces its expression. Other bacterial operons are repressible operons. They are usually active until a repressor turns them off.

Animation Combination of Switches: The *lac* Operon

Control of Gene Expression in Eukaryotes

In prokaryotes, a single promoter serves several genes that make up a transcription unit or operon. In eukaryotes, each gene has its own promoter where RNA polymerase binds. In contrast to the prokaryotes, eukaryotes employ a variety of mechanisms to regulate gene expression. These mechanisms affect whether the gene is expressed, the speed with which it is expressed, and how long it is expressed.

Animation Control of Gene Expression in Eukaryotes

Levels of Gene Control

Eukaryotic genes exhibit control of gene expression at five different levels: (1) pretranscriptional control, (2) transcriptional control, (3) posttranscriptional control, (4) translational control, and (5) posttranslational control.

Pretranscriptional Control Eukaryotes use DNA methylation and chromatin packing as a way to keep genes turned off. Genes within darkly staining, highly condensed portions of chromatin, called *heterochromatin,* are inactive. A dramatic example of this occurs with the X chromosome in mammalian females. Females have two X chromosomes, while males have only one X chromosome. This inequality is balanced by the inactivation of one X chromosome in every cell of the female body. Each inactivated X chromosome, called a Barr body in honor of its discoverer, can be seen as a small, darkly staining mass of condensed chromatin along the inner edge of the nuclear envelope. During early prenatal development, one or the other X chromosome in a cell is randomly shut off by chromatin condensation into heterochromatin. All of the cells that form from division of that cell will have the same X chromosome inactivated. Therefore, females have patches of tissue that differ in which X chromosome is being expressed. If that female is heterozygous for a particular X-linked gene, she will be a mosaic, containing patches of cells expressing different alleles. This can be clearly seen in calico and tortoiseshell cats (Fig. 25.17).

Animation X-inactivation

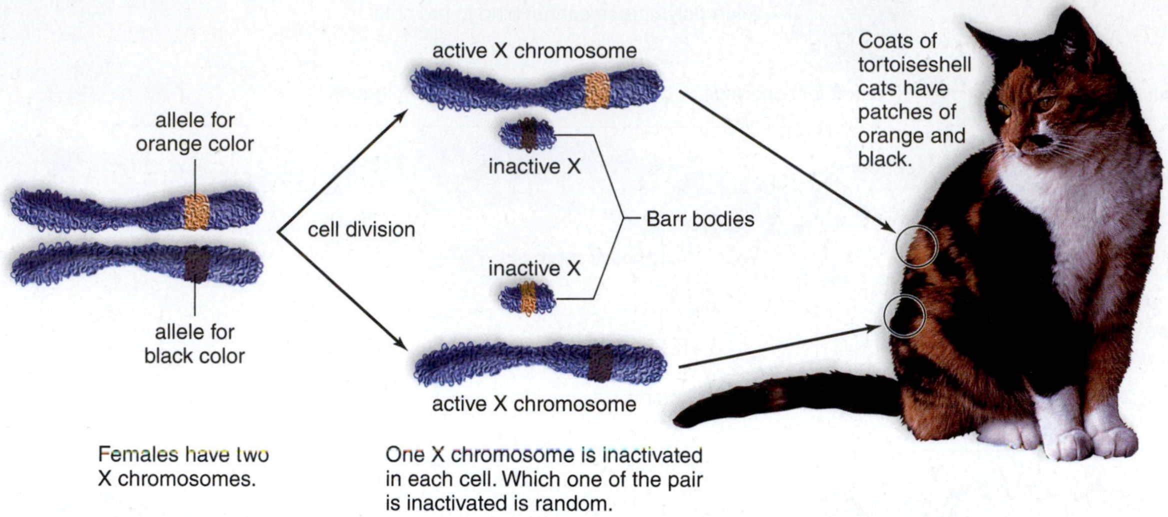

active X chromosome
allele for orange color
inactive X
cell division
Barr bodies
inactive X
allele for black color
active X chromosome
Coats of tortoiseshell cats have patches of orange and black.
Females have two X chromosomes.
One X chromosome is inactivated in each cell. Which one of the pair is inactivated is random.

Figure 25.17 X-inactivation. In cats, the alleles for black or orange coat color are carried on the X chromosome. Random X-inactivation occurs in females. Therefore, in heterozygous females, some of the cells express the allele for black coat color, while other cells express the allele for orange coat color. If the X-inactivation happened early in development, resulting in large patches, a calico pattern is found. If it occurred later in development, producing smaller patches of color, a tortoiseshell pattern is found. The white color on calico cats is provided by another gene.

Active genes in eukaryotic cells are associated with more loosely packed chromatin, called *euchromatin.* However, even euchromatin must be "unpacked" before it can be transcribed. The presence of nucleosomes limits access to the DNA by the transcription machinery. A chromatin remodeling complex pushes the nucleosomes aside to open up sections of DNA for expression:

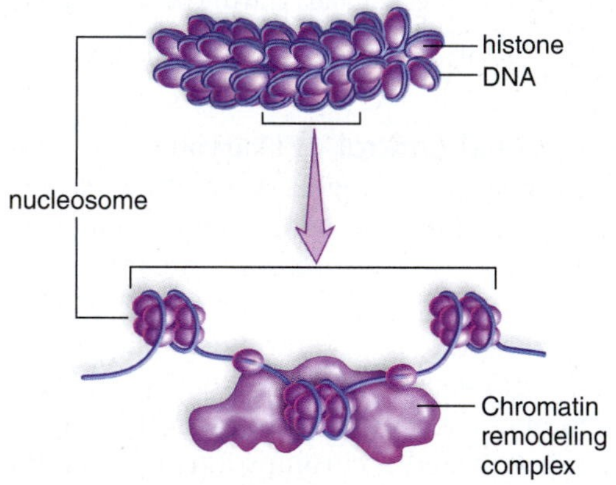

histone
DNA
nucleosome
Chromatin remodeling complex

Transcriptional Control As in prokaryotes, eukaryotic transcriptional control is dependent on the interaction of proteins with particular DNA sequences. The proteins are called transcription factors and activators, while the DNA sequences they are associated with are called *enhancers* and *promoters.* In eukaryotes, each gene has its own promoter. **Transcription factors** are proteins that help RNA polymerase bind to a promoter. Several transcription factors per gene form a *transcription initiation complex* that also helps pull double-stranded DNA apart and even acts to release RNA polymerase so that transcription can begin. The same transcription factors are used over again at other promoters, so it is easy to imagine that if one malfunctions, the result could be disastrous to the cell.

Animation Transcription Factors

In eukaryotes, *transcription activators* are proteins that speed transcription dramatically. In general, transcription activators are themselves activated in response to a signal received and transmitted by a signal transduction pathway as described later. They bind to enhancer regions in the DNA and stimulate transcription.

Posttranscriptional Control Following transcription, messenger RNA (mRNA) is processed before it leaves the nucleus and passes into the cytoplasm. The primary mRNA is converted to the mature mRNA by the addition of a poly-A tail and a guanine cap, and by the removal of the introns and splicing back together of the exons. The same mRNA can be spliced in different ways to make slightly different products in different tissues. For example, both the hypothalamus and the thyroid gland produce the hormone calcitonin, but the calcitonin mRNA that exits the nucleus contains different combinations of exons in the two tissues.

The speed of transport of mRNA from the nucleus into the cytoplasm can ultimately affect the amount of gene product following transcription. There is a difference in the length of time it takes various mRNA molecules to pass through a nuclear pore.

Translational Control The longer an mRNA remains in the cytoplasm before it is broken down, the more gene product can be translated. Differences in the poly-A tail or the guanine cap can determine how long a particular transcript remains

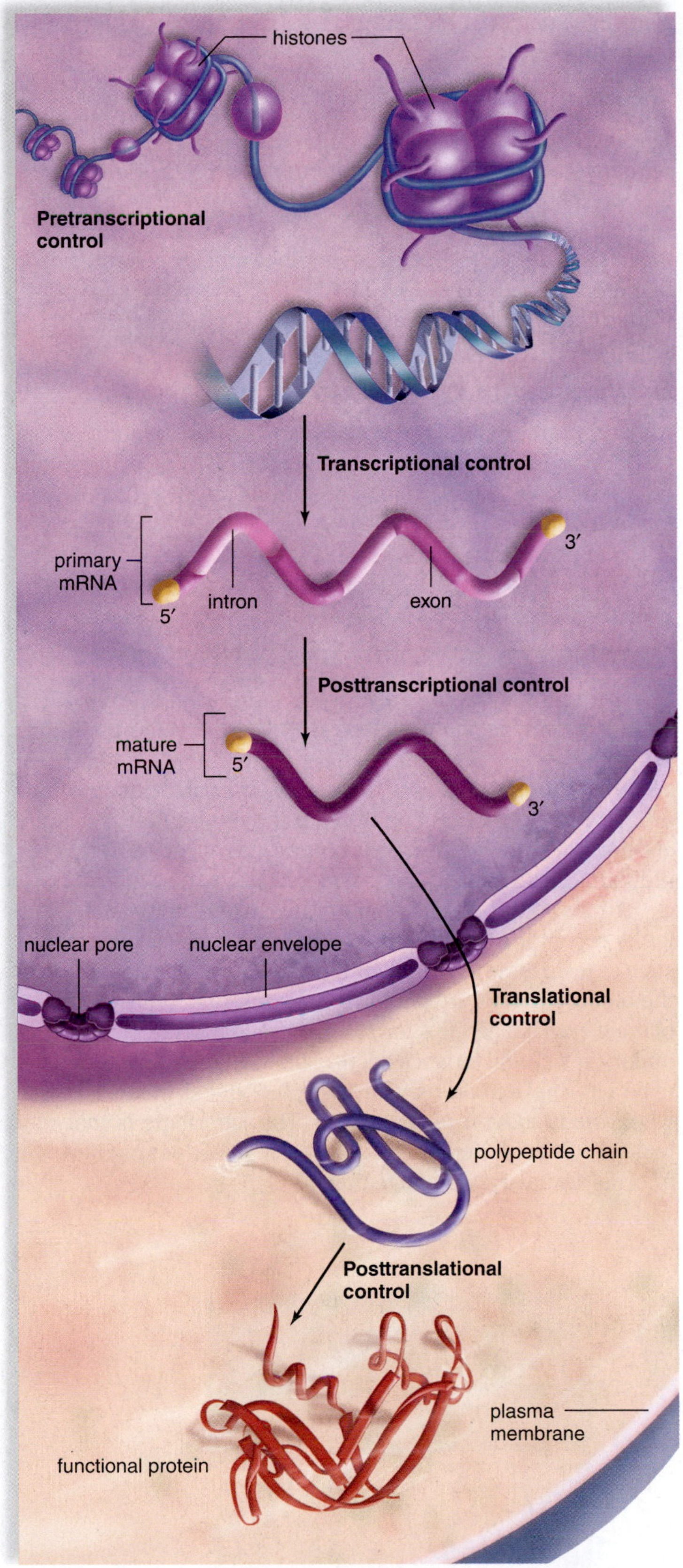
histones
Pretranscriptional control
Transcriptional control
primary mRNA
intron
exon
5′
3′
Posttranscriptional control
mature mRNA
5′
3′
nuclear pore
nuclear envelope
Translational control
polypeptide chain
Posttranslational control
plasma membrane
functional protein

Figure 25.18 Levels at which control of gene expression occurs in eukaryotic cells. The five levels of control are (1) pretranscriptional control, (2) transcriptional control, and (3) posttranscriptional control, which occur in the nucleus; and (4) translational and (5) posttranslational control, which occur in the cytoplasm.

active before it is destroyed by a ribonuclease associated with ribosomes. Hormones can cause the stabilization of certain mRNA transcripts. For example, the mRNA for vitelline, an egg membrane protein, can persist for three weeks if it is exposed to estrogen, as opposed to 15 hours without estrogen.

Posttranslational Control Some proteins are not active immediately after synthesis. After translation, insulin is folded into a three-dimensional structure that is inactive. Then a sequence of about 30 amino acids is enzymatically removed from the middle of the molecule, leaving two polypeptide chains that are bonded together by disulfide (S—S) bonds. This activates the protein. Other modifications, such as phosphorylation, also affect the activity of a protein. Many proteins only function a short time before they are degraded or destroyed by the cell.

The various levels of gene control are summarized in Figure 25.18.

Check Your Progress 25.4

1. Explain why gene expression would not be the same in all cells of an organism.
2. Describe how the *lac* operon represents an on/off switch for gene expression.
3. Summarize the different levels of gene regulation in eukaryotes.

25.5 Gene Mutations and Cancer

Learning Outcomes

Upon completion of this section, you should be able to

1. Summarize the causes of gene mutations.
2. Describe why cancer is a failure of genetic control.
3. Describe the characteristics of cancer cells.

A **gene mutation** is a permanent change in the sequence of bases in DNA. The effect of a DNA base sequence change on protein activity can range from no effect to complete inactivity. Germ-line mutations are those that originally occurred in sex cells and can be passed to subsequent generations. We now know that some germ-line mutations can result in cancer. Also, somatic mutations that are not passed on to future generations can sometimes lead to the development of cancer.

Causes of Mutations

Gene mutations may be caused by errors in replication, mutagens, and the activity of transposons.

Errors in Replication DNA replication errors are a rare source of mutations. DNA polymerase, the enzyme that carries out replication, proofreads the new strand against the old strand. Usually, mismatched pairs are then replaced with the correct nucleotides. In the end, there is typically only one mistake for every 1 billion nucleotide pairs replicated.

Mutagens Environmental influences called **mutagens** cause mutations in humans. Mutagens include radiation (e.g., radioactive elements, X rays, ultraviolet [UV] radiation) and certain organic chemicals (e.g., chemicals in cigarette smoke and certain pesticides). The rate of mutations resulting from mutagens is generally low because DNA repair enzymes constantly monitor and repair any irregularities.

Transposons **Transposons** are specific DNA sequences that have the remarkable ability to move within and between chromosomes. Their movement to a new location sometimes alters neighboring genes, particularly by increasing or decreasing their expression. Although "movable elements" in corn were described over 50 years ago by Barbara McClintock, their significance was only realized recently. Also called jumping genes, transposons have now been discovered in almost every species, including bacteria, fruit flies, and humans. McClintock described how the presence of white kernels in corn is due to a transposon located within a gene coding for a pigment-producing enzyme (Fig. 25.19*a*, *b*). "Indian corn" displays a variety of colors and patterns because of transposons (Fig. 25.19*c*). In a rare human neurological disorder called Charcot-Marie-Tooth disease, a transposon called Mariner causes the muscles and nerves of the legs and feet to gradually wither away.

Animation Transposons: Shifting Segments of the Genome

Effect of Mutations on Protein Activity

Point mutations involve a change in a single DNA nucleotide and, therefore, a possible change in a specific amino acid. The base change in the second row of Figure 25.20*a* has no effect on the resulting amino acid in hemoglobin. The change in the third row, however, codes for the amino acid glutamic acid instead of valine. This base change accounts for the genetic disorder sickle-cell disease because the incorporation of valine, instead of glutamic acid, causes hemoglobin molecules to form semirigid rods, and the red blood cells become sickle-shaped. (Compare Figure 25.20*b* to Figure 25.20*c*.) Sickle-shaped cells clog blood vessels and die off more quickly than normal-shaped cells. The base change in the fourth row of Figure 25.20*a* may also have drastic results because the DNA now codes for a stop codon.

Frameshift mutations occur most often because one or more nucleotides are either inserted or deleted from DNA. The result of a frameshift mutation can be a completely new sequence of codons and nonfunctional protein. Here is how this occurs: The sequence of codons is read from a specific starting point, as in this sentence, THE CAT ATE THE RAT. If the letter C is deleted from this sentence and the reading frame is shifted, we read THE ATA TET HER AT—something that doesn't make sense.

Animation Addition and Deletion Mutations

Nonfunctional Proteins

A single nonfunctioning protein can have a dramatic effect on the phenotype, because enzymes are often a part of metabolic pathways. One particular metabolic pathway in cells is as follows:

A (phenylalanine) $\xrightarrow{E_A}$ B (tyrosine) $\xrightarrow{E_B}$ C (melanin)

If a faulty code for enzyme E_A is inherited, a person is unable to convert the molecule A to B. Phenylalanine builds up in the system, and the excess causes mental impairment and the other symptoms of the genetic disorder phenylketonuria (PKU). In the same pathway, if a person inherits a faulty code for enzyme E_B, then B cannot be converted to C, and the individual is an albino.

A rare condition called androgen insensitivity is due to a faulty receptor for androgens, which are male sex hormones such as testosterone. Although there is plenty of testosterone in the blood, the cells are unable to respond to it. Female instead of male external genitals form, and female instead of male secondary sex characteristics occur. The individual, who appears to be a normal female, may be prompted to seek medical advice when menstruation never occurs. The karyotype is that of a male rather than a female, and the individual does not have the internal sexual organs of a female.

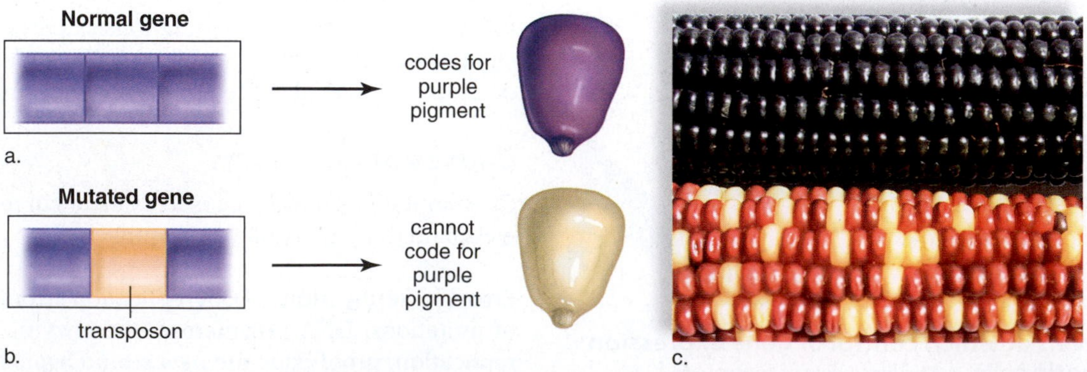
Normal gene
codes for purple pigment
a.
Mutated gene
cannot code for purple pigment
transposon
b.
c.

Figure 25.19 Transposon. **a.** A purple coding gene ordinarily codes for a purple pigment. **b.** A transposon "jumps" into the purple-coding gene. This mutated gene is unable to code for purple pigment and a white kernel results. **c.** Indian corn displays a variety of colors and patterns due to transposon activity.

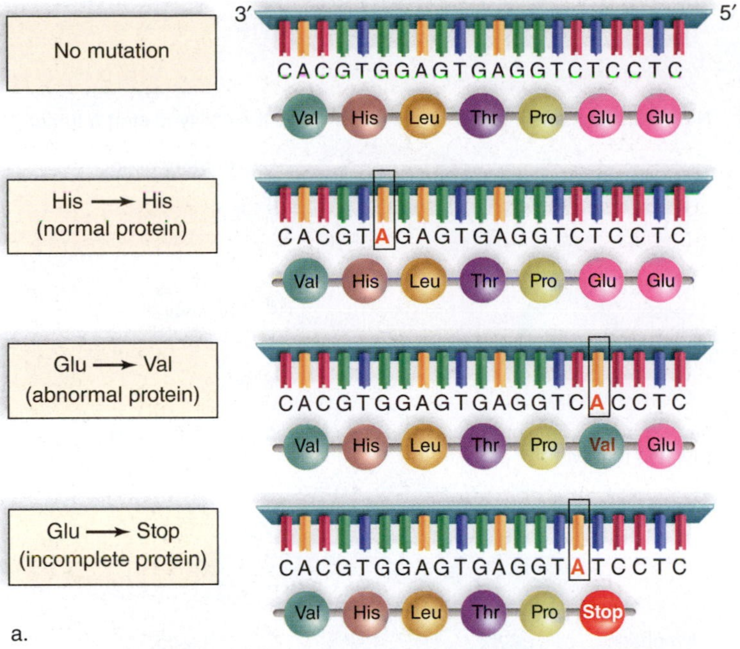
No mutation
3′
5′
CACGTGGAGTGAGGTCTCCTC
Val His Leu Thr Pro Glu Glu
His ⟶ His
(normal protein)
CACGTAGAGTGAGGTCTCCTC
Val His Leu Thr Pro Glu Glu
Glu ⟶ Val
(abnormal protein)
CACGTGGAGTGAGGTCACCTC
Val His Leu Thr Pro Val Glu
Glu ⟶ Stop
(incomplete protein)
CACGTGGAGTGAGGTATCCTC
Val His Leu Thr Pro Stop
a.

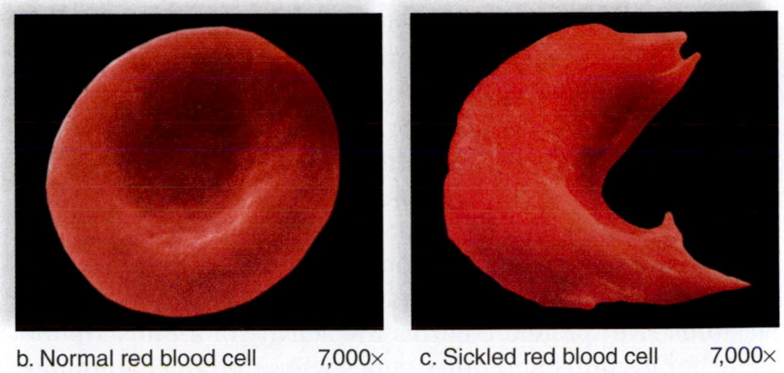
b. Normal red blood cell 7,000×
c. Sickled red blood cell 7,000×

Figure 25.20 Point mutations in hemoglobin. The effect of a point mutation can vary. **a.** Starting at the *top*: Normal sequence of bases in hemoglobin; next, the base change has no effect; next, due to base change, DNA now codes for valine instead of glutamic acid, and the result is that normal red blood cells (**b**) become sickle-shaped (**c**); next, base change will cause DNA to code for termination and the protein will be incomplete.

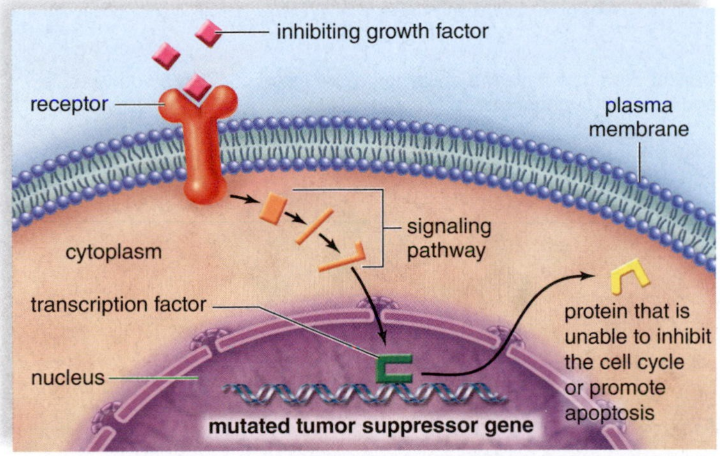
inhibiting growth factor
receptor
plasma membrane
signaling pathway
cytoplasm
transcription factor
protein that is unable to inhibit the cell cycle or promote apoptosis
nucleus
mutated tumor suppressor gene

Figure 25.21 Cell signaling pathway that no longer inhibits a mutated tumor suppressor gene. A mutated tumor suppressor gene codes for a product that directly or indirectly no longer inhibits the cell cycle.

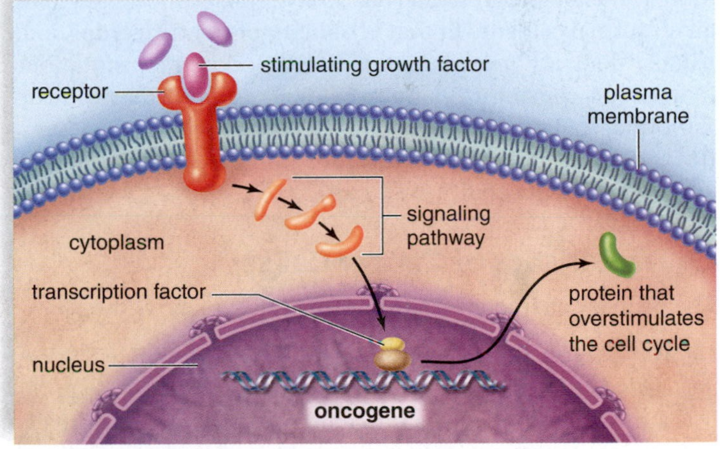
stimulating growth factor
receptor
plasma membrane
signaling pathway
cytoplasm
transcription factor
protein that overstimulates the cell cycle
nucleus
oncogene

Figure 25.22 Cell signaling pathway that stimulates an oncogene. An oncogene codes for a product that either directly or indirectly overstimulates the cell cycle.

Mutations Can Cause Cancer

It is estimated that one-third of us will develop cancer at some time in our lives. While the five-year survival rate for individuals with cancer has increased from 50% in 1977 to 68% today, around 1,500 people each day will die of the disease. In the United States, the three deadliest forms of cancer are lung cancer, colorectal cancer, and breast cancer.

The development of cancer involves a series of accumulating mutations that can be different for each type of cancer. As discussed in section 5.2, tumor suppressor genes ordinarily act as brakes on cell division, especially when it begins to occur abnormally. Proto-oncogenes stimulate cell division but are usually turned off in fully differentiated nondividing cells. When proto-oncogenes mutate, they become oncogenes that are active all the time. Carcinogenesis begins with the loss of tumor suppressor gene activity and/or the gain of oncogene activity. When tumor suppressor genes are inactive and oncogenes are active, cell division occurs uncontrollably because a cell signaling pathway that reaches from the plasma membrane to the nucleus no longer functions as it should (Figs. 25.21 and 25.22).

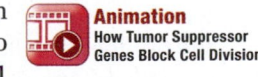

Animation How Tumor Suppressor Genes Block Cell Division

It often happens that tumor suppressor genes and proto-oncogenes code for transcription factors or proteins that control transcription factors. As we have seen, transcription factors are a part of the rich and diverse types of mechanisms that control gene expression in cells. They are of fundamental importance to DNA replication and repair, cell growth and division, control of apoptosis (programmed cell death), and cellular differentiation. Therefore, it is not surprising that inherited or acquired defects in transcription factor structure and function contribute to the development of cancer.

To take an example, a major tumor suppressor gene called *p*53 is more frequently mutated in human cancers than any other known gene. It has been found that the *p*53 protein acts as a transcription factor, and as such is involved in turning on the expression of genes whose products are cell cycle inhibitors. *p*53 also promotes apoptosis when it is needed. The retinoblastoma (RB) protein controls the activity of a transcription factor for cyclin D and other genes whose products promote entry into the S stage of the cell cycle. When the tumor suppressor gene *p*16 mutates, it is unable to control the cell cycle, the RB protein is continuously expressed, and the result is too much active cyclin D in the cell.

Mutations in many other genes also contribute to the development of cancer. Several proto-oncogenes code for ras proteins, which are needed for cells to grow, to make new DNA, and to not grow out of control. A point mutation is sufficient to turn a normally functioning *ras* proto-oncogene into an oncogene. Abnormal growth results.

Although cancers vary greatly, they usually follow a common multistep progression (Fig. 25.23). Most cancers begin as an abnormal cell growth that is **benign,** or not cancerous, and usually does not grow larger. However, additional mutations may occur, causing the abnormal cells to fail to respond to inhibiting signals that control the cell cycle. When this occurs, the growth becomes **malignant,** meaning that it is cancerous and possesses the ability to spread.

Characteristics of Cancer Cells

The primary characteristics of cancer cells are as follows:

Cancer cells are genetically unstable. Generation of cancer cells appears to be linked to mutagenesis. A cell acquires a mutation that allows it to continue to divide. Eventually one of the progeny cells will acquire another mutation and gain the ability to form a tumor. Further mutations occur, and the most aggressive cell becomes the dominant cell of the tumor. Tumor cells undergo multiple mutations and also tend to have chromosomal aberrations and rearrangements.

Video Melanoma Marker

Cancer cells do not correctly regulate the cell cycle. Cancer cells continue to cycle through the cell cycle. The normal controls of the cell cycle do not operate to stop the cycle and allow the cells to differentiate. Because of that, cancer cells tend to be nonspecialized. Both the rate of cell division and the number of cells increase.

Cancer cells escape the signals for cell death. A cell that has genetic damage or problems with the cell cycle will initiate apoptosis, or programmed cell death. However, cancer cells do not respond to internal signals to die, and they continue to divide even with genetic damage. Cells from the immune system, when they detect an abnormal cell, will send signals to that cell, inducing apoptosis. Cancer cells also ignore these signals.

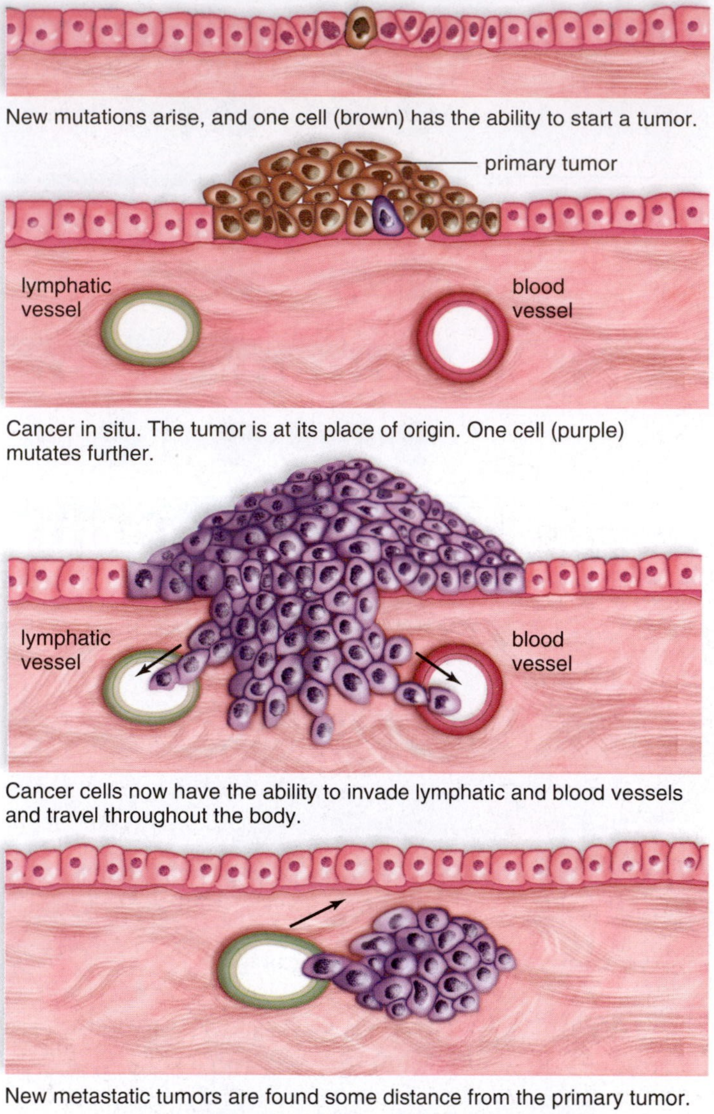
New mutations arise, and one cell (brown) has the ability to start a tumor.
primary tumor
lymphatic vessel
blood vessel
Cancer in situ. The tumor is at its place of origin. One cell (purple) mutates further.
lymphatic vessel
blood vessel
Cancer cells now have the ability to invade lymphatic and blood vessels and travel throughout the body.
New metastatic tumors are found some distance from the primary tumor.

Figure 25.23 Progression of cancer. A single abnormal cell begins the process, and the most aggressive cell, thereafter, becomes the one that divides the most and forms the tumor. Eventually, cancer cells gain the ability to invade underlying tissue and travel to other parts of the body, where they develop new tumors.

Most normal cells have a built-in limit to the number of times they can divide before they die. One of the reasons normal cells stop entering the cell cycle is that the telomeres become shortened. **Telomeres** are sequences at the ends of the chromosomes that keep them from fusing with each other. With each cell division, the telomeres shorten, eventually becoming short enough to signal apoptosis. Cancer cells turn on the gene that encodes the enzyme telomerase, which is capable of rebuilding and lengthening the telomeres. Cancer cells thus show characteristics of "immortality" in that they can enter the cell cycle repeatedly.

Animation Telomerase Function

SCIENCE IN YOUR LIFE ▶ HEALTH

Prevention of Cancer

Be tested for cancer.

Do a shower test for breast cancer or testicular cancer. Women should get a Pap smear for cervical cancer annually if they are over 21 or are sexually active. Have a friend or physician check your skin annually for any unusual moles. Have other exams done regularly by a physician.

Be aware of occupational hazards.

Exposure to several different industrial agents (nickel, chromate, asbestos, vinyl chloride, etc.) and/or radiation increases the risk of various cancers. Your employer should notify you of potential hazards in your workplace. The risk of several of these cancers is increased when combined with cigarette smoking.

Use sunscreen.

Almost all cases of basal cell and squamous cell skin cancers are sun-related. Use a sunscreen of at least SPF 30 (Fig. 25B) and wear protective clothing if you are going to be out during the brightest part of the day. Don't sunbathe on the beach or in a tanning salon.

Figure 25B UV protection. Use of sunscreen with SPF of 30 or higher can help prevent skin cancer, as can limiting midday exposure.

Check your home for radon.

Excessive radon exposure in homes increases the risk of lung cancer, especially in cigarette smokers. It is best to test your home and take the proper remedial actions.

Avoid unnecessary X rays.

Even though most medical and dental X rays are adjusted to deliver the lowest dose possible, unnecessary X rays should be avoided. Sensitive areas of the body that are not being X-rayed should be protected with lead screens.

Practice safe sex.

Human papillomaviruses (HPVs) cause genital warts and are associated with cervical cancer, as well as tumors of the vulva, vagina, anus, penis, and mouth. HPVs may be involved in 90–95% of all cases of cervical cancer, and 20 million people in the United States have an infection that can be transmitted to others.

Carefully consider hormone therapy.

Estrogen therapy to control menopausal symptoms increases the risk of endometrial cancer. However, including progesterone in estrogen replacement therapy helps minimize this risk.

Maintain a healthy weight.

The risk of cancer (especially colon, breast, and uterine cancers) is 55% greater among obese women and 33% greater among obese men, compared to people of normal weight. Eating a low-fat, healthy diet helps maintain your weight as well as reducing your risk of cancer.

Exercise regularly.

Regular activity can reduce the incidence of certain cancers, especially colon and breast cancer.

Increase consumption of foods that are rich in vitamins A and C.

Beta-carotene, a precursor of vitamin A, is found in dark-green, leafy vegetables, carrots, and various fruits. Vitamin C is present in citrus fruits. These vitamins are called antioxidants because they prevent the formation of chemicals called free radicals that can damage the DNA in the cell. Vitamin C also prevents the conversion of nitrates and nitrites into carcinogenic nitrosamines in the digestive tract.

Include vegetables from the cabbage family in your diet.

The cabbage family includes cabbage, broccoli, Brussels sprouts, kohlrabi, and cauliflower. These vegetables may reduce the risk of gastrointestinal and respiratory tract cancers.

Limit consumption of salt-cured, smoked, or nitrite-cured foods.

Salt-cured or pickled foods may increase the risk of stomach and esophageal cancer. Smoked foods, such as ham and sausage, contain chemical carcinogens similar to those in tobacco smoke. Nitrites are sometimes added to processed meats (e.g., hot dogs and cold cuts) and other foods to protect them from spoilage. These are converted to carcinogenic nitrosamines in the digestive tract.

Be moderate in the consumption of alcohol.

Cancers of the mouth, throat, esophagus, larynx, and liver occur more frequently among heavy drinkers, especially when accompanied by smoking cigarettes or chewing tobacco.

Don't smoke.

Cigarette smoking accounts for about 30% of all cancer deaths. Smoking is responsible for 90% of lung cancer cases among men and 79% among women—about 87% altogether. People who smoke two or more packs of cigarettes a day have lung cancer mortality rates 13 to 23 times greater than those of nonsmokers. Cigars and smokeless tobacco (chewing tobacco or snuff) increase the risk of cancers of the mouth, larynx, throat, and esophagus.

Questions to Consider

1. Which of these cancer-preventive measures are you currently practicing?
2. Why do you think tobacco use increases the risk of other types of cancer besides lung cancer?
3. Why does the use of tanning beds also increase the incidence of skin cancer?

Cancer cells can survive and proliferate elsewhere in the body. Many of the changes that must occur for cancer cells to form tumors elsewhere in the body are not understood. The cells apparently disrupt the normal adhesive mechanism and move to another place within the body. They travel through the blood and lymphatic vessels and then invade new tissues, where they form tumors. This process is known as **metastasis.** As a tumor grows, it must increase its blood supply by forming new blood vessels, a process called **angiogenesis.** Tumor cells switch on genes that code for the production of growth factors encouraging blood vessel formation. These new blood vessels supply the tumor cells with the nutrients and oxygen they require for rapid growth, but they also rob normal tissues of nutrients and oxygen.

Check Your Progress 25.5

1. Explain how gene mutations occur.
2. Distinguish between a point mutation and a frameshift mutation.
3. Describe the potential effects of a mutation in a tumor suppressor gene versus an oncogene.

Case Study Conclusion

Mutations are the result of changes in the DNA of an individual. In some cases, these changes affect specific proteins. For example, children who suffer from xeroderma pigmentosum are missing the enzymes that repair DNA after exposure to UV light. These enzymes constantly search the DNA for abnormalities and then either directly repair the damage, or remove a section of the DNA with the abnormality and resynthesize a new strand of DNA.

SCIENCE IN YOUR LIFE ▶ **BIOETHICAL**

Genetic Testing for Cancer Genes

Genetic tests have become increasingly available for certain cancer genes. If women test positive for defective *BRCA1* and *BRCA2* genes, they have an increased risk for early-onset breast and ovarian cancer. If individuals test positive for the *APC* gene, they are at greater risk for the development of colon cancer. Other genetic tests exist for rare cancers, including retinoblastoma and Wilms tumor.

Advocates of genetic testing say that it can alert those who test positive for these mutated genes to undergo more frequent mammograms or colonoscopies. Early detection of cancer clearly offers the best chance for successful treatment.

Others feel that genetic testing is unnecessary because nothing can presently be done to prevent the disease. Perhaps it is enough for those who have a family history of cancer to schedule more frequent checkups, beginning at a younger age.

Although the U.S. Equal Employment Opportunity Commission (EEOC) prohibits discrimination based on genetic information, recent surveys suggest that the majority of people remain concerned that being predisposed to cancer might threaten one's job or health insurance. Individuals opposed to genetic testing suggest that genetic testing be confined to a research setting, especially because it is not known which particular mutations in the genes predispose a person to cancer. They are afraid, for example, that a woman with a defective *BRCA1* or *BRCA2* gene might make the unnecessary decision to have a bilateral mastectomy. The lack of proper counseling also concerns many. In a study of 177 patients who underwent *APC* gene testing for susceptibility to colon cancer, less than 20% received counseling before the test. Moreover, physicians misinterpreted the test results in nearly one-third of the cases.

Another concern is that testing negative for a particular genetic mutation may give people the false impression that they are not at risk for cancer. Such a false sense of security can prevent them from having routine cancer screening. Regular testing and avoiding known causes of cancer—such as smoking, a high-fat diet, or too much sunlight—are important for everyone.

Questions to Consider

1. Should genetic testing for cancer be available for everyone, or should genetic testing be confined to a research setting? Explain.
2. If genetic testing for cancer were offered to you, would you take advantage of it? Why or why not?
3. Do you feel that everyone should do all they can to avoid having cancer, such as by not smoking, or is it the individual's choice to take that risk? Explain.

connect BIOLOGY Explore the concepts through a variety of multimedia assets, question types, and data interpretation.
www.mcgrawhillconnect.com

MEDIA STUDY TOOLS

www.mhhe.com/maderinquiry14

Enhance your study of this chapter with study tools and practice tests. Also ask your instructor about the resources available through ConnectPlus, including LearnSmart, the media-rich eBook, interactive learning tools, and animations.

Mc Graw Hill connect plus+ | BIOLOGY

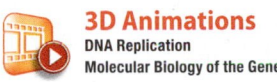

3D Animations
DNA Replication
Molecular Biology of the Gene

For an interactive examination of DNA replication and gene expression, review the 3D animations "DNA Replication" and "Molecular Biology of the Gene."

SUMMARIZE

25.1 DNA Structure

- In a series of experiments, researchers were able to determine that **DNA** (**deoxyribonucleic acid**) was the genetic material.
- DNA is structured as a **double helix,** with two sugar-phosphate backbones and paired nitrogen-containing bases (which may be either **purines** or **pyrimidines**). **Complementary base pairing** of A (adenine) with T (thymine), and G (guanine) with C (cytosine) occurs between the strands. The DNA strands of the double helix are oriented in an antiparallel configuration.

25.2 DNA Replication

- **DNA replication** is semiconservative, meaning that following replication each double helix contains one old strand and one new strand. Enzymes, including DNA polymerase, helicase, and DNA ligase are involved in the replication process.

25.3 Gene Expression

Genes contain the information for the synthesis of proteins or **RNA** (**ribonucleic acid**) molecules. RNA plays an active role in all stages of gene expression. Gene expression involves two processes: transcription and translation.

- **Transcription** produces an RNA sequence complementary to one of the DNA strands. Transcription uses an **RNA polymerase** enzyme, which binds to a region of the DNA called a **promoter. Messenger RNA** (**mRNA**) produced in eukaryotes must be processed by the addition of a cap and tail, and removal of **introns,** before being transported to the cytoplasm where the information in the **exons** can be translated.
- DNA specifies the synthesis of proteins using a **codon** consisting of three bases of RNA. Each codon codes for one amino acid.
- During **translation, transfer RNA** (**tRNA**) molecules, attached to their own particular amino acid, travel to a ribosome, and through complementary base pairing between **anticodons** of tRNA and codons of mRNA, the tRNAs, and the amino acids they carry, are sequenced in a predetermined way to form a polypeptide chain.
- Translation consists of three stages: **initiation, elongation,** and **termination.** A ribosome, which contains **ribosomal RNA** (**rRNA**), has a binding site for two tRNAs at a time. The tRNA at the P site passes a peptide to a newly arrived tRNA–amino acid at the A site. Then translocation occurs: the ribosome moves, and the tRNA with the peptide is at the P site. In this way, an amino acid grows one amino acid at a time until a polypeptide is formed.

25.4 Control of Gene Expression

- Each cell type in the body contains its own mix of proteins that make it different from all other cell types. Only certain genes are active in cells that perform specialized functions. In prokaryotes, genes are clustered into operons. An **operon** contains a group of genes, usually coding for proteins related to a particular metabolic pathway, along with the promoter, a DNA sequence where RNA polymerase binds to begin transcription, and an operator, where a repressor protein binds. In the case of the *lac* operon, the structural genes code for enzymes that metabolize lactose. When lactose is absent, the repressor binds to the operator and shuts off the operon. When lactose is present, it binds to the repressor so that it can no longer bind to the operator, and this allows RNA polymerase to transcribe the structural genes.
- Control of gene expression is more complicated in eukaryotes. Eukaryotes have five levels of control: pretranscriptional control, transcriptional control, posttranscriptional control, translational control, and posttranslational control.
- Genes that are found within highly condensed heterochromatin are inactive. A dramatic example of this is inactivated X chromosomes, called Barr bodies. Even genes that are found in looser-packed DNA, called euchromatin, must have the nucleosomes shifted in order to be expressed.
- The transcriptional level of control includes **transcription factors** that assist the RNA polymerase and transcriptional activators that greatly increase the rate of transcription.
- In posttranscriptional control, differences in mRNA processing affect gene expression.
- Translational controls occur in the cytoplasm and involve the length of time the mRNA is functional.
- Posttranslational controls also occur in the cytoplasm and involve the length of time the protein is functional. Proteins can be activated by a variety of methods.

25.5 Gene Mutations and Cancer

- A **gene mutation** is a permanent change in the sequence of bases in DNA. The effect of a DNA base sequence change on protein activity can range from no effect to complete inactivity.
- Gene mutations result in a change in the sequence of nucleotides in the DNA and can be caused by errors in replication, **mutagens,** and **transposons.**
- Point mutations involve a change in a single DNA nucleotide and, therefore, a possible change in a specific amino acid. Frameshift mutations occur most often because one or more nucleotides are either inserted or deleted from DNA.
- The development of cancer involves a series of accumulating mutations that can be different for each type of cancer. Carcinogenesis

begins with the loss of tumor suppressor gene activity and/or the gain of oncogene activity. Mutations in many other genes also contribute to the development of cancer. Although cancers vary greatly, they usually follow a common multistep progression. Cancer cells defy the normal regulation of the cell cycle and can invade and colonize other areas of the body. Cancer cells do not exhibit contact inhibition and thus form **benign** tumors. When tumor cells gain the ability to invade surrounding tissues, they are said to be **malignant.**

- In terms of their primary characteristics, cancer cells are genetically unstable, do not correctly regulate the cell cycle, cause **angiogenesis,** escape the signals for cell death, and can survive and proliferate elsewhere in the body, a process called **metastasis.** Multiple mutations and chromosomal aberrations are present in cancer cells. They continue to cycle through the cell cycle and tend to be nonspecialized. They do not respond to either internal or external signals for apoptosis, and they induce the expression of telomerase to repair their **telomeres.**

ASSESS

Testing Yourself

Choose the best answer for each question.

1. In the Hershey-Chase experiments, radioactive phosphorus was found
  a. outside the bacterial cells.
  b. inside the bacterial cells.
  c. both inside and outside the bacterial cells.

2. According to Chargaff's rules, in DNA, the amount of
  a. A = T and G = C.
  b. A = G and T = C.
  c. A = C and T = G.
  d. A = T = C = G.

3. During DNA replication, the parental strand ATTGGC would code for the daughter strand
  a. ATTGGC.
  b. CGGTTA.
  c. TAACCG.
  d. GCCAAT.

For questions 4–6, match the function with the type of RNA in the key.

Key:
  a. messenger RNA
  b. ribosomal RNA
  c. transfer RNA

4. Carries amino acids to ribosomes

5. Is a structural component of the organelle responsible for protein synthesis

6. Carries information from genes to the translation machinery

7. In contrast to DNA, RNA
  a. is a helix.
  b. contains uracil.
  c. is double stranded.
  d. contains deoxyribose sugar.

8. Using the information in Figure 25.8, determine the sequence of amino acids that would be produced following transcription and translation of the following DNA template strand:

TACGTTCCAACT

  a. tyrosine - valine - proline - threonine
  b. methionine - glutamic acid - glycine
  c. isoleucine - valine - proline
  d. methionine - glutamic acid - glycine - threonine

9. Mutation rates are typically low due to
  a. proofreading by DNA polymerase.
  b. DNA repair enzymes.
  c. silent mutations.
  d. None of the choices are correct.
  e. All of the choices are correct.

10. The process of converting the information contained in the nucleotide sequence of RNA into a sequence of amino acids is called
  a. transcription.
  b. translation.
  c. translocation.
  d. replication.

11. During protein synthesis, an anticodon of a transfer RNA (tRNA) pairs with
  a. amino acids in the polypeptide.
  b. DNA nucleotide bases.
  c. ribosomal RNA (rRNA) nucleotide bases.
  d. messenger RNA (mRNA) nucleotide bases.
  e. other tRNA nucleotide bases.

12. An operon is a short sequence of DNA
  a. that prevents RNA polymerase from binding to the promoter.
  b. that prevents transcription from occurring.
  c. and the sequences that control its transcription.
  d. that codes for the repressor protein.
  e. that functions to prevent the repressor from binding to the operator.

13. How is transcription directly controlled in eukaryotic cells?
  a. through the use of phosphorylation
  b. by means of apoptosis
  c. using transcription factors and activators
  d. when chromatin is packed to keep genes turned on
  e. None of the choices are correct.

14. Which of these is a characteristic of cancer cells?
  a. They do not exhibit contact inhibition.
  b. They lack specialization.
  c. They have abnormal chromosomes.
  d. They fail to undergo apoptosis.
  e. All of the choices are correct.

15. If the original DNA sequence is ACGCGT, which of the following represents a point mutation?
  a. ACGGGT
  b. TGCCGT
  c. GGCCCC
  d. ACGCGT

ENGAGE

Virtual Lab
DNA and Genes

The virtual lab "DNA and Genes" provides an interactive tutorial for understanding the influence of mutations on protein structure.

Thinking Critically

1. What kind of genes do you think would be included in the category of "housekeeping" genes?

2. When the enzyme telomerase was first discovered, some people thought it might be "the fountain of youth" because it could immortalize cells. Why was this not found to be true?

3. Why is it only the *risk* for cancer that is inherited?

Biotechnology and Genomics 26

CASE STUDY The Age of Biotechnology

Biotechnology is the study and application of living organisms and processes to manufacture products, or improve the characteristics of bacteria, plants, or animals. As a few examples, biotechnologists are actively investigating ways of making our world less dependent on fossil fuels, such as oil. Some bacteria produce biodegradable plastic granules inside of their cells. By introducing these genes into plants, such as the corn plants above, it may be possible someday to produce large quantities of biodegradable plastic. In addition to plastic-producing plants, biotechnology has made it possible for bioengineers and medical scientists to alter the genotype, and subsequently the phenotype, of other organisms—bacteria, animals, and humans. Genetically modified crops are resistant to disease and able to grow under stressful conditions. Farm animals can be made to grow larger than usual, and humans may be supplied with normal genes to make up for ones that do not function as they should.

But many people worry that genetically modified bacteria and plants might harm the environment and that the products produced by altered organisms might not be healthy for humans. Other ethical concerns abound. Is it ethical to give a cat a gene that makes it glow in the dark? To what extent would it be proper to improve the human genome? Everyone should be knowledgeable about modern genetics and biotechnology so they can participate in deciding these issues.

As you read through the chapter, think about the following questions:

1. What procedures are used to introduce a bacterial gene into a plant?
2. What is the difference between a genetically modified and transgenic organism?
3. How are animals being genetically modified?

CHAPTER OUTLINE

BEFORE YOU BEGIN

Before beginning this chapter, take a few moments to review the following discussions:

Figure 25.3 What are the structural characteristics of a DNA molecule?

Figure 25.5 What is the role of the DNA polymerase in DNA replication?

Section 25.3 What are the stages in gene expression?

26.1 DNA Cloning

Learning Outcomes

Upon completion of this section, you should be able to

1. Describe the steps in forming recombinant DNA.
2. Discuss how the polymerase chain reaction works.
3. Explain what is meant by a DNA "fingerprint."

Knowledge of DNA biology has led to an ability to manipulate the genes of organisms. We can clone genes and then use them to alter the **genome** (the complete genetic makeup of an organism) of viruses and cells, whether bacterial, plant, or animal cells. This practice, called **genetic engineering,** has innumerable uses, from producing a product to treating cancer and genetic disorders.

The Cloning of a Gene

We often think of **cloning** as the production of identical copies of an organism through some asexual means. The members of a bacterial colony on a petri dish are clones because they all came from the division of the same original cell. Human identical twins are also considered clones, because the first two cells of the embryo separated and each became a complete individual.

Another major biological application of cloning is **gene cloning,** which is the production of many identical copies of a single gene. Biologists clone genes for a number of reasons. They might want to produce large quantities of the gene's protein product, such as human insulin, learn how a cloned gene codes for a particular protein, or use the genes to alter the phenotypes of other organisms in a beneficial way. When cloned genes are used to modify a human, the process is called gene therapy (see section 26.3). Otherwise, organisms with foreign DNA or genes inserted into them are called **transgenic organisms,** which are frequently used to produce a product desired by humans. While a variety of techniques now exist to produce cloned DNA, most processes rely on recombinant DNA technology and the polymerase chain reaction (PCR).

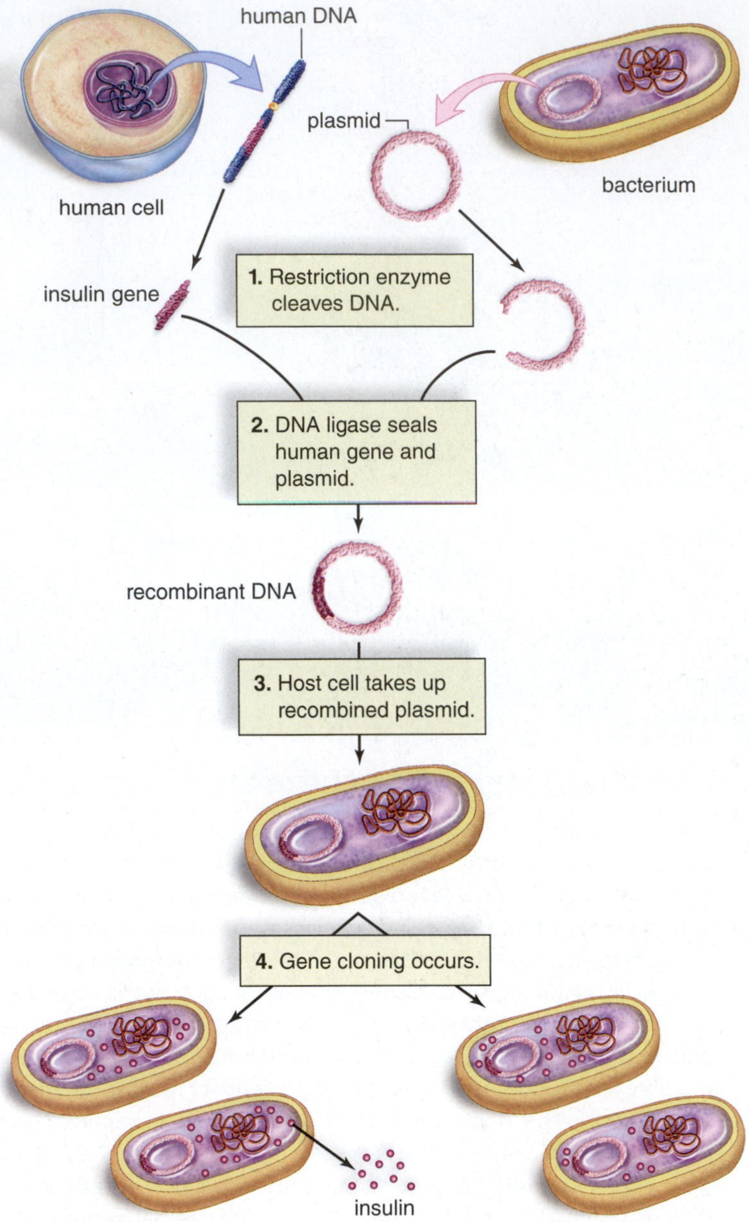
human DNA
plasmid
human cell
bacterium
insulin gene
1. Restriction enzyme cleaves DNA.
2. DNA ligase seals human gene and plasmid.
recombinant DNA
3. Host cell takes up recombined plasmid.
4. Gene cloning occurs.
insulin

Figure 26.1 Cloning a human gene. Human DNA and plasmid DNA are cleaved by a specific type of restriction enzyme. For example, human DNA containing the insulin gene is spliced into a plasmid by the enzyme DNA ligase. Gene cloning is achieved after a bacterium takes up the plasmid. If the gene functions normally as expected, the product (e.g., insulin) may also be retrieved.

Recombinant DNA Technology

Recombinant DNA (rDNA) contains DNA from two or more different sources, such as the human cell and the bacterial cell in Figure 26.1. To make rDNA, a researcher needs a **vector,** a piece of DNA that can be manipulated such that foreign DNA can be added to it. One common vector is a plasmid. **Plasmids** are small accessory rings of DNA from bacteria that are not part of the bacterial chromosome and are capable of self-replicating. Plasmids were discovered by investigators studying the bacterium *Escherichia coli.*

Two enzymes are needed to introduce foreign DNA into vector DNA: (1) a **restriction enzyme** to cleave the vector DNA and (2) **DNA ligase** to seal DNA into an opening created by the restriction enzyme. Hundreds of restriction enzymes occur naturally in bacteria, where they act as a primitive immune system by cutting up any viral DNA that enters the cell. They are called restriction enzymes because they *restrict* the growth of viruses, but they can also be used as molecular scissors to cut double-stranded

Animation Restriction Endonucleases

DNA at a specific site. For example, the restriction enzyme called *Eco*RI always recognizes and cuts double-stranded DNA in the following manner when DNA has the sequence of bases GAATTC:

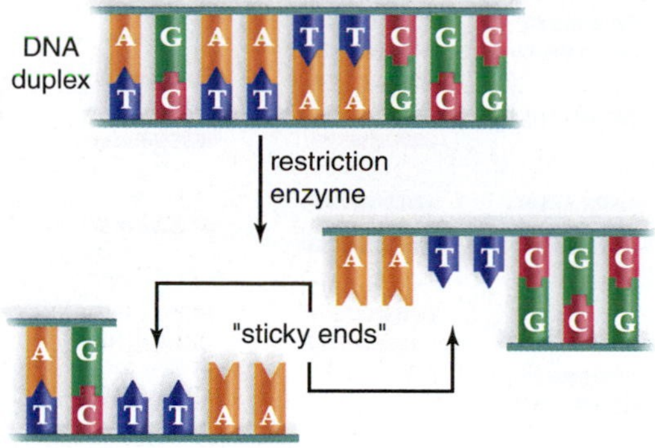
DNA duplex
A G A A T T C G C
T C T T A A G C G
restriction enzyme
"sticky ends"
A A T T C G C
G C G
A G
T C T T A A

Notice that a gap now exists into which a piece of foreign DNA can be placed if it ends in bases complementary to those exposed by the restriction enzyme. To ensure this, it is necessary only to cleave the foreign DNA with the same type of restriction enzyme. The single-stranded, but complementary, ends of the two DNA molecules are called "sticky ends" because they can bind a piece of foreign DNA by complementary base pairing. Sticky ends facilitate the insertion of foreign DNA into vector DNA.

DNA ligase, an enzyme that functions in DNA replication, is then used to seal the foreign piece of DNA into the vector. Bacterial cells take up recombinant plasmids, especially if the cells are treated to make them more permeable. Thereafter, as the plasmid replicates, so does the foreign DNA and thus the gene is cloned.

The Polymerase Chain Reaction

The **polymerase chain reaction (PCR)** can create billions of copies of a segment of DNA in a test tube in a matter of hours. PCR is very specific—it *amplifies* (makes copies of) a targeted DNA sequence, usually a few hundred bases in length. PCR requires the use of DNA polymerase, the enzyme that carries out DNA replication, and a supply of nucleotides for the new DNA strands. PCR involves three basic steps that occur repeatedly, usually for about 35 to 40 cycles: (1) a denaturation step at 95°C, where DNA is heated to become single stranded; (2) an annealing step at a temperature usually between 50° and 60°C, where an oligonucleotide primer hybridizes to each of the single DNA strands; and (3) an extension step at 72°C, where an engineered DNA polymerase adds complementary bases to each of the single DNA strands, creating double-stranded DNA.

PCR is a chain reaction because the targeted DNA is repeatedly replicated, much in the same way natural DNA replication occurs, as long as the process continues. Figure 26.2 uses color

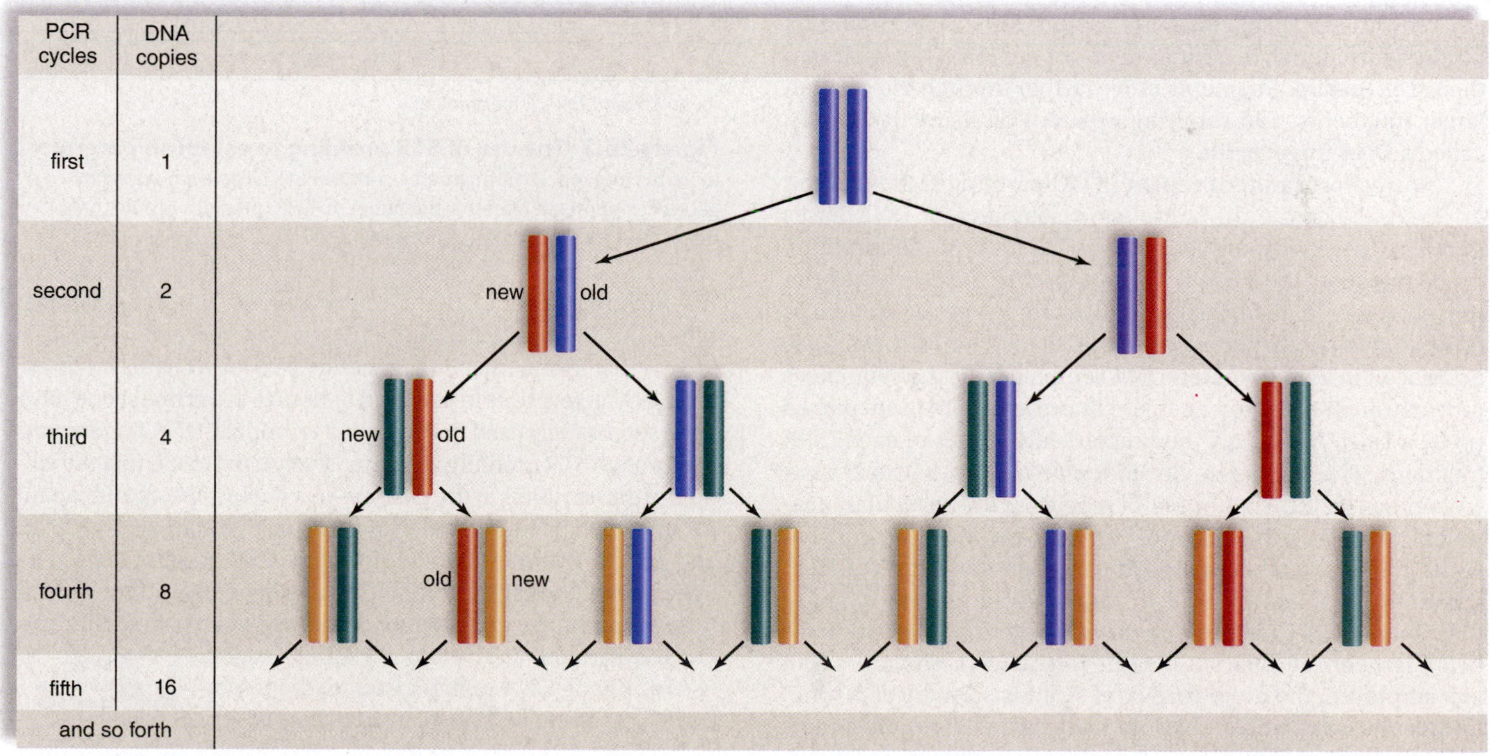
PCR cycles
DNA copies
first
1
second
2
third
4
fourth
8
fifth
16
and so forth
new
old
new
old
old
new

Figure 26.2 Polymerase chain reaction (PCR). PCR allows the production of many identical copies of DNA in a laboratory setting. Assuming you start with only one copy, you can create millions of copies with only 20 to 25 cycles. For this reason, only a tiny DNA sample is necessary for forensic genetics.

to distinguish the old strand from the new DNA strand. Notice that the amount of DNA doubles with each replication cycle. Thus, assuming you start with only one copy of DNA, after one cycle, you will have two copies, after two cycles four copies, and so on. PCR has been in use since its development in 1985 by Kary Banks Mullis, and now almost every laboratory has automated PCR machines to carry out the procedure. Automation became possible after a temperature-insensitive (thermostable) DNA polymerase was extracted from the bacterium *Thermus aquaticus,* which lives in hot springs. The enzyme can withstand the high temperature used to denature double-stranded DNA. Therefore, replication does not have to be interrupted by the need to add more enzyme.

Animation Polymerase Chain Reaction

DNA amplified by PCR is often analyzed for various purposes. For example, mitochondrial DNA base sequences have been used to decipher the evolutionary history of human populations. Because so little DNA is required for PCR to be effective, it is commonly used as a forensic method for analyzing DNA found at crime scenes—only a drop of semen, a flake of skin, or the root of a single hair is necessary!

DNA Analysis

Analysis of DNA following PCR has undergone improvements over the years. At first, the entire genome was treated with restriction enzymes, and because each person has their own restriction enzyme sites, they would have a unique collection of DNA fragment sizes. During a process called *gel electrophoresis,* whereby an electrical current is used to force DNA through a porous gel material, these fragments are separated according to their size. Smaller fragments move farther through the gel than larger fragments, and result in a pattern of distinctive bands, called a **DNA fingerprint.**

Now, **short tandem repeat** (**STR**) profiling is the method of choice. STRs are the same short sequence of DNA bases that recur several times, as in GATAGATAGATA. STR profiling is advantageous because it doesn't require the use of restriction enzymes. Instead, PCR is used to amplify target sequences of DNA, which are fluorescently labeled. The PCR products are placed in an automated DNA sequencer. As the sequences move through the sequencer, the fluorescent labels are picked up by a laser. A detector then records the length of each DNA fragment. The fragments are different lengths because each person has their own number of repeats at the particular location of the STR on the chromosome (i.e., each STR locus). That is, the greater the number of STRs at a locus, the longer the DNA fragment amplified by PCR. If individuals are homozygotes, they will have a single fragment, and heterozygotes will have two fragments of different lengths (Fig. 26.3*a*). The more STR loci employed, the more confident scientists can be of distinctive results for each person (Fig. 26.3*b*). The FBI is collecting a massive number of human STR profiles in its databases for current and future crime scene analyses, and genetic profiles may someday be on peoples' drivers' licenses in the United States. The Bioethical feature, "DNA Forensics," explores some of the pros and cons associated with DNA forensics that may result from these trends.

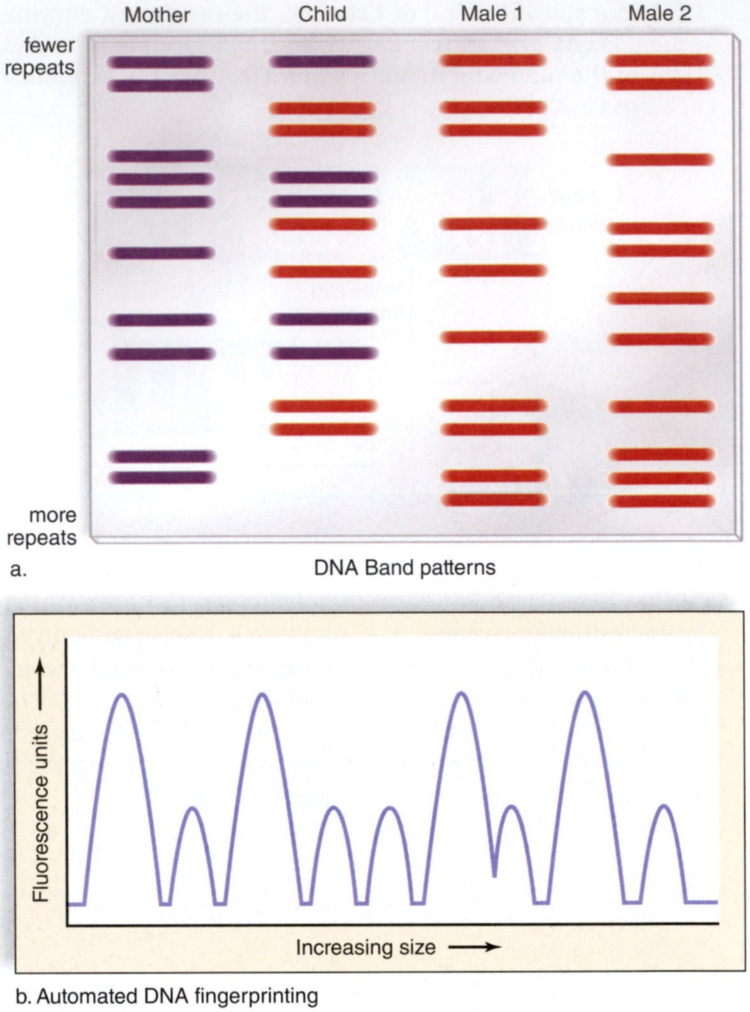
Mother
Child
Male 1
Male 2
fewer repeats
more repeats
a.
DNA Band patterns
Fluorescence units
Increasing size
b. Automated DNA fingerprinting

Figure 26.3 The use of STR profiling to establish paternity. **a.** In this method, DNA fragments containing STRs are separated by gel electrophoresis. Male 1 is the father. **b.** Each person's profile (only one shown) represents a unique pattern.

Applications of PCR are limited only by our imaginations. When the DNA matches that of a virus or mutated gene, it is known that a viral infection, genetic disorder, or cancer is present. DNA fingerprints from blood or tissues at a crime scene have been successfully used in convicting criminals. DNA fingerprinting through STR profiling was extensively used to identify the victims of the tsunamis in the past few years in Indonesia and Japan. Relatives can be found, paternity suits can be settled (Fig. 26.3), and genetic disorders can be detected. PCR has also shed new light on evolutionary studies by comparing DNA extracted from human mummies thousands of years old or animal fossils millions of years old. The National Football League uses synthetic DNA to mark each of the Super Bowl footballs to be able to authenticate them.

Video World Trade Center DNA

Check Your Progress 26.1

1. Summarize the two required steps for producing recombinant DNA.
2. Explain why STRs may be used for identification.

SCIENCE IN YOUR LIFE ▶ BIOETHICAL

DNA Forensics

Traditional fingerprinting has been used for years to identify criminals and to exonerate those wrongly accused of crimes. However, sometimes fingerprints are not left at the crime scene, whereas at other times fingerprints may be present but not connected to the particular crime at all. Forensic DNA analysis, on the other hand, does not depend on fingerprints. It can be used to identify suspects based on the unique DNA sequences in their genes, as obtained from biological samples found at the crime scene. One advantage of DNA profiling is that only a tiny sample is needed because PCR is used. So, the sample can come from a drop of blood, semen from a rape victim, or even a single hair root!

In DNA profiling, target sequences of DNA simple tandem repeats (also called *microsatellites*) are amplified using PCR and separated using automated equipment that can determine the length of the fragments because the DNA is labeled with fluorescent dyes that are picked up by a laser beam as they move through a gel contained in capillaries (see Fig. 26.3*b*). A person could have a single fragment at an STR locus, indicating he or she is a homozygote, or two fragments, indicating he or she is a heterozygote. The suspect's genotype is then compared to that found at the crime scene. A mismatch at one STR locus is enough to exclude a suspect. However, a match at a single locus is not enough to prove that the DNA sample at the crime scene belongs to a particular suspect. Why? The particular genotype is compared to the frequency of that genotype in the FBI database. Let's say that a particular genotype is present in 10% of the population of U.S. males, and that a suspect in a rape case has DNA that matches that in semen collected from the crime scene. This means there is a 10% chance of *accidental match.* Is this guilt beyond a reasonable doubt? Surely not. Therefore, the microsatellite genotyping procedure is done over several different locations in the genome (currently 13, plus a marker for sex) to ensure that the probability of accidental match is incredibly low. Thus, advocates of DNA profiling claim that identification is "beyond a reasonable doubt."

Opponents of this technology, however, point out that it is not without its problems. Police or laboratory negligence can invalidate the evidence. Problems involving sloppy laboratory procedures and the credibility of forensic experts have also been reported.

In addition to identifying criminals, DNA fingerprinting can be used to establish paternity, determine nationality for immigration purposes, and identify victims of a national disaster, such as a tsunami or earthquake.

Now that genomic sequencing is continually becoming cheaper and faster, it may not be in the too distant future that the genome of all humans can be sequenced inexpensively. Considering the usefulness of DNA fingerprints, perhaps everyone should be required to contribute blood to create a national DNA fingerprint databank. To this end, some have even argued for having a complete genome sequence of each person on their driver's license. Others say, however, this would constitute "search without cause," which is unconstitutional.

Questions to Consider

1. Would you be willing to provide your DNA for a national DNA databank? Why or why not?
2. If not everyone, do you think that convicted felons, at least, should be required to provide DNA for a databank?
3. Should all defendants have access to DNA fingerprinting (at government expense) to prove they didn't commit a crime? Should this include those already convicted of crimes who want to reopen their cases using new DNA evidence?

26.2 Biotechnology Products

Learning Outcomes

Upon completion of this section, you should be able to
1. Identify ways in which human society benefits from genetically modified bacteria, plants, and animals.
2. Describe the steps involved in the production of a transgenic animal.

Today, transgenic bacteria, plants, and animals are often called **genetically modified organisms (GMOs)**, and the products they produce are called **biotechnology** products (Fig. 26.4).

Figure 26.4 Biotechnology products. Products, such as the biodegradable plastic poncho shown here, or fuels like bioethanol, are becoming increasingly common.

Transgenic Bacteria

Recombinant DNA technology is used to produce transgenic bacteria, which are grown in huge vats called bioreactors. The bacteria express the cloned gene, and the gene product is usually collected from the medium in which the bacteria are grown. Biotechnology products produced by bacteria include insulin, human growth hormone, tPA (tissue plasminogen activator), and hepatitis B vaccine.

Animation Early Genetic Engineering Experiment

Transgenic bacteria have many other uses as well. Some have been produced to promote the health of plants. For example, bacteria that normally live on plants and encourage the formation of ice crystals have been changed from frost-plus to frost-minus bacteria. As a result, new crops such as frost-resistant strawberries are being developed. Also, a bacterium that normally colonizes the roots of corn plants has now been endowed with genes (from another bacterium) that code for an insect toxin. The toxin protects the roots from insects.

Bacteria can be selected for their ability to degrade a particular substance, and this ability can then be enhanced by bioengineering. For instance, naturally occurring bacteria that eat oil can be genetically engineered to do an even better job of cleaning up beaches after oil spills (Fig. 26.5), such as the 2010 *Deep Water Horizon* spill in the Gulf of Mexico. Bacteria can also remove sulfur from coal before it is burned, resulting in cleaner emissions. One bacterial strain was given genes that allowed it to clean up levels of toxins that would have killed other bacterial strains. Further, these bacteria were given

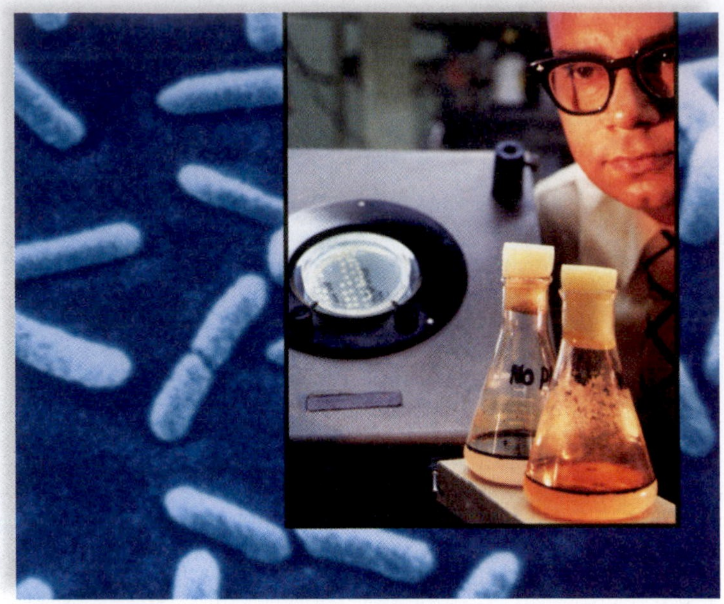

Figure 26.5 Bioremediation. These bacteria, which are capable of decomposing oil, were engineered and patented by the investigator Dr. Chakrabarty. The flask toward the rear contains oil and no bacteria. The flask toward the front contains the engineered bacteria and is almost clear of oil.

SCIENCE IN YOUR LIFE ▶ HEALTH

Are Genetically Engineered Foods Safe?

A series of focus groups conducted by the Food and Drug Administration (FDA) in 2000 showed that although most participants believed that genetically engineered foods, now also called GMOs, might offer benefits, they also feared possible unknown long-term health consequences. The discovery by activists that a type of genetically engineered corn called StarLink had inadvertently made it into the food supply triggered the recall of taco shells, tortillas, and many other corn-based foodstuffs from supermarkets. Further, the makers of StarLink were forced to buy back StarLink from farmers and to compensate food producers at an estimated cost of several hundred million dollars in late 2000.

StarLink is a type of "Bt" corn. It contains a foreign gene taken from a common soil organism, *Bacillus thuringiensis,* which makes a protein that is toxic to many insect pests. About a dozen Bt varieties, including corn, potato, and even a tomato, have now been approved for human consumption. These strains contain a gene for an insecticidal protein called CryIA. Instead, StarLink contained a gene for a related protein called Cry9C, which researchers thought might slow down the chances of pest resistance to Bt corn. In order to get FDA approval for use in foods, the makers of StarLink performed the required tests. Like the other now-approved strains, StarLink

a.

b.

c.

Figure 26A Genetically engineered crops. Genetically engineered (**a**) corn, (**b**) soybeans, and (**c**) cotton crops are increasingly being planted by today's farmers.

"suicide" genes that caused them to self-destruct when their job was done.

Organic chemicals are often synthesized by having catalysts act on precursor molecules or by using bacteria to carry out the synthesis. Today, it is possible to go one step further and manipulate the genes that code for these enzymes. For instance, biochemists discovered a strain of bacteria that is especially good at producing phenylalanine, an organic chemical needed to make aspartame, better known as NutraSweet. They isolated, altered, and cloned the appropriate genes so that various bacteria could be genetically engineered to produce phenylalanine.

Transgenic Plants

Techniques have been developed to introduce foreign genes into immature plant embryos or into plant cells called *protoplasts* that have had their cell wall removed. It is possible to treat protoplasts with an electric current while they are suspended in a liquid containing foreign DNA. The electric current makes tiny, self-sealing holes in the plasma membrane through which the desired genetic material can enter. Protoplasts go on to develop into mature plants containing and expressing the foreign DNA.

One altered plant known as the pomato is the result of these technologies. This plant produces potatoes belowground and tomatoes aboveground. Foreign genes transferred to cotton, corn, and potato strains have made these plants resistant to pests because their cells now produce an insect toxin. Similarly, soybeans have been made resistant to a common herbicide that is sprayed to kill weeds that compete with soybean growth. Some corn and cotton plants are both pest- and herbicide-resistant. These and other genetically engineered crops that are expected to have increased yields are now commonly sold commercially. However, as discussed in the Health feature, "Are Genetically Engineered Foods Safe?," the public is concerned about the possible effect of genetically modified organisms, also called GMOs, on human health, as well as the environment.

Video
GM Food Safety

Like bacteria, plants are also being engineered to produce human proteins, such as hormones, clotting factors, and antibodies, in their seeds. One type of antibody made by corn can deliver radioisotopes to tumor cells, and another made by soybeans can be used to treat genital herpes.

Video
Good Tobacco

Video
Potato Vaccine

wasn't poisonous to rodents, and its biochemical structure is not similar to those of most chemicals in food that commonly cause allergic reactions in humans (called allergens). But the Cry9C protein resisted digestion longer than the other Bt proteins when it was put in simulated stomach acid and subjected to heat. Because most food allergens resist digestion in a similar fashion, StarLink was not approved for human consumption.

The scientific community is now trying to devise more tests for allergens because it has not been possible to determine conclusively whether Cry9C is or is not an allergen. Also, at this point, it is unclear how resistant to digestion a protein must be in order to be an allergen, and it is also unclear what degree of amino acid sequence similarity a potential allergen must have to a known allergen to raise concern. Dean D. Metcalfe, chief of the Laboratory of Allergic Diseases at the National Institute of Allergy and Infectious Diseases, said, "We need to understand thresholds for sensitization to food allergens and thresholds for elicitation of a reaction with food allergens."

Other scientists are concerned about the following potential drawbacks to the planting of Bt corn: (1) resistance among populations of the target pest, (2) exchange of genetic material between the transgenic crop and related plant species, and (3) Bt crops' impact on nontarget species. They feel that many more studies are needed before stating for certain that Bt corn has no ecological drawbacks.

Despite controversies, the planting of genetically engineered corn has increased annually. The USDA reports that U.S. farmers planted genetically engineered corn on between 81% and 86% of all corn acres in 2011, up from 26% in 2001 (Fig. 26A). Today, almost 90% of soybean acreage, and 96% of cotton acreage is genetically modified. Some groups advocate that GMOs should be labeled as such, but this may not be easy to accomplish because, for example, most cornmeal is derived from both conventional and genetically engineered corn. So far, there has been no attempt to sort out one type of food product from the other. However, at many health food stores, foods that *do not* contain GMOs are now labeled.

Questions to Consider

1. Do you think genetically modified organisms (GMOs) should be labeled? Construct an argument both for and against labeling GMOs.
2. Some people are strongly advocating the complete removal of GMOs from the market. A few people, called ecoterrorists, are taking such drastic actions as burning crops or even setting biotechnology labs on fire. Do you think GMOs should be removed from the market? Why or why not? What further information would you need to make your decision?
3. Rice is a staple in the diet of millions of people worldwide, many of them living in less-developed countries. In some of those same countries, vitamin A deficiency is a major cause of blindness in small, malnourished children. Scientists have developed a new form of rice, called "golden rice," which is genetically modified to assist with the metabolism of vitamin A and might prevent millions of cases of blindness. The scientists who created golden rice claim it is safe for consumption, although critics say the health effects are not yet fully understood. Given its nutritional potential, should golden rice be planted and distributed on a wide scale? What information would you need to make your decision?

Transgenic Animals

Techniques have been developed to insert genes into the eggs of animals. It is possible to microinject foreign genes into eggs by hand, but another method uses vortex mixing. The eggs are placed in an agitator with DNA and silicon-carbide needles. The needles make tiny holes in the eggs through which the DNA can enter. When these eggs are fertilized, the resulting offspring are transgenic animals. Using this technique, many types of animal eggs have acquired the gene for bovine growth hormone (BGH). The procedure has been used to produce larger fishes, cows, pigs, rabbits, and sheep.

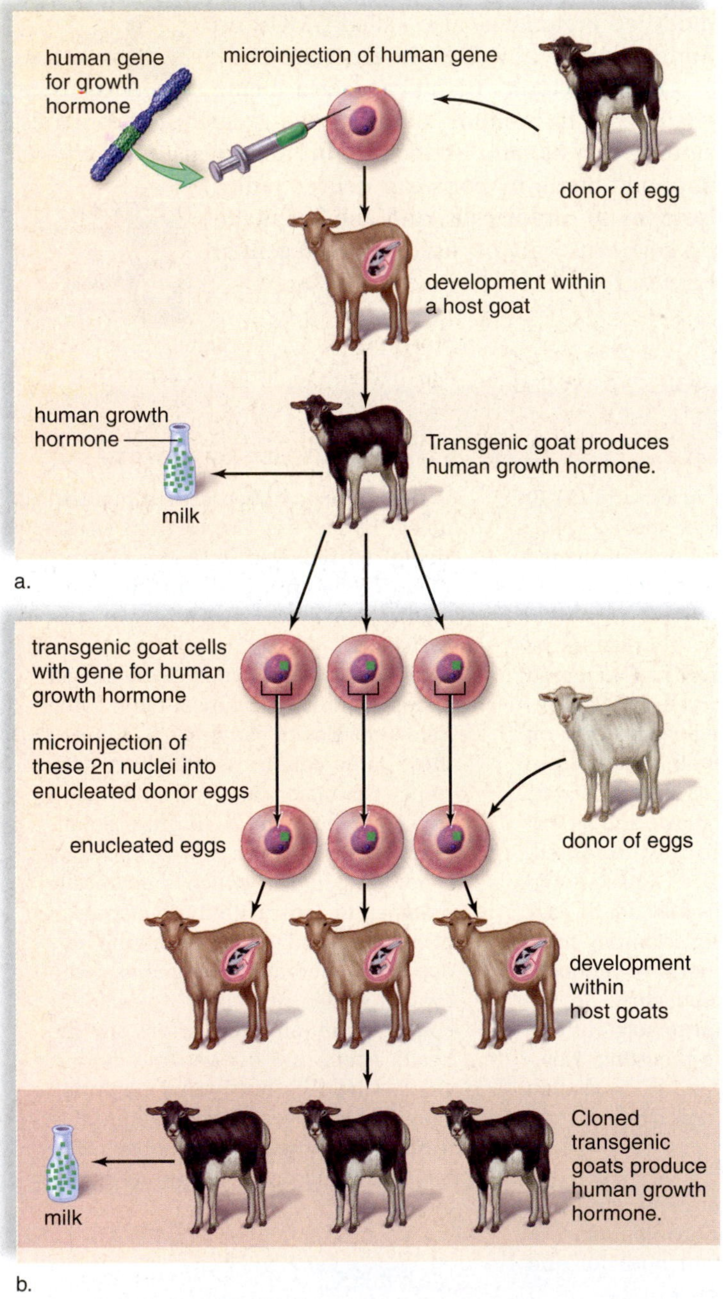
human gene for growth hormone
microinjection of human gene
donor of egg
development within a host goat
human growth hormone
Transgenic goat produces human growth hormone.
milk
a.
transgenic goat cells with gene for human growth hormone
microinjection of these 2n nuclei into enucleated donor eggs
enucleated eggs
donor of eggs
development within host goats
milk
Cloned transgenic goats produce human growth hormone.
b.

Figure 26.6 Production of transgenic animals. **a.** A genetically engineered egg develops in a host to create a transgenic goat that produces a biotechnology product in its milk. **b.** Nuclei from the transgenic goat are transferred into donor eggs, which develop into cloned transgenic goats.

Gene pharming, the use of transgenic farm animals to produce pharmaceuticals, is being pursued by a number of firms. Genes that code for therapeutic and diagnostic proteins are incorporated into an animal's DNA, and the proteins appear in the animal's milk. Plans are under way to produce drugs for the treatment of cystic fibrosis, cancer, blood diseases, and other disorders by this method. Figure 26.6 outlines the procedure for producing transgenic animals: DNA containing the gene of interest is injected into donor eggs. Following in vitro fertilization, the zygotes are placed in host females, where they develop. After female offspring mature, the product is secreted in their milk.

Video
Cloned Milk
Video
Firefly Rx

Eliminating a gene is another way to study a gene's function. A knockout mouse has had both alleles of a gene removed or made nonfunctional. For example, scientists have constructed a knockout mouse lacking the *CFTR* gene, the same gene mutated in cystic fibrosis patients. The mutant mouse has a phenotype similar to a human with cystic fibrosis and can be used to test new drugs for the treatment of the disease.

Check Your Progress 26.2

1. List some of the beneficial applications of transgenic bacteria, plants, and animals.
2. Distinguish between a transgenic organism and a cloned organism.

26.3 Gene Therapy

Learning Outcome

Upon completion of this section, you should be able to

1. Compare and contrast in vivo and ex vivo gene therapy.

Gene Therapy

Once a genetic disorder is detected, gene therapy is a potential course of treatment. **Gene therapy** is the insertion of genetic material into human cells for the treatment of genetic disorders and various other human illnesses, such as cardiovascular disease and cancer. Figure 26.7 shows regions of the body that have received copies of normal genes by various methods of gene transfer. Viruses genetically modified to be safe can be used to ferry a normal gene into the body, and so can liposomes, which are microscopic globules of lipids specially prepared to enclose the normal gene (i.e., ex vivo gene therapy). On the other hand, sometimes the gene is injected directly into a particular region of the body (i.e., in vivo gene therapy).

Note that despite its promise for treating disorders, gene therapy is still in its infancy and has had detrimental side effects for many patients, such as causing leukemia. Nonetheless, some strides are being made with particular diseases, such as those described in the following sections.

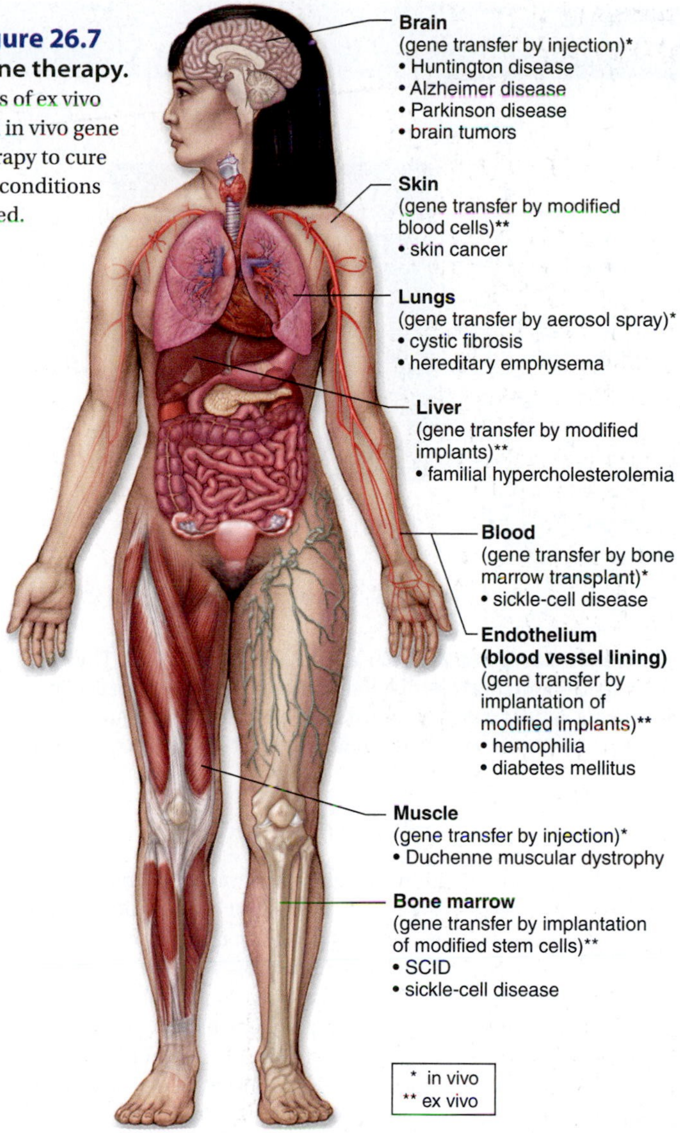
Brain
(gene transfer by injection)*
• Huntington disease
• Alzheimer disease
• Parkinson disease
• brain tumors
Skin
(gene transfer by modified blood cells)**
• skin cancer
Lungs
(gene transfer by aerosol spray)*
• cystic fibrosis
• hereditary emphysema
Liver
(gene transfer by modified implants)**
• familial hypercholesterolemia
Blood
(gene transfer by bone marrow transplant)*
• sickle-cell disease
Endothelium (blood vessel lining)
(gene transfer by implantation of modified implants)**
• hemophilia
• diabetes mellitus
Muscle
(gene transfer by injection)*
• Duchenne muscular dystrophy
Bone marrow
(gene transfer by implantation of modified stem cells)**
• SCID
• sickle-cell disease
* in vivo
** ex vivo

Figure 26.7 Gene therapy. Sites of ex vivo and in vivo gene therapy to cure the conditions noted.

Ex Vivo Gene Therapy

Figure 26.8 describes an ex vivo methodology for treating children who have SCID (severe combined immunodeficiency). These children lack the enzyme ADA (adenosine deaminase), which is involved in the maturation of T and B cells. Therefore, these children are prone to constant infections and may die without treatment. To carry out gene therapy, bone marrow stem cells are removed from the bone marrow of the patient and infected with a virus that carries a normal gene for the enzyme into their DNA. Then the cells are returned to the patient, where it is hoped they will divide to produce more blood cells with the same genes. Patients who have undergone this procedure show significantly improved immune function associated with a sustained rise in the level of ADA enzyme activity in the blood.

Another example of ex vivo gene therapy has been used to treat familial hypercholesterolemia, a condition that develops when liver cells lack a receptor protein for removing cholesterol from the blood. The high levels of blood cholesterol make the patient subject to fatal heart attacks at a young age. A small portion of the liver is surgically excised and then infected with a virus containing a normal gene for the receptor before being returned to the patient. Patients are expected to experience lowered serum cholesterol levels following this procedure.

In Vivo Gene Therapy

Cystic fibrosis patients lack a gene that codes for the transmembrane carrier of the chloride ion (for genetics of this disease see section 23.2). They often die due to numerous infections of the respiratory tract because a thick mucus forms in the lungs and attracts bacteria and other antigens. In gene therapy trials, the gene needed to cure cystic fibrosis is sprayed into the nose or delivered to the lower respiratory tract by an adenovirus vector or by using liposomes. So far, these treatments have met with limited success, but investigators are trying to improve uptake by using a combination of different vectors.

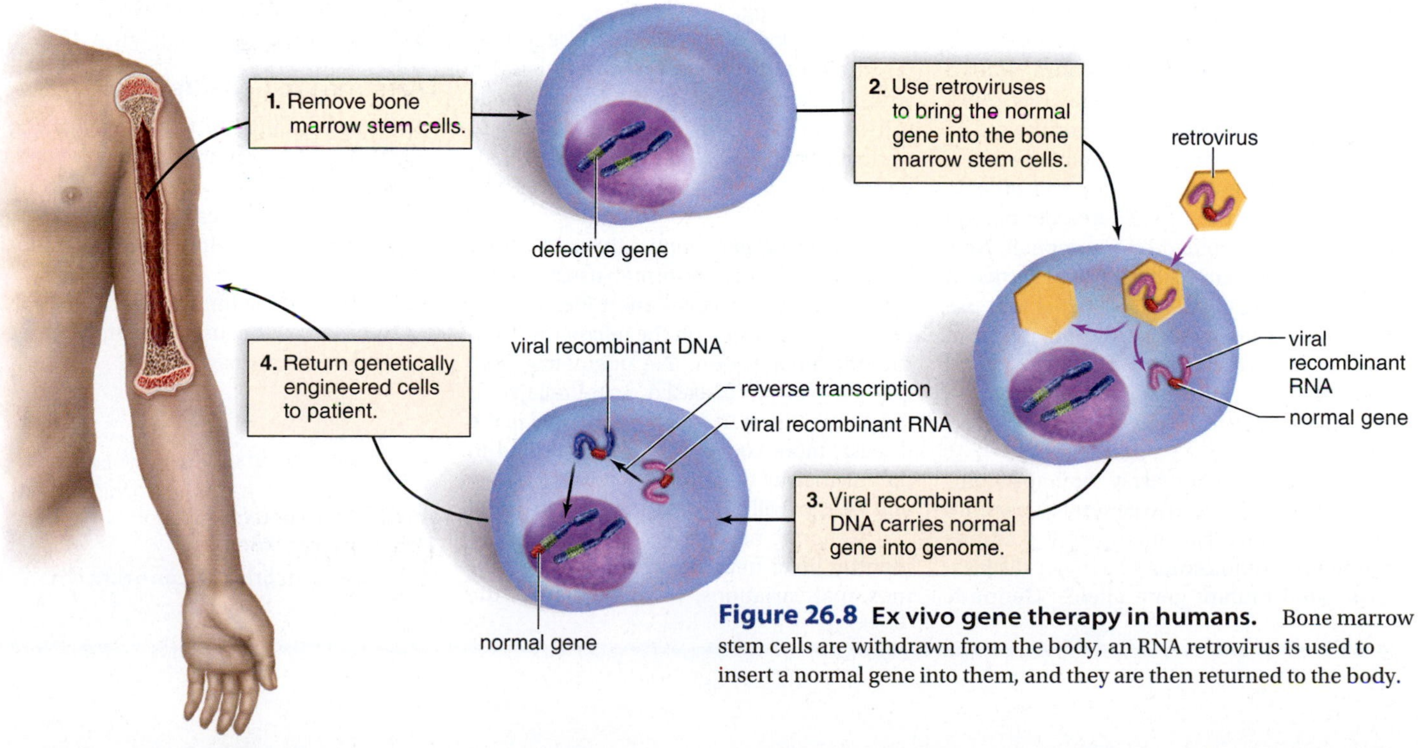
1. Remove bone marrow stem cells.
defective gene
2. Use retroviruses to bring the normal gene into the bone marrow stem cells.
retrovirus
viral recombinant RNA
normal gene
3. Viral recombinant DNA carries normal gene into genome.
viral recombinant DNA
reverse transcription
viral recombinant RNA
normal gene
4. Return genetically engineered cells to patient.

Figure 26.8 Ex vivo gene therapy in humans. Bone marrow stem cells are withdrawn from the body, an RNA retrovirus is used to insert a normal gene into them, and they are then returned to the body.

SCIENCE IN YOUR LIFE ▶ SCIENTIFIC INQUIRY

Testing for Genetic Disorders

Prospective parents know if either of them has an autosomal dominant disorder because the person will show it. However, genetic testing is required to detect if either is a carrier for an autosomal recessive disorder. If a woman is already pregnant, the parents may want to know if the unborn child has the disorder. If the woman is not pregnant, the parents may opt for testing of an embryo or egg before she does become pregnant. One way to detect genetic disorders is to test the DNA for mutated genes.

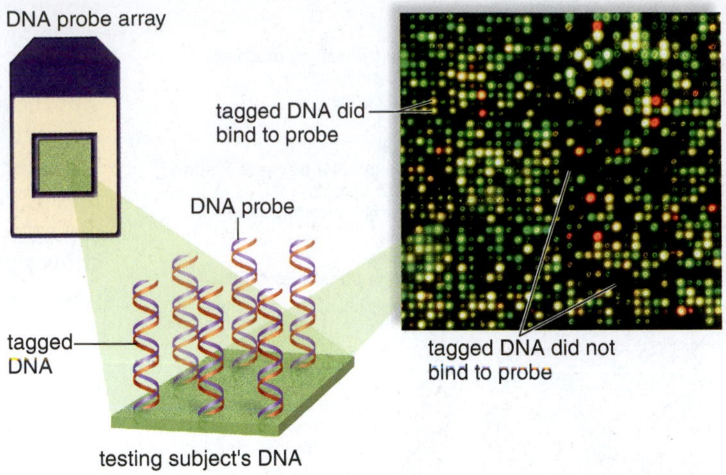
DNA probe array
tagged DNA did bind to probe
DNA probe
tagged DNA
tagged DNA did not bind to probe
testing subject's DNA

Figure 26B Use of a DNA microarray to test for a genetic disorder. This DNA chip contains rows of DNA sequences for mutations that indicate the presence of particular genetic disorders. If DNA fragments derived from an individual's DNA bind to a sequence representing a mutation on the DNA chip, that sequence fluoresces, and the individual has the mutation.

Testing the DNA

DNA testing typically uses procedures that test for a specific genetic marker, or probe the genome for sequences of interest using DNA microarrays.

Testing for a genetic marker is similar to the traditional procedure for DNA fingerprinting, as discussed earlier. As an example, consider that individuals with Huntington disease have an abnormality in the sequence of their bases at a particular location on a chromosome. This abnormality in sequence is a *genetic marker.* Huntington disease, specifically, results from a STR that is so long that it actually causes a frameshift mutation within a gene even though the STR itself occurs outside, but nearby the gene. In this and similar cases, the length of the STR can be detected with PCR and analysis on an automated DNA sequencer.

DNA Microarrays

With advances in robotic technology, it is now possible to place the entire human genome onto a single microarray (Fig. 26B). The mRNA from the organism or the cell to be tested is labeled with a fluorescent dye and added to the chip. When the mRNAs bind to the microarray, a fluorescent pattern results that is recorded by a computer. Now the investigator knows what DNA is active in that cell or organism. A researcher can use this method to determine the difference in gene expression between two different cell types, such as between liver cells and muscle cells.

Animation DNA Microarray

A mutation microarray, the most common type, can be used to generate a person's genetic profile. The microarray contains hundreds to thousands of known disease-associated mutant gene alleles. Genomic DNA from the individual to be tested is labeled with a fluorescent dye, and then added to the microarray. The spots on the microarray fluoresce if the individual's DNA binds to the mutant genes on the chip, indicating that the individual may have a particular disorder or is at risk for developing it later in life. This technique can generate a genetic profile much more quickly and inexpensively than older methods involving DNA sequencing.

DNA microarrays also promise to hasten the identification of genes associated with diseased tissues. In the first instance, mRNA derived from diseased tissue and normal tissue is labeled with different fluorescent dyes. The normal tissue serves as a control.

The investigator applies the mRNA from both normal and abnormal tissue to the microarray. The relative intensities of fluorescence from a spot on the microarray indicate the amount of mRNA originating from that gene in the diseased tissue relative to the normal tissue. If a gene is activated in the disease, more copies of mRNA will bind to the microarray than from the control tissue, and the spot will appear more red than green.

Genomic microarrays are also used to identify links between disease and chromosomal variations. In this instance, the chip contains genomic DNA that is cut into fragments. Each spot on the microarray corresponds to a known chromosomal location. Labeled genomic DNA from diseased tissues and control tissues bind to the DNA on the chip, and the relative fluorescence from both dyes is determined. If the number of copies of any particular target DNA has increased, more sample DNA will bind to that spot on the microarray relative to the control DNA, and a difference in fluorescence of the two dyes will be detected.

Questions to Consider

1. What benefits are there when using a DNA microarray over a genetic marker such as a STR?
2. Why might a researcher want to know what genes are being expressed in different cell types?
3. How might the information from a DNA microarray be used to develop new drugs to treat disease?

Gene therapy is increasingly relied upon as a part of cancer treatment. Genes are being used to make healthy cells more tolerant of chemotherapy, while making tumor cells more sensitive. Knowing that the tumor suppressor gene *p53* brings about apoptosis (cell death), researchers are interested in finding a way to selectively introduce *p53* into cancer cells, and in that way, kill them.

Check Your Progress 26.3

1. Summarize the methods that are being used to introduce genes into humans for gene therapy.
2. Provide examples of ex vivo and of in vivo gene therapy.

26.4 Genomics, Proteomics, and Bioinformatics

Learning Outcomes

Upon completion of this section, you should be able to

1. Discuss the implications of knowing the human genome sequence.
2. Describe a major insight from comparative genomics.
3. Compare and contrast functional genomics and proteomics.

In the preceding century, researchers discovered the structure of DNA, how DNA replicates, and how DNA and RNA are involved in the process of protein synthesis. Genetics in the twenty-first century largely concerns **genomics,** the study of the complete genetic sequences of humans and other organisms. Knowing the sequence of bases in genomes is the first step, and mapping their location on the chromosomes is the next step. The enormity of the task can be appreciated by knowing not only that we have approximately 25,000 genes that code for proteins, but also that nearly 99% of the 3.2 billion bases of our genome is noncoding and contains many repetitive sequences of unknown function. Many other organisms have even a larger number of protein-coding genes but fewer noncoding regions when compared to the human genome.

Sequencing the Genome

We now know the sequence of the roughly 3.2 billion pairs of DNA bases in our genome. Stretched out, the DNA in each of our cells is 6 feet long, and the nucleotides would make a book over a half million pages if printed as text. This feat, which has been compared to completion of the periodic table of the elements in chemistry, was accomplished by the **Human Genome Project** (**HGP**), a 13-year effort that involved both university and private laboratories around the world. How did they do it? First, investigators developed a laboratory procedure that would allow them to decipher a short sequence of base pairs, and then instruments became available that could carry out sequencing automatically. Over the 13-year span, DNA sequencers were constantly improved, and today's instruments can automatically analyze up to 120 million base pairs of DNA in a 24-hour period. So, new genomes are being sequenced all the time, and at a much faster rate than the human genome. For example, the genome of the African clawed frog, *Xenopus laevis,* which is roughly the same size as the human genome, was sequenced in under a year.

Completion of the human genome sequence has opened up great possibilities for biomedical research and treatment. These methods have been used to screen for particular diseases, as discussed earlier in this chapter. The HGP also led to the discovery of many small regions of DNA that vary among individuals (polymorphisms). Most of these are *single nucleotide polymorphisms* (*SNPs*), meaning that they have a difference of only one nucleotide. Many SNPs have no effect. Others may contribute to protein-coding differences affecting the phenotype. It's possible that certain SNP patterns change an individual's susceptibility to disease and alter their response to medical treatments. These discoveries have now led pharmaceutical companies to consider producing "designer drugs," which are tailored for an individual's genotype.

Determining that humans have approximately 25,000 genes required a number of techniques, many of which relied on identifying RNAs in cells and then working backward to find the DNA that can pair with that RNA. Structural genomics—knowing the sequence of the bases and how many genes we have—is now being followed by functional genomics. Most of the known human genes are expected to code for proteins. However, most of the human genome is noncoding because it does not specify the order of amino acids in a polypeptide. This noncoding DNA, once dismissed as "junk DNA," is now thought to serve many important functions, and has piqued the curiosity of many investigators.

Genome Architecture

As mentioned previously, researchers were somewhat surprised that nearly 99% of the human genome is DNA that does not directly code for amino acid sequences. Some of the DNA that does not specify polypeptides is transcribed into ribosomal RNA and transfer RNA, both structural molecules involved in protein assembly. The rest of the genome consists of transposable elements (or transposons), repetitive DNA elements, and sequences with unknown function. Transposable elements make up approximately 45% of the human genome. Transposable elements, originally discovered by Barbara McClintock in 1950 (who later won a Nobel Prize for this work), are short sequences of DNA that are able to jump from one location on a chromosome to another. Their movement to a new location sometimes alters neighboring genes, particularly decreasing their expression. In other words, a transposon sometimes acts like a regulator gene. The movement of transposons throughout the genome is thought to be a driving force in the evolution of life.

Animation Transposons: Shifting Segments of the Genome

Nearly half of the human genome is made up of repetitive elements, which occur when the same sequence of two or more nucleotides (e.g., CACACA) are repeated many times along the length of one or more chromosomes. Although many scientists still dismiss them as having no function, others point out that the centromeres and telomeres of chromosomes are composed of repetitive elements and, therefore, repetitive DNA elements may not be as useless as once thought. Telomeres are repetitive DNA sequences found near the ends of chromosomes and are thought to help maintain their structural stability. In addition, perhaps repetitive sequences in centromeres may help with segregating sister chromatids during cell division.

Redefining the Gene

Knowledge of the human genome sequences has changed the way researchers think about the concept of a "gene." Historically, a gene was thought of as a particular location (locus) of a chromosome.

While prokaryotes typically possess a single circular chromosome with genes that are tightly packed together, eukaryotic chromosomes are much more complex. The genes are seemingly randomly distributed along the length of a chromosome and are fragmented into exons, with intervening sequences called introns scattered throughout the length of the gene. In fact, 95% or more of most human genes is composed of introns. Recall that after transcription, introns need to be spliced out and exons joined together to form a functional mRNA transcript that will next be translated into a protein. Once regarded as merely intervening sequences, introns are now attracting attention as regulators of gene expression. The presence of introns allows exons to be put together in various sequences so that different mRNAs and proteins can result from a single gene. It could also be that introns function to regulate gene expression and help determine which genes are to be expressed and how they are to be spliced. In fact, entire genes have been found embedded within the introns of other genes.

Thus, perhaps the modern definition of a gene should take the emphasis away from the chromosome and place it on the results of transcription. Previously, molecular genetics considered a gene to be a nucleic acid sequence that codes for the sequence of amino acids in a protein. In contrast to this definition, we have known for some time that all three types of RNA (rRNA, mRNA, and tRNA) are transcribed from DNA and that these RNAs are useful products. We also know that protein-coding regions can be interrupted by regions that do not code for a protein but do produce RNAs with various functions. In light of these new findings, Mark Gerstein and associates suggested a new definition in 2007: "A gene is a genomic sequence (either DNA or RNA) directly encoding functional products, either RNA or protein." This definition takes into account three new things we have learned by investigating genomic sequences: (1) a gene product may not necessarily be a protein; (2) a gene may not be found at a particular locus on a chromosome; and (3) the genetic material need not be only DNA—some prokaryotes have RNA genes.

Functional and Comparative Genomics

In addition to the human genome, the genomes of many other organisms, including a common bacterium, a yeast, and a mouse, are also complete (Table 26.1). Because we now know the nucleotide sequence of many genomes, we can now focus on comparative and functional genomics.

Using **comparative genomics,** researchers have identified many similarities between the sequence of human bases and those of other organisms. Model organisms (e.g., those found in Table 26.1) can be used in these types of genetic analyses because they share many mechanisms and cellular pathways with other organisms, including humans. For example, scientists inserted a human gene associated with Parkinson disease into the fruit fly, *Drosophila melanogaster,* and the flies showed symptoms similar to those seen in humans with the disorder. These studies confirmed that the suspected gene is involved in Parkinson disease and suggested we might be able to use fruit flies to test potential therapies.

Comparative genomics also offers a way to study changes in the genome through time because some diseases, as well as model organisms, have a shorter generation time than humans. In this way, we have been able to track the evolution of the human immunodeficiency virus (HIV), the virus that causes AIDS, in individual patients. Tracking the genome sequence of the virus

TABLE 26.1 Comparison of Sequenced Genomes

Organism	 ***Homo sapiens*** **(human)**	 ***Mus musculus*** **(mouse)**	 ***Drosophila melanogaster*** **(fruit fly)**	 ***Arabidopsis thaliana*** **(flowering plant)**	 ***Caenorhabditis elegans*** **(roundworm)**	 ***Saccharomyces cerevisiae*** **(yeast)**
Estimated Size	3.2 billion bases	2.5 billion bases	180 million bases	125 million bases	97 million bases	12 million bases
Estimated Number of Genes	~25,000	~25,000	13,600	25,500	19,100	6,300
Chromosome Number	46	40	8	10	12	32

through time has allowed scientists and doctors to understand how the virus responds to different drug therapy regimens and, in some cases, modify treatment to improve a patient's life span.

Comparing genomes will also help us understand the evolutionary relationships among organisms. One surprising discovery is that the genomes of all vertebrates are similar. Researchers were not surprised to find that the genomes of humans and chimpanzees were approximately 98% alike, but they did not expect to find that the human and mouse sequence were 85% similar. Genomic comparisons will likely yield improved ability to reconstruct evolutionary relationships among organisms, as discussed in Chapter 27. Comparative genomics also allows researchers to infer function of unknown genes in one species through amino acid similarity of those genes in another species.

The aim of **functional genomics** is to understand the function of the various genes discovered within each genomic sequence and how these genes interact. In fact, functional genomics has utilized comparative genomics to assess similarities between the human genes and genes of other organisms to help deduce the probable function of many of our estimated 25,000 genes. Functional genomics also uses DNA microarrays to monitor the expression of thousands of genes simultaneously. The use of a microarray can tell what genes are turned on in a specific cell or tissue type in a particular organism at a particular point in time and under certain environmental circumstances. For example, we could compare gene expression of a patient in different stages of cancer growth to assist with treatment. As discussed earlier in this chapter, DNA microarrays can also be used to identify various mutations in a human's genome. This is called the person's **genetic profile,** which can be used to determine if various genetic diseases are likely, as well as to suggest which drug therapy may be most appropriate based on the individual's genotype.

Proteomics

Now that entire genomes are being published for different species, there is a race to sequence their proteomes, or a species' entire collection of proteins. **Proteomics** is the study of the structure, function, and interaction of cellular proteins, which differ depending on each cell type. Each cell produces hundreds of different proteins that can vary between cells and within the same cell, depending on conditions. Therefore, the goal of proteomics is an overwhelming endeavor.

Computer modeling of the three-dimensional shape of these proteins is an important part of proteomics. The study of protein shape and function is essential to the discovery of better drugs so that their chemical structure can match that of different protein shapes. One day, it may be possible to correlate drug treatment to the particular genome of the individual to increase efficiency and decrease side effects.

Bioinformatics

Bioinformatics is the application of computer technologies, specially developed software, and statistical techniques to the study of biological information, particularly databases that contain much genomic and proteomic information (Fig. 26.9). The new data produced by structural genomics and proteomics have produced literally terabytes of raw data stored in databases that are readily available to research scientists. It is called raw data because billions of base pairs of DNA nucleotide sequence have little meaning by themselves. Functional genomics and proteomics are dependent on computer analysis to find significant patterns in the raw data. For example, BLAST, which stands for *b*asic *l*ocal *a*lignment *s*earch *t*ool, is a computer program that can identify homologous genes among the genomic sequences of model organisms. Homologous genes are genes that code for the same proteins, although the base sequence may be slightly different. Finding these differences can help identify the putative function of genes as new organisms' genomes are sequenced, and also help to trace the history of evolution among a group of organisms. For example, researchers found the function of the protein that causes cystic fibrosis by using the computer to search for genes in model organisms that have the same sequence. Because they knew the function of this same gene in model organisms, they could deduce the function in humans. This was a necessary step toward possibly developing specific treatments for cystic fibrosis.

Figure 26.9 Bioinformatics. New computer programs are being developed to make sense out of the raw data generated by genomics and proteomics. Bioinformatics allows researchers to study both functional and comparative genomics in a meaningful way.

Bioinformatics has various applications in human genetics. The human genome has 3.2 billion known base pairs, and without the computer it would be almost impossible to make sense of these data. For example, it is now known that an individual's genome often contains multiple copies of a gene. But individuals may differ as to the number of copies—called copy number variations. Now it seems that the number of copies of a gene in a genome can be associated with specific diseases. The computer can help make correlations between genomic differences among large numbers of people and certain diseases.

It is safe to say that without bioinformatics, our progress in assembling DNA sequences into genomes; determining the function of DNA sequences; mapping genes on chromosomes; comparing our genome to model organisms; knowing how genes and proteins interact in cells; and so forth, would be extremely slow. Instead, with the help of bioinformatics, progress should proceed rapidly in these and other areas.

Check Your Progress 26.4

1. Explain the difference between genomics and proteomics.
2. Explain how comparative genomics can provide insights into gene function.
3. Discuss the importance of bioinformatics to the study of genomics and proteomics.

Case Study Conclusion

Through reading this chapter, it should be clear that advances in genomics and biotechnology will allow us to better understand the causes of human diseases, as well as to develop potential treatments and cures. One of the great challenges of the twenty-first century will be to learn how to assemble and interpret the massive quantities of data generated through genome sequencing projects, comparative genomics, and proteomics. An even greater challenge to society is to use these data and the technologies that result in an ethically responsible manner. There are many ethical dilemmas facing the use of biotechnology, and thus the need for everyone to understand the principles of the material presented in this chapter.

MEDIA STUDY TOOLS

www.mhhe.com/maderinquiry14

Enhance your study of this chapter with study tools and practice tests. Also ask your instructor about the resources available through ConnectPlus, including LearnSmart, the media-rich eBook, interactive learning tools, and animations.

SUMMARIZE

26.1 DNA Cloning

- Knowledge of the **genome** has enabled scientists to conduct **genetic engineering** and several types of **cloning. Gene cloning** can be used to isolate a gene and produce many copies of it. The gene can be studied in the laboratory or inserted into a bacterium, plant, or animal, producing a **transgenic organism.** Several methods are available for studying DNA molecules: **recombinant DNA** (**rDNA**) technology and the **polymerase chain reaction** (**PCR**). Recombinant DNA contains DNA from two different sources. A **restriction enzyme** cleaves both **vector** (often a **plasmid**) DNA and foreign DNA. The resulting "sticky ends" facilitate insertion of foreign DNA into vector DNA. The foreign gene is sealed into the vector DNA by **DNA ligase.**
- PCR uses a heat-resistant DNA polymerase to quickly make multiple copies of a specific piece (target) of DNA. PCR is a chain reaction because the targeted DNA is replicated over and over again. Analysis of DNA segments following PCR may generate a **DNA fingerprint** (or profile). These have multiple uses from assisting genomic research to DNA forensics studies. Increasingly, **short tandem repeat** (**STR**) profiling is being used to identify individuals or for forensic studies.

26.2 Biotechnology Products

- Transgenic organisms, also called **genetically modified organisms** (**GMOs**), have had a foreign gene inserted into them.
- Genetically modified bacteria, agricultural plants, and farm animals now produce **biotechnology** products of interest to humans, such as hormones and vaccines. Bacteria usually secrete the product, but the seeds of plants and the milk of animals contain the product.
- Transgenic bacteria have also been engineered to promote the health of plants, extract minerals, and produce medically important chemicals.
- Transgenic crops, engineered to resist herbicides and pests, are commercially available.
- Transgenic animals have been given various genes, in particular the one for bovine growth hormone (BGH). Cloning of whole animals is now possible.

26.3 Gene Therapy

- **Gene therapy,** by either ex vivo or in vivo methods, is used to correct the genotype of humans and to cure various human ills by giving the patient a foreign gene.
- During ex vivo therapy, cells are removed from the patient, treated, and returned to the patient. Ex vivo gene therapy has apparently helped children with SCID lead normal lives.

- In vivo therapy consists of directly giving the patient a foreign gene that will improve his or her health. Although it has limited success for treating cystic fibrosis, a number of in vivo therapies are being employed in the war against cancer and other human illnesses, such as cardiovascular disease.

26.4 Genomics, Proteomics, and Bioinformatics

- **Genomics** is the study of the complete genetic sequences of a species. Because of the **Human Genome Project (HGP)**, researchers now know the sequence of all the base pairs of the human genome. So far, between approximately 25,000 genes that code for proteins have been found; the rest of our DNA consists of noncoding regions.
- Noncoding regions include introns, transposable elements, and repetitive DNA. Noncoding DNA, once dismissed as "junk," is now thought to have important functions that may include maintaining structural integrity of chromosomes and gene regulation.
- Currently, researchers are placing an emphasis on functional and comparative genomics. **Functional genomics** aims to understand the function of protein-coding regions and noncoding regions of our genome. To that end, researchers are utilizing new tools such as DNA microarrays to generate **genetic profiles.**
- **Comparative genomics** has revealed little difference between the DNA sequence of our bases and those of many other organisms. Genome comparisons have revolutionized our understanding of evolutionary relations by revealing previously unknown relationships between organisms.
- **Proteomics** is the study of which genes are active in producing proteins in which cells and under which circumstances.
- **Bioinformatics** is the use of computers to assist with analysis of data from proteomics and functional and comparative genomics.

ASSESS

Testing Yourself

Choose the best answer for each question.

1. Restriction enzymes found in bacterial cells are ordinarily used
   a. during DNA replication.
   b. to degrade the bacterial cell's DNA.
   c. to degrade viral DNA that enters the cell.
   d. to attach pieces of DNA together.
2. Which of the following enzymes are needed to introduce foreign DNA into a vector?
   a. DNA gyrase and DNA ligase
   b. DNA ligase and DNA polymerase
   c. DNA gyrase and DNA polymerase
   d. restriction enzyme and DNA gyrase
   e. restriction enzyme and DNA ligase
3. A genetic profile can
   a. assist in maintaining good health.
   b. be accomplished utilizing bioinformatics.
   c. show how many genes are normal.
   d. be accomplished utilizing a microarray.
   e. Both a and d are correct.
4. The polymerase chain reaction
   a. uses RNA polymerase.
   b. takes place in huge bioreactors.
   c. uses a temperature-insensitive enzyme.
   d. makes many nonidentical copies of DNA.
   e. All of the choices are correct.
5. During the PCR reaction, the DNA sample is heated in order to
   a. separate it into single strands.
   b. allow primers to bind.
   c. allow DNA polymerase to work.
   d. synthesize RNA.
6. Gene therapy
   a. is sometimes used in medicine today.
   b. is always successful.
   c. is used only to cure genetic disorders, such as SCID and cystic fibrosis.
   d. makes use of viruses to carry foreign genes into human cells.
   e. Both a and d are correct.
7. Because of the Human Genome Project, we now know
   a. the sequence of the base pairs of our DNA.
   b. the sequence of all genes along the human chromosomes.
   c. all the mutations that lead to genetic disorders.
   d. All of the choices are correct.
   e. Only a and c are correct.
8. Bioinformatics can
   a. assist genomics and proteomics.
   b. compare our genome to that of a monkey.
   c. depend on computer technology.
   d. match up genes with proteins.
   e. All of the choices are correct.
9. Proteomics is used to discover
   a. which genes are active in which cells.
   b. which proteins are active in which cells.
   c. the structure and function of proteins.
   d. how proteins interact.
   e. All but a are correct.
10. Which of the following statements is incorrect?
    a. Bacteria usually secrete the biotechnology product into the medium.
    b. Plants are being engineered to have human proteins in their seeds.
    c. Animals are engineered to have a human protein in their milk.
    d. Animals can be cloned, but plants and bacteria cannot.
11. Which of these is a true statement?
    a. Plasmids can serve as vectors.
    b. Plasmids can carry recombinant DNA, but viruses cannot.
    c. Vectors carry only the foreign gene into the host cell.
    d. Only gene therapy uses vectors.
    e. Both a and d are correct.

12. Comparative genomics
  a. is the application of computer technologies to the study of the genome.
  b. is the study of the structure, function, and interaction of cellular proteins.
  c. can be used to understand human gene function by investigating genes in other species.
  d. involves studying all the genes that occur in a cell.
  e. is the study of a person's complete genotype, or genetic profile.
13. What is ex vivo gene therapy?
  a. when a person's genes are cloned using PCR
  b. when a person is infected with a virus that delivers a functional gene into a chromosome with a dysfunctional gene
  c. when cells containing a dysfunctional gene are removed from the patient, treated, and then returned to the patient
  d. when a person's gene is cloned into a plant, which then produces the correct gene product in its seeds
  e. None of the choices are correct.

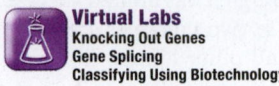

ENGAGE

Virtual Labs
Knocking Out Genes
Gene Splicing
Classifying Using Biotechnology

The virtual labs "Knocking Out Genes," "Gene Splicing," and "Classifying Using Biotechnology" all provide an interactive exploration of the use of recombinant DNA (rDNA) technology.

Thinking Critically

1. We can use transgenic viruses to infect humans and help treat genetic disorders. That is, the viruses are genetically modified to contain "normal" human genes to try to replace nonfunctional human genes. Using this type of gene therapy, a person is infected with a particular virus, which then delivers the "normal" human gene to cells by infecting them. What are some pros and cons of viral gene therapy?
2. In a genomic comparison between humans and yeast, what genes would you expect to be similar?

UNIT 6 Evolution and Diversity

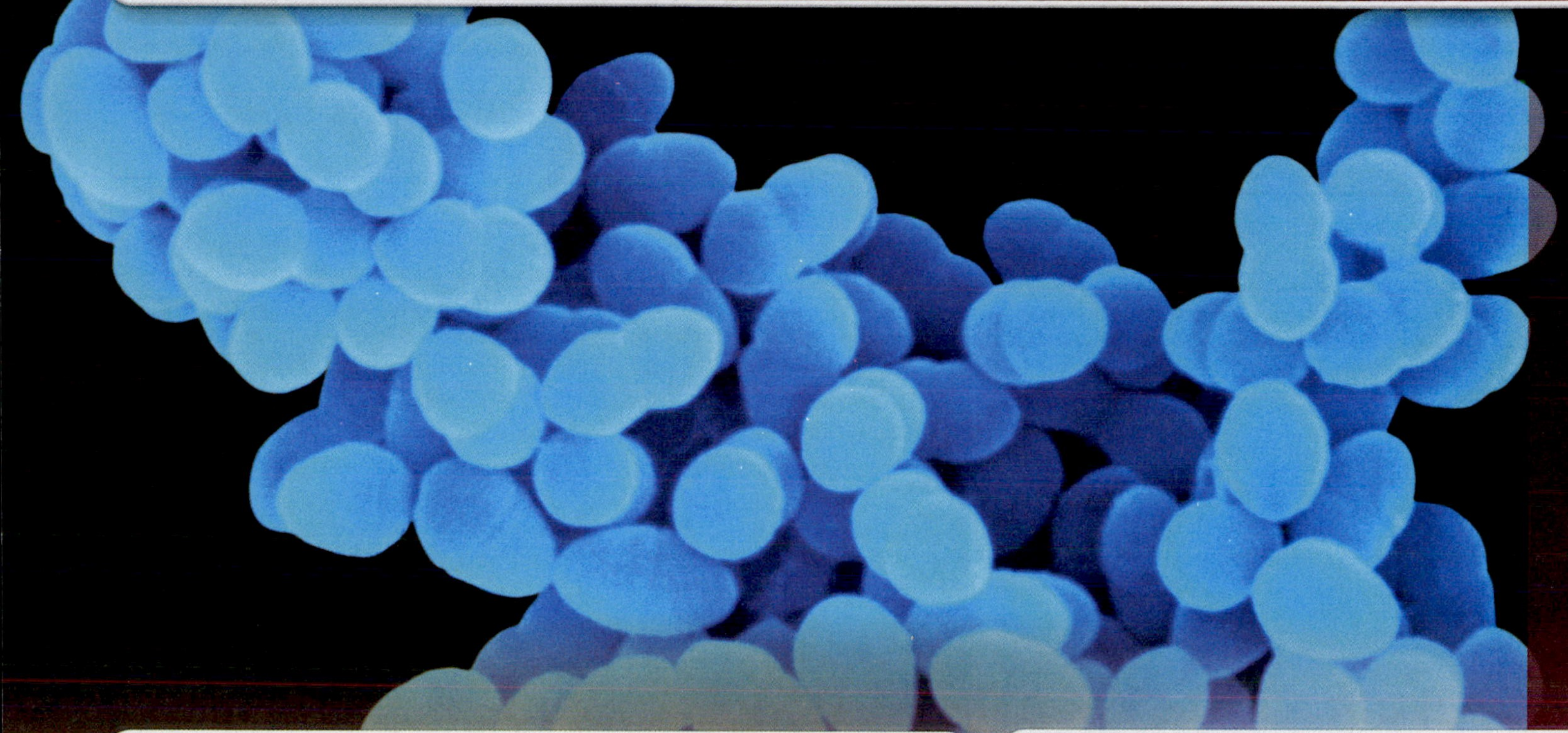

27 Evolution of Life

CASE STUDY Evolution of Antibiotic Resistance

MRSA (methicillin-resistant *Staphylococcus aureus*) is a strain of a common bacterium that can be deadly because of its resistance to several types of antibiotics. The bacterium is often found on the skin and in the noses of healthy people, but *S. aureus* can cause "staph" infections in scrapes, cuts, and open wounds. Normally, regular hand washing with soap and warm water prevents most staph infections. Until recently, severe staph infections were found primarily in hospitals. However, MRSA may now be found in prisons and schools, especially in gymnasiums and locker rooms.

Use and overuse of antibiotics has resulted in evolution of resistant bacterial strains, such as MRSA. Although we tend to think of evolution as happening over long timescales, human activities can accelerate the process of evolution quite rapidly. In fact, evolution of resistance to the antibiotic methicillin occurred in just one year! Antibiotic-resistant strains of bacteria are generally hard to treat, and treatment of infected patients can cost thousands of dollars.

Some scientists believe that "superbugs," or bacteria that have evolved antibiotic resistance, will be a far bigger threat to human health than emerging diseases such as H1N1 flu and AIDS. The good news is that our understanding of evolutionary biology has helped change human behavior to deal with superbugs. For example, doctors no longer prescribe antibiotics unless they are relatively certain a patient has a bacterial infection. Antibiotic resistance is an example of why evolution is important in people's everyday lives. In this chapter, you will learn about evidence that indicates evolution has occurred and about how the evolutionary process works.

As you read through the chapter, think about the following questions:

1. In the case of evolution of antibiotic resistance, what type of selection is operating?
2. Discuss some strategies doctors might use to help reduce the rate at which antibiotic-resistant bacteria evolve.

CHAPTER OUTLINE

BEFORE YOU BEGIN

Before beginning this chapter, take a few moments to review the following discussions:

Sections 3.2 and 3.3 What are the differences between eukaryotic and prokaryotic cells?

Section 23.1 What is the genetic basis of inheritance?

Section 25.1 How does the structure of DNA relate to its role as the genetic material?

27.1 Theory of Evolution

Learning Outcomes

Upon completion of this section, you should be able to

1. List two observations Darwin made on his voyage that influenced the development of his theory.
2. Identify how Darwin's and Lamarck's theories of evolution are different.
3. Summarize the four main observations that make up Darwin's theory of natural selection.

In December 1831, a new chapter in the history of biology had its humble origins. A 22-year-old naturalist, Charles Darwin, set sail on a journey of a lifetime aboard the British naval vessel HMS *Beagle* (Fig. 27.1). Darwin's primary mission on this voyage was to serve as the ship's naturalist—to collect and record the geological and biological diversity he saw during the voyage.

As Darwin set sail on the HMS *Beagle,* he was a supporter of the long-held idea that species had remained unchanged since the time of creation. During the five-year voyage of the *Beagle,* Darwin's observations challenged his own belief that species do not change over time. His observations of geological formations and species variation led him to propose a new process by which species arise and change. This process, called **evolution** (Latin for "an unrolling"), proposes that species arise, change, and become extinct due to natural, not supernatural, forces.

This new view was not readily accepted by Darwin's peers, but it gained gradual credibility as a result of a scientific and intellectual revolution that began in Europe in the late 1800s. Today, over 150 years since Darwin first published his idea of natural selection, the principle he proposed has been subjected to rigorous scientific tests, so much so that it is now considered one of the unifying theories of biology. Darwin's theory of evolution by natural selection explains both the unity and diversity of life on Earth, how all living organisms share a common ancestor, and how species adapt to various habitats and ways of life.

Although many have believed that Darwin forged this change in worldview by himself, several biologists during the preceding century and some of Darwin's contemporaries had a large influence on Darwin as he developed his theory. The European scientists of the eighteenth and nineteenth centuries were keenly interested in understanding the nature of biological diversity. This was a time of exploration and discovery as the natural history of new lands was mapped and

Charles Darwin
HMS *Beagle*

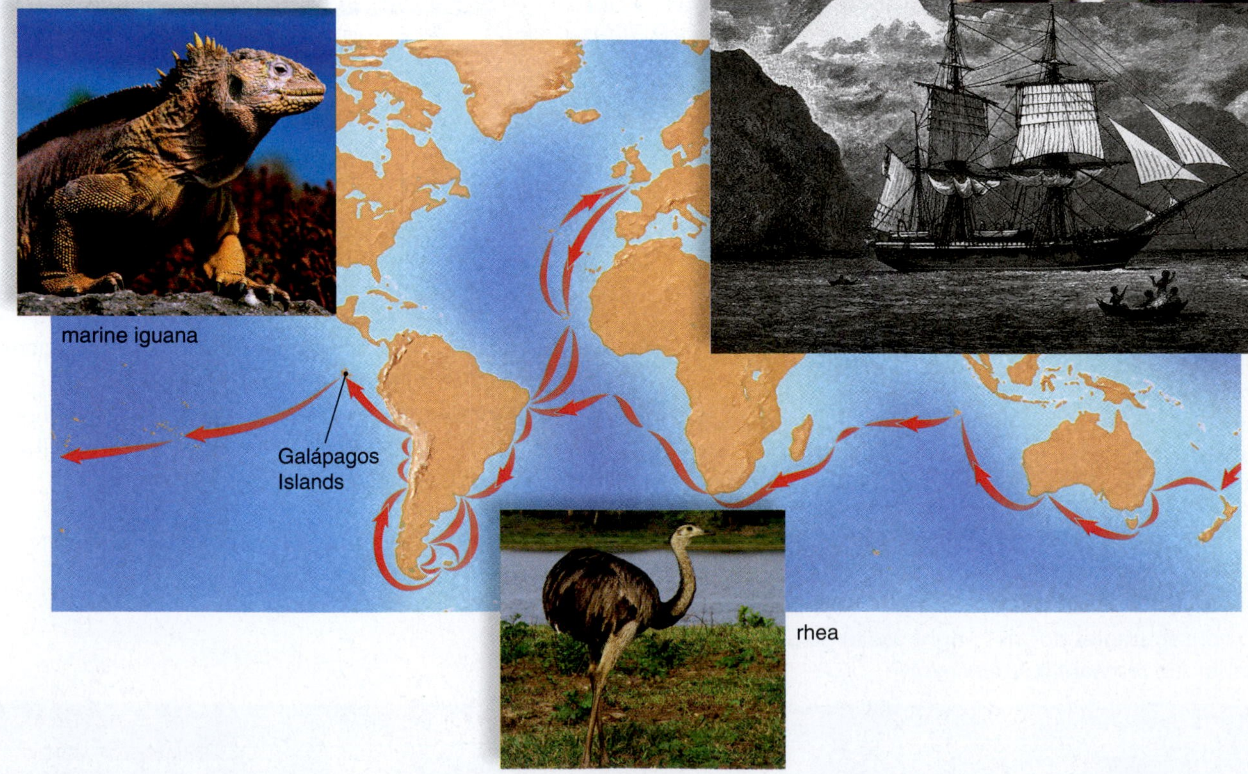
marine iguana
Galápagos Islands
rhea

Figure 27.1 Voyage of the HMS *Beagle*. The map shows the journey of the HMS *Beagle* around the world. As Darwin traveled along the east coast of South America, he noted that a bird called a rhea looked like the African ostrich. On the Galápagos Islands, marine iguanas, found no other place on Earth, use their large claws to cling to rocks and their blunt snouts for eating marine algae.

documented. Shipments of new plants and animals from newly explored regions were arriving in England to be identified and described by biologists—it was a time of rapid expansion of our understanding of the Earth's biological diversity. In this atmosphere of discovery, Darwin's theory first took root and grew.

One scientist that influenced Darwin as he developed his theory was Jean-Baptiste de Lamarck. Lamarck, a predecessor of Darwin, proposed one of the first hypotheses to explain how species evolve and adapt to the environment. Lamarck developed the theory of **inheritance of acquired characteristics**—that the environment can bring about inherited change. One example he gave, and the one for which he is most famous, is that the long neck of a giraffe developed over time because animals stretched their necks to reach food high in trees and then passed gradually longer necks to their offspring (Fig. 27.2*a*). This hypothesis for the inheritance of acquired characteristics has never been verified. The reason is due to the molecular mechanism of inheritance. Changes to an organism's visible characteristics, or **phenotype,** acquired during an organism's lifetime do not result in genetic changes that are heritable. As an example, consider tail cropping in Doberman pinschers. All Doberman puppies are born with tails, even though their parents' tails are most often cropped. That is, tail cropping is a phenotypic change that is not inherited in the DNA. We now know that Lamarck's ideas, although important for advancing ideas about evolution, were incorrect.

In contrast to Lamarck, Darwin proposed that traits that provided an advantage must be passed on to the next generation, that is, must be heritable for adaptation to occur (Fig. 27.2*b*). In the example of the giraffe, Darwin's theory would predict that giraffe populations, which have variation in the length of necks, would contain some individuals with longer necks than others. Those with the longer necks would be able to reach more leaves, acquire more energy, and potentially produce more offspring. These offspring would inherit the long neck trait from their parents. Over time, the longer neck trait would become more common in a population, resulting in an adaptation to the environment.

Early giraffes probably had short necks that they stretched to reach food.
Their offspring had longer necks that they stretched to reach food.
Eventually, the continued stretching of the neck led to today's giraffe.
a. Lamark
Early giraffes probably had necks of various lengths.
Natural selection due to competition led to survival of the longer-necked giraffes and their offspring.
Eventually, only long-necked giraffes survived the competition.
b. Darwin

Figure 27.2 A comparison of Lamarck's and Darwin's theories of evolution. **a.** Jean-Baptiste de Lamarck's proposal of the inheritance of acquired characteristics. **b.** Charles Darwin's theory of natural selection.

Darwin called this process of adaptation evolution by natural selection. Darwin's theory developed over a long period of time. During his voyage on the HMS *Beagle,* he made many observations that shaped the development of his theory of natural selection. Darwin observed that some animals, such as the African ostrich and the South American rhea, looked very similar, but could not be related because they lived on different continents. He reasoned that the similarity between these two large, flightless birds was due to adaptation to similar environments. While visiting the Galápagos Islands Darwin observed that the finches on each island were similar in overall appearance, but had differences in beak size and shape (Fig. 27.3). To explain these observations, Darwin speculated whether these different species of finches could have descended from a mainland finch species. In other words, he wondered if a

a. Large, ground-dwelling finch b. Warbler-finch c. Cactus-finch

Figure 27.3 Three Galápagos finches. Each of the 13 species of finches has a beak adapted to a particular way of life. For example: (**a**) the heavy beak of the large ground-dwelling finch (*Geospiza magnirostris*) is suited to a diet of large seeds; (**b**) the beak of the warbler-finch (*Certhidea olivacea*) is suited to feeding on insects found among ground vegetation or caught in the air; and (**c**) the longer beak, somewhat decurved, and the split tongue of the cactus-finch (*Cactornis scandens*) are suited to extracting the flesh of cactus fruit.

finch from South America was the common ancestor to all the types on the Galápagos Islands. Perhaps new species had arisen because the geographic distance between the islands isolated populations of birds long enough for them to evolve independently. And perhaps the present-day species had resulted from accumulated changes that occurred within each of these isolated populations.

Video Galápagos Finches

With this and other types of evidence, Darwin returned home from his voyage and spent the next 20 years developing and fine-tuning his **theory of natural selection.** His theory can be summarized as follows (see Fig. 27.2*b*):

- *Individual organisms within a species exhibit variation that can be passed from one generation to the next—that is, they have heritable variation.*
  Darwin emphasized that the members of a population vary in their functional, physical, and behavioral traits. Darwin emphasized that variation, abundant in natural populations, must be heritable for the process of natural selection to operate. We now know that genes, along with the environment, determine the phenotype of individuals and trait variation within populations.
- *Organisms compete for available resources.*
  Darwin read an essay by Thomas Malthus in which he proposed that human population growth is kept in check because an environment can support only a limited number of people. Darwin realized that if all offspring born to a population were to survive, there would not be sufficient resources to support it indefinitely. He calculated the reproductive potential of elephants, assuming an average life span of 100 years and that each female bears no fewer than six young over her lifetime. If all these young were to survive and reproduce, then after only 750 years, each breeding pair would have 19 million descendants! Obviously, no environment has the resources to support an elephant population of this magnitude, and no such elephant population has ever existed.
- *Individual organisms within a population differ in terms of their reproductive success.*
  Darwin observed that there are some individuals in a population that have favorable traits that enable them to better compete for limited resources, and thus devote more energy to reproduction. Darwin called this ability to have more offspring *differential reproductive success.* Natural selection occurs because certain members of a population happen to have an advantageous trait that allows them to survive and reproduce to a greater extent than do other members. For example, a mutation in a wild dog that increases its sense of smell may help it find prey.
- *Organisms become adapted to conditions as the environment changes.*
  Darwin's theory proposes that natural selection through differential reproductive success shapes traits, inherited from predecessors, in response to the environment. The result of this *descent with modification* is adaptation. An **adaptation** is any evolved trait that helps an organism be more suited to its environment. Adaptations are especially recognizable when unrelated organisms living in a particular environment display similar characteristics. For example, manatees, penguins, and sea turtles all have flippers, which help them move through the water.

Such adaptations to specific environments result from natural selection. Over time, selection can cause adaptive traits to be increasingly represented in the population. Evolution includes other processes in addition to natural selection (see sections 27.3 and 27.4), but natural selection is the only process that results in adaptation to the environment.

Check Your Progress 27.1

1. Describe the four critical elements of Darwin's theory.
2. Explain why Lamarck's "inheritance of acquired characteristics" does not explain evolutionary change.
3. Describe the role of adaptations in evolutionary change.

27.2 Evidence of Evolution

Learning Outcomes

Upon completion of this section, you should be able to

1. Define biological evolution.
2. Describe, with examples, four lines of evidence for evolution.

Biological evolution, or simply evolution, is all the changes that have occurred, due to differential reproductive success, in living organisms over geological time. As we have previously discussed, differential reproductive success indicates that some individuals reproduce more than others because they are better suited to their environment. Table 27.1 on page 539 indicates that Earth is about 4.6 billion years old and that prokaryotes, probably the first living organisms, evolved about 3.5 billion years ago. The eukaryotic cell arose about 2.1 billion years ago, but multicellularity did not begin until perhaps 700 million years ago. This means that only unicellular organisms were present for 80% of the time that life has existed on Earth. Consequently, most evolutionary events we will be discussing in the next few chapters occurred in less than 20% of the history of life!

Because of descent with modification, all living organisms share the same fundamental characteristics: they are made of cells, take chemicals and energy from the environment, respond to external stimuli, and reproduce (see section 1.1). Life is diverse because living organisms are adapted to different environments and the features that enable them to survive in those environments vary tremendously.

Many fields of biology provide evidence that evolution through descent with modification occurred in the past and is still occurring. Let us look at the various types of evidence for evolution.

Fossil Evidence

Fossils are the remains and traces of past life or any other direct evidence of past life. Most fossils consist only of hard parts of organisms, such as shells, bones, or teeth, because these are usually preserved after death. The soft parts of a dead organism are often consumed by scavengers or decomposed by bacteria. Occasionally, however, an organism is buried quickly and in such a way that decomposition is never completed or is completed so slowly that the soft parts leave an imprint of their structure. Traces include trails, footprints, burrows, worm casts, or even preserved droppings.

The great majority of fossils are found embedded in sedimentary rock. Sedimentation, a process that has been going on since Earth formed, can take place on land or in bodies of water. The weathering and erosion of rocks produces particles that vary in size and are called sediment. As such particles accumulate, the sediments form strata (sing., stratum), or recognizable layers of rock. Any given stratum is older than the one above it and younger than the one immediately below it, so that the relative age of fossils can be determined based on their depth.

Paleontologists are biologists who study the fossil record and from it draw conclusions about the history of life. When fossils are arranged from oldest to youngest, they can provide evidence of evolutionary change through time. A particularly good example of this is the horse, which evolved from a dog-sized, forest-dwelling tree browser with forward-looking eyes and teeth geared toward chewing leaves, to the animal we recognize as a horse today. Modern horses are adapted for an open field-type habitat, with eye sockets more to the sides of their head to allow for better peripheral vision and thereby detection of possible predators, as well as teeth geared toward grinding grasses.

Particularly interesting are the fossils that serve as **transitional links** between groups. A famous example are the fossils of *Archaeopteryx* that lived about 165 million years ago (Fig. 27.4). The fossil clearly seems to be an intermediate form between dinosaurs and birds. *Archaeopteryx* had dinosaur-like features including jaws with teeth and a long, jointed tail, but it also had feathers and wings similar to those of modern birds. The fossils of *Archaeopteryx* are similar to other transitional fossils in that they have some traits like their ancestors and others like their descendants, rather than expressing intermediate traits.

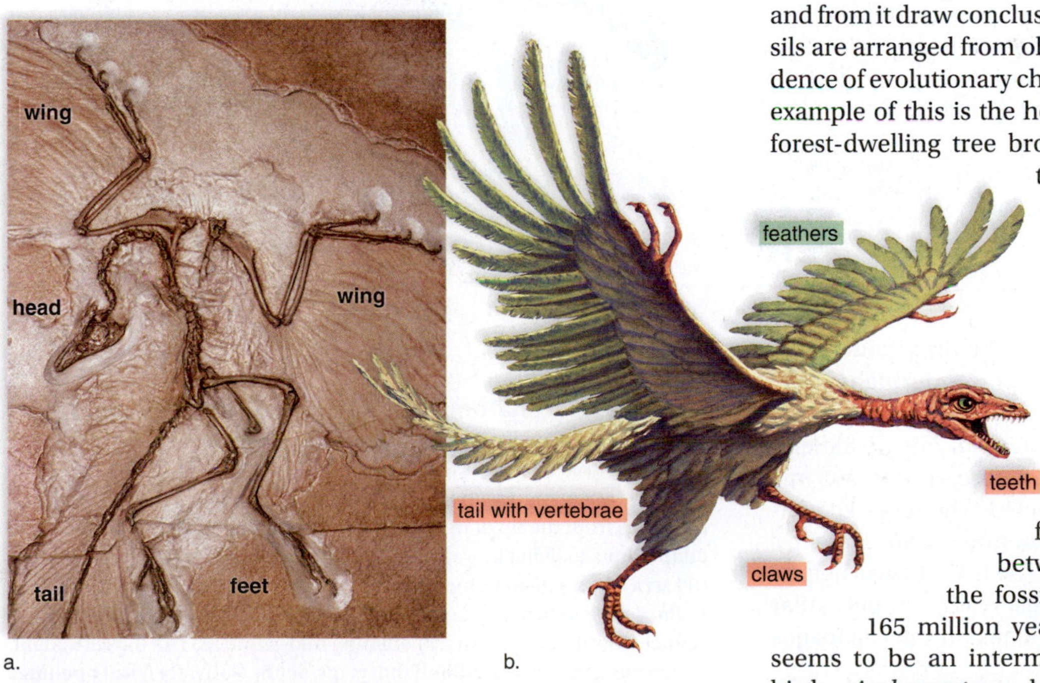
wing
head
wing
tail
feet
feathers
tail with vertebrae
teeth
claws
a.
b.

Figure 27.4 Transitional fossils. **a.** A fossil of *Archaeopteryx,* now considered the first bird, has features of both birds and dinosaurs. Fossils indicate it had feathers and wing claws. Most likely, it was a poor flier. Perhaps it ran over the ground on strong legs and climbed into trees with the assistance of these claws. **b.** *Archaeopteryx* also had a feather-covered, reptilelike tail that shows up well in this artist's representation.

Fossils have been discovered that support the hypothesis that whales had terrestrial ancestors. *Ambulocetus natans* (meaning the walking whale that swims) was the size of a large sea lion, with broad, webbed feet on its forelimbs and hindlimbs that enabled it to both walk and swim. It also had tiny hooves on its toes and the primitive skull and teeth of early whales (Fig. 27.5). Modern whales still have remnants of a hindlimb consisting of only a few bones that are very reduced in size. As the ancestors of whales adopted an increasingly aquatic lifestyle, the location of the nasal opening underwent a transition, from the tip of the snout as in *Ambulocetus,* to midway between the tip of the snout and the skull in *Basilosaurus,* to the very top of the head in modern whales (Fig. 27.5). Other transitional links among fossil vertebrates suggest that fishes evolved before amphibians, which evolved before reptiles, which evolved before mammals in the history of life.

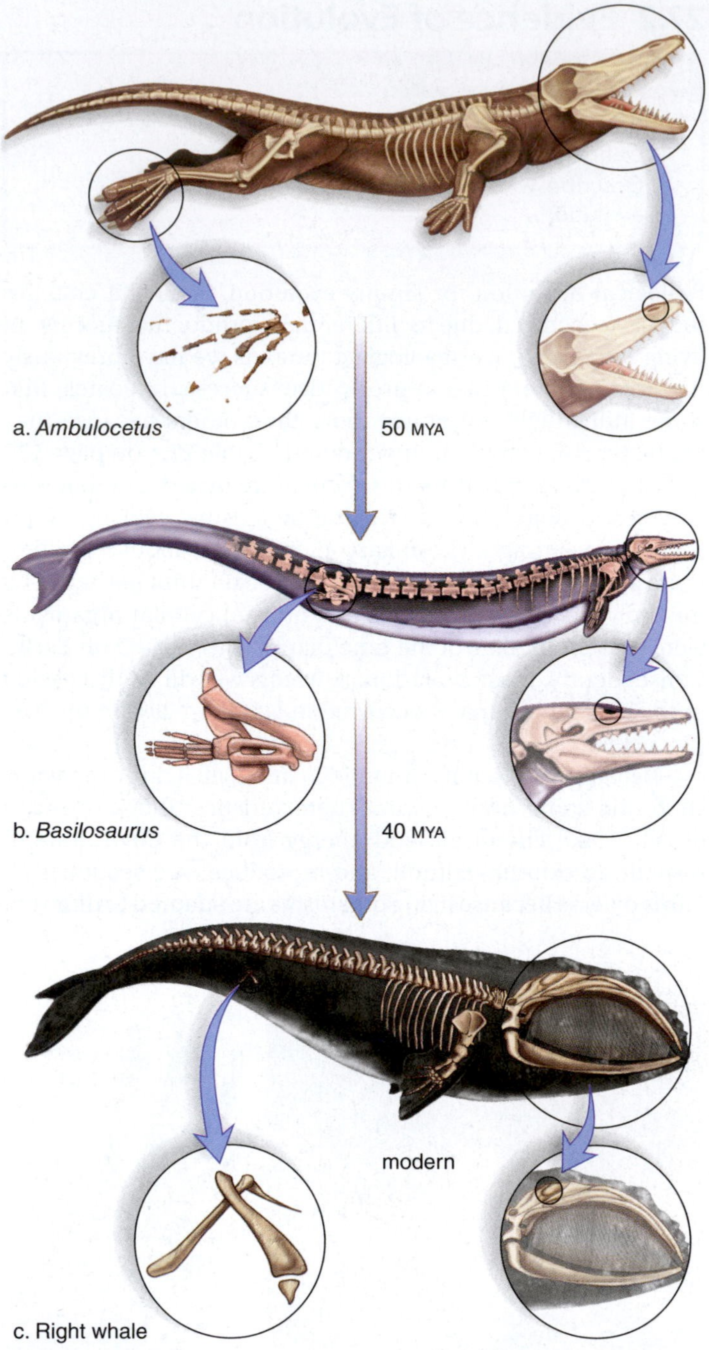
a. Ambulocetus
50 MYA
b. Basilosaurus
40 MYA
modern
c. Right whale

Figure 27.5. Anatomical transitions during the evolution of whales. Fossil evidence suggests that modern whales evolved from terrestrial ancestors that walked on four limbs. Transitional fossils show a gradual reduction in the hindlimb and a movement of the nasal openings from the tip of the nose to the top of the head—both are adaptations to living in water. **a.** *Ambulocetus,* an ancestor to whales (50 MYA), had a distinct hindlimb and a nose at the tip of its snout. **b.** *Basilosaurus,* a more recent ancestor of whales (40 MYA), had a greatly reduced hindlimb and a nasal opening mid-snout. **c.** The modern Right whale has a vestigial hindlimb that is inside the body. Its nasal opening is far back on the top of the head, from which it breathes as it surfaces from under the water.

Geological Timescale

As a result of studying strata, scientists have divided Earth's history into eras, and then periods and epochs (see Table 27.1). The fossil record has helped determine the dates given in the table. There are two ways to age fossils. The relative dating method determines the relative order of fossils and strata depending on the layer of rock in which they were found, but it does not determine the actual date they were formed.

The absolute dating method relies on radioactive dating techniques to assign an actual date to a fossil. All radioactive isotopes have a particular *half-life,* the length of time it takes for half of the radioactive isotope to change into another stable element. Carbon 14 (^{14}C) is the only radioactive isotope in organic matter. Assuming a fossil contains organic matter, half of the ^{14}C will have changed to nitrogen 14 (^{14}N) in 5,730 years. For example, if a fossil has one-fourth the amount of radioactive ^{14}C as a modern sample, then the fossil is approximately 11,460 years old (two half-lives).

Animation
Half-Life

Biogeographical Evidence

The field of **biogeography** is the study of the range and distribution of plants and animals in different places throughout the world. Such distributions are consistent with the hypothesis that, when forms are related, they evolved in one locale and then spread to accessible regions. Therefore, a different mix of plants and animals would be expected whenever geography separates continents, islands, seas, and so on. Darwin's observations of biogeography while on the HMS *Beagle* helped to form his theory of evolution. One observation in particular was that of the variation among the finches on the Galápagos Islands (see section 27.1).

TABLE 27.1 The Geological Timescale: Major Divisions of Geological Time and Some of the Major Evolutionary Events of Each Time Period

Era	Period	Epoch	Millions of Years Ago (MYA)	Plant Life	Animal Life
Cenozoic		Holocene	(0.01–0)	Humans influence plant life.	Age of *Homo sapiens* 
		Significant Mammalian Extinction			
	Quaternary	Pleistocene	(1.8–0.01)	Herbaceous plants spread and diversify. 	Presence of ice age mammals. Modern humans appear. 
		Pliocene	(5.33–1.8)	Herbaceous angiosperms flourish.	First hominids appear.
		Miocene	(23.03–5.33)	Grasslands spread as forests contract.	Apelike mammals and grazing mammals flourish; insects flourish. 
	Tertiary	Oligocene	(33.9–23.03)	Many modern families of flowering plants evolve; appearance of grasses.	Browsing mammals and monkeylike primates appear.
		Eocene	(55.8–33.9)	Subtropical forests with heavy rainfall thrive. 	All modern orders of mammals are represented. 
		Paleocene	(65.5–55.8)	Flowering plants continue to diversify.	Ancestral primates, herbivores, carnivores, and insectivores appear.
		Mass Extinction: **50% of all species, dinosaurs and most reptiles**			
Mesozoic	Cretaceous		(145.5–65.5)	Flowering plants spread; conifers persist. 	Placental mammals appear; modern insect groups appear.
	Jurassic		(199.6–145.5)	Flowering plants appear.	Dinosaurs flourish; birds appear.
		Mass Extinction: **48% of all species, including corals and ferns**			
	Triassic		(251–199.6)	Forests of conifers and cycads dominate. 	First mammals appear; first dinosaurs appear; corals and molluscs dominate seas.
		Mass Extinction **("The Great Dying"): 83% of all species on land and sea**			
Paleozoic	Permian		(299–251)	Gymnosperms diversify.	Reptiles diversify; amphibians decline.
	Carboniferous		(359.2–299)	Age of great coal-forming forests: ferns, club mosses, and horsetails flourish. 	Amphibians diversify; first reptiles appear; first great radiation of insects. 
		Mass Extinction**: Over 50% of coastal marine species, corals**			
	Devonian		(416–359.2)	First seed plants appear. Seedless vascular plants diversify.	First insects and first amphibians appear on land. 
	Silurian		(443.7–416)	Seedless vascular plants appear.	Jawed fishes diversify and dominate the seas. 
		Mass Extinction**: Over 57% of marine species**			
	Ordovician		(488.3–443.7)	Nonvascular land plants appear. 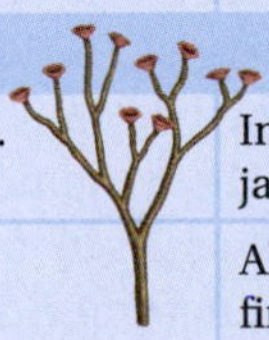	Invertebrates spread and diversify; first jawless and then jawed fishes appear.
	Cambrian		(542–488.3)	Marine algae flourish.	All invertebrate phyla present; first chordates appear. 
			630	First soft-bodied invertebrates evolve.	 
			1,000	Protists diversify.	
			2,100	First eukaryotic cells evolve. 	
			2,700	O_2 accumulates in atmosphere.	
			3,500	First prokaryotic cells evolve.	
			4,570	Earth forms.	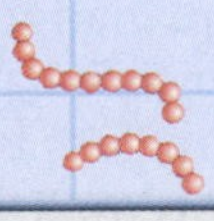 

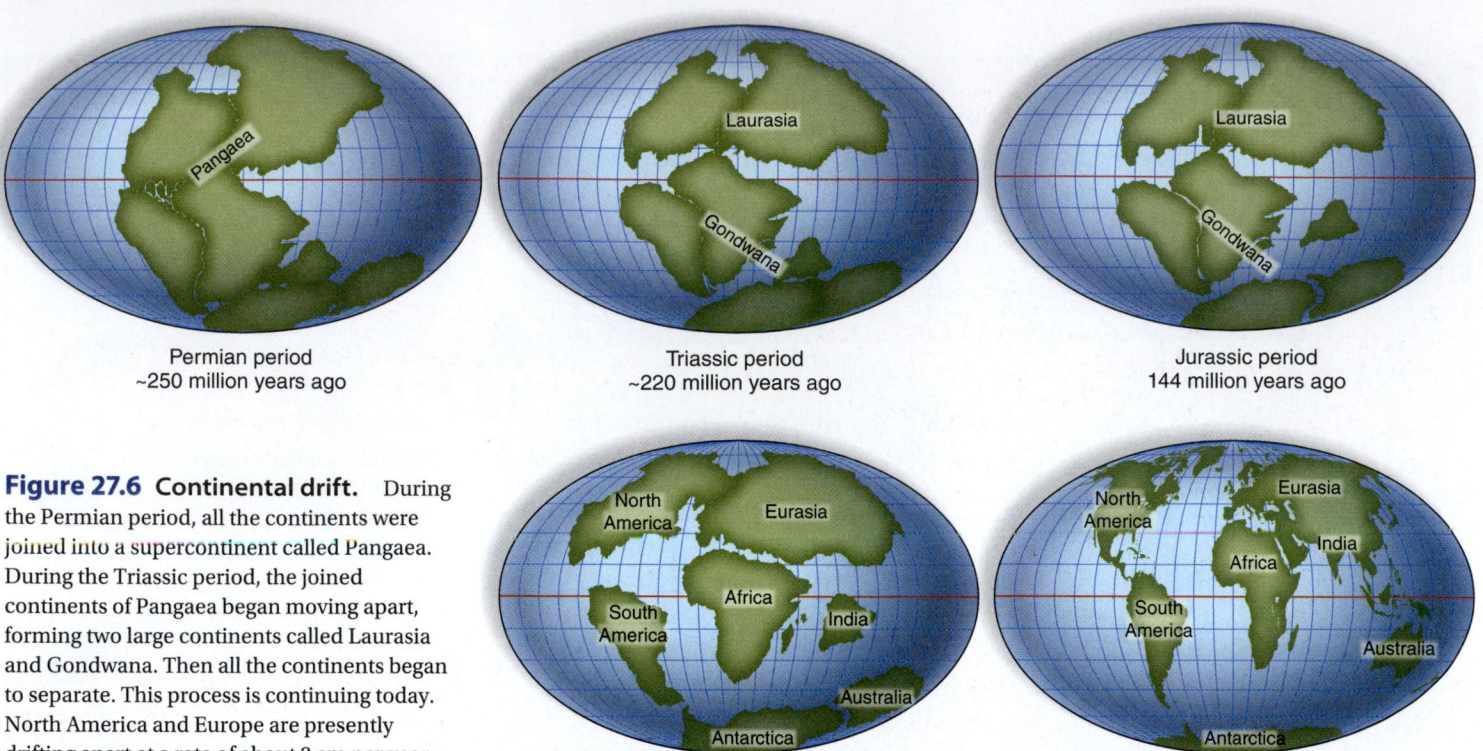

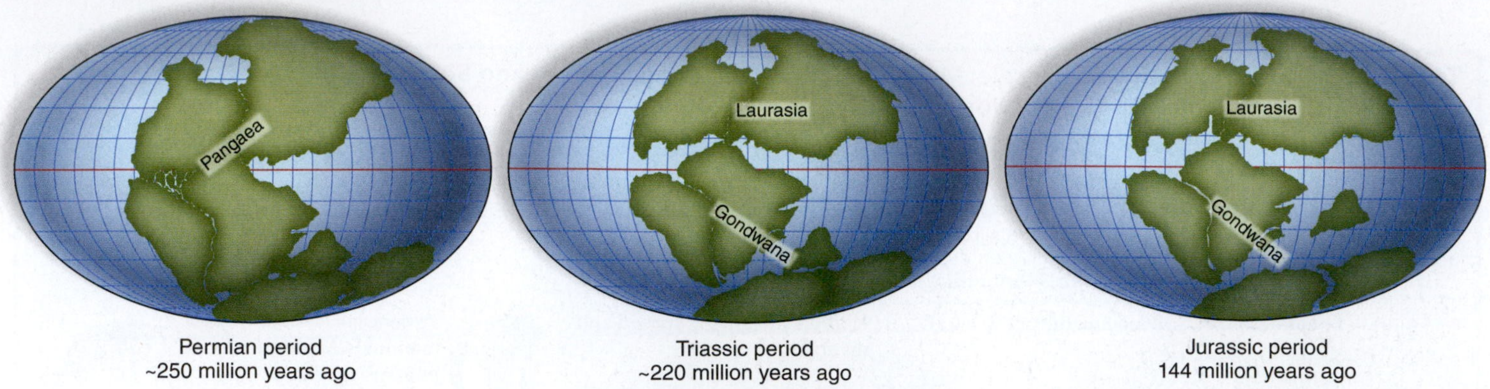
Pangaea
Permian period
~250 million years ago
Laurasia
Gondwana
Triassic period
~220 million years ago
Laurasia
Gondwana
Jurassic period
144 million years ago

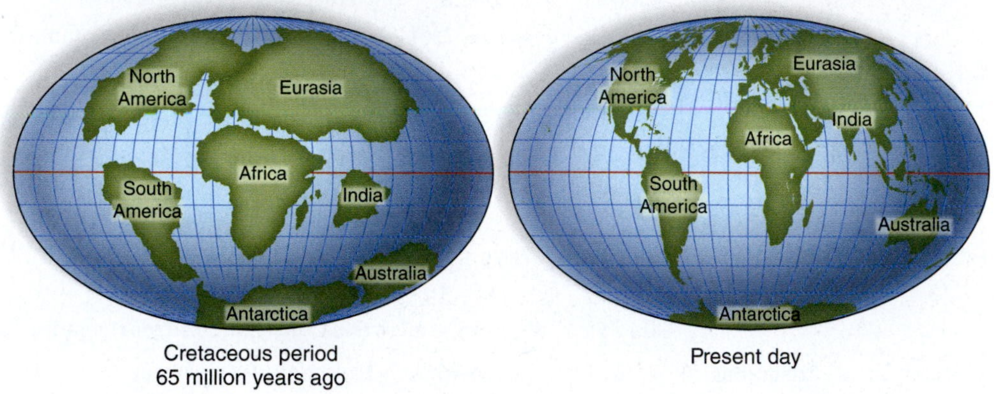
North America
Eurasia
Africa
South America
India
Australia
Antarctica
Cretaceous period
65 million years ago
North America
Eurasia
India
Africa
South America
Australia
Antarctica
Present day

Figure 27.6 Continental drift. During the Permian period, all the continents were joined into a supercontinent called Pangaea. During the Triassic period, the joined continents of Pangaea began moving apart, forming two large continents called Laurasia and Gondwana. Then all the continents began to separate. This process is continuing today. North America and Europe are presently drifting apart at a rate of about 2 cm per year.

Many of the barriers between landmasses arose through a process called **continental drift.** That is, the position of the continents has never been fixed. Rather, their positions and the positions of the oceans have changed over time (Fig. 27.6). During the Permian period, all the present landmasses belonged to one continent and then later drifted apart. As evidence of this, fossils of one species of seed fern (*Glossopteris*) have been found on all the southern continents separated by oceans. This species' presence on Antarctica is evidence that this continent was not always frozen. In contrast, many Australian species are restricted to that continent, including the majority of marsupials (pouched mammals such as the kangaroo). What is the explanation for these distributions? Some organisms must have evolved and spread out before the continents broke up. Then they became extinct.

The world's six biogeographical regions each have their own distinctive mix of living organisms. Darwin noted that South America lacked rabbits, even though the environment was quite suitable to them. He concluded there are no rabbits in South America because rabbits evolved somewhere else and had no means of reaching South America. To take another example, both cacti and spurges (*Euphorbia*) are plants adapted to a hot, dry environment—both are succulent, spiny, flowering plants. Why do cacti grow in the American deserts and most *Euphorbia* grow in African deserts when each would do well on the other continent? It seems obvious that they just happened to evolve on their respective continents. What is the best explanation for this phenomenon? Different mammals and flowering plants evolved separately in each biogeographical region, and barriers such as mountain ranges and oceans prevented them from migrating to other regions.

Mass Extinctions

Extinction is the death of every member of a species. During mass extinctions, a large percentage of species become extinct within a relatively short period of time. So far, there have been five major mass extinctions. These occurred at the ends of the Ordovician, Devonian, Permian, Triassic, and Cretaceous periods (see Table 27.1), and a sixth is likely occurring now, probably as a result of human activities (discussed in Chapter 37). Following mass extinctions, the remaining groups of organisms are likely to spread out and fill the habitats vacated by those that have become extinct.

It was proposed in 1977 that the Cretaceous extinction (or "Cretaceous crisis") was due to an asteroid that exploded, producing meteorites that fell to Earth. A large meteorite striking Earth could have produced a cloud of dust that mushroomed into the atmosphere, blocking out the sun and causing plants to freeze and die. In support of this hypothesis, a huge crater that could have been caused by a meteorite involved in the Cretaceous extinction was found in the Caribbean–Gulf of Mexico region on the Yucatán Peninsula. During the Cretaceous period, great herds of dinosaurs roamed the plains, but all dinosaur species went extinct near the end of the Cretaceous period.

Certainly, continental drift contributed to the Ordovician extinction. This extinction occurred after Gondwana arrived at the South Pole. Immense glaciers, which drew water from the oceans, chilled even once-tropical land. Marine invertebrates and coral reefs, which were especially hard hit, didn't recover until Gondwana drifted away from the pole and warmth returned. The mass extinction at the end of the Devonian period saw an end to 70% of marine invertebrates. Other scientists believe that this mass extinction could have been due to movement of Gondwana back to the South Pole.

Anatomical Evidence

The concept of common descent offers a plausible explanation for anatomical similarities among organisms. Vertebrate forelimbs are used for flight (birds and bats), orientation during swimming (whales and seals), running (horses), climbing (arboreal lizards), or swinging from tree branches (monkeys). Yet all vertebrate forelimbs contain the same sets of bones organized in similar ways, despite their dissimilar functions. The most plausible explanation for this unity is that the basic forelimb plan belonged to a common ancestor, and then the plan was modified in the succeeding groups as each continued along its own evolutionary pathway. Structures that are anatomically similar because they are inherited from a common ancestor are called **homologous structures** (Fig. 27.7). In contrast, **analogous structures** serve the same function, but are not constructed similarly, nor do they share a common ancestry. The wings of birds and insects and the eyes of octopi and humans are analogous structures and are similar due to a common environment, not common ancestry. The presence of homology, not analogy, is evidence that organisms are related.

Vestigial structures are anatomical features that are fully developed in one group of organisms but that are reduced and may have no function in similar groups. Most birds, for example, have well-developed wings for flight. However, some bird species (e.g., ostrich) have greatly reduced wings and do not fly. Similarly, snakes have no use for hindlimbs, and yet some have remnants of hindlimbs in a pelvic girdle and legs. The presence of vestigial structures can be explained by common descent. Vestigial structures occur because organisms inherit their anatomy from their ancestors. They are traces of an organism's evolutionary history.

The homology shared by vertebrates extends to their embryological development (Fig. 27.8). At some time during development, all vertebrates have a postanal tail and exhibit paired pharyngeal pouches. In fishes and amphibian larvae, these pouches develop into functioning gills. In humans, the first pair of pouches becomes the cavity of the middle ear and the auditory tube. The second pair becomes the tonsils, while the third and fourth pairs become the thymus and parathyroid glands. Why should terrestrial vertebrates develop and then

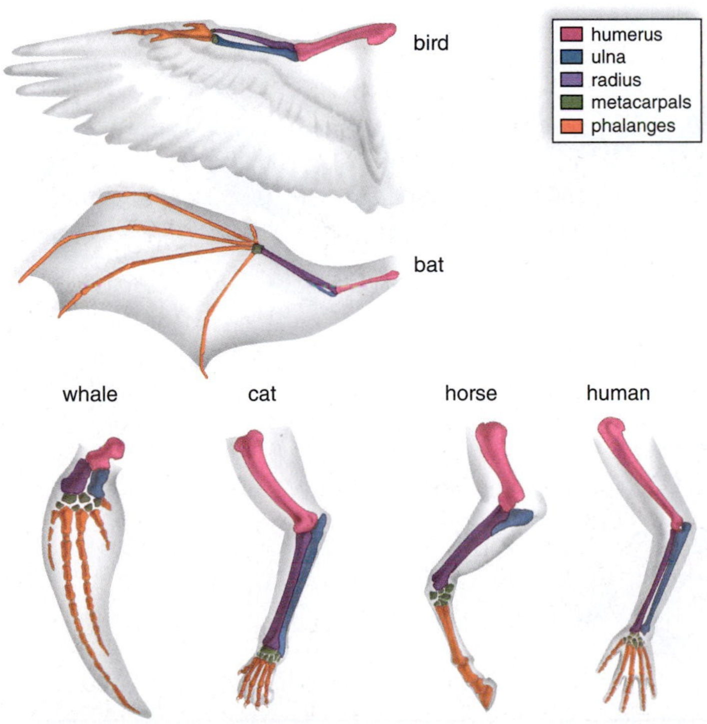
bird
humerus
ulna
radius
metacarpals
phalanges
bat
whale
cat
horse
human

Figure 27.7 Significance of homologous structures. Although the specific design details of vertebrate forelimbs are different, the same bones are present (note color-coding). Homologous structures provide evidence of a common ancestor.

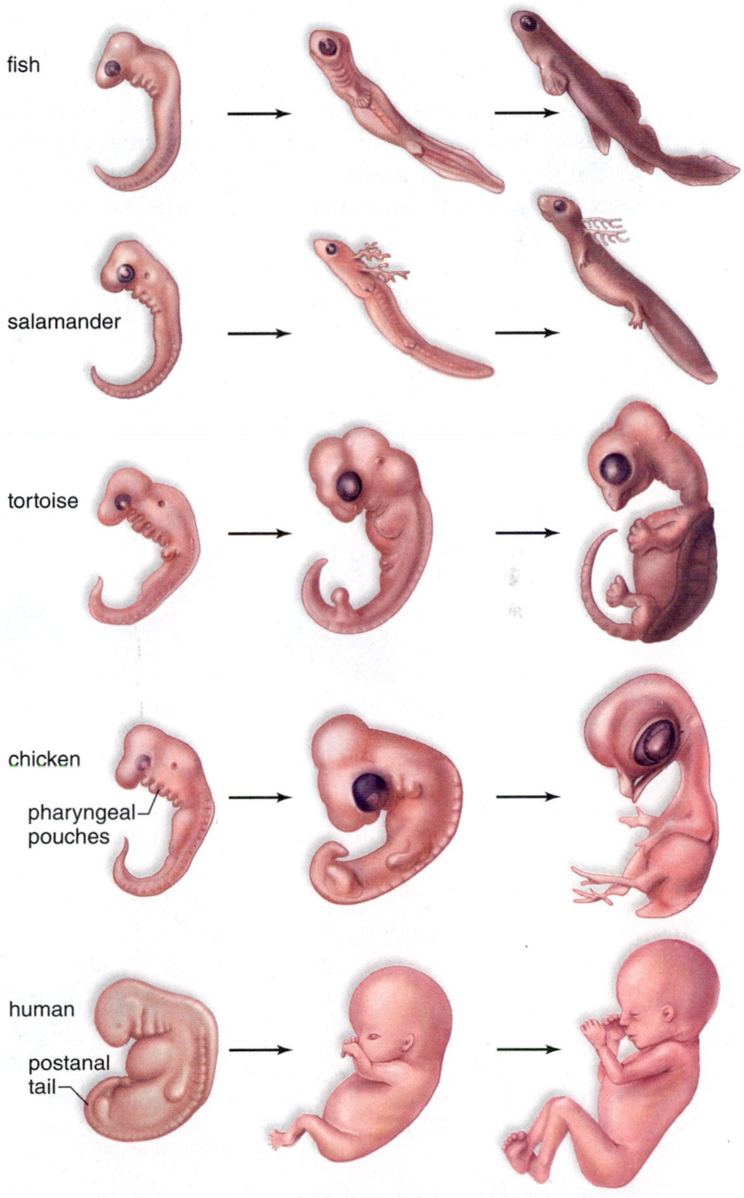
fish
salamander
tortoise
chicken
pharyngeal pouches
human
postanal tail

Figure 27.8 Significance of developmental similarities. At these comparable developmental stages, vertebrate embryos have many features in common, which suggests they evolved from a common ancestor. (These embryos are not drawn to scale.)

modify structures like pharyngeal pouches that have lost their original function? The most likely explanation is that fishes are ancestral to other vertebrate groups.

In 1859, Charles Darwin speculated that whales evolved from a land mammal. His hypothesis has now been substantiated. In recent years the fossil record has yielded an incredible parade of fossils that link modern whales and dolphins to land ancestors. The presence of a vestigial pelvic girdle and legs in modern whales is also significant evidence that ancestors of these creatures once walked on land. However, that these structures are severely reduced in modern whales is consistent with the fact that they are exclusively aquatic today (see Fig. 27.5).

Biochemical Evidence

Almost all living organisms use the same basic biochemical molecules, including DNA, ATP (adenosine triphosphate), and many identical or nearly identical enzymes. Further, organisms use the same DNA triplet code for the same 20 amino acids in their proteins. Because the sequences of DNA bases in the genomes of many organisms are now known, it has become clear that humans share a large number of genes with much simpler organisms. It appears that life's vast diversity has come about by only a slight difference in many of the same genes and regulatory genes often found in introns and other regions of the genome. The result has been widely divergent types of bodies.

When the degree of similarity in DNA nucleotide sequences or the degree of similarity in amino acid sequences of proteins is examined, the more similar the DNA sequences are, generally the more closely related the organisms are. For example, humans and chimpanzees are about 97% similar! Cytochrome *c* is a molecule that is used in the electron transport chain of all the organisms appearing in Figure 27.9. Data regarding differences in the amino acid sequence of cytochrome *c* show that the sequence in a human differs from that in a monkey by only one amino acid, from that in a duck by 11 amino acids, and from that in a yeast by 51 amino acids. These data are consistent with other data regarding the anatomical similarities of these organisms and, therefore, how closely they are related. In addition, we now know that chromosomal changes, such as translocations and inversions (see Chapter 24), play an important role in evolutionary change.

Evolution is no longer considered a hypothesis. It is one of the great unifying theories of biology. In science, the word *theory* is reserved for those conceptual schemes that are supported by a large number of observations and scientific experiments. The theory of evolution has the same status in biology that the theory of gravity has in physics.

We Can Observe Selection at Work

Table 27.1 outlines major events in the evolution of life over a timescale of millions and billions of years. But evolution does not necessarily happen over long periods of time. Researchers

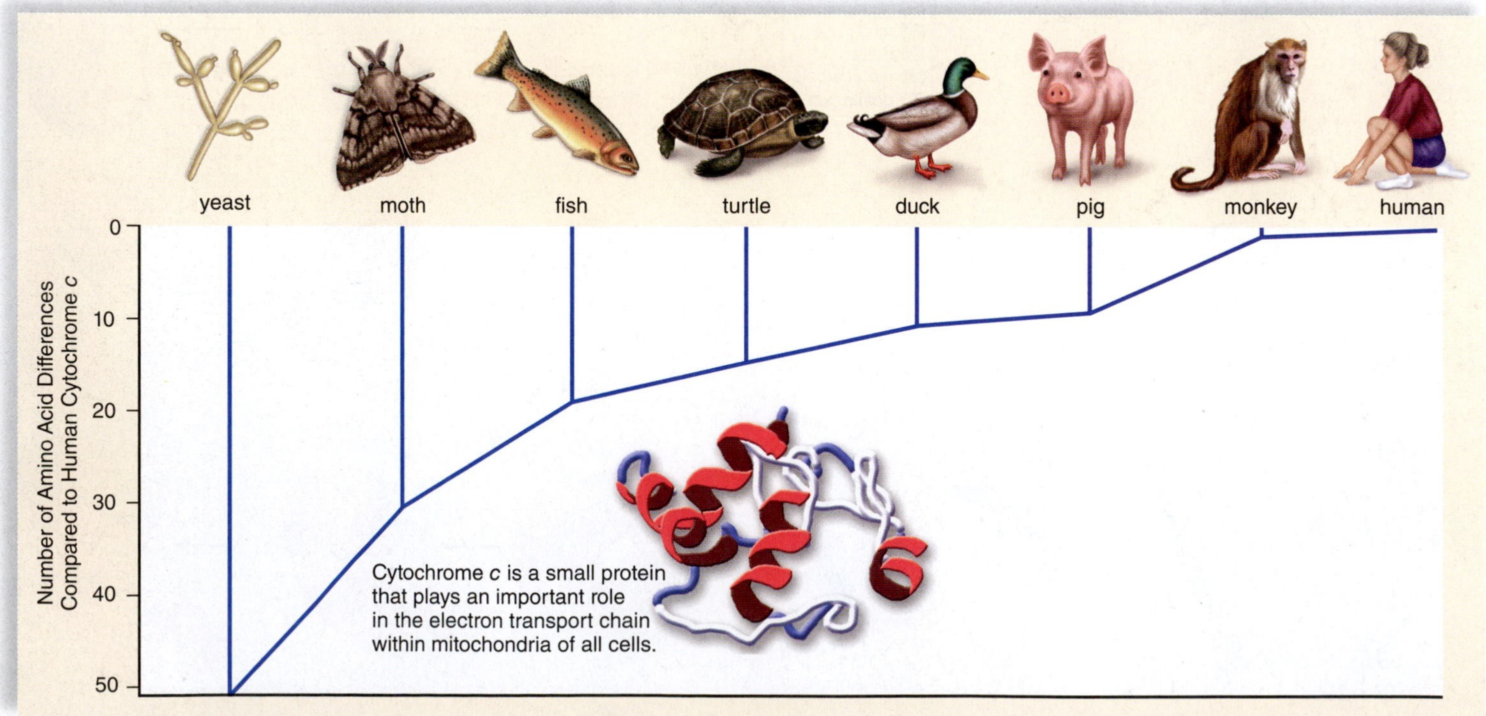
yeast
moth
fish
turtle
duck
pig
monkey
human
Number of Amino Acid Differences Compared to Human Cytochrome c
0
10
20
30
40
50
Cytochrome c is a small protein that plays an important role in the electron transport chain within mitochondria of all cells.

Figure 27.9 Significance of biochemical differences. The branch points in this diagram indicate the number of amino acids that differ between human cytochrome *c* and the organisms depicted. These biochemical data are consistent with those provided by a study of the fossil record and comparative anatomy.

Wolf
Boston terrier
Irish wolfhound

Figure 27.10 Artificial selection. All dogs, *Canis lupus familiaris*, are descended from the gray wolf, *Canis lupus*, which humans started to domesticate about 14,000 years ago. The process of selective breeding by humans has led to extreme differences among breeds.

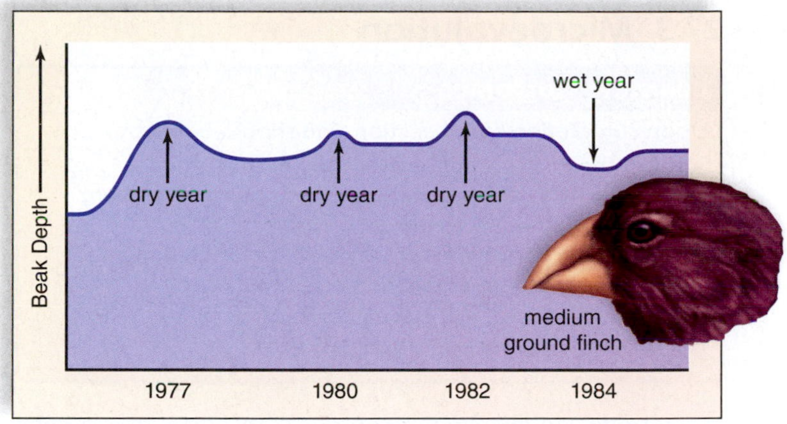
Beak Depth
wet year
dry year
dry year
dry year
medium ground finch
1977
1980
1982
1984

Figure 27.11 Evolution in action. The average beak depth of ground finches varies from generation to generation, according to the amount of rainfall. The amount of rainfall affects the hardness and size of seeds on the islands, and different beak depths were better suited to eating different types of seeds. Average beak features were observed to change many times over a period of a decade. This is one way in which evolution by natural selection has been observable over short periods of time.

have recorded the evolution of traits in natural populations over decades. Over the course of history, humans have used the power of evolution to shape traits of livestock, agricultural plants, and even our animal companions, over relatively short periods of time.

Humans as Agents of Evolution

Darwin noted that humans could artificially modify desired traits in plants and animals by selecting to breed individuals with preferred traits. For example, the diversity of domestic dogs has resulted from prehistoric humans selectively breeding wolves with particular traits, such as hair length, height, and guarding behavior. This type of human-controlled breeding to increase the frequency of desired traits is called **artificial selection.** Artificial selection, like natural selection, is possible only because the original population exhibits a variety of characteristics, allowing humans to select traits they prefer. For dogs, the result of artificial selection is the existence of many breeds of dogs all descended from the wolf (Fig. 27.10). Darwin surmised that if humans could create such a wide variety of organisms by artificial selection, then natural selection could also produce diversity. But in this case the environment, and not humans, is the force selecting for particular traits.

Evolution in Natural Populations

The Galápagos finches have beaks adapted to the food they eat, with different species of finches on each island (see Fig. 27.3). Today, many investigators, including Peter and Rosemary Grant of Princeton University, are documenting natural selection as it occurs on the Galápagos Islands. In 1973, the Grants began a study of the various finches on Daphne Major, an island near the center of the Galápagos Islands. The rainfall on this island varies widely from wet years to dry years. The Grants measured the beaks of each finch on the island and found that the beak depth of the medium ground finch, *Geospiza fortis,* changed between wet and dry years (Fig. 27.11). Medium ground finches like to eat small, tender seeds, but when the weather turns dry, they must eat larger, drier seeds that are harder to crush. During dry years the majority of finches died from starvation because their beaks were not well equipped to feed on the harder seeds.

But the few finches with deeper beaks were better able to crush larger, harder seeds, and these finches survived to reproduce. Therefore, among the next generation of *G. fortis* birds, the mean, or average, beak was deeper than the previous generation. The Grants' research demonstrates that evolutionary change can sometimes be observed within the timeframe of a human life span.

Video Finches Natural Selection

Check Your Progress 27.2

1. Summarize how the fossil record, geological timescale, and biogeography each provide evidence in support of the theory of evolution.
2. Explain how a breeder might produce a new breed of dog with long ears and a snub nose and how this new breed is evidence of evolution.
3. Determine which would have fewer protein and DNA base sequence differences with humans—chimpanzees or cows.

27.3 Microevolution

Learning Outcomes

Upon completion of this section, you should be able to

1. List the five conditions necessary for the allele frequencies of a population to be in Hardy-Weinberg equilibrium.
2. Calculate genotype frequencies of a population in Hardy-Weinberg equilibrium.
3. Identify an evolving population from a change in its allele frequencies over generations.

Many traits can change temporarily in response to a varying environment. For example, the color change in the fur of an Arctic fox from brown to white in winter, the increased thickness of your dog's fur in cold weather, or the bronzing of your skin when exposed to the sun last only for a season.

These are not evolutionary changes. Changes to traits over an individual's lifetime are not evidence that an individual has evolved, because these traits are not heritable (see Fig. 27.2). In order for traits to evolve, they must have the ability to be passed on to subsequent generations. Evolution is about change in a heritable trait within a population, not within individuals, over many generations.

Darwin observed that populations, not individuals, evolve, but he could not explain how traits change over time. Now we know that genes interact with the environment to determine traits. Because genes and traits are linked, evolution is really about genetic change—or more specifically, evolution is the *change in allele frequencies in a population over time.*

Microevolution pertains to evolutionary change within populations. In this chapter, we use the example of the peppered moth to examine how populations evolve.

Microevolution in the Peppered Moth

Population genetics, as its name implies, is the field of biology that studies the diversity of populations at the level of the gene. A **population** is all the members of a single species that occupy a particular area at the same time and that interbreed and exchange genes. Population geneticists are interested in how genetic diversity in populations changes over generations, and in the forces that cause populations to evolve. Population geneticists study microevolution by measuring the diversity of a population in terms of allele and genotype frequencies over generations.

You may recall from Chapter 23 that diploid organisms, such as moths, carry two copies of each chromosome, with one copy of each gene on each chromosome. A single gene can come in many forms, or alleles, that encode variations of a single trait. In the peppered moth, a single gene for body color has two alleles, *D* (dark color) and *d* (light color), with *D* dominant to *d* (Fig. 27.12).

We know that with only two alleles, there are three possible genotypes (the combinations of alleles in an individual) for the color gene in the peppered moth: *DD* (homozygous dominant), *Dd* (heterozygous), or *dd* (homozygous recessive). *DD* or *Dd* genotypes produce dark moths, and the *dd* genotype produces light moths (Fig. 27.12).

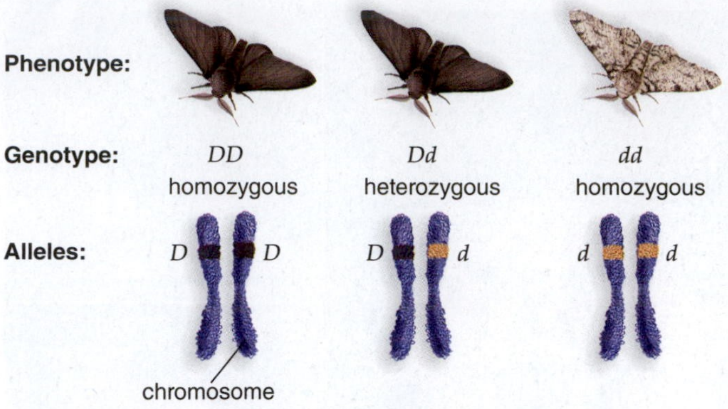
Phenotype:
Genotype:
DD
homozygous
Dd
heterozygous
dd
homozygous
Alleles:
D
D
D
d
d
d
chromosome

Figure 27.12 The genetic basis of body color in the peppered moth. Light or dark body color in the peppered moth is determined by a gene with two alleles, *D* and *d*. Genotypes *DD* and *Dd* produce dark body color, and *dd* produces light body color. The *D* and *d* alleles are variants of a gene at a particular locus on a chromosome.

Allele Frequencies

Suppose that a population geneticist collected a population of 25 peppered moths, some dark and some light, from a forest outside London (Fig. 27.13). In this population you would expect to find a mixture of *D* and *d* alleles in the **gene pool**—the alleles of all genes in all individuals in a population. The population geneticist ran tests to determine the alleles present in each moth. Of the 50 alleles in the peppered moth gene pool (2 alleles $\times$ 25 moths), 10 were *D* and 40 were *d*. Thus the frequency of the *D* and the *d* alleles would be 10/50, or 0.20, and 40/50, or 0.80. The allele frequency, as illustrated in this example, is the proportion of each allele in a population's gene pool (Fig. 27.13).

Notice that the frequencies of *D* and *d* add up to 1. This relationship is true of the sum of allele frequencies in a population for any gene of any diploid organism. This relationship is described by the expression $p + q = 1$, where p is the frequency of one allele, in this case *D*, and q is the frequency of the other allele, *d* (Fig. 27.13).

For the next three seasons, samples of 25 moths were collected from the same forest, and the allele frequencies were always the same: 0.20 *D*, and 0.80 *d*. Because allele frequencies in this population did not change over generations, we could conclude that this population has not evolved.

Hardy-Weinberg Equilibrium

A population in which allele frequencies do not change over time, such as in the moth population just described, is said to be in **genetic equilibrium,** or **Hardy-Weinberg equilibrium**—a stable, nonevolving state. Hardy-Weinberg equilibrium is derived from the work of British mathematician Godfrey H. Hardy and German physician Wilhelm Weinberg, who in 1908 developed a mathematical model to estimate genotype

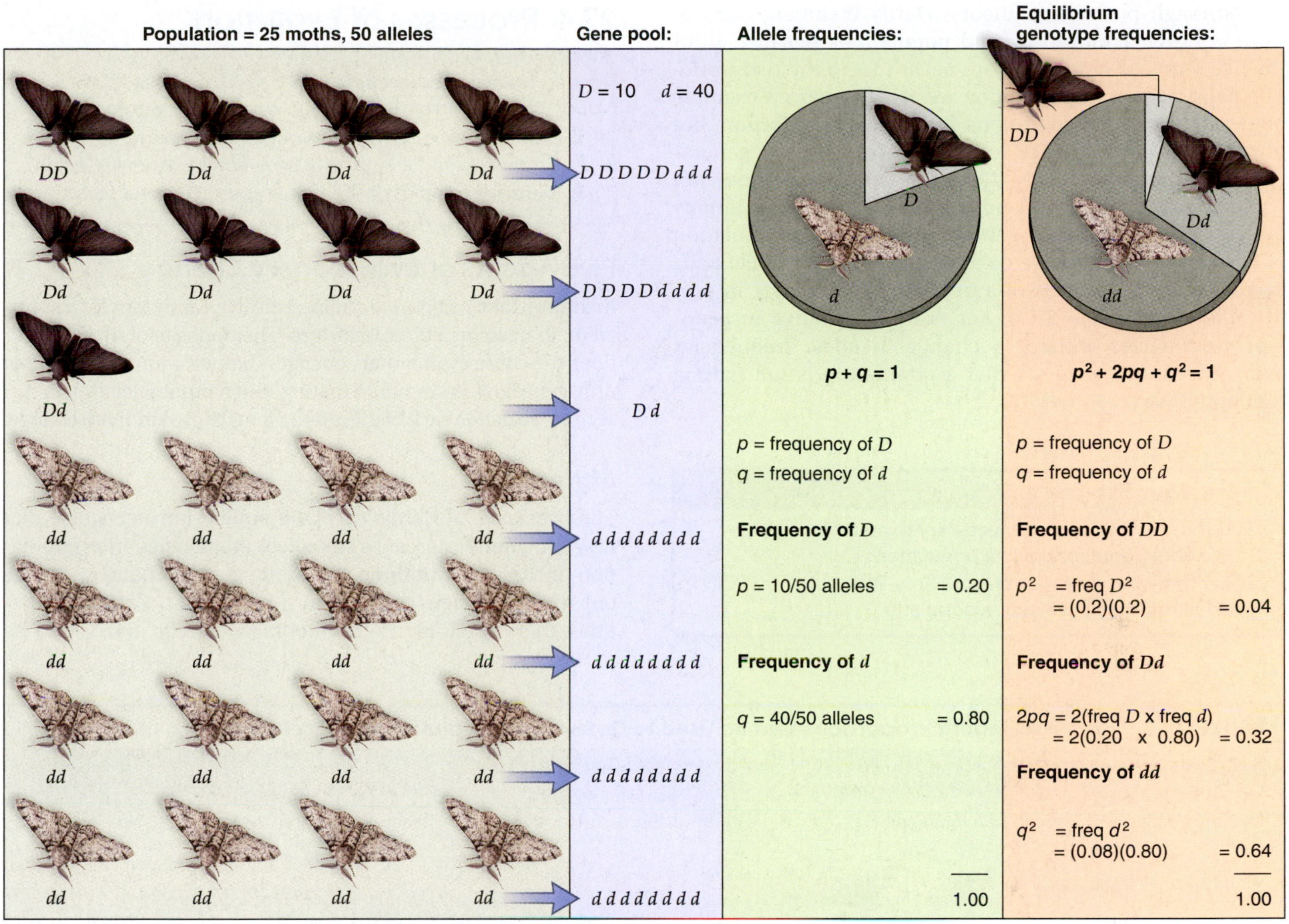

Population = 25 moths, 50 alleles
Gene pool:
Allele frequencies:
Equilibrium genotype frequencies:
DD Dd Dd Dd
Dd Dd Dd Dd
Dd
dd dd dd dd
dd dd dd dd
dd dd dd dd
dd dd dd dd
D = 10 d = 40
D D D D D d d d
D D D D d d d d
D d
d d d d d d d d
d d d d d d d d
d d d d d d d d
d d d d d d d d
D
d
p + q = 1
p = frequency of D
q = frequency of d
Frequency of D
p = 10/50 alleles = 0.20
Frequency of d
q = 40/50 alleles = 0.80
1.00
DD
Dd
dd
p² + 2pq + q² = 1
p = frequency of D
q = frequency of d
Frequency of DD
p² = freq D² = (0.2)(0.2) = 0.04
Frequency of Dd
2pq = 2(freq D x freq d) = 2(0.20 x 0.80) = 0.32
Frequency of dd
q² = freq d² = (0.08)(0.80) = 0.64
1.00

Figure 27.13 How Hardy-Weinberg equilibrium is estimated. A population of 25 moths contains a gene pool of *D* and *d* alleles. The frequencies of the *D* and *d* alleles can be estimated from the gene pool. Under Hardy-Weinberg equilibrium, the frequencies of *D* and *d* alleles should produce in the next generation a predictable frequency of genotypes that can be calculated with the Hardy-Weinberg equation.

frequencies of a population that is in genetic equilibrium. Their mathematical model proposes that the genotype frequencies of a nonevolving population can be described by the equation $p^2 + 2pq + q^2$, again with p and q representing the frequency of alleles D and d. Recall that in our moth population, $D = p$ and $d = q$, so that D^2 is the frequency of the *DD* genotype, $2Dd$ is the frequency of the *Dd* genotype, and d^2 is the frequency of the *dd* genotype (Fig. 27.13).

Conditions of Hardy-Weinberg Equilibrium

The principles of the Hardy-Weinberg equation indicate that allele frequencies in a gene pool will remain at equilibrium, and thus constant, after one generation of random mating in a large, sexually reproducing population as long as five conditions are met:

1. *No mutations.* Genetic mutations are an alteration in an allele, due to a change in DNA composition. Under Hardy-Weinberg assumptions, allele changes do not occur, or changes in one direction are balanced by changes in the opposite direction.
2. *No genetic drift.* Genetic drift is random changes in allele frequencies by chance. If a population is very large, changes in allele frequencies due to chance alone are insignificant.
3. *No gene flow.* Gene flow is the sharing of alleles between two populations through interbreeding. If there is no gene flow, migration of individuals, and therefore their genes, into or out of the population does not occur.
4. *Random mating.* Random mating occurs when individuals pair by chance, not according to their genotypes or phenotypes.
5. *No selection.* Often, the environment selects certain phenotypes to reproduce and have more offspring than other phenotypes. If selection does not occur, no phenotype is favored over another to reproduce.

Although possible in theory, Hardy-Weinberg equilibrium is never achieved in wild populations because all of the five required conditions are never met in the real world. The peppered moth population we are using as an example obeys all five of the conditions for genetic equilibrium, but in reality, populations are constantly evolving from generation to generation. Hardy-Weinberg equilibrium does not typically occur in natural populations, but it is an important tool for population geneticists because the violation of one or more of the five conditions causes the allele and/or genotype frequencies of a population to change in predictable ways (Table 27.2). For example, change in genotype frequencies without a change in allele frequencies over generations suggests that a population is not mating randomly.

Check Your Progress 27.3

1. List the five conditions necessary to maintain a Hardy-Weinberg equilibrium in a population.
2. Explain what deviations from Hardy-Weinberg equilibrium tell us about a population.

27.4 Processes of Evolution

Learning Outcomes

Upon completion of this section, you should be able to

1. Define the five agents of evolutionary change.
2. List the four requirements of evolution by natural selection.
3. Identify examples of the three types of natural selection.

Five Agents of Evolutionary Change

In the previous section, we outlined the five conditions for a population to be in genetic equilibrium. The opposite of these conditions can cause evolutionary change—namely mutations, genetic drift, gene flow, nonrandom mating, and natural selection. In this section we define each of these five agents of evolutionary change.

Mutations

The principles of Hardy-Weinberg equilibrium recognize that new mutations can cause the allele frequencies in a population to change. **Mutations,** which are genetic changes, are the only source of new variation in a population. Without mutations, there could be no new heritable genetic diversity to be

TABLE 27.2 Hardy-Weinberg Proportions Can Be Used to Determine if Evolution Has Occurred.

Hardy-Weinberg Condition	Deviation from Condition	Effect of Deviation on Population	Expected Deviation from HWE	Evolution Occurred?
Random mating	Nonrandom mating 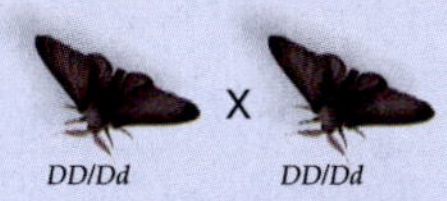 X DD/Dd DD/Dd	Alleles do not assort randomly	Change in genotype frequencies	No
No selection	Selection 	Certain alleles are selected for or against	Change in allele frequencies	Yes
No mutation	Mutation	Addition of new alleles	Change in allele frequencies	Yes
No migration	Immigration or emigration 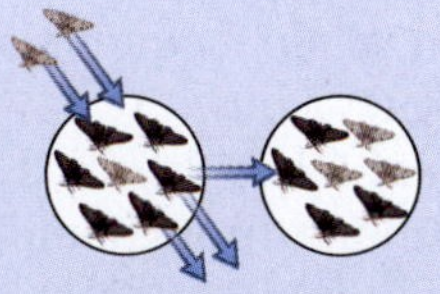	Individuals carry alleles into, or out of, the population	Change in allele frequencies	Yes
Large population (no genetic drift)	Small population (genetic drift) 1. bottleneck effect 2. founder effect 	Loss of allele diversity; some alleles may disappear	Change in allele frequencies	Yes

shaped by the forces of natural selection. It is important to realize that mutations are random events, they do not arise because the organism "needs" one. For example, a mutation that gives bacteria resistance to antibiotics appears by chance. It is because the mutation provides a survival advantage that it becomes increasingly common in the bacteria population over generations, not because the bacteria needed to evolve resistance to antibiotics. Not all mutations provide an advantage. Some mutations are harmful. But in reality most mutations have little to no effect on an organism's fitness. For example, one type of mutation, called a *point mutation,* is a change in a single nucleotide in a gene. A point mutation is called "silent" if it does not result in a change to the function of the protein encoded by the gene. Mutations that cause a malfunctioning protein can be harmful to an organism.

Animation Mutation by Base Substitution

Genetic Drift

Genetic drift refers to changes in the allele frequencies of a gene pool due to the random meeting of gametes during fertilization. Each time an organism reproduces, one in millions of sperm will fertilize one of many potential eggs. The random selection and assortment of gametes in a population causes allele frequencies to shift each generation. Thus populations are always undergoing genetic drift.

As you can imagine, genetic drift has greater effects in smaller populations because there are fewer gametes to assort. Removal of gametes from a population due to random events can have an effect on allele frequencies in the next generation. For example, the chance death of one individual in a population of a million will not have an appreciable effect on allele frequencies, but the chance death of one individual in a population of ten could change the frequency of an allele by 10% or even cause its loss altogether (if that individual was the only one with that allele) (Fig. 27.14). In nature, the *founder effect* and *bottleneck effect* occur when populations are drastically reduced in size.

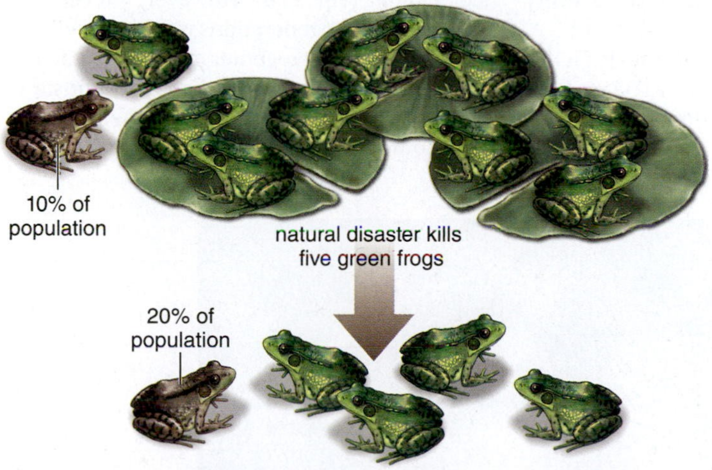
10% of population
natural disaster kills five green frogs
20% of population

Figure 27.14 Genetic drift. Genetic drift occurs when, by chance, only certain members of a population (in this case, green frogs) reproduce and pass on their alleles to the next generation. A natural disaster can cause the allele frequencies of the next generation's gene pool to be markedly different from those of the previous generation.

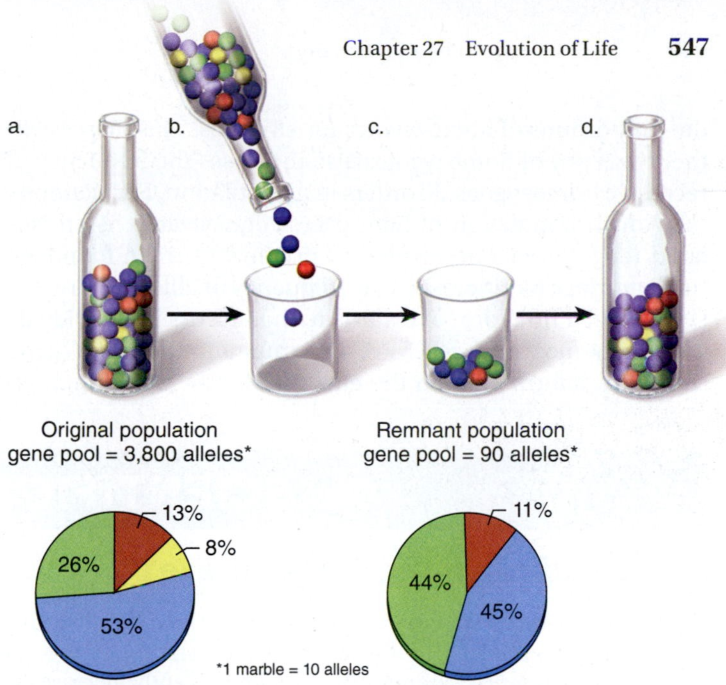
a.
b.
c.
d.
Original population gene pool = 3,800 alleles*
Remnant population gene pool = 90 alleles*
13%
8%
26%
53%
11%
44%
45%
*1 marble = 10 alleles

Figure 27.15 Bottleneck and founder effects change allele frequencies. **a.** The gene pool of a large population contains four different alleles, represented by colored marbles in a bottle, each with a different frequency. **b.** A population bottleneck occurs. The marbles, or alleles, that exit the bottle must pass through the narrow neck into the cup. The new gene pool will have a fraction of the alleles from the original population. **c.** The gene pool of the new population has changed from the original. Some alleles are in high frequency, while some are not present. **d.** A founder event is the same as a bottleneck, except that in the founder event, the original population still exists.

For example, a **bottleneck effect** can occur following a natural disaster that kills a large proportion of a population (Fig. 27.15). Both of these effects result in a severe reduction in the gene pool that can drastically affect allele frequencies.

The **founder effect** occurs when a few individuals form a new colony, and only a fraction of the total genetic diversity of the original gene pool is represented in these individuals (Fig. 27.15). The particular alleles carried by the founders are dictated by chance alone.

Animation Simulation of Genetic Drift

Gene Flow

Gene flow is the movement of alleles among populations when individuals or their gametes migrate from one population to another and breed in that new population. For example, adult plants are not able to migrate, but their pollen is often blown by the wind or carried by insects into different populations.

Gene flow mixes genetic diversity and keeps the gene pools of populations similar.

Nonrandom Mating

Nonrandom mating occurs when individuals are selective about choosing a mate with a preferred trait. Because most sexually reproducing organisms select their mates based on some trait, random mating is never observed in natural populations. *Inbreeding,* or mating between close relatives, is an example of nonrandom mating. Inbreeding changes genotype frequencies by decreasing the proportion of heterozygotes and increasing the proportions of homozygotes for all genes. This increase in the frequency of homozygotes also increases the frequency of recessive homozygous disorders in a population. For example, the Amish population of Lancaster, Pennsylvania, is an isolated religious sect descended from a few German founders. Intermarriage has increased the frequency of Ellis-van Creveld syndrome, a rare form of dwarfism, that occurs in individuals who are homozygous recessive for a mutation to the *EVC* gene (for more information on the effects of inbreeding in human populations, see the Scientific Inquiry feature, "Inbreeding in the Pingelapese."

Natural Selection

In section 27.1 we introduced natural selection as the process that allows some individuals, with an advantage over others, to produce more offspring. Here, we restate these steps in the context of modern evolutionary theory. Evolution by natural selection requires the following:

SCIENCE IN YOUR LIFE ▶ **SCIENTIFIC INQUIRY**

Inbreeding in the Pingelapese

One of the requirements of a population in Hardy-Weinberg equilibrium is that mates are chosen at random—that is, without preference for a particular trait. Most populations, however, do not meet this requirement because some traits are more attractive in a mate than others. Humans, for example, select mates based on a set of traits that we find appealing in a mate. This type of assortative mating based on trait preference is common in many species. Inbreeding is a unique form of nonrandom, or assortative, mating where individuals mate with close relatives, such as cousins.

One consequence of inbreeding is an increase in the frequency of homozygous genotypes in a population. For the most part, an increase in homozygotes does not have a large detrimental effect, especially if the population is very large. But when populations are small, inbreeding can have a large impact on the health of a population. Many human diseases are caused by the inheritance of two recessive alleles, such that the disease appears only in persons who are homozygous recessive for the disease-causing allele. In very small populations, the probability that individuals carrying the recessive allele will mate increases because the mating pool is very small. In this case, inbreeding significantly increases the frequency of homozygous recessive genotypes, and thus those afflicted with a disease.

Human cultures tend to have social rules that discourage inbreeding, but in very small populations, such as those following a bottleneck or founder event, inbreeding is sometimes unavoidable. One example of the effects of inbreeding on human populations is the occurrence of a rare form of non-sex-linked color blindness, achromatopsia, on the island of Pingelap, a small coral atoll in the Pacific Ocean. People who are achromatic are completely color blind (Fig. 27A). Normal human color vision is possible because of cells in the back of the eye called cones. Those with achromatopsia do not have any cone cells because they are homozygous for a rare recessive allele that prevents cone cells from developing. Complete color blindness is so common on Pingelap that it is considered part of everyday life for most families.

Experts propose that the high frequency of achromatopsia appeared following a severe population bottleneck in 1775 when a typhoon killed 90% of the inhabitants. Approximately 20 people survived the typhoon. Four generations after the typhoon struck, achromatopsia began to appear frequently in the population. How could this happen? Figure 27B shows how inbreeding in a population can produce homozygous recessive genotypes. Geneticists explain that the high frequency of this rare genetic disorder on Pingelap is consistent with a large degree of intermarriage among relatives, or inbreeding, following the typhoon.

Mwanenised, a male survivor of the typhoon, was a carrier, or a heterozygote, for the achromatopsia allele. He had 10 children, which was a large proportion of the first generation of Pingelapese after the typhoon. Geneticists would predict that on average 50% of his children would have been homozygous for the normal allele, and 50% would have been heterozygous carriers of the color-blind allele. Thus none of Mwanenised's children would have been color blind (Fig. 27B). However, intermarriage of the next generations would unavoidably have brought together men and women who were carriers for the achromatopsia allele. Thus, in subsequent generations the homozygous recessive genotype, and

Figure 27A Achromatopsia. Complete achromatopsia is a recessive genetic disorder that causes complete color blindness. People with complete achromatopsia see no color, only shades of gray. This image shows how a person with this form of color blindness would see the world.

1. *Individual variation.* The members of a population differ from one another.
2. *Inheritance.* Many of these differences are heritable genetic differences.
3. *Overproduction.* Individuals in a population are engaged in a struggle for existence because breeding individuals in a population tend to produce more offspring than the environment can support.
4. *Differential reproductive success.* Individuals that are better adapted to their environment produce more offspring than those that are not as well adapted, and consequently, their fertile offspring will make up a greater proportion of the next generation.

In biology, the **fitness** of an individual is measured by the number of fertile offspring produced throughout its lifetime. Gene mutations are the ultimate source of variation because

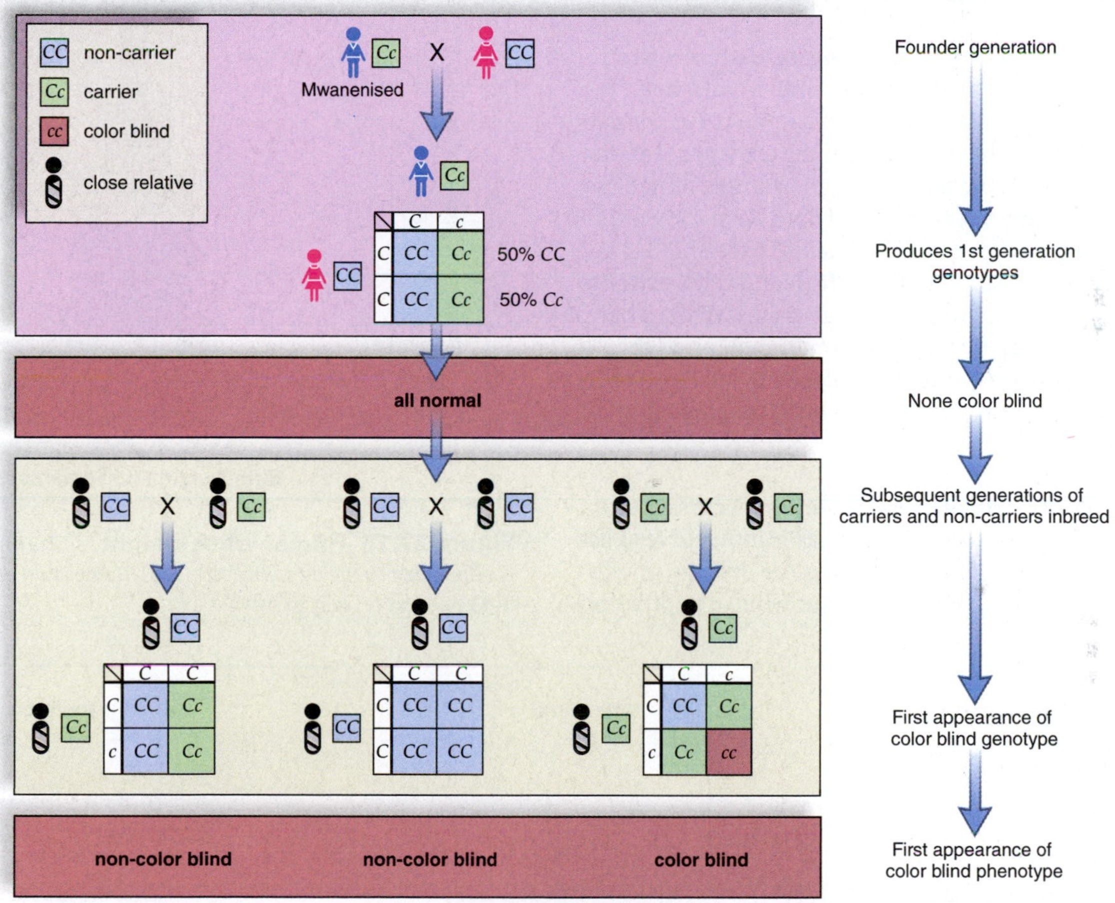
CC non-carrier
Cc carrier
cc color blind
close relative
Mwanenised
50% CC
50% Cc
all normal
non-color blind
non-color blind
color blind
Founder generation
Produces 1st generation genotypes
None color blind
Subsequent generations of carriers and non-carriers inbreed
First appearance of color blind genotype
First appearance of color blind phenotype

Figure 27B Inbreeding in small populations increases the frequency of recessive genetic disorders. On the island of Pingelap, a high level of incidental inbreeding in a small population has resulted in a higher than expected occurrence of complete achromatopsia.

complete color blindness, did appear due to inbreeding within a small population (Fig. 27B). On Pingelap, an increase in color blindness was observed by the fourth generation.[1]

The effect of inbreeding can be long-term, especially if a population remains relatively small and isolated. Today 1 in 12 Pingelapese suffers from achromatopsia, more than 3,000 times more frequent than in the United States, where it occurs in approximately 1 in 40,000 births.

Questions to Consider

1. What might have happened to the color-blind allele in the Pingelapese population following the typhoon if Mwanenised had not had any children? Only five children?
2. Would you predict that the Pingelapese population is in Hardy-Weinberg equilibrium? How would you measure this?
3. Has the Pingelapese population evolved? Explain. (*Hint:* Does inbreeding cause a change in allele frequencies?)

[1]See Sacks, Oliver (1998). *The Island of the Colorblind.* (Vintage/Anchor Books, New York). An interesting, nontechnical account of the Pingelapese.

they provide new alleles. Sometimes, these mutations result in positive fitness consequences for an organism perhaps providing variation for adaptation to a new environmental change. However, in sexually reproducing organisms, genetic variation can also result from crossing-over and independent assortment of chromosomes during meiosis and also fertilization when gametes are combined. A different combination of alleles can lead to a new and different phenotype.

In this context, consider that most of the traits on which natural selection acts are polygenic and thus controlled by more than one gene. Such traits have a range of phenotypes that often follow a bell-shaped curve.

The three main types of natural selection are stabilizing selection, directional selection, and disruptive selection.

Stabilizing Selection With **stabilizing selection,** extreme phenotypes are selected against, and individuals near the average are favored. Stabilizing selection can improve adaptation of the population to those aspects of the environment that remain constant. Human birth weight is an example of stabilizing selection. Over many years, hospital data have shown that human infants born with an intermediate birth weight (3–4 kg) have a better chance of survival than those at either extreme (either much less or much greater than average). When a baby is small, its systems may not be fully functional, and when a baby is large, it may have experienced a difficult delivery. Stabilizing selection reduces the variability in birth weight in human populations (Fig. 27.16).

Directional Selection **Directional selection** occurs when an extreme phenotype is favored and the distribution curve shifts in that direction (Fig. 27.17). This changes the average phenotype in a population. Such a shift can occur when a population is adapting to a changing environment. For example, the gradual increase in the size of the modern horse, *Equus,* can be correlated with a change in the environment from forest conditions to grassland conditions. *Hyracotherium,* the ancestor of the modern horse, was about the size of a dog and was adapted to the forestlike environment of the Eocene epoch of the Paleocene period. This animal could have hidden among the trees for protection, and its low-crowned teeth would have been appropriate for browsing on leaves. Later, in the Miocene and Pliocene epochs, grasslands began to replace the forests. Then

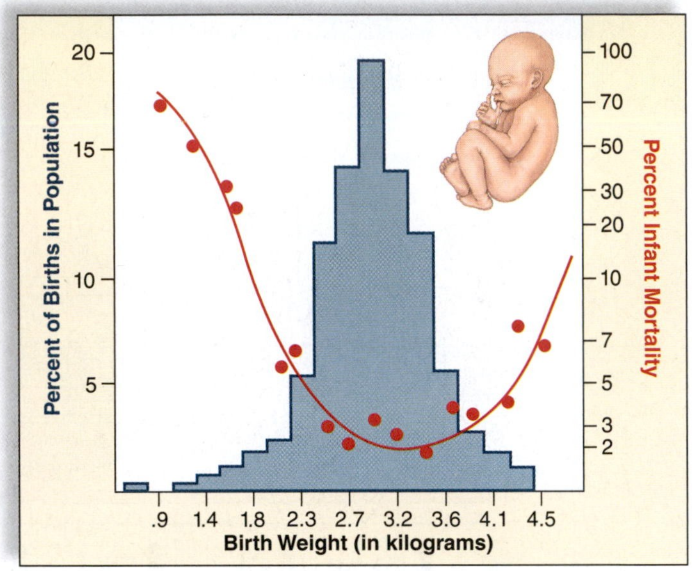

Percent of Births in Population
20
15
10
5
100
70
50
30
20
10
7
5
3
2
Percent Infant Mortality
.9
1.4
1.8
2.3
2.7
3.2
3.6
4.1
4.5
Birth Weight (in kilograms)

Figure 27.16 Human birth weight. The birth weight (blue) is influenced by the mortality rate (red). Babies that are born weighing 3–4 kg are more likely to survive.

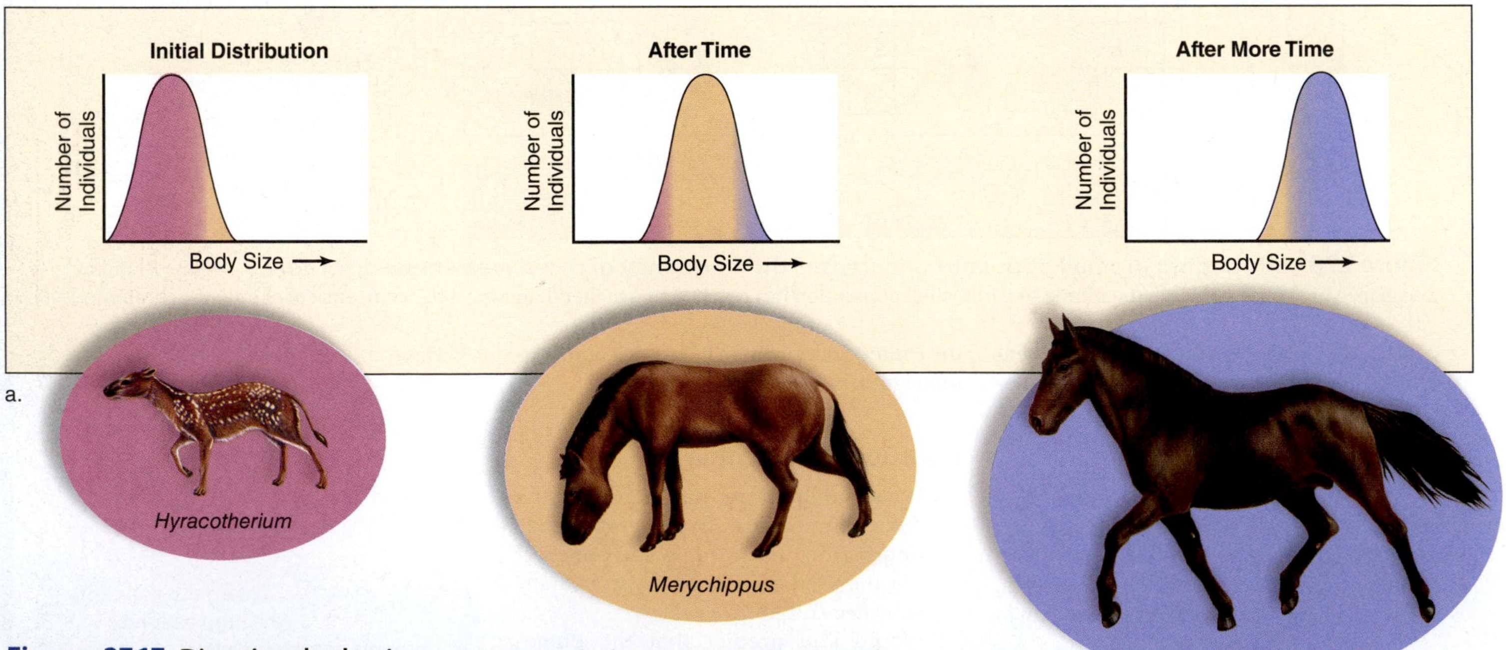

Initial Distribution
After Time
After More Time
Number of Individuals
Body Size
a.
Hyracotherium
Merychippus
Equus
b.

Figure 27.17 Directional selection. **a.** Directional selection occurs when natural selection favors one extreme phenotype, resulting in a shift in the distribution curve. **b.** For example, *Equus,* the modern-day horse, which is adapted to a grassland habitat, is much larger than its ancestor, *Hyracotherium,* which was adapted to a forest habitat.

Initial Distribution
Number of Individuals
Banding Pattern
After Time
Number of Individuals
Banding Pattern
After More Time
Number of Individuals
Banding Pattern
a.
b.

Figure 27.18 Disruptive selection. **a.** Disruptive selection favors two or more extreme phenotypes. **b.** Today, British land snails primarily possess one of two different phenotypes, each adapted to a different habitat. Snails with dark shells are more prevalent in forested areas, and light-banded snails are more prevalent in areas with low-lying vegetation.

the ancestors of *Equus* were subject to selective pressure for the development of strength, intelligence, speed, and durable grinding teeth. A larger size provided the strength needed for combat, elongated legs ending in hooves gave speed for escaping from enemies, and the durable grinding teeth enabled the animals to feed efficiently on grasses. Nevertheless, the evolution of the horse should not be viewed as a straight line of descent; there were many side branches that became extinct.

Disruptive Selection In **disruptive selection,** two or more extreme phenotypes are favored over any intermediate phenotype (Fig. 27.18). For example, British land snails (*Cepaea nemoralis*) have a wide habitat range that includes grass fields, hedgerows, and forested areas. In areas with low-lying vegetation, thrushes feed mainly on snails with dark shells that lack light bands, and in forested areas, they feed mainly on snails with light-banded shells. Therefore, the two different habitats have resulted in two different phenotypes in the population.

Maintenance of Variation

A population nearly always shows some genotypic variation. The maintenance of variation is beneficial because populations with limited variation may not be able to adapt to new conditions should the environment change and may become extinct. How can variation be maintained in spite of selection constantly working to reduce it? First, we must remember that the forces that promote variation are still at work: mutation still generates new alleles, recombination and independent assortment still shuffle the alleles during gametogenesis, and fertilization still creates new combinations of alleles from those present in the gene pool. Second, gene flow might still occur. If the receiving population is small and mostly homozygous, gene flow can be a significant source of new alleles. Finally, natural selection favors certain phenotypes, but the other types may still remain in reduced frequency. Disruptive selection even promotes polymorphism in a population. Genetic variation can be maintained in diploid organisms in heterozygote individuals that can maintain recessive alleles in the gene pool.

Diploidy and the Heterozygote

Only alleles that are expressed (cause a phenotypic difference) are subject to natural selection. In diploid organisms, this fact makes the heterozygote a potential protector of recessive alleles that might otherwise be weeded out of the gene pool by natural selection. Because the heterozygote remains in a population, so does the possibility of the recessive phenotype, which might have greater fitness in a changing environment. When natural selection favors the ratio of two or more phenotypes in generation after generation, it is called balanced polymorphism. Sickle cell disease offers an example of balanced polymorphism (Fig. 27.19).

Individuals with sickle cell disease have the genotype Hb^SHb^S (Hb = hemoglobin, the oxygen-carrying protein in red blood cells; s = sickle cell) and tend to die at an early age due to hemorrhaging and organ destruction. Those who are heterozygous and have sickle cell trait (Hb^AHb^S; A = normal) are better off because their red blood cells usually become sickle-shaped only when the oxygen content of the environment is

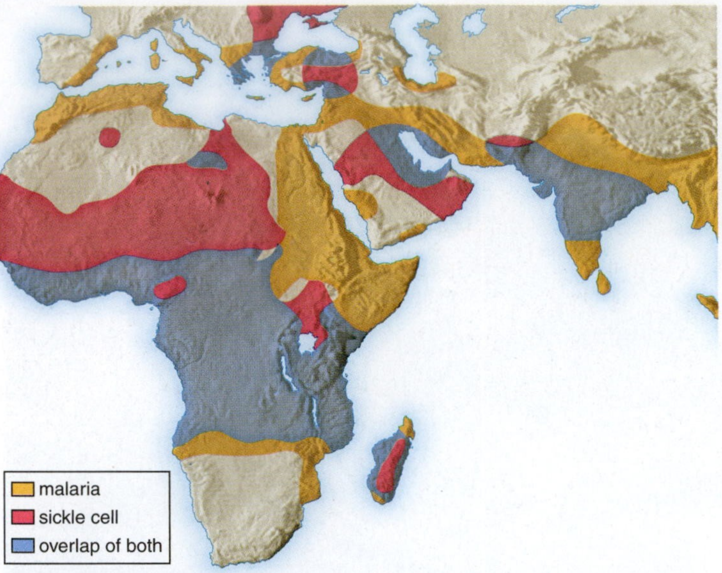
malaria
sickle cell
overlap of both

Genotype	Phenotype	Result
Hb^AHb^A	Normal	Dies due to malarial infection
Hb^AHb^S	Sickle cell trait	Lives due to protection from both
Hb^SHb^S	Sickle cell disease	Dies due to sickle cell disease

Figure 27.19 Sickle cell disease. **a.** Sickle cell disease is more prevalent in areas of Africa where malaria is more common. **b.** Whether or not a person dies from the sickle cell disease, malaria, or survives depends upon the genotype. Heterozygotes for the sickle cell trait are protected against both diseases.

low. Ordinarily, those with a normal genotype (Hb^AHb^A) are the most fit.

Geneticists studying the distribution of sickle cell disease in Africa have found that the recessive allele (Hb^S) has a higher frequency (0.2 to as high as 0.4 in a few areas) in regions where malaria is also prevalent. What is the connection between higher frequency of the recessive allele and malaria? Malaria is caused by a parasite that lives in and destroys the red blood cells of the normal homozygote (Hb^AHb^A). However, the parasite is unable to live in the red blood cells of the heterozygote (Hb^AHb^S) because the infection causes the red blood cells to become sickle-shaped. Sickle-shaped red blood cells lose potassium, and this causes the parasite to die. Thus, in an environment where malaria is prevalent, the heterozygote is favored. Each of the homozygotes is selected against but is maintained because the heterozygote is favored in parts of Africa subject to malaria. Figure 27.19 summarizes the effects of the three possible genotypes.

Check Your Progress 27.4

1. Summarize the five agents of evolutionary change.
2. Explain how the agents of evolutionary change relate to the process of natural selection.
3. Describe how malaria makes having a sickle cell allele advantageous and explain if having a sickle cell allele would be advantageous in all parts of the world.

27.5 Macroevolution and Speciation

Learning Outcomes

Upon completion of this section, you should be able to

1. Distinguish between prezygotic and postzygotic isolation mechanisms.
2. Explain how adaptive radiation can lead to speciation.
3. Compare and contrast phyletic gradualism with punctuated equilibrium.

In this section, we turn our attention to evolution on a large scale, that is, **macroevolution.** The history of life on Earth is a part of macroevolution. Macroevolution involves speciation, or the splitting of one species into two or more species. The same microevolutionary agents that are at play within populations—genetic drift, natural selection, mutation, and migration—also shape the evolution of new species. Thus microevolution and macroevolution are the result of the same agents, differing only at the scale at which they occur. Macroevolution is the result of the accumulation of microevolutionary change that results in the formation of new species. For our present discussion, **species** is defined as a group of organisms that are capable of interbreeding and are isolated reproductively from other organisms. The groups, or populations, of the same species can exchange genes, but different species do not exchange genes. Reproductive isolation of similar species is accomplished by the isolating mechanisms illustrated in Figure 27.20. **Prezygotic isolating mechanisms** are in place before fertilization, and thus reproduction is never attempted. **Postzygotic isolating mechanisms** are in place after fertilization, so reproduction may take place, but it does not produce fertile offspring.

Video Flirting Flies

The Process of Speciation

Speciation has occurred when one species gives rise to two species, each of which continues on its own evolutionary pathway. How can we recognize speciation? Whenever reproductive isolation develops between two formerly interbreeding groups or populations, speciation has occurred. One type of speciation, called **allopatric speciation,** usually occurs when populations become separated by a geographic barrier and gene flow is no longer possible. Figure 27.21 illustrates an example of allopatric speciation that has been extensively studied in California. Apparently, members of an ancestral population of *Ensatina* salamanders existing in the Pacific Northwest migrated southward, establishing a series of populations. Each population was exposed to its own selective pressures along the coastal and Sierra Nevada Mountains. Due to the presence of the Central Valley of California, which is largely dry and thus an unsuitable habitat for amphibians, gene flow rarely occurs between eastern and western populations of *Ensatina.* Genetic differences also increased from north to south, resulting in two distinct forms of *Ensatina* salamanders in Southern California that differ dramatically in color.

It is also possible that a single population could suddenly divide into two reproductively isolated groups without being geographically isolated. The best evidence for this type

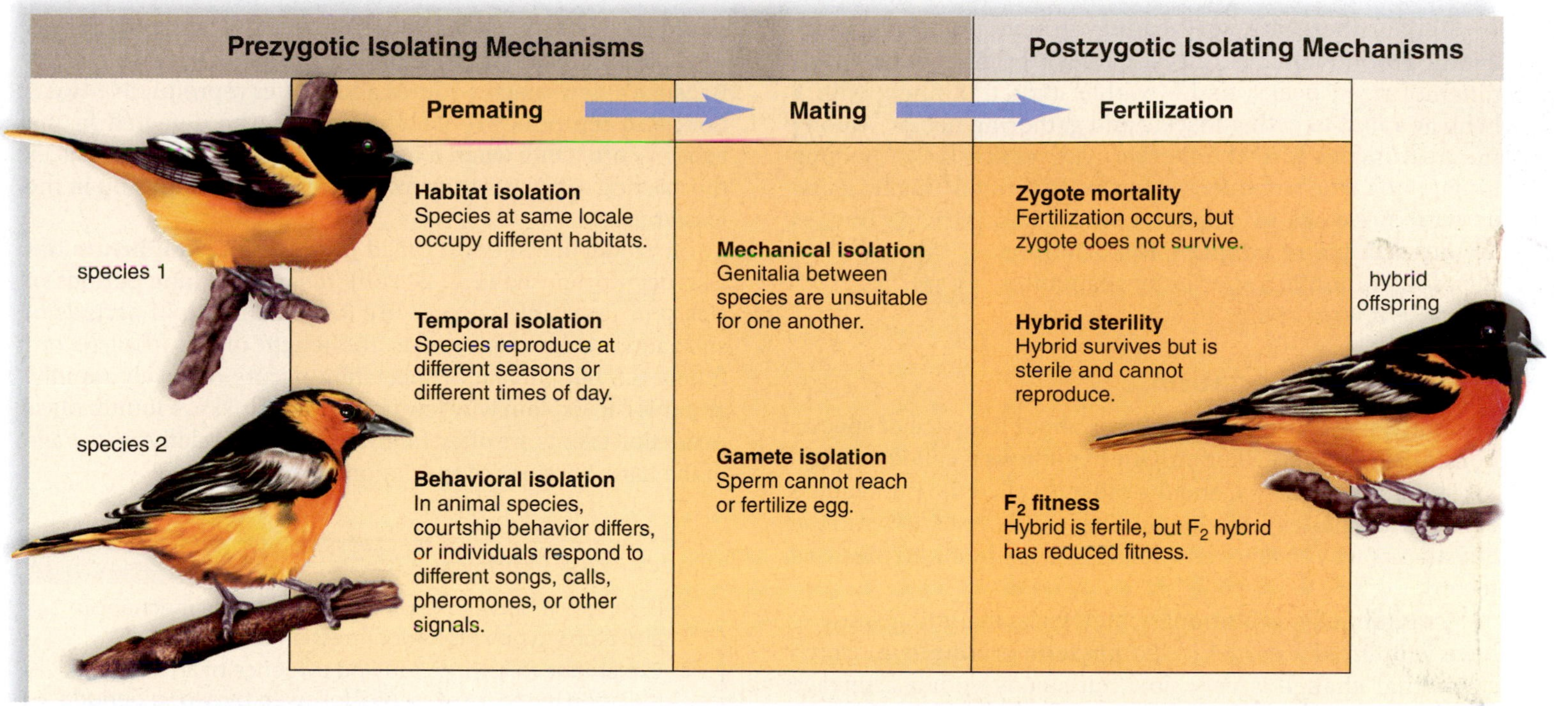
Prezygotic Isolating Mechanisms
Postzygotic Isolating Mechanisms
Premating
Mating
Fertilization
species 1
species 2
hybrid offspring
Habitat isolation
Species at same locale occupy different habitats.
Temporal isolation
Species reproduce at different seasons or different times of day.
Behavioral isolation
In animal species, courtship behavior differs, or individuals respond to different songs, calls, pheromones, or other signals.
Mechanical isolation
Genitalia between species are unsuitable for one another.
Gamete isolation
Sperm cannot reach or fertilize egg.
Zygote mortality
Fertilization occurs, but zygote does not survive.
Hybrid sterility
Hybrid survives but is sterile and cannot reproduce.
F_2 fitness
Hybrid is fertile, but F_2 hybrid has reduced fitness.

Figure 27.20 Reproductive barriers. Prezygotic isolating mechanisms prevent mating attempts or a successful outcome should mating take place. No zygote ever forms. Postzygotic isolating mechanisms prevent the zygote from developing—or should an offspring result, it is not fertile.

of speciation, called **sympatric speciation,** is found among plants, where multiplication of the chromosome number in one plant prevents it from successfully reproducing with others of its kind. Self-reproduction can maintain such a new plant species.

Adaptive Radiation

One of the best examples of speciation is provided by the finches on the Galápagos Islands, located 600 miles west of Ecuador, South America. As Darwin proposed, these 13 species of finches are descended from a species of mainland finch that migrated to one of the islands. It is likely that after the original species colonized a single island, some individuals dispersed to other islands where they adapted, via natural selection, to feed in various habitats on ecologically different islands. Remarkably, the shape of the beak of each finch species is well matched to the abundant food type on each island.

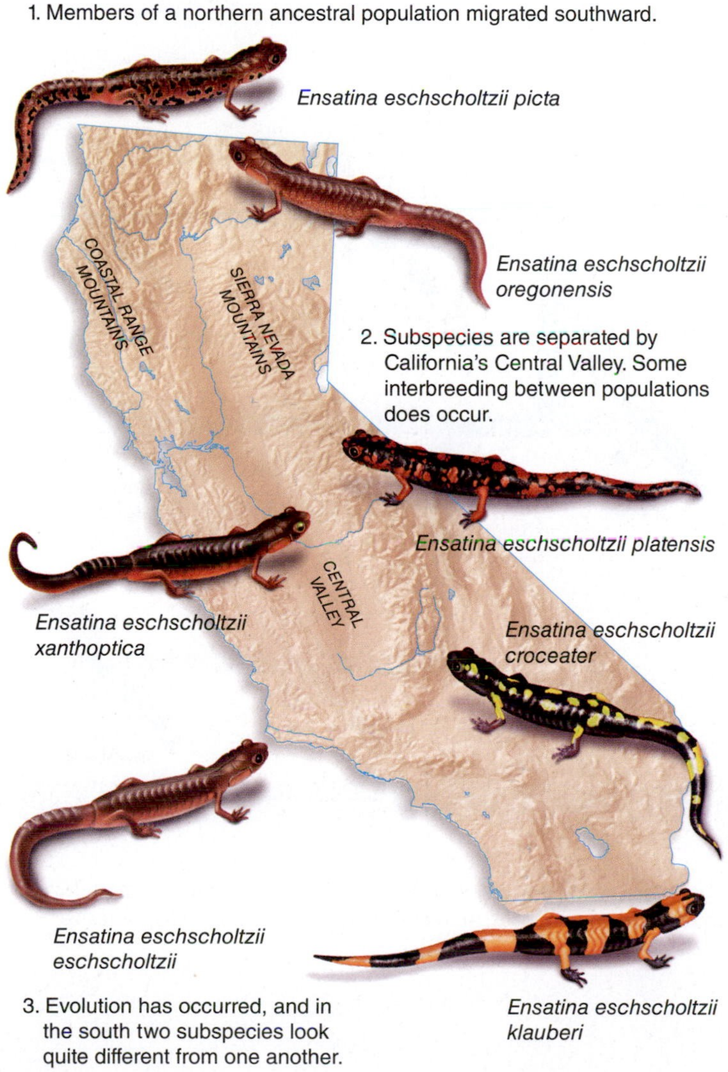
1. Members of a northern ancestral population migrated southward.
Ensatina eschscholtzii picta
Ensatina eschscholtzii oregonensis
COASTAL RANGE MOUNTAINS
SIERRA NEVADA MOUNTAINS
2. Subspecies are separated by California's Central Valley. Some interbreeding between populations does occur.
Ensatina eschscholtzii platensis
CENTRAL VALLEY
Ensatina eschscholtzii xanthoptica
Ensatina eschscholtzii croceater
Ensatina eschscholtzii eschscholtzii
Ensatina eschscholtzii klauberi
3. Evolution has occurred, and in the south two subspecies look quite different from one another.

Figure 27.21 Allopatric speciation. In this example of allopatric speciation, the Central Valley of California is separating a range of populations descended from the same northern ancestral species. Those to the west along the coastal mountains and those to the east along the Sierra Nevada Mountains experience gene flow, but gene flow is limited between the eastern populations and the western populations. Members of the most southerly eastern and western populations are quite different in color pattern.

Today, there are seed-eating ground finches, cactus-eating ground finches, insect-eating tree finches, all with different-sized beaks; and a warbler-type tree finch, with a beak adapted to eating insects and gathering nectar. Among the tree finches is a woodpecker type, which lacks the long tongue of a true woodpecker but makes up for this by using a cactus spine or a twig to ferret out insects. Darwin's finches are an example of **adaptive radiation,** or the proliferation of a species by adaptation to different ways of life.

Video Finches' Adaptive Radiation

The Pace of Speciation

Currently, there are two hypotheses about the pace of speciation and, therefore, evolution. One hypothesis is called the phyletic gradualism model, and the other is called the punctuated equilibrium model. Each model gives a different answer to the question of why so few transitional links are found in the fossil record.

Traditionally, evolutionary biologists, including Darwin, have supported a model called **phyletic gradualism,** which states that change is very slow but steady within a lineage before and after a divergence (splitting of the line of descent) (Fig. 27.22*a*). Therefore, it is not surprising that few transitional links such as *Archaeopteryx* (see Fig. 27.4) have been found. Indeed, the fossil record, even if it were complete, might be unable to show when speciation has occurred. Why? Because a new species comes about after reproductive isolation, and reproductive isolation cannot be detected in the fossil record! Only when a new species evolves and displaces the existing species is the new species likely to show up in the fossil record.

A model of evolution called **punctuated equilibrium** has also been proposed (Fig. 27.22*b*). It says that long periods of stasis, or no visible change, are followed by rapid periods of speciation. With reference to the length of the fossil record (about 3.5 billion years), speciation occurs relatively rapidly, and this can explain why few transitional links are found. Mass extinction events are often followed by rapid (relative to the age of the Earth) periods of speciation.

Check Your Progress 27.5

1. Distinguish between a prezygotic isolation mechanism and a postzygotic isolation mechanism.
2. Explain the following: You find evidence of an adaptive radiation in the fossil record, followed by a long period of little change. Which model of evolution would this support?

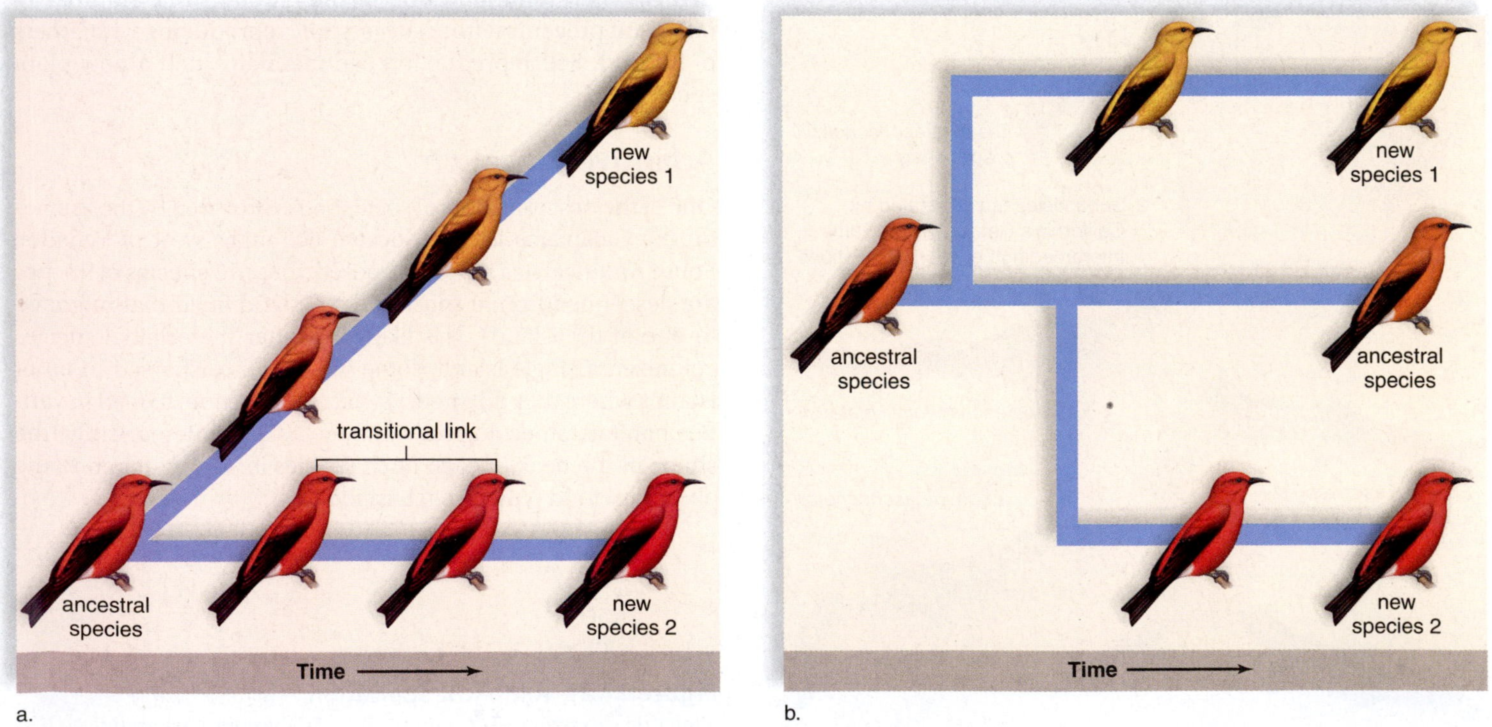
new
species 1
transitional link
ancestral
species
new
species 2
Time
a.
new
species 1
ancestral
species
ancestral
species
new
species 2
Time
b.

Figure 27.22 Phyletic gradualism compared to punctuated equilibrium. **a.** Supporters of the phyletic gradualism model hypothesize that speciation takes place gradually and many transitional links occur. **b.** Supporters of the more recent, punctuated equilibrium model hypothesize that speciation occurs rapidly, with no transitional links.

27.6 Systematics

Learning Outcomes

Upon completion of this section, you should be able to

1. Recognize how phylogenetics, the theory of evolution, and classification are interrelated.
2. Define the goal of systematic biology and its two branches of study: taxonomy and phylogenetics.
3. Interpret the information provided in both a cladogram and a phylogenetic tree.

In section 27.5 we examined macroevolution, or evolutionary change that results in the formation of new species. Macroevolution is the source of past and present biodiversity. **Systematic biology** (or **systematics**) is the study of the evolutionary history of biodiversity. Systematic biology is a quantitative science that uses characteristics of living and fossil organisms, or traits, to infer the evolutionary relationships among organisms, and to then organize biodiversity based upon these relationships.

Classification

Taxonomy is the branch of systematic biology concerned with identifying, naming, and classifying organisms. A **taxon** (pl., **taxa**) is the general name for a group containing an organism or group of organisms that exhibit a set of shared traits. **Classification** is the process of naming and assigning organisms or groups of organisms to a taxon. For example, the taxon Vertebrata contains organisms with a bony spinal column. As another example, the taxon Canidae contains the wolf (*Canis lupus*) and its close relative, the domestic dog (*Canis familiaris*). The binomial system of nomenclature is used to assign a two-part name (*genus* and *species*) to each taxon. For example, the scientific name for modern humans is *Homo sapiens.* Notice that the scientific name is in italics and only the genus is capitalized. The name of an organism usually tells you something about the organism. In this instance, the species name, *sapiens,* refers to a large brain.

Today, **taxonomists,** scientists that study taxonomy, use several categories of classification created by Swedish biologist Carl Linnaeus in the 1700s to show varying levels of similarity: species, genus, family, order, class, phylum, and kingdom. Recently, the domain, a higher taxonomic category, has been added to this list. There can be several species within a genus, several genera within a family, and so forth. In this hierarchy, the higher the category, the more inclusive it is. Therefore species in the same domain have general traits in common, whereas those in the same genus have quite specific traits in common.

Phylogeny

Phylogenetics is the branch of systematic biology that studies the evolutionary relatedness among groups of organisms. Phylogeneticists use characters from the fossil record, comparative anatomy and development, and the sequence, structure, and function of RNA and DNA molecules to construct a **phylogeny,** a hypothesis of evolutionary relatedness among taxa represented by a "family tree." In essence, systematic biology is the study of the evolutionary history of biodiversity, and a phylogeny is the estimation, and visual representation, of this history. Taxonomists can then use a phylogeny to classify organisms into taxa based upon shared evolutionary history. **Cladistics** is one method phylogeneticists use to construct a phylogeny called a **cladogram.** Branches in a cladogram are called **clades,** and each clade contains a most recent common ancestor and all its descendants. Cladistics uses only those traits that are shared among all individuals to define a clade. These **shared derived traits** are not found in organisms outside of the clade, called "outgroups." Figure 27.23 depicts a cladogram for seven groups of vertebrates. Only the lamprey, the "outgroup," lacks jaws, but the other six groups of vertebrates are in the same clade because they all have jaws, a shared derived trait. In the same way, the vertebrates in the clade without the lamprey and the shark all have the shared derived trait of lungs, and so forth. Figure 27.23 is somewhat misleading because, although single traits are noted on the tree, cladistics uses much more data to

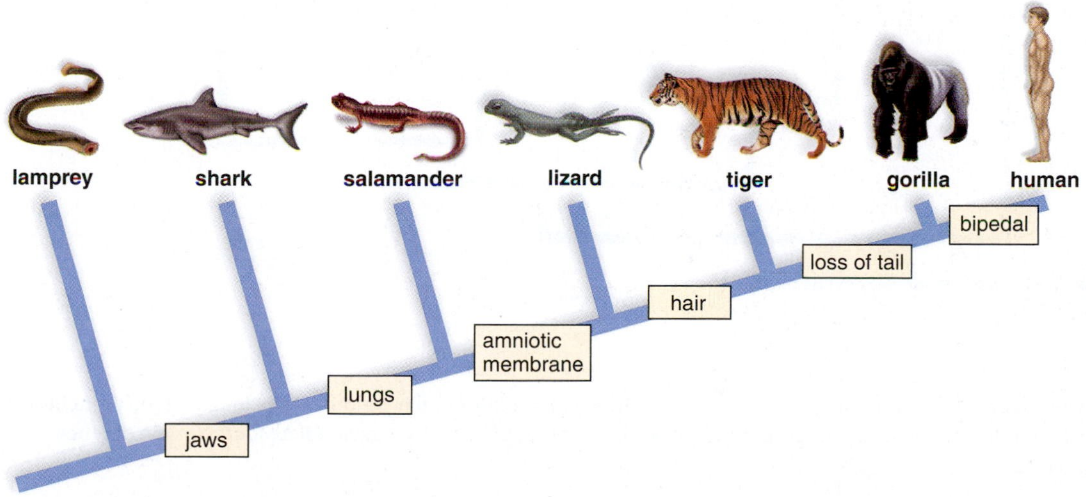
lamprey
shark
salamander
lizard
tiger
gorilla
human
jaws
lungs
amniotic membrane
hair
loss of tail
bipedal

Figure 27.23 Cladogram. A cladogram gives comparative information about relationships. Organisms in the same clade share the same derived traits. Humans and all the other vertebrates shown, except lampreys, are in the same clade as sharks because they all have jaws. However, humans are alone on a branch because only they are bipedal. This cladogram is simplified because phylogeneticists actually use a great deal more data to construct cladograms.

arrange groups of organisms into clades. Cladistics is based on the premise of **parsimony,** which considers the simplest explanation—that is, the cladogram, which requires the fewest number of evolutionary changes—to be the best hypothesis of evolutionary history.

Animation Phylogenetic Trees

It is important to note that a phylogeny (or in the case of cladistics, a cladogram), represents only evolutionary relatedness if homologous traits, or those that share a common ancestry (see Fig. 27.7), are used. Deciphering homology is sometimes difficult because of convergent evolution. **Convergent evolution** is the independent evolution of analogous traits, or the same or similar traits in distantly related lines of descent (see section 27.2). Analogous structures have the same function in different groups, but organisms with these structures do not share a recent common ancestor. Instead, analogous structures arise because of adaptations to similar environments. DNA nucleotide sequences are traits that are commonly used to construct a phylogeny. Modern computer systems and laboratory techniques have made acquiring DNA sequences easy and economical. Homology of nucleotides in a DNA sequence is determined by alignment, or matching sections of sequence from the same section of the genome among all the taxa under study. Figure 27.24 is a phylogeny of primates based on DNA data. If we assume that DNA differences, or mutations, occur at a relatively constant rate, the so-called "*molecular clock,*" the number of differences can be used to estimate the absolute age of each branch in the phylogeny.

Linnaean Classification Versus Phylogenetics

The Linnaean classification system was developed long before Darwin proposed his theory of evolution, and even longer before the field of phylogenetics was born. The Linnaean classification system still carries historical taxa that were assembled without an understanding of evolutionary relatedness. This can cause problems for the taxonomist, because in some cases analogous, and not homologous, traits were used to classify organisms. Figure 27.25 illustrates the types of problems that arise when trying to reconcile Linnaean classification with the principles of phylogenetics. The phylogeny in this figure proposes that birds are in a clade with crocodiles because they share a recent common ancestor. This ancestor had a gizzard among other shared derived traits of the skull. Yet Linnaean classification places birds in their own group separate from crocodiles and separate from reptiles in general. In many other instances, also, Linnaean classification is not consistent with new understandings about phylogenetic relationships. Therefore some phylogeneticists have proposed a different system

Galago
Capuchin
Green monkey
Rhesus monkey
Gibbon
Chimpanzee
Human
Millions of Years Ago (MYA)
0
10
20
30
40

Figure 27.24 Molecular data. The relationships of certain primate species are based on a study of their genomes. The length of the branches indicates the relative number of DNA base-pair differences between the groups. These data make it possible to suggest a date for the origin of other branches in the tree by using a molecular clock.

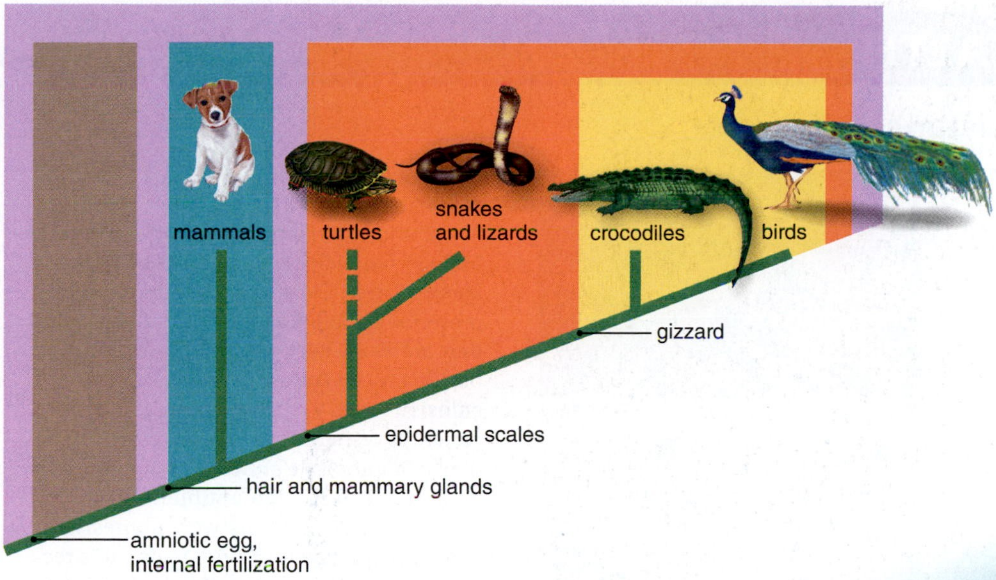
mammals
turtles
snakes and lizards
crocodiles
birds
gizzard
epidermal scales
hair and mammary glands
amniotic egg, internal fertilization

Figure 27.25 Cladistic classification. Taxonomic designations are based upon evolutionary history. Each taxon includes a common ancestor and all of its descendants.

of classification called the International Code of Phylogenetic Nomenclature, or PhyloCode, which sets forth rules to follow in naming of clades. Other biologists are hoping to modify Linnaean classification to be consistent with the principles of modern systematic biology.

Three-Domain Classification System

Classification systems change over time. Historically, most biologists have utilized the five-kingdom system of classification: plants, animals, fungi, protists, and monerans. Organisms were classified into these kingdoms based on type of cell (prokaryotic or eukaryotic), level of organization (unicellular or multicellular), and type of nutrition. In the five-kingdom system, the monerans were distinguished by their structure—they were prokaryotic (lack a membrane-bound nucleus)—whereas the organisms in the other kingdoms were eukaryotic (have a membrane-bound nucleus). The kingdom Monera contained all prokaryotes, which according to the fossil record evolved first (see section 28.1 for discussion about the origin of the first cell).

Ribosomal RNA (rRNA) genes and new data about cells challenged the five-kingdom system of classification. These data support that there are two, not one, groups of prokaryotes. Based on this information, a **three-domain system** of classification was developed (Fig. 27.26). A domain is a classification category higher than the kingdom. The two prokaryote groups were classified into the **domain Bacteria** and the **domain Archaea** because they are so fundamentally different from each other that they warrant being assigned to separate domains. A third domain, the eukaryotes, were classified into the **domain Eukarya,** which contains kingdoms for all eukaryotes, including protists, animals, fungi, and plants. Interestingly, the Archaea and Eukarya are more closely related to each other than either is to domain Bacteria.

Animation Three Domains

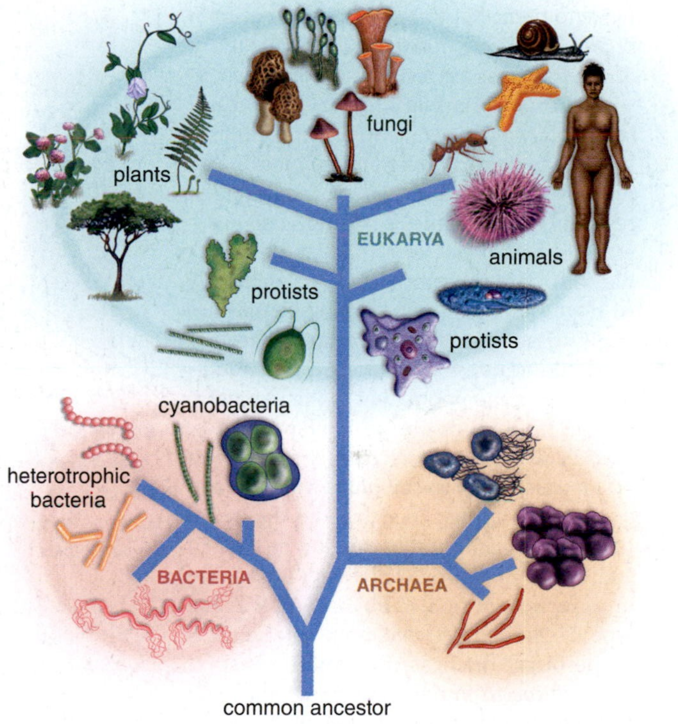
plants
fungi
EUKARYA
animals
protists
protists
cyanobacteria
heterotrophic bacteria
BACTERIA
ARCHAEA
common ancestor

Figure 27.26 The three-domain system of classification. Representatives of each domain are depicted. The phylogenetic tree shows that domain Archaea is more closely related to domain Eukarya than either is to domain Bacteria.

Check Your Progress 27.6

1. Explain the differences between the study of taxonomy and phylogenetics. How do they relate to the classification of organisms?
2. Describe the differences between a cladogram and a phylogenetic tree.
3. List the three domains of life.

SCIENCE IN YOUR LIFE ▶ HEALTH

Evolution of Antibiotic Resistance

The antibiotic penicillin was discovered by Alexander Fleming in 1928. On November 25, 1930, penicillin successfully cured a bacterial infection for the first time. Since the discovery of penicillin, modern science has produced an array of antibiotics effective against most bacterial infections. Prior to the discovery of penicillin, bacterial infections were a major cause of death worldwide, so much so that today it is difficult to imagine dying from a bacterial infection. Yet we are now coming full circle; many of the antibiotics that have saved millions of lives have evolved resistance to antibiotics (Fig. 27C). In 1945 the first resistance to penicillin was discovered in a strain of *Staphylococcus aureus.* Today, over 90% of *S. aureus* strains are resistant to penicillin.

Through Darwin's theory of natural selection, we have come to understand how bacteria become resistant to antibiotics. The process by which bacteria evolve antibiotic resistance is the same as natural selection—antibiotics kill bacteria that are susceptible, but bacterial populations are usually so large that some resistant individuals are likely to survive and reproduce. Through time, the frequency of resistant individuals increases in the population to the point that certain antibiotics are no longer effective. Overall, antibiotic resistance is evolving at an alarming rate, to the point where bacterial infections that just decades ago were treatable with antibiotics can now be fatal. In fact, resistance often evolves soon after the introduction of new antibiotics. This is the type of "accelerated evolution" that creates "superbugs," such as methicillin-resistant *Staphylococcus aureus,* or MRSA, which you learned about in the beginning of this chapter (Fig. 27C*a*). Resistant bacteria, such as MRSA, can cause abscessed lesions on the skin that are difficult to treat. In some cases resistant bacteria cause a systemic infection of the blood called *bacterial septicemia. Pseudomonas aeruginosa* (Fig. 27C*b*) produces a toxin that causes abscesses to erupt on the skin in patients who have *P. aeruginosa septicemia.* Bacterial resistance creates the need for even newer antibiotics. The development of a single new antibiotic is estimated to cost between $400 and $500 million. Antibiotic resistance adds over $30 billion to annual medical costs in the United States alone!

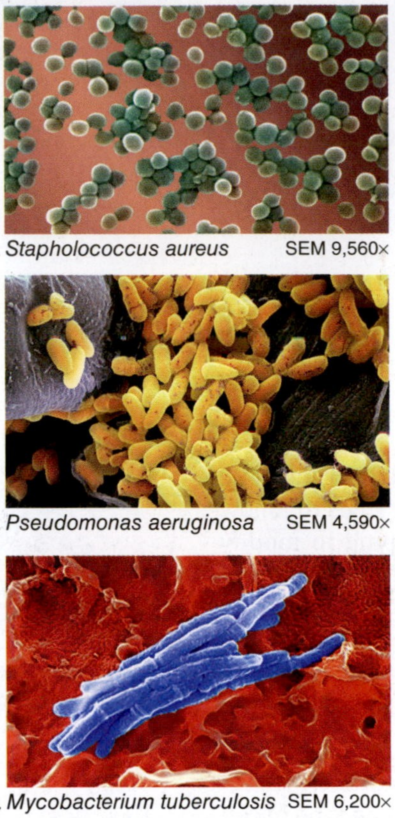
a. Stapholococcus aureus SEM 9,560×
b. Pseudomonas aeruginosa SEM 4,590×
c. Mycobacterium tuberculosis SEM 6,200×

Figure 27C Bacteria with antibiotic resistant strains. **a.** Methicillin-resistant *Staphylococcus aureus* (MRSA). **b.** *Pseudomonas aeruginosa.* **c.** *Mycobacterium tuberculosis.*

Our understanding of evolutionary biology has helped doctors treat patients appropriately. Doctors generally no longer prescribe antibiotics for colds or flu because they are viral infections. Antibiotics kill only bacteria, not viruses. In the case of tuberculosis (Fig. 27C*c*), the strain infecting a patient is tested for antibiotic resistance so that an appropriate antibiotic treatment can be administered. The antibiotic isoniazid is used to treat patients with nonresistant strains, whereas a four-drug regimen is recommended for treatment of resistant strains.

Questions to Consider

1. In New York City, state health officials have the power to quarantine tuberculosis patients who do not take their medicine. That is, they can essentially lock them up for as long as needed (often up to 18 months) to treat their illness. Note also that some of the medications can have serious side effects. What do you think about this policy?
2. Many people in less-developed countries die from tuberculosis, not because their disease is incurable, but simply because they do not have health insurance and cannot afford the medications. Why might it be advantageous for the United States to pay more for our medications so that pharmaceutical companies can provide them to lower-income people at a reduced cost or for free?

connect BIOLOGY Explore the concepts through a variety of multimedia assets, question types, and data interpretation.
www.mcgrawhillconnect.com

Case Study Conclusion

In this chapter, you have learned about both macroevolution and microevolution. Although people tend to think of evolution occurring over long timescales, it can occur quickly, particularly under human influence. The evolution of antibiotic resistance and pesticide resistance are examples of how such rapid evolution can be important in your everyday life. The evolution of antibiotic-resistant strains of bacteria such as methicillin-resistant *Staphylococcus aureus* is an increasing health problem, costing billions in increased medical costs and even resulting in deaths of some affected patients. By the same token, pesticide-resistant insects are costing billions in crop damage and decreasing yields. An evolutionary approach to treating bacterial infections involves careful use of antibiotics, thereby slowing the rate at which newly resistant strains evolve.

MEDIA STUDY TOOLS

www.mhhe.com/maderinquiry14

Enhance your study of this chapter with study tools and practice tests. Also ask your instructor about the resources available through ConnectPlus, including LearnSmart, the media-rich eBook, interactive learning tools, and animations.

Mc Graw Hill connect plus+ | BIOLOGY

SUMMARIZE

27.1 Theory of Evolution

Charles Darwin served as the naturalist aboard the HMS *Beagle* during its five-year voyage. Darwin's observations of geology and species variation led him to his theory of **evolution:** how species change over time.

- Jean-Baptiste de Lamarck proposed before Darwin a theory of evolution called **inheritance of acquired characteristics,** which proposed changes to the **phenotype** acquired during a lifetime could be passed on to subsequent generations.
- Darwin proposed the **theory of natural selection,** which proposes **adaptation** to the environment is due to heritable changes that, if advantageous, are passed to future generations differentially because of high reproductive success.

27.2 Evidence of Evolution

Darwin's theory of evolution by natural selection is supported by the following:

- The fossil record and biogeography, as well as studies of comparative anatomy and development, and biochemistry, all provide evidence for evolution. **Transitional links** are **fossils** that have traits of both ancestor and descendant and represent evolutionary shift from one trait to another.
- **Biogeography** shows that the distribution of organisms on Earth can be influenced by a combination of evolutionary and geological processes. **Continental drift** is one geological process that affected the biogeographical distribution of organisms.
- Comparing the anatomy and the development of organisms distinguishes **homologous structures** from **analogous structures.** Only homologous structures provide information about common ancestry. **Vestigial structures** are remnants of past adaptations, but which are no longer used in the current environment. **Artificial selection** by humans for preferred traits has led to evolution of many plants and animals. In natural populations, evolution can be observed to occur over periods as short as decades, such as the beak size of the Galápagos medium ground finch.
- All organisms have certain biochemical molecules in common, and these chemical similarities indicate the degree of relatedness.

27.3 Microevolution

- **Microevolution** is a process that involves a change in allele frequencies within the gene pool of a sexually reproducing population. **Population genetics** is the study of the change in allele frequencies in the **gene pool** of a **population.** The work of Hardy-Weinberg indicates that gene pool frequencies arrive at a **genetic equilibrium,** or **Hardy-Weinberg equilibrium,** that is maintained generation after generation unless disrupted by mutations, genetic drift, gene flow, nonrandom mating, or natural selection. Any change from the initial allele frequencies in the gene pool of a population signifies that evolution has occurred.

27.4 The Processes of Evolution

- The principles of the Hardy-Weinberg equation state that gene pool frequencies arrive at an equilibrium that is maintained generation after generation unless disrupted by **mutations, genetic drift, gene flow, nonrandom mating,** or **natural selection.**
- Mutations may influence the **fitness,** or reproductive success, of the individual. The **founder effect** and **bottleneck effect** are special types of genetic drift due to severe reduction in population size.
- **Stabilizing selection, directional selection,** and **disruptive selection** are different ways in which the distribution of a trait changes in response to natural selection.

27.5 Macroevolution and Speciation

- **Macroevolution** is the origin of new **species,** or **speciation.** A new species forms when the two populations can no longer interbreed because of either (or both) **prezygotic** or **postzygotic isolating mechanisms.**
- **Allopatric speciation,** the most common form of speciation, occurs when populations become reproductively isolated because of a geographic barrier.
- **Sympatric speciation** occurs when organisms overlap in their distributions.
- The evolution of several species of finches on the Galápagos Islands is an example of speciation caused by **adaptive radiation** because each one has a different way of life.
- Currently, there are two models about the pace of speciation. **Phyletic gradualism** is slow, steady change and **punctuated equilibrium,** long periods of no change interrupted by rapid change.

27.6 Systematics

- **Systematic biology,** or **systematics,** involves **phylogenetics,** the study of evolutionary relatedness (or **phylogeny**) among **taxa** (**taxon**), and **taxonomy,** the **classification** of taxa into a hierarchy of categories: domain, kingdom, phylum, class, order, family, genus, and species. **Taxonomists** are scientists who classify biodiversity.
- **Cladistics** is one method of phylogenetics that applies the principle of **parsimony** to **shared derived traits** to develop a **cladogram,** or a treelike hypothesis of evolutionary history. Cladistics strives to classify organisms based upon evolutionary relatedness. This may be complicated by **convergent evolution** and analogous traits.
- Each **clade** contains the most recent common ancestor and all its descendants. Modern phylogenetic methods most commonly analyze similarities and differences among nucleotide sequences, aided by computers, to reconstruct phylogenies.
- The **three-domain system** (**domain Bacteria, domain Archaea,** and **domain Eukarya**), based on molecular data, is currently preferred to the formerly used five-kingdom system (Monera, Protista, Fungi, Plantae, Animalia). Both bacteria and archaea are prokaryotes. Members of the kingdoms Protista, Fungi, Plantae, and Animalia are eukaryotes.

ASSESS

Testing Yourself

Choose the best answer for each question.

1. Inheritance of acquired characteristics is associated with
   a. Darwin.
   b. Malthus.
   c. Lamarck.
   d. None of these are correct.
2. A(n) _____ makes a species better suited to its environment.
   a. homologous trait
   b. analogous trait
   c. adaptation
   d. mutation
3. Fossils that serve as transitional links allow scientists to
   a. determine how prehistoric animals interacted with each other.
   b. deduce the order in which various groups of animals arose.
   c. relate climate change to evolutionary trends.
   d. determine why evolutionary changes occur.
4. The flipper of a dolphin and the fin of a tuna are
   a. homologous structures.
   b. homogeneous structures.
   c. analogous structures.
   d. reciprocal structures.
5. The frequency of a rare disorder expressed as an autosomal recessive trait is 0.0064. Using the Hardy-Weinberg equation, determine the frequency of carriers for the disease in this population.
   a. 0.147
   b. 0.080
   c. 0.920
   d. 0.020
   e. 0.846

For questions 6–10, match the description with the appropriate term in the key.

Key:

a. mutation
b. natural selection
c. founder effect
d. bottleneck
e. gene flow
f. nonrandom mating

6. The northern elephant seal went through a severe population decline as a result of hunting in the late 1800s. As a result of a hunting ban, the population has rebounded but is now homozygous for nearly every gene studied.
7. A small, reproductively isolated religious sect called the Dunkers was established by 27 families that came to the United States from Germany over 200 years ago. The frequencies for blood group alleles in this population differ significantly from those in the general U.S. population.
8. Turtles on a small island tend to mate with relatives more often than turtles on the mainland.
9. Within a population, plants that produce an insect toxin are more likely to survive and reproduce than plants that do not produce the toxin.
10. The gene pool of a population of bighorn sheep in the southwest United States is altered when several animals cross over a mountain pass and join the population.
11. People who are heterozygous for the cystic fibrosis gene are more likely than others to survive a cholera epidemic. This heterozygote advantage seems to explain why homozygotes are maintained in the human population, and is an example of
    a. disruptive selection.
    b. balanced polymorphism.
    c. high mutation rate.
    d. nonrandom mating.
12. The creation of new species due to geographic barriers is called
    a. isolation speciation.
    b. allopatric speciation.
    c. allelomorphic speciation.
    d. sympatric speciation.
    e. symbiotic speciation.
13. Which of the following models is supported by the observation that few transitional links are found in the fossil record?
    a. phyletic gradualism
    b. punctuated equilibrium
    c. Both a and b are correct.
    d. None of these are correct.
14. The three-domain classification system has recently been developed based on
    a. mitochondrial biochemistry and plasma membrane structure.
    b. cellular and rRNA sequence data.
    c. plasma membrane and cell wall structure.
    d. rRNA sequence data and plasma membrane structure.
    e. nuclear and mitochondrial biochemistry.
15. The branch of systematic biology that assigns organisms to specific classifications (domain, phyla, class, etc.) is called
    a. phylogenetics.
    b. systematics.
    c. taxonomy.
    d. cladistics.

ENGAGE

Thinking Critically

1. Viruses such as HIV are rapidly replicated and have very high mutation rates. Thus, evolution of the virus can be observed in a single infected person. Using what you have learned in this chapter, explain why HIV is so hard to treat, even though multiple drugs to treat HIV have been developed.
2. You observe a wasting disease in cattle that you know is genetically caused and thus heritable. The disease is fatal in young cattle. The allele frequency for the gene that causes the disease is 0.05 in the United States, but 0.35 in South America. Explain why such a difference in allele frequencies might exist.
3. If $p^2 = 0.36$, what percentage of the population has the recessive phenotype, assuming a Hardy-Weinberg equilibrium?
4. In a population of snails, 10 had no antennae (*aa*); 180 were heterozygous with antennae (*Aa*); and 810 were homozygous with antennae (*AA*). What is the frequency of the *a* allele in the population?

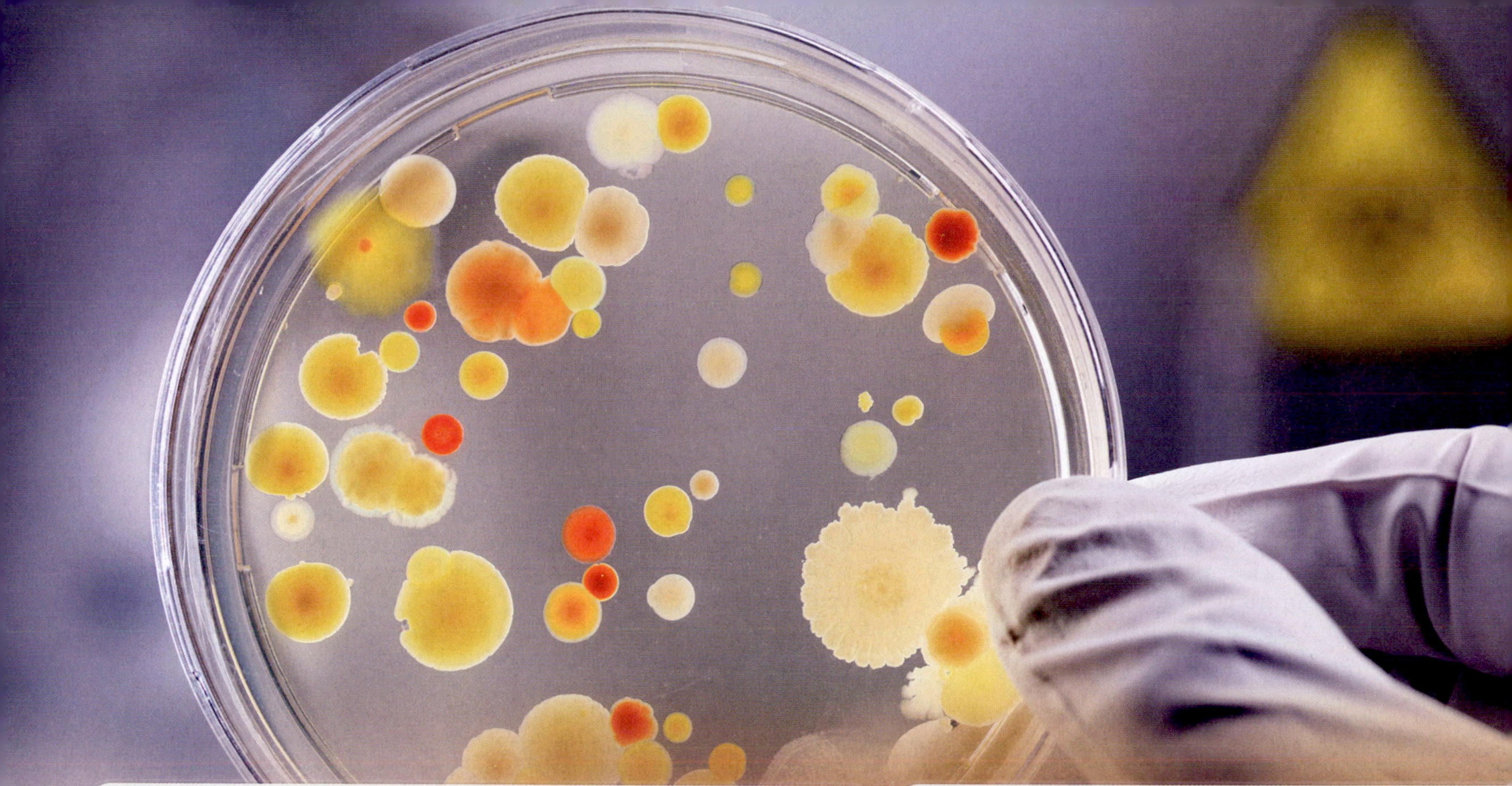

Microbiology 28

CASE STUDY The Germaphobic Billionaire

Howard Hughes was the world's richest man in the 1930s and 1940s. He built airplanes, produced movies, and romanced movie stars. But he was also terrified of "germs," and throughout his life he became increasingly fearful of contact with other people, whom he considered to be the source of most harmful microbes. According to the 2004 movie *The Aviator,* Hughes' mother instilled this fear of germs in him beginning when he was a small boy.

Even those of us who are not terrified of germs tend to become more aware of microbes when we are using a public restroom. However, because they aren't cleaned regularly, desks, not toilets, are among the most germ-filled indoor environments! It has been estimated that the average office worker's hands contact 10 million bacteria a day. Almost anything that people touch frequently, including ATM keypads, cell phones, and paper money, is teeming with microbial life.

Humans have always shared the Earth with bacteria, viruses, and other microbes, and there is no need to live in fear of most of these. In fact, many experts suggest that many children are receiving too little exposure to common, harmless microbes early in life, leading to decreased immunity to harmful varieties and, perhaps, to an increasing incidence of allergies and autoimmune disease later on. For this reason, and because we cannot avoid microbes anyway, it seems that while perhaps washing our hands or using a hand sanitizer a little more often, we might also want to be a little more accepting of the microscopic life-forms with which we share the planet.

As you read through the chapter, think about the following questions:

1. By what process do prokaryotes, such as bacteria, reproduce?
2. Considering the fact that they are structurally simple, what factors account for the tremendous diversity of bacteria?
3. What are some human diseases that are caused by bacteria and viruses?

CHAPTER OUTLINE

BEFORE YOU BEGIN

Before beginning this chapter, take a few moments to review the following discussions:

Section 1.2 To which two domains do the prokaryotes belong?

Section 3.2 What are some basic structural features of prokaryotic cells?

Table 27.1 When did prokaryotic cells first appear on Earth?

28.1 The Microbial World

Learning Outcomes

Upon completion of this section, you should be able to

1. Describe the contributions of Leeuwenhoek and Pasteur to the science of microbiology.
2. List the major groups of microbes studied in microbiology.
3. Provide several specific examples of the beneficial effects of microbes.

Antonie van Leeuwenhoek was an early eighteenth-century Dutch tradesman and scientist. He was apparently very skilled at working with glass lenses, which enabled him to greatly improve the microscopes that already existed in the late 1600s. Using these instruments, he was among the first to view microscopic life-forms in a drop of water, which he called "animalcules" and described in this way:

> *[They] were incredibly small, nay so small, in my sight, that I judged that even if 100 of these very wee animals lay stretched out one against another, they could not reach to the length of a grain of coarse sand.*

Leeuwenhoek and others after him believed that the "wee animals" he had observed could arise spontaneously from inanimate matter. For about 200 years, scientists carried out various experiments to determine the origin of microscopic organisms in laboratory cultures. Finally, in about 1859, Louis Pasteur devised the experiment shown in Figure 28.1. In the first experiment, when flasks containing sterilized broth were exposed to either outdoor or indoor air, they often became contaminated with microbial growth. In the second experiment, however, if the neck of the flask was curved so that microbes could not enter the broth from the air, no growth occurred. Thus, the bacteria were not arising spontaneously in the broth, but rather were coming from the air. In 1884, Pasteur also suggested that something even smaller than a bacterium was the cause of rabies, and it was he who chose the word *virus* from a Latin word meaning poison. Pasteur went on to develop a vaccine for rabies, which he used to save the life of a young French boy who had been bitten by a rabid dog.

Microbiology is the study of *microbes,* a term that includes the bacteria, archaea, protists, fungi, viruses, viroids, and prions. Most, but not all, of these organisms are so small that they require a microscope to be seen. Today we know that bacteria and other microbes are incredibly numerous in air, water, soil, and on objects. A single spoonful of soil can contain 10^{10} bacteria, and the total number of bacteria on Earth has been estimated at around 10^{30}, which clearly exceeds the numbers of any other type of living organism on Earth. It is a good thing bacteria are microscopic—if they were the size

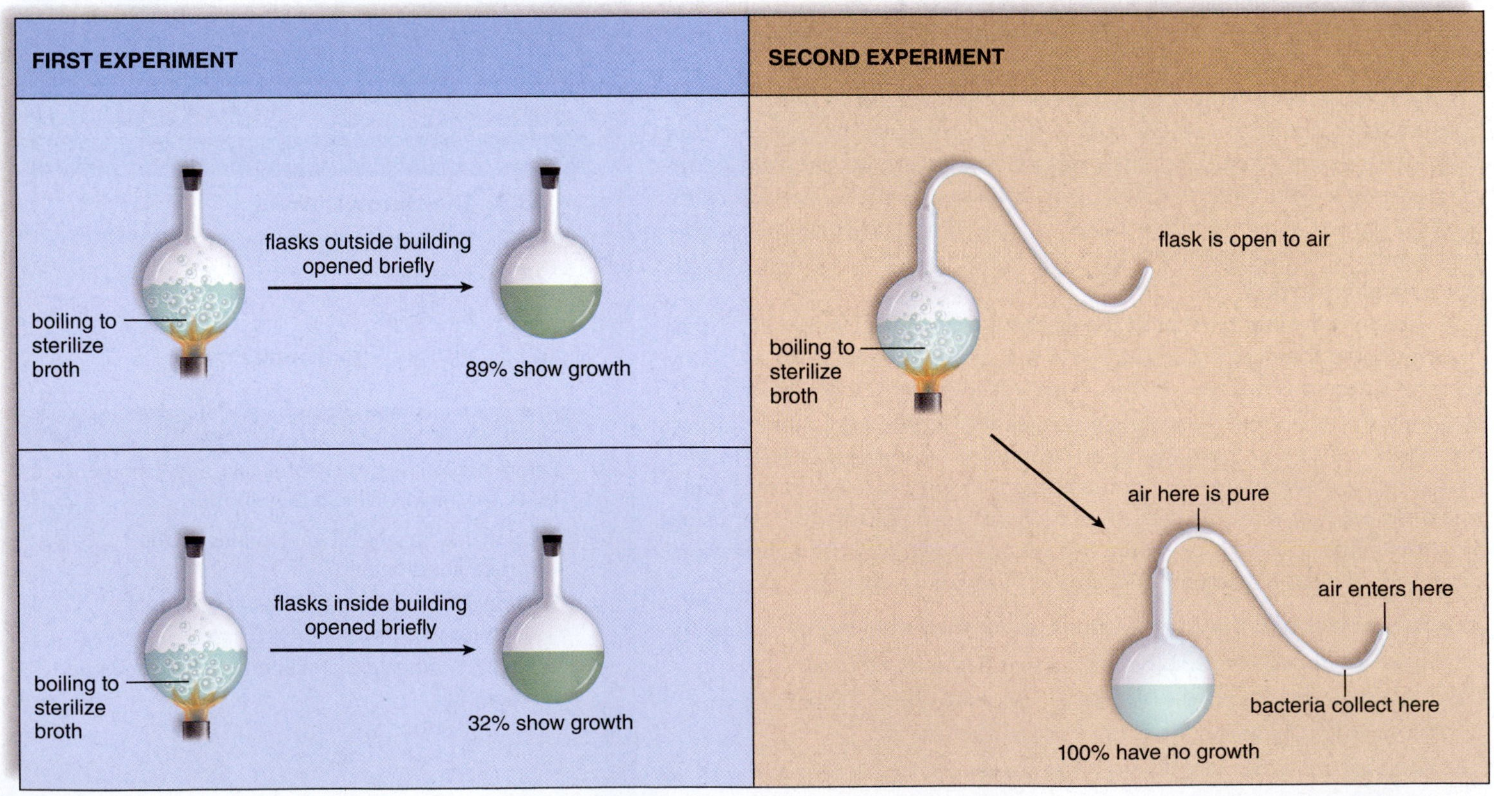
FIRST EXPERIMENT
flasks outside building opened briefly
boiling to sterilize broth
89% show growth
flasks inside building opened briefly
boiling to sterilize broth
32% show growth
SECOND EXPERIMENT
flask is open to air
boiling to sterilize broth
air here is pure
air enters here
bacteria collect here
100% have no growth

Figure 28.1 Pasteur's experiments. Pasteur disproved the theory of spontaneous generation of microbes by performing these types of experiments.

of beetles, the Earth would be covered in a layer of bacteria several miles deep!

We need only to turn on the nightly TV news to hear about diseases caused by microbes. H1N1 flu and diseases caused by antibiotic-resistant strains of the bacteria *Staphylococcus aureus* and *Mycobacterium tuberculosis* are recent examples of "emerging" infectious diseases. However, many microbes have beneficial, and sometimes essential, roles to play in human health as well as in the proper functioning of the biosphere. For example, you have approximately 10^{14} bacteria living on or in your body right now, even though your body is composed of only about 10^{13} human cells! In other words, you have ten times as many bacterial cells as human cells, although the bacterial cells are much smaller. Most of these microbes, also known as the **normal microbiota,** have beneficial effects. Bacteria that live on your skin help crowd out harmful microbes that might grow in those areas, and bacteria in the intestines aid in digestion and synthesize vitamin K and vitamin B_{12}. Our bodies are actually a finely balanced ecosystem for microbes.

Bacteria and fungi are the **decomposers** that play essential roles in various nutrient cycles on Earth by breaking down organic and inorganic materials that can be reused by plants and animals. Photosynthetic algae and other protists are *primary producers* that capture energy from the sun or inorganic material, providing nutrients for more complex organisms. Some bacteria also perform photosynthesis, often in ways very different from algae or plants. We have also learned to use bacteria to clean up our environment. Wastewater that leaves your bathroom and kitchen is probably cleaned at a treatment plant by processes that rely on the activity of bacteria. It seems that bacteria can consume virtually any substance found on Earth. Soil contaminated by oil spills or just about any other toxic compounds can be cleaned by encouraging the growth of bacteria that eat these compounds.

Not only do many microbes play essential roles in the environment, but they are also valuable for industrial processes, particularly food processing. Also, most antibiotics known today were first discovered in soil bacteria or fungi. Genetic techniques can be used to alter the products generated by bacterial cultures. A variety of valuable products can be made in this way, including insulin and vaccines. Although it might not seem so to an individual suffering from tuberculosis, AIDS, or another serious infectious disease, most microbes are far more beneficial than harmful—in fact, we could not live without them!

Check Your Progress 28.1

1. Describe the experiments Pasteur used to disprove the idea of spontaneous generation.
2. List the major groups of microbes that are studied in microbiology.
3. Define the term decomposers.

28.2 Origin of Microbial Life

Learning Outcomes

Upon completion of this section, you should be able to

1. Distinguish between chemical and biological evolution.
2. List and briefly describe the four stages that are thought to have led to the formation of the first cells.
3. Explain how the Miller-Urey experiment (among others) provided support for the "primordial soup" hypothesis.
4. Compare the implications of the protein-first, RNA-first, and membrane-first hypotheses for the evolution of the first living cell.

As noted in section 27.1, Darwin's theory of evolution by natural selection is guided by the principle of *common ancestry,* that all life on Earth can be traced back to a single ancestor. This ancestor, also called the **last universal common ancestor (LUCA)**, is common to all organisms that live, and have lived, on Earth since life began (Fig. 28.2). What were the properties of this common ancestor? How did life first begin on Earth? Darwin mused that perhaps the first living organism arose in a "warm little pond with all sorts of ammonia and phosphoric salts, light, heat, electricity, etc." With typical insight, Darwin was describing some of the actual conditions of ancient Earth that gave rise to the first forms of life.

In Chapter 1 we considered the characteristics shared by all organisms. Organisms acquire energy through *metabolism,* or the chemical reactions that occur within cells. They also respond and interact with their environment, self-replicate, and are subject to the forces of natural selection that drive adaptation to the environment. The molecules of organisms, called **biomolecules,** are organic molecules. The first organisms on Earth would have had all of these characteristics. Yet early Earth was very different from the Earth we know today, and consisted mainly of inorganic substances. How, then, did life get started in this inorganic "warm little pond"?

Advances in chemistry, evolutionary biology, paleontology, microbiology, and other branches of science have helped to develop new, and test old, hypotheses about the origin of life. These studies contribute to an ever-growing body of scientific evidence that life originated 3.5–4 billion years ago (BYA) from nonliving matter in a series of four stages:

Stage 1. *Organic monomers.* Simple organic molecules, called monomers, evolved from inorganic compounds prior to the existence of cells. Amino acids, the basis of proteins, and nucleotides, the building blocks of DNA and RNA, are examples of organic monomers.

Stage 2. *Organic polymers.* Organic monomers were joined, or polymerized, to form organic polymers, such as DNA, RNA, and proteins.

Stage 3. *Protobionts.* Organic polymers became enclosed in a membrane to form the first cell precursors, called protobionts (protocells).

Stage 4. *Living cells.* Protobionts acquired the ability to self-replicate, as well as other cellular properties.

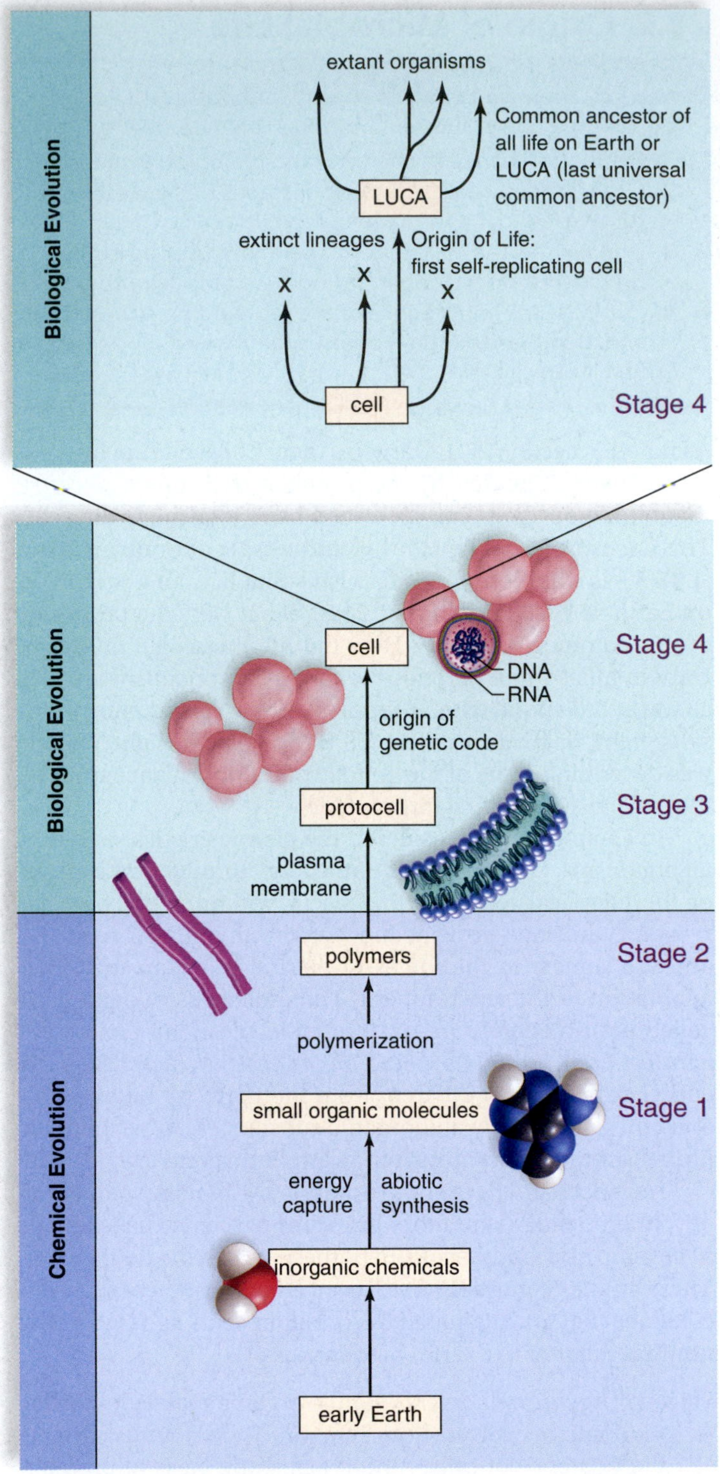
Biological Evolution
extant organisms
Common ancestor of all life on Earth or LUCA (last universal common ancestor)
LUCA
extinct lineages
Origin of Life: first self-replicating cell
X
X
X
cell
Stage 4
Biological Evolution
cell
Stage 4
DNA
RNA
origin of genetic code
protocell
Stage 3
plasma membrane
Chemical Evolution
polymers
Stage 2
polymerization
small organic molecules
Stage 1
energy capture
abiotic synthesis
inorganic chemicals
early Earth

Figure 28.2 Stages of the origin of life. The first organic molecules (*bottom*) originated from chemically altered inorganic molecules present on early Earth (Stage 1). More complex organic macromolecules were synthesized to create polymers (Stage 2) that were then enclosed in a plasma membrane to form the protobiont, or protocell (Stage 3). The protobiont underwent biological evolution to produce the first true, self-replicating, living cell (Stage 4). This first living cell underwent continued biological evolution (*top*), with a single surviving lineage, the LUCA, which became the common ancestor of all life on Earth.

Scientists have performed experiments to test hypotheses at each stage of the origin of life. Stages 1–3 involve the processes of a "chemical evolution," before the origin of life (see Fig. 28.2). Stage 4 is when life first evolved through the processes of "biological evolution."

Evolution of Monomers

In the 1920s, Russian biochemist Alexander Oparin and British geneticist J. B. S. Haldane independently hypothesized that the first stage in the origin of life was the evolution of simple organic molecules from the inorganic compounds that were present in the Earth's early atmosphere. The Oparin-Haldane hypothesis, sometimes called the *"primordial soup" hypothesis,* proposes that early Earth had very little oxygen (O_2), but instead was made up of water vapor (H_2O), hydrogen gas (H_2), methane (CH_4), and ammonia (NH_3).

Methane and ammonia are reducing agents because they readily donate their electrons. In the absence of oxygen, their reducing capability is powerful. Thus, the early Earth had a reducing atmosphere in which oxidation-reduction (redox) reactions could have driven the chemical evolution, or abiotic synthesis, of organic monomers from inorganic molecules in the presence of strong energy sources. Note that "oxidation" used in this context refers to a chemical redox reaction (see section 6.4), and not to oxygen gas.

In 1953, American chemist Stanley Miller performed a famous experiment to test the primordial soup hypothesis of early chemical evolution (Fig. 28.3). He reasoned that the energy sources on early Earth included heat from volcanoes and meteorites, radioactivity from isotopes, powerful electric discharges in lightning, and solar radiation. For his experiment, Miller placed a mixture of methane (CH_4), ammonia (NH_3), hydrogen (H_2), and water (H_2O), in a closed system, heated the mixture, and circulated it past an electric spark (simulating lightning). After a week's run, Miller discovered that a variety of amino acids and other organic acids had been produced.

Animation Miller-Urey

The *Miller-Urey experiment* has been repeatedly tested over the decades since it was first performed. In 2008, a group of scientists examined 11 vials of compounds produced from variations of the Miller-Urey experiment, and found a greater variety of organic molecules than Miller reported, including all 22 amino acids. If early atmospheric gases did react with one another to produce small organic compounds, neither oxidation (no free oxygen was present) nor decay (no decomposers existed) would have destroyed these molecules, and rainfall would have washed them into the ocean, where they would have accumulated for hundreds of millions of years. Therefore, the oceans would have been a thick, warm organic soup—much like Darwin's "warm little pond."

The *Oparin-Haldane hypothesis* was an important contribution to our understanding of the early stages of life's origins, but other hypotheses have also been proposed and tested. In the late 1980s, German chemist Günter Wächtershäuser proposed that thermal vents at the bottom of the Earth's oceans (Fig. 28.4) provided all the elements and conditions necessary

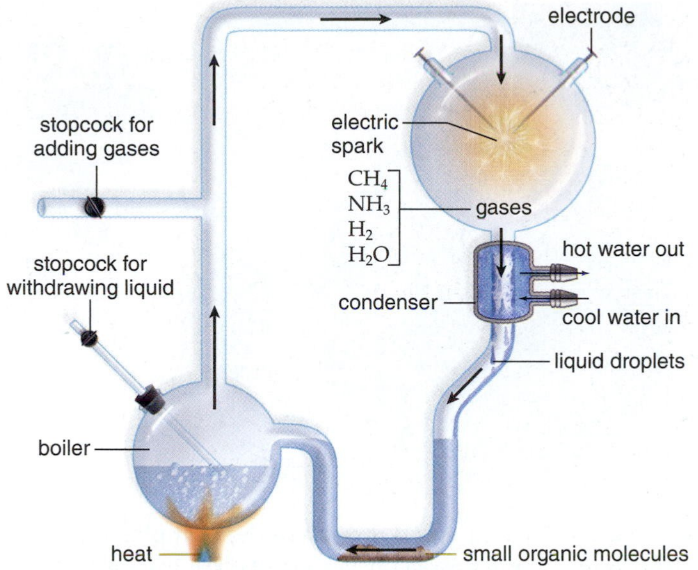
electrode
stopcock for adding gases
electric spark
CH_4
NH_3
H_2
H_2O
gases
hot water out
stopcock for withdrawing liquid
condenser
cool water in
liquid droplets
boiler
heat
small organic molecules

Figure 28.3 Stanley Miller's experiment. Gases that were thought to be present in the early Earth's atmosphere were admitted to the apparatus, circulated past an energy source (electric spark), and cooled to produce a liquid that could be withdrawn. Upon chemical analysis, the liquid was found to contain various small organic molecules, which could serve as monomers for large cellular polymers.

plume of hot water rich in iron-nickel sulfides
hydrothermal vent

Figure 28.4 Chemical evolution at hydrothermal vents. Minerals that form at deep-sea hydrothermal vents like this one can catalyze the formation of ammonia and even monomers of larger organic molecules that occur in cells.

to synthesize organic monomers. According to his *"iron-sulfur world" hypothesis,* dissolved gases emitted from thermal vents, such as carbon monoxide (CO), ammonia, and hydrogen sulfide, pass over iron and nickel sulfide minerals, also present at thermal vents. The iron and nickel sulfide molecules act as catalysts that drive the chemical evolution from inorganic to organic molecules.

A very different line of thinking involves the comets and meteorites that have pelted the Earth throughout history. In recent years, scientists have confirmed the presence of organic molecules in some meteorites. Many scientists feel that these organic molecules could have seeded the chemical origin of life on early Earth. Others even hypothesize that bacterium-like cells evolved first on another planet and then were carried to Earth, an idea known as *panspermia.* A meteorite from Mars labeled ALH84001 landed on Earth some 13,000 years ago. When examined, experts found tiny rods similar in shape to fossilized bacteria. This hypothesis continues to be investigated.

Evolution of Polymers

Within a cell's cytoplasm, organic monomers join to form polymers in the presence of enzymes—such as the synthesis of protein polymers from amino acids by ribosomes. Enzymes themselves are proteins, but presumably there were no proteins present on early Earth. How did the first organic polymers form if no enzymes were present?

Building on the Miller-Urey experiment, American biochemist Sidney Fox suggested that once amino acids were present in the oceans, they could have collected in shallow puddles along the rocky shore. Then, the heat of the sun could have caused them to form *proteinoids,* small polypeptides that have some catalytic properties.

The formation of proteinoids has been simulated in the laboratory. When placed in water, proteinoids form microspheres, structures composed only of protein that have many properties of a cell. According to the **protein-first hypothesis,** some of these newly formed polypeptides had enzymatic properties. Moreover, if a certain level of enzyme activity provided an advantage over others, this would have set the stage for natural selection to shape the evolution of these first organic polymers. Those that evolved to be part of the first cell would have had a selective advantage over those that did not become part of a cell.

Fox's protein-first hypothesis assumes that protein enzymes arose prior to the first DNA molecule. Thus the genes that encode proteins followed the evolution of the first polypeptides.

In contrast, the **RNA-first hypothesis** suggests that only the macromolecule RNA was needed to progress toward formation of the first cell or cells. Thomas Cech and Sidney Altman shared a Nobel Prize in 1989 for their discovery that RNA can be both a substrate and an enzyme. Some viruses today have RNA genes; therefore, the first genes could have been RNA. It would seem, then, that RNA could have carried out the processes of life commonly associated today with DNA and proteins. Those who support this hypothesis say that it was an "RNA world" some 4 BYA.

Evolution of Protobionts

Before the first true cell arose, a **protobiont** (also called a **protocell**) would have emerged. Protobionts are characterized as having an outer membrane. This membrane would have provided a boundary between the inside of the cell and its outside world, which was critical to the proper regulation and maintenance of cellular activities. Therefore, the evolution of a membrane was a critical step in the origin of life.

The plasma membrane of modern cells is made up of phospholipids assembled in a bilayer. The first plasma membranes were likely made up of fatty acids, which are smaller than phospholipids but like phospholipids have a hydrophobic "tail" and hydrophilic "head" (Fig. 28.5). Fatty acids are one of the organic polymers that could have formed from chemical reactions at deep-water thermal vents early in the history of life. In water, fatty acids assemble into small spheres called **micelles,** consisting of a single layer of fatty acids organized with their heads pointing out and tails pointing toward the center of the sphere.

Under appropriate conditions micelles can merge to form vesicles. **Vesicles** are larger than micelles and are surrounded by a bilayer (two layers) of fatty acids (Fig. 28.6), similar to the phospholipid bilayer of modern cell membranes. An important feature of a vesicle lipid bilayer is that the individual fatty acids can flip between the two layers, which helps to move select molecules, such as amino acids, from outside to the inside of the vesicle. The first protobiont would likely have been a type of vesicle with this type of fatty acid bilayer membrane. Interestingly, if lipids are made available to protein microspheres, lipids tend to become associated with microspheres, producing a lipid-protein membrane. Lipid-protein microspheres share some interesting properties with modern cells: they resemble bacteria, they have an electrical potential difference, and they divide and perhaps are subject to selection.

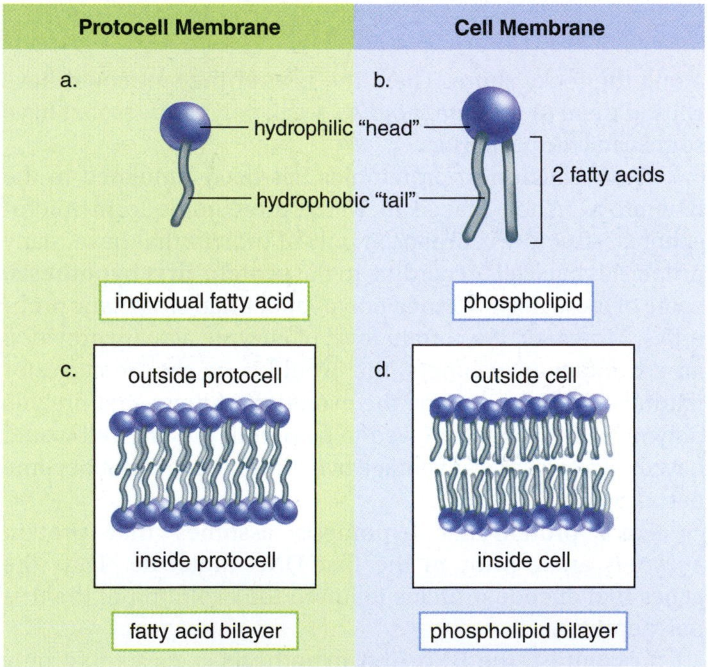
Protocell Membrane
Cell Membrane
a.
b.
hydrophilic "head"
hydrophobic "tail"
2 fatty acids
individual fatty acid
phospholipid
c.
d.
outside protocell
inside protocell
outside cell
inside cell
fatty acid bilayer
phospholipid bilayer

Figure 28.5 A comparison of protobiont and modern cell plasma membranes. The first lipid membrane was likely made of a single layer of fatty acids. These first protobiont (protocell) membranes and the modern cell membrane have features in common. **a.** Individual fatty acids have a single fatty acid chain with a hydrophilic head and hydrophobic tail. **b.** Phospholipids of modern cell membranes are made of two hydrophobic fatty acid chains (the "tails") attached to a hydrophilic head. **c.** Protobiont membranes were likely made up of a bilayer of fatty acids, with hydrophilic heads pointing outward, and hydrophobic tails pointing inward. **d.** Modern cells are organized in a similar fashion. Both bilayers create a semipermeable barrier between the inside and outside of a cell.

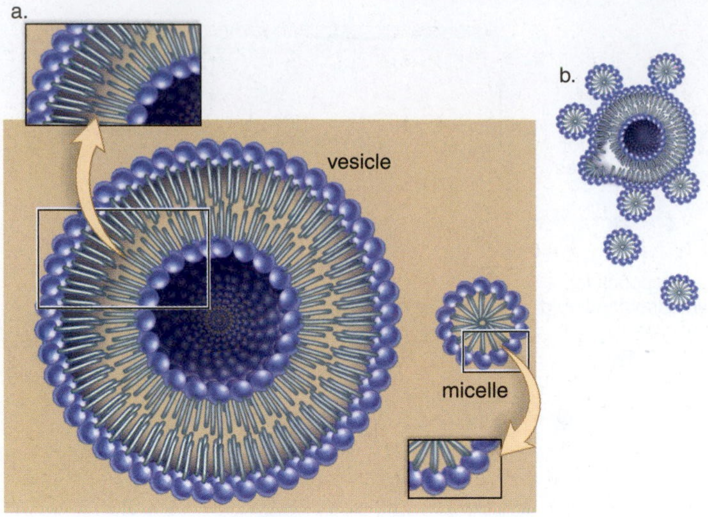
a.
b.
vesicle
micelle

Figure 28.6 Structure and growth of the first plasma membrane. The first plasma membrane was likely made of a fatty acid bilayer, similar to that seen in vesicles. Protobionts (protocells), the ancestor of modern cells, are thought to have had this type of membrane. **a.** A cross section of a vesicle reveals the fatty acid bilayer of a vesicle membrane. Micelles are spherical droplets formed by a single layer of fatty acids, and are much smaller than vesicles. **b.** Under proper conditions, micelles can merge to form vesicles. As micelles are added to the growing vesicle, individual fatty acids flip their hydrophilic heads toward the inside and outside of the vesicle, and their hydrophobic tails toward each other. This process forms a bilayer of fatty acids.

In the early 1960s, British biophysicist Alec Bangham discovered that when he extracted lipids from egg yolks and placed them in water, the lipids would naturally organize themselves into double-layered bubbles roughly the size of a cell. Bangham's bubbles soon became known as **liposomes.** Bangham and others soon realized that liposomes might have provided life's first membranous boundary. Perhaps liposomes with a phospholipid membrane engulfed early molecules that had enzymatic, even replicative, abilities. The liposomes would have protected the molecules from their surroundings and concentrated them so they could react (and evolve) quickly and efficiently. These investigators called this the **membrane-first hypothesis,** meaning that the first cell had to have a plasma membrane before any of its other parts.

A protobiont would have had to acquire nutrition, that is, other molecules, so that it could grow. One hypothesis suggests that protobionts were **heterotrophs,** organisms that consume preformed organic molecules. However, if the protobiont evolved at hydrothermal vents, it may have carried out *chemosynthesis*—the synthesis of organic molecules from inorganic molecules and nutrients. Indeed, many modern bacteria are **chemoautotrophs** that obtain energy by oxidizing inorganic compounds, such as hydrogen sulfide (H_2S), a molecule that is abundant at thermal vents. When hydrothermal vents in the deep and extremely dark ocean were first discovered in the 1970s, investigators were surprised to discover complex vent ecosystems supported by organic molecules formed by chemosynthesis, a process that does not require the energy of the sun.

ATP (adenosine triphosphate) is the most important energy-carrying molecule in living organisms. Because all life on Earth uses ATP to fuel cellular metabolism, the evolution of a means to synthesize ATP must have occurred very early in the history of life. The first protobionts evolved in the oxygen-poor environment of early Earth. Thus it is probable that ATP was likely synthesized first by fermentation (see section 7.3). The subsequent evolution of oxidative phosphorylation would have provided an advantage because it greatly increased the amount of ATP synthesized per unit of energy. Mitochondria share a common ancestor with a group of bacteria that synthesize ATP via an electron transport chain. Oxidative phosphorylation is possible in eukaryotes because mitochondria provide an electron transport chain ATP factory. At first, the protobiont must have had limited ability to break down organic molecules, and scientists speculate that it took millions of years for complex biochemical pathways to evolve completely.

Evolution of Living Cells

Today's cell is able to carry on protein synthesis in order to produce the enzymes that allow DNA to replicate. The **central dogma** of genetics states that DNA directs protein synthesis and that information flows from DNA to RNA to protein. It is possible that this sequence developed in stages.

According to the RNA-first hypothesis, RNA would have been the first to evolve, and the first true cell would have had RNA genes. These genes would have directed and enzymatically carried out protein synthesis. As noted, today we know a number of viruses have RNA genes. These viruses have a protein enzyme called *reverse transcriptase* that uses RNA as a template to form DNA. Perhaps with time, reverse transcription occurred within the protobiont, and this is how DNA-encoded genes arose. If so, RNA was responsible for both DNA and protein formation.

According to the protein-first hypothesis, proteins, or at least polypeptides, were the first of the three (DNA, RNA, and protein) to arise. Only after the protobiont developed a plasma membrane and sophisticated enzymes did it have the ability to synthesize DNA and RNA from small molecules provided by the ocean. Because a nucleic acid is a complicated molecule, it is unlikely that RNA arose spontaneously from simple chemicals. It seems more likely that enzymes were needed to guide the synthesis of nucleotides and then nucleic acids. Once there were DNA genes, protein synthesis would have been carried out in the manner dictated by the central dogma of genetics.

The Scottish chemist Alexander Cairns-Smith proposed that polypeptides and RNA evolved simultaneously. Therefore, the first true cell would have contained RNA genes that could have replicated because of the presence of proteins. This eliminates the baffling chicken-or-egg paradox: assuming a plasma membrane, which came first, proteins or RNA? It means, however, that two unlikely events would have had to happen at the same time.

After DNA formed, the genetic code had to evolve before DNA could store genetic information. The present genetic code is subject to fewer errors than a million other possible codes. Also, the present code is among the best at minimizing the effect of mutations. A single-base change in a present codon is likely to result in the substitution of a chemically similar amino acid and, therefore, minimal changes in the final protein. This evidence suggests that the genetic code did undergo a natural selection process before finalizing into today's code.

Check Your Progress 28.2

1. Explain the role of biomolecules in chemical and biological evolution.
2. List the four stages of the evolution of life, and explain why they likely occurred in a particular order.
3. Compare and contrast the "primordial soup" and the "iron-sulfur world" hypotheses.
4. Explain why the protein-first versus RNA-first debate can be compared to a chicken-or-egg paradox.

28.3 Archaea

Learning Outcomes

Upon completion of this section, you should be able to

1. List some of the major criteria that are used to distinguish the domains Archaea, Bacteria, and Eukarya.
2. Review the structural features of the plasma membranes and cell walls of archaea.
3. Distinguish between halophiles, thermoacidophiles, and methanogens.

As noted in section 1.2, the bacteria (domain Bacteria) and archaea (domain Archaea) are prokaryotes, but each is placed in its own domain because of molecular and cellular differences. **Prokaryotes** are single-celled organisms that lack the nuclei and membrane-bound cytoplasmic organelles found in eukaryotic cells. Archaea and bacteria are not close relatives, even though both are prokaryotes. Based on a number of criteria, including nucleic acid similarities, the eukarya are more closely related to archaea than to bacteria (Table 28.1).

TABLE 28.1 Comparison of Domains Archaea and Eukarya

	Domain	
Feature	**Archaea**	**Eukarya**
Nucleus	No	Yes
Organelles	No	Yes
Introns	Sometimes	Yes
Histones	Yes	Yes
RNA polymerase	Several types	Several types
Methionine is at start of protein synthesis	Yes	Yes

a.
b.
c.

Figure 28.7 Extreme habitats. Many archaea thrive in unusual environmental conditions. **a.** Halophilic archaea can live in salt lakes. **b.** Thermophilic archaea can live in the hot springs of Yellowstone National Park. **c.** Methanogens live in anaerobic swamps and in the guts of animals.

Archaeal Size and Structure

Archaea usually range from 0.1–15 μm in size. Their genome is a single, closed, circular DNA molecule, often smaller than a bacterial genome. Like bacteria, archaea reproduce asexually by *binary fission.*

Animation Binary Fission

The plasma membrane of archaea differs markedly from those of bacteria and eukaryotes. Rather than a lipid bilayer, archaea often have a monolayer of lipids with branched side chains. These chemical characteristics help many archaea tolerate acid and heat.

The cell walls of archaea may also facilitate their survival in extreme environments. The cell walls of archaea lack peptidoglycan, which is a distinguishing feature of bacterial cell walls (see next section). In some archaea, the cell wall is largely composed of polysaccharide, and in others, it is pure protein. A few have no cell wall.

Types of Archaea

Three main types of archaea are distinguished based on their unique habitats and metabolic activities: halophiles, thermoacidophiles, and methanogens (Fig. 28.7). Other archaea are found in moderate environments, such as lake sediments and soil, where they are likely involved in nutrient cycling. Several archaea have been found living in symbiotic relationships with animals, including sponges, sea cucumbers, and in the digestive tracts of humans and other animals (see next section on "Methanogens"). However, no parasitic archaea have yet been found—that is, they are not known to cause any infectious disease.

Halophiles

The **halophiles** (salt lovers) usually thrive at salt concentrations of around 12–15%; in contrast, ocean water is about 3.5% salt. Halophiles have been isolated from environments such as the Great Salt Lake in Utah, the Dead Sea, and hypersaline soils, where most other organisms cannot survive.

These archaea have evolved a number of mechanisms enabling them to survive in very salty environments. Their plasma membranes often have chloride pumps that contain a light-powered protein called *halorhodopsin* (related to the rhodopsin pigment found in the human retina). This unique protein pumps chloride and water into the cell, thereby preventing excessive dehydration.

Some halophiles also perform a type of photosynthesis that uses a pigment called *bacteriorhodopsin* instead of chlorophyll. When the pigment absorbs solar energy, it moves a hydrogen ion to outside the plasma membrane. This establishes an H^+ gradient that promotes ATP synthesis.

Thermoacidophiles

Another major type of archaea, the **thermoacidophiles,** are usually isolated from extremely hot, acidic, aquatic environments such as hot springs, geysers, and underwater volcanoes. Unlike the great majority of life-forms on Earth, the lipid membranes and proteins of these archaea have evolved to withstand and function optimally at temperatures as high as 80°C; some can even grow at 105°C (remember that water boils at 100°C)!

One thermoacidophilic species, *Picrophilus torridus,* has been particularly well studied for its ability to survive at a pH of less than 1, similar to that found in a 1.2-molar sulfuric acid solution (car battery acid is around 4.5 molar). Sequencing of the genome of this organism has shown that a relatively high number (12%) of its genes are involved in transport functions of its plasma membrane. Many of these transporter proteins are presumably involved in pumping excess H^+ ions out of the cell, as well as in taking advantage of the high density of H^+ outside the cell to transport other desired molecules into the cell. As one of the most acid-tolerant organisms known, *P. torridus* will likely provide enzymes for various biotechnology processes that require acidic conditions.

Methanogens

The **methanogens** (methane-makers) mostly use carbon dioxide and hydrogen as energy sources, producing methane as a by-product. Methanogens are found in anaerobic environments such as swamps, lake sediments, rice paddies, and the intestines of animals. Cows, which have large populations of methanogens in their digestive tracts, release a significant amount of methane into the environment. Because methane is a greenhouse gas that may contribute to global warming and climate change, some scientists are suggesting that the amount of methane produced by cattle should be reduced by changing the animals' diets, adding substances to their feed to alter the microbial populations, or developing a vaccine to specifically inhibit the growth of methanogens. However, the relative contribution of cattle, compared to other potential sources of methane, remains controversial.

Check Your Progress 28.3

1. List some key characteristics of archaea, bacteria, and eukarya.
2. Explain how the plasma membranes of archaea differ from those of bacteria and eukaryotes.
3. Describe adaptations that allow archaea to survive in extremely hot or salty environments.
4. Discuss why a dedicated environmentalist might give up eating beef.

28.4 Bacteria

Learning Outcomes

Upon completion of this section, you should be able to

1. Identify the major structural features of bacteria.
2. Describe bacterial reproduction, including three mechanisms by which bacteria increase variation.
3. Explain why bacteria are considered to be metabolically diverse.
4. List several major bacterial diseases of humans and describe how they are treated.

Bacteria (domain Bacteria) are the most common type of prokaryote on Earth. To date over 9,000 different species have been named, but the number of unnamed species is probably in the tens of millions. The late Harvard paleontologist and science writer Steven Jay Gould once labeled the Earth as "Planet of the Bacteria." Indeed, bacteria are found in virtually every environment on Earth.

Bacterial Size and Structure

Most bacteria are between 0.2–10 μm in size. A few, however, are quite large, including one species that is about the same size as the period at the end of this sentence. Bacteria have three basic shapes: rod (bacillus, pl., bacilli); spherical (coccus, pl., cocci); and spiral-shaped or helical (spirillum, pl., spirilla) (Fig. 28.8). A bacillus or coccus can occur singly or may occur in particular arrangements. For example, when cocci form a cluster, they are called staphylococci, whereas cocci that form chains are called streptococci.

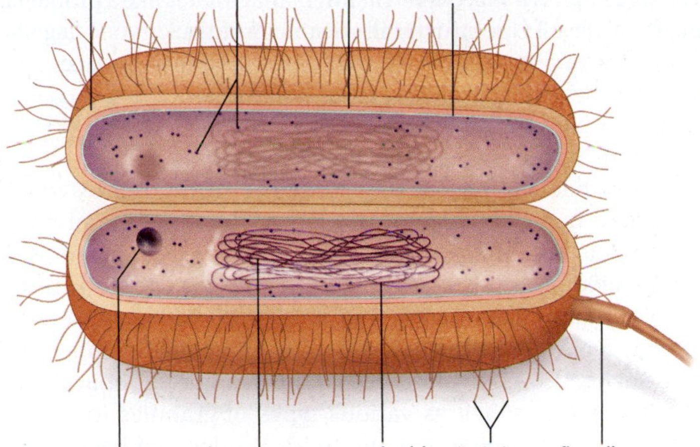
capsule
ribosomes
cell wall
plasma membrane
storage granule
chromosome (DNA)
nucleoid
fimbriae
flagellum

Figure 28.9 Typical bacterial cell. Bacteria are prokaryotes and thus have no membrane-bound nuclei or organelles.

The typical structure of a bacterium is shown in Figure 28.9. All bacterial cells have a plasma membrane, which is a lipid bilayer, similar to the plasma membrane in plant and animal cells. Most bacterial cells are further protected by a cell wall that contains the unique molecule peptidoglycan. Bacteria can be classified by differences in their cell walls, which are detected using a staining procedure devised more than 100 years ago by Hans Christian Gram. When you go to the doctor for a bacterial infection, one of the most common tests performed is the **Gram stain,** because different antibiotics tend to be more effective against Gram-positive or Gram-negative bacteria. Cell walls that have a thick layer of peptidoglycan outside the plasma membrane stain purple with the Gram stain procedure, and are called Gram-positive bacteria. If the peptidoglycan layer is either thin or lacking altogether, the cells stain pink and are considered Gram-negative. In addition to their plasma membrane, Gram-negative bacteria have an outer membrane that contains **lipopolysaccharide** molecules. When these Gram-negative cells are killed by your immune system, LPS molecules are released, stimulating inflammation and fever. Beyond the cell wall, some bacteria have a slimy polysaccharide capsule that can protect the cell from dehydration and the immune system.

Animation Gram Stain

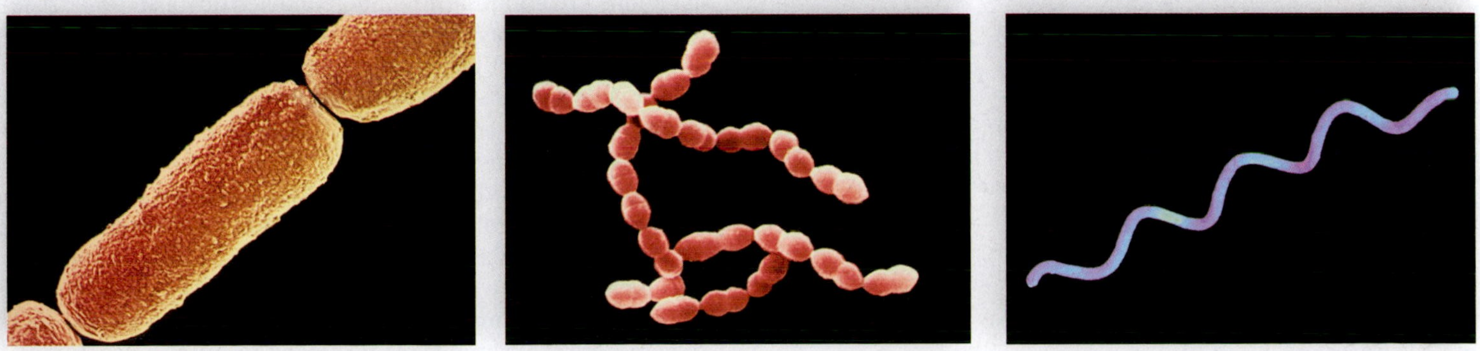

a. Bacilli (rod): *Bacillus anthracis* SEM 35,000×

b. Cocci (spherical): *Streptococcus thermophilus* SEM 3,520×

c. Spirillum (curved): *Spirillum volutans* SEM 6,250×

Figure 28.8 Typical shapes of bacteria. **a.** Bacilli, rod-shaped bacteria. **b.** Cocci, round bacteria. **c.** Spirillum, a spiral-shaped bacterium.

Motile bacteria have **flagella** for locomotion but never cilia. The flagellum is a stiff, curved filament that rotates like a propeller. Bacterial flagella are structurally distinct from eukaryotic flagella. Some bacteria have fimbriae that bind to various surfaces. For example, bacteria that cause urinary tract infections can bind to urinary tract cells. Drinking cranberry juice seems to inhibit this binding.

Video Cranberries vs Bacteria

Most bacteria have a single circular chromosome, which is located in a **nucleoid** region, rather than in a membrane-bound nucleus. Many bacteria also harbor accessory rings of DNA called plasmids that can carry genes for antibiotic resistance, among other things, and are commonly used to carry foreign DNA into other bacteria during genetic engineering (see section 26.1). Bacterial cells also contain an abundance of ribosomes, as well as various types of granules that store nutrients such as glycogen and lipids. Notice that bacteria do not have an endoplasmic reticulum, a Golgi apparatus, mitochondria, or chloroplasts. This diagram summarizes the major structural features of bacteria:

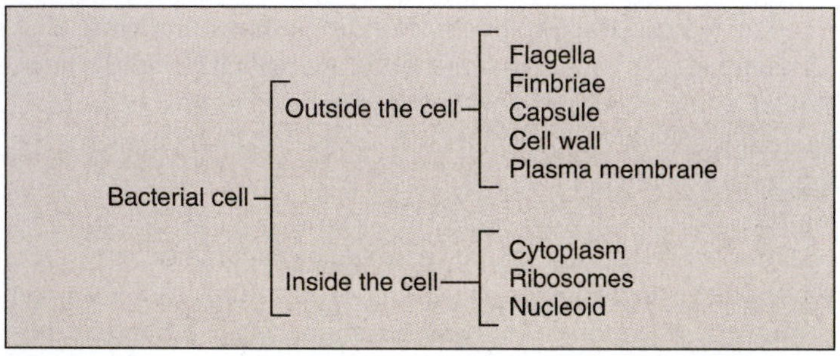
Bacterial cell
Outside the cell
Flagella
Fimbriae
Capsule
Cell wall
Plasma membrane
Inside the cell
Cytoplasm
Ribosomes
Nucleoid

Bacterial Reproduction and Gene Transfer

Bacteria reproduce asexually. After a period of sufficient growth, the bacterial cell simply divides into two new cells, with each cell getting a copy of the genome and approximately half of the cytoplasm. This process is known as **binary fission** (Fig. 28.10). Each daughter cell is a *clone,* or exact copy, of the parent cell.

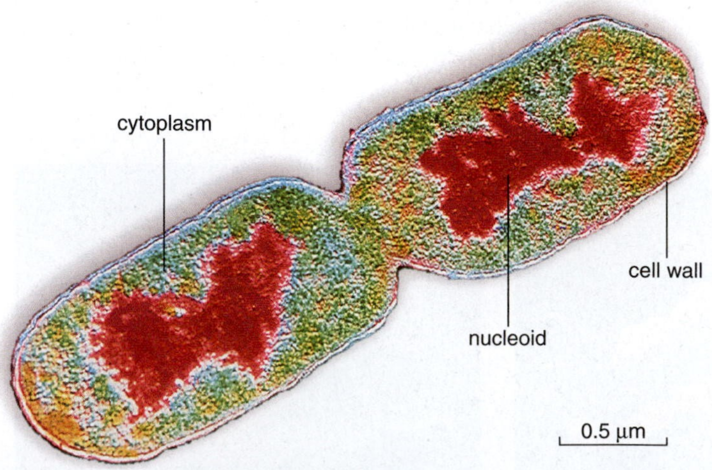
cytoplasm
cell wall
nucleoid
0.5 μm

Figure 28.10 Binary fission. When conditions are favorable for growth, prokaryotes divide to reproduce. Binary fission is a form of asexual reproduction because the daughter cells have exactly the same genetic material as the parent.

When bacteria are spread out and grown on agar plates, each individual cell can give rise to a colony of millions of cells. In the absence of genetic mutations, each of these cells is a clone of the initial cell. Some bacteria need only 20 minutes to reproduce, whereas others grow more slowly, with generation times of a day or more.

Animation Binary Fission

Under unfavorable conditions, some bacteria, such as *Clostridium tetani,* the cause of tetanus, can produce resistant structures called **endospores,** thick-walled, dehydrated structures capable of surviving the harshest conditions, perhaps even for thousands of years. Endospores are not for reproduction, but are simply a way to survive unfavorable conditions.

Sexual reproduction does not occur among prokaryotes, but at least three means of *horizontal gene transfer* have been observed. **Conjugation** takes place when a donor cell passes DNA to a recipient cell by way of a *sex pilus.* **Transformation** occurs when a bacterium takes up DNA released into the medium by dead bacteria. During **transduction,** viruses carry portions of bacterial DNA from one bacterium to another.

Animation Bacterial Conjugation

Animation Bacterial Transformation

Bacterial Metabolism

Although most bacteria are structurally similar to each other, they demonstrate a remarkable range of metabolic abilities. Most bacteria are heterotrophs that require an outside source of organic compounds in the same way that animals do. Some heterotrophic bacteria are anaerobic and cannot use oxygen to capture electrons at the end of their electron transport chain (which is physically located in their plasma membrane, as all prokaryotes lack mitochondria). Instead, they use a variety of substances as final electron acceptors. Sulfate-reducing bacteria transfer the electrons to sulfate, producing hydrogen sulfide, which smells like rotten eggs. Denitrifying bacteria use nitrate, whereas others use minerals, such as iron or manganese, as electron receivers.

Other bacteria are chemoautotrophs. They reduce carbon dioxide to an organic compound by using energetic electrons derived from chemicals, such as ammonia, hydrogen gas, and hydrogen sulfide. Electrons can also be extracted from certain minerals, such as iron.

Some bacteria are photosynthesizers that use solar energy to produce their own food. **Cyanobacteria** are believed to have arisen some 3.8 billion years ago and to have produced much of the oxygen in the atmosphere at that time. Sometimes erroneously called *blue-green algae,* these organisms contain green chlorophyll and have other pigments that give them a bluish-green color (Fig. 28.11). Other bacterial photosynthesizers split hydrogen sulfide, instead of water, in anaerobic environments. Therefore, they produce sulfur rather than oxygen as a by-product of photosynthesis.

Aside from producing oxygen, cyanobacteria are often the first colonizers of rocks. Many are capable of both carbon fixation and nitrogen fixation, needing only minerals, air, sunlight, and water for growth. Cyanobacteria are also notorious for forming toxic blooms in waters enriched by nutrients, which may require that some lakes be closed to human use. Cyanobacteria (and in some cases eukaryotic algae) form a symbiotic relationship with fungi in a *lichen.* The fungi provide a place for the cyanobacteria to grow and obtain water and mineral

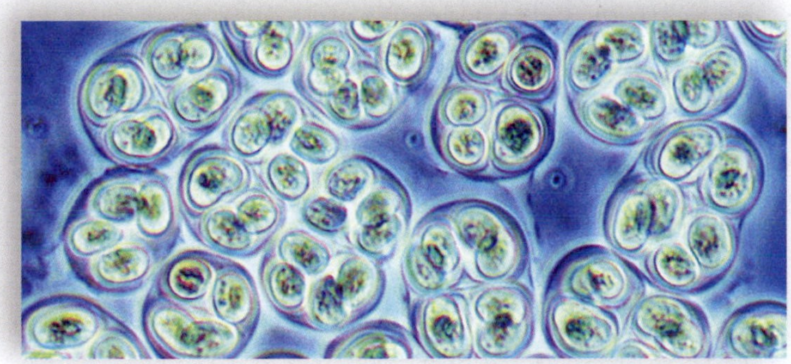

a.

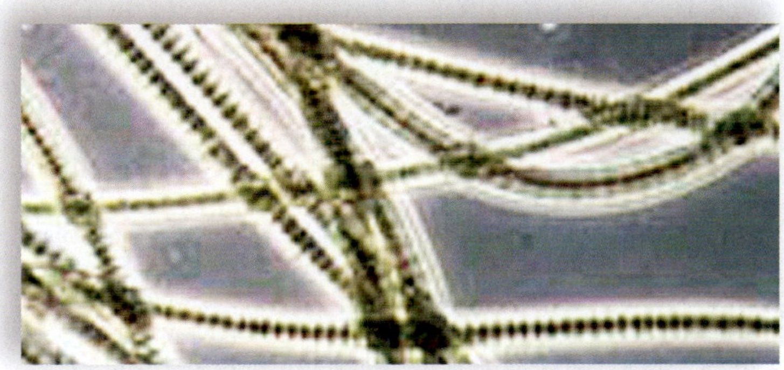

b.

Figure 28.11 Cyanobacteria. Cyanobacteria are photosynthetic bacteria that contain chlorophyll. **a.** In *Chroococcus,* single cells are grouped in a common gelatinous sheath. **b.** Filaments of cells occur in *Oscillatoria,* a common inhabitant in freshwater environments.

nutrients. Some cyanobacteria even appear to live in a symbiotic relationship in the fur of sloths, providing the animals with a form of camouflage in their dense green forest home!

Bacterial Diseases of Humans

Most types of bacteria do not cause disease, but a significant number do. Why does one microbe cause disease, whereas another, closely related species is completely harmless? Pathogenic microbes often carry genes that code for specific virulence factors that determine the type and extent of illness they are capable of causing. For example, right now you have billions of *Escherichia coli* living in your large intestine, without causing any problems. However, some strains of *E. coli* have acquired genes that make them dangerous invasive pathogens. The strain of *E. coli* called O157:H7 has the ability to generate a toxin that damages the lining of the intestine, resulting in a bloody diarrhea. It has also obtained other virulence factors that help it stick to the lining of the gut more efficiently. In 2011, around 60 people in ten states became ill from *E. coli* O157:H7 contamination in Romaine lettuce. That same year an outbreak of pathogenic *E. coli* in Germany sickened 852 people and caused 32 deaths.

Video *E. coli* Wars

By acquiring genes coding for virulence factors (i.e., via conjugation, transformation, and/or transduction) harmless bacteria may be converted into pathogens. Antibiotic resistance genes may pass between organisms in the same ways, creating antibiotic-resistant pathogens.

Some bacterial diseases that are transmitted sexually were discussed in section 21.5. Next we discuss a few more common types of bacterial diseases in humans.

Streptococcal Infections More different types of human disease are caused by bacteria from the genus *Streptococcus* than by any other type of bacterium. *Streptococcus pneumoniae* can cause pneumonia, meningitis, and middle ear infections. Up to 70% of apparently healthy adults are carriers of this potentially harmful bacterium, which causes disease mainly in children and the elderly. *S. mutans* is found on the teeth and contributes to tooth decay and the formation of dental caries.

The most common illness caused by streptococci is pharyngitis, commonly called strep throat, which is usually due to infection by *S. pyogenes* (Fig. 28.12*a*). *S. pyogenes* also causes relatively mild skin diseases, such as impetigo in infants (Fig. 28.12*b*). A number of more serious and invasive diseases can occur after or during infections with *S. pyogenes.* The nature and severity of these diseases depend on which virulence factors the pathogen carries. Scarlet fever is a strep infection caused by a *Streptococcus* strain that produces a toxin causing a red rash. An intense immune response to the infection can lead to rheumatic fever, which is characterized by a high temperature, swollen joints that form nodules, and heart damage. Although rarely seen in the United States today, rheumatic fever killed more school-age children than all other diseases combined in the early twentieth century.

By releasing enzymes that destroy connective and muscle tissues and kill cells, *S. pyogenes* infection can lead to large-scale cell lysis and tissue destruction. "Flesh-eating" bacteria cause necrotizing fasciitis (Fig. 28.12*c*), in which rapid tissue damage can require amputation of infected limbs or, in 40% of cases, rapid death.

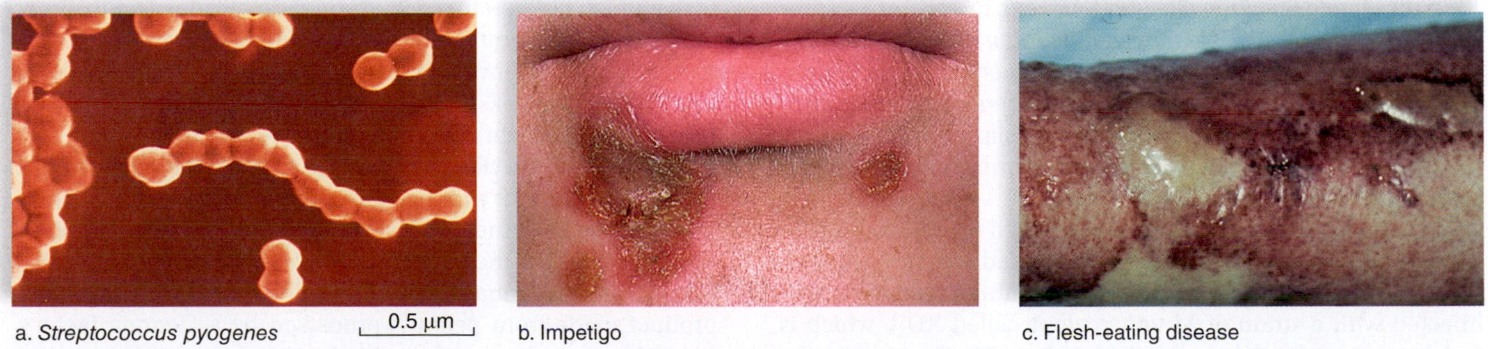
0.5 μm

a. *Streptococcus pyogenes* b. Impetigo c. Flesh-eating disease

Figure 28.12 *Streptococcus pyogenes.* **a.** *Streptococcus pyogenes,* shown here in an electron micrograph, often forms chains of cells, as seen here. **b.** Impetigo is a common mild skin infection. **c.** Flesh-eating disease is rare but life-threatening.

Staphylococcus aureus and MRSA *Staphylococcus aureus* is perhaps the bacterial species that has received the most media attention recently. About 20% of people are carriers of this bacterium (mostly on their skin or in their nostrils) without any symptoms. When *S. aureus* does cause disease, it is usually limited to skin infections. However, in people who are very young, very old, or immunocompromised for any other reason, it can invade the body and cause life-threatening disease. Moreover, a strain of *S. aureus* that is resistant to methicillin, called MRSA, is killing an increasing number of young, otherwise healthy individuals. Besides being resistant to many antibiotics, MRSA strains often possess genes coding for toxins not found in other *S. aureus* strains. Some of these toxins can be very damaging to tissues. In February 2009, a Texas jury awarded $17.5 million to a man who lost both arms and legs to a hospital-acquired MRSA infection, although the award was later reduced. Fortunately, most health-care facilities have initiated infection-control procedures that are now decreasing the rate of hospital-acquired MRSA infections.

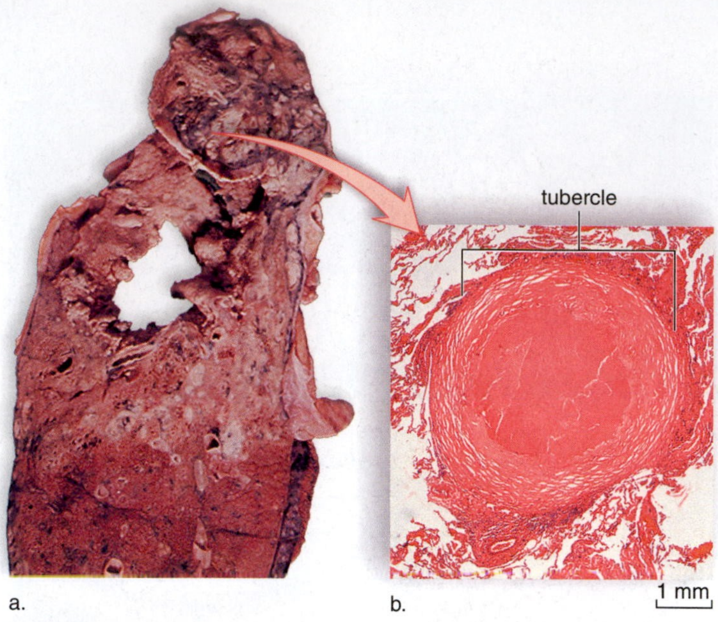
tubercle
a.
b.
1 mm

Figure 28.13 Tuberculosis. Tuberculosis, an infection caused by the bacterium *Mycobacterium tuberculosis,* usually settles in the lungs. **a.** Lung tissue with a large cavity surrounded by tubercles, which are hard, calcified nodules where the bacteria are trapped. **b.** Photomicrograph shows a cross section of one large tubercle, with a diseased center.

Tuberculosis

Tuberculosis (TB) is one of the leading worldwide causes of death due to infectious disease. When first described in 1882 by the father of medical microbiology, Robert Koch, TB caused one of every seven deaths in Europe, and one-third of all deaths of young adults. Today, it is estimated that one-third of the world's population is infected with the TB bacterium, causing approximately 2 million deaths each year. Tuberculosis is a chronic disease caused by *Mycobacterium tuberculosis,* a pathogen closely related to *M. leprae,* the causative agent of leprosy. Generally, *M. tuberculosis* infection occurs in the lungs, but it can occur elsewhere as well.

M. tuberculosis is very slow growing, and the extent of the disease is determined by host susceptibility. An immune response results in inflammation of the lungs. Active lesions can produce dense structures called *tubercles* (Fig. 28.13). These can persist for years, causing symptoms such as coughing and spreading the bacteria to other areas of the lungs and body as well as to other individuals. Eventually, the damaged lung tissue hardens and calcifies, leaving characteristic spots that can be observed with chest X rays. The number of TB cases in the United States surged in the late 1980s, mainly due to increased rates of infection with HIV, which damages helper T cells that are needed to fight the TB bacterium. Since 1992, however, the prevalence of TB in the United States has steadily declined, to a rate of 3.4 cases per 100,000 people in 2011. This reduction is mainly attributed to better testing and treatment of infected persons. This trend is also occurring worldwide, although numbers of TB cases continue to increase in Africa and Southeast Asia.

The 2007 story of a 31-year-old Atlanta lawyer with TB shows how easily an infectious agent can spread around the world. Despite knowing that he was infected with *M. tuberculosis,* he flew from Atlanta to Paris to get married, traveling through five countries in the process. Making matters worse, he was infected with a strain of *M. tuberculosis* called XDR, which is resistant to nearly all drugs available to treat TB. Upon returning to the United States, he was taken into forced quarantine for four weeks, the first time an American had been forcibly isolated by the Centers for Disease Control and Prevention (CDC) in several decades. He was released after it was determined that he was no longer infectious. It should be noted, however, that there is no way to know how many people infected with *M. tuberculosis* are traveling on airplanes every day.

Food Poisoning Whether the source was mishandled food at a salad bar, or potato salad that sat in the sun too long at a picnic, all of us have probably experienced the intestinal discomfort known as food poisoning. Two basic types of bacteria cause food poisoning: those that produce toxins while they are growing in food, and those that cause infections once they are in the intestine.

Several species of bacteria can produce toxins in foods, especially those containing dairy, eggs, or meat products. The symptoms, which consist mainly of vomiting and diarrhea, tend to appear suddenly within a few hours of ingestion, and are usually self-limiting. In contrast, *Clostridium botulinum,* the causative agent of *botulism,* produces one of the most toxic substances on Earth. When people are canning and don't heat foods to a high enough temperature, *Clostridium* can produce endospores that survive the canning process. These endospores then germinate in the airless environment of the can or bottle and become toxin-producing cells. If untreated, about 25% of people who ingest botulism toxin die from respiratory paralysis.

Salmonella is a classic example of a bacterial food poisoning agent that does not produce gastroenteritis until it has reproduced in the intestines for several days. A recent large outbreak of salmonellosis in the United States occurred between late summer of 2008 and January 2009, prompting the FDA to recall every product made from peanuts processed by a Georgia factory. Over 700 people became ill, and at least nine died after ingesting contaminated products.

Antibiotics and Probiotics

In 1928 at a London lab, Scottish scientist Alexander Fleming was researching substances that might inhibit bacterial growth when he took a two-week vacation. Upon returning, he noticed some petri dishes that had bacteria growing on them were contaminated with *Penicillium* fungus. In one of the most famous moments in medical history, Fleming realized that the fungus was producing a chemical that inhibited the growth of the bacteria (Fig. 28A*a*). Of course that substance turned out to be penicillin, and the practice of medicine was changed forever. Many bacterial infections that had previously meant serious illness or even death were now treatable.

As often happens, however, the widespread use of Fleming's new drug, and many others like it, had unintended consequences. Especially in patients who take antibiotics frequently or for a prolonged period, normal microbial populations in the body can be disrupted, resulting in diarrhea, yeast infections, or worse. A bacterium called *Clostridium difficile,* which is normally present in the intestines at relatively low levels, tends to overgrow in these patients, and may cause a severe, or even fatal, inflammation of the colon. Treatment for this disease is usually more antibiotics, but some authorities are suggesting that probiotics may be helpful in preventing this and many similar diseases.

Grocery store shelves and Internet health sites are full of products claiming to contain "probiotics," usually containing bacteria of the genera *Lactobacillus* or *Bifidobacterium.* According to the U.N. Food and Agricultural Organization, probiotics are "live microorganisms, which, when administered in adequate amounts, confer a health benefit on the host." By this definition, foods like yogurt, cheeses, and other fermented dairy products, are probiotics. But because of the increased interest in the health benefits of "friendly" bacteria, live bacteria are being added to an increasing variety of products (Fig. 28A*b*). Americans spent over $1 billion on probiotic products in 2011, and are expected to spend twice as much on probiotics by 2016.

How do probiotics work? As already mentioned in this chapter, the normal microflora of the body can produce vitamins and aid digestion. They also inhibit harmful microbes by simply taking up space and nutrients, and in some cases, by secreting chemicals that directly kill pathogens. But the beneficial effects of normal microflora go beyond competing with other microbes. Body surfaces such as the intestinal wall are fortified with large populations of immune cells that guard against invasion of the tissues by pathogens. An ongoing interaction between these cells and the normal microflora is necessary to maintain a healthy immune system. Experimentally, animals that are raised under gnotobiotic (germ-free) conditions have poorly developed immune systems, and replenishing their bodies with "good" bacteria restores healthy immune function, sort of like priming a pump. That is one reason why probiotics are also being tested for treatment of inflammatory conditions such as irritable bowel syndrome, ulcerative colitis, and Crohn disease. Future applications may also include treating urinary tract and vaginal infections, inhibiting food allergies, preventing tooth decay, and improving the effectiveness of vaccines!

Despite their potential, however, one aspect of probiotics to be aware of is quality control. Currently there is very little regulation of which bacterial strains, or even the number of beneficial organisms, a product must contain to be labeled as probiotic. And of course, it is extremely unlikely that all the potential benefits of probiotics will pan out. In a period of less than 100 years, however, we have gone from discovering a new way to inhibit or kill harmful bacteria that invade our bodies, to realizing how closely our health is linked to our relationship with harmless microflora.

Questions to Consider

1. If no antibiotics existed, how might your life be different?
2. Besides taking antibiotics, what other factors might influence the numbers and types of microflora in your body?
3. High numbers of "good" bacteria are found in the intestine and on the skin. The immune system needs to protect these areas from invading microbes, but cannot respond as strongly to the normal microflora without causing problems. What are some possible ways that immune cells could distinguish "good" from "bad" bacteria?

Figure 28A Changing attitudes about microbes.
a. When Fleming discovered the inhibition of bacterial growth by *Penicillium* (a fungal contaminant) in 1928, few would have predicted the (**b**) benefits of bacteria less than 100 years later.

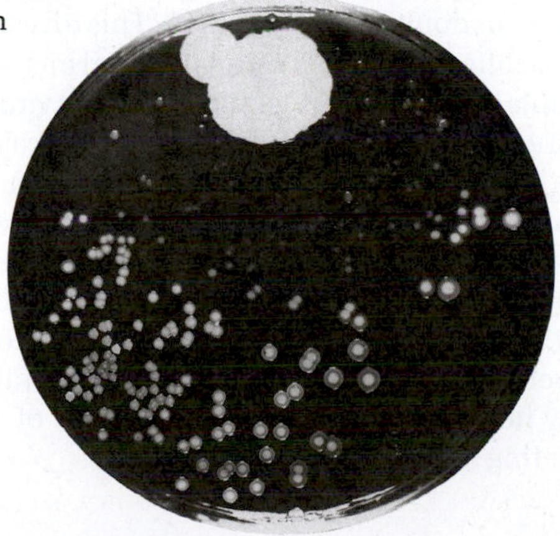

a.

CERTIFIED ORGANIC
MUGI MISO
Lifeway
Lowfat
Kefir
Stonyfield Farm
Smoothie
Naked
Probiotic
CULTURED
BROWN COW
ALL NATURAL
Vanilla
Fruits of Life
Naked
LIVING FOODS
Perfect Food
Rachel's

b.

Drug Control of Bacterial Diseases

Most **antibiotics** kill or inhibit bacteria by interfering with their unique metabolic pathways. Therefore, they are not expected to harm human cells. For example, erythromycin and tetracyclines inhibit bacterial protein synthesis by binding to bacterial ribosomes, whereas penicillins and cephalosporins inhibit bacterial cell wall synthesis.

Animation Antibiotic Inhibition of Protein Synthesis

There are problems associated with antibiotic therapy. Some individuals are allergic to certain antibiotics, and the reaction may even be fatal. Antibiotics not only kill off disease-causing bacteria, but they may also reduce the number of beneficial bacteria in the intestinal tract and vagina. This may allow the overgrowth of certain harmful bacteria or yeast, especially in the intestine, vagina, or mouth. See the Health feature "Antibiotics and Probiotics," on page 573, for a discussion of how probiotics can be used to replenish these beneficial organisms.

Most important perhaps is the growing resistance of bacteria to antibiotics, as discussed in the story that opened Chapter 27. Antibiotics were introduced in the 1940s, and for several decades they worked so well it appeared that infectious diseases had been brought under control. However, we now know that bacterial strains can mutate and become resistant to a particular antibiotic. Worse yet, when bacteria exchange genetic material, resistance can pass between different bacteria. Penicillin and tetracycline now have a failure rate of more than 22% against *Neisseria gonorrhoeae,* which causes gonorrhea. Bacteria with multiple drug resistances, such as MRSA and multidrug-resistant *M. tuberculosis,* are an especially serious threat in hospitals, prisons, and long-term care facilities. With the rise of antibiotic resistance, many companies are now working to develop innovative kinds of antibacterial therapies. The need for new, potent antibiotics is especially critical since the U.S. anthrax attacks of October 2001 raised the potential of bioterrorist activity (see the Bioethical feature, "Bioterrorism," on page 579).

Check Your Progress 28.4

1. Describe the three basic shapes of bacteria.
2. Explain how bacterial conjugation differs from transformation and transduction.
3. Compare the nutritional strategy of a heterotrophic bacterium with that of a chemoautotroph.
4. Consider how antibiotics work, and then describe two specific mechanisms bacteria could use to become resistant.

28.5 Viruses, Viroids, and Prions

Learning Outcomes

Upon completion of this section, you should be able to

1. Identify the major structural features of viruses.
2. Describe the steps in a typical viral reproductive cycle, and explain how some viruses become latent.
3. Describe several viral diseases of humans, and explain why it is difficult to produce vaccines against some viruses.
4. Distinguish between viruses, viroids, and prions.

As we learned in Chapter 1, all living organisms are composed of cells. **Viruses** are not composed of cells. They are acellular. Also, viruses are obligate parasites, meaning that they can reproduce only inside a living cell (called the host cell) by utilizing at least some of the machinery (ribosomes, certain enzymes, etc.) of that cell. Therefore, the question arises, "Are viruses alive?"

Scientists and philosophers have long argued this question. Some say that viruses are not alive. After all, not only are they acellular, but also some have been synthesized in the lab from chemicals! Moreover, when viruses are outside a host cell, they are totally quiescent, exhibiting no metabolic activity. Others argue that viruses are alive because they have a genome that directs their reproduction when they are inside the host cell.

If viruses are not considered alive, then certainly the simpler viroids and prions are not alive either. Viroids are strands of RNA that can reproduce inside a cell, and prions are protein molecules that cause other proteins to become prions.

Viral Size and Structure

Most viruses are much smaller than bacteria. Viruses typically measure between 0.03–0.2 μm, whereas most bacteria measure at least 0.5 μm. Interestingly, the largest virus discovered so far, called the Mimivirus, is about 0.4 μm in size, which is larger than the smallest known bacterium.

Viruses come in a variety of shapes, including helical, spheres, polyhedrons, and more complex forms. A virus always has at least two parts—an outer **capsid** composed of protein subunits, which protects an inner core of nucleic acid (Fig. 28.14). The viral genome can be single- or double-stranded DNA, or single- or double-stranded RNA. This diversity in genetic material is different from all cellular organisms, which always have a double-stranded DNA genome. Special viral enzymes that help a virus reproduce can also be inside the capsid.

In some viruses, especially those that infect animals, the capsid is surrounded by a membrane called an **envelope.** This envelope is made of lipid, and is usually derived from the host cell's plasma membrane when the virus buds from the host cell. Viral glycoproteins called spikes often extend from the envelope. These spikes are critical to viral infection because they help the virus bind to the surface of the host cell before entering it.

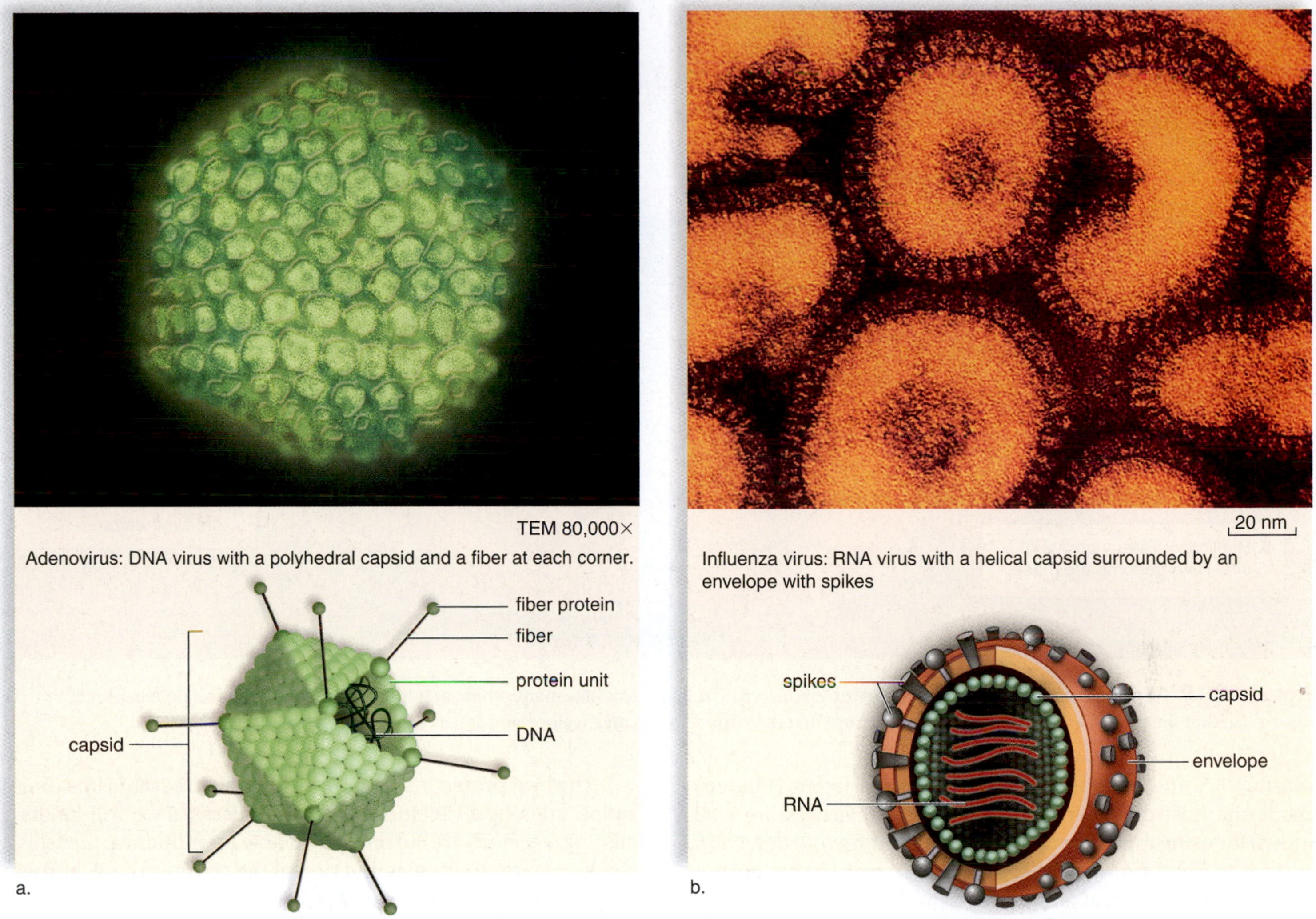
TEM 80,000×
Adenovirus: DNA virus with a polyhedral capsid and a fiber at each corner.
fiber protein
fiber
protein unit
capsid
DNA
a.
20 nm
Influenza virus: RNA virus with a helical capsid surrounded by an envelope with spikes
spikes
capsid
envelope
RNA
b.

Figure 28.14 Viruses. **a.** Despite their diversity, all viruses have an outer capsid composed of protein subunits and a nucleic acid core that is composed of either DNA or RNA. **b.** Some types of viruses also have a membranous envelope.

Viral Reproduction

Viruses infect almost every type of organism on Earth. Viruses called *bacteriophages* infect bacterial cells, usually just a particular group or species of bacteria. Some viruses infect only plants, whereas others infect only animals. Some viruses can infect and cause disease only in humans. The specificity of a virus for a host occurs because the spike of a virus and a receptor molecule on a cell's plasma membrane fit together like a hand fits a glove, which triggers events allowing viral entry into the cell.

Video How Viruses Attack

Animation Entry of an Animal Virus into Host Cells

Figure 28.15 illustrates the reproductive cycle of a typical enveloped animal RNA virus. This reproductive cycle has six steps:

1. During *attachment,* the spikes of the virus bind to a specific receptor molecule on the surface of a host cell. The host cell normally uses this receptor for another purpose, but it is effectively "hijacked" by the virus for its own purposes.
2. During *entry,* also called penetration, the viral envelope fuses with the host's plasma membrane, and the rest of the virus (capsid and viral genome) enters the cell. The genome is freed when cellular enzymes remove the capsid, a process called *uncoating.*
3. *Replication* occurs when a viral enzyme makes complementary copies of the genome, which is RNA in this case.
4. During *biosynthesis,* some of these RNA molecules serve as mRNA for the production of more capsid and spike proteins, using host ribosomes.
5. At *assembly,* a mature capsid forms around a copy of the viral genome.
6. During *budding,* new viruses are released from the cell surface. During this process, they acquire a portion of the host cell's plasma membrane and spikes, which were specified by viral genes during biosynthesis. The enveloped viruses are now free to spread the infection to other cells.

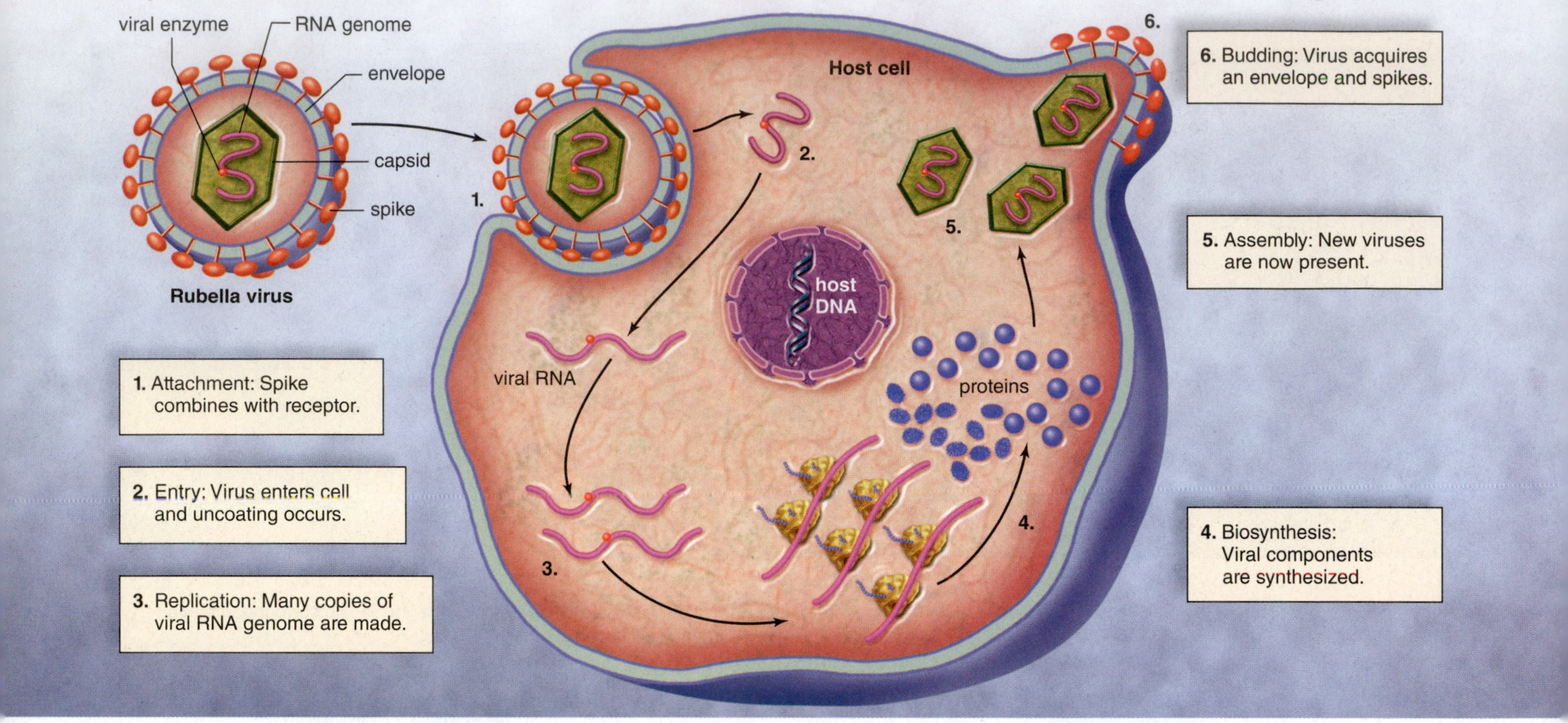

viral enzyme
RNA genome
envelope
capsid
spike
Rubella virus
Host cell
1.
2.
3.
4.
5.
6.
host DNA
viral RNA
proteins
6. Budding: Virus acquires an envelope and spikes.
5. Assembly: New viruses are now present.
4. Biosynthesis: Viral components are synthesized.
1. Attachment: Spike combines with receptor.
2. Entry: Virus enters cell and uncoating occurs.
3. Replication: Many copies of viral RNA genome are made.

Figure 28.15 Reproductive cycle of an animal virus. The rubella virus, like many others that infect animal cells, has an RNA genome. Notice also that the mode of entry requires uncoating and that the virus acquires an envelope and spikes when it buds from the host cell.

Latency Some animal viruses can become latent (hidden) inside the host cell. Herpesviruses and retroviruses are well known for using this strategy, which helps them avoid detection by the host immune system. During latency, new viruses are not produced, but the viral genome is reproduced along with the host cell. Environmental stresses, such as ultraviolet radiation, can induce the latent virus to enter the biosynthesis stage, leading to the production of new virus particles. Borrowing terms that were first defined in bacteriophages, latent viruses are sometimes said to be "lysogenic," whereas viruses that are actively reproducing are known as "lytic."

Video Virus Lytic Cycle

Animation Lambda Phage Replication

Retroviruses add an interesting twist to the story of viral reproduction. The genome of a retrovirus is RNA, but these viruses are able to convert their genome into DNA because they contain an enzyme called reverse transcriptase. This enzyme, which is not found in host cells, is so named because it catalyzes a process that is the reverse of normal transcription, which goes from DNA to RNA. The DNA copy (cDNA) of the retroviral genome can be integrated into the host DNA, where it is called a *provirus* (see Fig. 21.14). Not only is the integrated provirus resistant to antiviral medications taken by the host, but it is also able to escape detection by the host immune system.

Animation Replication Cycle of a Retrovirus

Viral Diseases of Humans

Viruses cause many important human diseases, but we have room to discuss only a few of them: colds, influenza, measles, and four herpes viral infections. Some viral diseases that are transmitted sexually, including HIV/AIDS, were discussed in section 21.5.

The best protection against most viral diseases is immunization utilizing a vaccine. Of the viral diseases we will be discussing, vaccines are currently available for influenza, measles, chickenpox, and shingles, but not for the common cold, herpes simplex, or Epstein-Barr viruses.

The Common Cold and Influenza

Colds are most commonly caused by rhinoviruses, and the symptoms usually include a runny nose, mild fever, and fatigue. The flu, caused by the influenza virus, is characterized by more severe symptoms, such as a high fever, body aches, and severe fatigue. Cold symptoms tend to subside within a week, but the flu may last for two or three weeks, and can result in death, especially in elderly patients or those with weakened immune systems.

While most common cold viruses are endemic (always present) in the human population, flu outbreaks tend to be epidemic, affecting large numbers of people in more limited geographic areas at any one time. Why can you get several colds or the flu year after year? There are over 100 different strains of rhinoviruses, plus several other viruses that cause very similar symptoms. In the case of colds, you become immune only to those strains you have contracted before. However, the reason you may get the flu each year is that the influenza virus can change rapidly. Small changes called *antigenic drift,* especially those affecting the surface spikes, may be enough to make the virus capable of temporarily evading the immune response of individuals who were immune to the original virus (Fig. 28.16*a*). Therefore, manufacturers of influenza vaccines try to incorporate the strains that are predicted to cause the highest number of cases each year. Still, antigenic drift can lead to local epidemics.

Video Killer Flu Recreated

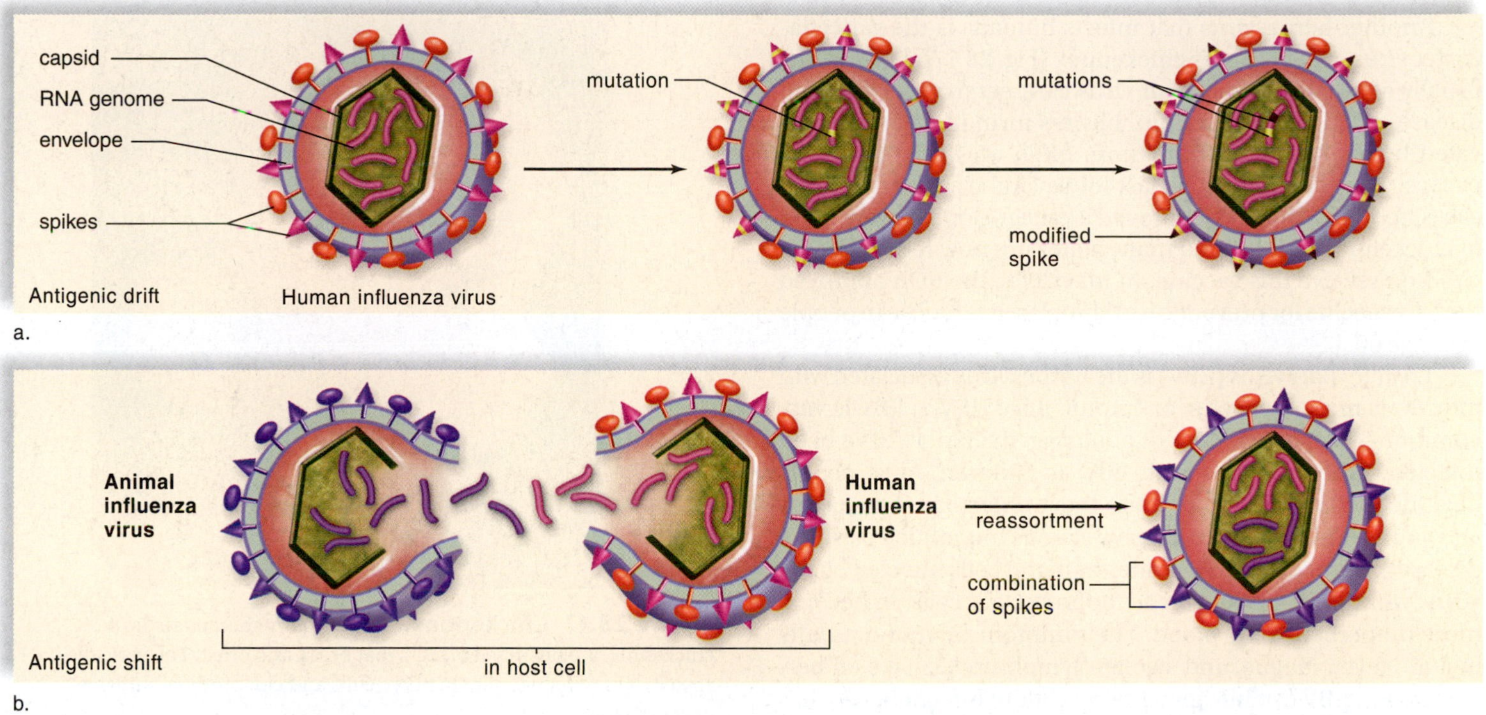
capsid
RNA genome
envelope
spikes
mutation
mutations
modified spike
Antigenic drift
Human influenza virus
a.
Animal influenza virus
Human influenza virus
reassortment
combination of spikes
Antigenic shift
in host cell
b.

Figure 28.16 Antigenic drift and shift. **a.** In antigenic drift, small mutations gradually change surface antigens so that antibodies to the original virus become less effective. As time goes by, people are more likely to become ill upon exposure to the virus. **b.** In antigenic shift, even more serious major changes take place on surface antigens as genome segments are reassorted between two influenza viruses that infect the same cell. Now, antibodies are more likely to be ineffective, and most people will become ill when they are exposed to the virus.

The genome of the influenza virus is composed of eight segments of RNA. When two different influenza viruses infect the same cell, these RNA segments can get mixed up as the viruses reproduce. When new virus particles are assembled that contain RNA from both original viruses, this reassortment event, called an *antigenic shift,* may lead to new combinations of surface spikes (Fig. 28.16*b*). Because the human population has not previously been exposed to this combination of antigens, the result can be a pandemic, or worldwide epidemic. The "avian flu" virus, also called H5N1, which arose in southern China in 2005, contains genes from a bird virus, as well as a human virus, and thus may have resulted from a reassortment event. Fortunately, the H5N1 virus doesn't seem to be efficiently transmitted to humans. However, the most recent "swine flu" outbreak of 2009 serves to remind us it is impossible to predict what future reassortment events may occur, so vaccine manufacturers must play catch-up whenever a new virus evolves.

Video Virus Crisis

Measles

Measles is one of the most contagious human diseases. An unvaccinated person can become infected simply by breathing viral particles in the air of a room where an infected person has been, even two hours later. After exposure, there is a seven- to twelve-day incubation period before the onset of flulike symptoms, including fever. A red rash develops on the face and moves to the trunk and limbs. The rash lasts for several days. In the more-developed nations, the fatality rate is about 1 in 3,000. However, in the less-developed countries, the fatality rate is 10–15%. The number of deaths is decreasing in many areas, however, due to efforts to vaccinate children, who still make up the majority of measles fatalities. In the United States, the number of measles cases dropped from 5 million each year to a few thousand after the introduction of the measles vaccine in 1963. Between 2000 and 2007, measles deaths worldwide fell by 74%, from an estimated 750,000 to 197,000. The measles vaccine is usually given to infants in combination with vaccines against mumps and rubella, collectively called the MMR vaccine.

Herpesviruses

Herpesviruses cause chronic infections that remain latent for much of the time. Eight herpesviruses capable of infecting humans have been described. Some of these infect epithelial cells and neurons, whereas others infect blood cells. Some herpesviruses do not seem to cause any pathology, but we will discuss four herpesviruses that cause disease in humans.

Herpes simplex virus type 1 (HSV-1) is usually associated with cold sores and fever blisters around the mouth. It is estimated that 70–90% of the adult population is infected with HSV-1. Herpes infections of the genitals are generally caused by HSV-2 and are transmitted through sexual contact. Painful blisters fill with clear fluid that is very infectious (see Fig. 21.15). Genital herpes infections are widespread in the United States, affecting 20–25% of the adult population. Most individuals infected with genital herpes will experience recurrences of symptoms induced by various stresses.

Another herpesvirus that infects humans is the varicella-zoster virus, which causes chickenpox (Fig. 28.17). Later in life, usually after age 60, the latent virus can reemerge as a related disease called shingles. Painful blisters form in an area innervated by a single sensory neuron, often on the upper chest or face. The symptoms may last for several weeks or, in some cases, months. The FDA approved a vaccine for the prevention of chickenpox in 1995, and many states are now requiring that children receive this vaccine. In May 2006, the FDA approved the first vaccine for prevention of shingles. It is for use in people over age 60.

Epstein-Barr virus (EBV) is the herpesvirus associated with infectious mononucleosis, or "mono." Like HSV-1, EBV is very common—as many as 95% of adults aged 35 to 40 have been infected. When infected with EBV as children, most people have no symptoms. In contrast, those infected as teenagers and young adults often develop infectious mononucleosis. The disease is named after the tendency of the cells infected by the virus, which are also known as mononuclear cells, to become more numerous in the blood. The symptoms of mono usually include fever, fatigue, and swollen lymph nodes. Like all herpesviruses, EBV establishes a lifelong latent infection. EBV has also been associated with other conditions, such as chronic fatigue syndrome and certain rare cancers.

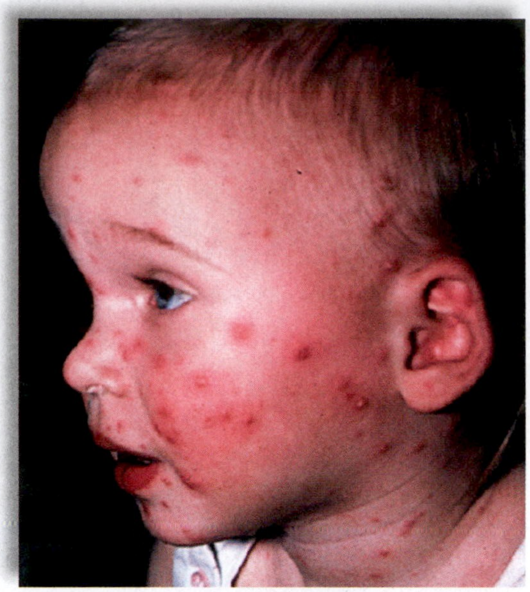

Figure 28.17 Chickenpox. The lesions characteristic of chickenpox are pus-filled blisters that break and crust. The varicella-zoster virus remains latent in nerve cells and can produce shingles decades later.

Antiviral Drugs

Because viruses use the machinery of host cells for viral replication, it is difficult to develop drugs that affect viral replication without harming host cells. As noted earlier, however, many viruses use their own enzymes to copy their genetic material, and many antiviral drugs inhibit these enzymes. As discussed in section 21.5, a variety of antiretroviral agents have been developed to inhibit the reverse transcriptase and protease enzymes used by HIV, which can delay the onset of AIDS. Other drugs, such as acyclovir and valacyclovir (or Valtrex), are commonly used to control recurrences of HSV-1 and HSV-2 infections. These compounds mimic nucleotides, and thus can inhibit viral genome synthesis. Other antiviral drugs may affect virus attachment, entry, or assembly. If no effective antiviral drugs are available for a particular viral infection, patients are often told to simply let the virus run its course, because antibiotics are not effective against viral infections.

Animation Antiviral Agents

Viroids and Prions

Viroids and prions are also acellular pathogens. A **viroid** consists of a circular piece of naked RNA. The viroid RNA is ten times shorter than that of most viral genomes, and apparently it does not code for any proteins. Viroid replication causes diseases in plants, the only known hosts. The mechanism of viroid diseases, such as potato spindle tuber and apple scar skin, are not known.

Prions are proteinaceous infectious particles that cause degenerative diseases of the nervous system in humans and other animals. They are derived from normal proteins of unknown function in the brains of healthy individuals. Disease occurs when the normal proteins change into the abnormal, prion shape. This forms more prion proteins, which go on to convert other normal proteins into the wrong shape. These abnormal proteins seem to build up in the brain, causing a loss of neurons and what appear to be holes in the tissue (see Fig. 17.22).

Animation How Prions Arise

The first prion disease to be described was *scrapie,* which occurs in sheep. It is called scrapie because, as the disease affects the animal's brain, an affected sheep begins to scrape off much of its wool. Scrapie is not capable of causing disease in humans. However, another prion disease in cattle, called *bovine spongiform encephalopathy* (*BSE*), or mad cow disease, does appear capable of infecting humans and has caused over 150 human deaths in Great Britain and a few other countries. This disease is now called *variant Creutzfeldt-Jakob disease* (*vCJD*), to distinguish it from the "classic" form of CJD, which occurs spontaneously in a very low percentage of people. In April 2012, a California dairy cow was diagnosed with BSE, which was the fourth case of BSE detected in U.S. cattle since 2003. Three human cases of vCJD have been documented in the United States, but all three victims were thought to have acquired the disease in other countries prior to moving to the United States.

Chronic wasting disease is another prion disease that occurs in deer, elk, and moose, but there have been no documented cases of transmission to humans. One prion disease that passes directly from human to human is kuru, which was transmitted among a cannibalistic tribe in New Guinea due to their traditional practice of consuming their dead relatives. Even though prion diseases are as frightening as they are fascinating, their incidence in humans remains very low.

Animation Prion Diseases

Check Your Progress 28.5

1. Describe the basic structure of an enveloped virus, including capsid, nucleic acid, envelope, and spikes.
2. Explain the six stages of a typical animal virus reproductive cycle.
3. List three viral diseases for which a vaccine is available, and three for which there is no vaccine.
4. Compare how viruses differ from viroids and prions.

SCIENCE IN YOUR LIFE ▶ **BIOETHICAL**

Bioterrorism

The events of September 11, 2001, caused scientists and ethicists to consider the possibility of bioterrorism—the use of a biological agent, such as a microbe, to kill thousands of people at once. For a few weeks after 9/11, this possibility became even more real as several people died when anthrax spores were sent through the U.S. mail. Unfortunately, the person(s) responsible for those deaths have not been identified, although a leading suspect, microbiologist Bruce Ivins, committed suicide in July 2008 as the government was preparing to charge him for the attacks.

Bioterrorism has been employed throughout the history of the civilized world, so much so, that the United States, the United Kingdom, and the USSR signed the 1972 Biological Weapons Convention banning research and the production of offensive biological agents. However, the possibility of using microbes and inexpensive nerve agents as weapons is particularly attractive to countries that lack the means to produce conventional weapons of war. When a group of Japanese extremists used sarin, a nerve agent, to kill many innocent civilians in the Tokyo subway system in 1995, it became apparent that individuals, as well as hostile countries, could obtain and use these agents as weapons.

Numerous ethical questions surround the process of researching biological weapons. For example, should scientists be allowed to perfect and publish their findings regarding such agents? Many biologists say not publishing these findings would be detrimental to the progress of science. "How do doctors talk about research if we don't publish it?" asks Dr. Ariella Rosengard of the University of Pennsylvania. In 2002, her laboratory published a paper on the variola virus, which causes smallpox, showing that a viral protein is 100 times more effective at inhibiting immune system proteins called complement (see section 13.2) than is a similar protein from a related virus. This knowledge may help to explain why variola virus is so lethal to humans, but it could also be useful to bioterrorists attempting to produce even more lethal forms of the virus. While it may seem that potentially harmful information like this shouldn't be published, such research is important for developing suitable vaccines.

Other scientists feel the need to be more socially responsible. "When you wield power equivalent to nuclear weapons, you need some oversight," says John Steinbruner of the Center for International and Security Studies in Maryland. This organization and others are attempting to define what types of research should be allowed, and what types should possibly be banned. The main difficulty is that a better understanding of what makes these biological agents so deadly could save lives, as well as destroy them.

Questions to Consider

1. Should the United States maintain an active research program to develop biological weapons? Why or why not? Is the use of these kinds of weapons more objectionable than conventional warfare? Than nuclear warfare?
2. Many experts believe that Russia still possesses large stockpiles of smallpox virus produced during the Cold War. The United States also retains some smallpox virus, reportedly for research purposes. Ideally, if both countries could agree, would it be best to destroy all samples of this deadly virus?
3. Since September 11, 2001, the U.S. government has increased funding for research into infectious agents that bioterrorists could use. However, this also means that more scientists will be working with these agents and developing knowledge that could be used for detrimental purposes. Should scientists working, for example, with the bacterium that causes anthrax, be required to submit to extensive background checks? To periodic questioning by the FBI? If so, might this discourage people with good intentions from doing this work?

connect BIOLOGY Explore the concepts through a variety of multimedia assets, question types, and data interpretation.
www.mcgrawhillconnect.com

Case Study Conclusion

In his later years, Howard Hughes' fear of germs became so intense that he became a recluse, living in a darkened hotel room that he considered to be a germ-free zone. He wore tissue boxes on his feet to protect them, and burned his clothing if someone on his staff of caretakers became ill. When serving his food, he ordered his staff to wash their hands multiple times, then wrap their hands in paper towels to avoid touching his food.

Hughes believed that germs came only from outside him, while he remained seemingly unaware of the vast populations of microscopic life, most of it beneficial, that thrives inside the mouth, nasopharynx, intestinal and genital tracts, as well as the entire surface of the skin. Ironically, as his obsession grew, Hughes neglected his own hygiene, rarely bathing or brushing his teeth.

Today we know that the great majority of microbes in our bodies, as well as in the environment, are beneficial to us. We also have developed effective treatment for most of those that mean us harm. And yet, a relatively low number of microbes, like HIV, influenza, and various antibiotic-resistant bacteria, continue to cause tremendous human suffering and death. One of modern medicine's greatest challenges is to continue winning the war against these ancient enemies.

MEDIA STUDY TOOLS

www.mhhe.com/maderinquiry14

Enhance your study of this chapter with study tools and practice tests. Also ask your instructor about the resources available through ConnectPlus, including LearnSmart, the media-rich eBook, interactive learning tools, and animations.

SUMMARIZE

28.1 The Microbial World

Microbiology is primarily the study of bacteria, archaea, protists, fungi, and viruses. Most microbes are microscopic, and some are extremely numerous.

- Two of the most significant contributors to the early discipline of microbiology were Antonie van Leeuwenhoek, who was the first to visualize microbes using his microscopes, and Louis Pasteur, whose many accomplishments include disproving the idea that microbes could arise spontaneously.
- Microbes are perhaps best known for causing infectious diseases, which can range from the merely annoying to the swiftly fatal. However, most microorganisms are far more beneficial than harmful, not only to humans, but to Earth's ecosystem. These include the **normal microbiota** that reside on and in the human body, as well as **decomposers** that break down organic and inorganic matter.

28.2 Origin of Microbial Life

A central principle of evolutionary theory is that all life on Earth has arisen from a **last universal common ancestor (LUCA)**. Chemical reactions are hypothesized to have led to the formation of **biomolecules** as well as the first true cells in the following stages:

- In stage 1, simple organic molecules, or monomers, arise from inorganic chemicals. The "primordial soup" hypothesis suggested by Oparin and Haldane is an influential idea about this stage, which is supported by the work of Miller, Urey, and others.
- In stage 2, monomers join to form polymers such as proteins, RNA, or membranes. The **protein-first hypothesis** suggests that some early proteins were enzymes that could synthesize other molecules. The **RNA-first hypothesis** holds that RNA was the first macromolecule.
- In stage 3, the evolution of a **probiont** (**protocell**) very likely involved the formation of **micelles,** which would have fused to form larger **vesicles.** The **membrane-first hypothesis** focuses on the need for an outer membrane to contain other biochemical reactions. The first membranes may have been similar to **liposomes,** which form spontaneously from lipids in water. The probiont may have been a **heterotroph** that consumed preformed organic molecules, or a **chemoautotroph** that required only inorganic nutrients.
- Stage 4 produced living cells that were able to reproduce. According to the **central dogma** of genetics, information flows from DNA to RNA to protein. However, this may not have been the case in the earliest cells.

28.3 Archaea

Archaea and bacteria are **prokaryotes,** single-celled microbes that lack a nucleus or membrane-bound organelles. However, the two groups are classified into different domains of life due to nucleic acid sequence and biochemical differences.

- Structurally, archaeal plasma membranes have unique lipids that help some to survive in extreme environments.
- Archaea are often associated with extreme habitats, but actually are widespread in the environment. Three major types of archaea are the **halophiles, thermoacidophiles,** and **methanogens.**

28.4 Bacteria

Bacteria are the most widespread, and in many ways successful, type of organism on Earth.

- Most bacteria have a cell wall that contains peptidoglycan, and most bacteria can be classified as either Gram-positive or Gram-negative using the **Gram stain** procedure. The outer membrane of Gram-negative bacteria contains a toxic substance called **lipopolysaccharide.** Many bacteria have **flagella** for motility.

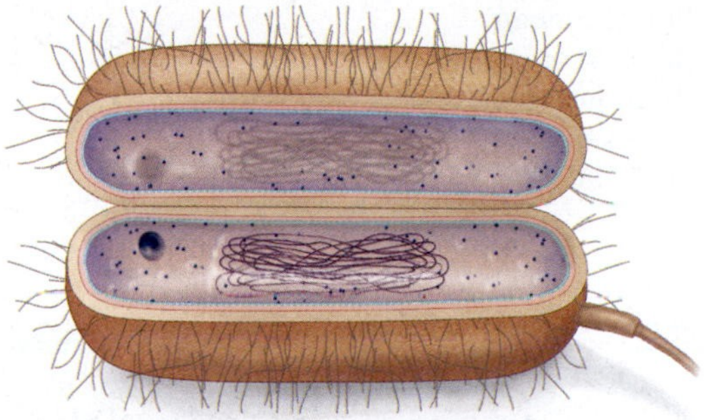

- All bacteria have a DNA genome, usually found on a single, circular chromosome in the **nucleoid** region. Bacteria reproduce by **binary fission,** producing identical daughter cells. Horizontal gene transfer also occurs, by **conjugation, transformation,** or **transduction.** Some bacteria can survive for long periods by producing **endospores.**
- Most bacteria are heterotrophs, requiring carbon in the form of organic molecules. Some are chemoautotrophs, which acquire carbon from carbon dioxide and energy from chemicals. **Cyanobacteria** are an important group of photosynthetic bacteria.
- Bacterial species that cause important diseases of humans include members of the genera *Staphylococcus, Streptococcus, Mycobacterium,* and *Salmonella*. **Antibiotics** are chemicals that either kill or inhibit the growth of bacteria.

28.5 Viruses, Viroids, and Prions

Viruses are minute, acellular pathogens that reproduce as obligate intracellular parasites.

- Structural features shared by all viruses include an outer **capsid** composed of protein and an inner core of nucleic acid. The viral genome can be single- or double-stranded DNA or RNA. Many animal viruses also have an **envelope** with spikes.
- The reproductive cycle of an animal virus typically has six steps: attachment, entry, genome replication, biosynthesis, assembly, and budding. During budding, enveloped viruses usually acquire a portion of the host cell's plasma membrane. **Retroviruses** (e.g., HIV) have RNA genomes that are converted to DNA by reverse transcriptase, before integrating into the host cell's chromosome as a provirus.
- Viruses infect almost all types of organisms and cause many important diseases. Viral diseases of humans include the common cold, influenza, measles, and chickenpox. Antiviral drugs usually inhibit viral enzymes.

Like viruses, viroids and prions are acellular pathogens.

- **Viroids** are obligate, intracellular plant pathogens that are autonomously replicating short RNA molecules. They contain no protein.
- **Prions** contain no nucleic acid. They are proteinaceous, infectious particles that convert a normal cellular protein into the abnormal form, which accumulates and damages brain tissue, inducing dementia. Diseases caused by prions include scrapie, kuru, and mad cow disease.

ASSESS

Testing Yourself

Choose the best answer for each question.

1. Decomposers
   a. break down dead organic matter in the environment by secreting digestive enzymes.
   b. break down living organic matter by secreting digestive enzymes.
   c. destroy living cells and then break them down with digestive enzymes.
   d. live in close association with another species.

2. The RNA-first hypothesis for the origin of cells is supported by the discovery of
   a. ribozymes.
   b. proteinoids.
   c. polypeptides.
   d. nucleic acid polymerization.

3. The Oparin-Haldane hypothesis is also known as the
   a. "iron-sulfur world" hypothesis.
   b. "prehistoric stew" hypothesis.
   c. "primordial soup" hypothesis.
   d. "warm little pond" hypothesis.

4. Which is not true of prokaryotes? They
   a. are living cells.
   b. lack a nucleus.
   c. all are parasitic.
   d. include both archaea and bacteria.
   e. evolved early in the history of life.

5. Archaea differ from bacteria in that they
   a. have a nucleus.
   b. have membrane-bound organelles.
   c. have peptidoglycan in their cell walls.
   d. are often photosynthetic.
   e. None of these are correct.

6. Which bacterium is spherical in shape?
   a. bacillus
   b. coccus
   c. spirillum

7. Bacterial endospores function in
   a. reproduction.
   b. survival.
   c. protein synthesis.
   d. storage.

8. Scarlet fever and rheumatic fever are both caused by
   a. *Clostridium tetani.*
   b. *Salmonella* sp.
   c. *Staphylococcus aureus.*
   d. *Streptococcus pyogenes.*

9. About one-third of the world's population is infected by
   a. *Escherichia coli* O157:H7.
   b. *Mycobacterium tuberculosis.*
   c. *Mycobacterium leprae.*
   d. *Streptococcus pyogenes.*

10. Label the reproductive cycle of an RNA animal virus in the following diagram.

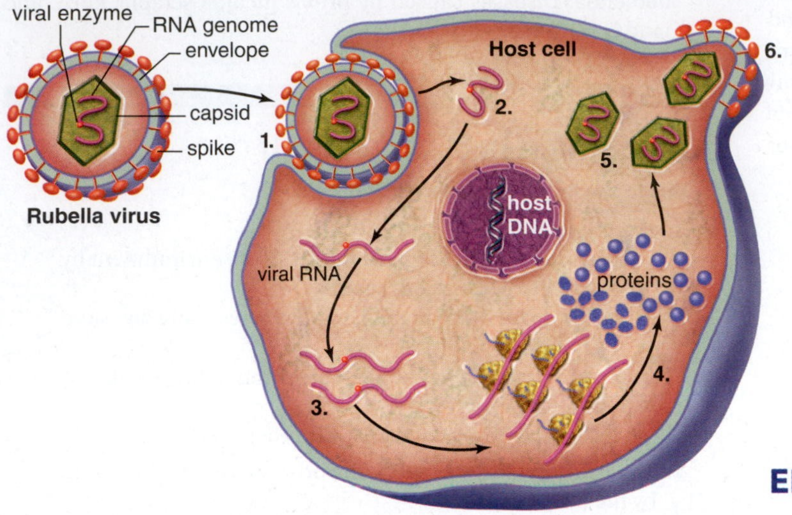
viral enzyme
RNA genome
envelope
capsid
spike
Rubella virus
Host cell
1.
2.
3.
4.
5.
6.
host DNA
viral RNA
proteins

11. The envelope of an animal virus is usually derived from the _____ of its host cell.
- **a.** cell wall
- **b.** plasma membrane
- **c.** capsule
- **d.** receptors

12. Capsid proteins are synthesized during which phase of viral replication?
- **a.** replication
- **b.** biosynthesis
- **c.** assembly
- **d.** proteination
- **e.** All of these are correct.

13. Retroviruses have a unique enzyme that
- **a.** disintegrates host DNA.
- **b.** polymerizes host DNA.
- **c.** transcribes viral RNA to cDNA.
- **d.** translates host DNA.
- **e.** repairs viral DNA.

14. Small changes in influenza surface antigens lead to
- **a.** antigenic drift.
- **b.** antigenic shift.
- **c.** antigenic recombination.
- **d.** antigenic schism.

15. Shingles is a disease observed mainly in the elderly that is a reactivation of which of the following?
- **a.** herpes
- **b.** polio
- **c.** chickenpox
- **d.** AIDS

16. Prions contain
- **a.** DNA only.
- **b.** protein only.
- **c.** RNA only.
- **d.** DNA, RNA, and protein.

ENGAGE

Thinking Critically

1. Model organisms are those widely used by researchers who wish to understand basic processes that are common to many species. Bacteria such as *Escherichia coli* are model organisms for modern geneticists. Give three reasons why bacteria would be useful in genetic experiments.
2. Many viruses contain their own enzymes for replicating their genetic material, whereas others use the host cell's enzymes. Would DNA viruses or RNA viruses be more likely to produce their own enzyme(s) for this purpose, and why?

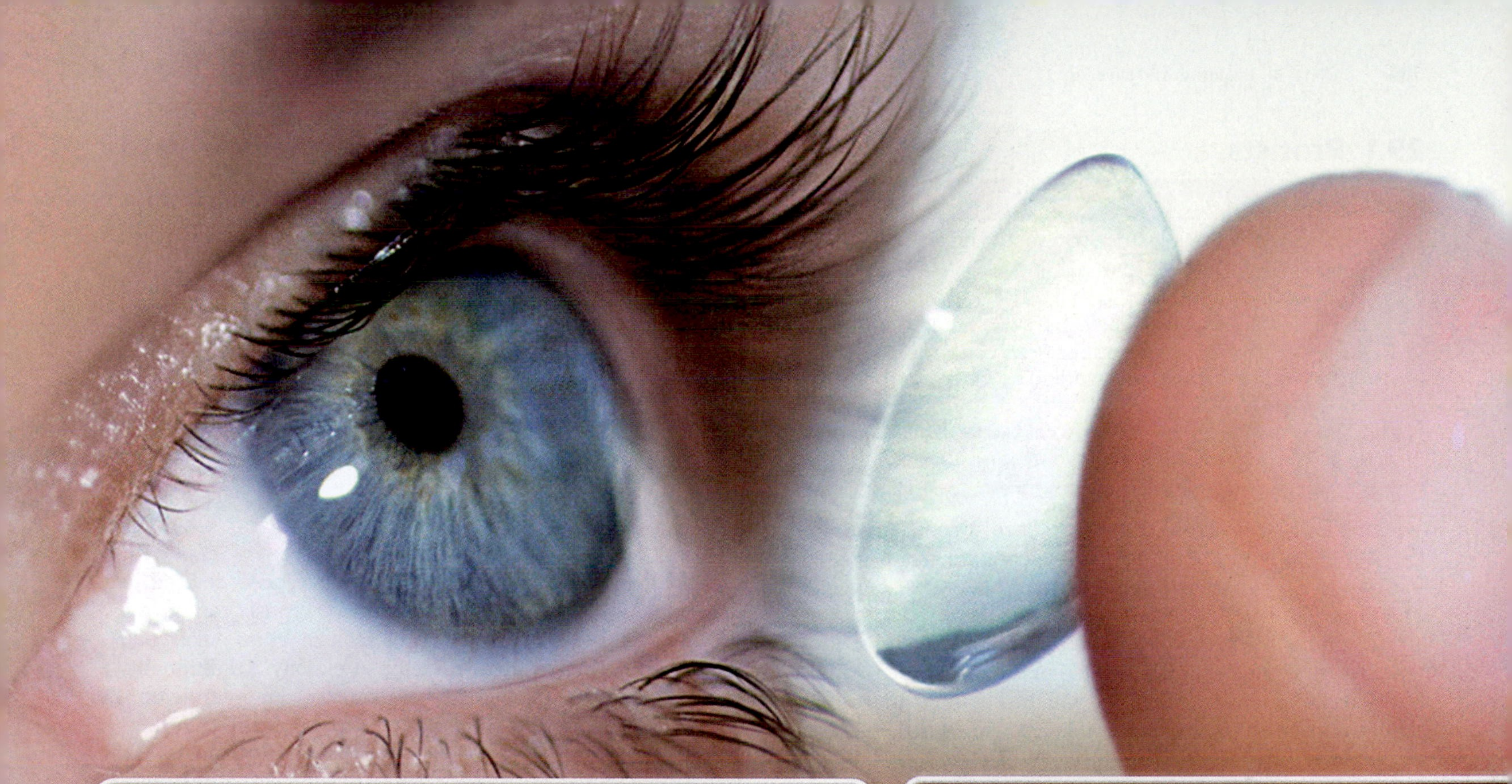

29 Protists and Fungi

CASE STUDY Keeping Your Eye on the Ball

Josh enjoyed playing shortstop on his small-college baseball team. Halfway through his sophomore season, however, he began to feel a burning pain in his left eye. At first he thought he must have just gotten some dirt under his contact lens, so when the pain persisted, he switched back to wearing his glasses in an attempt to keep playing. When the pain continued to get worse and his vision became blurred, he knew he needed to see his eye doctor ASAP. When an initial round of testing at his optometrist's office turned up no apparent cause of his problem, Josh was referred to an ophthalmologist, who diagnosed Josh with keratitis (corneal inflammation) caused by a single-celled parasite of the genus *Acanthamoeba.*

Protists, such as *Acanthamoeba,* are a very diverse group of single-celled eukaryotes. Although rare, eye infections caused by *Acanthamoeba* can be quite serious, especially if they are not diagnosed and treated early. In some cases, the infection may damage the cornea so severely that a corneal transplant is required.

During the course of his treatment, Josh learned that *Acanthamoeba* organisms are commonly found in water sources, like tap water, well water, hot tubs, and freshwater lakes and rivers. About 90% of cases involve contact lens wearers. He wondered if he may have acquired the infection when, after receiving a face full of dirt after sliding into a base, he had rinsed his contact lenses out in a drinking fountain near the team dugout. He also felt somewhat lucky when he learned that, especially in people whose immune system isn't functioning properly, *Acanthamoeba* infections can result in serious infections involving the brain, spinal cord, and other organs.

As you read through this chapter, think about the following questions:

1. Are all protistans parasites?
2. How are protists related to other eukaryotes, such as humans?
3. How do protists and fungi impact human heath and welfare?

CHAPTER OUTLINE

29.1 Protists
29.2 Fungi

BEFORE YOU BEGIN

Before beginning this chapter, take a few moments to review the following discussions:

Figure 1.5 Protists and fungi belong to which domain of life?

Section 3.5 How does the endosymbiotic theory explain the origin of energy-producing organelles in the eukaryotic cell?

Section 5.4 What is the difference between a haploid and a diploid eukaryotic cell?

29.1 Protists

Learning Outcomes

Upon completion of this section, you should be able to

1. Summarize what is known regarding the evolution of the first unicellular protist.
2. Describe the general characteristics of a protist.
3. Summarize the unique and specific structural features of each group of protists, including how they reproduce and obtain food.
4. Identify several human diseases caused by protists, and explain why many are difficult to treat.

The first eukaryotic cell is thought to have arisen from a prokaryotic cell around 1.7 billion years ago. As we first discussed in Chapter 1, all **eukaryotic cells** (**eukaryotes**) have a nuclear membrane and other membranous organelles.

The domain Eukarya is divided into four kingdoms: Protista, Fungi, Plantae, and Animalia. For the most part, **protists** are unicellular and microscopic, which sets them apart from other eukaryotes. They are such a diverse group, however, that perhaps the best definition of a protist is any eukaryotic organism that is not a plant, animal, or fungus.

Protists offer us a glimpse into the past because they are most likely related to the first eukaryotic cell to have evolved. According to the **endosymbiotic theory,** mitochondria may have resulted when a nucleated cell engulfed aerobic bacteria, and chloroplasts may have originated when a nucleated cell with mitochondria engulfed cyanobacteria (see Fig. 3.16). Protists also bridge the gap between the first eukaryotic cells and multicellular organisms—the other three types of eukaryotes (fungi, plants, and animals) all trace their ancestry to a protist.

Animation Endosymbiosis

Biology of Protists

Protists can be unicellular, colonial, or multicellular. They are structurally diverse, having many complex shapes. Some have an outer cell wall, while others have external shells or tests made of minerals, cellulose, or even glass! Some protists have more than one nucleus. In some cases they have organelles that are not found in other eukaryotes. For example, **contractile vacuoles** are responsible for osmoregulation, particularly in freshwater environments. They pump out the excess water that tends to enter the cell. The cell membrane of many protists is supported by a thin, often flexible **pellicle** that maintains cell shape and allows for better movement in water. Some motile, photosynthetic protists have an **eyespot apparatus** that allows them to sense light intensity and move to areas for optimum photosynthesis.

The variability of protists also extends to their lifestyles and modes of reproduction. Most carry out asexual reproduction, with sexual reproduction as an option, particularly under unfavorable conditions. Sexual reproduction often results in a **spore** or **cyst** that can survive until environmental conditions improve. Spore formation may also promote dispersal to areas better suited for growth. This involves the development of a cell with a thick cell wall and low metabolic rate, analogous to the endospores produced by some bacteria. Some protists, especially parasitic forms such as *Plasmodium* that cause malaria, have complicated life cycles in their different hosts.

Another major difference between groups of protists is their mode of locomotion. Some protists are nonmotile, and other groups use flagella, cilia, or pseudopods for getting about. The term **protozoan** is a common term used to describe a heterotrophic, usually motile, unicellular protist. Protozoans are distributed in great number in many habitats. For example, in aquatic environments, they are part of the **zooplankton,** microscopic suspended organisms that feed on other organisms. A number of protozoans are parasitic, often causing diseases of the blood. In many cases, their complex life cycles hinder the development of suitable treatments.

Diversity of Protists

Many different classification schemes have been devised to define relationships between the protists. Protists can have a combination of characteristics not seen in other eukaryotic groups, which make them difficult to classify. Traditionally, protists have been classified by their source of energy and nutrients: the algae are photosynthetic (but use a variety of pigments); the protozoans are heterotrophic by ingestion; and the water molds and slime molds are heterotrophic by absorption. Newer data, especially genetic sequence information, have resulted in some controversy, however. One system that has gained popularity divides all eukaryotes, including the protists, into six **supergroups,** high-level taxonomic groups just below the domain level. Each of these supergroups represents a separate evolutionary lineage (Table 29.1).

As biologists continue to learn more about these organisms, more agreement will likely emerge on how to classify the protists. In this chapter, we will group the protists into six groups (Fig. 29.1) according to some of their major shared characteristics: (1) photosynthetic protists, (2) flagellates, (3) ciliates, (4) amoeboids, (5) sporozoans, and (6) water molds and slime molds.

Photosynthetic Protists

Algae are photosynthetic eukaryotes that can be unicellular or form colonies or filaments. The unicellular and colonial forms may have *flagella,* which permit motility. Some types of algae are multicellular seaweeds, such as kelp. Unicellular types can be as small as bacteria (about 1 μm), while multicellular

TABLE 29.1 Eukaryotic Supergroups

Supergroup	Types of Organisms
Archaeplastids	Land plants as well as green and red algae
Chromalveolates	Brown algae, diatoms, ciliates, sporozoans, and water molds
Excavates	Euglenids and certain other flagellates
Amoebozoans	Amoeboids, as well as plasmodial and cellular slime molds
Rhizarians	Foraminiferans and radiolarians
Opisthokonts	Animals, fungi, and certain flagellates

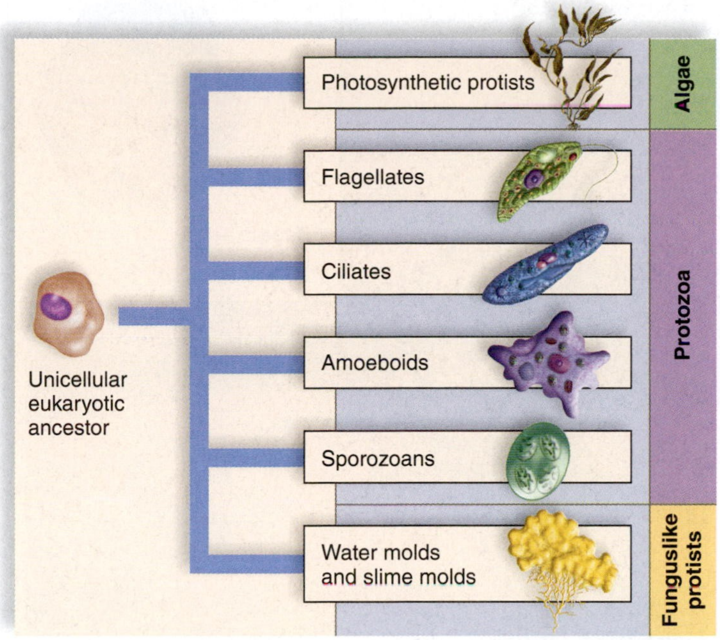
Photosynthetic protists
Flagellates
Ciliates
Amoeboids
Sporozoans
Water molds
and slime molds
Unicellular
eukaryotic
ancestor
Algae
Protozoa
Funguslike
protists

Figure 29.1 Major groups of protists. Because the precise evolutionary relationships between these groups are not yet known, they are grouped here (and discussed in the text) by major shared characteristics.

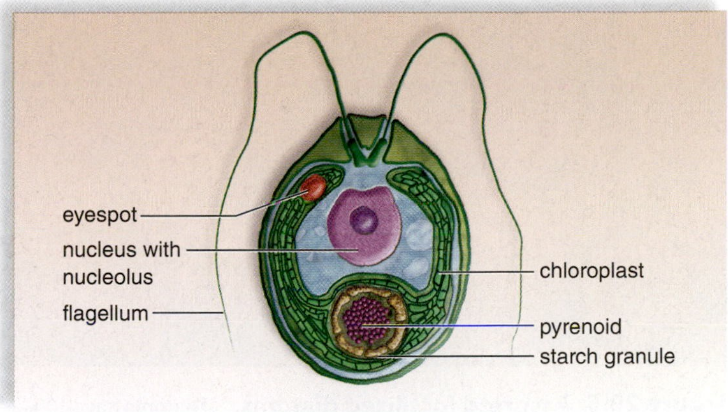
eyespot
nucleus with
nucleolus
flagellum
chloroplast
pyrenoid
starch granule

Figure 29.2 *Chlamydomonas.* *Chlamydomonas* is a motile, unicellular green alga.

Figure 29.3 *Volvox.* *Volvox* is a colony of flagellated cells.

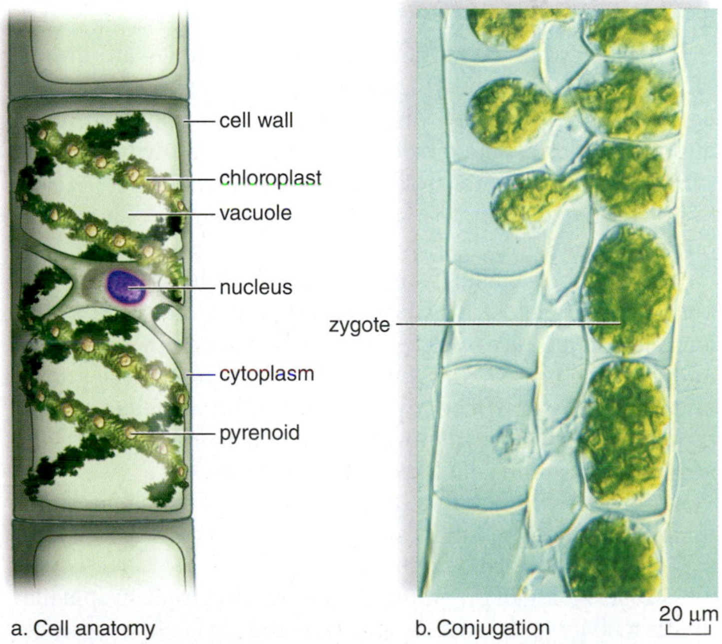
cell wall
chloroplast
vacuole
nucleus
cytoplasm
pyrenoid
zygote
a. Cell anatomy
b. Conjugation
20 µm

Figure 29.4 *Spirogyra.* **a.** *Spirogyra* is a filamentous green alga in which each cell has a ribbonlike chloroplast. **b.** During conjugation, the cell contents of one filament enter the cells of another filament. Zygote formation follows.

forms can grow as large as hundreds of feet. In aqueous environments, small algae are often a component of suspended, photosynthesizing organisms termed **phytoplankton.** Algae are important in terrestrial ecosystems as well, being found in soils, on rocks, and on trees. In addition, algae contribute to the formation of coral reefs and partner with fungi in lichens.

Algae have chloroplasts that contain chlorophyll and sometimes other pigments. Algae perform plantlike photosynthesis. Oxygen is released from water, and electrons energized by the sun are used to reduce carbon dioxide to carbohydrate. Thus, algae are important primary producers of food for heterotrophs and provide oxygen to the atmosphere. *Pyrenoids* are organelles found in algae that are active in starch storage and metabolism.

Algae generally have a rigid cell wall, sometimes made of cellulose. Beyond the cell wall, many algae produce a *slime layer* that can be harvested and used in foods. Some algae also produce oils that can be burned, which has stimulated much interest in algae as an alternative to petroleum fuel.

Green algae are believed to be closely related to the first plants, because both groups have some of the same characteristics: (1) a cell wall that contains cellulose, (2) chlorophylls *a* and *b,* and (3) storage of reserve food as starch inside the chloroplast. The many different types of green algae include those that are unicellular, colonial, or multicellular. *Chlamydomonas* is a unicellular green alga with a cup-shaped chloroplast and two whiplike flagella (Fig. 29.2). *Volvox* is a colony (loose association of cells) in which thousands of flagellated cells are arranged in a single layer surrounding a watery interior (Fig. 29.3). *Spirogyra* is filamentous, consisting of end-to-end chains of cells. A ribbonlike chloroplast is arranged in a spiral within each cell. *Spirogyra* can reproduce by conjugation, a process whereby physical contact between cells allows for the transmission of genetic material (Fig. 29.4). **Diatoms** are the

Figure 29.5 Assorted fossilized diatoms. Diatoms have overlapping shells made of silica. Scientists use their delicate markings to identify the particular species.

most numerous unicellular algae in the oceans, and they are also common in fresh water. In addition to chlorophyll, they utilize a brown pigment to capture solar energy. Diatoms constitute a large fraction of the phytoplankton and are important *primary producers* at the base of marine food chains. Diatoms are easy to recognize under the microscope because they have a wide variety of elaborate shells made of silica (Fig. 29.5). The two overlapping shells have intricately shaped depressions, pores, and passageways that bring the diatom's plasma membrane in contact with the environment. The fossilized remains of diatoms, also called *diatomaceous earth,* accumulate on the ocean floor and can be mined for use as filtering agents and abrasives.

Dinoflagellates are best known for the **red tide** they cause when they greatly increase in number, an event called an algal bloom. *Gonyaulax,* one species implicated in red tides, produces a very potent toxin (Fig. 29.6). This can be harmful by itself, but it also accumulates in shellfish. The shellfish are not damaged, but people can become quite ill when they eat them.

Despite some harmful effects, dinoflagellates are important members of the phytoplankton in marine and freshwater ecosystems. Generally, dinoflagellates are photosynthetic, although colorless heterotrophic forms are known to live as symbionts inside other organisms. Corals, which build coral reefs, contain large numbers of symbiotic dinoflagellates.

Video Good Poison

Many dinoflagellates have protective cellulose plates that become encrusted with silica, turning them into hard shells. Dinoflagellates have two flagella. One is located in a groove that encircles the protist, and the other is in a longitudinal groove and has a free end. The arrangement of the flagella makes a dinoflagellate whirl as it moves. In fact, the name dinoflagellate is derived from the Greek *dino-,* for whirling.

It is interesting to note that luminous dinoflagellates produce a twinkling light that can give seas a phosphorescent glow at night, particularly in the tropics.

Most **red algae** are multicellular, ranging from simple filaments to leafy structures. As *seaweeds,* red algae often resemble the brown algae seaweeds, although red algae can be more delicate (Fig. 29.7).

In addition to chlorophyll, the red algae contain red and blue pigments that give them characteristic colors. Coralline

cellulose plate
flagella

Figure 29.6 Dinoflagellates. Dinoflagellates have cellulose plates. These belong to *Gonyaulax,* a dinoflagellate that contains a reddish-brown pigment and is responsible for occasional "red tides."

Figure 29.7 Red algae. Red algae contain accessory pigments in addition to chlorophyll. Represented here by *Chondrus crispus,* they are smaller and more delicate than brown algae.

algae have calcium carbonate in their cell walls and contribute to the formation of coral reefs.

Red algae produce a number of useful gelling agents. *Agar* is commonly used in microbiology laboratories to solidify culture media and commercially to encapsulate vitamins or drugs. Agar is also a gelatin substitute for vegetarians, an antidrying agent in baked goods, and an additive in cosmetics, jellies, and desserts. *Carrageenan* is a related product used in cosmetics and chocolate manufacture. The species *Porphyra,* a red seaweed, is popular as a sushi wrap in Japan.

Brown algae are the conspicuous multicellular seaweeds that dominate rocky shores along cold and temperate coasts. The color of brown algae is due to accessory pigments that actually range from pale beige to yellow-brown to almost black. These pigments allow the brown algae to extend their range

Figure 29.8 Brown algae. Brown algae have accessory pigments that allow them to obtain their light energy while living in deeper waters compared to other algae. This is a type of brown algae known as bull kelp, *Nereocystis luetkeana*. The photo on the upper right shows flotation bladders that brown algae form to keep their blades close to the surface.

down into deeper waters because the pigments are more efficient than green chlorophyll in absorbing the sunlight away from the ocean surface. The alga produces a slimy matrix that retains water when the tide is out and the seaweed is exposed. This gelatinous material, algin, is used in ice cream, cream cheese, and some cosmetics.

The Sargasso Sea in the North Atlantic Ocean commonly has large floating mats of brown algae called *sargassum.* The most familiar brown algae are kelp (genus *Laminaria*) in coastal regions and giant marine kelp (genus *Macrocystis*) in deeper waters (Fig. 29.8). Tissue differentiation in kelp results in blades, stalks, and holdfasts, which are analogous to the leaves, stems, and roots of plants.

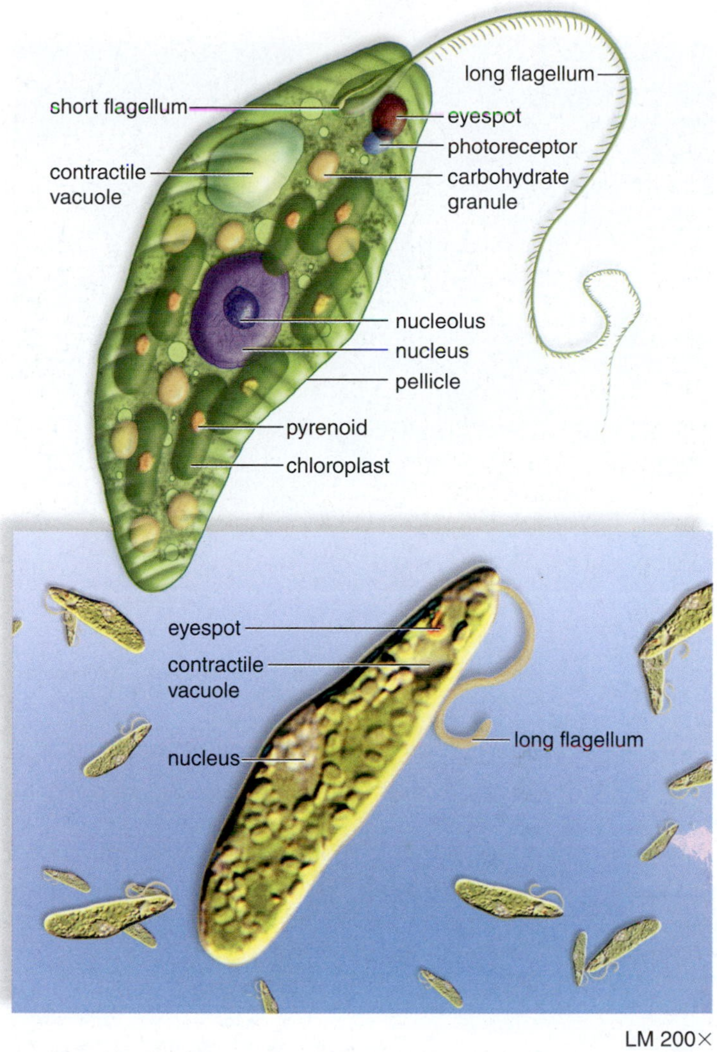

short flagellum
long flagellum
contractile vacuole
eyespot
photoreceptor
carbohydrate granule
nucleolus
nucleus
pellicle
pyrenoid
chloroplast
eyespot
contractile vacuole
nucleus
long flagellum
LM 200×

Figure 29.9 *Euglena*. Euglenoids have a flexible pellicle and a long flagellum that propels the body.

Flagellates

Flagellates are heterotrophic protists that propel themselves using one or more flagella. Most are symbiotic, and many are parasitic. Several of these are discussed in the upcoming section on human diseases caused by protozoans.

Euglenids are freshwater unicellular organisms that typify the problem of classifying protists. Many euglenids have chloroplasts, but some do not. Those that lack chloroplasts ingest or absorb their food. Those that have chloroplasts are believed to have originally acquired them by ingestion and subsequent endosymbiosis of a green algal cell. Three, rather than two, membranes surround these chloroplasts. The outermost membrane is believed to represent the plasma membrane of an original host cell that engulfed a green alga.

Euglenids have two flagella, one of which is typically much longer than the other and projects out of the anterior, vase-shaped invagination (Fig. 29.9). Near the base of this flagellum is an eyespot apparatus, which is a photoreceptive organelle for detecting light. Because euglenids are bounded by a flexible pellicle composed of protein strips lying side by side, they can assume different shapes as the underlying cytoplasm undulates and contracts.

Ciliates

Ciliates are the largest group of protozoans. All of them have cilia, hairlike structures that rhythmically beat, moving the cell forward or in reverse. Typically, the cell rotates as it moves. Cilia also help capture prey and particles and then move them toward the mouthparts. After phagocytic vacuoles engulf food, they combine with lysosomes, which supply the enzymes needed for digestion.

Some ciliates are up to 3 mm long, which is large enough to be seen with the naked eye. Most are freely motile, but some can be anchored like members of the genus *Stentor,* which form a stalk, attaching to a surface and collecting food with cilia (Fig. 29.10*a*). *Paramecium* is the most widely known ciliate, and it is commonly used for research and teaching. It is shaped like a slipper and has visible contractile vacuoles (Fig. 29.10*b*). Members of the genus *Paramecium* have a large *macronucleus* and a small *micronucleus.* The macronucleus produces mRNA and directs metabolic functions. The micronucleus is important during sexual reproduction. *Paramecium* has been important for studying ciliate sexual reproduction, which involves *conjugation,* with interactions between micronuclei and macronuclei (Fig. 29.10*c*).

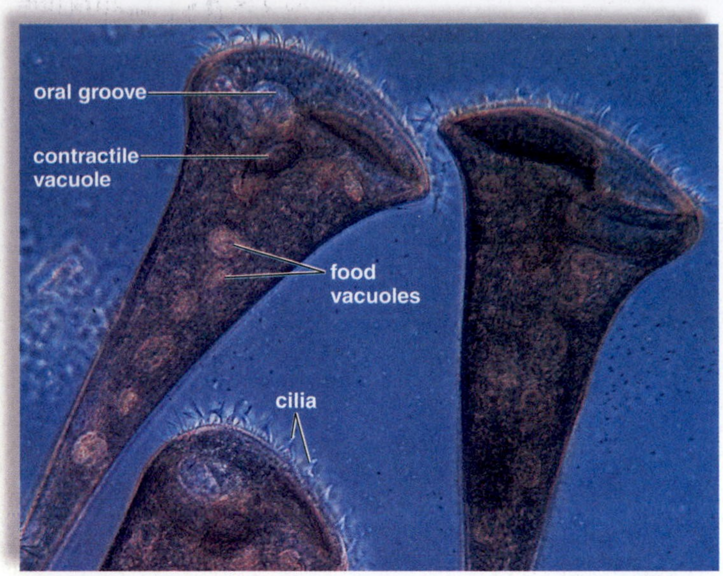
oral groove
contractile vacuole
food vacuoles
cilia
200 µm

a. *Stentor*

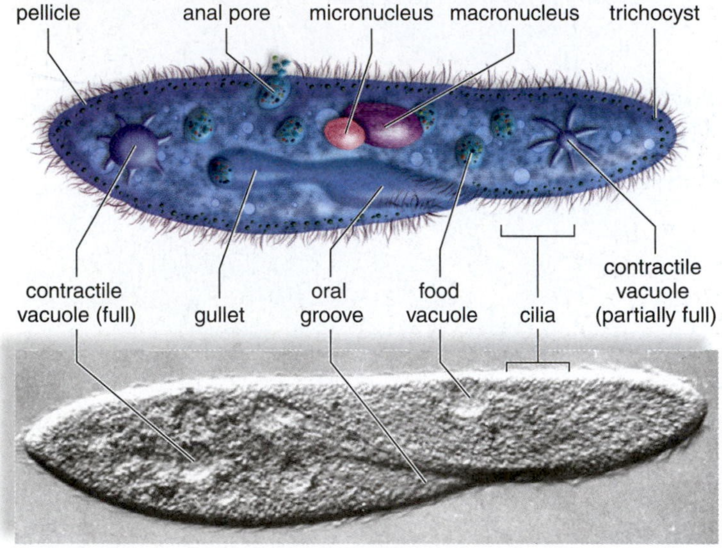
pellicle
anal pore
micronucleus
macronucleus
trichocyst
contractile vacuole (full)
gullet
oral groove
food vacuole
cilia
contractile vacuole (partially full)

b. *Paramecium*

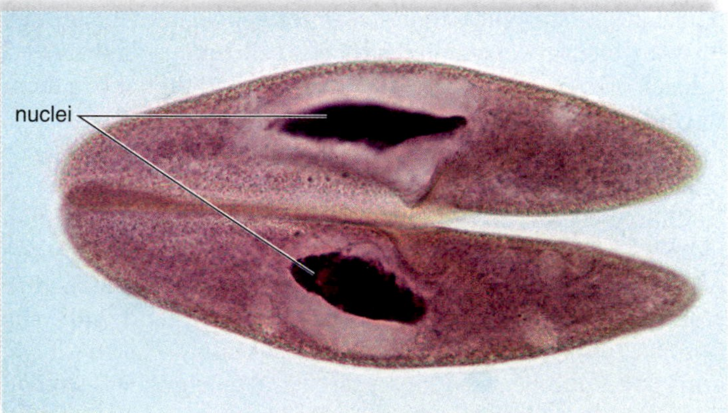
nuclei

c. During conjugation, two paramecia first unite at oral areas.

Figure 29.10 Ciliates. Ciliates are motile by means of cilia. **a.** *Stentor,* a large, vase-shaped, freshwater ciliate. **b.** Structure of *Paramecium,* adjacent to an electron micrograph. Note the oral groove and the gullet and anal pore. **c.** A form of sexual reproduction called conjugation occurs periodically in paramecia.

Amoeboids

Amoeboids move by *pseudopods,* processes that form when cytoplasm streams forward in a particular direction (Fig. 29.11*a*). They usually live in aquatic environments, such as oceans and freshwater lakes and ponds, where they are a part of the zooplankton. When amoeboids feed, their pseudopods surround and engulf their prey, which may be algae, bacteria, or other protists. Digestion then occurs within a food vacuole.

Video Amoeba Locomotion

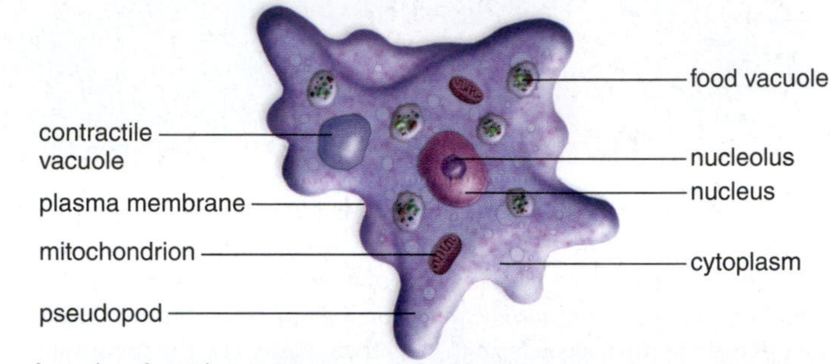
food vacuole
contractile vacuole
nucleolus
nucleus
plasma membrane
mitochondrion
cytoplasm
pseudopod

a. Amoeba, *Amoeba proteus*

25 µm

b. Foraminiferan, *Globigerina,* and the White Cliffs of Dover, England

c. Radiolarian tests SEM 200×

Figure 29.11 Amoeboids. Amoeboids are motile by way of pseudopods. **a.** Structure of *Amoeba proteus,* an amoeboid common in freshwater ponds. **b.** Pseudopods of a live foraminiferan project through holes in the calcium carbonate shell. Fossilized shells were so numerous they became a large part of the White Cliffs of Dover when a geologic upheaval occurred. **c.** Skeletal tests of radiolarians.

Foraminiferans and radiolarians are related amoeboid groups that have an external skeleton (test). In **foraminiferans,** pseudopods extend through openings in the test, which covers the plasma membrane. Because each geological time period has a distinctive form of foraminiferan, they can be used to date sedimentary rock. Depositions over millions of years followed by geological upheaval formed the White Cliffs of Dover along the southern coast of England (Fig. 29.11*b*). Also, the great Egyptian pyramids are built of foraminiferous limestone. In radiolarians, the test is internal (Fig. 29.11*c*). The tests of dead foraminiferans and radiolarians form a deep layer of sediment on the ocean floor. Their presence is used as an indicator of oil deposits on land or sea.

Sporozoans

Members of the phylum Apicomplexa are commonly called **sporozoans** because they produce spores. All phases of the life cycle are generally nonmotile, with the exception of male gametes and zygotes. All sporozoans are either intercellular or extracellular parasites. An apical complex composed of fibrils, microtubules, and organelles is found at one end of the single cell. Enzymes for attacking host cells are secreted through a pore at the end of the apical complex. Specific diseases of humans caused by sporozoans are discussed in the next section.

Water Molds and Slime Molds

We tend to think of molds as fungi, but the water molds and slime molds are classified as protists.

Water Molds Most **water molds** are **saprotrophs,** meaning that they feed on dead organic matter. They usually live in water, where they decompose remains and form furry growths when they parasitize fish (Fig. 29.12). In spite of their common name, some water molds live on land and parasitize insects and plants. The water mold *Phytophthora* was responsible for the 1840s potato famine in Ireland.

Water molds have a filamentous body, as do fungi, but the cell walls of water molds are largely composed of cellulose, whereas fungi have cell walls made of chitin. During asexual reproduction, water molds produce flagellated spores. During sexual reproduction, they produce eggs and sperm. Their phylum name, Oomycota, refers to the enlarged tips (oogonia), where eggs are produced.

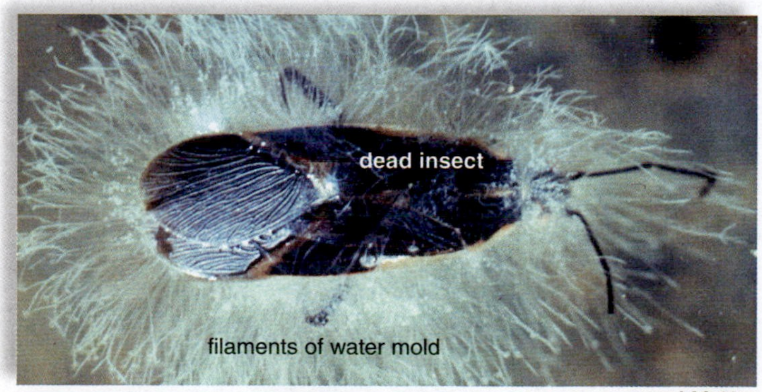
dead insect
filaments of water mold
LM 10×

Figure 29.12 Water mold. *Saprolegnia,* a water mold, feeding on a dead insect. Water molds are not fungi.

Slime Molds In forests and woodlands, slime molds feed on dead plant material. They also feed on bacteria, keeping their population under control. The vegetative cells of slime molds are mobile and amoeboid. They ingest their food by phagocytosis.

Usually, **plasmodial slime molds** exist as a *plasmodium,* a diploid, multinucleated, cytoplasmic mass enveloped by a slime sheath, that creeps along, phagocytizing decaying plant material in a forest or agricultural field (Fig. 29.13). At times

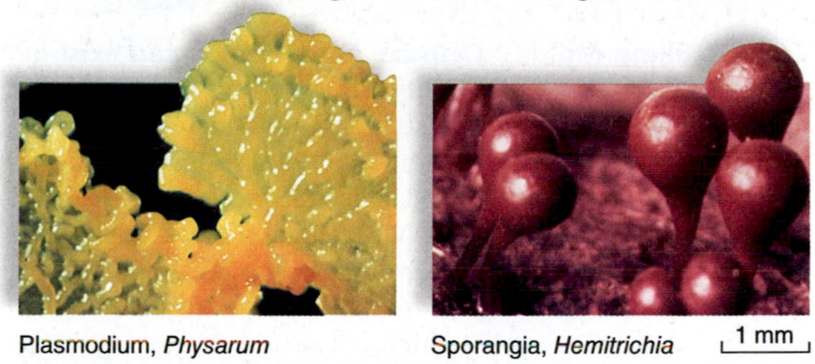
Plasmodium, *Physarum*
Sporangia, *Hemitrichia*
1 mm

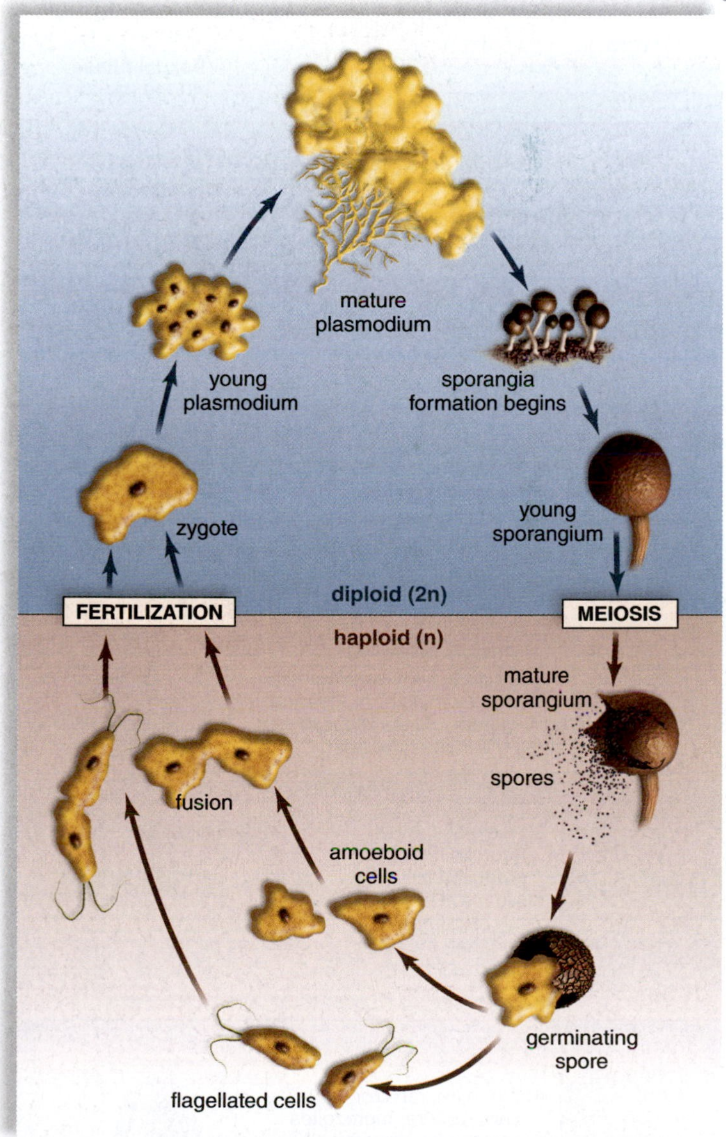
mature plasmodium
young plasmodium
sporangia formation begins
young sporangium
zygote
diploid (2n)
FERTILIZATION
MEIOSIS
haploid (n)
mature sporangium
spores
fusion
amoeboid cells
germinating spore
flagellated cells

Figure 29.13 Plasmodial slime molds. The diploid adult forms sporangia during sexual reproduction, when conditions are unfavorable to growth. Haploid spores germinate, releasing haploid amoeboid or flagellated cells that fuse.

unfavorable to growth, such as during a drought, the plasmodium develops many *sporangia,* reproductive structures that produce spores. The spores produced by a sporangium can survive until moisture is sufficient for them to germinate. In plasmodial slime molds, spores release a haploid flagellated cell or an amoeboid cell. Eventually, two of these fuse to form a zygote that feeds and grows, producing a multinucleated plasmodium once again.

Cellular Slime Molds **Cellular slime molds** can exist as individual amoeboid cells. They are common in soil, where they feed on bacteria and yeasts. *Dictyostelium discoideum* is the commonly researched example. As food supplies dwindle, the cells release a chemical that causes them to aggregate into a sluglike *pseudoplasmodium,* which eventually gives rise to a fruiting body that produces spores. When favorable conditions return, the spores germinate, releasing haploid amoeboid cells, and this cycle, which is asexual, begins again. A sexual cycle is known to occur under very moist conditions.

Genetic analyses of several slime mold species has revealed that these may have been some of the earliest eukaryotes to adapt to life on land, perhaps a billion years ago. Recent research has also shown that slime mold behavior can be surprisingly complex. Japanese researchers showed that when placed into a "maze" constructed on an agar plate, slime molds could find the shortest path to food. When presented with different types of food, the slime molds chose the most nutritious type.

Protozoal Diseases of Humans

Compared to the number of diseases caused by bacteria and viruses, a relatively few protists cause human diseases, and all of these are caused by heterotrophic protozoans. However, some cause significant sickness and mortality.

Malaria

The most widespread and dangerous protozoan disease is **malaria.** According to the World Health Organization (WHO), about half of the world's population lives in areas where malaria is endemic. In 2010, an estimated 216 million people were infected with malaria worldwide, and 655,000 of these people died. As significant as those numbers are, they are actually decreasing, at least in part due to the concerted efforts of many governments and private organizations.

Malaria can be caused by several sporozoan parasites in the genus *Plasmodium.* All of these have a complex life cycle that involves transmission by a mosquito vector (Fig. 29.14).

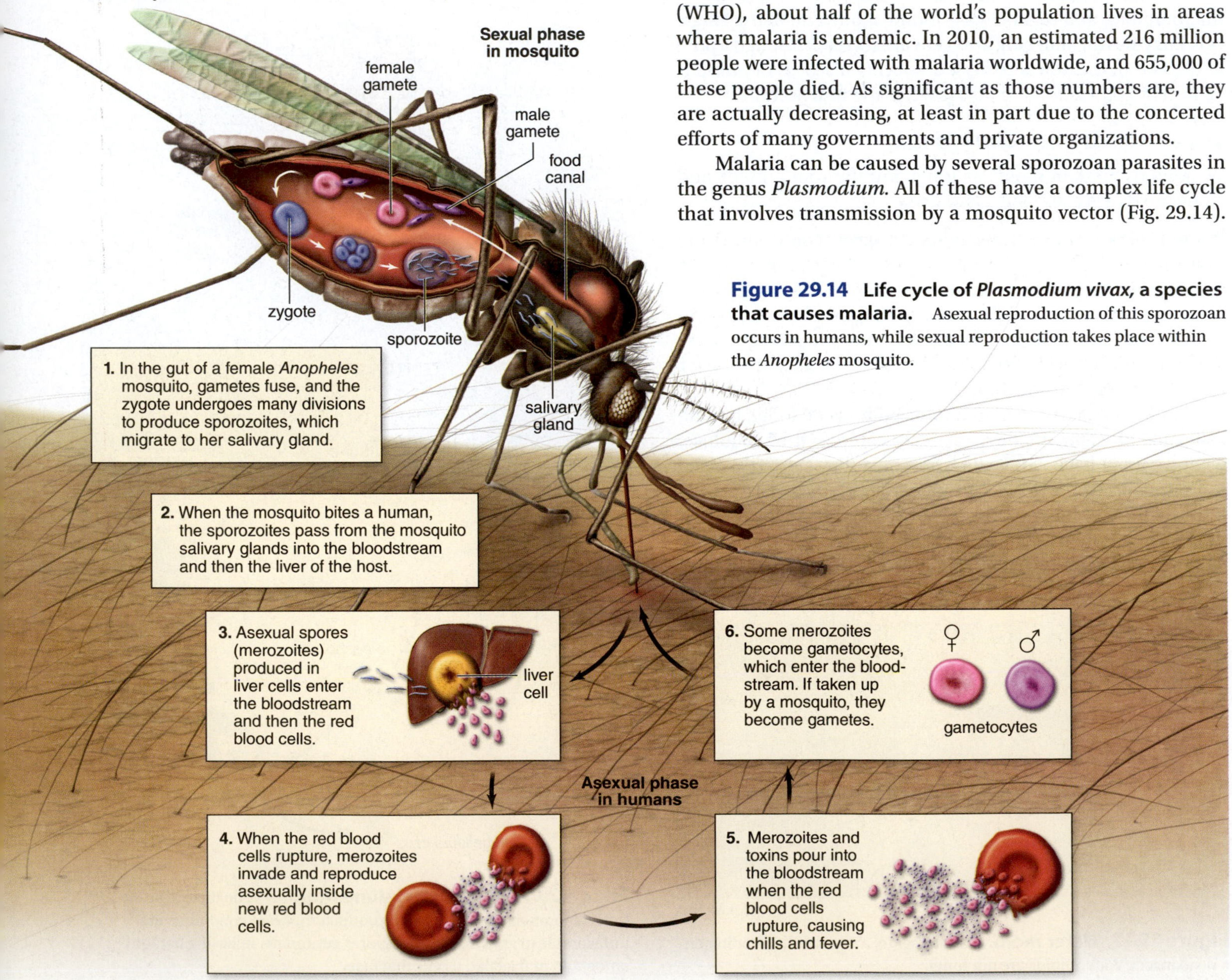

Sexual phase in mosquito
female gamete
male gamete
food canal
zygote
sporozoite
salivary gland
1. In the gut of a female *Anopheles* mosquito, gametes fuse, and the zygote undergoes many divisions to produce sporozoites, which migrate to her salivary gland.
2. When the mosquito bites a human, the sporozoites pass from the mosquito salivary glands into the bloodstream and then the liver of the host.
3. Asexual spores (merozoites) produced in liver cells enter the bloodstream and then the red blood cells.
liver cell
4. When the red blood cells rupture, merozoites invade and reproduce asexually inside new red blood cells.
Asexual phase in humans
5. Merozoites and toxins pour into the bloodstream when the red blood cells rupture, causing chills and fever.
6. Some merozoites become gametocytes, which enter the bloodstream. If taken up by a mosquito, they become gametes.
♀ ♂
gametocytes

Figure 29.14 Life cycle of *Plasmodium vivax,* a species that causes malaria. Asexual reproduction of this sporozoan occurs in humans, while sexual reproduction takes place within the *Anopheles* mosquito.

The sexual reproduction phase in the mosquito ends with the migration of a developmental stage called sporozoites into the salivary glands. These sporozoites are then transmitted by the mosquito's bite to a human, where they reproduce asexually in the liver and red blood cells, forming merozoites. The infected red blood cells frequently rupture, releasing merozoites and toxins, and causing the person to experience the chills and fever that are characteristic of malaria. Some of these merozoites become gametocytes, which, if ingested by another mosquito, can start the sexual reproduction phase again. People who survive malaria have only a limited immunity to reinfection. Interestingly, the sickle cell trait found in up to one-third of sub-Saharan Africans provides some protection against malaria. Efforts to control mosquito populations have been successful in the United States, where malaria is rarely seen. The control of mosquitoes worldwide has now been hampered by the rise of resistance to pesticides like DDT. Scientists are working to find new ways to inhibit mosquitoes, such as genetically engineering them to be resistant to *Plasmodium* infection. A Seattle company is even designing a laser system that could protect villages by zapping mosquitoes in flight!

An intense research effort is under way to develop a malaria vaccine, funded largely by the Bill and Melinda Gates Foundation, which has contributed about $1.2 billion to malaria research since 2000. A 2011 study published in the *New England Journal of Medicine* reported the results of a malaria vaccine trial conducted in seven countries and involving 15,460 children. Overall, the vaccine reduced the incidence of malaria in children by approximately 50% during the 12 months after vaccination.

Other Sporozoan Diseases

The sporozoan called *Toxoplasma gondii* is commonly transmitted by cat feces. The disease toxoplasmosis generally causes no appreciable symptoms, but the parasite can be harmful to a developing fetus. Thus, pregnant women are advised not to empty cat litter boxes and to avoid working in gardens or other locations where cats may defecate. A related protozoal disease is caused by *Cryptosporidium parvum.* The organism and its cysts are common in surface waters and in the feces of animals and birds, and can pass through sand filters in water treatment plants, while being unaffected by chlorine treatment. Cryptosporidiosis usually causes a self-limiting gastroenteritis, but in some cases can lead to a fatal watery diarrhea.

Diseases Caused by Flagellates

Trypanosoma brucei, the cause of **African sleeping sickness,** is transmitted by the tsetse fly (see the Health feature, "African Sleeping Sickness," on page 592). It attacks the patient's blood, causing inflammation that decreases oxygen flow to the brain. Chagas disease, caused by *T. cruzi,* is transmitted by the kissing bug, so called because it tends to bite lips. *Giardia lamblia* is well known as a human pathogen that can cause epidemics of diarrhea, known as giardiasis. The parasite is unusual in that

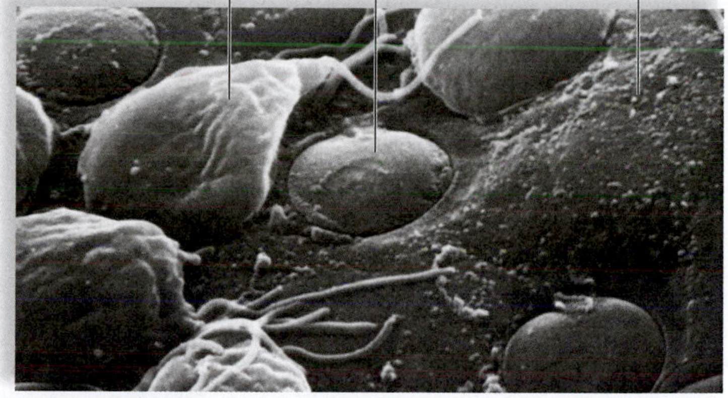

Giardia
circular marking
surface

Figure 29.15 ***Giardia lamblia.*** This protist adheres to any surface, including epithelial cells, by means of a sucking disk. Characteristic markings can be seen on the host cell after the disk detaches.

it has two nuclei and two sets of flagella (Fig. 29.15). *Giardia* is common in environments contaminated with fecal materials, such as day-care centers, and can be found in natural surface waters. Finally, a very common sexually transmitted disease is caused by *Trichomonas vaginalis,* a flagellate that mainly causes vaginitis in women.

Diseases Caused by Amoeboids

Parasitic amoeboids in the genus *Entamoeba* cause **amoebic dysentery.** Complications arise when the parasite invades the intestinal lining and reproduces there. If the parasite enters the body proper, liver and brain involvement can be fatal. As noted in the chapter-opening story, an amoeboid in the genus *Acanthamoeba* can cause corneal inflammation as well as serious infections. Even more foreboding, the amoeboid protist *Naegleria fowleri* can invade and attack the human nervous system, nearly always resulting in the death of the victim. It is usually acquired by swimming in warm bodies of fresh water, such as ponds, lakes, rivers, and unchlorinated swimming pools. Fortunately, this scenario is extremely rare.

Check Your Progress 29.1

1. Describe the endosymbiotic theory.
2. Describe the circumstances by which some protists produce spores or cysts, and explain the benefit this provides to the organism.
3. Compare the type of nutrition that is typically required by an alga, a protozoan, a water mold, and a cellular slime mold.
4. Identify the type of protist that causes each of the following diseases: malaria, African sleeping sickness, and amoebic dysentery.

SCIENCE IN YOUR LIFE ▶ HEALTH

African Sleeping Sickness

The World Health Organization (WHO) recognizes several neglected tropical diseases (NTDs) that affect more than 1 billion people exclusively in the most impoverished communities. Several of these NTDs, including African sleeping sickness, are caused by parasitic protists that are transmitted to humans from the bite of an infected insect.

African sleeping sickness, also called African human trypanosomiasis, is the deadliest of all of the NTDs. It is caused by a type of parasitic, single-celled protist, *Trypanosoma brucei,* which is transmitted to humans in the bite of the blood sucking tsetse fly (*Glossina*). The tsetse fly is the transmission vector of the disease because it carries the trypanosomes and transmits the long, blade-shaped protists (Fig. 29A) into the human bloodstream while feeding. The flies first become infected with the parasite when they bite either an infected human or other mammal, such as domestic cattle and pigs, which carry *T. brucei.*

Figure 29B illustrates the relationship between humans, the tsetse fly, and domestic livestock. Unlike humans, livestock can carry *T. brucei* without getting sick. Therefore, they act as a reservoir for the disease. Humans also can serve as a reservoir for new tsetse fly infections because the parasite can be picked up from humans by a fly bite and transmitted to others.

It is estimated that as many as 50,000 people are plagued by this disease. In the later stages of infection, *T. brucei* attacks the brain, causing behavioral changes and a shift in sleeping patterns. If caught early enough, it can be cured with medication, but because it is most common in very poor nations, medication and treatment are difficult to come by. Without treatment, the disease is fatal.

The tsetse fly prefers shaded, cool, moist habitats, such as the sandy edges of pools and rivers in central Africa. Regions that are plagued by sleeping sickness have no running water, so daily trips to local rivers and watering holes are necessary to collect water. The people who live and work in tsetse habitat are the most vulnerable to the disease. As a result, many communities have abandoned the areas infested with the tsetse fly. The abandonment of these fertile valleys has had a huge impact on the ecology of surrounding areas and the economy of communities and nations on a large scale.

Recently, an effort to eliminate the tsetse fly from these lands has been successful. The WHO reports that the number of new cases of sleeping sickness has dropped to the lowest levels in 50 years. The treatment and control of sleeping sickness has allowed the resettlement of more than 25 million hectares of prime agricultural land.

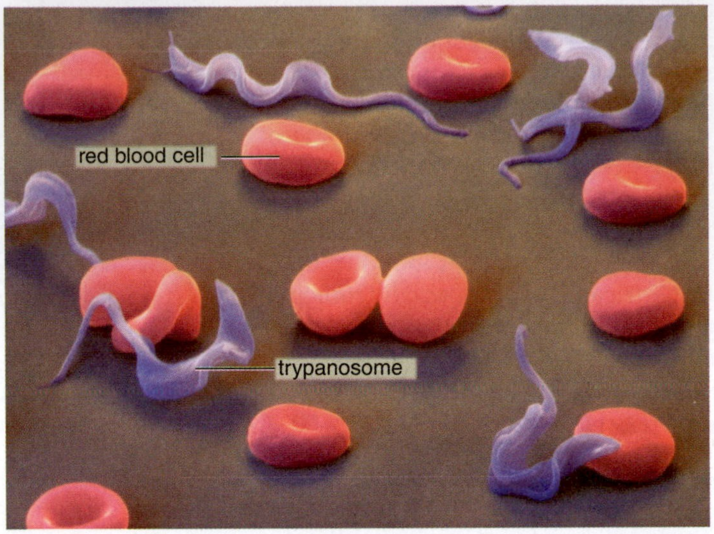
red blood cell
trypanosome

Figure 29A ***Trypanosoma brucei.*** A micrograph showing *T. brucei,* a causal agent of African sleeping sickness, among red blood cells.

Questions to Consider

1. How does poverty reinforce a high occurrence of African sleeping sickness?
2. How are the economy, ecology, and disease biology of African sleeping sickness interdependent?

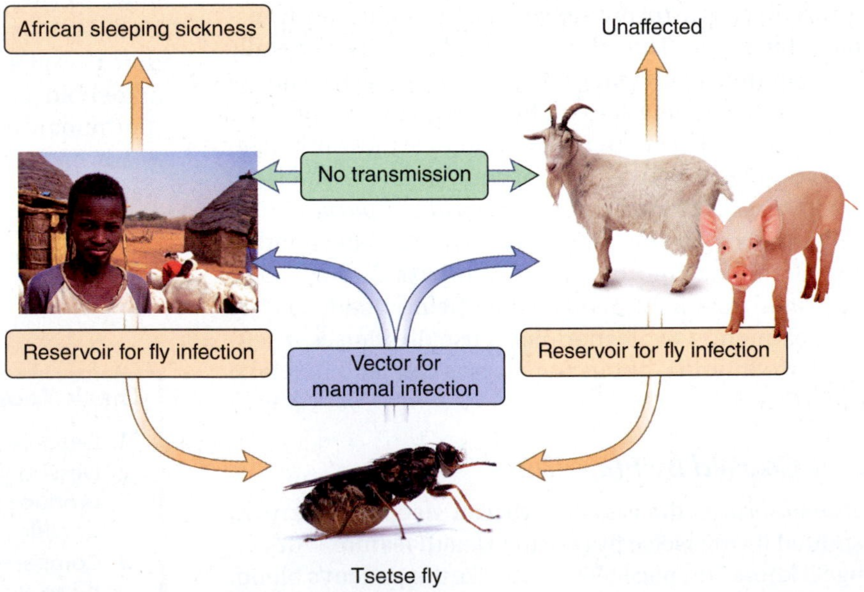
African sleeping sickness
Unaffected
No transmission
Reservoir for fly infection
Vector for mammal infection
Reservoir for fly infection
Tsetse fly

Figure 29B The transmission pattern of *Trypanosoma brucei.* Tsetse flies can become infected with *T. brucei* by feeding on infected humans or livestock. The flies can then transmit the parasitic protist to other humans, causing African sleeping sickness.

29.2 Fungi

Learning Outcomes

Upon completion of this section, you should be able to

1. Explain how most fungi obtain their nutrients.
2. Describe the structural components of a fungal body.
3. Compare the reproductive strategies of the major groups of fungi.
4. Identify several fungal diseases of humans, and explain why many can be difficult to treat.

Fungi (domain Eukarya, kingdom Fungi) are a structurally diverse group of eukaryotes that are strict heterotrophs. Unlike animals, fungi (sing., fungus) release digestive enzymes into their external environment and digest their food outside the body, while animals ingest their food and digest it internally. A few fungi are parasitic, but most are saprotrophs that decompose dead plants, animals, and microbes. Along with bacteria, fungi play an important role in ecosystems by breaking down complex organic molecules and returning inorganic nutrients to producers of food—that is, photosynthesizers. Fungi can degrade even cellulose and lignin in the woody parts of trees. It is common to see fungi (brown rot or white rot) on the trunks of fallen trees. The body of a fungus can become large enough to cover acres of land.

Biology of Fungi

The body of a fungus is composed of a mass of individual filaments called **hyphae** (sing., **hypha**). Collectively, the mass of filaments is called a **mycelium** (pl., mycelia) (Fig. 29.16). Some fungi have cross walls that divide a hypha into a chain of cells. These hyphae are termed *septate.* Septa have pores that allow cytoplasm and even organelles to pass from one cell to the other along the length of the hypha. *Nonseptate* fungi have no cross walls, and their hyphae are multinucleated.

Hyphae give the mycelium quite a large surface area per volume of cytoplasm, which facilitates absorption of nutrients into the body of a fungus. Hyphae grow at their tips, and the mycelium absorbs and then passes nutrients on to the growing tips. Individual hyphae can grow quite rapidly, as much as 18 ft per day. Altogether, a single mycelium may add as much as a kilometer of new hyphae in a single day!

Fungal cells are quite different from plant cells, not only because they lack chloroplasts, but also because their cell walls contain chitin, not cellulose. **Chitin,** like cellulose, is a polymer of glucose, but in chitin, each glucose molecule has a nitrogen-containing amino group attached to it. Chitin is also the major structural component of the exoskeleton of arthropods, such as insects, lobsters, and crabs. Unlike plants, the energy reserve of fungi is not starch, but glycogen, as in animals. Fungi are nonmotile and most do not have flagella at any stage in their life cycle. They move toward a food source by growing toward it.

Although a few fungal species are aquatic, most have adapted to life on land by producing windblown spores during both asexual and sexual reproduction (Fig. 29.17). A spore is a haploid reproductive cell that develops into a new organism without the need to fuse with another reproductive cell. In fungi, spores germinate into new mycelia. Sexual reproduction in fungi involves conjugation of hyphae from two different mating types (usually designated + and –). Often, the haploid nuclei from the two hyphae do not immediately fuse to form a zygote. The hyphae contain + and – nuclei for long periods of time. Eventually, the nuclei fuse to form a zygote that undergoes meiosis, followed by spore formation.

Video Spore Dispersal

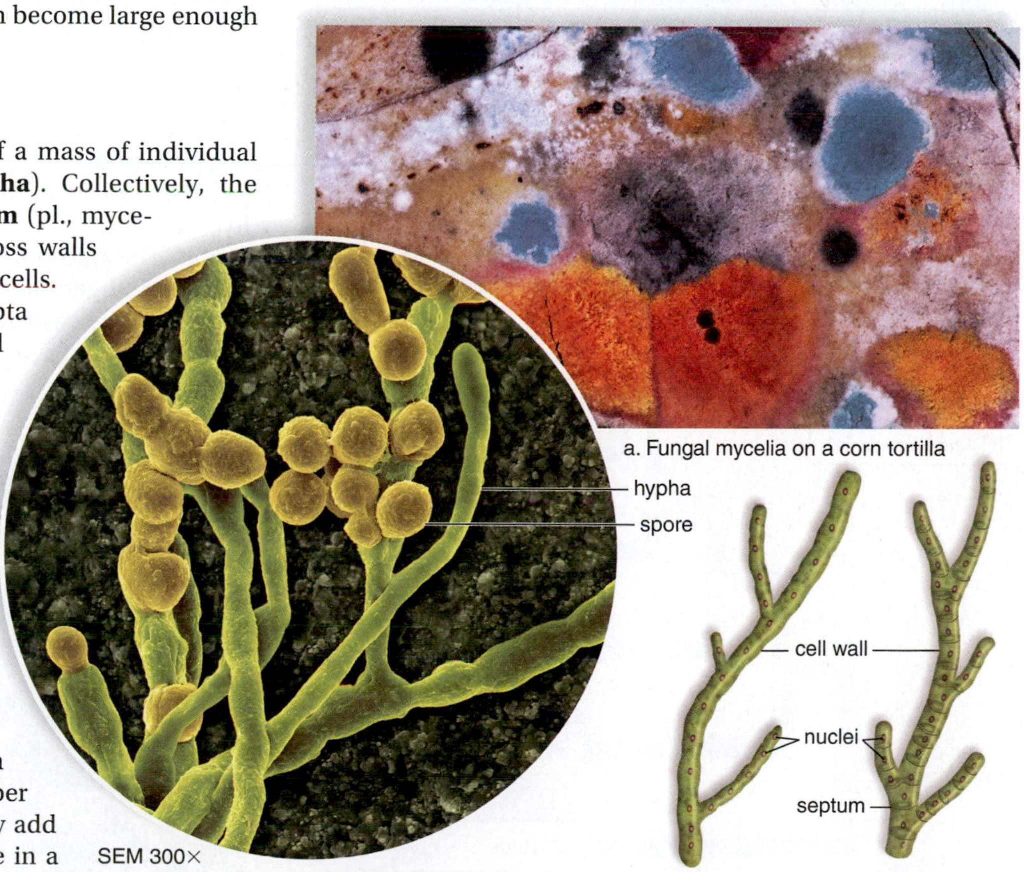
hypha
spore
SEM 300×
cell wall
nuclei
septum

a. Fungal mycelia on a corn tortilla
b. Specialized fungal hyphae that bear spores
c. Nonseptate hypha
d. Septate hypha

Figure 29.16 Fungal mycelia and hyphae. **a.** A colorful variety of fungal mycelia growing on a corn tortilla. **b.** Scanning electron micrograph of specialized aerial fungal hyphae that bear spores. Hyphae are either (**c**) nonseptate (do not have cross walls) or (**d**) septate (have cross walls).

b.
3,000×
a.

Figure 29.17 Fungi reproduce by spore formation. **a.** Here a pear-shaped puffball, *Lycoperdon pyriforme,* is releasing perhaps billions of spores. **b.** The spores of a puffball fungus as viewed under an electron microscope.

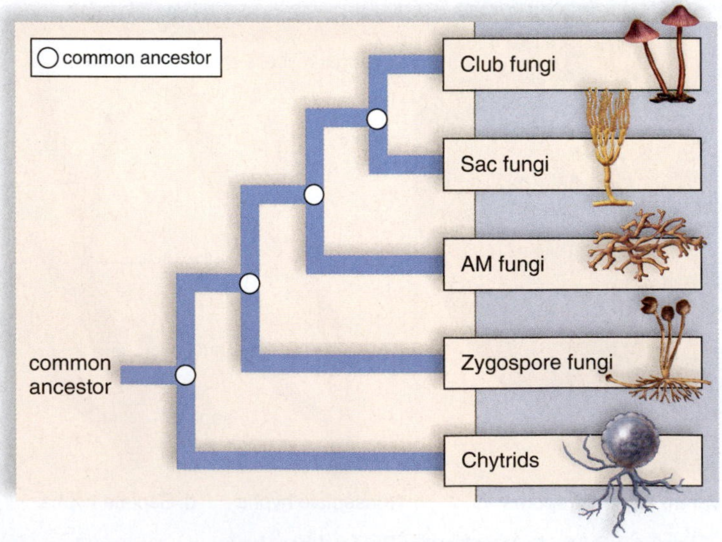
common ancestor
Club fungi
Sac fungi
AM fungi
Zygospore fungi
Chytrids
common ancestor

Figure 29.18 Evolutionary relationships among the fungi. The common ancestor for the fungi was a flagellated saprotroph with chitin in the cell walls. All but chytrids lost the flagella at some point. They also became multicellular.

Diversity of Fungi

Fungi are traditionally classified based on their mode of sexual reproduction. Figure 29.18 is an evolutionary tree showing how the five major groups of fungi that will be discussed in this section are believed to be related. Major fungal groups (phyla) include the chytrids, zygospore fungi, sac fungi, club fungi, and AM fungi.

Chytrid Fungi

The **chytrid fungi** (phylum Chytridomycota) may have been the first type of fungi to evolve. They are unique among the fungi because (1) they are aquatic, and (2) they produce flagellated reproductive cells. Most chytrids reproduce asexually through the production of zoospores, which grow into new chytrids. Many chytrids play a role in the decay and digestion of dead aquatic organisms, but some are parasitic on plants, animals, and protists (Fig. 29.19).

Zygospore Fungi

The **zygospore fungi** (phylum Zygomycota) are mainly saprotrophs, but some are parasites of small soil protists or worms and even insects, such as the housefly. *Rhizopus stolonifer* (Fig. 29.20) is well known to many of us as the mold that appears on old bread, even when it has been refrigerated. In *Rhizopus,* the hyphae are specialized: some are horizontal and exist on the surface of the bread; others grow into the bread to anchor the mycelium and carry out digestion; and still others are stalks that bear sporangia. A **sporangium** is a capsule that produces spores.

When *Rhizopus* reproduces sexually, the ends of + and − hyphae join, haploid nuclei fuse, and a thick-walled **zygospore** results. The zygospore undergoes a period of dormancy before meiosis and germination take place. Following germination, aerial hyphae, with sporangia at their tips, produce many spores. The spores are dispersed by air currents and give rise to new mycelia.

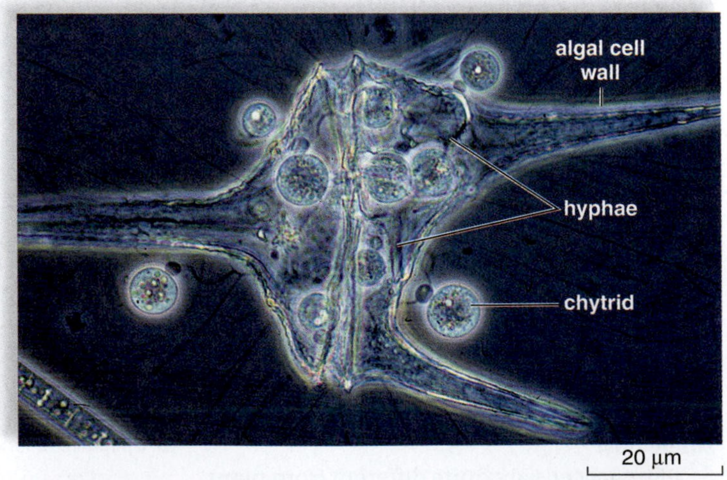
algal cell wall
hyphae
chytrid
20 µm

Figure 29.19 Chytrids parasitizing a protist. These aquatic chytrids *(Chytriomyces hyalinus)* have penetrated the cell walls of this dinoflagellate and are absorbing nutrients meant for their host. They will produce flagellated zoospores that will go on to parasitize other protists.

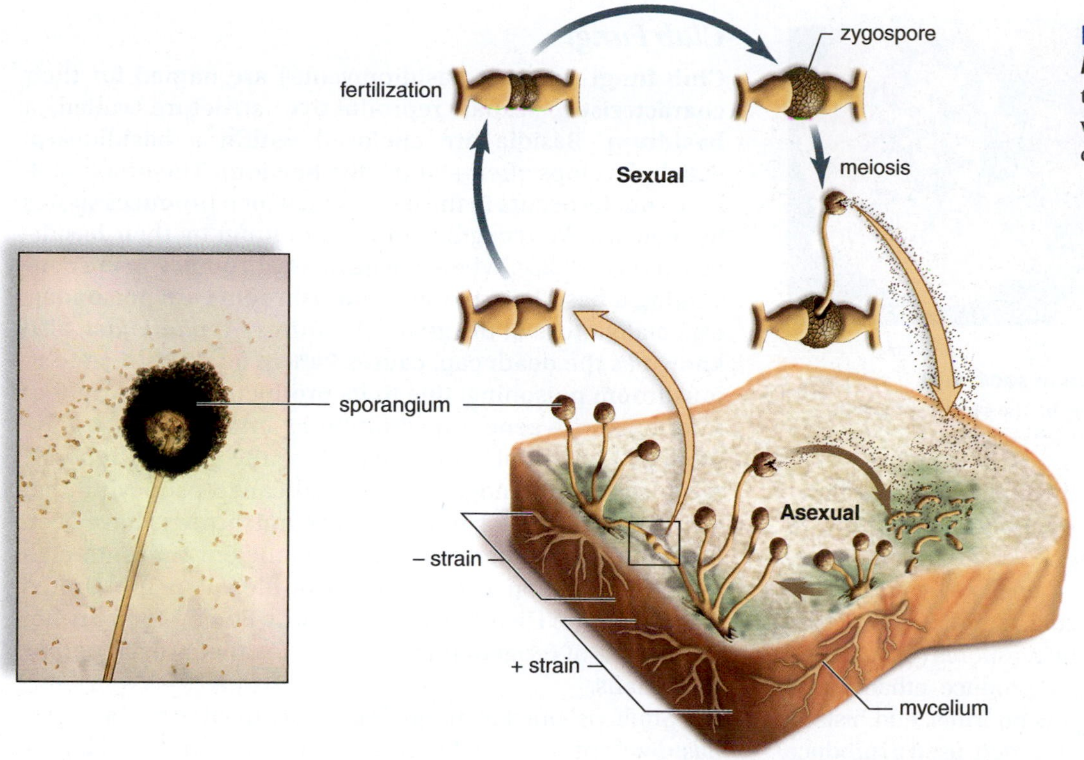
zygospore
fertilization
Sexual
meiosis
sporangium
Asexual
– strain
+ strain
mycelium

Figure 29.20 Black bread mold, *Rhizopus stolonifer.* The mycelium of this mold utilizes sporangia to produce windblown spores. A zygospore forms during the sexual life cycle.

Sac Fungi

Approximately 75% of all known fungi are **sac fungi** (phylum Ascomycota), named for their characteristic cuplike sexual reproductive structure called an **ascocarp.** Many sac fungi reproduce by producing chains of sexual spores called **conidia** (sing., conidium). Cup fungi, morels, and truffles have conspicuous ascocarps (Fig. 29.21). Truffles, underground symbionts of hazelnut and oak trees, are highly prized as gourmet delights. Pigs and dogs are trained to sniff out truffles in the woods, but they are also cultivated on the roots of seedlings.

Morels (Fig. 29.21*c*) are often collected by chefs and other devotees, because of their unique and delicate flavor. Collectors must learn, however, to avoid "false morels," which can be poisonous. Interestingly, even true morels should not be eaten raw. As saprophytes, these organisms secrete digestive enzymes capable of digesting wood or leaves. If consumed prior to being inactivated by cooking, these enzymes can cause a digestive upset.

Most fungal plant pathogens are sac fungi. Examples of these pathogenic fungi include powdery mildews that grow on plant leaves, as well as chestnut blight and Dutch elm disease that destroy trees. *Ergot,* a parasitic sac fungus that infects rye, produces hallucinogenic compounds similar to LSD. Psychoses resulting from ingesting grain contaminated with rye ergot may have led to the Salem, Massachusetts, witch trials in 1692.

Some of the most familiar fungi were formerly considered "imperfect fungi," because their means of sexual reproduction is unknown. However, based on DNA sequencing and other recent information, these fungi are now classified as sac fungi. Examples include *Penicillium,* the original source of penicillin, a breakthrough antibiotic that led to the important class of "cillin" antibiotics that have saved millions of lives. Other species of *Penicillium—P. roquefortii* and *P. camemberti*—are necessary to the production of blue cheeses. *Aspergillus* is used widely for its ability to produce citric acid and various enzymes. Certain species also cause serious infections in people with compromised immune systems, as discussed later in this chapter.

ascospores
mature ascus
a. Ascocarp
ascocarp
b. Cup fungi
ascocarp
c. Morels

Figure 29.21 Sexual reproduction in sac fungi. Some sac fungi are known to us by their ascocarp, the structure that produces spores as a part of sexual reproduction. **a.** Diagram of an ascocarp containing asci, where spores are produced. **b.** The ascocarps of cup fungi. **c.** The ascocarp of a morel, which is a prized delicacy.

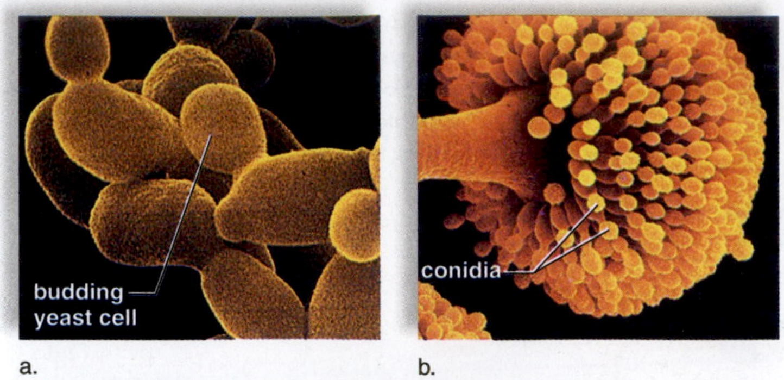
budding
yeast cell
conidia
a.
b.

Figure 29.22 Asexual reproduction in sac fungi. **a.** Yeasts, unique among fungi, reproduce by budding. **b.** The sac fungi usually reproduce asexually by producing spores called conidia.

Yeasts The term **yeasts** is generally applied to unicellular fungi, and many of these organisms are sac fungi. *Saccharomyces cerevisiae,* brewer's yeast, is representative of budding yeasts. When unequal binary fission occurs, a small cell gets pinched off and then grows to full size. Asexual reproduction occurs when the food supply runs out, producing spores (Fig. 29.22).

When some yeasts ferment, they produce ethanol and carbon dioxide. In the wild, yeasts grow on fruits, and historically, the yeasts already present on grapes were used to produce wine. Today, selected yeasts are added to relatively sterile grape juice in order to make wine. Also, yeasts are added to grains to make beer and liquor. Both the ethanol and carbon dioxide are retained in beers and sparkling wines, while carbon dioxide is released from wines. In breadmaking, the carbon dioxide produced by yeasts causes the dough to rise, and the ethanol quickly evaporates. The gas pockets are preserved as the bread bakes.

Club Fungi

Club fungi (phylum Basidiomycota) are named for their characteristic sexual reproductive structure called a **basidium.** Basidia are enclosed within a basidiocarp, which develops after + and – hyphae join. The union of + and – nuclei occurs in the basidium, which produces spores by meiosis. We recognize most club fungi by their basidiocarp (Fig. 29.23). When you eat a mushroom, you are consuming a basidiocarp. Certain mushrooms are poisonous, especially those of the genus *Amanitas*. *A. phalloides,* also known as the death cap, causes 90% of fatalities related to mushroom poisoning, due to its production of a toxin that interferes with gene transcription by inhibiting RNA polymerase. Other club fungi, mainly of the genus *Psilocybe,* produce a hallucinogenic chemical called psilocybin that is a structural analog of LSD. These mushrooms have been used in religious ceremonies since ancient times, though their recreational use is currently illegal in the United States. See the Health feature, "Deadly Fungi," for a further discussion of fungi that produce toxins that are dangerous to humans.

Shelf or bracket fungi found on dead trees are also basidiocarps. Less well known are puffballs and stinkhorns. In *puffballs,* spores are produced inside parchmentlike membranes, and the spores are released through a pore or when the membrane breaks down. *Stinkhorns* resemble a mushroom, but they emit a very disagreeable odor. Flies are attracted by the odor. When they linger to feed on the sweet jelly, they pick up spores and later distribute them.

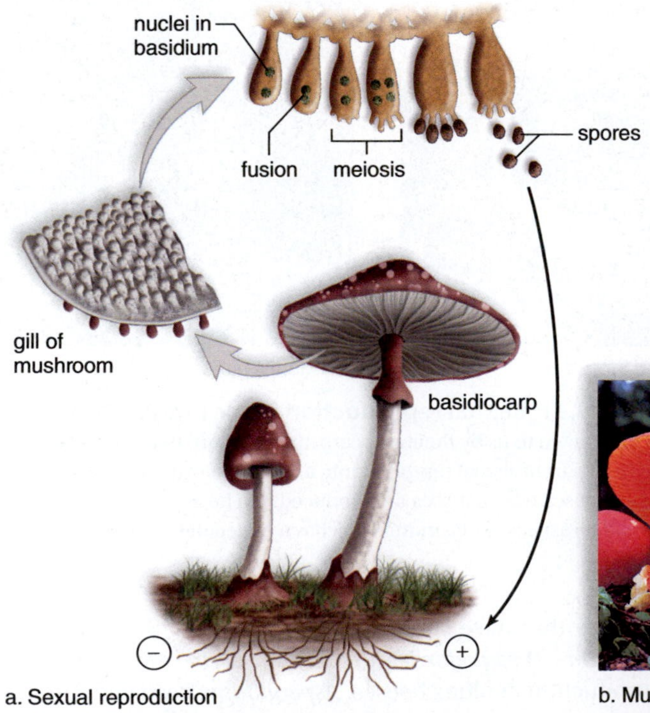
nuclei in
basidium
fusion
meiosis
spores
gill of
mushroom
basidiocarp
−
+
a. Sexual reproduction

b. Mushroom
c. Shelf fungi
d. Giant puffball

Figure 29.23 Sexual reproduction in club fungi. Most club fungi are known to us by their basidiocarp, the structure that produces spores as a part of sexual reproduction. **a.** Sexual life cycle of a basidiocarp and a basidium (pl., basidia) where spores are produced. Also shown are basidiocarps of (**b**) scarlet hood mushroom, (**c**) chicken-of-the-woods shelf fungi, and (**d**) a giant puffball. Puffballs contain many spores, and their basidiocarps tend to be rounded. When they are mature, any pressure from outside, such as a raindrop or the kick of a shoe, ejects the spores through a hole as a cloud of dust.

SCIENCE IN YOUR LIFE ▶ HEALTH

Deadly Fungi

It is unwise for amateurs to collect mushrooms in the wild because certain mushroom species are poisonous. The red and yellow *Amanita* mushrooms are especially dangerous. These species are also known as fly agaric because they were once thought to kill flies (the mushrooms were gathered and then sprinkled with sugar to attract flies). Its toxins include muscarine and muscaridine, which produce symptoms similar to those of acute alcoholic intoxication. In one to six hours, the victim staggers, loses consciousness, and becomes delirious, sometimes suffering from hallucinations, manic conditions, and stupor. Luckily, it also causes vomiting, which rids the system of the poison, so death occurs in less than 1% of cases. The death angel mushroom (*Amanita phalloides*, Fig. 29C) causes 90% of the fatalities attributed to mushroom poisoning. When this mushroom is eaten, symptoms don't begin until 10 to 12 hours later. Abdominal pain, vomiting, delirium, and hallucinations are not the real problem; rather, a poison interferes with RNA (ribonucleic acid) transcription by inhibiting RNA polymerase, and the victim dies from liver and kidney damage.

Some hallucinogenic mushrooms are used in religious ceremonies, particularly among Mexican Indians. *Psilocybe mexicana* contains a chemical called psilocybin that is a structural analogue of LSD and mescaline. It produces a dreamlike state in which visions of colorful patterns and objects seem to fill up space and dance past in endless succession. Other senses are also sharpened to produce a feeling of intense reality.

The only reliable way to tell a nonpoisonous mushroom from a poisonous one is to be able to correctly identify the species. Poisonous mushrooms cannot be identified with simple tests, such as whether they peel easily, have a bad odor, or blacken a silver coin during cooking. Only consume mushrooms identified by an expert!

Like club fungi, some sac fungi also contain chemicals that can be dangerous to people. *Claviceps purpurea,* the ergot fungus, infects rye and replaces the grain with ergot—hard, purple-black bodies consisting of tightly cemented hyphae (Fig. 29D). When ground with the rye and made into bread, the fungus releases toxic alkaloids that cause the disease ergotism. In humans, vomiting, feelings of intense heat or cold, muscle pain, a yellow face, and lesions on the hands and feet are accompanied by hysteria and hallucinations. Ergotism was common in Europe during the Middle Ages. During this period, it was known as St. Anthony's Fire and was responsible for 40,000 deaths in an epidemic in A.D. 994. We now know that ergot contains lysergic acid, from which LSD is easily synthesized. Based on recorded symptoms, historians believe that those individuals who were accused of practicing witchcraft in Salem, Massachusetts, during the seventeenth century were actually suffering from ergotism. It is also speculated that ergotism is to blame for supposed demonic possessions throughout the centuries. As recently as 1951, an epidemic of ergotism occurred in Pont-Saint-Esprit, France. Over 150 persons became hysterical, and four died.

Because the alkaloids that cause ergotism stimulate smooth muscle and selectively block the sympathetic nervous system, they can be used in medicine to cause uterine contractions and to treat certain circulatory disorders, including migraine headaches. Although the ergot fungus can be cultured in petri dishes, no one has succeeded in inducing it to form ergot in the laboratory. So far, the only way to obtain ergot, even for medical purposes, is to collect it in an infected field of rye.

ergot

Figure 29D Ergot infection of rye, caused by *Claviceps purpurea.*

Figure 29C Poisonous mushroom, *Amanita phalloides.*

Questions to Consider

1. Why do you think that some species of fungi contain chemicals that are harmful to people?
2. How might an understanding of fungal poisons aid in the development of new drugs for humans?

connect BIOLOGY Explore the concepts through a variety of multimedia assets, question types, and data interpretation.
www.mcgrawhillconnect.com

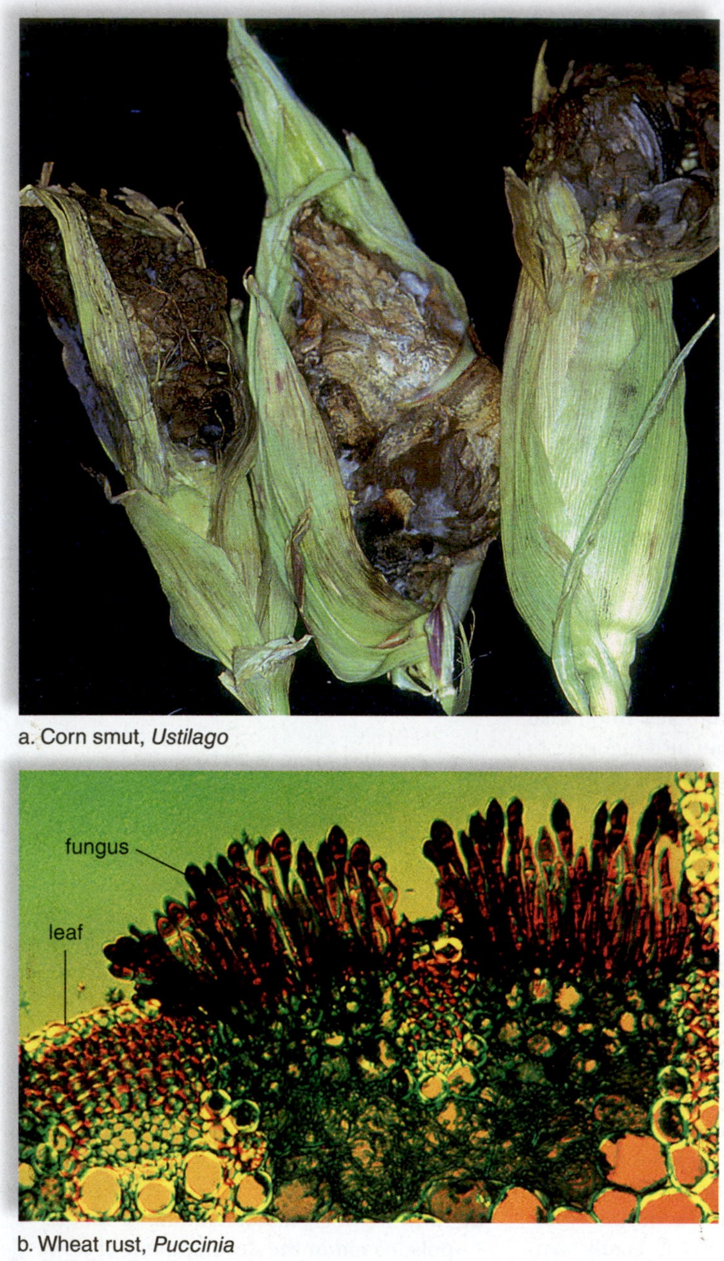
a. Corn smut, *Ustilago*
fungus
leaf
b. Wheat rust, *Puccinia*

Figure 29.24 Smuts and rusts. **a.** Corn smut. **b.** Wheat rust. Both are important fungal diseases affecting agricultural crops in the United States and other countries.

Smuts and *rusts* are club fungi that parasitize cereal crops, such as corn, wheat, oats, and rye. They are of great economic importance because of the crops they destroy each year. The corn smut mycelium grows between corn kernels and secretes substances that cause tumorlike swellings to develop (Fig. 29.24*a*). The life cycle of rusts often requires two different plant host species to complete the cycle, and one way to keep them in check is to eradicate the alternate host. Wheat rust (Fig. 29.24*b*) can also be controlled by producing new and resistant strains of wheat.

Symbiotic Relationships of Fungi

Several instances in which fungi are parasites of plants have already been mentioned. Fungi also commonly form *symbiotic* relationships with other organisms, in which both partners benefit.

Lichens

Lichens are associations between fungi and cyanobacteria or green algae. The different lichen species are identified according to the fungal partner. Lichens are efficient at acquiring nutrients and moisture, and therefore, they can survive in poor soils, as well as on rocks with no soil. These primary colonizers produce organic matter and create new soil, allowing plants to invade the area. Lichens exhibit three structures: compact *crustose* lichens, often seen on bare rocks or tree bark; shrublike *fruticose* lichens; and leaflike *foliose* lichens (Fig. 29.25).

The body of a lichen has three layers. The fungus forms a thin, tough upper layer and a loosely packed lower layer. These shield the photosynthetic cells in the middle layer. Specialized fungal hyphae, which penetrate or envelop the photosynthetic cells, transfer organic nutrients directly to the fungus. The fungus protects the algae from predation and desiccation and provides them with minerals and water. Lichens can reproduce asexually by releasing fragments that contain hyphae and an algal cell. As with many symbioses, the relationship between fungi and algae was likely a pathogen-and-host interaction originally, but became mutually beneficial over evolutionary time.

Mycorrhizae

Mycorrhizae are mutualistic relationships between soil fungi and the roots of most plants. Plants whose roots are invaded by mycorrhizae grow more successfully in dry or poor soils, particularly those deficient in phosphates. The relationship is very ancient, as it is seen in early plant fossils. Perhaps it helped plants adapt to life on dry land. Mycorrhizal fungi generally go unnoticed, except for the truffle, discussed previously.

Mycorrhizal fungi may live on the outside of roots, enter the cortex of roots, or penetrate root cells. The fungus and plant cells can easily exchange nutrients, with the plant providing organic nutrients to the fungus and the fungus bringing water and minerals to the plant. The fungal hyphae greatly increase the surface area from which the plant can absorb water and nutrients.

AM fungi (phylum Glomeromycota) are a group whose name stands for *a*rbuscular *m*ycorrhizal fungi. Arbuscules are branching invaginations the fungus makes when it invades plant roots. AM fungi are one type of mycorrhiza, fungi that form mutually beneficial relationships with the roots of plants.

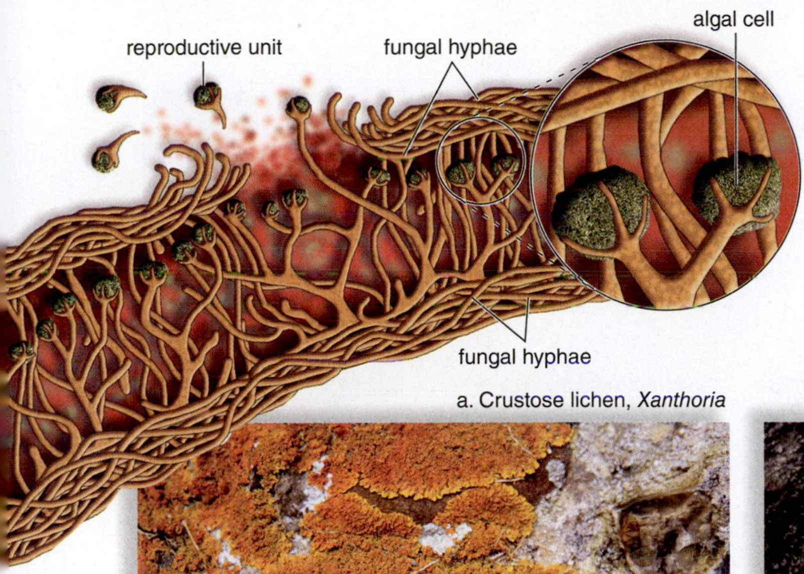
reproductive unit
fungal hyphae
algal cell
fungal hyphae

Figure 29.25 Lichen morphology. **a.** A section of a compact crustose lichen shows the placement of the algal cells and the fungal hyphae, which encircle and penetrate the algal cells. **b.** Fruticose lichens are shrublike. **c.** Foliose lichens are leaflike.

a. Crustose lichen, *Xanthoria*

b. Fruticose lichen, *Cladonia*

sac fungi
reproductive
cups

c. Foliose lichen, *Xanthoparmelia*

Fungal Diseases of Humans

For reasons that aren't completely understood, fungi tend to cause disease mainly in people whose immune system isn't working properly. **Mycoses,** fungal diseases of animals, vary in levels of seriousness.

Superficial Mycoses

Most *superficial mycoses* are confined to the outer layers of the skin, hair, or nails. Fungi called *dermatophytes* cause infections of the skin called *tineas.* Athlete's foot is a tinea characterized by itching and peeling of the skin between the toes (Fig. 29.26*a*). In **ringworm,** which is not caused by a worm but instead by several different fungi, the fungus releases enzymes that degrade keratin and collagen in skin. The area of infection becomes red and inflamed, and the fungal colony grows outward, forming a ring of inflammation. As the center of the lesion begins to heal, ringworm acquires its characteristic appearance, a red ring surrounding an area of healed skin (Fig. 29.26*b*). Some people can harbor and spread the fungus, but show no signs of disease. At any one time, an estimated 3–8% of the U.S. population is infected with scalp ringworm.

Candida albicans causes a wide variety of fungal infections. Superficial *C. albicans* infections ("yeast infections") tend to occur when the balance of the normal microbiota in an area of the body is disturbed, or the immune system is suppressed. For example, resident bacteria of the genus *Lactobacillus* normally produce organic acids that lower the pH of the vagina, and this inhibits *C. albicans,* a normal vaginal inhabitant, from proliferating. When lactobacilli are killed

Figure 29.26 Human fungal diseases. **a.** Athlete's foot and (**b**) ringworm are caused by fungi called *dermatophytes.* **c.** Thrush, or oral candidiasis, is characterized by the formation of white patches on the tongue.

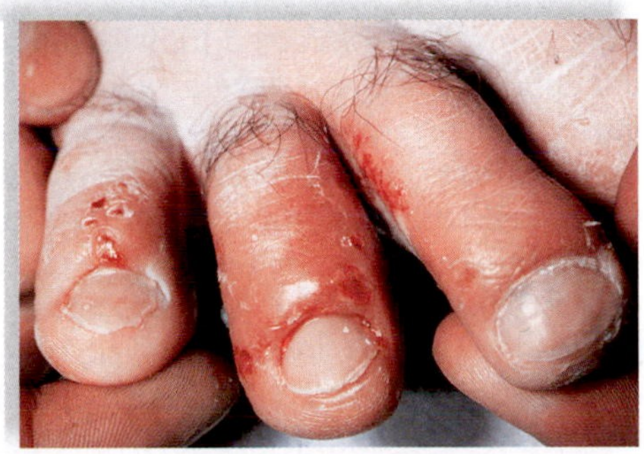

a.

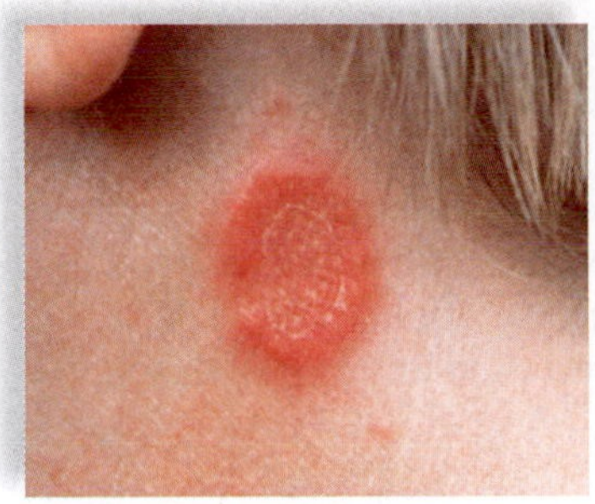

b.

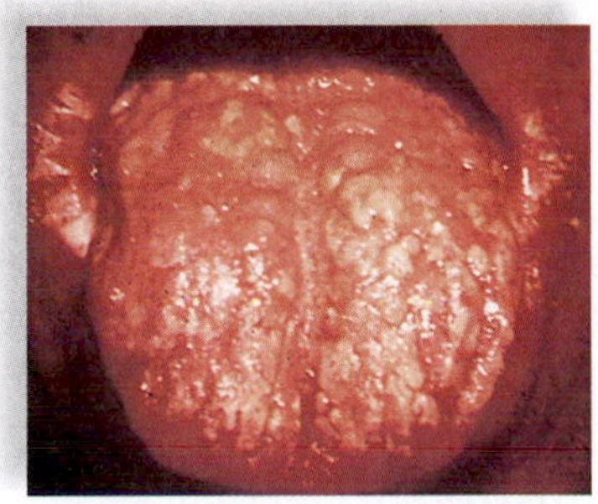

c.

off by antibiotics, the yeast proliferates, resulting in inflammation, itching, and discharge. **Thrush,** a *Candida* infection of the mouth, is common in newborns and AIDS patients (Fig. 29.26*c*). In immunosuppressed individuals, *Candida* may cause an invasive infection that can damage the heart, the brain, and other organs.

Systemic Mycoses

Systemic mycoses refer to fungal infections affecting the internal organs, mainly the lungs. Especially in immunocompromised individuals, the fungi can then spread through the bloodstream to multiple organs, eventually resulting in the death of the patient. This is well known in HIV infection leading to AIDS, and several fungal diseases are considered to be AIDS-defining illnesses (see the Health feature, "Opportunistic Infections and HIV," in Chapter 13).

Occasionally, systemic fungal infections occur in otherwise healthy patients. Due to an effective immune response, most of these infections are either asymptomatic, or cause only mild disease. In rare situations, however, an invasive and progressive infection occurs. Most serious systemic fungal infections start with respiratory symptoms such as coughing, chest pain, hoarseness, and perhaps blood in the sputum. Depending on which fungal species is involved, different organ systems like the skin, central nervous system, bones and joints, and even the heart can become involved.

The majority of people living in the Midwest and eastern United States have been infected with *Histoplasma capsulatum,* a common soil fungus often associated with bird droppings. Only about 5% of infected individuals notice any symptoms, but about 3,000 a year develop a lung disease called *histoplasmosis,* which resembles tuberculosis. Similarly, *Cryptococcus neoformans* occurs throughout the world, mainly in soils contaminated with the droppings of pigeons and chickens. *Coccidioidomycosis,* also called valley fever, can range from benign to severe infections. The causative agent, *Coccidioides immitis,* is found mainly in desert areas, such as in the southwestern United States. Research has shown that 80% of people become infected with *C. immitis* within five years of moving to an area where the organism is prevalent.

Also known as black mold, *Stachybotrys chartarum* grows well on building materials—especially those containing cellulose—that become moist (Fig. 29.27). It grows best in dark, unventilated locations. *S. chartarum* is thought to play a role in "sick building syndrome," in which individuals exposed to toxins produced by the fungus may experience allergies, flulike symptoms, headaches, fatigue, and dermatitis.

Figure 29.27 Sick building syndrome. Spores released by *Stachybotrys chartarum,* or black mold, has been associated with "sick building syndrome."

Antifungal Drugs

As eukaryotes, fungal cells contain a nucleus, organelles, and ribosomes that are like those of human cells. These similarities make it a challenge to design antimicrobials against fungi that do not also harm humans. It is generally safer to treat fungal skin infections with a topical medication, which is not absorbed. For systemic fungal infections, medication must be taken into the body, and thus side effects are more of a problem. To minimize this, researchers exploit any biochemical differences they can discover. The biosynthesis of membrane sterols differs somewhat between fungi and humans. Therefore, a variety of fungicides are directed against sterol biosynthesis.

Check Your Progress 29.2

1. Explain how the type of nutrition required by fungal cells differs from that required by algae or amoeboids.
2. Define hypha, mycelium, and chitin.
3. Compare the modes of reproduction of the zygospore fungi, sac fungi, yeasts, and club fungi.
4. List two types of superficial mycoses, and two systemic mycoses.

Case Study Conclusion

Though rare, infection of the eye with the protist *Acanthamoeba* can be serious and difficult to treat. Josh soon discovered this, as the pain in his left eye became almost unbearable, and his vision became blurred to the point where he could barely recognize how many fingers his doctor was holding up a few inches in front of his face.

Once diagnosed, Josh was prescribed eyedrops every hour (while awake), an oral antiprotozoal agent, and other medications were used to decrease pain. Over a period of several weeks, the eye slowly improved. Although Josh had to miss the remainder of that baseball season, by the following year his eye was healed enough that he was back on the field. Even though Josh knows that his situation was very unusual and is unlikely to reoccur, he now chooses to wear glasses, instead of contacts.

MEDIA STUDY TOOLS

www.mhhe.com/maderinquiry14

Enhance your study of this chapter with study tools and practice tests. Also ask your instructor about the resources available through ConnectPlus, including LearnSmart, the media-rich eBook, interactive learning tools, and animations.

SUMMARIZE

29.1 Protists

Protists are a diverse group of mostly unicellular **eukaryotic cells** (**eukaryotes**) that are widespread in aquatic and moist environments. It is likely that the first eukaryotic cells to evolve were protists, and as stated by the **endosymbiotic theory,** that mitochondria and chloroplasts first arose when a nucleated cell engulfed an aerobic or photosynthetic bacterium, respectively. The biology of protists can be quite complex.

- Many protists have unique organelles not found in other eukaryotes. Examples include **contractile vacuoles** found in many freshwater protists, and the **eyespot apparatus** of some.
- Most protists carry out asexual reproduction, but sexual reproduction involving the production of **spores** or **cysts** is an option.
- Modes of motility are also variable. The term **protozoan** generally describes a motile, heterotrophic protist. **Zooplankton** are heterotrophic protists found suspended in water, which make up an important part of many food chains.

Protist classification schemes are still being debated. Although some experts place protists into **supergroups** that include all eukaryotes, the protists have more traditionally been divided into the following groups:

- **Algae** are photosynthetic protists that derive energy from the sun. These protists are an important component of **phytoplankton** in various bodies of water. Various types of algae can be unicellular, colonial, or multicellular. **Green algae** possess chlorophyll, store reserve food as starch, and have cell walls of cellulose, as do plants. **Diatoms,** which have an outer layer of silica, are extremely numerous in both marine and freshwater ecosystems. **Dinoflagellates** usually have cellulose plates and two flagella. They are extremely numerous in the ocean and can produce a toxin when they form **red tides. Red algae** and **brown algae,** also called seaweeds, are economically important. Red algae are often more delicate than brown algae and are usually found in warmer waters.
- **Flagellates,** which use flagella for motility, are often symbiotic, and some are parasitic. **Euglenids** are flagellated protists with a **pellicle** instead of a cell wall. Their chloroplasts are most likely derived from a green alga, through endosymbiosis.
- **Ciliates** move by means of their many cilia. They are remarkably diverse in form—and as exemplified by *Paramecium,* they show how complex a protist can be, despite being a single cell.
- **Amoeboids** move and feed by forming pseudopods. Parasitic amoeboids, in the genus *Entamoeba,* cause amoebic dysentery.
- **Sporozoans** are nonmotile parasites that form spores.
- Water molds, slime molds, and cellular slime molds are actually classified as protists. **Water molds,** which ordinarily are **saprotrophs,** are sometimes parasitic. **Plasmodial slime molds** are amoeboid and ingest food by phagocytosis. **Cellular slime molds** exist as individual amoeboid cells until nutrients become limited. The cells then fuse into a pseudoplasmodium that produces spores.

A few protists cause some important human diseases, as follows:

- Sporozoans (apicomplexans) cause **malaria,** one of the most significant diseases in the world. Toxoplasmosis and cryptosporidiosis are other diseases caused by sporozoans.
- The most important diseases caused by flagellates are **African sleeping sickness** and **Chagas disease.** Infections with *Giardia lamblia* and *Trichomonas vaginalis* are also very common.
- The most common human disease caused by an amoeboid is **amoebic dysentery.** Rarely, amoeboids can cause a severe keratitis, and even a fatal brain infection.

29.2 Fungi

Fungi are unicellular or multicellular eukaryotes that are strict heterotrophs.

- After external digestion, fungi absorb the resulting nutrient molecules. Most fungi are saprotrophs that aid the cycling of chemicals in ecosystems by decomposing dead remains.
- The body of a fungus is composed of thin filaments, each of which is called a **hypha.** A mass of **hyphae** is termed a **mycelium.** Nonseptate hyphae have no cross walls, forming multinucleate cells. Septate hyphae have cross walls, but pores allow for exchange of cytoplasm. The cell wall contains **chitin.**
- Most fungi produce nonmotile, and often windblown, spores during both asexual and sexual reproduction. During sexual reproduction, hyphae tips fuse so that hyphae containing both + and − nuclei usually result. Following nuclear fusion, meiosis occurs during the production of the sexual spores.

Fungi can be classified into the following five major groups:

- **Chytrid fungi** are aquatic and have flagella at some stages.
- **Zygospore fungi,** of which the common bread mold *Rhizopus stolonifer* is a member, reproduce by forming a capsule called a **sporangium** that produces thick-walled **zygospores.**
- **Sac fungi** are named for their production of a cuplike **ascocarp,** which produces spores called **conidia.** Most known fungi are sac fungi, including morels, truffles, *Penicillium, Aspergillus,* and plant pathogens like Dutch Elm disease and ergot. Some sac fungi, like the brewer's yeast *Saccharomyces cerevisiae,* produce unicellular forms called **yeasts.**
- **Club fungi** are named for their production of a club-shaped reproductive structure called a **basidium.** Some club fungi produce toxins; those in the genus *Psilocybe* produce an LSD-like substance.

Fungi often form the following symbiotic relationships with other organisms:

- **Lichens** are symbiotic associations between fungi and cyanobacteria, or green algae. The algae provide organic nutrients to the fungi, while the fungi provide protection and enhanced absorption of water and minerals.

- **Mycorrhizae** enjoy a mutualistic relationship with plant roots, such that the fungus helps the plant absorb minerals and water, while the plant supplies the fungus with organic nutrients. **AM fungi** are a type of mycorrhiza that invades plant roots.

A few fungi are parasites of animals, causing fungal infections or **mycoses.** Most of these fungi are opportunists that mainly infect those with a compromised immune system.

- Superficial mycoses affect the outermost layers of skin, or sometimes the nails or hair. **Ringworm** is a type of tinea, or fungal infection of the skin. *Candida albicans,* a normal inhabitant of several body areas, can overgrow and cause "yeast infections" of the vagina, or **thrush** in the oral cavity.
- Systemic mycoses affect other body systems, especially the lungs. The most common fungi to be involved are *Histoplasma capsulatum, Cryptococcus neoformans,* and *Coccidioides immitis. Stachybotrys chartarum,* also known as "black mold," grows well in building materials that have gotten wet, and it is thought to play a role in "sick building syndrome." Because fungal cells are eukaryotic and share many similarities with animal cells, antifungal drugs can have many side effects, and systemic mycoses can be difficult to treat.

ASSESS

Testing Yourself

Choose the best answer for each question.

1. Which description does NOT apply to protists?
   a. All protists are eukaryotic.
   b. Some protists are photosynthetic.
   c. All protists are unicellular.
   d. Some protists are parasitic.
2. The best general description of an alga is
   a. a usually motile, unicellular protist.
   b. photosynthetic eukaryotes that can be unicellular or form colonies or filaments.
   c. photosynthetic eukaryotes that have a wide variety of shells made of silica.
   d. a group of parasitic protists that produce spores.
3. Dinoflagellates
   a. usually reproduce sexually.
   b. often have protective cellulose plates.
   c. are insignificant producers of food and oxygen.
   d. have cilia instead of flagella.
   e. tend to be larger than brown algae.
4. Ciliates
   a. can move by pseudopods.
   b. are not as varied as other protists.
   c. have a gullet for food gathering.
   d. are closely related to the radiolarians.
5. Which are found in slime molds but not in fungi?
   a. nonmotile spores
   b. amoeboid vegetative cells
   c. zygote formation
   d. photosynthesis
   e. All of the choices are correct.
6. Sporozoan parasites of the genus *Plasmodium* cause
   a. African sleeping sickness. **c.** malaria.
   b. Chagas disease. **d.** thrush.
7. Toxoplasmosis is mostly spread by
   a. cats. **c.** standing water.
   b. mosquitoes. **d.** tsetse flies.
8. Fungi are classified according to
   a. sexual reproductive structures.
   b. shape of their hyphae.
   c. mode of nutrition.
   d. type of cell wall.
   e. All of the choices are correct.
9. Which feature is best associated with hyphae?
   a. strong, impermeable walls **d.** pigmented cells
   b. rapid growth **e.** Both b and c are correct.
   c. large surface area
10. Conidia are formed
    a. asexually at the tips of special hyphae.
    b. during sexual reproduction.
    c. by all types of fungi except water molds.
    d. only when it is windy and dry.
    e. as a way to survive a harsh environment.
11. About 75% of all known fungi are
    a. chylotrichids. **c.** sac fungi.
    b. club fungi. **d.** zygospore fungi.
12. Lichens
    a. cannot reproduce.
    b. need a nitrogen source to live.
    c. are parasitic on trees.
    d. are able to live in extreme environments.
13. Mycorrhizal fungi
    a. are a type of lichen.
    b. are in a mutualistic relationship.
    c. help plants gather solar energy.
    d. help plants gather inorganic nutrients.
    e. Both b and d are correct.
14. Which of the following diseases is caused by *Candida*?
    a. oral thrush **c.** histoplasmosis
    b. tinea **d.** ringworm
15. Systemic fungal infections most frequently affect the
    a. heart. **c.** liver.
    b. immune system. **d.** lungs.

ENGAGE

Thinking Critically

1. Sickle cell disease occurs mostly in people of African ancestry. When a person inherits two copies of the sickle cell gene, his or her hemoglobin is abnormal, causing red blood cells to be more fragile and sticky than usual. This leads to anemia and circulatory problems. About 70,000 Americans have sickle cell disease, and about 2 million carry one copy of the gene. Even though it can cause a serious disease, the sickle cell gene is thought to have been favored by natural selection, because it gives some protection against malaria. Considering the type of cells infected by *Plasmodium* species, how might this protection work? What do you think may happen to the prevalence of the sickle cell disease gene if an effective vaccine against malaria is ever developed and widely used?
2. Many fungal infections of humans are considered to be opportunistic, meaning that fungi that are normally free-living (usually in soil) can sometimes survive, and even thrive, on or inside the human body. From the fungal "point of view," what unique challenges would be encountered when trying to survive on human skin? What about inside human lungs?

Plants

30

CASE STUDY The Importance of Plants

Many scientists believe that if all the plants on Earth were suddenly to die, all animals, including humans, would soon perish. Yet, if animals were to die, many plants would still survive and flourish. Plants provide important food and nesting sites for many animals, trees provide oxygen to the air, and even many medicines come directly from plants or plant products. The fossil record indicates that plants moved onto dry land from an aquatic environment prior to animals. The first plants to conquer land were the bryophytes (mosses and liverworts), which evolved around 450 million years ago (MYA). Their lack of vascular tissue and the inability to conduct water and nutrients limited their size. Seedless vascular plants like ferns evolved about 350 MYA. Vascular tissue allowed ferns and their allies to grow very tall. Gymnosperms were the first seed plants to evolve during the Devonian period around 365 MYA. The ability to produce seeds is a highly successful trait that increases reproductive success by protecting the embryo and allowing dispersal. Gymnosperms, such as cycads and evergreen trees, were quite large and dominated the land during the Mesozoic era when the dinosaurs ruled. In fact, botanists refer to the Mesozoic as the "age of the cycads" instead of the "age of the dinosaurs." Later, between 200 and 135 MYA, angiosperms evolved. Today, angiosperms are the most widespread of the plants because of their ability to produce flowers and seeds enclosed by fruit. Overall, plants display an astonishing diversity on every continent, as well as amazing abilities to adapt to varied environments.

As you read through the chapter, think about the following questions:

1. What environmental challenges did plants have to overcome in order to evolve to a terrestrial existence?
2. Which characteristic proved to be the most important during the evolution of each of the plant groups?
3. Which features enabled the angiosperms to become the most dominant group of plants on Earth today?

CHAPTER OUTLINE

BEFORE YOU BEGIN

Before beginning this chapter, take a few moments to review the following discussions:

Figure 3.5 What cellular structures are unique to plants?

Section 5.4 What is the role of meiosis in a cell?

Section 9.2 How does vascular tissue contribute to the success of the various groups of plants?

30.1 Evolutionary History of Plants

Learning Outcomes

Upon completion of this section, you should be able to

1. Identify the features present in the ancestor of land plants that helped give rise to the land plants.
2. List the five major events in the evolutionary history of plants.
3. Describe alternation of generations in plants.

Plants (domain Eukarya, kingdom Plantae) are vital to human survival. As explained by the Ecology feature, "Plants and Humans," on page 606, our dependence on them is nothing less than absolute. Although a land environment does offer advantages such as plentiful light, it also has certain challenges such as the constant threat of desiccation (drying out). Most important, all stages of reproduction—gametes, zygote, and embryo—must be protected from the drying effects of air. To keep the internal environment of cells moist, a land plant must acquire water, and transport it to all parts of the body. The evolution of an internal vascular system has made certain groups of plants incredibly successful. The evolutionary history of plants begins in the water. Most likely land plants evolved from a type of freshwater green algae some 450 MYA (million years ago): both green algae and plants (1) contain chlorophylls *a* and *b* and various accessory pigments, (2) store excess carbohydrates as starch, and (3) have cellulose in their cell walls. A comparison of DNA and RNA base sequences suggests that land plants are most closely related to a group of freshwater green algae known as charophytes. Although the common ancestor of modern charophytes and land plants no longer exists, if it did, it would have features resembling those of the *Chara* and *Coleochaete* (Fig. 30.1).

Chara
Coleochaete

Figure 30.1 The Charophytes. The charophytes (represented here by *Chara* and *Coleochaete*) are the green algae most closely related to the land plants.

Chara are commonly known as stoneworts because they are encrusted with calcium carbonate deposits. The body consists of a single file of very long cells anchored in mud by thin filaments. Whorls of branches occur at regions called nodes, located between the cells of the main axis. Male and female reproductive structures grow at the nodes. A *Coleochaete* looks flat like a pancake, but the body is actually composed of long, branched filaments of cells. A key feature in the evolution of plants was the protection of the zygote. Land plants not only protect the zygote, they also protect and nourish the resulting embryo. This may be the first derived feature that separates land plants from green algae.

To illustrate the evolution of land plants, it is possible to associate each group with one of the major evolutionary events representing adaptations to a land existence.

- Mosses were the first group to have some form of protection for the embryo to prevent it from drying out.
- Lycophytes were the first group to evolve **vascular tissue,** which transports water and nutrients throughout the plant body.
- Ferns evolved **megaphylls,** large leaves that increase the surface area for photosynthesis.
- Gymnosperms were the first plants with **seeds.** Seeds contain the embryo and supportive nutrients within a protective coat.
- Angiosperms are the only group to have flowers. Flowers attract a variety of pollinators and give rise to fruits.

Figure 30.2 traces the evolutionary history of land plants and will serve as a backdrop as we discuss major groups of plants in the pages that follow.

Alternation of Generations

All plants have a life cycle that includes an **alternation of generations.** In this life cycle, two multicellular individuals alternate, each producing the other (Fig. 30.3). The two individuals are (1) a sporophyte, which represents the diploid (2n) generation, and (2) a gametophyte, which represents the haploid (n) generation.

The **sporophyte** (2n) is the structure that produces spores by meiosis. A spore is a haploid reproductive cell that develops into a new organism without needing to fuse with another reproductive cell. In the plant life cycle, a spore undergoes mitosis and becomes a gametophyte.

The **gametophyte** (n) is so named for its production of gametes. In plants, eggs and sperm are produced by mitotic cell division. A sperm and egg fuse, forming a diploid zygote that undergoes mitosis and becomes the sporophyte.

Two observations are in order. First, meiosis produces haploid spores. This is consistent with the realization that the sporophyte is the diploid generation and spores are haploid reproductive cells. Second, mitosis occurs as a spore becomes a gametophyte, and mitosis also occurs as a zygote becomes a sporophyte. Indeed, it is the occurrence of mitosis at these times that results in two generations.

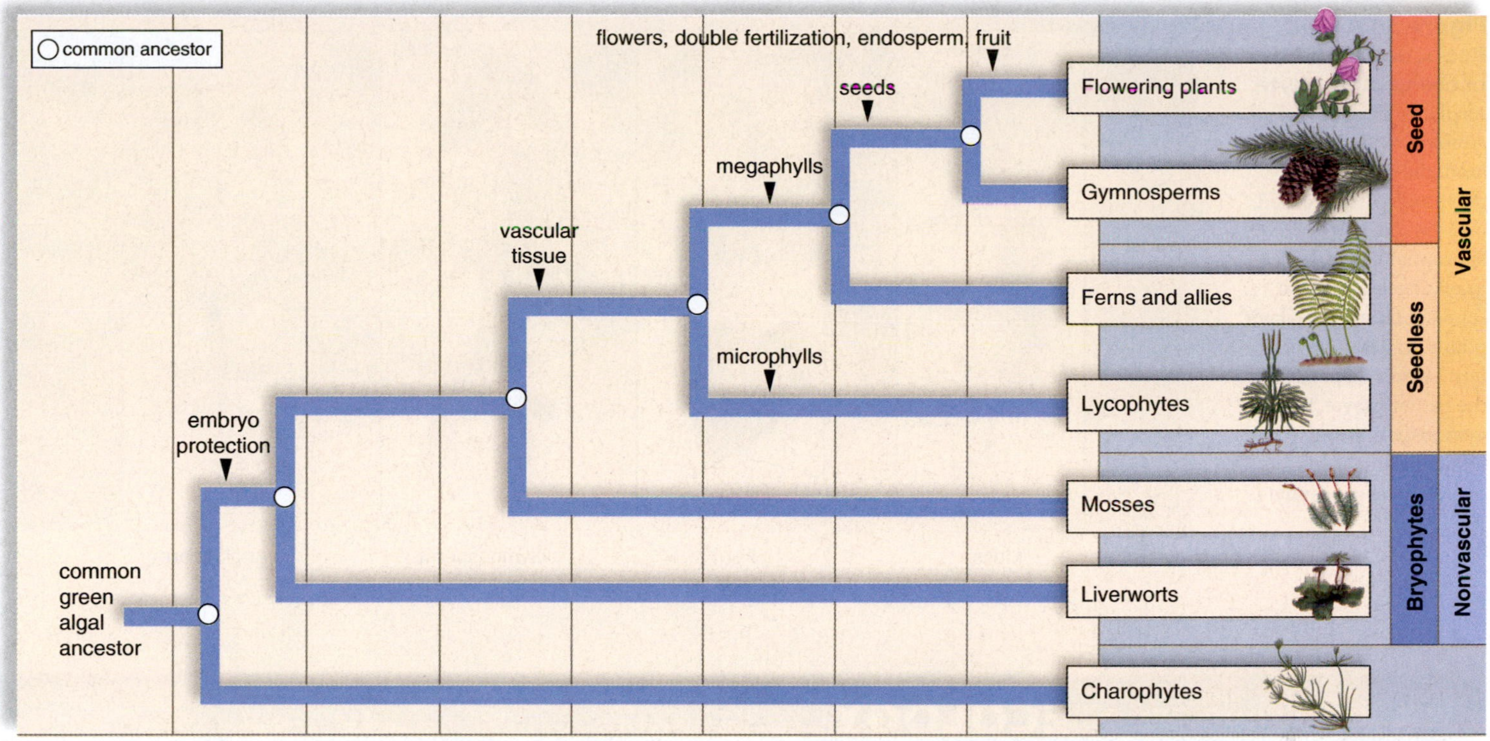
common ancestor
flowers, double fertilization, endosperm, fruit
seeds
megaphylls
vascular tissue
microphylls
embryo protection
common green algal ancestor
Flowering plants
Gymnosperms
Ferns and allies
Lycophytes
Mosses
Liverworts
Charophytes
Seed
Seedless
Vascular
Bryophytes
Nonvascular

Figure 30.2 Evolutionary history of plants. The evolution of plants involves five significant innovations: (1) Protection of a multicellular embryo was seen in the first plants to live on land. (2) The evolution of vascular tissue was another important adaptation for life on land. (3) The evolution of megaphylls aided dispersion of nutrients into leaves by veination. (4) The evolution of the seed increased the chance of survival for the next generation. (5) The evolution of the flower fostered the use of animals as pollinators and the use of fruits to aid in the dispersal of seeds.

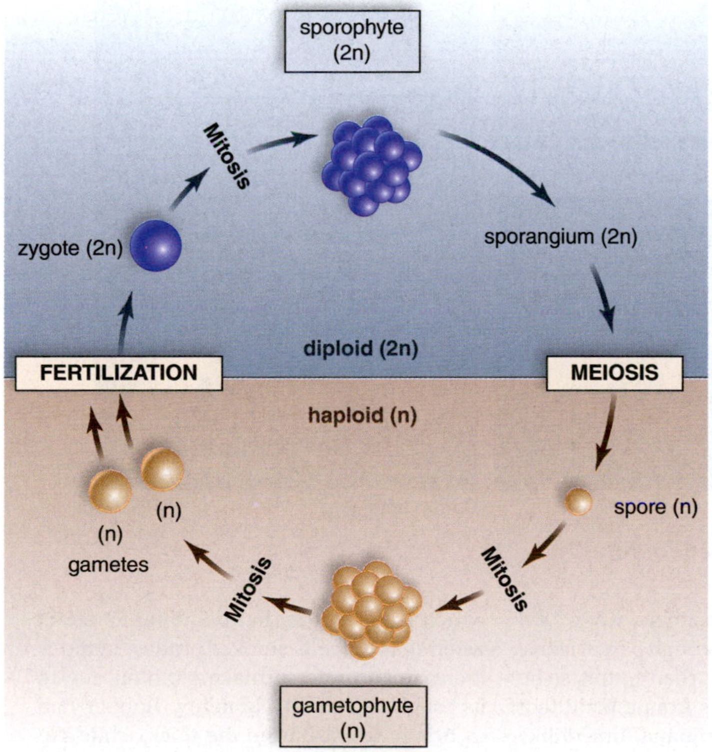
sporophyte (2n)
Mitosis
zygote (2n)
sporangium (2n)
diploid (2n)
FERTILIZATION
MEIOSIS
haploid (n)
(n)
(n)
gametes
spore (n)
Mitosis
Mitosis
gametophyte (n)

Figure 30.3 Alternation of generations. Plants alternate between a diploid sporophyte stage and a haploid gametophyte stage.

Plants differ as to which generation is the dominant one. In plants, the dominant generation carries out the majority of photosynthesis. In nonvascular plants, the gametophyte is dominant, but in the other three groups of plants, the sporophyte is dominant. In the history of plants, only the sporophyte evolves vascular tissue. Therefore, the shift to sporophyte dominance is an adaptation to life on land. Notice that as the sporophyte gains in dominance, the gametophyte becomes microscopic (Fig. 30.4) and dependent upon the sporophyte.

All plants have an alternation-of-generations life cycle, but the appearance of the generations varies widely. In ferns, the gametophyte is a small, independent, heart-shaped structure. Archegonia are female reproductive organs that produce eggs. Antheridia are male reproductive organs that produce motile sperm. The eggs are fertilized by flagellated sperm, which swim to the archegonia in a film of water. In contrast, the female gametophyte of an angiosperm, called the embryo sac, is retained within the body of the sporophyte plant and consists of seven cells within a structure called an ovule. Following fertilization, the ovule becomes a seed. In seed plants, pollen grains are mature, sperm-bearing male gametophytes. Pollen grains can be transported by wind, insects, bats, or birds, and therefore, they do not need free water to reach the egg. In seed plants, reproductive cells are protected from drying out in the land environment.

spores
spores
seed
seed
Gametophyte (n)
Sporophyte (2n)
roots
roots
roots
rhizoids
rhizoids
Moss
Fern
Gymnosperm
Angiosperm

Figure 30.4 Reduction in the size of the gametophyte. As plants became adapted to life on land, the size of the gametophyte decreased, and the size of the sporophyte increased, as evidenced by these representatives of today's plants. In the moss and fern, spores disperse the gametophyte. In gymnosperms and angiosperms, seeds disperse the sporophyte.

SCIENCE IN YOUR LIFE ▶ **ECOLOGY**

Plants and Humans

Humans derive most of their nourishment from three flowering plants: wheat, corn, and rice. All three of these plants are members of the grass family and are collectively called grains (Fig. 30A). Wheat is commonly used to produce flour and bread. It was first cultivated in the Middle East about 8000 B.C., and is thought to be one of the earliest cultivated plants. Early settlers brought wheat to North America, where many varieties of corn, more properly called maize, were already in existence. Rice originated in southeastern Asia several thousand years ago, where it grew in swamps. Many foods of both plant and animal origin are considered bland or tasteless without seasonings or spices. The Americas were discovered when Columbus was seeking a new route to the Far East to acquire spices, such as nutmeg, oregano, rosemary, and sage. In addition, our primary sweetener, sugar, comes almost exclusively from two plants—sugarcane (grown in South America, Africa, Asia, and the Caribbean) and sugar beets (grown mostly in Europe and North America). Our most popular drinks—coffee, tea, and cola—also come from flowering plants. Coffee originated in Ethiopia, whereas tea is thought to have been first used somewhere in central Asia. Cotton and rubber are two plants that still have many uses today (Fig. 30B). Until a few decades ago, cotton and other natural fibers were our only source of clothing. Interestingly, when Levi Strauss wanted to make a tough pair of jeans, he needed a stronger fiber than cotton, so he used hemp. Hemp comes from a plant that is closely related to marijuana, but unlike marijuana, has extremely small amounts of the chemical tetrahydrocannabinol (THC), which causes the hallucinogenic effect when marijuana is smoked. Today, hemp is increasingly used to make clothing due to its toughness and wearability. Rubber had its origin in Brazil from the thick, white sap of the rubber tree. To produce a stronger rubber, such as that in tires, sulfur is added,

grain head
Wheat plants, *Triticum*

grain head
Rice plants, *Oryza*

ear
Corn plants, *Zea*

Figure 30A Plants used for food.

Check Your Progress 30.1

1. Explain the benefits of a land existence for plants.
2. Identify the role of each generation in the alternation-of-generations life cycle.

30.2 Nonvascular Plants

Learning Outcomes

Upon completion of this section, you should be able to

1. Identify the traits of nonvascular plants that enable them to survive on land.
2. Compare and contrast the features of liverworts and mosses.

The **nonvascular plants** lack vascular tissue. Although the nonvascular plants often have a "leafy" appearance, they do not have true roots, stems, or leaves—which, by definition, must contain true vascular tissue. Therefore, the nonvascular plants are said to have rootlike, stemlike, and leaflike structures.

The gametophyte is the dominant generation in nonvascular plants. The flagellated sperm swim in a continuous film of water to the vicinity of the egg. The sporophyte, which develops from the zygote, is attached to, and derives its nourishment from the gametophyte.

The nonvascular plants consist of three divisions—one for hornworts, liverworts, and mosses, which collectively are known as bryophytes. The placement of these plants in separate divisions reflects current thinking that they are not that closely related. In this section, we discuss the liverworts and mosses.

Rubber tree, *Hevea*

Cotton, *Gossypium*

Figure 30B Plants used commercially.

and the sap is heated in a process called vulcanization. This produces a flexible material less sensitive to temperature changes. However, at present, most rubber is synthetically produced.

Plants have also been used for centuries for a number of important household items, including the house itself. We are most familiar with lumber as the major structural portion in buildings. This wood comes mainly from a variety of conifers: pine, fir, and spruce, among others. In the tropics, trees and even herbs provide important components for houses. In rural parts of Central and South America, palm leaves are preferable to tin for roofs, because they last as long as ten years and are quieter during rainstorms. In the Middle East, numerous houses along rivers are made entirely of reeds.

Today, plants are increasingly researched for their use in medicinal products. Currently, about 50% of all pharmaceutical drugs originate from plants or are derived from plant products. Some cancers are even treatable with medicinal plants, such as the rosy periwinkle, extracts from which have shown some success in treating childhood leukemia. Indeed, the National Cancer Institute and most pharmaceutical companies have spent millions of dollars to send botanists out to collect and test plant samples from around the world. Tribal medicine men, or shamans, of South America and Africa have already been of great importance in developing numerous drugs. However, plant extracts also continue to be misused for their hallucinogenic or other effects on the human body; examples are coca, the source of cocaine and crack, and the opium poppy, the source of heroin.

In addition to all these uses of plants, we should not forget their aesthetic value. Flowers brighten any yard, ornamental plants accent landscaping, and trees provide cooling shade during the summer as well as shelter from harsh winds during the winter.

Questions to Consider

1. Why do you think plants are such a good source of drugs for human use? What advantage does this give the plants?
2. With over 250,000 species of angiosperms alone, why do you think we get most of our nourishment from only three plants (wheat, corn, and rice)?
3. What are some uses of plants, other than those just discussed?

male gametophyte
female gametophyte
gemma cup
thallus
rhizoids
gemma
a. Gemma cup
Thallus with gemmae cups
b. Male gametophytes bear antheridia
c. Female gametophytes bear archegonia

Figure 30.5 Liverwort, *Marchantia*. **a.** A gemma is a group of cells that can detach and start a new plant. **b.** Antheridia are present in disk-shaped stalks. **c.** Archegonia are present in umbrella-shaped stalks.

Liverworts

Liverworts (phylum Hepaticophyta) come in two types: those that have a flat, lobed body called a *thallus* (pl., thalli) and those that are leafy. *Marchantia* is a familiar liverwort (Fig. 30.5) that has a smooth upper surface. The lower surface bears numerous *rhizoids* (rootlike hairs) that project into the soil. *Marchantia* reproduces both asexually and sexually. Gemmae cups on the upper surface of the thallus contain gemmae, groups of cells that detach from the thallus and can asexually start a new plant. Sexual reproduction depends on disk-headed stalks that bear **antheridia** (sing., antheridium), where flagellated sperm are produced, and on umbrella-headed stalks that bear **archegonia** (sing., archegonium), where eggs are produced. Following fertilization, tiny sporophytes composed of a foot, a short stalk, and a capsule begin growing within archegonia. Windblown spores are produced within the capsule.

Mosses

Mosses (phylum Bryophyta) can be found from the Arctic through the tropics to parts of the Antarctic. Although most live in damp, shaded locations in the temperate zone, some survive in deserts, whereas others inhabit bogs and streams. In forests, mosses frequently form a mat that covers the ground or the surfaces of rotting logs. Mosses can store large quantities of water in their cells, but if a dry spell continues for long, they become dormant until it rains.

Most mosses can reproduce asexually by fragmentation. Just about any part of the plant is able to grow and eventually produce leaflike thalli. Figure 30.6 describes the life cycle of a typical temperate-zone moss. The gametophyte of mosses has two stages. First, there is the algalike *protonema,* a branching filament of cells. Then, after about three days of favorable growing conditions, upright leafy thalli appear at intervals along the protonema. Rhizoids anchor the thalli, which bear antheridia and archegonia. An antheridium consists of a short stalk, an outer layer of sterile cells, and an inner mass of cells that become the flagellated sperm. An archegonium, which looks like a vase with a long neck, has a single egg located inside its base.

The dependent sporophyte consists of a *foot,* which grows down into the gametophyte tissue; a *stalk;* and an upper *capsule,* or *sporangium,* where windblown spores are produced. At first, the sporophyte is green and photosynthetic. At maturity, it is brown and nonphotosynthetic. Because the gametophyte is the dominant generation, it seems consistent for spores to be dispersal agents—that is, when the haploid spores germinate, the gametophyte is in a new location.

Adaptations and Uses of Nonvascular Plants

Although mosses are nonvascular, they are better than flowering plants at living on stone walls, fences and even in the shady cracks of hot, exposed rocks. For these particular microhabitats, being small and simple seems to be a selective advantage. When bryophytes colonize bare rock, they help convert the rocks to soil that can be used for the growth of other organisms.

In areas such as bogs, where the ground is wet and acidic, dead mosses, especially *Sphagnum,* do not decay. The accumulated moss, called peat or bog moss, can be used as fuel. Peat moss is also commercially important because it has special nonliving cells that can absorb moisture. Thus, peat moss is often used in gardens to improve the water-holding capacity of the soil.

Check Your Progress 30.2

1. List two features limiting the adaptation of nonvascular plants.
2. Describe two features of nonvascular plants that indicate they are adapted to the land environment.
3. Identify a nonvascular plant as a liverwort or a moss.

Figure 30.6 Moss life cycle, *Polytrichum* sp.

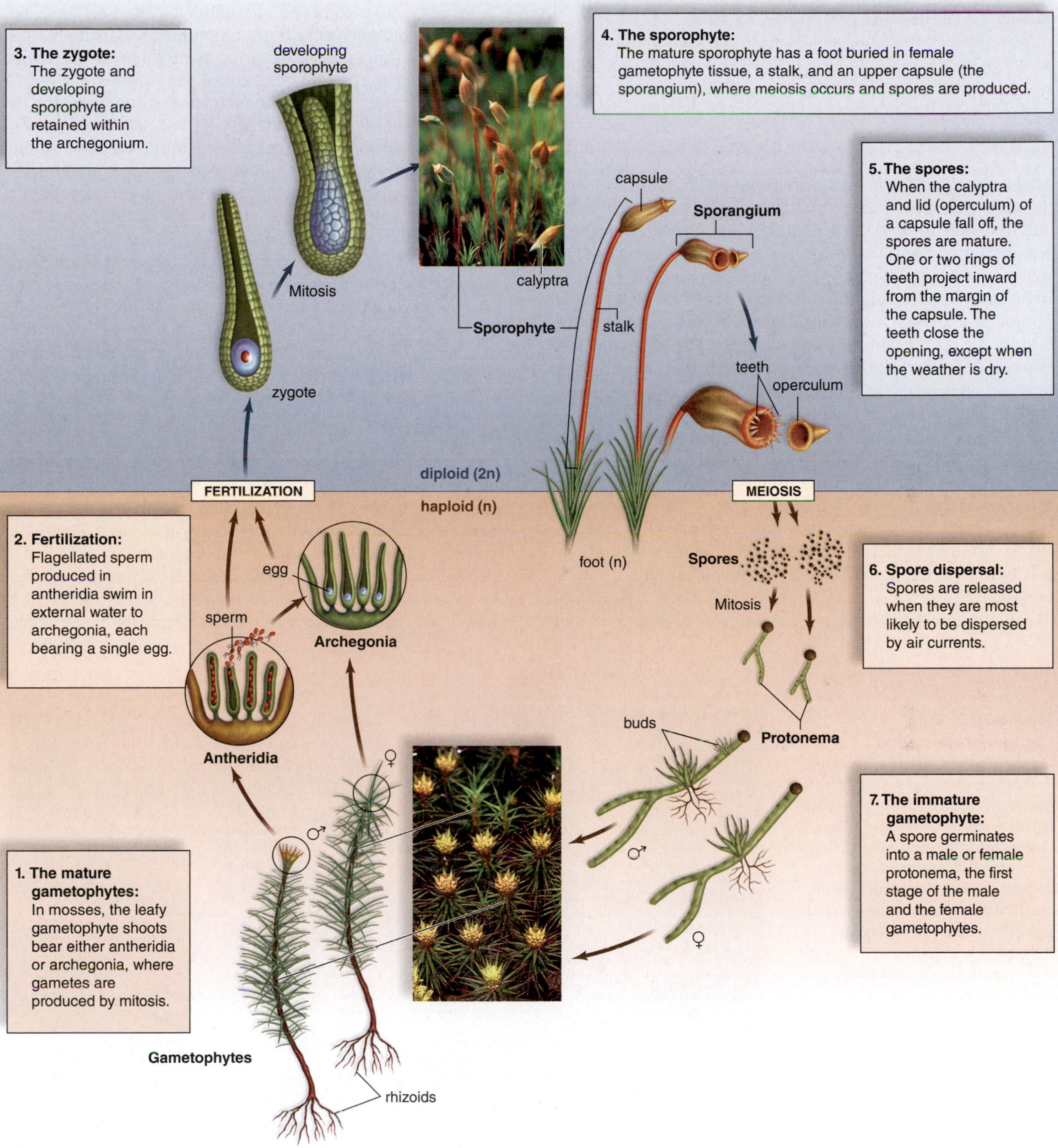

3. The zygote:
The zygote and developing sporophyte are retained within the archegonium.
developing sporophyte
Mitosis
zygote
4. The sporophyte:
The mature sporophyte has a foot buried in female gametophyte tissue, a stalk, and an upper capsule (the sporangium), where meiosis occurs and spores are produced.
calyptra
Sporophyte
capsule
stalk
Sporangium
teeth
operculum
5. The spores:
When the calyptra and lid (operculum) of a capsule fall off, the spores are mature. One or two rings of teeth project inward from the margin of the capsule. The teeth close the opening, except when the weather is dry.
diploid (2n)
FERTILIZATION
haploid (n)
MEIOSIS
foot (n)
Spores
Mitosis
6. Spore dispersal:
Spores are released when they are most likely to be dispersed by air currents.
2. Fertilization:
Flagellated sperm produced in antheridia swim in external water to archegonia, each bearing a single egg.
egg
sperm
Archegonia
Antheridia
buds
Protonema
♀
♂
7. The immature gametophyte:
A spore germinates into a male or female protonema, the first stage of the male and the female gametophytes.
1. The mature gametophytes:
In mosses, the leafy gametophyte shoots bear either antheridia or archegonia, where gametes are produced by mitosis.
Gametophytes
rhizoids

30.3 Seedless Vascular Plants

Learning Outcomes

Upon completion of this section, you should be able to

1. List two reasons why ferns are the most plentiful group of seedless vascular plants.
2. Describe the similarities and differences between the fern and the moss life cycle.

All the other plants we will study are **vascular plants.** Vascular tissue in these plants consists of **xylem,** which conducts water and minerals up from the soil, and **phloem,** which transports organic nutrients from one part of the plant to another. The vascular plants usually have true roots, stems, and leaves. The roots absorb water from the soil, and the stem conducts water to the leaves. Xylem, with its strong-walled cells, supports the body of the plant against the pull of gravity. The leaves are fully covered by a waxy cuticle except where it is interrupted by stomata, little pores whose size can be regulated to control water loss.

Animation Vascular Tissue

The sporophyte is the dominant generation in vascular plants. Seedless vascular plants include two groups: lycophytes and ferns and their allies. Both of these groups disperse their offspring by producing wind-blown spores. This is advantageous because the sporophyte is the generation with vascular tissue. Another advantage of having a dominant sporophyte relates to its being diploid. If a faulty gene is present, it can be masked by a functional gene. Then, too, the greater the amount of genetic material, the greater the possibility of mutations that will lead to increased variety and complexity. Indeed, vascular plants are complex, extremely varied, and widely distributed.

Some vascular plants do not produce seeds. Seedless vascular plants include whisk ferns, club mosses, horsetails, and ferns, which disperse their offspring by producing windblown spores. When the spores germinate, they produce a small gametophyte that is independent of the sporophyte for its nutrition. In these plants, antheridia release flagellated sperm, which swim in a film of external water to the archegonia, where fertilization occurs. Because spores are dispersal agents and the nonvascular gametophyte is independent of the sporophyte, these plants cannot wholly benefit from the adaptations of the sporophyte to a terrestrial environment.

The seedless vascular plants formed the great swamp forests of the Carboniferous period (Fig. 30.7). A large number of these plants died but did not decompose completely. Instead, they were compressed to form the coal that we still mine and burn today. Oil has a similar origin, but it typically forms in marine sedimentary rocks and includes animal remains.

Lycophytes

Lycophytes (phylum Lycophyta) are also called **club mosses** and were among the first land plants to have vascular tissue. Unlike true mosses (bryophytes), the lycophytes have well-developed vascular tissue in roots, stems, and leaves (Fig. 30.8). Typically, a fleshy underground and horizontal stem, called a rhizome, sends up upright aerial stems. Tightly packed, scale-like leaves cover the stems and branches, giving the plant a mossy look. The small leaves, termed microphylls, have a single

Figure 30.7 The Carboniferous period. Growing in the swamp forests of the Carboniferous period were treelike club mosses (*left*), treelike horsetails (*right*), and shorter, fernlike foliage (*lower left*). When the trees fell, they were covered by water and did not decompose well. Sediment built up and turned to rock, and the resulting compression caused the organic material to become coal, a fossil fuel that helps run our industrialized society today.

strobili
leaves
branches
aerial stem
rhizome
root

Figure 30.8 Ground pine, *Lycopodium*. Lycophytes such as *Lycopodium* have vascular tissue and thus true roots, stems, and leaves. The *Lycopodium* sporophyte develops an underground rhizome system. A rhizome is an underground stem; this rhizome produces roots along its length.

vein composed of xylem and phloem. The sporangia are borne on terminal clusters of leaves, called strobili, which are club-shaped. The spores are sometimes harvested and sold as lycopodium powder, or vegetable sulfur, for use in pharmaceuticals and in fireworks because it is highly flammable. Members of the genus *Lycopodium* featured in Figure 30.8 are common in moist woodlands in temperate climates, where they are called ground pines; they are also abundant in the tropics and subtropics.

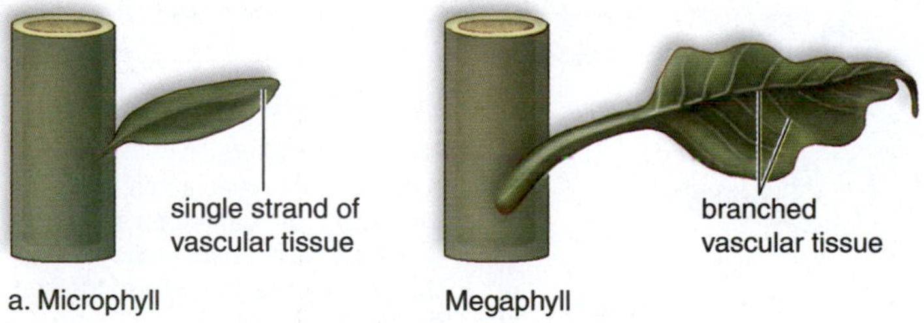
single strand of vascular tissue
a. Microphyll
branched vascular tissue
Megaphyll

Ferns and Their Allies

In this section, we will discuss the characteristics of ferns, whisk ferns, and horsetails.

Ferns

Ferns (phylum Polypodiophyta) are the largest group of plants other than the flowering plants, and they display great diversity in form and habitat. Ferns are most abundant in warm, moist, tropical regions, but can also be found in northern regions and in relatively dry, rocky, places. They range in size from low-growing plants resembling mosses to tall trees. The *fronds* (leaves) that grow from a rhizome can vary. The royal fern has fronds that stand 6 ft tall; those of the maidenhair fern are branched, with broad leaflets; and those of the hart's tongue fern are straplike and leathery. Figure 30.9 provides some examples of ferns.

Life Cycle, Adaptations, and Uses of Ferns Ferns have true roots, stems, and leaves that contain vascular tissue. The well-developed leaves fan out, capture solar energy, and photosynthesize. The life cycle of a typical temperate fern is shown in Figure 30.10. The dominant sporophyte produces windblown spores. When the spores germinate, a tiny green and independent gametophyte develops. The gametophyte is water dependent because it lacks vascular tissue. Also, flagellated sperm produced within antheridia require an outside source of moisture to swim to the eggs in the archegonia. Upon fertilization the zygote develops into a sporophyte. In nearly all ferns, the leaves of the sporophyte first appear in a curled-up form called a fiddlehead, which unrolls as it grows. Once established, some ferns, such as the bracken fern *Pteridium aquilinum,* can spread into drier areas by means of vegetative (asexual) reproduction. For example, ferns can spread when their rhizomes grow horizontally in the soil, producing fiddleheads that become new fronds.

In terms of economic value, people use ferns in decorative bouquets and as ornamental plants in the home and garden. Wood from tropical tree ferns often serves as a building material because it resists decay, as well as damage by termites. Ferns, especially the ostrich fern, are eaten as food, and in the Northeast, many restaurants feature fiddleheads as a special treat, although studies have shown that some are carcinogenic.

frond (undivided)
spores on fertile frond
Cinnamon fern, *Osmunda cinnamomea*
Hart's tongue fern, *Campyloneurum scolopendrium*

Figure 30.9 Fern diversity. The many different types of ferns grow in places that offer moisture. These photos show the dominant sporophyte. The separate, delicate, and almost microscopic gametophyte is water dependent and produces sperm that require moisture to swim to the egg.

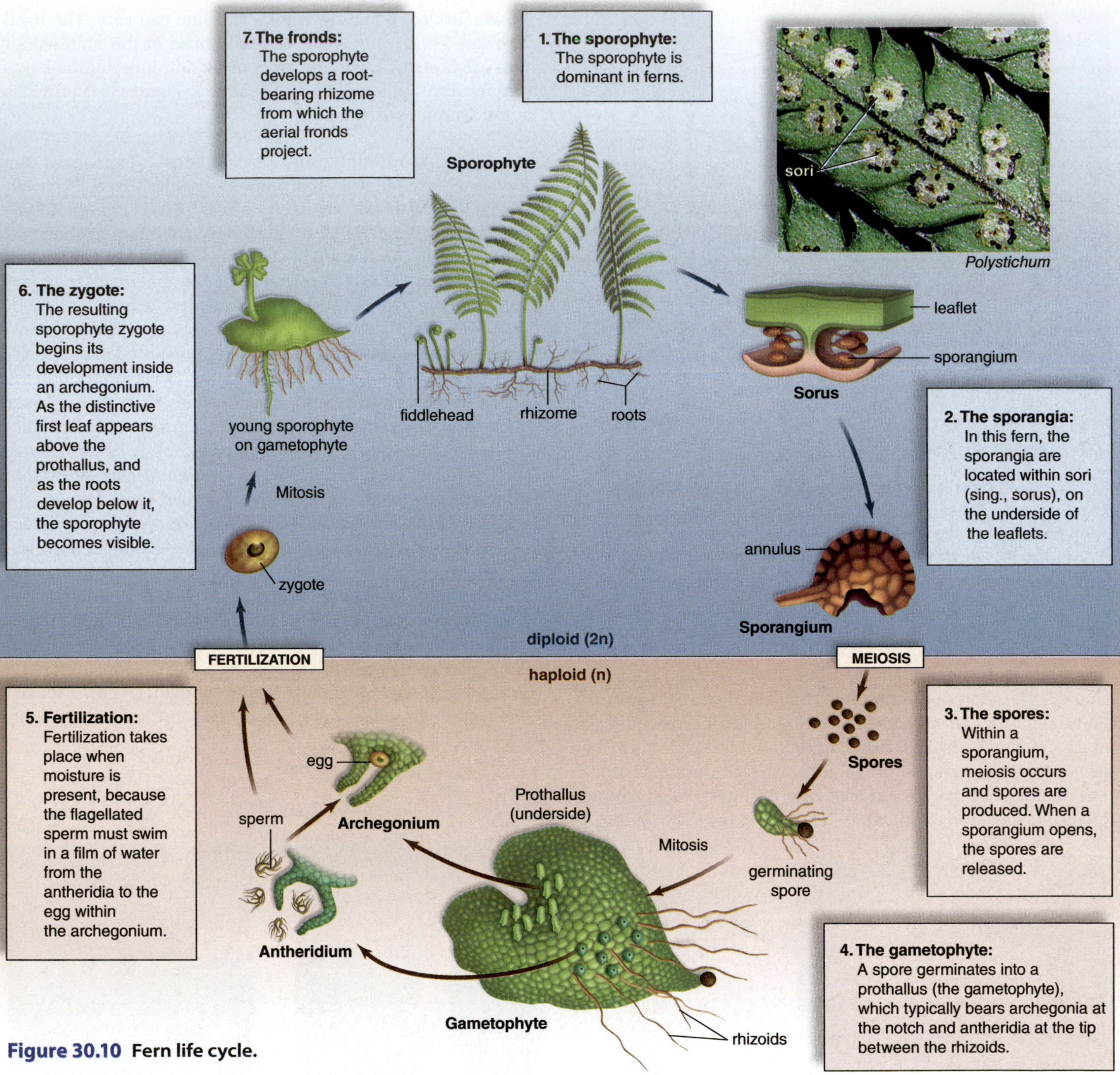
7. The fronds:
The sporophyte develops a root-bearing rhizome from which the aerial fronds project.
1. The sporophyte:
The sporophyte is dominant in ferns.
sori
Polystichum
Sporophyte
6. The zygote:
The resulting sporophyte zygote begins its development inside an archegonium. As the distinctive first leaf appears above the prothallus, and as the roots develop below it, the sporophyte becomes visible.
young sporophyte on gametophyte
fiddlehead
rhizome
roots
leaflet
sporangium
Sorus
2. The sporangia:
In this fern, the sporangia are located within sori (sing., sorus), on the underside of the leaflets.
Mitosis
zygote
annulus
Sporangium
diploid (2n)
haploid (n)
FERTILIZATION
MEIOSIS
5. Fertilization:
Fertilization takes place when moisture is present, because the flagellated sperm must swim in a film of water from the antheridia to the egg within the archegonium.
egg
sperm
Archegonium
Antheridium
Prothallus (underside)
Mitosis
Spores
germinating spore
3. The spores:
Within a sporangium, meiosis occurs and spores are produced. When a sporangium opens, the spores are released.
Gametophyte
rhizoids
4. The gametophyte:
A spore germinates into a prothallus (the gametophyte), which typically bears archegonia at the notch and antheridia at the tip between the rhizoids.

Figure 30.10 Fern life cycle.

Whisk Ferns and Horsetails

Whisk ferns (phylum Psilotophyta), named for their resemblance to whisk brooms, are found in Arizona, Texas, Louisiana, and Florida, as well as Hawaii and Puerto Rico. *Psilotum* (Fig. 30.11) looks like a rhyniophyte, an ancient vascular plant that is known only from the fossil record. Its aerial stem forks repeatedly and is attached to a *rhizome,* a fleshy horizontal stem that lies underground. Whisk ferns have no leaves, so the branches carry on photosynthesis. Sporangia, located at the ends of short branches, produce spores that are dispersal agents.

The *Psilotum* gametophyte is independent, small (less than a centimeter), and found underground associated with mutualistic mycorrhizal fungi. The resemblance of its life cycle to that of a fern suggests that *Psilotum* is actually a fern devoid of leaves and roots. If so, it could be said to be a vestigial fern.

Horsetails (phylum Equisetophyta), which thrive in moist habitats around the globe, are represented by *Equisetum,* the only genus in existence today (Fig. 30.12). A perennial rhizome produces roots and photosynthetic aerial stems that can stand about 1.3 m high. In some species, the whorls of slender

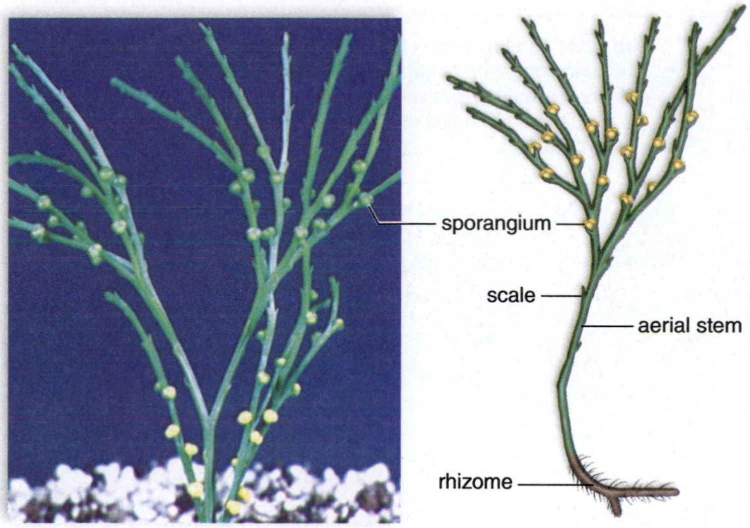
sporangium
scale
aerial stem
rhizome

Figure 30.11 Whisk fern, *Psilotum*. Whisk ferns have no roots or leaves—the branches carry on photosynthesis. The sporangia are yellow.

strobilus
branches
node
leaves
rhizome
root

Figure 30.12 Horsetail, *Equisetum*. Whorls of branches and tiny leaves are at the nodes of the stem. Spore-producing sporangia are borne in strobili (cones).

green side branches at the joints (nodes) of the stem bear a fanciful resemblance to a horse's tail. The small, scalelike leaves are whorled at the nodes and are nonphotosynthetic. Many horsetails have strobili at the tips of branch-bearing stems. Others send up special buff-colored, naked stems that bear the strobili.

The stems of horsetails are tough and rigid because of silica deposited in their cell walls. Early Americans, in particular, used horsetails to scour pots and called them "scouring rushes." Today, they are still used as ingredients in a few abrasive powders. *Equisetum* is sometimes considered a weed in the garden, and it is difficult to control because any portion of the rhizome left in the soil will regenerate.

Check Your Progress 30.3

1. Explain what makes lycophytes significant in plant evolution.
2. Describe two significant differences between the fern and moss life cycle.
3. Identify why ferns are the more plentiful group of seedless vascular plants.

30.4 Seed Plants

Learning Outcomes

Upon completion of this section, you should be able to

1. List the features of seed plants that make them fully adapted to life on land.
2. Compare and contrast gymnosperms with angiosperms, explaining why angiosperms are more species-rich today.
3. List the key features of monocots and eudicots.

Gymnosperms and angiosperms are categorized as seed plants and are the most plentiful plants on the Earth today. Seeds contain a sporophyte embryo and stored food within a protective seed coat. The seed coat and stored food allow an embryo to survive harsh conditions during long periods of dormancy (arrested state) until environmental conditions become favorable for growth. Seeds can even remain dormant for hundreds of years. When a seed germinates, the stored food is a source of nutrients for the growing seedling. The survival value of seeds largely accounts for the dominance of seed plants today.

Seed plants, such as the gymnosperms, are *heterosporous,* meaning that they have two types of spores and produce a male and female gametophyte during the course of their life cycle (Fig. 30.13). **Pollen grains,** which resist drying out, become the multicellular male gametophyte. **Pollination** occurs when a pollen grain is brought to the vicinity of a female gametophyte by wind or, in the case of flowering plants, by either wind or a pollinator. A **pollinator** is an animal, often an insect, that carries pollen from flower to flower. After a pollen grain germinates, sperm move toward the female gametophyte through a growing **pollen tube.** External water is not needed to accomplish fertilization. The female gametophyte develops within an **ovule,** eventually becoming a seed that disperses the offspring.

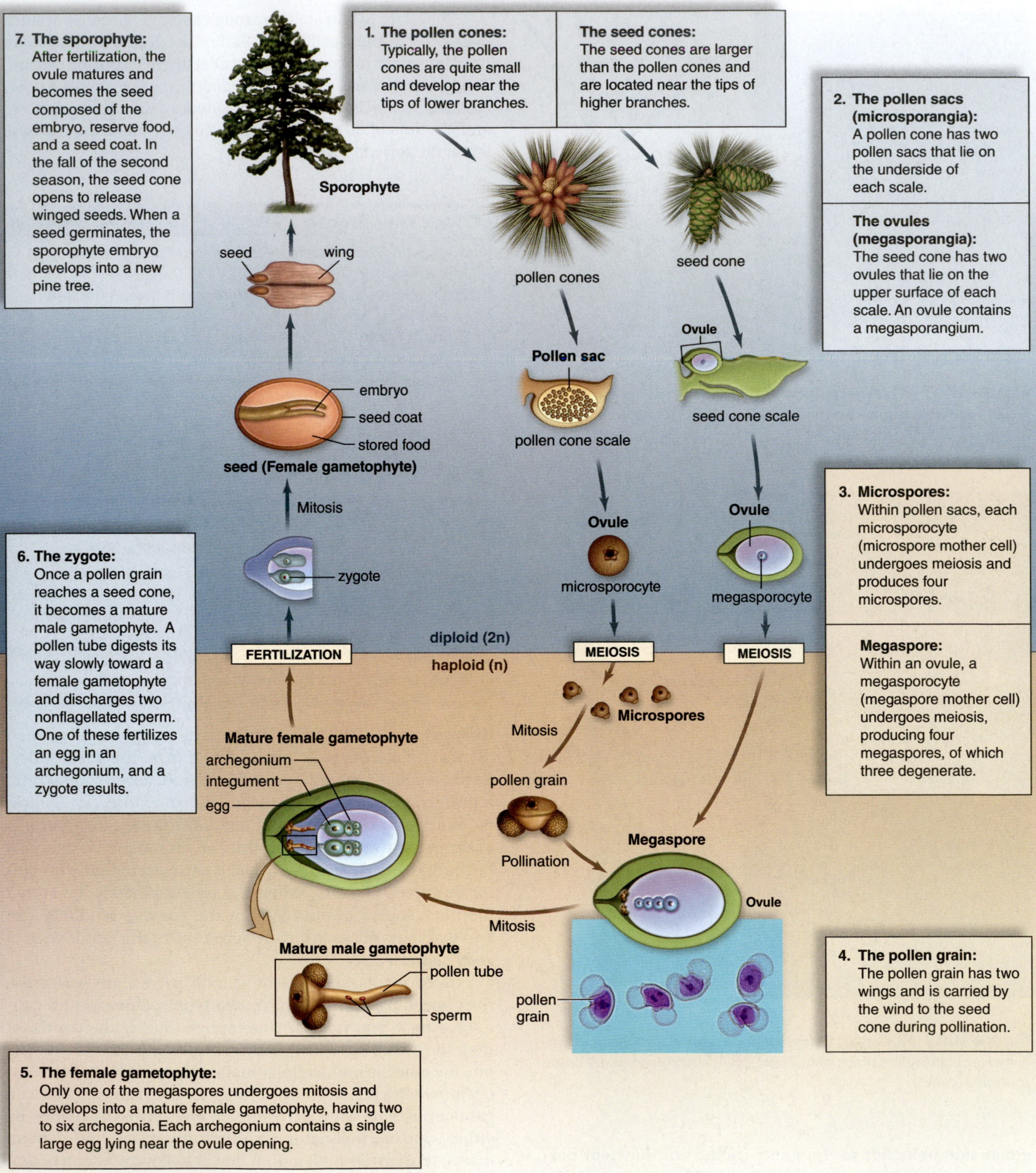
7. The sporophyte: After fertilization, the ovule matures and becomes the seed composed of the embryo, reserve food, and a seed coat. In the fall of the second season, the seed cone opens to release winged seeds. When a seed germinates, the sporophyte embryo develops into a new pine tree.
1. The pollen cones: Typically, the pollen cones are quite small and develop near the tips of lower branches.
The seed cones: The seed cones are larger than the pollen cones and are located near the tips of higher branches.
2. The pollen sacs (microsporangia): A pollen cone has two pollen sacs that lie on the underside of each scale.
The ovules (megasporangia): The seed cone has two ovules that lie on the upper surface of each scale. An ovule contains a megasporangium.
Sporophyte
seed
wing
pollen cones
seed cone
Ovule
Pollen sac
embryo
seed coat
stored food
seed (Female gametophyte)
pollen cone scale
seed cone scale
Mitosis
Ovule
Ovule
zygote
microsporocyte
megasporocyte
3. Microspores: Within pollen sacs, each microsporocyte (microspore mother cell) undergoes meiosis and produces four microspores.
Megaspore: Within an ovule, a megasporocyte (megaspore mother cell) undergoes meiosis, producing four megaspores, of which three degenerate.
6. The zygote: Once a pollen grain reaches a seed cone, it becomes a mature male gametophyte. A pollen tube digests its way slowly toward a female gametophyte and discharges two nonflagellated sperm. One of these fertilizes an egg in an archegonium, and a zygote results.
diploid (2n)
FERTILIZATION
MEIOSIS
MEIOSIS
haploid (n)
Microspores
Mitosis
Mature female gametophyte
archegonium
integument
egg
pollen grain
Pollination
Megaspore
Ovule
Mitosis
Mature male gametophyte
pollen tube
sperm
pollen grain
4. The pollen grain: The pollen grain has two wings and is carried by the wind to the seed cone during pollination.
5. The female gametophyte: Only one of the megaspores undergoes mitosis and develops into a mature female gametophyte, having two to six archegonia. Each archegonium contains a single large egg lying near the ovule opening.

Figure 30.13 Pine life cycle. The sporophyte is dominant, and its sporangia are borne in cones. There are two types of cones: pollen cones (microstrobili) and seed cones (megastrobili).

Gymnosperms

Most of the approximately 800 to 1,000 species of **gymnosperms** are cone-bearing plants. On the surfaces of their cone scales are ovules, which later become seeds. These ovules are said to be naked because they are not completely enclosed by diploid tissue (as the ovules of angiosperms are). This characteristic gives the gymnosperms their name; *gymnos* is a Greek word meaning naked, and *sperma* means seed. Botanists now divide the gymnosperms into four groups: conifers, cycads, ginkgoes, and gnetophytes.

Conifers

The better-known gymnosperms are evergreen, cone-bearing trees called **conifers** (phylum Pinophyta). The 575 species of conifers include the familiar pine, spruce, fir, cedar, hemlock, redwood, and cypress. Perhaps the oldest and largest trees in the world are conifers. Bristlecone pines in the Nevada mountains are known to be more than 4,900 years old, and a number of redwood trees in California are 2,000 years old and more than 100 meters tall.

Adaptations and Uses of Conifers Conifers are adapted to cold, dry weather. Thus, vast areas of northern temperate regions are covered in coniferous forests (Fig. 30.14*a*). The tough, needlelike leaves of pines conserve water because they have a thick cuticle and recessed stomata.

Note in the life cycle of the pine (see Fig. 30.13) that the dominant sporophyte produces two kinds of cones. **Pollen cones** (microstrobili) produce windblown pollen, and **seed cones** (megastrobili) produce windblown seeds (Fig. 30.14*b, c*). These cones are adaptations to a land environment.

Conifers supply much of the wood used to construct buildings and to manufacture paper. They also produce many valuable chemicals, such as those extracted from resin (pitch), a substance that protects conifers from attack by fungi and insects, and is used for waterproofing and in varnish and paints.

Other Gymnosperms

Cycads (phylum Cycadophyta, approximately 130 species) have large, finely divided leaves growing in clusters at the top of the stem. Depending on their height, they resemble palms or ferns.

a. A northern coniferous forest

seed cones
pollen cones

b. Cones of lodgepole pine, *Pinus contorta*

seed cones

c. Fleshy seed cones of juniper, *Juniperus*

Figure 30.14 Conifers. **a.** Conifers are adapted to living in cold climates. **b.** In pine trees, the seed cones become woody, but the pollen cones do not. **c.** The seed cones of a juniper have fleshy scales that fuse into a berrylike structure.

pollen cone   seed cone

a.

b.

c.

Figure 30.15 Gymnosperm diversity. **a.** Cycad cones. **b.** The female ginkgo trees are known for their malodorous seeds. **c.** *Welwitschia* is known for its enormous straplike leaves.

Pollen cones or seed cones, which grow at the top of the stem surrounded by the leaves, can be huge—more than a meter long and weighing 40 kg (Fig. 30.15*a*). Cycads were very plentiful in the Mesozoic era at the time of the dinosaurs, and it's likely that dinosaurs fed on cycad seeds. Now, cycads are in danger of extinction because they grow very slowly, a distinct disadvantage.

Ginkgos (phylum Ginkgophyta), although plentiful in the fossil record, are represented today by only one surviving species, *Ginkgo biloba,* the maidenhair tree, so named because its leaves resemble those of the southern maidenhair fern, *Adiantum capillus-veneris.* Female trees produce fleshy seeds, which ripen in the fall, and give off such a foul odor that male trees are usually preferred for planting (Fig. 30.15*b*). Ginkgo trees are resistant to pollution and do well along city streets and in city parks.

The three living genera of **gnetophytes** don't resemble one another. *Gnetum,* which occurs in the tropics, consists of trees or climbing vines with broad, leathery leaves arranged in pairs. *Ephedra,* occurring only in southwestern North America and Southeast Asia, is a shrub with small, scalelike leaves. *Welwitschia,* living in the deserts of southwestern Africa, has only two enormous, straplike leaves, which split lengthwise as the plant ages (Fig. 30.15*c*).

Angiosperms

Angiosperms, the flowering plants (phylum Magnoliophyta), are an exceptionally large and successful group of plants. There are 250,000 known species of angiosperms. The flowers of the Angiosperms vary widely in appearance (Fig. 30.16). Angiosperms live in all sorts of habitats, from fresh water to desert, and from the frigid north to the torrid tropics. They range in size from the tiny, almost microscopic duck weed to *Eucalyptus* trees over 100 m tall. It would be impossible to exaggerate the importance of angiosperms in our everyday lives. As discussed in the Ecology feature, "Plants and Humans," on page 606, they provide us with clothing, food, medicines, and other commercially valuable products.

The name angiosperm is derived from the Greek words *angio,* meaning vessel, and *sperma,* meaning seed. The seed develops from an ovule within an ovary, which becomes a fruit. Therefore, angiosperms produce covered seeds (in contrast to the exposed seeds of gymnosperms). Although the oldest fossils of angiosperms date back no more than about 135 million years, the angiosperms probably arose much earlier. Indirect evidence suggests the possible ancestors of angiosperms may have originated as long as 200 million years ago. Scientists hypothesize that angiosperms coevolved with the insects that act as pollinators. As angiosperms became diverse, flying insects also enjoyed an adaptive radiation increasing in morphological and ecological diversity.

Monocots and Eudicots

Most flowering plants belong to one of two classes. The **Monocotyledones** (Liliopsida), often shortened to **monocots,** are composed of about 65,000 species. The **Eudicotyledones,**

TABLE 30.1 Monocots and Eudicots

Monocots	Eudicots
One cotyledon	Two cotyledons
Flower parts in threes or multiples of three	Flower parts in fours or fives or multiples of four or five
Usually herbaceous	Woody or herbaceous
Usually parallel venation	Usually net venation
Scattered bundles in stem	Vascular bundles in a ring
Fibrous root system	Taproot system

Barbary fig cactus, *Opuntia ficus-indica*

Water lily, *Nymphaea odorata*

Blue flag iris, *Iris versicolor*

Amborella, *Amborella trichopoda*

Snow trillium, *Trillium nivale*

Apple blossom, *Malus domestica*

Butterfly weed, *Asclepias tuberosa*

Figure 30.16 Flower diversity. Angiosperm flowers vary widely in their appearance, but their function is similar.

shortened to **eudicots** (formerly known as dicots), are composed of about 185,000 species. Table 30.1 lists the differences between monocots and eudicots.

Monocots are named because they have only one cotyledon in their seeds, whereas eudicots have two cotyledons. Cotyledons are the seed leaves—they contain nutrients that nourish the developing embryo.

The Flower

The flowers of Angiosperms have certain structures in common (Fig. 30.17). The flower stalk expands slightly at the tip into a **receptacle,** which bears the other flower parts. These parts, called **sepals, petals, stamens,** and **carpels,** are attached to the receptacle in whorls (circles). The sepals, collectively called the **calyx,** protect the flower bud before it opens. The sepals may drop off or may be colored like the petals. Usually, however, sepals are green and remain attached to the receptacle. The petals, collectively called the **corolla,** are quite diverse in size, shape, and color. The corolla often attracts a particular pollinator.

Each stamen consists of a saclike **anther,** where pollen is produced, and a stalk called a **filament.** In most flowers, the anther is positioned where the pollen can be carried away by wind or a pollinator. One or more carpels is at the center of a flower. A carpel has three major regions: ovary, style, and stigma. The combination of the ovary, style, stigma, and carpels is sometimes called a *pistil.* The swollen base is the **ovary,** which contains from one to hundreds of ovules. The **style** elevates the **stigma,** which is adapted to receive pollen grains. Glands in the region of the ovary produce nectar, a nutrient that pollinators gather as they go from flower to flower. The flower contributes to the success of the angiosperms.

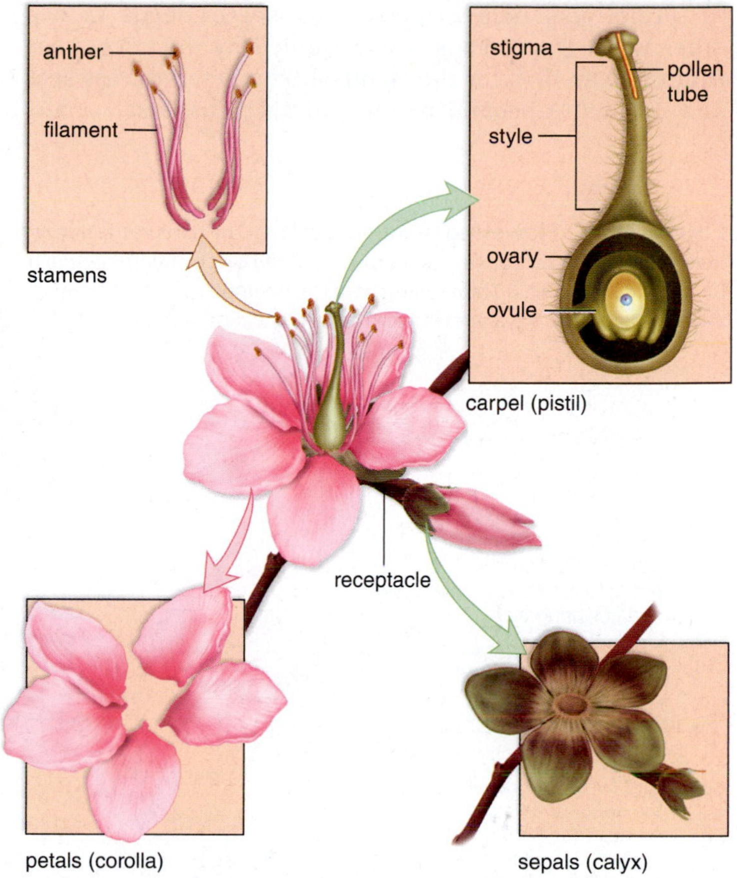
anther
filament
stamens
stigma
pollen tube
style
ovary
ovule
carpel (pistil)
receptacle
petals (corolla)
sepals (calyx)

Figure 30.17 Generalized flower. A flower has four main parts: sepals, petals, stamens, and carpels. Each stamen has an anther and a filament. A carpel has a stigma, a style, and an ovary. The ovary contains ovules.

Life Cycle, Adaptations, and Uses of Flowering Plants

Figure 30.18 depicts the life cycle of a typical flowering plant. Sexual reproduction in flowering plants is dependent on the flower, which produces both pollen and seeds. A gymnosperm, such as a pine tree, depends wholly on the wind to disperse both pollen and seeds. In contrast, among angiosperms, some species have windblown pollen, whereas others rely on a pollinator to carry pollen to another member of the same species. Bees, wasps, flies, butterflies, moths, beetles, and even bats can act as pollinators. The flower provides the pollinator with nectar, a sugary substance that, in the case of bees, will become honey in the hive. Several species of bees (such as honeybees and bumblebees) also have two "pollen baskets," one on each of their hind legs. When bees make a brief stop on a flower, they suck up nectar with specialized mouthparts and gather pollen grains, which are stored in their pollen baskets. Honey and pollen are the main diets of immature bees (called larvae). When a bee visits another flower, some of the pollen grains are rubbed off, and in this way the bee delivers pollen from one flower to another. Other types of pollinators also carry pollen between flowers of the same species, because pollinators and flowers are adapted to one another.

Video Pollinators

Flowering plants that have windblown pollen are usually not very showy, whereas insect-pollinated flowers and bird-pollinated flowers are often colorful. Many flowers are adapted to attract specific pollinators. For example, bee-pollinated flowers are usually blue or yellow and have ultraviolet shadings that lead the pollinator to the location of nectar at the base of the flower. Night-blooming flowers are usually aromatic and white or cream-colored, so that their smell alone, rather than color, can attract nocturnal pollinators such as bats.

Fruits, the final product of a flower, aid in the dispersal of seeds (Fig. 30.19). The production of fruit is another advantage angiosperms have over gymnosperms. Dry fruits, such as pods, assist the dispersal of windblown seeds. Mature pods sometimes explode and shoot out the seeds. Other pods, such as those of peas and beans, simply break open and scatter the seeds. The infamous dandelion produces a one-seeded fruit with an umbrella-like crown of intricately branched hairs. The slightest gust of wind catches the hairs, raising and propelling the fruit into the air like a parachute.

Fruits also help disperse angiosperm seeds in two other ways. One means is exemplified by the coconut, a fruit that can drift for thousands of kilometers across seas and oceans. A second method of seed dispersal occurs because the seeds of some fruits have a fleshy covering that animals eat as a source of food. Only the fleshy part is digested, so the stones or seeds pass through the animal's digestive system and are deposited perhaps far away from

Figure 30.18 Flowering plant life cycle. The parts of the flower involved in reproduction are the stamens and the carpel. Reproduction has been divided into significant stages: female gametophyte development, male gametophyte development, and sporophyte development.

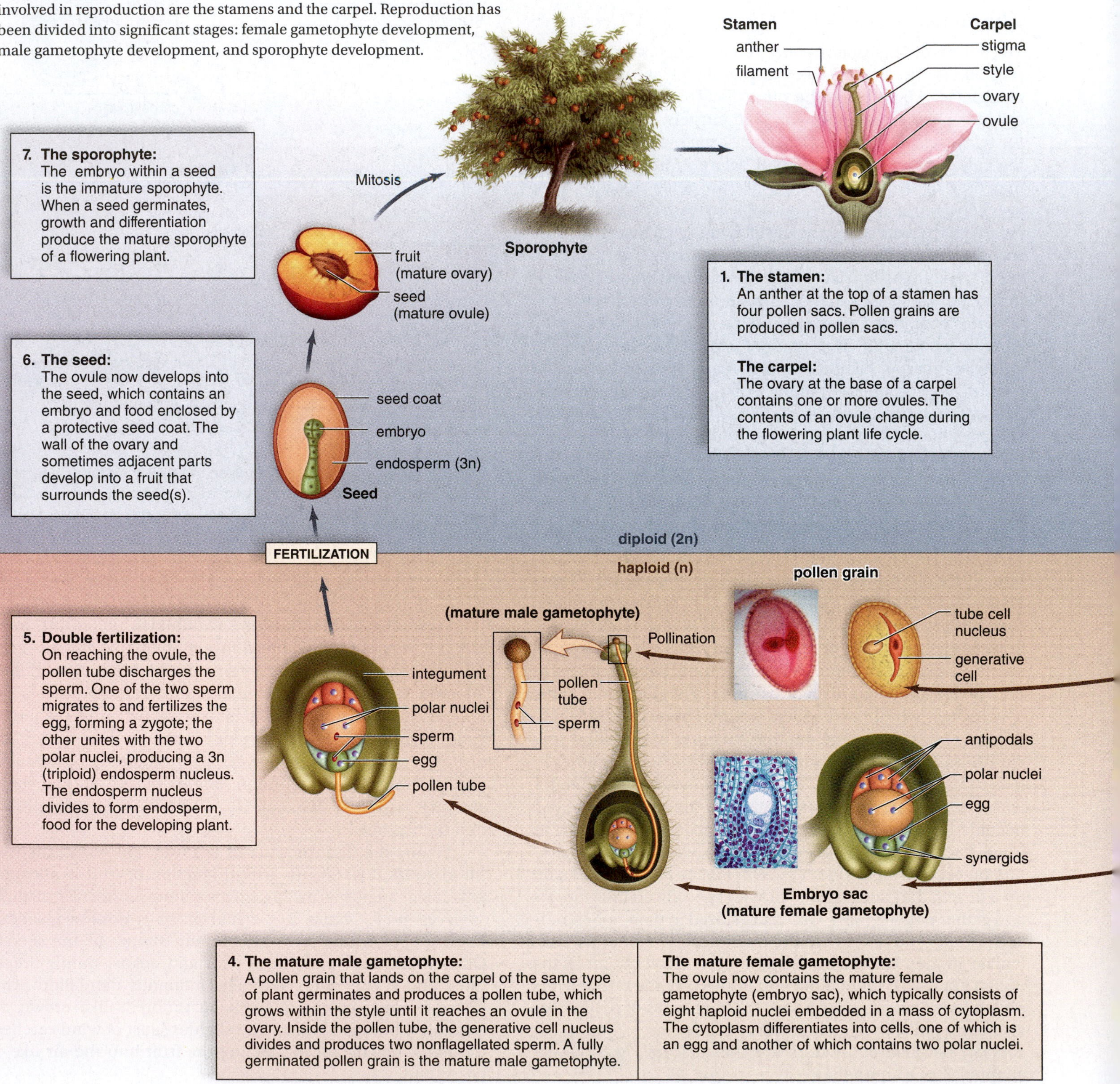
Stamen
anther
filament
Carpel
stigma
style
ovary
ovule
Sporophyte
Mitosis
fruit (mature ovary)
seed (mature ovule)
seed coat
embryo
endosperm (3n)
Seed
FERTILIZATION
diploid (2n)
haploid (n)
pollen grain
tube cell nucleus
generative cell
Pollination
(mature male gametophyte)
pollen tube
sperm
integument
polar nuclei
sperm
egg
pollen tube
antipodals
polar nuclei
egg
synergids
Embryo sac (mature female gametophyte)
1. The stamen: An anther at the top of a stamen has four pollen sacs. Pollen grains are produced in pollen sacs.
The carpel: The ovary at the base of a carpel contains one or more ovules. The contents of an ovule change during the flowering plant life cycle.
4. The mature male gametophyte: A pollen grain that lands on the carpel of the same type of plant germinates and produces a pollen tube, which grows within the style until it reaches an ovule in the ovary. Inside the pollen tube, the generative cell nucleus divides and produces two nonflagellated sperm. A fully germinated pollen grain is the mature male gametophyte.
The mature female gametophyte: The ovule now contains the mature female gametophyte (embryo sac), which typically consists of eight haploid nuclei embedded in a mass of cytoplasm. The cytoplasm differentiates into cells, one of which is an egg and another of which contains two polar nuclei.
5. Double fertilization: On reaching the ovule, the pollen tube discharges the sperm. One of the two sperm migrates to and fertilizes the egg, forming a zygote; the other unites with the two polar nuclei, producing a 3n (triploid) endosperm nucleus. The endosperm nucleus divides to form endosperm, food for the developing plant.
6. The seed: The ovule now develops into the seed, which contains an embryo and food enclosed by a protective seed coat. The wall of the ovary and sometimes adjacent parts develop into a fruit that surrounds the seed(s).
7. The sporophyte: The embryo within a seed is the immature sporophyte. When a seed germinates, growth and differentiation produce the mature sporophyte of a flowering plant.

the parent plant. Still other fruits, like those of the cocklebur, have seeds with hooks that catch on the fur of animals and are carried away.

Video Fruit Bat Seed Dispersal

Video Dung Seed Dispersal

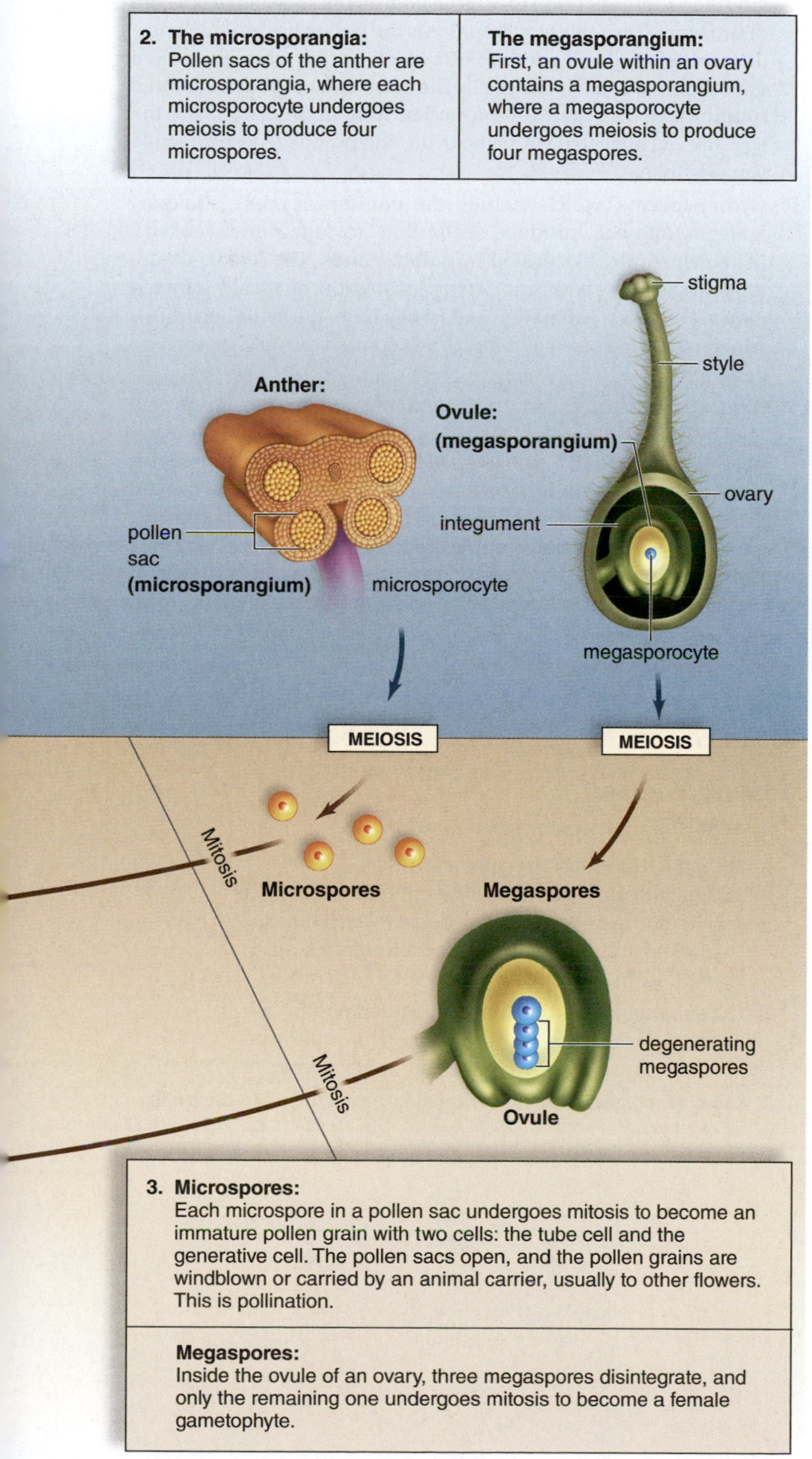
2. The microsporangia:
Pollen sacs of the anther are microsporangia, where each microsporocyte undergoes meiosis to produce four microspores.
The megasporangium:
First, an ovule within an ovary contains a megasporangium, where a megasporocyte undergoes meiosis to produce four megaspores.
stigma
style
Anther:
Ovule: (megasporangium)
ovary
integument
pollen sac (microsporangium)
microsporocyte
megasporocyte
MEIOSIS
MEIOSIS
Mitosis
Microspores
Megaspores
degenerating megaspores
Mitosis
Ovule
3. Microspores:
Each microspore in a pollen sac undergoes mitosis to become an immature pollen grain with two cells: the tube cell and the generative cell. The pollen sacs open, and the pollen grains are windblown or carried by an animal carrier, usually to other flowers. This is pollination.
Megaspores:
Inside the ovule of an ovary, three megaspores disintegrate, and only the remaining one undergoes mitosis to become a female gametophyte.

a.
b.
c.
d.
e.
f.

Figure 30.19 Fruits. Angiosperms have diverse fruits. **a.** A tomato is technically a berry. **b.** A strawberry is an aggregate fruit formed from many carpels within a single flower. **c.** Apples have seeds inside a fleshy fruit. **d.** A corn cob carries many fruits (kernels). **e.** Peas have seeds in pods. **f.** Dandelion fruits are carried by wind.

Check Your Progress 30.4

1. Recognize the two types of spores that occur in the seed plant life cycle, and describe where they develop, and what role they play in the life cycle.
2. List the features of the flowering plant life cycle that are not found in any other group of plants.
3. Distinguish between a monocot and a eudicot plant.

Case Study Conclusion

If we suddenly lost all of the plants in the world, the animals would soon follow. Plants, and particularly angiosperms, provide us important products, such as food, clothing, and medicines. Land plants also provide shelter for many animals, oxygen to the air, and help buffer noise in cities. The earliest plants to colonize land were the bryophytes, low-lying nonvascular plants, approximately 450 MYA. Seedless vascular plants evolved about 350 MYA and possessed vascular tissue. Gymnosperms were the first seed plants to have evolved approximately 365 MYA followed by the angiosperms between 200 and 135 MYA. Each group of plants evolved unique features that provided them an advantage over the previous groups. Currently, the angiosperms are the dominant group of plants on the planet.

MEDIA STUDY TOOLS

www.mhhe.com/maderinquiry14

Enhance your study of this chapter with study tools and practice tests. Also ask your instructor about the resources available through ConnectPlus, including LearnSmart, the media-rich eBook, interactive learning tools, and animations.

Mc Graw Hill connect plus+ BIOLOGY

SUMMARIZE

30.1 Evolutionary History of Plants

- Plants are multicellular, photosynthetic organisms adapted to a land existence that probably evolved from a multicellular, freshwater green alga about 500 MYA.
- Five significant events associated with plant adaptation to a land existence are (1) embryo protection, (2) **vascular tissue,** (3) **megaphylls,** (4) **seeds,** and (5) the flower.
- The life cycle of plants is characterized by **alternation of generations**—some have a dominant **gametophyte** (haploid) generation and others have a dominant **sporophyte** (diploid) generation.
- Land plants are divided into nonvascular plants, seedless vascular plants, and seed plants based on presence/absence of vascular tissue and their dispersal mechanism.

30.2 Nonvascular Plants

- **Nonvascular plants,** including **liverworts** and **mosses,** lack vascular tissue—and therefore, they do not have true roots, stems, or leaves.
- The gametophyte is dominant. **Antheridia** produce swimming sperm that use external water to reach eggs in the **archegonia.** Following fertilization, the dependent moss sporophyte consists of a foot, a stalk, and a capsule within which windblown spores are produced by meiosis.
- Use of windblown spores to disperse the gametophyte is nonvascular plants' chief adaptation for reproducing on land. Each spore germinates to produce a gametophyte.

30.3 Seedless Vascular Plants

- **Vascular plants** have vascular tissue—**xylem** and **phloem.** Xylem transports water and minerals and also supports plants against the pull of gravity. Phloem transports organic nutrients from one part of the plant to another.
- Lycophytes are often called **club mosses.** They contain true roots, stems, and leaves because they possess vascular tissue. Lycophytes have narrow leaves called microphylls.
- Lycophytes and **ferns** and their allies (whisk ferns and **horsetails**) were prominent in swamp forests of the Carboniferous period.
- The sporophyte is dominant; in the life cycle of seedless vascular plants, windborne spores disperse the gametophyte, and the separate gametophyte produces flagellated sperm within antheridia and eggs within archegonia.
- Vegetative (asexual) reproduction is sometimes used to disperse ferns in dry habitats.

30.4 Seed Plants

- Seed plants have a life cycle that is fully adapted to existence on land. The spores develop inside the body of the sporophyte and are of two types. The male gametophyte is the **pollen grain,** which produces nonflagellated sperm. **Pollination** occurs when the pollen grain is transferred to the egg by the wind or carried by an animal (**pollinator**). The sperm use a **pollen tube** in order to reach the egg. The egg is located within an ovule and produced by the female gametophyte.
- **Gymnosperms,** which include the **conifers, cycads, ginkgos,** and **gnetophytes,** produce seeds that are uncovered (naked). Male gametophytes develop in **pollen cones.** The female gametophyte develops within an ovule located on the scales of **seed cones.** Following pollination and fertilization, the ovule becomes a winged seed dispersed by wind. Seeds contain the next sporophyte generation.
- **Angiosperms,** also called the flowering plants, produce seeds covered by fruits. The **petals** of flowers attract pollinators, and the ovary develops into a fruit, which aids seed dispersal. **Monocotyledones (monocots)** and **Eudicotyledones (eudicots)** are the two main classes of flowering plants. A flower consists of a **corolla** (petals), **calyx (sepals), stamens (anther, filament),** and **carpel (stigma, style, ovary,** and **ovule**). Angiosperms provide most of the food that sustains terrestrial animals, and they are the source of many products used by humans.

ASSESS

Testing Yourself

Choose the best answer for each question.

1. The spores produced by a plant are
   a. haploid and genetically different from each other.
   b. haploid and genetically identical to each other.
   c. diploid and genetically different from each other.
   d. diploid and genetically identical to each other.
2. The gametophyte is the dominant generation in
   a. ferns.
   b. mosses.
   c. gymnosperms.
   d. angiosperms.
3. Label the stages in the following diagram of the plant life cycle.

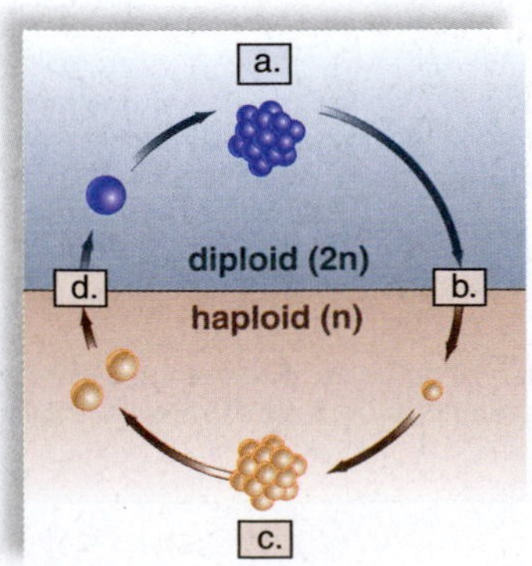
a.
diploid (2n)
haploid (n)
b.
c.
d.

4. In mosses, meiosis occurs in
   a. antheridia.
   b. archegonia.
   c. capsules.
   d. protonema.

For questions 5–10, match the description with the appropriate plant group in the key. Some answers can be used more than once. Some questions will have more than one answer.

Key:

a. nonvascular plants
b. seedless vascular plants
c. gymnosperms
d. angiosperms

5. Contain xylem and phloem
6. Produce seeds
7. Produce flagellated sperm
8. Gametophyte is separate from sporophyte
9. Produce seeds on scales
10. Produce flowers
11. In contrast to seed plants, seedless vascular plants
    a. are not heterosporous.
    b. produce gametophytes.
    c. have male gametophytes that travel to the female gametophytes.
    d. produce ovules.
12. Which statement about the conifer life cycle is false?
    a. Meiosis produces spores.
    b. The gametophyte is the dominant generation.
    c. The seed is a dispersal agent.
    d. Male gametophytes are carried by the wind.
13. Cycads are in danger of extinction because
    a. they occupy very specialized niches.
    b. they have unusual mineral nutrient requirements.
    c. they grow very slowly.
    d. their natural habitats are becoming too warm.
14. Label the parts of the flower in the following diagram.

a.
b.
c.
d.
e.

15. Unlike eudicots, monocots have
    a. woody tissue.
    b. two seed leaves.
    c. scattered vascular bundles in their stems.
    d. flower parts in multiples of fours or fives.

ENGAGE

Virtual Lab
Classifying Using Biotechnology

The virtual lab "Classifying Using Biotechnology" provides for an interactive exploration of how scientists analyze molecular data to establish evolutionary relationships.

Thinking Critically

1. What characteristics of angiosperms have allowed this group of plants to dominate Earth?
2. Compare and contrast the plant alternation-of-generations life cycle with the human life cycle.

31 Animals: The Invertebrates

CHAPTER OUTLINE

BEFORE YOU BEGIN

Before beginning this chapter, take a few moments to review the following discussions:

Figure 1.5 What is the basis for classifying an organism as a member of kingdom Animalia?

Table 3.2 Which cellular features are unique to animals?

Table 27.1 When did animals first appear in the tree of life?

CASE STUDY Neglected Tropical Diseases

A fad diet called the "tapeworm diet" proposes that people intentionally ingest human tapeworms, a parasitic invertebrate, to decrease their body's absorption of nutrients, thus promoting weight loss. The side effects of tapeworm infection, however, are potentially severe. In underdeveloped nations the World Health Organization (WHO) lists cysticercosis, an advanced tapeworm infection, as a serious neglected tropical disease (NTD) that threatens more than 50 million people worldwide. Cysticercosis causes severe malnutrition in growing children, as well as anemia and organ damage in children and adults. This fad diet has been banned by the FDA in the United States. Public education about the risks of parasite infection is our greatest defense against such a potentially life-threatening pursuit of an ideal body type. The invertebrates—animals that lack a backbone—are far more diverse and numerous than the vertebrates. Many invertebrates enhance the quality of our lives. For example, various types of insects act as essential pollinators of many food crops. However, not all invertebrates are beneficial. Mosquitoes can be vectors of malaria whereas some parasitic worms are responsible for other NTDs such as elephantiasis and schistosomiasis.

In this chapter, you will read about the evolution and diversity of the major groups of invertebrates as well as the traits that have enabled them to successfully adapt to a wide range of habitats.

As you read through the chapter, think about the following questions:

1. Where do invertebrates fit into the evolutionary history of animals?
2. What evolutionary advantages are there for invertebrates to have a variety of developmental stages, body plans, and lifestyles?

31.1 Evolutionary Trends Among Animals

Learning Outcomes

Upon completion of this section, you should be able to

1. List the general characteristics of animals.
2. Compare and contrast the types of symmetry found in animals.
3. Describe the differences between protostomes and deuterostomes.

Animals are members of the domain Eukarya and the kingdom Animalia. While animals are extremely diverse (Fig. 31.1), they share some important differences from the other multicellular eukaryotes, the plants and fungi. Unlike plants, which make their food through photosynthesis, animals are heterotrophs, and must acquire nutrients from an external source. Unlike fungi, which digest their food externally and absorb the breakdown products, animals ingest (eat) whole food and digest it internally. In general, animals share the following characteristics:

- Typically have the power of movement or locomotion by means of muscle fibers
- Multicellular; most have specialized cells that form tissues and organs
- Have a life cycle in which the adult is typically diploid
- Usually undergo sexual reproduction and produce an embryo that goes through developmental stages
- Heterotrophic; usually acquire food by ingestion followed by digestion

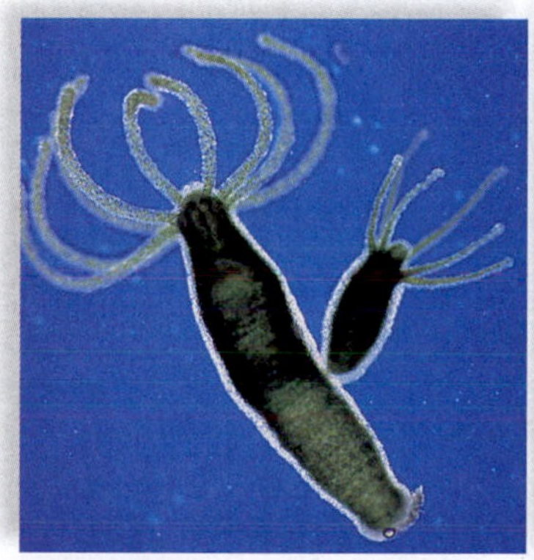

Hydra, *Hydra*

Green crayfish, *Barbi*

Humans, *Homo sapiens*

Figure 31.1 Animal diversity. Hydras, crayfish, and humans are all multicellular heterotrophic animals, but they differ in certain fundamental ways. In terms of level of organization, hydras have only tissues, but crayfish and humans have organ systems. Hydras do not have an identifiable head, whereas crayfish and humans do. Hydras do not have a complete digestive system, whereas crayfish and humans have a complete digestive tract surrounded by a body wall. Crayfish and humans share other features, even though one is an invertebrate and the other is a vertebrate. They both have a skeleton with jointed appendages, but the crayfish skeleton is external, while the human skeleton is internal.

The animal kingdom is currently divided into 35 groups, or phyla. These are recognized to have evolved from a protistan ancestor approximately 600 million years ago. This chapter will explore animals called the invertebrates. **Invertebrates** lack an internal skeleton, or endoskeleton, of bone or cartilage. The invertebrates evolved first, and far outnumber the **vertebrates** (animals with an endoskeleton). Because early animals were simple invertebrates, the fossil record is sparse regarding the early evolution of animals. Therefore, the evolutionary tree shown in Figure 31.2 is primarily based on molecular and anatomical data (Table 31.1). For molecular comparisons, it is assumed that the more closely related two organisms are, the more nucleotide sequences they will have in common (see section 26.4).

Anatomical Data

Although many anatomical criteria are used to classify animals (Fig. 31.2), we will focus on the general characteristics of symmetry and embryonic development.

Type of Symmetry

Animals can be asymmetrical, radially symmetrical, or bilaterally symmetrical. **Asymmetrical** animals have no particular symmetry, such as some species of sponges. **Radial symmetry** means that the animal is organized circularly, similar to a wheel, so that no matter how the animal is sliced longitudinally, mirror images are obtained. **Bilateral symmetry** means that the animal has definite right and left halves; only a longitudinal cut down the center of the animal will produce a mirror image.

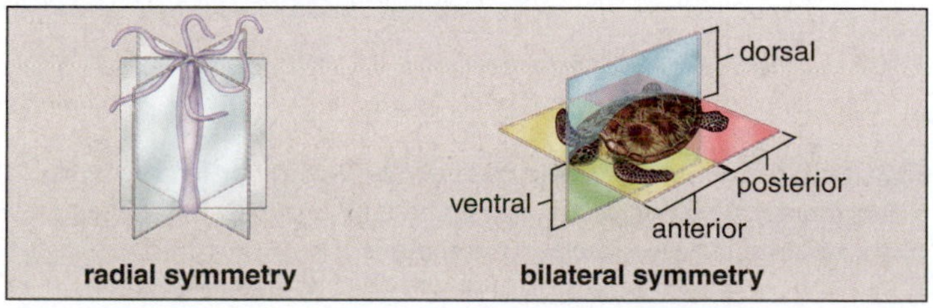
dorsal
posterior
ventral
anterior
radial symmetry
bilateral symmetry

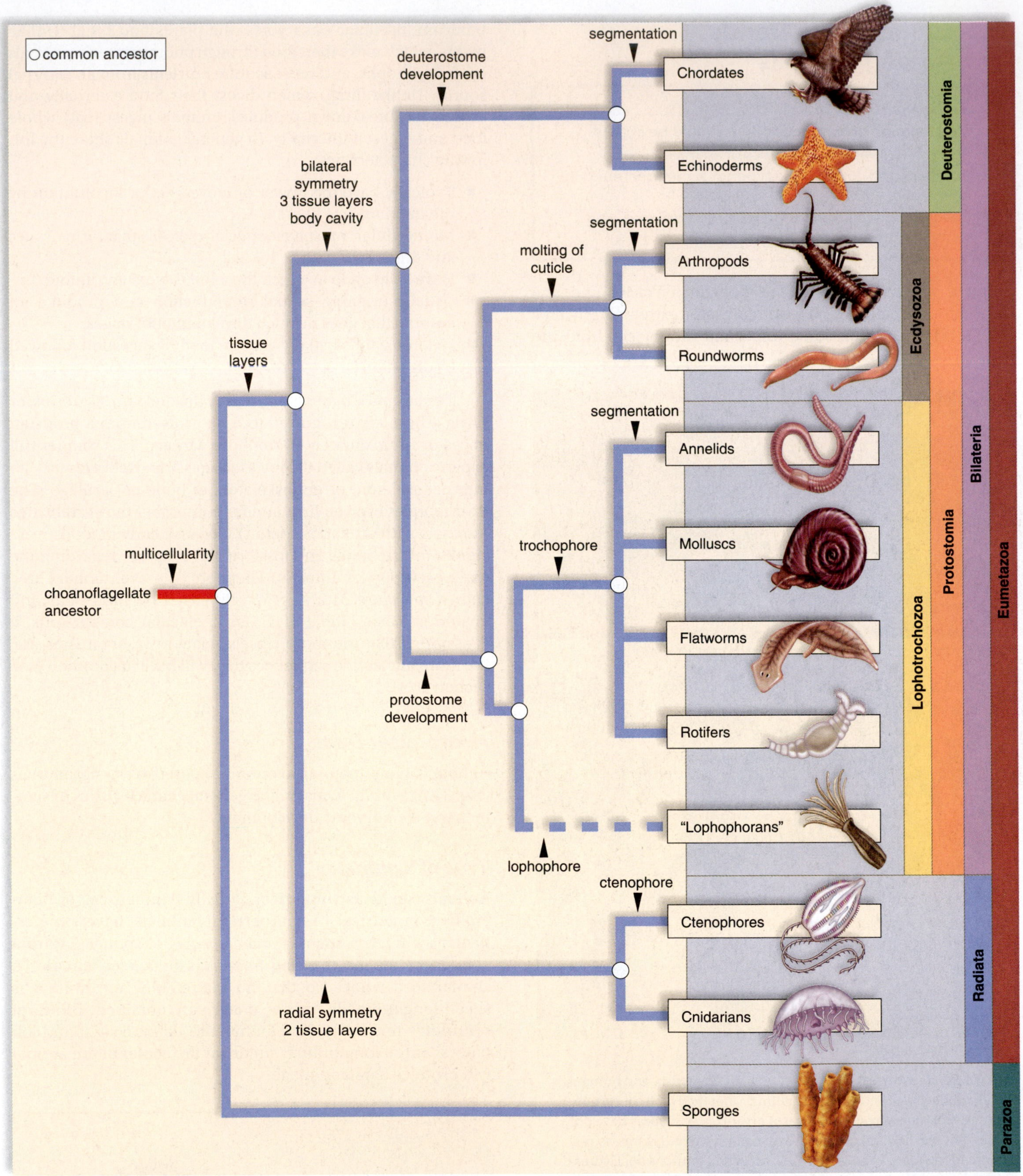

common ancestor
segmentation
deuterostome development
Chordates
Echinoderms
bilateral symmetry
3 tissue layers
body cavity
segmentation
molting of cuticle
Arthropods
Roundworms
tissue layers
segmentation
Annelids
Molluscs
trochophore
multicellularity
choanoflagellate ancestor
Flatworms
protostome development
Rotifers
"Lophophorans"
lophophore
ctenophore
Ctenophores
radial symmetry
2 tissue layers
Cnidarians
Sponges
Deuterostomia
Ecdysozoa
Lophotrochozoa
Protostomia
Bilateria
Eumetazoa
Radiata
Parazoa

Figure 31.2 Phylogenetic tree of the major animal phyla. All animal phyla living today are most likely descended from an ancestral protist living about 600 million years ago. This evolutionary tree is based on molecular data (nucleotide sequences) and anatomical data used to indicate which phyla are most closely related to one another.

TABLE 31.1 Animal Characteristics

DOMAIN: Eukarya
KINGDOM: Animalia

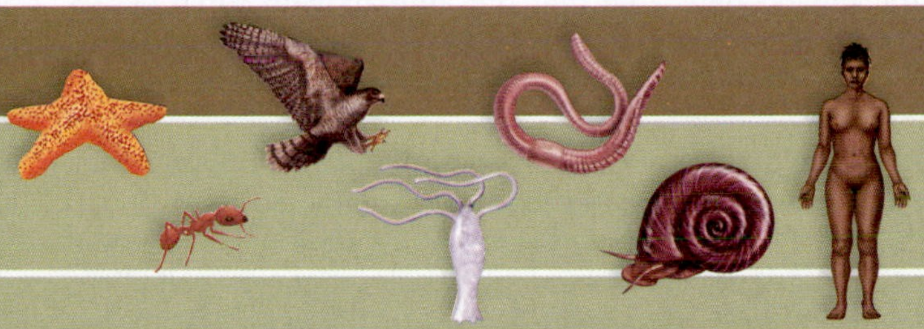

CHARACTERISTICS
Multicellular, usually with specialized tissues; ingest or absorb food; diploid life cycle

INVERTEBRATES

Sponges (bony, glass, spongin): *Asymmetrical, saclike body perforated by pores; internal cavity lined by choanocytes; spicules serve as internal skeleton. 5,150**

Radiata

Comb jellies: Possess eight longitudinal rows of cilia (combs) that aid in locomotion; lack nematocysts. 150

Cnidarians (hydra, jellyfish, corals, sea anemones): Radially symmetrical with two tissue layers; incomplete digestive tract; tentacles with nematocysts. 10,000**

Protostomia (Lophotrochozoa)

Lophophorates (bryozoans): Filter feeders with a mouth surrounded by a ciliated tentacle-like structure. 5,935

Flatworms (planarians, tapeworms, flukes): Bilateral symmetry with cephalization; three tissue layers and organ systems; acoelomate with incomplete digestive tract that can be lost in parasites; hermaphroditic. 20,000**

Rotifers (wheel animals): Microscopic animals with a corona (crown of cilia) that looks like a spinning wheel when in motion. 2,000**

Molluscs (chitons, clams, snails, squids): Coelom, all have a foot, mantle, and visceral mass; foot is variously modified; in many, the mantle secretes a calcium carbonate shell as an exoskeleton; true coelom and all organ systems. 110,000**

Annelids (polychaetes, earthworms, leeches): Segmented with body rings and setae; cephalization in some polychaetes; hydroskeleton; closed circulatory system. 16,000**

Protostomia (Ecdysozoa)

Roundworms (*Ascaris*, pinworms, hookworms, filarial worms): Pseudocoelom and hydroskeleton; complete digestive tract; free-living forms in soil and water; parasites common. 25,000**

Arthropods (crustaceans, spiders, scorpions, centipedes, millipedes, insects): Chitinous exoskeleton with jointed appendages undergoes molting; insects—most have wings—are most numerous of all animals. 1,000,000**

Deuterostomes

Echinoderms (sea stars, sea urchins, sand dollars, sea cucumbers): Radial symmetry as adults; unique water-vascular system and tube feet; endoskeleton of calcium plates. 7,000**

Chordates (tunicates, lancelets, vertebrates): All have notochord, dorsal tubular nerve cord, pharyngeal pouches, and postanal tail at some time; contains mostly vertebrates in which notochord is replaced by vertebral column. 56,000**

VERTEBRATES

Fishes (jawless, cartilaginous, bony): Endoskeleton, jaws, and paired appendages in most; internal gills; single-loop circulation; usually scales. 28,000**

Amphibians (frogs, toads, salamanders): Most have jointed limbs; lungs; three-chambered heart with double-loop circulation; moist, thin skin. 6,900**

Reptiles (snakes, turtles, crocodiles): Amniotic egg; rib cage in addition to lungs; three- or four-chambered heart typical; scaly, dry skin; copulatory organ in males and internal fertilization. 8,000**

Birds*** (songbirds, waterfowl, parrots, ostriches): Endothermy, feathers, and skeletal modifications for flying; lungs with air sacs; four-chambered heart. 10,000**

Mammals (monotremes, marsupials, placental): Hair and mammary glands. 4,800**

*After a character is listed, it is present in the rest, unless stated otherwise.
**Estimated number of species.
***Birds are commonly classified with the reptiles.

Radially symmetrical animals have the advantage of being able to reach out in all directions from one center. Radially symmetrical animals may be permanently attached to a substrate (**sessile**) or free floating, such as jellyfish. Bilaterally symmetrical animals tend to be more active, with movement directed toward an anterior end. During the evolution of animals, bilateral symmetry is accompanied by **cephalization,** localization of a brain and specialized sensory organs at the anterior end (head).

Embryonic Development

Sponges are multicellular animals, but unlike other animals, they do not have true tissues. Therefore, sponges have the *cellular level of organization.* Because of this, sponges are often classified as a separate group, called the parazoans (Fig. 31.2), from the other animals. *True tissues* appear in the other animals, the eumetazoans, as they undergo embryological development. The first three tissue layers are often called germ layers because they give rise to the organs and organ systems of complex animals. Animals such as the cnidarians, are diploblastic, meaning that they have only two tissue layers (ectoderm and endoderm) as embryos. These animals diplay the tissue level of organization (see Fig. 1.2). Those animals that develop further and have all three tissue layers (ectoderm, mesoderm, and endoderm) as embryos are triploblastic and have the organ level of organization. Notice in the phylogenetic tree that the animals with three tissue layers are either **protostomes** (*proto* is Greek for first; *stoma* is Greek for mouth) or **deuterostomes** (*deuter* is Greek for second).

Figure 31.3 shows that protostome and deuterostome development are differentiated by three major events:

1. Cleavage, the first event of development, is cell division without cell growth. In protostomes, spiral cleavage occurs, and daughter cells sit in grooves formed by the previous cleavages. The fate of these cells is fixed and determinate in protostomes; each can contribute to development in only one particular way. In deuterostomes, radial cleavage occurs, and the daughter cells sit right on top of the previous cells. The fate of these cells is indeterminate—that is, if they are separated from one another, each cell can go on to become a complete organism.
2. As development proceeds, a hollow sphere of cells, or blastula, forms and the indentation that follows develops into an opening called the blastopore. In protostomes, the mouth appears at or near the blastopore, hence the origin of their name. In deuterostomes, the anus appears at or near the blastopore, and only later does a second opening form the mouth, hence the origin of their name.

Video Blastocyst Formation

3. Certain members of the protostomes and all deuterostomes have a body cavity completely lined by mesoderm, called a true coelom. However, a true coelom develops differently in the two groups. In protostomes, the mesoderm arises from cells located near the embryonic blastopore, and a splitting occurs that produces the coelom. In deuterostomes, the coelom arises as a pair of mesodermal pouches from the wall of the primitive gut. The pouches enlarge until they meet and fuse.

The deuterostomes include the echinoderms and the chordates, two groups of animals that will be examined in detail in Chapter 32. The protostomes are divided into two groups: the **ecdysozoa** and the **lophotrochozoa.** The ecdysozoans include the roundworms and arthropods. Both of these types of animals molt; they shed their outer covering as they grow. Ecdysozoa means molting animals. The lophotrochozoa contain the **lophophorans,** which have a mouth surrounded by ciliated tentacle-like structures, and the **trochophores,** which either have presently, or their ancestors had, a trochophore larva.

Check Your Progress 31.1

1. Distinguish an animal from a plant or fungi.
2. List the general characteristics that animals have in common.
3. Using Figure 31.2, identify the general characteristics of the ecdysozoa.

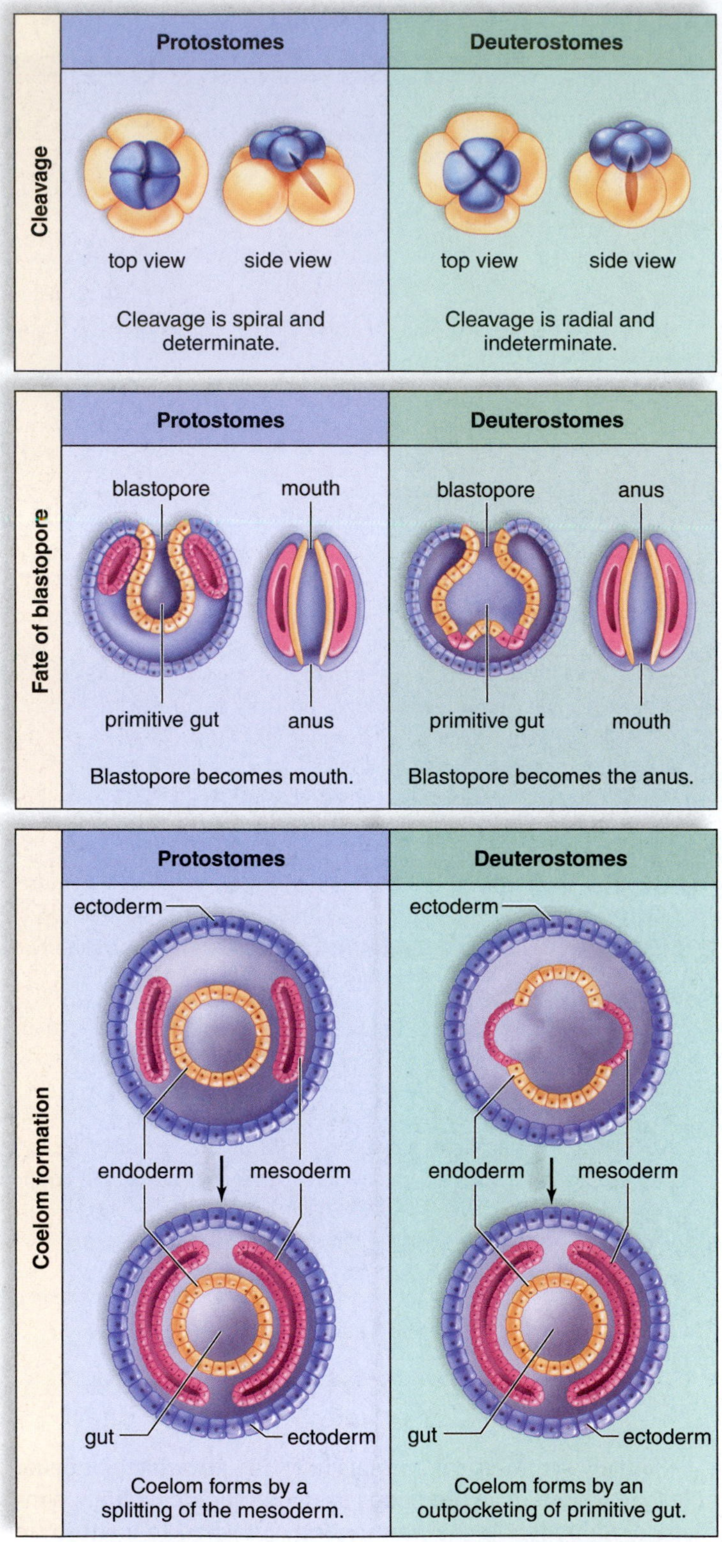
Cleavage
Protostomes
Deuterostomes
top view
side view
top view
side view
Cleavage is spiral and determinate.
Cleavage is radial and indeterminate.
Fate of blastopore
Protostomes
Deuterostomes
blastopore
mouth
blastopore
anus
primitive gut
anus
primitive gut
mouth
Blastopore becomes mouth.
Blastopore becomes the anus.
Coelom formation
Protostomes
Deuterostomes
ectoderm
ectoderm
endoderm
mesoderm
endoderm
mesoderm
gut
ectoderm
gut
ectoderm
Coelom forms by a splitting of the mesoderm.
Coelom forms by an outpocketing of primitive gut.

Figure 31.3 Protostomes compared to deuterostomes. *Left:* In the embryo of protostomes, cleavage is spiral—new cells are at an angle to old cells—and each cell has limited potential and cannot develop into a complete embryo; the blastopore is associated with the mouth; and the coelom, if present, develops by a splitting of the mesoderm. *Right:* In deuterostomes, cleavage is radial—new cells sit on top of old cells—and each one can develop into a complete embryo; the blastopore is associated with the anus; and the coelom, if present, develops by an outpocketing of the primitive gut.

31.2 The Simplest Invertebrates

Learning Outcomes

Upon completion of this section, you should be able to

1. Describe the anatomy and life cycle of sponges and cnidarians.
2. Contrast the way of life of a sponge with that of a cnidarian.
3. Identify the basic morphology of a hydra.

Sponges

Sponges are placed in phylum Porifera because their saclike bodies are perforated by many pores (Fig. 31.4). Sponges are aquatic, largely marine animals that vary greatly in size, shape, and color. Sponges are multicellular, although they have few cell types and no nerve or muscle cells or organized tissues. However, despite their apparent simplicity, the molecular data supports the fact that they are at the base of the evolutionary tree of animals. In sponges, the outer layer of the body wall contains flattened epidermal cells, some of which have contractile fibers; the middle layer is a semifluid matrix with wandering amoeboid cells; and the inner layer is composed of flagellated cells called **collar cells,** or choanocytes (Fig. 31.4). The beating of the flagella produces water currents that flow through the pores into the central cavity and out through the osculum, the upper opening of the body. Even a simple sponge only 10 cm tall is estimated to filter as much as 100 liters of water each day. It takes this much water to supply the needs of the sponge. A sponge is a sessile **filter feeder,** an organism that filters its food from the water by means of a straining device—in this case, the pores of the walls and the microvilli making up the collar of collar cells. Microscopic food particles that pass between the microvilli are engulfed by the collar cells and digested by them in food vacuoles or are passed to the amoeboid cells for digestion. The amoeboid cells also act as a circulatory device to transport nutrients from cell to cell, and they produce the sex cells (the egg and the sperm) and spicules.

Sponges can reproduce both asexually and sexually. They reproduce asexually by fragmentation followed by regeneration, gemmule formation, or budding. A gemmule is like a spore and is resistant to drying out, freezing, or the lack of oxygen. Gemmule formation occurs in freshwater sponges, and when conditions are favorable, a gemmule gives rise to an adult sponge. During sexual reproduction, sperm are released through the osculum and drawn into other sponges through the pores, where they fertilize eggs within the body. Most sponges are **hermaphroditic,** meaning they possess both male and female sex organs. However, they usually do not self-fertilize. After fertilization, the zygote develops into a flagellated larva that may swim to a new location. Sponges are capable of regeneration—if the cells of a sponge are separated, they are capable of reassembling and regenerating into a complete and functioning organism!

Sponges are classified on the basis of their skeleton. Some sponges have an internal skeleton composed of **spicules** (Fig. 31.4), small needle-shaped structures with one to six rays. Chalk sponges have spicules made of calcium carbonate; glass sponges

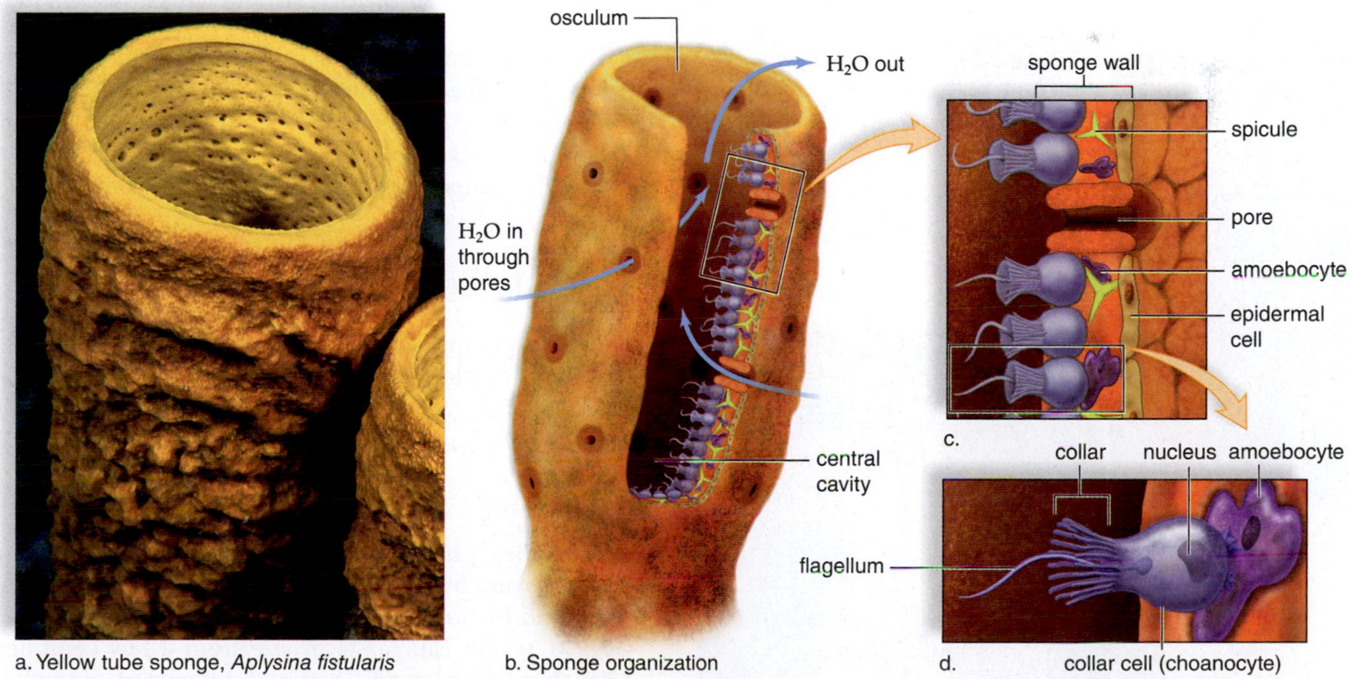
osculum
H_2O out
H_2O in through pores
central cavity
flagellum
sponge wall
spicule
pore
amoebocyte
epidermal cell
c.
collar
nucleus
amoebocyte
d.
collar cell (choanocyte)
a. Yellow tube sponge, *Aplysina fistularis*
b. Sponge organization

Figure 31.4 Simple sponge anatomy. **a.** A simple sponge. **b.** The wall of a sponge contains two layers of cells, the outer epidermal cells and the inner collar cells as shown in (**c**). The collar cells (**d,** enlarged) have flagella that beat, moving the water through pores as indicated by the blue arrows in (**b**). Food particles in the water are trapped by the collar cells and digested within their food vacuoles. Amoebocytes transport nutrients from cell to cell. Spicules form an internal skeleton in some sponges.

SCIENCE IN YOUR LIFE ▶ ECOLOGY

Destruction of the Coral Reefs

Coral reefs are among the most biologically diverse and productive communities on Earth. Coral reefs tend to be found in warm, clear, and shallow tropical waters worldwide and are typically formed by reef-building corals, which are cnidarians. Aside from being beautiful and giving shelter to many colorful species of fishes, coral reefs help generate economic income from tourism, protect ocean shores from erosion, and may serve as the source of medicines derived from antimicrobial compounds that reef-dwelling organisms produce. However, coral reefs around the globe are being destroyed for a variety of reasons, most of them linked to human development. Deforestation, for example, causes tons of soil to settle on the top of coral reefs. This sediment prevents photosynthesis of symbiotic algae that provide food for the corals. When the algae die, so do the corals, which then turn white. This so-called coral "bleaching" has been seen in the Pacific Ocean and the Caribbean Sea. Evidence suggests that climate change is one of the factors that is leading to coral bleaching and death because corals can tolerate only a narrow range of temperatures. As global temperatures rise, so do water temperatures, and corals can die as a result. Global warming also contributes to favorable conditions for various pathogens that can kill corals, such as those similar to pathogens that cause cholera in humans. Increases in aquatic nutrients from fertilizers that wash into the ocean also make corals more susceptible to diseases, which can also kill them. Marine scientist Edgardo Gomez estimates that 90% of coral reefs in the Philippines are dead or deteriorating due to human activities such as pollution and, especially, overfishing. Fishing methods that employ dynamite or cyanide to kill or stun the fish for food or the pet trade can easily kill corals. Paleobiologist Jeremy Jackson of the Smithsonian Tropical Research Institute in Panama estimates that we may lose 60% of the world's coral reefs by the year 2050.

Figure 31A Coral bleaching.

Video Coral Reef Ecosystems

Questions to Consider

1. What features of coral reefs help explain why they are so biologically diverse?
2. Considering what is causing the loss of coral reefs, is it possible to save them? How?

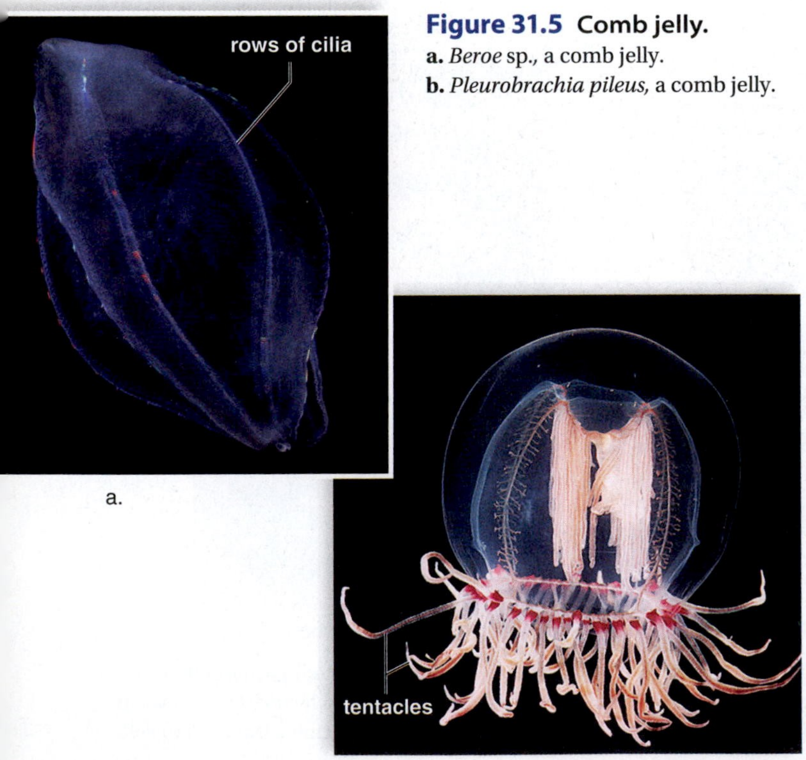
rows of cilia
tentacles
a.
b.

Figure 31.5 Comb jelly.
a. *Beroe* sp., a comb jelly.
b. *Pleurobrachia pileus,* a comb jelly.

have spicules that contain silica. Most sponges have a skeleton composed of fibers of spongin, a soft protein. A bath sponge is the dried spongin skeleton. Today, however, commercial "sponges" are usually synthetic. Note that the popular "loofah" is not really a sponge; it is actually a spongy plant related to cucumbers.

Comb Jellies and Cnidarians

These two groups of animals have true tissues and, as embryos, they have ectoderm and endoderm as their germ layers. They are radially symmetrical as adults, which offers the advantage of being able to reach out in all directions for food.

Comb Jellies

Comb jellies (phylum Ctenophora) are solitary, free-swimming marine invertebrates that are found primarily in warm waters. Ctenophores propel themselves along by beating their cilia (Fig. 31.5) and can range in size from a few centimeters to 1.5 meters (m) in length. Their body is made up of a transparent jellylike substance called **mesoglea.** Most ctenophores capture their prey by using sticky adhesive cells called colloblasts. Some ctenophores are bioluminescent, enabling them to produce their own light.

Video Ctenophores

Cnidarians

Cnidarians (phylum Cnidaria) are tubular or bell-shaped animals that reside mainly in shallow coastal waters. However, there are some freshwater, brackish, and oceanic forms. The term cnidaria is derived from the presence of specialized stinging cells called **cnidocytes** (see Fig. 31.7). Each cnidocyte has a toxin-filled capsule called a **nematocyst** that contains a long, spirally coiled hollow thread. When the trigger of the cnidocyte is touched, the nematocyst is discharged. Some threads merely trap prey, and others have spines that penetrate and inject paralyzing venom.

The body of a cnidarian is a two-layered sac. The outer tissue layer is a protective epidermis derived from ectoderm. The inner tissue layer, which is derived from endoderm, secretes digestive juices into the internal cavity, called the **gastrovascular cavity,** which is involved in the digestion of food and circulation of nutrients. The fluid-filled gastrovascular cavity also serves as a supportive hydrostatic skeleton. This type of skeleton offers some resistance to the contraction of muscle but permits flexibility. The two tissue layers are separated by mesoglea.

Two basic body forms are seen among cnidarians (Fig. 31.6*a*). The mouth of a **polyp** is directed upward, while the mouth of a jellyfish, or medusa, is directed downward. The bell-shaped medusa has more mesoglea than a polyp, and the tentacles are concentrated on the margin of the bell. At one time, both body forms may have been a part of the life cycle of all cnidarians. When both are present, the animal is dimorphic: the sessile polyp stage produces medusa by asexual budding, and the motile **medusa** stage produces egg and sperm. In some cnidarians, one stage is dominant and the other is reduced; in other species, one form is absent altogether.

Cnidarian Diversity

Cnidarians are quite diverse (Fig. 31.6*b*–*e*). Sea anemones (Fig. 31.6*b*) are sessile polyps that live attached to a submerged substrate. Most sea anemones range in size from 0.5–20 cm in length and 0.5–10 cm in diameter and are often colorful. Their upward-turned oral disk that contains the mouth is surrounded by a large number of hollow tentacles containing nematocysts.

Corals (Fig. 31.6*c*) resemble sea anemones encased in a calcium carbonate (limestone) house. The coral polyp can extend into the water to feed on microorganisms and retreat into the house for safety. Some corals can be solitary, but the vast majority live in colonies that vary in shape from rounded to branching. Many corals exhibit elaborate geometric designs and stunning colors and are responsible for the building of coral reefs. The slow accumulation of limestone can result in massive structures, such as the Belize Barrier Reef along the eastern coast of Belize. Coral reef ecosystems are very productive, and an extremely diverse group of marine life exists there.

Video
Coral Reef
Ecosystems

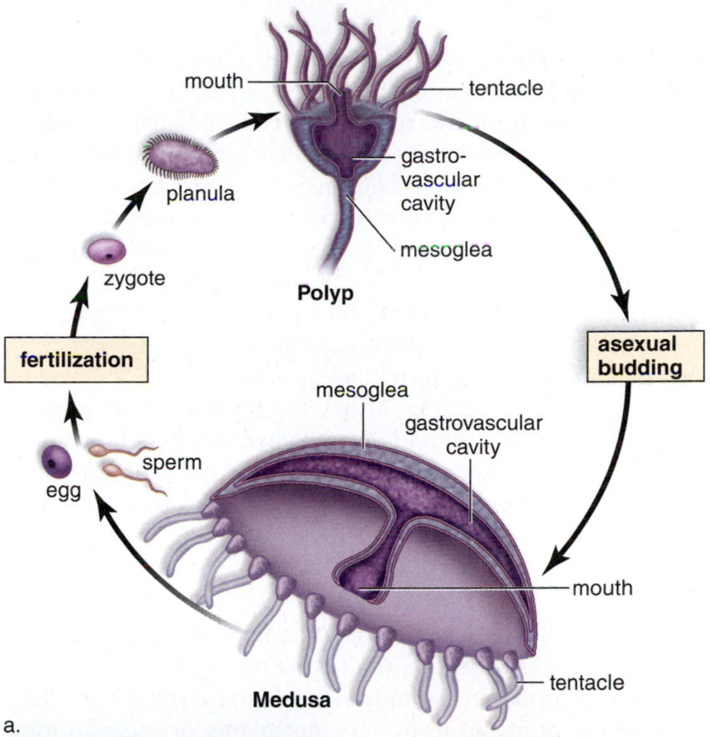
mouth
tentacle
planula
gastro-
vascular
cavity
zygote
mesoglea
Polyp
fertilization
asexual
budding
mesoglea
gastrovascular
cavity
sperm
egg
mouth
tentacle
Medusa

a.

b. Sea anemone, *Corynactis*

c. Cup coral, *Tubastrea*

d. Portuguese man-of-war, *Physalia*

e. Jellyfish, *Aurelia*

Figure 31.6 Cnidarian diversity. **a.** The life cycle of a cnidarian. Some cnidarians have both a polyp stage and a medusa stage. In others, one stage may be dominant or absent altogether. **b.** The anemone, which is sometimes called the flower of the sea, is a solitary polyp. **c.** Corals are colonial polyps residing in a calcium carbonate or proteinaceous skeleton. **d.** The Portuguese man-of-war is a colony of modified polyps and medusae. **e.** True jellyfishes undergo the complete life cycle. This is the medusa stage. The polyp is small.

The hydrozoans have a dominant polyp. Both hydra (Fig. 31.6) and the Portuguese man-of war (Fig. 31.6*d*) are hydrozoans. You might think the Portuguese man-of-war is an odd-shaped medusa, but actually it is a colony of polyps. The original polyp becomes a gas-filled float that provides buoyancy, keeping the colony afloat. Other polyps, which bud from this one, are specialized for feeding or for reproduction. A long, single tentacle armed with numerous nematocysts arises from the base of each feeding polyp. Swimmers who accidentally come upon a Portuguese man-of-war can receive painful, even serious, injuries from these stinging tentacles.

Video Portuguese Man-of-War

In true jellyfishes (Fig. 31.6*e*), the medusa is the primary stage, and the polyp remains small. Jellyfishes depend on tides and currents for their primary means of movement. They feed on a variety of invertebrates and fishes and are themselves food for marine animals.

Hydra

The body of a hydra (Fig. 31.7 *top*) is a small tubular polyp about one-quarter inch in length. Hydras are often studied in biology classes and labs as an example of a cnidarian. Hydras are likely to be found attached to underwater plants or rocks in most lakes and ponds. The only opening (the mouth) is in a raised area surrounded by four to six tentacles that contain a large number of nematocysts.

The outer tissue layer is a protective epidermis derived from ectoderm. The inner tissue layer, derived from endoderm, is called a gastrodermis. The two tissue layers are separated by mesoglea. There are both circular and longitudinal muscle fibers. Nerve cells located below the epidermis, near the mesoglea, interconnect and form a nerve net that communicates with sensory cells throughout the body. The nerve net allows transmission of impulses in several directions at once. Because they have both muscle fibers and nerve fibers, cnidarians are capable of directional movement.

The body of a hydra can contract or extend, and the tentacles that ring the mouth can reach out and grasp prey and discharge nematocysts (Fig. 31.7). Digestion begins within the central cavity but is completed within the food vacuoles of gastrodermal cells. Nutrient molecules are passed to the other cells of the body by diffusion. The large gastrovascular cavity allows digestion and gastrodermal cells to exchange gases directly with a watery medium.

Although hydras exist only as polyps (there are no medusae), they can reproduce either sexually or asexually. When sexual reproduction is going to occur, an ovary or a testis develops in the body wall. Like the sponges, cnidarians have great regenerative powers, and hydras can grow an entire organism from a small piece. When conditions are favorable, hydras reproduce asexually by making small outgrowths, or buds, that pinch off and begin to live independently.

Check Your Progress 31.2

1. Describe three main characteristics of sponges.
2. Explain how cnidarians are more anatomically complex than sponges.
3. Explain the basic anatomical characteristics of a hydra.

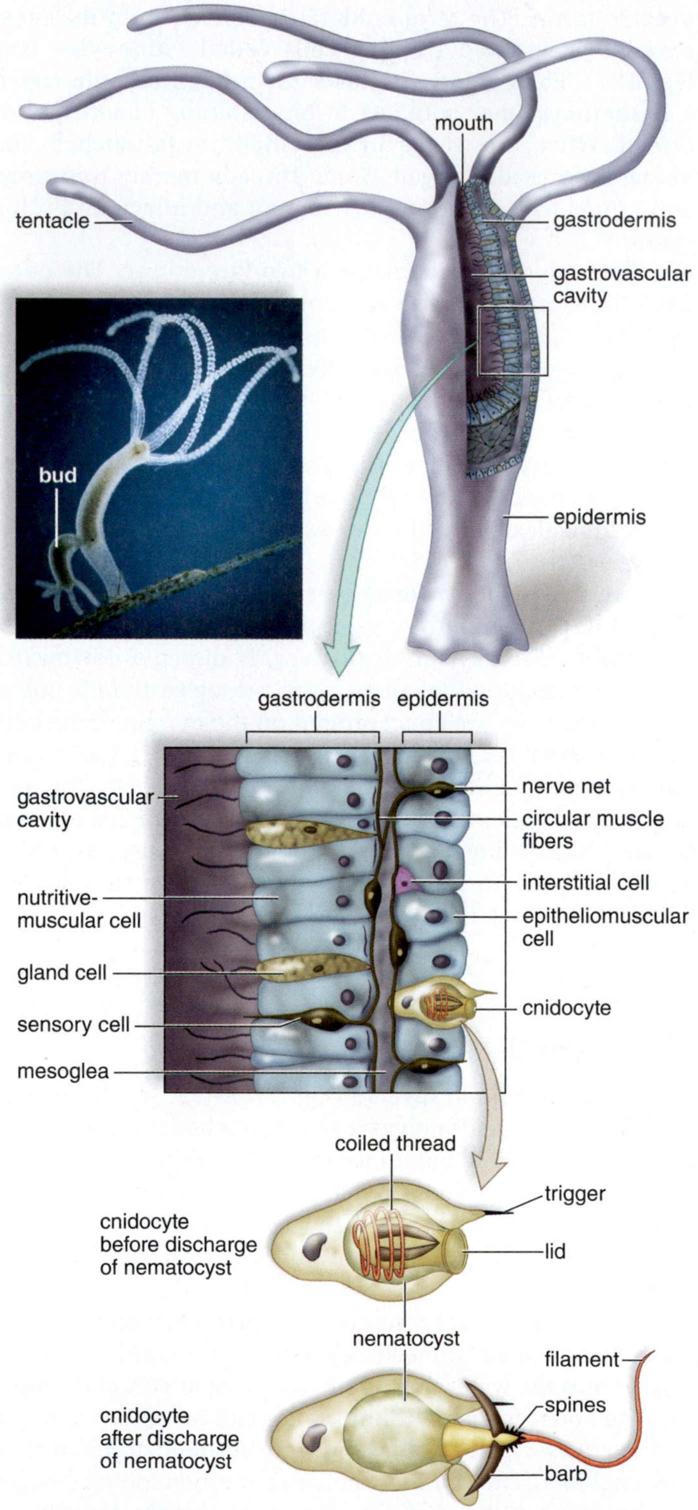
mouth
tentacle
gastrodermis
gastrovascular cavity
bud
epidermis
gastrodermis
epidermis
gastrovascular cavity
nutritive-muscular cell
gland cell
sensory cell
mesoglea
nerve net
circular muscle fibers
interstitial cell
epitheliomuscular cell
cnidocyte
coiled thread
trigger
cnidocyte before discharge of nematocyst
lid
nematocyst
filament
spines
cnidocyte after discharge of nematocyst
barb

Figure 31.7 Anatomy of *Hydra*. *Top:* The body of *Hydra* is a small, tubular polyp whose wall contains two tissue layers. *Hydra* reproduces asexually by forming outgrowths called buds (see photo) that develop into a complete animal. *Middle:* Various types of cells in the body wall. *Bottom:* Cnidocytes are cells that contain nematocysts.

31.3 The Lophotrochozoa

Learning Outcomes

Upon completion of this section, you should be able to

1. Describe the basic features of the lophotrochozoa.
2. Identify the unique features of the molluscs.
3. Contrast the differences between the flatworms and annelids.

The lophotrochozoa are bilaterally symmetrical, at least in some stage of their development. As embryos, they have three germ layers, and as adults, they have the organ level of organization. Lophotrochozoans are protostomes and include the lophophorans (bryozoans, phoronids, and brachiopods) and the **trochozoans** (flatworms, rotifers, molluscs, and annelids). Lophophorans are aquatic and have a feeding apparatus called the lophophore, which is a mouth surrounded by a ciliated tentacle-like structure (Fig. 31.8). The trochophores either have a trochophore larva today (molluscs and annelids), or an ancestor had one in the past (flatworms and rotifers). We will focus our attention on the trochozoans, which encompass the majority of the diversity of this group.

Flatworms

Flatworms (phylum Platyhelminthes) are aptly named because they have an extremely flat body. Like the cnidarians, flatworms have an incomplete digestive tract and only one opening, the mouth. When one opening is present, the digestive tract is said to be *incomplete,* and when two openings are present, the digestive tract is *complete.* Also, flatworms have no body cavity, and instead the third germ layer, mesoderm, fills the space between their organs.

Among flatworms, planarians are free-living, whereas flukes and tapeworms are parasitic. Free-living flatworms have muscles and excretory, reproductive, and digestive systems. The worms lack respiratory and circulatory systems, however. Because the body is flat and thin, diffusion alone can pass needed oxygen and other substances from cell to cell.

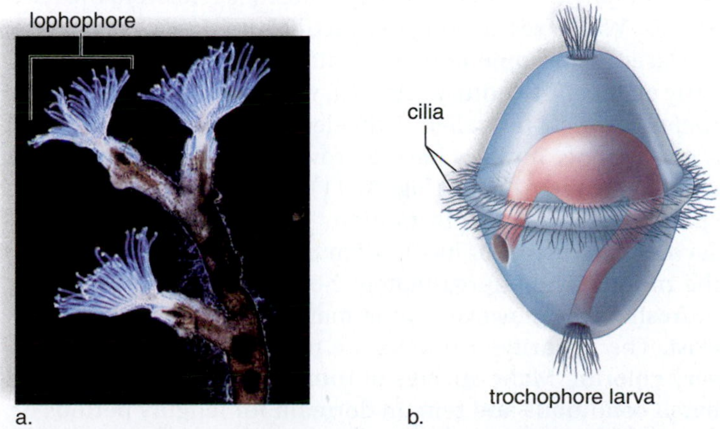
lophophore
cilia
trochophore larva
a.
b.

Figure 31.8 Lophotrochozoan characteristics. **a.** The lophophore feeding apparatus. **b.** A trochophore larva.

Free-Living Flatworms

Freshwater planarians, shown in Figure 31.9, are small (several millimeters to several centimeters) worms that live in lakes, ponds, streams, and springs. Some tend to be colorless; others have brown or black pigmentation. They feed on small living or dead organisms, such as worms and crustaceans.

Planarians have an excretory system that consists of a network of interconnecting canals extending through much of the body. The beating of cilia in the **flame cells** (so named because the beating of the cilia reminded early investigators of the flickering of a flame) keeps the water moving toward the excretory pores.

Planarians have a ladderlike nervous system. They possess a small anterior brain and two lateral nerve cords that are joined by cross-branches. Planarians exhibit cephalization; in addition to a brain, they have light-sensitive organs (eyespots) and chemosensitive organs located on the auricles. Their three muscle layers—outer circular, inner longitudinal, and diagonal—allow for varied movement. A ciliated epidermis allows planarians to glide along a film of mucus.

Planarians capture food by wrapping around the prey, entangling it in slime, and pinning it down. Then it extends a muscular pharynx, and by a sucking motion, tears up and swallows the food. The pharynx leads into a three-branched gastrovascular cavity within which digestion is completed. The digestive tract is considered incomplete because it has only one opening.

Planarians can reproduce both sexually and asexually. Although *hermaphroditic* (possessing both male and female reproductive organs), they typically practice cross-fertilization, in which the penis of one is inserted into the genital pore of the other (and vice versa), and reciprocal transfer of sperm takes place. Fertilized eggs hatch in two to three weeks as tiny worms. Asexual reproduction occurs by regeneration—if you slice a planarian in half, two new planarians will grow. You can even make a two-headed planarian by slicing the head in half!

Parasitic Flatworms

There are several classes of parasitic flatworms, two of which are discussed here: the tapeworms (class Cestoda) and the flukes (class Trematoda).

Tapeworms As adults, tapeworms are endoparasites (internal parasites) of various vertebrates, including humans. They vary in length from a few millimeters to nearly 20 m.

Tapeworms have a tough integument, a specialized body covering resistant to the host's digestive juices. Their excretory, muscular, and nervous systems are similar to those of other flatworms. Tapeworms have a well-developed anterior region, called the **scolex,** which bears hooks for attachment to the intestinal wall of the host and suckers for feeding. Behind the scolex are **proglottids,** a series of reproductive units with a full set of male and female sex organs. Each proglottid fertilizes its own eggs, which number in the thousands. Immature proglottids are located closer to the scolex, while mature (or gravid) proglottids are farther away. The gravids, which

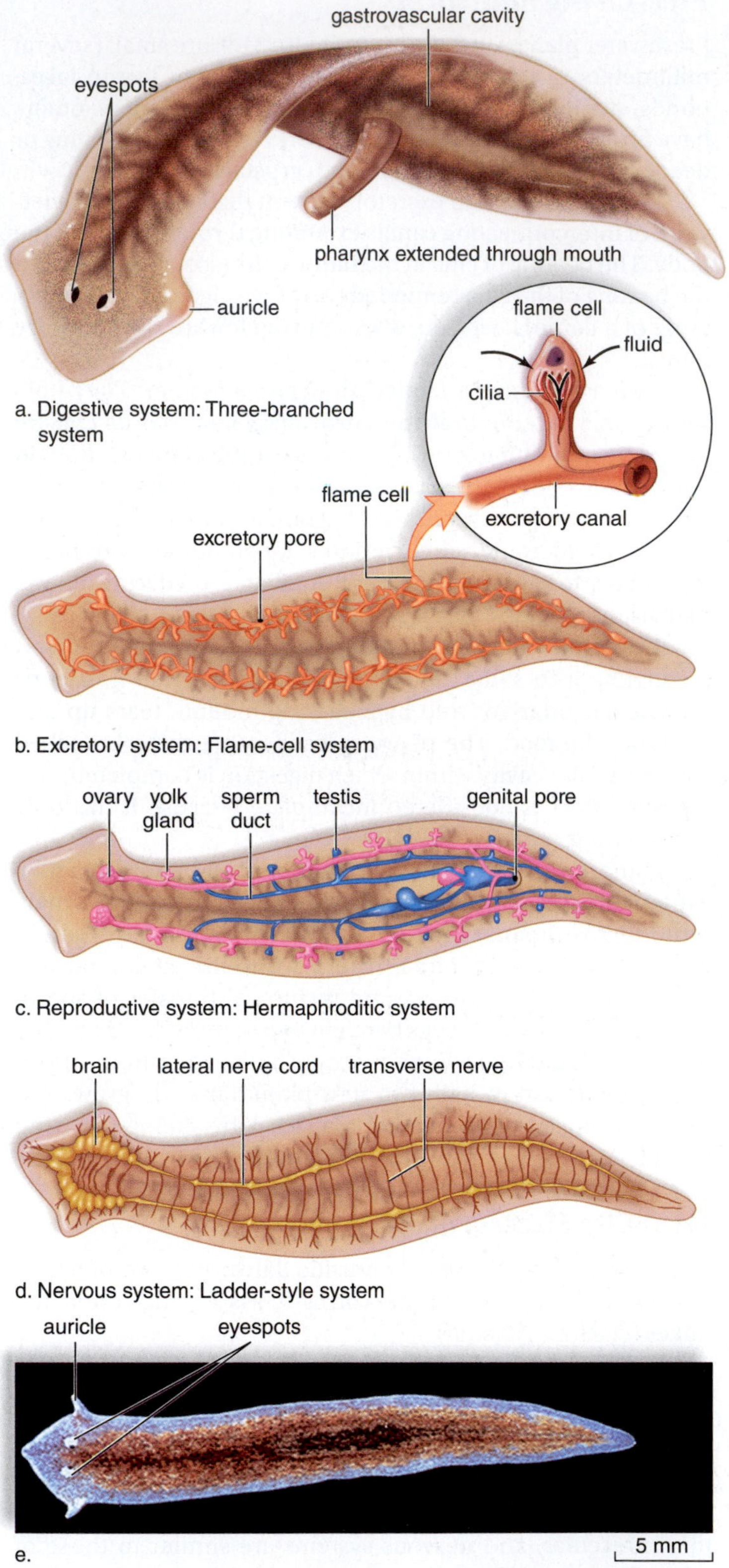
gastrovascular cavity
eyespots
pharynx extended through mouth
auricle
a. Digestive system: Three-branched system
flame cell
fluid
cilia
excretory canal
flame cell
excretory pore
b. Excretory system: Flame-cell system
ovary
yolk gland
sperm duct
testis
genital pore
c. Reproductive system: Hermaphroditic system
brain
lateral nerve cord
transverse nerve
d. Nervous system: Ladder-style system
auricle
eyespots
e.
5 mm

Figure 31.9 Planarian anatomy. **a.** A planarian extends its pharynx to suck food into a gastrovascular cavity that branches throughout the body. **b.** The excretory system includes flame cells. **c.** The reproductive system (shown in pink and blue) has both male and female organs. **d.** The nervous system has a ladderlike appearance. **e.** A flatworm, such as *Dugesia,* is bilaterally symmetrical and has a head region with eyespots.

contain fertilized eggs, break away and are eliminated in the host's feces. A secondary host must ingest the eggs for the life cycle to continue. Figure 31.10 illustrates the life cycle of the pork tapeworm, *Taenia solium,* where the human is the primary host and the pig is the secondary host. The larvae burrow through the intestinal wall and travel in the bloodstream to finally lodge and encyst in muscle. The cyst is a small, hard-walled structure that contains a larva called a bladder worm. When humans eat infected meat that has not been thoroughly cooked, the bladder worm breaks out of the cyst, attaches itself to the intestinal wall, and grows to adulthood. Then the life cycle begins again.

Flukes Flukes are endoparasites of various vertebrates. Their flattened and oval-to-elongated body is covered by a nonciliated integument. At the anterior end of these animals, an oral sucker is surrounded by sensory papillae, and there is at least one other sucker used for attachment to the host. Although the digestive system is reduced compared to that of free-living flatworms, the alimentary canal is well developed. The excretory and muscular systems are similar to those of free-living flatworms, but the nervous system is reduced, with poorly developed sense organs. Most flukes are hermaphroditic.

Flukes are usually named for the type of vertebrate organ they inhabit; for example, there are blood, liver, and lung flukes. The blood fluke (*Schistosoma* spp.) occurs predominantly in the Middle East, Asia, and Africa. Nearly 800,000 infected persons die each year from schistosomiasis. Adult flukes are small (approximately 2.5 cm long) and may live for years in their human hosts. The Chinese liver fluke, *Clonorchis sinensis,* is a major parasite of humans, cats, dogs, and pigs. This 20 mm-long fluke is commonly found in many regions of the Orient. It requires two intermediate hosts, a snail and a fish. Eggs are shed into the water in feces of the human or other mammal host and enter the body of a snail, where they undergo development. Larvae escape into the water and bore into the muscles of a fish. When humans eat infected fish, the juveniles migrate into the bile duct, where they mature. A heavy infection can cause destruction of the liver and death.

Rotifers

Rotifers (phylum Rotifera) are trochozoans related to the flatworms. With sizes ranging between 0.5 mm and 3 mm, rotifers are microscopic animals. Anton van Leeuwenhoek, an early eighteenth-century scientist, was one of the first to view rotifers (which he called "wheeled animalcules") through a microscope. Rotifers have a crown of cilia, known as the corona, on their heads (Fig. 31.11). The corona looks like a spinning wheel when in motion. This disc of beating cilia serves as an organ of locomotion and also directs food into the mouth. The approximately 2,000 species primarily live in fresh water; however, some marine and terrestrial forms exist. The majority of rotifers are transparent, but some are very colorful. Many species of rotifers can desiccate during harsh conditions and remain dormant for lengthy periods of time. This characteristic has earned them the title "resurrection animalcules."

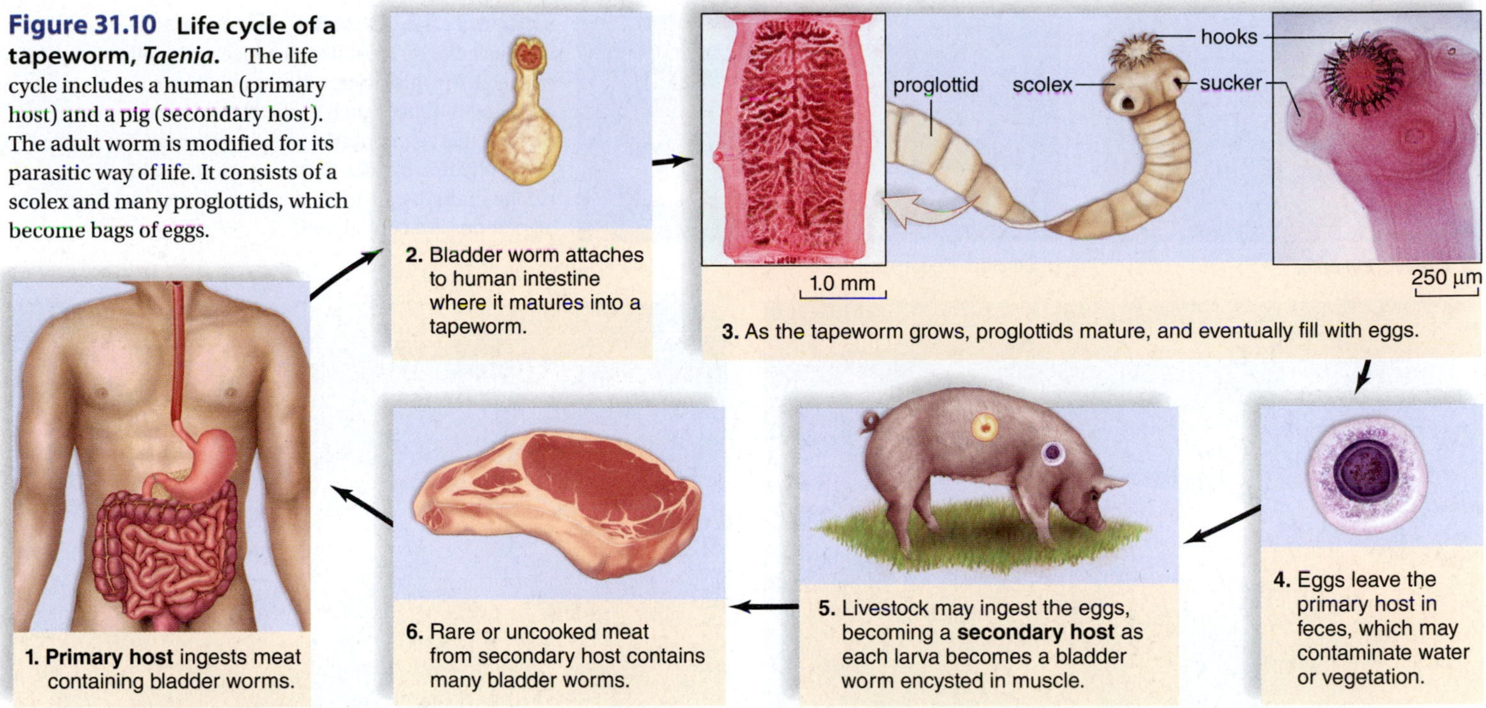
2. Bladder worm attaches to human intestine where it matures into a tapeworm.
proglottid
scolex
hooks
sucker
1.0 mm
250 µm
3. As the tapeworm grows, proglottids mature, and eventually fill with eggs.
4. Eggs leave the primary host in feces, which may contaminate water or vegetation.
5. Livestock may ingest the eggs, becoming a **secondary host** as each larva becomes a bladder worm encysted in muscle.
6. Rare or uncooked meat from secondary host contains many bladder worms.
1. **Primary host** ingests meat containing bladder worms.

Figure 31.10 Life cycle of a tapeworm, *Taenia*. The life cycle includes a human (primary host) and a pig (secondary host). The adult worm is modified for its parasitic way of life. It consists of a scolex and many proglottids, which become bags of eggs.

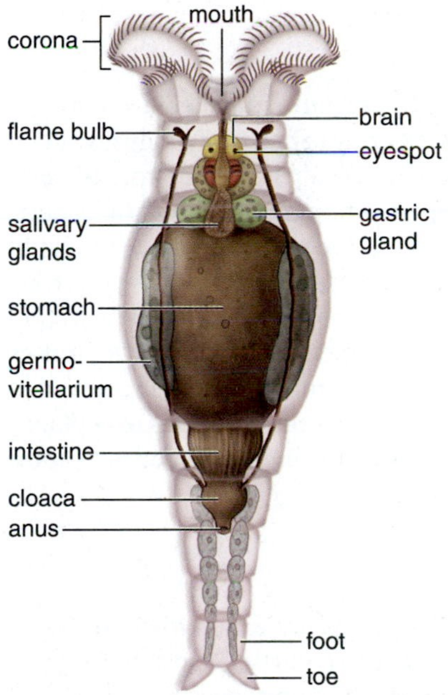
mouth
corona
flame bulb
brain
eyespot
salivary glands
gastric gland
stomach
germo-vitellarium
intestine
cloaca
anus
foot
toe

Figure 31.11 Rotifer. Rotifers are microscopic animals only 0.1–3 mm in length. The beating of cilia on two lobes at the anterior end of the animal gives the impression of a pair of spinning wheels.

Molluscs

The **molluscs** (phylum Mollusca) are the second most numerous group of animals, numbering over 110,000 species. They inhabit a variety of environments, including marine, freshwater, and terrestrial habitats. Although almost everyone enjoys looking at the intricate patterns and beauty of seashells, few people are aware of the tremendous diversity found in the phylum Mollusca (Fig. 30.12). Molluscs include chitons, limpets, slugs, snails, abalones, conchs, nudibranchs, clams, scallops, squid, and octopuses. Molluscs have a true coelom, and all coelomates have bilateral symmetry, three germ layers, the organ level of organization, and a complete digestive tract. The advantages of having a coelom include freer body movements, space for development of complex organs, greater surface area for absorption of nutrients, and protection of internal organs from damage.

The Unique Characteristics of Molluscs

Despite being a very large and diversified group, all molluscs have a body composed of at least three distinct parts:

1. The **visceral mass** is the soft-bodied portion that contains internal organs, including a highly specialized digestive tract, paired kidneys, and reproductive organs. The nervous system of a mollusc consists of several ganglia connected by nerve cords. The amount of cephalization and sensory organs varies from nonexistent in clams to complex in squid and octopuses.
2. The **foot** is a strong, muscular portion used for locomotion. Molluscan groups can be distinguished by modifications of the foot. Molluscs exhibit varying amounts of mobility. Oysters are sessile; snails are extremely slow moving; and squid are fast-moving, active predators.
3. The **mantle,** a membranous or sometimes muscular covering, envelops but does not completely enclose the visceral mass. The *mantle cavity* is the space between the two folds of the mantle. The mantle may secrete a *shell,* which is an exoskeleton. Some molluscs have a univalve (single shell) while others are bivalves (possessing two shells).

Another feature often present is a rasping, tonguelike *radula,* an organ that bears many rows of teeth and is used to obtain food.

a. Chiton, *Tonicella*

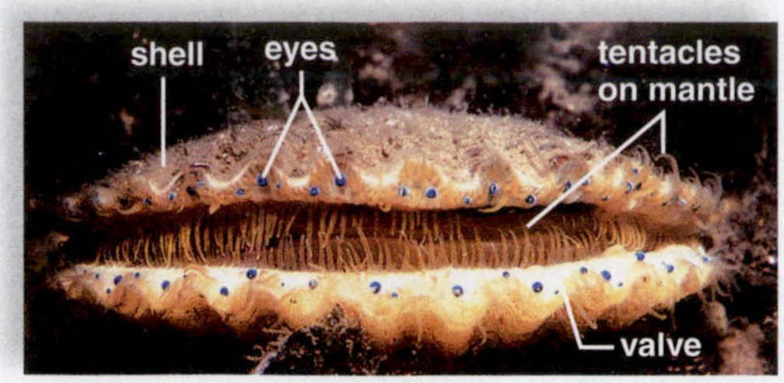
shell
eyes
tentacles on mantle
valve

b. Scallop, *Pecten* sp.

Figure 31.12 Diversity of molluscs. **a.** A chiton is believed to be most like the ancestral mollusc that gave rise to all the other types. Recent evidence suggests that ancestral molluscs such as the chiton were segmented, but that this segmentation was lost throughout much of the phylum. **b.** Like clams, a scallop is a bivalve. **c.** Like snails, a nudibranch is a gastropod. **d.** Like squids, an octopus is a cephalopod.

tentacle
gills
mantle
foot

c. Nudibranch, *Glossodoris macfarlandi*

eye
arm
suckers

d. Two-spotted octopus, *Octopus bimaculoides*

Gastropods

The **gastropods** (class Gastropoda) include nudibranchs, conchs, and snails. In gastropods, whose name means "stomach-footed," the foot is ventrally flattened, and the animal moves by muscle contractions that pass along the foot. Many are herbivores that use their radulas to scrape algae from surfaces. Others are carnivores, using their radulas to bore through surfaces, such as bivalve shells, to obtain food.

While nudibranchs, also called sea slugs, lack a shell, conchs and snails have a univalve coiled shell in which the visceral mass spirals. Land snails, such as *Helix aspera,* have a head with two pairs of tentacles; one pair bears eyes at the tips (Fig. 30.13*a*). The shell not only offers protection, but also prevents desiccation (drying out). While aquatic gastropods have gills, land snails have a mantle that is richly supplied with blood vessels and functions as a lung. Reproduction is also adapted to a land existence. Land snails are hermaphroditic. When two snails meet, each inserts its penis into the vagina of the other, and following fertilization and the deposit of eggs externally, development proceeds directly without the formation of a swimming larvae.

Video Snail's Pace

Cephalopods

In **cephalopods** (class Cephalopoda, meaning head-footed), including octopuses, squid, and nautiluses, the foot has evolved into a funnel or siphon about the head (Fig. 31.13*b*). Aside from the tentacles, which seize prey, cephalopods have a powerful beak and a radula (toothy tongue) to tear prey apart. Cephalization is apparent. The eyes have a lens and a retina with photoreceptors similar to those of vertebrates. However, the eye is constructed so differently from the vertebrate eye that the so-called camera-type eye must have evolved independently in both the molluscs and in the vertebrates. In cephalopods, the brain is formed from a fusion of ganglia, with nerves leaving the brain that supply various parts of the body. An especially large pair of nerves controls the rapid contraction of the mantle, allowing these animals to move quickly by a jet propulsion of water. Rapid movement and the secretion of a dark ink help cephalopods escape their enemies. Octopuses have no shell, and squid have only a remnant of a shell concealed beneath the skin. Octopuses, like some other species of cephalopods, are thought to be among the most intelligent invertebrates and are even capable of learning!

Bivalves

Clams, mussels, oysters, and scallops are called **bivalves** (class Bivalvia) because their shells have two parts. In a clam, the shell, which is secreted by the mantle, is composed of protein and calcium carbonate, with an inner layer of mother-of-pearl. If a foreign body is placed between the mantle and the shell, pearls form as concentric layers of shell are deposited about the particle.

Figure 31.14 shows the internal anatomy of the freshwater clam, *Anodonta.* The adductor muscles hold the valves of the shell together. Within the mantle cavity, the gills, which are the organs for gas exchange in aquatic forms, hang down on either side of the visceral mass, which lies above the foot.

The clam is a filter feeder. Food particles and water enter the mantle cavity by way of the incurrent siphon, a posterior opening between the two valves. Mucous secretions cause smaller particles to adhere to the gills, and ciliary action sweeps them toward the mouth.

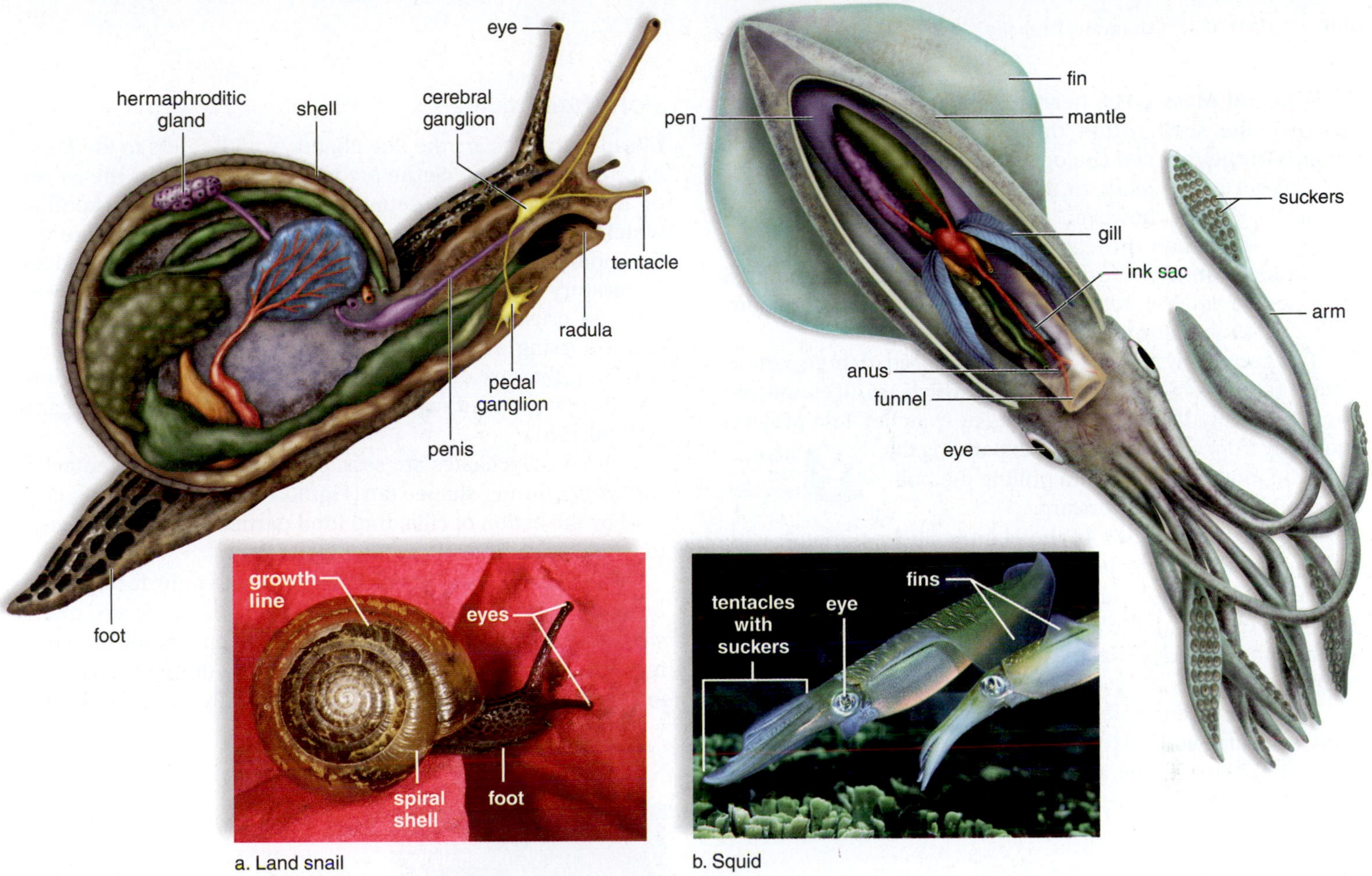
eye
hermaphroditic gland
shell
cerebral ganglion
tentacle
radula
pedal ganglion
penis
foot
pen
fin
mantle
suckers
gill
ink sac
arm
anus
funnel
eye
growth line
eyes
spiral shell
foot
fins
tentacles with suckers
eye
a. Land snail
b. Squid

Figure 31.13 **Gastropod and cephalopod anatomy.** **a.** Snails have a long muscular foot that allows them to creep along slowly by way of muscular contraction. Note the absence of lungs in this land snail. **b.** Squids are torpedo-shaped and adapted for fast swimming by jet propulsion.

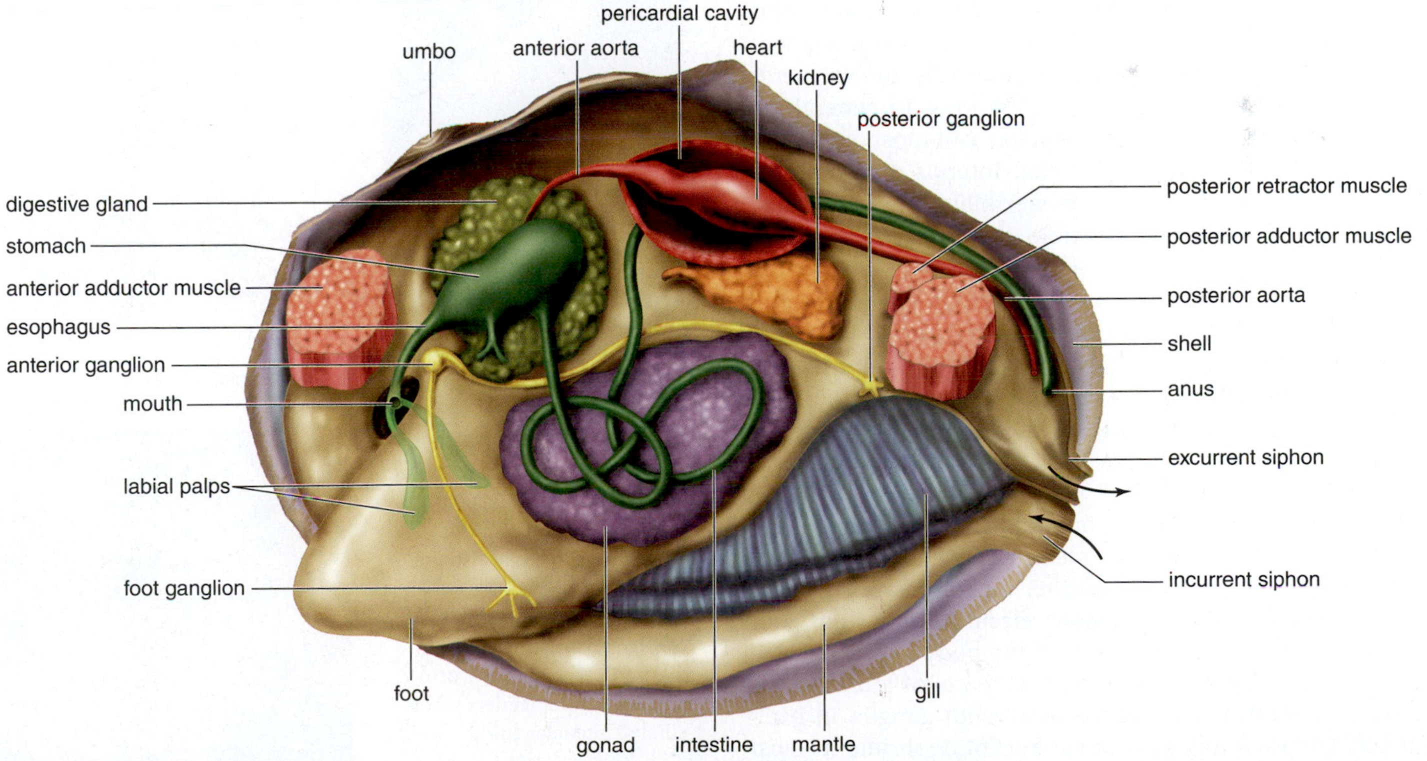
pericardial cavity
umbo
anterior aorta
heart
kidney
posterior ganglion
digestive gland
stomach
anterior adductor muscle
esophagus
anterior ganglion
mouth
labial palps
foot ganglion
foot
gonad
intestine
mantle
gill
posterior retractor muscle
posterior adductor muscle
posterior aorta
shell
anus
excurrent siphon
incurrent siphon

Figure 31.14 **Clam, *Anodonta*.** In this drawing of a clam, one of the bivalve shells and the mantle have been removed from one side. Follow the path of food from the incurrent siphon to the gills, the mouth, the stomach, the intestine, the anus, and the excurrent siphon. Locate the three ganglia: anterior, foot, and posterior. The heart lies in the reduced coelom.

The Visceral Mass The heart of a clam lies just below the hump of the shell within the pericardial cavity, the only remains of the reduced coelom. The heart pumps blood into a dorsal aorta that leads to the various organs of the body. Within the organs, however, blood flows through spaces, or sinuses, rather than through vessels. This is called an **open circulatory system** because the blood is not entirely contained within blood vessels.This type of circulatory system functions well in a slow-moving animal.

The nervous system of a clam is composed of three pairs of ganglia (anterior, foot, and posterior), which are all connected by nerves. Clams lack cephalization. The "hatchet" foot projects anteriorly from the shell, and by expanding the tip of the foot with blood and pulling the body after it, the clam moves forward.

Video Clam Locomotion

The digestive system of the clam includes a mouth with labial palps, an esophagus, a stomach, and an intestine, which coils about in the visceral mass and then is surrounded by the heart as it extends to the anus. Two excretory kidneys remove waste from the pericardial cavity for excretion into the mantle cavity.

The sexes are usually separate. The gonad (i.e., ovary or testis) is located around the coils of the intestine. All clams have some type of larval stage, and marine clams have a trochophore larva.

Annelids

Annelids (phylum Annelida, about 12,000 species) are *segmented,* as evidenced by the rings encircling the outside of their bodies. Segmentation is also seen in arthropods and chordates, although annelids are the only trochozoan with segmentation and a well-developed coelom. Internally, the segments of annelids are partitioned by septa. Worms, in general, do not have an internal or external skeleton, but most often have a **hydrostatic skeleton,** a fluid-filled interior that supports muscle contraction and enhances flexibility. Along with the partitioning of the fluid-filled coelom, this hydroskeleton permits each body segment to move independently. Locomotion occurs by contraction and expansion of each body segment, propelling the animal forward. Thus, a terrestrial annelid is capable of crawling on the surface in addition to burrowing in the mud. Although the most familiar members of this phylum are leeches and earthworms, the majority of annelids are marine. Annelids vary in size from microscopic to tropical earthworms as long as 4 m.

The body plan in annelids has led to specialization of the digestive tract. The digestive system may include a pharynx, esophagus, crop, gizzard, intestine, and accessory glands. Annelids have an extensive **closed circulatory system** with blood vessels that run the length of the body and branch to every segment. The nervous system consists of a brain connected to a ventral solid **nerve cord,** with ganglia in each segment. The excretory system consists of nephridia in most segments. A **nephridium** (pl., nephridia) is a tubule that collects waste material and excretes it through an opening in the body wall.

Polychaetes

Marine annelids are the Polychaeta, which refers to the presence of many setae. **Setae** are bristles that anchor the worm or help it move. The setae are in bundles on parapodia, which are paddlelike appendages found on most segments. Parapodia are used in swimming, but can also be used as respiratory organs. Clam worms, such as *Nereis* (Fig. 31.15*a*), prey on crustaceans and other small animals, which they capture using a pair of strong, chitinous jaws that extend with the pharynx. Associated with its way of life, *Nereis* has a well-defined head region with eyes and other sense organs (Fig. 30.15*b*).

Other polychaetes are sessile tube worms, with tentacles that form a funnel-shaped fan (Fig. 30.15*c*). Water currents, created by the action of cilia, trap food particles that are directed toward the mouth of these filter feeders.

Polychaetes have breeding seasons, and only during these times do they possess functional sex organs. In *Nereis,* many worms concurrently shed a portion of their bodies containing either eggs or sperm, and these float to the surface, where fertilization takes place. The zygote rapidly develops into a

sensory projections
parapodia

a. Clam worm, *Nereis*

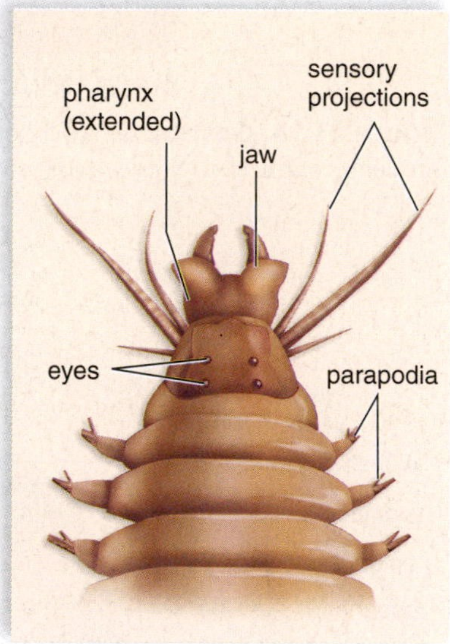
pharynx (extended)
jaw
sensory projections
eyes
parapodia

b. Head region of *Nereis*

Figure 31.15 Annelid diversity. **a.** Clam worms are predatory polychaetes that have a well-defined head region. **b.** Note also the parapodia, which are used for swimming and as respiratory organs. **c.** Christmas tree worms (a type of tube worm) are sessile filter feeders whose ciliated tentacles spiral.

spiraled tentacles

c. Christmas tree worms, *Spirobranchus*

trochophore larva, just as in marine clams. The existence of this larval form in both the annelids and molluscs is evidence that these two groups of animals are evolutionarily related.

Oligochaetes

The oligochaetes (class Oligochaeta), which include earthworms, have few setae per segment. Earthworms (e.g., *Lumbricus*) (Fig. 31.16) do not have a well-developed head or parapodia. Their setae protrude in pairs directly from the surface of the body. Locomotion, which is accomplished section by section, utilizes muscle contraction and the setae. When longitudinal muscles contract, segments bulge, and their setae protrude into the soil. Then, when circular muscles contract, the setae are withdrawn, and these segments move forward in sequence, pushing the whole animal forward.

Earthworms reside in soil where there is adequate moisture to keep the body wall moist for gas exchange. They are scavengers that do not have an obvious head and feed on leaves or any other organic matter. Food drawn into the mouth by the action of the muscular pharynx is stored in a crop and ground up in a thick, muscular gizzard. Digestion and absorption occur in a long intestine, whose dorsal surface has an expanded region, called a typhlosole, that increases the surface for absorption.

Segmentation Earthworm segmentation is visible internally by the presence of septa. The long, ventral solid nerve cord leading from the brain has ganglionic swellings and lateral nerves in each segment. The paired nephridia in most segments have two openings: one is a ciliated funnel that collects coelomic fluid, and the other is an exit in the body wall. Between the two openings is a convoluted region where waste material is removed from the blood vessels about the tubule. Red blood moves anteriorly in the dorsal blood vessel, which connects to the ventral blood vessel by five pairs of connectives called "hearts." Pulsations of the dorsal blood vessel and the five pairs of hearts are responsible for blood flow. As the ventral vessel takes the blood toward the posterior regions of the worm's body, it gives off branches in every segment. Altogether, segmentation is evidenced by (1) body rings, (2) coelom divided by septa, (3) setae on most segments, (4) ganglia and lateral nerves in each segment, (5) nephridia in most segments, and (6) branch blood vessels in each segment.

Reproduction Earthworms are hermaphroditic: the male organs are the testes, the seminal vesicles, and the sperm ducts, and the female organs are the ovaries, the oviducts, and the seminal receptacles. When mating, two worms lie parallel to each other facing in opposite directions. The fused midbody segment, called a clitellum, secretes mucus that protects the sperm from drying out as they pass between the worms. After the worms separate, the clitellum of each produces a slime tube, which is moved along over the anterior end by muscular contractions. As it passes, eggs and the sperm received earlier are deposited, and fertilization occurs. The slime tube then forms a cocoon to protect the worms as they develop. There is no larval stage.

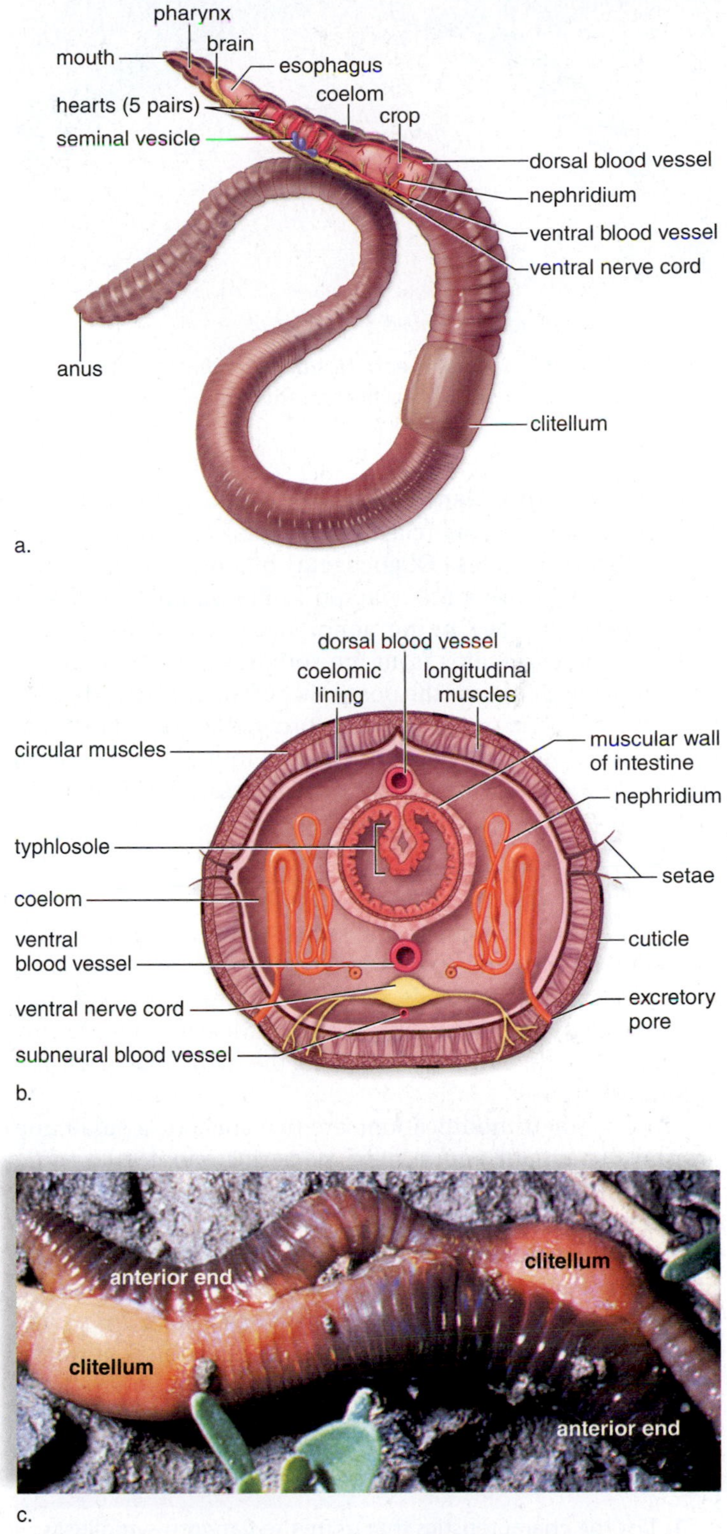
pharynx
brain
mouth
esophagus
coelom
hearts (5 pairs)
crop
seminal vesicle
dorsal blood vessel
nephridium
ventral blood vessel
ventral nerve cord
anus
clitellum
a.
dorsal blood vessel
coelomic lining
longitudinal muscles
circular muscles
muscular wall of intestine
nephridium
typhlosole
setae
coelom
ventral blood vessel
cuticle
ventral nerve cord
excretory pore
subneural blood vessel
b.
anterior end
clitellum
clitellum
anterior end
c.

Figure 31.16 Earthworm, *Lumbricus*. **a.** Internal anatomy of the anterior part of an earthworm. Notice that each body segment bears a pair of setae and that internal septa divide the coelom into compartments. **b.** Cross section of an earthworm. **c.** When earthworms mate, they are held in place by a mucus secreted by the clitellum. The worms are hermaphroditic, and when mating, sperm pass from the seminal vesicles of each to the seminal receptacles of the other.

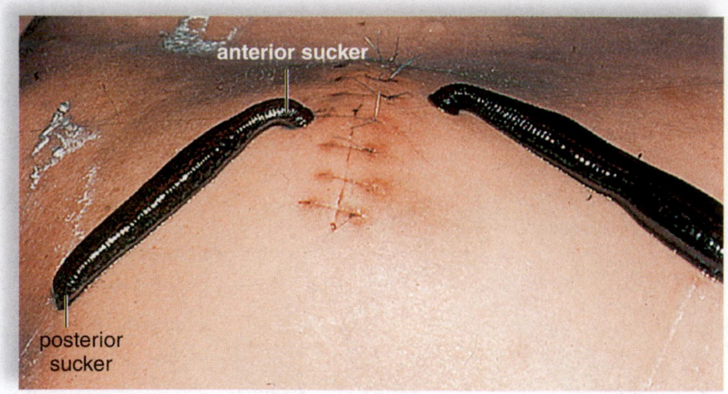
anterior sucker
posterior sucker
Medicinal leech, *Hirudo medicinalis*

Figure 31.17 Medicinal leech, *Hirudo medicinalis*. Leeches have been used in medicinal practices for thousands of years.

Comparison with Clam Worms Comparing the anatomy of marine clam worms (class Polychaeta) to that of terrestrial earthworms (class Oligochaeta) highlights the manner in which earthworms are adapted to life on land. A lack of cephalization is seen in the nonpredatory earthworms that extract organic remains from the soil they eat. Their lack of parapodia helps reduce the possibility of water loss and facilitates burrowing in soil. The clam worm makes use of external water, while the earthworm provides a mucous secretion to aid fertilization. It is the aquatic form that has the swimming, or trochophore larva, not the land form.

Leeches

Leeches (class Hirudinea) are usually found in fresh water, but some are marine or even terrestrial. They have the same body plan as other annelids, but they have no setae, and each body ring has several transverse grooves. Most leeches are only 2–6 cm in length, but some, including the medicinal leech, are as long as 20 cm.

Among their modifications are two suckers, a small one around the mouth and a large, posterior one. While some leeches are free-living, feeding on plant material or invertebrates, most are fluid feeders that attach themselves to open wounds. Some bloodsuckers, such as the medicinal leech (Fig. 31.17), can cut through tissue. Leeches are able to keep blood flowing and prevent clotting by means of a substance in their saliva known as hirudin, a powerful anticoagulant.

Check Your Progress 31.3

1. List the characteristics that unite the flatworms, molluscs, and annelids.
2. Compare features of the flatworm, mollusc, and annelid body cavity, digestive tract, and circulatory system.
3. Identify three characteristics found in all molluscs.

31.4 The Ecdysozoa

Learning Outcomes

Upon completion of this section, you should be able to

1. Identify the characteristics unique to ecdysozoans.
2. Describe the characteristics contributing to the success and diversity of arthropods.
3. List the major features that distinguish the crustaceans, insects, and arachnids.

Both the ecdysozoans and trochozoa are protostomes. The term *ecdysis* means molting, and both roundworms and arthropods, which belong to this group, periodically shed their outer covering.

Roundworms

Roundworms (phylum Nematoda, about 90,000 species) are nonsegmented worms that are prevalent in almost any environment. Generally, nematodes are colorless and range in size from microscopic to exceeding 1 m in length. The internal organs, including the tubular reproductive organs, lie within the pseudocoelom. A **pseudocoelom** is a body cavity that is incompletely lined by mesoderm. In other words, mesoderm occurs inside the body wall but not around the digestive cavity (gut). The fluid-filled pseudocoelom provides space for the development of organs, substitutes for a circulatory system by allowing easy passage of molecules, and provides a type of skeleton.

Nematodes have developed a variety of lifestyles from free-living to parasitic. One species, *Caenorhabditis elegans,* a free-living nematode, is a model animal used in genetics and developmental biology as it was one of the first species to have its genome sequenced.

Ascaris

In the roundworm, *Ascaris lumbricoides* (Fig. 31.18*a*) females tend to be larger (20–35 cm in length) than males (Fig. 31.18*b*). Both sexes move by a characteristic whiplike motion because they have only longitudinal muscles and no circular muscles next to the body wall. *Ascaris* species are most commonly parasites of humans and pigs.

A female *Ascaris* is very reproductively prolific, producing over 200,000 eggs daily. The eggs are passed with host feces and, under the right conditions, can develop into a new worm within two weeks. The eggs need to enter a new host's body via uncooked vegetables, soiled fingers, or ingested fecal material and hatch in the intestines. The juveniles make their way into the veins and lymphatic vessels and are carried to the heart and lungs. From the lungs, the larvae travel up the trachea, where they are swallowed and eventually reach the intestines. There, the larvae mature and begin feeding on intestinal contents. The symptoms of an *Ascaris* infection depend upon the site and the stage of infection.

Figure 31.18 Roundworm anatomy. **a.** The roundworm *Ascaris.* **b.** Roundworms such as *Ascaris* have a pseudocoelom and a complete digestive tract with a mouth and an anus. **c.** A filarial worm infection causes elephantiasis, which is characterized by a swollen body part when the worms block lymphatic vessels.

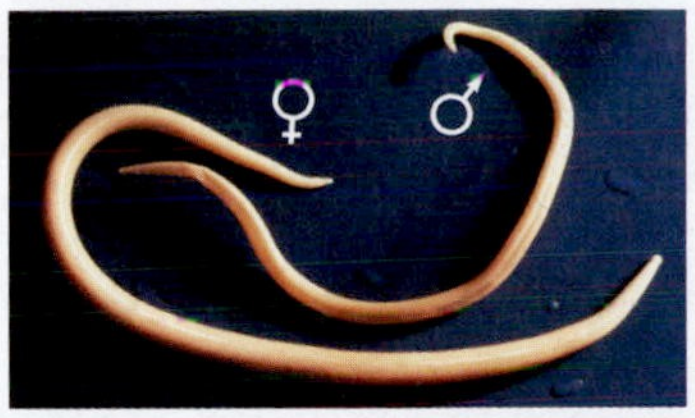
♀
♂

a. *Ascaris*

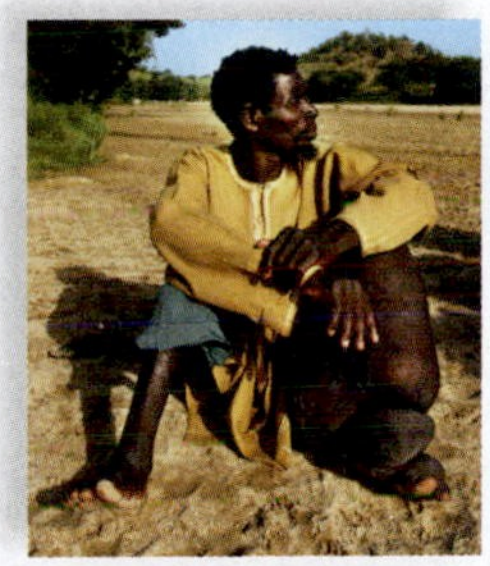

c.

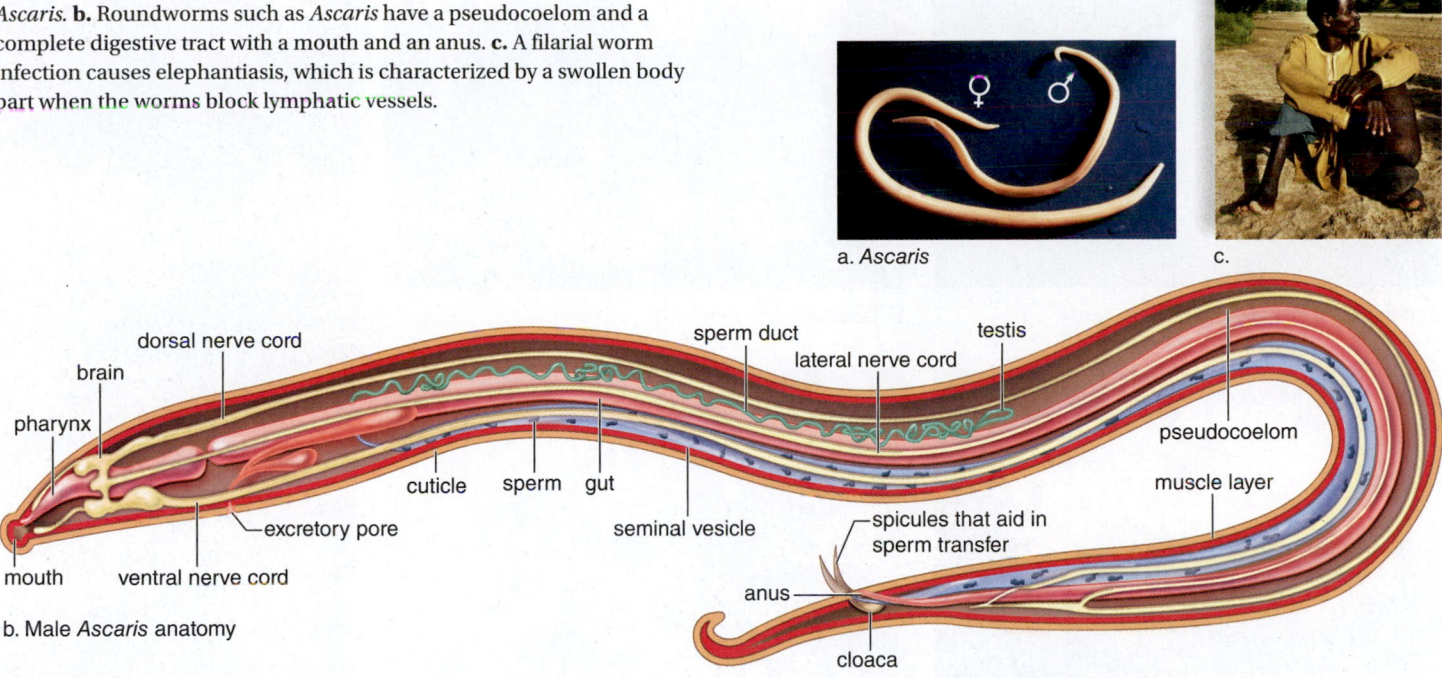

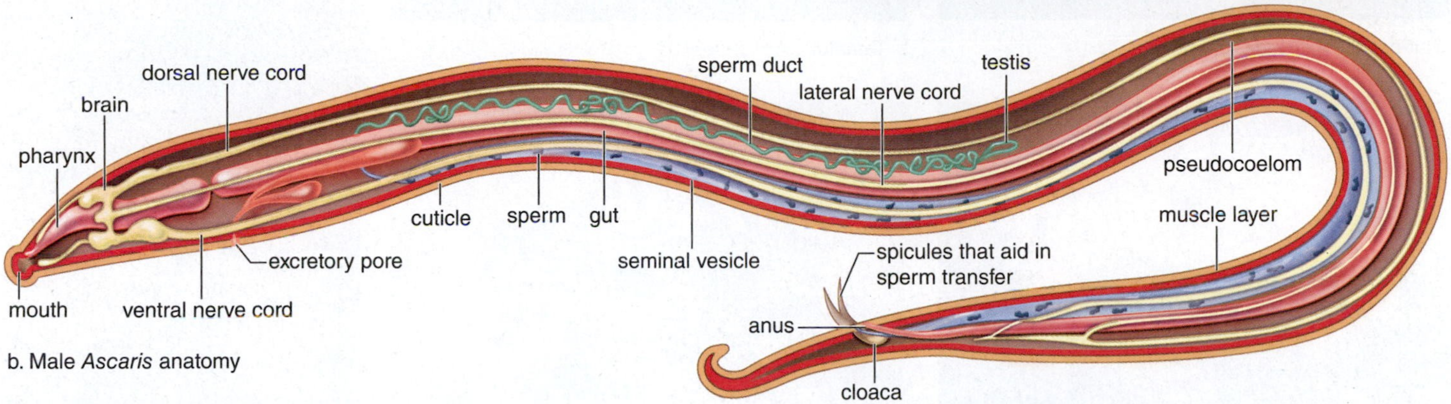
dorsal nerve cord
brain
pharynx
mouth
ventral nerve cord
excretory pore
cuticle
sperm
gut
seminal vesicle
sperm duct
lateral nerve cord
testis
pseudocoelom
muscle layer
spicules that aid in sperm transfer
anus
cloaca

b. Male *Ascaris* anatomy

Other Roundworms

Trichinosis is a fairly serious infection caused by *Trichinella spiralis,* a roundworm that rarely infects humans in the United States. Humans contract the disease when they eat undercooked pork that contains encysted larvae. After maturation, the female adult burrows into the wall of the host's small intestine, where she deposits live larvae that are then carried by the bloodstream and encyst in the skeletal muscles. When the adults are in the small intestine, digestive disorders, fatigue, and fever occur. When the larvae encyst, the symptoms include aching joints, muscle pain, and itchy skin.

Elephantiasis is caused by a roundworm called the filarial worm (Fig. 31.18*c*), which utilizes the mosquito as an intermediate host. When a mosquito bites an infected person, it transports larvae to a new host. Because the adult worms reside in lymphatic vessels, fluid return is impeded, and the limbs of an infected human can swell to an enormous size, even resembling those of an elephant.

Other roundworm infections are more common in the United States. Children frequently acquire a pinworm infection, and hookworm is seen in the southern states, as well as worldwide. Good hygiene, proper disposal of sewage, and cooking meat thoroughly usually protect people from parasitic roundworms.

Arthropods

Arthropods (phylum Arthropoda) are extremely diverse (Fig. 31.19), ranging in size from less than 0.1 mm (mites) to 4 m (Japanese crab) in length. Over 1 million species have been discovered and described, but some experts suggest that as many as 30 million arthropod species may exist—most of them insects. Arthropods, which also occupy every type of habitat, are considered the most successful group of all the animals. The remarkable success of arthropods is dependent on five characteristics:

1. *A rigid but jointed exoskeleton* (Fig. 31.20*a, b*). The exoskeleton is composed primarily of chitin, a strong, flexible, nitrogenous polysaccharide. The exoskeleton serves many functions, including protection, attachment for muscles, locomotion, and prevention of desiccation. However, because it is hard and nonexpandable, arthropods must molt, or shed, the exoskeleton in order to grow larger (Fig. 31.20*c*).
2. *Segmentation.* Segmentation is readily apparent because each segment has a pair of jointed appendages, even though certain segments are fused into a head, thorax, and abdomen. The jointed appendages of arthropods are basically hollow tubes moved by muscles. Typically, the appendages are highly adapted for a particular function, such as food gathering, reproduction, and locomotion. In addition, many appendages are associated with sensory structures and used for tactile purposes.
3. *Well-developed nervous system.* Arthropods have a brain and a ventral nerve cord. The head bears various types of sense organs, including eyes of two types—simple and compound. The compound eye is composed of many complete visual units, each of which operates independently (Fig. 31.20*d*). The lens of each visual unit focuses an image on a small number of photoreceptors within that unit. The simple eye, like that of vertebrates, has a single lens that brings the image to focus onto many receptors, each of which receives only a portion of the image. In addition to sight, many arthropods have well-developed touch, smell, taste, balance, and hearing.

a. Flat-backed millipede, *Sigmoria*
b. Tarantula, *Aphonopelma*
c. Dungeness crab, *Cancer*
d. Paper wasp, *Polistes*
e. Stone centipede, *Lithobius*

Figure 31.19 Arthropod diversity. **a.** A millipede has only one pair of antennae, and the head is followed by a series of segments, each with two pairs of appendages. **b.** The hairy tarantulas of the genus *Aphonopelma* are dark in color and move carefully and steadily. Their bite is harmless to people. **c.** A crab is a crustacean with a calcified exoskeleton, one pair of claws, and four other pairs of walking legs. **d.** A wasp is an insect with two pairs of wings, both used for flying, and three pairs of walking legs. **e.** A centipede has only one pair of antennae, and the head is followed by a series of segments, each with a single pair of appendages.

Figure 31.20 Arthropod skeleton and eye. **a.** The joint in an arthropod skeleton is a region where the cuticle is thinner and not as hard as the rest of the cuticle. The direction of movement is toward the flexor muscle or the extensor muscle, whichever one has contracted. **b.** The exoskeleton is secreted by the epidermis and consists of the endocuticle; the exocuticle, hardened by the deposition of calcium carbonate; and the epicuticle, a waxy layer. Chitin makes up the bulk of the exo- and endocuticles. **c.** Because the exoskeleton is nonliving, it must be shed through a process called molting for the arthropod to grow. **d.** Arthropods have a compound eye that contains many individual units, each with its own lens and photoreceptors.

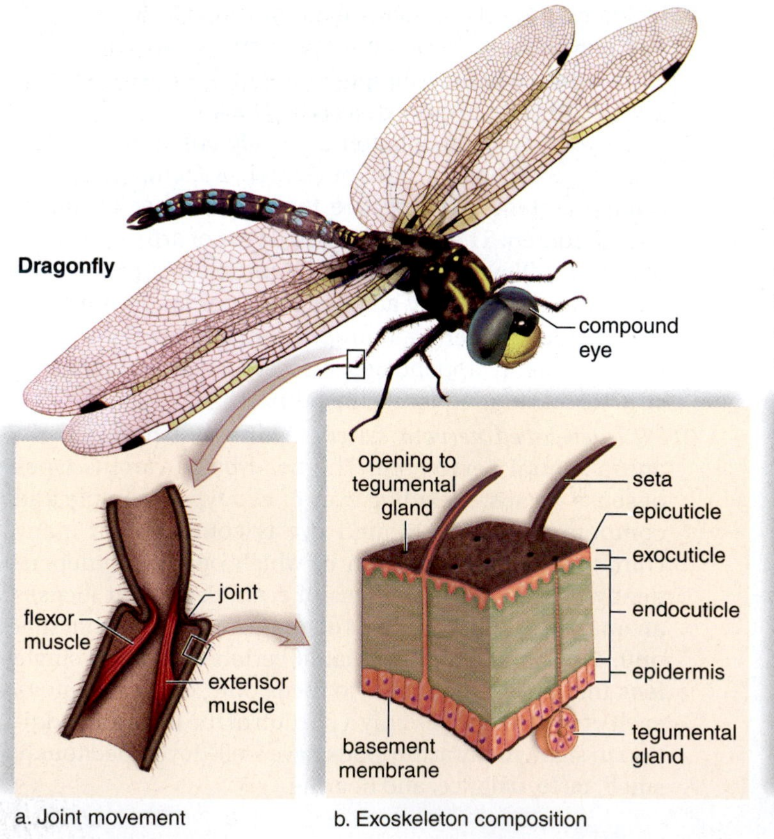
Dragonfly
compound eye
flexor muscle
joint
extensor muscle
a. Joint movement
opening to tegumental gland
seta
epicuticle
exocuticle
endocuticle
epidermis
basement membrane
tegumental gland
b. Exoskeleton composition

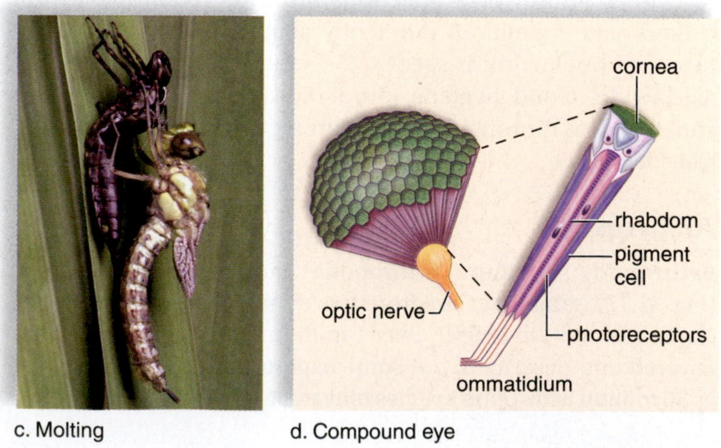
c. Molting
cornea
rhabdom
pigment cell
optic nerve
photoreceptors
ommatidium
d. Compound eye

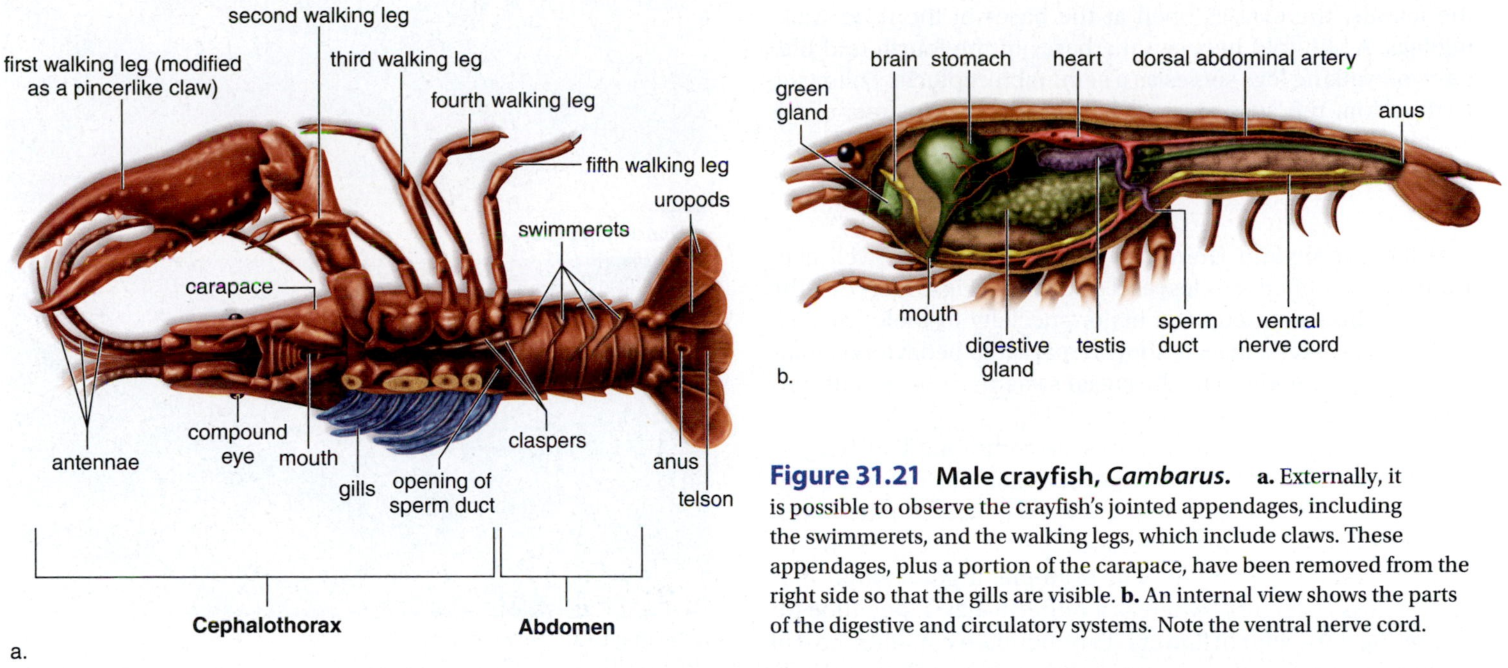
second walking leg
first walking leg (modified as a pincerlike claw)
third walking leg
fourth walking leg
fifth walking leg
uropods
swimmerets
carapace
antennae
compound eye
mouth
gills
opening of sperm duct
claspers
anus
telson
Cephalothorax
Abdomen
a.
green gland
brain
stomach
heart
dorsal abdominal artery
anus
mouth
digestive gland
testis
sperm duct
ventral nerve cord
b.

Figure 31.21 Male crayfish, *Cambarus*. **a.** Externally, it is possible to observe the crayfish's jointed appendages, including the swimmerets, and the walking legs, which include claws. These appendages, plus a portion of the carapace, have been removed from the right side so that the gills are visible. **b.** An internal view shows the parts of the digestive and circulatory systems. Note the ventral nerve cord.

4. *Variety of respiratory organs.* Marine forms use gills, which are vascularized, highly convoluted, thin-walled tissue specialized for gas exchange. Terrestrial forms have book lungs (e.g., spiders) or air tubes called tracheae. Tracheae serve as a rapid way to transport oxygen directly to the cells.
5. *Reduced competition through metamorphosis.* Many arthropods undergo a drastic change in form and physiology that occurs as an immature stage, called a larva, becomes an adult. Among arthropods, the larva eats different food and lives in a different environment than the adult. For example, larval crabs live among and feed on plankton, while adult crabs are bottom dwellers that catch live prey or scavenge dead organic matter. Among insects, such as butterflies, the caterpillar feeds on leafy vegetation, while the adult feeds on nectar.

Crustaceans

Video Voice of the Lobster

Video Barnacle-Free Boats

Crustaceans (subphylum Crustacea) are a group of largely marine arthropods that include barnacles, shrimps, lobsters, and crabs. There are also some freshwater crustaceans, including the crayfish, and some terrestrial ones, including the sowbug, or pillbug.

Crustaceans are named for their hard shells. The exoskeleton is calcified to a greater degree in some forms than in others. Although crustacean anatomy is extremely diverse, the head usually bears a pair of compound eyes and five pairs of appendages. The first two pairs, called antennae, lie in front of the mouth and have sensory functions. The other three pairs are mouthparts used in feeding.

In crayfish, such as *Cambarus,* the thorax bears five pairs of walking legs. The first walking leg is a pinching claw (Fig. 31.21). The gills are situated above the walking legs. The head and thorax are fused into a cephalothorax, which is covered on the top and sides by a nonsegmented carapace. The abdominal segments, which contain much musculature, are equipped with small paddlelike structures called swimmerets. The last two segments bear the uropods and the telson, which make up a fan-shaped tail to propel the crayfish backward.

Internal Organs The digestive system includes a stomach, which is divided into two main regions. The anterior portion consists of a gastric mill, a special grinding apparatus equipped with chitinous teeth. The posterior region acts as a filter to prevent coarse particles from entering the digestive glands where absorption takes place. *Green glands* lying in the head region, anterior to the esophagus, excrete metabolic wastes through a duct that opens externally at the base of the antennae. The coelom is reduced to a space around the reproductive system. A heart within a pericardial cavity pumps blood containing the respiratory pigment hemocyanin into a **hemocoel** consisting of sinuses (open spaces). Hemocyanin contains copper (compared to hemoglobin—the red, iron-containing pigment of humans), causing the blood to have a blueish color. These sinuses are common in an open circulatory system, where the hemolymph flows freely about the organs and is not typically contained within blood vessels.

The nervous system of the crayfish is very similar to that of the earthworm. The crayfish has a brain and a ventral nerve cord that passes posteriorly. Along the length of the nerve cord, segmental ganglia give off 8 to 19 paired lateral nerves.

The sexes are separate in the crayfish with the gonads located just ventral to the pericardial cavity. In the male, a coiled sperm duct opens to the outside at the base of the fifth walking leg. Sperm transfer is accomplished by the first two pairs of swimmerets, which are enlarged and quite strong. In the female, the ovaries open at the bases of the third walking legs. A stiff fold between the bases of the fourth and fifth pairs of walking legs serves as a seminal receptacle. Following fertilization, the eggs are attached to the swimmerets of the female.

Insects

Insects (subphylum Uniramia) are so numerous (well over 1 million identified species) and so diverse (Fig. 31.22) that the study of this one group is a major specialty in biology called entomology. Some insects show remarkable behavioral adaptations, as exemplified by the social systems of bees, ants, termites, and other colonial insects.

Insects have certain features in common. The body is divided into a head, a thorax, and an abdomen. The head usually bears a pair of sensory antennae, a pair of compound eyes, and several simple eyes. The mouthparts are adapted to the specific insect's way of life. For example, a grasshopper has mouthparts that chew, whereas a butterfly has a long tube for siphoning the nectar of flowers. The abdomen contains most of the internal organs. The thorax bears three pairs of legs and zero to three pairs of wings. Wings, if present, enhance an insect's ability to survive by providing a way of escaping enemies, finding food, facilitating mating, and dispersing the offspring. The exoskeleton of an insect is lighter and contains less chitin than that of many other arthropods.

Here, we will discuss the features of a grasshopper as a representative example of insect form and function. In the grasshopper (Fig. 31.23), the third pair of legs is suited to jumping. There are two pairs of wings. The forewings are tough and leathery, and when folded back at rest, they protect the broad, thin hindwings. On the lateral surface, the first abdominal segment bears a large tympanum on each side for the reception of sound waves.

Internal Organs The digestive system of a grasshopper is suitable for a herbivorous diet. The mouthparts chew the food, which is temporarily stored in the crop before passing into a gastric mill, where it is finely ground before digestion is completed in the stomach. Nutrients are absorbed into the hemocoel from outpockets called gastric ceca. The stomach leads into an intestine and a rectum, which empties by way of an anus. The excretory system consists of **Malpighian tubules,** which extend into the hemocoel and collect nitrogenous wastes that are excreted into the digestive tract. The formation of a solid nitrogenous waste, namely uric acid, conserves water.

The respiratory system begins with openings in the exoskeleton called spiracles. From here, the air enters small tubules called **tracheae** (Fig. 31.23*a*). The tracheae branch and rebranch until they end intracellularly, where the actual exchange of gases takes place. The movement of air through this complex of tubules is not passive. Rather, air is actively pumped by alternate contraction and relaxation of the body wall. Breathing by tracheae is suitable to the small

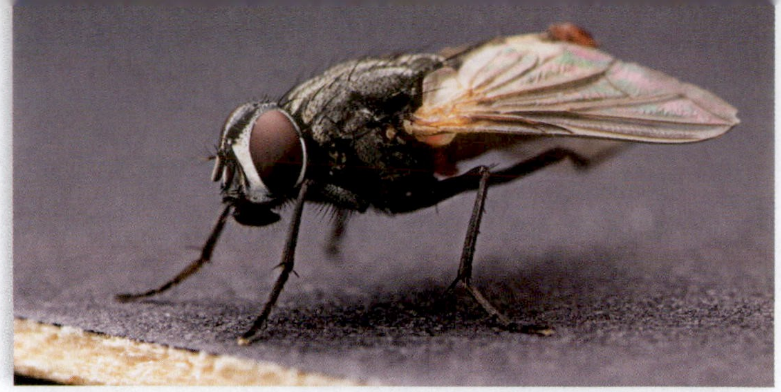

a. Housefly, *Musca*

b. Walking stick, *Diapheromera*

c. Bee, *Apis*

d. Dragonfly, *Aeshna*

e. Birdwing butterfly, *Troides*

Figure 31.22 Insect diversity. **a.** Flies have a single pair of wings and lapping mouthparts. **b.** Walking sticks are herbivorous, with biting and chewing mouthparts. **c.** Bees have four translucent wings and a thorax separated from the abdomen by a narrow waist. **d.** Dragonflies have two pairs of similar wings. They catch and eat other insects while flying. **e.** Butterflies have forewings larger than their hindwings. Their mouthparts form a long tube for siphoning up nectar from flowers.

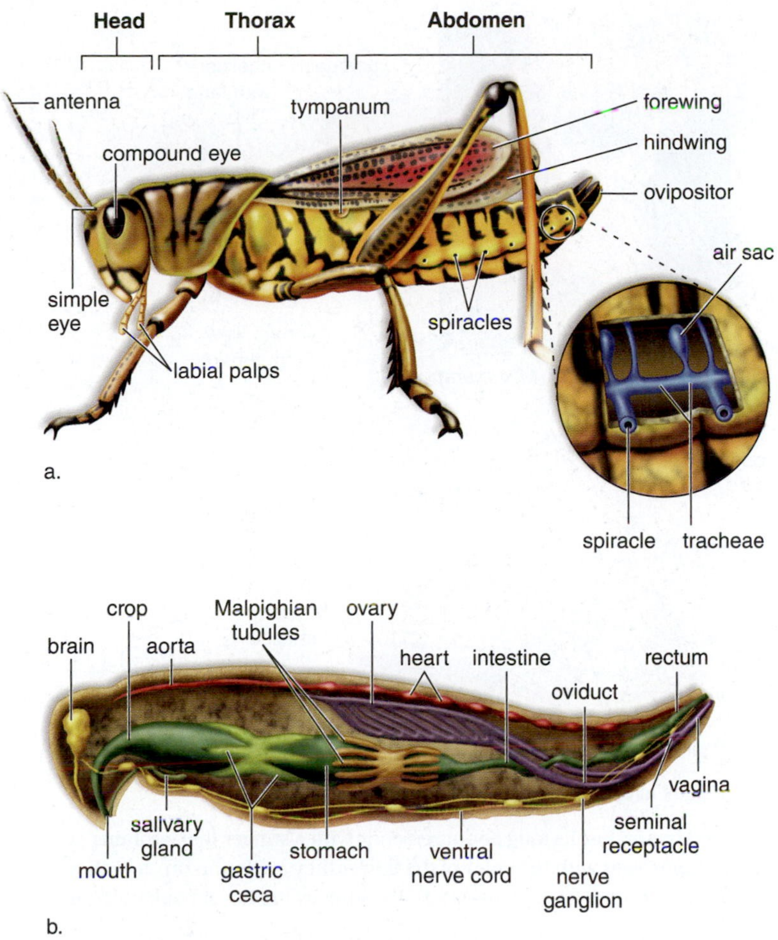
Head
Thorax
Abdomen
antenna
tympanum
forewing
hindwing
compound eye
ovipositor
air sac
simple eye
spiracles
labial palps
a.
spiracle
tracheae
crop
Malpighian tubules
ovary
brain
aorta
heart
intestine
rectum
oviduct
vagina
salivary gland
seminal receptacle
stomach
ventral nerve cord
mouth
gastric ceca
nerve ganglion
b.

Figure 31.23 Female grasshopper, *Romalea*. **a.** Externally, the body of a grasshopper is divided into three sections and has three pairs of legs. The tympanum receives sound waves, and the jumping legs and the wings are for locomotion. **b.** Internally, the digestive system is specialized. The Malpighian tubules excrete a solid nitrogenous waste (uric acid). A seminal receptacle receives sperm from the male, which has a penis.

size of insects (most are less than 60 mm in length), as the tracheae would be crushed by any significant amount of weight.

The circulatory system contains a slender, tubular heart that lies against the dorsal wall of the abdominal exoskeleton and pumps hemolymph into an aorta that leads to a hemocoel, where it circulates before returning to the heart again. The hemolymph is colorless and lacks a respiratory pigment because the tracheal system is responsible for gas exchange.

Reproduction and Development Grasshopper reproduction is adapted to life on land. The male grasshopper has a penis, and sperm passed to the female are stored in a seminal receptacle. Internal fertilization protects both gametes and zygotes from drying out. The female deposits the fertilized eggs in the ground with her ovipositor.

Metamorphosis is a change in form and physiology that occurs as an immature stage, called a larva, becomes an adult. Grasshoppers undergo gradual metamorphosis as they mature. The immature grasshopper, called a nymph, looks like an adult grasshopper, even though it differs somewhat in shape and form. Other insects, such as butterflies, undergo complete metamorphosis, involving drastic changes in form. At first, the animal is a wormlike larva (caterpillar) with chewing mouthparts. It then forms a case, or cocoon, about itself and becomes a pupa. During this stage, the body parts are completely reorganized. The adult then emerges from the cocoon. This life cycle allows the larvae and adults to use different food sources. Most eating by insects occurs during the larval stage. The field of forensic science has found a way to put our knowledge of insect life cycle stages to good use, as presented in the Scientific Inquiry feature, "Maggots: A Surprising Tool for Crime Scene Investigation," on page 644.

Comparison with Crayfish The grasshopper and the crayfish share a common ancestor, but they have diverged in their morphology to adapt to aquatic versus terrestrial environments. In crayfish, gills take up oxygen from water, while in the grasshopper, tracheae allow oxygen-laden air to enter the body. Appropriately, the crayfish has an oxygen-carrying pigment, but a grasshopper has no such pigment in its blood. The crayfish excretes a liquid nitrogenous waste (ammonia), while the grasshopper excretes a solid nitrogenous waste (uric acid). Only grasshoppers have (1) a tympanum for the reception of sound waves, and (2) a penis in males for passing sperm to females without possible desiccation, and an ovipositor in females for laying eggs in soil. Crayfish utilize their uropods when they swim; a grasshopper has legs for hopping and wings for flying.

Arachnids

The **arachnids** (subphylum Chelicerata) include scorpions, spiders, ticks, and mites (Fig. 31.24). In this group, the cephalothorax bears six pairs of appendages: the chelicerae and the pedipalps and four pairs of walking legs. The cephalothorax is followed by an abdomen that contains internal organs.

Scorpions are the oldest terrestrial arthropods. Today, they are widely distributed in the tropics, subtropics, and temperate regions, with a notable absence from New Zealand. They are nocturnal and spend most of the day hidden under a log or a rock. In scorpions, the pedipalps are large pincers, and the long abdomen ends with a stinger that contains venom.

Ticks and mites are parasites. Ticks suck the blood of vertebrates and sometimes transmit diseases, such as Rocky Mountain spotted fever or Lyme disease. Chiggers, the larvae of certain mites, feed on the skin of vertebrates.

Video Lyme Disease

Spiders, the most familiar arachnids, have a narrow waist that separates the cephalothorax from the abdomen. Each chelicera consists of a basal segment and a fang that delivers venom to paralyze or kill prey. Spiders use silk threads for all sorts of purposes, from lining their nests to catching prey. Biologists have used web-building behavior to discover how spider families are related.

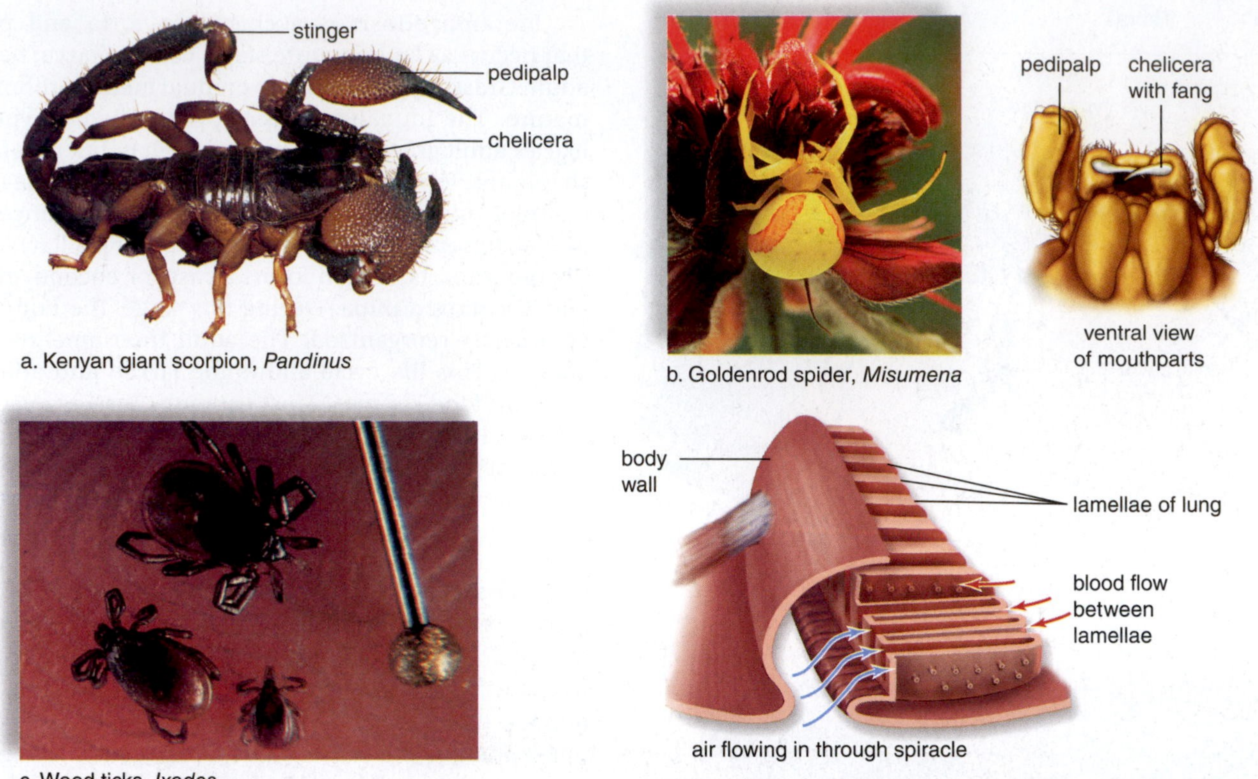
stinger
pedipalp
chelicera
pedipalp
chelicera with fang
ventral view of mouthparts
body wall
lamellae of lung
blood flow between lamellae
air flowing in through spiracle

a. Kenyan giant scorpion, *Pandinus*

b. Goldenrod spider, *Misumena*

c. Wood ticks, *Ixodes*

d. Book lung anatomy

Figure 31.24 Arachnid diversity. **a.** A scorpion has pincerlike chelicerae and pedipalps. Its long abdomen ends with a stinger that contains venom. **b.** Goldenrod spiders can change colors from yellow to white to camouflage themselves with the color of the flower they are found on. **c.** In the western United States, the wood tick carries a debilitating disease called Rocky Mountain spotted fever. **d.** Arachnids breathe by means of book lungs, in which the "pages" are double sheets of thin tissue (lamellae).

SCIENCE IN YOUR LIFE ▶ **SCIENTIFIC INQUIRY**

Maggots: A Surprising Tool for Crime Scene Investigation

A fresh human corpse is lying on the floor. Within 10 to 15 minutes, blowflies with blue and green metallic bodies arrive. An animal corpse is a perfect site for egg production. The flies lay eggs in the mouth, nostrils, ears, eyes, and any open wounds. Depending on the temperature, the eggs will hatch in the next 12 to 24 hours. The small hatchling larval flies, more commonly known as maggots, feed on fat and collagen in the corpse in order to grow. Their development occurs in three stages called instars, during which the larvae shed skin and grow. When the third instar grows, it leaves the corpse and forms a pupa that takes about 14 days (depending on the temperature) to emerge as an adult blowfly. The adult has to wait a day or two until it can fly because the wings need to expand and the body has to harden. Once capable of flight, the adult leaves to find a mate as well as another corpse in which to lay eggs, thereby completing the life cycle.

What do blowflies have to do with crime scene investigation? If the corpse involved is a human, the timing of the various stages of the life cycle of the flies (e.g., newly hatched eggs, maggot size, maggot weight, or pupa remains) can help forensic scientists determine the postmortem interval (or PMI) since the time of death. Forensic entomology uses insects such as blowflies to determine the time of death. A forensic entomologist can even determine if a human corpse has been moved based on the fact that different species of blowflies prefer to lay eggs in different environments. For example, some species lay their eggs in shade, whereas others lay eggs in sunny habitats; similarly, some species inhabit urban areas, whereas others inhabit rural areas. Thus, the combination of life stage(s) and species of blowfly found on a corpse can help a forensic entomologist determine the time and possible location of a person's death.

Questions to Consider

1. Given that the hatching of eggs and the growth rate of blowfly larvae depend on temperature, how sure do you think forensic entomologists can be about time of death?
2. As a juror, how would you feel about forensic entomology evidence entered into court?
3. When might forensic entomology evidence be useful in a murder case?

The internal organs of spiders also show how they are adapted to a terrestrial way of life. Malpighian tubules work in conjunction with rectal glands to reabsorb ions and water before a relatively dry nitrogenous waste (uric acid) is excreted. Invaginations of the inner body wall form the lamellae ("pages") of spiders' so-called book lungs. Air flowing into the folded lamellae on one side exchanges gases with blood flowing in the opposite direction on the other side.

Check Your Progress 31.4

1. Name two ways in which the roundworms are anatomically similar to the arthropods.
2. Explain the features that account for the great success of arthropods.
3. Identify the major anatomical characteristics of a crustacean.

31.5 Invertebrate Deuterostomes

Learning Outcomes

Upon completion of this section, you should be able to

1. Recognize the basic morphological characteristics of echinoderms.
2. Describe how sea stars move, feed, and reproduce.

Among animals, chordates are most closely related to the echinoderms (phylum Echinodermata) as witnessed by their similar embryological development, even though echinoderms are invertebrates. Both echinoderms and chordates are deuterostomes, in which the second embryonic opening becomes the mouth, and the coelom forms by outpocketing the primitive gut (see Fig. 31.3). Protostomes, in contrast, are characterized by a mouth that forms from the first embryonic opening (the blastopore).

Characteristics of Echinoderms

Echinoderms are a diverse group of marine animals. There are no terrestrial echinoderms. They have an endoskeleton (internal skeleton) consisting of spine-bearing, calcium-rich plates. The spines, which stick out through their delicate skin, account for their name.

It may seem surprising that echinoderms, although related to chordates, lack those features we associate with vertebrates such as humans. For example, echinoderms are often radially, not bilaterally, symmetrical. Their larva is a free-swimming filter feeder with bilateral symmetry, but it typically metamorphoses into a radially symmetrical adult. Recall that with radial symmetry, it is possible to obtain two mirror images, no matter how the animal is sliced longitudinally, whereas in bilaterally symmetrical animals, only a longitudinal cut gives two mirror images.

Echinoderm Diversity

Echinoderms are quite diverse (Fig. 31.25). They include sea lilies, motile feather stars, brittle stars, and sea cucumbers. Sea cucumbers actually look like a cucumber, except they have feeding tentacles surrounding their mouth. You may be more familiar with sea urchins and sand dollars in the class Echinoidea, which have spines for locomotion, defense, and burrowing. Sea urchins are food for sea otters off the coast of California.

Video Sea Urchin Reproduction

Sea Stars

Sea stars (starfish) are commonly found along rocky coasts, where they feed on clams, oysters, and other bivalve molluscs. The five-rayed body has an oral, or mouth, side (the underside) and an aboral, or anus, side (the upper side) (Fig. 31.26). Various structures project through the body wall: (1) spines from the endoskeletal plates offer some protection; (2) pincerlike structures called pedicellariae keep the surface free of small particles; and (3) skin gills, tiny fingerlike extensions of the skin, are used for gas exchange. On the oral surface, each arm has a groove lined by small tube feet.

To feed, a sea star positions itself over a bivalve such as a clam and attaches some of its tube feet to each side of the shell (Fig. 31.27). By working its tube feet in alternation, it pulls the shell open. A very small crack is enough for the sea star to evert its cardiac stomach and push it through the crack to contact the soft parts of the bivalve. The stomach secretes enzymes, and digestion begins even while the bivalve is attempting to close its shell. Later, partly digested food is taken into the sea star's body, where digestion continues in the pyloric stomach, using enzymes from digestive glands found in each arm. A short intestine opens at the anus on the aboral side.

In each arm, the well-developed coelomic cavity contains not only a pair of digestive glands, but also gonads (either male or female), which open on the aboral surface by very small pores. The nervous system consists of a central nerve ring from which radial nerves extend into each arm. A light-sensitive eyespot is at the tip of each arm. Sea stars are capable of coordinated but slow responses and body movements.

Locomotion depends on their water vascular system. Water enters this system through a structure on the aboral side called the madreporite, or sieve plate. From there it passes down a stone canal into a ring canal, which surrounds the mouth. A radial canal in each arm branches off from the central ring canal. For locomotion, water enters the ampullae from the radial canals. Contraction of ampullae forces water into the tube foot, expanding it. When the foot touches a surface, the center is withdrawn, giving it suction so that it can adhere to the surface. By alternating the expansion and contraction of the tube feet, a sea star moves slowly along.

Echinoderms don't have a respiratory, excretory, or circulatory system. Fluids within the coelomic cavity and the water vascular system carry out many of these functions. For example, gas exchange occurs across the skin gills and the tube feet. Nitrogenous wastes diffuse through the coelomic fluid and the

a. Brittle star, *Ophiothrix*

b. Sea cucumber, *Cucumaria*

c. Sea urchin, *Strongylocentrotus*

d. Feather stars, *Antedon*

Figure 31.25 Echinoderm diversity. **a.** Brittle stars have long, slender arms, which make them the most mobile echinoderms. **b.** Sea cucumbers look like a cucumber. They lack arms but have tentacle-like tube feet with suckers around the mouth. **c.** Sea urchins have large, colored, external spines for protection. **d.** A feather star extends its arms to filter suspended food particles from the sea.

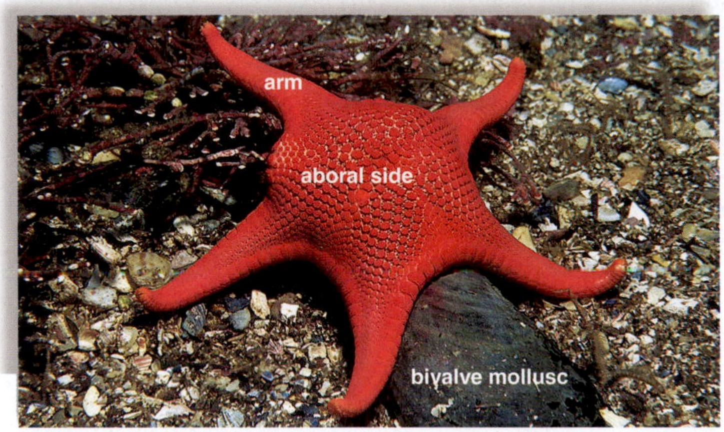
arm
aboral side
bivalve mollusc

a.

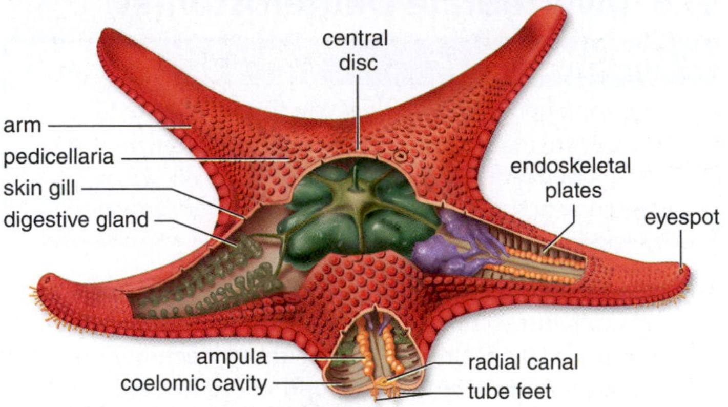
central disc
arm
pedicellaria
skin gill
digestive gland
endoskeletal plates
eyespot
ampula
coelomic cavity
radial canal
tube feet

b. Aboral side showing ray cross-section

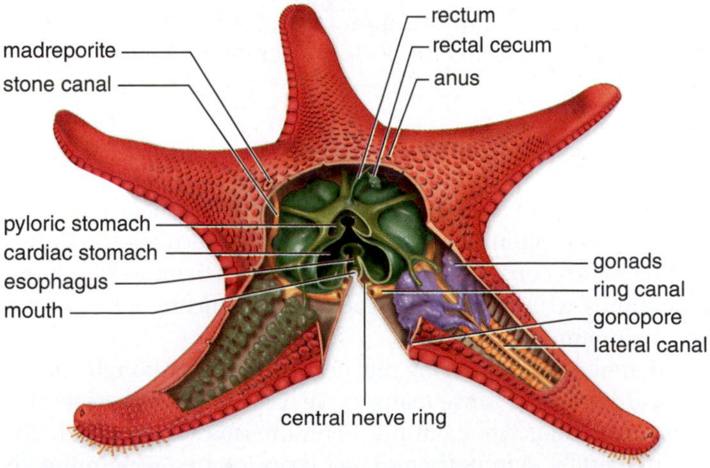
rectum
rectal cecum
anus
madreporite
stone canal
pyloric stomach
cardiac stomach
esophagus
mouth
gonads
ring canal
gonopore
lateral canal
central nerve ring

c. Aboral side showing internal cross-section

Figure 31.26 Sea star anatomy and behavior. **a.** A sea star uses the suction of its tube feet to open a bivalve mollusc, its primary source of food. **b.** Each arm of a sea star contains digestive glands, gonads, and portions of the water vascular system. This system (colored orange) terminates in tube feet. **c.** A different cross section shows the orientation of mouth, digestive system, and anus.

Figure 31.27 Sea star attacking a mollusc. Tube feet pry open the mollusc shell enough to allow the sea star to insert its stomach, allowing the digestive juices to kill the mollusc.

body wall. Cilia on the peritoneum lining the coelom keep the coelomic fluid moving.

Sea stars reproduce both asexually and sexually. If the body is fragmented, each fragment can regenerate a whole animal as long as the fragment contains part of the central disk. Fishermen who try to get rid of sea stars by cutting them up and tossing them overboard are merely propagating more sea stars! Sea stars also spawn, releasing either eggs or sperm. The larva is bilaterally symmetrical and metamorphoses to become the radially symmetrical adult.

Check Your Progress 31.5

1. Explain why echinoderms are more closely related to chordates than invertebrates.
2. Identify the basic structures of an echinoderm.

Case Study Conclusion

A fad diet called the "tapeworm diet" has been banned by the FDA due to the harmful side effects that parasitic tapeworms can cause to their hosts. More than 50 million people in less developed nations around the world are threatened with cysticercosis, a parasitic tapeworm infection that can lead to weight loss, anemia, and severe malnutrition.

The invertebrates—animals that lack a backbone—are far more diverse than the vertebrates. While various invertebrates are associated with disease and health problems, many of them are very beneficial to the survival of the human species. Various insects act as pollinators for a large number of commercially grown crops. Invertebrates possess a wide variety of traits that have allowed them to successfully adapt to a wide range of lifestyles and habitats.

MEDIA STUDY TOOLS

www.mhhe.com/maderinquiry14

Enhance your study of this chapter with study tools and practice tests. Also ask your instructor about the resources available through ConnectPlus, including LearnSmart, the media-rich eBook, interactive learning tools, and animations.

Mc Graw Hill connect plus+ | BIOLOGY

SUMMARIZE

31.1 Evolutionary Trends Among Animals

Animals are motile, multicellular heterotrophs that ingest their food. They exhibit a diploid life cycle.

- Animals are divided into two categories—**invertebrates,** those that lack an endoskeleton—and **vertebrates,** those that possess an endoskeleton.
- Symmetry varies among animals. Some are **asymmetrical,** whereas others have **radial symmetry** or **bilateral symmetry.** Some animals live portions of their life as **sessile** individuals. **Cephalization** is often associated with bilateral symmetry.
- Animals are further divided into the **protostomes** or **deuterostomes** based upon their developmental features and processes. **Lophotrochozoa** are examples of protostomes. This group is divided into the lophophores, which contain a **lophophoran,** or mouth surrounded by ciliated tentacles, as well as the **trochophores,** or individuals that either have presently, or their ancestors had, a trochophore larva.

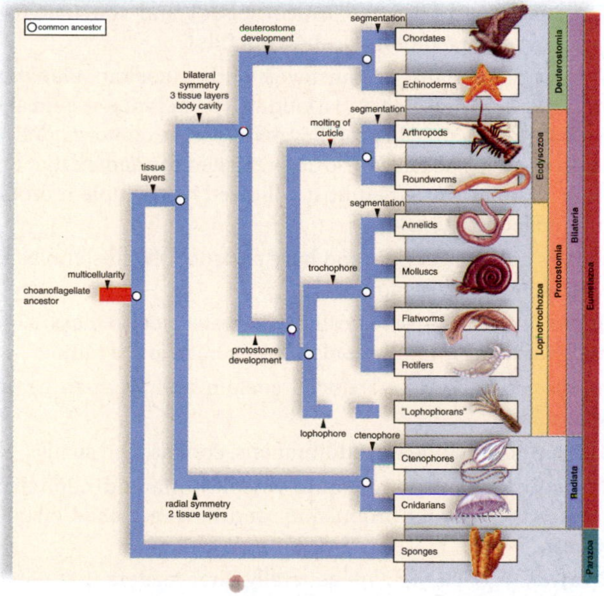

common ancestor
deuterostome development
segmentation
Chordates
Echinoderms
bilateral symmetry 3 tissue layers body cavity
segmentation
molting of cuticle
Arthropods
Roundworms
tissue layers
segmentation
Annelids
multicellularity
choanoflagellate ancestor
trochophore
Molluscs
Flatworms
protostome development
Rotifers
"Lophophorans"
lophophore
ctenophore
Ctenophores
radial symmetry 2 tissue layers
Cnidarians
Sponges
Deuterostomia
Ecdysozoa
Lophotrochozoa
Protostomia
Bilateria
Eumetazoa
Radiata
Parazoa

31.2 The Simplest Invertebrates

Table 31.2 contrasts selected anatomical features used to help classify animals.

- **Sponges** have the cellular level of organization, lack true tissues, and are typically asymmetrical.
- They use **collar cells** to create a water current, enabling them to act as **filter feeders.**
- Most sponges are **hermaphroditic,** meaning they have both male and female sex organs.

TABLE 31.2 Comparison of Invertebrates

	Sponges	Cnidarians	Flatworms	Roundworms	Other Phyla*
Level of organization	Cell	Tissue	Organs	Organs	Organs
Type of body plan	—	Incomplete digestive system	Incomplete digestive system	Complete digestive system	Complete digestive system
Type of symmetry	Usually asymmetrical	Radial	Bilateral	Bilateral	Bilateral except echinoderms
Type of body cavity	—	—	None	Pseudocoelom	Coelom
Segmentation	—	—	—	—	Only annelids, arthropods, and chordates
Jointed appendages	—	—	—	—	Only arthropods and chordates

*Molluscs, annelids, arthropods, echinoderms, chordates

- The internal skeleton is composed of **spicules,** which is often used to classify them. **Comb jellies** and **cnidarians** are radially symmetrical and have two tissue layers derived from the germ layers ectoderm and endoderm. The body of comb jellies is made up of a transparent jellylike substance called mesoglea.
- The **gastrovascular cavity** serves as a location for digestion and circulation of nutrients.
- The defining feature of the cnidarians is the presence of stinging cells called **cnidocytes** that contain a toxin-filled capsule called a **nematocyst.**
- Cnidarians exist as either a **polyp** or a **medusa,** or they can alternate between the two.

31.3 The Lophotrochozoans

Lophotrochozoans are protostomes and include the lophophorans (bryozoans, phoronids, and brachiopods) and the **trochozoans** (flatworms, rotifers, molluscs, and annelids).

- **Flatworms** have three tissue layers and no coelom. Planaria have cephalization, muscles, and a ladder-type nervous system and also use **flame cells** to help propel water toward excretory pores. Tapeworms have a specialized structure called the **scolex** that enables them to attach onto their host, followed by multiple reproductive units called **proglottids.**
- Rotifers are microscopic and have a corona that resembles a spinning wheel when in motion.
- The body of a **mollusc** typically contains a **visceral mass,** a **mantle,** and a **foot.** Many molluscs also have a head and a radula.
- Molluscs often have a reduced coelom and an open circulatory system.
- **Gastropods** include the nudibranchs, conchs, and snails.
- **Cephalopods,** such as squid, display marked cephalization, move rapidly by jet propulsion, and have a **closed circulatory system.**
- **Bivalves** (e.g., clams), are generally filter feeders that possess an **open circulatory system,** a radula, and a hatchet foot.
- **Annelids** are segmented both externally and internally. Their **hydrostatic skeleton** is a fluid-filled interior that supports muscle contraction and enhances flexibility. The nervous system consists of a ventral solid **nerve cord** and brain.
- The **nephridium** functions as a waste collection and excretory organ.
- Polychaetes are marine worms that have parapodia that contain tiny bristlelike hairs called **setae.**
- Earthworms are oligochaetes that are segmented and hermaphroditic scavengers that feed on organic matter in the soil.
- **Leeches** lack the setae found in other annelids. They contain suckers on each end of the body that enable attachment and feeding.

31.4 The Ecdysozoa

Roundworms are usually small, very diverse, and are present almost everywhere in great numbers. Roundworms have a pseudocoelom.

- **Trichinosis** and **elephantiasis** are two diseases associated with roundworm infections.
- **Arthropods** are the most diverse group of animals. Their success is largely attributable to a flexible exoskeleton, segmentation, a well-developed nervous system, respiratory organs, and **metamorphosis.** Like many other arthropods, **crustaceans** have a head that bears compound eyes, antennae, and mouthparts.
- Blood is pumped into the **hemocoel,** where hemolymph will flow about the organs.
- Like many other **insects,** grasshoppers have wings and three pairs of legs attached to the thorax. Grasshoppers also have several adaptations to a terrestrial life, including respiration by **tracheae** and an excretory system consisting of **Malpighian tubules,** which collect and excrete nitrogenous wastes.
- Spiders are **arachnids** with chelicerae, pedipalps, and four pairs of walking legs attached to a cephalothorax. Spiders, too, are adapted to life on land, and they spin silk that is used in various ways.
- The crayfish, a crustacean, also has an open circulatory system, respiration by gills, and a ventral solid nerve cord.

31.5 Invertebrate Deuterostomes

Echinoderms and chordates are deuterostomes, whereby the second embryonic opening becomes the mouth and the coelom forms by outpocketing of the primitive gut. Echinoderms (e.g., sea stars, sea urchins, sea cucumbers, sea lilies) have radial symmetry as adults (not as larvae) and spines from endoskeletal plates.

- Typical of echinoderms, sea stars have tiny skin gills, a central nerve ring with branches, and a water vascular system for locomotion. Each arm of a sea star contains components of the nervous, digestive, and reproductive systems.

ASSESS

Testing Yourself

Choose the best answer for each question.

1. Animals with a cellular level of organization produce
   a. cells only.
   b. cells and tissues.
   c. cells and organs.
   d. cells, tissues, and organs.
2. Which of the following has an incomplete digestive tract?
   a. earthworm
   b. hydra
   c. leech
   d. ant
   e. mosquito

For problems 3–7, indicate the type of symmetry exhibited by the animal listed. Each answer can be used more than once.

Key:

a. asymmetry
b. radial symmetry
c. bilateral symmetry

3. Lobster
4. Jellyfish
5. Sponge
6. Planarian
7. Octopus

For problems 8–10, match the feature with the type of flatworm. Each answer may be used more than once. Each problem may have more than one answer.

Key:

a. planarian
b. tapeworm
c. fluke

8. Endoparasite
9. Light-sensitive organs
10. Digestive tract lacks an anus
11. Compared to an animal species that has a coelom, one that lacks a coelom
    a. is more flexible.
    b. has more complex organs.
    c. is more likely to tolerate temperature variations.
    d. is less able to store excretory wastes.
12. Which of the following is not a feature of most coelomates?
    a. radial symmetry
    b. three germ layers
    c. complete digestive tract
    d. organ level of organization
13. Label the parts of an earthworm in the following illustration.

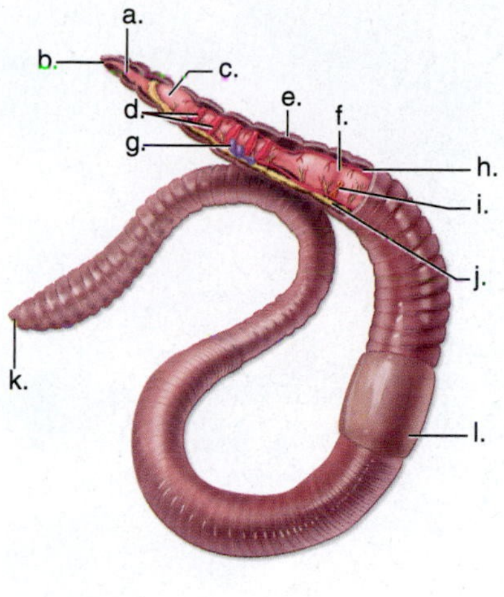
a.
b.
c.
d.
e.
f.
g.
h.
i.
j.
k.
l.

14. Which is a feature of an echinoderm?
    a. endoskeleton
    b. exoskeleton
    c. radial symmetry as a larva
    d. bilateral symmetry as an adult
15. Which echinoderm feature enables them to keep the surface of their skin free of small particles?
    a. pedicellariae
    b. skin gills
    c. tube feet
    d. endoskeleton plates

ENGAGE

Virtual Labs
Earthworm Dissection
Classifying Arthropods

The virtual lab "Earthworm Dissection" provides an interactive look at the anatomy of an earthworm. In addition, the virtual lab "Classifying Arthropods" provides a review of the key characteristics used to identify the major groups of arthropods.

Thinking Critically

1. Cnidarians are radially symmetrical, whereas flatworms, roundworms, molluscs, annelids, and arthropods are bilaterally symmetrical. How does the type of symmetry relate to the lifestyle of each type of animal?
2. What features of arthropods make them the most diverse animal phylum on Earth?
3. Why do you think coelomates far outnumber acoelomates or pseudocoelomates in terms of species diversity?

32 Animals: Chordates and Vertebrates

CHAPTER OUTLINE

BEFORE YOU BEGIN

Before beginning this chapter, take a few moments to review the following discussions:

Section 27.2 How does the fossil record support evolutionary theory?

Section 27.6 How does a phylogenetic tree indicate the evolutionary relatedness of species?

Figure 31.2 What phyla of animals are most closely related to the chordates?

CASE STUDY World's Smallest Vertebrate

There are almost 58,000 species of vertebrates (animals with a backbone) known to science. Vertebrates first evolved approximately 500 million years ago (MYA) in the marine environment. The earliest forms were jawless fishes. From this original ancestor the jawed fish, amphibians, reptiles (and birds), and mammals (and humans) would eventually arise. While many species of vertebrates are well known to science, by many estimates there may be as many as 18,000 unidentified vertebrate species waiting to be discovered.

In the past few years, scientists have discovered two new species of frogs in New Guinea—one of which, *Paedophryne amanuensis,* averages 7.7 mm in size and is believed to be the world's smallest vertebrate. These miniature frogs primarily inhabit the leaf litter on the floor of tropical environments, occupying a unique ecological niche. Adaptation to living in leaf litter and moss appears to be a common feature among miniaturized frogs, allowing them to exploit novel food sources in their environment. The small body size also reduces the number of eggs carried by the females to just two. Scientists are still uncertain as to the advantage this may provide *P. amanuensis.*

In Chapter 31, we learned about invertebrates, new species of which are being discovered all the time. While scientists are discovering new species of vertebrates, approximately 20% of the known species of mammals, 12% of birds, and 33% of amphibians are threatened with extinction. The future of many vertebrates may, in large part, depend on our ability to protect them. Amidst such discouraging messages, there are rays of hope. Large-scale cooperative efforts to conserve critical habitats, which will help to protect a wide range of species that inhabit these areas, is an example of how people can work together to protect biodiversity.

As you read through the chapter, think about the following questions:

1. What features need to be present in order to consider *P. amanuensis* a vertebrate?
2. Where do amphibians fit in the vertebrate evolutionary history?

32.1 Chordates

Learning Outcomes

Upon completion of this section, you should be able to

1. Identify the four morphological characteristics unique to the chordates.
2. Describe the features of the two groups of nonvertebrate chordates.

To be classified as a **chordate** (phylum Chordata, around 60,000 species), an animal must have, at some time during its life history, the following characteristics (Fig. 32.1):

1. A dorsal supporting rod called a **notochord.** The notochord is located just below the nerve cord toward the back (i.e., dorsal). Vertebrates have an embryonic notochord that is replaced by the vertebral column during development.
2. A dorsal tubular **nerve cord.** Tubular means that the cord contains a canal filled with fluid. In vertebrates, the nerve cord is protected by the vertebrae. Therefore, it is called the spinal cord because the vertebrae form the spine.
3. **Pharyngeal pouches.** Most vertebrates have pharyngeal pouches only during embryonic development. In the nonvertebrate chordates, the fishes, and some amphibian larvae, the pharyngeal pouches develop into functioning gills. Water passing into the mouth and the pharynx goes through the gill slits, which are supported by gill arches. In terrestrial vertebrates that breathe by lungs, the pouches are modified for various purposes. In humans, the first pair of pouches becomes the auditory tubes. The second pair becomes the tonsils, while the third and fourth pairs become the thymus and the parathyroids.
4. **Postanal tail.** The tail extends beyond the anus, hence the term postanal.

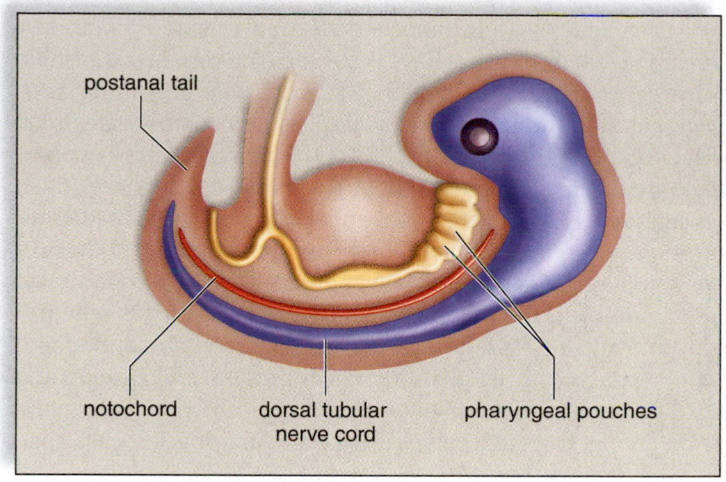
postanal tail
notochord
dorsal tubular nerve cord
pharyngeal pouches

Figure 32.1 Chordate characteristics.

Evolutionary Trends Among the Chordates

Figure 32.2 depicts the phylogenetic tree of the chordates, and lists at least one main evolutionary trend that distinguishes each group of animals from the preceding group.

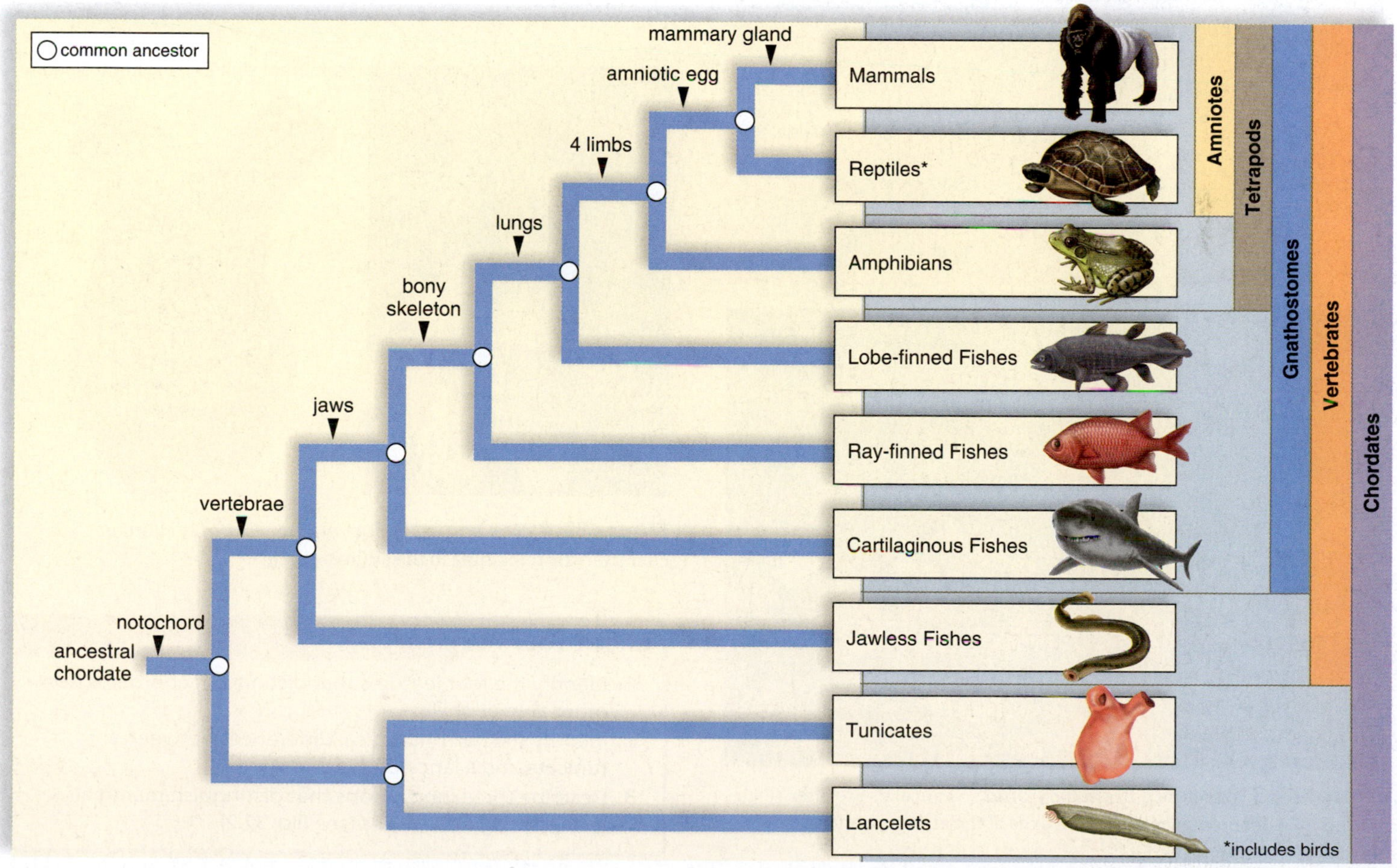
common ancestor
mammary gland
amniotic egg
4 limbs
lungs
bony skeleton
jaws
vertebrae
notochord
ancestral chordate
Mammals
Reptiles*
Amphibians
Lobe-finned Fishes
Ray-finned Fishes
Cartilaginous Fishes
Jawless Fishes
Tunicates
Lancelets
Amniotes
Tetrapods
Gnathostomes
Vertebrates
Chordates
*includes birds

Figure 32.2 Phylogenetic tree of chordates. Each of the innovations listed is an evolutionary trend shared by the classes beyond the branch point.

The tunicates and lancelets are nonvertebrate chordates, meaning that they do not have vertebrae. The vertebrates include the fishes, amphibians, reptiles, birds, and mammals. The cartilaginous fishes were the first to have jaws, and some early bony fishes had lungs. However, not all bony fishes have lungs. Amphibians are the first group to clearly have jointed appendages and to invade land. However, the fleshy appendages of lobe-finned fishes from the Devonian era contained bones homologous to those of terrestrial vertebrates. These fishes are believed to be ancestral to the amphibians. Reptiles and mammals have a means of reproduction suitable for land. During development, an amnion and other extraembryonic membranes are present. The amnion and extraembryonic membranes support the embryo and prevent it from drying out as it develops into a particular species' offspring.

Nonvertebrate Chordates

Nonvertebrate chordates are characterized by the fact that the notochord never becomes a vertebral column.

Lancelets (subphylum Cephalochordata, about 30 species), formerly referred to as *Amphioxus,* are marine chordates only a few centimeters long. They are named for their resemblance to a lancet—a small, two-edged surgical knife (Fig. 32.3). Lancelets are found in shallow water along most coasts, where they usually lie partly buried in sandy or muddy substrates with only their anterior mouth and gill apparatus exposed. They feed on microscopic particles filtered out of a constant stream of water that enters the mouth and exits through the gill slits.

Lancelets retain the four chordate characteristics as adults. In addition, segmentation is present, as evidenced by the segmental arrangement of the muscles and the periodic branches of the dorsal tubular nerve cord.

Tunicates (subphylum Urochordata, about 1,250 species) live on the ocean floor and take their name from a tunic that makes the adults look like thick-walled, squat sacs (Fig. 32.4). They are also called sea squirts because they squirt water from one of their siphons when disturbed. The tunicate larva is bilaterally symmetrical and has the four chordate characteristics. Metamorphosis produces the sessile adult, which has incurrent and excurrent siphons.

The pharynx is lined by numerous cilia. Beating of these cilia creates a current of water that moves into the pharynx and out the numerous gill slits, the only chordate characteristic that remains in the adult. Microscopic particles adhere to a mucous secretion and are digested.

Some biologists have suggested that tunicates are directly related to vertebrates. They hypothesize that a larva with the four chordate characteristics may have become sexually mature, and evolution thereafter produced a fishlike vertebrate.

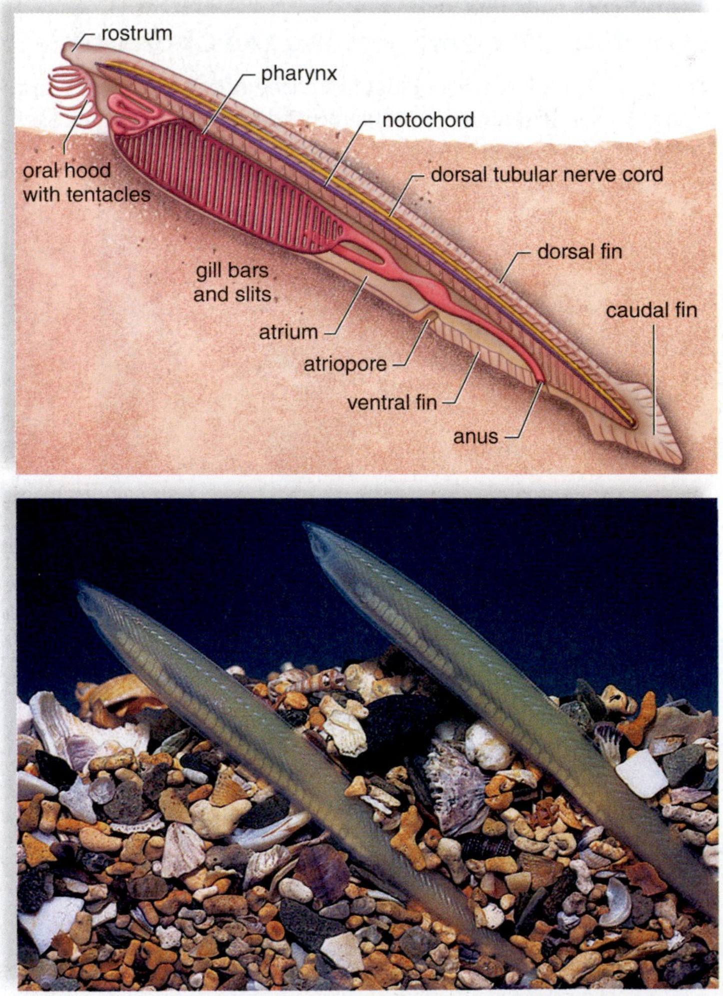
rostrum
pharynx
notochord
oral hood
with tentacles
dorsal tubular nerve cord
dorsal fin
gill bars
and slits
caudal fin
atrium
atriopore
ventral fin
anus

Figure 32.3 Lancelet, *Branchiostoma.* Lancelets are filter feeders. Water enters the mouth and exits at the atriopore after passing through the gill slits.

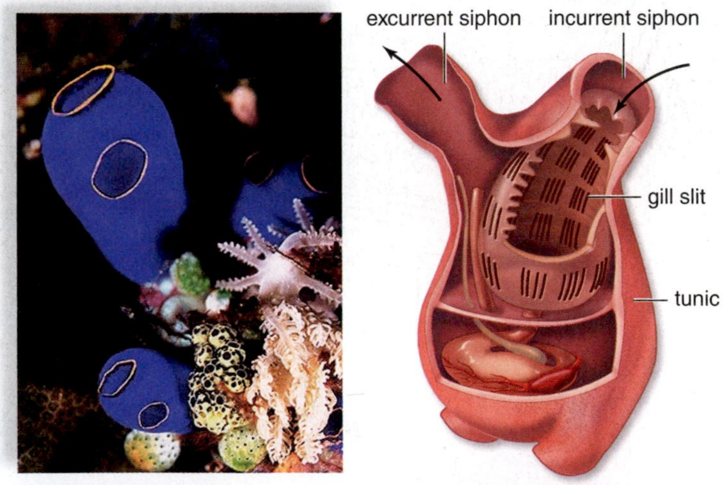
excurrent siphon
incurrent siphon
gill slit
tunic

Figure 32.4 Sea squirt, *Halocynthia.* The only chordate characteristic remaining in the adult is gill slits.

Check Your Progress 32.1

1. Identify the four features that distinguish chordates from other animal phyla.
2. Identify the morphological differences between a tunicate and a lancelet.
3. Describe three innovations that distinguish mammals from cartilaginous fishes (see Fig. 32.2).

32.2 Vertebrates: Fish and Amphibians

Learning Outcomes

Upon completion of this section, you should be able to

1. Compare and contrast the characteristics of a jawless fish to those of a jawed fish.
2. Describe the major evolutionary innovations that distinguish the fishes from the amphibians.
3. Identify the features of the three groups of living amphibians.

At some time in their life history, **vertebrates** (subphylum Vertebrata, about 58,000 species) have all four chordate characteristics. The embryonic notochord, however, is generally replaced by a vertebral column composed of individual vertebrae. The vertebral column, which is part of the flexible but strong jointed endoskeleton, provides evidence that vertebrates are segmented. The skeleton protects the internal organs and serves as a place of attachment for muscles. Together, the skeleton and muscles form a system that permits rapid and a more efficient movement. Two pairs of appendages are typical. The pectoral and pelvic fins of fishes evolved into the jointed appendages that allowed vertebrates to move onto land.

The main axis of the endoskeleton consists of not only the vertebral column, but also a skull that encloses and protects the brain. The high degree of cephalization is accompanied by complex sense organs. The eyes develop as outgrowths of the brain. The ears are primarily equilibrium devices in aquatic vertebrates, but they also function as sound-wave receivers in land vertebrates.

The evolution of jaws in vertebrates has allowed them to diversify into a variety of biological roles. Vertebrates have a complete digestive tract and a large coelom. Their circulatory system is closed, meaning that the blood is contained entirely within blood vessels. Vertebrates have an efficient means of obtaining oxygen from water or air, as appropriate. The kidneys are important excretory and water-regulating organs. The sexes are generally separate, and reproduction is usually sexual. The evolution of the amnion allowed reproduction to take place on land. Many species of reptiles, and a few species of mammals, lay a shelled egg. In placental mammals, development takes place within the uterus of the female. Although use of vertebrates, as well as other animals in laboratory research, has been very beneficial in many ways, it also generates controversy for some. The Bioethical feature, "Vertebrates and Human Medicine," on page 654, discusses issues related to use of animals in scientific research.

A strong, jointed endoskeleton, a vertebral column composed of vertebrae, a closed circulatory system, efficient respiration and excretion, and a high degree of cephalization are characteristics demonstrating vertebrates are adapted to an active lifestyle. Figure 32.5 shows major milestones in the history of vertebrates: the evolution of jaws, limbs, and the amnion, an extraembryonic membrane that is first seen in the shelled **amniotic egg** of reptiles.

dorsal fin
gill slits
jaw with teeth
pelvic fin
pectoral fin

a. Sand tiger shark, *Carcharias taurus*

b. Spotted salamander, *Ambystoma*

c. Veiled chameleons, *Chamaeleo*

Figure 32.5 Milestones in vertebrate evolution. **a.** The evolution of jaws in fishes allows animals to be predators and to feed off other animals. **b.** The evolution of limbs in most amphibians is adaptive for locomotion on land. **c.** The evolution of an amnion and a shelled egg in reptiles is adaptive for reproduction on land.

SCIENCE IN YOUR LIFE ▶ BIOETHICAL

Vertebrates and Human Medicine

Hundreds of pharmaceutical products come from other vertebrates, and even those that produce poisons and toxins provide medicines that benefit us.

Natural Products with Medical Applications

The Thailand cobra paralyzes its victim's nerves and muscles with a potent venom that eventually leads to respiratory arrest. However, that venom is also the source of the drug tetrachlorodecaoxide (Immunokine), which has been used for over a decade in multiple sclerosis patients. Immunokine, which is almost without side effects, actually protects the patient's nerve cells from destruction by their immune system.

A compound known as ABT-594, derived from the skin of the poison-dart frog (Fig. 32A*a*), is approximately 50 times more powerful than morphine in relieving chronic and acute pain without the addictive properties.

The southern copperhead snake and the fer-de-lance pit viper are two of the unlikely vertebrates that either serve as the source of pharmaceuticals or provide a chemical model for the synthesis of effective drugs in the laboratory. These drugs include anticoagulants ("clot busters"), painkillers, antibiotics, and anticancer drugs.

A variety of friendlier vertebrates produce proteins that are similar enough to human proteins to be used for medical treatment. Until 1978, when recombinant DNA human insulin was produced, diabetics injected purified insulin from pigs. Currently, the flu vaccine is produced in fertilized chicken eggs. The production of these drugs, however, is often time-consuming, labor intensive, and expensive. Approximately 600 million chicken eggs are used each year in vaccine production.

Animal Pharming

Some of the most powerful applications of genetic engineering can be found in the development of drugs and therapies for human diseases. In fact, this new biotechnology has actually led to a new industry: animal pharming. Animal pharming uses genetically altered vertebrates, such as mice, sheep, goats, cows, pigs, and chickens, to produce medically useful pharmaceutical products.

The human gene for some useful product is inserted into the embryo of the vertebrate. That embryo is implanted into a foster mother, which gives birth to the transgenic animal, which contains genes from the two sources. An adult transgenic vertebrate produces large quantities of the pharmed product in its blood, eggs, or milk, from which the product can be easily harvested and purified.

An example of a pharmed product advanced in development and approved by the FDA for high-risk situations is ATIII, a bioengineered form of human antithrombin. This medication is important in the treatment of individuals who have a hereditary deficiency of this protein and are at high risk for life-threatening blood clots during surgery, childbirth procedures, or the treatment of thromboembolism.

Xenotransplantation

Xenotransplantation, the transplantation of nonhuman vertebrate tissues and organs into humans, is another benefit of genetically altered animals. There is an alarming shortage of human donor organs to fill the need for hearts, kidneys, and livers. In 1963, doctors transplanted chimpanzee kidneys into 13 human patients, 12 of which lived for between 9 and 60 days. One patient survived for nine months. In the late 1990s, two patients were kept alive using a pig liver outside of their body to filter their blood until a human organ was available for transplantation.

Although baboons are phylogenetically closer to humans than pigs, pigs are generally healthier, produce more offspring in a shorter time, and are already farmed for food (Fig. 32A*b*). Despite the fears of some, scientists think that viruses unique to pigs are unlikely to cross the species barrier and infect the human recipient. Currently, pig heart valves and skin are routinely used for treatment of humans (Fig. 32A*c*). Miniature pigs, whose heart size is similar to humans, are being genetically engineered to make their tissues less foreign to the human immune system, in order to avoid rejection.

Questions to Consider

1. Is it ethical to change the genetic makeup of vertebrates in order to use them as drug or organ factories?
2. What are some of the health concerns that may arise due to xenotransplantation?

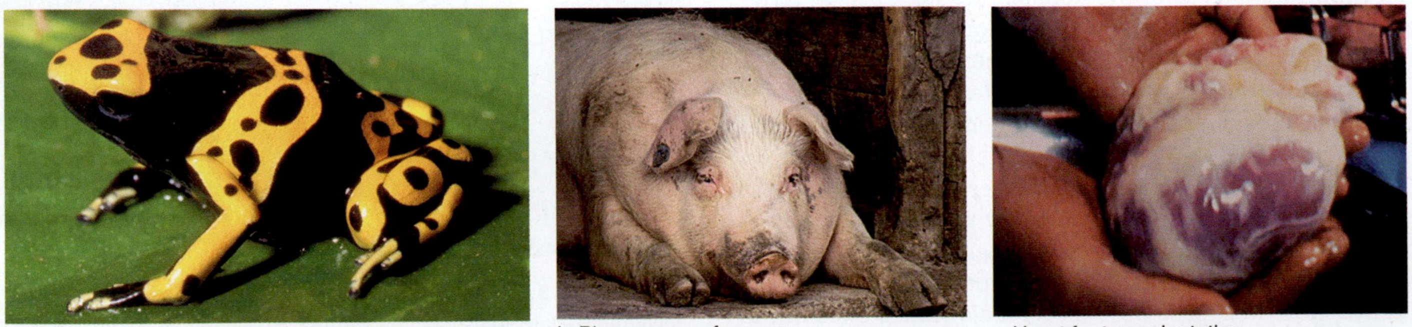

a. Poison-dart frogs, source of a medicine b. Pigs, source of organs c. Heart for transplantation

Figure 32A Medical uses of animals. **a.** The poison-dart frog is the source of a pain medication. **b.** Pigs are now being genetically altered to provide a supply of hearts for (**c**) heart transplant operations.

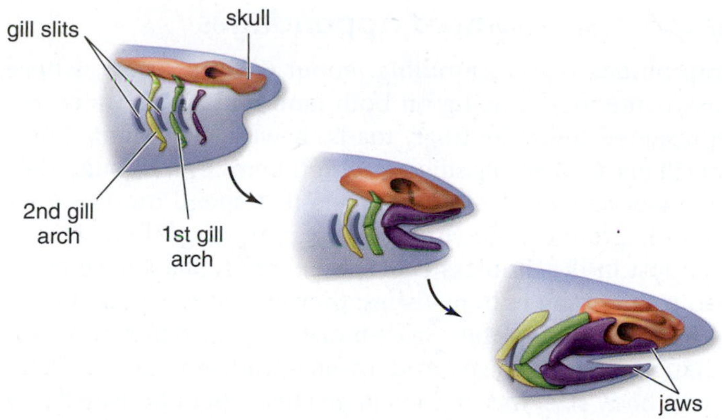
gill slits
skull
2nd gill arch
1st gill arch
jaws

Figure 32.6 Evolution of the jaw. The first jaw evolved from the first and second gill arches of fishes.

Fishes: First Jaws, Then Lungs

The first vertebrates were jawless fishes that wiggled through the water and sucked up food from the ocean floor. Today, there are three living groups of fishes: jawless fishes, cartilaginous fishes, and bony fishes. The two latter groups have *jaws,* tooth-bearing structures in the head. Jaws evolved from the first pair of gill arches, structures that ordinarily support gills (Fig. 32.6). The second pair of gill arches became support structures for the jaws, instead of gills. The presence of jaws permits a predatory way of life in many species of fishes.

Fishes are adapted to life in the water. Usually, they shed their sperm and eggs into the water, where fertilization occurs. The zygote develops into a swimming larva, which must fend for itself until it develops into the adult form.

Jawless Fishes

Living representatives of the **jawless fishes** (about 65 species) are cylindrical and up to a meter long. They have smooth, scaleless skin and no jaws or paired fins. The two groups of living jawless fishes are *hagfishes* (class Myxini) and *lampreys* (class Cephalaspidomorphi). Although both are jawless, there are distinct differences. Hagfishes are an ancient group of fish. They possess a skull, but lack the vertebrae found in the other classes of vertebrates. Hagfishes are scavengers, feeding mainly on dead fishes. Lampreys possess a true vertebral column. Most are parasites which use their round mouth as a sucker to attach itself to another fish and tap into its circulatory system. Unlike other fishes, the lamprey cannot take in water through its mouth. Instead, water moves in and out through the gill openings.

Cartilaginous Fishes

Cartilaginous fishes (class Chondrichthyes, about 750 species) are the sharks (see Fig. 32.5*a*), the rays (Fig. 32.7*a*), and the skates. This group of fishes is so named because they have skeletons of cartilage instead of bone. The small dogfish shark is often dissected in biology laboratories. One of the most dangerous sharks, and one that is capable of living in both fresh and salt water, is the bullshark, although, in general, people are

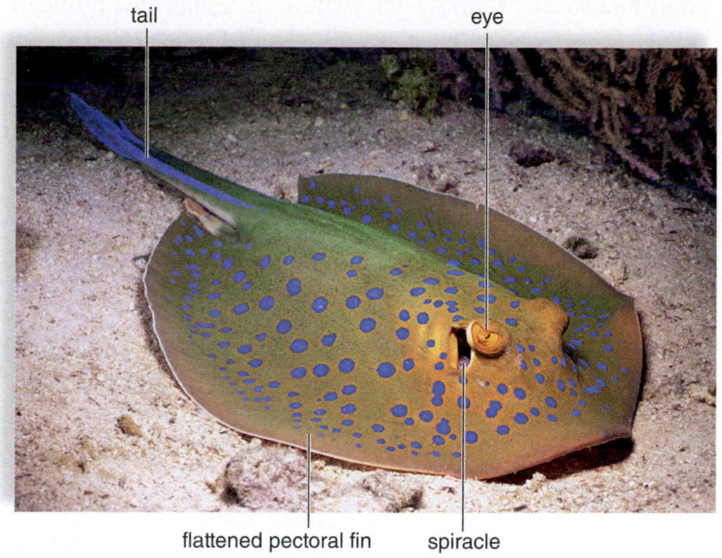
tail
eye
flattened pectoral fin
spiracle

a. Blue-spotted stingray, *Taeniura lymma*

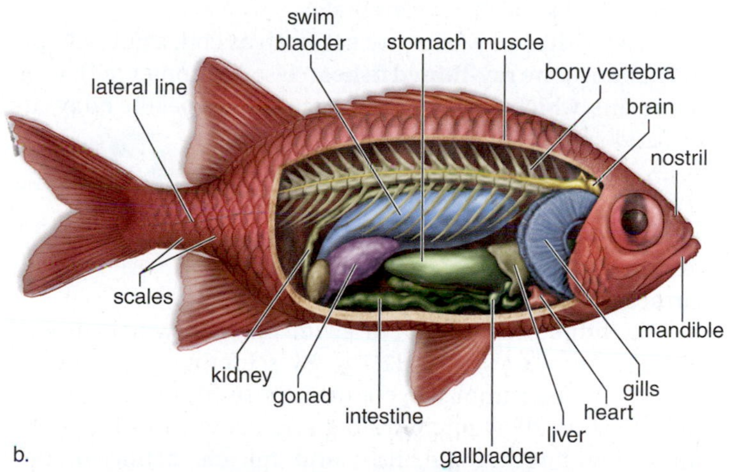
swim bladder
stomach
muscle
lateral line
bony vertebra
brain
nostril
scales
mandible
kidney
gonad
intestine
gallbladder
liver
heart
gills

b.

lobed fins

c. Coelacanth, *Latimeria chalumnae*

Figure 32.7 Jawed fishes. **a.** Rays, such as this blue-spotted stingray, are cartilaginous fishes. **b.** A soldierfish has the typical appearance and anatomy of a ray-finned fish, the most common type of bony fish. **c.** A coelacanth is a lobe-finned fish once thought to be extinct.

rarely attacked by sharks. Each year more people are injured or killed by dog attacks than shark attacks. The largest sharks, the whale sharks, feed on small fishes and marine invertebrates and do not attack humans. Skates and rays are rather flat fishes that live partly buried in the sand and feed on mussels, clams, and various crustaceans.

Three well-developed senses enable sharks and rays to detect their prey: (1) They have the ability to sense electric currents in water—even those generated by the muscle movements of animals. (2) They have a lateral line system, a series of pressure-sensitive cells along both sides of the body that can sense pressure caused by nearby movement in the water. (3) They have a keen sense of smell. The part of the brain associated with this sense is enlarged relative to the other parts. Sharks can detect some scents from about a third of a mile away in the open ocean.

Bony Fishes

Bony fishes (about 30,000 species) are by far the most numerous and diverse of all the vertebrates.

Most of the bony fishes we eat, such as cod, trout, salmon, and haddock, are **ray-finned fishes** (class Actinopterygii). Their paired fins, which they use to balance and propel the body, are thin and supported by bony rays. Ray-finned fishes have various ways of life. Some, such as herring, are filter feeders; others, such as trout, are opportunists; and still others are predaceous carnivores, such as piranhas and barracudas.

Video Cichlid Specialization

Animation Bony Fish

Ray-finned fishes have a swim bladder, which usually regulates buoyancy (Fig. 32.7*b*). By secreting gases into or absorbing gases from the bladder, these fishes can change their density, allowing them to go up or down in the water. The streamlined shape, fins, and muscle action of ray-finned fishes are all suited for locomotion in the water. Their skin is covered by bony scales that protect the body but do not prevent water loss. The gills of ray-finned fish are kept continuously moist by the passage of water through the mouth and out the gill slits during respiration. As the water passes over the gills, oxygen is absorbed and carbon dioxide is given off. Ray-finned fishes have a single-circuit circulatory system. The heart is a simple pump, and the blood flows through the chambers, including a nondivided atrium and ventricle, to the gills. Oxygenated blood leaves the gills and is circulated throughout the body by the heart (see Fig. 32.11*c*).

Another type of bony fish, called the **lobe-finned fishes** (class Sarcopterygii), are the ancestors of the amphibians. These fishes not only had fleshy appendages that could be adapted to land locomotion, but most also had a **lung** that was used for respiration. One type of lobe-finned fish, the coelacanth, was thought to have gone extinct about 20,000 years ago. However, some coelacanths have been discovered off the coasts of Eastern Africa and Indonesia, making them the only "living fossil" among these fishes (Fig. 32.7*c*).

Animation Early Vertebrates

Amphibians: Jointed Appendages

Amphibians (class Amphibia, about 6,000 species), whose class name means living on both land and in the water, are represented today by frogs, toads, newts, and salamanders. Caecilians are also classified as amphibians even though they are fossorial, wormlike amphibians that spend most of their life underground. Aside from *jointed appendages* (which have been lost in the limbless caecilians), amphibians have other features not seen in bony fishes: they usually have four limbs, eyelids for keeping their eyes moist, ears (a tympanum) for picking up sound waves, and a voice-producing larynx. Relative to body size, the brain is larger than that of a fish. Adult amphibians usually have small lungs, but some species respire entirely through their skin. Air enters the mouth by way of nostrils, and when the floor of the mouth is raised, air is forced into the relatively small lungs. Respiration is supplemented by gas exchange through the smooth, moist, and glandular skin. The amphibian heart has a divided atrium but only a single ventricle. The right atrium receives deoxygenated blood from the body, and the left atrium receives oxygenated blood from the lungs. These two types of blood are partially mixed in the single ventricle (see Fig. 32.11*c*). Mixed blood is then sent to all parts of the body. Some is sent to the skin, where it is further oxygenated.

Most members of this group lead a lifestyle that includes a larval stage that lives in the water, and an adult stage that lives on land. However, the adult usually returns to the water to reproduce. Figure 32.8 illustrates how a frog tadpole undergoes metamorphosis into a tetrapod (four-limbed) adult before taking up life on land. Currently, there is widespread

1. Tadpoles hatch.
2. Tadpole respires with gills.
3. Front and hind legs are present.
4. Frog respires with lungs.

Figure 32.8 Frog metamorphosis. During metamorphosis, the animal changes from an aquatic to a terrestrial organism.

concern about the future of amphibians because over 40% of all amphibian species are threatened with extinction. Because of their permeable skin and a life cycle that often depends on both water and land, they are thought to be "indicator species" of environmental quality. That is, they are the first to respond to environmental degradation, and disappearances of populations and species thus generate concern for other species—even humans.

Video Tadpole Development

Video Frog Reproduction

Check Your Progress 32.2

1. Recognize characteristics that distinguish fish from other vertebrates.
2. List the features that distinguish amphibians from fish.
3. Describe the features that allow most amphibians to have life stages in water and on land.

32.3 Vertebrates: Reptiles and Mammals

Learning Outcomes

Upon completion of this section, you should be able to

1. Compare the major evolutionary innovations that distinguish the reptiles and mammals.
2. List the features that distinguish birds from the rest of the reptile lineage.
3. Identify the unique features that define the three living lineages of mammals.

Reptiles: Amniotic Egg

Reptiles (class Reptilia, about 17,000 species, including birds) diversified and were most abundant between 245 and 65 million years ago. These reptiles included the mammal-like reptiles (now extinct), the ancestors of today's living reptiles, and the dinosaurs (now extinct). Some dinosaurs are remembered for their great size. *Brachiosaurus,* an herbivore, was about 23 m (75 ft) long and about 17 m (56 ft) tall. *Tyrannosaurus rex,* a carnivore, was 5 m (16 ft) tall when standing on its hind legs. The bipedal stance of some reptiles preceded the evolution of wings in birds. In fact, some say birds are actually living dinosaurs. A wide array of molecular, morphological, and behavioral studies support this hypothesis.

The reptiles living today are mainly turtles, alligators, snakes, lizards, and birds. The body is covered with hard, keratinized scales, which protect the animal from desiccation and from predators. Another adaptation for a land existence is the manner in which snakes typically use their tongue as a sense organ (Fig. 32.9). Reptiles have well-developed lungs enclosed by a protective rib cage. When the rib cage expands, a partial vacuum establishes a negative pressure, which causes air to rush into the lungs. The atrium of the heart is always separated into right and left chambers, but division of the ventricle varies. An interventricular septum is incomplete in certain species. Therefore, some oxygenated and deoxygenated blood is mixed between the ventricles.

Video Basilisk Lizard

Ducts to Jacobson's organ

Figure 32.9 The tongue as a sense organ. Snakes wave a forked tongue in the air to collect chemical molecules, which are then brought back into the mouth and delivered to an organ in the roof of the mouth called the Jacobson's organ. Analyzed chemicals help the snake trail a prey animal, recognize a predator, or find a mate.

Video Leaf-Tailed Geckos

Perhaps the most outstanding evolutionary innovation that first appeared in the reptiles is their means of reproduction that is suitable to a land existence. The penis of the male passes sperm directly to the female. Fertilization is internal, and the female typically possesses leathery, flexible, shelled eggs. Some snakes lay eggs, while other snakes actually give birth to live young. This *amniotic egg* made development on land possible and eliminated the need for a swimming-larval stage during development. This egg provides the developing embryo with atmospheric oxygen, food, and water; it removes nitrogenous wastes; and it protects the embryo from drying out and from mechanical injury. This is accomplished by the presence of extraembryonic membranes such as the chorion (Fig. 32.10).

Fishes, amphibians, and living reptiles other than birds are **ectothermic,** meaning that their body temperature matches the temperature of the external environment. If it is cold externally, they are cold internally; if it is hot externally, they are hot internally. Reptiles try to regulate their body temperatures by basking in the sun if they need warmth or by hiding in the shadows if they need cooling off.

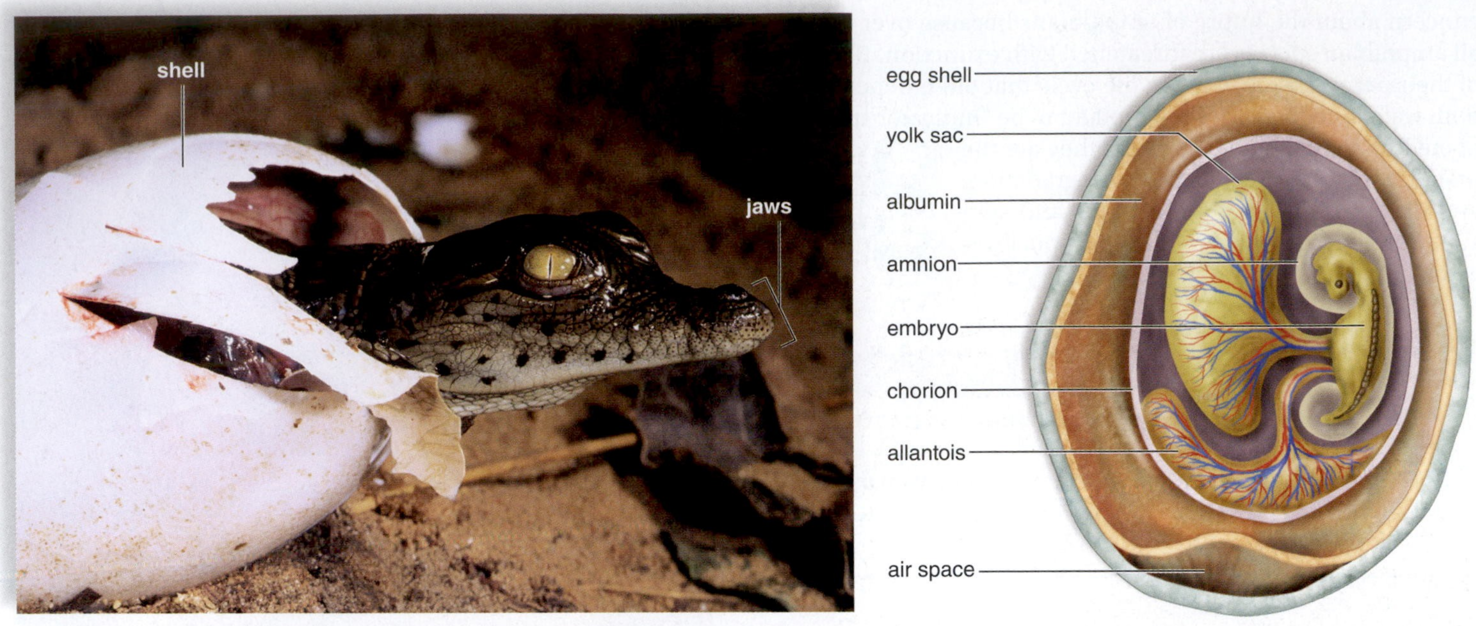
shell
jaws
egg shell
yolk sac
albumin
amnion
embryo
chorion
allantois
air space

a. American crocodile, *Crocodylus acutus*

b.

Figure 32.10 The reptilian egg allows reproduction on land. **a.** Baby American crocodile, *Crocodylus,* hatching out of its shell. Note that the shell is leathery and flexible, not brittle like birds' eggs. **b.** Inside the egg, the embryo is surrounded by extraembryonic membranes. The chorion aids gas exchange, the yolk sac provides nutrients, the allantois stores waste, and the amnion encloses a fluid that prevents drying out and provides protection.

Feathered Reptiles

To many people, the **birds** (class Aves) are the most conspicuous, melodic, beautiful, and fascinating group of vertebrates. Over 9,000 species of birds have been described. Birds range in size from the tiny "bee" hummingbird at 1.8 g to the ostrich at a maximum weight of 160 kg and a height of 2.7 m. The fossil record of birds, such as *Archaeopteryx* and newly described fossils from Mongolia and Spain, provides evidence that birds evolved from reptiles. Combined with molecular studies, birds are now considered part of the Reptilia. Their legs have scales and their feathers (Fig. 32.11*a*) are modified reptilian scales. However, birds typically lay a hard-shelled amniotic egg, rather than the leathery egg of reptiles.

Diversity of Birds

The majority of birds, including eagles, geese, and mockingbirds, have the ability to fly. However, some birds, such as emus, penguins, and ostriches, are flightless. Traditionally, birds have been classified based on beak and foot type (Fig. 32.12) and, to some extent, on their habitat and behavior. The various orders include birds of prey with notched beaks and sharp talons; shorebirds with long, slender, probing beaks and long, stilt-like legs; woodpeckers with sharp, chisel-like beaks and grasping feet; waterfowl with broad beaks and webbed toes; penguins with wings modified as paddles; and songbirds with perching feet.

Video Harris Hawks

Video Finches Adaptive Radiation

Anatomy and Physiology of Birds

Nearly every anatomical feature of a bird can be related to its ability to fly (see Fig. 32.11*a*). The forelimbs are modified as wings. The hollow, very light bones are laced with air cavities. A horny beak has replaced jaws equipped with teeth, and a slender neck connects the head to a rounded, compact torso. The sternum is enlarged and has a keel, to which strong muscles are attached for flying. Respiration is efficient because the lobular lungs form anterior and posterior air sacs. The presence of these sacs means that the air circulates one way through the lungs and that gases are continuously exchanged across respiratory tissues (see Fig. 32.11*b*). Another benefit of air sacs is that they lighten the body for more efficient flying.

Birds have a four-chambered heart that completely separates O_2-rich blood from O_2-poor blood. The left ventricle pumps O_2-rich blood under pressure to the muscles (see Fig. 32.11*c*). Birds are **endothermic.** Like mammals, their internal temperature is constant because they generate and maintain metabolic heat. This may be associated with their efficient nervous, respiratory, and circulatory systems. Also, their feathers provide insulation. Birds have no bladder and excrete uric acid in a semi-dry state.

Flight requires acute sense organs and an intricate nervous system. Birds have particularly good vision and well-developed brains. Their muscle reflexes are excellent. An enlarged portion of the brain seems to be the area responsible for instinctive behavior. A ritualized courtship often precedes mating. Many newly hatched birds require parental care before they are able to fly away and seek food for themselves. A remarkable aspect of bird behavior is the seasonal migration of many species over very long distances. Birds navigate by day and night, whether it's sunny or cloudy, by using the sun and stars and even Earth's magnetic field to guide them.

Video Bird Radar

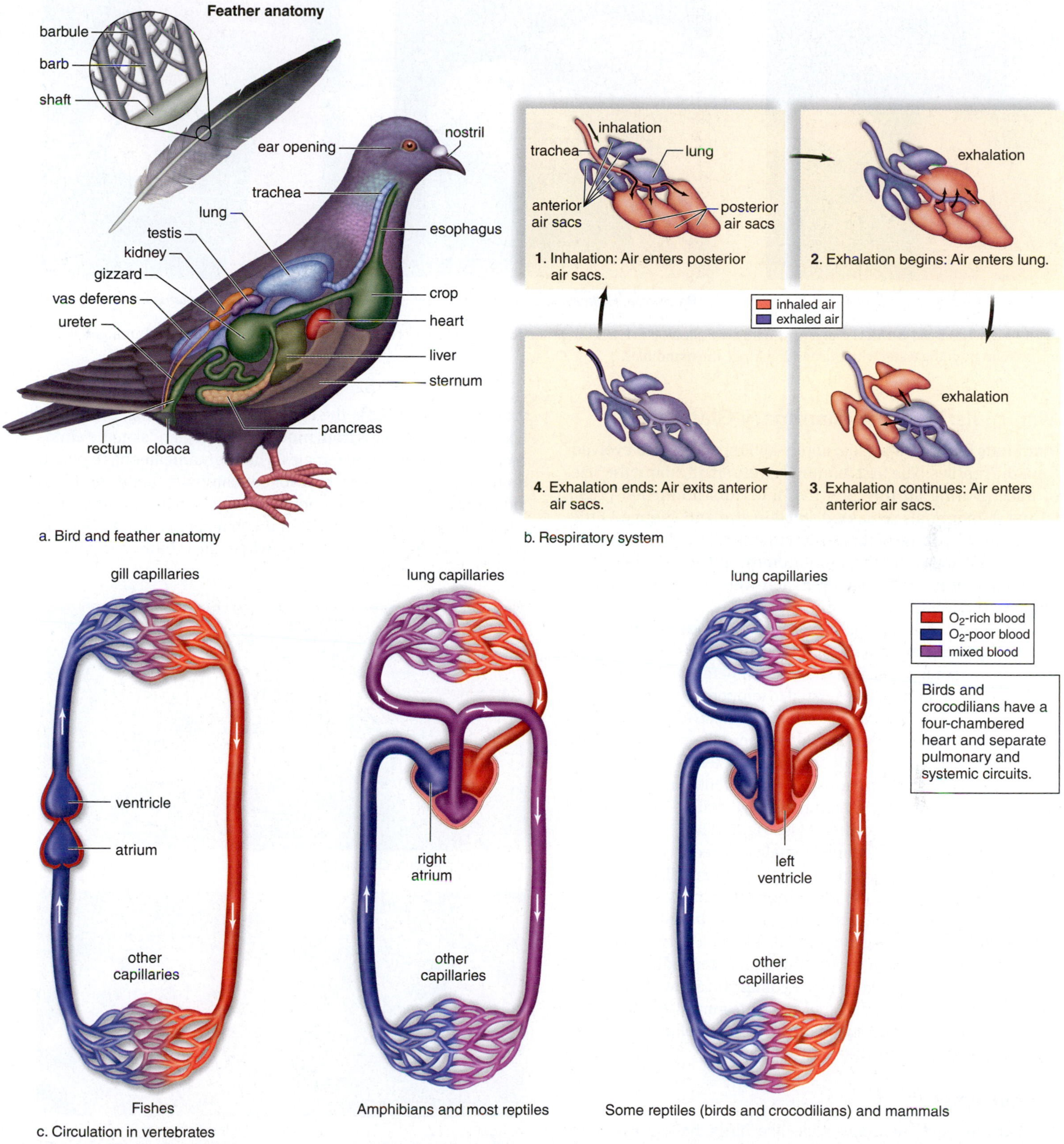
Feather anatomy
barbule
barb
shaft
ear opening
nostril
trachea
lung
testis
kidney
gizzard
vas deferens
ureter
esophagus
crop
heart
liver
sternum
pancreas
rectum
cloaca
a. Bird and feather anatomy
inhalation
trachea
lung
anterior air sacs
posterior air sacs
1. Inhalation: Air enters posterior air sacs.
exhalation
2. Exhalation begins: Air enters lung.
inhaled air
exhaled air
exhalation
3. Exhalation continues: Air enters anterior air sacs.
4. Exhalation ends: Air exits anterior air sacs.
b. Respiratory system
gill capillaries
ventricle
atrium
other capillaries
Fishes
lung capillaries
right atrium
other capillaries
Amphibians and most reptiles
lung capillaries
left ventricle
other capillaries
Some reptiles (birds and crocodilians) and mammals
O_2-rich blood
O_2-poor blood
mixed blood
Birds and crocodilians have a four-chambered heart and separate pulmonary and systemic circuits.
c. Circulation in vertebrates

Figure 32.11 Bird anatomy and physiology. **a.** The anatomy of a pigeon is representative of bird anatomy. **b.** In birds, air passes one way through the lungs. **c.** In fishes, a single-loop system utilizes a two-chambered heart. In amphibians and most reptiles, the heart has three chambers, and some mixing of O_2-rich and O_2-poor blood takes place. In birds, mammals, and some reptiles, the heart has four chambers and sends only O_2-poor blood to the lungs and O_2-rich blood to the body.

a. Cardinal, *Cardinalis cardinalis* b. Bald eagle, *Haliaetus leucocephalus* c. Flamingo, *Phoenicopterus ruber*

Figure 32.12 Bird beaks. **a.** A cardinal's beak allows it to crack tough seeds. **b.** A bald eagle's beak allows it to tear prey apart. **c.** A flamingo's beak strains food from the water with bristles that fringe the mandibles.

Mammals: Hair and Mammary Glands

Mammals (class Mammalia, about 4,600 species) also evolved from the reptiles. The first mammals were small, about the size of mice. During all the time the dinosaurs flourished (165–65 MYA), mammals remained small in size and changed little evolutionarily. Some of the earliest mammalian groups are still represented today by the monotremes and marsupials. The placental mammals that evolved later went on to live in many habitats, including air, land, and sea.

The chief characteristics of mammals are body hair and milk-producing mammary glands. Almost all mammals are endothermic and maintain a constant internal temperature. Many of the adaptations of mammals are related to temperature control. *Hair,* for example, provides insulation against heat loss and allows mammals to be active, even in cold weather. Like birds, mammals have efficient respiratory and circulatory systems, which ensure a ready supply of oxygen to muscles whose contraction produces body heat. Also, like birds, mammals have double-loop circulation and a four-chambered heart (see Fig. 32.11*c*).

Mammary glands enable females to feed (nurse) their young a nutrient-rich source of food. Nursing also creates a bond between mother and offspring that helps ensure parental care while the young are helpless. In all mammals (except monotremes), the young are born alive after a period of development in the uterus, a part of the female reproductive tract. Internal development shelters the young and allows the female to move actively about while the young are maturing. Mammals are classified as monotremes, marsupials, or placental mammals, according to how they reproduce.

Monotremes

Monotremes are mammals that, like birds, have a *cloaca,* a terminal region of the digestive tract that serves as a common chamber for feces, excretory wastes, and sex cells. They also lay hard-shelled amniotic eggs. They are represented by only five species: four species of spiny anteater and the duckbill platypus. All are found in Australia and New Guinea (Fig. 32.13*a*). The female duckbill platypus lays her eggs in a burrow in the ground. She incubates the eggs, and after hatching, the young lick up milk that seeps from modified sweat glands on the abdomens of both males and females. The spiny anteater has a pouch on the belly side formed by swollen mammary glands and longitudinal muscle. The egg moves from the cloaca to this pouch, where hatching takes place and the young remain for about 53 days. Then they stay in a burrow, where the mother periodically visits and nurses them.

a. Duckbill platypus, *Ornithorhynchus anatinus*

b. Koala, *Phascolarctos cinereus* c. Virginia opossum, *Didelphis virginianus*

Figure 32.13 Monotremes and marsupials. **a.** The duckbill platypus is a monotreme that inhabits Australian streams. **b.** The koala is an Australian marsupial that lives in trees. **c.** The opossum is the only marsupial in the Americas. The Virginia opossum is found in a variety of habitats.

Marsupials

The young of **marsupials** begin their development inside the female's body, but they are born in a very immature condition. Newborns crawl up into a pouch on their mother's abdomen. Inside the pouch, they attach to the nipples of mammary glands and continue to develop. Frequently, more are born than can be accommodated by the number of nipples, and it's "first come, first served."

Marsupial mammals are found mainly in Australia, where they underwent adaptive radiation for several million years without competition from placental mammals. Among the herbivorous marsupials in Australia, koalas are tree-climbing browsers (Fig. 32.13*b*), and kangaroos are grazers. The Virginia opossum (*Didelphis virginiana*) is the only marsupial that occurs north of Mexico (Fig. 32.13*c*). Other species of marsupials live in Central and South America. The Tasmanian devil, a carnivorous marsupial about the size of a small to medium-sized dog, is now threatened with extinction due to a transmissible tumor disease.

Placental Mammals

The vast majority of living mammals are **placental mammals** (Fig. 32.14). In these mammals, the extraembryonic membranes of the reptilian egg (see Fig. 32.10) have been modified for internal development within the uterus of the female. The chorion contributes to the fetal portion of the placenta, while a part of the uterine wall contributes to the maternal portion. Here, nutrients, oxygen, and waste are exchanged between fetal and maternal blood.

Mammals, with the exception of marine mammals, are adapted to life on land and have limbs that allow them to move rapidly. In fact, an evaluation of mammalian features leads us to the conclusion that they lead active lives. The lungs are expanded not only by the action of the rib cage, but also by the contraction of the diaphragm, a horizontal muscle that separates the thoracic cavity from the abdominal cavity. The heart has four chambers. The internal temperature is constant, and hair, when present, helps insulate the body.

The mammalian brain is well developed and enlarged due to the expansion of the cerebral hemispheres, which control the rest of the brain. The brain is not fully developed until after birth, and the young learn to take care of themselves during a period of dependency on their parents.

Mammals have differentiated teeth. Typically, in the front, the incisors and canine teeth have cutting edges for capturing and killing prey. On the sides, the premolars and molars chew food. The specific shape and size of the teeth may be associated with whether the mammal is an herbivore (eats vegetation), a carnivore (eats meat), or an omnivore (eats both meat and vegetation). For example, mice (order Rodentia) have continuously growing incisors; horses have large, grinding molars; and dogs (order Carnivora) have long canine teeth. Placental mammals are classified based on their methods of obtaining food and their mode of locomotion. For example, bats (order Chiroptera) have membranous wings supported by digits; horses (order Perissodactyla) have long, hoofed legs; and whales (order Cetacea) have paddlelike forelimbs.

Video Mom Grizzly Teaches Her Cubs

Video Bat Ecolocation

a. White-tailed deer, *Odocoileus virginianus*

b. African lioness, *Panthera leo*

c. Squirrel monkey, *Saimiri sciureus*

d. Killer whale, *Orcinus orca*

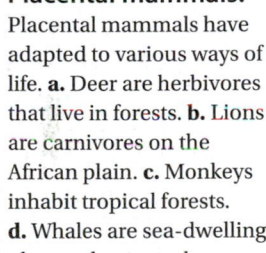

Figure 32.14 Placental mammals. Placental mammals have adapted to various ways of life. **a.** Deer are herbivores that live in forests. **b.** Lions are carnivores on the African plain. **c.** Monkeys inhabit tropical forests. **d.** Whales are sea-dwelling placental mammals.

Primates

Primata are members of the order **Primates.** In contrast to the other orders of placental mammals, most primates are adapted to an arboreal life—that is, for living in trees. Primate limbs are mobile, and the hands and feet both have five digits each. Many primates have a big toe and a thumb that are both opposable, meaning the big toe or thumb can touch each of the other toes or fingers. (Humans don't have an opposable big toe, but the thumb is opposable, and this results in a powerful and precise grip.) The opposable thumb allows a primate to easily reach out and bring food to the mouth. When locomoting, primates grasp and release tree limbs freely because they have nails instead of the claws of their evolutionary ancestors.

In primates, the snout is shortened considerably, allowing the eyes to move to the front of the head. The stereoscopic vision (or depth perception) that results permits primates to make accurate judgments about the distance and position of adjoining tree limbs. Some primates that are active during the day, such as humans, have color vision and strong visual acuity because the retina contains cone cells in addition to rod cells. Cone cells require bright light, but the image is sharp and in color. The lens of the eye focuses light directly on the fovea, a region of the retina where cone cells are concentrated.

The evolutionary trend among primates is toward a larger and more complex brain. The brain size is smallest in prosimians and largest in modern humans. Increases in size of the cerebral cortex, the brain region associated with learning, memory, thought, and awareness, is primarily responsible for the trend toward increased brain size. The portion of the brain devoted to smell gets smaller, and the portions devoted to sight increase in size and complexity during primate evolution. Also, more and more of the brain is involved in controlling and processing information received from the hands and the thumb. The result is good hand-eye coordination in humans.

It is difficult to care for several offspring while moving from limb to limb, and so one offspring per birth interval tends to be the norm in most primates. The juvenile period of dependency on parental care is extended, and there is an emphasis on learned behavior and complex social interactions.

The order Primates has two suborders: the **strepsirhini** (lemurs, aye ayes, bush babies, and lorises) and the **haplorhini** (monkeys, apes, and humans). Thus, humans are more closely related to the monkeys and apes than they are to the strepsirhini.

Check Your Progress 32.3

1. List three features that distinguish reptiles from the mammals.
2. Recognize the features that make placental mammals unique.
3. Identify the characteristics that separate primates from the other mammals.

32.4 Evolution of the Hominins

Learning Outcomes

1. Explain the significance of bipedalism in hominin evolution.
2. Summarize the importance of both ardipithecines and australopithecines in hominin evolution.
3. Arrange the early species of *Homo* in evolutionary order.

The evolutionary tree in Figure 32.15 shows that all primates share a common ancestor and that over time they diverged, producing the various lines of descent, or **lineage,** we see today. Notice that prosimians (also called strepsirhini), represented by lemurs, aye ayes, bush babies, and lorises, were the first types of primates to diverge from the primate line of descent, and African apes were the last group to diverge. Also note that the tree suggests that humans are most closely related to African apes. One of the most unfortunate misconceptions concerning human evolution is the belief that Darwin and others suggested that humans evolved from apes. On the contrary, humans and apes shared a common apelike ancestor. Today's apes are our distant cousins, and we couldn't have evolved from our cousins because we are contemporaries—living on Earth at the same time.

Molecular data have been used to estimate the date of the split between the human lineage and that of apes. When two lines of descent first diverge from a common ancestor, the genes of the two lineages are nearly identical. But as time goes by, each lineage accumulates genetic changes. Many genetic changes are neutral (not tied to adaptation) and accumulate at a fairly constant rate. Such changes can be used as a kind of **molecular clock** to infer how closely two groups are related and the point at which they diverged. Molecular data suggest that the split between the ape and human lineages occurred about 6 to 8 MYA. Currently, humans and chimpanzees are probably the most closely related, sharing more than 90% of our DNA. Hominins (the designation that includes humans and species very closely related to humans) first evolved about 5 MYA. Molecular data show that hominins, chimpanzees, and gorillas are all closely related and these groups must have shared a common ancestor sometime during the Miocene. Hominins, chimpanzees, and gorillas are now grouped together as hominines. The hominids include the hominines and the orangutan. The hominoids include the gibbon and the hominids. The hominoid common ancestor first evolved at the beginning of the Miocene about 23 MYA.

Evolution of Humanlike Hominins

The classification of humans is provided in the box on page 664 . The tribe hominini is found within the hominins. The biggest derived characteristic that separates modern humans is **bipedalism,** an anatomy suitable for standing erect and walking on two feet.

Until recently, many scientists thought that bipedalism evolved because of a dramatic change in climate that caused

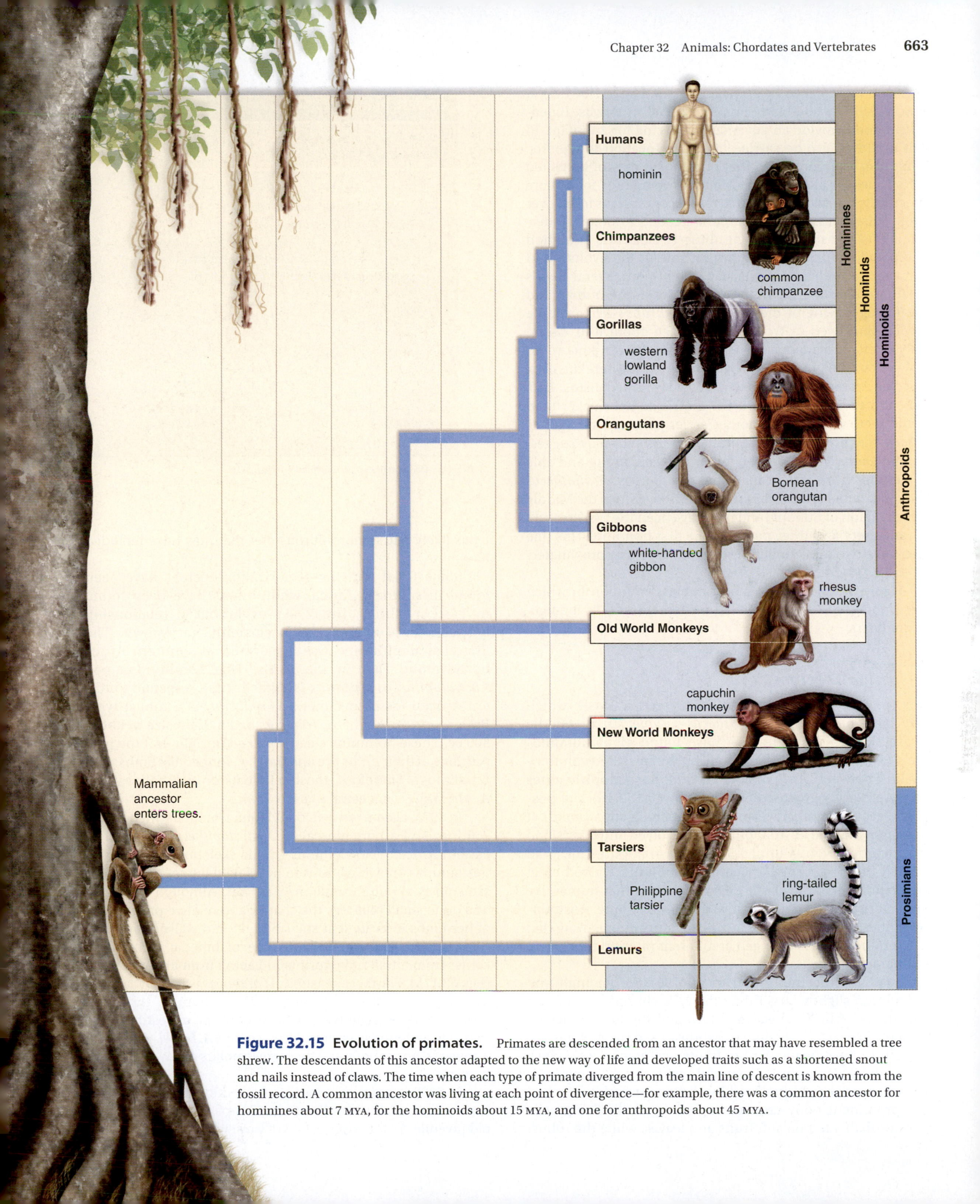

Humans
hominin
Chimpanzees
common chimpanzee
Gorillas
western lowland gorilla
Orangutans
Bornean orangutan
Gibbons
white-handed gibbon
Old World Monkeys
rhesus monkey
New World Monkeys
capuchin monkey
Tarsiers
Philippine tarsier
Lemurs
ring-tailed lemur
Mammalian ancestor enters trees.
Hominines
Hominids
Hominoids
Anthropoids
Prosimians

Figure 32.15 Evolution of primates. Primates are descended from an ancestor that may have resembled a tree shrew. The descendants of this ancestor adapted to the new way of life and developed traits such as a shortened snout and nails instead of claws. The time when each type of primate diverged from the main line of descent is known from the fossil record. A common ancestor was living at each point of divergence—for example, there was a common ancestor for hominines about 7 MYA, for the hominoids about 15 MYA, and one for anthropoids about 45 MYA.

forests to be replaced by grassland about 4 MYA. However, it is now believed that early hominids lived in forests. Bipedealism does provide an advantage in caring for an infant, in that it allows it to be carried from one location to another. It has also been proposed that bipedalism may of benefited the species by allowing it not only to forage for food more effectively, but also carry scarce resources (such as food) over longer distances. More research is needed to better understand this transition, but one thing is clear, the evolution of bipedalism played an important role in the evolution of our species. Paleontologists have identified several fossils dated around the time of the split between the human and ape lineages. One of these is *Sahelanthropus tchadensis,* a species that lived in West Central Africa about 6 to 7 MYA. Only cranial fragments of this species have been uncovered to date, but the point on the back of the skull where the neck muscles would have attached suggests bipedalism. Also, the canines are smaller and the tooth enamel thicker than in the teeth of an ape. Another series of leg fossils (genus *Orrorin*), which are about 6 million years old, may suggest upright walking because the femur appears strong and able to support more weight than that of the apes. *Ardipithecus kadabba,* dated around 5.5 MYA, was discovered in the Middle Awash region of Ethiopia and may have been the first creature to walk erect and thus, possibly, the first hominin. This line then gave rise to *Ardipithecus ramidus,* dated approximately 5.5 to 4.5 MYA.

Figure 32.16 is a timeline of human evolution in which these potential early hominids are represented by orange-colored bars.

Australopithecines

Most recently, scientists have discovered a fossil, which they have named *Australopithecus anamensis,* in Middle Awash, Ethiopia, where they have found eight hominid-like species so far. This fossil may be the missing evolutionary link between *Ardipithecus* and the **australopithecines** (represented by green-colored bars in Fig. 32.16), the possible direct hominin ancestors for humans (genus *Homo*). It is thought that the australopithecines evolved and diversified in Africa from 4 MYA until 1.5 MYA. Australopithecines had a small brain (an apelike characteristic) and walked erect (a humanlike characteristic). Thus, it seems that humanlike characteristics did not all evolve at the same time. Australopithecines give evidence of **mosaic evolution,** meaning that different body parts changed at different rates and therefore at different times.

Australopithecines stood about 100–115 cm high and had a brain slightly larger than that of a chimpanzee (about 370–515 cc). The forehead was low, and the face projected forward. There is no evidence that they used tools. Some are thought to have been slight of frame, or "gracile" (slender), while others were robust (powerful), with strong upper bodies and massive jaws powered by chewing muscles attached to a prominent bony crest on top of the skull. The gracile types probably fed on soft fruits and leaves, while the robust types likely had a more fibrous diet that may have included hard nuts.

Some fossil remains of australopithecines have been found in southern Africa, while others have been found in eastern Africa. The exact evolutionary relationship between these two groups is currently not known. The first australopithecine was discovered in southern Africa by Raymond Dart in the 1920s. This hominin, named *Australopithecus africanus,* is a gracile type. A second southern African fossil specimen, *A. robustus,* is a robust type. Both *A. africanus* and *A. robustus* had a brain size of about 500 cc. These hominins clearly walked upright, but the proportions of their limbs are apelike (i.e., longer forelimbs than hindlimbs). Therefore, most scientists do not believe that *A. africanus* is ancestral to early *Homo.*

In 1973, a team led by Donald Johanson unearthed nearly 250 fossils of a hominin called *Australopithecus afarensis* in eastern Africa. A now-famous female skeleton discovered in this group is known worldwide by its field name, "Lucy." This skeleton is about 3.2 million years old. Although her brain was quite small (400 cc), the shapes and relative proportions of her limbs indicate that she stood up and walked bipedally (Fig. 32.17*a*). Even better evidence of bipedal locomotion comes from a trail of footprints in Laetoli from about 3.7 MYA (Fig. 32.17*b*). For these reasons, *A. afarensis* is usually favored as being more directly related to *Homo* than *A. africanus. A. afarensis,* a gracile type, is believed to be ancestral to the robust types found in eastern Africa such as *A. boisei,* which had a powerful upper body and the largest molars of any humanlike hominin.

In 2000, a team of scientists from the Max Planck Institute unearthed the fossilized remains of a 3.3-million-year-old juvenile *A. afarensis* just 4 km from where Lucy had been

CLASSIFICATION

ORDER: Primates

- Adapted to an arboreal life
- Prosimians, Anthropoids

FAMILY: Hominidae (hominids)
SUBFAMILY: Homininae (hominines)
TRIBE*: Hominini (hominins)
Early Humanlike Hominins → *Sahelanthropus,* ardipithecines
Later Humanlike Hominins → australopithecines

GENUS: *Homo* (humans) → *Homo habilis, Homo rudolfensis, Homo ergaster, Homo erectus*
Early *Homo*
Brain size greater than 600 cc; tool use and culture

Later *Homo* → *Homo heidelbergensis, Homo neandertalensis, Homo sapiens*
Brain size greater than 1,000 cc; tool use and culture

**A taxonomic level that lies between subfamily and genus.*

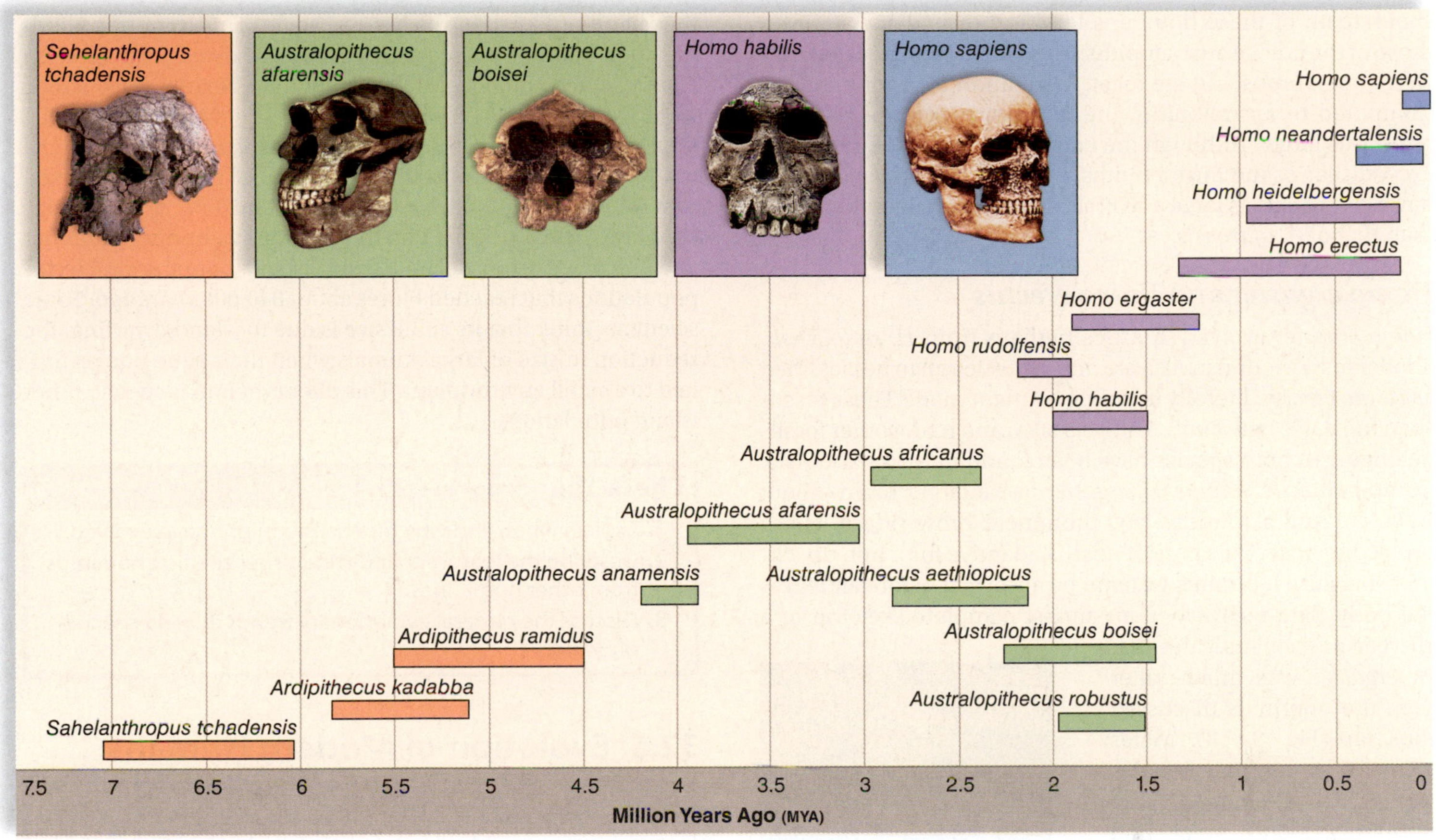
Sehelanthropus tchadensis
Australopithecus afarensis
Australopithecus boisei
Homo habilis
Homo sapiens
Homo sapiens
Homo neandertalensis
Homo heidelbergensis
Homo erectus
Homo ergaster
Homo rudolfensis
Homo habilis
Australopithecus africanus
Australopithecus afarensis
Australopithecus anamensis
Australopithecus aethiopicus
Ardipithecus ramidus
Australopithecus boisei
Ardipithecus kadabba
Australopithecus robustus
Sahelanthropus tchadensis
7.5 7 6.5 6 5.5 5 4.5 4 3.5 3 2.5 2 1.5 1 0.5 0
Million Years Ago (MYA)

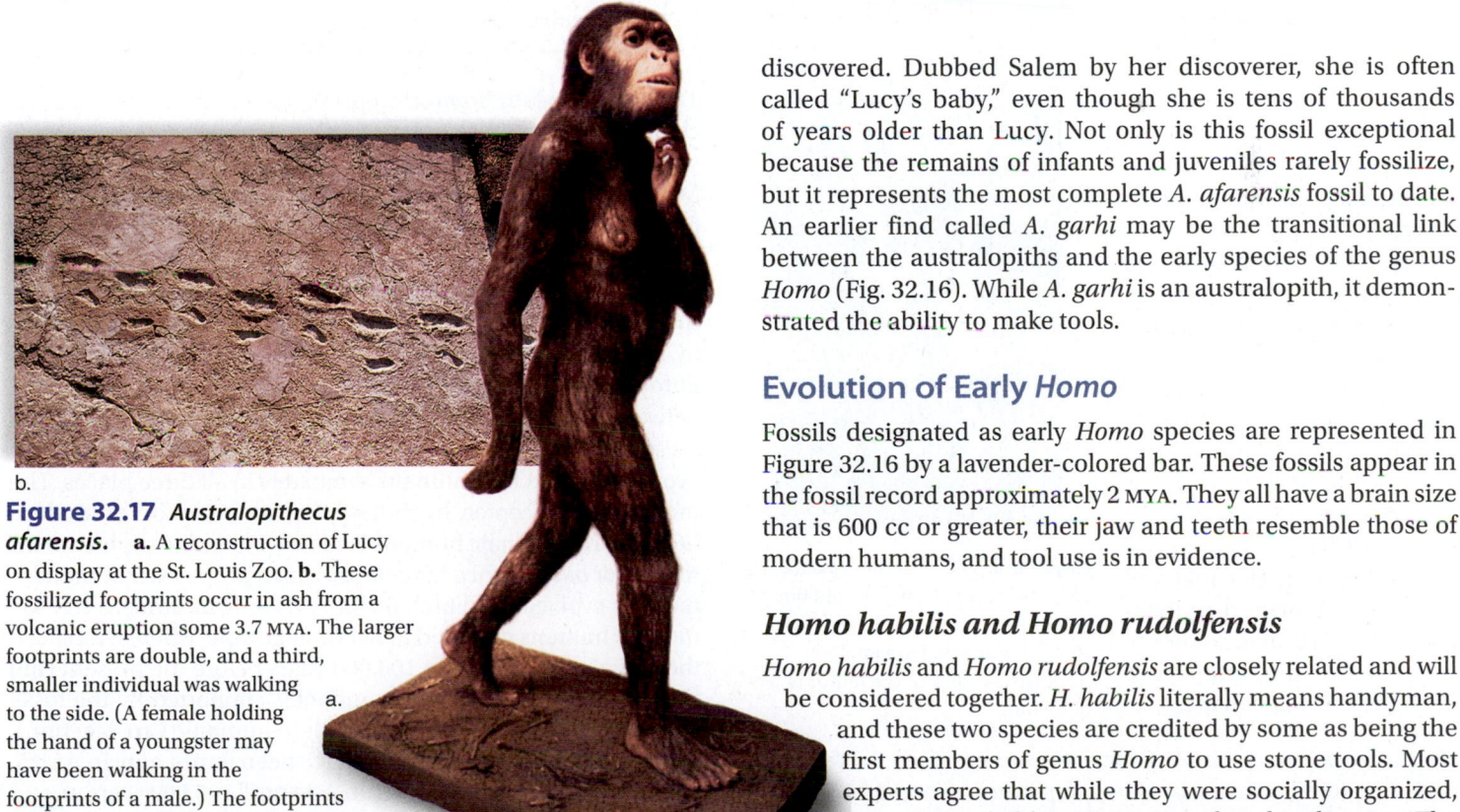

Figure 32.16 Human evolution. Several groups of extinct hominins preceded the evolution of modern humans. These groups have been divided into the early humanlike hominins (orange), later humanlike hominins (green), early *Homo* species (lavender), and finally the later *Homo* species (blue). Only modern humans are classified as *Homo sapiens.*

b.
a.

Figure 32.17 Australopithecus afarensis. **a.** A reconstruction of Lucy on display at the St. Louis Zoo. **b.** These fossilized footprints occur in ash from a volcanic eruption some 3.7 MYA. The larger footprints are double, and a third, smaller individual was walking to the side. (A female holding the hand of a youngster may have been walking in the footprints of a male.) The footprints suggest that *A. afarensis* walked bipedally.

discovered. Dubbed Salem by her discoverer, she is often called "Lucy's baby," even though she is tens of thousands of years older than Lucy. Not only is this fossil exceptional because the remains of infants and juveniles rarely fossilize, but it represents the most complete *A. afarensis* fossil to date. An earlier find called *A. garhi* may be the transitional link between the australopiths and the early species of the genus *Homo* (Fig. 32.16). While *A. garhi* is an australopith, it demonstrated the ability to make tools.

Evolution of Early *Homo*

Fossils designated as early *Homo* species are represented in Figure 32.16 by a lavender-colored bar. These fossils appear in the fossil record approximately 2 MYA. They all have a brain size that is 600 cc or greater, their jaw and teeth resemble those of modern humans, and tool use is in evidence.

Homo habilis and Homo rudolfensis

Homo habilis and *Homo rudolfensis* are closely related and will be considered together. *H. habilis* literally means handyman, and these two species are credited by some as being the first members of genus *Homo* to use stone tools. Most experts agree that while they were socially organized, they were probably scavengers rather than hunters. The cheek teeth of these hominins tend to be smaller than even those of the gracile australopiths. This is also evidence that they were omnivorous and ate meat, in addition to plant material. Compared to australopiths, the face protruded less, and the brain was larger. Although the height of *H. rudolfensis* did not exceed that of the australopiths, some of this species' fossils have a brain size as large as 800 cc, which is considerably larger than that of *A. afarensis*.

Homo ergaster and *Homo erectus*

Homo ergaster evolved in Africa perhaps from *H. rudolfensis.* Similar fossils found in Asia are different enough to be classified as *Homo erectus,* literally meaning "upright man." These fossils span the dates between 1.9 and 0.3 MYA, and many other fossils belonging to both species have been found in Africa and Asia. Compared to *H. habilis, H. ergaster* had a larger brain (about 1,000 cc) and a rounder jaw, prominent brow ridges, and a projecting nose. This type of nose is adaptive for a hot, dry climate because it permits water to be absorbed before air leaves the body. The recovery of an almost complete skeleton of a 10-year-old boy indicates that *H. ergaster* was much taller than the hominids discussed thus far (Fig. 32.18). Males were 1.8 m tall, and females were 1.55 m tall. Indeed, these hominids stood erect and most likely had a *striding gait* similar to modern humans. The robust and most likely heavily muscled skeleton still retained some australopithecine features. Even so, the size of the birth canal indicates that infants were born in an immature state that required an extended period of parental care.

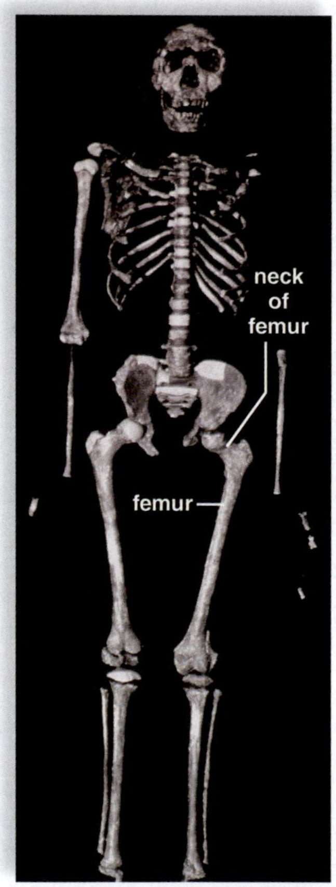
neck
of
femur
femur
Figure 32.18 ***Homo ergaster.*** This skeleton of a 10-year-old boy who lived 1.6 MYA in eastern Africa shows femurs that are angled because the neck of the femur is quite long.

H. ergaster first appeared in Africa but then migrated into Europe and Asia sometime between 1 and 2 MYA. Most likely, *H. erectus* evolved from *H. ergaster* after *H. ergaster* arrived in Asia. In any case, such an extensive population movement is a first in the history of humankind and a tribute to the intellectual and physical skills of these individuals. They also had a knowledge of fire and may have been the first to cook meat.

Homo floresiensis In 2004, scientists announced the discovery of the fossil remains of *Homo floresiensis.* The 18,000-year-old fossil of a 1-m tall, 25-kg adult female was discovered on the island of Flores in the South Pacific. This important finding suggests that a species of *Homo* coexisted with modern *Homo sapiens* much more recently than Neandertals, which went extinct about 28,000 years ago. The specimen was the size of a three-year-old *Homo sapiens sapiens* but possessed a braincase only one-third the size of that of a modern human. Apparently, *H. floresiensis* used tools and fire. A 2007 study supports the hypothesis that this diminutive hominin and her peers evolved from normal-sized, island-hopping *H. erectus* populations that reached Flores about 840,000 years ago. Some scientists think that its small size is due to island dwarfing, the reduction in size of large animals when their gene pool is limited to a small environment. This phenomenon is seen in other island populations.

Check Your Progress 32.4

1. Explain when and why bipedalism might have evolved.
2. Describe the feature that distinguishes the first hominids from other hominines.
3. Discuss the general evolutionary trends in early species of *Homo.*

32.5 Evolution of Modern Humans

Learning Outcomes

Upon completion of this section, you should be able to

1. Explain how the replacement model relates to evolution of later species of the genus *Homo.* .
2. Discuss the significance of increased tool use, language, and agriculture in Cro-Magnons.

The later species of *Homo* are represented in Figure 32.16 by blue-colored bars. Most researchers think that modern humans *Homo sapiens* evolved from *H. ergaster,* but they differ as to the details. Many disparate early *Homo* species in Europe are now classified as *Homo heidelbergensis.* Just as *H. erectus* is hypothesized to have evolved from *H. ergaster* in Asia, so *H. heidelbergensis* is hypothesized to have evolved from *H. ergaster* in Europe. Further, for the sake of discussion, *H. ergaster* in Africa, *H. erectus* in Asia, and *H. heidelbergensis* (and *H. neandertalensis*) in Europe can be grouped together as archaic humans who lived between 1.5 and 0.25 MYA. The presence of archaic humans at these different locations suggests to some that modern humans evolved from archaic humans separately in all three places. The most widely accepted hypothesis for the evolution of modern humans from archaic humans is referred to as the **replacement model,** or *out-of-Africa hypothesis,* which proposes that modern humans evolved from archaic humans only in Africa, and then modern humans migrated to Asia and Europe, where it replaced the archaic species about 100,000 years before the present (BP) (Fig. 32.19). The replacement model is supported by the fossil record. The earliest remains of modern humans (Cro-Magnon), dating at least 130,000 years BP, have been found only in Africa. Modern humans are not found in Asia until 100,000 years BP and not in Europe until 60,000 years BP. Until earlier modern human fossils are found in Asia and Europe, the replacement model is supported. The replacement model is also supported by DNA data. Several years ago, a study showed that the mitochondrial DNA of Africans is more diverse than the DNA of the people in Europe (and the world). This is significant because if mitochondrial DNA has a constant rate of mutation, Africans should show the greatest diversity, since modern humans have existed the longest in Africa. Similarly, Europeans are quite similar genetically, consistent with recent colonization of Europe under the out-of-Africa model.

An opposing hypothesis to the replacement model does exist. This hypothesis, called the *multiregional continuity hypothesis,* proposes that modern humans arose from archaic humans in essentially the same manner in Africa, Asia, and Europe. The hypothesis is multiregional because it applies equally to Africa, Asia, and Europe, and it supposes that in these regions, genetic continuity will be found between modern populations and archaic populations. This contrasting hypothesis has sparked many innovative studies to test which hypothesis is correct.

Neandertals

The **Neandertals,** which are classified as *Homo neandertalensis,* are a species of archaic humans that lived between 200,000 and 28,000 years ago. Neandertal fossils have been found from the Middle East throughout Europe. Neandertals take their name from Germany's Neander Valley, where one of the first Neandertal skeletons, estimated to be about 200,000 years old, was discovered.

According to the replacement model, Neandertals were supplanted by modern humans. Surprisingly, however, the Neandertal brain was, on the average, slightly larger than that of *Homo sapiens* (1,400 cc, compared with 1,360 cc in most modern humans). The Neandertals had massive brow ridges and wide, flat noses. They also had a forward-sloping forehead and a receding lower jaw. Their nose, jaws, and teeth protruded far forward. Physically, the Neandertals were powerful and heavily muscled, especially in the shoulders and neck. The bones of Neandertals were shorter and thicker than those of modern humans. New fossils show that the pubic bone was long compared to that of modern humans. The Neandertals lived in Europe and Asia during the last Ice Age, and their sturdy build could have helped conserve heat.

Archaeological evidence suggests that Neandertals were culturally advanced. Some Neandertals lived in caves. However, others probably constructed shelters. They manufactured a variety of stone tools, including spear points, which they could have used for hunting, and scrapers and knives, which would have helped in food preparation. They most likely hunted bears, woolly mammoths, rhinoceroses, reindeer, and other contemporary animals. They used and could control fire, which probably helped in cooking frozen meat and in keeping warm. They even buried their dead with flowers and tools and may have had a religion.

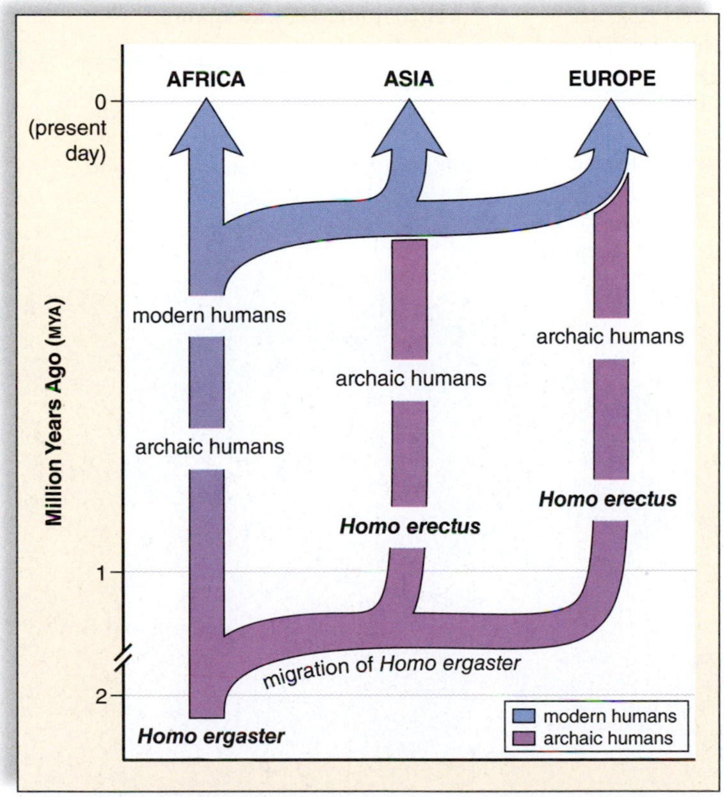
AFRICA
ASIA
EUROPE
0
(present day)
1
2
Million Years Ago (MYA)
modern humans
archaic humans
archaic humans
archaic humans
Homo erectus
Homo erectus
migration of Homo ergaster
Homo ergaster
modern humans
archaic humans

Figure 32.19 Evolution of modern humans. The replacement model proposes that *Homo sapiens* evolved in Africa and then migrated out of Africa, as *H. ergaster* did many thousands of years before. *H. sapiens* would have supplanted populations of ancestral *Homo* in Asia and Europe about 100,000 years ago.

Cro-Magnons

Cro-Magnons are the oldest fossils to be designated *Homo sapiens.* In keeping with the replacement model, the Cro-Magnons are named after a fossil location in France, where modern humans entered Asia and Europe from Africa approximately 100,000 years ago. They probably reached western Europe about 40,000 years ago. Cro-Magnons had a thoroughly modern appearance (Fig. 32.20). They had lighter bones, flat

Figure 32.20 Cro-Magnons. Cro-Magnon people are the first to be designated *Homo sapiens.* Their tool-making ability and other cultural attributes, such as their artistic talents, are legendary.

SCIENCE IN YOUR LIFE ▶ **SCIENTIFIC INQUIRY**

Biocultural Evolution Began with *Homo*

The term culture encompasses human activities and products that are passed on from one generation to another outside of direct biological inheritance. *Homo habilis* (and *Homo rudolfensis*) could make the simplest of stone tools, called Oldowan tools after a location in Africa where the tools were first found. The main tool could have been used for hammering, chopping, and digging. A flake tool was a type of knife sharp enough to scrape away hide and remove meat from bones. The diet of *H. habilis* most likely consisted of collected plants, but they probably had the opportunity to eat meat scavenged from kills abandoned by lions, leopards, and other large predators in Africa.

Homo erectus, who lived in Eurasia, also made stone tools, but the flakes were sharper and had straighter edges. They are called Acheulian tools for a location in France where they were first found. Their "multipurpose" hand axes were large flakes with an elongated oval shape, a pointed end, and sharp edges on the sides. Supposedly they were handheld, but no one knows for sure. *H. erectus* also made the same core and flake tools as *H. habilis.* In addition, *H. erectus* could have also made many other implements out of wood or bone, and even grass, which can be twisted together to make string and rope. Excavation of *H. erectus* campsites dated 400,000 years ago have uncovered literally tens of thousands of tools.

H. erectus, like *H. habilis,* also gathered plants as food. However, *H. erectus* may have also harvested large fields of wild plants. The members of this species were not master hunters, but they gained some meat through scavenging and hunting. The bones of all sorts of animals litter the areas where they lived. Apparently, they ate pigs, sheep, rhinoceroses, buffalo, deer, and many other smaller animals. *H. erectus* lived during the last Ice Age, but even so, moved northward. No wonder *H. erectus* is believed to have used fire. A campfire would have protected them from wild beasts and kept them warm at night. And the ability to cook would have made meat easier to eat.

For early humans to survive during the winter in northern climates, meat must have become a substantial part of the diet because plant sources are not available in the dead of winter. It is even possible that the campsites of *H. erectus* were "home bases" to which the group of individuals returned after the day's search for food. If so, these people may have been the first hunter-gatherers (Fig. 32B)—that is, they hunted animals and gathered plant products. This was a successful way of life that allowed the hominin populations to increase from a few thousand australopiths in Africa 2 MYA to hundreds of thousands of *H. erectus* by 300,000 years ago.

The hunter-gatherer lifestyle most likely encouraged the development and spread of culture between individuals and generations. Those who could speak a language would have been able to cooperate better as they hunted and sought places to gather food. Among animals, only humans have a complex language that allows them to communicate their experiences symbolically. Words stand for objects and events that can be pictured in the mind.

The cultural achievements of *H. erectus* essentially began a new phase of human evolution, called biocultural evolution, in which natural selection is influenced by cultural achievements rather than by anatomic phenotype. *H. erectus* succeeded in new, colder environments because these individuals occupied caves, used fire, and became more capable of obtaining and eating meat as a substantial part of their diet.

Questions to Consider

1. Explain why the development of culture plays such an important role in human evolution.
2. Give an example of how natural selection may interact with culture.

connect BIOLOGY Explore the concepts through a variety of multimedia assets, question types, and data interpretation.
www.mcgrawhillconnect.com

Figure 32B ***Homo erectus.*** The *H. erectus* people may have been hunter-gatherers.

high foreheads, domed skulls housing brains of 1,350 cc, small teeth, and a distinct chin. They made advanced stone tools, including compound tools, such as by fitting stone flakes to a wooden handle. They may have been the first to make knife-like blades and to throw spears, enabling them to kill animals from a distance. They were such accomplished hunters that some researchers hypothesize they may have been responsible for the extinction of many larger mammals, such as the giant sloth, the mammoth, the saber-toothed tiger, and the giant ox, during the late Pleistocene epoch. This event is known as the Pleistocene overkill.

Cro-Magnons hunted cooperatively, and perhaps they were the first to have language. They are believed to have lived in small groups, with the men hunting by day while the women remained at home with the children. It's quite possible that this hunting way of life among prehistoric people influences our behavior even today. The Cro-Magnon culture included art. They sculpted small figurines out of reindeer bones and antlers. They also painted beautiful drawings of animals on cave walls in Spain and France (Fig. 32.20).

Check Your Progress 32.5

1. Discuss the importance of the replacement model in the evolution of the genus *Homo*. .
2. Compare and contrast Neandertals and Cro-Magnons.

Case Study Conclusion

Vertebrates have been on the Earth for the past 500 million years. Scientists have identified approximately 58,000 species at this point. It is believed that there is an estimated 18,000 species still to be discovered. *Paedophryne amanuensis* represents the smallest vertebrate species known to science at this time. These miniature frogs inhabit the leaf litter of tropical forests and have evolved to a unique ecological niche. While new species are being discovered every year, approximately 20% of all mammals, 12% of birds, and 33% of amphibians are threatened with extinction. It will require a large-scale conservation effort of critical habitat to prevent the extinction of these vertebrates.

MEDIA STUDY TOOLS

www.mhhe.com/maderinquiry14

Enhance your study of this chapter with study tools and practice tests. Also ask your instructor about the resources available through ConnectPlus, including LearnSmart, the media-rich eBook, interactive learning tools, and animations.

SUMMARIZE

32.1 Chordates

- Organisms that are classified as **chordates** (tunicates, lancelets, and vertebrates) have a dorsal supporting rod known as a **notochord,** a dorsal tubular **nerve cord, pharyngeal pouches,** and a **postanal tail** at some time in their life history.
- **Lancelets** and **tunicates** are the nonvertebrate chordates. Adult tunicates lack chordate characteristics except gill slits, but adult lancelets have the four chordate characteristics.

32.2 Vertebrates: Fish and Amphibians

- **Vertebrates** have the four chordate characteristics as embryos, but the notochord is replaced by the vertebral column.
- Internal organs are well developed, and cephalization is apparent. The **amniotic egg** was a major evolutionary milestone as well.
- The vertebrate classes trace their evolutionary history as follows:
  - **Jawless fishes** were the first vertebrates and are represented by hagfishes and lampreys that lacked jaws and fins.
  - **Cartilaginous fishes** include the sharks, rays, and skates. Their skeleton is composed of cartilage. They possess jaws and fins.
  - **Bony fishes** have jaws and two pairs of fins. The bony fishes include the **ray-finned fishes** and the **lobe-finned fishes** (some of which have **lungs**).
  - **Amphibians** (e.g., frogs and salamanders) evolved from lobe-finned fishes and have two pairs of limbs. Most amphibian species return to the water to reproduce, and their tadpoles or larvae then metamorphose into terrestrial adults.

32.3 Vertebrates: Reptiles and Mammals

- **Reptiles** most often possess shelled eggs, which contain extraembryonic membranes, including an amnion that allows them to reproduce on land. They are **ectothermic,** meaning their body temperature matches the environment.
- **Birds** are feathered reptiles; feathers help maintain a constant body temperature. They are **endothermic,** meaning that they can maintain an internal temperature. Hollow bones, lungs with air sacs that penetrate bones and allow one-way ventilation, and a keeled breastbone adapt birds for flight.
- **Mammals** have hair to help them maintain a constant body temperature, and mammary glands for nursing of young. **Monotremes** lay eggs; **marsupials** have a pouch in which the newborn matures; and **placental mammals,** which are far more varied and numerous, retain offspring inside the uterus until birth.
- Primates are mammals adapted to living in trees. The order **Primates** includes the strepsirhini (lemurs, aye ayes, bush babies and lorises) and the haplorhini (monkeys, apes, and humans).
- Strepsirhini diverged first, followed by the monkeys and then the apes.

32.4 Evolution of the Hominins

- The hominin lineage diverged from the same primate line as the rest of the apes. A **molecular clock** can be used to infer the relatedness between the two groups. Molecular evidence suggests humans are most closely related to African apes, whose ancestry split from ours about 6 to 10 MYA.
- Hominin evolution began in eastern Africa with the rise of ardipithecines and the **australopithecines. Bipedalism,** or walking upright, is one of the main features that separates modern humans. The most famous australopithecine is Lucy (3.2 MYA), who walked bipedally and had a small brain. A **mosaic evolution** occurred in the australopithecines, showing that different body parts changed at different rates and times.
- *Homo habilis,* present about 2 MYA, is certain to have made tools.
- *Homo erectus,* with a brain capacity of 1,000 cc and a striding gait, was the first to migrate out of Africa.
- The recent discovery of the fossil, *Homo floresiensis,* however, suggests a species of *Homo* coexisted with modern humans as recently as 18,000 years ago.

32.5 Evolution of Modern Humans

- The **replacement model** suggests that modern humans (**Cro-Magnon**) originated in Africa and, after migrating into Europe and Asia, replaced the other *Homo* species (including **Neandertals**) found there. Various lines of evidence support this hypothesis.
- The Neandertals lived approximately 200,000 to 28,000 years ago. They were physically adapted to the last Ice Age. Cro-Magnons first appeared approximately 130,000 years ago. They had advanced culture and technology.

ASSESS

Testing Yourself

Choose the best answer for each question.

For questions 1–5, match the feature with the animal group in the key. Each answer may be used more than once. Each question may have more than one answer.

Key:

a. vertebrate chordates **b.** nonvertebrate chordates

1. Closed circulatory system
2. Notochord
3. Vertebral column
4. Lancelets and tunicates
5. Paired appendages

For questions 6–9, match the feature with the animal group in the key. Each answer may be used more than once. Each question may have more than one answer.

Key:

a. fishes
b. amphibians
c. reptiles
d. birds

6. Internal fertilization
7. Jointed appendages
8. Internal skeleton
9. Ectothermic

10. Label the parts of a bony fish in the following diagram.

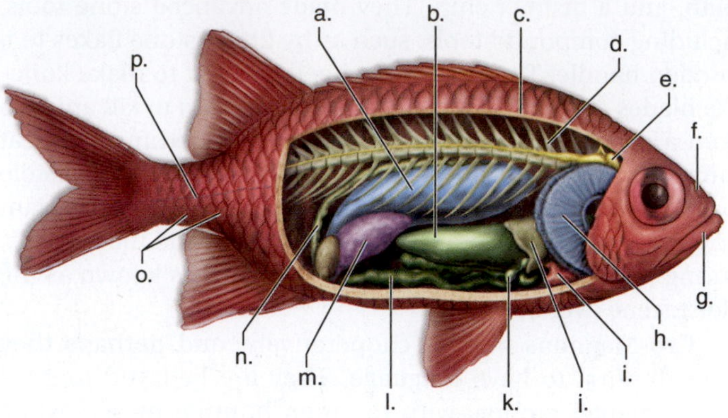
a.
b.
c.
d.
e.
f.
g.
h.
i.
j.
k.
l.
m.
n.
o.
p.

11. Which of the following is not an adaptation for flight in birds?
    a. air sacs
    b. modified forelimbs
    c. bones with air cavities
    d. enlarged sternum
    e. well-developed bladder
12. There are more marsupials in Australia than in North America because
    a. they did not need to compete with placental mammals until recently.
    b. their environment is more favorable for their survival in Australia.
    c. volcanoes led to the mass extinction of marsupials in North America.
    d. food sources are more diverse in Australia.
13. Stereoscopic vision is possible in primates due to
    a. the presence of cone cells.
    b. an enlarged brain.
    c. a shortened snout.
    d. None of these are correct.
14. The first humanlike feature to evolve in the hominids was
    a. a large brain.
    b. massive jaws.
    c. a slender body.
    d. bipedal locomotion.
15. In *H. habilis,* enlargement of the portions of the brain associated with speech probably led to
    a. cooperative hunting.
    b. the sharing of food.
    c. the development of culture.
    d. All of these are correct.

ENGAGE

Thinking Critically

1. Researchers have recently sequenced the genome of several vertebrate species, including the North American green anole and the African clawed frog. What information might these studies give us that could be applied to the development of new medical treatments for humans?
2. Bipedalism has many selective advantages, including the increased ability to spot predators and prey. However, bipedalism has one particular disadvantage—upright posture leads to a smaller pelvic opening, which makes giving birth to an offspring with a large head very difficult. This situation results in a higher percentage of deaths (of both mother and child) during birth in humans compared to other primates. How can you explain the selection for a trait, such as bipedalism, that has both positive and negative consequences for fitness?
3. In studying recent fossils of the genus *Homo,* such as Cro-Magnon, biologists have determined that modern humans have not undergone much biological evolution in the past 50,000 years. Rather, cultural anthropologists argue that cultural evolution has been far more important than biological evolution in the recent history of modern humans. What do they mean by this? Support your argument with some examples.

33 Behavioral Ecology

CASE STUDY The Benefits of Living in a Society

African lions often live in a group, or pride, containing between 2 and 18 closely related females. Living in a social unit increases the females' chances of survival. As a group they can overcome large prey, which would not be possible by an individual. However, although the pride increases the success of hunting large animals, the food must be shared by the entire pride. The advantages and disadvantages of living in a society was a focus of the documentary *African Cats,* which helped to reveal the secrets of the lives of lions.

Within a pride, all of the females tend to be genetically related (mothers, daughters, sisters, aunts). They work together to protect their territory while helping raise each other's young. Female cubs will stay with the pride, while male cubs leave when they reach two to four years of age. These young males often form coalitions with related males. Once they reach maturity, the male coalition establishes a territory that overlaps with a pride.

Sometimes a pride contains young cubs that are not related to the local male coalition. In this case, the cubs are often killed. This triggers the females to enter estrus, allowing each of the new males to have mating opportunities with the females of the pride. Killing the original cubs and mating with the females increases the likelihood that the males will be able to produce cubs with which they will share their genes.

In this chapter, you will learn about animal behavior in general and gain a better understanding of the complexity of living in a social unit. We will explore whether behavior has a genetic or environmental basis. Then we will examine how behavior and communication are necessary components of the struggle to survive.

As you read through the chapter, think about the following questions:

1. Which lion behaviors appear to have a genetic basis to them and which ones appear to have an environmental basis?
2. Which methods of communication are most likely employed by the lions to defend their territory?

CHAPTER OUTLINE

BEFORE YOU BEGIN

Before beginning this chapter, take a few moments to review the following discussions:

Table 20.1 What hormones influence behavior?

Section 23.3 How can the environment influence gene expression?

Section 27.3 What role does sexual selection and male competition play in the evolution of a species?

33.1 Nature Versus Nurture: Genetic Influences

Learning Outcomes

Upon completion of this section, you should be able to

1. Describe an experiment that shows behavior can be genetically based.
2. Describe the body systems that play a role in influencing behavior.

The "nature versus nurture" question asks to what degree our genes (nature) and environmental influences (nurture) affect our behavior. **Behavior** encompasses any action that can be observed and described. Because both the anatomy and physiology of animals, including humans, determine what types of behavior are possible, we immediately know that the genes, to a degree, control behavior. A variety of experiments have been conducted to help determine the degree that genetics controls behavior.

Nest-Building Behavior in Lovebirds

Lovebirds are small, green and pink African parrots that nest in tree hollows. There are several closely related species of lovebirds in the genus *Agapornis,* whose behavior differs by the way they build nests. Fischer lovebirds, *Agapornis fischeri,* pick up a large leaf (or in the laboratory, a piece of paper) with their bills, perforate it with a series of longitudinal bites, and then tear out long strips. They carry the strips in their bills to the nest (Fig. 33.1*a*) and weave them with others to make a deep cup. Peach-faced lovebirds, *Agapornis roseicollis,* cut somewhat shorter strips in a similar manner, but then carry them to the nest by inserting them deep into their rump feathers (Fig. 33.1*b*). In this way, they can carry several short strips during each trip to the nest, whereas Fischer lovebirds can carry only one longer strip at a time.

Researchers hypothesized that if the behavior for obtaining and carrying nesting material is inherited, then hybrids might show intermediate behavior. When the two species of birds were crossed, the hybrid offspring had difficulty carrying nesting materials. They cut strips of intermediate length and then attempted to tuck the strips into their rump feathers. They did not push the strips far enough into the feathers, however, and when they walked or flew, the strips always fell out. Hybrid birds eventually (after about three years) learned to carry the cut strips in their bills, but still briefly turned their heads toward their rumps before flying off. The intermediate behavior exhibited by the hybrids supports the hypothesis that behavior has a genetic basis.

Food Choice in Garter Snakes

Several experiments have been conducted using the garter snake, *Thamnophis elegans,* to determine if food preference has a genetic basis. In California, inland garter snake populations are aquatic and commonly feed underwater on frogs and fish,

a. Fischer lovebird with nesting material in its beak.

b. Peach-faced lovebird with nesting material in its rump feathers.

Figure 33.1 Nest building behavior in lovebirds. **a.** Fischer lovebirds carry strips of nesting material in their bills, as do most other birds. **b.** Peach-faced lovebirds tuck strips of nesting material into their rump feathers before flying back to the nest.

whereas coastal populations are terrestrial and feed mainly on slugs. In the laboratory, inland adult snakes refused to eat slugs, while coastal snakes readily did so. Newborns resulting from matings between snakes from the two populations (inland and coastal) have an intermediate preference for slugs, suggesting a genetic basis for food choice.

Differences between slug acceptors and slug rejecters appear to be inherited, but what physiological difference is there between the two populations? A clever experiment answered this question. Snakes use tongue flicks to recognize

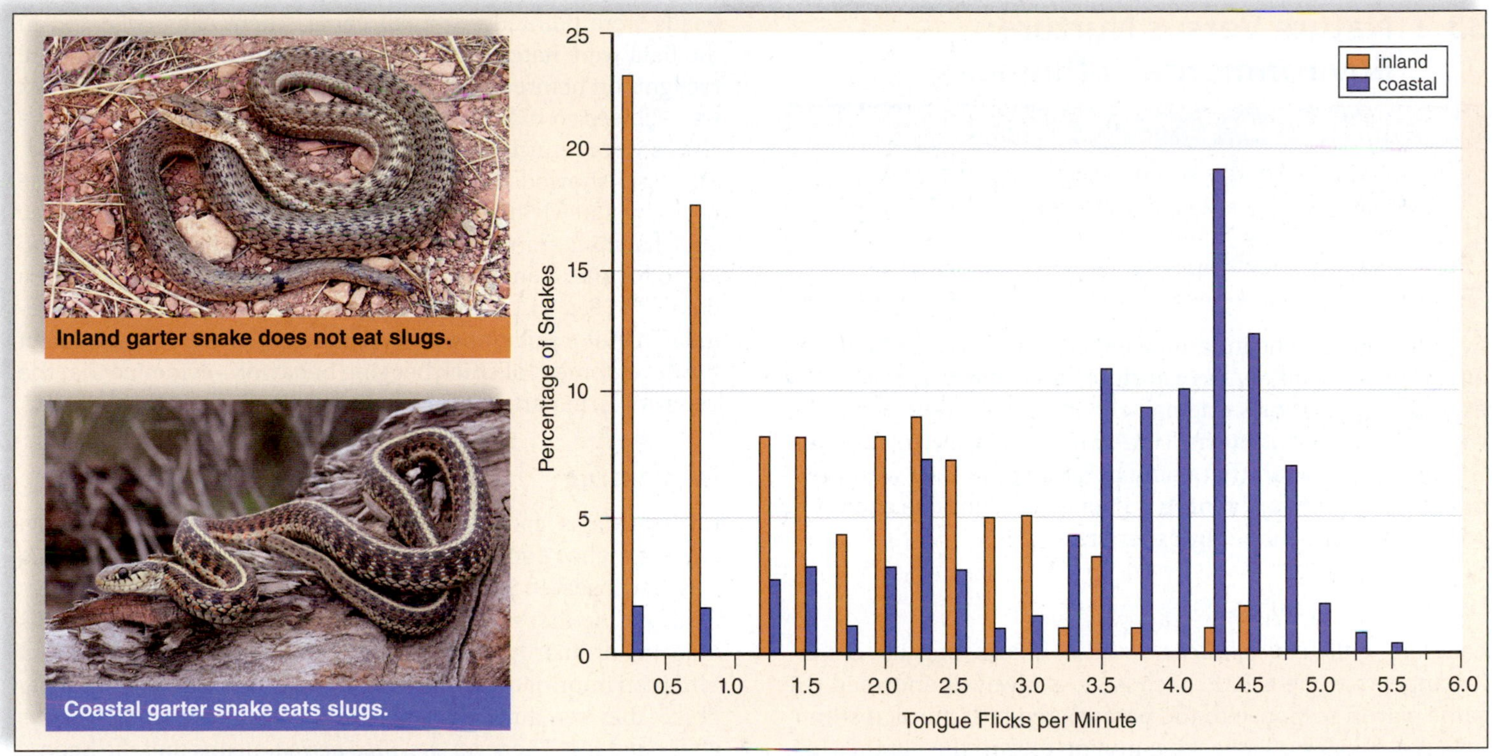
Inland garter snake does not eat slugs.
Coastal garter snake eats slugs.
inland
coastal
Percentage of Snakes
25
20
15
10
5
0
0.5 1.0 1.5 2.0 2.5 3.0 3.5 4.0 4.5 5.0 5.5 6.0
Tongue Flicks per Minute

Figure 33.2 Feeding behavior in garter snakes. The number of tongue flicks by inland and coastal garter snakes is measured in terms of their response to slug extract on cotton swabs. Coastal snakes tongue-flicked more than inland snakes, indicating that coastal snakes are more sensitive to the smell of slugs than inland snakes.

their prey, and their tongues carry chemicals to an odor receptor in their mouth. Newborns will even flick their tongues at cotton swabs dipped in fluids of their prey. Swabs were dipped in slug extract, and the number of tongue flicks were counted for newborn inland and coastal snakes. Coastal snakes had a higher number of tongue flicks (indicating more interest in the smell) than inland snakes (Fig. 33.2). Thus, inland snakes apparently do not eat slugs because they are not sensitive to their smell. A genetic difference between the two populations of snakes has resulted in a physiological difference in their nervous systems. Although hybrids showed a great deal of variation in the number of tongue flicks, they were generally intermediate, as predicted by the genetic hypothesis.

Egg-Laying Behavior of Marine Snails

The nervous and endocrine systems are both responsible for the coordination of body systems. Is the endocrine system also involved in behavior? Experiments support endocrine system involvement. For example, the egg-laying behavior in the marine snail *Aplysia* involves a set sequence of movements. Following copulation, the animal extrudes long strings of more than a million eggs per egg case. It then takes the egg case string in its mouth, covers it with mucus, waves its head back and forth to wind the string into an irregular mass, and attaches the mass to a solid object, such as a rock. Several years ago, scientists isolated and analyzed an egg-laying hormone (ELH) that causes the snail to lay eggs even if it has not mated. ELH was found to be a small protein of 36 amino acids that diffuses into the circulatory system and causes the smooth muscle cells of the reproductive duct to contract and expel the egg string. Using recombinant DNA techniques, the investigators isolated the entire ELH gene. The gene's product turned out to be a protein with 271 amino acids. The protein can be cleaved into as many as 11 possible products, and the ELH hormone is one of these. The hormone alone, or in conjunction with the gene's other products, is thought to control all the components of egg-laying behavior in *Aplysia*.

Twin Studies in Humans

Human twins, on occasion, have been separated at birth and raised under different environmental conditions. Studies of separated twins show that they have similar food preferences and activity patterns, and even select mates with similar characteristics. These twin studies lend support to the hypothesis that at least certain types of behavior are primarily influenced by nature (i.e., genes).

Check Your Progress 33.1

1. Identify how studies of animals suggest that behavior can be genetically based.
2. Summarize how the endocrine system influences behavior.

33.2 Nature Versus Nurture: Environmental Influences

Learning Outcomes

Upon completion of this section, you should be able to

1. Identify an experiment that shows how behavior can be environmentally influenced.
2. Compare three different types of learned behavior.

Even though genetic inheritance serves as a basis for behavior, environmental influences (nurture) also affect behavior. For example, behaviorists originally believed that some behaviors were **fixed action patterns** (**FAPs**) elicited by a sign stimulus. But then they found that many behaviors improve with practice. In this context, **learning** is defined as a durable change in behavior brought about by experience.

Learning in Birds

Laughing gull (*Leucophaeus atricilla*) chicks' begging behavior appears to be a FAP, because it is always performed the same way in response to the parent's red bill (the sign stimulus). A chick directs a pecking motion toward the parent's bill, grasps it, and strokes it downward (Fig. 33.3*a*). Parents bring about the begging behavior by swinging their bill gently from side to side. After the chick responds, the parent regurgitates food onto the nest floor. If need be, the parent then encourages the chick to eat. This interaction between the chicks and their parents suggests that the begging behavior involves learning. To test this hypothesis, diagrammatic pictures of gull heads were painted on small cards. Then, eggs collected in the field were hatched in a dark incubator to eliminate visual recognition before the test. On the day of hatching, each chick was allowed to make about a dozen pecks at the model. The chicks were returned to the nest, and then each was retested. The tests showed that on the average, only one-third of the pecks by a newly hatched chick strike the model. But one day after hatching, more than half of the pecks are accurate, and two days after hatching, the accuracy reaches a level of more than 75% (Fig. 33.3*b, c*). Investigators concluded that improvement in motor skills, as well as visual experience, strongly affect the development of chick begging behavior—evidence that the behavior is, in part, learned.

Imprinting

Imprinting is considered a very simple form of learning, although it has a strong genetic component as well. Imprinting was first observed in birds when chicks, ducklings, and goslings followed the first moving object they saw after hatching. This object is ordinarily their mother, but investigators found that birds can imprint on any object, as long as it is the first moving object they see during a sensitive period of two to three days after hatching. The term *sensitive period* means that the behavior develops only during this time.

A chick imprinted on a red ball follows it around and chirps whenever the ball is moved out of sight. Social interactions between parent and offspring during the sensitive period seem key to normal imprinting. For example, female mallards cluck during the entire time imprinting is occurring, and it could be that vocalization before and after hatching is necessary for normal imprinting.

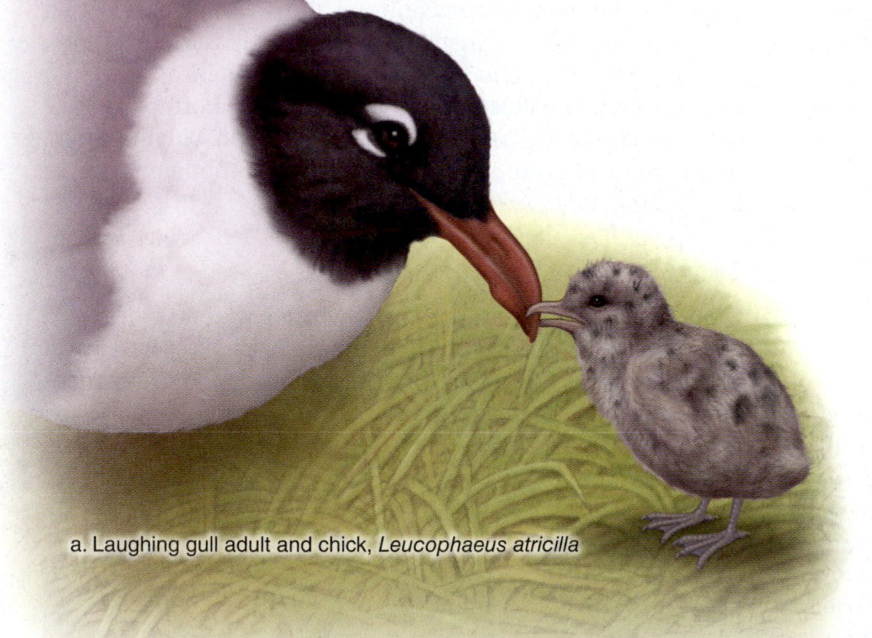
a. Laughing gull adult and chick, *Leucophaeus atricilla*

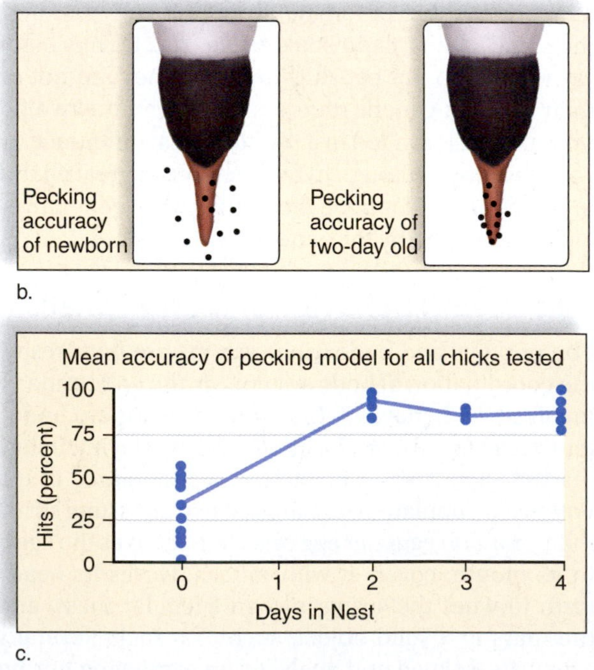
Pecking accuracy of newborn
Pecking accuracy of two-day old
b.
Mean accuracy of pecking model for all chicks tested
Hits (percent)
100
75
50
25
0
0
1
2
3
4
Days in Nest
c.

Figure 33.3 Begging behavior in laughing gull chicks. **a.** At about three days of age, a laughing gull chick grasps the red bill of a parent, stroking it downward, and the parent then regurgitates food. **b.** The accuracy of a chick striking a test probe, painted red. **c.** Chick-pecking accuracy illustrated graphically. Note from these diagrams that a chick markedly improves its ability (within only two days) to peck a bill, a behavior that normally causes a parent to regurgitate food.

Social Interactions and Learning

White-crowned sparrows sing a species-specific song, but males of a particular region have their own dialect. Birds were caged to test the hypothesis that young white-crowned sparrows learn how to sing from older members of their species.

Three groups of birds were tested. Birds in the first group *heard no songs at all.* When grown, these birds sang a song, but it was not fully developed. Birds in the second group *heard tapes of white-crowns singing.* When grown, they sang in that dialect, as long as the tapes had been played during a sensitive period from about age 10 to 50 days. White-crowned sparrows' dialects (or other species' songs) played before or after this sensitive period had no effect on their song. Birds in a third group did not hear tapes and instead were *given an adult tutor of a different species.* These birds sang the song of a still different species—no matter when the tutoring began—showing that social interactions apparently assist learning in birds.

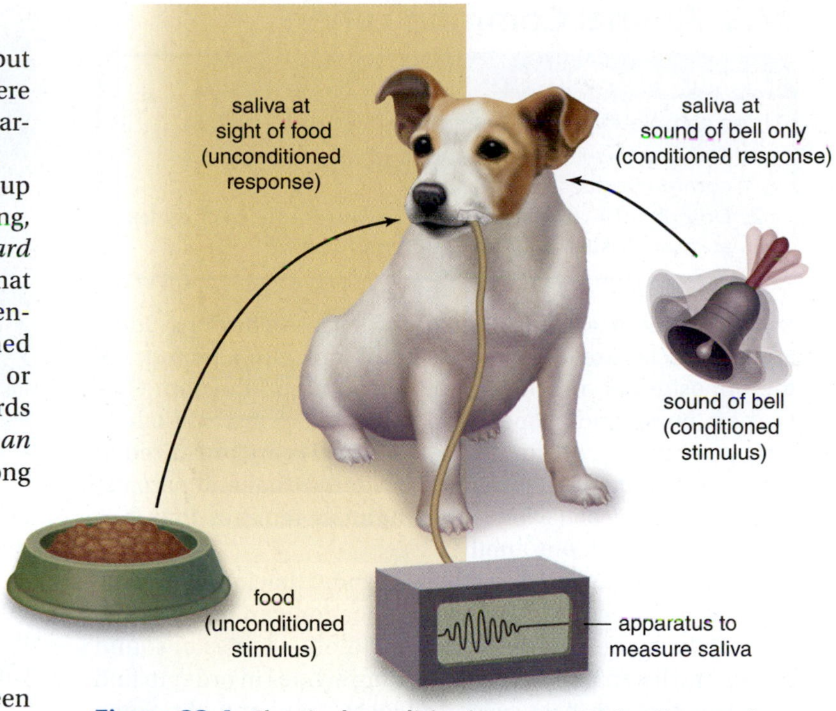
saliva at
sight of food
(unconditioned
response)
saliva at
sound of bell only
(conditioned response)
sound of bell
(conditioned
stimulus)
food
(unconditioned
stimulus)
apparatus to
measure saliva

Figure 33.4 Classical conditioning. Ivan Pavlov discovered classical conditioning by performing this experiment with dogs. A bell is rung when a dog is fed. Salivation is noted. Eventually, the dog salivates when the bell is rung even though no food is presented. Food is an unconditioned stimulus, and the sound of the bell is a conditioned stimulus that brings about the response—that is, salivation.

Associative Learning

A change in behavior that involves an association between two events is termed **associative learning.** For example, birds that get sick after eating a monarch butterfly no longer prey on monarch butterflies, even though they may be readily available. Or the smell of fresh-baked bread may entice you to eat even though you may have just eaten. Because you really enjoyed bread in the past, you associate the smell with those past experiences, which makes you hungry. Both classical conditioning and operant conditioning are examples of associative learning.

Classical Conditioning

In **classical conditioning,** the presentation of two different types of stimuli at the same time causes an animal to form an association between them. The best-known laboratory example of classical conditioning is an experiment done by the Russian psychologist Ivan Pavlov. First, Pavlov observed that dogs salivate when presented with food. Then he rang a bell whenever the dogs were fed. Eventually, the dogs would salivate whenever the bell was rung, regardless of whether food was present (Fig. 33.4).

Classical conditioning suggests that an organism can be trained or conditioned to associate a specific response to a specific stimulus. Unconditioned responses are those that occur naturally, as when salivation follows the presentation of food. Conditioned responses are those that are learned, as when a dog learns to salivate when it hears a bell. Advertisements attempt to use classical conditioning. For example, commercials pair attractive people with a particular product in the hope that viewers will associate attractiveness with that product. This pleasant association may cause them to buy the product.

In a similar way, it's been suggested that holding children on your lap when reading to them facilitates an interest in reading. The belief is that they will associate a pleasant feeling with reading.

Operant Conditioning

During **operant conditioning,** a stimulus-response connection is strengthened. Most people know that it is helpful to give an animal a reward, such as food or affection, when teaching them a trick. When we go to an animal show, it is quite obvious that trainers use operant conditioning. They present a stimulus, such as a hoop, and then give a reward (food) for the proper response (jumping through the hoop).

B. F. Skinner is well known for studying this type of learning in the laboratory. In Skinner's simplest type of experiment, a caged rat happens to press a lever and is rewarded with sugar pellets. Thereafter, the rat regularly presses the lever whenever it is hungry. In more sophisticated experiments, Skinner even taught pigeons to play ping-pong by rewarding desired responses to stimuli.

When operant conditioning is applied to child rearing, it's been suggested that parents who give a positive reinforcement for good behavior will be more successful than parents who punish behaviors they believe are undesirable.

Check Your Progress 33.2

1. Identify examples of various types of learning.
2. Contrast the different types of associative learning.

33.3 Animal Communication

Learning Outcomes

Upon completion of this section, you should be able to

1. Summarize the various ways in which animals communicate.
2. Describe the advantages and disadvantages of chemical, auditory, visual, and tactile communication.

Animals exhibit a wide diversity of social behaviors. Some animal species are largely solitary and join with a member of the opposite sex only for the purpose of reproduction. Others pair, bond, and cooperate in raising offspring. Still others form a **society** in which members of species are organized in a cooperative manner, extending beyond sexual and parental behavior. Social behavior among animals requires that they communicate with one another.

Communication is a signal by a sender that influences the behavior of a receiver. The communication can be purposeful, but does not have to be. Bats send out a series of sound pulses and listen for the corresponding echoes in order to find their way through dark caves and locate food at night. Some moths have an ability to hear these sound pulses, and they begin evasive tactics when they sense that a bat is near. The bats are not purposefully communicating with the moths. The bat sounds are simply a cue to the moths that danger is near.

The types of communication signals used by animals include chemical, auditory, visual, and tactile.

Chemical Communication

Chemical signals have the advantage of being effective both night and day. The term **pheromone** designates chemical signals in low concentration that are generally passed between members of the same species. For example, female moths secrete chemicals from special abdominal glands, which are detected by receptors on male antennae. The antennae are especially sensitive, and this ensures that only male moths of the same species will be able to detect them.

Video Bug Speak

Cheetahs and other cats mark their territories by depositing urine, feces, and anal gland secretions at the boundaries (Fig. 33.5). Klipspringers (small antelope) use secretions from a gland below the eye to mark twigs and grasses of their territory.

Researchers are studying to what degree pheromones and hormones affect the behavior of mammals. Some researchers maintain that human behavior is influenced by undetectable pheromones wafting through the air. They have discovered that like the mouse, humans have an organ in the nose, called the vomeronasal organ (VNO), which can detect not only odors, but also pheromones. The neurons from this organ lead to the hypothalamus, the part of the brain that controls the release of various hormones in the body. Perhaps this organ is even involved in how people choose their mates (see the Health feature, "Mate Choice and Smelly T-Shirts," on page 683).

Video Sex and the Senses

Figure 33.5 Use of a pheromone. This male cheetah is spraying urine onto a tree to mark its territory.

Auditory Communication

Auditory (sound) communication has various advantages over other kinds of communication. It is faster than chemical communication, and is effective both night and day. Further, auditory communication can be modified not only by loudness but also by pattern, duration, and repetition. In an experiment with rats, a researcher discovered that an intruder can avoid attack by increasing the frequency with which it makes an appeasement sound.

Male birds have songs for a number of different occasions, such as one song for distress, another for courting, and still another for marking territories. Sailors have long heard the songs of humpback whales transmitted through the hull of a ship. However, only recently has it been shown that the song has six basic themes, each with its own phrases, which can vary in length, be interspersed with sundry cries and chirps, and be heard for many miles. Interestingly, humpbacks' songs change from year to year, but all the humpbacks in an area learn and converge on singing the same song. The purpose of the song is probably sexual, serving to advertise the availability of the singer.

Language is the ultimate auditory communication. Only humans have the biological ability to produce a large number of different sounds and to put them together in many different ways. Nonhuman primates have different vocalizations, each having a definite meaning, such as when vervet monkeys give alarm calls (Fig. 33.6). Although chimpanzees can be taught to use an artificial language, they never progress beyond the capability level of a two-year-old child. It also has been difficult to prove that chimps understand the concept of grammar or can use their language to reason. As such, most anthropologists argue that humans possess a communication ability unparalleled by most animals.

Video Meerkat Warning Calls

a.
b.
Frequency (kilocycles per second)
0
1
2
3
4
5
6
7
8
0.5 Seconds
Eagle
Frequency (kilocycles per second)
0.5 Seconds
Leopard

Figure 33.6 Auditory communication. **a.** Vervet monkeys, *Cercopithecus aethiops,* are responding to an alarm call. Vervet monkeys can give different alarm calls according to whether a troop member sights an eagle or a leopard, for example. **b.** The frequency per second of the sound differs for each type of call.

Visual Communication

Visual signals are most often used by species that are active during the day. Contests between males make use of threat postures to help prevent outright fighting, a behavior that might result in reduced fitness. A male baboon displaying full threat is an awesome sight that establishes his dominance and keeps peace within the baboon troop (Fig. 33.7).

Many animals use complex courtship behaviors and displays. The plumage of a male Raggiana Bird of Paradise allows him to put on a spectacular courtship dance to attract a female, giving her a basis on which to select a mate. Defense and courtship displays are exaggerated and always performed in the same way so that their meaning is clear. Male fireflies use a flash pattern to signal females of the same species (Fig. 33.8).

Video
Flirting Flies

Figure 33.7 A male olive baboon displaying full threat. In olive baboons, males are larger than females and have enlarged canines. Competition between males establishes a dominance hierarchy for the distribution of resources.

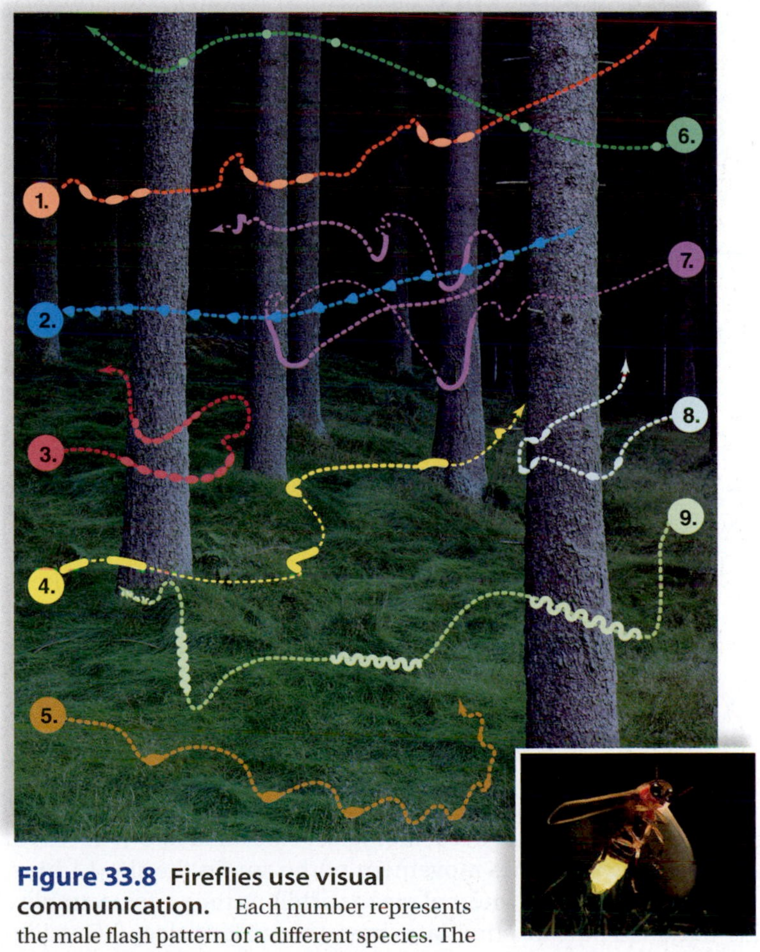
1.
2.
3.
4.
5.
6.
7.
8.
9.

Figure 33.8 Fireflies use visual communication. Each number represents the male flash pattern of a different species. The patterns are a behavioral reproductive isolation mechanism.

Visual communication allows animals to signal others of their intentions without the need to provide any auditory or chemical messages. The body language of students during a lecture provides an example. Some students lean forward in their seats and make eye contact with the instructor. They want the instructor to know they are interested and find the material of value. Others lean back in their chairs, look at the floor, or send text messages on their cell phones. These students indicate they are not interested in the material. Teachers can use students' body language to determine if they are effectively presenting the material and make changes accordingly.

Other human behaviors also send visual clues. The hairstyle and dress of a person or the way he or she walks and talks are ways to send messages to others. Psychologists have long tried to understand how visual clues can be used to better understand human emotions and behavior. Some studies have suggested that women are apt to dress in an appealing manner and be sexually inviting when they are ovulating. People who dress in black, move slowly, fail to make eye contact, and sit alone may be telling others that they are unhappy. Similarly, body language in animals is being used to suggest that they too have emotions, as discussed in the Bioethical feature, "Do Animals Have Emotions?"

Tactile Communication

Tactile communication occurs when one animal touches another. For example, laughing gull chicks peck at the parent's beak to induce the parent to feed them (see Fig. 33.3). A male leopard nuzzles the female's neck to calm her and stimulate her willingness to mate. In primates, grooming—one animal cleaning the coat and skin of another—helps cement social bonds within a group.

Honeybees use a combination of communication methods, including tactile ones, to convey information about their environment. When a foraging bee returns to the hive, it performs a waggle dance that communicates the distance and direction to the food source (Fig. 33.9). Inside the hive, other foragers crowd around her and touch the forager with their antennae, and as the bee moves between the two loops of a figure 8, it buzzes noisily and shakes its entire body in so-called waggles. The angle of the straight run of the figure 8 to that of the direction of gravity is the same as that of the angle between the sun and the food source. In other words, a 40-degree angle to the left of vertical means that food is 40 degrees to the left of the sun. Bees can use the sun as a compass because they have a biological clock that allows them to compensate for the movement of the sun in the sky. (A biological clock is an internal means of telling time—for example, darkness outside stimulates many animals to sleep.) Outside the hive, the dance is done on a horizontal surface, and the straight-run part of the figure 8 indicates the direction of the food.

a.

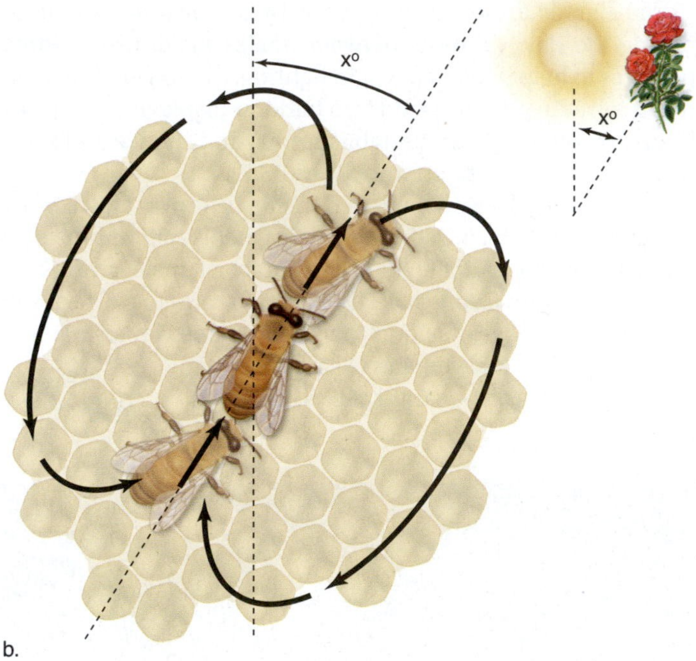
x°
x°

b.

Figure 33.9 Communication among bees. **a.** Honeybees do a waggle dance to indicate the direction of food. **b.** If the dance is done outside the hive on a horizontal surface, the straight run of the dance will point to the food source. If the dance is done inside the hive on a vertical surface, the angle of the straightaway to that of the direction of gravity is the same as the angle of the food source to the sun.

Check Your Progress 33.3

1. Identify ways in which communication is meant to affect the behavior of a receiver.
2. Describe the types of communication that would be the most effective in a dense forest.

SCIENCE IN YOUR LIFE ▶ BIOETHICAL

Do Animals Have Emotions?

a.

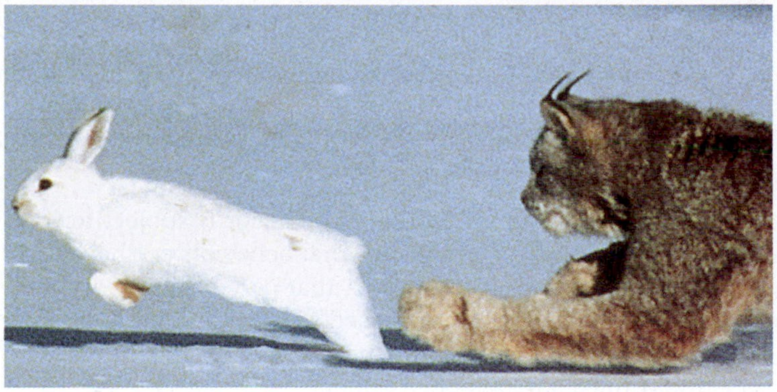

b.

Figure 33A Emotions in animals. **a.** Do chimpanzees feel motherly love, as this photograph seems to suggest? **b.** Does the hare feel fear as the lynx closes in?

Recently, investigators have become interested in determining whether animals have emotions. The body language of animals suggests that animals do have feelings (Fig. 33A). When wolves reunite, they wag their tails to and fro, whine, and jump up and down. Elephants vocalize—emit their "greeting rumble"—flap their ears, and spin about. Many young animals play with one another, or even by themselves, as when dogs chase their own tails. On the other hand, upon the death of a friend or parent, chimps are apt to sulk, stop eating, and even die. It seems reasonable to hypothesize that animals are "happy" when they reunite, "enjoy" themselves when they play, and are "depressed" over the loss of a troop member or relative. Even people who rarely observe animals usually agree about what the animal appears to be feeling.

In the past, scientists found it expedient to collect data only about observable behavior and to ignore the possible mental state of the animal. Why? Because emotions are personal and no one can ever know exactly how another animal is feeling. B. F. Skinner, whose research method is described in section 33.2, regarded animals as robots that become conditioned to respond automatically to various stimuli. He and others never considered that animals might have feelings. But now, some scientists argue that sufficient data exist to suggest that at least other vertebrates and/or mammals do have feelings, including fear, joy, embarrassment, jealousy, anger, love, sadness, and grief. And they believe that those who hypothesize otherwise should have to present the opposing data.

Perhaps it would be reasonable to consider the suggestion of Charles Darwin, the father of evolutionary biology, who said that animals are different in degree rather than in kind. This means that all animals can, say, feel love, but perhaps not to the degree that humans can. Researcher B. Würsig watched the courtship of two baleen whales. They touched, caressed, rolled side-by-side, and eventually swam off together. He wondered if their behavior indicated they felt love for one another. When you think about it, it seems unlikely that emotions first appeared in humans with no evolutionary homologies in animals.

Neurobiological data support the hypothesis that animals other than humans are capable of enjoying themselves when they perform an activity, such as playing. Researchers have found a high level of dopamine in the brain when rats play, and the dopamine level increases when the rats anticipate the opportunity to play. Certainly even the staunchest critic is aware that many different species of animals have limbic systems and are capable of fight-or-flight responses in dangerous situations. Can we go further and suggest that animals feel fear even when no physiological response has yet occurred?

Laboratory animals may be too stressed to provide convincing data on emotions, which makes field research more useful. Then, too, we have to consider that emotions evolved under an animal's normal environmental conditions. It is possible to fit animals with devices that transmit information on heart rate, body temperature, and eye movements as they go about their daily routine. Such information will help researchers learn how animal emotions might correlate with their behavior. In humans, emotions influence behavior, and the same may be true of other animals. One possible definition of emotion is a psychological phenomenon that helps animals direct and manage their behavior.

M. Bekoff, who is prominent in studying the basis of animal behavior, encourages us to be open to the possibility that animals have emotions. He states:

> By remaining open to the idea that many animals have rich emotional lives, even if we are wrong in some cases, little truly is lost. By closing the door on the possibility that many animals have rich emotional lives, even if they are very different from our own or from those of animals with whom we are most familiar, we will lose great opportunities to learn about the lives of animals with whom we share this wondrous planet.[1]

Questions to Consider

1. Do you believe animals have emotions? Why or why not? If not, what experiment(s) would convince you that animals do have emotions?
2. Pet psychology, an emerging field, is based on the premise that pets have feelings and emotions. Would you spend $50–100 per hour to take your pet to a psychologist?
3. If it can be shown that animals have emotions, should animal testing be allowed to continue?

[1]Bekoff, M. "Animal emotions: Exploring passionate natures," *Bioscience* 50:10, page 869 (October 2000).

33.4 Behaviors That Affect Fitness

Learning Outcomes

Upon completion of this section, you should be able to

1. Discuss why territoriality evolved in certain animal groups.
2. Explain why females should be choosier than males in selecting a mate.
3. Summarize the costs and benefits of living in a society.

Behavioral ecology assumes that most behavior is subject to natural selection. We have established that behaviors can have a genetic basis, and we would expect that certain behaviors more than others will lead to increased survival and number of offspring. Therefore, much of the behavior of organisms we observe today must have adaptive value.

Figure 33.10 Male and female gibbons. Siamang gibbons, *Hylobates syndactylus,* are monogamous, and they both share the task of raising offspring. They also share the task of marking their territory by singing. As is often the case in monogamous relationships, the sexes are similar in appearance. Male is above and female is below.

Territoriality and Fitness

In order to gather food, animals often have a particular home range where they can be found during the course of the day. Animals may actively defend a portion of their home range for their exclusive use as a food source or as a mating area. This portion of the home range is called their territory and the behavior is called territoriality.

Territoriality is more likely to occur during times of reproduction. For example, gibbons live in the tropical rain forest of South and Southeast Asia, and territories are maintained by loud singing (Fig. 33.10). Males sing just before sunrise, and mated pairs sing duets during the morning. Males, but not females, show evidence of fighting to defend their territory in the form of broken teeth and scars. Obviously, defense of a territory has a certain cost; it takes energy to sing and fight off others. Also, you might get hurt. In order for territoriality to persist, it must have an adaptive value. Chief among the benefits of territoriality are to ensure a source of food, exclusive rights to one or more females, and to have a place to rear young and possibly to protect yourself from predators.

The territory has to be the right size for the animal. Too large a territory cannot be defended, and too small a territory may not contain enough food. Cheetahs require a large territory to hunt for their prey and, therefore, they use urine to mark their territory (see Fig. 33.5). Hummingbirds are known to defend a very small territory because they depend on only a small patch of flowers as their food source.

Video Cichlid Territoriality

Foraging for Food

Animals need to ingest food that will provide more energy than the effort expended acquiring the food. In one study, it was shown that shore crabs eat intermediate-sized mussels because the net energy gain was more than if they ate larger-sized mussels (Fig. 33.11). The large mussels take too much energy to open per the amount of energy they provide. The optimal foraging model states that it is adaptive for foraging behavior (i.e., searching for food) and food choice to be as energetically efficient as possible. Even though it can be shown that animals that take in more energy are more likely to have more offspring, animals often have to consider other factors, such as escaping from predation. If an animal is killed and eaten, it has no chance at all of having offspring in the future. Thus, animals may have to avoid foraging when the threat of predation is high. Animals often face these types of trade-offs that cause them to modify their behavior. As another example, many animal species stop eating during peak reproductive periods because competition for mates is often fierce.

Reproductive Strategies and Fitness

Usually, primates are polygamous, and males monopolize multiple females. Because of gestation and lactation, females invest more in their offspring than do males and may not always be available for mating. Under these circumstances, it is adaptive for females to be concerned with a good food source. When food sources are clumped, females congregate in small groups. Because only a few females are expected to be receptive at a

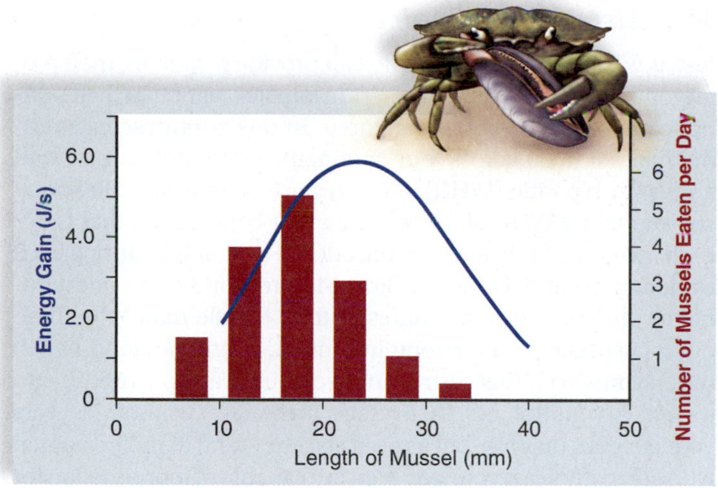
Energy Gain (J/s)
6.0
4.0
2.0
0
Number of Mussels Eaten per Day
6
5
4
3
2
1
0
10
20
30
40
50
Length of Mussel (mm)

Figure 33.11 Foraging for food. When offered a choice of an equal number of each size of mussel, the shore crab, *Carcinus maenas,* prefers the intermediate size. This size provides the highest rate of net energy return. Net energy is determined by the energetic yield of the crab minus the energetic costs of breaking open the shell and digesting the crab.

Figure 33.12 Hamadryas baboons. Among Hamadryas baboons, *Papio hamadryas,* a male, which is silver-white and twice the size of a female, keeps and guards a harem of females with whom he mates exclusively.

time, males will likely be able to defend these few from other males. Males are expected to compete with other males for the limited number of receptive females available (Fig. 33.12).

A limited number of primates are polyandrous. Tamarins are squirrel-sized New World monkeys that live in Central or South America. Tamarins live together in groups of one or more families in which one female mates with more than one male. The female normally gives birth to twins of such a large size that the father, and not the mother, carry them about. This may be the reason these animals are polyandrous. Polyandry also occurs when the environment does not have sufficient resources to support several young at a time.

Gibbons, as previously mentioned, are monogamous, which means that they pair bond. Subsequently, both the male and female help with the rearing of the young. Males are active fathers, frequently grooming and handling infants. Monogamy is relatively rare in primates, which includes prosimians, monkeys, and apes (only about 18% are monogamous). In primates, monogamy occurs when males have limited mating opportunities, territoriality exists, and the male is fairly certain the offspring are his. In gibbons, females are evenly distributed in the environment, most likely because they are aggressive to one another. Studies have shown that females do attack a speaker when it plays female sounds in their territory.

Sexual Selection

Sexual selection is a form of natural selection that favors features that increase an animal's chances of mating. These features are adaptive in the sense that they lead to increased fitness. Sexual selection often results in female choice and male competition. Because females produce a limited number of eggs in their lifetime and generally provide the majority of the parental care, it is adaptive for them to be choosy about their mate. If they choose a mate that passes on features to a male offspring that will cause him to be chosen by females, their fitness has increased. Whether females choose features that are adaptive to the environment is in question. For example, peahens are likely to choose peacocks that have the most elaborate tails. Such a fancy tail could otherwise be detrimental to the male and make him more likely to be captured by a predator. In one study, an extra ornament was attached to a father zebra finch and the daughters of this bird underwent the process of imprinting. Now, these females were more likely to choose a mate that also had the same artificial ornament.

While females can always be sure an offspring is theirs, males do not have this certainty. However, males produce a plentiful supply of sperm. The best strategy for males to increase their fitness, therefore, is to have as many offspring as possible. Competition may be required for them to gain access to females, and ornaments, such as antlers, can enhance a male's ability to fight (Fig. 33.13).

Figure 33.13 Competition. During the mating season, bull elk, *Cervus elaphus,* males may find it necessary to engage in antler wrestling in order to have sole access to females in a territory.

When bull elk compete, they issue a loud number of screams that gives way to a series of grunts. If still necessary, the two bulls walk in parallel to show each other their physique. If this doesn't convince one or the other to back off, the pair resorts to ramming each other with their antlers. Rarely is either bull actually hurt. Whereas a peacock cannot shed his tail, elk shed their antlers as soon as mating season is over.

Mating in Humans

A study of human mating behavior shows that the concepts of female choice and male competition apply to humans as well as to other animals. That is, mate choice behavior in men and women in most human cultures seems to be influenced by its fitness consequences. Of course, applying the principles of evolutionary biology to human behavior is not without controversy because we can't be reduced to being "preprogrammed" and still have the ability to make conscious choices. That said, understanding the evolutionary basis of animal behavior can provide some interesting insights into why people behave the way they do.

Human Males Compete

Consider that women, by nature, must invest more in having a child than men. After all, it takes nine months to have a child, and pregnancy is followed by lactation when a woman may nurse her infant. Men, on the other hand, need only contribute sperm during a sex act that may require only a few minutes. The result is that men are generally more available for mating than are women. Because more men are available, they necessarily have to compete with others for the opportunity to mate.

Like many other animals, humans are dimorphic. Males tend to be larger and more aggressive than females, perhaps as a result of past sexual selection by females. As in other animals, males may pay a price due to high levels of testosterone, the energetic costs of male-male competition, and increased stress associated with finding mates. Human men tend to live on average four to seven years less than females.

Females Choose

David Buss, an evolutionary psychologist at the University of Texas, conducted studies of female preference of a male mate across cultures in over 20 countries. His research, although somewhat controversial, suggests that the number-one trait females prefer in a male mate is his ability to obtain (financial) resources. Financial success means that men are more likely to provide females with the resources they need to raise their children, and thus increase the fitness (reproductive success) of the female.

Other recent studies have shown that symmetry in facial and body features is also important in female mate choice, possibly related to the "good genes" hypothesis. That is, females may choose males with "good genes" so that their offspring have good genes and a better chance at survival. As an example, symmetry in other animals is often a sign of good health and a strong immune system (e.g., animals with high parasite loads are often asymmetrical). Females may thus choose symmetrical males to pass these traits on to their offspring.

Men Also Have a Choice

Just as women choose men who can provide resources, men prefer youthfulness and attractiveness in females, signs that their partner can provide them with children. In one controversial study, men ages 8 to 80 across four continents preferred women with a waist-to-hip ratio (WHR) of 0.7, regardless of weight. Research showed that a WHR of 0.7 is optimal for conception; that is, for each increase of 10% in WHR, the odds of conception during each ovulation event decrease by 30%. Men responding to questionnaires prefer physical attributes in their female mates that biologists associate with a strong immune system and good health (e.g., symmetry), high estrogen levels (indicating fertility), and especially youthfulness. On average, men marry women 2.5 years younger than they are, but as men age, they tend to prefer women who are many years younger. Men are capable of reproduction for many more years than women. Therefore, by choosing younger women, older men can increase their fitness. Other factors are also involved in human mate choice, as explained in the Health feature, "Mate Choice and Smelly T-Shirts."

Societies and Fitness

The principles of evolutionary biology can be applied to the study of social behavior in animals. Sociobiologists hypothesize that societies form when living in a society has a greater reproductive benefit than reproductive cost. A **cost-benefit analysis** can help determine if this hypothesis is supported. Group living does have its benefits. It can help an animal avoid predators, rear offspring, and find food. A group of impalas is more likely to hear an approaching predator than a solitary one. Many fish moving rapidly in different directions might distract a would-be predator. Weaver birds form giant colonies that help protect them from predators, but the birds may also share information about food sources. Primate members of the same baboon troop signal to one another when they have found an especially bountiful fruit tree. Having a larger number of individuals looking for food increases the chances of finding it.

Video
Harris Hawks

Group living does have its disadvantages. When animals are crowded together into a small area, disputes can arise over access to the best feeding places and sleeping sites. Dominance hierarchies are one way to divide resources, but this puts subordinates at a disadvantage. Among red deer, sons increase the potential for the harem master to have a greater number of grandchildren. However, sons, being larger than daughters, need to be nursed more frequently and for a longer period of time. Subordinate females do not have access to enough food resources to adequately nurse sons and, therefore, they tend to rear daughters, not sons. Still, like the subordinate males in a baboon troop, subordinate females in a red deer harem may be better off in terms of fitness if they stay with a group instead of trying to survive on their own.

Living in close quarters exposes individuals to illness and parasites that can easily pass from one animal to another. Social behavior helps to offset some of the proximity disadvantages. For example, baboons and other types of social primates invest much time in grooming one another, and this most likely helps them remain healthy. Humans use extensive medical care to help offset the health problems that arise from living in the densely populated cities around the world.

SCIENCE IN YOUR LIFE ▶ HEALTH

Mate Choice and Smelly T-Shirts

Mate choice has been studied extensively in humans, largely because of our curiosity about how evolutionary biology and behavior in other animals may give us insight into human behavior. And, of course, mate choice is one of the more fascinating human behaviors. In one unusual and recent study, scientists had several men wear T-shirts for a few nights while they slept, thereby infusing the T-shirts with their unique smell. Then they asked women to rate the attractiveness of the male without seeing the person—that is, based only on the smell of the T-shirt. It turns out that women tended to choose men whose major histocompatibility complex (MHC) alleles were different from their own. Recall from Chapter 13 that the MHC is associated with recognition of foreign antigens (e.g., viruses or bacteria) and enhances the immune system's ability to fight off infections. In choosing men with different (complementary) MHC alleles, females are maximizing the chance that their offspring have highly heterozygous MHC and potentially a more robust immune system. A more heterozygous MHC is associated with a better innate ability to recognize a more diverse group of antigens. In fact, females in this study said unattractive shirt smells reminded them of their fathers, who, of course, have very similar MHC alleles. Where did scientists get the idea for such a strange experiment? It has already been well documented that mice preferentially choose mates with complementary MHC alleles based on smell.

Of course, while evolutionary and behavioral biology can provide insights into human mate choice, the process is clearly complex. Although males prefer young faithful females with physical attributes that indicate fertility, they also look for symmetry, intelligence, and a sense of humor. And while females prefer males with financial resources, they also list such qualities as dependability and emotional stability (indicators of good parenting), physical attractiveness (indicated by symmetry and testosterone markers, such as above-average strength and height, possibly to defend resources), sense of humor, and intelligence. The way a person smells also could play a factor, as indicated by the T-shirt study. Then there are unknowns—the context under which people meet, their professions, their likes/dislikes, etc.

Questions to Consider

1. Scientists have found a biological basis for infanticide. In lions, for example, males who kill the offspring of other males when they take over a pride then induce females to come into estrus, so they can mate with them. This increases male fitness. Should a biological basis for infanticide be used as a defense in human court cases where, say, a stepfather kills a stepchild?
2. As discussed in section 33.4, some studies suggest that men prefer women with a waist-to-hip ratio of 0.7. Women of different weights can have this ratio because it is based on bone structure, rather than weight. Nonetheless, women may misinterpret this information, which can contribute to the epidemic of eating disorders, such as bulimia and anorexia. How can we better educate people about the scientific information on mate choice and the dangers of eating disorders?
3. Studies of human mate choice are generally complex, and often controversial. Should federal agencies fund these studies? Why or why not?

connect BIOLOGY Explore the concepts through a variety of multimedia assets, question types, and data interpretation.
www.mcgrawhillconnect.com

Sociobiology and Human Culture

Humans today rely on living in organized societies. Clearly, the benefits of social living must outweigh the costs because we have organized governments, with laws that tend to increase the potential for human survival. The **culture** of a human society involves a wide spectrum of customs, ranging from how to dress to forms of entertainment, marriage rituals, and types of food. Language and the use of tools are essential to human culture. Language is used to socialize children, teach them how to use tools, educate them, and train them in skills that will increase their chances of success. Technological skills, such as knowing how to use a computer or how to fix plumbing, are taught from an early age. Certainly cultural evolution has surpassed biological evolution in the past few centuries. For example, medical advances, which are passed on through language, have increased the human life expectancy from 40 to 45 years a century ago to nearly 80 years today.

Why did human societies originate? Perhaps the earliest organized societies were composed of "hunter-gatherers." Which of our ancestors first became hunters is debated among paleontologists (biologists who study fossils) and sociologists. Nonetheless, scientists speculate that a predatory lifestyle may have encouraged the evolution of intelligence and the development of language. That is, cooperation, communication, and tools (such as weapons) are necessary for hunting large animals. In hunter-gatherer societies, men tended to hunt, while women utilized wild and cultivated plants as a source of food. Sometimes, when animal food was scarce, hunter-gatherer societies relied on plant food cultivated by women. Cooperation increased their chances for survival.

Altruism Versus Self-Interest

Altruism can be generally described as a self-sacrificing behavior for the good of another member of the society. In an evolutionary sense, altruism may compromise the fitness of the altruist, while benefiting the fitness of the recipient. Are animals truly altruists, given that natural selection should eventually rid populations of altruists due to their decreased reproductive success? In humans, we can clearly think of examples—a volunteer firefighter dying while trying to save a stranger, a soldier losing his life for his country, a woman diving in front of a car to save an elderly person from being hit. But what about other animals?

Figure 33.14 Altruism and army ants. A queen army ant has a large abdomen for egg production and is cared for by small ants, called nurses. The idea of inclusive fitness suggests that relatives, in addition to offspring, increase an individual's reproductive success. Therefore, sterile nurses are being altruistic when they help the queen produce offspring to whom they are closely related.

In general, altruistic behavior in animals is explained by the concept of **kin selection.** Because close relatives share many of your genes, it may make sense to self-sacrifice to save them. For example, your siblings share 50% of your genes, so sacrificing your life for two brothers or sisters makes sense evolutionarily. **Inclusive fitness** refers to an individual's personal reproductive success, as well as that of his or her relatives, and thus to an individual's total genetic contribution to the next generation. The concepts of kin selection and inclusive fitness are well supported by research in complex animal societies. For example, in a colony of army ants (Fig. 33.14), the queen is inseminated only during her nuptial flight. Thereafter, she spends the rest of her life reproducing constantly, laying up to 30,000 eggs per day! The eggs hatch into three different sizes of sterile female workers whose jobs contribute to the benefit of the entire colony. The smallest workers (3 mm), called nurses, take care of the queen and the larvae by feeding and cleaning them. The intermediate-sized workers (3–12 mm), which constitute most of the population, go out on raids to collect food. The soldiers (14 mm), which have relatively huge heads and powerful jaws, surround raiding parties to protect them and the colony from attack by intruders. The worker ants engage in this apparently altruistic behavior because it has a reproductive advantage.

The queen ant is diploid (2n), but her mate is haploid. Thus, assuming the queen has had only one mate, her sister workers are more closely related to each other than normal siblings (75% versus 50%). Therefore, a worker can achieve higher inclusive fitness by helping her mother (the queen) produce additional sisters than by producing her own offspring, which would only share 50% of her genes. Thus, this behavior is not really altruistic. Rather, it is adaptive because it is likely due to the increase in inclusive fitness of the helpers relative to those that hypothetically might defect and breed on their own. Similar social patterns are observed in certain species of bees and wasps.

Reciprocal Altruism

In some bird species, offspring from a previous clutch of eggs may stay at the nest to help parents rear the next batch of offspring. In a study of Florida scrub jays, the number of fledglings produced by an adult pair doubled when they had helpers. Mammalian offspring are also observed to help their parents (Fig. 33.15). Among jackals in Africa, solitary pairs managed to rear an average of 1.4 pups, whereas pairs with helpers reared 3.6 pups. What are the benefits of staying behind to help? First, a helper is contributing to the survival of its own kin. Therefore, the helper actually gains a fitness benefit. Second, a helper is more likely than a nonhelper to inherit a parental territory—including other helpers. Helping, then, involves making a minimal, short-term reproductive sacrifice in order to maximize future reproductive potential. Therefore, helpers at the nest are also practicing a form of reciprocal altruism. Reciprocal altruism also occurs in animals that are not necessarily closely related. In this event, an animal helps or cooperates with another animal with no immediate benefit. However, the animal that was helped will repay the debt at some later time. Reciprocal altruism usually occurs in groups of animals that are mutually dependent. Cheaters

Figure 33.15 Inclusive fitness. A meerkat is acting as a babysitter for its young sisters and brothers while their mother is away. Researchers point out that the helpful behavior of the older meerkat can lead to increased inclusive fitness.

in reciprocal altruism are recognized and not reciprocated in future events. Reciprocal altruism occurs in vampire bats that live in the tropics. Bats returning to the roost after a feeding activity share their blood meal with other bats in the roost. If a bat fails to share blood with one that had previously shared blood with it, the cheater bat will be excluded from future blood sharing.

Check Your Progress 33.4

1. Explain how territoriality is related to foraging for food.
2. Compare and contrast reproductive strategies and forms of sexual selection.
3. Describe the advantages and disadvantages of social living.

Case Study Conclusion

Female African lions form a society, called a pride, which provides benefits in foraging for food and protection of the young. The members of the pride are all closely related females (mothers, daughters, sisters, and aunts), which promotes altruism. However, this is not without a cost. While female cubs tend to stay in the pride their entire life, the young males leave and as they mature they will form small groups of genetically related individuals. Upon reaching maturity, they will establish a territory of their own that will coincide with that of a pride or prides of female lions. The males will often kill cubs that are not of their genetic line in order to get the females to enter estrus. They will then mate with the females in an attempt to pass on their genetics. While group life does have some disadvantages, the benefits must outweigh them in order for it to persist.

MEDIA STUDY TOOLS

www.mhhe.com/maderinquiry14

Enhance your study of this chapter with study tools and practice tests. Also ask your instructor about the resources available through ConnectPlus, including LearnSmart, the media-rich eBook, interactive learning tools, and animations.

Mc Graw Hill connect plus+ BIOLOGY

SUMMARIZE

33.1 Nature Versus Nurture: Genetic Influences

- Investigators have long been interested in the degree to which nature (genetics) or nurture (environment) influences **behavior.**
- Hybrid studies with lovebirds and human twin studies are consistent with the hypothesis that behavior has a genetic basis. Garter snake experiments indicate that the nervous system controls behavior. Egg-laying behavior in *Aplysia* has endocrine system involvement.

33.2 Nature Versus Nurture: Environmental Influences

- Even some behaviors formerly thought to be **fixed action patterns (FAPs)** can sometimes be modified by **learning.** A red spot on the bill of laughing gulls initiates chick begging behavior. However, with experience, chick begging accuracy improves.
- Other studies also support the involvement of learning in behaviors. **Imprinting** during a sensitive period causes birds to follow the first moving object they see.
- Song learning in birds involves various elements—including learning during the sensitive period, as well as the effect of social interactions outside the period.
- **Associative learning** includes **classical conditioning** and **operant conditioning.** In classical conditioning, the pairing of two different types of stimuli causes an animal to form an association between them (e.g., dogs salivating at the sound of a bell). In operant conditioning, animals learn behaviors because they are rewarded when they perform them.

33.3 Animal Communication

- **Communication** is an action by a sender that affects the behavior of a receiver.
- **Pheromones** are chemical signals passed between members of the same species.
- Auditory communication includes language, which occurs in humans and other animals.
- Visual communication allows signaling without auditory or chemical messages.
- Tactile communication is especially associated with social behavior.

33.4 Behaviors That Affect Fitness

- Traits that promote reproductive success are expected to have fitness advantages that outweigh their disadvantages.
- Some animals are territorial and defend a territory that may have high-quality food and/or nesting sites.
- **Sexual selection** is a form of natural selection that selects for traits that increase an animal's fitness.
- Males, who produce many sperm, are expected to compete to inseminate as many females as possible.
- Females, who produce few eggs, are expected to be selective about their mates. Females may choose mates based on increased survival of offspring or on traits that make their sons more attractive to other females.
- **Culture** in human society has enabled us to dramatically increase the success of our species.

- It is likely that mating behavior and mate choice in humans have also been shaped by evolutionary forces.
- Living in a social group can have its advantages (e.g., avoiding predators, raising young, and finding food). It also has disadvantages (e.g., competition between members, spread of illness and parasites, and reduced reproductive potential).
- When animals live in groups, the fitness (number of fertile offspring produced) benefits must outweigh the costs, or social behavior would not exist.
- Sometimes animals exhibit **altruism** in order to increase their reproductive success. However, it is necessary to consider **inclusive fitness,** which includes personal reproductive success, as well as the reproductive success of relatives. **Kin selection** is important in social species because of increasing the number of individuals who survive that share your genes. For example, social insects help their mother reproduce, but this behavior seems reasonable when we consider that siblings share 75% of their genes. Parental helpers in mammals and birds often inherit the parent's territory.

ASSESS

Testing Yourself

Choose the best answer for each question.

1. In lovebirds, if carrying strips in the bill is controlled by a single dominant gene, then hybrids between peach-faced lovebirds and Fischer lovebirds would
   a. carry strips in their bills.
   b. carry strips in their rump feathers.
   c. not carry strips.
   d. carry strips in both their bills and rump feathers.
   e. have decreased fitness.
2. Which of the following is not an example of a genetically based behavior?
   a. Inland garter snakes do not eat slugs, while coastal populations do.
   b. One species of lovebird carries nesting strips one at a time, while another carries several.
   c. One species of warbler migrates, while another one does not.
   d. Snails lay eggs in response to egg-laying hormone.
   e. Wild foxes raised in captivity are not capable of hunting for food.
3. How would the following graph differ if begging behavior in laughing gulls was a fixed action pattern?

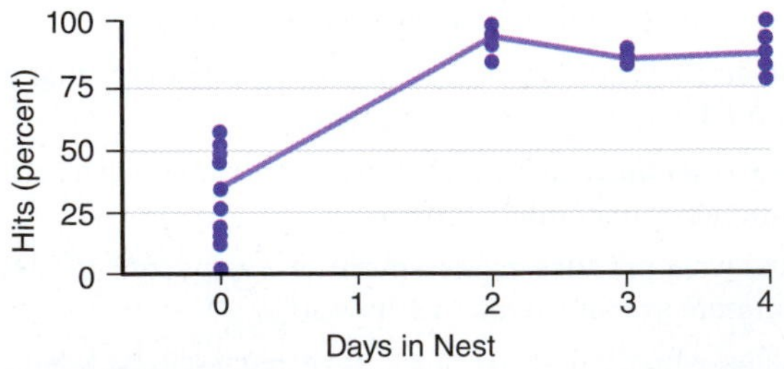
100
75
50
25
0
Hits (percent)
0
1
2
3
4
Days in Nest

   a. It would be a diagonal line with an upward incline.
   b. It would be a diagonal line with a downward incline.
   c. It would be a horizontal line.
   d. It would be a vertical line.
   e. None of these are correct.
4. Using treats to train a dog to do a trick is an example of
   a. imprinting.
   b. tutoring.
   c. vocalization.
   d. operant conditioning.
5. In white-crowned sparrows, social experience exhibits a very strong influence over the development of singing patterns. What observation led to this conclusion?
   a. Birds learned to sing only when they were trained by other birds.
   b. The window in which birds learn from other birds is wider than that in which birds learn from tape recordings.
   c. Birds could learn different dialects only from other birds.
   d. Birds that learned to sing from a tape recorder could change their song when they listened to another bird.
6. Which reaction best provides physiological evidence that animals have emotions?
   a. Adrenaline levels increase when a deer senses danger.
   b. Dopamine levels in the brains of rats increase when they play.
   c. Chimpanzees experience decreases in acetylcholine levels when they are content.
   d. The blood pressure of wolves drops when they return to the pack.
7. Which of the following best describes classical conditioning?
   a. the gradual strengthening of stimulus-response connections that are seemingly unrelated
   b. a type of associative learning in which there is no contingency between response and reinforcer
   c. the learning behavior in which an organism follows the first moving object it encounters
   d. the learning behavior in which an organism exhibits a fixed action pattern from the time of birth
8. Bees that do a waggle dance are teaching other bees
   a. how to dance.
   b. where to find food.
   c. how to find and use the sun for navigation.
   d. how to use auditory communication.
9. The benefits of imprinting generally outweigh the costs because
   a. an animal that has been imprinted on the wrong object can be reimprinted on the mother.
   b. imprinting behavior never lasts more than a few months.
   c. animals in the wild rarely imprint on anything other than their mother.
   d. animals that imprint on the wrong object generally die before they pass on their genes.
10. Which of the following are costs that a dominant male baboon must pay in order to gain a reproductive benefit?
   a. He requires more food and must travel longer distances.
   b. He requires more food and must care for his young.
   c. He is more prone to injury and requires more food.
   d. He is more prone to injury and must care for his young.
   e. He must care for his young and travel longer distances.

11. A red deer harem master typically dies earlier than other males because he is
   a. likely to get expelled from the herd and cannot survive alone.
   b. more prone to disease because he interacts with so many animals.
   c. likely to starve to death.
   d. apt to place himself between a predator and the herd to protect the herd.

For problems 12–15, match the type of communication in the key with its description. Answers may be used more than once.

Key:

   a. chemical communication
   b. auditory communication
   c. visual communication
   d. tactile communication

12. Aphids (insects) release an alarm pheromone when they sense they are in danger.
13. Male peacocks exhibit an elaborate display of feathers to attract females.
14. Ground squirrels give an alarm call to warn others of the approach of a predator.
15. Male silk moths are attracted to females by a sex attractant released by the female moth.

ENGAGE

Virtual Lab
Mealworm Behavior

The virtual lab "Mealworm Behavior" provides an interactive means of studying the factors that influence behavior in animals.

Thinking Critically

1. You are studying two populations of rats—one in New York and the other in Florida. Adult New York rats seem to prefer Swiss cheese, whereas Florida rats seem to prefer cheddar. Design an experiment to determine the extent to which this behavior may be genetically controlled versus environmentally influenced.
2. Until about 15 years ago, it was long thought that most bird species were monogamous—that is, at least within a breeding season, males and females paired and mated only with each other. However, with advances in DNA technology, researchers began to find out that offspring in a nest were often sired by males in neighboring territories. When females mate with neighboring males, it is called "extra pair copulation" or EPC. Why do you think EPC behavior might have evolved in many bird species?
3. If altruistic behavior has usually evolved to improve the altruist's inclusive fitness by helping relatives, then why do you think humans perform altruistic acts for completely unrelated individuals (e.g., jumping into a pool to save a drowning person)?

34 Population and Community Ecology

CHAPTER OUTLINE

BEFORE YOU BEGIN

Before beginning this chapter, take a few moments to review the following discussions:

Section 6.1 What are the forms of energy?

Section 8.1 How does photosynthesis capture solar energy?

Section 33.4 How does a species' behavior determine the type of relationship it has with other species?

CASE STUDY Exotic Species

In the early 1970s, several species of Asian carp were imported into Arkansas for the biocontrol of algal blooms in aquaculture facilities. Shortly thereafter, they escaped into the middle and lower Mississippi drainage. Over time, in some areas they have become the most abundant species and have now spread throughout the majority of the Mississippi River system.

Asian carp mainly eat the microscopic algae and zooplankton in freshwater ecosystems. They can achieve weights up to a hundred pounds, grow to a length of more than 4 ft, and live up to 30 years. Because of their voracious appetite they have the potential to reduce the population of native fish species such as gizzard shad, bigmouth buffalo, and native mussels due to competition for the same food sources. The main fear is that the Asian carp will put extreme pressure on the zooplankton populations, which can lead to a dense planktonic algae bloom. Ultimately they can produce a significant disruption of the freshwater ecosystems that they invade.

These fish also pose an economic threat by fouling the nets of commercial fisherman. Another major concern is the impact these fish may have upon the Great Lakes $7 billion fishing industry. Our knowledge of population growth and regulation has led to controls on fishing, including size limits and catch limits, to try and sustain the world's commercially valuable fish population. This knowledge can also be applied to management of species that can be considered harmful to a specific ecosystem.

As you read through the chapter, think about the following questions:

1. Does the old saying "90% of the fish are found in 10% of the lake" have any truth to it?
2. Do population growth models apply to humans the same way they apply to other species?
3. Do more-developed or less-developed countries have a greater impact upon the Earth's resources?

34.1 The Scope of Ecology

Learning Outcomes

Upon completion of this section, you should be able to

1. Identify what aspects of biology the study of ecology encompasses.
2. Recognize the different levels of ecological study.

Ecology is the study of the interactions of organisms with each other and with their physical environment. Ecology, like so many biological disciplines, is wide-ranging and involves several levels of study (Fig. 34.1).

At the most basic level, ecologists study how organisms are adapted to their environment. For example, they might study why a particular species of fish in a coral reef lives only within a narrow temperature range in warm tropical waters. Most organisms do not exist singly. Rather, they are part of a **population,** which is defined as all the organisms of the same species interacting with the environment in a particular area. At the population level of study, ecologists describe the size of populations over time. For example, an ecologist might study the relative sizes of parrotfishes' populations within a coral reef over time. A **community** consists of all the various populations at a particular locale. A coral reef community contains numerous populations of fish species, crustaceans, corals, and so forth. At the community level, ecologists study how various extrinsic factors (e.g., weather) and intrinsic factors (e.g., species' competition for resources) affect the size of these populations. An **ecosystem** encompasses a community of populations, as well as the nonliving environment. For example, energy flow and chemical cycling in a coral reef can affect the success of the organisms that inhabit it. Finally, the **biosphere** is that portion of the entire Earth's surface—air, water, land—where living organisms exist. Knowing the composition and diversity of an ecosystem, such as a coral reef, is important to the dynamics of the biosphere. Table 34.1 defines and summarizes the levels commonly studied in ecology.

Video Coral Reef Ecosystem

TABLE 34.1 Ecological Terms

Term	Definition
Ecology	Study of the interactions of organisms with each other and with the physical environment
Population	All the members of the same species that inhabit a particular area
Community	All the populations found in a particular area
Ecosystem	A community and its physical environment, including both nonliving (abiotic) and living (biotic) components
Biosphere	All of the ecosystems on Earth, including their biotic communities

Figure 34.1 Levels of ecological organization. The study of ecology encompasses levels of organization that proceed from the individual organism to the population, to the community, and to an ecosystem.

Individual
Population
Community
Ecosystem

Coral reef community

Modern ecology is not just descriptive; it primarily focuses on developing testable hypotheses. A central goal of modern ecology is to develop models that explain and predict the distribution and abundance of populations and species. Ultimately, ecology considers not just one particular area, but species' distributions throughout the biosphere. In this chapter, we particularly concentrate on the patterns of population growth, and how population growth is regulated. Then, at the community level, we examine the interactions between populations and how they change through time during the process of ecological succession.

Check Your Progress 34.1

1. Distinguish between a population and a community.
2. Describe the central goal of modern ecology.

34.2 Patterns of Population Growth

Learning Outcomes

Upon completion of this section, you should be able to

1. Recognize how the per capita rate of increase in a population is calculated.
2. Compare exponential and logistic population growth curves.
3. Recognize how the proportion of individuals at varying reproductive stages determines a population's age distribution.

Each population has a particular pattern of growth. The population size can change according to a *per capita rate of increase* or growth rate. Suppose, for example, a human population presently has a size of 1,000 individuals, the birthrate is 30 per year, and the death rate is 10 per year. The growth rate per year will be as follows:

$$30 - 10 / 1{,}000 = 0.02 = 2.0\% \text{ per year}$$

Note that this per capita rate of increase disregards both immigration and emigration, which for this example can be assumed to be equal and thus to cancel each other out. The highest possible per capita rate of increase for a population is called its **biotic potential** (Fig. 34.2). Whether the biotic potential is high or low depends on a variety of factors such as (1) reproductive potential of the current population, (2) availability of food, (3) presence or absence of disease, and (4) presence or absence of predators.

Suppose we are studying the growth of an insect population that is capable of infesting and taking over an area. Under these circumstances, **exponential growth** is expected. An exponential pattern of population growth results in a J-shaped curve (Fig. 34.3*a*). This pattern of population growth can be likened to compound interest at the bank: the more your money increases, the more interest you earn. If the insect population has 2,000 individuals and the per capita rate of increase is 20% per month, there will be 2,400 insects after one month, 2,880 after two months, 3,456 after three months, and so forth.

Video Exponential Growth

Notice that a J-shaped curve has these phases:

Lag phase. Growth is slow because the population is small.

Exponential growth phase. Growth is accelerating, and the population is exhibiting its biotic potential.

Usually, exponential growth cannot continue for long because of environmental resistance. **Environmental resistance** encompasses all those environmental conditions—such as limited food supply, accumulation of waste products, increased competition, or predation—that prevent populations from achieving their biotic potential. The amount of environmental resistance will increase as the population grows larger. This eventually causes the population growth to level off, resulting in an S-shaped pattern called **logistic growth** (Fig. 34.3*b*).

Notice that an S-shaped curve has these phases:

Lag phase. Growth is slow because the population is small.

Exponential growth phase. Growth is accelerating, due to biotic potential.

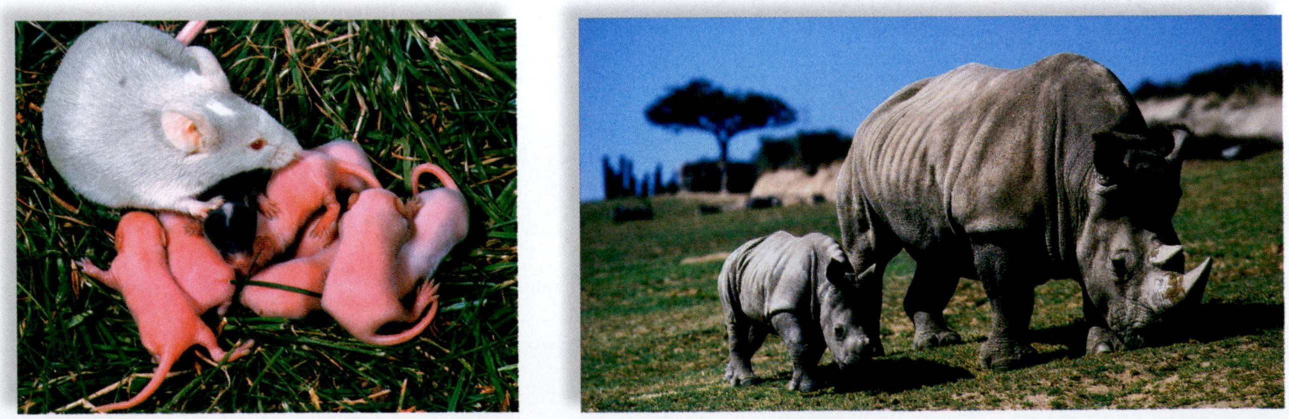

a. b.

Figure 34.2 Biotic potential. A population's maximum growth rate under ideal conditions—that is, its biotic potential—is greatly influenced by the number of offspring produced in each reproductive event. **a.** Mice, which produce many offspring that quickly mature to produce more offspring, have a much higher biotic potential than (**b**) the rhinoceros, which produces only one or two offspring during infrequent reproductive events.

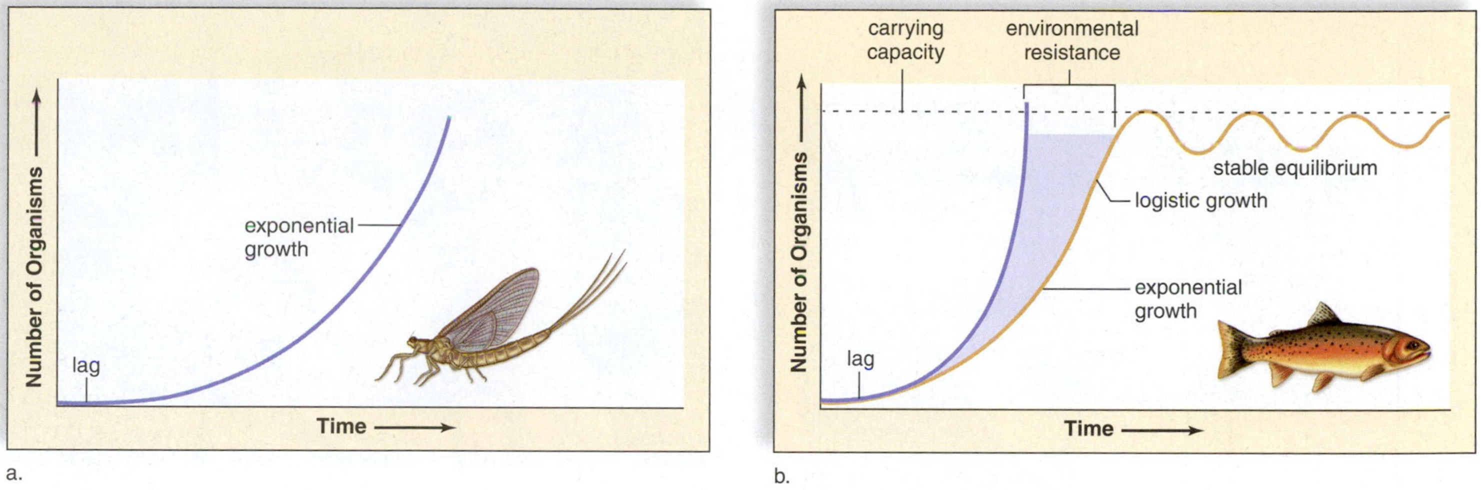
Number of Organisms
exponential growth
lag
Time
a.
carrying capacity
environmental resistance
stable equilibrium
logistic growth
exponential growth
Number of Organisms
lag
Time
b.

Figure 34.3 Patterns of population growth. **a.** Exponential growth results in a J-shaped growth curve because the growth rate is positive, and increasing, as in insects. **b.** Logistic growth results in an S-shaped growth curve because environmental resistance causes the population size to level off and be in a steady state at the carrying capacity of the environment, as in fish.

Logistic growth phase. The rate of population growth slows down.
Stable equilibrium phase. Little if any growth takes place because births and deaths are about equal.

The stable equilibrium phase is said to occur at the **carrying capacity** of the environment. The carrying capacity is the number of individuals of a species that a particular environment can support.

Our knowledge of logistic growth has practical implications. The model predicts that exponential growth occurs only when population size is much lower than the carrying capacity. So, for example, if humans are using a fish population as a continuous food source, it would be best to maintain that population size in the exponential phase of growth when the birthrate is the highest. If we overharvest, the fish population will sink into the lag phase, and it will be years before exponential growth recurs. On the other hand, if we are trying to limit the growth of a pest, it is best to reduce the carrying capacity. Simply reducing the population size only encourages exponential growth to begin again. Farmers can reduce the carrying capacity for a pest by alternating rows of different crops within their fields instead of growing one type of crop throughout the entire field.

Survivorship

Population growth patterns assume that populations are made up of identically aged individuals. In real populations, however, individuals are in different stages of their life spans. Let us consider how many members of an original group of individuals born at the same time, called a **cohort,** are still alive after certain intervals of time. The result is a survivorship curve.

For the sake of discussion, three types of idealized survivorship curves are recognized (Fig. 34.4). The type I curve is characteristic of a population in which most individuals survive well past the midpoint, and death comes near the end of the maximum life span. On the other hand, the type III curve is typical for a population in which most individuals die very young. In the type II curve, survivorship decreases at a constant rate throughout the life span. Some species, however, do not fit any of these curves exactly.

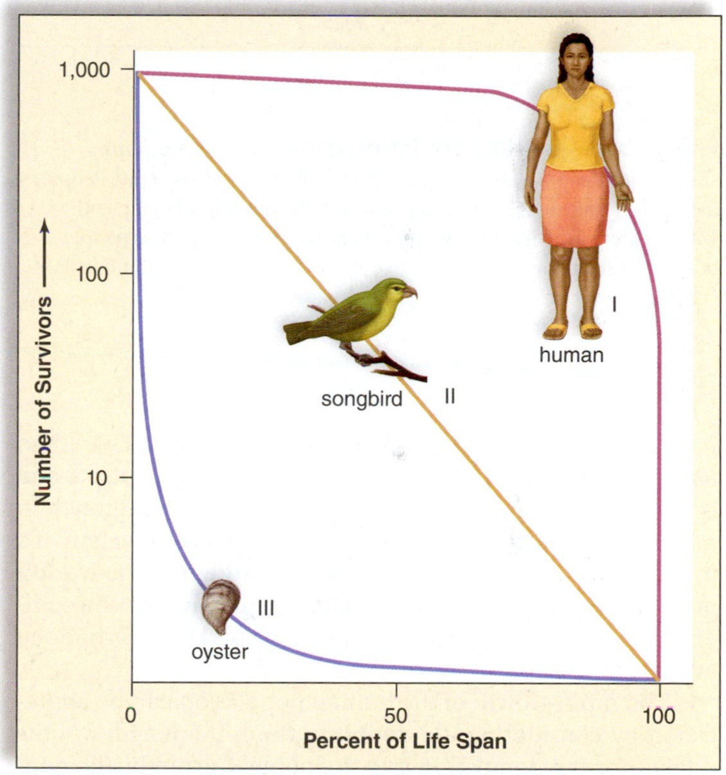
1,000
100
10
Number of Survivors
human
I
songbird
II
oyster
III
0
50
100
Percent of Life Span

Figure 34.4 Survivorship curves. Humans have a type I survivorship curve: the individual usually lives a normal life span, and then death is increasingly expected. Songbirds have a type II curve: the chances of surviving are the same for any particular age. Oysters have a type III curve: most deaths occur during the free-swimming larva stage, but oysters that survive to adulthood usually live a normal life span.

Much can be learned about the life history of a species by studying its survivorship curve. Only a few members of a population with a type III survivorship curve will actually contribute offspring to the next generation. Because death comes early for most members, only a few live long enough to reproduce.

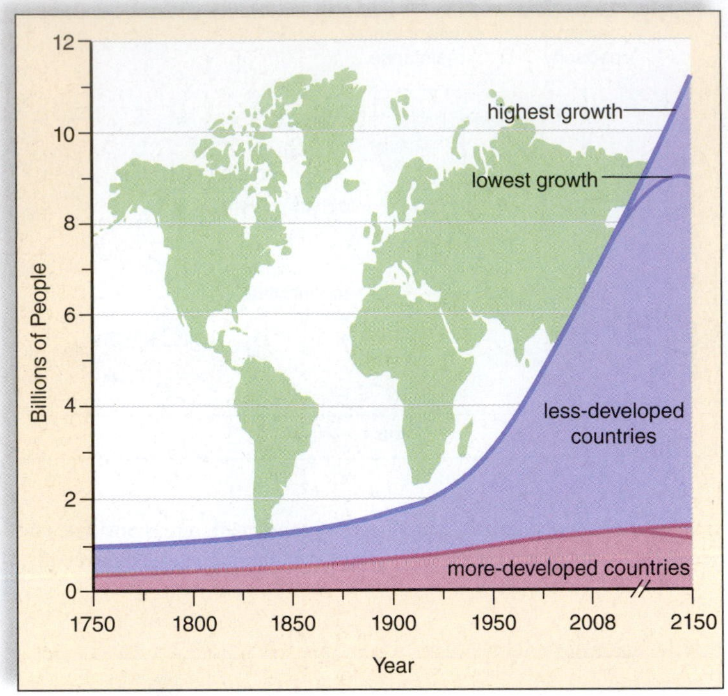

highest growth
lowest growth
less-developed countries
more-developed countries
Billions of People
0
2
4
6
8
10
12
1750
1800
1850
1900
1950
2008
2150
Year

a.

b.

c.

Figure 34.5 World population growth. **a.** The graph shows world population size in the past, with estimates to 2150. People in (**b**) more-developed countries have a high standard of living and will contribute the least to world population growth, whereas people in (**c**) less-developed countries have a low standard of living and will contribute the most to world population growth.

Human Population Growth

Figure 34.5*a* illustrates human population growth. Growth in *less-developed countries* is still in the exponential phase and is not expected to level off until about 2040, while growth in *more-developed countries* has leveled off. The equivalent of a medium-sized city (roughly 220,000) is added to the world's population every day, and 79 million (equivalent to the combined populations of Argentina, Ecuador, and Peru) are added every year.

The rapid growth of the human population can be appreciated by considering the **doubling time,** the length of time it takes for the population size to double. Currently, the doubling time for the human population is estimated to be between 50 and 60 years. Such an increase in population size will put extreme demands on our ability to produce and distribute resources. In the relatively near future, the world will need to double the amount of food, jobs, water, energy, and so on just to maintain the present standard of living.

Many people are gravely concerned that the amount of time it takes to add each additional billion persons to the world population has become shorter. The first billion was reached in 1800; the second billion in 1930; the third billion in 1960; and today there are over 7.0 billion. By contrast, **zero population growth,** in which the birthrate equals the death rate and population size remains steady, can be achieved only if the per capita rate of increase declines. The world's population may level off at 10, 12, or 14.2 billion, depending on the speed with which the per capita rate of increase declines.

More-Developed Versus Less-Developed Countries

The **more-developed countries** (**MDCs**), typified by countries in North America and Europe, are those in which population growth is low and people enjoy a good standard of living (Fig. 34.5*b*). The **less-developed countries** (**LDCs**), such as countries in Latin America, Africa, and Asia, are those in which population growth is expanding rapidly and the majority of people live in poverty (Fig. 34.5*c*).

The MDCs doubled their populations between 1850 and 1950 due to a decline in the death rate, the development of modern medicine, and improved socioeconomic conditions. However, the transition from an agricultural society to an urban society reduced the incentive for people to have large families (e.g., more people to help on the farm) because maintaining large families in urban areas is more costly than in rural areas, causing birthrates to decline. These factors, combined with the decreased death rate, caused populations in MDCs to experience only modest growth between 1950 and 1975. This sequence of events (i.e., decreased death rate followed by decreased birthrate) is termed a **demographic transition.**

Yearly growth of the MDCs as a whole has now stabilized at about 0.1%. The populations of several of the MDCs, including Germany, Greece, Italy, Hungary, and Japan, are not growing or are actually decreasing in size. In contrast, there is no leveling off in U.S. population growth due to the number of individuals that immigrate to the United States each year. In addition, an unusually large number of babies were born between 1947 and 1964 (called a baby boom). This means a large number of women are still of reproductive age.

Although death rates began to decline steeply in the LDCs following World War II with the importation of modern medicine from the MDCs, birthrates remained high. The yearly growth of the LDCs peaked at 2.5% between 1960 and 1965. Since that time, a demographic transition has occurred: the decline in the death rates slowed, and birthrates fell. Yearly growth now averages 1.9%. Due to exponential growth, the population of LDCs may explode from 4.9 billion today to 11 billion by 2100. Most of this growth will occur in Africa, Asia, and Latin America. Suggestions for greatly reducing the expected increase include the following:

1. Establish and/or strengthen family planning programs. A decline in growth is seen in countries with good family planning programs supported by community leaders. Currently, 25% of women in sub-Saharan Africa say they would like to delay or stop childbearing, and yet they lack access to birth control. Similarly, 15% of women in Asia and Latin America have an unmet need for birth control.
2. Use social progress to reduce the desire for large families. Many couples in LDCs presently desire as many as four to six children. But providing education, raising the status of women, reducing child mortality, and improving economic stability are desirable social improvements that could help decrease population growth.
3. Delay the onset of childbearing. A delay in the onset of childbearing and wider spacing of births could cause a temporary decline in the birthrate and reduce the present reproductive rate.

Age Distributions

The **age-structure diagram** divides the population into three age groups: prereproductive, reproductive, and postreproductive (Fig. 34.6). The LDCs are experiencing a population momentum because they have more women entering the reproductive years than older women leaving them.

It is a misconception that if each couple has two children, referred to as **replacement reproduction,** that zero population growth will occur. The population will continue to grow as long as there are more young women entering their reproductive years than there are older women leaving them, which is common among most of the LDCs. Living many years after reproduction also contributes to an increase in the population. This type of population growth results in *unstable age structure* (Fig. 34.6*b*). In LDCs, the more quickly replacement reproduction is achieved, the sooner zero population growth will occur. In MDCs, on the other hand, it is more typical that the number of people in the prereproductive class roughly equals the number in the reproductive age class. This causes a *stable age structure,* and population numbers will remain about the same for the foreseeable future (Fig. 34.6*a*).

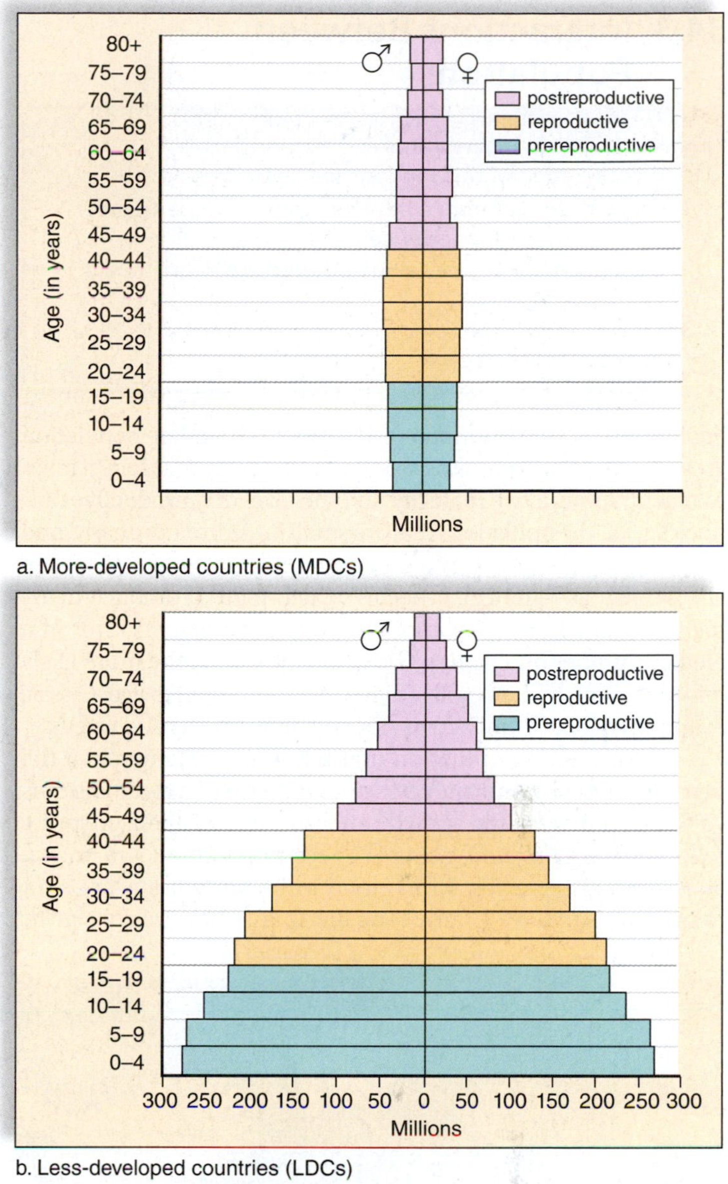

Age (in years)
80+
75–79
70–74
65–69
60–64
55–59
50–54
45–49
40–44
35–39
30–34
25–29
20–24
15–19
10–14
5–9
0–4
♂
♀
postreproductive
reproductive
prereproductive
Millions
a. More-developed countries (MDCs)
300 250 200 150 100 50 0 50 100 150 200 250 300
b. Less-developed countries (LDCs)

Figure 34.6 Age structure diagrams. The diagrams illustrate that (**a**) the MDCs are approaching stabilization, whereas (**b**) the LCDs will expand rapidly, resulting in unstable age structure.

Check Your Progress 34.2

1. Explain why populations don't grow to their biotic potential.
2. Describe the three types of idealized population survivorship curves.
3. Recognize how differences in age structure explain differences in population growth rates between LDCs and MDCs.

34.3 Interactions Between Populations

Learning Outcomes

Upon completion of this section, you should be able to

1. Identify the difference between density dependence and density independence.
2. Describe how competition and predation can impact population growth.
3. List the three types of symbiosis, and discuss the characteristics of each.

Ecologists wish to determine the factors that regulate population growth. Life history patterns are characterized by how long it takes to reach reproductive maturity and the level of reproductive output (Fig. 34.7). Populations that are small in size, mature early, and have a short life span exhibit an *opportunistic pattern*. They are also known as *r*-strategist species, which tend to produce many relatively small offspring and forego parental care in favor of a greater number of offspring. The more offspring, the more likely it is that some of them will survive to a reproductive age. Classic examples of opportunistic species are many insects and weeds.

In contrast, a population that remains pretty much at the carrying capacity and allocates energy to their growth and survival as well as to the growth and survival of their offspring follows an *equilibrium pattern*. These populations are known as *K*-strategist species, which tend to be fairly large, are slow to mature, and have a fairly long life span. They are specialists rather than generalists and tend to become extinct when their normal way of life is altered. The best examples of equilibrium species are found among birds and mammals. For example, the Florida panther, the largest mammal in the Florida Everglades, requires a very large range, and produces few offspring, which must be cared for. Currently, the Florida panther is unable to compensate for a reduction in its range due to human destruction of its habitat and is therefore on the verge of extinction.

Animation *r* and *K* Strategies

It is recognized that the environment contains both abiotic and biotic components. Abiotic factors, such as weather and natural disasters, are **density-independent factors,** meaning that the effects of the factor are the same for all sizes of populations. For example, fires don't necessarily kill a larger percentage of individuals as the population increases in size. On the other hand, biotic factors, such as competition, predation, and parasitism, are called **density-dependent factors.** The effects of a density-dependent factor depend on the size of the population. The denser a population is, the faster a disease might spread, for example. Populations that have the opportunistic life history pattern tend to be regulated by density-independent factors, and those that have the equilibrium life history pattern tend to be regulated by density-dependent factors.

Video Hungry Reindeer

Competition

Competition, a density-dependent factor, occurs when members of different species try to utilize a resource (such as light, space, or nutrients) that is in limited supply. Understanding

Opportunistic Species
(r-strategist)
• Small individuals
• Short life span
• Fast to mature
• Many offspring
• Little or no care of offspring
• Many offspring die before reproducing
• Early reproductive age
Equilibrium Species
(K-strategist)
• Large individuals
• Long life span
• Slow to mature
• Few and large offspring
• Much care of offspring
• Most young survive to reproductive age
• Adapted to stable environment

Figure 34.7 Life history patterns. Dandelions are an opportunistic species (*r*-strategist) with the characteristics noted, and bears are an equilibrium species (*K*-strategist) with the characteristics noted. Often the distinctions between these two possible life history patterns are not as clear-cut as they may seem.

the effects of competition requires a grasp of the concept of ecological niche. The **ecological niche** is the role a species plays in the community, including the **habitat** it requires and its interactions with other organisms. The niche is essentially a species' total way of life, and thus includes the resources needed to meet its energy, nutrient, survival, and reproductive demands.

Video Cichlid Specialization

According to the **competitive exclusion principle,** no two species can occupy the same ecological niche at the same time if resources are limited. This principle is exemplified by an experiment involving two species of paramecia. When grown together in one test tube with limited resources, only one species will survive (Fig. 34.8). In another laboratory experiment, two species of paramecia can occupy the same tube if one species feeds on bacteria at the bottom of the tube and the other feeds on bacteria suspended in solution. The division of feeding niches, called **resource partitioning,** decreases competition between the two species and allows occupancy of different niches and therefore survival.

Video Finches Adaptive Radiation

On the Scottish coast, a small barnacle (*Chthamalus stellatus*) lives on the high part of the intertidal zone, and a large barnacle (*Balanus balanoides*) lives on the lower part (Fig. 34.9). Free-swimming larvae of both species attach themselves to rocks at any point in the intertidal zone, where they develop into the sessile adult forms. In the lower zone, the large *Balanus* barnacles seem to either force the smaller *Chthamalus* individuals off the rocks or grow over them. Competition is therefore restricting the range of *Chthamalus* on the rocks. *Chthamalus* is more resistant to drying out than is *Balanus*. Therefore, it has an advantage that permits it to grow in the upper intertidal zone.

It should be noted, however, that the greatest competition for resources will always come from within your own species, an occurrence known as *intraspecific competition.* Two individuals of the same species will have nearly identical requirements for survival, thus producing competition when resources are limited.

Video Booby Chick Competition

Predation

Predation occurs when one organism, called the **predator,** feeds on another, called the **prey.** In the broadest sense, predators include not only animals such as lions that kill zebras, but also filter-feeding blue whales that strain krill from ocean waters, and even white-tailed deer that feed upon a farmer's cornfield.

Video Harris Hawks

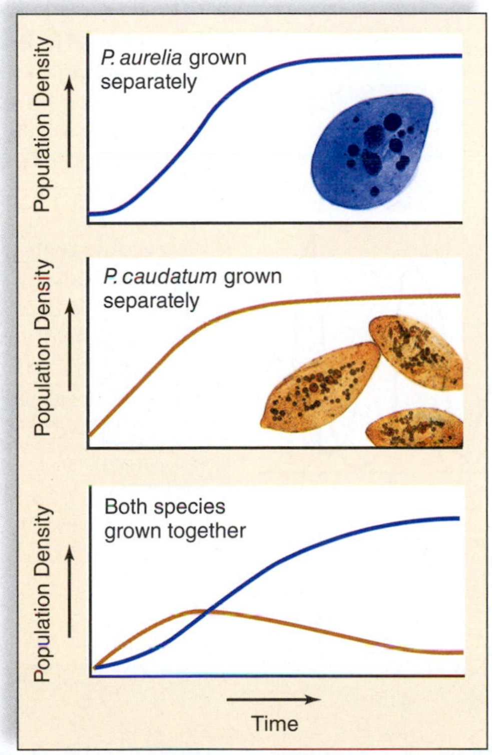
Population Density
P. aurelia grown separately
Population Density
P. caudatum grown separately
Population Density
Both species grown together
Time

Figure 34.8 Competition between two laboratory populations of paramecia. When grown alone in pure culture (top two graphs), *Paramecium aurelia* and *Paramecium caudatum* exhibit logistic growth. When the two species are grown together in mixed culture (bottom graph), *P. aurelia* is the better competitor, and *P. caudatum* dies out. This experiment illustrates the competitive exclusion principle.

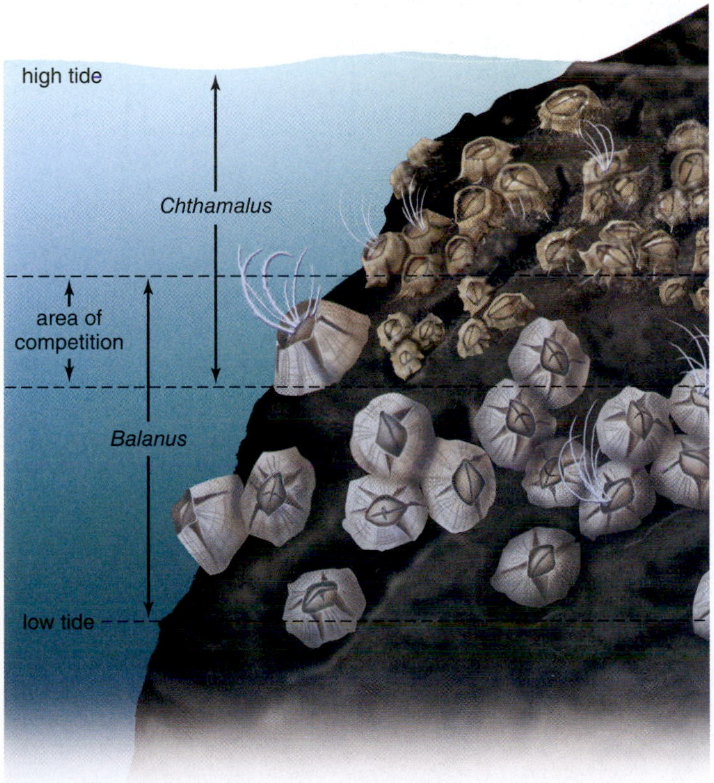
high tide
Chthamalus
area of competition
Balanus
low tide

Figure 34.9 Competition between two species of barnacles. Competition prevents two species of barnacles from occupying as much of the intertidal zone as possible. Both exist in the area of competition between *Chthamalus* and *Balanus.* Above this area, only *Chthamalus* survives, and below it only *Balanus* survives.

Predator-Prey Population Dynamics

Predators reduce the population density of prey, as shown by a laboratory study in which the protozoans *Paramecium caudatum* (prey) and *Didinium nasutum* (predator) were grown together in a culture medium. *Didinium* ate all the *Paramecium* and then died of starvation. In nature, we can find a similar example. When a gardener brought prickly-pear cactus to Australia from South America, the cactus spread out of control until millions of acres were covered with nothing but cacti. The cacti were brought under control when a moth from South America, whose caterpillar feeds only on the cactus, was introduced. Now, both cactus and moth are found at greatly reduced densities in Australia.

Mathematical formulas predict that predator and prey populations tend to cycle instead of maintaining a steady state. Cycling can occur when (1) the predator population overkills the prey, causing the predator population to decline in number, or (2) the prey population overshoots the carrying capacity and suffers a crash, causing the predator population to crash due to the lack of food. In either case, the result would be a series of peaks and valleys in population density of both species, with the predator population lagging slightly behind the prey.

A famous case of predator-prey cycles occurs between the snowshoe hare and the Canadian lynx, a type of small, predatory cat (Fig. 34.10). The snowshoe hare is a common herbivore in the coniferous forests of North America, where it feeds on terminal twigs of various shrubs and small trees. The Canadian lynx feeds on snowshoe hares but also on ruffed grouse and spruce grouse, two types of birds. Investigators at first assumed that the lynx had brought about the decline of the hare population. But others noted that the decline in snowshoe hare abundance was accompanied by low growth and reproductive rates that could be signs of a food shortage. It appears that both explanations apply to the data. In other words, both a predator-hare cycle and a hare-food cycle have combined to produce an overall effect, which is observed in Figure 34.10. Although not shown on the graph, densities of the grouse populations also cycle, perhaps because the lynx switches to this food source when the hare population declines. Predators and prey do not normally exist as simple, two-species systems, and therefore, abundance patterns should be viewed with the complete community in mind.

Antipredator Defenses

While predators have evolved strategies to secure the maximum amount of food with minimal expenditure of energy, prey organisms have evolved strategies to escape predation. **Coevolution** is present when two species adapt in response to selective pressure imposed by the other. This phenomenon applies to predation, as well as to symbiotic interactions.

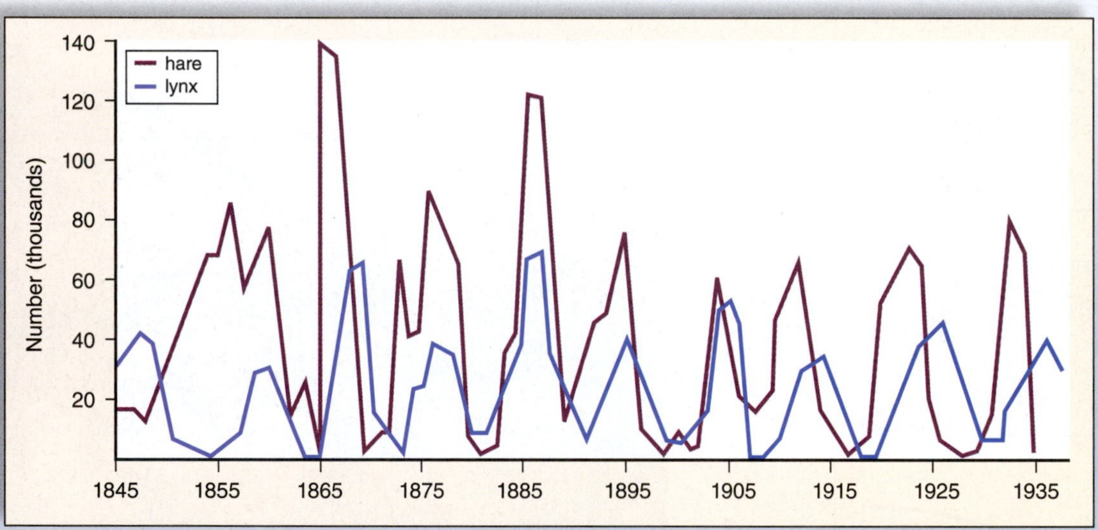

hare
lynx
Number (thousands)
140
120
100
80
60
40
20
1845
1855
1865
1875
1885
1895
1905
1915
1925
1935

snowshoe hare
lynx

Figure 34.10 Predator-prey cycling of a lynx and a snowshoe hare. The number of pelts received yearly by the Hudson Bay Company for almost 100 years shows a pattern of ten-year cycles in population densities. The snowshoe hare (prey) population reaches a peak before that of the lynx (predator) by a year or more.

Data from D. A. MacLulich, *Fluctuations in the Numbers of the Varying Hare (Lepus americanus)*, University of Toronto Press, Toronto, 1937; reprinted 1974, p. 543.

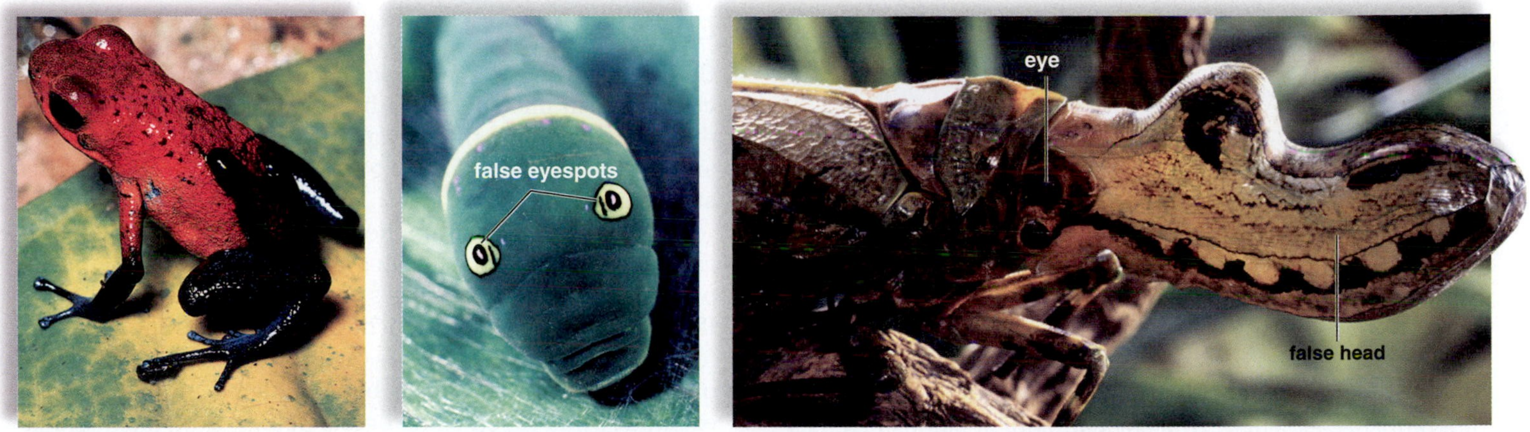
false eyespots
eye
false head

a. b. c.

Figure 34.11 Antipredator defenses. **a.** The skin secretions of poison-dart frogs are so poisonous that they were used by natives to make their arrows instant lethal weapons. The bright coloration of these frogs is called "warning coloration" because brightly colored species are generally toxic. **b.** The caterpillar of the eastern tiger swallowtail butterfly has false eyespots used to confuse a predator. **c.** The South American lantern fly has a large false head that resembles that of an alligator. This may frighten a predator into thinking it is facing a dangerous animal.

In plants, the sharp spines of the cactus, the pointed leaves of holly, and the tough, leathery leaves of the oak tree all discourage predation by insects. Plants even produce poisonous chemical compounds to deter predation. Animals have varied antipredator defenses such as poisonous secretions, concealment, fright, flocking together, and warning coloration (Fig. 34.11), as well as mimicry. A good example of concealment occurs with caddisfly larvae, which build protective cases out of sticks and sand, making themselves look like the stream bottom where they reside.

Video Flounder Camouflage

Video Leaf-Tailed Gecko

Mimicry **Mimicry** occurs when one species resembles another species that has evolved to defend against predators or resembles an object in the environment. Mimicry can help a predator capture food or a prey avoid capture. For example, angler fishes have lures that resemble worms for the purpose of bringing fish within reach. To avoid capture, some inchworms resemble twigs, and some caterpillars can transform themselves into shapes resembling snakes.

Batesian mimicry (named for Henry Bates, who discovered it) occurs when a prey species that is not harmful mimics another species that has a successful antipredator defense. Many examples of Batesian mimicry involve warning coloration. For example, stinging insects all tend to have black and yellow bands, like those of the wasp in Figure 34.12*a*. Once a predator has experienced the defense of the wasp, it remembers the coloration and tends to avoid animals that look similar, such as the nonstinging flower fly and longhorn beetle in Figure 34.12*b, c,* which can be considered Batesian mimics.

Another type of mimicry occurs when species that resemble each other all have successful defenses. For example, many coral snake species have brilliant red, black, and yellow body rings and are venomous. Mimics that share the same protective defense are called Müllerian mimics, named after Fritz Müller, who discovered the phenomenon. The bumblebee in Figure 34.12*d* is a Müllerian mimic of the wasp in Figure 34.12*a*—that is, both insects have the same appearance and mode of defense.

a. b. c. d.

Figure 34.12 Mimicry among insects. **a.** A yellow jacket wasp uses stinging as its defense. **b.** A flower fly and (**c**) a longhorn beetle are Batesian mimics because they are incapable of stinging another animal, and yet they have the same appearance as the yellow jacket wasp. **d.** A bumblebee and the yellow jacket wasp are Müllerian mimics because they have a similar appearance, and both use stinging as a defense.

Just as with other antipredator defenses, behavior plays a role in mimicry. Mimicry works better if the mimic acts like the model (the species it is mimicking) in addition to looking like it. For example, beetles that phenotypically resemble a wasp actively fly from place to place and spend most of their time in the same habitat as the wasp model.

TABLE 34.2 Symbiotic Relationships

	Species 1	Species 2
Parasitism*	Benefited	Harmed
Commensalism	Benefited	No effect
Mutualism	Benefited	Benefited

*Can be considered a type of predation.

Symbiosis

Symbiosis refers to close interactions between members of different species. Three types of symbiotic relationships have traditionally been defined—parasitism, commensalism, and mutualism. Table 34.2 lists these three categories in terms of their benefits to one species or the other. Of the three types, only mutualism may increase the population size of both species. However, some ecologists now consider it difficult to try to classify symbiotic relationships into these categories, because the amount of harm or good the individuals of two species do to one another depends on what the investigator chooses to measure. Although the following discussion describes the traditional classification system, bear in mind that symbiotic relationships do not always fall neatly into these three categories.

Parasitism

Parasitism is a symbiotic relationship in which the **parasite** derives nourishment from another organism, called the *host.* Therefore, the parasite benefits and the host is harmed. Parasites occur in all kingdoms of life. Bacteria (e.g., strep infection), protists (e.g., malaria), fungi (e.g., athlete's foot), plants (e.g., mistletoe), and animals (e.g., tapeworm) all include parasitic species. The effects of parasites on the health of the host can range from slightly weakening them to eventually killing them over time.

In addition to providing nourishment, some host organisms also provide their parasites with a place to live and reproduce, as well as a mechanism for dispersing offspring to new hosts. Many parasites have both a primary and secondary host. The secondary host may be a vector that transmits the parasite to the next primary host. As an example, consider the deer ticks *Ixodes dammini* and *I. ricinus* in the eastern and western United States, respectively. Deer ticks are arthropods that go through a number of stages (egg, larva, nymph, adult). They are so named because adults feed and mate on white-tailed deer in the fall. The female lays her eggs on the ground, and when the eggs hatch in the spring, they become larvae that feed primarily on white-footed mice. If a mouse is infected with the bacterium *Borrelia burgdorferi,* the larvae become infected also. The larvae overwinter and molt the next spring to become nymphs that can, by chance, take a blood meal from a human. At this time, the tick may pass the bacterium on to a human, who subsequently comes down with Lyme disease, characterized by arthritic-like symptoms and often a "bull's-eye" rash around the site of the tick bite. The nymphs develop into adults, and the cycle begins again.

Commensalism

Commensalism is a symbiotic relationship between two species in which one species is benefited and the other is neither benefited nor harmed. Often one species provides a home and/or transportation for the other species, as when barnacles attach themselves to whales.

Clownfishes live within the waving mass of tentacles of sea anemones. Because most fishes avoid the venomous tentacles of the anemones, clownfishes are protected from predators. If clownfishes attract other fishes on which the anemone can feed, this relationship borders on mutualism. Other examples of commensalism may also be considered mutualistic. For example, birds called cattle egrets benefit from feeding near cattle because the cattle flush insects and other animals from the vegetation as they graze (Fig. 34.13).

Mutualism

Mutualism is a symbiotic relationship in which both members of the association benefit, though not necessarily equally. An example is the relationship between plants and their animal pollinators. When herbivores, such as insects, feed on pollen they gain a meal while the plants increase their reproductive chances. Ants form mutualistic relationships with both plants and insects. In tropical America, the bullhorn acacia tree is adapted to provide a home for ants of the species *Pseudomyrmex ferruginea.* Unlike other acacias, this species has swollen thorns with a hollow interior, where ant larvae can grow and develop. In addition to housing the ants, the acacias provide them with food. The ants constantly protect the plant from the caterpillars of moths and butterflies by swarming and stinging them, and from other

Video Pollinators

Figure 34.13 Egret symbiosis. Cattle egrets eat insects off and around various animals, such as this African cape buffalo.

plants that might shade the plant. When the ants on experimental trees were poisoned, the trees died.

Video Thorn Tree Ants

Cleaning symbiosis is a relationship in which the individuals being cleaned are often vertebrates. Crustaceans, fish, and birds act as cleaners and are associated with a variety of vertebrate clients. Large fish in coral reefs line up at cleaning stations and wait their turn to be cleaned by small fish that often enter the mouths of the large fish (Fig. 34.14). It's been suggested that cleaners may be exploiting the relationship by feeding on host tissues as well as on ectoparasites. On the other hand, cleaning could ultimately lead to net gains in client fitness by ridding them of parasites.

Figure 34.14 Cleaning symbiosis. A cleaner wrasse, *Labroides dimidiatus*, in the mouth of a spotted sweetlip, *Plectorhincus chaetodontoides*, is feeding off parasites. Does this association improve the health of the sweetlip, or is the sweetlip being exploited? Investigation is under way.

Check Your Progress 34.3

1. Describe an example of mutualism.
2. Identify examples of populations that would exhibit opportunistic growth.
3. Explain how resource partitioning can allow for the coexistence of species that have similar niches.
4. Recognize the various interactions that will exist between various species within a community.

34.4 Ecological Succession

Learning Outcomes

Upon completion of this section, you should be able to

1. Compare the various types of succession.
2. Choose the correct sequence of events that occur during ecological succession.

Communities are composed of all of the interacting populations in an area that are regulated by biotic interactions and ecological disturbances. **Ecological succession** is a change in a community's composition that is directional and follows a continuous pattern of colonization by new species. On land, *primary succession* is the establishment of a plant community in a newly formed area where there is no soil formation. Primary succession

SCIENCE IN YOUR LIFE ▶ ECOLOGY

Ecological Niche of Top-Level Predators

General fear of wolves and social pressure from ranchers in the 1920s provided justification for the elimination of wolves from Yellowstone National Park by 1926. The extinction of the wolves produced a multitude of ecological effects that would ripple through the entire park for the next 70 years. Fewer wolves meant less predation upon the elk and other large herbivores. The increase in the large herbivore populations was greeted with enthusiasm by hunters. Inevitably, an increase in herbivores led to an overbrowsing of streamside vegetation, producing a ripple effect throughout the entire Yellowstone ecosystem.

With the decreased stability of the stream bank, erosion and sediment load in the streams increased. Fish, insects, birds, and all of the wildlife associated with the streams were negatively impacted.

Against the objections of many ranchers, wolves have been slowly reintroduced into Yellowstone beginning in 1995. Since then, biologists have seen signs of the stream bank stabilizing due to the decreased browsing by elk and other herbivores. Many of the problems that were caused by the eradication of wolves from Yellowstone have begun to be repaired as a direct result of their reintroduction.

Questions to Consider

1. Do you think individual wolves should be killed if they are caught killing cattle?
2. Now that the ecosystem is stabilizing itself, do you think wolves should ever be eliminated from Yellowstone National Park again?
3. What is more important, a healthy ecosystem or the financial gain of the ranching industry?

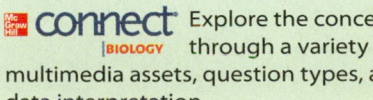

connect BIOLOGY Explore the concepts through a variety of multimedia assets, question types, and data interpretation.

www.mcgrawhillconnect.com

typically follows a major disturbance such as a volcanic eruption or a glacial retreat. *Secondary succession* is the return of a community to its natural vegetation following a disturbance, as when a cultivated cornfield returns to its natural state.

Pioneer species are the first species to begin the process of secondary succession of disturbed habitats. Succession progresses through a series of stages, as shown in Figure 34.15. Note that, in this case, the process began with grasses followed by shrubs and then to a mixture of shrubs and trees, until finally there were only trees.

Succession can also occur in aquatic communities, as when lakes and ponds sometimes undergo stages whereby they become filled in and eventually disappear.

Models of Succession

Ecologists have developed various hypotheses to explain succession and predict future events. The *climax-pattern model* of succession says that particular areas will always lead to the same type of community, called a **climax community.** This model is based on the observation that climate plays a significant role in determining whether a desert, a grassland, or a particular type of forest results. Therefore, coniferous forests are expected to occur in northern latitudes, deciduous forests in temperate zones, and tropical rain forests in the tropics. The climax-pattern model of succession now recognizes that the exact composition of a community need not always be the same. That is, while we might expect to see a coniferous forest as opposed to a tropical rain forest in northern latitudes, the exact mix of plants and animals after each stage of succession can vary.

The *facilitation model* of succession can be observed as shown in the example in Figure 34.15. It's possible that each successive community prepares the way for the next, and this is why grass-shrub-forest development occurs sequentially.

On the other hand, the *inhibition model* says that colonists hold on to their space and inhibit the growth of other plants until the colonists die or are damaged. Still another possible model, the *tolerance model,* predicts that different types of plants can colonize an area at the same time. Sheer chance determines which seeds arrive first, and successional stages may simply reflect the length of time it takes individuals to mature. This alone could account for the grass-shrub-forest development seen in Figure 34.15. In reality, the models we have mentioned are not mutually exclusive, and succession is recognized as a complex process.

Check Your Progress 34.4

1. Identify the events that would lead to succession.
2. Compare and contrast the various models used to explain succession.

Case Study Conclusion

Asian carp were initially brought into the United States to help with algae control of aquaculture facilities, but soon thereafter they escaped and have spread through the majority of the Mississippi River system. These fish are voracious eaters of zooplankton, which is in direct competition with many of the native fish and mussels. Due to the extreme size, weight, and life span of the Asian carp, they have the potential to completely disrupt the Mississippi River ecosystem.

Other concerns include their fouling the nets of commercial fishermen as well as the impact they may have upon the $7 billion fishing industry of the Great Lakes. Understanding population ecology is critical to determining the impact these fish may have upon the other species that live in the same community.

Figure 34.15 Secondary succession in a forest. Secondary succession is a process that begins in areas that already have soil (not on bare rock as primary succession does). This example of secondary succession occurred in a large conifer plantation in central New York State. Notice that certain species are common to the particular stages. However, the process of regrowth shows approximately the same series of stages as secondary succession would in a former cornfield. (Arrows indicate the passage of time.)

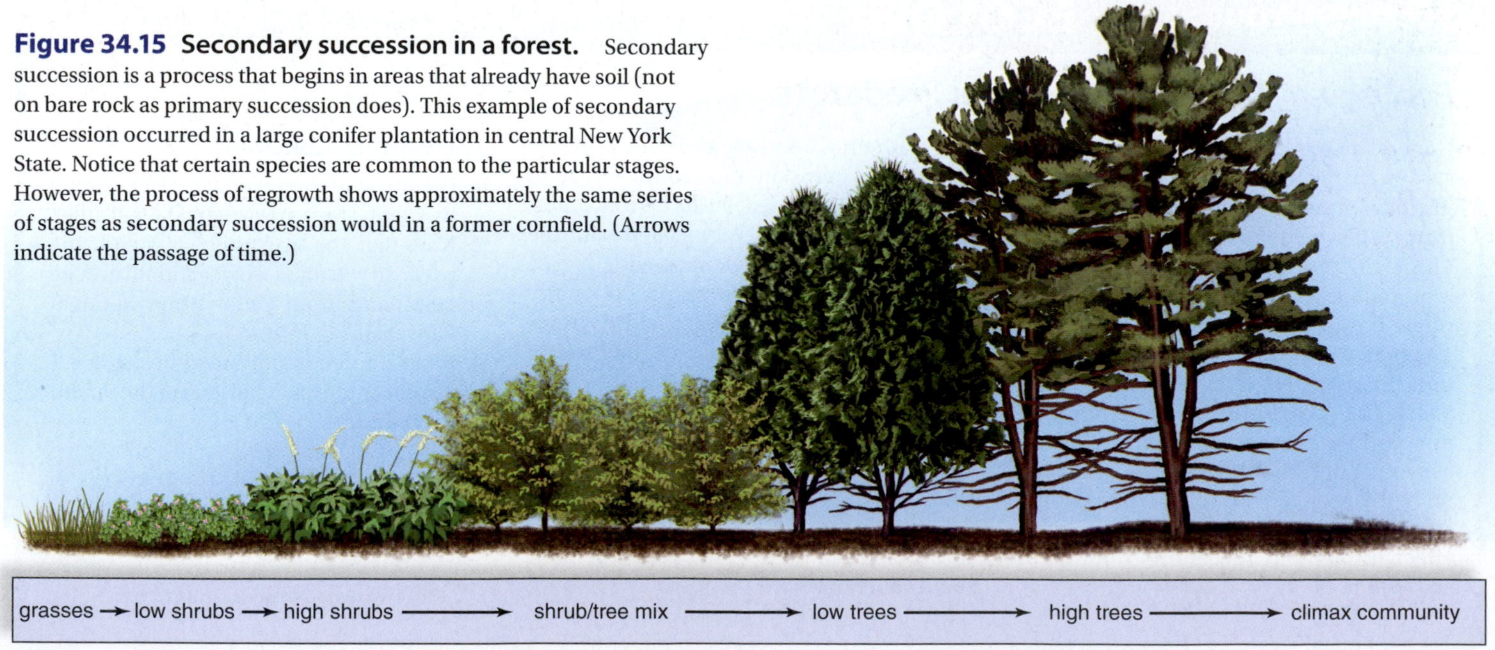
grasses → low shrubs → high shrubs → shrub/tree mix → low trees → high trees → climax community

MEDIA STUDY TOOLS

www.mhhe.com/maderinquiry14

Enhance your study of this chapter with study tools and practice tests. Also ask your instructor about the resources available through ConnectPlus, including LearnSmart, the media-rich eBook, interactive learning tools, and animations.

SUMMARIZE

34.1 The Scope of Ecology

- Ecologists study biotic and abiotic interactions of organisms at several levels: individual, **population, community, ecosystem,** and **biosphere. Ecology** is the study of the interactions between organisms and their physical environment.

34.2 Patterns of Population Growth

- The per capita rate of increase is calculated by subtracting the number of deaths from the number of births and dividing by the number of individuals in the population.
- Two patterns of population growth are possible. Exponential growth results in a J-shaped curve because, as the population increases in size, so does the number of new members.
- Under ideal circumstances, exponential growth occurs due to a population's **biotic potential.** However, exponential growth cannot continue indefinitely because **environmental resistance** opposes biotic potential, and logistic population growth occurs.
- Under **logistic growth,** an S-shaped growth curve results when the population stops growing near the **carrying capacity.**
- Populations tend to have one of three types of survivorship curves, depending on whether most individuals live out the normal life span (type I), die at a constant rate regardless of age (type II), or die early (type III). To best follow population growth a **cohort,** or group of individuals born at the same time, is tracked throughout their life.
- Human population can be appreciated by looking at the **doubling time** or the length of time it takes for the population size to double. **Zero population growth** can be achieved when the birthrate equals the death rate.
- **More-developed countries** (**MDCs**), such as the United States and most of western Europe, are those in which the population growth is low and people have a good standard of living. A **demographic transition** occurs when a decreased death rate is followed by a decreased birthrate, causing the population growth of countries to slow.
- The human population is expanding exponentially, mostly within **less-developed countries** (**LDCs**) in Africa, Asia, and Latin America. Support for family planning, human development, and delayed childbearing could help lessen the expected increase.
- **Age-structure diagrams** help divide the population into three age groups: prereproductive, reproductive, and postreproductive. **Replacement reproduction** occurs when each couple has two children, but it does not lead to zero population growth.

34.3 Interactions Between Populations

- Opportunistic species rapidly produce many young, generally lack parental care, and rely on rapid dispersal to new, unoccupied environments. Their population size is regulated by **density-independent factors.**
- Equilibrium species produce few young, which often require parental care, and their population size is typically regulated by **density-dependent factors,** such as **competition** and **predation.**
- The **competitive exclusion principle** states that no two species can occupy the same **ecological niche** at the same time when resources are limiting. With limited resources, either **resource partitioning** occurs or one species goes locally extinct. Each species will require a specific **habitat** in order to meet its basic needs.
- **Coevolution** occurs when two species adapt in response to selective pressure imposed by the other.
- Predator-prey interactions can cause **prey** populations to decline and remain at relatively low densities, cause decline of **predator** populations, or cycling of both predator and prey population densities.
- Prey defenses take many forms, such as concealment, use of fright, and warning coloration. Batesian **mimicry** occurs when one species has the warning coloration but lacks the defense of another species. Müllerian mimicry occurs when two species with the same warning coloration have a similar defense.
- In **parasitism,** the **parasite** benefits and the host is harmed. Parasites often utilize more than one host. In **commensalism,** neither party is harmed, and one species often provides a home and/or transportation for another species. In **mutualism,** both partners benefit. Examples of mutualistic relationships include flowers and their pollinators, ants and acacia trees, and cleaning **symbiosis.**

34.4 Ecological Succession

- A change in community composition over time is called **ecological succession.**
- The climax-pattern model of succession says that particular areas will always lead to the same type of community.
- The facilitation model implies that some community members prepare the landscape for new species during succession. The inhibition model suggests that some species prevent colonization, and the tolerance model suggests that species colonize simultaneously.
- A **climax community** is stable and associated with a particular geographic area.

ASSESS

Testing Yourself

Choose the best answer for each question.

1. Place the following levels of organization in order, from lowest to highest.
   a. population, organism, community, ecosystem
   b. community, ecosystem, population, organism
   c. organism, community, population, ecosystem
   d. population, ecosystem, organism, community
   e. organism, population, community, ecosystem

2. Assume that a deer population contains 500 animals. The birthrate is 105 animals per year, and the death rate is 100 animals per year. The per capita rate of increase per year is
a. 5%. **b.** 4%. **c.** 3%. **d.** 2%. **e.** 1%.

3. What type of survivorship curve would you expect for a plant species in which most seedlings die?
a. type I **b.** type II **c.** type III

4. Label this logistic growth curve:

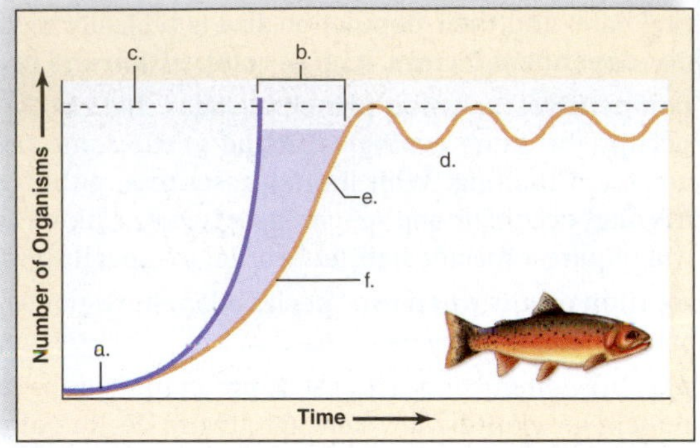
c.
b.
d.
e.
f.
a.
Number of Organisms
Time

5. If an age-structure diagram indicates that there are more people in the prereproductive stage of life than in the reproductive stage, then replacement reproduction, over the long run, will result in
a. zero population growth.
b. an increase in population size.
c. a decrease in population size.

6. An ecological niche includes an organism's
a. interactions with other organisms.
b. habitat.
c. resources.
d. Both a and b are correct.
e. All of the choices are correct.

7. A colorful, nonpoisonous frog that mimics a poison-dart frog is exhibiting
a. Batesian mimicry. **b.** Müllerian mimicry.

For problems 8–10, indicate the type of symbiosis illustrated by each example. Some answers may be used more than once.

Key:

a. parasitism **b.** commensalism **c.** mutualism

8. Unicellular algae live in the tissues of coral animals. The algae provide food for the coral, while the coral provides a stable home for the algae.

9. Flowering plants reward visiting insects with nectar. The insects, in turn, carry pollen to other flowers.

10. Small wasps lay eggs on other insects. The eggs hatch into larvae that feed on the insects and kill them.

11. If a population has a type I survivorship curve (most of its members live the entire life span), which of these would you also expect?
a. a single reproductive event per adult
b. most individuals reproduce
c. sporadic reproductive events
d. reproduction occurring near the end of the life span
e. None of the choices are correct.

12. Which of these is a density-independent regulating factor?
a. competition
b. predation
c. size of population
d. weather
e. resource availability

13. Six species of monkeys are found in a tropical forest. Most likely, they
a. occupy the same ecological niche.
b. eat different foods and occupy different ranges.
c. spend much time fighting each other.
d. are from different stages of succession.
e. All of the choices are correct.

14. Mosses growing on bare rock will eventually help to create soil. These mosses are involved in _____ succession.
a. primary **b.** secondary **c.** tertiary

15. Assume that a farm field is allowed to return to its natural state. By chance, the field is first colonized by native grasses, which begin the succession process. This is an example of which model of succession?
a. climax pattern
b. tolerance
c. facilitation
d. inhibition

ENGAGE

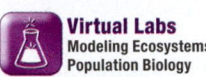

Virtual Labs
Modeling Ecosystems
Population Biology

The virtual lab "Modeling Ecosystems" allows you to explore the dynamics of energy within an ecosystem. The virtual lab "Population Biology" provides an investigative look at the factors that influence population size.

Thinking Critically

1. You are a farmer fighting off a crop insect pest. You know from evolutionary biology and natural selection that using the same pesticide causes resistance to evolve in the insect population. From what you have learned about population growth and regulation, what are some strategies you might use to control the insect population?

2. You find two species of insects, both using bright coloration and similar color patterns as an antipredator defense. Design an experiment to tell whether this is an example of Müllerian or Batesian mimicry.

3. If predators tend to reduce the population densities of prey, what prevents predators from reducing prey populations to such low levels that they drive themselves extinct?

Nature of Ecosystems 35

CASE STUDY Earth's Potential Future

In the year 2100, unspoiled ecosystems, such as the one pictured here, may be a thing of the past. If current trends continue, scientists predict that the northeastern United States will no longer have winters that consistently get below freezing temperatures. Corn will be able to be grown in the middle latitudes of Canada and many coastal areas will be under water. Skin cancer will reach an all-time high due to an increase in solar exposure. Nearly all bodies of water will be overgrown with algae and pond scum, and freshwater fish will be on the verge of extinction. Fresh water, once plentiful, will become a limited resource, and will come at a high price. Although this may seem extreme, this scenario is possible given the influence of modern humans on ecosystems and chemical cycling. Ecosystems are characterized by the interactions of biotic (living) and abiotic (nonliving) components. The biotic community depends on cycling of chemicals, such as carbon, nitrogen, and phosphorus. As we continue to burn carbon-rich fossil fuels such as oil, carbon dioxide is released into the atmosphere and acts like a blanket, trapping solar radiation. Steady atmospheric increases in carbon dioxide contribute to global climate change. In turn, sea levels are predicted to rise and extreme weather events, like hurricanes and cyclones, are predicted to increase in frequency. Use of detergents with phosphorus and nitrogen-containing fertilizers add excess amounts of these nutrients to water bodies, such as ponds and lakes. The result is cultural eutrophication, which causes massive algal blooms. When these blooms decompose, oxygen levels in water bodies plummet, causing widespread fish kills.

In this chapter, you will learn about how energy flows through the members of biotic communities and how abiotic chemicals and nutrients cycle through ecosystems. This knowledge will help you understand why phenomena such as climate change are of major concern.

As you read through the chapter, think about the following questions:

1. What happens to the concentration of pollutants in animals as you move from lower to higher trophic levels?
2. How do we alter biogeochemical cycles, and what are some of the environmental consequences?

CHAPTER OUTLINE

BEFORE YOU BEGIN

Before beginning this chapter, take a few moments to review the following discussions:

Section 2.1 Which elements are basic to all life?

Section 2.3 How do the properties of water influence the biogeochemical cycles?

Figure 34.10 What role does the predator-prey cycle play in the energy flow of an ecosystem?

35.1 The Biotic Components of Ecosystems

Learning Outcomes

Upon completion of this section, you should be able to

1. Identify the ways that autotrophs and heterotrophs obtain nutrients.
2. Interpret the energy flow and biogeochemical cycling within and among ecosystems.

An **ecosystem** possesses both abiotic and biotic components. The **abiotic** components include the nonliving aspects, such as sunlight, inorganic nutrients, and water availability, and conditions, such as type of soil, average temperature, and wind speed. The **biotic** components of an ecosystem are the various populations of species that form the community.

Populations Within an Ecosystem

The populations within an ecosystem are categorized according to their food source. **Autotrophs** (i.e., "self-feeders") require only inorganic nutrients and an outside energy source to produce organic nutrients for their own use as food. Because they produce food for themselves and other members of the community, they are called **producers** (Fig. 35.1*a*). Photosynthetic organisms such as algae and plants produce most of the organic nutrients for the biosphere.

Video Plants

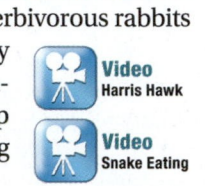

Some bacteria are chemoautotrophs. They reduce carbon dioxide to an organic compound by using energetic electrons derived from inorganic compounds, such as hydrogen sulfides, hydrogen gas, or ammonium ions. Chemoautotrophs have been found to support communities at hydrothermal vents along deep-sea oceanic ridges.

Heterotrophs (i.e., "other feeders") need an outside source of organic nutrients. Because they consume food, they are called **consumers. Herbivores** are animals that feed directly on plants or algae (Fig. 35.1*b*). **Carnivores** feed on other animals. Snakes that feed on rabbits are carnivores, and so are the hawks that feed on these snakes (Fig. 35.1*c*). In this example, herbivorous rabbits are primary consumers, snakes are secondary consumers, and the hawks are tertiary consumers. Sometimes tertiary consumers are called top predators. **Omnivores** are animals (including humans) that feed on both plants and animals.

Video Harris Hawk

Video Snake Eating

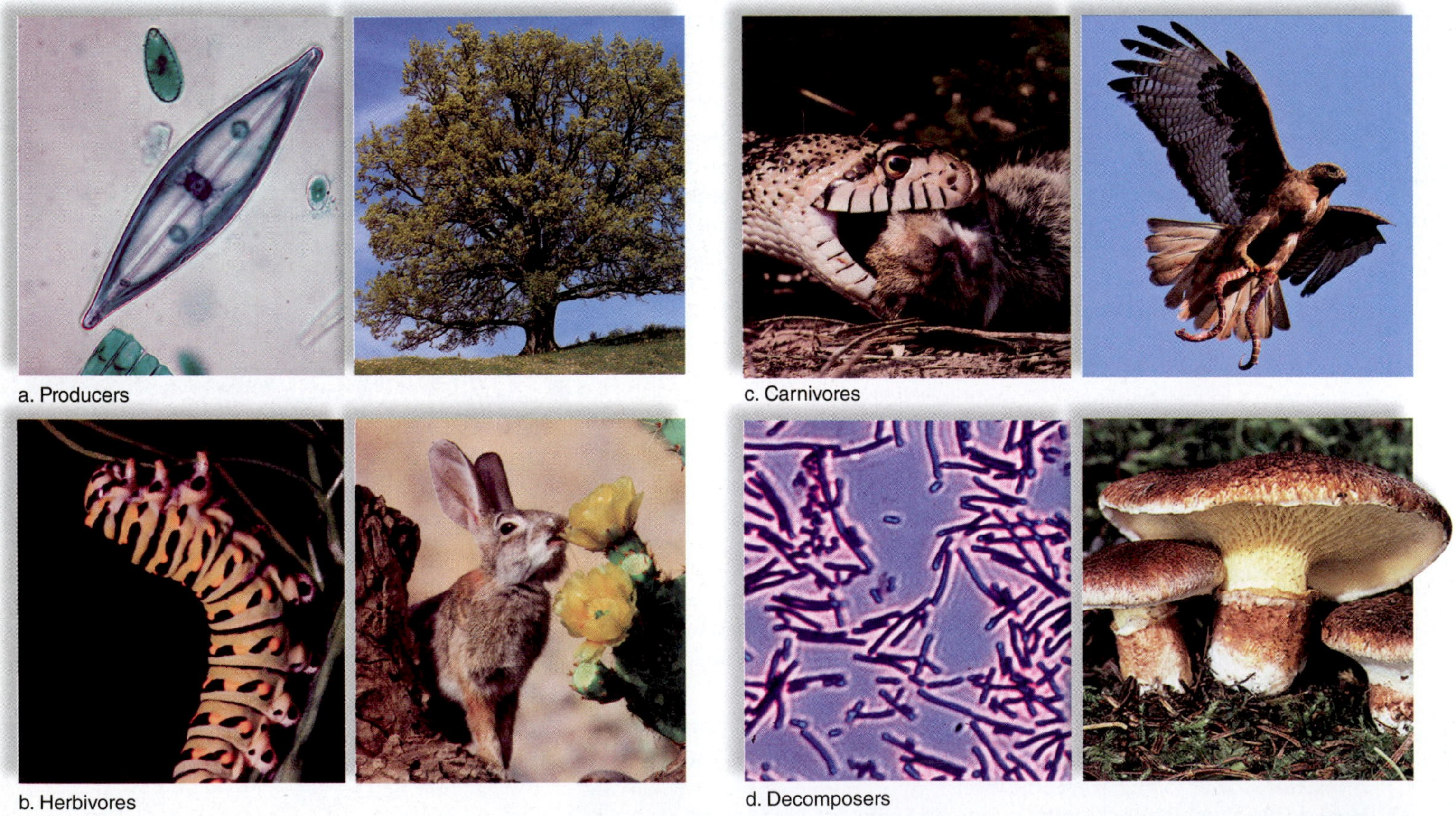
a. Producers
c. Carnivores
b. Herbivores
d. Decomposers

Figure 35.1 Biotic components. **a.** Diatoms, a type of algae, and green plants are producers. **b.** Caterpillars and rabbits are herbivores. **c.** Snakes and hawks are carnivores. **d.** Bacteria and mushrooms are decomposers.

Decomposers are heterotrophic bacteria, some species of protists and fungi, such as molds and mushrooms, that break down nonliving organic matter (Fig. 35.1*d*). They perform a very valuable service because they release inorganic nutrients that are taken up by plants back into the environment. **Detritus** is partially decomposed matter in the water or soil. Detritivores are found in all ecosystems. Earthworms and some beetles, termites, and maggots are soil detritivores.

Animation Decomposers

Energy Flow and Chemical Cycling

Every ecosystem is characterized by two fundamental phenomena: energy flow and chemical cycling. Energy flow begins when producers absorb solar energy, and chemical cycling begins when producers take in inorganic nutrients from the physical environment (Fig. 35.2). Thereafter, producers make organic nutrients (food) directly for themselves and indirectly for the other populations of the ecosystem. Most ecosystems cannot exist without a continual supply of solar energy. Chemicals cycle when inorganic nutrients are returned to the producers from the atmosphere or soil.

Only a portion of the organic nutrients made by autotrophs is passed on to heterotrophs because plants use some of the organic molecules to fuel their own cellular respiration. Similarly, only a small percentage of nutrients taken in by heterotrophs is available to higher-level consumers. Figure 35.3 shows why. A certain amount of the food eaten by a herbivore is eliminated as feces that are recycled by decomposers. Metabolic wastes are excreted as urine. Of the assimilated energy, a large portion is utilized during cellular respiration to provide energy to the organism and thereafter becomes heat. Only the remaining food, which is converted into increased body weight (or additional offspring), becomes available to carnivores.

In Chapter 6, you learned that energy flow is described by the laws of thermodynamics: (1) energy cannot be created or destroyed; and (2) in every energy transformation, some energy is lost as heat. These principles explain why ecosystems are dependent on a continual outside source of energy, such as sunlight, and why only a part of the original energy from producers is available to consumers.

Check Your Progress 35.1

1. Identify the nutritional differences between producers, consumers, and decomposers.
2. Explain why most energy fails to be converted to a usable form when one organism eats another.

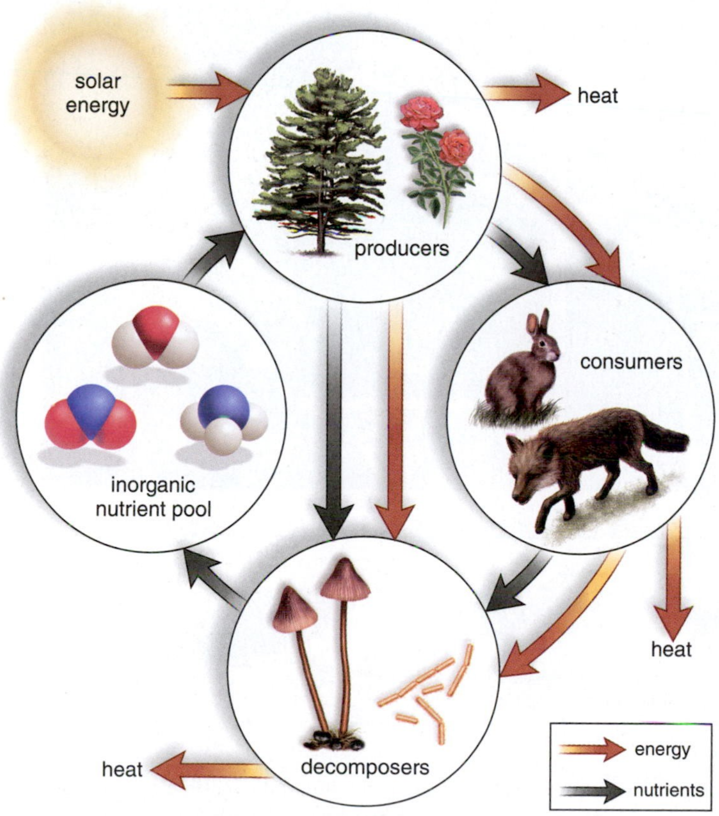
solar energy
heat
producers
consumers
inorganic nutrient pool
heat
heat
decomposers
energy
nutrients

Figure 35.2 Energy flow and nutrient cycling. Nutrients cycle, but energy flows through an ecosystem. As energy transformations repeatedly occur, all the energy derived from the sun eventually dissipates as heat.

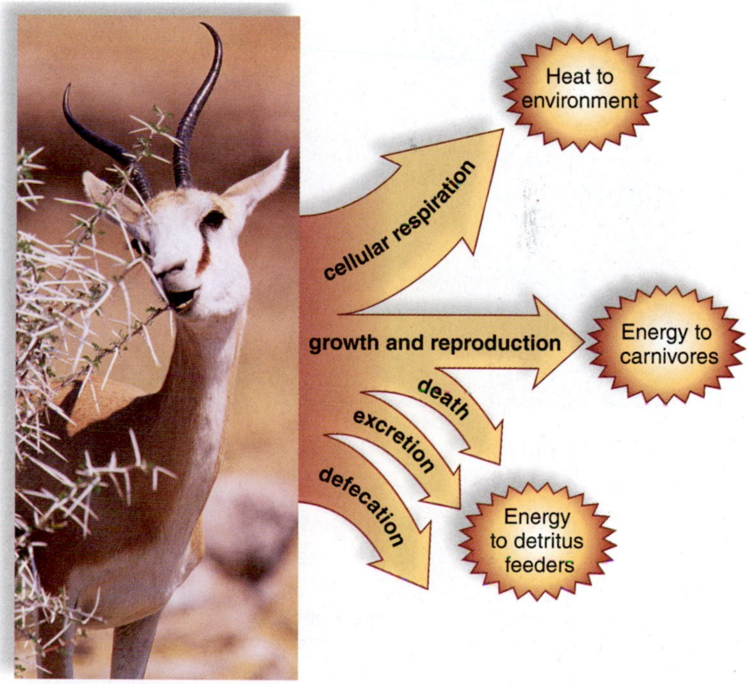
Heat to environment
cellular respiration
growth and reproduction
Energy to carnivores
death
excretion
defecation
Energy to detritus feeders

Figure 35.3 Energy balances. Only about 10% of the food energy taken in by a herbivore is passed on to carnivores and much of the rest is lost as heat to the environment. A large portion goes to detritivores via defecation, excretion, and death, and another large portion is used for cellular respiration.

35.2 Energy Flow

Learning Outcomes

Upon completion of this section, you should be able to

1. Recognize the difference between a food chain and a food web.
2. Explain the energy flow among populations through food webs and ecological pyramids.

The interconnecting paths of energy flow within ecosystems are represented by diagrams called **food webs.** Figure 35.4*a* is a **grazing food web.** Grazing food webs do not begin with grazers; instead they begin with producers like the trees in a forest. Caterpillars feed on leaves in the trees; and mice, rabbits, and deer feed on leaves at or near the ground. Birds, chipmunks, and mice eat fruits and nuts, but are omnivores because they also feed on caterpillars. These herbivores and omnivores all provide food for a number of different carnivores.

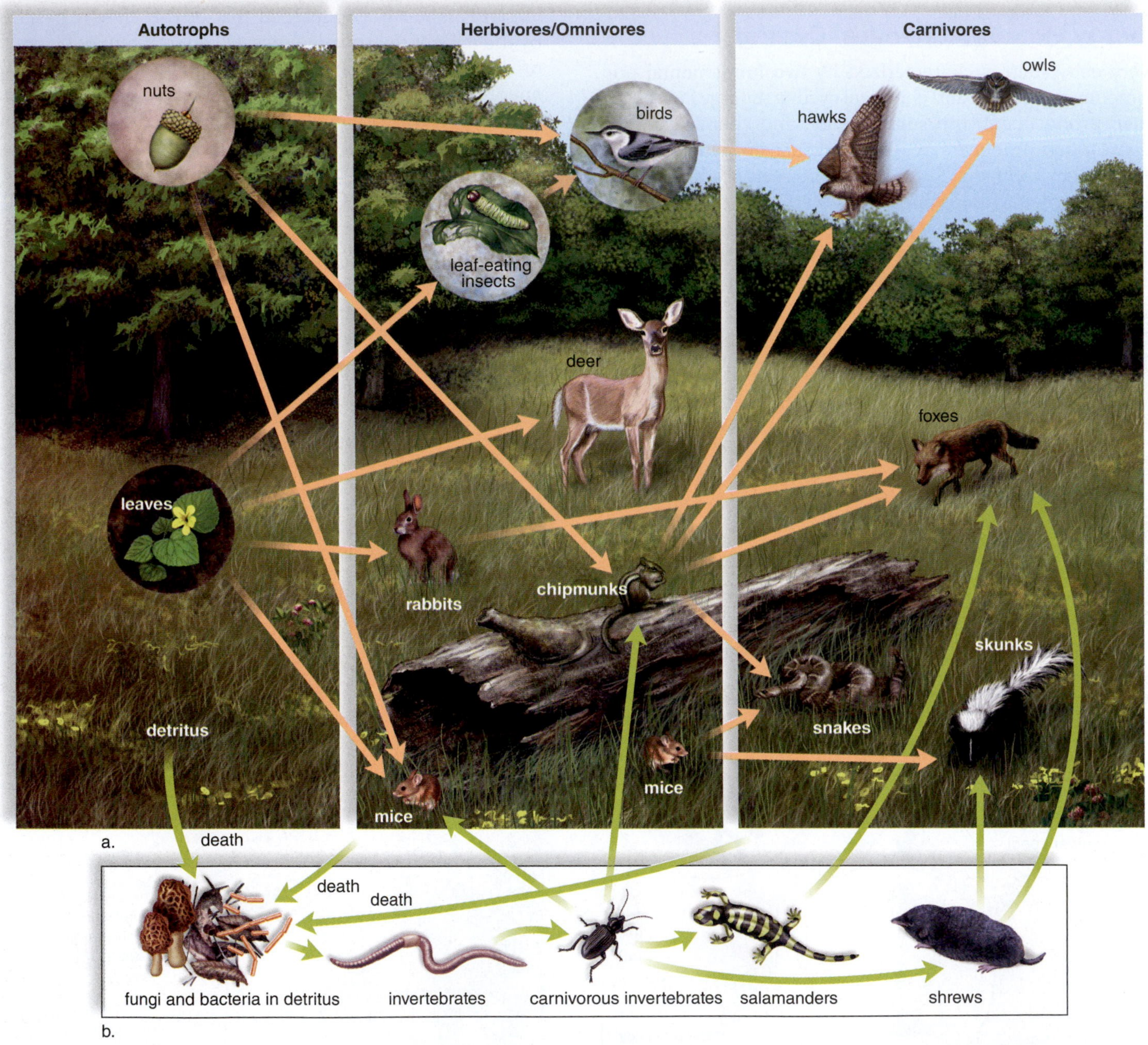
Autotrophs
Herbivores/Omnivores
Carnivores
nuts
birds
owls
hawks
leaf-eating insects
deer
foxes
leaves
rabbits
chipmunks
skunks
snakes
detritus
mice
mice
a.
death
death
death
fungi and bacteria in detritus
invertebrates
carnivorous invertebrates
salamanders
shrews
b.

Figure 35.4 Grazing and detrital food webs. Food webs are descriptions of who eats whom. **a.** Tan arrows illustrate possible grazing food webs. For example, birds, which feed on nuts, may be eaten by a hawk. Autotrophs such as the tree are producers (first trophic level); the first series of animals are primary consumers (second trophic level); and the next group of animals are secondary consumers (third trophic level). **b.** Green arrows illustrate possible detrital food webs, which begin with detritus, the waste products and remains of dead organisms. A large portion of these remains are from the grazing food web illustrated in (**a**). The organisms in the detrital food web are sometimes consumed by animals in the grazing food web, as when robins feed on earthworms. Thus, the grazing food web and the detrital food web are connected.

Figure 35.4*b* is a **detrital food web,** which begins with detritus. Detritus is food for soil organisms such as earthworms and beetles. These animals are, in turn, fed on by salamanders and shrews. Because the members of the detrital food web may become food for aboveground carnivores, the detrital and grazing food webs are joined.

In this particular forest, the organic matter lying on the forest floor and mixed into the soil contains over twice as much energy as the leaves of living trees. Therefore, more energy in a forest may be funneling through the detrital food web than through the grazing food web.

Trophic Levels

You can see that Figure 35.4*a* would allow us to link organisms one to another in a straight line, according to who eats whom. Diagrams that show a single path of energy flow such as this are called **food chains.** For example, in the grazing food web, we could find this **grazing food chain:**

leaves ⟶ caterpillars ⟶ sparrows ⟶ hawks

And in the detrital food web (Fig. 35.4*b*), we could find this **detrital food chain:**

detritus ⟶ earthworms ⟶ shrews

A **trophic level** is composed of all the organisms that feed at a particular link in a food chain. In the grazing food web in Figure 35.4*a,* going from left to right, the trees are primary producers (first trophic level). The first series of animals, the herbivores, are primary consumers (second trophic level). The group of animals at the far right, the carnivores, are secondary consumers (third trophic level).

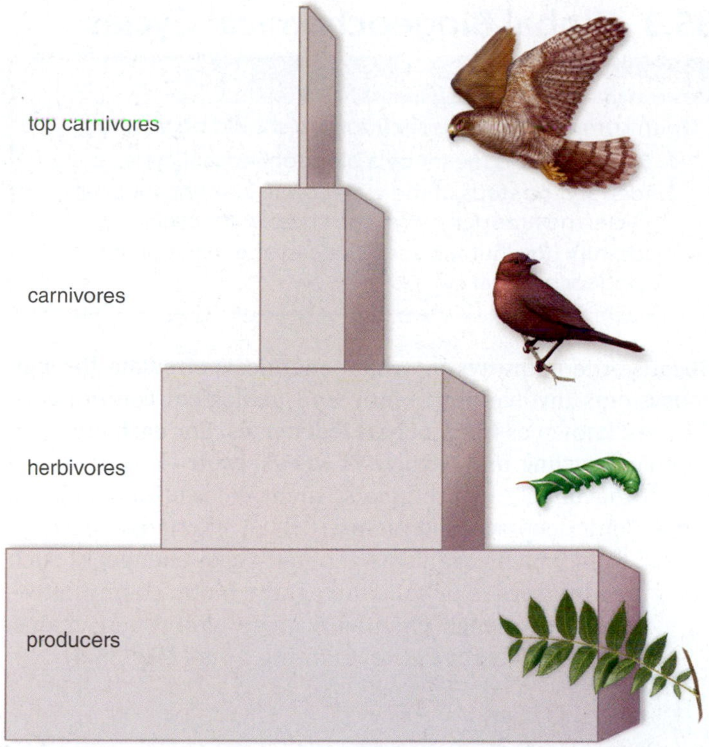
top carnivores
carnivores
herbivores
producers

Figure 35.5 Ecological pyramid. An ecological pyramid reflects the loss of energy from one trophic level to the next. Energy is lost as herbivores feed on producers, carnivores feed on herbivores, and top carnivores feed on carnivores. This loss of energy results in decreasing biomass and numbers of organisms at higher trophic levels relative to lower trophic levels (levels not to scale).

Ecological Pyramids

The shortness of food chains can be attributed to the loss of energy between trophic levels. In general, only about 10% of the energy of one trophic level is available to the next trophic level. Therefore, if a herbivore population consumes 1,000 kg of plant material, only about 100 kg is converted to herbivore tissue, 10 kg to first-level carnivores, and 1 kg to second-level carnivores. This 10% rule explains why few carnivores can be supported in a food web. The large energy losses that occur between successive trophic levels are sometimes depicted as an **ecological pyramid** (Fig. 35.5).

Biomass is the number of organisms multiplied by their weight. Thinking in terms of biomass rather than simply the number of organisms can be useful, because in terms of only numbers, one oak tree may have hundreds of caterpillars. Although it appears that there are more herbivores than autotrophs, the biomass of autotrophs is much greater. Similarly, the biomass of herbivores is greater than that of carnivores.

However, in aquatic ecosystems, such as lakes and open seas where algae are the main producers, the herbivores may have a greater biomass than the producers. The reason is that, over time, the algae reproduce rapidly, but they are also consumed at a high rate. Any pyramids like this one, which have more herbivores than producers, are called inverted pyramids:

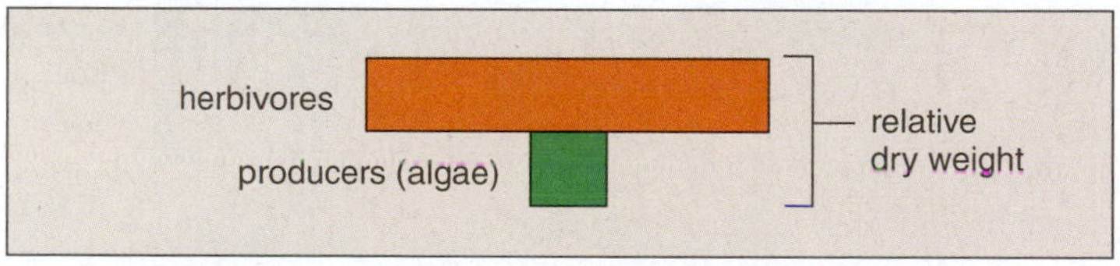
herbivores
producers (algae)
relative
dry weight

Ecologists are hesitant to use pyramids to describe ecological relationships for all situations. Detrital food chains are rarely included in pyramids, but they may represent a large portion of energy in many ecosystems, making conclusions based on aboveground pyramids potentially inaccurate.

Check Your Progress 35.2

1. Explain why ecosystems generally support few carnivores.
2. Explain how biomass relates to the structure of ecological pyramids.

35.3 Global Biogeochemical Cycles

Learning Outcomes

Upon completion of this section, you should be able to

1. Define what is meant by a biogeochemical cycle.
2. Identify the steps of the water cycle, the phosphorus cycle, the nitrogen cycle, and the carbon cycle.
3. Identify how human activities can alter each of the biogeochemical cycles.

Because the pathways by which chemicals circulate through ecosystems involve both biotic and geological components, they are known as **biogeochemical cycles.** For each element, chemical cycling may involve (1) a **reservoir**—a source normally unavailable to producers, such as fossilized remains, rocks, and deep-sea sediments; (2) an **exchange pool**—a source from which organisms generally take chemicals, such as the atmosphere or soil; and (3) the biotic community—through which chemicals move along food chains, perhaps never entering a pool (Fig. 35.6).

With the exception of water, which exists in gas, liquid, and solid forms, there are two types of biogeochemical cycles. In a gaseous cycle, exemplified by the carbon and nitrogen cycles, the element is withdrawn from and returns to the atmosphere as a gas. In a sedimentary cycle, exemplified by the phosphorus cycle, the element is absorbed from soil by plant roots, passed to heterotrophs, and eventually returned to the soil by decomposers.

The diagrams on the next few pages show how nutrients flow between terrestrial and aquatic ecosystems. In the nitrogen and phosphorus cycles, these nutrients run off from a terrestrial to an aquatic ecosystem and in that way enrich the aquatic ecosystem. However, too much of these nutrients can lead to excessive growth of algae. Decaying organic matter in aquatic ecosystems can be a source of nutrients for intertidal inhabitants such as fiddler crabs. Seabirds feed on fish but deposit guano (droppings) on land, and in that way phosphorus from the water is deposited on land. Anything put into the environment in one ecosystem can find its way to another ecosystem. As evidence of this, scientists find the soot from urban areas and pesticides from agricultural fields in the snow and animals of the Arctic.

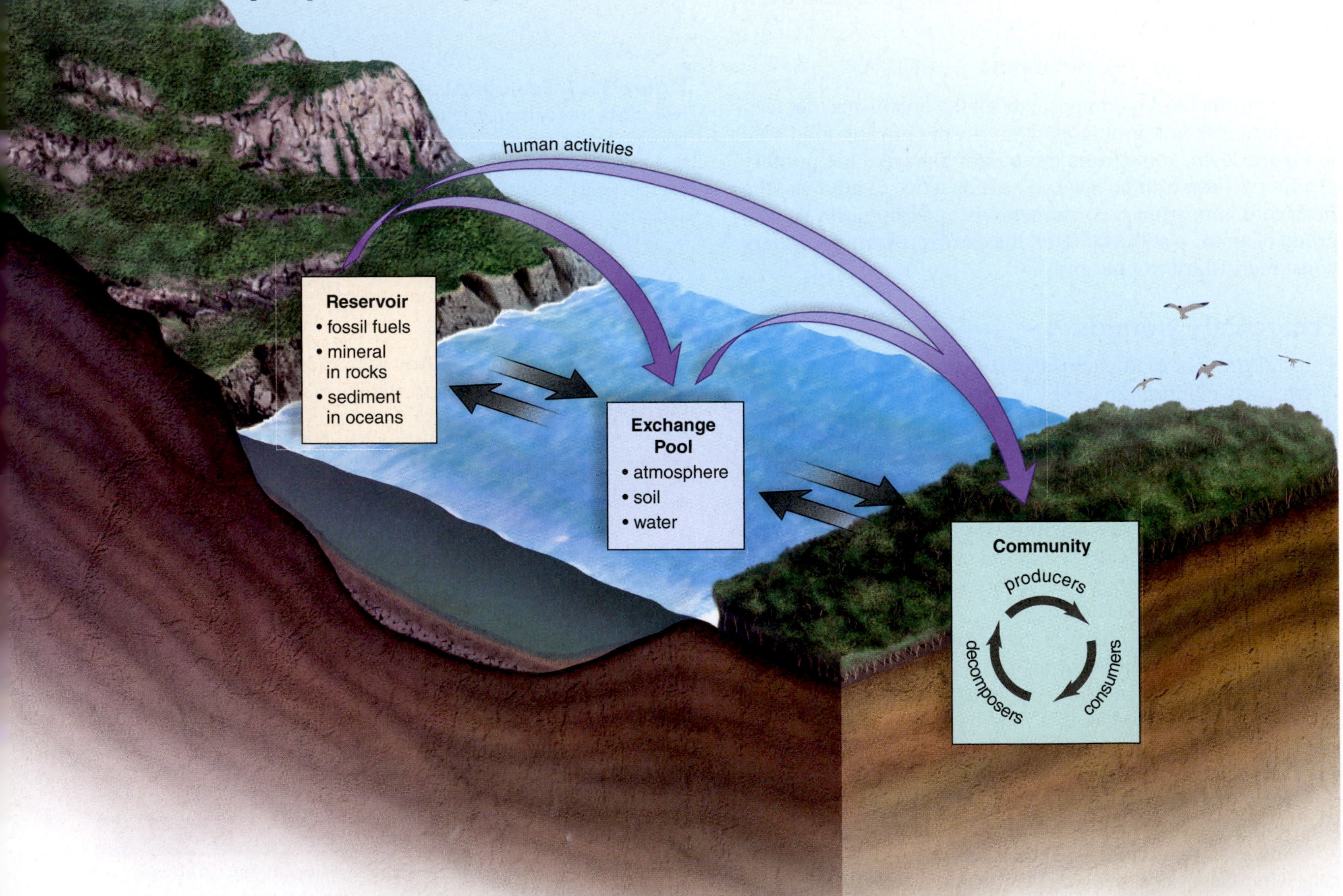

human activities
Reservoir
• fossil fuels
• mineral in rocks
• sediment in oceans
Exchange Pool
• atmosphere
• soil
• water
Community
producers
consumers
decomposers

Figure 35.6 Model for chemical cycling. Chemical nutrients cycle between these components of ecosystems. Reservoirs, which include fossil fuels, minerals in rocks, and sediments in oceans, are normally relatively unavailable sources. However, exchange pools, such as those in the atmosphere, soil, and water, are available sources of chemicals for the biotic community. When human activities (purple arrows) remove chemicals from reservoirs and pools and make them available to the biotic community, pollution can result.

The Water Cycle

The **water (hydrologic) cycle** is described in Figure 35.7. During the water cycle, fresh water is distilled from salt water. First, evaporation occurs. During **evaporation,** a liquid (in this case, water) changes from a liquid state to a gaseous state. The sun's rays cause fresh water to evaporate from seawater, and the salts are left behind. Next, condensation occurs. During **condensation,** a gas is changed to a liquid. The amount of water evaporating from the oceans exceeds the amount of precipitation that falls back into the ocean; often this excess moves over and falls on land. Water also evaporates from land, from plants (transpiration), and from bodies of fresh water. Because land lies above sea level, gravity eventually returns all fresh water to the sea. In the meantime, water is contained within standing waters (lakes and ponds), flowing water (streams and rivers), and groundwater.

Some of the water from **precipitation** (e.g., rain, snow, sleet, hail, and fog) sinks, or percolates, into the ground and saturates the earth to a certain level. The top of the saturation zone is called the groundwater table, or simply, the water table. Because water infiltrates through the soil and rock layers, sometimes groundwater is also located in **aquifers,** rock layers that contain water and release it in appreciable quantities to wells or springs. Aquifers are recharged when rainfall and melted snow percolate into the soil.

Human Activities

In some parts of the United States, especially the arid West and southern Florida, withdrawals from aquifers exceed any possibility of recharge. This is called "groundwater mining." In these locations, the groundwater level is dropping, and residents may run out of groundwater, at least for irrigation purposes, within a few years.

Fresh water, which makes up only about 3% of the world's supply of water, is called a renewable resource because a new supply is always being produced as a result of the water cycle. But it is possible to run out of fresh water when the available supply is not adequate or has become polluted so that it is not usable. Notice that the price of bottled water has gone up steadily over the past several years.

Video Thames River

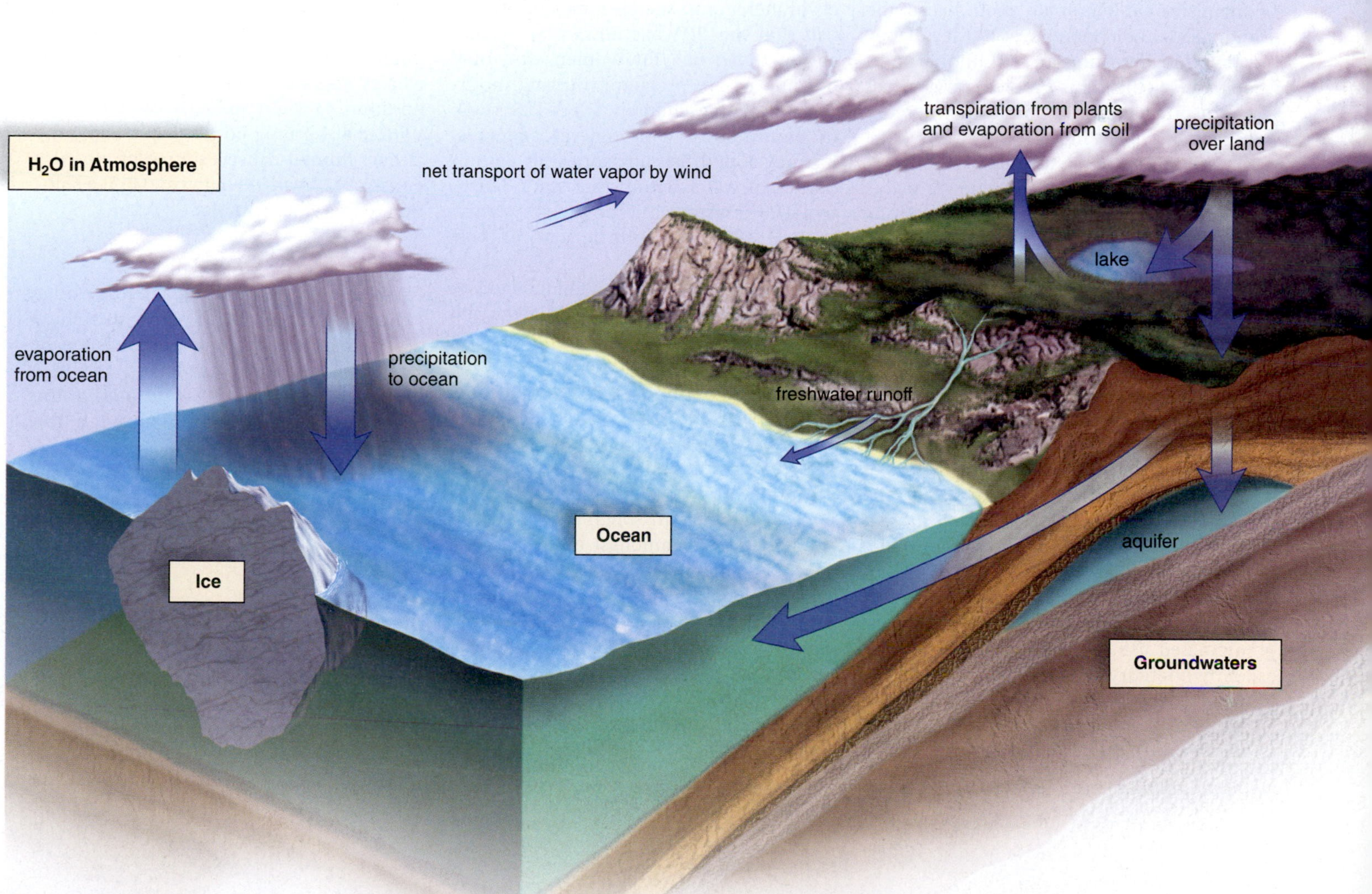

H_2O in Atmosphere
net transport of water vapor by wind
transpiration from plants and evaporation from soil
precipitation over land
lake
evaporation from ocean
precipitation to ocean
freshwater runoff
Ocean
Ice
aquifer
Groundwaters

Figure 35.7 The water cycle. Evaporation from the ocean exceeds precipitation, so there is a net movement of water vapor onto land, where precipitation results in the eventual flow of surface water and groundwater back to the sea. On land, transpiration by plants contributes to evaporation.

SCIENCE IN YOUR LIFE ▶ HEALTH

Association of Mercury with the Hydrologic Cycle

Figure 35A Mercury in the hydrologic cycle. Mercury, released by human activities such as coal-fired power plants, waste incineration, and metal processing can enter the hydrologic cycle and eventually wind up in the bodies of various plants and animals.

Scientists have known since the 1950s that the emission of mercury into the environment (Fig. 35A) can lead to serious health effects for humans. Studies show that fish and wildlife exposed to mercury emissions are negatively impacted. Humans are then impacted when they come into contact with affected fish and wildlife. Recent fish studies have shown a widespread contamination of mercury in streams, wetlands, reservoirs, and lakes throughout the majority of the United States because of the movement of mercury though the hydrologic cycle.

Environmental risk increases when mercury increases in concentration within an organism's body. This increase occurs when an organism accumulates a contaminant faster than it can eliminate it. Most organisms can eliminate about half of the mercury in their bodies every 70 days if they can avoid ingesting any additional mercury during this time. Problems arise when organisms cannot eliminate the mercury before they ingest more.

Mercury tends to enter an ecosystem when it is released from power plants and chemical facilities. The emissions can directly enter the hydrologic cycle through discharge into a nearby stream or river. Mercury can also enter an ecosystem through coal combustion, waste incineration, and metal processing. Once discharged into the atmosphere, it can become associated with rainfall and enter the aquatic ecosystem. Once in the aquatic ecosystem, mercury enters a complex cycle in which it is converted into a variety of different forms. It can then settle into the sediments of the body of water it is in, or it can be released back into the atmosphere through volatilization.

Exposure to mercury for humans generally happens when people eat contaminated fish or breathe mercury vapor. Methylmercury is the form that leads to health problems such as sterility in men, damage to the central nervous system, and in severe cases birth defects in infants. Developing fetuses and children can have health consequences from intake levels five to ten times lower than adult intake.

Studies have shown such elevated levels of mercury in sharks, tuna, and swordfish that the EPA has advisories against eating these fish for women who may become or are pregnant, nursing mothers, and young children. Every U.S. state, in conjunction with federal agencies, has developed fish advisories for certain bodies of water within that state. Forty-five U.S. states warn pregnant women to limit fish consumption from their waters.

Mercury poisoning isn't limited to just aquatic species. Research conducted in the northeastern United States and Canada showed the presence of mercury in a variety of birds, ranging from thrushes to loons to bald eagles. It is no surprise that loons and bald eagles can have high levels of mercury accumulation due to their consumption of contaminated fish. The presence of mercury in songbirds has raised serious concerns amongst ecologists. It is speculated that northeastern songbirds are ingesting mercury when they feed upon insects that have picked up the toxin from eating smaller insects that ingested it from vegetation. This raised concerns about mercury's ability to enter food webs and increase in concentration in previously unknown ways.

Ultimately, the blame for mercury pollution falls squarely on the shoulders of humans. Every ecosystem on the planet will have some degree of exposure to this pollutant, and with this exposure comes the risk of mercury contamination. Species ranging from polar bears and bald eagles to whales and sharks show us that there are no limits to where mercury can be found. With coal-burning power plants being the largest human-caused source of mercury emissions, it will be up to us to find a solution to this global problem.

Questions to Consider

1. Would you support tighter regulations on coal-fired power plants to reduce their mercury emissions even if it meant an increase in energy costs?
2. What is the easiest way to prevent the movement of mercury through the hydrologic cycle?
3. Are there any species in an ecosystem that are not impacted by exposure to mercury?

The Phosphorus Cycle

In the phosphorus cycle, phosphorus moves from rocks on land to the oceans, where it gets trapped in sediments. Then phosphorus moves back onto land following a geological upheaval. You can verify this by following the appropriate arrows in Figure 35.8. However, on land, the very slow weathering of rocks makes phosphate ions (PO_4^{3-} and HPO_4^{2-}) available to plants, which take up phosphate from the soil.

Producers use phosphate to form a variety of molecules, including phospholipids, ATP, and the nucleotides that become a part of DNA and RNA. Animals incorporate some of the phosphate into teeth, bones, and shells that take many years to decompose. Eventually, however, phosphate ions become available to producers once again. Because the available amount of phosphate is already being utilized within food chains, phosphate is usually a limiting inorganic nutrient for plants—that is, their growth is limited by the amount of available phosphorus.

Some phosphate runs off into aquatic ecosystems, where algae acquire phosphate before the excess becomes trapped in sediments. Phosphate in marine sediments does not become available to producers on land again until a geological upheaval exposes sedimentary rocks to weathering once more. Phosphorus does not enter the atmosphere, therefore, the phosphorus cycle is called a sedimentary cycle.

Human Activities

Humans boost the amount of available phosphate by mining phosphate ores and using them to make fertilizers, animal feed supplements, and detergents. Fertilizers usually contain three basic ingredients: nitrogen, phosphorus, and potassium. Some laundry detergents still contain approximately 35–75% sodium triphosphate, but many companies are phasing out the phosphates altogether. The amount of phosphate in animal feed varies.

Animal wastes from livestock feedlots, fertilizers from lawns and cropland, and untreated and treated sewage discharged from cities all add excess phosphate to nearby waters. The end result is **cultural eutrophication** (over-enrichment), which can lead to an algal bloom, as indicated by green scum floating on the water or excessive mats of filamentous algae. When the algae die off, decomposers use up all available oxygen for cellular respiration. The result can be a massive fish kill.

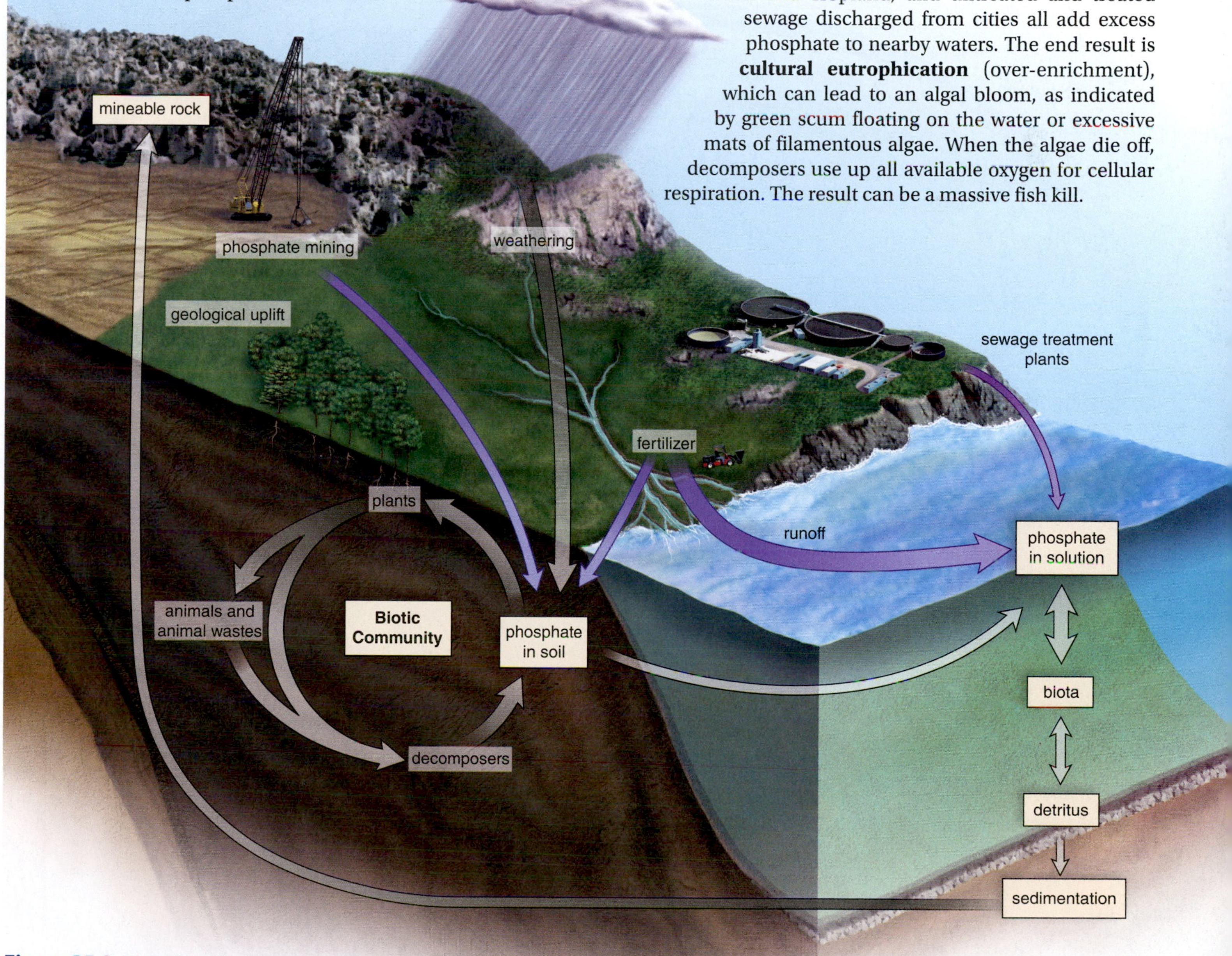
mineable rock
phosphate mining
weathering
geological uplift
sewage treatment plants
fertilizer
plants
runoff
phosphate in solution
animals and animal wastes
Biotic Community
phosphate in soil
biota
decomposers
detritus
sedimentation

Figure 35.8 The phosphorus cycle.
Purple arrows represent human activities; gray arrows represent natural events.

The Nitrogen Cycle

Even though nitrogen gas (N_2) makes up about 78% of the atmosphere, it is unavailable for plants to use. Therefore, nitrogen is also a limiting inorganic nutrient for plants.

Nitrogen Fixation

In the nitrogen cycle, **nitrogen fixation** occurs when nitrogen gas (N_2) is converted to ammonium ions (NH_4^+), a form plants can use (Fig. 35.9). Some cyanobacteria in aquatic ecosystems and some free-living bacteria in soil are able to fix atmospheric nitrogen in this way. Other nitrogen-fixing bacteria live in nodules on the roots of legumes (see Chapter 9). They make organic compounds containing nitrogen available to the host plants so that the plant can form proteins and nucleic acids.

Video Dung Beetles

Nitrification

Plants can also use nitrates as a source of nitrogen. The production of nitrates during the nitrogen cycle is called nitrification. Nitrification can occur in two ways: (1) Nitrogen gas is converted to nitrate (NO_3^-) in the atmosphere when cosmic radiation, meteor trails, and lightning provide the high energy needed for nitrogen to react with oxygen. (2) Ammonium ions in the soil from decomposition are converted to nitrate by soil bacteria in a two-step process. First, nitrite-producing bacteria convert ammonium to nitrite (NO_2^-), and then nitrate-producing bacteria convert nitrite to nitrate. These two groups of bacteria, called the nitrifying bacteria, are chemoautotrophs.

Denitrification

Denitrification is the conversion of nitrate back to nitrogen gas, which will enter the atmosphere. Denitrifying bacteria living in

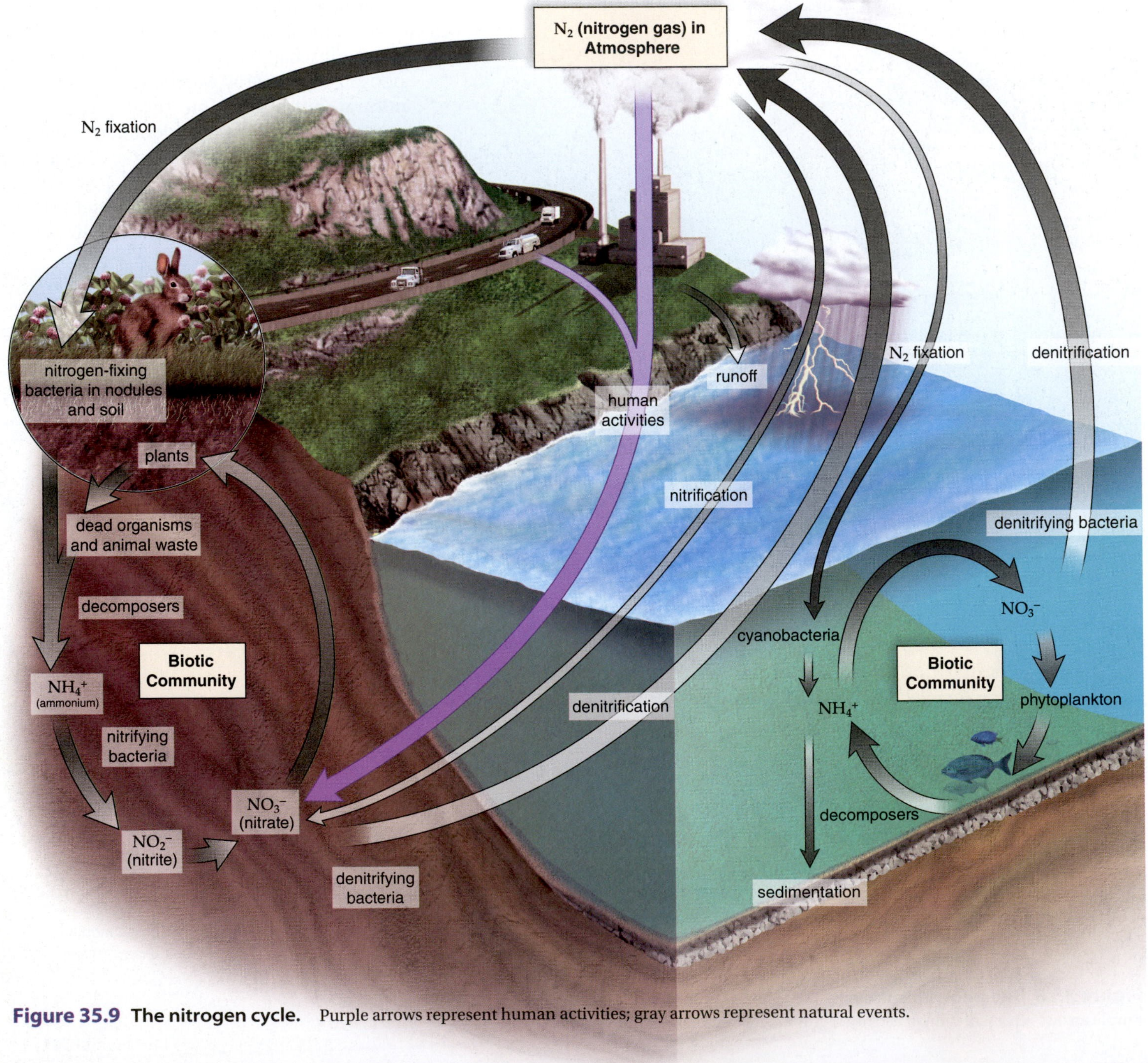

N₂ (nitrogen gas) in Atmosphere
N₂ fixation
nitrogen-fixing bacteria in nodules and soil
plants
dead organisms and animal waste
decomposers
Biotic Community
NH₄⁺ (ammonium)
nitrifying bacteria
NO₂⁻ (nitrite)
NO₃⁻ (nitrate)
denitrifying bacteria
human activities
runoff
nitrification
denitrification
N₂ fixation
denitrification
denitrifying bacteria
NO₃⁻
cyanobacteria
NH₄⁺
Biotic Community
phytoplankton
decomposers
sedimentation

Figure 35.9 The nitrogen cycle. Purple arrows represent human activities; gray arrows represent natural events.

SCIENCE IN YOUR LIFE ▶ ECOLOGY

Photochemical Smog

Smog is well known to most people, especially those who live in large cities, as a type of air pollution that is a mixture of smoke and fog. While breathing any type of air pollution can be damaging to the respiratory system, a particular type of smog that may be even more harmful is called photochemical smog. Photochemical smog arises when primary pollutants react with one another under the influence of sunlight to form a more deadly combination of chemicals. For example, two primary pollutants, nitrogen oxides (NO_x) and volatile organic compounds (VOCs), including hydrocarbons from fossil fuel use, react with one another in the presence of sunlight to produce nitrogen dioxide (NO_2), ozone (O_3), and PAN (peroxyacetylnitrate). Exposure to these chemicals can result in respiratory distress, eye irritation, headache, and certain types of cancer.

Large, industrialized cities with warm, sunny climates—such as Los Angeles; Sydney, Australia; Mexico City; and Buenos Aires, Argentina—are particularly susceptible to photochemical smog. If the city is surrounded by hills, a thermal inversion may aggravate the situation. Normally, warm air near the ground rises, so that pollutants are dispersed and carried away by air currents. But sometimes during a thermal inversion, smog gets trapped near the Earth by a blanket of warm air (Fig. 35B). This may occur when a cold front brings in cold air, which settles beneath a warm layer. The trapped pollutants cannot disperse, and the results are dangerous to a person's respiratory health. Even healthy adults experience a reduction in lung capacity when exposed to photochemical smog for long periods or during vigorous outdoor activities. Repeated exposures to high concentrations of ozone are associated with respiratory problems, such as an increased rate of lung infections and permanent lung damage. Children, the elderly, asthmatics, and individuals with emphysema or other similar disorders are particularly at risk.

Even though federal legislation is in place to bring air pollution under control, more than half the people in the United States live in cities polluted by too much smog. In the long run, pollution prevention is usually easier and cheaper than pollution cleanup. Some prevention suggestions are as follows:

- Encourage use of public transportation and burn fuels that do not produce pollutants.
- Increase recycling in order to reduce the amount of waste that is incinerated.
- Reduce energy use so that power plants need to provide less.
- Use renewable energy sources, such as solar, wind, or water power.
- Require industries to meet clean-air standards.

Questions to Consider

1. One of the most significant sources of the pollutants that contribute to photochemical smog is exhaust from automobiles. In June 2008, Honda introduced a car that runs on hydrogen and electricity and emits only water. How much extra would you be willing to pay for a car that doesn't emit any toxic exhaust fumes? Do you think that the government should provide strong incentives, such as tax breaks, for people who pay extra for these cars?
2. Why do you think exposure to ozone causes more serious respiratory problems in the young and the elderly than in healthy adults?
3. How often do you use public transportation? Whether you live in a big city or not, would you be willing to use public transportation if it added, for example, an hour to the total time you spend each day traveling to school or work? Why or why not?

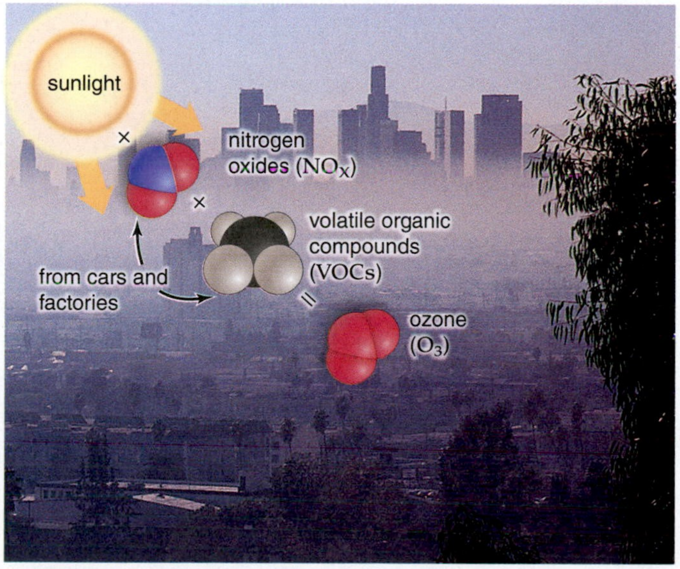
sunlight
nitrogen oxides (NO_x)
volatile organic compounds (VOCs)
from cars and factories
ozone (O_3)

a. Ground-level ozone formation

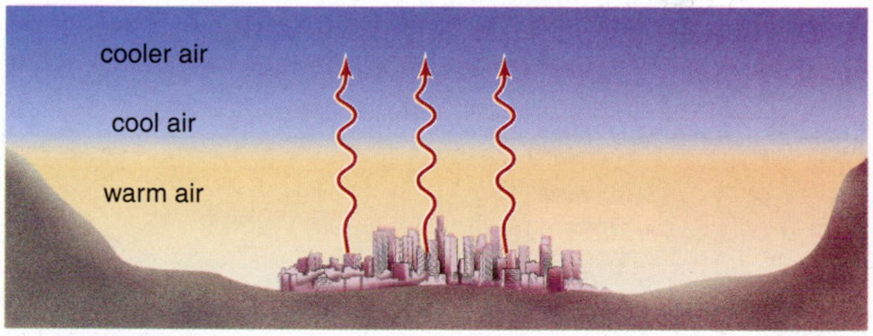
cooler air
cool air
warm air

b. Normal pattern

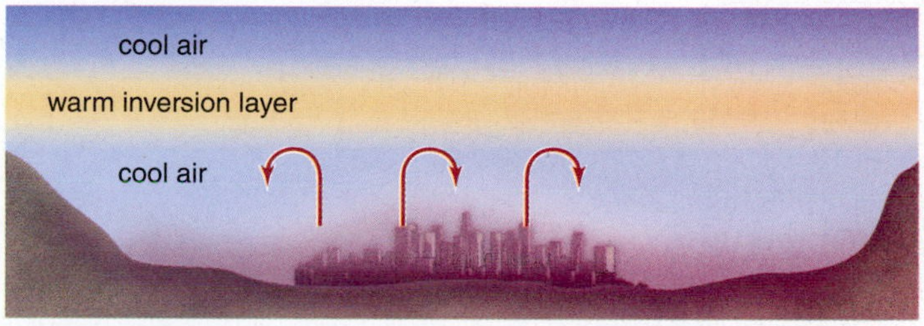
cool air
warm inversion layer
cool air

c. Thermal inversion

Figure 35B Thermal inversion. **a.** Los Angeles is the "air pollution capital" of the world. Its millions of cars and thousands of factories make this city particularly susceptible to photochemical smog, which contains ozone due to the chemical reaction shown here. **b.** Normally, pollutants escape into the atmosphere when warm air rises. **c.** During a thermal inversion, a layer of warm air (warm inversion layer) overlies and traps pollutants in cool air below.

the anaerobic mud of lakes, bogs, and estuaries carry out this process as a part of their own metabolism. In the nitrogen cycle, denitrification would counterbalance nitrogen fixation except for human activities.

Human Activities

Human activities nearly double the nitrogen fixation rate when fertilizers are produced from N_2. Fertilizer, which also contains phosphate, runs off into lakes and rivers and results in algal overgrowth and fish kill.

Fertilizer use also results in the release of nitrous oxide (N_2O), a greenhouse gas, component of acid rain, and contributor to ozone shield depletion.

The Carbon Cycle

In the carbon cycle, plants in both terrestrial and aquatic ecosystems take up carbon dioxide (CO_2) from the air through photosynthesis. They incorporate carbon into food that is used by autotrophs and heterotrophs alike (Fig. 35.10). When all organisms, including plants, respire, a portion of this carbon is returned to the atmosphere as carbon dioxide.

In aquatic ecosystems, the exchange of carbon dioxide with the atmosphere is primarily indirect. However, there is a small quantity of free carbon dioxide in the water. Carbon dioxide from the air combines with water to produce bicarbonate ions (HCO_3^-), a source of carbon for algae that make up the base of most aquatic ecosystems. Similarly, when aquatic organisms respire, the carbon dioxide they give off becomes bicarbonate ion. The amount of bicarbonate in the water is in equilibrium

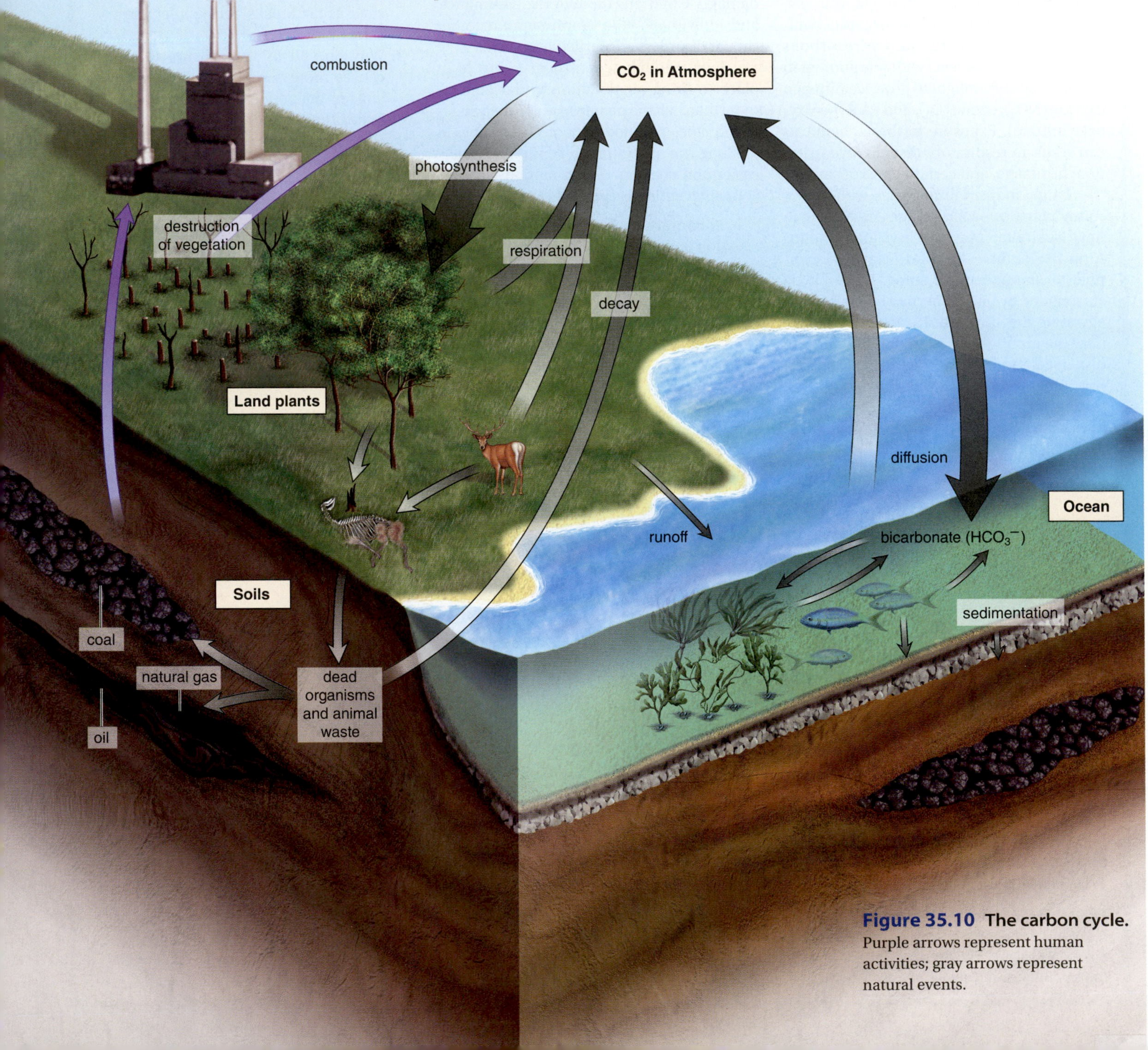
combustion
CO₂ in Atmosphere
photosynthesis
destruction of vegetation
respiration
decay
Land plants
diffusion
Ocean
runoff
bicarbonate (HCO₃⁻)
sedimentation
Soils
coal
natural gas
dead organisms and animal waste
oil

Figure 35.10 **The carbon cycle.** Purple arrows represent human activities; gray arrows represent natural events.

with the amount of carbon dioxide in the air. Thus far, the ocean has acted as a buffer absorbing much of the excess CO_2, but scientists are concerned about how much the ocean can actually hold.

Reservoirs Hold Carbon

Living and dead organisms contain organic carbon and serve as one of the reservoirs for the carbon cycle. The world's biotic components, particularly trees, contain 800 billion tons of organic carbon, and an additional 1,000–3,000 billion metric tons are estimated to be held in the remains of plants and animals in the soil. Before decomposition can occur, some of these remains are subjected to physical processes that transform them into coal, oil, and natural gas. We call these materials the **fossil fuels.** Most of the fossil fuels were formed during the Carboniferous period, 286 to 360 million years ago, when an exceptionally large amount of organic matter was buried before decomposing. Another reservoir is the inorganic carbonate that accumulates in limestone and in calcium carbonate shells of many marine organisms.

Human Activities

The transfer rates of carbon dioxide due to photosynthesis and cellular respiration are just about even. However, due to human activities, more carbon dioxide is being released into the atmosphere than is being removed. In 1850, atmospheric CO_2 was at about 280 parts per million (ppm); today, it is over 390 ppm. This increase is largely due to the burning of fossil fuels and the destruction of forests to make way for farmland and pasture. Today, the amount of carbon dioxide released into the atmosphere is about twice the amount that remains in the atmosphere. It's believed that most of this dissolves in the ocean.

Video Global Warming

Video Warming Hurts Rice

Video Karoo Global Warming

The increased amount of carbon dioxide (and other gases) in the atmosphere is causing a rise in temperature called **global warming.** These gases allow the sun's rays to pass through, but they absorb and radiate heat back to Earth, a phenomenon called the **greenhouse effect.** Global warming is contributing to **climate change,** which is causing significant changes to the Earth, as discussed in Chapter 36. The Ecology feature, "Global Climate Change," discusses worldwide efforts to cut back on greenhouse gases.

Check Your Progress 35.3

1. Recognize the steps of each of the biogeochemical cycles.
2. Provide examples of how human activities can disrupt the biogeochemical cycles.

SCIENCE IN YOUR LIFE ▶ ECOLOGY

Global Climate Change

Scientists around the world are working on collecting and interpreting environmental indicators that will help us understand how and why the Earth's climate is changing. Changes in the average temperature of a region, precipitation patterns, sea levels, and greenhouse gas concentrations are all indicators that our climate is changing.

Since 1901, the average temperature across the United States has risen, with 2000–2009 being the warmest decade on record worldwide with 30–60% of the United States experiencing drought conditions. Average precipitation rates have also increased by 6% over the past century. Since 1990, the United States has experienced eight of the top ten years of extreme precipitation events.

Sea levels worldwide have risen an average of 1 in per decade due to the overall increase in the surface temperature of the world's oceans. Climate models suggest that we will see a rise in sea levels of 3–4 ft over the next century. Even if the rising waters don't produce flooding, many coastal areas will be exposed to increasingly severe storms and storm surges that could lead to significant economic losses.

Between 1990 and 2008, the U.S.-produced greenhouse gas emissions have increased by 14%. Generation of electricity is the largest producer of greenhouse gas emissions in the United States, followed by transportation. Carbon dioxide and methane are two of the most common greenhouse gases. These gases allow the sun's rays to pass through, but then trap the heat and prevent it from escaping. The amount of carbon being released into the atmosphere is exceeding the amount of carbon that is being sequestered by the producers on Earth.

The Kyoto Protocol was initially adopted on December 11, 1997, and entered into force on February 16, 2005. The goal of the protocol was to achieve stabilization and reduction of the greenhouse gas concentrations in the atmosphere. Unfortunately, the Copenhagen conference in 2009 ended without any type of binding agreement for long-term action against climate change. It did produce a collective commitment by many developed nations to raise $30 billion to be used to help poor nations cope with the effects of, and combat, climate change.

The 2010 Cancun summit on climate change helped to solidify the agreement to raise funds. Because deforestation produces about 15% of the global carbon emissions, many developing countries will be able to receive incentives to prevent the destruction of their rainforests.

The main concern of the conferences is that the hard decisions of making significant changes to our greenhouse emissions continues to be pushed off into the future. The longer we delay in making changes, the higher the risks become.

Questions to Consider

1. Should the United States and other developed nations pay developing nations to preserve their forests?
2. Do individuals have a personal responsibility to help prevent climate change, or is it a governmental responsibility?

connect BIOLOGY Explore the concepts through a variety of multimedia assets, question types, and data interpretation.

www.mcgrawhillconnect.com

Case Study Conclusion

Scientists predict that the Earth's ecosystems will dramatically change within the next 100 years due to human actions. Ecosystems are characterized by the interactions of living (biotic) and nonliving (abiotic) components. The biotic components of an ecosystem depend on the cycling of nutrients such as carbon, nitrogen, and phosphorus. The burning of fossil fuels and the release of excessive carbon dioxide into the atmosphere is playing a significant role in global climate change. Sea levels and extreme weather events such as hurricanes and tornados are expected to increase in frequency. As we enrich aquatic ecosystems with additional amounts of phosphates and nitrates, we are causing cultural eutrophication to occur. The flow of energy through the members of biotic communities and how abiotic chemicals and nutrients cycle through ecosystems are critical to the future health of the Earth.

MEDIA STUDY TOOLS

www.mhhe.com/maderinquiry14

Enhance your study of this chapter with study tools and practice tests. Also ask your instructor about the resources available through ConnectPlus, including LearnSmart, the media-rich eBook, interactive learning tools, and animations.

Mc Graw Hill connect plus+ BIOLOGY

SUMMARIZE

5.1 The Biotic Components of Ecosystems

- An ecosystem is composed of populations of organisms (**biotic** component) plus their physical environment (**abiotic** component).
- **Producers** are **autotrophs** that transform solar energy into food for themselves and all consumers. **Consumers** are **heterotrophs** that take in organic food. As **herbivores** feed on plants or algae and **carnivores** feed on herbivores, energy is converted to heat. **Omnivores** are organisms that feed on both plants and animals. Feces, urine, and dead bodies become food for **decomposers.** Decomposers return some proportion of inorganic nutrients to autotrophs, and other portions are imported or exported among **ecosystems** in global cycles. **Detritus** is the partially decomposed matter in the soil and water.
- Ecosystems are characterized by energy flow and chemical cycling. Energy is lost from the biosphere, but inorganic nutrients are not. They recycle within and among ecosystems. Eventually, all the solar energy that enters an ecosystem is converted to heat, and thus ecosystems require a continual supply of solar energy.

35.2 Energy Flow

- Ecosystems contain **food webs** in which the various organisms are connected by trophic relationships. In a **grazing food web, food chains** begin with a producer. In a **detrital food web,** food chains begin with detritus. The two food webs are joined when the same consumer links both a **grazing food chain** and a **detrital food chain.**
- A **trophic level** contains all the organisms that feed at a particular link in a food chain. **Ecological pyramids** show trophic levels stacked one on top of the other like building blocks. They are shaped like pyramids because most energy is lost from one trophic level to the next, and thus the number of species that can be sustained decreases. **Biomass** is the number of organisms multiplied by their weight.

35.3 Global Biogeochemical Cycles

- **Biogeochemical cycles** consist of reservoirs, exchange pools, and biotic communities. A **reservoir** pool contains elements available on a limited basis to living organisms, such as **fossil fuels,** sediments, and rocks. An **exchange pool,** such as the atmosphere, soil, and water, is a ready source of nutrients for living organisms.
- In the **water (hydrologic) cycle, evaporation** over the ocean is not compensated for by rainfall. Evaporation from terrestrial ecosystems includes transpiration from plants. **Condensation** occurs when a gas is changed into a liquid. **Precipitation** returns water to the land in the form of rain, snow, sleet and fog. Rainfall over land results in bodies of fresh water plus groundwater, including **aquifers.** Eventually, all water returns to the oceans.
- In the phosphorus cycle, the biotic community recycles phosphorus back to the producers, and only limited quantities are made available by the weathering of rocks. Phosphates are mined for fertilizer production. When phosphates and nitrates enter lakes, ponds, and eventually the ocean, pollution and **cultural eutrophication,** or overenrichment, occurs.
- In the nitrogen cycle, certain bacteria in water, soil, and within root nodules undergo **nitrogen fixation. Denitrification** is when various bacteria return nitrogen to the atmosphere. Human activities convert atmospheric nitrogen to fertilizer, which is broken down by soil bacteria. Humans also burn fossil fuels. In this way, nitrogen oxides can contribute to the formation of smog and acid rain, both of which are detrimental to animal and plant life.
- In the carbon cycle, organisms add as much carbon dioxide to the atmosphere as they remove. Shells in ocean sediments, organic compounds in living and dead organisms, and fossil fuels are reservoirs for carbon. Human activities, such as burning fossil fuels and trees, add carbon dioxide and other gases to the atmosphere. A buildup of these "greenhouse gases" allows the sun's rays to pass through but they radiate the heat back to the Earth creating a **greenhouse effect.** This is leading to **global warming** and causing **climate change.** A rise in sea level and a change in climate patterns is expected to follow.

ASSESS

Testing Yourself

Choose the best answer for each question.

1. Which of the following would be a primary consumer in a vegetable garden?
   a. aphid sucking sap from cucumber leaves
   b. lady beetle eating aphids
   c. songbird eating lady beetles
   d. fox eating songbirds
   e. All of the choices are correct.
2. Label the populations in the ecosystem diagram to the right.

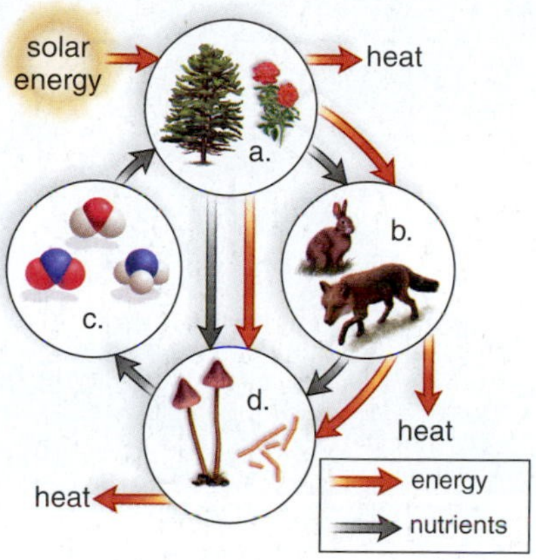
solar energy
heat
a.
b.
c.
d.
heat
heat
energy
nutrients

3. During chemical cycling, inorganic nutrients are typically returned to the soil by
   a. autotrophs.
   b. detritivores.
   c. decomposers.
   d. tertiary consumers.
4. When a heterotroph takes in food, only a small percentage of the energy in that food is used for growth. The remainder is
   a. not digested and eliminated as feces.
   b. excreted as urine.
   c. given off as heat.
   d. All of the choices are correct.
   e. None of the choices are correct.
5. The first trophic level on a food web is composed of the
   a. producers.
   b. primary consumers.
   c. secondary consumers.
   d. tertiary consumers.
6. Which of the following is a grazing food chain?
   a. leaves ⟶ detritivores ⟶ deer ⟶ owls
   b. birds ⟶ mice ⟶ snakes
   c. nuts ⟶ leaf-eating insects ⟶ chipmunks ⟶ hawks
   d. leaves ⟶ leaf-eating insects ⟶ mice ⟶ snakes
7. In this diagram, label the trophic levels (blanks a–d), using two of these terms for each level: producers, top carnivores, secondary consumers, autotrophs, primary consumers, tertiary consumers, carnivores, herbivores.

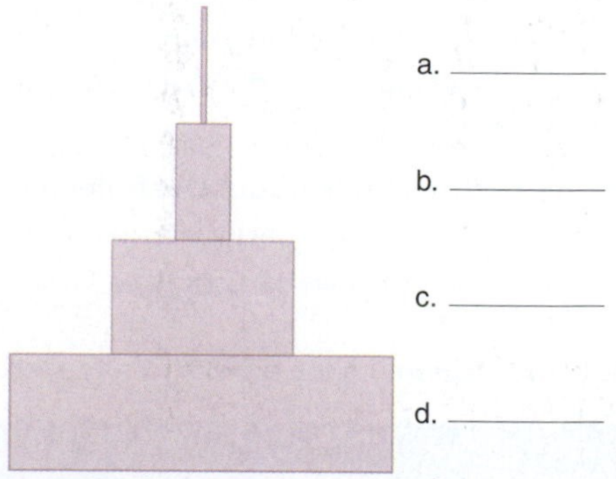
a. ________
b. ________
c. ________
d. ________

8. In a grazing food web, carnivores that eat herbivores are
   a. producers.
   b. primary consumers.
   c. secondary consumers.
   d. tertiary consumers.
9. Few carnivores can be supported in a food web because
   a. mineral nutrients do not cycle quickly enough.
   b. they produce only a few young per generation.
   c. their generation time is very long.
   d. a large amount of energy is lost at each trophic level.
10. Which of the following is a sedimentary biogeochemical cycle?
    a. carbon
    b. nitrogen
    c. phosphorus
    d. water
11. Which of the following is not a component of the water cycle?
    a. evaporation
    b. filling of aquifers
    c. runoff
    d. photosynthesis
    e. condensation

For questions 12–14, match each type of nitrogen conversion with a process in the key.

Key:

a. nitrogen fixation b. nitrification c. denitrification

12. Nitrate to nitrogen gas
13. Nitrogen gas to nitrate
14. Nitrogen gas to ammonium
15. Choose the statement that is true concerning this food chain: grass ⟶ rabbits ⟶ snakes ⟶ hawks
    a. Each predator population has a greater biomass than its prey population.
    b. Each prey population has a greater biomass than its predator population.
    c. Each population is omnivorous.
    d. Each population returns inorganic nutrients and energy to the producer.
    e. Both a and c are correct.

ENGAGE

Virtual Lab
Model Ecosystems

The virtual lab "Model Ecosystems" allows you to explore the factors that influence the movement of energy through an ecosystem.

Thinking Critically

1. A large forest has been removed by timber harvest (clear-cutting) and the land has not been replanted. After several years, humidity and rainfall in the area seem to have decreased. Use your knowledge of the water cycle to explain these observations.
2. In mountainous regions of the western United States, large predators (e.g., wolves in Yellowstone National Park), which were previously driven out of the area, are being reintroduced. People living in these areas are concerned that livestock may be lost if there is not enough wild food for the predators. What types of data are needed concerning food webs and ecological pyramids to ensure the successful reintroduction of these predators and minimal impact on livestock?
3. Based on your knowledge of the carbon cycle, why are fossil fuels such as oil so limited?

36 Major Ecosystems of the Biosphere

CHAPTER OUTLINE

BEFORE YOU BEGIN

Before beginning this chapter, take a few moments to review the following discussions:

Section 8.4 How do cellular processes like photosynthesis adapt to climate?

Section 34.3 What types of symbiotic relationships may allow species to survive in harsh environments?

Section 35.3 What are the major reservoirs for water and carbon?

CASE STUDY DDT in the Water

DDT was originally developed as a broad-spectrum pesticide to help control insect populations on South Pacific islands during World War II. Due to concerns about environmental effects, DDT has been banned in the United States since 1972, although it is still used in other parts of the world in agricultural and disease-control programs. However, we are still finding traces of it in various locations across the country. DDT's persistence in the environment causes it to be a global environmental and health problem.

Due to its stability (it takes 15 years to break down), and its ability to accumulate in fatty tissue, it has been found in organisms ranging from Adelie penguins in Antarctica to bald eagles in the United States and even in humans. The harmful effects of DDT include liver damage, a potential link to various cancers, decreased reproductive success, and temporary damage to the nervous system. Human exposure to DDT tends to come in the form of eating plants and animals that are contaminated. The Great Lakes and many other waterways have fish consumption advisories due to the presence of DDT in these systems. Once DDT enters an aquatic ecosystem, it can increase in concentration as it moves up the food chain.

The use of DDT in many developing nations around the world enables it to become a part of the global ecosystem. Soil and sediment runoff provides a means for DDT to enter into the aquatic ecosystems. As the water evaporates, the DDT may enter into the atmosphere. When the water reaches cooler latitudes, it condenses, forming rain. The rain carries the DDT particles back to the surface. The global wind circulation patterns allow DDT to be carried to every region of the planet.

As you read through the chapter, think about the following questions:

1. Should DDT use be permitted in any part of the world?
2. Is there any biome in which DDT can be used that would not impact another biome?
3. Is it possible to isolate the actions in one biome from the rest of the planet?

36.1 Climate and the Biosphere

Learning Outcomes

Upon completion of this section, you should be able to

1. Describe how solar radiation produces variations in the Earth's climate.
2. Explain how global air circulation patterns and physical geographic features are associated with the Earth's temperature and rainfall.

Climate refers to the prevailing weather conditions in a particular region. Climate is dictated by temperature and rainfall, which in turn are influenced by the following factors: (1) variations in the distribution of solar radiation due to the tilt of the Earth as it orbits about the sun; and (2) other effects, such as topography and proximity to water bodies.

Effect of Solar Radiation

Because the Earth is spherically shaped, the sun's rays are more direct at the equator and more spread out progressing toward the poles. Therefore, the tropical regions nearest the equator are warmer than the temperate regions farther away from the equator. In addition, Earth does not face the sun directly. Rather, it is on a slight tilt (about 23°) away from the sun. As the Earth orbits around the sun throughout the year, different parts of the planet are tilted toward or away from the sun, which determines the seasons. For example, during winter, the Northern Hemisphere is tilted away from the sun (Fig. 36.1). At the same time, the Southern Hemisphere is tilted toward the sun and it experiences summer.

Because the Earth completes one rotation on its axis per day and its surface consists of continents and oceans, the flow of warm and cold air form three large circulation patterns in each hemisphere. The direction in which the air rises and cools determines the direction of the wind (Fig. 36.2). At the equator, the sun heats the air and evaporates water. The warm, moist air rises and loses most of its moisture as rain, resulting in the greatest amounts of rainfall occurring nearest to the

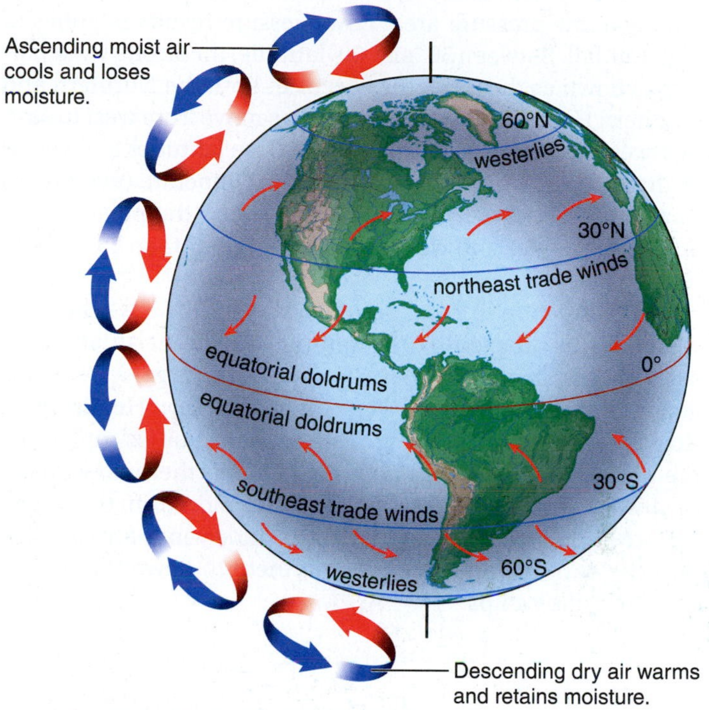

Ascending moist air cools and loses moisture.
60°N
westerlies
30°N
northeast trade winds
equatorial doldrums
0°
equatorial doldrums
30°S
southeast trade winds
60°S
westerlies
Descending dry air warms and retains moisture.

Figure 36.2 Global wind circulation. Air ascends and descends as shown because Earth rotates on its axis. Also, the trade winds blow from the northeast to the west in the Northern Hemisphere, and blow from the southeast to the west in the Southern Hemisphere. The westerlies blow toward the east.

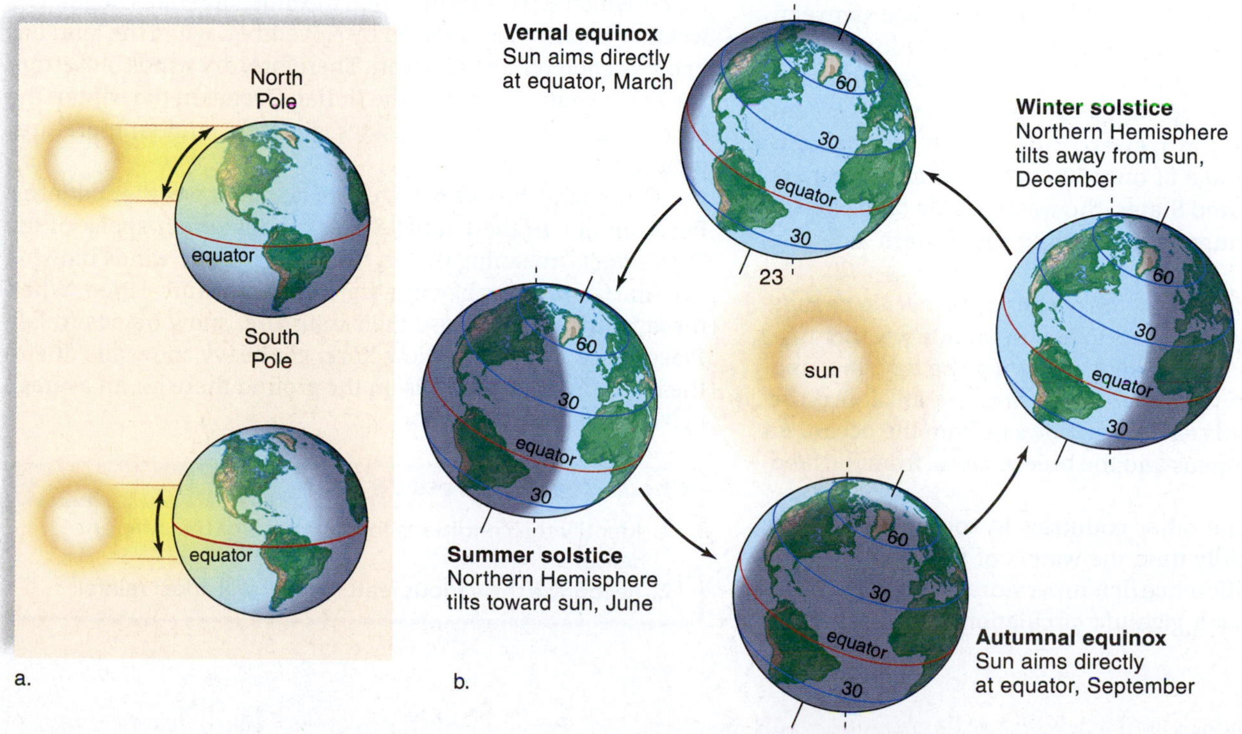

North Pole
equator
South Pole
equator
a.
Vernal equinox
Sun aims directly at equator, March
60
30
equator
30
23°
Winter solstice
Northern Hemisphere tilts away from sun, December
60
30
equator
30
sun
60
30
equator
30
Summer solstice
Northern Hemisphere tilts toward sun, June
60
30
equator
30
Autumnal equinox
Sun aims directly at equator, September
b.

Figure 36.1 Distribution of solar energy. **a.** Because Earth is a sphere, beams of solar energy striking the planet near one of the poles are spread over a wider area than similar beams striking Earth at the equator. **b.** The seasons of the Northern and Southern Hemispheres are due to the tilt of Earth on its axis as it rotates around the sun.

equator. The rising air flows toward the poles, but at about 30° north and south latitude, it sinks toward Earth's surface and reheats. As the dry air descends and warms once again, areas of high pressure are generated, which results in low rainfall. As a result, the great deserts of Africa, Australia, and the Americas occur at these latitudes. At about 60° north and south latitude, the warm air rises and cools as it does near the equator, producing a low-pressure area. Low pressure results in zones of high rainfall. Between 30° and 60° latitude, the strong wind patterns known as the westerlies occur in both the Northern and Southern Hemispheres. The westerlies move from west to east. The west coasts of the continents at these latitudes are wet, as in the U.S. Pacific Northwest, where a temperate (evergreen) rain forest is located. Weaker winds, called the polar easterlies, blow from east to west at latitudes higher than 60° in both hemispheres.

The direction of wind patterns, such as the easterlies and westerlies, is affected by the spinning of the Earth about its axis. That is, in the Northern Hemisphere, large-scale winds generally move clockwise, while in the Southern Hemisphere, they move counterclockwise. This explains, for example, why the northeast trade winds blow from the northeast toward the southwest and the southeast trade winds blow from the southeast toward the northwest (Fig. 36.2). Trade winds are so called because early sailors depended on them to power the movement of sailing ships.

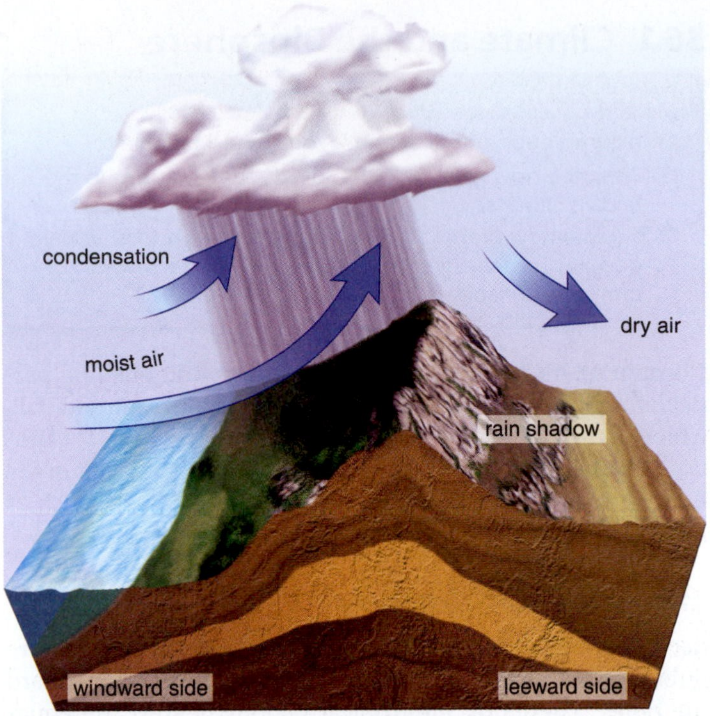
condensation
moist air
dry air
rain shadow
windward side
leeward side

Figure 36.3 Formation of a rain shadow. When winds from the sea cross a coastal mountain range, they rise and release their moisture as they cool the windward side of the mountain. The leeward side of a mountain is warmer and receives relatively little rain. Therefore, it is said to lie in a "rain shadow."

Other Effects

The term topography refers to the physical features, or "the lay," of the land. One physical feature that affects climate is the presence of mountains. As air blows up and over a mountain range, it rises and cools, causing condensation to occur. One side of the mountain, called the windward side, receives more rainfall than the other side, called the leeward side. On the leeward side, the dry air descends, often producing a dry arid environment (Fig. 36.3). The difference between the windward side and the leeward side can be quite dramatic. In the Hawaiian Islands, for example, the windward side of the mountains typically receives more than 750 cm of rain a year while the leeward side, which is in a **rain shadow,** gets an average of only 50 cm of rain and is generally sunny. In the United States, the western side of the Sierra Nevada mountain range is lush, while the eastern side is a semidesert.

In contrast to landmasses, the oceans are slower to change temperature. This causes coasts to have a unique weather pattern that is not seen farther inland. During the day, the land warms more quickly than the ocean, and the air above the land rises. Then a cool sea breeze blows in from the ocean. At night, the reverse happens and the breeze blows from the land to the sea.

In India and some other countries in southern Asia, the land heats more rapidly than the waters of the Indian Ocean during spring. The difference in temperature between the land and the ocean causes a gigantic circulation of air: warm air rises over the land, and cooler air comes in off the ocean to replace it. As the warm air rises, it loses its moisture, creating a **monsoon** climate in which wet ocean winds blow onshore for almost half the year. During the monsoon season, rainfall is particularly heavy on the windward side of hills. Cherrapunji in northern India receives an annual average of 1,090 cm of rain a year because of its high altitude. The chief crop of India is rice, which starts to grow when the monsoon rains begin. The weather pattern has reversed by November, when the land has become cooler than the ocean. Therefore, dry winds blow from the Asian continent across the Indian Ocean. In the winter, the air over the land is dry, the skies cloudless, and temperatures pleasant.

Other large bodies of water create major weather patterns. For example, in the United States, people often speak of the "lake effect," meaning that in the winter, arctic winds blowing over the Great Lakes become warm and moisture-laden. When these winds rise and lose their moisture, snow begins to fall. Places such as Buffalo, New York, get heavy snowfalls due to the lake effect, and snow is on the ground there for an average of 90 to 140 days every year.

Check Your Progress 36.1

1. Identify the conditions that account for the different seasons.
2. Recognize the various features that will affect rainfall.

36.2 Terrestrial Ecosystems

Learning Outcomes

Upon completion of this section, you should be able to

1. Identify the geographical distribution of the major terrestrial ecosystems.
2. Summarize the key characteristics of the major terrestrial biomes.

A major type of terrestrial ecosystem is called a **biome.** Biomes are characterized by particular climatic conditions, as well as the plants and animals living there. When terrestrial biomes are plotted according to their mean annual temperature and mean annual precipitation, a particular pattern results (Fig. 36.4*a*). The distribution of biomes is shown in Figure 36.4*b*. Even though this figure shows what appear to be clear demarcations, note that biomes gradually change from one type to another. To simplify our discussion in this chapter, we will group the

Figure 36.4 Pattern of biome distribution. **a.** Pattern of world biomes in relation to temperature and moisture. The dashed line encloses a wide range of environments in which either grasses or woody plants can dominate the area, depending on the soil type. **b.** The same type of biome can occur in different regions of the world, as shown on this global map.

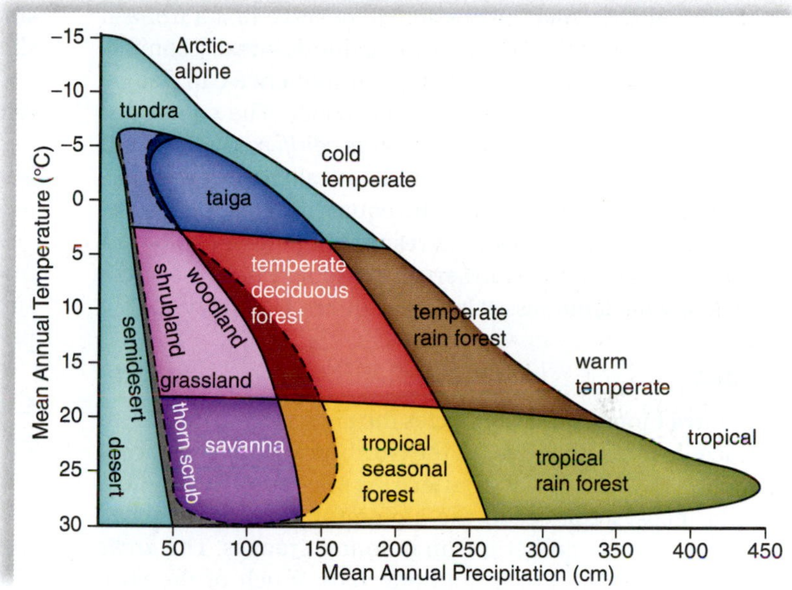
Arctic-alpine
tundra
cold temperate
taiga
temperate deciduous forest
woodland
shrubland
temperate rain forest
warm temperate
grassland
semidesert
desert
thorn scrub
savanna
tropical seasonal forest
tropical rain forest
tropical
Mean Annual Temperature (°C)
-15
-10
-5
0
5
10
15
20
25
30
50
100
150
200
250
300
350
400
450
Mean Annual Precipitation (cm)

a. Biome pattern of temperature and precipitation

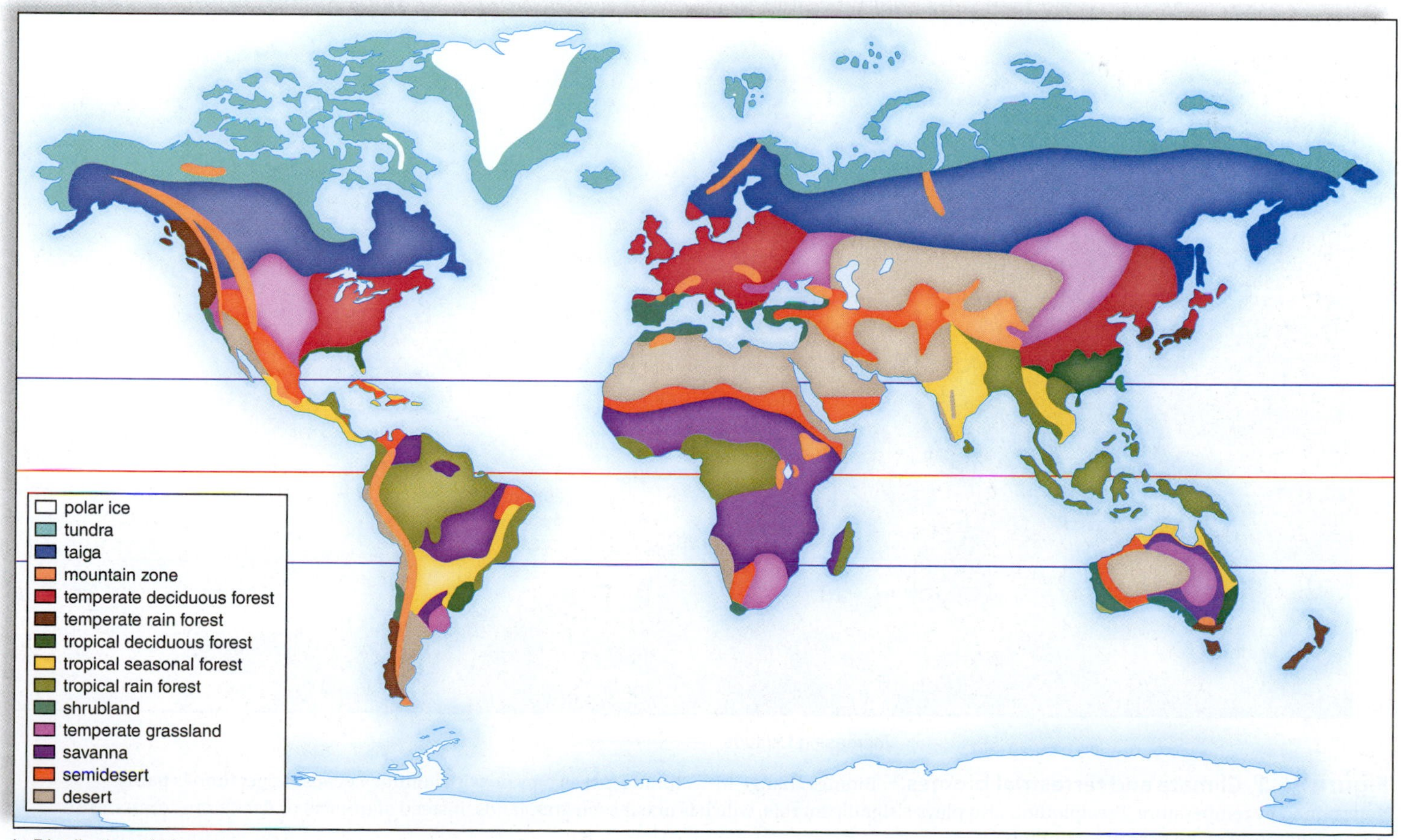
polar ice
tundra
taiga
mountain zone
temperate deciduous forest
temperate rain forest
tropical deciduous forest
tropical seasonal forest
tropical rain forest
shrubland
temperate grassland
savanna
semidesert
desert

b. Distribution of biomes

biomes into general categories. Although we will be discussing each type of biome separately, keep in mind that all biomes are linked to form the biosphere, with each biome having connections to all the other terrestrial and aquatic ecosystems.

Animation Biomes

The distribution of the biomes and their corresponding communities of organisms is determined principally by differences in climate due to solar radiation, water, and defining topographical features. Both latitude and altitude are responsible for temperature gradients. When traveling from the equator to the North Pole, it would be possible to observe first a tropical rain forest, followed by a temperate deciduous forest, a coniferous forest, and tundra. A similar type of sequence can also be seen when ascending a mountain (Fig. 36.5). The coniferous forest of a mountain is called a *montane coniferous forest,* and the tundra near the peak of a mountain is called an *alpine tundra.* However, when going from the equator to the South Pole, one would not reach a region corresponding to the coniferous forest and tundra of the Northern Hemisphere because of the absence of large landmasses in the Southern Hemisphere.

Tundra

The **tundra** biome, which encircles the arctic region just south of the ice-covered polar seas in the Northern Hemisphere, covers about 20% of Earth's land surface. As previously mentioned, a similar ecosystem, called the alpine tundra, occurs above the timberline on mountain ranges. The *arctic tundra* is cold and dark much of the year. Because rainfall amounts to only about 20 cm a year, the tundra could possibly be considered a desert. However, melting snow creates a landscape of pools and bogs in the summer. This is because only the topmost layer of earth thaws. The **permafrost** beneath this layer is always frozen, and therefore, drainage is minimal.

Trees are not found in the tundra because the growing season is too short, their roots cannot penetrate the permafrost, and they cannot become anchored in the shallow, boggy soil during the brief summer. Instead, the ground is covered with short grasses and sedges, as well as numerous patches of lichens and mosses in summer (Fig. 36.6*a*). Dwarf woody shrubs, such as dwarf birch, flower and seed quickly during the short growing season.

Few animals live in the tundra year-round, but nearly all species have adaptations for living in extreme cold and short growing seasons. In winter, for example, the ptarmigan (a grouse) burrows in the snow during storms, and the musk ox conserves heat because of its thick coat and short, squat body. Other species, such as caribou (Fig. 36.6*b*) and reindeer, migrate to and from the tundra, as do the wolves that prey upon them. In the summer, the tundra is alive with numerous insects and birds, particularly shore-birds and waterfowl that migrate inland.

Coniferous Forests

Coniferous forests are found in three locations: in the **taiga,** which extends around the world in the northern part of North America and Eurasia; near mountaintops (where it is called a montane coniferous forest); and along the Pacific coast of North America, as far south as northern California.

Taiga typifies the coniferous forest with its cone-bearing trees, such as spruce, fir, and pine (Fig. 36.7*a*). These trees are

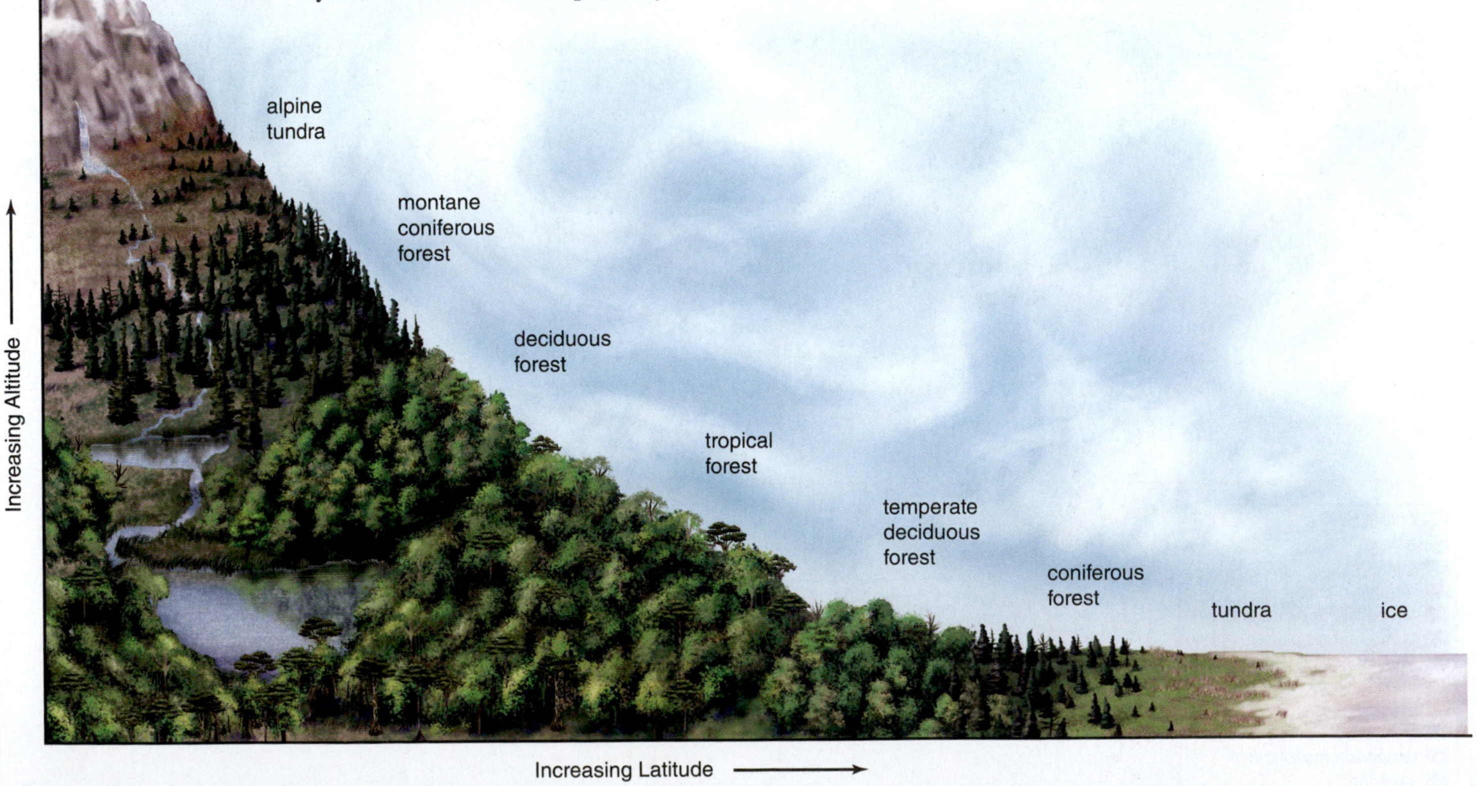
ice
alpine tundra
montane coniferous forest
deciduous forest
tropical forest
temperate deciduous forest
coniferous forest
tundra
ice
Increasing Altitude
Increasing Latitude

Figure 36.5 Climate and terrestrial biomes. Biomes change with altitude just as they do with latitude because vegetation is partly determined by temperature. Precipitation also plays a significant role, which is one reason grasslands, instead of tropical or deciduous forests, are sometimes found at the base of mountains.

a. Tundra

b. Bull caribou

Figure 36.6 Tundra. **a.** Vegetation consists principally of lichens, mosses, grasses, and low-growing shrubs. **b.** Caribou feed on the vegetation in the summer.

Figure 36.7 Coniferous forest. **a.** The taiga, which means swampland, is a coniferous forest that spans northern Europe, Asia, and North America. **b.** The appellation "spruce-moose" refers to the dominant presence of spruce trees and moose, which frequent the ponds.

a. Spruce trees in the taiga biome

b. Bull moose

well adapted to the cold because both the leaves (reduced to needles) and bark have thick coverings. Also, the needle-like leaves can withstand the weight of heavy snow. There is a limited understory of plants, but the floor is covered by low-lying mosses and lichens. Birds harvest the seeds of the conifers, and bears, deer, moose, beaver, and muskrat live around the cool lakes and along the streams (Fig. 36.7*b*). Wolves prey on these larger mammals. A montane coniferous forest also harbors the wolverine and the mountain lion.

The coniferous forest that runs along the west coast of Canada and the United States is sometimes called a *temperate rain forest.* The plentiful rainfall and rich soil have produced some of the tallest conifer trees ever known, including the coastal redwoods. Small sections of this forest are considered old-growth forest because trees average over 150 years old, with some trees even as old as 800 years. It truly is an evergreen forest because mosses, ferns, and other plants often grow on tree trunks. Whether the limited portion of old-growth forest that remains should be protected from logging is an important and controversial conservation issue.

Temperate Deciduous Forests

Temperate deciduous forests are found south of the taiga in eastern North America, eastern Asia, and much of Europe. The climate in these areas is moderate, with relatively high rainfall (75–150 cm per year). The seasons are well defined, and the growing season ranges between 140 and 300 days. The trees, such as oak, beech, and maple, have broad leaves and are deciduous because they lose their leaves in fall and regrow them in spring.

The tallest trees form a canopy, an upper layer of leaves that are the first to receive sunlight and thereby create shade below (Fig. 36.8). Even so, enough sunlight penetrates to provide energy for another layer of trees, called understory trees. Beneath these trees are shrubs and herbaceous plants that may flower in the spring before the trees have put forth their leaves. Mosses, lichens, and ferns can reside beneath the shrub layer. This stratification provides a variety of habitats for insects and birds. Ground life is also plentiful. Squirrels, cottontail rabbits, shrews, skunks, woodchucks, and chipmunks are small herbivores. These and ground birds such as turkeys, pheasants, and grouse are preyed on by red foxes. White-tail deer and black bears have recently increased in number. In contrast to

a.

Figure 36.8 Temperate deciduous forest. **a.** The Shawnee National Forest in Illinois is home to many varied plants and animals. **b.** Marsh marigolds may be found in wetland areas, chipmunks feed on acorns, and bobcats prey on these and other small mammals.

b.

marsh marigolds

eastern chipmunk

bobcat

the taiga, a greater diversity of amphibians and reptiles occur in this biome because the winters are not as cold. Frogs and turtles generally prefer an aquatic existence, as do the beaver and muskrat.

Abundant fruits, nuts, and berries provide a supply of food for the winter. The leaves, after turning brilliant colors and falling to the ground, contribute to the rich layer of humus after decomposition. The minerals within the rich soil are washed far into the ground by spring rains, but the deep tree roots capture these nutrients and cycle them back through the forest system.

Tropical Forests

The most common type of **tropical forest** is the *tropical rain forest,* which is found in areas of South America, Africa, and the Indo-Malayan region near the equator. The temperature in a tropical rain forest is always warm (between 20° and 25°C), and rainfall is plentiful (a minimum of 190 cm per year). As a result of these favorable climate conditions, this is the richest land biome in terms of species diversity. The diversity of species is enormous—a 10-km^2 area of tropical rain forest may contain 750 species of trees and 1,500 species of flowering plants.

A tropical rain forest has a complex structure, with many levels of life. Some broadleaf evergreen trees grow to heights of 15–50 m or more. These tall trees often have trunks buttressed at ground level to prevent them from toppling over. Lianas, or woody vines, often encircle rainforest trees as they grow. Although some animals live on the ground (e.g., pacas, agoutis, peccaries, and armadillos), many also live in the trees (Fig. 36.9). Insect life is so abundant that the majority of species have not yet been identified. Termites play a vital role in the decomposition of woody plant material, and ants are found everywhere. The various birds, such as hummingbirds, parakeets, parrots, and toucans, are often beautifully colored. Amphibians and reptiles are well represented by many types of frogs, snakes, and lizards. Lemurs, sloths, and monkeys are well-known primates that feed on the fruits of the trees. The largest carnivores are the big cats—the jaguars in South America and the leopards in Africa and Asia.

Many animals spend their entire life in the canopy, as do some plants. **Epiphytes** are plants that grow on other plants but usually have roots of their own that absorb moisture and minerals leached from the canopy. Others catch rain and debris

Figure 36.9 Representative animals and plants of tropical rain forests. Some plants and animals of the rain forest are shown here.

poison-dart frog

spike-headed katydid

panther

blue and gold macaw

blue morpho butterfly

orchid

chameleon

in hollows produced by overlapping leaf bases. The most common epiphytes are related to pineapples, orchids, and ferns.

Whereas the soil of a temperate deciduous forest biome is rich enough for agricultural purposes, the soil of a tropical rain forest biome is not. Nutrients are cycled directly from the litter to the plants again. Productivity is high because of warm and consistent temperatures, a year-long growing season, and the rapid recycling of nutrients from litter decomposition. To make up for the soil's low fertility, trees are sometimes cut and burned, so that the resulting ashes can provide enough nutrients for several harvests. This practice, called swidden agriculture, or slash-and-burn agriculture, can be successful if done on a small scale. Once the crops are harvested, the soil nutrients become depleted, consequently removing nutrients from the system. Erosion is high due to lack of tree roots and the heavy rainfall causing additional nutrients to be washed away.

While we usually think of tropical forests as nonseasonal rain forests, *tropical deciduous forests* that have wet and dry seasons are found in India, Southeast Asia, West Africa, South and Central America, the West Indies, and northern Australia. They also have deciduous trees, and some of these forests also contain elephants, tigers, and hippopotamuses in addition to the animals mentioned previously.

Shrublands

Shrublands tend to occur along coasts that have dry summers and receive most of their rainfall during their winter. They are characterized by shrubs that have small but thick evergreen leaves, often coated with a waxy material that prevents loss of moisture. These shrubs are adapted to withstand arid conditions and can also quickly regrow after a fire. In fact, the seeds of many species require the heat and scarring action of fire to induce germination. Other shrubs sprout from the roots after a fire. Typical shrubland species include coyotes, jackrabbits, gophers, and other rodents, as well as fire-adapted plant species such as chemise (Fig. 36.10). The dense shrubland that occurs in California is known as **chaparral.** This type of shrubland, called Mediterranean, lacks an understory and ground litter, and is highly flammable.

Grasslands

Grasslands occur where rainfall is greater than 25 cm but generally insufficient to support trees. For example, it is too dry for forests and too wet for deserts to form in temperate areas where rainfall is between 25 to 75 cm. Natural grasslands once covered more than 40% of Earth's land surface, but most areas that were once grasslands are now used to grow crops such as wheat, corn, and soybeans.

Grasses are well adapted to a changing environment and can tolerate some grazing, as well as flooding, drought, and sometimes fire. Where rainfall is high, large tall grasses that reach more than 2 m in height (e.g., pampas grass) can flourish. In drier areas, shorter grasses between 5 and 10 cm are dominant (e.g., grama grass). The growth of grasses is also seasonal. As a result, some grassland animals such as bison migrate, whereas ground squirrels hibernate, when there is little grass for them to eat.

a. Shrubland overview

b. Chemise

Figure 36.10 Shrublands. Shrublands **(a),** are subject to raging fires, but the shrubs, such as chemise **(b),** are adapted to quickly regrow.

a. Tall-grass prairie

b. American bison

Figure 36.11 The prairie. **a.** Tall-grass prairies are seas of grasses dotted by pines and junipers. **b.** Bison, once abundant, are now being reintroduced into certain areas.

The temperate grasslands include the Russian *steppes,* the South American *pampas,* and the North American *prairies* (Fig. 36.11). Large herds of bison—estimated at hundreds of thousands—once roamed the prairies, as did herds of pronghorn antelope. Now, small mammals, such as mice, prairie dogs, and rabbits, typically live below ground but usually feed above ground. Hawks, snakes, badgers, coyotes, and foxes feed on these mammals.

Video Tallgrass Prairie Ecology

Video Dung Beetles

Savannas

Savannas, which are grasslands that contain some trees, occur in regions where a relatively cool dry season is followed by a hot, rainy season. One tree that can survive the severe dry season is the flat-topped acacia, which sheds its leaves during a drought. The African savanna supports a tremendous variety and number of large herbivores (Fig. 36.12). Elephants and giraffes are browsers that feed on tree vegetation. Antelopes, zebras, wildebeests, water buffalo, and rhinoceroses are grazers that feed on grasses. Any plant litter that is not consumed by grazers is attacked by a variety of smaller organisms, among them termites. Termites build towering mounds in which they tend fungal gardens that they use for food. The herbivores support a large population of carnivores. Lions and hyenas sometimes hunt in packs, cheetahs hunt singly by day, and leopards hunt singly by night.

Video Thorn Tree Ant

a. Zebra (in foreground)

b. Wildebeest

Figure 36.12 The savanna. The African savanna varies from grassland to widely spaced shrubs and trees. This biome supports a large assemblage of herbivores, including zebras **(a)**, wildebeests **(b)**, and giraffes **(c)**. Carnivores, such as the cheetah **(d)**, prey on these.

d. Cheetah

c. Giraffe

Deserts

As discussed in section 36.1, **deserts** are usually found at latitudes of about 30°, in both the Northern and Southern Hemispheres. The winds that descend in these regions lack moisture, and annual rainfall is less than 25 cm. Days are hot because a lack of cloud cover allows the sun's rays to penetrate easily, but nights are cold because heat escapes easily into the atmosphere.

The Sahara—which stretches all the way from the Atlantic coast of Africa to the Arabian Peninsula—and a few other deserts have little or no vegetation. But most deserts have a variety of plants (Fig. 36.13). The best-known desert perennials in North America are the succulent, spiny-leafed cacti, which have stems that store water and carry on photosynthesis. Also common are nonsucculent shrubs, such as the many-branched sagebrush and the spiny-branched ocotillo that produces leaves during wet periods and sheds them during dry periods.

Some animals are adapted to the desert environment. For example, many desert animals are nocturnal. They are active at night when it is cooler. Reptiles and insects have waterproof outer coverings that conserve water. A desert has numerous insects, which pass through the stages of development from pupa to adult while there is rain. Reptiles, especially lizards and snakes, are a characteristic group of vertebrates found in deserts, but running birds (e.g., the roadrunner) and rodents (e.g., the kangaroo rat) are also well known (Fig. 36.13). Coyotes and hawks prey on the rodents.

Check Your Progress 36.2

1. Identify the features that would enable one biome to have a greater biodiversity than another.
2. Compare and contrast the biodiversity of the various land biomes.

36.3 Aquatic Ecosystems

Learning Outcomes

Upon completion of this section, you should be able to

1. Compare the characteristics of freshwater and saltwater ecosystems.
2. Recognize the differences between the pelagic and benthic divisions of the ocean.

Aquatic ecosystems are generally classified as freshwater (inland) or saltwater (usually marine). Brackish water, however, is a mixture of fresh and salt water. In this section, we describe lakes as examples of freshwater ecosystems and oceans as examples of saltwater ecosystems. Coastal ecosystems will represent areas of brackish water.

Video Thames River

a.

Figure 36.13 Deserts. Plants and animals that live in the Mojave Desert **(a)** are adapted to arid conditions. The plants are either succulents, which retain moisture, or shrubs with woody stems and small leaves, which lose little moisture. **b.** The kangaroo rat feeds on seeds and other vegetation. **c.** The roadrunner preys on insects, lizards, and snakes. **d.** The kit fox is a desert carnivore.

b. Bannertail kangaroo rat

c. Greater roadrunner

d. Kit fox

Lakes

Lakes are bodies of fresh water often classified by their nutrient abundance. Oligotrophic (nutrient-poor) lakes are characterized by a small amount of organic matter and low productivity (Fig. 36.14*a*). Eutrophic (nutrient-rich) lakes typically have plentiful organic matter and high productivity (Fig. 36.14*b*). Such lakes are usually situated in naturally nutrient-rich regions or are enriched by agricultural or urban and suburban runoff. Oligotrophic lakes can become eutrophic through large inputs of nutrients, a process called **eutrophication.** Excess nutrient inputs, such as nitrogen fertilizer, can cause eutrophication, which can lead to fish kills. See the Ecology feature, "Water Pollution," on page 731, for a discussion of water pollution issues.

In temperate environments, deep lakes are stratified during the summer and winter. In summer, lakes in the temperate zone have three layers of water that differ in temperature (Fig. 36.15). The surface layer, the epilimnion, is warm due to solar radiation; the thermocline is the middle layer that decreases 1°C per meter of depth; and the lowest layer, the hypolimnion, is cold. These differences in temperature prevent mixing. The warmer, less dense water of the epilimnion "floats" on top of the colder, more dense water of the hypolimnion.

As the season progresses, the epilimnion becomes nutrient-poor, while the hypolimnion begins to be depleted of oxygen. Phytoplankton (see Chapter 29) found in the sunlit epilimnion use up nutrients during photosynthesis and in turn release oxygen. Detritus naturally falls by gravity to the bottom of the lake where oxygen is used up as decomposition occurs. Decomposition in turn releases nutrients.

In the fall, as the epilimnion cools, and in the spring, as it warms, an overturn occurs. In the fall, the upper epilimnion waters become cooler than the hypolimnion waters. This causes the surface water to sink and the deep water to rise. The **fall overturn** continues until the temperature is uniform throughout the lake. At this point, wind helps circulate the water so that mixing occurs.

As winter approaches, the water cools further. Ice formation begins at the top, and the ice floats because it is less dense than cold water. Ice has an insulating effect, preventing further cooling of the water below. This permits aquatic organisms to live through the winter in the water beneath the surface of the ice.

In the spring, as the ice melts, the cooler water on top sinks below the warmer water below it. The **spring overturn**

a. Oligotrophic lake

b. Eutrophic lake

Figure 36.14 Types of lakes. Lakes can be classified according to whether they are **(a)** oligotrophic (nutrient-poor) or **(b)** eutrophic (nutrient-rich). Eutrophic lakes tend to have large populations of algae and rooted plants, resulting in a large population of decomposers that use up much of the oxygen and leave little oxygen for fishes.

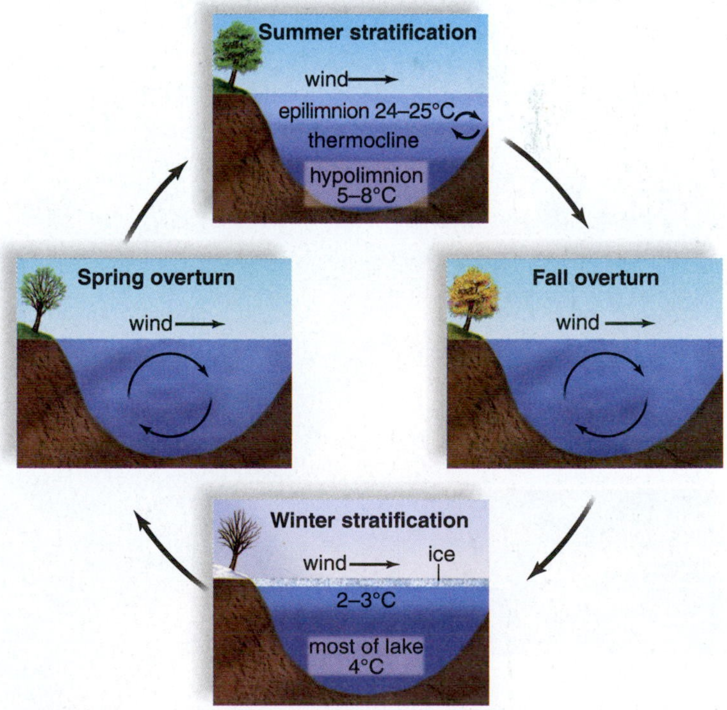

Summer stratification
wind
epilimnion 24–25°C
thermocline
hypolimnion
5–8°C
Spring overturn
wind
Fall overturn
wind
Winter stratification
wind
ice
2–3°C
most of lake
4°C

Figure 36.15 Lake stratification in a temperate region. Temperature profiles of a large oligotrophic lake in a temperate region vary with the season. During the spring and fall overturns, the deep waters receive oxygen from surface waters, and surface waters receive inorganic nutrients from deep waters.

continues until the temperature is uniform through the lake. At this point, wind aids in the circulation of water as before. When the surface waters absorb solar radiation, thermal stratification occurs once more.

This vertical stratification and seasonal change of temperatures in a lake basin influence the seasonal distribution of fish and other aquatic life. For example, cold-water fish move to the deeper water in summer and inhabit the upper water in winter. In the fall and spring just after mixing occurs, phytoplankton growth at the surface is most abundant.

Life Zones

In both fresh and salt water, microscopic **plankton,** which includes phytoplankton and also zooplankton (see Chapter 29), play important roles in aquatic ecosystems. Lakes and ponds can be divided into several life zones, as shown in Figure 36.16. Aquatic plants are rooted in the shallow littoral zone of a lake, and various microscopic organisms cling to these plants and to rocks. Some organisms, such as the water strider, live at the water-air interface and can literally walk on water. In the limnetic zone, small fishes, such as minnows and killifish, feed on plankton and also serve as food for large fishes. In the **profundal zone,** zooplankton and fishes, such as whitefish, feed on debris that falls from above. Pike species are an example of "lurking predators." They wait among vegetation around the margins of lakes and surge out to catch passing prey.

Video Plankton Diversity

A few insect larvae are in the limnetic zone, but they are far more abundant in both the littoral and profundal zones. Midge larvae and ghost worms are common members of the benthos, animals that live on the bottom in the benthic zone. In a lake, benthos organisms include crayfishes, snails, clams, and various types of worms and insect larvae.

Coastal Ecosystems

Estuaries

An **estuary** is where fresh water and salt water meet and mix (Fig. 36.17). Coastal bays, tidal marshes, fjords (an inlet of water between high cliffs), some deltas (a triangular-shaped area of land at the mouth of a river), and lagoons (a body of water

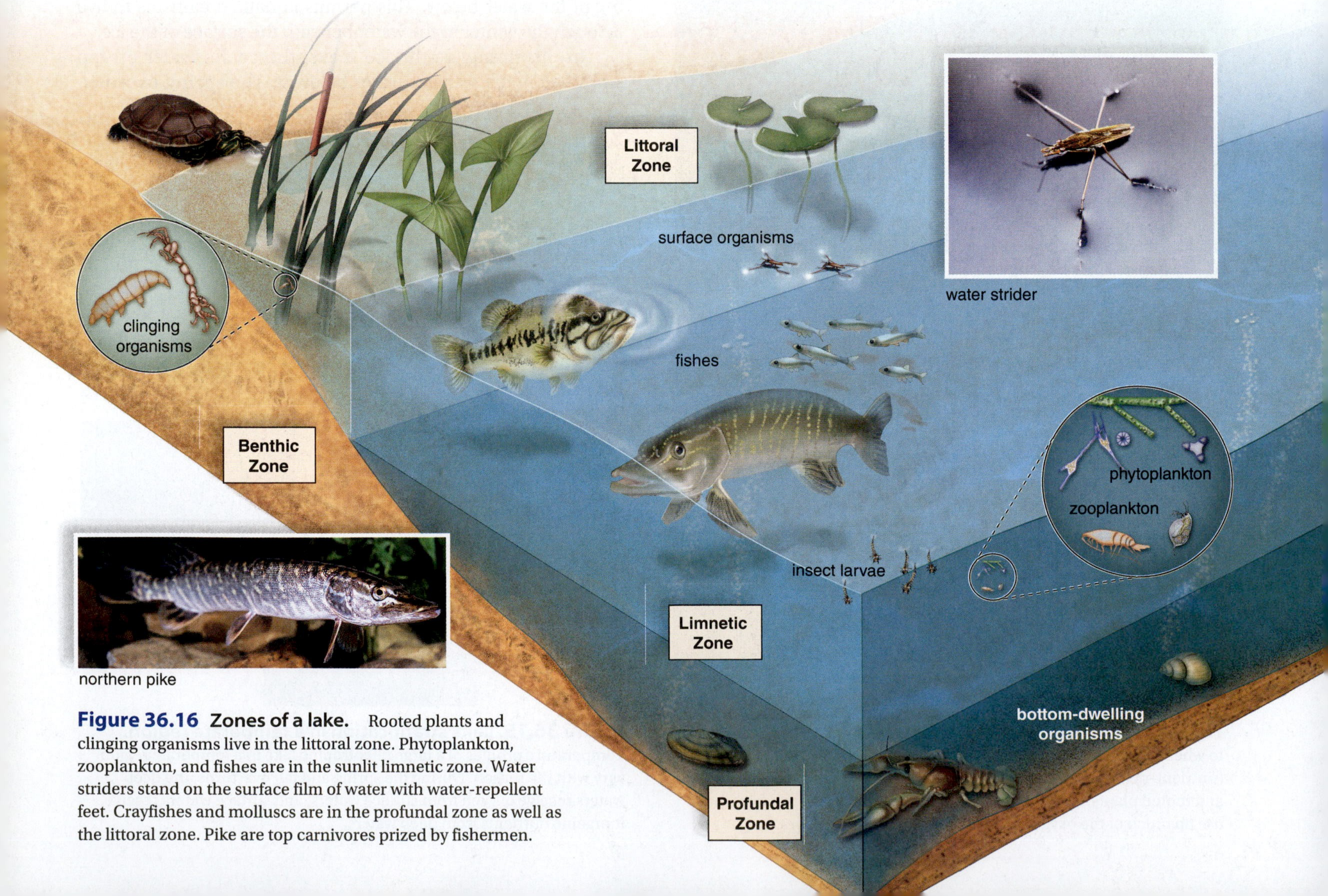
Littoral Zone
surface organisms
water strider
clinging organisms
fishes
Benthic Zone
phytoplankton
zooplankton
insect larvae
northern pike
Limnetic Zone
bottom-dwelling organisms
Profundal Zone

Figure 36.16 Zones of a lake. Rooted plants and clinging organisms live in the littoral zone. Phytoplankton, zooplankton, and fishes are in the sunlit limnetic zone. Water striders stand on the surface film of water with water-repellent feet. Crayfishes and molluscs are in the profundal zone as well as the littoral zone. Pike are top carnivores prized by fishermen.

SCIENCE IN YOUR LIFE ▶ ECOLOGY

Water Pollution

Agricultural fertilizers are the chief cause of nitrate contamination of drinking water. Excessive nitrates in a baby's bloodstream can lead to slow suffocation, known as "blue baby" syndrome. Agricultural herbicides are suspected carcinogens in the tap water of scattered ecosystems coast to coast.

Some farmers are already using irrigation methods that deliver water directly to plant roots, no-till agriculture that reduces the loss of topsoil and cuts back on herbicide use, and integrated pest management, which relies heavily on predatory bugs to kill bugs that prey upon the crops. Ideally, more farmers would adopt various methods of agriculture that help decrease water pollution. Encouraged, and in some cases compelled by state and federal agents, dairy farmers have built sheds, concrete containments, and underground liquid storage tanks to hold wastes when it rains. Later, the manure is trucked to fields and spread as fertilizer. Residential buildings, like golf courses and ski resorts, also contribute to the problem. Lawn care, motor vehicles, and the construction and use of roads and buildings all add contaminants to aquatic ecosystems.

Citizens around Grand Traverse Bay on the eastern shore of Lake Michigan have also gotten the message, especially because they want to keep enjoying water-dependent activities such as boating, swimming, and fishing. James Haverman, a concerned member of the Traverse Bay Watershed Initiative says, "If we can't change the way people live their everyday lives, we are not going to be able to make a difference." Builders in Traverse County are already required to control soil erosion with filter fences, steer rainwater away from exposed soil, build sediment basins, and plant protective buffers. Presently, homeowners must have a 25-ft setback from wetlands and a 50-ft setback from lakes and creeks. They are also encouraged to pump out their septic systems every two years.

Questions to Consider

1. Should farmers and homeowners be required by law to protect freshwater supplies?
2. Would you hire an environmentally friendly builder to construct your new home if he or she charged 20% more than a builder who did not employ environmentally friendly techniques?

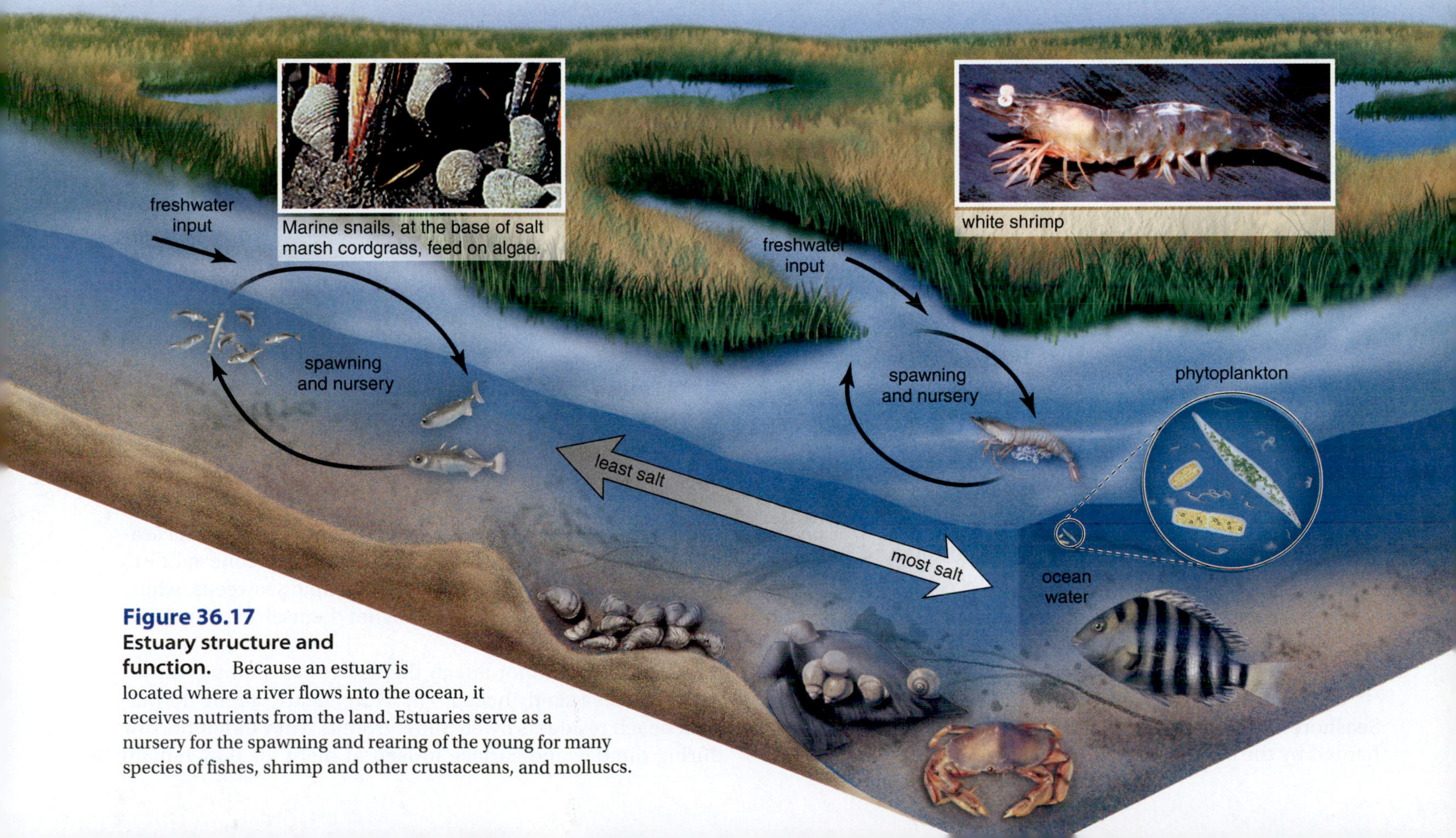
freshwater input
Marine snails, at the base of salt marsh cordgrass, feed on algae.
white shrimp
freshwater input
spawning and nursery
spawning and nursery
phytoplankton
least salt
most salt
ocean water

Figure 36.17
Estuary structure and function. Because an estuary is located where a river flows into the ocean, it receives nutrients from the land. Estuaries serve as a nursery for the spawning and rearing of the young for many species of fishes, shrimp and other crustaceans, and molluscs.

a. Mudflats

b. Mangrove swamp

Figure 36.18 Types of estuaries. Types of estuaries include **(a)** mudflats, which form at the mouth of rivers in temperate zones and are frequented by migrant birds, and **(b)** mangrove swamps, which skirt the coastlines of many tropical and subtropical lands. The tangled roots of mangrove trees trap sediments and nutrients that sustain many immature forms of sea life. The salt marsh depicted in Figure 36.17 is yet another type of estuary-associated system.

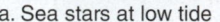

a. Sea stars at low tide

b. Shorebirds

Figure 36.19 Seashores. **a.** Some organisms of a rocky coast live in tidal pools. **b.** A sandy shore looks devoid of life except for the birds that feed there.

separated from the sea by a narrow strip of land) are all examples of estuaries (Fig. 36.18).

Organisms living in an estuary must be able to withstand constant mixing of waters and rapid changes in salinity. Not many organisms are adapted to this environment, but those that can survive here find an abundance of nutrients. An estuary acts as a nutrient trap because the sea prevents the rapid escape of nutrients brought by a river. It has been estimated that well over half of all marine fishes develop in the protective environment of an estuary, which explains why estuaries are called the nurseries of the sea. Estuaries are also feeding grounds for many birds, fish, and shellfish because they offer a ready supply of food.

Salt marshes, dominated by salt marsh cordgrass, are often associated with estuaries. So are mudflats and mangrove swamps, where sediment and nutrients from the land collect.

Seashores

Seashores, which may be rocky or sandy, are constantly bombarded by the sea as the tides roll in and out (Fig. 36.19). The **littoral zone** lies between the high- and low-tide marks. The littoral zone of a rocky beach is divided into subzones. In the upper portion of the littoral zone, barnacles are attached so tightly to stone by their own secretions that their calcareous outer plates remain in place even after the enclosed shrimplike animal dies. In the midportion of the littoral zone, brown algae known as rockweed may overlie the barnacles. In the lower portions of the littoral zone, oysters and mussels attach themselves to the rocks by filaments called *byssal threads.* Also present are snails called limpets and periwinkles. Periwinkles have a coiled shell and secure themselves by hiding in crevices or under seaweeds, while limpets press their single, flattened cone tightly to a rock. Below the littoral zone, macroscopic seaweeds, which are the main photosynthesizers, anchor themselves to the rocks by holdfasts.

Organisms cannot attach themselves to shifting, unstable sands on a sandy beach. Therefore, nearly all the permanent beach residents dwell underground. They either burrow during the day and surface to feed at night, or they remain permanently within their burrows and tubes. Ghost crabs and sandhoppers (amphipods) burrow themselves above the high-tide mark and feed at night when the tide is out. Sandworms and sand (ghost) shrimp remain within their burrows in the littoral zone and feed on detritus whenever possible. Still lower in the sand, clams, cockles, and sand dollars can be found.

Oceans

Oceans cover approximately three-quarters of our planet. Climate is driven by the sun, but the oceans play a major role in redistributing heat in the biosphere. Water tends to be warm at the equator and much cooler at the poles because of the distribution of the sun's rays, as discussed earlier in this chapter (see Fig. 36.1*a*). Air takes on the temperature of the water below, and then warm air moves from the equator to the poles. In other words, the oceans have a major influence on wind patterns. (The landmasses also play a role, but water holds heat longer and remains cool longer during periods of changing temperature than land.)

When the wind blows strongly and steadily across a great expanse of ocean for a long time, friction from the moving air begins to drag the water along with it. Once the water has been set in motion, its momentum, aided by the wind, keeps it moving in a steady flow called a current. Because the ocean currents eventually strike land, they move in a circular path—clockwise in the Northern Hemisphere and counterclockwise in the Southern Hemisphere. As the currents flow, they move warm water from the equator to the poles. One such current, called the Gulf Stream, brings tropical Caribbean water to the east coast of North America and the higher latitudes of western Europe. Without the Gulf Stream, Great Britain, which has a relatively warm temperature, would be as cold as Greenland. In the Southern Hemisphere, another major ocean current warms the eastern coast of South America.

Also, in the Southern Hemisphere, the Humboldt Current flows toward the equator. The Humboldt Current carries phosphorus-rich cold water northward along the west coast of South America. During a process called **upwelling,** cold offshore winds cause cold nutrient-rich waters to rise and take the place of warm nutrient-poor waters. In South America, the enriched waters nourish an abundance of marine life that supports the fisheries of Peru and northern Chile. When the Humboldt Current is not as cool as usual, upwelling does not occur, stagnation results, the fisheries decline, and climate patterns change globally. This phenomenon is discussed later in the Ecology feature, "El Niño–Southern Oscillation," on page 734.

Pelagic Division

It is customary to place the organisms of the oceans into either the pelagic division (open waters) or the benthic division (ocean floor). The geographic areas and zones of an ocean are shown in Figure 36.20, and the animals that inhabit the different zones are shown in Figure 36.21. The **pelagic division** includes the neritic province and the oceanic province.

Neritic Province The abundant sunlight and inorganic nutrients provide for a large concentration of organisms in the neritic province. Phytoplankton, consisting of suspended algae, is food not only for zooplankton but also for small fishes. These small fishes, in turn, are food for larger commercially valuable fishes.

Video Phytoplankton Diversity

Coral reefs are areas of high biological abundance and productivity found in shallow, nutrient-poor, tropical waters. Their chief constituents are stony corals, animals that have a calcium carbonate (limestone) exoskeleton, and calcareous red and green algae. Corals tend to form colonies derived from individual corals that have reproduced by means of budding. Corals provide a home for a microscopic alga called zooxanthella. The corals, which feed at night, and the algae, which photosynthesize during the day, are mutualistic and share materials and nutrients. The close relationship of corals and zooxanthellae is likely the reason coral reefs form only in shallow sunlit water—the algae need sunlight for photosynthesis.

Video Coral Reef Ecosystems

Figure 36.20 Marine environment. Organisms reside in the pelagic division (blue) where waters are divided as indicated. Organisms also reside in the benthic division (brown) in the surfaces and zones indicated.

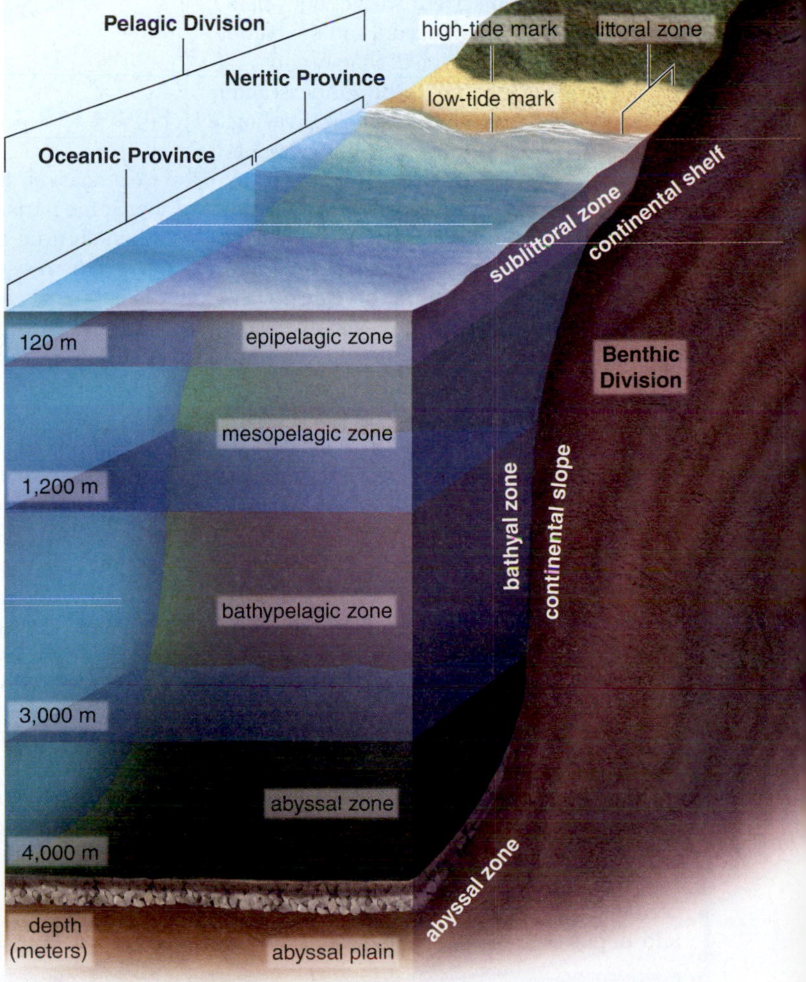
Pelagic Division
Neritic Province
Oceanic Province
high-tide mark
littoral zone
low-tide mark
sublittoral zone
continental shelf
Benthic Division
120 m
epipelagic zone
mesopelagic zone
1,200 m
bathyal zone
continental slope
bathypelagic zone
3,000 m
abyssal zone
4,000 m
abyssal zone
depth (meters)
abyssal plain

El Niño–Southern Oscillation

Climate largely determines the distribution of life on Earth. Short-term variations in climate, which we call weather, also have a pronounced effect on living organisms. There is no better example than an El Niño. Originally, El Niño referred to a warming of the seas off the coast of Peru around Christmas—hence, the name El Niño, "the boy child," after Jesus.

Now scientists prefer the term El Niño–Southern Oscillation (ENSO) to denote a severe weather change brought on by an interaction between the atmosphere and ocean currents. Ordinarily, the southeast trade winds move along the coast of South America and turn west because of Earth's daily rotation on its axis. As the winds drag warm ocean waters from east to west, there is an upwelling of nutrient-rich cold water from the ocean's depths, resulting in a bountiful Peruvian harvest of anchovies. When the warm ocean waters reach their western destination, the monsoons bring rain to India and Indonesia. Scientists have noted that these events correlate with a difference in the barometric pressure over the Indian Ocean and the southeastern Pacific—that is, the barometric pressure is low over the Indian Ocean and high over the southeastern Pacific. But when a "southern oscillation" occurs and the barometric pressures switch, an El Niño begins.

During an El Niño, both the northeast and the southeast trade winds slacken. Upwelling no longer occurs, and the anchovy catch off the coast of Peru plummets. During a severe El Niño, waters from the east never reach the west, and the winds lose their moisture in the middle of the Pacific instead of over the Indian Ocean. The monsoons fail, and drought occurs in India, Indonesia, Africa, and Australia. Harvests decline, cattle must be slaughtered, and famine is likely in highly populated India and Africa, where funds to import replacement supplies of food are limited. In 2009, severe drought caused fires that killed hundreds of people and caused billions of dollars in damage in the state of Victoria, Australia.

A backward movement of winds and ocean currents may even occur so that the waters warm to more than 14° above normal along the west coast of the Americas. This is a sign that a severe El Niño has occurred, and the weather changes are dramatic in the Americas also. Southern California is hit by storms and even hurricanes, and the deserts of Peru and Chile receive so much rain that flooding occurs. A jet stream (strong wind currents) can carry moisture into Texas, Louisiana, and Florida, with flooding a near certainty. Or the winds can turn northward and deposit snow in the mountains along the west coast so that flooding occurs in the spring. Some parts of the United States, however, benefit from an El Niño. The Northeast is warmer than usual, few if any hurricanes hit the east coast, and there is a lull in tornadoes throughout the Midwest. Altogether, a severe El Niño affects the weather over three-quarters of the globe.

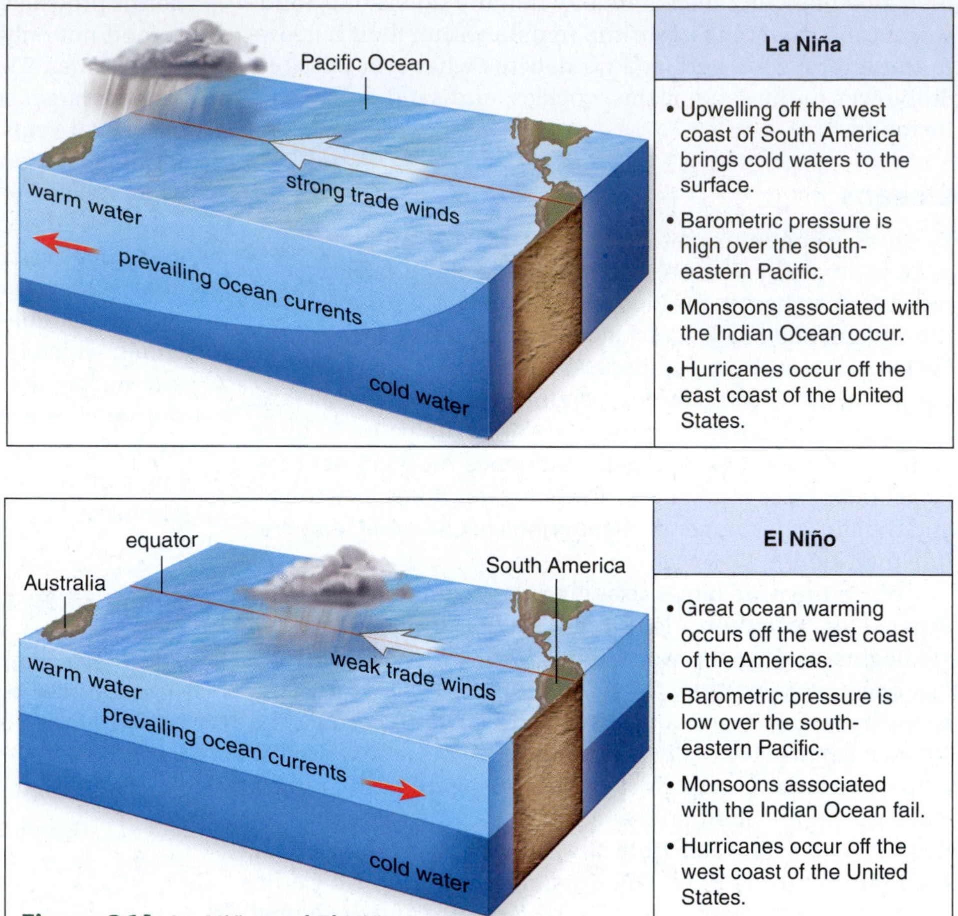

Pacific Ocean
strong trade winds
warm water
prevailing ocean currents
cold water
La Niña
• Upwelling off the west coast of South America brings cold waters to the surface.
• Barometric pressure is high over the south-eastern Pacific.
• Monsoons associated with the Indian Ocean occur.
• Hurricanes occur off the east coast of the United States.
equator
Australia
South America
weak trade winds
warm water
prevailing ocean currents
cold water
El Niño
• Great ocean warming occurs off the west coast of the Americas.
• Barometric pressure is low over the south-eastern Pacific.
• Monsoons associated with the Indian Ocean fail.
• Hurricanes occur off the west coast of the United States.

Figure 36A La Niña and El Niño.

Eventually, an El Niño dies out, and normal conditions return. The normal cold-water state off the coast of Peru is known as La Niña (the girl). Figure 36A contrasts the weather conditions of a La Niña with those of an El Niño. Since 1991, the sea surface has been almost continuously warm, and two record-breaking El Niños have occurred. What could be causing more of the El Niño state than the La Niña state? Some scientists argue that this environmental change is related to global warming, a rise in environmental temperature due to greenhouse gases in the atmosphere, as discussed in Chapter 35. Greenhouse gases, including carbon dioxide, are released by humans in mass quantities into the atmosphere by, for example, burning fossil fuels. Like the glass of a greenhouse, these gases allow the sun's rays to pass through, but they trap the heat.

Questions to Consider

1. Recent research has suggested that global warming is contributing to the increased frequency and severity of El Niño events. This can cause severe drought in India and Africa, as well as greatly reduce fish catch in parts of the Southern Hemisphere. Should the responsibility for decreasing greenhouse gas emissions be greater for more-developed countries (MDCs), because they produce a greater amount of greenhouse gases per capita than less-developed countries (LDCs), even though El Niños have more severe consequences for LDCs?
2. Aside from the recent increases in gasoline prices, would you consider buying a more fuel-efficient car or spend more on plane flights to purchase "carbon credits" to do your part in decreasing the emission of greenhouse gases such as carbon dioxide? Why or why not?

A reef is densely populated with life. The large number of crevices and caves provide shelter for filter feeders (e.g., sponges, sea squirts, and fan worms) and for scavengers (e.g., crabs and sea urchins). There are many types of small, beautifully colored fishes. Parrotfishes feed directly on corals, and others feed on plankton or detritus. Small fishes become food for larger fishes, including snappers and other fish that are caught for human consumption. The barracuda, moray eel, and various shark species are top predators in coral reefs.

Video Reef Balls

Video Coral Reef Spawning

Oceanic Province The oceanic province lacks the inorganic nutrients of the neritic province and, therefore, does not have as high a concentration of phytoplankton, even though it is sunlit. Photosynthesizers are food for a large assembly of zooplankton, which then become food for herrings and bluefishes. These, in turn, are eaten by larger mackerels, tunas, and sharks. Flying fishes, which glide above the surface, are preyed upon by dolphins. Whales are other mammals found in the epipelagic zone (Fig. 36.21). Baleen whales strain krill (small crustaceans) from the water, and toothed sperm whales feed primarily on the common squid.

Video Ocean Fishing Ban

Animals in the mesopelagic zone are carnivores that are adapted to the absence of light, and tend to be translucent, red, or luminescent. There are luminescent shrimps, squids, and fishes, such as lantern and hatchet fishes.

The bathypelagic zone is in complete darkness except for an occasional flash of bioluminescent light. Carnivores and scavengers are found in this zone. Strange-looking fishes with distensible mouths and abdomens and small, tubular eyes feed on infrequent prey.

Benthic Division

The **benthic division** includes organisms that live on or in the soil of the continental shelf (sublittoral zone), the continental slope (bathyal zone), and the abyssal plain (abyssal zone) (see Fig. 36.20).

Seaweed grows in the sublittoral zone, and can be found in batches on outcroppings as the water gets deeper. More diversity of life exists in the sublittoral and bathyal zones than in the abyssal zone. In the first two zones, clams, worms, and sea urchins are preyed upon by starfishes, lobsters, crabs, and brittle stars. Photosynthesizing algae

Figure 36.21 Ocean inhabitants. Different organisms are characteristic of the epipelagic, mesopelagic, and bathypelagic zones of the pelagic division compared to the abyssal zone of the benthic division.

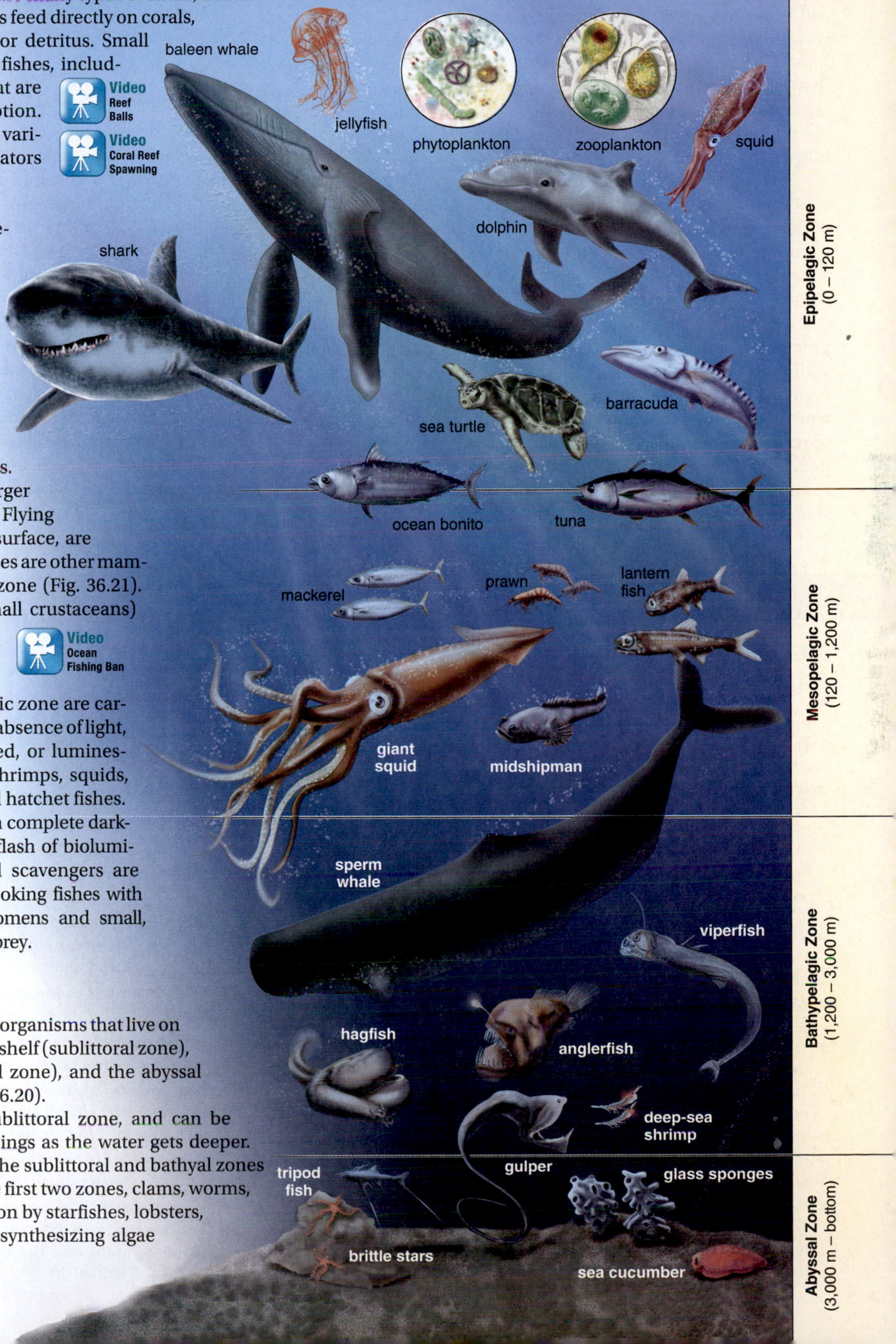
baleen whale
jellyfish
phytoplankton
zooplankton
squid
dolphin
shark
sea turtle
barracuda
ocean bonito
tuna
mackerel
prawn
lantern fish
giant squid
midshipman
sperm whale
viperfish
hagfish
anglerfish
deep-sea shrimp
gulper
tripod fish
glass sponges
brittle stars
sea cucumber
Epipelagic Zone (0 – 120 m)
Mesopelagic Zone (120 – 1,200 m)
Bathypelagic Zone (1,200 – 3,000 m)
Abyssal Zone (3,000 m – bottom)

occur in the sunlit sublittoral zone, but benthic organisms in the bathyal zone are dependent on the detritus that falls from the waters above.

The abyssal zone is inhabited by animals that live at the soil-water interface of the abyssal plain (Fig. 36.21). It once was thought that few animals exist in this zone because of the intense pressure and the extreme cold. Yet many invertebrates live there by feeding on debris floating down from the mesopelagic zone. Sea lilies rise above the seafloor; sea cucumbers and sea urchins crawl around on the sea bottom; and tube worms burrow in the mud.

The flat abyssal plain is interrupted by enormous underwater mountain chains called oceanic ridges. Along the axes of the ridges, crustal plates spread apart, and molten magma from Earth's core rises to fill the gap. At **hydrothermal vents,** seawater percolates through cracks and is heated to about 350°C, causing sulfate to react with water and form hydrogen sulfide (H_2S). Free-living or mutualistic chemoautotrophic bacteria use energetic electrons derived from hydrogen sulfide to reduce bicarbonate to organic compounds. These compounds support an ecosystem that includes huge tube worms and clams. It was a surprise to find such communities of organisms living so deep in the ocean, where light never penetrates. Life is possible so deep because unlike photosynthesizers, chemoautotrophs are producers that do not require solar energy.

Check Your Progress 36.3

1. Compare and contrast the various freshwater ecosystems.
2. Explain the major characteristics of the two oceanic divisions.

Case Study Conclusion

DDT was developed in the 1940s to be used as a broad-spectrum pesticide. Due to the ecological damage it can lead to, DDT has been banned in the United States. Various countries throughout the world still use it to help combat insects. DDT can persist in the environment for up to 15 years, enabling it to find its way into a variety of ecosystems. When released, it will eventually run off into nearby aquatic ecosystems. This can lead to an accumulation of DDT in the organisms that live in these biomes. It can also evaporate into the atmosphere and get carried across the planet by global air currents. DDT can then be released into communities thousands of miles away from the original source. Health officials are concerned with the harmful effects it can have upon people, including reproductive issues and damage to the liver and nervous system. Because of its persistence, DDT has been able to reach every corner of the planet.

MEDIA STUDY TOOLS

www.mhhe.com/maderinquiry14

Enhance your study of this chapter with study tools and practice tests. Also ask your instructor about the resources available through ConnectPlus, including LearnSmart, the media-rich eBook, interactive learning tools, and animations.

Mc Graw Hill connect plus+ | BIOLOGY

SUMMARIZE

36.1 Climate and the Biosphere

- **Climate** refers to the prevailing weather conditions in a particular region.
- The sun's rays are spread out over a larger area at the poles than the vertical rays at the equator because the Earth is spherical. Thus, surface temperatures decrease from the equator toward each pole.
- The planet is tilted on its axis, and the seasons change depending on which hemisphere is closest to the sun as Earth completes its annual rotation.
- Warm air rises near the equator, loses its moisture, and then descends at about 30° north and south latitude, and continuing toward the poles. When the air descends, it retains moisture, and therefore, the great deserts of the world are formed at 30° latitudes.
- Because Earth rotates on its axis daily, the winds blow in opposite directions above and below the equator.
- Air rising over coastal regions loses its moisture on the windward side as the elevation increases. This may cause **monsoons,** or extended periods of rain. Less rain falls on the leeward side, creating a **rain shadow.**

36.2 Terrestrial Ecosystems

- A **biome** is a major type of terrestrial ecosystem.
- Temperature and rainfall influence the distribution of biomes around the world.
- The **tundra** is the northernmost biome and consists largely of short grasses and sedges and dwarf woody plants. Most of the water in the soil is frozen year-round (i.e., **permafrost**) because of cold winters and short summers.
- The **taiga,** a **coniferous forest,** has less rainfall than other types of forests.
- The **temperate deciduous forest** has trees that gain and lose their leaves during spring and fall, respectively.
- **Tropical forests** are the most complex and productive of all biomes. Many of the plants in the tropics are **epiphytes,** meaning that they grow on other plants.
- **Shrublands** usually occur along coasts that have dry summers and receive most of their rainfall in the winter.
- Among **grasslands,** the savanna, a tropical grassland, supports the greatest number of different types of large herbivores. Prairies, such as those in the mid United States, have a limited variety of vegetation and animal life.

- **Deserts** are characterized by rainfall less than 25 cm a year. Plants, such as cacti, are succulents, and others are shrubs with thick leaves—both of which are adaptations to conserve water.

36.3 Aquatic Ecosystems

- In deep temperate zone lakes, water depth determines the temperature and the concentrations of nutrients and gases. A large input of nutrients into a body of water is called **eutrophication.** During **spring overturn** and **fall overturn,** nutrients from the bottom lake layers are redistributed while the entire body of water is cycled.
- **Lakes** and ponds have three life zones. Rooted plants and clinging organisms live in the **littoral zone; plankton** and fishes live in the sunlit **limnetic zone;** and bottom-dwelling organisms, such as crayfishes and molluscs, live in the **profundal zone.**
- Oceans play a major role in determining the climate of the land biomes. Ocean currents help move warm water from the equator toward the poles. During an **upwelling,** cold offshore winds cause cold nutrient-rich waters to rise and take the place of warm nutrient-poor waters.
- Marine ecosystems are divided into coastal ecosystems and the oceans.
- The coastal ecosystems, especially **estuaries,** are more productive than the oceans. Estuaries (e.g., salt marshes, mudflats, and mangrove forests) are near the mouth of a river and are called the nurseries of the sea. **Coral reefs** contain a tremendous diversity of species and play important roles in the health of the ocean.
- An ocean is divided into the **pelagic division** and the **benthic division.** The pelagic division (open waters) has three zones. The epipelagic zone receives adequate sunlight and supports the most life. The mesopelagic and bathypelagic zones contain organisms adapted to minimum light and no light, respectively. The benthic division (ocean floor) includes organisms living on the continental shelf in the sublittoral zone, the continental slope in the bathyal zone, and the abyssal plain in the abyssal zone. Underwater **hydrothermal vents** are formed when seawater percolates through cracks and is heated by molten magma from the Earth's core. Life here is based upon the sulfate reacting with water to form hydrogen sulfide.

ASSESS

Testing Yourself

Choose the best answer for each question.

1. The Northern Hemisphere is tilted toward the sun during the
   a. winter solstice. **c.** summer solstice.
   b. autumnal equinox. **d.** vernal equinox.
2. Which side of a mountain range lies in a rain shadow?
   a. windward **b.** leeward
3. Whether you are ascending a mountain or traveling from the equator to the North Pole, you would observe terrestrial biomes in which order?
   a. coniferous forest, tropical rain forest, temperate deciduous forest, tundra
   b. tropical rain forest, coniferous forest, temperate deciduous forest, tundra
   c. tropical rain forest, temperate deciduous forest, coniferous forest, tundra
   d. tropical rain forest, coniferous forest, tundra, temperate deciduous forest
   e. tundra, temperate deciduous forest, tropical rain forest, coniferous forest

For questions 4–10, match the description to the biome in the key.

Key:

a. tundra
b. taiga
c. temperate deciduous forests
d. tropical forests
e. shrublands
f. grasslands
g. deserts

4. Dominated by coniferous trees
5. Chaparral in California
6. Contains permafrost
7. Typically found at latitudes of about 30°
8. Has a complex structure, with many levels of life, from the canopy to the floor
9. Moderate climate, with relatively high rainfall and well-defined seasons
10. Includes savannas
11. Compared to the hypolimnion layer, the epilimnion layer of a lake in late summer is
    a. nutrient-rich and oxygen-rich.
    b. nutrient-rich and oxygen-poor.
    c. nutrient-poor and oxygen-rich.
    d. nutrient-poor and oxygen-poor.

For questions 12–15, indicate the life zone that each lake organism is most likely to inhabit. A question may have more than one answer.

Key:

a. littoral **c.** profundal
b. limnetic **d.** benthic

12. Predatory fish
13. Water lily
14. Clams
15. Unicellular algae

ENGAGE

Thinking Critically

1. Climate change is predicted to have the least severe consequences in terms of temperature change nearest the equator and more severe consequences progressing into the more temperate zones. Why do you think this is?
2. Scientists have documented a phenomenon called "biodiversity gradients," which shows that the greatest amounts of species diversity exist nearest the equator and that diversity tends to diminish moving through increasing latitudes. Based on your knowledge of climate, topography, and general characteristics of terrestrial biomes, explain the phenomenon of biodiversity gradients.
3. Based on what you know about tropical forests versus temperate forests, why do you think temperate forest habitats are more suitable for long-term agriculture, despite the fact that they have a shorter growing season than tropical forests?

37 Conservation Biology

CHAPTER OUTLINE

BEFORE YOU BEGIN

Before beginning this chapter, take a few moments to review the following discussions:

Table 27.1 What is the cause of the current extinction crisis?

Section 34.2 What factors are influencing human population growth?

Section 35.2 What role does biodiversity play in the energy flow of ecosystems?

CASE STUDY Gills Onions' Waste-to-Energy Project

Gills Onions is one of the largest onion producer/processors in the United States. Over 1 million pounds (lb) of onions are processed every day. During processing, Gills Onions generates approximately 300,000 lb of onion waste daily. Previously, the waste was composted and then hauled to local farm fields and spread as fertilizer. The hauling and spreading of the onion compost led to a multitude of problems, including odor, runoff, acidification of the soil, and a significantly large carbon footprint. Nearly $400,000 was spent annually on the disposal of the onion waste.

Gills Onions' Advanced Energy Recovery System (AERS) has enabled them to become the first food-processing facility in the world to produce ultra-clean energy from their own waste. They are able to convert 100% of their daily onion waste into renewable energy and cattle feed.

The AERS extracts the juice from the onion peels and uses an anaerobic reactor to produce methane-rich biogas that powers two fuel cells. This electricity is then used to power the onion-processing plant, saving Gills Onions an estimated $700,000 in annual electrical cost. The remaining onion pulp is further processed into cattle feed and sold to local ranchers. Additional savings come from the elimination of $400,000 in annual cost associated with the disposal of the onion waste. Greenhouse emissions have also been reduced by the elimination of the truck traffic required to haul the compost to local fields, thus reducing Gills Onions' carbon footprint by an estimated 14,500 metric tons (mt) of CO_2e (equivalent) emissions per year. The AERS is expected to produce a full payback of the $10.8 million spent in less than six years.

As you read through the chapter, think about the following questions:

1. Should all major food processing plants be required to develop and use systems similar to Gills Onions' AERS?
2. What are the potential negative consequences associated with Gills Onions' development of the AERS?
3. Can the indirect values associated with the AERS actually be measured in direct value terms?

37.1 Conservation Biology and Biodiversity

Learning Outcomes

Upon completion of this section, you should be able to

1. Summarize the goals of conservation biology.
2. Recognize the spectrum of biodiversity.

Conservation biology is an interdisciplinary science with the explicit goal of protecting biodiversity and the Earth's natural resources. Conservation biology is considered *interdisciplinary* because it relies on many subdisciplines of biology for the development of basic scientific concepts to describe biodiversity. The application of these concepts, so that biodiversity can be sustainably managed and conserved for future generations, is critical to the future of our planet.

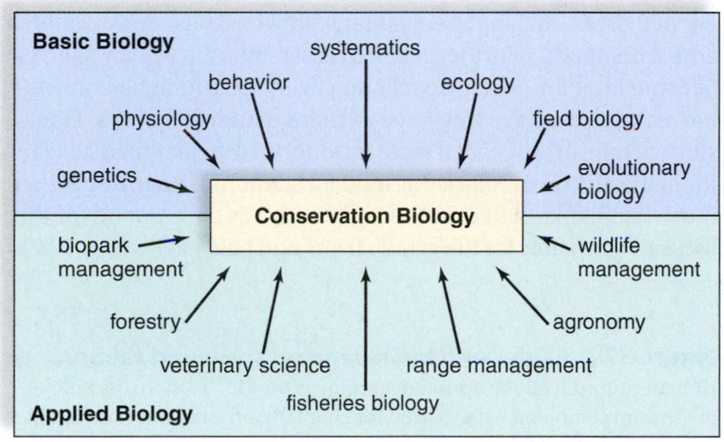
Basic Biology
systematics
behavior
ecology
physiology
field biology
genetics
evolutionary biology
Conservation Biology
biopark management
wildlife management
forestry
agronomy
veterinary science
range management
fisheries biology
Applied Biology

The application of basic scientific principles for conservation and management means that it is often necessary for conservation biologists to work with government officials at both the local and federal levels. Social scientists and economists are also involved because many conservation decisions have socioeconomic impacts in addition to biological ones. Public education is also critical for the success of conservation biology—educating people facilitates better-informed consumer decisions, such as product choices that minimize environmental impact.

Conservation biology is unique among the life sciences in that it embodies the following ethical principles: (1) biodiversity is desirable for the biosphere and therefore for humans; (2) human-induced extinctions are therefore undesirable; (3) complex interactions within ecosystems and communities support biodiversity and the maintenance of such interactions is therefore desirable; and (4) biodiversity generated by evolutionary change has intrinsic value, regardless of any practical benefit. Because species within communities often interact in subtle and complex ways, disruption to the ecosystems in which they live can have unpredictable and potentially dire consequences. For example, loss of a single, ecologically important species could result in large numbers of secondary extinctions. Thus, disruption of ecosystems or ecosystem functioning needs to be avoided.

Conservation biology has often been called a "crisis discipline" or a "discipline with a deadline" because we are facing a sixth mass extinction. Unlike the five previous mass extinctions that have occurred naturally throughout Earth's history, the current mass extinction is much more rapid. It is estimated that as many as 10–20% of all species may be extinct in the next 30 to 50 years, and some researchers suggest that we may lose as many as 50% of all species by the year 2100. Thus, it is urgent that we take steps to ensure that everyone understands the concept of biodiversity, the value of biodiversity, the causes of biodiversity loss, and the potential consequences of reduced biodiversity.

Biodiversity

At its simplest level, **biodiversity** is the variety of life on Earth. Biodiversity is commonly described in terms of the number of species found in a given area or ecosystem. To date, approximately 1.9 million species have been described across the globe (Fig. 37.1). This may only be a small fraction of Earth's species, however. Estimates are that there may be between 10 and 50 million species in all, leaving many species yet to be discovered and described.

Of those that are described, nearly 1,200 species in the United States and 30,000 identified species worldwide are threatened with extinction. An **endangered species** is one that faces immediate extinction throughout all or most of its range. Examples of endangered species include the hawksbill sea turtle, California condor, giant panda, and the snow leopard.

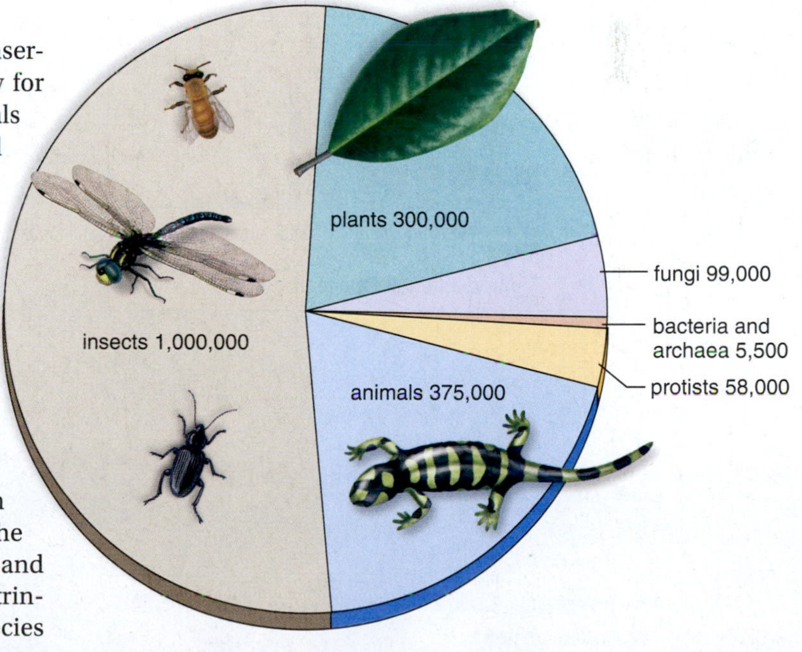
plants 300,000
fungi 99,000
bacteria and archaea 5,500
protists 58,000
animals 375,000
insects 1,000,000

Figure 37.1 Number of described species. There are only about 1.9 million described species; insects are far more prevalent than organisms in other groups. Undescribed species probably number far more than those species that have been described.

Threatened species are likely to become endangered in the foreseeable future. Examples of **threatened species** include the Navaho sedge, northern spotted owl, and coho salmon.

To develop a meaningful understanding of life on Earth, we need to know more about species than just their total number. Although we most often think at the species level, ecologists and conservation biologists necessarily describe biodiversity at three other levels of biological organization: genetic diversity, ecosystem diversity, and landscape diversity.

Genetic diversity includes the number of different alleles, as well as the relative frequencies of those alleles in populations and species. It is genetic diversity that underlies the evolutionary potential of populations, or their capacity to adapt to future environmental change. Thus, populations with high genetic diversity are more likely to have some individuals that can survive changes in their community or ecosystem. Low genetic diversity made the 1846 potato blight in Ireland and the 1984 outbreak of citrus canker in Florida worse than they would have been if the populations were more genetically diverse. Higher genetic diversity would have likely meant that there would have been a higher proportion of individuals that, by chance, were resistant to the pathogens that decimated these crops.

Video Cloning Endangered Species

Ecosystem diversity is dependent on the interactions between species and their abiotic environment in a particular area. An ecosystem may have one or several types of communities within it. The composition of species within a community may differ dramatically among different community types. Consider the species you would find in a temperate rain forest versus a savannah. Variation in community composition increases the levels of biodiversity in the biosphere. Although past conservation efforts frequently focused on saving particular charismatic species, such as the California condor, the black-footed ferret, or the spotted owl, such species-specific methods may not be the most beneficial for all species in the ecosystem. Instead, a more effective approach is to conserve species that play critical roles in their ecosystem, such as those whose loss may result in many secondary extinctions.

Video Coral Reef Ecosystems

Video Tallgrass Prairie Ecology

The reintroduction of wolves in Yellowstone National Park has had a dramatic effect upon the health of the park. Without wolves the elk population significantly increased, which then caused overgrazing of the aspen groves. Wolves have helped reduce the number of elk, thereby increasing the number of aspen and other trees found in floodplains that, in turn, increase the habitat for songbirds and beavers. Beavers have then built dams that have flooded areas, making them suitable for muskrats, ducks, otters, and amphibians. Further, elk carcasses that are left by wolves become food for grizzly bears and eagles. Disrupting a community can threaten the existence of more than one species. Opossum shrimp, *Mysis relicta,* were introduced into Flathead Lake in Montana and its tributaries as food for salmon. The shrimp ate so much zooplankton that, in the end, there was far less food for the fish and ultimately for the grizzly bears and bald eagles (Fig. 37.2).

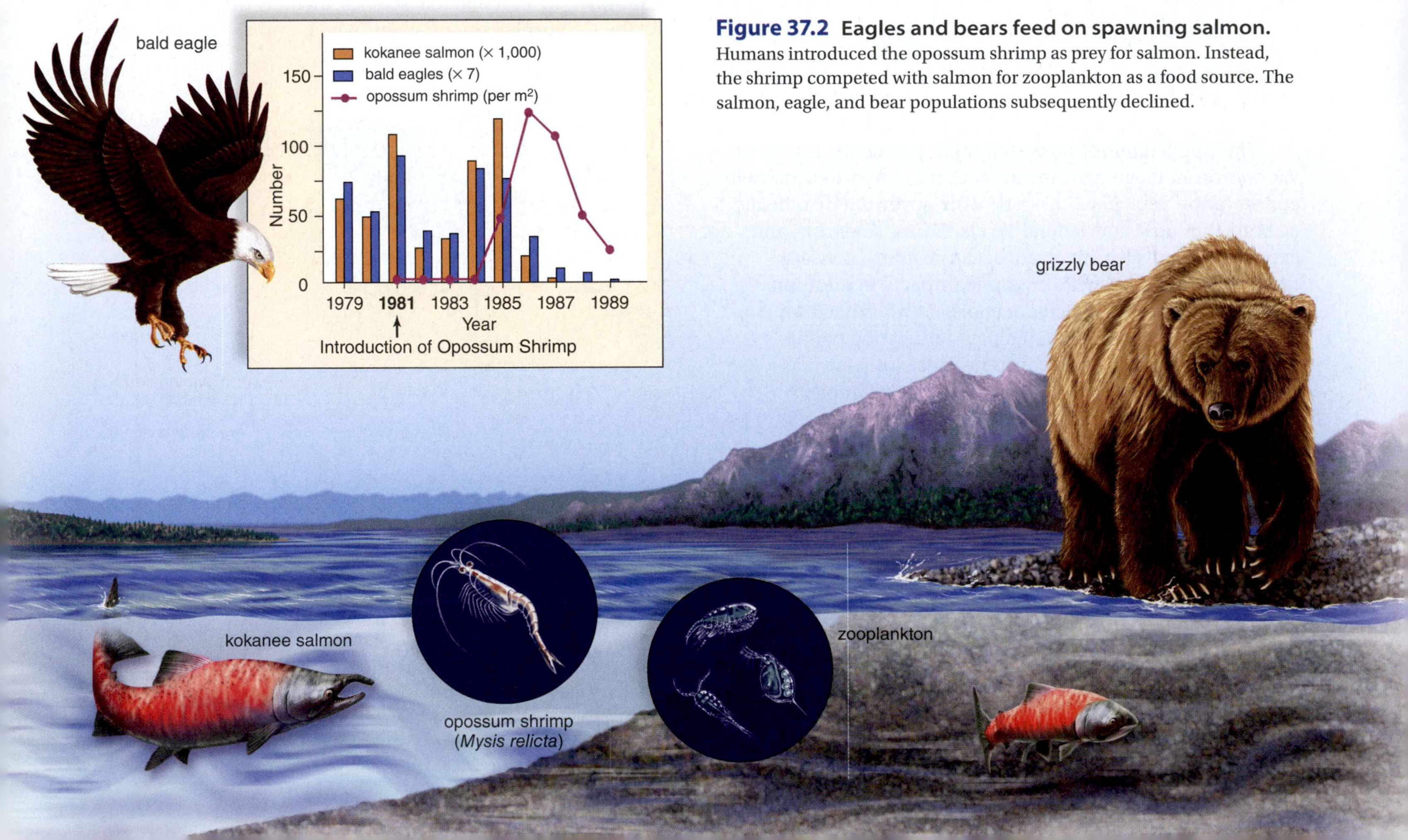
bald eagle
kokanee salmon (× 1,000)
bald eagles (× 7)
opossum shrimp (per m²)
150
100
50
0
Number
1979
1981
1983
1985
1987
1989
Year
Introduction of Opossum Shrimp
grizzly bear
kokanee salmon
opossum shrimp
(Mysis relicta)
zooplankton

Figure 37.2 Eagles and bears feed on spawning salmon. Humans introduced the opossum shrimp as prey for salmon. Instead, the shrimp competed with salmon for zooplankton as a food source. The salmon, eagle, and bear populations subsequently declined.

Landscape diversity involves a group of interacting ecosystems within a single landscape. For example, plains, mountains, and rivers may be fragmented to the point that they are connected only by patches (or corridors) that allow organisms to move from one ecosystem to the other. Fragmentation of the landscape may reduce reproductive capacity by increasing predation risk during dispersal between suitable habitat patches or creating difficulty in finding mates.

Check Your Progress 37.1

1. Describe how conservation biology is supported by a variety of disciplines.
2. Explain why ecosystem-level conservation may be more important than species-level conservation.

37.2 Value of Biodiversity

Learning Outcomes

Upon completion of this section, you should be able to

1. Compare the direct and indirect values of biodiversity.
2. Describe the role that biodiversity plays in a natural ecosystem.

Conservation biology strives to reverse the trend toward the extinction of thousands of plants and animals. To achieve this goal, it is necessary to make all people aware that biodiversity is a resource of immense value.

Direct Value

Various individual species perform services for humans and contribute greatly to the value we should place on biodiversity. Figure 37.3 gives examples of the direct value of wildlife.

Medicinal Value

Most of the prescription drugs used in the United States, valued at over $200 billion, were originally derived from living organisms. The rosy periwinkle from Madagascar is an excellent example of a tropical plant that has provided us with useful medicines. Potent chemicals from this plant are now used to treat two forms of cancer: leukemia and Hodgkin disease. Because of these drugs, the survival rate for childhood leukemia has improved from 10 to 90%, and Hodgkin disease is usually curable. Although the value of saving a life cannot be calculated, it is still sometimes easier for us to appreciate the worth of a resource if it is explained in monetary terms. Thus, researchers estimate that, judging from the success rate in the past, hundreds of additional types of drugs are yet to be found in tropical rain forests, and the value of this resource to society is probably $147 billion.

You may already know that the antibiotic penicillin is derived from a fungus and that certain species of bacteria produce the antibiotics tetracycline and streptomycin. These drugs have proven to be indispensable in the treatment of various sexually transmitted diseases (STDs). Leprosy is among those diseases for which, at this time, there is no cure. The bacterium that causes leprosy will not grow in the laboratory, but scientists discovered that it grows naturally in the nine-banded armadillo. Having a source of the bacterium may make it possible to find a cure for leprosy. The blood of horseshoe crabs contains a substance called limulus amoebocyte lysate, which is used to ensure that medical devices, such as pacemakers, surgical implants, and prosthetic devices, are free of bacteria. Blood is taken from 250,000 crabs a year, which are then returned to the sea unharmed.

Video Indigenous Medicine

Video Good Poison

Agricultural Value

Crops such as wheat, corn, and rice are derived from wild plants that have been modified to be high producers. The same high-yield, genetically similar strains tend to be grown worldwide. When rice crops in Africa were being devastated by a virus, researchers grew wild rice plants from thousands of seed samples until they found one that contained a gene for resistance to the virus. These wild plants were then used in a breeding program to transfer the gene into high-yield rice plants. If this variety of wild rice had become extinct before it could be discovered, rice cultivation in Africa might have collapsed.

Biological pest controls—natural predators and parasites—are often preferable to using chemical pesticides. When a rice pest called the brown planthopper became resistant to pesticides, farmers began to use natural brown planthopper enemies instead. The economic savings were calculated at well over $1 billion. Similarly, cotton growers in Cañete Valley, Peru, found that pesticides were no longer working against the cotton aphid because they had evolved resistance. Research identified natural predators that are now being used to an ever-greater degree by cotton farmers. The savings have been enormous.

Most flowering plants are pollinated by animals such as bees, wasps, butterflies, beetles, birds, and bats. The honeybee, *Apis mellifera,* has been domesticated, and it is estimated that they pollinate between $10 and $15 billion worth of food crops annually in the United States. The danger of this dependency on a single species is exemplified by the fact that mites have now wiped out more than 20% of the commercial U.S. honeybee population. Our main source of resistant bees will be to obtain them from the wild. The value of wild pollinators to the U.S. agricultural economy has been calculated between $4.1 and $6.7 billion a year.

Video Pollinators

Consumptive Use Value

We have experienced much success in cultivating crops, keeping domesticated animals, growing trees on plantations, and so forth. But so far, aquaculture, the growing of fish and shellfish for human consumption, has contributed only minimally to human welfare. Instead, most freshwater and marine harvests depend on wild animals, such as fishes (e.g., trout, cod,

Figure 37.3 Direct value of wildlife. The direct services of wild species benefit humans immensely. It is sometimes possible to calculate the monetary value, which is always surprisingly large.

Wild species, like the rosy periwinkle, *Catharanthus roseus,* are sources of many medicines.

Wild species, like these fish, provide us with food.

Wild species, like the nine-banded armadillo, *Dasypus novemcinctus,* play a role in medical research.

Wild species, like the lesser long-nosed bat, are pollinators of agricultural and other plants.

Wild species, like rubber trees, can provide a product indefinitely if the forest is not destroyed.

Wild species, like ladybugs, play a role in biological control of agricultural pests.

and tuna) and crustaceans (e.g., lobsters, shrimps, and crabs). Obviously, these aquatic organisms are an invaluable biodiversity resource.

The environment provides all sorts of other products that are sold in the marketplace worldwide, including wild fruits and vegetables, skins, fibers, beeswax, and seaweed. Also, some people obtain their meat directly from the wild. In one study, researchers calculated that the economic value of wild pig in the diet of native hunters in Sarawak, East Malaysia, was approximately $40 million per year.

Similarly, many trees are still felled for their wood. Researchers have calculated that a species-rich forest in the Peruvian Amazon region is worth far more if the forest is used for fruit and rubber production than for timber production. Whereas fruit and the latex needed to produce rubber can be brought to market indefinitely, no more timber or other resulting products can be produced when trees are cut down.

Indirect Value

The wild species we have been discussing live in ecosystems. If we want to preserve them, many scientists argue it is more economical to save the ecosystems than the individual species. Ecosystems perform many services for modern humans. These services are said to be indirect because they are pervasive and it is not easy to associate a direct dollar value to them (Fig. 37.4). Human survival depends on the functions that natural ecosystems perform for us. Studies have suggested that the indirect value of ecosystem services surpasses that of the global gross national product—about $33 trillion per year!

biogeochemical cycles
waste disposal
provision of fresh water
prevention of soil erosion
regulation of climate
ecotourism

Figure 37.4 Indirect value of ecosystems. Natural ecosystems provide many physiological and psychological benefits to humans.

Biogeochemical Cycles

You'll recall from Chapter 35 that ecosystems are characterized by energy flow and chemical cycling. The biodiversity within ecosystems contributes to the workings of the water, carbon, nitrogen, phosphorus, and other biogeochemical cycles. We are dependent on these cycles for fresh water, removal of carbon dioxide from the atmosphere, uptake of excess soil nitrogen, and provision of phosphate. When human activities upset the usual workings of biogeochemical cycles, the dire environmental consequences include the release of excess pollutants that are harmful to us. Unfortunately, technology is unable to artificially replicate any of the biogeochemical cycles.

Waste Disposal

Decomposers break down dead organic matter and other types of wastes to inorganic nutrients that are used by the producers within ecosystems. This function aids humans immensely because we dump millions of tons of waste material into natural ecosystems each year. If it were not for decomposition, waste would soon cover the entire surface of our planet. We can build sewage treatment plants, but they are expensive. In addition, few of them break down solid wastes completely to inorganic nutrients. It is less expensive and more efficient to water plants and trees with partially treated wastewater and let soil bacteria cleanse it completely.

Video Decomposers

Biological communities are also capable of breaking down and immobilizing pollutants like heavy metals and pesticides that humans release into the environment. A review of wetland functions in Canada assigned a value of $50,000 per hectare (2.5 acres or 10,000 m^2) per year to the ability of natural areas to purify water and take up pollutants.

Provision of Fresh Water

Few terrestrial organisms are adapted to living in a salty environment—they need fresh water. The water cycle continually supplies fresh water to terrestrial ecosystems. Humans use fresh water in innumerable ways, including drinking and irrigation of their crops. Freshwater ecosystems, such as rivers and lakes, also provide us with fish and other types of edible organisms. Unlike other commodities, there is no substitute for fresh water. We can remove salt from seawater to obtain fresh water, but the cost of desalinization is about four to eight times the average cost of fresh water acquired via the water cycle.

Video Thames River

Flood Prevention

Forests and other natural ecosystems exert a "sponge effect," thereby reducing flooding. They soak up water and then release it at a regular rate. When rain falls in a natural area, plant foliage and dead leaves lessen its impact, and the soil slowly absorbs it, especially if the soil has been aerated by organisms. The water-holding capacity of forests reduces the possibility of flooding. The value of a marshland outside Boston, Massachusetts, has been estimated at $72,000 per hectare per year based solely on its ability to reduce floods. Flooding in New Orleans by Hurricane Katrina in 2005 would likely have been much less severe if the natural wetlands and marshlands around the Gulf of Mexico had still been intact.

Prevention of Soil Erosion

Intact ecosystems naturally retain soil and prevent soil erosion. The importance of this ecosystem attribute is especially observed following deforestation. In Pakistan, the world's largest dam, the Tarbela Dam, is losing its storage capacity of 12 billion cubic meters (m^3) many years sooner than expected because silt is building up behind the dam due to deforestation. At one time, the Philippines exported $100 million worth of oysters, mussels, and clams each year. Now, silt carried down rivers following deforestation is smothering the mangrove ecosystem that serves as a nursery for the sea. In general, most coastal ecosystems are not as bountiful as they once were because of deforestation, coastal development, and aquatic pollution.

Regulation of Climate

At the local level, trees provide shade, reduce the need for fans and air conditioners during the summer, and provide buffers from noise. In tropical rain forests, trees maintain regional rainfall, and without them the forests become arid.

Globally, forests ameliorate the climate because they take up carbon dioxide and release oxygen. The leaves of trees use carbon dioxide when they photosynthesize, and the bodies of the trees store carbon. When trees are cut and burned, carbon dioxide is released into the atmosphere. Carbon dioxide makes a significant contribution to global warming, which is expected to be stressful for many plants and animals.

Ecotourism

Many people prefer to vacation in natural areas. In the United States, nearly 100 million people spend nearly $5 billion each year on fees, travel, lodging, and food associated with ecotourism. Tourists are often interested in activities such as sport fishing, whale watching, boat riding, hiking, birdwatching, and the like. Some merely want to immerse themselves in the beauty and serenity of nature. Many underdeveloped countries in tropical regions, such as Belize and Costa Rica, are taking advantage of this by offering "ecotours" of the local biodiversity. Providing guided tours of natural ecosystems, such as tropical rain forests, is often more profitable than destroying them. Ecotourism also provides indirect value for wildlife. As an example, the tusks of an elephant are worth around $10,000 for the ivory, but the Kenyan government estimates that a single elephant is worth as much as $1.2 million in tourist dollars.

Check Your Progress 37.2

1. Identify the differences between consumptive use value and agricultural value of biodiversity.
2. Explain the difference between direct and indirect values of biodiversity.

37.3 Threats to Biodiversity

Learning Outcomes

Upon completion of this section, you should be able to

1. Classify the five causes of extinction.
2. Compare natural and human influenced causes of extinction.

We are presently in a biodiversity crisis—the number of extinctions (loss of species) expected to occur in the near future is unparalleled in Earth's history. To identify the role that humans are playing in modern extinctions, researchers examined the records of 1,880 threatened and endangered wild species in the United States. Habitat loss is the most significant factor that was involved in 85% of the cases (Fig. 37.5*a*). Exotic species had a hand in nearly 50%, pollution was a factor in 24%, overexploitation in 17%, and disease in 3%. Note that the percentages add up to more than 100% because most of these species are imperiled for more than one reason. These five causes reflect the relative importance of the five leading causes for species' extinctions worldwide. Macaws illustrate that a combination of factors can lead to a species decline (Fig. 37.5*b*). Not only has their habitat been reduced by encroaching timber and mining companies, but macaws are also hunted for food and collected for the pet trade. DNA technology may now help enforce laws to protect wildlife collected for the pet trade, as described in the Scientific Inquiry feature, "Wildlife Conservation and DNA," on page 746.

Habitat Loss

Habitat loss has occurred in all ecosystems, and human disruption of natural habitats is the most influential factor in biodiversity loss. Concern has now centered on tropical rain forests and coral reefs because they are particularly rich in species diversity. A sequence of events in Brazil offers a fairly typical example of the manner in which rain forest is converted to land uninhabitable for wildlife. The construction of a major highway into the forest first provided a way to reach the interior of the forest (Fig. 37.5*c*). Small towns and industries sprang up along the highway, and roads branching off the main highway gave rise to even more roads. The result was fragmentation of the once immense forest. The government offered subsidies to anyone willing to take up residence in the forest, and the people who came cut and burned trees in patches to facilitate grazing (Fig. 37.5*c*). Tropical soils contain limited nutrients, and burning of trees releases additional nutrients that support cattle grazing for only about three years. Once the land was degraded, the farmers moved on to another portion of the forest to start over again.

Figure 37.5 Habitat loss. **a.** In a study that examined records of imperiled U.S. plants and animals, habitat loss emerged as the greatest threat to wildlife. **b.** Macaws that reside in South American tropical rain forests are endangered for some of the reasons listed in the graph in (a). **c.** Habitat loss due to road construction in Brazil. (*top*) Road construction opened up the rain forest and subjected it to fragmentation. (*middle*) The result was patches of forest and degraded land. (*bottom*) Wildlife could not live in destroyed portions of the forest.

Habitat Loss
Exotic Species
Pollution
Overexploitation
Disease
0 20 40 60 80 100
% Species Affected by Threat

a. Threats to wildlife

b. Macaws on salt lick

Roads cut through forest

Forest occurs in patches

Destroyed areas

c. Wildlife habitat is reduced

Loss of habitat also affects freshwater and marine biodiversity. Coastal degradation is mainly due to the large concentration of people living on or near the coast. At least 40% of the world's population lives within 100 km (60 mi) of a coastline, and this number is expected to increase. In the United States, over one-half of the population lives within 80 km (50 mi) of the coasts (including the Great Lakes). Coastal habitation leads to beach erosion and direct inputs of pollutants into freshwater or marine systems. Already, 60% of coral reefs have been destroyed or nearly so, with siltation resulting from beach erosion as a leading cause. It is possible that all coral reefs may disappear during the next 40 years unless our behaviors drastically change. Mangrove forest destruction is also a problem. Indonesia, with the most mangrove acreage, has lost 45% of its mangroves, and the percentage is even higher for other tropical countries. Wetland areas, estuaries, and seagrass beds are also being rapidly destroyed by the actions of humans.

Exotic Species

Exotic species, sometimes also called alien species, are nonnative members of a community. Communities around the world are characterized by unique assemblages of species that have evolved together in an area. Introduction of exotic species can disrupt this balance by changing the interactions between species in a food web, as shown in Figure 37.2. In this example, opossum shrimp introduced in a lake in Montana added a trophic level that, in the end, meant less food for bald eagles and grizzly bears. Often, exotic species can directly compete with or prey upon native species, thereby reducing their abundance or causing localized extinction. For these reasons, exotics are the second most important reason for biodiversity loss. Humans have introduced exotic species into new ecosystems by the following means:

Video Alien Invasion

Colonization. Europeans, in particular, brought various familiar species with them when they colonized new places. For example, the pilgrims brought the dandelion to the United States as a familiar salad green. The British brought foxes and stoats (a type of weasel) to New Zealand, and these predators have resulted in the loss of nearly 40% of all of New Zealand's bird species. These birds are easy prey because they evolved without mammalian predators, which were completely absent from New Zealand.

Horticulture and agriculture. Some exotics now taking over vast tracts of land have escaped from cultivated areas. Kudzu is a vine from Japan that the U.S. Department of Agriculture thought would help prevent soil erosion. The plant now covers much of the landscape in the South, including walnut, magnolia, and gum trees (Fig. 37.6*a*). Similarly, the water hyacinth was introduced to the

a. Kudzu

b. Mongoose

Figure 37.6 Exotic species. **a.** Kudzu, a vine from Japan, was introduced in several southern states to control erosion. Today, kudzu has taken over and displaced many native plants. **b.** Mongooses were introduced into Hawaii to control rats, but they also prey on native birds.

SCIENCE IN YOUR LIFE ▶ SCIENTIFIC INQUIRY

Wildlife Conservation and DNA

After DNA analysis, scientists were amazed to find that some 60% of loggerhead turtles found drowned in Mediterranean fishing nets were from beaches in the southeastern United States. Because the unlucky creatures were a good representative sample of the turtles in the area, that meant more than half of the young turtles living in the Mediterranean Sea had hatched from nests on beaches in Florida, Georgia, and South Carolina (Fig. 37A*a*). Some 20,000 to 50,000 loggerheads die each year due to the Mediterranean fisheries, which may partly explain the decline in loggerheads nesting on southeastern U.S. beaches for the last 25 years.

The sequencing of DNA from Alaskan brown bears allowed Sandra Talbot (a graduate student at the University of Alaska's Institute of Arctic Biology) and wildlife geneticist Gerald Shields to conclude that there are two types of brown bears in Alaska. One type resides only on southeastern Alaska's Admiralty, Baranof, and Chichagof islands, known as the ABC Islands. The other brown bear in Alaska is found throughout the rest of the state, as well as in Siberia and western Asia (Fig. 37A*b*).

a. Loggerhead turtle

b. Brown bears

Figure 37A DNA studies. **a.** Many loggerhead turtles found in the Mediterranean Sea are from the southeastern United States. **b.** These two brown bears appear similar, but one type, known as an ABC bear, resides only on southeastern Alaska's Admiralty, Baranof, and Chichagof islands.

A third distinct type of brown bear, known as the Montana grizzly, resides in other parts of North America. These three types make up all of the known brown bears in the New World.

The ABC bears' uniqueness may be bad news for the timber industry, which has expressed interest in logging parts of the ABC Islands. Says Shields, "Studies show that when roads are built and the habitat is fragmented, the population of brown bears declines. Our genetic observations suggest they are truly unique, and we should consider their heritage. They could never be replaced by transplants."

In what will become a classic example of how DNA analysis might be used to protect endangered species from future ruin, scientists from the United States and New Zealand carried out discreet experiments in a Japanese hotel room on whale sushi bought in local markets. Sushi, a staple of the Japanese diet, is rice and meat wrapped in seaweed. Armed with a miniature DNA sampling machine, the scientists found that, of the 16 pieces of whale sushi they examined, many were from whales that are endangered or protected under an international moratorium on whaling.

United States from South America because of its beautiful flowers. Today, it clogs up waterways and diminishes natural diversity.

Accidental transport. Global trade and travel accidentally bring many new species from one country to another. The zebra mussel from the Caspian Sea was accidentally introduced into the Great Lakes in 1988. It now forms dense beds that reduce biodiversity by outcompeting native mussels. Zebra mussels also decrease the amount of food available for higher levels in the food chain because they are such effective filter feeders. Zebra mussels cause millions in damage per year by clogging sewage and other pipes. Other exotics introduced into the United States include the Argentinian fire ant and the nutria, a large rodent found throughout the Southeast.

Exotics on Islands

Islands are particularly susceptible to environmental discord caused by the introduction of exotic species. Islands have unique assemblages of native species that are closely adapted to one another and cannot compete well against exotics. Myrtle trees, *Myrica faya,* introduced into the Hawaiian Islands from the Canary Islands, are symbiotic with a type of bacterium that is capable of nitrogen fixation. This feature allows the species to establish itself on nutrient-poor volcanic soil such as that found in Hawaii. Once established, myrtle trees halt normal ecological succession by outcompeting native plants on volcanic soil.

The brown tree snake has been introduced onto a number of islands in the Pacific Ocean. The snake eats eggs, nestlings, and adult birds. On Guam, it has reduced ten native bird species to the point of extinction. Mongooses introduced into the Hawaiian Islands to control rats have increased dramatically in abundance and also prey on native birds, causing some population declines (Fig. 37.6*b*).

Pollution

In the present context, **pollution** can be defined as any environmental change that adversely affects the lives and health of living organisms. Pollution has been identified as the third main cause of extinction. Pollution can also weaken organisms

"Their findings demonstrated the true power of DNA studies," says David Woodruff, a conservation biologist at the University of California, San Diego.

One sample was from an endangered humpback, four were from fin whales, one was from a northern minke, and another from a beaked whale. Stephen Palumbi of Harvard University says the technique could be used for monitoring and verifying catches. Until then, he says, "no species of whale can be considered safe."

Meanwhile, Ken Goddard, director of the unique U.S. Fish and Wildlife Service Forensics Laboratory in Ashland, Oregon, is already on the watch for wildlife crimes in the United States and 122 other countries that send samples to him for analysis. "DNA is one of the most powerful tools we've got," says Goddard, a former California police crime-lab director.

The lab has blood samples, for example, for all of the wolves being released into Yellowstone Park—"for the obvious reason that we can match those samples to a crime scene," says Goddard. The lab has many cases currently pending in court that he cannot discuss. But he likes to tell the story of the lab's first DNA-matching case. Shortly after the lab opened in 1989, California wildlife authorities contacted Goddard. They had seized the carcass of a trophy-sized deer from a hunter. They believed the deer had been shot illegally on a 3,000-acre preserve owned by actor Clint Eastwood. The agents found a gut pile on the property but had no way to match it to the carcass. The hunter had two witnesses to deny the deer had been shot on the preserve.

Goddard's lab analysis made a perfect match between tissue from the gut pile and tissue from the carcass. Says Goddard: "We now have a cardboard cutout of Clint Eastwood at the lab saying 'Go ahead: Make my DNA.'"

Questions to Consider

1. In some situations, distinctiveness, in terms of genetic DNA sequence, is being used to set policy. That is, if researchers find a population of a species, such as ABC brown bears, to be genetically distinct from other populations of the same species, then that population may be afforded special protection, such as threatened or endangered status under the U.S. Endangered Species Act. This policy is then used to make major environmental decisions, such as halting logging in that area. Should DNA distinctness be used to make such major policy decisions?
2. Forensic DNA evidence can be used to prosecute hunters who illegally hunt animals, such as deer, in areas where they do not have a proper hunting license. Occasionally, such hunters face harsh sentences that may even include jail time. Does DNA evidence provide enough proof to convict people of illegal hunting?
3. DNA studies, such as those conducted in Japan, opened our eyes to the fact that illegal whale hunting still occurs. Some people have recommended we screen the DNA of much of the imported fish meat coming from different countries. However, this can be quite expensive. Do you support genetic testing of imported fish meat, even if it increases the cost of the meat to the consumer?

and make them more susceptible to disease. Biodiversity is particularly threatened by the following types of environmental pollution:

Acid deposition. Sulfur dioxide from power plants and nitrogen oxides from auto exhaust are converted to acids when combined with atmospheric water vapor. These acids return to Earth during precipitation (rain or snow) or via dry deposition. This acid deposition decimates forests because it weakens trees and increases their susceptibility to disease and insects. In addition, because acid deposition can make the pH of water too low for organisms to survive, many lakes in northern states are now lifeless.

Eutrophication. Lakes are also under stress due to overenrichment. When lakes receive excess nutrients due to runoff from agricultural fields and wastewater from sewage treatment, algae begin to grow in abundance. Death of algae leads to an abundance of decomposers that decreases the available oxygen and often leads to fish kills.

Ozone depletion. The ozone shield protects Earth from harmful ultraviolet (UV) radiation. The release of chlorofluorocarbons (CFCs), in particular, into the atmosphere causes the shield to break down, leading to impaired growth of crops and trees and the death of plankton that sustain oceanic life. The immune systems and ability of all organisms to resist disease will most likely be weakened. Ozone depletion has also led to dramatic increases in skin cancer.

Organic chemicals. Our modern society uses organic chemicals in all sorts of ways. Organic chemicals called nonylphenols are used in products ranging from pesticides to dishwashing detergents, cosmetics, plastics, and spermicides. Many of these chemicals can mimic the effects of hormones, and if so, they are called endocrine-disruptors. For example, investigators exposed young salmon to nonylphenol. Although these fish are born in fresh water and mature in salt water, 20–30% of fish exposed to nonylphenol were unable to make the transition from fresh to salt water. Nonylphenols cause the pituitary to produce prolactin, a hormone that may prevent saltwater adaptation. Other endocrine-disrupting contaminants can possibly affect the endocrine system and thereby the reproductive potential of other animals, including humans.

Climate change. The term **climate change** refers to recent changes in the Earth's climate. The major contributor to climate change is the phenomenon of global warming, which is an increase in Earth's temperature due to the increase of greenhouse gases, such as carbon dioxide and methane, in the atmosphere. Climate change is expected to have many detrimental effects, including the destruction of coastal wetlands due to a rise in sea levels, a shift in suitable temperatures to locations where various species cannot live, and the death of coral reefs (Fig. 37.7). An upward shift in temperatures could influence everything from growing seasons in plants to migratory patterns in animals. As temperatures rise, regions of suitable climate for various terrestrial species may shift toward the poles and higher elevations. Extinctions are expected because the present assemblages of species in ecosystems will be disrupted as some species migrate northward (or southward in the Southern Hemisphere) following the environmental changes. Other species with limited mobility may not be able to migrate quickly enough. To remain in a favorable habitat, it's been calculated that the rate of beech tree migration would have to be 40 times faster than has ever been observed.

Video Global Warming

Video Warming Hurts Rice

Video Karoo Global Warming

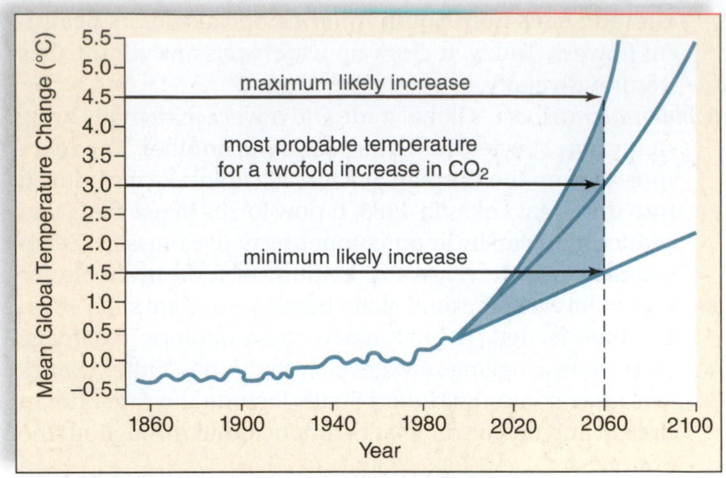
Mean Global Temperature Change (°C)
5.5
5.0
4.5
4.0
3.5
3.0
2.5
2.0
1.5
1.0
0.5
0.0
−0.5
maximum likely increase
most probable temperature
for a twofold increase in CO_2
minimum likely increase
1860
1900
1940
1980
2020
2060
2100
Year

a.

b.

Figure 37.7 Climate change and global warming.
a. Mean global temperature is expected to rise due to the introduction of greenhouse gases into the atmosphere. **b.** Global warming has the potential to significantly affect the world's biodiversity distribution. A temperature rise of only a few degrees causes coral reefs to "bleach" and become lifeless.

Overexploitation

Overexploitation occurs when the number of individuals taken from a wild population is so great that the population becomes severely reduced in numbers. A positive feedback cycle explains overexploitation: the smaller the population, the more commercially valuable its members, and the greater the incentive to capture the few remaining organisms. Poachers are very active in the collecting and sale of endangered and threatened species because it has become so lucrative. The overall international value of trading wildlife species is $20 billion, of which $8 billion is attributed to the illegal sale of rare species.

Markets for rare plants and exotic pets support both legal and illegal trade in wild species. Rustlers dig up rare cacti, such as the crested saguaros, and sell them to gardeners for as much as $15,000 each. Parrots are among birds taken from the wild for sale to pet owners. For every bird delivered alive, many more have died in the process. The same holds true for tropical fish, which often come from the coral reefs of Indonesia and the Philippines. Divers dynamite reefs or use plastic squeeze-bottles of cyanide to stun the fish within them. In the process, many fish and corals die.

The Convention on International Trade in Endangered Species (CITES) was an agreement established in 1973 to ensure that international trade of species does not threaten their survival. Today, over 30,000 species of plants and animals receive some level of protection from over 172 countries worldwide.

Poachers still hunt for hides, claws, tusks, horns, or bones of many endangered mammals. Because of its rarity, a single Siberian tiger is now worth more than $500,000—its bones are pulverized and used as a medicinal powder. The horns of rhinoceroses become ornate carved daggers, and their bones are also ground up to sell as a medicine. The ivory of an elephant's tusk is used to make art objects, jewelry, or piano keys. The fur of a Bengal tiger sells for as much as $100,000 in Tokyo.

The Food and Agricultural Organization of the United Nations tells us that we have now overexploited 11 of 15 major oceanic fishing areas. Fish are a renewable resource if harvesting does not exceed the ability of the population to reproduce. Our society uses larger and more efficient fishing fleets to decimate fishing stocks. Pelagic species such as tuna are captured by purse seine fishing, in which a very large net surrounds a school of fish, and then the net is closed in the same manner as a drawstring purse. Up to thousands of dolphins that swim above schools of tuna are often captured and then killed in this type of net. However, many tuna suppliers advertise their product as "dolphin safe." Other fishing boats drag huge trawling nets, large enough to accommodate 12 jumbo jets, along the seafloor to capture bottom-dwelling fish (Fig. 37.8*a*). Only large fish are kept; undesirable small fish and sea turtles are discarded, dying, back into the ocean. Trawling has been called the marine equivalent of clear-cutting trees because after the net goes by, the sea bottom is devastated (Fig. 37.8*b*). Today's fishing practices don't allow fisheries to recover. Cod and haddock, once the most abundant bottom-dwelling fish along the northeast coast of the United States, are now often outnumbered by dogfish and skate.

Entire marine ecosystems can be disrupted by overfishing, as exemplified on the U.S. west coast. When sea otters began to decline in numbers, investigators found that they were being eaten by orcas (killer whales). Usually orcas prefer seals and sea lions to sea otters, but they began eating sea otters when few seals and sea lions could be found. The decline in the seal and sea lion population was due to the decline in the perch and herring populations as a result of overfishing. Ordinarily, sea otters keep the population of sea urchins, which feed on kelp, under control. But with fewer sea otters around, the sea urchin population exploded and decimated the kelp beds. Thus, overfishing set in motion a chain of events that detrimentally altered the food web of an ecosystem.

a. Fishing by use of a drag net

b. Result of drag net fishing

Figure 37.8 Trawling. **a.** These Alaskan pollock were caught by dragging a net along the seafloor. **b.** Appearance of the seafloor after the net passed.

Video Ocean Fishing Ban

Disease

Wildlife is subject to emerging diseases just as humans are. Due to the encroachment of humans on their habitat, wildlife have been exposed to new pathogens from domestic animals living nearby. For example, canine distemper was spread from domesticated dogs to lions in the African Serengeti, causing population declines. Avian influenza likely emerged from domesticated fowl (e.g., chicken) populations and has led to the deaths of millions of wild birds.

Figure 37.9 Amphibian at risk. The harlequin toad is nearly extinct due to infections by a fungal pathogen. Diseases are implicated in global population declines and extinctions of amphibians.

The significant effect of diseases on biodiversity is underscored by National Wildlife Health Center findings that almost half of sea otter deaths along the coast of California are due to infectious diseases. Scientists tell us the number of pathogens that cause disease is on the rise, and just as human health is threatened, so is that of wildlife. Extinctions due simply to disease may occur, and are currently implicated in the worldwide decline of amphibians (Fig. 37.9).

Check Your Progress 37.3

1. Discuss two ways in which exotic species are introduced into new areas.
2. Identify the five main causes of extinction.
3. Recognize various reasons why pollution can impact biodiversity.

37.4 Habitat Conservation and Restoration

Learning Outcomes

Upon completion of this section, you should be able to

1. Describe the value of preserving biodiversity hotspots.
2. Distinguish between keystone species and flagship species.
3. Summarize the goals of habitat restoration.

Habitat Conservation

Because habitat loss is the leading cause of species' extinctions, conservation of habitat is of primary concern. One way to prioritize which habitats to conserve is to focus on those that contain the highest levels of biodiversity. Generally, biodiversity is highest at the tropics, and it declines toward each pole. Tropical rain forests and coral reefs are examples of ecosystems known for possessing high levels of biodiversity.

Some regions of the world are called **biodiversity hotspots** because they contain unusually large concentrations of species. Biodiversity in these hotspots accounts for about 44% of all known higher plant species and 35% of all terrestrial vertebrate species but covers only about half of 1.4% of Earth's land area. Hotspots may also contain large numbers of endemic species, or those not found anywhere else. An example of such a hotspot is the tropical forests of Madagascar, within which 93% of the primate species, 99% of the amphibian species, and over 80% of the plant species are endemic. The Cape region of South Africa, Indonesia, the coast of California, and the Great Barrier Reef of Australia are also considered biodiversity hotspots.

Video Coral Reef Ecosystems

We can also focus our efforts on conserving habitat for **keystone species,** or those whose loss would result in a great number of secondary extinctions. Keystone species need not be the most abundant. Although they are relatively low in numbers, wolves can be considered a keystone species in Yellowstone National Park, as discussed earlier. Bats have also been designated keystone species in Old World tropical forests. Bats are extremely important pollinators and dispersers of tree seeds. The loss of bats results in failure of many tree species to reproduce and can lead to a loss of biodiversity. Other keystone species include grizzly bears (Fig. 37.10*a*), beavers in wetlands, and elephants in grasslands and forests.

Keystone species should not be confused with **flagship species,** or those that evoke an emotional response from humans. Flagship species are considered charismatic and are valued for their beauty, regal nature, or similarity to people's pets. Flagship species such as lions, tigers, dolphins, and the giant panda can motivate the public to conserve biodiversity.

Landscape Conservation and Reserve Design

Conservation often has to occur at the landscape level because sufficient habitat may not be available in a single place to sustain a viable population of a particular species. Grizzly bears, once numbering between 50,000 and 100,000 south of Canada, are now estimated at about 1,000 individuals in six small, subdivided populations. Grizzly bear conservation entails maintenance of a number of different types of ecosystems, including plains, mountains, and rivers. Other species also rely on multiple ecosystem types, such as amphibians, that rely on wetlands and terrestrial habitat. Conserving a single one of these ecosystems would not be sufficient, nor would conservation of several ecosystems in isolation of one another. It is thus necessary to establish **conservation corridors** that allow animals to move safely between habitats that would otherwise be isolated. Corridors are often necessary because landscapes are subdivided due to urbanization, agriculture, and other aspects of human development. A

a. Grizzly bear, *Ursus arctos horribilis*

b. Old-growth forest; northern spotted owl, *Strix occidentalis caurina* (inset)

Figure 37.10
Habitat preservation.
When particular species are protected, other wildlife benefits. **a.** The Greater Yellowstone Ecosystem has been delineated in an effort to save grizzly bears, which need a very large habitat. **b.** Currently, the remaining portions of old-growth forests in the Pacific Northwest are not being logged in order to save the northern spotted owl (inset).

corridor may be as small as an overpass for wildlife to cross a highway or a strip of forested habitat that is maintained along a stream after timber harvest. They can be as large as a patch of land several kilometers wide to allow for animals to follow migration routes.

Landscape conservation for a single species is often beneficial for other species that share the same habitats. For example, conservation of the northern spotted owl (Fig. 37.10*b*) helps protect many other species that inhabit temperate old growth forests. The last of the contiguous 48 states' harlequin ducks, westslope cutthroat trout, lynx, pine martins, wolverines, and great gray owls are found in areas occupied by grizzly bears. The geographic range of grizzly bears also overlaps with 40% of Montana's vascular plants of conservation concern.

Edge Effects

When conserving landscapes, it is necessary to consider **edge effects.** An edge reduces the amount of habitat typical of an ecosystem because the edges around a patch have a habitat slightly different from the interior of the patch. For example, forest edges are brighter, warmer, drier, and windier, with more vines, shrubs, and weeds than the forest interior. Also, Figure 37.11*a* shows that a small and a large patch of habitat have the same amount of edge. Therefore, the effective habitat shrinks as a patch gets smaller.

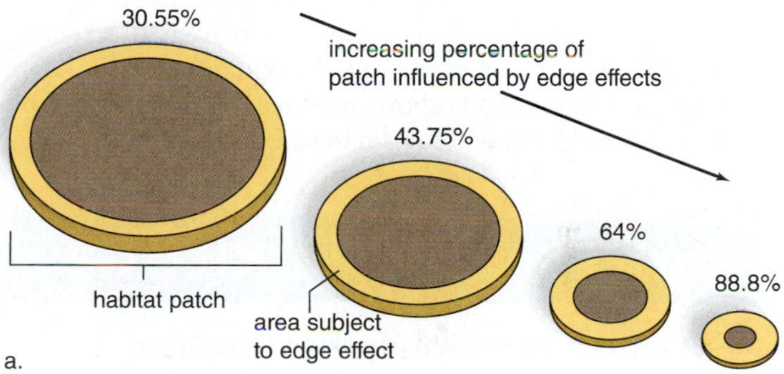
30.55%
increasing percentage of
patch influenced by edge effects
43.75%
64%
88.8%
habitat patch
area subject
to edge effect

a.

brown-headed
cowbird chick
yellow warbler
chick

b.

Figure 37.11 Edge effect. **a.** The smaller the patch, the greater the proportion that is subject to the edge effect. **b.** Cowbirds lay their eggs in the nests of songbirds (yellow warblers). A cowbird is bigger than a warbler nestling and will be able to acquire most of the food brought by the warbler parent.

Many popular game animals, such as turkeys and white-tail deer, are more plentiful in the edge region of a particular area. However, today it is known that creating edges can be detrimental to wildlife because of habitat fragmentation.

Edge effects can also have a serious impact on population size. Songbird populations west of the Mississippi have been declining of late, and ornithologists have noticed that the nesting success of songbirds is quite low at the edge of a forest. The cause turns out to be the brown-headed cowbird, a social parasite of songbirds. Adult cowbirds prefer to feed in open agricultural areas, and they only briefly enter the forest when searching for a host nest in which to lay their eggs (Fig. 37.11*b*). Cowbirds are therefore benefited, while songbirds are disadvantaged, by the edge effect.

Reserve Design

Conservation reserves are those areas that are set aside with the primary goal of protecting biodiversity within them. As such, reserves should be largely protected from human activities, except perhaps ecotourism or limited scientific research. When considering the arguments above, reserves should contain sufficient amounts of habitat to sustain the biodiversity within them. This space should include multiple ecosystems that are connected by corridors to allow movement of animals and dispersal of plant species between habitat types. This space should also include enough area to account for edge effects, meaning that the absolute amount of land included in each reserve is likely higher than the usable habitat within it.

Globally, a network of more than 480 biosphere reserves has been designated by the United Nations. The maintenance of biosphere reserves is in the hands of the countries and territories in which they are located, and compliance is voluntary. Because the goals of biosphere reserves include preservation of local cultural values, as well as maintenance of biodiversity and the ability to foster sustainable human development, there has been more cooperation than was initially expected.

Each biosphere reserve is divided into three areas. First, there is the central core reserve, which allows only research, light tourism, and limited sustainable use for cultural purposes. Next, there is a buffer zone that surrounds the core where only low-impact human activities are allowed. A transition area then surrounds the buffer zone. This area is meant to support sustainable human development, tourism, and agriculture.

SCIENCE IN YOUR LIFE ▶ ECOLOGY

Emiquon Floodplain Restoration

Emiquon was once a vital component of the Illinois River system, which is a major tributary of the Mississippi River. It included a complex system of backwater wetlands and lakes that originally covered an area of approximately 14,000 acres (Fig. 37B). The vast diversity of native plants and animals found within Emiquon have supported human civilizations for over 10,000 years. Early cultures accessed the abundant inland fisheries and a variety of mussel species that could be found there. River otters, mink, beaver, and migrating birds, as well as a variety of prairie grasses on the floodplains would have been used by native peoples (Fig. 37C).

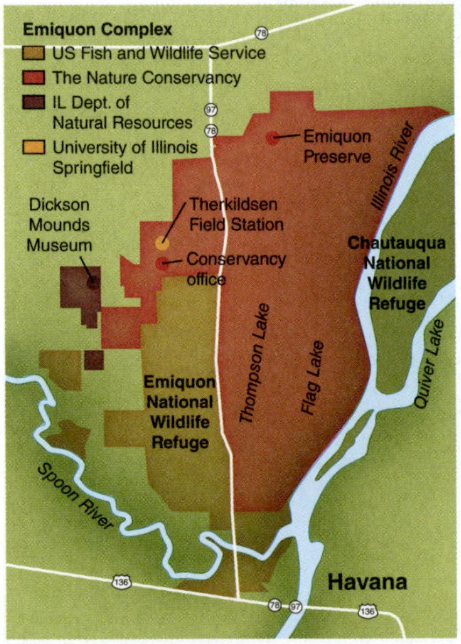
Emiquon Complex
US Fish and Wildlife Service
The Nature Conservancy
IL Dept. of Natural Resources
University of Illinois Springfield
78
97
Emiquon Preserve
Illinois River
Dickson Mounds Museum
Therkildsen Field Station
Conservancy office
Chautauqua National Wildlife Refuge
Thompson Lake
Flag Lake
Quiver Lake
Emiquon National Wildlife Refuge
Spoon River
136
Havana

Figure 37B The Emiquon Restoration Project. Restoration plans call for allowing the river to return to some of its former range.

In the early 1900s, the Illinois River was one of the most economically significant river ecosystems in the United States. It boasted the highest mussel abundance and productive commercial fisheries per mile of any stream in the United States. Each year during the seasonal flooding, the river would deposit nutrient-laden sediment onto the floodplain. This created a highly productive floodplain ecosystem. With the abundance of nutrients, a wide diversity of plant life would have been supported. This broad producer base would have been able to support a number of trophic levels as well as a wide diversity of food webs.

Throughout the twentieth century, modern cultures altered the natural ebb and flow of the Illinois River to better suit their needs. Levees were constructed, resulting in the isolation of nearly half of the original floodplain from the river. Native habitats were converted to agricultural land. Corn and soybeans soon replaced the variety of prairie and wetland vegetation that were once a hallmark of the Emiquon floodplain. The river itself was also altered to better facilitate the export and import of a variety of commercial goods by boat. With the isolation of these areas from the seasonal flooding, a decrease in the amount of nutrients being deposited also occurred. This resulted in a continuous decline in the natural productivity of the floodplain.

Restoration Plan

The National Research Council has identified Emiquon as one of three large-floodplain river ecosystems in the United States that can be restored to some semblance of its original diversity and function. The floodplain restoration work at Emiquon is a key part of The Nature Conservancy's efforts to conserve the Illinois River.

The Emiquon Science Advisory Council is composed of over 40 scientists and managers who provide guidance for the restoration and managed connection between Emiquon's restored floodplain and the Illinois River. Once established, this connection would allow for the seasonal flooding that it previously experienced. This seasonal flooding would help naturalize the flow of the river and restore the cyclical process of flooding and drying out. Flooding would provide more nutrients for wetland plants as well as help to improve the water quality.

Although there has been widespread cooperation with the establishment of biosphere reserves, there are few that strictly follow the United Nations' model. People are often unwilling to modify their lifestyles to the extent necessary to achieve sustainability, and funding for biosphere maintenance is often limited. For example, funding levels may be insufficient for compensating landowners for adhering to sustainable development practices in buffer zones or transition areas.

Habitat Restoration

In cases where habitat has already been modified in an area to the extent that conservation and reserve formation may not be viable, or to reverse existing damage, habitat restoration is an alternative. **Restoration ecology** is a subdiscipline of conservation biology that seeks scientific ways to return ecosystems to their state prior to habitat degradation. Although habitat restoration is perceived as beneficial, there is some concern that the restored areas may not be as functionally equivalent to the natural regions that were once there. Nonetheless, habitat restoration is increasing, and in the process, three principles have emerged thus far. First, it is best to begin as soon as possible before remaining fragments of the original habitat are lost. These fragments are sources of wildlife and seeds from which to restock the restored habitat. Second, once the natural histories of the species in the habitat are understood, it is best to use biological techniques that mimic natural processes to bring about restoration. This might take the form of using controlled burns to bring back grassland habitats, biological pest controls to rid the area of exotic species, or bioremediation techniques to clean up pollutants. Third, the goal is sustainable development, the ability of an ecosystem to maintain itself while providing services to humans. The Ecology feature, "Emiquon Floodplain Restoration," presents an example of a successful conservation project in west-central Illinois.

Check Your Progress 37.4

1. Identify ways in which landscape preservation is more valuable than ecosystem preservation.
2. Explain how edge effects influence nature reserve design.
3. List and explain the three principles of habitat restoration.

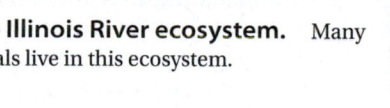

Figure 37C Wildlife of the Illinois River ecosystem. Many species of both plants and animals live in this ecosystem.

Access between the river and floodplain would once again be restored for various aquatic species, such as paddlefish and gar. Both species need a variety of habitats for reproduction and survival. The wetlands could in turn provide vital nutrients to the species that reside in the river. The Conservancy is working with the Illinois Natural History Survey, the University of Illinois, Springfield, and others to determine the baseline community diversity. These groups will monitor the community diversity to provide vital information about the progress of the restoration plan.

Local communities are investigating the potential economic benefits that can be gained as a result of the restoration. Several communities are exploring the potential for increasing the amount of nature-based tourism that can be offered at Emiquon and along the Illinois River.

The knowledge gained from the research at Emiquon and The Nature Conservancy's work on the Illinois River and within the Upper Mississippi River system will potentially guide large floodplain-river restoration efforts around the world.

Questions to Consider

1. In what ways might scientists assess whether the restoration plan was successful in its goal?
2. How might the Emiquon Floodplain Restoration plan be used as a model for other restoration projects?

connect BIOLOGY Explore the concepts through a variety of multimedia assets, question types, and data interpretation.

www.mcgrawhillconnect.com

37.5 Working Toward a Sustainable Society

Learning Outcomes

Upon completion of this section, you should be able to

1. Compare renewable versus nonrenewable resources.
2. List two ways to transition to renewable energy resources.
3. Discuss how modern agriculture can be changed to minimize environmental impacts.

The majority of biodiversity loss is the result of human consumption of resources (Fig. 37.12). Some resources are nonrenewable, and some are renewable. **Nonrenewable resources** are limited in supply. For example, the amount of land, fossil fuels, and minerals is finite and can be exhausted. Even though better extraction methods can make more fossil fuels and minerals available, and efficient use and recycling can make the supply last longer, eventually these resources will run out. **Renewable resources** are not limited in supply. For example, certain forms of energy, such as solar and wind energy, can be replenished.

The following characteristics indicate that human society in its current form is not sustainable (Fig. 37.13*a*):

1. As discussed previously, a considerable proportion of land, and therefore of natural ecosystems, is being altered for human purposes.
2. Our society primarily utilizes nonrenewable fossil fuel energy, which leads to global warming, acid precipitation, and smog.
3. Even though fresh water is a renewable resource, we are using it faster than it can be replenished in aquifers and other sources.
4. Agriculture requires large inputs of nonrenewable fossil fuel energy, fertilizer, and pesticides, which create much pollution.
5. At least half of the agricultural yield in the United States goes toward feeding animals; it takes 10 lb of grain to produce 1 lb of meat.
6. Minerals are nonrenewable, and the mining, manufacture, and use of mineral products is responsible for a significant amount of environmental pollution.

Human population
land
water
food
energy
minerals

Figure 37.12 Resources. Humans use land, water, food, energy, and minerals to meet their basic needs, such as a place to live, food to eat, and products that make their lives easier.

To move toward a more sustainable society, we should move away from the use of nonrenewable resources to renewable ones. A **sustainable** society, like a sustainable ecosystem, should be able to provide the same goods and services for future generations that it provides for the current one. At the same time, biodiversity would be conserved.

A natural ecosystem can offer clues as to what a sustainable human society would be like. A natural ecosystem uses only renewable solar energy, and its materials cycle through herbivores, carnivores, and detritivores, and back to producers once again. It is clear that if we want to develop a sustainable society, we too should use renewable energy sources and recycle materials (Fig. 37.13*b*). Section 37.4 discusses ways to conserve land and habitat for species. In addition to recycling and reuse of many of Earth's minerals (e.g., aluminum, copper, iron, and gold), which are nonrenewable, we should also consider alternative energy use, water conservation, and modifications to the way we conduct modern agriculture.

Energy

Presently, about 6% of the world's energy supply comes from nuclear power, and 81% comes from fossil fuels; both of these are finite, nonrenewable sources. Fossil fuels (oil, natural gas, and coal) are so named because they are derived from the compressed remains of plants and animals that died millions of years ago. Currently, shortage of fossil fuels such as oil has contributed to the dramatic increases in heating and gasoline costs. Comparatively speaking, each person in a more-developed country (MDC) uses approximately as much energy in one day as a person in a less-developed country (LDC) uses in one year. Increasing our reliance on renewable energy resources is a major step toward becoming a sustainable society.

Renewable Energy Sources

Renewable types of energy include hydropower, geothermal energy, wind power, and solar energy. Hydroelectric plants convert the energy of falling water into electricity (Fig. 37.14). Worldwide, hydropower presently generates 20% of all electricity utilized. In the United States, hydropower accounts for approximately 6% of total energy produced and 67% of the renewable energy used. Brazil, New Zealand, and Switzerland produce at least 75% of their electricity with hydropower, but Canada is the Western world's leading hydropower producer, accounting for 97% of its renewable energy generation. One way to reduce our reliance on nonrenewable resources such as fossil fuels is to increase our reliance on hydropower. However, much of the recent hydropower development has been due to construction of enormous dams that have detrimental environmental effects. Small-scale dams that generate less power per dam, but do not have the same environmental impact, are believed to be the more environmentally responsible choice.

Geothermal energy is produced because Earth has an internal source of heat. Elements such as uranium, thorium, radium, and plutonium undergo radioactive decay underground and then heat the surrounding rocks to hundreds of degrees Celsius. When the rocks are in contact with underground streams or

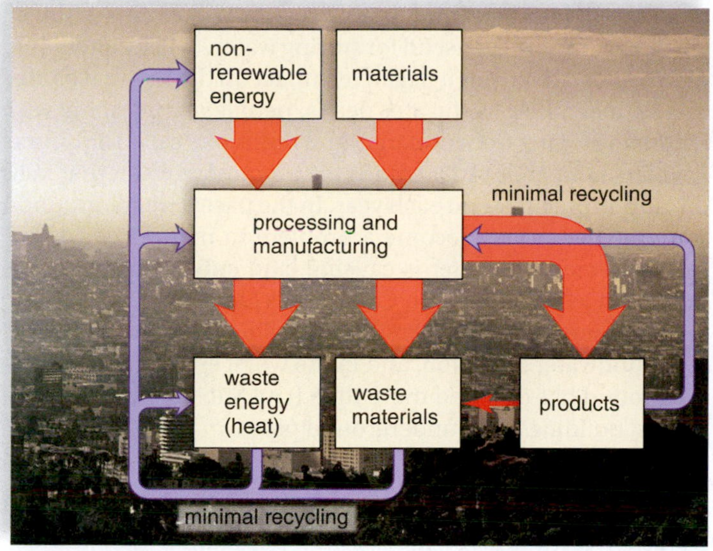
non-renewable energy
materials
processing and manufacturing
minimal recycling
waste energy (heat)
waste materials
products
minimal recycling
a. Human society at present

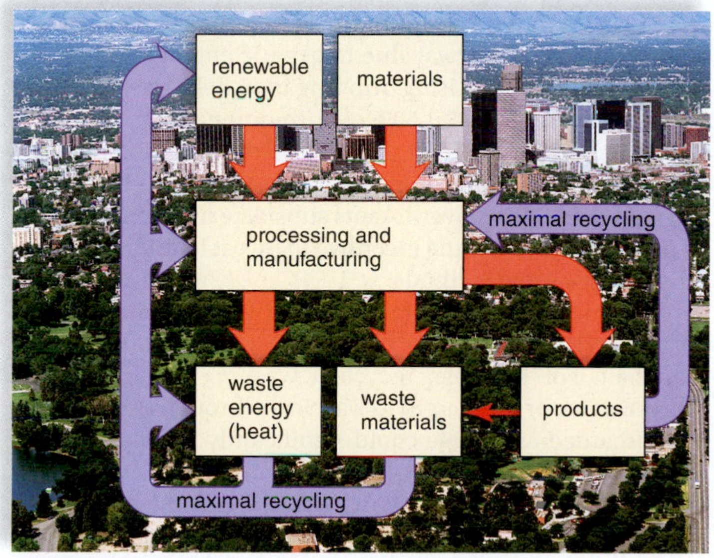
renewable energy
materials
processing and manufacturing
maximal recycling
waste energy (heat)
waste materials
products
maximal recycling
b. Sustainable society

Figure 37.13 Current human society versus a sustainable society. **a.** At present, our "throwaway" society is characterized by high input of energy and raw materials, large output of waste materials and energy in the form of heat (red arrows), and minimal recycling (purple arrows). **b.** A sustainable society would be characterized by the use of only renewable energy sources, reuse of heat and waste materials, and maximal recycling of products (purple arrows).

Figure 37.14 Hydropower. Hydropower dams provide a clean form of energy but can be ecologically disastrous in other ways.

a. Wind

b. Solar

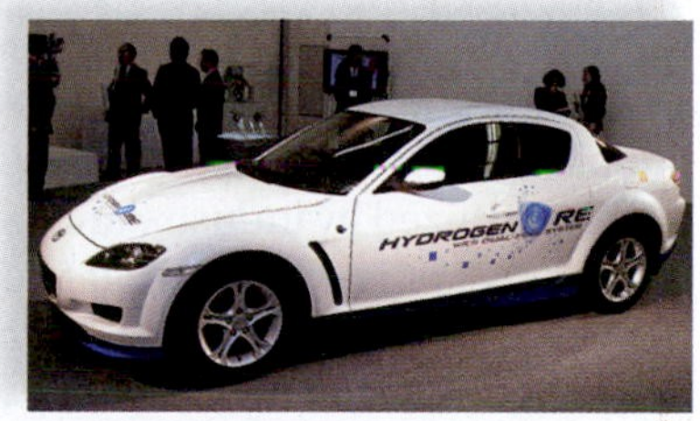
c. Hydrogen

Figure 37.15 Other renewable energy sources. **a.** Wind power requires land on which to place enough windmills to generate energy. **b.** Photovoltaic cells on rooftops and **(c)** the use of hydrogen cars will reduce air pollution and dependence on fossil fuels.

lakes, huge amounts of steam and hot water are produced. This steam can be piped up to the surface to supply hot water for home heating or to run steam-driven turbogenerators. The California Geysers Project, for example, is one of the world's largest geothermal electricity generating complexes.

Wind power is expected to account for a significant percentage of our energy needs in the future (Fig. 37.15*a*). Despite the common belief that a huge amount of land is required for "wind farms" that produce commercial electricity, the actual amount of space for a wind farm compares favorably to the amount of land required by a coal-fired power plant or a solar thermal energy system. A community that generates its own electricity by using wind power can solve the problem of uneven energy production by selling electricity to a local public utility when an excess is available and buying electricity from the same facility when wind power is in short supply.

Solar energy is diffuse energy that must be (1) collected, (2) converted to another form, and (3) stored if it is to compete with other available forms of energy. Passive solar heating of a house is successful when the windows of the house face the sun, the building is well insulated, and heat can be stored in water tanks, rocks, bricks, or some other suitable material.

In a **photovoltaic (solar) cell,** a wafer of an electron-emitting metal is in contact with another metal that collects the electrons and passes them along into wires in a steady stream. Spurred by the oil shocks of the 1970s and in 2008, the U.S. government has been supporting the development of photovoltaic cells. As a result, the price of buying one has dropped from about $100 per watt to around $2.43. The photovoltaic cells placed on roofs, for example, generate electricity that can be used inside a building and/or sold back to a power company (Fig. 37.15*b*).

Traditional cars have internal combustion engines that run on gasoline, but hybrid cars that are increasing in popularity due to rising fuel costs, run on both gasoline and electricity, increasing mileage per gallon. Now that we have better ways to capture solar energy, scientists are investigating the possibility of using solar energy to extract hydrogen from water via electrolysis. The hydrogen can then be used as a clean-burning fuel that produces water as a waste product.

Increasingly, vehicles are powered by fuel cells that use hydrogen to produce electricity. Fuel cells are now powering buses in Vancouver and Chicago. Hydrogen cars are already in limited production (Fig. 37.15*c*). Hydrogen fuel can be produced locally or in central locations, using energy from photovoltaic cells. If in central locations, hydrogen can be piped to filling stations using the natural gas pipes already in place in the United States. The advantages of such a solar-hydrogen revolution would be at least twofold: (1) the world would no longer be dependent on its limited oil reserves, and (2) environmental problems, such as global warming, acid rain, and smog, would begin to lessen.

Water

In some areas of the world, people do not have ready access to clean and safe drinking water. It's considered a human right for people to have clean drinking water, but actually, most fresh water is utilized by industry and agriculture. Worldwide, 60% of all fresh water is used to irrigate crops! Domestically in MDCs, more water is usually used for bathing, flushing toilets, and watering lawns than for drinking and cooking.

Although the needs of the human population overall do not exceed the renewable supply of water, this is not the case in certain regions of the United States and the world. When needed, humans increase the supply of fresh water by damming rivers and withdrawing water from aquifers. Dams have drawbacks: (1) Reservoirs behind the dam lose water due to evaporation and seepage into underlying rock beds. The amount of water lost sometimes equals the amount dams make available! (2) The salt left behind by evaporation and agricultural runoff increases salinity and can make a river's water unusable farther downstream. (3) Over time, dams hold back less water because of sediment buildup. Sometimes a reservoir becomes so full of silt that it is no longer useful for storing water. (4) The alteration of a river or stream has a negative impact on the native wildlife.

To meet their freshwater needs, people are pumping vast amounts of water from underground reservoirs called **aquifers.** Aquifers hold about 1,000 times the amount of water that falls on land as precipitation each year. In the past 50 years, groundwater depletion has become a problem in many areas of the world. Removal of water is causing land **subsidence,** settling of the soil as it dries out. In California's San Joaquin valley, an area of more than 13,000 km^2 has subsided at least 30 cm due to groundwater depletion, and in the worst spot, the surface of the ground has dropped more than 9 m! Subsidence damages canals, buildings, and underground pipes.

Conservation of Water

By 2025, two-thirds of the world's population may be facing serious water shortages. Some solutions for expanding water supplies have been suggested. Planting drought- and salt-tolerant crops would help a lot. Development of many such crops is already underway due to genetic engineering, as discussed in Chapter 26. Using drip irrigation delivers more water to crops and saves about 50% over traditional methods while increasing crop yields. Although the first drip systems were developed in 1960, they are currently used on less than 1% of irrigated land. Most governments subsidize irrigation so heavily that farmers have little incentive to invest in drip systems or other water-saving methods.

Reusing water and adopting conservation measures could help the world's industries cut their water demands by more than half. For example, recycling washing machine water, shower water, or water used to wash cars through a filter before it is discarded as sewage could significantly reduce domestic water usage. Home yard irrigation should occur during dusk and dawn hours, as opposed to in the middle of the day when evaporation is at its highest. Purchasing and using dual-flush toilets can also save millions of gallons of water per year.

Agriculture

In 1950, the global human population numbered 2.5 billion, and at that time, there was only enough food to provide less than 2,000 calories per person per day. Today, current agricultural practices provide enough food to provide everyone on Earth a healthy diet consisting of 2,500 calories per day. However, one-sixth of the world's population (over 1 billion people) are currently considered malnourished due to lack of proper distribution and the redirection of grain to feed livestock. In addition, modern farming methods are environmentally destructive in the following ways. First, planting only a few genetic varieties, or monocultures (a genetically identical crop), means that a single destructive parasite or pathogen can cause huge crop losses. Second, modern farming relies on heavy use of fertilizers, pesticides, and herbicides. Pesticides reduce soil fertility because they kill off beneficial soil organisms as well as pests. Pesticides also select for artificial resistance in insects, increasing eradication costs via increased pesticide use. In addition, fertilizers, pesticides, and herbicides all contribute to pollution. Third, modern agriculture uses significant amounts of fresh water through irrigation. Fourth, modern agriculture uses large amounts of fuels. Fertilizer production is energy intensive, irrigation pumps require energy to remove water from aquifers, large tractors and even airplanes are used to spread fertilizers, pesticides, and herbicides, and large machines are often used to harvest crops.

Several alternatives exist to employing modern agricultural methods. Polyculture, or planting of several varieties of crop in the field simultaneously can reduce the susceptibility of crops to pests or diseases (Fig. 37.16*a*). Polyculture also reduces the amount of herbicides necessary to kill competing weeds and can be used to replenish nutrients to topsoil. Crop rotation—where, for example, nitrogen-fixing crops, such as legumes, are alternated across harvest years with crops that utilize soil nitrogen, such as corn—can help reduce the use of nitrogen-containing fertilizers. Such multiuse farming techniques generally help increase the amount of organic matter and nutrients in the soil.

Organic farms are also increasing in number. Organic farms, as mandated by the U.S. Department of Agriculture, are those in which synthetic pesticides and herbicides are avoided or largely excluded from use. Organic farming has become increasingly profitable in recent years because people are more willing to purchase organic produce, despite the increased cost relative to nonorganic produce. Health concerns surrounding pesticide content in nonorganic produce, as well as better tasting food, have helped this trend.

One way that organic farmers have eliminated the need for pesticides is by using integrated pest management. This technique encourages the growth of competitive beneficial insects and uses biological pest control methods (also called "biocontrol"). Biocontrol has also helped reduce pesticide use in traditional farms as well. Such methods include using natural predators, such as spiders and ladybird beetles, to reduce the numbers of pests (Fig. 37.16*b*). Use of natural or engineered pathogens has also been suggested for biocontrol of pests, but there is concern that such diseases will escape crop areas and affect natural plants or wildlife.

Several types of farming now exist that reduce erosion and help to minimize topsoil loss, such as contour farming (Fig. 37.16*c*). Terrace farming involves converting steep slopes into steplike hills to minimize erosion, and some farmers plant "natural fences," such as rows of trees, around crops to prevent topsoil loss due to wind or other factors. These trees can also be used as other products; mature rubber trees provide us with rubber, and tagua nuts are an excellent substitute for ivory, for example. Cover crops, which are often a mixture of legumes and grasses, also help stabilize soil between rows of cash crops. Finally, avoiding farming on steep slopes also helps reduce erosion. Soil nutrients can be increased through composting, organic farming techniques, or other self-renewable methods. In general, we should consider using precision farming (PF) techniques that rely on accumulated knowledge to reduce habitat destruction, while improving crop yields.

Urban Growth

More and more people are moving to cities. Growth of cities increases pollution via many sources, including automobile exhaust, runoff of pollutants on impervious surfaces (e.g., roads) and into waterways, noise pollution, and industrial

a. Polyculture b. Biological pest control c. Contour farming

Figure 37.16 Agricultural conservation methods. **a.** Polyculture reduces the potential damaging effects of parasites and lessens the need for herbicides. Here, alfalfa has been planted between strips of corn. Alfalfa, a legume, has root nodules that contain nitrogen-fixing bacteria, thus helping to replenish the nitrogen content of the soil and reducing the amount of fertilizers needed. **b.** Instead of pesticides, it is sometimes possible to use a natural predator. Here, ladybird beetles are feeding on cottony-cushion scale insects on citrus trees. **c.** Contour farming conserves topsoil in gently sloping areas because water has less tendency to run off.

and domestic wastes. Thus, the growth of cities involves careful planning to serve the needs of new arrivals, without overexpansion of the city.

Energy use in cities can be curtailed by providing a good public transportation system, preferably one that is energy efficient. Many U.S. cities are now encouraging carpooling by having HOV (high-occupancy vehicle) lanes on highways, as well as reduced toll costs for cars occupied by more than one person. Portland, Oregon, even has short-term electric car rental stations around the city, whereby people can rent cars for an evening and easily drop them off when they are finished. Maintaining a network of safe bicycle lanes also encourages people to ride their bikes to work.

A way to curtail urban sprawl, or extensive expansion of cities outward, is to build them upward. Several cities, such as Vancouver, Canada, are building many high-rise apartment buildings to accommodate more people in a smaller area. As new buildings are built, they should be as "green" as possible—by using solar or geothermal energy for heating, as well as being constructed out of sustainable or recycled materials. Space on top of buildings could be used to make "green roofs," whereby a garden of grasses, herbs, and vegetables is planted to assist temperature control, supply food, reduce rainwater runoff, and be visually appealing. As more people move into cities, city officials can focus on renovating older parts of the city, as opposed to spreading out further.

Modern cities can also have more planned green spaces, including plentiful walking and bicycle paths. In city parks, native species that attract bees and butterflies and require less water and fertilizers could be planted, as opposed to traditional grasses.

Sustainable cities can also improve storm-water management by using sediment traps for storm drains, artificial wetlands, and holding ponds. As new development occurs, cities can increase the use of porous surfaces for walking paths, parking lots, and roads. These surfaces reflect less heat, while soaking up rainwater runoff.

Check Your Progress 37.5

1. Identify which nonrenewable energy use is potentially environmentally harmful.
2. List ways in which we can conserve water to minimize negative environmental impacts.
3. Discuss two alternative agricultural practices that are environmentally friendly.

Case Study Conclusion

Gills Onions processes over 1 million lb of onions every day, producing approximately 300,000 lb of waste. Previously, the waste was composted and hauled to local farm fields and spread as fertilizer. This technique resulted in a variety of environmental concerns as well as a significant amount of financial loss. To help resolve the problem, Gills Onions developed an Advanced Energy Recovery System that would turn the onion waste into a methane-rich biogas that powers two fuel cells. The fuel cells in turn supply energy to the processing plant, saving an estimated $700,000 in electricity costs per year. The remaining onion waste is then sold as livestock feed, eliminating the $400,000 in expenses associated with the waste disposal. Gills Onions has become the first food-processing facility in the world to produce ultraclean energy from their own waste. They have also reduced their carbon footprint by an estimated 14,500 metric tons (mt) of CO_2e (equivalent) emissions per year.

MEDIA STUDY TOOLS

www.mhhe.com/maderinquiry14

Enhance your study of this chapter with study tools and practice tests. Also ask your instructor about the resources available through ConnectPlus, including LearnSmart, the media-rich eBook, interactive learning tools, and animations.

SUMMARIZE

37.1 Conservation Biology and Biodiversity

- **Conservation biology** is an interdisciplinary science with the goal of conserving Earth's biodiversity for the good of all species that requires input from a variety of subdisciplines.
- **Biodiversity** is the variety of life that exists on Earth. Of the estimated 10 to 50 million species that live on Earth, nearly 75,000 of the described species are classified as **endangered species.** This means they face immediate extinction throughout all or most of their range. Thousands of species are listed as **threatened species,** meaning that they may become endangered in the foreseeable future. Although biodiversity is often described as the total number of species, it is also considered at the genetic, ecosystem, and landscape levels.
- **Genetic diversity** helps ensure that populations can evolve in the face of future environmental change.
- **Ecosystem diversity** considers the functioning of communities within ecosystems and the interconnectedness of species within such communities.
- **Landscape diversity** includes all of the ecosystems that may interact in a particular area.

37.2 Value of Biodiversity

- A direct value of biodiversity includes the use of wild species for our medical needs, agricultural value, and consumptive purposes.
- The indirect value of biodiversity includes the workings of biogeochemical cycles, waste disposal, provision of fresh water, prevention of soil erosion, and regulation of climate.

- Ecotourism is an indirect value of biodiversity that is increasing as an important source of income for many less-developed countries (LDCs).

37.3 Threats to Biodiversity

- The five major causes of extinction, in decreasing order of importance, are habitat loss, introduction of **exotic species, pollution, overexploitation,** and disease.
- Habitat loss has occurred in all parts of the biosphere, but concern has now centered on tropical rain forests and coral reefs, where biodiversity is especially high.
- Exotic species have been introduced into foreign ecosystems because of colonization, horticulture or agriculture, and accidental transport.
- Among the various causes of pollution (acid rain, eutrophication, ozone depletion, and organic chemicals), **climate change** and global warming are expected to cause the most instances of extinction.
- Overexploitation is exemplified by commercial fishing, which is so efficient that fisheries of the world are collapsing.
- Wildlife species are increasingly threatened by disease-causing pathogens.

37.4 Habitat Conservation and Restoration

- **Biodiversity hotspots,** such as tropical rain forests and coral reefs, are often high-priority conservation areas because they contain unusually large concentrations of species.
- **Keystone species** are often a target for conservation because their loss would result in many secondary extinctions. **Flagship species** are those that elicit an emotional response and often motivate the public to support conservation.
- Habitat conservation should be considered at the landscape level, with the formation of **conservation corridors** that allow species to travel safely between ecosystem types.
- **Edge effects** reduce the amount of suitable habitat and should be considered in reserve design.
- **Conservation reserves** are areas set aside with the primary goal of protecting the biodiversity within them.
- Biosphere reserves should consist of a core area with nearly no disturbance, a surrounding buffer area where only low-impact human activities are allowed, and an outer transition area meant to support sustainable human development, tourism, and agriculture.
- **Restoration ecology** is the field of biology that is used to help return ecosystems to their state prior to habitat degradation.
- Restoration ecology has three basic principles: (1) begin as soon as possible, (2) use biological techniques that mimic natural processes to bring about restoration, and (3) maintain ability of an ecosystem to maintain itself while providing services to humans.

37.5 Working Toward a Sustainable Society

- The current biodiversity crisis is ultimately caused by human actions.
- Today's society is considered unsustainable because we rely heavily on **nonrenewable resources** to maintain our existence.
- A shift of focus toward **renewable resources** (wind, solar, geothermal, and hydropower) and **sustainable** use of water and agricultural practices will help reduce human impact. The use of a **photovoltaic (solar) cell** is a form of renewable energy.
- Globally, many people do not have access to clean drinking water. In the southwestern United States, people are rapidly draining the **aquifers** to meet the growing demand for water. The removal of the water from the aquifers has led to land **subsidence,** settling of the soil as it dries out. In order to prevent problems, people need to become more aware of water conservation methods.
- These methods include drip irrigation, planting drought- and salt-tolerant crops, and household water reuse and recycling methods.
- Modern-day agriculture can employ new methods, such as biocontrol to reduce reliance on pesticides, and new forms of farming, such as contour farming, to reduce topsoil erosion.

ASSESS

Testing Yourself

Choose the best answer for each question.

1. Which type of resource is renewable and readily available?
   a. oil
   b. coal
   c. solar energy
   d. underground water supplies
   e. copper
2. Most fresh water is used for
   a. domestic purposes, such as bathing, flushing toilets, and watering lawns.
   b. domestic purposes, such as cooking and drinking.
   c. agriculture.
   d. industry.
3. Removal of groundwater from aquifers may cause
   a. pollution.
   b. subsidence.
   c. mineral depletion.
   d. soil erosion.
   e. All of the choices are correct.
4. The most significant cause of the loss of biodiversity is
   a. habitat loss.
   b. pollution.
   c. exotic species.
   d. disease.
   e. overexploitation.
5. Exotic species are introduced into new areas by which method?
   a. colonization
   b. agriculture
   c. accidental transport
   d. All of the choices are correct.
6. Edge effects
   a. result from species crowding at edge of habitats.
   b. mean that no plants can grow well at the edge of a mountain.
   c. may increase wind and temperatures around the outside of habitats.
   d. increase competition between species at the edge of their territories.
   e. All of these are correct.
7. Agricultural runoff of nutrients can result in a fish kill due to
   a. toxic effects of the high nutrient levels on the fish.
   b. toxic effects of the high nutrient levels of the food sources of the fish.
   c. low oxygen levels in the water as decomposers break down algae.
   d. the cooling of the water as algae shade the bottom.

8. A transition to hydrogen fuel technology will
   a. be long in coming and probably not of major significance.
   b. lessen many current environmental problems.
   c. not be likely because it will always be expensive and as polluting as natural gas.
   d. be of major consequence, but resource limitations for obtaining hydrogen will hinder its progress.
9. Global warming does not bring about
   a. the inability of species to migrate to cooler climates as environmental temperatures rise.
   b. the bleaching and drowning of coral reefs.
   c. a rise in sea levels and loss of wetlands.
   d. preservation of species because cold weather causes hardships.
   e. All of the choices are correct.
10. Biodiversity hotspots
   a. have few populations because the temperature is too hot.
   b. contain about 20% of Earth's species even though their area is small.
   c. are always found in tropical rain forests and coral reefs.
   d. are sources of species for the ecosystems of the world.
   e. All except a are correct.
11. Consumptive use value
   a. means we should think of conservation in terms of the long run.
   b. means we are placing too much emphasis on living organisms that are useful to us.
   c. means some organisms, other than crops and farm animals, are valuable as products.
   d. is a type of direct value.
   e. Both c and d are correct.
12. Which of these is not an indirect value of species?
   a. participates in biogeochemical cycles
   b. participates in waste disposal
   c. helps provide fresh water
   d. prevents soil erosion
   e. All of the choices are correct.
13. The services provided to us by ecosystems are unseen. This means
   a. they are not valuable.
   b. they are noticed particularly when the service is disrupted.
   c. biodiversity is not needed for ecosystems to keep functioning as before.
   d. we should be knowledgeable about them and protect them.
   e. Both b and d are correct.
14. Eagles and bears feed on spawning salmon. If shrimp are introduced that compete with salmon for food,
   a. the salmon population will decline.
   b. the eagle and bear populations will decline.
   c. only the shrimp population will decline.
   d. all populations will increase in size.
   e. Both a and b are correct.
15. Which of the following should not be considered when designing nature reserves?
   a. edge effects that may require an increase in size
   b. corridors to facilitate wildlife movement among habitat patches
   c. primarily the needs of a single, important species
   d. inclusion of multiple ecosystem types

ENGAGE

Thinking Critically

1. What are some ways we can reduce our reliance on nonrenewable resources and move toward increased reliance on renewable resources?
2. Why should people be generally more concerned about pollutants in our meat than in our vegetables?
3. Discuss how removal of a keystone species can disrupt ecosystem functioning. Give an example.

APPENDIX A

Answer Key

CHAPTER 1

Case Study

1. Levels of ^{137}C present throughout the environment must be measured over time to determine potential sources and magnitude of exposure. The atmosphere, sources of water, plants, and soil need to be monitored to track ^{137}C available to be inhaled or ingested by humans and other organisms. **2.** Exposure to radioactivity influences all levels of biological organization from molecules upward. Radioactivity can alter DNA present in cells which make up tissues, organs, organ systems, and organisms. Populations made up of affected organisms combine to form a community that is present in an ecosystem in the Earth's biosphere. Japan and the United States are a part of this biosphere.

Check Your Progress

1.1: 1. Characteristics are: (1): organization; (2) ways to acquire materials and energy; (3) Reproduction; (4) responses to stimuli; (5) homeostasis; (6) growth and development; (7) capacity for adaptation to their environment. **2.** Organisms share these characteristics because at some point in the very early history of life, a common ancestor to all species was the first cell or cells. Evolution over approximately the last 3.5 billion years has since shaped the characteristics of species seen today. **1.2: 1.** Domains are the largest, most inclusive classification category. The three domains are Archaea, Bacteria, and Eukarya. The first two are composed of species with prokaryotic cells (no membrane-bound nucleus), and the third has organisms with eukaryotic cells. Archaea can live in extreme aquatic environments, and Eukarya and Bacteria live almost everywhere else. Four kingdoms have been classified within the Eukarya: Protista, Fungi, Plantae and Animalia. **2.** Domain, Kingdom, Phylum, Class, Order, Family, Genus, Species **3.** A hierarchical classification system allows increasing specificity about the organisms under study which helps define their evolutionary relationships. **1.3: 1.** An ecosystem is made up of the physical environment and all the organisms found there. All organisms in an ecosystem are grouped as a community. All individuals of one species found in a community in that ecosystem comprise a population. **2.** Humans have actively changed most ecosystems including forests, grasslands, wetlands, and oceans for their own short term benefit. Human demands for energy and food influence the atmosphere, land, and oceans. These activities, while improving the condition of humans, disrupt basic ecosystem processes. **3.** Ecosystems with high biodiversity include species of importance to humans. Many of these species are rare. Harvesting them disturbs the balance between all of the species within the ecosystem, thus increasing their vulnerability to collapse. **1.4: 1.** The control group in an experiment is included to determine if the experiment is sensitive to the effect of the variable of interest. The test group experiences the variable and the control group does not. **2.** An experiment can be structured so that one group is exposed to or includes something that another group does not experience. This variation in exposure will or will not produce different outcomes. The experiment can then determine the impact of the variable on the outcomes. **3.** See Fig. 1.8. **1.5: 1.** Examples include modern agricultural practices that increase crop yields but use fertilizers that pollute, bioengineered organisms that produce chemicals of interest but can damage the environment, and gene therapy techniques with unknown long term outcomes. **2.** Technology advances the survivability and comfort of humans as well as other selected organisms. Without technology, humans and others would not have clean water, vaccinations, available fuel, pharmaceuticals, or other tools that help us control the world we live in.

Science in Your Life

How Safe is Your Cell Phone? 1. Establishing that cell phones are safe to use depends not on the number of studies but the quality of the experimental design, the significance of the data collected, and whether others were able to repeat the experiment with the same results. Even then, proof is not achieved. Rather, evidence is gathered supporting or refuting the safety of cell phone use. **2.** Double blind, controlled, experiment first using animals. **3.** Students will answer yes or no. The impact could be first on molecules but result in changes carried up through all levels to the biosphere.

Testing Yourself

1. c; **2.** d; **3.** b; **4.** a; **5.** b; **6.** b; **7.** d; **8.** a; **9.** c; **10.** c; **11.** a; **12.** e; **13.** d; **14.** c; **15.** b.

Thinking Critically

1. Evolution explains the unity of life based on all organisms sharing a common ancestor, which is the first cell or cells that existed almost 4 billion years ago. Evolution also explains life's diversity which results from the environment always changing. Because there is variation among individuals, populations, and species, there is variation in responses to environmental pressures. Some individuals have had adaptations or features to make them more suited to the new environment and these individuals tend to live to produce more offspring than less suited individuals. **2.** Life is defined as organisms that have the following basic characteristics: (1) organization; (2) ways to acquire materials and energy; (3) reproduction; (4) responses to stimuli; (5) homeostasis; (6) growth and development; (7) capacity for adaptation to their environment. Computers are not living because they cannot acquire energy on their own and are not self-replicating. **3.** The experimental chemical should be studied in a model species. The experiment should have both a control group (that receives no drug, or the inert ingredients used in the drug's formulation) and a treatment group (that receives the drug). Both groups should experience identical conditions and have the same type of cancer. Data is collected on tumor development in the treatment group relative to the control group. If the treatment group shows greater tumor reduction than the control group, the drug may then be approved for trials on humans.

CHAPTER 2

Case Study

1. High heat capacity, high heat of vaporization, effective solvent, cohesive, adhesive, high surface tension, less dense in frozen state compared to liquid state **2.** Carbon, hydrogen, nitrogen, oxygen, phosphorus, and sulfur (CHNOPS)

Check Your Progress

2.1: 1. Earth's crust has a different function and is constructed differently than living organisms. It is not a carbon-based life form that exhibits the characteristics of life. **2.** The drawing would be similar to Fig. 2.2, with twenty protons and twenty neutrons in the nucleus, two electrons in the first orbital, eight electrons in the second orbital, eight electrons in the third orbital, and two electrons in the fourth orbital. **3.** High levels of radiation can cause burns, harm cells, and change DNA structure with cancer being a possible result of exposure. However, radiation can be used beneficially to kill cancer cells. Also, radioisotopes are used extensively in medicine to image the body through X-rays, positron emission tomography, and computerized tomography. Bacteria and fungi present on medical equipment, in mail, and in food can be killed by exposure to radiation. **4.** Oxygen 16 has 8 protons, 8 electrons, and 8 neutrons. Oxygen 18 differs by having 10 neutrons (two more). **2.2: 1.** Nitrogen gas (N_2) is a molecule because two atoms of nitrogen share a covalent bond. Carbon dioxide (CO_2) is a compound because two different elements are bonded together. However CO_2 is also a molecule because if you broke apart the carbon and oxygen atoms the properties of carbon dioxide would be lost. **2.** Carbon has six electrons, two in the first orbital, and four in the second orbital. Its incomplete outer orbital can accept or share four electrons, forming four covalent bonds. **3.** Hydrogen tends to form polar covalent bonds because the larger element with which it is bonding attracts hydrogen's electron more strongly, becoming more negatively charged than the hydrogen. Since the electrons are shared unequally the charge difference is partial. **2.3: 1.** Heat capacity of water is the amount of energy needed to raise the temperature of one gram one degree Celsius while heat of vaporization is the amount of energy needed to convert one gram of the hottest water to a gas. **2.** A solution at pH 6 contains 1×10^{-6} moles per liter H^+ ions while a solution at pH 8 contains 1×10^{-8} moles per liter H^+ ions. 1×10^{-6} is a hundred times larger than 1×10^{-8}. **3.** Because it minimally dissociates and is present in dissociated and associated forms. Hydrogen ions can react with the dissociated acid and hydroxide ions can react with the associated acid. **2.4: 1.** Because they contain carbon and hydrogen and are present in living organisms. **2.** Hydrolysis reactions involve the addition of a molecule of water across a chemical bond, splitting the molecule in two. Dehydration reactions remove a molecule of water, joining two molecules. **2.5: 1.** A ring structure of carbons with each bonded to a hydrogen and an OH group. **2.** Because we are able to break the bonds between the glucose molecules in starch but not between the glucose molecules in cellulose. **2.6: 1.** Phospholipids and cholesterol, which is a steroid. **2.** A double bond tends to make a fatty acid liquid since it introduces a "kink" in the chain that prevents the fatty acids from packing as tightly together. **2.7: 1.** Enzymes, structural proteins, hormones, transport molecules, and membrane channels. **2.** Each one has an amino group, an acidic group, and an R group bound to a central carbon atom. The R group varies in structure between one amino acid and another. **3.** primary—linear sequence of amino acids; secondary—orientation of polypeptide chain, coiled or folded determined by hydrogen bonding; tertiary—three dimensional structure maintained by covalent, ionic or hydrogen bonding; quaternary—structure of two or more polypeptides joined in a protein. **2.8:1.** In the sequence of the nucleotide bases A, C, T, and G. **2.** In the chemical bonds of the phosphate groups.

Science in Your Life

The Fate of Prescription Medicine: 1. If a community take-back program exists, then people should be encouraged to participate through public service announcements and educational programs. **2.** Access to clean water is a basic human need and is considered a right that society as a whole should support and protect. **3.** Those contaminants that should be eliminated first are those that are having the highest current and long term impact on the ecosystem. **A Balanced Diet: 1.** Fifty percent should be fruits and vegetables. Each person will answer according to their eating habits. **2.** Good fats provide energy and have important biological roles. Bad fats contribute to plaque formation, leading to cardiovascular disease. **3.** Whole-grain carbohydrates contain more fiber, vitamins, and minerals than processed simple carbohydrates.

Testing Yourself

1. d; **2.** d; **3.** d; **4.** d; **5.** c; **6.** d; **7.** a; **8.** a; **9.** b; **10.** a; **11.** a; **12.** a; **13.** d; **14.** a.

Thinking Critically

1. Potatoes contain starch while wood contains cellulose. Our bodies are capable of breaking down the bonds holding the glucose molecules of starch together but not of breaking the bonds that hold the glucose molecules of cellulose together. **2.** The 20 different amino acids form an alphabet. Think of the number of words that could be formed with 20 different letters. That just accounts for the primary structure of the proteins. When you add the secondary and tertiary structures, this adds even more variety to the different types of proteins possible. **3.** These answers can be verified with the periodic table in the appendix. **a.** K loses 1 electron to become +1. Therefore it must have 1 electron in its outer shell. **b.** Ca loses 2 electrons to become +2. Therefore it must have 2 electrons in its outer shell. **c.** F gains 1 electron to become -1. Therefore it must have 7 electrons in its outer shell. **d.** N gains 3 electrons to become -3. Therefore it must have 5 electrons in its outer shell.

CHAPTER 3

Case Study

1. Peroxisomes, present in animals and plants, have enzymes that break down fats. In plant leaves, peroxisomes can counteract photosynthesis by generating carbon dioxide and using up oxygen. **2.** The cell relies on lysosomes to dispose of faulty organelles. The cell lacks a mechanism for disposing of faulty lysozomes.

Check Your Progress

3.1: 1. If humans were just one cell the surface-area-to-volume ratio would be too small to allow for efficient movement of nutrients and oxygen into the cell and wastes, including carbon dioxide, out of the cell. **2.** Eukaryotic. **3.2: 1.** To form a barrier between inside and outside of the cell and regulate what crosses that barrier. **2.** See Fig. 3.3. **3.** A bacterium contains many structures that allow it to live independently as a single cell. It moves by flagella, regulates what moves in and out with the plasma membrane, makes proteins in the cytoplasm, protects itself with a cell wall, and reproduces using DNA in the nucleoid. **3.3: 1.** Protecting the cell while remaining permeable. **2.** See Table 3.2. **3.** The different compartments of the cell allow for separation of cellular functions which increases the efficiency of the whole. **3.4: 1.** All three are part of the cytoskeleton and are dynamic. Intermediate filaments play a structural role while both actin filaments and microtubules have a structural and a movement role. Actin filaments are composed of actin monomers, microtubules are composed of tubulin monomers, and intermediate filaments are composed of various types of fibrous polypeptides. **2.** Cilia and flagella have a 9 + 2 pattern of microtubule doublets while centrioles have a 9 + 0 pattern of microtubule triplets. **3.** The dynein side arms on the microtubule doublets slide past each other using the energy of ATP. **3.5: 1.** They are similar to bacteria in size and structure, are bounded by a double membrane, contain a limited amount of genetic material, divide by splitting, have their own ribosomes (similar to prokaryotic ribosomes) and do produce some proteins, and have ribosomal RNA sequences similar to prokaryotic rRNA sequences. **2.** Mitochondria were originally aerobic heterotrophic bacteria taken up by a eukaryotic cell and chloroplasts were cyanobacteria similarly taken up. Neither the bacteria nor cyanobacteria were destroyed but instead successfully lived inside the host cell, contributing their metabolic functions to the whole.

Science in Your Life

Microscopy Today: 1. Processing the sample to prepare it for viewing can change its structure and appearance. **2.** The image becomes blurred and detail is lost. **3.** A stream of electrons is used rather than visible light which humans see as color.

Testing Yourself

1. a; **2.** c; **3.** d; **4.** c; **5.** d; **6.** c; **7.** a; **8.** b; **9.** e; **10.** d; **11.** d; **12. a.** RER, folding, processing of proteins, **b.** nucleus, information in the form of DNA to make the proteins, **c.** nucleolus, where rRNA is made and ribosomes are assembled, **d.** Golgi apparatus, proteins are transported from here; **13.** c; **14.** c.

Thinking Critically

1. The Blob would not have an adequate surface area for exchanging nutrients and wastes and it would not be able to survive for long. **2.** The surface area of a 1 mm cube is 1mm × 1mm × 6 = 6 mm^2. The volume of a 1 mm cube is 1mm × 1mm × 1mm = 1 mm^3. The ratio of surface area to volume is 6:1. The surface area of a 2 mm cube is 2mm × 2mm × 6 = 24 mm^2. The volume of a 2 mm cube is 2mm × 2mm × 2mm = 8 mm^3. The ratio of surface area to volume is 24:8 or 3:1. The 1 mm cube has a greater surface-area-to-volume ratio. **3.** Since plants do not have centrioles but do assemble and disassemble microtubules, this indicates that centrioles are not required. Why animals have centrioles is still a mystery.

CHAPTER 4

Case Study

1. The roles of membrane proteins include: stabilize and shape the membrane; create channels through which molecules pass; interact with molecules and carry them across the membrane; identify cells as being foreign or self; bind to specific molecules at the cell surface and bring about a cellular response; serve as enzymes, catalyzing biological reactions. **2.** facilitated diffusion.

Check Your Progress

4.1: 1. Phospholipids compose a bilayer that separates the inside from the outside of the cell. Steroids in the bilayer regulate the fluidity of the membrane. Proteins present in the membrane contribute to its structure, the passage of molecules across the membrane, signaling pathways, cell recognition, and enzyme reactions. **2.** See Fig. 4.2. **4.2: 1.** During diffusion, molecules move from an area of high concentration to an area of low concentration. Facilitated transport promotes the movement of these molecules down a concentration gradient across a plasma membrane. Carrier proteins reversibly bind to the molecule and speed up their passage. **2.** A hypertonic environment has a lower concentration of water and a higher concentration of solutes than a hypotonic environment. Water will move from the hypotonic environment where it is present at a higher concentration, to the hypertonic environment across a semipermeable membrane by osmosis. **3.** Both move molecules across the plasma membrane and require a carrier molecule. Facilitated transport does not use energy while active transport does. Facilitated transport moves items down their concentration gradient while active transport moves against the concentration gradient. **4.** It is specific for a particular solute. **4.3: 1.** Collagen and elastin fibers provide structure to the ECM. Fibronectin binds to integrin in the membrane and can signal the cell's cytoskeleton. Proteoglycans, shaped as bottle brushes, assist in cell signaling by regulating the passage molecules through the ECM. **2.** Adhesion junction—mechanically attaches adjacent cells; gap junction—allows cells to communicate through channels; tight junction—connects plasma membranes creating a tight barrier. **3.** The ECM is a meshwork of proteins and polysaccharides that can vary in amount and flexibility depending on the cell type. The plant cell wall, external to the plasma membrane, is composed of cellulose fibrils and pectin that allows flexibility during growth and eventually hardens into a rigid structure.

Science in Your Life

How Cells Talk to One Another: 1. The signaling molecule will have no effect on the cell. **2.** Because the transduction pathway is a series of proteins, each of which can bring about a response in another protein. **3.** That part of the signal transduction pathway involved with changes in the movement of the cell. This might control the spread of the cancer to other parts of the body.

Testing Yourself

1. c; **2.** a; **3.** d; **4.** a; **5.** b; **6.** b; **7.** b; **8.** c; **9.** a; **10.** d; **11.** a; **12.** c; **13.** b; **14.** b; **15.** c.

Thinking Critically

1. The receptor molecule changes its shape and that activates an internal signaling pathway, usually involving peripheral proteins and secondary messengers. **2.** If the cell wall is weakened, it may not be able to withstand the osmotic pressure as the cell swells due to osmosis. The bacterium may lyse in hypotonic solutions. **3.** Channel proteins often have gates that require the binding of a specific molecule to activate them. The gate is opened only if that molecule is present.

CHAPTER 5

Case Study

1. G_1 checkpoint—determines if DNA is damaged and if growth signals and nutrients necessary for cell division are available. G_2 checkpoint—determines if DNA has not finished replicating or if it is damaged, thus preventing start of M stage of cell cycle. M checkpoint—determines if chromosomes are properly aligned by the spindle assembly, assuring proper distribution to the daughter cells. **2.** G_1 checkpoint **3.** If the checkpoints fail, the cell cycle will proceed even if the DNA is damaged, not replicated completely, or not distributed evenly between the daughter cells. Cancer, which is unregulated cell growth, can result.

Check Your Progress

5.1: 1. See Fig. 5.1. **2.** It removes unwanted tissue and abnormal cells. **5.2: 1.** Cells stop at the G_1 checkpoint if conditions are not favorable for cell division and/or there is DNA damage. Cells stop at the G_2 checkpoint if the DNA has not finished replicating and/or there is DNA damage. Cells stop at the M checkpoint if the cells are not lined up correctly at the metaphase plate. **2.** Oncogenes encode proteins that continuously promote the cell cycle, leading to unregulated cell division. Tumor suppressor genes function to inhibit the cell cycle. If mutated the proteins that they express would not be active and the cell cycle would continue, possibly leading to cancer. **5.3: 1.** One chromatid before replication; two chromatids after. **2.** See Fig. 5.5. **3.** Both result in two daughter cells with identical genetic material. Animal cells form a cleavage furrow between the daughter nuclei which is constricted by the action of a band of actin filaments. Plant cells build a new cell wall between the daughter cells by fusing together vesicles produced by the Golgi apparatus. **5.4: 1.** In order to divide the DNA content in half. Because the DNA was already replicated before meiosis began. **2.** See Figs. 5.9 and 5.12. **3.** During crossing-over genetic material is exchanged between non-sister chromatids. This produces a different combination of genes. Independent assortment mixes the whole chromosomes donated by both parents. The end result of both processes is cells that do not have genetic material identical to either parent. **5.5: 1.** See Table 5.1. **2.** See Table 5.2. **5.6: 1.** Sperm and egg cells are haploid. They function in sexual reproduction. **2.** Four sperm; one egg.

Science in Your Life

Benign Versus Malignant Tumors: 1. To determine if the cancer has spread which would indicate a malignancy. **2.** Malignant tumor cells divide more rapidly than benign tumor cells. They display greater alterations in shape and function. The malignant cell cycle is under less regulation than that of benign cells.

Testing Yourself

1. a; **2.** b; **3.** a; **4.** d; **5.** c; **6.** a; **7.** d; **8.** a; **9.** f; **10.** c; **11.** e; **12.** d; **13.** d; **14.** b; **15.** b.

Thinking Critically

1. The homologous chromosomes are aligned at the metaphase plate in meiosis I and crossing over between non-identical sister chromatids is more likely. **2.** If a cell cycle checkpoint is nonfunctional, then the cell proceeds through the cell cycle even if it is damaged. It can replicate unstopped, leading to cancer. **3.** Sexual reproduction introduces genetic variation that may be an adaptive advantage to organisms responding to changing environments.

CHAPTER 6

Case Study

1. To act as a catalyst, increasing the rate of the conversion of reactant to product in a series of linked reactions. **2.** Temperature, pH, concentration of substrate, presence of activators, inhibitors, and cofactors.

Check Your Progress

6.1: 1. Once energy is released as heat it cannot be recaptured for recycling. **2.** With digestion you are converting chemical energy to kinetic energy with a loss of heat according to the 1st law which says that energy cannot be created or destroyed, but can be changed from one form to another and the 2nd law that says that there will be a loss of usable energy every time energy is converted from one type to another. **6.2: 1.** Endergonic **2.** ATP with its three phosphate groups, is analogous to a charged battery which can provide energy via a coupled reaction when one phosphate is cleaved off. ADP that is produced is like a discharged battery which requires an input of energy to become ATP again. **6.3: 1.** Enzymes are needed to reduce that activation energy of the reactions present in a biochemical reaction, thus allowing reactions to occur under conditions of the cell. A cell can convert an enzyme from an inactive form to an active one by the addition or removal of phosphate groups. Cellular enzymes are subject to feedback inhibition. The presence of cofactors and coenzymes in the cell regulates enzyme activity. **2.** Denaturation changes the shape of the enzyme and its active site, altering the fit of the reactant with the enzyme. **3.** How well the enzyme interacts with the reactants and the rate at which product is formed is determined by the three-dimensional shape of the protein molecule and its active site. **6.4: 1.** They both involve the gain and loss of hydrogen atoms. In photosynthesis, water loses hydrogen atoms and carbon dioxide gains them. In cellular respiration, glucose loses hydrogen atoms and oxygen gains them. **2.** See Fig. 6.11. **3.** Carbohydrates, fats, proteins.

Science in Your Life

Enzyme Inhibitors Can Spell Death: 1. The arguments for using chemicals in medicine even if they are dangerous are based on the benefit they can provide versus the risk they represent. **2.** The search for and harvesting of species can be detrimental to the existence of the organism and its environment. However, the chemicals may benefit the health and wellbeing of humans and other organisms.

Testing Yourself

1. b; **2.** c; **3.** a; **4.** d; **5.** b; **6.** d; **7.** a; **8.** a; **9.** b; **10.** c; **11.** e.

Thinking Critically

1. The energy of activation must still be overcome. Enzymes enable the reactions to occur at the conditions present within the body. **2.** Glycogen must be synthesized from individual glucose molecules. This type of reaction (anabolism) requires the input of energy. **3.** The chemical reactions of photosynthesis and cellular respiration are very different. Mitochondria do not have the appropriate enzymes, structure, or light harvesting pigments to carry out photosynthesis.

CHAPTER 7

Case Study

1. Aerobic metabolism in the cell uses oxygen to break down glucose to form ATP. Without oxygen under anaerobic conditions, glucose is broken down just to pyruvate with much less ATP production. **2.** As glucose is broken down in a series of controlled reactions in the cytoplasm and mitochondria, some of its chemical energy is released and used to make ATP. **3.** Proteins are deaminated by the liver and their carbon skeletons can enter glycolysis, be converted to acetyl groups, or enter the citric acid cycle. Fats are broken down to glycerol and fatty acids. The glycerol can be converted to pyruvate to be converted to acetyl CoA and the fatty acids are broken down to acetyl CoA which enter the citric acid cycle.

Check Your Progress

7.1: 1. NAD^+ and FAD serve as coenzymes for the enzymes involved in cellular respiration. Both molecules receive electrons and become reduced (NADH and $FADH_2$) and transport the electrons to inside the mitochondria, where they function in the electron transport chain. **2.** Glycolysis is an anaerobic process. The preparatory reaction, citric acid cycle, and electron transport chain are dependent on the presence of oxygen. **3.** See Fig. 7.2. **7.2: 1.** Glucose must be activated by the investment of 2 ATPs before the cascade of reactions oxidizing glyceraldehyde 3-phosphate forming 4 ATPs can happen. **2.** For each molecule of glucose the inputs are 2 ATPs and $2NAD^+$s. The outputs are 2 molecules of pyruvate, 2 NADHs, 2 ADP, and 4 ATPs. There is a net gain of 2 ATPs. **7.3: 1.** When muscles are working hard in a burst of activity, using up available oxygen. **2.** During fermentation pyruvate accepts electrons from NADH formed during glycolysis and lactate is formed. The oxidized NAD^+ is available to pick up more hydrogen atoms from glycolysis. **7.4: 1.** See Fig. 7.2. **2.** See Fig. 7.9. **3.** The NADH formed outside the mitochondria during glycolysis sometimes cannot cross into the mitochondria. Instead NADH delivers its electrons to the electron transport chain using a process that costs one ATP/2 electrons. Since two NADH are formed during glycolysis the total ATP count is reduced to 4 from 6.

Science in Your Life

Metabolic Fate of Pizza: 1. The cheeseburger and fries would be digested into molecules of fat, protein, and carbohydrate. Each of these molecules would enter the cellular respiration pathways as depicted in Figure 7A. **2.** Water is used in the hydrolysis reactions involved in the breakdown of carbohydrates, proteins, and lipids.

Testing Yourself

1. b; **2.** d; **3.** a; **4.** c; **5.** e; **6.** a; **7.** b; **8.** c; **9.** b; **10.** e; **11.** b; **12.** b; **13.** d; **14.** c; **15.** a.

Thinking Critically

1. Plasma membrane. **2.** If the electron transport chain is the same in insects and humans, then Rotenone would inhibit the human chain also. In fact, rotenone is very toxic to humans. **3.** The faster fatty acids are converted to acetyl CoA and enter the citric acid cycle, the greater the rate of respiration. These compounds may convert fatty acids to acetyl CoA more rapidly.

CHAPTER 8

Case Study

1. Chlorophyll. **2.** Green chlorophyll is present in the spring and summer but it degrades in the fall and the red or yellow carotenoids become visible.

Check Your Progress

8.1: 1. The sun. **2.** See Fig. 8.2. **3.** $6CO_2 + 6H_2O \rightarrow C_6H_{12}O_6 + 6O_2$ with the input of solar energy and presence of pigments. **4.** Light reactions require the input of light energy. They result in the generation of NADPH and ATP which function in the Calvin cycle. The Calvin cycle reactions are independent of the input of light energy. They reduce CO_2 to produce carbohydrates. **8.2: 1.** They reflect and do not absorb wavelengths of light that humans see as green. **2.** In the noncyclic electron pathway, the excited electron from photosystem II passes down an electron transport chain to photosystem I, producing ATP. The excited electron from photosystem I is passed to $NADP^+$ making NADPH. Both ATP and NADPH are used in the Calvin cycle reactions. **3.** See Fig. 8.8. **4.** H^+ ions present in the thylakoid space are at a higher concentration than in the stroma. The tendency is for the ions to move down this electrochemical gradient, releasing the energy to fuel ATP synthesis. **8.3: 1.** Carbon dioxide is attached to RuBP by the enzyme RuBP carboxylase, and then converted to G3P in a two-step reaction involving ATP as the energy source and NADPH + H^+ as the source of the electrons for the reduction step. **2.** It takes 3 turns to make one G3P but for every 3 turns, 5 molecules of G3P are used to re-form 3 molecules of RuBP. **3.** G3P can be converted into glucose, sucrose, starch, cellulose, fatty acids, or amino acids. **8.4: 1.** Sugarcane, corn, succulent plants such as cacti. **2.** It allows the stomata to close during the day so there is less loss of water without affecting the plant's ability to fix carbon dioxide. **8.5: 1.** See Fig. 8.12. **2.** Because the chemical pathways are not the reverse of each other. Different enzymes, cellular structures, and pigments are involved in the two processes.

Science in Your Life

The New Rice: 1. Reducing meat and fuel consumption frees up disproportionately large amounts of resources that could be directed toward grain production. **2.** No. Sustainable farming practices could help but GMOs solve some of the challenges of flooding and climate change directly. **Diesel Power from Algae: 1.** The production and distribution of the biofuel must be cheaper than foreign oil sources and the margin must be great enough to win over the consumer. **2.** Transportation, distribution, availability, ease of acquisition by the consumer, consumer confidence in the product.

Testing Yourself

1. d; **2.** c; **3.** d; **4.** b; **5.** a; **6.** a; **7.** b; **8.** a; **9.** d; **10.** e; **11. a.** thylakoid, **b.** oxygen, **c.** stroma, **d.** Calvin cycle, **e.** granum; **12.** a; **13.** e.

Thinking Critically

1. Leaves that are broad allow maximum surface area for exposure to sunlight. Leaves that are thin allow sunlight to penetrate and reach the chloroplasts. **2.** White reflects all of the wavelengths of visible light, resulting in less heat absorption. **3.** A carrier at the start of the electron transport chain in the thylakoid membrane pumps hydrogen ions from the stroma into the thylakoid space.

CHAPTER 9

Case Study

1. Stems, roots, and leaves; epidermis with a cuticle **2.** The structural differences have supported eudicots becoming the larger group of familiar flowering plants. The monocots have succeeded at becoming significant food crops. **3.** Plants have adaptations that allow them to succeed on land including a cuticle, stems, roots, and leaves, and specific reproductive strategies. The adaptations against predators include epidermal hairs and antiherbivory chemicals.

Check Your Progress

9.1: 1. Because all stems have nodes and internodes while not all stems are aboveground. **2.** See Fig. 9.1 for structure comparison. Roots: the root hairs with their large surface area allow for uptake of water and minerals. The branching of the roots allows them to stabilize the aboveground portion of the plant. Stems allow for continued growth and contain xylem and phloem for transportation of water and nutrients through the plant. Leaves are thin so that sunlight can reach the chloroplasts and gases can diffuse easily. The petiole allows for maximum exposure to the sun. **9.2: 1.** Meristematic tissue. **2.** Epidermal tissue: protection, conservation of water; Ground tissue: photosynthesis and carbohydrate storage; Vascular tissue: transport of water and nutrients. **3.** Ground tissue: parenchyma (progenitor cells), collenchyma (flexible support), sclerenchyma (support, some transport water); Vascular tissue: xylem (transport water and minerals from roots to leaves), phloem (transports sugar and other organic compounds throughout the plant). **9.3: 1.** See Fig. 9.7. **2.** Monocots: corn, grass, palms; Dicots: dandelions, oak trees, potatoes, kale. **9.4: 1.** Via a Casparian strip which requires water and minerals to pass through the cells to enter the xylem. **2.** The monocot root has a vascular cylinder surrounding a central pith, while the eudicot root has the phloem and xylem in the center of the root. **3.** Adventitious roots (stabilizing the shoot system), epiphytes (absorption of water), haustoria (absorption of water and nutrients), root nodules (obtaining nutrients), mycorrhizae (absorption of water and nutrients). **9.5: 1.** See Fig. 9.15. **2.** In the spring when water is abundant, secondary xylem contains wide vessel elements with thin walls. When water becomes restricted in the summer, the wood has fewer vessels and contains thick-walled vessels. **3.** Stolons in aboveground horizontal stems produce new plants when touching the soil. Cacti have aboveground vertical stems that store water. Many vines have stems with tendrils that twine around structures. Rhizomes are underground horizontal stems. Corms are bulbous underground stems which are dormant in winter. **9.6: 1. 1.** Palisade mesophyll (elongated cells packed tightly) and spongy mesophyll (irregular cells bounded by air spaces). **2.** Shade plants have broad, wide leaves; desert plants have reduced leaves with sunken stomata. Leaves tend to be thin and flat to allow maximum sunlight absorption. **9.7: 1.** Cohesion-water molecules tend to cling together; adhesion-water molecules interact with the walls of the xylem vessels; both are required to keep the column of water together as it ascends the xylem channel. **2.** Evaporation of water from a leaf causes the whole channel of water to move upward, drawing in more water from the roots. **3.** After sugar is actively transported into the phloem sieve tubes, water follows by osmosis. This creates a pressure that causes the phloem contents to flow from the source to the sink.

Science in Your Life

The Many Uses of Bamboo: 1. Whether bamboo products are worth the extra cost depends on the price, how much money the buyer has, and the value the buyer puts on the product and its lesser environmental impact. **2.** Farmers could be offered economic incentives and subsidies to switch to growing bamboo. **3.** Competition with other markets that are established and employ many, and displacement of other habitats to grow bamboo. **Using Plants to Clean up Toxic Messes: 1.** If the plant has converted the pollutant into a form that is not a hazard then the dead plants can be treated as usual. If the plant has collected and concentrated the pollutant then the dead plant material must be collected and treated as hazardous waste. **2.** Each plant has a unique physiology which allows it to handle toxic materials differently. **3.** If farmers have pollutants in their fields then they should be required to prevent their release into the environment. If phytoremediation is a way to do this, then regulations requiring its use seem appropriate.

Testing Yourself

1. c; **2.** c; **3.** c; **4.** c; **5.** c; **6.** d; **7.** c; **8.** d; **9.** d; **10.** b; **11.** a; **12.** d;**13.** b; **14.** a; **15.** c.

Thinking Critically

1. See Table 9.1. **2.** Woody plants can increase in girth and grow taller to reach sunlight, but this costs energy. **3.** Both use active transport and osmosis for transporting substances into the vessels. Plants use passive means of transport whereas humans use a pump. Plants separate water and minerals from organic nutrients while humans do not. Plants have two separate sets of vessels (xylem and phloem) with the xylem being arranged in a one way pathway for transport, while humans have two connecting sets of vessels (arteries and veins) arranged in a circular pathway. Some of the vessels in plants are dead, while all of the vessels are alive in humans.

CHAPTER 10

Case Study

1. Auxins, gibberellins, cytokinins, abscisic acid, and ethylene **2.** Selective breeding; creation of transgenic organisms often called genetically modified organisms; introduction of genetic traits that offer resistance to insects, herbicides, salty environments, potato blight, or lead to production of monounsaturated fatty acids in soybeans or increase productivity by altering C_3 plants to C_4 plants.

Check Your Progress

10.1:1. The sporophyte generation produces microspores and megaspores. A microspore undergoes mitosis and becomes a pollen grain (male gametophyte), and the megaspore undergoes mitosis to become a microscopic embryo sac (female gametophyte). **2.** Anther: development of pollen grains. Filament: holds up anther. Stamen: male part of flower consisting of anther and filament. Pistil (or carpel): female part of flower, consisting of stigma, style, ovary, and ovule. Stigma: sticky knob for holding pollen. Style: holds up stigma. Ovary: enlarged base that holds one or more ovules. Ovule: house megaspores and when fertilized, is chamber where seed develops. **3.** Heterosporous means that a plant produces two types of spores: microspores which become a pollen grain (male gametophyte) and megaspores which become the embryo sac (female gametophyte). **4.** See Fig. 10.5. **10.2: 1.** In monocots food molecules are absorbed from the endosperm. In eudicots the nutrients are stored in the cotyledons and the endosperm disappears. **2.** Mechanisms for seed dispersal include hooks and spines on seeds for attachment to animals, seeds which survive passage through digestive systems and are deposited in feces, coevolution with animals that harvest and bury seeds, seeds that can float in water, hairs or wings that carry seeds in the wind, seed pods that explode upon ripening. **3.** Regulation of the germination process assures that seeds are in the best environment for survival. **4.** Compare Fig. 10.9 with Fig. 10.10. **10.3: 1.** Vegetative propagation and tissue culture. **2.** See Fig. 10.11. **3.** Corn, cotton, soybean, and potato plants all have been genetically engineered to be resistant to herbicides. Tomatoes have been engineered to be salt tolerant. **10.4: 1.** Auxins—promote shoot growth from the apical meristem or from the ancillary buds when the terminal bud is removed. Gibberellins cause elongation of cells. Cytokinins promote cell division. Abscisic acid initiates and maintains seed and bud dormancy while controlling the opening of stomata under water stress. Ethylene functions in the process of abscission. **2.** Plant growth tropisms include growth toward or away from a stimulus such as light, gravity, and touch. **3.** Short day plants flower during shorter daylight periods while long day plants flower during longer light days. **4.** Since phytochrome conversion from inactive to active forms is the first step in a signaling pathway that results in flowering, regulating the conversion could lead to controlling plant flowering commercially.

Science in Your Life

Pollinators and You: 1. They have a symbiotic relationship in which both partners benefit. **2.** Flowers that are not colorful and give off strong odors most likely have pollinators that are nocturnal who are not attracted by color. **3.** Reduce pesticide use, reduce exposure to parasites, promote health of the hive, provide good food sources, reduce stress, and avoid monoculture hives. **Transgenic Crops: 1.** A risk exists if the pollinator is not absolutely specific to its plant that it is pollinating. **2.** Advantages: faster development time, cost savings, more specific product formation. Disadvantages: unknown outcomes from GMOs in the environment. **3.** Some people will not purchase GMO products out of concern for their health and the health of the environment.

Testing Yourself

1. b; **2.** c; **3.** b; **4.** c; **5.** e; **6.** a; **7.** d; **8.** c; **9.** d; **10.** d; **11. a.** stigma, **b.** style, **c.** ovary, **d.** ovule, **e.** receptacle, **f.** peduncle, **g.** sepal, **h.** petal, **i.** filament, **j.** anther; **12. a.** diploid, **b.** anther, **c.** ovule, **d.** ovary, **e.** haploid, **f.** megaspore, **g.** male gametophyte, **h.** female gametophyte, **i.** sperm, **j.** seed; **13.** b; **14.** b; **15.** c.

Thinking Critically

1. In both humans and plants the developing embryo is protected during its development from desiccation and exposure to the land environment. **2.** A single, specialized pollinator species will ensure a greater probability of pollen transfer and fertilization to the correct species. **3.** Plants are totipotent.

CHAPTER 11

Case Study

1. More engaged systems: respiratory, cardiovascular, skeletal, muscular, nervous, integumentary. Less engaged: lymphatic and immune, digestive, urinary, endocrine, reproductive **2.** Epithelial, connective, muscular, nervous **3.** Organ systems interact within the body to maintain homeostasis, transport substances, support function, and control disease.

Check Your Progress

11.1: 1. See Fig. 11.1. **2.** See Fig. 11.2. **3.** See Fig. 11.4. **4.** Cell body, dendrites, axon. A nerve fiber is a myelinated axon which conducts nerve impulses to the cell body. **11.2: 1.** Thoracic, abdominal. **2.** Mucus protects the digestive, respiratory, urinary, and reproductive system from bacteria and viruses. Serous fluid lubricates membranes that support internal organs and divide body cavities. Synovial fluid lubricates cartilage at the end of bones. **11.3: 1.** Integumentary—skin; cardiovascular—heart; lymphatic and immune—lymph nodes; digestive—small intestine; respiratory—lungs; urinary—kidneys; nervous—spinal cord;—musculoskeletal—skeleton; endocrine—pancreas; reproductive—ovaries. **2.** Integumentary, and lymphatic and immune. **11.4: 1.** See Fig. 11.8. **2.** They might not absorb enough UVB rays to produce adequate levels of Vitamin D which helps to keep bones strong. **3.** Nails—protect the distal parts of digits; hair follicles—produce hair; oil glands secrete sebum. **11.5: 1.** In negative feedback, a sensor detects a change in internal conditions, resulting in a response that brings the conditions back to normal. Positive feedback involves an ever-greater change in the same direction until the initial stimulus stops. **2.** The respiratory, digestive, and urinary systems interact to take in O_2 and nutrients needed by the body, and eliminate unused or waste products such as CO_2 and nitrates. They assure that blood pH, blood volume, and glucose levels are maintained within strict ranges. **3.** Failure of the immune system results in infectious diseases such as shingles and candidiasis. Failure of the cardiovascular system results in heart failure or strokes. Failure of endocrine organs can result in diabetes or hypothyroidism.

Science in Your Life

UV Rays: 1. Long term health outcomes are of more value than appearance based on social expectations. Healthy skin is most important. **2.** Being unaware of what melanoma looks like or that it is a serious health threat; not being able to see the skin on your back; having darkly pigmented skin that makes moles harder to see. Barriers to accessing health care services such as poverty, lack of insurance and adequate transportation also contribute. **3.** There is disagreement as to what is a desirable level of vitamin D in the bloodstream and uncertainty about the effect of maintaining any specific level. There are benefits for having enough vitamin D and risks with ingesting too much. **Testing Hair for Drugs: 1.** Procedures and policies for testing for drugs must comply with the Fourth Amendment. However, an employer has the right to deny an applicant on the basis of drug evidence. **2.** Anytime drug use interferes with job performance or the safety of others, drug testing is warranted. In competitive sports, performance enhancing drugs alter the balance between players. An organization certifying athletes' performance has the right to set the policies covering testing for drugs.

Testing Yourself

1. c; **2.** d; **3.** b; **4.** a; **5.** d; **6.** b; **7. a.** ventral cavity, **b.** thoracic cavity, **c.** abdominal cavity, **d.** pelvic cavity, **e.** dorsal cavity, **f.** cranial cavity, **g.** vertebral canal, **h.** diaphragm; **8.** e; **9.** c; **10.** c; **11.** c; **12. a.** hair shaft, **b.** epidermis, **c.** dermis, **d.** subcutaneous layer, **e.** adipose tissue, **f.** sweat gland, **g.** hair follicle, **h.** arrector pili muscle, **i.** oil gland, **j.** sensory receptor, **k.** basal cells, **l.** sweat pore; **13.** a; **14.** b; **15.** d.

Thinking Critically

1. Blood is like a tissue because it contains similarly specialized cells that perform a common body function. Blood cells are specialized to circulate throughout the body, bringing O_2 to the tissues (red blood cells), preventing the blood from being lost from the body (platelets), and protecting the body from infection (white blood cells). **2. a.** Negative feedback, since secretion of the pituitary gland hormone is inhibited by epinephrine. **b.** Positive feedback, because the stronger the signal becomes, the stronger the stimulus to respond, until urination occurs. **c.** Negative feedback, because the various sensory receptors are detecting a change in a set point (blood volume), and the kidney is responding by increasing urine production. **3.** The cardiovascular and lymphatic systems could both be placed in the transport, maintenance, and support systems. The respiratory, urinary, and digestive systems could be considered as maintenance or support. Other examples exist.

CHAPTER 12

Case Study

1. Lack of oxygen to tissues **2.** Delivery of oxygen and nutrients to the tissues, removal of CO_2 and other wastes from the tissues. **3.** Major preventable risk factors for cardiovascular disease include smoking, obesity, poor diet, lack of exercise, drug abuse, and excessive alcohol consumption.

Check Your Progress

12.1:1. Blood flow is controlled in arteries by contraction of smooth muscle in artery walls. Capillary blood flow is affected by the pressure of arterial supply plus the contraction of pre-capillary sphincters. Venous blood flow is affected by arterial and capillary blood flow, and valves that prevent blood from flowing backward. **2.** O_2, CO_2, glucose, amino acids. **12.2: 1.** See Fig. 12.3. **2.** See Fig. 12.6. **3.** Diabetes, heart disease, liver disease, Alzheimer, and Parkinson diseases. **4.** If the protein content of the blood is reduced then the osmotic pressure of the blood is lowered and less water moves from the tissue fluid into the blood. **12.3: 1.** Blood enters the right atrium and right ventricle, then travels through the pulmonary arteries to the lungs, and through the pulmonary veins back to the left atrium and left ventricle. **2.** If the left ventricle is not working properly, blood would back up into the lung, increasing blood pressure there, resulting in increased fluid leakage from blood vessels in the lung. **3.** The heartbeat "lub" sound is caused by the closure of the atrioventricular valves and the "dub" sound is caused by the semilunar valves closing. **4.** Intrinsic control of the heartbeat is needed to maintain regular contractions, while extrinsic control allows the heartbeat to change in response to different physiological demands. **12.4: 1.** Pulmonary arteries. **2.** Blood flows to the lungs via the pulmonary arteries, and returns to the heart via the pulmonary veins. Blood flows to the body via the aorta, and returns via the superior and inferior vena cava. **3.** Blood flows in one direction in arteries due to the pumping of the heart, and in veins mainly due to muscle contraction and the presence of valves. **12.5: 1.** A thrombus is a blood clot that remains stationary, an embolus is a clot that moves with the blood, and an aneurysm is the ballooning of the side of a weakened blood vessel. **2.** Stents can directly activate the blood clotting system, or irritate blood vessels in which they are placed, causing inflammation. **3.** Failing hearts can be replaced with a transplant from a human donor, or with an artificial heart. A left ventricular assist device can help a failing heart function until it can be replaced. Potential problems include rejection of a transplanted heart by the recipient's immune system, and the mechanical failure of an artificial pump.

Science in Your Life

Medicinal Leeches: 1. The response to this question depends on one's discomfort with leeches as a medical device. **2.** Diseases were once thought to occur because of an imbalance between the "humors" in the body. It was thought that blood letting would reestablish the correct balance. **3.** The introduction of yeast and bacteria to reestablish a healthful gut flora; therapy pets to help people cope with physical and mental challenges. **Prevention of Cardiovascular Disease: 1.** The response to this will depend on the individual. **2.** Infection with Group B Streptococcus can impact heart valve health. Infections in general increase inflammation which can lead to increased plaque formation. Infections of the gut can lead to decreased absorption of nutrients and lead to lipid imbalances and vitamin deficiencies, both of which can influence cardiovascular health.

Testing Yourself

1. b; **2.** a; **3.** d; **4.** e; **5. a.** P wave, **b.** QRS complex, **c.** T wave; **6.** b; **7** c; **8. a.** internal jugular vein, **b.** common carotid artery, **c.** inferior vena cava, **d.** common iliac artery, **e.** aorta, **f.** femoral artery; **9.** b; **10.** b; **11.** d; **12.** b; **13.** c; **14.** b; **15.** b.

Thinking Critically

1. Tissues that do not contain blood vessels include the corneas of the eyes, the outermost layer of the skin, and the cartilage that lines joints. These tissues presumably either receive their required nutrients by diffusion from nearby vessels, and/or have lower metabolic requirements than more highly vascularized tissues. **2.** The answer will vary depending on age, but for someone with an average heart rate of 70 beats per minute, on their 20th birthday their heart will have beaten 70 × 60 minutes/hour × 24 hours/day × 365 days/year × 20 years, or 735,840,000 times. In terms of volume, 5.25 liters/minute × 60 minutes/hour × 24 hours/day × 365 days/year × 20 years = 55,188,000 liters (over 14 million gallons). **3.** Tracing "a" shows too many QRS complexes, indicating atrial fibrillation. The complexes in tracing "b" are missing the "T" waves, and the heartbeat is too slow, suggesting a problem with electrical transmission in the heart. A pacemaker is an electrical device that can be implanted in a patient's heart muscle, which takes the place of a malfunctioning SA node, to deliver a regular electrical impulse to the heart muscle. **4.** A total artificial heart (TAH) would have to be as incredibly durable as the heart is, to last even 20 years. The TAH also needs to be made of a material that does not stimulate the body to react against it in any way, including the immune and clotting systems.

CHAPTER 13

Case Study

1. Up until 6 months children benefit from the passive immunity received from their mothers.
2. Without sufficient levels of antibodies, XLA patients cannot effectively remove bacteria which are outside of the cells in the body. NK cells and cytotoxic T cells function to remove virus infected cells.

Check Your Progress

13.1: 1. Excess tissue fluid moves into the lymph vessels and is returned to the blood stream. Fats from the small intestine move into the lymph vessels and are transported into the bloodstream. White blood cells in the lymph system function in protecting the body from disease. **2.** Edema is the accumulation of excess tissue fluid, caused by too much fluid produced (i.e., by a tumor), or not enough drained away (i.e., because of low blood protein content or blockage in a lymphatic vessel). **3.** In the red bone marrow and the thymus, which are primary lymphoid organs, lymphocytes develop and mature. In the lymph nodes and the spleen, lymphocytes become activated. **13.2: 1.** The skin and mucous membranes are physical barriers; a chemical barrier is the acid of the stomach. **2.** Redness, heat, swelling, and pain are the signs of inflammation. Mast cells release histamine, which causes capillaries to dilate and become permeable to fluids which move into the tissue, producing swelling and pain. Increased blood flow results in redness and heat. **3.** Macrophages kill pathogens by engulfing them into a vesicle that has an acid pH, hydrolytic enzymes, and reactive oxygen compounds. NK cells induce cells that lack self-MHC-I molecules to undergo apoptosis (cell suicide). **4.** The complement system enhances phagocytosis of pathogens, activates inflammation, and kills pathogens by forming a membrane attack complex. **13.3: 1.** Diversity in antigen receptor occurs because: 1. genes for T and B cell receptors containing segments that code for many parts of the antigen receptor. 2. enzymes in the T and B cells cut out these segments and combine them differently. 3. mutations may be introduced as these segments are combined. 4. the two protein chains, containing the variation of steps 1–3, combine to produce a new receptor. **2.** The clonal selection theory states that B cells and T cells have cell surface receptors for only one specific antigen. When the cell contacts that specific antigen, it is selected to undergo clonal expansion (divide) and differentiate into either memory cells or cells that actively fight infection. **3.** The five classes are IgG (activates complement, crosses the placenta), IgM (early response, activates complement), IgA (found in body secretions), IgD (found only on immature B cells), and IgE (protects against parasites). **4.** Helper T cells recognize antigen fragments in combination with MCH molecules presented by antigen-presenting cells that have MHC class II proteins on their surface. Cytotoxic T cells recognize antigen-presenting cells with MHC class I proteins. Suppressor T cells inhibit these responses and act to control adaptive immunity. **13.4: 1.** Three types include: attenuated (live but non-virulent), genetically engineered (also called subunit vaccines), and DNA vaccines. **2.** The immune system of the newborn is immature, plus the newborn has not been exposed to any infectious agents, so it would be very susceptible to infectious diseases if it did not receive antibodies from its mother. **3.** The interaction of cytokines with their receptors could be blocked using antibodies that react with the cytokine, or possibly with the receptor. **4.** Monoclonal antibodies are extremely specific because they react with only one antigen. Because of their specificity and purity, reaction against the preparation is avoided. **13.5: 1.** Immediate allergic responses are caused by IgE antibodies that have already been produced against the offending allergen; delayed-type responses take longer because they require sensitized T cells to secrete cytokines that cause inflammation. **2.** If a mother is Rh-positive, she will not produce anti-Rh antibodies that would react with the antigens of the fetus. **3.** Drugs that inhibit cytokine production may inhibit certain desirable immune responses that protect the individual. **13.6: 1.** Muscular weakness occurs in patients with myasthenia gravis because antibodies attach to and disrupt neuromuscular junctions. In multiple sclerosis, T cells attack the myelin surrounding nerves and hamper nerve conduction. Antibody–antigen complexes form in people with systemic lupus erythematosus and these get deposited in organs, including the kidneys. Antibody–antigen complexes form in rheumatoid arthritis and these primarily affect the joints. **2.** A congenital immunodeficiency disease is severe combined immunodeficiency. AIDS is an acquired immunodeficiency disease.

Science in Your Life

Opportunistic Infections and HIV: 1. Deficiencies in immune function can occur in healthy people undergoing medical challenges. Once over the challenge, immune function is restored. With HIV-infected people the virus remains, constantly compromising their immune function. **2.** The number of helper T cells which would help to combat herpes virus 8 is reduced in HIV positive people. This increases their susceptibility. **3.** Almost all helper T cells are gone. The individual has little defense against the virus. **Requirement of Children Vaccinations 1.** From a public health perspective the benefits of vaccination clearly outweigh the risks. But this is assessed on a population level. Individuals will perceive the level of risk for any disease based on their personal experience. **2.** To respond to this question the importance of religious beliefs must be weighed against good practices of health care for the whole population. **3.** Pediatricians have belief sets that are implemented in their practice of medicine. In choosing not to see unvaccinated children they are following their belief that it is dangerous to the community to not vaccinate children. Perhaps a better approach would be to see unvaccinated children and attempt to educate them and their parents on the benefits of vaccination. **Immediate Allergic Responses: 1.** People previously exposed to an antigen can develop hypersensitivity leading to immediate allergic responses. **2.** One explanation is that children in developed countries are exposed to reduced kinds and levels of allergens. Their ability to cope with the allergens when encountered later in life is compromised. **3.** Epinephrine constricts blood vessels, increasing blood pressure, and reducing the possibility of anaphylactic shock.

Testing Yourself

1. b; **2. a.** thymus (primary), **b.** lymph node (secondary), **c.** bone marrow (primary), **d.** spleen (secondary); **3.** a; **4.** c; **5.** d; **6.** d; **7.** e; **8.** d; **9.** c; **10.** c; **11.** d; **12.** a; **13. a.** Ag-binding sites, **b.** light chain, **c.** heavy chain, V = variable, C = constant; **14.** b; **15.** c.

Thinking Critically

1. Innate immune mechanisms are valuable mainly because they can react very quickly to a wide variety of pathogens. However, organisms lacking adaptive immunity are not able to mount specific antibody and T cell responses targeted at one specific organism, nor are they able to develop memory cells that are able to respond faster to subsequent exposures to the same organism. **2.** NK cells recognize target cells that lack MHC-I molecules; Cytotoxic T cells recognize target cells that have pieces of foreign antigens presented on MHC-I molecules on their surface. If a virus is able to interfere with the normal expression of MHC-I molecules on the cells they infect (and many do this), then NK cells can kill these infected cells. **3.** ABO incompatibility between mother and fetus usually causes no problems because the antibodies the mother has against the nonself A or B antigen are normally of the IgM type, which is too large to cross the placenta. Because around 85% of the population is Rh-positive, most females do not produce anti-Rh antibodies that could damage or kill the fetus; therefore, there is not a strong evolutionary selective pressure to eliminate the Rh molecule.

CHAPTER 14

Case Study

1. The mouth and stomach mechanically and chemically break down food. The stomach enzymatically breaks down some proteins. With the exception of alcohol and glucose, nutrients are not absorbed until food reaches the small intestine. Enzymatic digestion occurs in the small intestine. Once absorbed into the bloodstream nutrients are distributed to the liver or tissues for processing or utilization. **2.** Genetic factors influencing the breakdown and absorption of food as well as the rate at which it is metabolized influence a person's weight. Also, one's disposition toward physical activity or other behaviors requiring energy can impact weight.

Check Your Progress

14.1: 1. Food passes through the mouth, pharynx, esophagus, stomach, small intestine, and large intestine. Mechanical digestion occurs in the mouth and stomach. Chemical digestion occurs in the stomach and small intestine. **2.** Mucosa, submucosa, muscularis, serosa **3.** The small intestine processes food for digestion and absorbs nutrients. Its lower part, the ileum, is involved in immune responses. The large intestine absorbs water, salts, and some vitamins, and eliminates indigestible material as feces. **4.** Hormones that increase digestive activity include gastrin, secretin, and CCK. Inhibiting these might regulate weight but they can cause difficulties in digesting food, leading to excess gas or constipation. **14.2: 1.** The pancreas secretes hormones such as insulin and glucagon into the blood, and also secretes digestive enzymes into the small intestine by way of ducts. **2.** The liver performs functions that are essential for life including removing toxins from the blood, processing some wastes, and synthesizing some essential plasma proteins such as albumin. The main function of the gallbladder is to store bile, the absence of which is not life threatening. **14.3: 1.** Carbohydrate digestion begins in the mouth, with salivary amylase, and continues in the small intestine, due to maltase and pancreatic amylase. Protein digestion begins in the stomach, where pepsin breaks proteins into peptides, and continues with trypsin and peptidases in the small intestine. Fats are mainly digested by pancreatic lipase in the small intestine. **2.** Carbohydrates are broken down to glucose; proteins to amino acids; fats to glycerol and fatty acids. **3.** The digestive system is divided into many compartments, each of which has a distinctive internal environment and pH. These differences contribute to regulation of digestive enzymes. **14.4: 1.** The glycemic index (GI) of foods is an indicator of whether the blood glucose response to their ingestion is high or low. The advantage of consuming foods with a high GI is that they can provide a quick burst of energy, but that energy is usually short-lived and over time, consuming too many high-GI foods can lead to type 2 diabetes and heart disease. **2.** A diet deficient in protein does not supply adequate amounts of essential amino acids used to synthesize necessary proteins. Too much protein must be deaminated in the liver which requires abundant water for excretion, calcium loss in urine. **3.** Saturated fat intake leads to increased levels of LDL. LDL carries cholesterol from the liver to the cells. This is related to increased plaque formation and greater risk for cardiovascular disease (CVD). HDL carries cholesterol to the liver where it is converted to bile salts. High HDL levels are correlated with reduced risk of CVD. Lack of exercise, diet deficient in fruits and vegetables, smoking, and excessive alcohol intake contribute to risk of CVD caused by plaque. **4.** Vitamins are organic compounds that the body needs for metabolic processes, usually functioning as coenzymes. However Vitamin A and D are different. Vitamin A is a precursor to a visual pigment necessary for night vision, and vitamin D which is formed in the skin, is converted to calcitriol which promotes the absorption of calcium in the intestines. **14.5: 1.** People with anorexia nervosa restrict their food intake, thinking of themselves as overweight no matter what they weigh. Those with bulimia nervosa eat more than they think is appropriate and rid themselves of the food by vomiting or the use of laxatives. Binge eating involves ingesting more food than needed, usually for emotional reasons. This most often leads to obesity. People with pica eat unusual substances for reasons that are not understood. **2.** Abnormally increased or decreased levels of hormones that influence appetite, such as gastrin, gastric inhibitory peptide (GIP), or leptin, can predispose a person to eating disorders. **14.6: 1.** Most stomach ulcers are caused by the bacterium *Helicobacter pylori*. **2.** The pancreas is essential in regulating serum glucose levels by the secretion of insulin and glucagon. The pancreatic juice it secretes neutralizes stomach acid and provides digestive enzymes. The liver synthesizes bile and essential proteins, regulates glucose and cholesterol levels, metabolizes many waste products, and stores iron and many vitamins. Both organs are essential to life.

Science in Your Life

Vegetarians: Where Do You Get Your Protein? 1. If these animals have grown big then it indicates that a vegetarian diet is adequate. However, their digestive systems are structured differently and the way they digest food varies from humans. What works for cattle, horses, and elephants may not work for humans. **2.** If amino acids are not stored, it seems logical that you must eat a full complement of all 20 essential amino acids at one time. Eating different foods that supply the full complement of amino acids is desirable. **3.** If a specific amino acid is not available then elongation of a peptide chain on the ribosome during translation will not occur. **Stomach Ulcers: 1.** To survive *H. pylori* must either have enzymes that operate with acidic optimal pHs or it must be able to isolate itself from the acidic environment. **2.** People are reluctant to change their beliefs and thinking about scientific principles. **3.** The argument could be made that it was ethical but not prudent behavior. Better to have used animal models to test the impact of *H. pylori.*

Testing Yourself

1. d; **2.** d; **3.** d; **4.** c; **5.** e; **6.** d; **7. a.** large intestine, **b.** small intestine (ileum), **c.** cecum, **d.** vermiform appendix **8.** d; **9.** a; **10.** b; **11.** c; **12.** a; **13.** d.

Thinking Critically

1. A loss of mechanical digestion could be replaced by eating cooked and pureed foods. A lack of chemical digestion, however, would prevent the breakdown of food into smaller, absorbable molecules which would necessitate direct introduction of nutrients into the bloodstream. **2.** Digestion of complex carbohydrates begins in the mouth with salivary amylase, and continues in the small intestine with pancreatic amylase. Ultimately, monosaccharides such as glucose are absorbed from the small intestine into the hepatic portal vein, where some travel to the liver and some enter the general circulation via the hepatic veins and inferior vena cava. Cells transport glucose molecules across their plasma membranes and utilize them to produce ATP. **4.** Only tube 4 containing pepsin, HCl, and water should result in the breakdown of the egg white. The other tubes lack a necessary component of this reaction mix.

CHAPTER 15

Case Study

1. Narrowing of the airways due to swelling would restrict how much air could pass through. **2.** Treatments for asthma often dilate bronchi. They address the symptoms rather than the causes of asthma. **3.** Asthma is a response to the body's exposure to allergens. Regulating the immune system's response to allergens is extremely difficult.

Check Your Progress

15.1: 1. The upper respiratory tract consists of the nasal cavities, pharynx, and larynx. The lower respiratory tract consists of the trachea, bronchial tree, and lungs.

2. The trachea lies ventral to the esophagus. When food is swallowed, the epiglottis covers the tracheal opening (glottis) so the food slides over the epiglottis and into the esophagus. **3.** O_2 travels through the nasal cavity, pharynx, larynx, trachea, bronchial tree, and lungs, where it diffuses across the alveolar and capillary endothelial cells to the bloodstream. **4.** Pulmonary surfactants lower the surface tension of the coating of the alveoli, preventing their sides from collapsing upon themselves. **15.2: 1.** Tidal volume is the amount of air that normally moves in and out of the lungs with each breath. Vital capacity is the maximum volume of air that can be moved in and out during a single breath. Expiratory reserve volume is the air that can be forcibly exhaled beyond the tidal volume. Residual volume is the air left in the lungs after a forced exhalation. **2.** Inspiration requires the contraction of the diaphragm and external intercostal muscles; expiration is passive because it requires no muscle contractions, just the elastic recoil of the thoracic wall and lungs. **3.** The respiratory center in the brain automatically sends nerve impulses to the diaphragm and intercostal muscles. The vagus nerve carries inhibitory impulses from the lungs to the brain to stop the lungs from overstretching. The carotid bodies have chemoreceptors that monitor levels of O_2 in the blood. **15.3: 1.** Hemoglobin's most essential function is to carry O_2 (as oxyhemoglobin) to the tissues, but it also carries a small amount of CO_2 (as carbaminohemoglobin) back to the lungs. Excess H^+ ions can be taken up by hemoglobin, which then becomes reduced hemoglobin. **2.** Arterial blood is brighter red than venous blood because the red blood cells in arterial blood contain oxyhemoglobin, which is bright red, compared to the red blood cells in venous blood that contain reduced hemoglobin, which is darker. Blood from a cut appears bright red because hemoglobin becomes oxyhemoglobin upon exposure to O_2 in the air. **3.** Carbon monoxide has a stronger binding affinity to hemoglobin than oxygen and remains bound for a much longer time which makes the hemoglobin involved unavailable to transport oxygen. **15.4: 1.** Asthma causes narrowing of the airways. Pulmonary fibrosis results in the buildup of fibrous connective tissue in the lungs, resulting in a loss of elasticity of the lungs during inhalation. **2.** Smoking damages the tissues of the respiratory track directly. Infections from bacteria, fungi, protozoans, and viruses become more likely if the lung tissue is damaged. **3.** Three common disorders of the upper respiratory tract are the common cold, usually caused by rhinoviruses; strep throat, caused by the bacterium *Streptococcus pyogenes;* and otitis media (middle ear infections), caused by various bacteria, viruses, and allergens. **4.** Infections can easily spread from the respiratory system (pharynx) to the middle ear via the auditory (eustachian) tubes.

Science in Your Life

FAQs about Tobacco and Health: 1. Answers will vary according to different experiences and perceptions about smoking. **2.** Responses to this question will vary but respect must be given to the fact that nicotine is a highly addictive substance. **3.** Breaking addictions whether it is to nicotine or something else, involves conviction and also benefits from any aid that is proven effective. **Artificial Lung Technology: 1.** Cost of the procedure as well as expanded time frames for development would restrict its distribution to the general public. **2.** Lungs have a complex structure which includes surfaces for gas exchange and an association with a circulatory system for transportation. **3.** Tissues grown in the lab can be engineered to not display surface antigens which would identify the tissue as nonself. This would reduce tissue rejection. Timing of the transplant could be controlled rather than waiting for a donor to appear. **Bans on Smoking: 1.** Public education efforts aimed at communicating the dangers of tobacco smoking, as well as increased taxes on cigarettes, have played a role in reducing smoking. **2.** and **3.** Answers will vary according to people's own perceptions, experiences, and opinions.

Testing Yourself

1. a. nasal cavity, **b.** nostril, **c.** pharynx, **d.** epiglottis, **e.** glottis, **f.** larynx, **g.** trachea, **h.** bronchus, **i.** bronchiole; **2.** a; **3.** b; **4.** d; **5.** c; **6. a.** sinus, **b.** hard palate, **c.** epiglottis, **d.** larynx, **e.** uvula, **f.** glottis; **7.** b; **8.** b; **9.** a; **10.** b; **11.** e; **12.** c; **13.** c; **14.** d.

Thinking Critically

1. The respiratory system of birds is more efficient than that of mammals where inspired air passes into a "dead end" (the alveoli) and must be moved back out. Some amount of air always remains in the lungs. In the bird system, air passes in one direction through the air sacs. **2.** Since CO binds to hemoglobin much more strongly than O_2 does, much less O_2 is delivered to the tissues and respiration is compromised. **3.** When a person breathes through a tracheostomy, the air does not pass through the upper respiratory tract, which normally warms and humidifies the air, as well as removing many impurities. Therefore, tracheostomy patients are prone to lower respiratory tract infections.

CHAPTER 16

Case Study

1. Excretion of metabolic wastes, osmoregulation, maintaining acid base balance, secretion of hormones. **2.** If collecting ducts are diseased or obstructed then urine cannot reach the renal pelvis and will back up. **3.** Sickled red blood cells are not as flexible as normal red blood cells and get clogged up in capillary beds, obstructing the flow of blood. Proper kidney function is dependent on many capillary beds allowing the exchange of gases, ions, and nutrients.

Check Your Progress

16.1: 1. Excretion is the removal of metabolic wastes from the body. **2.** The four major functions are excretion of wastes, maintenance of water-salt balance, maintenance of acid–base balance, and secretion of hormones. **3.** Excretion is mainly performed by the kidneys. Water-salt balance is also affected by the digestive system as well as the integumentary system (sweating). Acid–base balance is also affected by the digestive system and the respiratory system. Hormones are mainly produced by the endocrine system, which involves kidneys. **16.2: 1.** The renal cortex contains the glomeruli, proximal convoluted tubules, and distal convoluted tubules. The renal medulla contains the loops of the nephron. **2.** The epithelium of the PCT has many microvilli which increase the surface area for reabsorption. The DCT is composed of cuboidal epithelial cells that are not designed for reabsorption. **3.** All these processes involve the movement of a substance across biological membranes. Diffusion and passive transport are dependent on concentration; active transport is not but it requires energy. Active and passive transport usually require carrier molecules. **4.** Glucose is normally returned from the glomerular filtrate to the blood by reabsorption via carrier proteins at the proximal convoluted tubule. **16.3: 1.** See Fig. 16.7. Reabsorption of salt, establishment of a solute gradient, reabsorption of water. **2.** When blood volume and pressure is low, the JG apparatus secretes renin, which converts angiotensin I, which is then converted to angiotensin II that causes aldosterone secretion. **3.** Diuretics can be used medically to reduce blood volume and lower blood pressure. They are used illicitly to lose weight quickly or to dilute urine, making it more difficult to detect drugs. **4.** Kidneys regulate H^+, HCO_3^-, OH^- and NH_4^+ to maintain blood pH 7.4. **16.4: 1.** If kidneys cannot produce urine then excretion of metabolic wastes is blocked, regulation of blood volume is compromised, maintenance of blood pH at 7.4 is impaired, and hormone secretion including renin, aldosterone and erythropoietin is hindered. The lack of any of these functions is life threatening. **2.** Smoking increases the risk of bladder cancer 5-fold due to the toxic byproducts of cigarettes being secreted into the bladder. **3.** Females have a shorter, broader urethra that is closer to the anal opening and is more easily contaminated with bacteria. An enlarged prostate may cause difficult urination.

Science in Your Life

Matching Organs for Transplantation: 1. The B antigen which appears on microbes surrounding our bodies is viewed as foreign and an antigenic response, which produces antibodies, happens. **2.** There are more antigens than the MHC molecules that are displayed on our cells which define one person versus another. **3.** Since available organs are rare and the costs involved are considerable, consideration must be made of the probable success of a transplant. Factors could include the age and health of the recipient as well as their capacity to cope with the process of receiving a transplant and the long-term health maintenance.

Testing Yourself

1. c; **2.** d; **3.** c; **4.** b; **5.** d; **6. a.** renal artery; **b.** renal vein; **c.** aorta; **d.** inferior vena cava; **e.** kidney; **f.** ureter; **g.** urinary bladder; **h.** urethra; **7. a.** renal cortex; **b.** renal vein; **c.** ureter; **d.** renal pelvis; **8.** d; **9.** c, b, d, a; **10.** a; **11.** b; **12.** c; **13.** b; **14.** a; **15.** c; **16.** d; **17.** c; **18.** b; **19.** a; **20.** b.

Thinking Critically

1. If one is not eating large amounts of salt then the regulation of sodium levels in the blood and the symptom of high blood pressure would be dependent on kidney function rather than diet. If the release of aldosterone is being successfully regulated by the kidney's release of renin then blood pressure should be under control. **2.** Angiotensin II is a powerful vasoconstrictor in addition to its stimulating the adrenal cortex to release aldosterone, which leads to the reabsorption of sodium ions. Blocking angiotensin II's production would lower blood pressure.

CHAPTER 17

Case Study

1. Protection and insulation; speed up nerve conduction. **2.** In the PNS Schwann cells wrap themselves around axons forming myelin. In the CNS, oligodentrocytes, a type of neuroglial cell, form the myelin. With MS, oligodendrocytes are targeted by the immune system. **3.** White matter, which contains myelinated axons, is disrupted by MS.

Check Your Progress

17.1: 1. Three classes of neurons are sensory neurons, which take messages to the CNS; interneurons, which sum up messages from sensory neurons and other interneurons and communicate with motor neurons; and motor neurons, which take messages away from the CNS to effector organs, muscles, or glands. **2.** Most neurons contain dendrites, a cell body, and an axon. See Fig. 17.2. **3.** Myelin is formed by Schwann cells in the PNS, and by oligodendroglial cells in the CNS. **4.** Gray matter contains nonmyelinated nerve fibers; in white matter the fibers are myelinated. The brain has gray matter on the surface and white matter in deeper tissue. That pattern is reversed in the spinal cord. **17.2: 1.** The sodium-potassium pumps in neurons are always transporting Na^+ to the outside, and K^+ to the inside of the cell. **2.** During an action potential, the Na^+ gates open and Na^+ enters the cell, causing a depolarization. Then the Na^+ gates close and the K^+ gates open, causing a repolarization (even a slight hyperpolarization). See Fig. 17.4. **3.** A node of Ranvier is a gap between Schwann cells that make up the myelin sheath of an axon of a PNS neuron. Saltatory conduction is the "jumping" of action potentials as they spread from node to node. After the action potential passes one part of an axon, the sodium gates in that part are unable to open for a period of time called the refractory period. This prevents action potentials from moving backward. Synaptic integration is the summing up of all incoming excitatory and inhibitory messages by a neuron. **4.** AChE normally breaks down acetylcholine in the synaptic cleft. The listed effects of black widow spider venom suggest that acetylcholine normally transmits impulses that contract certain muscles and increase salivation, heart rate, and blood pressure. **17.3: 1.** The spinal cord is structured to provide a way for the brain to communicate to the peripheral nerves. This communication involves sensory and motor functions. **2.** Cerebrum: receives sensory information, integrates it, and commands voluntary motor responses. Diencephalon has two parts: hypothalamus serves as a link between the nervous and endocrine system maintaining homeostasis; thalamus receives sensory inputs except for smell. Cerebellum: processes information about body position and maintains posture and balance. Brain stem: acts as a relay between the cerebrum and spinal cord or cerebellum; regulates breathing through the pons; regulates vital functions through the medulla oblongata; receives and sends signals between the higher brain centers and the spinal cord. **3.** Cerebellum: poor balance and uncoordinated voluntary muscle movements. Medulla oblongata: erratic heartbeat, breathing, and blood pressure; uncoordinated reflexes for vomiting, coughing, sneezing, hiccuping, and swallowing. RAS: inability to be alert to sensory stimuli. **17.4: 1.** Because the amygdala influences responses like anger and fear, a tumor in this location could cause a person to be excessively angry or fearful. **2.** S.S. had his hippocampus destroyed by a viral infection, and he could not convert short-term to long-term memories. **3.** Mice that lack a glutamate receptor in their hippocampus were unable to learn to run mazes. **4.** Wernicke's area is involved in the comprehension of speech; Broca's area is involved in the actual motor function of speech. **17.5: 1.** See Fig. 17.15. **2.** When you touch a hot stove your hand withdraws before you feel the pain, because the nerve pathway for a reflex arc travels directly through the spinal cord, without input from the brain. **3.** See Table 17.1. **4.** The neurological explanation for a stomachache after jogging might be that your brain directed more activity to the sympathetic system, and less to the parasympathetic. This could decrease peristalsis and direct blood away from the digestive system, leading to some discomfort. **17.6: 1.** Continued exposure to a drug over time might lead to compensatory responses at synaptic clefts that result in a decreased sensitivity to the drug. For example, if the drug prevents release of the NT (see Fig. 17.18 part 2) the cell might synthesize more of the NT. **2.** Cocaine prevents synaptic uptake of dopamine. Heroin binds to endorphin receptors. Methamphetamine results in high amounts of dopamine released in the brain. Bath Salts inhibit the reuptake of several neurotransmitters. **17.7: 1.** Major symptoms of AD include loss of short-term memory, progressing to an inability to carry out basic functions. These symptoms can begin before age 50, but usually not before 65. **2.** L-dopa is converted to dopamine, which helps to compensate for the loss of dopamine-producing cells in the brain with Parkinson's disease. **3.** Autoimmune diseases occur when the immune system attacks self-cells, tissues, or organs. In MS, the immune system attacks myelin, oligodendrocytes, and neurons. **4.** *H. influenza* can cause disease if the immune status of the carrier is compromised by fatigue, poor health, viral infection, lack of exercise, or risky behaviors.

Science in Your Life

Why Do We Sleep? 1. Sleep is a period of inactivity involving a suspension of consciousness, lack of responsiveness, and changes in brain wave activity. It is essential to health. **2.** The difficulty in measuring the processes and actions of sleep contribute to the mystery. **3.** Responses will reflect individual perceptions of the role of sleep in one's life. **Caffeine: Good or Bad for You? 1.** Individual experiences with caffeine will be reported here. **2.** Because caffeine is a drug it should be reported, just like other ingredients are listed in the Nutrition Facts labels of foods. **3.** Arguments for or against this should recognize the large effects caffeine can have on human physiology.

Testing Yourself

1. a; **2.** b; **3.** c; **4.** a; **5. a.** depolarization, **b.** repolarization, **c.** action potential, **d.** resting potential; **6.** b; **7.** c; **8.** b; **9.** c; **10. a.** skull, **b.** meninges, **c.** corpus callosum, **d.** pituitary gland, **e.** midbrain, **f.** pons, **g.** medulla oblongata, **h.** thalamus, **i.** hypothalamus, **j.** pineal gland, **k.** spinal cord; **11.** d; **12.** d; **13.** d; **14.** d; **15.** d; **16.** d; **17.** d; **18.** d; **19.** d; **20.** d.

Thinking Critically

1. A way scientists study the functions of specific brain areas is to note which brain functions are disrupted by brain tumors or other disorders like strokes that affect only a specific part of the brain. Other data has been obtained from brain surgeries in which the patient is kept awake, and different areas of the brain can be electrically stimulated to see what parts of the body (or higher functions) they control. **2.** The first reason that nicotine, cocaine, and methamphetamine would be poor choices as therapeutic drugs is the high incidence of addiction and other undesirable side effects. A second problem would be the adaptation of the body over time, so that less dopamine is produced in the brain, which would counteract any temporary beneficial effects for Parkinson disease. **3.** In order to develop rational treatments for any disease it is very helpful, if not essential, to understand what causes the disease. In many brain diseases these factors are not completely understood. In addition, the brain is less accessible to investigation than other organs, due to its protected location inside the skull, the blood-brain barrier, and its essential function for life.

CHAPTER 18

Case Study

1. The cochlea in the inner ear houses mechanoreceptors that sense fluid-borne sound waves and produce nerve impulses that travel to the brain and are interpreted as sound. **2.** Cochlear implants sense sound waves and convert them into neural impulses that are transmitted to the brain by auditory nerves. Cochlear implants are dependent on functioning auditory nerves. **3.** Sound is detected by mechanoreceptors and light by photoreceptors. If technology existed that converted light signals to electrical impulses then an implant could be created to help people with vision problems.

Check Your Progress

18.1: 1. Interoceptors are involved in regulating critical body functions like blood pressure, blood volume, and blood pH. **2.** Chemoreceptors—taste and smell; photoreceptors—vision; mechanoreceptors—hearing and balance; thermoreceptors—heat and cold. **3.** Without sensory adaptation, we would constantly be stimulated by all sorts of inputs, which the brain could not interpret. **18.2: 1.** Somatic sensory receptors that are interoceptors include proprioceptors (such as muscle spindles and Golgi tendon organs), and pain receptors in the skin and internal organs. Somatic sensory receptors that are exteroceptors include cutaneous receptors like Meissner corpuscles, Krause end bulbs, Merkel disks, Pacinian corpuscles, Ruffini endings, and free nerve endings sensitive to temperature. **2.** See Fig. 18.3. **3.** Someone who lacks muscle spindles would be susceptible to injuring their muscles and joints due to a lack of awareness of limb position and degree of muscle contraction. A person who lacked pain receptors would be without the benefit of the protective function of pain and would be very prone to injury. **18.3: 1.** The tongue has taste cells and the nasal cavity has olfactory cells. Both are chemoreceptors to which molecules bind and nerve impulses are sent to the brain. The tongue has 5 types of taste cells distributed over the tongue while there are about 1,000 different types of olfactory cells in the olfactory epithelium. **2.** Adaptations that allow some animals to have more sensitive sense of smell would include having a greater number of olfactory receptors (and increased space in the nasal cavity for these receptors) and having more sensory nerve axons to receive signals from these receptors. **3.** The olfactory bulbs in the brain have direct connections to centers for memory in the limbic system. **18.4: 1.** Conjunctiva, cornea, aqueous humor in the anterior compartment, through the iris, lens, vitreous humor in the posterior compartment, retina including the ganglion cell layer, bipolar cell layer, and finally the rod cell and cone cell layer where light is converted to nerve impulses. **2.** When the ciliary muscle is relaxed, the suspensory ligaments that attach to the lens are taut, which causes the lens to become flatter. When the ciliary muscle contracts, the suspensory ligaments are relaxed, allowing the lens to become more round. **3.** As many as 150 rod cells, but perhaps only one cone cell, synapse(s) on an individual ganglion cell, so that more information can be sent to the brain for an individual cone cell versus a rod cell. **4.** The optic chiasma is X-shaped, formed by the optic nerve fibers crossing over. The image sent to the brain is split because one side of the optic tract carries information about the opposite side of the visual field. The two sides of the visual field must communicate with each other in the brain for perception of the entire visual field. **18.5: 1.** See Fig. 18.1. Sound waves travel down the auditory canal to the tympanic membrane and then to the bones of the middle ear which are connected to the cochlea in the inner ear. It is in the cochlea that the sound waves are converted to neural impulses that are transmitted to the brain. **2.** The mechanoreceptors responsible for transducing sound waves into nerve impulses are located on hair cells on the basilar membrane of the spiral organ in the cochlea of the inner ear. **3.** Sound waves transmitted via the oval window cause the basilar membrane to vibrate. Near the tip of the spiral organ, the basilar membrane vibrates more in response to higher pitched sound, and at the base it responds to lower pitches. Different sensory nerves supply each part of the basilar membrane, and take information to different parts of the auditory cortex, where the pitch is interpreted. **4.** Loud sounds cause the basilar membrane to vibrate more rapidly, which is interpreted by the brain as volume. **18.6: 1.** See Fig. 18.13. **2.** The ampulla and cupula are associated with the semicircular canals; the otoliths, saccule, and utricle are associated with the vestibule. **3.** Having two systems for equilibrium allows the brain to receive information about rotational movement from the semicircular canals, and about the body's position at rest from the vestibule. Although it might be possible to distinguish between these with just one system, it would not be as efficient or sensitive. **18.7: 1.** The sense of smell can protect us from dangerous chemicals and situations. Without it, one is more vulnerable to injury from such sources. **2.** Nearsightedness results when the eyeball is elongated; hence the lens focuses in front of the retina. Farsightedness results when the eyeball is too short. Astigmatism is caused by a misshapen cornea or lens. **3.** In macular degeneration the choroid vessels thicken and no longer function, which leads to the destruction of cone cells, resulting in blindness. The drainage system in the eyes fails and fluid builds up in patients with glaucoma. The increased pressure in the eye destroys nerves first in the peripheral vision but can lead eventually to complete blindness. With cataracts the lens of the eye becomes cloudy and eventually opaque. No light can get through and total lack of vision can result.

Testing Yourself

1. e; **2.** b; **3.** a; **4.** b; **5. a.** retina, **b.** choroid, **c.** sclera, **d.** optic nerve, **e.** fovea centralis, **f.** ciliary body, **g.** lens, **h.** iris, **i.** pupil, **j.** cornea; **6.** d; **7.** e; **8. a.** stapes, **b.** incus, **c.** malleus, **d.** tympanic membrane, **e.** semicircular canals, **f.** cochlea, **g.** auditory tube; **9.** c; **10.** c; **11.** a; **12.** b; **13.** d; **14.** d; **15.** a.

Thinking Critically

1. Without sensory adaptation to constant everyday stimuli, we would constantly be bombarded with and distracted by a huge amount of sensory data (about touch, temperature, pressure, etc.). In contrast, there are large risks for not responding to pain, as shown by the damage that occurs to people who lack normal pain perception. **2.** Once it reaches the retina, light must pass through layers of axons, ganglion cells, and bipolar cells before it reaches the rod or cone cells. From the right side of the left eyeball, nerve impulses travel via the optic nerve, through the optic chiasma, then through the right optic tract, until they synapse with neurons in the thalamus. Axons from thalamic neurons transmit this information to the visual area of the occipital cortex. **3.** The null hypothesis would be that there is no association between the density of taste buds on the tongue and obesity. The prediction based on this hypothesis is that obese people will have the same density of taste buds as normal weight people.

CHAPTER 19

Case Study

1. She is relatively young, in good health, and remains active. **2.** It provides a smooth surface for the end of the bones to move against each other and gives support. **3.** The knee is a simpler joint that involves hinge movement, not rotational.

Check Your Progress

19.1: 1. Compact bone is composed of tubular units and is highly organized, with central canals providing blood, nerves, and lymph. Spongy bone appears at the end of the long bones and is composed of thin plates arranged to form spaces. **2.** Hyaline cartilage is found at the ends of long bones (in joints), in the nose, ribs, larynx, and trachea. The disks between the vertebrae and in the knee are made of fibrocartilage and elastic cartilage is found in the ear and epiglottis. **3.** Both ligaments and tendons are composed of dense fibrous connective tissue, but ligaments connect bone to bone, while tendons connect muscle to bone. **4.** During bone remodeling, osteoclasts break down bone, and osteoblasts rebuild it. **19.2: 1.** Axial skeleton consists of skull, hyoid bone, 7 cervical, 12 thoracic, and 5 lumbar vertebrae, sacrum, coccyx, rib cage, and ossicles. **2.** The four major bones of the cranium are the frontal, parietal, occipital, and temporal bones. The most prominent facial bones are the mandible, maxillae, zygomatic bones, and nasal bones. **3.** True ribs are attached to the sternum by costal cartilage. False ribs connect to the sternum via a common cartilage. Floating ribs do not attach to the sternum. **4.** The knee is a typical hinge joint, the elbow is a pivot joint, and the hip is a ball-and-socket joint. **19.3: 1.** Smooth, cardiac, and skeletal. **2.** The origin of a muscle is its bone attachment that remains stationary, while the insertion is on the bone that moves. **3.** See Tables 19.2 and 19.3. **4.** Flexion involves movement that decreases the angle between bones; extension is the opposite of flexion; abduction moves a bone away from the body; adduction brings a limb toward the midline; rotation moves the part around an axis. **19.4: 1.** Three unique components of muscle cells are the T tubules, the myofibrils, and myoglobin. **2.** During muscle contraction, myosin filaments use energy from ATP to pull actin filaments toward the center of the sarcomere. **3.** Neurons control muscle contraction by releasing neurotransmitters at the neuromuscular junction. These chemicals bind to receptors on the sarcolemma, stimulating impulses that spread via the T tubules. **4.** Creatine phosphate breakdown is the fastest way to provide ATP for muscle contraction but it only supplies enough energy for a few seconds of intense activity. Cellular respiration provides most ATP used by a muscle cell by breaking down glycogen and fatty acids. Fermentation can supply ATP without requiring oxygen. **19.5: 1.** Three stages of a muscle twitch are the latent, contraction, and relaxation periods. **2.** The index finger requires fine motor control and would have a higher ratio of motor nerve axons per muscle fibers than a muscle in your back. **3.** Domesticated birds no longer migrate by flying long distances, so they have no need for the endurance associated with slow-twitch fibers. **19.6: 1.** Untreated bone fractures can heal, especially if the individual is able to prevent the bone from moving, or it occurs in a bone that does not move a lot (i.e., a rib). However, if two broken ends of a bone continue to move, usually an exaggerated bone callus forms, and the fracture may not heal well. **2.** Astronauts experience bone loss because the bones are not bearing weight. They would be remodeled over time by over activity of the osteoclasts. **3.** Rheumatoid arthritis is an autoimmune disease in which the immune system attacks and damages the joints; osteoarthritis is usually secondary to traumatic damage to a joint (either from an injury or over the lifetime of an older person).

Science in Your Life

Dead on the "Farm": 1. Concerns include whether bodies were being treated ethically, whether bodies were secure from predators, whether the site would attract the thrill seekers, and contamination of ground water. **2.** Osteoarthritis, rheumatoid arthritis, rickets, scurvy, tuberculosis **3.** Botanists might study the age and variety of plant growth over a burial site to determine how long the remains had been there. Chemists might study the composition of the remains searching for toxins. Geologists could assess the impact of local soil and rock conditions on the deterioration rate of the remains. Psychologists could interpret evidence to determine the motivation and habits of a criminal. **What Constitutes an Unfair Advantage in Sports? 1.** Yes, similarities exist but aspirin is legal and available to all. It is taken to reduce inflammation. Anabolic steroids are illegal and available to only a few. They enhance performance and provide an advantage over others. **2.** Opinions will vary and distinguishing one enhancer to performance from the other is difficult. **3.** Such a show would be fascinating but also repulsive due to the inevitable destructiveness of drugs and surgical modification on usual body structure and function. Under such unregulated conditions competitors would go to extremes to win.

Testing Yourself

1. d; **2.** c; **3. a.** frontal bone, **b.** maxilla, **c.** mandible, **d.** occipital bone, **e.** temporal bone, **f.** parietal bone; **4.** e; **5.** a; **6.** b; **7.** e; **8** d; **9** d; **10.** T tubule; **11.** sarcoplasmic reticulum; **12.** myofibril; **13.** Z line; **14.** sarcomere; **15.** Sarcolemma.

Thinking Critically

1. The elbow, neck, and phalanges are capable of flexion, extension, and rotation. The knee is capable of flexion and extension only. The hip is capable of flexion, extension, abduction, adduction, and rotation. **2.** Aerobic respiration is preferable to fermentation because it generates much more ATP per molecule of glucose. Oxygen serves as a final electron acceptor in the electron transport chain present in mitochondria. **3.** Animal tissues contain high levels of amino acids, and the consumption of excess amino acids causes the urine to become more acidic, which may cause increased loss of calcium in the urine. The level of daily exercise may be another important factor, if Chinese people are more active than Americans.

CHAPTER 20

Case Study

1. Insulin and glucagon. **2.** See Fig. 20.9. **3.** The brain requires a constant supply of glucose and without it unconsciousness can result.

Check Your Progress

20.1: 1. Hypothalamus, pituitary gland, thyroid gland, adrenal glands, thymus, pancreas. **2.** When an endocrine gland is controlled by negative feedback, it is sensitive to the condition it is regulating or the blood level of the hormone it is producing. **3.** Peptide hormones bind to receptors in the plasma membrane and do not enter the cell. A second messenger inside the cell is needed to initiate an enzyme cascade. **4.** It has been found that women prefer axillary odors from men with different MHC than themselves. **20.2: 1.** Neurons in the hypothalamus produce hormones that pass through axons into the posterior pituitary where they are stored. The hypothalamus controls the anterior pituitary by producing hormones that travel to the anterior pituitary through a portal system. **2.** See Table 20.1. **3.** The release of oxytocin from the posterior pituitary is controlled by positive feedback in which the amount of oxytocin released increases as the stimulus continues. **4.** Diagnosis is difficult because of the many hormones controlled by the pituitary and the many organs and tissues affected. **20.3: 1.** Levels of T_3 and T_4 are controlled by negative feedback. When levels of T_3 and T_4 rise, the anterior pituitary stops producing thyroid-stimulating hormone. **2.** Calcitonin decreases blood calcium by increasing calcium deposition in bones. A person with a calcitonin-producing thyroid tumor would be expected to have low blood calcium levels leading to disrupted nerve conduction, muscle contraction, and blood clotting, as well as increased bone density. **3.** Parathyroid hormone increases the activity of osteoclasts, promotes reabsorption of calcium by the kidneys, and activates vitamin D, so removal of the parathyroid glands would tend to cause decreased blood calcium and increased blood phosphate. **20.4: 1.** Secretion of hormones by the adrenal medulla is under direct nervous control; secretions of the adrenal cortex are controlled either by release of ACTH from the anterior pituitary (for glucocorticoids) or by the renin-angiotensin system (for mineralcorticoids). **2.** Epinephrine from the adrenal medulla helps the body deal with stress by activating the fight-or-flight response (increased heart rate, blood pressure, and energy level); glucocorticoids from the adrenal cortex raise blood glucose levels and inhibit the body's inflammatory response. **3.** Aldosterone causes the kidneys to absorb sodium and excrete potassium, which generally raises blood pressure. **20.5: 1.** The exocrine tissue of the pancreas produces and secretes juices that help in digestion in the small intestine. The endocrine cells of the pancreas produce the hormones glucagon, insulin, and somatostatin. **2.** See Fig. 20.9. **3.** Somatostatin inhibits the release of growth hormone and suppresses glucagon and insulin secretion by negative feedback mechanism. **20.6: 1.** Testosterone and estrogen stimulate the development of male and female sex characteristics and the production of gametes. **2.** Some of the worst side effects from taking anabolic steroids are liver dysfunction and cancer, kidney disease, heart damage, stunted growth, infertility, and impotency. **3.** Leptin signals satiety, a sense of not being hungry. **4.** Prostaglandins are not distributed in the blood, but instead act close to where they are produced within cells. **20.7: 1.** ADH causes more water to be reabsorbed in the kidneys, decreasing urine volume. Inhibition of ADH secretion by alcohol results in increased amounts of urine and frequency of urination. **2.** In iodine deficiency, the thyroid gland cannot synthesize enough T_3 and T_4. The anterior pituitary responds by releasing large amounts of TSH, which stimulates the thyroid gland to enlarge. **3.** In Addison disease, antibodies play a role in destruction of the adrenal cortex by the immune system. In Graves disease, antibodies bind to the TSH receptor, stimulating production of T_3 and T_4 rather than destroying the gland. Type 1 diabetes is due to an insulin shortage, whereas type 2 results from an insensitivity to insulin, often specifically due to insufficient numbers of insulin receptors on cells.

Science in Your Life

Melatonin: 1. Depression, whatever the cause, is often viewed as a weakness that people should just get over. Explaining SAD as a chemical imbalance can evoke more sympathy for those suffering from it. **2.** If melatonin levels are peeking within your body at a time that you need to be awake, it will be more difficult to function then. **3.** Responses to this question should address the potential harm that might result from an unregulated market of melatonin, which can be classified as a drug. **Human Growth Hormone: 1, 2,** and **3.** Here people are asked to present their opinions about HGH, supplements, and the responsibility of celebrities in our society. They should be able to defend their position with logical arguments on these issues.

Testing Yourself

1. b; **2. a.** pineal gland, **b.** hypothalamus, **c.** pituitary gland, **d.** thymus, **e.** pancreas, **f.** adrenal glands, **g.** thyroid gland; **3.** c; **4.** c; **5.** b; **6.** e; **7.** d; **8.** f; **9.** b; **10.** c; **11.** a. **12.** e; **13. a.** ANH, **b.** renin, **c.** aldosterone; **14.** d; **15.** d.

Thinking Critically

1. The nervous system allows for quick, almost immediate responses to stress or danger, and fine-tuning of all other bodily functions (plus all the other benefits of higher brain functions), while the endocrine system provides longer-lasting responses. **2.** If Cushing syndrome is caused by a problem (such as cancer) of the adrenal gland itself, ACTH levels would be very low, because of negative feedback action of cortisol on the hypothalamus/pituitary. **3.** Pheromones received as odors, may affect sexual attraction, aggression, love, or almost any emotion, and we would not necessarily be aware of these affects.

CHAPTER 21

Case Study

1. The uterus can accommodate a limited number of embryos. Space, ability to provide nutrients, and efficient gas exchange are limitations. **2.** It is generally agreed that bringing children into the world is an ethical choice dependent on the parents' ability to rear them and the world's ability to supply them with resources for a healthful life. The burden on society must be considered. **3.** AID, GIFT, ICSI, and surrogate mothers.

Check Your Progress

21.1: 1. The seminiferous tubules in the testes produce sperm. The epididymis stores the sperm, and the vasa deferentia carry the sperm to the female via the urethra. The urethra also functions in urination. **2.** Seminal vesicles, prostate gland, bulbourethral glands **3.** GnRH stimulates the anterior pituitary to secrete the two gonadotropic hormones, FSH and LSH. FSH promotes the production of sperm. ICSH controls the production of testosterone by interstitial cells. Testosterone controls the development and function of the testes and associated organs. It also brings about male secondary sex characteristics.

Check Your Progress

21.2: 1. The uterine tubes are severed and tied off. Movement of eggs down the uterine tubes is prevented. **2.** In males, the urethra and external genitalia (the penis) function in both reproduction and urination. In contrast, these two systems are separate in females. **3.** Contractions of the uterus may help to move sperm from the uterus to the fallopian tubes. **21.3: 1.** During the follicular phase, FSH from the anterior pituitary promotes the development of an ovarian follicle. Ovulation signals the end of the follicular phase. The major feature of the luteal phase is secretion of progesterone by the corpus luteum, which causes the uterine lining to thicken in order to be ready to receive a fertilized oocyte. **2.** Estrogen from an ovarian follicle stimulates the proliferative phase of the uterine cycle, during which the endometrium begins to thicken. Progesterone from the corpus luteum leads to the secretory phase of the uterine cycle, in which the endometrium thickens and uterine glands produce thick mucus. **3.** In pregnancy the placenta produces human chorionic gonadotropin, which maintains progesterone production by the corpus luteum until the placenta can produce progesterone and estrogen, which inhibit the ovarian cycle and maintain the endometrium. See Fig. 21.9 for an explanation of the cycle of a nonpregnant woman. During menopause no eggs develop and the ovarian and uterine cycles cease. **21.4:1.** Methods that prevent ovulation: oral contraception, contraceptive implants, contraceptive injections/patch, vaginal ring. Methods that prevent fertilization: diaphragm, cervical cap, male condom, female condom, coitus interruptus, spermicidal jellies, creams or foams, natural family planning, vaginal film. **2.** The following methods physically block sperm from entering the uterus: condoms, diaphragm, and cervical cap. **3.** Birth control methods that prevent the contact of male ejaculate with female tissues reduce the transmission of STDs. **4.** Emergency contraception includes methods of birth control that prevent pregnancy after unprotected sex. **21.5: 1.** STDs caused by bacteria: Chlamydia, gonorrhea, syphilis; STDs caused by viruses: genital herpes, HPV, hepatitis **2.** HIV can destroy $CD4^+$ T cells by directly infecting them, by inducing apoptosis in infected and uninfected cells, and by causing cytotoxic T cells to kill HIV-infected cells. **3.** Reverse transcriptase inhibitors block production of viral DNA from RNA; protease inhibitors prevent cleavage of long viral proteins; fusion inhibitors interfere with HIV entry into cells; integrase inhibitors prevent insertion of viral DNA into host DNA. **4.** Viruses are not affected by antibiotics. They can remain inside cells protected from antibodies produced by the immune system. **21.6: 1.** Viagra, Levitra, and Cialis increase the blood flow to the penis during sexual intercourse. **2.** Endometriosis, or inflammation of the uterine lining, can cause infertility by preventing implantation of a fertilized oocyte. **3.** Ovarian cancer is more frequently fatal than testicular cancer mainly because the ovaries are inside a woman's body, hidden from detection until the cancer has spread to other organs. **4.** Artificial insemination by donor requires the least medical intervention, since sperm is simply collected, concentrated, and placed into the vagina (or uterus) by a physician. In vitro fertilization mixes oocytes and sperm outside the body and the embryos are transferred to the uterus of the female. Gamete intrafallopian transfer requires more intervention, since the oocytes and sperm are brought together in vitro, then introduced into one of the woman's uterine tubes. Intracytoplasmic sperm injection requires the most intervention, as a single sperm must be injected into an oocyte.

Science in Your Life

Preventing Transmission of STDs: 1. You cannot be totally certain that your partner does not have a STD even if they have been tested multiple times. Many people are unaware that they are carriers of STDs. **2.** A negative HIV test results from no or low levels of antibodies to the virus being present in the blood. This may result during the earliest stages of infection. **3.** Abstinence or barrier methods of birth control are most effective at preventing the transmission of HIV. Least effective methods are those that allow the ejaculate to enter the female reproductive system. **Endocrine-Disrupting Contaminants: 1.** Responses to this question will be based on one's own experiences and perceptions of risk and benefits of endocrine-disrupting chemicals in food. **2.** Mechanisms of disruption of endocrine pathways include blocking binding sites at the plasma membrane, interference of the movement of the hormone through the plasma membrane, and interruption of transmission of the signal by a second messenger. **3.** Agriculture and manufacturing practices today are dependent on many chemicals, some of which are potential endocrine disruptors. Without the chemicals, the cost of production would rise and those costs would be passed on to the consumer. Whether you would be willing to pay the extra costs is dependent on the individual.

Testing Yourself

1. d; **2.** e; **3. a.** seminal vesicle, **b.** ejaculatory duct, **c.** prostate gland, **d.** bulbourethral gland, **e.** anus, **f.** vas deferens, **g.** epididymis, **h.** testis, **i.** scrotum, **j.** foreskin, **k.** glans penis, **l.** penis, **m.** urethra, **n.** vas deferens, **o.** urinary bladder; **4.** c; **5.** c; **6.** d; **7. a.** uterine tube, **b.** ovary, **c.** uterus, **d.** urethra, **e.** clitoris, **f.** anus, **g.** vagina, **h.** cervix; **8.** d. **9.** d; **10.** e; **11.** b; **12.** f; **13.** a. **14.** g; **15.** c; **16.** b.

Thinking Critically

1. Current exposure levels to chemicals that potentially could reduce sperm production in males could be measured and compared to previous levels. **2.** The main advantage of a monthly menstrual cycle is that the human female ovulates every month, which increases the number of opportunities for becoming pregnant. **3.** Although BPH may cause frequent or difficult urination, it usually does not block urination completely. Untreated prostate cancer, however, often spreads to other tissues, such as the bone marrow, where the cancer cells may interfere with normal functions.

CHAPTER 22

Case Study

1. Trophoblast. **2.** Weight gain; lower blood pressure leading to increased blood volume; increased RBC count; increased cardiac output; increased blood flow to the kidneys, placenta, skin, and breasts; increased symptoms of heartburn; constipation.

Check Your Progress

22.1: 1. See Fig. 22.1. **2.** The oocyte's membrane depolarizes and lifts off of the oocyte, preventing any further sperm from binding. Enzymes released by the cortical granules lead to the formation of the fertilization membrane which results in the lifting of the zona pellucida from the surface of the oocyte. This prevents other sperm from binding to the zona pellucida. **3.** Cellular: All three have cleavage resulting in the formation of a multicellular embryo and the formation of a blastula. In the lancelet, the cleavage produces a ball of cells of equal size. In the frog, the cells are not of equal size, and in the chicken, cleavage is restricted to a layer of cells over the yolk. Tissue: All develop three germ layers: ectoderm, mesoderm, and endoderm. In the lancelet, mesoderm forms by out pocketing of the archenteron, while in the frog and chicken, mesoderm forms by migration of cells between the ectoderm and endoderm. **4.** See Table 22.1. **22.2: 1.** Cellular differentiation results in cells becoming specialized in structure and function due to differential gene expression. Morphogenesis produces the shape and form of the body first through the migration of cell types to produce the germ layers. **2.** A morphogen is the product of a morphogen gene and is present usually as a gradient within a developing organism. The gradients determine the shape of the organism. **3.** The homeobox sequence in a homeotic gene codes for a homeodomain composed of 60 amino acids. The homeodomain protein binds to DNA and controls gene expression. **22.3. 1.** Umbilical blood vessels from the allantois; first blood cells from the yolk sac; fetal half of the placenta from the chorion **2.** Months 1 and 2: week 1—fertilization, cleavage, blastocyst formation; week 2—implantation, yolk sac and amnion form, gastrulation occurs; week 3—the nervous system appears, development of the heart begins; weeks 4–5—development of the chorion and allantois, umbilical cord forms, limb buds appear, head enlarges, developing eyes, ears, and nose; weeks 6–8—head assumes normal position relative to body, all organ systems are developed. Months 3 & 4: cartilage replaces bone, head growth slows, eyelashes, eyebrows, head hair, fingernails, and nipples appear. Months 5 through 7: lanugo covered by vernix caseosa appears, eyelids are open. **3.** See Fig 22.15. The unique fetal structures are: foramen ovale, ductus arteriosus, umbilical arteries, umbilical vein, ductus venosus. **4.** The placenta is a structure of the developing embryo that is attached to the uterine wall that allows the exchange of gases, nutrients, and wastes between the maternal and embryo circulatory systems. **22.4: 1.** Progesterone causes smooth muscle to relax and arteries to expand, leading to low blood pressure. The renin-angiotensin-aldosterone mechanism leads to sodium and water retention. Blood volume, number of RBCs, and cardiac output increase. Blood flow to the kidneys, placenta, skin, and breasts rises. **2.** Stage 2. **3.** Colostrum is a thin, yellow, milky fluid, rich in proteins including antibodies, that is produced by breasts during the first two days after birth, before milk production begins. **22.5: 1.** Changes include: physical, social, psychological, growth rates, puberty, development of language, sensory development, skill development, learning processes. **2.** With an involuted thymus, older people are less able to generate T-cell responses to new antigens and since most B-cell responses are dependent on T cells, antibody responses to vaccinations are less robust. **3.** Menopause is defined as the absence of menstruation for a year. It is the result of the unresponsiveness of the ovaries to gonadotropic hormones, resulting in the lack of estrogen and progesterone secretion. **4.** The preprogrammed theory of aging proposes that aging is genetically determined and is based on the fact that longevity runs in families. The damage accumulation theory of aging is based on the observation that over time, harmful DNA mutations and harmful metabolites increase within our bodies. Increased cross-linking of collagen fibers observed in older people is an example of this.

Science in Your Life

Cloning: 1. This question will be answered based on how the ethical questions of therapeutic cloning are perceived by the individual. **2.** The allocation of funding sources is based on a value system that reflects one's perception of the contribution to society that will be made. The answers presented will be based on the personal values. **3.** This is a question of ethics as it involves the balance of monetary power. Students will choose either side of the argument. **4.** The funding of scientific investigations is always a matter of ethics and ethical decisions are based on value systems. Reference should be made to the fact that funding sources are limited and if one project is funded another will not be. **Preventing and Testing for Birth Defects: 1.** Many birth defects are not immediately detectable and the identification of their existence may not be made. Also some birth defects, such as reduced mental capacity, are difficult to analyze and correlate to interuterine conditions. **2.** Environmental exposure to toxins, unbalanced diet, lack of exercise, stress, depression, long hours standing on your feet at work, etc. **3.** This is a matter of intense debate that involves the rights and freedoms of the mother versus those of the developing fetus. Prosecution of the mother may not be the most productive means of promoting healthy babies.

Testing Yourself

1. d; **2.** a; **3.** b; **4.** c; **5.** c; **6.** c; **7.** b; **8.** c; **9.** b; **10. a.** zygote, **b.** morula, **c.** blastula, **d.** early gastrula, **e.** late gastrula; **11. a.** neural tube, **b.** somite, **c.** notochord, **d.** gut, **e.** coelom; **12.** c; **13.** d; **14.** d; **15.** b.

Thinking Critically

1. The egg is what contributes the cytoplasm and the organelles to the developing embryo. Therefore only the mother's mitochondria are present in the embryo. **2.** The chorionic villi develop from the chorion, which is fetal tissue. Therefore, the chorionic villi have the same genetic makeup as the fetus. **3.** It is a good idea to be healthy as a mother so as to be able to have a healthy baby. Also, much of the important embryonic development begins before a woman even knows she is pregnant. Therefore, if a woman is planning to become pregnant, she should begin exercising the same caution she would if she knew she was pregnant.

CHAPTER 23

Case Study

1. The sperm or oocytes produced during meiosis carry just one copy of each gene which can then combine to create a variety of diploid zygotes, each unique in its genetic make-up and phenotype. **2.** See Figs. 23.9 and 23.11.

Check Your Progress

23.1: 1. The phenotype of an individual is its appearance, which is determined by the genes it carries for traits carried in its genome. The genes present in the genome comprise its genotype. **2.** A testcross is conducted to determine if an individual that is expressing a dominant allele is heterozygous or homozygous for that gene. **3.** The Law of Segregation states that each individual has two factors for each trait and these factors separate during gamete formation such that each gamete contains only one factor. At fertilization the zygote receives two factors for each trait. The Law of Independent Assortment states that during gamete formation, each pair of factors separates independently of other factors and that all possible combinations of factors can occur. **4.** A phenotypic ratio of 3:1 results from a one-trait cross and a ratio of 9:3:3:1 results from a two-trait cross. **23.2: 1.** See Figs. 23.9 and 23.11. **2.** Autosomal recessive disorders result from the inheritance of two recessive alleles. The child is affected and is homozygous recessive, but neither parent is because they are heterozygous. In autosomal dominant disorders, both parents are affected but the child can be unaffected. **23.3: 1.** With incomplete dominance, the heterozygote exhibits a phenotype that is intermediate between the homozygous recessive and the homozygous dominant. **2.** The mother has to be heterozygous for type A blood and the second allele for type O blood would have come from the father. He could have been heterozygous for type A, type B, or type O blood. **23.4: 1.** Environmental factors include: nutritional status, diet, temperature, exposure to sunlight. **2.** Identical twins have identical genomes and have developed from identical oocytes containing the same cytoplasmic composition. However, the expression of their genes is influenced by environmental inputs. Variation between identical twins is attributed to the inputs from the environment on their development.

Science in Your Life

Investigations of Gregor Mendel: 1. Large numbers are needed to determine if the event has happened by chance alone or as a result of the variable being studied. **2.** There are serious ethical considerations that must be addressed in human studies that are not pertinent to plant studies. **3.** Incomplete dominance, codominance, multiple allele inheritance, polygenic inheritance. **Genetics of Taste: 1.** Organisms in the environment that sense bitterness by taste will avoid bitter tasting food. Those bitter plants or animals will survive to propagate and their numbers and the frequency of the alleles of their genes will increase in the population. **2.** If plants taste bitter they will be more resistant to being eaten. More plants will survive to reproduce.

Testing Yourself

1. e; **2.** c; **3.** c; **4.** d; **5.** a; **6.** c; **7.** c; **8.** a; **9.** a; **10.** d; **11.** c; **12.** b.

Additional Genetics Problems

1. a. W, **b.** WS, Ws. **c.** T, t. **d.** Tg, tg. **e.** AB, Ab, aB, ab. **2. a.** gamete, **b.** genotype, **c.** gamete, **d.** genotype. **3.** 75%. **4.** Both are Ee. **5.** father DD, mother dd, children Dd. **6.** WwEe. **7.** 1/16 or 6.25%. **8.** father DdFf, mother ddff, child ddff.

Thinking Critically

1. Cross the fruit fly with the trait to a true breeding fruit fly without the trait and determine the phenotypic ratios. If the trait does not appear in the offspring or appears 25% of the time then it is recessive. If the trait appears 75% of the time then it is dominant. **2.** Track the appearance of that trait in identical twins raised in different environments. **3.** Even identical twins have differences in environmental influences. Beginning in the womb, each twin experiences their environment in different ways.

CHAPTER 24

Case Study

1. Nondisjunction during meiosis I or during meiosis II. **2.** Inheritance of three copies of chromosome 21. **3.** By amniocentesis or chorionic villus sampling.

Check Your Progress

24.1: 1. Linked alleles are all on one chromosome and they tend to be inherited all together, thus they do not exhibit independent assortment. **2.** Two genes that are 10 map units apart exhibit a 10% chance of crossing-over. **24.2: 1.** Sex linked alleles appear on the X-chromosome and more rarely on the Y-chromosome. Their inheritance is dependent on the distribution of the X and Y chromosomes in males and females. Recessive alleles on the X-chromosome are expressed in males, while they are not in heterozygous females. **2.** 50% of female offspring and 50% of male offspring would express the recessive allele. **3.** Color blindness: 8% of Caucasians perceive red and green colors differently. Duchenne muscular dystrophy: caused by the absence of the protein dystrophin which leads to calcium entering the cell, which promotes an enzyme that dissolves muscle fibers. Fragile X syndrome: an abnormally high number of repeat sequences of CGG in the X-chromosome. Hemophilia: either due to the absence or low level of clotting factor VIII or factor IX, resulting in a reduced ability of the blood to clot. **24.3: 1.** In normal XX females one of the X-chromosomes becomes inactive and forms a Barr body, which indicates that only one functional X-chromosome is needed in females just like in males. Extra X-chromosomes all become Barr bodies. In males extra X-chromosomes also become Barr bodies. Extra Y-chromosomes do not seem to interfere with normal development, perhaps because there are so few genes on the Y-chromosome. **2.** Turner syndrome and Kleinfelter syndrome can be caused by nondisjunction during oogenesis at meiosis I or at meiosis II. **24.4: 1.** When chromosomes break and do not rejoin, the deletion from one chromosome can be added to the corresponding homologous chromosome, creating a duplication. **2.** If in the process of translocation the allele of interest remains whole, then it can be that there is no loss of function. If after an inversion, the alignment of the two homologous chromosomes can be achieved by forming a loop, then it is possible that the individual can be phenotypically normal.

Science in Your Life

TREDS: 1. Polymerase slippage along unstable triplet expansions during meiosis is greater than in mitotic divisions. **2.** Mother, because the syndrome is X-linked. **3.** Huntington disease results in the alteration of the huntingtin protein by the inclusion of many glutamines. The disease is apparent by the degeneration of brain cells. With fragile X syndrome, there are a range of symptoms the severity of which depends on the number of CGG repeats present. **Viewing Chromosomes: 1.** The chromosomes are most highly condensed and compacted during metaphase. **2.** The shape and structure of the chromosomes can be assessed as well as their numbers.

Testing Yourself

1. b; **2.** c; **3.** c; **4.** a; **5.** a; **6.** b; **7.** d; **8.** c; **9.** a; **10.** e; **11.** d; **12.** c; **13.** a; **14.** b; **15.** e.

Additional Genetics Problems

1. Females all have the dominant trait, half of the males have the dominant trait, half of the males have the recessive trait. **2.** Mother, mother X^HX^h, father X^HY, son X^hY. **3.** 100%, 0%, 100%. **4.** The father must be color blind; unless there is a genetic abnormality in the daughter (i.e., she has only one X chromosome), the husband is not the father.

Thinking Critically

1. Parents may need to come to terms emotionally with having a handicapped child. In addition, they will need to prepare for the necessary added responsibilities of a child with a handicap. They may need to arrange the finances for additional caregivers in the home and extensive medical care. **2.** Chromosome 1 is so large and carries so many genes that three copies are lethal. **3.** As long as the translocation is balanced, meaning that the entire amount of DNA is still present, and no alleles are disrupted, the person can express all of the genes normally and be phenotypically normal. However, when the translocated chromosomes go through meiosis and crossing-over, gametes without the entire amount of DNA present will be produced. This would result in infertility.

CHAPTER 25

Case Study

1. The information in DNA is transcribed into RNA molecules, which through their various functions translate that information into proteins. **2.** An alteration in the DNA leads to a change in the information transcribed to RNA and translated to proteins. The function of those proteins is altered, allowing unregulated cell growth and reduced apoptosis.

Check Your Progress

25.1: 1. See Fig 25.1. Griffith concluded that some substance was passed from the dead S strain bacteria to the living R strain and this substance had the capacity to transform the R strain. This was indicated to be genetic material. Later Avery and others identified the genetic material as DNA. **2.** See Fig. 25.2. **3.** See Fig. 25.3. **25.2: 1.** A new double strand of DNA is composed of one old strand and one newly synthesized strand. **2.** See Fig. 25.4 and Fig. 25.5. **25.3: 1.** mRNA carries the genetic information encoded in DNA from the nucleus to the ribosomes in the cytoplasm. At the ribosome composed of rRNA, mRNA assembles amino acids carried by tRNA into chains, which are eventually released to form functional proteins. **2.** The information stored in DNA is transcribed into mRNA which then, with the help of tRNA at the ribosomes made of rRNA, translates the information into protein structure. **3.** The genetic code contains more than one triplet codon for each amino acid. **25.4: 1.**The control of gene expression in eukaryotes involves 5 levels: pretranscriptional, transcriptional, posttranscriptional, translational, posttranslational. Each level of control results in variation between one cell and another. **2.** In the absence of lactose a repressor is bound to the promoter/operator complex and no lactase is produced. Lactose will bind to the repressor and prevent it from binding to the operator. The gene for lactase can then be expressed. **3.** See Fig. 25.18. **25.5: 1.** Gene mutations can happen when DNA replication produces errors, when a mutagen causes physical damage to DNA, and when transposons jump into a new location disrupting the neighboring genes. **2.** A point mutation occurs when one DNA nucleotide is changed. A frameshift mutation occurs when a nucleotide is inserted or deleted. **3.** A mutation in a tumor suppressor gene means that its function as a brake to cell division will be altered. A mutation in a proto-oncogene that results in an active oncogene results in constantly stimulated cell division.

Science in Your Life

Finding the Structure of DNA: 1. 30% **2.** Knowing the structure of DNA has elucidated its function, allowing humans to regulate gene expression, analyze genetic material's presence, and recognize its relatedness from species to species. The advances have been enormous. **3.** DNA has two strands that are composed of specific pairs of nucleotides. Separate the strands and you have two templates for replication. **Prevention of Cancer: 1.** The responses to this question will reflect people's personal habits. **2.** Tobacco smoke contains harmful chemicals that can be transmitted throughout the body to cause cancer at multiple sites. **3.** Tanning beds increase exposure to UV radiation—a cause of skin cancer. **Genetic Testing for Cancer Genes: 1.** Genetic testing is already becoming increasingly available based on people's ability to pay for it. Whether it should be financed by insurance will be determined by the limits of health care plans to supply services. **2.** The use of this type of service would depend on one's commitment to its value and the ability to pay for it. **3.** Maintaining one's health is a personal responsibility because not maintaining it affects others negatively in our society, both emotionally and physically, as well as increasing the cost of health care delivery. It is likely that the role of personal freedom and personal responsibility will be addressed in this question.

Testing Yourself

1. b; **2.** a; **3.** c; **4.** c; **5.** b; **6.** a; **7.** b; **8.** b; **9.** e; **10.** b; **11.** d; **12.** c; **13.** c; **14.** e; **15.** a.

Thinking Critically

1. Housekeeping genes are those that all cells require such as those that encode for rRNA, genes encoding enzymes needed for energy metabolism, and genes for membrane transport proteins. **2.** Cancer cells express telomerase, allowing the cells that have damaged DNA to continue to divide. The cell loses the mechanism for cell death. **3.** In cases where a copy of a gene inherited from a parent is defective, the second gene inherited from the other parent must be altered before cancer develops. Multiple environmental exposures and habits can influence whether cancer develops or not.

CHAPTER 26

Case Study

1. Treat protoplasts with electric current while they are suspended in a solution containing foreign DNA. **2.** A GMO has had its genetic material altered by techniques used in recombinant DNA technology. It contains various DNA molecules that have been combined to create a new genome. Transgenic organisms are a subset of GMOs. They have a gene from another species inserted into their genome. **3.** Examples of animals that have been genetically modified include: 1—fishes, cows, pigs, rabbits, and sheep carrying bovine growth hormone which leads to larger animals; 2—goats which produce human growth hormone in their milk; 3—animals to produce drugs for cystic fibrosis in their milk.

Check Your Progress

26.1: 1. Using a restriction enzyme to cut open plasmid or other DNA followed by use of DNA ligase to seal the foreign gene into the plasmid. **2.** Each person has a unique number of short tandem repeats (STRs) at multiple loci. Separating these STRs by gel electrophoresis or by the technique of STR profiling allows the identification of each individual. **26.2: 1.** Improved agriculture, production of pharmaceuticals, production of hormones, clean-up of environmental pollution. **2.** Cloned organism has the identical genetic make-up. Transgenic organisms carry a gene that has been inserted from another organism. **26.3: 1.** In vivo and ex vivo methods using viral vectors are used to infect cells and introduce corrected genes. They also infect people with viruses that help fight off cancer. **2.** An example of in vivo gene therapy is use of an adenovirus vector as an inhalant to treat cystic fibrosis patients. An example of ex vivo gene therapy is removal of dysfunctional genes from an SCID patient, using a virus to "inject" the correct gene into these cells, and re-inoculating the patient with the corrected cells. **26.4: 1.** Genomics is the study of the complete genomic sequences of organisms. Proteomics studies the proteins that are produced from the genome within a cell. **2.** Comparative genomics can allow us to better understand gene function because unknown function in a focal organism that is close in genetic sequence to known genes in related organisms can often have a similar function. **3.** Bioinformatics, which involves computer technologies, special software, and statistics, is used to study the enormous amount of information obtained from genomics and proteomics.

Science in Your Life

DNA Forensics: 1. A valid fear for contributing your DNA information to a national databank is that at some point the release of that information will cause harm through some discriminatory process. **2.** The legal system is always striving to achieve their goals. The balance of personal freedoms versus society's need of a legal system should be discussed. Regulation of the DNA databank to insure proper sample collection, storage, processing, and interpretation must be insured. **3.** Access of defendants to DNA fingerprinting data for their own defense seems like a personal right. Funding the cost of this is a societal concern. **Are Genetically Engineered Foods Safe? 1.** Labeling is an effective way to be transparent about production of food. It is an easy way to communicate to the consumer what is in their food. **2.** Opinion one way or the other on this issue will depend on the perceived risk of GMOs versus their measured benefit to the world. **3.** Respondents would need to get more information confirming the safety of golden rice and its benefit to people eating it. Unfortunately this type of information may take many years to accumulate. **Testing for Genetic Disorders: 1.** DNA microarrays allow for the detection of what DNA is being expressed in a cell or organism. Previous techniques just determined that an aberrant sequence was present in the genome. **2.** Comparison between diseased tissue gene expression and normal tissue can be made. **3.** Cells from an individual with heart disease would have chromosomal variations that could be detected using genomic microarrays, when they were compared to cells from healthy individuals. Drugs could then be developed to target these chromosomal variations.

Testing Yourself

1. c; **2.** e; **3.** b; **4.** c; **5.** a; **6.** e; **7.** a; **8.** e; **9.** a; **10.** d; **11.** a; **12.** c; **13.** c.

Thinking Critically

1. Pros of viral gene therapy include treating or correcting potentially fatal or severely debilitating genetic disorders such as SCID. Gene therapy can also be used to treat difficult cancers. Cons include safety of using viruses and the fact that many patients have developed leukemia as a result of gene therapy. Many consequences of gene therapy are unknown. **2.** Genes for basic function, such as DNA replication, protein synthesis, and basic cellular processes such as respiration, should be similar.

CHAPTER 27

Case Study

1. Directional selection. **2.** Isolate the bacteria in question and determine its susceptibility to antibiotics before treatment; treat the bacteria with the most effective antibiotic; follow the patient to encourage completion of the full regime of antibiotics.

Check Your Progress

27.1: 1. The four elements of Darwin's theory of natural selection are: 1) Individuals within a species exhibit genetically inheritable variations. 2) Organisms compete for limited resources. 3) Individual organisms have different reproductive success. 4) Organisms adapt to their environment over generations through natural selection. **2.** Changes to an organism's phenotype acquired during its lifetime are not inheritable and therefore are not passed on to the next generation. **3.** Adaptations are evolved traits that result in an organism being better suited to its environment. **27.2: 1.** Fossils found in the strata of sedimentary rocks of different geological eras, periods, and epochs show evolutionary links between modern organisms and ancient organisms. The range and distribution of plants and animals is studied in the field of biogeography and this information helps to understand how different communities begin whenever landmasses or water environments are separated. **2.** Breeders use artificial selection to increase the frequency of desired traits in a breed of dogs. The breed is evolving over time due to this. **3.** Chimpanzees, because they are more closely related evolutionarily. **27.3: 1.** There must be no mutations, selection, genetic drift, or gene flow. The population must also have random mating. **2.** Allele frequencies will change if the Hardy-Weinberg conditions are not met. As an example, directional selection will decrease the frequency of the allele that is not favored. **27.4: 1.** Mutations create genetic variation in a population. Genetic drift is the change in allelic frequency due to the random meeting of gametes during fertilization. Small populations are particularly susceptible to this process. Gene flow is the movement of alleles between populations and results in increased genetic diversity. Nonrandom mating, in which individuals select a mate based on desired traits, results in increased allelic frequencies of those traits. Natural selection results in individuals that are better adapted to the environment having more offspring than those that are less well-adapted. **2.** With natural selection, there must be variation in phenotypic traits. This variation must be heritable. There must be over-reproduction and a struggle for survival due to limited resources. There must be differential reproductive success, where some individuals produce more offspring than others. **3.** The parasite that causes malaria cannot live within the red blood cells of the heterozygote because upon infection the cell sickles and loses K^+, resulting in the death of the parasite. Carrying the trait costs the individual. They are less fit and in areas of the world without the malaria parasite this would represent a disadvantage. **27.5: 1.** See Fig. 27.20. **2.** Punctuated equilibrium. **27.6: 1.** Taxonomy involves identifying, naming, and classifying organisms into taxa. Phylogenetics studies the evolutionary relatedness of organisms and classifies organisms into clades which reflect their evolutionary descent. **2.** A phylogenetic tree is an estimation and visual representation of the evolutionary history of biodiversity. A cladogram is composed of branches which are clades. Each clade contains the most recent ancestor and its descendants who share a derived trait. **3.** Domain Bacteria, domain Archaea, domain Eukarya.

Science in Your Life

Inbreeding in the Pingelapese: 1. The trait would have been removed from the population once Mwanenised died. If he had only five children the distribution of the trait would have been smaller and the impact on the population would have been less. **2.** The Pingelapese population is not in Hardy-Weinberg equilibrium because the measured frequency of the allele for achromatopsia is changing in the small and isolated population. **Evolution of Antibiotic Resistance: 1.** Responses to this question will be based on the person's view of individual rights versus the good of the community. The rights of the individual are being compromised for the greater good of all the individuals in the city. **2.** Tuberculosis is a world-wide disease that can move from country to country. Efforts that the USA takes to control tuberculosis internationally will ultimately benefit all humanity as well as the U.S. population.

Testing Yourself

1. c; **2.** c; **3.** b; **4.** a; **5.** a; **6.** d; **7.** c; **8.** f; **9.** b; **10.** e; **11.** b; **12.** b; **13.** b; **14.** b; **15.** c.

Thinking Critically

1. HIV is hard to treat because it evolves resistance to drugs used to treat it very quickly. HIV's mutation rates are so high that millions of new strains evolve daily, some of which are resistant to drugs that an infected patient may be being treated with. Eventually these strains will increase in frequency within a patient such that a drug is no longer effective. The somewhat good news is that this process often takes years, and with proper, and early treatment HIV-infected patients can live up to 15 to 20 years or more before the virus becomes fatal. **2.** The disease may be in higher allele frequency in South America because there is balancing selection. That is, heterozygote carriers of the disease may be resistant to some pathogen that is present in South America but not North America. **3.** 16%. **4.** 10%.

CHAPTER 28

Case Study

1. Binary fission. **2.** Bacteria are the most common prokaryote because they have adapted to every environment on this planet. **3.** Bacterial caused diseases include: food poisoning, dental caries, pneumonia, meningitis, middle ear infections, strep throat, necrotizing fasciitis, tuberculosis. Viral caused diseases include: common cold, flu, avian flu, measles, mumps, rubella, cold sores, genital warts, chickenpox, mononucleosis, AIDS. **28.1: 1.** Pasteur demonstrated that sterile broth enclosed in a flask that was not open to the air did not spontaneously generate microbes. **2.** Microbes include bacteria, archaea, protists, fungi, and viruses. **3.** Decomposers break down and recycle organic and inorganic material. Without decomposers, dead organisms and organic material would accumulate, and eventually the supply of many nutrients would run out. **28.2: 1.** The generation of biomolecules from inorganic compounds was the first step in biological evolution. **2.** Biomolecules, such as amino acids and nucleotides, had to be generated before polymers of these (DNA, RNA, proteins) could be constructed. Once polymers appeared, protobionts could evolve, leading eventually to the appearance of living cells. **3.** The "primordial soup" hypothesis explains the evolution of organic molecules from inorganic molecules by simulating the atmospheric environment containing CH_4, NH_3, H_2, and H_2O exposed to heat and electrical sparks. The "iron-sulfur world" hypothesis explains the appearance of organic molecules at thermal vents at the bottom of the oceans. Ammonia, CO, and hydrogen sulfide pass over iron and nickel sulfide minerals which act as catalysts to form organic molecules. **3.** Which comes first, the proteins or the genetic material? In the protein-first hypothesis, polypeptides formed and some of them had enzymatic activity which eventually lead to DNA formation. In the RNA-first hypothesis, RNA appeared first and could act as a substrate and an enzyme. DNA and proteins appeared later. Another hypothesis is that polypeptides and RNA evolved simultaneously, thus eliminating the paradox. **28.3: 1.** Archaea and Bacteria lack a nucleus and membrane-bound organelles in the cytoplasm. Archaea shares features with Eukarya listed in Table 28.1. **2.** The plasma membranes of Archaea have a monolayer of lipids with branched side chains, compared to the lipid bilayer of bacterial and eukaryote membranes. This contributes to the ability of many archaea to grow at very high temperatures. **3.** In addition to the membrane structural variations described above, archaeal enzymes must be able to function at high temperatures and/or salt concentrations that would inactivate the enzymes of most other organisms. **4.** Archaea living in the digestive tracts of cattle produce methane that contributes to global warming. Giving up beef in your diet would help to reduce global warming. **28.4: 1.** The three basic bacterial shapes are bacillus (rod), coccus (spherical), and spirillum (spiral-shaped). **2.** In bacterial conjugation, actual physical contact (via a sex pilus) is required for two bacteria to share DNA. Transformation is the uptake of DNA from the environment by a bacterial cell, and in transduction DNA is transferred from one bacterium to another by viruses. **3.** Heterotrophic bacteria use organic compounds from their environment for energy. Chemoautotrophs reduce CO_2 to an organic compound using electrons derived from ammonia, hydrogen gas, and hydrogen sulfide, and sometimes minerals such as iron. **4.** Antibiotics work by interfering with normal bacterial metabolism such as protein synthesis and cell wall synthesis. Bacteria can become resistant to antibiotics if a mutation occurs that alters the target of the drug (often an enzyme). Alternatively, some bacteria may acquire the ability to alter the antibiotic itself, so the drug is destroyed or inactivated. **28.5: 1.** A typical enveloped virus has a protein capsid that contains its nucleic acid (DNA or RNA) and is surrounded by a lipid envelope in which protein spikes are embedded. **2.** Attachment: viral spikes bind to receptor molecules on host cell. Entry: viral envelope fuses with plasma membrane and capsid and viral genome enters cell. Replication: viral enzyme makes copies of its genome. Biosynthesis: more capsid and spike proteins are made on host ribosomes. Assembly: capsid forms around viral genome. Budding: new viruses containing some host plasma membrane and spikes are released from cell. **3.** Vaccines are available for measles, mumps, and rubella. Vaccines are not available for HIV, hepatitis C, and mononucleosis. **4.** Viruses, viroids, and prions are all acellular pathogens. Viroids' genomes are a single strand of circular RNA that is 1/10 the size of a viral genome. It codes for no proteins. Viroids cause disease in plants and not animals. Prions are proteinaceous molecules that are derived from normal proteins that have changed shape. They lead to neurological wasting diseases in animals, including humans.

Science in Your Life

Antibiotics and Probiotics: 1. Without antibiotics, a tool to fight infections of pathogenic bacteria would be removed. Each of us would be more vulnerable to bacterial infections. **2.** Microflora can be influenced by the host's exposure to chemicals, diet, health, exercise, and the functioning of the immune system. **3.** Just as our immune system can discern between self and nonself cells, it has the potential to recognize one type of bacteria from another through cell surface receptors. **Bioterrorism: 1.** The use of biological weapons is held in disfavor by many. Just like nuclear warfare, it has the potential to affect entire populations, not just military personnel. People have few defense mechanisms to protect themselves and their families from biological weapons. **2.** One could argue that studying smallpox virus would lead to advances in managing such destructive viruses but since stockpiles are kept as a potential bioterrorist tool, this argument seems weak. Stockpiles should be destroyed. **3.** Yes, background checks and monitoring of those working with infectious agents that could be used for bioterrorism should be required. Self-regulation for such a sensitive issue is unrealistic. Such regulations would discourage some investigators.

Testing Yourself

1. a; **2.** a; **3.** c; **4.** c; **5.** e; **6.** b; **7.** b; **8.** d; **9.** b; **10.** 1. attachment, 2. entry, 3. replication, 4. biosynthesis, 5. assembly, 6. budding; **11.** b; **12.** b; **13** c; **14.** a; **15.** c; **16.** b.

Thinking Critically

1. Three reasons why bacteria are useful as a model genetic organism include: fast generation time, size, and ease of generating genetic mutations. **2.** Viruses that use RNA as their genetic material would be more likely to contain their own enzyme than those that use DNA, because DNA viruses can use the enzymes that their host cells normally use to copy their cellular DNA. Since host cells do not have enzymes to make copies of RNA, RNA viruses must produce their own RNA-replicating enzymes.

CHAPTER 29

Case Study

1. No. **2.** Protists which are eukaryotes and for the most part unicellular and microscopic, are likely related to the first eukaryotic cell to have evolved. They represent a bridge between the first eukaryotic cells and multicellular organisms. **3.** Protists and fungi both benefit human health and welfare and severely negatively impact it. Some provide useful products while others are parasitic or disease causing.

Check Your Progress

29.1: 1. The endosymbiotic theory proposes that mitochondria were the result of a nucleated cell engulfing an aerobic bacterial cell. Under the theory chloroplasts originated when a nucleated cell with a mitochondria engulfed a cyanobacteria. **2.** Protists carry out sexual reproduction when experiencing unfavorable conditions. This often results in cysts or spores which can survive in the environment until conditions improve, thus promoting dispersion to environments better suited for growth. **3.** Algae are typically photosynthetic, converting CO_2 into carbohydrates which are used as a food source. Protozoa are heterotrophs. Water molds and cellular slime molds are saprotrophs that feed on dead plant material. Cellular slime molds can also eat bacteria. **4.** Plasmodium causes malaria. Trypanosoma causes African sleeping sickness. Members of the genus Entamoeba cause amoebic dysentery. **29.2: 1.** Fungi secrete enzymes and digest available food externally. The broken down food is then absorbed. Algae are primarily photosynthetic and require carbon dioxide, water, and a source of energy. Amoeboids surround algae, bacteria, or other protists which they use as food with their pseudopods, and digest the food in a food vacuole. **2.** Hyphae—individual filaments of a fungus. Mycelium—mass of hyphae composing the body of a fungus. Chitin—a component of the fungal cell wall. **3.** See Fig. 29.20 for zygospore fungi reproduction; Fig. 29.21 for sac fungi; Fig. 29.22 for yeast; Fig. 29.23 for club fungi. **4.** Superficial mycoses include athlete's foot and ring worm. Systemic mycoses include histoplasmosis and coccidioidomycosis.

Science in Your Life

African Sleeping Sickness: 1. Poor nations have fewer resources that can be directed to controlling tsetse fly populations and providing clean running water. Thus their populations experience increased exposure to fly bites. Also, the access to health care in poor nations is limited so treatment of the disease is not assured. **2.** African sleeping sickness is most prevalent in poor nations where tsetse flies thrive and livestock live in close connection with humans. The number of flies is large and the routes of transmission between unaffected domestic animals and humans are well established. **Deadly Fungi: 1.** Toxins help fungi avoid predation by animals, including humans who eat them. They serve as a defense mechanism. **2.** Understanding the mechanism of action of fungal toxins could lead to modeling new drugs with similar but modified effects.

Testing Yourself

1. c; **2.** b; **3.** b; **4.** c; **5.** b; **6.** c; **7.** a; **8.** a; **9.** e; **10.** a; **11.** c; **12.** d; **13.** e; **14.** a; **15.** d.

Thinking Critically

1. The *Plasmodium* parasites complete part of their life cycle inside red blood cells. Two basic mechanisms have been proposed to explain malaria resistance in people who carry the sickle cell gene: 1) the abnormal hemoglobin inhibits the parasite from completing some part of its life cycle, or 2) infected RBCs from sickle cell carriers are more susceptible to being destroyed by the host immune system than infected RBCs from normal individuals. If an effective malaria vaccine becomes widely implemented, the natural resistance to malaria provided by the sickle cell gene would become a less significant factor. Over time, the sickle cell gene would become less prevalent in the population as the advantage of having only one copy of the gene diminished. **2.** The outer layer of skin is composed of dead epithelial cells which are constantly being shed. Human skin is often exposed to soap and water. In order to attach and grow on the skin, the fungi need to be able to penetrate to the lower layers of the dermis and be able to withstand the usual methods of washing. The lungs are more hospitable to fungi, due to the lining being moist and protected.

CHAPTER 30

Case Study

1. Land plants need to support themselves against gravity, move nutrients and water throughout their structures, and devise strategies for reproduction in dry environments. **2.** Mosses developed structures for protecting embryos from desiccation. Lycophytes had vascular tissue. Ferns had large leaves called megaphylls. Gymnosperms developed seeds. Angiosperms developed flowers and fruits. **3.** The development of flowers and their coevolution with insect pollinators afforded angiosperms high reproductive success which promoted their adaptive radiation to most parts of the world.

Check Your Progress

30.1: 1. Land plants benefit from greater exposure to light and greatly expanded habitats. **2.** The sporophyte produces haploid spores by meiosis. Spores are reproductive cells that can produce gametophytes by mitosis. The gametophytes then produce gametes by mitosis that can fuse to form the diploid zygote which develops into the sporophyte by mitosis. **30.2: 1.** Nonvascular plants are limited by not having vascular tissue to transport water and nutrients. They are restricted to moist environments to grow because the gametophye requires a film of water for the flagellated sperm to swim in to reach the egg. **2.** Nonvascular plants have developed rootlike, stemlike, and leaflike structures which help them to live on land. Because they are small and simple they can inhabit microhabitats that other plants cannot. **3.** A liverwort has a flat lobed thallus, whereas a moss has leafy green shoots. **30.3: 1.** Lycophytes were the first plants to have vascular tissue and true roots, stems, and leaves. **2.** The sporophyte is dominant in ferns (gametophyte is dominant in mosses); the gametophyte is separate from the sporophyte in ferns, whereas the two life stages are attached in mosses. **3.** Ferns have large leaves that can absorb more sunshine. They are very diverse in structure and can live in a variety of habitats including dry areas. **30.4: 1.** Microspores develop in pollen sacs and they become pollen grains (male gametophyes). Megaspores develop in the ovule and one out of four becomes an embryo sac (female gametophyte). **2.** Double fertilization in flowering plants results in presence of endosperm in addition to a zygote. Presence of an ovary leads to production of seeds enclosed by a fruit. Animals are often used as pollinators. **3.** Monocots have one cotyledon (seed leaves that nourish the embryo) and eudicots have two.

Science in Your Life

Plants and Humans: 1. Plants contain chemicals that they use for defense against viruses and organisms that invade and harm them. Isolating and understanding the mechanism of action of these chemicals could lead to the development of potentially beneficial drugs for humans. **2.** Wheat, corn, and rice have been selectively bred for many years to increase ease of cultivation and crop yields. The standardization of plant types that are successfully grown commercially has led to limitations on the types of plants used for food. **3.** Other uses include fuel, soil stabilization and enrichment, control of wetlands, windbreaks, and shade providers.

Testing Yourself

1. b; **2.** b; **3. a.** sporophyte, **b.** meiosis, **c.** gametophyte, **d.** fertilization; **4.** c; **5.** b, c, d; **6.** c, d; **7.** a; **8.** b; **9.** c; **10.** d; **11.** a; **12.** b; **13.** c; **14. a.** stamen, **b.** carpel, **c.** receptacle, **d.** petals, **e.** sepals; **15.** c.

Thinking Critically

1. Angiosperms have protected seeds that can lay dormant before germination. They also have double fertilization, which provides endosperm as nourishment for the embryo. Angiosperms have flowers, which attract animal pollinators and allow for dispersal of pollen and fertilization of other plants. Finally, angiosperms have fruits, which are also attractive to animals and can allow long-distance dispersal of seeds to new habitats. **2.** Humans do not reach reproductive maturity until puberty. All cells, except eggs and sperm, are diploid. Haploid eggs or sperm are produced by meiosis in ovaries or testes. Fertilization occurs when a sperm cell meets an egg cell, forming a diploid zygote which grows into a fetus. In contrast, plants have an alternation between a diploid sporophyte generation and a haploid gametophyte generation. The sporophyte produces haploid, reproductive spores by meiosis that can develop into a gametophyte by mitosis. The gametophyte produces gametes by mitosis. Sperm and egg fuse, forming a diploid zygote that undergoes mitosis and becomes the sporophyte.

CHAPTER 31

Case Study

1. Invertebrates evolved from a choanoflagellate ancestor and outnumber the vertebrates by a large margin. See Fig. 31.2 for the phylogenetic tree of the major animal phyla and notice the placement of the Chordates which include the vertebrates. **2.** With a variety of developmental stages, body plans, and lifestyles, invertebrates have been able to radiate to occupy innumerable habitats on Earth.

Check Your Progress

31.1: 1. Unlike plants, animals are heterotrophs and must digest their food internally. Fungi secrete enzymes to break down food externally and then absorb the nutrients. **2.** Five characteristics animals have in common are: 1. movement or locomotion via muscle fibers; 2. multicellularity; 3. typically diploid adults; 4. usually undergo sexual reproduction and experience developmental stages; 5. usually heterotrophic. **2.** Ecdysozoa are multicellular, have three tissue layers, are bilaterally symmetrical, have a body cavity, exhibit protostome development, and molt their cuticle. **31.2: 1.** Sponges are multicellular sessile filter feeders. They have an outer layer of flattened epidermal cells, a middle semifluid matrix with amoeboid cells, and an inner layer composed of collar cells. They are capable of reassembling from single cells into a complete organism. **2.** Cnidarians have true tissues and have an ectoderm and endoderm as embryonic germ layers. They exhibit radial symmetry as adults and demonstrate two body forms, a polyp and a medusa. **3.** See Fig. 31.7. **31.3: 1.** Flatworms, molluscs, and annelids all are protostomes that are bilaterally symmetrical, have three tissue layers, and are multicellular. **2.** Flatworms have no body cavity, an incomplete digestive tract, and no circulatory system. Molluscs have a true coelom, a complete digestive tract, and an open circulatory system. Annelids have a true coelom, a complete digestive tract, and a closed circulatory system. **3.** Molluscs have a visceral mass, a strong muscular foot, and a mantle. **31.4: 1.** Both roundworms and arthropods periodically shed their outer covering and they both have a complete digestive tract. **2.** Rigid, but jointed exoskeleton, segmentation, well developed nervous system, wide variety of respiratory organs, and reduced competition through metamorphosis. **3.** Crustaceans have hard, calcified exoskeletons and typically have heads with compound eyes and five pairs of appendages. **31.5: 1.** Echinoderms and chordates are closely related because they are both deuterostomes. They share similar embryological development and the coelom forms from an outpocketing of the primitive gut. **2.** Echinoderms have endoskeletons composed of calcium, containing plates which have spines that protrude through their skin. They are most often radially symmetrical and have free-swimming larvae.

Science in Your Life

Destruction of the Coral Reefs: 1. Coral reefs are found in shallow, warm, clear water through which light can pass. They provide a variety of types of shelter for many different organisms. **2.** Deforestation, increased sedimentation, global warming, emergence of pathogens, and increased levels of nutrients are all reversible causes of coral reef decline if there is sufficient political capital to make the needed changes. **Maggots: A Surprising Tool for Crime Scene Investigation: 1.** Forensic entomologists can correlate the growth rate of larvae they observe with the temperature records for the area where the body was found and can pinpoint the time of death accurately. **2.** Responses will vary but entomology evidence has been shown to be quite accurate. **3.** When a body has been exposed to insects and has been dead for a day or more.

Testing Yourself

1. a; **2.** b; **3.** c; **4.** b; **5.** a; **6.** c; **7.** c; **8.** b, c; **9.** a; **10.** a, b, c; **11.** d; **12.** a; **13. a.** pharynx, **b.** mouth, **c.** esophagus,

d. hearts, **e.** coelom, **f.** crop, **g.** seminal vesicle, **h.** dorsal blood vessel, **i.** nephridium, **j.** ventral nerve cord, **k.** anus, **l.** clitellum; **14.** a; **15.** a.

Thinking Critically

1. Cniderians are aquatic and usually have a sessile polyp and motile medusa life stage. Cniderians can reproduce both asexually and sexually. Bilaterally symmetrical species are typically motile and most often reproduce sexually. **2.** Arthropods have a rigid, but jointed exoskeleton, segmentation, well-developed nervous system, wide variety of respiratory organs and reduced competition through metamorphosis. Many species also have flight, which gives them many advantages. **3.** Relative to acoelomates or pseudocoelomates, coelomates have: freer body movements, space for development of complex organs, greater surface area for absorption of nutrients, and protection of internal organs from damage.

CHAPTER 32

Case Study

1. Vertebrates have all four chordate characteristics but the notochord is replaced by a column of vertebrae. Usually there are two pairs of appendages along with a jointed endoskeleton. A skull and a high degree of cephalization as well as jaws, eyes, and ears are present. **2.** Amphibians, compared to lobe-finned fishes, have four limbs with joints, eyelids, ears, and larynx. The larval stage lives in water but the adult can live on land.

Check Your Progress

32.1: 1. Chordates have a notochord, nerve cord, pharyngeal pouches, and a postanal tail. **2.** Lancelots are shaped like a two-edged knife while tunicates look like a sac with thick walls. **3.** Mammals have a bony skeleton, lungs, four limbs, an amniotic egg, and mammary glands. **32.2: 1.** Fish are adapted to living in water, undergo external fertilization, and have a zygote that is a swimming larva. **2.** Amphibians have jointed appendages, four limbs, eyelids, ears, and a larynx. **3.** The amphibian larval stage requires a water environment for survival. Metamorphosis of the larvae to the tetrapod adult that can breathe through lungs and its skin allows survival on land. **32.2: 1.** Reptiles compared to mammals are covered by keratinized scales, do not have a diaphragm, and have an interventricular septum that varies as to its completeness. **2.** Placental mammals are unique because the extraembryonic membranes have been modified for internal development within the uterus. **3.** Primates are adapted to an arboreal life, have mobile limbs, and the hands and feet have five digits. **32.4: 1.** Bipedalism may have evolved because a dramatic change in climate caused forests to be replaced by grassland about 4 mya. It is thought that hominids began to stand upright to keep the sun off their backs and to have better vision for detecting predators or prey in grasslands. However, recent fossil findings suggest bipedalism might have evolved as early as 7 mya, while hominids still lived in forests. The first hominids may have walked upright on large branches as they collected fruit from overhead. The upright stance later made it easier for traveling on the ground and foraging among bushes. **2.** Bipedalism. **3.** See Fig. 32.16. **32.5: 1.** The replacement model or out-of-Africa hypothesis, proposes that modern humans evolved from archaic humans only in Africa, and then modern humans migrated to Asia and Europe, where they replaced the archaic species about 100,000 years bp. In contrast, the multiregional continuity hypothesis proposes that modern humans arose from archaic humans in Africa, Asia, and Europe at roughly the same time. **2.** Neandertals lived between 200,000 and 28,000 years ago and were likely supplanted by modern humans (Cro-Magnons at the time). Neandertal brains were slightly larger (1,400 cc). They had massive brow ridges, a forward-sloping forehead, a receding lower jaw, and wide, flat noses. The bones of Neandertals were shorter and thicker than those of Cro-Magnons. Cro-Magnons had had lighter bones, flat high foreheads, domed skulls housing brains of 1,350 cc, small teeth, and a distinct chin.

Science in Your Life

Vertebrates and Human Medicine: 1. How this question is answered will depend on people's opinions of the status of humans within the whole of the Earth's organisms. Whether animals are here to serve humans should be addressed. **2.** Receiving an organ from another species demands that drugs to prevent organ rejection be taken forever. Other species may harbor viruses to which humans are susceptible. Organs from other species can vary as to size and function from humans and challenge compatibility within the human body. **Biocultural Evolution Began with Homo: 1.** Cultural skills such as using tools, food gathering practices, and language changed the relationship of humans to their environment, allowing them to survive and reproduce under more varied conditions. **2.** A human who can gather food by hunting and gathering will be better fed than a human who cannot. This is an advantage that will support their reproductive success.

Testing Yourself

1. a, b; **2.** b; **3.** a; **4.** b; **5.** a; **6.** c, d; **7.** b, c, d; **8.** a, b, c, d; **9.** a, b, c; **10. a.** swim bladder, **b.** stomach, **c.** muscle, **d.** vertebra(e), **e.** brain, **f.** nostril, **g.** mandible, **h.** gills, **i.** heart, **j.** liver, **k.** gallbladder, **l.** intestine, **m.** gonad, **n.** kidney, **o.** scales, **p.** lateral line; **11.** e; **12.** a; **13.** c; **14.** d; **15.** c.

Thinking Critically

1. Because we have a distant common ancestry with all vertebrates, we share many genes in common with them. Continued study of processes such as regeneration or immune response in amphibians could give us clues as to which human genes may be similar and potentially "turned on" or stimulated in such a way to benefit people by curing diseases or regenerating limbs. In general, genomic studies benefit humans because we are better understanding genes associated with developmental processes and disease resistance. **2.** Bipedalism must have provided more of a fitness advantage than the cost of a smaller pelvis for birthing. Upright stance may have given hominids advantages when forests shifted to grasslands about 4 mya. These advantages included: keeping the sun off their backs and better vision to detect predators and prey. These advantages, which included better avoiding starvation and/or being eaten, the cost of a more difficult birthing process than other primates. **3.** Cultural evolution encompasses human behavior and products, including technology. Language is also critical for cultural evolution because products and ideas are passed on from generation to generation not by genetic inheritance, but by verbal transmission. We learn about history from others, which plants are good for us and which ones are poisonous, and most recently, about medicines that can save our lives. In the past 100 years, the human life span has increased by more than 50% due to technology and culture.

CHAPTER 33

Case Study

1. Genetic based lion behaviors include living in a pride of females, female cubs staying with the pride, male cubs forming coalitions with related males, and local males killing cubs they are not related to. Environmentally based behaviors include hunting patterns that reflect the available resources and defined territories determined by the habitat. **2.** Chemical and visual.

Check Your Progress

33.1: 1. Identical twins that are separated at birth and exposed to different environments growing up will show similarities, simply because they are genetically identical. Similar food preferences or other behaviors suggest that there is a genetic basis, since their environments were mostly different. **2.** Endocrine systems can coordinate body systems just like the nervous system. By coordination with other gene products, hormones can control behaviors. **33.2: 1.** Examples of learning include laughing gull chicks learning begging behaviors, ducklings imprinting on their mother, and white-crowned sparrows learning a species specific song. **2.** Classical conditioning presents a subject with two different stimuli and leads them to make an association between them (e.g. dogs and food and the sound of a bell). Operant conditioning involves strengthening a stimulus-response connection through such methods as rewards. **33.3: 1.** Communication could warn a receiver, show receptivity to mating, mark a territory, tell a receiver where food is, and help in navigation. **2.** Auditory communication would be most effective in a forest. Visual would be least effective. **33.4: 1.** Animals may defend a portion of their home range that has particularly good food resources. **2.** Females produce a limited number of eggs in their lifetime, whereas males produce essentially limitless amounts of sperm. Females are also usually the sex that provides more parental care than males, so they are choosy to ensure the highest quality males in terms of fitness or parental care that the males may provide for the offspring. **3.** Advantages: avoiding predators, finding food, cooperative offspring rearing. Disadvantages: higher competition for mates and food, spread of diseases and parasites.

Science in Your Life

Do Animals Have Emotions? 1. It appears that animals can be happy, can enjoy themselves, and even get sad at times. Those who respond should be able to support their position about animal emotions by specific examples and experimental evidence that is convincing. **2.** The answer to this question will be based on a person's perception of cost and return relative to one's available resources and value system. **3.** Whether animals having emotions determines their status as research subjects could be argued either way, especially if they differ in the degree of their emotional response. **Mate Choice and Smelly T-shirts: 1.** Definitely not. Humans have societal roles that are more defined than those operating in a lion pride. We as a group work to insure the survival of offspring. **2.** It is necessary to communicate that health is not defined by absolute weight but instead by one's ability to live well and reproduce. **3.** Research into the behaviors of humans in their choice of mates could have a positive impact on our society's ability to understand and manage human societies. If money is available then research in mate choice is warranted.

Testing Yourself

1. a; **2.** e; **3.** c; **4.** d; **5.** b; **6.** b; **7.** b; **8.** b; **9.** c; **10.** c; **11.** d; **12.** a; **13.** c; **14.** b; **15.** a.

Thinking Critically

1. One experiment would be to catch rats from New York and Florida, breed them in the lab, and see if hand-raised newborns (without their parents) have the same preferences as the adults in natural populations. If offspring have no preference, an environmental or learned component would be suggested. This could further be tested by rearing one offspring group (composed of both New York and Florida rats) on one type of cheese and another offspring group on another type of cheese. A cheese preference test could be conducted later in life, and if rats that showed no preference at birth prefer the cheese they were raised on later in life, an environmental ("nurture") component is suggested. **2.** For males, it is fairly straightforward, since sperm are energetically inexpensive and EPC behavior would tend to increase male fitness in absence of parental care (since the paired male would care for the offspring), then EPC behavior would be favored by natural selection for males of many bird species. For females, it is a bit more complicated. Females may choose to mate with neighboring males that have "good genes"—that is, perhaps they appear more healthy (e.g., lower parasite load) than the male with which they formed a pair bond. Or, perhaps they mate with neighboring males to have "sexier sons." Neighboring males may have sexually-selected characteristics, such as bright plumage, that may make females have sexier sons (who would tend to mate with more females than males with duller plumage). By choosing "sexier" mates, females would have sexier sons, and those females would tend to have more grandchildren, thereby increasing their inclusive fitness. **3.** It is not exactly clear why altruistic behavior exists in humans. In the context of culture, someone might save someone else's life because it makes them feel good to perform such an act. Perhaps people do it so others would be indebted to them or for monetary gain. Perhaps some people perform altruistic behavior so that their societal status would increase (e.g., to be a "hero")—this may afford them more opportunities (e.g., jobs) that may allow them to improve their ability to obtain resources. Nonetheless, altruism in humans is a complicated issue, but since we are a social species, there may be some benefit in the context of living in a society.

CHAPTER 34

Case Study

1. The distribution of fish within a lake will vary according to the availability of suitable habitats and food for the species in the community. That may be only 10% of the lake. **2.** To some extent population growth models apply to humans. However, humans have been able to increase the carrying capacity of the Earth for their species through technology, science, and use of the Earth's resources. They also use methods of birth control. **3.** More developed countries use more resources per capita and therefore have a greater impact.

Check Your Progress

34.1: 1. A population is all of the organisms of the same species within a specific area. A community is all of the populations present in one location. **2.** Ecology seeks to better understand how organisms interact with other organisms within their physical environment. **34.2: 1.** To reach their biotic potential a population would need to experience no environmental resistance. It would have no limitations on its reproductive potential, have unlimited food, be disease free, and have no predators. This does not happen. **2.** Type I individuals have a normal life span and then die at an increasing rate after that. Type II individuals experience a constant rate of survival throughout their lives. Type III individuals have a decreased rate of survival early in their life and the rate of survival increases during their life. **3.** LDCs tend to have a larger percentage of their population in the pre-reproductive group than MDCs. This produces an unstable age structure because that group will reach reproductive age and have babies, increasing the population. **34.3: 1.** Mutualism is exemplified by insect pollinators that receive nectar from flowers and spread pollen to other flowers. **2.** Opportunistic growth patterns are exhibited by organisms such as insects and weeds that produce many small offspring with little parental input. **3.** By partitioning resources such as food into different niches, competition between species is reduced and survival is increased. **4.** Species within a community can experience many types of interactions, including predator-prey, parasitism, commensalism, and mutualism. **34.4: 1.** The directional changes in a community's composition that is called succession occurs when the community is newly established and there is no soil, or when a major disturbance has disrupted the structure of the community. **2.** Climax-pattern model—areas will always lead to the climax community. Facilitation model—each successive community helps the development of the next community. Inhibition model—each species in a community resists being replaced by a successive species. Tolerance model—chance determines which species colonize an area. The order of succession may reflect the time it takes the species to mature.

Science in Your Life

Ecological Niche of Top-Level Predators: 1. If possible, relocation of the wolf to another area would be a better solution than killing. **2.** The value of wolves has been established and they should remain unless some other imbalance occurs within the park. **3.** A healthy ecosystem is a long term goal that is sustainable. It is of more value than short term financial rewards obtained by the ranching industry.

Testing Yourself

1. e; **2.** e; **3.** c; **4. a.** lag; **b.** environmental resistance; **c.** carrying capacity; **d.** stable equilibrium; **e.** logistic growth; **f.** exponential growth; **5.** b; **6.** e; **7.** a; **8.** c; **9.** c; **10.** a; 11. e; **12.** d; **13.** b; **14.** a; **15.** c.

Thinking Critically

1. Rotate the use of different pesticides. Use biological controls. Grow different crops year to year. **2.** You would need to test if the two species have the same mode of defense. If so it would be Mullerian mimicry. If only one species was harmful then it would be Batesian mimicry. **3.** As a predator reduces the population of prey, it will become increasingly difficult to survive on the limited resource of that prey and the predator will seek out different sources of food to survive. This will happen before the prey becomes extinct.

CHAPTER 35

Case Study

1. Some pollutants, such as mercury, become increasingly concentrated in animals moving from lower to higher trophic levels. **2.** Biogeochemical cycles include the reservoir and exchange pools of the chemicals involved. Human activity can alter the balance between these and the communities of producers, consumers, and decomposers.

Check Your Progress

35.1: 1. Producers are at the base of the food chain and are plants, and thus autotrophs. Consumers need to eat other organisms to survive and are heterotrophs. Decomposers consume dead organic matter and break it down. **2.** Most energy (90%) is lost as heat to the environment due to inefficiencies in consumption and digestion. **35.2: 1.** Because most energy is lost as heat when one organism eats another, little biomass, in the form of carnivores, remains at the top trophic levels in an ecosystem. **2.** Usually the biomass of autotrophs is greater than that of herbivores which is greater than the carnivores. But in aquatic ecosystems herbivores can have higher biomass than producers, which can be consumed at a high rate. **35.3: 1.** See Figs. 35.7, 35.8, 35.9, and 35.10. **2.** Human activities burn fossil fuels, releasing carbon dioxide into the atmosphere, thus causing rises in global temperatures. Humans mine phosphorus and distribute it into the environment as fertilizers, animal feed supplements, and detergents, where it becomes available to the biotic community.

Science in Your Life

Association of Mercury with the Hydrologic Cycle: 1. Reducing mercury emissions even if energy costs rise, is essential for the Earth's health. **2.** Prevent the entry of mercury into the hydrologic cycle by tighter regulations on power plants and chemical facilities. Also reduce the amount entering the atmosphere by controlling coal burning, waste incineration, and metal processing. **3.** Because mercury moves through all levels of an ecosystem it has an impact on all species. **Photochemical Smog: 1.** How much extra people are willing to pay for a fuel efficient car is a combination of how important a person feels it is and a person's financial capacity. The government's role should be to support the development of reasonably priced fuel efficient vehicles and also the development of public transportation systems that obviate the need for cars. **2.** The young and elderly are less able to fight off lung infections as a result of ozone-related lung damage. **3.** The willingness to use public transportation depends on the balance between the level of inconvenience of using the system versus the personal benefits received. Such benefits include satisfaction at conserving energy. **Global Climate Change: 1.** Developing nations often have limited resources that can be devoted to conservation measures. Financial support from developed nations is one possible strategy that would help to preserve forests, which would benefit all countries. **2.** Yes, individuals have a personal responsibility to help prevent climate change.

Testing Yourself

1. a; **2. a.** producers, **b.** consumers, **c.** inorganic nutrient pool, **d.** decomposers; **3.** c; **4.** d; **5.** a; **6.** d; **7. a.** tertiary consumers, top carnivores, **b.** secondary consumers, carnivores, **c.** primary consumers, herbivores, **d.** producers, autotrophs; **8.** c; **9.** d; **10.** c; **11.** d; **12.** c; **13.** b; **14.** a; **15.** b.

Thinking Critically

1. Removal of trees changes the balance of evaporation, condensation, and precipitation in an ecosystem. Removal of trees removes shade, causing increases in evaporation relative to condensation and precipitation. In addition, removal of trees causes erosion (due to reduction of tree roots that hold the soil together), which causes more water to run off into the ocean. This causes precipitation patterns to shift from overland, where the forest was, to over the ocean. **2.** You would need information on the amount of biomass produced by primary producers and the proportion of that biomass eaten by primary consumers, secondary consumers, and so forth, because energy is lost at each trophic level in the food chain. You would need to know the trophic level that your predator would be on the food chain in order to know if there is enough energy in the ecosystem to support the proposed predator you plan to introduce. Based on estimates of available energy at that trophic level, scientists can determine appropriate numbers of animals to introduce to minimize impacts on domestic livestock. **3.** Degradation of organic matter to the point where it forms fossil fuels occurs very, very slowly (thousands to millions of years). Because modern humans are using fossil fuels at such a rapid rate, there is not enough time to replenish them in the environment.

CHAPTER 36

Case Study

1. The use of DDT should be controlled and restricted to extreme situations that warrant it. **2.** DDT moves throughout all biomes and impacts them all eventually. **3.** All biomes are interconnected and cannot be isolated from each other.

Check Your Progress

36.1: 1. The proximity of the portion of the Earth to the sun determines the different seasons. Because the Earth is tilted on its axis, during part of the year the southern hemisphere is further from the sun than the northern hemisphere and vice versa. **2.** Solar energy heats up air and causes water to evaporate. As this air rises, it cools and the moisture condenses. It falls out as rain. Mountains affect rainfall. As air blows up and over a mountain range, it rises and cools and the moisture condenses as rain. **36.2: 1.** Features in a biome that promote biodiversity include warm temperatures, plenty of rain, and complex structure with many levels of living organisms. **2.** The tropical rain forest has the highest species diversity, while the tundra has the lowest. **36.3: 1.** Lakes can range from nutrient poor (oligotrophic) to nutrient rich (eutrophic). Deep lakes are stratified during summer and winter into epilimnion, thermocline, and hypolimnion layers. Estuaries are where fresh water and saltwater meet and mix. **2.**The pelagic division, which is divided into the neritic and oceanic province, is characterized by organisms that live in the open waters under varied levels of sunlight and nutrient availability. The benthic division is characterized by organisms that live on or in the ocean floor.

Science in Your Life

Water Pollution: 1. Yes, because freshwater supplies must be preserved for the benefit of everyone. Protection of freshwater supplies is not an individual decision. **2.** Whether you would pay 20% more depends on how important you feel environmentally friendly techniques are and if you have the financial capability to afford the investment. **El Nino—Southern Oscillation: 1.** Yes MDCs have the responsibility to reduce their emissions because their habits have a worldwide impact and they should be accountable to all people. **2.** Yes, if one believed reducing greenhouse gases was important, that buying a fuel-efficient car or paying "carbon credits" would be an effective measure to take, and if one had enough money to support such efforts.

Testing Yourself

1. c; **2.** b; **3.** c; **4.** b; **5.** e; **6.** a; **7.** g; **8.** d; **9.** c; **10.** f; **11.** c; **12.** b, c; **13.** a; **14.** c, d; **15.** b.

Thinking Critically

1. Because the sun aims directly at the equator, it is already relatively warm and temperatures are relatively constant in tropical areas near the equator. As you move into the temperate zones (both north and south) away from the equator, the sun's rays get dispersed. Rising air flows toward the poles, sinks to the Earth's surface and reheats. With greenhouse gases trapping solar radiation, the warmer air will move toward the poles and away from the equator, making temperature increases in temperate areas larger than those in tropical areas nearest to the equator. **2.** Equatorial climates have the highest amount of rainfall, which supports high primary productivity. They also have the most stable climate in terms of temperature, so they can support many plant and animal species that have narrow temperature tolerances. As one moves more north or south from the equator, temperatures get hotter in the deserts, with large temperature fluctuations and little rainfall. This requires plant and animal species adapted for water conservation and those that can deal with large temperature fluctuations. Further from the equator are taiga and tundra. Taiga has a very short growing season, as it is cold and the ground only thaws for a few months of the year. This makes primary productivity lower and thus, fewer animals can be supported in the food web. Tundra is extremely cold and has permafrost, making ground cover ice and plant growth virtually impossible. Thus, there is very low diversity in areas of tundra—for example, marine life such as penguins that are aquatic for part of the year where they can feed. **3.** Although tropical forests have basically a year-round growing season, versus a growing season that ranges from 140–300 days in temperate forest habitats, temperate forests have more nutrient rich soil. Decomposition rates are slower in temperate areas because of colder temperatures and thus, nutrients are recycled more slowly. In tropical forests, nutrients are cycled rapidly from the soil directly back into the plants. When tropical forests are slashed and burned to initiate agriculture, nutrients are released into the soil, providing nutrients for several harvests. However, the soils become nutrient-poor rather quickly, requiring forest regrowth to occur before further harvesting can occur.

CHAPTER 37

Case Study

1. Establishing systems such as AERS should be strongly encouraged through technical support, economic incentives, and backing for the capital outlay necessary for implementation. **2.** While the AERS system appears to be a win-win situation it is somewhat vulnerable to shifts in market pricing. It represents a very large capital investment that needs to be flexible enough to survive. **3.** To some degree indirect costs can be estimated and translated into direct value terms.

Check Your Progress

37.1: 1. Conservation biology relies on multiple disciplines in basic (e.g., physiology, genetics, ecology) and applied (e.g., forestry, fisheries biology, agronomy) biology to achieve its goal of protecting biodiversity and Earth's natural resources. **2.** Ecosystem-level conservation will protect habitats for many communities and the species they contain, whereas species-level conservation may protect one species while not accounting for the needs of the whole community. **37.2: 1.** Consumptive use value describes the direct value of a natural product, such as timber. Agricultural use value is the value of those products produced by humans for agricultural purposes. **2.** Direct value is the dollar value of product in the market. Indirect values, which are difficult to measure, are those benefits incurred by as a result of the existence and functioning of ecosystems. **37.3: 1.** Exotic species can be introduced by humans when they colonize a new area, by accidental transport, or by horticulture and agriculture. **2.** The five main causes of extinction are: habitat loss, introduction of exotic species, pollution, overexploitation, and diseases. **3.** Pollution can weaken immunity and lead to increased susceptibility to disease. Pollution leads to global warming, which can have negative impacts on species' geographic ranges. **37.4: 1.** Landscape preservation maintains multiple ecosystem types that are connected by corridors. This supports the populations of species such as grizzly bears, that require different types of ecosystems for survival. Maintaining just one ecosystem often cannot provide adequate habitat for species. **2.** Because edge effects essentially remove the amount of suitable habitat in an area, nature reserves should account for this and generally be designed to be larger. **3.** Three principles of habitat restoration are: 1) begin as soon as possible to protect remaining habitat; 2) use biological techniques to mimic natural processes; and, 3) focus on sustainable development, allowing the ecosystem to maintain itself while providing services to humans. **37.5: 1.** Most nonrenewable energy use is concentrated on fossil fuel burning, which releases greenhouse gases into the environment and contributes to global warming. **2.** We can plant salt and drought tolerant crops, use drip irrigation, recycle home use water, limit yard irrigation to dawn and dusk hours, and use water efficient toilets. **3.** Crop rotation helps maintain nutrients in the soil (e.g., by alternating nitrogen fixing crops, such as legumes). Both organic farming and biological pest control remove or reduce the use of pesticides and herbicides. Contour farming, terrace farming, planting cover crops, and natural fences all help reduce erosion.

Applying the Concepts Revisited

Testing Yourself

1. c; **2.** c; **3.** b; **4.** a; **5.** d; **6.** c; **7.** c; **8.** b; **9.** d; **10.** b; **11.** e; **12.** e; **13.** e; **14.** e; **15.** c.

Thinking Critically

1. One simple way to reduce our reliance on nonrenewable resources is to reduce consumption by conserving electricity, water, and gas for example. Another way is through recycling, which reuses nonrenewable resources. We can also switch from using nonrenewable energy resources, such as fossil fuels, to renewable energy sources, such as hydropower, geothermal energy, wind power, solar power, and hydrogen fuel. **2.** Biological magnification is the process by which pollutants increase in their concentration up the food chain. While an herbivore may eat a certain amount of biomass of plants, a carnivore eats several herbivores, thereby magnifying the concentration of pesticides in the plants. The further an organism is up the food chain, the more the chemicals are magnified in the food. Thus, people who often are tertiary or even quaternary consumers should be more concerned about pesticide concentrations in their food than herbivores, since they eat plants directly. **3.** Extinction of a keystone species may cause secondary extinctions or change the food web dynamics of an ecosystem, as seen with wolves in Yellowstone National Park. Without wolves, elk are numerous and alter aspen groves, their favorite food. Wolves have helped reduce the number of elk, and their carcasses become food for grizzly bears and eagles. Reduction of elk also results in increases in the number of aspen and other trees found in floodplains that, in turn, increases habitat for songbirds and beavers. Beavers' dams flood areas, making them suitable for muskrats, ducks, otters, and amphibians.

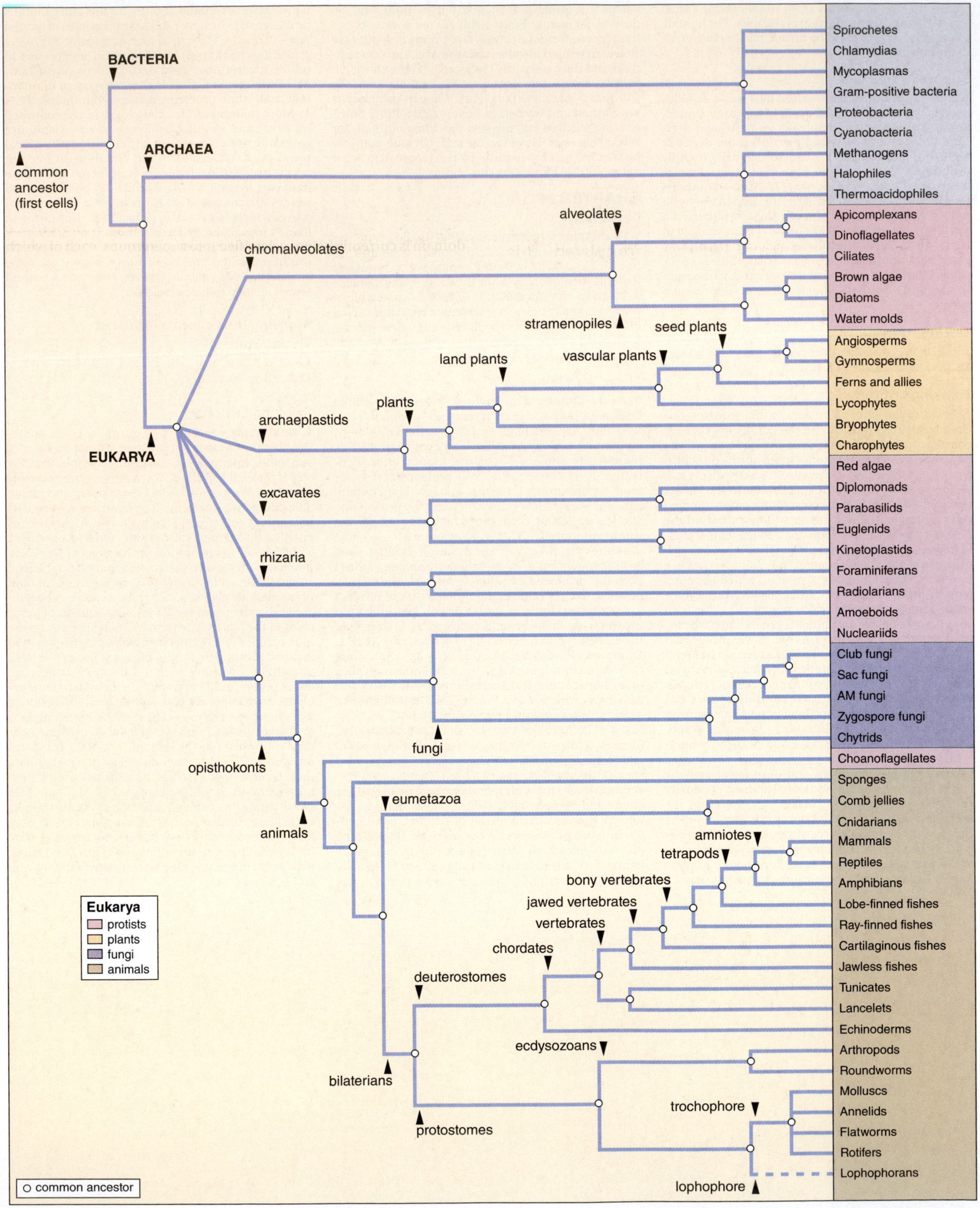
BACTERIA
ARCHAEA
EUKARYA
common ancestor (first cells)
alveolates
chromalveolates
stramenopiles
seed plants
vascular plants
land plants
plants
archaeplastids
excavates
rhizaria
opisthokonts
fungi
animals
eumetazoa
amniotes
tetrapods
bony vertebrates
jawed vertebrates
vertebrates
chordates
deuterostomes
ecdysozoans
bilaterians
protostomes
trochophore
lophophore
Spirochetes
Chlamydias
Mycoplasmas
Gram-positive bacteria
Proteobacteria
Cyanobacteria
Methanogens
Halophiles
Thermoacidophiles
Apicomplexans
Dinoflagellates
Ciliates
Brown algae
Diatoms
Water molds
Angiosperms
Gymnosperms
Ferns and allies
Lycophytes
Bryophytes
Charophytes
Red algae
Diplomonads
Parabasilids
Euglenids
Kinetoplastids
Foraminiferans
Radiolarians
Amoeboids
Nucleariids
Club fungi
Sac fungi
AM fungi
Zygospore fungi
Chytrids
Choanoflagellates
Sponges
Comb jellies
Cnidarians
Mammals
Reptiles
Amphibians
Lobe-finned fishes
Ray-finned fishes
Cartilaginous fishes
Jawless fishes
Tunicates
Lancelets
Echinoderms
Arthropods
Roundworms
Molluscs
Annelids
Flatworms
Rotifers
Lophophorans
Eukarya
protists
plants
fungi
animals
common ancestor

APPENDIX B

Tree of Life

The tree of life depicted in this appendix is based on the phylogenetic (evolutionary) trees presented in the text. Section 1.2 describes how the three domains of life—Bacteria, Archaea, and Eukarya—are related. This relationship is also apparent in the tree of life, which combines the individual trees given in the text for eukaryotic protists, plants, fungi, and animals. In combining these trees, we show how all organisms may be related to one another through the evolutionary process.

PROKARYOTES

Domains Bacteria and Archaea constitute the prokaryotic organisms that are characterized by their simple structure but a complex metabolism. The chromosome of a prokaryote is not bounded by a nuclear envelope and, therefore, these organisms do not have a nucleus. Prokaryotes carry out all the metabolic processes performed by eukaryotes and many others besides. However, they do not have organelles, except for multiple ribosomes.

Domain Bacteria

Bacteria (Chapter 28) are the most plentiful of all organisms, capable of living in most habitats, and they can carry out many different metabolic processes. Most bacteria are aerobic heterotrophs, but some are photosynthetic, and some are chemosynthetic. Motile forms move by flagella consisting of a single filament. Their cell wall contains peptidoglycan, and they have distinctive RNA sequences.

Domain Archaea

Archaean (Chapter 28) cell walls lack peptidoglycan, their lipids have a unique branched structure, and their ribosomal RNA sequences are distinctive. Examples are methanogens and the extremophiles.

EUKARYOTES

Eukaryotes have a complex cell structure with a nucleus and several types of organelles that compartmentalize the cell. Mitochondria that produce ATP and chloroplasts that produce carbohydrates are derived from prokaryotes that took up residence in a larger, nucleated cell. They may be unicellular (the majority of the protists), or multicellular (the plants, fungi, animals). Each multicellular group is characterized by a particular mode of nutrition. Flagella, if present, have a 9 + 2 organization.

Domain Eukarya

From a classical perspective, this domain has been divided into four kingdoms: the protists, plants, animals, and fungi. However, this domain is currently being reclassified into supergroups, each of which represents a major eukaryotic group. The six proposed supergroups will encompass all members of the domain Eukarya, including all protists, plants, fungi, and animals.

Kingdom Protista

The protists (Chapter 29) represent a group containing those eukaryotes that are not classified as a plant, fungus, or animal. Examples of protists include amoebas, the brown algae, and the paramecia.

Kingdom Plantae

Plants (Chapter 30) are multicellular photosynthetic eukaryotes that became adapted to living on land. This group includes aquatic green algae called charophytes, which have a haploid life cycle and share certain significant traits with the land plants.The land plants exhibit an alternation-of-generations life cycle; protect a multicellular sporophyte embryo; produce gametes in gametangia; have apical meristem that produces differentiated tissues; and have a waxy cuticle that prevents water loss. Examples include mosses, ferns, conifers, and flowering plants.

Kingdom Fungi

Fungi (Chapter 29) have multicellular bodies composed of hyphae that absorb nutrients from the environment. They have a haploid life cycle and usually produce nonmotile spores during both asexual and sexual reproduction. Chytrids are aquatic fungi with flagellated spores and gametes. Examples include the mushrooms, cup fungi, and molds.

Kingdom Animalia

Animals (Chapters 31 and 32) are multicellular and have a diploid life cycle. Most have specialized tissues and ingest their food in a digestive cavity. Examples include invertebrate organisms such as sponges, earthworms, molluscs, and insects, and vertebrates such as fishes, reptiles, and humans.

APPENDIX C

History of Biology

Year	Name	Contribution
1628	William Harvey	Demonstrates that the blood circulates and the heart is a pump.
1665	Robert Hooke	Uses the word cell to describe compartments he sees in cork under the microscope.
1668	Francesco Redi	Shows that decaying meat protected from flies does not spontaneously produce maggots.
1673	Antonie van Leeuwenhoek	Uses microscope to view living microorganisms.
1735	Carolus Linnaeus	Initiates the binomial system of naming organisms.
1809	Jean B. Lamarck	Supports the idea of evolution by the inheritance of acquired characteristics.
1825	Georges Cuvier	Founds the science of paleontology and shows that fossils are related to living forms.
1828	Karl E. von Baer	Establishes the germ layer theory of development.
1838	Matthias Schleiden	States that plants are multicellular organisms.
1839	Theodor Schwann	States that animals are multicellular organisms.
1851	Claude Bernard	Concludes that a relatively constant internal environment allows organisms to survive under varying conditions.
1858	Rudolf Virchow	States that cells come only from preexisting cells.
1858	Charles Darwin, Alfred Wallace	Independently present evidence that natural selection guides the evolutionary process.
1865	Louis Pasteur	Disproves the theory of spontaneous generation for bacteria; shows that infections are caused by bacteria.
1866	Gregor Mendel	Proposes basic laws of genetics based on his experiments with garden peas.
1882	Robert Koch	Establishes the germ theory of disease and develops many techniques used in bacteriology.
1902	Walter S. Sutton, Theodor Boveri	Suggest that genes are on the chromosomes, after noting the similar behavior of genes and chromosomes.
1904	Ivan Pavlov	Shows that conditioned reflexes affect behavior, based on experiments with dogs.
1910	Thomas H. Morgan	States that each gene has a locus on a particular chromosome, based on experiments with Drosophila.
1924	Hans Spemann, Hilde Mangold	Show that induction occurs during development, based on experiments with frog embryos.
1929	Sir Alexander Fleming	Discovers the toxic effect of a mold product he called penicillin on certain bacteria.
1937	Konrad Z. Lorenz	Founds the study of ethology and shows the importance of imprinting in learning.
1937	Sir Hans A. Krebs	Discovers the reactions of a cycle that produces carbon dioxide during cellular respiration.
1940	George Beadle, Edward Tatum	Develop the one gene—one enzyme theory, based on red bread mold studies.
1944	O. T. Avery, Colin MacLeod, Maclyn McCarty	Demonstrate that DNA alone from virulent bacteria can transform nonvirulent bacteria.
1945	Melvin Calvin, Andrew A. Benson	Discover the individual reactions of a cycle that reduces carbon dioxide during photosynthesis.
1950	Barbara McClintock	Discovers transposons (jumping genes) while doing experiments with corn.
1952	Alfred D. Hershey, Martha Chase	Find that only DNA from viruses enters cells and directs the reproduction of new viruses.
1953	James Watson, Francis Crick, Rosalind Franklin	Establish that the molecular structure for DNA is a double helix.
1953	Harold Urey, Stanley Miller	Demonstrate that the first organic molecules may have arisen from the gases of the primitive atmosphere.
1954	Linus Pauling	States that disease-causing abnormal hemoglobins are due to mutations.
1954	Jonas Salk	Develops a vaccine that protects against polio.
1958	Matthew S. Meselson, Franklin W. Stahl	Demonstrate that DNA replication is semiconservative.
1961	Francois Jacob, Jacques Monod	Discover that genetic expression is controlled by regulatory genes.
1964	Marshall W. Nirenberg, Philip Leder	Produce synthetic RNA, enabling them to break the DNA code.
1973	Stanley Cohen	Uses recombinant DNA technique (genetic engineering) to place plant and animal genes in *Escherichia coli.*
1977	Carl Woese	Based on differences in ribosomal RNA sequences, proposes the three domain system of classifications.
1978	Peter Mitchell	Determines chemiosmotic mechanism by which ATP is produced in chloroplasts and mitochondria.
1989	Sidney Altman, Thomas R. Check	Independently discover that some RNA molecules can act as enzymes.
1990	R. Michael Blaese, W. French Anderson, Kenneth W. Culver	Develop procedure to infuse genetically engineered blood cells for treatment of immune system disorder—first gene therapy used in a human.
1997	Ian Wilmut	Clones an adult mammal for the first time.
2003	Human Genome Project	Complete human genome sequenced, creating push to discover genomic links to health and disease.
2010	Craig Venter	Leads team that develops the first synthetic life form.

GLOSSARY

A

abiotic Nonliving; examples are water, gases, and minerals.

abscisic acid (ABA) Plant hormone that causes stomata to close and initiates and maintains dormancy.

abscission Dropping of leaves, fruits, or flowers from a land plant.

acid Molecules tending to raise the hydrogen ion concentration in a solution and thus to lower its pH numerically.

acquired immunodeficiency syndrome (AIDS) Disease caused by the HIV virus that destroys helper T cells and macrophages of the immune system, thus preventing an immune response to pathogens; caused by sexual contact with an infected person, intravenous drug use, and transfusions of contaminated blood (rare).

acromegaly Condition caused by an overproduction of growth hormones as an adult.

actin One of two major proteins of muscle; makes up thin filaments in myofibrils of muscle fibers. *See also* myosin.

actin filament Component of the cytoskeleton; plays a role in the movement of the cell and its organelles; a protein filament in a sarcomere of a muscle, its movement shortens the sarcomere, yielding muscle contraction.

action potential Electrochemical changes that take place across the axon membrane; the nerve impulse.

active immunity Ability to produce antibodies due to the immune system's response to a microorganism or a vaccine.

active site Region of an enzyme where the substrate binds and where the chemical reaction occurs.

active transport Use of a plasma membrane carrier protein to move a molecule or ion from a region of lower concentration to one of higher concentration; it opposes equilibrium and requires energy.

acute bronchitis Inflammation of the bronchi in the lungs; often caused by a viral or bacterial infection.

adaptation Species' modification in structure, function, or behavior that makes a species more suitable to its environment.

adaptive immunity Type of immunity that is characterized by the action of lymphocytes to specific antigens.

adaptive radiation Rapid evolution of several species from a common ancestor into new ecological or geographical zones.

Addison disease Condition caused by an insufficient production of cortisol by the adrenal glands.

adenine (A) One of four nitrogen-containing bases in nucleotides composing the structure of DNA and RNA. Pairs with uracil (U) and thymine (T).

adenosine Portion of ATP and ADP that is composed of the base adenine and the sugar ribose.

adhesion junction Junction between cells in which the adjacent plasma membranes do not touch but are held together by intercellular filaments attached to buttonlike thickenings.

adipose tissue Connective tissue in which fat is stored.

ADP (adenosine diphosphate) Nucleotide with two phosphate groups that can accept another phosphate group and become ATP.

adrenal cortex Outer portion of the adrenal gland; secretes mineralocorticoids, such as aldosterone, and glucocorticoids, such as cortisol.

adrenal gland Gland that lies atop a kidney; the *adrenal medulla* produces the hormones epinephrine and norepinephrine, and the *adrenal cortex* produces the glucocorticoid and mineralocorticoid hormones.

adrenal medulla Inner portion of the adrenal gland; secretes the hormones epinephrine and norepinephrine.

adrenocorticotropic hormone (ACTH) Hormone secreted by the anterior lobe of the pituitary gland that stimulates activity in the adrenal cortex.

aerobic A chemical process that requires air (oxygen); phase of cellular respiration that requires oxygen.

African sleeping sickness Disease caused by the protist *Trypanosoma brucei*, which is transmitted by the tsetse fly.

afterbirth Another term for the placenta; the expulsion of the afterbirth represents the third stage of parturition.

age structure diagram In demographics, a display of the age groups of a population; a growing population has a pyramid-shaped diagram.

aging General term for the progressive physiological changes that occur to the human body over time.

aldosterone Hormone secreted by the adrenal cortex that regulates the sodium and potassium ion balance of the blood.

algae Photosynthetic protists that may be unicellular or live in colonies or filaments; historically have been classified according to color.

allantois Extraembryonic membrane that accumulates nitrogenous wastes in the eggs of reptiles, including birds; contributes to the formation of umbilical blood vessels in mammals.

allele Alternative form of a gene; alleles occur at the same locus on homologous chromosomes.

allergy Immune response to substances that usually are not recognized as foreign.

allopatric speciation Model that proposes that new species arise due to an interruption of gene flow between populations that are separated geographically.

alternation of generations Life cycle, typical of land plants, in which a diploid sporophyte alternates with a haploid gametophyte.

altruism Social interaction that has the potential to decrease the lifetime reproductive success of the member exhibiting the behavior.

alveoli (sing., alveolus) In humans, terminal, microscopic, grapelike air sac found in lungs.

Alzheimer disease Disease of the central nervous system (brain) that is characterized by an accumulation of beta amyloid protein and neurofibrillary tangles in the hippocampus and amygdala.

AM fungi Fungi with branching invaginations (arbuscular mycorrhizae or AM) used to invade plant roots.

amino acid Organic molecule composed of an amino group and an acid group; covalently bonds to produce peptide molecules.

ammonia Nitrogenous end product that takes a limited amount of energy to produce but requires much water to excrete because it is toxic.

amnion Extraembryonic membrane of birds, reptiles, and mammals that forms an enclosing, fluid-filled sac.

amniotic egg Egg that has an amnion, as seen during the development of reptiles (including birds) and mammals.

amoebic dysentery Condition caused by protistans belonging to the genus *Entamoeba*; characterized by severe diarrhea and fluid loss.

amoeboid Cell that moves and engulfs debris with pseudopods.

amphibian Member of vertebrate class Amphibia that includes frogs, toads, and salamanders; they are tied to a watery environment for reproduction.

amygdala Part of the limbic system of the brain; associated with emotional experiences.

amyotrophic lateral sclerosis (ALS) Neurodegenerative disease affecting the motor neurons in the brain and spinal cord, resulting in paralysis and death; also known as Lou Gehrig's disease.

anabolic steroid Synthetic steroid that mimics the effect of testosterone.

anabolism Chemical reaction in which smaller molecules (monomers) are combined to form larger molecules (polymers); anabolic metabolism.

anaerobic A chemical reaction that occurs in the absence of oxygen; an example is the fermentation reactions.

analogous, analogous structure Structure that has a similar function in separate lineages but differs in anatomy and ancestry.

anaphase Fourth phase of mitosis; chromosomes move toward the poles of the spindle.

anaphylactic shock Severe systemic form of anaphylaxis involving bronchiolar constriction, impaired breathing, vasodilation, and a rapid drop in blood pressure with a threat of circulatory failure.

androgen Male sex hormone (e.g., testosterone).

anemia Disease that is characterized by a decrease in the number of red blood cells (erythrocytes); may be genetic in origin or the result of another disease or condition in the body.

aneurysm Expansion in the walls of the blood vessel; often caused by atherosclerosis or hypertension.

angina pectoris Condition characterized by thoracic pain resulting from occluded coronary arteries; may precede a heart attack.

angiogenesis Formation of new blood vessels; rapid angiogenesis is a characteristic of cancer cells.

angiosperm Flowering land plant; the seeds are borne within a fruit.

animal Multicellular, heterotrophic eukaryote that undergoes development to achieve its final form. In general, animals are mobile organisms, characterized by the presence of muscular and nervous tissue.

annelid The segmented worms, such as the earthworm and the clam worm.

annual ring Layer of wood (secondary xylem) usually produced during one growing season.

anorexia nervosa Eating disorder characterized by a severe psychological fear of gaining weight.

anterior pituitary Portion of the pituitary gland that is controlled by the hypothalamus and produces six types of hormones, some of which control other endocrine glands.

anther Male structure of a flowering plant where pollen is produced; attached to the flower by the filament.

antheridium (pl., **antheridia**) Sperm-producing structures, as in the moss life cycle.

anthropoid Group of primates that includes monkeys, apes, and humans.

antibiotics Chemicals that interfere with the physiological activities, or structure, of bacteria.

antibody Protein produced in response to the presence of an antigen; each antibody combines with a specific antigen.

antibody-mediated immunity Specific mechanism of defense in which plasma cells derived from B cells produce antibodies that combine with antigens.

anticodon Three-base sequence in a transfer RNA molecule base that pairs with a complementary codon in mRNA.

antidiuretic hormone (ADH) Hormone secreted by the posterior pituitary that increases the permeability of the collecting ducts in a kidney.

antigen Foreign substance, usually a protein or a polysaccharide, that stimulates the immune system to react, such as to produce antibodies.

antigen-presenting cell (APC) Cell that displays an antigen to certain cells of the immune system so they can defend the body against that particular antigen.

anus Outlet of the digestive tube.

aorta In humans, the major systemic artery that takes blood from the heart to the tissues.

apical dominance Influence of a terminal bud in suppressing the growth of axillary buds.

apical meristem In vascular land plants, masses of cells in the root and shoot that reproduce and elongate as primary growth occurs.

apoptosis Programmed cell death; involves a cascade of specific cellular events leading to death and destruction of the cell.

appendicular skeleton Part of the vertebrate skeleton comprising the appendages, shoulder girdle, and hip girdle.

appendix (vermiform appendix) In humans, small, tubular appendage that extends outward from the cecum of the large intestine.

aquaporin Channel protein through which water can diffuse across a membrane.

aquifer Rock layers that contain water and release it in appreciable quantities to wells or springs.

arachnid Group of arthropods that includes spiders and scorpions.

arboreal Living in trees.

archaeans, archaea Prokaryotic organisms that are members of the domain Archaea.

archegonium Egg-producing structures, as in the moss life cycle.

arteriole Vessel that takes blood from an artery to capillaries.

artery Blood vessel that transports blood away from the heart.

arthropod Invertebrates, with an exoskeleton and jointed appendages, such as crustaceans and insects.

articular cartilage Form of hyaline cartilage that occurs at joints where two bones meet.

artificial selection Intentional breeding of certain traits, or combinations of traits, over others to produce a desirable outcome.

ascocarp Reproductive structure of the sac fungi (phylum Ascomycota).

associative learning Acquired ability to associate two stimuli or between a stimulus and a response.

asthma Condition in which bronchioles constrict and cause difficulty in breathing.

astigmatism Uneven shape of the cornea or lens of the eye; causes a distortion in the light reaching the retina of the eye.

asymmetrical Lack of any symmetrical relationship in the morphology of an organism.

atherosclerosis Form of cardiovascular disease characterized by the accumulation of fatty materials (usually cholesterol) in the arteries.

atom Smallest particle of an element that displays the properties of the element.

atomic mass Average of atom mass units for all the isotopes of an atom.

atomic number Number of protons within the nucleus of an atom.

ATP (adenosine triphosphate) Nucleotide with three phosphate groups. The breakdown of ATP into ADP + P makes energy available for energy-requiring processes in cells.

ATP synthase Complex of proteins in the cristae of mitochondria and thylakoid membrane of chloroplasts that produces ATP from the diffusion of hydrogen ions across a membrane.

atrial natriuretic hormone (ANH) Hormone secreted by the heart that increases sodium excretion.

atrioventricular valve Heart valve located between an atrium and a ventricle.

atrium Chamber; particularly an upper chamber of the heart, lying above a ventricle.

auditory canal Structure that funnels sounds from the outer ear to the middle ear.

auditory tube Also called the eustachian tube; connects the middle ear to the nasopharynx for the equalization of pressure.

australopithecine One of several species of *Australopithecus*, a genus that contains the first generally recognized humanlike hominins.

autoimmune disease Disease that results when the immune system mistakenly attacks the body's own tissues.

autonomic system Portion of the peripheral nervous system that regulates internal organs.

autosome Chromosome pairs that are the same between the sexes; in humans, all but the X and Y chromosomes.

autotroph Organism that can capture energy and synthesize organic molecules from inorganic nutrients.

auxin Plant hormone regulating growth, particularly cell elongation; also called indoleacetic acid (IAA).

axial skeleton Part of the vertebrate skeleton forming the vertical support or axis, including the skull, the rib cage, and the vertebral column.

axillary bud Bud located in the axil of a leaf.

axon Elongated portion of a neuron that conducts nerve impulses, typically from the cell body to the synapse.

B

bacteriophage Virus that infects bacteria.

bacterium (pl., **bacteria**) Member of the domain Bacteria.

bark External part of a tree, containing cork, cork cambium, and phloem.

Barr body Dark-staining body in the cell nuclei of female mammals that contains a condensed, inactive X chromosome; named after its discoverer, Murray Barr.

basal nuclei Subcortical nuclei deep within the white matter that serve as relay stations for motor impulses and produce dopamine to help control skeletal muscle activities.

base Molecules tending to lower the hydrogen ion concentration in a solution and thus raise the pH numerically.

basidium Clublike structure in which nuclear fusion, meiosis, and basidiospore production occur during sexual reproduction of club fungi.

basophil White blood cell with a granular cytoplasm; able to be stained with a basic dye.

B cell Lymphocyte that matures in the bone marrow and, when stimulated by the presence of a specific antigen, gives rise to antibody-producing plasma cells.

B-cell receptor (BCR) Molecule on the surface of a B cell that binds to a specific antigen.

behavior Observable, coordinated responses to environmental stimuli.

benign Mass of cells derived from a single mutated cell that has repeatedly undergone cell division but has remained at the site of origin.

benign prostatic hyperplasia (BPH) Enlargement of the prostate gland that may cause irritation of the bladder and an urge to urinate more frequently.

benthic division Ocean or lake floor, extending from the high-tide mark to the deepest depths.

bicarbonate ion Ion that participates in buffering the blood, and the form in which carbon dioxide is transported in the bloodstream.

bilateral symmetry Body plan having two corresponding or complementary halves.

bile Secretion of the liver that is temporarily stored and concentrated in the gallbladder before being released into the small intestine, where it emulsifies fat.

binary fission Splitting of a parent cell into two daughter cells; serves as an asexual form of reproduction in bacteria.

biodiversity Total number of species, the variability of their genes, and the communities in which they live.

biodiversity hotspot Region of the world that contains unusually large concentrations of species.

biogeochemical cycle Circulating pathway of elements such as carbon and nitrogen involving exchange pools, storage areas, and biotic communities.

biogeography Study of the geographical distribution of organisms.

bioinformatics Area of scientific study that utilizes computer technologies to analyze large sets of data, typically in the study of genomics and proteomics.

biology The branch of science that is concerned with the study of life and living organisms.

biomagnification The accumulation of pollutants as they move up the food web.

biomass The number of organisms multiplied by their weight.

biome One of the biosphere's major communities, characterized in particular by certain climatic conditions and particular types of plants.

biomolecule Organic molecules such as proteins, nucleic acids, carbohydrates, and fats.

biosphere Zone of air, land, and water at the surface of the Earth in which living organisms are found.

biotechnology Scientific research that involves researching and producing commercial or agricultural products that are made with or derived from transgenic organisms.

biotic Term that indicates life; a cell is an example of a biotic factor in the environment.

biotic potential Maximum population growth rate under ideal conditions.

bipedalism Walking erect on two feet; distinguishing characteristic of primates.

bird Endothermic reptile that has feathers and wings, is often adapted for flight, and lays hard-shelled eggs.

birth control pill Form of oral contraception that typically contains a combination of estrogen and progesterone.

bivalve Type of mollusc with a shell composed of two valves; includes clams, oysters, and scallops.

bladder stones Condition that causes inflammation of the bladder; may be caused by kidney stones, infections, or other conditions that restrict the flow of urine from the bladder.

blade Broad, expanded portion of a land plant leaf that may be single or compound leaflets.

blastocoel Fluid-filled cavity of a blastula.

blastocyst Early stage of human embryonic development that consists of a hollow, fluid-filled ball of cells.

blastopore Opening into the primitive gut formed at gastrulation.

blastula Hollow, fluid-filled ball of cells occurring during animal development prior to gastrula formation.

blind spot Region of the retina, lacking rods or cones, where the optic nerve leaves the eye.

blood Fluid circulated by the heart through a closed system of vessels; type of connective tissue.

blood pressure Force of blood pushing against the inside wall of blood vessels.

body cavity In vertebrates, defined regions of the body in which organs reside.

bone Connective tissue having protein fibers and a hard matrix of inorganic salts, notably calcium salts.

bony fishes Vertebrates belonging to the class of fish called Osteichthyes that have a bony, rather than cartilaginous, skeleton.

bottleneck effect Type of genetic drift; occurs when a majority of genotypes are prevented from participating in the production of the next generation as a result of a natural disaster or human interference.

brain Ganglionic mass at the anterior end of the nerve cord; in vertebrates, the brain is located in the cranial cavity of the skull.

brain stem In mammals, portion of the brain consisting of the medulla oblongata, pons, and midbrain.

bronchi (sing., **bronchus**) In terrestrial vertebrates, branch of the trachea that leads to the lungs.

bronchiole In terrestrial vertebrates, small tube that conducts air from a bronchus to the alveoli.

brown algae Marine photosynthetic protists with a notable abundance of xanthophyll pigments; this group includes well-known seaweeds of northern rocky shores.

buffer Substance or group of substances that tend to resist pH changes of a solution, thus stabilizing its relative acidity and basicity.

bulbourethral glands Male sex glands that produce pre-ejaculate fluid that neutralizes acid in the urethra.

bulimia nervosa Eating disorder characterized by binge eating followed by purging of food (vomiting).

bundle sheath Layer of cells surrounding the vascular tissue in the leaf of a plant.

bursae (sing., bursa) Fluid-filled sacs that reduce friction between tendons and ligaments; often found near synovial joints such as the elbow and knee.

bursitis Inflammation of the bursae.

C

C_3 plant Plant that fixes carbon dioxide via the Calvin cycle; the first stable product of C_3 photosynthesis is a 3-carbon compound.

C_4 plant Plant that fixes carbon dioxide to produce a C_4 molecule that releases carbon dioxide to the Calvin cycle.

calcitonin Hormone secreted by the thyroid gland that increases the blood calcium level.

calorie Amount of heat energy required to raise the temperature of one gram of water 1°C.

Calvin cycle reaction Portion of photosynthesis that takes place in the stroma of chloroplasts and can occur in the dark; it uses the products of the light reactions to reduce CO_2 to a carbohydrate.

calyx The sepals collectively; the outermost flower whorl.

CAM Crassulacean-acid metabolism; a form of photosynthesis in succulent plants that separates the light-dependent and Calvin reactions by time.

cancer Malignant tumor whose nondifferentiated cells exhibit loss of contact inhibition, uncontrolled growth, and the ability to invade tissue and metastasize.

capillary Microscopic blood vessel; gases and other substances are exchanged across the walls of a capillary between blood and tissue fluid.

capsid Protective protein containing the genetic material of a virus.

capsule A form of glycocalyx that consists of a gelatinous layer; found in blue-green algae and certain bacteria.

carbaminohemoglobin ($HBCO_2$) Hemoglobin carrying carbon dioxide.

carbohydrate Class of organic compounds that typically contain carbon, hydrogen, and oxygen in a 1:2:1 ratio; includes the monosaccharides, disaccharides, and polysaccharides.

carbon dioxide fixation Process by which carbon dioxide gas is attached to an organic compound; in photosynthesis, this occurs in the Calvin cycle reactions.

carbonic anhydrase Enzyme in red blood cells that speeds the formation of carbonic acid from water and carbon dioxide.

carcinogen Environmental agent that causes mutations leading to the development of cancer.

cardiac cycle One complete cycle of systole and diastole for all heart chambers.

cardiac muscle Striated, involuntary muscle tissue found only in the heart.

cardiac output Blood volume pumped by each ventricle per minute (not total output pumped by both ventricles).

cardiovascular system Organ system of humans that consists of the heart and blood vessels; transports blood, nutrients, gases and wastes; assists in the defense against disease; helps control homeostasis.

carnivore Consumer in a food chain that eats other animals.

carotenoid An accessory photosynthetic pigment of plants and algae that are often yellow or orange in color; consists of two classes—the xanthophylls and the carotenes.

carotid body Structure located at the branching of the carotid arteries; contains chemoreceptors sensitive to the O_2, CO_2, and H^+ content in blood.

carpel Ovule-bearing unit that is a part of a pistil.

carrier protein Protein in the plasma membrane that combines with and transports a molecule or ion across the plasma membrane.

carrying capacity (K) Largest number of organisms of a particular species that can be maintained indefinitely by a given environment.

cartilage Connective tissue in which the cells lie within lacunae embedded in a flexible, proteinaceous matrix.

cartilaginous fishes Vertebrates that belong to the class of fish called the Chondrichthyes: possess a cartilaginous, rather than bony, skeleton; includes sharks, rays, and skates.

cartilaginous joint Form of joint that is connected by hyaline cartilage.

Casparian strip Layer of impermeable lignin and suberin bordering four sides of root endodermal cells; prevents water and solute transport between adjacent cells.

catabolism Metabolic process that breaks down large molecules into smaller ones; catabolic metabolism.

cataracts Condition where the lens of the eye becomes opaque, preventing the transmission of light to the retina.

cecum Region of the large intestine just below the small intestine; also called the appendix.

cell The smallest unit of life that displays all the properties of life; composed of cytoplasm surrounded by a plasma membrane.

cell body Portion of a neuron that contains a nucleus and from which dendrites and an axon extend.

cell cycle An ordered sequence of events in eukaryotes that involves cell growth and nuclear division; consists of the stages G_1, S, G_2, and M.

cell-mediated immunity Specific mechanism of defense in which T cells destroy antigen-bearing cells.

cell plate Structure across a dividing plant cell that signals the location of new plasma membranes and cell walls.

cell recognition protein Glycoproteins in the plasma membrane that identify self and help the body defend itself against pathogens.

cell theory One of the major theories of biology, which states that all organisms are made up of cells; cells are capable of self-reproduction and come only from preexisting cells.

cellular differentiation Process and developmental stages by which a cell becomes specialized for a particular function.

cellular respiration Metabolic reactions that use the energy from carbohydrate, fatty acid, or amino acid breakdown to produce ATP molecules.

cellular slime mold Free-living amoeboid cells that feed on bacteria and yeasts by phagocytosis and aggregate to form a plasmodium that produces spores.

cellulose Polysaccharide that is the major complex carbohydrate in plant cell walls.

cell wall Cellular structure that surrounds a plant, protistan, fungal, or bacterial cell and maintains the cell's shape and rigidity; composed of polysaccharides.

central dogma Processes that dictate the flow of information from the DNA to RNA to protein in a cell.

central nervous system (CNS) Portion of the nervous system consisting of the brain and spinal cord.

centriole Cell structure, existing in pairs, that occurs in the centrosome and may help organize a mitotic spindle for chromosome movement during animal cell division.

centromere Constriction where sister chromatids of a chromosome are held together.

centrosome Central microtubule organizing center of cells. In animal cells, it contains two centrioles.

cephalization Having a well-recognized anterior head with a brain and sensory receptors.

cephalopod Type of mollusc in which the head is prominent and the foot is modified to form two arms and several tentacles; includes squids, cuttlefish, octopuses, and nautiluses.

cerebellum In terrestrial vertebrates, portion of the brain that coordinates skeletal muscles to produce smooth, graceful motions.

cerebral cortex Outer layer of cerebral hemispheres; receives sensory information and controls motor activities.

cerebral hemisphere Either of the two lobes of the cerebrum in vertebrates.

cerebrospinal fluid Fluid found in the ventricles of the brain, in the central canal of the spinal cord, and in association with the meninges.

cerebrum Largest part of the brain in mammals.

cervical cancer Form of cancer that is often caused by the human papillomavirus (HPV); vaccinations are available to reduce risk.

cervix Narrow end of the uterus, which leads into the vagina.

channel protein Protein that forms a channel to allow a particular molecule or ion to cross the plasma membrane.

chemical energy Energy associated with the interaction of atoms in a molecule.

chemical signal Molecule that brings about a change in a cell, tissue, organ, or individual when it binds to a specific receptor.

chemiosmosis Process by which mitochondria and chloroplasts use the energy of an electron transport chain to create a hydrogen ion gradient that drives ATP formation.

chemoautotroph Organism able to synthesize organic molecules by using carbon dioxide as the carbon source and the oxidation of an inorganic substance (such as hydrogen sulfide) as the energy source.

chemoreceptor Sensory receptor that is sensitive to chemical stimulation—for example, receptors for taste and smell.

chitin Strong but flexible nitrogenous polysaccharide found in the exoskeleton of arthropods and in the cell walls of fungi.

chlamydia Sexually-transmitted disease caused by *Chlamydia trachomatis;* one of the more common forms of STDs.

chlorophyll Green photosynthetic pigment of algae and plants that absorbs solar energy; occurs as chlorophyll *a* and chlorophyll *b*.

chloroplast Membrane-bounded organelle in algae and plants with chlorophyll-containing membranous thylakoids; where photosynthesis takes place.

cholesterol A steroid found in the plasma membrane of animal cells and from which other types of steroids are derived.

chordate Animals that have a dorsal tubular nerve cord, a notochord, pharyngeal gill pouches, and a postanal tail at some point in their life cycle; includes a few types of invertebrates (e.g., sea squirts and lancelets) and the vertebrates.

chorion Extraembryonic membrane functioning for respiratory exchange in birds and reptiles; contributes to placenta formation in mammals.

choroid Vascular, pigmented middle layer of the eyeball.

chromalveolate Supergroup of eukaryotes that includes alveolates and stramenopiles.

chromatid Following replication, a chromosome consists of a pair of sister chromatids, held together at the centromere; each chromatid is comprised of a single DNA helix.

chromatin Network of DNA strands and associated proteins observed within a nucleus of a cell.

chromoplast Plastid in land plants responsible for orange, yellow, and red color of plants, including the autumn colors in leaves.

chromosomal mutations Changes in the physical structure of a chromosome; includes deletions, duplications, inversions, and translocations.

chromosome The structure that transmits the genetic material from one generation to the next; composed of condensed chromatin; each species has a particular number of chromosomes that is passed on to the next generation.

chronic bronchitis Inflammation of the bronchi in the lungs that is usually caused by smoking or exposure to environmental contaminants.

chyme Thick, semiliquid food material that passes from the stomach to the small intestine.

chytrid Mostly aquatic fungi with flagellated spores that may represent the most ancestral fungal lineage.

cilia (sing., cilium) Short, hairlike projections from the plasma membrane, occurring usually in larger numbers.

ciliary muscle Within the ciliary body of the vertebrate eye, the ciliary muscle that controls the shape of the lens.

ciliate Complex unicellular protist that moves by means of cilia and digests food in food vacuoles.

circadian rhythm Biological rhythm with a 24-hour cycle.

circulatory system In animals, an organ system that moves substances to and from cells, usually via a heart, blood, and blood vessels.

circumcision Surgical removal of the foreskin from the penis; female circumcision involves the partial or total removal of some part of the female genitalia.

cirrhosis Chronic, irreversible injury to liver tissue; commonly caused by frequent alcohol consumption.

citric acid cycle Cycle of reactions in mitochondria that begins with citric acid. This cycle breaks down an acetyl group and produces CO_2, ATP, NADH, and $FADH_2$; also called the Krebs cycle.

cladistics Method of systematics that uses derived characters to determine monophyletic groups and construct cladograms.

cladogram In cladistics, a branching diagram that shows the relationship among species in regard to their shared derived characters.

class One of the categories, or taxa, used to group species; the taxon above the order level.

classical conditioning Type of learning whereby an unconditioned stimulus that elicits a specific response is paired with a neutral stimulus so that the response becomes conditioned.

classification Process of naming organisms and assigning them to taxonomic groups (taxa).

cleavage Cell division without cytoplasmic addition or enlargement; occurs during the first stage of animal development.

cleavage furrow Indentation in the plasma membrane of animal cells during cell division; formation marks the start of cytokinesis.

climate Generalized weather patterns of an area, primarily determined by temperature and average rainfall.

climate change Recent changes in the Earth's climate; evidence suggests that this is primarily due to human influence, including the increased release of greenhouse gases.

climax community In ecology, community that results when succession has come to an end.

clitoris Organ of sexual arousal in a female.

clonal selection theory States that the antigen selects which lymphocyte will undergo clonal expansion and produce more lymphocytes bearing the same type of receptor.

cloning Production of identical copies. In organisms, the production of organisms with the same genes; in genetic engineering, the production of many identical copies of a gene.

closed circulatory system A type of circulatory system where blood is confined to vessels and is kept separate from the interstitial fluid.

clotting Also called coagulation, the response of the body to an injury in the vessels of the circulatory system; involves platelets and clotting proteins.

club fungi Fungi that produce spores in club-shaped basidia within a fruiting body; includes mushrooms, shelf fungi, and puffballs.

club moss Type of seedless vascular plant; also called ground pines.

cnidarian Invertebrates existing as either a polyp or medusa with two tissue layers and radial symmetry.

cnidocytes Specialized stinging cells of a cnidarian; contain a toxin-filled capsule called a nematocyst.

cochlea Spiral-shaped structure of the vertebrate inner ear containing the sensory receptors for hearing.

codominance Inheritance pattern in which both alleles of a gene are equally expressed in a heterozygote.

codon Three-base sequence in messenger RNA that during translation directs the addition of a particular amino acid into a protein or directs termination of the process.

coelom Body cavity of an animal; the method by which the coelom is formed (or lack of formation) is an identifying characteristic in animal classification.

coenzyme Nonprotein organic molecule that aids the action of the enzyme to which it is loosely bound.

coevolution Mutual evolution in which two species exert selective pressures on the other species.

cofactor Nonprotein assistant required by an enzyme in order to function; many cofactors are metal ions, others are coenzymes.

cohesion-tension model Explanation for upward transport of water in xylem based upon transpiration-created tension and the cohesive properties of water molecules.

cohort Group of individuals having a statistical factor in common, such as year of birth, in a population study.

coleoptile Protective sheath that covers the young leaves of a seedling.

collagen fiber White fiber in the matrix of connective tissue giving flexibility and strength.

collar cells Choanocytes of poriferans (sponges); inner layer of flagellated cells.

collecting duct Duct within the kidney that receives fluid from several nephrons; the reabsorption of water occurs here.

collenchyma Plant tissue composed of cells with unevenly thickened walls; supports growth of stems and petioles.

color blindness Inability to detect specific wavelengths of light associated with color; red-green color blindness is the most common type.

color vision Ability of the eye to detect specific wavelengths of light; involves specialized cone cells that detect blue, green, and red wavelengths.

colostrum Substance produced during the early days of lactation; rich in proteins and antibodies.

columnar epithelium Type of epithelial tissue with cylindrical cells.

comb jelly Invertebrates that resemble jelly fishes and are the largest animals to be propelled by beating cilia.

commensalism Symbiotic relationship in which one species is benefited, and the other is neither harmed nor benefited.

common ancestor Ancestor common to at least two lines of descent.

communication Signal by a sender that influences the behavior of a receiver.

community Assemblage of species interacting with one another within the same environment.

compact bone Type of bone that contains osteons consisting of concentric layers of matrix and osteocytes in lacunae.

companion cell Plant cell in the vascular tissue of plants that metabolically supports the conducting cells of the phloem.

comparative genomics Study of genomes through the direct comparison of their genes and DNA sequences from multiple species.

competition Results when members of a species attempt to use a resource that is in limited supply.

competitive exclusion principle Theory that no two species can occupy the same niche in the same place and at the same time.

complement Collective name for a series of enzymes and activators in the blood, some of which may bind to antibody and may lead to rupture of a foreign cell.

complementary base pairing Hydrogen bonding between particular purines and pyrimidines; responsible for the structure of DNA, and some RNA, molecules.

compound Substance having two or more different elements in a fixed ratio.

compound eye Type of eye found in arthropods; it is composed of many independent visual units.

concentration gradient Gradual change in chemical concentration between two areas of differing concentrations.

conclusion Statement made following an experiment as to whether or not the results support the hypothesis.

condensation The process by which water vapor turns into liquid; at the cellular level, the removal of water between two molecules for the purpose of forming a chemical bond.

cone cell Photoreceptor in vertebrate eyes that responds to bright light and makes color vision possible.

conifer Member of a group of cone-bearing gymnosperm land plants that includes pine, cedar, and spruce trees.

coniferous forest Forested area generally found in areas with temperate climate and characterized by cone-bearing trees that are mostly evergreens with needle-shaped or scalelike leaves.

conjugation Transfer of genetic material from one cell to another.

conjunctiva Delicate membrane that lines the eyelid, protecting the sclera.

connective tissue Type of animal tissue that binds structures together, provides support and protection, fills spaces, stores fat, and forms blood cells; adipose tissue, cartilage, bone, and blood are types of connective tissue; living cells in a nonliving matrix.

conservation biology Discipline that seeks to understand the effects of human activities on species, communities, and ecosystems and to develop practical approaches to preventing the extinction of species and the destruction of ecosystems.

conservation corridors Strips of land that allow animals to move safely between habitats that would otherwise be isolated.

conservation reserves Areas that are set aside with the primary goal of protecting biodiversity within them.

constipation Changes in the frequency of bowel movements that leads to the formation of hard, dry fecal material.

consumer Organism that feeds on another organism in a food chain generally; primary consumers eat plants, and secondary consumers eat animals.

continental drift The movement of the Earth's crust by plate tectonics resulting in the movement of continents with respect to one another.

contraceptive Medications and devices that reduce the chance of pregnancy.

contraceptive implants Synthetic forms of progesterone that prevent ovulation by disrupting the ovarian cycle.

contraceptive injection This form of contraception contains synthetic forms of progesterone (or a combination of progesterone and estrogen) that prevent ovulation by disrupting the ovarian cycle.

contraceptive patch Applied to the skin, this form of contraception contains synthetic forms of progesterone that prevent ovulation by disrupting the ovarian cycle.

contraceptive vaccine Under development, this birth control method immunizes against the hormone HCG, crucial to maintaining implantation of the embryo.

contractile vacuoles Structure found in some protistan cells, usually freshwater species, that are involved in osmoregulation.

control Sample that goes through all the steps of an experiment but does not contain the variable being tested; a standard against which the results of an experiment are checked.

convergent evolution Similarity in structure in distantly related groups generally due to similiar selective pressures in like environments.

coral reef Coral formations in shallow tropical waters that support an abundance of diversity.

cork Outer covering of the bark of trees; made of dead cells that may be sloughed off.

cork cambium Lateral meristem that produces cork.

cornea Transparent, anterior portion of the outer layer of the eyeball.

corolla The petals, collectively; usually the conspicuously colored flower whorl.

corpus luteum Follicle that has released an egg and increases its secretion of progesterone.

cortex In plants, ground tissue bounded by the epidermis and vascular tissue in stems and roots; in animals, outer layer of an organ, such as the cortex of the kidney or adrenal gland.

cortisol Glucocorticoid secreted by the adrenal cortex that responds to stress on a long-term basis; reduces inflammation and promotes protein and fat metabolism.

cotyledon Seed leaf for embryo of a flowering plant; provides nutrient molecules for the developing plant before photosynthesis begins.

coupled reactions Reactions that occur simultaneously; one is an exergonic reaction that releases energy, and the other is an endergonic reaction that requires an input of energy in order to occur.

covalent bond Chemical bond in which atoms share one pair of electrons.

cranial nerve Nerve that arises from the brain.

creatine phosphate High energy compound found in muscular tissue; represents the fastest source of energy for muscle contraction.

cristae (sing., **crista**) Short, fingerlike projections formed by the folding of the inner membrane of mitochondria.

Cro-Magnon Common name for the first fossils to be designated *Homo sapiens.*

crossing-over Exchange of segments between nonsister chromatids of a bivalent during meiosis.

crustacean Member of a group of aquatic arthropods that contains, among others, shrimps, crabs, crayfish, and lobsters.

cuboidal epithelium Type of epithelial tissue with cube-shaped cells.

cultural eutrophication Overenrichment caused by human activities leading to excessive bacterial growth and oxygen depletion of a body of water.

culture Total pattern of human behavior; includes technology and the arts, and is dependent upon the capacity to speak and transmit knowledge.

Cushing syndrome Condition caused by the overproduction of ACTH, resulting in too much cortisol in the body.

cutaneous receptor Sensory receptors of the dermis that are activated by touch, pain, pressure, and temperature.

cuticle Waxy layer covering the epidermis of plants that protects the plant against water loss and disease-causing organisms.

cyanobacterium (pl., **cyanobacteria**) Photosynthetic bacterium that contains chlorophyll and releases oxygen; formerly called a blue-green alga.

cycad Type of gymnosperm with palmate leaves and massive cones; cycads are most often found in the tropics and subtropics.

cyclic adenosine monophosphate (cAMP) ATP-related compound that acts as the second messenger in peptide hormone transduction; it initiates activity of the metabolic machinery.

cyclic electron pathway Alternate form of the light-dependent reactions in photosynthesis that does not produce NADPH.

cyclin Protein that cycles in quantity as the cell cycle progresses; combines with and activates the kinases that function to promote the events of the cycle.

cyst In protists and invertebrates, resting structure that contains reproductive bodies or embryos.

cystic fibrosis (CF) Genetic disease caused by a defect in the *CFTR* gene, which is responsible for the formation of a transmembrane chloride ion transporter; causes the mucus of the body to be viscous.

cystitis Inflammation of the bladder.

cytokine Type of protein secreted by a T lymphocyte that attacks viruses, virally infected cells, and cancer cells.

cytokinesis Division of the cytoplasm following mitosis or meiosis.

cytokinin Plant hormone that promotes cell division; often works in combination with auxin during organ development in plant embryos.

cytoplasm Region of a cell between the nucleus, or the nucleoid region of a bacterium, and the plasma membrane; contains the organelles of the cell.

cytoplasmic segregation Process that parcels out the maternal determinants, which play a role in development, during mitosis.

cytosine (C) One of four nitrogen-containing bases in the nucleotides composing the structure of DNA and RNA; pairs with guanine.

cytoskeleton Internal framework of the cell, consisting of microtubules, actin filaments, and intermediate filaments.

cytotoxic T cell T lymphocyte that attacks and kills antigen-bearing cells.

D

data Facts or information collected through observation and/or experimentation.

daughter chromosomes Following anaphase in cell division, the structures produced by the separation of the sister chromatids.

day-neutral plant Plant whose flowering is not dependent on day length—e.g., tomato and cucumber.

deciduous Land plant that sheds its leaves annually.

decomposer Organism, usually a bacterium or fungus, that breaks down organic matter into inorganic nutrients that can be recycled in the environment.

deductive reasoning The use of general principles to predict specific outcomes. Often uses "if . . . then" statements.

dehydration reaction Chemical reaction in which a water molecule is released during the formation of a covalent bond.

delayed allergic response Allergic response initiated at the site of the allergen by sensitized T cells, involving macrophages and regulated by cytokines.

deletion Change in chromosome structure in which the end of a chromosome breaks off or two simultaneous breaks lead to the loss of an internal segment; often causes abnormalities—e.g., cri du chat syndrome.

demographic transition Due to industrialization, a decline in the birthrate following a reduction in the death rate so that the population growth rate is lowered.

denatured Loss of a protein's or enzyme's normal shape so that it no longer functions; usually caused by a less than optimal pH and temperature.

dendrite Part of a neuron that sends signals toward the cell body.

dendritic cell Antigen-presenting cell of the epidermis and mucous membranes.

denitrification Conversion of nitrate or nitrite to nitrogen gas by bacteria in soil.

dense fibrous connective tissue Type of connective tissue containing many collagen fibers packed together; found in tendons and ligaments, for example.

density-dependent factor Biotic factor, such as disease or competition, that affects population size in a direct relationship to the population's density.

density-independent factor Abiotic factor, such as fire or flood, that affects population size independent of the population's density.

dermis In mammals, thick layer of the skin underlying the epidermis.

desert Ecological biome characterized by a limited amount of rainfall; deserts have hot days and cool nights.

detection Stimulation of sensory receptors by environmental stimuli.

detrital food chain Straight-line linking of organisms according to who eats whom, beginning with detritus.

detrital food web Complex pattern of interlocking and crisscrossing food chains, beginning with detritus.

detritivore Any organism that obtains most of its nutrients from the detritus in an ecosystem.

detritus Partially decomposed material; may be found in either soil or water.

deuterostome Group of coelomate animals in which the second embryonic opening is associated with the mouth; the first embryonic opening, the blastopore, is associated with the anus.

development Process of regulated growth and differentiation of cells and tissues.

diabetes insipidus Condition caused by a lack of antidiuretic hormone (ADH); characterized by excessive thirst and over-production of urine.

diabetes mellitus Condition caused by an insulin imbalance in the body; Type I diabetes results from not enough insulin being produced; Type II diabetes is caused by the body (specifically adipose tissue) not responding to insulin in the blood.

diabetic retinopathy Complication of diabetes that causes the capillaries in the retina to become damaged, potentially causing blindness.

diaphragm In mammals, dome-shaped muscularized sheet separating the thoracic cavity from the abdominal cavity; contraceptive device that prevents sperm from reaching the egg.

diarrhea Excessively frequent and watery bowel movements.

diastole Relaxation period of a heart chamber during the cardiac cycle.

diatom Golden-brown alga with a cell wall in two parts, or valves; significant component of phytoplankton.

diffusion Movement of molecules or ions from a region of higher to lower concentration; it requires no energy and tends to lead to an equal distribution (equilibrium).

digestive system Organ system that consists of digestive organs (stomach, intestine, etc.) and accessory organs (liver, etc.); ingests and digests food; absorbs nutrients and eliminates wastes.

dihybrid cross Cross between parents that differ in two traits.

dinoflagellate Photosynthetic unicellular protist with two flagella, one whiplash and the other located within a groove between protective cellulose plates; significant part of phytoplankton.

diploid (2n) Cell condition in which two of each type of chromosome are present.

directional selection Outcome of natural selection in which an extreme phenotype is favored, usually in a changing environment.

disaccharide Sugar that contains two monosaccharide units; e.g., maltose.

disease Abnormality in the body's physiology that typically upsets homeostasis.

disruptive selection Outcome of natural selection in which the two extreme phenotypes are favored over the average phenotype, leading to more than one distinct form.

distal convoluted tubule Final portion of a nephron that joins with a collecting duct; associated with tubular secretion.

diuretic Chemical that increases the flow of urine; an example is caffeine.

DNA (deoxyribonucleic acid) Nucleic acid polymer produced from covalent bonding of nucleotide monomers that contain the sugar deoxyribose; the genetic material of nearly all organisms.

DNA ligase Enzyme that links DNA fragments; used during production of recombinant DNA to join foreign DNA to vector DNA.

DNA polymerase During replication, an enzyme that joins the nucleotides complementary to a DNA template.

DNA replication Synthesis of a new DNA double helix prior to mitosis and meiosis in eukaryotic cells and during prokaryotic fission in prokaryotic cells.

domain Largest of the categories, or taxa, used to group species; the three domains are Archaea, Bacteria, and Eukarya.

domain Archaea One of the three domains of life; contains prokaryotic cells that often live in extreme habitats and have unique genetic, biochemical, and physiological characteristics; its members are sometimes referred to as *archaea*.

domain Bacteria One of the three domains of life; contains prokaryotic cells that differ from archaea because they have their own unique genetic, biochemical, and physiological characteristics.

domain Eukarya One of the three domains of life, consisting of organisms with eukaryotic cells; includes protists, fungi, plants, and animals.

dominant allele Allele that exerts its phenotypic effect in the heterozygote; it masks the expression of the recessive allele.

dormancy In plants, a cessation of growth under conditions that seem appropriate for growth.

dorsal cavity One of two main body cavities in humans; contains the cranial cavity and vertebral canal.

dorsal root ganglion Mass of sensory neuron cell bodies located in the dorsal root of a spinal nerve.

double fertilization In flowering plants, one sperm nucleus unites with the egg nucleus, and a second sperm nucleus unites with the polar nuclei of an embryo sac.

double helix Double spiral; describes the three-dimensional shape of DNA.

doubling time Number of years it takes for a population to double in size.

dryopithecine Tree dwelling primate existing 12–9 MYA; ancestral to apes.

Duchenne muscular dystrophy Genetic disorder that is characterized by a wasting of muscle tissue; displays an X-linked recessive pattern of inheritance.

ductus arteriosus Structure of the heart associated with fetal circulation; provides a pathway for returning any blood that enters the right ventricle back to the aorta.

duodenum First part of the small intestine, where chyme enters from the stomach.

duplication Change in chromosome structure in which a particular segment is present more than once in the same chromosome.

E

ecdysozoan Protostome characterized by periodic molting of the exoskeleton. Includes the roundworms and arthropods.

echinoderm Invertebrates such as sea stars, sea urchins, and sand dollars; characterized by radial symmetry and a water vascular system.

ecological niche Role an organism plays in its community, including its habitat and its interactions with other organisms.

ecological pyramid Visual depiction of the biomass, number of organisms, or energy content of various trophic levels in a food web—from the producer to the final consumer populations.

ecological succession The gradual replacement of communities in an area following a disturbance (secondary succession) or the creation of new soil (primary succession).

ecology Study of the interactions of organisms with other organisms and with the physical and chemical environment.

ecosystem Biological community together with the associated abiotic environment; characterized by a flow of energy and a cycling of inorganic nutrients.

ecosystem diversity Variety of species in a particular locale, dependent on the species interactions.

ectoderm Outermost primary tissue layer of an animal embryo; gives rise to the nervous system and the outer layer of the integument.

ectothermic Body temperature that varies according to the environmental temperature.

edge effect Phenomenon in which the edges around a landscape patch provide a slightly different habitat than the favorable habitat in the interior of the patch.

egg Also called an ovum; haploid cell that is usually fertilized by a sperm to form a diploid zygote.

elastic cartilage Type of cartilage composed of elastic fibers, allowing greater flexibility.

elastic fiber Yellow fiber in the matrix of connective tissue, providing flexibility.

electrocardiogram (ECG) Recording of the electrical activity associated with the heartbeat.

electron Negative subatomic particle, moving about in an energy level around the nucleus of an atom.

electronegativity The ability of an atom to attract electrons toward itself in a chemical bond.

electron shell The average location, or energy level, of an electron in an atom. Often drawn as concentric circles around the nucleus.

electron transport chain (ETC) Process in a cell that involves the passage of electrons along a series of membrane-bound electron carrier molecules from a higher to lower energy level; the energy released is used for the synthesis of ATP.

elements Substances that cannot be broken down into substances with different properties; composed of only one type atom.

elephantiasis Disease caused by a roundworm called the filarial worm.

elongation Middle stage of translation in which additional amino acids specified by the mRNA are added to the growing polypeptide.

embryo Stage of a multicellular organism that develops from a zygote before it becomes free-living; in seed plants, the embryo is part of the seed.

embryo sac Female gametophyte (megagametophyte) of flowering plants.

embryonic development In humans, the first two months of development following fertilization during which the major organs are formed.

embryonic disk During human development, flattened area during gastrulation from which the embryo arises.

emergency contraception Form of contraception that is taken following unprotected intercourse.

emphysema Disorder of the respiratory system, specifically the lungs, that is characterized by damage to the alveoli, thus reducing the ability to exchange gases with the external environment.

endangered species A species that is in peril of immediate extinction throughout all or most of its range (e.g., California condor, snow leopard).

endergonic reaction Chemical reaction that requires an input of energy; opposite of exergonic reaction.

endocrine gland Ductless organ that secretes hormone(s) into the bloodstream.

endocrine system Organ system involved in the coordination of body activities; uses hormones as chemical signals secreted into the bloodstream.

endocytosis Process by which substances are moved into the cell from the environment; includes phagocytosis, pinocytosis, and receptor-mediated endocytosis.

endoderm Innermost primary tissue layer of an animal embryo that gives rise to the linings of the digestive tract and associated structures.

endodermis Internal plant root tissue forming a boundary between the cortex and the vascular cylinder.

endometrioisis Condition in which endometrial tissue is located outside of the uterine cavity, causing pain and discomfort.

endometrium Mucous membrane lining the interior surface of the uterus.

endoplasmic reticulum (ER) System of membranous saccules and channels in the cytoplasm, often with attached ribosomes.

endoskeleton Protective internal skeleton, as in vertebrates.

endosperm In flowering plants, nutritive storage tissue that is derived from the union of a sperm nucleus and polar nuclei in the embryo sac.

endospore Spore formed within a cell; certain bacteria form endospores.

endosymbiosis *See also* endosymbiotic theory.

endosymbiotic theory Explanation of the evolution of eukaryotic organelles by phagocytosis of prokaryotes.

endothermic Maintenance of a constant body temperature independent of the environmental temperature.

energy Capacity to do work and bring about change; occurs in a variety of forms.

energy of activation Energy that must be added in order for molecules to react with one another.

entropy Measure of disorder or randomness in a system.

envelope Lipid covering of some viruses; located outside of the capsid.

environmental resistance Total of factors in the environment that limit the numerical increase of a population in a particular region.

enzymatic protein Protein that catalyzes a specific reaction; may be found in the plasma membrane or the cytoplasm of the cell.

enzyme Organic catalyst, usually a protein, that speeds a reaction in cells due to its particular shape.

enzyme inhibition Means by which cells regulate enzyme activity; may be competitive or noncompetitive inhibition.

eosinophil White blood cell containing cytoplasmic granules that stain with acidic dye.

epidermal tissue Exterior tissue, usually one cell thick, of leaves, young stems, roots, and other parts of plants.

epidermis In mammals, the outer, protective layer of the skin; in plants, tissue that covers roots, leaves, and stems of nonwoody organisms.

epididymis Location of sperm maturation in an adult human; located in the testis.

epigenetic inheritance An inheritance pattern in which a nuclear gene has been modified but the changed expression of the gene is not permanent over many generations; the transmission of genetic information by means that are not based on the coding sequences of a gene.

epiglottis Structure that covers the glottis, the air-tract opening, during the process of swallowing.

epinephrine Hormone secreted by the adrenal medulla in times of stress; adrenaline.

epiphyte Plant that takes its nourishment from the air because its placement in other plants gives it an aerial position.

episiotomy Surgical procedure during the process of birth (parturition) that involves a small incision to enlarge the vagina; allows for passage of the baby's head.

epithelial tissue Tissue that lines hollow organs and covers surfaces.

erectile disfunction (ED) Also called impotence, this represents an inability to produce or maintain an erection.

erythropoietin (EPO) Hormone produced by the kidneys that speeds red blood cell formation.

esophagus Muscular tube for moving swallowed food from the pharynx to the stomach.

essential amino acid One of eight amino acids that are unable to be produced by the human body and therefore must be obtained in the diet.

essential fatty acid Forms of fatty acids that are unable to be produced by the human body and therefore must be obtained in the diet.

estrogen Female sex hormone that helps maintain sexual organs and secondary sex characteristics.

estuary Portion of the ocean located where a river enters and fresh water mixes with salt water.

ethylene Plant hormone that causes ripening of fruit and is also involved in abscission.

euchromatin Chromatin with a lower level of compaction and therefore accessible for transcription.

eudicot (Eudicotyledone) Flowering plant group; members have two embryonic leaves (cotyledons), net-veined leaves, vascular bundles in a ring, flower parts in fours or fives and their multiples, and other characteristics.

euglenoid Flagellated and flexible freshwater unicellular protist that usually contains chloroplasts and has a semirigid cell wall.

eukaryotic cell (eukaryote) Type of cell that has a membrane-bound nucleus and membranous organelles; found in organisms within the domain Eukarya.

eutrophication Enrichment of water by inorganic nutrients used by phytoplankton. Often, overenrichment caused by human activities leads to excessive bacterial growth and oxygen depletion.

evaporation The process by which liquid water turns into gaseous form (water vapor).

evergreen Land plant that sheds leaves over a long period, so some leaves are always present.

evolution Genetic change in a species over time resulting in the development of genetic and phenotypic differences that are the basis of natural selection; descent of organisms from a common ancestor.

exchange pool Location in a biogeochemical cycle from which organisms generally take a specific chemical (i.e., phosphorus)

excretion Elimination of metabolic wastes by an organism at exchange boundaries such as the plasma membrane of unicellular organisms and excretory tubules of multicellular animals.

exergonic reaction Chemical reaction that releases energy; opposite of endergonic reaction.

exocrine gland Gland that secretes its product to an epithelial surface directly or through ducts.

exocytosis Process in which an intracellular vesicle fuses with the plasma membrane so that the vesicle's contents are released outside the cell.

exon Segment of mRNA containing the protein-coding portion of a gene that remains within the mRNA after splicing has occurred.

exotic species Nonnative species that migrate or are introduced by humans into a new ecosystem; also called alien species.

experiment A test of a hypothesis that examines the influence of a single variable. Often involves both control and test groups.

experimental design Methodology by which an experiment will seek to support the hypothesis.

experimental variable Factor of the experiment being tested.

expiration Act of expelling air from the lungs; exhalation.

exponential growth Growth, particularly of a population, in which the increase occurs in the same manner as compound interest.

external respiration Exchange of oxygen and carbon dioxide between alveoli of the lungs and blood.

exteroceptor Sensory receptors of the peripheral nervous system that detect stimuli from outside the body.

extinct; extinction Total disappearance of a species or higher group.

extracellular matrix (ECM) Nonliving substance secreted by some animal cells; is composed of protein and polysaccharides.

extraembryonic membrane Membrane that is not a part of the embryo but is necessary to the continued existence and health of the embryo.

ex vivo gene therapy Gene therapy in which cells are removed from an organism, and DNA is injected to correct a genetic defect; the cells are returned to the organism to treat a disease or disorder.

eyespot apparatus Structure found in some species of protists that is used to sense light and light intensity.

F

facilitated transport Passive transfer of a substance into or out of a cell along a concentration gradient by a process that requires a protein carrier.

FAD Flavin adenine dinucleotide; a coenzyme of oxidation-reduction that becomes $FADH_2$ as oxidation of substrates occurs in the mitochondria during cellular respiration, FAD then delivers electrons to the electron transport chain.

fall overturn Mixing process that occurs in fall in stratified lakes, whereby oxygen-rich top waters mix with nutrient-rich bottom waters.

familial hypercholesterolemia Genetic disorder that causes an accumulation of cholesterol in the blood due to defects in the LDL-receptors on the cell surface.

family One of the categories, or taxa, used to group species; the taxon located above the genus level.

farsighted Condition in which an individual cannot focus on objects closely; caused by the focusing of the image behind the retina of the eye.

fat Organic molecule that contains glycerol and three fatty acids; energy storage molecule.

fate maps Diagram used by developmental geneticists that tracks the differentiation of a cell in an embryo.

fatty acid Molecule that contains a hydrocarbon chain and ends with an acid group.

female condom Contraceptive device, usually made from polyurethane, the fits into the female's cervix.

fermentation Anaerobic breakdown of glucose that results in a gain of two ATP and end products such as alcohol and lactate; occurs in the cytoplasm of cells.

fern Member of a group of land plants that have large fronds; in the sexual life cycle, the independent gametophyte produces flagellated sperm, and the vascular sporophyte produces windblown spores.

fertilization Fusion of sperm and egg nuclei, producing a zygote that develops into a new individual.

fiber General term for indigestible plant material, typically complex carbohydrates such as cellulose.

fibrin Protein involved in blood clotting; acts to trap cells to seal wounds in the blood vessels.

fibroblast Cell found in loose connective tissue that synthesizes collagen and elastic fibers in the matrix.

fibrocartilage Cartilage with a matrix of strong collagenous fibers.

fibromyalgia Disorder associated with chronic, widespread pain; cause is not currently known.

fibrous joint Type of joint between two bones that is immovable; sutures are examples of fibrous joints.

fibrous root system In most monocots, a mass of similarly sized roots that cling to the soil.

filament End-to-end chains of cells that form as cell division occurs in only one plane; in plants, the elongated stalk of a stamen.

fimbria (pl., **fimbriae**) Small, bristlelike fiber on the surface of a bacterial cell, which attaches bacteria to a surface; also fingerlike extension from the oviduct near the ovary.

first messenger Chemical signal such as a peptide hormone that binds to a plasma membrane receptor protein and alters the metabolism of a cell because a second messenger is activated.

fish Aquatic, gill-breathing vertebrate that usually has fins and skin covered with scales; fishes were among the earliest vertebrates that evolved.

fitness Ability of an organism to reproduce and pass its genes to the next fertile generation; measured against the ability of other organisms to reproduce in the same environment.

fixed action pattern (FAP) Innate behavior pattern that is stereotyped, spontaneous, independent of immediate control, genetically encoded, and independent of individual learning.

flagellates Heterotrophic protozoans that propel themselves using one or more flagella.

flagellum (pl., **flagella**) Long, slender extension used for locomotion by some bacteria, protozoans, and sperm.

flagship species Species that evoke a strong emotional response in humans; charismatic, cute, regal (e.g., lions, tigers, dolphins, pandas).

flame cells Cells in the excretory system of flatworms (and other animals) that contain cilia; function is to keep water/fluids moving in the excretory system.

flatworm Invertebrates such as planarians and tapeworms with extremely thin bodies, a three-branched gastrovascular cavity, and a ladder type nervous system.

flower Reproductive organ of a flowering plant, consisting of several kinds of modified leaves arranged in concentric rings and attached to a modified stem called the receptacle.

fluid-mosaic model Model for the plasma membrane based on the changing location and pattern of protein molecules in a fluid phospholipid bilayer.

follicle Structure in the ovary of animals that contains an oocyte; site of oocyte production.

follicle-stimulating hormone (FSH) Hormone released by the anterior pituitary; in males it promotes the production of sperm; in females it promotes the development of the follicle in the ovary.

follicular phase First half of the ovarian cycle, during which the follicle matures and much estrogen (and some progesterone) is produced.

fontanels Membranous regions found in the skulls of infants; permit the skull to move through the birth canal.

food chain The order in which one population feeds on another in an ecosystem, thereby showing the flow of energy from a detritivore (detrital food chain) or a producer (grazing food chain) to the final consumer.

food web In ecosystems, a complex pattern of interlocking and crisscrossing food chains.

foot Animal structure involved in locomotion.

foramen magnum Large opening at the base of the skull that allows for a connection between the spinal cord and brain stem.

foramen ovale Structural feature of the fetal heart that allows for blood to move directly between the right and left atrium.

foraminiferan A protist bearing a calcium carbonate test with many openings through which pseudopods extend.

formed elements Portion of the blood that consists of erythrocytes, leukocytes, and platelets (thrombocytes).

formula A group of symbols and numbers used to express the composition of a compound.

fossil Any past evidence of an organism that has been preserved in the Earth's crust.

fossil fuel Remains of once-living organisms that are burned to release energy, such as coal, oil, and natural gas.

founder effect Cause of genetic drift due to colonization by a limited number of individuals who, by chance, have different genotype and allele frequencies than the parent population.

fovea centralis Region of the retina consisting of densely packed cones; responsible for the greatest visual acuity.

fragile X syndrome Genetic condition caused by an abnormal number of nucleotide repeats; named after the appearance, and not physical characteristics, of the X chromosome.

frameshift mutation Insertion or deletion of at least one base so that the reading frame of the corresponding mRNA changes.

free energy Energy in a system that is capable of performing work.

fruit Flowering plant structure consisting of one or more ripened ovaries that usually contain seeds.

functional genomics Study of gene function at the genome level. It involves the study of many genes simultaneously and the use of DNA microarrays.

functional group Specific cluster of atoms attached to the carbon skeleton of organic molecules that enters into reactions and behaves in a predictable way.

fungi (sing., **fungus**) Eukaryotic saprotrophic decomposer; the body is made up of filaments called hyphae that form a mass called a mycelium.

G

gallbladder Organ attached to the liver that serves to store and concentrate bile.

gallstones Hardened form of bile that may block the bile ducts and cause pancreatitis.

gamete Haploid sex cell; e.g., egg or sperm.

gametogenesis Development of the male and female sex gametes.

gametophyte Haploid generation of the alternation of generations life cycle of a plant; produces gametes that unite to form a diploid zygote.

ganglia (sing., **ganglion**) Collection or bundle of neuron cell bodies usually outside the central nervous system.

gap junction Junction between cells formed by the joining of two adjacent plasma membranes; it lends strength and allows ions, sugars, and small molecules to pass between cells.

gastropod Mollusc with a broad, flat foot for crawling (e.g., snails and slugs).

gastrovascular cavity Blind digestive cavity in animals that have a sac body plan.

gastrula Stage of animal development during which the germ layers form, at least in part, by invagination.

gastrulation Formation of a gastrula from a blastula; characterized by an invagination to form cell layers of a caplike structure.

gene Unit of heredity existing as alleles on the chromosomes; in diploid organisms, typically two alleles are inherited—one from each parent.

gene cloning DNA cloning to produce many identical copies of the same gene.

gene flow Sharing of genes between two populations through interbreeding.

gene mutation Altered gene whose sequence of bases differs from the original sequence.

gene pool Total of the alleles of all the individuals in a population.

gene therapy Correction of a detrimental mutation by the insertion of DNA sequences into the genome of a cell.

genetically modified organisms (GMO) Organisms whose genetic material has been altered or enhanced using DNA technology.

genetic code Universal code that has existed for eons and allows for conversion of DNA and RNA's chemical code to a sequence of amino acids in a protein. Each codon consists of three bases that stand for one of the 20 amino acids found in proteins or directs the termination of translation.

genetic diversity Variety among members of a population.

genetic drift Mechanism of evolution due to random changes in the allelic frequencies of a population; more likely to occur in small populations or when only a few individuals of a large population reproduce.

genetic engineering Human-induced changes in the genome of an organism, often performed for the benefit of humans.

genetic equilibrium Description for a population in which the frequency of alleles for a given trait is not changing over time.

genetic profile An individual's genome, including any possible mutations.

genetic recombination Process in which chromosomes are broken and rejoined to form novel combinations; in this way offspring receive alleles in combinations different from their parents.

genetic variation Differences in the sequences of nucleotides in the genes of an individual. Introduced during the process of meiosis, or by mutation.

genital herpes Sexually-transmitted disease caused by the herpes simplex virus (both type 1 and type 2).

genital warts Sexually-transmitted disease caused by the human papillomavirus (HPV).

genome Sum of all of the genetic information in a cell or organism.

genomics Area of study that examines the genome of a species or group of species.

genotype Genes of an organism for a particular trait or traits; often designated by letters—for example, *BB* or *Aa*.

genus One of the categories, or taxa, used to group species; contains those species that are most closely related through evolution.

geologic timescale History of the Earth based on the fossil record and divided into eras, periods, and epochs.

germ cell During zygote development, cells that are set aside from the somatic cells and that will eventually undergo meiosis to produce gametes.

germinate Beginning of growth of a seed, spore, or zygote, especially after a period of dormancy.

germ layer Primary tissue layer of a vertebrate embryo—namely, ectoderm, mesoderm, or endoderm.

gerontology Study of aging and the aging processes.

gibberellin Plant hormone promoting increased stem growth; also involved in flowering and seed germination.

gigantism Condition caused by the overproduction of growth hormone during childhood; individuals often also have diabetes mellitus and other health problems.

gills Respiratory organ in most aquatic animals; in fish, an outward extension of the pharynx.

ginkgo Member of phylum Ginkgophyte; maidenhair tree.

girdling Removing a strip of bark from around a tree.

gland Epithelial cell or group of epithelial cells that are specialized to secrete a substance.

glaucoma Condition in which the fluid in the eye (aqueous humor) accumulates, increasing pressure in the eye and damaging nerve fibers.

global warming Predicted increase in the Earth's temperature due to human activities that promote the greenhouse effect.

glomerular capsule Cuplike structure that is the initial portion of a nephron.

glomerular filtration Movement of small molecules from the glomerulus into the glomerular capsule due to the action of blood pressure.

glomerulus Capillary network within the glomerular capsule of a nephron.

glottis Opening for airflow in the larynx.

glucagon Hormone produced by the pancreas that stimulates the liver to break down glycogen, thus raising blood glucose levels.

glucocorticoid Type of hormone secreted by the adrenal cortex that influences carbohydrate, fat, and protein metabolism; *See also* cortisol.

glucose Six-carbon monosaccharide; used as an energy source during cellular respiration and as a monomer of the structural polysaccharides.

glycemic index Nutritional measurement of how the body responds to the glucose level of a food. Foods with a high glycemic index result in the release of high amounts of insulin.

glycogen Storage polysaccharide found in animals; composed of glucose molecules joined in a linear fashion but having numerous branches.

glycolipid Lipid in plasma membranes that contains an attached carbohydrate chain; assembled in the Golgi apparatus.

glycolysis Anaerobic breakdown of glucose that results in a gain of two ATP and the production of pyruvate; occurs in the cytoplasm of cells.

glycoprotein Protein in plasma membranes that has an attached carbohydrate chain; assembled in the Golgi apparatus.

gnetophyte Member of one of the four phyla of gymnosperms; Gnetophyta has only three living genera, which differ greatly from one another—e.g., *Welwitschia* and *Ephedra*.

Golgi apparatus Organelle consisting of sacs and vesicles that processes, packages, and distributes molecules about or from the cell.

gonad Organ that produces gametes; the ovary produces eggs, and the testis produces sperm.

gonadotropic hormone Substance secreted by the anterior pituitary that regulates the activity of the ovaries and testes; principally, follicle-stimulating hormone (FSH) and luteinizing hormone (LH).

gonadotropin-releasing hormone (GnRH) Hormone released by the hypothalamus that influences the secretion of hormones by the anterior pituitary, thus regulating the activity of the testes.

gonorrhea Sexually-transmitted disease caused by the bacterium *Neisseria gonorrhoeae.*

gout Condition caused by the inability of the urinary system to remove uric acid from the body.

Gram stain Method of determining whether a culture of bacteria possess a layer of exposed peptidoglycan outside of their plasma membrane; historically used to classify bacteria as Gram-positive or Gram-negative, now often used to determine the effective treatment of antibiotics.

granum (pl., **grana**) Stack of chlorophyll-containing thylakoids in a chloroplast.

grassland Biome characterized by rainfall greater than 25 cm/yr, grazing animals, and warm summers; includes the prairie of the U.S. midwest and the African savanna.

Graves disease Disease associated with the overproduction of T_3 and T_4 by the thyroid; caused by antibodies that interact with TSH receptors in the thyroid.

gravitational equilibrium Maintenance of balance when the head and body are motionless.

gravitropism Growth response of roots and stems of plants to the Earth's gravity; roots demonstrate positive gravitropism, and stems demonstrate negative gravitropism.

gray crescent Region that forms during early cellular differentiation of the embryo; believed to contain unique chemical signals that are important during development.

gray matter Nonmyelinated axons and cell bodies in the central nervous system.

grazing food chain A flow of energy to a straight-line linking of organisms according to who eats whom.

grazing food web Complex pattern of interlocking and crisscrossing food chains that begins with populations of autotrophs serving as producers.

green algae Members of a diverse group of photosynthetic protists; contain chlorophylls *a* and *b* and have other biochemical characteristics like those of plants.

greenhouse effect Reradiation of solar heat toward the Earth, caused by an atmosphere that allows the sun's rays to pass through but traps the heat in the same manner as the glass of a greenhouse.

greenhouse gases Gases in the atmosphere such as carbon dioxide, methane, water vapor, ozone, and nitrous oxide that are involved in the greenhouse effect.

ground tissue Tissue that constitutes most of the body of a plant; consists of parenchyma, collenchyma, and sclerenchyma cells that function in storage, basic metabolism, and support.

growth factor A hormone or chemical, secreted by one cell, that may stimulate or inhibit growth of another cell or cells.

growth hormone (GH) Substance secreted by the anterior pituitary; controls size of an individual by promoting cell division, protein synthesis, and bone growth.

growth plate Area of cartilage within a long bone that permits the bone to increase in length.

guanine (G) One of four nitrogen-containing bases in nucleotides composing the structure of DNA and RNA; pairs with cytosine.

guard cell One of two cells that surround a leaf stoma; changes in the turgor pressure of these cells cause the stoma to open or close.

Guillain-Barré syndrome Disease caused by inflammation in the nervous system, resulting in demyelination of the nerve axons in the peripheral nervous system.

gymnosperm Type of woody seed plant in which the seeds are not enclosed by fruit and are usually borne in cones, such as those of the conifers.

H

habitat Place where an organism lives and is able to survive and reproduce.

hair cells Cells located in the inner ear that function as mechanoreceptors for hearing.

hair follicle Tubelike depression in the skin in which a hair develops.

halophile Type of archaean that lives in extremely salty habitats.

haploid (n) Cell condition in which only one of each type of chromosome is present.

Hardy-Weinberg equilibrium Mathematical law stating that the gene frequencies in a population remain stable if evolution does not occur due to nonrandom mating, selection, migration, and genetic drift.

heart Muscular organ whose contraction causes blood to circulate in the body of an animal.

heart attack Damage to the myocardium due to blocked circulation in the coronary arteries; myocardial infarction.

heat Type of kinetic energy associated with the random motion of molecules.

helper T cell Secretes lymphokines, which stimulate all kinds of immune cells.

hemocoel Residual coelom found in arthropods, which is filled with hemolymph.

hemodialysis Treatment for kidney failure in which the patient's blood is passed through an artificial kidney to remove waste material and toxins.

hemoglobin (Hb) Iron-containing respiratory pigment occurring in vertebrate red blood cells and in the blood plasma of some invertebrates.

hemophilia Genetic disorder that results in a deficiency of a clotting factor

hepatitis Inflammation of the liver. Viral hepatitis occurs in several forms.

herbaceous stem Nonwoody stem; herbaceous plants tend to die back to ground level at the end of the growing season.

herbivore Primary consumer in a grazing food chain; a plant eater.

hermaphroditic Type of animal that has both male and female sex organs.

heterochromatin Highly compacted chromatin that is not accessible for transcription.

heterotrophy Organism that cannot synthesize needed organic compounds from inorganic substances and therefore must take in organic food.

heterozygote advantage Situation in which individuals heterozygous for a trait have a selective advantage over those who are homozygous dominant or recessive; an example is sickle cell disease.

heterozygous Possessing unlike alleles for a particular trait.

hexose Any monosaccharide that contains six carbons; examples are glucose and galactose.

hippocampus Region of the central nervous system associated with learning and memory; part of the limbic system.

histamine Substance, produced by basophils in blood and mast cells in connective tissue, that causes capillaries to dilate.

histone A group of proteins involved in forming the nucleosome structure of eukaryote chromatin.

homeodomain Conserved DNA-binding region of transcription factors encoded by the homeobox of homeotic genes.

homeostasis Maintenance of normal internal conditions in a cell or an organism by means of self-regulating mechanisms.

homeotic genes Genes that control the overall body plan by controlling the fate of groups of cells during development.

hominid Taxon that includes humans, chimpanzees, gorillas, and orangutans.

hominin Taxon that includes humans and species very closely related to humans plus chimpanzees.

hominine Taxon that includes humans, chimpanzees, and gorillas.

hominoid Taxon that includes the hominids plus the gibbons.

homologous chromosome Member of a pair of chromosomes that are alike and come together in synapsis during prophase of the first meiotic division; a *homologue*.

homologous gene Gene that codes for the same protein, even if the base sequence may be different.

homologous, homologous structure A structure that is similar in different types of organisms because these organisms descended from a common ancestor.

homologue Member of a homologous pair of chromosomes.

homology Similarity of parts or organs of different organisms caused by evolutionary derivation from a corresponding part or organ in a remote ancestor, and usually having a similar embryonic origin.

homozygous Possessing two identical alleles for a particular trait.

hormone Chemical messenger produced in one part of the body that controls the activity of other parts.

horsetail A seedless vascular plant having only one genus (*Equisetum*) in existence today; characterized by rhizomes, scalelike leaves, strobili, and tough, rigid stems.

host Organism that provides nourishment and/or shelter for a parasite.

human chorionic gonadotropin (HCG) Gonadotropic hormone produced by the chorion that functions to maintain the uterine lining.

Human Genome Project (HGP) Initiative to determine the complete sequence of the human genome and to analyze this information.

human immunodeficiency virus (HIV) Virus responsible for the disease AIDS.

hunter-gatherer A hominin that hunted animals and gathered plants for food.

Huntington disease Autosomal dominant genetic disorder that affects the nervous system; results in a progressive loss of neurons in the brain.

hyaline cartilage Cartilage whose cells lie in lacunae separated by a white translucent matrix containing very fine collagen fibers.

hybridization Interbreeding between two different species; typically prevented by prezygotic isolation mechanisms.

hydrogen bond Weak bond that arises between a slightly positive hydrogen atom of one molecule and a slightly negative atom of another molecule or between parts of the same molecule.

hydrolysis reaction Splitting of a chemical bond by the addition of water, with the H^+ going to one molecule and the OH^- going to the other.

hydrophilic Type of molecule, often polar, that interacts with water by dissolving in water and/or by forming hydrogen bonds with water molecules.

hydrophobic Type of molecule, that is typically nonpolar, and therefore does not interact easily with water.

hydrostatic skeleton Fluid-filled body compartment that provides support for muscle contraction resulting in movement; seen in cnidarians, flatworms, roundworms, and segmented worms.

hydrothermal vent Hot springs in the seafloor along ocean ridges where heated sea water and sulfate react to produce hydrogen sulfide; here, chemosynthetic bacteria support a community of varied organisms.

hyperthyroidism Caused by the over-secretion of hormones from the thyroid gland; symptoms include hyperactivity, nervousness, and insomnia.

hypertension Form of cardiovascular disease characterized by blood pressure over 140/95 (over 45 years of age) or 130/90 (under 45 years of age).

hypertonic solution Higher solute concentration (less water) than the cytoplasm of a cell; causes cell to lose water by osmosis.

hypha Filament of the vegetative body of a fungus.

hypothalamic-inhibiting hormone One of many hormones produced by the hypothalamus that inhibits the secretion of an anterior pituitary hormone.

hypothalamic-releasing hormone One of many hormones produced by the hypothalamus that stimulates the secretion of an anterior pituitary hormone.

hypothalamus In vertebrates, part of the brain that helps regulate the internal environment of the body—for example, heart rate, body temperature, and water balance.

hypothesis Supposition established by reasoning after consideration of available evidence; it can be tested by obtaining more data, often by experimentation.

hypothyroidism Caused by the under-secretion of hormones from the thyroid gland; symptoms include weight gain, lethargic behavior, and depression.

hypotonic solution Solution that contains a lower solute (more water) concentration than the cytoplasm of a cell; causes cell to gain water by osmosis.

I

ileum Region of the small intestine; connects the jejunum and large intestine.

immediate allergic response Allergic response that occurs within seconds of contact with an allergen; caused by the attachment of the allergen to IgE antibodies.

immune system System associated with protection against pathogens, toxins, and some cancerous cells; in humans this is an organ system.

immunity Ability of the body to protect itself from foreign substances and cells, including disease-causing agents.

immunization Strategy for achieving immunity to the effects of specific disease-causing agents.

immunodeficiency disease Condition that is the result of the immune system's inability to protect the body.

immunoglobulin (Ig) Globular plasma protein that functions as an antibody.

implantation In placental mammals, the embedding of an embryo at the blastocyst stage into the endometrium of the uterus.

imprinting Learning to make a particular response to only one type of animal or object.

inbreeding Mating between closely related individuals; influences the genotype ratios of the gene pool.

inclusive fitness Fitness that results from personal reproduction and from helping nondescendant relatives reproduce.

incomplete dominance Inheritance pattern in which an offspring has an intermediate phenotype, as when a red-flowered plant and a white-flowered plant produce pink-flowered offspring.

incomplete penetrance Dominant alleles that are either not always, or partially expressed.

incus Bone found in the middle ear that assists in the transmission of sound to the inner ear; also called the anvil.

independent assortment Alleles of unlinked genes segregate independently of each other during meiosis so that the gametes can contain all possible combinations of alleles.

induced fit model Change in the shape of an enzyme's active site that enhances the fit between the active site and its substrate(s).

induction Ability of a chemical or a tissue to influence the development of another tissue.

inductive reasoning Using specific observations and the process of logic and reasoning to arrive at general scientific principles.

infertility Inability to have as many children as desired.

inflammatory response Tissue response to injury that is characterized by redness, swelling, pain, and heat.

inheritance of acquired characteristics Lamarckian belief that characteristics acquired during the lifetime of an organism can be passed on to offspring.

initiation First stage of translation in which the translational machinery binds an mRNA and assembles.

innate immunity An immune response that does not require a previous exposure to the pathogen.

inner ear Portion of the ear consisting of a vestibule, semicircular canals, and the cochlea where equilibrium is maintained and sound is transmitted.

insect Type of arthropod. The head has antennae, compound eyes, and simple eyes; the thorax has three pairs of legs and often wings; and the abdomen has internal organs.

insertion When referring to a muscle, the connection of the muscle to the bone that moves.

inspiration Act of taking air into the lungs; inhalation.

insulin Hormone released by the pancreas that serves to lower blood glucose levels by stimulating the uptake of glucose by cells, especially muscle and liver cells.

integration Summing up of excitatory and inhibitory signals by a neuron or by some part of the brain.

integumentary system Organ system of humans that contains the skin, protects the body, synthesizes sensory input, assists in temperature regulation, and synthesizes vitamin D.

interferon Antiviral agent produced by an infected cell that blocks the infection of another cell.

interkinesis Period of time between meiosis I and meiosis II during which no DNA replication takes place.

intermediate filament Ropelike assemblies of fibrous polypeptides in the cytoskeleton that provide support and strength to cells; so called because they are intermediate in size between actin filaments and microtubules.

internal respiration Exchange of oxygen and carbon dioxide between blood and tissue fluid.

interneuron Neuron located within the central nervous system that conveys messages between parts of the central nervous system.

internode In vascular plants, the region of a stem between two successive nodes.

interoceptor Sensory receptors of the peripheral nervous system that detect stimuli from inside the body.

interphase Stages of the cell cycle (G_1, S, G_2) during which growth and DNA synthesis occur when the nucleus is not actively dividing.

interstitial cells Cells of the male reproductive system that secrete the androgens (such as testosterone).

interstitial cell-stimulating hormone (ICSH) Another name for luteinizing hormone in males; regulates the production of testosterone by the interstitial cells.

intervertebral disks Fibrocartilage located between the vertebrae that act as shock absorbers.

intrauterine device (IUD) Contraceptive device that is inserted into the female's uterus; believed to work by altering the environment of the uterus.

intron Intervening sequence found between exons in mRNA; removed by RNA processing before translation.

inversion Change in chromosome structure in which a segment of a chromosome is turned around 180°; this reversed sequence of genes can lead to altered gene activity and abnormalities.

invertebrate Animal without a vertebral column or backbone.

ion Charged particle that carries a negative or positive charge.

ionic bond Chemical bond in which ions are attracted to one another by opposite charges.

iris Muscular ring that surrounds the pupil and regulates the passage of light through this opening.

isomer Molecules with the same molecular formula but a different structure, and therefore a different shape.

isotonic solution Solution that is equal in solute concentration to that of the cytoplasm of a cell; causes cell to neither lose nor gain water by osmosis.

isotope Atoms of the same element having the same atomic number but a different mass number due to a variation in the number of neutrons.

J

jaundice Yellowish tint to the skin caused by an abnormal amount of bilirubin (bile pigment) in the blood, indicating liver malfunction.

jawless fishes Type of fishes that lack jaws (agnathan); includes hagfishes and lampreys.

jejunum Region of the small intestine located between the duodenum and ileum.

joint Articulation between two bones of a skeleton.

junction protein(s) Proteins in the cell membrane that assist in cell-to-cell communication.

K

karyotype Chromosomes arranged by pairs according to their size, shape, and general appearance in mitotic metaphase.

keystone species Species whose activities significantly affect community structure.

kidney stones Hardened granules that may form on the renal pelvis of the kidney; caused by factors such as diet, infections, or pH imbalance.

kidneys Paired organs of the vertebrate urinary system that regulate the chemical composition of the blood and produce a waste product called urine.

kinetic energy Energy associated with motion.

kinetochore An assembly of proteins that attaches to the centromere of a chromosome during mitosis.

kingdom One of the categories, or taxa, used to group species; the taxon above phylum.

kin selection Indirect selection; adaptation to the environment due to the reproductive success of an individual's relatives.

***K*-selection** Favorable life-history strategy under stable environmental conditions characterized by the production of a few offspring with much attention given to offspring survival.

L

lactation Secretion of milk by mammary glands, usually for the nourishment of an infant.

lacteal Lymphatic vessel in an intestinal villus; aids in the absorption of fats.

lacuna Small pit or hollow cavity, as in bone or cartilage, where a cell or cells are located.

lake Body of fresh water, often classified by nutrient status, such as oligotrophic (nutrient- poor) or eutrophic (nutrient-rich).

lancelet Invertebrate chordate with a body that resembles a lancet and has the four chordate characteristics as an adult.

landscape diversity Variety of habitat elements within an ecosystem (e.g., plains, mountains, and rivers).

lanugo Fine, down-like covering of the fetus.

large intestine In vertebrates, portion of the digestive tract that follows the small intestine; in humans, consists of the cecum, colon, rectum, and anal canal.

laryngitis Inflammation of the larynx, usually resulting in an inability or difficulty in speaking.

larynx Cartilaginous organ located between the pharynx and the trachea; in humans, contains the vocal cords; sometimes called the voice box.

last universal common ancestor (LUCA) The first living organism on the planet, from which all life evolved.

law Universal principle that describes the basic functions of the natural world.

law of independent assortment Mendelian principle that explains how combinations of traits appear in gametes; *see also* independent assortment.

law of segregation Mendelian principle that explains how, in a diploid organism, alleles separate during the formation of the gametes.

laws of thermodynamics Two laws explaining energy and its relationships and exchanges. The first, also called the "law of conservation," says that energy cannot be created or destroyed but can only be changed from one form to another; the second says that energy cannot be changed from one form to another without a loss of usable energy.

leaf Lateral appendage of a stem, highly variable in structure, often containing cells that carry out photosynthesis.

leaf vein Vascular tissue within a leaf.

learning Relatively permanent change in an animal's behavior that results from practice and experience.

leech Blood-sucking annelid, usually found in fresh water, with a sucker at each end of a segmented body.

lens Transparent, "disk" structure found in the vertebrate eye behind the iris; brings objects into focus on the retina.

leptin Hormone produced by adipose tissue that acts on the hypothalamus to signal satiety (fullness).

less-developed country (LDC) Country that is becoming industrialized; typically, population growth is expanding rapidly, and the majority of people live in poverty.

leukocyte Another name for a white blood cell (WBC).

lichen Symbiotic relationship between certain fungi and either cyanobacteria or algae, in which the fungi possibly provide inorganic food or water and the algae or cyanobacteria provide organic food.

ligament Tough cord or band of dense fibrous tissue that binds bone to bone at a joint.

light reaction Portion of photosynthesis that captures solar energy and takes place in thylakoid membranes of chloroplasts; it produces ATP and NADPH.

lignin Chemical that hardens the cell walls of land plants.

limbic system In humans, functional association of various brain centers, including the amygdala and hippocampus; governs learning and memory and various emotions such as pleasure, fear, and happiness.

limiting factor Resource or environmental condition that restricts the abundance and distribution of an organism.

lineage Line of descent represented by a branch in a phylogenetic tree.

linkage group Term used to identify groups of alleles on a chromosome that tend to be inherited together.

lipase Fat-digesting enzyme secreted by the pancreas.

lipid Class of organic compounds that tends to be soluble in nonpolar solvents; includes fats and oils.

lipopolysaccharide Lipid-sugar molecules that are found on the exterior of some species of bacteria; presence indicates that the species is Gram-negative.

liposome Droplet of phospholipid molecules formed in a liquid environment.

littoral zone Shore zone between high-tide mark and low-tide mark; also, shallow water of a lake where light penetrates to the bottom.

liver Large, dark red internal organ that produces urea and bile, detoxifies the blood, stores glycogen, and produces the plasma proteins, among other functions.

liverwort Bryophyte with a lobed or leafy gametophyte and a sporophyte composed of a stalk and capsule.

lobe-finned fishes Type of fishes with limblike fins; also called the sarcopterygians.

locus Physical location of a trait (or gene) on a chromosome.

logistic growth Population increase that results in an S-shaped curve; growth is slow at first, steepens, and then levels off due to environmental resistance.

long-day plant Plant that flowers when day length is longer than a critical length; e.g., wheat, barley, clover, and spinach.

loop of the nephron Portion of a nephron between the proximal and distal convoluted tubules; functions in water reabsorption.

loose fibrous connective tissue Tissue composed mainly of fibroblasts widely separated by a matrix containing collagen and elastic fibers.

lophophoran A general term to describe several groups of lophotrochozoans that have a feeding structure called a lophophore.

lophotrochozoa Main group of protostomes; widely diverse. Includes the flatworms, rotifers, annelids, and molluscs.

lumen Cavity inside any tubular structure, such as the lumen of the digestive tract.

lung cancer Uncontrolled cell growth that affects any component of the respiratory system.

lungs Internal respiratory organ containing moist surfaces for gas exchange.

luteal phase Second half of the ovarian cycle, during which the corpus luteum develops and much progesterone (and some estrogen) is produced.

luteinizing hormone (LH) Hormone released by the anterior pituitary; in males it regulates the production of testosterone by the interstitial cells; in females it promotes the development of the corpus luteum in the ovary; also called interstitial cell-stimulating hormone (ICSH) in males.

lymph Fluid, derived from tissue fluid, that is carried in lymphatic vessels.

lymphatic capillary Smallest vessels of the lymphatic system; closed-ended; responsible for the uptake of fluids from the surrounding tissues.

lymphatic system Organ system consisting of lymphatic vessels and lymphatic organs; transports lymph and lipids, and aids the immune system.

lymphatic vessel Vessel that carries lymph.

lymph node Mass of lymphatic tissue located along the course of a lymphatic vessel.

lymphocyte Specialized white blood cell that functions in specific defense; occurs in two forms—T lymphocytes and B lymphocytes.

lymphoid organ Organ other than a lymphatic vessel that is part of the lymphatic system; the lymphatic organs are the lymph nodes, tonsils, spleen, thymus gland, and bone marrow.

lysogenic cycle Bacteriophage life cycle in which the virus incorporates its DNA into that of a bacterium; occurs preliminary to the lytic cycle.

lysosome Membrane-bound vesicle that contains hydrolytic enzymes for digesting macromolecules and bacteria; used to recycle worn-out cellular organelles.

lytic cycle Bacteriophage life cycle in which the virus takes over the operation of the bacterium immediately upon entering it and subsequently destroys the bacterium.

M

macroevolution Large-scale evolutionary change, such as the formation of new species.

macrophage In vertebrates, large phagocytic cell derived from a monocyte that ingests microbes and debris.

macular degeneration Condition in which the capillaries supplying the retina of the eye become damaged, resulting in reduced vision and blindness.

malaria Disease caused by the protozoan *Plasmodium vivax*; transmitted by the *Anopheles* mosquito.

male condom Contraceptive device, usually made of latex, that fits over the male penis.

malignant The power to threaten life; cancerous.

malleus Bone found in the middle ear that assists in the transmission of sound to the inner ear; also called the hammer.

Malpighian tubule Blind, threadlike excretory tubule attached to the gut of an insect.

maltase Enzyme produced in small intestine that breaks down maltose to two glucose molecules.

mammal Endothermic vertebrate characterized especially by the presence of hair and mammary glands.

mantle In molluscs, an extension of the body wall that covers the visceral mass and may secrete a shell.

Marfan syndrome Autosomal dominant genetic disorder of the connective tissue, specifically the fibrillin protein.

marsupial Member of a group of mammals bearing immature young nursed in a marsupium, or pouch—for example, kangaroo and opossum.

mass extinction Episode of large-scale extinction in which large numbers of species disappear in a few million years or less.

mass number Mass of an atom equal to the number of protons plus the number of neutrons within the nucleus.

mast cell Connective tissue cell that releases histamine in allergic reactions.

matrix Unstructured semifluid substance that fills the space between cells in connective tissues or inside organelles.

matter Anything that takes up space and has mass.

mechanical energy A type of kinetic energy associated with the position, or motion (such as walking or running) of an object.

mechanoreceptor Sensory receptor that responds to mechanical stimuli, such as pressure, sound waves, or gravity.

medulla oblongata In vertebrates, part of the brain stem that is continuous with the spinal cord; controls heartbeat, blood pressure, breathing, and other vital functions.

medusa Among cnidarians, bell-shaped body form that is directed downward and contains much mesoglea.

megaphyll Large leaf with several to many veins.

megaspore One of the two types of spores produced by seed plants; develops into a female gametophyte (embryo sac).

megaosporocyte Female cell of a plant that undergoes meiosis to produce four haploid megaspores.

meiosis Type of nuclear division that reduces the chromosome number from 2n to n; daughter cells receive the haploid number of chromosomes in varied combinations; also called reduction division.

melanocyte Specialized cell in the epidermis that produces melanin, the pigment responsible for skin color.

melanocyte-stimulating hormone (MSH) Substance that causes melanocytes to secrete melanin in most vertebrates.

melatonin Hormone, secreted by the pineal gland, that is involved in biorhythms.

membrane-first hypothesis Proposes that the plasma membrane was the first component of the early cells to evolve.

memory Capacity of the brain to store and retrieve information about past sensations and perceptions; essential to learning.

memory B cell Forms during a primary immune response but enters a resting phase until a secondary immune response occurs.

memory T cell T cell that differentiates during an initial infection and responds rapidly during subsequent exposure to the same antigen.

meninges Protective membranous coverings around the central nervous system.

meningitis A condition that refers to inflammation of the brain or spinal cord meninges (membranes).

menopause Termination of the ovarian and uterine cycles in older women.

menstruation Periodic shedding of tissue and blood from the inner lining of the uterus in primates.

meristem Undifferentiated embryonic tissue in the active growth regions of plants.

mesoderm Middle primary tissue layer of an animal embryo that gives rise to muscle, several internal organs, and connective tissue layers.

mesoglea Transparent jellylike substance located between the endoderm and ectoderm of some sponges and cnidarians.

mesophyll Inner, thickest layer of a leaf consisting of palisade and spongy mesophyll; the site of most of photosynthesis.

messenger RNA (mRNA) Type of RNA formed from a DNA template and bearing coded information for the amino acid sequence of a polypeptide.

metabolic pathway Series of linked reactions, beginning with a particular reactant and terminating with an end product.

metabolic pool Metabolites that are the products of and/or the substrates for key reactions in cells, allowing one type of molecule to be changed into another type, such as carbohydrates converted to fats.

metabolism The sum of the chemical reactions that occur in a cell.

metamorphosis Change in shape and form that some animals, such as insects, undergo during development.

metaphase Third phase of mitosis; chromosomes are aligned at the metaphase plate.

metaphase plate A disk formed during metaphase in which all of a cell's chromosomes lie in a single plane at right angles to the spindle fibers.

metastasis Spread of cancer from the place of origin throughout the body; caused by the ability of cancer cells to migrate and invade tissues.

methanogen Type of archaean that lives in oxygen-free habitats, such as swamps, and releases methane gas.

MHC (major histocompatibility complex) protein Protein marker that is a part of cell-surface markers anchored in the plasma membrane, which the immune system uses to identify "self."

micelle Single layer of fatty acids (or phospholipids) that orientate themselves in an aqueous environment.

microevolution Change in gene frequencies between populations of a species over time.

micronutrient Essential element needed in small amounts for plant growth, such as boron, copper, and zinc.

microRNA (miRNA) Short sequences of RNA, usually less than 22 nucleotides, that are involved in posttranscriptional regulation of gene expression. These molecules either inhibit, or reduce, the expression of specific genes.

microspore One of the two types of spores produced by seed plants; develops into a male gametophyte (pollen grain).

microsporocyte Male cells of a plant that undergo meiosis to produce four haploid microspores.

microtubule Small, cylindrical organelle composed of tubulin protein around an empty central core; present in the cytoplasm, centrioles, cilia, and flagella.

midbrain In mammals, the part of the brain located below the thalamus and above the pons.

middle ear Portion of the ear consisting of the tympanic membrane, the oval and round windows, and the ossicles, where sound is amplified.

migration Regular back-and-forth movement of animals between two geographic areas at particular times of the year.

mimicry Superficial resemblance of two or more species; a survival mechanism that avoids predation by appearing to be noxious.

mineral Naturally occurring inorganic substance containing two or more elements; certain minerals are needed in the diet.

mineralocorticoid Hormone secreted by the adrenal cortex that regulate salt and water balance, leading to increases in blood volume and blood pressure.

mitochondria (sing., **mitochondrion**) Membrane-bounded organelle in which ATP molecules are produced during the process of cellular respiration.

mitosis The stage of cellular reproduction in which nuclear division occurs; process in which a parent nucleus produces two daughter nuclei, each having the same number and kinds of chromosomes as the parent nucleus.

model Simulation of a process that aids conceptual understanding until the process can be studied firsthand; a hypothesis that describes how a particular process could possibly be carried out.

mold Various fungi whose body consists of a mass of hyphae (filaments) that grow on and receive nourishment from organic matter such as human food and clothing.

mole The molecular weight of a molecule expressed in grams; contains 6.023 X 10^{23} molecules.

molecular clock Idea that the rate at which mutational changes accumulate in certain genes is constant over time and is not involved in adaptation to the environment.

molecule Union of two or more atoms of the same element; also, the smallest part of a compound that retains the properties of the compound.

mollusc Invertebrates including squids, clams, snails, and chitons; characterized by a visceral mass, a mantle, and a foot.

monoclonal antibody One of many antibodies produced by a clone of hybridoma cells that all bind to the same antigen.

monocot (Monocotyledone) Flowering plant group; members have one embryonic leaf (cotyledon), parallel-veined leaves, scattered vascular bundles, flower parts in threes or multiples of three, and other characteristics.

monocyte Type of agranular leukocyte that functions as a phagocyte, particularly after it becomes a macrophage, which is also an antigen-presenting cell.

monohybrid cross Cross between parents that differ in only one trait.

monomer Small molecule that is a subunit of a polymer—e.g., glucose is a monomer of starch.

monosaccharide Simple sugar; a carbohydrate that cannot be broken down by hydrolysis—e.g., glucose; also, any monomer of the polysaccharides.

monosomy Chromosome condition in which a diploid cell has one less chromosome than normal; designated as 2n-1.

monotreme Egg-laying mammal—e.g., duckbill platypus or spiny anteater.

monsoon Climate in India and southern Asia caused by wet ocean winds that blow onshore for almost half the year.

more-developed country (MDC) Country that is industrialized; typically, population growth is low, and the people enjoy a good standard of living overall.

morphogen genes Specific sequences of DNA that determine the relationship of structures during development; often expressed in a gradient across the embryo.

morphogenesis Emergence of shape in tissues, organs, or entire embryo during development.

morphology Physical characteristics that contribute to the appearance of an organism.

morula Spherical mass of cells resulting from cleavage during animal development prior to the blastula stage.

mosaic evolution Concept that human characteristics did not evolve at the same rate; for example, some body parts are more humanlike than others in early hominins.

moss Bryophyte that is typically found in moist habitats.

motor molecule Protein that moves along either actin filaments or microtubules and translocates organelles.

motor (efferent) neuron Nerve cell that conducts nerve impulses away from the central nervous system and innervates effectors (muscle and glands).

motor unit Combination of nerve fibers and muscles fibers in a muscle.

mouth In humans, organ of the digestive tract where food is chewed and mixed with saliva.

mRNA transcript mRNA molecule formed during transcription that has a sequence of bases complementary to a gene.

mucosa Epithelial membrane containing cells that secrete mucus; found in the inner cell layers of the digestive (first layer) and respiratory tracts.

mucous membrane Body membrane that lines the tubes of several organ systems in humans; secretes mucus that helps protect the body from infection.

muscle tone Physiological condition by which, at any given time, some muscle fibers are always contracting in a muscle, even when the muscle is at rest.

muscle twitch Single contraction of a muscle fiber; typically lasts only a fraction of a second.

musculoskeletal system Name for the combined muscular and skeletal systems of humans; involved in movement and posture.

multicellular Organism composed of many cells; usually has organized tissues, organs, and organ systems.

multiple alleles Inheritance pattern in which there are more than two alleles for a particular trait; each individual has only two of all possible alleles.

multiple sclerosis (MS) Disease of the central nervous system characterized by the breakdown of myelin in the neurons; considered to be an autoimmune disease.

muscular dystrophy (MD) Disease that causes a weakening of the muscles; has a strong genetic component.

muscular tissue Type of animal tissue composed of fibers that shorten and lengthen to produce movements.

muscularis Smooth muscle layer found in the digestive tract.

mutagen Chemical or physical agent that increases the chance of mutation.

mutation Change in the nucleotide structure of an organism's DNA.

mutualism Symbiotic relationship in which both species benefit in terms of growth and reproduction.

myasthenia gravis (MG) Autoimmune disease in which antibodies interfere with neuromuscular junctions, causing muscle weakness.

mycelium Tangled mass of hyphal filaments composing the vegetative body of a fungus.

mycorrhizae (sing., **mycorrhiza**) Mutualistic relationship between fungal hyphae and roots of vascular plants.

mycoses General name for fungal diseases of animals.

myelin sheath White, fatty material—derived from the membrane of neurolemmocytes—that forms a covering for nerve fibers.

myofibril Specific muscle cell organelle containing a linear arrangement of sarcomeres, which shorten to produce muscle contraction.

myosin Muscle protein making up the thick filaments in a sarcomere; it pulls actin to shorten the sarcomere, yielding muscle contraction.

N

NAD^+ (nicotinamide adenine dinucleotide) Coenzyme in oxidation-reduction reactions that accepts electrons and hydrogen ions to become NADH + H^+ as oxidation of substrates occurs. During cellular respiration, NADH carries electrons to the electron transport chain in mitochondria.

$NADP^+$ (nicotinamide adenine dinucleotide phosphate) Coenzyme in oxidation-reduction reactions that accepts electrons and hydrogen ions to become NADPH + H^+. During photosynthesis, NADPH participates in the reduction of carbon dioxide to a carbohydrate.

nail Flattened epithelial tissue from the stratum lucidum of the skin; located on the tips of fingers and toes.

natural killer (NK) cell Lymphocyte that causes an infected or cancerous cell to burst.

natural selection Mechanism of evolutionary change caused by environmental selection of organisms most fit to reproduce; results in adaptation to the environment.

Neandertal Hominin with a sturdy build that lived during the last Ice Age in Europe and the Middle East; hunted large game and left evidence of being culturally advanced.

nearsighted (myopic) Condition in which an individual cannot focus on objects at a distance; caused by the focusing of the image in front of the retina of the eye.

negative feedback Mechanism of homeostatic response by which the output of a system suppresses or inhibits activity of the system.

nematocyst In cnidarians, a capsule that contains a threadlike fiber, the release of which aids in the capture of prey.

nephridium (pl., **nephridia**) Segmentally arranged, paired excretory tubules of many invertebrates, as in the earthworm.

nephron Microscopic kidney unit that regulates blood composition by glomerular filtration, tubular reabsorption, and tubular secretion.

nerve Bundle of long axons outside the central nervous system.

nerve cord Component of an invertebrate's nervous system; usually located ventrally in invertebrates; commonly called the spinal cord in vertebrates.

nerve fiber Axon; conducts nerve impulses away from the cell. Axons are classified as either myelinated or unmyelinated based on the presence or absence of a myelin sheath.

nerve impulse Electrical signal that conveys information along the length of a neuron.

nerve net Diffuse, noncentralized arrangement of nerve cells in cnidarians.

nervous system Organ system of humans that includes the brain, spinal cord, sense organs (eyes, ears, etc.) and associated nerves. Receives, integrates, and stores sensory input; coordinates activity of other organ systems.

nervous tissue Tissue that contains nerve cells (neurons), which conduct impulses, and neuroglia, which support, protect, and provide nutrients to neurons.

neurodegenerative disease Disease, usually caused by a prion, virus, or bacterium, that damages or impairs the function of nervous tissue.

neuroglia Nonconducting nerve cells that are intimately associated with neurons and function in a supportive capacity.

neuromuscular junction Region where an axon bulb approaches a muscle fiber; contains a presynaptic membrane, a synaptic cleft, and a postsynaptic membrane.

neuron Nerve cell that characteristically has three parts: dendrites, cell body, and an axon.

neurotransmitter Chemical stored at the ends of axons that is responsible for transmission across a synapse.

neutron Neutral subatomic particle, located in the nucleus and assigned one atomic mass unit.

neutrophil Granular leukocyte that is the most abundant of the white blood cells; first to respond to infection.

nitrification Process by which nitrogen in ammonia and organic compounds is oxidized to nitrites and nitrates by soil bacteria.

nitrogen fixation Process whereby free atmospheric nitrogen is converted into compounds, such as ammonium and nitrates, usually by bacteria.

node In plants, the place where one or more leaves attach to a stem.

nodes of Ranvier Gaps in the myelin sheath around a nerve fiber.

noncyclic electron pathway Light-dependent photosynthetic pathway that is used to generate ATP and NADPH; because the pathway is noncyclic, the electrons must be replaced by the splitting of water (photolysis).

nondisjunction Failure of the homologous chromosomes or sister chromatids to separate during either mitosis or meiosis; produces cells with abnormal chromosome numbers.

nonpolar covalent bond Bond in which the sharing of electrons between atoms is fairly equal.

nonrandom mating Mating among individuals on the basis of their phenotypic similarities or differences, rather than mating on a random basis.

nonrenewable resource Minerals, fossil fuels, and other materials present in essentially fixed amounts (within the human timescale) in our environment.

nonvascular plants Bryophytes, such as mosses and liverworts, that have no vascular tissue and either occur in moist locations or have special adaptations for living in dry locations.

norepinephrine Neurotransmitter of the postganglionic fibers in the sympathetic division of the autonomic system; also, a hormone produced by the adrenal medulla.

normal microbiota Typical complement of bacterial species found on the human body; includes species found both internally and externally.

notochord Cartilage-like supportive dorsal rod in all chordates at some time in their life cycle; replaced by vertebrae in vertebrates.

nuclear envelope Double membrane that surrounds the nucleus in eukaryotic cells and is connected to the endoplasmic reticulum; has pores that allow substances to pass between the nucleus and the cytoplasm.

nuclear pore Opening in the nuclear envelope that permits the passage of proteins into the nucleus and ribosomal subunits out of the nucleus.

nucleic acid Polymer of nucleotides; both DNA and RNA are nucleic acids.

nucleoid Region of prokaryotic cells where DNA is located; it is not bounded by a nuclear envelope.

nucleolus Dark-staining, spherical body in the nucleus that produces ribosomal subunits.

nucleoplasm Semifluid medium of the nucleus containing chromatin.

nucleosome In the nucleus of a eukaryotic cell, a unit composed of DNA wound around a core of eight histone proteins, giving the appearance of a string of beads.

nucleotide Monomer of DNA and RNA consisting of a 5-carbon sugar bonded to a nitrogenous base and a phosphate group.

nucleus Membrane-bound organelle within a eukaryotic cell that contains chromosomes and controls the structure and function of the cell.

O

obesity Overweight condition characterized as having a body mass index (BMI) greater than 30.

observation Initial step in the scientific method that often involves the recording of data from an experiment or natural event.

octet rule The observation that an atom is most stable when its outer shell is complete and contains eight electrons; an exception is hydrogen which requires only two electrons in its outer shell to have a completed shell.

oil Triglyceride, usually of plant origin, that is composed of glycerol and three fatty acids and is liquid in consistency due to many unsaturated bonds in the hydrocarbon chains of the fatty acids.

oil gland Gland of the skin, associated with a hair follicle, that secretes sebum; sebaceous gland.

olfactory cell Modified neuron that is a sensory receptor for the sense of smell.

oligodendrocyte Type of glial cell that forms myelin sheaths around neurons in the CNS.

omnivore Organism in a food chain that feeds on both plants and animals.

oncogene Cancer-causing gene formed by a mutation in a proto-oncogene; codes for proteins that stimulate the cell cycle and inhibit apoptosis.

oocyte Immature egg that is undergoing meiosis; upon completion of meiosis, the oocyte becomes an egg.

oogenesis Production of eggs in females by the process of meiosis and maturation.

open circulatory system Arrangement of internal transport in which blood bathes the organs directly, and there is no distinction between blood and interstitial fluid.

operant conditioning Learning that results from rewarding or reinforcing a particular behavior.

operon Group of structural and regulating genes that function as a single unit.

opportunistic infections Diseases caused by pathogens that are normally suppressed by the immune system but may become pathogenic if the immune system is compromised, as is the case with HIV/AIDS.

order One of the categories, or taxa, used to group species; the taxon located above the family level.

organ Combination of two or more different tissues performing a common function.

organelle Small, membranous structures in the cytoplasm having a specific structure and function.

organic chemistry Branch of science that deals with organic molecules including those that are unique to living things.

organic molecule Molecule that always contains carbon and hydrogen, and often contains oxygen as well; organic molecules are associated with living things.

organ of Corti Structure in the vertebrate inner ear that contains auditory receptors (also called spiral organ).

organ system Group of related organs working together; examples are the digestive and endocrine systems.

orientation In birds, the ability to know present location by tracking stimuli in the environment.

origin When referring to a muscle, the connection on the stationary bone.

osmoregulation Regulation of the water–salt balance to maintain a normal balance within internal fluids.

osmosis Diffusion of water through a selectively permeable membrane.

osmotic pressure Measure of the tendency of water to move across a selectively permeable membrane; visible as an increase in liquid on the side of the membrane with higher solute concentration.

ossicle One of the small bones of the vertebrate middle ear—malleus, incus, and stapes.

osteoarthritis (OA) Form of degenerative joint disease which is the result of the loss of cartilage in a synovial joint.

osteoblast Bone-forming cell.

osteoclast Cell that is responsible for bone resorption.

osteocyte Mature bone cell located within the lacunae of bone.

osteoporosis Condition characterized by a loss of bone density; associated with levels of sex hormones and diet.

otitis media Inflammation of the middle ear.

otolith Calcium carbonate granule associated with sensory receptors for detecting movement of the head; in vertebrates, located in the utricle and saccule.

outer ear Portion of the ear consisting of the pinna and the auditory canal.

oval window Structure of the middle ear that conducts sound from the middle ear to the inner ear.

ovarian cancer Form of cancer that affects the uterus; often difficult to diagnose due to a lack of symptoms.

ovarian cycle Monthly changes occurring in the ovary that determine the level of sex hormones in the blood.

ovary In flowering plants, the enlarged, ovule-bearing portion of the carpel that develops into a fruit; female gonad in animals that produces an egg and female sex hormones.

overexploitation When the number of individuals taken from a wild population is so great that the population becomes severely reduced in numbers.

ovulation Bursting of a follicle when a secondary oocyte is released from the ovary; if fertilization occurs, the secondary oocyte becomes an egg.

ovule In seed plants, a structure that contains the female gametophyte and has the potential to develop into a seed.

oxidation Loss of one or more electrons from an atom or molecule; in biological systems, generally the loss of hydrogen atoms.

oxidation-reduction reaction A paired set of chemical reactions in which one molecule gives up electrons (oxidized) while another molecule accepts electrons (reduced); commonly called a redox reaction.

oxygen debt Amount of oxygen required to oxidize lactic acid produced anaerobically during strenuous muscle activity.

oxyhemoglobin Compound formed when oxygen combines with hemoglobin.

oxytocin Hormone released by the posterior pituitary that causes contraction of the uterus and milk letdown.

ozone shield Accumulation of O_3, formed from oxygen in the upper atmosphere; a filtering layer that protects the Earth from ultraviolet radiation.

P

p53 The protein produced from a tumor suppressor gene that (1) attempts to repair DNA damage, or (2) stops the cell cycle, or (3) initiates apoptosis.

pacemaker Cells of the sinoatrial node of the heart; electrical device designed to mimic the normal electrical patterns of the heart.

pain receptors Sensory receptors that are sensitive to chemicals released by damaged cells; also called nociceptors.

paleontology Study of fossils that results in knowledge about the history of life.

pancreas Internal organ that produces digestive enzymes and the hormones insulin and glucagon.

pancreatic amylase Enzyme that digests starch to maltose.

pancreatic cancer Form of cancer that originates in the pancreas; one of the more fatal forms of cancer.

pancreatic islet Masses of cells that constitute the endocrine portion of the pancreas.

pancreatitis Inflammation of the pancreas; may be caused by infection, gallstones, or excessive use of alcohol.

Pap test Procedure that removes a few cells from the cervix to look for evidence of cervical cancer.

parasite Species that is dependent on a host species for survival, usually to the detriment of the host species.

parasitism Symbiotic relationship in which one species (the *parasite*) benefits in terms of growth and reproduction to the detriment of the other species (the *host*).

parasympathetic division Division of the autonomic system that is active under normal conditions; uses acetylcholine as a neurotransmitter.

parathyroid gland Gland embedded in the posterior surface of the thyroid gland; it produces parathyroid hormone.

parathyroid hormone (PTH) Hormone secreted by the four parathyroid glands that increases the blood calcium level and decreases the phosphate level.

parenchyma Plant tissue composed of the least-specialized of all plant cells; found in all organs of a plant.

Parkinson disease (PD) Progressive deterioration of the central nervous system due to a deficiency in the neurotransmitter dopamine.

parsimony In systematics, the simplest solution in the analysis of evolutionary relationships.

partial pressure Pressure exerted by each gas in a mixture of gases.

parturition Process of giving birth; divided into three stages.

passive immunity Protection against infection acquired by transfer of antibodies to a susceptible individual.

pathogen Disease-causing agent such as viruses, parasitic bacteria, fungi, and animals.

pattern formation Positioning of cells during development that determines the final shape of an organism.

pectoral girdle Portion of the vertebrate skeleton that provides support and attachment for the upper (fore) limbs; consists of the scapula and clavicle on each side of the body.

pedigree Chart of genetic relationship of family individuals across generations.

pelagic division Open portion of the ocean.

pellicle Structure that supports the cell membrane of some species of protists; used to maintain structure in a water environment.

pelvic girdle Portion of the vertebrate skeleton to which the lower (hind) limbs are attached; consists of the coxal bones.

pelvic inflammatory disease (PID) Condition in which the uterine tubes become inflamed; often caused by chlamydial infections.

penis Male copulatory organ; in humans, the male organ of sexual intercourse.

pentose Five-carbon monosaccharide. Examples are deoxyribose found in DNA and ribose found in RNA.

pepsin Enzyme secreted by gastric glands that digests proteins to peptides.

peptidase Intestinal enzyme that breaks down short chains of amino acids to individual amino acids that are absorbed across the intestinal wall.

peptide Two or more amino acids joined together by covalent bonding.

peptide bond Type of covalent bond that joins two amino acids.

peptide hormone Type of hormone that is a protein, a peptide, or derived from an amino acid.

peptidoglycan Polysaccharide that contains short chains of amino acids; found in bacterial cell walls.

perception Processing of sensory stimuli that occurs when the brain interprets information being received from the sensory receptors.

perennial Flowering plant that lives more than one growing season because the underground parts regrow each season.

pericycle Layer of cells surrounding the vascular tissue of roots; produces branch roots.

periosteum Fibrous connective tissue that covers the surface of a long bone; contains blood vessels that service the cells within the bone.

peripheral nervous system (PNS) Nerves and ganglia that lie outside the central nervous system.

peristalsis Wavelike contractions that propel substances along a tubular structure such as the esophagus.

permafrost Permanently frozen ground, usually occurring in the tundra, a biome of Arctic regions.

peroxisome Enzyme-filled vesicle in which fatty acids and amino acids are metabolized to hydrogen peroxide that is broken down to harmless products.

petal A flower part that occurs just inside the sepals; often conspicuously colored to attract pollinators.

petiole The part of a plant leaf that connects the blade to the stem.

phagocytes Type of white blood cells that destroy pathogens using phagocytosis.

phagocytosis Process by which cells engulf large substances, forming an intracellular vacuole.

pharyngeal pouches Developmental characteristic of a chordate; these may develop into gills, or other structures such as auditory tubes and glands.

pharyngitis Inflammation of the pharynx; often caused by viruses or bacteria.

pharynx In vertebrates, common passageway for both food intake and air movement; located between the mouth and the esophagus.

phenomena Observable natural event or fact.

phenotype Visible expression of a genotype—e.g., brown eyes or attached earlobes.

phenylketonuria (PKU) Autosomal recessive genetic disorder that causes a lack of the enzyme that metabolizes phenylalanine; the accumulation of phenylalanine causes problems with nervous system development and function.

pheromone Chemical messenger that works at a distance and alters the behavior of another member of the same species.

phloem Vascular tissue that conducts organic solutes in plants; contains sieve-tube members and companion cells.

phospholipid Molecule that forms the bilayer of the cell's membranes; has a polar, hydrophilic head bonded to two nonpolar, hydrophobic tails.

photoautotroph Organism able to synthesize organic molecules by using carbon dioxide as the carbon source and sunlight as the energy source.

photoperiod (photoperiodism) Relative lengths of daylight and darkness that affect the physiology and behavior of an organism.

photoreceptor Sensory receptor that responds to light stimuli.

photosynthesis Process, usually occurring within chloroplasts, that uses solar energy to reduce carbon dioxide to carbohydrate.

photosystem Photosynthetic unit where solar energy is absorbed and high-energy electrons are generated; contains a pigment complex and an electron acceptor; occurs as PS (photosystem) I and PS II.

phototropism Growth response of plant stems to light; stems demonstrate positive phototropism.

pH scale Measurement scale for hydrogen ion concentration. Based on the formula $-\log[H^+]$.

phylogenetic tree Diagram that indicates common ancestors and lines of descent among a group of organisms.

phylogenetics Area of systematic biology that uses evidence to construct phylogeny, or a hypothesis of the evolutionary relationships between species.

phylogeny Evolutionary history of a group of organisms.

phyletic gradualism Model of genetic change in a species which suggests that change occurs at a slow, steady pace over time.

phylum One of the categories, or taxa, used to group species; the taxon located above the class level.

phytochrome Photoreversible plant pigment that is involved in photoperiodism and other responses of plants, such as etiolation.

phytoplankton Part of plankton containing organisms that photosynthesize, releasing oxygen to the atmosphere and serving as food producers in aquatic ecosystems.

pineal gland Gland—either at the skin surface (fish, amphibians) or in the third ventricle of the brain (mammals)—that produces melatonin.

pinna External flap of the ear; serves to collect and funnel sound toward the middle ear.

pinocytosis Process by which vesicle formation brings macromolecules into the cell.

pith Parenchyma tissue in the center of some stems and roots.

pituitary dwarfism Condition caused when too little growth hormone is produced during childhood; individuals with this condition have a small stature, but body parts in normal proportions.

pituitary gland Small gland that lies just inferior to the hypothalamus; consists of the anterior and posterior pituitary, both of which produce hormones.

placenta Organ formed during the development of placental mammals from the chorion and the uterine wall; allows the embryo, and then the fetus, to acquire nutrients and rid itself of wastes; produces hormones that regulate pregnancy.

placental mammal Also called the eutherians; species that rely on internal development whereby the fetus exchanges nutrients and wastes with its mother via a placenta.

plankton Freshwater and marine organisms that are suspended on or near the surface of the water; includes phytoplankton and zooplankton.

plants Multicellular, photosynthetic eukaryotes that increasingly became adapted to live on land.

plasma In vertebrates, the liquid portion of blood; contains nutrients, wastes, salts, and proteins.

plasma cell Mature B cell that mass-produces antibodies.

plasma membrane Membrane surrounding the cytoplasm that consists of a phospholipid bilayer with embedded proteins; functions to regulate the entrance and exit of molecules from cell.

plasmid Extrachromosomal ring of accessory DNA in the cytoplasm of prokaryotes.

plasmodesmata In plants, cytoplasmic connections in the cell wall that connect two adjacent cells.

plasmodial slime mold Free-living mass of cytoplasm that moves by pseudopods on a forest floor or in a field, feeding on decaying plant material by phagocytosis; reproduces by spore formation.

plasmolysis Contraction of the cell contents due to the loss of water.

plate tectonics Concept that the Earth's crust is divided into a number of fairly rigid plates whose movements account for continental drift.

platelet Component of blood that is necessary to blood clotting.

pleiotropy Inheritance pattern in which one gene affects many phenotypic characteristics of the individual.

pneumonia Condition of the respiratory system characterized by the filling of the bronchi and alveoli with fluid; caused by a viral, fungal, or bacterial pathogen.

point mutation Change of only one base in the sequence of bases in a gene.

polar body Nonfunctional product of oogenesis produced by the unequal division of cytoplasm in females during meiosis; in humans three of the four cells produced by meiosis are polar bodies.

polar covalent bond Bond in which the sharing of electrons between atoms is unequal.

pollen cone Reproductive structure of a gymnosperm that produces windblown pollen; microstrobili.

pollen grain In seed plants, structure that is derived from a microspore and develops into a male gametophyte.

pollen tube In seed plants, a tube that forms when a pollen grain lands on the stigma and germinates. The tube grows, passing between the cells of the stigma and the style to reach the egg inside an ovule, where fertilization occurs.

pollination In gymnosperms, the transfer of pollen from pollen cone to seed cone; in angiosperms, the transfer of pollen from anther to stigma.

pollinator Animal, typically an insect or bird, that transfers pollen between plants.

pollution Any environmental change that adversely affects the lives and health of living organisms.

polyandrous Female animals that have several male mates; found in the New World monkeys where the males help in rearing the offspring.

polygamous Male animals that have several female mates.

polygenic inheritance Pattern of inheritance in which a trait is controlled by several allelic pairs.

polygenic trait Traits that are under the control of multiple genes as opposed to monogenic (single-gene) traits.

polymer Macromolecule consisting of covalently bonded monomers; for example, a polypeptide is a polymer of monomers called amino acids.

polymerase chain reaction (PCR) Technique that uses the enzyme DNA polymerase to produce millions of copies of a particular piece of DNA.

polyp Among cnidarians, body form that is directed upward and contains much mesoglea; in anatomy: small, abnormal growth that arises from the epithelial lining.

polypeptide Polymer of many amino acids linked by peptide bonds.

polyribosome String of ribosomes simultaneously translating regions of the same mRNA strand during protein synthesis.

polysaccharide Polymer made from carbohydrate monomers; the polysaccharides starch and glycogen are polymers of glucose monomers.

pons Portion of the brain stem above the medulla oblongata and below the midbrain; assists the medulla oblongata in regulating the breathing rate.

population Group of organisms of the same species occupying a certain area and sharing a common gene pool.

population genetics The study of gene frequencies and their changes within a population.

portal system Pathway of blood flow that begins and ends in capillaries, such as the portal system located between the small intestine and liver.

positive feedback Mechanism of homeostatic response in which the output of the system intensifies and increases the activity of the system.

postanal tail Characteristic of a chordate; tail extends past the anus of the digestive system.

posterior pituitary Portion of the pituitary gland that stores and secretes oxytocin and antidiuretic hormone produced by the hypothalamus.

posttranscriptional control Gene expression following transcription that regulates the way mRNA transcripts are processed.

posttranslational control Alternation of gene expression by changing a protein's activity after it is translated.

potential energy Stored energy in a potentially usable form, as a result of location or spatial arrangement.

precipitation The process by which water leaves the atmosphere and falls to the ground as water, ice, or snow.

predation Interaction in which one organism (the predator) uses another (the prey) as a food source.

predator Organism that practices predation.

prediction Step of the scientific process that follows the formulation of a hypothesis and assists in creating the experimental design.

preparatory (prep) reaction Reaction that oxidizes pyruvate with the release of carbon dioxide; results in acetyl CoA and connects glycolysis to the citric acid cycle.

pressure-flow model Explanation for phloem transport; osmotic pressure following active transport of sugar into phloem produces a flow of sap from a source to a sink.

prey Organism that provides nourishment for a predator.

prezygotic isolating mechanism Anatomical, physiological, or behavioral difference between two species that prevents the possibility of mating.

primary lymphoid organ Location in the lymphatic system where lymphocytes develop and mature; e.g., bone marrow and thymus.

primary root Original root that grows straight down and remains the dominant root of the plant; contrasts with fibrous root system.

primate Member of the order Primates; includes prosimians, monkeys, apes, and hominins, all of whom have adaptations for living in trees.

principle Theory that is generally accepted by an overwhelming number of scientists; also called a law.

prion Infectious particle consisting of protein only and no nucleic acid.

producer Photosynthetic organism at the start of a grazing food chain that makes its own food—e.g., green plants on land and algae in water.

product Substance that forms as a result of a reaction.

profundal zone Underwater region of a lake (or ocean) that is just below where light penetrates; often very cold.

progesterone Female sex hormone that helps maintain sexual organs and secondary sex characteristics.

proglottid Segment of a tapeworm that contains both male and female sex organs and becomes a bag of eggs.

prokaryote Organism that lacks the membrane-bound nucleus and the membranous organelles typical of eukaryotes.

prokaryotic cell Cells that generally lack a membrane-bound nucleus and organelles; the cell type within the domains Bacteria and Archaea.

prolactin (PRL) Hormone secreted by the anterior pituitary that stimulates the production of milk from the mammary glands.

promoter In an operon, a sequence of DNA where RNA polymerase binds prior to transcription.

prophase First phase of mitosis; characterized by the condensation of the chromatin; chromosomes are visible, but scattered in the nucleus.

proprioceptor Class of mechanoreceptors responsible for maintaining the body's equilibrium and posture; involved in reflex actions.

prosimian Group of primates that includes lemurs and tarsiers, and may resemble the first primates to have evolved.

prostaglandin Hormone that has various and powerful local effects.

prostate cancer Cancer of the prostate gland; may be detected using a PSA test.

prostate gland Gland in male humans that secretes an alkaline, cloudy, fluid that increases the motility of sperm.

protease Enzyme that breaks the peptide bonds between amino acids in proteins, polypeptides, and peptides.

protein Polymer of amino acids; often consisting of one or more polypeptides and having a complex three-dimensional shape.

protein-first hypothesis In chemical evolution, the proposal that protein originated before other macromolecules and made possible the formation of protocells.

proteome Sum of the expressed proteins in a cell.

proteomics Study of the complete collection of proteins that a cell or organism expresses.

protists The group of eukaryotic organisms that are not a plant, fungus, or animal. Protists are generally a microscopic complex single cell; they evolved before other types of eukaryotes in the history of Earth.

protobiont (protocell) In biological evolution, a possible cell forerunner that became a cell once it acquired genes.

proton Positive subatomic particle located in the nucleus and assigned one atomic mass unit.

proto-oncogene Gene that promotes the cell cycle and prevents apoptosis; may become an oncogene through mutation.

protostome Group of coelomate animals in which the first embryonic opening (the blastopore) is associated with the mouth.

protozoan Heterotrophic, unicellular protist that moves by flagella, cilia, or pseudopodia.

proximal convoluted tubule Portion of a nephron following the glomerular capsule where tubular reabsorption of filtrate occurs.

pseudocoelom Body cavity lying between the digestive tract and body wall that is incompletely lined by mesoderm.

pseudopod Cytoplasmic extension of amoeboid protists; used for locomotion and engulfing food.

pteridophyte Ferns and their allies (horsetail and whisk ferns).

pulmonary artery Blood vessel that transports oxygen-poor blood from the heart to the lungs.

pulmonary circuit Circulatory pathway between the lungs and the heart.

pulmonary fibrosis Respiratory condition characterized by the buildup of connective tissue in the lungs; typically caused by inhalation of coal dust, silica, or asbestos.

pulmonary tuberculosis Respiratory infection caused by the bacterium *Mycobacterium tuberculosis*.

pulmonary vein Blood vessel that transports oxygen-rich blood from the lungs to the heart.

pulse Vibration felt in arterial walls due to expansion of the aorta following ventricle contraction.

punctuated equilibrium Model of evolutionary change in a species which suggests that there are long periods of little or no change, followed by brief periods of rapid speciation.

Punnett square Visual representation developed by Reginald Punnett that is used to calculate the expected results of simple genetic crosses.

pupil Opening in the center of the iris of the vertebrate eye.

purine Nucleotides with a double-ring structure; examples are adenine and guanine.

pyelonephritis Infection of the kidneys, often caused by an initial infection in one of the ureters.

pyrimidine Nucleotides with a single ring in their structure; examples are thymine, cytosine, and uracil.

R

radial symmetry Body plan in which similar parts are arranged around a central axis, like spokes of a wheel.

radioactive isotope Atoms of the same element having the same atomic number but a different mass number due to a variation in the number of neutrons. If the nucleus is unstable, and emits particles, it is considered to be radioactive.

radiocarbon dating Process of radiometric dating that measures the decay of ^{14}C to ^{14}N.

rain shadow Leeward side (side sheltered from the wind) of a mountainous barrier, which receives much less precipitation than the windward side.

ray-finned bony fishes Group of bony fishes with fins supported by parallel bony rays connected by webs of thin tissue.

RB The protein of a tumor suppressor gene; interprets growth signals and nutrient availability before allowing the cell cycle to proceed.

reactant Substance that participates in a reaction.

receptacle Area where a flower attaches to a floral stalk.

receptor Type of membrane protein that binds to specific molecules in the environment, providing a mechanism for the cell to sense and adjust to its surroundings.

receptor-mediated endocytosis Selective uptake of molecules into a cell by vacuole formation after they bind to specific receptor proteins in the plasma membrane.

receptor protein Protein located in the plasma membrane or within the cell; binds to a substance that alters some metabolic aspect of the cell.

recessive allele Allele that exerts its phenotypic effect only in the homozygote; its expression is masked by a dominant allele.

reciprocal altruism The trading of helpful or cooperative acts, such as helping at the nest, by individuals—the animal that was helped will repay the debt at some later time.

recombinant DNA (rDNA) DNA that contains genes from more than one source.

red algae Marine photosynthetic protists with a notable abundance of phycobilin pigments; includes coralline algae of coral reefs.

red blood cell Erythrocyte; contains hemoglobin and carries oxygen from the lungs or gills to the tissues in vertebrates.

red bone marrow Vascularized, modified connective tissue that is sometimes found in the cavities of spongy bone; site of blood cell formation.

red tide A population bloom of dinoflagellates that causes coastal waters to turn red. Releases a toxin that can lead to paralytic shellfish poisoning.

redox reaction A paired set of chemical reactions in which one molecule gives up electrons (oxidized) while another molecule accepts electrons (reduced); also called an oxidation-reduction reaction.

reduced hemoglobin (HHb) Globin chains within the hemoglobin molecule that have combined with hydrogen ions (H^+).

reduction Gain of electrons by an atom or molecule with a concurrent storage of energy; in biological systems, the electrons are accompanied by hydrogen ions.

reflex action Automatic, involuntary response of an organism to a stimulus.

refractory period Time following an action potential when a neuron is unable to conduct another nerve impulse.

renewable resource Resource normally replaced or replenished by natural processes and not depleted by moderate use.

renin Enzyme released by the kidneys that leads to the secretion of aldosterone and a rise in blood pressure.

replacement model Hypothesis of the evolution of modern humans from archaic species; also called the Out-of-Africa hypothesis.

replacement reproduction Population in which each person is replaced by only one child.

replication fork In eukaryotic DNA replication, the location where the two parental DNA strands separate.

repressor In an operon, protein molecule that binds to an operator, preventing transcription of structural genes.

reproduce To produce a new individual of the same kind.

reproduction The process of producing a new individual of the same kind.

reproductive cloning Used to create an organism that is genetically identical to the original individual.

reproductive isolation Model by which new species arise when gene flow is disrupted between two populations, genetic changes accumulate, and the populations are subsequently unable to mate and produce viable offspring.

reproductive system Organ system in humans that includes the sex-specific organs (testes, ovaries, etc.); produces, transports gametes; in females, nutures and gives birth to offspring.

reproductively isolated Descriptive term that indicates that a population is incapable of interbreeding with another population.

reptile Terrestrial vertebrate with internal fertilization, scaly skin, and an egg with a leathery shell; includes snakes, lizards, turtles, crocodiles, and birds.

reservoir Location in a biogeochemical cycle where a chemical or resource is stored for long periods of time and is typically unavailable to living organisms.

resource Abiotic and biotic components of an environment that support or are needed by living organisms.

resource partitioning Mechanism that increases the number of niches by apportioning the supply of a resource such as food or living space between species.

respiration Sequence of events that results in gas exchange between the cells of the body and the environment.

respiratory center Group of nerve cells in the medulla oblongata that send out nerve impulses on a rhythmic basis, resulting in involuntary inspiration on an ongoing basis.

respiratory system Organ system of humans that includes the lungs and associated structures; involved in the exchange of gases; helps control pH.

responding variable Result or change that occurs when an experimental variable is utilized in an experiment.

resting potential Membrane potential of an inactive neuron.

restriction enzyme Bacterial enzyme that stops viral reproduction by cleaving viral DNA; used to cut DNA at specific points during production of recombinant DNA.

reticular activating system (RAS) Area of the brain that contains the reticular formation; acts as a relay for information to and from the peripheral nervous system and higher processing centers of the brain.

reticular connective tissue Form of connective tissue that supports the lymph nodes, spleen, thymus, and bone marrow.

reticular fiber Very thin collagen fiber in the matrix of connective tissue, highly branched and forming delicate supporting networks.

retina Innermost layer of the vertebrate eyeball containing the photoreceptors—rod cells and cone cells.

retinal Light-absorbing molecule found in the photoreceptors of the eye; usually combined with opsin to form rhodopsin.

retinal detachment Condition characterized by the separation of the retina from the choroid layer of the eye.

restoration ecology Seeks scientific ways to return ecosystems to their state prior to habitat degradation.

retrovirus RNA virus containing the enzyme reverse transcriptase that carries out RNA/DNA transcription.

reverse transcriptase Viral enzyme found in retroviruses that is capable of converting their RNA genome into a DNA copy.

rheumatoid arthritis (RA) Autoimmune disease that causes inflammation of the joints.

rhizome Rootlike underground stem.

rhodopsin Light-absorbing molecule in rod cells and cone cells that contains a pigment and the protein opsin.

ribose Pentose sugar found in RNA.

ribosomal RNA (rRNA) Structural form of RNA found in the ribosomes.

ribosome Site of protein synthesis in a cell; composed of proteins and ribosomal RNA (rRNA).

ribozyme RNA molecule that functions as an enzyme that can catalyze chemical reactions.

ringworm Condition of the skin caused by a fungal infection.

RNA (ribonucleic acid) Nucleic acid produced from covalent bonding of nucleotide monomers that contain the sugar ribose; occurs in many forms, including: messenger RNA, ribosomal RNA, and transfer RNA.

RNA-first hypothesis In chemical evolution, the proposal that RNA originated before other macromolecules and allowed the formation of the first cell(s).

RNA interference Cellular process that utilizes miRNA and siRNA molecules to reduce, or inhibit, the expression of specific genes.

RNA polymerase During transcription, an enzyme that creates an mRNA transcript by joining nucleotides complementary to a DNA template.

rod cell Photoreceptor in vertebrate eyes that responds to dim light.

root Organ system of a plant that is responsible for anchoring the plant, absorbing water and minerals, and storing carbohydrates.

root cap Protective cover of the root tip, whose cells are constantly replaced as they are ground off when the root pushes through rough soil particles.

root hair Extension of a root epidermal cell that collectively increases the surface area for the absorption of water and minerals.

root nodule Structure on plant root that contains nitrogen-fixing bacteria.

root system Includes the main root and all of its lateral (side) branches.

rotational equilibrium Maintenance of balance when the head and body are suddenly moved or rotated.

rotifer Microscopic invertebrates characterized by ciliated corona that when beating looks like a rotating wheel.

rough ER (endoplasmic reticulum) Membranous system of tubules, vesicles, and sacs in cells; has attached ribosomes.

round window Structure of the middle ear that assists in the transmission of sound from the middle ear to the inner ear.

roundworm Invertebrates with nonsegmented cylindrical body covered by a cuticle that molts; some forms are free-living in water and soil, and many are parasitic.

***r*-selection** Favorable life history strategy under certain environmental conditions; characterized by a high reproductive rate with little or no attention given to offspring survival.

RuBP carboxylase An enzyme that starts the Calvin cycle reactions by catalyzing attachment of the carbon atom from CO_2 to RuBP.

S

saccule Saclike cavity in the vestibule of the vertebrate inner ear; contains sensory receptors for gravitational equilibrium.

sac fungi Fungi that produce spores in fingerlike sacs called asci within a fuiting body; includes morels, truffles, yeasts, and molds.

salivary amylase In humans, enzyme in saliva that digests starch to maltose.

salivary gland In humans, gland associated with the mouth that secretes saliva.

salt Solid substances formed by ionic bonds that usually dissociate into individual ions in water.

saltatory conduction Movement of nerve impulses from one node to another along a myelinated axon.

saprotroph Organism that secretes digestive enzymes and absorbs the resulting nutrients back across the plasma membrane.

sarcolemma Plasma membrane of a muscle fiber; also forms the tubules of the T system involved in muscular contraction.

sarcomere One of many units, arranged linearly within a myofibril, whose contraction produces muscle contraction.

sarcoplasmic reticulum Smooth endoplasmic reticulum of skeletal muscle cells; surrounds the myofibrils and stores calcium ions.

saturated fatty acid Fatty acid molecule that lacks double bonds between the carbons of its hydrocarbon chain. The chain bears the maximum number of hydrogens possible.

savanna Terrestrial biome that is a grassland in Africa, characterized by few trees and a severe dry season.

Schwann cell Cell that surrounds a fiber of a peripheral nerve and forms the myelin sheath.

scientific method Process by which scientists formulate a hypothesis, gather data by observation and experimentation, and come to a conclusion.

scientific theory Concept, or a collection of concepts, widely supported by a broad range of observations, experiments, and data.

sclera White, fibrous, outer layer of the eyeball.

sclerenchyma Plant tissue composed of cells with heavily lignified cell walls; functions in support.

scolex Tapeworm head region; contains hooks and suckers for attachment to host.

scrotum Saclike structures of a male that houses the testes.

second messenger Chemical signal such as cyclic AMP that causes the cell to respond to the first messenger—a hormone bound to plasma membrane receptor protein.

secondary lymphoid organ Location in the lymphatic system where lymphocytes are activated by antigens; example is the lymph nodes.

secretion Release of a substance by exocytosis from a cell.

sedimentation Process by which particulate material accumulates and forms a stratum.

seed Mature ovule that contains an embryo, with stored food enclosed in a protective coat.

seed cone Reproductive structure of a gymnosperm that produces windblown seeds; megastrobili.

seed plant Vascular plant that disperses seeds; the gymnosperms and angiosperms.

seedless vascular plant Collective name for club mosses and ferns; characterized by windblown spores.

segmentation Repetition of body units as seen in the earthworm.

selectively permeable Property of the plasma membrane that allows some substances to pass, but prohibits the movement of others.

semen (seminal fluid) Thick, whitish fluid consisting of sperm and secretions from several glands of the male reproductive tract.

semicircular canal One of three half-circle-shaped canals of the vertebrate inner ear; contains sensory receptors for rotational equilibrium.

semiconservative replication Process of DNA replication that results in two double helix molecules, each having one parental and one new strand.

semilunar valve Valve resembling a half moon located between the ventricles and their attached vessels.

seminal vesicles Glands in male humans that secrete a viscous fluid that provides nutrition to the sperm cells.

seminiferous tubule Long, coiled structure contained within chambers of the testis where sperm are produced.

senescence Sum of the processes involving aging, decline, and eventual death of a plant or plant part.

sensation Processing of sensory stimuli that involves detection of nerve impulses by the cerebral cortex of the brain.

sensory (afferent) neuron Nerve cell that transmits nerve impulses to the central nervous system after a sensory receptor has been stimulated.

sensory receptor Structure that receives either external or internal environmental stimuli and is a part of a sensory neuron or transmits signals to a sensory neuron.

sensory transduction Process by which a sensory receptor converts an input to a nerve impulse.

sepal Outermost, leaflike covering of the flower; usually green in color.

septate Having cell walls; some fungal species have hyphae that are septate.

septum Partition or wall that divides two areas; the septum in the heart separates the right half from the left half.

serosa Outer embryonic membrane of birds and reptiles; chorion.

serous membrane Body membrane in the thoracic and abdominal cavities that secretes a watery lubricant.

Sertoli cell Male reproductive cells that support, nourish and regulate the sperm-producing cells.

sessile Organism that is permanently attached to a substrate, such as rock.

seta (pl., setae) A needlelike, chitinous bristle in annelids, arthropods, and others.

severe combined immunodeficiency (SCID) Immune disease characterized by the impairment, or lack of, T or B cells in the body.

sex chromosome Chromosomes that differ between the sexes; in humans, these represent the X and Y chromosomes.

sex-linked Trait controlled by a gene on a sex chromosome; often described as either X-linked or Y-linked.

sexual reproduction Reproduction involving meiosis, gamete formation, and fertilization; produces offspring with chromosomes inherited from each parent with a unique combination of genes.

sexual selection Changes in males and females, often due to male competition and female selectivity, leading to increased fitness.

shared derived traits Used in classification systems, such as cladistics, to determine the evolutionary relationship between two species.

shoot system Aboveground portion of a plant consisting of the stem, leaves, and flowers.

short-day plant Plant that flowers when day length is shorter than a critical length—e.g., cocklebur, poinsettia, and chrysanthemum.

short tandem repeat (STR) profiling Procedure of analyzing DNA in which PCR and gel electrophoresis are used to create a banding pattern; these are usually unique for each individual; process used in DNA barcoding.

shrubland Arid terrestrial biome characterized by shrubs and tending to occur along coasts that have dry summers and receive most of their rainfall in the winter.

sickle cell disease Autosomal recessive genetic disorder that causes a malformation of hemoglobin molecules, causing red blood cells to form a sickle shape; also sometimes called sickle cell anemia due to the symptoms of the disease.

sieve-tube member Member that joins with others in the phloem tissue of plants as a means of transport for nutrient sap.

signal transduction Process that occurs within a cell when a molecular signal (protein, hormone, etc.) initiates a response within the interior of the cell.

sink In the pressure-flow model of phloem transport, the location (roots) from which sugar is constantly being removed. Sugar will flow to the roots from the source.

sinus Spaces within the cranium that reduce the overall weight of the skull.

sinusitis Inflammation of the cranial sinuses in the head.

sister chromatid One of two genetically identical chromosomal units that are the result of DNA replication and are attached to each other at the centromere.

skeletal muscle Striated, voluntary muscle tissue that comprises skeletal muscles; also called striated muscle.

skin Outer covering of the body; can be called the integumentary system because it contains organs such as sense organs.

skull Common name for facial bones and bones of the cranium.

sliding filament model An explanation for muscle contraction based on the movement of actin filaments in relation to myosin filaments.

small interfering RNAs (siRNA) Short sequences of RNA, typically less than 25 nucleotides, that are involved in posttranscriptional control of gene expression through a process called RNA interference.

small intestine In vertebrates, the portion of the digestive tract that precedes the large intestine; in humans, consists of the duodenum, jejunum, and ileum.

smooth ER (endoplasmic reticulum) Membranous system of tubules, vesicles, and sacs in eukaryotic cells; site of lipid synthesis; lacks attached ribosomes.

smooth muscle Nonstriated, involuntary muscles found in the walls of internal organs.

sodium-potassium pump Carrier protein in the plasma membrane that moves sodium ions out of and potassium ions into cells; important in the function of nerve and muscle cells in animals.

soil Accumulation of inorganic rock material and organic matter that is capable of supporting the growth of vegetation.

solute Substance that is dissolved in a solvent, forming a solution.

solution Fluid (the solvent) that contains a dissolved solid (the solute).

solvent Liquid portion of a solution that serves to dissolve a solute.

somatic cell Body cell; excludes cells that undergo meiosis and become sperm or eggs.

somatic system Portion of the peripheral nervous system containing motor neurons that control skeletal muscles.

somatostatin Hormone produced by the pancreas, stomach, and small intestine that inhibits the effects of growth hormones; also suppresses activity of hormones such as insulin and glucagon.

source In the pressure-flow model of phloem transport, the location (leaves) of sugar production. Sugar will flow from the leaves to the sink.

speciation Origin of new species due to the evolutionary process of descent with modification.

species Group of similarly constructed organisms capable of interbreeding and producing fertile offspring; organisms that share a common gene pool; the taxon at the lowest level of classification.

specific epithet In the binomial system of taxonomy, the second part of an organism's name; it may be descriptive.

sperm Male gamete having a haploid number of chromosomes and the ability to fertilize an egg, the female gamete.

spermatogenesis Production of sperm in males by the process of meiosis and maturation.

spicule Skeletal structure of sponges composed of calcium carbonate or silicate.

spinal cord In vertebrates, the nerve cord that is continuous with the base of the brain and housed within the vertebral column.

spinal nerve Nerve that arises from the spinal cord.

spindle Collection of microtubules that assist in the orderly distribution of chromosomes during cell division.

spleen Large, glandular organ located in the upper left region of the abdomen; stores and filters blood.

sponge Invertebrates that are pore-bearing filter feeders whose inner body wall is lined by collar cells that resemble a unicellular choanoflagellate.

spongy bone Type of bone that has an irregular, meshlike arrangement of thin plates of bone.

sporangium (pl., sporangia) Structure that produces spores.

spore Asexual reproductive or resting cell capable of developing into a new organism without fusion with another cell, in contrast to a gamete.

sporophyte Diploid generation of the alternation-of-generations life cycle of a plant; produces haploid spores that develop into the haploid generation.

sporozoan Spore-forming protist that has no means of locomotion and is typically a parasite with a complex life cycle; usually has both sexual and asexual phases.

spring overturn Mixing process that occurs in spring in stratified lakes whereby oxygen-rich top waters mix with nutrient-rich bottom waters.

squamous epithelium Type of epithelial tissue that contains flat cells.

stabilizing selection Outcome of natural selection in which extreme phenotypes are eliminated and the average phenotype is conserved.

standard deviation A statistical analysis of data from an observation or experiment; measures how much the data varies.

stamen In flowering plants, the portion of the flower that consists of a filament and an anther containing pollen sacs where pollen is produced.

stapes Bone found in the middle ear that assists in the transmission of sound to the inner ear; also called the stirrup.

starch Storage polysaccharide found in plants that is composed of glucose molecules joined in a linear fashion with few side chains.

stem Usually the upright, vertical portion of a plant that transports substances to and from the leaves.

stem cell Type of cell that acts as a source for other types of cells; capable of continuously dividing.

stereoscopic vision Vision characterized by depth perception and three-dimensionality.

steroid Type of lipid molecule having a complex of four carbon rings—e.g., cholesterol, estrogen, progesterone, and testosterone.

steroid hormone Type of hormone that has a complex of four carbon rings, but each one has different side chains.

stigma In flowering plants, portion of the carpel where pollen grains adhere and germinate before fertilization can occur.

stolon Stem that grows horizontally along the ground and may give rise to new plants where it contacts the soil—e.g., the runners of a strawberry plant.

stomach In vertebrates, muscular sac that mixes food with gastric juices to form chyme, which enters the small intestine.

stomach ulcer Open sore in the lining of the stomach; frequently caused by the bacteria *Helicobacter pylori.*

stomata (sing., stoma) Small openings between two guard cells on the underside of leaf epidermis through which gases pass.

strata (sing., stratum) Ancient layer of sedimentary rock; results from slow deposition of silt, volcanic ash, and other materials.

striated Having bands; in cardiac and skeletal muscle, alternating light and dark bands produced by the distribution of contractile proteins.

stroke Condition resulting when an arteriole in the brain bursts or becomes blocked by an embolism; cerebrovascular accident.

stroma Region within a chloroplast that surrounds the grana; contains enzymes involved in the synthesis of carbohydrates during the Calvin cycle of photosynthesis.

style Elongated, central portion of the carpel between the ovary and stigma.

submucosa Tissue layer just under the epithelial lining of the lumen of the digestive tract (second layer).

subsidence Occurs when a portion of Earth's surface gradually settles downward.

substrate Reactant in an enzyme-controlled reaction.

substrate-level ATP synthesis Process in which ATP is formed by transferring a phosphate from a metabolic substrate to ADP.

supergroup Systematic term that refers to the major groups of eukaryotes.

surface-area-to-volume ratio Ratio of a cell's outside area to its internal volume; the relationship limits the maximum size of a cell.

surface tension Force that holds moist membranes together due to the attraction of water molecules through hydrogen bonds.

survivorship Probability of newborn individuals of a cohort surviving to particular ages.

sustainable Ability of a society or ecosystem to maintain itself while also providing services to human beings.

suture Line of union between two nonarticulating bones, as in the skull.

sweat gland Skin gland that secretes a fluid substance for evaporative cooling; sudoriferous gland.

symbiosis Relationship that occurs when two different species live together in a unique way; it may be beneficial, neutral, or detrimental to one or both species.

symbiotic relationship *See also* symbiosis.

symmetry Pattern of similarity in an object.

sympathetic division Division of the autonomic system that is active when an organism is under stress; uses norepinephrine as a neurotransmitter.

sympatric speciation Origin of new species in populations that overlap geographically.

synapse Junction between neurons consisting of the presynaptic (axon) membrane, the synaptic cleft, and the postsynaptic (usually dendrite) membrane.

synapsis Pairing of homologous chromosomes during meiosis I.

synaptic cleft Small gap between presynaptic and postsynaptic cells of a synapse.

synaptonemal complex Protein structure that forms between the homologous chromosomes of prophase I of meiosis; promotes the process of crossing-over.

synovial joint Freely moving joint in which two bones are separated by a cavity.

synovial membrane Body membrane that lines synovial joints; secretes synovial fluid that lubricates the joint.

syphillus Sexually-transmitted disease caused by the bacterium *Treponema pallidum*.

systematics Study of the diversity of life for the purpose of understanding the evolutionary relationships between species.

systematic biology Another name for the study of systematics, or the study of the evolutionary relationships between species.

systemic circuit Circulatory pathway of blood flow between the tissues and the heart.

systemic lupus erythematosus (SLE) An autoimmune disease that eventually causes death through kidney failure.

systole Contraction period of the heart during the cardiac cycle.

T

taiga Terrestrial biome that is a coniferous forest extending in a broad belt across northern Eurasia and North America.

taproot Main axis of a root that penetrates deeply and is used by certain plants (such as carrots) for food storage.

taste bud Structure in the vertebrate mouth containing sensory receptors for taste; in humans, most taste buds are on the tongue.

taxon (pl., taxa) Group of organisms that fills a particular classification category.

taxonomist Scientist that investigates the identification and naming of new organisms.

taxonomy Branch of biology concerned with identifying, describing, and naming organisms.

Tay-Sachs disease Autosomal recessive genetic disorder that results in a deficiency in the enzyme hexosaminidase A; causes an accumulation of glycolipids in the lysosomes resulting in a progressive loss of psychomotor functions.

T cell Lymphocyte that matures in the thymus and exists in four varieties, one of which kills antigen-bearing cells outright.

T-cell receptor (TCR) Molecule on the surface of a T cell that can bind to a specific antigen fragment in combination with an MHC molecule.

telomere Tip of the end of a chromosome that shortens with each cell division and may thereby regulate the number of times a cell can divide.

telophase Final phase of mitosis; daughter cells are located at each pole.

temperate deciduous forest Forest found south of the taiga; characterized by deciduous trees such as oak, beech, and maple, moderate climate, relatively high rainfall, stratified plant growth, and plentiful ground life.

template Parental strand of DNA that serves as a guide for the complementary daughter strand produced during DNA replication.

tendon Strap of fibrous connective tissue that connects skeletal muscle to bone.

terminal bud Bud that develops at the apex of a shoot.

termination End of translation that occurs when a ribosome reaches a stop codon on the mRNA that it is translating, causing release of the completed protein.

testcross Cross between an individual with a dominant phenotype and an individual with a recessive phenotype to determine whether the dominant individual is homozygous or heterozygous.

testes Male gonads that produce sperm and the male sex hormones.

testicular cancer One of several forms of cancer that affect the testes of males; usually characterized by abnormal tenderness or lumps in one of the testicles.

testosterone Male sex hormone that helps maintain sexual organs and secondary sex characteristics.

tetanus Summation of muscle contractions to a level of maximum sustainability.

tetany Severe spasm caused by involuntary contraction of the skeletal muscles due to a calcium imbalance.

tetrapod Four-footed vertebrate; includes amphibians, reptiles, birds, and mammals.

thalamus In vertebrates, the portion of the diencephalon that passes on selected sensory information to the cerebrum.

therapeutic cloning Used to create mature cells of various cell types. Facilitates study of specialization of cells and provide cells and tissue to treat human illnesses.

thermoacidophile Type of archaean that lives in hot, acidic, aquatic habitats, such as hot springs or near hydrothermal vents.

thermoreceptor Sensory receptor that detects heat.

threatened species Species that is likely to become an endangered species in the foreseeable future (e.g., bald eagle, gray wolf, Louisiana black bear).

three-domain system Classification system that places all organisms into one of three large domains—Bacteria, Archaea, and Eukarya.

thrombin Enzyme that is involved in blood clotting; acts on fibrinogen molecules to produce fibrin.

thrush Condition caused by an infection of the oral cavity with the fungi *Candida*.

thylakoid Flattened sac within a granum of a chloroplast; membrane contains chlorophyll; location where the light reactions of photosynthesis occur.

thymine (T) One of four nitrogen-containing bases in nucleotides composing the structure of DNA; pairs with adenine.

thymus Lymphoid organ involved in the development and functioning of the immune system; T lymphocytes mature in the thymus.

thyroid gland Large gland in the neck that produces several important hormones, including thyroxine, triiodothyronine, and calcitonin.

thyroid-stimulating hormone (TSH) Substance produced by the anterior pituitary that causes the thyroid to secrete thyroxine and triiodothyronine.

thyroxine (T_4) Hormone secreted from the thyroid gland that promotes growth and development; in general, it increases the metabolic rate in cells.

tight junction Junction between cells when adjacent plasma membrane proteins join to form an impermeable barrier.

tissue Group of similar cells combined to perform a common function.

tissue culture Process of growing tissue artificially, usually in a liquid medium in laboratory glassware.

tissue fluid Fluid that surrounds the body's cells; consists of dissolved substances that leave the blood capillaries by filtration and diffusion.

tone Continuous, partial contraction of muscle.

tonicity The solute concentration (osmolarity) of a solution compared to that of a cell. If the solution is isotonic to the cell, there is no net movement of water; if the solution is hypotonic, the cell gains water; and if the solution is hypertonic, the cell loses water.

tonsillitis Inflammation of the tonsils—lymphoid tissue located in the pharynx.

totipotent Cell that has the full genetic potential of the organism, including the potential to develop into a complete organism.

toxin Poisonous substance produced by living cells or organisms. Toxins are often proteins that are capable of causing disease on contact with or absorption by body tissues.

trachea (pl., tracheae) In insects, air tube located between the spiracles and the tracheoles. In tetrapod vertebrates, air tube (windpipe) that runs between the larynx and the bronchi.

tracheid In vascular plants, type of cell in xylem that has tapered ends and pits through which water and minerals flow.

tract Bundle of myelinated axons in the central nervous system.

trait A characteristic of an organism; may be based on the physiology, morphology, or the genetics of the organism.

transcription First stage of gene expression; process whereby a DNA strand serves as a template for the formation of mRNA.

transcription factor In eukaryotes, protein required for the initiation of transcription by RNA polymerase.

transcriptional control Control of gene expression by the use of transcription factors, and other proteins, that regulate either the initiation of transcription or the rate at which it occurs.

transduction Exchange of DNA between bacteria by means of a bacteriophage.

transduction pathway Series of proteins or enzymes that change a signal to one understood by the cell.

trans-fats Unsaturated fatty acid chains in which the configuration of the carbon-carbon double bonds is such that the hydrogen atoms are across from each other, as opposed to being on the same side (cis).

transfer RNA (tRNA) Type of RNA that transfers a particular amino acid to a ribosome during protein synthesis; at one end, it binds to the amino acid, and at the other end it has an anticodon that binds to an mRNA codon.

transformation Taking up of extraneous genetic material from the environment by bacteria.

transgenic organism An organism whose genome has been altered by the insertion of genes from another species.

transitional fossil Fossil that bears a resemblance to two groups that in the present day are classified separately.

transitional links Evidence of evolution, typically fossils, that bears a resemblance to two groups that in the present day are classified separately.

translation During gene expression, the process whereby ribosomes use the sequence of codons in mRNA to produce a polypeptide with a particular sequence of amino acids.

translational control Gene expression regulated by influencing the interaction of the mRNA transcripts with the ribosome.

translocation Movement of a chromosomal segment from one chromosome to another nonhomologous chromosome, leading to abnormalities—e.g., Down syndrome.

transpiration Plant's loss of water to the atmosphere, mainly through evaporation at leaf stomata.

transposon DNA sequence capable of randomly moving from one site to another in the genome.

trichinosis Infection caused by the roundworm *Trichinella spiralis*.

triglyceride Neutral fat composed of glycerol and three fatty acids; typically involved in energy storage.

triplet code During gene expression, each sequence of three nucleotide bases stands for a particular amino acid.

trisomy Chromosome condition in which a diploid cell has one more chromosome than normal; designated as 2n+1.

trochozoan Type of protostome that produces a trochophore larva; also has two bands of cilia around its middle.

trophic level Feeding level of one or more populations in a food web.

trophoblast Outer membrane surrounding the embryo in mammals; when thickened by a layer of mesoderm, it becomes the chorion, an extraembryonic membrane.

tropical forest Can be wet or dry. Tropical rain forests and tropical deciduous forests are characterized by high annual precipitation, lush vegetation, an annual mean temperature of 20–25°C, high humidity, and both a wet and dry season. The difference between the two forest types is in the length of the dry season. Tropical deciduous forests have a longer dry season and consequently, trees lose their leaves.

tropism In plants, a growth response toward or away from a directional stimulus.

trypsin Protein-digesting enzyme secreted by the pancreas.

T (traverse) tubules Portion of the sarcolemma (plasma membrane) of a muscle cell that interact with the sarcoplastic reticulum of the cell.

tubal ligation Surgical form of reproductive sterilization that cuts and seals the uterine tubes in a female.

tubular reabsorption Movement of primarily nutrient molecules and water from the contents of the nephron into blood at the proximal convoluted tubule.

tubular secretion Movement of certain molecules from blood into the distal convoluted tubule of a nephron so that they are added to urine.

tumor Cells derived from a single mutated cell that has repeatedly undergone cell division; benign tumors remain at the site of origin, while malignant tumors metastasize.

tumor suppressor gene Gene that codes for a protein that ordinarily suppresses the cell cycle; inactivity due to a mutation can lead to a tumor.

tundra Biome characterized by permanently frozen subsoil found between the ice cap and the tree line of regions of the Arctic, just south of the ice-covered polar seas in the Northern Hemisphere; also known as the arctic tundra. Alpine tundra, having similar characteristics, is found near the peak of a mountain.

tunicate Type of primitive invertebrate chordate.

turgor pressure Pressure of the cell contents against the cell wall; in plant cells, determined by the water content of the vacuole; provides internal support.

tympanic membrane Membranous region that receives air vibrations in an auditory organ; in humans, the eardrum.

U

umbilical cord Cord connecting the fetus to the placenta through which blood vessels pass.

unicellular An organism comprised of a single cell, as in the Bacteria.

unsaturated fatty acid Fatty acid molecule that contains double bonds between some carbons of its hydrocarbon chain; thus contains fewer hydrogens than a saturated hydrocarbon chain.

upwelling Upward movement of deep, nutrient-rich water along coasts; it replaces surface waters that move away from shore when the direction of prevailing wind shifts.

uracil (U) Pyrimidine base that occurs in RNA, replacing thymine.

urea Main nitrogenous waste of terrestrial amphibians and most mammals.

ureter Tubular structure conducting urine from the kidney to the urinary bladder.

urethra Tubular structure that receives urine from the bladder and carries it to the outside of the body.

uric acid Main nitrogenous waste of insects, reptiles, and birds.

urinary bladder Organ where urine is stored.

urinary system Organ system of humans that includes the kidney, urinary bladder and associated structures; excretes metabolic wastes; maintains fluid balance; helps control pH.

urine Liquid waste product made by the nephrons of the vertebrate kidney through the processes of glomerular filtration, tubular reabsorption, and tubular secretion.

uterine cycle Cycle that runs concurrently with the ovarian cycle; it prepares the uterus to receive a developing zygote.

uterine tubes Transport the oocyte (egg) from the ovary to the uterus; location of fertilization; also called the oviducts or fallopian tubes.

uterus In mammals, expanded portion of the female reproductive tract through which eggs pass to the environment or in which an embryo develops and is nourished before birth.

utricle Cavity in the vestibule of the vertebrate inner ear; contains sensory receptors for gravitational equilibrium.

V

Vaccine Preventative measure that uses antigens prepared in such a way so as to promote immunity without causing the disease.

vacuole Membrane-bound sac, larger than a vesicle; usually functions in storage and can contain a variety of substances. In plants, the central vacuole fills much of the interior of the cell.

vagina Component of the female reproductive system that serves as the birth canal; receives the penis during copulation.

vaginal contraceptive ring Form of contraceptive containing estrogen and progesterone that is inserted into the vagina following menstruation.

valence shell The outer electron shell of an atom. Contains the valence electrons, which determine the chemical reactivity of the atom.

vas deferens Also called the ductus deferens; storage location for mature sperm before they pass into the ejaculatory duct.

vascular bundle In plants, primary phloem and primary xylem enclosed by a bundle sheath.

vascular cambium In plants, lateral meristem that produces secondary phloem and secondary xylem.

vascular cylinder In eudicots, the tissues in the middle of a root, consisting of the pericycle and vascular tissues.

vascular plant Plant that has xylem and phloem.

vascular tissue Transport tissue in plants, consisting of xylem and phloem.

vasectomy Surgical form of reproductive sterilization that cuts and seals the vas deferens in a male.

vector In genetic engineering, a means to transfer foreign genetic material into a cell—e.g., a plasmid.

vein Blood vessel that arises from venules and transports blood toward the heart.

vena cava Large systemic vein that returns blood to the right atrium of the heart in tetrapods; either the superior or inferior vena cava.

ventilation Process of moving air into and out of the lungs; breathing.

ventral cavity One of two main body cavities in humans; contains the thoracic, abdominal, and pelvic cavities.

ventricle Cavity in an organ, such as a lower chamber of the heart or the ventricles of the brain.

venule Vessel that takes blood from capillaries to a vein.

vernix caseosa White, greasy covering of the developing fetus; protects the fetus' skin from the amniotic fluid.

vertebral column Portion of the vertebrate endoskeleton that houses the spinal cord; consists of many vertebrae separated by intervertebral disks.

vertebrate Chordate in which the notochord is replaced by a vertebral column.

vertigo Equilibrium disorder that is often associated with problems in the receptors of the semicircular canals in the ear.

vesicle Small, membrane-bound sac that stores substances within a cell.

vessel element Cell that joins with others to form a major conducting tube found in xylem.

vestibule Space or cavity at the entrance to a canal, such as the cavity that lies between the semicircular canals and the cochlea.

vestigial structure Remnant of a structure that was functional in some ancestor but is no longer functional in the organism in question.

villus Small, fingerlike projection of the inner small intestinal wall.

viroid Infectious strand of RNA devoid of a capsid and much smaller than a virus.

virus Noncellular parasitic agent consisting of an outer capsid and an inner core of nucleic acid.

visceral mass Internal organs; in a mollusc, the tissues between the mantle.

visual accommodation Ability of the eye to focus at different distances by changing the curvature of the lens.

vitamin Organic nutrient that is required in small amounts for metabolic functions. Vitamins are often part of coenzymes.

viviparous Animal that gives birth after partial development of offspring within mother.

vocal cord In humans, fold of tissue within the larynx; creates vocal sounds when it vibrates.

vulva Common name for the external genital organs of a female.

W

water (hydrologic) cycle Interdependent and continuous circulation of water from the ocean, to the atmosphere, to the land, and back to the ocean.

water mold Filamentous organisms having cell walls made of cellulose; typically decomposers of dead freshwater organisms, but some are parasites of aquatic or terrestrial organisms.

water vascular system Series of canals that takes water to the tube feet of an echinoderm, allowing them to expand.

wax Sticky, solid, water-repellent lipid consisting of many long-chain fatty acids usually linked to long-chain alcohols.

wetland Area that is covered by water at some point in the year. *See also* bog, marsh, or swamp.

white blood cell Leukocyte, of which there are several types, each having a specific function in protecting the body from invasion by foreign substances and organisms.

white matter Myelinated axons in the central nervous system.

whorl Cluster of branches, or other plant structures, that occurs in a circular pattern.

wobble hypothesis Ability of the tRNAs to recognize more than one codon; the codons differ in their third nucleotide.

wood Secondary xylem that builds up year after year in woody plants and becomes the annual rings.

X

X-linked Allele that is located on an X chromosome; not all X-linked genes code for sexual characteristics.

xylem Vascular tissue that transports water and mineral solutes upward through the plant body; it contains vessel elements and tracheids.

Y

yeast Unicellular fungus that has a single nucleus and reproduces asexually by budding or fission, or sexually through spore formation.

yolk Dense nutrient material in the egg of a bird or reptile.

yolk sac One of the extraembryonic membranes that, in shelled vertebrates, contains yolk for the nourishment of the embryo, and in placental mammals is the first site for blood cell formation.

Z

zero population growth No growth in population size.

zooplankton Part of plankton containing protozoans and other types of microscopic animals.

zoospore Spore that is motile by means of one or more flagella.

zygospore Thick-walled resting cell formed during sexual reproduction of zygospore fungi.

zygospore fungi Fungi such as black bread mold that reproduce by forming windblown spores in sporangia; sexual reproduction involves a thick-walled zygospore.

zygote Diploid cell formed by the union of two gametes; the product of fertilization.

CREDITS

Chapter 1 Opener: © AFP/TEPCO; 1.1 (giraffes): © Sylvia S. Mader; 1.1 (butterfly): © Creatas/PunchStock RF; 1.1 (sequoia): © Robert Glusic/Getty RF; 1.1 (Earth): © Ingram Publishing/Alamy RF; 1.1 (mushroom): © IT Stock/Age Fotostock RF; 1.1 (humans): © Heath Korvola/UpperCut Images/Getty RF; 1.1 (peacock): © Brand X Pictures/PunchStock RF; 1.3: © BananaStock/PunchStock RF; 1.4 (main, mature tree): © PhotoDisc Website RF; 1.4 (sapling): Courtesy Paul Wray, Iowa State University; 1.4 (seedling): © Herman Eisenbeiss/Photo Researchers, Inc.; 1.5a: © Ralph Robinson/Visuals Unlimited; 1.5b: © A.B. Dowsett/SPL/Photo Researchers, Inc.; 1.8: © Photo by Ron Nichols, USDA Natural Resources Conservation Service; 1.9: © Dr. Jeremy Burgess/Photo Researchers, Inc.; 1.10a (all): Courtesy Jim Bidlack; 1A: © Purestock/Superstock RF.

Chapter 2 Opener: © Tobias Titz/Getty RF; 2.1: © Tom Mareschal/Alamy RF; 2.4a: © Biomed Commun./Custom Medical Stock Photo; 2.4b (left): © Mazzlota et al./Photo Researchers, Inc; 2.4b (right): Courtesy National Institutes of Health; 2.5a: © Tony Freeman/PhotoEdit; 2.5b: © Mark Kostich/Getty RF; 2.7b (salting food): © PM Images/Getty RF; 2.7b (crystals): © Evelyn Jo Johnson; 2.10: © PNC/Brand X/Corbis RF; 2.11: © Steve Mason/Getty RF; 2A (left): © The McGraw-Hill Companies, Inc. Mark A. Dierker, photographer; 2A (right): © Courtesy Pacific Northwest National Laboratory; 2.16: © Jeremy Burgess/SPL/Photo Researchers, Inc.; 2.17: © Don W. Fawcett/Photo Researchers, Inc.; 2.18: © Science VU/Visuals Unlimited; 2B: Courtesy USDA, ChooseMyPlate.gov website; 2.22 (both): © Purestock/Superstock RF; 2.27a: © Photodisc Red/Getty Images.

Chapter 3 Opener: © Bruce Dale/National Geographic/Getty Images; 3.3: © Sercomi/Photo Researchers, Inc.; 3.4: © Dr. Dennis Kunkel/Visuals Unlimited; 3.5: © E.H. Newcomb/W.P. Wergin/Biological Photo Service; 3.6 (top right): Courtesy Ron Milligan/Scripps Research Institute; 3.6 (bottom blow up): Courtesy E.G. Pollock; 3Aa (amoeba): © Stephen Durr; 3Ab (mitochondrion): © Dr. Don W. Fawcett/Visuals Unlimited; 3Ac (dinoflagellate): © Biophoto Assoc./Photo Researchers, Inc.; 3.7: © R. Bolender & D. Fawcett/Visuals Unlimited; 3.9: © EM Research Services, Newcastle University; 3.11a: Courtesy Herbert W. Israel, Cornell University; 3.12a: Courtesy Dr. Keith Porter; 3.13a (actin): © M. Schliwa/Visuals Unlimited; 3.13a *(Chara)*: © The McGraw-Hill Companies, Inc. Dennis Strete and Darrell Vodopich, photographers; 3.13b (intermediate): © K.G. Murti/Visuals Unlimited; 3.13b (girls): © Amos Morgan/Getty RF; 3.13c (microtubules): © K.G. Murti/Visuals Unlimited; 3.13c (chameleon): © Photodisc/Vol. 6/Getty RF; 3.15(sperm): © David M. Phillips/Photo Researchers, Inc.; 3.15 (flagellum): © William L. Dentler/Biological Photo Service; 3.15 (cilia): © Dr. G. Moscoso/Photo Researchers, Inc.

Chapter 4 Opener: © Tooga Productions, Inc./Getty Images; 4.7 (top left, center, and right): © David M. Phillips/Photo Researchers, Inc.; 4.7 (bottom left and center): © Dwight Kuhn; 4.7 (bottom right): © Ed Reschke/Getty Images; 4.11a: © Eric Grave/Phototake; 4.11b: © Don W. Fawcett/Photo Researchers, Inc.; 4.11c (both): Courtesy Mark Bretscher; 4.13a: © Kelly, 1966. Originally published in The Journal of Cell Biology, 28: 51–72; 4.13b: © David M. Phillips/Visuals Unlimited; 4.13c: Courtesy Camillo Peracchia, M.D.

Chapter 5 Opener: © Kevin Mazur/Contributor/WireImage/Getty Images; 5.2: Courtesy Douglas R. Green/LaJolla Institute for Allergy and Immunology; 5A: © Dr. P. Marazzi/Photo Researchers, Inc.; 5.5 Animal cell (early prophase, prophase, metaphase, anaphase, telophase): © Ed Reschke; 5.5 (prometaphase): © Michael Abbey/Photo Researchers, Inc.; 5.5 Plant cell (early prophase, prometaphase): © Ed Reschke; 5.5 (prophase, metaphase, anaphase, telophase): © R. Calentine/Visuals Unlimited; 5.6 (top): © Thomas Deerinck/Visuals Unlimited; 5.6 (bottom): © Steve Gschmeissner/SPL/Getty RF; 5.7: © Biophoto Associates/Photo Researchers, Inc.

Chapter 6 Opener: © Michael Fogden/Animals Animals; 6.2c (both): © Keith Eng, 2008; 6.8b: © Brand X Pictures/PunchStock RF; 6.8c: © Digital Vision/PunchStock RF; 6.11 (leaves): © Comstock/PunchStock RF; 6.11(runner): © PhotoDisc/Getty RF; 6.12: © Getty Images/SW Productions RF; 6A: © AP Photo/Chikumo Chiaki.

Chapter 7 Opener: © Ronald Martinez/Getty Images; 7A: © C Squared Studios/Getty RF.

Chapter 8 Openers (trees): © Image Plan/Corbis RF and (fruit): © The McGraw-Hill Companies, Inc. Evelyn Jo Johnson, photographer; 8.1a: © Tom Adams/Visuals Unlimited/Getty Images; 8.1b: © Chuck Davis/Stone/Getty Images; 8.1c: © Dynamic Graphics Group/Creatas/Alamy RF; 8.2d: © Dr. George Chapman/Visuals Unlimited; 8.10a: © The McGraw-Hill Companies, Inc. Evelyn Jo Johnson, photographer; 8.10b: © Corbis RF; 8.10c: © S. Alden/PhotoLink/Getty RF; 8A: © Doable/A.collection/Getty Images.

Chapter 9 Opener: © The McGraw-Hill Companies, Inc. Evelyn Jo Johnson, photographer; 9.2a: © Dorling Kindersley/Getty Images; 9.2b,c: © Dwight Kuhn; .4a: © Evelyn Jo Johnson; 9.4b: © J.Robert Waaland/Biological Photo Service; 9.4c: © Kingsley Stern; 9.5a: © Dr. Ken Wagner/Visuals Unlimited; 9.5b: © George Wilder/Visuals Unlimited; 9.5c: © Biophoto Assoc./Photo Researchers, Inc.; 9.6a: © N.C. Brown Center for Ultrastructure Studies, SUNY, College of Environmental Science & Forestry, Syracuse, NY; 9.6b: © Dr. Michael Clayton, University of Wisconsin, Madison, Dept. of Botany; 9.8a (root tip): Courtesy Ray F. Evert/University of Wisconsin Madison; 9.8b: © CABISCO/Phototake; 9.9: © Steven P. Lynch; 9.10a: © John D. Cunningham/Visuals Unlimited; 9.10b: Courtesy George Ellmore, Tufts University; 9.11a: © Dr. Robert Calentine/Visuals Unlimited; 9.11b: © The McGraw-Hill Companies, Inc. Evelyn Jo Johnson, photographer; 9.11c: © Brad Mogen/Visuals Unlimited/Getty Images; 9.11d: © Alan and Linda Detrick/Photo Researchers, Inc.; 9.11e: © David Sieren/Visuals Unlimited; 9.11f: © Professor David F. Cox, Lincoln Land Community College; 9.12a: © Dwight Kuhn; 9.12b: © Dana Richter/Visuals Unlimited; 9.12c: © Science VU/R. Roncadori/Visuals Unlimited; 9.15 (top): © Ed Reschke; 9.15 (bottom): Courtesy Ray F. Evert/University of Wisconsin Madison; 9.16 (top): © CABISCO/Phototake; 9.16 (bottom): © Kingsley Stern; 9.18 (circular cross section): © Ed Reschke/Getty Images; 9A (flooring): © Roman Borodaev/Alamy RF; 9A (shoots): © Koki Iino/Getty RF; 9A (bamboo grove): © Michele Westmorland/Getty Images; 9.19a: © Ardea London Limited; 9.20a: © The McGraw-Hill Companies, Inc. Evelyn Jo Johnson, photographer; 9.20b: © Science Pictures Limited/Photo Researchers, Inc.; 9.20c, d: © The McGraw-Hill Companies, Inc. Carlyn Iverson, photographer; 9.21: © Jeremy Burgess/SPL/Photo Researchers, Inc.; 9.23a: © Patti Murray Animals Animals; 9.23b: © Gerald & Buff Corsi/Visuals Unlimited; 9.23c: © Steven P. Lynch; 9.25 (both): © Jeremy Burgess/SPL/Photo Researchers, Inc.; 9.26a: © M. H. Zimmermann, Courtesy Dr. P. B. Tomlinson, Harvard University; 9.26b: © Steven P. Lynch; 9B: © Yann Arthus-Bertrand/Corbis.

Chapter 10 Opener: © Mitch Hrdlicka/Getty RF; 10.3a: © Farley Bridges; 10.3b: © Pat Pendarvis; 10.4a: © Adam Hart-Davis/SPL/Photo Researchers, Inc.; 10.4b: © Design Pics Inc./Alamy RF; 10.5 (pollen grain): Courtesy Graham Kent; 10.5 (embryo sac): © Ed Reschke; 10.6a: © George Bernard/Animals Animals; 10.6b: © Simko/Visuals Unlimited; 10.6c: © Dwight Kuhn; 10Aa: © Steven P. Lynch; 10Ab: © IT Stock Free/Alamy RF; 10.8a: © James Mauseth; 10.8b: © MIXA/Getty RF; 10.8c: © David Marsden/Getty Images; 10.8d: © Dinodia Photos/Alamy; 10.9: © Ed Reschke; 10.10 (top left): © Ed Reschke/Getty Images; 10.10 (top right): © Scott Sinklier/AgStock Images/Corbis; 10.11(all): Courtesy Prof. Dr. Hans-Ulrich Koop, from Plant Cell Reports, 17:601-604; 10.12 (all): Courtesy Monsanto; 10.13b: Courtesy Eduardo Blumwald; 10.15 (both): Courtesy Prof. Malcolm B. Wilkins; 10.16: © Sylvan Wittwer/Visuals Unlimited; 10.18a–b (both): © Kingsley Stern; 10.18c: © Kent Knudson/PhotoLink/Getty RF; 10.19a: © Dorling Kindersley/Getty Images; 10.19b: © Kingsley Stern; 10.19c: © John D. Cunningham/Visuals Unlimited; 10.21b (both): © Nigel Cattlin/Visuals Unlimited/Getty Images.

Chapter 11 Opener: © Christopher Polk/Staff/Getty Images; 11.1 (all) and 11.2a: © Ed Reschke; 11.2b, e: © The McGraw-Hill Companies, Inc. Dennis Strete, photographer; 11.2c: © The McGraw-Hill Companies, Inc. Al Telser, photographer; 11.2d: © Dr. Fred Hossler/Visuals Unlimited/Getty Images; 11.4a: © Ed Reschke; 11.4b: © The McGraw-Hill Companies, Inc. Dennis Strete, photographer; 11.4c (both) and 11.5b: © Ed Reschke; 11Aa, b: © Dr. P. Marazzi/Photo Researchers, Inc.; 11Ac: © James Stevenson/SPL/Photo Researchers, Inc.; 11B: © MCT via Getty Images.

Chapter 12 Opener: © Mark Sullivan/WireImage/Getty Images; 12.1d and 12.3c: © Ed Reschke; 12.4a: © Ed Reschke/Getty Images; 12.4b: © Andrew Syred/Photo Researchers, Inc.; 12.5: © Dennis Kunkel/Phototake; 12.6b: © Eye of Science/Photo Researchers, Inc.; 12A: © Astrid & Hanns-Frieder Michler/Photo Researchers, Inc.; 12.13b–c (both): © Ed Reschke; 12B: © Biophoto Associates/Photo Researchers, Inc.; 12.18: © CCN/Phototake; 12.20: © ABIOMED, Inc.; p. 229 (ECG): © Ed Reschke.

Chapter 13 Opener: © Dr. Richard Kessel & Dr. Randy Kardon/Tissues & Organs/Visuals Unlimited; 13.2a: © The McGraw-Hill Companies, Inc. Dennis Strete, photographer; 13.2b: © R. Calentine/Visuals Unlimited; 13.2c: © Fred E. Hossler/Visuals Unlimited; 13.2d: © The McGraw-Hill Companies, Inc. Al Telser, photographer; 13.6b: Courtesy Dr. Arthur J. Olson, Scripps Institute; 13.8b: © Steve Gschmeissner/Photo Researchers, Inc.; 13.9a (top): © The McGraw-Hill Companies, Inc. Jill Braaten, photographer; 13.10b: © Digital Vision/Getty RF; 13.10c: © Aaron Haupt/Photo Researchers, Inc.; 13.12: © Bart's Medical Library; 13B (right): © Damien Lovegrove/SPL/Photo Researchers, Inc.; 13B (left): © David Scharf/SPL/Photo Researchers, Inc.; 13.1 (both): © J. C. Revy/Phototake.

Chapter 14 Figure 14.4b: © Biophoto Associates/Photo Researchers, Inc.; 14.5b: © Ed Reschke/Getty Images; 14.6 (villi): © Kage Mikrofotografie/Phototake; 14.6 (microvilli): This photo was published in Medical Cell Biology, Charles Flickinger, photo by Susumu Ito, Copyright Elsevier, 1979.; 14.13: Courtesy USDA, ChooseMyPlate.gov website; 14A: © The McGraw-Hill Companies, Inc. Andrew Resek, photographer; 14.17a, c: © Biophoto Associates/Photo Researchers, Inc.; 14.17b: © Ken Greer/Visuals Unlimited; 14.19a: © Ted Foxx/Alamy RF; 14.19b: © Donna Day/Stone/Getty Images; 14.19c: © The McGraw-Hill Companies, Inc. Mark Dierker, photographer; 14B: © J. James/Photo Researchers, Inc.

Chapter 15 Opener: © Michael Keller/Corbis; 15.3 (left): © CNRI/Phototake; 15.4: © Dr. Kessel & Dr. Kardon/Tissues & Organs/Visuals Unlimited; 15.6: © Yoav Levy/Phototake; 15.10: © Dr. P. Marazzi/Photo Researchers, Inc.; 15.11: © Clinica Claros/Phototake; 15.13a: © Matt Meadows/Getty Images; 15.13b: © Biophoto Associates/Photo Researchers, Inc.; 15Aa: Courtesy The Wyss Institute, Harvard University.

Chapter 16 Openers (polycystic): © Biophoto Associates/Photo Researchers, Inc. and (normal): © SIU/Visuals Unlimited; 16.3d: © Kage Mikrofotografie/Phototake; 16.4b: © Science Photo Library RF/Getty RF; 16.4c, d: © Ed Reschke/Getty Images; 16A: © Will & Deni McIntyre/Photo Researchers, Inc.; 16.11: Image reprinted with permission from eMedicine.com, 2008. Available at http://emedicine.medscape.com/article/440657-overview.; 16.12: © AP Images/Brian Walker.

Chapter 17 Opener: © Peter Duddek/Visum/The Image Works; 17.2 (myelin): © M.B. Bunge/Biological Photo Service; 17.2 (cell body): © doc-stock/Visuals Unlimited; 17.3b: © The McGraw-Hill Companies, Inc. Dr. Dennis Emery, Dept. of Zoology and Genetics, Iowa State University, photographer; 17.6a: © Science VU/Lewis-Everhart-Zeevi/Visuals Unlimited; 17.8c: © Karl E. Deckart/Phototake; 17.9b: © Colin Chumbley/Science Source/Photo Researchers, Inc.; 17A (student): © Photodisc/Alamy RF; 17A (brain waves): © L. Birmington/Custom Medical Stock Photo; 17.14 (all): © Marcus Raichle; 17.19 (both): © Science VU/Visuals Unlimited; 17B: © The McGraw-Hill Companies, Inc. Mark Dierker, photographer; 17.21: © ISM/Phototake; 17.22: © Ralph C. Eagle, Jr./Photo Researchers, Inc.; 17.23: © Topham/The Image Works.

Chapter 18 Openers, both: © AP Images/Gene J. Puskar; 18.4b, c (all): © Omikron/SPL/Photo Researchers, Inc.; 18.8: © Lennart Nilsson/Scanpix; 18.9b: © Biophoto Associates/Photo Researchers, Inc.; 18.12: © P. Motta/SPL/Photo Researchers, Inc.; 18A (both): Courtesy Robert S. Preston and Joseph E. Hawkins, Kresge Hearing Research Institute, University of Michigan; 18.14: © Steve Allen/Getty RF; 18.16: © Sue Ford/Photo Researchers, Inc.

Chapter 19 Opener: © Scott Camazine/Photo Researchers, Inc.; 19.1(hyaline, compact bone): © Ed Reschke; 19.1(osteocyte): © Biophoto Associates/Photo Researchers, Inc.; 19.5b: © Corbis RF; 19A (both): Courtesy Melissa Connor; 19.10a: © Gerard Vandystadt/Photo Researchers, Inc.; 19.13: © EM Research Services, Newcastle University; 19.14a (top): © Victor B. Eichler, Ph.D.; 19.17 (woman): © Rubberball/Getty RF; 19.17 (fibers): © G.W. Willis/Visuals Unlimited; 19.17 (man): © Corbis RF; 19.18: © Scott Camazine/Phototake; 19.19 (left): © Alfred Pasieka/Photo Researchers, Inc.; 19.19 (right): © Yoav Levy/Phototake USA.com; 19.20: © Dr. P. Marazzi/Photo Researchers, Inc.; 19B: © AFP/Getty Images.

Chapter 20 Opener: © Michael Krasowitz/Getty Images; 20.11: © General Photographic Agency/Getty Images; 20A: © James Darell/Stone/Getty Images; 20.12(all): Reprinted from Clinical Pathological Conference, American Journal of Medicine, Vol. 20, page 133, "Acromegaly, Diabetes, Hypermetabolism, Proteinuria and Heart Failure," copyright 1956, with permission from Elsevier; 20.13 (both): Courtesy Shannon Halverson; 20B: © Angela J. Cesere/AnnArbor.com; 20.14a: © Medical-on-Line/Alamy; 20.14b: © Biophoto Associates/Photo Researchers, Inc.; 20.14c: © Dr. P. Marazzi/SPL/Photo Researchers, Inc.; 20.15a: © Custom Medical Stock Photo; 20.15b: © NMSB/Custom Medical Stock Photo; 20.17a–b(both): Courtesy of Insulet Corporation; 20C: © Ramey Photo Agency.

Chapter 21 Opener: © Stockbyte/Getty RF; 21.3b (top): © CNRI/Photo Researchers, Inc.; 21.3b (bottom): © Ed Reschke; 21.7 (oocyte): © Ed Reschke/Getty Images; 21.10b: © Bettman/Corbis; 21.11a: © The McGraw-Hill Companies, Inc. Bob Coyle, photographer; 21.11b: © The McGraw-Hill Companies, Inc. Vincent Ho, photographer; 21.11c: © The McGraw-Hill Companies, Inc. Bob Coyle, photographer; 21.11d: © Scott Camazone/Photo Researchers, Inc.; 21.11e (left): Courtesy Population Council, Karen-Tweedy Holmes.; 21.11e (right): © Bob Pardue/Alamy; 21.11f: © Michaela Begsteiger Image Broker/Newscom; 21.12: © Scott Camazine/Photo Researchers, Inc.; 21Ba(both): © Corbis RF; 21Bb: © Brand X Pictures RF; 21.13 (all): © Nicholas Nixon; 21B (woman): © Steven Peters/Stone/Getty Images; 21B (owls): © Art Wolfe/Stone/Getty Images; 21B (fish): © Georgette Douwma/Getty RF; 21.15a: Courtesy of the Centers for Disease Control, Atlanta, GA. (Crooks, R. and Baur, K. "Our Sexuality, 8/e", Wadsworth 200d 17.4, pg. 489; 21.15b: Courtesy of the Centers for Disease Control, Susan Lindsley, photographer; 21.16: Courtesy Seattle STD Training Center/HIV Prevention, Seattle, WA; 21.17a: © Science VU/CDC/Visuals Unlimited; 21.17b: Courtesy of the Centers for Disease Control, Dr. N.J. Fiumara, Dr. Gavin Hart; 21.17c: © Collection CNRI/Phototake; 21.17d: © Science VU/Visuals Unlimited; 21.19: © Laparoscopic Appearance of Endometriosis (copyright 1991), www.memfert.com, Dan Martin, M. D.; 21.20: © CC Studio/SPL/Photo Researchers, Inc.

Chapter 22 Opener: © John Slater/Getty RF; 22.1: © David M. Phillips/Visuals Unlimited; 22.2a: © William Jorgensen/Visuals Unlimited; p. 438 (chick): © Photodisc/Getty Images; 22A: © Associated Press; 22.4b: Courtesy Kathryn Tosney; 22.8 (both): Courtesy Steve Paddock, Howard Hughes Medical Research Institute; 22.9a: Courtesy E.B. Lewis; 22.13a: © Lennart Nilsson/Scanpix; 22.14: © John Watney/Photo Researchers, Inc.; 22.18: © Jose Luis Pelaez, Inc./Corbis; 22.19: © Corbis RF.

Chapter 23 Opener: © Sports Illustrated/Getty Images; 23.1: © Ned M. Seidler/National Geographic Image Collection; 23.3a: © HFPA, 63rd Golden Globe Awards; 23.3b: © Dynamic Graphics/PictureQuest RF; 23.3c–d(both): © The McGraw-Hill Companies, Inc. Bob Coyle, photographer; 23.10a: Courtesy Robert D. Terry/University of California, San Diego School of Medicine; 23.10c: © Eye of Science/Photo Researchers, Inc.; 23.12 (both): Courtesy Dr. Hemachandra Reddy, The Neurological Science Institute, Oregon Health & Science University; 23.13 (man): © Vol. 88/PhotoDisc RF; 23.13(woman): © Larry Williams/Corbis; 23.14: © Mediscan/Medical-On-Line; 23.17: Courtesy University of Connecticut, Peter Morenus, photographer; 23.18: © Jane Burton/Bruce Coleman, Inc.

Chapter 24 Opener: © moodboard/Alamy; 24.5 (left, right): Courtesy Dr. Rabi Tawil, Director, Neuromuscular Pathology Laboratory, University of Rochester Medical Center; 24.5 (center): Courtesy Muscular Dystrophy; 24Aa (both): From R. Simensen and R. Curtis Rogers, "Fragile X Syndrome," American Family Physician 39(5):186, May 1989, © American Academy of Family Physicians; 24Ab: © David M. Phillips/Visuals Unlimited; 24.6 (queen): © Stapleton Collection/Corbis; 24.6(prince): © Huton Archive/Getty Images; 24.8a: © Jose Carrilo/PhotoEdit; 24.8b, 24Bc: © CNRI/SPL/Photo Researchers, Inc.; 24.9b: Courtesy The Williams Syndrome Association; 24.10b: Courtesy Kathy Wise; 24.11b: © Fred, Herbert, and Hendrik van Dijk. "Images of Memorable Cases, 50 Years at the Bedside: Case 133."

Chapter 25 Opener: © AP Images/Jim McKnight; 25Aa: © Photo Researchers, Inc.; 25Ac: © Science Source/Photo Researchers, Inc.; 25.11d: Courtesy Alexander Rich; 25.17: © Photodisc/Getty RF; 25.19c: © Mondae Leigh Baker; 25.20 (both): © Dr. Stanley Flegler/Visuals Unlimited; 25B: © Jeff Maloney/Getty Images.

Chapter 26 Opener: © Zacarias Pereira da Mata/Shutterstock; 26.4: © Europics/Newscom; 26.5 (both): Courtesy General Electric Research & Development.; 26Aa: Courtesy USDA/Doug Wilson, photographer; 26Ab: © Norm Thomas/Photo Researchers, Inc.; 26Ac: Courtesy Carolyn Merchant/UC Berkeley; 26B (array, right): © Deco/Alamy RF; 26.1 (human): © Image Source/JupiterImages; 26.1(mouse): © David A. Northcott/Corbis; 26.1(fly): © Herman Eisenbeiss/Photo Researchers, Inc.; 26.1 (plant): Courtesy USDA/Peggy Greb, photographer; 26.1 (roundworm): © CMSP/J.L. Carson/Getty Images; 26.1 (yeast): © G. Wanner/Getty Images; 26.9: © Geno Zhou/FeatureChina/Newscom.

Chapter 27 Opener: © Dr. Dennis Kunkel/Visuals Unlimited; 27.1 (Darwin): Transparency #326697 American Museum of Natural History Library; 27.1 (rhea): © Nicole Duplaix/National Geographic/Getty Images; 27.1 (ship): © Mary Evans Picture Library/The Image Works; 27.1 (iguana): © FAN Travelstock/Alamy RF; 27.3a: © Adrienne T. Gibson/Animals Animals; 27.3b: © Joe McDonald/Animals Animals; 27.3c: © Leonard Lee Rue/Animals Animals; 27.4a (fossil): © Jason Edwards/Getty RF; 27.4b: © Joe Tucciarone; 27.5a (fossil foot): © J.G.M. Thewissen, Northeastern Ohio Universities College of Medicine; 27.10 (Boston terrier): © Robert Dowling/Corbis; 27.10 (wolf): © DLILLC/Corbis RF; 27.10 (Irish wolfhound): © Ralph Reinhold/Index Stock Imagery/Photolibrary RF; 27A (both): © Ingram Publishing/SuperStock RF; 27.18b (both): © John Cancalosi/Getty Images; 27Ca (MRSA): © Science Source/Photo Researchers, Inc.; 27Cb *(P. aeruginosa):* © Steve Gschmeissner/Photo Researchers, Inc.; 27Cc *(M. tuberculosis):* © SPL/Photo Researchers, Inc.

Chapter 28 Opener: © Andreas Reh/Getty RF; 28.4: © Ralph White/Corbis; 28.7a (main): © John Sohlden/Visuals Unlimited; 28.7a (inset): From J.T. Staley, et al., Bergey's Manual of Systematic Bacteriology, Vol. 13, © 1989 Williams & Wilkins Co., Baltimore, prepared by A.L. Usted, photography by Dept. of Biophysics, Norwegian Institute of Technology; 28.7b (main): © Jeff Lepore/Photo Researchers, Inc.; 28.7b (inset): Courtesy Dennis W. Grogan, Univ. of Cincinnati; 28.7c (main): © Susan Rosenthal/Corbis; 28.7c (inset): © Ralph Robinson/Visuals Unlimited; 28.8a: © Gary Gaugler/Visuals Unlimited; 28.8b: © SciMAT/Photo Researchers, Inc.; 28.8c: © Dr. Richard Kessel & Dr. Gene Shih/Visuals Unlimited; 28.10: © CNRI/SPL/Photo Researchers, Inc.; 28.11a: © R. Knauft/Biology Media/Photo Researchers, Inc.; 28.11b: © Eric Grave/Photo Researchers, Inc.; 28.12a: © Fred Hossler/Visuals Unlimited; 28.12b: © SPL/Photo Researchers, Inc.; 28.12c: © Kenneth E. Greer/Visuals Unlimited; 28.1 (both): © © Leonard V. Crowley, *An Introduction to Human Disease: Pathology and Pathophysiology Correlations* 5/e, page 383 right and page 385 left; 2001: Jones and Bartlett Publishers, Sudbury, MA, www.jblearning.com, reprinted with permission; 28Aa: © Bettmann/Corbis; 28Ab: © The McGraw-Hill Companies, Inc. Mark Dierker, photographer; 28.14a (top): © Dr. Hans Gelderblom/Visuals Unlimited; 28.14b (top): © K.G. Murti/Visuals Unlimited; 28.17: © John D. Cunningham/Visuals Unlimited.

Chapter 29 Opener: © Laurence Monneret/Getty Images; 29.3: © Stephen Durr; 29.4b: © M.I. Walker/Science Source/Photo Researchers, Inc.; 29.5: © Darlyne A. Murawski/Getty Images; 29.6(top): © Kevinschafer.com; 29.7–29.8 (all): © Dan Ippolito; 29.9: © Michael Abbey/Visuals Unlimited; 29.10a: © Eric Grave/Photo Researchers, Inc.; 29.10b: © CABISCO/Phototake; 29.10c: © Ed Reschke/Getty Images; 29.11b (foraminiferan): © NHPA/Superstock; 29.11b (cliffs): © Stockbyte/Getty RF; 29.11c: © Dr. Richard Kessel & Dr. Gene Shih/Visuals Unlimited; 29.12: © James Richardson/Visuals Unlimited; 29.13 (plasmodium): © CABISCO/Visuals Unlimited; 29.13 (sporangia): © V. Duran/Visuals Unlimited; 29.15: © Stanley Erlandsen; 29A: © Eye of Science/Photo Researchers, Inc.; 29B (boy): © Author's Image/PunchStock RF; 29B (tsetse fly): © Frank Greenaway/Getty Images; 29B (pig): © G.K. & Vikki Hart/Getty RF; 29B (goat): © Eureka/Alamy RF; 29.16a: © Gary R. Robinson/Visuals Unlimited; 29.16b: © Dennis Kunkel/Visuals Unlimited; 29.17a: © Richard Packwood/Getty Images; 29.17b: © Steve Gschmeissner/Getty RF; 29.19: Reproduced with permission of the Freshwater Biological Association on behalf of The Estate of Dr. Hilda Canter-Lund. (c) The Freshwater Biological Association.; 29.20: © Garry DeLong/Getty Images; 29.21b: © Robert Calentine/Visuals Unlimited; 29.21c: © Bear Dancer Studios/Mark Dierker; 29.2 (both): © David Philips/Visuals Unlimited; 29.23b: © Biophoto Assoc./Photo Researchers, Inc.; 29.23c: © Inga Spence/Photo Researchers, Inc.; 29.23d: © L. West/Photo Researchers, Inc.; 29C: © Patrick Endres/Visuals Unlimited; 29D: © R. Calentine/Visuals Unlimited; 29.24a: © Steven P. Lynch; 29.24b: © Arthur M. Siegelman/Visuals Unlimited; 29.25a: © Dr. Jeremy Burgess/SPL/Photo Researchers, Inc.; 29.25b: © Stephen Sharnoff/Visuals Unlimited; 29.25c: © Kerry T. Givens; 29.26a: © P. Marazzi/SPL/Photo Researchers, Inc.; 29.26b: © John Hadfield/SPL/Photo Researchers, Inc.; 29.26c: © Everett S. Beneke/Visuals Unlimited; 29.27: © AP Photo/Kevork Djansezian.

Chapter 30 Opener: © Design Pics/Jack Goldfarb/Getty RF; 30.1 *(Coleochaete):* © T. Mellichamp/Visuals Unlimited; 30.1 *(Chara):* © Heather Angel /Natural Visions; 30A (wheat): © Pixtal /age fotostock; 30A (corn plants): © Corbis RF; 30A (corn ear): © Nigel Cattlin/Photo Researchers, Inc.; 30A (rice plants): © Corbis RF; 30A (rice grain heads): © Dex Image/Getty RF; 30B (rubber): © moodboard/Corbis RF; 30B (cotton): © Dale Jackson/Visuals Unlimited; 30.5a: © Ed Reschke; 30.5b: © J.M. Conrarder/Nat'l Audubon Society/Photo Researchers, Inc.; 30.5c: © R. Calentine/Visuals Unlimited; 30.6 (sporophyte): © Heather Angel/Natural Visions; 30.6(gametophyte): © Steven P.

Lynch; 30.7: © The Field Museum. #CSGEO75400c; 30.8: © Steve Solum/Bruce Coleman, Inc.; 30.9 (cinnamon): © Laszlo Podor Photography/Getty RF; 30.9 (hart's tongue): © Organics image library/Alamy RF; 30.10: © Steven P. Lynch; 30.11: © CABISCO/Phototake; 30.12: © Robert P. Carr/Bruce Coleman, Inc.; 30.13 (pollen grains): © Carolina Biological Supply Co./Visuals Unlimited; 30.14a: © Pete Ryan/National Geographic/Getty RF; 30.14b: © Walt Anderson/Visuals Unlimited; 30.14c: © The McGraw-Hill Companies, Inc. Evelyn Jo Johnson, photographer; 30.15a (both): © Steven P. Lynch; 30.15b: © Kathy Merrifield/Photo Researchers, Inc.; 30.15c: Courtesy Fiona Norris; 30.16 (cactus): © Author's Image/PunchStock RF; 30.16 (waterlily): © Pat Pendarvis; 30.16 (iris): © Steffen Hauser/botanikfoto/Alamy; 30.16 *(Amborella):* Courtesy Stephen McCabe/Arboretum at University of California Santa Cruz; 30.16 *(trillium):* © Adam Jones/Photo Researchers, Inc.; 30.16 (apple blossoms): © Inga Spence/Photo Researchers, Inc.; 30.16 (butterfly weed): © Evelyn Jo Johnson; 30.18 (pollen grain): Courtesy Graham Kent; 30.18 (embryo sac): © Ed Reschke; 30.19a: © Stockbyte/PunchStock; 30.19b, c: © Mondae Leigh Baker; 30.19d: © Photolink/Getty RF; 30.19e: © Corbis RF; 30.19f: © Brand X Pictures RF.

Chapter 31 Opener: © James H. Robinson/Photo Researchers, Inc.; 31.1 (hydra): © Biophoto Associates/Photo Researchers, Inc.; 31.1 (crayfish): © David M. Dennis; 31.1 (humans): © Justin Kase z10z/Alamy RF; 31.4a: © Andrew J. Martinez/Photo Researchers, Inc.; 31A: © Secret Sea Visions/Getty Images; 31.5a: © NHPA/Image Quest 3-D RF; 31.5b: © J. McCollugh/Visuals Unlimited; 31.6b: © Azure Computer & Photo Services/Animals Animals; 31.6c: © Corbis RF; 31.6d: © NHPA/Charles Hood/Photoshot RF; 31.6e: © Amos Nachoum/Corbis; 31.7: © CABISCO/Visuals Unlimited; 31.8a: © Robert Brons/Biological Photo Service; 31.9e: © Tom E. Adams/Visuals Unlimited; 31.10 (#3 proglottid): © John D. Cunningham/Visuals Unlimited; 31.10 (#3 scolex): © James Webb/Phototake NYC; 31.12a: © Fred Bavendam/Minden Pictures; 31.12b: Courtesy Larry S. Roberts; 31.12c: © Kenneth W. Fink/Bruce Coleman, Inc.; 31.12d: © Ken Lucas/Visuals Unlimited; 31.13a: © Farley Bridges; 31.13b: © Georgette Douwma/Photo Researchers, Inc.; 31.15a: © Heather Angel/Natural Visions; 31.15c: © Diane R. Nelson; 31.16c: © Roger K. Burnard/Biological Photo Service; 31.17: © St. Bartholomews Hospital/SPL/Photo Researchers, Inc.; 31.18a: © Lauritz Jensen/Visuals Unlimited; 31.18c: © Vanessa Vick/The New York Times/Redux; 31.19a: © John MacGregor/Getty Images; 31.19b: © G.C. Kelley/Photo Researchers, Inc.; 31.19c: © Bob Evans/Getty Images; 31.19d: © James Robinson/Photo Researchers, Inc.; 31.19e: © Dwight Kuhn; 31.20c: © Sinclair Stammers/Photo Researchers, Inc.; 31.22a: © Joel Sartore/Getty RF; 31.22b: © Creatas Images/PictureQuest; 31.22c: © MedioImages/Photodisc/Getty RF; 31.22d: © Photos.comSelect/Index Stock Imagery; 31.22e: © Darlyne A. Murawski/Getty Images; 31.24a: © Tom McHugh/Photo Researchers, Inc.; 31.24b: © Robert Lubeck/Animals Animals; 31.24c: © Scott Camazine/Photo Researchers, Inc.; 31.25a: © Diane R. Nelson; 31.25b, c: © David Wrobel/Getty Images; 31.25d: © Philippe Bourseiller/Getty Images; 31.26a: © Randy Morse, GoldenStateImages.com; 31.27: © Andy Long/Painet.

Chapter 32 Opener: © Christopher Austin/AFP/Getty Images/Newscom; 32.3: © Heather Angel/Natural Visions; 32.4: © Corbis RF; 32.5a: © James Watt/Animals Animals; 32.5b: © Dwight Kuhn; 32.5c: © Zig Leszczynski/Animals Animals; 32Aa: © MedioImages/SuperStock RF; 32Ab: © Allan Friedlander/SuperStock; 32Ac: © Account Phototake/Phototake; 32.7a: © Comstock Images/PictureQuest; 32.7c: © Peter Scoones/SPL/Photo Researchers, Inc.; 32.8(1): © Jane Burton/Bruce Coleman, Inc.; 32.8(2, 3): © Michael Redmer/Visuals Unlimited; 32.8(4): © Joe McDonald/Visuals Unlimited; 32.9: © Papilio/Alamy; 32.10a: © Martin Harvey/Gallo Images/Corbis; 32.12a: © Gary W. Carter/Corbis; 32.12b: © Dale DeGabriele/Getty Images; 32.12c: © Medford Taylor/Getty Images; 32.13a: © D. Parer & E. Parer-Cook/Ardea; 32.13b: © Fritz Prenzel/Animals Animals; 32.13c: © Leonard Lee Rue/Photo Researchers, Inc.; 32.14a: © Stephen J. Krasemann/Photo Researchers, Inc.; 32.14b: © John Downer/Getty Images; 32.14c: © Gerald Lacz/Animals Animals; 32.14d: © Mike Bacon; 32.16 *(S. tchadensis):* © HO/AFP/Newscom; 32.16 (A. afarensis): © Friedrich Saurer/Alamy; 32.16 *(A. boisei):* © Science VU/NMK/Visuals Unlimited; 32.16 *(H. habilis):* © Herve Conge/ISM/Phototake; 32.16 *(H. sapiens):* © 2007 Educational Images Ltd./Custom Medical Stock Photo; 32.17a: © Dan Dreyfus and Associates; 32.17b: © John Reader/Photo Researchers, Inc.; 32.18: © Science VU/Visuals Unlimited; 32.20: Image # 608, courtesy American Museum of Natural History Library.

Chapter 33 Opener: © Getty RF; 33.1a: © Joe McDonald; 33.1b: Courtesy Jeff and Wendy Martin, Refuge for Saving the Wildlife, Inc.; 33.2 (coastal): © Gerold & Cynthia Merker/Visuals Unlimited; 33.2 (inland): © Matt Jeppson/Shutterstock; 33.5: © Gregory G. Dimijian/Photo Researchers, Inc.; 33.6a (main): © Arco Images/GmbH/Alamy; 33.6a (inset): © Fritz Polking/Visuals Unlimited; 32.7: © Raul Arboleda/AFP/Getty Images; 33.8 (trees): © PhotoLink RF; 33.8 (firefly): © Phil Degginger/Alamy; 33.9a: © OSF/Animals Animals; 33Aa: © Tom McHugh/Photo Researchers, Inc.; 33Ab: © Alan Carey/Photo Researchers, Inc.; 33.10: © Nicole Duplaix/Getty Images; 33.12: © Thomas Dobner 2006/Alamy RF; 33.13: © D. Robert & Lorri Franz/Corbis; 33.14: © Mark Moffett/Minden Pictures; 33.15: © J & B Photo/Animals Animals.

Chapter 34 Opener © AP Photo/Illinois River Biological Station via the Detroit Free Press, Nerissa Michaels; 34.1: © David Hall/Photo Researchers, Inc.; 34.2a: © E. R. Degginger/Photo Researchers, Inc.; 34.2b: © Tracey Thompson/Corbis RF; 34.5b: © Image State/Alamy; 34.5c: © Michael Coyne/Getty Images 34.7 (dandelions): © Ted Levin/Animals Animals; 34.7 (bears): © Michio Hoshino/Minden Pictures; 34.10: © Alan Carey/Photo Researchers, Inc.; 34.11a: © Michael Fogden/Animals Animals; 34.11b: © Scott Camazine/Photo Researchers, Inc.; 34.11c: © National Audubon Society/A. Cosmos Blank/Photo Researchers, Inc.; 34.12a–c: © Edward S. Ross; 34.12d: © James H. Robinson/Photo Researchers, Inc.; 34.13: © James Hager/Getty Images; 34.14: © Bill Wood/Bruce Coleman, Inc.

Chapter 35 Opener: © Jeremy Woodhouse 2008/Getty Images; 35.1a (diatom): © Ed Reschke; 35.1a (tree): © Herman Eisenbeiss/Photo Researchers, Inc.; 35.1b (caterpillar): © Corbis RF; 35.1b (rabbit): © Gerald C. Kelley/Photo Researchers, Inc.; 35.1c (snake eating rabbit): © Derrick Hamrick/imagebroker/Corbis; 35.1c (hawk and snake): © Tze-hsin Woo/Getty RF; 35.1d (bacteria): © David M. Phillips/Visuals Unlimited; 35.1d (mushrooms): © Michael Beug; 35.3: © George D. Lepp/Photo Researchers, Inc.; 35A: © Harrison Shull/Getty Images; 35Ba: © Bill Aron/PhotoEdit.

Chapter 36 Openers (Earth): Courtesy NASA, (eagle): © Frank Leung/Getty RF, and (penguins): © Brand X Pictures/Getty RF; 36.6a: © Johnny Johnson/Getty Images; 36.6b: © John Shaw/Bruce Coleman, Inc.; 36.7a: © Creatas/Jupiterimages RF; 36.7b: © Bill Silliker, Jr./Animals Animals; 36.8 (Shawnee National Forest): © David S. Rice/Photo Researchers, Inc.; 36.8 (bobcat): © Tom McHugh/Photo Researchers, Inc.; 36.8 (marigolds): © Virginia Neefus/Animals Animals; 36.8 (chipmunk): © Zig Lesczynski/Animals Animals; 36.9 (frog): © Art Wolfe/Photo Researchers, Inc.; 36.9 (katydid): © M. Fogden/OSF/Animals Animals; 36.9 (jaguar, orchids): © Professor David F. Cox, Lincoln Land Community College; 36.9 (macaw): © PhotoLink/Getty RF; 36.9 (butterfly): © Kevin Schafer/Getty Images; 36.9 (lizard): © J.H. Pete Carmichael/Getty Images; 36.10a: © Walt Anderson/Visuals Unlimited; 36.10b: © Kathy Merrifield/Photo Researchers, Inc.; 36.11a: © Jim Steinberg/Photo Researchers, Inc.; 36.11b: © Steven Fuller/Animals Animals; 36.12a, b: © Darla G. Cox; 36.12c: © George W. Cox; 36.12d: © Digital Vision/Getty RF; 36.13a: © Corbis RF; 36.13b: © Bob Calhoun/Bruce Coleman, Inc.; 36.13c: © Jack Wilburn/Animals Animals; 36.13d: © Kevin Schafer/Getty Images; 36.14a: © Roger Evans/Photo Researchers, Inc.; 36.14b: © The McGraw-Hill Companies, Inc. Pat Watson, photographer; 36.16(pike): © Robert Maier/Animals Animals; 36.16 (water strider): © Frank Greenaway/Getty Images; 36.17 (shrimp): © Ken Lucas/Ardea London Limited; 36.17 (snails): © Fred Whitehead/Animals Animals; 36.18a: © John Eastcott/Yva Momatiuk/Animals Animals; 36.18b: © Mark Lewis/Getty Images; 36.19a: © Steve Bower/Shutterstock; 36.19b: © Jeff Greenburg/Photo Researchers, Inc.

Chapter 37 Opener: © J. Emilio Flores/La Opinion/Newscom; 37.3 (periwinkle): © Steven P. Lynch; 37.3 (armadillo): © Photodisc/Getty RF; 37.3 (fishing): © Herve Donnezan/Photo Researchers, Inc.; 37.3 (bat): © Merlin D. Tuttle/Bat Conservation International; 37.3 (ladybug): © D. Hurst/Alamy RF; 37.3 (rubber tree): © Bryn Campbell/Stone/Getty Images; 37.4 (mountains): © William Smithey Jr./Getty Images; 37.4 (seashore): © Vol. 121/Corbis RF; 37.5b: © IT Stock/PunchStock RF; 37.5c (top): Courtesy Woods Hole Research Center; 37.5c (center): Courtesy R.O. Bierregaard, Jr.; 37.5c (bottom): Courtesy Thomas Stone, Woods Hole Research Center; 37.6a: © Chuck Pratt/Bruce Coleman, Inc.; 37.6b: © Chris Johns/National Geographic Image Collection; 37Aa: © Porterfield/Chickering/Photo Researchers, Inc.; 37Ab: © Michio Hoshino/Minden Pictures; 37.7b: Courtesy Walter C. Jaap/Florida Fish & Wildlife Conservation Commission; 37.8a: © Shane Moore/Animals Animals; 37.8b: © Peter Auster/University of Connecticut; 37.9: © Dr. Paul A. Zahl/Photo Researchers, Inc; 37.10a: © Design Pics/Deb Garside; 37.10b (forest): © Buddy Mays/Corbis RF; 37.10b (owl): © Pat & Tom Leeson/Photo Researchers, Inc.; 37.11b: © Jeff Foott Productions; 37C (researcher in water, vial with sample): © Marilyn Kok, University of Illinois Springfield; 37C (birds): Courtesy Deborah M. Berman; 37C (paddlefish): © Kinne, Russ/Animals Animals; 37C (otter): © Getty RF; 37.12 (land): © Vol. 39/PhotoDisc/Getty RF; 37.12 (water): © Evelyn Jo Johnson; 37.12(food): © The McGraw-Hill Companies, Inc. John Thoeming, photographer; 37.12 (energy): © Gerald and Buff Corsi/Visuals Unlimited; 37.12 (minerals): © James P. Blair/National Geographic Image Collection; 37.13a: © Kent Knudson/PhotoLink/Getty RF; 37.13b: © Scenics of America/PhotoLink/Getty RF; 37.14: © David L. Pearson/Visuals Unlimited; 37.15a: © S.K. Patrick/Visuals Unlimited; 37.15b: © Danita Delimont/Getty Images; 37.15c: © 2009 The Associated Press; 37.16a: © Laish Bristol/Visuals Unlimited; 37.16b: Courtesy V. Jane Windsor, Division of Plant Industry, Florida Department of Agriculture & Consumer Services; 37.16c: © Inga Spence/Visuals Unlimited.

INDEX

Note: Page numbers in *italics* indicate material presented in figures and tables.

A

B

E

F

G

H

P

Q

R

S

Periodic Table of Elements

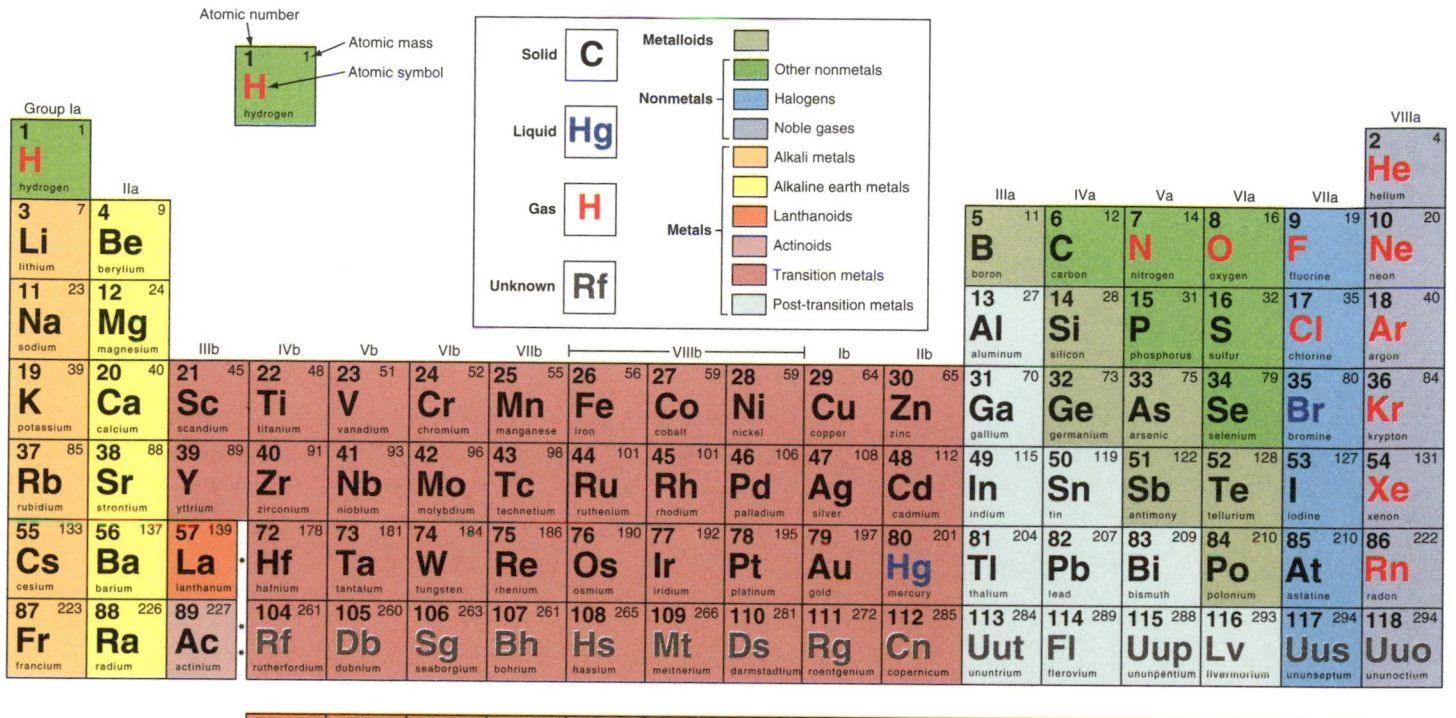

Key: Atomic number — 1; Atomic mass — 1; Atomic symbol — H; hydrogen

Solid	C
Liquid	Hg
Gas	H
Unknown	Rf

Metalloids	
Nonmetals	Other nonmetals
	Halogens
	Noble gases
Metals	Alkali metals
	Alkaline earth metals
	Lanthanoids
	Actinoids
	Transition metals
	Post-transition metals

Group Ia	IIa	IIIb	IVb	Vb	VIb	VIIb	VIIIb			Ib	IIb	IIIa	IVa	Va	VIa	VIIa	VIIIa
1 1 H hydrogen																	2 4 He helium
3 7 Li lithium	4 9 Be beryllium											5 11 B boron	6 12 C carbon	7 14 N nitrogen	8 16 O oxygen	9 19 F fluorine	10 20 Ne neon
11 23 Na sodium	12 24 Mg magnesium											13 27 Al aluminum	14 28 Si silicon	15 31 P phosphorus	16 32 S sulfur	17 35 Cl chlorine	18 40 Ar argon
19 39 K potassium	20 40 Ca calcium	21 45 Sc scandium	22 48 Ti titanium	23 51 V vanadium	24 52 Cr chromium	25 55 Mn manganese	26 56 Fe iron	27 59 Co cobalt	28 59 Ni nickel	29 64 Cu copper	30 65 Zn zinc	31 70 Ga gallium	32 73 Ge germanium	33 75 As arsenic	34 79 Se selenium	35 80 Br bromine	36 84 Kr krypton
37 85 Rb rubidium	38 88 Sr strontium	39 89 Y yttrium	40 91 Zr zirconium	41 93 Nb niobium	42 96 Mo molybdium	43 98 Tc technetium	44 101 Ru ruthenium	45 101 Rh rhodium	46 106 Pd palladium	47 108 Ag silver	48 112 Cd cadmium	49 115 In indium	50 119 Sn tin	51 122 Sb antimony	52 128 Te tellurium	53 127 I iodine	54 131 Xe xenon
55 133 Cs cesium	56 137 Ba barium	57 139 La lanthanum	72 178 Hf hafnium	73 181 Ta tantalum	74 184 W tungsten	75 186 Re rhenium	76 190 Os osmium	77 192 Ir iridium	78 195 Pt platinum	79 197 Au gold	80 201 Hg mercury	81 204 Tl thalium	82 207 Pb lead	83 209 Bi bismuth	84 210 Po polonium	85 210 At astatine	86 222 Rn radon
87 223 Fr francium	88 226 Ra radium	89 227 Ac actinium	104 261 Rf rutherfordium	105 260 Db dubnium	106 263 Sg seaborgium	107 261 Bh bohrium	108 265 Hs hassium	109 266 Mt meitnerium	110 281 Ds darmstadtium	111 272 Rg roentgenium	112 285 Cn copernicum	113 284 Uut ununtrium	114 289 Fl flerovium	115 288 Uup ununpentium	116 293 Lv livermorium	117 294 Uus ununseptum	118 294 Uuo ununoctium
			58 140 Ce cerium	59 141 Pr praseodymium	60 144 Nd neodymium	61 147 Pm promethium	62 150 Sm samarium	63 152 Eu europium	64 157 Gd gadolinium	65 159 Tb terbium	66 163 Dy dysprosium	67 165 Ho holmium	68 167 Er erbium	69 169 Tm thulium	70 173 Yb ytterbium	71 175 Lu lutetium	
			90 232 Th thorium	91 231 Pa protactinium	92 238 U uranium	93 237 Np neptunium	94 242 Pu plutonium	95 243 Am americium	96 247 Cm curium	97 247 Bk berkelium	98 249 Cf californium	99 254 Es einsteinium	100 253 Fm fermium	101 256 Md mendelevium	102 254 No nobelium	103 257 Lr lawrencium	

Metric System

Unit and Abbreviation	Metric Equivalent	Approximate English-to-Metric Equivalents
Length		
nanometer (nm)	$= 10^{-9}$ m (10^{-3} µm)	
micrometer (µm)	$= 10^{-6}$ m (10^{-3} mm)	
millimeter (mm)	= 0.001 (10^{-3}) m	
centimeter (cm)	= 0.01 (10^{-2}) m	1 inch = 2.54 cm 1 foot = 30.5 cm
meter (m)	= 100 (10^{2}) cm = 1,000 mm	1 foot = 0.30 m 1 yard = 0.91 m
kilometer (km)	= 1,000 (10^{3}) m	1 mi = 1.6 km
Weight (mass)		
nanogram (ng)	$= 10^{-9}$ g	
microgram (µg)	$= 10^{-6}$ g	
milligram (mg)	$= 10^{-3}$ g	
gram (g)	= 1,000 mg	1 ounce = 28.3 g 1 pound = 454 g
kilogram (kg)	= 1,000 (10^{3}) g	= 0.45 kg
metric ton (t)	= 1,000 kg	1 ton = 0.91 t
Volume		
microliter (µl)	$= 10^{-6}$ l (10^{-3} ml)	
milliliter (ml)	$= 10^{-3}$ liter = 1 cm^3 (cc) = 1,000 mm^3	1 tsp = 5 ml 1 fl oz = 30 ml
liter (l)	= 1,000 ml	1 pint = 0.47 liter 1 quart = 0.95 liter 1 gallon = 3.79 liter
kiloliter (kl)	= 1,000 liter	

°C	°F	
100	212	Water boils at standard temperature and pressure.
71	160	Flash pasteurization of milk
57	134	Highest recorded temperature in the United States, Death Valley, July 10, 1913
41	105.8	Average body temperature of a marathon runner in hot weather
37	98.6	Human body temperature
13.7	56.66	Human survival is still possible at this temperature.
0	32.0	Water freezes at standard temperature and pressure.

Units of Temperature

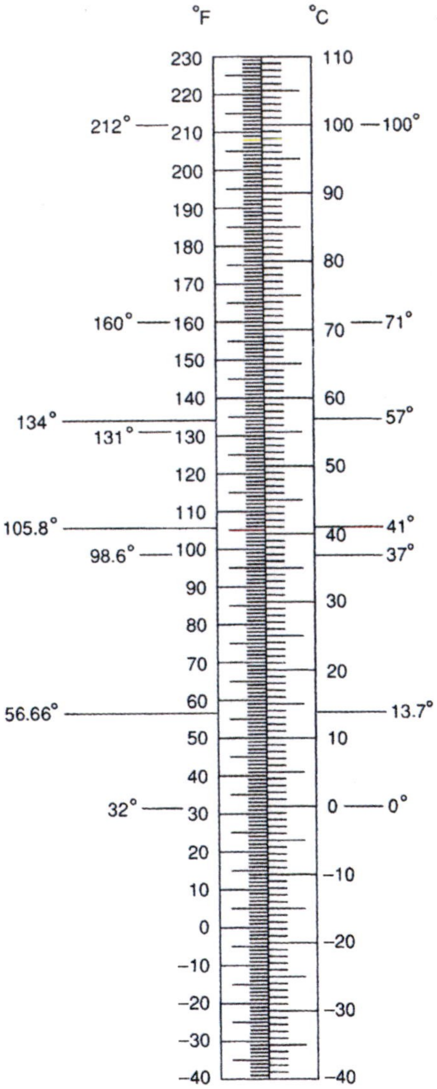

°F
°C
230
220
210
200
190
180
170
160
150
140
130
120
110
100
90
80
70
60
50
40
30
20
10
0
–10
–20
–30
–40
110
100
90
80
70
60
50
40
30
20
10
0
–10
–20
–30
–40
212°
160°
134°
131°
105.8°
98.6°
56.66°
32°
100°
71°
57°
41°
37°
13.7°
0°

To convert temperature scales:

$$°C = \frac{(°F - 32)}{1.8}$$

$$°F = 1.8\,(°C) + 32$$